FOURTH EDITION

FOOD CHEMICALS CODEX

Effective July 1, 1996

COMMITTEE ON FOOD CHEMICALS CODEX

Food and Nutrition Board
Institute of Medicine
National Academy of Sciences

NATIONAL ACADEMY PRESS
Washington, D.C.   1996

NATIONAL ACADEMY PRESS • 2101 CONSTITUTION AVENUE, NW • WASHINGTON, DC 20418

NOTICE   The project that is the subject of this report was approved by the Governing Board of the National Research Council, whose members are drawn from the Councils of the National Academy of Sciences, the National Academy of Engineering, and the Institute of Medicine. The members of the Committee responsible for the report were chosen for their special competences and with regard for appropriate balance.

INSTITUTE OF MEDICINE   The Institute of Medicine was chartered in 1970 by the National Academy of Sciences to enlist distinguished members of the appropriate professions in the examination of policy matters pertaining to the health of the public. In this, the Institute acts under both the Academy's 1863 congressional charter responsibility to be an adviser to the federal government and its own initiative in identifying issues of medical care, research, and education. Dr. Kenneth I. Shine is President of the Institute of Medicine.

FOOD AND NUTRITION BOARD   The Food and Nutrition Board (FNB) was established in 1940 to address issues of national importance that pertain to the safety and adequacy of the nation's food supply; to establish principles and guidelines for adequate nutrition; and to render authoritative judgment on the relationships among food intake, nutrition, and health. The FNB is a multidisciplinary group of scientists with expertise in various aspects of nutrition, nutritional biochemistry, food science and technology, epidemiology, food toxicology, food safety, public health, and food and nutrition policy. These scientists respond to requests from federal agencies about issues concerning food and nutrition, initiate studies that are later assigned to standing or ad hoc FNB committees, and oversee the work of these committees. Through members of its liaison panels, technical input in aspects of nutrition, food safety, food technology, and food processing is provided.

This study is supported by U.S. Food and Drug Administration Contract No. 223-92-2250.

COMPLIANCE WITH FEDERAL STATUTES   The fact that an article appears in the *Food Chemicals Codex* or its supplements does not exempt it from compliance with requirements of acts of Congress, with regulations and rulings issued by agencies of the United States Government under authority of these acts, or with requirements and regulations of governments in other countries that have adopted the *Food Chemicals Codex*. Revisions of the federal requirements that affect the *Codex* specifications will be included in *Codex* supplements as promptly as practicable.

EFFECTIVE DATE   The specifications in this edition become effective July 1, 1996.

Library of Congress Cataloging-in-Publication Data

Food chemicals codex / Committee on Food Chemicals Codex, Food and
    Nutrition Board, Institute of Medicine, National Academy of
    Sciences. — 4th ed.
        p.   cm.
    "Effective July 1, 1996."
    Includes bibliographical references and index.
    ISBN 0-309-05394-3 (alk. paper)
    1. Food additives—Standards—United States.   2. Food additives-
-Analysis.   I. Institute of Medicine (U.S.).   Committee on Food
Chemicals Codex.
TP455.F66   1996   95-49093
664'.06'021873—dc20   CIP

Copyright 1996 by the National Academy of Sciences. All rights reserved.

Additional copies of this report are available from:

National Academy Press
2101 Constitution Avenue, N.W.
Lockbox 285
Washington, DC 20055

Call 1-800-624-6242 or 202-334-3313 (in the Washington metropolitan area).

No part of this publication may be reproduced by any mechanical, photographic, or electronic process, or in the form of a phonographic recording, nor may it be stored in a retrieval system, transmitted, or otherwise copied for public or private use, without written permission from the publisher, except for official use by the United States Government or by governments in other countries that have adopted the Food Chemicals Codex.

Printed in the United States of America

The serpent has been a symbol of long life, healing, and knowledge among almost all cultures and religions since the beginning of recorded history. The image adopted as a logotype by the Institute of Medicine is based on a relief carving from ancient Greece, now held by the Stätlichemuseen in Berlin.

Contents

Organization of the *Food Chemicals Codex*, 1981–1995 ... v
Participants in Committee Activities and Other Programs ... viii
Others Who Provided Assistance, 1981–1995 ... x

PREFACE ... xiii

GENERAL INFORMATION ... xix
Operating Procedures of the *Food Chemicals Codex* ... xix
Validation of Codex Methods ... xxii
General Good Manufacturing Practices Guidelines for Food Chemicals ... xxvii
Lists of New Monographs and Former and Current Titles ... xxxii

Section

1 GENERAL PROVISIONS AND REQUIREMENTS APPLYING TO SPECIFICATIONS, TESTS, AND ASSAYS OF THE *FOOD CHEMICALS CODEX* ... 1

2 MONOGRAPHS ... 9

3 FLAVOR CHEMICALS ... 447
Specifications for Flavor Chemicals (table) ... 449
Test Methods for Flavor Chemicals ... 564
Gas Chromatographic (GC) Assay of Flavor Chemicals ... 569

4 INFRARED SPECTRA ... 571
Essential Oils ... 572
Flavor Chemicals ... 601
Other Substances ... 713

5 GENERAL TESTS AND ASSAYS ... 723

Appendix I: Apparatus for Tests and Assays ... 727
Appendix II: Physical Tests and Determinations ... 729
    A. Chromatography ... 729
    B. Physicochemical Properties ... 737
    C. Others ... 748
Appendix III: Chemical Tests and Determinations ... 753
    A. Identification Tests ... 753
    B. Limit Tests ... 755
    C. Others ... 768
Appendix IV: Chewing Gum Base ... 782
Appendix V: Enzyme Assays, 786
Appendix VI: Essential Oils and Flavors ... 815
Appendix VII: Fats and Related Substances ... 820
Appendix VIII: Oleoresins ... 829
Appendix IX: Rosins and Related Substances ... 832
Appendix X: Carbohydrates (Starches, Sugars, and Related Substances) ... 836

Solutions and Indicators ... 848

INDEX ... 863

Organization of the Food Chemicals Codex, 1981–1995

COMMITTEE ON FOOD CHEMICALS CODEX 1992–1995

Steve L. Taylor (*Chair*), Department of Food Science and Technology and Food Processing Center, University of Nebraska, Lincoln, NE (since 1989)
Samuel M. Tuthill (*Vice-Chair*), Mallinckrodt Chemical, Inc., St. Louis, MO (since 1973)
Herbert Blumenthal, Silver Spring, MD (since 1989)
Grady W. Chism, III, Department of Food Science and Nutrition, Ohio State University, Columbus, OH
Andrew G. Ebert, The Robert H. Kellen Company, Atlanta, GA (since 1988)
Nancy Higley, Tastemaker, Cincinnati, OH
Joseph H. Hotchkiss, Department of Food Science, Cornell University, Ithaca, NY (since 1989)
Bev L. Huston, Health Canada, Ottawa, Canada (since 1988)
John C. Kirschman, Emmaus, PA
Francis P. Mahn, Hoffman-La Roche, Inc., Nutley, NJ (since 1988)
Robert L. Wade, The Procter and Gamble Company, Cincinnati, OH (since 1993)
Connie M. Weaver, Department of Foods and Nutrition, Purdue University, West Lafayette, IN

Fatima N. Johnson, *Project Director*
Marcia S. Lewis, *Research Assistant* (since 1994)

WORKING GROUP ON GENERAL ANALYTICAL METHODS 1993–1995

Samuel M. Tuthill, *Chair*
John P. Fletcher
Douglas L. Terry

FORMER MEMBERS OF THE COMMITTEE ON FOOD CHEMICALS CODEX 1981–1992

Frank L. Boyd, 1981–1983
Joseph T. Brady, 1982–1991
Rhys Bryant, 1986–1987
Bruce H. Campbell, 1981–1984
Durward F. Dodgen, 1987
John P. Fletcher, 1981–1984
Sol W. Gunner, 1981–1987
Susan K. Harlander, 1986, 1989–1991
Jenny C. Hunter-Cevera, 1992–1993
James R. Kirk, 1981–1983
Marvin Legator, 1984–1987
Harold M. McNair, 1983–1989

Thomas Medwick, 1981–1986
*Fred A. Morecombe, 1981–1986
Ian C. Munro, 1983–1987
*Jessie M. Norris, 1981–1983
*Andrew J. Schmitz, 1987–1991
Stephen G. Schulman, 1988–1991
Jane C. Sheridan, 1981–1986
James T. Stewart, 1985–1989
Jan Stofberg, 1981–1991
Douglas L. Terry, 1987–1989
*Edgar Theimer, 1986–1987

---

*Deceased

MEMBERS OF THE FOOD AND NUTRITION BOARD

Cutberto Garza (*Chair*), Division of Nutritional Sciences, Cornell University, Ithaca, NY
John W. Erdman, Jr. (*Vice-Chair*), Division of Nutritional Sciences, College of Agriculture, University of Illinois at Urbana-Champaign
Perry L. Adkisson, Department of Entomology, Texas A&M University, College Station
Lindsay H. Allen, Department of Nutrition, University of California, Davis
Dennis M. Bier, Children's Nutrition Research Center, Baylor College of Medicine, Houston, TX
Fergus M. Clydesdale, Department of Food Science and Nutrition, University of Massachusetts, Amherst
Michael P. Doyle, Center for Food Safety and Quality Enhancement, Department of Food Science and Technology, The University of Georgia, Griffin
Johanna T. Dwyer, Tufts University School of Medicine and Frances Stern Nutrition Center, New England Medical Center, Boston, MA
Scott M. Grundy, Center for Human Nutrition, University of Texas Southwestern Medical Center at Dallas
K. Michael Hambidge, Center for Human Nutrition, University of Colorado Health Sciences Center, Denver
Janet C. King, University of California, Berkeley, and USDA Western Human Nutrition Research Center, Presidio of San Francisco
Sanford A. Miller, Graduate School of Biomedical Sciences, University of Texas Health Science Center, San Antonio
Alfred Sommer, School of Hygiene and Public Health, The Johns Hopkins University, Baltimore, MD
Vernon R. Young, Laboratory of Human Nutrition, School of Science, Massachusetts Institute of Technology, Cambridge
Steve L. Taylor (*Ex-Officio Member*), Department of Food Science and Technology and Food Processing Center, University of Nebraska, Lincoln

Allison A. Yates, *Director* (1994–present)
Catherine E. Woteki, *Director* (1990–1993)
Bernadette M. Marriott, *Deputy Director* (1993–1995)
Gail E. Spears, *Administrative Assistant* (1994–present)

Marcia S. Lewis, *Administrative Assistant* (1991–1994)
Jamaine L. Tinker, *Financial Associate* (1994–present)
Sue M. Wyatt, *Financial Associate* (1991–1994)

FORMER FOOD AND NUTRITION BOARD CHAIRS AND VICE-CHAIRS 1981–1995

Chairs

Janet C. King, 1994–1995
M. R. C. Greenwood, 1990–1993
Richard J. Havel, 1987–1990

Kurt J. Isselbacher, 1983–1987
Irwin Rosenberg, 1981–1983

Vice-Chairs

Edwin L. Bierman, 1992–1995
Donald B. McCormick, 1989–1992
*Hamish N. Munro, 1988–1989

Kurt J. Isselbacher, 1987–1988
Richard J. Havel, 1985–1987
Richard L. Hall, 1982–1985

---

*Deceased

FORMER FOOD CHEMICALS CODEX STAFF

Sanford W. Bigelow, *Project Director*, 1989–1991

Durward F. Dodgen, *Project Director*, 1988–1989

*Robert E. Rehwoldt, *Interim Project Director*, 1988

Robert A. Mathews, *Project Director*, 1981–1987

Sheila A. Moats, *Research Associate*, 1990–1993

Patricia A. Takach, *Project Assistant*, 1994–1995

Marilyn Mitchell, *Project Assistant*, 1993–1994

Geraldine Kennedo, *Project Assistant*, 1992–1993

Marcia S. Lewis, *Project Assistant*, 1989–1991

Talitha D. Evans, *Secretary*, 1986

Betty C. Guyot, *Secretary*, 1981–1983, 1986

---

*Deceased

Participants in Committee Activities and Other Programs*

FCC OPEN SESSION ON CARRAGEENAN
(FCC Committee Meeting, July 26, 1994)

Harris "Pete" Bixler
James Carr
Donald H. Combs
Eunice M. Cuirle

Rodney J. H. Gray
Paul M. Kuznesof
Denis LaSota

Robert Mayer
Scott Rangus
Peter Salling

WORKSHOP ON ANALYTICAL METHODS FOR FOOD INGREDIENTS
(In conjunction with AOAC International 107th Annual Meeting, July 29–30, 1993)

Charles H. Barnstein
Stephen Capar
Jonathan W. DeVries
Michael J. DiNovi
James T. Elfstrum

William Horwitz
Allen W. Matthys
Stan Omaye
Denis Page
Steve L. Taylor

Samuel M. Tuthill
Harriet Wallin
Charles Warner

WORKSHOP ON THE ALLERGENICITY OF FOOD–USE PROTEIN HYDROLYSATES
(July 31, 1991)

Fred Atkins
A. Wesley Burks, Jr.
Robert K. Bush
Christopher Cordle
Cutberto Garza
Walter Glinsmann

Susan K. Harlander
Rolf Jost
John C. Kirschman
Zdenek Kratky
Paul M. Kuznesof
H. Lee Leary

Charles Manley
Lanny J. Rosenwasser
Hugh A. Sampson
R. Grant Smith
James A. Whitten

WORKING GROUP ON MICROBIOLOGICAL SPECIFICATIONS FOR FOOD CHEMICALS
(August 30, 1991)

Cleve Denny
Maurice Fagan
Rodney J. H. Gray
John Humber

George Jackson
John C. Kirschman
Paul M. Kuznesof
Aubrey S. Outschoorn

Steve L. Taylor
Kay Wachsmuth

---

*The following lists comprise invited speakers, including committee members, and participants outside of the committee.

WORKSHOP ON LEAD SPECIFICATIONS FOR FOOD INGREDIENTS
(May 2, 1991)

Michael Adams	Lloyd J. Filer	Peter Method
Herbert Blumenthal	Dee Graham	Richard Ronk
Michael Bolger	Clark Hartford	*Andrew J. Schmitz
Stephen Capar	Joseph H. Hotchkiss	Stephen G. Schulman
Margaret A. Clarke	Paul M. Kuznesof	Steve L. Taylor
Janet Dudek	Kathryn Mahaffey	Samuel M. Tuthill

FCC FORUM ON FUNCTIONS AND ACTIVITIES FOR DEVELOPMENT FOR FCC IV
(November 17–18, 1987)

Joseph T. Brady	Paul F. Hopper	*Roger Middlekauf
Rhys Bryant	Julia C. Howell	Richard Ronk
Durward F. Dodgen	John C. Kirschman	Daniel Rosenfield
James T. Elfstrum	Harold M. McNair	Jan Stofberg
F. J. Francis		

FOOD CHEMICALS CODEX WORKSHOP ON LEAD, HELD IN CONJUNCTION WITH THE HEALTH PROTECTION BRANCH OF HEALTH AND WELFARE CANADA
(August 14–15, 1986)

Michael Bolger	Don Grant	Kathryn Mahaffey
Kenneth W. Boyer	Sol W. Gunner	Deborah C. Rice
Harry Conacher	Bev L. Huston	John W. Salminen
Robert W. Dabeka	Diane Kirkpatrick	Jacqueline Sitwell
Steve Gilbert		

TASK FORCE ON HYDROCHLORIC ACID 1981–1984

1—June 1, 1981; 2—February 28, 1983; 3—August 1, 1983; 4—February 13, 1984

Samuel M. Tuthill *Chair*	John P. Fletcher *Member*	*Jessie M. Norris *Member*
Bruce H. Campbell *Member*	*Fred A. Morecombe *Member*	Robert A. Mathews *FCC Staff*

---

*Deceased

Others Who Provided Assistance, 1981–1995*

Rebecca Allen
Don Ayerlee

Charles Baker
James K. Baker
William Balke
Rasma B. Balodis
Stephen Barker
Gerrit Bekendam
David L. Berner
Bruce M. Bertram
Joseph Bertucci
Anthony P. Bimbo
Gary Blair
Louis Blecher
B. D. Bolt
Edwin Bontenbal
Ben Borsje
Marion Bradford
Kyd D. Brenner
Simon Brooke-Taylor
Franta J. Broulik

Michael F. Campbell
J. C. Carlaw
Paul Cassaletto
Duane Chase
James P. Clark
Warren S. Clark
Richard E. Cristol
Ray Croes

Roger Dabbah
David E. Dalsis
Christopher C. DeMerlis
R. A. DePalma
Mario Diaz-Cruz
Gontran Dondain
Doug Drogosh

Lee Elwell
Joseph E. Englesberg
Elizabeth Erman
Robert Evans

Roger W. Fenstermaker

George Ford
Joseph Fordham
Larry Fosdick
Geoffrey G. Fowler
John V. Fratus
James Freeman
Carl Frey
William J. Frost
John Fry

A. A. Gaballa
James A. Gall
James E. Geyer
Curtis E. Gidding
Jo Gilbert
Stanley P. Gorak
J. A. Gosselin
Elwood E. Graham Jr.
Vivian A. Gray
Karl H. Griessmann

Martin J. Hahn
John B. Hallagan
Jerome A. Halperin
J. C. Henze
A. Heron
David H. Hickman
Jerry Hjelle
Connie Horn
Julia C. Howell

Kenji Ishii
Glen Ishikawa

Walter S. Jeffery

Isabelle Kamishlian
Sam Kennedy
Warren W. Kindt
Charles L. King
Donald L. Kiser
Lorie Klopf
Catherine Knoka
Willem Kohl
Kohei Kojima
John Kropiewnicki

Robert Kuna

Patricia L. Lawson
Leonard A. Levy
J. H. Lichtenbelt
C. I. Luckhoo

Janine Manhart
Alton E. Martin
Allen W. Matthys
Duane McDowell
Roger A. McKenna
William McMullen
Carolyn Merkel
Inge Meyland
Nancy Miller-Ihli
John P. Modderman
G. Müller

Donald D. Naragon
June M. Neades
Lyle Nehls
Robert E. Norland

Edgar Odom
Owen J. Olesen
Philip M. Olinger

Juhani Paakkanen
Barbara Pagliocca
G. R. Parikh
Robert Patti
W. Penning
Glyn O. Phillips
Walter Pilnik
Robert T. Plutnick
Rinske Potjewijd

B. Quock

Arthur Raczynski
Robert M. Reeves
Joe M. Regenstein
Jill Rickman
Marion Riordan
Larry Roberts

*The listing of individuals in this section, which includes those not listed in the preceding sections, does not necessarily indicate that their contributions were less significant than those made by individuals listed in the preceding sections.

J. A. Robertson
Daniel A. Roman
Louis Rothschild, Jr.
John Rotruck

Wayne Sander
Jean Savigny
Patricia Scarbinsky
David R. Schoneker
Mark A. Seese
Richard Servon
Sheldon Silbiger
Elisabeth A. Snipes
Charles Sokol
Joan M. Stapleton

Jim Steinke
Hope Stewart
W. M. Strauss

Ralph J. Tenney
J. Thomsen
Helen R. Thorsheim
Frances Turnak
Jean Turner
William C. Twieg

Richard Ungvarsky

Joseph G. Valentino
John A. van Velthuysen

Florian M. Ward
Alan Warren
Jerry Weigel
Jerry Wertz
Brian Whitehouse
Wayne Wolf
John T. Woodward

Tony Yates
Gary L. Yingling

James Zawecki
Patricia S. Zawislak

Preface

The fourth edition of the *Food Chemicals Codex* is the culmination of efforts of the many members, past and present, of the Committee on Food Chemicals Codex (FCC). The current committee, formed in the fall of 1992 at the request of the U.S. Food and Drug Administration, has brought all these efforts to fruition with this edition. The charge to the committee states that "the committee shall (1) provide information on matters related to the purity of food ingredients used in the United States and shall be knowledgeable of the purity of food ingredients used in other countries; (2) provide information on food-grade specifications for food additives, GRAS substances, and any other food substances used as ingredients; and (3) publish specification monographs in a fourth edition of the *Food Chemicals Codex*. To provide such information, the committee shall review proposals from industry, government, or any other source."

The FCC project, currently under the administrative supervision of the Food and Nutrition Board of the Institute of Medicine/National Academy of Sciences, began in 1961, soon after the passage of the 1958 Food Additives Amendment to the federal Food, Drug, and Cosmetic Act. Although the U.S. Food and Drug Administration had by regulations and informal statements defined in general terms quality requirements for food chemicals generally recognized as safe and other food additive chemicals, these requirements were not sufficiently specific to serve as release, procurement, and acceptance specifications for manufacturers and users of food chemicals. Therefore, regulators and other interested parties believed that the publication of a book of standards designed especially for food chemicals would promote uniformity of quality and added assurance of safety for such chemicals. For these reasons, the Food Protection Committee of the National Academy of Sciences/National Research Council received requests in 1958 from its Industry Liaison Panel and other sources to undertake a project to produce a *Food Chemicals Codex* comparable in many respects to the *United States Pharmacopeia* and the *National Formulary* for drugs. As a result of these requests, representatives of industry and government agencies agreed that there was a definite need for such a *Codex* and that the Food Protection Committee was a suitable agency to undertake the project.

The first edition, published in 1966, was supported by a Public Health Service grant and more than 100 supplementary grants from industry, associations, and foundations. Its goal, which is still that of the *Food Chemicals Codex*, was to define the quality of food-grade chemicals in terms of identity, strength, and purity based on the elements of safety and good manufacturing practice. Later editions were supported by direct contracts with the U.S. Food and Drug Administration. Although sponsorship has not been continuous, it has been sufficient to have supported the publication of 3 earlier editions and 11 supplements in a 35-year span.

SCOPE

The scope of the *Food Chemicals Codex* has expanded with each new edition. Substances included in the first edition were limited to chemicals added directly to foods to achieve a desired function. Succeeding editions included these substances as well as those that come into contact with foods, such as processing aids (including enzymes, extraction solvents, filter media, and packaging materials and ingredients), and those that are regarded as foods, such as fructose and dextrose, rather than as additives. This fourth edition includes 773 monographs from the third edition; 142 monographs, including those for 69 flavor chemicals, added in the four supplements to the third edition; and 52 new monographs, including 33 for flavor chemicals, new to this fourth edition, bringing the total to 967.

Three monographs included in the third edition have been deleted from the fourth edition. These are Carrageenan, Cinnamyl Anthranilate, and Methyl Formate (see list of deletions, page xxxii). Since the publication of the third edition in 1981, the following events provided the impetus for the deletions: (1) A second type of Carrageenan, a semi-refined grade, was introduced into the marketplace. Its regulatory status in terms of nomenclature, specifications, and labeling was, at press time, unresolved by the U.S. Food and Drug Administration, with several food additive petitions pending. Therefore, the Committee on Food Chemicals Codex decided to wait for the resolution of the regulatory issues before including one or more Carrageenan monographs in the fourth edition of the Codex. (2) The use of Cinnamyl Anthranilate in foods has been banned by the U.S. Food and Drug Administration (see 50 FR 42932, October 23, 1985). (3) The regulatory authority for fumigants/insecticides, including those for food use, such as Methyl Formate used on raisins, was transferred from the U.S. Food and Drug Administration to the Environmental Protection Agency.

UPDATING AND DEVELOPING SPECIFICATIONS

Substantive changes in the fourth edition resulted from changes in committee policies (see the *General Provisions*) and scientific advances during the 15 years since the appearance of the third edition in 1981. Symposia and various meetings with the affected industry (see pages viii–ix) were sponsored by the Committee on Food Chemicals Codex to guide the committee in its deliberations.

The committee has invariably sought to define, using physicochemical and microbiological parameters, ingredients prepared under good manufacturing practices as safe for human consumption. Special emphasis has been placed on reducing contaminants, including trace elements, particularly lead. The need for practical, sensitive analytical methods to achieve this goal was recognized, and industry participation was enlisted. As a result, several quantitative procedures to measure low lead levels are new to this edition. This effort also led to collaboration with the Agricultural Research Service of the U.S. Department of Agriculture and the International Life Sciences Institute in developing a program to validate an atomic absorption graphite furnace method to quantitate lead in water-soluble sweeteners at levels below 0.1 mg/kg.

Limits on contaminants, specifically lead and heavy metals, have been reduced by a minimum of one-half in more than 71 and more than 111 monographs, respectively, in this edition. This trend is expected to continue. Manufacturers and suppliers of food ingredients are encouraged to inform the committee of their ability to supply food ingredients with lead and heavy metals limits lower than those specified in this edition. The arsenic specification remains in only about 110 articles in this edition where (1) the ingredient or additive is a high-volume consumption item (greater than 25 million pounds a year), (2) the ingredient or additive is derived from a natural (mineral) source where arsenic may be an intrinsic contaminant, and/or (3) there is reason to believe that arsenic constitutes a significant part of the total heavy metals content.

The committee has been cognizant of the need for international harmonization of specifications in today's world. Efforts were made, where feasible, to harmonize the specifications in this edition with those of other standards-setting organizations, in particular with those in the *Compendium of Food Additive Specifications* published by the FAO/WHO Joint Expert Committee on Food Additives.

FORMAT

Generally the presentation follows that of the third edition, but a number of significant changes and additions have been made. As expected, the passage of 15 years since the appearance of the third edition has been accompanied by changes, not only substantive in character, but also of common, accepted scientific terminology.

- The *General Provisions* has been expanded and retitled to *General Provisions and Requirements* to more accurately reflect the material contained therein.

- For easy reference, most of the material in Chapter 8 of the third edition has been moved to an information section in the front of this edition (pages xix–xxxii) preceding the requirements sections.

- A new section on *Validation of Codex Methods* has been added to the information section to serve as a guide to all interested parties who want to suggest revising existing FCC analytical methods and limits or to suggest new methods for committee consideration.

- The analytical methods for flavor chemicals, especially for quantitative measurements, have, in keeping with present common laboratory practices, been revised significantly from mostly wet-chemical methods to gas chromatographic procedures and are presented with the tabular specifications for flavor chemicals.

- The general analytical test procedures listed alphabetically in Chapter 6 of the third edition have been recast, by virtue of their utility or purpose, into ten distinctive appendixes that appear at the end of the book. This eliminates the need to provide repetitive page references to those procedures referenced in the monographs that are needed to demonstrate compliance with specific limits and identification tests.

- Additional information, in terms of CAS (Chemical Abstracts Services) Registry and INS (International Numbering System of the FAO/WHO Codex Alimentarius) numbers, has been added to specific monographs where definitive numbers could be identified for the referenced article.

- In this edition, all *cis* and *trans* isomers have been changed to (Z) and (E), zusammen and entgegen, respectively.

- The designation "molecular weight" has been revised to "formula weight," a term that more broadly defines the sum of the atomic weights of the referenced article without regard to its intrinsic nature.

- Odor tests have been deleted, especially where the substance generated, for example, acrolein, may pose a hazard to the analyst.

- The mention of highly toxic solvents, for example, carbon tetrachloride and benzene, has been eliminated where such reference has been purely informational and replaced where a suitable substitute could be identified.

FUTURE REVISIONS

The introduction of new food additives as well as constant changes and advances in manufacturing processes and analytical sciences lead to a need for continued revision of this compendium. The first supplement to this edition may be expected 18 months after the edition's release.

Users of this edition are requested and encouraged to submit suggestions for updating the specifications as well as the general analytical methods. Constructive criticism and notification of errors should also be brought to the attention of the Food Chemicals Codex, National Academy of Sciences, 2101 Constitution Avenue, N.W., Washington, D.C. 20418. The worth of the *Food Chemicals Codex* to its users (food additive manufacturers, food processors, and national and international regulatory bodies) would be enhanced by the continuity of the project and revision of this edition through appropriate and timely support.

LEGAL STATUS

The *Food Chemicals Codex* enjoys international recognition by manufacturers, vendors, and users of food chemicals. The specifications herein serve as the basis for many buyer and seller contractual agreements.

In the United States, the first edition was given quasi-legal recognition in July 1966 by means of a letter of endorsement from FDA Commissioner James L. Goddard, which was reprinted in the book. The letter stated that "the FDA will regard the specifications in the *Food Chemicals Codex* as defining an 'appropriate food grade' within the meaning of Sec. 121.101(b)(3) and Sec. 121.1000(a)(2) of the food additive regulations, subject to the following qualification: this endorsement is not construed to exempt any food chemical appearing in the *Food Chemicals Codex* from compliance with requirements of Acts of Congress or with regulations and rulings issued by the Food and Drug Administration under authority of such Acts."

Subsequently, the specifications in the second edition, followed by those in the third edition, were cited, by reference, in the U.S. *Code of Federal Regulations* to define specific safe ingredients under title 21, in various parts of sections 172, 173, and 184.

In Canada, the current edition of the *Food Chemicals Codex*, including its supplements, is officially recognized in the *Canadian Food and Drug Regulations* under Section B.01.045(b) as the reference for specifications for food additives. The New Zealand government, under its food regulation 245(6)(a), defines a food additive as being of appropriate quality "if it complies with the monograph for that food additive (if any) in the current edition of the *Food Chemicals Codex* published by the National Academy of Sciences and the National Research Council of the United States of America in Washington, D.C." Similarly, the national food authority of Australia frequently refers to the *Food Chemicals Codex* specifications to define food additives.

ACKNOWLEDGMENTS

A compendium of this breadth can only result from the cooperation of many individuals and organizations. Underlying this, the support provided by U.S. Food and Drug Administration contract number 223-92-2250, monitored by project officer Paul M. Kuznesof, is gratefully acknowledged.

Several monographs and various sections in this edition have extensive portions based on other publications, and are used with permission granted by the parent organizations: United States Pharmacopeial Convention, Inc.; the American Oil Chemists Society; AOAC International; and the American Society for Testing and Materials. This edition of the *Food Chemicals Codex* directly references the procedures in the seventh edition of the *FDA Bacteriological Analytical Manual* (BAM) for its microbial limit tests. Where the sample size is not defined in the limit, the results are based on the sampling procedures described in BAM.

While participating individuals have been listed on pages viii–xi, the following organizations have also been active participants:

American Dairy Products Institute
American Oil Chemists Association
American Spice Trade Association
Corn Refiners Association
Enzyme Technical Association
European Association of the Chewing Gum Industry
Flavor and Extract Manufacturers Association
Gelatin Manufacturers of Europe
Gelatin Manufacturers Institute of America
International Food Additives Council
International Glutamate Technical Committee
International Hydrolyzed Protein Council
International Life Sciences Institute
International Pectin Producers Association
International Pharmaceutical Excipients Council

International Technical Caramel Association
Japan Food Additives Association
National Association of Chewing Gum Manufacturers
National Association of Color Manufacturers
National Soft Drink Association
Salt Institute
Seaweed Industry Association of the Philippines
Synthetic Amorphous Silica and Silicate Industry Association

Members of the National Academy Press, namely James M. Gormley, Sally S. Stanfield, Estelle H. Miller, Dawn M. Eichenlaub, and William B. Mason, and IOM Reports and Information Office staffers Claudia M. Carl and Michael A. Edington provided valuable support to the FCC staff toward the publication of this edition.

Success in the complex task of completing the fourth edition is due to the dedication and determination of the members of the Committee on Food Chemicals Codex under the focused leadership of its chair, Steve L. Taylor, and its vice-chair, Samuel M. Tuthill, during the past 30 months.

Washington, D.C.
September 1995

FATIMA N. JOHNSON
Project Director

General Information

OPERATING PROCEDURES OF THE *FOOD CHEMICALS CODEX*

Organization

The Food Chemicals Codex (FCC) project is an activity of the Food and Nutrition Board, a unit of the Institute of Medicine of the National Academy of Sciences. The immediate responsibility for developing the *Food Chemicals Codex* lies with the Board's Committee on Food Chemicals Codex. The committee consists of 12 to 15 members, chosen for their expertise in the various aspects of the committee's work, who are appointed, upon recommendation of the Food and Nutrition Board and the President of the Institute of Medicine, by the Chairman of the National Research Council. Committee members are paid no consulting fees or honoraria and are reimbursed only for expenses incurred while attending meetings and other activities of the committee.

Functions of the Committee on Food Chemicals Codex

The committee's principal functions are as follows:

- To establish the general policies and guidelines by which FCC specifications are prepared.

- To evaluate comments submitted by interested parties on any aspect of the specifications and test procedures.

- To propose means by which the specifications may be kept current in reflecting food-grade quality on the basis of product safety and good manufacturing practice.

- To provide information on issues dealing with specifications for particular substances and analytical test procedures.

- To seek the advice of specialists when additional expert opinion is needed in making decisions regarding the appropriateness of specifications.

- To establish working groups consisting of committee members and other experts to address specific issues relevant to monograph development and to report their findings and recommendations to the full committee.

- To consider and act on any other matter concerning the development and publication of specifications and test procedures for food-grade ingredients.

- To approve the final manuscript for National Academy of Sciences review before the publication of any edition of the FCC or its supplements.

The business of the committee is conducted through a central office at the National Academy of Sciences in Washington, D.C. The appointed Academy staff officer (Project Director) coordinates all committee activities. The committee meets in regular session, usually twice a year, to discuss the project's progress, including technical and policy issues relevant to the FCC. One or more members of the committee as well as the Project Director conduct ad hoc meetings on short-term projects as needed. The committee and Project Director also organize workshops and symposia as appropriate to exchange information with interested parties on key issues, whether of broad or limited scope.

Requirements for Listing Substances in the *Food Chemicals Codex*

The requirements are as follows: (1) the substance is permitted for use in food or in food processing by the U.S. Food and Drug Administration (or, in certain cases, by other countries in which FCC specifications are recognized), (2) it is commercially available, and (3) suitable specifications and analytical test procedures are available to determine its purity.

Criteria for *Food Chemicals Codex* Grade

The specifications published in the FCC are based primarily on the criteria of safety and good manufacturing practices (GMP). An FCC-grade substance is one that is prepared under GMP (discussed in detail in this section) and is of such purity as to ensure that potentially harmful or objectionable contaminants are not present at levels that would represent a hazard to the consumer of the foods in which the substance is intended to be used. Thus, FCC specifications define substances of a level of quality sufficiently high to represent a reasonable certainty of safety when they are used under customary conditions of intentional use in food or in food processing. The specifications generally represent acceptable levels of quality and purity of food-grade substances available in the United States and in other countries in which FCC specifications are recognized. Because the different types of ingredients are diverse and complex, few general criteria can be established that will apply to all substances for which FCC specifications are prepared. The committee recognizes that limits and tests cannot be provided to cover all possible unusual or unexpected impurities, the presence of which would be inconsistent with GMP. This matter is discussed further under *Trace Impurities*, in the *General Provisions*, and in *General Good Manufacturing Practices Guidelines for Food Chemicals*.

In addition to impurity limits, specifications, where applicable, must include the following: empirical formula, structural formula, and formula weight; description of the substance, including physical form, odor, and solubility (see the descriptive terms for solubility in the *General Provisions*); identification; assay (or a quantitative test to serve as an assay); physicochemical characteristics such as specific rotation, melting range or solidification point, viscosity, specific gravity, refractive index, pH, etc.; loss on drying or water content; residual solvents; limits for mycotoxins and microbiological contaminants; and limits for by-products and other adventitious constituents usually occurring in, or arising from the manufacture of, the substance. Furthermore, the data provided, taken together, represent a complete compositional analysis of the substance. Additional information items include how the substance is to be packaged and stored to maintain its integrity and its functional use(s) in foods. If the substance contains an "added substance," mentioning this fact enables the committee to judge whether the specifications should include it (see *Added Substances* in the *General Provisions*).

Procedures for Submission and Development of Specifications

The committee will consider suggested specifications submitted with supporting data, such as elaborated above, by any interested party, including food ingredient manufacturers and suppliers, food processors, and industry associations. Suggested specifications should be submitted, in duplicate, to Food Chemicals Codex, Food and Nutrition Board, National Academy of Sciences, 2101 Constitution Avenue, N.W., Washington, D.C. 20418. The committee and/or the project staff critically examine suggested specifications and often expand them to meet the general criteria the committee requires. Because committee discussions involving quality characteristics of substances used in food or food processing might result in sharing privileged or proprietary information, contributors may request that such discussions be held in closed sessions. The final outcome of such discussions must be openly shared with all manufacturers, users, and parties interested in the substance discussed; therefore, open discussions are the norm, except during unusual circumstances. Where privileged or proprietary information is concerned, the project staff can put such information in a format so that the end results are not associated with particular manufacturers or users. The National Academy of Sciences–Institute of Medicine is not bound by the Freedom of Information Act; thus, information shared with the staff or the committee need not be made accessible to others. If the submitter wishes, information can be labeled confidential, and the project staff will exercise due discretion. The committee and/or the staff draft a new monograph and send it to the originator for comment (and to any other manufacturers of that substance that can be identified). After the draft has gone through this process and all necessary revisions have been made, the committee votes by mail ballot whether to propose these specifications for public comment. If the committee finds deficiencies, or if any questions are raised, the draft is returned to the originator and other interested parties with the committee's comments and recommendations for improvement. Once a draft has gained committee acceptance, availability of the proposed specification for comment is announced in the *Federal Register* or through notices in trade journals. This notification allows the public and other interested parties as well as manufacturers and users that may be inadvertently overlooked to provide their comments to the committee. Once the public comments are considered and any necessary changes made, the committee votes to determine whether the monograph is suitable for publication. Monographs are then reviewed by the National Academy of Sciences and, if approved as official monographs, are announced or are published in the next edition of the FCC or a supplement.

Procedure for Revising Specifications

FCC specifications are subject to revision at any time, and suggested revisions may be initiated by regulatory bodies, manufacturers, suppliers, or users of the ingredients; by the committee itself; or by any other interested parties. All suggestions for revision must be accompanied by supporting data. In the case of revisions of test procedures and analytical methods, comparative data for both the existing and suggested procedures must be submitted. Where changes in limits or other tolerances are suggested, supporting data should be presented on representative production batches. Suggestions for changing the limits of certain impurities (e.g., arsenic, heavy metals, lead, fluoride, and mercury) may require the submission of safety data and information concerning the daily intake of the substance. All suggestions for revision, together with the supporting data, are reviewed by the Committee on Food Chemicals Codex and/or by the staff. If other manufacturers are involved (and can be identified), they are also asked to comment. If the committee finds deficiencies, or if any questions arise, the suggested revised specifications are returned to the originator (and other manufacturers, where appropriate) with the committee's comments or questions. If agreement cannot be reached at this point between the committee and the originator, or among manufacturers and other interested parties, a special meeting may be held to discuss the matter, or the parties involved may be invited to one of the committee's regular meetings to examine the question in depth. Approved revisions are published in either the next edition of the FCC or in a supplement by the same procedure described above in *Procedures for Submission and Development of Specifications*.

Further Information

Users of the FCC should become thoroughly familiar with the *General Provisions* pertaining to this edition. Inquiries regarding any aspect of the operation of the FCC project may be directed to Food Chemicals Codex, Food and Nutrition Board, National Academy of Sciences, 2101 Constitution Avenue, N.W., Washington, D.C. 20418 (telephone 202-334-2580; facsimile 202-334-2316).

VALIDATION OF CODEX METHODS

Submissions to the Codex

Submissions for new or revised specifications and analytical methods must contain sufficient information to enable committee members to evaluate the proposals. In most cases, evaluations involve assessment of the clarity and completeness of the description of the analytical methods, determination of the need for the methods, and documentation that the methods have been appropriately validated. Information may vary depending on the type of test method involved. However, in most cases a submission will consist of the following sections:

Rationale This section should identify the need for the method and describe the capability of the specific method proposed and why it is preferred over other types of determinations. For revised methods, a comparison should be provided of limitations of the existing FCC analytical method and advantages offered by the suggested method.

Suggested Analytical Method This section should contain a complete description of the analytical method sufficiently detailed to enable persons "skilled in the art" to replicate it. The write-up should include all important operational parameters and specific instructions such as preparation of reagents, performance of systems suitability tests, description of blanks used, precautions, and explicit formulas for calculation of test results.

Data Elements This section should provide thorough and complete documentation of the validation of the analytical method. It should include summaries of experimental data and calculations substantiating each of the applicable analytical performance parameters. These parameters are described in the following section.

Validation

Validation of an analytical method is the process by which it is established, by laboratory studies, that the performance characteristics of the method meet the requirements for the intended analytical applications. Performance characteristics are expressed in terms of analytical parameters. Typical analytical parameters that should be considered in the validation of the types of tests described in this document are listed in Table 1. Each of the parameters is defined in the next section of this chapter along with a delineation of a typical method by which it may be measured.

Table 1. Typical Analytical Parameters Used in Assay Validation

Accuracy	Limit of Quantitation
Precision	Linearity
Specificity	Range
Limit of Detection	Ruggedness

Analytical Performance Parameters

Accuracy

Definition The accuracy of an analytical method is the closeness of test results obtained by that method to the true value. Accuracy may often be expressed as percent recovery by the assay of known, added amounts of analyte.

Determination The accuracy of an analytical method may be determined by applying that method to samples to which known amounts of analyte have been added both above and below the normal levels expected in the samples. The accuracy is then calculated from the test results as the percentage of analyte recovered by the assay.

Precision

Definition The precision of an analytical method is the degree of agreement among individual test results when the procedure is applied repeatedly to multiple samplings of a homogeneous sample. The precision of an analytical method is usually expressed as the standard deviation or relative standard deviation (coefficient of variation). Precision may be a measure of the degree of either reproducibility or repeatability of the analytical method under normal operating conditions. In this context, reproducibility refers to the use of the analytical procedure in different laboratories. Intermediate precision expresses within laboratory variation, as on different days, or with different analysts or equipment within the same laboratory. Repeatability refers to the use of the analytical procedure within a laboratory over a short period of time, using the same analyst with the same equipment.

Determination The precision of an analytical method is determined by assaying a sufficient number of aliquots of a homogeneous sample to be able to calculate statistically valid estimates of standard deviation or relative standard deviation (coefficient of variation). Assays in this context are independent analyses of samples that have been carried through the complete analytical procedure from sample preparation to final test result.

Specificity

Definition The specificity of an analytical method is its ability to measure accurately and specifically the analyte in the presence of components that may be expected to be present in the sample matrix. Specificity may often be expressed as the degree of bias of test results obtained by analysis of samples containing added impurities, degradation products, or related chemical compounds when compared with test results from samples without added substances. The bias may be expressed as the difference in assay results between the two groups of samples. Specificity is a measure of the degree of interference (or absence thereof) in the analysis of complex sample mixtures.

Determination The specificity of an analytical method is determined by comparing test results from the analysis of samples containing impurities, degradation products, or related chemical compounds with those obtained from the analysis of samples without these elements. The bias of the assay, if any, is the difference in test results between the two groups of samples.

When impurities or degradation products are unidentified or unavailable, specificity may be demonstrated by analysis by the method in question of samples containing impurities or degradation products and comparing the results to those from additional purity assays (e.g., chromatographic assay). The degree of agreement of test results is a measure of the specificity.

Limit of Detection

Definition The limit of detection is a parameter of limit tests. It is the lowest concentration of analyte in a sample that can be detected, but not necessarily quantitated, under the stated experimental conditions. Thus, limit tests merely substantiate that the analyte concentration is above or below a certain level. The limit of detection is usually expressed as the concentration of analyte (e.g., percentage, parts per billion) in the sample.

Determination Determination of the limit of detection of an analytical method will vary depending on whether it is an instrumental or a noninstrumental procedure. For instrumental procedures, different techniques may be used. Some investigators determine the signal-to-noise ratio by comparing test results from samples with known concentrations of analyte with those of blank samples and establish the minimum level at which the analyte can be reliably detected. A signal-to-noise ratio of 2:1 or 3:1 is generally accepted. Other investigators measure the magnitude of analytical background response by analyzing a number of blank samples and calculating the standard deviation of this response. The standard deviation multiplied by a factor, usually 2 or 3, provides an estimate of the limit of detection. This limit is subsequently validated by the analysis of a suitable number of samples known to be near or prepared at the limit of detection.

For noninstrumental methods, the limit of detection is generally determined by the analysis of samples with known concentrations of analyte and by establishing the minimum level at which the analyte can be reliably detected.

Limit of Quantitation

Definition Limit of quantitation is a parameter of quantitative assays for low levels of compounds in sample matrices, such as impurities and degradation products in food additives and processing aids. It is the lowest concentration of analyte in a sample that can be determined with acceptable precision and accuracy under the stated experimental conditions. The limit of quantitation is expressed as the concentration of analyte (e.g., percentage, parts per billion) in the sample.

Determination Determination of the limit of quantitation of an analytical method may vary depending on whether it is an instrumental or a noninstrumental procedure. For instrumental procedures, a common approach is to measure the magnitude of analytical background response by analyzing a number of blank samples and calculating the standard deviation of this response The standard deviation multiplied by a factor, usually 10, provides an estimate of the limit of quantitation. This limit is subsequently validated by the analysis of a suitable number of samples known to be near or prepared at the limit of quantitation.

For noninstrumental methods, the limit of quantitation is generally determined by the analysis of samples with known concentrations of analyte and by establishing the minimum level at which the analyte can be detected with acceptable accuracy and precision.

Linearity and Range

Definition of Linearity The linearity of an analytical method is its ability (within a given range) to elicit test results that are directly, or by a well-defined mathematical transformation, proportional to the concentration of analyte in samples within a given range. Linearity is usually expressed in terms of the variance around the slope of the regression line (correlation coefficient), calculated according to an established mathematical relationship from test results obtained by the analysis of samples with varying concentrations of analyte.

Definition of Range The range of an analytical method is the interval between and including the upper and lower levels of analyte that have been demonstrated to be determined with precision, accuracy, and linearity using the method as written. The range is normally expressed in the same units as test results (e.g., percent, parts per million) obtained by the analytical method.

Determination of Linearity and Range The linearity of an analytical method is determined by mathematical treatment of test results obtained by analysis of samples with analyte concentrations across the claimed range of the method. The treatment is normally a calculation of a regression line by the method of least squares of test results versus analyte concentrations. In some cases, to obtain proportionality between assays and sample concentrations, the test data may have to be subjected to a mathematical transformation before the regression analysis. The slope of the regression line and its variance (correlation coefficient) provide a mathematical measure of linearity; the y-intercept is a measure of the potential assay bias.

The range of the method is validated by verifying that the analytical method provides acceptable precision, accuracy, and linearity when applied to samples containing analyte at the extremes of the range as well as within the range.

Ruggedness

Definition The ruggedness of an analytical method is the degree of reproducibility of test results obtained by the analysis of the same samples under a variety of normal test conditions, such as different laboratories, analysts, instruments, lots of reagents, elapsed assay times, assay temperatures, and days. Ruggedness is normally expressed as the lack of influence on test results of operational and environmental variables of the analytical method. Ruggedness is a measure of reproducibility of test results under normal, expected operational conditions from laboratory to laboratory and from analyst to analyst.

Determination The ruggedness of an analytical method is determined by analysis of aliquots from homogeneous lots in different laboratories, by different analysts, using operational and environmental conditions that may differ but are still within the specified parameters of the assay. The degree of reproducibility of test results is then determined as a function of the assay variables. This reproducibility may be compared to the precision of the assay under normal conditions to obtain a measure of the ruggedness of the analytical method.

Robustness

Definition The robustness of an analytical method is a measure of the method's capacity to remain unaffected by small, but deliberate, variations in method parameters, and it provides an indication of the method's reliability during normal use.

Data Elements Required for Assay Validation

FCC assay procedures vary from highly exacting analytical determinations to subjective evaluation of attributes. Considering this variety of assays, it is only logical that different test methods require different validation schemes. This section covers only the most common categories of assays for which validation data should be required. These categories are as follows:

Category I Analytical methods for quantitation of major components of food additives or processing aids (including preservatives).

Category II Analytical methods for determination of impurities in food additives or processing aids. These methods include quantitative assays and limit tests.

Category III Analytical methods for determination of performance characteristics (e.g., solubility, melting point).

For each assay category, different analytical information is needed. Listed in Table 2 are data elements that are normally required for each assay category.

Already established general assays and tests (e.g., titrimetric method of water determination, identification test) should also be validated to verify their accuracy (and absence of possible interference) when used for a new product or raw material.

The validity of an analytical method can be verified only by laboratory studies. Therefore, documentation of the successful completion of such studies is a basic requirement for determining whether a method is suitable for its intended applications. Appropriate documentation should accompany any proposal for new or revised compendial analytical procedures.

Table 2. Data Elements Required for Assay Validation

Analytical Performance Parameter	Assay Category I	Assay Category II		Assay Category III
		Quantitative	Limit Tests	
Accuracy	Yes	Yes	*	*
Precision	Yes	Yes	No	Yes
Specificity	No	Yes	No	*
Limit of Detection	Yes	Yes	Yes	*
Limit of Quantitation	No	No	Yes	*
Linearity	Yes	Yes	No	*
Range	Yes	Yes	*	*
Ruggedness	Yes	Yes	Yes	Yes

*May be required, depending on the nature of the specific test.

GENERAL GOOD MANUFACTURING PRACTICES GUIDELINES FOR FOOD CHEMICALS[1]

Food chemicals and other substances employed as adjuncts in foods and as aids in food processing must meet recognized standards of performance and quality for their intended uses and applications. The requirements contained in the monographs of the FCC pertain to the characteristics of food chemical articles at the time of their use.

It is not sufficient, however, for an end product merely to meet the FCC requirements. Production of food-quality chemicals is best achieved by implementing procedures that place primary emphasis on preventing defects and deficiencies. Thus, a product must be made and handled in a sanitary manner, in a way designed either to preclude the formation of undesirable by-products, or to ensure their adequate removal, as well as to prevent contamination, deterioration, mix-up and mislabeling, and the introduction of unusual or unexpected impurities.

---

[1]These guidelines are presented for information only and are not intended to be mandatory in any sense as regards compliance with FCC specifications.

Food chemicals are subject to applicable regulations promulgated by the responsible government agencies in countries in which FCC specifications are recognized. In the United States, for example, the pertinent regulations, which deal primarily with sanitation, are "Current Good Manufacturing Practice in Manufacturing, Packing, or Holding Human Food."[2]

Beyond requirements related to sanitation, however, manufacturers, processors, packers, and distributors should establish and exercise other appropriate systems of controls throughout their operations. These controls, together with the regulations cited above, constitute "good manufacturing practice." While the details of the application of the principles of good manufacturing practice to the manufacturing, processing, packing, and distribution of food chemical substances will vary, the fundamental relevance of such principles at all stages of an operation should be recognized.

The principles of good manufacturing practice encompass such considerations as

- Systems of quality control and assurance, including self-auditing procedures.

- Clearly defined responsibilities of supervisory and other personnel, all of whom must be qualified and adequately trained.

- Design, operation and maintenance of buildings and equipment, with attention to housekeeping, sanitation, pest control, prevention of contamination of product, cleaning of equipment, a calibration program for all instruments and gauges, and environmentally satisfactory methods of waste disposal.

- Documentation of validation studies pertaining to the manufacturing process, laboratory test methods, and equipment and computer applications, when any such studies are appropriate.

- Written operational instructions that should include such items as
  — General instructions and hazards.
  — Master manufacturing instructions.
  — Master packaging instructions.
  — Master specifications for raw materials, in-process materials, packaging materials, labels, and finished products.
  — Laboratory test methods.
  — Control instrumentation and computer applications.
  — Labeling, holding, and distribution instructions.

- Handling and control, including the testing and approval, of raw materials, process aids, intermediates, and finished products.

- Product containers, closures, and labeling (including the control of labels and labeling).

- Laboratory and inspection controls and records (including the effect of process changes).

---

[2]*Code of Federal Regulations,* Title 21, Part 110, which may be obtained from the Superintendent of Documents, U.S. Government Printing Office, Washington, D.C. 20402. Also, Parts 113 and 114 are of interest, particularly with regard to record keeping.

- Reserve samples of raw materials and products.

- Written records that contain essential operational data for each individual lot of food chemical, and that permit tracing the lot history from the raw materials through manufacture, packaging, holding, and distribution of the product.

- Product stability and lifespan.

- Systems for holding, evaluating, and disposing of rejected products and returned materials.

- Procedures for investigating complaints and taking appropriate corrective action.

Some food chemicals have uses other than as food chemicals—in fact, the food-grade material may be only a small part of production for industrial or other uses. In such situations, the principles of good manufacturing practice, of necessity, must apply, and particular attention must be paid to the suitability of the raw materials used; prevention of cross-contamination; and the segregation of food chemicals from nonfood chemicals, including material in process, final product, and product in storage. The necessary controls to ensure the above must be developed and implemented.

> **Note:** Depending on the process involved, it frequently is possible to divert the food-grade product from the main product stream, as the final steps in arriving at the food-grade product are approached, and to complete the processing under conditions suitable for food-grade substances. In such cases, if the diverted material can be adequately characterized by a knowledge of its history, and/or by appropriate analytical testing, it may be considered to be the raw material for the food-grade product.

Biotechnology (processes involving the use of biological systems) is an important source of chemicals, enzymes, and other substances used in foods and in food processing. Some food ingredients have long been made by fermentation and by enzymatic processes, but now processes involving genetically modified organisms have become a prominent emerging source of such substances.

The manufacture of food chemicals, whether it involves chemical or biological synthesis and purification, or recovery from natural materials, has a number of characteristics that must be taken into account in establishing a system of good manufacturing practice. For example, in the production of many chemicals, recycling of process liquors and recovery from waste streams are necessary for reasons of quality, economics, and environmental protection. In addition, the production of some food chemicals involves processes in which chemical and biochemical mechanisms have not been fully elucidated, and thus the methods and procedures for materials accountability usually will differ from those applicable to the manufacture of other classes of materials.

Another aspect of good manufacturing practice for food chemicals relates to the possible presence of objectionable impurities. While the limits and tests provided in the FCC are consistent with the information available to the FCC concerning current methods of manufacture and common impurities that may be present, it obviously is impossible to provide limits and tests in each FCC monograph for the detection of all possible impurities, since these may vary with the raw materials and the method of processing used in making the chemical. Thus, to evaluate whether other undesirable impurities may be present, the manufacturer should understand to the best degree possible for the process at hand, the factors that contribute to the presence of impurities. Solvents, and impurities in the raw materials and processing aids, all of which might carry into the final product, need to be considered. In synthetic processes, it is necessary similarly to consider intermediates and the products of side reactions, as well as the possible formation of isomeric compounds, including epimers and enantiomorphs.

The same general considerations apply to biological processes, be they traditional or based on newer biotechnology. For products of biotechnology, a review that includes adequate characterization and documentation of the genetic origins of the starting materials, and the characteristics of the process, provides the necessary guidance for identifying and setting levels of undesirable impurities, which need to be controlled to suitable levels or to be absent altogether. Therefore, the active genetic components in the process (e.g., culture, recombinant DNA) should be known, well characterized, and free from any potential for introducing biologically significant levels of undesirable constituents (e.g., toxins, antibiotics, antinutrients) that cannot be kept out of the final product by preliminary processing. As in the production of all food chemicals, testing is appropriate when possible and particularly to demonstrate the absence of certain toxins and/or certain specific DNA sequences.

Because of the necessity to maintain the purity and integrity of the genetic materials associated with the biotechnological process, containment[3] is a particularly important consideration in preventing cross-contamination as well as the inadvertent release of biologically active materials.

Exposure of all products used in foods and food processing to foreign material contamination must be prevented. If objectionable impurities from any source, other than those covered by FCC requirements, are suspected to be present, good manufacturing practice requires that additional tests and limits be applied by the manufacturer to ensure that the substance is suitable for its intended applications as a food chemical. Current analytical technology should be applied wherever possible.

---

[3]See *NIH Guidelines for Research Involving Recombinant DNA Molecules, Federal Register*, Vol. 51, No. 88, pages 16957–16985, May 7, 1986. Copies, which include later revisions of the Guidelines, may be obtained from the Office of Recombinant DNA Activities, National Institutes of Health, Building 31, Room 4B11, Bethesda, MD 20892.

MONOGRAPHS ADDED TO THE *FOOD CHEMICALS CODEX*, THIRD EDITION, IN SUPPLEMENTS 1 THROUGH 4

Supplement 1:
Acid Hydrolyzed Proteins
Autolyzed Yeast Extract
Grape Skin Extract

Flavors
Allyl Heptanoate
Benzodihydropyrone
Butyl Isovalerate
Dibenzyl Ether
Ethyl Isobutyrate
Ethyl Myristate
n-Hexyl Acetate
4-(*p*-Hydroxyphenyl)-2-butanone

Supplement 2:
Annatto Extracts
Casein and Caseinate Salts
Coconut Oil (Unhydrogenated)
Corn Oil (Unhydrogenated)
Cottonseed Oil (Unhydrogenated)
F D & C Blue No. 1
F D & C Blue No. 2
F D & C Green No. 3
F D & C Red No. 3
F D & C Red No. 40
F D & C Yellow No. 5
F D & C Yellow No. 6
Hexanes
High-Fructose Corn Syrup
Invert Sugar
Lard (Unhydrogenated)
Palm Kernel Oil (Unhydrogenated)
Palm Oil (Unhydrogenated)
Peanut Oil (Unhydrogenated)
Polydextrose
Polydextrose Solution
Safflower Oil (Unhydrogenated)
Soybean Oil (Unhydrogenated)
Sunflower Oil (Unhydrogenated)
Tallow

Flavors
2-Acetyl Pyrazine
Amyl Heptanoate
Benzaldehyde Glyceryl Acetal
Butyl Phenylacetate
p-Cymene
2,6-Dimethyl-5-heptenal
Fusel Oil Refined
Isoamyl Benzoate
4-*p*-Methoxyphenyl-2-butanone
6-Methylcoumarin
Methyl Hexyl Ketone
δ-Nonalactone
δ-Octalactone

Supplement 3:
Butane
Calcium Sorbate
Canola Oil
Copper Sulfate
Gellan Gum
Glucose Syrup
Glucose Syrup, Dried
Glutaraldehyde
Glyceryl Behenate
Glyceryl Monostearate
Gum Ghatti
Helium
Isobutane
Konjac Flour
Maltodextrin
Natamycin
Nitrogen
Nitrogen Enriched Air
Nitrous Oxide
Ox Bile Extract
Ozone
Potassium Benzoate
Potassium Lactate Solution
Propane
Rapeseed Oil, Fully Hydrogenated
Rapeseed Oil, Superglycerinated
Sodium Lactate Solution

Sodium Magnesium Aluminosilicate
Starter Distillate
Sucrose
Urea
Zein

Flavors
Aconitic Acid
Anisyl Formate
Benzyl Formate
Butyl Stearate
D-Camphor
L-Carveol
L-Carvyl Acetate
Cinnamyl Butyrate
Cinnamyl Cinnamate
Cinnamyl Isobutyrate
γ-Decalactone
Dimethyl Sulfide
γ-Dodecalactone
Ethylene Brassylate
4-Ethylguaiacol
Ethyl-3-Methylthiopropionate
Ethyl Oleate
Glyceryl Tripropanoate
γ-Hexalactone
(*Z*)-3-Hexanyl Acetate
(*E*)-2-Hexenyl Acetate
4-Hydroxy-2,5-dimethyl-3(2H)furanone
Isoborneol
2-Methylpentanoic Acid
2-Methyl-2-pentenoic Acid
4-Methylpentanoic Acid
3-Methylthiopropionaldehyde
Myristaldehyde
Neryl Acetate
2-Nonanone
1-Octene-3-ol
Octyl Isobutyrate
Phenylethyl Anthranilate
Phenylethyl Butyrate
Piperidine
Tetrahydrofurfuryl Alcohol
Thymol

Tolualdehyde
para-Tolualdehyde
Trimethylamine
2-Undecanone
Valeraldehyde

Supplement 4:
Ammoniated Glycyrrhizin
Cocoa Butter Substitute
Enzyme-Modified Milkfat

Lactose
Malt Syrup
Monoammonium
 Glycyrrhizinate
Morpholine
Mustard Oil
Nickel
Nisin Preparation
Sodium Metasilicate
Sucralose

Wheat Gluten

Flavors
Allyl Isovalerate
Ethyl 10-Undecenoate
2-Mercaptopropionic Acid
Methyl-3-methylthio-
 propionate
Myristyl Alcohol
δ-Undecalactone

NEW MONOGRAPHS IN *FOOD CHEMICALS CODEX*, FOURTH EDITION

Acesulfame Potassium
Bentonite
Calcium Acid Pyrophosphate
Carbon Dioxide
Dammar Gum
Ferrous Lactate
Food Starch, Unmodified
Furcelleran
Gelatin
Glycerol Ester of Gum Rosin
Glycerol Ester of Partially
 Hydrogenated Gum Rosin
Glyceryl Monooleate
Glyceryl Tristearate
Magnesium Gluconate
Maltitol Syrup
Partially Hydrolyzed Proteins
Sodium Potassium
 Tripolyphosphate
Whey

Dried Yeast

Flavors
1-Amyl Alcohol
Amyl Butyrate
Amyl Formate
Cyclohexyl Acetate
1,2-Di[(1′-ethoxy)-
 ethoxy]propane
Dimethyl Succinate
Ethone
Ethyl Benzoyl Acetate
Ethyl-(*E*)-2-butenoate
2-Ethyl Hexanol
Ethyl Levulinate
Ethyl-2-methylpentanoate
Ethyl Valerate
Isoamyl Alcohol
Isoamyl Phenyl Acetate
Levulinic Acid

Methyl Acetate
2-Methyl Butanal
3-Methyl Butanal
2-Methylbutyl Acetate
Methyl Butyrate
2-Methylbutyric Acid
Methyl Ionones
Methyl Isobutyrate
Methyl Propyl 3-Methyl
 Butyrate
4-Methyl-5-thiazole Ethanol
Nonanoic Acid
Propyl Acetate
Propyl Alcohol
Propyl Propionate
Terpinen-4-ol
3,5,5,-Trimethyl Hexanal
Zingerone

FORMER AND CURRENT TITLES OF *FOOD CHEMICALS CODEX* MONOGRAPHS

Third Edition Title	*Fourth Edition Title*
Acid Hydrolyzed Proteins	Acid Hydrolysates of Proteins
Autolyzed Yeast Extract	Yeast Extract
Calcium Chloride, Anhydrous	Calcium Chloride
Lactated Mono-Diglycerides	Glyceryl-Lacto Esters of Fatty Acids
Cellulose, Microcrystalline	Cellulose Gel
PVP	Polyvinylpyrrolidone

MONOGRAPHS DELETED

Carrageenan
Cinnamyl Anthranilate
Methyl Formate

FOOD CHEMICALS CODEX

# 1 / General Provisions and Requirements Applying to Specifications, Tests, and Assays of the Food Chemicals Codex

The General Provisions provide in summary form the basic policies and guidelines for the interpretation and application of the standards, tests, assays, and other specifications of the *Food Chemicals Codex*, and make it unnecessary to repeat throughout the book those requirements that are pertinent in numerous instances.

Where exceptions to the General Provisions are made, the wording in the individual monograph or general test section takes precedence and specifically indicates the directions or the intent.

TITLE OF BOOK

The title of this book, including supplements thereto issued separately, is the *Food Chemicals Codex*, fourth edition. It may be abbreviated to FCC IV.

Where the term "Codex" is used without further qualification in the text of this book, it applies to the *Food Chemicals Codex*, fourth edition.

INQUIRIES

Inquiries regarding any aspect of the operation of the Food Chemicals Codex may be directed to the Food Chemicals Codex, Institute of Medicine, National Academy of Sciences, 2101 Constitution Avenue, N.W., Washington, D.C. 20418.

CODEX SPECIFICATIONS

Food Chemicals Codex specifications, comprising the *Description*, *Requirements*, and *Tests*, are presented in monograph form (*Section 2*) or tabular form (*Section 3*) for each individual substance or group of substances and are designed to serve for ingredients of a quality level sufficiently high to ensure their safety under usual conditions of intentional use in foods, both directly or indirectly, or in food processing. Thus, FCC specifications generally represent acceptable levels of quality and purity of food-grade ingredients available in the United States (or in other countries in which FCC specifications are recognized).

The titles of FCC monographs are in most instances the common or usual names. The FCC specifications apply equally to substances bearing the main titles, or synonyms listed under the main titles, or names derived by transposition of definitive words in main titles.

The assays and tests described constitute methods upon which the specifications of the *Food Chemicals Codex* depend. The analyst is not prevented, however, from applying alternative methods if he or she is satisfied that the procedures used will produce results of equal or greater accuracy. In the event of doubt or disagreement concerning a substance purported to comply with the requirements of this Codex, only the methods described herein are applicable and authoritative.

POLICIES AND GUIDELINES

General Policy It shall be the policy of the Codex to set maximum limits for trace impurities wherever they are deemed to be important for a particular chemical, and they shall be set at levels consistent with safety and good manufacturing practice.

Maximum limits for inorganic trace impurities (e.g., arsenic, fluoride, heavy metals, lead, mercury, selenium) will be included in any monographs where safety or manufacturing experience indicates their desirability.

Where, in a given monograph, a heavy metals limit between 5 and 10 mg/kg has been established and no separate lead limit is included, the lead limit, *de facto*, is the same as the heavy metals limit.

All requests to increase limits shall be considered on the basis of the toxicological risk involved, the principles of good manufacturing practice, and the availability from other sources of substances meeting the FCC limits that are requested to be increased.

Added Substances (Policy) FCC specifications are intended for application to individual substances (single entities) and not to proprietary blends or other mixtures. Some specifications, however, provide for "added substances" (i.e., functional secondary ingredients such as anti-caking agents, antioxidants, diluents, emulsifiers, and preservatives) intentionally added when necessary to ensure the integrity, stability, utility, or functionality of the primary substance in commercial use.

When an FCC monograph provides for such additions, the added substance(s) must meet the following requirements: (a) it is approved for use in foods by the U.S. Food and Drug Administration, or by the responsible government agency in other countries in which FCC specifications are recognized; (b) it is of appropriate food-grade quality and meets the requirements of the *Food Chemicals Codex*, if listed therein; (c) it is used in an amount not to exceed the minimum required to impart its intended technical effect or function in the primary substance; (d) its use will not result in concentrations exceeding permitted levels in any food as a consequence of the subsequent use in foods of the FCC primary substance(s) to which it has been added; and (e) it does not interfere with the assay and tests prescribed for determining compliance with the FCC requirements for the primary substance, unless the monograph for the primary substance has provided for such interferences.

Where added substances are specifically permitted in an FCC substance, the label shall state the name(s) and amount(s) of any added substance(s).

An FCC substance to which are added substances not specifically provided for and mentioned by name or function in its monograph shall not be designated as an FCC substance. Such a combination is a mixture to be described by disclosure of its ingredients, including any that are not FCC substances.

Arsenic Specifications (Policy) The General Policy for arsenic specifications in *Food Chemicals Codex* monographs is that such specifications will be included in monographs only when there is specific reason for the committee to believe that arsenic constitutes a significant part of the heavy metals limit.

Flavor Chemicals (Policy) (a) For volatile oils prepared by distillation, the only requirement is a simple test for heavy metals using a 1:1 acidified mixture of the oil and water, through which hydrogen sulfide is passed. The oil passes the test, which is sensitive to 10 mg/kg of Pb, if there is no darkening in color in either the oil or the water. (When this test is required, it will be included in the applicable monograph, with a limit of "Passes test.") (b) For cold-pressed oils (i.e., those not purified by distillation), the limits for arsenic, heavy metals, and lead shall be consistent with the established policies for these impurities, taking into account the usage levels of the oils. However, when they are used in foods at levels of 0.01% or less, the heavy metals limit shall be as low as possible, but not greater than 0.004%, and the requirements for arsenic and lead may be omitted. (c) For flavor chemicals that are liquids at or near room temperature and that are prepared and/or purified by distillation, no limits for arsenic, heavy metals, or lead are required, but for flavor chemicals that are crystalline materials, or for other solids not prepared by distillation, the limits for arsenic, heavy metals, and lead shall be consistent with the established policies for these impurities, taking into account the usage levels of the chemicals. However, when they are used in food at levels of 0.01% or less, the heavy metals limit shall be as low as possible, but not greater then 0.004%, and the requirements for arsenic and lead may be omitted.

Fluoride Limits (Guideline) These guidelines were established because of frequent requests by manufacturers to increase the fluoride limits for certain substances, specifically phosphates, and compounds containing calcium and magnesium. In developing these guidelines, it was recognized that the uncontrolled addition of fluoride to the diet via food additives is inadvisable, especially in view of the widely variable consumption of fluoride-containing food additives and of the highly irregular dietary intake of fluoride from sources other than processed foods. (a) Fluoride limits shall be as low as possible, and shall exceed 0.005% only under the most unusual circumstances. (b) Restraint shall be exercised in increasing existing fluoride requirements (up to a possible maximum of 0.005%), and requests to increase such limits will not be granted unless the increases are judged to be essential. (c) All requests to increase fluoride limits will be considered in light of available information concerning the amounts of fluoride contributed to the total dietary intake by public water supplies, by foods containing fluorides, and by special food products (e.g., fish protein concentrate), as well as by processed foods to which fluoride-containing substances are added intentionally. (d) Decisions to grant or deny such requests shall be based upon the toxicological risk involved, upon the principles of good manufacturing practice, and upon the availability from other sources of substances meeting the Codex limits current at the time of the request.

FCC Substances Containing Sulfur Dioxide (Policy) If an FCC substance is found to contain 10 mg/kg or more of sulfur dioxide (SO_2), the presence of sulfur dioxide shall be indicated on the labeling.

Labeling (Policy) For purposes of compliance with *Food Chemicals Codex* monographs, "labeling" means all labels and other written, printed, or graphic matter (1) upon any article or any of its containers or wrappers or (2) accompanying such article, or otherwise provided by vendors to purchasers for purposes of product identification.

Substances included in the *Food Chemicals Codex* are subject to compliance with such labeling requirements as may be promulgated by government bodies, in addition to the FCC requirements. Such substances are intended for use in foods, either directly or indirectly, and in food processing. The name of the substance on a container label, plus the designation "Food Chemicals Codex Grade," "FCC Grade," or simply "FCC," is a representation by the manufacturer, vendor, or user of the substance that at the time of shipment, it conforms to the specifications in FCC IV, including its supplements, that are current at that time.

When an FCC substance is available commercially in solution form or as a component of a mixture, and there is no provision in the *Food Chemicals Codex* for such solution or mixture, the manufacturer, vendor, or user may indicate on the label that the product contains substances meeting FCC specifications by use of the initials "FCC" after the name of the component(s) that meets the FCC requirements.

For the labeling of FCC substances in which added substances are permitted, see *Added Substances (Policy)*, above.

For the labeling of FCC substances that contain 10 mg/kg or more of sulfur dioxide, see *FCC Substances Containing Sulfur Dioxide (Policy)*, above.

Lead (and Heavy Metals) Limits (Policy) The Committee on Food Chemicals Codex recognizes the desirability of lowering lead exposure, especially in the case of infants and children. Overall exposure to lead is a public health concern. Whereas diet is not the largest source of lead exposure, it is a significant one. While ingestion of FCC substances does not represent the major source of dietary lead, it is desirable to lower the lead limits for all FCC substances, particularly for those substances consumed in high amounts. Therefore, the committee's policy is to reduce lead limits (as well as heavy metals limits because of their interrelated nature) to the lowest extent feasible for FCC substances, especially given that more recent evidence shows deleterious neurobehavioral effects occurring in children exposed to lead at levels below those previously considered acceptable.

In setting heavy metals and lead limits, the committee considers the amount of a food chemical consumed, the feasibility of manufacturing a product within these limits, and the availability of analytical methods to ensure compliance. The constraints of good manufacturing practice and the availability of reliable analytical methods are often limiting factors in setting lower lead and heavy metals limits.

The committee regards as one of its goals the assurance of the safety of properly used food chemicals. This means that FCC specifications will respond to advances in knowledge about new manufacturing methods, analytical techniques, or toxicology and safety issues.

Microbiological Attributes (Policy) Manufacturers, vendors, and/or users of FCC substances are expected to exercise good manufacturing practices (see page xxvii) and to establish microbiological controls in production processes as necessary to ensure that FCC substances are not contaminated with pathogenic or other objectional organisms or unwanted microbial metabolites and that the FCC substance is otherwise suitable for its intended use. Where the FCC recognizes a specific need for microbiological limits for an individual substance, such a requirement will be included in the monograph.

Among the criteria considered by the committee, on a case-by-case basis, for developing microbial limits for food chemicals, are the following:

• the origin of the food chemical (plant, animal, microbially derived via fermentation, and natural mineral sources);

• evidence of health hazard or potential hazard based on epidemiological data, hazard analysis, or known user-specific populations that may be at risk;

• the nature of the natural and commonly acquired microflora of the food chemical and the ability of the food chemical to support its growth;

• the effect of processing on the microflora of the food chemical;

• the potential for microbial contamination and/or growth in a food chemical during its procurement, processing, handling, storage, and distribution;

• the state in which the food chemical is packaged, stored, and distributed (e.g., frozen, refrigerated, heat processed, etc.); and

• the potential for direct consumer use.

Any one of several of these criteria will act to flag the substance for consideration for the establishment of microbiological specifications. However, the expertise of the committee, and the expertise available to it from other sources, as well as the experience of producers and users, may provide additional input for the committee to consider in arriving at a final decision.

Mg/Kg and Percent (Policy) Beginning with the Second Supplement to FCC III, in order to bring the Codex into concordance with current nomenclature, as used in analytical chemistry, the term "ppm" [parts per million (by weight)] was replaced by "mg/kg" (milligrams per kilogram).

The term "mg/kg" is used for expressing the concentrations of trace amounts of substances, such as impurities, up to 10 mg/kg. Above 10 mg/kg, percent (by weight) is used. For example, a monograph requirement equivalent to 20 mg/kg is expressed as 0.002%, or 0.0020%, depending on the number of significant figures justified by the test specified for use in conjunction with the requirement.

ATOMIC WEIGHTS AND CHEMICAL FORMULAS

The atomic weights used in computing formula weights and volumetric and gravimetric factors stated in tests and assays are those recommended in 1991 by the IUPAC Commission on Atomic Weights and Isotopic Abundances.

Molecular and structural formulas and formula weights immediately following titles are included for the purpose of information and are not to be considered an indication of the purity of the substance. Molecular formulas given in specifications, tests, and assays, however, denote the pure chemical entity.

ASSAYS AND TESTS

Every FCC substance in commerce, when tested in accord with these assays and tests, meets all of the requirements in the monograph defining it.

Analytical Samples In the description of assays and tests, the approximate quantity of the analytical sample to be used is usually indicated. The quantity actually used, however, should not deviate by more than 10% from that stated.

Some substances are directed to be dried before a sample is taken for an assay or test. When a *Loss on Drying* or *Water* test is specified, the undried substance may be used and the results calculated on the dried basis, provided that any moisture or other volatile matter in the undried sample does not interfere with the specified assay and test procedures.

The word "accurately," used in connection with gravimetric or volumetric measurements, means that the operation should be carried out within the limits of error prescribed under *Volumetric Apparatus* or under *Weights and Balances*, Appendix I. The same significance also applies to the term "exactly" or quantitative expressions such as "100.0 mL" or "50.0 mg."

The word "transfer," when used in describing assays and tests, means that the procedure should be carried out quantitatively.

Apparatus With the exception of volumetric flasks and other exact measuring or weighing devices, directions to use a definite size or type container or other laboratory apparatus are intended only as recommendations, unless otherwise specified.

Where an instrument for physical measurement, such as a thermometer, spectrophotometer, gas chromatograph, etc., is designated by its distinctive name or tradename in a test or assay, a similar instrument of equivalent or greater sensitivity or accuracy may be employed.

Where low-actinic or light-resistant containers are specified, clear glass containers that have been rendered opaque by application of a suitable coating or wrapping may be used.

Blank Tests Where a blank determination is specified in a test or assay, it is to be conducted by using the same quantities of the same reagents and by the same procedure repeated in every detail except that the substance being tested is omitted.

A *residual blank titration* may be stipulated in assays and tests involving a back titration in which a volume of a volumetric solution larger than is required to react with the sample is added, and the excess of this solution is then titrated with a second volumetric solution. Where a residual blank titration is specified or where the procedure involves such a titration, a blank is run as directed in the preceding paragraph. The volume of the titrant consumed in the back titration is then subtracted from the volume required for the blank. The difference between the two, equivalent to the actual volume consumed by the sample, is the corrected volume of the volumetric solution to be used in calculating the quantity of the substance being determined.

Centrifuge Where the use of a centrifuge is indicated, unless otherwise specified, the directions are predicated upon the use of apparatus having an effective radius of about 20 cm (8 in.) and driven at a speed sufficient to clarify the supernatant layer within 15 min.

Desiccators and Desiccants The expression "in a desiccator" means using a tightly closed container of appropriate design in which a low moisture content can be maintained by means of a suitable desiccant. Preferred desiccants include anhydrous calcium chloride, magnesium perchlorate, phosphorus pentoxide, and silica gel.

Filtration Where it is directed to "filter," without further qualification, the intent is that the liquid be filtered through suitable filter paper or an equivalent device until the filtrate is clear.

Identification The tests described under this heading in monographs are designed for application to substances taken from labeled containers and are provided only as an aid to substantiate identification. These tests, regardless of their specificity, are not necessarily sufficient to establish proof of identity, but failure of a substance taken from a labeled container to meet the requirements of a prescribed identification test means that it does not conform to the requirements of the monograph.

Indicators The quantity of an indicator solution used should be 0.2 mL (approximately 3 drops) unless otherwise directed in an assay or test.

Negligible The term "negligible," as used in some *Residue on Ignition* specifications, indicates a quantity not exceeding 0.5 mg.

Odorless This term, when used in describing a substance, applies to the examination, after exposure to air for 15 min, of about 25 g of the substance that has been transferred quickly from the original container to an open evaporating dish of about 100-mL capacity. If the package contains 25 g or less, the entire contents should be examined.

Pressure Measurements The term "mm of mercury" used with respect to pressure within an apparatus, or atmospheric pressure, refers to the use of a suitable manometer or barometer calibrated in terms of the pressure exerted by a column of mercury of the stated height.

Reagents Specifications for reagents are not included in the *Food Chemicals Codex*. Unless otherwise specified, reagents required in tests and assays should conform to the specifications of the current edition of *Reagent Chemicals—American Chemical Society Specifications* or in the section on Reagent Specifications in the *United States Pharmacopeia*. Reagents not covered by any of these specifications should be of a grade suitable to the proper performance of the method of assay or test involved.

Acids and Ammonium Hydroxide Where ammonium hydroxide, glacial acetic acid, hydrochloric acid, hydrofluoric acid, nitric acid, phosphoric acid, or sulfuric acid are called for in tests and assays, reagents of ACS grade and strengths are to be used. (These reagents sometimes are called "concentrated," but this term is not used in the *Food Chemicals Codex*.)

Alcohol, Ethyl Alcohol, Ethanol Where one of the foregoing is called for in tests and assays, ACS-grade Ethyl Alcohol (95%) is to be used.

Alcohol Absolute, Anhydrous Alcohol, Dehydrated Alcohol Where one of the foregoing is called for in tests and assays, ACS-grade Ethyl Alcohol, Absolute, is to be used.

Water Where water is called for in tests and assays and in the preparation of solutions, it shall have been prepared by distillation, ion-exchange treatment, or reverse osmosis. ACS Reagent Water and Purified Water USP meet these requirements.

Water, Carbon Dioxide-Free Where this type of water is called for, it shall have been boiled vigorously for 5 min or more, and allowed to cool while protected from absorption of carbon dioxide from the atmosphere. "Deaerated water" is water that has been treated to reduce the content of dissolved air by suitable means, such as by boiling vigorously for 5 min and cooling or by the application of ultrasonic vibration.

Note: It is recognized that certain chemical reagents specified in FCC test procedures may be considered to be hazardous or toxic by the Occupational Safety and Health Administration, by the Environmental Protection Agency (under provisions of the Toxic Substances Control Act), or by health authorities in other countries in which the *Food Chemicals Codex* is recognized. In preparing this edition, the Committee on Food Chemicals Codex has attempted to specify use of different reagents where suitable substitutes are known. For some procedures, however, the original chemicals have been retained due to the lack of information on suitable substitutes. In such cases, the analyst is encouraged to investigate the use of suitable substitute reagents, as appropriate, and to inform the committee of the results so obtained.

The methods and analytical procedures described in the Codex are designed to be carried out by properly trained personnel in a suitably equipped laboratory. In common with many laboratory procedures, the methods quoted frequently involve hazardous materials.

In performing the assay or test procedures in the Codex, it is expected that safe laboratory practices will be followed. This includes the utilization of precautionary measures, protective equipment, and work practices consistent with the chemicals and procedures utilized. Prior to undertaking any assay or procedures described in this compendium, the individual should be aware of the hazards associated with the chemicals and the procedures and means of protecting against them. Material Safety Data Sheets, which contain precautionary information related to safety and health concerns, are available from manufacturers and distributors of many chemicals, and should provide helpful information about such chemicals.

Reference Standards Some instrumental and chromatographic tests and assays specify the use of a reference standard. Where a reference standard is designated as FCC or USP, it may be obtained from the United States Pharmacopeial Convention, Inc., 12601 Twinbrook Parkway, Rockville, MD 20852. Where a reference standard is designated as a NIST Standard Reference Material, it may be obtained from the National Institute of Standards and Technology, Office of Standard Reference Materials, Building 202, Room 204, Gaithersburg, MD 20899.

To serve its intended purpose, each reference standard must be properly stored, handled, and used. Generally, reference standards should be stored in their original containers away from heat and protected from light. Follow any special instructions accompanying the containers.

Assay and test results are determined on the basis of comparison of the test sample with the reference standard that has been freed from or corrected for volatile residues or water content as instructed on the reference standard label. Where a reference standard is required to be dried before using, transfer a sufficient amount to a clean, dry vessel. Do not use the original container as the drying vessel, and do not dry a reference standard repeatedly at temperatures above 25°. Where the titrimetric determination of water is required at the time a reference standard is to be used, proceed as directed for the *Karl Fischer Titrimetric Method,* Appendix IIB.

Unless a reference standard label bears a specific potency or content, the reference standard is taken as being 100.0% pure.

Significant Figures Where tolerance limits are expressed numerically, the values are considered to be significant to the number of digits indicated. The observed or calculated analytical result is recorded, with only one digit included in the decimal place to the right of the last place in the limit expression. If this digit is smaller than 5, it is eliminated and the preceding digit is unchanged. If this digit is greater than 5, it is eliminated and the preceding digit is increased by one. If this digit equals 5, the 5 is eliminated, and the preceding digit is increased by one. For example, a requirement of not less than 96.0% would not be met by a result of 95.94%, but would be met by results of 95.96% or 95.95%, both of which would be rounded to 96.0%. When a range is stated, the upper and lower limits are inclusive so that the range consists of the two values themselves, properly rounded, and all intermediate values between them.

Solutions All solutions, unless otherwise specified, are to be prepared with *ACS Reagent Water or Purified Water USP* (See *Water,* under Reagents).

Such expressions as "1 in 10" or "10%" mean that *1 part by volume* of a liquid or *1 part by weight* of a solid is to be dissolved in a volume of the diluent or solvent sufficient to make the finished solution 10 parts by volume. Directions for the preparation of colorimetric solutions (CS), test solutions (TS), and volumetric solutions (VS), are provided in the section on *Solutions and Indicators.*

A volumetric solution should be prepared to have a normality (molarity) within 10% of the stated value and should be standardized to four significant figures. When volumetric equivalence factors are provided in tests and assays, the term "0.X N (M)" is understood to mean a VS having a normality (molarity) of exactly 0.X000 N (M). If the normality (molarity) of the VS employed in a particular procedure differs from 0.X000, an appropriate correction factor must be applied.

Specific Gravity Numerical values for specific gravity, unless otherwise noted, refer to the ratio of the weight of a substance in air at 25° to that of an equal volume of water at the same temperature. Specific gravity may be determined by any reliable method.

Temperatures Unless otherwise specified, temperatures are expressed in centigrade (Celsius) degrees, and all measurements are to be made at 25°.

Test Solutions See *Solutions and Indicators.*

Time Limits Unless otherwise specified, 5 min is to be allowed for a reaction to take place in conducting limit tests for trace impurities such as chloride, iron, etc.

Expressions such as "exactly 5 min" mean that the stated period should be accurately timed.

Tolerances The minimum purity tolerances specified for FCC items have been established with the expectation that the substances to which they apply will be used as direct or indirect food additives, ingredients, or food-processing aids. These tolerance limits should neither bar the use of lots of articles that more nearly approach 100% purity nor constitute a basis for a claim that such lots exceed the quality prescribed by the *Food Chemicals Codex*.

When a maximum assay tolerance is not given, the assay should show the equivalent of not more than 100.5%.

Trace Impurities Tests for inherent trace impurities are provided to limit such substances to levels consistent with good manufacturing practice and that are safe and otherwise unobjectionable under conditions in which the food additive or ingredient is customarily employed.

It obviously is impossible to provide limits and tests in each monograph for the detection of all possible unusual or unexpected impurities, the presence of which would be inconsistent with good manufacturing practice. The limits and tests provided are those considered to be necessary according to currently recognized methods of manufacture and are based on information available to or provided to the Committee on Food Chemicals Codex. If other methods of manufacture or other than the usual raw materials are used, or if other possible impurities may be present, additional tests may be required and should be applied, as necessary, by the manufacturer, vendor, or user to demonstrate that the substance is suitable for its intended application.

In instances where a lead limit is specified in a monograph at levels not less than 5 mg/kg and not greater than 10 mg/kg, and the heavy metals content is found to be equal to, or less than, the lead limit, the lead content need not be determined.

Vacuum The unqualified use of the term "in vacuum" means a pressure at least as low as that obtainable by an efficient aspirating water pump (not higher than 20 mm of Hg).

Water and Loss on Drying In general, a limit test, to be determined by the *Karl Fischer Titrimetric Method*, is provided under the heading *Water* for compounds containing water of crystallization or adsorbed water. Limit tests under the heading *Loss on Drying*, determined by other methods, are designed for compounds in which the loss on drying may not necessarily be attributable to water.

Weighing Practices

Constant Weight A direction that a substance is to be "dried to constant weight" means that the drying should be continued until two consecutive weighings differ by not more than 0.5 mg/g of sample taken, the second weighing to follow an additional h of drying.

The direction "ignite to constant weight" means that the ignition should be continued at 800° ± 25°, unless otherwise specified, until two consecutive weighings do not differ by more than 0.5 mg/g of sample taken, the second weighing to follow an additional 15-min ignition period.

Tared Container When a tared container, such as a glass filtering crucible, a porcelain crucible, or a platinum dish, is called for in an analytical procedure, it shall be treated in the same manner as is specified after it has been used in the procedure. For example, dried or ignited for a specified time, or to constant weight, cooled in a desiccator as necessary, and weighed accurately.

Weights and Measures, Symbols and Abbreviations The International System of Units (SI), to the extent possible, is used in most specifications, assays, and tests in this *Food Chemicals Codex*. The SI metric units, and other units and abbreviations commonly employed, are as follows:

kg	= kilogram
g	= gram
mg	= milligram
μg	= microgram
ng	= nanogram
pg	= picogram
L	= liter
mL	= milliliter
μL	= microliter
m	= meter
cm	= centimeter
dm	= decimeter
mm	= millimeter
μm	= micrometer (0.001 mm)
nm	= nanometer
C	= coulomb
A	= ampere
V	= volt
mV	= millivolt
W	= watt
dc	= direct current
ft	= foot
in.	= inch
in.^{3}	= cubic inch
gal	= gallon
lb	= pound
oz	= ounce
ppm	= parts per million (10^6) parts
mg/kg	= parts per million (by weight)
ppb	= parts per billion (10^9) parts
ng/kg	= parts per billion (by weight)
psi	= pounds per square inch
sp. gr.	= specific gravity
b.p.	= boiling point
m.p.	= melting point
id	= inside diameter
od	= outside diameter
h	= hour
min	= minute
s	= second

% = percent
N = normality
M = molarity
μ*M* = micromolar
μmol = micromole
CFU = colony-forming unit(s)

GENERAL SPECIFICATIONS AND STATEMENTS

Certain specifications and statements in the monographs of the *Food Chemicals Codex* are not amenable to precise description and accurate determination within narrow limiting ranges. Because of the subjective or general nature of these specifications, good judgment, based upon experience, must be used in interpreting and attaching significance to them. Specifications or statements that are most likely to cause doubt are discussed in the subsequent paragraphs.

Description The material given under this heading in monographs is provided for general information. It includes a description of physical characteristics such as color, odor, taste, form, etc., and information on stability under certain conditions of exposure to air and light. Statements in this section may also cover approximate indications of properties such as solubility (see below) in various solvents, pH, melting point, and boiling point, with numerical values modified by "about," "approximately," "usually," and other comparable nonspecific terms. These characteristics and statements are not requirements, but are provided as information that may assist with the overall evaluation of a food chemical.

Solubility Statements included in the *Requirements* section in monographs under headings such as *Solubility in Alcohol* express exact requirements and constitute quality specifications.

Statements relating to solubility given under the heading *Description*, however, are intended as information regarding approximate solubilities only and are not to be considered as Codex-quality requirements. Such statements are considered to be of minor significance as a means of identification or determination of purity. For those purposes, dependence must be placed upon other specifications.

Approximate solubilities are indicated by the following descriptive terms:

Descriptive Term	Parts of Solvent Required for 1 Part of Solute
Very Soluble	less than 1
Freely Soluble	from 1 to 10
Soluble	from 10 to 30
Sparingly Soluble	from 30 to 100
Slightly Soluble	from 100 to 1000
Very Slightly Soluble	from 1000 to 10,000
Practically Insoluble or Insoluble	more than 10,000

Soluble substances, when brought into solution, may show slight physical impurities, such as fragments of filter paper, fibers, and dust particles, unless excluded by definite tests or other requirements; however, it is a requirement that significant amounts of black specks, metallic chips, glass fragments, or other insoluble matter are not permitted.

Functional Use in Foods A statement of functional classification is provided in each monograph as useful information to indicate the principal applications or technical effect of the substance in foods or in food processing. The statement is not intended to limit in any way the choice or use of the substance or to indicate that it has no other utility.

Packaging and Storage Statements in monographs relating to packaging are advisory in character and are intended only as general information to emphasize instances where deterioration may be accelerated under adverse packaging and storage conditions, such as exposure to air, light, or extremes of temperature, or where safety hazards are involved.

Cool Place A cool place is one where the temperature is between 8° and 15° (46° and 59°F). Alternatively, it may be a refrigerator, unless otherwise specified in the original monograph.

Excessive Heat Any temperature above 40° (104°F).

Storage Under Nonspecific Conditions Where no specific storage directions or limitations are provided in the individual monograph, it is to be understood that the conditions of storage and distribution include protection from moisture, freezing, and excessive heat.

Containers The container is the device that holds the substance and that is or may be in direct contact with it. The immediate container is in direct contact with the substance at all times. The closure is a part of the container.

The container should not interact physically or chemically with the material that it holds so as to alter its strength, quality, or purity, and the food (additive) contact surface of the container should comply with the food additive regulations promulgated under the Food, Drug, and Cosmetic Act (or with applicable laws and regulations in other countries in which FCC specifications are recognized).

Light-Resistant Container A light-resistant container is designed to prevent deterioration of the contents beyond the prescribed limits of strength, quality, or purity under the ordinary or customary conditions of handling, shipment, storage, and sale. A colorless container may be made light-resistant by enclosing it in an opaque carton or wrapper (see also *Apparatus*, above).

Well-Closed Container A well-closed container protects the contents from extraneous solids and from loss of the chemical under the ordinary or customary conditions of handling, shipment, storage, and sale.

Tight Container A tight container protects the contents from contamination by extraneous liquids, solids, or vapors, from loss of the chemical, and from efflorescence, deliquescence, or evaporation under the ordinary or customary conditions of handling, shipment, storage, and sale, and is capable of tight reclosure.

2 / Monograph Specifications

Acacia

Gum Arabic

INS: 414 CAS: [9000-01-5]

DESCRIPTION

A dried gummy exudation obtained from the stems and branches of *Acacia senegal* (L.) Willd. or of related species of *Acacia* (Fam. *Leguminosae*). Unground Acacia occurs as white or yellowish white spheroidal tears of varying size or in angular fragments. It is also available commercially in the form of white to yellowish white flakes, granules, or powder. One g dissolves in 2 mL of water, forming a solution that flows readily and is acid to litmus. It is insoluble in alcohol.

Functional Use in Foods Stabilizer; thickener; emulsifier.

REQUIREMENTS

Identification To 10 mL of a cold 1 in 50 solution of Acacia add 0.2 mL of diluted lead subacetate TS. A flocculent, or curdy, white precipitate is formed immediately.
Arsenic (as As) Not more than 3 mg/kg.
Ash (Acid-Insoluble) Not more than 0.5%.
Ash (Total) Not more than 4.0%.
Heavy Metals (as Pb) Not more than 0.002%.
Insoluble Matter Not more than 1.0%.
Lead Not more than 5 mg/kg.
Loss on Drying Not more than 15.0%.
Starch or Dextrin Passes test.
Tannin-Bearing Gums Passes test.

TESTS

Arsenic A *Sample Solution* prepared as directed for organic compounds meets the requirements of the *Arsenic Test*, Appendix IIIB.
Ash (Acid-Insoluble) Determine as directed in the general method, Appendix IIC.
Ash (Total) Determine as directed in the general method, Appendix IIC.
Heavy Metals Prepare and test a 1-g sample as directed in *Method II* under the *Heavy Metals Test*, Appendix IIIB, using 20 µg of lead ion (Pb) in the control (*Solution A*).
Insoluble Matter Dissolve a 5-g sample in about 100 mL of water contained in a 250-mL Erlenmeyer flask, add 10 mL of 2.7 *N* hydrochloric acid, and boil gently for 15 min. Filter the hot solution by suction through a tared filtering crucible, wash thoroughly with hot water, dry at 105° for 2 h, and weigh.
Lead A *Sample Solution* prepared as directed for organic compounds meets the requirements of the *Lead Limit Test*, Appendix IIIB, using 5 µg of lead ion (Pb) in the control.
Loss on Drying, Appendix IIC Dry at 105° for 5 h. Unground samples should be powdered to pass through a No. 40 sieve and mixed well before weighing.
Starch or Dextrin Boil a 1 in 50 solution, cool, and add a few drops of iodine TS. No bluish or reddish color is produced.
Tannin-Bearing Gums To 10 mL of a 1 in 50 solution add about 0.1 mL of ferric chloride TS. No blackish coloration or blackish precipitate is formed.

Packaging and Storage Store in well-closed containers.

Acesulfame Potassium

Acesulfame K; 6-Methyl-1,2,3-oxathiazine-4(3H)-one-2,2 Dioxide; Potassium Salt

$$\text{CH}_3\text{-C=CH-C(=O)-N(K)-SO}_2\text{-O}$$ (ring structure)

$C_4H_4KNO_4S$	Formula wt 201.24
INS: 950	CAS: [55589-62-3]

DESCRIPTION

A white, odorless, free-flowing crystalline powder having a sweet taste. It is freely soluble in water and very slightly soluble in ethanol.

Functional Use in Foods Nonnutritive sweetener.

REQUIREMENTS

Identification
   A. Add a few drops of sodium cobaltinitrite TS to a solution of 0.3 g of the sample in 1 mL of glacial acetic acid and 5 mL of water. A yellow precipitate is produced.
   B. Dissolve 10 mg of the sample in 1000 mL of water. The solution shows an absorption maximum at 227 ± 2 nm.

Assay Not less than 99.0% and not more than 101.0% of $C_4H_4KNO_4S$ after drying.
Fluoride Not more than 0.003%.
Heavy Metals (as Pb) Not more than 10 mg/kg.
Lead Not more than 1 mg/kg.
Loss on Drying Not more than 1.0%.
pH of a 1 in 100 Solution Between 6.5 and 7.5.
Potassium Not less than 17.0% and not more than 21.0%.

TESTS

Assay Weigh accurately between 200 and 300 mg of the sample, previously dried at 105° for 2 h, and dissolve in 50 mL of glacial acetic acid in a 250-mL flask.

   Note: Dissolution may be slow.

To the flask, add 2 or 3 drops of crystal violet TS, and titrate with 0.1 N perchloric acid to a blue-green endpoint that persists for at least 30 s. Perform a blank determination (see *General Provisions*), and make any necessary correction. Each mL of 0.1 N perchloric acid is equivalent to 20.12 mg of $C_4H_4KNO_4S$ (Acesulfame Potassium).
Fluoride Determine as directed in *Method III* of the *Fluoride Limit Test*, Appendix IIIB.
Heavy Metals A solution of 2 g in 20 mL of water meets the requirements of the *Heavy Metals Test*, Appendix IIIB, using 20 μg of lead ion (Pb) in the control (*Solution A*).
Lead A solution of 2 g in 20 mL of water meets the requirements of the *Lead Limit Test*, Appendix IIIB, using 2 μg of lead ion (Pb) in the control.
Loss on Drying, Appendix IIC Dry at 105° for 2 h.
pH of a 1 in 100 Solution Determine by the *Potentiometric Method*, Appendix IIB.
Potassium

   Standard Solutions Transfer 190.7 mg of potassium chloride, previously dried at 105° for 2 h, to a 1000-mL volumetric flask, dilute with water to volume, and mix. Transfer 100.0 mL of this solution to a second 1000-mL volumetric flask, dilute with water to volume, and mix to obtain a *Stock Solution* containing 10 μg of potassium per mL (equivalent to 19.07 μg of potassium chloride). Into separate 100-mL volumetric flasks, pipet 10.0-, 15.0- and 20.0-mL aliquots of the *Stock Solution*, add 2.0 mL of a 1 in 5 sodium chloride solution and 1.0 mL of hydrochloric acid to each, dilute with water to volume, and mix. The *Standard Solutions* obtained contain, respectively, 1.0, 1.5, and 2.0 μg of potassium per mL.

   Test Solution Transfer about 160 mg of the sample, accurately weighed, to a 1000-mL volumetric flask, dilute to volume with water, and mix. Filter 50 mL of this solution. Transfer 5.0 mL of the filtrate to a 100-mL volumetric flask, add 2.0 mL of a 1 in 5 sodium chloride solution and 1.0 mL of hydrochloric acid, dilute with water to volume, and mix.

   Procedure Concomitantly determine the absorbances of the *Standard Solutions* and the *Test Solution* at the potassium emission line of 766.5 nm, with a suitable atomic absorption spectrophotometer equipped with a potassium hollow-cathode lamp and an air–acetylene flame, using water as the blank. Plot the absorbance of the *Standard Solutions* versus concentration, in μg/mL, of potassium, and draw the straight line best fitting the plotted points. From the graph so obtained, determine the concentration, C, in μg/mL, of potassium in the *Test Solution*. Calculate the percent potassium in the sample of Acesulfame Potassium taken by the formula

$$2000C/W,$$

in which W is the quantity, in mg, of Acesulfame Potassium taken to prepare the *Test Solution*.

Packaging and Storage Store in well-closed containers in a cool, dry place.

Acetic Acid, Glacial

$$\text{CH}_3\text{COOH}$$

$C_2H_4O_2$	Formula wt 60.05
INS: 260	CAS: [64-19-7]

DESCRIPTION

A clear, colorless liquid having a pungent, characteristic odor and, when well diluted with water, an acid taste. It boils at about 118° and has a specific gravity of about 1.049. It is miscible with water, with alcohol, and with glycerin.

Functional Use in Foods Acidifier; flavoring agent.

REQUIREMENTS

Identification A 1 in 3 solution gives positive tests for *Acetate*, Appendix IIIA.
Assay Not less than 99.5% and not more than 100.5%, by weight, of $C_2H_4O_2$.
Heavy Metals (as Pb) Not more than 10 mg/kg.
Nonvolatile Residue Not more than 0.005%.
Readily Oxidizable Substances Passes test.
Solidification Point Not lower than 15.6°.

TESTS

Assay Measure about 2 mL into a tared, glass-stoppered flask, and weigh accurately. Add 40 mL of water, then add phenolphthalein TS, and titrate with 1 N sodium hydroxide. Each mL of 1 N sodium hydroxide is equivalent to 60.05 mg of $C_2H_4O_2$.
Heavy Metals To the residue obtained in the test for *Nonvolatile Residue* add 8 mL of 0.1 N hydrochloric acid, warm gently until solution is complete, and dilute to 100 mL with water. A 10-mL portion of this solution diluted to 25 mL meets the requirements of the *Heavy Metals Test*, Appendix IIIB, using 20 μg of lead ion (Pb) in the control (*Solution A*).
Nonvolatile Residue Evaporate 19 mL (20 g), accurately measured, in a tared dish on a steam bath, and dry at 105° for 1 h.
Readily Oxidizable Substances Dilute 2 mL in a glass-stoppered container with 10 mL of water, and add 0.1 mL of 0.1 N potassium permanganate. The pink color is not changed to brown within 2 h.
Solidification Point Determine as directed in the general method, Appendix IIB.

Packaging and Storage Store in tight containers.

Acetone

2-Propanone; Dimethyl Ketone

$$CH_3COCH_3$$

C_3H_6O \hfill Formula wt 58.08

\hfill CAS: [67-64-1]

DESCRIPTION

A clear, colorless, volatile liquid having a characteristic odor. It is miscible with water, with alcohol, with ether, with chloroform, and with most volatile oils. Its refractive index is about 1.356.

Caution: Acetone is highly flammable.

Functional Use in Foods Extraction solvent.

REQUIREMENTS

Identification Mix 0.1 mL of the sample with 10 mL of water, add 5 mL of 1 N sodium hydroxide, warm, and add 5 mL of iodine TS. A yellow precipitate of iodoform is produced.
Assay Not less than 99.5% and not more than 100.5% of C_3H_6O, by weight.
Acidity (as acetic acid) Not more than 0.002%.
Aldehydes (as formaldehyde) Not more than 0.002%.
Alkalinity (as NH_3) Not more than 10 mg/kg.
Distillation Range Within a range of 1°, including 56.1°.
Heavy Metals (as Pb) Not more than 1 mg/kg.
Methanol Not more than 0.05%.
Nonvolatile Residue Not more than 10 mg/kg.
Phenols Passes test.
Solubility in Water Passes test.
Substances Reducing Permanganate Passes test.
Water Not more than 0.5%.

TESTS

Assay Its specific gravity, determined by any reliable method (see *General Provisions*), is not greater than 0.7880 at 25°/25° (equivalent to 0.7930 at 20°/20°).
Acidity Mix 38 mL (about 30 g) of the sample with an equal volume of carbon dioxide-free water, add 0.1 mL of phenolphthalein TS, and titrate with 0.1 N sodium hydroxide. Not more than 0.1 mL is required to produce a pink color.
Aldehydes Dilute 2.5 mL (about 2 g) of the sample with 7.5 mL of water. Prepare a standard solution containing 40 μg of formaldehyde in 10 mL of water. To each solution add 0.15 mL of a 5% solution of 5,5-dimethyl-1,3-cyclohexanedione in alcohol, and evaporate on a steam bath until the Acetone is volatilized. Dilute to 10 mL with water, and cool quickly in an ice bath while stirring vigorously. Any turbidity in the sample solution does not exceed that produced in the standard.
Alkalinity Add 1 drop of methyl red TS to 25 mL of water, add 0.1 N sulfuric acid until a red color just appears, then add 23 mL (about 18 g) of the sample, and mix. Not more than 0.1 mL of 0.1 N sulfuric acid is required to restore the red color.
Distillation Range Proceed as directed in the general method, Appendix IIB.
Heavy Metals Evaporate 25 mL (about 20 g) of the sample to dryness on a steam bath in a glass evaporating dish. Cool, add 2 mL of hydrochloric acid, and slowly evaporate to dryness again on the steam bath. Moisten the residue with 1 drop of hydrochloric acid, add 10 mL of hot water, and digest for 2 min. Cool, and dilute to 25 mL with water. This solution meets the requirements of the *Heavy Metals Test*, Appendix IIIB, using 20 μg of lead ion (Pb) in the control (*Solution A*).
Methanol Dilute 10 mL of the sample to 100 mL with water. Prepare a standard solution in water containing 40 μg of methanol in each mL. To 1 mL of each solution add 0.2 mL of 10% phosphoric acid and 0.25 mL of potassium permanganate solution (1 in 20). Allow to stand for 15 min, then add 0.3 mL

of sodium bisulfite solution (1 in 10), and shake until colorless. Slowly add 5 mL of ice-cold 80% sulfuric acid, keeping the mixture cold during the addition. Add 0.1 mL of chromotropic acid solution (1 in 100), mix, and digest on a steam bath for 20 min. Any violet color produced in the sample solution does not exceed that produced in the standard.

Nonvolatile Residue Evaporate 125 mL (about 100 g) of the sample to dryness in a tared dish on a steam bath, dry the residue at 105° for 30 min, cool, and weigh.

Phenols Evaporate 3 mL of the sample to dryness at 60°. To the residue add 3 drops of a solution of 100 mg of sodium nitrite in 5 mL of sulfuric acid, allow to stand for about 3 min, and then carefully add 3 mL of 2 N sodium hydroxide. No color is produced.

Solubility in Water Mix 38 mL (about 30 g) of the sample with an equal volume of carbon dioxide-free water. The solution remains clear for at least 30 min.

Substances Reducing Permanganate Transfer 10 mL of the sample into a glass-stoppered cylinder, add 0.05 mL of 0.1 N potassium permanganate, mix, and allow to stand for 15 min. The pink color is not entirely discharged.

Water Determine by the *Karl Fischer Titrimetric Method*, Appendix IIB, using freshly distilled pyridine instead of methanol as the solvent.

Packaging and Storage Store in tight containers remote from fire.

Acetone Peroxides

INS: 929   CAS: [1336-17-0]

DESCRIPTION

A mixture of monomeric and linear dimeric acetone peroxides (mainly 2,2-hydroperoxypropane), with minor proportions of higher polymers, usually mixed with an edible carrier such as cornstarch. The cornstarch mixture is a fine, white, free-flowing powder having a sharp, acrid odor similar to that of hydrogen peroxide when the container is first opened.

Caution: Acetone Peroxides are strong oxidizing agents. Exposure to the skin and eyes should be avoided.

Functional Use in Foods Bleaching agent; maturing agent; dough conditioner.

REQUIREMENTS

Identification Dissolve about 20 mg of the sample in 5 mL of dilute sulfuric acid (1 in 10), allow to stand for a few min, and add a drop of potassium permanganate TS. The pink color is discharged.

Assay It yields an amount of hydrogen peroxide equivalent to not less than 16.0% of Acetone Peroxides.

Heavy Metals (as Pb) Not more than 10 mg/kg.

TESTS

Assay Transfer about 200 mg, accurately weighed, into a 250-mL beaker, add 50 mL of dilute sulfuric acid (1 in 10), allow to stand for at least 3 min, stirring occasionally, and titrate with 0.1 N potassium permanganate to a light pink color that persists for at least 20 s. Calculate the total peroxides, P, as g of hydrogen peroxide equivalents per 100 g of the sample, by the formula

$$V \times N \times 0.017 \times 100/W,$$

in which V and N are the volume and exact normality, respectively, of the potassium permanganate; 0.017 is the milliequivalent weight of H_2O_2; and W is the weight, in g, of the sample taken. Multiply the value P so obtained by 1.6 to convert to percentage of Acetone Peroxides.

Heavy Metals Mix 10 g with 100 mL of dilute sulfuric acid (1 in 10), allow to stand for 5 min, stirring occasionally, and filter. Heat the filtrate on a steam bath for 15 min, then boil for 1 min, cool, and dilute to 100 mL with water. Dilute 20 mL of the resulting solution to 25 mL with water. This solution meets the requirements of the *Heavy Metals Test*, Appendix IIIB, using 20 µg of lead ion (Pb) in the control (*Solution A*).

Packaging and Storage Store in a cool, dry place, preferably below 24°.

Acetylated Monoglycerides

Acetylated Mono- and Diglycerides

DESCRIPTION

Acetylated Monoglycerides consist of partial or complete esters of glycerin with a mixture of acetic acid and edible fat-forming fatty acids. They may be manufactured by the interesterification of edible fats with triacetin and glycerin in the presence of catalytic agents, followed by molecular distillation, or by the direct acetylation of edible monoglycerides with acetic anhydride without the use of catalyst or molecular distillation. They vary in consistency from clear, thin liquids to solids, and are from white to pale yellow in color. They may have an acetic acid odor, but are practically bland in taste. They are insoluble in water, but are soluble in alcohol, in acetone, and in other organic solvents, the extent of solubility depending upon the degree of esterification and the melting range.

Functional Use in Foods Emulsifier; coating agent; texture-modifying agent; solvent; lubricant.

REQUIREMENTS

Acid Value Not more than 6.
Heavy Metals (as Pb) Not more than 10 mg/kg.
Reichert-Meissl Value Between 75 and 150.

The following specifications should conform to the representations of the vendor: **Free Glycerin, Iodine Value,** and **Saponification Value.**

TESTS

Acid Value Determine as directed for *Acid Value, Method II,* under *Fats and Related Substances,* Appendix VII.
Free Glycerin Determine as directed in the general method, Appendix VII.
Heavy Metals Prepare and test a 2-g sample as directed in *Method II* under the *Heavy Metals Test,* Appendix IIIB, using 20 μg of lead ion (Pb) in the control (*Solution A*).
Iodine Value Determine by the *Modified Wijs Method,* Appendix VII.
Reichert-Meissl Value Determine as directed in the general method, Appendix VII.
Saponification Value Determine as directed for *Saponification Value* under *Fats and Related Substances,* Appendix VII.

Packaging and Storage Store in well-closed containers.

N-Acetyl-L-Methionine

N-Acetyl-L-2-amino-4-(methylthio)butyric Acid

$$CH_3SCH_2CH_2\underset{\underset{HNCOCH_3}{|}}{CH}COOH$$

$C_7H_{13}NO_3S$ Formula wt 191.25

CAS: [65-82-7]

DESCRIPTION

A colorless or lustrous white crystalline solid, or a white powder. It is odorless or practically odorless. It is soluble in water, in alcohol, in alkali solutions, and in dilute mineral acids, but is practically insoluble in ether.

Functional Use in Foods Nutrient; dietary supplement.

REQUIREMENTS

Identification Dissolve 250 mg of the sample in 2.5 mL of isopropyl alcohol, dilute to 25 mL with water, and then dilute 10 mL of this solution to 100 mL with water. Spot 0.5, 30, and 50 μL of the solution 2 cm from the bottom of a suitable thin-layer chromatographic plate coated with silica gel (e.g., 20- × 20-cm Brinkman Silica Gel 60, 250 μm, or equivalent), and allow the plate to develop for a distance of 10 cm in a sealed equilibrated thin-layer chromatographic chamber, using a solvent composed of 75 volumes of *n*-butanol, 20 volumes of glacial acetic acid, and 20 volumes of water. Dry the plate overnight, and spray with iodoplatinate solution prepared fresh before use by mixing 3 mL of 10% hexachloroplatinic (IV) acid with 97 mL of water and 100 mL of 6% potassium iodide solution. The sample forms a single colorless spot having an R_f value of 0.67 ± 0.1.
Assay Not less than 98.5% and not more than 101.5% of $C_7H_{13}NO_3S$, calculated on the dried basis.
Heavy Metals (as Pb) Not more than 0.002%.
Lead Not more than 10 mg/kg.
Loss on Drying Not more than 0.5%.
Residue on Ignition Not more than 0.1%.
Specific Rotation $[\alpha]_D^{20°}$: Between −18.0° and −22.0°, after drying.

TESTS

Assay Transfer about 250 mg of the sample, accurately weighed, into a glass-stoppered flask, and add 100 mL of water, 5 g of dibasic potassium phosphate, 2 g of monobasic potassium phosphate, and 2 g of potassium iodide. Mix well to dissolve, add 50.0 mL of 0.1 *N* iodine, stopper the flask, and mix. Allow to stand for 30 min, and then titrate the excess iodine with 0.1 *N* sodium thiosulfate. Perform a residual blank titration. Each mL of 0.1 *N* iodine is equivalent to 9.563 mg of $C_7H_{13}NO_3S$.
Heavy Metals Prepare and test a 1-g sample as directed in *Method II* under the *Heavy Metals Test,* Appendix IIIB, using 20 μg of lead ion (Pb) in the control (*Solution A*).
Lead A *Sample Solution* prepared as directed for organic compounds meets the requirements of the *Lead Limit Test,* Appendix IIIB, using 10 μg of lead ion (Pb) in the control.
Loss on Drying Dry at 105° for 2 h.
Residue on Ignition, Appendix IIC Ignite 1 g as directed in the general method.
Specific Rotation, Appendix IIB Determine in a solution containing 2 g of the previously dried sample in sufficient water to make 100 mL.

Packaging and Storage Store in well-closed, light-resistant containers.

Acid Hydrolysates of Proteins

Acid-Hydrolyzed Proteins; Hydrolyzed Vegetable Protein (HVP); Hydrolyzed Plant Protein (HPP); Hydrolyzed (Source) Protein Extract; Acid-Hydrolyzed Milk Protein

DESCRIPTION

Acid Hydrolysates of Proteins are composed primarily of amino acids, small peptides (peptide chains of 5 or fewer amino acids), and salts resulting from the essentially complete hydrolysis of peptide bonds in edible proteinaceous materials catalyzed by heat and/or food-grade acids. Cleavage of peptide bonds typically ranges from a low of 85% to essentially 100%. In processing, the protein hydrolysates may be treated with safe and suitable alkaline materials. The edible proteinaceous materials used as raw materials are derived from corn, soy, wheat, yeast, peanut, rice, or other safe and suitable vegetable or plant sources, or from milk. Individual products may be in liquid, paste, powder, or granular form.

Functional Use in Foods Flavoring agent; flavor enhancer; adjuvant.

REQUIREMENTS

Calculate all analyses on the dried basis. In a suitable tared container, evaporate liquid and paste samples to dryness on a steam bath, then, as for the powdered and granular forms, dry to constant weight at 105° (see the *General Provisions*).
Assay (Total Nitrogen; TN) Not less than 4.0% total nitrogen.
α-Amino Nitrogen (AN) Not less than 3.0%.
α-Amino Nitrogen/Total Nitrogen (AN/TN) Percent Ratio Not less than 62.0% and not more than 85.0%, when calculated on an ammonia nitrogen-free basis.
Ammonia Nitrogen (NH_3-N) Not more than 1.5%.
Glutamic Acid Not more than 20.0% as $C_5H_9NO_4$ and not more than 35.0% of the total amino acids.
Heavy Metals (as Pb) Not more than 10 mg/kg.
Insoluble Matter Not more than 0.5%.
Lead Not more than 5 mg/kg.
Potassium Not more than 30.0%.
Sodium Not more than 20.0%.

TESTS

Assay (Total Nitrogen) Proceed as directed under *Nitrogen Determination*, Appendix IIIC.
α-Amino Nitrogen Transfer 7 to 25 g, accurately weighed, into a 500-mL volumetric flask with the aid of several 50-mL portions of warm ammonia-free water, dilute to volume with water, and mix. Neutralize 20.0 mL of the solution with 0.2 N barium hydroxide or 0.2 N sodium hydroxide, using phenolphthalein TS as indicator, and add 10 mL of freshly prepared phenolphthalein–formol solution (50 mL of 40% formaldehyde containing 1 mL of 0.05% phenolphthalein in 50% alcohol neutralized exactly to pH 7 with 0.2 N barium hydroxide or 0.2 N sodium hydroxide). Titrate with 0.2 N barium hydroxide or 0.2 N sodium hydroxide to a distinct red color, add a small but accurately measured volume of 0.2 N barium hydroxide or 0.2 N sodium hydroxide in excess, and back titrate to neutrality with 0.2 N hydrochloric acid. Conduct a blank titration using the same reagents, with 20 mL of water in place of the test solution. Each mL of 0.2 N barium hydroxide or 0.2 N sodium hydroxide is equivalent to 2.8 mg of α-amino nitrogen.
α-Amino Nitrogen/Total Nitrogen (AN/TN) Percent Ratio Calculate by dividing the percent α-amino nitrogen (*AN*) by the percent total nitrogen (*TN*) as corrected for ammonia nitrogen (*NH_3-N*) according to the formula

$$100[(AN - NH_3\text{-}N)/(TN - NH_3\text{-}N)].$$

Ammonia Nitrogen (**Caution**: Provide adequate ventilation.) (**Note**: All reagents should be nitrogen free, where available, or otherwise very low in nitrogen content.) Transfer from 700 mg to 2.2 g of the sample to a 500- to 800-mL Kjeldahl digestion flask of hard, moderately thick, well-annealed glass, wrapping the sample, if solid or semi-solid, in nitrogen-free filter paper to facilitate the transfer if desired.

Add about 200 mL of water, and mix. Add a few granules of zinc to prevent bumping, tilt the flask, and cautiously pour sodium hydroxide pellets, or a 2 in 5 sodium hydroxide solution, down the inside of the flask so that it forms a layer under the solution, using a sufficient amount (usually about 25 g of solid NaOH) to make the mixture strongly alkaline. Immediately connect the flask to a distillation apparatus consisting of a Kjeldahl connecting bulb and a condenser, the delivery tube of which extends well beneath the surface of a measured excess of 0.5 N hydrochloric or sulfuric acid contained in a 500-mL flask. Add from 5 to 7 drops of methyl red indicator (1 g of methyl red in 200 mL of alcohol) to the receiver flask. Rotate the Kjeldahl flask to mix its contents thoroughly, and then heat until all of the ammonia has distilled, collecting at least 150 mL of distillate. Wash the tip of the delivery tube, collecting the washings in the receiving flask, and titrate the excess acid with 0.5 N sodium hydroxide. Perform a blank determination, substituting 2 g of sucrose for the sample, and make any necessary correction (see the *General Provisions*). Each mL of 0.5 N acid consumed is equivalent to 7.003 mg of ammonia nitrogen.

Note: If it is known that the substance to be determined has a low nitrogen content, 0.1 N acid and alkali may be used, in which case each mL of 0.1 N acid consumed is equivalent to 1.401 mg of nitrogen.

Calculate the percent ammonia nitrogen by dividing the weight of ammonia nitrogen, in mg, by the weight of the sample, in mg, times 100.
Glutamic Acid
Apparatus Use an ion-exchange amino acid analyzer, equipped with sulfonated polystyrene columns, in which the effluent from the sample is mixed with ninhydrin reagent and the absorbance of the resultant color is measured continuously and automatically at 570 and 440 nm by a recording photometer.
Standard Solution Weigh 1250 ± 2 mg of glutamic acid, reagent grade, and place in a 500-mL volumetric flask. Fill the flask half-full with water, and add 5 mL of hydrochloric acid to help dissolve the amino acid, dilute to volume with water, and mix. Prepare the standard for analysis by diluting 1 mL of this solution with 4 mL of 0.2 N sodium citrate, pH 2.2, buffer. This *Standard Solution* contains 0.5 mg of glutamic acid per mL (C_S).
Sample Preparation Accurately weigh 5 mg of the sample and dilute to exactly 5 mL with 0.2 N sodium citrate, pH 2.2, buffer. Remove any insoluble material by centrifugation or filtration.
Procedure Using 2-mL aliquots of the *Standard Solution* and *Sample Preparation*, proceed as directed according to the apparatus manufacturer's instructions. From the chromatograms thus obtained, match the retention times produced by the *Standard Solution* with those produced by the *Sample Solution*, and identify the peak produced by glutamic acid. Record the area of the glutamic acid peak from the sample as A_U, and that from the standards as A_S.
Calculations Calculate the concentration, C_A, in mg/mL, of glutamic acid in the *Sample Preparation* by the formula

$A_U \times C_S/A_S$,

in which C_S is the concentration, in mg/mL, of the glutamic acid in the *Standard Solution*.

Calculate the percentage of glutamic acid, on the basis of total amino acids, by the formula

$$100C_A/6.25N_T,$$

in which N_T is the percentage of total nitrogen determined in the *Assay*.

Calculate the percentage of glutamic acid in the sample by the formula

$$100 \times C_A/S_W,$$

in which S_W is the weight of the sample taken, in mg.

Heavy Metals Prepare and test a 2-g sample as directed in *Method II* under the *Heavy Metals Test*, Appendix IIIB, using 20 μg of lead ion (Pb) in the control (*Solution A*).

Insoluble Matter Transfer about 5 g, accurately weighed, into a 250-mL Erlenmeyer flask, add 75 mL of water, cover the flask with a watch glass, and boil gently for 2 min. Filter the solution through a tared filtering crucible, dry at 105° for 1 h, cool, and weigh.

Lead A *Sample Solution* prepared as directed for organic compounds meets the requirements of the *Lead Limit Test*, Appendix IIIB, using 5 μg of lead ion (Pb) in the control.

Potassium

Spectrophotometer Use any suitable atomic absorption spectrophotometer.

Standard Solution Transfer 38.20 mg of reagent-grade potassium chloride, accurately weighed, into a 100-mL volumetric flask, dissolve in and dilute to volume with deionized water, and mix. Transfer 5.0 mL of this solution to a 1000-mL volumetric flask, dilute to volume with deionized water, and mix. Each mL contains 1.0 μg of K.

Sample Solution Transfer 1.00 ± 0.05 g of previously dried sample, accurately weighed, into a silica or porcelain dish. Ash in a muffle furnace at 246° to 260° for 2 to 4 h. Allow the ash to cool, and dissolve in 5 mL of 20% hydrochloric acid, warming the solution if necessary to complete solution of the residue. Filter the solution through acid-washed filter paper into a 500-mL volumetric flask. Wash the filter paper with hot water, dilute to volume, and mix.

Procedure Determine the absorbance of each solution at 766.5 nm, following the manufacturer's instruction for optimum operation of the spectrophotometer. The absorbance produced by the *Sample Solution* does not exceed that of the *Standard Solution*.

Sodium

Spectrophotometer Use any suitable atomic absorption spectrophotometer.

Standard Solution Transfer 25.42 mg of reagent-grade sodium chloride, accurately weighed, into a 100-mL volumetric flask, dissolve in and dilute to volume with deionized water, and mix. Transfer 5.0 mL of this solution to a 1000-mL volumetric flask, dilute to volume with deionized water, and mix. Each mL contains 0.5 mg of Na. Using water as the solvent, prepare a 1 in 100 dilution of this solution to obtain the final working *Standard Solution*.

Sample Solution Transfer 1.00 ± 0.05 g of previously dried sample, accurately weighed, into a silica or porcelain dish. Ash in a muffle furnace at 246° to 260° for 2 to 4 h. Allow the ash to cool, and dissolve in 5 mL of 20% hydrochloric acid, warming the solution if necessary to complete solution of the residue. Filter the solution through acid-washed filter paper into a 500-mL volumetric flask. Wash the filter paper with hot water, dilute to volume, and mix. Using water as the solvent, prepare a 1 in 100 dilution of this solution to obtain the final *Sample Solution*.

Procedure Determine the absorbance of each solution at 589.0 nm, following the manufacturer's instructions for optimum operation of the spectrophotometer. The absorbance produced by the *Sample Solution* does not exceed that of the *Standard Solution*.

Packaging and Storage Store in well-closed containers.

Aconitic Acid

Equisetic Acid; Citridic Acid; Achilleic Acid

$C_6H_6O_6$　　　　　　　　　　　　Formula wt 174.11

　　　　　　　　　　　　　　　　　CAS [499-12-7]

DESCRIPTION

Aconitic Acid (1-propene-1,2,3-tricarboxylic acid) occurs in the leaves and tubers of *Ranunculaceae Aconitum napellus* L. and various species of *Achillea* and *Equisetum*, in beet root, and in sugar cane. It may be synthesized by the dehydration of citric acid by sulfuric or methanesulfonic acid. Aconitic Acid from the above sources has the "*trans*" configuration. It has a melting point of 195° to 200°, with decomposition. It is soluble in water and in alcohol and slightly soluble in ether.

Functional Use in Foods Flavoring substance and adjuvant.

REQUIREMENTS

Identification The substance as a potassium bromide dispersion exhibits infrared absorption bands at 3030, 2630, and 1720 cm^{-1}. An aqueous solution of the substance exhibits major absorption peaks at 411 and 432 nm, with little or no absorption at 389 nm.

Assay Not less than 98.0% and not more than 100.5% of $C_3H_3(COOH)_3$, calculated on the anhydrous basis.

Heavy Metals (as Pb) Not more than 10 mg/kg.

Oxalate Passes test.

Readily Carbonizable Substances Passes test.
Residue on Ignition Not more than 0.1%.
Tridodecylamine Not more than 0.1 mg/kg.
Ultraviolet Absorbance For the specified spectral ranges: 280 to 289 nm, not more than 0.25 AU; 290 to 299 nm, not more than 0.20 AU; 300 to 359 nm, not more than 0.13 AU; 360 to 400 nm, not more than 0.03 AU.
Water Not more than 0.5%.

TESTS

Assay Proceed as directed under *Assay* in the monograph for *Citric Acid*. Each mL of 1 N sodium hydroxide is equivalent to 58.04 mg of $C_3H_3(COOH)_3$.
Heavy Metals A solution of 2 g in 25 mL of water meets the requirements of the *Heavy Metals Test*, Appendix IIIB, using 20 μg of lead ion (Pb) in the control (*Solution A*).
Oxalate Neutralize 10 mL of a 1 in 10 solution with 6 N ammonium hydroxide, add 5 drops of 2.7 N hydrochloric acid, cool, and add 2 mL of calcium chloride TS. No turbidity is produced.
Readily Carbonizable Substances Transfer 1.0 g, finely powdered, to a 22-mm × 175-mm test tube, previously rinsed with 10 mL of 95% sulfuric acid and allowed to drain for 10 min. Add 10 mL of 95% sulfuric acid, agitate the tube until solution is complete, and immerse the tube in a water bath at 90° ± 1° for 60 ± 0.5 min, keeping the level of the acid below the level of the water during the heating period. Cool the tube in a stream of water, and transfer the acid solution to a color-comparison tube. The color of the acid solution is not darker than that of the same volume of *Matching Fluid K* in a similar matching tube, viewing the tubes vertically against a white background.
Residue on Ignition Ignite 4 g as directed in the general method, Appendix IIC.
Tridodecylamine

Indicator Buffer Solution Prepare a mixture consisting of 700 mL of 0.1 M citric acid (anhydrous, reagent grade), 200 mL of 0.2 M disodium phosphate, and 50 mL each of 0.2% bromophenol blue and 0.2% bromocresol green in spectrograde methanol.

No-Indicator Buffer Solution Prepare a mixture consisting of 700 mL of 0.1 M citric acid (anhydrous, reagent grade), 200 mL of 0.2 M disodium phosphate, and 100 mL of spectrograde methanol.

Amine Stock Solution Transfer between 40 and 45 mg of tridodecylamine (trilaurylamine), accurately weighed, into a 500-mL volumetric flask, dilute to volume with isopropyl alcohol, and mix. Discard after 3 weeks.

Standard Amine Solution Prepare the solution fresh daily. Using a graduated 5-mL pipet, transfer into a 100-mL volumetric flask an amount of *Amine Stock Solution* equivalent to 400 μg of tridodecylamine, dilute to volume with isopropyl alcohol, and mix.

Procedure Dissolve 160 g of anhydrous reagent-grade citric acid (not the sample to be tested) in 320 mL of water, and divide the solution equally between two 250-mL separators, S_1 and S_2. To S_1 add 5 mL of *No-Indicator Buffer Solution*. To S_2 add 2.0 mL of *Standard Amine Solution* and 5 mL of *Indicator Buffer Solution*.

To prepare solutions of the sample being tested, dissolve 145 g of the anhydrous Aconitic Acid sample in 320 mL of water. Divide the test solution equally between two 250-mL separators, S_3 and S_4. Add 5 mL of *No-Indicator Buffer Solution* to S_3 and 5 mL of *Indicator Buffer Solution* to S_4.

To each of the four separators, add 20 mL of a 1:1 mixture (v/v) prepared from spectrograde chloroform and *n*-heptane, shake for 15 min on a mechanical shaker, and allow the phases to separate for 45 min. Drain all except the last few drops of the lower (aqueous) phases and discard. Hand-shake the organic phases with 25 mL each of 0.05 N sulfuric acid for 30 s, and allow the phases to separate for 30 min. Drain all except the last few drops of the lower (organic phase) through dry Whatman No. 40 (or equivalent) paper, and collect the aqueous filtrates in separate, small, glass-stoppered containers.

Determine the absorbance of each solution in a 5-cm cell at 400 nm, with a suitable spectrophotometer standardized before analysis, against chloroform–heptane (1:1 v/v). The net absorbance of the sample ($S_4 - S_3$) is not greater than that of the standard ($S_2 - S_1$).
Ultraviolet Absorbance Determine as directed for *Ultraviolet Absorbance* in the monograph for *Citric Acid*.
Water Determine under *Water Determination* by the *Karl Fischer Titrimetric Method*, Appendix IIB.

Packaging and Storage Store in tight containers.

Adipic Acid

Hexanedioic Acid; 1,4-Butanedicarboxylic Acid

$$HOOC(CH_2)_4COOH$$

$C_6H_{10}O_4$	Formula wt 146.14
INS: 355	CAS: [124-04-9]

DESCRIPTION

White crystals or crystalline powder. It is soluble in acetone, freely soluble in alcohol, and slightly soluble in water. It is not hygroscopic.

Functional Use in Foods Buffer; neutralizing agent.

REQUIREMENTS

Assay Not less than 99.6% and not more than 101.0% of $C_6H_{10}O_4$, calculated on the anhydrous basis.
Heavy Metals (as Pb) Not more than 10 mg/kg.
Melting Range Between 151.5° and 154°.
Residue on Ignition Not more than 0.002%,
Water Not more than 0.2%.

TESTS

Assay Mix about 1.5 g, accurately weighed, with 75 mL of recently boiled and cooled water in a 250-mL glass-stoppered Erlenmeyer flask, add phenolphthalein TS, and titrate with 0.5 N sodium hydroxide to the first appearance of a faint pink endpoint that persists for at least 30 s, shaking the flask as the endpoint is approached. Each mL of 0.5 N sodium hydroxide is equivalent to 36.54 mg of $C_6H_{10}O_4$.

Heavy Metals Prepare and test a 2-g sample as directed in *Method II* under the *Heavy Metals Test*, Appendix IIIB, using 20 μg of lead ion (Pb) in the control (*Solution A*).

Melting Range Determine as directed for *Melting Range or Temperature*, Appendix IIB.

Residue on Ignition Transfer 100.0 g to a tared 125-mL platinum dish that has been previously cleaned by fusing with 5 g of potassium pyrosulfate or bisulfate, followed by boiling in 2 N sulfuric acid and rinsing with water. Melt the sample completely over a gas burner, then ignite the melt with the burner. After ignition starts, lower or remove the flame in order to prevent the sample from boiling and to keep it burning slowly until it is completely carbonized. Ignite at 850° in a muffle furnace for 30 min or until the carbon is completely removed, cool, and weigh.

Water Determine by the *Karl Fischer Titrimetric Method*, Appendix IIB.

Packaging and Storage Store in well-closed containers.

Agar

INS: 406   CAS: [9002-18-0]

DESCRIPTION

A dried hydrophilic, colloidal polygalactoside extracted from *Gelidium cartilagineum* (L.) Gaillon (Fam. *Gelidiaceae*), *Gracilaria confervoides* (L.) Greville (Fam. *Sphaerococcaceae*), and related red algae (Class *Rhodophyceae*). It is commercially available in bundles consisting of thin, membranous agglutinated strips, or in cut, flaked, granulated, or powdered forms. It is white to pale yellow in color, is either odorless or has a slight characteristic odor, and has a mucilaginous taste. Agar is insoluble in cold water, but is soluble in boiling water.

Functional Use in Foods Stabilizer; emulsifier; thickener.

REQUIREMENTS

Identification
A. Place a few fragments of unground Agar or a small amount of the powder on a slide, add a few drops of water, and examine microscopically. The Agar appears granular and somewhat filamentous. A few fragments of the spicules of sponges and a few frustules of diatoms may be present.

B. Boil 1 g with 65 mL of water for 10 min with continuous stirring, and adjust to a concentration of 1.5%, by weight, with hot water. A clear liquid is obtained that congeals between 32° and 39° to form a firm, resilient gel that does not liquefy below 85°.

Arsenic (as As)   Not more than 3 mg/kg.
Ash (Acid-Insoluble)   Not more than 0.5%, calculated on the dried basis.
Ash (Total)   Not more than 6.5%, calculated on the dried basis.
Gelatin   Passes test.
Heavy Metals (as Pb)   Not more than 10 mg/kg.
Insoluble Matter   Not more than 1.0%.
Lead   Not more than 5 mg/kg.
Loss on Drying   Not more than 20.0%.
Starch   Passes test.
Water Absorption   Passes test.

TESTS

Arsenic A *Sample Solution* prepared as directed for organic compounds meets the requirements of the *Arsenic Test*, Appendix IIIB.

Ash (Acid-Insoluble) Determine as directed in the general method, Appendix IIC.

Ash (Total) Determine as directed in the general method, Appendix IIC.

Gelatin Dissolve about 1 g in 100 mL of boiling water, and allow to cool to about 50°. To 5 mL of the solution add 5 mL of trinitrophenol TS. No turbidity appears within 10 min.

Heavy Metals Prepare and test a 2-g sample as directed in *Method II* under the *Heavy Metals Test*, Appendix IIIB, using 20 μg of lead ion (Pb) in the control (*Solution A*).

Insoluble Matter To 7.5 g add sufficient water to make 500 g, boil for 15 min, and readjust to the original weight. To 100 g of the mixture add hot water to make 200 mL, heat almost to boiling, filter while hot through a tared filtering crucible, rinse the container with several portions of hot water, and pass the rinsings through the crucible. Dry the crucible and its contents at 105° to constant weight, cool, and weigh. The weight of the residue does not exceed 15 mg.

Lead A *Sample Solution* prepared as directed for organic compounds meets the requirements of the *Lead Limit Test*, Appendix IIIB, using 5 μg of lead ion (Pb) in the control.

Loss on Drying, Appendix IIC   Dry at 105° for 5 h. Cut unground Agar into pieces from 2 to 5 mm square before drying.

Starch Boil 100 mg in 100 mL of water, cool, and add a few drops of iodine TS. No blue color is produced.

Water Absorption Place 5 g in a 100-mL graduated cylinder, fill to the 100-mL mark with water, mix, and allow to stand at about 25° for 24 h. Pour the contents of the cylinder through moistened glass wool, allowing the water to drain into another 100-mL graduated cylinder. Not more than 75 mL of water is obtained.

Packaging and Storage Store in well-closed containers.

DL-Alanine

DL-2-Aminopropanoic Acid

$$CH_3CH(NH_2)COOH$$

$C_3H_7NO_2$ Formula wt 89.09

CAS: [302-72-7]

DESCRIPTION

A white, odorless, crystalline powder having a sweetish taste. It is freely soluble in water, but sparingly soluble in alcohol. It is optically inactive. The pH of a 1 in 20 solution is between 5.5 and 7.0. It melts with decomposition at about 198°.

Functional Use in Foods   Nutrient; dietary supplement; flavor enhancer.

REQUIREMENTS

Identification

A. Heat 5 mL of a 1 in 1000 solution with 1 mL of triketohydrindene hydrate TS for 3 min. A violet color is produced.

B. Dissolve 200 mg in 10 mL of water, add 100 mg of potassium permanganate, and heat to boiling. The odor of acetaldehyde is detected.

Assay   Not less than 98.5% and not more than 101.5% of $C_3H_7NO_2$, calculated on the dried basis.
Heavy Metals (as Pb)   Not more than 0.002%.
Lead   Not more than 10 mg/kg.
Loss on Drying   Not more than 0.3%.
Residue on Ignition   Not more than 0.2%.

TESTS

Assay   Dissolve about 200 mg, accurately weighed, in 3 mL of formic acid and 50 mL of glacial acetic acid, add 2 drops of crystal violet TS, and titrate with 0.1 N perchloric acid to a bluish green endpoint. Perform a blank determination, and make any necessary correction. Each mL of 0.1 N perchloric acid is equivalent to 8.909 mg of $C_3H_7NO_2$.
Heavy Metals   Prepare and test a 1-g sample as directed in *Method II* under the *Heavy Metals Test*, Appendix IIIB, using 20 µg of lead ion (Pb) in the control (*Solution A*).
Lead   A *Sample Solution* prepared as directed for organic compounds meets the requirements of the *Lead Limit Test*, Appendix IIIB, using 10 µg of lead ion (Pb) in the control.
Loss on Drying, Appendix IIC   Dry at 105° for 3 h.
Residue on Ignition, Appendix IIC   Ignite 1 g as directed in the general method.

Packaging and Storage   Store in well-closed, light-resistant containers.

L-Alanine

L-2-Aminopropanoic Acid

$$\begin{array}{c} CH_3CCOOH \\ |\backslash \\ HNH_2 \end{array}$$

$C_3H_7NO_2$ Formula wt 89.09

CAS: [56-41-7]

DESCRIPTION

A white, odorless, crystalline powder having a sweetish taste. It is freely soluble in water, sparingly soluble in alcohol, and insoluble in ether. The pH of a 1 in 20 solution is between 5.5 and 7.0.

Functional Use in Foods   Nutrient; dietary supplement.

REQUIREMENTS

Identification

A. Heat 5 mL of a 1 in 1000 solution with 1 mL of triketohydrindene hydrate TS for 3 min. A violet color is produced.

B. Dissolve 200 mg in 10 mL of water, add 100 mg of potassium permanganate, and heat to boiling. The odor of acetaldehyde is detected.

Assay   Not less than 98.5% and not more than 101.5% of $C_3H_7NO_2$, calculated on the dried basis.
Heavy Metals (as Pb)   Not more than 0.002%.
Lead   Not more than 10 mg/kg.
Loss on Drying   Not more than 0.3%.
Residue on Ignition   Not more than 0.2%.
Specific Rotation $[\alpha]_D^{20°}$: Between +13.5° and +15.5° after drying; or $[\alpha]_D^{25°}$: Between +13.2° and +15.2° after drying.

TESTS

Assay   Dissolve about 200 mg, accurately weighed, in 3 mL of formic acid and 50 mL of glacial acetic acid, add 2 drops of crystal violet TS, and titrate with 0.1 N perchloric acid to a bluish green endpoint. Perform a blank determination (see *General Provisions*), and make any necessary correction. Each mL of 0.1 N perchloric acid is equivalent to 8.909 mg of $C_3H_7NO_2$.
Heavy Metals   Prepare and test a 1-g sample as directed in *Method II* under the *Heavy Metals Test*, Appendix IIIB, using 20 µg of lead ion (Pb) in the control (*Solution A*).
Lead   A *Sample Solution* prepared as directed for organic compounds meets the requirements of the *Lead Limit Test*, Appendix IIIB, using 10 µg of lead ion (Pb) in the control.
Loss on Drying, Appendix IIC   Dry at 105° for 3 h.
Residue on Ignition, Appendix IIC   Ignite 1 g as directed in the general method.
Specific Rotation, Appendix IIB   Determine in a solution containing 10 g of a previously dried sample in sufficient 6 N hydrochloric acid to make 100 mL.

Packaging and Storage Store in well-closed, light-resistant containers.

Alginic Acid

$(C_6H_8O_6)_n$ Formula wt, *Calculated*, 176.13
              Formula wt, *Actual* (Avg), 200.00

INS: 400   CAS: [9005-32-7]

DESCRIPTION

Alginic Acid is a hydrophilic colloidal carbohydrate extracted by the use of dilute alkali from various species of brown seaweeds (*Phaeophyceae*). It may be described chemically as a linear glycuronoglycan consisting mainly of β-(1 → 4) linked D-mannuronic and L-guluronic acid units in the pyranose ring form. It occurs as a white to yellowish white, fibrous powder. It is odorless and tasteless. Alginic Acid is insoluble in water, readily soluble in alkaline solutions, and insoluble in organic solvents. The pH of a 3 in 100 suspension in water is between 2.0 and 3.4.

Functional Use in Foods Stabilizer; thickener; emulsifier.

REQUIREMENTS

Identification
  A. To 5 mL of a 1 in 150 solution in 0.1 *N* sodium hydroxide add 1 mL of calcium chloride TS. A voluminous gelatinous precipitate is formed.
  B. To 5 mL of the solution prepared for *Identification Test A* add 1 mL of 2 *N* sulfuric acid. A heavy gelatinous precipitate is formed.
  C. To about 5 mg, contained in a test tube, add 5 mL of water, 1 mL of a freshly prepared 1 in 100 solution of naphtholresorcinol in ethanol, and 5 mL of hydrochloric acid. Heat the mixture to boiling, boil gently for about 3 min, and then cool to about 15°. Transfer the contents of the test tube to a 30-mL separator with the aid of 5 mL of water and extract with 15 mL of isopropyl ether. Perform a blank test (see *General Provisions*). The isopropyl ether extract from the sample exhibits a deeper purplish hue than that from the blank.
Assay It yields not less than 20% and not more than 23% of carbon dioxide (CO_2), corresponding to between 91.0% and 104.5% of Alginic Acid (Equiv wt 200.00), calculated on the dried basis.
Arsenic (as As) Not more than 3 mg/kg.
Ash Not more than 4.0% after drying.
Heavy Metals (as Pb) Not more than 0.002%.
Lead Not more than 5 mg/kg.
Loss on Drying Not more than 15.0%.

TESTS

Assay Proceed as directed under *Alginates Assay*, Appendix IIIC. Each mL of 0.25 *N* sodium hydroxide consumed in the assay is equivalent to 25 mg of Alginic Acid (equiv wt 200.00).
Arsenic A *Sample Solution* prepared as directed for organic compounds meets the requirements of the *Arsenic Test*, Appendix IIIB.
Ash Weigh accurately about 3 g in a tared crucible, and incinerate at about 650° until free from carbon. Cool the crucible and its contents in a desiccator, weigh, and determine the weight of the ash.
Heavy Metals Prepare and test a 1-g sample as directed in *Method II* under the *Heavy Metals Test*, Appendix IIIB, but use nitric acid instead of sulfuric acid to wet the sample prior to ignition, and cautiously ignite in a platinum crucible. Any color does not exceed that produced in a control (*Solution A*) containing 20 μg of lead ion (Pb).
Lead A *Sample Solution* prepared as directed for organic compounds meets the requirements of the *Lead Limit Test*, Appendix IIIB, using 5 μg of lead ion (Pb) in the control.
Loss on Drying, Appendix IIC Dry at 105° for 4 h.

Packaging and Storage Store in well-closed containers.

Almond Oil, Bitter, FFPA

Bitter Almond Oil Free from Prussic Acid

                                CAS: [8013-76-1]

DESCRIPTION

A volatile oil obtained from *Prunus amygdalus* Batsch var. *amara* (De Candolle) Focke (Fam. *Rosaceae*), apricot kernel (*Prunus armeniaca* L.), and other fruit kernels containing amygdalin. It is prepared by steam distillation of a water-macerated, powdered, and pressed cake that has been specially treated and redistilled to remove hydrocyanic acid. It is a colorless to slightly yellow liquid having a strong almondlike aroma and a slightly astringent, mild taste. It is soluble in most fixed oils and in propylene glycol, and it is slightly soluble in mineral oil. It is insoluble in glycerin.

Functional Use in Foods Flavoring agent.

REQUIREMENTS

Identification The infrared absorption spectrum of the sample exhibits relative maxima (that may vary in intensity) at the same wavelengths (or frequencies) as those shown in the respective spectrum in the section on *Infrared Spectra* (*Series A: Essential Oils*), using the same test conditions as specified therein.
Assay Not less than 95.0% of aldehydes, calculated as benzaldehyde (C_7H_6O).

Acid Value Not more than 8.0.
Angular Rotation Optically inactive, or not more than ±0.15°.
Chlorinated Compounds Passes test.
Heavy Metals (as Pb) Passes test.
Hydrocyanic Acid Passes test (about 0.15%).
Refractive Index Between 1.541 and 1.546 at 20°.
Solubility in Alcohol Passes test.
Specific Gravity Between 1.040 and 1.050.

TESTS

Assay Weigh accurately about 1 mL, and proceed as directed under *Aldehydes*, Appendix VI, using 53.05 as the equivalence factor (*e*) in the calculation.
Acid Value Determine as directed for *Acid Value* under *Essential Oils and Flavors*, Appendix VI.
Angular Rotation Determine in a 100-mm tube as directed under *Optical (Specific) Rotation*, Appendix IIB.
Chlorinated Compounds Proceed as directed in the general method, Appendix VI.
Heavy Metals Shake 10 mL of the oil with an equal volume of water to which 1 drop of hydrochloric acid has been added, and pass hydrogen sulfide through the mixture until it is saturated. No darkening in color is produced in either the oil or the water.
Hydrocyanic Acid To a 1-mL sample in a test tube, add 1 mL of water, 5 drops of a 1 in 10 sodium hydroxide solution, and 5 drops of a 1 in 10 ferrous sulfate solution. Shake thoroughly and acidify with 0.5 *N* hydrochloric acid. No blue precipitate or color forms.
Refractive Index, Appendix IIB Determine with an Abbé or other refractometer of equal or greater accuracy.
Solubility in Alcohol Proceed as directed in the general method, Appendix VI. One mL dissolves to form a clear solution in 2 mL of 70% alcohol.
Specific Gravity Determine by any reliable method (see *General Provisions*).

Packaging and Storage Store in full, tight, glass, aluminum, tin-lined, or other suitably lined containers in a cool place protected from light.

Aluminum Ammonium Sulfate

Ammonium Alum

$AlNH_4(SO_4)_2 \cdot 12H_2O$ Formula wt 453.33

INS: 523   CAS: [7784-25-0]

DESCRIPTION

Large, colorless crystals, white granules, or a powder. It is odorless and has a sweetish, strongly astringent taste. One g dissolves in 7 mL of water at 25° and in about 0.3 mL of boiling water. It is insoluble in alcohol, and is freely, but slowly, soluble in glycerin. Its solutions are acid to litmus.

Functional Use in Foods Buffer; neutralizing agent.

REQUIREMENTS

Identification A 1 in 20 solution gives positive tests for *Aluminum*, for *Ammonium*, and for *Sulfate*, Appendix IIIA.
Assay Not less than 99.5% and not more than 100.5% of $AlNH_4(SO_4)_2 \cdot 12H_2O$.
Alkalies and Alkaline Earths Passes test.
Fluoride Not more than 0.003%.
Heavy Metals (as Pb) Not more than 0.002%.
Lead Not more than 10 mg/kg.
Selenium Not more than 0.003%.

TESTS

Assay Weigh accurately about 1 g of sample, dissolve in 50 mL of water, add 50.0 mL of 0.05 *M* disodium EDTA and 20 mL of pH 4.5 buffer solution (77.1 g of ammonium acetate and 57 mL of glacial acetic acid in 1000 mL of solution), and boil gently for 5 min. Cool, and add 50 mL of alcohol, and 2 mL of dithizone TS. Titrate with 0.05 *M* zinc sulfate to a bright rose-pink color, and perform a blank determination (see *General Provisions*). Each mL of 0.05 *M* disodium EDTA is equivalent to 22.67 mg of $AlNH_4(SO_4)_2 \cdot 12H_2O$.
Alkalies and Alkaline Earths Completely precipitate the aluminum from a boiling solution of 1 g of the sample in 100 mL of water by the addition of enough 6 *N* ammonium hydroxide to render the solution distinctly alkaline to methyl red TS, and filter. Evaporate the filtrate to dryness, and ignite. The weight of the residue does not exceed 5 mg.
Fluoride

Lime Suspension Carefully slake about 56 g of low-fluorine calcium oxide (about 2 mg/kg F) with 250 mL of water, and add 250 mL of 60% perchloric acid slowly and with stirring. Add a few glass beads, and boil to copious fumes of perchloric acid, then cool, add 200 mL of water, and boil again. Repeat the dilution and boiling once more, cool, dilute considerably, and filter through a fritted-glass filter if precipitated silicon dioxide appears. Pour the clear solution, with stirring, into 1000 mL of sodium hydroxide solution (1 in 10), allow the precipitate to settle, and siphon off the supernatant liquid. Remove the sodium salts from the precipitate by washing five times in large centrifuge bottles, shaking the mass thoroughly each time. Finally, shake the precipitate into a suspension and dilute to 2000 mL. Store in paraffin-lined bottles and shake well before use.

Note: 100 mL of this suspension should give no appreciable fluoride blank when evaporated, distilled, and titrated as directed in the *Fluoride Limit Test*, Appendix IIIB.

Procedure Assemble the distilling apparatus as described in the *Fluoride Limit Test*, Appendix IIIB, and add to the distilling flask 1.67 g of the sample, accurately weighed, and 25 mL of dilute sulfuric acid (1 in 2). Distill until the temperature reaches 160°, then maintain at 160° to 165° by adding water from the funnel, collecting 300 mL of distillate. Oxidize the distillate by the cautious addition of 2 or 3 mL of fluorine-free 30% hydrogen peroxide (to remove sulfates), allow to stand for a few min, and evaporate in a platinum dish with an excess of *Lime Suspension*. Ignite briefly at 600°, then cool and wet the ash with about 10 mL of water. Cover the dish with a watch glass, and cautiously introduce under cover just sufficient 60% perchloric acid to dissolve the ash. Add the contents of the dish through the dropping funnel of a freshly prepared distilling apparatus (the distilling flask should contain a few glass beads), using a total of 20 mL of the 60% perchloric acid for dissolving the ash and transferring the solution. Add 10 mL of water and a few drops of silver perchlorate solution (1 in 2) through the dropping funnel, and continue as directed in the *Fluoride Limit Test*, Appendix IIIB, beginning with "Distill until the temperature reaches 135°. . . ."

Heavy Metals Dissolve 1 g in 20 mL of water, add a few drops of 2.7 N hydrochloric acid, and evaporate to dryness in a porcelain dish. Treat the residue with 20 mL of water, and add 50 mg of hydroxylamine hydrochloride. Heat on a steam bath for 10 min, cool, and dilute to 25 mL with water. This solution meets the requirements of the *Heavy Metals Test*, Appendix IIIB, using 20 μg of lead ion (Pb) and 50 mg of hydroxylamine hydrochloride in the control (*Solution A*).

Lead A solution of 1 g in 10 mL of water meets the requirements of the *Lead Limit Test*, Appendix IIIB, using 10 μg of lead ion (Pb) in the control.

Selenium Determine as directed in *Method II* under the *Selenium Limit Test*, Appendix IIIB, using 200 mg of sample.

Packaging and Storage Store in well-closed containers.

Aluminum Potassium Sulfate

Potassium Alum

AlK(SO$_4$)$_2$·12H$_2$O   Formula wt 474.39

INS: 522   CAS: [7784-24-9]

DESCRIPTION

Large, transparent crystals or crystalline fragments, or a white crystalline powder. It is odorless and has a sweetish, astringent taste. One g dissolves in 7.5 mL of water at 25° and in about 0.3 mL of boiling water. It is insoluble in alcohol, but is freely soluble in glycerin. Its solutions are acid to litmus.

Functional Use in Foods Buffer; neutralizing agent; firming agent.

REQUIREMENTS

Identification A 1 in 20 solution gives positive tests for *Aluminum*, for *Potassium*, and for *Sulfate*, Appendix IIIA.

Assay Not less than 99.5% and not more than 100.5% of AlK(SO$_4$)$_2$·12H$_2$O.
Ammonium Salts Passes test.
Fluoride Not more than 0.003%.
Heavy Metals (as Pb) Not more than 10 mg/kg.
Selenium Not more than 0.003%.

TESTS

Assay Weigh accurately about 1 g of sample, dissolve in 50 mL of water, add 50.0 mL of 0.05 M disodium EDTA and 20 mL of pH 4.5 buffer solution (77.1 g of ammonium acetate and 57 mL of glacial acetic acid in 1000 mL of solution), and boil gently for 5 min. Cool, and add 50 mL of alcohol, and 2 mL of dithizone TS. Titrate with 0.05 M zinc sulfate to a bright rose-pink color, and perform a blank determination (see *General Provisions*). Each mL of 0.05 M disodium EDTA is equivalent to 23.72 mg of AlK(SO$_4$)$_2$·12H$_2$O.

Ammonium Salts Heat 1 g with 10 mL of 1 N sodium hydroxide on a steam bath for 1 min. The odor of ammonia is not perceptible.

Fluoride Determine as directed in the test for *Fluoride* in the monograph for *Aluminum Ammonium Sulfate*.

Heavy Metals Dissolve 2 g in 20 mL of water, add a few drops of 2.7 N hydrochloric acid, and evaporate to dryness in a porcelain dish. Treat the residue with 20 mL of water, and add 50 mg of hydroxylamine hydrochloride. Heat on a steam bath for 10 min, cool, and dilute to 25 mL with water. This solution meets the requirements of the *Heavy Metals Test*, Appendix IIIB, using 20 μg of lead ion (Pb) and 50 mg of hydroxylamine hydrochloride in the control (*Solution A*).

Selenium Determine as directed in *Method II* under the *Selenium Limit Test*, Appendix IIIB, using 200 mg of sample.

Packaging and Storage Store in well-closed containers.

Aluminum Sodium Sulfate

Soda Alum; Sodium Alum

AlNa(SO$_4$)$_2$ Formula wt 242.10

INS: 521   CAS: [10102-71-3]

DESCRIPTION

Aluminum Sodium Sulfate is anhydrous or may contain up to 12 molecules of water of hydration. It occurs as colorless crystals, white granules, or a powder. It is odorless and has a saline, astringent taste. The anhydrous form is slowly soluble in water. The dodecahydrate is freely soluble in water, and it effloresces in air. Both forms are insoluble in alcohol.

Functional Use in Foods Buffer; neutralizing agent; firming agent.

REQUIREMENTS

Identification It responds to the flame test for *Sodium*, Appendix IIIA, and gives positive tests for *Aluminum* and for *Sulfate*, Appendix IIIA.

Assay *Anhydrous form*: Not less than 99.0% and not more than 104.0% of $AlNa(SO_4)_2$ after drying; *dodecahydrate*: not less than 99.5% of $AlNa(SO_4)_2$ after drying.

Ammonium Salts Passes test.

Fluoride Not more than 0.003%.

Heavy Metals (as Pb) Not more than 0.002%.

Lead Not more than 10 mg/kg.

Loss on Drying *Anhydrous form*: Not more than 10%; *dodecahydrate*: not more than 47.2%.

Neutralizing Value *Anhydrous form*: Between 104 and 108.

Selenium Not more than 0.003%.

TESTS

Assay Weigh accurately about 500 mg of sample previously dried as directed in the test for *Loss on Drying*, moisten with 1 mL of glacial acetic acid, and dissolve in 50 mL of water, warming gently on a steam bath until solution is complete. Cool, neutralize with 6 N ammonium hydroxide, add 50.0 mL of 0.05 M disodium EDTA and 20 mL of pH 4.5 buffer solution (77.1 g of ammonium acetate and 57 mL of glacial acetic acid in 1000 mL of solution), and boil gently for 5 min. Cool, and add 50 mL of alcohol and 2 mL of dithizone TS. Titrate with 0.05 M zinc sulfate to a bright rose-pink color, and perform a blank determination (see *General Provisions*). Each mL of 0.05 M disodium EDTA is equivalent to 12.10 mg of $AlNa(SO_4)_2$.

Ammonium Salts Heat 1 g with 10 mL of 1 N sodium hydroxide on a steam bath for 1 min. The odor of ammonia is not perceptible.

Fluoride Determine as directed in the test for *Fluoride* in the monograph for *Aluminum Ammonium Sulfate*.

Heavy Metals Dissolve 1 g in 20 mL of water, add a few drops of 2.7 N hydrochloric acid, and evaporate to dryness in a porcelain dish. Treat the residue with 20 mL of water, and add 50 mg of hydroxylamine hydrochloride. Heat on a steam bath for 10 min, cool, and dilute to 25 mL with water. This solution meets the requirements of the *Heavy Metals Test*, Appendix IIIB, using 20 µg of lead ion (Pb) and 50 mg of hydroxylamine hydrochloride in the control (*Solution A*).

Lead A solution of 1 g in 10 mL of water meets the requirements of the *Lead Limit Test*, Appendix IIIB, using 10 µg of lead ion (Pb) in the control.

Loss on Drying, Appendix IIC *Anhydrous form*: dry at 200° for 16 h; *dodecahydrate*: dry first at 50° to 55° for 1 h, then at 200° for 16 h.

Neutralizing Value Weigh accurately 500 mg of the anhydrous form into a 200-mL Erlenmeyer flask, add 30 mL of water and 4 drops of phenolphthalein TS, and boil until the sample dissolves. Add 13.0 mL of 0.5 N sodium hydroxide, boil for a few s, and titrate with 0.5 N hydrochloric acid to the disappearance of the pink color, adding the acid dropwise and agitating vigorously after each addition. Calculate the neutralizing value, as parts of $NaHCO_3$ equivalent to 100 parts of the sample, by the formula

$$8.4V,$$

in which V is the volume, in mL, of 0.5 N sodium hydroxide consumed by the sample.

Selenium Determine as directed in *Method II* under the *Selenium Limit Test*, Appendix IIIB, using 200 mg of sample.

Packaging and Storage Store in tight containers.

Aluminum Sulfate

$Al_2(SO_4)_3 \cdot xH_2O$ Formula wt, anhydrous 342.15

INS: 520   CAS: anhydrous [10043-01-3]

DESCRIPTION

Aluminum Sulfate is anhydrous or contains 18 molecules of water of crystallization. Due to efflorescence, the hydrate may have a composition approximating the formula $Al_2(SO_4)_3 \cdot 14H_2O$. It occurs as a white powder, as shining plates, or as crystalline fragments. It is odorless and has a sweet taste, becoming mildly astringent. One g of the hydrate dissolves in about 2 mL of water. The anhydrous product approaches the same solubility, but the rate of solution is so slow that it initially appears to be relatively insoluble. The pH of a 1 in 20 solution is 2.9 or above.

Functional Use in Foods Firming agent.

REQUIREMENTS

Identification A 1 in 10 solution gives positive tests for *Aluminum* and for *Sulfate*, Appendix IIIA.

Assay $Al_2(SO_4)_3$ (anhydrous): not less than 99.5% of $Al_2(SO_4)_3$, calculated on the ignited basis; $Al_2(SO_4)_3 \cdot 18H_2O$ (hydrate): not less than 99.5% and not more than 114.0% of $Al_2(SO_4)_3 \cdot 18H_2O$, corresponding to not more than approximately 101.7% of $Al_2(SO_4)_3 \cdot 14H_2O$.

Alkalies and Alkaline Earths Passes test (about 0.4%).

Ammonium Salts Passes test.

Fluoride Not more than 0.003%.

Heavy Metals (as Pb) Not more than 0.002%.

Lead Not more than 10 mg/kg.

Loss on Ignition $Al_2(SO_4)_3$ (anhydrous): not more than 5%.

> Note: This **REQUIREMENT** does not apply to $Al_2(SO_4)_3 \cdot 18H_2O$.

Selenium Not more than 0.003%.

TESTS

Assay Weigh accurately an amount of sample equivalent to about 4 g of $Al_2(SO_4)_3$, transfer into a 250-mL volumetric flask, dissolve in water, dilute to volume, and mix. Pipet 10 mL of this solution into a 250-mL beaker, add 25.0 mL of 0.05 M

disodium EDTA and 20 mL of pH 4.5 buffer solution (77.1 g of ammonium acetate and 57 mL of glacial acetic acid in 1000 mL of solution), and boil gently for 5 min. Cool, and add 50 mL of alcohol and 2 mL of dithizone TS. Titrate with 0.05 M zinc sulfate to a bright rose-pink color, and perform a blank determination (see *General Provisions*). Each mL of 0.05 M disodium EDTA is equivalent to 8.554 mg of $Al_2(SO_4)_3$ or to 16.66 mg of $Al_2(SO_4)_3 \cdot 18H_2O$.

Alkalies and Alkaline Earths To a boiling solution of 2 g in 150 mL of water add a few drops of methyl red TS, and then add 6 N ammonium hydroxide until the color of the solution just changes to a distinct yellow. Add hot water to restore the original volume, and filter while hot. Evaporate 75 mL of the filtrate to dryness, and ignite to constant weight. Not more than 4 mg of residue remains.

Ammonium Salts Heat 1 g with 10 mL of 1 N sodium hydroxide on a steam bath for 1 min. The odor of ammonia is not perceptible.

Fluoride Determine as directed in the test for *Fluoride* in the monograph for *Aluminum Ammonium Sulfate*.

Heavy Metals Dissolve 1 g in 20 mL of water, add a few drops of 2.7 N hydrochloric acid, and evaporate to dryness in a porcelain dish. Treat the residue with 20 mL of water, and add 100 mg of hydroxylamine hydrochloride. Heat on a steam bath for 10 min, cool, and dilute to 25 mL with water. This solution meets the requirements of the *Heavy Metals Test*, Appendix IIIB, using 20 μg of lead ion (Pb) and 100 mg of hydroxylamine hydrochloride in the control (*Solution A*).

Lead A solution of 1 g in 10 mL of water meets the requirements of the *Lead Limit Test*, Appendix IIIB, using 10 μg of lead ion (Pb) in the control.

Loss on Ignition Weigh accurately about 2 g of $Al_2(SO_4)_3$ (anhydrous), and ignite, preferably in a muffle furnace, at about 500° for 3 h.

Note: This TEST does not apply to $Al_2(SO_4)_3 \cdot 18H_2O$.

Selenium Determine as directed in *Method II* under the *Selenium Limit Test*, Appendix IIIB, using 200 mg of sample.

Packaging and Storage Store in well-closed containers.

Ambrette Seed Oil

Ambrette Seed Liquid

CAS: [8015-62-1]

DESCRIPTION

The volatile oil obtained by steam distillation from the partially dried and crushed seeds of the plant *Abelmoschus moschatus* Moench, syn. *Hibiscus Abelmoschus* L. (Fam. *Malvaceae*). It is refined by solvent extraction to remove fatty acids, or precipitation of the fatty acid salts. It is a clear yellow to amber liquid having the strong musky odor of ambrettolide. It is soluble in most fixed oils and in mineral oil, often with cloudiness. It is relatively insoluble in glycerin and in propylene glycol.

Functional Use in Foods Flavoring agent.

REQUIREMENTS

Identification The infrared absorption spectrum of the sample exhibits relative maxima (that may vary in intensity) at the same wavelengths (or frequencies) as those shown in the respective spectrum in the section on *Infrared Spectra* (*Series A: Essential Oils*), using the same test conditions as specified therein.
Acid Value Not more than 3.0.
Angular Rotation Between −2.5° and +3°.
Heavy Metals (as Pb) Passes test.
Refractive Index Between 1.468 and 1.485 at 20°.
Saponification Value Between 140 and 200.
Specific Gravity Between 0.898 and 0.920.

TESTS

Acid Value Determine as directed for *Acid Value* under *Essential Oils and Flavors*, Appendix VI.

Angular Rotation Determine in a 100-mm tube as directed under *Optical (Specific) Rotation*, Appendix IIB.

Heavy Metals Shake 10 mL of the oil with an equal volume of water to which 1 drop of hydrochloric acid has been added, and pass hydrogen sulfide through the mixture until it is saturated. No darkening in color is produced in either the oil or the water.

Refractive Index, Appendix IIB Determine with an Abbé or other refractometer of equal or greater accuracy.

Saponification Value Determine as directed for *Saponification Value* under *Essential Oils and Flavors*, Appendix VI, using about 1 g, accurately weighed.

Specific Gravity Determine by any reliable method (see *General Provisions*).

Packaging and Storage Store in full, preferably glass, aluminum, tin-lined, or other suitably lined containers in a cool place protected from light.

Ammoniated Glycyrrhizin

DESCRIPTION

It is a brown powder. It is precipitated by acid from the water extract of dried and ground rhizomes and roots of *Glycyrrhiza glabra* or related *Glycyrrhiza* (licorice root) and neutralized with dilute ammonia. Suitable diluents may be added.

Functional Use in Foods Flavoring agent; flavor enhancer.

REQUIREMENTS

Identification It gives positive tests for *Ammonium*, Appendix IIIA.

Assay Not less than 22.0% and not more than 32.0% of monoammonium glycyrrhizinate, $C_{42}H_{65}NO_{16}$, calculated on the dried basis.

Ash (Total) Not more than 2.5%.

Heavy Metals (as Pb) Not more than 10 mg/kg.

Loss on Drying Not more than 6.0%.

TESTS

Assay (Based on AOAC method 982.19.)

Apparatus Use a high-pressure liquid chromatograph, operated at room temperature, and containing a 10-μm particle size, 30-cm × 4-mm id, C_{18} reverse-phase column (μBondapak C_{18} or equivalent). Maintain the *Mobile Phase* at a pressure and flow rate (typically 2.0 mL/min) capable of giving the required elution time (see *System Suitability*). Use an ultraviolet detector that monitors absorption at 254 nm (0.2 to 0.1 AUFS range).

Mobile Phase Add 380 mL of acetonitrile and 10 mL of glacial acetic acid to 610 mL of glass-distilled water filtered through a 0.45-μm filter (Millipore or equivalent). Mix, and de-gas thoroughly.

Standard Solution Weigh accurately about 10 mg of Monoammonium Glycyrrhizinate Standard for analytical use (available from MacAndrews & Forbes Company[1]) in 20 mL of a 1:1 acetonitrile–water solution. Filter the solution through a 0.45-μm Millipore filter or equivalent. Prepare fresh daily.

Note: Correct the weight of Monoammonium Glycyrrhizinate Standard taken for the percent loss on drying shown on its label.

Assay Solution Weigh accurately about 40 mg of the sample, and dissolve in 20 mL of water. Filter the solution through a 0.45-μm Millipore filter or equivalent.

System Suitability Inject duplicate 10-μL portions of the *Standard Solution* into the chromatograph. The retention time of the monoammonium glycyrrhizinate is approximately 6 min. Adjust the operating conditions if necessary. The mean standard deviation for replicate injections is not more than 2.0%.

Procedure Separately inject, in duplicate, 10-μL volumes of the *Standard Solution* and the *Assay Solution* into the chromatograph, and determine the mean peak area for each solution. Calculate the percentage of monoammonium glycyrrhizinate, equivalent to $C_{42}H_{65}NO_{16}$, in the portion of Ammoniated Glycyrrhizin taken by the formula

$$100(20C_S/W_U)(A_U/A_S),$$

in which C_S is the concentration, in mg/mL, of the *Standard Solution*; A_S and A_U are the peak areas of the *Standard Solution* and the *Assay Solution*, respectively; and W_U is the weight, in mg, of the sample taken.

Ash (Total) Proceed as directed in the general method, Appendix IIC.

---
[1] Third Street and Jefferson Avenue, Camden, NJ 08104.

Heavy Metals A 2-g sample meets the requirements of the *Heavy Metals Test, Method II*, Appendix IIIB, using 20 μg of lead ion (Pb) in the control (*Solution A*).

Loss on Drying, Appendix IIC Dry 1 g at 105° for 1 h.

Packaging and Storage Store in a tight container in a cool, dry place.

Ammonium Alginate

Algin

$(C_6H_7O_6NH_4)_n$ Formula wt, Calculated 193.16
                        Formula wt, Actual (Avg) 217.00

INS: 403   CAS: [9005-34-9]

DESCRIPTION

The ammonium salt of alginic acid (see the monograph for *Alginic Acid*) occurs as a white to yellowish, fibrous or granular powder. It dissolves in water to form a viscous, colloidal solution. It is insoluble in alcohol and in hydroalcoholic solutions in which the alcohol content is greater than about 30% by weight. It is insoluble in chloroform, in ether, and in acids having a pH lower than about 3.

Functional Use in Foods Stabilizer; thickener; emulsifier.

REQUIREMENTS

Identification

A. To 5 mL of a 1 in 100 solution add 1 mL of calcium chloride TS. A voluminous, gelatinous precipitate is formed.

B. To 10 mL of a 1 in 100 solution add 1 mL of 2.7 N sulfuric acid. A heavy gelatinous precipitate is formed.

C. Ammonium Alginate meets the requirements of *Identification Test C* in the monograph for *Alginic Acid*.

D. To about 1 g of Ammonium Alginate contained in a test tube add 5 mL of 1 N sodium hydroxide, and shake the mixture briefly. The odor of ammonia is evolved.

Assay It yields not less than 18% and not more than 21% of carbon dioxide (CO_2), corresponding to between 88.7% and 103.6% of Ammonium Alginate (equiv wt 217.00), calculated on the dried basis.

Arsenic (as As) Not more than 3 mg/kg.

Ash Not more than 4.0% after drying.

Heavy Metals (as Pb) Not more than 0.002%.

Lead Not more than 5 mg/kg.

Loss on Drying Not more than 15.0%.

TESTS

Assay Proceed as directed under *Alginates Assay*, Appendix IIIC. Each mL of 0.25 N sodium hydroxide consumed in the assay is equivalent to 27.12 mg of Ammonium Alginate (equiv wt 217.00).

Arsenic A *Sample Solution* prepared as directed for organic compounds meets the requirements of the *Arsenic Test*, Appendix IIIB.

Ash Determine as directed under *Ash* in the monograph for *Alginic Acid*.

Heavy Metals Prepare and test a 1-g sample as directed in *Method II* under the *Heavy Metals Test*, Appendix IIIB, but use nitric acid instead of sulfuric acid to wet the sample prior to ignition, and cautiously ignite in a platinum crucible. Any color does not exceed that produced in a control (*Solution A*) containing 20 μg of lead ion (Pb).

Lead A *Sample Solution* prepared as directed for organic compounds meets the requirements of the *Lead Limit Test*, Appendix IIIB, using 5 μg of lead ion (Pb) in the control.

Loss on Drying, Appendix IIC Dry at 105° for 4 h.

Packaging and Storage Store in well-closed containers.

Chloride, Appendix IIIB Any turbidity produced by a 500-mg sample does not exceed that shown in a control containing 15 μg of chloride ion (Cl).

Heavy Metals Dissolve the residue from the test for *Nonvolatile Residue* in 1 mL of 2.7 N hydrochloric acid, evaporate to dryness, and dissolve the residue in 25 mL of water. This solution meets the requirements of the *Heavy Metals Test*, Appendix IIIB, using 20 μg of lead ion (Pb) in the control (*Solution A*).

Nonvolatile Residue Transfer 4 g into a tared dish, add 10 mL of water, and evaporate to dryness on a steam bath. Heat the dish at 105° for 1 h, cool in a desiccator, and weigh. Retain the residue for the *Heavy Metals Test*.

Sulfur Compounds Dissolve 4 g in 40 mL of water, add about 10 mg of sodium carbonate and 1 mL of 30% hydrogen peroxide, and evaporate the solution to dryness on a steam bath. Treat the residue as directed in the *Sulfate Limit Test*, Appendix IIIB. Any turbidity produced does not exceed that shown by 280 μg of sulfate ion (SO_4).

Packaging and Storage Store in well-closed containers.

Ammonium Bicarbonate

NH_4HCO_3 Formula wt 79.06

INS: 503(ii) CAS: [1066-33-7]

DESCRIPTION

White crystals or a crystalline powder having a slight odor of ammonia. At a temperature of 60° or above it volatilizes rapidly, dissociating into ammonia, carbon dioxide, and water, but at room temperature it is quite stable. One g dissolves in about 6 mL of water. It is insoluble in alcohol.

Functional Use in Foods Alkali; leavening agent.

REQUIREMENTS

Identification It gives positive tests for *Ammonium* and for *Bicarbonate*, Appendix IIIA.

Assay Not less than 99.0% and not more than 100.5% of NH_4HCO_3.

Chloride Not more than 0.003%.

Heavy Metals (as Pb) Not more than 5 mg/kg.

Nonvolatile Residue Not more than 0.05% (0.55% for products containing a suitable anticaking agent).

Sulfur Compounds (as SO_4) Not more than 0.007%.

TESTS

Assay Weigh accurately about 3 g of sample, and dissolve it in 40 mL of water. Add 2 drops of methyl red TS, and titrate with 1 N hydrochloric acid. Add the acid slowly, with constant stirring, until the solution becomes faintly pink. Heat the solution to boiling, cool, and continue the titration until the faint pink color no longer fades after boiling. Each mL of 1 N hydrochloric acid is equivalent to 79.06 mg of NH_4HCO_3.

Ammonium Carbonate

INS: 503(i) CAS: [10361-29-2]

DESCRIPTION

Ammonium Carbonate consists of ammonium bicarbonate (NH_4HCO_3) and ammonium carbamate ($NH_2.COONH_4$) in varying proportions. It occurs as a white powder or as hard, white or translucent masses. Its solutions are alkaline to litmus. On exposure to air it becomes opaque and is finally converted into porous lumps or a white powder of ammonium bicarbonate due to the loss of ammonia and carbon dioxide. One g dissolves slowly in about 4 mL of water.

Functional Use in Foods Buffer; leavening agent; neutralizing agent.

REQUIREMENTS

Identification When heated, it volatilizes without charring and the vapor is alkaline to moistened litmus paper. A 1 in 20 solution effervesces upon the addition of an acid.

Assay Not less than 30.0% and not more than 34.0% of NH_3.

Chloride Not more than 0.003%.

Heavy Metals (as Pb) Not more than 5 mg/kg.

Nonvolatile Residue Not more than 0.05%.

Sulfur Compounds (as SO_4) Not more than 0.005%.

TESTS

Assay Place about 10 mL of water in a weighing bottle, tare the bottle and its contents, add about 2 g of Ammonium Carbonate, and weigh accurately. Transfer the contents of the bottle to a 250-mL flask, and slowly add, with mixing, 50.0 mL of 1 N sulfuric acid, allowing for the release of carbon dioxide. When solution has been effected, wash down the sides of the flask with a few mL of water, add methyl orange TS, and titrate the excess acid with 1 N sodium hydroxide. Each mL of 1 N sulfuric acid is equivalent to 17.03 mg of NH_3.

Chloride Dissolve 500 mg in 10 mL of hot water, add about 5 mg of sodium carbonate, and evaporate to dryness on a steam bath. Treat the residue as directed in the *Chloride Limit Test*, Appendix IIIB. Any turbidity produced does not exceed that shown in a control containing 15 μg of chloride ion (Cl).

Heavy Metals Dissolve the residue from the test for *Nonvolatile Residue* in 1 mL of 2.7 N hydrochloric acid, and evaporate to dryness. Dissolve the residue in 25 mL of water. This solution meets the requirements of the *Heavy Metals Test*, Appendix IIIB, using 20 μg of lead ion (Pb) in the control (*Solution A*).

Nonvolatile Residue Transfer 4 g into a tared dish, add 10 mL of water, evaporate on a steam bath. Heat the dish at 105° for 1 h, cool in a desiccator, and weigh. Retain the residue for the *Heavy Metals Test*.

Sulfur Compounds Dissolve 4 g in 40 mL of water, add about 10 mg of sodium carbonate and 1 mL of 30% hydrogen peroxide, and evaporate the solution to dryness on a steam bath. Treat the residue as directed in the *Sulfate Limit Test*, Appendix IIIB. Any turbidity produced does not exceed that shown in a control containing 200 μg of sulfate (SO_4).

Packaging and Storage Store in tight, light-resistant containers, preferably at a temperature not exceeding 30°.

Ammonium Chloride

NH_4Cl Formula wt 53.49

INS: 510 CAS: [12125-02-9]

DESCRIPTION

Colorless crystals, or a white, fine or coarse crystalline powder. It has a cool, saline taste and is somewhat hygroscopic. One g dissolves in 2.6 mL of water at 25°, in 1.4 mL of boiling water, in about 100 mL of alcohol, and in about 8 mL of glycerin. The pH of a 1 in 20 solution is between 4.5 and 6.0.

Functional Use in Foods Yeast food; dough conditioner.

REQUIREMENTS

Identification A 1 in 10 solution gives positive tests for *Ammonium* and for *Chloride*, Appendix IIIA.

Assay Not less than 99.0% of NH_4Cl after drying.

Heavy Metals (as Pb) Not more than 10 mg/kg.

Loss on Drying Not more than 0.5%.

TESTS

Assay Dry about 200 mg over silica gel for 4 h, weigh accurately, and dissolve it in about 40 mL of water in a glass-stoppered flask. Add, while agitating, 3 mL of nitric acid, 5 mL of nitrobenzene, 50.0 mL of 0.1 N silver nitrate, shake vigorously, then add 2 mL of ferric ammonium sulfate TS, and titrate the excess silver nitrate with 0.1 N ammonium thiocyanate. Each mL of 0.1 N silver nitrate is equivalent to 5.349 mg of NH_4Cl.

Heavy Metals A solution of 2 g in 25 mL of water meets the requirements of the *Heavy Metals Test*, Appendix IIIB, using 20 μg of lead ion (Pb) in the control (*Solution A*).

Loss on Drying, Appendix IIC Dry over silica gel for 4 h.

Packaging and Storage Store in tight containers.

Ammonium Hydroxide

Strong Ammonia Solution; Stronger Ammonia Water

NH_4OH Formula wt 35.05

INS: 527 CAS: [7664-41-7]

DESCRIPTION

A clear, colorless solution of NH_3 having an exceedingly pungent, characteristic odor. Upon exposure to air it loses ammonia rapidly. Its specific gravity is about 0.90.

Functional Use in Foods Alkali.

REQUIREMENTS

Identification Dense, white fumes are produced when a glass rod wet with hydrochloric acid is held near the surface of the liquid.

Assay Not less than 27.0% and not more than 30.0%, by weight, of NH_3.

Heavy Metals (as Pb) Not more than 5 mg/kg.

Nonvolatile Residue Not more than 0.02%.

Readily Oxidizable Substances Passes test.

TESTS

Assay Tare accurately a 125-mL glass-stoppered Erlenmeyer flask containing 35.0 mL of 1 N sulfuric acid. Partially fill a 10-mL graduated pipet from near the bottom of a sample, previously cooled in the original sample bottle to 10° or lower. (Do not use vacuum for drawing up the sample.) Wipe off any liquid adhering to the outside of the pipet, and discard the first mL. Hold the pipet just above the surface of the acid, and transfer 2 mL into the flask, leaving at least 1 mL in the pipet. Stopper the flask, mix, and weigh again to obtain the weight of the sample. Add methyl red TS, and titrate the excess acid with 1

N sodium hydroxide. Each mL of 1 N sulfuric acid is equivalent to 17.03 mg of NH_3.
Heavy Metals Transfer 22 mL (20-g sample) to a beaker, add about 5 mg of sodium chloride, evaporate to dryness on a steam bath, and dissolve the residue in 2 mL of 1 N acetic acid and sufficient water to make 50 mL. A 10-mL portion of this solution, diluted to 25 mL with water, meets the requirements of the *Heavy Metals Test*, Appendix IIIB, using 20 μg of lead ion (Pb) in the control (*Solution A*).
Nonvolatile Residue Evaporate 11 mL (10-g sample) in a tared platinum or porcelain dish to dryness, dry at 105° for 1 h, cool, and weigh.
Readily Oxidizable Substances Dilute 4 mL with 6 mL of water, and add a slight excess of 2 N sulfuric acid and 0.1 mL of 0.1 N potassium permanganate. The pink color does not completely disappear within 10 min.

Packaging and Storage Store in tight containers, preferably at a temperature not exceeding 25°.

Ammonium Phosphate, Dibasic

Diammonium Phosphate

$(NH_4)_2HPO_4$ Formula wt 132.06

INS: 342(ii)   CAS: [7783-28-0]

DESCRIPTION

White, odorless crystals, crystalline powder, or granules having a cooling, saline taste. It is freely soluble in water. The pH of a 1 in 100 solution is between 7.6 and 8.2.

Functional Use in Foods Buffer; dough conditioner; leavening agent; yeast food.

REQUIREMENTS

Identification A 1 in 20 solution gives positive tests for *Ammonium* and for *Phosphate*, Appendix IIIA.
Assay Not less than 96.0% and not more than 102.0% of $(NH_4)_2HPO_4$.
Arsenic (as As) Not more than 3 mg/kg.
Fluoride Not more than 10 mg/kg.
Heavy Metals (as Pb) Not more than 10 mg/kg.

TESTS

Assay Dissolve about 600 mg, accurately weighed, in 40 mL of water, and titrate to a pH of 4.6 with 0.1 N sulfuric acid. Each mL of 0.1 N sulfuric acid is equivalent to 13.21 mg of $(NH_4)_2HPO_4$.
Arsenic A solution of 1 g in 35 mL of water meets the requirements of the *Arsenic Test*, Appendix IIIB.
Fluoride Determine on a 2-g sample as directed in *Method IV* under the *Fluoride Limit Test*, Appendix IIIB, using *Buffer Solution A* and 0.1 mL of *Fluoride Standard Solution*.
Heavy Metals A solution of 2 g in 25 mL of water meets the requirements of the *Heavy Metals Test*, Appendix IIIB, using 20 μg of lead ion (Pb) in the control (*Solution A*).

Note: Use glacial acetic acid in making the pH adjustment.

Packaging and Storage Store in tight containers.

Ammonium Phosphate, Monobasic

Monoammonium Phosphate

$NH_4H_2PO_4$ Formula wt 115.03

INS: 342(i)   CAS: [7722-76-1]

DESCRIPTION

White, odorless crystals, crystalline powder, or granules. It is freely soluble in water. The pH of a 1 in 100 solution is between 4.3 and 5.0.

Functional Use in Foods Buffer; dough conditioner; leavening agent; yeast food.

REQUIREMENTS

Identification A 1 in 20 solution gives positive tests for *Ammonium* and for *Phosphate*, Appendix IIIA.
Assay Not less than 96.0% and not more than 102.0% of $NH_4H_2PO_4$.
Arsenic (as As) Not more than 3 mg/kg.
Fluoride Not more than 10 mg/kg.
Heavy Metals (as Pb) Not more than 10 mg/kg.

TESTS

Assay Dissolve about 500 mg, accurately weighed, in 50 mL of water, and titrate to a pH of 8.0 with 0.1 N sodium hydroxide. Each mL of 0.1 N sodium hydroxide is equivalent to 11.50 mg of $NH_4H_2PO_4$.
Arsenic A solution of 1 g in 35 mL of water meets the requirements of the *Arsenic Test*, Appendix IIIB.
Fluoride Determine on a 2-g sample as directed in *Method IV* under the *Fluoride Limit Test*, Appendix IIIB, using *Buffer Solution B* and 0.1 mL of *Fluoride Standard Solution*.
Heavy Metals A solution of 2 g in 25 mL of water meets the requirements of the *Heavy Metals Test*, Appendix IIIB, using 20 μg of lead ion (Pb) in the control (*Solution A*).

Packaging and Storage Store in tight containers.

Ammonium Saccharin

1,2-Benzisothiazolin-3-one 1,1-Dioxide Ammonium Salt

$C_7H_8N_2O_3S$ Formula wt 200.22

DESCRIPTION

White crystals or a white crystalline powder. It is freely soluble in water. The pH of a 1 in 3 solution is between 5 and 6. It is intensely sweet.

Functional Use in Foods   Nonnutritive sweetener.

REQUIREMENTS

Identification

A. Dissolve about 100 mg in 5 mL of sodium hydroxide solution (1 in 20), evaporate to dryness, and gently fuse the residue over a small flame until it no longer evolves ammonia. After the residue has cooled, dissolve it in 20 mL of water, neutralize the solution with 2.7 N hydrochloric acid, and filter. The addition of a drop of ferric chloride TS to the filtrate produces a violet color.

B. Mix 20 mg with 40 mg of resorcinol, cautiously add 10 drops of sulfuric acid, and heat the mixture in a liquid bath at 200° for 3 min. After cooling, add 10 mL of water and an excess of 1 N sodium hydroxide. A fluorescent green liquid results.

C. A 1 in 10 solution gives positive tests for *Ammonium*, Appendix IIIA.

D. To 10 mL of a 1 in 10 solution add 1 mL of hydrochloric acid. A crystalline precipitate of saccharin is formed. Wash the precipitate well with cold water and dry at 105° for 2 h. It melts between 226° and 230° (*Class Ia*, Appendix IIB).

Assay   Not less than 98.0% and not more than 101.0% of $C_7H_8N_2O_3S$, calculated on the anhydrous basis.
Benzoate and Salicylate   Passes test.
Heavy Metals (as Pb)   Not more than 10 mg/kg.
Readily Carbonizable Substances   Passes test.
Selenium   Not more than 0.003%.
Toluenesulfonamides   Not more than 25 mg/kg.
Water   Not more than 0.3%.

TESTS

Assay   Weigh accurately about 500 mg, and transfer it quantitatively to a separator with the aid of 10 mL of water. Add 2 mL of 2.7 N hydrochloric acid, and extract the precipitated saccharin first with 30 mL, then with five 20-mL portions of a solvent composed of 9 volumes of chloroform and 1 volume of alcohol. Filter each extract through a small filter paper moistened with the solvent mixture, and evaporate the combined filtrates on a steam bath to dryness with the aid of a current of air. Dissolve the residue in 75 mL of hot water, cool, add phenolphthalein TS, and titrate with 0.1 N sodium hydroxide. Perform a blank determination, and make any necessary correction (see *General Provisions*). Each mL of 0.1 N sodium hydroxide is equivalent to 20.02 mg of $C_7H_8N_2O_3S$.

Benzoate and Salicylate   To 10 mL of a 1 in 20 solution previously acidified with 5 drops of glacial acetic acid, add 3 drops of ferric chloride TS. No precipitate or violet color appears.

Heavy Metals   Prepare and test a 2-g sample as directed in *Method II* under the *Heavy Metals Test*, Appendix IIIB, using 20 μg of lead ion (Pb) in the control (*Solution A*).

Readily Carbonizable Substances, Appendix IIB   Dissolve 200 mg in 5 mL of 95% sulfuric acid, and keep at a temperature of 48° to 50° for 10 min. The color is no darker than *Matching Fluid A*.

Selenium   Determine as directed in *Method I* under the *Selenium Limit Test*, Appendix IIIB, using 200 mg of sample.

Toluenesulfonamides   Determine as directed in the monograph for *Sodium Saccharin*.

Water   Determine by the *Karl Fischer Titrimetric Method*, Appendix IIB.

Packaging and Storage   Store in well-closed containers.

Ammonium Sulfate

$(NH_4)_2SO_4$ Formula wt 132.14

INS: 517   CAS: [7783-20-2]

DESCRIPTION

Colorless or white, odorless crystals or granules that decompose at temperatures above 280°. One g is soluble in about 1.5 mL of water, and is insoluble in alcohol. The pH of a 0.1 M solution is about 4.5 to 6.0.

Functional Use in Foods   Dough conditioner; yeast nutrient.

REQUIREMENTS

Identification   It gives positive tests for *Ammonium* and for *Sulfate*, Appendix IIIA.
Assay   Not less than 99.0% and not more than 100.5% of $(NH_4)_2SO_4$.
Heavy Metals (as Pb)   Not more than 10 mg/kg.
Residue on Ignition   Not more than 0.25%.
Selenium   Not more than 0.003%.

TESTS

Assay   Transfer about 2 g, accurately weighed, into a 250-mL flask and dissolve it in 100 mL of water. To the solution add 40 mL of a mixture of equal volumes of formaldehyde and water, previously neutralized to phenolphthalein TS with 1 N sodium hydroxide. Mix, allow to stand for 30 min, and titrate the mixture with 1 N sodium hydroxide to a pink endpoint that persists for 5 min. Each mL of 1 N sodium hydroxide is equivalent to 66.06 mg of (NH$_4$)$_2$SO$_4$.

Heavy Metals A solution of 2 g in 25 mL of water meets the requirements of the *Heavy Metals Test*, Appendix IIIB, using 20 µg of lead ion (Pb) in the control (*Solution A*).

Residue on Ignition, Appendix IIC Ignite 1 g as directed in the general method.

Selenium Determine as directed in *Method II* under the *Selenium Limit Test*, Appendix IIIB, using 200 mg of sample.

Packaging and Storage Store in well-closed containers.

Amyris Oil, West Indian Type

Sandalwood Oil, West Indian Type

DESCRIPTION

The volatile oil obtained by steam distillation from the wood of *Amyris balsamifera* L. (Fam. *Rutaceae*). It is a clear, pale yellow, viscous liquid having a distinct odor suggestive of sandalwood. It is soluble in most fixed oils and usually in mineral oil. It is soluble in an equal volume of propylene glycol, the solution often becoming opalescent on further dilution. It is practically insoluble in glycerin.

Functional Use in Foods Flavoring agent.

REQUIREMENTS

Identification The infrared absorption spectrum of the sample exhibits relative maxima (that may vary in intensity) at the same wavelengths (or frequencies) as those shown in the respective spectrum in the section on *Infrared Spectra (Series A: Essential Oils)*, using the same test conditions as specified therein.
Acid Value Not more than 3.0.
Angular Rotation Between +10° and +53°.
Ester Value Not more than 7.
Ester Value after Acetylation Between 115 and 165.
Heavy Metals (as Pb) Passes test.
Refractive Index Between 1.503 and 1.512 at 20°.
Solubility in Alcohol Passes test.
Specific Gravity Between 0.943 and 0.976.

TESTS

Acid Value Determine as directed for *Acid Value* under *Essential Oils and Flavors*, Appendix VI.
Angular Rotation Determine in a 100-mm tube as directed under *Optical (Specific) Rotation*, Appendix IIB.
Ester Value Determine as directed in the general method, Appendix VI, using about 5 g, accurately weighed.
Ester Value after Acetylation Proceed as directed under *Total Alcohols*, Appendix VI, using about 2 g of the dried acetylated oil, accurately weighed. Reflux for a period of 2 h. Calculate the *Ester Value after Acetylation* by the formula

$$A \times 28.05/B,$$

in which A is the number of mL of 0.5 N alcoholic potassium hydroxide consumed in the saponification, and B is the weight, in g, of the acetylated oil used in the test.

Heavy Metals Shake 10 mL of the oil with an equal volume of water to which 1 drop of hydrochloric acid has been added, and pass hydrogen sulfide through the mixture until it is saturated. No darkening in color is produced in either the oil or the water.

Refractive Index, Appendix IIB Determine with an Abbé or other refractometer of equal or greater accuracy.

Solubility in Alcohol Proceed as directed in the general method, Appendix VI. One mL dissolves in 3 mL of 80% alcohol, often with opalescence.

Specific Gravity Determine by any reliable method (see *General Provisions*).

Packaging and Storage Store in full, tight, preferably aluminum, glass, or tin-lined containers in a cool place protected from light.

Angelica Root Oil

CAS: [8015-64-3]

DESCRIPTION

Angelica Root Oil is obtained by steam distillation of the dried slender rootlets of *Angelica archangelica* L. It is a pale yellow to deep amber liquid having a warm pungent odor and bittersweet taste. It is soluble in most fixed oils, slightly soluble in mineral oil, but relatively insoluble in glycerin and in propylene glycol.

Functional Use in Foods Flavoring agent.

REQUIREMENTS

Identification The infrared absorption spectrum of the sample exhibits relative maxima (that may vary in intensity) at the same wavelengths (or frequencies) as those shown in the respective spectrum in the section on *Infrared Spectra (Series A: Essential Oils)*, using the same test conditions as specified therein.
Acid Value Not more than 7.0.
Angular Rotation Optically inactive, or not more than +46.0°.
Ester Value Between 10 and 65.
Heavy Metals (as Pb) Passes test.
Refractive Index Between 1.473 and 1.487 at 20°.
Solubility in Alcohol Passes test.
Specific Gravity Between 0.850 and 0.880.

TESTS

Acid Value Determine as directed for *Acid Value* under *Essential Oils and Flavors*, Appendix VI.

Angular Rotation Determine in a 100-mm tube as directed under *Optical (Specific) Rotation*, Appendix IIB.

Ester Value Determine as directed in the general method, Appendix VI, using about 5 g, accurately weighed.

Heavy Metals Shake 10 mL of the oil with an equal volume of water to which 1 drop of hydrochloric acid has been added, and pass hydrogen sulfide through the mixture until it is saturated. No darkening in color is produced in either the oil or the water.

Refractive Index, Appendix IIB Determine with an Abbé or other refractometer of equal or greater accuracy.

Solubility in Alcohol Proceed as directed in the general method, Appendix VI. One mL dissolves in 1 mL of 90% alcohol, often with turbidity, and remains in solution on further addition of alcohol to a total of 10 mL.

Specific Gravity Determine by any reliable method (see *General Provisions*).

Packaging and Storage Store in full, tight, preferably dark glass bottles, or aluminum or tin-lined containers, in a cool place protected from light. The oils increase in specific gravity and viscosity on storage.

TESTS

Acid Value Determine as directed for *Acid Value* under *Essential Oils and Flavors*, Appendix VI.

Angular Rotation Determine in a 100-mm tube as directed under *Optical (Specific) Rotation*, Appendix IIB.

Ester Value Determine as directed in the general method, Appendix VI, using about 5 g, accurately weighed.

Heavy Metals Shake 10 mL of the oil with an equal volume of water to which 1 drop of hydrochloric acid has been added, and pass hydrogen sulfide through the mixture until it is saturated. No darkening in color is produced in either the oil or the water.

Refractive Index, Appendix IIB Determine with an Abbé or other refractometer of equal or greater accuracy.

Solubility in Alcohol Proceed as directed in the general method, Appendix VI. One mL dissolves in 4 mL of 90% alcohol, often with considerable turbidity, and remains in solution on further addition of alcohol to a total of 10 mL.

Specific Gravity Determine by any reliable method (see *General Provisions*).

Packaging and Storage Store in full, tight, preferably dark glass bottles, or aluminum or tin-lined containers, in a cool place protected from light.

Angelica Seed Oil

DESCRIPTION

Angelica Seed Oil is obtained by steam distillation of the fresh seeds of *Angelica archangelica* L. It is a light yellow liquid having a sweeter and more delicate aroma than the root oil. It is soluble in most fixed oils, slightly soluble in mineral oil, but relatively insoluble in glycerin and in propylene glycol.

Functional Use in Foods Flavoring agent.

REQUIREMENTS

Identification The infrared absorption spectrum of the sample exhibits relative maxima (that may vary in intensity) at the same wavelengths (or frequencies) as those shown in the respective spectrum in the section on *Infrared Spectra (Series A: Essential Oils)*, using the same test conditions as specified therein.

Acid Value Not more than 3.0.

Angular Rotation Between +4° and +16°.

Ester Value Between 14.0 and 32.0.

Heavy Metals (as Pb) Passes test.

Refractive Index Between 1.480 and 1.488 at 20°.

Solubility in Alcohol Passes test.

Specific Gravity Between 0.853 and 0.876.

Anise Oil

CAS: [8007-70-3]

DESCRIPTION

Anise Oil is obtained by steam distillation of the dried ripe fruit of *Pimpinella anisum* L. (Fam. *Umbelliferae*) or *Illicium verum* Hooker filius (Fam. *Magnoliaceae*). It is a colorless to pale yellow, strongly refractive liquid having the characteristic odor and taste of anise.

Note: If solid material has separated, carefully warm the Anise Oil until it is completely liquefied, and mix before using it.

Functional Use in Foods Flavoring agent.

REQUIREMENTS

Identification The infrared absorption spectrum of the sample exhibits relative maxima (that may vary in intensity) at the same wavelengths (or frequencies) as those shown in the respective spectrum in the section on *Infrared Spectra (Series A: Essential Oils)*, using the same test conditions as specified therein.

Angular Rotation Between −2° and +1°.

Heavy Metals (as Pb) Passes test.

Phenols Passes test.

Refractive Index Between 1.553 and 1.560 at 20°.

Solidification Point Not lower than 15°.
Solubility in Alcohol Passes test.
Specific Gravity Between 0.978 and 0.988.

TESTS

Angular Rotation Determine in a 100-mm tube as directed under *Optical (Specific) Rotation*, Appendix IIB.
Heavy Metals Shake 10 mL of the oil with an equal volume of water to which 1 drop of hydrochloric acid has been added, and pass hydrogen sulfide through the mixture until it is saturated. No darkening in color is produced in either the oil or the water.
Phenols Prepare a 1 in 3 solution of recently distilled Anise Oil in 90% alcohol. It is neutral to moistened litmus paper, and the mixture develops no blue or brownish color upon the addition of 1 drop of ferric chloride TS to 5 mL of the solution.
Refractive Index, Appendix IIB Determine with an Abbé or other refractometer of equal or greater accuracy.
Solidification Point Determine as directed in the general method, Appendix IIB.
Solubility in Alcohol Proceed as directed in the general method, Appendix VI. One mL dissolves in 3 mL of 90% alcohol.
Specific Gravity Determine by any reliable method (see *General Provisions*).

Packaging and Storage Store in full, tight containers. Avoid exposure to excessive heat.

Annatto Extracts

INS: 160b   CAS: [1393-63-1]

DESCRIPTION

The extract prepared from annatto seed, *Bixa orellana* (L.), using a food-grade extraction solvent. Annatto Extracts occur as dark red solutions, emulsions, or suspensions in water or oil or as dark red powders. Bixin is the principal pigment of oil-soluble Annatto Extracts. Norbixin is the principal pigment of alkaline water-soluble Annatto Extracts. Commercial preparations are usually mixtures of bixin, norbixin, and other carotenoids.

Functional Use in Foods Color.

REQUIREMENTS

Identification

A. Oil-soluble Annatto Extract diluted with acetone exhibits absorbance maxima at 439, 470, and 501 nm. Water-soluble Annatto Extract diluted with water exhibits absorbance maxima at 451 to 455 nm and 480 to 484 nm.

B. *Carr-Price Reaction* Prepare a small chromatography column by filling a glass tube (e.g., 7 × 200 mm), stoppered with glass wool, with alumina (800–200 mesh) slurried in toluene so that the settled alumina fills about $2/3$ of the tube. Using a rubber outlet tube and clamp, adjust the flow rate to about 30 drops/min.

Oil-Soluble Annatto Add to the top of the alumina column 3 mL of a solution containing sufficient sample, in toluene, to impart a color equivalent to a 0.1% potassium dichromate solution. Elute with toluene until a pale yellow fraction is washed from the column. Wash the column with 3 10-mL volumes of dry acetone, add 5 mL of *Carr-Price Reagent* (see *Solutions and Indicators*), and allow it to run onto the top of the column. The orange-red zone (bixin) at the top of the column immediately becomes blue-green.

Water-Soluble Annatto Transfer 2 mL or 2 g to a 50-mL separatory funnel, and add sufficient 2 N sulfuric acid to make the solution acidic to pH test paper (pH 1 to 2). Dissolve the red precipitate of norbixin by mixing the solution with 50 mL of toluene. Discard the water layer, and wash the toluene phase with water until it no longer gives an acid reaction. Remove any undissolved norbixin by centrifugation or filtration, and dry the solution over anhydrous sodium sulfate. Transfer 3 to 5 mL of the dry solution to the top of an alumina column prepared as described above. Elute the column with toluene, then 3 10-mL volumes of dry acetone, followed by 5 mL of *Carr-Price Reagent* added to the top of the column. The orange-red band of norbixin immediately becomes blue-green.

Arsenic (as As) Not more than 3 mg/kg.
Color Intensity Meets the representations of the vendor.
Lead Not more than 10 mg/kg.
Residual Solvent *Acetone*: not more than 0.003%; *hexanes*: not more than 0.0025%; *isopropyl alcohol*: not more than 0.005%; *methyl alcohol*: not more than 0.005% in excess of that produced naturally; *trichloroethylene and dichloromethane*: not more than 0.003% individually or in combination.

TESTS

Arsenic A *Sample Solution* prepared as directed for organic compounds meets the requirements of the *Arsenic Test*, Appendix IIIB.
Color Intensity
Oil-Soluble Extracts Transfer an accurately weighed sample to a solution of 1% glacial acetic acid in acetone, and dilute to a suitable volume (absorbance of 0.5 to 1.0). Filter the sample to clarify if necessary. Measure the absorbance at 454 nm, and calculate the color intensity (I) by the formula

$$I = A/(b \times c),$$

in which A is the absorbance; b is the cell length, in cm; and c is the concentration, in g/L.

Water-Soluble Extracts Proceed as directed under *Oil-Soluble Extracts*, above, but dissolve the sample in 0.1 M sodium hydroxide and measure the absorbance at 453 nm.
Lead A *Sample Solution* prepared as directed for organic compounds meets the requirements of the *Lead Limit Test*, Appendix IIIB, using 10 μg of lead ion (Pb) in the control.

Residual Solvent Proceed as directed under *Residual Solvent*, Appendix VIII.

Packaging and Storage Store under refrigeration in well-closed containers.

β-Apo-8′-Carotenal

Apocarotenal; APO

$C_{30}H_{40}O$ Formula wt 416.65
INS: 160e CAS: [1107-26-2]

DESCRIPTION

A fine crystalline powder with a dark metallic sheen. It is freely soluble in chloroform and sparingly soluble in acetone, but is insoluble in water. It melts at about 136° to 142° with decomposition.

Functional Use in Foods Color.

REQUIREMENTS

Identification
A. Determine the absorbance of *Sample Solution B*, prepared as directed in the *Assay*, at 488 nm and at 460 nm. The ratio A_{488}/A_{460} is between 0.77 and 0.85.
B. Determine the absorbance of *Sample Solution B* at 460 nm, and that of *Sample Solution A*, prepared as directed in the *Assay*, at 332 nm. The ratio $A_{332}/(10 \times A_{460})$ is between 0.063 and 0.075.

Assay Not less than 96.0% and not more than 101.0% of $C_{30}H_{40}O$.
Heavy Metals (as Pb) Not more than 10 mg/kg.
Residue on Ignition Not more than 0.2%.

TESTS

Assay (Note: Carry out all work in low-actinic glassware and in subdued light.)
Sample Solution A Transfer about 40 mg of the sample, accurately weighed, into a 100-mL volumetric flask, dissolve in 10 mL of acid-free chloroform, dilute to volume with cyclohexane, and mix. Pipet 2 mL of this solution into a 50-mL volumetric flask, dilute to volume with cyclohexane, and mix.
Sample Solution B Pipet 5 mL of *Sample Solution A* into a 50-mL volumetric flask, dilute to volume with cyclohexane, and mix.

Procedure Determine the absorbance of *Sample Solution B* in a 1-cm cell at the wavelength of maximum absorption at about 460 nm, with a suitable spectrophotometer, using cyclohexane as the blank. Calculate the quantity, in mg, of $C_{30}H_{40}O$ in the sample taken by the formula

$$25,000A/264,$$

in which A is the absorbance of the solution and 264 is the absorptivity of pure β-Apo-8′-Carotenal.

Heavy Metals Prepare and test a 2-g sample as directed in *Method II* under the *Heavy Metals Test*, Appendix IIIB, using 20 μg of lead ion (Pb) in the control (*Solution A*).

Residue on Ignition Ignite 2 g as directed in the general method, Appendix IIC.

Packaging and Storage Store in tight light-resistant containers under inert gas.

L-Arginine

L-2-Amino-5-guanidinovaleric Acid

$C_6H_{14}N_4O_2$ Formula wt 174.20
 CAS: [74-79-3]

DESCRIPTION

White crystals or a white crystalline powder. It is soluble in water, but insoluble in ether and sparingly soluble in alcohol. It is strongly alkaline, and its water solutions absorb carbon dioxide from the air.

Functional Use in Foods Nutrient; dietary supplement.

REQUIREMENTS

Identification To 5 mL of a 1 in 1000 solution add 1 mL of triketohydrindene hydrate TS. A reddish purple color appears.
Assay Not less than 98.5% and not more than 101.5% of $C_6H_{14}N_4O_2$, calculated on the dried basis.
Heavy Metals (as Pb) Not more than 0.002%.
Lead Not more than 10 mg/kg.
Loss on Drying Not more than 1.0%.
Residue on Ignition Not more than 0.2%.
Specific Rotation $[\alpha]_D^{20°}$: Between +26.0° and +27.9°, after drying; or $[\alpha]_D^{25°}$: Between +25.8° and +27.7°, after drying.

TESTS

Assay   Dissolve about 200 mg, accurately weighed, in 3 mL of formic acid and 50 mL of glacial acetic acid, add 2 drops of crystal violet TS, and titrate with 0.1 N perchloric acid to a green endpoint or until the blue color disappears completely. Each mL of 0.1 N perchloric acid is equivalent to 8.710 mg of $C_6H_{14}N_4O_2$.

Heavy Metals   Prepare and test a 1-g sample as directed in *Method II* under the *Heavy Metals Test*, Appendix IIIB, using 20 µg of lead ion (Pb) in the control (*Solution A*).

Lead   A *Sample Solution* prepared as directed for organic compounds meets the requirements of the *Lead Limit Test*, Appendix IIIB, using 10 µg of lead ion (Pb) in the control.

Loss on Drying, Appendix IIC   Dry at 105° for 3 h.

Residue on Ignition, Appendix IIC   Ignite 1 g as directed in the general method.

Specific Rotation, Appendix IIB   Determine in a solution containing 8 g of a previously dried sample in sufficient 6 N hydrochloric acid to make 100 mL.

Packaging and Storage   Store in well-closed, light-resistant containers.

L-Arginine Monohydrochloride

L-2-Amino-5-guanidinovaleric Acid Monohydrochloride

$$\begin{array}{c} NHCH_2CH_2CH_2CCOOH \\ | \quad\quad\quad\quad\quad\quad | \\ C=NH \quad\quad H\ NH_2 \\ | \\ NH_2\cdot HCl \end{array}$$

$C_6H_{14}N_4O_2\cdot HCl$ Formula wt 210.66

CAS: [1119-34-2]

DESCRIPTION

A white or nearly white, practically odorless crystalline powder. It is soluble in water, slightly soluble in hot alcohol, and insoluble in ether. It is acidic and melts with decomposition at about 235°.

Functional Use in Foods   Nutrient; dietary supplement.

REQUIREMENTS

Identification

A. Heat 5 mL of a 1 in 1000 solution with 1 mL of triketohydrindene hydrate TS. A reddish purple color is produced.

B. A 1 in 1000 solution gives positive tests for *Chloride*, Appendix IIIA.

Assay   Not less than 98.5% and not more than 101.5% of $C_6H_{14}N_4O_2\cdot HCl$ after drying.

Heavy Metals (as Pb)   Not more than 0.002%.
Lead   Not more than 10 mg/kg.
Loss on Drying   Not more than 0.3%.
Residue on Ignition   Not more than 0.1%.
Specific Rotation $[\alpha]_D^{20°}$: Between +21.3° and +23.5° after drying; or $[\alpha]_D^{25°}$: Between +21.2° and +23.4° after drying.

TESTS

Assay   Dissolve about 100 mg of the sample, previously dried at 105° for 3 h and accurately weighed, in 2 mL of formic acid, add exactly 15.0 mL of 0.1 N perchloric acid, and heat on a water bath for 30 min. After cooling, add 45 mL of glacial acetic acid, and titrate the excess perchloric acid with 0.1 N sodium acetate, determining the endpoint potentiometrically. Perform a blank determination, and make any necessary correction. Each mL of 0.1 N perchloric acid is equivalent to 10.53 mg of $C_6H_{14}N_4O_2\cdot HCl$.

Heavy Metals   Prepare and test a 1-g sample as directed in *Method II* under the *Heavy Metals Test*, Appendix IIIB, using 20 µg of lead ion (Pb) in the control (*Solution A*).

Lead   A *Sample Solution* prepared as directed for organic compounds meets the requirements of the *Lead Limit Test*, Appendix IIIB, using 10 µg of lead ion (Pb) in the control.

Loss on Drying, Appendix IIC   Dry at 105° for 3 h.

Residue on Ignition, Appendix IIC   Ignite 1 g as directed in the general method.

Specific Rotation, Appendix IIB   Determine in a solution containing 8 g of a previously dried sample in sufficient 6 N hydrochloric acid to make 100 mL.

Packaging and Storage   Store in well-closed, light-resistant containers.

Ascorbic Acid

Vitamin C; L-Ascorbic Acid

$C_6H_8O_6$ Formula wt 176.13

INS: 300   CAS: [50-81-7]

DESCRIPTION

White or slightly yellow crystals or powder, melting at about 190°. It gradually darkens on exposure to light, is reasonably stable in air when dry, but rapidly deteriorates in solution in the presence of air. One g is soluble in about 3 mL of water and in about 30 mL of alcohol. It is insoluble in chloroform and in ether.

Functional Use in Foods Antioxidant; meat-curing aid; nutrient; dietary supplement.

REQUIREMENTS

Identification
A. A 1 in 50 solution slowly reduces alkaline cupric tartrate TS at 25°, but more readily upon heating.
B. The infrared absorption spectrum of a potassium bromide dispersion of the sample exhibits maxima at the same wavelengths as that of a similar preparation of USP Ascorbic Acid Reference Standard.

Assay Not less than 99.0% and not more than 100.5% of $C_6H_8O_6$.
Heavy Metals (as Pb) Not more than 10 mg/kg.
Residue on Ignition Not more than 0.1%.
Specific Rotation $[\alpha]_D^{25°}$: Between +20.5° and +21.5°.

TESTS

Assay Dissolve about 400 mg, accurately weighed, in a mixture of 100 mL of water, recently boiled and cooled, and 25 mL of 2 N sulfuric acid. Titrate the solution immediately with 0.1 N iodine, adding starch TS near the endpoint. Each mL of 0.1 N iodine is equivalent to 8.806 mg of $C_6H_8O_6$.
Heavy Metals A solution of 2 g in 25 mL of water meets the requirements of the *Heavy Metals Test*, Appendix IIIB, using 20 µg of lead ion (Pb) in the control (*Solution A*).
Residue on Ignition, Appendix IIC Ignite 2 g as directed in the general method.
Specific Rotation, Appendix IIB Determine in a solution containing 1 g in 10 mL of carbon dioxide-free water.

Packaging and Storage Store in tight, light-resistant containers.

Ascorbyl Palmitate

Palmitoyl L-Ascorbic Acid

$CH_3(CH_2)_{14}COOCH_2-\overset{OH}{\underset{H}{C}}-\text{(ascorbic ring with OH, OH, O, O)}$

$C_{22}H_{38}O_7$ Formula wt 414.54
INS: 304

DESCRIPTION

A white or yellowish white powder having a slight odor. It is very slightly soluble in water and in vegetable oils. One g dissolves in about 4.5 mL of alcohol.

Functional Use in Foods Antioxidant.

REQUIREMENTS

Identification A 1 in 10 solution in alcohol decolorizes dichlorophenol–indophenol TS.
Assay Not less than 95.0% of $C_{22}H_{38}O_7$, calculated on the dried basis.
Heavy Metals (as Pb) Not more than 10 mg/kg.
Loss on Drying Not more than 2%.
Melting Range Between 107° and 117°.
Residue on Ignition Not more than 0.1%.
Specific Rotation $[\alpha]_D^{25°}$: Between +21° and +24°, calculated on the dried basis.

TESTS

Assay Dissolve about 300 mg, accurately weighed, in 50 mL of alcohol in a 250-mL Erlenmeyer flask, add 30 mL of water, and immediately titrate with 0.1 N iodine to a yellow color that persists for at least 30 s. Each mL of 0.1 N iodine is equivalent to 20.73 mg of $C_{22}H_{38}O_7$.
Heavy Metals Prepare and test a 2-g sample as directed in *Method II* under the *Heavy Metals Test*, Appendix IIIB, using 20 µg of lead ion (Pb) in the control (*Solution A*).
Loss on Drying, Appendix IIC Dry in a vacuum oven at 56° to 60° for 1 h.
Melting Range Determine as directed in *Melting Range or Temperature*, *Procedure for Class Ia*, Appendix IIB.
Residue on Ignition Ignite 2 g as directed in the general method, Appendix IIC.
Specific Rotation, Appendix IIB Determine in a solution containing 1 g in 10 mL of methanol.

Packaging and Storage Store in tight containers, preferably in a cool, dry place.

L-Asparagine

L-α-Aminosuccinamic Acid

$$\text{H}_2\text{NCOCH}_2\overset{\text{NH}_2}{\underset{\text{H}}{\text{C}}}\text{COOH}$$

$C_4H_8N_2O_3 \cdot H_2O$ Formula wt 150.13

CAS: anhydrous [70-47-3]
CAS: monohydrate [5794-13-8]

DESCRIPTION

White crystals or crystalline powder having a slightly sweet taste. It is soluble in water and practically insoluble in alcohol and in ether. Its solutions are acid to litmus. It melts at about 234°.

Functional Use in Foods   Nutrient; dietary supplement.

REQUIREMENTS

Identification   To 100 mg of the sample add 5 mL of 1 N sodium hydroxide, heat on a water bath for 1 h, adjust the pH to 5.0 with 2.7 N hydrochloric acid, and add 100 mg of triketohydrindene hydrate. The vapor evolved changes the color of acetaldehyde test paper to blue.
Assay   Not less than 98.0% and not more than 101.5% of $C_4H_8N_2O_3$ after drying.
Heavy Metals (as Pb)   Not more than 10 mg/kg.
Lead   Not more than 5 mg/kg.
Loss on Drying   Between 11.5% and 12.5%.
Residue on Ignition   Not more than 0.1%.
Specific Rotation $[\alpha]_D^{20°}$: Between +33.0° and +36.5° after drying.

TESTS

Assay   Dissolve about 130 mg of the sample, previously dried at 130° for 3 h and accurately weighed, in 3 mL of formic acid and 50 mL of glacial acetic acid, and titrate with 0.1 N perchloric acid, determining the endpoint potentiometrically. Perform a blank determination (see *General Provisions*), and make any necessary correction. Each mL of 0.1 N perchloric acid is equivalent to 13.21 mg of $C_4H_8N_2O_3$.
Heavy Metals   Prepare and test a 2-g sample as directed in *Method II* under the *Heavy Metals Test*, Appendix IIIB, using 20 μg of lead ion (Pb) in the control (*Solution A*).
Lead   A *Sample Solution* prepared as directed for organic compounds meets the requirements of the *Lead Limit Test*, Appendix IIIB, using 5 μg of lead ion (Pb) in the control.
Loss on Drying, Appendix IIC   Dry at 130° for 3 h.
Residue on Ignition, Appendix IIC   Ignite 1 g as directed in the general method.
Specific Rotation, Appendix IIB   Determine in a solution containing 10 g of a previously dried sample in sufficient 6 N hydrochloric acid to make 100 mL.

Packaging and Storage   Store in well-closed, light-resistant containers.

Aspartame

N-L-α-Aspartyl-L-phenylalanine 1-Methyl Ester; APM

$$\text{H}_2\text{NCHCONHCHCH}_2-\text{C}_6\text{H}_5$$
with COOCH$_3$ and CH$_2$COOH substituents

$C_{14}H_{18}N_2O_5$ Formula wt 294.31
INS: 951   CAS: [22839-47-0]

DESCRIPTION

A white, odorless, crystalline powder having a sweet taste. It is sparingly soluble in water and slightly soluble in alcohol. The pH of a 0.8% solution is between about 4.5 and 6.0.

Functional Use in Foods   Sweetener; sugar substitute; flavor enhancer.

REQUIREMENTS

Identification
The infrared absorption spectrum of a potassium bromide dispersion of Aspartame, previously dried, exhibits maxima only at the same wavelengths as that of a similar preparation of USP Aspartame Reference Standard.
Assay   Not less than 98.0% and not more than 102.0% of $C_{14}H_{18}N_2O_5$, calculated on the dried basis.
5-Benzyl-3,6-dioxo-2-piperazineacetic Acid   Not more than 1.5%.
Heavy Metals (as Pb)   Not more than 10 mg/kg.
Loss on Drying   Not more than 4.5%.
Other Related Substances   Not more than 2.0%.
Residue on Ignition   Not more than 0.2%.
Specific Rotation $[\alpha]_D^{20°}$: Between +14.5° and +16.5°, calculated on the dried basis.
Transmittance   Passes test.

TESTS

Assay   Transfer about 300 mg of the sample, accurately weighed, to a 150-mL beaker, dissolve in 1.5 mL of formic acid (96%), and add 60 mL of glacial acetic acid. Add crystal violet TS, and titrate immediately with 0.1 N perchloric acid to a green endpoint. Perform a blank determination, and make any necessary correction. Each mL of 0.1 N perchloric acid is equivalent to 29.43 mg of $C_{14}H_{18}N_2O_5$.

Note: Use 0.1 N perchloric acid previously standardized to a green endpoint. A blank titration exceeding 0.1 mL may be due to excessive water content and may cause loss of visual endpoint sensitivity.

5-Benzyl-3,6-dioxo-2-piperazineacetic Acid

Mobile Phase Weigh 5.6 g of potassium phosphate monobasic into a 1-L flask, add 820 mL of water, and dissolve. Adjust the pH to 4.3 using phosphoric acid, add 180 mL of methanol, and mix. Filter through a 0.45-μm disk, and de-gas.

Diluting Solvent Add 200 mL of methanol to 1800 mL of water, and mix.

Impurity Standard Preparation Transfer about 25 mg of USP 5-Benzyl-3,6-dioxo-2-piperazineacetic Acid Reference Standard, accurately weighed, into a 100-mL volumetric flask. Add 10 mL of methanol, and dissolve. Dilute to volume with water, and mix. Pipet 15 mL of this solution into a 50-mL volumetric flask, dilute to volume with *Diluting Solvent*, and mix. Use a freshly prepared solution.

Sample Preparation Transfer about 50 mg of the Aspartame sample, accurately weighed, to a 10-mL volumetric flask. Dilute to volume with *Diluting Solvent*, and mix. Use a freshly prepared solution.

Chromatographic System Use a suitable high-pressure liquid chromatograph equipped with a detector measuring at 210 nm and a 250- × 4.6-mm column packed with octadecyl silanized silica (10-μm Partisil ODS-3, or equivalent) and operated under isocratic conditions at 40°. The flow rate of the *Mobile Phase* is about 2 mL/min.

System Suitability The area responses of three replicate injections of the *Impurity Standard Preparation* show a relative standard deviation of not more than 2.0%.

Procedure Separately inject equal 20-μL portions of the *Impurity Standard Preparation* and the *Sample Preparation* into the chromatograph, and record the chromatograms (the approximate retention time of 5-benzyl-3,6-dioxo-2-piperazineacetic acid is 4 min, and the approximate retention time of Aspartame is 11 min). Measure the peak area response of 5-benzyl-3,6-dioxo-2-piperazineacetic acid in each chromatogram. Calculate the percentage of 5-benzyl-3,6-dioxo-2-piperazineacetic acid in the sample by the formula

$$1000(A_U C_S)/(A_S W_U),$$

in which A_U and A_S are the peak area responses of 5-benzyl-3,6-dioxo-2-piperazineacetic acid in the *Sample Preparation* and in the *Impurity Standard Preparation*, respectively; C_S is the concentration, in mg/mL, of 5-benzyl-3,6-dioxo-2-piperazineacetic acid in the *Impurity Standard Preparation*; and W_U is the weight, in mg, of Aspartame taken for the *Sample Preparation*.

Heavy Metals Prepare and test a 2-g sample as directed in *Method II* under the *Heavy Metals Test*, Appendix IIIB, using 20 μg of lead ion (Pb) in the control (*Solution A*).

Other Related Impurities Proceed as directed in the test for *5-Benzyl-3,6-dioxo-2-piperazineacetic Acid*, except to use the following in place of the *Standard Preparation*:

Other Related Substances Standard Preparation Pipet 2 mL of the *Sample Preparation* from the test for *5-Benzyl-3,6-dioxo-2-piperazineacetic Acid* into a 100-mL volumetric flask, dilute to volume with the *Diluting Solvent*, and mix.

Procedure Inject about 20-μL portions of the *Other Related Substances Standard Preparation* and the *Sample Preparation* into the chromatograph, and record the chromatogram for a time equal to twice the retention time of Aspartame. In the chromatogram obtained from the *Sample Preparation*, the sum of the responses of all secondary peaks, other than that for 5-benzyl-3,6-dioxo-2-piperazineacetic acid, is not more than the response of the Aspartame peak obtained in the chromatogram from the *Other Related Substances Standard Preparation*.

Loss on Drying, Appendix IIC Dry at 105° for 4 h.

Residue on Ignition Ignite a 1-g sample as directed in the general method, Appendix IIC.

Specific Rotation, Appendix IIB Determine in a solution containing 4 g of sample in sufficient 15 N formic acid to make 100 mL. Make the determination within 30 min of preparation of the sample solution.

Transmittance The transmittance, T, of a 1% solution, in 2 N hydrochloric acid, determined in a 1-cm cell at 430 nm with a suitable spectrophotometer, using 2 N hydrochloric acid as the blank, is not less than 0.95, equivalent to an absorbance, A, of not more than approximately 0.022.

Packaging and Storage Store in well-closed containers in a cool, dry place.

DL-Aspartic Acid

DL-Aminosuccinic Acid

$$HOOCCH_2CH(NH_2)COOH$$

$C_4H_7NO_4$ Formula wt 133.11

CAS: [617-45-8]

DESCRIPTION

Colorless or white, odorless crystals having an acid taste. It is slightly soluble in water, but insoluble in alcohol and in ether. It is optically inactive and melts with decomposition at about 280°.

Functional Use in Foods Nutrient; dietary supplement.

REQUIREMENTS

Identification To 5 mL of a 1 in 1000 solution add 1 mL of triketohydrindene hydrate TS. A bluish purple color is produced.

Assay Not less than 98.5% and not more than 101.5% of $C_4H_7NO_4$, calculated on the dried basis.

Heavy Metals (as Pb) Not more than 0.002%.

Lead Not more than 10 mg/kg.

Loss on Drying Not more than 0.3%.
Residue on Ignition Not more than 0.1%.

TESTS

Assay Dissolve about 200 mg, accurately weighed, in 3 mL of formic acid and 50 mL of glacial acetic acid, add 2 drops of crystal violet TS, and titrate with 0.1 N perchloric acid to a green endpoint or until the blue color disappears completely. Perform a blank determination (see *General Provisions*), and make any necessary correction. Each mL of 0.1 N perchloric acid is equivalent to 13.31 mg of $C_4H_7NO_4$.
Heavy Metals Prepare and test a 1-g sample as directed in *Method II* under the *Heavy Metals Test*, Appendix IIIB, using 20 μg of lead ion (Pb) in the control (*Solution A*).
Lead A *Sample Solution* prepared as directed for organic compounds meets the requirements of the *Lead Limit Test*, Appendix IIIB, using 10 μg of lead ion (Pb) in the control.
Loss on Drying, Appendix IIC   Dry at 105° for 3 h.
Residue on Ignition, Appendix IIC   Ignite 1 g as directed in the general method.

Packaging and Storage   Store in well-closed, light-resistant containers.

L-Aspartic Acid

L-Aminosuccinic Acid

$$HOOCCH_2\overset{\overset{H}{|}}{\underset{NH_2}{C}}COOH$$

$C_4H_7NO_4$ Formula wt 133.10

CAS: [56-84-8]

DESCRIPTION

White, odorless crystals or crystalline powder having a slightly acid taste. It is slightly soluble in water but insoluble in alcohol and in ether. It melts at about 270°.

Functional Use in Foods   Nutrient; dietary supplement.

REQUIREMENTS

Identification   To 5 mL of a 1 in 1000 solution add 1 mL of triketohydrindene hydrate TS. A bluish purple color appears.
Assay   Not less than 98.5% and not more than 101.5% of $C_4H_7NO_4$, calculated on the dried basis.
Heavy Metals (as Pb)   Not more than 10 mg/kg.
Lead   Not more than 5 mg/kg.
Loss on Drying   Not more than 0.25%.
Residue on Ignition   Not more than 0.1%.
Specific Rotation $[\alpha]_D^{20°}$: Between +24.5° and +26.0° after drying.

TESTS

Assay Dissolve about 200 mg, accurately weighed, in 3 mL of formic acid and 50 mL of glacial acetic acid, add 2 drops of crystal violet TS, and titrate with 0.1 N perchloric acid to a green endpoint or until the blue color disappears completely. Perform a blank determination (see *General Provisions*), and make any necessary correction. Each mL of 0.1 N perchloric acid is equivalent to 13.31 mg of $C_4H_7NO_4$.
Heavy Metals Prepare and test a 2-g sample as directed in *Method II* under the *Heavy Metals Test*, Appendix IIIB, using 20 μg of lead ion (Pb) in the control (*Solution A*).
Lead A *Sample Solution* prepared as directed for organic compounds meets the requirements of the *Lead Limit Test*, Appendix IIIB, using 5 μg of lead ion (Pb) in the control.
Loss on Drying, Appendix IIC   Dry at 105° for 3 h.
Residue on Ignition, Appendix IIC   Ignite 1 g as directed in the general method.
Specific Rotation, Appendix IIB   Determine in a solution containing 8 g of a previously dried sample in sufficient 6 N hydrochloric acid to make 100 mL.

Packaging and Storage   Store in well-closed, light-resistant containers.

Azodicarbonamide

Azodicarboxylic Acid Diamide

$$H_2N-\overset{\overset{O}{\|}}{C}-N=N-\overset{\overset{O}{\|}}{C}-NH_2$$

$C_2H_4N_4O_2$ Formula wt 116.08

INS: 927a   CAS: [123-77-3]

DESCRIPTION

A yellow to orange red, odorless, crystalline powder. It is practically insoluble in water and in most organic solvents. It is slightly soluble in dimethyl sulfoxide. It melts above 180° with decomposition.

Functional Use in Foods   Maturing agent for flour.

REQUIREMENTS

Identification   A solution of 35 mg of the sample in 1000 mL of water exhibits an ultraviolet absorption maximum at about 245 nm.
Assay   Not less than 98.6% and not more than 100.5% of $C_2H_4N_4O_2$ after drying.
Heavy Metals (as Pb)   Not more than 0.002%.
Lead   Not more than 10 mg/kg.
Loss on Drying   Not more than 0.5%.
Nitrogen   Between 47.2% and 48.7%.

pH of a 2% Suspension Not less than 5.0.
Residue on Ignition Not more than 0.15%.

TESTS

Assay Transfer about 225 mg of the sample, previously dried in a vacuum oven at 50° for 2 h and accurately weighed, into a 250-mL glass-stoppered iodine flask. Add about 23 mL of dimethyl sulfoxide to the flask, washing any adhered sample down with the solvent, then stopper the flask, and place about 2 mL of the solvent in the cup or lip of the flask. Swirl occasionally until complete solution of the sample is effected, and then loosen the stopper to drain the remainder of solvent into the flask and to rinse down any dissolved sample into the solution. Add 5.0 g of potassium iodide followed by 15 mL of water, then immediately pipet 10 mL of 0.5 N hydrochloric acid into the flask, and stopper rapidly. Swirl until the potassium iodide dissolves, and allow to stand for 20 to 25 min protected from light. Titrate the liberated iodine with 0.1 N sodium thiosulfate to the disappearance of the yellow color. Titrate with additional thiosulfate if any yellow color appears within 15 min. Perform a blank determination on a solution consisting of 25 mL of dimethyl sulfoxide, 5.0 g of potassium iodide, 15 mL of water, and 5 mL of 0.5 N hydrochloric acid, and make any necessary correction. Each mL of 0.1 N sodium thiosulfate is equivalent to 5.804 mg of $C_2H_4N_4O_2$.

Heavy Metals Prepare and test a 1-g sample as directed in *Method II* under the *Heavy Metals Test*, Appendix IIIB, using 20 µg of lead ion (Pb) in the control (*Solution A*).

Lead A *Sample Solution* prepared as directed for organic compounds meets the requirements of the *Lead Limit Test*, Appendix IIIB, using 10 µg of lead ion (Pb) in the control.

Loss on Drying, Appendix IIC   Dry in a vacuum oven at 50° for 2 h.

Nitrogen Transfer about 50 mg into a 100-mL Kjeldahl flask, add 3 mL of concentrated hydriodic acid solution (57% freshly assayed), and digest the mixture with gentle heating for 1.25 h, adding sufficient water, when necessary, to maintain the original volume. Increase the heat at the end of the digestion period, and continue heating until the volume is reduced by about one-half. Cool to room temperature, add 1.5 g of potassium sulfate, 3 mL of water, and 4.5 mL of sulfuric acid, and heat until iodine fumes are no longer evolved. Allow the mixture to cool, wash down the sides of the flask with water, heat until charring occurs, and again cool to room temperature. To the charred material add 40 mg of mercuric oxide, heat until the color of the solution is pale yellow, then cool, wash down the sides of the flask with a few mL of water, and digest the mixture for 3 additional h. Cool the digest, add 20 mL of ammonia-free water, 16 mL of a 50% sodium hydroxide solution, and 5 mL of a 44% sodium thiosulfate solution. Immediately connect the flask to a distillation apparatus as directed under *Nitrogen Determination*, Appendix IIIC, and distill, collecting the distillate in 10 mL of a 4% boric acid solution. Add a few drops of methyl red–methylene blue TS to the distillate and titrate with 0.05 N sulfuric acid. Perform a blank determination (see *General Provisions*). Each mL of 0.05 N sulfuric acid is equivalent to 0.7004 mg of N.

pH of a 2% Suspension Add 2 g to 100 mL of water, agitate the mixture with a power stirrer for 5 min, and determine the pH of the resulting suspension potentiometrically (Appendix IIB).

Residue on Ignition Ignite 1.5 g as directed in the general method, Appendix IIC.

Packaging and Storage Store in well-closed, light-resistant containers.

Balsam Peru Oil

DESCRIPTION

The oil obtained by extraction or distillation from Peruvian Balsam obtained from *Myroxylon pereirae* Royle Klotzsche (Fam. *Leguminosae*). It is a yellow to pale brown, slightly viscous liquid having a sweet balsamic odor. Occasionally, crystals may separate from the liquid. It is soluble in most fixed oils, and is soluble, with turbidity, in mineral oil. It is partly soluble in propylene glycol, but it is practically insoluble in glycerin.

Functional Use in Foods   Flavoring agent.

REQUIREMENTS

Identification The infrared absorption spectrum of the sample exhibits relative maxima (that may vary in intensity) at the same wavelengths (or frequencies) as those shown in the respective spectrum in the section on *Infrared Spectra (Series A: Essential Oils)*, using the same test conditions as specified therein.

Acid Value   Between 30 and 60.
Angular Rotation   Between −1° and +2°.
Ester Value   Between 200 and 225.
Heavy Metals (as Pb)   Passes test.
Refractive Index   Between 1.567 and 1.579 at 20°.
Solubility in Alcohol   Passes test.
Specific Gravity   Between 1.095 and 1.110.

TESTS

Acid Value Determine as directed for *Acid Value* under *Essential Oils and Flavors*, Appendix VI.

Angular Rotation Determine in a 100-mm tube as directed under *Optical (Specific) Rotation*, Appendix IIB.

Ester Value Proceed as directed in the general method, Appendix VI, using about 1 g, accurately weighed.

Heavy Metals Shake 10 mL of the oil with an equal volume of water to which 1 drop of hydrochloric acid has been added, and pass hydrogen sulfide through the mixture until it is saturated. No darkening in color is produced in either the oil or the water.

Refractive Index, Appendix IIB   Determine with an Abbé or other refractometer of equal or greater accuracy.

Solubility in Alcohol Proceed as directed in the general method, Appendix VI. One mL dissolves in 0.5 mL of 90% alcohol and remains in solution upon dilution to 10 mL.
Specific Gravity Determine by any reliable method (see *General Provisions*).

Packaging and Storage Store in full, tight, preferably glass, tin-lined, or aluminum containers in a cool place protected from light.

Basil Oil, Comoros Type

Basil Oil Exotic; Basil Oil, Réunion Type

DESCRIPTION

Basil Oil, Comoros Type, is obtained by steam distillation of the flowering tops or the entire plant of *Ocimum basilicum* L. It may be distinguished from other types, such as basil oil, European type, by its camphoraceous odor and physicochemical constants. It is a light yellow liquid with a spicy odor. It is soluble in most fixed oils and, with turbidity, in mineral oil. One mL is soluble in 20 mL of propylene glycol with slight haziness, but it is insoluble in glycerin.

Functional Use in Foods Flavoring agent.

REQUIREMENTS

Identification The infrared absorption spectrum of the sample exhibits relative maxima (that may vary in intensity) at the same wavelengths (or frequencies) as those shown in the respective spectrum in the section on *Infrared Spectra (Series A: Essential Oils)*, using the same test conditions as specified therein.
Acid Value Not more than 1.0.
Angular Rotation Between −2° and +2°.
Ester Value after Acetylation Between 25 and 45.
Heavy Metals (as Pb) Passes test.
Refractive Index Between 1.512 and 1.520 at 20°.
Saponification Value Between 4 and 10.
Solubility in Alcohol Passes test.
Specific Gravity Between 0.952 and 0.973.

TESTS

Acid Value Determine as directed for *Acid Value* under *Essential Oils and Flavors*, Appendix VI.
Angular Rotation Determine in a 100-mm tube as directed under *Optical (Specific) Rotation*, Appendix IIB.
Ester Value after Acetylation Proceed as directed under *Linalool Determination*, Appendix VI, using 2.5 g of the dry acetylated oil, accurately weighed, for the saponification. Calculate the *Ester Value after Acetylation* by the formula

$$a \times 28.05/b,$$

in which a is the number of mL of 0.5 N alcoholic potassium hydroxide consumed in the saponification, and b is the weight of the acetylated oil, in g, used in the test.
Heavy Metals Shake 10 mL of the oil with an equal volume of water to which 1 drop of hydrochloric acid has been added, and pass hydrogen sulfide through the mixture until it is saturated. No darkening in color is produced in either the oil or the water.
Refractive Index, Appendix IIB Determine with an Abbé or other refractometer of equal or greater accuracy.
Saponification Value Determine as directed for *Saponification Value* under *Essential Oils and Flavors*, Appendix VI, using about 5 g of sample, accurately weighed.
Solubility in Alcohol Proceed as directed in the general method, Appendix VI. One mL dissolves in 4 mL of 80% alcohol.
Specific Gravity Determine by any reliable method (see *General Provisions*).

Packaging and Storage Store in full, tight, preferably glass or aluminum containers in a cool place protected from light.

Basil Oil, European Type

Basil Oil, Italian Type; Sweet Basil Oil

DESCRIPTION

Basil Oil, European Type, is obtained by steam distillation of the flowering tops or the entire plant of *Ocimum basilicum* L. It may be distinguished from other types, such as basil oil, Comoros type, or basil oil, Réunion type, by its more floral odor and its physicochemical constants. It is a pale yellow to yellow liquid with a floral-spicy odor. It is soluble in most fixed oils and, with turbidity, in mineral oil. One mL is soluble in 20 mL of propylene glycol with slight haziness, but it is insoluble in glycerin and in mineral oil.

Functional Use in Foods Flavoring agent.

REQUIREMENTS

Identification The infrared absorption spectrum of the sample exhibits relative maxima (that may vary in intensity) at the same wavelengths (or frequencies) as those shown in the respective spectrum in the section on *Infrared Spectra (Series A: Essential Oils)*, using the same test conditions as specified therein.
Acid Value Not more than 2.5.
Angular Rotation Between −5° and −15°.
Ester Value after Acetylation Between 140 and 180.
Heavy Metals (as Pb) Passes test.
Refractive Index Between 1.483 and 1.493 at 20°.
Solubility in Alcohol Passes test.
Specific Gravity Between 0.900 and 0.920.

TESTS

Acid Value Determine as directed for *Acid Value* under *Essential Oils and Flavors*, Appendix VI.

Angular Rotation Determine in a 100-mm tube as directed under *Optical (Specific) Rotation*, Appendix IIB.

Ester Value after Acetylation Proceed as directed under *Linalool Determination*, Appendix VI, using 2.5 g of the dry acetylated oil, accurately weighed, for the saponification. Calculate the *Ester Value after Acetylation* by the formula

$$a \times 28.05/b,$$

in which a is the number of mL of 0.5 N alcoholic potassium hydroxide consumed in the saponification, and b is the weight of the acetylated oil, in g, used in the test.

Heavy Metals Shake 10 mL of the oil with an equal volume of water to which 1 drop of hydrochloric acid has been added, and pass hydrogen sulfide through the mixture until it is saturated. No darkening in color is produced in either the oil or the water.

Refractive Index, Appendix IIB Determine with an Abbé or other refractometer of equal or greater accuracy.

Solubility in Alcohol Proceed as directed in the general method, Appendix VI. One mL dissolves in 4 mL of 80% alcohol.

Specific Gravity Determine by any reliable method (see *General Provisions*).

Packaging and Storage Store in full, tight, preferably glass or aluminum containers in a cool place protected from light.

Bay Oil

Myrcia Oil

DESCRIPTION

The volatile oil distilled from the leaves of *Pimenta acris* Kostel. It occurs as a yellow or brownish yellow liquid with a pleasant aromatic odor and a pungent, spicy taste. It is soluble in alcohol and in glacial acetic acid. Its solutions in alcohol are acid to litmus.

Functional Use in Foods Flavoring agent.

REQUIREMENTS

Identification

A. The infrared absorption spectrum of the sample exhibits relative maxima (that may vary in intensity) at the same wavelengths (or frequencies) as those shown in the respective spectrum in the section on *Infrared Spectra (Series A: Essential Oils)*, using the same test conditions as specified therein.

B. Shake 1 mL with 20 mL of hot water and filter. The filtrate gives not more than a slight acid reaction with litmus, and on the addition of 1 drop of ferric chloride TS yields only a transient grayish green, not a blue or purple color.

Assay Not less than 50% and not more than 65%, by volume, of phenols.
Angular Rotation Levorotatory, but not more than –3°.
Heavy Metals (as Pb) Passes test.
Refractive Index Between 1.507 and 1.516 at 20°.
Specific Gravity Between 0.950 and 0.990.

TESTS

Assay Proceed as directed under *Phenols*, Appendix VI.

Angular Rotation Determine in a 100-mm tube as directed under *Optical (Specific) Rotation*, Appendix IIB.

Heavy Metals Shake 10 mL of the oil with an equal volume of water to which 1 drop of hydrochloric acid has been added, and pass hydrogen sulfide through the mixture until it is saturated. No darkening in color is produced in either the oil or the water.

Refractive Index, Appendix IIB Determine with an Abbé or other refractometer of equal or greater accuracy.

Specific Gravity Determine by any reliable method (see *General Provisions*).

Packaging and Storage Store in full, tight containers in a cool place protected from light.

Beeswax, White

White Wax

INS: 901

DESCRIPTION

The bleached, purified wax from the honeycomb of the bee *Apis mellifera* L. (Fam. *Apidae*). It is a yellowish white solid, somewhat translucent in thin layers, having a faint characteristic odor, free from rancidity. Its specific gravity is about 0.95. White Beeswax is insoluble in water and sparingly soluble in cold alcohol. Boiling alcohol dissolves cerotic acid and part of the myricin, which are constituents of the wax. It is completely soluble in chloroform, in ether, and in fixed and volatile oils. It is partly soluble in cold carbon disulfide and is completely soluble at temperatures of 30° or above.

Functional Use in Foods Candy glaze and polish; flavoring agent.

REQUIREMENTS

Acid Value Between 17 and 24.
Carnauba Wax Passes test.
Ester Value Between 72 and 79.
Fats, Japan Wax, Rosin, and Soap Passes test.
Heavy Metals (as Pb) Not more than 0.002%.
Lead Not more than 10 mg/kg.
Melting Range Between 62° and 65°.
Saponification Cloud Test Passes test.

TESTS

Acid Value, Appendix VII (Fats and Related Substances) Warm about 3 g, accurately weighed, in a 200-mL flask with 25 mL of absolute alcohol, previously neutralized to phenolphthalein with potassium hydroxide, until the sample is melted. Shake the mixture, add 1 mL of phenolphthalein TS, and titrate the warm solution with 0.5 N alcoholic potassium hydroxide to a permanent, faint pink color.

Carnauba Wax Place 100 mg in a test tube, and add 20 mL of n-butanol. Immerse the test tube in boiling water, and shake the mixture gently until solution is complete. Transfer the test tube into a beaker of water at 60°, and allow it to cool to room temperature. A loose mass of fine, needle-like crystals separate from a clear mother liquor. Under the microscope the crystals appear as loose needles or stellate clusters, and no amorphous masses are observed (*absence of carnauba wax*).

Ester Value To the solution resulting from the determination of *Acid Value* add 25.0 mL of 0.5 N alcoholic potassium hydroxide and 50 mL of alcohol, heat the mixture under a reflux condenser for 4 h, and titrate the excess alkali with 0.5 N hydrochloric acid. Perform a residual blank titration, and calculate the *Ester Value* as the number of mg of potassium hydroxide required for each g of the sample taken for the test.

Fats, Japan Wax, Rosin, and Soap Boil 1 g for 30 min with 35 mL of a 1 in 7 solution of sodium hydroxide, maintaining the volume by the occasional addition of water, and cool the mixture. The wax separates and the liquid remains clear. Filter the cold mixture and acidify the filtrate with hydrochloric acid. No precipitate is formed.

Heavy Metals Prepare and test a 1-g sample as directed in *Method II* under the *Heavy Metals Test*, Appendix IIIB, using 20 μg of lead ion (Pb) in the control (*Solution A*).

Lead A *Sample Solution* prepared as directed for organic compounds meets the requirements of the *Lead Limit Test*, Appendix IIIB, using 10 μg of lead ion (Pb) in the control.

Melting Range Determine as directed for *Melting Range or Temperature*, *Procedure for Class II*, Appendix IIB.

Saponification Cloud Test

Saponifying Solution Dissolve 40 g of potassium hydroxide in about 900 mL of aldehyde-free alcohol maintained at a temperature of 15° until solution is complete, then warm to room temperature, and add sufficient aldehyde-free alcohol to make 1000 mL.

Procedure Transfer 3.00 g into a round-bottom, 100-mL boiling flask provided with a ground-glass joint, add 30 mL of the *Saponifying Solution*, attach a reflux condenser to the flask, and heat the mixture gently on a steam bath for 2 h. At the end of this period, remove the reflux condenser, insert a thermometer into the solution, and place the flask in a water bath at a temperature of 80°. Rotate the flask while both the bath and the solution cool to 65°. The solution shows no cloudiness or globule formation before this temperature is reached.

Packaging and Storage Store in well-closed containers.

Beeswax, Yellow

Yellow Wax

INS: 901 CAS: [8012-89-3]

DESCRIPTION

The purified wax from the honeycomb of the bee *Apis mellifera* L. (Fam. *Apidae*). It is a yellowish to grayish brown solid having an agreeable, honeylike odor. It is somewhat brittle when cold, and presents a dull, granular, noncrystalline fracture when broken. It becomes pliable at a temperature of about 35°. Its specific gravity is about 0.95. Yellow Beeswax is insoluble in water and sparingly soluble in cold alcohol. Boiling alcohol dissolves cerotic acid and part of the myricin, which are constituents of the wax. It is completely soluble in chloroform, in ether, and in fixed and volatile oils. It is partly soluble in cold carbon disulfide and is completely soluble at temperatures of 30° or above.

Functional Use in Foods Candy glaze and polish; flavoring agent.

REQUIREMENTS

Acid Value Between 18 and 24.
Carnauba Wax Passes test.
Ester Value Between 72 and 77.
Fats, Japan Wax, Rosin, and Soap Passes test.
Heavy Metals (as Pb) Not more than 0.002%.
Lead Not more than 10 mg/kg.
Melting Range Between 62° and 65°.
Saponification Cloud Test Passes test.

TESTS

For the determination of *Acid Value*; *Carnauba Wax*; *Ester Value*; *Fats, Japan Wax, Rosin, and Soap*; *Heavy Metals*; *Lead*; *Melting Range*; and *Saponification Cloud Test*, proceed as directed in the monograph for *White Beeswax*.

Packaging and Storage Store in well-closed containers.

Bentonite

Smectite; Aluminum Magnesium Silicate

$Al_2O_3 \cdot 4SiO_2 \cdot nH_2O$

INS: 558 CAS: [1302-78-9]

DESCRIPTION

Natural smectite clays consisting primarily of colloidal hydrated aluminum magnesium silicates of the montmorillonite or hectorite type of minerals with varying quantities of alkalies, alkaline earths and iron. Materials of commerce are powders ranging in colors and tints from off-white to pale brown to gray depending on the cations present in natural deposits. When moistened, Bentonite exhibits a distinct earthy or claylike odor. It is insoluble in water, alcohol, dilute acids or alkalies.

Functional Use in Foods   Processing aid as clarifying, filter agents.

REQUIREMENTS

Identification

A. Add 2 g in small portions to 100 mL of water, with intense agitation. Allow to stand for 12 h to ensure complete hydration. Place 2 mL of the mixture so obtained on a suitable glass slide, and allow to air-dry at room temperature to produce an oriented film. Place the slide in a vacuum desiccator over a free surface of ethylene glycol. Evacuate the desiccator, and close the stopcock so that ethylene glycol saturates the desiccator chamber. Allow to stand for 12 h. Record the X-ray diffraction pattern using a copper source, and calculate the d values: the largest peak corresponds to a d value between 15.0 and 17.2 Å. Prepare a random powder specimen of Bentonite, and determine the d values in the region between 1.48 and 1.54 Å: the peak is between 1.492 and 1.504 Å or between 1.510 and 1.540 Å.

B. To 0.5 g in a metal crucible add 1 g of potassium nitrate and 3 g of anhydrous sodium carbonate, heat until the mixture has melted and allow to cool. To the residue add 20 mL of boiling water, mix, filter and wash the residue with 50 mL of water. To the residue add 1 mL of hydrochloric acid and 5 mL of water and filter. To the filtrate add 1 mL of 10 N sodium hydroxide and filter. To the filtrate add 3 mL of 2 M ammonium chloride. A gelatinous, white precipitate is produced.

Arsenic (as As)   Not more than 5 mg/kg.
Coarse Particles   Not more than 0.5% is retained on a 75-μm sieve.
Gel Formation   Passes test.
Lead   Not more than 0.004%.
Loss on Drying   Not more than 8.0%.
Microbial Limits:
  Aerobic Plate Count   Not more than 1000 per g.
  E. coli   Negative in 25 g.
pH of a 1 in 50 Dispersion   Between 8.5 and 10.5.

TESTS

Arsenic   Transfer 8.0 g of a dried sample to a 250-mL beaker containing 100 mL of dilute hydrochloric acid (1 in 25), mix, cover with a watch glass, and boil gently, with occasional stirring, for 15 min without allowing excessive foaming. Filter the hot supernatant liquid through a rapid-flow filter paper into a 200-mL volumetric flask, and wash with four 25-mL portions of hot dilute hydrochloric acid (1 in 25), collecting the washings in the volumetric flask. Cool the combined filtrates to room temperature, add dilute hydrochloric acid (1 in 25) to volume, and mix. A 25-mL aliquot of this solution meets the requirements of the *Arsenic Test*, Appendix IIIB, using 5.0 mL of *Standard Arsenic Solution* (5 μg of As).

Coarse Particles   To 20 g add 100 mL of water and mix for 15 min at not less than 5000 revolutions per min. Transfer to a wet sieve of nominal mesh aperture 75 μm, previously dried at 100° to 105° and weighed, and wash with three 500-mL quantities of water, ensuring than any agglomerates are dispersed. Dry at 100° to 105° and weigh.

Gel Formation   Mix a 6 g sample with 300 mg of magnesium oxide. Add the mixture, in several divided portions, to 200 mL of water contained in a blender of approximately 500-mL capacity. Blend thoroughly for 5 min at high speed, transfer 100 mL of the mixture to a 100-mL graduated cylinder, and allow to remain undisturbed for 24 h: Not more than 2 mL of supernatant liquid appears on the surface.

Lead   (**Note**: The *Standard Preparation* and the *Test Preparation* may be modified, if necessary, to obtain solutions of suitable concentrations, adaptable to the linear or working range of the instrument.)

Standard Preparation   On the day of use, dilute 3.0 mL of *Lead Nitrate Stock Solution* (*Heavy Metals Test*, Appendix IIIB) with water to 100 mL. Each mL of the *Standard Preparation* contains the equivalent of 3 μg of lead.

Test Preparation   Transfer 3.75 g of a dried sample to a 250-mL beaker containing 100 mL of dilute hydrochloric acid (1 in 25), stir, cover with a watch glass, and boil for 15 min. Cool to room temperature, and allow the insoluble matter to settle. Decant the supernatant liquid through a rapid-flow filter paper into a 400-mL beaker. Wash the filter with four 25-mL portions of hot water, collecting the filtrate in the 400-mL beaker. Concentrate the combined extracts by gentle boiling to approximately 20 mL. If a precipitate appears, add 2 to 3 drops of nitric acid, heat to boiling, and cool to room temperature. Filter the concentrated extracts through a rapid-flow filter paper into a 50-mL volumetric flask. Transfer the remaining contents of the 400-mL beaker through the filter paper and into the flask with water. Dilute with water to volume, and mix.

Procedure   Determine the absorbances of the *Test Preparation* and the *Standard Preparation* at 284 nm in a suitable atomic absorption spectrophotometer equipped with a lead hollow-cathode lamp, deuterium arc background correction, and a single-slot burner, using an oxidizing flame of air and acetylene. The absorbance of the *Test Preparation* is not greater than that of the *Standard Preparation*.

Loss on Drying, Appendix IIC   Dry at 105° for 2 h.
Microbial Limits:
  Aerobic Plate Count   Proceed as directed in chapter 3 of the *FDA Bacteriological Analytical Manual*, Seventh Edition, 1992.
  E. coli   Proceed as directed in chapter 4 of the *FDA Bacteriological Analytical Manual*, Seventh Edition, 1992.
pH   Disperse a 4.0-g sample in 200 mL of water, mixing vigorously to facilitate wetting, and determine by the *Potentiometric Method*, Appendix IIB.

Packaging and Storage   Store in tight containers.

Benzoic Acid

⌬—COOH

C$_7$H$_6$O$_2$ Formula wt 122.12
INS: 210 CAS: [65-85-0]

DESCRIPTION

White crystals, scales, or needles. It is odorless or has a slightly benzoin-like or benzaldehyde-like odor. It begins to sublime at about 100° and is volatile with steam. One g is soluble in 275 mL of water at 25°, in 20 mL of boiling water, in 3 mL of alcohol, in 5 mL of chloroform, and in 3 mL of ether. It is soluble in fixed and in volatile oils, and is sparingly soluble in solvent hexane.

Functional Use in Foods Preservative; antimicrobial agent.

REQUIREMENTS

Identification Dissolve 1 g in a mixture of 20 mL of water and 1 mL of 1 N sodium hydroxide, filter the solution, and add about 1 mL of ferric chloride TS. A buff-colored precipitate is formed.
Assay Not less than 99.5% and not more than 100.5% of C$_7$H$_6$O$_2$, calculated on the anhydrous basis.
Heavy Metals (as Pb) Not more than 10 mg/kg.
Readily Carbonizable Substances Passes test.
Readily Oxidizable Substances Passes test.
Residue on Ignition Not more than 0.05%.
Solidification Point Between 121° and 123°.
Water Not more than 0.7%.

TESTS

Assay Dissolve about 500 mg of the sample, accurately weighed, in 25 mL of 50% alcohol previously neutralized with 0.1 N sodium hydroxide, add phenolphthalein TS, and titrate with 0.1 N sodium hydroxide. Each mL of 0.1 N sodium hydroxide is equivalent to 12.21 mg of C$_7$H$_6$O$_2$.
Heavy Metals Volatilize 2 g over a low flame. To the residue add 2 mL of nitric acid and about 10 mg of sodium carbonate, and evaporate to dryness on a steam bath. Dissolve the residue in a mixture of 1 mL of 1 N acetic acid and 24 mL of water. This solution meets the requirements of the *Heavy Metals Test*, Appendix IIIB, using 20 μg of lead ion (Pb) in the control (*Solution A*).
Readily Carbonizable Substances, Appendix IIB Dissolve 500 mg in 5 mL of 95% sulfuric acid. The color is no darker than *Matching Fluid Q*.
Readily Oxidizable Substances To a mixture of 100 mL of water and 1.5 mL of sulfuric acid heated to 100° add dropwise 0.1 N potassium permanganate until a pink color persists for 30 s. Dissolve 1.0 g of the Benzoic Acid in the hot solution, and titrate with 0.1 N potassium permanganate to a pink color that persists for 15 s. The volume of 0.1 N potassium permanganate consumed does not exceed 0.5 mL.
Residue on Ignition, Appendix IIC Ignite 2 g as directed in the general method.
Solidification Point Determine as directed in the general method, Appendix IIB.
Water Determine by the *Karl Fischer Titrimetric Method*, Appendix IIB, using methanol in pyridine (1 in 2) as the solvent.

Packaging and Storage Store in well-closed containers.

Benzoyl Peroxide

C$_{14}$H$_{10}$O$_4$ Formula wt 242.23
INS: 928 CAS [94-36-0]

DESCRIPTION

A colorless, crystalline solid having a faint odor of benzaldehyde. It is insoluble in water, slightly soluble in alcohol, and soluble in chloroform and ether. One g dissolves in 40 mL of carbon disulfide. It melts between 103° and 106° with decomposition.

> **Caution**: Benzoyl Peroxide, especially in the dry form, is a dangerous, highly reactive, oxidizing material and has been known to explode spontaneously. Observe safety precautions printed on the label of the container.

Functional Use in Foods Bleaching agent.

REQUIREMENTS

Identification To 500 mg of the sample add 50 mL of 0.5 N alcoholic potassium hydroxide, heat gradually to boiling, and continue boiling for 15 min. Cool, dilute to 200 mL with water, and make the solution strongly acid with 0.5 N hydrochloric acid. Extract with ether, dry the extract with anhydrous sodium sulfate, and then evaporate to dryness on a steam bath. The residue of benzoic acid so obtained melts between 121° and 123° (see Appendix IIB).
Assay Not less than 96.0% of C$_{14}$H$_{10}$O$_4$.
Heavy Metals (as Pb) Not more than 0.002%.
Lead Not more than 10 mg/kg.

TESTS

Assay Dissolve about 250 mg, accurately weighed, in 15 mL of acetone in a 100-mL glass-stoppered bottle, and add 3 mL of potassium iodide solution (1 in 2). Swirl for 1 min, then immediately titrate with 0.1 N sodium thiosulfate (without the addition of starch TS). Each mL of 0.1 N sodium thiosulfate is equivalent to 12.11 mg of $C_{14}H_{10}O_4$.

Heavy Metals Mix 1 g with 5 mL of 1 N sodium hydroxide, slowly evaporate to dryness on a steam bath, cool, and dissolve the residue in 25 mL of water. This solution meets the requirements of the *Heavy Metals Test*, Appendix IIIB, using 20 μg of lead ion (Pb) and 5 mL of 1 N sodium hydroxide in the control (*Solution A*).

Lead Mix 1 g with 10 mL of 1 N sodium hydroxide, slowly evaporate to dryness on a steam bath, and cool. A *Sample Solution*, prepared as directed for organic compounds from the residue so obtained, meets the requirements of the *Lead Limit Test*, Appendix IIIB, using 10 μg of lead ion (Pb) in the control.

Packaging and Storage Store in the original container and observe the safety precautions printed on the label.

Bergamot Oil, Coldpressed

DESCRIPTION

A volatile oil obtained by expression, without the aid of heat, from the fresh peel of the fruit of *Citrus bergamia* Risso et Poiteau (Fam. *Rutaceae*). It is a green to yellowish green or yellowish brown liquid having a fragrant, sweet-fruity odor. It is miscible with alcohol and with glacial acetic acid. It is soluble in most fixed oils, but is insoluble in glycerin and in propylene glycol. It may contain a suitable antioxidant.

Functional Use in Foods Flavoring agent.

REQUIREMENTS

Identification The infrared absorption spectrum of the sample exhibits relative maxima (that may vary in intensity) at the same wavelengths (or frequencies) as those shown in the respective spectrum in this section on *Infrared Spectra* (*Series A: Essential Oils*), using the same test conditions as specified therein.

Assay Not less than 36.0% of esters, calculated as linalyl acetate ($C_{12}H_{20}O_2$).

Angular Rotation Between +8° and +24°.

Heavy Metals (as Pb) Not more than 0.004%.

Lead Not more than 10 mg/kg.

Refractive Index Between 1.465 and 1.468 at 20°.

Residue on Evaporation Not more than 6.0%.

Solubility in Alcohol Passes test.

Specific Gravity Between 0.875 and 0.880.

Ultraviolet Absorbance Not less than 0.32.

TESTS

Assay Weigh accurately about 2 g, and proceed as directed under *Ester Determination*, Appendix VI, but heat the mixture for 30 min on the steam bath. Use 98.15 as the equivalence factor (*e*) in the calculation.

Angular Rotation Determine in a 100-mm tube as directed under *Optical (Specific) Rotation*, Appendix IIB.

Heavy Metals Prepare and test a 500-mg sample as directed in *Method II* under the *Heavy Metals Test*, Appendix IIIB, using 20 μg of lead ion (Pb) in the control (*Solution A*).

Lead A *Sample Solution* prepared as directed for organic compounds meets the requirements of the *Lead Limit Test*, Appendix IIIB, using 10 μg of lead ion (Pb) in the control.

Refractive Index, Appendix IIB Determine with an Abbé or other refractometer of equal or greater accuracy.

Residue on Evaporation Proceed as directed in the general method, Appendix VI, heating for 5 h.

Solubility in Alcohol Proceed as directed in the general method, Appendix VI. One mL dissolves in 2 mL of 90% alcohol.

Specific Gravity Determine by any reliable method (see *General Provisions*).

Ultraviolet Absorbance Proceed as directed under *Ultraviolet Absorbance of Citrus Oils*, Appendix VI, using about 50 mg of sample, accurately weighed. The absorbance maximum occurs at 315 ± 3 nm.

Packaging and Storage Store in full, tight, preferably glass, aluminum, tin-lined, or other suitably lined containers in a cool place protected from light.

BHA

Butylated Hydroxyanisole

$C_{11}H_{16}O_2$ Formula wt 180.25

INS: 320 CAS: [25013-16-5]

DESCRIPTION

BHA is predominately 3-*tert*-butyl-4-hydroxyanisole (3-BHA), with varying amounts of 2-*tert*-butyl-4-hydroxyanisole (2-BHA). It occurs as a white or slightly yellow, waxy solid having a faint characteristic odor. It is insoluble in water, but is freely soluble in alcohol and in propylene glycol. It melts between 48° and 63°.

Functional Use in Foods Antioxidant.

REQUIREMENTS

Identification To 5 mL of a 1 in 10,000 solution of the sample in 72% alcohol add 2 mL of sodium borate TS and 1 mL of a 1 in 10,000 solution of 2,6-dichloroquinonechlorimide in absolute alcohol, and mix. A blue color develops.
Assay Not less than 98.5% of $C_{11}H_{16}O_2$.
Heavy Metals (as Pb) Not more than 10 mg/kg.
Residue on Ignition Not more than 0.05%.

TESTS

Assay

Internal Standard Solution Dissolve about 500 mg of 4-*tert*-butylphenol, accurately weighed, in acetone in a 100-mL volumetric flask, add acetone to volume, and mix.

Standard Preparation Dissolve together, accurately weighed quantities of USP reference standards 3-*tert*-butyl-4-hydroxyanisole and 2-*tert*-butyl-4-hydroxyanisole to final concentrations of 9 mg/mL and 1 mg/mL, respectively, in *Internal Standard Solution*.

Assay Preparation Dissolve about 100 mg of Butylated Hydroxyanisole, accurately weighed, in the *Internal Standard Solution* in a 10-mL volumetric flask, dilute with the *Internal Standard Solution* to volume, and mix.

Chromatographic System The gas chromatograph is equipped with a flame-ionization detector and contains a 1.8-m × 2-mm (id) stainless-steel column packed with 10% silicone GE XE-60; the column is maintained isothermally at a temperature between 175° and 185°, and helium is used as the carrier gas, at a flow rate of 30 mL/min. Chromatograph a sufficient number of injections of the *Standard Preparation*, and record the areas as directed under *Procedure*, to ensure that the relative standard deviation does not exceed 2.0% for the 3-*tert*-butyl-4-hydroxyanisole isomer and 6.0% for the 2-*tert*-butyl-4-hydroxyanisole isomer; the resolution between the isomers is not less than 1.3, and the tailing factor does not exceed 2.0. (See *Chromatography*, Appendix IIA.)

Procedure Separately inject suitable portions (about 5 µL) of the *Standard Preparation* and the *Assay Preparation* into the gas chromatograph, and record the chromatograms. Measure the areas under the peaks for each isomer and the internal standard in each chromatogram, and calculate the quantity, (*I*), in mg, of each isomer in the Butylated Hydroxyanisole by the formula

$$I = 10\ C_S\ (R_U/R_S),$$

in which C_S is the concentration, in mg/mL, of the isomer in the *Standard Preparation*; R_S is the ratio of the area of the isomer to that of the internal standard in the chromatogram from the *Standard Preparation*; and R_U is the ratio of the area of the isomer to that of the internal standard in the chromatogram from the *Assay Preparation*. Calculate the weight, in mg, of $C_{11}H_{16}O_2$ in the Butylated Hydroxyanisole by adding the quantities of the two isomers.

Heavy Metals Prepare and test a 2-g sample as directed in *Method II* under the *Heavy Metals Test*, Appendix IIIB, using 20 µg of lead ion (Pb) in the control (*Solution A*).

Residue on Ignition Ignite 10 g as directed in the general method, Appendix IIC.

Packaging and Storage Store in well-closed containers.

BHT

Butylated Hydroxytoluene; 2,6-Di-*tert*-butyl-*p*-cresol

$C_{15}H_{24}O$ Formula wt 220.35
INS: 321 CAS: [128-37-0]

DESCRIPTION

A white crystalline solid having a faint characteristic odor. It is insoluble in water and in propylene glycol, but is freely soluble in alcohol.

Functional Use in Foods Antioxidant.

REQUIREMENTS

Identification To 10 mL of a 1 in 100,000 solution of the sample in methanol add 10 mL of water, 2 mL of sodium nitrite solution (3 in 1000), and 5 mL of dianisidine solution (200 mg of 3,3'-dimethoxybenzidine dihydrochloride dissolved in a mixture of 40 mL of methanol and 60 mL of 1 N hydrochloric acid). An orange-red color develops within 3 min. Add 5 mL of chloroform, and shake. The chloroform layer exhibits a purple or magenta color that fades when exposed to light.
Assay Not less than 99.0 weight % of $C_{15}H_{24}O$.
Heavy Metals (as Pb) Not more than 10 mg/kg.
Residue on Ignition Not more than 0.002%.

TESTS

Assay Its solidification point (see Appendix IIB) is not lower than 69.2°, indicating a purity of not less than 99.0% of $C_{15}H_{24}O$.
Heavy Metals Prepare and test a 2-g sample as directed in *Method II* under the *Heavy Metals Test*, Appendix IIIB, using 20 µg of lead ion (Pb) in the control (*Solution A*).
Residue on Ignition Transfer a 50-g sample into a tared crucible, ignite until thoroughly charred, and cool. Moisten the ash

with 1 mL of sulfuric acid, and complete the ignition by heating for 15-min periods at 800° ± 25° to constant weight.

Packaging and Storage Store in well-closed containers.

Biotin

cis-Hexahydro-2-oxo-1H-thieno[3,4]imidazole-4-valeric Acid; *d*-Biotin

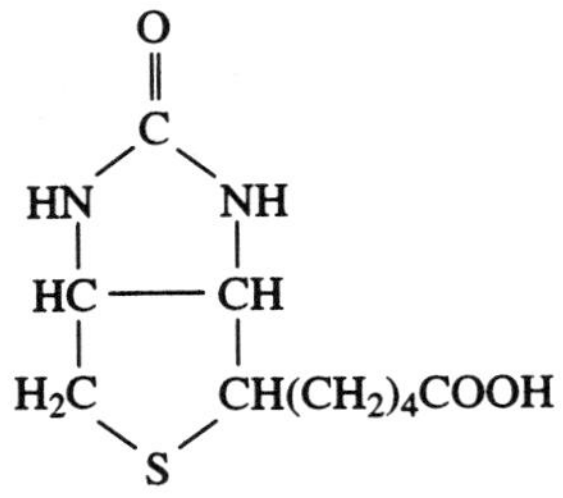

$C_{10}H_{16}N_2O_3S$ Formula wt 244.31

CAS: [58-85-5]

DESCRIPTION

A practically white, crystalline powder. It is stable to air and heat. One g dissolves in about 5000 mL of water at 25° and in about 1300 mL of alcohol; it is more soluble in hot water and in dilute alkali, and is insoluble in other common organic solvents.

Functional Use in Foods Nutrient; dietary supplement.

REQUIREMENTS

Identification

A. The infrared absorption spectrum of a potassium bromide dispersion of the sample exhibits maxima only at the same wavelengths as that of a similar preparation of USP Biotin Reference Standard.

B. A saturated solution in warm water decolorizes bromine TS, added dropwise.

Assay Not less than 97.5% and not more than 100.5% of $C_{10}H_{16}N_2O_3S$.
Heavy Metals (as Pb) Not more than 10 mg/kg.
Melting Range Between 229° and 232° with decomposition.
Specific Rotation $[\alpha]_D^{25°}$: Between +89° and +93°.

TESTS

Assay Mix about 500 mg, accurately weighed, with 100 mL of water, add phenolphthalein TS, and titrate the suspension slowly, while heating and stirring continuously, with 0.1 *N* sodium hydroxide to a pink color. Each mL of 0.1 *N* sodium hydroxide is equivalent to 24.43 mg of $C_{10}H_{16}N_2O_3S$.
Heavy Metals Prepare and test a 2-g sample as directed in *Method II* under the *Heavy Metals Test*, Appendix IIIB, using 20 μg of lead ion (Pb) in the control (*Solution A*).
Melting Range Determine as directed for *Melting Range or Temperature*, Appendix IIB.
Specific Rotation, Appendix IIB Determine in a solution in 0.1 *N* sodium hydroxide containing 500 mg in each 25 mL.

Packaging and Storage Store in tight containers.

Birch Tar Oil, Rectified

DESCRIPTION

The pyroligneous oil obtained by dry distillation of the bark and the wood of *Betula pendula* Roth and related species of *Betula* (Fam. *Betulaceae*) and rectified by steam distillation. It is a clear, dark brown liquid having a strong leatherlike odor. It is soluble in most fixed oils, but it is insoluble in glycerin, in mineral oil, and in propylene glycol.

Functional Use in Foods Flavoring agent.

REQUIREMENTS

Identification The infrared absorption spectrum of the sample exhibits relative maxima (that may vary in intensity) at the same wavelengths (or frequencies) as those shown in the respective spectrum in the section on *Infrared Spectra* (*Series A: Essential Oils*), using the same test conditions as specified therein.
Heavy Metals (as Pb) Passes test.
Solubility in Alcohol Passes test.
Specific Gravity Between 0.886 and 0.950.

TESTS

Heavy Metals Shake 10 mL of the oil with an equal volume of water to which 1 drop of hydrochloric acid has been added, and pass hydrogen sulfide through the mixture until it is saturated. No darkening in color is produced in either the oil or the water.
Solubility in Alcohol Proceed as directed in the general method, Appendix VI. One mL dissolves in 3 mL of absolute alcohol.
Specific Gravity Determine by any reliable method (see Appendix VII).

Packaging and Storage Store in full, tight, preferably glass containers in a cool place protected from light.

Black Pepper Oil

CAS: [8006-82-4]

DESCRIPTION

The volatile oil obtained by steam distillation from the dried, unripened fruit of the plant *Piper nigrum* L. (Fam. *Piperaceae*). It is an almost colorless to slightly greenish liquid having the characteristic odor of pepper and a relatively mild taste. It is soluble in most fixed oils, in mineral oil, and in propylene glycol. It is sparingly soluble in glycerin.

Functional Use in Foods Flavoring agent.

REQUIREMENTS

Identification The infrared absorption spectrum of the sample exhibits relative maxima (that may vary in intensity) at the same wavelengths (or frequencies) as those shown in the respective spectrum in the section on *Infrared Spectra (Series A: Essential Oils)*, using the same test conditions as specified therein.
Heavy Metals (as Pb) Passes test.
Refractive Index Between 1.479 and 1.488 at 20°.
Solubility in Alcohol Passes test.
Specific Gravity Between 0.864 and 0.884.
Specific Rotation $[\alpha]_D^{20°}$: Between +1° and −33.5°.

TESTS

Heavy Metals Shake 10 mL of the oil with an equal volume of water to which 1 drop of hydrochloric acid has been added, and pass hydrogen sulfide through the mixture until it is saturated. No darkening in color is produced in either the oil or the water.
Refractive Index, Appendix IIB Determine with an Abbé or other refractometer of equal or greater accuracy.
Solubility in Alcohol Proceed as directed in the general method, Appendix VI. One mL dissolves in 3 mL of 95% alcohol.
Specific Gravity Determine by any reliable method (see Appendix VII).
Specific Rotation, Appendix IIB Determine in a solution containing 5 g in sufficient 2 *N* hydrochloric acid to make 100 mL.

Packaging and Storage Store in full, tight, glass containers in a cool place protected from light.

Bois de Rose Oil

DESCRIPTION

The volatile oil obtained by steam distillation from the chipped wood of *Aniba rosaeodora* var. *amazonica* Ducke, (Fam. *Lauraceae*). The oils from the coastal region of Brazil and the Amazon valley tend to differ in odor and in linalool content from that produced in the Loreto province of Peru. The oil is a colorless to pale yellow liquid having a slightly camphoraceous, pleasant floral odor. It is soluble in most fixed oils and in propylene glycol. It is soluble in mineral oil, occasionally with turbidity, but only slightly soluble in glycerin.

Functional Use in Foods Flavoring agent.

REQUIREMENTS

Identification The infrared absorption spectrum of the sample exhibits relative maxima (that may vary in intensity) at the same wavelengths (or frequencies) as those shown in the respective spectrum in the section on *Infrared Spectra (Series A: Essential Oils)*, using the same test conditions as specified therein.
Assay Not less than 82.0% and not more than 92.0% of total alcohols, calculated as linalool ($C_{10}H_{18}O$).
Angular Rotation Between −4° and +6°.
Distillation Range Not less than 70% distills between 195° and 205°.
Heavy Metals (as Pb) Passes test.
Refractive Index Between 1.462 and 1.470 at 20°.
Solubility in Alcohol Passes test.
Specific Gravity Between 0.868 and 0.889.

TESTS

Assay Proceed as directed under *Linalool Determination*, Appendix VI, using about 1.2 g of the acetylated oil, accurately weighed.
Angular Rotation Determine in a 100-mm tube as directed under *Optical (Specific) Rotation*, Appendix IIB.
Distillation Range Proceed as directed in the general method, Appendix IIB, using 50 mL of the sample, previously dried over anhydrous sodium sulfate, and employing a 125-mL flask.
Heavy Metals Shake 10 mL of the oil with an equal volume of water to which 1 drop of hydrochloric acid has been added, and pass hydrogen sulfide through the mixture until it is saturated. No darkening in color is produced in either the oil or the water.
Refractive Index, Appendix IIB Determine with an Abbé or other refractometer of equal or greater accuracy.
Solubility in Alcohol Proceed as directed in the general method, Appendix VI. One mL dissolves in 6 mL of 60% alcohol.
Specific Gravity Determine by any reliable method (see Appendix VII).

Packaging and Storage Store in full, tight, preferably glass, aluminum, tin-lined, or other suitably lined containers in a cool place protected from light.

Brominated Vegetable Oil

DESCRIPTION

Brominated Vegetable Oil is a bromine addition product of vegetable oil or oils. It is a pale yellow to dark brown, viscous, oily liquid having a bland or fruity odor and a bland taste. It is insoluble in water, but is soluble in chloroform, in ether, in hexane, and in fixed oils. Brominated Vegetable Oil may contain a suitable stabilizer.

Functional Use in Foods Flavoring agent; beverage stabilizer.

REQUIREMENTS

Identification Mix about 0.2 mL of the sample with 1 g of anhydrous sodium carbonate in a suitable crucible, cover the mixture with an additional 1 g of sodium carbonate, compact the mixture by gentle tapping, and heat the crucible rapidly and strongly for 10 min. Cool the crucible and its contents, dissolve the residue in 20 mL of hot water, and filter. To the filtrate add 1.7 N nitric acid until effervescence ceases, then add 1 mL of silver nitrate TS. A curdy, yellowish precipitate, which is insoluble in nitric acid but soluble in an excess of stronger ammonia water, is formed.
Free Bromine Passes test.
Free Fatty Acids (as oleic) Not more than 2.5%.
Heavy Metals (as Pb) Not more than 10 mg/kg.
Iodine Value Not more than 16.
Specific Gravity Within the range specified by the vendor.

TESTS

Free Bromine Dissolve 1 g in 20 mL of acetone, add 1 g of sodium iodide, and allow to stand in a stoppered flask in the dark for 30 min, with occasional shaking. Add 25 mL of water and 1 mL of starch TS. No blue color is produced.
Free Fatty Acids Determine as directed in the general method, Appendix VII, using 28.2 as the equivalence factor (e) in the calculation for oleic acid. Titrate with the appropriate normality of sodium hydroxide solution, shaking vigorously, to the first permanent pink color of the same intensity as that of the neutralized alcohol, or, if the color of the sample interferes, titrate to a pH of 8.5, determined with a suitable instrument.
Heavy Metals Prepare and test a 2-g sample as directed in *Method II* under the *Heavy Metals Test*, Appendix IIIB, using 20 µg of lead ion (Pb) in the control (*Solution A*).
Iodine Value Determine by the *Modified Wijs Method*, Appendix VII.
Specific Gravity Determine as directed in the general method, in *General Provisions*, at the temperature specified by the vendor.

Packaging and Storage Store in well-closed containers.

Butadiene-Styrene 75/25 Rubber

DESCRIPTION

Butadiene-Styrene 75/25 Rubber is available as a liquid latex or solid rubber that is produced by the emulsion polymerization of butadiene and styrene, using fatty acid soaps as emulsifiers, a persulfate catalyst, a suitable molecular weight regulator (if required), and a suitable shortstop. It is also available as a solid rubber produced by the solution-copolymerization of butadiene and styrene in a hexane solution, using butyl lithium as a catalyst.

The latex has a pH of 9.5 to 11.0 and a solids content of 26% to 42%. It is coagulated with or without other food-grade ingredients in a heated kettle. The coagulated mass is squeezed to drain off serums, then the coagulum is washed with hot water (with or without alkali), and it is rinsed with water until the batch is neutral. Finally, the coagulum is dried to remove residual volatiles. When butadiene-styrene rubber is purchased in the latex form, it must be washed by the preceding or an equivalent procedure.

In the case of the solvent-polymerized product, solvent and volatiles are removed by processing with hot water or by drum-drying. Both of the solid forms are supplied by the manufacturer either in slab form or as a uniform, free-flowing crumb and may contain a suitable food-grade antioxidant. The crumb form, in addition, may contain a suitable food-grade partitioning agent.

Functional Use in Foods Masticatory substance in chewing gum base.

REQUIREMENTS

Note: The following **REQUIREMENTS** apply to the solid rubber as supplied by the manufacturer, or to the washed and dried coagulum obtained from the latex as described above.

Identification Identify emulsion-polymerized Butadiene-Styrene 75/25 Rubber latex and solid by comparing their infrared absorption spectra with the respective typical spectra as shown in the section on *Infrared Spectra* (*Series C: Other Substances*). Prepare latex samples by first drying at 105° for 4 h, then by dissolving in hot toluene and evaporating on a potassium bromide plate. Prepare solid samples by dissolving them in hot toluene and evaporating on a potassium bromide plate.
Arsenic (as As) Not more than 3 mg/kg.
Bound Styrene Between 22.0% and 26.0%.
Heavy Metals (as Pb) Not more than 10 mg/kg.
Lead Not more than 3 mg/kg.
Lithium Not more than 0.0075%.
Quinones Not more than 0.002%.
Residual Hexane Not more than 0.01%.
Residual Styrene Not more than 0.002%.

TESTS

Arsenic Prepare a *Sample Solution* as directed in the general method under *Chewing Gum Base*, Appendix IV. This solution meets the requirements of the *Arsenic Test*, Appendix IIIB.

Bound Styrene Determine as directed in the general method, Appendix IV.

Heavy Metals Prepare and test a 2-g sample as directed in *Method II* under the *Heavy Metals Test*, Appendix IIIB, using 20 µg of lead ion (Pb) in the control (*Solution A*).

Lead Prepare a *Sample Solution* as directed in the general method under *Chewing Gum Base*, Appendix IV. This solution meets the requirements of the *Lead Limit Test*, Appendix IIIB, using 10 µg of lead ion (Pb) in the control.

Lithium

Atomic Absorption Spectrophotometer Use a suitable instrument, equipped with a lithium hollow cathode lamp, capable of measuring the radiation absorbed by lithium in the 6707-nm spectral band.

Standard Solution Transfer 399.3 mg of ACS reagent-grade lithium carbonate to a 1000-mL volumetric flask, dissolve in a minimum amount of 1:1 hydrochloric acid–water, dilute to volume with water, and mix. Transfer 10.0 mL of this solution to a 100-mL volumetric flask, dilute to volume with water, and mix. Finally, transfer 10.0 mL of this solution to a second 100-mL volumetric flask, add 1.0 mL of hydrochloric acid, dilute to volume with water, and mix. This solution contains 75 µg of Li per 100 mL. *Sample Solution* Weigh accurately 1 g of a solid rubber sample, wrap it tightly in ashless filter paper, and place in a tared platinum crucible. Heat in an oven at 100° for 15 min, and then transfer to a muffle furnace programmed to reach 500° within 1 to 3 h after introduction of the sample. Remove the crucible from the furnace 15 to 20 min after 500° has been reached, and cool in a desiccator. Quantitatively transfer the contents of the crucible to a 100-mL volumetric flask, using 1 mL of hydrochloric acid and water, dilute to volume with water, and mix.

Procedure Following the manufacturer's instructions for operating the atomic absorption spectrophotometer, aspirate a suitable portion of the *Standard Solution* through the flame. In a similar manner, aspirate a suitable portion of the *Sample Solution*. Any absorbance produced by the *Sample Solution* does not exceed that produced by the *Standard Solution*.

Quinones Determine as directed in the general method, Appendix IV.

Residual Hexane (Note: The isooctane, 2,2,4-trimethylpentane, used in this test should be of chromatographic-grade quality.)

Internal Standard Stock Solution Transfer 150 mg of *n*-nonane, accurately weighed, to a 50-mL volumetric flask, dilute to volume with isooctane, and mix.

Dilute Internal Standard Solution Pipet 10.0 mL of *Internal Standard Stock Solution* into a 100-mL volumetric flask, dilute to volume with isooctane, and mix. Pipet 5.0 mL of this solution into a 250-mL volumetric flask, dilute to volume with isooctane, and mix. Each mL of the final solution contains 6 µg of *n*-nonane.

Hexane Standard Solution Transfer 150 mg of *n*-hexane, accurately weighed, to a 50-mL volumetric flask, dilute to volume with isooctane, and mix. Pipet 1.0 mL of this solution into a 100-mL volumetric flask, dilute to volume with isooctane, and mix. Finally, pipet 10.0 mL of this solution and 10.0 mL of *Internal Standard Stock Solution* into a 50-mL volumetric flask, dilute to volume with isooctane, and mix.

Sample Preparation Weigh accurately 1.5 g of a solid rubber sample, transfer it into a 4-oz bottle, and pipet 25.0 mL of the *Dilute Internal Standard Solution* into the bottle. Stopper the bottle, and shake mechanically overnight to dissolve the rubber. Add 50 mL of methanol to precipitate out the polymer, and shake vigorously for 15 min. Allow the mixture to settle, and decant the liquid phase into a 250-mL separator. Wash the polymer with 25 mL of methanol, and add the wash to the separator. Add 50 to 75 mL of cold water to the separator, and shake vigorously for 1 min, venting periodically to release any pressure. Allow the phases to separate, drain off the bottom (aqueous) phase, and rewash the isooctane phase with a second 50-mL portion of cold water. Shake again, allow to separate, and drain off the bottom layer. Transfer 10 mL of the isooctane phase to a 20-mL vial for the analysis.

Procedure Use a gas chromatograph equipped with a flame-ionization detector and a column capable of separating hexane, isooctane, and *n*-nonane. Under typical conditions, the instrument contains a 3-m × 3-mm stainless steel column packed with 60- to 80-mesh Chromosorb P containing 15% didecyl phthalate. The column is maintained isothermally at 120°, the injection port at 240°, and the detector at 250°. Helium is the carrier gas, flowing at a rate of 30 mL/min. A digital integrator or computer is recommended for data acquisition, although any mode (other than triangulation and planimetry) that gives accurate and reliable measurement of the peak areas is satisfactory.

Chromatograph duplicate 5-µL portions of the *Hexane Standard Solution*, and measure the areas under the hexane and nonane peaks. In a similar manner, chromatograph duplicate 5-µL portions of the *Sample Preparation*, and measure the areas under the hexane and nonane peaks. The peak area ratio of hexane to nonane (i.e., hexane divided by nonane) produced by the *Sample Preparation* does not exceed that produced by the *Hexane Standard Solution*.

Residual Styrene Determine as directed in the general method, Appendix IV.

Packaging and Storage Store in well-closed containers.

Butadiene-Styrene 50/50 Rubber

DESCRIPTION

A synthetic liquid latex (SBR 2000 Type) or solid rubber (SBR 1028 Type) produced by the emulsion copolymerization of butadiene and styrene, using rosin acid soaps or fatty acid soaps as emulsifiers, a persulfate catalyst, a suitable molecular weight regulator (if required), and a suitable shortstop. It is also available as a solid rubber produced by the solution copolymerization of butadiene and styrene in a hexane solution, using butyl lithium as a catalyst. The latex, which has a pH between 10.0 and 11.5 and a solids content of 41% to 63%, is coagulated with or without other food-grade ingredients in a heated kettle, the coagulated mass is squeezed to drain off serums, and the coagulum is washed with hot water (with or without alkali) and rinsed with water until the batch is neutral. Finally, the coagulum is dried to remove residual volatiles. When butadiene-styrene rubber is purchased in the latex form, it must be washed by the preceding or an equivalent procedure. The solid form is supplied by the manufacturer either in slab form or as a uniform, free-flowing crumb and may contain a suitable food-grade antioxidant. The crumb form, in addition, may contain a suitable food-grade partitioning agent.

Functional Use in Foods Masticatory substance in chewing gum base.

REQUIREMENTS

Note: The following **REQUIREMENTS** apply to the solid rubber as supplied by the manufacturer, or to the washed and dried coagulum obtained from the latex as described above.

Identification Identify Butadiene-Styrene 50/50 Rubber latex and solid by comparing their infrared absorption spectra with the respective typical spectra as shown in the section on *Infrared Spectra*, (*Series C: Other Substances*). Prepare latex samples by first drying at 105° for 4 h, then by dissolving in hot toluene and evaporating on a potassium bromide plate. Prepare solid samples by dissolving them in hot toluene and evaporating on a potassium bromide plate.
Arsenic (as As) Not more than 3 mg/kg.
Bound Styrene Between 45.0% and 50.0%.
Heavy Metals (as Pb) Not more than 10 mg/kg.
Lead Not more than 3 mg/kg.
Lithium Not more than 0.0075%.
Quinones Not more than 0.002%.
Residual Hexane Not more than 0.01%.
Residual Styrene Not more than 0.003%.

TESTS

Proceed as directed in the monograph for *Butadiene-Styrene 75/25 Rubber*.

Packaging and Storage Store in well-closed containers.

Butane

n-Butane

$$CH_3CH_2CH_2CH_3$$

C_4H_{10} Formula wt 58.12

CAS: [106-97-8]

DESCRIPTION

A colorless, flammable gas with a characteristic odor (boiling temperature is −0.5°). One volume of water dissolves 0.15 volume, and 1 volume of alcohol dissolves 18 volumes at 17° and 770 mm of mercury; 1 volume of ether or chloroform at 17° dissolves 25 or 30 volumes, respectively. Vapor pressure at 21° is about 1620 mm of mercury (17 psi).

Functional Use in Foods Propellant; aerating agent.

REQUIREMENTS

Caution: Butane is highly flammable and explosive. Observe precautions and perform sampling and analytical operations in a well-ventilated fume hood.

Identification
  A. The infrared absorption spectrum of Butane exhibits maxima, among others, at about the following wavelengths, in μm: 3.4 (vs), 6.8 (s), 7.2 (m), and 10.4 (m).
  B. The vapor pressure of a test specimen, obtained as directed in the *Sampling Procedure* and determined at 21° by means of a suitable pressure gauge, is between 205 and 235 kPa absolute (30 and 34 psia, respectively).
Assay Not less than 97.0% of C_4H_{10}.
Acidity of Residue Passes test.
High-Boiling Residue Not more than 5 mg/kg.
Sulfur Compounds Passes test.
Water Not more than 10 mg/kg.

TESTS

Sampling Procedure Use a stainless steel specimen cylinder equipped with a stainless steel valve and having a capacity of not less than 200 mL and a pressure rating of 240 psi or more. Dry the cylinder with the valve open at 110° for 2 h, and evacuate the hot cylinder to less than 1 mm of mercury. Close the valve, and cool and weigh the cylinder. Tightly connect one end of a charging line to the Butane container, and loosely connect the other end to the specimen cylinder. Carefully open the Butane container, and allow the Butane to flush out the charging line through the loose connection. Avoid excessive flushing that causes moisture to freeze in the charging line and connections. Tighten the fitting on the specimen cylinder, and open the specimen cylinder valve, allowing the Butane to flow into the evacuated cylinder. Continue sampling until the desired amount of specimen is obtained, then close the Butane container valve, and finally, close the specimen cylinder valve.

Caution: Do not overload the specimen cylinder.

Again weigh the charged specimen cylinder, and calculate the specimen weight.

Assay

Chromatographic System   Under typical conditions, the gas chromatograph is equipped with a thermal-conductivity detector and contains a 6-m × 3-mm aluminum column packed with 10 weight percent tetraethylene glycol dimethyl ether liquid phase on a support of crushed firebrick (GasChrom R or equivalent), which has been calcined or burned with a clay binder above 900° and silanized. Helium is used as the carrier gas at a flow rate of 50 mL/min, and the temperature of the column is maintained at 33°.

System Suitability   The peak responses obtained for Butane in the chromatograms from duplicate determinations agree within 1%.

Procedure   Connect one Butane cylinder to the chromatograph through a suitable sampling valve and a flow control valve downstream from the sampling valve. Flush the liquid specimen through the sampling valve, taking care to avoid trapping gas or air in the valve. Inject a suitable volume, typically 2 μL, of Butane into the chromatograph, and record the chromatogram. Calculate the percentage purity by dividing 100 times the Butane response by the sum of all the responses in the chromatogram.

Acidity of Residue   Add 10 mL of water to the residue obtained in *High-Boiling Residue* (see below), mix by swirling for about 30 s, add 2 drops of methyl orange TS, insert the stopper in the tube, and shake vigorously. No pink or red color appears in the aqueous layer.

High-Boiling Residue   Prepare a cooling coil from copper tubing (about 6 mm od × about 6.1 m long) to fit into a suitable vacuum-jacketed flask. Immerse the cooling coil in a mixture of dry ice and acetone in a vacuum-jacketed flask, and connect one end of the tubing to a specimen cylinder (see *Sampling Procedure*). Carefully open the specimen cylinder valve, flush the cooling coil with about 50 mL of the liquified Butane, and discard this portion of liquid. Continue delivering liquid from the cooling coil, and collect it in a previously chilled 1000-mL sedimentation cone until the cone is filled to the 1000-mL mark (approximately 600 g). Allow the liquid to evaporate, using a warm water bath maintained at about 40° to reduce evaporating time. When all of the liquid has evaporated, rinse the sedimentation cone with two 50-mL portions of pentane, and combine the rinsings in a tared 150-mL evaporating dish. Transfer 100 mL of the pentane solvent to a second tared 150-mL evaporating dish, place both evaporating dishes on a water bath, evaporate to dryness, and heat the dishes in an oven at 100° for 60 min. Cool the dishes in a desiccator, and weigh. Repeat the heating for 15-min periods until successive weighings are within 0.1 mg. The weight of the residue obtained from the specimen is the difference between the weights of the residues in the two evaporating dishes. Calculate the mg/kg of high-boiling residue based on a sample weight of 600 g.

Sulfur Compounds   Carefully open the container valve to produce a moderate flow of gas. Do not direct the gas stream toward the face, but deflect a portion of the stream toward the nose. The gas is free from the characteristic odor of sulfur compounds.

Water   Determine by the *Karl Fischer Titrimetric Method*, Appendix IIB. Proceed, using the following modifications: (a) Provide the closed-system titrating vessel with an opening through which passes a coarse-porosity gas dispersion tube connected to a sampling cylinder. (b) Dilute the reagent with anhydrous methanol to give a water equivalence factor of between 0.2 and 1.0 mg/mL; age this diluted solution for not less than 16 h before standardization. (c) Obtain a 100-g specimen as directed in the *Sampling Procedure*, and introduce the specimen into the titration vessel through the gas dispersion tube at a rate of about 100 mL of gas per min; if necessary, heat the specimen cylinder gently to maintain this flow rate.

Packaging and Storage   Store in tight cylinders protected from excessive heat.

Butylated Hydroxymethylphenol

$(CH_3)_3C$ — ring with OH (top), $C(CH_3)_3$, and CH_2OH (bottom)

$C_{15}H_{24}O_2$ Formula wt 236.35

DESCRIPTION

A nearly white crystalline solid having a faint characteristic odor. It is insoluble in water and in propylene glycol, but is freely soluble in alcohol.

Functional Use in Foods   Antioxidant.

REQUIREMENTS

Identification   Butylated Hydroxymethylphenol may be identified by its solidification point, as determined in the *Assay*.
Assay   Not less than 98.0% of $C_{15}H_{24}O_2$.
Heavy Metals (as Pb)   Not more than 10 mg/kg.

TESTS

Assay   Its solidification point (see Appendix IIB) is not lower than 140°, indicating a purity of not less than 98.0%, by weight, of $C_{15}H_{24}O_2$.
Heavy Metals   Prepare and test a 2-g sample as directed in *Method II* under the *Heavy Metals Test*, Appendix IIIB, using 20 μg of lead ion (Pb) in the control (*Solution A*).

Packaging and Storage   Store in well-closed containers.

1,3-Butylene Glycol

CH$_2$OHCH$_2$CHOHCH$_3$

C$_4$H$_{10}$O$_2$　　　　　　　　　　Formula wt 90.12

　　　　　　　　　　　　　　　　CAS: [107-88-0]

DESCRIPTION

A clear, colorless, hygroscopic, viscous liquid having a slight, characteristic taste. It is practically odorless. It is miscible with water, with acetone, and with ether in all proportions, but it is immiscible with fixed oils. It dissolves most essential oils and synthetic flavoring substances.

Functional Use in Foods　Solvent for flavoring agents.

REQUIREMENTS

Assay　Not less than 99.0% of C$_4$H$_{10}$O$_2$.
Distillation Range　Between 200° and 215°.
Heavy Metals (as Pb)　Not more than 5 mg/kg.
Specific Gravity　Between 1.004 and 1.006 at 20°.

TESTS

Assay　Prepare an acetylating reagent, within one week of use, by mixing 3.4 mL of water and 130 mL of acetic anhydride with 1000 mL of anhydrous pyridine. Pipet 20 mL of this reagent into a 250-mL iodine flask, and add about 1 g of the sample, accurately weighed. Attach a dry reflux condenser to the flask, and reflux for 1 h. Allow the flask to cool to room temperature, then rinse the condenser with 50 mL of chilled (10°) carbon dioxide-free water, allowing the water to drain into the flask. Stopper the flask, cool to below 20°, add phenolphthalein TS, and titrate with 0.5 N sodium hydroxide, swirling the contents of the flask continuously during the titration. Perform a blank determination (see *General Provisions*). Each mL of 0.5 N sodium hydroxide is equivalent to 2.253 mg of C$_4$H$_{10}$O$_2$.
Distillation Range　Proceed as directed in the general method, Appendix IIB.
Heavy Metals　Prepare and test a 4-g sample as directed in *Method II* under the *Heavy Metals Test*, Appendix IIIB, using 20 μg of lead ion (Pb) in the control (*Solution A*).
Specific Gravity　Determine by any reliable method (see *General Provisions*).

Packaging and Storage　Store in well-closed containers.

Caffeine

1,3,7-Trimethylxanthine

C$_8$H$_{10}$N$_4$O$_2$　　　　　　　　　　Formula wt 194.19

　　　　　　　　　　　　　　　　CAS: [58-08-2]

DESCRIPTION

A white powder, or white, glistening needles, usually matted together. It may be compacted or compressed into free-flowing granules or pellets. Caffeine is anhydrous or contains one molecule of water of hydration. It is odorless and has a bitter taste. Its solutions are neutral to litmus. The hydrate is efflorescent in air. One g of hydrated Caffeine is soluble in about 50 mL of water, in 75 mL of alcohol, in about 6 mL of chloroform, and in 600 mL of ether.

Functional Use in Foods　Flavoring agent.

REQUIREMENTS

Labeling　Indicate whether it is anhydrous or hydrous.
Identification
　A.　Dissolve about 5 mg in 1 mL of hydrochloric acid in a porcelain dish, add 50 mg of potassium chlorate, and evaporate on a steam bath to dryness. Invert the dish over a vessel containing a few drops of 6 N ammonium hydroxide. The residue acquires a purple color, which disappears upon the addition of a solution of a fixed alkali.
　B.　The infrared absorption spectrum of a mineral oil dispersion of the sample, previously dried at 80° for 4 h, exhibits maxima only at the same wavelengths as that of a similar preparation of USP Caffeine Reference Standard.
Assay　Not less than 98.5% and not more than 101.0% of C$_8$H$_{10}$N$_4$O$_2$, calculated on the anhydrous basis.
Heavy Metals (as Pb)　Not more than 10 mg/kg.
Lead　Not more than 1 mg/kg.
Melting Range　Between 235° and 237.5°.
Other Alkaloids　Passes test.
Readily Carbonizable Substances　Passes test.
Residue on Ignition　Not more than 0.1%.
Water　*Anhydrous caffeine*: not more than 0.5%; *hydrous caffeine*: not more than 8.5%.

TESTS

Assay　Dissolve about 170 mg, accurately weighed, of finely powdered Caffeine, with warming, in 5 mL of glacial acetic acid. Cool, add 10 mL of acetic anhydride and 20 mL of toluene, and titrate with 0.1 N perchloric acid, determining the endpoint potentiometrically. Each mL of 0.1 N perchloric acid is equivalent to 19.42 mg of $C_8H_{10}N_4O_2$.

Heavy Metals A solution of 2 g in 5 mL of 0.1 N hydrochloric acid and 23 mL of water. It may be necessary to warm gently to effect solution and then cool to room temperature. This solution meets the requirements of the *Heavy Metals Test*, Appendix IIIB, using 20 µg of lead ion (Pb) in the control (*Solution A*).

Lead A *Sample Solution* prepared as directed for organic compounds using a 4-g sample meets the requirements of the *Lead Limit Test*, Appendix IIIB, using 4 µg of lead ion (Pb) in the control.

Melting Range Dry at 80° for 4 h and then determine as directed for *Melting Range or Temperature*, Appendix IIB.

Other Alkaloids Add a few drops of mercuric–potassium iodide TS to 5 mL of a 1 in 50 solution of the sample. No precipitate forms.

Readily Carbonizable Substances, Appendix IIB Dissolve 500 mg in 5 mL of 95% sulfuric acid. The color is no darker than *Matching Fluid D*.

Residue on Ignition, Appendix IIC Ignite 2 g as directed in the general method.

Water Determine the water content by the *Karl Fischer Titrimetric Method*, Appendix IIB.

Packaging and Storage Store hydrous caffeine in tight containers and anhydrous caffeine in well-closed containers.

Calcium Acetate

$Ca(C_2H_3O_2)_2$ Formula wt 158.17

INS: 263 CAS: [62-54-4]

DESCRIPTION

A fine, white, bulky, odorless powder. It is freely soluble in water, and is slightly soluble in alcohol.

Functional Use in Foods Sequestrant.

REQUIREMENTS

Identification A 1 in 10 solution gives positive tests for *Calcium* and for *Acetate*, Appendix IIIA.

Assay Not less than 99.0% and not more than 100.5% of $Ca(C_2H_3O_2)_2$, calculated on the anhydrous basis.

Chloride Not more than 0.05%.

Fluoride Not more than 0.005%.

Heavy Metals (as Pb) Not more than 10 mg/kg.

Sulfate Not more than 0.1%.

Water Not more than 7.0%.

TESTS

Assay Dissolve about 300 mg, accurately weighed, in 150 mL of water containing 2 mL of 2.7 N hydrochloric acid. While stirring, preferably with a magnetic stirrer, add about 30 mL of 0.05 M disodium EDTA from a 50-mL buret, then add 15 mL of 1 N sodium hydroxide and 300 mg of hydroxy naphthol blue indicator, and continue the titration to a blue endpoint. Each mL of 0.05 M disodium EDTA is equivalent to 7.909 mg of $Ca(C_2H_3O_2)_2$.

Chloride, Appendix IIIB Any turbidity produced by a 40-mg sample does not exceed that shown in a control containing 20 µg of chloride ion (Cl).

Fluoride Determine as directed in *Method III* under the *Fluoride Limit Test*, Appendix IIIB, except in the *Procedure* use 10 mL of 1 N hydrochloric acid to dissolve the sample.

Heavy Metals A solution of 2 g in 25 mL of water meets the requirements of the *Heavy Metals Test*, Appendix IIIB, using 20 µg of lead ion (Pb) in the control (*Solution A*).

Sulfate, Appendix IIIB Any turbidity produced by a 200-mg sample does not exceed that shown in a control containing 200 µg of sulfate (SO_4).

Water Determine by the *Karl Fischer Titrimetric Method*, Appendix IIB.

Packaging and Storage Store in well-closed containers.

Calcium Acid Pyrophosphate

$CaH_2P_2O_7$ Formula wt 216.04

DESCRIPTION

A fine, white, colorless and acidic powder. It is insoluble in water, but is soluble in dilute hydrochloric and nitric acids.

Functional Use in Foods Leavening agent; nutrient; dietary supplement.

REQUIREMENTS

Identification

A. Dissolve about 100 mg by warming with a mixture of 5 mL of 2.7 N hydrochloric acid and 5 mL of water, add 2.5 mL of 6 N ammonium hydroxide, dropwise, with shaking, and then add 5 mL of ammonium oxalate TS. A white precipitate is formed.

B. Dissolve 100 mg of the sample in 100 mL of 1.7 N nitric acid. Add 0.5 mL of this solution to 30 mL of quimociac TS. A yellow precipitate does not form. Heat the remaining portion of the sample solution for 10 min at 95°, and then add 0.5 mL of the solution to 30 mL of quimociac TS. A yellow precipitate forms immediately.

Assay Not less than 95.0% and not more than 100.5% of $CaH_2P_2O_7$.

Arsenic (as As) Not more than 3 mg/kg.

Fluoride Not more than 0.005%.

Heavy Metals (as Pb) Not more than 0.0015%.

Lead Not more than 2 mg/kg.

Loss on Ignition Not more than 10.0%.

TESTS

Assay Dissolve about 300 mg, accurately weighed, in 10 mL of 2.7 N hydrochloric acid, add about 120 mL of water and a few drops of methyl orange TS, and boil for 30 min, keeping the volume and pH of the solution constant during the boiling period by adding hydrochloric acid or water, if necessary. Add 2 drops of methyl red TS and 30 mL of ammonium oxalate TS, then add dropwise, with constant stirring, a mixture of equal volumes of 6 N ammonium hydroxide and water until the pink color of the indicator just disappears. Digest on a steam bath for 30 min, cool to room temperature, allow the precipitate to settle, and filter the supernatant liquid through a sintered-glass filter crucible, using gentle suction. Wash the precipitate in the beaker with about 30 mL of cold (below 20°) wash solution, prepared by diluting 10 mL of ammonium oxalate TS to 1000 mL. Allow the precipitate to settle, and pour the supernatant liquid through the filter. Repeat this washing by decantation three more times. Using the wash solution, transfer the precipitate as completely as possible to the filter. Finally, wash the beaker and the filter with two 10-mL portions of cold (below 20°) water. Place the sintered-glass filter crucible in the beaker, and add 10 mL of water and 50 mL of cold, dilute sulfuric acid (1 in 6). Add from a buret 35 mL of 0.1 N potassium permanganate, and stir until the color disappears. Heat to about 70°, and complete the titration with 0.1 N potassium permanganate. Each mL of 0.1 N potassium permanganate is equivalent to 5.40 mg of $CaH_2P_2O_7$.

Arsenic A solution of 1 g in 5 mL of 2.7 N hydrochloric acid meets the requirements of the *Arsenic Test*, Appendix IIIB.

Fluoride Weigh accurately 1.0 g, and proceed as directed in the *Fluoride Limit Test*, Appendix IIIB.

Heavy Metals Warm 2.66 g with 7 mL of 2.7 N hydrochloric acid until no more dissolves, dilute to 50 mL with water, and filter. A 25-mL portion of the filtrate meets the requirements of the *Heavy Metals Test*, Appendix IIIB, using 20 μg of lead ion (Pb) in the control (*Solution A*).

Lead A 10-g sample meets the requirements of the *APDC Extraction Method for Lead*, Appendix IIIB.

Loss on Ignition Transfer about 1 g, accurately weighed, to a suitable tared crucible, ignite at 800° ± 25° for 30 min, cool in a desiccator, and weigh.

Packaging and Storage Store in well-closed containers.

Calcium Alginate

Algin

$[(C_6H_7O_6)_2Ca]_n$ Formula wt, *Calculated* 195.16
Formula wt, *Actual* (Avg) 219.00

INS: 404

DESCRIPTION

The calcium salt of alginic acid (see the monograph for *Alginic Acid*) occurs as a white to yellowish, fibrous or granular powder. It is nearly odorless and tasteless. It is insoluble in water, but is soluble in alkaline solutions or in solutions of substances that combine with the calcium. It is insoluble in organic solvents.

Functional Use in Foods Stabilizer; thickener; emulsifier.

REQUIREMENTS

Identification Calcium Alginate meets the requirements of *Identification Test C* in the monograph for *Alginic Acid*.

Assay It yields not less than 18% and not more than 21% of carbon dioxide (CO_2), corresponding to between 89.6% and 104.5% of Calcium Alginate (equiv wt 219.00), calculated on the dried basis.

Arsenic (as As)   Not more than 3 mg/kg.

Heavy Metals (as Pb)   Not more than 0.002%.

Lead   Not more than 5 mg/kg.

Loss on Drying   Not more than 15.0%.

TESTS

Assay Proceed as directed under *Alginates Assay*, Appendix IIIC. Each mL of 0.25 N sodium hydroxide consumed in the assay is equivalent to 27.38 mg of Calcium Alginate (equiv wt 219.00).

Arsenic A *Sample Solution* prepared as directed for organic compounds meets the requirements of the *Arsenic Test*, Appendix IIIB.

Heavy Metals Determine as directed in the test for *Heavy Metals* in the monograph for *Alginic Acid*.

Lead A *Sample Solution* prepared as directed for organic compounds meets the requirements of the *Lead Limit Test*, Appendix IIIB, using 5 μg of lead ion (Pb) in the control.

Loss on Drying, Appendix IIC   Dry at 105° for 4 h.

Packaging and Storage Store in well-closed containers.

Calcium Ascorbate

$C_{12}H_{14}CaO_{12} \cdot 2H_2O$ Formula wt 426.34

INS: 302   CAS: [5743-27-1]

DESCRIPTION

A white to slightly yellow, odorless, crystalline powder. It is soluble in water, slightly soluble in alcohol, and insoluble in ether. The pH of a 1 in 10 solution is between 6.8 and 7.4.

Functional Use in Foods Antioxidant.

REQUIREMENTS

Identification A 1 in 10 solution gives positive tests for *Calcium*, Appendix IIIA, and it decolorizes dichlorophenol–indophenol TS.
Assay Not less than 98.0% and not more than 100.5% of $C_{12}H_{14}CaO_{12}\cdot 2H_2O$.
Arsenic (as As) Not more than 3 mg/kg.
Fluoride Not more than 10 mg/kg.
Heavy Metals (as Pb) Not more than 10 mg/kg.
Oxalate Passes test.
Specific Rotation $[\alpha]_D^{25°}$: Between +95° and +97°.

TESTS

Assay Dissolve about 300 mg, accurately weighed, in 50 mL of water in a 250-mL Erlenmeyer flask, and immediately titrate with 0.1 N iodine to a pale yellow color that persists for at least 30 s. Each mL of 0.1 N iodine is equivalent to 10.66 mg of $C_{12}H_{14}CaO_{12}\cdot 2H_2O$.
Arsenic A *Sample Solution* prepared as directed for organic compounds meets the requirements of the *Arsenic Test*, Appendix IIIB.
Fluoride Proceed as directed in the *Fluoride Limit Test*, Appendix IIIB.
Heavy Metals A solution of 2 g in 25 mL of water meets the requirements of the *Heavy Metals Test*, Appendix IIIB, using 20 μg of lead ion (Pb) in the control (*Solution A*).
Oxalate To a solution of 1 g in 10 mL of water add 2 drops of glacial acetic acid and 5 mL of a 1 in 10 solution of calcium acetate. The solution remains clear after standing for 5 min.
Specific Rotation, Appendix IIB Determine in a solution containing 1 g in each 20 mL.

Packaging and Storage Store in tight containers, preferably in a cool, dry place.

Calcium Bromate

$Ca(BrO_3)_2\cdot H_2O$ Formula wt 313.90
INS: 924b

DESCRIPTION

A white crystalline powder. It is very soluble in water.

Functional Use in Foods Maturing agent; oxidizing agent.

REQUIREMENTS

Identification
   A. A 1 in 20 solution in 2.7 N hydrochloric acid imparts a transient yellowish red color to a nonluminous flame.
   B. To a 1 in 20 solution add sulfurous acid dropwise. A yellow color is produced that disappears upon the addition of an excess of sulfurous acid.
Assay Not less than 99.8% and not more than 100.5% of $Ca(BrO_3)_2\cdot H_2O$.
Heavy Metals (as Pb) Not more than 0.002%.
Lead Not more than 10 mg/kg.

TESTS

Assay Dissolve about 900 mg, accurately weighed, in 50 mL of water in a 250-mL glass-stoppered Erlenmeyer flask. Add 3 g of potassium iodide, followed by 3 mL of hydrochloric acid. Allow the mixture to stand for 5 min, add 100 mL of cold water, and titrate the liberated iodine with 0.1 N sodium thiosulfate, adding starch TS near the endpoint. Perform a blank determination (see *General Provisions*). Each mL of 0.1 N sodium thiosulfate is equivalent to 26.16 mg of $Ca(BrO_3)_2\cdot H_2O$.

Sample Solution for the Determination of Heavy Metals and Lead Dissolve 2 g in 10 mL of water, add 10 mL of hydrochloric acid, and evaporate to dryness on a steam bath. Dissolve the residue in 5 mL of hydrochloric acid, again evaporate to dryness, and then dissolve the residue in 40 mL of water.

Heavy Metals A 20-mL portion of the *Sample Solution*, diluted to 25 mL with water, meets the requirements of the *Heavy Metals Test*, Appendix IIIB, using 20 μg of lead ion (Pb) in the control (*Solution A*).
Lead A 20-mL portion of the *Sample Solution* meets the requirements of the *Lead Limit Test*, Appendix IIIB, using 10 μg of lead ion (Pb) in the control.

Packaging and Storage Store in well-closed containers.

Calcium Carbonate

$CaCO_3$ Formula wt 100.09
INS: 170(i)   CAS: [471-34-1]

DESCRIPTION

A fine, white microcrystalline powder. It is colorless and tasteless, and is stable in air. It is practically insoluble in water and in alcohol. The presence of any ammonium salt or carbon dioxide increases its solubility in water, but the presence of any alkali hydroxide reduces the solubility.

Functional Use in Foods Alkali; nutrient; dietary supplement; dough conditioner; firming agent; yeast nutrient.

REQUIREMENTS

Identification It dissolves with effervescence in 1 N acetic acid, in 2.7 N hydrochloric acid, and in 1.7 N nitric acid, and the resulting solutions, after boiling, give positive tests for *Calcium*, Appendix IIIA.
Assay Not less than 98.0% and not more than 100.5% of $CaCO_3$ after drying.
Acid-Insoluble Substances Not more than 0.2%.
Arsenic (as As) Not more than 3 mg/kg.
Fluoride Not more than 0.005%.
Heavy Metals (as Pb) Not more than 0.002%.
Lead Not more than 3 mg/kg.
Loss on Drying Not more than 2%.
Magnesium and Alkali Salts Not more than 1%.

TESTS

Assay Transfer about 200 mg, previously dried at 200° for 4 h and accurately weighed, into a 400-mL beaker, add 10 mL of water, and swirl to form a slurry. Cover the beaker with a watch glass, and introduce 2 mL of 2 N hydrochloric acid from a pipet inserted between the lip of the beaker and the edge of the watch glass. Swirl the contents of the beaker to dissolve the sample. Wash down the sides of the beaker, the outer surface of the pipet, and the watch glass, and dilute to about 100 mL with water. While stirring, preferably with a magnetic stirrer, add about 30 mL of 0.05 M disodium EDTA from a 50-mL buret, then add 15 mL of 1 N sodium hydroxide and 300 mg of hydroxy naphthol blue indicator, and continue the titration to a blue endpoint. Each mL of 0.05 M disodium EDTA is equivalent to 5.004 mg of $CaCO_3$.
Acid-Insoluble Substances Suspend 5 g in 25 mL of water, cautiously add with agitation 25 mL of dilute hydrochloric acid (1 in 2), then add water to make a volume of about 200 mL. Heat the solution to boiling, cover, digest on a steam bath 1 h, cool, and filter. Wash the precipitate with water until the last washing shows no chloride with silver nitrate TS, and then ignite it. The weight of the residue does not exceed 10 mg.
Arsenic A solution of 1 g in 10 mL of 2.7 N hydrochloric acid meets the requirements of the *Arsenic Test*, Appendix IIIB.
Fluoride Determine as directed in *Method III* under the *Fluoride Limit Test*, Appendix IIIB.

> **Sample Solution for the Determination of Heavy Metals and Lead** Cautiously dissolve 5 g in 25 mL of dilute hydrochloric acid (1 in 2), and evaporate to dryness on a steam bath. Dissolve the residue in about 15 mL of water, and dilute to 25 mL (1 mL = 200 mg).

Heavy Metals Neutralize 5.0 mL (1 g) of the *Sample Solution* with 1 N sodium hydroxide, using phenolphthalein as the indicator, and dilute to 25 mL. This solution meets the requirements of the *Heavy Metals Test*, Appendix IIIB, using 20 μg of lead ion (Pb) in the control (*Solution A*).
Lead A 20-mL portion of the *Sample Solution* meets the requirements of the *Lead Limit Test*, Appendix IIIB, using 12 μg of lead ion (Pb) in the control.
Loss on Drying, Appendix IIC Dry at 200° for 4 h.

Magnesium and Alkali Salts Mix 1 g with 40 mL of water, carefully add 5 mL of hydrochloric acid, mix, and boil for 1 min. Rapidly add 40 mL of oxalic acid TS, and stir vigorously until precipitation is well established. Immediately add 2 drops of methyl red TS, then add 6 N ammonium hydroxide, dropwise, until the mixture is just alkaline, and cool. Transfer the mixture to a 100-mL cylinder, dilute with water to 100 mL, let it stand for 4 h or overnight, then decant the clear, supernatant liquid through a dry filter paper. To 50 mL of the clear filtrate in a platinum dish add 0.5 mL of sulfuric acid, and evaporate the mixture on a steam bath to a small volume. Carefully evaporate the remaining liquid to dryness over a free flame, and continue heating until the ammonium salts have been completely decomposed and volatilized. Finally, ignite the residue to constant weight. The weight of the residue does not exceed 5 mg.

Packaging and Storage Store in well-closed containers.

Calcium Chloride

$CaCl_2 \cdot 2H_2O$	Formula wt, dihydrate 147.01
$CaCl_2$	Formula wt, anhydrous 110.98
INS: 509	CAS: dihydrate [10035-04-8]
	CAS: anhydrous [10043-52-4]

DESCRIPTION

White, hard, odorless fragments, granules, or powder. It is deliquescent. It is anhydrous or the dihydrate. It is soluble in water and slightly soluble in alcohol. The pH of a 1 in 20 solution is between 4.5 and 9.5.

Functional Use in Foods Sequestrant; firming agent.

REQUIREMENTS

Labeling Indicate whether it is the dihydrate or anhydrous.
Identification A 1 in 10 solution gives positive tests for *Calcium* and for *Chloride*, Appendix IIIA.
Assay For the dihydrate, not less than 99.0% and not more than 107.0% of $CaCl_2 \cdot 2H_2O$; for the anhydrous, not less than 93.0% and not more than 100.5% of $CaCl_2$.
Acid-Insoluble Matter (for the anhydrous) Not more than 0.02%; no particles per kg of sample greater than 2 mm in any dimension.
Arsenic (as As) Not more than 3 mg/kg.
Fluoride Not more than 0.004%.
Heavy Metals (as Pb) Not more than 0.002%.
Lead Not more than 5 mg/kg.
Magnesium and Alkali Salts Not more than 4.0% for the dihydrate, and not more than 5.0% for the anhydrous.

TESTS

Assay Transfer about 1.5 g, accurately weighed, into a 250-mL volumetric flask, dissolve it in a mixture of 100 mL of water and 5 mL of 2.7 N hydrochloric acid, dilute to volume with water, and mix. Transfer 50.0 mL of this solution into a suitable container, and add 50 mL of water. While stirring, preferably with a magnetic stirrer, add about 30 mL of 0.05 M disodium EDTA from a 50-mL buret, then add 15 mL of 1 N sodium hydroxide and 300 mg of hydroxy naphthol blue indicator, and continue the titration to a blue endpoint. Each mL of 0.05 M disodium EDTA is equivalent to 5.55 mg of $CaCl_2$ or 7.35 mg of $CaCl_2 \cdot 2H_2O$.

Acid-Insoluble Matter (Note: The following test is intended for Anhydrous Calcium Chloride.) Place a 32-mm od lintine disc filter[1] in a suitable filter assembly comprising a 2.5-L screw-cap bottle cut in half horizontally and fitted with a rubber washer that is 35 mm od and 25 mm id, followed by the lintine disc, a 20-mesh stainless steel screen 35 mm od, and a bottle cap with a 25-mm hole in the top. With the filter at the bottom, wash the assembly with 100 mL of diluted acetic acid (1 in 300), followed by 100 mL of water. Remove the disc from the assembly, place on a watch glass, and dry the combination at 105° for 2 h.

Dissolve 1 kg of sample in 3 L of water containing 10 mL of glacial acetic acid. Allow to cool, and filter through the lintine disc. Rinse the walls of the filter assembly so that all insoluble matter is transferred to the disc, and wash with 100 mL of water. Place the disc on the same watch glass mentioned above, and dry at 105° for 2 h, being careful at all times not to lose any particles that may be on the disc. The difference in the two weights is the weight of the acid-insoluble matter.

Place the disc under a low-power magnifier (4× to 10× magnification). Using a millimeter rule, measure the largest dimension of each particle (or as many as may be necessary) on the disc. No particles greater than 2 mm in any dimension are present.

Arsenic A solution of 1 g in 10 mL of water meets the requirements of the *Arsenic Test*, Appendix IIIB.

Fluoride Determine as directed in *Method III* under the *Fluoride Limit Test*, Appendix IIIB.

Heavy Metals Dissolve 1 g in 2 mL of 1 N acetic acid, and add water to make 25 mL. This solution meets the requirements of the *Heavy Metals Test*, Appendix IIIB, using 20 μg of lead ion (Pb) in the control (*Solution A*).

Lead A solution of 1 g in 20 mL of water meets the requirements of the *Lead Limit Test*, Appendix IIIB, using 5 μg of lead ion (Pb) in the control.

Magnesium and Alkali Salts Dissolve 1.0 g in about 50 mL of water, add 500 mg of ammonium chloride, mix, and boil for 1 min. Rapidly add 40 mL of oxalic acid TS, and stir vigorously until precipitation is well established. Immediately add 2 drops of methyl red TS, then add 6 N ammonium hydroxide, dropwise, until the mixture is just alkaline, and cool. Transfer the mixture to a 100-mL cylinder, dilute with water to 100 mL, let it stand for 4 h or overnight, then decant the clear, supernatant liquid through a dry filter paper. To 50 mL of the clear filtrate in a platinum dish add 0.5 mL of sulfuric acid, and evaporate the mixture on a steam bath to a small volume. Carefully evaporate the remaining liquid to dryness over a free flame, and continue heating until the ammonium salts have been completely decomposed and volatilized. Finally, ignite the residue to constant weight. The weight of the residue does not exceed 20 mg for the dihydrate or 25 mg for the anhydrous.

Packaging and Storage Store in tight containers.

Calcium Chloride Solution

DESCRIPTION

Calcium Chloride Solution occurs as a clear to slightly turbid, colorless or slightly colored liquid at room temperature. It is nominally available in a concentration range of about 35% to 45% of $CaCl_2$.

Functional Use in Foods Sequestrant; firming agent.

REQUIREMENTS

Identification When diluted to a concentration of about 1 to 10 ($CaCl_2$ basis), it gives positive tests for *Calcium* and for *Chloride*, Appendix IIIA.

Assay Not less than 90.0% and not more than 110.0%, by weight, of the labeled amount of calcium chloride, expressed as $CaCl_2$.

Alkalinity [as $Ca(OH)_2$] Not more than 0.3%.

Fluoride Not more than 0.004%, calculated on the $CaCl_2$ determined in the *Assay*.

Heavy Metals (as Pb) Not more than 0.002%, calculated on the $CaCl_2$ determined in the *Assay*.

Lead Not more than 10 mg/kg, calculated on the $CaCl_2$ determined in the *Assay*.

Magnesium and Alkali Salts Not more than 5.0%, calculated on the $CaCl_2$ determined in the *Assay*.

TESTS

Assay Transfer an accurately weighed amount of the solution, equivalent to about 1 g of $CaCl_2$, into a 250-mL volumetric flask, dissolve it in a mixture of 100 mL of water and 5 mL of 2.7 N hydrochloric acid, dilute to volume with water, and mix. Transfer 50.0 mL of this solution into a suitable container, and add 50 mL of water. While stirring, preferably with a magnetic stirrer, add about 30 mL of 0.05 M disodium EDTA from a 50-mL buret, then add 15 mL of 1 N sodium hydroxide and 300 mg of hydroxy naphthol blue indicator, and continue the titration to a blue endpoint. Each mL of 0.05 M disodium EDTA is equivalent to 5.550 mg of $CaCl_2$.

Alkalinity Dilute an accurately weighed amount of the solution, equivalent to about 5 g of $CaCl_2$, to 50 mL with water, add phenolphthalein TS, and titrate with 0.1 N hydrochloric acid. Each mL of 0.1 N hydrochloric acid is equivalent to 3.71 mg of $Ca(OH)_2$.

---

[1] Available from Filter Fabrics, Inc., 814 E. Jefferson, Goshen, IN 46526; 219/533-3114.

Fluoride   Determine as directed in *Method III* under the *Fluoride Limit Test*, Appendix IIIB, using as the sample an accurately weighed amount of the solution equivalent to 1 g of $CaCl_2$.

Heavy Metals   Dilute an accurately weighed amount of the solution, equivalent to 1 g of $CaCl_2$, to 25 mL with water. This solution meets the requirements of the *Heavy Metals Test*, Appendix IIIB, using 20 µg of lead ion (Pb) in the control (*Solution A*).

Lead   Dilute an accurately weighed amount of the solution, equivalent to 1 g of $CaCl_2$, to 10 mL with water. The resulting solution meets the requirements of the *Lead Limit Test*, Appendix IIIB, using 10 µg of lead ion (Pb) in the control.

Magnesium and Alkali Salts   Dilute an accurately weighed amount of the solution, equivalent to 1.0 g of $CaCl_2$, to 50 mL with water, add 500 mg of ammonium chloride, mix, and boil for about 1 min. Rapidly add 40 mL of oxalic acid TS, and stir vigorously until precipitation is well established. Immediately add 2 drops of methyl red TS, then add 6 N ammonium hydroxide, dropwise, until the mixture is just alkaline, and cool. Transfer the mixture into a 100-mL cylinder, dilute with water to 100 mL, let it stand for 4 h or overnight, and then decant the clear, supernatant liquid through a dry filter paper. To 50 mL of the clear filtrate in a platinum dish add 0.5 mL of sulfuric acid, and evaporate the mixture on a steam bath to a small volume. Carefully evaporate the remaining liquid to dryness over a free flame, and continue heating until the ammonium salts have been completely decomposed and volatilized. Finally, ignite the residue to constant weight. The weight of the residue does not exceed 25 mg.

Packaging and Storage   Store in tight containers.

Calcium Citrate

$[CH_2(COO^-)C(OH)(COO^-)CH_2COO^-]_2Ca_3$

$Ca_3(C_6H_5O_7)_2 \cdot 4H_2O$ Formula wt 570.50

INS: 333   CAS: [813-94-5]

DESCRIPTION

A fine, white, odorless powder. It is very slightly soluble in water and insoluble in alcohol.

Functional Use in Foods   Sequestrant; buffer; firming agent.

REQUIREMENTS

Identification

A. Dissolve 500 mg in 10 mL of water and 2.5 mL of 1.7 N nitric acid, add 1 mL of mercuric sulfate TS, heat to boiling, and then add potassium permanganate TS. A white precipitate is formed.

B. Ignite 500 mg completely at as low a temperature as possible, cool, and dissolve the residue in 10 mL of water and 1 mL of glacial acetic acid. Filter and add 10 mL of ammonium oxalate TS to the filtrate. A voluminous white precipitate appears that is soluble in hydrochloric acid.

Assay   Not less than 97.5% and not more than 100.5% of $Ca_3(C_6H_5O_7)_2$ after drying.
Fluoride   Not more than 0.003%.
Heavy Metals (as Pb)   Not more than 0.002%.
Lead   Not more than 10 mg/kg.
Loss on Drying   Between 10.0% and 14.0%.

TESTS

Assay   Dissolve about 350 mg, previously dried at 150° for 4 h and accurately weighed, in a mixture of 10 mL of water and 2 mL of 2.7 N hydrochloric acid, and dilute to about 100 mL with water. While stirring, preferably with a magnetic stirrer, add about 30 mL of 0.05 M disodium EDTA from a 50-mL buret, then add 15 mL of 1 N sodium hydroxide and 300 mg of hydroxy naphthol blue indicator, and continue the titration to a blue endpoint. Each mL of 0.05 M disodium EDTA is equivalent to 8.300 mg of $Ca_3(C_6H_5O_7)_2$.

Fluoride   Weigh accurately 1.67 g, and proceed as directed in the *Fluoride Limit Test*, Appendix IIIB.

Heavy Metals   Dissolve 1 g in 20 mL of water and 2 mL of hydrochloric acid, add 1.5 mL of ammonium hydroxide, and dilute to 25 mL with water. This solution meets the requirements of the *Heavy Metals Test*, Appendix IIIB, using 20 µg of lead ion (Pb) in the control (*Solution A*).

Lead   Dissolve 1 g in 10 mL of water and 1 mL of hydrochloric acid. This solution meets the requirements of the *Lead Limit Test*, Appendix IIIB, using 10 µg of lead ion (Pb) in the control.

Loss on Drying, Appendix IIC   Dry for 4 h at 150°.

Packaging and Storage   Store in well-closed containers.

Calcium Disodium EDTA

Calcium Disodium Ethylenediaminetetraacetate; Calcium Disodium (Ethylenedinitrilo)tetraacetate; Calcium Disodium Edetate

$$\left[\begin{array}{c} NaOOCCH_2 \quad\quad CH_2COONa \\ \quad\backslash\quad CH_2CH_2 \quad / \\ N \quad\quad\quad\quad N \\ / \quad\quad\quad\quad\quad\quad \backslash \\ CH_2 \quad\quad Ca \quad\quad CH_2 \\ | \quad\quad\quad\quad\quad\quad | \\ COO \quad\quad\quad\quad OOC \end{array}\right] \cdot 2H_2O$$

$C_{10}H_{12}CaN_2Na_2O_8 \cdot 2H_2O$ Formula wt 410.30
INS: 385 CAS: [23411-34-9]

DESCRIPTION

White, odorless crystalline granules or a white to off-white powder. It is slightly hygroscopic, has a faint saline taste, and is stable in air. It is freely soluble in water.

Functional Use in Foods Preservative; sequestrant.

REQUIREMENTS

Identification
   A. A 1 in 20 solution responds to the oxalate test for *Calcium* and to the flame test for *Sodium*, Appendix IIIA.
   B. The infrared absorption spectrum of a mineral oil dispersion of the sample exhibits maxima only at the same wavelengths as that of a similar preparation of USP Edetate Calcium Disodium Reference Standard.
   C. To 5 mL of water in a test tube add 2 drops of ammonium thiocyanate TS and 2 drops of ferric chloride TS. To the deep red solution so obtained add about 50 mg of the sample, and mix. The deep red color disappears.

Assay Not less than 97.0% and not more than 102.0% of $C_{10}H_{12}CaN_2Na_2O_8$, calculated on the anhydrous basis.
Heavy Metals (as Pb) Not more than 0.002%.
Lead Not more than 10 mg/kg.
Magnesium-Chelating Substances Passes test.
Nitrilotriacetic Acid Not more than 0.1%.
pH of a 1 in 100 Solution Between 6.5 and 7.5.
Water Not more than 13.0%.

TESTS

Assay Transfer about 1.2 g of the sample, accurately weighed, into a 250-mL beaker, and dissolve in 75 mL of water. Add 25 mL of 1 N acetic acid and 1.0 mL of diphenylcarbazone TS, and titrate slowly with 0.1 M mercuric nitrate to the first appearance of a purplish color. Each mL of 0.1 M mercuric nitrate is equivalent to 37.43 mg of $C_{10}H_{12}CaN_2Na_2O_8$.

Heavy Metals Prepare and test a 1-g sample as directed in *Method II* under the *Heavy Metals Test*, Appendix IIIB, using 20 µg of lead ion (Pb) in the control (*Solution A*).

Lead Prepare a *Sample Solution* as directed for organic compounds on Appendix IIIB, but use 70% perchloric acid instead of 30% hydrogen peroxide in the decomposition of the sample. The resulting solution meets the requirements of the *Lead Limit Test*, Appendix IIIB.

Magnesium-Chelating Substances Transfer a 1-g sample, accurately weighed, to a small beaker, and dissolve it in 5 mL of water. Add 5 mL of a buffer solution prepared by dissolving 67.5 g of ammonium chloride in 200 mL of water, adding 570 mL of ammonium hydroxide, and diluting with water to 1000 mL. Then to the buffered solution add 5 drops of eriochrome black TS, and titrate with 0.1 M magnesium acetate to the appearance of a deep wine-red color. Not more than 2.0 mL is required.

Nitrilotriacetic Acid
   Mobile Phase, Cupric Nitrate Solution, Stock Standard Solution, and *Chromatographic System* Prepare as directed in the test for *Nitrilotriacetic Acid* in the monograph for *Disodium EDTA*.
   Standard Preparation Transfer 1.0 g of the sample to a 100-mL volumetric flask, add 100 µL of *Stock Standard Solution*, dilute with *Cupric Nitrate Solution* to volume, and mix. Sonicate, if necessary, to achieve complete solution.
   Test Preparation Transfer 1.0 g of the sample to a 100-mL volumetric flask, dilute with *Cupric Nitrate Solution* to volume, and mix. Sonicate, if necessary, to achieve complete solution.
   Procedure Proceed as directed for *Procedure* in the test for *Nitrilotriacetic Acid* in the monograph for *Disodium EDTA*: The response of the nitrilotriacetic acid peak of the *Test Preparation* does not exceed the difference between the nitrilotriacetic acid peak responses obtained from the *Standard Preparation* and the *Test Preparation*.

pH of a 1 in 100 Solution Determine by the *Potentiometric Method*, Appendix IIB.

Water Determine by the *Karl Fischer Titrimetric Method*, Appendix IIB.

Packaging and Storage Store in well-closed containers.

Calcium Gluconate

$$[CH_2OH(CHOH)_4COO]_2Ca$$

$C_{12}H_{22}CaO_{14}$ Formula wt, anhydrous 430.38
$C_{12}H_{22}CaO_{14} \cdot H_2O$ Formula wt, monohydrate 448.39

INS: 578

CAS: anhydrous [18016-24-5]
CAS: monohydrate [299-28-5]

DESCRIPTION

Calcium Gluconate is anhydrous or contains one molecule of water of hydration. It is in the form of white, crystalline granules or powder. It is odorless, tasteless, and stable in air. Its solutions are neutral to litmus. One g dissolves slowly in about 30 mL of water at 25° and in about 5 mL of boiling water. It is insoluble in alcohol and in many other organic solvents.

Functional Use in Foods   Firming agent; formulation aid; sequestrant; stabilizer; thickener; texturizer.

REQUIREMENTS

Labeling   Indicate whether it is anhydrous or the monohydrate.
Identification
     A. A 1 in 50 solution gives positive tests for *Calcium*, Appendix IIIA.
     B. Dissolve a quantity of the sample in water to obtain a test solution containing 10 mg/mL, heating in a water bath at 60° if necessary. Similarly, prepare a standard solution of USP Potassium Gluconate Reference Standard in water containing 10 mg/mL. Apply separate 5-µL portions of the test solution and the standard solution on a suitable thin-layer chromatographic plate (see *Thin-Layer Chromatography*, Appendix IIA) coated with a 0.25-mm layer of chromatographic silica gel, and allow to dry. Develop the chromatogram in a solvent system consisting of a mixture of alcohol, water, ammonium hydroxide, and ethyl acetate (50:30:10:10) until the solvent front has moved about three-fourths of the length of the plate. Remove the plate from the chamber, and dry at 110° for 20 min. Allow to cool, and spray with a spray reagent prepared as follows: dissolve 2.5 g of ammonium molybdate in about 50 mL of 2 *N* sulfuric acid in a 100-mL volumetric flask, add 1.0 g of ceric sulfate, swirl to dissolve, dilute with 2 *N* sulfuric acid to volume, and mix. Heat the plate at 110° for about 10 min: the principal spot obtained from the test solution corresponds in color, size, and R_f value to that obtained from the standard solution.
Assay   *Anhydrous form*: not less than 98.0% and not more than 102.0% of $C_{12}H_{22}CaO_{14}$, calculated on the dried basis; *monohydrate*: not less than 98.0% and not more than 102.0% of $C_{12}H_{22}CaO_{14} \cdot H_2O$, calculated on the as-is basis.
Heavy Metals (as Pb)   Not more than 10 mg/kg.
Lead   Not more than 5 mg/kg.
Loss on Drying   *Anhydrous form*: not more than 3.0%; *monohydrate*: not more than 2.0%.
Sucrose and Reducing Sugars   Not more than 1.0%.

TESTS

Assay   Dissolve about 800 mg, accurately weighed, in 100 mL of water containing 2 mL of 2.7 *N* hydrochloric acid. While stirring, preferably with a magnetic stirrer, add about 30 mL of 0.05 *M* disodium EDTA from a 50-mL buret, then add 15 mL of 1 *N* sodium hydroxide and 300 mg of hydroxy naphthol blue indicator, and continue the titration to a blue endpoint. Each mL of 0.05 *M* disodium EDTA is equivalent to 21.52 mg of $C_{12}H_{22}CaO_{14}$ or 22.42 mg of $C_{12}H_{22}CaO_{14} \cdot H_2O$.
Heavy Metals   Prepare and test a 2-g sample as directed in *Method II* under the *Heavy Metals Test*, Appendix IIIB, using 20 µg of lead ion (Pb) in the control (*Solution A*).
Lead   A *Sample Solution* prepared as directed for organic compounds meets the requirements of the *Lead Limit Test*, Appendix IIIB, using 5 µg of lead ion (Pb) in the control.
Loss on Drying, Appendix IIC   Dry at 105° for 16 h.
Sucrose and Reducing Sugars   Transfer 1.0 g of the sample to a 250-mL conical flask, add, and dissolve in, 20 mL of hot water. Cool, add 25 mL of alkaline cupric citrate TS, cover the flask, and boil gently for 5 min, accurately timed. Cool rapidly to room temperature, add 25 mL of 0.6 *N* acetic acid, 10.0 mL of 0.1 *N* iodine, 10 mL of 2.7 *N* hydrochloric acid. Immediately titrate with 0.1 *N* sodium thiosulfate using starch TS as the indicator. Perform a blank determination and make any necessary correction. Each mL of 0.1 *N* sodium thiosulfate consumed is equivalent to 2.7 mg of reducing substances (as dextrose).

Packaging and Storage   Store in well-closed containers.

Calcium Glycerophosphate

$C_3H_7CaO_6P$ Formula wt 210.14

INS: 383   CAS: [27214-00-2]

DESCRIPTION

A fine, white, odorless, almost tasteless powder. It is somewhat hygroscopic. One g dissolves in about 50 mL of water at 25°. It is more soluble in water at a lower temperature, and citric acid increases its solubility in water. It is insoluble in alcohol.

Functional Use in Foods   Nutrient; dietary supplement.

REQUIREMENTS

Identification   A saturated solution gives positive tests for *Calcium*, Appendix IIIA.
Assay   Not less than 98.0% and not more than 100.5% of $C_3H_7CaO_6P$ after drying.
Alkalinity   Passes test.
Heavy Metals (as Pb)   Not more than 0.002%.
Lead   Not more than 10 mg/kg.
Loss on Drying   Not more than 12.0%.

TESTS

Assay Weigh accurately about 2 g, previously dried at 150° for 4 h, and dissolve in 100 mL of water and 5 mL of 2.7 N hydrochloric acid. Transfer the solution to a 250-mL volumetric flask, dilute to volume with water, and mix well. Pipet 50.0 mL of this solution into a suitable container, and add 50 mL of water. While stirring, preferably with a magnetic stirrer, add about 30 mL of 0.05 M disodium EDTA from a 50-mL buret, then add 15 mL of 1 N sodium hydroxide and 300 mg of hydroxy naphthol blue indicator, and continue the titration to a blue endpoint. Each mL of 0.05 M disodium EDTA is equivalent to 10.51 mg of $C_3H_7CaO_6P$.

Alkalinity A solution of 1 g in 60 mL of water requires not more than 1.5 mL of 0.1 N sulfuric acid for neutralization, using 3 drops of phenolphthalein TS as indicator.

Heavy Metals Dissolve 1 g in 3 mL of 1 N acetic acid, and dilute to 25 mL with water. This solution meets the requirements of the *Heavy Metals Test*, Appendix IIIB, using 20 μg of lead ion (Pb) in the control (*Solution A*).

Lead A *Sample Solution* prepared as directed for organic compounds meets the requirements of the *Lead Limit Test*, Appendix IIIB, using 10 μg of lead ion (Pb) in the control.

Loss on Drying, Appendix IIC Dry at 150° for 4 h.

Packaging and Storage Store in tight containers.

Calcium Hydroxide

Slaked Lime

$Ca(OH)_2$ Formula wt 74.09

INS: 526   CAS: [1305-62-0]

DESCRIPTION

A white powder, possessing an alkaline, slightly bitter taste. One g dissolves in 630 mL of water at 25°, and in 1300 mL of boiling water. It is soluble in glycerin and in a saturated solution of sucrose, but is insoluble in alcohol.

Functional Use in Foods Miscellaneous and general purpose; buffer; neutralizing agent; firming agent.

REQUIREMENTS

Identification
A. When mixed with from 3 to 4 times its weight of water, it forms a smooth magma. The clear, supernatant liquid from the magma is alkaline to litmus.
B. Mix 1 g with 20 mL of water, and add sufficient glacial acetic acid to effect solution. The resulting solution gives positive tests for *Calcium*, Appendix IIIA.

Assay Not less than 95.0% and not more than 100.5% of $Ca(OH)_2$.
Acid-Insoluble Substances Not more than 0.5%.
Arsenic (as As) Not more than 3 mg/kg.
Carbonate Passes test.
Fluoride Not more than 0.005%.
Heavy Metals (as Pb) Not more than 0.003%.
Lead Not more than 10 mg/kg.
Magnesium and Alkali Salts Not more than 4.8%.

TESTS

Assay Weigh accurately about 1.5 g, transfer to a beaker, and gradually add 30 mL of 2.7 N hydrochloric acid. When solution is complete, transfer it to a 500-mL volumetric flask, rinse the beaker thoroughly, adding the rinsings to the flask, dilute with water to volume, and mix. Transfer 50.0 mL of this solution into a suitable container, and add 50 mL of water. While stirring, preferably with a magnetic stirrer, add about 30 mL of 0.05 M disodium EDTA from a 50-mL buret, then add 15 mL of 1 N sodium hydroxide and 300 mg of hydroxy naphthol blue indicator, and continue the titration to a blue endpoint. Each mL of 0.05 M disodium EDTA is equivalent to 3.705 mg of $Ca(OH)_2$.

Acid-Insoluble Substances Dissolve 2 g in 30 mL of dilute hydrochloric acid (1 in 3), and heat to boiling. Filter the mixture through a suitable tared porous bottom porcelain crucible, wash the residue with hot water until the last washing is free from chloride, ignite at 800° ± 25° for 45 min, cool, and weigh.

Note: Avoid exposing the crucible to sudden temperature changes.

Arsenic A solution of 1 g in 15 mL of 2.7 N hydrochloric acid meets the requirements of the *Arsenic Test*, Appendix IIIB.

Carbonate Mix 2 g with 50 mL of water, and add an excess of 2.7 N hydrochloric acid. No more than a slight effervescence is produced.

Fluoride Weigh accurately 1.0 g, and proceed as directed in the *Fluoride Limit Test*, Appendix IIIB.

Heavy Metals Dissolve 667 mg in 15 mL of 2.7 N hydrochloric acid, and evaporate to dryness on a steam bath. Dissolve the residue in 25 mL of water, and filter. The filtrate meets the requirements of the *Heavy Metals Test*, Appendix IIIB, using 20 μg of lead ion (Pb) in the control (*Solution A*).

Lead A solution of 1 g in 15 mL of 2.7 N hydrochloric acid meets the requirements of the *Lead Limit Test*, Appendix IIIB, using 10 μg of lead ion (Pb) in the control.

Magnesium and Alkali Salts Dissolve 500 mg in a mixture of 30 mL of water and 10 mL of 2.7 N hydrochloric acid, and boil for 1 min. Rapidly add 40 mL of oxalic acid TS, and stir vigorously until precipitation is well established. Immediately add 2 drops of methyl red TS, then add 6 N ammonium hydroxide, dropwise, until the mixture is just alkaline, and cool. Transfer the mixture to a 100-mL cylinder, dilute with water to 100 mL, let it stand for 4 h or overnight, then decant the clear, supernatant liquid through a dry filter paper. To 50 mL of the clear filtrate in a tared platinum dish add 0.5 mL of sulfuric acid, and evaporate the mixture on a steam bath to a small volume. Carefully evaporate the remaining liquid to dryness over a free flame, and continue heating until the ammonium salts have been completely decomposed and volatilized. Finally, ignite the residue at 800° ± 25° to constant weight.

Packaging and Storage Store in tight containers.

Calcium Iodate

$Ca(IO_3)_2 \cdot H_2O$ Formula wt 407.90
INS: 916 CAS: [7789-80-2]

DESCRIPTION

A white powder. It is odorless or has a slight odor. It is slightly soluble in water, and is insoluble in alcohol.

Functional Use in Foods Maturing agent; dough conditioner.

REQUIREMENTS

Identification To 5 mL of a saturated solution of the sample add 1 drop of starch TS and a few drops of 20% hypophosphorous acid. A transient blue color appears.
Assay Not less than 99.0% and not more than 101.0% of $Ca(IO_3)_2 \cdot H_2O$.
Heavy Metals (as Pb) Not more than 10 mg/kg.

TESTS

Assay Weigh accurately about 600 mg, dissolve it in 10 mL of 70% perchloric acid and 10 mL of water, heating gently if necessary, and dilute with water to 250.0 mL. Transfer 50.0 mL to a 250-mL glass-stoppered Erlenmeyer flask, add 1 mL of 70% perchloric acid and 5 g of potassium iodide, stopper the flask, and swirl briefly. Let stand for 5 min, then titrate with 0.1 N sodium thiosulfate, adding starch TS just before the endpoint is reached. Each mL of 0.1 N sodium thiosulfate is equivalent to 3.398 mg of $Ca(IO_3)_2 \cdot H_2O$.
Heavy Metals Mix 5 mL of hydrochloric acid with a 2-g sample, evaporate to dryness on a steam bath, and cool. Add 5 mL of hydrochloric acid, and again evaporate to dryness. Dissolve the residue in 15 mL of water, heat nearly to boiling, and add just enough hydrazine sulfate to discharge any yellow color. Cool, and dilute to 25 mL with water. This solution meets the requirements of the *Heavy Metals Test*, Appendix IIIB, using 20 μg of lead ion (Pb) in the control (*Solution A*).

Packaging and Storage Store in well-closed containers.

Calcium Lactate

$[CH_3CH(OH)COO]_2Ca \cdot xH_2O$

$C_6H_{10}CaO_6 \cdot xH_2O$ Formula wt, anhydrous 218.22
INS: 327 CAS: [814-80-2]

DESCRIPTION

White to cream-colored, almost odorless, crystalline powder or granules, containing up to five molecules of water of crystallization. The pentahydrate is somewhat efflorescent and at 120° becomes anhydrous. It is soluble in water and practically insoluble in alcohol.

Functional Use in Foods Buffer; dough conditioner; yeast nutrient.

REQUIREMENTS

Identification A 1 in 20 solution gives positive tests for *Calcium* and for *Lactate*, Appendix IIIA.
Assay Not less than 98.0% and not more than 101.0% of $C_6H_{10}CaO_6$, calculated on the dried basis.
Acidity Passes test (about 0.45%, as lactic acid).
Fluoride Not more than 0.0015%.
Heavy Metals (as Pb) Not more than 0.002%.
Lead Not more than 10 mg/kg.
Loss on Drying *Pentahydrate*: between 22.0% and 27.0%; *trihydrate*: between 15.0% and 20.0%; *monohydrate*: between 5.0% and 8.0%; *dried form*: not more than 3.0%.
Magnesium and Alkali Salts Not more than 1%.
Volatile Fatty Acids Passes test.

TESTS

Assay Dissolve an accurately weighed amount of the sample, equivalent to about 350 mg of $C_6H_{10}CaO_6$, in 150 mL of water containing 2 mL of 2.7 N hydrochloric acid. While stirring, preferably with a magnetic stirrer, add about 30 mL of 0.05 M disodium EDTA from a 50-mL buret, then add 15 mL of 1 N sodium hydroxide and 300 mg of hydroxy naphthol blue indicator, and continue the titration to a blue endpoint. Each mL of 0.05 M disodium EDTA is equivalent to 10.91 mg of $C_6H_{10}CaO_6$.
Acidity Dissolve 1 g in 20 mL of water, add 3 drops of phenolphthalein TS, and titrate with 0.1 N sodium hydroxide. Not more than 0.5 mL is required.
Fluoride Proceed as directed in the *Fluoride Limit Test*, Appendix IIIB, using *Method I* (3.3-g sample) or *Method III* (1.0-g sample).

Heavy Metals A solution of 1 g in 25 mL of water meets the requirements of the *Heavy Metals Test*, Appendix IIIB, using 20 μg of lead ion (Pb) in the control (*Solution A*).

Lead Dissolve 1 g in 3 mL of dilute nitric acid (1 in 2), boil for 1 min, cool, and dilute to 20 mL with water. This solution meets the requirements of the *Lead Limit Test*, Appendix IIIB, using 10 μg of lead ion (Pb) in the control.

Loss on Drying, Appendix IIC Distribute a sample of about 1.5 g evenly in a suitable weighing dish to a depth not exceeding 3 mm, and dry at 120° for 4 h.

Magnesium and Alkali Salts Mix 1 g with 40 mL of water, carefully add 1 mL of hydrochloric acid, boil for 1 min, and add rapidly 40 mL of oxalic acid TS. Add immediately to the warm mixture 2 drops of methyl red TS, then add 6 N ammonium hydroxide, dropwise, from a buret until the mixture is just alkaline, and cool to room temperature. Transfer the mixture into a 100-mL graduate, dilute with water to 100 mL, mix, and allow to stand for 4 h or overnight. Decant the clear, supernatant liquid through a dry filter paper, transfer 50 mL of the clear filtrate to a tared platinum dish, and add 0.5 mL of sulfuric acid. Evaporate to a small volume on a steam bath, then carefully heat over a free flame to dryness, and continue heating to complete decomposition and volatilization of the ammonium salts. Finally, ignite the residue to constant weight. The weight of the residue does not exceed 5 mg.

Volatile Fatty Acids Stir about 500 mg of the sample with 1 mL of sulfuric acid, and warm. The mixture does not emit an odor of volatile fatty acids.

Packaging and Storage Store in tight containers.

Calcium Lactobionate

Calcium 4-(β,D-Galactosido)-D-gluconate

$C_{24}H_{42}CaO_{24}$ Formula wt, anhydrous 754.66

INS: 399 CAS: [5001-51-4]

DESCRIPTION

A white to cream-colored, odorless, free-flowing powder. It is freely soluble in water but is insoluble in alcohol and in ether. It has a bland taste, and it readily forms double salts, such as the chloride, bromide, and gluconate. It is anhydrous when obtained by spray-drying or the dihydrate when obtained by crystallization. It decomposes at about 120°. The pH of a 1 in 10 solution is between 6.5 and 7.5.

Functional Use in Foods Firming agent in dry pudding mixes; nutrient.

REQUIREMENTS

Labeling Indicate whether it has been obtained through spray-drying or from crystallization.

Identification

A. The infrared absorption spectrum of a potassium bromide dispersion of the sample, previously dried at 105° for 8 h, exhibits maxima only at the same wavelengths as that of a similar preparation of USP Calcium Lactobionate Reference Standard.

B. It gives positive tests for *Calcium*, Appendix IIIA.

Calcium Content Not less than 5.05% and not more than 5.55% of Ca, calculated on the dried basis.

Halides Not more than 0.04%.

Heavy Metals (as Pb) Not more than 0.002%.

Lead Not more than 10 mg/kg.

Loss on Drying Not more than 8.0%.

Reducing Substances Not more than 1.0%.

Specific Rotation $[\alpha]_D^{20°}$: Between +23° and +25°.

Sulfate Not more than 0.7%.

TESTS

Calcium Content Weigh accurately about 1.5 g, and dissolve it in 100 mL of water containing 2 mL of 2.7 N hydrochloric acid. While stirring, preferably with a magnetic stirrer, add about 30 mL of 0.05 M disodium EDTA from a 50-mL buret, then add 15 mL of 1 N sodium hydroxide and 300 mg of hydroxy naphthol blue indicator, and continue the titration to a blue endpoint. Each mL of 0.05 M disodium EDTA is equivalent to 2.004 mg of Ca.

Halides A 1.2-g sample tested as directed under *Chloride Test*, Appendix IIIB, shows no more turbidity than corresponds to 0.7 mL of 0.020 N hydrochloric acid.

Heavy Metals Prepare and test a 1-g sample as directed in *Method II* under the *Heavy Metals Test*, Appendix IIIB, using 20 μg of lead ion (Pb) in the control (*Solution A*).

Lead A *Sample Solution* prepared as directed for organic compounds meets the requirements of the *Lead Limit Test*, Appendix IIIB, using 10 μg of lead ion (Pb) in the control.

Loss on Drying, Appendix IIC Dry at 105° for 8 h.

Reducing Substances Transfer 1.0 g to a 250-mL conical flask, dissolve in 20 mL of water, and add 25 mL of alkaline cupric citrate TS. Cover the flask, boil gently for 5 min, accurately timed, and cool rapidly to room temperature. Add 25 mL of 0.6 N acetic acid, 10.0 mL of 0.1 N iodine, and 10 mL of 3 N hydrochloric acid, and titrate with 0.1 N sodium thiosulfate, adding 3 mL of starch TS as the endpoint is approached. Perform a blank determination, omitting the specimen, and note the difference in volumes required. Each mL of the difference in volume of 0.1 N sodium thiosulfate consumed is equivalent to 2.7 mg of reducing substances (as dextrose).

Specific Rotation, Appendix IIB Determine in a solution containing 500 mg, calculated on the anhydrous basis, in each 10 mL.

Sulfate Transfer about 25 g, accurately weighed, into a 600-mL beaker, dissolve it in 200 mL of water, adjust the solution to a pH between 4.5 and 6.5 with 2.7 N hydrochloric acid, and filter, if necessary. Heat the filtrate or clear solution to just below the boiling point, then add 10 mL of barium chloride TS, stirring vigorously, boil gently for 5 min, and allow to stand for at least 2 h, or preferably overnight. Collect the precipitate of barium sulfate on a suitable, tared crucible, wash until free from chloride, dry, and ignite at 600° to constant weight. The weight of barium sulfate so obtained, multiplied by 0.412, represents the weight of SO_4 in the sample taken.

Packaging and Storage Store in well-closed containers.

Calcium Oxide

Lime

CaO   Formula wt 56.08

INS: 529   CAS: [1305-78-8]

DESCRIPTION

Hard, white or grayish white masses or granules, or a white to grayish white powder. It is odorless. One g dissolves in about 840 mL of water at 25°, and in about 1740 mL of boiling water. It is soluble in glycerin, but is insoluble in alcohol.

Functional Use in Foods Alkali; nutrient; dietary supplement; dough conditioner; yeast food.

REQUIREMENTS

Identification Slake 1 g with 20 mL of water, and add glacial acetic acid until the sample is dissolved. The resulting solution gives positive tests for *Calcium*, Appendix IIIA.
Assay Not less than 95.0% and not more than 100.5% of CaO after ignition.
Acid-Insoluble Substances Not more than 1%.
Alkalies or Magnesium Not more than 3.6%.
Arsenic (as As) Not more than 3 mg/kg.
Fluoride Not more than 0.015%.
Heavy Metals (as Pb) Not more than 0.003%.
Lead Not more than 5 mg/kg.
Loss on Ignition Not more than 10.0%.

TESTS

Assay Ignite about 1 g to constant weight, and dissolve the ignited sample, accurately weighed, in 20 mL of 2.7 N hydrochloric acid. Cool the solution, dilute with water to 500.0 mL, and mix. Pipet 50.0 mL of this solution into a suitable container, and add 50 mL of water. While stirring, preferably with a magnetic stirrer, add about 30 mL of 0.05 M disodium EDTA from a 50-mL buret, then add 15 mL of 1 N sodium hydroxide and 300 mg of hydroxy naphthol blue indicator, and continue the titration to a blue endpoint. Each mL of 0.05 M disodium EDTA is equivalent to 2.804 mg of CaO.

Acid-Insoluble Substances Slake a 5-g sample, then mix it with 100 mL of water and sufficient hydrochloric acid, added dropwise, to effect solution. Boil the solution, cool, add hydrochloric acid, if necessary, to make the solution distinctly acid, and filter through a tared glass filter crucible. Wash the residue with water until free of chlorides, dry at 105° for 1 h, cool, and weigh.

Alkalies or Magnesium Dissolve 500 mg in 30 mL of water and 15 mL of 2.7 N hydrochloric acid. Heat the solution and boil for 1 min. Add rapidly 40 mL of oxalic acid TS, and stir vigorously. Add 2 drops of methyl red TS, and neutralize the solution with 6 N ammonium hydroxide to precipitate the calcium completely. Heat the mixture on a steam bath for 1 h, cool, dilute to 100 mL with water, mix well, and filter. To 50 mL of the filtrate add 0.5 mL of sulfuric acid, then evaporate to dryness, and ignite to constant weight in a tared platinum crucible at 800° ± 25°.

Arsenic A solution of 1 g in 15 mL of 2.7 N hydrochloric acid meets the requirements of the *Arsenic Test*, Appendix IIIB.

Fluoride Weigh accurately 1.0 g, and proceed as directed in the *Fluoride Limit Test*, Appendix IIIB.

Heavy Metals Mix 2 g with 25 mL of water, cautiously add 7 mL of hydrochloric acid, followed by 3 mL of nitric acid, and evaporate to dryness on a steam bath. Dissolve the residue in 1 mL of 2.7 N hydrochloric acid and 25 mL of hot water, filter, wash with a few mL of water, and dilute the filtrate to 100 mL with water. A 33.4-mL portion of this solution, to which has been added 1.0 mL of 10% hydroxylamine hydrochloride solution, meets the requirements of the *Heavy Metals Test*, Appendix IIIB, using 20 µg of lead ion (Pb) in the control (*Solution A*).

Lead A solution of 1 g in 15 mL of 2.7 N hydrochloric acid meets the requirements of the *Lead Limit Test*, Appendix IIIB, using 5 µg of lead ion (Pb) in the control.

Loss on Ignition Ignite 1 g to constant weight in a tared platinum crucible at 1100° ± 50°.

Packaging and Storage Store in tight containers.

Calcium Pantothenate

d-Calcium Pantothenate; Dextro Calcium Pantothenate

$[HOCH_2C(CH_3)_2CH(OH)CONH(CH_2)_2COO]_2Ca$

$C_{18}H_{32}CaN_2O_{10}$ Formula wt 476.54

CAS: [137-08-6]

DESCRIPTION

The calcium salt of the dextrorotatory isomer of pantothenic acid occurs as a slightly hygroscopic, white powder. It is odorless and has a bitter taste. It is stable in air. One g dissolves in about 3 mL of water. It is soluble in glycerin, but is practically insoluble in alcohol, in chloroform, and in ether.

Functional Use in Foods   Nutrient; dietary supplement.

REQUIREMENTS

Identification

A. A 1 in 20 solution gives positive tests for *Calcium*, Appendix IIIA.

B. The infrared absorption spectrum of a potassium bromide dispersion of the sample, previously dried at 105° for 3 h, exhibits maxima only at the same wavelengths as that of a similar preparation of USP Calcium Pantothenate Reference Standard.

C. Boil 50 mg in 5 mL of 1 N sodium hydroxide for 1 min, cool, and add 5 mL of 1 N hydrochloric acid and 2 drops of ferric chloride TS. A strong yellow color is produced.

Assay   Not less than 97.0% and not more than 103.0% of Dextrorotatory Calcium Pantothenate ($C_{18}H_{32}CaN_2O_{10}$) after drying.

Alkalinity   Passes test.

Alkaloids   Passes test.

Calcium Content   Not less than 8.2% and not more than 8.6% of Ca after drying.

Heavy Metals (as Pb)   Not more than 10 mg/kg.

Loss on Drying   Not more than 5.0%.

Specific Rotation $[\alpha]_D^{25°}$: Between +25.0° and +27.5° after drying.

TESTS

Assay   (Use low-actinic glassware throughout this procedure.)

Mobile Phase   Transfer 2.0 mL of phosphoric acid into a 2-L volumetric flask, and dilute to volume with water. Filter the solution through a 0.45-μm pore-size disk.

Internal Standard Preparation   Transfer about 80 mg of *p*-hydroxybenzoic acid, accurately weighed, to a 1000-mL volumetric flask, dissolve in 5 mL of alcohol, dilute with *Mobile Phase* to volume, and mix.

Standard Preparation   Transfer about 15 mg of USP Calcium Pantothenate Reference Standard, previously dried at 105° for 3 h and accurately weighed, into a 25-mL volumetric flask. Dilute to volume with *Internal Standard Preparation*, and mix.

Sample Preparation   Proceed as directed for the *Standard Preparation*, using an accurately weighed amount of the sample equivalent to about 15 mg of Calcium Pantothenate previously dried at 105° for 3 h.

Chromatographic System (see *Chromatography*, Appendix IIA)   Use a high-pressure liquid chromatograph equipped with an ultraviolet detector that measures at 210 nm. Under typical conditions, the instrument contains a 15-cm × 3.9-mm column packed with octadecylsilanized silica (10-μm μBondapak C 18, or equivalent). The flow rate is about 1.5 mL/min.

System Suitability   Three replicate injections of the *Standard Preparation* show a relative standard deviation of not more than 2.0%.

Procedure   Separately inject equal volumes (about 10 μL) of the *Standard Preparation* and the *Sample Preparation* into the chromatograph, record the chromatograms, and measure the peak responses obtained for Calcium Pantothenate and the *Internal Standard Preparation*. The relative retention times are 0.5 for Calcium Pantothenate and 1.0 for *p*-hydroxybenzoic acid. Calculate the quantity, in mg, of $C_{18}H_{32}CaN_2O_{10}$ in the portion of Calcium Pantothenate taken by the formula

$$25C(R_U/R_S),$$

in which C is the concentration, in mg/mL, of USP Calcium Pantothenate Reference Standard in the *Standard Preparation*, and R_U and R_S are the ratios of the peak responses obtained for Calcium Pantothenate and *p*-hydroxybenzoic acid from the *Sample Preparation* and the *Standard Preparation*, respectively.

Alkalinity   Dissolve 1 g in 15 mL of recently boiled and cooled water in a small flask. As soon as solution is complete, add 1.0 mL of 0.1 N hydrochloric acid, then add 0.05 mL of phenolphthalein TS, and mix. No pink color is produced within 5 s.

Alkaloids   Dissolve 200 mg in 5 mL of water, and add 1 mL of 2.7 N hydrochloric acid and 2 drops of mercuric-potassium iodide TS. No turbidity is produced in 1 min.

Calcium Content   Weigh accurately about 950 mg, previously dried at 105° for 3 h, and dissolve it in 100 mL of water containing 2 mL of 2.7 N hydrochloric acid. While stirring, preferably with a magnetic stirrer, add about 30 mL of 0.05 M disodium EDTA from a 50-mL buret, then add 15 mL of 1 N sodium hydroxide and 300 mg of hydroxy naphthol blue indicator, and continue the titration to a blue endpoint. Each mL of 0.05 M disodium EDTA is equivalent to 2.004 mg of Ca.

Heavy Metals   A solution of 2 g in 25 mL of water meets the requirements of the *Heavy Metals Test*, Appendix IIIB, using 20 μg of lead ion (Pb) in the control (*Solution A*).

Loss on Drying, Appendix IIC   Dry at 105° for 3 h.

Specific Rotation, Appendix IIB   Determine in a solution containing 500 mg of the sample, previously dried at 105° for 3 h.

Packaging and Storage   Store in tight containers.

Calcium Pantothenate, Racemic

$C_{18}H_{32}CaN_2O_{10}$ Formula wt 476.54

CAS: [6381-63-1]

DESCRIPTION

A mixture of the calcium salts of the dextrorotatory and levorotatory isomers of pantothenic acid. It occurs as a white, slightly hygroscopic powder. It is odorless, has a bitter taste, and is stable in air. Its solutions are neutral or alkaline to litmus. It is optically inactive. It is freely soluble in water. It is soluble in glycerin, and is practically insoluble in alcohol, in chloroform, and in ether.

Note: The physiological activity of Racemic Calcium Pantothenate is approximately one-half that of the dextrorotatory isomer.

Functional Use in Foods Nutrient; dietary supplement.

REQUIREMENTS

Identification

A. A 1 in 20 solution gives positive tests for *Calcium*, Appendix IIIA.

B. The infrared absorption spectrum of a potassium bromide dispersion of the sample, previously dried at 105° for 3 h, exhibits maxima only at the same wavelengths as that of a similar preparation of USP Calcium Pantothenate Reference Standard.

C. Boil 50 mg in 5 mL of 1 N sodium hydroxide for 1 min, cool, and add 5 mL of 1 N hydrochloric acid and 2 drops of ferric chloride TS. A strong yellow color is produced.

Assay Not less than 97.0% and not more than 103.0% of Calcium Pantothenate ($C_{18}H_{32}CaN_2O_{10}$) after drying.
Alkalinity Passes test.
Alkaloids Passes test.
Calcium Content Not less than 8.2% and not more than 8.6% of Ca after drying.
Heavy Metals (as Pb) Not more than 10 mg/kg.
Loss on Drying Not more than 5.0%.
Specific Rotation $[\alpha]_D^{25°}$: Between −0.05° and +0.05° after drying.

TESTS

Assay Proceed as directed for *Assay* in the monograph for *Calcium Pantothenate*.
Alkalinity Dissolve 1 g in 15 mL of recently boiled and cooled water in a small flask. As soon as solution is complete, add 1.6 mL of 0.1 N hydrochloric acid, then add 0.05 mL of phenolphthalein TS, and mix. No pink color is produced within 5 s.
Alkaloids Dissolve 200 mg in 5 mL of water, and add 1 mL of 2.7 N hydrochloric acid and 2 drops of mercuric-potassium iodide TS. No turbidity is produced in 1 min.
Calcium Content Weigh accurately about 950 mg, previously dried at 105° for 3 h, and dissolve it in 100 mL of water containing 2 mL of 2.7 N hydrochloric acid. While stirring, preferably with a magnetic stirrer, add about 30 mL of 0.05 M disodium EDTA from a 50-mL buret, then add 15 mL of 1 N sodium hydroxide and 300 mg of hydroxy naphthol blue indicator, and continue the titration to a blue endpoint. Each mL of 0.05 M disodium EDTA is equivalent to 2.004 mg of Ca.
Heavy Metals A solution of 2 g in 25 mL of water meets the requirements of the *Heavy Metals Test*, Appendix IIIB, using 20 μg of lead ion (Pb) in the control (*Solution A*).
Loss on Drying, Appendix IIC Dry at 105° for 3 h.
Specific Rotation, Appendix IIB Determine in a solution containing 500 mg of the sample, previously dried at 105° for 3 h, in each 10-mL portion.

Packaging and Storage Store in tight containers.

Calcium Pantothenate, Calcium Chloride Double Salt

Calcium Chloride Double Salt of DL- or D-Calcium Pantothenate

$C_{18}H_{32}CaN_2O_{10} \cdot CaCl_2$ Formula wt 587.52

CAS: [6363-38-8]]

DESCRIPTION

A chemical complex composed of approximately equimolecular quantities of dextrorotatory (D) or racemic (DL) calcium pantothenate and calcium chloride. It occurs as a white, odorless, free-flowing, fine powder having a bitter taste. It is freely soluble in water, but insoluble in alcohol. Its solutions in water are alkaline to litmus.

Functional Use in Foods Nutrient; dietary supplement.

REQUIREMENTS

Identification

A. A 1 in 20 solution gives positive tests for *Calcium*, Appendix IIIA.

B. Dissolve 50 mg in 5 mL of 1 N sodium hydroxide and filter. To the filtrate add 1 drop of cupric sulfate TS. A deep blue color develops.

C. Stir 1.0 g of a dried sample with 15 mL of dimethylformamide for 5 min. Centrifuge the mixture, then transfer 2.0 mL of the clear supernatant liquid to a weighing dish, evaporate it under vacuum on a steam bath, and dry the residue in an oven at 105° for 1 h. The weight of the residue, composed of uncombined calcium pantothenate and calcium chloride, in g, multiplied by 750 equals the percentage of uncomplexed material in the sample. It does not exceed 10.0% of the weight of the sample.

Assay Not less than 45.0% and not more than 55.0% of Calcium Pantothenate ($C_{18}H_{32}Ca_2N_2O_{10}C_{l4}$) after drying.
Arsenic (as As) Not more than 3 mg/kg.
Calcium Content Between 12.4% and 13.6% of Ca after drying.
Chloride Between 10.5% and 12.1% of Cl after drying.
Heavy Metals (as Pb) Not more than 0.002%.
Lead Not more than 10 mg/kg.
Loss on Drying Not more than 5.0%.

TESTS

Assay Proceed as directed for *Assay* in the monograph for *Calcium Pantothenate*.
Arsenic A solution of 1 g in 25 mL of water meets the requirements of the *Arsenic Test*, Appendix IIIB.
Calcium Content Proceed as directed for *Calcium Content* in the monograph for *Calcium Pantothenate*.
Chloride Transfer about 1 g, previously dried in vacuum for 1 h and accurately weighed, into a 250-mL beaker, and add sufficient water to make 100 mL. Equip a pH meter with glass and silver electrodes, and set it on the "+ millivolt" scale. Insert

the electrodes and a motor-driven glass stirring rod into the sample beaker. Add 1 to 2 drops of methyl orange TS. Stir and add, dropwise, 10% nitric acid until a pink color is obtained, then add 10 mL excess. Titrate the solution with 0.1 N silver nitrate to a reading of +1.0 millivolt on the pH meter. Each mL of 0.1 N silver nitrate is equivalent to 3.545 mg of Cl.

Heavy Metals A solution of 1 g in 25 mL of water meets the requirements of the *Heavy Metals Test*, Appendix IIIB, using 20 μg of lead ion (Pb) in the control (*Solution A*).

Lead A solution of 1 g in 25 mL of water meets the requirements of the *Lead Limit Test*, Appendix IIIB, using 10 μg of lead ion (Pb) in the control.

Loss on Drying, Appendix IIC Dry in vacuum at 100° for 1 h.

Packaging and Storage Store in tight containers.

Calcium Peroxide

CaO_2 Formula wt 72.08

INS: 930 CAS: [1305-79-9]

DESCRIPTION

A white or yellowish, odorless, almost tasteless powder or granular material. It decomposes in moist air. It is practically insoluble in water. It dissolves in acids, forming hydrogen peroxide. A 1 in 100 aqueous slurry has a pH of about 12.

Functional Use in Foods Dough conditioner; oxidizing agent.

REQUIREMENTS

Identification Cautiously dissolve 250 mg in 5 mL of glacial acetic acid, and add a few drops of a saturated solution of potassium iodide. Iodine is liberated. Add 20 mL of water and sufficient sodium thiosulfate TS to remove the iodine color. The resulting solution gives positive tests for *Calcium*, Appendix IIIA.
Assay Not less than 60.0% of CaO_2.
Fluoride Not more than 0.005%.
Heavy Metals (as Pb) Not more than 0.002%.
Lead Not more than 10 mg/kg.

TESTS

Assay Transfer about 1 g of the sample, accurately weighed, into an Erlenmeyer flask, add 30 mL of water and 30 mL of 85% phosphoric acid diluted 1 to 1 with water, and titrate immediately with 0.5 N potassium permanganate to the first faint pink color that persists for 1 min. Each mL of 0.5 N potassium permanganate is equivalent to 18.02 mg of CaO_2.
Fluoride Weigh accurately 1.0 g, and proceed as directed in the *Fluoride Limit Test*, Appendix IIIB.

Sample Solution for the Determination of Heavy Metals and Lead Weigh accurately 4.0 g of the sample into a 250-mL beaker, cautiously add 50 mL of nitric acid, and evaporate just to dryness on a steam bath. Add 20 mL of nitric acid, repeat the evaporation, cool, and dissolve the residue in sufficient water, containing 4 drops of nitric acid, to make 40.0 mL.

Heavy Metals A 10-mL portion of the *Sample Solution* meets the requirements of the *Heavy Metals Test*, Appendix IIIB, using 20 μg of lead ion (Pb) in the control (*Solution A*), and adjusting the solutions to a pH of 2.0, instead of between 3.0 and 4.0.
Lead A 10-mL portion of the *Sample Solution* meets the requirements of the *Lead Limit Test*, Appendix IIIB, using 10 μg of lead ion (Pb) in the control.

Packaging and Storage Store in tight containers, and avoid contact with readily oxidizable materials. Observe safety precautions printed on the label of the original container.

Calcium Phosphate, Dibasic

Dicalcium Phosphate

$CaHPO_4 \cdot 2H_2O$ Formula wt, anhydrous 136.06
Formula wt, dihydrate 172.09

INS: 341(ii) CAS: anhydrous [7757-93-9]
CAS: dihydrate [7789-77-7]

DESCRIPTION

Dibasic Calcium Phosphate is anhydrous or contains two molecules of water of hydration. It occurs as a white, odorless, tasteless powder that is stable in air. It is practically insoluble in water, but is readily soluble in dilute hydrochloric and nitric acids. It is insoluble in alcohol.

Functional Use in Foods Leavening agent; dough conditioner; nutrient; dietary supplement; yeast food.

REQUIREMENTS

Labeling Indicate whether it is anhydrous or the dihydrate.
Identification
 A. Dissolve about 100 mg by warming with a mixture of 5 mL of 2.7 N hydrochloric acid and 5 mL of water, add 2.5 mL of 6 N ammonium hydroxide, dropwise, with shaking, and then add 5 mL of ammonium oxalate TS. A white precipitate is formed.
 B. To 10 mL of a warm solution (1 in 100) in a slight excess of nitric acid add 10 mL of ammonium molybdate TS. A yellow precipitate of ammonium phosphomolybdate is formed.
Assay Not less than 97.0% and not more than 105.0% of Dibasic Calcium Phosphate ($CaHPO_4$) or of Dibasic Calcium Phosphate, Dihydrate, ($CaHPO_4 \cdot 2H_2O$).

Arsenic (as As) Not more than 3 mg/kg.
Fluoride Not more than 0.005%.
Heavy Metals (as Pb) Not more than 0.0015%.
Lead Not more than 5 mg/kg.
Loss on Ignition $CaHPO_4$ (anhydrous): between 7.0% and 8.5%; $CaHPO_4.2H_2O$ (dihydrate): between 24.5% and 26.5%.

TESTS

Assay Dissolve about 250 mg of Dibasic Calcium Phosphate, accurately weighed, with the aid of gentle heat if necessary, in a mixture of 5 mL of hydrochloric acid and 3 mL of water contained in a 250-mL beaker equipped with a magnetic stirrer, and cautiously add 125 mL of water. With constant stirring, add, in the order named, 0.5 mL of triethanolamine, 300 mg of hydroxy naphthol blue indicator, and from a 50-mL buret, about 23 mL of 0.05 M disodium ethylenediaminetetraacetate. Add sodium hydroxide solution (45 in 100) until the initial red color changes to clear blue, then continue to add it dropwise until the color changes to violet, then add an additional 0.5 mL. The pH is between 12.3 and 12.5. Continue the titration dropwise with the 0.05 M disodium ethylenediaminetetraacetate to the appearance of a clear blue endpoint that persists for not less than 60 s. Each mL of 0.05 M disodium ethylenediaminetetraacetate is equivalent to 6.803 mg of $CaHPO_4$ or to 8.604 mg of $CaHPO_4.2H_2O$.

Arsenic A solution of 1 g in 5 mL of 2.7 N hydrochloric acid meets the requirements of the *Arsenic Test*, Appendix IIIB.

Fluoride (Note: Prepare and store all solutions in plastic containers.)

Buffer Solution Dissolve 73.5 g of sodium citrate in water to make 250 mL of solution.

Standard Solution Dissolve an accurately weighed quantity of USP Sodium Fluoride RS quantitatively in water to obtain a solution containing 1.1052 mg/mL. Transfer 20.0 mL of the resulting solution to a 100-mL volumetric flask containing 50 mL of *Buffer Solution*, dilute with water to volume, and mix. Each mL of this solution contains 100 μg of fluoride ion.

Electrode System Use a fluoride-specific, ion-indicating electrode and a silver–silver chloride reference electrode connected to a pH meter capable of measuring potentials with a minimum reproducibility of ± 0.2 mV.

Standard Response Line Transfer 50.0 mL of *Buffer Solution* and 2.0 mL of hydrochloric acid to a beaker, and add water to make 100 mL. Add a plastic-coated stirring bar, insert the electrodes into the solution, stir for 15 min, and read the potential, in mV. Continue stirring, and at 5-min intervals, add 100 μL, 100 μL, 300 μL, and 500 μL of *Standard Solution*, reading the potential 5 min after each addition. Plot the logarithms of the cumulative fluoride ion concentrations (0.1, 0.2, 0.5, and 1.0 μg/mL) versus potential, in mV.

Procedure Transfer 2.0 g of the specimen under test to a beaker containing a plastic-coated stirring bar, add 20 mL of water and 2.0 mL of hydrochloric acid, and stir until dissolved. Add 50.0 mL of *Buffer Solution* and sufficient water to make 100 mL of test solution. Rinse and dry the electrodes, insert them into the test solution, stir for 5 min, and read the potential, in mV. From the measured potential and the *Standard Response Line* determine the concentration, C, in μg/mL, of fluoride ion in the test solution. Calculate the percentage of fluoride in the specimen taken by the formula

$$C \times 0.005.$$

Heavy Metals Warm 2.66 g with 5 mL of 2.7 N hydrochloric acid until no more dissolves, dilute to 50 mL with water, and filter. A 25-mL portion of the filtrate meets the requirements of the *Heavy Metals Test*, Appendix IIIB, using 20 μg of lead ion (Pb) in the control (*Solution A*).

Lead A 10-g sample using a 5-μg/mL *Standard Lead Solution* meets the requirements of the *APDC Extraction Method for Lead*, Appendix IIIB.

Loss on Ignition Weigh accurately about 3 g, and ignite, preferably in a muffle furnace, at 800° to 825° to constant weight.

Packaging and Storage Store in well-closed containers.

Calcium Phosphate, Monobasic

Monocalcium Phosphate; Calcium Biphosphate; Acid Calcium Phosphate

$Ca(H_2PO_4)_2.H_2O$ Formula wt, anhydrous 234.05
                             Formula wt, monohydrate 252.07

INS: 341(i)   CAS: anhydrous [7758-23-8]
                             CAS: monohydrate [10031-30-8]

DESCRIPTION

Monobasic Calcium Phosphate is anhydrous or contains one molecule of water of hydration, but, due to its deliquescent nature, more than the calculated amount of water may be present. It occurs as white crystals or granules, or as a granular powder. It is sparingly soluble in water and is insoluble in alcohol.

Functional Use in Foods Buffer; dough conditioner; firming agent; leavening agent; nutrient; dietary supplement; yeast food; sequestrant.

REQUIREMENTS

Labeling Indicate the state of hydration.
Identification

A. Dissolve 100 mg by warming in a mixture of 2 mL of 2.7 N hydrochloric acid and 8 mL of water, and add 5 mL of ammonium oxalate TS. A white precipitate forms.

B. To a warm solution of the sample in a slight excess of nitric acid add ammonium molybdate TS. A yellow precipitate forms.

Assay $Ca(H_2PO_4)_2$ (anhydrous): not less than 16.8% and not more than 18.3% of Ca; $Ca(H_2PO_4)_2.H_2O$ (monohydrate): not less than 15.9% and not more than 17.7% of Ca.
Arsenic (as As) Not more than 3 mg/kg.
Fluoride Not more than 0.005%.
Heavy Metals (as Pb) Not more than 0.0015%.

Lead Not more than 5 mg/kg.
Loss on Drying Ca(H$_2$PO$_4$)$_2$.H$_2$O (monohydrate): not more than 1%.
Loss on Ignition Ca(H$_2$PO$_4$)$_2$ (anhydrous): between 14.0% and 15.5%.

TESTS

Assay Weigh accurately a portion of the sample equivalent to about 475 mg of Ca(H$_2$PO$_4$)$_2$, dissolve it in 10 mL of 2.7 N hydrochloric acid, add a few drops of methyl orange TS, and boil for 5 min, keeping the volume and pH of the solution constant during the boiling period by adding hydrochloric acid or water, if necessary. Add 2 drops of methyl red TS and 30 mL of ammonium oxalate TS, then add dropwise, with constant stirring, a mixture of equal volumes of 6 N ammonium hydroxide and water until the pink color of the indicator just disappears. Digest on a steam bath for 30 min, cool to room temperature, allow the precipitate to settle, and filter the supernatant liquid through a sintered-glass crucible, using gentle suction. Wash the precipitate in the beaker with about 30 mL of cold (below 20°) wash solution, prepared by diluting 10 mL of ammonium oxalate TS to 1000 mL. Allow the precipitate to settle, and pour the supernatant liquid through the filter. Repeat this washing by decantation three more times. Using the wash solution, transfer the precipitate as completely as possible to the filter. Finally, wash the beaker and the filter with two 10-mL portions of cold (below 20°) water. Place the sintered-glass crucible in the beaker, and add 100 mL of water and 50 mL of cold dilute sulfuric acid (1 in 6). Add from a buret 35 mL of 0.1 N potassium permanganate, and stir until the color disappears. Heat to about 70°, and complete the titration with 0.1 N potassium permanganate. Each mL of 0.1 N potassium permanganate is equivalent to 2.004 mg of Ca.
Arsenic A solution of 1 g in 5 mL of 2.7 N hydrochloric acid meets the requirements of the *Arsenic Test*, Appendix IIIB.
Fluoride (anhydrous): determine as directed in *Method II* under the *Fluoride Limit Test*, Appendix IIIB; (monohydrate): proceed as directed under *Fluoride* in the monograph for *Calcium Phosphate, Dibasic*.
Heavy Metals Warm 2.66 g with 5 mL of 2.7 N hydrochloric acid until no more dissolves, dilute to 50 mL with water, and filter. A 25-mL portion of the filtrate meets the requirements of the *Heavy Metals Test*, Appendix IIIB, using 20 µg of lead ion (Pb) in the control (*Solution A*).
Lead A 10-g sample using a 5-µg/mL *Standard Lead Solution* meets the requirements of the *APDC Extraction Method for Lead*, Appendix IIIB.
Loss on Drying, Appendix IIC Dry Ca(H$_2$PO$_4$)$_2$.H$_2$O (monohydrate) at 60° for 3 h.
Loss on Ignition Weigh accurately about 3 g of Ca(H$_2$PO$_4$)$_2$ (anhydrous), and ignite, preferably in a muffle furnace, at 800° for 30 min.

Packaging and Storage Store in well-closed containers.

Calcium Phosphate, Tribasic

Tricalcium Phosphate; Precipitated Calcium Phosphate; Calcium Hydroxyapatite

INS: 341(iii)   CAS: [7758-87-4]

DESCRIPTION

Tribasic Calcium Phosphate consists of a variable mixture of calcium phosphates having the approximate composition of 10CaO.3P$_2$O$_5$.H$_2$O. It occurs as a white, odorless, tasteless powder that is stable in air. It is insoluble in alcohol and almost insoluble in water, but it dissolves readily in dilute hydrochloric and nitric acids.

Functional Use in Foods Anticaking agent; buffer; nutrient; dietary supplement; clouding agent.

REQUIREMENTS

Identification
A. To a warm solution of the sample in a slight excess of nitric acid add ammonium molybdate TS. A yellow precipitate forms.
B. Dissolve about 100 mg by warming with 5 mL of 2.7 N hydrochloric acid and 5 mL of water, add 1 mL of 6 N ammonium hydroxide, dropwise, with shaking, and then add 5 mL of ammonium oxalate TS. A white precipitate forms.
Assay Not less than 34.0% and not more than 40.0% of calcium (Ca).
Arsenic (as As) Not more than 3 mg/kg.
Fluoride Not more than 0.0075%.
Heavy Metals (as Pb) Not more than 0.0015%.
Lead Not more than 5 mg/kg.
Loss on Ignition Not more than 10.0%.

TESTS

Assay Proceed as directed in the *Assay* in the monograph for *Dibasic Calcium Phosphate*, using a 150-mg sample, accurately weighed. Each mL of 0.05 M disodium EDTA is equivalent to 2.004 mg of Ca.
Arsenic A solution of 1 g in 25 mL of 2.7 N hydrochloric acid meets the requirements of the *Arsenic Test*, Appendix IIIB.
Fluoride (Note: Prepare and store all solutions in plastic containers.)
Buffer Solution, *Standard Solution*, and *Electrode System* Proceed as directed under *Fluoride* in the monograph for *Dibasic Calcium Phosphate*.
Standard Response Line Proceed as directed in the test for *Fluoride* in the monograph for *Calcium Phosphate, Dibasic*, except use 3.0 mL of hydrochloric acid instead of 2.0 mL.
Procedure Proceed as directed under *Fluoride* in the monograph for *Dibasic Calcium Phosphate*, except use 3.0 mL of hydrochloric acid instead of 2.0 mL.
Heavy Metals Warm 2.66 g with 7 mL of 2.7 N hydrochloric acid until no more dissolves, dilute to 50 mL with water, and filter. A 25-mL portion of the filtrate meets the requirements

of the *Heavy Metals Test*, Appendix IIIB, using 20 μg of lead ion (Pb) in the control (*Solution A*).

Note: Filter the mixture after pH adjustment.

Lead A 10-g sample meets the requirements of the *APDC Extraction Method for Lead*, Appendix IIIB.

Loss on Ignition Weigh accurately about 3 g, and ignite, preferably in a muffle furnace, at 800° to 825° to constant weight.

Packaging and Storage Store in well-closed containers.

Calcium Propionate

$(CH_3CH_2COO)_2Ca$

$C_6H_{10}CaO_4$ Formula wt 186.22

INS: 282 CAS: [4075-81-4]

DESCRIPTION

White crystals or crystalline solid, possessing not more than a faint odor of propionic acid. One g dissolves in about 3 mL of water. The pH of a 1 in 10 solution is between 8 and 10.

Functional Use in Foods Preservative; mold inhibitor.

REQUIREMENTS

Identification
  A. A 1 in 20 solution gives positive tests for *Calcium*, Appendix IIIA.
  B. Upon ignition at a relatively low temperature, it yields an alkaline residue that effervesces with acids.

Assay Not less than 98.0% and not more than 100.5% of $C_6H_{10}CaO_4$, calculated on the anhydrous basis.

Fluoride Not more than 0.003%.

Heavy Metals (as Pb) Not more than 10 mg/kg.

Insoluble Substances Not more than 0.2%.

Magnesium (as MgO) Passes test (about 0.4%).

Water Not more than 5.0%.

TESTS

Assay Dissolve about 400 mg, accurately weighed, in 100 mL of water. While stirring, preferably with a magnetic stirrer, add about 30 mL of 0.05 M disodium EDTA from a 50-mL buret, then add 15 mL of 1 N sodium hydroxide and 300 mg of hydroxy naphthol blue indicator, and continue the titration to a blue endpoint. Each mL of 0.05 M disodium EDTA is equivalent to 9.311 mg of $C_6H_{10}CaO_4$.

Fluoride Proceed as directed in the *Fluoride Limit Test*, Appendix IIIB, using *Method III* (1.0-g sample).

Heavy Metals A solution of 2 g in 25 mL of water meets the requirements of the *Heavy Metals Test*, Appendix IIIB, using 20 μg of lead ion (Pb) in the control (*Solution A*).

Insoluble Substances Dissolve 10 g in 100 mL of hot water, filter through a tared filtering crucible, wash the insoluble residue with hot water, and dry at 105° to constant weight.

Magnesium Place 400.0 mg of the sample, 5 mL of 2.7 N hydrochloric acid, and about 10 mL of water in a small beaker, and dissolve the sample by heating on a hot plate. Evaporate the solution to a volume of about 2 mL, and cool. Transfer the residual liquid into a 100-mL volumetric flask, dilute to volume with water, and mix. Dilute 7.5 mL of this solution to 20 mL with water, add 2 mL of 1 N sodium hydroxide and 0.05 mL of a 1 in 1000 solution of Titan yellow (Clayton yellow), mix, allow to stand for 10 min, and shake. Any color does not exceed that produced by 1.0 mL of *Magnesium Standard Solution* (50 μg Mg ion) in the same volume of a control containing 2.5 mL of the sample solution (10-mg sample) and the quantities of the reagents used in the test.

Water Determine by the *Karl Fischer Titrimetric Method*, Appendix IIB.

Packaging and Storage Store in tight containers.

Calcium Pyrophosphate

$Ca_2P_2O_7$ Formula wt 254.10

INS: 450(vi) CAS: [7790-76-3]

DESCRIPTION

A fine, white, odorless and tasteless powder. It is insoluble in water, but is soluble in dilute hydrochloric and nitric acids.

Functional Use in Foods Buffer; neutralizing agent; nutrient; dietary supplement.

REQUIREMENTS

Identification
  A. Dissolve about 100 mg by warming with a mixture of 5 mL of 2.7 N hydrochloric acid and 5 mL of water, add 2.5 mL of 6 N ammonium hydroxide, dropwise, with shaking, and then add 5 mL of ammonium oxalate TS. A white precipitate is formed.
  B. Dissolve 100 mg of the sample in 100 mL of 1.7 N nitric acid. Add 0.5 mL of this solution to 30 mL of quimociac TS. A yellow precipitate does not form. Heat the remaining portion of the sample solution for 10 min at 95°, and then add 0.5 mL of the solution to 30 mL of quimociac TS. A yellow precipitate forms immediately.

Assay Not less than 96.0% of $Ca_2P_2O_7$.

Arsenic (as As) Not more than 3 mg/kg.

Fluoride Not more than 0.005%.

Heavy Metals (as Pb)   Not more than 0.0015%.
Lead   Not more than 2 mg/kg.
Loss on Ignition   Not more than 1%.

TESTS

Assay   Dissolve about 300 mg, accurately weighed, in 10 mL of 2.7 N hydrochloric acid, add about 120 mL of water and a few drops of methyl orange TS, and boil for 30 min, keeping the volume and pH of the solution constant during the boiling period by adding hydrochloric acid or water, if necessary. Add 2 drops of methyl red TS and 30 mL of ammonium oxalate TS, then add dropwise, with constant stirring, a mixture of equal volumes of 6 N ammonium hydroxide and water until the pink color of the indicator just disappears. Digest on a steam bath for 30 min, cool to room temperature, allow the precipitate to settle, and filter the supernatant liquid through a sintered-glass crucible, using gentle suction. Wash the precipitate in the beaker with about 30 mL of cold (below 20°) wash solution, prepared by diluting 10 mL of ammonium oxalate TS to 1000 mL. Allow the precipitate to settle, and pour the supernatant liquid through the filter. Repeat this washing by decantation three more times. Using the wash solution, transfer the precipitate as completely as possible to the filter. Finally, wash the beaker and the filter with two 10-mL portions of cold (below 20°) water. Place the sintered-glass crucible in the beaker, and add 100 mL of water and 50 mL of cold dilute sulfuric acid (1 in 6). Add from a buret 35 mL of 0.1 N potassium permanganate, and stir until the color disappears. Heat to about 70°, and complete the titration with 0.1 N potassium permanganate. Each mL of 0.1 N potassium permanganate is equivalent to 6.35 mg of $Ca_2P_2O_7$.

Arsenic   A solution of 1 g in 5 mL of 2.7 N hydrochloric acid meets the requirements of the *Arsenic Test*, Appendix IIIB.

Fluoride   Weigh accurately 1.0 g, and proceed as directed in the *Fluoride Limit Test*, Appendix IIIB.

Heavy Metals   Warm 2.66 g with 7 mL of 2.7 N hydrochloric acid until no more dissolves, dilute to 50 mL with water, and filter. A 25-mL portion of the filtrate meets the requirements of the *Heavy Metals Test*, Appendix IIIB, using 20 µg of lead ion (Pb) in the control (*Solution A*).

Lead   A 10-g sample meets the requirements of the *APDC Extraction Method for Lead*, Appendix IIIB.

Loss on Ignition   Weigh accurately about 1 g, and ignite, preferably in a muffle furnace, at 800° to 825° for 30 min.

Packaging and Storage   Store in well-closed containers.

Calcium Saccharin

1,2-Benzisothiazolin-3-one 1,1-Dioxide Calcium Salt

$$\left[\begin{array}{c}\text{benzisothiazoline structure with } N^-, SO_2\end{array}\right]_2 Ca \cdot 3\frac{1}{2} H_2O$$

$C_{14}H_8CaN_2O_6S_2 \cdot 3\frac{1}{2}H_2O$ Formula wt 467.49

INS: 954   CAS: anhydrous [6485-34-3]

DESCRIPTION

White crystals or a white, crystalline powder. It is odorless or has a faint, aromatic odor. It is intensely sweet even in dilute solutions. One g is soluble in 1.5 mL of water.

Functional Use in Foods   Nonnutritive sweetener.

REQUIREMENTS

Identification

A. Dissolve about 100 mg in 5 mL of sodium hydroxide solution (1 in 20), evaporate to dryness, and gently fuse the residue over a small flame until it no longer evolves ammonia. After the residue has cooled, dissolve it in 20 mL of water, neutralize the solution with 2.7 N hydrochloric acid, and filter. The addition of a drop of ferric chloride TS to the filtrate produces a violet color.

B. Mix 20 mg with 40 mg of resorcinol, add 10 drops of sulfuric acid, and heat the mixture in a liquid bath at 200° for 3 min. After cooling, add 10 mL of water and an excess of 1 N sodium hydroxide. A fluorescent green liquid results.

C. A 1 in 10 solution gives positive tests for *Calcium*, Appendix IIIA.

D. To 10 mL of a 1 in 10 solution add 1 mL of hydrochloric acid. A crystalline precipitate of saccharin is formed. Wash the precipitate well with cold water, and dry at 105° for 2 hr. It melts between 226° and 230° (*Class Ia*, Appendix IIB).

Assay   Not less than 98.0% and not more than 101.0% of $C_{14}H_8CaN_2O_6S_2$, calculated on the anhydrous basis.

Benzoate and Salicylate   Passes test.

Heavy Metals (as Pb)   Not more than 10 mg/kg.

Readily Carbonizable Substances   Passes test.

Selenium   Not more than 0.003%.

Toluenesulfonamides   Not more than 0.0025%.

Water   Not more than 15.0%.

TESTS

Assay   Weigh accurately about 500 mg, and transfer it quantitatively to a separator with the aid of 10 mL of water. Add 2 mL of 2.7 N hydrochloric acid, and extract the precipitated saccharin first with 30 mL, then with five 20-mL portions, of a solvent composed of 9 volumes of chloroform and 1 volume of alcohol. Filter each extract through a small filter paper moistened with the solvent mixture, and evaporate the combined filtrates on a steam bath to dryness with the aid of a current of air. Dissolve the residue in 75 mL of hot water, cool, add phenolphthalein TS, and titrate with 0.1 N sodium hydroxide. Perform a blank determination, and make any necessary correction (see *General Provisions*). Each mL of 0.1 N sodium hydroxide is equivalent to 20.22 mg of $C_{14}H_8CaN_2O_6S_2$.

Benzoate and Salicylate To 10 mL of a 1 in 20 solution previously acidified with 5 drops of glacial acetic acid, add 3 drops of ferric chloride TS. No precipitate or violet color appears.

Heavy Metals Prepare and test a 2-g sample as directed in *Method II* under the *Heavy Metals Test*, Appendix IIIB, using 20 µg of lead ion (Pb) in the control (*Solution A*).

Readily Carbonizable Substances, Appendix IIB Dissolve 200 mg in 5 mL of 95% sulfuric acid, and keep at a temperature of 48° to 50° for 10 min. The color is no darker than *Matching Fluid A*.

Selenium Determine as directed in *Method I* under the *Selenium Limit Test*, Appendix IIIB, using 200 mg of sample.

Toluenesulfonamides Determine as directed in the monograph for *Sodium Saccharin*.

Water Determine by the *Karl Fischer Titrimetric Method*, Appendix IIB.

Packaging and Storage Store in well-closed containers.

Calcium Silicate

INS: 552 CAS: [1344-95-2]

DESCRIPTION

A hydrous or anhydrous silicate with varying proportions of CaO and SiO_2. It occurs as a white to off-white free-flowing powder that remains so after absorbing relatively large amounts of water or other liquids. It is insoluble in water, but forms a gel with mineral acids. The pH of a 1 in 20 aqueous slurry is between 8.4 and 12.5.

Functional Use in Foods Anticaking agent; filter aid.

REQUIREMENTS

Identification
A. Mix about 500 mg with 10 mL of 2.7 N hydrochloric acid, filter, and neutralize the filtrate to litmus paper with 6 N ammonium hydroxide. The neutralized filtrate gives positive tests for *Calcium*, Appendix IIIA.

B. Prepare a bead by fusing a few crystals of sodium ammonium phosphate on a platinum loop in the flame of a Bunsen burner. Place the hot, transparent bead in contact with a sample, and again fuse. Silica floats about in the bead, producing, upon cooling, an opaque bead with a weblike structure.

Assay for Calcium Oxide and Silicon Dioxide Not less than the percentages stated or within the range claimed by the vendor.
Fluoride Not more than 10 mg/kg.
Heavy Metals (as Pb) Not more than 0.002%.
Lead Not more than 5 mg/kg.
Loss on Drying and **Loss on Ignition** Not more than the percentages stated or within the range claimed by the vendor.

TESTS

Assay for Silicon Dioxide Transfer about 400 mg of the sample, accurately weighed, into a beaker, add 5 mL of water and 10 mL of perchloric acid, and heat until dense, white fumes of perchloric acid are evolved. Cover the beaker with a watch glass, and continue to heat for 15 min longer. Allow to cool, add 30 mL of water, filter, and wash the precipitate with 200 mL of hot water. Retain the combined filtrate and washings for use in the *Assay for Calcium Oxide*. Transfer the filter paper and its contents to a platinum crucible, heat slowly to dryness, and then heat sufficiently to char the filter paper. After cooling, add a few drops of sulfuric acid, and then ignite at about 1300° to constant weight. Moisten the residue with 5 drops of sulfuric acid, add 15 mL of hydrofluoric acid, heat cautiously on a hot plate until all of the acid is driven off, and ignite to constant weight at a temperature not lower than 1000°. Cool in a desiccator and weigh. The loss in weight is equivalent to the SiO_2 in the sample taken.

Assay for Calcium Oxide Using 1 N sodium hydroxide, neutralize to litmus the combined filtrate and washings retained in the *Assay for Silicon Dioxide*, and add, while stirring, about 30 mL of 0.05 M disodium EDTA from a 50-mL buret. Add 15 mL of 1 N sodium hydroxide and 300 mg of hydroxy naphthol blue indicator, and continue the titration to a blue endpoint. Each mL of 0.05 M disodium EDTA is equivalent to 2.804 mg of CaO.

Fluoride Prepare a slurry consisting of 5 g of the sample and 45 mL of 0.1 N hydrochloric acid, stir for 15 min at room temperature, and filter through a 0.45-µm membrane filter into a 50-mL volumetric flask. Wash the filter with five 1-mL portions of 0.1 N hydrochloric acid, collecting the washings in the flask, then dilute to volume with 0.1 N hydrochloric acid, and mix. Transfer 5.0 mL of this solution into a 25-mL volumetric flask, add 5.0 mL of a 10% solution of Amadac-F[1] in 60% isopropanol, dilute to volume with water, mix, and allow to stand for 1 h in diffuse light at room temperature. Determine the absorbance of this solution in a 1-cm cell with a suitable spectrophotometer, at the wavelength of maximum absorption at about 620 nm, against a blank consisting of 5.0 mL of 0.1 N hydrochloric acid, 5.0 mL of the Amadac indicator solution, and 15.0 mL of water. The absorbance is not greater than that produced by 5.0 mL of a solution containing 2.21 µg of NaF per mL of 0.1 N hydrochloric acid when treated in the same manner as the sample.

---
[1]Amadac-F is a product of Burdick & Jackson Laboratories, Inc., Muskegon, MI 49442, consisting of a blended solid mixture of partially hydrated sodium acetate, acetic acid, stabilizers, lanthanum nitrate, and 3-amino-methylalizarin-N,N-diacetate (alizarin complexan), the lanthanum and complexan being equimolar.

Sample Solution for the Determination of Heavy Metals and Lead   Transfer 5.0 g of the sample into a 250-mL beaker, add 50 mL of 0.5 N hydrochloric acid, cover with a watch glass, and heat slowly to boiling. Boil gently for 15 min, cool, and let the undissolved material settle. Decant the supernatant liquid through Whatman No. 4, or equivalent, filter paper into a 100-mL volumetric flask, retaining as much as possible of the insoluble material in the beaker. Wash the slurry and beaker with three 10-mL portions of hot water, decanting each washing through the filter paper into the flask. Finally, wash the filter paper with 15 mL of hot water, cool the filtrate to room temperature, dilute to volume with water, and mix.

Heavy Metals   A 20-mL portion of the *Sample Solution* meets the requirements of the *Heavy Metals Test*, Appendix IIIB, using 20 μg of lead ion (Pb) in the control (*Solution A*).

Lead

Lead Nitrate Stock Solution   Dissolve 159.8 mg of ACS reagent-grade lead nitrate, $Pb(NO_3)_2$, in 100 mL of water containing 1 mL of nitric acid, dilute with water to 1000.0 mL, and mix. Each mL of this solution contains 100 μg of lead ion. Prepare and store this solution in glass containers that are free from lead salts.

Standard Lead Solution   On the day of use, dilute stepwise and quantitatively an accurately measured volume of *Lead Nitrate Stock Solution* with water to obtain the *Standard Lead Solution* containing 0.25 μg/mL of lead ion (Pb).

Procedure   Set a suitable atomic absorption spectrophotometer to a wavelength of 217 nm. Adjust the instrument to zero absorbance against water. Read the absorbance of the *Standard Lead Solution* containing 0.25 μg/mL of lead ion.

Aspirate the *Sample Solution* prepared as directed above into the spectrophotometer and measure the absorbance in the same manner. The absorbance obtained from the *Sample Solution* is not greater than that obtained from the *Standard Lead Solution*.

Loss on Drying, Appendix IIC   Dry at 105° for 2 h.

Loss on Ignition   Transfer about 1 g, previously dried at 105° for 2 h and accurately weighed, into a suitable tared crucible, and ignite at 900° to constant weight.

Packaging and Storage   Store in well-closed containers.

Calcium Sorbate

2,4-Hexadienoic Acid, Calcium Salt

$$[CH_3CH=CHCH=CHCOO]_2Ca$$

$C_{12}H_{14}CaO_4$ Formula wt 262.32

INS: 203   CAS: [7492-55-9]

DESCRIPTION

White, fine crystalline powder. It decomposes at about 400°. It is sparingly soluble in water and practically insoluble in organic solvents as well as in fats and in oils.

Functional Use in Foods   Preservative.

REQUIREMENTS

Identification

A. Ignite 1 g at 800°. Cool, and slake with 10 mL of water. Add glacial acetic acid until the sample is dissolved, and filter if necessary. The resultant solution gives positive tests for *Calcium*, Appendix IIIA.

B. Place 200 mg of the sample in 5 mL of methanol. Add 0.1 mL of 1 N sodium hydroxide, and dissolve in 95 mL of water. After addition of a few drops of bromine TS, the color is discharged.

Assay   Not less than 98.0% and not more than 101.0% of $C_{12}H_{14}CaO_4$, calculated on the dried basis.

Acidity (as sorbic acid)   Passes test (approximately 1%).

Alkalinity [as $Ca(OH)_2$]   Passes test (approximately 0.5%).

Heavy Metals (as Pb)   Not more than 10 mg/kg.

Loss on Drying   Not more than 1.0%.

TESTS

Assay   Dissolve about 150 mg of the sample, accurately weighed, in 50 mL of glacial acetic acid in a 250-mL glass-stoppered Erlenmeyer flask, warming if necessary to effect solution. Cool to room temperature, add 2 drops of crystal violet TS, and titrate with 0.1 N perchloric acid in glacial acetic acid to a blue-green endpoint that persists for at least 30 s. Perform a blank determination (see *General Provisions*) and make any necessary correction. Two mL of 0.1 N perchloric acid is equivalent to 26.23 mg of $C_{12}H_{14}CaO_4$.

Acidity or Alkalinity   Add some drops of methanol to 1 g of the sample. Add 30 mL of water and several drops of phenolphthalein TS. If the mixture is colorless, titrate with 0.1 N sodium hydroxide to a pink color that persists for 15 s. Not more than 1.0 mL is required. If the mixture is pink, titrate with 0.1 N hydrochloric acid. Not more than 1.35 mL is required to discharge the pink color.

Heavy Metals   Prepare and test a 2-g sample as directed in *Method II* under the *Heavy Metals Test*, Appendix IIIB, using 20 μg of lead ion (Pb) in the control (*Solution A*).

Loss on Drying, Appendix IIC Dry at 105° for 3 h.

Packaging and Storage Store in tight containers.

Calcium Stearate

CAS: [1592-23-0]

DESCRIPTION

Calcium Stearate is a compound of calcium with a mixture of solid organic acids obtained from edible sources, and consists chiefly of variable proportions of Calcium Stearate and calcium palmitate. It occurs as a fine, white to yellowish white, bulky powder having a slight, characteristic odor. It is unctuous, and is free from grittiness. It is insoluble in water, in alcohol, and in ether.

Functional Use in Foods Anticaking agent; binder; emulsifier.

REQUIREMENTS

Identification
 A. Heat 1 g with a mixture of 25 mL of water and 5 mL of hydrochloric acid. Fatty acids are liberated, floating as an oily layer on the surface of the liquid. The water layer gives positive tests for *Calcium*, Appendix IIIA.
 B. Mix 25 g of the sample with 200 mL of hot water, then add 60 mL of 2 N sulfuric acid, and heat the mixture, with frequent stirring, until the fatty acids separate cleanly as a transparent layer. Wash the fatty acids with boiling water until free from sulfate, collect them in a small beaker, and warm on a steam bath until the water has separated and the fatty acids are clear. Allow the acids to cool, pour off the water layer, then melt the acids, filter into a dry beaker, and dry at 105° for 20 min. The solidification point of the fatty acids so obtained is not below 54° (see Appendix IIB).
Assay Not less than 9.0% and not more than 10.5% of CaO, calculated on the dried basis.
Free Fatty Acid (as stearic acid) Not more than 3.0%.
Heavy Metals (as Pb) Not more than 10 mg/kg.
Loss on Drying Not more than 4.0%.

TESTS

Assay Boil about 1.2 g, accurately weighed, with 50 mL of 0.1 N hydrochloric acid for 10 min, or until the fatty acid layer is clear, adding water if necessary to maintain the original volume. Cool, filter, and wash the filter and flask thoroughly with water until the last washing is not acid to litmus. Neutralize the filtrate to litmus with 1 N sodium hydroxide. While stirring, preferably with a magnetic stirrer, add about 30 mL of 0.05 M disodium EDTA from a 50-mL buret, then add 15 mL of 1 N sodium hydroxide and 300 mg of hydroxy naphthol blue indicator, and continue the titration to a blue endpoint. Each mL of 0.05 M disodium EDTA is equivalent to 2.804 mg of CaO.
Free Fatty Acid Transfer 2 g of the sample, accurately weighed, into a dry 125-mL Erlenmeyer flask containing 50 mL of acetone, fit an air-cooled reflux condenser onto the neck of the flask, boil the mixture on a steam bath for 10 min, and cool. Filter through two layers of Whatman No. 42, or equivalent, filter paper, and wash the flask, residue, and filter with 50 mL of acetone. Add phenolphthalein TS and 5 mL of water to the filtrate, and titrate with 0.1 N sodium hydroxide. Perform a blank determination, using 100 mL of acetone and 5 mL of water (see *General Provisions*). Each mL of 0.1 N sodium hydroxide is equivalent to 28.45 mg of stearic acid ($C_{18}H_{36}O_2$).
Heavy Metals, Appendix IIIB Place 2.5 g of the sample in a porcelain dish, place 500 mg of the sample in a second dish for the control, and to each add 5 mL of a 1 in 4 solution of magnesium nitrate in alcohol. Cover the dishes with 3-in. short-stem funnels so that the stems are straight up. Heat for 30 min on a hot plate at the low setting, then heat for 30 min at the medium setting, and cool. Remove the funnels, add 20 μg of lead ion (Pb) to the control, and heat each dish over an Argand burner until most of the carbon is burned off. Cool, add 10 mL of nitric acid, and transfer the solutions into 250-mL beakers. Add 5 mL of 70% perchloric acid, evaporate to dryness, then add 2 mL of hydrochloric acid to the residues, and wash down the inside of the beakers with water. Evaporate carefully to dryness again, swirling near the dry point to avoid spattering. Repeat the hydrochloric acid treatment, then cool, and dissolve the residues in about 10 mL of water. To each solution add 1 drop of phenolphthalein TS and sufficient 1 N sodium hydroxide until the solutions just turn pink, and then add 2.7 N hydrochloric acid until the solutions become colorless. Add 1 mL of 1 N acetic acid and a small amount of charcoal to each solution, and filter through Whatman No. 2, or equivalent, filter paper into 50-mL Nessler tubes. Wash with water, dilute to 40 mL, and add 10 mL of hydrogen sulfide TS to each tube. The color in the solution of the sample does not exceed that produced in the control.
Loss on Drying, Appendix IIC Dry at 105° to constant weight, using 2-h increments of heating.

Packaging and Storage Store in well-closed containers.

Calcium Stearoyl Lactylate

INS: 482(i) CAS: [5793-94-2]

DESCRIPTION

A mixture of calcium salts of stearoyl lactic acid, with minor proportions of other salts of related acids. It occurs as a cream-colored powder having a mild, caramel-like odor. It is slightly soluble in hot water.

Functional Use in Foods Dough conditioner; stabilizer; whipping agent.

REQUIREMENTS

Identification Calcium Stearoyl Lactylate responds to the tests for *Identification* in the monograph for *Calcium Stearate*.
Acid Value Between 50 and 86.
Calcium Content Between 4.2% and 5.2%.
Ester Value Between 125 and 164.
Heavy Metals (as Pb) Not more than 10 mg/kg.
Total Lactic Acid Between 32.0% and 38.0%.

TESTS

Acid Value Transfer about 1 g, accurately weighed, to a 125-mL volumetric flask, add 25 mL of alcohol, previously neutralized in phenolphthalein TS, and heat on a hot plate until the sample is dissolved. Cool, add 5 drops of phenolphthalein TS, and titrate rapidly with 0.1 N sodium hydroxide to the first pink color that persists for at least 30 s. Calculate the acid value by the formula

$$56.1 V \times N/W,$$

in which V is the volume, in mL, and N is the normality of the sodium hydroxide solution, and W is the weight, in g, of the sample taken. Retain the neutralized solution for the determination of *Ester Value*.

Calcium Content

Stock Lanthanum Solution Transfer 5.86 g of lanthanum oxide, La_2O_3, into a 100-mL volumetric flask, wet with a few mL of water, slowly add 25 mL of hydrochloric acid, and swirl until the material is completely dissolved. Dilute to volume with water, and mix.

Stock Calcium Solution Use a solution containing 0.5 mg of Ca in each mL (500 mg/kg Ca). The solution may be obtained commercially or prepared as follows: Transfer 124.8 mg of calcium carbonate, $CaCO_3$, previously dried at 200° for 4 h, into a 100-mL volumetric flask, carefully dissolve in 2 mL of 2.7 N hydrochloric acid, dilute to volume with water, and mix.

Standard Preparations Transfer 10.0 mL of the *Stock Lanthanum Solution* into each of three 50-mL volumetric flasks. Using a microliter syringe, transfer 0.20 mL of the *Stock Calcium Solution* into the first flask, 0.40 mL into the second flask, and 0.50 mL into the third flask. Dilute each flask to volume with water, and mix. The flasks contain 2.0, 4.0, and 5.0 µg of Ca per mL, respectively. Prepare these solutions fresh daily.

Sample Preparation Transfer about 250 mg of the sample, accurately weighed, into a 30-mL beaker, dissolve with heating in 10 mL of alcohol, and quantitatively transfer the solution into a 25-mL volumetric flask. Wash the beaker with two 5-mL portions of alcohol, adding the washings to the flask, dilute to volume with alcohol, and mix. Transfer 5.0 mL of the *Stock Lanthanum Solution* to a second 25-mL volumetric flask. Using a microliter syringe, transfer 0.25 mL of the alcoholic solution of the sample to the second flask, dilute to volume with water, and mix.

Procedure Concomitantly determine the absorbance of each *Standard Preparation* and of the *Sample Preparation* at 422.7 nm, with a suitable atomic absorption spectrophotometer, following the operating parameters as recommended by the manufacturer of the instrument. Plot the absorbance of the *Standard Preparations* versus concentration of Ca, in µg/mL, and from the curve so obtained determine the concentration, C, in µg/mL, of Ca in the *Sample Preparation*. Calculate the quantity, in mg, of Ca in the sample taken by the formula $2.5C$.

Ester Value To the neutralized solution retained in the test for *Acid Value* add 10.0 mL of alcoholic potassium hydroxide solution prepared by dissolving 11.2 g of potassium hydroxide in 250 mL of alcohol and diluting with 25 mL of water. Add 5 drops of phenolphthalein TS, connect a suitable condenser, and reflux for 2 h. Cool, add 5 additional drops of phenolphthalein TS, and titrate the excess alkali with 0.1 N sulfuric acid. Perform a blank determination using 10.0 mL of the alcoholic potassium hydroxide solution. Calculate the ester value by the formula

$$56.1(B - S)N/W,$$

in which $B - S$ represents the difference between the volumes of 0.1 N sulfuric acid required for the blank and the sample, respectively, N is the normality of the sulfuric acid, and W is the weight, in g, of the sample taken.

Heavy Metals Prepare and test a 2-g sample as directed in *Method II* under the *Heavy Metals Test*, Appendix IIIB, using 20 µg of lead ion (Pb) in the control (*Solution A*).

Total Lactic Acid

Standard Curve Dissolve 1.067 g of lithium lactate in sufficient water to make 1000.0 mL. Transfer 10.0 mL of this solution into a 100-mL volumetric flask, dilute to volume with water, and mix. Transfer 1.0, 2.0, 4.0, 6.0, and 8.0 mL of the diluted standard solution into separate 100-mL volumetric flasks, dilute each flask to volume with water, and mix. These standards represent 1, 2, 4, 6, and 8 µg of lactic acid per mL, respectively. Transfer 1.0 mL of each solution into separate test tubes, and continue as directed in the *Procedure*, beginning with "Add 1 drop of cupric sulfate TS ...." After color development and reading the absorbance values, construct a *Standard Curve* by plotting absorbance versus µg of lactic acid.

Test Preparation Transfer about 200 mg of the sample, accurately weighed, into a 125-mL Erlenmeyer flask, add 10 mL of 0.5 N alcoholic potassium hydroxide and 10 mL of water, attach an air condenser, and reflux gently for 45 min. Wash the sides of the flask and the condenser with about 40 mL of water, and heat on a steam bath until no odor of alcohol remains. Add 6 mL of dilute sulfuric acid (1 in 2), heat until the fatty acids are melted, then cool to about 60°, and add 25 mL of petroleum ether. Swirl the mixture gently, and transfer quantitatively to a separator. Collect the water layer in a 100-mL volumetric flask, and wash the petroleum ether layer with two 20-mL portions of water, adding the washings to the volumetric flask. Dilute to volume with water, and mix. Transfer 1.0 mL of this solution into a second 100-mL volumetric flask, dilute to volume with water, and mix.

Procedure Transfer 1.0 mL of the *Test Preparation* into a test tube, and transfer 1.0 mL of water to a second test tube to serve as the blank. Treat each tube as follows: Add 1 drop of cupric sulfate TS, swirl gently, and then add rapidly from a buret 9.0 mL of sulfuric acid. Loosely stopper the tube, and heat in a water bath at 90° for exactly 5 min. Cool immediately to below 20° in an ice bath for 5 min, add 3 drops of *p*-phenylphenol TS, shake immediately, and heat in a water bath at 30° for 30 min, shaking the tube twice during this time to disperse the reagent. Heat the tube in a water bath at 90° for exactly 90 s, and then cool immediately to room temperature in an ice water bath. Determine the absorbance of the solution in a 1-cm cell, at 570 nm, with a suitable spectrophotometer, using the blank to set the instrument. Obtain the weight, in μg, of lactic acid in the portion of the *Test Preparation* taken for the *Procedure* by means of the *Standard Curve*.

Packaging and Storage   Store in tight containers in a cool, dry place.

Calcium Sulfate

$CaSO_4 \cdot xH_2O$ Formula wt, anhydrous 136.14
INS: 516   CAS: anhydrous [7778-18-9]

DESCRIPTION

Calcium Sulfate is anhydrous or contains two molecules of water of hydration. It occurs as a fine, white to slightly yellow-white, odorless powder.

Functional Use in Foods   Nutrient; dietary supplement; yeast food; dough conditioner; firming agent; sequestrant.

REQUIREMENTS

Identification   Dissolve about 200 mg by warming with a mixture of 4 mL of 2.7 *N* hydrochloric acid and 16 mL of water. A white precipitate forms when 5 mL of ammonium oxalate TS is added to 10 mL of the solution. Upon the addition of barium chloride TS to the remaining 10 mL, a white precipitate forms that is insoluble in hydrochloric and nitric acids.
Assay   Not less than 98.0% of $CaSO_4$, calculated on the dried basis.
Fluoride   Not more than 0.003%.
Heavy Metals (as Pb)   Not more than 10 mg/kg.
Loss on Drying $CaSO_4$ (anhydrous): not more than 1.5%; $CaSO_4 \cdot 2H_2O$ (dihydrate): between 19.0% and 23.0%.
Selenium   Not more than 0.003%.

TESTS

Assay   Dissolve 250 mg, accurately weighed, in 100 mL of water and 4 mL of 2.7 *N* hydrochloric acid, boil to effect solution, and cool. While stirring, preferably with a magnetic stirrer, add about 30 mL of 0.05 *M* disodium EDTA from a 50-mL buret, then add 25 mL of 1 *N* sodium hydroxide and 300 mg of hydroxy naphthol blue indicator, and continue the titration to a blue endpoint. Each mL of 0.05 *M* disodium EDTA is equivalent to 6.807 mg of $CaSO_4$.

Fluoride   Weigh accurately 1.67 g, and proceed as directed in the *Fluoride Limit Test*, Appendix IIIB.
Heavy Metals   Mix 2 g with 20 mL of water, add 25 mL of 2.7 *N* hydrochloric acid, and heat to boiling to dissolve the sample. Cool, and add ammonium hydroxide to a pH of 7. Filter, evaporate to a volume of about 25 mL, and refilter if necessary to obtain a clear solution. This solution meets the requirements of the *Heavy Metals Test*, Appendix IIIB, using 20 μg of lead ion (Pb) in the control (*Solution A*).
Loss on Drying, Appendix IIC   Dry at 250° to constant weight.
Selenium   Determine as directed in *Method II* under the *Selenium Limit Test*, Appendix IIIB, using 200 mg of sample.

Packaging and Storage   Store in well-closed containers.

Cananga Oil

CAS: [68606-83-7]

DESCRIPTION

The oil obtained by distillation from the flowers of the tree *Cananga odorata* Hook f. et Thoms., (Fam. *Anonaceae*). It is a light to deep yellow liquid having a harsh floral odor suggestive of ylang ylang. It is soluble in most fixed oils and in mineral oil, but it is practically insoluble in glycerin and in propylene glycol.

Functional Use in Foods   Flavoring agent.

REQUIREMENTS

Identification   The infrared absorption spectrum of the sample exhibits relative maxima (that may vary in intensity) at the same wavelengths (or frequencies) as those shown in the respective spectrum in the section on *Infrared Spectra*, (*Series A: Essential Oils*), using the same test conditions as specified therein.
Angular Rotation   Between −15° and −30°.
Heavy Metals (as Pb)   Passes test.
Refractive Index   Between 1.495 and 1.505 at 20°.
Saponification Value   Between 10 and 40.
Solubility in Alcohol   Passes test.
Specific Gravity   Between 0.904 and 0.920.

TESTS

Angular Rotation   Determine in a 100-mm tube as directed under *Optical (Specific) Rotation*, Appendix IIB.
Heavy Metals   Shake 10 mL of the oil with an equal volume of water to which 1 drop of hydrochloric acid has been added, and pass hydrogen sulfide through the mixture until it is saturated. No darkening in color is produced in either the oil or the water.
Refractive Index, Appendix IIB   Determine with an Abbé or other refractometer of equal or greater accuracy.

Saponification Value Determine as directed for *Saponification Value* under *Essential Oils and Flavors*, Appendix VI, using about 5 g, accurately weighed.
Solubility in Alcohol Proceed as directed in the general method, Appendix VI. One mL dissolves in 0.5 mL of 95% alcohol, usually becoming cloudy on further dilution.
Specific Gravity Determine by any reliable method (see *General Provisions*).

Packaging and Storage Store in full, tight, preferably glass, aluminum, tin-lined, or other suitably lined containers in a cool place protected from light.

Candelilla Wax

INS: 902　　　　　　　　　　　　　　　CAS: [8006-44-8]

DESCRIPTION

A purified wax obtained from the leaves of the candelilla plant, *Euphorbia antisyphilitica*. It is a hard, yellowish brown, opaque to translucent wax. Its specific gravity is about 0.983. It is soluble in chloroform and in toluene, but is insoluble in water.

Functional Use in Foods Masticatory substance in chewing gum base; surface-finishing agent.

REQUIREMENTS

Identification Identify Candelilla Wax by comparing its infrared absorption spectrum with a typical spectrum as shown in the Section on *Infrared Spectra (Series C: Other Substances)*. The sample is melted and prepared for analysis on a potassium bromide plate.
Acid Value Between 12 and 22.
Heavy Metals (as Pb) Not more than 0.002%.
Lead Not more than 3 mg/kg.
Melting Range Between 68.5° and 72.5°.
Saponification Value Between 43 and 65.

TESTS

Acid Value Determine as directed for *Acid Value, Method I*, under *Fats and Related Substances*, Appendix VII.
Heavy Metals Prepare and test a 1-g sample as directed in *Method II* under the *Heavy Metals Test*, Appendix IIIB, using 20 µg of lead ion (Pb) in the control (*Solution A*).
Lead Prepare a *Sample Solution* as directed in the general method under *Chewing Gum Base*, Appendix IV. This solution meets the requirements of the *Lead Limit Test*, Appendix IIIB, using 10 µg of lead ion (Pb) in the control.
Melting Range Determine as directed for *Melting Range or Temperature, Procedure for Class II*, Appendix IIB.
Saponification Value Determine as directed for *Saponification Value* under *Fats and Related Substances*, Appendix VII.

Packaging and Storage Store in well-closed containers.

Canola Oil

Low Erucic Acid Rapeseed Oil; LEAR

DESCRIPTION

A light yellow oil, typically obtained by a combination of mechanical expression followed by *n*-hexane extraction, from the seed of the plant *Brassica napus* or *Brassica campestris* of the family *Cruciferae*. The plant varieties are those producing oil-bearing seeds with a low erucic acid ($C_{22:1}$) content. It is a mixture of triglycerides composed of both saturated and unsaturated fatty acids. It is refined, bleached, and deodorized to substantially remove free fatty acids; phospholipids; color; odor and flavor components; and miscellaneous, other non-oil materials. It can be hydrogenated to reduce the level of unsaturated fatty acids for functional purposes in foods. It is a liquid at 0° and above.

Functional Use in Foods Coating agent; emulsifying agent; formulation aid; texturizer.

REQUIREMENTS

Labeling Hydrogenated Canola Oil less than fully hydrogenated must be labeled as Partially Hydrogenated Canola Oil.
Identification Unhydrogenated Canola Oil exhibits the following composition profile of fatty acids as determined under *Fatty Acid Composition*, Appendix VII.

Fatty Acid: Weight % (Range):	<14	14:0	16:0	16:1	18:0	18:1	18:2
	<0.1	<0.2	<6.0	<1.0	<2.5	>50	<40.0

Fatty Acid: Weight % (Range):	18:3	20:0	20:1	22:0	22:1	24:0	24:1
	<14	<1.0	<2.0	<0.5	<2.0	<0.2	≤0.2

Acid Value Not more than 6.
Cold Test Passes test.
Color (AOCS-Wesson) Not more than 1.5 red/15 yellow.
Erucic Acid Not more than 2.0%.
Free Fatty Acids (as oleic acid) Not more than 0.05%.
Heavy Metals (as Pb) Not more than 5 mg/kg.
Iodine Value Between 110 and 126.
Lead Not more than 0.1 mg/kg.
Linolenic Acid Not more than 14.0%.
Peroxide Value Not more than 10 meq/kg.
Refractive Index Between 1.465 and 1.467 at 40°.
Saponifiable Value Between 178 and 193.

Stability Not less than 7 h.
Sulfur Not more than 10 mg/kg.
Unsaponifiable Matter Not more than 1.5%.
Water Not more than 0.1%.

TESTS

Acid Value Determine as directed for *Acid Value, Method II*, under *Fats and Related Substances*, Appendix VII.
Cold Test Proceed as directed under *Cold Test*, Appendix VII.
Color Proceed as directed for *Color* (AOCS-Wesson) under *Fats and Related Substances*, Appendix VII. Use a 133.4-mm cell.
Erucic Acid Determine as part of *Fatty Acid Composition*, Appendix VII.
Free Fatty Acids Proceed as directed under *Free Fatty Acids*, Appendix VII, using the following equivalence factor (*e*) in the formula given in the procedure:

Free fatty acids as oleic acid, $e = 28.2$.

Heavy Metals Prepare and test a 2-g sample as directed in *Method II* under the *Heavy Metals Test*, Appendix IIIB, using 10 μg of lead ion (Pb) in the control (*Solution A*).
Iodine Value Proceed as directed under *Modified Wijs Method*, Appendix VII.
Lead Determine as directed under *Method II* in the *Atomic Absorption Spectrophotometric Graphite Furnace Method* under the *Lead Limit Test*, Appendix IIIB, using a 1-g sample.
Linolenic Acid Proceed as directed under *Fatty Acid Composition*, Appendix VII.
Peroxide Value Proceed as directed under *Peroxide Value* in the monograph for *Hydroxylated Lecithin*. However, after the addition of saturated potassium iodide and mixing, mix the solution for only 1 min and begin the titration immediately instead of allowing the solution to stand for 10 min.
Refractive Index, Appendix IIB Determine with an Abbé or other refractometer of equal or greater accuracy.
Saponifiable Value Determine as directed under the general method, Appendix VII.
Stability Proceed as directed under *Stability*, Appendix VII.
Sulfur Organosulfur compounds present in the sample react with Raney nickel to produce nickel sulfides. Nickel sulfides are treated with a strong acid to produce hydrogen sulfide, which is trapped and titrated with mercuric acetate using a dithizone indicator.

Caution: This test requires the use of the following hazardous substances: mercuric acetate, spongy nickel, and dibenzyl disulfide. Conduct the test in a fume hood.

Apparatus Fit a 125-mL round-bottom boiling flask with a cylindrical filling funnel (20 mL with open top), an ST PTFE metering valve stopcock, and a gas inlet tube (see the figure for Raney Nickel Reduction Apparatus in Appendix IIIC under *Sulfur (by Oxidative Microcoulometry)*). On top of the boiling flask, fit a water-jacketed distillation column with hooks. To the distillation column, fit a piece of glass tubing with ground ST inner joints with hooks, and connect the distillation column and a gas dispersion tube with ST outer joints with hooks.

Dibenzyl Disulfide Solution Accurately weigh 0.75 g of dibenzyl disulfide and place in a 250-mL volumetric flask. Dilute to volume with methyl isobutyl ketone, and mix.

Sulfur Standard Accurately weigh five 250.0-g samples of food-grade peanut oil. Transfer 0.0, 1.0, 2.0, 3.0, and 4.0 mL of the *Dibenzyl Disulfide Solution* into the peanut oil samples; the samples contain 0, 3, 6, 9, and 12 mg/kg sulfur, respectively.

Raney Nickel Preparation (**Caution**: Raney nickel is pyrophoric when dry.) Raney nickel is produced by reacting nickel–aluminum alloy with sodium hydroxide. Weigh accurately 1 g of nickel–aluminum alloy powder (50% Ni, 50% Al), place it in a 50-mL centrifuge tube, and chill it in an ice bath. Each pellet is enough catalyst for one determination. Slowly add 5 mL of water per tube, and let the tube stand for 10 min. Then, slowly add 10 mL of 2.5 N sodium hydroxide, and allow the mixture to react for 30 min. Cap the tubes, and place them in a 50° water bath for 2 h. Centrifuge the mixture at 1000 rpm for 10 min, and discard the supernatant liquid. Wash the pellets twice with 15 mL of water and twice with 15 mL of isopropanol, centrifuging between each wash. The catalyst may be stored under isopropanol for a period no longer than 2 weeks.

Note: Properly dispose of the unused *Raney Nickel Preparation* by transferring it to a 250-mL Erlenmeyer flask, and placing it in a fume hood. Add 20 mL of 60% (w/v) hydrochloric acid, and allow complete digestion of the catalyst.

Caution: Hydrogen gas is evolved during the digestion process.

Dithizone Indicator Solution Dissolve 10 mg of dithizone (diphenylthiocarbazone) in a 10-mL volumetric flask with acetone.

Mercuric Acetate Titrant (**Note**: Mercuric acetate is a strong irritant when ingested or inhaled or upon dermal exposure.) Transfer 3.82 g of mercuric acetate into a 1000-mL volumetric flask containing 950 mL of water. Add 12.2 mL of glacial acetic acid, dilute to volume with water, and mix. Transfer 10.0 mL of this solution into a 100-mL volumetric flask, dilute to volume with water, and mix. The titrant solution contains 0.0012 M mercuric acetate.

Titration Reagent Blank Add 50.0 mL of 1 N sodium hydroxide and 50.0 mL of acetone to a 250-mL beaker, and mix. Add 0.5 mL of the *Dithizone Indicator Solution*, and titrate with *Mercuric Acetate Titrant* until the color changes from bright amber to strawberry red. Record the volume of titrant used.

Procedure Test a representative portion of the sample. Accurately weigh 15 to 20 g of the sample, and place it on the bottom of the boiling flask. Discard the isopropanol from the *Raney Nickel Preparation*, add 10 mL of 95% isopropanol, mix, and place the mixture in the sample oil. Attach the water condenser and the nitrogen line to the boiling flask and adjust the gas flow to 4 psi through the sample. Place a heating mantle under the flask. Immerse the bubbler in a 250-mL beaker containing 50.0 mL of 1 N sodium hydroxide, and stir slowly. Boil the sample for 90 min. Add 50 mL of acetone and 0.5 mL of *Dithizone Indicator Solution* to the 250-mL beaker. Add 20 mL of 60% hydrochloric acid into the filling funnel fitted onto the boiling flask, and adjust the nitrogen flow to 2 to 3 psi. Position the stir bar directly under the bubbler for maximum dispersion of the hydrogen sulfide bubbles. Slowly add the solution of 60% hydrochloric acid to the boiling flask. Begin the titration with *Mercuric Acetate Titrant* until the bright amber color changes to strawberry red. Add enough hydrochloric acid to turn the solution in the boiling flask green, and then let it boil for 15 min. Continue the titration throughout the boiling stage, making sure to rinse the inside of the bubbler with the solution in the beaker by turning off the nitrogen flow until the solution rises to the top of the vertical tube. Rinse the tube a second time (the solution usually returns to amber during the first rinse). Continue the titration and record the volume of titrant used to the nearest 0.01 mL.

Calculation The concentration of sulfur in the sample, in mg/kg, is calculated by the following formula:

$$(V_S - V_B) \times K/W,$$

in which V_S is the volume, in mL, of titrant to the endpoint for the sample; V_B is the volume, in mL, of titrant to the endpoint for the blank (usually about 0.10 mL); K is a constant determined from the calibration of the *Sulfur Standard* (expressed as μg sulfur per mL titrant); and W is the weight, in g, of the sample.

The *Sulfur Standards* are analyzed, in duplicate, to determine the constant, K, and are calculated by the following formula:

$$K = W \times C/(V_S - V_B),$$

where W is the weight, in g, of the *Sulfur Standard*; C is the concentration, in mg/kg, of the *Sulfur Standard*; V_S is the volume, in mL, of titrant for the *Sulfur Standard*; and V_B is the volume, in mL, of titrant for the *Titration Reagent Blank*.

Unsaponifiable Matter Proceed as directed under *Unsaponifiable Matter*, Appendix VII.

Water Proceed as directed under *Water Determination* using the *Karl Fischer Titrimetric Method*, Appendix IIB. However, in place of 35 to 40 mL of methanol, use 50 mL of a 1:1 chloroform–methanol mixture to dissolve the sample.

Packaging and Storage Store in well-closed containers.

Canthaxanthin

4,4′-Diketo-β-carotene; Cantha

$C_{40}H_{52}O_2$ Formula wt 564.85
INS: 161g   CAS: [514-78-3]

DESCRIPTION

A dark crystalline powder. It is soluble in chloroform, and is very slightly soluble in acetone, but is insoluble in water. It melts at about 207° to 212° with decomposition.

Functional Use in Foods Color.

REQUIREMENTS

Identification The absorbance spectrum of *Sample Solution B*, prepared as directed in the *Assay*, exhibits a maximum between 468 nm and 472 nm.
Assay Not less than 96.0% and not more than 101.0% of $C_{40}H_{52}O_2$.
Heavy Metals (as Pb) Not more than 10 mg/kg.
Residue on Ignition Not more than 0.2%.

TESTS

Assay (Note: Carry out all work in low-actinic glassware and in subdued light.)
Sample Solution A Transfer about 50 mg of the sample, accurately weighed, into a 100-mL volumetric flask, dissolve in 10 mL of acid-free chloroform, immediately dilute to volume with cyclohexane, and mix. Pipet 5 mL of this solution into a second 100-mL volumetric flask, dilute to volume with cyclohexane, and mix.
Sample Solution B Pipet 5 mL of *Sample Solution A* into a 50-mL volumetric flask, dilute to volume with cyclohexane, and mix.
Procedure Determine the absorbance of *Sample Solution B* in a 1-cm cell at the wavelength of maximum absorption at about 470 nm, with a suitable spectrophotometer, using cyclohexane as the blank. Calculate the quantity, in mg, of $C_{40}H_{52}O_2$ in the sample taken by the formula

$$20{,}000A/220,$$

in which A is the absorbance of the solution and 220 is the absorptivity of pure canthaxanthin.
Heavy Metals Prepare and test a 2-g sample as directed in *Method II* under the *Heavy Metals Test*, Appendix IIIB, using 20 μg of lead ion (Pb) in the control (*Solution A*).

Residue on Ignition Ignite 1 g as directed in the general method, Appendix IIC, using a silica crucible and moistening the residue with 2 mL of nitric acid and 1 mL of sulfuric acid.

Packaging and Storage Store in tight light-resistant containers under inert gas.

Caramel

Caramel Color

INS: 150   CAS: [8028-89-5]

DESCRIPTION

Caramel is a complex mixture of compounds, some of which are in the form of colloidal aggregates. Caramel is manufactured by heating carbohydrates either alone or in the presence of food-grade acids, alkalies, and/or salts. Caramel usually is a dark brown to black liquid or solid having an odor of burnt sugar and a somewhat bitter taste. Caramel is produced from commercially available food-grade nutritive sweeteners consisting of fructose, dextrose (glucose), invert sugar, sucrose, and/or starch hydrolysates and fractions thereof. The acids that may be used are food-grade sulfuric, sulfurous, phosphoric, acetic, and citric acids, and the alkalies are ammonium, sodium, potassium, and calcium hydroxides. The salts that may be used are ammonium, sodium, and potassium carbonate, bicarbonate, phosphate (including mono- and dibasic), sulfate, and bisulfite. Food-grade antifoaming agents such as polyglycerol esters of fatty acids may be used as processing aids during its manufacture. Caramel is soluble in water.

Four distinct classes of Caramel can be distinguished by the reactants used in their manufacture and by specific identification tests:

Class I (Plain Caramel, Caustic Caramel) Prepared by heating carbohydrates with or without acids or alkalies; no ammonium or sulfite compounds are used.

Class II (Caustic Sulfite Caramel) Prepared by heating carbohydrates with or without acids or alkalies in the presence of sulfite compounds; no ammonium compounds are used.

Class III (Ammonia Caramel) Prepared by heating carbohydrates with or without acids or alkalies in the presence of ammonium compounds; no sulfite compounds are used.

Class IV (Sulfite Ammonia Caramel) Prepared by heating carbohydrates with or without acids or alkalies in the presence of both sulfite and ammonium compounds.

All of these Caramels shall meet the criteria established for Caramel in this monograph.

Functional Use in Foods Color.

REQUIREMENTS

Ammoniacal Nitrogen[1] Not more than 0.6%, calculated on an equivalent color basis.
Arsenic[2] (as As) Not more than 1 mg/kg.
Color Intensity[3] Between 0.01 and 0.6 absorbance units (a.u.).
Heavy Metals Not more than 25 mg/kg.
Lead[2] Not More than 2 mg/kg.
Mercury[2] Not more than 0.1 mg/kg.
4-Methylimidazole[1] Not more than 0.025%, calculated on an equivalent color basis.
Sulfur Dioxide[1] Not more than 0.2%, calculated on an equivalent color basis.
Total Nitrogen[1] Not more than 3.3%, calculated on an equivalent color basis.
Total Sulfur[1] Not more than 3.5%, calculated on an equivalent color basis.

TESTS

Ammoniacal Nitrogen Transfer 25.0 mL of 0.1 N sulfuric acid to a 500-mL receiving flask, and connect it to a distillation apparatus consisting of a Kjeldahl connecting bulb and a condenser so that the condenser delivery tube is immersed beneath the surface of the acid solution in the receiving flask. Transfer about 2 g of the sample, accurately weighed, into an 800-mL, long-neck Kjeldahl digestion flask, and to the flask add 2 g of magnesium oxide (carbonate-free), 200 mL of water, and several boiling chips. Swirl the digestion flask to mix the contents, and quickly connect it to the distillation apparatus. Heat the digestion flask to boiling, and collect about 100 mL of distillate in the receiving flask. Wash the tip of the delivery tube with a few mL of water, collecting the washings in the receiving flask, then add 4 or 5 drops of methyl red TS, titrate with 0.1 N sodium hydroxide, and record the volume, in mL, as S. Conduct a blank determination, and record the mL of 0.1 N sodium hydroxide required as B. Calculate the percent ammoniacal nitrogen (equivalent color basis) by the formula

$$[(B - S) \times 0.0014 \times 100/W] \times 0.1/A_{610},$$

in which W is the weight, in g, of the sample taken; A_{610} is the color intensity as is; and 0.1 is the basis of color equivalency.

Arsenic A *Sample Solution* prepared as directed for organic compounds meets the requirements of the *Arsenic Test*, Appendix IIIB, using 1.0 mL of the *Standard Arsenic Solution* (1 g As) in the control.

Color Intensity For the purposes of this monograph, *Color Intensity* is defined as the absorbance of a 0.1% (w/v) solution of Caramel solids in water in a 1-cm cell at 610 nm. Since *Color Intensity* is expressed on a solids basis, *Total Solids* must be determined in order to determine *Color Intensity*.

---

[1] These tests are calculated on an equivalent color basis that permits the values to be expressed in terms of a Caramel having a color intensity standardized to 0.1 a.u.

[2] These tests are calculated on an as-is basis.

[3] *Color Intensity* is defined as the absorbance of a 0.1% (w/v) solution of Caramel in water measured in a 1-cm cell at 610 nm and is expressed on a *Total Solids* basis.

Procedure for Total Solids   For liquid samples, *Total Solids* is determined by drying a sample on a carrier composed of fine quartz sand that passes a No. 40, but not a No. 60, sieve and that has been prepared by digestion with hydrochloric acid, washed acid free, dried, and ignited. Mix 30.0 g of prepared sand, accurately weighed, with 1.5 to 2.0 g of the sample, accurately weighed, and dry to constant weight at 60° under reduced pressure (50 mm Hg). Record the final weight of the sand plus the Caramel solids. Calculate the percent total solids, S, as follows:

$$[(W_F - W_S)/W_C] \times 100,$$

in which W_F is the final weight of the sand plus Caramel solids, W_S is the weight of the prepared sand taken, and W_C is the weight of the Caramel sample taken.

For solid samples (powdered or granular), *Total Solids* is determined by the *Loss on Drying* test, Appendix IIC. Dry at 60° under reduced pressure (50 mm Hg), to constant weight. Calculate the percent total solids by the formula

$$[(W_D - W_B)/(W_S - W_B)] \times 100,$$

in which W_D is the weight of the bottle and sample after drying, W_B is the weight of the empty bottle, and W_S is the weight of the bottle and sample before drying.

To determine *Color Intensity*, transfer 100 mg of Caramel into a 100-mL volumetric flask, dilute to volume with water, and mix. Centrifuge if the solution is cloudy. Determine the absorbance (A_{610}) of the clear solution in a 1-cm cell at 610 nm with a suitable spectrophotometer previously standardized using water as the reference. Calculate the *Color Intensity* by the formula

$$(A_{610} \times 100)/S,$$

in which S is the percent total solids.

Heavy Metals   Proceed as directed in the *Heavy Metals Test*, Method II, Appendix IIIB, using an 800-mg sample and 20 μg of lead ion in the control (*Solution A*).

Lead   (**Note**: For this test, use reagent-grade chemicals with as low a lead content as is practicable, as well as high-purity water and gases. Before use in this analysis, rinse all glassware and plasticware twice with 10% nitric acid and twice with 10% hydrochloric acid, and then rinse them thoroughly with high-purity water, preferably obtained from a mixed-bed strong-acid, strong-base ion-exchange cartridge capable of producing water with an electrical resistivity of 12 to 15 megohms.)

Lead Nitrate Stock Solution   Prepare as directed in the *Heavy Metals Test*, Appendix IIIB. Each mL of this solution contains 100 μg of lead ion (Pb).

Standard Lead Solution   On the day of use, transfer 50.0 mL of *Lead Nitrate Stock Solution* to a 500-mL volumetric flask containing 50 mL of water, add 5 mL of nitric acid, dilute to volume with water, and mix. Each mL of *Standard Lead Solution* contains the equivalent of 10 μg of lead ion (Pb).

Standard Solutions   Prepare a series of lead standard solutions serially diluted from the *Standard Lead Solution*. Into separate 100-mL volumetric flasks, pipet 2, 5, 10, and 20 mL, respectively, of *Standard Lead Solution*, add 1 mL of nitric acid, dilute to volume, and mix. The *Standard Solutions* contain, respectively, 0.20, 0.50, 1.00 and 2.00 μg of lead per mL.

Sample Solution   Transfer about 25 g of the sample, accurately weighed, into an ashing vessel. Suitable ashing vessels are approximately of 100 mL capacity and are flat-bottom platinum crucibles or dishes, Vycor or quartz tall-form beakers, or evaporating dishes (Corning Glass Works No. 13180, or equivalent). Discard Vycor vessels when the inner surfaces become etched. Dry the sample overnight at 120° in a forced-draft oven. (The sample must be absolutely dry to prevent flowing or spattering in the furnace.) Place the sample in a furnace set at 250°. The furnace should be equipped with a pyrometer to control the temperature over a range of 260° to 600°, with a variation of less than 10°. Slowly, in 50° increments, raise the temperature to 350°, and hold at this temperature until smoking ceases. Increase the temperature to 500° in approximately 75° increments (the sample must not ignite). Ash for 16 h (overnight) at 500°. Remove the sample from the furnace, and allow it to cool. The ash should be white and essentially carbon free. If the ash still contains excess carbon particles (i.e., the ash is gray rather than white), proceed as follows: Wet with a minimal amount of water followed by the dropwise addition of nitric acid (0.5 to 3 mL). Dry on a hot plate. Transfer the ash to a furnace set at 250°, slowly increase the temperature to 500°, and continue heating for 1 to 2 h. Repeat the nitric acid treatment and ashing, if necessary, to obtain a carbon-free residue.

Note: Local overheating or deflagration may result if the sample still contains much intermingled carbon and especially if much potassium is present in the ash.

Dissolve the residue in 5 mL of 1 N nitric acid, warming on a steam bath or hot plate for 2 to 3 min to aid solution. Filter, if necessary, and decant through S&S 589 Black Ribbon paper, or equivalent, into a 50-mL volumetric flask. Repeat with two 5-mL portions of 1 N nitric acid, filter, and add the washings to the original filtrate. Dilute to volume with 1 N nitric acid, and mix to prepare the *Sample Solution*.

Similarly, prepare duplicate reagent blanks for each *Standard Solution* and *Sample Solution*, including any additional water and nitric acid if used for sample ashing.

Note: Do not ash nitric acid in a furnace because the lead contaminant will be lost.

Evaporate nitric acid to dryness in an ashing vessel on a steam bath or hot plate, and then proceed as above, beginning at "Dissolve the residue in 5 mL of 1 N nitric acid, warming on a steam bath. . . ."

Extraction   (Complete analysis on the same day.)

Butyl Acetate, Aqueous   Use spectral-grade butyl acetate, and saturate it with water.

APDC Solution   Transfer 2.00 g of APDC (ammonium 1-pyrrolidinedithicarbamate) (Aldrich Chemical, or equivalent) into a 100-mL volumetric flask, dilute to volume with water, and mix. Remove insoluble free acid and other impurities normally present by two to three extractions with 10-mL portions of *Butyl Acetate, Aqueous*.

Citric Acid Solution, Lead-Free   Dissolve 10 g of citric acid in 30 mL of water. Add ammonium hydroxide slowly with stirring until the pH is between 8.0 and 8.5, using short-range pH paper as an external indicator. Transfer the solution to a separatory funnel, and extract with 10-mL portions of *Dithizone*

Extraction Solution, Lead Limit Test, Appendix IIIB, until the dithizone solution retains its green color or remains unchanged. Drain the final dithizone layer, plus about 1 mL of the aqueous layer, into a beaker, and add nitric acid (1 in 1) slowly with stirring until the pH is between 3.5 and 4, again using short-range pH paper as an external indicator. Transfer this solution to a 100-mL volumetric flask (through a filter, if necessary), dilute to volume with water, and mix thoroughly.

Pipet 20 mL each of the *Standard Solutions*, the *Sample Solution*, and the appropriate reagent blanks into separate 60-mL separatory funnels. Treat each solution as follows: Add 4 mL of *Citric Acid Solution* and 2 to 3 drops of bromocresol green TS. The solution should be yellow. Adjust the pH to about 5.4, using ammonium hydroxide initially and then ammonium hydroxide diluted with 4 volumes of water in the vicinity of the color change (the first permanent appearance of light blue). Add 4 mL of *APDC Solution*, stopper, and shake for 30 to 60 s. Add by pipet 5.0 mL of *Butyl Acetate, Aqueous*, stopper the separatory funnel, and shake vigorously for 30 to 60 s. Let stand until the layers separate clearly, and drain and discard the lower aqueous phase. If an emulsion forms or the solvent layer is cloudy, drain the solvent layer into a 15-mL centrifuge tube, cover with aluminum foil or Parafilm (or equivalent), and centrifuge for about 1 min at 2000 rpm. The lead content is determined using these *Standard Solutions* and the *Sample Solution* (containing butyl acetate) by atomic absorption spectrophotometry.

Procedure Use an atomic absorption spectrophotometer equipped with a 4-in., single-slot burner head. Set the instrument to previously determined optimum conditions for organic solvent aspiration (3 to 5 mL/min) and at a wavelength of 283.3 nm. Use an air–acetylene flame adjusted for maximum lead absorption with a fuel-lean flame. Aspirate the blanks, the *Standard Solutions*, and the *Sample Solution*, flushing with water and then with *Butyl Acetate, Aqueous* between measurements. Record the absorbance of the *Standard Solutions* and the *Sample Solution* (containing butyl acetate), and correct for the blanks. Prepare the *Standard Curve* by plotting the absorbance of each *Standard Solution* against its concentration, in μg of lead per mL. (This concentration, in butyl acetate, is four times that in the aqueous standard.) From the *Standard Curve*, determine the concentration, C, in μg/mL, of the *Sample Solution*. Calculate the quantity, in mg/kg, of lead in the sample by the formula

$$12.5 \times C/W,$$

in which W is the weight, in g, of the sample taken.

Mercury

Standard Preparation Prepare as directed in the *Mercury Limit Test*, Appendix IIIB, using 1.0 mL of the stock solution, equivalent to 1 μg of Hg, instead of the 2.0 mL specified therein.

Sample Preparation Transfer 5 g of the sample into a 250-mL Erlenmeyer flask, and continue as directed in the second full paragraph under *Sample Solution* in the *Arsenic Test*, Appendix IIIB, beginning with "... add 5 mL of sulfuric acid and a few glass beads ...." After the sample has been digested and the solution diluted to 35 mL, as directed therein, add 1 mL of potassium permanganate solution (1 in 25), and mix.

Procedure Continue as directed for *Procedure* in the *Mercury Limit Test*, Appendix IIIB. Any absorbance produced by the *Sample Preparation* is not more than half that produced by the *Standard Preparation*, indicating not more than 0.1 mg of mercury per kg of Caramel.

4-Methylimidazole (4-MeI)

4-Methylimidazole Stock Solution Purify reagent-grade 4-methylimidazole[4] by redistillation (b.p. 92° to 93°, 0.05 mm Hg), and then prepare a stock solution by transferring 50 mg of the distillate, accurately weighed, into a 50-mL volumetric flask and diluting to volume with tetrahydrofuran (acetone is also an acceptable solvent). Mix thoroughly, and store in a refrigerator.

To prepare the standard solutions, pipet 1.0-, 1.5-, 2.0-, 2.5-, 3.0-, 3.5-, 4.0-, and 5.0-mL portions of the *4-Methylimidazole Stock Solution* into separate 10-mL volumetric flasks, dilute each to volume with the same solvent used to prepare the stock solution, and mix. The standards thus prepared represent 4-methylimidazole concentrations (w/v) of 100, 150, 200, 250, 300, 350, 400, and 500 mg/L, respectively. Store the standard solutions in a refrigerator, and use within 1 month.

Sample Preparation Place a plug of fine glass wool in the base of a 300- × 22-mm chromatographic tube having a Teflon stopcock. Mix thoroughly 10.0 g of Caramel, accurately weighed, and 5.0 g 3 N sodium hydroxide in a 250-mL polypropylene beaker. The pH of the mixture should exceed 12. Add 20.0 g chromatographic siliceous earth (Johns-Manville Celite 545, or equivalent) to the beaker, and thoroughly mix with a wide-blade, stainless-steel spatula until a homogeneous, semi-dry mixture is obtained. Homogeneity is obtained when the color is uniform and no dark clumps are seen. Quantitatively transfer the mixture to the column. Place a plug of glass wool on top of the column, and then allow the column to fall a short distance vertically to help settle the contents. The column bed, approximately 150-mm in height, should be of uniform consistency, yet open enough to allow elution to occur readily.

Rinse the sample beaker with methylene chloride, and pour the contents into the column with the stopcock open. Allow the methylene chloride to pass down the column until it reaches the stopcock. Close the stopcock and allow the methylene chloride to remain in contact with the bed for 5 min. Open the stopcock, and pass methylene chloride through the column at a rate of 5 mL/min. Collect 200 mL of eluate in a 300-mL round-bottom flask. Remove the bulk of the solvent from the eluate by rotary vacuum evaporation (350 to 390 mm Hg) while maintaining the flask at a temperature of 35° in a water bath. Reduce the volume to about 1 mL. During the concentration step, watch the flask carefully to ensure that no loss of sample occurs by bumping. Quantitatively transfer the extract residue to a 5-mL volumetric flask using a disposable Pasteur pipet, by rinsing the flask several times with small (ca. 0.7 mL) portions of the same solvent used to prepare the original solutions (tetrahydrofuran or acetone) until the dilution mark is reached. Mix thoroughly.

Apparatus Use a suitable gas chromatograph equipped with a hydrogen flame-ionization detector, and a silanized 1-m × 4-mm (id) glass column packed with 90/100-mesh Anakrom ABS,

---

[4] A suitable grade may be obtained from Aldrich Chemical Co., P.O. Box 355, Milwaukee, WI 53201, Catalog No. 19,988-5, or equivalent.

or equivalent, containing 7.5% Carbowax 20M and 2% potassium hydroxide (available from Supelco[5]).

Operating Conditions The operating conditions may vary, depending on the particular instrument used, but a suitable chromatogram may be obtained by using the following conditions: *column temperature*: 190° (isothermal); *injection port temperature*: 200°; *carrier gas*: nitrogen, flowing at a rate of 50 mL/min; *detector temperature*: 250°.

Procedure To prepare a calibration curve using an integrator, inject 5.0 μL of each *Standard Solution* using an autosampler, and determine the concentration of each *Standard Solution*.

Note: If manual injections are used, to avoid fractionation in the syringe needle and to ensure that 5.0 μL is injected, use the solvent-flush technique, with the solvent used to prepare the standard solutions.

For each standard chromatogram, obtain the corrected peak area. If an integrator is not used, calculate the corrected peak area by multiplying the peak height, in mm, by the peak width at one-half height, in mm, by the proper attenuation and range factors, depending on the particular apparatus and operating parameters used. Plot each corrected peak area thus obtained versus its respective concentration of 4-methylimidazole to obtain the standard curve. In the same manner, chromatograph a 5.0-μL portion of the *Sample Preparation*, calculate the peak area corresponding to any 4-methylimidazole contained in the sample, and by reference to the standard curve, obtain the content of the 4-methylimidazole in the sample.

Sulfur Dioxide Proceed as directed under *Sulfur Dioxide Determination*, Appendix X, using 0.5 g of the sample, accurately weighed.

Total Nitrogen Determine as directed under *Nitrogen Determination* (Kjeldahl Method) using *Method II*, Appendix IIIC.

Total Sulfur In the largest casserole available that fits in an electric muffle furnace, place 1 to 3 g of magnesium oxide or an equivalent quantity of $Mg(NO_3)_2 6H_2O$ (6.4 to 19.2 g), 1 g of powdered sucrose, and 50 mL of nitric acid. Transfer into the casserole about 5 g of sample, accurately weighed, when the expected amount of sulfur is 2.5% or less, or 1 g of sample when the expected amount is greater than 2.5%. Place the same quantities of reagents in another casserole for the blank. Evaporate on a steam bath to the consistency of paste. Place the casserole in a cold electric muffle furnace, gradually heat to 525°, and hold at that temperature until all nitrogen dioxide fumes are driven off. Cool the casserole, add 100 mL of water to dissolve the sample, and neutralize to pH 7 with hydrochloric acid, using short-range pH indicator paper as an external indicator. Add an additional 2 mL of hydrochloric acid, filter the solution into a suitable beaker, heat to boiling, and while stirring, slowly add 20 mL of barium chloride TS to the hot solution. Boil the contents of the beaker for 5 min, and allow to stand overnight. Filter the contents of the beaker through a tight, ashless filter paper, and quantitatively transfer the precipitate to the paper. Thoroughly wash the paper and the precipitate with hot water, and then transfer the paper to a tared crucible previously ignited for 1 h at 800° in a muffle furnace. Dry the paper in a crucible for 1 h at 105°, and then carefully char it, with free access of air, at low heat, over a burner. Gradually increase the heat to burn away the paper, and finally, ignite the crucible and contents for 1 h at 800°. Cool and weigh, and calculate the percent sulfur by the formula

$$[(W_S - W_B)/S] \times 13.74 \times 0.1/A_{610},$$

in which W_S is the weight, in g, of the ignited residue of barium sulfate from the sample determination; W_B is the weight, in g, of the ignited residue from the blank determination; and S is the weight, in g, of the sample taken.

Packaging and Storage Store in well-closed containers and avoid exposure to excessive heating and, for solid products, excessive humidity.

ADDITIONAL INFORMATION

Identification of Classes The four classes of Caramel may be distinguished from each other by the following methods:

Class I Not more than 50% of the color is bound by DEAE (diethylaminoethyl) cellulose, and not more than 50% of the color is bound by phosphoryl cellulose.

Class II More than 50% of the color is bound by DEAE cellulose, and it exhibits an absorbance ratio (at equal concentrations) of more than 50.

Class III Not more than 50% of the color is bound by DEAE cellulose, and more than 50% of the color is bound by phosphoryl cellulose.

Class IV More than 50% of the color is bound by DEAE cellulose, and it exhibits an absorbance ratio (at equal concentrations) of not more than 50.

Identification Tests for Classes

Absorbance Ratio For the purposes of this test, *Absorbance Ratio* is defined as the absorbance of Caramel, at equal concentrations, at 280 nm divided by the absorbance at 560 nm.

Procedure Transfer 100 mg of the sample into a 100-mL volumetric flask with the aid of water, dilute to volume, mix, and centrifuge if solution is cloudy. Pipet a 5.0-mL portion of the clear solution into a 100-mL volumetric flask, dilute to volume with water, and mix. Determine the absorbance of the 0.1% solution in a 1-cm cell at 560 nm and that of the 1:20 diluted solution at 280 nm with a suitable spectrophotometer previously standardized using water as a reference. (A suitable spectrophotometer is one equipped with a monochromator to provide a band width of 2 nm or less and of such quality that the stray-light characteristic is 0.5% or less.) Calculate the absorbance ratio of the Caramel by the formula

$$(A_{C1} \times 20)/A_{C2},$$

in which A_{C1} is the absorbance of Caramel at 280 nm, 20 is the dilution factor, and A_{C2} is the absorbance of Caramel at 560 nm.

Color Bound by DEAE Cellulose For the purposes of this monograph, *Color Bound by DEAE Cellulose* is defined as the percent decrease in absorbance of a Caramel solution at 560 nm after treatment with DEAE cellulose.

Special Reagent DEAE cellulose of 0.7 meq/g capacity (e.g., Cellex D from Bio-Rad, or equivalent); DEAE cellulose

---
[5]Supelco, Supelco Park, Bellefonte, PA 16823-0048.

of higher or lower capacities may be used in proportionately higher or lower quantities.

Procedure Prepare a Caramel solution of approximately 0.5 absorbance unit at 560 nm by transferring an appropriate amount of sample into a 100-mL volumetric flask with the aid of 0.025 N hydrochloric acid. Dilute to volume with 0.025 N hydrochloric acid, and centrifuge or filter if the solution is cloudy. To a 20-mL aliquot of the Caramel solution, add 200 mg of DEAE cellulose, mix thoroughly for several min, centrifuge or filter, and collect the clear supernatant liquid. Determine the absorbance of the Caramel solution and the supernatant liquid in a 1-cm cell at 560 nm, with a suitable spectrophotometer previously standardized using 0.025 N hydrochloric acid as a reference. Calculate the percent color bound by DEAE cellulose by the formula

$$[(X_1 - X_2)/X_1]100,$$

in which X_1 is the absorbance of the Caramel solution at 560 nm, and X_2 is the absorbance of the supernatant liquid at 560 nm after DEAE cellulose treatment.

Color Bound by Phosphoryl Cellulose For the purposes of this monograph, *Color Bound by Phosphoryl Cellulose* is defined as the percent decrease in absorbance of a Caramel solution at 560 nm after treatment with phosphoryl cellulose.

Special Reagent Phosphoryl cellulose of 0.85 meq/g capacity (e.g., Cellex P from Bio-Rad, or equivalent); phosphoryl cellulose of higher or lower capacities may be used in proportionately higher or lower quantities.

Procedure Transfer 200 to 300 mg of the sample into a 100-mL volumetric flask, dilute to volume with 0.025 N hydrochloric acid, and centrifuge or filter if the solution is cloudy. To a 40-mL aliquot of the Caramel solution, add 2.0 g of phosphoryl cellulose, mix thoroughly for several min, centrifuge or filter, and collect the clear supernatant liquid. Determine the absorbance of the Caramel solution and the supernatant liquid in a 1-cm cell at 560 nm with a suitable spectrophotometer previously standardized using 0.025 N hydrochloric acid as a reference. Calculate the percent color bound by phosphoryl cellulose by the formula

$$[(X_1 - X_2)/X_1]100,$$

in which X_1 is the absorbance of the Caramel solution at 560 nm, and X_2 is the absorbance of the supernatant liquid at 560 nm after phosphoryl cellulose treatment.

2-Acetyl-4-(5)-tetrahydroxybutylimidazole (THI) Class III (Ammonia Caramel) is the only class of Caramel color found to contain the impurity THI (2-Acetyl-4-(5)-tetrahydroxybutylimidazole), which in some countries has a limit of 25 ppm on an Equivalent Color basis; therefore, a method for the determination of THI is provided as information.

Materials

2,4-Dinitrophenylhydrazine (DNPH) Hydrochloride Add 5 g of reagent-grade 2,4-dinitrophenylhydrazine to 10 mL of hydrochloric acid in a 100-mL Erlenmeyer flask, and gently shake the latter until the free base (red) is converted to the hydrochloride (yellow). Add 100 mL of ethanol, and heat the mixture on a steam bath until all the solid has dissolved. Cool to room temperature, and after the solution has crystallized, filter off the hydrochloride. Wash with ether, dry at room temperature, and store in a desiccator. Upon storage, the hydrochloride slowly converts to the free base. The latter can be removed by washing with purified (peroxide-free) dimethoxyethane. Prepare the reagent by mixing 0.5 g of *2,4-Dinitrophenylhydrazine (DNPH) Hydrochloride* in 15 mL of 5% methanol in dimethoxyethane for 30 min. Store in a refrigerator at 4°. When properly prepared and stored, this reagent is stable for at least 3 months.

THI–DNPH Standard Using THI[6], prepare the *THI–DNPH Standard* as follows. Add 0.5 g of *2,4-Dinitrophenylhydrazine (DNPH) Hydrochloride* to 1 mL of HCl, followed by 10 mL of ethyl alcohol, and heat on a steam bath until in solution. Add 100 mg of THI to the hot solution. Crystallization begins in a few min, and the THI–DNPH is filtered off when the suspension reaches room temperature. The *THI–DNPH Standard* is obtained by recrystallization of the THI–DNPH from ethyl alcohol (use 1 drop of HCl per 5 mL of ethyl alcohol). The yield is 70% to 80% based on the THI. When stored in the refrigerator, the *THI–DNPH Standard* is stable for at least 1 year.

Stock THI–DNPH Solution Dissolve about 10 mg of *THI–DNPH Standard*, accurately weighed, in a 100-mL volumetric flask, and dilute to volume with absolute, carbonyl-free methanol. Dilute a portion of this solution tenfold with methanol. The THI concentration, in mg/L, of the *Stock THI–DNPH Solution* is 0.47 times the mg of the THI–DNPH. *Stock THI–DNPH Solution* is stable for at least 20 weeks when stored in the refrigerator.

Cation-Exchange Resin (Strong) Dowex 50 AG x 8, proton form, 100- to 200-mesh.

Cation-Exchange Resin (Weak) Amberlite CG AG 50 I, proton form, 100- to 200-mesh. (Sediment two or three times before use.)

Methanol, Carbonyl-Free To 500 mL of methanol, add 5 g of Girard's Reagent P (Aldrich, or equivalent) and 0.2 mL of hydrochloric acid, and reflux for 2 h. Distill the refluxed methanol through a short Vigreux column, and store in tightly closed bottles.

Dimethoxyethane Purify dimethoxyethane by distillation from 2,4-dinitrophenylhydrazine in the presence of acid, and redistill it from sodium hydroxide. Immediately before use, pass it through a column of neutral alumina to remove peroxides.

Apparatus

Combination Columns Two connected columns are used, one above the other, as described in *J. Agr. Fd. Chem.*, (1974) 22, 110. Fill the upper column (150 × 12.5 mm, filling height max 9 cm; or 200 × 10 mm, filling height max 14 cm, with capillary outlet of 1 mm (id) with *Cation-Exchange Resin (Weak)*; bed height, approximately 50 to 60 mm, or 80 to 90 mm, respectively. Fill the lower column (total length 175 mm, 10 mm id, with capillary outlet and Teflon stopcock) with *Cation-Exchange Resin (Strong)* to a bed-height of 60 mm. Use a dropping funnel (100 mL) with Teflon stopcock as a solvent reservoir. All parts are connected by standard ground-glass joints (14.5 mm).

Sample Preparation Accurately weigh 200 to 250 mg of the sample, and dissolve it in 3 mL of water. Quantitatively

---

[6]THI may be obtained from the International Technical Caramel Association, 1575 Eye Street, N.W., Suite 800, Washington, D.C. 20005.

transfer the solution to the upper part of the combination column. Elute with water until a total of about 100 mL of water has passed through the column. Disconnect the upper column, and elute the lower column with 0.5 N HCl. Discard the first 10.0 mL of eluate, and subsequently collect a volume of 35 mL. Concentrate the solution to dryness at 40° and 15 mm of Hg, then dissolve the syrup residue in 250 μL of carbonyl-free methanol, and add 250 μL of the *2,4-Dinitrophenylhydrazine Hydrochloride Reagent*. Transfer the reaction mixture (sample) to a septum-capped vial, and store for 5 h at room temperature.

Procedure Prepare a series of *THI–DNPH Standard Solutions* serially diluted from the *Stock THI–DNPH Solution*. Into separate 10-mL volumetric flasks, pipet 1, 2, and 5 mL, respectively, of the *Stock THI–DNPH Solution*, and dilute to volume with absolute, carbonyl-free methanol. Prepare a standard curve by injecting 5 μL of the *Stock THI–DNPH Solution*, and the serially diluted *THI–DNPH Standard Solutions* into a 250-mm × 4-mm, 10-μm LiChrosorb RP-8 HPLC column (Alltech Associates, Inc., or equivalent) fitted with an ultraviolet detector set at 385 nm. The mobile phase is 50:50 methanol:0.1 M phosphoric acid (v/v). Inject 5 μL of sample into the column. Adjustments in the mobile phase composition may be needed as column characteristics vary among manufacturers. At a mobile phase flow rate of 2 mL/min and column dimensions of 250 × 4.6 mm, elute THI–DNPH at about 6.3 ± 0.1 min. Measure the peak areas. Calculate the amount of THI in the sample from the standard curve. (For THI limits greater than 25 mg/kg, prepare a series of *Standard THI–DNPH Solutions* in a range encompassing the expected THI concentration in the sample.)

Caraway Oil

CAS: [8000-42-8]

DESCRIPTION

A volatile oil distilled from the dried, ripe fruit of *Carum carvi* L. (Fam. *Umbelliferae*). It is a colorless to pale yellow liquid having the characteristic odor and taste of caraway.

Functional Use in Foods Flavoring agent.

REQUIREMENTS

Identification The infrared absorption spectrum of the sample exhibits relative maxima (that may vary in intensity) at the same wavelengths (or frequencies) as those shown in the respective spectrum in the section on *Infrared Spectra* (*Series A: Essential Oils*), using the same test conditions as specified therein.
Assay Not less than 50.0%, by volume, of ketones as carvone.
Angular Rotation Between +70° and +80°.
Heavy Metals (as Pb) Passes tests.
Refractive Index Between 1.484 and 1.488 at 20°.
Solubility in Alcohol Passes test.
Specific Gravity Between 0.900 and 0.910.

TESTS

Assay Proceed as directed under *Aldehydes and Ketones—Neutral Sulfite Method*, Appendix VI.
Angular Rotation Determine in a 100-mm tube as directed under *Optical (Specific) Rotation*, Appendix IIB.
Heavy Metals Shake 10 mL of the oil with an equal volume of water to which 1 drop of hydrochloric acid has been added, and pass hydrogen sulfide through the mixture until it is saturated. No darkening in color is produced in either the oil or the water.
Refractive Index, Appendix IIB Determine with an Abbé or other refractometer of equal or greater accuracy.
Solubility in Alcohol Proceed as directed in the general method, Appendix VI. One mL dissolves in 8 mL of 80% alcohol.
Specific Gravity Determine by any reliable method (see *General Provisions*).

Packaging and Storage Store in full, tight containers in a cool place protected from light.

Carbon, Activated

DESCRIPTION

A solid, porous, carbonaceous material prepared by carbonizing and activating organic substances. The raw materials, which include sawdust, peat, lignite, coal, cellulose residues, coconut shells, petroleum coke, etc., may be carbonized and activated at a high temperature with or without the addition of inorganic salts in a stream of activating gases such as steam or carbon dioxide. Alternatively, carbonaceous matter may be treated with a chemical activating agent such as phosphoric acid or zinc chloride and the mixture carbonized at an elevated temperature, followed by removal of the chemical activating agent by water washing. Activated Carbon occurs as a black, tasteless substance, varying in particle size from coarse granules to a fine powder. It is insoluble in water and in organic solvents.

Functional Use in Foods Decolorizing agent; taste- and odor-removing agent; purification agent in food processing.

REQUIREMENTS
Identification
A. Place about 3 g of powdered sample in a glass-stoppered Erlenmeyer flask containing 10 mL of dilute hydrochloric acid (5%), boil for 30 s, and cool to room temperature. Add 100 mL of iodine TS, stopper, and shake vigorously for 30 s. Filter through Whatman No. 12 filter paper, or equivalent, discarding

the first portion of filtrate. Compare 50 mL of the subsequent filtrate with a reference solution prepared by diluting 10 mL of iodine TS to 50 mL with water, but not treated with carbon. The color of the carbon-treated iodine solution is no darker than that of the reference solution, indicating the adsorptivity of the sample.

B. Ignite a portion of the sample in air. Carbon monoxide and carbon dioxide are produced, and an ash remains.

Cyanogen Compounds   Passes test.
Heavy Metals (as Pb)   Not more than 0.004%.
Higher Aromatic Hydrocarbons   Passes test.
Iodine Value   Not less than 400.
Lead   Not more than 10 mg/kg.
Water Extractables   Not more than 4.0%.

The following additional **REQUIREMENTS** should conform to the representations of the vendor: **Loss on Drying** and **Residue on Ignition**.

TESTS

Cyanogen Compounds   Mix 5 g of the sample with 50 mL of water and 2 g of tartaric acid, and distill the mixture, collecting 25 mL of distillate below the surface of a mixture of 2 mL of 1 N sodium hydroxide and 10 mL of water contained in a small flask placed in an ice bath. Dilute the distillate to 50 mL with water, and mix. Add 12 drops of ferrous sulfate TS to 25 mL of the diluted distillate, heat almost to boiling, cool, and add 1 mL of hydrochloric acid. No blue color is produced.

Heavy Metals   A 10-mL portion of the filtrate obtained in the test for *Water Extractables* meets the requirements of the *Heavy Metals Test*, Appendix IIIB, using 20 µg of lead ion (Pb) in the control (*Solution A*).

Higher Aromatic Hydrocarbons   Extract 1 g of the sample with 12 mL of cyclohexane in a continuous-extraction apparatus for 2 h. Using matched Nessler tubes, the extract shows no more color or fluorescence than does a solution of 100 µg of quinine sulfate in 1000 mL of 0.1 N sulfuric acid when observed in ultraviolet light.

Iodine Value

Standard Iodine Solution   Using a wide-mouth funnel, transfer 805 g of potassium iodide to a 2-L volumetric flask, and add enough water to the flask to cover the sample. Prepare a solution of 50 g of potassium iodide in 150 mL of water in a 250-mL beaker. Weigh 120 g of molecular iodine in a glass-stoppered weighing bottle. Pour the iodine into the funnel fitted to the 2-L volumetric flask, and then immediately restopper the weighing bottle, add an additional 285 g of potassium iodide to the funnel, and wash the funnel clean with a stream of water. With the potassium iodide solution, rinse the remaining iodine from the weighing bottle until the washings are colorless. Pour the remaining potassium iodide solution into the volumetric flask, and rinse all glassware with water into the volumetric flask. Grease the stopper, and insert it into the flask.

Gently shake the volumetric flask on a mechanical shaker for 30 min, add about 300 mL of water, and repeat the shaking until the flask has cooled to room temperature. Add progressively smaller amounts of water to the flask so that the final quantity required to dilute to volume is small enough that no further heat of solution is detectable. Allow the solution to stand overnight.

Transfer 100 mL of the iodine solution to a 1-L volumetric flask, and fill with water to the bottom of the flask neck. Allow the solution to stand 30 min before standardizing.

Standardization   Transfer approximately 125 mg of primary standard-grade barium thiosulfate (dried at 40°), accurately weighed, to a 125-mL Erlenmeyer flask. Cover the sample with about 50 mL of starch TS, and continue the titration until 1 drop of iodine solution produces a distinct, light blue color. The normality (N) of the iodine solution, which is approximately 0.047, is given by $V \times 0.4673$, in which V is the volume, in mL, of titrant.

Procedure   Transfer approximately 50 to 60 mg of sample, previously dried at 105° for 30 min and accurately weighed, to a glass vial or bottle. Pipet 25 mL of *Standard Iodine Solution*, stopper the container, and shake mechanically at about 240 strokes per min for 2 min. Transfer the mixture to a centrifuge tube, and centrifuge until the sample forms a pellet firm enough to permit decanting of the supernatant solution. Pipet 20 mL of the supernatant solution into a 250-mL Erlenmeyer flask, and titrate with sodium thiosulfate (*Volumetric Solutions, Solutions and Indicators*), diluted 1 in 2.5 with water, until the yellow iodine color becomes pale. Add 1 mL of starch TS, and continue titrating until the blue color is discharged. Record the volume of titrant as S. Titrate a 25-mL aliquot of *Standard Iodine Solution* with the sodium thiosulfate, and record the volume of titrant required as B.

Calculation   Calculate the iodine value (I), in mg of iodine adsorbed per g of Carbon, by the formula

$$I = [N(B - S)(126.91)]/W,$$

in which $(B - S)$ is the difference in volumes of sodium thiosulfate required for the blank and the sample, respectively; N is the exact normality of the sodium thiosulfate; and W is the weight, in g, of the sample.

Lead   A 20-mL portion of the filtrate obtained in the test for *Water Extractables* meets the requirements of the *Lead Limit Test*, Appendix IIIB, using 10 µg of lead ion (Pb) in the control.

Loss on Drying, Appendix IIC   Dry at 120° for 4 h.

Residue on Ignition   Ignite 500 mg as directed in the general method, Appendix IIC.

Water Extractables   Transfer 5.00 g of the sample into a 250-mL flask provided with a reflux condenser and a Bunsen valve. Add 100 mL of water and several glass beads, and reflux for 1 h. Cool slightly, and filter through Whatman No. 12 or equivalent filter paper, discarding the first 10 mL of filtrate. Cool the subsequent filtrate to room temperature, and pipet 25.0 mL into a tared crystallization dish.

Note: Retain the remainder of the filtrate for the *Heavy Metals* and *Lead* tests.

Evaporate the filtrate in the dish to incipient dryness on a hot plate, never allowing the solution to boil. Dry for 1 h at 100° in a vacuum oven, cool, and weigh.

Packaging and Storage   Store in well-closed containers.

Carbon Dioxide

CO_2 Formula wt 44.01

INS: 290   CAS: [124-38-9]

DESCRIPTION

A colorless, odorless gas, 1 L of which weighs about 1.98 g at 0° and a pressure of 760 mm of mercury. Under a pressure of about 59 atmospheres it may be condensed to a liquid, a portion of which forms a white solid ("dry ice") upon rapid vaporization. Solid Carbon Dioxide evaporates without melting upon exposure to air. One volume of the gas dissolves in about 1 volume of water, forming a solution that is acid to litmus.

Functional Use in Foods   Propellant and aerating agent; direct-contact freezing agent.

REQUIREMENTS

Identification

A. Pass 100 ± 5 mL, released from the vapor phase of the contents of the container, through a carbon dioxide detector tube (see *Detector Tubes* under *Solutions and Indicators*) at the rate specified for the tube: The indicator change extends throughout the entire indicating range of the tube.

B. The gas, when passed through barium hydroxide TS, forms a precipitate that dissolves with effervescence in acetic acid.

Assay   Not less than 99.5% of CO_2, by volume.

Carbonyl Sulfide   Not more than 0.5 ppm, by volume.

Hydrogen Sulfide   Not more than 0.5 ppm, by volume.

Nitric Oxide and Nitrogen Dioxide   Not more than 2.5 ppm each, by volume.

Nonvolatile Hydrocarbons   Not more than 10 mg/kg.

Sulfur Dioxide   Not more than 5 ppm, by volume.

Volatile Hydrocarbons (as methane)   Not more than 0.005%, by volume.

Water   Passes test.

TESTS

Note: The following **TESTS** are designed to reflect the quality of Carbon Dioxide in both its vapor and liquid phases, which are present in previously unopened cylinders. Reduce the container pressure by means of a regulator.

Withdraw the specimens for the **TESTS** with the least possible release of Carbon Dioxide consistent with proper purging of the sampling apparatus. Measure the gases with a gas volume meter downstream from the detector tubes to minimize contamination of or changes to the specimens. Perform **TESTS** in the sequence in which they are listed.

The various detector tubes called for in the respective **TESTS** are listed under *Detector Tubes* in *Solutions and Indicators*.

Assay   (**Note**: Sampling for this *Assay* may be done from the vapor phase for convenience, but this results in more residual volume. If the specification of 0.5 mL is exceeded from the vapor phase, a liquid specimen may be taken.) Assemble a 100-mL gas buret provided with a leveling bulb and a two-way stopcock to a gas absorption pipet of suitable capacity by connecting the pipet to one of the buret outlets. Fill the buret with slightly acidified water (turned pink with methyl orange), and fill the pipet with potassium hydroxide solution (1 in 2). By manipulating the leveling bulb and leveling water, draw the potassium hydroxide solution to fill the pipet and capillary connection up to the stopcock, and then fill the buret with the leveling water, and draw it through the other stopcock opening in such a manner that all gas bubbles are eliminated from the system. Draw into the buret 100.0 mL of specimen taken from the liquid phase as directed below in the test for *Nitric Oxide and Nitrogen Dioxide*. By raising the leveling bottle, force the measured specimen into the pipet. The absorption may be facilitated by rocking the pipet or by flowing the specimen between pipet and buret. Draw any residual gas into the buret, and measure its volume: Not more than 0.5 mL of gas remains.

Carbonyl Sulfide

Standard Preparation   Flush a 500-mL, glass, septum-equipped sampling bulb with helium, and inject into the bulb a 0.25-mL sample of pure carbonyl sulfide. Allow the bulb to stand for 15 min to permit the gases to mix, and then inject 0.50 mL of the mixture into a second 500-mL sampling bulb, also flushed with helium, and allow this tube to stand for 15 min to permit the gases to mix. This mixture is a nominal 0.5 ppm v/v standard. Determine the exact concentration from the exact volumes of the gas-sampling bulbs. To determine these volumes, weigh the empty tubes, fill them with water, and reweigh. From the weight of the water, and its temperature, calculate the volumes of the tubes.

Chromatographic System   The gas chromatograph is equipped with a Sievers 350 (or equivalent)[1] Chemiluminescence Detector (SCD) and a Supelco (or equivalent) 30-m × 0.53-mm id, 5-μm DB-5 capillary column. The carrier gas is helium set at a head-pressure of 5 psig. The split/splitless injection port is set at 100°, and the split ratio is set at 1:1. The column temperature is set at 30°. The retention time for carbonyl sulfide is approximately 3 min. The SCD is operated with 190 mL/min of hydrogen and 396 mL/min of air. The gas flows and probe position of the SCD are optimized for maximum sensitivity.

Procedure   Inject, in triplicate, 5.00 mL of the *Standard Preparation* into the gas chromatograph, record the chromatograms, and average the peak area responses. The relative standard deviation does not exceed 5.0%.

Similarly, inject, in triplicate, 5.00 mL of the sample, average the peak area responses, and calculate the ppm v/v in the sample by the formula

$$\text{ppm} = S(A_U/A_S),$$

in which S is the calculated ppm of carbonyl sulfide in the *Standard Preparation* (approximately 0.5 ppm), A_U is the aver-

---

[1] Any sulfur-selective detector may be used; e.g., electrolytic conductivity, flame photometric, or sulfur chemiluminescence. The detector must be capable of detecting less than 0.1 ppm v/v of carbonyl sulfide with a signal-to-noise ratio of 10:1.

age of the sample peak area responses, and A_S is the average area of the *Standard Preparation* area responses.

Hydrogen Sulfide   Pass 10,050 ± 50 mL, released from the vapor phase, through a hydrogen sulfide detector tube at the rate specified for the tube: The indicator change corresponds to not more than 5 ppm, which is not more than 0.5 ppm, for the volume of carbon dioxide specified in this test.

Nitrogen Oxide and Nitrogen Dioxide   Position the sample container so that when its valve is opened, the liquid phase can be sampled (generally this requires that the cylinder be inverted). Attach a section of tubing long enough to act as a vaporizer for the small quantity of liquid to be sampled. Connect one end of a nitric oxide–nitrogen dioxide detector tube to the tubing and the other end to a gas flow meter. Pass 500 mL of the liquid sample through the tube at a suitable rate. No frost should reach the tube inlet from the expanding sample. The indicator change corresponds to not more than 2.5 ppm. Repeat the test, using another identical detector tube but taking the sample from the gaseous phase. The indicator change from this second test corresponds to not more than 2.5 ppm.

Nonvolatile Hydrocarbons   Pass a sample of liquid Carbon Dioxide from a storage container or sample cylinder through a commercial carbon dioxide snow horn directly into an open, clean container. Collect the resulting Carbon Dioxide snow in this container. Weigh 500 g of this sample into a clean beaker. Allow the Carbon Dioxide solid to sublime completely, with a watch-glass placed over the beaker to prevent ambient contamination. Wash the beaker with a residue-free solvent, and transfer the solvent from the beaker to a clean, tared watch-glass or petri dish with two additional rinses of the beaker with the solvent. Allow the solvent to evaporate, using heating to 104°, until the watch-glass or petri dish is at a constant weight. Determine the weight of the residue by difference. The weight of the residue does not exceed 5 mg (10 mg/kg).

Sulfur Dioxide   Pass 1050 ± 50 mL, taken from the liquid phase as described in the test for *Nitrogen Oxide and Nitrogen Dioxide*, through a sulfur dioxide detector tube at the rate specified for the tube: The indicator change corresponds to not more than 5 ppm.

Volatile Hydrocarbons

Standard Preparation   Flush a 500-mL, glass, septum-equipped sampling bulb with helium, and inject into the bulb a 5.00-mL sample of methane. Allow the bulb to stand for 15 min to permit the gases to mix, and then inject 2.50 mL of the mixture into a second 500-mL sampling bulb, also flushed with helium, and allow this tube to stand for 15 min to permit the gases to mix. This mixture is a nominal 50 ppm v/v standard. Determine the exact concentration from the exact volumes of the gas-sampling bulbs. To determine these volumes, weigh the empty tubes, fill them with water, and reweigh. From the weight of the water and its temperature, calculate the volumes of the tubes.

Chromatographic System   The gas chromatograph is equipped with a flame ionization detector and a 1.8-m × 3-mm od metal column packed with 80- to 100-mesh Hayesep Q (or equivalent). The carrier gas is helium at a flow rate of 30 mL/min. The injector temperature and the detector temperatures both are maintained at 230°. The column temperature is programmed according to the following steps: It is held at 70° for 1 min, then increased to 200° at a rate of 20°/min, and then held at 200° for 10 min. The parameters for the detector are sensitivity range: 10^{-12} A/mV; attenuation: 32. The concentration of volatile hydrocarbons is reported in methane equivalents. The various gas chromatographic responses, excluding the Carbon Dioxide response, are summed to yield the total volatile hydrocarbon concentration. The composition of hydrocarbons present will vary from sample to sample. Typical retention times are methane: 0.4 min; carbon dioxide: 0.8 min; hexane: 14.4 min.

Procedure   Inject in triplicate 1.00 mL of the *Standard Preparation* into the gas chromatograph, and average the peak area responses. The relative standard deviation should not exceed 5.0%. Similarly, inject in triplicate 1.00 mL of sample, sum the average peak areas of the individual peaks, *except for the carbon dioxide peaks*, and calculate the ppm v/v in the sample by the formula

$$\text{ppm} = S(A_U/A_S),$$

in which S is the calculated ppm of methane in the *Standard Preparation* (approximately 50 ppm), A_U is the sum of the averages of the individual peak area responses in the sample, and A_S is the average area of the *Standard Preparation* area responses.

Water   Pass 24,000 mL of the gas sample through a suitable water-absorption tube not less than 100 mm in length, which previously has been flushed with about 500 mL of the sample and weighed. Regulate the flow so that about 60 min will be required for passage of the gas. The gain in weight of the absorption tube does not exceed 1.0 mg.

Packaging and Storage   Store in metal cylinders.

Cardamom Oil

CAS: [8000-66-6]

DESCRIPTION

The volatile oil distilled from the seed of *Elettaria cardamomum* (L.) Maton (Fam. *Zingiberaceae*). It is a colorless or very pale yellow liquid with the aromatic, penetrating, and somewhat camphoraceous odor of cardamom and a pungent, strongly aromatic taste. It is affected by light. It is miscible with alcohol.

Functional Use in Foods   Flavoring agent.

REQUIREMENTS

Identification   The infrared absorption spectrum of the sample exhibits relative maxima (that may vary in intensity) at the same wavelengths (or frequencies) as those shown in the respective spectrum in the section on *Infrared Spectra (Series A: Essential Oils)*, using the same test conditions as specified therein.

Angular Rotation   Between +22° and +44°.

Heavy Metals (as Pb)   Passes test.
Refractive Index   Between 1.462 and 1.466 at 20°.
Solubility in Alcohol   Passes test.
Specific Gravity   Between 0.917 and 0.947.

TESTS

Angular Rotation   Determine in a 100-mm tube as directed under *Optical (Specific) Rotation*, Appendix IIB.
Heavy Metals   Shake 10 mL of the oil with an equal volume of water to which 1 drop of hydrochloric acid has been added, and pass hydrogen sulfide through the mixture until it is saturated. No darkening in color is produced in either the oil or the water.
Refractive Index, Appendix IIB   Determine with an Abbé or other refractometer of equal or greater accuracy.
Solubility in Alcohol   Proceed as directed in the general method, Appendix VI. One mL dissolves in 5 mL of 70% alcohol. The solution may be clear or hazy.
Specific Gravity   Determine by any reliable method (see *General Provisions*).

Packaging and Storage   Store in full, tight containers in a cool place protected from light.

Carmine

Carminic Acid

$C_{22}H_{20}O_{13}$ Formula wt 492.39
INS: 120   CAS: [1390-65-4]

DESCRIPTION

Carmine is the aluminum or the calcium–aluminum lake, on an aluminum hydroxide substrate, of the coloring principles obtained by an aqueous extraction of cochineal. Cochineal consists of the dried female insects *Dactylopius coccus costa* (*Coccus cacti L.*), enclosing young larvae; the coloring principles derived therefrom consist chiefly of carminic acid ($C_{22}H_{20}O_{13}$).

Carminic acid crystallizes from water as bright red crystals that darken at 130° and decompose at 250°; it is freely soluble in water, in alcohol, in ether, in concentrated sulfuric acid, and in solutions of alkali hydroxides; it is insoluble in petroleum benzin and in chloroform; its solutions at pH 4.8 are red-orange to yellow, and at 6.2 are dark red to violet.

Carmine occurs as bright red, friable pieces or as a dark red powder. It is soluble in alkali solutions, slightly soluble in hot water, and practically insoluble in cold water and in dilute acids.

Before use in food, Carmine must have been pasteurized or otherwise treated to destroy all viable *Salmonella* microorganisms. According to the pertinent U.S. color additive regulation (**21** *CFR* 73.100), pasteurization or such other treatment is deemed to permit the addition of safe and suitable substances (other than chemical preservatives) that are essential to the method of pasteurization or other treatment used.

Functional Use in Foods   Color.

REQUIREMENTS

Identification   Mix 333 mg of Carmine with 44 mL of water, 0.15 mL of sodium hydroxide solution (1 in 10), and 0.2 mL of ammonium hydroxide, warm to dissolve, and dilute to volume with water in a 500-mL volumetric flask. Pipet 10.0 mL of this solution into a 250-mL volumetric flask, dilute to volume with water, and mix. The resulting solution exhibits absorption maxima at 520 nm and 550 nm, when determined in a 1-cm cell with a suitable spectrophotometer against a water blank, and the absorbance at 520 nm is not less than 0.30.
Assay   Not less than 42.0%[1] of carminic acid ($C_{22}H_{20}O_{13}$), calculated on the dried basis.
Arsenic (as As)   Not more than 1 mg/kg.
Ash   Not more than 12.0%.
Lead   Not more than 10 mg/kg.
Loss on Drying   Not more than 20.0%.
Microbial Limits:
   Salmonella   Negative in 25 g.

TESTS

Assay   Accurately weigh about 81 mg of Carmine, dissolve in 30 mL of 2 *N* hydrochloric acid, and heat to a boil for 30 s. After cooling, dilute to a volume of 1000 mL. Determine the absorbance of this solution in a 1-cm cell at the wavelength of maximum absorbance at about 494 nm, with a suitable spectrophotometer, using a 1 in 3 dilution of 2 *N* hydrochloric acid as the blank. Calculate the percentage of carminic acid in the sample of Carmine taken by the formula

$$0.1 A/13.9 W,$$

in which *A* is the absorbance of the sample solution and *W* is the weight, in mg, of the sample taken.
Arsenic, Appendix IIIB   Transfer 3.0 g of the sample into a 500-mL Kjeldahl flask equipped with a steam trap, add 5 g of ferrous sulfate and 75 mL of hydrochloric acid, and mix. Connect the flask with the steam trap and with a condenser, the delivery tube of which consists of a large-size straight adapter and extends

---

[1]The 42.0% minimum content of carminic acid specified herein does not indicate that the FCC-grade product is of any lower quality than that described in the second supplement to FCC II (page 18), which specified 50.0%. Rather, the revised *Assay* procedure used for this edition gives a more accurate indication of the true carminic acid content.

to slightly above the bottom of a 500-mL Erlenmeyer flask containing 100 mL of water. Begin heating the Kjeldahl flask and collect about 40 mL of distillate in the Erlenmeyer flask. Pour the distillate mixture into a 600-mL beaker, add 20 mL of bromine water, and heat on a hot plate until the volume is reduced to about 2 mL. Transfer the residual liquid into a 125-mL arsine generator flask (see Appendix IIIB, Figure 11) with the aid of 35 mL of water, and continue as directed in the *Procedure* under *Arsenic Test*, Appendix IIIB, beginning with "Add 20 mL of dilute sulfuric acid (1 in 5). . . ."

Ash Transfer about 1 g of the sample into a tared, previously ignited and cooled porcelain crucible, and ignite with a Meker burner (red hot) to constant weight.

Lead A *Sample Solution* prepared as directed for organic compounds meets the requirements of the *Lead Limit Test*, Appendix IIIB, using 10 μg of lead ion (Pb) in the control.

Loss on Drying, Appendix IIC Dry a 1-g sample at 135° for 3 h.

Microbial Limits:
  Salmonella Proceed as directed in chapter 5 of the *FDA Bacteriological Analytical Manual*, Seventh Edition, 1992.

Packaging and Storage Store in well-closed containers in a cool, dry place.

Carnauba Wax

INS: 903 CAS: [8015-86-9]

DESCRIPTION

A purified wax obtained from the leaf buds and leaves of the Brazilian wax palm *Copernicia cereferia* (Arruda) Mart. It is hard and brittle, has a resinous fracture, and ranges in color from light brown to pale yellow. Its specific gravity is about 0.997. It is partially soluble in boiling alcohol, is soluble in chloroform and in ether, but is insoluble in water.

Functional Use in Foods Candy glaze and polish.

REQUIREMENTS

Acid Value Between 2 and 7.
Ester Value Between 71 and 88.
Heavy Metals (as Pb) Not more than 0.002%.
Lead Not more than 10 mg/kg.
Melting Range Between 80° and 86°.
Residue on Ignition Not more than 0.25%.
Saponification Value Between 78 and 95.
Unsaponifiable Matter Between 50.0% and 55.0%.

TESTS

Acid Value Determine as directed for *Acid Value, Method I*, under *Fats and Related Substances*, Appendix VII.

Ester Value Subtract the *Acid Value* from the *Saponification Value* to obtain the *Ester Value*.

Heavy Metals Prepare and test a 1-g sample as directed for *Method II* under the *Heavy Metals Test*, Appendix IIIB, using 20 μg of lead ion (Pb) in the control (*Solution A*).

Lead A *Sample Solution* prepared as directed for organic compounds meets the requirements of the *Lead Limit Test*, Appendix IIIB, using 10 μg of lead ion (Pb) in the control.

Melting Range Determine as directed for *Melting Range or Temperature, Procedure for Class II*, Appendix IIB.

Residue on Ignition Heat a 2-g sample in a tared, open, porcelain or platinum dish over an open flame. It volatilizes without emitting an acrid odor. Ignite as directed in the general method, Appendix IIC.

Saponification Value Weigh accurately about 5 g of the sample, and determine as directed for *Saponification Value* under *Fats and Related Substances*, Appendix VII.

Unsaponifiable Matter Determine as directed in the general method, Appendix VII.

Packaging and Storage Store in well-closed containers.

β-Carotene

Carotene

$C_{40}H_{56}$ Formula wt 536.88
INS: 160a(i) CAS: [7235-40-7]

DESCRIPTION

Red crystals or crystalline powder. It is insoluble in water and in acids and alkalies, but is soluble in carbon disulfide and in chloroform. It is sparingly soluble in ether, in solvent hexane, and in vegetable oils, and is practically insoluble in methanol and in ethanol. It melts between 176° and 182°, with decomposition.

Functional Use in Foods Nutrient; dietary supplement; color.

REQUIREMENTS

Identification
A. Determine the absorbance of *Sample Solution B* (prepared for the *Assay*) at 455 nm and at 483 nm. The ratio A_{455}/A_{483} is between 1.14 and 1.18.

B. Determine the absorbance of *Sample Solution B* at 455 nm and that of *Sample Solution A* at 340 nm. The ratio A_{455}/A_{340} is not lower than 1.5.

Assay Not less than 96.0% and not more than 101.0% of $C_{40}H_{56}$, calculated on the dried basis.
Heavy Metals (as Pb) Not more than 10 mg/kg.
Loss on Drying Not more than 0.2%.
Residue on Ignition Not more than 0.2%.

TESTS

Assay (**Note**: Carry out all work in low-actinic glassware and in subdued light.)

Sample Solution A Transfer about 50 mg, accurately weighed, into a 100-mL volumetric flask, dissolve in 10 mL of acid-free chloroform, immediately dilute to volume with cyclohexane, and mix. Pipet 5 mL of this solution into a second 100-mL volumetric flask, dilute to volume with cyclohexane, and mix.

Sample Solution B Pipet 5 mL of *Sample Solution A* into a 50-mL volumetric flask, dilute to volume with cyclohexane, and mix.

Procedure Determine the absorbance of *Sample Solution B* in a 1-cm cell at the wavelength of maximum absorption at about 455 nm, with a suitable spectrophotometer, using cyclohexane as the blank. Calculate the quantity, in mg, of $C_{40}H_{56}$ in the sample taken by the formula

$$20{,}000A/250,$$

in which A is the absorbance of the solution, and 250 is the absorptivity of pure β-Carotene.

Heavy Metals Prepare and test a 2-g sample as directed in *Method II* under the *Heavy Metals Test*, Appendix IIIB, using 20 μg of lead ion (Pb) in the control (*Solution A*).

Loss on Drying, Appendix IIC Dry in a vacuum over phosphorus pentoxide at 40° for 4 h.

Residue on Ignition Ignite 2 g as directed in the general method, Appendix IIC.

Packaging and Storage Store in a cool place in tight, light-resistant containers under inert gas.

Carrot Seed Oil

DESCRIPTION

The volatile oil obtained by steam distillation from the crushed seeds of *Daucus carota* L. (Fam. *Umbelliferae*). It is a light yellow to amber liquid having a pleasant aromatic odor. It is soluble in most fixed oils, and is soluble, with opalescence, in mineral oil. It is practically insoluble in glycerin and in propylene glycol.

Functional Use in Foods Flavoring agent.

REQUIREMENTS

Identification The infrared absorption spectrum of the sample exhibits relative maxima (that may vary in intensity) at the same wavelengths (or frequencies) as those shown in the respective spectrum in the section on *Infrared Spectra* (*Series A: Essential Oils*), using the same test conditions as specified therein.
Acid Value Not more than 5.0.
Angular Rotation Between −4° and −30°.
Heavy Metals (as Pb) Passes test.
Refractive Index Between 1.483 and 1.493 at 20°.
Saponification Value Between 9 and 58.
Solubility in Alcohol Passes test.
Specific Gravity Between 0.900 and 0.943.

TESTS

Acid Value Determine as directed for *Acid Value* under *Essential Oils and Flavors*, Appendix VI.

Angular Rotation Determine in a 100-mm tube as directed under *Optical (Specific) Rotation*, Appendix IIB.

Heavy Metals Shake 10 mL of the oil with an equal volume of water to which 1 drop of hydrochloric acid has been added, and pass hydrogen sulfide through the mixture until it is saturated. No darkening in color is produced in either the oil or the water.

Refractive Index, Appendix IIB Determine with an Abbé or other refractometer of equal or greater accuracy.

Saponification Value Determine as directed for *Saponification Value* under *Essential Oils and Flavors*, Appendix VI, using about 5 g, accurately weighed.

Solubility in Alcohol Proceed as directed in the general method, Appendix VI. One mL dissolves in 0.5 mL of 90% alcohol. The solution may become opalescent upon further dilution up to 10 mL.

Specific Gravity Determine by any reliable method (see *General Provisions*).

Packaging and Storage Store in full, tight, preferably glass, aluminum, tin-lined, or other suitably lined containers in a cool place protected from light.

Cascarilla Oil

Sweetwood Bark Oil

CAS: [8007-06-5]

DESCRIPTION

The volatile oil obtained by steam distillation of the dried bark of *Croton cascarilla* Benn. and of *Croton eluteria* Benn. (Fam. *Euphorbiaceae*). It is a light yellow to brown amber liquid having a pleasant spicy odor. It is soluble in most fixed oils and in mineral oil, but it is practically insoluble in glycerin and in propylene glycol.

Functional Use in Foods Flavoring agent.

REQUIREMENTS

Identification The infrared absorption spectrum of the sample exhibits relative maxima (that may vary in intensity) at the same wavelengths (or frequencies) as those shown in the respective spectrum in the section on *Infrared Spectra (Series A: Essential Oils)*, using the same test conditions as specified therein.
Acid Value Between 3 and 10.
Angular Rotation Between −1° and +8°.
Ester Value after Acetylation Between 62 and 88.
Heavy Metals (as Pb) Passes test.
Refractive Index Between 1.488 and 1.494 at 20°.
Saponification Value Between 8 and 20.
Solubility in Alcohol Passes test.
Specific Gravity Between 0.892 and 0.914.

TESTS

Acid Value Determine as directed for *Acid Value* under *Essential Oils and Flavors*, Appendix VI.
Angular Rotation Determine in a 100-mm tube as directed under *Optical (Specific) Rotation*, Appendix IIB.
Ester Value after Acetylation Proceed as directed under *Total Alcohols*, Appendix VI, using about 2 g of the dried acetylated oil, accurately weighed. Calculate the *Ester Value after Acetylation* by the formula

$$A \times 28.05/B,$$

in which A is the number of mL of 0.5 N alcoholic potassium hydroxide consumed in the saponification, and B is the weight of the sample of acetylized oil, in g.
Heavy Metals Shake 10 mL of the oil with an equal volume of water to which 1 drop of hydrochloric acid has been added, and pass hydrogen sulfide through the mixture until it is saturated. No darkening in color is produced in either the oil or the water.
Refractive Index, Appendix IIB Determine with an Abbé or other refractometer of equal or greater accuracy.
Saponification Value Determine as directed for *Saponification Value* under *Essential Oils and Flavors*, Appendix VI, using 5 g, accurately weighed.
Solubility in Alcohol Proceed as directed in the general method, Appendix VI. One mL dissolves in 0.5 mL of 90% alcohol and remains in solution on dilution to 10 mL.
Specific Gravity Determine by any reliable method (see *General Requirements*).

Packaging and Storage Store in full, tight, preferably glass, aluminum, or tin-lined containers in a cool place protected from light.

Casein and Caseinate Salts

CAS: [9000-71-9]

DESCRIPTION

Casein is an off-white to cream-colored granular or fine powder derived from the coagulum formed by treating skim milk with a food-grade acid (acid Casein), enzyme (rennet Casein), or other food-grade precipitating agent. After the precipitation, Casein is separated from the soluble milk fraction, washed, and dried. Chemically, Casein is a mixture of at least 20 electrophoretically distinct phosphoproteins. The main fractions—designated α-casein, β-casein, and κ-casein—are known to be mixtures, rather than single proteins. Casein contains all the amino acids known to be essential for human nutrition. It is insoluble in water and alcohol, but can be dissolved by aqueous alkalies to form Caseinate Salts. Caseinate Salts are white- to cream-colored granules or powders soluble or dispersible in water. They are prepared by treatment of Casein with food-grade alkalies, neutralizing agents, enzymes, buffers, or sequestrants. Common counter ions are NH_4^+, Ca^{++}, Mg^{++}, K^+, and Na^+.

Functional Use in Foods Binder; extender; clarifying agent; emulsifier; stabilizer.

REQUIREMENTS

Assay Not less than 90.0% protein for acid Casein; not less than 86.0% protein for rennet Casein; not less than 84.0% for Caseinate Salts, calculated on the dried basis.
Fat Not more than 2.25%.
Free Acid Passes test.
Heavy Metals (as Pb) Not more than 0.002%.
Lactose Not more than 2.0%.
Lead Not more than 5 mg/kg.
Loss on Drying Not more than 12.0%.
Microbial Limits:
  Aerobic Plate Count Not more than 100,000 per g.
  Coliforms Not more than 2/0.1 g.
  Salmonella Negative in 25 g.

TESTS

Assay Proceed as directed under *Nitrogen Determination*, Appendix IIIC. Calculate the percent protein (P) by the formula

$$P = N \times 6.38,$$

in which N is the percent nitrogen.
Fat Transfer to a fat-extraction flask 1 g, accurately weighed; add 10 mL of water, and shake until homogeneous (warm if necessary). Add approximately 1 mL of ammonium hydroxide and heat in a water bath for 15 min at 60° to 70°, shaking occasionally. Add 10 mL of alcohol and mix well. Add 25 mL of peroxide-free ether, stopper, and shake vigorously for 1 min; allow to cool if necessary; add 25 mL of petroleum ether and repeat vigorous shaking. Allow the layers to separate and clarify or centrifuge at 600 rpm to expedite the process. Decant the

organic layer into a suitable flask or dish and repeat the extraction twice with 15 mL each of ether and petroleum ether for each extraction. Evaporate the combined ether extractions on a steam bath and dry the residue to a constant weight at 102°, or 70° to 75° at less than 50 mm Hg. Calculate the percent fat (F) by the formula

$$F = (R \times 100)/S,$$

in which R is the weight of the residue and S is the weight of the sample.

Free Acid Accurately weigh a 10-g portion of the finely ground sample, and transfer to a 500-mL conical flask. Add 200 mL of freshly boiled water maintained at 60°, swirl, and stopper. Place the flask in an 80° water bath, and hold for 30 min. Shake at 10-min intervals. Cool to room temperature, and filter. Accurately transfer a 100.0-mL portion of the clear filtrate to a 250-mL conical flask, add 0.5 mL of phenolphthalein TS, and titrate with 0.1 N sodium hydroxide to a pink endpoint that persists for 30 s. Not more than 2.7 mL of 0.1 N sodium hydroxide is consumed.

Heavy Metals Prepare and test a 1-g sample as directed in *Method II* under the *Heavy Metals Test*, Appendix IIIB, using 20 µg of lead ion (Pb) in the control (*Solution A*).

Lactose

Apparatus Use a suitable absorption spectrophotometer capable of operating in the visible range.

Phenol Reagent Heat a mixture of 8 g of phenol and 2 g of water until the crystals dissolve.

Lactose Solution Transfer approximately 2 g of lactose monohydrate, accurately weighed, to a 100-mL volumetric flask; dissolve in and dilute to volume with water.

Sample Solution Transfer approximately 1 g of sample, accurately weighed, to a 150-mL beaker. If the sample is acid Casein, add 0.10 g of sodium hydrogen carbonate. If the sample is rennet Casein, add 0.10 g sodium tripolyphosphate. Add 25 mL of water, and dissolve the sample by gently swirling while warming to 60° to 70° on a hot plate. Cool the solution to ambient temperature and add 15 mL water, 8 mL of 0.1 N hydrochloric acid, and 1 mL of a 10% solution of acetic acid. Mix well by swirling, and after 5 min, add 1 mL of 1 M sodium acetate; mix well.

After the precipitate has settled, filter and discard the first 5 mL of filtrate. Pipet 2 mL of the remaining filtrate into a test tube, add 0.2 mL *Phenol Reagent*, and mix well. Add 5 mL of sulfuric acid using an automatic dispenser or other means that permits mixing within 1 to 2 s. Ensure that the solution has been thoroughly mixed, and allow it to stand for 15 min, then cool to 20° in a water bath for 5 min.

Standard Solutions Transfer 10 mL of *Lactose Solution* to a 100-mL volumetric flask; dissolve in and dilute to volume with water (diluted *Lactose Solution*). Transfer respectively, 1, 2, 3, and 4 mL of diluted *Lactose Solution* to four 100-mL volumetric flasks; dilute to volume with water. These dilutions (standard dilutions) contain 20, 40, 60, and 80 µg of lactose per mL of solution, respectively. Into each of five test tubes add, in sequence, 2 mL of water and, respectively, 3 mL each of the standard dilutions of lactose. Then to each test tube add *Phenol Reagent* and sulfuric acid as described under *Sample Solution*.

Calibration Determine the absorbance of each *Standard Solution* at 490 nm against the water blank. Calculate the slope of the curve obtained by plotting absorbance versus µg/mL of lactose. The slope of the curve is the absorptivity (a) of the lactose–reagent product, presuming a cell of 1-cm pathlength is used for absorbance readings.

Procedure Determine the absorbance of the *Sample Solution* at 490 nm against a blank prepared using identical reagents.

Calculation Calculate the percent lactose (L) in the Casein sample by the formula

$$L = (A \times 0.00475)/(a \times m),$$

in which A is the absorbance of the *Sample Solution* at 490 nm; a is the absorptivity calculated under *Calibration*; m is the sample weight, in g; and the numerical factor accounts for dilution and conversion to percent from µg/mL.

Lead A *Sample Solution* prepared as directed for organic compounds meets the requirements of the *Lead Limit Test*, Appendix IIIB, using 5 µg of lead ion in the control.

Loss on Drying, Appendix IIC Dry at 102° for 3 h.

Microbial Limits:

Aerobic Plate Count Proceed as directed in chapter 3, *FDA Bacteriological Analytical Manual*, Seventh Edition, Food and Drug Administration, 1992.

Coliforms Proceed as directed in chapter 4, *FDA Bacteriological Analytical Manual*, Seventh Edition, Food and Drug Administration, 1992.

Salmonella Proceed as directed in chapter 5, *FDA Bacteriological Analytical Manual*, Seventh Edition, Food and Drug Administration, 1992.

Packaging and Storage Store in well-closed containers.

Cassia Oil

Cinnamon Oil

DESCRIPTION

The volatile oil obtained by steam distillation from the leaves and twigs of *Cinnamomum cassia* Blume (Fam. *Lauraceae*), rectified by distillation. It is a yellowish or brownish liquid having the characteristic odor and taste of cassia cinnamon. Upon aging or exposure to air it darkens and thickens. It is soluble in glacial acetic acid and in alcohol.

Functional Use in Foods Flavoring agent.

REQUIREMENTS

Identification The infrared absorption spectrum of the sample exhibits relative maxima (that may vary in intensity) at the same wavelengths (or frequencies) as those shown in the respective spectrum in the section on *Infrared Spectra* (*Series A: Essential Oils*), using the same test conditions as specified therein.

Assay Not less than 80.0%, by volume, of total aldehydes.
Angular Rotation Between −1° and +1°.
Chlorinated Compounds Passes test.
Heavy Metals (as Pb) Passes test.
Refractive Index Between 1.602 and 1.614 at 20°.
Rosin or Rosin Oils Passes test.
Solubility in Alcohol Passes test.
Specific Gravity Between 1.045 and 1.063.

TESTS

Assay Proceed as directed under *Aldehydes and Ketones—Neutral Sulfite Method*, Appendix VI.
Angular Rotation Determine in a 100-mm tube as directed under *Optical (Specific) Rotation*, Appendix IIB.
Chlorinated Compounds Proceed as directed in the general method, Appendix VI.
Heavy Metals Shake 10 mL of the oil with an equal volume of water to which 1 drop of hydrochloric acid has been added, and pass hydrogen sulfide through the mixture until it is saturated. No darkening in color is produced in either the oil or the water.
Refractive Index, Appendix IIB Determine with an Abbé or other refractometer of equal or greater accuracy.
Rosin or Rosin Oils Shake a 2-mL sample in a test tube with 5 to 10 mL of solvent hexane, allow the liquids to separate, decant the hexane layer, which is but slightly colored, into another test tube, and shake it with an equal volume of cupric acetate solution (1 in 1000). The mixture does not assume a green color.
Solubility in Alcohol Proceed as directed in the general method, Appendix VI. One mL dissolves in 2 mL of 70% alcohol.
Specific Gravity Determine by any reliable method (see *General Provisions*).

Packaging and Storage Store in full, tight, light-resistant containers. Avoid exposure to excessive heat.

Castor Oil

INS: 1503   CAS: [8001-79-4]

DESCRIPTION

The fixed oil obtained from the seed of *Ricinus communis* L. (Fam. *Euphorbiaceae*). It is a pale yellowish or almost colorless, transparent, viscous liquid and has a faint, mild odor and a bland, characteristic taste. It is soluble in alcohol, and is miscible with absolute alcohol, with glacial acetic acid, with chloroform, and with ether.

Functional Use in Foods Antisticking agent; release agent; component of protective coatings.

REQUIREMENTS

Identification It is only partly soluble in solvent hexane (distinction from *most other fixed oils*), but it yields a clear liquid with an equal volume of alcohol (*foreign fixed oils*).
Free Fatty Acids Passes test.
Heavy Metals (as Pb) Not more than 10 mg/kg.
Hydroxyl Value Between 160 and 168.
Iodine Value Between 83 and 88.
Saponification Value Between 176 and 185.
Specific Gravity Between 0.952 and 0.966.

TESTS

Free Fatty Acids Dissolve about 10 g, accurately weighed, in 50 mL of a mixture of equal volumes of alcohol and ether (which has been neutralized to phenolphthalein with 0.1 N sodium hydroxide) contained in a flask. Add 1 mL of phenolphthalein TS, and titrate with 0.1 N sodium hydroxide until the solution remains pink after shaking for 30 s. Not more than 7 mL of 0.1 N sodium hydroxide is required for a 10.0-g sample.
Heavy Metals Prepare and test a 2-g sample as directed in *Method II* under the *Heavy Metals Test*, Appendix IIIB, using 20 μg of lead ion (Pb) in the control (*Solution A*).
Hydroxyl Value Determine as directed under *Method II* in the general method, Appendix VII.
Iodine Value Determine by the *Modified Wijs Method*, Appendix VII, using about 300 mg, accurately weighed.
Saponification Value Determine as directed for *Saponification Value* under *Fats and Related Substances*, Appendix VII, using about 3 g, accurately weighed.
Specific Gravity Determine as directed in the general method, Appendix VII.

Packaging and Storage Store in tight containers, and avoid exposure to excessive heat.

Cedar Leaf Oil

Thuja Oil; White Cedar Leaf Oil

                                                       CAS: [8007-20-3]

DESCRIPTION

The volatile oil obtained by steam distillation from the fresh leaves and branch ends of the eastern arborvitae, *Thuja occidentalis* L. (Fam. *Cupressaceae*). It is a colorless to yellow liquid having a strong camphoraceous and sagelike odor. It is soluble in most fixed oils, in mineral oil, and in propylene glycol. It is practically insoluble in glycerin.

Functional Use in Foods Flavoring agent.

REQUIREMENTS

Identification The infrared absorption spectrum of the sample exhibits relative maxima (that may vary in intensity) at the same wavelengths (or frequencies) as those shown in the respective spectrum in the section on *Infrared Spectra* (*Series A: Essential Oils*), using the same test conditions as specified therein.
Assay Not less than 60.0% of ketones, calculated as thujone ($C_{10}H_{16}O$).
Angular Rotation Between −10° and −14°.
Heavy Metals (as Pb) Passes test.
Refractive Index Between 1.456 and 1.459 at 20°.
Solubility in Alcohol Passes test.
Specific Gravity Between 0.910 and 0.920.

TESTS

Assay Weigh accurately about 1 g, and proceed as directed under *Aldehydes and Ketones—Hydroxylamine Method*, Appendix VI, using 76.10 as the equivalence factor (*e*) in the calculation.
Angular Rotation Determine in a 100-mm tube as directed under *Optical (Specific) Rotation*, Appendix IIB.
Heavy Metals Shake 10 mL of the oil with an equal volume of water to which 1 drop of hydrochloric acid has been added, and pass hydrogen sulfide through the mixture until it is saturated. No darkening in color is produced in either the oil or the water.
Refractive Index, Appendix IIB Determine with an Abbé or other refractometer of equal or greater accuracy.
Solubility in Alcohol Proceed as directed in the general method, Appendix VI. One mL dissolves in 3 mL of 70% alcohol, occasionally becoming cloudy on dilution to 10 mL.
Specific Gravity Determine by any reliable method (see *General Provisions*).

Packaging and Storage Store in full, tight, preferably glass or tin-lined containers in a cool place protected from light.

Celery Seed Oil

DESCRIPTION

The volatile oil obtained by steam distillation of the fruit or seed of *Apium graveolens* L. It is a yellow to greenish brown liquid having a pleasant aromatic odor. It is soluble in most fixed oils with the formation of a flocculent precipitate, and in mineral oil with turbidity. It is partly soluble in propylene glycol, and it is insoluble in glycerin.

Functional Use in Foods Flavoring agent.

REQUIREMENTS

Identification The infrared absorption spectrum of the sample exhibits relative maxima (that may vary in intensity) at the same wavelengths (or frequencies) as those shown in the respective spectrum in the section on *Infrared Spectra* (*Series A: Essential Oils*), using the same test conditions as specified therein.
Acid Value Not more than 4.5.
Angular Rotation Between +48° and +78°.
Heavy Metals (as Pb) Passes test.
Refractive Index Between 1.480 and 1.490 at 20°.
Saponification Value Between 25 and 65.
Solubility in Alcohol Passes test.
Specific Gravity Between 0.870 and 0.910.

TESTS

Acid Value Determine as directed for *Acid Value* under *Essential Oils and Flavors*, Appendix VI.
Angular Rotation Determine in a 100-mm tube as directed under *Optical (Specific) Rotation*, Appendix IIB.
Heavy Metals Shake 10 mL of the oil with an equal volume of water to which 1 drop of hydrochloric acid has been added, and pass hydrogen sulfide through the mixture until it is saturated. No darkening in color is produced in either the oil or the water.
Refractive Index, Appendix IIB Determine with an Abbé or other refractometer of equal or greater accuracy.
Saponification Value Determine as directed for *Saponification Value* under *Essential Oils and Flavors*, Appendix VI, using 5 g, accurately weighed.
Solubility in Alcohol Proceed as directed in the general method, Appendix VI. One mL dissolves in 8 mL of 90% alcohol, usually with turbidity.
Specific Gravity Determine by any reliable method (see *General Provisions*).

Packaging and Storage Store in full, tight, glass, tin-lined, or aluminum containers in a cool place protected from light.

Cellulose Gel

Cellulose, Microcrystalline

INS: 460(i)

DESCRIPTION

Cellulose Gel is purified, partially depolymerized cellulose prepared by treating alpha cellulose, obtained as a pulp from fibrous plant material, with mineral acids. It occurs as a fine, white, or almost white, powder. It consists of free-flowing, nonfibrous particles that may be compressed into self-binding tablets that disintegrate rapidly in water. It is insoluble in water, in dilute acids, in dilute sodium hydroxide solutions, and in most organic solvents.

Functional Use in Foods Anticaking agent; binding agent; disintegrating agent; dispersing agent; tableting aid.

REQUIREMENTS

Identification

A. Sieve 20 g for 5 min on an air-jet sieve equipped with a screen having 38-μm openings. If more than 5% is retained on the screen, mix 30 g of Cellulose Gel with 270 mL of water; otherwise, mix 45 g with 255 mL of water. Perform the mixing for 5 min in a single-speed, high-speed (equal to or greater than 18,000 rpm) power blender (use a Waring Blender, Model 700G, available from Fisher Scientific, or equivalent) that has a clover-shaped jar design. The jar and blades meet the following specifications: The jar has an inside diameter of 7.0 cm at the bottom and 9.2 cm at the top and an overall height of 21.9 cm, and the four blades are arranged so that two of the blades are pointed up and two are pointed down. Transfer 100 mL of the dispersion to a 100-mL graduated cylinder, and allow to stand for 3 h: A white, opaque, bubble-free dispersion, which does not form a supernatant liquid at the surface, is obtained.

B. To 20 mL of the dispersion obtained in *Identification Test A* add a few drops of iodine TS, and mix. No purplish to blue or blue color is produced.

Assay Not less than 97.0% and not more than 102.0% of carbohydrate, calculated as cellulose on the dried basis.
Heavy Metals (as Pb) Not more than 10 mg/kg.
Loss on Drying Not more than 7.0%.
pH Between 5.0 and 7.5.
Residue on Ignition Not more than 0.05%.
Water-Soluble Substances Not more than 0.24%.

TESTS

Assay Transfer about 125 mg of the sample, accurately weighed, to a 300-mL Erlenmeyer flask, using about 25 mL of water. Add 50.0 mL of 0.5 N potassium dichromate, mix, then carefully add 100 mL of sulfuric acid, and heat to boiling. Remove from heat, allow to stand at room temperature for 15 min, cool in a water bath, and transfer into a 250-mL volumetric flask. Dilute with water almost to volume, cool to 25°, then dilute to volume with water, and mix. Titrate a 50.0-mL aliquot with 0.1 N ferrous ammonium sulfate, using 2 or 3 drops of orthophenanthroline TS as the indicator, and record the volume required as *S*, in mL. Perform a blank determination, and record the volume of 0.1 N ferrous ammonium sulfate required as *B*, in mL. Calculate the percentage of cellulose in the sample by the formula

$$(B - S) \times 338/W,$$

in which *W* is the weight of sample taken, in mg, corrected for *Loss on Drying*.

Heavy Metals Prepare and test a 2-g sample as directed in *Method II* under the *Heavy Metals Test*, Appendix IIIB, using 20 μg of lead ion (Pb) in the control (*Solution A*).

Loss on Drying, Appendix IIC Dry to constant weight at 105°.

pH Shake about 5 g with 40 mL of water for 20 min, centrifuge, and determine the pH of the supernatant liquid by the *Potentiometric Method*, Appendix IIB.

Residue on Ignition, Appendix IIC Ignite 2 g as directed in the general method.

Water-Soluble Substances Shake 5 g with 80 mL of water for 10 min. Filter the mixture through Whatman No. 42 or equivalent filter paper into a tared beaker, evaporate the filtrate to dryness on a steam bath, dry at 105° for 1 h, cool, and weigh.

Packaging and Storage Store in well-closed containers.

Cellulose, Powdered

INS: 460(ii)   CAS: [9004-34-6]

DESCRIPTION

Powdered Cellulose is purified, mechanically disintegrated cellulose prepared by processing bleached cellulose obtained as a pulp from such fibrous materials as wood or cotton. It occurs as a white, odorless substance and consists of fibrous particles that may be compressed into self-binding tablets that disintegrate rapidly in water. It exists in various grades, exhibiting degrees of fineness ranging from a dense, free-flowing powder to a coarse, fluffy, nonflowing material. It is insoluble in water, in dilute acids, and in nearly all organic solvents. It is slightly soluble in 1 N sodium hydroxide.

Functional Use in Foods Anticaking agent; binding agent; bulking agent; disintegrating agent; dispersing agent; filter aid; texturizing agent; thickening agent.

REQUIREMENTS

Identification

A. Mix approximately 30 g of the sample with 270 mL of water in a high-speed (approximately 12,000 rpm) power blender for 5 min. The mixture will be either a free-flowing suspension or a heavy, lumpy suspension that flows poorly (if at all), settles only slightly, and contains many trapped air bubbles. The mixture is not slimy. If a free-flowing suspension is obtained, transfer 100 mL of it into a 100-mL graduate, and allow to settle for 1 h: The solids settle in the cylinder and a supernatant liquid appears above the layer of the Cellulose.

B. Boil 10 g of the sample with 90 mL of water for 5 min, filter while hot through ashless fine-quantitative paper (S & S 589 Blue Ribbon, or equivalent), and add 2 drops of iodine TS to the filtrate. No change in color from the yellow-red is produced.

C. To 20 mL of a 0.1% solution of anthrone in 75% sulfuric acid add from 2 to 5 mg of the sample, and heat on a steam bath. The solution turns blue green within 5 min.

D. Place a few drops of the mixture from *Identification Test A* on a microscope slide, and insert a coverglass. Observe at 100 magnifications with a microscope. Fibers and fiber fragments are visible, regardless of the degree of fineness of the sample.

E. Dilute 10 mL of the mixture from *Identification Test A* to 1000 mL with water, and filter 125 mL of the dilution through a Büchner funnel. Rinse the pad with 25 mL of acetone, and dry (paper included) at 105°. Transfer the powder to a tared weighing bottle, weigh, then transfer to a 50-mL Erlenmeyer flask, and seal with a rubber stopper. Record the weight of the sample as w, in mg. Dissolve the sample in 0.167 M and 1.0 M solutions of cupriethylenediamine (CED), the volumes of which are determined as follows: $0.12 \times w$ equals the mL of 0.167 M CED, and $0.08 \times w$ equals the mL of 1.0 M CED. Add a few 3-mm glass beads and the calculated volume of 0.167 M CED, blow nitrogen over the surface of the solution, and shake for 2 min. Add the calculated volume of 1.0 M CED, again introduce the nitrogen, and shake vigorously for at least 3 min. A dark blue solution, clear under microscopic examination, is produced.

Assay Not less than 97.0% and not more than 102.0% of carbohydrate, calculated as Cellulose.
Ash (Total) Not more than 0.3%.
Chloride Not more than 0.05%.
Heavy Metals (as Pb) Not more than 10 mg/kg.
Loss on Drying Not more than 7.0%.
pH Between 5.0 and 7.5.
Sulfur (Total) Not more than 0.01%.
Water-Soluble Substances Not more than 1.5%.

TESTS

Assay Transfer about 125 mg of the sample, accurately weighed, to a 300-mL Erlenmeyer flask, using about 25 mL of water. Add 50.0 mL of 0.5 N potassium dichromate, mix, then carefully add 100 mL of sulfuric acid, and heat to boiling. Remove from heat, allow to stand at room temperature for 15 min, then cool in a water bath, and transfer the solution to a 250-mL volumetric flask. Dilute with water almost to volume, cool to 25°, dilute to volume with water, and mix. Titrate a 50-mL aliquot with 0.1 N ferrous ammonium sulfate, using 2 or 3 drops of orthophenanthroline TS. Perform a blank determination, and calculate the normality, N, of the ferrous ammonium sulfate solution by the formula

$$(0.1 \times 50)/B,$$

in which B is the volume, in mL, of ferrous ammonium sulfate solution required in the blank titration. Calculate the percentage of Cellulose in the sample by the formula

$$6.75(B - S) \times N/2W,$$

in which S is the volume, in mL, of ferrous ammonium sulfate solution used in the sample titration, and W is the weight of the sample taken, in g, corrected for moisture content (see *Loss on Drying*).

Ash (Total) Heat 3 g at 550° ± 50° until completely charred, then ignite at 800° ± 25° until free from carbon, cool in a desiccator, and weigh.

Chloride Transfer about 5 g of the sample, accurately weighed, to a 500-mL conical flask, add 250 mL of water, and reflux the mixture for 1 h. Filter through paper, and again reflux the sample with 200 mL of water for 30 min. Filter and combine the filtrates and hot water rinses. Add 1 mL of nitric acid, heat to boiling, and slowly add 5 mL of a 5% solution of silver nitrate. After the precipitate has coagulated, cool, and filter through a sintered-glass filtering funnel. Wash with nitric acid solution (1 in 100) until free from silver nitrate, then rinse with water, dry at 130°, and weigh. Perform a blank determination to obtain the corrected weight of the sample precipitate, each mg of which is equivalent to 0.247 mg of chloride.

Heavy Metals Prepare and test a 2-g sample as directed in *Method II* under the *Heavy Metals Test*, Appendix IIIB, using 20 μg of lead ion (Pb) in the control (*Solution A*).

Loss on Drying, Appendix IIC Dry to constant weight at 105°.

pH Mix 10 g of the sample with 90 mL of water, allow to stand with occasional stirring for 1 h, and determine the pH of the supernatant liquid by the *Potentiometric Method*, Appendix IIB.

Sulfur (Total) Transfer about 5 g of the sample, previously dried at 105° to constant weight and accurately weighed, to a 300-mL conical flask, and add 50 mL of a 2:3 mixture, v/v, of perchloric acid and nitric acid. Heat on a hot plate under a hood, and boil until all organic matter has been destroyed and copious fumes of perchloric acid are evolved. If the organic matter chars and cannot be destroyed quickly by further heating for a short time, add 10 to 20 mL of the acid mixture and continue the treatment until a clear, syrupy residue is obtained.

> Note: It is absolutely necessary that all of the nitric acid be driven from the flask, as it will form a double salt with the barium sulfate formed later.

Allow the mixture to cool for a few min, then add 200 mL of hot water, and heat again to boiling. (If the solution is cloudy, filter and rinse the filter with a small amount of hot water before boiling.) As soon as the mixture is boiling gently, carefully run in 20 mL of barium chloride TS, boil for a few min longer, and allow to stand for at least 12 h on a steam bath. Filter any barium sulfate onto an ashless filter paper, and rinse with five portions of boiling water to remove traces of perchloric acid. Place the paper in a tared platinum dish, dry in an oven at 105°, and ignite at 800° ± 25° for 1 h. Perform a blank determination to obtain the corrected weight of the sample precipitate, each mg of which is equivalent to 0.137 mg of sulfur.

Water-Soluble Substances Mix 6 g of the sample with 90 mL of recently boiled and cooled water, and allow to stand with occasional stirring for 10 min. Filter, discard the first 10 mL of filtrate, and pass the filtrate through the same filter a second time, if necessary, to obtain a clear filtrate. Evaporate a 15-mL portion of the filtrate to dryness in a tared evaporating dish on a steam bath, dry at 105° for 1 h, cool in a desiccator, and weigh.

Packaging and Storage Store in well-closed containers.

Chamomile Oil, English Type

CAS: [8015-92-7]

DESCRIPTION

The oil obtained by steam distillation of the dried flowers of the so-called English or Roman Chamomile, *Anthemis nobilis* L. It is a light blue or light greenish blue liquid with a strong aromatic odor, characteristic of the flowers. The color may change with age to greenish yellow or yellow brown. It is soluble in most fixed oils, and it is almost completely soluble in mineral oil. It is soluble, with slight haziness, in propylene glycol, but it is insoluble in glycerin.

Functional Use in Foods Flavoring agent.

REQUIREMENTS

Identification The infrared absorption spectrum of the sample exhibits relative maxima (that may vary in intensity) at the same wavelengths (or frequencies) as those shown in the respective spectrum in the section on *Infrared Spectra (Series A: Essential Oils)*, using the same test conditions as specified therein.
Acid Value Not more than 15.0.
Ester Value Between 250 and 310.
Heavy Metals (as Pb) Passes test.
Refractive Index Between 1.440 and 1.450 at 20°.
Solubility in Alcohol Passes test.
Specific Gravity Between 0.892 and 0.910.

TESTS

Acid Value Determine as directed for *Acid Value* under *Essential Oils and Flavors*, Appendix VI.
Ester Value Determine as directed in the general method, Appendix VI, using about 1 g, accurately weighed.
Heavy Metals Shake 10 mL of the oil with an equal volume of water to which 1 drop of hydrochloric acid has been added, and pass hydrogen sulfide through the mixture until it is saturated. No darkening in color is produced in either the oil or the water.
Refractive Index, Appendix IIB Determine with an Abbé or other refractometer of equal or greater accuracy.
Solubility in Alcohol Proceed as directed in the general method, Appendix VI. One mL dissolves in 2 mL of 80% alcohol, sometimes with a slight precipitate.
Specific Gravity Determine by any reliable method (see *General Provisions*).

Packaging and Storage Store in full, tight, glass or aluminum containers in a cool place protected from light.

Chamomile Oil, German Type

Chamomile Oil, Hungarian Type

DESCRIPTION

The oil obtained by steam distillation of the flowers and stalks of *Matricaria chamomilla* L. It is a deep blue or bluish green liquid with a strong and characteristic odor and a bitter aromatic taste. When exposed to light or air, the blue color changes to green and finally to brown. Upon cooling, the oil may become viscous. It is soluble in most fixed oils and in propylene glycol. It is insoluble in glycerin and in mineral oil.

Functional Use in Foods Flavoring agent.

REQUIREMENTS

Identification The infrared absorption spectrum of the sample exhibits relative maxima (that may vary in intensity) at the same wavelengths (or frequencies) as those shown in the respective spectrum in the section on *Infrared Spectra (Series A: Essential Oils)*, using the same test conditions as specified therein.
Acid Value Between 5 and 50.
Ester Value Not more than 40.
Ester Value after Acetylation Between 65 and 155.
Heavy Metals (as Pb) Passes test.
Solubility in Alcohol Passes test.
Specific Gravity Between 0.910 and 0.950.

TESTS

Acid Value Determine as directed for *Acid Value* under *Essential Oils and Flavors*, Appendix VI.
Ester Value Determine as directed in the general method, Appendix VI, using about 5 g, accurately weighed.
Ester Value after Acetylation Acetylate a 10-mL sample as directed under *Total Alcohols*, Appendix VI. Weigh accurately about 1.5 g of the dried, acetylated oil, and proceed as directed under *Ester Value*, Appendix VI.
Heavy Metals Shake 10 mL of the oil with an equal volume of water to which 1 drop of hydrochloric acid has been added, and pass hydrogen sulfide through the mixture until it is saturated. No darkening in color is produced in either the oil or the water.
Solubility in Alcohol Proceed as directed in the general method, Appendix VI. The oil does not usually dissolve clearly in 95% alcohol.
Specific Gravity Determine by any reliable method (see *General Provisions*).

Packaging and Storage Store in full, tight, glass or aluminum containers in a cool place protected from light.

Chlorine

Cl₂ — Formula wt 70.91
INS: 925 — CAS: [7782-50-5]

DESCRIPTION

A greenish yellow gas, normally packaged as a liquid under pressure in containers approved by the U.S. Department of Transportation. At 60°F, it has a vapor pressure of 70.91 psig. Its vapor density is about 2.5 times that of air. About 0.8 lb (0.362 kg) is soluble in 100 lb (45.4 kg) of water at 60°F under atmospheric pressure.

Caution: Chlorine gas is a respiratory irritant. Large amounts cause coughing, labored breathing, and irritation of the eyes. In extreme cases, the difficulty in breathing may cause death due to suffocation. Liquid Chlorine causes skin and eye burns on contact. (Safety precautions to be observed in handling the material are specified in the *Chlorine Manual*, available from the Chlorine Institute, Suite 506, 2001 L Street, N.W., Washington, D.C. 20036.)

Functional Use in Foods Antimicrobial agent; bleaching agent; oxidizing agent.

REQUIREMENTS

Identification Cautiously pass a few mL of Chlorine gas through 10 mL of 1 N sodium hydroxide that has been previously chilled in an ice bath. The resulting solution gives positive tests for *Chloride*, Appendix IIIA, and it darkens starch iodide paper.
Assay Not less than 99.5%, by volume.
Heavy Metals (as Pb) Not more than 0.002%, by weight.
Lead Not more than 10 mg/kg.
Mercury Not more than 1 mg/kg.
Moisture Not more than 0.015%, by weight.
Residue Not more than 0.015%, by weight, of nonvolatile matter.

TESTS

Assay Determine by ASTM Method E 412-93, "Assay of Liquid Chlorine (Zinc Amalgam Method)."

Sample Solution for the Determination of Heavy Metals, Lead, and Mercury Dissolve the residue, obtained in the test for *Residue*, in 2.5 mL of freshly prepared aqua regia, and dilute with water to a volume, in mL, equivalent to the weight, in g, of the initial Chlorine sample, so that 1 mL of the final dilution is equivalent to 1 g of Chlorine.

Heavy Metals A 1.0-mL portion of the *Sample Solution*, diluted to 25 mL with water, meets the requirements of the *Heavy Metals Test*, Appendix IIIB, using 20 μg of lead ion (Pb) in the control (*Solution A*).
Lead A 1.0-mL portion of the *Sample Solution*, mixed with 5 mL of water and 11 mL of 2.7 N hydrochloric acid, meets the requirements of the *Lead Limit Test*, Appendix IIIB, using 10 μg of lead ion (Pb) in the control.
Mercury Transfer 2.0 mL of the *Sample Solution* into a 50-mL beaker, add 10 mL of water, 1 mL of dilute sulfuric acid (1 in 5), and 1 mL of potassium permanganate solution (1 in 25), cover with a watch glass, boil for a few s, and cool. Use the resulting solution as the *Sample Preparation* as directed under the *Mercury Limit Test*, Appendix IIIB.
Moisture and **Residue** Determine by ASTM Method E 410-92, "Moisture and Residue in Liquid Chlorine."

Packaging and Storage Store in suitable pressure containers, observing applicable U.S. regulations pertaining to shipping containers.

Cholic Acid

Cholalic Acid; 3,7,12-Trihydroxycholanic Acid

C₂₄H₄₀O₅ — Formula wt 408.58
INS: 1000 — CAS: [81-25-4]

DESCRIPTION

Colorless plates or a white, crystalline powder having a bitter taste with a sweetish aftertaste. One g dissolves in about 30 mL of alcohol or acetone and in about 7 mL of glacial acetic acid. It is very slightly soluble in water.

Functional Use in Foods Emulsifier.

REQUIREMENTS

Identification To 1 mL of a 1 in 5000 solution in 50% acetic acid add 1 mL of a solution of furfural (1 in 100). Cool in an ice bath for 5 min, add 15 mL of dilute sulfuric acid (1 in 2), mix, and warm in a water bath at 70° for 10 min. Immediately cool in an ice bath and stir for 2 min. A blue color develops.
Assay Not less than 98.0% of C₂₄H₄₀O₅, calculated on the dried basis.
Heavy Metals (as Pb) Not more than 0.002%.
Lead Not more than 10 mg/kg.
Loss on Drying Not more than 0.5%.

Melting Range Between 197° and 202°.
Residue on Ignition Not more than 0.1%.
Specific Rotation $[\alpha]_D^{25°}$: Not less than +37°, calculated on the dried basis.

TESTS

Assay Transfer about 400 mg, accurately weighed, into a 250-mL Erlenmeyer flask, add 20 mL of water and 40 mL of alcohol, cover with a watch glass, heat gently on a steam bath until dissolved, and cool. Add 5 drops of phenolphthalein TS, and titrate with 0.1 N sodium hydroxide, using a 10-mL microburet, to the first pink color that persists for 15 s. Perform a blank determination (see *General Provisions*) and make any necessary correction. Each mL of 0.1 N sodium hydroxide is equivalent to 40.86 mg of $C_{24}H_{40}O_5$.

Heavy Metals Prepare and test a 1-g sample as directed in *Method II* under the *Heavy Metals Test*, Appendix IIIB, using 20 μg of lead ion (Pb) in the control (*Solution A*).

Lead A *Sample Solution* prepared as directed for organic compounds meets the requirements of the *Lead Limit Test*, Appendix IIIB, using 10 μg of lead ion (Pb) in the control.

Loss on Drying, Appendix IIC Dry at 140° under a vacuum of not more than 5 mm of Hg for 4 h.

Melting Range Determine as directed for *Melting Range or Temperature*, Appendix IIB.

Residue on Ignition, Appendix IIC Ignite 2 g as directed in the general method.

Specific Rotation, Appendix IIB Determine in a solution in alcohol containing 200 mg in each 10 mL.

Packaging and Storage Store in tight containers.

Choline Bitartrate

(2-Hydroxyethyl)trimethylammonium Bitartrate

$[HOCH_2CH_2N^+(CH_3)_3]C_4H_4O_6^-$

$C_9H_{19}NO_7$ Formula wt 253.25
INS: 1001(v)   CAS: [87-67-2]

DESCRIPTION

A white, hygroscopic, crystalline powder having an acidic taste. It is odorless or may have a faint trimethylamine-like odor. It is freely soluble in water, slightly soluble in alcohol, and insoluble in ether and in chloroform.

Functional Use in Foods Nutrient; dietary supplement.

REQUIREMENTS

Identification

A. Dissolve 500 mg in 2 mL of water, add 3 mL of 1 N sodium hydroxide, and heat to boiling. The odor of trimethylamine is detectable.

B. Dissolve 500 mg in 2 mL of iodine TS. A reddish brown precipitate is immediately formed. Add 5 mL of 1 N sodium hydroxide. The precipitate dissolves and the solution becomes clear yellow. Heat the solution. A pale yellow precipitate forms and the odor of iodoform may be detected.

C. To 2 mL of cobaltous chloride TS add 1 mL of a 1 in 100 solution of the sample and 2 mL of potassium ferrocyanide solution (1 in 50). An emerald green color develops immediately.

Assay Not less than 98.0% of $C_9H_{19}NO_7$, calculated on the anhydrous basis.
1,4-Dioxane Passes test.
Heavy Metals (as Pb) Not more than 10 mg/kg.
Residue on Ignition Not more than 0.1%.
Water Not more than 0.5%.

TESTS

Assay Transfer about 500 mg, accurately weighed, into a 250-mL Erlenmeyer flask, add 50 mL of glacial acetic acid, and warm on a steam bath until solution is complete. Cool, add 2 drops of crystal violet TS, and titrate with 0.1 N perchloric acid in glacial acetic acid to a green endpoint. Perform a blank determination (see *General Provisions*), and make any necessary correction. Each mL of 0.1 N perchloric acid is equivalent to 25.36 mg of $C_9H_{19}NO_7$.

1,4-Dioxane Determine as directed in the general method, Appendix IIIB.

Heavy Metals A solution of 2 g in 25 mL of water meets the requirements of the *Heavy Metals Test*, Appendix IIIB, using 20 μg of lead ion (Pb) in the control (*Solution A*).

Residue on Ignition Ignite 2 g as directed in the general method, Appendix IIC.

Water Determine by drying in a vacuum desiccator over phosphorus pentoxide for 4 h or by the *Karl Fischer Titrimetric Method*, Appendix IIB, using a 2-g sample dissolved in 50 mL of methanol.

Packaging and Storage Store in tight containers.

Choline Chloride

(2-Hydroxyethyl)trimethylammonium Chloride

$$[HOCH_2CH_2N^+(CH_3)_3]Cl^-$$

$C_5H_{14}ClNO$ Formula wt 139.62
INS: 1001(iii)   CAS: [67-48-1]

DESCRIPTION

Colorless or white crystals or crystalline powder, usually having a slight odor of trimethylamine. It is hygroscopic, and is very soluble in water and in alcohol.

Functional Use in Foods   Nutrient; dietary supplement.

REQUIREMENTS

Identification
    A. It responds to *Identification Tests A, B*, and *C* in the monograph for *Choline Bitartrate*.
    B. A 1 in 20 solution gives positive tests for *Chloride*, Appendix IIIA.
Assay   Not less than 98.0% and not more than 100.5% of $C_5H_{14}ClNO$, calculated on the anhydrous basis.
1,4-Dioxane   Passes test.
Heavy Metals (as Pb)   Not more than 10 mg/kg.
Residue on Ignition   Not more than 0.05%.
Water   Not more than 0.5%.

TESTS

Assay   Transfer about 300 mg, accurately weighed, into a 250-mL Erlenmeyer flask, add 50 mL of glacial acetic acid, and warm on a steam bath until solution is complete. Cool, add 10 mL of mercuric acetate TS and 2 drops of crystal violet TS, and titrate with 0.1 *N* perchloric acid in glacial acetic acid to a green endpoint. Perform a blank determination (see *General Provisions*), and make any necessary correction. Each mL of 0.1 *N* perchloric acid is equivalent to 13.96 mg of $C_5H_{14}ClNO$.
1,4-Dioxane   Determine as directed in the general method, Appendix IIIB.
Heavy Metals   A solution of 2 g in 25 mL of water meets the requirements of the *Heavy Metals Test*, Appendix IIIB, using 20 μg of lead ion (Pb) in the control (*Solution A*).
Residue on Ignition   Ignite 4 g as directed in the general method, Appendix IIC.
Water   Determine by drying in a vacuum desiccator for 4 h over phosphorus pentoxide or by the *Karl Fischer Titrimetric Method*, Appendix IIB.

Packaging and Storage   Store in tight containers.

Cinnamon Bark Oil, Ceylon Type

DESCRIPTION

The volatile oil obtained by steam distillation from the dried inner bark of the clipped cinnamon shrub *Cinnamomum zeylanicum* Nees (Fam. *Lauraceae*). It is a yellow liquid with an odor of cinnamon and a spicy burning taste. It is soluble in most fixed oils and in propylene glycol. It is insoluble in glycerin and in mineral oil.

Functional Use in Foods   Flavoring agent.

REQUIREMENTS

Identification   The infrared absorption spectrum of the sample exhibits relative maxima (that may vary in intensity) at the same wavelengths (or frequencies) as those shown in the respective spectrum in the section on *Infrared Spectra (Series A: Essential Oils)*, using the same test conditions as specified therein.
Assay   Not less than 55.0% and not more than 78.0% of aldehydes, calculated as cinnamic aldehyde (C_9H_8O).
Angular Rotation   Between −2° and 0°.
Heavy Metals (as Pb)   Passes test.
Refractive Index   Between 1.573 and 1.591 at 20°.
Solubility in Alcohol   Passes test.
Specific Gravity   Between 1.010 and 1.030.

TESTS

Assay   Weigh accurately about 2.5 g, and proceed as directed under *Aldehydes*, Appendix VI, using 66.10 as the equivalence factor (*e*) in the calculation.
Angular Rotation   Determine in a 100-mm tube as directed under *Optical (Specific) Rotation*, Appendix IIB.
Heavy Metals   Shake 10 mL of the oil with an equal volume of water to which 1 drop of hydrochloric acid has been added, and pass hydrogen sulfide through the mixture until it is saturated. No darkening in color is produced in either the oil or the water.
Refractive Index, Appendix IIB   Determine with an Abbé or other refractometer of equal or greater accuracy.
Solubility in Alcohol   Proceed as directed in the general method, Appendix VI. One mL dissolves in 3 mL of 70% alcohol.
Specific Gravity   Determine by any reliable method (see *General Provisions*).

Packaging and Storage   Store in full, tight, light-resistant glass, aluminum, or tin-lined containers in a cool place protected from light.

Cinnamon Leaf Oil

CAS: [8015-96-1]

DESCRIPTION

The volatile oil obtained by steam distillation from the leaves and twigs of the true cinnamon shrub *Cinnamomum zeylanicum* Nees. The commercial oils, according to the geographical origin, are designated as either Cinnamon Leaf Oil, Ceylon, or Cinnamon Leaf Oil, Seychelles, and the two types differ in physical and chemical properties. The oil is a light to dark brown liquid having a spicy cinnamon, clovelike odor and taste. It is soluble in most fixed oils and in propylene glycol. It is soluble, with cloudiness, in mineral oil, but it is insoluble in glycerin.

Functional Use in Foods Flavoring agent.

REQUIREMENTS

Labeling Indicate whether it is the Ceylon or Seychelles type.
Identification The infrared absorption spectrum of the sample exhibits relative maxima (that may vary in intensity) at the same wavelengths (or frequencies) as those shown in the respective spectrum in the section on *Infrared Spectra (Series A: Essential Oils)*, using the same test conditions as specified therein.
Assay *Ceylon type*: not less than 80.0% and not more than 88.0%, by volume, of phenols as eugenol; *Seychelles type*: not less than 87.0% and not more than 96.0%, by volume, of phenols as eugenol.
Angular Rotation *Ceylon type*: between −2° and +1°; *Seychelles type*: between −2° and 0°.
Heavy Metals (as Pb) Passes test.
Refractive Index *Ceylon type*: between 1.529 and 1.537; *Seychelles type*: between 1.533 and 1.540 at 20°.
Solubility in Alcohol Passes test.
Specific Gravity *Ceylon type*: between 1.030 and 1.050; *Seychelles type*: between 1.040 and 1.060.

TESTS

Assay Shake a suitable quantity of the oil with about 2% of powdered tartaric acid, and filter. Proceed with a sample of the filtered oil as directed under *Phenols*, Appendix VI.
Angular Rotation Determine in a 100-mm tube as directed under *Optical (Specific) Rotation*, Appendix IIB.
Heavy Metals Shake 10 mL of the oil with an equal volume of water to which 1 drop of hydrochloric acid has been added, and pass hydrogen sulfide through the mixture until it is saturated. No darkening in color is produced in either the oil or the water.
Refractive Index, Appendix IIB Determine with an Abbé or other refractometer of equal or greater accuracy.
Solubility in Alcohol Proceed as directed in the general method, Appendix VI. One mL of the Ceylon type oil dissolves in 1.5 mL of 70% alcohol. One mL of the Seychelles type oil dissolves in 1 mL of 70% alcohol. The solutions may cloud upon further dilution.
Specific Gravity Determine by any reliable method (see *General Provisions*).

Packaging and Storage Store in full, tight, light-resistant, glass, aluminum, or tin-lined containers in a cool place protected from light.

Citric Acid

$C_6H_8O_7$ Formula wt 192.13
INS: 330 CAS: [77-92-9]

DESCRIPTION

Citric Acid is anhydrous or contains one molecule of water of hydration. It occurs as colorless, translucent crystals or as a white, granular to fine crystalline powder. It is odorless and has a strongly acid taste, and the hydrous form is efflorescent in dry air. One g is soluble in about 0.5 mL of water, in about 2 mL of alcohol, and in about 30 mL of ether.

Functional Use in Foods Sequestrant; dispersing agent; acidifier; flavoring agent.

REQUIREMENTS

Labeling Indicate whether it is anhydrous or hydrous.
Identification A 1 in 10 solution gives positive tests for *Citrate*, Appendix IIIA.
Assay Not less than 99.5% and not more than 100.5% of $C_6H_8O_7$, calculated on the anhydrous basis.
Heavy Metals (as Pb) Not more than 5 mg/kg.
Lead Not more than 0.5 mg/kg.
Oxalate Passes test.
Readily Carbonizable Substances Passes test.
Residue on Ignition Not more than 0.05%.
Tridodecylamine Not more than 0.1 mg/kg.
Ultraviolet Absorbance (polycyclic aromatic hydrocarbons) 280 to 289 nm, not more than 0.25; 290 to 299 nm, not more than 0.20; 300 to 359 nm, not more than 0.13; 360 to 400 nm, not more than 0.03.
Water *Anhydrous form*: not more than 0.5%; *hydrous form*: not more than 8.8%.

TESTS

Assay Dissolve about 3 g, accurately weighed, in 40 mL of water, add phenolphthalein TS, and titrate with 1 *N* sodium hydroxide. Each mL of 1 N sodium hydroxide is equivalent to 64.04 mg of $C_6H_8O_7$.

Heavy Metals A solution of 4 g in 25 mL of water meets the requirements of the *Heavy Metals Test*, Appendix IIIB, using 20 μg of lead ion (Pb) in the control (*Solution A*).

Lead Determine as directed under *Method I* in the *Atomic Absorption Spectrophotometric Graphite Furnace Method* under the *Lead Limit Test*, Appendix IIIB.

Oxalate Neutralize 10 mL of a 1 in 10 solution with 6 N ammonium hydroxide, add 5 drops of 2.7 N hydrochloric acid, cool, and add 2 mL of calcium chloride TS. No turbidity is produced.

Readily Carbonizable Substances, Appendix IIB Transfer 1.0 g, finely powdered, to a 22- × 175-mm test tube, previously rinsed with 10 mL of 95% sulfuric acid and allowed to drain for 10 min. Add 10 mL of 95% sulfuric acid, agitate the tube until solution is complete, and immerse the tube in a water bath at 90° ± 1° for 60 ± 0.5 min, keeping the level of the acid below the level of the water during the heating period. Cool the tube in a stream of water, and transfer the acid solution to a color-comparison tube. The color of the acid solution is not darker than that of the same volume of *Matching Fluid K* in a similar matching tube, viewing the tubes vertically against a white background.

Residue on Ignition, Appendix IIC Ignite 4 g as directed in the general method.

Tridodecylamine

Buffered Indicator Solution Prepare a mixture consisting of 700 mL of 0.1 M Citric Acid (anhydrous, reagent grade), 200 mL of 0.2 M disodium phosphate, and 50 mL each of 0.2% bromophenol blue and of 0.2% bromocresol green in spectrograde methanol.

No-Indicator Buffer Solution Prepare a mixture consisting of 700 mL of 0.1 M Citric Acid (anhydrous, reagent grade), 200 mL of 0.2 M disodium phosphate, and 100 mL of spectrograde methanol.

Amine Stock Solution Transfer between 40 and 45 mg of tridodecyl(trilauryl)amine, accurately weighed, into a 500-mL volumetric flask, dilute to volume with isopropyl alcohol, and mix. Discard after 3 weeks.

Standard Amine Solution Using a graduated 5-mL pipet, transfer into a 100-mL volumetric flask an amount of *Amine Stock Solution* equivalent to 400 μg of tridodecylamine, dilute to volume with isopropyl alcohol, and mix. Prepare this solution fresh on the day of use.

Procedure Dissolve 160 g of anhydrous reagent-grade Citric Acid (not the sample to be tested) in 320 mL of water, and divide the solution equally between two 250-mL separators, S_1 and S_2. To S_1 add 5 mL of *No-Indicator Buffer Solution*. To S_2 add 2.0 mL of *Standard Amine Solution* and 5 mL of *Buffered Indicator Solution*.

To prepare solutions of the sample being tested, dissolve 160 g of anhydrous Citric Acid sample in 320 mL of water (or 174 g of Citric Acid monohydrate sample in 306 mL of water). Divide the test solution equally between two 250-mL separators, S_3 and S_4. Add 5 mL of *No-Indicator Buffer Solution* to S_3, and 5 mL of *Buffered Indicator Solution* to S_4.

To each of the four separators add 20 mL of a 1 to 1 mixture (v/v) prepared from spectro-grade chloroform and *n*-heptane, shake for 15 min on a mechanical shaker, and allow the phases to separate for 45 min. Drain all except the last few drops of the lower (aqueous) phases, and discard. Hand-shake the organic phases with 25 mL each of 0.05 N sulfuric acid for 30 s, and allow the phases to separate for 30 min. Drain all except the last few drops of the lower (organic) phases through dry Whatman No. 40 (or equivalent) paper, and collect the aqueous filtrates in separate small glass-stoppered containers.

Determine the absorbance of each solution in a 5-cm cell at 400 nm, with a suitable spectrophotometer standardized prior to analysis, against chloroform–heptane (1:1 v/v). The net absorbance of the sample ($S_4 - S_3$) is not greater than that of the standard ($S_2 - S_1$).

Ultraviolet Absorbance Determine as directed in the U.S. food additive regulation pertaining to the use of *Candida lipolytica* in the production of Citric Acid (**21** *CFR* 173.165).

Water Determine by the *Karl Fischer Titrimetric Method*, Appendix IIB.

Packaging and Storage Store in tight containers.

Clary Oil

Clary Sage Oil; Oil of Muscatel

DESCRIPTION

The oil obtained by steam distillation from the flowering tops and leaves of the clary sage plant, *Salvia sclarea* L. (Fam. *Labiatae*). It is a pale yellow to yellow liquid having a herbaceous odor and a winy bouquet. It is soluble in most fixed oils, and in mineral oil up to 3 volumes, but becomes opalescent on further dilution. It is insoluble in glycerin and in propylene glycol.

Functional Use in Foods Flavoring agent.

REQUIREMENTS

Identification The infrared absorption spectrum of the sample exhibits relative maxima (that may vary in intensity) at the same wavelengths (or frequencies) as those shown in the respective spectrum in the section on *Infrared Spectra* (*Series A: Essential Oils*), using the same test conditions as specified therein.

Assay Not less than 48.0% and not more than 75.0% of esters, calculated as linalyl acetate ($C_{12}H_{20}O_2$).

Acid Value Not more than 2.5.

Angular Rotation Between −6° and −20°.

Heavy Metals (as Pb) Passes test.

Refractive Index Between 1.458 and 1.473 at 20°.

Solubility in Alcohol Passes test.

Specific Gravity Between 0.886 and 0.929.

TESTS

Assay Weigh accurately about 2 g, and proceed as directed under *Ester Determination*, Appendix VI, using 98.15 as the equivalence factor (*e*) in the calculation.

Acid Value Determine as directed for *Acid Value* under *Essential Oils and Flavors*, Appendix VI.

Angular Rotation Determine in a 100-mm tube as directed under *Optical (Specific) Rotation*, Appendix IIB.

Heavy Metals Shake 10 mL of the oil with an equal volume of water to which 1 drop of hydrochloric acid has been added, and pass hydrogen sulfide through the mixture until it is saturated. No darkening in color is produced in either the oil or the water.

Refractive Index, Appendix IIB Determine with an Abbé or other refractometer of equal or greater accuracy.

Solubility in Alcohol Proceed as directed in the general method, Appendix VI. One mL dissolves in 3 mL of 90% alcohol, becoming opalescent upon further dilution.

Specific Gravity Determine by any reliable method (see *General Provisions*).

Packaging and Storage Store in full, tight, preferably glass, aluminum, tin-lined, or galvanized iron containers in a cool place protected from light.

Clove Leaf Oil

CAS: [8015-97-2]

DESCRIPTION

The volatile oil obtained by steam distillation of the leaves of *Eugenia caryophyllata* Thunberg (*Eugenia aromatica* L. Baill.) (Fam. *Myrtaceae*). It is a pale yellow liquid. It is soluble in propylene glycol, and in most fixed oils with slight opalescence, and it is relatively insoluble in glycerin and in mineral oil.

Functional Use in Foods Flavoring agent.

REQUIREMENTS

Identification The infrared absorption spectrum of the sample exhibits relative maxima (that may vary in intensity) at the same wavelengths (or frequencies) as those shown in the respective spectrum in the section on *Infrared Spectra (Series A: Essential Oils)*, using the same test conditions as specified therein.

Assay Not less than 84.0% and not more than 88.0%, by volume, of phenols as eugenol.

Angular Rotation Between −2° and 0°.

Heavy Metals (as Pb) Passes test.

Refractive Index Between 1.531 and 1.535 at 20°.

Solubility in Alcohol Passes test.

Specific Gravity Between 1.036 and 1.046.

TESTS

Assay Shake a suitable quantity of the oil with 2% of powdered tartaric acid for about 2 min, and filter. Then, using a sample of the filtered oil, proceed as directed under *Phenols*, Appendix VI, modified by heating the flask in a boiling water bath for 10 min, after shaking the oil with 1 N potassium hydroxide. Remove from the boiling water bath, cool, and proceed as directed.

Angular Rotation Determine in a 100-mm tube as directed under *Optical (Specific) Rotation*, Appendix IIB.

Heavy Metals Shake 10 mL of the oil with an equal volume of water to which 1 drop of hydrochloric acid has been added, and pass hydrogen sulfide through the mixture until it is saturated. No darkening in color is produced in either the oil or the water.

Refractive Index, Appendix IIB Determine with an Abbé or other refractometer of equal or greater accuracy.

Solubility in Alcohol Proceed as directed in the general method, Appendix VI. One mL dissolves in 2 mL of 70% alcohol. A slight opalescence may occur when additional solvent is added.

Specific Gravity Determine by any reliable method (see *General Provisions*).

Packaging and Storage Store in full, tight, light-resistant, glass, tin-lined, stainless, or aluminum containers in a cool place protected from light.

Clove Oil

Clove Bud Oil

DESCRIPTION

The volatile oil obtained by steam distillation from the dried flowerbuds of *Eugenia caryophyllata* Thunberg (*Eugenia aromatica* L. Baill.) (Fam. *Myrtaceae*). It is a colorless or pale yellow liquid having the characteristic clove odor and taste. It darkens and thickens upon aging or exposure to air.

Functional Use in Foods Flavoring agent.

REQUIREMENTS

Identification The infrared absorption spectrum of the sample exhibits relative maxima (that may vary in intensity) at the same wavelengths (or frequencies) as those shown in the respective spectrum in the section on *Infrared Spectra (Series A: Essential Oils)*, using the same test conditions as specified therein.

Assay Not less than 85.0%, by volume, of phenols as eugenol.

Angular Rotation Between −1.5° and 0°.

Heavy Metals (as Pb) Passes test.

Phenol Passes test.

Refractive Index Between 1.527 and 1.535 at 20°.

Solubility in Alcohol Passes test.
Specific Gravity Between 1.038 and 1.060.

TESTS

Assay Proceed as directed under *Phenols*, Appendix VI.
Angular Rotation Determine in a 100-mm tube as directed under *Optical (Specific) Rotation*, Appendix IIB.
Heavy Metals Shake 10 mL of the oil with an equal volume of water to which 1 drop of hydrochloric acid has been added, and pass hydrogen sulfide through the mixture until it is saturated. No darkening in color is produced in either the oil or the water.
Phenol Shake 1 mL of sample with 20 mL of hot water. The water shows no more than a scarcely perceptible acid reaction with blue litmus paper. Cool the mixture, pass the water layer through a wetted filter, and treat the clear filtrate with 1 drop of ferric chloride TS. The mixture has only a transient grayish green color, but not a blue or violet color.
Refractive Index, Appendix IIB Determine with an Abbé or other refractometer of equal or greater accuracy.
Solubility in Alcohol Proceed as directed in the general method, Appendix VI. One mL dissolves in 2 mL of 70% alcohol.
Specific Gravity Determine by any reliable method (see *General Provisions*).

Packaging and Storage Store in full, tight, light-resistant containers and avoid exposure to excessive heat.

Clove Stem Oil

CAS: [8015-98-3]

DESCRIPTION

The volatile oil obtained by steam distillation from the dried stems of the buds of *Eugenia caryophyllata* Thunberg (*Eugenia aromatica* L. Baill.) (Fam. *Myrtaceae*). It is a yellow to light brown liquid with a characteristic odor and taste. It is soluble in fixed oils and in propylene glycol, but it is relatively insoluble in glycerin and in mineral oil.

Functional Use in Foods Flavoring agent.

REQUIREMENTS

Identification The infrared absorption spectrum of the sample exhibits relative maxima (that may vary in intensity) at the same wavelengths (or frequencies) as those shown in the respective spectrum in the section on *Infrared Spectra (Series A: Essential Oils)*, using the same test conditions as specified therein.
Assay Not less than 89.0% and not more than 95.0%, by volume, of phenols as eugenol.
Angular Rotation Between −1.5° and 0°.

Heavy Metals (as Pb) Passes test.
Refractive Index Between 1.534 and 1.538 at 20°.
Solubility in Alcohol Passes test.
Specific Gravity Between 1.048 and 1.056.

TESTS

Assay Shake a suitable quantity of the oil with about 2% of powdered tartaric acid for about 2 min, and filter. Then, using a sample of the filtered oil, proceed as directed under *Phenols*, Appendix VI, modified by heating the flask in a boiling water bath for 10 min, after shaking the oil with 1 N potassium hydroxide. Remove from the boiling water bath, cool, and proceed as directed.
Angular Rotation Determine in a 100-mm tube as directed under *Optical (Specific) Rotation*, Appendix IIB.
Heavy Metals Shake 10 mL of the oil with an equal volume of water to which 1 drop of hydrochloric acid has been added, and pass hydrogen sulfide through the mixture until it is saturated. No darkening in color is produced in either the oil or the water.
Refractive Index, Appendix IIB Determine with an Abbé or other refractometer of equal or greater accuracy.
Solubility in Alcohol Proceed as directed in the general method, Appendix VI. One mL dissolves in 2 mL of 70% alcohol.
Specific Gravity Determine by any reliable method (see *General Provisions*).

Packaging and Storage Store preferably in full, tight, light-resistant glass, aluminum, or tin-lined containers in a cool place protected from light.

Cocoa Butter Substitute

DESCRIPTION

A white, waxy, odorless solid that is predominantly a mixture of triglycerides derived primarily from palm oils. Cocoa Butter Substitute is the common name for the triglyceride consisting mainly of 1-palmitoyl-2-oleoyl-3-stearin. It may be synthesized by the esterification of fully saturated 1,3-diglycerides with the anhydride of food-grade oleic acid in the presence of a catalyst (trifluoromethane sulfonic acid) or by transesterification of partially saturated 1,2,3-triglycerides with ethyl stearate in the presence of a suitable food-grade lipase enzyme preparation approved for such use. The resulting product may be used directly or with cocoa butter in all proportions for the preparation of coatings. In contrast to many edible oils and hard butters, Cocoa Butter Substitute has an abrupt melting range, changing from a rather firm, plastic solid below 32° to a liquid at about 33.8° to 35.5°. Cocoa Butter Substitute is free from any rancid odor and taste.

Functional Use in Foods Coating agent; formulation aid; texturizer.

REQUIREMENTS

Identification Cocoa Butter Substitute exhibits the following typical composition profile of fatty acids as determined under *Fatty Acid Composition*, Appendix VII:

Fatty Acid:	≤12	12:0	14:0	16:0	16:1
Weight % (Range):	0.0	0.0	0.0	21–24	0.0

Fatty Acid:	18:0	18:1	18:2	≥20
Weight % (Range):	40–44	31–35	0.5–1.5	0.3–0.7

Color (AOCS-Wesson) Not more than 2.5 red.
Free Fatty Acids (as oleic acid) Not more than 1.0%.
Glycerides Not less than 98.0% of total.
  Diglycerides Not more than 7.0%.
  Monoglycerides Not more than 1.0%.
  Triglycerides Not less than 90.0%.
Heavy Metals (as Pb) Not more than 10 mg/kg.
Hexane Not more than 5 mg/kg.
Iodine Value Between 30 and 33.
Lead Not more than 0.1 mg/kg.
Peroxide Value Not more than 10 meq/kg.
Residual Catalyst (as F) Not more than 0.5 mg/kg.
Unsaponifiable Matter Not more than 1.0%.
Water Not more than 0.1%.

TESTS

Color Proceed as directed for *Color* (AOCS-Wesson) under *Fats and Related Substances*, Appendix VII.

Free Fatty Acids Using the diglyceride fraction under *Glycerides*, proceed as directed under *Free Fatty Acids*, Appendix VII, except add 2 mL of phenolphthalein TS, and titrate with the appropriate normality of sodium hydroxide. Use the following equivalence factor (e) in the formula given in the procedure:

Free fatty acids as oleic acid, $e = 28.2$.

Glycerides Proceed as directed under *Total Monoglycerides*, Appendix VII, except save all three elution fractions to determine the percentages of *Monoglycerides*, *Diglycerides*, and *Triglycerides*.

Note: Use toluene instead of benzene.

The diglyceride fraction also contains free fatty acids, the percentage of which is determined under *Free Fatty Acids*. Calculate the percentage of *Glycerides*, which is the sum of *Monoglycerides*, *Diglycerides*, and *Triglycerides*, by the following formulas:

$$TG = T + D + M,$$
$$T = W_T 100/W_U,$$
$$D = (W_D 100/W_U) - F,$$
$$M = W_M 100/W_U,$$

where TG is the percent of total glycerides; T is the percent of triglycerides; D is the percent of diglycerides; M is the percent of monoglycerides; W_T is the weight, in g, of triglycerides; W_U is the weight, in g, of the sample taken; W_D is the weight, in g, of diglycerides; F is the percent of free fatty acids; and W_M is the weight, in g, of monoglycerides.

Heavy Metals A 2-g sample meets the requirements of the *Heavy Metals Test, Method II*, Appendix IIIB, using 20 µg of lead ion (Pb) in the control (*Solution A*).

Hexane

Standard Preparation Using a micropipet, transfer and dissolve 34 µL of hexane in 45 g of cold-pressed cottonseed oil (that has not been extracted with hexane). As directed under *Procedure*, analyze aliquots of 0.1, 0.25, 0.5, and 5.0 mg; the aliquots correspond to 2, 5, 10, and 100 mg/kg, respectively, of residual hexane in a 25-mg sample.

Assay Preparation Pack the lower half of 8.5-cm × 9.5-mm (od) borosilicate glass tubing (inlet liner) with glass wool that has been heated at 200° for 16 h to expel volatiles. Transfer a 25-mg sample, accurately weighed, into the glass tubing, and cover it with a small plug of treated glass wool.

Chromatographic System Use a suitable gas chromatograph that is equipped with independent dual flame-ionization detectors and contains a 0.6-m × 6.35-mm (od) stainless-steel U-tube packed with Porapak P or equivalent. After the *Assay Preparation* is inserted into the chromatograph, subject it to the following operating conditions: Maintain the inlet temperature at 110°, maintain the detectors at 200°, and hold the column oven initially at 70° for 2 min followed by a linear temperature gradient at 5°/min for 22 min at a temperature between 70° and 180° and a final hold at 180° for 10 min or until the column is clean. Use helium as the carrier gas at a flow rate of 60 mL/min, hydrogen as the fuel gas at a flow rate of 52 mL/min for each flame, and air as the scavenger gas for both flames at a flow rate of 500 mL/min. To ensure that the relative standard deviation does not exceed 2.0%, chromatograph a sufficient number of replicates of each *Standard Preparation*, and record the areas as directed under *Procedure*. (See *Chromatography*, Appendix IIA.)

Standard Curve Chromatograph aliquots of each *Standard Preparation* as directed under *Procedure*. Measure the peak areas for each *Standard Preparation*. Plot a standard curve using the concentration, in mg/kg, of each *Standard Preparation* versus its corresponding peak area, and draw the best straight line.

Procedure Insert the *Assay Preparation* into the inlet liner of the gas chromatograph, immediately sealing the base of the inlet and the lower lip of the glass tubing with a silicone O-ring (Applied Science Laboratories, Inc., or equivalent) previously heated at 200° for 2 h to remove volatile impurities. Immediately close the inlet liner with the septum and septum liner. Allow the carrier gas to flow through the *Assay Preparation*, chromatograph as directed under *Chromatographic System*, and record the chromatograms. Using the peak area of hexane eluting from the *Assay Preparation* at the same time as the *Standard Preparation*, read directly from the *Standard Curve* the concentration, C, of hexane in mg/kg, of the *Assay Preparation*. Calculate the quantity, in mg/kg, of hexane in the sample by the formula

$25C/W$,

in which W is the weight, in mg, of the sample introduced into the gas chromatograph.

Iodine Value Proceed as directed under *Iodine Value*, Appendix VII.

Lead Determine as directed under *Method II* in the *Atomic Absorption Spectrophotometric Graphite Furnace Method* under the *Lead Limit Test*, Appendix IIIB, using a 5-g sample.

Peroxide Value Proceed as directed under *Peroxide Value* in the monograph for *Hydroxylated Lecithin*. However, after the addition of saturated potassium iodide and mixing, instead of allowing the solution to stand for 10 min, mix the solution for 1 min, and begin the titration immediately.

Residual Catalyst Transfer a 30-g sample, accurately weighed, into a 250-mL distillation flask having a side arm and a trap. Connect the flask with a condenser, and fit it with a thermometer and a capillary tube. Both of these should reach nearly to the bottom of the flask so that they extend into the liquid during the distillation. Add 0.2 g of silver sulfate, three boiling beads, and 25 mL of 1:1 sulfuric acid to the flask. Connect a dropping funnel or a steam generator to the capillary tube. Distill until the temperature reaches 135°. Then, through the capillary, add water from the funnel, or introduce steam, as necessary, to maintain the temperature as close to 135° as possible, until 250 mL of distillate has been collected in a beaker. Cool the distillate. Add 3 mL of 30% hydrogen peroxide to remove any sulfites, let stand for 5 min, and evaporate the distillate in a dish containing 15 mL of saturated calcium hydroxide suspension. Ash the residue at 600° for 4 h. Using the ashed residue as the sample, proceed as directed for *Method I* under *Fluoride Limit Test*, Appendix IIIB, beginning with ". . . and 30 mL of water in a 125-mL distillation flask having a side arm and trap." The total volume of sodium fluoride TS required for the solutions from both *Distillate A* and *Distillate B* should not exceed 0.75 mL.

Unsaponifiable Matter Proceed as directed under *Unsaponifiable Matter*, Appendix VII.

Water Proceed as directed under *Water Determination*, Appendix IIB. However, in place of 35 to 40 mL of methanol, use 50 mL of a 1:1 chloroform–methanol mixture to dissolve the sample.

Packaging and Storage Store in well-closed containers.

Coconut Oil (Unhydrogenated)

CAS: [8001-31-8]

DESCRIPTION

A fat with a sweet nutty flavor obtained from the kernel of the fruit of the coconut palm *Cocos nucifera*. The crude oil is obtained by mechanically expressing dried coconut meat (copra) and is refined, bleached, and deodorized to substantially remove free fatty acids, phospholipids, color, odor and flavor components, and miscellaneous other non-oil materials. Compared with many natural fats, Coconut Oil has an abrupt melting range, changing from a rather firm, plastic solid at about 21° or below to a liquid at about 27°.

Functional Use in Foods Coating agent; emulsifying agent; formulation aid; texturizer.

REQUIREMENTS

Identification Coconut Oil exhibits the following composition profile of fatty acids as determined under *Fatty Acid Composition*, Appendix VII:

Fatty Acid:	6:0	8:0	10:0	12:0	14:0	16:0	16:1
Weight % (Range):	0–0.8	5–9	4–8	44–52	15–21	8–11	0–1

Fatty Acid:	18:0	18:1	18:2	20:0
Weight % (Range):	1–4	5–8	0–2.5	0–0.4

Arsenic (as As) Not more than 0.5 mg/kg.
Color (AOCS-Wesson) Not more than 20 yellow/2.0 red.
Free Fatty Acids Not more than 0.1% (as oleic acid); Not more than 0.07% (as lauric acid).
Iodine Value Between 6 and 11.
Lead Not more than 0.1 mg/kg.
Melting Range Between 23.5° and 27°.
Peroxide Value Not more than 10 meq/kg.
Unsaponifiable Matter Not more than 1.5%.
Water Not more than 0.1%.

TESTS

Arsenic A *Sample Solution* prepared using 2 g of sample, accurately weighed, meets the requirements of the *Arsenic Test*, Appendix IIIB. The absorbance due to any red color from the solution of the sample does not exceed that produced by 1.0 mL of *Standard Arsenic Solution* (1 μg As) when treated in the same manner and under the same conditions as the sample.

Color Proceed as directed for *Color* (AOCS-Wesson) under *Fats and Related Substances*, Appendix VII.

Free Fatty Acids Proceed as directed under *Free Fatty Acids*, Appendix VII, using the following equivalence factors (e) in the formula given in the procedure:

Free fatty acids as oleic acid, $e = 28.2$.
Free fatty acids as lauric acid, $e = 20.0$.

Iodine Value Proceed as directed under *Iodine Value*, Appendix VII.

Lead Determine as directed under *Method II* in the *Atomic Absorption Spectrophotometric Graphite Furnace Method* under the *Lead Limit Test*, Appendix IIIB, using a 3-g sample.

Melting Range Proceed as directed for *Melting Range* under *Fats and Related Substances*, Appendix VII.

Peroxide Value Proceed as directed under *Peroxide Value* in the monograph for *Hydroxylated Lecithin*. However, after the addition of saturated potassium iodide and mixing, instead of allowing the solution to stand for 10 min, mix the solution for 1 min and begin the titration immediately.

Unsaponifiable Matter Proceed as directed under *Unsaponifiable Matter*, Appendix VII.

Water Proceed as directed under *Water Determination*, Appendix IIB. However, in place of 35 to 40 mL of methanol, use 50 mL of chloroform to dissolve the sample.

Packaging and Storage Store in well-closed containers.

Cognac Oil, Green

Wine Yeast Oil

CAS: [8016-21-5]

DESCRIPTION

The volatile oil obtained by steam distillation from wine lees. It is a green to bluish green liquid with the characteristic aroma of cognac. It is soluble in most fixed oils and in mineral oil. It is very slightly soluble in propylene glycol, and it is insoluble in glycerin.

Functional Use in Foods Flavoring agent.

REQUIREMENTS

Identification The infrared absorption spectrum of the sample exhibits relative maxima (that may vary in intensity) at the same wavelengths (or frequencies) as those shown in the respective spectrum in the section on *Infrared Spectra (Series A: Essential Oils)*, using the same test conditions as specified therein.

Acid Value Between 32 and 70.
Angular Rotation Between −1° and +2°.
Ester Value Between 200 and 245.
Heavy Metals (as Pb) Passes test.
Refractive Index Between 1.427 and 1.430 at 20°.
Solubility in Alcohol Passes test.
Specific Gravity Between 0.864 and 0.870.

TESTS

Acid Value Determine as directed for *Acid Value* under *Essential Oils and Flavors*, Appendix VI.

Angular Rotation Determine in a 100-mm tube as directed under *Optical (Specific) Rotation*, Appendix IIB.

Ester Value Proceed as directed in the general method, Appendix VI, using about 1 g, accurately weighed.

Heavy Metals Shake 10 mL of the oil with an equal volume of water to which 1 drop of hydrochloric acid has been added, and pass hydrogen sulfide through the mixture until it is saturated. No darkening in color is produced in either the oil or the water.

Refractive Index, Appendix IIB Determine with an Abbé or other refractometer of equal or greater accuracy.

Solubility in Alcohol Proceed as directed in the general method, Appendix VI. One mL dissolves in 2 mL of 80% alcohol.

Specific Gravity Determine by any reliable method (see *General Provisions*).

Packaging and Storage Store in full, tight containers in a cool place protected from light.

Copaiba Oil

CAS: [8013-97-6]

DESCRIPTION

The volatile oil obtained by steam distillation of copaiba balsam, an exudate from the trunk of various South American species of *Copaifera* L. (Fam. *Leguminosae*). It is a colorless to slightly yellow liquid having the characteristic odor of copaiba balsam and an aromatic, slightly bitter, and pungent taste. It is soluble in alcohol, in most fixed oils, and in mineral oil. It is insoluble in glycerin and practically insoluble in propylene glycol.

Functional Use in Foods Flavoring agent.

REQUIREMENTS

Identification The infrared absorption spectrum of the sample exhibits relative maxima (that may vary in intensity) at the same wavelengths (or frequencies) as those shown in the respective spectrum in the section on *Infrared Spectra (Series A: Essential Oils)*, using the same test conditions as specified therein.

Angular Rotation Between −7° and −33°.
Gurjun Oil Passes test.
Heavy Metals (as Pb) Passes test.
Refractive Index Between 1.493 and 1.500 at 20°.
Specific Gravity Between 0.880 and 0.907.

TESTS

Angular Rotation Determine in a 100-mm tube as directed under *Optical (Specific) Rotation*, Appendix IIB.

Gurjun Oil Add 5 or 6 drops of the sample to 10 mL of glacial acetic acid containing 5 drops of nitric acid. No purple color develops in 2 min, indicating the absence of gurjun oil.

Heavy Metals Shake 10 mL of the oil with an equal volume of water to which 1 drop of hydrochloric acid has been added, and pass hydrogen sulfide through the mixture until it is saturated. No darkening in color is produced in either the oil or the water.

Refractive Index, Appendix IIB   Determine with an Abbé or other refractometer of equal or greater accuracy.
Specific Gravity   Determine by any reliable method (see *General Provisions*).

Packaging and Storage   Store in full, tight, preferably glass, tin, aluminum, or other suitably lined containers in a cool place protected from light.

Copper Gluconate

$$[CH_2OH(CHOH)_4COO]_2Cu$$

$C_{12}H_{22}CuO_{14}$ Formula wt 453.84

                               CAS: [527-09-3]

DESCRIPTION

A fine, light blue powder. It is very soluble in water, and is very slightly soluble in alcohol.

Functional Use in Foods   Nutrient; dietary supplement.

REQUIREMENTS

Identification
  A. A 1 in 20 solution gives positive tests for *Copper*, Appendix IIIA.
  B. It meets the requirements of *Identification Test B* in the monograph for *Calcium Gluconate*.
Assay   Not less than 98.0% and not more than 102.0% of $C_{12}H_{22}CuO_{14}$.
Lead   Not more than 10 mg/kg.
Reducing Substances   Not more than 1.0%.

TESTS

Assay   Dissolve about 1.5 g, accurately weighed, in 100 mL of water in a 250-mL Erlenmeyer flask, add 2 mL of glacial acetic acid and 5 g of potassium iodide, mix well, and titrate with 0.1 N sodium thiosulfate to a light yellow color. Add 2 g of ammonium thiocyanate, mix, then add 3 mL of starch TS and continue titrating to a milk-white endpoint. Each mL of 0.1 N sodium thiosulfate is equivalent to 45.38 mg of $C_{12}H_{22}CuO_{14}$.
Lead   A solution of 1 g in 25 mL of water meets the requirements of the *Lead Limit Test*, Appendix IIIB.
Reducing Substances   Transfer about 1 g of the sample, accurately weighed, into a 250-mL Erlenmeyer flask, dissolve in 10 mL of water, add 25 mL of alkaline cupric citrate TS, and cover the flask with a small beaker. Boil gently for exactly 5 min and cool rapidly to room temperature. Add 25 mL of a 1 in 10 solution of acetic acid, 10.0 mL of 0.1 N iodine, 10 mL of 2.7 N hydrochloric acid, and 3 mL of starch TS, and titrate with 0.1 N sodium thiosulfate to the disappearance of the blue color. Calculate the weight, in mg, of reducing substances (as D-glucose) by the formula

$$(V_1N_1 - V_2N_2)27,$$

in which V_1 and N_1 are the volume and normality, respectively, of the iodine solution, V_2 and N_2 are the volume and normality, respectively, of the sodium thiosulfate solution, and 27 is an empirically determined equivalence factor for D-glucose.

Packaging and Storage   Store in well-closed containers.

Copper Sulfate
Cupric Sulfate

$CuSO_4.5H_2O$ Formula wt 249.68

INS: 519   CAS: [7758-98-7]

DESCRIPTION

Blue crystals, crystalline granules, or powder. It effloresces slowly in dry air and is freely soluble in water, soluble in glycerin, and slightly soluble in alcohol.

Functional Use in Foods   Nutrient supplement; processing aid.

REQUIREMENTS

Identification   A 1 in 20 solution gives positive tests for *Copper* and for *Sulfate*, Appendix IIIA.
Assay   Not less than 98.0% and not more than 102.0% of $CuSO_4.5H_2O$.
Iron   Not more than 0.01%.
Lead   Not more than 10 mg/kg.
Substances Not Precipitated by Hydrogen Sulfide   Not more than 0.3%.

TESTS

Assay   Dissolve about 1 g of the sample, accurately weighed, in 50 mL of water, add 4 mL of glacial acetic acid and 3 g of potassium iodide, mix well, and titrate with 0.1 N sodium thiosulfate to a light yellow color. Add 2 g of ammonium thiocyanate, mix, and then add 3 mL of starch TS, and continue titrating to a milky white endpoint. Perform a blank titration, and make any necessary correction. Each mL of 0.1 N sodium thiosulfate is equivalent to 24.97 mg of $CuSO_4.5H_2O$.
Iron   To the residue from the test entitled *Substances Not Precipitated by Hydrogen Sulfide*, add 2 mL of hydrochloric acid and 0.1 mL of nitric acid, cover with a watch glass, and digest on a steam bath for 20 min. Remove the watch glass, and evaporate to dryness. Dissolve the residue in 1 mL of hydrochloric acid, and dilute to 60 mL. Dilute 5 mL of this

solution to 40 mL, add 2 mL of hydrochloric acid, and dilute to 50 mL. Add 40 mg of ammonium peroxydisulfate crystals and 10 mL of ammonium thiocyanate TS, and mix thoroughly. Any red color produced within 1 h shall not exceed that produced by 0.033 mg of Fe in an equal volume of solution containing the reagents used in the test.

Lead   A solution of 1 g in 25 mL of water meets the requirements of the *Lead Limit Test*, Appendix IIIB, using 10 µg of lead ion (Pb) in the control.

Substances Not Precipitated by Hydrogen Sulfide   Dissolve 5 g in 200 mL of sulfuric acid (1 in 100), heat to 70°, and pass hydrogen sulfide through the solution until the copper is completely precipitated. Dilute to 250 mL, mix thoroughly, allow the precipitate to settle, and filter. Evaporate 200 mL of the filtrate to dryness in a tared dish, ignite at 800° ± 25° for 15 min, cool, and weigh.

Packaging and Storage   Store in tight containers.

Coriander Oil

CAS: [8008-52-4]

DESCRIPTION

The volatile oil obtained by steam distillation from the dried ripe fruit of *Coriandrum sativum* L. (Fam. *Umbelliferae*). It is a colorless or pale yellow liquid having the characteristic odor and taste of coriander.

Functional Use in Foods   Flavoring agent.

REQUIREMENTS

Identification   The infrared absorption spectrum of the sample exhibits relative maxima (that may vary in intensity) at the same wavelengths (or frequencies) as those shown in the respective spectrum in the section on *Infrared Spectra (Series A: Essential Oils)*, using the same test conditions as specified therein.
Angular Rotation   Between +8° and +15°.
Heavy Metals (as Pb)   Passes test.
Refractive Index   Between 1.462 and 1.472 at 20°.
Solubility in Alcohol   Passes test.
Specific Gravity   Between 0.863 and 0.875.

TESTS

Angular Rotation   Determine in a 100-mm tube as directed under *Optical (Specific) Rotation*, Appendix IIB.
Heavy Metals   Shake 10 mL of the oil with an equal volume of water to which 1 drop of hydrochloric acid has been added, and pass hydrogen sulfide through the mixture until it is saturated. No darkening in color is produced in either the oil or the water.

Refractive Index, Appendix IIB   Determine with an Abbé or other refractometer of equal or greater accuracy.
Solubility in Alcohol   Proceed as directed in the general method, Appendix VI. One mL dissolves in 3 mL of 70% alcohol.
Specific Gravity   Determine by any reliable method (see *General Provisions*).

Packaging and Storage   Store in full, tight containers protected from light. Avoid exposure to excessive heat.

Corn Oil (Unhydrogenated)

CAS: [8001-30-7]

DESCRIPTION

An amber-colored oil with a characteristic slight corn flavor obtained from the corn plant *Zea mays*, usually by solvent extraction of the corn germ. It is refined, bleached, and deodorized to substantially remove free fatty acids, phospholipids, color, odor and flavor components, and miscellaneous other non-oil materials. It is a liquid at 21° to 27°, but traces of wax may cause the oil to cloud when cooled to low temperature, unless they are removed by winterization. It is free from visible foreign material (other than wax) at 21° to 27°.

Functional Use in Foods   Coating agent; emulsifying agent; formulation aid; texturizer.

REQUIREMENTS

Identification   Corn Oil exhibits the following composition profile of fatty acids as determined under *Fatty Acid Composition*, Appendix VII.

Fatty Acid:	<14	14:0	16:0	16:1	18:0	18:1
Weight % (Range):	<0.1	<1.0	8.0–19	<0.5	0.5–4.0	19–50

Fatty Acid:	18:2	18:3	20:0	20:1	22:0	22:1	24:0
Weight % (Range):	38–65	<2.0	<1.0	<0.5	<0.3	<0.1	<0.4

Arsenic (as As)   Not more than 0.5 mg/kg.
Color (AOCS-Wesson)   Not more than 5.0 red.
Free Fatty Acids (as oleic acid)   Not more than 0.1%.
Iodine Value   Between 120 and 130.
Lead   Not more than 0.1 mg/kg.
Linolenic Acid   Not more than 2.0%.
Peroxide Value   Not more than 10 meq/kg.
Unsaponifiable Matter   Not more than 1.5%.
Water   Not more than 0.1%.

TESTS

Arsenic A *Sample Solution* prepared as directed for organic compounds using 4 g of sample, accurately weighed, meets the requirements of the *Arsenic Test*, Appendix IIIB. The absorbance due to any red color from the solution of the sample does not exceed that produced by 2.0 mL of *Standard Arsenic Solution* (2 μg As) when treated in the same manner and under the same conditions as the sample.

Color Proceed as directed for *Color* (AOCS-Wesson) under *Fats and Related Substances*, Appendix VII.

Free Fatty Acids Proceed as directed under *Free Fatty Acids*, Appendix VII, using the following equivalence factors (*e*) in the formula given in the procedure:

Free fatty acids as oleic acid, $e = 28.2$.

Iodine Value Proceed as directed under *Iodine Value*, Appendix VII.

Lead Determine as directed under *Method II* in the *Atomic Absorption Spectrophotometric Graphite Furnace Method* under the *Lead Limit Test*, Appendix IIIB, using a 3-g sample.

Linolenic Acid Proceed as directed under *Fatty Acid Composition*, Appendix VII.

Peroxide Value Proceed as directed under *Peroxide Value* in the monograph for *Hydroxylated Lecithin*. However, after the addition of saturated potassium iodide and mixing, instead of allowing the solution to stand for 10 min, mix the solution for 1 min and begin the titration immediately.

Unsaponifiable Matter Proceed as directed under *Unsaponifiable Matter*, Appendix VII.

Water Proceed as directed under *Water Determination*, Appendix IIB. However, in place of 35 to 40 mL of methanol use 50 mL of chloroform to dissolve the sample.

Packaging and Storage Store in well closed containers.

Costus Root Oil

CAS: [8023-88-9]

DESCRIPTION

The volatile oil obtained by steam distillation from the dried, triturated roots of the herbaceous perennial plant *Saussurea lappa* Clarke (Fam. *Compositae*), or by a solvent extraction procedure followed by vacuum distillation of the resinoid extract. It is a light yellow to brown, viscous liquid having a peculiar, persistent odor reminiscent of violet, orris, and vetivert. It is soluble in most fixed oils and in mineral oil. It is insoluble in glycerin and in propylene glycol.

Functional Use in Foods Flavoring agent.

REQUIREMENTS

Identification The infrared absorption spectrum of the sample exhibits relative maxima (that may vary in intensity) at the same wavelengths (or frequencies) as those shown in the respective spectrum in the section on *Infrared Spectra* (*Series A: Essential Oils*), using the same test conditions as specified therein.

Acid Value Not more than 42.
Angular Rotation Between +10° and +36°.
Ester Value Between 90 and 150.
Heavy Metals (as Pb) Passes test.
Refractive Index Between 1.512 and 1.523 at 20°.
Solubility in Alcohol Passes test.
Specific Gravity Between 0.995 and 1.039.

TESTS

Acid Value Determine as directed for *Acid Value* under *Essential Oils and Flavors*, Appendix VI.

Angular Rotation Determine in a 100-mm tube as directed under *Optical (Specific) Rotation*, Appendix IIB.

Ester Value Determine as directed in the general method, Appendix VI, using about 1 g, accurately weighed.

Heavy Metals Shake 10 mL of the oil with an equal volume of water to which 1 drop of hydrochloric acid has been added, and pass hydrogen sulfide through the mixture until it is saturated. No darkening in color is produced in either the oil or the water.

Refractive Index, Appendix IIB Determine with an Abbé or other refractometer of equal or greater accuracy.

Solubility in Alcohol Proceed as directed in the general method, Appendix VI. One mL dissolves in 0.5 mL of 90% alcohol, but the solution becomes cloudy upon further dilution and occasionally paraffin crystals may separate.

Specific Gravity Determine by any reliable method (see *General Provisions*).

Packaging and Storage Store in full, tight, preferably glass or aluminum containers in a cool place protected from light.

Cottonseed Oil (Unhydrogenated)

CAS: [8001-29-4]

DESCRIPTION

A dark, reddish brown oil with a slight nutty flavor obtained from the seed of the cotton plant *Gossypium hirsutum* (American) or *Gossypium barbadense* (Egyptian) by mechanical expression or solvent extraction. It is refined, bleached, and deodorized to substantially remove free fatty acids, phospholipids, color, odor and flavor components, and miscellaneous other non-oil materials. It is a liquid at 21° to 27°, will cloud at 21°, but partially solidifies at storage temperatures below 10° to 16°. It is free from visible foreign material at 23° to 27°.

Functional Use in Foods Coating agent; emulsifying agent; formulation aid; texturizer.

REQUIREMENTS

Identification Unhydrogenated Cottonseed Oil exhibits the following composition profile of fatty acids as determined under *Fatty Acid Composition*, Appendix VII:

Fatty Acid:	<14:0	14:0	16:0	16:1	18:0	18:1
Weight % (Range):	<0.1	0.5–2.0	17–29	<1.5	1.0–4.0	13–44

Fatty Acid:	18:2	18:3	20:0	20:1	22:0	22:1	24:0
Weight % (Range):	40–63	0.1–2.1	<0.5	<0.5	<0.5	<0.5	<0.5

Arsenic (as As) Not more than 0.5 mg/kg.
Color (AOCS-Wesson) Not more than 70 yellow/4.5 red.
Free Fatty Acids (as oleic acid) Not more than 0.1%
Iodine Value Between 99 and 119.
Lead Not more than 0.1 mg/kg.
Linolenic Acid Not more than 2.1%.
Peroxide Value Not more than 10 meq/kg.
Unsaponifiable Matter Not more than 1.5%.
Water Not more than 0.1%.

TESTS

Arsenic A *Sample Solution* prepared using 4 g of sample, accurately weighed, meets the requirements of the *Arsenic Test*, Appendix IIIB, using 2 mL of the *Standard Arsenic Solution* in the control (2 μg As).
Color Proceed as directed for *Color* (AOCS-Wesson) under *Fats and Related Substances*, Appendix VII.
Free Fatty Acids Proceed as directed under *Free Fatty Acids*, Appendix VII, using the following equivalence factor (*e*) in the formula given in the procedure:

Free fatty acids as oleic acid, *e* = 28.2.

Iodine Value Proceed as directed under *Iodine Value*, Appendix VII.
Lead Determine as directed under *Method II* in the *Atomic Absorption Spectrophotometric Graphite Furnace Method* under the *Lead Limit Test*, Appendix IIIB, using a 3-g sample.
Linoleic Acid Proceed as directed under *Fatty Acid Composition*, Appendix VII.
Peroxide Value Proceed as directed under *Peroxide Value* in the monograph for *Hydroxylated Lecithin*. However, after the addition of saturated potassium iodide and mixing, instead of allowing the solution to stand for 10 min, mix the solution for 1 min and begin the titration immediately.
Unsaponifiable Matter Proceed as directed under *Unsaponifiable Matter*, Appendix VII.
Water Proceed as directed under *Water Determination*, Appendix IIB. However, in place of 35 to 40 mL of methanol use 50 mL of chloroform to dissolve the sample.

Packaging and Storage Store in well-closed containers.

Cubeb Oil

CAS: [8007-87-2]

DESCRIPTION

The volatile oil obtained by steam distillation from the mature, unripe, sun-dried fruit of the perennial vine *Piper cubeba* L. (Fam. *Piperaceae*). It is a colorless or light green to bluish green liquid having a spicy odor and a slightly acrid taste. It is soluble in most fixed oils and in mineral oil, but it is insoluble in glycerin and propylene glycol.

Functional Use in Foods Flavoring agent.

REQUIREMENTS

Identification The infrared absorption spectrum of the sample exhibits relative maxima (that may vary in intensity) at the same wavelengths (or frequencies) as those shown in the respective spectrum in the section on *Infrared Spectra* (Series A: *Essential Oils*), using the same test conditions as specified therein.
Acid Value Not more than 2.0.
Angular Rotation Between −12° and −43°.
Heavy Metals (as Pb) Passes test.
Refractive Index Between 1.492 and 1.502 at 20°.
Saponification Value Not more than 8.
Solubility in Alcohol Passes test.
Specific Gravity Between 0.898 and 0.928.

TESTS

Acid Value Determine as directed for *Acid Value* under *Essential Oils and Flavors*, Appendix VI.
Angular Rotation Determine in a 100-mm tube as directed under *Optical (Specific) Rotation*, Appendix IIB.
Heavy Metals Shake 10 mL of the oil with an equal volume of water to which 1 drop of hydrochloric acid has been added, and pass hydrogen sulfide through the mixture until it is saturated. No darkening in color is produced in either the oil or the water.
Refractive Index, Appendix IIB Determine with an Abbé or other refractometer of equal or greater accuracy.
Saponification Value Determine as directed for *Saponification Value* under *Essential Oils and Flavors*, Appendix VI, using about 5 g, accurately weighed.
Solubility in Alcohol Proceed as directed in the general method, Appendix VI. One mL dissolves in 10 mL of 90% alcohol.
Specific Gravity Determine by any reliable method (see *General Provisions*).

Packaging and Storage Store in full, tight, preferably glass, aluminum, or tin-lined containers in a cool place protected from light.

Cumin Oil

CAS: [8014-13-9]

DESCRIPTION

The volatile oil obtained by steam distillation from the plant *Cuminum cyminum* L. It is a light yellow to brown liquid having a strong and somewhat disagreeable odor. It is relatively soluble in most fixed oils and in mineral oil. It is very soluble in glycerin and in propylene glycol.

Functional Use in Foods Flavoring agent.

REQUIREMENTS

Identification The infrared absorption spectrum of the sample exhibits relative maxima (that may vary in intensity) at the same wavelengths (or frequencies) as those shown in the respective spectrum in the section on *Infrared Spectra (Series A: Essential Oils)*, using the same test conditions as specified therein.
Assay Not less than 45.0% and not more than 54.0% of aldehydes, calculated as cuminaldehyde ($C_{10}H_{12}O$).
Angular Rotation Between +3° and +8°.
Heavy Metals (as Pb) Passes test.
Refractive Index Between 1.500 and 1.506 at 20°.
Solubility in Alcohol Passes test.
Specific Gravity Between 0.905 and 0.925.

TESTS

Assay Weigh accurately about 1 g, and proceed as directed under *Aldehydes*, Appendix VI, using 74.10 as the equivalence factor (*e*) in the calculation. Allow the mixture to stand for 30 min at room temperature before titrating.
Angular Rotation Determine in a 100-mm tube as directed under *Optical (Specific) Rotation*, Appendix IIB.
Heavy Metals Shake 10 mL of the oil with an equal volume of water to which 1 drop of hydrochloric acid has been added, and pass hydrogen sulfide through the mixture until it is saturated. No darkening in color is produced in either the oil or the water.
Refractive Index, Appendix IIB Determine with an Abbé or other refractometer of equal or greater accuracy.
Solubility in Alcohol Proceed as directed in the general method, Appendix VI. One mL dissolves in 8 mL of 80% alcohol. The solution may become hazy upon the addition of more alcohol.
Specific Gravity Determine by any reliable method (see *General Provisions*).

Packaging and Storage Store in full, tight, preferably glass, tin, or suitably lined containers in a cool place protected from light.

L-Cysteine Monohydrochloride

L-2-Amino-3-mercaptopropanoic Acid Monohydrochloride

$$\underset{NH_2 \cdot HCl}{HSCH_2CHCOOH \cdot H_2O}$$

$C_3H_7NO_2S \cdot HCl \cdot H_2O$ Formula wt 175.64
INS: 920 CAS: monohydrate [7048-04-6]
CAS: anhydrous [52-89-1]

DESCRIPTION

A white, crystalline powder having a characteristic odor and acidic taste. It is freely soluble in water and in alcohol. The anhydrous form melts with decomposition at about 175°.

Functional Use in Foods Nutrient; dietary supplement.

REQUIREMENTS

Identification
 A. Dissolve 100 mg in 5 mL of water and add 10 mL of cupric nitrate TS. A bluish gray precipitate is formed.
 B. A 1 in 20 solution gives positive tests for *Chloride*, Appendix IIIA.
Assay Not less than 98.0% and not more than 101.5% of $C_3H_7NO_2S \cdot HCl$ after drying.
Heavy Metals (as Pb) Not more than 0.002%.
Lead Not more than 10 mg/kg.
Loss on Drying Not less than 8.0% and not more than 12.0%.
Residue on Ignition Not more than 0.1%.
Specific Rotation $[\alpha]_D^{20°}$: Between +5.0° and +8.0°; or $[\alpha]_D^{25°}$: Between +4.9° and +7.9°.

TESTS

Assay Transfer about 300 mg, previously dried as directed under *Loss on Drying* and accurately weighed, into a 250-mL glass-stoppered flask. Add 20 mL of water, 4 g of potassium iodide, 5 mL of 2.7 N hydrochloric acid, and 25.0 mL of 0.1 N iodine. Stopper the flask, allow the mixture to stand for 30 min in a dark place, and titrate the excess iodine with 0.1 N sodium thiosulfate. Perform a blank determination (see *General Provisions*), and make any necessary correction. Each mL of 0.1 N iodine is equivalent to 15.76 mg of $C_3H_7NO_2S \cdot HCl$.
Heavy Metals Prepare and test a 1-g sample as directed in *Method II* under the *Heavy Metals Test*, Appendix IIIB, using 20 μg of lead ion (Pb) in the control (*Solution A*).
Lead A *Sample Solution* prepared as directed for organic compounds meets the requirements of the *Lead Limit Test*, Appendix IIIB, using 10 μg of lead ion (Pb) in the control.
Loss on Drying, Appendix IIC Dry at room temperature for 24 h in a vacuum desiccator using a suitable desiccant and maintaining a pressure of not more than 5 mm of Hg.

Residue on Ignition, Appendix IIC   Ignite 1 g as directed in the general method.
Specific Rotation, Appendix IIB   Determine in a solution containing 8 g of undried sample in sufficient 1 N hydrochloric acid to make 100 mL.

Packaging and Storage   Store in well-closed, light-resistant containers.

L-Cystine

3,3′-Dithiobis(2-aminopropanoic acid)

HOOCCH(NH$_2$)CH$_2$SSCH$_2$CH(NH$_2$)COOH

C$_6$H$_{12}$N$_2$O$_4$S$_2$	Formula wt 240.30
INS: 921	CAS: [56-89-3]

DESCRIPTION

Colorless, practically odorless, white crystals. It is soluble in diluted mineral acids and in alkaline solutions. It is very slightly soluble in water and in alcohol.

Functional Use in Foods   Nutrient; dietary supplement.

REQUIREMENTS

Identification   The infrared absorption spectrum of a potassium bromide dispersion of L-Cystine, previously dried, exhibits maxima only at the same wavelengths as those of a similar preparation of NIST Cystine Standard Reference Material.
Assay   Not less than 98.5% and not more than 101.5% of C$_6$H$_{12}$N$_2$O$_4$S$_2$, calculated on the dried basis.
Heavy Metals (as Pb)   Not more than 0.002%.
Lead   Not more than 10 mg/kg.
Loss on Drying   Not more than 0.2%.
Residue on Ignition   Not more than 0.1%.
Specific Rotation   [α]$_D^{20°}$: Between −215° and −225°, after drying.

TESTS

Assay   Determine as directed under *Nitrogen Determination*, Appendix IIIC, using a 200-mg sample. Percent L-Cystine equals percent N × 8.58.
Heavy Metals   Prepare and test a 1-g sample as directed in *Method II* under the *Heavy Metals Test*, Appendix IIIB, using 20 μg of lead ion (Pb) in the control (*Solution A*).
Lead   A *Sample Solution* prepared as directed for organic compounds meets the requirements of the *Lead Limit Test*, Appendix IIIB, using 10 μg of lead ion (Pb) in the control.
Loss on Drying, Appendix IIC   Dry at 105° for 3 h.

Residue on Ignition   Ignite 2 g as directed in the general method, Appendix IIC.
Specific Rotation, Appendix IIB   Determine in a solution containing 2 g of a previously dried sample in sufficient 1 N hydrochloric acid to make 100 mL.

Packaging and Storage   Store in well-closed containers.

Dammar Gum

Dammar Resin; Damar Gum; Damar Resin; Dammar

CAS: [9000-16-2]

DESCRIPTION

Dammar Gum is the dried exudate from trees of the *Agathis*, *Hopea*, or *Shorea* species. It consists of a complex mixture of acidic and neutral terpenoid compounds together with polysaccharide material. Crude Dammar Gum occurs as irregular, white to yellowish to brownish tears, fragments, or powder, sometimes admixed with fragments of bark. Refined grades are white to yellowish and are free of fragments of ligneous matter. It is insoluble in water and in ethanol and is freely soluble in toluene and in limonene. Dammar Gum is practically odorless, although refined grades may carry an odor of the essential oils used in the cleaning process. A chloroform solution of Dammar Gum is dextrorotatory.

Functional Use in Foods   Stabilizer; glazing agent.

REQUIREMENTS

Identification   Prepare a 10% solution of the sample in chloroform, and spot 20 μL on a thin-layer plate coated with a 0.2-mm layer of silica (Merck F254 or equivalent) in a previously equilibrated chamber. Elute with a mixture of 30 parts of diethyl ether and 25 parts of heptane. In a suitable fume hood, spray the plate with sulfuric acid, and dry at 180° for 3 min. Two dark spots are observed at R$_f$ values of 0.8 and 0.7, with the ratio of the faster-moving spot to the second spot being about 1.1.
Acid Number   Between 20 and 40.
Ash (Total)   Not more than 0.5%.
Heavy Metals (as Pb)   Not more than 0.002%.
Iodine Value   Between 10 and 40.
Lead   Not more than 5 mg/kg.
Loss on Drying   Not more than 6.0%.
Melting Range   Between 90° and 95°.
Microbial Limits:
  E. coli   Negative in 25 g.
  Salmonella   Negative in 25 g.
Softening Point   Between 86° and 90°.

TESTS

Acid Number Determine as directed in the general method, Appendix IX, modified as follows: To an accurately weighed 5-g sample add 30 mL of toluene and 30 mL of neutral ethanol. Titrate with 0.5 N alcoholic potassium hydroxide using phenolphthalein TS as the indicator.

Ash (Total) Determine as directed in the general method, Appendix IIC.

Heavy Metals Prepare and test a 1-g sample as directed in *Method II*, Appendix IIIB, using 20 µg of lead ion (Pb) in the control (Solution A).

Iodine Value Determine by the *Modified Wijs Method*, Appendix VII.

Lead A sample solution prepared as directed for organic compounds meets the requirements of the *Lead Limit Test*, Appendix IIIB, using 5 µg of lead ion (Pb) in the control.

Loss on Drying, Appendix IIC Dry at 105° for 18 h.

Melting Range Proceed as directed for *Melting Range or Temperature*, Appendix IIB.

Microbial Limits:
  E. coli Proceed as directed in chapter 4 of the *FDA Bacteriological Analytical Manual*, Seventh Edition, 1992.
  Salmonella Proceed as directed in chapter 5 of the *FDA Bacteriological Analytical Manual*, Seventh Edition, 1992.

Softening Point Determine as directed in the general method, Appendix IX, using the *Ring-and-Ball Method*.

Packaging and Storage Store in well-closed containers.

Decanoic Acid

Capric Acid

$$CH_3(CH_2)_8COOH$$

$C_{10}H_{20}O_2$ Formula wt 172.27

CAS: [334-48-5]

DESCRIPTION

White crystals having a characteristic, unpleasant, rancid odor. It is soluble in most organic solvents and practically insoluble in water.

Functional Use in Foods Component in the manufacture of other food-grade additives; defoaming agent.

REQUIREMENTS

Acid Value Between 320 and 329.
Heavy Metals (as Pb) Not more than 10 mg/kg.
Iodine Value Not more than 0.6.
Residue on Ignition Not more than 0.1%.
Saponification Value Between 320 and 331.

Titer (Solidification Point) Between 27° and 32°.
Unsaponifiable Matter Not more than 0.2%.
Water Not more than 0.2%.

TESTS

Acid Value Determine as directed for *Acid Value, Method I*, under *Fats and Related Substances*, Appendix VII.

Heavy Metals Prepare and test a 2-g sample as directed in *Method II* under the *Heavy Metals Test*, Appendix IIIB, using 20 µg of lead ion (Pb) in the control (Solution A).

Iodine Value Determine by the *Modified Wijs Method*, Appendix VII.

Residue on Ignition, Appendix IIC Ignite 10 g as directed in the general method.

Saponification Value Determine as directed for *Saponification Value* under *Fats and Related Substances*, Appendix VII, using about 2 g, accurately weighed.

Titer (Solidification Point) Determine as directed under *Solidification Point*, Appendix IIB.

Unsaponifiable Matter Determine as directed in the general method, Appendix VII.

Water Determine by the *Karl Fischer Titrimetric Method*, Appendix IIB.

Packaging and Storage Store in well-closed containers.

Dehydroacetic Acid

3-Acetyl-6-methyl-1,2-pyran-2,4(3H)-dione; Methylacetopyronone; DHA

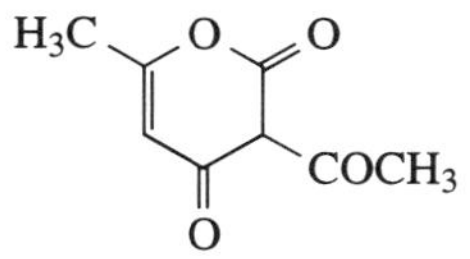

$C_8H_8O_4$ Formula wt 168.15

CAS: [520-45-6]

DESCRIPTION

A white or nearly white crystalline powder. It is odorless or almost odorless, and has a faint, acid taste. It is soluble in aqueous solutions of fixed alkalies, and is very slightly soluble in water. One g dissolves in about 35 mL of alcohol and in 5 mL of acetone.

Functional Use in Foods Preservative.

REQUIREMENTS

Identification The infrared absorption spectrum of a potassium bromide dispersion of the sample exhibits maxima only at the same wavelengths as that of USP Dehydroacetic Acid Reference Standard.

Assay Not less than 98.0% and not more than 100.5% of $C_8H_8O_4$, calculated on the dried basis.

Heavy Metals (as Pb) Not more than 10 mg/kg.

Loss on Drying Not more than 1%.

Melting Range Between 109° and 111°.

Residue on Ignition Not more than 0.1%.

TESTS

Assay Transfer about 500 mg, accurately weighed, to a 250-mL Erlenmeyer flask, dissolve it in 75 mL of neutral alcohol, add phenolphthalein TS, and titrate with 0.1 N sodium hydroxide to a pink endpoint that persists for at least 30 s. Each mL of 0.1 N sodium hydroxide is equivalent to 16.82 mg of $C_8H_8O_4$.

Heavy Metals Prepare and test a 2-g sample as directed in *Method II* under the *Heavy Metals Test*, Appendix IIIB, using 20 μg of lead ion (Pb) in the control (*Solution A*).

Loss on Drying, Appendix IIC Dry at 80° for 4 h.

Melting Range Determine as directed for *Melting Range or Temperature*, Appendix IIB.

Residue on Ignition Ignite 2 g as directed in the general method, Appendix IIC.

Packaging and Storage Store in well-closed containers.

Desoxycholic Acid

Deoxycholic Acid; 13α,12α-Dihydroxycholanic Acid

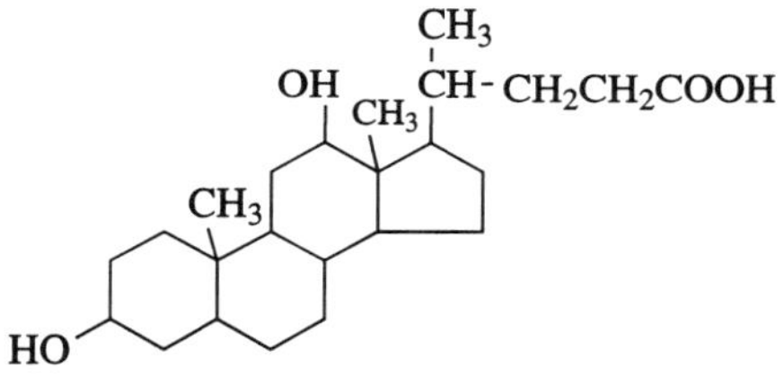

$C_{24}H_{40}O_4$ Formula wt 392.58

CAS: [83-44-3]

DESCRIPTION

A white crystalline powder. It is practically insoluble in water, slightly soluble in chloroform and in ether, soluble in acetone and in solutions of alkali hydroxides and carbonates, and freely soluble in alcohol.

Functional Use in Foods Emulsifier.

REQUIREMENTS

Identification To about 10 mg of the sample add 2 drops of benzaldehyde and 3 drops of 75% sulfuric acid, heat at 50° for 5 min, and then add 10 mL of glacial acetic acid. A green color is produced. (Cholic acid produces a brown color.)

Assay Not less than 98.0% and not more than 102.0% of $C_{24}H_{40}O_4$, calculated on the dried basis.

Heavy Metals (as Pb) Not more than 0.002%.

Lead Not more than 10 mg/kg.

Loss on Drying Not more than 1%.

Melting Range Between 172° and 175°.

Residue on Ignition Not more than 0.2%.

TESTS

Assay Transfer about 500 mg, accurately weighed, into a 250-mL Erlenmeyer flask, and add 20 mL of water and 40 mL of alcohol. Cover the flask with a watch glass, heat the mixture gently on a steam bath until the sample is dissolved, and allow to cool to room temperature. To the solution add a few drops of phenolphthalein TS, and titrate with 0.1 N sodium hydroxide to a pink endpoint that persists for 15 s. Each mL of 0.1 N sodium hydroxide is equivalent to 39.26 mg of $C_{24}H_{40}O_4$.

Heavy Metals Prepare and test a 1-g sample as directed in *Method II* under the *Heavy Metals Test*, Appendix IIIB, using 20 μg of lead ion (Pb) in the control (*Solution A*).

Lead A *Sample Solution* prepared as directed for organic compounds meets the requirements of the *Lead Limit Test*, Appendix IIIB, using 10 μg of lead ion (Pb) in the control.

Loss on Drying, Appendix IIC Dry at 140° under a vacuum of not more than 5 mm of Hg for 4 h.

Melting Range Determine as directed for *Melting Range or Temperature*, Appendix IIB.

Residue on Ignition, Appendix IIC Ignite 1 g as directed in the general method.

Packaging and Storage Store in tight containers.

Dexpanthenol

D(+)-Pantothenyl Alcohol; Panthenol

$HOCH_2C(CH_3)_2CH(OH)CONH(CH_2)_2CH_2OH$

$C_9H_{19}NO_4$ Formula wt 205.25

CAS: [81-13-0]

DESCRIPTION

The dextrorotatory isomer of the alcohol analogue of pantothenic acid. It occurs as a clear, viscous, somewhat hygroscopic liquid having a slight characteristic odor. Some crystallization may occur on standing. Its solutions are alkaline to litmus. It is freely soluble in water, in alcohol, in methanol, and in propylene glycol. It is soluble in chloroform and in ether, and is slightly soluble in glycerin.

Note: The physiological activity of 1.0 g of Dexpanthenol is equivalent to 1.16 g of dextro-calcium pantothenate.

Functional Use in Foods Nutrient; dietary supplement.

REQUIREMENTS

Identification
A. To 1 mL of a 10% solution of the sample add 5 mL of 1 N sodium hydroxide and 1 drop of cupric sulfate TS, and shake vigorously. A deep blue color develops.
B. To 1 mL of a 1% solution of the sample add 1 mL of 1 N hydrochloric acid, and heat on a steam bath for about 30 min. Cool, add 100 mg of hydroxylamine hydrochloride, mix, and add 5 mL of 1 N sodium hydroxide. Allow to stand for 5 min, then adjust the pH to 2.5 to 3.0 with 1 N hydrochloric acid, and add 1 drop of ferric chloride TS. A purplish red color develops.
C. The infrared absorption spectrum of a film of the sample exhibits maxima only at the same wavelengths as that of a similar preparation of USP Dexpanthenol Reference Standard.
Assay Not less than 98.0% and not more than 102.0% of $C_9H_{19}NO_4$ (Dexpanthenol), calculated on the anhydrous basis.
Aminopropanol Not more than 1%.
Heavy Metals (as Pb) Not more than 10 mg/kg.
Refractive Index Between 1.495 and 1.502 at 20°.
Residue on Ignition Not more than 0.1%.
Specific Rotation $[\alpha]_D^{25°}$: Between +29.0° and +31.5°, calculated on the anhydrous basis.
Water Not more than 1%.

TESTS

Assay Transfer about 400 mg, accurately weighed, into a 300-mL reflux flask fitted with a standard-taper glass joint, add 50.0 mL of 0.1 N perchloric acid in glacial acetic acid, and reflux for 5 h. Cool, covering the condenser with foil to prevent contamination by moisture, and rinse the condenser with glacial acetic acid. Add 5 drops of crystal violet TS, and titrate with 0.1 N potassium acid phthalate in glacial acetic acid to a blue green endpoint. Perform a blank determination and make any necessary correction (see *General Provisions*). Each mL of 0.1 N perchloric acid is equivalent to 20.53 mg of $C_9H_{19}NO_4$.
Aminopropanol Transfer about 5 g of the sample, accurately weighed, into a 50-mL flask, and dissolve in 10 mL of water. Add bromothymol blue TS, and titrate with 0.1 N sulfuric acid from a microburet to a yellow endpoint. Each mL of 0.1 N sulfuric acid is equivalent to 7.5 mg of aminopropanol.
Heavy Metals Prepare and test a 2-g sample as directed in *Method II* under the *Heavy Metals Test*, Appendix IIIB, using 20 μg of lead ion (Pb) in the control (*Solution A*).
Refractive Index, Appendix IIB Determine with an Abbé or other refractometer of equal or greater accuracy.
Residue on Ignition Ignite a 1-g sample as directed in the general method, Appendix IIC.

Specific Rotation, Appendix IIB Determine in a solution containing 500 mg, calculated on the anhydrous basis, in each 10 mL of water.
Water Determine by the *Karl Fischer Titrimetric Method*, Appendix IIB.

Packaging and Storage Store in tight containers.

Dextrin

INS: 1400   CAS: [9004-53-9]

DESCRIPTION

Dextrin is partially hydrolyzed starch converted by heat alone, or by heating in the presence of suitable food-grade acids and buffers, from any of several grain- or root-based unmodified native starches (e.g., corn, waxy maize, high amylose, milo, waxy milo, potato, arrowroot, wheat, rice, tapioca, sago, etc.). The products thus obtained occur as free-flowing white, yellow, or brown powders and consist chiefly of polygonal, rounded, or oblong or truncated granules. They are partially to completely soluble in water.

Functional Use in Foods Thickener; colloidal stabilizer; binder; surface-finishing agent.

REQUIREMENTS

Labeling Indicate the presence of sulfur dioxide if the residual concentration is greater than 10 mg/kg.
Identification Suspend about 1 g of the sample in 20 mL of water, and add a few drops of iodine TS. A dark blue to reddish brown color is produced.
Chloride Not more than 0.2%.
Crude Fat Not more than 1.0%.
Heavy Metals (as Pb) Not more than 0.002%.
Lead Not more than 1 mg/kg.
Loss on Drying Not more than 13.0%.
Protein Not more than 1.0%.
Reducing Sugars Not more than 18.0% (expressed as D-glucose), calculated on the dried basis.
Residue on Ignition Not more than 0.5%.

TESTS

Heavy Metals, Lead, Loss on Drying, and **Protein** Determine as directed in the monograph for *Food Starch, Modified*.
Chloride, Appendix IIIB Dissolve 1 g in 25 mL of boiling water, cool, dilute to 100 mL with water, and filter. To 1 mL of the filtrate add 24 mL of water, 2 mL of nitric acid, and 1 mL of silver nitrate TS. Any turbidity produced does not exceed that shown in a control containing 20 μg of chloride ion.
Crude Fat Determine as directed in the general method, Appendix X.

Reducing Sugars Transfer about 10 g of the sample, accurately weighed, into a 200-mL collecting flask, dilute to volume with water, shake for 30 min, and filter through Whatman No. 1 filter paper, or equivalent, collecting the filtrate in a clean, dry flask. Pipet 10 mL each of *Fehling's Solution A* and of *Fehling's Solution B* (see *Cupric Tartrate TS, Alkaline*, in the section on *General Tests and Assays, Solutions and Indicators*) into a 250-mL Erlenmeyer flask, add 20.0 mL of the sample filtrate and 10 mL of water, and mix. Add two small glass beads, cover the mouth of the flask with a small glass funnel or glass bulb, and heat on a hot plate adjusted to bring the solution to a boil in 3 min. Continue boiling for exactly 2 min (total heating time, 5 min), and then quickly cool to room temperature in an ice bath or in a cold running-water bath. Add 10 mL each of 30% potassium iodide solution and of 28% sulfuric acid, and titrate immediately with 0.1 N sodium thiosulfate. Near the endpoint add 1 mL of starch TS, and continue titrating carefully, while agitating the solution continuously, until the blue color is discharged. Record the volume, in mL, of 0.1 N sodium thiosulfate required as S. Conduct two reagent blank determinations in the same manner, substituting water for the sample filtrate, and record the average volume, in mL, of the blanks as B. Obtain the *Titer Difference*, expressed as mL of 0.1 N sodium thiosulfate, by subtracting S from B. Determine the weight, in mg, of reducing sugars, expressed as D-glucose (dextrose), by reference to the table entitled *Conversion of Titer Difference to Reducing Sugars Content*, and record this value as R. Calculate the percentage of reducing sugars, as D-glucose, on the dried basis, by the formula

$$(R \times 200 \times 100)/(W \times 20 \times 1000),$$

in which W is the weight of sample taken, in g, corrected for *Loss on Drying*.

Residue on Ignition Ignite 5 g as directed in the general method, Appendix IIC.

Packaging and Storage Store in well-closed containers.

Conversion of Titer Difference to Reducing Sugars Content[a]

Titer Difference (mL)	0.0	0.1	0.2	0.3	0.4	0.5	0.6	0.7	0.8	0.9
	Reducing Sugar (as Dextrose) (mg)									
0.0	0.0	0.3	0.7	1.0	1.3	1.6	1.9	2.2	2.5	2.8
1.0	3.2	3.5	3.8	4.1	4.4	4.7	5.0	5.3	5.6	5.9
2.0	6.4	6.6	6.9	7.2	7.5	7.8	8.1	8.5	8.8	9.1
3.0	9.4	9.8	10.1	10.4	10.7	11.0	11.4	11.7	12.0	12.3
4.0	12.6	13.0	13.3	13.6	14.0	14.3	14.6	15.0	15.3	15.6
5.0	15.9	16.3	16.6	16.9	17.2	17.6	17.9	18.2	18.5	18.9
6.0	19.2	19.5	19.8	20.1	20.5	20.8	21.1	21.4	21.8	22.1
7.0	22.4	22.7	23.0	23.3	23.7	24.0	24.3	24.6	24.9	25.2
8.0	25.6	25.9	26.2	26.6	26.9	27.3	27.6	28.0	28.3	28.6
9.0	28.9	29.3	29.6	30.0	30.3	30.6	31.0	31.3	31.6	31.9
10.0	32.3	32.7	33.0	33.3	33.7	34.0	34.3	34.6	35.0	35.3
11.0	35.7	36.0	36.3	36.7	37.0	37.3	37.6	38.0	38.3	38.7
12.0	39.0	39.3	39.6	40.0	40.3	40.6	41.0	41.3	41.7	42.0
13.0	42.4	42.8	43.1	43.4	43.7	44.1	44.4	44.8	45.2	45.5
14.0	45.8	46.2	46.5	46.9	47.2	47.6	47.9	48.3	48.6	48.9
15.0	49.3	49.6	49.9	50.3	50.7	51.1	51.4	51.7	52.1	52.4
16.0	52.8	53.2	53.5	53.9	54.2	54.5	54.9	55.3	55.6	56.0
17.0	56.3	56.7	57.0	57.3	57.7	58.1	58.4	58.8	59.1	59.5
18.0	59.8	60.1	60.5	60.9	61.2	61.5	61.9	62.3	62.6	63.0
19.0	63.3	63.6	64.0	64.3	64.7	65.0	65.4	65.8	66.1	66.5
20.0	66.9	67.2	67.6	68.0	68.4	68.8	69.1	69.5	69.9	70.3
21.0	70.7	71.1	71.5	71.9	72.2	72.6	73.0	73.4	73.7	74.1
22.0	74.5	74.9	75.3	75.7	76.1	76.5	76.9	77.3	77.7	78.1
23.0	78.5	78.9	79.3	79.7	80.1	80.5	80.9	81.3	81.7	82.1
24.0	82.6	83.0	83.4	83.8	84.2	84.6	85.0	85.4	85.8	86.2
25.0	86.6	87.0	87.4	87.8	88.2	88.6	89.0	89.4	89.8	90.2
26.0	90.7	91.1	91.5	91.9	92.3	92.7	93.1	93.5	93.9	94.3
27.0	94.8									

[a]Use of this table presumes the ability of the analyst to duplicate exactly the conditions under which the data were developed. The risk of error can be avoided by careful duplicate standardization with known quantities of pure dextrose (5 samples, ranging from 10 to 70 mg). A plot of *Titer Difference* versus mg of dextrose is slightly curvilinear, passing through the origin. If use of a standardization curve is adopted, the thiosulfate solution need not be standardized. Some additional increase in accuracy results from use of a 0.065 N sodium thiosulfate solution, which increases the blank titer to about 44 to 45 mL.

Dextrose

D-Glucose; Glucose; Corn Sugar

$C_6H_{12}O_6$ Formula wt 180.16

CAS: [492-62-6]

DESCRIPTION

Dextrose is purified and crystallized D-glucose. It is anhydrous or contains one molecule of water of crystallization. It occurs as white, odorless, crystalline granules or as a granular powder having a bland, sweet taste. It is freely soluble in water, very soluble in boiling water, and slightly soluble in alcohol.

Functional Use in Foods Nutritive sweetener; humectant; texturizing agent; formulation and processing aid.

REQUIREMENTS

Labeling Indicate the presence of sulfur dioxide if the residual concentration is greater than 10 mg/kg.
Identification Add a few drops of a 1 in 20 solution of the sample to 5 mL of hot alkaline cupric tartrate TS. A copious red precipitate of cuprous oxide is formed.
Assay Not less than 99.5% and not more than 100.5% of reducing sugar content (dextrose equivalent), expressed as D-glucose, calculated on the dried basis.
Arsenic (as As) Not more than 1 mg/kg.
Chloride Not more than 0.018%.
Heavy Metals (as Pb) Not more than 5 mg/kg.
Lead Not more than 0.1 mg/kg.
Loss on Drying *Anhydrous form*: not more than 2.0%; *monohydrate*: not more than 10.0%.
Residue on Ignition Not more than 0.1%.
Specific Rotation $[\alpha]_D^{25°}$: Between +52.6° and +53.2° after drying.
Starch Passes test.
Sulfur Dioxide Not more than 0.002%.

TESTS

Assay Determine as directed under the *Reducing Sugars Assay*, Appendix X.
Arsenic A sample solution prepared using a 1-g sample meets the requirements of the *Arsenic Test*, Appendix IIIB, using 1 mL of *Standard Arsenic Solution* in the control (1 μg As).
Chloride A 2.0-g sample shows no more chloride than corresponds to 0.50 mL of 0.020 N hydrochloric acid.
Heavy Metals Prepare and test a 4-g sample as directed under the *Heavy Metals Limit Test*, Appendix IIIB, using 20 μg of lead (Pb) in the control (*Solution A*).
Lead Determine as directed under *Method I* in the *Atomic Absorption Spectrophotometric Graphite Furnace Method* under the *Lead Limit Test*, Appendix IIIB, using a 5-g sample.
Loss on Drying, Appendix IIC Dry 10 g of anhydrous Dextrose, or 5 g of Dextrose monohydrate, at 70° in a vacuum oven not exceeding 50 mm of Hg for 2 h, cool in a desiccator for 30 min, and weigh. Dry for successive 1-h intervals until the weight change is less than 2 mg.
Residue on Ignition Ignite 10 g as directed in the general method, Appendix IIC.
Specific Rotation, Appendix IIB Determine in a solution containing 10 g of a previously dried sample and 0.2 mL of 6 N ammonium hydroxide in sufficient water to make 100 mL.
Starch To 1 g dissolved in 10 mL of water add 1 drop of iodine TS. A yellow color indicates the absence of soluble starch.
Sulfur Dioxide Determine as directed in the general method, Appendix X, using a 75-g sample.

Packaging and Storage Store in tight containers in a dry place.

Diacetyl Tartaric Acid Esters of Mono- and Diglycerides

DESCRIPTION

The reaction product of partial glycerides of edible oils, fats, or fat-forming fatty acids with diacetyl tartaric anhydride. The esters range in appearance from sticky, viscous liquids through a fatlike consistency to a waxy solid, depending upon the iodine value of the oils or fats used in their manufacture. The diacetyl tartaroyl esters have a faint acid odor and are miscible in all proportions with oils and fats. They are soluble in most common fat solvents, in methanol, in acetone, and in ethyl acetate, but are insoluble in other alcohols, in acetic acid, and in water. They are dispersible in water and resistant to hydrolysis for moderate periods of time. The pH of a 3% dispersion in water is between 2 and 3.

Functional Use in Foods Emulsifier.

REQUIREMENTS

Identification To a solution of 500 mg in 10 mL of methanol add, dropwise, lead acetate TS. A white, flocculent, practically insoluble precipitate forms.
Assay for Tartaric Acid Between 17.0 and 20.0 g of tartaric acid ($C_4H_6O_6$) per 100 g after saponification.
Acetic Acid Between 14.0 and 17.0 g of CH_3COOH per 100 g after saponification.
Acid Value Between 62 and 76.
Fatty Acids, Total Not less than 56.0 g of total fatty acids per 100 g after saponification.
Glycerin Not less than 12.0 g of $C_3H_8O_3$ per 100 g after saponification.
Heavy Metals (as Pb) Not more than 10 mg/kg.
Residue on Ignition Not more than 0.01%.
Saponification Value Between 380 and 425.

TESTS

Assay for Tartaric Acid

Standard Reference Curve Transfer 100 mg of reagent-grade tartaric acid, accurately weighed, into a 100-mL volumetric flask, dissolve it in about 90 mL of water, add water to volume, and mix well. Transfer 3.0-, 4.0-, 5.0-, and 6.0-mL portions into separate 19- × 150-mm matched cuvettes, and add sufficient water to make 10.0 mL. To each cuvette add 4.0 mL of a freshly prepared 1 in 20 solution of sodium metavanadate and 1.0 mL of glacial acetic acid.

Note: Use these solutions within 10 min after color development.

Prepare a blank in the same manner, using 10 mL of water in place of the tartaric acid solutions. Set the instrument at zero with the blank, and then determine the absorbance of the four solutions of tartaric acid at 520 nm with a suitable spectrophotometer or a photoelectric colorimeter equipped with a 520-nm filter. From the data thus obtained, prepare a reference curve by plotting the absorbances on the ordinate against the corresponding quantities, in mg, of the tartaric acid on the abscissa.

Assay Preparation Transfer about 4 g of the sample, accurately weighed, into a 250-mL Erlenmeyer flask, and add 80 mL of approximately 0.5 N potassium hydroxide and 0.5 mL of phenolphthalein TS. Connect an air condenser at least 65 cm in length to the flask, and heat the mixture on a hot plate for about 2.5 h. Add to the hot mixture approximately 10% phosphoric acid until it is definitely acid to congo red test paper. Reconnect the air condenser, and heat until the fatty acids are liquified and clear. Cool and then transfer the mixture into a 250-mL separator with the aid of small portions of water and chloroform. Extract the liberated fatty acids with three successive 25-mL portions of chloroform, and collect the extracts in a second separator. Wash the combined chloroform extracts with two 25-mL portions of water, and add the washings to the separator containing the water layer. Retain the combined chloroform extracts for the determination of total fatty acids. Transfer the contents of the first separator to a 250-mL beaker, heat on a steam bath to remove traces of chloroform, filter through acid-washed, fine-texture filter paper into a 500-mL volumetric flask, and finally dilute to volume with water (*Solution I*). Pipet 25.0 mL of this solution into a 100-mL volumetric flask, and dilute to volume with water (*Solution II*). Retain the rest of *Solution I* for the determination of *Glycerin*.

Procedure Transfer 10.0 mL of *Solution II* prepared under *Assay Preparation* into a 19- × 150-mm cuvette, and continue as directed under *Standard Reference Curve*, beginning with "... add 4.0 mL of a freshly prepared 1 in 20 solution of sodium metavanadate...." From the reference curve determine the weight, in mg, of tartaric acid in the final dilution, multiply this by 20, and divide the result by the weight of the original sample to obtain the percentage of tartaric acid.

Acetic Acid Determine as directed under *Volatile Acidity*, Appendix VII, using a 4-g sample, accurately weighed, and 30.03 as the equivalence factor (*e*).

Acid Value Transfer about 1 g, accurately weighed, into a 125-mL Erlenmeyer flask. Prepare a solvent by mixing 1 volume of benzene with 4 volumes of methanol, adding phenol red TS, and neutralizing, if necessary. Dissolve the sample in about 25 mL of this solvent by warming gently, if necessary. Titrate the solution with 0.1 N methanolic potassium hydroxide to a light red endpoint. Perform a blank determination on a 25-mL portion of the solvent, and make any necessary correction (see *General Provisions*). Calculate the acid value by the formula

$$56.1V \times N/W,$$

in which V is the volume, in mL, N is the normality of the methanolic potassium hydroxide, and W is the weight, in g, of the sample taken.

Fatty Acids, Total Dry the combined chloroform extracts of fatty acids obtained in the *Assay for Tartaric Acid* by shaking with a few g of anhydrous sodium sulfate. Filter the solution into a tared 250-mL beaker, evaporate the chloroform on a steam bath, cool, and weigh.

Glycerin Transfer 5.0 mL of *Solution I* prepared in the *Assay for Tartaric Acid* into a 250-mL glass-stoppered Erlenmeyer or iodine flask. Add to the flask 15 mL of glacial acetic acid and 25.0 mL of periodic acid solution, prepared by dissolving 2.7 g of periodic acid (H_5IO_6) in 50 mL of water, adding 950 mL of glacial acetic acid, and mixing thoroughly; protect this solution from light. Shake the mixture for 1 or 2 min, allow it to stand for 15 min, add 15 mL of potassium iodide solution (15 in 100) and 15 mL of water, swirl, let stand 1 min, and then titrate the liberated iodine with 0.1 N sodium thiosulfate, using starch TS as the indicator. Perform a *Residual Blank Titration* (see *General Provisions*) using water in place of the sample. The corrected volume is the number of mL of 0.1 N sodium thiosulfate required for the glycerin and the tartaric acid in the sample represented by the 5 mL of *Solution I*. From the percentage determined in the *Assay for Tartaric Acid*, calculate the volume of 0.1 N sodium thiosulfate required for the tartaric acid in the titration. The difference between the corrected volume and the calculated volume required for the tartaric acid is the number of mL of 0.1 N sodium thiosulfate consumed due to the glycerin in the sample. One mL of 0.1 N sodium thiosulfate is equivalent to 2.303 mg of glycerin and to 7.505 mg of tartaric acid.

Heavy Metals Prepare and test a 2-g sample as directed in *Method II* under the *Heavy Metals Test*, Appendix IIIB, using 20 µg of lead ion (Pb) in the control (*Solution A*).

Residue on Ignition Ignite 10 g as directed in the general method, Appendix IIC.

Saponification Value Determine as directed for *Saponification Value* under *Fats and Related Substances*, Appendix VII, using about 2 g, accurately weighed. Add 5 to 10 mL of water to samples and blanks before saponification; otherwise sufficient salts precipitate during saponification to cause serious bumping and spattering.

Packaging and Storage Store in well-closed containers.

Diatomaceous Earth

Diatomaceous Silica; Diatomite; D.E.

CAS: [61790-53-2]

DESCRIPTION

A white to gray or buff-colored powder consisting of processed siliceous skeletons of diatoms. It is insoluble in water, in acids (except hydrofluoric), and in dilute alkalies. The *natural* powder (gray to off-white) is air dried and classified by particle size; the *calcined* powder (pink to buff-colored) is air dried, classified, calcined at a high temperature (1500° to 1800°F), and again classified; and the *flux-calcined* powder (white) is air dried, classified, calcined in the presence of a suitable flux (generally soda ash or other alkaline salt), and again classified.

Functional Use in Foods Filter aid in food processing.

REQUIREMENTS

Identification   When examined with a 100- to 200-power microscope, typical diatom shapes are observed.
Arsenic (as As)   Not more than 10 mg/kg.
Lead   Not more than 10 mg/kg.
Loss on Drying   *Natural powders*: not more than 10.0%; *calcined* and *flux-calcined powders*: not more than 3.0%.
Loss on Ignition   *Natural powders*: not more than 7.0%, calculated on the dried basis; *calcined* and *flux-calcined powders*: not more than 0.5%, calculated on the dried basis.
Nonsiliceous Substances   Not more than 25.0%, calculated on the dried basis.
pH   Passes test.

TESTS

Arsenic   Transfer 10.0 g of the sample into a 250-mL beaker, add 50 mL of 0.5 N hydrochloric acid, cover with a watch glass, and heat at 70° for 15 min. Cool, and decant through a Whatman No. 3 filter paper into a 100-mL volumetric flask. Wash the slurry with three 10-mL portions of hot water and the filter paper with 15 mL of hot water, dilute to volume with water, and mix. A 3.0-mL portion of this solution meets the requirements of the *Arsenic Test*, Appendix IIIB.
Lead   A 10.0-mL portion of the solution prepared in the *Arsenic Test* meets the requirements of the *Lead Limit Test*, Appendix IIIB, using 10 μg of lead ion (Pb) in the control.
Loss on Drying, Appendix IIC   Dry at 105° for 2 h.
Loss on Ignition   Weigh accurately about 1 g, and ignite at 800° to constant weight in a suitable tared crucible.
Nonsiliceous Substances   Transfer about 200 mg, accurately weighed, into a tared platinum crucible, add 5 mL of hydrofluoric acid and 2 drops of sulfuric acid (1 in 2), and evaporate gently to dryness. Cool, add 5 mL of hydrofluoric acid, evaporate again to dryness, and then ignite to constant weight.
pH, Appendix IIB   Boil 10 g with 100 mL of water for 30 min, make up to 100 mL with water, and filter through a fine-porosity sintered-glass funnel. The pH of the filtrate prepared with *natural* or *calcined* powders is between 5.0 and 10.0, and of that prepared with *flux-calcined* powders is between 8.0 and 11.0.

Packaging and Storage   Store in well-closed containers.

Dilauryl Thiodipropionate

$(C_{12}H_{25}OOCCH_2CH_2)_2S$

$C_{30}H_{58}O_4S$ Formula wt 514.85
INS: 389   CAS: [123-28-4]

DESCRIPTION

White crystalline flakes having a characteristic sweetish, esterlike odor. It is insoluble in water, but is soluble in most organic solvents.

Functional Use in Foods   Antioxidant.

REQUIREMENTS

Identification   Dilauryl Thiodipropionate may be identified by its solidification point (see below).
Assay   Not less than 99.0% and not more than 100.5% of $C_{30}H_{58}O_4S$.
Acidity (as thiodipropionic acid)   Not more than 0.2% of $C_6H_{10}O_4S$.
Heavy Metals (as Pb)   Not more than 0.002%.
Lead   Not more than 10 mg/kg.
Solidification Point   Not below 40°.

TESTS

Assay   Transfer about 700 mg, accurately weighed, into a 250-mL Erlenmeyer flask, and add 100 mL of glacial acetic acid and 50 mL of alcohol. Heat the mixture at a temperature of about 40° until the sample is completely dissolved, then add 3 mL of hydrochloric acid and 4 drops of *p*-ethoxychrysoidin TS, and immediately titrate the solution with 0.1 N bromine. When the endpoint is approached (pink color), add 4 more drops of the indicator solution and continue the titration, dropwise, to a color change from red to pale yellow. Perform a blank determination (see *General Provisions*) and make any necessary correction. Each mL of 0.1 N bromine is equivalent to 25.74 mg of $C_{30}H_{58}O_4S$. Multiply the percentage of thiodipropionic acid, determined in the *Acidity* test, by 2.89, and subtract this value from the percentage of Dilauryl Thiodipropionate calculated from the titration. The difference is the percentage purity of $C_{30}H_{58}O_4S$.
Acidity (as thiodipropionic acid)   Transfer about 2 g, accurately weighed, into a 250-Erlenmeyer flask. Dissolve the sample in 50 mL of a mixture composed of 1 part of methyl alcohol and 3 parts of benzene, add 5 drops of phenolphthalein TS, and titrate with 0.1 N alcoholic potassium hydroxide. Each mL of 0.1 N alcoholic potassium hydroxide is equivalent to 8.91 mg of $C_6H_{10}O_4S$.
Heavy Metals   Prepare and test a 1-g sample as directed in *Method II* under the *Heavy Metals Test*, Appendix IIIB, using 20 μg of lead ion (Pb) in the control (*Solution A*).
Lead   A *Sample Solution* prepared as directed for organic compounds meets the requirements of the *Lead Limit Test*, Appendix IIIB, using 10 μg of lead ion (Pb) in the control.
Solidification Point   Determine as directed in the general method, Appendix IIB.

Packaging and Storage   Store in well-closed containers.

Dill Seed Oil, European Type

DESCRIPTION

The volatile oil obtained by steam distillation from the crushed, dried fruit (or seeds) of *Anethum graveolens* L. (Fam. *Umbelliferae*). It is a slightly yellowish to light yellow liquid with a caraway-like odor and flavor. It is soluble in most fixed oils and in mineral oil. It is soluble, with slight opalescence, in propylene glycol, but it is practically insoluble in glycerin.

Functional Use in Foods Flavoring agent.

REQUIREMENTS

Identification The infrared absorption spectrum of the sample exhibits relative maxima (that may vary in intensity) at the same wavelengths (or frequencies) as those shown in the respective spectrum in the section on *Infrared Spectra (Series A: Essential Oils)*, using the same test conditions as specified therein.
Assay Not less than 42.0% and not more than 60.0%, by volume, of ketones as carvone.
Angular Rotation Between +70° and +82°.
Heavy Metals (as Pb) Passes test.
Refractive Index Between 1.483 and 1.490 at 20°.
Solubility in Alcohol Passes test.
Specific Gravity Between 0.890 and 0.915.

TESTS

Assay Proceed as directed under *Aldehydes and Ketones—Neutral Sulfite Method*, Appendix VI.
Angular Rotation Determine in a 100-mm tube as directed under *Optical (Specific) Rotation*, Appendix IIB.
Heavy Metals Shake 10 mL of the oil with an equal volume of water to which 1 drop of hydrochloric acid has been added, and pass hydrogen sulfide through the mixture until it is saturated. No darkening in color is produced in either the oil or the water.
Refractive Index, Appendix IIB Determine with an Abbé or other refractometer of equal or greater accuracy.
Solubility in Alcohol Proceed as directed in the general method, Appendix VI. One mL dissolves in 2 mL of 80% alcohol, with slight opalescence that may not disappear on dilution to as much as 10 mL.
Specific Gravity Determine by any reliable method (see *General Provisions*).

Packaging and Storage Store in full, tight, preferably glass, aluminum, or other suitably lined containers in a cool place protected from light.

Dill Seed Oil, Indian Type

Dill Seed Oil, Indian; Dill Oil, Indian Type

DESCRIPTION

The volatile oil obtained by steam distillation from the crushed mature fruit of Indian dill, *Anethum sowa* D.C. (Fam. *Umbelliferae*). It is a light yellow to light brown liquid with a rather harsh caraway-like odor and flavor. It is soluble in most fixed oils and in mineral oil, occasionally with slight opalescence. It is sparingly soluble in propylene glycol and practically insoluble in glycerin.

Functional Use in Foods Flavoring agent.

REQUIREMENTS

Identification The infrared absorption spectrum of the sample exhibits relative maxima (that may vary in intensity) at the same wavelengths (or frequencies) as those shown in the respective spectrum in the section on *Infrared Spectra (Series A: Essential Oils)*, using the same test conditions as specified therein.
Assay Not less than 20.0% and not more than 30.0%, by volume, of ketones as carvone.
Angular Rotation Between +40° and +58°.
Heavy Metals (as Pb) Passes test.
Refractive Index Between 1.486 and 1.495 at 20°.
Solubility in Alcohol Passes test.
Specific Gravity Between 0.925 and 0.980.

TESTS

Assay Proceed as directed under *Aldehydes and Ketones—Neutral Sulfite Method*, Appendix VI.
Angular Rotation Determine in a 100-mm tube as directed under *Optical (Specific) Rotation*, Appendix IIB.
Heavy Metals Shake 10 mL of the oil with an equal volume of water to which 1 drop of hydrochloric acid has been added, and pass hydrogen sulfide through the mixture until it is saturated. No darkening in color is produced in either the oil or the water.
Refractive Index, Appendix IIB Determine with an Abbé or other refractometer of equal or greater accuracy.
Solubility in Alcohol Proceed as directed in the general method, Appendix VI. One mL dissolves in 0.5 mL of 90% alcohol and remains clear on dilution.
Specific Gravity Determine by any reliable method (see *General Provisions*).

Packaging and Storage Store in full, tight, preferably glass, aluminum, or other suitably lined containers protected from light.

Dillweed Oil, American Type

Dill Oil; Dill Herb Oil, American Type

CAS: [8006-75-5]

DESCRIPTION

The volatile oil obtained by steam distillation from the freshly cut stalks, leaves, and seeds of the plant *Anethum graveolens* L. It is a light yellow to yellow liquid. It is soluble in most fixed oils and in mineral oil. It is soluble, usually with opalescence or turbidity, in propylene glycol, but it is practically insoluble in glycerin.

Functional Use in Foods   Flavoring agent.

REQUIREMENTS

Identification   The infrared absorption spectrum of the sample exhibits relative maxima (that may vary in intensity) at the same wavelengths (or frequencies) as those shown in the respective spectrum in the section on *Infrared Spectra (Series A: Essential Oils)*, using the same test conditions as specified therein.
Assay   Usually not less than 28.0% and not more than 45.0%, by volume, of ketones as carvone.

Note: Oil obtained from early season distillation may show a carvone content as low as 25.0% and a correspondingly lower specific gravity, lower refractive index, and higher angular rotation.

Angular Rotation   Between +84° and +95°.
Heavy Metals (as Pb)   Passes test.
Refractive Index   Between 1.480 and 1.485 at 20°.
Solubility in Alcohol   Passes test.
Specific Gravity   Between 0.884 and 0.900.

TESTS

Assay   Proceed as directed under *Aldehydes and Ketones—Neutral Sulfite Method*, Appendix VI.
Angular Rotation   Determine in a 100-mm tube as directed under *Optical (Specific) Rotation*, Appendix IIB.
Heavy Metals   Shake 10 mL of the oil with an equal volume of water to which 1 drop of hydrochloric acid has been added, and pass hydrogen sulfide through the mixture until it is saturated. No darkening in color is produced in either the oil or the water.
Refractive Index, Appendix IIB   Determine with an Abbé or other refractometer of equal or greater accuracy.
Solubility in Alcohol   Proceed as directed in the general method, Appendix VI. One mL dissolves in 1 mL of 90% alcohol, frequently with opalescence that may not disappear on dilution to as much as 10 mL.
Specific Gravity   Determine by any reliable method (see *General Provisions*).

Packaging and Storage   Store in full, tight, preferably glass, aluminum, or tin-lined containers in a cool place protected from light.

Dimethylpolysiloxane

Dimethyl Silicone; Polydimethylsiloxane

CAS: [9006-65-9]

DESCRIPTION

Dimethylpolysiloxane is a mixture of fully methylated linear siloxane polymers containing repeating units of the formula $[(CH_3)_2SiO]$ and stabilized with trimethylsiloxy end-blocking units of the formula $[(CH_3)_3SiO—]$. It occurs as a clear, colorless, viscous liquid that is soluble in most aliphatic and aromatic hydrocarbon solvents but insoluble in water.

Note: Dimethylpolysiloxane is frequently used in commerce as such, or as a liquid containing silica (usually 4% to 5%), which must be removed by high-speed centrifugation (about 20,000 rpm) before testing the Dimethylpolysiloxane for *Identification, Refractive Index, Specific Gravity*, and *Viscosity*. This monograph does not apply to aqueous emulsions containing emulsifying agents and preservatives, in addition to silica.

Functional Use in Foods   Defoaming agent.

REQUIREMENTS

Identification   The infrared absorption spectrum, determined in a 0.5-mm cell, of a solution of 1 mg of the sample in 25 mL of carbon tetrachloride exhibits maxima only at the same wavelengths as that of a similar preparation of USP Polydimethylsiloxane Reference Standard.
Heavy Metals (as Pb)   Not more than 10 mg/kg.
Loss on Heating   Not more than 18.0%.
Refractive Index   Between 1.4000 and 1.4050.
Specific Gravity   Between 0.96 and 0.98.
Viscosity   Between 300 and 1500 centistokes.

TESTS

Heavy Metals   Prepare and test a 2-g sample as directed in *Method II* under the *Heavy Metals Test*, Appendix IIIB, using 20 µg of lead ion (Pb) in the control (*Solution A*).

Note: If silica is present, it must be removed by filtration before the pH is adjusted.

Loss on Heating   Heat 15 g of the sample in an open tared aluminum cup, having an internal surface of about 30 cm^2, for 4 h at 200° in a circulating air oven, cool, and weigh.

Refractive Index, Appendix IIB Determine with an Abbé or other refractometer of equal or greater accuracy.

Specific Gravity Determine by any reliable method (see *General Provisions*).

Viscosity Determine as directed in the general method, Appendix IIB.

Packaging and Storage Store in tight containers.

Dioctyl Sodium Sulfosuccinate

DSS

```
            C2H5
            |
COO—CH2—CH—(CH2)3—CH3
|
CH2
|
CH—SO3Na
|
COO—CH2—CH—(CH2)3—CH3
            |
            C2H5
```

$C_{20}H_{37}NaO_7S$ Formula wt 444.56

CAS: [577-11-7]

DESCRIPTION

A white, waxlike, plastic solid having a characteristic odor suggestive of octyl alcohol. It is free from the odor of other solvents. One g dissolves slowly in about 70 mL of water. It is freely soluble in alcohol and in glycerin, and is very soluble in solvent hexane.

Functional Use in Foods Emulsifier; wetting agent.

REQUIREMENTS

Identification Dry 50 mg of the sample at 105° for 2 h, cool, and immediately dissolve in 2 mL of carbon tetrachloride. The infrared absorption spectrum of the solution, measured in 0.1-mm cells, exhibits maxima only at the same wavelengths as that of a similar preparation of USP Dioctyl Sodium Sulfosuccinate Reference Standard, previously dried in the same manner.

Assay Not less than 98.5% of $C_{20}H_{37}NaO_7S$, calculated on the dried basis.

Bis(2-ethylhexyl)maleate Not more than 0.4%.

Clarity of Solution Passes test.

Heavy Metals (as Pb) Not more than 10 mg/kg.

Loss on Drying Not more than 2.0%.

Residue on Ignition Between 15.5% and 16.2%.

TESTS

Assay

Sample Solution Transfer about 3.8 g of the sample, accurately weighed, into a 500-mL volumetric flask, dissolve in chloroform, dilute to volume with the same solvent, and mix.

Tetra-n-butylammonium Iodide Solution Transfer 1.250 g of tetra-*n*-butylammonium iodide to a 500-mL volumetric flask, dilute to volume with water, and mix.

Salt Solution Dissolve 100 g of anhydrous sodium sulfate and 10 g of sodium carbonate in sufficient water to make 1000.0 mL.

Procedure Pipet 10.0 mL of the *Sample Solution* into a 250-mL flask, and add 40 mL of chloroform, 50 mL of *Salt Solution*, and 10 drops of bromophenol blue TS. Titrate with *Tetra-n-butylammonium Iodide Solution* to the first appearance of a blue color in the chloroform layer after vigorous shaking. Calculate the percentage of $C_{20}H_{37}NaO_7S$ by the formula

$$(V \times 1.250 \times 444.6 \times 10)/(W \times 369.4),$$

in which V is the volume, in mL, of *Tetra-n-butylammonium Iodide Solution* required; 444.6 is the molecular weight of Dioctyl Sodium Sulfosuccinate; W is the weight, in g, of the sample taken; and 369.4 is the molecular weight of tetra-*n*-butylammonium iodide.

Bis(2-ethylhexyl)maleate

Supporting Electrolyte Dissolve 21.2 g of anhydrous lithium perchlorate ($LiClO_4$) in 175 mL of water in a 250-mL beaker. Adjust the pH of this solution to 3.0 by the dropwise addition of glacial acetic acid (usually 1 or 2 drops is sufficient), using a suitable pH meter. Quantitatively transfer the solution into a 200-mL volumetric flask, dilute to volume with water, and mix.

Standard Solution Transfer 100 to 110 mg of USP Bis(2-ethylhexyl)maleate Reference Standard, accurately weighed, into a 100-mL volumetric flask. Record the exact weight, to the nearest 0.1 mg, as W_A. Add 60 to 70 mL of isopropyl alcohol, swirl to dissolve, then dilute to volume with water, and mix.

Sample Stock Solution Transfer 12.5 g of the sample, accurately weighed, into a 150-mL beaker. Record the exact weight, to the nearest 10 mg, as W_S. Add 80 to 90 mL of isopropyl alcohol, and stir with a glass stirring rod until the sample is dissolved. Quantitatively transfer this solution, with the aid of isopropyl alcohol, into a 250-mL volumetric flask, then dilute to volume with isopropyl alcohol, and mix.

Test Solution A Pipet 50.0 mL of the *Sample Stock Solution* and 20.0 mL of the *Supporting Electrolyte* into a 100-mL volumetric flask. Dilute to within 15 mm of the graduated volume line with isopropyl alcohol, stopper, shake to facilitate solution, and set aside for 2 min. Dilute to volume with isopropyl alcohol, and mix. A completely clear solution should be obtained.

Test Solution B Pipet 50.0 mL of the *Sample Stock Solution*, 10.0 mL of the *Standard Solution*, and 20.0 mL of the *Supporting Electrolyte* into a 100-mL volumetric flask, and complete the preparation as described for *Test Solution A*.

Blank Pipet 20.0 mL of the *Supporting Electrolyte* into a 100-mL volumetric flask, dilute to volume with isopropyl alcohol, and mix.

Procedure Rinse a polarographic H-cell several times with small portions of *Test Solution A*, then fill the cell half full with the solution, place a paper tissue in the top of the sample side of the cell, and pass a moderate stream of nitrogen through the solution for 15 min.

Note: The nitrogen should first be saturated by passing it through a suitable scrubber containing isopropyl alcohol.

After 15 min, divert the nitrogen stream over the surface of the solution, and remove the tissue from the cell.

Set the polarizing voltage of a suitable, previously calibrated polarograph (Metrohm Polarocord E-261 or equivalent) at −1.3 V, adjust the current sensitivity to the lowest range (most sensitive) at which the current oscillations will remain on scale, and record the polarogram, scanning a voltage range of −0.9 V to −1.5 V at this sensitivity and using a saturated calomel electrode as the reference electrode. Record the average oscillations, in mm, at −1.3 V as A, and those at −1.0 V as B.

Note: If a manual polarograph is used, record the average oscillations of the solutions at −1.3 V and −1.0 V, respectively.

Repeat the entire procedure using *Test Solution B*, recording the average oscillations at −1.3 V as D, and those at −1.0 V as E. Similarly, repeat the entire procedure using the *Blank*, recording the average oscillations at −1.3 V as G, and those at −1.0 V as H.

Calculation Make the following preliminary calculations (in mA) to obtain C (net diffusion current of *Test Solution A*); F (net diffusion current of *Test Solution B*); I (net current introduced by the *Blank*); J (diffusion current due to added maleate); and K (diffusion current due to originally present maleate):

$$C = (A - B) \times S_1;$$
$$F = (D - E) \times S_2;$$
$$I = (G - H) \times S_3;$$
$$J = F - C;$$
$$K = C - I;$$

in which S_1, S_2, and S_3 represent the current sensitivities used for *Test Solution A*, *Test Solution B*, and the *Blank*, respectively.

Finally, calculate the percentage of bis(2-ethylhexyl)maleate in the original sample taken by the formula

$$(K \times 50W_A)/(J \times W_S).$$

Clarity of Solution Dissolve 25 g in 94 mL of alcohol. The solution does not develop a haze within 24 h.

Heavy Metals Ignite 2 g in a platinum crucible until free from carbon, cool, moisten the residue with 1 mL of hydrochloric acid, and evaporate to dryness on a steam bath. Add 2 mL of 1 N acetic acid, digest on a steam bath for 5 min, filter into a 50-mL Nessler tube, and wash the residue with sufficient water to make 25 mL. This solution meets the requirements of the *Heavy Metals Test*, Appendix IIIB, using 20 μg of lead ion (Pb) in the control (*Solution A*).

Loss on Drying, Appendix IIC Dry at 105° for 2 h.

Residue on Ignition Ignite 1 g as directed in the general method, Appendix IIC.

Packaging and Storage Store in well-closed containers.

Disodium EDTA

Disodium Ethylenediaminetetraacetate; Disodium (Ethylenedinitrilo)tetraacetate; Disodium Edetate

$$\left[\begin{array}{c} \text{NaOOCCH}_2 \quad\quad\quad \text{CH}_2\text{COONa} \\ \quad | \quad\quad \text{CH}_2\text{CH}_2 \quad | \\ \text{CH}_2 \diagdown \text{N} \quad\quad\quad\quad \text{N} \diagdown \text{CH}_2 \\ \quad | \quad\quad\quad\quad\quad\quad\quad\quad | \\ \text{COOH} \quad\quad\quad\quad \text{HOOC} \end{array}\right] \cdot 2\text{H}_2\text{O}$$

$C_{10}H_{14}N_2Na_2O_8 \cdot 2H_2O$ Formula wt 372.24

INS: 386   CAS: [6381-92-6]

DESCRIPTION

A white, crystalline powder. It is soluble in water.

Functional Use in Foods Preservative; sequestrant; stabilizer.

REQUIREMENTS

Identification

A. A 1 in 20 solution responds to the flame test for *Sodium*, Appendix IIIA.

B. To 5 mL of water in a test tube add 2 drops of ammonium thiocyanate TS and 2 drops of ferric chloride TS. To the deep red solution so obtained add about 50 mg of the sample, and mix. The deep red color disappears.

C. The infrared absorption spectrum of a potassium bromide dispersion of the sample exhibits maxima only at the same wavelengths as that of a similar preparation of USP Edetate Disodium Reference Standard.

Assay Not less than 99.0% and not more than 101.0% of $C_{10}H_{14}N_2Na_2O_8 \cdot 2H_2O$.

Calcium Negative.

Heavy Metals (as Pb) Not more than 0.002%.

Lead Not more than 10 mg/kg.

Nitrilotriacetic Acid Not more than 0.1%.

pH of a 1 in 100 Solution Between 4.3 and 4.7.

TESTS

Assay

Assay Preparation Transfer about 5 g of the sample, accurately weighed, into a 250-mL volumetric flask, dissolve in water, dilute to volume, and mix.

Procedure Place about 200 mg of chelometric standard calcium carbonate, accurately weighed, in a 400-mL beaker, add 10 mL of water, and swirl to form a slurry. Cover the beaker with a watch glass, and introduce 2 mL of 2.7 N hydrochloric acid from a pipet inserted between the lip of the beaker and the edge of the watch glass. Swirl the contents of the beaker to dissolve the calcium carbonate. Wash down the sides of the beaker, the outer surface of the pipet, and the watch glass, and dilute to about 100 mL with water. While stirring, preferably with a magnetic stirrer, add about 30 mL of the *Assay Preparation* from a 50-mL buret, then add 15 mL of 1 N sodium hydroxide and 300 mg of hydroxy naphthol blue indicator, and continue the titration to a blue endpoint. Calculate the weight, in mg, of $C_{10}H_{14}N_2Na_2O_8 \cdot 2H_2O$ in the sample taken by the formula

$$929.8(W/V),$$

in which W is the weight, in mg, of calcium carbonate, and V is the volume, in mL, of the *Assay Preparation* consumed in the titration.

Calcium To a 1 in 20 aqueous solution of the sample, add 2 drops of methyl red TS, and neutralize with 6 N ammonium hydroxide. Add 3 N hydrochloric acid dropwise until the solution is just acid, and then add 1 mL of ammonium oxalate TS. No precipitate is formed.

Heavy Metals Prepare and test a 1-g sample as directed in *Method II* under the *Heavy Metals Test*, Appendix IIIB, using 20 μg of lead ion (Pb) in the control.

Lead A *Sample Solution* prepared as directed for organic compounds meets the requirements of the *Lead Limit Test*, Appendix IIIB, using 10 μg of lead ion (Pb) in the control.

Nitrilotriacetic Acid

Mobile Phase Add 10 mL of a 1 in 4 solution of tetrabutylammonium hydroxide in methanol to 200 mL of water, and adjust with 1 M phosphoric acid to a pH of 7.5 ± 0.1. Transfer the solution to a 1000-mL volumetric flask, add 90 mL of methanol, dilute with water to volume, mix, filter through a membrane filter (0.5-μm or finer porosity), and degas.

Cupric Nitrate Solution Prepare a solution containing about 10 mg/mL.

Stock Standard Solution Transfer about 100 mg of nitrilotriacetic acid, accurately weighed, to a 10-mL volumetric flask, add 0.5 mL of ammonium hydroxide, and mix. Dilute to volume, and mix.

Standard Preparation Transfer 1.0 g of the sample to a 100-mL volumetric flask, add 100 μL of *Stock Standard Solution*, dilute with *Cupric Nitrate Solution* to volume, and mix. Sonicate, if necessary, to achieve complete solution.

Test Preparation Transfer 1.0 g of the sample to a 100-mL volumetric flask, dilute with *Cupric Nitrate Solution* to volume, and mix. Sonicate, if necessary, to achieve complete solution.

Chromatographic System (see *High-Pressure Liquid Chromatography*, Appendix IIA) The chromatograph is equipped with a 254-nm detector and a 4.6-mm × 15-cm column that contains 5- to 10-μm porous microparticles of silica to which is bonded octylsilane (Zorbax 8 or equivalent). The flow rate is about 2 mL/min. Chromatograph three replicate injections of the *Standard Preparation*, and record the peak responses as directed under *Procedure*: The relative standard deviation is not more than 2.0%, and the resolution factor between nitrilotriacetic acid and Disodium EDTA is not less than 4.0.

Procedure Separately inject equal volumes (about 50 μL) of the *Standard Preparation* and the *Test Preparation* into the chromatograph, record the chromatograms, and measure the responses for the major peaks. The retention times are about 3.5 min for nitrilotriacetic acid and 9 min for Disodium EDTA. The response of the nitrilotriacetic acid peak of the *Test Preparation* does not exceed the difference between the nitrilotriacetic acid peak responses obtained from the *Standard Preparation* and the *Test Preparation*.

pH of a 1 in 100 Solution Determine by the *Potentiometric Method*, Appendix IIB.

Packaging and Storage Store in well-closed containers.

Disodium Guanylate

Sodium 5′-Guanylate; Disodium Guanosine-5′-monophosphate

$C_{10}H_{12}N_5Na_2O_8P \cdot xH_2O$ Formula wt, anhydrous 407.19

DESCRIPTION

Disodium Guanylate contains approximately seven molecules of water of crystallization. It occurs as colorless or white crystals, or as a white, crystalline powder, having a characteristic taste. It is soluble in water, sparingly soluble in alcohol, and practically insoluble in ether.

Functional Use in Foods Flavor enhancer.

REQUIREMENTS

Identification The ultraviolet absorption spectrum of a 1 in 50,000 solution of the sample in 0.01 N hydrochloric acid exhibits an absorbance maximum at 256 ± 2 nm.

Assay Not less than 97.0% and not more than 102.0% of $C_{10}H_{12}N_5Na_2O_8P$, calculated on the dried basis.

Amino Acids   Passes test.
Ammonium Salts   Passes test.
Clarity and Color of Solution   Passes test.
Heavy Metals (as Pb)   Not more than 0.002%.
Lead   Not more than 10 mg/kg.
Loss on Drying   Not more than 25.0%.
Other Nucleotides   Passes test.
pH of a 1 in 20 Solution   Between 7.0 and 8.5.

TESTS

Assay   Transfer about 500 mg of the sample, accurately weighed, into a 1000-mL volumetric flask, dissolve in 0.01 N hydrochloric acid, dilute to volume with 0.01 N hydrochloric acid, and mix. Transfer 10.0 mL of this solution into a 250-mL volumetric flask, dilute to volume with 0.01 N hydrochloric acid, and mix. Determine the absorbance of this solution and of a similar solution of FCC Disodium Guanylate Reference Standard, at a concentration of 20 μg/mL, in 1-cm cells, at the maximum at about 260 nm, with a suitable spectrophotometer, using 0.01 N hydrochloric acid as the blank. Calculate the quantity, in mg, of $C_{10}H_{12}N_5Na_2O_8P$ in the sample taken by the formula

$$25C \times A_U/A_S,$$

in which C is the exact concentration of the Reference Standard solution, in μg/mL; A_U is the absorbance of the sample solution; and A_S is the absorbance of the Reference Standard solution.
Amino Acids   To 5 mL of a 1 in 1000 solution of the sample add 1 mL of ninhydrin TS, and heat for 3 min. No color is produced.
Ammonium Salts   Transfer about 100 mg of the sample into a small test tube, and add 50 mg of magnesium oxide and 1 mL of water. Moisten a piece of red litmus paper with water, suspend it in the tube, cover the mouth of the tube, and heat in a water bath for 5 min. The litmus paper does not change to blue.
Clarity and Color of Solution   A 100-mg portion of the sample dissolved in 10 mL of water is colorless and shows no more than a trace of turbidity.
Heavy Metals   Prepare and test a 1-g sample as directed in *Method II* under the *Heavy Metals Test*, Appendix IIIB, using 20 μg of lead ion (Pb) in the control (*Solution A*).
Lead   A *Sample Solution* prepared as directed for organic compounds meets the requirements of the *Lead Limit Test*, Appendix IIIB, using 10 μg of lead ion (Pb) in the control.
Other Nucleotides   Prepare a strip of Whatman No. 2 or equivalent filter paper about 20 × 40 cm, and draw a line across the narrow dimension about 5 cm from one end. Using a micropipet, apply on the center of the line 10 μL of a 1 in 100 solution of the sample in water, and dry the paper in air. Fill the trough of an apparatus suitable for descending chromatography (see Appendix IIA) with a 160:3:40 mixture of saturated ammonium sulfate solution, *tert*-butyl alcohol, and 0.025 N ammonia, respectively, and suspend the strip in the chamber, placing the end of the strip in the trough at a distance about 1 cm from the pencil line. Seal the chamber, and allow the chromatogram to develop until the solvent front descends to a distance about 30 cm from the starting line. Remove the strip from the chamber, dry in air, and observe under shortwave (254 nm) ultraviolet light in the dark. Only one spot is visible.
Loss on Drying, Appendix IIC   Dry at 120° for 4 h.
pH of a 1 in 20 Solution   Determine by the *Potentiometric Method*, Appendix IIB.

Packaging and Storage   Store in well-closed containers.

Disodium Inosinate

Sodium 5′-Inosinate; Disodium Inosine-5′-monophosphate

$C_{10}H_{11}N_4Na_2O_8P \cdot xH_2O$ Formula wt, anhydrous 392.17

DESCRIPTION

Disodium Inosinate contains approximately 7.5 molecules of water of crystallization. It occurs as colorless or white crystals, or as a white, crystalline powder, having a characteristic taste. It is soluble in water, sparingly soluble in alcohol, and practically insoluble in ether.

Functional Use in Foods   Flavor enhancer.

REQUIREMENTS

Identification   The ultraviolet absorption spectrum of a 1 in 50,000 solution of the sample in 0.01 N hydrochloric acid exhibits an absorbance maximum at 250 ± 2 nm. The ratio A_{250}/A_{260} is between 1.55 and 1.65, and the ratio A_{280}/A_{260} is between 0.20 and 0.30.
Assay   Not less than 97.0% and not more than 102.0% of $C_{10}H_{11}N_4Na_2O_8P$, calculated on the anhydrous basis.
Amino Acids   Passes test.
Ammonium Salts   Passes test.
Barium   Not more than 0.015%.
Clarity and Color of Solution   Passes test.
Heavy Metals (as Pb)   Not more than 0.002%.
Lead   Not more than 10 mg/kg.
Other Nucleotides   Passes test.
pH of a 1 in 20 Solution   Between 7.0 and 8.5.
Water   Not more than 28.5%.

TESTS

Assay Transfer about 500 mg of the sample, accurately weighed, into a 1000-mL volumetric flask, dissolve in 0.01 N hydrochloric acid, dilute to volume with 0.01 N hydrochloric acid, and mix. Transfer 10.0 mL of this solution into a 250-mL volumetric flask, dilute to volume with 0.01 N hydrochloric acid, and mix. Determine the absorbance of this solution and of a similar solution of FCC Disodium Inosinate Reference Standard, at a concentration of 20 μg/mL, in 1-cm cells, at the maximum at about 250 nm, with a suitable spectrophotometer, using 0.01 N hydrochloric acid as the blank. Calculate the quantity, in mg, of $C_{10}H_{11}N_4Na_2O_8P$ in the sample taken by the formula

$$25C \times A_U/A_S,$$

in which C is the exact concentration of the Reference Standard solution, in μg/mL; A_U is the absorbance of the sample solution; and A_S is the absorbance of the Reference Standard solution.

Amino Acids To 5 mL of a 1 in 1000 solution of the sample add 1 mL of ninhydrin TS. No color is produced.

Ammonium Salts Transfer about 100 mg of the sample into a small test tube, and add 50 mg of magnesium oxide and 1 mL of water. Moisten a piece of red litmus paper with water, suspend it in the tube, cover the mouth of the tube, and heat in a water bath for 5 min. The litmus paper does not change to blue.

Barium Dissolve 1 g of the sample in 100 mL of water, filter, and add 5 mL of 2 N sulfuric acid to the filtrate. Any turbidity is not greater than that produced in a similar solution containing 1.5 mL of *Barium Standard Solution* (150 μg Ba).

Clarity and Color of Solution A 500-mg portion of the sample dissolved in 10 mL of water is colorless and shows no more than a trace of turbidity.

Heavy Metals Prepare and test a 1-g sample as directed in *Method II* under the *Heavy Metals Test*, Appendix IIIB, using 20 μg of lead ion (Pb) in the control (*Solution A*).

Lead A *Sample Solution* prepared as directed for organic compounds meets the requirements of the *Lead Limit Test*, Appendix IIIB, using 10 μg of lead ion (Pb) in the control.

Other Nucleotides Prepare a strip of Whatman No. 2 or equivalent filter paper about 20 × 40 cm, and draw a line across the narrow dimension about 5 cm from one end. Using a micropipet, apply on the center of the line 10 μL of a 1 in 100 solution of the sample in water, and dry the paper in air. Fill the trough of an apparatus suitable for descending chromatography (see Appendix IIA) with a 160:3:40 mixture of saturated ammonium sulfate solution, *tert*-butyl alcohol, and 0.025 N ammonia, respectively, and suspend the strip in the chamber, placing the end of the strip in the trough at a distance about 1 cm from the pencil line. Seal the chamber, and allow the chromatogram to develop until the solvent front descends to a distance about 30 cm from the starting line. Remove the strip from the chamber, dry in air, and observe under shortwave (254 nm) ultraviolet light in the dark. Only one spot is visible.

pH of a 1 in 20 Solution Determine by the *Potentiometric Method*, Appendix IIB.

Water Determine by the *Karl Fischer Titrimetric Method*, Appendix IIB.

Packaging and Storage Store in well-closed containers.

Enzyme-Modified Milkfat

Enzyme-Modified Fat

DESCRIPTION

Light- to medium-tan liquid, paste, or powder with strong fatty acid odor and flavor. Produced by enzyme lipolysis, using suitable food-grade enzymes, of milkfat obtained from the following sources: milk, concentrated milk, dry whole milk, cream, concentrated cream(s), dry cream, butter, butter oil, dried butter, or anhydrous milkfat. Milkfat emulsions are reacted with suitable food-grade enzymes under controlled conditions to increase the flavor components. Optional dairy ingredients such as skim milk, concentrated skim milk, nonfat dry milk, buttermilk, concentrated buttermilk, dried buttermilk, liquid whey, concentrated whey, and dried whey may be used to adjust the concentration of the flavors. Thermoprocessing is then used to destroy the enzyme activity and provide acceptable microbiological quality. Suitable preservatives, emulsifiers, buffers, stabilizers, and antioxidants as well as sodium chloride may be added.

Functional Use in Food Flavoring agent.

REQUIREMENTS

Labeling Indicate the *Acid Value*.
Identification Very strong fatty acid odor.
Acid Value Not less than 98.0% and not more than 102.0% of the labeled value.
Heavy Metals (as Pb) Not more than 10 mg/kg.
Lead Not more than 1 mg/kg.
Loss on Drying Not more than 4.0%.
Microbial Limits:
    Aerobic Plate Count Not more than 10,000 per g.
    Coliforms Not more than 10 per g.
    Salmonella Negative in 25 g.
    Staphylococcus aureus Not more than 100 per g.
    Yeasts and Molds Not more than 10 per g.

TESTS

Acid Value Determine as directed for *Acid Value, Method II*, under *Fats and Related Substances*, Appendix VII, using a 5-g sample.

Heavy Metals Prepare and test a 2-g sample as directed in *Method II* under the *Heavy Metals Test*, Appendix IIIB, using 20 μg of lead ion (Pb) in the control (*Solution A*).

Lead Determine as directed under *Method II* in the *Atomic Absorption Spectrophotometric Graphite Furnace Method* under the *Lead Limit Test*, Appendix IIIB.

Loss on Drying, Appendix IIC Dry at 105° for 48 h.

Microbial Limits:

Aerobic Plate Count   Proceed as directed in chapter 3 of the *FDA Bacteriological Analytical Manual*, Seventh Edition, 1992.

Coliforms   Proceed as directed in chapter 4 of the *FDA Bacteriological Analytical Manual*, Seventh Edition, 1992.

Salmonella   Proceed as directed in chapter 5 of the *FDA Bacteriological Analytical Manual*, Seventh Edition, 1992.

Staphylococcus aureus   Proceed as directed in chapter 12 of the *FDA Bacteriological Analytical Manual*, Seventh Edition, 1992.

Yeasts and Molds   Proceed as directed in chapter 18 of the *FDA Bacteriological Analytical Manual*, Seventh Edition, 1992.

Packaging and Storage   Store in tight containers in a cool place.

Enzyme Preparations

DESCRIPTION

Enzyme Preparations used in food processing are derived from animal, plant, or microbial sources (see **CLASSIFICATION**, below). They may consist of whole cells, parts of cells, or cell-free extracts of the source used, and they may contain one active component or, more commonly, a mixture of several, as well as food-grade diluents, preservatives, antioxidants, and other substances consistent with good manufacturing practice.

The individual preparations usually are named according to the substance to which they are applied, such as *Protease* or *Amylase*; such traditional names as *Malt*, *Pepsin*, and *Rennet* also are used, however.

The color of the preparations—which may be liquid, semiliquid, or dry—may vary from virtually colorless to dark brown. The active components consist of the biologically active proteins, which are sometimes conjugated with metals, carbohydrates, and/or lipids. Known molecular weights of the active components range from approximately 12,000 to several hundred thousand.

The activity of enzyme preparations is measured according to the reaction catalyzed by individual enzymes (see below) and is usually expressed in activity units per unit weight of the preparation. In commercial practice (but not for *Food Chemicals Codex* purposes), the activity of the product is sometimes also given as the quantity of the preparation to be added to a given quantity of food to achieve the desired effect.

Additional information relating to the nomenclature and the sources from which the active components are derived is provided in the *General Tests* section under *Enzyme Assays*, Appendix V.

Functional Use in Foods   Enzyme (see discussion under **CLASSIFICATION**, below).

CLASSIFICATION

Animal-Derived Preparations

Catalase (bovine liver)   Partially purified liquid or powdered extracts from bovine liver. Major active principle: *catalase*. Typical application: manufacture of certain cheeses.

Chymotrypsin   Obtained from purified extracts of bovine or porcine pancreatic tissue. White to tan, amorphous powders soluble in water but practically insoluble in alcohol, in chloroform, and in ether. Major active principle: *chymotrypsin*. Typical application: hydrolysis of protein.

Lipase, Animal   Obtained from two primary sources: (1) edible forestomach tissue of calves, kids, or lambs, and (2) animal pancreatic tissue. Produced as purified edible tissue preparations or as aqueous extracts. Dispersible in water; insoluble in alcohol. Major active principle: *lipase*. Typical applications: manufacture of cheese; modification of lipids.

Pancreatin   Obtained from porcine or bovine (ox) pancreatic tissue. White to tan, water-soluble powder. Major active principles: (1) α-*amylase*, (2) *protease*, and (3) *lipase*. Typical applications: preparation of precooked cereals, infant foods, protein hydrolysates.

Pepsin   Obtained from the glandular layer of hog stomach. White to light tan, water-soluble powder; amber paste; or clear amber to brown aqueous liquids. Major active principle: *pepsin*. Typical applications: preparation of fish meal and other protein hydrolysates; clotting of milk in manufacture of cheese (in combination with rennet).

Phospholipase A$_2$ Obtained from porcine pancreatic tissue. Produced as a white to tan powder or pale- to dark-yellow liquid. Major active principle: *phospholipase A$_2$*. Typical application: hydrolysis of lecithins.

Rennet, Bovine   Aqueous extracts made from the fourth stomach of bovine animals. Clear, amber to dark-brown liquid or white to tan powder. Major active principle: *protease* (pepsin). Typical application: manufacture of cheese. Similar preparations can be made from the fourth stomach of sheep or goats.

Rennet, Calf   Aqueous extracts made from the fourth stomach of calves. Clear, amber to dark-brown liquid or white to tan powder. Major active principle: *protease* (chymosin). Typical application: manufacture of cheese. Similar preparations can be made from the fourth stomach of lambs or kids.

Trypsin   Obtained from purified extracts of porcine or bovine pancreas. White to tan, amorphous powders soluble in water but practically insoluble in alcohol, in chloroform, and in ether. Major active principle: *trypsin*. Typical applications: baking; tenderizing of meat; production of protein hydrolysates.

Plant-Derived Preparations

Amylase Obtained from extraction of ungerminated barley. Clear, amber to dark-brown liquid or white to tan powder. Major active principle: β-*amylase*. Typical applications: production of alcoholic beverages and sugar syrups.

Bromelain The purified proteolytic substance derived from the pineapples *Ananas comosus* and *Ananas bracteatus* L. White to light-tan, amorphous powder. Soluble in water (the solution is usually colorless to light yellow and somewhat opalescent) but practically insoluble in alcohol, in chloroform, and in ether. Major active principle: *bromelain*. Typical applications: chillproofing of beer; tenderizing of meat; preparation of precooked cereals; production of protein hydrolysates; baking.

Ficin The purified proteolytic substance derived from the latex of *Ficus* sp., which include a variety of tropical fig trees. White to off-white powders completely soluble in water. (Liquid fig latex concentrates are light brown to dark brown.) Major active principle: *ficin*. Typical applications: chillproofing of beer; tenderizing of meat; conditioner of dough in baking.

Malt The product of the controlled germination of barley. Clear, amber to dark-brown liquid preparations or white to tan powder. Major active principles: (1) α-*amylase* and (2) β-*amylase*. Typical applications: baking; manufacture of alcoholic beverages; manufacture of syrups.

Papain The purified proteolytic substance derived from the fruit of the papaya *Carica papaya* L. (Fam. *Caricaceae*). Produced as a white to light-tan, amorphous powder or a liquid. Soluble in water (the solution is usually colorless or light yellow and somewhat opalescent) but practically insoluble in alcohol, in chloroform, and in ether. Major active principles: (1) *papain* and (2) *chymopapain*. Typical applications: chillproofing of beer; tenderizing of meat; preparation of precooked cereals; production of protein hydrolysates.

Microbially Derived Preparations

Carbohydrase (*Aspergillus niger* var., including *Aspergillus aculeatus*) Produced by the controlled fermentation of *Aspergillus niger* var. (including *Aspergillus aculeatus*) as an off-white to tan powder or a tan to dark-brown liquid. Soluble in water (the solution is usually light yellow to dark brown) but practically insoluble in alcohol, in chloroform, and in ether. Major active principles: (1) α-*amylase*, (2) *pectinase* (a mixture of enzymes, including *pectin depolymerase, pectin methyl esterase, pectin lyase,* and *pectate lyase*), (3) *cellulase*, (4) *glucoamylase* (amyloglucosidase), (5) *amylo 1,6-glucosidase*, (6) *hemicellulase* (a mixture of enzymes, including *poly(galacturonate) hydrolase, arabinosidase, mannosidase, mannanase,* and *xylanase*), (7) *lactase*, (8) β-*glucanase*, (9) β-D-*glucosidase*, (10) *pentosanase*, and (11) α-*galactosidase*. Typical applications: preparation of starch syrups and dextrose, alcohol, beer, ale, fruit juices, chocolate syrups, bakery products, liquid coffee, wine, dairy products, cereals, and spice and flavor extracts.

Carbohydrase (*Aspergillus oryzae* var.) Produced by the controlled fermentation of *Aspergillus oryzae* var. as an off-white to tan, amorphous powder or a liquid. Soluble in water (the solution is usually light yellow to dark brown) but practically insoluble in alcohol, in chloroform, and in ether. Major active principles: (1) α-*amylase*, (2) *glucoamylase* (amyloglucosidase), and (3) *lactase*. Typical applications: preparation of starch syrups, alcohol, beer, ale, bakery products, and dairy products.

Carbohydrase (*Bacillus acidopullulyticus*) Produced by the controlled fermentation of *Bacillus acidopullulyticus* as an off-white to brown, amorphous powder or a liquid. Soluble in water (the solution is usually light yellow to dark brown) but practically insoluble in alcohol, in chloroform, and in ether. Major active principle: *pullulanase*. Typical applications: hydrolysis of amylopectins and other branched polysaccharides.

Carbohydrase (*Bacillus stearothermophilus*) Produced by the controlled fermentation of *Bacillus stearothermophilus* as an off-white to tan powder or a light-yellow to dark-brown liquid. Soluble in water but practically insoluble in alcohol, in ether, and in chloroform. Major active principle: α-*amylase*. Typical applications: preparation of starch syrups, alcohol, beer, dextrose, and bakery products.

Carbohydrase (*Candida pseudotropicalis*) Produced by the controlled fermentation of *Candida pseudotropicalis* as an off-white to tan, amorphous powder or a liquid. Soluble in water (the solution is usually light yellow to dark brown) but insoluble in alcohol, in chloroform, and in ether. Major active principle: *lactase*. Typical applications: manufacture of candy and ice cream; modification of dairy products.

Carbohydrase (*Kluyveromyces marxianus* var. *lactis*) Produced by the controlled fermentation of *Kluyveromyces marxianus* var. *lactis* as an off-white to tan, amorphous powder or a liquid. Soluble in water (the solution is usually light yellow to dark brown) but insoluble in alcohol, in chloroform, and in ether. Major active principle: *lactase*. Typical applications: manufacture of candy and ice cream; modification of dairy products.

Carbohydrase (*Mortierella vinaceae* var. *raffinoseutilizer*) Produced by the controlled fermentation of *Mortierella vinaceae* var. *raffinoseutilizer* as an off-white to tan powder or as pellets. Soluble in water (pellets may be insoluble in water) but practically insoluble in alcohol, in chloroform, and in ether. Major active principle: α-*galactosidase*. Typical application: production of sugar from sugar beets.

Carbohydrase (*Rhizopus niveus*) Produced by the controlled fermentation of *Rhizopus niveus* as an off-white to brown, amorphous powder or a liquid. Soluble in water (the solution is usually light yellow to dark brown) but practically insoluble in alcohol, in chloroform, and in ether. Major active principles: (1) α-*amylase* and (2) *glucoamylase*. Typical application: hydrolysis of starch.

Carbohydrase (*Rhizopus oryzae* var.) Produced by the controlled fermentation of *Rhizopus oryzae* var. as a powder or a liquid. Soluble in water but practically insoluble in alcohol, in chloroform, and in ether. Major active principles: (1) α-*amylase*, (2) *pectinase*, and (3) *glucoamylase* (amyloglucosidase). Typical applications: preparation of starch syrups and fruit juices, vegetable purees, and juices; manufacture of cheese.

Carbohydrase (*Saccharomyces* species) Produced by the controlled fermentation of a number of species of *Saccharomyces* traditionally used in the manufacture of food. White to tan, amorphous powder. Soluble in water (the solution is usually light yellow) but practically insoluble in alcohol, in chloroform, and in ether. Major active principles: (1) *invertase* and (2) *lactase*. Typical applications: manufacture of candy and ice cream; modifications of dairy products.

Carbohydrase [(*Trichoderma longibrachiatum* var.) (formerly *reesei*)] Produced by the controlled fermentation of *Trichoderma longibrachiatum* var. as an off-white to tan, amorphous powder or a liquid. Soluble in water (the solution is usually tan to brown) but practically insoluble in alcohol, in chloroform, and in ether. Major active principles: (1) *cellulase* (2) β-*glucanase*, (3) β-D-*glucosidase*,(4) *hemicellulase*, and (5) *pentosanase*. Typical applications: preparation of fruit juices, wine, vegetable oils, beer, and baked goods.

Carbohydrase (*Bacillus subtilis* containing a *Bacillus megaterium* α-*amylase* gene) Produced by controlled fermentation of the modified *Bacillus subtilis* as an off-white to brown, amorphous powder or liquid. Soluble in water (the solution is usually light yellow to dark brown) but practically insoluble in alcohol, in chloroform, and in ether. Major active principle: α-*amylase*. Typical applications: preparation of starch syrups, alcohol, beer, and dextrose.

Carbohydrase (*Bacillus subtilis* containing a *Bacillus stearothermophilus* α-*amylase* gene) Produced by controlled fermentation of the modified *Bacillus subtilis* as an off-white to brown, amorphous powder or a liquid. Soluble in water (the solution is usually light yellow to dark brown) but practically insoluble in alcohol, in chloroform, and in ether. Major active principle: maltogenic *amylase*. Typical applications: preparation of starch syrups, dextrose, alcohol, beer, and baked goods.

Carbohydrase and Protease, Mixed (*Bacillus licheniformis* var.) Produced by the controlled fermentation of *Bacillus licheniformis* var. as an off-white to brown, amorphous powder or a liquid. Soluble in water (the solution is usually light yellow to dark brown) but practically insoluble in alcohol, in chloroform, and in ether. Major active principles: (1) α-*amylase* and (2) *protease*. Typical applications: preparation of starch syrups, alcohol, beer, dextrose, fish meal, and protein hydrolysates.

Carbohydrase and Protease, Mixed (*Bacillus subtilis* var. including *Bacillus amyloliquefaciens*) Produced by the controlled fermentation of *Bacillus subtilis* var. as an off-white to tan, amorphous powder or a liquid. Soluble in water (the solution is usually light yellow to dark brown) but practically insoluble in alcohol, in chloroform, and in ether. Major active principles: (1) α-*amylase*, (2) β-*glucanase*, (3) *protease*, and (4) *pentosanase*. Typical applications: preparation of starch syrups, alcohol, beer, dextrose, bakery products, fish meal; tenderizing of meat; preparation of protein hydrolysates.

Catalase (*Aspergillus niger* var.) Produced by the controlled fermentation of *Aspergillus niger* var. as an off-white to tan, amorphous powder or a liquid. Soluble in water (the solution is usually tan to brown) but practically insoluble in alcohol, in chloroform, and in ether. Major active principle: *catalase*. Typical applications: manufacture of cheese, egg products, and soft drinks.

Catalase (*Micrococcus lysodeikticus*) Produced by the controlled fermentation of *Micrococcus lysodeikticus*. Soluble in water (the solution is usually light yellow to dark brown) but practically insoluble in alcohol, in chloroform, and in ether. Major active principle: *catalase*. Typical application: manufacture of cheese, egg products, and soft drinks.

Chymosin (*Aspergillus niger* var. *awamori*, *Escherichia coli* K-12, and *Kluyveromyces marxianus*, each microorganism containing a calf *prochymosin* gene) Produced by the controlled fermentation of the above-named genetically modified microorganisms as a white to tan, amorphous powder or as a light-yellow to brown liquid. The powder is soluble in water but practically insoluble in alcohol, in chloroform, and in ether. Major active principle: *chymosin*. Typical application: manufacture of cheese; preparation of milk-based desserts.

Glucose Isomerase (*Actinoplanes missouriensis*, *Bacillus coagulans*, *Streptomyces olivaceus*, *Streptomyces olivochromogenes*, *Microbacteruim arborescens*, *Streptomyces rubiginosus* var., or *Streptomyces murinus*) Produced by the controlled fermentation of any of the above-named organisms as an off-white to tan, brown, or pink, amorphous powder, granules, or a liquid. The products may be soluble in water but practically insoluble in alcohol, in chloroform, and in ether, or if immobilized, may be insoluble in water and partially soluble in alcohol, in chloroform, and in ether. Major active principle: *glucose* (or *xylose isomerase*). Typical applications: manufacture of high-fructose corn syrup and other fructose starch syrups.

Glucose Oxidase (*Aspergillus niger* var.) Produced by the controlled fermentation of *Aspergillus niger* var. as a yellow to brown solution or as a yellow to tan or off-white powder. Soluble in water (the solution is usually light-yellow to brown) but practically insoluble in alcohol, in chloroform, and in ether. Major active principles: (1) *glucose oxidase* and (2) *catalase*. Typical applications: removal of sugar from liquid eggs; deoxygenation of citrus beverages.

Lipase (*Aspergillus niger* var.) Produced by the controlled fermentation of *Aspergillus niger* var. as an off-white to tan, amorphous powder. Soluble in water (the solution is usually light yellow) but practically insoluble in alcohol, in chloroform, and in ether. Major active principle: *lipase*. Typical application:

hydrolysis of lipids (e.g., fish oil concentrates and cereal derived lipids).

Lipase (*Aspergillus oryzae* var.) Produced by the controlled fermentation of *Aspergillus oryzae* var. as an off-white to tan, amorphous powder or a liquid. Soluble in water (the solution is usually light yellow) but practically insoluble in alcohol, in chloroform, and in ether. Major active principle: *lipase*. Typical applications: hydrolysis of lipids (e.g., fish oil concentrates); manufacture of cheese and cheese flavors.

Lipase (*Candida rugosa*; formerly *Candida cylindracea*) Produced by the controlled fermentation of *Candida rugosa* as an off-white to tan powder. Soluble in water but practically insoluble in alcohol, in chloroform, and in ether. Major active principle: lipase. Typical applications: hydrolysis of lipids, manufacture of dairy products and confectionery goods, development of flavor in processed foods.

Lipase [*Rhizomucor (Mucor) miehei*] Produced by the controlled fermentation of *Rhizomucor miehei* as an off-white to tan powder or a liquid. Soluble in water (the solution is usually light yellow to dark brown) but practically insoluble in alcohol, in chloroform, and in ether. Major active principle: *lipase*. Typical applications: hydrolysis of lipids; manufacture of cheese; removal of haze in fruit juices.

Phytase (*Aspergillus niger* var.) Produced by the controlled fermentation of *Aspergillus niger* var. as an off-white to brown powder or as a tan to dark-brown liquid. Soluble in water but practically insoluble in alcohol, in chloroform, and in ether. Major active principle: (1) *3-phytase*, and (2) *acid phosphatase*. Typical applications: soy protein isolate production; phytic acid removal from plant materials.

Protease (*Aspergillus niger* var.) Produced by the controlled fermentation of *Aspergillus niger* var. The purified enzyme occurs as an off-white to tan, amorphous powder. Soluble in water (the solution is usually light yellow) but practically insoluble in alcohol, in chloroform, and in ether. Major active principle: *protease*. Typical application: production of protein hydrolysates.

Protease (*Aspergillus oryzae* var.) Produced by the controlled fermentation of *Aspergillus oryzae* var. The purified enzyme occurs as an off-white to tan, amorphous powder. Soluble in water (the solution is usually light yellow) but practically insoluble in alcohol, in chloroform, and in ether. Major active principle: *protease*. Typical applications: chillproofing of beer; production of bakery products; tenderizing of meat; production of protein hydrolysates; development of flavor in processed foods.

Rennet, Microbial (nonpathogenic strain of *Bacillus cereus*) Produced by the controlled fermentation of *Bacillus cereus* as a white to tan, amorphous powder or a light-yellow to dark-brown liquid. Soluble in water but practically insoluble in alcohol, in chloroform, and in ether. Major active principle: *protease*. Typical application: manufacture of cheese.

Rennet, Microbial (*Endothia parasitica*) Produced by the controlled fermentation of nonpathogenic strains of *Endothia parasitica* as an off-white to tan, amorphous powder or a liquid. The powder is soluble in water (the solution is usually tan to dark brown) but practically insoluble in alcohol, in chloroform, and in ether. Major active principle: *protease*. Typical application: manufacture of cheese.

Rennet, Microbial [*Rhizomucor (Mucor)* sp.] Produced by the controlled fermentation of *Rhizomucor miehei*, or *pusillus* var. Lindt as a white to tan, amorphous powder. The powder is soluble in water (the solution is usually light yellow) but practically insoluble in alcohol, in chloroform, and in ether. Major active principle: *protease*. Typical application: manufacture of cheese.

REACTIONS CATALYZED

Note: The reactions catalyzed by any given active component are essentially the same, regardless of the source from which that component is derived.

α-Amylase Endohydrolysis of α-1,4-glucan bonds in polysaccharides (starch, glycogen, etc.), yielding dextrins and oligo- and monosaccharides.

β-Amylase Hydrolysis of α-1,4-glucan bonds in polysaccharides (starch, glycogen, etc.), yielding maltose and beta limit dextrins.

Bromelain Hydrolysis of polypeptides, amides, and esters (especially at bonds involving basic amino acids, leucine, or glycine), yielding peptides of lower molecular weight.

Catalase $2H_2O_2 \rightarrow O_2 + 2H_2O$.

Cellulase Hydrolysis of β-1,4-glucan bonds in such polysaccharides as cellulose, yielding β-dextrins.

Chymosin (calf and fermentation derived) Cleaves a single bond in *kappa* casein.

Ficin Hydrolysis of polypeptides, amides, and esters (especially at bonds involving basic amino acids, leucine, or glycine), yielding peptides of lower molecular weight.

α-Galactosidase Hydrolysis of terminal nonreducing α-D-galactose residues in α-D-galactosides.

β-Glucanase Hydrolysis of β-1,3- and β-1,4-linkages in β-D-glucans, yielding oligosaccharides and glucose.

Glucoamylase (*Amyloglucosidase*) Hydrolysis of terminal α-1,4- and α-1,6-glucan bonds in polysaccharides (starch, glycogen, etc.), yielding glucose (dextrose).

Glucose Isomerase (*Xylose isomerase*) Isomerization of glucose to fructose, and xylose to xylulose.

Glucose Oxidase β-D-glucose + O$_2$ → D-glucono-δ-lactone + H$_2$O$_2$.

β-D-Glucosidase Hydrolysis of terminal, nonreducing β-D-glucose residues with the release of β-D-glucose.

Hemicellulase Hydrolysis of β-1,4-glucans, α-L arabinosides, β-D-mannosides, 1,3-β-D-xylans, and other polysaccharides, yielding polysaccharides of lower molecular weight.

Invertase (β-fructofuranosidase) Hydrolysis of sucrose to a mixture of glucose and fructose (invert sugar).

Lactase (β-galactosidase) Hydrolysis of lactose to a mixture of glucose and galactose.

Lipase Hydrolysis of triglycerides of simple fatty acids, yielding mono- and diglycerides, glycerol, and free fatty acids.

Maltogenic Amylase Hydrolysis of 1,4-α-glucan bonds.

Pancreatin
  α-amylase Hydrolysis of α-1,4 glucan bonds.
  Protease Hydrolysis of proteins and polypeptides.
  Lipase Hydrolysis of triglycerides of simple fatty acids.

Pectinase
  Pectate lyase Hydrolysis of pectate to oligosaccharides.
  Pectin depolymerase Hydrolysis of 1,4-galacturonide bonds.
  Pectin lyase Hydrolysis of oligosaccharides formed by pectate lyase.
  Pectinesterase Demethylation of pectin.

Pepsin Hydrolysis of polypeptides, including those with bonds adjacent to aromatic or dicarboxylic L-amino acid residues, yielding peptides of lower molecular weight.

Phospholipase A$_2$ Hydrolysis of lecithins and phosphatidylcholine, producing fatty acid anions.

Phytase
  3-phytase myo-Inositol hexakisphosphate + H$_2$O → 1,2,4,5,6-pentakisphosphate + orthophosphate.
  Acid Phosphatase Orthophosphate monoester + H$_2$O → an alcohol + orthophosphate.

Protease (generic) Hydrolysis of polypeptides, yielding peptides of lower molecular weight.

Pullulanase Hydrolysis of 1,6-α-D-glycosidic bonds on amylopectin and glycogen and in α- and β-limit dextrins, yielding linear polysaccharides.

Rennet (Bovine and calf) Hydrolysis of polypeptides; specificity may be similar to *pepsin*.

Trypsin Hydrolysis of polypeptides, amides, and esters at bonds involving the carboxyl groups of L-arginine and L-lysine, yielding peptides of lower molecular weight.

GENERAL REQUIREMENTS

Enzyme preparations are produced in accordance with good manufacturing practices. Regardless of the source of derivation, they should cause no increase in the total microbial count in the treated food over the level accepted for the respective food.

Animal tissues used to produce enzymes must comply with the applicable U.S. meat inspection requirements and must be handled in accordance with good hygienic practices.

Plant material used to produce enzymes or culture media used to grow microorganisms consist of components that leave no residues harmful to health in the finished food under normal conditions of use.

Preparations derived from microbial sources are produced by methods and under culture conditions that ensure a controlled fermentation, thus preventing the introduction of microorganisms that could be the source of toxic materials and other undesirable substances.

The carriers, diluents, and processing aids used to produce the enzyme preparations shall be substances that are acceptable for general use in foods, including water and substances that are insoluble in foods but removed from the foods after processing.

Although tolerances have not been established for mycotoxins, appropriate measures should be taken to ensure that the products do not contain such contaminants.

ADDITIONAL REQUIREMENTS

Assay Not less than 85.0% and not more than 115.0% of the declared units of enzyme activity.
Heavy Metals (as Pb) Not more than 30 mg/kg.
Lead Not more than 5 mg/kg.
Microbial Limits:
  Coliforms Not more than 30 per g.
  Salmonella sp. Negative in 25 g.

TESTS

Assay The following procedures, which are included in the *General Tests* under *Enzyme Assays*, Appendix V, are provided for application as necessary in determining compliance with the declared representations for enzyme activity:[1] Acid Phosphatase Activity, α-Amylase Activity (Nonbacterial); Bacterial α-Amylase Activity (BAU); Catalase Activity; Cellulase Activity; Chymotrypsin Activity; Diastase Activity (Diastatic Power); α-Galactosidase Activity, β-Glucanase Activity; Glucoamylase Activity (Amyloglucosidase Activity); Glucose Isomerase Activity; Glucose Oxidase Activity; β-D-Glucosidase Activity;

---

[1] Because of the varied conditions under which pectinases are employed, and because laboratory hydrolysis of a purified pectin substrate does not correlate with results observed with the natural substrates under use conditions, pectinase suppliers and users should develop their own assay procedures that would relate to the specific application under consideration.

Hemicellulase Activity; Invertase Activity; Lactase (Neutral) (β-Galactosidase) Activity; Lactase (Acid)(β-Galactosidase) Activity; Lipase Activity; Lipase/Esterase (Forestomach) Activity; Maltogenic Amylase Activity; Milk-Clotting Activity; Pancreatin Activity; Pepsin Activity; Phospholipase Activity; Phytase Activity; Plant Proteolytic Activity; Proteolytic Activity, Bacterial (PC); Proteolytic Activity, Fungal (HUT); Proteolytic Activity, Fungal (SAP); Pullulanase Activity; and Trypsin Activity.

Heavy Metals Prepare and test a 667-mg sample as directed in *Method II* under the *Heavy Metals Test*, Appendix IIIB, using 20 μg of lead ion (Pb) in the control (*Solution A*).

Lead A *Sample Solution* prepared as directed for organic compounds meets the requirements of the *Lead Limit Test*, Appendix IIIB, using 5 μg of lead ion (Pb) in the control.

Microbial Limits:
  Coliforms Proceed as directed in chapter 4 of the *FDA Bacteriological Analytical Manual*, Seventh Edition, 1992.
  Salmonella sp. Proceed as directed in chapter 5 of the *FDA Bacteriological Analytical Manual*, Seventh Edition, 1992.

Packaging and Storage Store in tight containers in a cool, dry place.

Erythorbic Acid

D-Araboascorbic Acid

$C_6H_8O_6$ Formula wt 176.13
INS: 315 CAS: [89-65-6]

DESCRIPTION

White or slightly yellow crystals or powder. On exposure to light it gradually darkens. In the dry state it is reasonably stable in air, but in solution it rapidly deteriorates in the presence of air. It melts between 164° and 171° with decomposition. One g is soluble in about 2.5 mL of water and in about 20 mL of alcohol. It is slightly soluble in glycerin.

Functional Use in Foods Preservative; antioxidant.

REQUIREMENTS

Identification
  A. A 1 in 50 solution slowly reduces alkaline cupric tartrate TS at 25°, but more readily upon heating.
  B. To 2 mL of a 1 in 50 solution add a few drops of sodium nitroferricyanide TS, followed by 1 mL of approximately 0.1 N sodium hydroxide. A transient blue color is produced immediately.
  C. Dissolve about 15 mg in 15 mL of a trichloroacetic acid solution (1 in 20), add about 200 mg of activated charcoal, and shake the mixture vigorously for 1 min. Filter through a small fluted filter, refiltering if necessary to obtain a clear filtrate. To 5 mL of the clear filtrate add 1 drop of pyrrole, agitate the mixture until the pyrrole is dissolved, then heat in a water bath at 50°. A blue color develops.

Assay Not less than 99.0% and not more than 100.5% of $C_6H_8O_6$, calculated on the dried basis.
Heavy Metals (as Pb) Not more than 10 mg/kg.
Lead Not more than 5 mg/kg.
Loss on Drying Not more than 0.4%.
Residue on Ignition Not more than 0.3%.
Specific Rotation $[\alpha]_D^{25°}$: Between −16.5° and −18.0°.

TESTS

Assay Dissolve about 400 mg, accurately weighed, in a mixture of 100 mL of water, recently boiled and cooled, and 25 mL of 2 N sulfuric acid. Titrate the solution immediately with 0.1 N iodine, adding starch TS near the endpoint. Each mL of 0.1 N iodine is equivalent to 8.806 mg of $C_6H_8O_6$.

Heavy Metals A solution of 2 g in 25 mL of water meets the requirements of the *Heavy Metals Test*, Appendix IIIB, using 20 μg of lead ion (Pb) in the control (*Solution A*).

Lead A *Sample Solution* prepared from a 2-g sample as directed for organic compounds meets the requirements of the *Lead Limit Test*, Appendix IIIB, using 10 μg of lead ion (Pb) in the control.

Loss on Drying, Appendix IIC Dry under reduced pressure over silica gel for 3 h.

Residue on Ignition, Appendix IIC Ignite 1 g as directed in the general method.

Specific Rotation Transfer about 2.5 g, accurately weighed, into a 25-mL volumetric flask, dissolve it in about 20 mL of water, and dilute to volume. Determine the specific rotation as directed under *Optical (Specific) Rotation*, Appendix IIB.

Packaging and Storage Store in tight, light-resistant containers.

Ethoxylated Mono- and Diglycerides

Polyoxyethylene (20) Mono- and Diglycerides of Fatty Acids; Polyglycerate (60)

INS: 488

DESCRIPTION

A mixture of stearate, palmitate, and lesser amounts of myristate partial esters of glycerin condensed with approximately 20 moles of ethylene oxide per mole of alpha-monoglyceride reaction mixture, having an average molecular weight of 535 (± 10%). It occurs as a pale, slightly yellow colored, oily liquid or semigel having a faint, characteristic odor and a mildly bitter taste. It is soluble in water, in alcohol, and in xylene. It is partially soluble in mineral oil and in vegetable oils.

Note: If the product is manufactured by direct esterification of glycerin with a mixture of primary stearic, palmitic, and myristic acids, the intermediate product (before reaction with ethylene oxide) has an acid value of not greater than 0.3 and a water content not greater than 0.2%.

Functional Use in Foods Dough conditioner; emulsifier.

REQUIREMENTS

Identification
A. To 5 mL of a 1 in 20 solution in water add 5 mL of 1 N sodium hydroxide, boil for a few min, cool, and acidify with 2.7 N hydrochloric acid. The solution is strongly opalescent.
B. A mixture of 46 volumes of the sample with 54 volumes of water at 40° or below yields a gelatinous mass.
Acid Value Not more than 2.
1,4-Dioxane Passes test.
Heavy Metals (as Pb) Not more than 10 mg/kg.
Hydroxyl Value Between 65 and 80.
Oxyethylene Content (apparent) Not less than 60.5% and not more than 65.0%, calculated as ethylene oxide (C_2H_4O), on the anhydrous basis.
Saponification Value Between 65 and 75.
Stearic, Palmitic, and Myristic Acids Between 31 and 33 g per 100 g of sample.
Water Not more than 1%.

TESTS

Acid Value Determine as directed for *Acid Value, Method II*, under *Fats and Related Substances*, Appendix VII.
1,4-Dioxane Determine as directed in the general method, Appendix IIIB.
Heavy Metals Prepare and test a 2-g sample as directed in *Method II* under the *Heavy Metals Test*, Appendix IIIB, using 20 μg of lead ion (Pb) in the control (*Solution A*).
Hydroxyl Value Determine as directed under *Method II* in the general method, Appendix VII.
Oxyethylene Content (apparent) Weigh accurately about 70 mg, and proceed as directed under *Oxyethylene Determination*, Appendix VII.
Saponification Value Determine as directed for *Saponification Value* under *Fats and Related Substances*, Appendix VII, using about 6 g, accurately weighed.
Stearic, Palmitic, and Myristic Acids Isolate the fatty acids as directed in the test for *Lauric Acid* in the monograph for *Polysorbate 20*, and determine the weight of the acids. The product so obtained has an *Acid Value* between 199 and 211 (*Method I*, Appendix VII) and a *Solidification Point*, Appendix IIB, not below 50°.

Water Determine by the *Karl Fischer Titrimetric Method*, Appendix IIB.

Packaging and Storage Store in well-closed containers.

Ethoxyquin

6-Ethoxy-1,2-dihydro-2,2,4-trimethylquinoline

$C_{14}H_{19}NO$ Formula wt, monomer 217.31
INS: 324 CAS: [91-53-2]

DESCRIPTION

Ethoxyquin is a mixture consisting predominantly of the monomer ($C_{14}H_{19}NO$). It occurs as a clear liquid that may darken with age without affecting its antioxidant activity. Its specific gravity is about 1.02, and its refractive index is about 1.57.

Functional Use in Foods Antioxidant.

REQUIREMENTS

Identification A solution of 1 mg of the sample in 10 mL of acetonitrile exhibits a strong fluorescence when viewed under short-wavelength ultraviolet light.
Assay Not less than 90.0% of $C_{14}H_{19}NO$.
Heavy Metals (as Pb) Not more than 10 mg/kg.

TESTS

Assay Transfer about 200 mg of the sample, accurately weighed, into a 150-mL beaker containing 50 mL of glacial acetic acid, and immediately titrate with 0.1 N perchloric acid in glacial acetic acid, determining the endpoint potentiometrically. Perform a blank determination and make any necessary correction (see *General Provisions*). Each mL of 0.1 N perchloric acid is equivalent to 21.73 mg of $C_{14}H_{19}NO$ (monomer).
Heavy Metals Prepare and test a 2-g sample as directed in *Method II* under the *Heavy Metals Test*, Appendix IIIB, using 20 μg of lead ion (Pb) in the control (*Solution A*).

Packaging and Storage Store in tightly closed carbon steel or black iron (not rubber, neoprene, or nylon) containers in a cool, dark place. Prolonged exposure to sunlight causes polymerization.

Ethyl Alcohol

Alcohol; Ethanol

C_2H_6O Formula wt 46.07

CAS: [64-17-5]

DESCRIPTION

A clear, colorless, mobile liquid having a slight, characteristic odor and a burning taste. It is miscible with water, with ether, and with chloroform. It boils at about 78° and is flammable. Its refractive index at 20° is about 1.364.

Note: This monograph applies only to undenatured alcohol.

Functional Use in Foods   Extraction solvent; vehicle.

REQUIREMENTS

Assay   Not less than 94.9% by volume (92.3% by weight) of C_2H_6O.
Acidity (as acetic acid)   Not more than 0.003%.
Alkalinity (as NH_3)   Not more than 3 mg/kg.
Fusel Oil   Passes test.
Heavy Metals (as Pb)   Not more than 1 mg/kg.
Ketones, Isopropyl Alcohol   Passes test.
Methanol   Passes test.
Nonvolatile Residue   Not more than 0.003%.
Solubility in Water   Passes test.
Substances Darkened by Sulfuric Acid   Passes test.
Substances Reducing Permanganate   Passes test.

TESTS

Assay   Its specific gravity, determined by any reliable method (see *General Provisions*), is not greater than 0.8096 at 25°/25° (equivalent to 0.8161 at 15.56°/15.56°).
Acidity   Transfer 10 mL of the sample to a glass-stoppered flask containing 25 mL of water, add 0.5 mL of phenolphthalein TS, and add 0.02 N sodium hydroxide to the first appearance of a pink color that persists after shaking for 30 s. Add 25 mL (about 20 g) of the sample, mix, and titrate with 0.02 N sodium hydroxide until the pink color is restored. Not more than 0.5 mL is required.
Alkalinity   Add 2 drops of methyl red TS to 25 mL of water, add 0.02 N sulfuric acid until a red color just appears, then add 25 mL (about 20 g) of the sample, and mix. Not more than 0.2 mL of 0.02 N sulfuric acid is required to restore the red color.
Fusel Oil   Mix 10 mL of the sample with 1 mL of glycerin and 1 mL of water, and allow to evaporate from a piece of clean, odorless absorbent paper. No foreign odor is perceptible when the last traces of alcohol leave the paper.
Heavy Metals   Evaporate 25 mL (about 20 g) of the sample to dryness on a steam bath in a glass evaporating dish. Cool, add 2 mL of hydrochloric acid, and slowly evaporate to dryness again on the steam bath. Moisten the residue with 1 drop of hydrochloric acid, add 10 mL of hot water, and digest for 2 min. Cool, and dilute to 25 mL with water. This solution meets the requirements of the *Heavy Metals Test*, Appendix IIIB, using 20 µg of lead ion (Pb) in the control (*Solution A*).
Ketones, Isopropyl Alcohol   Transfer 1 mL of the sample, 3 mL of water, and 10 mL of mercuric sulfate TS to a test tube, mix, and heat in a boiling water bath. No precipitate forms within 3 min.
Methanol   Transfer 1 drop of the sample to a test tube, add 1 drop of dilute phosphoric acid (1 in 20) and 1 drop of potassium permanganate solution (1 in 20), mix, and allow to stand for 1 min. Add sodium bisulfite solution (1 in 10), dropwise, until the permanganate color is discharged. If a brown color remains, add 1 drop of the phosphoric acid solution. To the colorless solution add 5 mL of freshly prepared chromotropic acid TS, and heat on a water bath at 60° for 10 min. No violet color develops.
Nonvolatile Residue   Evaporate 125 mL (about 100 g) of the sample to dryness in a tared dish on a steam bath, dry the residue at 105° for 30 min, cool, and weigh.
Solubility in Water   Transfer 50 mL of the sample to a 100-mL glass-stoppered graduate, dilute to 100 mL with water, and mix. Place the graduate in a water bath maintained at 10°, and allow to stand for 30 min. No haze or turbidity develops.
Substances Darkened by Sulfuric Acid   Transfer 10 mL of sulfuric acid into a small Erlenmeyer flask, cool to 10°, and add 10 mL of the sample, dropwise, with constant agitation. The mixture is colorless or has no more color than either the acid or sample before mixing.
Substances Reducing Permanganate   Transfer 20 mL of the sample, previously cooled to 15°, to a glass-stoppered cylinder, add 0.1 mL of 0.1 N potassium permanganate, mix, and allow to stand for 5 min. The pink color is not entirely discharged.

Packaging and Storage   Store in tight containers, remote from fire.

Ethyl Cellulose

Modified Cellulose, EC

INS: 462   CAS: [9004-57-3]

DESCRIPTION

Ethyl Cellulose is the ethyl ether of cellulose in the form of a free-flowing, white to light tan powder. It is heat-labile, and exposure to high temperatures (240°) causes color degradation and loss of properties. It is practically insoluble in water, in glycerin, and in propylene glycol, but is soluble in varying proportions in certain organic solvents, depending upon the ethoxyl content. Ethyl Cellulose containing less than 46% to 48% of ethoxyl groups is freely soluble in tetrahydrofuran, in methyl acetate, in chloroform, and in aromatic hydrocarbon–alcohol mixtures. Ethyl Cellulose containing 46% to 48% or more of ethoxyl groups is freely soluble in alcohol, in methanol, in toluene, in chloroform, and in ethyl acetate. A 1 in 20 aqueous suspension of the sample is neutral to litmus.

Functional Use in Foods Protective coating component for vitamin and mineral tablets; binder and filler in dry vitamin preparations.

REQUIREMENTS

Identification Dissolve 5 g of the sample in 95 g of an 80:20 (w/w) mixture of toluene–ethanol. A clear, stable, slightly yellow solution is formed. Pour a few mL of the solution onto a glass plate, and allow the solvent to evaporate. A thick, tough, continuous, clear film remains. The film is flammable.
Assay Not less than 44.0% and not more than 50.0% of ethoxyl groups (—OC_2H_5), after drying (equivalent to not more than 2.6 ethoxyl groups per anhydroglucose unit).
Heavy Metals (as Pb) Not more than 10 mg/kg.
Lead Not more than 3 mg/kg.
Loss on Drying Not more than 3%.
Residue on Ignition Not more than 0.4%.
Viscosity Not less than 90% and not more than 110% of that stated on the label for a labeled viscosity of 10 centipoises or more; not less than 80% and not more than 120% of that stated on the label for a labeled viscosity of less than 10 centipoises.

TESTS

Assay Place about 50 mg of the sample, previously dried at 105° for 2 h, in a tared gelatin capsule, weigh accurately, transfer the capsule and its contents into the boiling flask of a methoxyl determination apparatus, and proceed as directed under *Methoxyl Determination*, Appendix IIIC. Each mL of 0.1 N sodium thiosulfate is equivalent to 751 μg of ethoxyl groups (—OC_2H_5).
Heavy Metals Prepare and test a 2-g sample as directed in *Method II* under the *Heavy Metals Test*, Appendix IIIB, using 20 μg of lead ion (Pb) in the control (*Solution A*).
Lead A *Sample Solution* prepared from a 2-g sample as directed for organic compounds meets the requirements of the *Lead Limit Test*, Appendix IIIB, using 6 μg of lead ion (Pb) in the control.
Loss on Drying, Appendix IIC Dry at 105° for 2 h.
Residue on Ignition Ignite 1 g as directed in the general method, Appendix IIC.
Viscosity
Solvent Systems For Ethyl Cellulose containing less than 46% to 48% of ethoxyl groups, prepare a solvent consisting of a 60:40 (w/w) mixture of toluene–alcohol; for Ethyl Cellulose containing 46% to 48% or more of ethoxyl groups, prepare a solvent consisting of an 80:20 (w/w) mixture of toluene–alcohol.
Procedure Transfer 5.0 g of the sample, previously dried at 105° for 2 h and accurately weighed, into a bottle containing 95 ± 0.5 g of the appropriate solvent system. Shake or tumble the bottle until the sample is completely dissolved, and then adjust the temperature of the solution to 25° ± 0.1°. Determine the viscosity as directed under *Viscosity of Methylcellulose*, Appendix IIB, but make all determinations at 25° instead of 20° as directed therein.

Packaging and Storage Store in well-closed containers.

Ethylene Dichloride

1,2-Dichloroethane

$$Cl-\underset{H}{\overset{H}{C}}-\underset{H}{\overset{H}{C}}-Cl$$

$C_2H_4Cl_2$ Formula wt 98.96

CAS: [107-06-2]

DESCRIPTION

A clear, colorless, flammable, oily liquid having a chloroform-like odor. It is slightly soluble in water, and is soluble in alcohol, in ether, and in acetone. Its refractive index at 20° is about 1.445.

Functional Use in Foods Extraction solvent.

REQUIREMENTS

Acidity (as HCl) Not more than 10 mg/kg.
Distillation Range Between 82° and 85°.
Free Halogens Passes test.
Heavy Metals (as Pb) Not more than 1 mg/kg.
Nonvolatile Residue Not more than 0.002%.
Specific Gravity Between 1.245 and 1.255.
Water Not more than 0.03%.

TESTS

Acidity Transfer 25 mL of alcohol to a 100-mL glass-stoppered flask, add 2 drops of phenolphthalein TS, and titrate with 0.01 N sodium hydroxide to the first appearance of a slight pink color. Add 25 mL (about 31 g) of the sample, mix, and titrate with 0.01 N sodium hydroxide until the faint pink color is restored. Not more than 0.85 mL is required.
Distillation Range Determine as directed in the general method, Appendix IIB.
Free Halogens Shake 10 mL of the sample vigorously for 2 min with 10 mL of 10% potassium iodide solution and 1 mL of starch TS. A blue color does not appear in the water layer.
Heavy Metals Evaporate 16 mL (about 20 g) of the sample to dryness on a steam bath in a glass evaporating dish.

Caution: Use hood.

Cool, add 2 mL of hydrochloric acid, and slowly evaporate to dryness again on the steam bath. Moisten the residue with 1 drop of hydrochloric acid, add 10 mL of hot water, and digest for 2 min. Filter if necessary through a small filter, wash the evaporating dish and the filter with about 10 mL of water, and dilute to 25 mL with water. This solution meets the requirements of the *Heavy Metals Test*, Appendix IIIB, using 20 μg of lead ion (Pb) in the control (*Solution A*).

Nonvolatile Residue Evaporate 80 mL (about 100 g) of the sample to dryness in a tared dish on a steam bath, dry the residue at 105° for 30 min, cool, and weigh.

Caution: Use hood.

Specific Gravity Determine by any reliable method (see *General Provisions*).

Water Determine by the *Karl Fischer Titrimetric Method*, Appendix IIB.

Packaging and Storage Store in tight containers.

Ethyl Maltol

3-Hydroxy-2-ethyl-4-pyrone

$C_7H_8O_3$ Formula wt 140.14

INS: 637

DESCRIPTION

A white, crystalline powder having a characteristic odor and a sweet, fruitlike flavor in dilute solution. One g dissolves in about 55 mL of water, 10 mL of alcohol, 17 mL of propylene glycol, and 5 mL of chloroform. It melts at about 90°.

Functional Use in Foods Flavoring agent.

REQUIREMENTS

Identification The infrared absorption spectrum of a 1 in 50 solution of Ethyl Maltol in chloroform, determined in a 0.1-mm cell, exhibits maxima only at the same wavelengths as that of FCC Ethyl Maltol Reference Standard, similarly measured.

Assay Not less than 99.0% of $C_7H_8O_3$, calculated on the anhydrous basis.

Heavy Metals (as Pb) Not more than 0.002%.

Lead Not more than 10 mg/kg.

Residue on Ignition Not more than 0.2%.

Water Not more than 0.5%.

TESTS

Assay

Standard Solution Weigh accurately about 50 mg of FCC Ethyl Maltol Reference Standard, dissolve it in sufficient 0.1 *N* hydrochloric acid to make 250.0 mL, and mix. Transfer 5.0 mL of this solution into a 100-mL volumetric flask, dilute to volume with 0.1 *N* hydrochloric acid, and mix.

Assay Solution Weigh accurately about 50 mg of the sample, dissolve it in sufficient 0.1 *N* hydrochloric acid to make 250.0 mL, and mix. Transfer 5.0 mL of this solution to a 100-mL volumetric flask, dilute to volume with 0.1 *N* hydrochloric acid, and mix.

Procedure Determine the absorbance of each solution in a 1-cm cell at the wavelength of maximum absorption at about 276 nm, with a suitable spectrophotometer, using 0.1 *N* hydrochloric acid as the blank. Calculate the quantity, in mg, of $C_7H_8O_3$ in the sample taken by the formula

$$5C(A_U/A_S),$$

in which C is the concentration, in μg/mL, of FCC Ethyl Maltol Reference Standard in the *Standard Solution*, A_U is the absorbance of the *Assay Solution*, and A_S is the absorbance of the *Standard Solution*.

Heavy Metals Prepare and test a 1-g sample as directed in *Method II* under the *Heavy Metals Test*, Appendix IIIB, using 20 μg of lead ion (Pb) in the control (*Solution A*).

Lead A *Sample Solution* prepared as directed for organic compounds meets the requirements of the *Lead Limit Test*, Appendix IIIB, using 10 μg of lead ion (Pb) in the control.

Residue on Ignition Ignite a 1-g sample as directed in the general method, Appendix IIC.

Water Determine by the *Karl Fischer Titrimetric Method*, Appendix IIB.

Packaging and Storage Store in tight containers.

Eucalyptus Oil

CAS: [8000-48-4]

DESCRIPTION

The volatile oil obtained by steam distillation from the fresh leaves of *Eucalyptus globulus* Labillardiere and other species of *Eucalyptus* L'Heritier (Fam. *Myrtaceae*). It is a colorless or pale yellow liquid having a characteristic aromatic, somewhat camphoraceous odor and a pungent, spicy, cooling taste.

Functional Use in Foods Flavoring agent.

REQUIREMENTS

Identification The infrared absorption spectrum of the sample exhibits relative maxima (that may vary in intensity) at the same wavelengths (or frequencies) as those shown in the respective spectrum in the section on *Infrared Spectra* (*Series A: Essential Oils*), using the same test conditions as specified therein.

Assay Not less than 70.0% of cineole ($C_{10}H_{18}O$).
Heavy Metals (as Pb) Passes test.
Phellandrene Passes test.
Refractive Index Between 1.458 and 1.470 at 20°.
Solubility in Alcohol Passes test.
Specific Gravity Between 0.905 and 0.925.

TESTS

Assay Transfer about 3 g, previously dried with anhydrous sodium sulfate and accurately weighed, into a 25- × 150-mm test tube. Add to the sample 2.100 g of melted *o*-cresol that is pure and dry, having a solidification point of 30° or higher.

Note: Moisture in the *o*-cresol may cause low results.

Stir the mixture with the thermometer to induce crystallization, and note the highest temperature reading obtained. Warm the tube gently until the contents are completely melted, then insert the test tube into the apparatus assembled as directed under *Solidification Point*, Appendix IIB. Allow the mixture to cool slowly until crystallization starts, or until the temperature has fallen to the point previously noted. Stir the mixture vigorously with the thermometer, rubbing the sides of the test tube with an up and down motion to induce crystallization. Continue the stirring and rubbing so long as the temperature rises. Take the highest temperature obtained as the solidification point. Repeat the procedure until two results agreeing within 0.1° are obtained. Calculate the percentage of cineole from the *Percentage of Cineole* table, Appendix VI.

Heavy Metals Shake 10 mL of the oil with an equal volume of water to which 1 drop of hydrochloric acid has been added, and pass hydrogen sulfide through the mixture until it is saturated. No darkening in color is produced in either the oil or the water.

Phellandrene Mix 2.5 mL of sample with 5 mL of solvent hexane, add 5 mL of a solution of sodium nitrite made by dissolving 5 g of sodium nitrite in 8 mL of water, then gradually add 5 mL of glacial acetic acid. No crystals form in the mixture within 10 min.

Refractive Index, Appendix IIB Determine with an Abbé or other refractometer of equal or greater accuracy.

Solubility in Alcohol Proceed as directed in the general method, Appendix VI. One mL dissolves in 5 mL of 70% alcohol.

Specific Gravity Determine by any reliable method (see *General Provisions*).

Packaging and Storage Store in well-filled, tight containers in a cool place protected from light.

F D & C Blue No. 1[1]

Brilliant Blue FCF; CI 42090; Class: Triphenylmethane

$C_{37}H_{34}N_2O_9S_3Na_2$ Formula wt 792.86
INS: 133   CAS: [3844-45-9]

DESCRIPTION

F D & C Blue No. 1 is principally the disodium salt of ethyl[4-[*p*-[ethyl(*m*-sulfobenzyl)amino]-α-(*o*-sulfophenyl)benzylidene]-2,5-cyclohexadien-1-ylidene](*m*-sulfobenzyl) ammonium hydroxide inner salt.

The colorant is a dark purple to bronze powder that dissolves in water to give a solution green-blue at neutrality, green in weak acid, and yellow in stronger acid. Addition of base to its neutral solution produces a violet color only on boiling. When dissolved in concentrated sulfuric acid, it yields a yellow solution that turns green when diluted with water. It is soluble in 95% ethanol.

Functional Use in Foods Color.

REQUIREMENTS

Identification A freshly prepared aqueous solution containing 10 mg/L exhibits absorbance intensities (*A*) and wavelength maxima as follows: at pH 7, *A* = 1.11 at 630 nm; at pH 1, *A* = 0.95 at 629 nm, and *A* = 0.2 at 410 nm; and at pH 13, *A* = 1.29 at 630 nm, and *A* = 0.15 at 408 nm.
Arsenic (as As) Not more than 3 mg/kg.
Chromium (as Cr) Not more than 0.005%.
Ether Extracts[2] (combined) Not more than 0.4%.
Lead Not more than 10 mg/kg.
Leuco Base Not more than 5.0%.
Subsidiary Colors Not more than 6.0%, combined, of isomeric disodium salts of ethyl[4-[*p*-[ethyl(*p*-sulfobenzyl)amino]-α-(*o*-sulfophenyl)benzylidene]-2,5-cyclohexadien-1-ylidene](*p*-sulfobenzyl) ammonium hydroxide inner salt and ethyl[4-[*p*-[ethyl(*o*-sulfobenzyl)amino]-α-(*o*-sulfophenyl)benzylidene]-2,5-cyclohexadien-1-ylidene](*o*-sulfobenzyl) ammonium hydroxide inner salt.
Total Color Not less than 85.0%.
Uncombined Intermediates and Products of Side Reactions
  o-, m-, and p-*sulfobenzaldehydes* Not more than 1.5%, combined.
  N-*ethyl,N-(m-sulfobenzyl)sulfanilic acid* Not more than 0.3%.
Volatile Matter at 135°, **Chloride**, and **Sulfate** (as sodium salts) Not more than 15.0% in combination.
Water-Insoluble Matter Not more than 0.2%.

---

[1]To be used or sold in the United States, this colorant must be certified by the U.S. Food and Drug Administration.

[2]Not required for certification in the United States.

TESTS

Arsenic A sample solution prepared as directed for organic compounds meets the requirements of the *Arsenic Test*, Appendix IIIB.

Chloride Proceed as directed for *Sodium Chloride* under *F D & C Colors*, Appendix IIIC.

Chromium Proceed as directed for *Chromium* under *F D & C Colors*, Appendix IIIC.

Ether Extracts Proceed as directed for *Ether Extracts* under *F D & C Colors*, Appendix IIIC.

Lead A *Sample Solution* prepared as directed for organic compounds meets the requirements of the *Lead Limit Test*, Appendix IIIB, using 10 μg of lead ion (Pb) in the control.

Leuco Base Transfer approximately 120 mg of colorant, accurately weighed, to a 1-L volumetric flask; dissolve in and dilute to volume with water. Proceed as directed for *Leuco Base* under *F D & C Colors*, Appendix IIIC.

Subsidiary Colors

Apparatus Use a 20- × 20-cm glass plate coated with a 0.25-mm layer of Silica Gel G. The solvent system is composed of acetonitrile (50 mL), isoamyl alcohol (50 mL), 2-butanone (15 mL), water (5 mL), and ammonium hydroxide (5 mL).

Sample Solution Transfer approximately 1 g of colorant, accurately weighed, to a 100-mL volumetric flask. Fill the flask about $\frac{3}{4}$ full with water and place in the dark for 1 h; then dilute to volume and mix well.

Procedure Spot 0.1 mL of *Sample Solution* in a line across the plate, approximately 3 cm from the bottom edge. Allow the plate to dry for about 20 min, in the dark; then develop with the solvent system in an unlined tank equilibrated for at least 20 min before the plate is inserted. Allow the solvent front to reach within about 3 cm of the top of the plate. Dry the developed plate in the dark.

When the plate has dried, scrape off all the colored bands above the F D & C Blue No. 1, which remains close to the origin, into a 30-mL beaker. Extract the subsidiary colors with three 6-mL portions of 95% ethanol, or until no color remains on the gel by visual inspection. Record the volume of ethanol used and record the spectrum of the solution between 400 and 700 nm. Calculate the percent of subsidiary colors (P) by the formula

$$P = (A \times V \times 100)/(a \times W \times b),$$

in which A is the absorbance at the wavelength maximum; V is the volume, in mL, of the solution; a is the absorptivity (0.126 mg/L/cm); W is the sample weight, in mg; and b is the pathlength of the cell.

Sulfate Proceed as directed for *Sodium Sulfate* under *F D & C Colors*, Appendix IIIC.

Total Color Determine the total color strength as the weight percent of colorant using *Methods I and II* under *F D & C Colors*, Appendix IIIC. Express the *Total Color* as the average of the two results.

Method I (Sample Preparation) Transfer 50 to 75 mg of colorant, accurately weighed, to a 1-L volumetric flask; dissolve in and dilute to volume with water. The absorptivity (a) for F D & C Blue No. 1 is 0.164 mg/L/cm at 630 nm.

Method II (Sample Preparation) Transfer approximately 0.5 g of colorant, accurately weighed, to the titration flask. The stoichiometric factor (F_s) for F D & C Blue No. 1 is 2.52.

Uncombined Intermediates and Products of Side Reactions Proceed as directed for *Uncombined Intermediates and Products of Side Reactions, Method I* under *F D & C Colors*, Appendix IIIC. Calculate the concentrations of m-sulfobenzaldehyde and N-ethyl-N-(3-sulfobenzyl)-sulfanilic acid using the following absorptivities:

m-Sulfobenzaldehyde, a = 0.0495 mg/L/cm at 246 nm (acid solution).

N-Ethyl-N-(3-sulfobenzyl)-sulfanilic acid, a = 0.078 mg/L/cm at 277 nm (alkaline solution).

Volatile Matter at 135° Proceed as directed for *Volatile Matter* under *F D & C Colors*, Appendix IIIC.

Water-Insoluble Matter Proceed as directed for *Water-Insoluble Matter* under *F D & C Colors*, Appendix IIIC.

Packaging and Storage Store in well-closed containers.

F D & C Blue No. 2[1]

Indigotine; Indigotine Disulfonate; Indigo Carmine; CI 73015; Class: Indigoid

$C_{16}H_8N_2O_8S_2Na_2$ Formula wt 466.36

INS: 132 CAS: [860-22-0]

DESCRIPTION

F D & C Blue No. 2 is principally the disodium salt of 2-(1,3-dihydro-3-oxo-5-sulfo-2H-indol-2-ylidene)-2,3-dihydro-3-oxo-1H-indole-5-sulfonic acid.

The colorant is a blue-brown to red-brown powder that dissolves in water to give a solution that is blue at neutrality, blue-violet in acid, and green to yellow-green in base. When dissolved in concentrated sulfuric acid, it yields a blue-violet solution that turns blue when diluted with water. It is sparingly soluble in 95% ethanol.

Functional Use in Foods Color.

REQUIREMENTS

Identification A freshly prepared solution containing 20 mg/L exhibits absorbance intensities (A) and wavelength maxima as follows: at pH 7, A = 0.82 at 610 nm; at pH 1, A = 0.81 at 610 nm; and at pH 13, A = 0.2 at 610 nm, and A = 0.31 at 442 nm.

Arsenic (as As) Not more than 3 mg/kg.

Ether Extracts[2] (combined) Not more than 0.4%.

---

[1]To be used or sold in the United States, this colorant must be certified by the U.S. Food and Drug Administration.

[2]Not required for certification in the United States.

Lead Not more than 10 mg/kg.
Mercury (as Hg) Not more than 1 mg/kg.
Subsidiary and Isomeric Colors
Disodium salt of 2-(1,3-dihydro-3-oxo-7-sulfo-2H-indole-2-ylidene)-2,3-dihydro-3-oxo-1H-indole-5-sulfonic acid Not more than 18.0%.
Sodium salt of 2-(1,3-dihydro-3-oxo-2H-indole-2-ylidene)-2,3-dihydro-3-oxo-1H-indole-5-sulfonic acid Not more than 2.0%.
Total Color Not less than 85.0%.
Uncombined Intermediates and Products of Side Reactions
Isatin-5-sulfonic acid Not more than 0.4%.
5-Sulfoanthranilic acid Not more than 0.2%.
Volatile Matter at 135°, Chlorides, and **Sulfates** (as sodium salts) Not more than 15.0% in combination.
Water-Insoluble Matter Not more than 0.4%.

TESTS

Arsenic A sample solution prepared as directed for organic compounds meets the requirements of the *Arsenic Test*, Appendix IIIB.
Chloride Proceed as directed for *Sodium Chloride* under *F D & C Colors*, Appendix IIIC.
Ether Extracts Proceed as directed for *Ether Extracts* under *F D & C Colors*, Appendix IIIC.
Lead A *Sample Solution* prepared as directed for organic compounds meets the requirements of the *Lead Limit Test*, Appendix IIIB, using 10 μg of lead ion (Pb) in the control.
Mercury Proceed as directed for *Mercury* under *F D & C Colors*, Appendix IIIC.
Subsidiary and Isomeric Colors
Apparatus Use a 2.5- × 40-cm glass column packed with Celite (Johns Mansville No. 595, or the equivalent) prepared as described under *Procedure*. Dissolve 20 g of hydroxylamine hydrochloride in 500 mL of water, place the solution in a 2-L separatory funnel, and add 450 mL of butanol, 450 mL of chloroform, 300 mL of water, and 100 mL of hydrochloric acid. Agitate the mixture well, periodically venting the funnel. After settling, the bottom layer (organic), called the *Mobile Phase*, and the top layer (aqueous), called the *Stationary Phase*, should be separated and stored.
Sample Solution Dissolve approximately 100 mg of colorant, accurately weighed, in 100 mL of *Stationary Phase*; warm on a steam bath, if necessary, to dissolve the sample.
Procedure Slurry 12 g of Celite with 7 mL of *Stationary Phase* and pour into the column. Mix 5 mL of *Sample Solution* with 10 g of Celite, and pour into the column over the slurry, ensuring that the sample is quantitatively transferred to the column.
Elute the column with the *Mobile Phase*. Collect the monosulfonated derivative, the first band eluting, in a 25-mL graduated cylinder, and note the volume. Collect the next band, the isomeric (unsulfonated) derivative, in a similar manner.
Mix each aliquot collected with an equal volume of hexane, and transfer to a separatory funnel. Extract this mixture with three 15-mL aliquots of water; combine the water extracts, and calculate the percent concentration (*P*) of the monosulfonated derivative (a = 0.0513 mg/L/cm at 615 nm) and the isomeric derivative (a = 0.0478 mg/L/cm at 610 nm) by the formula

$$P = (A \times V)/(a \times W \times 10),$$

in which *A* is the absorbance; *V* is the volume of extract; *a* is the absorptivity, in mg/L/cm; and *W* is the sample weight, in mg.
Sulfate Proceed as directed for *Sodium Sulfate* under *F D & C Colors*, Appendix IIIC.
Total Color Determine the total color strength as the weight percent of colorant using *Methods I and II* under *F D & C Colors*, Appendix IIIC. Express the *Total Color* as the average of the two results.
Method I (Sample Preparation) Transfer 175 to 225 mg of colorant, accurately weighed, to a 1-L volumetric flask; dissolve in and dilute to volume with water. The absorptivity (*a*) for F D & C Blue No. 2 is 0.0478 mg/L/cm at 610 nm.
Method II (Sample Preparation) Transfer approximately 0.3 g of colorant, accurately weighed, to the titration flask. The stoichiometric factor (F_s) for F D & C Blue No. 2 is 4.29.
Uncombined Intermediates and Products of Side Reactions Proceed as directed for *Uncombined Intermediates and Products of Side Reactions, Method I* under *F D & C Colors*, Appendix IIIC. Calculate the concentration of isatin-5-sulfonic acid using an absorptivity of 0.089 mg/L/cm at 245 nm.
Volatile Matter at 135° Proceed as directed for *Volatile Matter* under *F D & C Colors*, Appendix IIIC.
Water-Insoluble Matter Proceed as directed for *Water-Insoluble Matter* under *F D & C Colors*, Appendix IIIC.

Packaging and Storage Store in well-closed containers.

F D & C Green No. 3[1]

Fast Green FCF; CI 42053; Class: Triphenylmethane

$C_{37}H_{34}N_2O_{10}S_3Na_2$ Formula wt 808.86

INS: 143   CAS: [2353-45-9]

DESCRIPTION

F D & C Green No. 3 is principally the inner salt disodium salt of *N*-ethyl-*N*-[4-[[4-[ethyl[(3-sulfophenyl)methyl]amino]phenyl](4-hydroxy-2-sulfophenyl)methylene]-2,5-cyclohexadien-1-ylidene]-3-sulfobenzenemethanaminium hydroxide.
The colorant is a red to brown-violet powder or collection of crystals that dissolves in water to give a solution blue-green at neutrality, green in acid, and blue-violet in base. When dissolved in concentrated sulfuric acid, it yields a brown-orange solution that turns green when diluted with water. When heated to 130° with glycerine triacetate and an excess of acetic anhy-

---

[1] To be used or sold in the United States, this colorant must be certified by the U.S. Food and Drug Administration.

dride, acetylation of its phenolic hydroxyl group causes a color change from green to light blue.

Functional Use in Foods Color.

REQUIREMENTS

Identification A freshly prepared aqueous solution containing 5 mg/L exhibits absorbance intensities (A) and wavelength maxima as follows: at pH 7, $A = 0.80$ at 624 nm, and $A = 0.08$ at 423 nm; at pH 1, $A = 0.83$ at 625 nm, and $A = 0.09$ at 423 nm; and at pH 13, $A = 0.74$ at 610 nm.
Arsenic (as As) Not more than 3 mg/kg.
Chromium (as Cr) Not more than 0.005%.
Ether Extracts[2] (combined) Not more than 0.4%.
Lead Not more than 10 mg/kg.
Leuco Base Not more than 5.0%.
Mercury (as Hg) Not more than 1 mg/kg.
Subsidiary Colors Not more than 6.0%, total, of isomeric inner salt disodium salt of N-ethyl-N-[4-[[4-[ethyl[(3-sulfophenyl)methyl]amino]phenyl](4-hydroxy-2-sulfophenyl)methylene]-2,5-cyclohexadien-1-ylidene]-4-sulfobenzenemethanaminium hydroxide, N-ethyl-N-[4-[[4-[ethyl[(4-sulfophenyl)methyl]amino]phenyl](4-hydroxy-2-sulfophenyl)methylene]-2,5-cyclohexadien-1-ylidene]-4-sulfobenzenemethanaminium hydroxide, and N-ethyl-N-[4-[[4-[ethyl[(2-sulfophenyl)methyl]amino]phenyl](4-hydroxy-2-sulfophenyl)methylene]-2,5-cyclohexadiene-1-ylidene]-3-sulfobenzenemethanaminium hydroxide.
Total Color Not less than 85.0%.
Uncombined Intermediates and Products of Side Reactions
  Sum of 3- and 4-[[ethyl(4-sulfophenyl)amino]methyl]benzenesulfonic acid disodium salts Not more than 0.3%.
  Sum of 2-, 3-, and 4-formyl benzenesulfonic acids sodium salts Not more than 0.5%.
  2-Formyl-5-hydroxybenzenesulfonic acid Not more than 0.5%.
Volatile Matter at 135°, Chlorides and **Sulfates** (as sodium salts) Not more than 15.0% in combination.
Water-Insoluble Matter Not more than 0.2%.

TESTS

Arsenic A sample solution prepared as directed for organic compounds meets the requirements of the *Arsenic Test*, Appendix IIIB.
Chloride Proceed as directed for *Sodium Chloride* in *F D & C Colors*, Appendix IIIC.
Chromium Proceed as directed for *Chromium* under *F D & C Colors*, Appendix IIIC.
Ether Extracts Proceed as directed for *Ether Extracts* under *F D & C Colors*, Appendix IIIC.
Lead A *Sample Solution* prepared as directed for organic compounds meets the requirements of the *Lead Limit Test*, Appendix IIIB, using 10 μg of lead ion (Pb) in the control.

Leuco Base Transfer approximately 130 mg of colorant, accurately weighed, to a 1-L volumetric flask; dissolve in and dilute to volume with water. Proceed as directed for *Leuco Base* under *F D & C Colors*, Appendix IIIC.
Mercury Proceed as directed for *Mercury* under *F D & C Colors*, Appendix IIIC.
Subsidiary Colors
  Apparatus Use a 20- × 20-cm glass plate coated with a 0.25-mm layer of Silica Gel G. The solvent system is composed of acetonitrile (50 mL), isoamyl alcohol (50 mL), 2-butanone (15 mL), water (10 mL), and ammonium hydroxide (5 mL).
  Sample Solution Transfer approximately 1 g of colorant, accurately weighed, to a 100-mL volumetric flask. Fill the flask about ¾ full with water, and incubate in the dark for 1 h; dilute to volume, and mix well.
  Procedure Proceed as directed under *Subsidiary Colors* in the monograph for *F D & C Blue No. 1*.
Sulfate Proceed as directed for *Sodium Sulfate* under *F D & C Colors*, Appendix IIIC.
Total Color Determine the total color strength as the weight percent of colorant using *Methods I and II* under *F D & C Colors*, Appendix IIIC. Express the *Total Color* as the average of the two results.
  Method I (Sample Preparation) Transfer 50 to 75 mg of colorant, accurately weighed, to a 1-L volumetric flask; dissolve in and dilute to volume with water. The absorptivity (a) for F D & C Green No. 3 is 0.156 mg/L/cm at 625 nm.
  Method II (Sample Preparation) Transfer approximately 0.5 g of colorant, accurately weighed, to the titration flask. The stoichiometric factor (F_s) for F D & C Green No. 3 is 2.47.
Uncombined Intermediates and Products of Side Reactions Transfer approximately 2 g of colorant, accurately weighed, to a 100-mL volumetric flask; dissolve in and dilute to volume with water. Proceed as directed under *Uncombined Intermediates and Products of Side Reactions, Method I* under *F D & C Colors*, Appendix IIIC. Calculate the amounts of intermediates and other products present using the following absorptivities after identifying the unknowns by comparing their spectra with standards:
  4-Hydroxy-2-sulfobenzaldehyde, $a = 0.080$ mg/L/cm at 335 nm (alkaline solution).
  m-Sulfobenzaldehyde, $a = 0.495$ mg/L/cm at 246 nm (acid solution).
  N-Ethyl-N-(3-sulfobenzyl)-sulfanilic acid, $a = 0.078$ mg/L/cm at 277 nm (alkaline solution).
Volatile Matter at 135° Proceed as directed for *Volatile Matter* under *F D & C Colors*, Appendix IIIC.
Water-Insoluble Matter Proceed as directed for *Water-Insoluble Matter* under *F D & C Colors*, Appendix IIIC.

Packaging and Storage Store in well-closed containers.

---

[2] Not required for certification in the United States.

F D & C Red No. 3[1]

Erythrosine; CI 45430; Class: Xanthene

$C_{20}H_6O_5I_4Na_2$ Formula wt 879.86

INS: 127   CAS: [16423-68-0]

DESCRIPTION

F D & C Red No. 3 is principally the disodium salt of the monohydrate of 9(o-carboxyphenyl)-6-hydroxy-2,4,5,7-tetraiodo-3H-xanthen-3-one.

The colorant is a brown powder that dissolves in water to give a solution red at neutrality, with a yellow-brown precipitate in acid, and with a red precipitate in base. When dissolved in concentrated sulfuric acid, it yields a brown-yellow solution that evolves iodine and a precipitate of the free acid when heated.

Functional Use in Foods   Color.

REQUIREMENTS

Identification   A solution containing 2.8 mg/L exhibits absorbance intensities (A) and wavelength maxima as follows: in neutral (pH = 7) and alkaline (pH = 13) solutions, A = 0.32 at 527 nm with a shoulder at 490 nm. In acid solution, a yellow-brown precipitate forms.

Arsenic (as As)   Not more than 3 mg/kg.

Ether Extracts[2] (combined)   Not more than 0.2%.

Lead   Not more than 10 mg/kg.

Subsidiary Colors

Monoiodofluoresceins   Not more than 1.0%.

Other lower iodinated fluoresceins   Not more than 9.0%.

Total Color   Not less than 87.0%.

Uncombined Intermediates and Products of Side Reactions

2-(2′,4′-Dihydroxy-3′,5′-diiodobenzoyl)benzoic acid   Not more than 0.2%.

Sodium iodide   Not more than 0.4%.

Triiodoresorcinol   Not more than 0.2%.

Unhalogenated intermediates   Total not more than 0.1%.

Volatile Matter at 135°, Chlorides, and **Sulfates** (as sodium salts)   Not more than 13.0% in combination.

Water-Insoluble Matter   Not more than 0.2%.

TESTS

Arsenic   A sample solution prepared as directed for organic compounds meets the requirements of the *Arsenic Test*, Appendix IIIB.

Chloride   Proceed as directed for *Sodium Chloride* under *F D & C Colors*, Appendix IIIC.

Ether Extracts   Proceed as directed for *Ether Extracts* under *F D & C Colors*, Appendix IIIC.

Lead   A *Sample Solution* prepared as directed for organic compounds meets the requirements of the *Lead Limit Test*, Appendix IIIB, using 10 μg of lead ion (Pb) in the control.

Subsidiary Colors

Apparatus   Use a 20- × 20-cm glass plate coated with a 0.25-mm layer of Silica Gel G. The solvent system is composed of acetone (95 mL), chloroform (25 mL), butylamine (10 mL), and water (10 mL).

Sample Solution   Transfer approximately 2 g of colorant, accurately weighed, to a 100-mL volumetric flask. Fill the flask about $3/4$ full with water, and incubate in the dark for 1 h; dilute to volume, and mix well.

Procedure   Spot 0.1 mL of *Sample Solution* in a line across the plate, approximately 3 cm from the bottom edge. Allow the plate to dry for about 20 min in the dark, then develop with the solvent system in an unlined tank equilibrated for at least 20 min before inserting the plate. Allow the solvent front to reach to within about 3 cm of the top of the plate. Dry the developed plate in the dark.

Scrape off each subsidiary color and extract with 3- to 5-mL portions of 50% aqueous ethanol until no color remains on the gel by visual inspection. Dilute each sample to 13 to 15 mL, add a few drops of ammonium hydroxide, and record the final volume. Repeat this procedure for the band of F D & C Red No. 3 using 10- to 20-mL portions of 50% ethanol, and dilute the eluant to 250 mL in a volumetric flask after adding enough ammonium hydroxide to make the solution slightly alkaline. The approximate band positions (R_f), wavelengths of maximal absorbance (λ), and absorptivities (a) are as follows:

Color	R_f	λ	a
Unknown	0.84	524	0.110
Red-3	0.84	526	0.110
2,4,7	0.76	521	0.140
2,4,5	0.67	521	0.116
2,4/2,5	0.45	513	0.145
Unknown	0.45	524	0.110

Record the spectrum of each solution between 400 and 600 nm and calculate the quantity in percent (P) of each subsidiary color by the formula

$$P = (A \times V \times 100)/(a \times W \times b),$$

in which A is the absorbance at the wavelength maximum, V is the volume, in mL, of the solution; a is the absorptivity, in mg/L/cm, as given above; W is the sample weight, in mg; and b is the path length, in cm, of the cell.

Sulfate   Proceed as directed for *Sodium Sulfate* under *F D & C Colors*, Appendix IIIC.

Total Color   Determine the total color strength as the weight percent of colorant using *Methods I and III* under *F D & C Colors*, Appendix IIIC. Express the *Total Color* as the average of the two results.

Method I (Sample Preparation)   Transfer 75 to 100 mg of colorant, accurately weighed, to a 1-L volumetric flask; dissolve

---

[1] To be used or sold in the United States, this colorant must be certified by the U.S. Food and Drug Administration.

[2] Not required for certification in the United States.

in and dilute to volume with water. The absorptivity (*a*) for F D & C Red No. 3 is 0.110 mg/L/cm at 527 nm.

Method III (Sample Preparation) Proceed as directed for *Total Color, Method III* under *F D & C Colors*, Appendix IIIC. The gravimetric conversion factor (*F*) for F D & C Red No. 3 is 1.074.

Uncombined Intermediates and Products of Side Reactions
Transfer 2 g of colorant, accurately weighed, to a 100-mL volumetric flask; dissolve in and dilute to volume with water. Proceed as directed for *Uncombined Intermediates and Products of Side Reactions, Method I* under *F D & C Colors*, Appendix IIIC. Calculate the concentrations of 2-(2,4-dihydroxy-3,5-diiodobenzoyl)benzoic acid, iodine, phthalic acid, sodium iodide, and triiodoresorcinol, using the following absorptivities:

2-(2,4-Dihydroxy-3,5-diiodobenzoyl)benzoic acid, $a = 0.047$ mg/L/cm at 348 nm (alkaline).

Iodine, $a = 0.082$ mg/L/cm at 245 nm (acidic).

Phthalic acid, $a = 0.045$ mg/L/cm at 228 nm (acidic).

Sodium iodide, $a = 0.091$ mg/L/cm at 220 nm (acidic).

Triiodoresorcinol, $a = 0.079$ mg/L/cm at 223 nm (acidic).

Volatile Matter at 135° Proceed as directed for *Volatile Matter* under *F D & C Colors*, Appendix IIIC.

Water-Insoluble Matter Proceed as directed for *Water-Insoluble Matter* under *F D & C Colors*, Appendix IIIC.

Packaging and Storage Store in well-closed containers.

F D & C Red No. 40[1]

Allura Red AC; CI 16035; Class: Monoazo

$C_{18}H_{14}N_2O_8S_2Na_2$ Formula wt 496.43

INS: 129 CAS: [25956-17-6]

DESCRIPTION

F D & C Red No. 40 is principally the disodium salt of 6-hydroxy-5-[(2-methoxy-5-methyl-4-sulfophenyl)azo]-2-naphthalenesulfonic acid.

The colorant is a red powder that dissolves in water to give a solution red at neutrality and in acid and dark red in base. It is slightly soluble in 95% ethanol.

Functional Use in Foods Color.

REQUIREMENTS

Identification A solution containing 16.4 mg/L exhibits absorbance intensities (*A*) and wavelength maxima as follows: at pH 7, $A = 0.87$ at 500 nm; at pH 1, $A = 0.83$ at 490 nm (both neutral and acid solutions exhibit a shoulder at about 410 nm); and at pH 13, $A = 0.37$ at 500 nm, and $A = 0.41$ at 450 nm.

---
[1] To be used or sold in the United States, this colorant must be certified by the U.S. Food and Drug Administration.

Arsenic (as As) Not more than 3 mg/kg.
Lead Not more than 10 mg/kg.
Subsidiary Colors

Disodium 6-hydroxy-5-[(2-methoxy-5-methyl-4-sulfophenyl)azo]-8-(2-methoxy-5-methyl-4-sulfophenoxy)-2-naphthalenesulfonic acid Not more than 1.0%.

Higher and lower sulfonated subsidiary colors (as sodium salts) Not more than 1.0% each.

Total Color Not less than 85.0%.
Uncombined Intermediates and Products of Side Reactions

4-Amino-5-methoxy-o-toluenesulfonic acid Not more than 0.2%.

Disodium 6,6'-oxybis(2-naphthalenesulfonic acid) Not more than 1%.

Sodium 6-hydroxy-2-naphthalenesulfonic acid Not more than 0.3%.

Volatile Matter at 135°, **Chlorides**, and **Sulfates** (as sodium salts) Not more than 14.0% in combination.

Water-Insoluble Matter Not more than 0.2%.

TESTS

Arsenic A sample solution prepared as directed for organic compounds meets the requirements of the *Arsenic Test*, Appendix IIIB.

Chloride Proceed as directed for *Sodium Chloride* under *F D & C Colors*, Appendix IIIC.

Lead A *Sample Solution* prepared as directed for organic compounds meets the requirements of the *Lead Limit Test*, Appendix IIIB, using 10 μg of lead ion (Pb) in the control.

Subsidiary Colors

Apparatus Use a 20- × 20-cm glass plate coated with a 0.25-mm layer of Silica Gel G. The solvent system is composed of acetonitrile (20 mL), dioxane (20 mL), ethyl acetate (20 mL), isoamyl alcohol (20 mL), water (20 mL), and ammonium hydroxide (4 mL).

Standard Solution Transfer approximately 1 g, accurately weighed, of purified F D & C Red No. 40, free of subsidiary colors, to a 50-mL volumetric flask. Add 50 mg each of lower and higher sulfonated subsidiary colors, accurately weighed; dissolve in and dilute to volume with water. Store in the dark.

Sample Solution Transfer approximately 2 g of colorant, accurately weighed, to a 100-mL volumetric flask; dissolve in and dilute to volume with water.

Procedure Spot 3-μL aliquots of *Sample Solution* and *Standard Solution* side by side 3 cm from the bottom of the plate. Up to seven samples and standards may be run simultaneously.

When the plate has air dried for 15 min, develop it in an unlined tank equilibrated with the solvent system for at least 20 min. Allow the solvent front to reach to within about 3 cm from the top of the plate.

Allow the plate to dry in a fume hood, and by visual inspection, compare the intensities of the lower and higher sulfonated subsidiary colors with those in the *Standard Solution*. If the subsidiary colors in the *Sample Solution* appear more concentrated than those in the *Standard Solution*, determine the quantity of each, using a densitometer set to monitor the absorbance maximum of each. Calculate the concentrations of the subsidiary colors in percent (*P*), if present above 0.1%, by the formula

$$P = (A \times p)/A_S,$$

in which A is the area of the densitometer curve, p is the percent of subsidiary color in the *Standard Solution*, and A_S is the area of the densitometer curve for the subsidiary color in the *Standard Solution*.

Sulfate Proceed as directed for *Sodium Sulfate* under *F D & C Colors*, Appendix IIIC.

Total Color Determine the total color strength as the weight percent of colorant using *Methods I and II* under *F D & C Colors*, Appendix IIIC. Express the *Total Color* as the average of the two results.

Method I (Sample Preparation) Transfer 175 to 225 mg of colorant, accurately weighed, to a 1-L volumetric flask; dissolve in and dilute to volume with water. The absorptivity (a) for F D & C Red No. 40 is 0.052 mg/L/cm at 502 nm.

Method II (Sample Preparation) Transfer approximately 0.2 g of colorant, accurately weighed, to the titration flask. The stoichiometric factor (F_s) for F D & C Red No. 40 is 8.06.

Uncombined Intermediates and Products of Side Reactions

Sample Preparation Transfer 0.25 g of colorant, accurately weighed, to a 100-mL volumetric flask. Dissolve in and dilute to volume with 0.1 M $Na_2B_4O_7$. Inject 20 µL of this solution as directed for *Uncombined Intermediates and Products of Side Reactions, Method II* under *F D & C Colors*, Appendix IIIC.

Volatile Matter at 135° Proceed as directed for *Volatile Matter* under *F D & C Colors*, Appendix IIIC.

Water-Insoluble Matter Proceed as directed for *Water-Insoluble Matter* under *F D & C Colors*, Appendix IIIC.

Packaging and Storage Store in well-closed containers.

F D & C Yellow No. 5[1]

Tartrazine; CI 19140; Class: Pyrazalone

$C_{16}H_9N_4O_9S_2Na_3$ Formula wt 534.37

INS: 102 CAS: [1934-21-0]

DESCRIPTION

F D & C Yellow No. 5 is principally the trisodium salt of 4,5-dihydro-5-oxo-1-(4-sulfophenyl)-4-[4-sulfophenyl-azo]-1*H*-pyrazole-3-carboxylic acid.

The colorant is a yellow-orange powder that dissolves in water to give a solution golden yellow at neutrality and in acid. When dissolved in concentrated sulfuric acid, it yields an orange-yellow solution that turns yellow when diluted with water.

Functional Use in Foods Color.

REQUIREMENTS

Identification A solution (*A*) containing 19.9 mg/L exhibits absorbance intensities and wavelength maxima as follows: at pH 7, $A = 1.4$ at 425 nm; at pH 1, $A = 1.1$ at 426 nm; and at pH 13, the absorbance maximum is shifted below 400 nm.

Arsenic (as As) Not more than 3 mg/kg.

Ether Extracts[2] (combined) Not more than 0.2%.

Lead Not more than 10 mg/kg.

Mercury (as Hg) Not more than 1 mg/kg.

Total Color Not less than 87.0%.

Uncombined Intermediates and Products of Side Reactions

4,4'-[4,5-Dihydro-5-oxo-4-[(4-sulfophenyl)-hydrazono]-1H-pyrazol-1,3-diyl]bis[benzenesulfonic acid], trisodium salt Not more than 1%.

4-[(4',5-Disulfo[1,1'-biphenyl]-2-yl)hydrazono]-4,5-dihydro-5-oxo-1-(4-sulfophenyl)-1H-pyrazole-3-carboxylic acid, tetrasodium salt Not more than 1%.

Ethyl or methyl 4,5-dihydro-5-oxo-1-(4-sulfophenyl)-4-[(4-sulfophenyl)hydrazono]-1H-pyrazole-3-carboxylate, disodium salt Not more than 1%.

Sum of 4,5-dihydro-5-oxo-1-phenyl-4-[(4-sulfophenyl)azo]-1H-pyrazole-3-carboxylic acid, disodium salt, and 4,5-dihydro-5-oxo-4-(phenylazo)-1-(4-sulfophenyl)-1H-pyrazole-3-carboxylic acid, disodium salt Not more than 0.5%.

4-Aminobenzenesulfonic acid, sodium salt Not more than 0.2%.

4,5-Dihydro-5-oxo-1-(4-sulfophenyl)-1H-pyrazole-3-carboxylic acid, disodium salt Not more than 0.2%.

Ethyl or methyl 4,5-dihydro-5-oxo-1-(4-sulfophenyl)-1H-pyrazole-3-carboxylate, sodium salt Not more than 0.1%.

4,4'-(1-Triazene-1,3-diyl)bis[benzenesulfonic acid], disodium salt Not more than 0.05%.

4-Aminoazobenzene Not more than 0.075 mg/kg.

4-Aminophenyl Not more than 0.005 mg/kg.

Aniline Not more than 0.1 mg/kg.

Azobenzene Not more than 0.04 mg/kg.

Benzidine Not more than 0.001 mg/kg.

1,3-Diphenyltriazene Not more than 0.04 mg/kg.

Volatile Matter at 135°, Chlorides, and **Sulfates** (as sodium salts) Not more than 13.0% in combination.

Water-Insoluble Matter Not more than 0.2%.

TESTS

Arsenic A sample solution prepared as directed for organic compounds meets the requirements of the *Arsenic Test*, Appendix IIIB.

Chloride Proceed as directed for *Sodium Chloride* under *F D & C Colors*, Appendix IIIC.

Ether Extracts Proceed as directed for *Ether Extracts* under *F D & C Colors*, Appendix IIIC.

Lead A *Sample Solution* prepared as directed for organic compounds meets the requirements of the *Lead Limit Test*, Appendix IIIB, using 10 µg of lead ion (Pb) in the control.

Mercury Proceed as directed for *Mercury* under *F D & C Colors*, Appendix IIIC.

---

[1] To be used or sold in the United States, this colorant must be certified by the U.S. Food and Drug Administration.

[2] Not required for certification in the United States.

Sulfate Proceed as directed for *Sodium Sulfate* under *F D & C Colors*, Appendix IIIC.

Total Color Determine the total color strength as the weight percent of colorant, using *Methods I and II* under *F D & C Colors*, Appendix IIIC. Express the *Total Color* as the average of the two results.

Method I (Sample Preparation) Transfer 175 to 250 mg of colorant, accurately weighed, to a 1-L volumetric flask; dissolve in and dilute to volume with water. The absorptivity (*a*) for F D & C Yellow No. 5 is 0.053 mg/L/cm at 428 nm.

Method II (Sample Preparation) Transfer approximately 0.2 g of colorant, accurately weighed, to the titration flask. The stoichiometric factor (F_s) for F D & C Yellow No. 5 is 7.49.

Uncombined Intermediates and Products of Side Reactions

Sample Preparation Transfer 0.15 g of colorant, accurately weighed, to a 100-mL volumetric flask. Dissolve in and dilute to volume with 0.1 M $Na_2B_4O_7$. Inject 50 µL of this solution as directed for *Uncombined Intermediates and Products of Side Reactions, Method II* under *F D & C Colors*, Appendix IIIC.

Volatile Matter at 135° Proceed as directed for *Volatile Matter* under *F D & C Colors*, Appendix IIIC.

Water-Insoluble Matter Proceed as directed for *Water-Insoluble Matter* under *F D & C Colors*, Appendix IIIC.

Packaging and Storage Store in well-closed containers.

F D & C Yellow No. 6[1]

Sunset Yellow FCF; CI 15985; Class: Monoazo

$C_{16}H_{10}N_2O_7S_2Na_2$ Formula wt 452.38

INS: 110   CAS: [2783-94-0]

DESCRIPTION

F D & C Yellow No. 6 is principally the disodium salt of 6-hydroxy-5-[(4-sulfophenyl)azo]-2-naphthalenesulfonic acid. The trisodium salt of 3-hydroxy-4-[(4-sulfophenyl)azo]-2,7-naphthalenedisulfonic acid may be added in small amounts.

The colorant is an orange powder that dissolves in water to give a solution yellow-orange at neutrality or in acid and red-brown in base. When dissolved in concentrated sulfuric acid, it yields an orange solution that turns yellow when diluted with water. It is slightly soluble in 95% ethanol.

Functional Use in Foods Color.

REQUIREMENTS

Identification A solution (*A*) containing 18.5 mg/L exhibits absorbance intensities and wavelength maxima as follows: at pH 1, A = 1.1 at 480 nm; and at pH 13, A = 0.46 at 443 nm, with a shoulder at about 500 nm.

Arsenic (as As)   Not more than 3 mg/kg.

Ether Extracts[2] (combined)   Not more than 0.2%.

Lead   Not more than 10 mg/kg.

Mercury (as Hg)   Not more than 1 mg/kg.

Total Color   Not less than 87.0%.

Uncombined Intermediates and Products of Side Reactions

4-Aminoazobenzene   Not more than 0.05%.

4-Aminobiphenyl   Not more than 0.015%.

Aniline   Not more than 0.25%.

Azobenzene   Not more than 0.2%.

Sodium salt of 4-aminobenzenesulfonic acid   Not more than 0.2%.

Sodium salt of 6-hydroxy-2-naphthalenesulfonic acid   Not more than 0.3%.

Disodium salt of 6,6'-oxybis[2-naphthalenesulfonic acid]   Not more than 1%.

Disodium salt of 4,4'-(1-triazene-1,3-diyl)bis[benzenesulfonic acid]   Not more than 0.1%.

Sum of the sodium salt of 6-hydroxy-5-(phenylazo)-2-naphthalenesulfonic acid and the sodium salt of 4-[(2-hydroxy-1-naphthylenyl)azo]benzenesulfonic acid   Not more than 1%.

Sum of the trisodium salt of 3-hydroxy-4-[(4-sulfophenyl)azo-2,7-naphthalenedisulfonic acid and other, higher sulfonated subsidiaries   Not more than 5%.

Volatile Matter at 135°, Chlorides, and **Sulfates** (as sodium salts)   Not more than 13.0% in combination.

Water-Insoluble Matter   Not more than 0.2%.

TESTS

Arsenic A sample solution prepared as directed for organic compounds meets the requirements of the *Arsenic Test*, Appendix IIIB.

Chloride Proceed as directed for *Sodium Chloride* under *F D & C Colors*, Appendix IIIC.

Ether Extracts Proceed as directed for *Ether Extracts* under *F D & C Colors*, Appendix IIIC.

Lead A *Sample Solution* prepared as directed for organic compounds meets the requirements of the *Lead Limit Test*, Appendix IIIB, using 10 µg of lead ion (Pb) in the control.

Mercury Proceed as directed for *Mercury* under *F D & C Colors*, Appendix IIIC.

Total Color Determine the total color strength as the weight percent of colorant using *Methods I and II* under *F D & C Colors*, Appendix IIIC. Express the *Total Color* as the average of the two results.

Method I (Sample Preparation) Transfer 200 to 225 mg of colorant, accurately weighed, to a 1-L volumetric flask; dissolve in and dilute to volume with water. The absorptivity (*a*) for F D & C Yellow No. 6 is 0.054 mg/L/cm at 484 nm.

Method II (Sample Preparation) Transfer approximately 0.2 g of colorant, accurately weighed, to the titration flask. The stoichiometric factor (F_s) for F D & C Yellow No. 6 is 8.84.

---

[1]To be used or sold in the United States, this colorant must be certified by the U.S. Food and Drug Administration.

[2]Not required for certification in the United States.

Uncombined Intermediates and Products of Side Reactions
Sample Preparation Transfer 0.25 g of colorant, accurately weighed, to a 100-mL volumetric flask. Dissolve in and dilute to volume with 0.1 M $Na_2B_4O_7$. Inject 20 μL of this solution as directed for *Uncombined Intermediates and Products of Side Reactions, Method II* under *F D & C Colors*, Appendix IIIC.
Volatile Matter at 135° Proceed as directed for *Volatile Matter* under *F D & C Colors*, Appendix IIIC.
Water-Insoluble Matter Proceed as directed for *Water-Insoluble Matter* under *F D & C Colors*, Appendix IIIC.

Packaging and Storage Store in well-closed containers.

Fennel Oil

CAS: [8006-84-6]

DESCRIPTION

The volatile oil obtained by steam distillation from the dried ripe fruit of *Foeniculum vulgare* Miller (Fam. *Umbelliferae*). It is a colorless or pale yellow liquid having the characteristic odor and taste of fennel.

Note: If solid material has separated, carefully warm the sample until it is completely liquefied, and mix it before using.

Functional Use in Foods Flavoring agent.

REQUIREMENTS

Identification The infrared absorption spectrum of the sample exhibits relative maxima (that may vary in intensity) at the same wavelengths (or frequencies) as those shown in the respective spectrum in the section on *Infrared Spectra (Series A: Essential Oils)*, using the same test conditions as specified therein.
Angular Rotation Between +12° and +24°.
Heavy Metals (as Pb) Passes test.
Refractive Index Between 1.532 and 1.543 at 20°.
Solidification Point Not lower than 3°.
Solubility in Alcohol Passes test.
Specific Gravity Between 0.953 and 0.973.

TESTS

Angular Rotation Determine in a 100-mm tube as directed under *Optical (Specific) Rotation*, Appendix IIB.
Heavy Metals Shake 10 mL of the oil with an equal volume of water to which 1 drop of hydrochloric acid has been added, and pass hydrogen sulfide through the mixture until it is saturated. No darkening in color is produced in either the oil or the water.
Refractive Index, Appendix IIB Determine with an Abbé or other refractometer of equal or greater accuracy.

Solidification Point Determine as directed in the general method, Appendix IIB.
Solubility in Alcohol Proceed as directed in the general method, Appendix VI. One mL dissolves in 1 mL of 90% alcohol.
Specific Gravity Determine by any reliable method (see *General Provisions*).

Packaging and Storage Store in full, tight containers in a cool place protected from light.

Ferric Ammonium Citrate, Brown

Iron Ammonium Citrate

INS: 381

DESCRIPTION

A complex salt of undetermined structure, composed of iron, ammonia, and citric acid and occurring as thin, transparent brown, reddish brown, or garnet red scales or granules, or as a brownish yellow powder. It is odorless or has a slight ammoniacal odor and has a mild iron-metallic taste. It is very soluble in water but is insoluble in alcohol. The pH of a 1 in 20 solution is about 5.0 to 8.0. It is deliquescent in air and is affected by light.

Functional Use in Foods Nutrient; dietary supplement.

REQUIREMENTS

Identification
    A. A 500-mg sample, when ignited, chars and leaves a residue of iron oxide.
    B. To 5 mL of a 1 in 10 solution of the sample add 0.3 mL of potassium permanganate TS and 4 mL of mercuric sulfate TS, and then heat the mixture to boiling. A white precipitate forms.
    C. Dissolve about 500 mg of the sample in 5 mL of water, and add 5 mL of 1 N sodium hydroxide. A reddish brown precipitate forms, and ammonia is evolved when the mixture is heated.
Assay Not less than 16.5% and not more than 18.5% of iron (Fe).
Ferric Citrate Passes test.
Lead Not more than 10 mg/kg.
Mercury Not more than 1 mg/kg.
Oxalate Passes test.
Sulfate Not more than 0.3%.

TESTS

Assay Transfer about 1 g, accurately weighed, into a 250-mL glass-stoppered Erlenmeyer flask, and dissolve in 25 mL of water and 5 mL of hydrochloric acid. Add 4 g of potassium iodide, stopper, and allow to stand protected from light for 15 min. Add 100 mL of water, and titrate the liberated iodine with 0.1 N sodium thiosulfate, using starch TS as the indicator. Perform a blank determination (see *General Provisions*) and make any necessary correction. Each mL of 0.1 N sodium thiosulfate is equivalent to 5.585 mg of iron (Fe).

Ferric Citrate Add potassium ferrocyanide TS to a 1 in 100 solution of the sample. No blue precipitate forms.

Lead (Note: The following method has been found to be satisfactory when the particular atomic absorption spectrophotometer specified is used. The method may be modified as necessary for use with other suitable atomic absorption spectrophotometers capable of determining lead in the sample at the limit specified.)

Standard Preparation Transfer 10.0 mL of *Lead Nitrate Stock Solution* (see *Heavy Metals Test*, Appendix IIIB) into a 500-mL volumetric flask, dilute to volume with water, and mix. This solution should be prepared on the day of use. Each mL contains the equivalent of 2 μg of lead ion (Pb).

Sample Preparation Transfer about 15 g of the sample, accurately weighed, into a 100-mL volumetric flask (previously rinsed with nitric acid and water), dissolve in a mixture of 50 mL of water and 1 mL of nitric acid, dilute to volume with water, and mix.

Procedure Using a Perkin-Elmer 403 atomic absorption spectrophotometer equipped with a deuterium arc background corrector, a digital readout device, and a burner head capable of handling 15% solids content, blank the instrument with water following the manufacturer's operating instructions. Aspirate a portion of the *Standard Preparation*, and record the absorbance as A_S; then aspirate a portion of the *Sample Preparation*, and record the absorbance as A_U. Calculate the lead content, in mg/kg, of the sample taken by the formula

$$100 \times (C/W) \times (A_U/A_S),$$

in which C is the concentration of Pb in the *Standard Preparation*, in μg/mL, and W is the weight of the sample taken, in g.

Mercury

Standard Preparations Prepare a solution containing 1 μg of mercury (Hg) per mL as directed for *Standard Preparation* under *Mercury Limit Test*, Appendix IIIB. Pipet 0.25, 0.50, 1.0, and 3.5 mL of this solution into each of four glass-stoppered bottles of about 300-mL capacity, such as BOD (biological oxygen demand) bottles. Dilute the contents of each bottle to 100 mL with water, and mix. These solutions contain the equivalent of 0.25, 0.50, 1.0, and 3.5 mg/kg of Hg, respectively.

Sample Preparation Transfer 1.000 g of the sample into a 200-mL, screw-cap centrifuge bottle, and add 5 mL of nitric acid and 5 mL of hydrochloric acid. Close the bottle tightly with a Teflon-lined screw-cap, digest on a steam bath for 1 h, and cool. Quantitatively transfer into a suitable glass-stoppered bottle (see *Standard Preparations*), dilute to 100 mL with water, and bubble air through the sample for 2 min. Prepare a reagent blank in the same manner.

Apparatus Use a suitable atomic absorption spectrophotometer assembly designed for mercury analysis, such as the Coleman MAS-50 Mercury Analyzer.

Note: The *Apparatus* and *Procedure* described under *Mercury Limit Test*, Appendix IIIB, may be suitably modified for this determination.

Procedure Add 5 mL of a 10% stannous chloride solution (prepared fresh each week by dissolving 20 g of $SnCl_2 \cdot 2H_2O$ in 40 mL of warm hydrochloric acid and diluting with 160 mL of water) to the solution to be tested, and immediately insert the bubbler of the mercury analysis apparatus. Obtain the absorbance reading by following the instrument manufacturer's operating instructions. Correct the sample readings for the reagent blank, and determine the mercury concentration of the *Sample Preparation* from a standard curve prepared by plotting the readings obtained with the *Standard Preparations* against mercury concentration, in mg/kg.

Oxalate Transfer 1 g of the sample into a 125-mL separator, dissolve in 10 mL of water, add 2 mL of hydrochloric acid, and extract successively with one 50-mL portion and one 20-mL portion of ether. Transfer the combined ether extracts to a 150-mL beaker, add 10 mL of water, and remove the ether by evaporation on a steam bath. Add 1 drop of glacial acetic acid and 1 mL of calcium acetate solution (1 in 20) to the residual aqueous solution. No turbidity is produced within 5 min.

Sulfate Dissolve a 100-mg sample of 1 mL in 2.7 N hydrochloric acid, and dilute to 30 to 40 mL with water. Proceed as directed in the *Sulfate Limit Test*, Appendix IIIB, beginning with the addition of 3 mL of barium chloride TS. Any turbidity produced does not exceed that shown in a control containing 300 μg of sulfate (SO_4).

Packaging and Storage Store in tight, light-resistant containers in a cool place.

Ferric Ammonium Citrate, Green

Iron Ammonium Citrate

INS: 381

DESCRIPTION

A complex salt of undetermined structure, composed of iron, ammonia, and citric acid and occurring as thin, transparent green scales, as granules, as a powder, or as transparent green crystals. It is odorless and has a mild iron-metallic taste. It is very soluble in water but is insoluble in alcohol. Its solutions are acid to litmus. It may deliquesce in air and is affected by light.

Functional Use in Foods Nutrient; dietary supplement; anticaking agent for sodium chloride.

REQUIREMENTS

Identification It responds to the *Identification Tests* in the monograph for *Ferric Ammonium Citrate, Brown*.

Assay Not less than 14.5% and not more than 16.0% of iron (Fe).
Ferric Citrate Passes test.
Lead Not more than 10 mg/kg.
Mercury Not more than 1 mg/kg.
Oxalate Passes test.
Sulfate Not more than 0.3%.

TESTS

Determine as directed for the respective TESTS in the monograph for *Ferric Ammonium Citrate, Brown*.

Packaging and Storage Store in tight, light-resistant containers in a cool place.

Ferric Phosphate

Iron Phosphate; Ferric Orthophosphate

$FePO_4 \cdot xH_2O$ Formula wt, anhydrous 150.82

CAS: [10045-86-0]

DESCRIPTION

Ferric Phosphate contains from one to four molecules of water of hydration. It occurs as an odorless, yellowish white- to buff-colored powder. It is insoluble in water and in acetic acid, but is soluble in mineral acids.

Functional Use in Foods Nutrient; dietary supplement.

REQUIREMENTS

Identification Dissolve 1 g in 5 mL of dilute hydrochloric acid (1 in 2), and add an excess of 1 N sodium hydroxide. A reddish brown precipitate forms. Boil the mixture, filter to remove the iron, and strongly acidify a portion of the filtrate with hydrochloric acid. Cool, mix with an equal volume of magnesia mixture TS, and treat with a slight excess of 6 N ammonium oxide. An abundant white precipitate forms. This precipitate, after being washed, turns greenish yellow when treated with a few drops of silver nitrate TS.
Assay Not less than 26.0% and not more than 32.0% of Fe.
Arsenic (as As) Not more than 3 mg/kg.
Fluoride Not more than 0.005%.
Lead Not more than 10 mg/kg.
Loss on Ignition Not more than 32.5%.
Mercury Not more than 3 mg/kg.

TESTS

Assay Dissolve about 3.5 g of the sample, accurately weighed, in 75 mL of dilute hydrochloric acid (1 in 2), heat to boiling, and boil for about 5 min. Cool, transfer into a 100-mL volumetric flask, dilute to volume with the dilute hydrochloric acid, and mix. To 25.0 mL of this solution add 100 mL of the dilute hydrochloric acid, boil again for 5 min, and to the boiling solution add stannous chloride TS, dropwise, with stirring, until the iron is just reduced as indicated by the disappearance of the yellow color. Add 2 drops in excess (but no more) of the stannous chloride TS, dilute with about 50 mL of water, and cool to room temperature. While stirring vigorously, add 15 mL of a saturated solution of mercuric chloride, and then allow to stand for 5 min. Add 15 mL of a sulfuric acid–phosphoric acid mixture, prepared by slowly adding 75 mL of sulfuric acid to 300 mL of water, cooling, adding 75 mL of phosphoric acid, and then diluting to 500 mL with water, and mix. Add 0.5 mL of barium diphenylamine sulfonate TS, and titrate with 0.1 N potassium dichromate to a reddish violet endpoint. Each mL of 0.1 N potassium dichromate is equivalent to 5.585 mg of Fe.
Arsenic Assemble the special distillation apparatus as shown in Fig. 13 of the general test in Appendix IIIB. Transfer 2 g of the sample, 50 mL of hydrochloric acid, and 5 g of cuprous chloride into the distilling flask (*B*). Reassemble the distillation apparatus and apply gentle suction to flask *F* to produce a continuous stream of bubbles. Heat the solution in flask *B* to boiling and distill until between 30 and 35 mL of distillate has been collected in flask *D*. Quantitatively transfer the distillate to a 100-mL volumetric flask with the aid of water, dilute to volume with water, and mix (*Sample Solution*). Prepare *Standard* and *Blank Solutions* in the same manner, using 6.0 mL of *Standard Arsenic Solution* (Appendix IIIB) in place of the sample in the *Standard Solution*, and 6.0 mL of water in the *Blank Solution*. Transfer 50.0 mL of the *Sample Solution* into the generator flask (Fig. 13, Appendix IIIB), add 2 mL of potassium iodide solution (15 in 100), and continue as directed in the *Procedure* under *Arsenic Test*, Appendix IIIB, beginning with "[add] 0.5 mL of *Stannous Chloride Solution*, and mix...." Modify the *Procedure* by using 5.0 g of Devarda's metal in place of the 3.0 g of 20-mesh granular zinc, and maintain the temperature of the reaction mixture in the generator flask between 25° and 27°. Treat 50.0 mL each of the *Standard Solution* and of the *Blank Solution* in the same manner and under the same conditions. Determine the absorbance at 525 nm produced by each solution as directed under *Procedure*. Calculate the arsenic content (in mg/kg) of the sample by the formula

$$3 \times (A_U - A_B)/(A_S - A_B),$$

in which A_U is the absorbance produced by the *Sample Solution*, A_S is the absorbance produced by the *Standard Solution*, and A_B is the absorbance produced by the *Blank Solution*.

Note: If A_B exceeds 0.300, different samples of reagent-grade cuprous chloride and Devarda's metal should be tested for arsenic content by the procedure described herein, and lots of these reagents should be selected that will give blank readings that do not exceed 0.300.

Fluoride Weigh accurately 1.0 g, and proceed as directed in the *Fluoride Limit Test*, Appendix IIIB.
Lead Proceed as directed under *Lead* in the monograph for *Ferrous Gluconate*.
Loss on Ignition Ignite at 800° for 1 h.

Mercury

Standard Preparations Dissolve 338.5 mg of mercuric chloride, $HgCl_2$, in about 200 mL of water in a 250-mL volumetric flask, add 14 mL of dilute sulfuric acid (1 in 2), dilute to volume with water, and mix. Pipet 10.0 mL of this solution into a 1000-mL volumetric flask containing about 800 mL of water and 56 mL of dilute sulfuric acid (1 in 2), dilute to volume with water, and mix. Pipet 10.0 mL of the second solution into a second 1000-mL volumetric flask containing 800 mL of water and 56 mL of dilute sulfuric acid (1 in 2), dilute to volume with water, and mix. Each mL of this diluted stock solution contains 0.1 µg of Hg. Pipet 1.25, 2.50, 5.00, 7.50, and 10.00 mL of the last solution (equivalent to 0.125, 0.250, 0.500, 0.750, and 1.00 µg of Hg, respectively) into each of five 150-mL beakers. To each add 25 mL of aqua regia, cover with watch glasses, heat just to boiling, simmer for about 5 min, and cool to room temperature. Transfer the solutions into separate 250-mL volumetric flasks, dilute to volume with water, and mix. Transfer a 50.0-mL aliquot from each solution into respective 150-mL beakers, and to each add 1.0 mL of dilute sulfuric acid (1 in 5) and 1.0 mL of a filtered solution of potassium permanganate (1 in 25). Heat the solutions just to boiling, simmer for about 5 min, and cool.

Sample Preparation Transfer 5.00 g of the sample into a 150-mL beaker, add 25 mL of aqua regia, cover with a watch glass, and allow to stand at room temperature for about 5 min. Heat just to boiling, allow to simmer for about 5 min, and cool. Transfer the solution into a 250-mL volumetric flask, dilute to volume with water, and mix.

Note: Disregard any undissolved material that may be present.

Transfer a 50.0-mL aliquot of this solution into a 150-mL beaker, and add 1.0 mL of dilute sulfuric acid (1 in 5) and 1.0 mL of a filtered solution of potassium permanganate (1 in 25). Heat the solution just to boiling, simmer for about 5 min, and cool. Prepare a reagent blank in the same manner.

Apparatus Use a *Mercury Detection Instrument* and *Aeration Apparatus* as described under the *Mercury Limit Test*, Appendix IIIB. For the purposes of the test described herein, the Techtron AA-1000 atomic absorption spectrophotometer, equipped with a 10-cm silica absorption cell (Beckman Part No. 75144, or equivalent) and coupled with a strip chart recorder (Varian Series A-25, or equivalent), is satisfactory.

Procedure Assemble the *Aerating Apparatus* as shown in the figure, *Aeration Apparatus for Mercury Limit Test*, under *Mercury Limit Test*, Appendix IIIB. Use magnesium perchlorate as the absorbent in the absorption cell (*e*), fill gas washing bottle *c* with 60 mL of water, and place stopcock *b* in the bypass position. Connect the assembly to the 10-cm absorption cell (analogous to *f* in the figure) of the spectrophotometer, and adjust the air or nitrogen flow rate so that, in the following procedure, maximum absorption and reproducibility are obtained without excessive foaming in the test solution. Obtain a baseline reading at 253.7 nm by following the manufacturer's instructions for operating the particular *Mercury Detection Instrument* in use. Using the Techtron AA-1000 spectrophotometer, the following conditions are suitable: *slit width*: 2 Å; *lamp current*: 3 mA; and *scale expansion*: × 1. With the Varian A-25 recorder, set the *chart speed* at 25 in./h and the *span* at 2 mV. Precondition the apparatus by an appropriate modification of the procedures described below for treatment of the test solutions.

Note: The fritted bubbler in gas washing bottle *c* should be kept immersed in water between determinations. After each determination, wash the bubbler with a stream of water.

Treat the blank, each of the *Standard Preparations*, and the *Sample Preparation* as follows: Transfer the solution to be tested to a 125-mL gas washing bottle, using a few drops of hydroxylamine hydrochloride solution (1 in 10) to remove any manganese hydroxide from the beaker. Dilute to about 55 mL with water, and add a magnetic stirring bar. Discharge the permanganate color by the dropwise addition of the hydroxylamine hydrochloride solution, swirling after each drop is added. Add 15.0 mL of 10% stannous chloride solution (prepared by dissolving 20 g of $SnCl_2.2H_2O$ in 40 mL of warm hydrochloric acid and diluting with 160 mL of water), and immediately connect the gas washing bottle to the *Aeration Assembly*. Switch on the magnetic stirrer, turn stopcock *b* from the bypass to the aerating position, and obtain the absorbance reading. Disconnect bottle *c* from the *Aerating Apparatus*, discard the solution just tested, wash bottle *c* with water, wash the fritted bubbler with water, and repeat the foregoing procedure with the remaining solutions. Correct the sample readings for the reagent blank, and determine the mercury concentration of the *Sample Preparation* from a standard curve prepared by plotting the readings obtained with the *Standard Preparations* against mercury concentration, in mg/kg, suitable adjustments being made for dilution factors.

Packaging and Storage Store in well-closed containers.

Ferric Pyrophosphate

Iron Pyrophosphate

$Fe_4(P_2O_7)_3 \cdot xH_2O$　　　　　　Formula wt, anhydrous 745.22

CAS: [10058-44-3]

DESCRIPTION

A tan or yellowish white, odorless powder. It is insoluble in water but is soluble in mineral acids.

Functional Use in Foods Nutrient; dietary supplement.

REQUIREMENTS

Identification Dissolve 500 mg in 5 mL of dilute hydrochloric acid (1 in 2), and add an excess of 1 *N* sodium hydroxide. A reddish brown precipitate forms. Allow the solution to stand for several min, and then filter, discarding the first few mL. To 5 mL of the clear filtrate add 1 drop of bromophenol blue TS, and titrate with 1 N hydrochloric acid to a green color. Add 10 mL of a 1 in 8 solution of zinc sulfate, and readjust the pH to 3.8 (green color). A white precipitate forms (distinction from *orthophosphates*).

Assay   Not less than 24.0% and not more than 26.0% of Fe.
Arsenic (as As)   Not more than 3 mg/kg.
Lead   Not more than 10 mg/kg.
Loss on Ignition   Not more than 20.0%.
Mercury   Not more than 3 mg/kg.

TESTS

Assay   Proceed as directed under *Assay* in the monograph for *Ferric Phosphate*.
Arsenic   Prepare and test a 2-g sample as directed in the test for *Arsenic* in the monograph for *Ferric Phosphate*.
Lead   Proceed as directed under *Lead* in the monograph for *Ferrous Gluconate*.
Loss on Ignition   Ignite at 800° for 1 h.
Mercury   Determine as directed under *Mercury* in the monograph for *Ferric Phosphate*.

Packaging and Storage   Store in well-closed containers.

Ferrous Fumarate

$C_4H_2FeO_4$ Formula wt 169.90

                                              CAS: [141-01-5]

DESCRIPTION

An odorless, reddish orange to red-brown powder. It may contain soft lumps that produce a yellow streak when crushed.

Functional Use in Foods   Nutrient; dietary supplement.

REQUIREMENTS

Identification

A. To about 1.5 g add 25 mL of dilute hydrochloric acid (1 in 2), and dilute to 50 mL with water. Heat to effect complete solution, then cool, filter on a fine-porosity sintered-glass crucible, wash the precipitate with dilute hydrochloric acid (3 in 100), saving the filtrate for *Identification Test B*, and dry the precipitate at 105°. To 400 mg of the dried precipitate add 3 mL of water and 7 mL of 1 N sodium hydroxide, and stir until solution is complete. Add 2.7 N hydrochloric acid, dropwise, until the solution is just acid to litmus, add 1 g of *p*-nitrobenzyl bromide and 10 mL of alcohol, and reflux the mixture for 2 h. Cool, filter, and wash the precipitate with two small portions of a mixture of 2 parts of alcohol and 1 part of water, followed by two small portions of water. The precipitate, recrystallized from hot alcohol and dried at 105°, melts at about 152° (see Appendix IIB).

B. A portion of the filtrate obtained in the preceding test gives positive tests for *Iron*, Appendix IIIA.

Assay   Not less than 97.0% and not more than 101.0% of $C_4H_2FeO_4$, calculated on the dried basis.
Ferric Iron   Not more than 2.0%.
Lead   Not more than 10 mg/kg.
Loss on Drying   Not more than 1.5%.
Mercury   Not more than 3 mg/kg.
Sulfate   Not more than 0.2%.

TESTS

Assay   Transfer about 500 mg, accurately weighed, into a 500-mL Erlenmeyer flask, add 25 mL of dilute hydrochloric acid (2 in 5), and heat to boiling. Add, dropwise, a solution of 5.6 g of stannous chloride in 50 mL of dilute hydrochloric acid (3 in 10) until the yellow color disappears, and then add 2 drops in excess. Cool the solution in an ice bath to room temperature, add 8 mL of mercuric chloride TS, and allow to stand for 5 min. Add 200 mL of water, 25 mL of dilute sulfuric acid (1 in 2), and 4 mL of phosphoric acid, then add orthophenanthroline TS, and titrate with 0.1 N ceric sulfate. Each mL of 0.1 N ceric sulfate is equivalent to 16.99 mg of $C_4H_2FeO_4$.
Ferric Iron   Transfer 2 g of the sample into a 250-mL glass-stoppered Erlenmeyer flask, add 25 mL of water and 4 mL of hydrochloric acid, and heat on a hot plate until solution is complete. Stopper the flask, and cool to room temperature. Add 3 g of potassium iodide, stopper, swirl to mix, and allow to stand in the dark for 5 min. Remove the stopper, add 75 mL of water, and titrate with 0.1 N sodium thiosulfate, adding starch TS near the endpoint. Not more than 7.16 mL of 0.1 N sodium thiosulfate is consumed.
Lead   (**Note:** For the preparation of all aqueous solutions and for the rinsing of glassware before use, employ water that has been passed through a strong-acid, strong-base, mixed-bed ion-exchange resin before use. Select all reagents to have as low a content of lead as practicable, and store all reagent solutions in containers of borosilicate glass. Clean glassware before use by soaking in warm 8 N nitric acid for 30 min and by rinsing with deionized water.)

Ascorbic Acid–Sodium Iodide Solution   Dissolve 20 g of ascorbic acid and 38.5 g of sodium iodide in water in a 200-mL volumetric flask, dilute with water to volume, and mix.

Trioctylphosphine Oxide Solution   (**Caution:** This solution causes irritation. Avoid contact with eyes, skin, and clothing. Take special precautions in disposing of unused portions of solutions to which this reagent is added.) Dissolve 5.0 g of trioctylphosphine oxide in 4-methyl-2-pentanone in a 100-mL volumetric flask, dilute with the same solvent to volume, and mix.

Standard Preparation and Blank   Transfer 5.0 mL of *Lead Nitrate Stock Solution*, prepared as directed in the test for *Heavy Metals*, Appendix IIIB, to a 100-mL volumetric flask, dilute with water to volume, and mix. Transfer 2.0 mL of the resulting solution to a 50-mL beaker. To this beaker and to a second, empty, beaker (*Blank*) add 6 mL of nitric acid and 10 mL of perchloric acid, and evaporate in a hood to dryness.

Caution: Use perchloric acid in a well-ventilated fume hood with proper precautions.

Cool, dissolve the residues in 10 mL of 9 *N* hydrochloric acid, and transfer with the aid of about 10 mL of water to separate 50-mL volumetric flasks. To each flask add 20 mL of *Ascorbic Acid–Sodium Iodide Solution* and 5.0 mL of *Trioctylphosphine Oxide Solution*, shake for 30 s, and allow to separate. Add water to bring the organic solvent layer into the neck of each flask, shake again, and allow to separate. The organic solvent layers are the *Blank* and the *Standard Preparation*, and they contain 0.0 and 2.0 μg of lead per mL, respectively.

Test Preparation   Add 1.0 g of Ferrous Fumarate to a 50-mL beaker, and add 6 mL of nitric acid and 10 mL of perchloric acid.

Caution: Use perchloric acid in a well-ventilated fume hood with proper precautions.

Cover with a ribbed watch glass, and heat in a hood until completely dry. Cool, dissolve the residue in 10 mL of 9 *N* hydrochloric acid, and transfer with the aid of about 10 mL of water to a 50-mL volumetric flask. Add 20 mL of *Ascorbic Acid–Sodium Iodide Solution* and 5.0 mL of *Trioctylphosphine Oxide Solution*, shake for 30 s, and allow to separate. Add water to bring the organic solvent layer into the neck of the flask, shake again, and allow to separate. The organic solvent layer is the *Test Preparation*.

Procedure   Concomitantly determine the absorbances of the *Blank*, the *Standard Preparation*, and the *Test Preparation* at the lead emission line at 283.3 nm with a suitable atomic absorption spectrophotometer equipped with a lead hollow-cathode lamp and an air–acetylene flame, using 4-methyl-2-pentanone to set the instrument to zero. In a suitable analysis, the absorbance of the *Blank* is not greater than 20% of the difference between the absorbance of the *Standard Preparation* and the absorbance of the *Blank*. The absorbance of the *Test Preparation* does not exceed that of the *Standard Preparation*.

Loss on Drying, Appendix IIC   Dry at 105° for 16 h.
Mercury

Dithizone Extraction Solution   Dissolve 30 mg of dithizone in 1000 mL of chloroform, add 5 mL of alcohol, and mix. Store in a refrigerator. Before use, shake a suitable volume with about half its volume of dilute nitric acid (1 in 100), discarding the nitric acid. Do not use if more than a month old.

Diluted Dithizone Extraction Solution   Just prior to use, dilute 5 mL of *Dithizone Extraction Solution* with 25 mL of chloroform.

Hydroxylamine Hydrochloride Solution   Dissolve 20 g of hydroxylamine hydrochloride in sufficient water to make about 65 mL, transfer the solution into a separator, add a few drops of thymol blue TS, and then add ammonium hydroxide until the solution assumes a yellow color. Add 10 mL of sodium diethyldithiocarbamate solution (1 in 25), mix, and allow to stand for 5 min. Extract the solution with successive 10- to 15-mL portions of chloroform until a 5-mL test portion of the chloroform extract does not assume a yellow color when shaken with a dilute cupric sulfate solution. Add 2.7 *N* hydrochloric acid until the extracted solution is pink, adding 1 or 2 drops more of thymol blue TS, if necessary, then dilute to 100 mL with water, and mix.

Mercury Stock Solution   Transfer 135.4 mg, accurately weighed, of mercuric chloride into a 100-mL volumetric flask, dissolve in 1 *N* sulfuric acid, dilute to volume with the acid, and mix. Dilute 5.0 mL of this solution to 500.0 mL with 1 *N* sulfuric acid. Each mL contains the equivalent of 10 μg of Hg.

Diluted Standard Mercury Solution   On the day of use, transfer 10.0 mL of *Mercury Stock Solution* into a 100-mL volumetric flask, dilute to volume with 1 *N* sulfuric acid, and mix. Each mL contains the equivalent of 1 μg of Hg.

Sodium Citrate Solution   Dissolve 250 g of sodium citrate dihydrate in 1000 mL of water.

Sample Solution   Dissolve 1 g of the sample in 30 mL of 1.7 *N* nitric acid by heating on a steam bath. Cool to room temperature in an ice bath, stir, and filter through S and S No. 589, or equivalent, filter paper that has been previously washed with 1.7 *N* nitric acid, followed by water. To the filtrate add 20 mL of *Sodium Citrate Solution* and 1 mL of *Hydroxylamine Hydrochloride Solution*.

Procedure   (**Note:** Because mercuric dithizonate is light sensitive, this procedure should be performed in subdued light.) Prepare a control containing 3.0 mL of *Diluted Standard Mercury Solution* (3 μg Hg), 30 mL of 1.7 *N* nitric acid, 5 mL of *Sodium Citrate Solution*, and 1 mL of *Hydroxylamine Hydrochloride Solution*. Treat the control and the *Sample Solution* as follows: Using a pH meter, adjust the pH of each solution to 1.8 with ammonium hydroxide and transfer the solutions into different separators. Extract with two 5-mL portions of *Dithizone Extraction Solution*, and then extract again with 5 mL of chloroform, discarding the aqueous solutions. Transfer the combined extracts from each separator into different separators, add 10 mL of dilute hydrochloric acid (1 in 2) to each, shake well, and discard the chloroform layers. Extract the acid solutions with about 3 mL of chloroform, shake well, and discard the chloroform layers. Add to each separator 0.1 mL of 0.05 *M* disodium EDTA and 2 mL of 6 *N* acetic acid, mix, and then slowly add 5 mL of ammonium hydroxide. Stopper the separators, and cool under the stream of cold water. Dry the outside of the separators, pour the solutions (through the top of the separators) carefully, to avoid loss, into separate beakers, and using a pH meter, adjust the pH of both solutions to 1.8 with 6 *N* ammonium hydroxide. Return the sample and control solutions to their original separators, add 5.0 mL of *Diluted Dithizone Extraction Solution*, and shake vigorously. Any color developed in the *Sample Solution* does not exceed that in the control.

Sulfate   Mix 1 g of the sample with 100 mL of water in a 250-mL beaker, and heat on a steam bath, adding hydrochloric acid, dropwise, until complete solution is effected (about 2 mL of the acid will be required). Filter the solution, if necessary, and dilute the clear solution or filtrate to 100 mL with water. Heat to boiling, add 10 mL of barium chloride TS, warm on a steam bath for 2 h, cover, and allow to stand overnight. If crystals of ferrous fumarate form, warm on a steam bath to dissolve them, then filter through paper, wash the residue with hot water, and transfer the paper containing the residue to a tared crucible. Char the paper, without burning, and ignite the crucible and its contents at 600° to constant weight. Each mg of the residue is equivalent to 0.412 mg (412 µg) of SO$_4$.

Packaging and Storage   Store in well-closed containers.

Ferrous Gluconate

HO-CH(OH)-CH(OH)-CH(OH)-CH(OH)-C(=O)-O-Fe-O-C(=O)-CH(OH)-CH(OH)-CH(OH)-CH(OH)-CH$_2$OH · 2H$_2$O

C$_{12}$H$_{22}$FeO$_{14}$·2H$_2$O   Formula wt 482.18
INS: 579   CAS: [299-29-6]

DESCRIPTION

Fine, yellowish gray or pale greenish yellow powder or granules having a slight odor resembling that of burned sugar. One g dissolves in about 10 mL of water with slight heating. It is practically insoluble in alcohol. A 1 in 20 solution is acid to litmus.

Functional Use in Foods   Nutrient; dietary supplement; color adjunct.

REQUIREMENTS

Identification
A. It meets the requirements of *Identification Test B* in the monograph for *Calcium Gluconate*.
B. A 1 in 20 solution gives positive tests for *Ferrous Salts* (Iron), Appendix IIIA.
Assay   Not less than 97.0% and not more than 102.0% of C$_{12}$H$_{22}$FeO$_{14}$, calculated on the dried basis.
Chloride   Not more than 0.07%.
Ferric Iron   Not more than 2.0%.
Lead   Not more than 10 mg/kg.
Loss on Drying   Between 6.5% and 10.0%.
Mercury   Not more than 3 mg/kg.
Oxalic Acid   Passes test.
Reducing Sugars   Passes test.
Sulfate   Not more than 0.1%.

TESTS

Assay   Dissolve about 1.5 g, accurately weighed, in a mixture of 75 mL of water and 15 mL of 2 N sulfuric acid in a 300-mL Erlenmeyer flask, and add 250 mg of zinc dust. Close the flask with a stopper containing a Bunsen valve, allow to stand at room temperature for 20 min, then filter through a sintered-glass filter crucible containing a thin layer of zinc dust, and wash the crucible and contents with 10 mL of 2 N sulfuric acid, followed by 10 mL of water. Add orthophenanthroline TS, and titrate the filtrate in the suction flask immediately with 0.1 N ceric sulfate. Perform a blank determination (see *General Provisions*) and make any necessary correction. Each mL of 0.1 N ceric sulfate is equivalent to 44.62 mg of C$_{12}$H$_{22}$FeO$_{14}$.
Chloride, Appendix IIIA   Dissolve 1 g in 100 mL of water. Any turbidity produced by a 10-mL portion of this solution does not exceed that shown in a control containing 70 µg of chloride ion (Cl).
Ferric Iron   Dissolve about 5 g, accurately weighed, in a mixture of 100 mL of water and 10 mL of hydrochloric acid in a 250-mL glass-stoppered flask, add 3 g of potassium iodide, shake well, and allow to stand in the dark for 5 min. Titrate any liberated iodine with 0.1 N sodium thiosulfate, using starch TS as the indicator. Each mL of 0.1 N sodium thiosulfate is equivalent to 5.585 mg of ferric iron.
Lead   (**Note**: For the preparation of all aqueous solutions and for the rinsing of glassware before use, employ water that has been passed through a strong-acid, strong-base, mixed-bed ion-exchange resin before use. Select all reagents to have as low a content of lead as practicable, and store all reagent solutions in containers of borosilicate glass. Clean glassware before use by soaking in warm 8 N nitric acid for 30 min and by rinsing with deionized water.)

Ascorbic Acid–Sodium Iodide Solution   Dissolve 20 g of ascorbic acid and 38.5 g of sodium iodide in water in a 200-mL volumetric flask, dilute with water to volume, and mix.

Trioctylphosphine Oxide Solution   (**Caution**: This solution causes irritation. Avoid contact with eyes, skin, and clothing. Take special precautions in disposing of unused portions of solutions to which this reagent is added.) Dissolve 5.0 g of trioctylphosphine oxide in 4-methyl-2-pentanone in a 100-mL volumetric flask, dilute with the same solvent to volume, and mix.

Standard Preparation and Blank   Transfer 5.0 mL of *Lead Nitrate Stock Solution*, prepared as directed in the test for *Heavy Metals*, Appendix IIIB, to a 100-mL volumetric flask, dilute with water to volume, and mix. Transfer 2.0 mL of the resulting solution to a 50-mL volumetric flask. To this volumetric flask and to a second, empty 50-mL volumetric flask (*Blank*) add 10 mL of 9 N hydrochloric acid and about 10 mL of water. To each flask add 20 mL of *Ascorbic Acid–Sodium Iodide Solution* and 5.0 mL of *Trioctylphosphine Oxide Solution*, shake for 30 s, and allow to separate. Add water to bring the organic solvent layer into the neck of each flask, shake again, and allow to separate. The organic solvent layers are the *Blank* and the *Standard Preparation*, and they contain 0.0 and 2.0 µg of lead per mL, respectively.

Test Preparation   Add 1.0 g of the sample, 10 mL of 9 N hydrochloric acid, about 10 mL of water, 20 mL of *Ascorbic Acid–Sodium Iodide Solution*, and 5.0 mL of *Trioctylphosphine Oxide Solution* to a 50-mL volumetric flask, shake for 30 s, and allow to separate. Add water to bring the organic solvent layer into the neck of the flask, shake again and allow to separate. The organic solvent layer is the *Test Preparation*.

Procedure   Concomitantly determine the absorbance of the *Blank*, the *Standard Preparation*, and the *Test Preparation* at the lead emission line at 283.3 nm, with a suitable atomic absorption spectrophotometer equipped with a lead hollow-cathode lamp and an air–acetylene flame, using 4-methyl-2-

pentanone to set the instrument to zero. In a suitable analysis, the absorbance of the *Blank* is not greater than 20% of the difference between the absorbance of the *Standard Preparation* and the absorbance of the *Blank*: The absorbance of the *Test Preparation* does not exceed that of the *Standard Preparation*.

Loss on Drying, Appendix IIC   Dry at 105° for 16 h.

Mercury   Determine as directed in the test for *Mercury* in the monograph for *Ferrous Fumarate*.

Oxalic Acid   Dissolve 1 g in 10 mL of water, add 2 mL of hydrochloric acid, transfer to a separator, and extract successively with 50 and 20 mL of ether. Combine the ether extracts, add 10 mL of water, and evaporate the ether on a steam bath. Add 1 drop of acetic acid (36%) and 1 mL of calcium acetate solution (1 in 20). No turbidity is produced within 5 min.

Reducing Sugars   Dissolve 500 mg in 10 mL of water, warm, and make the solution alkaline with 1 mL of 6 N ammonium hydroxide. Pass hydrogen sulfide gas into the solution to precipitate the iron, and allow the mixture to stand for 30 min to coagulate the precipitate. Filter, and wash the precipitate with two successive 5-mL portions of water. Acidify the combined filtrate and washings with hydrochloric acid, and add 2 mL of 2.7 N hydrochloric acid in excess. Boil the solution until the vapors no longer darken lead acetate paper, and continue to boil, if necessary, until it has been concentrated to about 10 mL. Cool, add 5 mL of sodium carbonate TS and 20 mL of water, filter, and adjust the volume of the filtrate to 100 mL. To 5 mL of the filtrate add 2 mL of alkaline cupric tartrate TS, and boil for 1 min. No red precipitate is formed within 1 min.

Sulfate, Appendix IIIB   Any turbidity produced by a 200-mg sample does not exceed that shown in a control containing 200 μg of sulfate (SO_4).

Packaging and Storage   Store in tight containers.

Ferrous Lactate

Iron (II) Lactate; Iron (II) 2-Hydroxypropionate

$[CH_3CH(OH)COO]_2Fe$

$C_6H_{10}FeO_6 \cdot xH_2O$ Formula wt, anhydrous 233.99

INS: 585   CAS: [5905-52-2]

DESCRIPTION

Greenish-white powder or crystals, having a characteristic odor. The levo enantiomer occurs as the dihydrate and the racemic mixture occurs as the trihydrate. It is sparingly soluble in water and practically insoluble in ethanol. A 1 in 50 solution has a pH between 5 and 6.

Functional Use in Foods   Nutrient; dietary supplement.

REQUIREMENTS

Labeling   Indicate the state of hydration.

Identification   A 1 in 50 solution of the sample gives positive tests for *Lactate* and for *Iron (Ferrous Salts)* as described in Appendix IIIA.

Assay   Not less than 97.0% and not more than 100.5% of $C_6H_{10}FeO_6$, calculated on the anhydrous basis.

Chloride   Not more than 0.1%.

Ferric Iron   Not more than 0.2%.

Lead   Not more than 1 mg/kg.

Oxalic Acid   Passes test.

Specific Rotation   For the dihydrate: $[\alpha]_D^{20°}$: Between +6.0° and +11.0°, calculated on the anhydrous basis.

Sulfate   Not more than 0.1%.

Water   Between 12.0% and 14.0% for the dihydrate and between 18.0% and 20.0% for the trihydrate.

TESTS

Assay   Dissolve about 800 mg, accurately weighed, of the sample in a mixture of 150 mL of water and 10 mL of sulfuric acid in a 300-mL flask. Add 5 mL of phosphoric acid and cool to room temperature if necessary. Add 1 drop of orthophenanthroline TS, and immediately titrate with 0.1 N ceric sulfate. Perform a blank determination and make any necessary correction. Each mL of 0.1 N ceric sulfate is equivalent to 23.40 mg of $C_6H_{10}FeO_6$.

Chloride, Appendix IIIA   Dissolve 1 g of the sample, accurately weighed, in 100 mL of water. Any turbidity produced by a 10-mL portion of this solution does not exceed that shown in a control containing 100 μg of chloride ion (Cl). (Save the remaining sample solution for the *Sulfate Test*.)

Ferric Iron   Dissolve about 5 g, accurately weighed, in a mixture of 100 mL of water and 10 mL of hydrochloric acid in a 250-mL glass-stoppered flask, add 3 g of potassium iodide, shake well, and allow to stand in the dark for 5 min. Titrate any liberated iodine with 0.1 N sodium thiosulfate, using starch TS as the indicator. Each mL of 0.1 N sodium thiosulfate is equivalent to 5.585 mg of ferric iron.

Lead   (**Note**: For the preparation of all aqueous solutions and for the rinsing of glassware before use, employ water that has been passed through a strong-acid, strong-base, mixed-bed ion-exchange resin before use. Select all reagents to have as low a content of lead as practicable, and store all reagent solutions in containers of borosilicate glass. Clean glassware before use by soaking in warm 8 N nitric acid for 30 min and by rinsing with deionized water.)

Standard Solutions   Prepare all lead solutions in 0.1% nitric acid. Use a single-element 1000 μg/mL lead stock solution to prepare (weekly) an intermediate stock solution (1 μg/mL). Prepare (daily) a *Lead Standard Solution* (10 ng/mL) by diluting the intermediate stock solution 1:100 with 0.1 N nitric acid.

Modifier Stock Solution   Weigh an amount of palladium nitrate equivalent to 1 g of palladium, and dilute to 100 mL with 15% nitric acid. Just before use, prepare a *Modifier Working Solution* by diluting the stock solution 1:10 with water.

Blank Solution   Use 0.1% nitric acid.

Sample Preparation Dissolve an accurately weighed 2-g sample in 5 mL of water and 10 mL of 10% nitric acid. Dilute to 100.0 mL with water.

Procedure Use the following furnace program with a suitable graphite furnace atomic absorption spectrophotometer set at 283.3 nm and equipped with an autosampler:

Temperature (°C)	Time (s)	Gas flow (argon) (L/min)
85	5.0	3.0
95	40.0	3.0
120	10.0	3.0
300	30.0	3.0
900	5.0	3.0
900	1.0	3.0
900	2.0	0.0
2100	0.6	0.0
2100	2.0	0.0
2800	3.0	3.0

Separately inject the following solution mixtures: Solution A: 30 µL of the *Blank Solution* and 5 µL of the *Modifier Working Solution*; Solution B: 10 µL of the *Standard Lead Solution*, 10 µL of the *Sample Preparation*, 10 µL of the *Blank Solution*, and 5 µL of the *Modifier Working Solution*; Solution C: 20 µL of the *Standard Lead Solution*, 10 µL of the *Sample Preparation*, and 5 µL of the *Modifier Working Solution*; Solution D: 10 µL of the *Sample Preparation*, 20 µL of the *Blank Solution*, and 5 µL of the *Modifier Working Solution*. Calculate the blank-corrected absorbances of Solutions B, C, and D by subtracting the absorbance measured for Solution A. Plot the blank-corrected absorbances of Solutions B, C, and D (y-axis) versus the quantity of lead, in ng, added to each solution (x-axis). These are equal to 0.1, 0.2, and 0 ng, respectively. Draw the best straight line through the points. Extrapolate the line to the x-axis intercept to obtain the quantity C, in ng, of lead in 10 µL of the *Sample Solution*. Calculate the concentration, in mg/kg, of lead in the sample taken by the formula

$$10C/W$$

in which C is defined above and W is the weight, in g, of the sample taken.

Oxalic Acid Dissolve 1 g of the sample in 10 mL of water and 2 mL of hydrochloric acid. Transfer to a separator and extract with two 35-mL portions of ether. Evaporate the combined ether extracts on a rotary evaporator or on a steam bath. Dissolve any residue in 10 mL of water, add 1 mL of glacial acetic acid and add 1 mL of a 1 in 20 calcium acetate solution. No turbidity is produced in 5 min.

Specific Rotation, Appendix IIB Determine in a solution containing 2 g in 100 mL of oxygen-free water.

Sulfate, Appendix IIIA Any turbidity produced by a 20-mL portion of the solution prepared for the *Chloride Test* does not exceed that shown in a control containing 200 µg of sulfate (SO_4).

Water Determine by the *Karl Fischer Titrimetric Method*, Appendix IIB, at 50° using 100 mg of the sample dissolved in a freshly prepared mixture of 20 mL of methanol and 20 mL of formamide.

Packaging and Storage Store in tight containers.

Ferrous Sulfate

$FeSO_4 \cdot 7H_2O$ Formula wt 278.02

CAS: [7720-78-7]

DESCRIPTION

Pale, bluish green crystals or granules. It is odorless; has a saline, styptic taste; and is efflorescent in dry air. In moist air it oxidizes readily to form brownish yellow basic ferric sulfate. Its 1 in 10 solution is acid to litmus, having a pH of about 3.7. One g dissolves in 1.5 mL of water at 25° and in 0.5 mL of boiling water. It is insoluble in alcohol.

Functional Use in Foods Nutrient; dietary supplement.

REQUIREMENTS

Identification It gives positive tests for *Ferrous Salts* (Iron) and for *Sulfate*, Appendix IIIA.
Assay Not less than 99.5% and not more than 104.5% of $FeSO_4 \cdot 7H_2O$.
Lead Not more than 10 mg/kg.
Mercury Not more than 3 mg/kg.

TESTS

Assay Dissolve about 1 g, accurately weighed, in a mixture of 25 mL of 2 N sulfuric acid and 25 mL of recently boiled and cooled water, add orthophenanthroline TS, and immediately titrate with 0.1 N ceric sulfate. Perform a blank determination (see *General Provisions*), and make any necessary correction. Each mL of 0.1 N ceric sulfate is equivalent to 27.80 mg of $FeSO_4 \cdot 7H_2O$.
Lead Determine as directed under *Lead* in the monograph for *Ferric Gluconate*.
Mercury Determine as directed under *Mercury* in the monograph for *Ferrous Fumarate*.

Packaging and Storage Store in tight containers.

Ferrous Sulfate, Dried

$FeSO_4 \cdot xH_2O$ Formula wt, anhydrous 151.91

DESCRIPTION

A grayish white- to buff-colored powder consisting primarily of $FeSO_4 \cdot H_2O$, with varying amounts of $FeSO_4 \cdot 4H_2O$. It dissolves slowly in water but is insoluble in alcohol.

Functional Use in Foods Nutrient; dietary supplement.

REQUIREMENTS

Identification It gives positive tests for *Ferrous Salts* (Iron) and for *Sulfate*, Appendix IIIA.
Assay Not less than 86.0% and not more than 89.0% of $FeSO_4$.
Insoluble Substances Not more than 0.05%.
Lead Not more than 10 mg/kg.
Mercury Not more than 3 mg/kg.

TESTS

Assay Proceed as directed under *Assay* in the monograph for *Ferrous Sulfate*. Each mL of 0.1 N ceric sulfate is equivalent to 15.19 mg of $FeSO_4$.
Insoluble Residue Dissolve 2 g in 20 mL of freshly boiled dilute sulfuric acid (1 in 100), heat to boiling, and then digest in a covered beaker on a steam bath for 1 h. Filter through a tared filtering crucible, wash thoroughly, and dry at 105°. The weight of the insoluble residue does not exceed 1 mg.
Lead Determine as directed under *Lead* in the monograph for *Ferrous Gluconate*.
Mercury Determine as directed under *Mercury* in the monograph for *Ferrous Fumarate*.

Packaging and Storage Store in tight containers.

Fir Needle Oil, Canadian Type

Balsam Fir Oil

DESCRIPTION

The volatile oil obtained by steam distillation from needles and twigs of *Abies balsamea* L., Mill (Fam. *Pinaceae*). It is a colorless to faintly yellow liquid having a pleasant balsamlike odor. It is soluble in most fixed oils and in mineral oil. It is slightly soluble in propylene glycol, but it is insoluble in glycerin.

Functional Use in Foods Flavoring agent.

REQUIREMENTS

Identification The infrared absorption spectrum of the sample exhibits relative maxima (that may vary in intensity) at the same wavelengths (or frequencies) as those shown in the respective spectrum in the section on *Infrared Spectra (Series A: Essential Oils)*, using the same test conditions as specified therein.
Assay Not less than 8.0% and not more than 16.0% of esters, calculated as bornyl acetate ($C_{12}H_{20}O_2$).
Angular Rotation Between −19° and −24°.
Heavy Metals (as Pb) Passes test.
Refractive Index Between 1.473 and 1.476 at 20°.
Solubility in Alcohol Passes test.
Specific Gravity Between 0.872 and 0.878.

TESTS

Assay Weigh accurately about 5 g, and proceed as directed under *Ester Determination*, Appendix VI, using 98.15 as the equivalence factor (e) in the calculation.
Angular Rotation Determine in a 100-mm tube as directed under *Optical (Specific) Rotation*, Appendix IIB.
Heavy Metals Shake 10 mL of the oil with an equal volume of water to which 1 drop of hydrochloric acid has been added, and pass hydrogen sulfide through the mixture until it is saturated. No darkening in color is produced in either the oil or the water.
Refractive Index, Appendix IIB Determine with an Abbé or other refractometer of equal or greater accuracy.
Solubility in Alcohol Proceed as directed in the general method, Appendix VI. One mL dissolves in 4 mL of 90% alcohol, occasionally with haziness.
Specific Gravity Determine by any reliable method (see *General Provisions*).

Packaging and Storage Store in full, tight, preferably glass, aluminum, tin-lined, or other suitable containers in a cool place protected from light.

Fir Needle Oil, Siberian Type

Pine Needle Oil

DESCRIPTION

The volatile oil obtained by steam distillation from needles and twigs of *Abies sibirica* Lebed. (Fam. *Pinaceae*). It is an almost colorless or faintly yellow liquid. It is soluble in most fixed oils and in mineral oil. It is insoluble in glycerin and in propylene glycol.

Functional Use in Foods Flavoring agent.

REQUIREMENTS

Identification The infrared absorption spectrum of the sample exhibits relative maxima (that may vary in intensity) at the same wavelengths (or frequencies) as those shown in the respective spectrum in the section on *Infrared Spectra (Series A: Essential Oils)*, using the same test conditions as specified therein.

Assay Not less than 32.0% and not more than 44.0% of esters, calculated as bornyl acetate ($C_{12}H_{20}O_2$).
Angular Rotation Between −33° and −45°.
Heavy Metals (as Pb) Passes test.
Refractive Index Between 1.468 and 1.473 at 20°.
Solubility in Alcohol Passes test.
Specific Gravity Between 0.898 and 0.912.

TESTS

Assay Weigh accurately about 2 g, and proceed as directed under *Ester Determination*, Appendix VI, using 98.15 as the equivalence factor (*e*) in the calculation.
Angular Rotation Determine in a 100-mm tube as directed under *Optical (Specific) Rotation*, Appendix IIB.
Heavy Metals Shake 10 mL of the oil with an equal volume of water to which 1 drop of hydrochloric acid has been added, and pass hydrogen sulfide through the mixture until it is saturated. No darkening in color is produced in either the oil or the water.
Refractive Index, Appendix IIB Determine with an Abbé or other refractometer of equal or greater accuracy.
Solubility in Alcohol Proceed as directed in the general method, Appendix VI. One mL dissolves in 1 mL of 90% alcohol. Occasionally the solution may become hazy upon further dilution.
Specific Gravity Determine by any reliable method (see *General Provisions*).

Packaging and Storage Store in full, tight, preferably glass, aluminum, or tin-lined containers in a cool place protected from light.

Folic Acid

N-[4-[[(2-Amino-1,4-dihydro-4-oxo-6-pteridinyl)methyl]-amino]benzoyl]-L-glutamic Acid; *N*-[*p*-[[(2-Amino-4-hydroxy-6-pteridinyl)methyl]amino]benzoyl]glutamic Acid; Pteroylglutamic Acid; Folacin

$C_{19}H_{19}N_7O_6$ Formula wt 441.40

CAS: [59-30-3]

DESCRIPTION

Yellow or yellowish orange, odorless crystals or crystalline powder. About 1.6 mg dissolves in 1 mL of water. It is insoluble in acetone, in alcohol, in chloroform, and in ether, but dissolves in solutions of alkali hydroxides and carbonates. The pH of a suspension of 1 g in 10 mL of water is between about 4.0 and 4.8.

Functional Use in Foods Nutrient; dietary supplement.

REQUIREMENTS

Identification The ultraviolet absorption spectrum of a 1 in 100,000 solution of the sample in sodium hydroxide solution (1 in 250) exhibits maxima and minima at the same wavelengths as that of a similar solution of USP Folic Acid Reference Standard, concomitantly measured. The ratio A_{256}/A_{365} is between 2.80 and 3.00.
Assay Not less than 95.0% and not more than 102.0% of $C_{19}H_{19}N_7O_6$, calculated on the anhydrous basis.
Residue on Ignition Not more than 0.3%.
Water Not more than 8.5%.

TESTS

Assay

Standard Solution Weigh accurately about 30 mg of USP Folic Acid Reference Standard, corrected for water content, and dissolve in an aqueous solvent containing 2 mL of ammonium hydroxide and 1 g of sodium perchlorate per 100 mL. Using the same solvent, adjust the volume quantitatively, according to the injection size to be used in the *Procedure*, so that between 5 and 20 μg of Folic Acid is chromatographed.

Sample Solution Prepare as directed for the *Standard Solution*, using an accurately weighed quantity of the sample, and adjust to the same volume as the *Standard Solution*.

Mobile Phase Transfer 35.1 g of sodium perchlorate, 1.40 g of monobasic potassium phosphate, 7.0 mL of 1 *N* potassium hydroxide, and 40 mL of methanol to a 1000-mL volumetric flask, dilute with water to volume, and mix. Adjust the pH to 7.2 with 1 *N* potassium hydroxide.

Note: The methanol concentration may be varied to meet system suitability requirements and to provide a suitable elution time for Folic Acid.

Chromatographic System Typically, a high-pressure liquid chromatograph, operated at room temperature, is fitted with a 25- to 30-cm × 4-mm stainless steel column packed with octadecyl silane chemically bonded to porous silica or ceramic microparticles 5 to 10 μm in diameter. The mobile phase is maintained at a pressure and flow rate capable of giving the required resolution (see *System Suitability Test*) and a suitable elution time. An ultraviolet detector that monitors absorption at 254 nm is used.

System Suitability Solution Prepare a solution containing about 1 mg/mL each of USP Folic Acid Reference Standard and USP Calcium Formyltetrahydrofolate Authentic Substance in an aqueous solvent containing 2 mL of ammonium hydroxide and 1 g of sodium perchlorate per 100 mL. Filter the solution before use through a membrane filter of 1-μm porosity or finer.

System Suitability Test Chromatograph five injections of equal volume, up to 25 μL, of the *Standard Solution*, and measure the peak response as directed in the *Procedure*. The relative standard deviation, calculated by the formula 100 × (standard deviation/mean peak response), for the peak response does not exceed 2%. Inject a volume, up to 25 μL, of the *System Suitability Solution* in a similar manner. The resolution factor, R, between calcium formyltetrahydrofolate and Folic Acid, calculated by the formula given for R under *Chromatography*, Appendix IIA, is not less than 3.6.

Note: For a particular column, resolution may be increased by decreasing the amount of methanol in the mobile phase.

Procedure Introduce equal volumes, up to 25 μL, of the *Sample Solution* and *Standard Solution* into the chromatograph by means of a suitable sampling valve or high-pressure microsyringe. Measure the responses for the major peaks obtained with the *Sample Solution* and the *Standard Solution*. Calculate the quantity, in mg, of $C_{19}H_{19}N_7O_6$ in the sample taken by the formula

$$VC \times (P_U/P_S),$$

in which V is the volume of the *Sample Solution*, in mL; C is the concentration of USP Folic Acid Reference in the *Standard Solution*, in mg/mL; and P_U and P_S are the peak responses of the solutions from the *Sample Solution* and the *Standard Solution*, respectively.

Residue on Ignition Ignite 1 g as directed in the general method, Appendix IIC.

Water Determine by the *Karl Fischer Titrimetric Method*, Appendix IIB, using a 200-mg sample.

Packaging and Storage Store in well-closed, light-resistant containers.

Food Starch, Modified

Modified Food Starch; Food Starch–Modified

DESCRIPTION

Modified food starches are products of the treatment of any of several grain- or root-based native starches (e.g., corn, sorghum, wheat, potato, tapioca, sago, etc.) with small amounts of certain chemical agents, which modify the physical characteristics of the native starches to produce desirable properties.

Starch molecules are polymers of anhydroglucose and occur in both linear and branched form. The degree of polymerization and, accordingly, the molecular weight of the naturally occurring starch molecules vary radically. Furthermore, they vary in the ratio of branched chain polymers (amylopectin) to linear chain polymers (amylose), both within a given type of starch and from one type to another. These factors, in addition to any type of chemical modification used, affect the viscosity, texture, and stability of the starch sols significantly.

Starch is chemically modified by mild degradation reactions or by reactions between the hydroxyl groups of the native starch and the reactant selected. One or more of the following processes are used: mild oxidation (bleaching), moderate oxidation, acid and/or enzyme depolymerization, monofunctional esterification, polyfunctional esterification (cross-linking), monofunctional etherification, alkaline gelatinization, and certain combinations of these treatments. These methods of preparation can be used as a basis for classifying the starches thus produced (see **ADDITIONAL REQUIREMENTS** below). Generally, however, the products are called Modified Food Starch, or Food Starch–Modified.

Modified food starches are usually produced as white or nearly white, tasteless, odorless powders; as intact granules; and if pregelatinized (i.e., subjected to heat treatment in the presence of water), as flakes, amorphous powders, or coarse particles. Modified food starches are insoluble in alcohol, in ether, and in chloroform. If not pregelatinized, they are practically insoluble in cold water. Upon heating in water, the granules usually begin to swell at temperatures between 45° and 80°, depending on the botanical origin and the degree of modification. They gelatinize completely at higher temperatures. Pregelatinized starches hydrate in cold water.

Functional Use of Foods Thickener; colloidal stabilizer; binder.

REQUIREMENTS

Labeling Indicate the presence of sulfur dioxide if the residual concentration is greater than 10 mg/kg.

Identification

A. Suspend about 1 g of the sample in 20 mL of water, and add a few drops of iodine TS. A dark blue to red color is produced.

B. Place about 2.5 g of the sample in a boiling flask, add 10 mL of dilute hydrochloric acid (3%) and 70 mL of water, mix, reflux for about 3 h, and cool. Add 0.5 mL of the resulting solution to 5 mL of hot alkaline cupric tartrate TS. A copious red precipitate is produced.

C. Examine a portion of the sample with a polarizing microscope in polarized light under crossed Nicol prisms. The typical polarization cross is observed, except in the case of pregelatinized starches.

Arsenic (as As) Not more than 3 mg/kg.

Crude Fat Not more than 0.15%.

Heavy Metals (as Pb) Not more than 0.002%.

Lead Not more than 1 mg/kg.

Loss on Drying *Cereal starch*: not more than 15.0%; *potato starch*: not more than 21.0%; *sago and tapioca starch*: not more than 18.0%.

pH of Dispersions Between 3.0 and 9.0.

Protein Not more than 0.5%, except not more than 1% in modified high-amylose starches.

Sulfur Dioxide Not more than 0.005%.

ADDITIONAL REQUIREMENTS

The modified food starches listed below according to method of preparation must meet all of the above **REQUIREMENTS** in addition to the specified methods of *Treatment* (the reagent for which, if not specifically limited, should not exceed the amount reasonably required to accomplish the intended modification) and any requirements for *Residuals Limitation*.

Alkaline Gelatinization (Gelatinized Starch)

Treatment to Produce Gelatinized Starch	Residuals Limitation
Sodium hydroxide, not to exceed 1%	—

Depolymerization (Thin-Boiling, or Acid-Modified Starch) This treatment results in partial depolymerization, causing a reduction in viscosity. The treatments may be used in combination with the other treatments that follow.

Treatment to Produce Thin-Boiling Starch	Residuals Limitation
Hydrochloric acid and/or sulfuric acid	—
Alpha-amylase enzyme	The enzyme must be generally recognized as safe or approved as a food additive for this purpose. The resulting nonsweet nutritive saccharide polymer has a dextrose equivalent of less than 20.

Etherification and Esterification (Starch Ether-Esters)

Treatment to Produce Hydroxypropyl Distarch Phosphate	Residuals Limitation
Phosphorus oxychloride, not to exceed 0.1%, and propylene oxide, not to exceed 10%	Not more than 3 mg/kg of residual propylene chlorohydrin

Etherification with Oxidation (Oxidized Starch Ethers)

Treatment to Produce Oxidized Hydroxypropyl Starch	Residuals Limitation
Chlorine, as sodium hypochlorite, not to exceed 0.055 lb (25 g) of chlorine per lb (454 g) of dry starch; active oxygen obtained from hydrogen peroxide, not to exceed 0.45%; and propylene oxide, not to exceed 25%	Not more than 1 mg/kg of residual propylene chlorohydrin

Mild Oxidation (Bleached Starch) The starches resulting from mild oxidation are not altered chemically; in all cases, extraneous color bodies are oxidized, solubilized, and removed by washing and filtration. These treatments may be used in combination with the other forms of treatment listed herein.

Treatment to Produce Bleached Starch	Residuals Limitation
Active oxygen obtained from hydrogen peroxide, and/or peracetic acid, not to exceed 0.45% of active oxygen	—
Ammonium persulfate, not to exceed 0.075%, and sulfur dioxide, not to exceed 0.05%	—
Chlorine, as sodium hypochlorite, not to exceed 0.0082 lb (3.72 g) of chlorine per lb (454 g) of dry starch	—
Chlorine, as calcium hypochlorite, not to exceed 0.036% of dry starch	—
Potassium permanganate, not to exceed 0.2%	Not more than 0.005% of residual manganese (as Mn)
Sodium chlorite, not to exceed 0.5%	—

Moderate Oxidation (Oxidized Starch) The maximum specified treatment introduces about 1 carboxyl group per 28 anhydroglucose units. The starch is whitened, and its molecular weight and viscosity are reduced.

Treatment to Produce Oxidized Starch	Residuals Limitation
Chlorine, as sodium hypochlorite, not to exceed 0.055 lb (25 g) of chlorine per lb (454 g) of dry starch	—

Monofunctional and/or Polyfunctional Esterification (Starch Esters) The starch esters are named individually, depending on the method of preparation.

Treatment to Produce Starch Acetate	Residuals Limitation
Acetic anhydride or vinyl acetate	Not more than 2.5% of acetyl groups introduced into finished product

Treatment to Produce Acetylated Distarch Adipate	Residuals Limitation
Adipic anhydride, not to exceed 0.12%, and acetic anhydride	Not more than 2.5% of acetyl groups introduced into finished product

Treatment to Produce Starch Phosphate	Residuals Limitation
Monosodium orthophosphate	Not more than 0.4% of residual phosphate (calculated as P)

Treatment to Produce Starch Octenyl Succinate	Residuals Limitation
Octenyl succinic anhydride, not to exceed 3%, followed by treatment with alpha-amylase enzyme	—

Treatment to Produce Starch Sodium Octenyl Succinate	Residuals Limitation
Octenyl succinic anhydride, not to exceed 3%	—

Treatment to Produce Starch Aluminum Octenyl Succinate	*Residuals Limitation*
Octenyl succinic anhydride, not to exceed 2%, and aluminum sulfate, not to exceed 2%	—

Treatment to Produce Distarch Phosphate	*Residuals Limitation*
Phosphorus oxychloride, not to exceed 0.1%	—
Sodium trimetaphosphate	Not more than 0.04% of residual phosphate (calculated as P)

Treatment to Produce Phosphated Distarch Phosphate	*Residuals Limitation*
Sodium tripolyphosphate and sodium trimetaphosphate	Not more than 0.4% of residual phosphate (calculated as P)

Treatment to Produce Acetylated Distarch Phosphate	*Residuals Limitation*
Phosphorus oxychloride, not to exceed 0.1%, followed by either acetic anhydride, not to exceed 8%, or vinyl acetate, not to exceed 7.5%	Not more than 2.5% of acetyl groups introduced into finished product

Treatment to Produce Starch Sodium Succinate	*Residuals Limitation*
Succinic anhydride, not to exceed 4%	—

Monofunctional Etherification

Treatment to Produce Hydroxypropyl Starch	*Residuals Limitation*
Propylene oxide, not to exceed 25%	Not more than 1 mg/kg of residual propylene chlorohydrin

TESTS

Arsenic A *Sample Solution* prepared as directed for organic compounds meets the requirements of the *Arsenic Test*, Appendix IIIB.

Crude Fat Determine as directed in the general method, Appendix X.

Heavy Metals Prepare and test a 1-g sample as directed in *Method II* under the *Heavy Metals Test*, Appendix IIIB, using 20 µg of lead ion (Pb) in the control (*Solution A*) and 500° as the ignition temperature.

Lead Transfer 4.0 g of the sample to an evaporating dish, add 4 mL of sulfuric acid solution (1 in 4), distributing it evenly through the sample, and evaporate most of the water on a steam bath. Char and dehydrate the sample by heating on a hot plate, while heating at the same time with an infrared lamp from above, and then heat in a muffle furnace at 500° until the residue is free from carbon. Remove the dish from the furnace, cool, and cautiously wash down the inside of the dish with water. Add 1 mL of 1 N hydrochloric acid, evaporate to dryness on a steam bath, then add 2 mL of 1 N hydrochloric acid, and heat briefly, while stirring, on a steam bath. Quantitatively transfer the solution into a separator with the aid of small quantities of water, and neutralize with 1 N ammonium hydroxide. This *Sample Solution* meets the requirements of the *Lead Limit Test*, Appendix IIIB, using 4 µg of lead ion (Pb) in the control.

Loss on Drying, Appendix IIC Dry a 5-g sample in a vacuum oven, not exceeding 100 mm of Hg, at 120° for 4 h.

pH of Dispersions Mix 20 g of the sample with 80 mL of water, and agitate continuously at a moderate rate for 5 min. (In the case of pregelatinized starches, 3 g should be suspended in 97 mL of water.) Determine the pH of the resulting suspension by the *Potentiometric Method*, Appendix IIB.

Note: The distilled water used for sample dispersion should require not more than 0.05 mL of 0.1 N acid or alkali per 200 mL to obtain the methyl red or phenolphthalein endpoint, respectively.

Protein Transfer about 10 g of the sample, accurately weighed, into an 800-mL Kjeldahl flask, and add 10 g of anhydrous potassium or sodium sulfate, 300 mg of copper selenite or mercuric oxide, and 60 mL of sulfuric acid. Gently heat the mixture, keeping the flask inclined at about a 45° angle, and after frothing has ceased, boil briskly until the solution has remained clear for about 1 h. Cool, add 30 mL of water, mix, and cool again. Cautiously pour about 75 mL (or enough to make the mixture strongly alkaline) of sodium hydroxide solution (2 in 5) down the inside of the flask so that it forms a layer under the acid solution, and then add a few pieces of granular zinc. Immediately connect the flask to a distillation apparatus consisting of a Kjeldahl connecting bulb and a condenser, the delivery tube of which extends well beneath the surface of an accurately measured excess of 0.1 N sulfuric acid contained in a 50-mL flask. Gently rotate the contents of the Kjeldahl flask to mix, and distill until all ammonia has passed into the absorbing acid solution (about 250 mL of distillate). To the receiving flask add 0.25 mL of methyl red–methylene blue TS, and titrate the excess acid with 0.1 N sodium hydroxide. Perform a blank determination, substituting pure sucrose or dextrose for the sample, and make any necessary correction (see *General Provisions*). Each mL of 0.1 N sulfuric acid consumed is equivalent to 1.401 mg of nitrogen (N). Calculate the percent N in the sample, and then calculate the percent protein by multiplying the percent N by 6.25, in the case of starches obtained from corn, or by 5.7, in the case of starches obtained from wheat. Other factors may be applied as necessary for starches obtained from other sources.

Sulfur Dioxide Determine as directed in the general method, Appendix X.

TESTS (ADDITIONAL REQUIREMENTS)

Acetyl Groups Determine the content of acetyl groups in starch acetate, acetylated distarch adipate, and acetylated distarch phosphate as directed in the general method, Appendix X.

Manganese Determine the residual manganese in bleached starch prepared with potassium permanganate as directed in the general method, Appendix X.
Phosphate Determine the residual phosphate (calculated as P) in starch phosphate, distarch phosphate, and phosphated distarch phosphate as directed in the general method, Appendix X.
Propylene Chlorohydrin Determine the residual propylene chlorohydrin in hydroxypropyl starch, hydroxypropyl starch phosphate, and oxidized hydroxypropyl starch as directed in the general method, Appendix X.

Packaging and Storage Store in well-closed containers.

Food Starch, Unmodified

DESCRIPTION

Food starches are extracted from any of several grain or root crops, including corn (maize), sorghum, wheat, potato, tapioca, sago, and arrowroot and hybrids of these crops such as waxy maize, high amylose maize, etc. They are chemically composed of either one, or a mixture of two, glucose polysaccharides (amylose and amylopectin), the composition and relative proportions of which are characteristic of the plant source. Food starches are generally produced by extraction from the plant source using wet milling processes in which the starch is liberated by grinding aqueous slurries of the raw material. The extracted starch may be subjected to other nonchemical treatments such as purification, extraction, physical treatments, dehydration, heating, and minor pH adjustment during further processing steps. Food starch may be pregelatinized by heat treatment in the presence of water or made cold-water soluble.

Food starches are usually produced as white or nearly white, tasteless, odorless powders; as intact granules; and if pregelatinized, as flakes, powders, or coarse particles. Food starches are insoluble in alcohol, in ether, and in chloroform. If not treated to be pregelatinized or cold-water swelling, they are practically insoluble in cold water. Upon heating in water, the granules usually begin to swell at temperatures between 45° and 80°, depending on the botanical origin of the starch. They gelatinize completely at higher temperatures. Pregelatinized and cold-water swelling starches hydrate in cold water.

Functional Use of Foods Thickener; colloidal stabilizer; binder.

REQUIREMENTS

Labeling Indicate the presence of sulfur dioxide if the residual concentration is greater than 10 mg/kg.
Identification
  A. Suspend about 1 g of the sample in 20 mL of water, and add a few drops of iodine TS. A dark blue to red color is produced.
  B. Place about 2.5 g of the sample in a boiling flask, add 10 mL of dilute hydrochloric acid (3%) and 70 mL of water, mix, reflux for about 3 h, and cool. Add 0.5 mL of the resulting solution to 5 mL of hot alkaline cupric tartrate TS. A copious red precipitate is produced.
  C. Examine a portion of the sample with a polarizing microscope in polarized light under crossed Nicol prisms. The typical polarization cross is observed, except in the case of pregelatinized starches.
Crude Fat Not more than 0.15%.
Heavy Metals Not more than 0.002%.
Lead Not more than 1 mg/kg.
Loss on Drying *Cereal starch*: not more than 15.0%; *potato starch*: not more than 21.0%; *sago and tapioca starch*: not more than 18.0%.
pH of Dispersions Between 3.0 and 9.0.
Protein Not more than 0.5%, except not more than 1% in high-amylose and other hybrid starches.
Sulfur Dioxide Not more than 0.005%.

TESTS

Crude Fat Determine as directed in the general method, Appendix X.
Heavy Metals Prepare and test a 1-g sample as directed in *Method II* under the *Heavy Metals Test*, Appendix IIIB, using 20 μg of lead ion (Pb) in the control (*Solution A*) and 500° as the ignition temperature.
Lead Transfer 4.0 g of the sample to an evaporating dish, add 4 mL of sulfuric acid solution (1 in 4), distributing it evenly through the sample, and evaporate most of the water on a steam bath. Char and dehydrate the sample by heating on a hot plate, while heating at the same time with an infrared lamp from above, and then heat in a muffle furnace at 500° until the residue is free from carbon. Remove the dish from the furnace, cool, and cautiously wash down the inside of the dish with water. Add 1 mL of 1 *N* hydrochloric acid, evaporate to dryness on a steam bath, then add 2 mL of 1 *N* hydrochloric acid, and heat briefly, while stirring, on a steam bath. Quantitatively transfer the solution into a separator with the aid of small quantities of water, and neutralize with 1 *N* ammonium hydroxide. This *Sample Solution* meets the requirements of the *Lead Limit Test*, Appendix IIIB, using 4 μg of lead ion (Pb) in the control.
Loss on Drying, Appendix IIC Dry a 5-g sample in a vacuum oven, not exceeding 100 mm of Hg, at 120° for 4 h.
pH of Dispersions Mix 20 g of the sample with 80 mL of water, and agitate continuously at a moderate rate for 5 min. (In the case of pregelatinized starches, 3 g should be suspended in 97 mL of water.) Determine the pH of the resulting suspension by the *Potentiometric Method*, Appendix IIB.

  Note: The distilled water used for sample dispersion should require not more than 0.05 mL of 0.1 *N* acid or alkali per 200 mL to obtain the methyl red or phenolphthalein endpoint, respectively.

Protein Transfer about 10 g of the sample, accurately weighed, into an 800-mL Kjeldahl flask, and add 10 g of anhydrous potassium or sodium sulfate, 300 mg of copper selenite or mercuric oxide, and 60 mL of sulfuric acid. Gently heat the mixture, keeping the flask inclined at about a 45° angle, and after frothing has ceased, boil briskly until the solution has remained clear for about 1 h. Cool, add 30 mL of water, mix, and cool again. Cautiously pour about 75 mL (or enough to make the mixture strongly alkaline) of sodium hydroxide solution (2 in 5) down the inside of the flask so that it forms a layer under the acid solution, and then add a few pieces of granular zinc. Immediately connect the flask to a distillation apparatus consisting of a Kjeldahl connecting bulb and a condenser, the delivery tube of which extends well beneath the surface of an accurately measured excess of 0.1 N sulfuric acid contained in a 50-mL flask. Gently rotate the contents of the Kjeldahl flask to mix, and distill until all ammonia has passed into the absorbing acid solution (about 250 mL of distillate). To the receiving flask add 0.25 mL of methyl red–methylene blue TS, and titrate the excess acid with 0.1 N sodium hydroxide. Perform a blank determination, substituting pure sucrose or dextrose for the sample, and make any necessary correction (see *General Provisions*). Each mL of 0.1 N sulfuric acid consumed is equivalent to 1.401 mg of nitrogen (N). Calculate the percent N in the sample, and then calculate the percent protein by multiplying the percent N by 6.25, in the case of starches obtained from corn, or by 5.7, in the case of starches obtained from wheat. Other factors may be applied as necessary for starches obtained from other sources.

Sulfur Dioxide Determine as directed in the general method, Appendix X, using a 25-g sample.

Packaging and Storage Store in well-closed containers.

Formic Acid

HCOOH

CH_2O_2 Formula wt 46.03

INS: 236

DESCRIPTION

A colorless, *highly corrosive* liquid having a characteristic pungent odor. It is miscible with water, with alcohol, with glycerin, and with ether. Its specific gravity is about 1.20.

Functional Use in Foods Flavoring adjunct.

REQUIREMENTS

Identification To 5 mL add 2 mL of mercuric chloride TS, and warm the mixture. A white precipitate of mercurous chloride forms.
Assay Not less than 85.0% of CH_2O_2.
Acetic Acid Not more than 0.4%.
Dilution Test Passes test.
Heavy Metals (as Pb) Not more than 10 mg/kg.
Sulfate Not more than 0.004%.

TESTS

Assay Tare a small glass-stoppered Erlenmeyer flask containing about 15 mL of water. Transfer about 1.5 mL of the sample into the flask and weigh. Dilute the solution of the sample to 50 mL, add phenolphthalein TS, and titrate with 1 N sodium hydroxide. Each mL of 1 N sodium hydroxide is equivalent to 46.03 mg of CH_2O_2.
Acetic Acid Dilute 1 mL (1.2 g) to 100 mL with water, transfer 50 mL of this solution into a 250-mL boiling flask, and add 5 g of yellow mercuric oxide. Boil the mixture under a reflux condenser with continuous stirring for 2 h, cool, filter, and wash the residue with about 25 mL of water. To the combined filtrate and washings add phenolphthalein TS and titrate with 0.02 N sodium hydroxide. Not more than 2.0 mL of 0.02 N sodium hydroxide is required to produce a pink color.
Dilution Test Dilute 1 volume of the sample with 3 volumes of water. No turbidity is observed within 1 h.
Heavy Metals To 1.7 mL (2 g) in a beaker add about 10 mg of sodium carbonate, evaporate to dryness on a steam bath, and dissolve the residue in 25 mL of water. This solution meets the requirements of the *Heavy Metals Test*, Appendix IIIB, using 20 µg of lead ion (Pb) in the control (*Solution A*).
Sulfate, Appendix IIIB To 2.1 mL (2.5 g) in a beaker add about 10 mg of sodium carbonate, and evaporate to dryness on a steam bath. Any turbidity produced by the residue does not exceed that shown in a control containing 100 µg of sulfate (SO_4).

Packaging and Storage Store in tight containers.

Fructose

D-Fructose; Levulose; Fruit Sugar

$C_6H_{12}O_6$ Formula wt 180.16

CAS: [57-48-7]

DESCRIPTION

Fructose occurs as white, hygroscopic, odorless, purified crystals or as a purified crystalline powder having a sweet taste. It is a natural constituent of fruit (hence, the name "fruit sugar") and is obtained from glucose in corn syrup by the use of glucose isomerase. Its density is about 1.6. It is soluble in methanol and in ethanol, freely soluble in water, and insoluble in ether.

Functional Use in Foods Nutritive sweetener; processing aid; formulation aid.

REQUIREMENTS

Identification

A. To 5 mL of hot alkaline cupric tartrate TS add a few drops of a 1 in 10 solution of the sample. A copious red precipitate of cuprous oxide is formed.

B. The infrared absorption spectrum of a potassium bromide dispersion of the sample, previously dried, exhibits maxima only at the same wavelengths as those of a similar preparation of the USP Fructose Reference Standard.

Assay Not less than 98.0% and not more than 102.0% of Fructose ($C_6H_{12}O_6$), after drying.
Arsenic (as As) Not more than 1 mg/kg.
Chloride Not more than 0.018%.
Glucose Not more than 0.5%.
Heavy Metals (as Pb) Not more than 5 mg/kg.
Hydroxymethylfurfural Not more than 0.1%, calculated on the dried basis.
Lead Not more than 0.1 mg/kg.
Loss on Drying Not more than 0.5%.
Residue on Ignition Not more than 0.5%.
Sulfate Not more than 0.025%.

TESTS

Assay Transfer about 10 g of the sample, previously dried in vacuum at 70° for 4 h and accurately weighed, into a 100-mL volumetric flask, dissolve in 50 mL of water, add 0.2 mL of 15.2 N ammonium hydroxide, dilute to volume with water, and mix. After 30 min, determine the angular rotation (see Appendix IIB) in a 100-mm (or 200-mm) tube at 25° with the sodium D line. The observed rotation, in degrees (absolute value), multiplied by 1.124 (or 0.562 for the 200-mm tube), represents the weight, in g, of Fructose ($C_6H_{12}O_6$) in the sample taken.

Arsenic A *Sample Solution* prepared as directed for organic compounds meets the requirements of the *Arsenic Test*, Appendix IIIB, using 1 mL of *Standard Arsenic Solution* (1 µg As).

Chloride, Appendix IIIB Any turbidity produced by a 2-g sample does not exceed that shown in a control containing 0.5 mL of 0.02 N hydrochloric acid.

Glucose

Reagent Solution Dissolve 40 mg of o-dianisidine dihydrochloride, 40 mg of horseradish peroxidase (Worthington Biochemical Co., Freehold, NJ, or equivalent), and 0.4 mL of purified glucose oxidase (1000 glucose oxidase units per mL, Miles Laboratories, Inc., or equivalent) in 0.1 M acetate buffer (prepared as follows: dissolve 13.608 g of sodium acetate trihydrate in sufficient water to make 1000 mL, add 2.7 mL of acetic acid, and adjust the pH to 5.5 with acetic acid or sodium acetate), and dilute to 100 mL with the buffer solution.

Note: Commercially available preparations containing the reagents in the proper proportions may also be used.

Glucose Standard Solution Transfer about 300 mg, accurately weighed, of USP Dextrose Reference Standard, previously dried in vacuum at 70° for 4 h, into a 1000-mL volumetric flask, dissolve in and dilute to volume with water, and mix. Allow to stand for 2 h to allow mutarotation to occur, then transfer 20.0 mL to a 100-mL volumetric flask, dilute to volume with water, and mix. Prepare fresh on the day of use.

Sample Preparation Transfer 14 g of sample, accurately weighed, into a 100-mL volumetric flask, dissolve in and dilute to volume with water, and mix. Transfer 20.0 mL into a second 100-mL volumetric flask, dilute to volume with water, and mix.

Procedure Pipet 2 mL each of the *Sample Preparation*, the *Glucose Standard Solution*, and water into separate 18- × 150-mm test tubes. Heat the tubes in a water bath maintained at 30° for 5 min, and at zero time start the reaction by adding 1.0 mL of the *Reagent Solution* to the first tube. Allow 30- and 60-s intervals between the addition of *Reagent Solution* to each tube, and add 1.0 mL of the *Reagent Solution* to each of the remaining tubes. Mix the content of the tubes, and allow them to react for exactly 30 min from zero time. Immediately stop the reaction in the first tube by adding 10.0 mL of 25% sulfuric acid. Similarly, add 10.0 mL of 25% sulfuric acid to the remaining tubes after they have reacted for exactly 30 min. Mix the contents of the tubes, and cool them to room temperature. Using a suitable spectrophotometer, determine the absorbances of the sample and standard mixtures at 540 nm against the reagent mixture in the reference cell. Calculate the percentage of glucose in the sample by the formula

$$(50C/W) \times A_U/A_S,$$

in which C is the exact concentration of the *Glucose Standard Solution*, in mg/mL; W is the weight of sample taken, in g; A_U is the absorbance of the mixture obtained from the *Sample Preparation*; and A_S is the absorbance of the mixture obtained from the *Glucose Standard Solution*.

Heavy Metals Prepare and test a 4-g sample as directed under the *Heavy Metals Limit Test*, Appendix IIIB, using 20 µg of lead (Pb) in the control (*Solution A*).

Hydroxymethylfurfural Transfer approximately 1 g of the sample, accurately weighed, to a 100-mL volumetric flask, dilute to volume with water, and mix. Read the absorbance of this solution against a water blank at 283 nm in a 1-cm quartz cell in a spectrophotometer. Calculate the percentage of 5-hydroxymethylfurfural (HMF) by the following formula:

$$\% \text{ HMF} = (0.749 \times A)/C,$$

in which A is the absorbance of the sample solution, and C is the concentration, in mg/mL, of the sample solution corrected for ash and moisture.

Lead Determine as directed under *Method I* in the *Atomic Absorption Spectrophotometric Graphite Furnace Method* under the *Lead Limit Test*, Appendix IIIB, using a 5-g sample.

Loss on Drying, Appendix IIC Dry in vacuum at 70° for 4 h.

Residue on Ignition, Appendix IIC Ignite 2 g as directed in the general method.

Sulfate, Appendix IIIB Any turbidity produced by a 2-g sample does not exceed that shown in a control containing 0.5 mL of 0.02 N sulfuric acid.

Packaging and Storage Store in tight containers protected from humidity.

Fumaric Acid

(*E*)-Butenedioic Acid; *trans*-1,2-Ethylenedicarboxylic Acid

$$\begin{array}{c} \text{HOOCCH} \\ \| \\ \text{HCCOOH} \end{array}$$

$C_4H_4O_4$ Formula wt 116.07
INS: 297   CAS: [110-17-8]

DESCRIPTION

White, odorless granules or crystalline powder. It is soluble in alcohol, slightly soluble in water and in ether, and very slightly soluble in chloroform.

Functional Use in Foods   Acidifier; flavoring agent.

REQUIREMENTS

Identification   The infrared absorption spectrum of a potassium bromide dispersion of the sample exhibits maxima only at the same wavelengths as that of a similar preparation of USP Fumaric Acid Reference Standard.
Assay   Not less than 99.5% and not more than 100.5% of $C_4H_4O_4$, calculated on the anhydrous basis.
Heavy Metals (as Pb)   Not more than 10 mg/kg.
Maleic Acid   Not more than 0.1%.
Residue on Ignition   Not more than 0.1%.
Water   Not more than 0.5%.

TESTS

Assay   Transfer about 1 g, accurately weighed, into a 250-mL Erlenmeyer flask, add 50 mL of methanol, and dissolve the sample by warming gently on a steam bath. Cool, add phenolphthalein TS, and titrate with 0.5 *N* sodium hydroxide to the first appearance of a pink color that persists for at least 30 s. Perform a blank determination (see *General Provisions*), and make any necessary correction. Each mL of 0.5 *N* sodium hydroxide is equivalent to 29.02 mg of $C_4H_4O_4$.
Heavy Metals   Prepare and test a 2-g sample as directed in *Method II* under the *Heavy Metals Test*, Appendix IIIB, using 20 μg of lead ion (Pb) in the control (*Solution A*).
Maleic Acid
  Mobile Phase   Prepare 0.005 *N* sulfuric acid that has been suitably filtered and degassed.
  Standard Preparation   Transfer about 1 mg of USP Maleic Acid Reference Standard, accurately weighed, to a 1000-mL volumetric flask, dilute to volume with *Mobile Phase*, and mix.
  Test Preparation   Transfer about 100 mg of sample, accurately weighed, to a 100-mL volumetric flask, dilute to volume with *Mobile Phase*, and mix.
  Resolution Solution   Transfer 1 mg of USP Fumaric Acid Reference Standard and 0.5 mg of USP Maleic Acid Reference Standard into a 100-mL volumetric flask, dilute to volume with *Mobile Phase*, and mix.

  Chromatographic System   The liquid chromatograph is equipped with a 210-nm detector and 4.6-mm × 22-cm column packed with a strong cation exchange resin consisting of sulfonated cross-linked styrene–divinylbenzene copolymer in the hydrogen form (Polypore H from Brownlee Lab, or equivalent). The flow rate is about 0.3 mL/min. Chromatograph the *Resolution Solution*, and record the peak responses: The resolution, *R*, between the maleic acid and fumaric acid peaks is not less than 2.5, and the relative standard deviation of the maleic acid peak for replicate injections is not more than 2.0%.
  Procedure   Separately inject equal volumes (about 5 μL) of the *Standard Preparation* and the *Test Preparation* into the chromatograph, record the chromatograms, and measure the peak responses. The relative retention times are about 0.5 for maleic acid and 1.0 for Fumaric Acid. Calculate the quantity, in mg, of maleic acid in the total weight of the sample taken by the formula

$$100C(R_U/R_S),$$

in which *C* is the concentration, in mg/mL, of USP Maleic Acid Reference Standard in the *Standard Preparation*, and R_U and R_S are the responses of the maleic acid peaks obtained from the *Test Preparation* and the *Standard Preparation*, respectively.
Residue on Ignition   Ignite 2 g as directed in the general method, Appendix IIC.
Water   Determine by the *Karl Fischer Titrimetric Method*, Appendix IIB.

Packaging and Storage   Store in well-closed containers.

Furcelleran

Danish Agar

CAS: [9000-21-9]

DESCRIPTION

Furcelleran is a hydrocolloid obtained from *Furcellaria fastigiata* of the class *Rhodophyceae* (red seaweeds) by extraction with water or aqueous alkali. It consists mainly of the potassium, sodium, magnesium, calcium, and ammonium sulfate esters of galactose and 3,6-anhydrogalactose copolymers. These hexoses are alternately linked α-1,3 and β-1,4 in the polymer. The relative proportion of cations existing in Furcelleran may be changed during processing to the extent that one may become predominant.

  The ester sulfate content of Furcelleran ranges from 8% to 20% (see **REQUIREMENTS**). In addition, it contains inorganic salts that originate from the seaweed and the process of recovery from the extract. Furcelleran is recovered by alcohol precipitation, potassium precipitation, or by freezing. The alcohols used during recovery and purification are restricted to methanol, ethanol, and isopropanol.

Furcelleran is a brownish or tan to white, coarse to fine powder that is practically odorless and has a mucilaginous taste. It is soluble in water at a temperature of about 80°, forming a viscous, clear or slightly opalescent solution that flows readily. It disperses in water more readily if first moistened with alcohol, glycerin, or a saturated solution of sucrose in water.

Functional Use in Foods Emulsifier; stabilizer; thickener; gelling agent.

REQUIREMENTS

Identification

A. Add 4 g sample to 200 mL of water, and heat the mixture in a water bath at 80°, with constant stirring, until dissolved. Replace any water lost by evaporation, and allow the solution to cool to room temperature. It becomes viscous and may form a gel.

B. To 50 mL of the solution or gel obtained in *Identification Test A* add 200 mg of potassium chloride, then reheat, mix well, and cool. A short-textured ("brittle") gel will be formed.

C. To 5 mL of the solution obtained in *Identification Test A* add 1 drop of a 1 in 100 solution of methylene blue. A fibrous precipitate forms.

D. Obtain the infrared absorption spectrum of the sample by the following procedure: Prepare a 0.2% aqueous solution, cast films 0.0005 cm thick (when dry) on a suitable nonsticking surface such as Teflon, and obtain the infrared absorption spectrum. (Alternatively, the spectrum may be obtained on potassium bromide pellets if care is taken to avoid moisture).

Furcelleran has strong, broad absorption bands in the 1000 to 1100 cm^{-1} region. Absorption maximum is 1065. Other characteristic absorption bands and their intensities relative to the absorbance at 1050 cm^{-1} are as follows:

Wave Number (cm^{-1})	Molecular Assignment	Absorbance Relative to 1050 cm^{-1}
1220–1260	ester sulfate	0.2–0.6
928–933	3,6-anhydrogalactose	0.2–0.3
840–850	galactose-4-sulfate	0.1–0.3

Arsenic (as As) Not more than 3 mg/kg.
Ash (Acid-Insoluble) Not more than 1.0%.
Ash (Total) Not more than 35.0%.
Heavy Metals (as Pb) Not more than 0.002%.
Lead Not more than 5 mg/kg.
Loss on Drying Not more than 12.0%.
Sulfate Between 8.0% and 20.0% on the dry weight basis.
Viscosity of a 1.5% Solution Not less than 5 centipoises at 75°.
Acid-Insoluble Matter Not more than 1.0%.
Solubility in Water Not more than 30 mL of water is required to completely dissolve 1 g at a temperature of 80°.

TESTS

Arsenic A *Sample Solution* prepared as directed for organic compounds meets the requirements of the *Arsenic Test*, Appendix IIIB.

Ash (Acid-Insoluble) Proceed as directed in the general method, Appendix IIC.

Ash (Total) Transfer about 2 g, accurately weighed, into a previously ignited, tared, silica or platinum crucible. Heat the sample with a suitable infrared heat lamp, increasing the intensity gradually, until it is completely charred, and then continue for an additional 30 min. Transfer the crucible and charred sample into a muffle furnace and ignite at about 550° for 1 h, then cool in a desiccator and weigh. Repeat the ignition in the muffle furnace until a constant weight is attained. If a carbon-free ash is not obtained after the first ignition, moisten the charred spots with a 1 in 10 solution of ammonium nitrate and dry under an infrared heat lamp before reigniting.

Heavy Metals Prepare and test a 1-g sample as directed in *Method II* under the *Heavy Metals Test*, Appendix IIIB, using 20 μg of lead ion (Pb) in the control *(Solution A)*.

Lead A *Sample Solution* prepared as directed for organic compounds meets the requirements of the *Lead Limit Test*, Appendix IIIB, using 5 μg of lead ion (Pb) in the control.

Loss on Drying, Appendix IIC Dry at 105° for 4 h.

Sulfate Transfer about 500 mg, previously dried at 105° for 12 h and accurately weighed, into a 100-mL Kjeldahl flask. Add 10 mL of nitric acid, and heat gently for 30 min, adding more of the acid, if necessary, to prevent evaporation to dryness, and to yield a volume of about 3 mL at the end of the heating. Cool the mixture to room temperature, and decompose the excess nitric acid by the addition of formaldehyde TS, dropwise, heating, if necessary, until no brown fumes continue to be evolved. Continue the heating until the volume of the reaction mixture is reduced to about 5 mL, and then cool. Transfer the residue quantitatively, with the aid of water, into a 400-mL beaker, dilute it to about 100 mL, and filter, if necessary, to produce a clear solution. Dilute the solution to about 200 mL, and add 1 mL of hydrochloric acid. Heat to boiling and add, dropwise, with constant stirring, an excess (about 6 mL) of hot barium chloride TS. Heat the mixture for 1 h on a steam bath, collect the precipitate of barium sulfate on a filter, wash it until free from chloride, dry, ignite, and weigh. The weight of the barium sulfate so obtained, multiplied by 0.4116, gives the equivalent of sulfate (SO_4).

Viscosity of a 1.5% Solution Transfer 7.5 g of the sample into a tared, 600-mL tall-form (Berzelius) beaker, and disperse with agitation for 10 to 20 min in 450-mL of deionized water. Add sufficient water to bring the final weight to 500 g, and heat in a water bath, with continuous agitation, until a temperature of 80° is reached (20 to 30 min). Add water to adjust for loss by evaporation, cool to 76° to 77°, and place in a constant-temperature bath at 75°. Preheat the bob and guard of a Brookfield LVF or LVT viscometer to approximately 75° in water, then dry the bob and guard and attach them to the viscometer, which should be equipped with a No. 1 spindle (19 mm in diameter, approximately 65 mm in length) and capable of rotating at 30 rpm. Adjust the height of the bob in the sample solution, start the viscometer rotating at 30 rpm, and, after six complete revolutions, take the reading on the 0 to 100 scale. Record the results in centipoises by multiplying the reading by 2.

Note: Some samples of Furcelleran may be too viscous to be read when a No. 1 spindle is used. Such samples obviously pass the specification, but if a viscosity reading is desired for other reasons, use a No. 2 spindle, take the reading on the 0 to 100 scale, and multiply the reading by 10 to obtain the viscosity in centipoises, or read on the 0 to 500 scale and multiply by 2. If the viscosity is very low, increased precision may be obtained by using the Brookfield UL (ultra low) adapter, in which case the viscometer reading on the 0 to 100 scale should be multiplied by 0.2 to obtain the viscosity in centipoises.

Acid-Insoluble Matter Transfer about 2 g, accurately weighed, to a 250-mL beaker containing 150 mL of water and 1.5 mL of sulfuric acid. Cover with a watch glass, and heat on a steam bath for 6 h, rubbing down the wall of the beaker frequently with a rubber-tipped stirring rod, and replacing any water lost by evaporation. Transfer about 500 mg of a suitable filtering aid, accurately weighed, to the beaker, and filter through a tared filtering crucible containing a 2.4-cm glass fiber filter. Wash the residue several times with hot water, dry at 105° for 3 h, cool in a desiccator, and weigh. The difference between the total weight and the sum of the weights of the filter aid, crucible, and glass fiber filter is the weight of the acid-insoluble matter.

Solubility in Water Add 1 g of Furcelleran to 30 mL of cold water, stir well, and heat to a temperature of 80°. The resulting solution, when maintained at 80°, is uniformly viscous and clear or slightly opalescent.

Packaging and Storage Store in a well-closed container.

Garlic Oil

CAS: [8000-78-0]

DESCRIPTION

The volatile oil obtained by steam distillation from the crushed bulbs or cloves of the common garlic plant, *Allium sativum* L. (Fam. *Liliaceae*). It is a clear yellow to reddish orange liquid having a strong, pungent odor and a flavor characteristic of garlic. It is soluble in most fixed oils and in mineral oil. It may be incompletely soluble in alcohol. It is insoluble in glycerin and in propylene glycol.

Functional Use in Foods Flavoring agent.

REQUIREMENTS

Identification The infrared absorption spectrum of the sample exhibits relative maxima (that may vary in intensity) at the same wavelengths (or frequencies) as those shown in the respective spectrum in the section on *Infrared Spectra (Series A: Essential Oils)*, using the same test conditions as specified therein.
Heavy Metals (as Pb) Passes test.
Refractive Index Between 1.550 and 1.580 at 20°.
Specific Gravity Between 1.050 and 1.095.

TESTS

Heavy Metals Shake 10 mL of the oil with an equal volume of water to which 1 drop of hydrochloric acid has been added, and pass hydrogen sulfide through the mixture until it is saturated. No darkening in color is produced in either the oil or the water.
Refractive Index, Appendix IIB Determine with an Abbé or other refractometer of equal or greater accuracy.
Specific Gravity Determine by any reliable method (see *General Provisions*).

Packaging and Storage Store in full, tight, preferably glass or aluminum containers in a cool place protected from light.

Gelatin

Food-Grade Gelatin; Edible Gelatin

CAS: [9000-70-8]

DESCRIPTION

Gelatin is the product obtained from the acid, alkaline, or enzymatic hydrolysis of collagen, the chief protein component of the skin, bones, and connective tissues of animals. These animal sources shall not have been exposed to pentachlorophenol.

Type A Gelatin is produced by the acid processing of collagenous raw materials and exhibits an isoelectric point between pH 7 and pH 9. *Type B* Gelatin is produced by the alkaline or lime processing of collagenous raw materials and exhibits an isoelectric point between pH 4.6 and pH 5.2. Mixtures of *Types A and B* as well as Gelatins produced by modifications of the above mentioned processes may exhibit isoelectric points outside of the stated ranges.

Gelatin is nearly tasteless and odorless. It is a vitreous, brittle solid that is faintly yellow. When Gelatin granules are immersed in cold water, they hydrate into discrete, swollen particles. On being warmed, Gelatin disperses into the water, resulting in a stable suspension. Water solutions of Gelatin will form a reversible gel if cooled below the specific gel point of Gelatin. The gel point is dependent on the source of the raw material. Gelatin extracted from the tissues of warm-blooded animals will have a gel point in the range of 30° to 35°. Gelatin extracted from the skin of cold-water ocean fish will have a gel point in the range of 5° to 10°. Gelatin is soluble in aqueous solutions of polyhydric alcohols such as glycerin and propylene glycol. It is insoluble in most organic solvents.

Functional Use in Foods Firming agent; formulation and processing aid; stabilizer and thickener; surface-active agent; surface-finishing agent.

REQUIREMENTS

Identification
A. Gelatin forms a reversible gel when tested as follows: Dissolve 10 g in 100 mL of hot water in a suitable flask, and cool in a refrigerator at 2° for 24 h. A gel forms. Transfer the flask to a water bath heated to 60°. Within 30 min, upon stirring, the gel reverts to the original liquid state.

B. To a 1 in 100 solution of the sample add trinitrophenol TS or a 1 in 1.5 solution of potassium dichromate previously mixed with about one-fourth its volume of 3 N hydrochloric acid. A yellow precipitate forms.

Ash Not more than 3.0%.
Chromium Not more than 10 mg/kg.
Fluoride Not more than 0.005%.
Heavy Metals (as Pb) Not more than 0.002%.
Lead (as Pb) Not more than 1.5 mg/kg.
Loss on Drying Not more than 15.0%.
Microbial Limits:
   E. coli Negative in 25 g.
   Salmonella Negative in 25 g.
Pentachlorophenol Limit Not more than 0.3 mg/kg.
Protein The specification conforms to the representations of the vendor.
Sulfur Dioxide Not more than 0.005%.

TESTS

Ash Proceed as directed under *Ash (Total)*, Appendix IIC.

Chromium

Test Solution Transfer 10 g of the sample, accurately weighed, to a 100-mL silica dish. Using a very low flame, heat the dish over a Bunsen burner, taking care that the sample does not swell over the lip of the dish or catch fire. Gradually increase the flame until the sample is completely charred, transfer to a muffle furnace at 550°, and ash overnight. Cool to room temperature, add 10 mL of hydrochloric acid and 10 mL of nitric acid, and heat on a steam bath for 10 min. Cool and transfer to a 25-mL volumetric flask, dilute to volume, and mix.

Chromium Stock Solution Transfer 192.3 mg of chromium trioxide, accurately weighed, to a 1000-mL volumetric flask, dissolve in 100 mL of water and 10 mL of nitric acid, dilute to volume, and mix. This solution contains 0.1 mg of chromium per mL. Transfer 100.0 mL to a 1000-mL volumetric flask, dilute to volume, and mix. This solution contains 10 µg of chromium per mL.

Standard Preparations To separate 100-mL volumetric flasks, transfer 10.0, 30.0, 50.0, and 70.0 mL, respectively, of *Chromium Stock Solution*, dilute to volume, and mix. The *Standard Preparations* contain, respectively, 1.0, 3.0, 5.0, and 7.0 µg of chromium per mL.

Procedure Concomitantly determine the absorbances of the *Standard Preparations* and the *Test Solution* at the chromium emission line of 357.9 nm, with a suitable atomic absorption spectrophotometer equipped with a chromium hollow-cathode lamp and a slightly reducing air–acetylene flame, using water as the blank. Plot the absorbances of the *Standard Preparations* versus concentration, in µg/mL, of chromium, and draw the straight line best fitting the four plotted points. From the graph so obtained, determine the concentration, C_S, in µg/mL, of chromium in the *Test Solution*. Calculate the concentration of chromium, in mg/kg, in the portion of Gelatin taken, by the formula

$$25 C_S / W,$$

in which W is the quantity, in g, of the sample taken.

Fluoride Fluoride is separated from the sample by distillation from a solution containing sulfuric acid. Depending on the amount of chloride in the sample, silver sulfate solution is added to the solution being distilled. To ascertain how much silver sulfate solution is necessary, the concentration of chloride in the sample must be determined, as described below, under *Procedure for Chloride*.

The temperature of the distillation must be carefully controlled to prevent excess sulfate (as sulfur trioxide) from carrying over into the fluoride-containing distillate. Since excess sulfate (SO_4) may interfere with the colorimetric determination of fluoride, it is necessary, after measuring the fluoride, to determine sulfate in the distillate, and if the result so indicates, to repeat the distillation, paying closer attention to controlling the temperature of the solution being distilled.

Apparatus The distillation apparatus consists of a 2-L round-bottom borosilicate flask, a connector with a thermometer opening, a thermometer reading to 200° and long enough to reach almost to the bottom of the flask, a condenser with an inner spiral tube, and an electric heating mantle for the flask. All glassware shall be fitted together with standard-taper joints.

Standard Fluoride Solution Accurately weigh 221 mg of sodium fluoride, previously dried at 200° for 4 h, dissolve in water in a 1000-mL volumetric flask, dilute to volume, and mix. Transfer a 10.0-mL aliquot into a separate 1000-mL volumetric flask, dilute to volume, and mix. This solution contains 1 µg of fluoride per mL.

SPADNS Solution Dissolve 958 mg of SPADNS (sodium 2-(parasulfophenylazo)-1,8-dihydroxy-3,6-naphthalene disulfonate) in water, dilute to 500 mL, and mix. Protect from direct sunlight.

Zirconyl–Acid Reagent Dissolve 133 mg of zirconyl chloride octahydrate ($ZrOCl_2 \cdot 8H_2O$) in 25 mL of water, add 350 mL of hydrochloric acid, dilute to 500 mL, and mix.

Zirconyl–Acid/SPADNS Reagent Mix equal volumes of *SPADNS Solution* and *Zirconyl–Acid Reagent*.

Reference Solution Add 10 mL of *SPADNS Solution* to 100 mL of water, add 10 mL of hydrochloric acid (7 in 10), and mix.

Sodium Arsenite Solution Dissolve 5.0 g of sodium arsenite in 1 L of water, and mix.

Silver Sulfate Solution Dissolve 500 mg of silver sulfate in 100 mL of water, and mix.

Calibration Standards Transfer the following volumes of the *Standard Fluoride Solution* into separate 50-mL volumetric flasks, dilute each to volume, and mix:

Volume	Concentration
0.0 mL Reagent Blank (water)	0.0 μg/mL
5.0 mL	0.1 μg/mL
25.0 mL	0.5 μg/mL
50.0 mL	1.0 μg/mL

Procedure for Chloride Transfer 5 g of the sample, accurately weighed, into a 400-mL beaker. Add 150 mL of water, stir to hydrate the sample, and warm to dissolve. Add 25.0 mL of 0.1 N silver nitrate and 15 mL of nitric acid, mix well, and carefully heat the contents of the beaker to boiling. Slowly add saturated potassium permanganate solution in 5-drop increments, boiling the sample solution after each addition, until it becomes colorless. Cool the solution to room temperature, add 3 mL of ferric ammonium sulfate TS, and while agitating the solution vigorously, titrate with 0.1 N ammonium thiocyanate to a pale rose-colored endpoint that persists for 30 s.

Note: If the solution becomes pink upon the addition of the first few drops of 0.1 N ammonium thiocyanate, the determination must be repeated using a smaller sample.

Each mL of 0.1 N silver nitrate consumed is equivalent to 3.55 mg of chloride ion. Calculate the chloride, in mg/g, by the formula

$$C/S,$$

in which C is the weight, in mg, of chloride found, and S is the quantity, in g, of the sample taken.

Procedure for Fluoride Place 400 mL of water in the 2-L flask of the distillation apparatus, and cautiously add 200 mL of sulfuric acid. Swirl to mix, add 25 to 30 glass beads, and assemble the apparatus, making certain that all joints are tight. Distill into a receiver until the temperature of the solution reaches 180°, or until 250 mL of distillate have been collected in the receiver. Discontinue the distillation, and discard the distillate. Bring the contents of the distilling flask to 120°, or below. Then add 5 g of the sample, accurately weighed and dissolved in 300 mL of water, followed by 1 mL of silver sulfate per mg of chloride in the sample, as determined above. Again, assemble the apparatus and distill, collecting the distillate in a 250-mL volumetric flask. Continue the distillation until the temperature in the distilling flask reaches exactly 180° and no higher. If fewer than 250 mL have distilled, dilute the distillate to volume, and mix. If more than 250 mL have distilled, determine and record the volume, and mix.

Measure fluoride in the distillate as follows: Cool the distillate to a temperature of 25°. Transfer the standards, the reagent blank, and a 50-mL aliquot of the sample into separate beakers, being certain that the temperatures of all of the solutions fall within a range of 1°. Add 0.05 mL of *Sodium Arsenite Solution* to each beaker, followed by 10 mL of *Zirconyl–Acid/SPADNS Reagent*. Mix, allow to stand for 10 min, and measure the absorption of each solution against the reference solution at 570 nm in a suitable spectrophotometer.

Plot the absorbances of the *Standard Solutions* versus concentration, in μg/mL, of fluoride, and draw the straight line best fitting the data points. From the graph so obtained, determine the concentration, F, in μg/mL, of fluoride in the distillate, and calculate the percent fluoride by the formula

$$FV/10,000\ W,$$

in which V is the volume, in mL, of the distillate, and W is the quantity, in g, of the sample taken for the distillation.

After measuring the fluoride, determine sulfate in the distillate by the procedure that follows. If the concentration of sulfate is greater than 200 mg/L, repeat the distillation, being even more careful to control the temperature of the solution in the distillation flask so that it does not exceed 180°.

Procedure for Sulfate in the Fluoride Distillate Accurately determine the volume of distillate remaining from the fluoride distillation, and transfer it to a suitable beaker. Determine the pH with a pH meter, and if necessary, add dilute hydrochloric acid (1:3) to bring the pH to 4.5. Add an additional 2 mL of the hydrochloric acid, heat the solution to boiling, and while stirring, slowly add 30 mL of barium chloride TS to the hot solution. Boil the solution for 5 min, and allow it to stand until any precipitate settles out. If any precipitate has formed, filter the contents of the beaker through a tight, ashless filter paper, and quantitatively transfer the precipitate to the paper. Thoroughly wash the precipitate on the paper with hot water, and then transfer the paper to a tared crucible, previously ignited for 2 h at 1000° in a muffle furnace. Dry the paper in the crucible for 1 h at 105°, and then carefully char it at low heat. Gradually increase the heat to burn away the paper, and finally ignite the crucible and its contents for 2 h at 1000°. Cool and weigh, and calculate, in mg/L, the sulfate in the distillate by the formula

$$(B \times 0.4115 \times 1,000,000)/M,$$

in which B is the weight, in g, of the barium sulfate residue, and M is the mL of distillate taken.

Heavy Metals Prepare and test a 1-g sample as directed in *Method II* under the *Heavy Metals Test*, Appendix IIIB, using 20 μg of lead ion (Pb) in the control (*Solution A*).

Lead A Sample Solution containing 2 g of the sample prepared as directed for organic compounds meets the requirements of the *Lead Limit Test*, Appendix IIIB, using 3 μg lead ion (Pb) in the control.

Loss on Drying, Appendix IIC Dry at 105° to constant weight.

Microbial Limits:

  E. coli Proceed as directed in chapter 4 of the *FDA Bacteriological Analytical Manual*, Seventh Edition, 1992.

  Salmonella Proceed as directed in chapter 5 of the *FDA Bacteriological Analytical Manual*, Seventh Edition, 1992.

Pentachlorophenol

Pentachlorophenol (PCP) Stock Solution Transfer about 4.0 mg of PCP Reference Standard (Standard No. 5260, Pesticide Reference Standards Section, Environmental Protection Agency, Research Triangle Park, NC 27711, or equivalent, available from Aldrich Chemical Co.), accurately weighed, to a 1000-mL volumetric flask, dissolve in pesticide-grade benzene, dilute with benzene to volume, and mix. Each mL of this solution contains 4.0 μg of PCP.

PCP Standard Solutions Prepare a series of *PCP Standard Solutions* by serially diluting the *PCP Stock Solution* in hexane. Into separate 1000-mL volumetric flasks, pipet 0.0, 1.0, 5.0, 10.0, 25.0, 50.0, and 100.0 mL, respectively, of *PCP Stock*

Solution, dilute to volume with hexane, and mix. The *PCP Standard Solutions* contain in each mL 0.0, 0.004, 0.020, 0.040, 0.100, 0.200, and 0.400 μg of PCP, respectively.

Sample Preparation Transfer about 2 g of the Gelatin sample, accurately weighed, and 2.0 mL of water (to serve as the blank) into separate 25- × 150-mm screw-cap test tubes equipped with Teflon-lined caps. Treat each in the following manner: Add 10 mL of 12 N sulfuric acid. Close the tube, tighten the cap, and heat for 1 h in a water bath maintained at 100° in a fume hood, removing the tube periodically and mixing the sample by shaking. Remove the tube from the bath and allow to cool to room temperature. Add 10 mL of hexane:isopropanol (4:1 v/v) to the tube, and shake vigorously. Centrifuge for 2 min at 1000 × g in a suitable centrifuge (International Equipment Co., Needham Heights, MA 02194) with a head equipped to accommodate 25- × 150-mm test tubes. Transfer the upper hexane layer to a second 25- × 150-mm test tube with a Pasteur pipet. Repeat the extraction and centrifugation two additional times, and combine the hexane extracts in the second test tube. To the combined extracts, add 5.0 mL of 1.0 N potassium hydroxide, tighten the cap, shake the test tube vigorously, and centrifuge for 2 min at 1000 × g as before. Remove the upper layer with a Pasteur pipet, and discard. Add 10 mL of hexane to the test tube, tighten the cap, shake the test tube vigorously, and centrifuge as before. Remove the upper layer with a Pasteur pipet, and discard. Add 5.0 mL of 12 N sulfuric acid to the test tube, tighten the cap, and mix by carefully swirling the tube. Add 5.0 mL of hexane, tighten the cap, shake the test tube vigorously, centrifuge as before, and transfer the upper layer to a 10-mL volumetric flask. Repeat twice, using 2.0 mL of hexane each time, and transfer the upper layer to the 10-mL volumetric flask. Dilute to volume with hexane.

Chromatographic System Use a suitable gas chromatograph equipped with a ^{63}Ni electron capture detector and a 1.8-m × 4-mm id glass column containing 1% SP-1240DA on 100- to 120-mesh Supelcoport (Supelco Inc.) or equivalent. Place a small plug (2 to 3 mm) of phosphoric acid-washed glass wool in the detector end of the column. The carrier gas is 5% methane in argon at a flow rate of 60 mL/min. Condition the column by purging with carrier gas at ambient temperature for 10 to 15 min; program the column oven to increase from 70° to 190° at 1°/min, and hold the temperature at 190° for 8 h while continuing to purge with carrier gas.

Caution: Use only recently prepared and thoroughly conditioned columns; the appearance of ghost PCP peaks may be noted following the injection of samples containing high levels of PCP; repeated injections of solvent may be necessary until ghost PCP peaks disappear.

For sample analyses, the temperature of the column oven, injector port, and detector are maintained at 180°, 250°, and 350°, respectively. The electrometer should be adjusted to provide about half of the full-scale deflection when 0.1 ng of PCP is injected.

Procedure Inject 5-μL portions of each of the *PCP Standard Solutions* (0.0, 0.02, 0.10, 0.20, 1.0, and 2.0 ng, respectively) and the reagent blank into the gas chromatograph sequentially, and record the chromatograms. Measure the areas under the PCP peaks and the peak heights for each of the *PCP Standard Solutions* (retention time for PCP should be about 10 min), corrected for the reagent blank. The maximum acceptable reagent blank for satisfactory performance of the method is 0.01 μg/g. Similarly, inject 5 μL of the *Sample Preparation* into the gas chromatograph, and record the chromatogram. Measure the area under the PCP peak and the peak height, corrected for the reagent blank. Determine the amount of PCP in the *Sample Preparation* by comparing the peak area and height to the peak area and height obtained from injection of known amounts of *PCP Standard Solutions*; to ensure valid measurement of PCP in the *Sample Preparation*, the size of the PCP peak from the *Sample Preparation* and the standards should be within ±10%. The *Sample Preparation* may require further dilution. Designate as A_S the amount of PCP, expressed in ng, in the aliquot of the *Sample Preparation*. Each *PCP Standard Solution* and *Sample Preparation* should be injected twice to ensure that consistent responses are obtained. Following each injection of *PCP Standard Solutions* or *Sample Preparation*, rinse the syringe 10 times with hexane. After each injection of *PCP Standard Solutions* or *Sample Preparation*, inject 5 μL of hexane onto the gas chromatograph, and record the chromatogram. If peaks are observed at the retention time for PCP, repeat the hexane injection until such peaks are no longer encountered. Calculate the concentration of PCP, in μg/g, in the sample by the formula

$$5A_S.$$

Protein Transfer about 1 g, accurately weighed, to a 500-mL Kjeldahl flask, and proceed as directed under *Nitrogen Determination*, Appendix IIIC. Percent protein equals percent $N \times 5.55$.

Sulfur Dioxide Determine as directed in the general method, Appendix X. Instead of using a 50-g sample, dissolve 20.0 g in 100 mL of a 5% alcohol in water mixture, and proceed as directed under *Sample Introduction and Distillation*.

Packaging and Storage Store in tight containers.

Gellan Gum

CAS: [71010-52-1]

DESCRIPTION

A high-molecular-weight polysaccharide gum produced by a pure-culture fermentation of a carbohydrate with *Pseudomonas elodea*, and purified by recovery with isopropyl alcohol, dried, and milled. It is a heteropolysaccharide comprising a tetrasaccharide repeating unit of one rhamnose, one glucuronic acid, and two glucose units. The glucuronic acid is neutralized to mixed potassium, sodium, calcium, and magnesium salts. It may contain acyl (glyceryl and acetyl) groups as the *O*-glycosidically linked esters. It occurs as an off-white powder that is soluble in hot or cold deionized water.

Functional Use in Foods Stabilizer; thickener.

REQUIREMENTS

Identification

A. A 1% solution is made by hydrating a 1-g sample in 99 mL of deionized water. The mixture is stirred for about 2 h, using a motorized stirrer and a propeller-type stirring blade. Draw a small amount of the above solution into a wide-bore pipet, and transfer it into a solution of 10% calcium chloride. A tough, wormlike gel will form instantly.

B. To the 1% deionized water solution prepared for *Identification Test A*, add 0.5 g of sodium chloride, heat the solution to 80°, stirring constantly, and hold the temperature at 80° for 1 min. Stop heating and stirring the solution, and allow it to cool to room temperature. A firm gel will form.

Assay It yields not less than 3.3% and not more than 6.8% of carbon dioxide (CO_2), calculated on the dried basis.
Arsenic (as As) Not more than 3 mg/kg.
Heavy Metals (as Pb) Not more than 0.002%.
Isopropyl Alcohol Not more than 0.075%.
Lead Not more than 2 mg/kg.
Loss on Drying Not more than 15.0%.

TESTS

Assay Proceed as directed under *Alginates Assay*, Appendix IIIC, but use an undried sample of about 1.2 g, accurately weighed.
Arsenic A *Sample Solution* prepared as directed for organic compounds meets the requirements of the *Arsenic Test*, Appendix IIIB.
Heavy Metals Prepare and test a 1-g sample as directed in the *Heavy Metals Test, Method II*, Appendix IIIB, using a platinum crucible for the ignition and 20 µg of lead ion (Pb) in the control (*Solution A*).
Isopropyl Alcohol Proceed as directed for *Isopropyl Alcohol* in the monograph for *Xanthan Gum*, using a sample of about 5 g, accurately weighed.
Lead A *Sample Solution* prepared with a 2-g sample as directed for organic compounds meets the requirements of the *Lead Limit Test*, Appendix IIIB, using 4 µg of lead ion (Pb) in the control.
Loss on Drying, Appendix IIC Dry at 105° for 2.5 h.

Packaging and Storage Store in well-closed containers.

Geranium Oil, Algerian Type

Rose Geranium Oil, Algerian Type

DESCRIPTION

The oil obtained by steam distillation from the leaves of *Pelargonium graveolens* L'Her. (Fam. *Geraniaceae*). It is a light yellow to deep yellow liquid having a characteristic odor resembling rose and geraniol. It is soluble in most fixed oils, and it is soluble, usually with opalescence, in mineral oil and in propylene glycol. It is practically insoluble in glycerin.

Functional Use in Foods Flavoring agent.

REQUIREMENTS

Identification The infrared absorption spectrum of the sample exhibits relative maxima (that may vary in intensity) at the same wavelengths (or frequencies) as those shown in the respective spectrum in the section on *Infrared Spectra* (*Series A: Essential Oils*), using the same test conditions as specified therein.
Assay Not less than 13.0% and not more than 29.5% of esters, calculated as geranyl tiglate ($C_{15}H_{24}O_2$).
Acid Value Between 1.5 and 9.5.
Angular Rotation Between −7° and −13°.
Ester Value after Acetylation Between 203 and 234.
Heavy Metals (as Pb) Passes test.
Refractive Index Between 1.464 and 1.472 at 20°.
Solubility in Alcohol Passes test.
Specific Gravity Between 0.886 and 0.898.

TESTS

Assay Proceed as directed under *Ester Value*, Appendix VI, using about 6 g, accurately weighed. The ester value multiplied by 0.422 equals the percentage of geranyl tiglate ($C_{15}H_{24}O_2$).
Acid Value Determine as directed for *Acid Value* under *Essential Oils and Flavors*, Appendix VI, using about 5 g, accurately weighed, and modifying the procedure by using 15 mL of water, instead of alcohol, as diluent and by agitating the mixture thoroughly during the titration to keep the oil in suspension.
Angular Rotation Determine in a 100-mm tube as directed under *Optical (Specific) Rotation*, Appendix IIB.
Ester Value after Acetylation Proceed as directed under *Total Alcohols*, Appendix VI, using about 1.9 g of the acetylated oil, accurately weighed, for saponification. Calculate the ester value after acetylation by the formula

$$A \times 28.05/B,$$

in which A is the number of mL of 0.5 N alcoholic potassium hydroxide consumed in the saponification, and B is the weight of acetylated oil, in g.
Heavy Metals Shake 10 mL of the oil with an equal volume of water to which 1 drop of hydrochloric acid has been added, and pass hydrogen sulfide through the mixture until it is saturated. No darkening in color is produced in either the oil or the water.
Refractive Index, Appendix IIB Determine with an Abbé or other refractometer of equal or greater accuracy.
Solubility in Alcohol Proceed as directed in the general method, Appendix VI. One mL dissolves in 3 mL of 70% alcohol, but on further dilution with the alcohol, opalescence may occur, sometimes followed by separation of paraffin particles.
Specific Gravity Determine by any reliable method (see *General Provisions*).

Packaging and Storage Store in full, tight, preferably glass, aluminum, or tin-lined containers in a cool place protected from light.

Gibberellic Acid

$C_{19}H_{22}O_6$ Formula wt 346.38

DESCRIPTION

A white to pale yellow, odorless or practically odorless, crystalline powder. It is slightly soluble in water and is soluble in alcohol and in acetone. It melts at about 234°.

Functional Use in Foods Enzyme activator.

REQUIREMENTS

Identification Dissolve a few mg of the sample in 2 mL of sulfuric acid. A reddish solution having a green fluorescence is formed.
Assay Not less than 90.0% of $C_{19}H_{22}O_6$, calculated on the dried basis.
Heavy Metals (as Pb) Not more than 0.002%.
Lead Not more than 10 mg/kg.
Loss on Drying Not more than 3.0%.
Specific Rotation $[\alpha]_D^{20°}$: Between +75.0° and +90.0°, calculated on the dried basis.

TESTS

Assay

Standard Preparation Transfer an accurately weighed quantity of FCC Gibberellic Acid Reference Standard, equivalent to about 25 mg of pure Gibberellic Acid (corrected for phase purity and volatiles content), into a 50-mL volumetric flask, dissolve in methanol, dilute to volume with methanol, and mix. Transfer 10.0 mL of this solution into a second 50-mL volumetric flask, dilute to volume with methanol, and mix.

Assay Preparation Transfer about 40 mg of the sample, accurately weighed, into a 50-mL volumetric flask, dissolve in methanol, dilute to volume with methanol, and mix. Transfer 10.0 mL of this solution into a 100-mL volumetric flask, dilute to volume with methanol, and mix.

Procedure Transfer 5.0 mL of the *Assay Preparation* into a 25- × 200-mm glass-stoppered tube, and transfer 4.0-mL and 5.0-mL portions of the *Standard Preparation* into separate, similar tubes. Place the tubes in a boiling water bath, evaporate to dryness, and then dry in an oven at 90° for 10 min. Remove the tubes from the oven, stopper, and allow to cool to room temperature. Dissolve the residue in each tube in 10.0 mL of dilute sulfuric acid (8 in 10), heat in a boiling water bath for 10 min, and then cool in a 10° water bath for 5 min. Determine the absorbance of the solutions in 1-cm cells at 535 nm with a suitable spectrophotometer, using the dilute sulfuric acid as the blank. Record the absorbance of the solution from the *Assay Preparation* as A_U. Note the absorbance of the two solutions prepared from the 4.0-mL and 5.0-mL aliquots of the *Standard Preparation*, and record the absorbance of the final solution giving the value nearest to that of the sample as A_S; record as V the volume of the aliquot used in preparing this solution. Calculate the quantity, in mg, of $C_{19}H_{22}O_6$ in the sample taken by the formula

$$500C \times (V/5) \times (A_U/A_S),$$

in which C is the exact concentration, in mg/mL, of the *Standard Preparation*.

Heavy Metals Prepare and test a 1-g sample as directed in *Method II* under the *Heavy Metals Test*, Appendix IIIB, using 20 µg of lead ion (Pb) in the control (*Solution A*).
Lead A *Sample Solution* prepared as directed for organic compounds meets the requirements of the *Lead Limit Test*, Appendix IIIB, using 10 µg of lead ion (Pb) in the control.
Loss on Drying, Appendix IIC Dry at 100° in vacuum for 7 h.
Specific Rotation, Appendix IIB Determine in a solution in alcohol containing 100 mg in each mL. Avoid the use of heat in preparing the solution.

Packaging and Storage Store in well-closed containers.

Ginger Oil

CAS: [8007-08-7]

DESCRIPTION

The volatile oil obtained by steam distillation of the dried ground rhizome of *Zingiber officianale*, Roscoe (Fam. *Zingiberaceae*). It is a light yellow to yellow liquid having the aromatic, characteristic odor of ginger. It is soluble in most fixed oils and in mineral oil. It is soluble, usually with turbidity, in alcohol, but it is insoluble in glycerin and in propylene glycol.

Functional Use in Foods Flavoring agent.

REQUIREMENTS

Identification The infrared absorption spectrum of the sample exhibits relative maxima (that may vary in intensity) at the same wavelengths (or frequencies) as those shown in the respective spectrum under *Infrared Spectra (Series A: Essential Oils)*, using the same test conditions as specified therein.

Angular Rotation Between −28° and −47°.
Heavy Metals (as Pb) Passes test.
Refractive Index Between 1.488 and 1.494 at 20°.
Saponification Value Not more than 20.
Specific Gravity Between 0.870 and 0.882.

TESTS

Angular Rotation Determine in a 100-mm tube as directed under *Optical (Specific) Rotation*, Appendix IIB.
Heavy Metals Shake 10 mL of the oil with an equal volume of water to which 1 drop of hydrochloric acid has been added, and pass hydrogen sulfide through the mixture until it is saturated. No darkening in color is produced in either the oil or the water.
Refractive Index, Appendix IIB Determine with an Abbé or other refractometer of equal or greater accuracy.
Saponification Value Determine as directed for *Saponification Value* under *Essential Oils and Flavors*, Appendix VI, using about 5 g, accurately weighed.
Specific Gravity Determine by any reliable method (see *General Provisions*).

Packaging and Storage Store in full, tight, glass, aluminum, or tin-lined containers in a cool place protected from light.

Glucono Delta-Lactone

$C_6H_{10}O_6$ Formula wt 178.14
INS: 575 CAS: [90-80-2]

DESCRIPTION

A fine, white, practically odorless, crystalline powder. It is freely soluble in water and is sparingly soluble in alcohol. It decomposes at about 153°.

Functional Use in Foods Acid; leavening agent; sequestrant.

REQUIREMENTS

Identification Dissolve a portion of the sample in water, heating at 60° if necessary, to obtain a test solution containing 10 mg/mL. Similarly prepare a standard solution with the same concentration using USP Potassium Gluconate Reference Standard. Separately apply 5-μL portions of both solutions on a thin-layer chromatographic plate (see *Thin-layer Chromatography*, Appendix IIA) coated with a 0.25-mm layer of chromatographic silica gel, and allow to dry. Develop the chromatogram in a solvent mixture consisting of ethanol, water, ammonium hydroxide, and ethyl acetate (5:3:1:1) until the solvent front has moved three-fourths of the length of the plate. Remove the plate from the chamber, dry at 110° for 20 min, and cool. Spray the plate with a reagent prepared as follows: Dissolve 2.5 g of ammonium molybdate in 50 mL of 2 N sulfuric acid in a 100-mL volumetric flask, add 1.0 g of ceric sulfate, swirl to dissolve, dilute to volume with 2 N sulfuric acid, and mix. Heat the plate at 110° for about 10 min: The principal spot obtained from the sample solution corresponds in color, size, and R_f value to that obtained from the standard solution.
Assay Not less than 99.0% and not more than 100.5% of $C_6H_{10}O_6$.
Heavy Metals (as Pb) Not more than 0.002%.
Lead Not more than 10 mg/kg.
Reducing Substances (as D-glucose) Not more than 0.5%.

TESTS

Assay Dissolve about 600 mg, accurately weighed, in 100 mL of water in a 300-mL Erlenmeyer flask, add 50.0 mL of 0.1 N sodium hydroxide, and allow to stand for 15 min. Add phenolphthalein TS, and titrate the excess alkali with 0.1 N hydrochloric acid. Perform a blank determination (see *General Provisions*). Each mL of 0.1 N hydrochloric acid is equivalent to 17.81 mg of $C_6H_{10}O_6$.
Heavy Metals A solution of 1 g in 25 mL of water meets the requirements of the *Heavy Metals Test*, Appendix IIIB, using 20 μg of lead ion (Pb) in the control (*Solution A*).
Lead A *Sample Solution* prepared as directed for organic compounds meets the requirements of the *Lead Limit Test*, Appendix IIIB, using 10 μg of lead ion (Pb) in the control.
Reducing Substances Weigh accurately 10.0 g into a 400-mL beaker, dissolve the sample in 40 mL of water, add phenolphthalein TS, and neutralize with sodium hydroxide solution (1 in 2). Dilute to 50.0 mL with water, and add 50 mL of alkaline cupric tartrate TS. Heat the mixture over a Bunsen burner, regulating the flame so that boiling begins in 4 min, and continue the boiling for exactly 2 min. Filter through a sintered-glass filter crucible, wash the filter with 3 or more small portions of water, and place the crucible in an upright position in the original beaker. Add 5 mL of water and 3 mL of nitric acid to the crucible, mix with a stirring rod to ensure complete solution of the cuprous oxide, and wash the solution into a beaker with several mL of water. To the beaker add bromine TS (5 to 10 mL) until the color becomes yellow, and dilute with water to about 75 mL. Add a few glass beads, boil over a Bunsen burner until the bromine is completely removed, and cool. Slowly add ammonium hydroxide until a deep blue color appears, then adjust the pH to approximately 4 with glacial acetic acid, and dilute to about 100 mL with water. Add 4 g of potassium iodide, and titrate with 0.1 N sodium thiosulfate, adding starch TS just before the endpoint is reached. Not more than 16.1 mL is required.

Packaging and Storage Store in well-closed containers.

Glucose Syrup

Corn Syrup

DESCRIPTION

A clarified, concentrated, aqueous solution of saccharides obtained by the partial hydrolysis of edible starch by food-grade acids and/or enzymes. Depending on the degree of hydrolysis, it contains varying amounts of D-glucose. When obtained from corn starch, it is commonly designated as corn syrup. It is a sweet, clear white to light yellow viscous liquid miscible in all proportions with water.

Functional Use in Foods Nutritive sweetener.

REQUIREMENTS

Labeling Indicate the presence of sulfur dioxide if the residual concentration is greater than 10 mg/kg.
Identification Add a few drops of a 1 in 20 solution of the sample to 5 mL of hot alkaline cupric tartrate TS. A red precipitate of cuprous oxide is formed.
Assay Not less than 20.0% reducing sugar content (dextrose equivalent) expressed as D-glucose, calculated on the dried basis.
Arsenic (as As) Not more than 1 mg/kg.
Heavy Metals (as Pb) Not more than 5 mg/kg.
Lead Not more than 0.5 mg/kg.
Residue on Ignition Not more than 0.50%.
Starch Passes test.
Sulfur Dioxide Not more than 0.004%.
Total Solids Not less than 70.0%.

TESTS

Assay Determine as directed in the *Reducing Sugars Assay*, Appendix X.
Arsenic A *Sample Solution* prepared using a 1-g sample meets the requirements of the *Arsenic Test*, Appendix IIIB, using 1 mL of *Standard Arsenic Solution* in the control (1 µg As).
Heavy Metals A solution of 4 g in 25 mL of water meets the requirements of the *Heavy Metals Test*, Appendix IIIB, using 20 µg of lead ion (Pb) in the control (*Solution A*).
Lead Determine as directed under *Method I* in the *Atomic Absorption Spectrophotometric Graphite Furnace Method* under the *Lead Limit Test*, Appendix IIIB, using a 5-g sample.
Residue on Ignition, Appendix IIC Ignite 20 g.
Starch To 1 g dissolved in 10 mL of water add 1 drop of iodine TS. A yellow color indicates the absence of soluble starch.
Sulfur Dioxide Determine as directed in the general method, Appendix X, using a 35-g sample.
Total Solids Determine the refractive index of a sample of Glucose Syrup at 20° or 45°, and use the appropriate Glucose Syrup table under *Glucose Syrup (Corn Syrup)*, Appendix X.

Packaging and Storage Store in tightly closed containers in a dry place.

Glucose Syrup, Dried

Dried Glucose Syrup; Glucose Syrup Solids

DESCRIPTION

Dried Glucose Syrup is a purified, concentrated mixture of nutritive saccharides obtained by the hydrolysis of edible starch and by partially drying the resulting solution (glucose syrup). When obtained from corn starch, it is commonly designated dried corn syrup or corn syrup solids. Depending on the degree of hydrolysis, it contains varying amounts of D-glucose. It is a sweet, white to light yellow powder or granules soluble in water.

Functional Use in Foods Nutritive sweetener.

REQUIREMENTS

Labeling Indicate the presence of sulfur dioxide if the residual concentration is greater than 10 mg/kg.
Identification Add a few drops of a 1 in 20 solution of the sample to 5 mL of hot alkaline cupric tartrate TS. A red precipitate of cuprous oxide is formed.
Assay Not less than 20.0% of reducing sugar content (dextrose equivalent) expressed as D-glucose, calculated on the dried basis.
Arsenic (as As) Not more than 1 mg/kg.
Heavy Metals (as Pb) Not more than 5 mg/kg.
Lead Not more than 0.5 mg/kg.
Residue on Ignition Not more than 0.5%.
Starch Passes test.
Sulfur Dioxide Not more than 0.004%.
Total Solids Not less than 90.0% when the reducing sugar content is 88.0% or greater; not less than 93.0% when the reducing sugar content is between 20.0% and 88.0%.

TESTS

Assay Determine as directed in the *Reducing Sugars Assay*, Appendix X.
Arsenic A solution of 1 g in 35 mL of water meets the requirements of the *Arsenic Test*, Appendix IIIB, using 1 mL of the *Standard Arsenic Solution* in the control (1 µg As).
Heavy Metals (as Pb) A solution of 4 g in 25 mL of water meets the requirements of the *Heavy Metals Test*, Appendix IIIB, using 20 µg of lead ion (Pb) in the control (*Solution A*).
Lead Determine as directed under *Method I* in the *Atomic Absorption Spectrophotometric Graphite Furnace Method* under the *Lead Limit Test*, Appendix IIIB, using a 5-g sample.
Residue on Ignition Ignite 1 g as directed in the general method, Appendix IIC, using an ignition temperature of 525° for 2 h.
Starch To a 1-g sample dissolved in 10 mL of water add 1 drop of iodine TS. A yellow color indicates the absence of soluble starch.
Sulfur Dioxide Determine as directed in the general method, Appendix X, using a 25-g sample.
Total Solids Determine the water content of an accurately weighed portion as directed under the *Karl Fischer Titrimetric*

Method, Appendix IIB. Calculate the percent *Total Solids* by the formula

$$(W_U - W_W)100/W_U,$$

in which W_U is the weight, in mg, of the sample taken, and W_W is the weight, in mg, of water determined.

Packaging and Storage Store in tightly closed containers in a dry environment.

L-Glutamic Acid

Glutamic Acid; L-2-Aminopentanedioic Acid

$$\text{HOOCCH}_2\text{CH}_2\underset{\underset{H\ NH_2}{|\ |}}{C}\text{COOH}$$

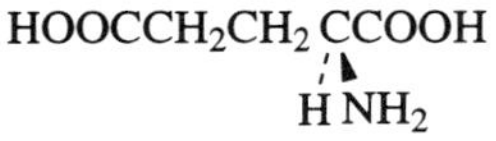

C$_5$H$_9$NO$_4$ Formula wt 147.13
INS: 620   CAS: [56-86-0]

DESCRIPTION

A white, practically odorless, free-flowing, crystalline powder. It is slightly soluble in water, forming acidic solutions. The pH of a saturated solution is about 3.2.

Functional Use in Foods Salt substitute; nutrient; dietary supplement.

REQUIREMENTS

Identification

A. Dissolve about 150 mg in a mixture of 4 mL of water and 1 mL of 1 N sodium hydroxide, add 1 mL of ninhydrin TS and 100 mg of sodium acetate, and heat in a boiling water bath for 10 min. An intense, violet blue color is formed.

B. The Glutamic Acid dissolves completely on stirring when either 5.6 mL of 1 N hydrochloric acid or 6.8 mL of 1 N sodium hydroxide is added to a suspension of 1 g of the sample in 9 mL of water.

Assay Not less than 98.5% and not more than 101.5% C$_5$H$_9$NO$_4$, calculated on the dried basis.
Heavy Metals (as Pb) Not more than 0.002%.
Lead Not more than 10 mg/kg.
Loss on Drying Not more than 0.1%.
Residue on Ignition Not more than 0.3%.
Specific Rotation $[\alpha]_D^{20°}$: Between +31.5° and +32.5° after drying.

TESTS

Assay Dissolve about 200 mg, accurately weighed, in 3 mL of formic acid and 50 mL of glacial acetic acid. Add 2 drops of crystal violet TS, and titrate with 0.1 N perchloric acid to a green endpoint or until the blue color disappears completely. Perform a blank determination (see *General Provisions*), and make any necessary correction. Each mL of 0.1 N perchloric acid is equivalent to 14.71 mg of C$_5$H$_9$NO$_4$.

Heavy Metals Prepare and test a 1-g sample as directed in *Method II* under the *Heavy Metals Test*, Appendix IIIB, using 20 µg of lead ion (Pb) in the control (*Solution A*).

Lead A *Sample Solution* prepared as directed for organic compounds meets the requirements of the *Lead Limit Test*, Appendix IIIB, using 10 µg of lead ion (Pb) in the control.

Loss on Drying, Appendix IIC Dry at 105° for 3 h.

Residue on Ignition, Appendix IIC Ignite 1 g as directed in the general method.

Specific Rotation, Appendix IIB $[\alpha]_D^{20°}$: Determine in a solution containing 10 g of a previously dried sample in sufficient 2 N hydrochloric acid to make 100 mL.

Packaging and Storage Store in well-closed containers.

L-Glutamic Acid Hydrochloride

2-Aminopentanedioic Acid Hydrochloride

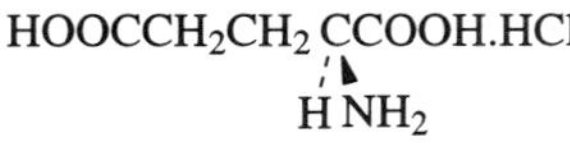

$$\text{HOOCCH}_2\text{CH}_2\underset{\underset{H\ NH_2}{|\ |}}{C}\text{COOH}\cdot\text{HCl}$$

C$_5$H$_9$NO$_4$·HCl   Formula wt 183.59
          CAS: [138-15-8]

DESCRIPTION

A white, crystalline powder. One g dissolves in about 3 mL of water. It is almost insoluble in alcohol and in ether. Its solutions are acid to litmus.

Functional Use in Foods Salt substitute; flavoring agent; nutrient; dietary supplement.

REQUIREMENTS

Identification

A. To 1 mL of a 1 in 3 solution add 1 mL of barium hydroxide solution (1 in 50), filter, and add 10 mL of alcohol. A white crystalline precipitate of barium glutamate forms on standing.

B. To 1 mL of a 1 in 30 solution add 1 mL of ninhydrin TS and 100 mg of sodium acetate, and boil for 10 min. An intense violet blue color is produced.

Assay Not less than 98.5% and not more than 101.5% of C$_5$H$_9$NO$_4$·HCl after drying.
Heavy Metals (as Pb) Not more than 0.002%.
Lead Not more than 10 mg/kg.
Loss on Drying Not more than 0.5%.
Residue on Ignition Not more than 0.25%.

Specific Rotation $[\alpha]_D^{20°}$: Between +25.2° and +25.8°, calculated on the dried basis.

TESTS

Assay Dissolve about 100 mg of the sample, previously dried at 80° for 4 h and accurately weighed, in 0.5 mL of water, add exactly 15.0 mL of 0.1 N perchloric acid, and heat on a water bath for 30 min. After cooling, add 45 mL of glacial acetic acid, and titrate the excess perchloric acid with 0.1 N sodium acetate, determining the endpoint potentiometrically. Perform a blank determination, and make any necessary correction. Each mL of 0.1 N perchloric acid is equivalent to 18.36 mg of $C_5H_9NO_4 \cdot HCl$.

Heavy Metals Prepare and test a 1-g sample as directed in *Method II* under the *Heavy Metals Test*, Appendix IIIB, using 20 μg of lead ion (Pb) in the control (*Solution A*).

Lead A *Sample Solution* prepared as directed for organic compounds meets the requirements of the *Lead Limit Test*, Appendix IIIB, using 10 μg of lead ion (Pb) in the control.

Loss on Drying, Appendix IIC   Dry at 80° for 4 h.

Residue on Ignition Ignite 1 g as directed in the general method, Appendix IIC.

Specific Rotation, Appendix IIB   Determine in a solution containing 10 g in sufficient 2 N hydrochloric acid to make 100 mL.

Packaging and Storage Store in well-closed, light-resistant containers.

L-Glutamine

L-2-Aminoglutaramic Acid

$$H_2NCOCH_2CH_2 \underset{H\ NH_2}{CCOOH}$$

$C_5H_{10}N_2O_3$ Formula wt 146.15

CAS: [56-85-9]

DESCRIPTION

White, odorless crystals or crystalline powder having a slightly sweet taste. It is soluble in water and practically insoluble in alcohol and in ether. Its solutions are acid to litmus. It melts with decomposition at about 185°.

Functional Use in Foods Nutrient; dietary supplement.

REQUIREMENTS

Identification Heat 5 mL of a 1 in 1000 solution with 1 mL of triketohydrindene hydrate TS in a water bath. A purple color appears.

Assay Not less than 98.5% and not more than 101.5% of $C_5H_{10}N_2O_3$ after drying.

Heavy Metals (as Pb)   Not more than 10 mg/kg.

Loss on Drying Not more than 0.3%.

Residue on Ignition Not more than 0.1%.

Specific Rotation $[\alpha]_D^{20°}$: Between +6.3° and +7.3° after drying.

TESTS

Assay Dissolve about 150 mg of the sample, previously dried at 105° for 3 h and accurately weighed, in 3 mL of formic acid and 50 mL of glacial acetic acid, and titrate with 0.1 N perchloric acid, determining the endpoint potentiometrically. Perform a blank determination (see *General Provisions*), and make any necessary correction. Each mL of 0.1 N perchloric acid is equivalent to 14.62 mg of $C_5H_{10}N_2O_3$.

Heavy Metals Prepare and test a 2-g sample as directed in *Method II* under the *Heavy Metals Test*, Appendix IIIB, using 20 μg of lead ion (Pb) in the control (*Solution A*).

Loss on Drying, Appendix IIC   Dry at 105° for 3 h.

Residue on Ignition, Appendix IIC   Ignite 1 g as directed in the general method.

Specific Rotation, Appendix IIB   Determine in solution containing 4 g of a previously dried sample in sufficient water to make 100 mL.

Packaging and Storage Store in well-closed, light-resistant containers.

Glutaraldehyde

Glutaral; 1,5-Pentanedial

$C_5H_8O_2$ Formula wt 100.12

CAS: [111-30-8]

DESCRIPTION

A clear, nearly colorless aqueous solution with a characteristic sharp odor. It is miscible with water. The grades of Glutaraldehyde suitable for food use usually have concentrations between 15% and 50%.

Functional Use in Foods Fixing agent in the immobilization of enzyme preparations; cross-linking agent for microencapsulating flavoring substances; antimicrobial, sugar-milling.

REQUIREMENTS

Labeling Indicate the concentration of Glutaraldehyde.

Identification

2,4-Dinitrophenylhydrazine Reagent Add 4 mL of sulfuric acid to 0.8 g of 2,4-dinitrophenylhydrazine, and then add 6 mL of water, dropwise, with swirling. When dissolution is essentially complete, add 20 mL of alcohol, mix, and filter. The filtrate is the reagent.

Procedure Add 0.4 mL of the sample to 20 mL of *2,4-Dinitrophenylhydrazine Reagent*. Mix by swirling, and allow the mixture to stand for 5 min. Collect the precipitate on a filter, and rinse thoroughly with alcohol. Dissolve the precipitate in 20 mL of hot ethylene dichloride, filter, and cool the filtrate in an ice bath until crystallization occurs. Collect the precipitate on a filter. Redissolve the precipitate by refluxing with 30 mL of acetone, filter, and cool the filtrate in an ice bath until crystallization occurs. Collect the precipitate on a filter: The 2,4-dinitrophenylhydrazone so obtained melts between 185° and 195° (see *Melting Range or Temperature*, Appendix IIB).

Assay Not less than 100.0% and not more than 105.0% of the labeled amount of $C_5H_8O_2$.

Heavy Metals (as Pb) Not more than 10 mg/kg.

pH Between 3.1 and 4.5.

TESTS

Assay

Hydroxylamine Hydrochloride Solution Prepare a 0.5 N solution by dissolving 35.0 g of hydroxylamine hydrochloride in water in a 1-L volumetric flask, dilute to volume with water, and mix.

Triethanolamine Solution Prepare a 0.5 N solution by transferring 65 mL (74 g) of 98% triethanolamine to a 1-L volumetric flask, dilute to volume with water, and mix.

Procedure Neutralize a volume of *Hydroxylamine Hydrochloride Solution* sufficient for analyzing both the blank and the sample to pH 3.60. Use a suitable autotitrator, and titrate with *Triethanolamine Solution*.

Caution: The stirring rate is critical throughout the neutralization and analysis. When stirring is required, ensure adequate mixing without whipping air bubbles into the solution. The stirring speed should be consistent for the sample and blank.

Transfer 65.0 mL of the neutralized *Hydroxylamine Hydrochloride Solution* to each of two titration cups. Add a Teflon (or equivalent) stirrer to each cup. Using the autotitrator, add 30.8 mL of the neutralized *Triethanolamine Solution* to each titration cup, cover, and mix. Using a weighing pipet, introduce into one of the cups a suitable portion of the sample equivalent to about 300 mg of Glutaraldehyde. Mix the solutions thoroughly, and allow the sample and blank to stand at room temperature for at least 60 min but not for more than 90 min.

Titrate the sample and the blank to pH 3.60 with 0.5 N hydrochloric acid, determining the endpoint potentiometrically. Calculate the percentage, by weight, of $C_5H_8O_2$ (w/w) in the sample by the formula

$$[N(B - A)(0.05006)/W]100,$$

in which B and A are the volumes, in mL, of 0.5 N hydrochloric acid consumed by the blank and sample solutions, respectively; N is the normality of the hydrochloric acid; 0.05006 is the milliequivalent weight, in g/milliequivalent, of Glutaraldehyde; and W is the weight, in g, of sample taken.

Heavy Metals Prepare and test 2.0 g of the solution as directed under the *Heavy Metals Test*, Appendix IIIB, using 20 µg of lead ion (Pb) in the control (*Solution A*).

pH, Appendix IIB Determine by the *Potentiometric Method*.

Packaging and Storage Store in tight, light-resistant containers protected from heat.

Glycerin

Glycerol

$$CH_2OHCHOHCH_2OH$$

$C_3H_8O_3$ \hfill Formula wt 92.09

\hfill CAS: [56-81-5]

DESCRIPTION

A clear, colorless, syrupy liquid having a sweet taste. It has not more than a slight characteristic odor, which is neither harsh nor disagreeable. It is hygroscopic and its solutions are neutral. Glycerin is miscible with water and with alcohol. It is insoluble in chloroform, in ether, and in fixed and volatile oils.

Functional Use in Foods Humectant; solvent; bodying agent; plasticizer.

REQUIREMENTS

Identification The infrared absorption spectrum of a thin film of the sample exhibits a very strong, broad band at 2.7 µm to 3.3 µm; a strong doublet at about 3.4 µm; a maximum at about 6.1 µm; a strong region of absorption between 6.7 µm and 8.3 µm, having maxima at about 7.1 µm, 7.6 µm, and 8.2 µm, and a very strong region of bands at about 9.0 µm, 9.6 µm, 10.1 µm, 10.9 µm, and 11.8 µm.

Note: Glycerin containing low content of water may not exhibit a maximum at about 6.1 µm.

Assay Not less than 95.0% and not more than 100.5% of $C_3H_8O_3$.

Chlorinated Compounds (as Cl) Not more than 0.003%.

Color Passes test.

Fatty Acids and Esters Passes test (limit about 0.1%, calculated as butyric acid).

Heavy Metals (as Pb) Not more than 5 mg/kg.

Readily Carbonizable Substances Passes test.

Residue on Ignition Not more than 0.01%.
Specific Gravity Not less than 1.249.

TESTS

Assay

Sodium Periodate Solution Dissolve 60 g of sodium metaperiodate ($NaIO_4$) in sufficient water containing 120 mL of 0.1 N sulfuric acid to make 1000 mL. Do not heat to dissolve the periodate. If the solution is not clear, filter through a sintered-glass filter. Store the solution in a glass-stoppered, light-resistant container. Test the suitability of this solution as follows: Pipet 10 mL into a 250-mL volumetric flask, dilute to volume with water, and mix. To about 550 mg of Glycerin dissolved in 50 mL of water add 50 mL of the diluted periodate solution with a pipet. For a blank, pipet 50 mL of the solution into a flask containing 50 mL of water. Allow the solutions to stand for 30 min, then add to each 5 mL of hydrochloric acid and 10 mL of potassium iodide TS, and rotate to mix. Allow to stand for 5 min, add 100 mL of water, and titrate with 0.1 N sodium thiosulfate, shaking continuously and adding starch TS near the endpoint. The ratio of the volume of 0.1 N sodium thiosulfate required for the Glycerin–periodate mixture to that required for the blank should be between 0.750 and 0.765.

Procedure Transfer about 400 mg of the sample, accurately weighed, into a 600-mL beaker, dilute with 50 mL of water, add bromothymol blue TS, and acidify with 0.2 N H_2SO_4 to a definite green or greenish yellow color. Neutralize with 0.05 N sodium hydroxide to a definite blue endpoint, free of green color. Prepare a blank containing 50 mL of water, and neutralize in the same manner. Pipet 50 mL of the *Sodium Periodate Solution* into each beaker, mix by swirling gently, cover with a watch glass, and allow to stand for 30 min at room temperature (not above 35°) in the dark or in subdued light. Add 10 mL of a mixture consisting of equal volumes of ethylene glycol and water, and allow to stand for 20 min. Dilute each solution to about 300 mL with water, and titrate with 0.1 N sodium hydroxide to a pH of 8.1 ± 0.1 for the sample and 6.5 ± 0.1 for the blank, using a pH meter previously calibrated with pH 4.0 Acid Phthalate Standard Buffer Solution (see *Solutions and Indicators*). Each mL of 0.1 N sodium hydroxide, after correction for the blank, is equivalent to 9.210 mg of Glycerin ($C_3H_8O_3$).

Chlorinated Compounds Transfer 5.0 g of Glycerin into a dry 100-mL round bottom, ground joint flask and add to it 15 mL of morpholine. Connect the flask with a ground joint reflux condenser, and reflux the mixture gently for 3 h. Rinse the condenser with 10 mL of water, receiving the washing into the flask, and cautiously acidify with nitric acid. Transfer the solution to a suitable comparison tube, add 0.5 mL of silver nitrate TS, dilute with water to 50.0 mL, and mix thoroughly. Any turbidity does not exceed that produced by 150 μg of chloride (Cl) in an equal volume of solution containing the quantities of reagents used in the test, omitting the refluxing.

Color The color of Glycerin, when viewed downward against a white surface in a 50-mL Nessler tube, is not darker than the color of a standard made by diluting 0.40 mL of ferric chloride CS with water to 50 mL and similarly viewed in a Nessler tube of approximately the same diameter and color as that containing the sample.

Fatty Acids and Esters Mix a 40.0-mL (50-g) sample with 50 mL of recently boiled water and 5.0 mL of 0.5 N sodium hydroxide. Boil the mixture for 5 min, cool, add phenolphthalein TS, and titrate the excess alkali with 0.5 N hydrochloric acid. Not more than 1 mL of 0.5 N sodium hydroxide is consumed.

Heavy Metals Mix a 4.0-mL (5-g) sample with 2 mL of 0.1 N hydrochloric acid, add water to make 25 mL, and proceed as directed under the *Heavy Metals Test*, Appendix IIIB. Any color does not exceed that produced in a control (*Solution A*) containing 25 μg of lead ion (Pb).

Readily Carbonizable Substances, Appendix IIB Rinse a glass-stoppered, 25-mL cylinder with 95% sulfuric acid, and allow it to drain for 10 min. Add 5 mL of Glycerin and 5 mL of 95% sulfuric acid, shake vigorously for 1 min, and allow to stand for 1 h. The mixture is no darker than *Matching Fluid H*.

Residue on Ignition Heat 50 g in a tared, open dish, and ignite the vapors, allowing them to burn until the sample has been completely consumed. After cooling, moisten the residue with 0.5 mL of sulfuric acid, and complete the ignition by heating for 15-min periods at 800° ± 25° to constant weight.

Specific Gravity Determine by any reliable method (see *General Provisions*).

Packaging and Storage Store in tight containers.

Glycerol Ester of Gum Rosin

DESCRIPTION

A hard, pale amber-colored resin (color N or paler as determined by ASTM Designation D 509) produced by the esterification of pale gum rosin with food-grade glycerin and purified by steam stripping. It is soluble in acetone and in toluene, but is insoluble in water.

Functional Use in Foods Masticatory substance in chewing gum base.

REQUIREMENTS

Identification Identify Glycerol Ester of Gum Rosin by comparing its infrared absorption spectrum with a typical spectrum as shown in the section on *Infrared Spectra (Series C: Other Substances)*. The sample is melted and prepared for analysis on a potassium bromide plate.
Acid Number Between 3 and 9.
Heavy Metals (as Pb) Not more than 10 mg/kg.
Lead Not more than 2 mg/kg.
Ring-and-Ball Softening Point Between 82° and 90°.

TESTS

Acid Number Determine as directed in the general method, Appendix IX.

Heavy Metals Prepare and test a 2-g sample as directed in *Method II* under the *Heavy Metals Test*, Appendix IIIB, using 20 µg of lead ion (Pb) in the control (*Solution A*).
Lead Prepare a *Sample Solution* from a 5-g sample as directed in the general method under *Chewing Gum Base*, Appendix IV. This solution meets the requirements of the *Lead Limit Test*, Appendix IIIB, using 10 µg of lead ion (Pb) in the control.
Ring-and-Ball Softening Point Determine as directed in the general method, Appendix IX.

Packaging and Storage Store in well-closed containers.

Glycerol Ester of Partially Dimerized Rosin

DESCRIPTION

A hard, pale amber-colored resin (color M or paler as determined by ASTM Designation D 509) produced by the esterification of partially dimerized rosin with food-grade glycerin and purified by steam stripping. It is soluble in acetone, but is insoluble in water.

Functional Use in Foods Masticatory substance in chewing gum base.

REQUIREMENTS

Identification Identify Glycerol Ester of Partially Dimerized Rosin by comparing its infrared absorption spectrum with a typical spectrum as shown in the section on *Infrared Spectra* (*Series C: Other Substances*). The sample is melted and prepared for analysis on a potassium bromide plate.
Acid Number Between 3 and 8.
Drop Softening Point Between 109° and 119°.
Heavy Metals (as Pb) Not more than 10 mg/kg.
Lead Not more than 1 mg/kg.

TESTS

Acid Number Determine as directed in the general method, Appendix IX.
Drop Softening Point Determine as directed in the general method, Appendix IX, using a bath temperature of 125°.
Heavy Metals Prepare and test a 2-g sample as directed in *Method II* under the *Heavy Metals Test*, Appendix IIIB, using 20 µg of lead ion (Pb) in the control (*Solution A*).
Lead Prepare a *Sample Solution* from a 5-g sample as directed in the general method under *Chewing Gum Base*, Appendix IV. This solution meets the requirements of the *Lead Limit Test*, Appendix IIIB, using 5 µg of lead ion (Pb) in the control.

Packaging and Storage Store in well-closed containers.

Glycerol Ester of Partially Hydrogenated Gum Rosin

DESCRIPTION

A medium-hard, pale amber-colored resin (color N or paler as determined by ASTM Designation D 509) produced by the esterification of partially hydrogenated gum rosin with food-grade glycerin and purified by steam stripping. It is soluble in acetone and in toluene, but is insoluble in water and in alcohol.

Functional Use in Foods Masticatory substance in chewing gum base.

REQUIREMENTS

Identification Identify glycerol ester of partially hydrogenated gum rosin by comparing its infrared absorption spectrum with a typical spectrum as shown in the section on *Infrared Spectra* (*Series C: Other Substances*). The sample is melted and prepared for analysis on a potassium bromide plate.
Acid Number Between 3 and 10.
Drop Softening Point Between 79° and 88°.
Heavy Metals (as Pb) Not more than 10 mg/kg.
Lead Not more than 2 mg/kg.

TESTS

Acid Number Determine as directed in the general method, Appendix IX.
Drop Softening Point Determine as directed in the general method, Appendix IX, using a bath temperature of 100°.
Heavy Metals Prepare and test a 2-g sample as directed in *Method II* under the *Heavy Metals Test*, Appendix IIIB, using 20 µg of lead ion (Pb) in the control (*Solution A*).
Lead Prepare a *Sample Solution* from a 5-g sample as directed in the general method under *Chewing Gum Base*, Appendix IV. This solution meets the requirements of the *Lead Limit Test*, Appendix IIIB, using 10 µg of lead ion (Pb) in the control.

Packaging and Storage Store in well-closed containers.

Glycerol Ester of Partially Hydrogenated Wood Rosin

DESCRIPTION

A medium-hard, pale amber-colored resin (color N or paler as determined by ASTM Designation D 509) produced by the esterification of partially hydrogenated wood rosin with food-grade glycerin and purified by steam stripping. It is soluble in acetone, but is insoluble in water and in alcohol.

Functional Use in Foods Masticatory substance in chewing gum base.

REQUIREMENTS

Identification   Identify Glycerol Ester of Partially Hydrogenated Wood Rosin by comparing its infrared absorption spectrum with a typical spectrum as shown in the section on *Infrared Spectra (Series C: Other Substances)*. The sample is melted and prepared for analysis on a potassium bromide plate.
Acid Number   Between 3 and 10.
Ring-and-Ball Softening Point   Between 68° and 78°.
Heavy Metals (as Pb)   Not more than 10 mg/kg.
Lead   Not more than 1 mg/kg.

TESTS

Acid Number   Determine as directed in the general method, Appendix IX.
Ring-and-Ball Softening Point   Determine as directed in the general method, Appendix IX, using a water bath.
Heavy Metals   Prepare and test a 2-g sample as directed in *Method II* under the *Heavy Metals Test*, Appendix IIIB, using 20 µg of lead ion (Pb) in the control (*Solution A*).
Lead   Prepare a *Sample Solution* from a 5-g sample as directed in the general method under *Chewing Gum Base*, Appendix IV. This solution meets the requirements of the *Lead Limit Test*, Appendix IIIB, using 5 µg of lead ion (Pb) in the control.

Packaging and Storage   Store in well-closed containers.

Glycerol Ester of Polymerized Rosin

DESCRIPTION

A hard, pale amber-colored resin (color M or paler as determined by ASTM Designation D 509) produced by the esterification of polymerized rosin with food-grade glycerin and purified by steam stripping. It is soluble in acetone, but is insoluble in water and in alcohol.

Functional Use in Foods   Masticatory substance in chewing gum base.

REQUIREMENTS

Identification   Identify Glycerol Ester of Polymerized Rosin by comparing its infrared absorption spectrum with a typical spectrum as shown in the section on *Infrared Spectra (Series C: Other Substances)*. The sample is melted and prepared for analysis on a potassium bromide plate.
Acid Number   Between 3 and 12.
Heavy Metals (as Pb)   Not more than 10 mg/kg.
Lead   Not more than 1 mg/kg.
Ring-and-Ball Softening Point   Between 80° and 126°.

TESTS

Acid Number   Determine as directed in the general method, Appendix IX.
Heavy Metals   Prepare and test a 2-g sample as directed in *Method II* under the *Heavy Metals Test*, Appendix IIIB, using 20 µg of lead ion (Pb) in the control (*Solution A*).
Lead   Prepare a *Sample Solution* from a 5-g sample as directed in the general method under *Chewing Gum Base*, Appendix IV. This solution meets the requirements of the *Lead Limit Test*, Appendix IIIB, using 5 µg of lead ion (Pb) in the control.
Ring-and-Ball Softening Point   Determine as directed in the general method, Appendix IX.

Packaging and Storage   Store in well-closed containers.

Glycerol Ester of Tall Oil Rosin

DESCRIPTION

A pale amber-colored resin (color N or paler as determined by ASTM Designation D 509) produced by the esterification of tall oil rosin with food-grade glycerin and purified by steam stripping. It is soluble in acetone, but is insoluble in water.

Functional Use in Foods   Masticatory substance in chewing gum base.

REQUIREMENTS

Identification   Identify Glycerol Ester of Tall Oil Rosin by comparing its infrared absorption spectrum with a typical spectrum as shown in the section on *Infrared Spectra (Series C: Other Substances)*. The sample is melted and prepared for analysis on a potassium bromide plate.
Acid Number   Between 2 and 12.
Heavy Metals (as Pb)   Not more than 10 mg/kg.
Lead   Not more than 1 mg/kg.
Ring-and-Ball Softening Point   Between 80° and 88°.

TESTS

Acid Number   Determine as directed in the general method, Appendix IX.
Heavy Metals   Prepare and test a 2-g sample as directed in *Method II* under the *Heavy Metals Test*, Appendix IIIB, using 20 µg of lead ion (Pb) in the control (*Solution A*).
Lead   Prepare a *Sample Solution* from a 5-g sample as directed in the general method under *Chewing Gum Base*, Appendix IV. This solution meets the requirements of the *Lead Limit Test*, Appendix IIIB, using 5 µg of lead ion (Pb) in the control.
Ring-and-Ball Softening Point   Determine as directed in the general method, Appendix IX.

Packaging and Storage   Store in well-closed containers.

Glycerol Ester of Wood Rosin

INS: 445 CAS: [8050-30-4]

DESCRIPTION

A hard, pale amber-colored resin (color N or paler as determined by ASTM Designation D 509) produced by the esterification of pale wood rosin with food-grade glycerin. When intended for use in chewing gum base, the product is usually purified by steam stripping, but when intended for use in adjusting the density of citrus oils for beverages, it is purified by countercurrent steam distillation. It is soluble in acetone, but is insoluble in water.

Functional Use in Foods Masticatory substance in chewing gum base; beverage stabilizer.

REQUIREMENTS

Identification Identify Glycerol Ester of Wood Rosin by comparing its infrared absorption spectrum with a typical spectrum as shown in the section on *Infrared Spectra (Series C: Other Substances)*. The sample is melted and prepared for analysis on a potassium bromide plate.
Acid Number Between 3 and 9.
Ring-and-Ball Softening Point Between 82° and 90°.
Heavy Metals (as Pb) Not more than 10 mg/kg.
Lead Not more than 1 mg/kg.

TESTS

Acid Number Determine as directed in the general method, Appendix IX.
Ring-and-Ball Softening Point Determine as directed in the general method, Appendix IX.
Heavy Metals Prepare and test a 2-g sample as directed in *Method II* under the *Heavy Metals Test*, Appendix IIIB, using 20 µg of lead ion (Pb) in the control (*Solution A*).
Lead Prepare a *Sample Solution* from a 5-g sample as directed in the general method under *Chewing Gum Base*, Appendix IV. This solution meets the requirements of the *Lead Limit Test*, Appendix IIIB, using 5 µg of lead ion (Pb) in the control.

Packaging and Storage Store in well-closed containers.

Glyceryl Behenate

Glyceryl Tribehenate; Glyceryl Tridocosanoate

DESCRIPTION

Glyceryl Behenate is a mixture of fatty acid glycerides, primarily glyceryl esters of behenic acid. It is a fine powder that melts at about 70°. It is soluble in chloroform and practically insoluble in water and in alcohol.

Functional Use in Foods Formulation aid.

REQUIREMENTS

Identification
A. (**Caution:** Ether is highly volatile and flammable. Its vapor, when mixed with air and ignited, may explode.)
Solvent Mixture Prepare a chloroform–acetone (96:4) mixture.
Chromatographic Plates Use suitable thin-layer chromatographic plates (see *Thin-Layer Chromatography,* Appendix IIA) coated with a 0.25-mm layer of chromatographic silica gel. Pretreat the plates by placing them in a chromatographic chamber saturated with ether. Remove the plates from the chamber, allow the ether to evaporate, and immerse them in a 2.5% solution of boric acid in alcohol. After about 1 min, withdraw the plates and allow them to dry at ambient temperature. Heat to 110° for 30 min to activate the plates, and then keep them in a desiccator.
Procedure Apply 10 µL of a 6% solution of the sample in chloroform and 10 µL of a 6% solution of USP Glyceryl Behenate Reference Standard in chloroform on one of the chromatographic plates. Develop the chromatogram in the solvent mixture until the solvent front has moved about 12 cm. Remove the plate from the developing chamber and allow the solvent to evaporate. Spray the chromatogram with a 0.02% solution of dichlorofluorescein in alcohol. Examine the spots under short-wavelength ultraviolet light: the R_f values of the spots obtained from the test solution correspond to those obtained from the standard solution.
B. Dissolve about 22 mg of the sample in 1 mL of toluene in a screw-cap vial with a Teflon-lined septum. Add about 0.4 mL of 0.2 N methanolic (*m*-trifluoromethylphenyl)trimethylammonium hydroxide, attach the cap, and mix. Allow the vial to stand at room temperature for not less than 30 min. Introduce a suitable volume into a gas chromatograph equipped with a flame-ionization detector and a column 1.8 m in length and 4 mm in internal diameter packed with a 10% coating of 50% 3-cyanopropyl–50% phenylmethylsilicone (SP 2300 or equivalent) on a silanized siliceous earth support (Supelcoport, or equivalent) maintained at a temperature of about 225°. The retention time of the main peak in the resulting chromatogram corresponds to that of the main peak in a similar preparation of USP Glyceryl Behenate Reference Standard chromatographed concomitantly. The ratio of the response of the main peak to the sum of all the responses is not less than 0.83.
Acid Value Not more than 4.
Heavy Metals (as Pb) Not more than 10 mg/kg.
Iodine Value Not more than 3.
1-Monoglycerides Content Not less than 12.0% and not more than 18.0%.
Free Glycerin Not more than 1.0%.
Residue on Ignition Not more than 0.1%.
Saponification Value Not less than 145 and not more than 165.

TESTS

Acid Value In a flask, suspend about 10 g of the sample, accurately weighed, in 50 mL of a 1:1 mixture of alcohol and ether that has been neutralized to phenolphthalein with 0.1 N sodium hydroxide. Connect the flask with a suitable condenser, and warm, with frequent shaking, for about 10 min. Add 1 mL of phenolphthalein TS, and titrate with 0.1 N sodium hydroxide until the solution remains faintly pink after shaking for 30 s. Calculate the acid value by the formula

$$56.1V \times N/W,$$

in which V is the volume, in mL, of the 0.1 N sodium hydroxide solution; N is the normality of the sodium hydroxide solution; and W is the weight, in g, of the sample taken.

Heavy Metals Proceed as directed under *Method II* of the *Heavy Metals Test*, Appendix IIIB, using a 2-g sample and 20 µg of lead ion (Pb) in the control solution (*Solution A*).

Iodine Value Proceed as directed under *Iodine Value*, Appendix VII, using the *Modified Wijs Method*.

1-Monoglycerides Content Melt the sample at a temperature not higher than 80°, and mix. Using about 1 g, accurately weighed, proceed as directed in the general method, Appendix VII.

Note: If the Glyceryl Behenate titration is less than 0.8 volumes of the blank titration, discard and repeat, using a smaller weight of Glyceryl Behenate.

Calculate the percentage of 1-monoglycerides, as glyceryl monobehenate, by the formula

$$20.73N(B - S)/W,$$

in which 20.73 is one-twentieth of the molecular weight of glyceryl monobehenate; N is the normality of the sodium thiosulfate; B and S are the volumes, in mL, of 0.1 N sodium thiosulfate consumed by the blank and the Glyceryl Behenate, respectively; and W is the weight, in g, of Glyceryl Behenate taken.

Free Glycerin Proceed as directed in the general method, Appendix VII.

Residue on Ignition Ignite 5 g of the sample as directed in the general method, Appendix IIC.

Saponification Value Determine as directed for *Saponification Value* under *Fats and Related Substances*, Appendix VII.

Packaging and Storage Store in tight containers at a temperature not higher than 35°.

Glyceryl-Lacto Esters of Fatty Acids

Lactated Mono-Diglycerides

DESCRIPTION

A mixture of partial lactic and fatty acid esters of glycerin. It varies in consistency from a soft to a hard, waxy solid. It is dispersible in hot water, and is moderately soluble in hot isopropanol, in xylene, and in cottonseed oil.

Functional Use in Foods Emulsifier; stabilizer.

REQUIREMENTS

Identification Transfer into a 25-mL glass-stoppered test tube 1 mL of the solution of the sample remaining after titrating with 0.1 N potassium hydroxide in the determination of *Total Lactic Acid*, add 0.1 mL of cupric sulfate solution (1 g of $CuSO_4.5H_2O$ in 25 mL of water) and 6 mL of sulfuric acid, and mix. Stopper loosely, heat in a boiling water bath for 5 min, and then cool in an ice bath for 5 min. Remove from the ice bath, add 0.1 mL of *p*-phenylphenol solution (75 mg dissolved in 5 mL of 1 N sodium hydroxide), and mix. Allow to stand at room temperature for 1 min, then heat in a boiling water bath for 1 min. A deep, blue-violet color indicates the presence of lactic acid.

Heavy Metals (as Pb) Not more than 5 mg/kg.
Lead Not more than 0.5 mg/kg.
Residue on Ignition Not more than 0.1%.
Unsaponifiable Matter Not more than 2.0%.

The following specifications should conform to the representations of the vendor: **Acid Value**, **Free Glycerin**, **1-Monoglyceride Content**, **Total Lactic Acid**, and **Water**.

TESTS

Acid Value Determine as directed for *Acid Value, Method II*, under *Fats and Related Substances*, Appendix VII.

Free Glycerin Determine as directed in the general method, Appendix VII.

Heavy Metals Prepare and test a 4-g sample as directed in *Method II* under the *Heavy Metals Test*, Appendix IIIB, using 20 µg of lead ion (Pb) in the control (*Solution A*).

Lead Prepare and test a 4-g sample as directed under the *Lead Limit Test*, Appendix IIIB, using 2 µg of lead ion (Pb) in the control.

1-Monoglyceride Content Determine as directed in the general method, Appendix VII.

Residue on Ignition, Appendix IIC Ignite 1 g as directed in the general method.

Total Lactic Acid Transfer an accurately weighed portion of the melted sample, equivalent to between 140 and 170 mg of lactic acid, into a 250-mL Erlenmeyer flask. Pipet 20 mL of 0.5 N alcoholic potassium hydroxide into the flask, connect an air condenser at least 65 cm in length, and reflux for 30 min. Run a blank determination using the same volume of alkali. Add 20 mL of water to each flask, then disconnect the condensers, evaporate to a volume of 20 mL, and cool to about 40°. Add methyl red TS to each flask, and titrate the blank with 0.5 N hydrochloric acid. Add exactly the same volume of 0.5 N hydrochloric acid to the sample flask, with swirling. To each flask add 50 mL of hexane, swirl vigorously to dissolve the fatty acids in the sample flask, then transfer quantitatively the contents of each flask into separate 250-mL separators and shake for 30 s. Collect the aqueous phases in 300-mL Erlenmeyer flasks, wash the hexane solutions with 50 mL of water, and combine the wash solutions with the original aqueous phases in the Erlenmeyer flasks, discarding the hexane solutions. Add phenolphthalein TS, and titrate with 0.1 N potassium hydroxide to a pink color that persists for at least 30 s. Each mL of 0.1 N potassium hydroxide is equivalent to 9.008 mg of lactic acid ($C_3H_6O_3$).

Unsaponifiable Matter, Appendix VII   Determine as directed in the general method.

Water   Determine by the *Karl Fischer Titrimetric Method*, Appendix IIB.

Packaging and Storage   Store in well-closed containers.

Glyceryl Monooleate

$C_{21}H_{40}O_4$ 　　　　　　　　　　　Formula wt 356.54

　　　　　　　　　　　　　　　　　　CAS: [25496-72-4]

DESCRIPTION

Glyceryl Monooleate is prepared by esterifying glycerin with food-grade oleic acid in the presence of a suitable catalyst such as aluminum oxide. It also occurs in many animal and vegetable fats such as tallow and cocoa butter. It is soluble in hot alcohol and in chloroform; very slightly soluble in cold alcohol, in ether, and in petroleum ether; and insoluble in water. It melts at around 15°. It may also contain tri- and diesters.

Functional Use in Foods   Moisture control agent; emulsifier; flavoring agent.

REQUIREMENTS

Identification   Glyceryl Monooleate exhibits the following typical composition profile of fatty acids as determined under *Fatty Acid Composition*, Appendix VII.

Fatty Acid:	≤12	12:0	14:0	16:0	16:1
Weight % (Range):	0	0	<4	1–5	<9

Fatty Acid:	18:0	18:1	18:2	≥20
Weight % (Range):	<3.0	≥82	3–7	<1.5

Assay   Not less than 35.0% monoglycerides, calculated on the anhydrous basis.
Acid Value   Not more than 6.
Free Glycerin   Not more than 6.0%.
Heavy Metals   Not more than 10 mg/kg.
Hydroxyl Value   Between 300 and 330.
Iodine Value   Between 58 and 80.
Lead   Not more than 1 mg/kg.
Residue on Ignition   Not more than 0.1%.
Saponification Value   Between 160 and 176.
Water   Not more than 1.0%

TESTS

Assay

Propionating Reagent   Mix 10 mL of pyridine and 20 mL of propionic anhydride.

Internal Standard Solution   Transfer about 400 mg of hexadecyl hexadecanoate, accurately weighed, to a 100-mL volumetric flask, dilute with chloroform to volume, and mix.

Standard Preparation   Transfer about 50 mg of USP Monoglycerides Reference Standard, accurately weighed, to a 25-mL flask, add by pipet 5 mL of *Internal Standard Solution*, and mix. When solution is complete, immerse the flask in a water bath maintained at a temperature between 45° and 50°, and volatilize the chloroform with the aid of a stream of nitrogen. Add 3.0 mL of *Propionating Reagent*, and heat on a hot plate at 75° for 30 min. Evaporate the reagents with the aid of a stream of nitrogen and gentle steam heat. Add 15 mL of chloroform, and swirl to dissolve the residue.

Assay Preparation   Transfer about 50 mg of the sample, accurately weighed, to a 25-mL conical flask, and proceed as directed for *Standard Preparation*, beginning with "... add by pipet 5 mL of *Internal Standard Solution*, ..."

Chromatographic System   Under typical conditions the gas chromatograph is equipped with a flame-ionization detector, and contains a 2.4-m × 4-mm borosilicate glass column packed with 2% liquid phase, 5% phenyl methyl silicone on 80- to 100-mesh support (Supelcoport, or equivalent). The column is maintained isothermally at a temperature between 270° and 280°, the injection port and detector block are maintained at about 310°, and helium is used as the carrier gas at a flow rate of about 70 mL/min.

System Suitability   Chromatograph 6 to 10 injections of the *Standard Preparation* as directed under *Procedure*: The resolution factor, R, between the peaks for the derivatized glyceryl hexadecanoate and glyceryl octadecanoate is not less than 2.0, and the relative standard deviation of the ratio of the peak area of the derivatized glyceryl *cis*-9-octadecanoate to that of the hexadecyl hexadecanoate is not more than 2.0%.

Procedure   Inject a suitable portion of the *Standard Preparation* into a suitable gas chromatograph, and record the chromatogram. Measure the areas under the peaks, and record the values of the sum of the areas under the derivatized monoglyceride peaks and of the area under the hexadecyl hexadecanoate peak as A_S and A_D, respectively. Calculate the response factor, F, taken by the formula

$$(A_S/A_D)(W_D/W_S),$$

in which W_D and W_S are the weights, in mg, of hexadecyl hexadecanoate and USP Monoglycerides Reference Standard, respectively, in the *Standard Preparation*. Similarly inject a suitable portion of the *Assay Preparation*, and record the chromatogram. Measure the areas under the peaks, and record the values of the sum of the areas under the derivatized monoglyceride peaks and of the area under the hexadecyl hexadecanoate peak as a_U and a_D, respectively. Calculate the quantity, in mg, of monoglycerides in the amount of Glyceryl Monooleate taken by the formula

$$(W_D/F)(a_U/a_D).$$

Acid Value Determine as directed for *Acid Value* under *Fats and Related Substances*, Appendix VII.

Free Glycerin

Propionating Reagent Mix 10 mL of pyridine with 20 mL of propionic anhydride.

Internal Standard Solution Dissolve a suitable quantity of tributyrin, accurately weighed, in chloroform and dilute quantitatively with chloroform to obtain a solution having a concentration of about 0.2 mg/mL.

Standard Preparation Transfer about 15 mg of glycerin and about 50 mg of tributyrin, both accurately weighed, to a glass-stoppered, 25-mL conical flask, add 3 mL of *Propionating Reagent*, and heat at 75° for 30 min. Volatilize the reagents with the aid of a stream of nitrogen at room temperature, add about 12 mL of chloroform, and mix. Dilute about 1 mL of this mixture with chloroform to about 20 mL and mix.

Test Preparation Transfer about 50 mg of the sample, accurately weighed, to a glass-stoppered, 25-mL conical flask, add by pipet 5 mL of *Internal Standard Solution*, and mix to dissolve. Immerse the flask in a water bath maintained at a temperature between 45° and 50°, and volatilize the chloroform with the aid of a stream of nitrogen. Add 3 mL of *Propionating Reagent*, and heat at 75° for 30 min. Volatilize the reagents with the aid of a stream of nitrogen at room temperature, add about 5 mL of chloroform, and mix.

Chromatographic System Under typical conditions, the gas chromatograph is equipped with a flame-ionization detector and contains a 2.4-m × 4-mm borosilicate glass column packed with 2% liquid phase consisting of a high-molecular-weight compound of polyethylene glycol and a diepoxide (Carbowax 20 M, or equivalent) on an 80- to 100-mesh siliceous earth support (Chromosorb W AW DMCS, or equivalent). The column is maintained isothermally at a temperature between 190° and 200°; the injection port and detector block are maintained at about 300° and 310°, respectively; and helium is used as the carrier gas at a flow rate of about 70 mL/min.

System Suitability Chromatograph 6 to 10 injections of the *Standard Preparation* as directed under *Procedure*: The resolution factor, R, between the peaks for the derivatized glycerin and tributyrin is not less than 4.0, and the relative standard deviation of the ratio of their peak areas is not more than 2.0%.

Procedure Inject a suitable portion of the *Standard Preparation* into a suitable gas chromatograph and record the chromatogram. Measure the areas under the peaks and record the values of the areas under the tripropionin and tributyrin peaks as A_S and A_D, respectively. Calculate the response factor, F, taken by the formula

$$(A_D/A_S)(W_S/W_D),$$

in which W_S and W_D are the weights, in mg, of glycerin and tributyrin, respectively, in the *Standard Preparation*. Similarly inject a suitable portion of the *Test Preparation*, and record the chromatogram. Measure the areas under the peaks and record the values of the areas under the tripropionin and tributyrin peaks as a_U and a_D, respectively. Calculate the percentage of glycerin by the formula

$$100F(a_U/a_D)(w_D/w_U),$$

in which w_D is the weight, in mg, of tributyrin in 5 mL of *Internal Standard Solution*, and w_U is the weight, in mg, of Glyceryl Monooleate in the *Test Preparation*.

Heavy Metals Proceed as directed using a 2-g sample under *Method II* of the *Heavy Metals Test*, Appendix IIIB, and 20 µg lead ion (Pb) in the control solution (*Solution A*).

Hydroxyl Value, Appendix VII Proceed as directed for *Method II*.

Iodine Value Proceed as directed under *Iodine Value*, Appendix VII, using the *Modified Wijs Method*.

Lead A *Sample Solution* prepared as directed for organic compounds using a 3-g sample meets the requirements of the *Lead Limit Test*, Appendix IIIB, using 3 µg of lead ion (Pb) in the control.

Residue on Ignition Ignite 5 g of the sample as directed in the general method, Appendix IIC.

Saponification Value Determine as directed for *Saponification Value* under *Fats and Related Substances*, Appendix VII.

Water Determine by the *Karl Fischer Titrimetric Method*, Appendix IIB. Use a 0.5-g sample and 20 mL of a 1:1 methanol-chloroform mixture.

Packaging and Storage Store in tight, light-resistant containers.

Glyceryl Monostearate

Monostearin; 1,2,3-Propanetriol Octadecanoate

DESCRIPTION

Glyceryl Monostearate occurs as a white, waxlike solid, as flakes, or as beads. It is a mixture of Glyceryl Monostearate and glyceryl monopalmitate. It may contain a suitable antioxidant. It is soluble in hot organic solvents such as acetone, alcohol, and ether and in mineral or fixed oils. It is dispersible in hot water with the aid of soap or suitable surfactants.

Functional Use in Foods Emulsifier.

REQUIREMENTS

Identification Heat the sample with 3 parts water to 2° to 5° above its melting point. An irreversible gel forms when the sample is held at this temperature.

Assay Not less than 90.0% monoglycerides of saturated fatty acids.

Acid Value Not more than 6.

Free Glycerin Not more than 1.2%.

Heavy Metals (as Pb) Not more than 10 mg/kg.

Hydroxyl Value Between 300 and 330.

Iodine Value Not more than 3.

Lead Not more than 1 mg/kg.

Melting Range Not below 65°.

Residue on Ignition Not more than 0.1%.

Saponification Value Not less than 150 and not more than 165.

TESTS

Assay

Propionating Reagent Mix 10 mL of pyridine and 20 mL of propionic anhydride.

Internal Standard Solution Transfer about 400 mg of hexadecyl hexadecanoate, accurately weighed, to a 100-mL volumetric flask, dissolve in chloroform, dilute with chloroform to volume, and mix.

Standard Preparation Transfer about 50 mg of USP Monoglycerides Reference Standard, accurately weighed, to a 25-mL conical flask, add by pipet 5 mL of *Internal Standard Solution*, and mix. When solution is complete, immerse the flask in a water bath maintained at a temperature between 45° and 50°, and volatilize the chloroform with the aid of a stream of nitrogen. Add 3.0 mL of *Propionating Reagent*, and heat the flask on a hot plate at 75° for 30 min. Evaporate the reagents with the aid of a stream of nitrogen and gentle steam heat. Add 15 mL of chloroform, and swirl to dissolve the residue.

Assay Preparation Transfer about 50 mg of the sample, accurately weighed, to a 25-mL conical flask, and proceed as directed for *Standard Preparation*, beginning with "... add by pipet 5 mL of *Internal Standard Solution*, ..."

Chromatographic System Under typical conditions, the gas chromatograph is equipped with a flame-ionization detector and contains a 2.4-m × 4-mm borosilicate glass column packed with 2% liquid phase 5% phenyl methyl silicone (SE 52, or equivalent) on an 80- to 100-mesh siliceous earth support (Diatoport S, or equivalent). The column is maintained isothermally at a temperature between 270° and 280°, the injection port and detector block are maintained at about 310°, and helium is used as the carrier gas at a flow rate of about 70 mL/min.

System Suitability Chromatograph 6 to 10 injections of the *Standard Preparation* as directed under *Procedure*: The resolution factor, R, between the peaks for the derivatized glyceryl hexadecanoate and glyceryl octadecanoate is not less than 2.0, and the relative standard deviation of the ratio of the peak area of the derivatized glyceryl octadecanoate to that of the hexadecyl hexadecanoate is not more than 2.0%.

Procedure Inject a suitable portion of the *Standard Preparation* into a suitable gas chromatograph, and record the chromatogram. Measure the areas under the peaks and record the values of the sum of the areas under the derivatized monoglyceride peaks and of the area under the hexadecyl hexadecanoate peak as A_S and A_D, respectively. Calculate the response factor, F, taken by the formula

$$(A_S/A_D)(W_D/W_S),$$

in which W_D and W_S are the weights, in mg, of hexadecyl hexadecanoate and USP Monoglycerides Reference Standard, respectively, in the *Standard Preparation*. Similarly, inject a suitable portion of the *Assay Preparation* and record the chromatogram. Measure the areas under the peaks, and record the values of the sum of the areas under the derivatized monoglyceride peaks and of the area under the hexadecyl hexadecanoate peak as a_U and a_D, respectively. Calculate the quantity, in mg, of monoglycerides in the amount of Glyceryl Monostearate taken by the formula

$$(W_D/F)(a_U/a_D).$$

Acid Value Determine as directed for *Acid Value* under *Fats and Related Substances*, Appendix VII.

Free Glycerin

Propionating Reagent Mix 10 mL of pyridine with 20 mL of propionic anhydride.

Internal Standard Solution Dissolve a suitable quantity of tributyrin, accurately weighed, in chloroform, and dilute quantitatively with chloroform to obtain a solution with a concentration of about 0.2 mg/mL.

Standard Preparation Transfer about 15 mg of glycerin and about 50 mg of tributyrin, both accurately weighed, to a glass-stoppered, 25-mL conical flask, add 3 mL of *Propionating Reagent*, and heat at 75° for 30 min. Volatilize the reagents with the aid of a stream of nitrogen at room temperature, add about 12 mL of chloroform, and mix. Dilute about 1 mL of this mixture with chloroform to about 20 mL, and mix.

Test Preparation Transfer about 50 mg of Glyceryl Monostearate, accurately weighed, to a glass-stoppered, 25-mL conical flask, add by pipet 5 mL of *Internal Standard Solution*, and mix to dissolve. Immerse the flask in a water bath maintained at a temperature between 45° and 50°, and volatilize the chloroform with the aid of a stream of nitrogen. Add 3 mL of *Propionating Reagent*, and heat at 75° for 30 min. Volatilize the reagents with the aid of a stream of nitrogen at room temperature, add about 5 mL of chloroform, and mix.

Chromatographic System Under typical conditions, the gas chromatograph is equipped with a flame-ionization detector and contains a 2.4-m × 4-mm borosilicate glass column packed with 2% liquid phase consisting of a high-molecular-weight compound of polyethylene glycol and a diepoxide (Carbowax 20 M, or equivalent) on an 80- to 100-mesh siliceous earth support (Chromosorb W AW DMCS, or equivalent). The column is maintained isothermally at a temperature between 190° and 200°; the injection port and detector block are maintained at about 300° and 310°, respectively; and helium is used as carrier gas at a flow rate of about 70 mL/min.

System Suitability Chromatograph 6 to 10 injections of the *Standard Preparation* as directed under *Procedure*: The resolution factor, R, between the peaks for the derivatized glycerin and tributyrin is not less than 4.0, and the relative standard deviation of the ratio of their peak areas is not more than 2.0%.

Procedure Inject a suitable portion of the *Standard Preparation* into a suitable gas chromatograph, and record the chromatogram. Measure the areas under the peaks, and record the values of the areas under the tripropionin and tributyrin peaks as A_S and A_D, respectively. Calculate the response factor, F, taken by the formula

$$(A_D/A_S)(W_S/W_D),$$

in which W_S and W_D are the weights, in mg, of glycerin and tributyrin, respectively, in the *Standard Preparation*. Similarly, inject a suitable portion of the *Test Preparation*, and record the chromatogram. Measure the areas under the peaks, and record the values of the areas under the tripropionin and tributyrin peaks as a_U and a_D, respectively. Calculate the percentage of glycerin by the formula

$$100F(a_U/a_D)(w_D/w_U),$$

in which w_D is the weight, in mg, of tributyrin in 5 mL of *Internal Standard Solution*, and w_U is the weight, in mg, of Glyceryl Monostearate in the *Test Preparation*.

Heavy Metals Proceed as directed under *Method II* of the *Heavy Metals Test*, Appendix IIIB, using a 2-g sample and 20 µg of lead ion (Pb) in the control solution (*Solution A*).

Hydroxyl Value, Appendix VII Proceed as directed for *Method II*.

Iodine Value Proceed as directed under *Iodine Value*, Appendix VII, using the *Modified Wijs Method*.

Lead Prepare and test a 3-g sample as directed under the *Lead Limit Test*, Appendix IIIB, using 3 µg of lead ion (Pb) in the control.

Melting Range Determine as directed for *Melting Range* under *Fats and Related Substances*, Appendix VII.

Residue on Ignition Ignite 5 g of the sample as directed in the general method, Appendix IIC.

Saponification Value Determine as directed for *Saponification Value* under *Fats and Related Substances*, Appendix VII.

Packaging and Storage Store in tight, light-resistant containers.

Glyceryl Tristearate

Tristearin; Stearin; Octadecanoic Acid, 1,2,3-Propane Triyl Ester

$C_{57}H_{110}O_6$ Formula wt 891.45

CAS: [555-43-1]

DESCRIPTION

Glyceryl Tristearate is prepared by reacting glycerin with stearic acid in the presence of a suitable catalyst such as aluminum oxide. It also occurs in many animal and vegetable fats such as tallow and cocoa butter. It is a white, microfine crystalline powder. It is soluble in hot alcohol and in chloroform; very slightly soluble in cold alcohol, in ether, and in petroleum ether; and insoluble in water.

Functional Use in Foods Crystallization accelerator; lubricant; surface-finishing agent; formulation aid.

REQUIREMENTS

Identification Glyceryl Tristearate exhibits the following typical composition profile of fatty acids as determined under *Fatty Acid Composition*, Appendix VII.

Fatty Acid: Weight % (Range):	≤12	12:0	14:0	16:0	16:1
	0.0–0.3	0.0–0.5	0.0–1.0	0.0	0.0

Fatty Acid: Weight % (Range):	18:0	18:1	18:2	≥20
	>95	0.0–0.5	0.0–0.5	0.0–0.5

Acid Value Not more than 1.0.
Free Glycerin Not more than 0.5%.
Heavy Metals (as Pb) Not more than 10 mg/kg.
Hydroxyl Value Not more than 5.0.
Iodine Number Not more than 1.0.
Lead Not more than 1 mg/kg.
Melting Range Between 69° and 73°.
Residue on Ignition Not more than 0.1%.
Saponification Value Between 186 and 192.
Unsaponifiable Matter Not more than 0.5%.

TESTS

Acid Value Determine as directed for *Acid Value* under *Fats and Related Substances*, Appendix VII.

Free Glycerin Determine as directed in the general method, Appendix VII.

Heavy Metals Proceed as directed under *Method II* of the *Heavy Metals Test*, Appendix IIIB, using a 2-g sample and 20-µg lead ion (Pb) in the control solution (*Solution A*).

Hydroxyl Value, Appendix VII Proceed as directed for *Method II*.

Iodine Value Proceed as directed under *Iodine Value*, Appendix VII, using the *Modified Wijs Method*.

Lead Prepare and test a 3-g sample as directed for organic compounds in the *Lead Limit Test*, Appendix IIIB, using 3 µg of lead ion (Pb) in the control.

Melting Range Determine as directed for *Melting Range or Temperature, Procedure for Class II*, Appendix IIB.

Residue on Ignition Ignite 5 g of the sample as directed in the general method, Appendix IIC.

Saponification Value Determine as directed for *Saponification Value* under *Fats and Related Substances*, Appendix VII.

Unsaponifiable Matter Determine as directed under *Unsaponifiable Matter*, Appendix VII.

Packaging and Storage Store in tight, light-resistant containers.

Glycine

Aminoacetic Acid; Glycocoll

$$H_2NCH_2COOH$$

$C_2H_5NO_2$ Formula wt 75.07
INS: 640 CAS: [56-40-6]

DESCRIPTION

A white, odorless, crystalline powder having a sweetish taste. Its solution is acid to litmus. One g dissolves in about 4 mL of water. It is very slightly soluble in alcohol and in ether.

Functional Use in Foods Nutrient; dietary supplement.

REQUIREMENTS

Identification
  A. To 5 mL of a 1 in 10 solution add 5 drops of 2.7 N hydrochloric acid and 5 drops of a solution of sodium nitrite (1 in 2). A vigorous evolution of a colorless gas is produced.
  B. Add 1 mL of ferric chloride TS to 2 mL of a 1 in 10 solution. A red color is produced that disappears upon the addition of an excess of 2.7 N hydrochloric acid, and that reappears upon the addition of an excess of ammonium hydroxide.
  C. To 2 mL of a 1 in 10 solution add 1 drop of liquefied phenol and 5 mL of sodium hypochlorite TS. A blue color is produced.
Assay Not less than 98.5% and not more than 101.5% of $C_2H_5NO_2$ after drying.
Heavy Metals (as Pb) Not more than 0.002%.
Lead Not more than 5 mg/kg.
Loss on Drying Not more than 0.2%.
Residue on Ignition Not more than 0.1%.

TESTS

Assay Transfer about 175 mg, previously dried at 105° for 3 h and accurately weighed, to a 250-mL flask. Dissolve the sample in 50 mL of glacial acetic acid, add 2 drops of crystal violet TS, and titrate with 0.1 N perchloric acid to a bluish green endpoint. Perform a blank determination (see *General Provisions*) and make any necessary correction. Each mL of 0.1 N perchloric acid is equivalent to 7.507 mg of $C_2H_5NO_2$.
Heavy Metals A solution of 1 g in 25 mL of water meets the requirements of the *Heavy Metals Test*, Appendix IIIB, using 20 µg of lead ion (Pb) in the control (*Solution A*).
Lead A *Sample Solution* prepared as directed for organic compounds meets the requirements of the *Lead Limit Test*, Appendix IIIB, using 5 µg of lead ion (Pb) in the control.
Loss on Drying, Appendix IIC Dry at 105° for 3 h.
Residue on Ignition, Appendix IIC Ignite 2 g as directed in the general method.

Packaging and Storage Store in well-closed containers.

Grapefruit Oil, Coldpressed

Grapefruit Oil, Expressed; Oil of Shaddock

DESCRIPTION

The oil obtained by expression from the fresh peel of the grapefruit *Citrus paradisi* Macfayden (*Citrus decumana* L.). It is a yellow, sometimes reddish liquid, often showing a flocculent separation of waxy material. It is soluble in most fixed oils and in mineral oil, often with opalescence or cloudiness. It is slightly soluble in propylene glycol and insoluble in glycerin. It may contain a suitable antioxidant.

Functional Use in Foods Flavoring agent.

REQUIREMENTS

Identification The infrared absorption spectrum of the sample exhibits relative maxima (that may vary in intensity) at the same wavelengths (or frequencies) as those shown in the respective spectrum under *Infrared Spectra* (*Series A: Essential Oils*), using the same test conditions as specified therein.
Angular Rotation Between +91° and +96°.
Heavy Metals (as Pb) Not more than 0.004%.
Lead Not more than 10 mg/kg.
Refractive Index Between 1.475 and 1.478 at 20°.
Residue on Evaporation Between 5.0% and 10.0%.
Specific Gravity Between 0.848 and 0.856.

TESTS

Angular Rotation Determine in a 100-mm tube as directed under *Optical (Specific) Rotation*, Appendix IIB.
Heavy Metals Prepare and test a 500-mg sample as directed in *Method II* under the *Heavy Metals Test*, Appendix IIIB, using 20 µg of lead ion (Pb) in the control (*Solution A*).
Lead A *Sample Solution* prepared as directed for organic compounds meets the requirements of the *Lead Limit Test*, Appendix IIIB, using 10 µg of lead ion (Pb) in the control.
Refractive Index, Appendix IIB Determine with an Abbé or other refractometer of equal or greater accuracy.
Residue on Evaporation Proceed as directed in the general method, Appendix VI, heating for 5 h.
Specific Gravity Determine by any reliable method (see *General Provisions*).

Packaging and Storage Store in full, tight, preferably glass or tin-lined containers in a cool place protected from light.

Grape Skin Extract

Enocianina

CAS: [11029-12-2]

DESCRIPTION

Grape Skin Extract is a red to purple powder or liquid concentrate prepared by aqueous extraction of grape marc remaining from the pressing of grapes to obtain juice. Extraction is effected with water containing sulfur dioxide. After concentration by vacuum evaporation, the sugar content is reduced by fermentation; further concentration removes most of the alcohol. The primary color components are anthocyanins such as the glucosides of malvidin, peonidin, petunidin, delphinidin, or cyanidin. Other components naturally present are sugars, tartrates, malates, tannins, and minerals. Alcohol or sulfur dioxide may be added. The powder may contain an added carrier such as maltodextrin, modified starch, or gum. In acid solution, Grape Skin Extract is red; in neutral to alkaline solution, it is violet to blue.

Functional Use in Foods Color.

REQUIREMENTS

Identification Transfer 1 g of sample and 1 g of potassium metabisulfite to a 100-mL volumetric flask, dissolve in about 50 mL of pH 3.0 *Citrate–Citric Acid Buffer* (see the **TEST** entitled *Assay*, below), and dilute to volume with the same buffer. The red color due to anthocyanins is bleached.
Assay The color strength (CS) expressed as the absorbance of a 1% solution in a cell of 1-cm pathlength at pH 3.0 shall not be less than 90% of the color strength as represented by the vendor.
Arsenic Not more than 1 mg/kg.
Lead Not more than 10 mg/kg.
Pesticides Pesticide levels shall conform with national regulations in the country of use.

TESTS

Assay Transfer about 0.2 g of Grape Skin Extract, accurately weighed, to a 100-mL volumetric flask, dissolve in about 25 mL of pH 3.0 *Citrate–Citric Acid Buffer*, and dilute to volume with the same buffer. Prepare the buffer by adding 0.1 M sodium citrate dropwise to 0.1 M citric acid until a pH of 3.0 is reached, as determined by a glass electrode. Remove any undissolved material by filtration or centrifugation. Adjust the pH to 3.0, and determine the absorbance of the clarified solution at 525 nm in a cell with a 1-cm pathlength. The color strength expressed as the absorbance of a 1% solution in a cell of 1-cm pathlength is calculated as

$$CS = \text{Absorbance at 525 nm/Sample Weight in g.}$$

Arsenic A *Sample Solution* from a 3-g sample prepared as directed for organic compounds meets the requirements of the *Arsenic Test*, Appendix IIIB.
Lead A *Sample Solution* prepared as directed for organic compounds meets the requirements of the *Lead Limit Test*, Appendix IIIB, using 10 µg of lead ion (Pb) in the control.

Packaging and Storage Store liquid Grape Skin Extract with aseptic packaging or in high-density polyethylene containers at 4° to 14°. Store powdered Grape Skin Extract in fiber drums at room temperature.

Guar Gum

CAS: [9000-30-0]

DESCRIPTION

A gum obtained from the ground endosperms of *Cyamopsis tetragonolobus* (L.) Taub (Fam. *Leguminosae*) (synonym *Cyamopsis psoraloides* [Lam.] D.C.). It consists chiefly of a high-molecular-weight hydrocolloidal polysaccharide, composed of galactose and mannose units combined through glycosidic linkages, which may be described chemically as a galactomannan. It is a white to yellowish white, nearly odorless powder. It is dispersible in either hot or cold water, forming a sol, having a pH between 5.4 and 7.0, that may be converted to a gel by the addition of small amounts of sodium borate.

Functional Use in Foods Stabilizer; thickener; emulsifier.

REQUIREMENTS

Identification
   A. Transfer a 2-g sample into a 400-mL beaker, moisten it thoroughly with about 4 mL of isopropyl alcohol, add with vigorous stirring 200 mL of cold water, and continue the stirring until the gum is completely and uniformly dispersed. An opalescent, viscous solution is formed.
   B. Transfer 100 mL of the solution prepared in *Identification Test A* into another 400-mL beaker, heat the mixture in a boiling water bath for about 10 min, and then cool to room temperature. No appreciable increase in viscosity is produced.
Acid-Insoluble Matter Not more than 7.0%.
Arsenic (as As) Not more than 3 mg/kg.
Ash (Total) Not more than 1.5%.
Galactomannans Not less than 70.0%.
Heavy Metals (as Pb) Not more than 0.002%.
Lead Not more than 5 mg/kg.
Loss on Drying Not more than 15.0%.
Protein Not more than 10.0%.
Starch Passes test.

TESTS

Acid-Insoluble Matter Transfer 1.5 g of the sample, accurately weighed, into a 250-mL beaker containing 150 mL of water and 1.5 mL of sulfuric acid. Cover the beaker with a watch glass, and heat the mixture on a steam bath for 6 h, rubbing down the wall of the beaker frequently with a rubber-tipped stirring rod and replacing any water lost by evaporation. At the end of the 6-h heating period, add about 500 mg of a suitable filter aid, previously dried for 3 h at 105° and accurately weighed, and filter through a tared, sintered-glass filter crucible. Wash the residue several times with hot water, dry the crucible and its contents at 105° for 3 h, cool in a desiccator, and weigh. The difference between the weight of the filter aid and that of the residue is the weight of *Acid-Insoluble Matter*.
Arsenic A *Sample Solution* prepared as directed for organic compounds meets the requirements of the *Arsenic Test*, Appendix IIIB.
Ash (Total) Determine as directed in the general method, Appendix IIC.
Galactomannans The difference between the sum of the percentages of *Acid-Soluble Matter*, *Total Ash*, *Loss on Drying*, and *Protein* and 100 represents the percentage of *Galactomannans*.
Heavy Metals Prepare and test a 1-g sample as directed in *Method II* under the *Heavy Metals Test*, Appendix IIIB, using 20 µg of lead ion (Pb) in the control (*Solution A*).
Lead A *Sample Solution* prepared as directed for organic compounds meets the requirements of the *Lead Limit Test*, Appendix IIIB, using 5 µg of lead ion (Pb) in the control.
Loss on Drying, Appendix IIC Dry at 105° for 5 h.
Protein Transfer about 3.5 g, accurately weighed, into a 500-mL Kjeldahl flask, and proceed as directed under *Nitrogen Determination*, Appendix IIIC. The percentage of nitrogen determined, multiplied by 6.25, gives the percentage of protein in the sample.
Starch To a 1 in 10 solution of the gum add a few drops of iodine TS. No blue color is produced.

Packaging and Storage Store in well-closed containers.

Gum Ghatti

Indian Gum

CAS: [9000-28-6]

DESCRIPTION

Gum Ghatti is the dried gummy exudate from the stems of *Anogeissus latifolia* Wall of the family *Combretaceae*. It is a complex, water-soluble polysaccharide composed of the calcium and magnesium salts of L-arabinose, D-galactose, D-mannose, D-xylose, and D-glucuronic acids in the approximate molar ratio of 10:6:2:1:2. It is light to dark tan and is insoluble in 90% alcohol.

Functional Use in Foods Emulsifier and emulsifier salt.

REQUIREMENTS

Labeling Indicate the viscosity, in centipoises.
Identification
   Lead Acetate Solution (**Caution**: Use gloves and goggles to avoid contact with skin and eyes. Use an effective fume-removal device or other respiratory protection.) Activate 5 to 60 g of lead (II) oxide by heating it for 2.5 to 3 h in a furnace at 650° to 670° (cooled product should have a lemon color). In a 500-mL Erlenmeyer flask provided with a reflux condenser, boil 80 g of lead acetate trihydrate and 40 g of freshly activated lead (II) oxide with 250 g of water for 45 min. Cool, filter off any residue, and dilute with recently boiled water to a density of 1.25 at 20°. Add 4 mL of water to 1 mL of the lead acetate solution, and filter.
   Procedure Add 0.2 mL of the *Lead Acetate Standard Solution* to 5 mL of a cold 1 in 100 aqueous solution of the gum. A slight precipitate or clear solution results that yields an opaque flocculent precipitate upon the addition of 1 mL of 3 *N* ammonium hydroxide.
Arsenic (as As) Not more than 3 mg/kg.
Ash (Acid-Insoluble) Not more than 1.75%.
Ash (Total) Not more than 6.0%.
Heavy Metals (as Pb) Not more than 0.002%.
Insoluble Matter Not more than 1.0%.
Lead Not more than 5 mg/kg.
Loss on Drying Not more than 14.0%.
Viscosity A 5% solution exhibits a viscosity measured in centipoises within the range stated on the label.

TESTS

Arsenic A *Sample Solution* prepared as directed for organic compounds meets the requirements of the *Arsenic Test*, Appendix IIIB.
Ash (Acid-Insoluble) Determine as directed in the general method, Appendix IIC.
Ash (Total) Determine as directed in the general method, Appendix IIC.
Heavy Metals Prepare and test a 1-g sample as directed in *Method II* under the *Heavy Metals Test*, Appendix IIIB, using 20 µg of lead ion (Pb) in the control (*Solution A*).
Insoluble Matter Dissolve a 5-g sample, accurately weighed, in about 100 mL of water contained in a 250-mL Erlenmeyer flask, add 10 mL of 2.7 *N* hydrochloric acid, and boil gently for 15 min. Filter the hot solution, using suction through a tared filtering crucible, wash thoroughly with hot water, dry at 105° for 2 h, and weigh.
Lead A *Sample Solution* prepared as directed for organic compounds meets the requirements of the *Lead Limit Test*, Appendix IIIB, using 5 µg of lead ion (Pb) in the control.
Loss on Drying, Appendix IIC Dry at 105° for 5 h. Before drying, unground samples should be powdered to pass through a No. 40 sieve and mixed well before weighing.

Viscosity Determine as directed under *Viscosity of Sodium Carboxymethylcellulose*, Appendix IIB, at 75°, using spindle No. 2 at 60 rpm.

Packaging and Storage Store in well-closed containers.

Gum Guaiac

Guaiac Resin

INS: 241　　　　　　　　　　　　　CAS: [9000-29-7]

DESCRIPTION

The resin of the wood of *Guajacum officinale* L. or of *Guajacum sanctum* L. (Fam. *Zygophyllaceae*). It occurs as irregular masses enclosing fragments of vegetable tissues, or in large, nearly homogeneous masses, and occasionally in more or less rounded or ovoid tears; it is externally brownish black to dusky brown, acquiring a greenish color on long exposure, the fractured surface having a glassy luster, the thin pieces being transparent and varying in color from brown to yellowish orange. The powder is moderate yellow brown, becoming olive brown on exposure to the air. It has a balsamic odor and a slightly acrid taste. Gum Guaiac dissolves incompletely but readily in alcohol, in ether, in chloroform, and in solutions of alkalies. It is slightly soluble in carbon disulfide.

Functional Use in Foods Preservative; antioxidant.

REQUIREMENTS

Identification

A. Add 1 drop of ferric chloride TS to 5 mL of an alcoholic solution of the sample (1 in 100). A blue color is produced that gradually changes to green, finally becoming greenish yellow.

B. A mixture of 5 mL of an alcoholic solution of the sample (1 in 100) and 5 mL of water becomes blue upon shaking with 20 mg of lead peroxide. Filter the solution, and boil a portion of the filtrate. The color disappears but may be restored by the addition of lead peroxide and shaking. Add a few drops of 2.7 N hydrochloric acid to a second portion of the filtrate. The color is immediately discharged.

Alcohol-Insoluble Residue Not more than 15.0%.
Ash (Acid-Insoluble) Not more than 2.0%.
Ash (Total) Not more than 5.0%.
Heavy Metals (as Pb) Not more than 0.002%.
Lead Not more than 10 mg/kg.
Melting Range Between 85° and 90°.
Rosin Passes test.

TESTS

Alcohol-Insoluble Residue Place 2 g of the sample, finely powdered and accurately weighed, in a dry, tared extraction thimble, and extract it with alcohol in a suitable continuous extraction apparatus for 3 h or until completely extracted. Dry the insoluble residue in the thimble for 4 h at 105°, and weigh.

Ash (Acid-Insoluble) Determine as directed in the general method, Appendix IIC.

Ash (Total) Determine as directed in the general method, Appendix IIC.

Heavy Metals Prepare and test a 1-g sample as directed in *Method II* under the *Heavy Metals Test*, Appendix IIIB, using 20 µg of lead ion (Pb) in the control (*Solution A*).

Lead A *Sample Solution* prepared as directed for organic compounds meets the requirements of the *Lead Limit Test*, Appendix IIIB, using 10 µg of lead ion (Pb) in the control.

Melting Range Determine as directed for *Melting Range or Temperature*, Appendix IIB.

Rosin A 1 in 10 solution of the sample in petroleum ether is colorless, and when shaken with an equal quantity of a fresh solution of cupric acetate (1 in 200), is not more green than a similar solution of cupric acetate in petroleum ether.

Packaging and Storage Store in well-closed containers.

Helium

He　　　　　　　　　　　　　　Formula wt 4.0026

　　　　　　　　　　　　　　　　CAS: [7440-59-7]

DESCRIPTION

A colorless, odorless gas that is not combustible and does not support combustion. Very slightly soluble in water. One L of the gas weighs about 180 mg at 0° and at a pressure of 760 mm of mercury.

Functional Use in Foods Processing aid.

REQUIREMENTS

Note: Reduce the container pressure by means of a regulator. Measure the gas with a gas volume meter downstream from the pertinent detector tube to minimize contamination of or change to the specimens.

Identification The flame of a burning splinter of wood is extinguished when inserted into an inverted test tube filled with Helium.

Note: Use caution.

A small balloon filled with Helium shows buoyancy.
Assay Not less than 99.0% of He, by volume.
Air Not more than 1.0%, by volume.
Carbon Monoxide Not more than 10 ppm, by volume.
Odor Passes test.

TESTS

Assay Introduce a specimen of Helium into a gas chromatograph by means of a gas sampling valve. Select the operating conditions of the gas chromatograph so that the standard peak signal resulting from the following procedure corresponds to not less than 70% of the full-scale reading. Preferably use an apparatus corresponding to the general type in which the column is 6 m in length and 4 mm in inside diameter and is packed with porous polymer beads (PoraPak Q, or equivalent), which permit complete separation of nitrogen and oxygen from Helium, although nitrogen and oxygen may not be separated from each other. Use industrial-grade Helium (99.99%) as the carrier gas, with a thermal-conductivity detector, and control the column temperature at 60°. The peak response produced by the assay specimen exhibits a retention time corresponding to that produced by an air–Helium certified standard (a mixture of 1.0% air in industrial-grade helium is available from most suppliers) and indicates not more than 1.0% air, by volume, when compared with the peak response of the air–Helium certified standard, and not less than 99.0% of He, by volume.

Air Determine as directed in the *Assay*.

Carbon Monoxide Pass 1050 ± 50 mL through a carbon monoxide detector tube (under *Solutions and Indicators*) at the rate specified for the tube. The indicator change corresponds to not more than 10 ppm, by volume.

Odor Carefully open the container valve to produce a moderate flow of gas. Do not direct the gas stream toward the face, but deflect a portion of the stream toward the nose. No appreciable odor is discernible.

Packaging and Storage Store in appropriate gas cylinders.

Heptylparaben

n-Heptyl-*p*-hydroxybenzoate

HO—⟨⟩—COO(CH$_2$)$_6$CH$_3$

$C_{14}H_{20}O_3$ Formula wt 236.31

CAS: [1085-12-7]

DESCRIPTION

Small, colorless crystals or a white crystalline powder. It is odorless or has a faint, characteristic odor and a slight burning taste. It is very slightly soluble in water but is freely soluble in alcohol and in ether.

Functional Use in Foods Preservative; antimicrobial agent.

REQUIREMENTS

Identification Dissolve 500 mg in 10 mL of 1 N sodium hydroxide, boil for 30 min, allow the solution to evaporate to a volume of about 5 mL, and cool. Acidify the solution with 2 N sulfuric acid, collect the crystals on a filter, wash several times with small portions of water, and dry in a desiccator over silica gel. The *p*-hydroxybenzoic acid so obtained melts between 212° and 217° (see Appendix IIB).

Assay Not less than 99.0% and not more than 100.5% of $C_{14}H_{20}O_3$, calculated on the dried basis.

Acidity Passes test.

Heavy Metals (as Pb) Not more than 10 mg/kg.

Loss on Drying Not more than 0.5%.

Melting Range Between 48° and 51°.

Residue on Ignition Not more than 0.05%.

TESTS

Assay Transfer into a flask about 3.5 g, accurately weighed, add 40.0 mL of 1 N sodium hydroxide, and rinse the sides of the flask with water. Cover with a watch glass, boil gently for 1 h, cool, and titrate the excess sodium hydroxide with 1 N sulfuric acid to pH 6.5. Perform a blank determination with the same quantities of the same reagents in the same manner, and make any necessary correction (see *General Provisions*). Each mL of 1 N sodium hydroxide is equivalent to 236.3 mg of $C_{14}H_{20}C_3$, calculated on the dried basis.

Acidity Heat 750 mg with 15 mL of water at 80° for 1 min, cool, and filter. The filtrate is acid or neutral to litmus. To 10 mL of the filtrate add 0.2 mL of 0.1 N sodium hydroxide and 2 drops of methyl red TS. The solution is yellow, without even a light cast of pink.

Heavy Metals, Appendix IIIB Dissolve 2 g in 23 mL of acetone, and add 2 mL of 1 N acetic acid, 2 mL of water, and 10 mL of hydrogen sulfide TS. Any color does not exceed that produced in a control (*Solution A*) made with 23 mL of acetone, 2 mL of 1 N acetic acid, 2 mL of *Standard Lead Solution* (20 μg Pb ion), and 10 mL of hydrogen sulfide TS.

Loss on Drying, Appendix IIC Dry in a desiccator over silica gel for 5 h.

Melting Range Determine as directed for *Melting Range or Temperature*, Appendix IIB.

Residue on Ignition Ignite 2 g as directed in the general method, Appendix IIC.

Packaging and Storage Store in tight containers.

Hexanes

Mixed Paraffinic Hydrocarbons

DESCRIPTION

Hexanes constitute a clear, colorless, flammable liquid composed predominantly of C_6, with some C_5 and C_7, isomeric paraffins. The relative proportion of isomers varies with the producer and the production lot. It is soluble in alcohol, in acetone, and in ether and is insoluble in water.

Functional Use in Foods Extraction solvent.

REQUIREMENTS

Benzene Not more than 0.05%.
Color (APHA) Not more than 10.
Distillation Range Between 64° and 71°.
Heavy Metals (as Pb) Not more than 1 mg/kg.
Nonvolatile Residue Not more than 10 mg/kg.
Specific Gravity Between 0.655 and 0.675.
Sulfur Not more than 5 mg/kg.

TESTS

Benzene Proceed as directed under *Benzene* (in Paraffinic Hydrocarbon Solvents), Appendix IIIC.
Color (APHA) Dilute 2.0 mL of platinum–cobalt stock solution (APHA No. 500) with water in a 100-mL volumetric flask. Compare this solution (APHA No. 10) with 100 mL of the sample in 100-mL Nessler tubes, viewed vertically over a white background.
Distillation Range Proceed as directed in the general method, Appendix IIB.
Heavy Metals Evaporate 30 mL (about 20 g) of the sample to dryness on a steam bath in a glass evaporating dish. Cool, add 2 mL of hydrochloric acid, and slowly evaporate to dryness again on the steam bath. Moisten the residue with one drop of hydrochloric acid, add 10 mL of hot water, and digest for 2 min. Cool, and dilute to 25 mL with water in a volumetric flask. This solution meets the requirements of the *Heavy Metals Test*, Appendix IIIB, using 20 μg of lead ion (Pb) in the control (*Solution A*).
Nonvolatile Residue Evaporate 150 mL (about 100 g) of the sample to dryness in a tared dish on a steam bath. Dry the residue at 105° for 30 min, cool, and weigh.
Specific Gravity Determine by any reliable method (see *General Provisions*).
Sulfur Proceed as directed under *Sulfur*, Appendix IIIC.

Packaging and Storage Store in tight containers, protected from fire.

High-Fructose Corn Syrup

DESCRIPTION

High-Fructose Corn Syrup (HFCS) is a sweet, nutritive saccharide mixture prepared as a clear, aqueous solution from high-dextrose-equivalent corn starch hydrolysate by the partial enzymatic conversion of glucose (dextrose) to fructose, using an insoluble glucose isomerase preparation that complies with **21 CFR** 184.1372 and that has been grown in a pure culture fermentation that produces no antibiotics.

It is a water-white to light yellow, somewhat viscous liquid that darkens at high temperatures. It is miscible in all proportions with water.

Functional Use in Foods Nutritive sweetener.

REQUIREMENTS

Labeling Indicate the color range and presence of sulfur dioxide if the residual concentration is greater than 10 mg/kg.
Identification To 5 mL of hot alkaline cupric tartrate TS add a few drops of a 1 in 10 solution of the sample. A copious red precipitate of cuprous oxide is formed.
Assay *42% HFCS*: Not less than 97.0% total saccharides, expressed as a percent of solids, of which not less than 42.0% consists of fructose, not less than 92.0% consists of monosaccharides, and not more than 8.0% consists of other saccharides. *55% HFCS*: Not less than 95.0% total saccharides, expressed as a percent of solids, of which not less than 55.0% consists of fructose, not less than 95.0% consists of monosaccharides, and not more than 5.0% consists of other saccharides.
Arsenic (as As) Not more than 1 mg/kg.
Color Within the range specified by the vendor.
Heavy Metals (as Pb) Not more than 5 mg/kg.
Lead Not more than 0.1 mg/kg.
Residue on Ignition Not more than 0.05%.
Sulfur Dioxide Not more than 0.003%.
Total Solids *42% HFCS*: Not less than 70.5%; *55% HFCS*: Not less than 76.5%.

TESTS

Assay

Apparatus Use a suitable high-performance liquid-chromatography (HPLC) system, such as described in Standard Analytical Methods of the Corn Refiners Association (also, see *Chromatography*, Appendix IIA) equipped with a 22- to 31-cm stainless-steel column, a strip-chart recorder, and a differential refractometer detector maintained at 45° ± 0.005°.

Stationary Phase Prepacked macroreticular polystyrene sulfonate divinylbenzene cation-exchange resin (2% to 8% cross-linked, 8- to 25-μm particle size), preferably in the calcium or silver form. Examples of acceptable resins are Bio-Rad Aminex HPX-87C (or the equivalent) for separating DP_1–DP_4 saccharides and Aminex HPX-42C and HPX-42A (or the equivalent) for separating DP_1–DP_7 saccharides. Maintain the column at 85° during operation.

Mobile Phase Degassed purified water passed through a 0.22-μm filter before use; maintain at 85° during operation of the chromatograph.

Standardization Prepare a standard solution containing a total of about 10% solids, using sugars of known purity (e.g., USP Fructose Reference Standard; USP Dextrose Reference Standard, or NIST Standard Reference Material; maltose, Aldrich Chemical Company; or any equivalent) that approximates, on the dry basis, the composition of the sample to be analyzed. Dissolve each standard sugar, accurately weighed, in 20 mL of purified water contained in a 50-mL beaker. Heat on a steam bath until all sugars are dissolved, then cool, and transfer to a 100-mL volumetric flask. Dilute to volume with water and mix. Freeze the solution if it is to be reused.

If a corn syrup or maltodextrin is used to supply a DP_{4+} fraction, take care to include all saccharides in the standard composition calculation.

Compute the dry-basis concentration (C), in percent, of each individual component in the standard solution by the formula

$$C = (W_C/\Sigma W_I) \times 100,$$

in which W_C is the weight of the sugar of interest and ΣW_I is the sum of all sugar components. Standardize by injecting 10 to 20 µL (about 1.0 to 2.0 mg solids) of the standard sugar solution. Integrate the peaks and normalize. Sum the individual DP_{4+} responses from the normalized printout to obtain the total DP_{4+} normalized response. Calculate the response factors as follows (see *Chromatography*, Appendix IIA):

$$R_I = \text{(known concentration, dry basis \%)/(measured concentration, normalized \%)},$$

in which R_I is the response factor for component i.

Compute the response factor for each component relative to glucose (R'_I) using the following equation:

$$R'_I = R_I/R_{G'}$$

in which R_G is the response factor for glucose. The R'_I for DP_{4+} should be programmed as a default value (if automated equipment is used) and used to compute the concentration of higher saccharides.

Sample Analysis Determine the solids content (see below) of the sample, and dilute to approximately 10% solids with water. Inject a volume (10 to 50 µL) appropriate for the specific solids content.

Calculation Calculate the concentration of each component as follows:

$$C_I = (A_I \times R_I \times 100)/\Sigma A_N R_N),$$

in which A_I is the area recorded for that component and $\Sigma A_N R_N$ is the sum of the product of the areas (A) and response factors (R) for all components detected.

Arsenic A *Sample Solution* prepared as directed for organic compounds meets the requirements of the *Arsenic Test* (Appendix IIIB), using 1 mL of *Standard Arsenic Solution* (1 µg As).

Color

Apparatus Use a suitable variable-wavelength spectrophotometer capable of measuring percent transmittance throughout the visible spectrum and designed to permit the use of sample and reference cells with pathlengths of 2 to 4 cm. The transmittance of all paired cells should agree within 0.5%.

Standard Solution Dissolve 0.10 g of reagent grade potassium dichromate ($K_2Cr_2O_7$) in 1 L of water, and mix thoroughly.

Procedure With water in sample and reference cells of 2-cm pathlength adjust the percent transmittance scale of the spectrophotometer to 100%. Leave the reference cell in place and replace the water in the sample cell with the *Standard Solution*; determine the wavelength at which it exhibits exactly 54.5 percent transmittance. This wavelength is defined as λ_c.

the corrected 450-nm wavelength. Remove the 2-cm cells from the spectrophotometer, and with water in the sample and reference cells of 4-cm pathlength, adjust the percent transmittance scale to 100% with the spectrophotometer set at λ_c. Leave the reference cell in place and replace the water in the sample cell with the sample of HFCS. Measure the percent transmittance (T_{450}). Remove the sample cell, set the wavelength at 600 nm, replace the sample with water, and adjust the percent transmittance scale to 100%; then determine the percent transmittance at 600 nm (T_{600}) with the same sample of HFCS in the sample cell. Calculate the *Color* (C) of the sample with the following formula:

$$C = (\log T_{600} - \log T_{450})/4,$$

in which T_{600} is the percent transmittance at 600 nm and T_{450} is the percent transmittance at 450 nm.

Heavy Metals Prepare and test a 4-g sample as directed in *Method II* under the *Heavy Metals Test*, Appendix IIIB, using 20 µg of lead ion (Pb) in the control (*Solution A*) and 500° as the ignition temperature.

Lead Determine as directed in *Method I* in the *Atomic Absorption Spectrophotometric Graphite Furnace Method* under the *Lead Limit Test*, Appendix IIIB, using a 5-g sample.

Residue on Ignition Ignite 10 g as directed under *Residue on Ignition*, Appendix IIC.

Sulfur Dioxide Determine as directed in the general method, Appendix X, using a 50-g sample.

Total Solids Determine the refractive index of a sample of HFCS at 20° or 45°, and use the tables for *High-Fructose Corn Syrup Solids*, Appendix X, to obtain the percent Total Solids.

Packaging and Storage Store in tight containers.

L-Histidine

L-α-Amino-4(or 5)-imidazolepropionic Acid

$$\begin{array}{c} \text{CH}_2\text{CCOOH} \\ \overset{\displaystyle \frown}{\text{N}\quad\text{NH}} \quad \text{H NH}_2 \end{array}$$

$C_6H_9N_3O_2$ Formula wt 155.16

CAS: [71-00-1]

DESCRIPTION

White, odorless crystals or crystalline powder having a slightly bitter taste. It is soluble in water, very slightly soluble in alcohol, and insoluble in ether. It melts with decomposition between about 277° and 288°.

Functional Use in Foods Nutrient; dietary supplement.

REQUIREMENTS

Identification   To 5 mL of a 1 in 100 solution of the sample add 2 mL of bromine TS. A yellow color is produced. When the solution is heated gently, it first becomes colorless, then reddish brown, and finally it forms a dark gray precipitate.
Assay   Not less than 98.5% and not more than 101.5% of $C_6H_9N_3O_2$ after drying.
Heavy Metals (as Pb)   Not more than 10 mg/kg.
Loss on Drying   Not more than 0.2%.
Residue on Ignition   Not more than 0.2%.
Specific Rotation $[\alpha]_D^{20°}$: Between +11.5° and +13.5° after drying; or $[\alpha]_D^{25°}$: Between +12.0° and +14.0° after drying.

TESTS

Assay   Dissolve about 150 mg of the sample, previously dried at 105° for 3 h and accurately weighed, in 3 mL of formic acid and 50 mL of glacial acetic acid, and titrate with 0.1 N perchloric acid, determining the endpoint potentiometrically. Perform a blank determination (see *General Provisions*), and make any necessary correction. Each mL of 0.1 N perchloric acid is equivalent to 15.52 mg of $C_6H_9N_3O_2$.
Heavy Metals   Prepare and test a 2-g sample as directed in *Method II* under the *Heavy Metals Test*, Appendix IIIB, using 20 µg of lead ion (Pb) in the control (*Solution A*).
Loss on Drying, Appendix IIC   Dry at 105° for 3 h.
Residue on Ignition, Appendix IIC   Ignite 1 g as directed in the general method.
Specific Rotation, Appendix IIB   Determine in a solution containing 11 g of a previously dried sample in sufficient 6 N hydrochloric acid to make 100 mL.

Packaging and Storage   Store in well-closed, light-resistant containers.

L-Histidine Monohydrochloride

L-α-Amino-4(or 5)-imidazolepropionic Acid Monohydrochloride

$C_6H_9N_3O_2 \cdot HCl \cdot H_2O$ Formula wt 209.63

CAS: monohydrate [5934-29-2]

DESCRIPTION

White, odorless crystals or crystalline powder having a slightly acid, bitter taste. It is soluble in water but insoluble in alcohol and in ether. It melts with decomposition at about 250° (after drying).

Functional Use in Foods   Nutrient; dietary supplement.

REQUIREMENTS

Identification
  A. Heat 5 mL of a 1 in 1000 solution with 1 mL of triketohydrindene hydrate TS. A reddish purple color is produced.
  B. A 1 in 1000 solution gives positive tests for *Chloride*, Appendix IIIA.
Assay   Not less than 98.5% and not more than 101.5% of $C_6H_9N_3O_2 \cdot HCl \cdot H_2O$ after drying.
Heavy Metals (as Pb)   Not more than 10 mg/kg.
Loss on Drying   Not more than 0.3%.
Residue on Ignition   Not more than 0.1%.
Specific Rotation $[\alpha]_D^{20°}$: Between +8.5° and +10.5° after drying.

TESTS

Assay   Dissolve about 100 mg of the sample, previously dried at 105° for 3 h and accurately weighed, in 3 mL of formic acid, add exactly 15.0 mL of 0.1 N perchloric acid, and heat on a water bath for 30 min. After cooling, add 45 mL of glacial acetic acid, and titrate the excess perchloric acid with 0.1 N sodium acetate, determining the endpoint potentiometrically. Perform a blank determination (see *General Provisions*), and make any necessary correction. Each mL of 0.1 N perchloric acid is equivalent to 10.48 mg of $C_6H_9N_3O_2 \cdot HCl \cdot H_2O$.
Heavy Metals   Prepare and test a 2-g sample as directed in *Method II* under the *Heavy Metals Test*, Appendix IIIB, using 20 µg of lead ion (Pb) in the control (*Solution A*).
Loss on Drying, Appendix IIC   Dry at 105° for 3 h.
Residue on Ignition, Appendix IIC   Ignite 1 g as directed in the general method.
Specific Rotation, Appendix IIB   Determine in a solution containing 11 g of a previously dried sample in sufficient 6 N hydrochloric acid to make 100 mL.

Packaging and Storage   Store in well-closed, light-resistant containers.

Hops Oil

CAS: [8007-04-3]

DESCRIPTION

The volatile oil obtained by steam distillation of the freshly dried membranous cones of the female plants of *Humulus lupulus* L. or *Humulus americanus* Nutt. (Fam. *Moraceae*). It is a light yellow to greenish yellow liquid having a characteristic aromatic odor. Age darkens the color, and the oil tends to become viscous.

It is soluble in most fixed oils and, sometimes with opalescence, in mineral oil. It is practically insoluble in glycerin and in propylene glycol.

Functional Use in Foods Flavoring agent.

REQUIREMENTS

Identification The infrared absorption spectrum of the sample exhibits relative maxima (that may vary in intensity) at the same wavelengths (or frequencies) as those shown in the respective spectrum under *Infrared Spectra (Series A: Essential Oils)*, using the same test conditions as specified therein.
Acid Value Not more than 11.0.
Angular Rotation Between −2° and +2°5′.
Heavy Metals (as Pb) Passes test.
Refractive Index Between 1.470 and 1.494 at 20°.
Saponification Value Between 14 and 69.
Solubility in Alcohol Passes test.
Specific Gravity Between 0.825 and 0.926.

TESTS

Acid Value Determine as directed for *Acid Value* under *Essential Oils and Flavors*, Appendix VI, using about 5 g, accurately weighed.
Angular Rotation Determine in a 100-mm tube as directed under *Optical (Specific) Rotation*, Appendix IIB.
Heavy Metals Shake 10 mL of the oil with an equal volume of water to which 1 drop of hydrochloric acid has been added, and pass hydrogen sulfide through the mixture until it is saturated. No darkening in color is produced in either the oil or the water.
Refractive Index, Appendix IIB Determine with an Abbé or other refractometer of equal or greater accuracy.
Saponification Value Determine as directed for *Saponification Value* under *Essential Oils and Flavors*, Appendix VI, using about 5 g, accurately weighed.
Solubility in Alcohol Proceed as directed in the general method, Appendix VI. One mL usually is not soluble in 95% alcohol. Old oils are less soluble than fresh oils.
Specific Gravity Determine by any reliable method (see *General Provisions*).

Packaging and Storage Store in full, tight, preferably glass, aluminum, tin-lined, or other suitably lined containers in a cool place protected from light.

Hydrochloric Acid

HCl Formula wt 36.46
INS: 507 CAS: [7647-01-0]

DESCRIPTION

A water solution of hydrogen chloride of varied concentrations. It is a clear, colorless or slightly yellowish, corrosive liquid having a pungent odor. It is miscible with water and with alcohol. Concentrations of Hydrochloric Acid are expressed in percent by weight, or may be expressed in Baumé degrees (Bé°) from which percentages of Hydrochloric Acid and specific gravities may readily be derived (see *Hydrochloric Acid Table*, Appendix IIC). The usually available concentrations are 18°, 20°, 22°, and 23° Bé. Concentrations above 13° Bé (19.6%) fume in moist air, lose hydrogen chloride, and create a corrosive atmosphere. Because of these characteristics, suitable precautions must be observed during sampling and analysis to prevent losses.

Note: Hydrochloric Acid is produced by various methods that might impart trace amounts of organic compounds as impurities. The manufacturer, vendor, or user is responsible for identifying the specific organic compounds that are present and for meeting the *Requirements* for *Organic Compounds*. Methods are provided for their determination. In applying the procedures any necessary standards should be used to quantitate the organic compounds present in each specific product.

The variety of organic impurities that might conceivably be found in Hydrochloric Acid is such that it is impossible to provide a comprehensive and accurate list here. Therefore, the manufacturer, vendor, or user is responsible for establishing the suitability of such Hydrochloric Acid for its intended application in foods or food processing in accordance with the provision of *Trace Impurities* (see *General Provisions*).

Functional Use in Foods Acid.

REQUIREMENTS

Labeling Indicate the content, by weight, of Hydrochloric Acid, HCl. Alternatively, indicate the range of HCl content, the range of Baumé degrees, and/or the specific gravity range.
Identification It gives positive tests for *Chloride*, Appendix IIIA.
Assay Not less than 97.0% and not more than 103.0% of the labeled amount of HCl, or within the range specified on the label.
Baumé Degrees Within the range shown on the label or claimed by the vendor.
Color Passes test.
Heavy Metals (as Pb) Not more than 1 mg/kg.
Iron Not more than 5 mg/kg.
Nonvolatile Residue Not more than 0.5%.
Organic Compounds
   Total Organic Compounds (Non-Fluorine-Containing) Not more than 5 mg/kg, including
      Benzene Not more than 0.05 mg/kg.
      Fluorinated Organic Compounds (total) Not more than 25 mg/kg.
Oxidizing Substances (as Cl_2) Not more than 0.003%.
Reducing Substances (as SO_3) Not more than 0.007%.
Specific Gravity Within the range specified or implied by the vendor.
Sulfate Not more than 0.5%.

TESTS

Assay Tare accurately a 125-mL glass-stoppered Erlenmeyer flask containing 35.0 mL of 1 N sodium hydroxide. Partially fill, without the use of vacuum, a 10-mL serological pipet from near the bottom of a representative sample, remove any acid adhering to the outside, and discard the first mL flowing from the pipet. Hold the tip of the pipet just above the surface of the sodium hydroxide solution, and transfer between 2.5 and 3 mL of the sample into the flask, leaving at least 1 mL in the pipet. Stopper the flask, mix the contents, and weigh accurately to obtain the weight of the sample. Add methyl orange TS, and titrate the excess sodium hydroxide with 1 N hydrochloric acid. Each mL of 1 N sodium hydroxide is equivalent to 36.46 mg of HCl.

Baumé Degrees Transfer about 200 mL of the sample, previously cooled to a temperature below 15° into a 250-mL hydrometer cylinder. Insert a suitable Baumé hydrometer graduated at 0.1° intervals, adjust the temperature to 15.6° (60°F), and note the reading at the bottom of the meniscus.

Color It shows no more color than *Matching Fluid A*, Appendix IIB (*Readily Carbonizable Substances*).

Heavy Metals Evaporate 20 g (17.5 mL) to dryness on a steam bath, dissolve the residue in 10 mL of 1 N acetic acid, and dilute to 25 mL with water. This solution meets the requirements of the *Heavy Metals Test*, Appendix IIIB, using 20 μg of lead ion (Pb) in the control (*Solution A*).

Iron Dilute 4.3 mL (5 g) to a volume of 40 mL, and add about 40 mg of ammonium persulfate and 10 mL of ammonium thiocyanate TS. Any red color does not exceed that produced by 2.5 mL of *Iron Standard Solution* (25 μg Fe) in an equal volume of solution containing the same quantities of ACS reagent-grade Hydrochloric Acid and the reagents used in the test.

Nonvolatile Residue Transfer 1 g into a tared glass dish, evaporate to dryness on a steam bath, and then dry at 110° for 1 h, cool in a desiccator, and weigh. The weight of the residue does not exceed 5 mg.

Organic Compounds Analyses are carried out by gas chromatography employing *Vapor Partitioning* or *Solvent Extraction*, depending upon the characteristics of the compound being determined. It is necessary, however, to use the *Vapor Partitioning* method for the determination of benzene.

Vapor Partitioning Method This method is suitable for the determination of extractable organic compounds at 0.05 to 100 mg/kg but is most appropriate for organic compounds with a vapor pressure greater than 10 mm Hg at 25°. Use a gas chromatograph equipped with a flame ionization detector and a 4-m × 2-mm (id) stainless-steel column packed with 15%, by weight, methyl trifluoropropyl silicone (DCFS 1265, or QF-1, or OV-210, or SP-2401) stationary phase on 80/100-mesh Gas Chrom R or the equivalent. A newly packed column should be conditioned at 120° and 30 mL/min helium flow for at least 2 h (preferably overnight) before it is attached to the detector. For analysis, the column is maintained isothermally at 105°; the injection port and detector are maintained at 250°; the carrier gas flow rate is set at 11 mL/min; fuel gas flows should be optimized for the gas chromatograph and detector in use. The experimental conditions may be changed as necessary for optimal resolution and sensitivity. The signal-to-noise ratio should be at least 10:1.

Preparation of Standard Solutions Prepare a standard solution of the organic compounds to be quantitated in Hydrochloric Acid (known to be free of interfering impurities) at approximate concentrations of 5 mg/kg, or within ±50% of the concentrations in the samples to be analyzed.

Place a stirring bar in a 1-L volumetric flask equipped with a ground-glass stopper, and tare the combination. Fill the flask with reagent-grade hydrochloric acid so that no air space is present when the flask is stoppered, and determine the weight of the Hydrochloric Acid. Calculate the volume (V) in μL of each organic component to be added from the formula

$$V = (C \times W)/(D \times 1000),$$

in which C is the desired concentration, in mg/kg; W is weight, in g, of the Hydrochloric Acid; D is the density, in mg/μL, of the organic compound; and 1000 is a conversion factor with the units g/kg. Add the calculated amount of each component to the Hydrochloric Acid with a syringe (ensure that the syringe tip is under the solution surface), stopper the flask, and stir the solution for at least 2 h using a magnetic stirrer.

Calibration Treat the standard in the same way as described for the sample under *Procedure* (below). Determine a blank for each lot of reagent-grade Hydrochloric Acid, and calculate a response factor (R) by dividing the concentration (C) in mg/kg for each component by the peak area (A) for that component (subtract any area obtained from the blank sample):

$$R = C/(A - \text{area of blank}).$$

Gaseous compounds present special problems in the preparation of standards. Therefore, to determine response factors for gaseous compounds use the following method, which will be referred to as the *Method of Multiple Extractions*. Dilute a sample of Hydrochloric Acid known to contain the gaseous compound of interest with an equal volume of water. Draw 20 mL of this solution into a 50-mL glass syringe; then draw 20 mL of air into the syringe, cap with a rubber septum, and place the syringe on a shaker for 5 min. Withdraw 1 mL of the vapor through the septum, and inject it into the chromatograph. Expel the vapor phase from the 50-mL syringe, draw in another 20 mL of air, repeat the extraction, and inject another 1-mL vapor sample into the chromatograph. Carry out the extraction and GC analysis on the same sample of acid a total of six times. For each impurity, plot the area (A_N) determined for extraction N against the difference between A_N and the area determined for extraction (N + 1); that is, plot A_N against $(A_N - A_{N+1})$. The slope of this line is the extraction efficiency (E) for that impurity into the air.

Inject into the chromatograph 1 mL of a 0.1% (by volume) standard gas sample of each impurity in air and determine the absolute factor (F_A), in g, per peak area (A) by the following formula:

$$F_A = (M \times 4.0816 \times 10^{-8})/A,$$

in which M is the molecular weight of the compound.

The concentration (C), in mg/kg, of the component in the original sample is calculated by the formula

$$C = (A \times F_A \times 1.6949 \times 10^6)/E,$$

in which A is the peak area corresponding to the compound (as above), F_A is the absolute factor, and E is extraction efficiency. The response factor is then calculated as

$$R = C/A.$$

Procedure Dilute a 10-mL sample of Hydrochloric Acid to be analyzed with an equal volume of water. Draw this solution into a 50-mL glass syringe. Then draw 20 mL of air into the syringe, cap with a rubber septum, and place the syringe on a shaker for 5 min. Draw 1 mL of the vapor through the septum, and inject it into the gas chromatograph. Approximate elution times in min for some specific organic compounds are as follows:

Methane and acetylene 1.70	1,1-Dichloroethane 4.53
Methyl chloride 2.21	Carbon tetrachloride ...4.86
Vinyl chloride 2.29	1,1,1-Trichloroethane 5.50
1,1,1-Trichlorofluoromethane ..2.62	Benzene 6.00
Ethyl chloride 2.90	Trichloroethylene 6.22
Vinylidene chloride 3.20	Ethylene dichloride 6.61
Methylene chloride 3.64	Propylene dichloride ...8.41
Chloroform 4.49	Perchloroethylene 9.73

Alternate columns may be required to resolve some combinations of components. Methyl chloride and vinyl chloride are resolved by a 3.7-m × 3-mm (id) squalane column at 45° and a helium flow of 10 mL/min. Chloroform and 1,1-dichloroethane are resolved by a 4-m × 3-mm (id) DC 550R column at 110° and a helium flow of 12 mL/min.

Calculation Calculate the concentration (C) in mg/kg of each compound by multiplying its corresponding peak area (A) by the appropriate response factor (R) determined in the *Calibration* protocol:

$$C = R \times A.$$

Precision The relative standard deviation at 5 mg/kg should not exceed 15% for five analyses.

Solvent Extraction Method The solvent extraction technique is suitable for the determination of extractable organic compounds at 0.3 to 100 mg/kg, but is most appropriate for organic compounds with vapor pressures less than 10 mm Hg at 25°. The conditions for the gas chromatograph are the same as for the *Vapor Partitioning* method, except that the column temperature is 120°, and the carrier-gas flow is 21 mL/min.

Preparation of Standards Prepare the *Standard Solution* as described under *Vapor Partitioning*.

Calibration Extract a sample of the *Standard Solution* as directed under *Procedure* (below) and inject it into the gas chromatograph. Determine a blank for each lot of reagent-grade Hydrochloric Acid and perchloroethylene by extracting the Hydrochloric Acid in the same way as the standard. Calculate a response factor (R) by dividing the concentration (C) in mg/kg for each component by the peak area (A) for that component (subtract any area obtained from the blank sample):

$$R = C/(A - \text{area of blank}).$$

Procedure Accurately transfer 90 mL of the Hydrochloric Acid sample and 10 mL of perchloroethylene (free of interfering impurities) into a narrow-mouth, 4-oz bottle. Place the bottle in a mechanical shaker for 30 min. Separate the two phases (perchloroethylene on the bottom) and inject 3 μL of the perchloroethylene extract into the gas chromatograph. Approximate elution times in min for some chlorinated organic compounds are as follows:

Vinylidene chloride 2.94	Ethylene dichloride 5.26
Methylene chloride 3.27	Propylene dichloride 6.36
Chloroform 3.83	Perchloroethylene 6.95
Carbon tetrachloride ...4.07	1,1,1,2-Tetrachloroethane 10.12
1,1,1-Trichloroethane 4.50	1,1,2,2-Tetrachloroethane 13.70
Trichloroethylene 4.97	Pentachloroethane 16.19

To determine perchloroethylene and higher-boiling impurities, substitute methylene chloride (free of interfering impurities) for perchloroethylene in the extraction step. For higher-boiling impurities such as monochlorobenzene and the three dichlorobenzenes, use a 2.74-m × 2.1-mm (id) stainless-steel column packed with 10% carbowax 20M/2% KOH on 80/100-mesh chromasorb W (acid washed) at 150° and a nitrogen flow of 35 mL/min.

Calculation Calculate the concentration (C), in mg/kg, of each compound by multiplying the corresponding peak area (A) (subtract any area obtained from a blank sample) by the appropriate response factor (R) determined in the *Calibration* protocol:

$$C = R \times (A - \text{area of blank}).$$

Precision The relative standard deviation at 5 mg/kg should not exceed 15% for five analyses.

Oxidizing Substances Transfer 1 mL into a 30-mL test tube, dilute to 20 mL with freshly boiled and cooled water, and add 1 mL of potassium iodide TS and 1 mL of starch TS. Stopper the test tube, and mix thoroughly. Any blue color does not exceed that produced in a control consisting of 1.0 mL of 0.001 N iodine in an equal volume of water containing the same quantities of the same reagents and 1 mL of ACS reagent-grade Hydrochloric Acid.

Reducing Substances Transfer 1 mL of ACS reagent-grade Hydrochloric Acid into a 30-mL test tube, dilute to 20 mL with recently boiled and cooled water, and add 1 mL of potassium iodide TS, 1 mL of starch TS, and 2.0 mL of 0.001 N iodine. Stopper the test tube, and mix thoroughly. The blue color produced is not discharged by 1 mL of the sample.

Specific Gravity Determine at 15.6° (60°F) with a hydrometer, or calculate it from the Baumé degrees observed in the **TEST** for *Baumé Degrees*.

Sulfate, Appendix IIIB Dilute a 1-g sample to 100.0 mL with water, transfer 5.0 mL of this dilution into a 50-mL, tall-form Nessler tube, and dilute to 20 mL with water. Add a drop of phenolphthalein TS, neutralize the solution with 6 N ammonium hydroxide, and then add 1 mL of 2.7 N hydrochloric acid. To the clear solution, previously filtered, if necessary, add 3 mL of barium chloride TS, dilute to 50 mL with water, and mix. Prepare a control consisting of 1 mL of ACS reagent-grade Hydrochloric Acid and 250 μg of sulfate (SO$_4$) and the same quantities of the reagents used for the sample. Any turbidity shown in the sample does not exceed that of the control.

Packaging and Storage Store in tight containers.

Hydrogen Peroxide

H_2O_2 Formula wt 34.01

CAS: [7722-84-1]

DESCRIPTION

A clear, colorless liquid having a slightly pungent odor. It is miscible with water. The grades of Hydrogen Peroxide suitable for food use usually have a concentration between 30% and 50%.

Note: Although Hydrogen Peroxide undergoes exothermic decomposition in the presence of dirt and other foreign materials, it is safe and stable under recommended conditions of handling and storage. Information on safe handling and use may be obtained from the supplier.

Functional Use in Foods Bleaching, oxidizing agent; starch modifier; preservative.

REQUIREMENTS

Identification Shake 1 mL of the sample with 10 mL of water containing 1 drop of 2 N sulfuric acid, and add 2 mL of ether. The subsequent addition of a drop of potassium dichromate TS produces an evanescent blue color in the water layer that upon agitation and standing passes into the ether layer.
Assay Not less than the labeled concentration or within the range stated on the label.
Acidity (as H_2SO_4) Not more than 0.03%.
Heavy Metals (as Pb) Not more than 10 mg/kg.
Iron Not more than 0.5 mg/kg.
Phosphate Not more than 0.005%.
Residue on Evaporation Not more than 0.006%.
Tin Not more than 10 mg/kg.

TESTS

Assay Accurately weigh a volume of the sample equivalent to about 300 mg of H_2O_2 into a 100-mL volumetric flask, dilute to volume with water, and mix thoroughly. To a 20.0-mL portion of this solution add 25 mL of 2 N sulfuric acid, and titrate with 0.1 N potassium permanganate. Each mL of 0.1 N potassium permanganate is equivalent to 1.701 mg of H_2O_2.
Acidity Dilute 9 mL (10 g) with 90 mL of carbon dioxide-free water, add methyl red TS, and titrate with 0.02 N sodium hydroxide. The volume of sodium hydroxide solution should not be more than 3 mL greater than the volume required for a blank test on 90 mL of the water used for dilution.

Heavy Metals Evaporate 1.8 mL (2 g) to dryness on a steam bath with 10 mg of sodium chloride, and dissolve the residue in 25 mL of water. The solution so obtained meets the requirements of the *Heavy Metals Test*, Appendix IIIB, using 20 μg of lead ion (Pb) in the control (*Solution A*).
Iron Evaporate 18 mL (20 g) to dryness on a steam bath with 10 mg of sodium chloride, dissolve the residue in 2 mL of hydrochloric acid, and dilute to 50 mL with water. Add about 40 mg of ammonium persulfate crystals and 10 mL of ammonium thiocyanate TS, and mix. Any red or pink color does not exceed that produced by 1.0 mL of *Iron Standard Solution* (10 μg Fe) in an equal volume of solution containing the quantities of the reagents used in the test.
Phosphate Evaporate 400 mg to dryness on a steam bath. Dissolve the residue in 25 mL of approximately 0.5 N sulfuric acid, add 1 mL of ammonium molybdate solution (500 mg of $(NH_4)_6Mo_7O_{24}.4H_2O$ in each 10 mL of water) and 1 mL of *p*-methylaminophenol sulfate TS, and allow to stand for 2 h. Any blue color does not exceed that produced by 2.0 mL of *Phosphate Standard Solution* (20 μg PO_4) in an equal volume of solution containing the quantities of the reagents used in the test.
Residue on Evaporation Evaporate 25 g to dryness in a tared porcelain or silica dish on a steam bath, and dry to constant weight at 105°. The weight of the residue does not exceed 1.5 mg.
Tin

Aluminum Chloride Solution Dissolve 8.93 g of aluminum chloride, $AlCl_3.6H_2O$, in sufficient water to make 1000 mL.

Gelatin Solution On the day of use, dissolve 100 mg of gelatin in 50 mL of boiled water that has been cooled to between 50° and 60°.

Tin Stock Solution Dissolve 250.0 mg of lead-free tin foil in 10 to 15 mL of hydrochloric acid, and dilute to 250.0 mL with dilute hydrochloric acid (1 in 2).

Standard Solution On the day of use, transfer 5.0 mL of *Tin Stock Solution* into a 100-mL volumetric flask, dilute to volume with water, and mix. Transfer 2.0 mL of this solution (100 μg Sn) into a 250-mL Erlenmeyer flask, and add 15 mL of water, 5 mL of nitric acid, and 2 mL of sulfuric acid. Place a small stemless funnel in the mouth of the flask, and heat until strong fumes of sulfuric acid are evolved. Cool, add 5 mL of water, evaporate again to strong fumes, and cool. Repeat the addition of water and heating to strong fumes, then add 15 mL of water, heat to boiling, and cool. Dilute to about 35 mL with water, add 1 drop of methyl red TS and 2.0 mL of the *Aluminum Chloride Solution*, and mix. Make the solution just alkaline by the dropwise addition of ammonium hydroxide, stirring gently, and then add 0.1 mL in excess.

Caution: To avoid dissolving the aluminum hydroxide precipitate, do not add more than 0.1 mL in excess of the ammonia solution.

Centrifuge for about 15 min at 4000 rpm, and then decant the supernatant liquid as completely as possible without disturbing the precipitate. Dissolve the precipitate in 5 mL of dilute hydrochloric acid (1 in 2), add 1.0 mL of the *Gelatin Solution*, and dilute to 20.0 mL with a saturated solution of aluminum chloride.

Sample Solution Transfer 9 mL (10 g) of the sample into a 250-mL Erlenmeyer flask, and add 15 mL of water, 5 mL of nitric acid, and 2 mL of sulfuric acid. Mix, and heat gently on a hot plate to initiate and maintain a vigorous decomposition. When decomposition is complete, place a small stemless funnel in the mouth of the flask, and continue as directed for the *Standard Solution*, beginning with ". . . and heat until strong fumes of sulfuric acid are evolved."

Procedure Rinse a polarographic cell or other vessel with a portion of the *Standard Solution*, then add a suitable volume to the cell, immerse it in a constant temperature bath maintained at 35° ± 0.2°, and deaerate by bubbling oxygen-free nitrogen or hydrogen through the solution for at least 10 min. Insert the dropping mercury electrode of a suitable polarograph, and record the polarogram from −0.2 to −0.7 V and at a sensitivity of 0.0003 μA/mm, using a saturated calomel reference electrode. In the same manner, polarograph a portion of the *Sample Solution* at the same current sensitivity. The height of the wave produced by the *Sample Solution* is not greater than that produced by the *Standard Solution* at the same half-wave potential.

Packaging and Storage Store in a cool place in containers with a vent in the stopper.

Hydroxylated Lecithin

CAS: [8029-76-3]

DESCRIPTION

Hydroxylated Lecithin is derived from a complex mixture of acetone-insoluble phosphatides from soybean and other plant lecithins, consisting chiefly of phosphatidyl choline, phosphatidyl ethanolamine, and phosphatidyl inositol, as well as other minor phospholipids and glycolipids mixed with varying amounts of triglycerides, fatty acids, sterols, and carbohydrates. The mixture is treated with hydrogen peroxide, benzoyl peroxide, lactic acid, and sodium hydroxide, or with hydrogen peroxide, acetic acid, and sodium hydroxide, to produce a hydroxylated product having an iodine value approximately 10% lower than that of the starting material. Hydroxylated Lecithin may vary in consistency from fluid to plastic, depending upon the content of free fatty acid and oil and whether it contains diluents. It is light yellow in color and has a characteristic "bleached" odor. It is partially soluble in water, but hydrates readily to form emulsions; it is more dispersible and hydrates more readily than crude lecithin.

Functional Use in Foods Emulsifier; clouding agent.

REQUIREMENTS

Acetone-Insoluble Matter (phosphatides) Not less than 50.0%.
Acid Value Not more than 70.
Heavy Metals (as Pb) Not more than 0.002%.
Hexane-Insoluble Matter Not more than 0.3%.
Iodine Value Between 85 and 95.
Lead Not more than 10 mg/kg.
Peroxide Value Not more than 100.
Water Not more than 1.5%.

TESTS

Acetone-Insoluble Matter (phosphatides)

Purification of Phosphatides Dissolve 5 g of phosphatides from previous *Acetone-Insoluble Matter* determinations in 10 mL of petroleum ether, and add 25 mL of acetone to the solution. Transfer approximately equal portions of the precipitate to each of two 40-mL centrifuge tubes using additional portions of acetone to facilitate the transfer. Stir thoroughly, dilute to 40 mL with acetone, stir again, chill for 15 min in an ice bath, stir again, and then centrifuge for 5 min. Decant the acetone, crush the solids with a stirring rod, refill the tube with acetone, stir, chill, centrifuge, and decant as before. The solids after the second centrifugation require no further purification and may be used for preparing the *Phosphatide–Acetone Solution*. Five g of the purified phosphatides are required to saturate about 16 L of acetone.

Phosphatide–Acetone Solution Add a quantity of purified phosphatides to sufficient acetone, previously cooled to a temperature of about 5°, to form a saturated solution, and maintain the mixture at this temperature for 2 h, shaking it vigorously at 15-min intervals. Decant the solution through a rapid filter paper, avoiding the transfer of any undissolved solids to the paper and conducting the filtration under refrigerated conditions (not above 5°).

Procedure If it is plastic or semisolid, soften a portion of the sample by warming it in a water bath at a temperature not exceeding 60° and then mixing it thoroughly. Transfer 2 g of a well-mixed sample, accurately weighed, into a 40-mL centrifuge tube, previously tared with a glass stirring rod, and add 15 mL of *Phosphatide–Acetone Solution* from a buret. Warm the mixture in a water bath until the sample melts, but avoid evaporation of the acetone. Stir until the sample is completely disintegrated and dispersed, and then transfer the tube into an ice bath, chill for 5 min, remove from the ice bath, and add about one-half of the required volume of *Phosphatide–Acetone Solution*, previously chilled for 5 min in an ice bath. Stir the mixture to complete dispersion of the sample, dilute to 40 mL with chilled *Phosphatide–Acetone Solution* (5°), again stir, and return the tube and contents to the ice bath for 15 min. At the end of the 15-min chilling period, stir again while still in the ice bath, remove the stirring rod, temporarily supporting it in a vertical upside-down position, and centrifuge the mixture immediately at about 2000 rpm for 5 min. Decant the supernatant liquid from the centrifuge tube, crush the centrifuged solids with the same stirring rod previously used, and refill the tube to the 40-mL mark with chilled (5°) *Phosphatide–Acetone Solution*, and repeat the chilling, stirring, centrifugation, and decantation procedure previously followed. After the second centrifugation and decantation of the supernatant acetone, again crush the solids with the assigned stirring rod, and place the tube and its contents in a horizontal position at room temperature until the excess acetone has evaporated. Mix the residue again, dry the centrifuge tube and its contents at 105° for 45 min in a forced-draft oven, cool, and weigh. Calculate the percentage of acetone-insoluble matter by the formula

$$(100R/S) - B,$$

in which R is the weight of residue, S is the weight of the sample, and B is the percentage of *Hexane-Insoluble Matter* determined as directed in this monograph.

Acid Value If it is plastic or semisolid, soften a portion of the sample by warming it in a water bath at a temperature not exceeding 60°, and then mix it thoroughly. Transfer about 2 g of a well-mixed sample into a 250-mL Erlenmeyer flask, and dissolve it in 50 mL of petroleum ether. To this solution add 50 mL of alcohol, previously neutralized to phenolphthalein with 0.1 N sodium hydroxide, and mix well. Add phenolphthalein TS, and titrate with 0.1 N sodium hydroxide to a pink endpoint that persists for 5 s. Calculate the number of mg of potassium hydroxide required to neutralize the acids in 1 g of the sample by multiplying the number of mL of 0.1 N sodium hydroxide consumed in the titration by 5.6 and dividing the result by the weight of the sample.

Heavy Metals Prepare and test a 1-g sample as directed in *Method II* under the *Heavy Metals Test*, Appendix IIIB, using 20 μg of lead ion (Pb) in the control (*Solution A*).

Hexane-Insoluble Matter If plastic or semisolid, soften a portion of the sample by warming it at a temperature not exceeding 60°, and then mix it thoroughly. Weigh 10 g of a previously well-mixed sample into a 250-mL wide-mouth Erlenmeyer flask, add 100 mL of solvent hexane, and shake until the sample is dissolved. Filter the solution through a 30-mL Corning "C" porosity or equivalent filtering funnel that previously has been dried at 105° for 1 h, cooled in a desiccator, and weighed. Wash the flask with two successive 25-mL portions of solvent hexane, and pass the washings through the filter. Dry the funnel at 105° for 1 h, cool to room temperature in a desiccator, and weigh. From the gain in weight of the funnel, calculate the percentage of the hexane-insoluble matter in the sample.

Iodine Value Determine by the *Modified Wijs Method*, Appendix VII.

Lead A *Sample Solution* prepared as directed for organic compounds meets the requirements of the *Lead Limit Test*, Appendix IIIB, using 10 μg of lead ion (Pb) in the control.

Peroxide Value Weigh accurately about 10 g of the sample, add 30 mL of a 3:2 mixture of glacial acetic acid–chloroform, and mix. Add 1 mL of a saturated solution of potassium iodide, mix, and allow to stand for 10 min. Add 100 mL of water, begin titrating with 0.05 N sodium thiosulfate, adding starch TS as the endpoint is approached, and continue the titration until the blue starch color has just disappeared. Perform a blank determination (see *General Provisions*), and make any necessary correction. Calculate the peroxide value, as meq of peroxide per kg of sample, by the formula

$$S \times N \times 1000/W,$$

in which S is the net volume, in mL, of sodium thiosulfate solution required for the sample; N is the exact normality of the sodium thiosulfate solution; and W is the weight, in g, of the sample taken.

Water Determine by the *Karl Fischer Titrimetric Method*, Appendix IIB.

Packaging and Storage Store in well-closed containers.

Hydroxypropyl Cellulose

Modified Cellulose

INS: 463 CAS: [9004-64-2]

DESCRIPTION

Hydroxypropyl Cellulose is a cellulose ether containing hydroxypropyl substitution. It occurs as a white powder. It is soluble in water and in certain organic solvents. It may contain a suitable anticaking agent.

Functional Use in Foods Emulsifier; film former; protective colloid; stabilizer; suspending agent; thickener.

REQUIREMENTS

Identification
  A. When a 0.1% solution of the sample is shaken, a layer of foam appears (distinction from sodium carboxymethylcellulose).
  B. To 5 mL of a 0.5% solution of the sample add 5 mL of a 5.0% solution of copper sulfate (or aluminum sulfate). No precipitate appears (distinction from sodium carboxymethylcellulose).

Assay Not more than 80.5% of hydroxypropoxyl groups ($-OCH_2CHOHCH_3$) after drying, equivalent to not more than 4.6 hydroxypropyl groups per anhydroglucose unit.

Heavy Metals (as Pb) Not more than 10 mg/kg.

Lead Not more than 3 mg/kg.

Loss on Drying Not more than 5.0%.

pH of a 1% Solution Between 5.0 and 8.0.

Residue on Ignition Not more than 0.5%.

Viscosity of a 10% Solution Not less than 145 centipoises.

TESTS

Assay Weigh accurately about 85 mg of the sample, previously dried at 105° for 3 h, and determine the hydroxypropoxyl content as directed under *Hydroxypropoxyl Determination*, Appendix IIIC.

Heavy Metals Prepare and test a 2-g sample as directed in *Method II* under the *Heavy Metals Test*, Appendix IIIB, adding 1 mL of hydroxylamine hydrochloride solution (1 in 5) to the solution of the residue. Any color does not exceed that produced in a control (*Solution A*) containing 20 μg of lead ion (Pb).

Lead A *Sample Solution* prepared from a 2-g sample as directed for organic compounds meets the requirements of the

Lead Limit Test, Appendix IIIB, using 6 μg of lead ion (Pb) in the control.

Loss on Drying, Appendix IIC   Dry at 105° for 3 h.

pH of a 1% Solution   Determine by the *Potentiometric Method*, Appendix IIB.

Residue on Ignition   Ignite 1 g as directed in the general method, Appendix IIC.

Viscosity of a 10% Solution   Transfer an accurately weighed sample, equivalent to 40 g of Hydroxypropyl Cellulose on the dried basis, into a tared sample container, and proceed as directed under *Viscosity of Sodium Carboxymethylcellulose*, Appendix IIB.

Packaging and Storage   Store in well-closed containers.

Hydroxypropyl Methylcellulose

Propylene Glycol Ether of Methylcellulose; Modified Cellulose, HPMC

INS: 464   CAS: [9004-65-3]

DESCRIPTION

Hydroxypropyl Methylcellulose is the propylene glycol ether of methylcellulose in which both the hydroxypropyl and methyl groups are attached to the anhydroglucose rings of cellulose by ether linkages. Several product types are available that are defined by varying combinations of methoxyl and hydroxypropoxyl content. It occurs as a white to off-white, fibrous powder or as granules. It is soluble in water and in certain organic solvent systems. Aqueous solutions are surface active, form films upon drying, and undergo a reversible transformation from sol to gel upon heating and cooling, respectively.

Functional Use in Foods   Thickening agent; stabilizer; emulsifier.

REQUIREMENTS

Identification

A. Add 1 g to 100 mL of water. It swells and disperses to form a clear to opalescent, mucilaginous solution, depending upon the intrinsic viscosity, which is stable in the presence of most electrolytes.

B. Add 1 g to 100 mL of boiling water and stir the mixture. A slurry is formed that, when cooled to 20°, dissolves to form a clear or opalescent mucilaginous solution.

C. Pour a few mL of the solution prepared for *Identification Test B* onto a glass plate, and allow the water to evaporate. A thin, self-sustaining film results.

Assay for Hydroxypropoxyl Groups   Within the range claimed by the vendor for any product type between a minimum of 3.0% and a maximum of 12.0% of hydroxypropoxyl groups (—OCH$_2$CHOHCH$_2$).

Assay for Methoxyl Groups   Within the range claimed by the vendor for any product type between a minimum of 19.0% and a maximum of 30.0% of methoxyl groups (—OCH$_3$).

Heavy Metals (as Pb)   Not more than 10 mg/kg.

Loss on Drying   Not more than 5.0%.

Residue on Ignition   Not more than 1.5% for products with viscosities of 50 centipoises or above; not more than 3.0% for products with viscosities below 50 centipoises.

Viscosity   The viscosity of a solution containing 2 g in each 100 g of solution is not less than 80.0% and not more than 120.0% of that stated on the label for viscosity types of 100 centipoises or less, and not less than 75.0% and not more than 140.0% of that stated on the label for viscosity types higher than 100 centipoises.

TESTS

Assay   (**Caution**: Perform all steps involving hydriodic acid carefully in a well-ventilated hood. Use goggles, acid-resistant gloves, and other appropriate safety equipment. Be extremely careful when handling the hot vials, since they are under pressure. In the event of hydriodic acid exposure, wash with copious amounts of water and seek medical attention at once.)

Internal Standard Solution   Transfer about 2.5 g of toluene, accurately weighed, into a 1000-mL volumetric flask containing 10 mL of *o*-xylene, dilute with *o*-xylene to volume, and mix.

Standard Preparation   Transfer about 135 mg of adipic acid into a suitable serum vial, add 4.0 mL of hydriodic acid followed by 4.0 mL of the *Internal Standard Solution*, and close the vial securely with a septum stopper. Weigh the vial and its contents accurately, add 30 μL of isopropyl iodide with a syringe through the septum, reweigh, and calculate the weight of isopropyl iodide added. Similarly, add 90 μL of methyl iodide, and calculate the weight added. Shake well, and allow the layers to separate.

Assay Preparation   Transfer about 0.065 g of the sample, accurately weighed, into a 5-mL vial equipped with a pressure-tight septum closure, add an amount of adipic acid equal to the weight of the sample, and pipet 2 mL of the *Internal Standard Solution* into the vial. Cautiously pipet 2 mL of hydriodic acid into the mixture, immediately secure the closure, and weigh accurately. Shake the vial for 30 s, heat at 150° for 20 min, remove from the heat, shake again, using extreme caution, and heat at 150° for 40 min. Allow the vial to cool for about 45 min, and then weigh. If the weight loss is greater than 10 mg, discard the mixture and prepare another *Assay Preparation*.

Chromatographic System   Use a suitable gas chromatograph equipped with a thermal conductivity detector. Under typical conditions, the instrument contains a 1.8-m × 4-mm glass column packed with 10% methylsilicone oil (UCW 982 or equivalent) on 100/120-mesh flux-calcined chromatographic siliceous earth (Chromosorb WHP, or equivalent). The column is maintained at 100°, and the injection port and detector at 200°, and helium is used as the carrier gas flowing at the rate of 20 mL/min.

Calibration   Inject about 2 μL of the upper layer of the *Standard Preparation* into the chromatograph, and record the chromatogram. The retention times for methyl iodide, isopropyl iodide, toluene, and *o*-xylene are approximately 3, 5, 7, and 13 min, respectively. Calculate the relative response factor, F, of equal weights of toluene and methyl iodide by the formula

$$Q/A,$$

in which Q is the quantity ratio of methyl iodide to toluene in the *Standard Preparation*, and A is the peak area ratio of methyl iodide to toluene obtained from the *Standard Preparation*. Similarly, calculate the relative response factor, F', of equal weights of toluene and isopropyl iodide by the formula

$$Q'/A',$$

in which Q' is the quantity ratio of isopropyl iodide to toluene in the *Standard Preparation*, and A' is the peak area ratio of isopropyl iodide to toluene obtained from the *Standard Preparation*.

Procedure Inject about 2 μL of the upper layer of the *Assay Preparation* into the chromatograph, and record the chromatogram. Calculate the percentage of methoxyl groups (—OCH$_3$) in the sample by the formula

$$2 \times (31/142) \times F \times a \times (W/w),$$

in which 31/142 is the ratio of the formula weights of methoxyl to methyl iodide; a is the ratio of the area of the methyl iodide peak to that of the toluene peak obtained from the *Assay Preparation*; W is the weight of toluene in the *Internal Standard Solution*, in g; and w is the weight of sample taken, in g. Similarly, calculate the percentage of hydroxypropoxyl groups (—OCH$_2$CHOHCH$_3$) in the sample by the formula

$$2 \times (75/170) \times F' \times a' \times (W/w),$$

in which a' is the ratio of the area of the isopropyl iodide peak to that of the toluene peak obtained from the *Assay Preparation*.

Heavy Metals Prepare and test a 2-g sample as directed in *Method II* under the *Heavy Metals Test*, Appendix IIIB, adding 1 mL of hydroxylamine hydrochloride solution (1 in 5) to the solution of the residue. Any color does not exceed that produced in a control (*Solution A*) containing 20 μg of lead ion (Pb).

Loss on Drying, Appendix IIC Dry a 3-g sample at 105° for 2 h.

Residue on Ignition Ignite 1 g as directed in the general method, Appendix IIC.

Viscosity Weigh accurately a sample, equivalent to 2 g of solids on the dried basis, transfer to a wide-mouth, 250-mL centrifuge bottle, and add 98 g of water previously heated to between 80° and 90°. Stir with a mechanical stirrer for 10 min, then place the bottle in an ice bath until solution is complete, adjust the weight of the solution to 100 g if necessary, and centrifuge it to expel any entrapped air. Adjust the temperature of the solution to 20° ± 0.1°, and determine the viscosity as directed under *Viscosity of Methylcellulose*, Appendix IIB.

Packaging and Storage Store in well-closed containers.

Inositol

1,2,3,5/4,6-Cyclohexanehexol; *i*-Inositol; *meso*-Inositol

$C_6H_{12}O_6$ Formula wt 180.16

CAS: [87-89-8]

DESCRIPTION

It occurs as fine, white crystals or as a white crystalline powder. It is odorless, has a sweet taste, and is stable in air. Its solutions are neutral to litmus. It is optically inactive. One g is soluble in 6 mL of water. It is slightly soluble in alcohol, and is insoluble in ether and in chloroform.

Functional Use in Foods Nutrient; dietary supplement.

REQUIREMENTS

Identification

A. To 1 mL of a 1 in 50 solution in a porcelain evaporating dish, add 6 mL of nitric acid, and evaporate to dryness on a water bath. Dissolve the residue in 1 mL of water, add 0.5 mL of a 1 in 10 solution of strontium acetate, and again evaporate to dryness on a steam bath. A violet color is produced.

B. The inositol hexaacetate obtained in the *Assay* melts between 212° and 216° (see Appendix IIB).

Assay Not less than 97.0% of $C_6H_{12}O_6$ after drying.
Calcium Passes test.
Chloride Not more than 0.005%.
Heavy Metals (as Pb) Not more than 0.002%.
Lead Not more than 10 mg/kg.
Loss on Drying Not more than 0.5%.
Melting Range Between 224° and 227°.
Residue on Ignition Not more than 0.1%.
Sulfate Not more than 0.006%.

TESTS

Assay Transfer about 200 mg, previously dried at 105° for 4 h and accurately weighed, to a 250-mL beaker, add 5 mL of a mixture consisting of 1 part of 2 N sulfuric acid in 50 parts of acetic anhydride, and cover the beaker with a watch glass. Heat on a steam bath for 20 min, then chill in an ice bath, and add 100 mL of water. Boil for 20 min, allow to cool, and transfer quantitatively, with the aid of a little water, to a 250-mL separator. Extract the solution with 6 successive 30-, 25-, 20-, 15-, 10-, and 10-mL portions of chloroform, using the solvent to rinse the original flask. Collect the chloroform extracts in a second 250-mL separator and wash the combined extracts with 10 mL of water. Transfer the chloroform solution through a funnel containing a pledget of cotton into a 150-mL tared Soxhlet flask. Wash the separator and funnel with 10 mL of chloroform and add to the combined extracts. Evaporate to dryness on a steam bath, dry in an oven at 105° for 1 h, cool in a desiccator, and weigh. The weight of the inositol hexaacetate obtained, multiplied by 0.4167, represents the equivalent of $C_6H_{12}O_6$.

Calcium To 10 mL of a 1 in 10 solution add 1 mL of ammonium oxalate TS. The solution remains clear for at least 1 min.

Chloride, Appendix IIIB   Any turbidity produced by a 400-mg sample does not exceed that shown in a control containing 20 μg of chloride ion (Cl).

Heavy Metals   A solution of 1 g in 25 mL of water meets the requirements of the *Heavy Metals Test*, Appendix IIIB, using 20 μg of lead ion (Pb) in the control (*Solution A*).

Lead   A *Sample Solution* prepared as directed for organic compounds meets the requirements of the *Lead Limit Test*, Appendix IIIB, using 10 μg of lead ion (Pb) in the control.

Loss on Drying, Appendix IIC   Dry at 105° for 4 h.

Melting Range   Determine as directed for *Melting Range or Temperature*, Appendix IIB.

Residue on Ignition, Appendix IIC   Ignite 2 g as directed in the general method.

Sulfate, Appendix IIIB   Any turbidity produced by a 5-g sample does not exceed that shown in a control containing 300 μg of sulfate (SO_4).

Packaging and Storage   Store in well-closed containers.

Invert Sugar

Invert Sugar Syrup

CAS: [8013-17-0]

DESCRIPTION

Invert Sugar is a mixture of glucose and fructose that results from the hydrolysis of sucrose in accordance with good manufacturing practices. Invert Sugar is marketed as Invert Sugar Syrup and also contains dextrose (glucose), fructose, and sucrose in various amounts as represented by the manufacturer.

Invert Sugar Syrup is a hygroscopic liquid that has a sweet taste; is very soluble in water, in glycerin, and in glycols; and is very sparingly soluble in acetone and in ethanol.

Functional Use in Foods   Nutritive sweetener.

REQUIREMENTS

Labeling   Indicate the percent sucrose and Invert Sugar.

Identification   Prepare a 10% solution in purified water and inject 7.5 μL into a high-performance liquid chromatographic system equipped with a cation exchange resin maintained at 85° and a differential refractometer detector; the liquid phase is purified water eluting at a flow rate of 0.7 mL/min. The chromatogram of the sample gives appropriate elution times for fructose, glucose, and sucrose when compared with a standard solution containing 1 g of each saccharide in 100 mL of purified water.

Assay   Not less than 90.0% and not more than 110.0% of the labeled amount of sucrose and of Invert Sugar.

Arsenic (as As)   Not more than 1 mg/kg.

Heavy Metals (as Pb)   Not more than 5 mg/kg.

Lead   Not more than 0.5 mg/kg.

pH   Not less than 3.0 and not more than 5.5.

Residue on Ignition   Not more than 0.2%.

Total Solids   As represented by the vendor.

Total Sugars   Not less than 99.5% of the total solids content.

TESTS

Assay   Determine as directed under *Invert Sugar*, Appendix X.

Arsenic   A *Sample Solution* prepared as directed using a 1-g sample meets the requirements of the *Arsenic Test*, Appendix IIIB, using 1 mL of the *Standard Arsenic Solution* in the control (1 μg As).

Heavy Metals   A solution of 4 g in 25 mL of water meets the requirements of the *Heavy Metals Test*, Appendix IIIB, using 20 μg of lead ion (Pb) in the control (*Solution A*).

Lead   Determine as directed under *Method I* in the *Atomic Absorption Spectrophotometric Graphite Furnace Method* under the *Lead Limit Test*, Appendix IIIB, using a 5-g sample.

pH   Determine by the *Potentiometric Method*, Appendix IIB.

Residue on Ignition   Determine as directed in *Method II* under *Residue on Ignition*, Appendix IIC.

Total Solids   Proceed as directed under *Invert Sugar, Total Solids*, Appendix X, using the table provided.

Total Sugars   Calculate the *Total Sugars* (T_S) as the sum of the percentages of Invert Sugar (P_I) and sucrose (P_S) determined under *Assay*:

$$T_S = P_I + P_S.$$

Packaging and Storage   Store in tight containers.

Iron, Carbonyl

Fe   At wt 55.85

CAS: [37220-42-1]

DESCRIPTION

Carbonyl Iron is elemental iron produced by the decomposition of iron pentacarbonyl as a dark gray powder. When viewed under a microscope having a magnifying power of 500 diameters or greater, it appears as spheres built up with concentric shells. It is stable in dry air.

Functional Use in Foods Nutrient; dietary supplement.

REQUIREMENTS

Identification It dissolves in dilute mineral acids with the evolution of hydrogen and the formation of solutions of the corresponding salts, which give positive tests for *Ferrous Salts (Iron)*, Appendix IIIA.
Assay Not less than 98.0% of Fe.
Acid-Insoluble Substances Not more than 0.2%.
Arsenic (as As) Not more than 3 mg/kg.
Lead Not more than 10 mg/kg.
Mercury Not more than 2 mg/kg.
Sieve Analysis Not less than 100% passes through a 200-mesh sieve; not less than 95% passes through a 325-mesh sieve.

TESTS

Assay Determine as directed in the monograph for *Iron, Reduced*.
Acid-Insoluble Substances Determine as directed in the monograph for *Iron, Electrolytic*.
Arsenic Dissolve 1.0 g in 25 mL of 2 N sulfuric acid, heat on a steam bath until the evolution of hydrogen ceases, cool, and dilute to 35 mL with water. This solution meets the requirements of the *Arsenic Test*, Appendix IIIB.
Lead Proceed as directed under *Lead* in the monograph for *Ferrous Gluconate*, but with the following modifications for *Standard Preparation and Blank* and *Test Preparation*.

Standard Preparation and Blank Transfer 5.0 mL of *Lead Nitrate Stock Solution*, prepared as directed in the test for *Heavy Metals*, Appendix IIIB, to a 100-mL volumetric flask, dilute with water to volume, and mix. Transfer to a 50-mL beaker 2.0 mL of the resulting solution. To this beaker, and to a separate 50-mL beaker (blank), add 8 mL of hydrochloric acid and 2 mL of nitric acid. Place a ribbed watch glass over each beaker, and evaporate to dryness on a steam bath. Add 10 mL of 9 N hydrochloric acid to each beaker, and transfer the resulting solutions, with the aid of about 10 mL of water, to separate 50-mL volumetric flasks. To each flask add 20 mL of *Ascorbic Acid–Sodium Iodide Solution* and 5.0 mL of *Trioctylphosphine Oxide Solution*, shake for 30 s, and allow to separate. Add water to bring the organic solvent layer into the neck of each flask, shake again, and allow to separate. The organic solvent layers are the *Blank* and the *Standard Preparation*, and they contain 0.0 and 2.0 μg of lead per mL, respectively.

Test Preparation Transfer 1.0 g of the sample to a 50-mL beaker, and cover it with a ribbed watch glass. Slowly add 8 mL of hydrochloric acid and 2 mL of nitric acid, keeping the beaker covered as much as possible. After the initial reaction subsides, evaporate to dryness on a steam bath, cool, and dissolve the residue in 10 mL of 9 N hydrochloric acid, warming if necessary to effect solution. Cool, and transfer the resultant solution, with the aid of about 10 mL of water, to a 50-mL volumetric flask. Add 20 mL of *Ascorbic Acid–Sodium Iodide Solution* and 5 mL of *Trioctylphosphine Oxide Solution*, shake for 30 s, and allow to separate. Add water to bring the organic solvent layer into the neck of the flask, shake again, and allow to separate. The organic solvent layer is the *Test Preparation*.
Mercury Determine as directed in the monograph for *Iron, Reduced*, but use 2 g of the sample and 40 mL of *Sodium Citrate Solution* in preparing the *Sample Solution*, and prepare the *Diluted Standard Mercury Solution* as follows: Transfer 4.0 mL of *Mercury Stock Solution* into a 250-mL volumetric flask, dilute to volume with 1 N hydrochloric acid, and mix (1 mL = 4 μg Hg). Modify the first sentence of the *Procedure* to read: "Prepare a control by treating 1.0 mL of *Diluted Standard Mercury Solution* (4 μg Hg) in the same manner. . . ."
Sieve Analysis Determine as directed under *Sieve Analysis of Granular Metal Powders*, Appendix IIC.

Packaging and Storage Store in well-closed containers.

Iron, Electrolytic

Fe   At wt 55.85

DESCRIPTION

Electrolytic Iron is elemental iron obtained by electrodeposition in the form of an amorphous, lusterless, grayish black powder. It is stable in dry air.

Functional Use in Foods Nutrient; dietary supplement.

REQUIREMENTS

Identification It dissolves in dilute mineral acids with the evolution of hydrogen and the formation of solutions of the corresponding salts, which give positive tests for *Ferrous Salts (Iron)*, Appendix IIIA.
Assay Not less than 97.0% of Fe.
Acid-Insoluble Substances Not more than 0.2%.
Arsenic (as As) Not more than 3 mg/kg.
Lead Not more than 10 mg/kg.
Mercury Not more than 2 mg/kg.
Sieve Analysis Not less than 100% passes through a 100-mesh sieve; not less than 95% passes through a 325-mesh sieve.

TESTS

Assay Determine as directed in the monograph for *Iron, Reduced*.
Acid-Insoluble Substances Dissolve 1 g in 25 mL of 2 N sulfuric acid, and heat on a steam bath until the evolution of hydrogen ceases. Filter through a tared filter crucible, wash with water until free from sulfate, dry at 105° for 1 h, cool, and weigh.
Arsenic Dissolve 1 g in 25 mL of 2 N sulfuric acid, heat on a steam bath until the evolution of hydrogen ceases, cool, and

dilute to 35 mL with water. This solution meets the requirements of the *Arsenic Test*, Appendix IIIB.

Lead Proceed as directed under *Lead* in the monograph for *Iron, Carbonyl*.

Mercury Determine as directed in the monograph for *Iron, Reduced*, but use 2 g of the sample and 40 mL of *Sodium Citrate Solution* in preparing the *Sample Solution*, and prepare the *Diluted Standard Mercury Solution* as follows: Transfer 4.0 mL of *Mercury Stock Solution* into a 250-mL volumetric flask, dilute to volume with 1 N hydrochloric acid, and mix (1 mL = 4 µg of Hg). Modify the first sentence of the *Procedure* to read: "Prepare a control by treating 1.0 mL of *Diluted Standard Mercury Solution* (4 µg Hg) in the same manner...."

Sieve Analysis Determine as directed under *Sieve Analysis of Granular Metal Powders*, Appendix IIC.

Packaging and Storage Store in well-closed containers.

Iron, Reduced

Fe   At wt 55.85

DESCRIPTION

Reduced Iron is elemental iron obtained by a chemical process in the form of a grayish black powder, all of which should pass through a 100-mesh sieve. It is lusterless or has not more than a slight luster. When viewed under a microscope having a magnifying power of 100 diameters, it appears as an amorphous powder, free from particles having a crystalline structure. It is stable in dry air.

Functional Use in Foods Nutrient; dietary supplement.

REQUIREMENTS

Identification It dissolves in dilute mineral acids with the evolution of hydrogen and the formation of solutions of the corresponding salts, which give positive tests for *Ferrous Salts* (Iron), Appendix IIIA.

Assay Not less than 96.0% of Fe.

Acid-Insoluble Substances Not more than 1.25%.

Arsenic (as As) Not more than 3 mg/kg.

Lead Not more than 10 mg/kg.

Mercury Not more than 5 mg/kg.

TESTS

Assay Transfer about 200 mg, accurately weighed, into a 300-mL Erlenmeyer flask, add 50 mL of 2 N sulfuric acid, and close the flask with a stopper containing a Bunsen valve, made by inserting a glass tube connected to a short piece of rubber tubing with a slit on the side and a glass rod inserted in the other end and arranged so that gases can escape but air cannot enter. Heat on a steam bath until the iron is dissolved, cool the solution, dilute it with 50 mL of recently boiled and cooled water, add 2 drops of orthophenanthroline TS, and titrate with 0.1 N ceric sulfate until the red color changes to a weak blue. Each mL of 0.1 N ceric sulfate is equivalent to 5.585 mg of Fe.

Acid-Insoluble Substances Dissolve 1.0 g in 25 mL of 2 N sulfuric acid, and heat on a steam bath until the evolution of hydrogen ceases. Filter through a tared filter crucible, wash with water until free from sulfate, and dry at 105° for 1 h. The weight of the residue does not exceed 12.5 mg.

Arsenic Dissolve 1.0 g in 25 mL of 2 N sulfuric acid, heat on a steam bath until the evolution of hydrogen ceases, cool, and dilute to 35 mL with water. This solution meets the requirements of the *Arsenic Test*, Appendix IIIB.

Lead Proceed as directed under *Lead* in the monograph for *Iron, Carbonyl*.

Mercury

Dithizone Stock Solution Dissolve 30 mg of dithizone in 1000 mL of chloroform, add 5 mL of alcohol, and mix. Store in a refrigerator in a dark bottle. Prepare fresh each month.

Dithizone Extraction Solution On the day of use, dilute 30 mL of *Dithizone Stock Solution* to 100 mL with chloroform.

Hydroxylamine Hydrochloride Solution Prepare as directed under *Mercury* in the monograph for *Ferrous Fumarate*.

Mercury Stock Solution Transfer 33.8 mg, accurately weighed, of mercuric chloride into a 100-mL volumetric flask, dissolve in 1 N hydrochloric acid, dilute to volume with the acid, and mix. This solution contains the equivalent of 250 µg of Hg in each mL.

Diluted Standard Mercury Solution Transfer 2.0 mL of *Mercury Stock Solution* into a 100-mL volumetric flask, dilute to volume with 1 N hydrochloric acid, and mix. Each mL contains the equivalent of 5 µg of Hg.

Sodium Citrate Solution Dissolve 250 g of sodium citrate dihydrate in 1000 mL of water.

Sample Solution Transfer 1 g of the sample into a 250-mL beaker, add 20 mL of dilute nitric acid (1 in 2), and digest on a steam bath for about 45 min. Add 5 mL of dilute hydrochloric acid (1 in 3), and continue heating on the steam bath until the sample is dissolved. Cool to room temperature, and filter, if necessary, through a medium-porosity filter paper. Wash with a few mL of water, add 20 mL of *Sodium Citrate Solution* and 1 mL of *Hydroxylamine Hydrochloride Solution* to the filtrate, and adjust the pH to 1.8 with ammonium hydroxide.

Procedure (Note: Because mercuric dithizonate is light sensitive, this procedure should be performed in subdued light.) Prepare a control by treating 1.0 mL of *Diluted Standard Mercury Solution* (5 µg Hg) in the same manner and with the same reagents as directed for the preparation of the *Sample Solution*. Transfer the control and the *Sample Solution* into separate 250-mL separators, and treat both solutions as follows: Extract with 5 mL of *Dithizone Extraction Solution*, shaking the mixtures vigorously for 1 min. Drain carefully, collecting the chloroform in another separator. If the chloroform does not show a pronounced green color due to excess reagent, add another 5 mL of the extraction solution, shake again, and drain into the separator. Continue the extraction with 5-mL portions, if necessary, collecting each successive extract in the second separator, until the final chloroform layer contains dithizone in marked excess. To the combined chloroform extracts add 15 mL of dilute hydrochloric acid (1 in 3), shake the mixture vigorously for 1 min, and discard the chloroform. Extract with 2 mL of chloroform, drain carefully, and discard the chloroform. Add 1 mL of 0.05 M disodium EDTA and 2 mL of 6 N acetic acid to the aqueous layer. Slowly add 5 mL of 6 N ammonium hydroxide, and cool the separator. Transfer the solution into a 150-mL beaker, adjust the pH to 1.8 with 6 N ammonium hydroxide or dilute nitric acid (1 in 10), using a pH meter, and return the solution to the separator. Add 5.0 mL of *Dithizone Extraction Solution*, and shake vigorously for 1 min. Allow the layers to separate, insert a plug of cotton into the stem of the separator, and collect the dithizone extract in a test tube. Determine the absorbance of each solution in 1-cm cells at 490 nm with a suitable spectrophotometer, using chloroform as the blank. The absorbance of the *Sample Solution* does not exceed that of the control.

Packaging and Storage Store in well-closed containers.

Isobutane

$$CH_3CH(CH_3)_2$$

C_4H_{10} \hspace{2cm} Formula wt 58.12

DESCRIPTION

A colorless, odorless, flammable gas (boiling temperature is about $-11°$). The vapor pressure at $21°$ is approximately 2950 mm of mercury (31 psi).

Functional Use in Foods Propellant; aerating agent.

REQUIREMENTS

Caution: Isobutane is highly flammable and explosive. Observe precautions, and perform sampling and analytical operations in a well-ventilated fume hood.

Identification
A. The infrared absorption spectrum exhibits maxima, among others, at about the following wavelengths, in μm: 3.4 (vs), 6.8 (s), 7.2 (m), and 10.9 (m).
B. The vapor pressure of a test specimen, obtained as directed in the *Sampling Procedure* and determined at $21°$ by means of a suitable pressure gauge, is between 303 and 331 kPa absolute (44 and 48 psia, respectively).
Assay Not less than 95.0% of C_4H_{10}.
Acidity of Residue Passes test.
High-Boiling Residue Not more than 5 mg/kg.
Sulfur Compounds Passes test.
Water Not more than 10 mg/kg.

TESTS

Sampling Procedure, Assay, Acidity of Residue, High-Boiling Residue, Sulfur Compounds, and **Water** Proceed as directed for these tests in the monograph for *Butane*, except substitute Isobutane wherever butane is specified in the text.

Packaging and Storage Store in tight cylinders protected from excessive heat.

Isobutylene–Isoprene Copolymer

Butyl Rubber

CAS: [9010-85-9]

DESCRIPTION

A synthetic copolymer containing from 0.5 to 2.0 molar percent of isoprene, the remainder, respectively, consisting of isobutylene. It is prepared by copolymerization of isobutylene and isoprene in methyl chloride solution, using aluminum chloride as the catalyst. After completion of polymerization, the rubber particles are treated with hot water containing a suitable food-grade deagglomerating agent, such as stearic acid. Finally, the coagulum is dried to remove residual volatiles.

Functional Use in Foods Masticatory substance in chewing gum base.

REQUIREMENTS

Identification Identify Isobutylene–Isoprene Copolymer by comparing its infrared absorption spectrum with a typical spectrum as shown in the section on *Infrared Spectra* (*Series C: Other Substances*). Prepare the sample by dissolving it in hot toluene and evaporating on a potassium bromide plate.
Heavy Metals (as Pb) Not more than 0.002%.
Lead Not more than 3 mg/kg.
Total Unsaturation Not more than 2.0%, as isoprene.

TESTS

Heavy Metals Prepare and test a 1-g sample as directed in *Method II* under the *Heavy Metals Test*, Appendix IIIB, using 20 μg of lead ion (Pb) in the control (*Solution A*).
Lead Prepare a *Sample Solution* as directed in the general method under *Chewing Gum Base*, Appendix IV. This solution meets the requirements of the *Lead Limit Test*, Appendix IIIB, using 10 μg of lead ion (Pb) in the control.
Total Unsaturation Determine as directed in the general method, Appendix IV.

Packaging and Storage Store in well-closed containers.

DL-Isoleucine

DL-2-Amino-3-methylvaleric Acid

$C_6H_{13}NO_2$ Formula wt 131.17

CAS: [443-79-8]

DESCRIPTION

A white, odorless, crystalline powder having a slightly bitter taste. It is soluble in water, but is practically insoluble in alcohol and in ether. It melts with decomposition at about 292°. The pH of a 1 in 100 solution is between 5.5 and 7.0. It is optically inactive.

Functional Use in Foods Nutrient; dietary supplement.

REQUIREMENTS

Identification To 5 mL of a 1 in 1000 solution add 1 mL of triketohydrindene hydrate TS. A bluish purple color is produced.
Assay Not less than 98.5% and not more than 101.5% of $C_6H_{13}NO_2$, calculated on the dried basis.
Heavy Metals (as Pb) Not more than 0.002%.
Lead Not more than 10 mg/kg.
Loss on Drying Not more than 0.3%.
Residue on Ignition Not more than 0.1%.

TESTS

Assay Dissolve about 250 mg, accurately weighed, in 3 mL of formic acid and 50 mL of glacial acetic acid, add 2 drops of crystal violet TS, and titrate with 0.1 N perchloric acid to the first appearance of a pure green color or until the blue color disappears completely. Perform a blank determination (see *General Provisions*), and make any necessary correction. Each mL of 0.1 N perchloric acid is equivalent to 13.12 mg of $C_6H_{13}NO_2$.
Heavy Metals Prepare and test a 1-g sample as directed in *Method II* under the *Heavy Metals Test*, Appendix IIIB, using 20 μg of lead ion (Pb) in the control (*Solution A*).
Lead A *Sample Solution* prepared as directed for organic compounds meets the requirements of the *Lead Limit Test*, Appendix IIIB, using 10 μg of lead ion (Pb) in the control.
Loss on Drying, Appendix IIC Dry at 105° for 3 h.
Residue on Ignition, Appendix IIC Ignite 1 g as directed in the general method.

Packaging and Storage Store in well-closed containers.

L-Isoleucine

L-2-Amino-3-methylvaleric Acid

$C_6H_{13}NO_2$ Formula wt 131.17

CAS: [73-32-5]

DESCRIPTION

Crystalline leaflets or a white crystalline powder having a bitter taste. It is soluble in 25 parts of water, slightly soluble in hot alcohol, and soluble in diluted mineral acids and in alkaline solutions. It sublimes at between 168° and 170°, and melts with decomposition at about 284°. The pH of a 1 in 100 solution is between 5.5 and 7.0.

Functional Use in Foods Nutrient; dietary supplement.

REQUIREMENTS

Identification To 5 mL of a 1 in 1000 solution add 1 mL of triketohydrindene hydrate TS. A reddish purple or bluish purple color is produced.
Assay Not less than 98.5% and not more than 101.5% of $C_6H_{13}NO_2$, calculated on the dried basis.
Heavy Metals (as Pb) Not more than 0.002%.
Lead Not more than 10 mg/kg.
Loss on Drying Not more than 0.3%.
Residue on Ignition Not more than 0.2%.
Specific Rotation $[\alpha]_D^{20°}$: Between +38.6° and +41.5° after drying; or $[\alpha]_D^{25°}$: Between +38.2° and +41.1° after drying.

TESTS

Assay Dissolve about 250 mg, accurately weighed, in 3 mL of formic acid and 50 mL of glacial acetic acid, add 2 drops of crystal violet TS, and titrate with 0.1 N perchloric acid to the first appearance of a pure green color or until the blue color disappears completely. Perform a blank determination (see *General Provisions*), and make any necessary correction. Each mL of 0.1 N perchloric acid is equivalent to 13.12 mg of $C_6H_{13}NO_2$.
Heavy Metals Prepare and test a 1-g sample as directed in *Method II* under the *Heavy Metals Test*, Appendix IIIB, using 20 μg of lead ion (Pb) in the control (*Solution A*).
Lead A *Sample Solution* prepared as directed for organic compounds meets the requirements of the *Lead Limit Test*, Appendix IIIB, using 10 μg of lead ion (Pb) in the control.
Loss on Drying, Appendix IIC Dry at 105° for 3 h.
Residue on Ignition, Appendix IIC Ignite 1 g as directed in the general method.

Specific Rotation, Appendix IIB   Determine in a solution containing 4 g of a previously dried sample in sufficient 6 N hydrochloric acid to make 100 mL.

Packaging and Storage   Store in well-closed containers.

Isopropyl Alcohol

2-Propanol; Isopropanol

$$CH_3CHOHCH_3$$

C_3H_8O Formula wt 60.10

CAS: [67-63-0]

DESCRIPTION

A clear, colorless, flammable liquid having a characteristic odor and a slightly bitter taste. It is miscible with water, with ethyl alcohol, with ether, and with many other organic solvents. Its refractive index at 20° is about 1.377.

Functional Use in Foods   Extraction solvent.

REQUIREMENTS

Identification   Add 3 mL of water and 1 mL of mercuric sulfate TS to 2 mL of the sample contained in a test tube, and warm gently. A white or yellowish precipitate is formed.
Assay   Not less than 99.7% of C_3H_8O, by weight.
Acidity (as acetic acid)   Not more than 10 mg/kg.
Distillation Range   Within a range of 1°, including 82.3°.
Heavy Metals (as Pb)   Not more than 1 mg/kg.
Nonvolatile Residue   Not more than 10 mg/kg.
Solubility in Water   Passes test.
Substances Reducing Permanganate   Passes test.
Water   Not more than 0.2%.

TESTS

Assay   Its specific gravity, determined by any reliable method (see *General Provisions*), is not greater than 0.7840 at 25°/25° (equivalent to 0.7870 at 20°/20°).
Acidity   Add 2 drops of phenolphthalein TS to 100 mL of water, add 0.01 N sodium hydroxide to the first pink color that persists for at least 30 s, then add 50 mL (about 39 g) of the sample, and mix. Not more than 0.7 mL of 0.01 N sodium hydroxide is required to restore the pink color.
Distillation Range   Proceed as directed in the general method, Appendix IIB.
Heavy Metals   Evaporate 25 mL (about 20 g) of the sample to dryness on a steam bath in a glass evaporating dish. Cool, add 2 mL of hydrochloric acid, and slowly evaporate to dryness again on the steam bath. Moisten the residue with 1 drop of hydrochloric acid, add 10 mL of hot water, and digest for 2 min. Cool, and dilute to 25 mL with water. This solution meets the requirements of the *Heavy Metals Test*, Appendix IIIB, using 20 μg of lead ion (Pb) in the control (*Solution A*).
Nonvolatile Residue   Evaporate 125 mL (about 100 g) of the sample to dryness in a tared dish on a steam bath, dry the residue at 105° for 30 min, cool, and weigh.
Solubility in Water   Mix 10 mL of the sample with 40 mL of water. After 1 h, the solution is as clear as an equal volume of water.
Substances Reducing Permanganate   Transfer 50 mL of the sample into a 50-mL glass-stoppered cylinder, add 0.25 mL of 0.1 N potassium permanganate, mix, and allow to stand for 10 min. The pink color is not entirely discharged.
Water   Determine by the *Karl Fischer Titrimetric Method*, Appendix IIB.

Packaging and Storage   Store in tight containers, remote from fire.

Juniper Berries Oil

DESCRIPTION

The volatile oil obtained by steam distillation from the dried ripe fruit of the plant *Juniperus communis* L. var. *erecta* Pursh (Fam. *Cupressaceae*). It is a colorless, faintly greenish, or yellowish liquid with a characteristic odor and an aromatic bitter taste. It is soluble in most fixed oils and in mineral oil. It is insoluble in glycerin and in propylene glycol. The oil tends to polymerize on long storage.

Functional Use in Foods   Flavoring agent.

REQUIREMENTS

Identification   The infrared absorption spectrum of the sample exhibits relative maxima (that may vary in intensity) at the same wavelengths (or frequencies) as those shown in the respective spectrum under *Infrared Spectra (Series A: Essential Oils)* using the same test conditions as specified therein.
Angular Rotation   Between −15° and 0°.
Heavy Metals (as Pb)   Passes test.
Refractive Index   Between 1.474 and 1.484 at 20°.
Specific Gravity   Between 0.854 and 0.879.

TESTS

Angular Rotation   Determine in a 100-mm tube as directed under *Optical (Specific) Rotation*, Appendix IIB.
Heavy Metals   Shake 10 mL of the oil with an equal volume of water to which 1 drop of hydrochloric acid has been added, and pass hydrogen sulfide through the mixture until it is saturated. No darkening in color is produced in either the oil or the water.

Refractive Index, Appendix IIB Determine with an Abbé or other refractometer of equal or greater accuracy.

Specific Gravity Determine by any reliable method (see *General Provisions*).

Packaging and Storage Store in full, tight, preferably glass, tin-lined, or galvanized containers in a cool place protected from light.

Kaolin

China Clay

DESCRIPTION

A purified clay consisting mainly of alumina, silica, and water. It occurs as a fine, white to yellowish white or grayish powder having an earthy taste. It becomes darker and has a distinct claylike odor when moistened. It is insoluble in water, in alcohol, in dilute acids, and in alkali solutions.

Functional Use in Foods Anticaking agent.

REQUIREMENTS

Identification Mix 1 g of the sample with 10 mL of water and 5 mL of sulfuric acid in a porcelain dish, and evaporate until the water is removed. Continue heating until dense white fumes of sulfur trioxide are evolved, then cool, and cautiously add 20 mL of water. Boil for a few min, and filter. A gray residue of silica remains on the filter. To a portion of the filtrate add 6 N ammonium hydroxide. A gelatinous, white precipitate of aluminum hydroxide is produced that is insoluble in an excess of 6 N ammonium hydroxide.

Acid-Soluble Substances Not more than 2.0%.
Arsenic (as As) Not more than 3 mg/kg.
Carbonate Passes test.
Heavy Metals (as Pb) Not more than 0.004%.
Iron Passes test.
Lead Not more than 10 mg/kg.
Loss on Ignition Not more than 15.0%.
Sulfide Passes test.

TESTS

Acid-Soluble Substances Digest a 1-g sample with 20 mL of 2.7 N hydrochloric acid for 15 min, and filter. Evaporate 10 mL of the filtrate to dryness in a tared dish, ignite gently, cool, and weigh.

Sample Solution for the Determination of Arsenic, Heavy Metals, and Lead Transfer 10.0 g of the sample into a 250-mL flask, and add 50 mL of 0.5 N hydrochloric acid. Attach a reflux condenser to the flask, heat on a steam bath for 30 min, cool, and let the undissolved material settle. Decant the supernatant liquid through Whatman No. 3 filter paper, or equivalent, into a 100-mL volumetric flask, retaining as much as possible of the insoluble material in the beaker. Wash the slurry and beaker with three 10-mL portions of hot water, decanting each washing through the filter into the flask. Finally, wash the filter paper with 15 mL of hot water, cool the filtrate to room temperature, dilute to volume with water, and mix.

Arsenic A 10-mL portion of the *Sample Solution* meets the requirements of the *Arsenic Test*, Appendix IIIB.

Carbonate Mix a 1-g sample with 10 mL of water, cool, and keep the mixture cool while adding 5 mL of sulfuric acid. No effervescence occurs during the addition of the acid.

Heavy Metals A 5-mL portion of the *Sample Solution* diluted to 25 mL with water meets the requirements of the *Heavy Metals Test*, Appendix IIIB, using 20 µg of lead ion (Pb) in the control (*Solution A*).

Iron Mix a 2-g sample with 10 mL of water in a mortar, and add 500 mg of sodium salicylate. No more than a light reddish tint is produced.

Lead A 10-mL portion of the *Sample Solution* meets the requirements of the *Lead Limit Test*, Appendix IIIB, using 10 µg of lead ion (Pb) in the control.

Loss on Ignition Ignite a 2-g sample, accurately weighed, in a tared crucible at 575° ± 25° to constant weight, cool, and weigh.

Sulfide Add a 1-g sample to 25 mL of water in a 250-mL flask, then add 15 mL of 2.7 N hydrochloric acid, and immediately cover the top of the flask with filter paper moistened with lead acetate TS. Heat to boiling, and boil for several min. The paper does not show any brown coloration.

Packaging and Storage Store in well-closed containers.

Karaya Gum

Sterculia Gum

CAS: [9000-36-6]

DESCRIPTION

A dried gummy exudation from *Sterculia urens* Roxburgh and other species of *Sterculia* (Fam. *Sterculiaceae*), or from *Cochlospermum gossypium* A. P. De Condolle or other species of *Cochlospermum* Kunth (Fam. *Bixaceae*). It occurs in tears of variable size or in broken irregular pieces having a somewhat crystalline appearance. It is pale yellow to pinkish brown, translucent, and horny, and is sometimes admixed with a few darker fragments and occasional pieces of bark. The gum has a slightly acetous odor and a mucilaginous and slightly acetous taste. In the powdered form it is light gray to pinkish gray. Karaya Gum is insoluble in alcohol, but it swells in water to form a gel.

Functional Use in Foods Stabilizer; thickener; emulsifier.

REQUIREMENTS

Identification
  A. Add 2 g to 50 mL of water. It swells to form a granular, stiff, slightly opalescent mucilage.
  B. Add a few drops of Millon's Reagent to a 1 in 100 solution of the gum. A white curdy precipitate forms.
Arsenic (as As) Not more than 3 mg/kg.
Ash (Acid-Insoluble) Not more than 1.0%.
Foreign Gums Passes test.
Heavy Metals (as Pb) Not more than 0.002%.
Insoluble Matter Not more than 3.0%.
Lead Not more than 5 mg/kg.
Loss on Drying Not more than 20.0%.
Starch Passes test.
Viscosity of a 1% Solution Not less than the minimum or within the range claimed by the vendor.

TESTS

Arsenic A *Sample Solution* prepared as directed for organic compounds meets the requirements of the *Arsenic Test*, Appendix IIIB.
Ash (Acid-Insoluble) Determine as directed in the general method, Appendix IIC.
Foreign Gums The gum swells in 60% alcohol (*distinction from other gums*).
Heavy Metals Prepare and test a 1-g sample as directed in *Method II* under the *Heavy Metals Test*, Appendix IIIB, using 20 µg of lead ion (Pb) in the control (*Solution A*).
Insoluble Matter Transfer about 5 g, accurately weighed, into a 250-mL Erlenmeyer flask, add a mixture of equal parts of 2.7 N hydrochloric acid and water, cover the flask with a watch glass, and boil gently until the mixture loses its viscosity. Filter the solution through a tared filtering crucible, wash the residue with water until the washings are free from acid, dry at 105° for 1 h, and weigh.
Lead A *Sample Solution* prepared as directed for organic compounds meets the requirements of the *Lead Limit Test*, Appendix IIIB, using 5 µg of lead ion (Pb) in the control.
Loss on Drying, Appendix IIC Powder an unground sample until it passes through a No. 40 sieve, mix well before weighing, and dry at 105° for 5 h.
Starch To a 1 in 10 solution of the gum add a few drops of iodine TS. No blue color is produced.
Viscosity Transfer a 4-g sample, finely powdered, into the container of a stirring apparatus equipped with blades capable of being adjusted to about 1000 rpm. Add 10 mL of alcohol to the sample, swirl to wet the gum uniformly, and then add 390 mL of water, avoiding the formation of lumps. Stir the mixture for 7 min, pour the resulting dispersion into a 500-mL bottle, insert a stopper, and allow to stand for about 12 h in a water bath at 25°. Determine the apparent viscosity at this temperature with a model LVF Brookfield or equivalent-type viscometer (see *Viscosity of Sodium Carboxymethylcellulose*, Appendix IIB) using a suitable spindle, speed, and factor.

Packaging and Storage Store in well-closed containers.

Kelp

DESCRIPTION

The dehydrated seaweed obtained from the class *Phaeophyceae* (brown algae) of the genera *Macrocystis* (including *M. pyrifera* and related species) and *Laminaria* (including *L. digitata, L. cloustoni,* and *L. saccharina*). The seaweed may be chopped to provide coarse particles and/or it may be ground to provide a fine powder. It is dark green to olive brown in color and has a salty, characteristic taste.

Functional Use in Foods Dietary supplement (source of iodine).

REQUIREMENTS

Arsenic (as As, inorganic) Not more than 3 mg/kg.
Ash (Total) Not more than 45.0%.
Heavy Metals (as Pb) Not more than 0.002%.
Iodine Content Between 0.1% and 0.5%.
Lead Not more than 10 mg/kg.
Loss on Drying Not more than 13.0%.

TESTS

Arsenic Assemble the special distillation apparatus for determination of inorganic arsenic as shown in the figure (Special Apparatus for the Determination of Inorganic Arsenic) in Appendix IIIB. Weigh accurately 2.00 g of Kelp, which has been previously ground to pass a 60-mesh screen, and transfer to the distillation flask (*A*) of the apparatus. To the flask add 50 mL of distillation-reducing solution (prepared on the day of use by dissolving 36 g of ACS low-arsenic ferrous chloride, $FeCl_2 \cdot 4H_2O$, in 500 mL of 6.6 N hydrochloric acid), connect the flask to the receiver chamber (*B*), then complete the assembly of the apparatus, and begin circulating tap water through the condenser (*C*). Half-fill the lower two bulbs of the splash head (*D*) with distilled water.

With the stopcock in position such that the receiver chamber drains into the distillation flask, heat the flask until the temperature above the solution reaches 106° to 108°, and continue refluxing at this temperature for 45 min. Close the stopcock, continue heating at 108° to 110°, and collect 30 to 33 mL of distillate in the receiver chamber. Remove the heating source and allow the temperature to drop to about 80°.

Drain the distillate from the receiver chamber into a 250-mL beaker that is contained in an ice-water bath. Close the stopcock, and add a second 50-mL portion of the distillation-reducing solution through the thermometer opening to the distillation flask. Replace the thermometer, increase the temperature to 108° to 110°, and collect a second 30- to 33-mL portion of distillate in the receiver chamber.

Drain the second distillate into the beaker containing the first portion, and continue cooling in the ice-water bath until the combined distillate cools to room temperature. Remove the splash head, and wash its contents into the beaker. Also, wash down the insides of the condenser and receiver chamber with water, collecting the washings in the beaker. Filter the distillate plus washings through a Whatman No. 40, or equivalent, filter paper, collecting the filtrate in a 300-mL Erlenmeyer flask having a 24/40 standard-taper joint, to be used later as the arsine generator flask. Wash the filter three times with water so that the final volume of filtrate measures 200 mL.

Add 2 mL of potassium iodide TS and 0.5 mL of *Stannous Chloride Solution*, and continue as directed in the *Procedure*, Appendix IIIB, under the *Arsenic Test*, beginning with "Allow the mixture to stand for 30 min at room temperature . . ." but using 6.0 mL, rather than 3.0 mL, of *Standard Arsenic Solution* in the preparation of the standard.

Ash (Total)   Determine as directed in the general method, Appendix IIC.

Heavy Metals   Prepare and test a 1-g sample as directed in *Method II* under the *Heavy Metals Test*, Appendix IIIB, using 20 µg of lead ion (Pb) in the control (*Solution A*).

Iodine Content   Transfer about 2 g of the sample, accurately weighed, into a large porcelain crucible, and mix thoroughly with 10 g of potassium carbonate. Ignite the sample in a muffle furnace, starting with low heat, and then ignite at 500° to 600° for 20 min or until combustion is complete. Dissolve the ash in about 200 mL of boiling water, filter, and wash the filter paper with two 15-mL portions of boiling water, adding the washings to the filtrate. Cool to room temperature, neutralize to methyl red TS with approximately 20 mL of 85% phosphoric acid diluted with 20 mL of water, and then add 5 mL in excess. Cool the reaction mixture on an ice bath, and add bromine TS (about 5 mL) until a permanent yellow color is obtained. Gently boil the solution to remove all free bromine, adding water if necessary to maintain a volume of 200 mL or more. Boil for an additional 5 min after the bromine color has been completely dissipated. Add a few mg of salicylic acid, stir, and cool to about 20°. Add 1 mL of the diluted phosphoric acid solution and 5 mL of potassium iodide TS, and titrate immediately with 0.01 N sodium thiosulfate, using starch TS as the indicator. Each mL of 0.01 N sodium thiosulfate is equivalent to 211.5 µg of I.

Lead   A *Sample Solution* prepared as directed for organic compounds meets the requirements of the *Lead Limit Test*, Appendix IIIB, using 10 µg of lead ion (Pb) in the control.

Loss on Drying, Appendix IIC   Dry at 105° for 4 h.

Packaging and Storage   Store in well-closed containers.

Konjac Flour

Konjac; Konnyaku; Konjac Gum; Yam Flour

DESCRIPTION

A hydrocolloidal polysaccharide obtained from the tubers of various species of *Amorphophallus*. Konjac Flour is a high molecular weight, nonionic glucomannan primarily consisting of mannose and glucose at a respective molar ratio of approximately 1.6:1.0. It is a slightly branched polysaccharide connected by β-1,4 linkages and has an average molecular weight of 200,000 to 2,000,000 daltons. Acetyl groups along the glucomannan backbone contribute to solubility properties and are located, on average, every 9 to 19 sugar units. The typical powder is cream to light tan in color.

Konjac Flour is dispersible in hot or cold water and forms a highly viscous solution with a pH between 4.0 and 7.0. Solubility is increased by heat and mechanical agitation. Addition of mild alkali to the solution results in the formation of a heat-stable gel that resists melting, even under extended heating conditions.

Functional Use in Foods   Gelling agent; thickener; film former; emulsifier; stabilizer.

REQUIREMENTS

Identification

A. *Microscopic Test*   Stain about 0.1 g of the sample with 0.01% methylene blue powder in 50% isopropyl alcohol, and observe microscopically. Konjac Flour may be identified by the presence of flattened elliptical particles, which are generally 100 to 500 µm in length along the long axis. Unground Konjac Flour is clearly distinguished from other hydrocolloids by the presence of saclike cells that contain glucomannan. The surface of these cells has a reticulated structure. Particles of Konjac Flour are also birefringent under polarized light. These visual characteristics may remain even if the sample is finely ground, but are less pronounced.

B. *Gel Test*   At room temperature, add 5 mL of a 4% sodium borate solution to a 1% solution of the sample in a test tube, and shake vigorously. If Konjac Flour is present, a gel forms. (Konjac Flour solutions gel in the presence of sodium borate, similar in reaction to that of other galactomannans such as guar gum and locust bean gum.) The *Heat-Stable Gel Test*, below, distinguishes Konjac Flour from guar and locust bean gums.

C. *Heat-Stable Gel Test*   Prepare a 2% solution of the sample by heating it in a boiling water bath for 30 min with continuous agitation and then cooling the solution to room temperature. For each g of the sample used to prepare the 2% solution, add 1 mL of 10% potassium carbonate solution to the fully hydrated sample at ambient temperature. Heat the mixture in a water bath to 85°, and hold quiescently for 2 h without agitation. Konjac Flour forms a thermally stable gel under these conditions. Related hydrocolloids such as guar gum and locust bean gum do not form thermally stable gels and are negative by this test.

Arsenic (as As)   Not more than 3 mg/kg.
Ash (Total)   Not more than 5.0%.
Carbohydrate (Total)   Not less than 75.0%.
Heavy Metals   Not more than 10 mg/kg.
Lead   Not more than 5 mg/kg.
Loss on Drying   Not more than 15.0%.
Protein   Not more than 8.0%.

TESTS

Arsenic   A *Sample Solution* prepared as directed for organic compounds meets the requirements of the *Arsenic Test*, Appendix IIIB.

Ash (Total)   Determine as directed in the general method, Appendix IIC.

Carbohydrate (Total)   The remainder, after subtracting from 100% the sum of the percentages of *Ash*, *Loss on Drying*, and *Protein*, represents the percentage of carbohydrates (glucomannans) in the sample.

Heavy Metals   Prepare and test a 2-g sample as directed in *Method II* under the *Heavy Metals Test*, Appendix IIIB, using 20 µg of lead ion (Pb) in the control (*Solution A*).

Lead   A 2-g sample prepared as directed for organic compounds meets the requirements of the *Lead Limit Test*, Appendix IIIB, using 10 µg of lead ion (Pb) in the control.

Loss on Drying, Appendix IIC   Dry at 105° for 5 h.

Protein   Transfer about 3.5 g, accurately weighed, into a 500-mL Kjeldahl flask, and proceed as directed under *Nitrogen Determination*, Appendix IIIC. Percent protein equals percent $N \times 5.7$.

Packaging and Storage   Store cool and dry in a closed container away from direct heat and sunlight.

Angular Rotation   Determine in a 100-mm tube as directed under *Optical (Specific) Rotation*, Appendix IIB.

Ester Value   Determine as directed in the general method, Appendix VI, using about 1 g, accurately weighed.

Heavy Metals   Shake 10 mL of the oil with an equal volume of water to which 1 drop of hydrochloric acid has been added, and pass hydrogen sulfide through the mixture until it is saturated. No darkening in color is produced in either the oil or the water.

Refractive Index, Appendix IIB   Determine with an Abbé or other refractometer of equal or greater accuracy.

Solubility in Alcohol   Proceed as directed in the general method, Appendix VI. One mL dissolves in 0.5 mL of 90% alcohol, but the solution usually becomes opalescent or turbid upon further dilution.

Specific Gravity   Determine by any reliable method (see *General Provisions*).

Packaging and Storage   Store in full, tight, preferably glass or aluminum containers in a cool place protected from light.

Labdanum Oil

DESCRIPTION

The volatile oil obtained by steam distillation from crude labdanum gum extracted from the perennial shrub *Cistus ladaniferus* L. (Fam. *Cistaceae*). It is a golden yellow, viscous liquid having a powerful balsamic odor, which on dilution, is reminiscent of ambergris. It turns dark brown on standing. It is soluble in most fixed oils and in mineral oil. It is insoluble in glycerin and in propylene glycol.

Functional Use in Foods   Flavoring agent.

REQUIREMENTS

Identification   The infrared absorption spectrum of the sample exhibits relative maxima (that may vary in intensity) at the same wavelengths (or frequencies) as those shown in the respective spectrum under *Infrared Spectra (Series A: Essential Oils)*, using the same test conditions as specified therein.

Acid Value   Between 18 and 86.

Angular Rotation   Between +0°15′ and +7°.

Ester Value   Between 31 and 86.

Heavy Metals (as Pb)   Passes test.

Refractive Index   Between 1.492 and 1.507 at 20°.

Solubility in Alcohol   Passes test.

Specific Gravity   Between 0.905 and 0.993.

TESTS

Acid Value   Determine as directed for *Acid Value* under *Essential Oils and Flavors*, Appendix VI.

Lactic Acid

α-Hydroxypropionic Acid; 2-Hydroxypropionic Acid

$C_3H_6O_3$ Formula wt 90.08

INS: 270   CAS: L(+)-Lactic Acid [79-33-4]
             CAS: DL-Lactic Acid [598-82-3]

DESCRIPTION

A colorless or yellowish, nearly odorless, syrupy liquid consisting of a mixture of lactic acid ($C_3H_6O_3$) and lactic acid lactate ($C_6H_{10}O_5$). It is obtained by the lactic fermentation of sugars or is prepared synthetically. It is usually available in solutions containing the equivalent of from 50% to 90% lactic acid. It is hygroscopic, and when concentrated by boiling, the acid condenses to form lactic acid lactate, 2-(lactoyloxy)propanoic acid, which on dilution and heating hydrolyzes to Lactic Acid. It is miscible with water and with alcohol.

Functional Use in Foods   Acid.

REQUIREMENTS

Labeling   Indicate the concentration of Lactic Acid.

Identification   It gives positive tests for *Lactate*, Appendix IIIA.

Assay   Not less than 95.0% and not more than 105.0% of the labeled concentration of $C_3H_6O_3$.

Chloride   Not more than 0.1%.

Citric, Oxalic, Phosphoric, or Tartaric Acid   Passes test.

Cyanide   Not more than 5 mg/kg.

Heavy Metals (as Pb)   Not more than 10 mg/kg.

Iron   Not more than 10 mg/kg.

Lead Not more than 5 mg/kg.
Residue on Ignition Not more than 0.1%.
Sugars Passes test.
Sulfate Not more than 0.25%.

TESTS

Assay Weigh accurately a portion of the sample equivalent to about 3 g of Lactic Acid, transfer to a 250-mL flask, add 50.0 mL of 1 N sodium hydroxide, mix, and boil for 20 min. Add phenolphthalein TS, titrate the excess alkali in the hot solution with 1 N sulfuric acid, and perform a blank determination (see *General Provisions*). Each mL of 1 N sodium hydroxide is equivalent to 90.08 mg of $C_3H_6O_3$.

Chloride, Appendix IIIB Dilute 20 g to 1000 mL, and mix thoroughly. Any turbidity produced by a 1.0-mL (20 mg of Lactic Acid) portion of this solution does not exceed that shown in a control containing 20 µg of chloride ion (Cl).

Citric, Oxalic, Phosphoric, or Tartaric Acid Dilute 1 g to 10 mL with water, add 40 mL of calcium hydroxide TS, and boil for 2 min. No turbidity is produced.

Cyanide

p-Phenylenediamine–Pyridine Mixed Reagent Dissolve 200 mg of p-phenylenediamine hydrochloride in 100 mL of water, warming to effect solution. Cool, allow the solids to settle, and use the supernatant liquid to make the mixed reagent. Dissolve 128 mL of pyridine in 365 mL of water, add 10 mL of hydrochloric acid, and mix. To prepare the mixed reagent, mix 30 mL of the p-phenylenediamine solution with all of the pyridine solution, and allow to stand for 24 h before using. The mixed reagent is stable for about 3 weeks when stored in an amber bottle.

Sample Solution Transfer an accurately weighed quantity of the sample, equivalent to 20.0 g of 100% Lactic Acid, into a 100-mL volumetric flask, dilute to volume with water, and mix.

Cyanide Standard Solution Dissolve 2.5 g of potassium cyanide in 1000 mL of 0.1 N sodium hydroxide. Transfer a 10-mL aliquot into a 1000-mL volumetric flask, dilute to volume with 0.1 N sodium hydroxide, and mix. Each mL of this solution contains 10 µg of CN.

Procedure Pipet a 10-mL aliquot of the *Sample Solution* into a 50-mL beaker. Into a second 50-mL beaker pipet 1.0 mL of the *Cyanide Standard Solution*, and add 10 mL of water. Place the beakers in an ice bath, and adjust the pH to between 9 and 10 with 20% sodium hydroxide, stirring slowly and adding the reagent slowly to avoid overheating. Allow the solutions to stand for 3 min, and then slowly add 10% phosphoric acid to a pH between 5 and 6. Transfer the solutions into 100-mL separators containing 25 mL of cold water, and rinse the beakers and pH meter electrodes with a few mL of cold water, collecting the washings in the respective separator. Add 2 mL of bromine TS, stopper, and mix. Add 2 mL of 2% sodium arsenite solution, stopper, and mix. To the clear solutions add 10 mL of n-butanol, stopper, and mix. Finally, add 5 mL of *p-Phenylenediamine–Pyridine Mixed Reagent*, mix, and allow to stand for 15 min. Remove and discard the aqueous phases, and filter the alcohol phases into 10-mm cells. The absorbance of the *Sample Solution*, determined at 480 nm with a suitable spectrophotometer, is no greater than that of the *Cyanide Standard Solution*.

Heavy Metals A solution of 2 g in 25 mL of water meets the requirements of the *Heavy Metals Test*, Appendix IIIB, using 20 µg of lead ion (Pb) in the control (*Solution A*).

Iron To the ash obtained in the test for *Residue on Ignition* add 2 mL of dilute hydrochloric acid (1 in 2), and evaporate to dryness on a steam bath. Dissolve the residue in 1 mL of hydrochloric acid, dilute to 40 mL with water, and add about 40 mg of ammonium persulfate crystals and 10 mL of ammonium thiocyanate TS. Any red or pink color does not exceed that produced by 20 mL of *Iron Standard Solution* (20 µg Fe) in an equal volume of solution containing the quantities of reagents used in the test.

Lead A 4-g sample prepared as directed for organic compounds meets the requirements of the *Lead Limit Test*, Appendix IIIB, using 20 µg of lead ion (Pb) in the control.

Residue on Ignition, Appendix IIC Ignite 2 g as directed in the general method.

Sugars Add 5 drops of the sample to 10 mL of hot alkaline cupric tartrate TS. No red precipitate is formed.

Sulfate, Appendix IIIB Dilute 10 g to 100 mL, and mix thoroughly. Any turbidity produced by a 1.6-mL (160 mg of Lactic Acid) portion of this solution does not exceed that shown in a control containing 400 µg of sulfate ion (SO_4).

Packaging and Storage Store in tight containers.

Lactose

4-*O*-β-Galactopyranosyl-D-glucose

(α-Lactose)

$C_{12}H_{22}O_{11}$ Formula wt, anhydrous 342.30
$C_{12}H_{22}O_{11} \cdot H_2O$ Formula wt, monohydrate 360.32

CAS: [63-42-3]

DESCRIPTION

A white to creamy white, crystalline powder, normally obtained from whey, possessing a mildly sweet taste; odorless to slight characteristic odor. It may be anhydrous, contain one molecule of water of hydration, or contain a mixture of both forms if it has been prepared by a spray-drying process. It is soluble in water, very slightly soluble in alcohol, and insoluble in chloroform and in ether.

Functional Use in Foods   Nutritive sweetener; formulation aid; processing aid; humectant (anhydrous form); texturizer.

REQUIREMENTS

Labeling   Indicate whether it is anhydrous or the monohydrate or a mixture of both forms if it has been prepared by a spray-drying process.
Identification   Add 5 mL of 1 N sodium hydroxide to 5 mL of a hot, saturated solution of Lactose, and gently warm the mixture. The liquid becomes yellow and, finally, brownish red. Cool to room temperature, and add a few drops of alkaline cupric tartrate TS. A red precipitate of cuprous oxide is formed.
Assay   Not less than 98.0% and not more than 100.5% of $C_{12}H_{22}O_{11}$, calculated on the dried basis.
Arsenic   Not more than 0.5 mg/kg.
Heavy Metals (as Pb)   Not more than 5 mg/kg.
Lead   Not more than 0.5 mg/kg.
Loss on Drying   *Monohydrate and spray-dried mixture*: Not less than 4.5% and not more than 5.5%; *anhydrous form*: Not more than 1.0%.
pH   Not less than 4.5 and not more than 7.5, in a 10% solution.
Residue on Ignition   Not more than 0.3%.
Microbial Limits:
  E. coli   Negative in 25 g.
  Salmonella   Negative in 25 g.

TESTS

Assay   Proceed as directed under *Lactose*, Appendix X. Use about 2 g of the sample, accurately weighed, and transfer it to a 100-mL volumetric flask. Add 10 mL of fructose internal standard solution, dilute to volume with water, and mix. Perform the analysis within 24 h.
Arsenic   A solution of 4 g in 35 mL of water meets the requirements of the *Arsenic Test*, Appendix IIIB, using 2.0 mL of the Standard Arsenic Solution in the control (2 μg As).
Heavy Metals   Prepare and test a 4-g sample as directed in *Method II* under the *Heavy Metals Test*, Appendix IIIB, using 20 μg of lead ion (Pb) in the control (*Solution A*).
Lead   Determine as directed under *Method I* in the *Atomic Absorption Spectrophotometric Graphite Furnace Method* under the *Lead Limit Test*, Appendix IIIB, using a 5-g sample.
Loss on Drying, Appendix IIC   Dry 2 g at 120° for 16 h.
pH   Transfer a 10-g sample, accurately weighed, into a clean, dry 100-mL Erlenmeyer flask, and add 90 mL of recently boiled water set at 25°. Shake until the particles are evenly suspended and the mixture is free of lumps. Heat the sample to boiling, and shake frequently to aid dissolution. Let the suspension stand for 10 min, decant the supernate into the hydrogen-ion vessel, and quickly cool to 25°. Immediately determine the pH by the *Potentiometric Method*, Appendix IIB, using pH 4.01 and 9.18 buffer solutions to standardize the pH meter.
Residue on Ignition, Appendix IIC   Ignite 2 g as directed in *Method I*.

Microbial Limits:
  E. coli   Proceed as directed in chapter 4 of the *FDA Bacteriological Analytical Manual*, Seventh Edition, 1992.
  Salmonella   Proceed as directed in chapter 5 of the *FDA Bacteriological Analytical Manual*, Seventh Edition, 1992.

Packaging and Storage   Store in well-closed containers protected from humidity.

Lactylated Fatty Acid Esters of Glycerol and Propylene Glycol

Propylene Glycol Lactostearate

INS: 478

DESCRIPTION

A mixture of partial lactic and fatty acid esters of propylene glycol and glycerin produced by the lactylation of a product obtained by reacting edible fats or oils with propylene glycol. It varies in consistency from a soft solid to a hard, waxy solid. It is dispersible in hot water, and is moderately soluble in hot isopropanol, in chloroform, and in soybean oil.

Functional Use in Foods   Emulsifier; stabilizer; whipping agent; plasticizer; surface-active agent.

REQUIREMENTS

Identification   Place about 150 mg of melted sample in a 16- × 125-mm tube equipped with a screw cap having a Teflon liner, and add 4 mL of absolute methanol, 4 drops of a 25% sodium methoxide solution in absolute methanol, and a boiling chip. Cap the tube, reflux for 15 min, and cool to room temperature. Add 8 drops of a 15% potassium acid sulfate solution, 4 mL of water, and 4 mL of *n*-hexane, cap the tube, shake for 1 min, and centrifuge for 30 to 60 s. Decant and discard the *n*-hexane layer, and repeat the extraction with three additional 4-mL portions of *n*-hexane, discarding each extract. Transfer the aqueous alcoholic phase from the tube to a 50-mL round bottom glass-stoppered flask, place the flask in a water bath at 50° to 55°, and evaporate to near dryness (about 0.5 mL) with a rotary film evaporator under full water aspirator vacuum.

Caution: Do not heat above 55°.

Remove the flask from the evaporator, add 1 mL of a 1:1 methanol–0.5 N hydrochloric acid solution, swirl for several min, and decant the clear solution into a small flask. Inject a portion of this solution into a suitable gas chromatograph (see Appendix IIA), and obtain the chromatogram, observing the following operating conditions or equivalent conditions: *detector*, flame ionization; *carrier gas*, helium, flowing at a rate of 50 mL/min; *column*, 1.8-m × 3-mm (id) packed with 80/100-mesh Porapak Q (ethylvinylbenzene–divinylbenzene polymer porous beads); *column temperature*, 175° to 210°, heated at a rate of 4°/min and holding at 210° until the glycerin is eluted; *inlet port temperature*, 310°; *detector temperature*, 385°; *recorder*, 0 to 1 mV range, with 1-s full-scale deflection at a chart speed of 6.5 mm/min; *sample size*, 2 to 3 μL. From the chromatogram so obtained, identify the peaks by their relative positions on the chart. The major peaks, representing propylene glycol, methyl lactate, lactic acid, and glycerin in the order listed, may be identified with suitable reference substances. Major peaks may also be identified by their relative retention times using a suitable internal standard.

Acid Value Not more than 12.0.
Heavy Metals (as Pb) Not more than 10 mg/kg.
Water-Insoluble Combined Lactic Acid Between 14.0% and 18.0%.

The following specifications should conform to the representations of the vendor: **1-Monoglyceride Content, Total Lactic Acid, Free Lactic Acid, Free Glycerin,** and **Water**.

TESTS

Acid Value Determine as directed for *Acid Value, Method II*, under *Fats and Related Substances*, Appendix VII.

Free Glycerin Determine as directed in the general method, Appendix VII.

Free Lactic Acid Transfer about 15 g of the sample, accurately weighed, into a beaker, dissolve in about 75 mL of benzene, and transfer the solution into a 500-mL glass-stoppered graduate. Wash the beaker with about 125 mL of benzene in divided portions, adding the washings to the graduate. Add 200 mL of water to the graduate, and shake vigorously for 1 min. After 125 mL or more of the aqueous phase has separated, pipet 100.0 mL of the aqueous phase into an Erlenmeyer flask, add 1 mL of phenolphthalein TS, and titrate with 0.5 N sodium hydroxide to the first appearance of a slight pink color. Calculate the percentage of free lactic acid in the sample by the formula

$$9.008 \times V \times N/(0.5 \times W),$$

in which V is the volume, in mL, of 0.5 N sodium hydroxide required; N is the exact normality of the sodium hydroxide solution; and W is the weight of the sample, in g.

Heavy Metals Prepare and test a 2-g sample as directed in *Method II* under the *Heavy Metals Test*, Appendix IIIB, using 20 μg of lead ion (Pb) in the control (*Solution A*).

1-Monoglyceride Content Determine as directed in the general method, Appendix VII.

Total Lactic Acid Transfer about 3 g of the sample, accurately weighed, into a 250-mL glass-stoppered flask, pipet 50.0 mL of 0.7 N alcoholic potassium hydroxide into the flask, attach an air condenser, and boil gently on a steam bath for 30 min or until the sample is completely saponified. Remove the flask from the steam bath, immediately remove the air condenser, and allow the solution to cool until it begins to jell. Add 75.0 mL of 0.5 N hydrochloric acid, mix, and transfer the solution to a 500-mL separator, washing the flask with two 15-mL portions of water. Cool to 35° or lower, and extract with 100 mL of diethyl ether. Transfer the aqueous layer to a second 500-mL separator, and wash the ether layer with two 20-mL portions of water, adding the wash water to the original aqueous phase in the second separator. Retain the ether solution. Extract the aqueous phase with a second 100-mL portion of diethyl ether, and transfer the aqueous phase to a 500-mL Erlenmeyer flask. Combine and wash the ether extracts with five 20-mL portions of water, and add the wash water to the flask. To the combined aqueous phases in the Erlenmeyer flask, add 1 mL of phenolphthalein TS, and titrate with 0.5 N sodium hydroxide to the first appearance of a slight pink color. Perform a blank determination (see *General Provisions*), and calculate the percentage of total lactic acid by the formula

$$9.008(S - B)(N)/W,$$

in which $S - B$ represents the difference, in mL, between the volumes of 0.5 N sodium hydroxide required for the sample and blank, respectively; N is the exact normality of the sodium hydroxide solution; and W is the weight of the sample, in g.

Water Determine by the *Karl Fischer Titrimetric Method*, Appendix IIB.

Water-Insoluble Combined Lactic Acid Transfer about 3 g of the sample, accurately weighed, into a 250-mL separator with the aid of 100 mL of benzene, and wash with three 30-mL portions of water, discarding the washings. Transfer the benzene layer to a 250-mL glass-stoppered Erlenmeyer flask, wash the separator with a few mL of benzene, and completely evaporate the combined benzene solution to dryness. Pipet 50.0 mL of 0.7 N alcoholic potassium hydroxide into the flask, attach an air condenser, boil gently on a steam bath for 30 min or until the sample is completely saponified, and remove the flask from the steam bath. Immediately remove the air condenser, and allow the solution to cool until it begins to jell. Add 75.0 mL of 0.5 N hydrochloric acid, mix, and transfer the solution into a 500-mL separator, washing the flask with two 15-mL portions of water. Cool to 35° or lower, and extract with 100 mL of diethyl ether. Transfer the water layer to a second 500-mL separator, and wash the diethyl ether with two 20-mL portions of water, adding the wash water to the original aqueous phase in the second separator. Retain the ether solution. Extract the aqueous phase with a second 100-mL portion of diethyl ether, and transfer the aqueous phase to a 500-mL Erlenmeyer flask. Combine and wash the ether extracts with five 20-mL portions of water, and add the wash water to the flask. To the combined aqueous phases in the flask add 1 mL of phenolphthalein TS, and titrate with 0.5 N sodium hydroxide to the first appearance of a slight pink color. Perform a blank determination (see *General Provisions*), and calculate the percentage of water-insoluble combined lactic acid in the sample by the formula

$$9.008(S - B)(N)/W,$$

in which $S - B$ represents the difference, in mL, between the volumes of 0.5 N sodium hydroxide required for the sample and blank, respectively; N is the exact normality of the sodium hydroxide solution; and W is the weight of the sample, in g.

Packaging and Storage Store in well-closed containers.

Lactylic Esters of Fatty Acids

DESCRIPTION

Lactylic Esters of Fatty Acids are mixed fatty acid esters of lactic acid and its polymers, with minor quantities of free lactic acid, polylactic acid, and fatty acids. They vary in consistency from liquids to hard, waxy solids. They are dispersible in hot water and are soluble in organic solvents and in vegetable oils.

Functional Use in Foods Emulsifier; surface-active agent.

REQUIREMENTS

Identification

A. Transfer into a 25-mL glass-stoppered test tube 1 mL of the solution obtained in the test for *Total Lactic Acid* after titrating with 0.1 N potassium hydroxide. Add 0.1 mL of cupric sulfate solution (1 g of $CuSO_4 \cdot 5H_2O$ in 25 mL of water) and 6 mL of sulfuric acid, and mix. Stopper loosely, heat in a boiling water bath for 5 min, then cool in an ice bath for 5 min, and remove from the bath. Add 0.1 mL of *p*-phenylphenol TS, mix, allow to stand at room temperature for 1 min, and then heat in a boiling water bath for 1 min. A deep, blue-violet color indicates the presence of lactic acid.

B. Assemble a suitable apparatus for ascending thin-layer chromatography. Prepare a slurry of chromatographic silica gel containing about 13% of calcium sulfate as the binder (use 1 g of $CaSO_4$ to each 2 mL of water), apply a uniformly thin layer to glass plates of convenient size, dry in the air for 10 min, and activate by drying at 100° for 1 h. Store the cool plates in a clean, dry place until ready for use.

Transfer 1 g of the sample into a 10-mL volumetric flask, dissolve, and dilute to volume with chloroform. Transfer 250 mg of stearic acid into another 10-mL volumetric flask, dissolve, and dilute to volume with chloroform.

Spot 2 µL of the sample solution and 1 µL of the stearic acid solution approximately 1.5 cm from the bottom of the plate, allow the spots to dry, and then place the plate in a suitable chromatographic chamber containing a mixture of 4 volumes of acetone, 4 volumes of acetic acid, and 92 volumes of hexane. Develop by ascending chromatography until the solvent front travels 15 cm beyond the sample spot. Remove the plate from the chamber, dry thoroughly in air, and spray evenly with a saturated solution of chromium trioxide in sulfuric acid. Immediately place the sprayed plate on a hot plate maintained at about 200° in a hood, char until white fumes of sulfur trioxide cease, and cool to room temperature. The spots from the sample are located according to the following R_f values: stearic acid, 1.00; fatty acid, 1.00; acylated monolactic acid, 0.84; acylated dilactic acid, 0.76; acylated trilactic acid, 0.68; and tetralactic acid, 0.62.

Heavy Metals (as Pb) Not more than 10 mg/kg.

The following specifications should conform to the representations of the vendor: **Acylated Monolactic Acid, Acylated Polylactic Acid, Free Fatty Acid, Total Lactic Acid, Acid Value, Saponification Value,** and **Water.**

TESTS

Assay for Acylated Monolactic Acid, Acylated Polylactic Acid, and Free Fatty Acid This assay is performed by gas–liquid chromatography (see Appendix IIA) using an instrument containing a thermal conductivity or flame ionization detector and helium as the carrier gas. The operating conditions of the apparatus may vary, depending upon the particular instrument used, but a suitable chromatogram is obtained with a 1.2-m × 6.3-mm column packed with 20% SE-30 or SE-52, or other comparable grades of silicone rubber gums, on Chromosorb P or W or Diatoport S, or other comparable grades of diatomaceous material. The column should be programmed between 150° and 310°, using a heating rate of 4°/min, the inlet port temperature should be 335°, and the detector temperature should be 315°. The recorder should be equipped with an attenuator switch and should be operated in the 0- to 1-mV range, with 1-s full-scale deflection at a chart speed of 12.7 mm/min. A constant gas flow rate of about 54 mL/min should be established and maintained throughout the determination.

Diazomethane Reagent (**Caution**: Diazomethane is both toxic and potentially explosive. Its preparation and use should be carried out in a hood.) Place 1 mL of potassium hydroxide solution (4 in 10), followed by 2.5 mL of methanol, in a 25-mL distilling flask fitted with a dropping funnel and an efficient spiral water-cooled condenser set downward for distillation. Connect the condenser to a 50-mL receiving flask that is cooled in ice and vented to the hood. Heat the distilling flask in a water bath to 65°, add 2 mL of ether through the dropping funnel, saturating the distillation apparatus with ether vapor, and close the stopcock. Place in the dropping funnel a solution containing 2.15 g of *N*-methyl-*N*-nitroso-*p*-toluene sulfonamide in 13 mL of ether, and adjust the stopcock so that the rate of distillation is about equal to the rate of addition from the funnel. When the funnel is empty, add another 2 mL of ether, and continue the distillation until the distillate is colorless. The ether–alcohol solution of diazomethane so obtained should be used immediately or stored at −10° until used.

Procedure To approximately 50 mg of the sample add the *Diazomethane Reagent* until a yellow color persists. Carefully evaporate the ether at 50° under a stream of clean, dry nitrogen. Inject 0.5 to 2.0 µL of the melted methyl esters so obtained into the gas chromatographic apparatus, using a 10-µL capacity Hamilton fixed needle or equivalent. The sample size should be adjusted so that the major peak is not attenuated more than ×8. From the chromatogram so obtained, identify the peaks by their relative position on the chart. The esters, appearing in the order of increasing number of carbon atoms in the fatty acid and in order of increasing length of the polymer, are eluted as follows: myristate, palmitate, stearate, palmitoyl lactylate (2-palmitoyloxypropionate), stearoyl lactylate (2-stearoyloxypropionate), palmitoyl lactoyl lactylate, stearoyl lactoyl lactylate, palmitoyl dilactoyl lactylate, stearoyl dilactoyl lactylate, palmitoyl trilactoyl lactylate, stearoyl trilactoyl lactylate, and palmitoyl tetralactoyl lactylate. Other esters may be determined by interpolation of a conventional carbon number-retention plot.

Determine the composition of the sample, using the area normalization method, by the formula

$$\%_i = 100A_i/\Sigma(A_i + ... + A_n),$$

in which i represents the component of interest, A_i is the equalized area for the component of interest, and $\Sigma(A_i + ... + A_n)$ is the sum of the equalized areas.

If free and polylactic acids are present, as determined below, the results should be corrected by multiplying $\%_i$ by [(100 − % free and polylactic acid)/100].

Acid Value Determine as directed for *Acid Value, Method II*, under *Fats and Related Substances*, Appendix VII.

Heavy Metals Prepare and test a 2-g sample as directed in *Method II* under the *Heavy Metals Test*, Appendix IIIB, using 20 µg of lead ion (Pb) in the control (*Solution A*).

Saponification Value Determine as directed for *Saponification Value* under *Fats and Related Substances*, Appendix VII.

Total Free and Polylactic Acids Weigh accurately about 500 mg of the sample, previously melted, transfer into a 50-mL glass-stoppered separator with the aid of 15 mL of benzene, and add 10 mL of water. Invert the funnel 10 times, and allow to stand until the layers have separated. Filter the aqueous layer through a plug of glass wool into a 125-mL flask, wash the benzene with two 10-mL portions of water, and combine the aqueous layers. To the flask add 5.0 mL of 0.1 N sodium hydroxide, and then heat on a steam bath for 15 min under a nitrogen atmosphere. Add phenolphthalein TS, and titrate with 0.1 N hydrochloric acid to the disappearance of the pink color. Conduct a blank determination, using 30 mL of water and 5.0 mL of 0.1 N sodium hydroxide, and calculate the percentage of free and polylactic acids in the sample by the formula

$$(B - S) \times 9.0/W,$$

in which $B - S$ represents the difference, in mL, between the volumes of 0.1 N hydrochloric acid required for the blank and the sample, respectively; 9.0 is an equivalence factor for the lactic acid; and W is the weight, in g, of the sample.

Total Lactic Acid Transfer an accurately weighed portion of the melted sample, equivalent to between 140 mg and 170 mg of lactic acid, into a 250-mL Erlenmeyer flask, and add to the flask 20.0 mL of 0.5 N alcoholic potassium hydroxide. Connect an air condenser, at least 65 cm in length, to the flask, and reflux for 30 min. Add 20 mL of water through the condenser, disconnect the condenser, and prepare a blank containing 20.0 mL of the alkali and 20 mL of water. Evaporate the contents of each flask to a volume of about 20 mL, cool to about 40°, add methyl red TS to the flask containing the blank, and titrate the blank with 0.5 N hydrochloric acid. Add exactly the same volume of the acid to the sample flask, with swirling, and then add to the sample flask 50 mL of hexane, swirling vigorously to dissolve the fatty acids. Transfer quantitatively the contents of the sample flask into a 250-mL separator, and shake for 30 s. Collect the aqueous phase in an Erlenmeyer flask, wash the hexane solution with 50 mL of water, and combine the wash solution with the original aqueous phase in the flask, discarding the hexane solution. Add phenolphthalein TS, and titrate with 0.1 N potassium hydroxide to a pink color that persists for at least 30 s. Each mL of 0.1 N potassium hydroxide is equivalent to 9.008 mg of lactic acid ($C_3H_6O_3$).

Water Determine by the *Karl Fischer Titrimetric Method*, Appendix IIB.

Packaging and Storage Store in tight, plastic-lined containers in a cool, dry place.

Lanolin, Anhydrous

Wool Fat

INS: 913

DESCRIPTION

A purified, yellowish white, semisolid, fatlike substance extracted from the wool of sheep. It is insoluble in water, but mixes with about twice its weight of water without separation. It is soluble in chloroform and in ether.

Functional Use in Foods Masticatory substance in chewing gum base.

REQUIREMENTS

Acid Value Not more than 1.12.
Arsenic (as As) Not more than 3 mg/kg.
Heavy Metals (as Pb) Not more than 0.002%.
Iodine Value Between 18 and 36.
Lead Not more than 3 mg/kg.
Loss on Heating Not more than 0.5%.
Melting Range Between 36° and 42°.

TESTS

Acid Value Determine as directed for *Acid Value, Method I*, under *Fats and Related Substances*, Appendix VII.

Arsenic Prepare a *Sample Solution* as directed in the general method under *Chewing Gum Base*, Appendix IV. This solution meets the requirements of the *Arsenic Test*, Appendix IIIB.

Heavy Metals Prepare and test a 1-g sample as directed in *Method II* under the *Heavy Metals Test*, Appendix IIIB, using 20 µg of lead ion (Pb) in the control (*Solution A*).

Iodine Value Determine by the *Modified Wijs Method*, Appendix VII.

Lead Prepare a *Sample Solution* as directed in the general method under *Chewing Gum Base*, Appendix IV. This solution meets the requirements of the *Lead Limit Test*, Appendix IIIB, using 10 µg of lead ion (Pb) in the control.

Loss on Heating Heat a 5-g sample on a steam bath, with frequent stirring, to constant weight.

Melting Range Determine as directed for *Melting Range or Temperature*, Appendix IIB.

Packaging and Storage Store in well-closed containers, preferably at a temperature not exceeding 30°.

Lard (Unhydrogenated)

DESCRIPTION

An off-white fat obtained by dry or wet (steam) rendering of fresh fatty porcine tissues (cuttings and trimmings) shortly after slaughtering. Rendered Lard may be bleached, or bleached and deodorized. It is soft to semisolid at 27° and melts completely at 42°.

Rendered, bleached, and bleached-deodorized Lard are off-white semisolids at 21° to 27°. Bleached, and bleached-deodorized Lard, which are pale yellow and clear at 54°, differ from rendered Lard, which is pale yellow, clear to hazy, and may contain extraneous matter.

Functional Use in Foods Coating agent; emulsifying agent; formulation aid; texturizer.

SPECIFIC REQUIREMENTS

	Rendered Lard	Bleached Lard	Bleached-Deodorized Lard
Color (AOCS-Wesson)	Not more than 3.0 red	Not more than 1.5 red	Not more than 1.5 red
Free Fatty Acids (as oleic acid)	Not more than 1.0%	Not more than 1.0%	Not more than 0.1%
Insoluble Matter	Not more than 0.1%	Not more than 0.05%	Not more than 0.05%
Iodine Value	Between 46 and 70	Between 46 and 70	Between 46 and 70
Water	Not more than 0.5%	Not more than 0.1%	Not more than 0.1%

GENERAL REQUIREMENTS

Identification Lard exhibits the following composition profile of fatty acids as determined under *Fatty Acid Composition*, Appendix VII.

Fatty Acid:	<14:0	14:0	14:1	15:0	16:0
Weight % (Range):	<0.5	0.5–2.5	0.2	<0.1	20–32

Fatty Acid:	16:1	17:0	17:1	18:0	18:1
Weight % (Range):	1.7–5	<1.0	<0.7	5.0–24	35–62

Fatty Acid:	18:2	18:3	20:0	20:1
Weight % (Range):	3.0–16	<2.0	<1.0	<1.0

Lead Not more than 0.1 mg/kg.
Peroxide Value Not more than 10 meq/kg.

TESTS

Color Proceed as directed for *Color* (AOCS-Wesson) under *Fats and Related Substances*, Appendix VII.

Free Fatty Acids Proceed as directed under *Free Fatty Acids*, Appendix VII, using the following equivalence factor (e) in the formula given in the procedure:

Free fatty acids as oleic acid, $e = 28.2$.

Iodine Value Proceed as directed under *Iodine Value*, Appendix VII.

Lead Determine as directed under *Method II* in the *Atomic Absorption Spectrophotometric Graphite Furnace Method* under the *Lead Limit Test*, Appendix IIIB, using a 3-g sample.

Peroxide Value Proceed as directed under *Peroxide Value* in the monograph for *Hydroxylated Lecithin*. However, after the addition of saturated potassium iodide and mixing, instead of allowing the solution to stand for 10 min, mix the solution for 1 min, and begin the titration immediately.

Unsaponifiable Matter Proceed as directed under *Unsaponifiable Matter*, Appendix VII.

Water Proceed as directed under *Water Determination*, Appendix IIB. However, in place of 35 to 40 mL of methanol, use 50 mL of chloroform to dissolve the sample.

Packaging and Storage Store in well-closed containers.

Laurel Leaf Oil

Bay Leaf Oil

DESCRIPTION

The oil obtained by steam distillation from the leaves of *Laurus nobilis* L. (Fam. *Lauraceae*). It is a light yellow to yellow liquid having an aromatic and spicy odor. It is soluble in most fixed oils, and is soluble with cloudiness in mineral oil and in propylene glycol. It is insoluble in glycerin.

Note: The oil from *Laurus nobilis* L. should not be confused with that of the West Indian bay tree, or California bay laurel.

Functional Use in Foods Flavoring agent.

REQUIREMENTS

Identification The infrared absorption spectrum of the sample exhibits relative maxima (that may vary in intensity) at the same wavelengths (or frequencies) as those shown in the respective spectrum under *Infrared Spectra (Series A: Essential Oils)*, using the same test conditions as specified therein.
Acid Value Not more than 3.0.
Angular Rotation Between −10° and −19°.
Heavy Metals (as Pb) Passes test.
Refractive Index Between 1.465 and 1.470 at 20°.

Saponification Value Between 15 and 45.
Saponification Value after Acetylation Between 36 and 85.
Solubility in Alcohol Passes test.
Specific Gravity Between 0.905 and 0.929.

TESTS

Acid Value Determine as directed for *Acid Value* under *Essential Oils and Flavors*, Appendix VI.
Angular Rotation Determine in a 100-mm tube as directed under *Optical (Specific) Rotation*, Appendix IIB.
Heavy Metals Shake 10 mL of the oil with an equal volume of water to which 1 drop of hydrochloric acid has been added, and pass hydrogen sulfide through the mixture until it is saturated. No darkening in color is produced in either the oil or the water.
Refractive Index, Appendix IIB Determine with an Abbé or other refractometer of equal or greater accuracy.
Saponification Value Determine as directed for *Saponification Value* under *Essential Oils and Flavors*, Appendix VI, using about 5 g, accurately weighed.
Saponification Value after Acetylation Proceed as directed under *Total Alcohols*, Appendix VI, using about 2.5 g of acetylated oil, accurately weighed. Calculate the saponification value by the formula

$$28.05 \times A/B,$$

in which A is the number of mL of 0.5 N alcoholic potassium hydroxide consumed in the titration and B is the weight of the acetylated oil, in g.
Solubility in Alcohol Proceed as directed in the general method, Appendix VI. One mL dissolves in 1 mL of 80% alcohol and remains in solution upon dilution to 10 mL.
Specific Gravity Determine by any reliable method (see *General Provisions*).

Packaging and Storage Store in full, tight, preferably glass, tin-lined, or aluminum containers in a cool place protected from light.

Lauric Acid

Dodecanoic Acid

$$CH_3(CH_2)_{10}COOH$$

$C_{12}H_{24}O_2$ Formula wt 200.32

DESCRIPTION

A solid organic acid obtained from coconut oil and other vegetable fats. It occurs as a white or faintly yellowish, somewhat glossy, crystalline solid or powder. It is practically insoluble in water, but is soluble in alcohol, in chloroform, and in ether.

Functional Use in Foods Component in the manufacture of other food-grade additives; defoaming agent.

REQUIREMENTS

Acid Value Between 252 and 287.
Heavy Metals (as Pb) Not more than 10 mg/kg.
Iodine Value Not more than 3.0.
Residue on Ignition Not more than 0.1%.
Saponification Value Between 253 and 287.
Titer (Solidification Point) Between 26° and 44°.
Unsaponifiable Matter Not more than 0.3%.
Water Not more than 0.2%.

TESTS

Acid Value Determine as directed for *Acid Value, Method I*, under *Fats and Related Substances*, Appendix VII.
Heavy Metals Prepare and test a 2-g sample as directed in *Method II* under the *Heavy Metals Test*, Appendix IIIB, using 20 µg of lead ion (Pb) in the control (*Solution A*).
Iodine Value Determine by the *Modified Wijs Method*, Appendix VII.
Residue on Ignition, Appendix IIC Ignite 10 g as directed in the general method.
Saponification Value Determine as directed for *Saponification Value* under *Fats and Related Substances*, Appendix VII, using about 3 g, accurately weighed.
Titer (Solidification Point) Determine as directed under *Solidification Point*, Appendix IIB.
Unsaponifiable Matter Determine as directed in the general method, Appendix VII.
Water Determine by the *Karl Fischer Titrimetric Method*, Appendix IIB.

Packaging and Storage Store in well-closed containers.

Lavandin Oil, Abrial Type

DESCRIPTION

An oil obtained by steam distillation of the fresh flowering tops of a hybrid, *Lavandula abrialis* unofficial (Fam. *Labiatae*), of true lavender, *Lavandula officinalis*, or of spike lavender, *Lavandula latifolia*. It is a pale yellow to yellow liquid having a slight camphoraceous odor that is strongly suggestive of lavender. It is soluble in most fixed oils and in propylene glycol. It is soluble with opalescence in mineral oil, but it is relatively insoluble in glycerin.

Functional Use in Foods Flavoring agent.

REQUIREMENTS

Identification The infrared absorption spectrum of the sample exhibits relative maxima (that may vary in intensity) at the same wavelengths (or frequencies) as those shown in the respective spectrum under *Infrared Spectra (Series A: Essential Oils)*, using the same test conditions as specified therein.
Assay Not less than 28.0% and not more than 35.0% of esters, calculated as linalyl acetate ($C_{12}H_{20}O_2$).
Angular Rotation Between $-2°$ and $-5°$.
Heavy Metals (as Pb) Passes test.
Refractive Index Between 1.460 and 1.464 at 20°.
Solubility in Alcohol Passes test.
Specific Gravity Between 0.885 and 0.893.

TESTS

Assay Weigh accurately about 3 g, and proceed as directed under *Ester Determination*, Appendix VI, using 98.15 as the equivalence factor (*e*) in the calculation.
Angular Rotation Determine in a 100-mm tube as directed under *Optical (Specific) Rotation*, Appendix IIB.
Heavy Metals Shake 10 mL of the oil with an equal volume of water to which 1 drop of hydrochloric acid has been added, and pass hydrogen sulfide through the mixture until it is saturated. No darkening in color is produced in either the oil or the water.
Refractive Index, Appendix IIB Determine with an Abbé or other refractometer of equal or greater accuracy.
Solubility in Alcohol Proceed as directed in the general method, Appendix VI. One mL dissolves in 2 mL of 70% alcohol. A slight opalescence sometimes develops on further dilution.
Specific Gravity Determine by any reliable method (see *General Provisions*).

Packaging and Storage Store in full, tight, preferably glass, tin-lined, or good-quality galvanized containers in a cool place protected from light.

Lavender Oil

DESCRIPTION

The volatile oil obtained by steam distillation from the fresh flowering tops of *Lavandula officinalis* Chaix ex Villars (*Lavandula vera* De Candolle) (Fam. *Labiatae*). It is a colorless or yellow liquid having the characteristic odor and taste of lavender flowers.

Functional Use in Foods Flavoring agent.

REQUIREMENTS

Identification The infrared absorption spectrum of the sample exhibits relative maxima (that may vary in intensity) at the same wavelengths (or frequencies) as those shown in the respective spectrum under *Infrared Spectra (Series A: Essential Oils)*, using the same test conditions as specified therein.
Assay Not less than 35.0% of esters, calculated as linalyl acetate ($C_{12}H_{20}O_2$).
Alcohol Passes test.
Angular Rotation Between $-3°$ and $-10°$.
Foreign Water-Soluble Esters Passes test.
Heavy Metals (as Pb) Passes test.
Refractive Index Between 1.459 and 1.470 at 20°.
Solubility in Alcohol Passes test.
Specific Gravity Between 0.875 and 0.888.

TESTS

Assay Weigh accurately about 5 g, and proceed as directed under *Ester Determination*, Appendix VI, using 98.15 as the equivalence factor (*e*) in the calculation.
Alcohol Transfer 5 mL to a narrow, graduated, glass-stoppered, 10-mL cylinder, add 5 mL of water, and shake. The volume of the oil does not diminish.
Angular Rotation Determine in a 100-mm tube as directed under *Optical (Specific) Rotation*, Appendix IIB.
Foreign Water-Soluble Esters Mix 20 mL of the sample with 40 mL of 5% alcohol in a glass-stoppered, 100-mL cylinder. When the mixture has cleared, pipet 30 mL of the alcohol layer into a 125-mL Erlenmeyer flask. Add phenolphthalein TS, and neutralize the solution with 0.5 *N* sodium hydroxide. Add 5.0 mL of 0.5 *N* sodium hydroxide, and heat the mixture on a boiling water bath under a reflux condenser for 1 h. Allow the mixture to cool, and titrate the excess alkali with 0.5 *N* hydrochloric acid. Not less than 4.7 mL of the acid is required to neutralize the mixture.
Heavy Metals Shake 10 mL of the oil with an equal volume of water to which 1 drop of hydrochloric acid has been added, and pass hydrogen sulfide through the mixture until it is saturated. No darkening in color is produced in either the oil or the water.
Refractive Index, Appendix IIB Determine with an Abbé or other refractometer of equal or greater accuracy.
Solubility in Alcohol Proceed as directed in the general method, Appendix VI. One mL dissolves in 4 mL of 70% alcohol.
Specific Gravity Determine by any reliable method (see *General Provisions*).

Packaging and Storage Store in full, tight containers in a cool place protected from light.

Lecithin

INS: 322   CAS: [8002-43-5]

DESCRIPTION

Food-grade Lecithin is obtained from soybeans and other plant sources. It is a complex mixture of acetone-insoluble phosphatides that consists chiefly of phosphatidyl choline, phosphatidyl ethanolamine, and phosphatidyl inositol, combined with various amounts of other substances such as triglycerides, fatty acids, and carbohydrates. Refined grades of Lecithin may contain any of these components in varying proportions and combinations depending on the type of fractionation used. In its oil-free form, the preponderance of triglycerides and fatty acids is removed and the product contains 90% or more of phosphatides representing all or certain fractions of the total phosphatide complex. The consistency of both natural grades and refined grades of Lecithin may vary from plastic to fluid, depending upon free fatty acid and oil content, and upon the presence or absence of other diluents. Its color varies from light yellow to brown, depending on the source, on crop variations, and on whether it is bleached or unbleached. It is odorless or has a characteristic, slight nutlike odor and a bland taste. Edible diluents, such as cocoa butter and vegetable oils, often replace soybean oil to improve functional and flavor characteristics. Lecithin is only partially soluble in water, but it readily hydrates to form emulsions. The oil-free phosphatides are soluble in fatty acids, but are practically insoluble in fixed oils. When all phosphatide fractions are present, Lecithin is partially soluble in alcohol and practically insoluble in acetone.

Functional Use in Foods   Antioxidant; emulsifier.

REQUIREMENTS

Acetone-Insoluble Matter (phosphatides)   Not less than 50.0%.
Acid Value   Not more than 36.
Heavy Metals (as Pb)   Not more than 0.002%.
Hexane-Insoluble Matter   Not more than 0.3%.
Lead   Not more than 10 mg/kg.
Peroxide Value   Not more than 100.
Water   Not more than 1.5%.

TESTS

Acetone-Insoluble Matter (phosphatides)

Purification of Phosphatides   Dissolve 5 g of phosphatides from previous *Acetone-Insoluble Matter* determinations in 10 mL of petroleum ether, and add 25 mL of acetone to the solution. Transfer approximately equal portions of the precipitate to each of two 40-mL centrifuge tubes using additional portions of acetone to facilitate the transfer. Stir thoroughly, dilute to 40 mL with acetone, stir again, chill for 15 min in an ice bath, stir again, and then centrifuge for 5 min. Decant the acetone, crush the solids with a stirring rod, refill the tube with acetone, stir, chill, centrifuge, and decant as before. The solids after the second centrifugation require no further purification and may be used for preparing the *Phosphatide–Acetone Solution*. Five g of the purified phosphatides are required to saturate about 16 L of acetone.

Phosphatide–Acetone Solution   Add a quantity of purified phosphatides to sufficient acetone, previously cooled to a temperature of about 5°, to form a saturated solution, and maintain the mixture at this temperature for 2 h, shaking it vigorously at 15-min intervals. Decant the solution through a rapid filter paper, avoiding the transfer of any undissolved solids to the paper and conducting the filtration under refrigerated conditions (not above 5°).

Procedure   If it is plastic or semisolid, soften a portion of the Lecithin by warming it in a water bath at a temperature not exceeding 60° and then mixing it thoroughly. Transfer 2 g of a well-mixed sample, accurately weighed, into a 40-mL centrifuge tube, previously tared with a glass stirring rod, and add 15 mL of *Phosphatide–Acetone Solution* from a buret. Warm the mixture in a water bath until the Lecithin melts, but avoid evaporation of the acetone. Stir until the sample is completely disintegrated and dispersed, and then transfer the tube into an ice bath, chill for 5 min, remove from the ice bath, and add about one-half of the required volume of *Phosphatide–Acetone Solution*, previously chilled for 5 min in an ice bath. Stir the mixture to complete dispersion of the sample, dilute to 40 mL with chilled *Phosphatide–Acetone Solution* (5°), again stir, and return the tube and contents to the ice bath for 15 min. At the end of the 15-min chilling period, stir again while still in the ice bath, remove the stirring rod, temporarily supporting it in a vertical upside-down position, and centrifuge the mixture immediately at about 2000 rpm for 5 min. Decant the supernatant liquid from the centrifuge tube, crush the centrifuged solids with the same stirring rod previously used, and refill the tube to the 40-mL mark with chilled (5°) *Phosphatide–Acetone Solution*, and repeat the chilling, stirring, centrifugation, and decantation procedure previously followed. After the second centrifugation and decantation of the supernatant acetone, again crush the solids with the assigned stirring rod, and place the tube and its contents in a horizontal position at room temperature until the excess acetone has evaporated. Mix the residue again, dry the centrifuge tube and its contents at 105° for 45 min in a forced-draft oven, cool, and weigh. Calculate the percentage of acetone-insoluble substances by the formula

$$(100R/S) - B,$$

in which R is the weight of residue, S is the weight of the sample, and B is the percentage of *Hexane-Insoluble Matter* determined as directed in this monograph.

Acid Value   If it is plastic or semisolid, soften a portion of the Lecithin by warming it in a water bath at a temperature not exceeding 60°, and then mix it thoroughly. Transfer about 2 g of a well-mixed sample into a 250-mL Erlenmeyer flask, and dissolve it in 50 mL of petroleum ether. To this solution add 50 mL of alcohol, previously neutralized to phenolphthalein with 0.1 N sodium hydroxide, and mix well. Add phenolphthalein TS, and titrate with 0.1 N sodium hydroxide to a pink endpoint that persists for 5 s. Calculate the number of mg of potassium hydroxide required to neutralize the acids in 1 g of the sample by multiplying the number of mL of 0.1 N sodium hydroxide consumed in the titration by 5.6 and dividing the result by the weight of the sample.

Heavy Metals Prepare and test a 1-g sample as directed in *Method II* under the *Heavy Metals Test*, Appendix IIIB, using 20 μg of lead ion (Pb) in the control (*Solution A*).

Hexane-Insoluble Matter If plastic or semisolid, soften a portion of the Lecithin by warming it at a temperature not exceeding 60°, and then mix it thoroughly. Weigh 10 g of a previously well-mixed sample into a 250-mL wide-mouth Erlenmeyer flask, add 100 mL of solvent hexane, and shake until the Lecithin is dissolved. Filter the solution through a 30-mL Corning "C" porosity or equivalent filtering funnel that previously has been dried at 105° for 1 h, cooled in a desiccator, and weighed. Wash the flask with two successive 25-mL portions of solvent hexane, and pass the washings through the filter. Dry the funnel at 105° for 1 h, cool to room temperature in a desiccator, and weigh. From the gain in weight of the funnel, calculate the percentage of the hexane-insoluble matter in the sample.

Lead A *Sample Solution* prepared as directed for organic compounds meets the requirements of the *Lead Limit Test*, Appendix IIIB, using 10 μg of lead ion (Pb) in the control.

Peroxide Value Determine as directed in the monograph for *Hydroxylated Lecithin*.

Water Determine by the *Karl Fischer Titrimetric Method*, Appendix IIB.

Packaging and Storage Store in well-closed containers.

Lemongrass Oil

DESCRIPTION

A volatile oil prepared by steam distillation of freshly cut and partially dried cymbopogon grasses indigenous to tropical and subtropical areas. Two types of Lemongrass Oil are commercially available. The *East Indian* type, also known as Cochin, Native, and British Indian Lemongrass Oil, is usually a dark yellow to light brownish red liquid having a pronounced heavy lemonlike odor. The *West Indian* type, also known as Madagascar, Guatemala, or other country of origin Lemongrass Oil, is light yellow to light brown in color and has a lemonlike odor of a lighter character than the East Indian type oil. Lemongrass Oils are soluble in mineral oil, freely soluble in propylene glycol, but practically insoluble in water and in glycerin. The East Indian variety dissolves readily in alcohol, but the West Indian oil yields cloudy solutions.

Functional Use in Foods Flavoring agent.

REQUIREMENTS

Labeling Indicate whether it is the East Indian or West Indian type.

Identification The infrared absorption spectrum of the sample exhibits relative maxima (that may vary in intensity) at the same wavelengths (or frequencies) as those shown in the respective spectrum under *Infrared Spectra* (*Series A: Essential Oils*), using the same test conditions as specified therein.

Assay Not less than 75.0%, by volume, of aldehydes as citral.

Angular Rotation Between −3° and +1°.

Heavy Metals (as Pb) Passes test.

Refractive Index Between 1.483 and 1.489.

Solubility in Alcohol Passes test.

Specific Gravity *East Indian type*: between 0.894 and 0.904; *West Indian type*: between 0.869 and 0.894.

Steam–Volatile Oil Not less than 93.0%, by volume.

TESTS

Assay Mix 50.0 mL of the sample with 500 mg of tartaric acid, shake for 5 min, and filter. Dry the filtered oil over anhydrous sodium sulfate, and then pipet 10.0 mL of the clear, treated oil into a 150-mL cassia flask.

> Note: Retain the remaining oil for the *Steam–Volatile Oil* test.

Add 75 mL of a 30% solution of sodium bisulfite, stopper the flask, and shake until a semisolid to solid sodium bisulfite addition product has formed. Allow the mixture to stand at room temperature for 5 min, then loosen the stopper, and immerse the flask in a water bath heated to between 85° and 90°. Maintain the water bath at this temperature, shaking the flask occasionally, until the addition product dissolves, and then continue the heating and intermittent shaking for another 30 min. When the liquids have separated completely, add enough of the sodium bisulfite solution to raise the lower level of the oily layer within the graduated portion of the neck of the flask. Calculate the percentage, by volume, of the citral by the formula

$$100 - (V \times 10),$$

in which V is the number of mL of separated oil in the graduated neck of the cassia flask.

Angular Rotation Determine in a 100-mm tube as directed under *Optical (Specific) Rotation*, Appendix IIB.

Heavy Metals Shake 10 mL of the oil with an equal volume of water to which 1 drop of hydrochloric acid has been added, and pass hydrogen sulfide through the mixture until it is saturated. No darkening in color is produced in either the oil or the water.

Refractive Index, Appendix IIB Determine with an Abbé or other refractometer of equal or greater accuracy.

Solubility in Alcohol Proceed as directed in the general method, Appendix VI. *East Indian type*: one mL dissolves in 3 mL of 70% alcohol, usually with slight turbidity; *West Indian type*: yields a cloudy solution with 70%, 80%, 90%, and 95% alcohol.

Specific Gravity Determine by any reliable method (see *General Provisions*).

Steam–Volatile Oil Proceed as directed for *Volatile Oil Content* under *Essential Oils and Flavors*, Appendix VI, using 25.0 mL of the oil prepared as directed in the *Assay*.

Packaging and Storage Store in a cool place in full, tight, preferably glass, aluminum, tin-lined, or other suitably lined containers, or in black iron unlined drums. If stored in glass containers, avoid exposure to light.

Lemon Oil, Coldpressed

Lemon Oil, Expressed

DESCRIPTION

A volatile oil obtained by expression, without the aid of heat, from the fresh peel of the fruit of *Citrus limon* L. Burmann filius (Fam. *Rutaceae*), with or without the previous separation of the pulp and the peel. It is a pale to deep yellow or greenish yellow liquid having the characteristic odor and taste of the outer part of fresh lemon peel. It is miscible with dehydrated alcohol and with glacial acetic acid. It may contain a suitable antioxidant.

Note: Do not use Lemon Oil that has a terebinthine odor.

Functional Use in Foods Flavoring agent.

REQUIREMENTS

Labeling Indicate whether it is the California type or the Italian type.
Identification The infrared absorption spectrum of the sample exhibits relative maxima (that may vary in intensity) at the same wavelengths (or frequencies) as those shown in the respective spectrum under *Infrared Spectra (Series A: Essential Oils)*, using the same test conditions as specified therein.
Assay *California type*: not less than 2.2% and not more than 3.8% of aldehydes, calculated as citral ($C_{10}H_{16}O$); *Italian type*: not less than 3.0% and not more than 5.5% of aldehydes, calculated as citral ($C_{10}H_{16}O$).
Angular Rotation Between +57° and +65.6°.
Foreign Oils Passes test.
Heavy Metals (as Pb) Not more than 0.004%.
Lead Not more than 10 mg/kg.
Refractive Index Between 1.473 and 1.476 at 20°.
Solubility in Alcohol Passes test.
Specific Gravity Between 0.849 and 0.855.
Ultraviolet Absorbance *California type*: not less than 0.2; *Italian type*: not less than 0.49.

TESTS

Assay Weigh accurately a 5-mL sample, and proceed as directed under *Aldehydes and Ketones—Hydroxylamine/Tert-Butyl Alcohol Method*, Appendix VI. Allow the mixture to stand for 15 min, with occasional shaking, before titrating, and use 76.12 as the equivalence factor (*e*) in the calculation.

Angular Rotation Determine in a 100-mm tube as directed under *Optical (Specific) Rotation*, Appendix IIB.
Foreign Oils Transfer 50 mL to a Ladenburg flask having four bulbs of the following approximate diameters: 6 cm, 3.5 cm, 3.0 cm, and 2.5 cm, respectively, in ascending order. The distance from the bottom of the flask to the side arm is 20 cm. Distill the oil at the rate of 1 drop per s until the distillate measures 5 mL. The angular rotation of the distillate is not more than 6° less than that of the original oil, and the refractive index is not less than 0.001 and not more than 0.003 lower than that of the original oil at 20°.
Heavy Metals Prepare and test a 500-mg sample as directed in *Method II* under the *Heavy Metals Test*, Appendix IIIB, using 20 µg of lead ion (Pb) in the control (*Solution A*).
Lead A *Sample Solution* prepared as directed for organic compounds meets the requirements of the *Lead Limit Test*, Appendix IIIB, using 10 µg of lead ion (Pb) in the control.
Refractive Index, Appendix IIB Determine with an Abbé or other refractometer of equal or greater accuracy.
Solubility in Alcohol Proceed as directed in the general method, Appendix VI. One mL dissolves in 3 mL of 95% alcohol, sometimes with a slight haze.
Specific Gravity Determine by any reliable method (see *General Provisions*).
Ultraviolet Absorbance Proceed as directed under *Ultraviolet Absorbance of Citrus Oils*, Appendix VI, using about 250 mg of sample, accurately weighed. The maximum absorbance occurs at 315 ± 3 nm.

Packaging and Storage Store in full, tight containers. Avoid exposure to excessive heat.

Lemon Oil, Distilled

DESCRIPTION

The volatile oil obtained by distillation from the fresh peel or juice of the fruit of *Citrus limon* L. Burmann filius (Fam. *Rutaceae*), with or without the previous separation of the juice, pulp, and peel. It is a colorless to pale yellow liquid having the characteristic odor of fresh lemon peel. It is soluble in most fixed oils, in mineral oil, and in alcohol (with haze). It is insoluble in glycerin and in propylene glycol. It may contain a suitable antioxidant.

Functional Use in Foods Flavoring agent.

REQUIREMENTS

Identification The infrared absorption spectrum of the sample exhibits relative maxima (that may vary in intensity) at the same wavelengths (or frequencies) as those shown in the respective spectrum under *Infrared Spectra (Series A: Essential Oils)*, using the same test conditions as specified therein.

Aldehydes Between 1.0% and 3.5% of aldehydes, calculated as citral ($C_{10}H_{16}O$).
Angular Rotation Between +55° and +75°.
Heavy Metals (as Pb) Passes test.
Refractive Index Between 1.470 and 1.475 at 20°.
Specific Gravity Between 0.842 and 0.856.
Ultraviolet Absorbance Not more than 0.01.

TESTS

Aldehydes Weigh accurately about 5 mL of the sample, and proceed as directed under *Aldehydes and Ketones—Hydroxylamine/Tert-Butyl Alcohol Method*, Appendix VI, using 76.12 as the equivalence factor (*e*) in the calculation. Allow the mixture to stand at room temperature for 1 h before titrating.
Angular Rotation Determine in a 100-mm tube as directed under *Optical (Specific) Rotation*, Appendix IIB.
Heavy Metals Shake 10 mL of the oil with an equal volume of water to which 1 drop of hydrochloric acid has been added, and pass hydrogen sulfide through the mixture until it is saturated. No darkening in color is produced in either the oil or the water.
Refractive Index, Appendix IIB Determine with an Abbé or other refractometer of equal or greater accuracy.
Specific Gravity Determine by any reliable method (see *General Provisions*).
Ultraviolet Absorbance Proceed as directed under *Ultraviolet Absorbance of Citrus Oils*, Appendix VI, using about 250 mg of sample, accurately weighed. The maximum absorbance occurs at 315 ± 5 nm.

Packaging and Storage Store in full, tight containers in a cool place protected from light.

Lemon Oil, Desert Type, Coldpressed

Lemon Oil Arizona

DESCRIPTION

The volatile oil obtained by expression, without the aid of heat, from the fresh peel of the fruit of *Citrus Limon* L. Burmann filius (Fam. *Rutaceae*), with or without the previous separation of the pulp and peel. It is a pale to deep yellow or greenish yellow liquid having the characteristic odor and taste of the outer part of fresh lemon peel. It is miscible with dehydrated alcohol and with glacial acetic acid. It may contain a suitable antioxidant.

  Note: Do not use if it has a terebinthine odor.

Functional Use in Foods Flavoring agent.

REQUIREMENTS

Identification The infrared absorption spectrum of the sample exhibits relative maxima (that may vary in intensity) at the same wavelengths (or frequencies) as those shown in the respective spectrum under *Infrared Spectra* (*Series A: Essential Oils*), using the same test conditions as specified therein.
Assay Not less than 1.7% of aldehydes, calculated as citral ($C_{10}H_{16}O$).
Angular Rotation Between +67° and +78°.
Heavy Metals (as Pb) Not more than 0.004%.
Lead Not more than 10 mg/kg.
Refractive Index Between 1.473 and 1.476.
Solubility in Alcohol Passes test.
Specific Gravity Between 0.846 and 0.851.
Ultraviolet Absorbance Not less than 0.20.

TESTS

Assay Weigh accurately about 5 mL of the sample, and proceed as directed under *Aldehydes and Ketones—Hydroxylamine/Tert-Butyl Alcohol Method*, Appendix VI. Allow the mixture to stand for 15 min, with occasional shaking, before titrating, and use 76.12 as the equivalence factor (*e*) in the calculation.
Angular Rotation Determine in a 100-mm tube as directed under *Optical (Specific) Rotation*, Appendix IIB.
Heavy Metals Prepare and test a 500-mg sample as directed in *Method II* under the *Heavy Metals Test*, Appendix IIIB, using 20 µg of lead ion (Pb) in the control (*Solution A*).
Lead A *Sample Solution* prepared as directed for organic compounds meets the requirements of the *Lead Limit Test*, Appendix IIIB, using 10 µg of lead ion (Pb) in the control.
Refractive Index, Appendix IIB Determine with an Abbé or other refractometer of equal or greater accuracy.
Solubility in Alcohol Proceed as directed in the general method, Appendix VI. One mL dissolves in 3 mL of alcohol, sometimes with a slight haze.
Specific Gravity Determine by any reliable method (see *General Provisions*).
Ultraviolet Absorbance Proceed as directed under *Ultraviolet Absorbance of Citrus Oils*, Appendix VI, using about 250 mg of sample, accurately weighed. The maximum absorbance occurs at 315 ± 3 nm.

Packaging and Storage Store in full, tight containers. Avoid exposure to excessive heat.

DL-Leucine

DL-2-Amino-4-methylvaleric Acid

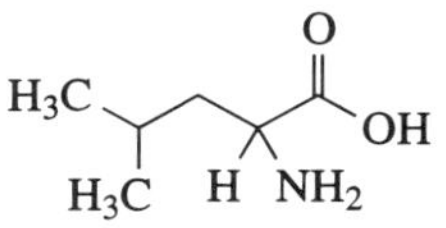

$C_6H_{13}NO_2$ — Formula wt 131.17

CAS: [328-39-2]

DESCRIPTION

Small, white crystals or a crystalline powder. It is odorless and has a slightly bitter taste. It is freely soluble in water, slightly soluble in alcohol, and insoluble in ether. It melts with decomposition at about 290°. The pH of a 1 in 100 solution is between 5.5 and 7.0. It is optically inactive.

Functional Use in Foods Nutrient; dietary supplement.

REQUIREMENTS

Identification To 5 mL of a 1 in 1000 solution add 1 mL of triketohydrindene hydrate TS. A reddish purple or bluish purple color is produced.
Assay Not less than 98.5% and not more than 101.5% of $C_6N_{13}NO_2$, calculated on the dried basis.
Heavy Metals (as Pb) Not more than 0.002%.
Lead Not more than 10 mg/kg.
Loss on Drying Not more than 0.3%.
Residue on Ignition Not more than 0.1%.

TESTS

Assay Dissolve about 400 mg, accurately weighed, in 3 mL of formic acid and 50 mL of glacial acetic acid. Add 2 drops of crystal violet TS, and titrate with 0.1 N perchloric acid to the first appearance of a pure green color or until the blue color disappears completely. Perform a blank determination (see *General Provisions*), and make any necessary correction. Each mL of 0.1 N perchloric acid is equivalent to 13.12 mg of $C_6H_{13}NO_2$.
Heavy Metals Prepare and test a 1-g sample as directed in *Method II* under the *Heavy Metals Test*, Appendix IIIB, using 20 µg of lead ion (Pb) in the control (*Solution A*).
Lead A *Sample Solution* prepared as directed for organic compounds meets the requirements of the *Lead Limit Test*, Appendix IIIB, using 10 µg of lead ion (Pb) in the control.
Loss on Drying, Appendix IIC Dry at 105° for 3 h.
Residue on Ignition, Appendix IIC Ignite 1 g as directed in the general method.

Packaging and Storage Store in well-closed containers.

L-Leucine

L-2-Amino-4-methylvaleric Acid

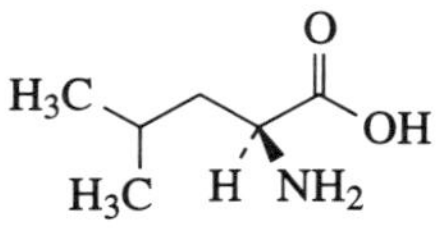

$C_6H_{13}NO_2$ — Formula wt 131.17

INS: 641

CAS: [61-90-5]

DESCRIPTION

Small, white, lustrous plates, or a white, crystalline powder. One g dissolves in about 40 mL of water and in about 100 mL of acetic acid. It is sparingly soluble in alcohol, but is soluble in dilute hydrochloric acid and in solutions of alkali hydroxides and carbonates.

Functional Use in Foods Nutrient; dietary supplement.

REQUIREMENTS

Identification It sublimes at about 150°.
Assay Not less than 98.5% and not more than 101.5% of $C_6H_{13}NO_2$ after drying.
Heavy Metals (as Pb) Not more than 0.002%.
Lead Not more than 10 mg/kg.
Loss on Drying Not more than 0.2%.
Residue on Ignition Not more than 0.1%.
Specific Rotation $[\alpha]_D^{20°}$: Between +14.5° and +16.5°, calculated on the dried basis; or $[\alpha]_D^{25°}$: Between +14.8° and +16.8°, calculated on the dried basis.

TESTS

Assay Transfer about 400 mg, previously dried at 105° for 3 h and accurately weighed, into a 250-mL flask. Dissolve the sample in 3 mL of formic acid and about 50 mL of glacial acetic acid, add 2 drops of crystal violet TS, and titrate with 0.1 N perchloric acid to a bluish green endpoint. Perform a blank determination (see *General Provisions*), and make any necessary correction. Each mL of 0.1 N perchloric acid is equivalent to 13.12 mg of $C_6H_{13}NO_2$.
Heavy Metals Prepare and test a 1-g sample as directed in *Method II* under the *Heavy Metals Test*, Appendix IIIB, using 20 µg of lead ion (Pb) in the control (*Solution A*).
Lead A *Sample Solution* prepared as directed for organic compounds meets the requirements of the *Lead Limit Test*, Appendix IIIB, using 10 µg of lead ion (Pb) in the control.
Loss on Drying, Appendix IIC Dry at 105° for 3 h.
Residue on Ignition Ignite 1 g as directed in the general method, Appendix IIC.

Specific Rotation, Appendix IIB   Determine in a solution containing 4 g in sufficient 6 N hydrochloric acid to make 100 mL.

Packaging and Storage   Store in well-closed containers.

Lime Oil, Coldpressed

Lime Oil, Expressed

DESCRIPTION

The volatile oil obtained by expression from the fresh peel or crushed whole fruit of *Citrus aurantifolia* Swingle (Mexican type) or *Citrus latifolia* (Tahitian type). It is a yellow to brownish green to green liquid and often shows a waxy separation. It is soluble in most fixed oils and in mineral oil. It is insoluble in glycerin and in propylene glycol. It may contain a suitable antioxidant.

Functional Use in Foods   Flavoring agent.

REQUIREMENTS

Labeling   Indicate whether it is the Mexican or Tahitian type.
Identification   The infrared absorption spectrum of the sample exhibits relative maxima (that may vary in intensity) at the same wavelengths (or frequencies) as those shown in the respective spectra under *Infrared Spectra (Series A: Essential Oils)*, using the same test conditions as specified therein.
Assay   *Mexican type*: not less than 4.5% and not more than 8.5% of aldehydes (as citral); *Tahitian type*: not less than 3.2% and not more than 7.5% of aldehydes (as citral).
Angular Rotation   *Mexican type*: between +35° and +41°; *Tahitian type*: between +38° and +53°.
Heavy Metals (as Pb)   Not more than 0.004%.
Lead   Not more than 10 mg/kg.
Refractive Index   *Mexican type*: between 1.482 and 1.486; *Tahitian type*: between 1.476 and 1.486.
Residue on Evaporation   *Mexican type*: between 10.0% and 14.5%; *Tahitian type*: between 5.0% and 12.0%.
Specific Gravity   *Mexican type*: between 0.872 and 0.881; *Tahitian type*: between 0.858 and 0.876.
Ultraviolet Absorbance   *Mexican type*: not less than 0.45; *Tahitian type*: not less than 0.24.

TESTS

Assay   Weigh accurately about 5 mL of the sample, and proceed as directed under *Aldehydes and Ketones—Hydroxylamine/Tert-Butyl Alcohol Method*, Appendix VI. Allow the mixture to stand for 1 h, with occasional shaking, before titrating, and use 76.12 as the equivalence factor (*e*) in the calculation.
Angular Rotation   Determine in a 100-mm tube as directed under *Optical (Specific) Rotation*, Appendix IIB.

Heavy Metals   Prepare and test a 500-mg sample as directed in *Method II* under the *Heavy Metals Test*, Appendix IIIB, using 20 µg of lead ion (Pb) in the control (*Solution A*).
Lead   A *Sample Solution* prepared as directed for organic compounds meets the requirements of the *Lead Limit Test*, Appendix IIIB, using 10 µg of lead ion (Pb) in the control.
Refractive Index, Appendix IIB   Determine with an Abbé or other refractometer of equal or greater accuracy.
Residue on Ignition   Proceed as directed in the general method, Appendix IIC, using a 3-g sample and heating for 6 h.
Specific Gravity   Determine by any reliable method (see *General Provisions*).
Ultraviolet Absorbance   Proceed as directed under *Ultraviolet Absorbance of Citrus Oils*, Appendix VI, using about 20 mg of sample, accurately weighed. The maximum absorbance occurs at 315 ± 3 nm.

Packaging and Storage   Store in full, tight containers. Avoid exposure to excessive heat.

Lime Oil, Distilled

DESCRIPTION

The volatile oil obtained by distillation from the juice or the whole crushed fruit of *Citrus aurantifolia* Swingle. It is a colorless to greenish yellow liquid. It is soluble in most fixed oils and in mineral oil. It is insoluble in glycerin and in propylene glycol. It may contain a suitable antioxidant.

Functional Use in Foods   Flavoring agent.

REQUIREMENTS

Identification   The infrared absorption spectrum of the sample exhibits relative maxima (that may vary in intensity) at the same wavelengths (or frequencies) as those shown in the respective spectrum under *Infrared Spectra (Series A: Essential Oils)*, using the same test conditions as specified therein.
Aldehydes   Between 0.5% and 2.5% of aldehydes, calculated as citral ($C_{10}H_{16}O$).
Angular Rotation   Between +34° and +47°.
Heavy Metals (as Pb)   Passes test.
Refractive Index   Between 1.474 and 1.477 at 20°.
Solubility in Alcohol   Passes test.
Specific Gravity   Between 0.855 and 0.863.

TESTS

Aldehydes   Weigh accurately about 5 g, and proceed as directed under *Aldehydes and Ketones—Hydroxylamine/Tert-Butyl Alcohol Method*, Appendix VI, using 76.12 as the equivalence factor (*e*) in the calculation. Allow the mixture to stand at room temperature for 15 min before titrating.

Angular Rotation Determine in a 100-mm tube as directed under *Optical (Specific) Rotation*, Appendix IIB.

Heavy Metals Shake 10 mL of the oil with an equal volume of water to which 1 drop of hydrochloric acid has been added, and pass hydrogen sulfide through the mixture until it is saturated. No darkening in color is produced in either the oil or the water.

Refractive Index, Appendix IIB Determine with an Abbé or other refractometer of equal or greater accuracy.

Solubility in Alcohol Proceed as directed in the general method, Appendix VI. One mL dissolves in 5 mL of 90% alcohol.

Specific Gravity Determine by any reliable method (see *General Provisions*).

Packaging and Storage Store in full, tight, preferably glass, tin-lined, galvanized, or black iron containers in a cool place protected from light.

Limestone, Ground

DESCRIPTION

Ground Limestone consists essentially of calcium carbonate. It is obtained by crushing, grinding, and classifying naturally occurring limestone benefited by flotation and/or air classification. It is produced as a fine, white to off-white, microcrystalline powder. It is odorless and tasteless and is stable in air. It is practically insoluble in water and in alcohol. The presence of any ammonium salt or carbon dioxide increases its solubility in water, but the presence of any alkali hydroxide reduces the solubility.

Functional Use in Foods Texturizing and release agent and modifier for chewing gum base and chewing gum.

REQUIREMENTS

Identification It dissolves with effervescence in 1 N acetic acid, in 2.7 N hydrochloric acid, and in 1.7 N nitric acid, and the resulting solutions, after boiling, give positive tests for *Calcium*, Appendix IIIA.

Assay Not less than 94.0% and not more than 100.5% of $CaCO_3$ after drying.

Acid-Insoluble Substances Not more than 2.5%.

Arsenic (as As) Not more than 3 mg/kg.

Fluoride Not more than 0.005%.

Heavy Metals (as Pb) Not more than 0.002%.

Lead Not more than 3 mg/kg.

Loss on Drying Not more than 2.0%.

Magnesium and Alkali Salts Not more than 3.5%.

TESTS

Assay Transfer about 200 mg, previously dried at 200° for 4 h and accurately weighed, into a 400-mL beaker, add 10 mL of water, and swirl to form a slurry. Cover the beaker with a watch glass, and introduce 2 mL of 2.7 N hydrochloric acid from a pipet inserted between the lip of the beaker and the edge of the watch glass. Swirl the contents of the beaker to dissolve the sample. Wash down the sides of the beaker, the outer surface of the pipet, and the watch glass, and dilute to about 100 mL with water. While stirring, preferably with a magnetic stirrer, add about 30 mL of 0.05 M disodium EDTA from a 50-mL buret, then add 15 mL of 1 N potassium hydroxide and 300 mg of hydroxy naphthol blue indicator, and continue the titration to a blue endpoint. Each mL of 0.05 M disodium EDTA is equivalent to 5.004 mg of $CaCO_3$.

Acid-Insoluble Substances Suspend 5 g in 25 mL of water, cautiously add with agitation 25 mL of dilute hydrochloric acid (1 in 2), then add water to make a volume of about 200 mL. Heat the solution to boiling, cover, digest on a steam bath for 1 h, cool, and filter. Wash the precipitate with water until the last washing shows no chloride with silver nitrate TS, and then ignite it. The weight of the residue does not exceed 125 mg.

Arsenic A solution of 1 g in 10 mL of 2.7 N hydrochloric acid meets the requirements of the *Arsenic Test*, Appendix IIIB.

Fluoride Determine as directed under *Method III* in the general method, Appendix IIIB.

Sample Solution for the Determination of Heavy Metals and Lead Cautiously dissolve 5 g in 25 mL of dilute hydrochloric acid (1 in 2), and evaporate to dryness on a steam bath. Dissolve the residue in about 15 mL of water, and dilute to 25.0 mL (1 mL = 200 mg).

Heavy Metals Neutralize 5.0 mL (1 g) of the *Sample Solution* with 1 N sodium hydroxide, using phenolphthalein as the indicator, and dilute to 25 mL. This solution meets the requirements of the *Heavy Metals Test*, Appendix IIIB, using 20 μg of lead ion (Pb) in the control (*Solution A*).

Lead A 5-mL portion of the *Sample Solution* (1 g) meets the requirements of the *Lead Limit Test*, Appendix IIIB, using 3 μg of lead ion (Pb) in the control.

Loss on Drying, Appendix IIC Dry at 200° for 4 h.

Magnesium and Alkali Salts Mix 1 g with 40 mL of water, carefully add 5 mL of hydrochloric acid, mix, and boil for 1 min. Rapidly add 40 mL of oxalic acid TS, and stir vigorously until precipitation is well established. Immediately add 2 drops of methyl red TS, then add 6 N ammonium hydroxide, dropwise, until the mixture is just alkaline, and cool. Transfer the mixture to a 100-mL cylinder, dilute with water to 100 mL, let it stand for 4 h or overnight, then decant the clear, supernatant liquid through a dry filter paper. To 50 mL of the clear filtrate in a platinum dish add 0.5 mL of sulfuric acid, and evaporate the mixture on a steam bath to a small volume. Carefully evaporate the remaining liquid to dryness over a free flame, and continue heating until the ammonium salts have been completely decomposed and volatilized. Finally, ignite the residue to constant weight. The weight of the residue does not exceed 17.5 mg.

Packaging and Storage Store in well-closed containers.

Linaloe Wood Oil

DESCRIPTION

The volatile oil obtained by steam distillation from the wood of *Bursera delpechiana* Poiss. (Fam. *Burseraceae*) and other *Bursera* species. It is a colorless to yellow liquid having a pleasant, flowery odor. It is soluble in most fixed oils and in propylene glycol. It is soluble in mineral oil, but becomes opalescent or turbid on dilution. It is insoluble in glycerin.

Functional Use in Foods Flavoring agent.

REQUIREMENTS

Identification The infrared absorption spectrum of the sample exhibits relative maxima (that may vary in intensity) at the same wavelengths (or frequencies) as those shown in the respective spectrum under *Infrared Spectra (Series A: Essential Oils)*, using the same test conditions as specified therein.
Assay Not less than 85.0% of alcohols, calculated as linalool ($C_{10}H_{18}O$).
Acid Value Not more than 3.0.
Angular Rotation Between −5° and −13°.
Ester Value Between 40 and 75.
Heavy Metals (as Pb) Passes test.
Refractive Index Between 1.459 and 1.463 at 20°.
Solubility in Alcohol Passes test.
Specific Gravity Between 0.876 and 0.883.

TESTS

Assay Proceed as directed under *Linalool Determination*, Appendix VI, using about 1.5 g of acetylated oil, accurately weighed, for the saponification.
Acid Value Determine as directed for *Acid Value* under *Essential Oils and Flavors*, Appendix VI.
Angular Rotation Determine in a 100-mm tube as directed under *Optical (Specific) Rotation*, Appendix IIB.
Ester Value Determine as directed in the general method, Appendix VI, using about 2.5 g, accurately weighed.
Heavy Metals Shake 10 mL of the oil with an equal volume of water to which 1 drop of hydrochloric acid has been added, and pass hydrogen sulfide through the mixture until it is saturated. No darkening in color is produced in either the oil or the water.
Refractive Index, Appendix IIB Determine with an Abbé or other refractometer of equal or greater accuracy.

Solubility in Alcohol Proceed as directed in the general method, Appendix VI. One mL dissolves in 5 mL of 60% alcohol.
Specific Gravity Determine by any reliable method (see *General Provisions*).

Packaging and Storage Store in full, tight, preferably glass, aluminum, or tin-lined containers in a cool place protected from light.

Linoleic Acid

(Z),(Z)-9,12-Octadecadienoic Acid

$$CH_3(CH_2)_4CH=CHCH_2CH=CH(CH_2)_7COOH$$

$C_{18}H_{32}O_2$ Formula wt 280.45

CAS: [60-33-3]

DESCRIPTION

An essential fatty acid and the major constituent of many vegetable oils, including cottonseed, soybean, peanut, corn, sunflower seed, safflower, poppy seed, and linseed. It is a colorless to pale-yellow, oily liquid that is easily oxidized by air. Its specific gravity is about 0.901, and its refractive index is about 1.4699. It has a boiling point ranging from 225° to 230° and a melting point around −5°. One mL dissolves in 10 mL of petroleum ether. It is freely soluble in ether; soluble in absolute alcohol and in chloroform; and miscible with dimethylformamide, fat solvents, and oils. It is insoluble in water.

Functional Use in Foods Flavoring adjuvant; dietary supplement.

REQUIREMENTS

Identification Linoleic Acid exhibits the following composition profile of fatty acids as determined under *Fatty Acid Composition*, Appendix VII.

Fatty Acid:	14:0	16:0	18:0	18:1	18:2	18:3
Weight % (Range):	<1.0	3–5	<1.0	<25.0	>60.0	<9.0

Assay Not less than 60.0% of fatty acid C18:2 equivalent to $C_{18}H_{32}O_2$, calculated on the anhydrous basis.
Acid Value Between 196 and 202.
Heavy Metals (as Pb) Not more than 10 mg/kg.
Iodine Value Between 145 and 160.
Residue on Ignition Not more than 0.01%.
Saponification Value Between 194 and 202.
Unsaponifiable Matter Not more than 2.0%.
Water Not more than 0.5%.

TESTS

Assay Proceed as directed under *Fatty Acid Composition*, Appendix VII.

Acid Value Determine as directed for *Acid Value, Method I*, under *Fats and Related Substances*, Appendix VII.

Heavy Metals Prepare and test a 2-g sample as directed in *Method II* under the *Heavy Metals Test*, Appendix IIIB, using 20 µg of lead ion (Pb) in the control (*Solution A*).

Iodine Value Determine by the *Modified Wijs Method*, Appendix VII.

Residue on Ignition Ignite 10 g as directed in the general method, Appendix IIC.

Saponification Value Determine as directed for *Saponification Value* under *Fats and Related Substances*, Appendix VII, using a 3-g sample, accurately weighed.

Unsaponifiable Matter Proceed as directed under the general method, Appendix VII, using a 10-g sample.

Water Determine by the *Karl Fischer Titrimetric Method*, Appendix IIB.

Packaging and Storage Store in tight containers.

Locust (Carob) Bean Gum

Locust Bean Gum; Carob Bean G

DESCRIPTION

A gum obtained from the ground endosperms of *Ceratonia siliqua* (L.) Taub. (Fam. *Leguminosae*). It consists chiefly of a high-molecular-weight hydrocolloidal polysaccharide, composed of galactose and mannose units combined through glycosidic linkages, which may be described chemically as a galactomannan. It is a white to yellowish white, nearly odorless powder. It is dispersible in either hot or cold water, forming a sol having a pH between 5.4 and 7.0, which may be converted to a gel by the addition of small amounts of sodium borate.

Functional Use in Foods Stabilizer; thickener; emulsifier.

REQUIREMENTS

Identification

A. Transfer a 2-g sample into a 400-mL beaker, moisten it with about 4 mL of isopropyl alcohol, add with vigorous stirring 200 mL of cold water, and continue the stirring until the gum is uniformly dispersed. An opalescent, slightly viscous solution is formed.

B. Transfer 100 mL of the solution prepared in *Identification Test A* into another 400-mL beaker, heat the mixture in a boiling water bath for about 10 min, and then cool to room temperature. An appreciable increase in viscosity is produced (*distinction from guar gum*), except for the flash-ground gum.

Acid-Insoluble Matter Not more than 4.0%.
Arsenic (as As) Not more than 3 mg/kg.
Ash (Total) Not more than 1.2%.
Galactomannans Not less than 75.0%.
Heavy Metals (as Pb) Not more than 0.002%.
Lead Not more than 5 mg/kg.
Loss on Drying Not more than 14.0%.
Protein Not more than 7.0%.
Starch Passes test.

TESTS

Acid-Insoluble Matter Transfer 1.5 g of the sample, accurately weighed, into a 250-mL beaker containing 150 mL of water and 1.5 mL of sulfuric acid. Cover the beaker with a watch glass, and heat the mixture on a steam bath for 6 h, rubbing down the wall of the beaker frequently with a rubber-tipped stirring rod and replacing any water lost by evaporation. Then add about 500 mg of a suitable filter aid, previously dried for 3 h at 105° and accurately weighed, and filter through a tared, sintered-glass filter crucible. Wash the residue several times with hot water, dry the crucible and its contents at 105° for 3 h, cool in a desiccator, and weigh. The difference between the weight of the filter aid and that of the residue is the weight of the *Acid-Insoluble Matter*.

Arsenic A *Sample Solution* prepared as directed for organic compounds meets the requirements of the *Arsenic Test*, Appendix IIIB.

Ash (Total) Determine as directed in the general method, Appendix IIC.

Galactomannans The remainder, after subtracting from 100% the sum of the percentages of *Acid-Insoluble Matter*, *Total Ash*, *Loss on Drying*, and *Protein*, represents the percentage of galactomannans in the sample.

Heavy Metals Prepare and test a 1-g sample as directed in *Method II* under the *Heavy Metals Test*, Appendix IIIB, using 20 µg of lead ion (Pb) in the control (*Solution A*).

Lead A *Sample Solution* prepared as directed for organic compounds meets the requirements of the *Lead Limit Test*, Appendix IIIB, using 5 µg of lead ion (Pb) in the control.

Loss on Drying, Appendix IIC Dry at 105° for 5 h.

Protein Transfer about 3.5 g, accurately weighed, into a 500-mL Kjeldahl flask, and proceed as directed under *Nitrogen Determination*, Appendix IIIC. The percentage of nitrogen determined, multiplied by 6.25, gives the percentage of protein in the sample.

Starch To a 1 in 10 solution of the gum add a few drops of iodine TS. No blue color is produced.

Packaging and Storage Store in well-closed containers.

Lovage Oil

DESCRIPTION

The volatile oil obtained by steam distillation of the fresh root of the plant *Levisticum officinale* L. Koch syn. *Angelica levisticum*, Baillon (Fam. *Umbelliferae*). It is a yellow greenish brown to deep-brown liquid having a strong characteristic aromatic odor and taste. It is soluble in most fixed oils and slightly soluble, with opalescence, in mineral oil, but it is relatively insoluble in glycerin and in propylene glycol.

Note: This oil becomes darker and more viscous under the influence of air and light.

Functional Use in Foods Flavoring agent.

REQUIREMENTS

Identification The infrared absorption spectrum of the sample exhibits relative maxima (that may vary in intensity) at the same wavelengths (or frequencies) as those shown in the respective spectrum under *Infrared Spectra (Series A: Essential Oils)*, using the same test conditions as specified therein.
Acid Value Between 2.0 and 16.0.
Angular Rotation Between −1° and +5°.
Heavy Metals (as Pb) Passes test.
Refractive Index Between 1.536 and 1.554 at 20°.
Saponification Value Between 238 and 258.
Solubility in Alcohol Passes test.
Specific Gravity Between 1.034 and 1.057.

TESTS

Acid Value Determine as directed for *Acid Value* under *Essential Oils and Flavors*, Appendix VI.
Angular Rotation Determine in a 100-mm tube as directed under *Optical (Specific) Rotation*, Appendix IIB.
Heavy Metals Shake 10 mL of the oil with an equal volume of water to which 1 drop of hydrochloric acid has been added, and pass hydrogen sulfide through the mixture until it is saturated. No darkening in color is produced in either the oil or the water.
Refractive Index, Appendix IIB Determine with an Abbé or other refractometer of equal or greater accuracy.
Saponification Value Determine as directed for *Saponification Value* under *Essential Oils and Flavors*, Appendix VI, using 1.5 g accurately weighed.
Solubility in Alcohol Proceed as directed in the general method, Appendix VI. One mL dissolves in 4 mL of 80% alcohol, sometimes with slight turbidity. The age of the oil has an adverse effect upon solubility.
Specific Gravity Determine by any reliable method (see *General Provisions*).

Packaging and Storage Store in full, tight, glass, aluminum, tin-lined, or other suitably lined containers in a cool place protected from light.

L-Lysine Monohydrochloride

2,6-Diaminohexanoic Acid Hydrochloride

$$NH_2(CH_2)_4 \underset{\underset{H}{|}}{\overset{\overset{NH_2}{|}}{C}}COOH \cdot HCl$$

$C_6H_{14}N_2O_2 \cdot HCl$ Formula wt 182.65

CAS: [657-27-2]

DESCRIPTION

A white or nearly white, practically odorless, free-flowing, crystalline powder. It is freely soluble in water, but is almost insoluble in alcohol and in ether. It melts at about 260° with decomposition.

Functional Use in Foods Nutrient; dietary supplement.

REQUIREMENTS

Identification
  A. Heat 5 mL of a 1 in 100 solution with 1 mL of ninhydrin TS. A violet color is produced.
  B. A 1 in 20 solution gives positive tests for *Chloride*, Appendix IIIA.
Assay Not less than 98.5% and not more than 101.5% of $C_6H_{14}N_2O_2 \cdot HCl$ after drying.
Heavy Metals (as Pb) Not more than 10 mg/kg.
Loss on Drying Not more than 1.0%.
Residue on Ignition Not more than 0.2%.
Specific Rotation $[\alpha]_D^{20°}$: Between +20.3° and +21.5° after drying; or $[\alpha]_D^{25°}$: Between +20.4° and +21.4° after drying.

TESTS

Assay Dissolve about 100 mg of the sample, previously dried at 105° for 3 h and accurately weighed, in 2 mL of formic acid, add exactly 15.0 mL of 0.1 N perchloric acid, and heat on a water bath for 30 min. After cooling, add 45 mL of glacial acetic acid, and titrate the excess perchloric acid with 0.1 N sodium acetate, determining the endpoint potentiometrically. Perform a blank determination, and make any necessary correction. Each mL of 0.1 N perchloric acid is equivalent to 9.133 mg of $C_6H_{14}N_2O_2 \cdot HCl$.
Heavy Metals Prepare and test a 2-g sample as directed in *Method II* under the *Heavy Metals Test*, Appendix IIIB, using 20 μg of lead ion (Pb) in the control (*Solution A*).
Loss on Drying, Appendix IIC Dry at 105° for 3 h.
Residue on Ignition Ignite 2 g as directed in the general method, Appendix IIC.
Specific Rotation, Appendix IIB Determine in a solution containing 8 g of a previously dried sample in sufficient 6 N hydrochloric acid to make 100 mL.

Packaging and Storage Store in well-closed containers.

Mace Oil

DESCRIPTION

The volatile oil obtained by steam distillation from the ground, dried arillode of the ripe seed of *Myristica fragrans* Houtt. (Fam. *Myristicaceae*). Two types of oil, the East Indian and the West Indian, are commercially available. It is a colorless to pale yellow liquid having the characteristic odor and taste of nutmeg. It is soluble in most fixed oils and in mineral oil, but it is insoluble in glycerin and in propylene glycol.

Functional Use in Foods Flavoring agent.

REQUIREMENTS

Labeling Indicate whether it is the East Indian or West Indian type.
Identification The infrared absorption spectrum of the sample exhibits relative maxima (that may vary in intensity) at the same wavelengths (or frequencies) as those shown in the respective spectrum in the section on *Infrared Spectra (Series A: Essential Oils)*, using the same test conditions as specified therein.
Angular Rotation *East Indian type*: between +2° and +30°; *West Indian type*: between +20° and +45°.
Heavy Metals (as Pb) Passes test.
Refractive Index *East Indian type*: between 1.474 and 1.488; *West Indian type*: between 1.469 and 1.480 at 20°.
Solubility in Alcohol Passes test.
Specific Gravity *East Indian type*: between 0.880 and 0.930; *West Indian type*: between 0.854 and 0.880.

TESTS

Angular Rotation Determine in a 100-mm tube as directed under *Optical (Specific) Rotation*, Appendix IIB.
Heavy Metals Shake 10 mL of the oil with an equal volume of water to which 1 drop of hydrochloric acid has been added, and pass hydrogen sulfide through the mixture until it is saturated. No darkening in color is produced in either the oil or the water.
Refractive Index, Appendix IIB Determine with an Abbé or other refractometer of equal or greater accuracy.
Solubility in Alcohol Proceed as directed in the general method, Appendix VI. One mL dissolves in 4 mL of 90% alcohol.
Specific Gravity Determine by any reliable method (see *General Provisions*).

Packaging and Storage Store in full, tight, glass, tin-lined, or other suitably lined containers in a cool place protected from light.

Magnesium Carbonate

INS: 504(i)   CAS: [39409-82-0]

DESCRIPTION

Magnesium Carbonate is a basic hydrated magnesium carbonate or a normal hydrated magnesium carbonate. It occurs as light, white, friable masses, or as a bulky, white powder. It is odorless, and is stable in air. It is practically insoluble in water, to which, however, it imparts a slightly alkaline reaction. It is insoluble in alcohol, but is dissolved by dilute acids with effervescence.

Functional Use in Foods Alkali; drying agent; color-retention agent; anticaking agent; carrier.

REQUIREMENTS

Identification When treated with 2.7 N hydrochloric acid, it dissolves with effervescence, and the resulting solution gives positive tests for *Magnesium*, Appendix IIIA.
Assay The equivalent of not less than 40.0% and not more than 43.5% of MgO.
Acid-Insoluble Substances Not more than 0.05%.
Calcium Oxide Not more than 0.6%.
Heavy Metals (as Pb) Not more than 0.002%.
Lead Not more than 10 mg/kg.
Soluble Salts Not more than 1%.

TESTS

Assay Dissolve about 1 g, accurately weighed, in 30.0 mL of 1 N sulfuric acid, add methyl orange TS, and titrate the excess acid with 1 N sodium hydroxide. From the volume of 1 N sulfuric acid consumed, deduct the volume of 1 N sulfuric acid corresponding to the content of calcium oxide in the weight of the sample taken for the assay. The difference is the volume of 1 N sulfuric acid equivalent to the magnesium oxide present. Each mL of 1 N sulfuric acid is equivalent to 20.16 mg of MgO and to 28.04 mg of CaO.
Acid-Insoluble Substances Mix 5.0 g with 75 mL of water, add hydrochloric acid in small portions, with agitation, until no more of the sample dissolves, and boil for 5 min. If an insoluble residue remains, filter through a suitable tared porous bottom porcelain crucible, wash well with water until the last washing is free from chloride, ignite at 800° ± 25° for 45 min, cool, and weigh.

Note: Avoid exposing the crucible to sudden temperature changes.

Calcium Oxide Dissolve about 1 g, accurately weighed, in a mixture of 3 mL of sulfuric acid and 22 mL of water. Add 50 mL of alcohol, and allow the mixture to stand overnight. If crystals of magnesium sulfate separate, warm the mixture to about 50° to dissolve them. Filter through a suitable tared porous bottom porcelain crucible, previously washed with 2 N sulfuric acid, water, and alcohol. Wash the crystals on the porous disk several times with a mixture of 2 volumes of alcohol and 1 volume of 2 N sulfuric acid. Ignite the crucible and contents at 450° ± 25° to constant weight. The weight of calcium sulfate so obtained, multiplied by 0.4119, gives the equivalent of calcium oxide in the sample taken for the test.

Note: Avoid exposing the crucible to sudden temperature changes.

Heavy Metals Dissolve 1 g in 15 mL of 2.7 N hydrochloric acid, and evaporate the solution to dryness on a steam bath. Toward the end of the evaporation, stir frequently to disintegrate the residue so that finally a dry powder is obtained. Dissolve the residue in 20 mL of water, and evaporate to dryness in the same manner as before. Redissolve the residue in 25 mL of water, and filter if necessary. This solution meets the requirements of the *Heavy Metals Test*, Appendix IIIB, using 20 μg of lead ion (Pb) in the control (*Solution A*).

Lead A solution of 1 g in 10 mL of 2.7 N hydrochloric acid meets the requirements of the *Lead Limit Test*, Appendix IIIB, using 10 μg of lead ion (Pb) in the control.

Soluble Salts Mix 2.0 g with 100 mL of a mixture of equal volumes of *n*-propyl alcohol and water. Heat the mixture to the boiling point with constant stirring, cool to room temperature, add water to make 100 mL, and filter. Evaporate 50 mL of the filtrate on a steam bath to dryness, and dry at 105° for 1 h. The weight of the residue does not exceed 10 mg.

Packaging and Storage Store in well-closed containers.

Heavy Metals (as Pb) Not more than 10 mg/kg.
Sulfate Not more than 0.03%.

TESTS

Assay Dissolve about 300 mg of the dihydrate, or about 450 mg of the hexahydrate, accurately weighed, in 25 mL of water, add 5 mL of ammonia–ammonium chloride buffer TS and 0.1 mL of eriochrome black TS, and titrate with 0.05 M disodium EDTA until the solution turns blue. Each mL of 0.05 M disodium EDTA is equivalent to 6.562 mg of $MgCl_2 \cdot 2H_2O$ or 10.16 mg of $MgCl_2 \cdot 6H_2O$.

Ammonium Dissolve 1 g in 90 mL of water, and slowly add 10 mL of a freshly boiled and cooled solution of sodium hydroxide (1 in 10). Allow to settle, then decant 20 mL of the supernatant liquid into a color comparison tube, dilute to 50 mL with water, and add 2 mL of Nessler's reagent. Any color does not exceed that produced by 10 μg of ammonium (NH_4) ion in 48 mL of water and 2 mL of the sodium hydroxide solution.

Heavy Metals A solution of 2 g in 25 mL of water meets the requirements of the *Heavy Metals Test*, Appendix IIIB, using 20 μg of lead ion (Pb) in the control (*Solution A*).

Sulfate, Appendix IIIB Any turbidity produced by a 1-g sample does not exceed that shown in a control containing 300 μg of sulfate (SO_4).

Packaging and Storage Store in tight containers.

Magnesium Chloride

$MgCl_2 \cdot xH_2O$ Formula wt, anhydrous 95.21

INS: 511 CAS: [7786-30-3]

DESCRIPTION

Magnesium Chloride contains two or six molecules of water of hydration. The dihydrate is in the form of small, white, hygroscopic granules. The hexahydrate is in the form of colorless, deliquescent flakes or crystals. Both forms are very soluble in water and freely soluble in alcohol.

Functional Use in Foods Color-retention agent; firming agent.

REQUIREMENTS

Labeling Indicate whether it is the dihydrate or the hexahydrate.

Identification A 1 in 10 solution gives positive tests for *Magnesium* and for *Chloride*, Appendix IIIA.

Assay *Dihydrate*: not less than 95.0% and not more than 100.5% of $MgCl_2 \cdot 2H_2O$; *hexahydrate*: not less than 99.0% and not more than 105.0% of $MgCl_2 \cdot 6H_2O$.

Ammonium Not more than 0.005%.

Magnesium Gluconate

$[CH_2OH(CHOH)_4COO]_2Mg$

$C_{12}H_{22}MgO_{14}$ Formula wt, anhydrous 414.60
$C_{12}H_{22}MgO_{14} \cdot 2H_2O$ Formula wt, dihydrate 450.63

INS: 580 CAS: anhydrous [3632-91-5]
  CAS: dihydrate [59625-89-7]

DESCRIPTION

A white to off-white powder or granulate. It is anhydrous, the dihydrate, or a mixture of both. It is very soluble in water and is sparingly soluble in alcohol. It is insoluble in ether.

Functional Use in Foods Nutrient; dietary supplement.

REQUIREMENTS

Identification

A. A 1 in 20 solution gives positive tests for *Magnesium*, Appendix IIIA.

B. It meets the requirements of *Identification Test B* in the monograph for *Calcium Gluconate*.

Assay Not less than 98.0% and not more than 102.0% of $C_{12}H_{22}MgO_{14}$, calculated on the anhydrous basis.
Heavy Metals Not more than 0.002%.
Lead Not more than 10 mg/kg.
Chloride Not more than 0.05%.
Sulfate Not more than 0.05%.
Reducing Substances Not more than 1.0%.
Water Between 3.0% and 12.0%.

TESTS

Assay Dissolve about 800 mg, accurately weighed, in 20 mL of water, add 5 mL of ammonia–ammonium chloride buffer TS and 0.1 mL of eriochrome black TS, and titrate with 0.05 M disodium EDTA to a blue endpoint. Each mL of 0.05 M disodium EDTA is equivalent to 20.73 mg of $C_{12}H_{22}MgO_{14}$.
Heavy Metals A solution of 1 g in 25 mL of water meets the requirements of the *Heavy Metals Test*, Appendix IIIB, using 20 μg of lead ion (Pb) in the control (*Solution A*).
Lead A solution of 1 g in 25 mL of water meets the requirements of the *Lead Limit Test*, Appendix IIIB, using 10 μg of lead ion (Pb) in the control.
Chloride Use the *Chloride and Sulfate Limit Test*, Appendix IIIB. Dissolve 1 g in 100 mL of water. Any turbidity produced by a 10-mL portion of this solution does not exceed that shown in a control containing 50 μg of chloride ion (Cl).
Sulfate Use the *Chloride and Sulfate Limit Test*, Appendix IIIB. Any turbidity produced by a 200-mg sample does not exceed that shown in a control containing 100 μg of sulfate (SO_4).
Reducing Substances Determine as directed under *Reducing Substances* in the monograph for *Copper Gluconate*.
Water Determine by the *Karl Fischer Titrimetric Method*, using *Method 1b (Residual Titration)*, Appendix IIB. Allow 30 min for solubilization of the sample, and perform a blank determination.

Packaging and Storage Store in well-closed containers.

Magnesium Hydroxide

$Mg(OH)_2$ Formula wt 58.32
INS: 528   CAS: [1309-42-8]

DESCRIPTION

A white, bulky powder. It dissolves in dilute acids, but is practically insoluble in water and in alcohol.

Functional Use in Foods Alkali; drying agent; color-retention agent.

REQUIREMENTS

Identification A 1 in 20 solution in 2.7 N hydrochloric acid gives positive tests for *Magnesium*, Appendix IIIA.
Assay Not less than 95.0% and not more than 100.5% of $Mg(OH)_2$ after drying.
Alkalies (Free) and Soluble Salts Passes test.
Calcium Oxide Not more than 1%.
Heavy Metals (as Pb) Not more than 0.002%.
Lead Not more than 10 mg/kg.
Loss on Drying Not more than 2.0%.
Loss on Ignition Between 30.0% and 33.0%.

TESTS

Assay Transfer about 400 mg, previously dried at 105° for 2 h and accurately weighed, into an Erlenmeyer flask. Add 25.0 mL of 1 N sulfuric acid, and after solution is complete, add methyl red TS and titrate the excess acid with 1 N sodium hydroxide. From the volume of 1 N sulfuric acid consumed, deduct the volume of 1 N sulfuric acid corresponding to the content of calcium oxide in the sample taken for the assay. The difference is the volume of 1 N sulfuric acid equivalent to the $Mg(OH)_2$ in the sample of Magnesium Hydroxide taken. Each mL of 1 N sulfuric acid is equivalent to 29.16 mg of $Mg(OH)_2$ and to 28.04 mg of CaO.
Alkalies (Free) and Soluble Salts Boil 2 g with 100 mL of water for 5 min in a covered beaker, then filter while hot. Titrate 50 mL of the cooled filtrate with 0.1 N sulfuric acid, using methyl red TS as the indicator. Not more than 2 mL of the acid is consumed. Evaporate 25 mL of the filtrate to dryness, and dry at 105° for 3 h. Not more than 10 mg of residue remains.
Calcium Oxide Dissolve about 500 mg, accurately weighed, in a mixture of 3 mL of sulfuric acid and 22 mL of water. Add 50 mL of alcohol, and allow the mixture to stand overnight. If crystals of magnesium sulfate separate, warm the mixture to about 50° to dissolve them. Filter through a suitable tared porous bottom porcelain crucible, previously washed with 2 N sulfuric acid, water, and alcohol. Wash the crystals on the porous disk several times with a mixture of 2 volumes of alcohol and 1 volume of 2 N sulfuric acid. Ignite the crucible and contents at 450° ± 25° to constant weight. The weight of calcium sulfate thus obtained, multiplied by 0.4119, gives the equivalent of calcium oxide in the sample taken for the test.

Note: Avoid exposing the crucible to sudden temperature changes.

Heavy Metals Dissolve 1 g in 10 mL of 2.7 N hydrochloric acid, and evaporate to dryness on a steam bath. Toward the end of the evaporation, stir the residue frequently, disintegrate it to obtain a dry powder, dissolve the powder in 25 mL of water, and filter. This solution meets the requirements of the *Heavy Metals Test*, Appendix IIIB, using 20 μg of lead ion (Pb) in the control (*Solution A*).
Lead A solution of 1 g in 20 mL of 2.7 N hydrochloric acid meets the requirements of the *Lead Limit Test*, Appendix IIIB, using 10 μg of lead ion (Pb) in the control.
Loss on Drying, Appendix IIC Dry at 105° for 2 h.

Loss on Ignition Transfer about 500 mg, accurately weighed, to a tared platinum crucible, and ignite, increasing the heat gradually, to constant weight at 800° ± 25°.

Packaging and Storage Store in tight containers.

Magnesium Oxide

MgO Formula wt 40.30
INS: 530 CAS: [1309-48-4]

DESCRIPTION

A very bulky, white powder known as Light Magnesium Oxide or a relatively dense, white powder known as Heavy Magnesium Oxide. Five g of Light Magnesium Oxide occupies a volume of approximately 40 to 50 mL, while 5 g of Heavy Magnesium Oxide occupies a volume of approximately 10 to 20 mL. It is practically insoluble in water and is insoluble in alcohol. It is soluble in dilute acids.

Functional Use in Foods Alkali; neutralizer.

REQUIREMENTS

Labeling Indicate whether it is Light Magnesium Oxide or Heavy Magnesium Oxide.
Identification A solution of Magnesium Oxide in 2.7 N hydrochloric acid gives positive tests for *Magnesium*, Appendix IIIA.
Assay Not less than 96.0% and not more than 100.5% of MgO after ignition.
Acid-Insoluble Substances Not more than 0.1%.
Alkalies (Free) and Soluble Salts Passes test.
Calcium Oxide Not more than 1.5%.
Heavy Metals (as Pb) Not more than 0.002%.
Lead Not more than 10 mg/kg.
Loss on Ignition Not more than 10.0%.

TESTS

Assay Ignite about 500 mg to constant weight at 800° ± 25° in a tared platinum crucible, weigh the residue accurately, dissolve it in 30.0 mL of 1 N sulfuric acid, boil gently to remove any carbon dioxide, cool, add methyl orange TS, and titrate the excess acid with 1 N sodium hydroxide. From the volume of 1 N sulfuric acid consumed deduct the volume of 1 N sulfuric acid corresponding to the content of calcium oxide in the Magnesium Oxide taken for the assay. The difference is the volume of 1 N sulfuric acid equivalent to the MgO in the portion of Magnesium Oxide taken. Each mL of 1 N sulfuric acid is equivalent to 20.15 mg of MgO and to 28.04 mg of CaO.
Acid-Insoluble Substances Mix 2.0 g with 75 mL of water, add hydrochloric acid in small portions, with agitation, until no more of the sample dissolves, and boil for 5 min. If an insoluble residue remains, filter through a suitable tared porous bottom porcelain crucible, wash well with water until the last washing is free from chloride, ignite at 800° ± 25° for 45 min, cool, and weigh.

Note: Avoid exposing the crucible to sudden temperature changes.

Alkalies (Free) and Soluble Salts Boil 2 g with 100 mL of water for 5 min in a covered beaker, and filter while hot. Add methyl red TS, and titrate 50 mL of the cooled filtrate with 0.1 N sulfuric acid. Not more than 2 mL of the acid is consumed. Evaporate 25 mL of the filtrate to dryness, and dry at 105° for 1 h. Not more than 10 mg of residue remains.
Calcium Oxide Dissolve about 400 mg, accurately weighed, in a mixture of 3 mL of sulfuric acid and 22 mL of water. Add 50 mL of alcohol, and allow the mixture to stand overnight. If crystals of magnesium sulfate separate, warm the mixture to about 50° to dissolve them. Filter through a suitable tared porous bottom porcelain crucible, previously washed with 2 N sulfuric acid, water, and alcohol. Wash the crystals on the porous disk several times with a mixture of 2 volumes of alcohol and 1 volume of 2 N sulfuric acid. Ignite the crucible and contents at 450° ± 25° to constant weight. The weight of calcium sulfate so obtained, multiplied by 0.4119, gives the equivalent of calcium oxide in the sample taken for the test.

Note: Avoid exposing the crucible to sudden temperature changes.

Heavy Metals Dissolve 1 g in 20 mL of 2.7 N hydrochloric acid, and evaporate the solution to dryness on a steam bath. Toward the end of the evaporation, stir frequently to disintegrate the residue so that finally a dry powder is obtained. Dissolve the residue in 20 mL of water and evaporate to dryness in the same manner as before. Redissolve the residue in 20 mL of water and filter if necessary. This solution meets the requirements of the *Heavy Metals Test*, Appendix IIIB, using 20 μg of lead ion (Pb) in the control (*Solution A*).
Lead A solution of 1 g in 20 mL of 2.7 N hydrochloric acid meets the requirements of the *Lead Limit Test*, Appendix IIIB, using 10 μg of lead ion (Pb) in the control.
Loss on Ignition Weigh accurately about 500 mg in a tared covered platinum crucible. Ignite at 800° ± 25° for 15 min, cool, and weigh.

Packaging and Storage Store in tight containers.

Magnesium Phosphate, Dibasic

Dimagnesium Phosphate

$MgHPO_4 \cdot 3H_2O$ Formula wt 174.33
INS: 343(ii) CAS: [7782-75-4]

DESCRIPTION

A white, odorless, crystalline powder. It is slightly soluble in water and insoluble in alcohol, but is soluble in dilute acids.

Functional Use in Foods Nutrient; dietary supplement; leavening agent; pH control agent.

REQUIREMENTS

Identification

A. Dissolve about 200 mg in 10 mL of 1.7 N nitric acid, and add, dropwise, ammonium molybdate TS. A greenish yellow precipitate of ammonium phosphomolybdate forms that is soluble in 6 N ammonium hydroxide.

B. Dissolve 100 mg in 0.5 mL of 1 N acetic acid and 20 mL of water. Add 1 mL of ferric chloride TS, let stand for 5 min, and filter. The filtrate gives a positive test for *Magnesium*, Appendix IIIA.

Assay Not less than 96.0% of $Mg_2P_2O_7$ after ignition.
Arsenic (as As) Not more than 3 mg/kg.
Fluoride Not more than 10 mg/kg.
Heavy Metals (as Pb) Not more than 15 mg/kg.
Lead Not more than 2 mg/kg.
Loss on Ignition Between 15.0% and 36.0%.

TESTS

Assay Weigh accurately about 500 mg of the residue obtained in the test for *Loss on Ignition*, and dissolve it by heating in a mixture of 50 mL of water and 2 mL of hydrochloric acid. Cool, dilute to 100.0 mL with water, and mix. Transfer 50.0 mL of this solution into a 400-mL beaker, add 100 mL of water, and heat to 55° to 60°. From a buret add 15 mL of 0.1 M disodium EDTA, add a magnetic stirring bar, and adjust with 1 N sodium hydroxide to pH 10 while stirring. Add 10 mL of ammonia–ammonium chloride buffer TS and 12 drops of eriochrome black TS, and continue the titration with 0.1 M disodium EDTA until the wine-red color changes to pure blue. Calculate the weight, in mg, of $Mg_2P_2O_7$ in the residue taken by the formula

$$2 \times 11.13 \times V,$$

in which V is the volume, in mL, of 0.1 M disodium EDTA required in the titration of the 50.0-mL aliquot.

Arsenic A solution of 1 g in 5 mL of 2.7 N hydrochloric acid meets the requirements of the *Arsenic Test*, Appendix IIIB.

Fluoride Determine as directed in *Method III* under *Fluoride Limit Test*, Appendix IIIB (except in the *Procedure* use 10 mL of 1 N hydrochloric acid to dissolve the sample), or use the following procedure: Transfer 5.0 g of the sample into a 200-mL distilling flask connected with a condenser and carrying a thermometer and a dropping funnel equipped with a stopcock. Dissolve the sample in 25 mL of dilute sulfuric acid (1 in 4), add 6 glass beads, and connect the apparatus for distillation, using a 600-mL beaker to collect the distillate. Add 40 mL of the dilute sulfuric acid to the flask through the dropping funnel, then fill the funnel with water, heat the solution to boiling, and continue heating until the temperature reaches 165°. Adjust the stopcock of the dropping funnel so that the temperature is maintained at 165° ± 5°, and continue the distillation until about 300 mL has been collected. Rinse the condenser and condenser arm with water, collecting the rinsings in the beaker. Add 1 N sodium hydroxide to the distillate to make it alkaline to litmus paper, and then add 5 mL in excess. Add 5 mL of 30% hydrogen peroxide and 6 glass beads to the beaker, boil until a volume of about 30 mL is reached, and cool. Transfer the condensed distillate, including the glass beads, into a 125-mL distilling flask connected with a condenser and carrying a thermometer and a capillary tube, both of which must extend into the liquid. Add 30 mL of perchloric acid, and continue as directed under the *Fluoride Limit Test, Method I*, Appendix IIIB, beginning with "Connect a small dropping funnel or a steam generator to the capillary tube. . . ."

Heavy Metals, Appendix IIIB Suspend 2.66 g in 20 mL of water, and add hydrochloric acid, dropwise, until the sample just dissolves. Adjust the pH to between 3 and 4, filter, and dilute the filtrate to 40 mL with water. For the control (*Solution A*), add 20 µg of lead ion (Pb) to 10 mL of the filtrate, and dilute to 40 mL. For the sample (*Solution B*), dilute the remaining 30 mL of the filtrate to 40 mL. Add 10 mL of hydrogen sulfide TS to each solution, and allow to stand for 5 min. *Solution B* is no darker than *Solution A*.

Lead A 10-g sample meets the requirements of the *APDC Extraction Method for Lead*, Appendix IIIB.

Loss on Ignition Weigh accurately about 1 g, and ignite, preferably in a muffle furnace, at 800° ± 25° to constant weight.

Packaging and Storage Store in well-closed containers.

Magnesium Phosphate, Tribasic

Trimagnesium Phosphate

$Mg_3(PO_4)_2 \cdot xH_2O$ Formula wt, anhydrous 262.86

INS: 334(iii)   CAS: [7757-87-1]

DESCRIPTION

Tribasic Magnesium Phosphate may contain four, five, or eight molecules of water of hydration. It occurs as a white, odorless, tasteless, crystalline powder. It is readily soluble in dilute mineral acids, but is almost insoluble in water.

Functional Use in Foods Nutrient; dietary supplement.

REQUIREMENTS

Identification

A. Dissolve about 200 mg in 10 mL of 1.7 N nitric acid, and add, dropwise, ammonium molybdate TS. A greenish yellow precipitate of ammonium phosphomolybdate forms that is soluble in 6 N ammonium hydroxide.

B. Dissolve 100 mg in 0.7 mL of 1 N acetic acid and 20 mL of water. Add 1 mL of ferric chloride TS, let stand for 5 min, and filter. The filtrate gives a positive test for *Magnesium*, Appendix IIIA.

Assay Not less than 98.0% and not more than 101.5% of $Mg_3(PO_4)_2$, calculated on the ignited basis.

Arsenic (as As) Not more than 3 mg/kg.
Fluoride Not more than 10 mg/kg.
Heavy Metals (as Pb) Not more than 0.003%.
Lead Not more than 5 mg/kg.
Loss on Heating $Mg_3(PO_4)_2.4H_2O$: between 15.0% and 23.0%; $Mg_3(PO_4)_2.5H_2O$: between 20.0% and 27.0%; $Mg_3(PO_4)_2.8H_2O$: between 30.0% and 37.0%.

TESTS

Assay Weigh accurately about 200 mg of the sample, and dissolve it in a mixture of 25 mL of water and 10 mL of 1.7 N nitric acid. Filter, if necessary, wash any precipitate, then dissolve the precipitate by the addition of 1 mL of 1.7 N nitric acid. Adjust the temperature to about 50°, add 75 mL of ammonium molybdate TS, and maintain the temperature at about 50° for 30 min, stirring occasionally. Allow to stand for 16 h or overnight at room temperature. Wash the precipitate once or twice with water by decantation, using from 30 to 40 mL each time, and pour these two washings through a filter. Transfer the precipitate to the same filter, and wash with potassium nitrate solution (1 in 100) until the last washing is not acid to litmus paper. Transfer the precipitate and filter to the precipitation vessel, add 50.0 mL of 1 N sodium hydroxide, agitate until the precipitate is dissolved, add 3 drops of phenolphthalein TS, and then titrate the excess alkali with 1 N sulfuric acid. Each mL of 1 N sodium hydroxide is equivalent to 5.714 mg of $Mg_3(PO_4)_2$.
Arsenic A solution of 1 g in 10 mL of 2.7 N hydrochloric acid meets the requirements of the *Arsenic Test*, Appendix IIIB.
Fluoride Determine as directed in the *Fluoride Limit Test* under *Fluoride* in the monograph for *Magnesium Phosphate, Dibasic.*
Heavy Metals, Appendix IIIB Suspend 1.33 g in 20 mL of water, and add hydrochloric acid, dropwise, until the sample just dissolves. Adjust the pH to between 3 and 4, filter, and dilute the filtrate to 40 mL with water. For the control (*Solution A*), add 20 μg of lead ion (Pb) to 10 mL of the filtrate, and dilute to 40 mL. For the sample (*Solution B*), dilute the remaining 30 mL of the filtrate to 40 mL. Add 10 mL of hydrogen sulfide TS to each solution, and allow to stand for 5 min. *Solution B* is no darker than *Solution A*.
Lead Dissolve 1 g in 20 mL of 2.7 N hydrochloric acid, evaporate the solution to a volume of about 10 mL on a steam bath, dilute to about 20 mL with water, and cool. This solution meets the requirements of the *Lead Limit Test*, Appendix IIIB, using 5 μg of lead ion (Pb) in the control.
Loss on Heating Weigh accurately about 1 g, and heat at about 425° to constant weight.

Packaging and Storage Store in well-closed containers.

Magnesium Silicate

Synthetic Magnesium Silicate

INS: 553   CAS: [1343-88-0]

DESCRIPTION

A synthetic, usually amorphous form of magnesium silicate in which the molar ratio of silicon dioxide to magnesium oxide is approximately 2.2 to 3.3. It occurs as a very fine, white, odorless, tasteless powder, free from grittiness. It is insoluble in water and in alcohol, but is readily decomposed by mineral acids. The pH of a 1 in 10 slurry is between 7.0 and 10.8.

Functional Use in Foods Anticaking agent; filter aid.

REQUIREMENTS

Identification
A. Mix about 500 mg with 10 mL of 2.7 N hydrochloric acid, filter, and neutralize the filtrate to litmus paper with 6 N ammonium hydroxide. The neutralized filtrate responds to the tests for *Magnesium*, Appendix IIIA.
B. Prepare a bead by fusing a few crystals of sodium ammonium phosphate on a platinum loop in the flame of a Bunsen burner. Place the hot, transparent bead in contact with a sample, and again fuse. Silica floats about in the bead, producing, upon cooling, an opaque bead with a weblike structure.
Assay Not less than 15.0% of MgO and not less than 67.0% of SiO_2, calculated on the ignited basis.
Fluoride Not more than 10 mg/kg.
Free Alkali (as NaOH) Not more than 1%.
Heavy Metals (as Pb) Not more than 0.002%.
Lead Not more than 10 mg/kg.
Loss on Drying Not more than 15.0%.
Loss on Ignition Not more than 15.0%, determined on a dried sample.
Soluble Salts Not more than 3.0%.

TESTS

Assay for Magnesium Oxide Weigh accurately about 1.5 g, and transfer it into a 250-mL conical flask. Add 50.0 mL of 1 N sulfuric acid, and digest on a steam bath for 1 h. Cool to room temperature, add methyl orange TS, and titrate the excess acid with 1 N sodium hydroxide. Each mL of 1 N sulfuric acid is equivalent to 20.15 mg of MgO.
Assay for Silicon Dioxide Transfer about 700 mg, accurately weighed, into a 150-mL beaker. Add 20 mL of 1 N sulfuric acid, and heat on a steam bath for 1 h and 30 min. Decant the supernatant liquid through an ashless filter paper, and wash the residue, by decantation, three times with hot water. Treat the residue with 25 mL of water and digest on a steam bath for 15 min. Finally, transfer the residue to the filter and wash thoroughly with hot water. Transfer the filter paper and its contents to a platinum crucible. Heat to dryness, incinerate, then ignite strongly for 30 min, cool, and weigh. Moisten the residue with

water, and add 6 mL of hydrofluoric acid and 3 drops of sulfuric acid. Evaporate to dryness, ignite for 5 min, cool, and weigh. The loss in weight represents the weight of SiO_2.

Fluoride Determine as directed under *Fluoride* in the monograph for *Calcium Silicate*.

Free Alkali Add 2 drops of phenolphthalein TS to 20 mL of diluted filtrate prepared in the test for *Soluble Salts*, representing 1 g of magnesium silicate. If a pink color is produced, not more than 2.5 mL of 0.1 N hydrochloric acid is required to discharge it.

Sample Solution for the Determination of Heavy Metals and Lead Transfer 10.0 g of the sample into a 250-mL flask, and add 50 mL of 0.5 N hydrochloric acid. Attach a reflux condenser to the flask, heat on a steam bath for 30 min, cool, and let the undissolved material settle. Decant the supernatant liquid through Whatman No. 3, or equivalent, filter paper, into a 100-mL volumetric flask, retaining as much as possible of the insoluble material in the beaker. Wash the slurry and beaker with three 10-mL portions of hot water, decanting each washing through the filter into the flask. Finally, wash the filter paper with 15 mL of hot water, cool the filtrate to room temperature, dilute to volume with water, and mix.

Heavy Metals A 10-mL portion of the *Sample Solution* diluted to 25 mL with water meets the requirements of the *Heavy Metals Test*, Appendix IIIB, using 20 μg of lead ion (Pb) in the control.

Lead A 10-mL portion of the *Sample Solution* meets the requirements of the *Lead Limit Test*, Appendix IIIB, using 10 μg of lead ion (Pb) in the control.

Loss on Drying, Appendix IIC Dry at 105° for 2 h. Retain the sample for determination of *Loss on Ignition*.

Loss on Ignition Ignite the sample, retained from the test for *Loss on Drying*, at 900° to 1000° for 20 min.

Soluble Salts Boil 10 g with 150 mL of water for 15 min. Cool to room temperature, and add water to restore the original volume. Allow the mixture to stand for 15 min, and filter until clear. To 75 mL of the clear filtrate add 25 mL of water. Evaporate 50 mL of this solution, representing 2.5 g of magnesium silicate, in a tared platinum dish on a steam bath to dryness, and ignite gently to constant weight. The weight of the residue does not exceed 75 mg.

Packaging and Storage Store in well-closed containers.

Magnesium Stearate

CAS: [557-04-0]

DESCRIPTION

Magnesium Stearate is a compound of magnesium with a mixture of solid organic acids obtained from edible sources and consists chiefly of variable proportions of Magnesium Stearate and magnesium palmitate. It occurs as a fine, white, bulky powder having a faint, characteristic odor. It is unctuous and is free from grittiness. It is insoluble in water, in alcohol, and in ether.

Functional Use in Foods Anticaking agent; binder; emulsifier.

REQUIREMENTS

Identification

A. Heat 1 g with a mixture of 25 mL of water and 5 mL of hydrochloric acid. Fatty acids are liberated, floating as an oily layer on the surface of the liquid. The water layer gives positive tests for *Magnesium*, Appendix IIIA.

B. Mix 25 g of the sample with 200 mL of hot water, then add 60 mL of 2 N sulfuric acid, and heat the mixture, with frequent stirring, until the fatty acids separate cleanly as a transparent layer. Wash the fatty acids with boiling water until free from sulfate, collect them in a small beaker, and warm on a steam bath until the water has separated and the fatty acids are clear. Allow the acids to cool, pour off the water layer, then melt the acids, filter into a dry beaker, and dry at 105° for 20 min. The solidification point of the fatty acids so obtained is not below 54° (see Appendix IIB).

Assay Not less than 6.8% and not more than 8.3% of MgO.
Heavy Metals (as Pb) Not more than 0.002%.
Lead Not more than 10 mg/kg.
Loss on Drying Not more than 4.0%.

TESTS

Assay Boil about 1 g of the sample, accurately weighed, with 50 mL of 0.1 N hydrochloric acid for about 30 min, or until the separated fatty acid layer is clear, adding water if necessary to maintain the original volume. Cool, filter, and wash the filter and the container thoroughly with water until the last washing is not acid to litmus. Neutralize the filtrate with 1 N sodium hydroxide to litmus. While stirring, preferably with a magnetic stirrer, titrate with 0.05 M disodium EDTA as follows: Add about 30 mL of the titrant from a 50-mL buret, then add 5 mL of ammonia–ammonium chloride buffer TS and 0.15 mL of eriochrome black TS, and continue the titration to a blue endpoint. Each mL of 0.05 M disodium EDTA is equivalent to 2.015 mg of MgO.

Heavy Metals, Appendix IIIB Place 1.5 g of the sample in a porcelain dish, place 500 mg of the sample in a second dish for the control, and to each add 5 mL of a 1 in 4 solution of magnesium nitrate in alcohol. Cover the dishes with 7.6-cm short-stem funnels so that the stems are straight up. Heat for 30 min on a hot plate at the low setting, then heat for 30 min at the medium setting, and cool. Remove the funnels, add 20 μg of lead ion (Pb) to the control, and heat each dish over an Argand burner until most of the carbon is burned off. Cool, add 10 mL of nitric acid, and transfer the solutions into 250-mL beakers. Add 5 mL of 70% perchloric acid, evaporate to dryness, then add 2 mL of hydrochloric acid to the residues, and wash down the inside of the beakers with water. Evaporate carefully to dryness again, swirling near the dry point to avoid spattering. Repeat the hydrochloric acid treatment, then cool, and dissolve the residues in about 10 mL of water. To each solution add 1 drop of phenolphthalein TS and sufficient 1 N sodium hydroxide until the solutions just turn pink, and then add 2.7 N hydrochloric acid until the solutions become colorless. Add 1 mL of 1 N acetic acid and a small amount of charcoal to each solution, and filter through Whatman No. 2, or equivalent, filter paper into 50-mL Nessler tubes. Wash with water, dilute to 40 mL, and add 10 mL of hydrogen sulfide TS to each tube. The color in the solution of the sample does not exceed that produced in the control.

Lead, Appendix IIIB Ignite 500 mg in a silica crucible in a muffle furnace at 475° to 500° for 15 to 20 min. Cool, add 3 drops of nitric acid, evaporate over a low flame to dryness, and re-ignite at 475° to 500° for 30 min. Dissolve the residue in 1 mL of a mixture of equal parts by volume of nitric acid and water, and wash into a separator with several successive portions of water. Add 3 mL of *Ammonium Citrate Solution* and 0.5 mL of *Hydroxylamine Hydrochloride Solution*, and make alkaline to phenol red TS with ammonium hydroxide. Add 10 mL of *Potassium Cyanide Solution*. Immediately extract the solution with successive 5-mL portions of *Dithizone Extraction Solution*, draining off each extract into another separator, until the last portion of dithizone solution retains its green color. Shake the combined extracts for 30 s with 20 mL of dilute nitric acid (1 in 100), and discard the chloroform layer. Add to the acid solution exactly 4 mL of *Ammonia–Cyanide Solution* and 2 drops of *Hydroxylamine Hydrochloride Solution*. Add 10 mL of *Standard Dithizone Solution*, and shake the mixture for 30 s. Filter the chloroform layer through an acid-washed filter paper into a Nessler tube, and compare the color with that of a standard prepared as follows: To 20 mL of dilute nitric acid (1 in 100), add 5 μg of lead ion (Pb), 4 mL of *Ammonia–Cyanide Solution*, and 2 drops of *Hydroxylamine Hydrochloride Solution*, and shake for 30 s with 10 mL of *Standard Dithizone Solution*. Filter through an acid-washed filter paper into a Nessler tube. The color of the sample solution does not exceed that in the control.

Loss on Drying, Appendix IIC Dry at 105° to constant weight, using 2-h increments of heating.

Packaging and Storage Store in well-closed containers.

Magnesium Sulfate

Epsom Salt

MgSO$_4 \cdot x$H$_2$O   Formula wt, anhydrous 120.36

INS: 518   CAS: [7487-88-9]

DESCRIPTION

Magnesium Sulfate is produced with one or seven molecules of water of hydration, or in a dried form containing the equivalent of about 2.3 waters of hydration. It is in the form of a colorless crystal or a granular crystalline powder. It is odorless. It is readily soluble in water, slowly soluble in glycerine, and sparingly soluble in alcohol.

Functional Use in Foods Nutrient; dietary supplement.

REQUIREMENTS

Labeling Indicate whether it is the monohydrate, the dried form, or the heptahydrate.
Identification A 1 in 20 solution gives positive tests for *Magnesium* and for *Sulfate*, Appendix IIIA.
Assay Not less than 99.5% of MgSO$_4$ after ignition.
Heavy Metals (as Pb) Not more than 10 mg/kg.
Loss on Ignition Between 13.0% and 16.0% for the monohydrate. Between 22.0% and 28.0% for the dried form. Between 40.0% and 52.0% for the heptahydrate.
Selenium Not more than 0.003%.

TESTS

Assay Weigh accurately about 500 mg of the residue obtained in the test for *Loss on Ignition*, dissolve it in a mixture of 50 mL of water and 1 mL of hydrochloric acid, dilute to 100.0 mL with water, and mix. Transfer 50.0 mL of this solution into a 250-mL Erlenmeyer flask, add 10 mL of ammonia–ammonium chloride buffer TS and 12 drops of eriochrome black TS, and titrate with 0.1 M disodium EDTA until the wine-red color changes to pure blue. Each mL of 0.1 M disodium EDTA is equivalent to 12.04 mg of MgSO$_4$.
Heavy Metals A solution of 2 g in 25 mL of water meets the requirements of the *Heavy Metals Test*, Appendix IIIB, using 20 μg of lead ion (Pb) in the control (*Solution A*).
Loss on Ignition Weigh accurately about 1 g in a crucible, heat at 105° for 2 h, then ignite in a muffle furnace at 450° ± 25° to constant weight.
Selenium Determine as directed in *Method II* under the *Selenium Limit Test*, Appendix IIIB, using 200 mg of sample.

Packaging and Storage Store in well-closed containers.

Malic Acid

DL-Malic Acid; Hydroxysuccinic Acid

$$\begin{array}{c} \text{HOCHCOOH} \\ | \\ \text{CH}_2\text{COOH} \end{array}$$

C$_4$H$_6$O$_5$ Formula wt 134.09

INS: 296   CAS: [6915-15-7]

DESCRIPTION

White or nearly white, crystalline powder or granules having a strongly acid taste. One g dissolves in 0.8 mL of water and in

1.4 mL of alcohol. Its solutions are optically inactive. It melts at about 130°.

Functional Use in Foods Acidifier; flavoring agent.

REQUIREMENTS

Identification The infrared absorption spectrum of a potassium bromide dispersion of the sample exhibits maxima only at the same wavelengths as that of a similar preparation of USP Malic Acid Reference Standard.
Assay Not less than 99.0% and not more than 100.5% of $C_4H_6O_5$.
Fumaric Acid Not more than 1.0%.
Heavy Metals (as Pb) Not more than 10 mg/kg.
Maleic Acid Not more than 0.05%.
Residue on Ignition Not more than 0.1%.
Specific Rotation $[\alpha]_D^{25°}$: Between −0.10° and +0.10°.
Water-Insoluble Matter Not more than 0.1%.

TESTS

Assay Dissolve about 2 g of the sample, accurately weighed, in 40 mL of recently boiled and cooled water, add phenolphthalein TS, and titrate with 1 N sodium hydroxide to the first appearance of a faint pink color that persists for at least 30 s. Each mL of 1 N sodium hydroxide is equivalent to 67.04 mg of $C_4H_6O_5$.

Fumaric and Maleic Acids

Mobile Phase Prepare a filtered, degassed solution of 0.01 N sulfuric acid in water solution.

Note: For all *Reference Standards*, do not dry before use, and keep the containers tightly closed and protected from light. Determine the water content titrimetrically of the *USP Fumaric Acid Reference Standard* before use, and make the necessary correction in preparing the *Standard Preparation*.

Standard Preparation Transfer about 5 mg of *USP Fumaric Acid Reference Standard* and about 2 mg of *USP Maleic Acid Reference Standard*, both accurately weighed, to a 1000-mL volumetric flask, dilute to volume with *Mobile Phase*, and mix.

Test Preparation Transfer about 100 mg of the sample, accurately weighed, to a 100-mL volumetric flask, dilute to volume with *Mobile Phase*, and mix.

Resolution Solution Transfer about 1 g of the sample, about 10 mg of *USP Fumaric Acid Reference Standard*, and about 4 mg of *USP Maleic Acid Reference Standard* to a 1000-mL volumetric flask, dilute to volume with *Mobile Phase*, and mix.

Chromatographic System Use a liquid chromatograph equipped with a 210-nm detector and a 6.5-mm × 30-cm column packed with a strong cation exchange resin consisting of sulfonated cross-linked styrene–divinylbenzene copolymer in the hydrogen form (Polypore H from Brownlee Lab, or equivalent). Maintain the temperature of the column at 37° ± 1°. Use a flow rate of about 0.6 mL/min. Chromatograph the *Resolution Solution*, and record the peak responses: The resolution, R, of the maleic acid and Malic Acid peaks is not less than 2.5; the resolution, R, of the Malic Acid and fumaric acid peaks is not less than 7.0; and the relative standard deviation of the sample solution peak for replicate injections is not more than 2.0%.

Procedure Separately inject about 20 μL of the *Standard Preparation* and then the *Test Preparation* into the chromatograph, record the chromatograms, and measure the peak responses. The relative retention times are approximately 0.6 for maleic acid, approximately 1.0 for Malic Acid, and approximately 1.5 for fumaric acid. Calculate the quantities, in mg, of maleic acid and fumaric acid, in the portion of the sample taken by the formula

$$100C(r_U/r_S),$$

in which C is the concentration, in mg/mL, of the corresponding *Reference Standard* in the *Standard Preparation*, and r_U and r_S are the responses of the corresponding peaks from the *Test Preparation* and the *Standard Preparation*, respectively.

Heavy Metals A solution of 2 g in 25 mL of water meets the requirements of the *Heavy Metals Test*, Appendix IIIB, using 20 μg of lead ion (Pb) in the control (*Solution A*).

Residue on Ignition Ignite 2 g as directed in the general method, Appendix IIC.

Specific Rotation, Appendix IIB Dissolve 850 mg of the sample in water, and dilute to 10 mL at 25°.

Water-Insoluble Matter Dissolve 25 g in 100 mL of water, and filter through a tared, sintered-glass filter crucible of suitable porosity. Wash the filter with hot water, dry at 100° to constant weight, cool, and weigh.

Packaging and Storage Store in well-closed containers.

Maltitol Syrup

Hydrogenated Glucose Syrup

INS: 965

DESCRIPTION

Maltitol Syrup is a water solution of a hydrogenated, partially hydrolyzed starch containing maltitol, sorbitol, and hydrogenated oligo- and polysaccharides. It is a clear, colorless, syrupy liquid having a sweet taste. It is very soluble in water and slightly soluble in alcohol.

Functional Use in Foods Humectant; texturizing agent; stabilizer; sweetener.

REQUIREMENTS

Identification Prepare a mixture of n-propyl alcohol, ethyl acetate, and water (70:20:10) as the *Developing Solvent*. Dissolve USP Maltitol Reference Standard in water to obtain a *Standard Solution* having a concentration of 2.5 mg/mL. Prepare the *Test Solution* by diluting Maltitol Syrup with water to obtain a solution containing, on the anhydrous basis, about 2.5 mg of Maltitol per mL. Apply separately 2 µL each of the *Standard Solution* and the *Test Solution* to a thin-layer chromatographic plate coated with a 0.25-mm layer of chromatographic silica gel. Allow the spots to dry, and develop the plate in a developing chamber containing the *Developing Solvent* until the solvent front has moved about 17 cm. Remove the plate from the chamber, mark the solvent front, and allow the solvent to evaporate. Spray the plate with sodium metaperiodate solution (1 in 500), air-dry for 15 min, and spray with a 1 in 50 solution of 4,4′-tetramethyldiaminodiphenylmethane in a mixture of acetone and glacial acetic acid (4:1): The principal spot obtained from the *Test Solution* corresponds in R_f value and color to that obtained from the *Standard Solution*.

Assay Not less than 50.0%, by weight, of D-maltitol as $C_{12}H_{24}O_{11}$, calculated on the anhydrous basis.

Chloride Not more than 0.005%.

Heavy Metals (as Pb) Not more than 10 mg/kg.

Lead Not more than 1 mg/kg.

Reducing Sugars Not more than 0.30% (as glucose).

Residue on Ignition Not more than 0.1%.

Sulfate Not more than 0.01%.

Water Not more than 31.0%.

TESTS

Assay

Mobile Phase Use degassed water.

Standard Preparation Dissolve an accurately weighed quantity of USP Maltitol Reference Standard in water, and dilute quantitatively with water to obtain a solution having a known concentration of about 10.0 mg/mL.

Assay Preparation Transfer 1 g of Maltitol Syrup, accurately weighed, to a 50-mL volumetric flask, dilute with water to volume, and mix.

Chromatographic System The liquid chromatograph is equipped with a refractive index detector that is maintained at a constant temperature and a 9-mm × 30-cm column packed with a strong cation-exchange resin, about 9 µm in diameter, consisting of sulfonated cross-linked styrene–divinylbenzene copolymer in the calcium form (Aminex HPX-87c or equivalent). The column temperature is maintained at 85° ± 0.5°, and the flow rate of the *Mobile Phase* is about 0.5 mL/min. Chromatograph the *Standard Preparation*, and record the peak responses as directed under *Procedure*. Replicate injections show a relative standard deviation not greater than 2.0%.

Procedure Separately inject suitable portions (about 20 µL) of the *Assay Preparation* and the *Standard Preparation* into the chromatograph, record the chromatograms, and measure the responses for the major peaks. The elution pattern includes the higher molecular weight hydrogenated polysaccharides, followed by three individual peaks representing maltotriitol, maltitol, and sorbitol. The principal peak is maltitol, which elutes at about twice the retention time of the void volume, and the retention time for sorbitol is about 1.7 relative to maltitol. Calculate the quantity, in mg, of maltitol in the portion of Syrup taken by the formula

$$50C(r_U/r_S),$$

in which C is the concentration, in mg/mL, of USP Maltitol Reference Standard in the *Standard Preparation*, and r_U and r_S are the peak responses of Maltitol obtained from the *Assay Preparation* and the *Standard Preparation*, respectively.

Chloride, Appendix IIIB Any turbidity produced by a 10.0-g sample does not exceed that shown in a control containing 1.5 mL of 0.01 N hydrochloric acid.

Heavy Metals A solution of 2 g in 25 mL of water meets the requirements of the *Heavy Metals Test*, Appendix IIIB, using 20 µg of lead ion (Pb) in the control (*Solution A*).

Lead Determine as directed for *Flame Atomic Absorption Spectrophotometric Method* under the *Lead Limit Test*, Appendix IIIB, using a 10-g sample.

Reducing Sugars Dissolve 7 g of the sample in 35 mL of water in a 400-mL beaker and mix. Add 25 mL of cupric sulfate TS and 25 mL of alkaline tartrate TS. Cover the beaker with a watch glass, heat the mixture at such a rate that it comes to a boil in approximately 4 min and boils for exactly 2 min. Filter the precipitated cuprous oxide through a tared, sintered-glass filter crucible previously washed with hot water, ethanol, and ether and dried at 100° for 30 min. Thoroughly wash the collected cuprous oxide on the filter with hot water, then with 10 mL of ethanol and finally with 10 mL of ether, and dry at 100° for 30 min. The weight of the cuprous oxide does not exceed 50 mg.

Residue on Ignition, Appendix IIC Ignite 2 g as directed under *Method II* (for liquids).

Sulfate, Appendix IIIB Any turbidity produced by a 10.0-g sample does not exceed that shown in a control containing 2.0 mL of 0.01 N sulfuric acid.

Water Determine as directed under *Karl Fischer Titrimetric Method*, Appendix IIB.

Packaging and Storage Store in well-closed containers.

Maltodextrin

CAS: [9050-36-6]

DESCRIPTION

Maltodextrin is a purified, concentrated, nonsweet, nutritive mixture of saccharide polymers obtained by the partial hydrolysis of edible starch. It occurs as a white, slightly hygroscopic powder, as granules of similar description, or as a clear to hazy solution in water. Powders or granules are freely soluble or readily dispersible in water.

Functional Use in Foods Anticaking and free-flowing agent; formulation aid; processing aid; bulking agent; stabilizer and thickener; surface-finishing agent.

REQUIREMENTS

Labeling Indicate the presence of sulfur dioxide if the residual concentration is greater than 10 mg/kg.

Identification To 5 mL of hot alkaline cupric tartrate TS add a few drops of a 1 in 10 solution of the sample. A red precipitate of cuprous oxide forms.

Assay Less than 20.0% reducing sugar content (dextrose equivalent) expressed as D-glucose.

Heavy Metals (as Pb) Not more than 5 mg/kg.

Lead Not more than 0.5 mg/kg.

Protein (Total) Not more than 0.5%, except not more than 1.0% in maltodextrins produced from high-amylose starches.

Residue on Ignition Not more than 0.5%.

Sulfur Dioxide Not more than 0.0025%.

Total Solids *Powders and Granules*: Not less than 90.0%; *Liquids*: Not less than 50.0%.

TESTS

Assay Determine as directed in the *Reducing Sugars Assay*, Appendix X.

Heavy Metals A solution of 2 g in 25 mL of water meets the requirements of the *Heavy Metals Test*, Appendix IIIB, using 10 μg of lead ion (Pb) in the control (*Solution A*).

Lead Determine as directed under *Method I* in the *Atomic Absorption Spectrophotometric Graphite Furnace Method* under the *Lead Limit Test*, Appendix IIIB, using a 5-g sample.

Protein (Total) Determine as directed under *Nitrogen Determination (Kjeldahl Method)*, Appendix IIIC. Protein content is $N \times 6.25$.

Residue on Ignition, Appendix IIC Ignite 1 g as directed in the general method.

Sulfur Dioxide Determine as directed in the general method, Appendix X.

Total Solids

Powders and Granules Prepare an approximate 60% solution by dissolving 60 g of Maltodextrin, accurately weighed, in water to a final total weight of 100 g. Heat the sample slightly, if necessary, to get the solution. Determine the refractive index of this solution at 20° or 45°, and use the tables for *Maltodextrin*, Appendix X, to obtain the percent *Total Solids* for the prepared solution. Calculate the *Total Solids* of the sample taken by the formula

$$(TS \times W_T)/W_S,$$

in which TS is the percent *Total Solids* for the prepared solution, W_T is the total weight of the prepared solution, and W_S is the weight of the sample taken.

Liquids Determine the refractive index of a sample of Maltodextrin at 20° or 45°, and use the tables for *Maltodextrin*, Appendix X, to obtain the percent *Total Solids*.

Packaging and Storage Keep dry, and store at ambient temperatures.

Maltol

3-Hydroxy-2-methyl-4-pyrone

$C_6H_6O_3$ Formula wt 126.12

INS: 636 CAS: [118-71-8]

DESCRIPTION

A white, crystalline powder having a characteristic caramel-butterscotch odor, and suggestive of a fruity-strawberry aroma in dilute solution. One g dissolves in about 82 mL of water, in 21 mL of alcohol, in 80 mL of glycerin, and in 28 mL of propylene glycol.

Functional Use in Foods Flavoring agent.

REQUIREMENTS

Identification A 1 in 100,000 solution in 0.1 N hydrochloric acid exhibits an absorbance maximum at 274 ± 2 nm.

Assay Not less than 99.0% of $C_6H_6O_3$, calculated on the anhydrous basis.

Heavy Metals (as Pb) Not more than 0.002%.

Lead Not more than 10 mg/kg.

Melting Range Between 160° and 164°.

Residue on Ignition Not more than 0.2%.

Water Not more than 0.5%.

TESTS

Assay

Standard Solution Weigh accurately about 50 mg of FCC Maltol Reference Standard, dissolve it in sufficient 0.1 N hydrochloric acid to make 250.0 mL, and mix. Transfer 5.0 mL of this solution into a 100-mL volumetric flask, dilute to volume with 0.1 N hydrochloric acid, and mix.

Assay Solution Weigh accurately about 50 mg of the sample, dissolve it in sufficient 0.1 N hydrochloric acid to make 250.0 mL, and mix. Transfer 5.0 mL of this solution into a 100-mL volumetric flask, dilute to volume with 0.1 N hydrochloric acid, and mix.

Procedure Determine the absorbance of each solution in a 1-cm quartz cell at the wavelength of maximum absorption at about 274 nm, with a suitable spectrophotometer, using 0.1 N hydrochloric acid as the blank. Calculate the quantity, in mg, of $C_6H_6O_3$ in the sample taken by the formula

$$5C(A_U/A_S),$$

in which C is the concentration, in μg/mL, of FCC Maltol Reference Standard in the *Standard Solution*; A_U is the absorbance of the *Assay Solution*; and A_S is the absorbance of the *Standard Solution*.

Heavy Metals Prepare and test a 1-g sample as directed in *Method II* under the *Heavy Metals Test*, Appendix IIIB, using 20 μg of lead ion (Pb) in the control (*Solution A*).

Lead A *Sample Solution* prepared as directed for organic compounds meets the requirements of the *Lead Limit Test*, Appendix IIIB, using 10 μg of lead ion (Pb) in the control.

Melting Range Determine as directed for *Melting Range or Temperature, Procedure for Class Ia*, Appendix IIB.

Residue on Ignition Ignite 1 g as directed in the general method, Appendix IIC.

Water Determine by the *Karl Fischer Titrimetric Method*, Appendix IIB.

Packaging and Storage Store in tight containers.

Malt Syrup

Malt Extract

CAS: [8002-48-0]

DESCRIPTION

Malt is the product of barley (*Hordeum vulgare L.*) germinated under controlled conditions. Malt Syrup and Malt Extract are interchangeable terms for a concentrate of the water extract of germinated barley grain with or without added food-grade preservatives. Malt Syrup is usually a yellow to brown, sweet, and viscous liquid containing varying amounts of amylolytic enzymes and plant constituents.

Functional Use in Foods Color; enzyme; flavoring agent; humectant; nutritive sweetener; stabilizer; thickener; and texturizer.

REQUIREMENTS

Identification To 5 mL of hot alkaline cupric tartrate TS add a few drops of a 1 in 10 solution of the sample. A red precipitate of cuprous oxide is formed.

Assay Not less than 40.0% and not more than 65.0% of reducing sugar content expressed as maltose.

Heavy Metals Not more than 5 mg/kg.

Lead Not more than 0.5 mg/kg.

N-Nitrosodimethylamine Not more than 0.005 mg/kg.

pH Between 4.5 and 5.5.

Protein Not more than 7.0%.

Sulfur Dioxide Not more than 10 mg/kg.

Total Solids Between 77.0% and 83.0%.

TESTS

Assay

Sample Solution Transfer about 5 g, accurately weighed, of the sample to a 500-mL volumetric flask, dilute with water, and mix. (It is not necessary to remove any protein before analysis.) Pipet 10.0 mL each of *Fehling Solutions A* and *B* into a 250-mL flask. Add 20.0 mL (choose the size of the aliquot so that the sample titration will be about half that of the blank titration) of the *Sample Solution*, add water to make a total volume of 50 mL, and mix the contents of the flask by gentle swirling. Add two small glass beads, and close the mouth of the flask with a small funnel or glass bulb. Heat the solution, preferably on a hot plate, at such a rate that the solution is brought to boiling within 3 min, and then continue boiling for exactly 2 min (total heating time, 5 min). Cool quickly to room temperature in an ice bath or with cold running water, and then rinse down the funnel (or bulb) and the walls of the flask with a few mL of water. Add 10 mL each of 30% potassium iodide solution and 28% sulfuric acid, and titrate rapidly with 0.1 *N* sodium thiosulfate until the iodine color almost disappears. Add 1 mL of starch TS, and titrate dropwise, with continuous agitation, to the disappearance of the blue color. Record the volume, in mL, of 0.1 *N* sodium thiosulfate required for the *Sample Solution* as *S*. Conduct two reagent blank determinations, substituting 20 mL (or the same volume as the aliquot of the *Sample Solution* taken) of water for the sample, and record the average volume, in mL, of the blanks as *B*. Obtain the *Titer Difference*, expressed as mL of 0.1 *N* sodium thiosulfate, for the sample by subtracting *S* from *B* and recording the value thus obtained as T_S.

Calculation By reference to the following table, determine the weight, in mg, of reducing sugars (as maltose) equivalent to the volume T_S, and record the value thus obtained as W_S. Calculate the total reducing sugars (as maltose), in mg, in the sample taken by the formula

$$25 W_S.$$

(If the aliquot of the *Sample Solution* taken for assay differs from 20.0 mL, adjust the factor accordingly.)

Conversion of Titer Differences to Reducing Sugars Content

Titer Diff. (mL)	.0	.1	.2	.3	.4	.5	.6	.7	.8	.9
				Reducing Sugars (as Maltose) (mg)						
5.0	27.0	27.6	28.1	28.7	29.2	29.8	30.3	30.9	31.4	32.0
6.0	32.5	33.1	33.6	34.2	34.7	35.3	35.8	36.3	36.9	37.5
7.0	38.0	38.6	39.1	39.7	40.2	40.8	41.3	41.9	42.4	43.0
8.0	43.5	44.1	44.6	45.2	45.7	46.3	46.8	47.4	47.9	48.5
9.0	49.0	49.6	50.2	50.8	51.4	52.0	52.6	53.2	53.8	54.4
10.0	55.0	55.6	56.1	56.7	57.2	57.8	58.3	58.9	59.4	60.0
11.0	60.5	61.1	61.6	62.2	62.7	63.3	63.8	64.4	64.9	65.5
12.0	66.0	66.6	67.2	67.8	68.4	69.0	69.6	70.2	70.8	71.4
13.0	72.0	72.6	73.2	73.8	74.4	75.0	75.6	76.2	76.8	77.4
14.0	78.0	78.6	79.1	79.7	80.2	80.8	81.3	81.9	82.4	83.0
15.0	83.5	84.1	84.6	85.2	85.7	86.3	86.8	87.4	87.9	88.5
16.0	89.0	89.6	90.2	90.8	91.4	92.0	92.6	93.2	93.8	94.4
17.0	95.0	95.6	96.2	96.8	97.4	98.0	98.6	99.2	99.8	100.4
18.0	101.0	101.6	102.2	102.8	103.4	104.0	104.6	105.2	105.8	106.4
19.0	107.0	107.6	108.1	108.7	109.2	109.8	110.3	110.9	111.4	112.0
20.0	112.5	113.1	113.7	114.3	114.9	115.5	116.1	116.7	117.3	117.9

Heavy Metals Prepare and test a 4-g sample as directed in *Method II* under the *Heavy Metals Test*, Appendix IIIB, using 20 μg of lead ion (Pb) in the control (*Solution A*).

Lead Determine as directed under *Method I* in the *Atomic Absorption Spectrophotometric Graphite Furnace Method* under the *Lead Limit Test*, Appendix IIIB, using a 2-g sample.

N-Nitrosodimethylamine (Based on AOAC method 982.12.) (**Caution**: *N*-Nitrosamines are potent carcinogens: Take adequate precaution to avoid exposure. Carry out all steps in a well-ventilated fume hood, and wear protective gloves while handling nitrosamine standards. Because these compounds are highly photolabile, carry out all procedures under subdued light. Do not pipet solutions by mouth, and do not use the same pipet for other reagents. Destroy all nitrosamine solutions by boiling with hydrochloric acid, potassium iodide, and sulfamic acid before disposal.) (**Note**: Thoroughly clean all glassware before use. After normal cleaning and washing, wash with chromic acid. If contamination still exists, rinse all glassware with dichloromethane before use. Let the charred residue in the distillation flask soak with diluted alkali, and then wash in a normal manner.)

Apparatus Set up the distillation apparatus consisting of a 1000-mL distillation flask, a heating mantle, an adapter, and a 200-mm Graham condenser (Kontes, or equivalent) so that the connecting adapter slopes downward toward the vertical Graham condenser. Loosely wrap glass wool around the distillation flask and connecting adapter. Set up a 100-mL graduate under the condenser to collect the distillate. The cooling water for the condenser should be ≤20°. Assemble a 250-mL Kuderna-Danish evaporative concentrator that has a 24/40 column connection and a 19/22 lower joint (Kontes, or equivalent) with a 4-mL Kuderna-Danish concentrator tube (Kontes, or equivalent) that has a 19/22 joint and 0.1-mL subdivisions from 0 to 2.0 mL at the bottom.

Use a gas chromatograph fitted with a 1.8-m × 3-mm (od) stainless steel column packed with 20% Carbowax 20M (or equivalent) and 2% sodium hydroxide on 80- to 100-mesh, acid-washed Chromosorb P (or equivalent). Set the temperature of the injector port and column to 220° and 170°, respectively, and use argon as the carrier gas at a flow rate of 25 to 30 mL/min. Use a Thermal Energy Analyzer detector (Thermo Electron Corporation, Waltham, MA, or equivalent). Operate according to the instrument manual and with a −110° to −130° slush bath. Adjust instrumental parameters, such as vacuum chamber pressure, oxygen flow, calibration knob, etc., to obtain the proper sensitivity.

NDMA Standard Solution Using an accurately weighed quantity of *N*-nitrosodimethylamine (NDMA) (Sigma Chemical Company, St. Louis, MO), prepare a stock solution in dichloromethane having a concentration of 1 mg/mL. By serial dilution with dichloromethane, prepare a series of solutions containing 500, 200, 100, 40, 20, 10, and 5 ng/mL. Store these solutions at −20°, and warm to room temperature before use. After 30 days, dispose of the *Standard Solutions*.

NDPA Standard Solution As described above, prepare a solution containing 250 ng of *N*-nitrosodi-*n*-propylamine (NDPA) per mL of anhydrous ethanol.

Sample Preparation Transfer 50 g of the sample, accurately weighed, into a 1000-mL distillation flask, and add 1.0 mL of 10% sulfamic acid, 1.0 mL of *NDPA Standard Solution*, 1.0 mL of 1 *N* hydrochloric acid, and 15 mL of water. Mix the contents by gentle swirling, and let the flask stand in the dark for 10 min. Add 10.0 mL of 3 *N* potassium hydroxide and two small boiling chips, mix, and connect the flask to the distillation apparatus.

Distillation During the initial 10 min of distillation, adjust the heating mantle so the mixture boils smoothly without too much frothing or bumping. Watch constantly for excessive foaming, and if necessary, turn off the heat for 1 to 2 min. After 10 min, increase the temperature, and continue distillation (watch for foaming) until approximately 55 mL of distillate is collected. Do not boil the distilling flask to complete dryness; this may give erroneous results. Total distillation time should be ≤1 h. If any portion of the sample foams over during distillation, discontinue the distillation, and start over with a fresh sample of Malt Syrup.

Add 2.0 mL of 10 *N* potassium hydroxide to the distillate in the graduate, and transfer to a 250-mL separatory funnel. Use the same cylinder for all subsequent measuring of dichloromethane. Rinse the condenser with 50 mL of dichloromethane, and collect the rinsing directly into the separatory funnel containing the distillate and potassium hydroxide. Extract the distillate with dichloromethane by shaking vigorously for 2 min. Drain off the dichloromethane layer (lower) into a second separatory funnel. Extract the aqueous layer with two additional 50-mL portions of dichloromethane, and combine all dichloromethane extracts in the second separatory funnel. Discard the aqueous layer.

Place 40 g of anhydrous sodium sulfate in a coarse, sintered-glass Büchner funnel, wash with about 20 mL of dichloromethane, and discard the washing. Dry the combined dichloromethane extract by passing it through a sodium sulfate bed on the Büchner funnel, and collect the extract directly in the Kuderna-Danish concentrator. Wash the sodium sulfate bed with an additional 20 mL of dichloromethane, and collect the washing in the Kuderna-Danish concentrator.

Add a 1- to 2-mm boiling chip to the contents of the Kuderna-Danish concentrator, attach a three-section Snyder column with three chambers and a 24/40 joint (Kontes, or equivalent), and concentrate the extract by heating the flask in a 50° to 60° water bath. Initially maintain the outside water level close to the level of dichloromethane inside the flask, and continue heating until the concentrated extract is about 4 mL (about 40 min). (If excessive boiling occurs during concentration, control it either by raising the flask slightly out of the water bath or by decreasing the bath temperature.) Finally, raise the flask above the water, and let condensed dichloromethane in the Snyder column drain into the flask. Add about 1 mL of dichloromethane to the top of the Snyder column, and let it drain into the flask. Disconnect the concentrator tube from the flask.

Add another boiling chip to the contents and attach a micro Snyder column with three chambers and a 19/22 joint (Kontes or equivalent) to the concentrator tube. Concentrate the extract to about 0.8 mL by heating the concentrator tube in a 50° to 60° water bath. Lift out or immerse the tube in water to control the boiling rate, but do not lift the tube completely out of the water bath; this will stop the action of the boiling chip. Avoid overheating and excessive accumulation of dichloromethane in the column chambers. Stop concentration when the dichloromethane level reaches 0.8 mL; do not concentrate to less than 0.8 mL. Carry out this final concentrating step slowly, taking at least 30 min. Raise the tube, and with the bottom still touching the water, let the liquid drain, and note the volume to see if it is around 0.8 mL. If it is greater than 0.8 mL, continue the concentration as above. Finally, rinse the micro Snyder column with a few drops of dichloromethane, let the rinsing drain to the tube, disconnect the column, and dilute the extract to 1.0 or 1.1 mL, but not greater than 1.1 mL. (Do not use a nitrogen stream for concentrating the extract at any stage.)

Stopper the tube, mix in a vortex mixer, and store at 4° in the dark until analysis. Let the extract warm to room temperature, and note its volume before analyzing it.

To ensure the absence of contamination, carry the reagent blank taken through all of the steps mentioned above, except use 50 mL of 4% alcohol in water instead of 50 g of the sample.

Procedure Set attenuation (usually 4) of the thermal energy analyzer detector so that injection of 30 pg of NDMA gives a definite peak with acceptable background. Using this attenuation, analyze 5- to 6-μL aliquots, in duplicate, of *NDMA Standard Solutions* of 5, 10, 20, and 40 ng/mL. (Note the volume injected.)

Next, choose a higher attenuation setting that gives an on-scale peak for 6 μL of *NDMA Standard Solution* at 500 ng/mL. Using this setting, analyze 6-μL aliquots, in duplicate, of *NDMA Standard Solutions* of 500, 200, 100, and 40 ng/mL.

Accurately measure the peak heights (±0.1 cm), and determine the average peak heights of two injections at each concentration. If exactly 6 μL are not injected, make appropriate corrections, and convert all peak heights equivalent to 5.0-μL injections. Draw two standard curves, one for each attenuation setting, of peak heights versus pg injection.

As above, inject a 6-μL aliquot of sample extract, in duplicate, using the lowest attenuation setting sensitive to 30 pg of NDMA. Measure and determine the average peak height. Compare the sample response with the standard curve that produces the closest peak height at the same attenuation. Choose the *NDMA Standard Solution* that gives the closest peak height, inject 6-μL aliquots, in duplicate, and determine the average peak height.

For samples giving off-scale peaks at an attenuation of 32, dilute the extracts with dichloromethane to 5.0 mL in a volumetric flask, and reanalyze. For accurate results, analyze the sample extract and corresponding standard under the same attenuation setting, and all within 60 min.

If the extract gives a negative result for NDMA, or if the peak is too small to measure, inject 10-μL aliquots, in duplicate, using a 25-μL syringe. Similarly, inject duplicate 10-μL aliquots of the 5 ng/mL *NDMA Standard Solution* for quantitation. To achieve a 0.1-μg/kg detection limit, 10-μL aliquots of *Sample Preparation* must be analyzed under an attenuation setting that gives a detectable peak corresponding to 30 pg of NDMA.

Note: If using a 25-μL syringe, which usually has a thick needle, watch for septum damage, and check for leaks. To be on the safe side, use a new septum daily.

Calculate the concentration of NDMA in the sample using the following formula:

Uncorrected NDMA, in μg/kg, in the sample = $(H_1 P V_2)/(H_2 G V_1)$,

in which H_1 is the average NDMA peak height, in cm, of the sample; H_2 is the average peak height, in cm, of the corresponding *NDMA Standard Solution*; P is the weight, in pg, of NDMA producing the H_2 peak height; V_1 is the volume, in μL, of *Sample Preparation* injected; V_2 is the final volume, in mL, of *Sample Preparation*; and G is the weight, in g, of the sample taken for analysis.

Correction for Percent Recovery of NDPA Accurately measure the peak height of the NDPA peak on each sample chromatogram and calculate the average peak height of two injections. Make appropriate corrections if the final volume of sample is not exactly 1.0 mL or the injection volume is not exactly 6.0 μL. Then inject, in duplicate, within 60 min, 6-μL of the *NDPA Standard Solution* under the same attenuation setting. Calculate the average peak height, and correct the value if exactly 6.0 μL is not injected. Calculate the percent recovery of NDPA for each sample. If recovery of NDPA is less than 80%, repeat the analysis from the beginning. Finally, correct the results as follows:

Corrected NDMA, in mg/kg, in the sample = (uncorrected μg/kg/% recovery of NDPA) × 0.1.

pH Determine by the potentiometric method, Appendix IIB, using a 1 in 10 aqueous solution.

Protein Using a 0.25-g sample, determine the percent of nitrogen as directed under *Nitrogen Determination*, Appendix IIIC. The percent of nitrogen, multiplied by 6.25, gives the percent of protein in the sample.

Sulfur Dioxide Determine as directed in the general method, Appendix X, using a 100-g sample.

Total Solids Determine the water content of an accurately weighed portion of the sample by the *Karl Fischer Titrimetric Method*, Appendix IIB. Calculate the percent of *Total Solids* by the formula

$$100[(W_U - W_W)/W_U],$$

in which W_U is the weight, in mg, of the sample, and W_W is the weight, in mg, of water determined.

Packaging and Storage Store in tight containers.

Mandarin Oil, Coldpressed

Mandarin Oil, Expressed

DESCRIPTION

The oil obtained by expression of the peels of the ripe fruit of the mandarin orange, *Citrus reticulata* Blanco var. *Mandarin*. It is a clear, dark orange to reddish yellow, or brownish orange liquid with a pleasant, orangelike odor. It often shows a bluish fluorescence in diffused light. Oils produced from unripe fruit often show a green color. It is soluble in most fixed oils and in mineral oil. It is slightly soluble in propylene glycol, but it is insoluble in glycerin. It may contain a suitable antioxidant.

Functional Use in Foods Flavoring agent.

REQUIREMENTS

Identification The infrared absorption spectrum of the sample exhibits relative maxima (that may vary in intensity) at the same wavelengths (or frequencies) as those shown in the respective spectrum in the section on *Infrared Spectra (Series A: Essential Oils)*, using the same test conditions as specified therein.
Assay Not less than 0.4% and not more than 1.8% of aldehydes, calculated as decyl aldehyde ($C_{10}H_{20}O$).
Angular Rotation Between +63° and +78°.
Heavy Metals (as Pb) Not more than 0.004%.
Lead Not more than 10 mg/kg.
Refractive Index Between 1.473 and 1.477 at 20°.
Residue on Evaporation Between 2.0% and 5.0%.
Specific Gravity Between 0.846 and 0.852.

TESTS

Assay Weigh accurately about 10 g, and proceed as directed under *Aldehydes and Ketones—Hydroxylamine/Tert-Butyl Alcohol Method*, Appendix VI, using 156.26 as the equivalence factor (*e*) in the calculation. Allow the mixture to stand for 30 min at room temperature before titrating.
Angular Rotation Determine in a 100-mm tube as directed under *Optical (Specific) Rotation*, Appendix IIB.
Heavy Metals Prepare and test a 500-mg sample as directed in *Method II* under the *Heavy Metals Test*, Appendix IIIB, using 20 µg of lead ion (Pb) in the control (*Solution A*).
Lead A *Sample Solution* prepared as directed for organic compounds meets the requirements of the *Lead Limit Test*, Appendix IIIB, using 10 µg of lead ion (Pb) in the control.
Refractive Index, Appendix IIB Determine with an Abbé or other refractometer of equal or greater accuracy.
Residue on Evaporation Proceed as directed in the general method, Appendix VI, using about 5 g, accurately weighed, and heating for 5 h.
Specific Gravity Determine by any reliable method (see *General Provisions*).

Packaging and Storage Store in full, tight, preferably glass, tin-lined, galvanized, or other suitably lined containers in a cool place protected from light.

Manganese Chloride

$MnCl_2 \cdot 4H_2O$ Formula wt 197.91

CAS: [7773-01-5]

DESCRIPTION

Large, irregular, pink, translucent crystals. It is freely soluble in water at room temperature and very soluble in hot water.

Functional Use in Foods Nutrient; dietary supplement.

REQUIREMENTS

Identification A 1 in 20 solution gives positive tests for *Manganese* and for *Chloride*, Appendix IIIA.
Assay Not less than 98.0% and not more than 102.0% of $MnCl_2 \cdot 4H_2O$.
Heavy Metals (as Pb) Not more than 10 mg/kg.
Insoluble Matter Not more than 0.005%.
Iron Not more than 5 mg/kg.
pH of a 5% Solution Between 4.0 and 6.0.
Substances Not Precipitated by Sulfide Not more than 0.2%, after ignition.
Sulfate Not more than 0.005%.

TESTS

Assay Transfer about 4 g, accurately weighed, into a 250-mL volumetric flask, dissolve in water, dilute to volume with water, and mix. Transfer 25.0 mL of this solution into a 400-mL beaker, and add 10 mL of a 1 in 10 solution of hydroxylamine hydrochloride, 25 mL of 0.05 *M* disodium EDTA measured from a buret, 25 mL of ammonia–ammonium chloride buffer TS, and 5 drops of eriochrome black TS. Heat the solution to between 55° and 65°, and titrate from the buret to a blue endpoint. Each mL of 0.05 *M* disodium EDTA is equivalent to 9.896 mg of $MnCl_2 \cdot 4H_2O$.
Heavy Metals A solution of 2 g in 25 mL of water meets the requirements of the *Heavy Metals Test*, Appendix IIIB, using 20 µg of lead ion (Pb) in the control (*Solution A*).
Insoluble Matter Dissolve about 20 g, accurately weighed, in 200 mL of water, and allow to stand on a steam bath for 1 h. Filter through a tared sintered-glass crucible, wash thoroughly with hot water, dry at 105° for 1 h, cool, and weigh.
Iron Dissolve 2.0 g in 20 mL of water, add 1 mL of hydrochloric acid, and dilute to 50 mL with water. Add about 40 mg of ammonium persulfate crystals and 3 mL of ammonium thiocyanate TS. Any red or pink color does not exceed that produced by 1.0 mL of *Iron Standard Solution* (10 µg Fe) in an equal volume of a solution containing the quantities of the reagents used in the test.
pH of a 5% Solution Determine by the *Potentiometric Method*, Appendix IIB.
Substances Not Precipitated by Sulfide Dissolve 2.0 g in about 90 mL of water, add 4 mL of ammonium hydroxide, heat to 80°, and pass hydrogen sulfide through the solution to completely precipitate the manganese. Dilute to 100 mL, mix, and allow the precipitate to settle. Decant the supernatant liquid through a filter, and evaporate 50 mL of the filtrate to dryness in a tared dish. Add 0.5 mL of sulfuric acid, ignite to constant weight, cool, and weigh.
Sulfate Dissolve 10.0 g in 100 mL of water, add 1 mL of 2.7 *N* hydrochloric acid, mix, and filter. Heat to boiling, then add 10 mL of barium chloride TS, and allow to stand overnight. Filter out any precipitate in a tared crucible, wash, ignite gently, cool, and weigh. The weight of the ignited precipitate should not be more than 1.2 mg greater than the weight obtained in a complete blank test.

Packaging and Storage Store in well-closed containers.

Manganese Gluconate

[CH$_2$OH(CHOH)$_4$COO]$_2$Mn.2H$_2$O

C$_{12}$H$_{22}$MnO$_{14}$.2H$_2$O Formula wt, dihydrate 481.27
Formula wt, anhydrous 445.24

CAS: [6485-39-8]

DESCRIPTION

A slightly pink-colored powder. Manganese Gluconate is anhydrous when obtained by spray-drying or the dihydrate when obtained by crystallization. It is very soluble in hot water and is very slightly soluble in alcohol.

Functional Use in Foods   Nutrient; dietary supplement.

REQUIREMENTS

Labeling   Indicate whether it has been obtained through spray-drying or from crystallization.
Identification
   A. A 1 in 20 solution gives positive tests for *Manganese*, Appendix IIIA.
   B. It meets the requirements of *Identification Test B* in the monograph for *Calcium Gluconate*.
Assay   Not less than 98.0% and not more than 102.0% of C$_{12}$H$_{22}$MnO$_{14}$, calculated on the anhydrous basis.
Heavy Metals (as Pb)   Not more than 0.002%.
Lead   Not more than 10 mg/kg.
Reducing Substances   Not more than 1.0%.
Water   Between 3.0% and 9.0% for dried (anhydrous) form obtained from spray-drying and between 6.0% and 9.0% for the dihydrate obtained by crystallization.

TESTS

Assay   Dissolve about 700 mg, accurately weighed, in 50 mL of water, add 1 g of ascorbic acid, 10 mL of ammonia–ammonium chloride buffer TS, and 5 drops of eriochrome black TS, and titrate with 0.05 M disodium EDTA to a deep blue color. Each mL of 0.05 M disodium EDTA is equivalent to 22.26 mg of C$_{12}$H$_{22}$MnO$_{14}$.
Heavy Metals   A solution of 1 g in 25 mL of water meets the requirements of the *Heavy Metals Test*, Appendix IIIB, using 20 μg of lead ion (Pb) in the control (*Solution A*).
Lead   A solution of 1 g in 25 mL of water meets the requirements of the *Lead Limit Test*, Appendix IIIB, using 10 μg of lead ion (Pb) in the control.
Reducing Substances   Determine as directed under *Reducing Substances* in the monograph for *Copper Gluconate*.
Water   Determine by the *Karl Fischer Titrimetric Method*, Appendix IIB, the determination being performed by stirring the mixture containing the test preparation, maintained at a temperature of 50°, for 30 min before titrating with the reagent.

Packaging and Storage   Store in well-closed containers.

Manganese Glycerophosphate

C$_3$H$_7$MnO$_6$P.xH$_2$O   Formula wt 225.00

CAS: [1320-46-3]

DESCRIPTION

A white or pinkish white powder. It is odorless and nearly tasteless. One g dissolves in about 5 mL of citric acid solution (1 in 4). It is slightly soluble in water, and is insoluble in alcohol.

Functional Use in Foods   Nutrient; dietary supplement.

REQUIREMENTS

Identification   A 1 in 20 solution in 2.7 N hydrochloric acid gives positive tests for *Manganese*, Appendix IIIA.
Assay   Not less than 98.0% and not more than 100.5% of C$_3$H$_7$MnO$_6$P after drying.
Heavy Metals (as Pb)   Not more than 0.002%.
Lead   Not more than 10 mg/kg.
Loss on Drying   Not more than 12.0%.

TESTS

Assay   Dissolve about 1 g, previously dried at 110° to constant weight and accurately weighed, in 1.5 mL of nitric acid and 5 mL of warm water. Dilute to 125 mL, add 2.0 g of dibasic ammonium phosphate and a few drops of methyl red TS, and heat to boiling. While the solution is boiling, slowly add ammonium hydroxide, dropwise and with constant stirring, until alkaline, and then add 2.0 mL in excess. Let stand 2 h at room temperature. Filter through a tared, porous-bottom porcelain filter crucible, and wash the precipitate with dilute ammonia (1 in 100). Dry at 105°, and ignite to constant weight at 800° ± 25°. Each g of manganese pyrophosphate so obtained is equivalent to 1.585 g of C$_3$H$_7$MnO$_6$P.

   Note: Avoid exposing the crucible to sudden temperature changes.

Heavy Metals   Dissolve 1 g in 5 mL of 2.7 N hydrochloric acid, and dilute to 25 mL with water. This solution meets the requirements of the *Heavy Metals Test*, Appendix IIIB, using 20 μg of lead ion (Pb) in the control (*Solution A*).
Lead   Mix 1 g with 3 mL of dilute nitric acid (1 in 2) and 10 mL of water, and boil until brown fumes appear. Add 10 mL of water, boil for 2 min, cool, and dilute to 100 mL with water. A 25-mL portion of this solution meets the requirements of the *Lead Limit Test*, Appendix IIIB, using 25 mL of *Ammonium Citrate Solution*, 1 mL of *Potassium Cyanide Solution*, 0.5 mL of *Hydroxylamine Hydrochloride Solution*, and 2.5 μg of lead ion (Pb).
Loss on Drying, Appendix IIC   Dry at 110° to constant weight.

Packaging and Storage   Store in well-closed containers.

Manganese Hypophosphite

Mn(H$_2$PO$_2$)$_2 \cdot x$H$_2$O   Formula wt 184.92

CAS: [10043-84-2]

DESCRIPTION

A pink, granular or crystalline powder that is stable in air. It is odorless and nearly tasteless. One g dissolves in about 6.5 mL of water at 25° or in about 6 mL of boiling water. It is soluble in alcohol.

Caution: Mix Manganese Hypophosphite with nitrates, chlorates, or other oxidizing agents very carefully, as an explosion may occur if it is triturated or heated.

Functional Use in Foods   Nutrient; dietary supplement.

REQUIREMENTS

Identification   A 1 in 20 solution gives positive tests for *Manganese* and for *Hypophosphite*, Appendix IIIA.
Assay   Not less than 97.0% and not more than 100.5% of Mn(PH$_2$O$_2$)$_2$ after drying.
Heavy Metals (as Pb)   Not more than 0.002%.
Lead   Not more than 10 mg/kg.
Loss on Drying   Not more than 9.0%.

TESTS

Assay   Transfer about 120 mg, previously dried at 105° for 1 h and accurately weighed, into a 100-mL volumetric flask, dissolve in water, and dilute with water to volume. Transfer 50.0 mL of this solution into a 250-mL glass-stoppered iodine flask, add 50.0 mL of 0.1 N bromine and 20 mL of 2 N sulfuric acid, and stopper the flask. Place a few mL of a saturated solution of potassium iodide in the lip around the stopper, shake the flask well, and allow to stand for 3 h. Place the flask in an ice bath for 5 min, then carefully remove the stopper and allow the potassium iodide solution to be drawn into the flask. Add 2 g of potassium iodide dissolved in 10 mL of recently boiled water, shake the flask, and titrate the liberated iodine with 0.1 N sodium thiosulfate, using starch TS as the indicator. Each mL of 0.1 N bromine is equivalent to 2.311 mg of Mn(PH$_2$O$_2$)$_2$.
Heavy Metals   Dissolve 1 g in 15 mL of water, add 3 mL of nitric acid, evaporate to dryness on a steam bath, and dissolve the residue in 25 mL of water. This solution meets the requirements of the *Heavy Metals Test*, Appendix IIIB, using 20 µg of lead ion (Pb) in the control (*Solution A*).
Lead   Dissolve 250 mg in 10 mL of water, add 2 mL of dilute nitric acid (1 in 2), and boil until brown fumes appear. Add 10 mL of water, boil for 2 min, then cool and dilute with water to about 25 mL. This solution meets the requirements of the *Lead Limit Test*, Appendix IIIB, using 25 mL of *Ammonium Citrate Solution*, 1 mL of *Potassium Cyanide Solution*, 0.5 mL of *Hydroxylamine Hydrochloride Solution*, and 2.5 µg of lead ion (Pb).

Loss on Drying, Appendix IIC   Dry at 105° for 1 h.

Packaging and Storage   Store in well-closed containers.

Manganese Sulfate

MnSO$_4 \cdot$H$_2$O   Formula wt 169.02

CAS: [7785-87-7]

DESCRIPTION

A pale pink, granular, odorless powder. It is freely soluble in water and is insoluble in alcohol.

Functional Use in Foods   Nutrient; dietary supplement.

REQUIREMENTS

Identification   A 1 in 10 solution gives positive tests for *Manganese* and for *Sulfate*, Appendix IIIA.
Assay   Not less than 98.0% and not more than 102.0% of MnSO$_4 \cdot$H$_2$O.
Heavy Metals (as Pb)   Not more than 0.002%.
Lead   Not more than 10 mg/kg.
Loss on Heating   Between 10.0% and 13.0%.
Selenium   Not more than 0.003%.

TESTS

Assay   Transfer about 4 g, accurately weighed, into a 250-mL volumetric flask, dissolve in water, dilute to volume with water, and mix. Transfer a 25.0-mL portion of this solution into a 400-mL beaker, and add 10 mL of a 1 in 10 solution of hydroxylamine hydrochloride, 25 mL of 0.05 M disodium EDTA measured from a buret, 25 mL of ammonia–ammonium chloride buffer TS, and 5 drops of eriochrome black TS. Heat the solution to between 55° and 65°, and titrate from the buret to a blue endpoint. Each mL of 0.05 M disodium EDTA is equivalent to 8.450 mg of MnSO$_4 \cdot$H$_2$O.
Heavy Metals   A solution of 1 g in 25 mL of water meets the requirements of the *Heavy Metals Test*, Appendix IIIB, using 20 µg of lead ion (Pb) in the control (*Solution A*).
Lead   Dissolve 1 g in 3 mL of dilute nitric acid (1 in 2) and 10 mL of water, and boil for 2 min. Cool, and dilute to 100 mL with water. A 25-mL portion of this solution meets the requirements of the *Lead Limit Test*, Appendix IIIB, using 25 mL of *Ammonium Citrate Solution*, 1 mL of *Potassium Cyanide Solution*, 0.5 mL of *Hydroxylamine Hydrochloride Solution*, and 2.5 µg of lead ion (Pb).
Loss on Heating   Heat about 1 g, accurately weighed, in a crucible tared in a stoppered weighing bottle, to constant weight at 400° to 500°. Cool in a desiccator, transfer to the stoppered weighing bottle, and weigh.

Selenium Determine as directed in *Method II* under the *Selenium Limit Test*, Appendix IIIB, using 200-mg of sample.

Packaging and Storage Store in well-closed containers.

Mannitol

D-Mannitol; Mannite; 1,2,3,4,5,6-Hexanehexol

$$\text{HO-CH}_2\text{-CHOH-CHOH-CHOH-CHOH-CH}_2\text{OH}$$

C$_6$H$_{14}$O$_6$ Formula wt 182.17
INS: 421   CAS: [69-65-8]

DESCRIPTION

A white, crystalline solid consisting of D-mannitol and a small quantity of sorbitol. It is odorless and has a sweet taste. It is soluble in water, very slightly soluble in alcohol, and practically insoluble in most other common organic solvents.

Functional Use in Foods Nutrient; dietary supplement; texturizing agent.

REQUIREMENTS

Identification The infrared absorption spectrum of a potassium bromide dispersion of the sample exhibits maxima only at the same wavelengths as that of a similar preparation of USP Mannitol Reference Standard.
Assay Not less than 96.0% and not more than 101.5% of C$_6$H$_{14}$O$_6$ (Mannitol), calculated on the dried basis.
Chloride Not more than 0.007%.
Heavy Metals (as Pb) Not more than 5 mg/kg.
Loss on Drying Not more than 0.3%.
Melting Range Between 164° and 168°.
Reducing Sugars Passes test.
Specific Rotation $[\alpha]_D^{25°}$: Between +137° and +145°.
Sulfate Not more than 0.01%.

TESTS

Assay
Mobile Phase Use degassed water.
Resolution Solution Dissolve sorbitol and USP Mannitol Reference Standard in water to obtain a solution having concentrations of about 4.8 mg/mL of each.
Standard Preparation Dissolve an accurately weighed quantity of USP Mannitol Reference Standard in water, and dilute quantitatively with water to obtain a solution having a known concentration of about 4.8 mg/mL.
Assay Preparation Transfer about 0.24 g of the sample, accurately weighed, to a 50-mL volumetric flask, dissolve in 10 mL of water, dilute with water to volume, and mix.
Chromatographic System (see *Chromatography*, Appendix IIA) The liquid chromatograph is equipped with a refractive index detector that is maintained at a constant temperature and a 4-mm × 25-cm column that contains Biorad Aminex HPX-87C (or the equivalent). The column is maintained at a temperature between 30° and 85°, controlled within ± 2° of the selected temperature, and the flow rate is about 0.5 mL/min. Chromatograph the *Standard Preparation*, and record the peak responses as directed under *Procedure*: The relative standard deviation for replicate injections is not more than 2.0%. In a similar manner, chromatograph the *Resolution Solution*: The resolution, R, between the sorbitol and Mannitol peaks is not less than 2.0.
Procedure Separately inject equal volumes (about 20 μL) of the *Assay Preparation* and the *Standard Preparation* into the chromatograph, record the chromatograms, and measure the responses for the major peaks. Calculate the quantity, in mg, of C$_6$H$_{14}$O$_6$ in the Mannitol taken by the formula

$$50C(r_u/r_s),$$

in which C is the concentration, in mg/mL, of USP Mannitol Reference Standard in the *Standard Preparation*, and r_u and r_s are the peak responses obtained from the *Assay Preparation* and the *Standard Preparation*, respectively.
Chloride, Appendix IIIB Any turbidity produced by a 200-mg sample does not exceed that shown in a control containing 14 μg of chloride ion (Cl).
Heavy Metals A solution of 4 g in 25 mL of water meets the requirements of the *Heavy Metals Test*, Appendix IIIB, using 20 μg of lead ion (Pb) in the control (*Solution A*).
Loss on Drying, Appendix IIC Dry at 105° for 4 h.
Melting Range Determine as directed for *Melting Range or Temperature*, Appendix IIB.
Reducing Sugars Add 1 mL of a saturated solution to 5 mL of alkaline cupric citrate TS, and heat for 5 min in a boiling water bath. No more than a very slight precipitation occurs.
Specific Rotation Transfer about 1 g of Mannitol, accurately weighed, to a 100-mL volumetric flask, and add 40 mL of ammonium molybdate solution (1 in 10), which previously has been filtered if necessary. Add 20 mL of 1 N sulfuric acid, dilute with water to volume, and mix. Determine the specific rotation of this solution as directed in the general method, Appendix IIB.
Sulfate, Appendix IIIB Any turbidity produced by a 2-g sample does not exceed that shown in a control containing 200 μg of sulfate (SO$_4$).

Packaging and Storage Store in well-closed containers.

Marjoram Oil, Spanish Type

DESCRIPTION

A volatile oil obtained by steam distillation from the flowering plant *Thymus mastichina* L. (Fam. *Labiatae*). It is a slightly yellow liquid having a camphoraceous note. It is soluble in most fixed oils, but it is insoluble in glycerin, in propylene glycol, and in mineral oil.

Functional Use in Foods Flavoring agent.

REQUIREMENTS

Identification The infrared absorption spectrum of the sample exhibits relative maxima (that may vary in intensity) at the same wavelengths (or frequencies) as those shown in the respective spectrum in the section on *Infrared Spectra (Series A: Essential Oils)*, using the same test conditions as specified therein.
Assay Not less than 49.0% and not more than 65.0% of cineole.
Acid Value Not more than 2.0.
Angular Rotation Between −5° and +10°.
Heavy Metals (as Pb) Passes test.
Refractive Index Between 1.463 and 1.468 at 20°.
Saponification Value Between 5 and 20.
Solubility in Alcohol Passes test.
Specific Gravity Between 0.904 and 0.920.

TESTS

Assay Dry a sample over anhydrous sodium sulfate, and transfer 3 g of the dried oil, accurately weighed, into a test tube as directed for *Solidification Point*, Appendix IIB. Add 2.1 g of melted *o*-cresol. The *o*-cresol must be pure and dry and have a solidification point not below 30°. Insert the thermometer, stir, and warm the tube gently until the mixture is completely melted. Proceed as directed under *Solidification Point*. Repeat the procedure until two successive readings agree within 0.1°. Compute the percentage of cineole from the *Percentage of Cineole* table, Appendix VI (*Essential Oils and Flavors*).
Acid Value Determine as directed for *Acid Value* under *Essential Oils and Flavors*, Appendix VI.
Angular Rotation Determine in a 100-mm tube as directed under *Optical (Specific) Rotation*, Appendix IIB.
Heavy Metals Shake 10 mL of the oil with an equal volume of water to which 1 drop of hydrochloric acid has been added, and pass hydrogen sulfide through the mixture until it is saturated. No darkening in color is produced in either the oil or the water.
Refractive Index, Appendix IIB Determine with an Abbé or other refractometer of equal or greater accuracy.
Saponification Value Determine as directed for *Saponification Value* under *Essential Oils and Flavors*, Appendix VI, using about 10 g, accurately weighed.
Solubility in Alcohol Proceed as directed in the general method, Appendix VI. One mL dissolves in 1 mL of 80% alcohol and remains in solution on further addition of the alcohol to a total volume of 10 mL.
Specific Gravity Determine by any reliable method (see *General Provisions*).

Packaging and Storage Store in full, tight, preferably glass, aluminum, tin-lined, or other suitably lined containers in a cool place protected from light.

Marjoram Oil, Sweet

DESCRIPTION

The volatile oil obtained by steam distillation of the dried herb of the marjoram shrub, *Marjoram hortensis* L. (Fam. *Labiatae*). It is a yellow or greenish yellow oil having a spicy or cardamom note. It is soluble in most fixed oils and in mineral oil (with turbidity). It is only partly soluble in propylene glycol, and is insoluble in glycerin.

Functional Use in Foods Flavoring agent.

REQUIREMENTS

Identification The infrared absorption spectrum of the sample exhibits relative maxima (that may vary in intensity) at the same wavelengths (or frequencies) as those shown in the respective spectrum in the section on *Infrared Spectra (Series A: Essential Oils)*, using the same test conditions as specified therein.
Acid Value Not more than 2.5.
Angular Rotation Between +14° and +24°.
Heavy Metals (as Pb) Passes test.
Refractive Index Between 1.470 and 1.475 at 20°.
Saponification Value Between 23 and 40.
Saponification Value after Acetylation Between 68 and 86.
Solubility in Alcohol Passes test.
Specific Gravity Between 0.890 and 0.906.

TESTS

Acid Value Determine as directed for *Acid Value* under *Essential Oils and Flavors*, Appendix VI.
Angular Rotation Determine in a 100-mm tube as directed under *Optical (Specific) Rotation*, Appendix IIB.
Heavy Metals Shake 10 mL of the oil with an equal volume of water to which 1 drop of hydrochloric acid has been added, and pass hydrogen sulfide through the mixture until it is saturated. No darkening in color is produced in either the oil or the water.
Refractive Index, Appendix IIB Determine with an Abbé or other refractometer of equal or greater accuracy.
Saponification Value Determine as directed for *Saponification Value* under *Essential Oils and Flavors*, Appendix VI, using about 5 g, accurately weighed.

Saponification Value after Acetylation Proceed as directed under *Total Alcohols*, Appendix VI, using about 2.5 g of acetylated oil, accurately weighed. Calculate the saponification value by the formula

$$28.05 \times A/B,$$

in which A is the number of mL of 0.5 N alcoholic potassium hydroxide consumed in the titration, and B is the weight of the acetylated oil, in g.

Solubility in Alcohol Proceed as directed in the general method, Appendix VI. One mL dissolves in 2 mL of 80% alcohol.

Specific Gravity Determine by any reliable method (see *General Provisions*).

Packaging and Storage Store in full, tight containers in a cool place protected from light.

Masticatory Substances, Natural

Coagulated or Concentrated Latices of Vegetable Origin

DESCRIPTION

The Masticatory Substances of vegetable origin consist of the gums from the trees of *Sapotaceae*, *Apocynaceae*, *Moraceae*, and *Euphorbiaceae* as listed below. The coagulated material varies in color from white to brown, depending on its moisture content and heat treatment during purification. The gums are purified by extensive treatment either alone or in combination with other gums or food-grade materials. They are heat treated and then clarified by centrifuging or by any other appropriate means of filtration.

Family	Genus and Species
Sapotaceae	
Chicle	*Manilkara zapotilla* Gilly and *Manilkara chicle* Gilly
Chiquibul	*Manilkara zapotilla* Gilly
Crown gum	*Manilkara zapotilla* Gilly and *Manilkara chicle* Gilly
Gutta hang kang	*Palaquium leiocarpum* Boerl. and *Palaquium oblongifolium* Burck
Gutta Katiau	*Palaquium ganua moteleyana* Clarke (also known as *Sideroxylon glabrescens*)
Massaranduba balata (and the solvent-free resin extract of Massaranduba balata)	*Manilkara huberi* (Ducke) Chevalier
Massaranduba chocolate	*Manilkara solimoesensis* Gilly
Nispero	*Manilkara zapotilla* Gilly and *Manilkara chicle* Gilly
Rosidinha (rosadinha)	*Micropholis* (also known as *Sideroxylon*) spp.
Venezuelan chicle	*Manilkara williamsii* Standley and related spp.
Apocynaceae	
Jelutong	*Dyera costulata* Hook. F. and *Dyera lowii* Hook. F.
Leche caspi (sorva)	*Couma macrocarpa* Barb. Rodr.
Pendare	*Couma macrocarpa* Barb. Rodr. and *Couma utilis* (Mart.) Muell. Arg.
Perillo	*Couma macrocarpa* Barb. Rodr. and *Couma utilis* (Mart.) Muell. Arg.
Moraceae	
Leche de vaca	*Brosimum utile* (H.B.K.) Pittier and *Poulsenia* ssp.; also *Lacmellea standleyi* (Woodson), Monachino (Apocynaceae)
Niger Gutta	*Ficus platyphylla* Del.
Tunu (tuno)	*Castilla fallax* Cook
Euphorbiaceae	
Chilte	*Cnidoscolus* (also known as *Jatropha*) *elasticus* Lundell and *Cnidoscolus tepiquensis* (Cost. and Gall.) McVaugh
Natural rubber (latex solids)	*Hevea brasiliensis*

Functional Use in Foods Masticatory substance in chewing gum base.

REQUIREMENTS

Arsenic (as As)   Not more than 3 mg/kg.
Heavy Metals (as Pb)   Not more than 0.002%.
Lead   Not more than 3 mg/kg.

The following specifications, where applicable, should conform to the representations of the vendor: cleanliness, color, texture, odor, and loss on drying.

TESTS

Arsenic Prepare a *Sample Solution* as directed in the general method under *Chewing Gum Base*, Appendix IV. This solution meets the requirements of the *Arsenic Test*, Appendix IIIB.

Heavy Metals Prepare and test a 1-g sample as directed in *Method II* under the *Heavy Metals Test*, Appendix IIIB, using 20 µg of lead ion (Pb) in the control (*Solution A*).

Lead Prepare a *Sample Solution* as directed in the general method under *Chewing Gum Base*, Appendix IV. This solution meets the requirements of the *Lead Limit Test*, Appendix IIIB, using 10 μg of lead ion (Pb) in the control.

Packaging and Storage Store in well-closed containers.

Mentha Arvensis Oil, Partially Dementholized

Cornmint Oil, Partially Dementholized

DESCRIPTION

The portion of oil remaining after the partial removal of menthol, by freezing operations only, from the oil of *Mentha arvensis* var. *piperascens* Holmes (forma piperascens Malinvaud). It is a colorless to yellow liquid having a characteristic minty odor. It is soluble in most fixed oils, in mineral oil, and in propylene glycol. It is insoluble in glycerin.

Functional Use in Foods Flavoring agent.

REQUIREMENTS

Identification The infrared absorption spectrum of the sample exhibits relative maxima (that may vary in intensity) at the same wavelengths (or frequencies) as those shown in the respective spectrum in the section on *Infrared Spectra (Series A: Essential Oils)*, using the same test conditions as specified therein.
Assay Not less than 40.0% and not more than 60.0% of total alcohols, calculated as menthol ($C_{10}H_{20}O$).
Angular Rotation Between −20° and −35°.
Heavy Metals (as Pb) Passes test.
Refractive Index Between 1.458 and 1.465 at 20°.
Solubility in Alcohol Passes test.
Specific Gravity Between 0.888 and 0.908.
Total Esters Between 5.0% and 20.0%, calculated as menthyl acetate ($C_{12}H_{22}O_2$).
Total Ketones Between 30.0% and 50.0%, calculated as menthone ($C_{10}H_{18}O$).

TESTS

Assay Proceed as directed under *Total Alcohols*, Appendix VI, using about 1.5 g of the acetylated oil, accurately weighed, for the saponification. Calculate the percentage of alcohol, as menthol, in the sample by the formula

$$A \times 7.813(1 - 0.0021E)/(B - 0.021A),$$

in which A is the number of mL of 0.5 N alcoholic potassium hydroxide consumed in the saponification; B is the weight, in g, of the acetylated oil taken; and E is the percentage of esters, as menthyl acetate, determined as directed under *Total Esters* below.
Angular Rotation Determine in a 100-mm tube as directed under *Optical (Specific) Rotation*, Appendix IIB.

Heavy Metals Shake 10 mL of the oil with an equal volume of water to which 1 drop of hydrochloric acid has been added, and pass hydrogen sulfide through the mixture until it is saturated. No darkening in color is produced in either the oil or the water.
Refractive Index, Appendix IIB Determine with an Abbé or other refractometer of equal or greater accuracy.
Solubility in Alcohol Proceed as directed in the general method. One mL dissolves in 2.5 to 4 mL of 80% alcohol and may become hazy upon further dilution.
Specific Gravity Determine by any reliable method (see *General Provisions*).
Total Esters Weigh accurately about 10 g, and proceed as directed under *Ester Determination*, Appendix VI, using 99.15 as the equivalence factor (*e*) in the calculation.
Total Ketones Weigh accurately about 1 g, and proceed as directed for ketones under *Aldehydes and Ketones—Hydroxylamine Method*, Appendix VI, using 77.12 as the equivalence factor (*e*) in the calculation.

Packaging and Storage Store in full, tight containers in a cool place protected from light.

DL-Methionine

DL-2-Amino-4-(methylthio)butyric Acid

$$CH_3SCH_2CH_2\underset{NH_2}{CHCOOH}$$

$C_5H_{11}NO_2S$ Formula wt 149.21

CAS: [59-51-8]

DESCRIPTION

White, crystalline platelets or powder having a characteristic odor. One g dissolves in about 30 mL of water. It is soluble in dilute acids and in solutions of alkali hydroxides. It is very slightly soluble in alcohol, and practically insoluble in ether. It is optically inactive. The pH of a 1 in 100 solution is between 5.6 and 6.1.

Functional Use in Foods Nutrient; dietary supplement.

REQUIREMENTS

Identification
A. Add 25 mg of a previously dried sample to 1 mL of a saturated solution of anhydrous cupric sulfate in sulfuric acid. A yellow color appears.
B. To 10 mL of a 1 in 1000 solution add successively, shaking after each addition, 1 mL of a 1 in 5 solution of sodium hydroxide, 1 mL of a 1 in 100 glycine solution, and 0.3 mL of a freshly prepared 1 in 10 solution of sodium nitroferricyanide. Keep the

mixture at about 40° for 10 min, cool in an ice bath for 2 min, then add 2 mL of 20% hydrochloric acid and shake the mixture. A red or orange red color appears.

C. To 1 mL of a 1 in 30 solution add 1 mL of ninhydrin TS and 100 mg of sodium acetate, and heat to boiling. An intense violet blue color is formed (distinction from hydroxy analog).

Assay Not less than 98.5% and not more than 101.5% of $C_5H_{11}NO_2S$ after drying.
Heavy Metals (as Pb) Not more than 0.002%.
Lead Not more than 10 mg/kg.
Loss on Drying Not more than 0.5%.
Residue on Ignition Not more than 0.1%.

TESTS

Assay Dissolve about 140 mg of the sample, previously dried at 105° for 3 h and accurately weighed, in 3 mL of formic acid and 50 mL of glacial acetic acid, and titrate with 0.1 N perchloric acid, determining the endpoint potentiometrically. Perform a blank determination, and make any necessary correction. Each mL of 0.1 N perchloric acid is equivalent to 14.92 mg of $C_5H_{11}NO_2S$.

Heavy Metals Prepare and test a 1-g sample as directed in *Method II* under the *Heavy Metals Test*, Appendix IIIB, using 20 μg of lead ion (Pb) in the control (*Solution A*).

Lead A *Sample Solution* prepared as directed for organic compounds meets the requirements of the *Lead Limit Test*, Appendix IIIB, using 10 μg of lead ion (Pb) in the control.

Loss on Drying, Appendix IIC Dry at 105° for 3 h.

Residue on Ignition, Appendix IIC Ignite 1 g as directed in the general method.

Packaging and Storage Store in well-closed, light-resistant containers.

L-Methionine

L-2-Amino-4-(methylthio)butyric Acid

$$CH_3SCH_2CH_2\underset{H}{\overset{NH_2}{C}}COOH$$

$C_5H_{11}NO_2S$ Formula wt 149.21

CAS: [63-68-3]

DESCRIPTION

Colorless or white, lustrous plates, or a white, crystalline powder. It has a slight, characteristic odor. It is soluble in water, in alkali solutions, and in dilute mineral acids. It is slightly soluble in alcohol and practically insoluble in ether.

Functional Use in Foods Nutrient; dietary supplement.

REQUIREMENTS

Identification L-Methionine responds to *Identification Tests A, B,* and *C* in the monograph for DL-*Methionine*.
Assay Not less than 98.5% and not more than 101.5% of $C_5H_{11}NO_2S$ after drying.
Heavy Metals (as Pb) Not more than 10 mg/kg.
Lead Not more than 10 mg/kg.
Loss on Drying Not more than 0.5%.
Residue on Ignition Not more than 0.1%.
Specific Rotation $[\alpha]_D^{20°}$: Between +21.0° and +25.0°, calculated on the dried basis; or $[\alpha]_D^{25°}$: Between +21.1° and +25.1°, calculated on the dried basis.

TESTS

Assay Dissolve about 140 mg of the sample, previously dried at 105° for 3 h and accurately weighed, in 3 mL of formic acid and 50 mL of glacial acetic acid, and titrate with 0.1 N perchloric acid, determining the endpoint potentiometrically. Perform a blank determination, and make any necessary correction. Each mL of 0.1 N perchloric acid is equivalent to 14.92 mg of $C_5H_{11}NO_2S$.

Heavy Metals Prepare and test a 2-g sample as directed in *Method II* under the *Heavy Metals Test*, Appendix IIIB, using 20 μg of lead ion (Pb) in the control (*Solution A*).

Lead A *Sample Solution* prepared as directed for organic compounds meets the requirements of the *Lead Limit Test*, Appendix IIIB, using 10 μg of lead ion (Pb) in the control.

Loss on Drying, Appendix IIC Dry at 105° for 3 h.

Residue on Ignition, Appendix IIC Ignite 1 g as directed in the general method.

Specific Rotation, Appendix IIB Determine in a solution containing 2 g in sufficient 6 N hydrochloric acid to make 100 mL.

Packaging and Storage Store in well-closed, light-resistant containers.

Methyl Alcohol

Methanol

CH_3OH Formula wt 32.04

CAS: [67-56-1]

DESCRIPTION

A clear, colorless, flammable liquid having a characteristic odor. It is miscible with water, with ethyl alcohol, and with ether. Its refractive index at 20° is about 1.329.

Functional Use in Foods Extraction solvent.

REQUIREMENTS

Assay Not less than 99.85% of CH$_3$OH, by weight.
Acetone and Aldehydes Not more than 0.003%.
Acidity (as formic acid) Not more than 0.0015%.
Alkalinity (as NH$_3$) Not more than 3 mg/kg.
Distillation Range Within a range of 1°, including 64.6° ± 0.1°.
Heavy Metals (as Pb) Not more than 1 mg/kg.
Nonvolatile Residue Not more than 10 mg/kg.
Readily Carbonizable Substances Passes test.
Solubility in Water Passes test.
Substances Reducing Permanganate Passes test.
Water Not more than 0.1%.

TESTS

Assay Its specific gravity, determined by any reliable method (see *General Provisions*), is not greater than 0.7893 at 25°/25° (equivalent to 0.7928 at 20°/20°).
Acetone and Aldehydes To 1.25 mL (about 1 g) of the sample add 3.75 mL of water and 5.0 mL of alkaline mercuric–potassium iodide TS. Any turbidity does not exceed that produced in a standard containing 30 µg of acetone.
Acidity To a mixture of 10 mL of alcohol and 25 mL of water add 0.5 mL of phenolphthalein TS, and titrate with 0.02 N sodium hydroxide to the first pink color that persists for at least 30 s. Add 19 mL (about 15 g) of the sample, mix, and titrate with 0.02 N sodium hydroxide until the pink color is restored. Not more than 0.25 mL is required.
Alkalinity Add 1 drop of methyl red TS to 25 mL of water, add 0.02 N sulfuric acid until a red color just appears, then add 29 mL (about 22.5 g) of the sample, and mix. Not more than 0.2 mL of 0.02 N sulfuric acid is required to restore the red color.
Distillation Range Proceed as directed in the general method, Appendix IIB.
Heavy Metals Evaporate 25 mL (about 20 g) of the sample to dryness on a steam bath in a glass evaporating dish. Cool, add 2 mL of hydrochloric acid, and slowly evaporate to dryness again on the steam bath. Moisten the residue with 1 drop of hydrochloric acid, add 10 mL of hot water, and digest for 2 min. Cool, and dilute to 25 mL with water. This solution meets the requirements of the *Heavy Metals Test*, Appendix IIIB, using 20 µg of lead ion (Pb) in the control (*Solution A*).
Nonvolatile Residue Evaporate 125 mL (about 100 g) of the sample to dryness in a tared dish on a steam bath, dry the residue at 105° for 30 min, cool, and weigh.
Readily Carbonizable Substances, Appendix IIB A mixture of 25 mL of 95% sulfuric acid (cooled to 10°) and 25 mL of the sample has no more color than 3.5 mL of platinum–cobalt CS, diluted to 50 mL with water (equivalent to not more than 35 APHA color units).
Solubility in Water Mix 15 mL of the sample with 45 mL of water. After 1 h, the solution is as clear as an equal volume of water.
Substances Reducing Permanganate Transfer 20 mL of the sample, previously cooled to 15°, to a glass-stoppered cylinder, add 0.1 mL of 0.1 N potassium permanganate, mix, and allow to stand for 5 min. The pink color is not entirely discharged.
Water Determine by the *Karl Fischer Titrimetric Method*, Appendix IIB.

Packaging and Storage Store in tight containers remote from heat, sparks, and open flames.

Methylcellulose

Modified Cellulose, MC

INS: 461 CAS: [9004-67-5]

DESCRIPTION

Methylcellulose is the methyl ether of cellulose in the form of a white, fibrous powder or granules. It is soluble in water and in a limited number of organic solvent systems. Aqueous solutions of Methylcellulose are surface active, form films upon drying, and undergo a reversible transformation from sol to gel upon heating and cooling, respectively.

Functional Use in Foods Thickener; stabilizer; emulsifier; bodying agent; bulking agent; binder; film former.

REQUIREMENTS

Identification
A. Add 1 g to 100 mL of water. It swells and disperses to form a clear to opalescent, mucilaginous solution, depending upon the intrinsic viscosity, which is stable in the presence of most electrolytes and alcohol in concentrations up to 40%.
B. Heat a few mL of the solution prepared for *Identification Test A*. The solution becomes cloudy, and a flaky precipitate appears that redissolves as the solution cools.
C. Pour a few mL of the solution prepared for *Identification Test A* onto a glass plate, and allow the water to evaporate. A thin, self-sustaining film results.
Assay Not less than 27.5% or more than 31.5% of methoxyl groups (—OCH$_3$), calculated on the dried basis.
Heavy Metals (as Pb) Not more than 10 mg/kg.
Loss on Drying Not more than 5.0%.
Residue on Ignition Not more than 1.5%.
Viscosity The apparent viscosity of a solution containing 2 g in each 100 mL is not less than 80% and not more than 120% of that stated on the label for viscosity types of 100 centipoises or less, and not less than 75% and not more than 140% of that stated on the label for viscosity types higher than 100 centipoises.

TESTS

Assay (**Caution:** Perform all steps involving hydriodic acid carefully, in a well-ventilated hood. Use goggles, acid-resistant gloves, and other appropriate safety equipment. Be extremely careful when handling the hot vials, since they are under pressure. In the event of hydriodic acid exposure, wash with copious amounts of water, and seek medical attention at once.)

Internal Standard Solution Transfer about 2.5 g of toluene, accurately weighed, into a 100-mL volumetric flask containing 10 mL of *o*-xylene, dilute with *o*-xylene to volume, and mix.

Standard Preparation Transfer about 135 mg of adipic acid into a suitable serum vial, add 4.0 mL of hydriodic acid followed by 4.0 mL of the *Internal Standard Solution*, and close the vial securely with a septum stopper. Weigh the vial and its contents accurately, add 90 μL of methyl iodide with a syringe through the septum, again weigh, and calculate the weight of methyl iodide added. Shake well, and allow the layers to separate.

Assay Preparation Transfer about 65 mg of the sample, accurately weighed, into a 5-mL vial equipped with a pressure-tight septum closure, add an amount of adipic acid equal to the weight of the sample, and pipet 2 mL of the *Internal Standard Solution* into the vial. Cautiously pipet 2 mL of hydriodic acid into the mixture, immediately secure the closure, and weigh accurately. Shake the vial for 30 s, heat at 150° for 20 min, remove from the heat, shake again, using extreme caution, and heat at 150° for 40 min. Allow the vial to cool for about 45 min, and then weigh. If the weight loss is greater than 10 mg, discard the mixture and prepare another *Assay Preparation*.

Chromatographic System Use a suitable gas chromatograph equipped with a thermal conductivity detector. Under typical conditions, the instrument contains a 1.8-m × 4-mm glass column packed with 10% methylsilicone oil (UCW 982 or equivalent) on 100/120-mesh flux-calcined chromatographic siliceous earth (Chromosorb WHP or equivalent). The column is maintained at 100°, and the injection port and detector at 200°, and helium is used as the carrier gas flowing at the rate of 20 mL/min.

Calibration Inject about 2 μL of the upper layer of the *Standard Preparation* into the chromatograph, and record the chromatogram. The retention times for methyl iodide, toluene, and *o*-xylene are approximately 3, 7, and 13 min, respectively. Calculate the relative response factor, F, of equal weights of toluene and methyl iodide by the formula

$$Q/A,$$

in which Q is the quantity ratio of methyl iodide to toluene in the *Standard Preparation*, and A is the peak area ratio of the methyl iodide to toluene obtained from the *Standard Preparation*.

Procedure Inject about 2 μL of the upper layer of the *Assay Preparation* into the chromatograph, and record the chromatogram. Calculate the percentage of methoxyl groups (—OCH$_3$) in the sample by the formula

$$2 \times (31/142) \times F \times a \times (W/w),$$

in which 31/142 is the ratio of the formula weights of methoxyl to methyl iodide; a is the ratio of the area of the methyl iodide peak to that of the toluene peak obtained from the *Assay Preparation*; W is the weight of the toluene in the *Internal Standard Solution*, in g; and w is the weight of sample taken, in g.

Heavy Metals Prepare and test a 2-g sample as directed in *Method II* under the *Heavy Metals Test*, Appendix IIIB, adding 1 mL of hydroxylamine hydrochloride solution (1 in 5) to the solution of the residue. Any color does not exceed that produced in a control (*Solution A*) containing 20 μg of lead ion (Pb) and 1 mL of the hydroxylamine hydrochloride solution.

Loss on Drying, Appendix IIC Dry a 3-g sample at 105° for 2 h.

Residue on Ignition Ignite 1 g as directed in the general method, Appendix IIC.

Viscosity Accurately weigh a sample, equivalent to 2 g of solids on the dried basis, transfer to a wide-mouth, 250-mL centrifuge bottle, and add 98 g of water previously heated to between 80° and 90°. Stir with a mechanical stirrer for 10 min, then place the bottle in an ice bath until solution is complete, adjust the weight of the solution to 100 g, if necessary, and centrifuge it to expel any entrapped air. Adjust the temperature of the solution to 20° ± 0.1°, and determine the viscosity as directed under *Viscosity of Methylcellulose*, Appendix IIB.

Packaging and Storage Store in well-closed containers.

Methylene Chloride

Dichloromethane

CH_2Cl_2 Formula wt 84.93

CAS: [75-09-2]

DESCRIPTION

A clear, colorless, nonflammable liquid having an odor resembling that of chloroform. It is soluble in about 50 parts of water, and is miscible with alcohol, with acetone, with chloroform, and with ether. Its refractive index at 20° is about 1.424.

Functional Use in Foods Extraction solvent.

REQUIREMENTS

Acidity (as HCl) Not more than 10 mg/kg.
Distillation Range Between 39.5° and 40.5°.
Free Halogens Passes test.
Heavy Metals (as Pb) Not more than 1 mg/kg.
Nonvolatile Residue Not more than 0.002%.
Specific Gravity Between 1.318 and 1.323.
Water Not more than 0.02%.

TESTS

Acidity Transfer 100 mL (about 132 g) of the sample into a separator, add 100 mL of neutralized water, and shake vigorously for 2 min. Allow the layers to separate, transfer the aqueous phase into an Erlenmeyer flask, add 4 drops of bromothymol blue TS, and titrate with 0.01 N sodium hydroxide. Not more than 3.6 mL is required.

Distillation Range Proceed as directed in the general method, Appendix IIB.

Free Halogens Transfer 10 mL of the sample to a separator, add 25 mL of water, and shake vigorously for 1 min. Allow the layers to separate completely, and then remove and discard the lower sample layer. To the aqueous phase add 1 mL of potassium iodide TS and a few drops of starch TS, and allow to stand for 5 min. A blue color does not develop.

Heavy Metals (Caution: use hood.) Evaporate 15 mL (about 20 g) of the sample to dryness in a glass evaporating dish on a steam bath. Cool, add 2 mL of hydrochloric acid, and slowly evaporate to dryness again on the steam bath. Moisten the residue with 1 drop of hydrochloric acid, add 10 mL of hot water, and digest for 2 min. Filter if necessary through a small filter, wash the evaporating dish and the filter with about 10 mL of water, and dilute to 25 mL with water. This solution meets the requirements of the *Heavy Metals Test*, Appendix IIIB, using 20 µg of lead ion (Pb) in the control (*Solution A*).

Nonvolatile Residue (Caution: use hood.) Evaporate 38 mL (about 50 g) of the sample to dryness in a tared dish on a steam bath, dry the residue at 105° for 30 min, cool, and weigh.

Specific Gravity Determine by any reliable method (see *General Provisions*).

Water Determine by the *Karl Fischer Titrimetric Method*, Appendix IIB.

Packaging and Storage Store in tight containers.

Methyl Ester of Rosin, Partially Hydrogenated

DESCRIPTION

A light amber-colored liquid resin produced by the esterification of rosin with methanol, followed by partial hydrogenation and purification by steam stripping. It is soluble in acetone, but is insoluble in water.

Functional Use in Foods Masticatory substance in chewing gum base.

REQUIREMENTS

Identification Identify Partially Hydrogenated Methyl Ester of Rosin by comparing its infrared absorption spectrum with a typical spectrum as shown in the section on *Infrared Spectra* (*Series C: Other Substances*). The sample is prepared for analysis neat (as is) on a potassium bromide plate.

Acid Number Between 4 and 8.

Arsenic (as As) Not more than 3 mg/kg.

Heavy Metals (as Pb) Not more than 10 mg/kg.

Lead Not more than 1 mg/kg.

Refractive Index Between 1.517 and 1.520 at 20°.

Viscosity Between 23 and 76 poises.

TESTS

Acid Number Determine as directed in the general method, Appendix IX.

Arsenic Prepare a *Sample Solution* as directed in the general method under *Chewing Gum Base*, Appendix IV. This solution meets the requirements of the *Arsenic Test*, Appendix IIIB.

Heavy Metals Prepare and test a 2-g sample as directed in *Method II* under the *Heavy Metals Test*, Appendix IIIB, using 20 µg of lead ion (Pb) in the control (*Solution A*).

Lead Prepare a *Sample Solution* using a 5-g sample as directed in the general method under *Chewing Gum Base*, Appendix IV. This solution meets the requirements of the *Lead Limit Test*, Appendix IIIB, using 5 µg of lead ion (Pb) in the control.

Refractive Index, Appendix IIB Determine with an Abbé or other refractometer of equal or greater accuracy.

Viscosity Determine as directed in the general method, Appendix IX.

Packaging and Storage Store in well-closed containers.

Methyl Ethyl Cellulose

Modified Cellulose

INS: 465 CAS: [9004-69-7]

DESCRIPTION

Methyl Ethyl Cellulose is the methyl ether of ethyl cellulose in which both the methyl and ethyl groups are attached to the anhydroglucose units by ether linkages. It occurs as a white or pale, cream-colored, fibrous solid or powder. It is practically odorless and disperses in cold water to form aqueous solutions that undergo a reversible transformation from sol to gel upon heating and cooling, respectively.

Functional Use in Foods Emulsifier; stabilizer; foaming agent.

REQUIREMENTS

Identification

A. Add 1 g to 100 mL of water. It disperses to form an opalescent, fibrous sol.

B. Heat a few mL of the sol prepared for *Identification Test A* to about 60°. The sol becomes cloudy, and a gelatinous precipitate forms that redissolves upon cooling.

C. The remaining sol from *Identification Test A*, when whipped as for egg white in a kitchen-type mixer, produces a stable air/liquid foam.

Assay Not less than 14.5% and not more than 19.0% of ethoxyl groups ($-OC_2H_5$) and not less than 3.5% and not more than 6.5% of methoxyl groups ($-OCH_3$).

Heavy Metals (as Pb) Not more than 10 mg/kg.

Lead Not more than 3 mg/kg.

Loss on Drying *Fibrous form*: not more than 15.0%; *powdered form*: not more than 10.0%.
Residue on Ignition Not more than 0.6%.
Viscosity The apparent viscosity of a solution containing the equivalent of 2.5 g of dry sample in each 100 g of solution is not less than 80% and not more than 120% of that stated on the label or otherwise represented by the vendor. The usual range of viscosity types is between 20 and 60 centipoises.

TESTS

Assay for Ethoxyl Groups Proceed as directed for *Hydroxypropoxyl Determination*, Appendix IIIC. Each mL of 0.02 N sodium hydroxide is equivalent to 0.9 mg of ethoxyl groups (—OC_2H_5).
Assay for Methoxyl Groups Place about 50 mg, previously dried at 105°, in a tared gelatin capsule, and weigh accurately. Proceed as directed for *Methoxyl Determination*, Appendix IIIC, but calculate the total alkoxyl content as ethoxyl groups (—OC_2H_5). Each mL of 0.1 N sodium thiosulfate is equivalent to 0.7510 mg of ethoxyl groups (—OC_2H_5). Now calculate the methoxyl groups (—OCH_3) by the formula

$$(A - B) \times 31/45,$$

in which A is the total alkoxyl content, calculated as —OC_2H_5; B is the —OC_2H_5 determined in the *Assay for Ethoxyl Groups*; 31 is the molecular weight of —OCH_3; and 45 is the molecular weight of —OC_2H_5.
Heavy Metals Prepare and test a 2-g sample as directed in *Method II* under the *Heavy Metals Test*, Appendix IIIB, adding 1 mL of a 1 in 5 hydroxylamine hydrochloride solution to the solution of the residue. Any color does not exceed that produced in a control (*Solution A*) containing 20 µg of lead ion (Pb) and 1 mL of the hydroxylamine hydrochloride solution.
Lead A *Sample Solution* prepared from a 2-g sample as directed for organic compounds meets the requirements of the *Lead Limit Test*, Appendix IIIB, using 6 µg of lead ion (Pb) in the control.
Loss on Drying, Appendix IIC Dry a 3-g sample at 105° for 4 h.
Residue on Ignition Ignite 1 g as directed in the general method, Appendix IIC.
Viscosity Transfer an accurately weighed sample, equivalent to 5.0 g on the dried basis, into a 250-mL beaker. Adjust the rotor of a variable-speed stirrer about 1 in. above the sample, add 195 mL of recently boiled and cooled water, and stir at a speed that will avoid undue aeration. Continue the stirring for about 1.5 h, then either set aside for 3 h or overnight, or centrifuge to expel any entrapped air. Adjust the temperature to 20° ± 0.1°, and determine as directed under *Viscosity of Methylcellulose*, Appendix IIB, using a *Viscometer for High Viscosity*.

Packaging and Storage Store in well-closed containers.

Methylparaben

Methyl *p*-Hydroxybenzoate

HO—⟨phenyl⟩—$COOCH_3$

$C_8H_8O_3$ Formula wt 152.15

CAS: [99-76-3]

DESCRIPTION

Small, colorless crystals or a white, crystalline powder. It is odorless or has a faint, characteristic odor and a slight, burning taste. One g dissolves in about 400 mL of water at 25°, in about 50 mL of water at 80°, in 2.5 mL of alcohol, in about 7 mL of ether, and in about 4 mL of propylene glycol. It is slightly soluble in glycerin and in fixed oils.

Functional Use in Foods Preservative; antimicrobial agent.

REQUIREMENTS

Identification Dissolve 500 mg in 10 mL of 1 N sodium hydroxide, boil for 30 min, allow the solution to evaporate to a volume of about 5 mL, and cool. Acidify the solution with 2 N sulfuric acid, collect the crystals on a filter, wash several times with small portions of water, and dry in a desiccator over silica gel. The *p*-hydroxybenzoic acid so obtained melts between 212° and 217° (see Appendix IIB, *Melting Range or Temperature*).
Assay Not less than 99.0% and not more than 100.5% of $C_8H_8O_3$, calculated on the dried basis.
Acidity Passes test.
Heavy Metals (as Pb) Not more than 10 mg/kg.
Loss on Drying Not more than 0.5%.
Melting Range Between 125° and 128°.
Residue on Ignition Not more than 0.05%.

TESTS

Assay Transfer about 2 g, accurately weighed, into a flask, add 40.0 mL of 1 N sodium hydroxide, and rinse the sides of the flask with water. Cover with a watch glass, boil gently for 1 h, cool, and titrate the excess sodium hydroxide with 1 N sulfuric acid to pH 6.5. Perform a blank determination with the same quantities of the same reagents in the same manner, and make any necessary correction (see *General Provisions*). Each mL of 1 N sodium hydroxide is equivalent to 152.2 mg of $C_8H_8O_3$, calculated on the dried basis.
Acidity Heat 750 mg with 15 mL of water at 80° for 1 min, cool, and filter. The filtrate is acid or neutral to litmus. To 10 mL of the filtrate add 0.2 mL of 0.1 N sodium hydroxide and 2 drops of methyl red TS. The solution is yellow, without even a light cast of pink.
Heavy Metals, Appendix IIIB Dissolve 2 g in 23 mL of acetone, and add 2 mL of 1 N acetic acid, 2 mL of water, and

10 mL of hydrogen sulfide TS. Any color does not exceed that produced in a control (*Solution A*) made with 23 mL of acetone, 2 mL of 1 *N* acetic acid, 2 mL of *Standard Lead Solution* (20 μg Pb ion), and 10 mL of hydrogen sulfide TS.

Loss on Drying, Appendix IIC   Dry over silica gel for 5 h.

Melting Range   Determine as directed for *Melting Range or Temperature*, Appendix IIB.

Residue on Ignition   Ignite 4 g as directed in the general method, Appendix IIC.

Packaging and Storage   Store in well-closed containers.

Mineral Oil, White

Liquid Petrolatum

INS: 905(a)

DESCRIPTION

A mixture of refined liquid hydrocarbons, essentially paraffinic and naphthenic in nature, obtained from petroleum. It occurs as a colorless, transparent, oily liquid, free or nearly free from fluorescence. It is odorless and tasteless when cold, and develops not more than a faint odor of petroleum when heated. It is insoluble in water and in alcohol, is soluble in volatile oils, and is miscible with most fixed oils, but not with castor oil. It may contain any antioxidant permitted in food by the U.S. Food and Drug Administration, in an amount not greater than that required to produce its intended effect.

Functional Use in Foods   Binder; defoaming agent; lubricant; release agent; fermentation aid; protective coating.

REQUIREMENTS

Readily Carbonizable Substances   Passes test.
Specific Gravity   Not less than that stated, or within the range claimed by the vendor.
Ultraviolet Absorbance (polynuclear hydrocarbons)   Passes test.
Viscosity   Not less than that stated, or within the range claimed by the vendor.

TESTS

Readily Carbonizable Substances   Place 5 mL of the sample in a glass-stoppered test tube that previously has been rinsed with chromic acid cleaning mixture (200 g of sodium dichromate dissolved in about 100 mL of water to which 1500 mL of sulfuric acid has been added slowly with stirring), then rinsed with water, and dried. Add 5 mL of sulfuric acid containing between 94.5% and 94.9% of H_2SO_4, and heat in a boiling water bath for 10 min. After the test tube has been in the bath for 30 s, remove it quickly, and while holding the stopper in place, give three vigorous vertical shakes over an amplitude of about 5 in. Repeat every 30 s. Do not keep the test tube out of the bath longer than 3 s for each shaking period. At the end of 10 min from the time when first placed in the water bath, remove the test tube. The sample remains unchanged in color, and the acid does not become darker than a standard color produced by mixing in a similar test tube 3 mL of ferric chloride CS, 1.5 mL of cobaltous chloride CS, and 0.5 mL of cupric sulfate CS, this mixture being overlaid with 5 mL of mineral oil.

Specific Gravity   Determine by any reliable method (see *General Provisions*).

Ultraviolet Absorbance (polynuclear hydrocarbons)   All measurements are made with an ultraviolet spectrophotometer in 1-cm cells and in the wavelength range of 260 to 350 nm, under the same instrumental conditions. The standard reference absorbance is the absorbance at 275 nm of a standard reference solution of naphthalene (National Institute for Standards and Technology Standard Material No. 577, or equivalent purity) containing a concentration of 7.0 mg/1000 mL, in purified isooctane, measured against isooctane of the same spectral purity in 1-cm cells. (The absorbance will be approximately 0.30.)

Hexane   Use a pure grade of hexanes (predominantly *n*-hexane and methylcyclopentane) having an ultraviolet absorbance not exceeding 0.10 down to 220 nm and not exceeding 0.02 down to 260 nm. The purity should be such that the "solvent control," as defined under *Procedure*, has an absorbance curve compared to water showing no extraneous impurity peaks and no absorbance exceeding that of dimethyl sulfoxide compared to water at any wavelength in the range 260 to 350 nm, inclusive. If necessary to obtain the prescribed purities, the hexane may be passed through activated silica gel.

Dimethyl Sulfoxide   Use a pure grade of dimethyl sulfoxide (99.9%, m.p. 18°) that is clear, water-white in appearance, having an absorbance curve compared to water not exceeding 1.0 at 264 nm and showing no extraneous impurity peaks in the wavelength range up to 350 nm. It should be stored in glass-stoppered bottles.

Apparatus   Use 125-mL glass-stoppered separators equipped with tetrafluoroethylene polymer stopcocks or other suitable stopcocks that will not contaminate the solvents.

Procedure   Transfer 25 mL of the sample and 25 mL of hexane to a separator, and mix. Add 5.0 mL of dimethyl sulfoxide, shake the mixture vigorously for at least 1 min, and allow to stand until the lower layer is clear. Completely transfer the lower layer to a second separator, add 2 mL of hexane, and shake the mixture vigorously. Allow to stand until the lower layer is clear, and then draw off the lower layer, designated as "mineral oil extract." Shake 5.0 mL of dimethyl sulfoxide with 25 mL of hexane vigorously for at least 1 min in a third separator, allow to stand until the lower layer is clear, and draw off this layer, designated as "solvent control." Determine the absorbance of the mineral oil extract in a 1-cm cell in the range 260 to 350 nm, inclusive, compared to the solvent control. The absorbance of the mineral oil extract does not exceed that of the solvent control at any wavelength in the specified range by more than one-third of the standard reference absorbance.

Note: Suitable corrections of the absorbance should be made when testing samples containing added antioxidants.

Viscosity Determine by any reliable method.

Packaging and Storage Store in tight containers.

Monoammonium L-Glutamate

Monoammonium Glutamate Monohydrate; Ammonium Glutamate

$$NH_4OOCCH_2CH_2CH(NH_2)COOH \cdot H_2O$$

$C_5H_{12}N_2O_4 \cdot H_2O$ Formula wt 182.18

INS: 624 CAS: monohydrate [7558-63-6]

DESCRIPTION

A white, practically odorless, free-flowing, crystalline powder. It is freely soluble in water, but is practically insoluble in common organic solvents. The pH of a 1 in 20 solution is between 6.0 and 7.0.

Functional Use in Foods Flavor enhancer; salt substitute.

REQUIREMENTS

Identification

A. To 1 mL of a 1 in 30 solution add 1 mL of triketohydrindene hydrate TS and 100 mg of sodium acetate, and heat in a boiling water bath for 10 min. An intense, violet blue color is formed.

B. To 10 mL of a 1 in 10 solution add 5.6 mL of 1 N hydrochloric acid. A white crystalline precipitate of glutamic acid forms on standing. When 6 mL of 1 N hydrochloric acid is added to the turbid solution, the glutamic acid dissolves on stirring.

C. A 1 in 10 solution gives positive tests for *Ammonium*, Appendix IIIA.

Assay Not less than 98.5% and not more than 101.5% of $C_5H_{12}N_2O_4 \cdot H_2O$, calculated on the dried basis.
Heavy Metals (as Pb) Not more than 0.002%.
Lead Not more than 10 mg/kg.
Loss on Drying Not more than 0.5%.
Residue on Ignition Not more than 0.1%.
Specific Rotation $[\alpha]_D^{20°}$: Not less than +25.4° and not more than +26.4°, calculated on the dried basis.

TESTS

Assay Dissolve about 250 mg, accurately weighed, in 100 mL of glacial acetic acid. A few drops of water may be added prior to the addition of the acetic acid to effect faster dissolution of the sample. Titrate with 0.1 N perchloric acid in glacial acetic acid, determining the endpoint potentiometrically. Perform a blank determination (see *General Provisions*), and make any necessary correction. Each mL of 0.1 N perchloric acid is equivalent to 9.109 mg of $C_5H_{12}N_2O_4 \cdot H_2O$.

Heavy Metals Prepare and test a 1-g sample as directed in *Method II* under the *Heavy Metals Test*, Appendix IIIB, using 20 µg of lead ion (Pb) in the control (*Solution A*).
Lead A *Sample Solution* prepared as directed for organic compounds meets the requirements of the *Lead Limit Test*, Appendix IIIB, using 10 µg of lead ion (Pb) in the control.
Loss on Drying, Appendix IIC Dry at 50° for 4 h.
Residue on Ignition, Appendix IIC Ignite 1 g as directed in the general method.
Specific Rotation, Appendix IIB Determine in a solution containing 10 g in sufficient 2 N hydrochloric acid to make 100 mL.

Packaging and Storage Store in tight containers.

Monoammonium Glycyrrhizinate

Ammonium Glycyrrhizinate, Pentahydrate; Ammonium Glycyrrhizinate

$$C_{42}H_{61}O_{16}NH_4 \cdot 5H_2O$$

$C_{42}H_{65}NO_{16} \cdot 5H_2O$ Formula wt, anhydrous 839.98

INS: 958

DESCRIPTION

It is a white powder with an intensely sweet taste and is obtained by extraction from ammoniated glycyrrhizin. It is soluble in ammonia water and is insoluble in glacial acetic acid.

Functional Use in Foods Flavoring agent.

REQUIREMENTS

Identification It gives positive tests for *Ammonium*, Appendix IIIA.
Assay Not less than 85.0% and not more than 102.0% of $C_{42}H_{65}NO_{16}$, calculated on the dried basis.
Ash (Total) Not more than 0.5%.
Heavy Metals (as Pb) Not more than 10 mg/kg.
Loss on Drying Not more than 6.0%.
Specific Rotation $[\alpha]_D^{20°}$: Between +45° and +53°, on the as-is basis.

TESTS

Assay (Based on AOAC method 982.19.)
Apparatus Fit a high-pressure liquid chromatograph, operated at room temperature, with a 10-µm particle size, 30-cm × 4-mm id, C_{18} reverse-phase column (µBondapak C_{18}, or equivalent). Maintain the mobile phase at a pressure and flow rate

(typically 2.0 mL/min) capable of giving the required elution time (see *System Suitability*). Use an ultraviolet detector that monitors absorption at 254 nm (0.2 to 0.1 AUFS range).

Mobile Phase   Add 380 mL of acetonitrile and 10 mL of acetic acid to 610 mL of glass-distilled water filtered through a 0.45-μm filter (Millipore or equivalent). Mix, and de-gas thoroughly.

Standard Solution   Weigh accurately about 10 mg of Monoammonium Glycyrrhizinate Standard for analytical use (available from MacAndrews & Forbes Company[1]) in 20 mL of a 1:1 acetonitrile–water solution. Filter the solution through a 0.45-μm Millipore filter, or equivalent. Prepare fresh daily.

Note: Correct the weight of Monoammonium Glycyrrhizinate Standard taken for the percent loss on drying shown on its label.

Assay Solution   Weigh accurately about 10 mg of the sample, and dissolve in 20 mL of a 1:1 acetonitrile–water solution. Filter the solution through a 0.45-μm Millipore filter, or equivalent.

System Suitability   Inject duplicate 10-μL portions of the *Standard Solution* into the chromatograph. The retention time of the Monoammonium Glycyrrhizinate is approximately 6 min. Adjust the operating conditions if necessary. The mean standard deviation for replicate injections is not more than 2.0%.

Procedure   Separately inject, in duplicate, 10-μL volumes of the *Standard Solution* and the *Assay Solution* into the chromatograph, and determine the mean peak area for each solution. Calculate the percentage of Monoammonium Glycyrrhizinate, equivalent to $C_{42}H_{65}NO_{16}$, in the portion of Monoammonium Glycyrrhizinate taken by the formula

$$100(20C_S/W_U)(A_U/A_S),$$

in which C_S is the concentration, in mg/mL, of the *Standard Solution*; A_S and A_U are the peak areas of the *Standard Solution* and the *Assay Solution*, respectively; and W_U is the weight, in mg, of the sample taken.

Ash (Total)   Proceed as directed in the general method, Appendix IIC.

Heavy Metals   A 2-g sample meets the requirements of the *Heavy Metals Test, Method II*, Appendix IIIB, using 20 μg of lead ion (Pb) in the control (*Solution A*).

Loss on Drying, Appendix IIC   Dry 1 g at 78° for 4 h under 1-mm Hg vacuum.

Specific Rotation, Appendix IIB   Determine in a solution containing a 1.5-g undried sample in sufficient 40% ethanol to make 100 mL.

Packaging and Storage   Store in a tight container in a cool, dry place.

Mono- and Diglycerides

DESCRIPTION

Mono- and Diglycerides consist of mixtures of glycerol mono- and diesters, with minor amounts of triesters, of edible fats or oils or edible fat-forming fatty acids. Those commercially available vary in consistency from yellow liquids through ivory-colored plastics to hard, ivory-colored solids having a bland odor and taste. They are insoluble in water, but are soluble in alcohol, in ethyl acetate, and in chloroform and other chlorinated hydrocarbons.

Functional Use in Foods   Emulsifier; stabilizer.

REQUIREMENTS

Acid Value   Not more than 6.
Arsenic (as As)   Not more than 3 mg/kg.
Free Glycerin   Not more than 7.0%.
Heavy Metals (as Pb)   Not more than 10 mg/kg.
Residue on Ignition   Not more than 0.5%.

The following specifications should conform to the representations of the vendor: **1-Monoglyceride Content, Total Monoglycerides, Hydroxyl Value, Iodine Value,** and **Saponification Value**.

TESTS

1-Monoglyceride Content   Determine as directed in the general method, Appendix VII.

Acid Value   Determine as directed for *Acid Value, Method II*, under *Fats and Related Substances*, Appendix VII.

Arsenic   A *Sample Solution* prepared as directed for organic compounds meets the requirements of the *Arsenic Test*, Appendix IIIB.

Free Glycerin   Determine as directed in the general method, Appendix VII.

Heavy Metals   Prepare and test a 2-g sample as directed in *Method II* under the *Heavy Metals Test*, Appendix IIIB, using 20 μg of lead ion (Pb) in the control (*Solution A*).

Hydroxyl Value   Determine as directed under *Method II* in the general method, Appendix VII.

Iodine Value   Determine by the *Modified Wijs Method*, Appendix VII.

Residue on Ignition   Ignite 5 g as directed in the general method, Appendix IIC.

Saponification Value   Determine as directed for *Saponification Value* under *Fats and Related Substances*, Appendix VII, using about 4 g, accurately weighed.

Total Monoglycerides   Determine as directed in the general method, Appendix VII.

Packaging and Storage   Store in well-closed containers.

---

[1] Third Street and Jefferson Avenue, Camden, NJ 08104.

Monoglyceride Citrate

CAS: [36291-32-4]

DESCRIPTION

A mixture of glyceryl monooleate and its citric acid monoester, manufactured by the reaction of glyceryl monooleate with citric acid under controlled conditions. It occurs as a viscous amber liquid. It is dispersible in most common fat solvents and in alcohol, but is insoluble in water.

Functional Use in Foods Synergist and solubilizer for antioxidants.

REQUIREMENTS

Acid Value Between 70 and 100.
Heavy Metals (as Pb) Not more than 10 mg/kg.
Residue on Ignition Not more than 0.3%.
Saponification Value Between 260 and 265.
Total Citric Acid Between 14.0% and 17.0%.
Water Not more than 0.2%.

TESTS

Acid Value Determine as directed for *Acid Value, Method II*, under *Fats and Related Substances*, Appendix VII.

Heavy Metals Prepare and test a 2-g sample as directed in *Method II* under the *Heavy Metals Test*, Appendix IIIB, using 20 μg of lead ion (Pb) in the control (*Solution A*).

Residue on Ignition Ignite 1 g as directed in the general method, Appendix IIC.

Saponification Value Determine as directed for *Saponification Value* under *Fats and Related Substances*, Appendix VII.

Total Citric Acid

Standard Solution Transfer about 35 mg of sodium citrate dihydrate, accurately weighed, into a 100-mL volumetric flask, dissolve and dilute to volume with water, and mix. Calculate the concentration (C), in μg/mL, of citric acid in the final solution by the formula

$$1000 \times 0.6533W/100,$$

in which W is the weight, in mg, of the sodium citrate dihydrate taken, and 0.6533 is a factor converting sodium citrate dihydrate to citric acid.

Sample Solution Transfer about 150 mg of the sample, accurately weighed, into a saponification flask, add 50 mL of 4% alcoholic potassium hydroxide solution, and reflux for 1 h. Acidify the reaction mixture with hydrochloric acid to a pH of 2.8 to 3.2, transfer to a 400-mL beaker, and evaporate to dryness on a steam bath. Quantitatively transfer the contents of the beaker into a separator, using no more than 50 mL of water, and then extract with three 50-mL portions of petroleum ether (b.p. 30° to 60°), discarding the extracts. Transfer the water layer to a 100-mL volumetric flask, dilute to volume with water, and mix.

Procedure Pipet 2.0 mL each of the *Standard Solution* and of the *Sample Solution* into separate 40-mL graduated centrifuge tubes, and to each tube add 2 mL of dilute sulfuric acid (1 in 2) and 11 mL of water. Boil for 3 min, cool, and add 5 mL of bromine TS to each tube. Dilute to the 20-mL mark, allow to stand for 10 min, and centrifuge. Transfer 4.0 mL of each solution into separate 19- × 110-mm test tubes, add 1 mL of water, 0.5 mL of dilute sulfuric acid (1 in 2), and 0.3 mL of 1 M potassium bromide, and shake. Add 0.3 mL of 1.5 N potassium permanganate, shake, and allow to stand for 2 min. Add 1 mL of a saturated solution of ferrous sulfate, shake, allow to stand for 2 min, and then dilute to 10 mL with water. Add 10.0 mL of n-hexane (previously washed with sulfuric acid, followed by a water wash, and then dried over anhydrous sodium sulfate), shake vigorously for 2 min, and then centrifuge at a low speed for 1 min. Transfer 5.0 mL of the hexane extract into a 20- × 145-mm tube containing 10.0 mL of sodium sulfide solution (4 g of $Na_2S.9H_2O$ in each 100 mL of water), and shake vigorously briefly (3 oscillations only). Centrifuge the mixture at low speed for 1 min. Immediately determine the absorbance of each aqueous layer in a 1-cm cell at 450 nm with a suitable spectrophotometer, using a reagent blank in the reference cell. Calculate the quantity, in mg, of citric acid in the sample taken by the formula

$$0.1C \times A_U/A_S,$$

in which C is as defined under *Standard Solution*, A_U is the absorbance of the final solution from the *Sample Solution*, and A_S is the absorbance of the final solution from the *Standard Solution*.

Water Determine by the *Karl Fischer Titrimetric Method*, Appendix IIB.

Packaging and Storage Store in well-closed containers.

Monopotassium L-Glutamate

Monopotassium Glutamate Monohydrate; Potassium Glutamate; MPG

$$KOOCCH_2CH_2CH(NH_2)COOH.H_2O$$

$C_5H_8KNO_4.H_2O$ Formula wt 203.24

INS: 622 CAS: anhydrous [19473-49-5]

DESCRIPTION

A white, practically odorless, free-flowing, crystalline powder. It is hygroscopic, is freely soluble in water, and is slightly soluble in alcohol. The pH of a 1 in 50 solution is between 6.7 and 7.3.

Functional Use in Foods Flavor enhancer; salt substitute.

REQUIREMENTS

Identification

A. To 1 mL of a 1 in 30 solution add 1 mL of triketohydrindene hydrate TS and 100 mg of sodium acetate, and heat in a boiling water bath for 10 min. An intense, violet blue color is formed.

B. To 10 mL of a 1 in 10 solution add 5.6 mL of 1 N hydrochloric acid. A white, crystalline precipitate of glutamic acid forms on standing. When 6 mL of 1 N hydrochloric acid is added to the turbid solution, the glutamic acid dissolves on stirring.

C. A 1 in 10 solution gives positive tests for *Potassium*, Appendix IIIA.

Assay Not less than 98.5% and not more than 101.5% of $C_5H_8KNO_4 \cdot H_2O$, calculated on the dried basis.

Heavy Metals (as Pb) Not more than 0.002%.

Lead Not more than 10 mg/kg.

Loss on Drying Not more than 0.2%.

Specific Rotation $[\alpha]_D^{20°}$: Not less than +22.5° and not more than +24.0°, calculated on the dried basis.

TESTS

Assay Dissolve about 250 mg, accurately weighed, in 100 mL of glacial acetic acid. A few drops of water may be added prior to the addition of the acetic acid to effect faster dissolution of the sample. Titrate with 0.1 N perchloric acid in glacial acetic acid, determining the endpoint potentiometrically. Perform a blank determination (see *General Provisions*), and make any necessary correction. Each mL of 0.1 N perchloric acid is equivalent to 10.16 mg of $C_5H_8KNO_4 \cdot H_2O$.

Heavy Metals Prepare and test a 1-g sample as directed in *Method II* under the *Heavy Metals Test*, Appendix IIIB, using 20 µg of lead ion (Pb) in the control (*Solution A*).

Lead A *Sample Solution* prepared as directed for organic compounds meets the requirements of the *Lead Limit Test*, Appendix IIIB, using 10 µg of lead ion (Pb) in the control.

Loss on Drying, Appendix IIC Dry at 80° in vacuum for 5 h.

Specific Rotation, Appendix IIB Determine in a solution containing 10 g in sufficient 2 N hydrochloric acid to make 100 mL.

Packaging and Storage Store in tight containers.

Monosodium L-Glutamate

Monosodium Glutamate Monohydrate; Sodium Glutamate; MSG

$$NaOOCCH_2CH_2CH(NH_2)COOH \cdot H_2O$$

$C_5H_8NNaO_4 \cdot H_2O$ Formula wt 187.13

INS: 621 CAS: monohydrate [6106-04-3]

DESCRIPTION

White, practically odorless, free-flowing crystals or crystalline powder. It is freely soluble in water, and is sparingly soluble in alcohol. It may have either a slightly sweet or a slightly salty taste. The pH of a 1 in 20 solution is between 6.7 and 7.2.

Functional Use in Foods Flavor enhancer.

REQUIREMENTS

Identification

A. To 1 mL of a 1 in 30 solution add 1 mL of triketohydrindene hydrate TS and 100 mg of sodium acetate and heat in a boiling water bath for 10 min. An intense, violet blue color is formed.

B. To 10 mL of a 1 in 10 solution add 5.6 mL of 1 N hydrochloric acid. A white crystalline precipitate of glutamic acid forms on standing. When 6 mL of 1 N hydrochloric acid is added to the turbid solution, the glutamic acid dissolves on stirring.

C. Prepare a 10% solution of the sample in 1 N hydrochloric acid. To 1 mL of this solution add 5 mL of cobalt–uranyl acetate TS, and agitate on a vortex mixer for 3 min. A golden yellow precipitate forms, indicating the presence of sodium.

Assay Not less than 98.5% and not more than 101.5% of $C_5H_8NNaO_4 \cdot H_2O$, calculated on the dried basis.

Chloride Not more than 0.2%.

Clarity and Color of Solution Passes test.

Heavy Metals (as Pb) Not more than 0.002%.

Lead Not more than 10 mg/kg.

Loss on Drying Not more than 0.5%.

Specific Rotation $[\alpha]_D^{20°}$: Between +24.8° and +25.3°, calculated on the dried basis; or $[\alpha]_{546.1\,nm}^{25°}$: Between +29.7° and +30.2°, calculated on the dried basis.

TESTS

Assay Dissolve about 250 mg, accurately weighed, in 100 mL of glacial acetic acid. A few drops of water may be added prior to the addition of the acetic acid to effect faster dissolution of the sample. Titrate with 0.1 N perchloric acid in glacial acetic acid, determining the endpoint potentiometrically. Perform a blank determination (see *General Provisions*), and make any necessary correction. Each mL of 0.1 N perchloric acid is equivalent to 9.356 mg of $C_5H_8NNaO_4 \cdot H_2O$.

Chloride, Appendix IIIB Any turbidity produced by a 10-mg sample does not exceed that shown in a control containing 20 μg of chloride ion (Cl).

Clarity and Color of Solution A solution of 1 g of the sample in 10 mL of water is colorless and has no more turbidity than a standard mixture prepared as follows: Dilute 0.2 mL of *Standard Chloride Solution* (see Appendix IIIB) to 20 mL with water, add 1 mL of dilute nitric acid (1 in 3), 0.2 mL of a 1 in 50 solution of dextrin, and 1 mL of a 1 in 50 solution of silver nitrate, mix, and allow to stand for 15 min.

Heavy Metals Prepare and test a 1-g sample as directed in *Method II* under the *Heavy Metals Test*, Appendix IIIB, using 20 μg of lead ion (Pb) in the control (*Solution A*).

Lead A *Sample Solution* prepared as directed for organic compounds meets the requirements of the *Lead Limit Test*, Appendix IIIB, using 10 μg of lead ion (Pb) in the control.

Loss on Drying, Appendix IIC Dry at 100° for 5 h.

Specific Rotation, Appendix IIB Determine in a solution containing 10 g in sufficient 2 N hydrochloric acid to make 100 mL.

Packaging and Storage Store in tight containers.

Morpholine

Tetrahydro-2H-1,4-oxazine; Diethylene Oximide; Diethylene Imidoxide

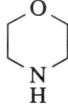

C$_4$H$_9$NO Formula wt 87.12
CAS: [110-91-8]

DESCRIPTION

A clear, colorless, mobile, hygroscopic liquid with a characteristic amine odor. It is miscible with water with the evolution of some heat. It is also miscible with acetone, ether, castor oil, methanol, and alcohol as well as in many oils such as linseed and pine.

Caution: Dermal irritant.

Functional Use in Foods Boiler water additive.

REQUIREMENTS

Identification The infrared absorption spectrum of a neat dispersion of the sample between two sodium chloride plates exhibits maxima at the same wavelengths as the typical spectrum, as shown in the section on *Infrared Spectra (Series C: Other Substances)*.

Assay Not less than 99.0%.
Distillation Range Between 126.0° and 130.0°.
Heavy Metals (as Pb) Not more than 1 mg/kg.
Refractive Index Between 1.454 and 1.455 at 20°.
Specific Gravity Between 0.994 and 0.997 at 25°.

TESTS

Assay

Mixed Indicator Solution Prepare separate 0.1% solutions of bromocresol green in methanol and the sodium salt of methyl red in water. Mix 5 parts by volume of the bromocresol green solution with one part of the methyl red solution.

Transfer 50 mL of water to a 250-mL flask. Add 0.4 mL of mixed indicator solution, and neutralize by the drop-wise addition of 0.1 N hydrochloric acid just to the disappearance of the green color. Weigh accurately 1.4 to 1.6 g of sample into the flask, and swirl to effect complete solution. Titrate with standard 0.5 N hydrochloric acid to the disappearance of the green color. Each mL of 0.5 N hydrochloric acid corresponds to 43.56 mg of C$_4$H$_9$NO.

Distillation Range Proceed as directed in the general method, Appendix IIB.

Heavy Metals Evaporate about 20.0 g of the sample to dryness on a steam bath in a glass evaporating dish. Cool, add 2 mL of hydrochloric acid, and slowly evaporate to dryness again on the steam bath. Moisten the residue with 1 drop of hydrochloric acid, add 10 mL of hot water, and digest for 2 min. Cool, and dilute to 25 mL with water. This solution meets the requirements of the *Heavy Metals Test*, Appendix IIIB, using 20 μg of lead ion (Pb) in the control (*Solution A*).

Refractive Index Determine as directed in the general method, Appendix IIB.

Specific Gravity Determine at 25° by any reliable method (see *General Provisions*).

Packaging and Storage Store in tight containers.

Mustard Oil

DESCRIPTION

The volatile oil obtained by the steam and water distillation of the comminuted presscakes of the seeds from *Brassica nigra* (Linnaeus) W.D.J. Koch or *Brassica juncea* (Linnaeus) Czernjajev. The essential oil forms upon maceration of the comminuted seeds in warm water that releases sinigrin, a β-glucopyranoside, which is subsequently enzymatically hydrolyzed to allyl isothiocyanate. The oil is a clear, pale yellow liquid with a sharp, pungent taste.

Caution: Mustard Oil is a lachrymator.

Functional Use in Foods Flavoring agent.

REQUIREMENTS

Identification The infrared absorption spectrum of the sample exhibits relative maxima (that may vary in intensity) at the same wavelengths (or frequencies) as those shown in the respective spectrum for allyl isothiocyanate in the section on *Infrared Spectra (Series B: Flavor Chemicals)*, using the same test conditions as specified therein.
Assay Not less than 93.0%, as C_3H_5NCS (allyl isothiocyanate).
Refractive Index Between 1.524 and 1.534 at 20°.
Specific Gravity Between 1.008 and 1.019.

TESTS

Assay Determine as directed under *Assay by Gas Chromatography, General Method, Polar Column M-1a* in Section 3.
Refractive Index Determine as directed under *Refractive Index*, Appendix IIB, using 20° as the determination temperature.
Specific Gravity Determine by any reliable method (see *General Provisions*).

Packaging and Storage Store in tight containers in a cool, dry place protected from light.

Myristic Acid

Tetradecanoic Acid

$$CH_3(CH_2)_{12}COOH$$

$C_{14}H_{28}O_2$ Formula wt 228.38

CAS: [544-63-8]

DESCRIPTION

A solid, organic acid obtained from coconut oil and other fats. It occurs as a hard, white or faintly yellowish, somewhat glossy, crystalline solid, or as a white or yellowish white powder. Myristic Acid is practically insoluble in water, but is soluble in alcohol, in chloroform, and in ether.

Functional Use in Foods Component in the manufacture of other food-grade additives; defoaming agent.

REQUIREMENTS

Acid Value Between 242 and 249.
Heavy Metals (as Pb) Not more than 10 mg/kg.
Iodine Value Not more than 1.0.
Residue on Ignition Not more than 0.1%.
Saponification Value Between 242 and 251.
Titer (Solidification Point) Between 48° and 55.5°.
Unsaponifiable Matter Not more than 1%.
Water Not more than 0.2%.

TESTS

Acid Value Determine as directed for *Acid Value, Method I*, under *Fats and Related Substances*, Appendix VII.
Heavy Metals Prepare and test a 2-g sample as directed in *Method II* under the *Heavy Metals Test*, Appendix IIIB, using 20 µg of lead ion (Pb) in the control (*Solution A*).
Iodine Value Determine by the *Modified Wijs Method*, Appendix VII.
Residue on Ignition, Appendix IIC Ignite 2 g as directed in the general method.
Saponification Value Determine as directed for *Saponification Value* under *Fats and Related Substances*, Appendix VII, using about 3 g, accurately weighed.
Titer (Solidification Point) Determine as directed under *Solidification Point*, Appendix IIB.
Unsaponifiable Matter, Appendix VII Determine as directed in the general method.
Water Determine by the *Karl Fischer Titrimetric Method*, Appendix IIB.

Packaging and Storage Store in well-closed containers.

Myrrh Oil

CAS: [8016-37-3]

DESCRIPTION

The volatile oil obtained by steam distillation from myrrh gum obtained from several species of *Commiphora* (Fam. *Burseraceae*). It is a light brown or green liquid having the characteristic odor of the gum. It is soluble in most fixed oils, but is only slightly soluble in mineral oil. It is insoluble in glycerin and in propylene glycol. It becomes darker in color and more viscous under the influence of air and light.

Functional Use in Foods Flavoring agent.

REQUIREMENTS

Identification The infrared absorption spectrum of the sample exhibits relative maxima (that may vary in intensity) at the same wavelengths (or frequencies) as those shown in the respective spectrum in the section on *Infrared Spectra (Series A: Essential Oils)*, using the same test conditions as specified therein.
Acid Value Between 2 and 13.
Angular Rotation Between −60° and −83°.
Heavy Metals (as Pb) Passes test.
Refractive Index Between 1.519 and 1.528 at 20°.
Saponification Value Between 9 and 35.
Solubility in Alcohol Passes test.
Specific Gravity Between 0.985 and 1.014.

TESTS

Acid Value Determine as directed for *Acid Value* under *Essential Oils and Flavors*, Appendix VI.

Angular Rotation Determine in a 100-mm tube as directed under *Optical (Specific) Rotation*, Appendix IIB.

Heavy Metals Shake 10 mL of the oil with an equal volume of water to which 1 drop of hydrochloric acid has been added, and pass hydrogen sulfide through the mixture until it is saturated. No darkening in color is produced in either the oil or the water.

Refractive Index, Appendix IIB Determine with an Abbé or other refractometer of equal or greater accuracy.

Saponification Value Determine as directed for *Saponification Value* under *Fats and Related Substances*, Appendix VII, using about 5 g, accurately weighed.

Solubility in Alcohol Proceed as directed in the general method, Appendix VI. One mL dissolves in 10 mL of 90% alcohol, occasionally with opalescence or turbidity.

Specific Gravity Determine by any reliable method (see *General Provisions*).

Packaging and Storage Store in full, tight, preferably glass containers in a cool place protected from light.

Natamycin

Pimaricin

$C_{33}H_{47}NO_{13}$ Formula wt 665.74

CAS: [7681-93-8]

DESCRIPTION

Off-white to cream-colored powder, which may contain up to 3 moles of water. It melts with decomposition at about 280°. Practically insoluble in water, slightly soluble in methanol, and soluble in glacial acetic acid and in dimethylformamide.

Functional Use in Foods Antimycotic.

REQUIREMENTS

Identification Transfer 50 mg, accurately weighed, to a 200-mL volumetric flask, add 5.0 mL of water, and moisten the specimen. Add 100 mL of a 1 in 1000 solution of glacial acetic acid in methanol, and shake by mechanical means in the dark until dissolved. Dilute with the acetic acid–methanol solution to volume, and mix. Transfer 2.0 mL of this solution to a 100-mL volumetric flask, dilute with the acetic acid–methanol solution to volume, and mix: The ultraviolet absorption spectrum of the solution so obtained exhibits maxima and minima at the same wavelengths as those of a similar solution of USP Natamycin Reference Standard, concomitantly measured.

Assay Not less than 97.0% and not more than 102.0% of $C_{33}H_{47}NO_{13}$, calculated on the anhydrous basis.

Heavy Metals (as Pb) Not more than 0.002%.

pH Between 5.0 and 7.5.

Specific Rotation $[\alpha]_D^{20°}$: Between +276° and +280°.

Water Between 6.0% and 9.0%.

TESTS

Assay (Note: Throughout this *Assay*, protect from direct light all solutions containing Natamycin.)

Mobile Phase Dissolve 3.0 g of ammonium acetate and 1.0 g of ammonium chloride in 760 mL of water, and mix. Add 5.0 mL of tetrahydrofuran and 240 mL of acetonitrile, mix, and filter through a 0.5-μm or finer porosity filter. Make adjustments if necessary to meet the system suitability requirements.

Standard Preparation Transfer about 20 mg of USP Natamycin Reference Standard, accurately weighed, to a 100-mL volumetric flask. Add 5.0 mL of tetrahydrofuran, and sonicate for 10 min. Add 60 mL of methanol, and swirl to dissolve. Add 25 mL of water, and mix. Allow to cool to room temperature. Dilute with water to volume, mix, and filter through a membrane filter of 5-μm or finer porosity.

Resolution Solution Dissolve 20 mg of Natamycin in a mixture of 99 mL of methanol and 1 mL of 0.1 N hydrochloric acid, and allow to stand for 2 h.

Note: Use this solution within 1 h.

Assay Preparation Transfer about 20 mg of Natamycin, accurately weighed, to a 100-mL volumetric flask. Proceed as directed under *Standard Preparation*, beginning with "add 5.0 mL of tetrahydrofuran. . . ."

Chromatographic System (see *Chromatography*, Appendix IIA) Use a high-pressure liquid chromatograph equipped with an ultraviolet detector measuring at 303 nm and a 4.6-mm × 25-cm column packed with octadecylsilanized silica (Supelcosil LC 18 or equivalent). The flow rate is about 3 mL/min. Chromatograph the *Standard Preparation*, and record the peak responses: The column efficiency is not less than 3000 theoretical plates, the tailing factor is between 0.8 and 1.3, and the relative standard deviation for three replicate injections is not more than 1.0%. Chromatograph the *Resolution Solution*. The resolution between Natamycin and its methyl ester is not less than 2.5. The relative retention times are about 0.7 for Natamycin and 1.0 for its methyl ester.

Procedure Separately inject about 20 μL each of the *Standard Preparation* and the *Assay Preparation* into the chromatograph, and record the peak areas of the major peaks. Calculate the percentage of Natamycin in the portion taken by the formula

$$0.1(W_s P_s / W_u)(r_u / r_s),$$

in which W_s is the weight, in mg, of USP Natamycin Reference Standard taken to prepare the *Standard Preparation*; P_s is the stated content, in μg/mg, of USP Natamycin Reference Standard; W_u is the weight, in mg, of Natamycin taken to prepare the *Assay Preparation*; and r_u and r_s are the peak area responses obtained with the *Assay Preparation* and the *Standard Preparation*, respectively.

Heavy Metals Prepare and test a 1-g sample as directed in *Method II* under the *Heavy Metals Test*, Appendix IIIB, using 20 μg of lead ion (Pb) in the control (*Solution A*).

pH Determine by the *Potentiometric Method*, Appendix IIB, using an aqueous suspension containing 10 mg/mL.

Specific Rotation, Appendix IIB Determine in a solution containing 100 mg in each 10 mL of glacial acetic acid.

Water Determine by the *Karl Fischer Titrimetric Method*, Appendix IIB.

Packaging and Storage Store in tight, light-resistant containers in a cool place.

Niacin

Nicotinic Acid; 3-Pyridinecarboxylic Acid

$C_6H_5NO_2$ Formula wt 123.11

CAS: [59-67-6]

DESCRIPTION

White or light yellow crystals or a crystalline powder. It is odorless or has a slight odor. One g dissolves in 60 mL of water. It is freely soluble in boiling water and in boiling alcohol and also in solutions of alkali hydroxides and carbonates. It is almost insoluble in ether.

Functional Use in Foods Nutrient; dietary supplement.

REQUIREMENTS

Identification

A. Triturate a sample with twice its weight of 2,4-dinitrochlorobenzene. Gently heat about 10 mg of the mixture in a test tube until melted, and continue the heating for a few s longer. Cool, and add 3 mL of 0.5 N alcoholic potassium hydroxide. A deep red to deep wine-red color is produced.

B. Dissolve about 50 mg in 20 mL of water, neutralize to litmus paper with 0.1 N sodium hydroxide, and add 3 mL of cupric sulfate TS. A blue precipitate gradually forms.

C. The infrared absorption spectrum of a mineral oil dispersion of the sample, previously dried at 105° for 1 h, exhibits maxima only at the same wavelengths as that of a similar preparation of USP Niacin Reference Standard.

D. Determine the absorbance of a solution of the sample containing 20 μg in each mL of water in a 1-cm cell at 237 nm and 262 nm, using water as the blank. The ratio A_{237}/A_{262} is between 0.35 and 0.39.

Assay Not less than 99.5% and not more than 101.0% of $C_6H_5NO_2$, calculated on the dried basis.

Heavy Metals (as Pb) Not more than 0.002%.

Loss on Drying Not more than 1.0%.

Melting Range Between 234° and 238°.

Residue on Ignition Not more than 0.1%.

TESTS

Assay Dissolve about 300 mg, accurately weighed, in about 50 mL of water, add phenolphthalein TS, and titrate with 0.1 N sodium hydroxide. Perform a blank determination (see *General Provisions*). Each mL of 0.1 N sodium hydroxide is equivalent to 12.31 mg of $C_6H_5NO_2$.

Heavy Metals Mix 1 g with 2 mL of 1 N acetic acid, add water to make 25 mL, heat gently until solution is complete, and cool. This solution meets the requirements of the *Heavy Metals Test*, Appendix IIIB, using 20 μg of lead ion (Pb) in the control (*Solution A*).

Loss on Drying, Appendix IIC Dry at 105° for 1 h.

Melting Range Determine as directed for *Melting Range or Temperature*, Appendix IIB.

Residue on Ignition, Appendix IIC Ignite 1 g as directed in the general method.

Packaging and Storage Store in well-closed containers.

Niacinamide

Nicotinamide

$C_6H_6N_2O$ Formula wt 122.13

CAS: [98-92-0]

DESCRIPTION

A white, crystalline powder. It is odorless or nearly so, and has a bitter taste. Its solutions are neutral to litmus. One g dissolves in about 1 mL of water, in about 1.5 mL of alcohol, and in about 10 mL of glycerin.

Functional Use in Foods Nutrient; dietary supplement.

REQUIREMENTS

Identification

A. Transfer about 20 mg to a 1000-mL volumetric flask, dissolve and dilute with water to volume, mix, and determine the absorbance of the solution in a 1-cm cell at 245 nm and at 262 nm with a suitable spectrophotometer, using water as the blank. The ratio A_{245}/A_{262} is 0.65 ± 0.02.

B. To 20 mg in a test tube add 1 pellet of 1 N sodium hydroxide, and heat gently over an open flame. The odor of ammonia is perceptible. Upon heating more vigorously, the odor of pyridine is perceptible.

C. The infrared absorption spectrum of a potassium bromide dispersion of the sample exhibits maxima at the same wavelengths as that of a similar preparation of USP Niacinamide Reference Standard.

Assay Not less than 98.5% and not more than 101.0% of $C_6H_6N_2O$ after drying.
Heavy Metals (as Pb) Not more than 0.002%.
Loss on Drying Not more than 0.5%.
Melting Range Between 128° and 131°.
Readily Carbonizable Substances Passes test.
Residue on Ignition Not more than 0.1%.

TESTS

Assay Dissolve about 300 mg, previously dried over silica gel for 4 h and accurately weighed, in 20 mL of glacial acetic acid, warming slightly if necessary to effect solution. Add 100 mL of benzene and 2 drops of crystal violet TS, and titrate with 0.1 N perchloric acid. Perform a blank determination (see *General Provisions*), and make any necessary correction. Each mL of 0.1 N perchloric acid is equivalent to 12.21 mg of $C_6H_6N_2O$.

Heavy Metals Prepare and test a 1-g sample as directed in *Method II* under the *Heavy Metals Test*, Appendix IIIB, using 20 µg of lead ion (Pb) in the control (*Solution A*).
Loss on Drying, Appendix IIC Dry over silica gel for 4 h.
Melting Range Determine as directed for *Melting Range or Temperature*, Appendix IIB.
Readily Carbonizable Substances, Appendix IIB Dissolve 200 mg in 5 mL of 95% sulfuric acid. The solution has no more color than *Matching Fluid A*.
Residue on Ignition, Appendix IIC Ignite 1 g as directed in the general method.

Packaging and Storage Store in tight containers.

Niacinamide Ascorbate

DESCRIPTION

A complex of ascorbic acid ($C_6H_8O_6$) and niacinamide ($C_6H_6N_2O$). Niacinamide Ascorbate is a lemon yellow-colored powder that is odorless or has a very slight odor. It may gradually darken upon exposure to air. One g is soluble in 3.5 mL of water and in about 20 mL of alcohol. It is very slightly soluble in chloroform and in ether and is sparingly soluble in glycerin.

Functional Use in Foods Nutrient; dietary supplement.

REQUIREMENTS

Identification

A. A 1 in 50 solution responds to the *Identification Tests* in the monograph for *Ascorbic Acid*.

B. An 80-mg sample responds to *Identification Test B* in the monograph for *Niacinamide*.

Assay Not less than 73.5% of ascorbic acid ($C_6H_8O_6$) and not less than 24.5% of niacinamide ($C_6H_6N_2O$), calculated on the anhydrous basis. The total of ascorbic acid and niacinamide is not less than 99.0%.
Heavy Metals (as Pb) Not more than 0.0025%.
Loss on Drying Not more than 0.5%.
Melting Range Between 141° and 145°.
Residue on Ignition Not more than 0.1%.

TESTS

Assay for Ascorbic Acid Proceed as directed under *Assay* in the monograph for *Ascorbic Acid*.
Assay for Niacinamide Proceed as directed under *Assay* in the monograph for *Niacinamide*, but use an undried sample and make the calculation on the anhydrous basis.
Heavy Metals Prepare and test an 800-mg sample as directed in *Method II* under the *Heavy Metals Test*, Appendix IIIB, using 20 µg of lead ion (Pb) in the control (*Solution A*).
Loss on Drying, Appendix IIC Dry to constant weight at 75°.
Melting Range Determine as directed for *Melting Range or Temperature*, *Procedure for Class Ia*, Appendix IIB.
Residue on Ignition, Appendix IIC Ignite 2 g as directed in the general method.

Packaging and Storage Store in tight, light-resistant containers.

Nickel

Nickel Catalysts

Ni   Atomic Wt 58.69
                                                 CAS: [7440-02-0]

DESCRIPTION

Nickel metal is a lustrous, white, hard, ferromagnetic, metallic solid with no characteristic odor. Nickel is commonly used as a catalyst for hydrogenation reactions for food chemicals. Depending on the use, Nickel Catalysts fall into two general categories: *Sponge Nickel Catalyst* and *Supported Nickel Catalyst*.

Sponge Nickel Catalyst is typically used in the manufacture of amines and polyols. It is prepared by chemically treating a Nickel–aluminum amalgam with sodium hydroxide to remove the majority of aluminum, thus leaving a highly porous (skeletal) Nickel solid. The resulting *Sponge Nickel Catalyst* is extremely pyrophoric in air and must be stored under an inert liquid.

Supported Nickel Catalyst is typically used in the manufacture of edible oils. It is prepared from a Nickel salt deposited onto an inert carrier consisting of various types of acceptable silicas, aluminas, or combinations thereof. The Nickel salt-carrier complex is not catalytically active and is converted to *Supported Nickel Catalyst* in a stream of hydrogen at elevated temperatures. After activation, *Supported Nickel Catalyst* is also pyrophoric and must be protected from air typically by suspending it in a food-grade stearine. It usually is supplied in the form of droplets or flakes.

Functional Use in Foods Catalyst for hydrogenation reactions.

REQUIREMENTS

Identification

Sponge Nickel Catalyst Dissolve approximately 100 mg in about 2 mL of hydrochloric acid, and dilute to about 20 mL with water. Place 5 mL of this solution in a test tube, add a few drops of bromine water, and make it slightly alkaline with ammonium hydroxide. Add 2 to 3 mL of a 1% solution of dimethylglyoxime in alcohol. An intense red color or precipitate develops.

Supported Nickel Catalyst Ash as described under *Assay*. Transfer a 5-mL aliquot of the ashed sample solution to a test tube, and complete the identification as described above.

Assay

Sponge Nickel Catalyst Not less than 83.0% of Ni on the as-is basis.

Caution: Handle with extreme care—Sponge Nickel is pyrophoric when dried.

Supported Nickel Catalyst Not less than 20.0% of Ni and not more than 27.0% of Ni on the as-is basis.

TESTS

Assay

Sponge Nickel Catalyst Weigh accurately about 500 mg of wet sample, place it into a 400-mL beaker, and dissolve it in 20 mL of 1:1 hydrochloric acid, heating if necessary. Dilute to 250 mL with water. Add 2 g of tartaric acid, heat to about 80°, and add 30 mL of a 1% solution of dimethylglyoxime in alcohol. Add ammonium hydroxide until the solution is slightly ammoniacal, and place the mixture on a steam bath for 20 min. Filter the precipitate into a tared, fritted-glass, medium-porosity filter crucible, and wash with hot water until the filtrate is free of chloride. Dry the precipitate at 120° for 2 h, and then to constant weight, and weigh. Calculate the percent Nickel by the following formula:

$$(W_P 20.32)/W_S,$$

in which W_P is the weight, in g, of the precipitate; 20.32 is the percent of Nickel in the precipitate; and W_S is the weight, in g, of the sample taken.

Supported Nickel Catalyst Fill a 100-mL porcelain crucible half-full of ashless filter paper pulp. Accurately weigh 2 g of the finished catalyst, in droplet or flake form, and place it on top of the paper pulp. Transfer the crucible to a muffle furnace set at room temperature, and slowly raise the temperature to 650° so that the stearine melts into the paper, and the organic mass burns and chars slowly. Continue heating at 650° for 2 h or until the carbon is burned off. Then cool, add 20 mL of hydrochloric acid, quantitatively transfer the solution or suspension to a 400-mL beaker, and carefully evaporate to dryness on a steam bath. Cool, add 20 mL of hydrochloric acid, warm to aid dissolution (catalysts containing silica will not dissolve completely), transfer to a 500-mL volumetric flask, dilute to volume with water, and mix. Allow any solids to settle, pipet a clear, 50-mL aliquot into a 400-mL beaker, and dilute to 250 mL with water. (If there is suspended matter in the volumetric flask, filter a portion through a dry, medium-speed filter paper into a dry receiver, and pipet from the receiver.) Continue as directed above, beginning with "Add 2 g of tartaric acid. . . ."

Calculate the percent Nickel by the following formula:

$$10(W_P 20.32)/W_S,$$

in which W_P is the weight, in g, of the precipitate; 20.32 is the percent of Nickel in the precipitate; and W_S is the weight, in g, of the sample taken.

Packaging and Storage

Sponge Nickel Catalyst Store under liquids such as water, alcohol, or methylcyclohexane, in a cool, dry place.

Supported Nickel Catalyst Store in tight containers in a cool, dry place.

Nisin Preparation

Contains 34 amino acids and has an approximate empirical formula of
$C_{143}H_{228}O_{37}N_{42}S_7$ Approximate Formula wt 3348

INS: 234   CAS: [1414-45-5]

DESCRIPTION

Nisin is a mixture of closely related polypeptides produced by strains of the *Lactococcus lactis* subsp. *lactis* Lancefield Group N in a sterilized milk-culture medium. Nisin in the fermentation broth can be recovered by various methods, such as injecting sterile, compressed air (froth concentration); acidification; salting out; and spray drying.

Nisin Preparation is a white, free-flowing powder comprising Nisin and sodium chloride that is adjusted to an activity level of not less than 900 IU/mg by the addition of sodium chloride and nonfat milk solids.

Functional Use in Foods Antimicrobial agent.

REQUIREMENTS

Assay Not less than 900 IU of Nisin per mg of Nisin Preparation.
Differentiation of Nisin from Other Antimicrobial Substances Passes tests.
Heavy Metals (as Pb) Not more than 10 mg/kg.
Loss on Drying Not more than 3.0%.
Microbial Limits:
  Aerobic Plate Count Not more than 10 per g.
  E. coli Negative in 25 g.
  Salmonella Negative in 25 g.
Sodium Chloride Content Not less than 50.0%.

TESTS

Assay

Assay Medium Dissolve 10 g of bacteriological peptone, 3 g of beef extract, 3 g of sodium chloride, 1.5 g of autolyzed yeast, 1 g of brown sugar, and 15 g of agar in distilled water to a final volume of 1000 mL. Sterilize in an autoclave at 121° for 15 min. The medium can be stored in a covered container at room temperature until use. At the time of use, melt the medium, and cool to approximately 50°. Add 2% of a 1:1 mixture of Tween 20 (polyoxyethylene sorbitan monolaurate) and distilled water, previously held for 20 to 30 min at 48°.

Assay Organism Maintain *Micrococcus luteus* (ATCC 10240[1], NCIMB 8166[2]) by subculturing on agar slants of the *Assay Medium* and incubating at 30° for 48 h. Prepared slants may be stored for a maximum of 14 days at 4° until required. When required for use, the growth on the slant cultures is suspended in 7 mL of sterilized normal saline solution.

Nisin Standard Solutions Suspend 100 mg of Nisin International Reference Preparation[3] (1000 IU of Nisin per mg), accurately weighed, in 80 mL of 0.02 N hydrochloric acid. Set aside at room temperature for 2 h. Dilute the suspension to a final volume of 100.0 mL with 0.02 N hydrochloric acid. This standard stock solution contains 1000 IU of Nisin per mL. From this solution, pipet 0.5, 1.0, 2.5, 5.0, and 10.0 mL into separate 1000-mL volumetric flasks. Dilute to volume with 0.02 N hydrochloric acid to obtain *Nisin Standard Solutions* with concentrations of 0.5, 1.0, 2.5, 5.0 and 10.0 IU/mL. Store the standard stock solution for up to 7 days at 4°, or prepare a fresh solution each day.

Preparation of the Standard Curve Using normal saline solution, dilute the suspension of the *Assay Organism* to 1 in 10, and mixing thoroughly, add 2 mL of this dilution to each 100 mL of melted *Assay Medium* held at 48°. Pour the inoculated medium to a depth of 3 to 4 mm into sterile, flat-bottomed Petri dishes, and allow to solidify. Invert the plates, and store at 4° for 1 h. Bore 64 8-mm holes on 30-mm centers in each plate of the agar medium with the aid of a sterile, hollow-steel borer, 7 to 9 mm in diameter, and discard the agar discs. Transfer, in quadruplicate, 0.20-mL volumes in the concentration range of 0.5 to 10.0 IU/mL of the *Nisin Standard Solutions* into the holes. Cover the plates, and incubate them overnight at 30°. Measure the zones of inhibition to the nearest 0.1 mm by means of calipers or other appropriate devices. Plot the log of the Nisin concentration in the critical range against the zone diameters, and draw the best straight line through the plotted points.

Procedure Suspend 100 mg of the sample in 80 mL of 0.02 N hydrochloric acid in a 100-mL volumetric flask, and set aside at room temperature for 2 h. Dilute the solution to volume by adding 0.02 N hydrochloric acid. Dilute to a 1 in 200 solution with 0.02 N hydrochloric acid. Proceed as described above for the *Standard Curve*, transferring in quadruplicate a measured volume of this solution into the holes of four agar discs. After incubation, measure the zones of inhibition. From the *Standard Curve*, determine the Nisin concentrations, and average the results.

Differentiation of Nisin from Other Antimicrobial Substances

Stability to Acid Suspend a 100-mg sample in 0.02 N hydrochloric acid as described in the preparation of the *Nisin Standard Solutions* under the *Assay*. Boil this solution for 5 min. Using the *Assay* method described above, determine the Nisin concentration. No significant loss of activity is noted following this heat treatment: The calculated Nisin concentration of the boiled sample is 100% ± 5% of the *Assay* value. Adjust the pH of the Nisin solution to 11.0 by adding 5 N sodium hydroxide. Heat the solution at 65° for 30 min, and then cool. Adjust the pH to 2.0 by adding hydrochloric acid dropwise. Again determine the Nisin concentration using the *Assay* method described above. Complete loss of the antimicrobial activity of Nisin is observed following this treatment.

Tolerance of Lactococcus lactis to High Concentrations of Nisin Prepare cultures of *Lactococcus lactis* (ATCC 11454, NCIMB 8586) in sterile, separated milk by incubating for 18 h at 30°. Prepare one or more flasks containing 100 mL of litmus milk, and sterilize at 121° for 15 min. Suspend 0.1 g of Nisin in the sterilized litmus milk, and allow to stand at room temperature for 2 h. Add 0.1 mL of the *L. lactis* culture, and incubate at 30° for 24 h. *L. lactis* will grow in this concentration of Nisin (about 1000 IU/mL); however, it will not grow in similar concentrations of other antimicrobial substances. This test will not differentiate Nisin from subtilin.

Heavy Metals Prepare and test a 2-g sample as directed for *Method II* under the *Heavy Metals Test*, Appendix IIIB, using 20 μg of lead ion (Pb) in the control (*Solution A*).

Loss on Drying, Appendix IIC Dry a 2-g sample at 105° for 2 h, and continue drying to a constant weight.

Microbial Limits:

  Aerobic Plate Count Proceed as directed in chapter 3 of the *FDA Bacteriological Analytical Manual*, Seventh Edition, 1992.

---

[1] ATCC is the American Type Culture Collection, 12301 Parklawn Drive, Rockville, MD 20852.

[2] NCIMB is the National Collection of Industrial and Marine Bacteria, NCIMB Ltd., 23 St. Machar Drive, Aberdeen, Scotland AB2 1RY.

[3] The Nisin International Reference Preparation is available from the WHO International Laboratory for Biological Standards, Ministry of Agriculture, Fisheries & Food, Central Veterinary Laboratory, New Haw, Weybridge, Surrey, England.

E. coli Proceed as directed in chapter 4 of the *FDA Bacteriological Analytical Manual*, Seventh Edition, 1992.

Salmonella Proceed as directed in chapter 5 of the *FDA Bacteriological Analytical Manual*, Seventh Edition, 1992.

Sodium Chloride Content Prepare and test a 250-mg sample as directed under *Assay* in the monograph for *Sodium Chloride*, except omit the drying. Calculate the percentage of sodium chloride in the sample.

Packaging and Storage Store in well-closed containers at temperatures not exceeding 22°.

Nitrogen

N_2 Formula wt 28.01

INS: 941 CAS: [7727-37-9]

DESCRIPTION

A colorless and odorless gas. It may be condensed to a colorless liquid boiling at −195.8° or to a white solid melting at −209.8°. It is nonflammable and does not support combustion. One L of the gas weighs about 1.25 g at 0° and a pressure of 760 mm of mercury. One volume of the gas dissolves in about 62 volumes of water and in about 8 volumes of alcohol at 20°, at a pressure of 760 mm of mercury.

Functional Use in Foods *Gas*: air and oxygen displacer; propellant and aerating agent; *liquid*: direct-contact freezing agent.

REQUIREMENTS

Note: Reduce the container pressure by means of a regulator. Measure the gas with a gas volume meter downstream from the pertinent detector tube to minimize contamination of or change to the specimens.

The detector tubes called for in certain **TESTS** are described under *Solutions and Indicators*.

Identification Insert a burning wood splinter into a test tube filled with the gas. The flame is extinguished.

Note: Use caution.

Assay Not less than 99.0% of N_2, by volume.
Carbon Dioxide Not more than 0.03%, by volume.
Carbon Monoxide Not more than 10 ppm, by volume.
Oxygen Not more than 1.0%, by volume.
Water Passes test.

TESTS

Assay Introduce a specimen of Nitrogen into a gas chromatograph by means of a gas sampling valve. Select the operating conditions of the gas chromatograph so that the standard peak signal resulting from the following procedure corresponds to not less than 70% of the full-scale reading. Preferably, use an apparatus corresponding to the general type in which the column is 3 m in length and 4 mm in inside diameter and is packed with a molecular sieve prepared from a synthetic alkali-metal aluminosilicate capable of absorbing molecules with diameters of up to 0.5 nm, which permits complete separation of oxygen from Nitrogen. Use industrial-grade helium (99.99%) as the carrier gas, with a thermal conductivity detector, and control the column temperature: The peak response produced by the assay specimen exhibits a retention time corresponding to that produced by the oxygen–helium certified standard (a mixture of 1.0% oxygen in industrial-grade helium is available from most suppliers) and is equivalent to not more than 1.0% of oxygen, by volume, when compared with the peak response of the oxygen–helium certified standard, indicating not less than 99.0% N_2, by volume.

Carbon Dioxide Pass 1050 ± 50 mL of the gas sample through a carbon dioxide detector tube at the rate specified for the tube. The indicator change corresponds to not more than 0.03%, by volume.

Carbon Monoxide Pass 1050 ± 50 mL of the gas sample through a carbon monoxide detector tube at the rate specified for the tube. The indicator change corresponds to not more than 10 ppm, by volume.

Oxygen Determine as directed under *Assay*.

Water Pass 24,000 mL of the gas sample through a suitable water-absorption tube no less than 100 mm in length, which previously has been flushed with about 500 mL of the sample and weighed. Regulate the flow so that about 60 min will be required for passage of the gas. The gain in weight of the absorption tube does not exceed 1.0 mg.

Packaging and Storage Preserve in cylinders.

Nitrogen Enriched Air

N_2 Formula wt 28.01

DESCRIPTION

Nitrogen Enriched Air is produced from air *in situ* by physical separation methods. It contains not less than 90% and not more than 99% nitrogen, by volume. The remaining components are noble gases and, primarily, oxygen.

Functional Use in Foods Air and oxygen displacer.

REQUIREMENTS

Note: Reduce the container pressure by means of a regulator. Measure the gas with a gas volume meter downstream from the detector tube to minimize contamination of or change to the specimens.

The detector tubes called for in certain **TESTS** are described under *Solutions and Indicators*.

Labeling Where the gas is piped from cylinders or directly from the collecting tank to the point of use, label each outlet "Nitrogen Enriched Air."

Identification Insert a burning wood splinter into a test tube filled with the gas. The flame is extinguished.

Note: Use caution.

Assay Not less than 90.0% and not more than 99.0% of N_2, by volume.

Carbon Dioxide Not more than 0.03%, by volume.

Carbon Monoxide Not more than 10 ppm, by volume.

Nitric Oxide and Nitrogen Dioxide Not more than 2.5 ppm, by volume.

Oxygen Not less than 1.0% and more than 10.0% of O_2, by volume.

Sulfur Dioxide Not more than 5 ppm, by volume.

Water Passes test.

TESTS

Assay Introduce a specimen of Nitrogen Enriched Air into a gas chromatograph by means of a gas sampling valve. Select the operating conditions of the gas chromatograph so that the peak signal of an oxygen–helium certified standard (a mixture of 5.0% oxygen, by volume, in industrial grade helium is available from most suppliers) resulting from the following procedure corresponds to approximately 45% of the full-scale reading. Preferably, use an apparatus corresponding to the general type in which the column is 3 m in length and 4 mm in inside diameter and is packed with a molecular sieve prepared from a synthetic alkali-metal aluminosilicate capable of absorbing molecules with diameters of up to 0.5 nm, which permits complete separation of oxygen from nitrogen. Use industrial-grade helium (99.99%) as the carrier gas, with a thermal conductivity detector, and control the column temperature: The peak response produced by the assay specimen exhibits a retention time corresponding to that produced by the 5.0% oxygen–helium standard and, when compared with the peak response of the standard, is equivalent to not less than 1.0% and not more than 10.0% of oxygen, indicating not less than 90.0% and not more than 99.0% N_2, by volume.

Carbon Dioxide Pass 1050 ± 50 mL of the gas sample through a carbon dioxide detector tube at the rate specified for the tube. The indicator change corresponds to not more than 0.03%, by volume.

Carbon Monoxide Pass 1050 ± 50 mL of the gas sample through a carbon monoxide detector tube at the rate specified for the tube. The indicator change corresponds to not more than 10 ppm, by volume.

Nitric Oxide and Nitrogen Dioxide Pass 550 ± 50 mL through a nitric oxide–nitrogen dioxide detector tube at the rate specified for the tube. The indicator change corresponds to not more than 2.5 ppm, by volume.

Oxygen Determine as directed under *Assay*.

Sulfur Dioxide Pass 1050 ± 50 mL through a sulfur dioxide detector tube at the rate specified for the tube. The indicator change corresponds to not more than 5 ppm, by volume.

Water Pass 24,000 mL of the gas sample through a suitable water-absorption tube no less than 100 mm in length, which previously has been flushed with about 500 mL of the sample and weighed. Regulate the flow so that about 60 min will be required for passage of the gas. The gain in weight of the absorption tube does not exceed 1.0 mg.

Packaging and Storage Preserve in metal cylinders or in a low-pressure collecting tank.

Nitrous Oxide

Nitrogen Oxide (N_2O)

N_2O Formula wt 44.01

INS: 942 CAS: [10024-97-2]

DESCRIPTION

A colorless gas without appreciable odor. One L at 0° and at a pressure of 760 mm of mercury weighs about 1.97 g. One volume dissolves in about 1.4 volumes of water at 20° and at a pressure of 760 mm of mercury. It is freely soluble in alcohol and soluble in ether and in oils.

Functional Use in Foods Propellant; aerating agent; gas.

REQUIREMENTS

Note: The following tests are designed to reflect the quality of Nitrous Oxide in both its vapor and liquid phases, which are present in previously unopened cylinders. Reduce the container pressure by means of a regulator. Withdraw the samples for the tests with the least possible release of Nitrous Oxide consistent with proper purging of the sample apparatus. Measure the gases with a gas volume meter downstream from the detector tubes to minimize contamination of or change to the specimens. Perform tests in the sequence in which they are listed under **REQUIREMENTS**.

The detector tubes called for in certain **TESTS** are described under *Solutions and Indicators*.

Identification

A. With the container temperatures the same and maintained between 15° and 25°, concomitantly read the pressure of the Nitrous Oxide container and of a container of 99.9% Nitrous Oxide certified standard (available from most suppliers).

Note: Do not use the Nitrous Oxide certified standard if it has been depleted to less than half of its full capacity.

The pressure of the Nitrous Oxide container is within 50 psi of that of the Nitrous Oxide certified standard.

B. Pass 100 ± 5 mL released from the vapor phase of the contents of the Nitrous Oxide container through a carbon dioxide detector tube (described under *Solutions and Indicators*) at the rate specified for the tube: No color change is observed (distinction from carbon dioxide).

C. Collect about 100 mL of the gas under test in a 100-mL tube fitted at the top with a stopcock. Open the stopcock, and quickly add a freshly prepared solution of 500 mg of pyrogallol in 2 mL of water and a freshly prepared solution of 12 g of potassium hydroxide in 8 mL of water. Immediately close the stopcock, and mix: The gas is not absorbed, and the solution does not become brown (distinction from oxygen).

Assay Not less than 99.0% N_2O, by volume.
Air Not more than 1.0%, by volume.
Ammonia Not more than 0.0025%, by volume.
Carbon Dioxide Not more than 0.03%, by volume.
Carbon Monoxide Not more than 10 ppm, by volume.
Halogens (as Cl) Not more than 1 ppm, by volume.
Nitric Oxide Not more than 1 ppm, by volume.
Nitrogen Dioxide Not more than 1 ppm, by volume.
Odor Passes test.
Water Not more than 150 mg/m^3.

TESTS

Assay Introduce a specimen of Nitrous Oxide taken from the liquid phase, as directed in the test for *Nitrogen Dioxide*, into a gas chromatograph by means of a gas sampling valve. Select the operating conditions of the gas chromatograph such that the peak response resulting from the following procedure corresponds to not less than 70% of the full-scale reading. Preferably, use an apparatus corresponding to the general type in which the column is 6 m in length and 4 mm in inside diameter and is packed with porous polymer beads, which permit complete separation of nitrogen and oxygen from Nitrous Oxide, although the nitrogen and oxygen may not be separated from each other. Use industrial-grade helium (99.99%) as the carrier gas, with a thermal-conductivity detector, and control the column temperature: The peak response produced by the assay specimen exhibits a retention time corresponding to that produced by an air–helium certified standard (described in the monograph for Helium) and is equivalent to not more than 1.0% of air, by volume, when compared with the peak response of the air–helium certified standard, indicating not less than 99.0% of N_2O, by volume.

Air Determine as directed in the *Assay*.

Ammonia Pass 1050 ± 50 mL, released from the vapor phase of the contents of the container, through an ammonia detector tube at the rate specified for the tube: The indicator change corresponds to not more than 0.0025%, by volume.

Carbon Dioxide Pass 1050 ± 50 mL, released from the vapor phase of the contents of the container, through a carbon dioxide detector tube at the rate specified for the tube: The indicator change corresponds to not more than 0.03%, by volume.

Carbon Monoxide Pass 1050 ± 50 mL, released from the vapor phase of the contents of the container, through a carbon monoxide detector tube at the rate specified for the tube: The indicator change corresponds to not more than 10 ppm, by volume.

Halogens Pass 1050 ± 50 mL, released from the vapor phase of the contents of the container, through a chlorine detector tube at the rate specified for the tube: The indicator change corresponds to not more than 1 ppm, by volume.

Nitric Oxide Pass 550 ± 50 mL, released from the vapor phase of the contents of the container, through a nitric oxide–nitrogen dioxide detector tube at the rate specified for the tube: The indicator change corresponds to not more than 1 ppm, by volume.

Nitrogen Dioxide Arrange a container so that when its valve is opened, a portion of the liquid phase of the contents is released through a piece of tubing of sufficient length to allow all of the liquid to vaporize during passage through it and to prevent frost from reaching the inlet of the detector tube. Release into the tubing a flow of liquid sufficient to provide 550 mL of the vaporized sample plus any excess necessary to ensure adequate flushing of air from the system. Pass 550 ± 50 mL of this gas through a nitric oxide–nitrogen dioxide detector tube at the rate specified for the tube: The indicator change corresponds to not more than 1 ppm, by volume.

Odor Carefully open the container valve to produce a moderate flow of gas. Do not direct the gas stream toward the face, but deflect a portion of the stream toward the nose: No appreciable odor is discernible.

Water Flush the regulator with 5 L or more of the gas specimen. Pass 50 ± 5 L, released from the vapor phase, through a water vapor detector tube connected to the regulator with a minimum length of metal or polyethylene tubing. Measure the gas passing through the detector tube with a gas flowmeter set at a flow rate of 2 L/min. The corrected indicator change corresponds to not more than 150 mg/m^3.

Packaging and Storage Preserve in cylinders.

Nutmeg Oil

Myristica Oil

CAS: [8008-45-5]

DESCRIPTION

The volatile oil obtained by steam distillation from the dried kernels of the ripe seed of *Myristica fragrans* Houttuyn (Fam. *Myristicaceae*). Two types of oil, the East Indian and the West Indian, are commercially available. It is a colorless or pale yellow liquid having the characteristic odor and taste of nutmeg. It is soluble in alcohol.

Functional Use in Foods Flavoring agent.

REQUIREMENTS

Labeling Indicate whether it is the East Indian or West Indian type.

Identification The infrared absorption spectrum of the sample exhibits relative maxima (that may vary in intensity) at the same wavelengths (or frequencies) as those shown in the respective spectrum in the section on *Infrared Spectra (Series A: Essential Oils)*, using the same test conditions as specified therein.
Angular Rotation *East Indian type*: between +8° and +30°; *West Indian type*: between +25° and +45°.
Heavy Metals (as Pb) Passes test.
Refractive Index *East Indian type*: between 1.474 and 1.488; *West Indian type*: between 1.469 and 1.476 at 20°.
Residue on Evaporation *East Indian type*: not more than 60 mg/3 mL; *West Indian type*: not more than 50 mg/3 mL.
Solubility in Alcohol Passes test.
Specific Gravity *East Indian type*: between 0.880 and 0.910; *West Indian type*: between 0.854 and 0.880.

TESTS

Angular Rotation Determine in a 100-mm tube as directed under *Optical (Specific) Rotation*, Appendix IIB.
Heavy Metals Shake 10 mL of the oil with an equal volume of water to which 1 drop of hydrochloric acid has been added, and pass hydrogen sulfide through the mixture until it is saturated. No darkening in color is produced in either the oil or the water.
Refractive Index, Appendix IIB Determine with an Abbé or other refractometer of equal or greater accuracy.
Residue on Evaporation Proceed as directed in the general method, Appendix VI, using 3 mL of sample. Heat on a steam bath for 5 h, and then heat at 105° for 1 h.
Solubility in Alcohol Proceed as directed in the general method, Appendix VI. One mL of East Indian oil dissolves in 3 mL of 90% alcohol. One mL of West Indian oil dissolves in 4 mL of 90% alcohol.
Specific Gravity Determine by any reliable method (see *General Provisions*).

Packaging and Storage Store in full, tight containers in a cool place protected from light.

Octanoic Acid

Caprylic Acid

$$CH_3(CH_2)_6COOH$$

$C_8H_{16}O_2$　　　　　　　　　　　Formula wt 144.21

CAS: [124-07-2]

DESCRIPTION

A colorless, oily liquid having a slight, unpleasant odor and a burning, rancid taste. It is slightly soluble in water and soluble in most organic solvents. Its specific gravity is about 0.910.

Functional Use in Foods Component in the manufacture of other food-grade additives; defoaming agent.

REQUIREMENTS

Acid Value Between 366 and 396.
Heavy Metals (as Pb) Not more than 10 mg/kg.
Iodine Value Not more than 2.0.
Residue on Ignition Not more than 0.1%.
Saponification Value Between 366 and 398.
Titer (Solidification Point) Between 8° and 15°.
Unsaponifiable Matter Not more than 0.2%.
Water Not more than 0.4%.

TESTS

Acid Value Determine as directed for *Acid Value, Method I*, under *Fats and Related Substances*, Appendix VII.
Heavy Metals Prepare and test a 2-g sample as directed in *Method II* under the *Heavy Metals Test*, Appendix IIIB, using 20 μg of lead ion (Pb) in the control (*Solution A*).
Iodine Value Determine by the *Modified Wijs Method*, Appendix VII.
Residue on Ignition, Appendix IIC Ignite 10 g as directed in the general method.
Saponification Value Determine as directed for *Saponification Value* under *Fats and Related Substances*, Appendix VII, using about 2 g, accurately weighed.
Titer (Solidification Point) Determine as directed under *Solidification Point*, Appendix IIB.
Unsaponifiable Matter Determine as directed in the general method, Appendix VII.
Water Determine by the *Karl Fischer Titrimetric Method*, Appendix IIB.

Packaging and Storage Store in tight containers.

Oleic Acid

(Z)-9-Octadecenoic Acid

$$CH_3(CH_2)_7CH = CH(CH_2)_7COOH$$

$C_{18}H_{34}O_2$　　　　　　　　　　　Formula wt 282.47

CAS: [112-80-1]

DESCRIPTION

An unsaturated acid obtained from fats. Oleic Acid is a colorless to pale-yellow, oily liquid when freshly prepared, but upon exposure to air it gradually absorbs oxygen and darkens. It has a characteristic lardlike odor and taste. When strongly heated in air, it is decomposed with the production of acrid vapors. Its specific gravity is about 0.895. It is practically insoluble in water, but is miscible with alcohol, with ether, and with fixed and volatile oils.

Functional Use in Foods Component in the manufacture of other food-grade additives; defoaming agent; lubricant; binder.

REQUIREMENTS

Acid Value Between 196 and 204.
Heavy Metals (as Pb) Not more than 10 mg/kg.
Iodine Value Between 83 and 103.
Residue on Ignition Not more than 0.01%.
Saponification Value Between 196 and 206.
Titer (Solidification Point) Not above 10°.
Unsaponifiable Matter Not more than 2.0%.
Water Not more than 0.4%.

TESTS

Acid Value Determine as directed for *Acid Value, Method I*, under *Fats and Related Substances*, Appendix VII.
Heavy Metals Prepare and test a 2-g sample as directed in *Method II* under the *Heavy Metals Test*, Appendix IIIB, using 20 µg of lead ion (Pb) in the control (*Solution A*).
Iodine Value Determine by the *Modified Wijs Method*, Appendix VII.
Residue on Ignition, Appendix IIC Ignite 10 g as directed in the general method.
Saponification Value Determine as directed for *Saponification Value* under *Fats and Related Substances*, Appendix VII, using about 3 g, accurately weighed.
Titer (Solidification Point) Determine as directed under *Solidification Point*, Appendix IIB.
Unsaponifiable Matter, Appendix VII Determine as directed in the general method.
Water Determine by the *Karl Fischer Titrimetric Method*, Appendix IIB.

Packaging and Storage Store in tight containers.

Olibanum Oil

Oil of Frankincense

CAS: [8016-36-2]

DESCRIPTION

The volatile oil distilled from a gum obtained from the trees *Boswellia carterii* Birdw. and other *Boswellia* species (Fam. *Burseraceae*). It is a pale yellow liquid having a balsamic odor with a faint lemon note. It is soluble in most fixed oils and, with a slight haze, in mineral oil. It is insoluble in glycerin and in propylene glycol.

Functional Use in Foods Flavoring agent.

REQUIREMENTS

Identification The infrared absorption spectrum of the sample exhibits relative maxima (that may vary in intensity) at the same wavelengths (or frequencies) as those shown in the respective spectrum in the section on *Infrared Spectra* (*Series A: Essential Oils*), using the same test conditions as specified therein.
Acid Value Not more than 4.0.
Angular Rotation Between −15° and +35°.
Ester Value Between 4 and 40.
Heavy Metals (as Pb) Passes test.
Refractive Index Between 1.465 and 1.482 at 20°.
Solubility in Alcohol Passes test.
Specific Gravity Between 0.862 and 0.889.

TESTS

Acid Value Determine as directed for *Acid Value* under *Essential Oils and Flavors*, Appendix VI.
Angular Rotation Determine in a 100-mm tube as directed under *Optical (Specific) Rotation*, Appendix IIB.
Ester Value Determine as directed in the general method, Appendix VI, using about 5 g, accurately weighed.
Heavy Metals Shake 10 mL of the oil with an equal volume of water to which 1 drop of hydrochloric acid has been added, and pass hydrogen sulfide through the mixture until it is saturated. No darkening in color is produced in either the oil or the water.
Refractive Index, Appendix IIB Determine with an Abbé or other refractometer of equal or greater accuracy.
Solubility in Alcohol Proceed as directed in the general method, Appendix VI. One mL dissolves in 6 mL of 90% alcohol, occasionally with opalescence.
Specific Gravity Determine by any reliable method (see *General Provisions*).

Packaging and Storage Store in full, tight, preferably glass, aluminum, or tin-lined containers in a cool place protected from light.

Onion Oil

CAS: [8002-72-0]

DESCRIPTION

A volatile oil obtained by steam distillation of the bulbs of *Allium cepa* L. (Fam. *Liliaceae*). It is a clear, amber yellow to amber orange liquid having a strong pungent odor and taste characteristic of onion. It is soluble in most fixed oils, in mineral oil, and in alcohol. It is insoluble in glycerin and in propylene glycol.

Note: Onion Oil is purchased mainly on the basis of its odor and flavor, which render definitive specifications of little value.

Functional Use in Foods Flavoring agent.

REQUIREMENTS

Identification The infrared absorption spectrum of the sample exhibits relative maxima (that may vary in intensity) at the same wavelengths (or frequencies) as those shown in the respective spectrum in the section on *Infrared Spectra* (*Series A: Essential Oils*), using the same test conditions as specified therein.
Heavy Metals (as Pb) Passes test.
Refractive Index Between 1.549 and 1.570 at 20°.
Specific Gravity Between 1.050 and 1.135.

TESTS

Heavy Metals Shake 10 mL of the oil with an equal volume of water to which 1 drop of hydrochloric acid has been added, and pass hydrogen sulfide through the mixture until it is saturated. No darkening in color is produced in either the oil or the water.
Refractive Index, Appendix IIB Determine with an Abbé or other refractometer of equal or greater accuracy.
Specific Gravity Determine by any reliable method (see *General Provisions*).

Packaging and Storage Store in full, tight, preferably glass or aluminum containers in a cool place protected from light.

Orange Oil, Coldpressed

Sweet Orange Oil

DESCRIPTION

The volatile oil obtained by expression, without the use of heat, from the fresh peel of the ripe fruit of *Citrus sinensis* L. Osbeck (Fam. *Rutaceae*). It is an intensely yellow, orange, or deep orange liquid having the characteristic odor and taste of the outer part of fresh, sweet orange peel. It is miscible with dehydrated alcohol and with carbon disulfide. It is soluble in glacial acetic acid. It may contain a suitable antioxidant.

Note: Do not use Sweet Orange Oil that has a terebinthine odor.

Functional Use in Foods Flavoring agent.

REQUIREMENTS

Labeling Indicate whether it is the California or Florida type.
Identification The infrared absorption spectrum of the sample exhibits relative maxima (that may vary in intensity) at the same wavelengths (or frequencies) as those shown in the respective spectrum in the section on *Infrared Spectra* (*Series A: Essential Oils*), using the same test conditions as specified therein.
Assay Not less than 1.2% and not more than 2.5% of aldehydes, calculated as decyl aldehyde ($C_{10}H_{20}O$).
Angular Rotation Between +94° and +99°.
Heavy Metals (as Pb) Not more than 0.004%.
Lead Not more than 10 mg/kg.
Refractive Index Between 1.472 and 1.474 at 20°.
Specific Gravity Between 0.842 and 0.846.
Ultraviolet Absorbance *California type*: not less than 0.130; *Florida type*: not less than 0.240.

TESTS

Assay Weigh accurately a 10-mL sample, and proceed as directed under *Aldehydes and Ketones—Hydroxylamine/Tert-Butyl Alcohol Method*, Appendix VI. Allow the mixture to stand for 15 min, with occasional shaking, before titrating, and use 78.14 as the equivalence factor (*e*) in the calculation.
Angular Rotation Determine in a 100-mm tube as directed under *Optical (Specific) Rotation*, Appendix IIB.
Heavy Metals Prepare and test a 500-mg sample as directed in *Method II* under the *Heavy Metals Test*, Appendix IIIB, using 20 µg of lead ion (Pb) in the control (*Solution A*).
Lead A *Sample Solution* prepared as directed for organic compounds meets the requirements of the *Lead Limit Test*, Appendix IIIB, using 10 µg of lead ion (Pb) in the control.
Refractive Index, Appendix IIB Determine with an Abbé or other refractometer of equal or greater accuracy.
Specific Gravity Determine by any reliable method (see *General Provisions*).
Ultraviolet Absorbance Proceed as directed under *Ultraviolet Absorbance of Citrus Oils*, Appendix VI, using about 250 mg of sample, accurately weighed. The maximum absorbance occurs at 330 ± 3 nm.

Packaging and Storage Store in full, tight containers. Avoid exposure to excessive heat.

Orange Oil, Bitter, Coldpressed

DESCRIPTION

The volatile oil obtained by expression, without the use of heat, from the fresh peel of the fruit of *Citrus aurantium* L. (Fam. *Rutaceae*). It is a pale yellow or yellowish brown liquid with the characteristic aromatic odor of the Seville orange and an aromatic, somewhat bitter, taste. It is miscible with absolute alcohol and with an equal volume of glacial acetic acid. It is soluble in fixed oils and in mineral oil. It is slightly soluble in propylene glycol, but it is relatively insoluble in glycerin. It is affected by light, and its alcohol solutions are neutral to litmus. It may contain a suitable antioxidant.

Functional Use in Foods Flavoring agent.

REQUIREMENTS

Identification The infrared absorption spectrum of the sample exhibits relative maxima (that may vary in intensity) at the same wavelengths (or frequencies) as those shown in the respective spectrum in the section on *Infrared Spectra (Series A: Essential Oils)*, using the same test conditions as specified therein.
Aldehydes Not less than 0.5% and not more than 1.0% of aldehydes, calculated as decyl aldehyde ($C_{10}H_{20}O$).
Angular Rotation Between +88° and +98°.
Heavy Metals (as Pb) Not more than 0.004%.
Lead Not more than 10 mg/kg.
Refractive Index Between 1.472 and 1.476 at 20°.
Residue on Evaporation Between 2.0% and 5.0%.
Specific Gravity Between 0.845 and 0.851.

TESTS

Aldehydes Weigh accurately about 10 g, and proceed as directed under *Aldehydes and Ketones—Hydroxylamine/Tert-Butyl Alcohol Method*, Appendix VI, using 78.14 as the equivalence factor (*e*) in the calculation. Allow the mixture to stand for 30 min at room temperature before titrating.
Angular Rotation Determine in a 100-mm tube as directed under *Optical (Specific) Rotation*, Appendix IIB.
Heavy Metals Prepare and test a 500-mg sample as directed in *Method II* under the *Heavy Metals Test*, Appendix IIIB, using 20 μg of lead ion (Pb) in the control (*Solution A*).
Lead A *Sample Solution* prepared as directed for organic compounds meets the requirements of the *Lead Limit Test*, Appendix IIIB, using 10 μg of lead ion (Pb) in the control.
Refractive Index, Appendix IIB Determine with an Abbé or other refractometer of equal or greater accuracy.
Residue on Evaporation Proceed as directed in the general method, Appendix VI, using 5 g of sample, and heat for 4.5 h.
Specific Gravity Determine by any reliable method (see *General Provisions*).

Packaging and Storage Store in full, tight, preferably glass or tin-lined containers in a cool place protected from light.

REQUIREMENTS

Identification The infrared absorption spectrum of the sample exhibits relative maxima (that may vary in intensity) at the same wavelengths (or frequencies) as those shown in the respective spectrum in the section on *Infrared Spectra (Series A: Essential Oils)*, using the same test conditions as specified therein.
Aldehydes Between 1.0% and 2.5% of aldehydes, calculated as decyl aldehyde ($C_{10}H_{20}O$).
Angular Rotation Between +94° and +99°.
Heavy Metals (as Pb) Passes test.
Refractive Index Between 1.471 and 1.474 at 20°.
Specific Gravity Between 0.840 and 0.844.
Ultraviolet Absorbance Not more than 0.01.

TESTS

Aldehydes Weigh accurately about 5 mL of the sample, and proceed as directed under *Aldehydes and Ketones—Hydroxylamine/Tert-Butyl Alcohol Method*, Appendix VI, using 78.14 as the equivalence factor (*e*) in the calculation. Allow the mixture to stand at room temperature for 1 h before titrating.
Angular Rotation Determine in a 100-mm tube as directed under *Optical (Specific) Rotation*, Appendix IIB.
Heavy Metals Shake 10 mL of the oil with an equal volume of water to which 1 drop of hydrochloric acid has been added, and pass hydrogen sulfide through the mixture until it is saturated. No darkening in color is produced in either the oil or the water.
Refractive Index, Appendix IIB Determine with an Abbé or other refractometer of equal or greater accuracy.
Specific Gravity Determine by any reliable method (see *General Provisions*).
Ultraviolet Absorbance Proceed as directed under *Ultraviolet Absorbance of Citrus Oils*, Appendix VI, using about 250 mg of sample, accurately weighed. The maximum absorbance occurs at 330 ± 3 nm.

Packaging and Storage Store in full, tight containers in a cool place protected from light.

Orange Oil, Distilled

DESCRIPTION

The volatile oil obtained by distillation from the fresh peel or juice of the fruit of *Citrus sinesis* L. Osbeck (Fam. *Rutaceae*), with or without the previous separation of the juice, pulp, or peel. It is a colorless to pale yellow liquid having the characteristic odor of fresh orange peel. It is soluble in most fixed oils, in mineral oil, and in alcohol (with haze). It is insoluble in glycerin and in propylene glycol. It may contain a suitable antioxidant.

Functional Use in Foods Flavoring agent.

Origanum Oil, Spanish Type

DESCRIPTION

The volatile oil obtained by steam distillation from the flowering herb *Thymus capitatus* Hoffm. et Link and various species of *Origanum*. It is a yellowish red to dark brownish red liquid having a pungent, spicy odor suggestive of thyme oil. It is soluble in most fixed oils and in propylene glycol. It is soluble, with turbidity, in mineral oil, but it is insoluble in glycerin.

Functional Use in Foods Flavoring agent.

REQUIREMENTS

Identification The infrared absorption spectrum of the sample exhibits relative maxima (that may vary in intensity) at the same wavelengths (or frequencies) as those shown in the respective spectrum in the section on *Infrared Spectra (Series A: Essential Oils)*, using the same test conditions as specified therein.
Assay Not less than 60.0% and not more than 75.0%, by volume, of phenols.
Angular Rotation Between −2° and +3°.
Heavy Metals (as Pb) Passes test.
Refractive Index Between 1.502 and 1.508 at 20°.
Solubility in Alcohol Passes test.
Specific Gravity Between 0.935 and 0.960.

TESTS

Assay Shake a suitable quantity of sample with about 2% of powdered tartaric acid, and filter. Proceed as directed under *Phenols*, Appendix VI.
Angular Rotation Determine in a 100-mm tube as directed under *Optical (Specific) Rotation*, Appendix IIB. Occasionally the oil is too dark to read in a 100-mm tube.
Heavy Metals Shake 10 mL of the oil with an equal volume of water to which 1 drop of hydrochloric acid has been added, and pass hydrogen sulfide through the mixture until it is saturated. No darkening in color is produced in either the oil or the water.
Refractive Index, Appendix IIB Determine with an Abbé or other refractometer of equal or greater accuracy.
Solubility in Alcohol Proceed as directed in the general method, Appendix VI. One mL is soluble in 2 mL of 70% alcohol. The solution may become cloudy on dilution.
Specific Gravity Determine by any reliable method (see *General Provisions*).

Packaging and Storage Store in full, tight, preferably glass, aluminum, or tin-lined containers in a cool place protected from light. A precipitate may form in galvanized containers, and the oil darkens in iron drums.

Orris Root Oil

DESCRIPTION

The volatile oil obtained by steam distillation from the peeled, dried, and aged rhizomes of *Iris pallida* Lam. (Fam. *Iridaceae*). At room temperature it is a light yellow to brown yellow mass, which melts between 38° and 50° to form a yellow to yellow brown liquid. It is soluble in most fixed oils, in mineral oil, and in propylene glycol. It is insoluble in glycerin.

Functional Use in Foods Flavoring agent.

REQUIREMENTS

Identification The infrared absorption spectrum of the sample exhibits relative maxima (that may vary in intensity) at the same wavelengths (or frequencies) as those shown in the respective spectrum in the section on *Infrared Spectra (Series A: Essential Oils)*, using the same test conditions as specified therein.
Assay Not less than 9.0% and not more than 20.0% of ketones, calculated as irone ($C_{14}H_{22}O$).
Acid Value Between 175 and 235.
Ester Value Between 4 and 35.
Heavy Metals (as Pb) Passes test.
Melting Range Between 38° and 50°.

TESTS

Assay Weigh accurately about 1 g, and proceed as directed under *Aldehydes*, Appendix VI, using 103.2 as the equivalence factor (*e*) in the calculation. Allow the mixture to stand 1 h at room temperature before titrating.
Acid Value Determine as directed for *Acid Value* under *Essential Oils and Flavors*, Appendix VI, using about 1 g, accurately weighed.
Ester Value Determine as directed in the general method, Appendix VI, using about 1 g, accurately weighed.
Heavy Metals Shake 10 mL of the oil with an equal volume of water to which 1 drop of hydrochloric acid has been added, and pass hydrogen sulfide through the mixture until it is saturated. No darkening in color is produced in either the oil or the water.
Melting Range Determine as directed for *Melting Range or Temperature, Procedure for Class II*, Appendix IIB.

Packaging and Storage Store in full, tight, preferably glass, tin-lined, or other suitably lined aluminum containers in a cool place protected from light.

Ox Bile Extract

Sodium Choleate; Purified Oxgall

$C_{24}H_{39}NaO_5$ Formula wt 430.56 (as sodium cholate)

DESCRIPTION

A yellowish green powder with a partly sweet, partly bitter, disagreeable taste. It is soluble in water and in alcohol. It contains ox bile acids, chiefly glycocholic and taurocholic, as sodium salts, equivalent to not less than 45.0% cholic acid ($C_{24}H_{40}O_5$). It is the purified portion of the bile of an ox, obtained by evaporating the alcohol extract of concentrated bile.

Functional Use in Foods Surfactant.

REQUIREMENTS

Identification
A. It gives a positive test for *Sodium*, Appendix IIIA.
B. A 1 in 10 solution in alcohol, when mixed with 0.5 mL of iodine TS, yields a blue color.

Assay Contains not less than 45.0% of cholic acid ($C_{24}H_{40}O_5$).
Ash (Total) Not more than 10.0%.
Loss on Drying Not more than 6.0%.
Microbial Limits:
  Aerobic Plate Count Not more than 20,000 per g.
  E. coli Not more than 3 per g.
  Salmonella sp. Negative in 25 g.
  Yeasts and Molds Not more than 10 per g.
pH Between 6.3 and 7.0.

TESTS

Assay
Standard Solution Using an accurately weighed quantity of USP Cholic Acid Reference Standard, prepare a solution in 60% acetic acid with a known concentration of about 0.5 mg/mL. When stored in a refrigerator, this solution may be used for several months.
Assay Preparation Dissolve about 50 mg of the sample, accurately weighed, in 100 mL of 60% acetic acid, and mix. Filter the solution, discarding the first 10 mL of the filtrate.
Procedure Transfer 1.0 mL each of the *Standard Solution* and the *Assay Preparation* into separate containers. To each container, add 1.0 mL of a freshly prepared 1 in 100 solution of furfural. Cool the containers in an ice bath for 5 min. Add 13 mL of a dilute solution of sulfuric acid, prepared by cautiously mixing 50 mL of sulfuric acid with 65 mL of water. Thoroughly mix the contents in each container, and place them in a water bath maintained at 70° for 10 min. Then immediately place the containers in an ice bath for 2 min. Determine the absorbance of each solution in a 1-cm cell at the wavelength of maximum absorbance at about 650 nm. Calculate the quantity, in mg, of cholic acid taken in the portion of Ox Bile Extract by the formula

$$100C(A_U/A_S),$$

in which C is the concentration, in mg/mL, of USP Cholic Acid Reference Standard in the *Standard Solution*, and A_U and A_S are the maximum absorbances at identical wavelengths of the *Assay Preparation* and the *Standard Solution*, respectively.
Ash (Total) Determine by the general method, Appendix IIC.
Loss on Drying, Appendix IIC Dry at 105° for 16 h.
Microbial Limits:
  Aerobic Plate Count Proceed as directed in chapter 3 of the *FDA Bacteriological Analytical Manual*, Seventh Edition, 1992.
  E. coli Proceed as directed in chapter 4 of the *FDA Bacteriological Analytical Manual*, Seventh Edition, 1992.
  Salmonella sp. Proceed as directed in chapter 5 of the *FDA Bacteriological Analytical Manual*, Seventh Edition, 1992.
  Yeasts and Molds Proceed as directed in chapter 18 of the *FDA Bacteriological Analytical Manual*, Seventh Edition, 1992.

pH Determine by the *Potentiometric Method*, Appendix IIB, of a 1 in 20 solution.

Packaging and Storage Store in tight containers.

Oxystearin

INS: 387 CAS: [8028-45-3]

DESCRIPTION

Oxystearin is a mixture of the glycerides of partially oxidized stearic and other fatty acids. It occurs as a tan to light brown, fatty or waxlike substance having a bland taste. It is soluble in ether, in solvent hexane, and in chloroform.

Functional Use in Foods Crystallization inhibitor in salad and cooking oils; sequestrant; defoaming agent.

REQUIREMENTS

Acid Value Not more than 15.
Heavy Metals (as Pb) Not more than 10 mg/kg.
Hydroxyl Value Between 30 and 45.
Iodine Value Not more than 15.
Refractive Index (butyro) Between 59 and 61 at 48° (equivalent to 1.465 to 1.467 on the Abbé scale).
Saponification Value Between 225 and 240.
Unsaponifiable Matter Not more than 0.8%.

TESTS

Acid Value Determine as directed for *Acid Value, Method II*, under *Fats and Related Substances*, Appendix VII.
Heavy Metals Prepare and test a 2-g sample as directed in *Method II* under the *Heavy Metals Test*, Appendix IIIB, using 20 µg of lead ion (Pb) in the control (*Solution A*).
Hydroxyl Value Determine as directed under *Method II* in the general method, Appendix VII, using about 5 g, accurately weighed.
Iodine Value Determine by the *Modified Wijs Method*, Appendix VII.
Refractive Index, Appendix IIB Melt the sample, filter through filter paper, and determine the refractive index at 48° with an Abbé or butyro refractometer.
Saponification Value Determine as directed for *Saponification Value* under *Fats and Related Substances*, Appendix VII, using about 3 g, accurately weighed.
Unsaponifiable Matter Determine as directed in the general method, Appendix VII.

Packaging and Storage Store in well-closed containers.

Ozone

Triatomic Oxygen

O_3 Formula wt 47.9982

CAS: [10028-15-6]

DESCRIPTION

Ozone is an unstable, colorless gas with a pungent, characteristic odor. It is produced *in situ* from oxygen either by ultraviolet irradiation of air or by passing a high-voltage discharge through air. It is a potent oxidizing agent that decomposes at ambient temperature to molecular oxygen.

Functional Use in Foods Antimicrobial and disinfectant for water to be used for direct consumption, such as for ice, or for indirect consumption, such as for water used in the treatment or display of fish, produce, and other perishable foods. It is also used in the treatment of wastewater.

REQUIREMENTS

Identification

Reagent Solution Disperse 124.5 mg of alizarin violet 3R in 500 mL of water in a 1-L volumetric flask. Mechanically stir overnight. Add 20 mg of sodium hexametaphosphate, 48.5 g of ammonium chloride, and 6.2 mL of ammonium hydroxide (equivalent to 1.6 g of NH_3). Dilute to volume with water, and stir overnight. A 10-fold dilution of this solution has an absorbance of 0.155 AU cm^{-1} at 548 nm; the pH of dilutions with sample waters is between 8.1 and 8.5.

Procedure Introduce 20 mL of the *Reagent Solution* into each of two 200-mL volumetric flasks. Fill one flask with Ozone-free water to serve as the blank. Fill the other with the sample by directly introducing the sample, with the aid of a long-stemmed funnel or pipet, below the surface of the *Reagent Solution* to prevent Ozone loss by degassing. Immediately measure the absorbance of both solutions at 548 nm, using 1- to 5-cm cells: The presence of Ozone is indicated if the sample solution has a lower absorbance than the blank.

Assay 0.01 to 0.5 mg of O_3 per L.

TESTS

Assay

Indigo Stock Solution In a 1-L volumetric flask, dissolve 0.770 g of potassium indigotrisulfonate in 500 mL of water and 1 mL of phosphoric acid, dilute to volume with water, and mix. A 1:100 dilution of this reagent has an absorbance of 0.20 ± 0.010 cm^{-1} at 600 nm.

Indigo Reagent I Just before use, transfer 20 mL of *Indigo Stock Solution*, 10 g of monobasic sodium phosphate, and 7 mL of phosphoric acid into a 1-L volumetric flask, dilute to volume with water, and mix.

Indigo Reagent II Proceed as directed for *Indigo Reagent I*, using 100 mL of *Indigo Stock Solution* instead of 20 mL.

Malonic Acid Reagent Dissolve 5 g of malonic acid in water and dilute to 100 mL.

Procedure (for a concentration range of 0.01 to 0.1 mg of Ozone per L) Add 10.0 mL of *Indigo Reagent I* to each of two 100-mL flasks. Fill one flask with Ozone-free water to serve as the blank. Fill the other with the sample by directly introducing the sample, with the aid of a long-stemmed funnel or pipet, below the surface of the dye solution to prevent Ozone loss by degassing. Without delay, mix and measure the absorbance of each solution at 600 nm, preferably in 10-cm cells. (For a concentration range of 0.05 to 0.5 mg of Ozone per L, use *Indigo Reagent II* and proceed as above.)

Control of Interferences In the presence of chlorine, add 1 mL of *Malonic Acid Reagent* to both flasks before adding the samples. Proceed as above, but measure absorbance immediately.

Calculate the concentration of Ozone, in mg/L, by the formula

$$100D/(f \times b \times V),$$

in which D is the difference in absorbance between the sample solution and blank solution; b is the path length, in cm; V is the volume of sample, in mL (normally 90 mL); and f is 0.42.

Palmarosa Oil

Geranium Oil, East Indian Type; Geranium Oil, Turkish Type

DESCRIPTION

The volatile oil obtained by steam distillation from the partially dried grass *Cymbopogon martini* Stapf. var. *motia*. It is a light yellow to yellow oil that is often hazy and brownish. It is soluble in most fixed oils and in propylene glycol. It is soluble, usually with opalescence or turbidity, in mineral oil. It is practically insoluble in glycerin.

Functional Use in Foods Flavoring agent.

REQUIREMENTS

Identification The infrared absorption spectrum of the sample exhibits relative maxima (that may vary in intensity) at the same wavelengths (or frequencies) as those shown in the respective spectrum in the section on *Infrared Spectra (Series A: Essential Oils)*, using the same test conditions as specified therein.
Assay for Alcohols Not less than 88.0% of total alcohols.
Assay for Esters Not less than 4.0% and not more than 18.0% of esters, calculated as geranyl acetate ($C_{12}H_{20}O_2$).
Angular Rotation Between −2° and +3°.
Heavy Metals (as Pb) Passes test.
Refractive Index Between 1.470 and 1.476 at 20°.
Solubility in Alcohol Passes test.
Specific Gravity Between 0.879 and 0.892.

TESTS

Assay for Alcohols Proceed as directed under *Total Alcohols*, Appendix VI. Weigh accurately about 1 g of the acetylated oil for the saponification, and use 77.13 as the equivalence factor (*e*) in the calculation.

Assay for Esters Weigh accurately about 5 g, and proceed as directed under *Ester Determination*, Appendix VI, using 98.15 as the equivalence factor (*e*) in the calculation.

Angular Rotation Determine in a 100-mm tube as directed under *Optical (Specific) Rotation*, Appendix IIB.

Heavy Metals Shake 10 mL of the oil with an equal volume of water to which 1 drop of hydrochloric acid has been added, and pass hydrogen sulfide through the mixture until it is saturated. No darkening in color is produced in either the oil or the water.

Refractive Index, Appendix IIB Determine with an Abbé or other refractometer of equal or greater accuracy.

Solubility in Alcohol Proceed as directed in the general method, Appendix VI. One mL dissolves in 2 mL of 70% alcohol.

Specific Gravity Determine by any reliable method (see *General Provisions*).

Packaging and Storage Store in full, tight, preferably glass or tin-lined containers in a cool place protected from light.

Palmitic Acid

Hexadecanoic Acid

$C_{16}H_{32}O_2$ Formula wt 256.43

                             CAS: [57-10-3]

DESCRIPTION

A mixture of solid organic acids obtained from fats consisting chiefly of Palmitic Acid ($C_{16}H_{32}O_2$) with varying amounts of stearic acid ($C_{16}H_{36}O_2$). It occurs as a hard, white or faintly yellowish, somewhat glossy crystalline solid, or as a white or yellowish white powder. It has a slight characteristic odor and taste. Palmitic Acid is practically insoluble in water. It is soluble in alcohol, in ether, and in chloroform.

Functional Use in Foods Component in the manufacture of other food-grade additives; defoaming agent.

REQUIREMENTS

Acid Value Between 204 and 220.
Heavy Metals (as Pb) Not more than 10 mg/kg.
Iodine Value Not more than 2.0.
Residue on Ignition Not more than 0.1%.
Saponification Value Between 205 and 221.
Titer (Solidification Point) Between 53.3° and 62°.
Unsaponifiable Matter Not more than 1.5%.
Water Not more than 0.2%.

TESTS

Acid Value Determine as directed for *Acid Value, Method I*, under *Fats and Related Substances*, Appendix VII.

Heavy Metals Prepare and test a 2-g sample as directed in *Method II* under the *Heavy Metals Test*, Appendix IIIB, using 20 μg of lead ion (Pb) in the control (*Solution A*).

Iodine Value Determine by the *Modified Wijs Method*, Appendix VII.

Residue on Ignition, Appendix IIC Ignite 2 g as directed in the general method.

Saponification Value Determine as directed for *Saponification Value* under *Fats and Related Substances*, Appendix VII, using about 3 g, accurately weighed.

Titer (Solidification Point) Determine as directed under *Solidification Point*, Appendix IIB.

Unsaponifiable Matter, Appendix VII Determine as directed in the general method.

Water Determine by the *Karl Fischer Titrimetric Method*, Appendix IIB.

Packaging and Storage Store in well-closed containers.

Palm Kernel Oil (Unhydrogenated)

DESCRIPTION

A fat with a slight, characteristic sweet nutty flavor obtained from the kernel of the fruit of the oil palm *Elaeis guineensis* by mechanical expression or solvent extraction. It is refined, bleached, and deodorized to substantially remove free fatty acids, phospholipids, color, odor and flavor components, and miscellaneous other non-oil materials. Like coconut oil, it has a more abrupt melting range than other fats and oils.

Functional Use in Foods Coating agent; emulsifying agent; formulation aid; texturizer.

REQUIREMENTS

Identification Palm Kernel Oil exhibits the following composition profile of fatty acids as determined under *Fatty Acid Composition*, Appendix VII.

Fatty Acid:	6:0	8:0	10:0	12:0	14:0	16:0
Weight % (Range):	0–1.5	3–5	2.5–6	40–52	14–18	7–10

Fatty Acid:	16:1	18:0	18:1	18:2	20:0
Weight % (Range):	0–1	1–3	11–19	0.5–4	tr.–1

Color (AOCS-Wesson) Not more than 20 yellow/2.0 red.

Free Fatty Acids *As oleic acid*: Not more than 0.1%; *As lauric acid*: Not more than 0.07%.
Iodine Value Between 13 and 23.
Lead Not more than 0.1 mg/kg.
Melting Range Between 27° and 29°.
Peroxide Value Not more than 10 meq/kg.
Unsaponifiable Matter Not more than 1.5%.
Water Not more than 0.1%.

TESTS

Color Proceed as directed for *Color* (AOCS-Wesson) under *Fats and Related Substances*, Appendix VII.
Free Fatty Acids Proceed as directed under *Free Fatty Acids*, Appendix VII, using the following equivalence factors (e) in the formula given in the procedure:

Free fatty acids as oleic acid, $e = 28.2$.
Free fatty acids as lauric acid, $e = 20.0$.

Iodine Value Proceed as directed under *Iodine Value*, Appendix VII.
Lead Determine as directed for *Method II* in the *Atomic Absorption Spectrophotometric Graphite Furnace Method* under the *Lead Limit Test*, Appendix IIIB, using a 3-g sample.
Melting Range Determine as directed for *Melting Range* under *Fats and Related Substances*, Appendix VII.
Peroxide Value Proceed as directed under *Peroxide Value* in the monograph for *Hydroxylated Lecithin*. However, after the addition of saturated potassium iodide and mixing, instead of allowing the solution to stand for 10 min, mix the solution for 1 min and begin the titration immediately.
Unsaponifiable Matter Proceed as directed under *Unsaponifiable Matter*, Appendix VII.
Water Proceed as directed under *Water Determination*, Appendix IIB. However, in place of 35 to 40 mL of methanol, use 50 mL of chloroform to dissolve the sample.

Packaging and Storage Store in well-closed containers.

Palm Oil (Unhydrogenated)

DESCRIPTION

A deep orange-red fat with a nutty flavor obtained from the pulp of the fruit of the oil palm *Elaeis guineensis* usually by boiling, centrifugation, and mechanical expression. It is refined, bleached, and deodorized to substantially remove free fatty acids, phospholipids, color, odor and flavor components, and miscellaneous other non-oil materials. It is a semisolid at 21° to 27°.

Functional Use in Foods Coating agent; emulsifying agent; formulation aid; texturizer.

REQUIREMENTS

Identification Palm Oil exhibits the following composition profile of fatty acids as determined under *Fatty Acid Composition*, Appendix VII.

Fatty Acid:	14:0	16:0	18:0	18:1	18:2
Weight % (Range):	0.5–5.9	32–47	2–8	34–44	7–12

Color (AOCS-Wesson) Not more than 35 yellow/5.0 red.
Free Fatty Acids *As oleic acid*: Not more than 0.1%; *As palmitic acid*: Not more than 0.09%.
Iodine Value Between 50 and 55.
Lead Not more than 0.1 mg/kg.
Peroxide Value Not more than 10 meq/kg.
Stability (AOM) Not less than 50 h.
Unsaponifiable Matter Not more than 1.5%.
Water Not more than 0.1%.

TESTS

Color Proceed as directed for *Color* (AOCS-Wesson) under *Fats and Related Substances*, Appendix VII.
Free Fatty Acids Proceed as directed under *Free Fatty Acids*, Appendix VII, using the following equivalence factors (e) in the formula given in the procedure:

Free fatty acids as oleic acid, $e = 28.2$.
Free fatty acids as palmitic acid, $e = 25.6$.

Iodine Value Proceed as directed under *Iodine Value*, Appendix VII.
Lead Determine as directed under *Method II* in the *Atomic Absorption Spectrophotometric Graphite Furnace Method* under the *Lead Limit Test*, Appendix IIIB, using a 3-g sample.
Peroxide Value Proceed as directed under *Peroxide Value* in the monograph for *Hydroxylated Lecithin*. However, after the addition of saturated potassium iodide and mixing, instead of allowing the solution to stand for 10 min, mix the solution for 1 min, and begin the titration immediately.
Stability (Active Oxygen Method) Proceed its directed under *Stability*, Appendix VII.
Unsaponifiable Matter Proceed as directed under *Unsaponifiable Matter*, Appendix VII.
Water Proceed as directed under *Water Determination*, Appendix IIB. However, in place of 35 to 40 mL of methanol, use 50 mL of chloroform to dissolve the sample.

Packaging and Storage Store in well-closed containers.

DL-Panthenol

DL-Pantothenyl Alcohol; Racemic Pantothenyl Alcohol

HOCH$_2$C(CH$_3$)$_2$CH(OH)CONH(CH$_2$)$_2$CH$_2$OH

C$_9$H$_{19}$NO$_4$ Formula wt 205.25

DESCRIPTION

A racemic mixture of the dextrorotatory (active) and levorotatory (inactive) isomers of panthenol, the alcohol analogue of pantothenic acid. It occurs as a white to creamy white, crystalline powder having a slight, characteristic odor. Its solutions are neutral or alkaline to litmus. It is freely soluble in water, in alcohol, and in propylene glycol. It is soluble in chloroform and in ether, and is slightly soluble in glycerin.

Note: The physiological activity of DL-Panthenol is one-half that of dexpanthenol (D-panthenol).

Functional Use in Foods   Nutrient; dietary supplement.

REQUIREMENTS

Identification   It responds to the *Identification Tests* in the monograph for *Dexpanthenol*.
Assay   Not less than 99.0% and not more than 102.0% of C$_9$H$_{19}$NO$_4$ (DL-Panthenol), calculated on the dried basis.
Aminopropanol   Not more than 0.1%.
Heavy Metals (as Pb)   Not more than 10 mg/kg.
Loss on Drying   Not more than 0.5%.
Melting Range   Between 64.5° and 68.5°.
Residue on Ignition   Not more than 0.1%.

TESTS

Assay   Transfer about 400 mg, accurately weighed, into a 300-mL reflux flask fitted with a standard-taper glass joint, add 50.0 mL of 0.1 N perchloric acid in glacial acetic acid, and reflux for 5 h. Cool, covering the condenser with foil to prevent contamination by moisture, and rinse the condenser with glacial acetic acid. Add 5 drops of crystal violet TS, and titrate with 0.1 N potassium acid phthalate in glacial acetic acid to a blue green endpoint. Perform a blank determination, and make any necessary correction (see *General Provisions*). Each mL of 0.1 N perchloric acid is equivalent to 20.53 mg of C$_9$H$_{19}$NO$_4$.
Aminopropanol   Transfer about 10 g of the sample, accurately weighed, into a 50-mL flask, and dissolve in 25 mL of water. Add bromothymol blue TS, and titrate with 0.01 N sulfuric acid from a microburet to a yellow endpoint. Each mL of 0.01 N sulfuric acid is equivalent to 0.75 mg (750 µg) of aminopropanol.
Heavy Metals   Prepare and test a 2-g sample as directed for *Method II* under the *Heavy Metals Test*, Appendix IIIB, using 20 µg of lead ion (Pb) in the control (*Solution A*).
Loss on Drying, Appendix IIC   Dry at 56° for 4 h in vacuum over phosphorus pentoxide.
Melting Range   Determine as directed for *Melting Range or Temperature*, Appendix IIB.
Residue on Ignition   Ignite a 1-g sample as directed in the general method, Appendix IIC.

Packaging and Storage   Store in tight containers.

Paraffin, Synthetic

Fischer-Tropsch Paraffin

CAS: [8002-74-2]

DESCRIPTION

A white, practically tasteless and odorless wax that is very hard at room temperature. It is synthesized by the Fischer-Tropsch process from carbon monoxide and hydrogen, which are catalytically converted to a mixture of paraffin hydrocarbons; the lower-molecular-weight fractions are removed by distillation, and the residue is hydrogenated and further treated by percolation through activated charcoal. It is soluble in hot hydrocarbon solvents.

Functional Use in Foods   Masticatory substance in chewing gum base.

REQUIREMENTS

Identification   The sample is melted and prepared for analysis on a potassium bromide plate. The infrared absorption spectrum of the sample exhibits maxima at the same wavelengths as the typical spectrum, as shown in the section on *Infrared Spectra* (*Series C: Other Substances*).
Absorptivity   Less than 0.01 at 290 nm, in decahydronaphthalene at 88°C (190°F).
Congealing Point   Between 200° and 210°F (93.3° and 98.9°C).
Heavy Metals (as Pb)   Not more than 0.002%.
Lead   Not more than 3 mg/kg.
Oil Content   Not more than 0.50%.

TESTS

Absorptivity   Transfer about 100 mg of the sample, accurately weighed, into a 100-mL volumetric flask, dissolve in decahydronaphthalene at 88°C (190°F), dilute to volume at this temperature, and mix. Determine the absorbance of this solution in a 10-cm cell at 290 nm with a suitable spectrophotometer, the cell holders of which shall maintain the temperature of the sample cell and the reference cell at 88°C. Use decahydronaphthalene at 88°C in a matched cell as the blank. Cell lengths should be known to within ± 0.5% or better of the nominal pathlength. Calculate the absorptivity (a) of the sample solution by the formula

A/bc,

in which A is the absorbance of the sample solution, corrected for the solvent blank; b is the exact pathlength of the sample cell, in cm; and c is the exact concentration of the sample solution, in g/L.

Note: A suitable spectrophotometer, as applied in this test, is an accurately calibrated instrument capable of measuring absorbance with a repeatability of ± 0.1% or better from an average of 0.4 absorbance level at 290 nm; it has a spectral bandwidth of 2 nm or less, and wavelength measurements made with it shall be repeatable within ± 0.2 nm.

Congealing Point

Definition The temperature at which the molten sample, when allowed to cool under the prescribed conditions, ceases to flow.

Thermometer Jacket Assembly Use an ASTM Congealing Point Thermometer having a range of 68° to 213°F and conforming to the requirements for an ASTM 54 F thermometer (see Appendix I). By means of a cork, fit the thermometer into a jacket consisting of a 1-oz glass vial, 25 mm in diameter and 55 mm in height, and adjust the thermometer so that the bottom of the bulb is 10 to 15 mm from the bottom of the vial.

Procedure Place a sample, of sufficient size to represent exactly the material under test, in a casserole or other suitable dish, and heat slowly in a water bath to a temperature approximately 15°F above the expected congealing point. Heat the *Thermometer Jacket Assembly* to approximately the same temperature as the prepared sample. When both the sample and the assembly have reached the required temperature, remove the assembly from the bath, then immediately remove the thermometer from its jacket, and immerse the thermometer bulb into the molten sample until the bulb is completely covered, taking care not to cover any part of the thermometer stem with the sample. As rapidly as possible, remove the thermometer and any adhering sample from the sample dish and place the thermometer in the jacket, holding both the thermometer and its jacket in a horizontal position during this operation. Rotate the thermometer in a horizontal position at the rate of approximately one revolution in 2 s, pausing momentarily at the completion of each revolution to inspect the drop of sample on the thermometer bulb. When the drop is observed to rotate with the bulb, read the thermometer, and record the reading as the congealing point, reported to the nearest 0.5°F. Repeat the determination. If the variation is greater than 1°F, make a third determination, and record the average of the three determinations as the *Congealing Point*.

Heavy Metals Prepare and test a 1-g sample as directed in *Method II* under the *Heavy Metals Test*, Appendix IIIB, using 20 μg of lead ion (Pb) in the control (*Solution A*).

Lead Prepare a *Sample Solution* as directed in the general method under *Chewing Gum Base*, Appendix IV. This solution meets the requirements of the *Lead Limit Test*, Appendix IIIB, using 10 μg of lead ion (Pb) in the control.

Oil Content Determine as directed in the general method, Appendix IIC.

Packaging and Storage Store in well-closed containers.

Parsley Herb Oil

DESCRIPTION

The oil obtained by steam distillation of the aboveground parts of the plant *Petroselinum crispum* (Fam. *Umbelliferae*), including the immature seed. It is a yellow to light brown liquid having the odor of parsley herb. It is soluble in most fixed oils, in mineral oil, and in alcohol (with opalescence). It is slightly soluble in propylene glycol, but it is insoluble in glycerin.

Functional Use in Foods Flavoring agent.

REQUIREMENTS

Identification The infrared absorption spectrum of the sample exhibits relative maxima (that may vary in intensity) at the same wavelengths (or frequencies) as those shown in the respective spectrum in the section on *Infrared Spectra* (*Series A: Essential Oils*), using the same test conditions as specified therein.
Acid Value Not more than 2.0.
Angular Rotation Between +1° and −9°.
Heavy Metals (as Pb) Passes test.
Refractive Index Between 1.503 and 1.530 at 20°.
Specific Gravity Between 0.908 and 0.940.

TESTS

Acid Value Determine as directed for *Acid Value* under *Essential Oils and Flavors*, Appendix VI.
Angular Rotation Determine in a 100-mm tube as directed under *Optical (Specific) Rotation*, Appendix IIB.
Heavy Metals Shake 10 mL of the oil with an equal volume of water to which 1 drop of hydrochloric acid has been added, and pass hydrogen sulfide through the mixture until it is saturated. No darkening in color is produced in either the oil or the water.
Refractive Index, Appendix IIB Determine with an Abbé or other refractometer of equal or greater accuracy.
Specific Gravity Determine by any reliable method (see *General Provisions*).

Packaging and Storage Store in full, tight containers in a cool place protected from light.

Parsley Seed Oil

DESCRIPTION

The oil obtained by steam distillation of the ripe seed of *Petroselinum crispum* (Fam. *Umbelliferae*). It is a yellow to light brown liquid having a rather harsh odor. It is soluble in most fixed oils and in mineral oil. It is slightly soluble in propylene glycol, but it is insoluble in glycerin.

Functional Use in Foods Flavoring agent.

REQUIREMENTS

Identification The infrared absorption spectrum of the sample exhibits relative maxima (that may vary in intensity) at the same wavelengths (or frequencies) as those shown in the respective spectrum in the section on *Infrared Spectra (Series A: Essential Oils)*, using the same test conditions as specified therein.
Acid Value Not more than 4.0.
Angular Rotation Between −4° and −10°.
Heavy Metals (as Pb) Passes test.
Refractive Index Between 1.513 and 1.522 at 20°.
Saponification Value Between 2 and 10.
Solubility in Alcohol Passes test.
Specific Gravity Between 1.040 and 1.080.

TESTS

Acid Value Determine as directed for *Acid Value* under *Fats and Related Substances*, Appendix VII.
Angular Rotation Determine in a 100-mm tube as directed under *Optical (Specific) Rotation*, Appendix IIB.
Heavy Metals Shake 10 mL of the oil with an equal volume of water to which 1 drop of hydrochloric acid has been added, and pass hydrogen sulfide through the mixture until it is saturated. No darkening in color is produced in either the oil or the water.
Refractive Index, Appendix IIB Determine with an Abbé or other refractometer of equal or greater accuracy.
Saponification Value Determine as directed for *Saponification Value* under *Essential Oils and Flavors*, Appendix VI, using about 5 g, accurately weighed.
Solubility in Alcohol Proceed as directed in the general method, Appendix VI. One mL dissolves in 6 mL of 80% alcohol, occasionally with slight haziness.
Specific Gravity Determine by any reliable method (see *General Provisions*).

Packaging and Storage Store in full, preferably glass, tin-lined, or other suitably lined containers in a cool place protected from light.

Partially Hydrolyzed Proteins

Enzyme-Hydrolyzed (Source) Protein; Partially Hydrolyzed (Source) Protein; (Source) Peptone; Enzyme-Modified (Source) Protein; Partial Enzymatic Digest of (Source) Protein; Partial Acid Digest of (Source) Protein

INS: 429

DESCRIPTION

Partially Hydrolyzed Proteins are composed of peptides and polypeptides resulting from the partial or incomplete hydrolysis (breakdown) of peptide bonds present in edible proteinaceous materials catalyzed by heat, food-grade proteolytic enzymes, and/or suitable food-grade acids. Their degree of hydrolysis typically range from 3% to 85% on the basis of peptide bond cleavage. During processing, the proteinaceous raw material may be treated with safe and suitable alkaline materials. The edible proteinaceous materials used as raw materials are derived from casein and other milk products such as whey protein; from animal tissue, including gelatin, defatted animal tissue, and egg albumen; and from soy protein products, yeast, wheat protein products, or other suitable and safe plant sources. Partially Hydrolyzed Proteins may be available in liquid, paste, powder, or granular form.

> Note: Depending on the protein source and the degree of hydrolysis, Partially Hydrolyzed Proteins may present an allergenic risk to sensitized individuals.

Functional Use in Foods Binder; dough conditioner; emulsifier and emulsifier salt; flavoring agent; flavor enhancer; nutrient supplement; fermentation aid; processing aid; surface-active agent; formulation aid; texturizer.

REQUIREMENTS

Calculate all analyses on the dried basis. In a suitable tared container, evaporate liquid and paste samples to dryness on a steam bath, then, as for the powdered and granular forms, dry to constant weight at 105° (see *General Provisions*).
Labeling Indicate the source of protein, including type.
Assay (Total Nitrogen; TN) Not less than 7.0%.
α-Amino Nitrogen (AN) Not less than 90.0% and not more than 110.0% of the amount claimed on the label.
α-Amino Nitrogen/Total Nitrogen (AN/TN) Percent Ratio Not less than 2.0% and not more than 62.0%, when calculated on an ammonia nitrogen-free basis.
Ammonia Nitrogen (NH_3-N) Not more than 1.5%.
Ash (Total) Not more than 40.0%.
Glutamic Acid Not more than 20.0% as $C_5H_9NO_4$ and not more than 35.0% of the total amino acids.
Heavy Metals (as Pb) Not more than 10 mg/kg.
Lead Not more than 3 mg/kg.

TESTS

Assay (Total Nitrogen) Proceed as directed under *Nitrogen Determination*, Appendix IIIC.

α-Amino Nitrogen Proceed as directed under *α-Amino Nitrogen* in the monograph for *Acid Hydrolysates of Proteins*.

α-Amino Nitrogen/Total Nitrogen (AN/TN) Percent Ratio Calculate by dividing the percent α-amino nitrogen (*AN*) by the percent total nitrogen (*TN*) as corrected for ammonia nitrogen (NH_3-N) according to the formula

$$100 \, [(AN - NH_3\text{-N})/(TN - NH_3\text{-N})].$$

Ammonia Nitrogen Proceed as directed under *Ammonia Nitrogen* in the monograph for *Acid Hydrolysates of Proteins*.

Ash (Total) Proceed as directed in the general method, Appendix IIC, using a 1-g sample.

Glutamic Acid Proceed as directed under *Glutamic Acid* in the monograph for *Acid Hydrolysates of Proteins*.

Heavy Metals Prepare and test a 2-g sample as directed in *Method II* under the *Heavy Metals Test*, Appendix IIIB, using 20 μg of lead ion (Pb) in the control (*Solution A*).

Lead A *Sample Solution* prepared as directed for organic compounds meets the requirements of the *Lead Limit Test*, Appendix IIIB, using 3 μg of lead ion (Pb) in the control.

Packaging and Storage Store in tight containers.

Peanut Oil (Unhydrogenated)

DESCRIPTION

A pale yellow oil with a bland flavor obtained from the kernel of the peanut plant *Arachis hypogaea* by mechanical expression or solvent extraction. It is refined, bleached, and deodorized to substantially remove free fatty acids, phospholipids, color, odor and flavor components, and miscellaneous other non-oil materials. It is a liquid at 21° to 27°, but solidifies to a gel-like consistency at refrigerator temperatures (2° to 4°). It is free from visible foreign matter at 21° to 27°, but sometimes clouds at temperatures above 21°.

Functional Use in Foods Coating agent; emulsifying agent; formulation aid; texturizer.

REQUIREMENTS

Identification Unhydrogenated Peanut Oil exhibits the following composition profile of fatty acids as determined under *Fatty Acid Composition*, Appendix VII:

Fatty Acid:	<14	14:0	16:0	16:1	18:0	18:1	18:2
Weight % (Range):	<0.1	<0.2	6–15	<1.0	1.3–6.5	36–72	13–45

Fatty Acid:	18:3	20:0	20:1	22:0	22:1	24:0
Weight % (Range):	<2.0	<1.0–2.5	0.5–2.1	1.5–4.8	<0.1	1.0–2.5

Color (AOCS-Wesson) Not more than 5.0 red.
Free Fatty Acids (as oleic acid) Not more than 0.1%.
Iodine Value Between 84 and 100.
Lead Not more than 0.1 mg/kg.
Linolenic Acid Not more than 2.0%.
Peroxide Value Not more than 10 meq/kg.
Unsaponifiable Matter Not more than 1.5%.
Water Not more than 0.1%.

TESTS

Color Proceed as directed for *Color* (AOCS-Wesson) under *Fats and Related Substances*, Appendix VII.

Free Fatty Acids Proceed as directed under *Free Fatty Acids*, Appendix VII, using the following equivalence factor (*e*) in the formula given in the procedure:

$$\text{Free fatty acids as oleic acid, } e = 28.2.$$

Iodine Value Proceed as directed under *Iodine Value*, Appendix VII.

Lead Determine as directed under *Method II* in the *Atomic Absorption Spectrophotometric Graphite Furnace Method* under *Lead Limit Test*, Appendix IIIB, using a 3-g sample.

Linolenic Acid Proceed as directed under *Fatty Acid Composition*, Appendix VII.

Peroxide Value Proceed as directed under *Peroxide Value* in the monograph for *Hydroxylated Lecithin*. However, after the addition of saturated potassium iodide and mixing, instead of allowing the solution to stand for 10 min, mix the solution for 1 min and begin the titration immediately.

Unsaponifiable Matter Proceed as directed under *Unsaponifiable Matter*, Appendix VII.

Water Proceed as directed under *Water Determination*, Appendix IIB. However, in place of 35 to 40 mL of methanol, use 50 mL of chloroform to dissolve the sample.

Packaging and Storage Store in well-closed containers.

Pectins

INS: 440   CAS: [9000-69-5]

DESCRIPTION

Pectins consist mainly of the partial methyl esters of polygalacturonic acid and their sodium, potassium, calcium, and ammonium salts. It is obtained by extraction in an aqueous medium of appropriate edible plant material, usually citrus fruits or apples. No organic precipitants shall be used other than methanol, ethanol, and isopropanol. In some types, a portion of the methyl esters may have been converted to primary amides by treatment with ammonia under alkaline conditions. It usually occurs as a white, yellowish, light grayish, or light brownish powder. The commercial product is normally diluted with sugars for standardization purposes. In addition to sugars, pectins may be mixed with suitable food-grade salts required for pH control and desirable setting characteristics.

Note: The following **REQUIREMENTS** and **TESTS** apply to the Pectins as supplied, whether standardized or not, except for specifications covering amide substitution and the weight percent of total galacturonic acid in the Pectin component, in which cases the test procedures provide for removing the sugars and soluble salts before analysis of the Pectin component.

Functional Use in Foods Gelling agent; thickener; stabilizer; emulsifier.

REQUIREMENTS

Labeling Indicate the presence of sulfur dioxide if the residual concentration is greater than 10 mg/kg.
Identification
 A. To a 1 in 100 aqueous solution of the sample add an equal volume of alcohol. A translucent, gelatinous precipitate is formed (most gums will not form such a precipitate).
 B. To 10 mL of a 1 in 100 aqueous solution of the sample add 1 mL of thorium nitrate solution (1 in 10), stir, and allow to stand for 2 min. A stable precipitate or gel forms (most gums will not form such a precipitate).
 C. To 5 mL of a 1 in 100 aqueous solution of the sample add 1 mL of 1 N sodium hydroxide, and allow to stand at room temperature for 15 min. A gel or semi-gel forms (tragacanth and other gums will not form such a precipitate).
 D. Acidify the gel from *Identification Test C* with 1 mL of 2.7 N hydrochloric acid, and shake well. If necessary, continue adding the acid dropwise until the mixture is acid to litmus. A voluminous, colorless, gelatinous precipitate forms, which, upon boiling, becomes white and flocculent (pectic acid).
Acid-Insoluble Ash Not more than 1.0%.
Arsenic (as As) Not more than 3 mg/kg.
Degree of Amide Substitution Not more than 25% total carboxylic groups.
Heavy Metals (as Pb) Not more than 0.002%.
Lead Not more than 5 mg/kg.
Loss on Drying Not more than 12.0%.
Methanol, Ethanol, and Isopropanol Not more than 1.0% total.
Sodium Methyl Sulfate Not more than 0.1%.
Sulfur Dioxide Not more than 0.005%.
Total Galacturonic Acid in Pectin Component Not less than 65.0%, calculated on the ash-free, dried basis.

TESTS

Acid-Insoluble Ash Determine as directed in the general method, Appendix IIC.
Arsenic A *Sample Solution* prepared as directed for organic compounds meets the requirements of the *Arsenic Test*, Appendix IIIB.
Degree of Amide Substitution and Total Galacturonic Acid in the Pectin Component Weigh 5 g of the sample to the nearest 0.1 mg, and transfer to a suitable beaker. Stir for 10 min with a mixture of 5 mL of 2.7 N hydrochloric acid, and 100 mL of 60% ethanol. Transfer to a fritted-glass filter tube (30- to 60-mL capacity), and wash with six 15-mL portions of the same hydrochloric acid–60% ethanol mixture, followed by 60% ethanol until the filtrate is free of chlorides. Finally, wash with 20 mL of ethanol, dry for 2.5 h in an oven at 105°, cool, and weigh. Transfer exactly one-tenth of the total net weight of the now ash-free, dried sample (representing 0.5 g of the original unwashed sample) to a 250-mL conical flask, and moisten the sample with 2 mL of ethanol. Add 100 mL of recently boiled and cooled distilled water, stopper, and swirl occasionally until a complete solution is formed. Add 5 drops of phenolphthalein TS, titrate with 0.1 N sodium hydroxide, and record the results as the initial titre (V_1).

Add exactly 20 mL of 0.5 N sodium hydroxide, stopper, shake vigorously, and let stand for 15 min. Add exactly 20 mL of 0.5 N hydrochloric acid, and shake until the pink color disappears. Titrate with 0.1 N sodium hydroxide to a faint pink color that persists after vigorous shaking; record this value as the saponification titre (V_2).

Quantitatively transfer the contents of the conical flask into a 500-mL distillation flask fitted with a Kjeldahl trap and a water-cooled condenser, the delivery tube of which extends well beneath the surface of a mixture of 150 mL of carbon dioxide-free water and 20.0 mL of 0.1 N hydrochloric acid in a receiving flask. To the distillation flask add 20 mL of a 1 in 10 sodium hydroxide solution, seal the connections, and then begin heating carefully to avoid excessive foaming. Continue heating until 80 to 120 mL of distillate has been collected. Add a few drops of methyl red TS to the receiving flask, titrate the excess acid with 0.1 N sodium hydroxide, and record the volume required, in mL, as S. Perform a blank determination on 20.0 mL of 0.1 N hydrochloric acid, and record the volume required, in mL, as B. Record the amide titre ($B - S$) as V_3.

Transfer exactly one-tenth of total net weight of the dried sample (representing 0.5 g of the original unwashed sample), and wet with about 2 mL of ethanol in a 50-mL beaker. Dissolve the Pectin in 25 mL of 0.125 M sodium hydroxide. Let the solution stand for 1 h, with agitation, at room temperature. Transfer quantitatively the saponified Pectin solution to a 50-mL volumetric flask, and dilute to volume with distilled water. Transfer 25.0 mL of the diluted Pectin solution to a distillation apparatus, and add 20 mL of Clark's solution, which consists of 100 g of magnesium sulfate heptahydrate and 0.8 mL of sulfuric acid and distilled water to a total of 180 mL. The distillation apparatus consists of a steam generator connected to a round-bottom flask to which a condenser is attached. Both steam generator and the round-bottom flask are equipped with heating mantles. Start the distillation by heating the round-bottom flask containing the sample. Collect the first 15 mL of distillate separately in a measuring cylinder. Then start the steam supply and continue distillation until 150 mL of distillate has been collected in a 200-mL beaker. Quantitatively combine the distillates, titrate with 0.05 M sodium hydroxide to pH 8.5, and record the volume required, in mL, as S.

Perform a blank determination using 20 mL of distilled water. Record the required volume, in mL, as B. Record acetate ester titre ($S - B$) as V_4.

Calculate the degree of amidation (as the percent of total carboxyl groups) by the formula

$$100 - [V_3/(V_1 + V_2 + V_3 - V_4)].$$

Calculate mg of galacturonic acid by the formula

$$19.41 (V_1 + V_2 + V_3 - V_4).$$

The mg of galacturonic acid obtained in this way is the content of one-tenth of the weight of the washed and dried sample. To calculate the percent galacturonic acid on a moisture- and ash-free basis, multiply the number of mg obtained by $1000/x$, in which x is the weight, in mg, of the washed and dried sample.

Note: If the Pectin is known to be of the nonamidated type, only V_1 and V_2 need to be determined, and V_3 may be regarded as zero.

Heavy Metals Prepare and test a 1-g sample as directed in *Method II* under the *Heavy Metals Test*, Appendix IIIB, using 20 μg of lead ion (Pb) in the control (*Solution A*).

Lead (Note: Use deionized water throughout this procedure.)

Diluted Standard Lead Solution (2 μg Pb/mL) Immediately before use, pipet 0.10 mL of a certified commercially available 1000 ppm (1000 μg/mL) Pb stock solution into a 50-mL volumetric flask containing 30 mL of water and 4 mL of 20% v/v hydrochloric acid and 4 mL of 0.1 M EDTA. Dilute to volume with water, and mix.

Control Lead Solution (0.4 μg Pb/mL) Pipet 5.0 mL of the *Diluted Standard Lead Solution* into a 25-mL volumetric flask containing 10 mL of water, 2 mL of 20% v/v hydrochloric acid and 2 mL of 0.1 M EDTA. Dilute to volume with water, and mix.

Standard Lead Blank Solution Add 30 mL of water, 4 mL of 20% v/v hydrochloric acid, and 4 mL of 0.1 M EDTA into a 50-mL volumetric flask. Dilute to volume with water, and mix.

Sample Preparation Transfer 2.0 g of the sample into a clean 100-mL glass beaker, add 25 mL of nitric acid (70% v/v), cover with a watch glass, and heat at low to moderate heat on a hot plate in a fume hood for 2 h. Remove watch glass, and continue to heat until the sample is dry with no visible fumes. Add 0.5 mL of nitric acid, and heat to dryness. Cool to room temperature and add 2 mL of 20% v/v hydrochloric acid and 2 mL of 0.1 M EDTA. Transfer quantitatively to a 25-mL volumetric flask, then dilute to volume with water, and mix.

Procedure Set up the inductively coupled plasma emission spectrometer according to manufacturer's instructions using the Pb emission line at 220.35 nm. Calibrate the instrument using the *Standard Lead Blank Solution* and the *Diluted Standard Lead Solution*. Then analyze the *Sample Preparation* and the *Control Lead Solution*. The sample passes the test if the lead concentration found in the *Sample Preparation* is equal to or less than that in the *Control Lead Solution*.

Loss on Drying, Appendix IIC Dry at 105° for 2 h.

Methanol, Ethanol, and Isopropanol The alcohols are converted to their nitrite esters, and their levels are determined by headspace gas chromatography.

Internal Standard Solution Dissolve 50 mg of *n*-propanol in 1 L of water.

Sample Solution Dissolve 100 mg of the sample in 10 mL of water, and as necessary, use sodium chloride as a dispersing agent.

Standard Alcohol Solution Using a micropipet, transfer 50 mg each of methanol (corresponding to 39.55 μL), ethanol (corresponding to 39.47 μL), and isopropanol (corresponding to 39.28 μL) into a 1000-mL volumetric flask, dilute to volume with water, and mix.

Sodium Nitrite Solution Dissolve 250 g of sodium nitrite in 1000 mL of water.

Chromatographic System Use a suitable gas chromatograph equipped with a flame-ionization detector. Use a 90-cm × 4-mm id glass column with the first 15 cm packed with Chromopack (or equivalent) and the remainder packed with Poropak R 120- to 150-mesh (or equivalent). The operating conditions of the gas chromatograph are as follows: the injection port temperature is 250°, and the column temperature is 150° isothermal. Use nitrogen as the carrier gas with a flow rate of 80 mL/min.

Procedure Weigh 200 mg of urea, and place it in a 25-mL amber-glass vial (Reacti-Flasks or equivalent). Purge with nitrogen for 5 min, add 1 mL of saturated oxalic acid solution, close with a rubber stopper, and swirl. Add 1 mL of *Sample Solution* and 1 mL of *Internal Standard Solution*, and simultaneously start a stopwatch ($t = 0$). Swirl the vial, and recap it with an open screw cap fitted with a silicone rubber septum. Swirl the vial until $t = 30$ s. At $t = 45$ s, inject through the septum 0.5 mL of *Sodium Nitrite Solution*. Swirl until $t = 70$ s, and at $t = 150$ s, withdraw through the septum 1.0 mL of the headspace using a pressure lock syringe (Precision Sampling Corporation, or equivalent). Inject the 1.0 mL into the injection port of the gas chromatograph. Repeat this procedure, but use 1 mL of the *Standard Alcohol Solution* instead of the *Sample Solution*.

Calculation Quantify the total methanol, ethanol, and isopropanol present in the sample by the following formula:

$$T = V_{MS} (R_{MU}/R_{MS})0.791 + V_{ES} (R_{EU}/R_{ES})0.7893 + V_{IS} (R_{IU}/R_{IS})0.7855,$$

in which T is the total amount, in mg, of methanol, ethanol, and isopropanol in the sample; the subscripts M, E, and I refer to methanol, ethanol and isopropanol, respectively; V_S is the volume, in mL, of the corresponding alcohol in the *Standard Alcohol Solution*; R_S is the ratio of the peak area of the corresponding alcohol in the *Standard Alcohol Solution* to that of *n*-propanol in *Internal Standard Solution*; R_U is the ratio of the peak area of the corresponding alcohol in the *Sample Solution* to that of *n*-propanol in the *Internal Standard Solution*; and 0.791, 0.7893, and 0.7855 are the densities, in g/mL, for methanol, ethanol, and isopropanol, respectively. Calculate the percent methanol, ethanol, and isopropanol present in the sample by the following formula:

$$(T100)/W,$$

in which W is the sample weight, in mg.

Sodium Methyl Sulfate

Mobile Phase Prepare a 0.04 M potassium hydrogen phthalate solution by transferring 16.4 g of potassium hydrogen phthalate into a 2-L volumetric flask, dilute to volume with water, and mix. Filter the solution through a 0.45-μm pore-size filter (Millipore, or equivalent).

Standard Preparation Transfer 10.0 mg of anhydrous sodium methyl sulfate into a 100-mL volumetric flask, dilute to volume with *Mobile Phase*, and mix.

Assay Preparation Suspend about 1 g of the sample, accurately weighed, in 10.0 mL of 50% (v/v) ethanol solution. Stir for 30 min using a Teflon-coated stirring bar. Allow the suspension to precipitate, and filter. Evaporate a 1.0-mL aliquot to dryness using reduced pressure (10 mm Hg), and heat at 60°. Redissolve the residue in 1.0 mL of the *Mobile Phase*.

Chromatographic System Use a high-performance liquid chromatograph equipped with a refractive index detector and a 25-cm × 4.6-mm column packed with Nucleosil 10SB (or equivalent) and maintained at 40°. Set the flow rate at 1 mL/min.

System Suitability Three replicate injections of the *Standard Preparation* show a relative standard deviation of not more than 4.0% for the response factor of the sodium methyl sulfate peak obtained using the formula

$$(A_S/C_S),$$

in which A_S is the peak area response of the *Standard Preparation*, and C_S is the concentration, in mg/mL, of sodium methyl sulfate in the *Standard Preparation*.

Procedure Inject 20 μL of the *Standard Preparation* followed by the *Assay Preparation*. Determine the peak area in the chromatograms for the *Standard Preparation* and *Assay Preparation*. Calculate the quantity in percent of sodium methyl sulfate in the sample by the formula

$$(C_S A_U)/(A_S W),$$

in which C_S is the concentration, in mg/mL, of sodium methyl sulfate in the *Standard Preparation*; A_U and A_S are the peak area responses obtained from the *Assay Preparation* and *Standard Preparation*, respectively; and W is the weight, in g, of the sample taken.

Sulfur Dioxide Determine as directed in the general method, Appendix X, using the following method under *Sample Introduction and Distillation*: Transfer about 20 g of the sample, accurately weighed, into flask C, and add 20 mL of ethanol to moisten the sample. Add 400 mL of water, swirling vigorously to disperse the sample. Reassemble the apparatus, making sure that the tapered joints are clean and greased with stopcock grease, and proceed as directed under *Sample Introduction and Distillation*, beginning with "the nitrogen flow through the *3% Hydrogen Peroxide Solution*. . . ."

Packaging and Storage Store in well-closed containers.

Pennyroyal Oil

DESCRIPTION

The volatile oil obtained by steam distillation from the fresh or partly dried plant *Mentha pulegium* L. (Fam. *Labiatae*). It is a light yellow to yellow, aromatic liquid having a mintlike odor. It is soluble in most fixed oils and in propylene glycol. It is soluble, with slight cloudiness, in mineral oil, but it is practically insoluble in glycerin.

Functional Use in Foods Flavoring agent.

REQUIREMENTS

Identification The infrared absorption spectrum of the sample exhibits relative maxima (that may vary in intensity) at the same wavelengths (or frequencies) as those shown in the respective spectrum in the section on *Infrared Spectra (Series A: Essential Oils)*, using the same test conditions as specified therein.
Assay Not less than 88.0% and not more than 96.0%, by volume, of ketones.
Angular Rotation Between +18° and +25°.
Heavy Metals (as Pb) Passes test.
Refractive Index Between 1.483 and 1.488 at 20°.
Solubility in Alcohol Passes test.
Specific Gravity Between 0.928 and 0.940.

TESTS

Assay Proceed as directed under *Aldehydes and Ketones—Neutral Sulfite Method*, Appendix VI.
Angular Rotation Determine in a 100-mm tube as directed under *Optical (Specific) Rotation*, Appendix IIB.
Heavy Metals Shake 10 mL of the oil with an equal volume of water to which 1 drop of hydrochloric acid has been added, and pass hydrogen sulfide through the mixture until it is saturated. No darkening in color is produced in either the oil or the water.
Refractive Index, Appendix IIB Determine with an Abbé or other refractometer of equal or greater accuracy.
Solubility in Alcohol Proceed as directed in the general method, Appendix VI. One mL dissolves in 2 mL of 70% alcohol.
Specific Gravity Determine by any reliable method (see *General Provisions*).

Packaging and Storage Store in full, tight, preferably glass, tin-lined, or suitably galvanized containers in a cool place protected from light.

Pentaerythritol Ester of Partially Hydrogenated Wood Rosin

DESCRIPTION

A hard, amber-colored resin (color K or paler as determined by ASTM Designation D 509) produced by the esterification of partially hydrogenated wood rosin with pentaerythritol and purified by steam stripping. It is soluble in acetone, but is insoluble in water.

Functional Use in Foods Masticatory substance in chewing gum base.

REQUIREMENTS

Identification The sample is melted and prepared for analysis on a potassium bromide plate. The infrared absorption spectrum of the sample exhibits maxima at the same wavelengths as the typical spectrum, as shown in the section on *Infrared Spectra* (*Series C: Other Substances*).
Acid Number Between 7 and 18.
Arsenic (as As) Not more than 3 mg/kg.
Heavy Metals (as Pb) Not more than 10 mg/kg.
Lead Not more than 1 mg/kg.
Ring-and-Ball Softening Point Between 94° and 102°.

TESTS

Acid Number Determine as directed in the general procedure, Appendix IX.
Arsenic Prepare a *Sample Solution* as directed in the general method under *Chewing Gum Base*, Appendix IV. This solution meets the requirements of the *Arsenic Test*, Appendix IIIB.
Heavy Metals Prepare and test a 2-g sample as directed in *Method II* under the *Heavy Metals Test*, Appendix IIIB, using 20 µg of lead ion (Pb) in the control (*Solution A*).
Lead Prepare a *Sample Solution* using a 5-g sample as directed in the general method under *Chewing Gum Base*, Appendix IV. This solution meets the requirements of the *Lead Limit Test*, Appendix IIIB, using 5 µg of lead ion (Pb) in the control.
Ring-and-Ball Softening Point Determine as directed in the general procedure, Appendix IX, using the *Ring-and-Ball Method*.

Packaging and Storage Store in well-closed containers.

Pentaerythritol Ester of Wood Rosin

DESCRIPTION

A hard, pale, amber-colored resin (color M or paler as determined by ASTM Designation D 509) produced by the esterification of pale wood rosin with pentaerythritol and purified by steam stripping. It is soluble in acetone, but is insoluble in water and in alcohol.

Functional Use in Foods Masticatory substance in chewing gum base.

REQUIREMENTS

Identification The sample is melted and prepared for analysis on a potassium bromide plate. The infrared absorption spectrum of the sample exhibits maxima at the same wavelengths as the typical spectrum, as shown in the section on *Infrared Spectra* (*Series C: Other Substances*).
Acid Number Between 6 and 16.
Arsenic (as As) Not more than 3 mg/kg.
Heavy Metals (as Pb) Not more than 10 mg/kg.
Lead Not more than 1 mg/kg.
Ring-and-Ball Softening Point Between 100° and 108°.

TESTS

Acid Number Determine as directed in the general procedure, Appendix IX.
Arsenic Prepare a *Sample Solution* as directed in the general method under *Chewing Gum Base*, Appendix IV. This solution meets the requirements of the *Arsenic Test*, Appendix IIIB.
Heavy Metals Prepare and test a 2-g sample as directed in *Method II* under the *Heavy Metals Test*, Appendix IIIB, using 20 µg of lead ion (Pb) in the control (*Solution A*).
Lead Prepare a *Sample Solution* using a 5-g sample as directed in the general method under *Chewing Gum Base*, Appendix IV. This solution meets the requirements of the *Lead Limit Test*, Appendix IIIB, using 5 µg of lead ion (Pb) in the control.
Ring-and-Ball Softening Point Determine as directed in the general procedure, Appendix IX, using the *Ring-and-Ball Method*.

Packaging and Storage Store in well-closed containers.

Peppermint Oil

CAS: [8006-90-4]

DESCRIPTION

An essential oil obtained by steam distillation from the fresh overground parts of the flowering plant of *Mentha piperita* L. (Fam. *Labiatae*); it may be rectified by distillation, but is neither partially nor wholly dementholized. It is a colorless or pale yellow liquid having a strong, penetrating odor of peppermint and a pungent taste, followed by a sensation of coldness when air is drawn into the mouth.

Functional Use in Foods Flavoring agent.

REQUIREMENTS

Labeling Indicate whether it is natural or rectified.
Identification
 A. The infrared absorption spectrum of the sample exhibits relative maxima (that may vary in intensity) at the same wavelengths (or frequencies) as those shown in the respective spectrum in the section on *Infrared Spectra* (*Series A: Essential Oils*), using the same test conditions as specified therein.
 B. Mix in a dry test tube 3 drops of the oil with 5 mL of a solution of nitric acid in glacial acetic acid (1 in 300), and place

the tube in a beaker of boiling water. A blue color develops within 5 min, which, on continued heating, deepens and shows a copper-colored fluorescence, and then fades, leaving a golden yellow solution.

Assay for Total Esters Not less than 5.0% of esters, calculated as menthyl acetate ($C_{12}H_{22}O_2$).
Assay for Total Menthol Not less than 50.0% of menthol ($C_{10}H_{20}O$).
Angular Rotation Between −18° and −32°.
Dimethyl Sulfide Passes test (rectified); fails test (natural).
Heavy Metals (as Pb) Passes test.
Refractive Index Between 1.459 and 1.465 at 20°.
Solubility in Alcohol Passes test.
Specific Gravity Between 0.896 and 0.908.

TESTS

Assay for Total Esters Weigh accurately about 10 g, and proceed as directed under *Ester Determination*, Appendix VI, using 99.16 as the equivalence factor (*e*) in the calculation.
Assay for Total Menthol Proceed as directed under *Total Alcohols*, Appendix VI, using a 2.5-g sample of the acetylated oil. Calculate the percentage of total menthol by the formula

$$7.814A(1 - 0.0021E)/(B - 0.021A),$$

in which *A* is the difference between the number of mL of 0.5 *N* hydrochloric acid required in the titration and the number of mL of 0.5 *N* hydrochloric acid required in the residual blank titration, *B* is the weight of the sample of the acetylated oil, and *E* is the percentage of total esters determined and calculated as menthyl acetate ($C_{12}H_{22}O_2$).
Angular Rotation Determine in a 100-mm tube as directed under *Optical (Specific) Rotation*, Appendix IIB.
Dimethyl Sulfide Distill 1 mL from a sample of 25 mL, and carefully superimpose the distillate on 5 mL of mercuric chloride TS in a test tube. A white film does not form at the zone of contact within 1 min if the sample is rectified.
Heavy Metals Shake 10 mL of the oil with an equal volume of water to which 1 drop of hydrochloric acid has been added, and pass hydrogen sulfide through the mixture until it is saturated. No darkening in color is produced in either the oil or the water.
Refractive Index, Appendix IIB Determine with an Abbé or other refractometer of equal or greater accuracy.
Solubility in Alcohol Proceed as directed in the general method, Appendix VI. One mL dissolves in 3 mL of 70% alcohol.
Specific Gravity Determine by any reliable method (see *General Provisions*).

Packaging and Storage Store in full, tight containers in a cool place protected from light.

Perlite

Expanded Perlite

DESCRIPTION

In its natural state, Perlite occurs as a dense, gray to brown, glassy volcanic rock consisting essentially of fused sodium potassium aluminum silicate plus 3% to 5% water. When fractured and heated at high temperature (900° to 1100°) under proper conditions, it pops like popcorn (due to the presence of the occluded water), expanding to 20 or more times its original volume. The expanded material is crushed to yield a white, nonhygroscopic powder having a bulk density of 32 to 400 kg/m^3 (2 to 25 lb/ft^3) and a particle size ranging from less than one to several hundred μm. It is in this latter expanded and powdered state that Perlite is used as a filter aid in food processing. Acceptable food-grade free-flowing agents such as sodium carbonate and sodium silicate may be added. The powder is slightly soluble in water and sparingly soluble in dilute acids and alkalies.

Functional Use in Foods Filter aid in food processing.

REQUIREMENTS

Identification
  A. Mix about 1 g of the sample with 25 mL of 2.7 *N* hydrochloric acid in a beaker, cover with a watch glass, heat on a steam bath for 15 min, and cool. Filter, and neutralize the filtrate to litmus paper with 6 *N* ammonium hydroxide. The neutralized filtrate gives positive tests for *Aluminum*, for *Potassium*, and for *Sodium*, Appendix IIIA.
  B. Prepare a bead by fusing a few crystals of sodium ammonium phosphate on a platinum loop in the flame of a burner. Place the hot, transparent bead in contact with a sample, and again fuse. Silica floats about in the bead, producing, upon cooling, an opaque bead with a weblike structure.
Arsenic (as As) Not more than 10 mg/kg.
Lead Not more than 10 mg/kg.
Loss on Drying Not more than 3.0% (powdered form).
Loss on Ignition Not more than 7.0% (glassy form).
pH Between 5 and 11 (filtrate from a 10% suspension).

TESTS

Arsenic Transfer 10.0 g of the sample into a 250-mL beaker, add 50 mL of 0.5 *N* hydrochloric acid, cover with a watch glass, and heat at 70° for 15 min. Cool, and decant through a Whatman No. 3, or equivalent, filter paper into a 100-mL volumetric flask. Wash the slurry with three 10-mL portions of hot water and the filter paper with 15 mL of hot water, dilute the solution to volume with water, and mix. A 3.0-mL portion of this solution meets the requirements of the *Arsenic Test*, Appendix IIIB.
Lead A 10.0-mL portion of the solution prepared in the *Arsenic Test* meets the requirements of the *Lead Limit Test*, Appendix IIIB, using 10 μg of lead ion (Pb) in the control.

Loss on Drying, Appendix IIC   Dry the powder at 105° for 2 h.

Loss on Ignition   Ignite a 250-mg crushed sample of the glassy form to constant weight at 1000°.

pH, Appendix IIB   Boil 10 g with 100 mL of water for 30 min, make up to 100 mL with water, and filter through a fine-pore, sintered-glass funnel. Use the resulting filtrate for the determination of pH.

Packaging and Storage   Store in well-closed containers.

Petitgrain Oil, Paraguay Type

DESCRIPTION

The volatile oil obtained by steam distillation from the leaves and small twigs of the bitter orange tree, *Citrus aurantium* L. subspecies *amara*. It is a yellow to brownish yellow liquid having a somewhat harsh, bitter-sweet, floral odor. It is soluble in most fixed oils and is soluble, with opalescence or turbidity, in mineral oil and in propylene glycol. It is relatively insoluble in glycerin.

Functional Use in Foods   Flavoring agent.

REQUIREMENTS

Identification   The infrared absorption spectrum of the sample exhibits relative maxima (that may vary in intensity) at the same wavelengths (or frequencies) as those shown in the respective spectrum in the section on *Infrared Spectra (Series A: Essential Oils)*, using the same test conditions as specified therein.

Assay   Not less than 45.0% and not more than 60.0% of esters, calculated as linalyl acetate ($C_{12}H_{20}O_2$).

Angular Rotation   Between −4° and +1°.

Heavy Metals (as Pb)   Passes test.

Refractive Index   Between 1.455 and 1.462 at 20°.

Solubility in Alcohol   Passes test.

Specific Gravity   Between 0.878 and 0.889.

TESTS

Assay   Weigh accurately about 2 g, and proceed as directed under *Ester Determination*, Appendix VI, using 98.15 as the equivalence factor (*e*) in the calculation.

Angular Rotation   Determine in a 100-mm tube as directed under *Optical (Specific) Rotation*, Appendix IIB.

Heavy Metals   Shake 10 mL of the oil with an equal volume of water to which 1 drop of hydrochloric acid has been added, and pass hydrogen sulfide through the mixture until it is saturated. No darkening in color is produced in either the oil or the water.

Refractive Index, Appendix IIB   Determine with an Abbé or other refractometer of equal or greater accuracy.

Solubility in Alcohol   Proceed as directed in the general method, Appendix VI. One mL dissolves in 4 mL of 70% alcohol. The solution usually develops opalescence or turbidity upon further dilution.

Specific Gravity   Determine by any reliable method (see *General Provisions*).

Packaging and Storage   Store in full, tight, preferably glass, tin-lined, or aluminum containers in a cool place protected from light.

Petrolatum

White Petrolatum; Yellow Petrolatum

INS: 905b   CAS: [8009-03-8]

DESCRIPTION

A purified mixture of semisolid hydrocarbons obtained from petroleum, occurring as an unctuous mass, and varying in color from white to yellowish or light amber. It is transparent in thin layers, and has not more than a slight fluorescence, even after being melted. It is insoluble in water, and is almost insoluble in cold or hot alcohol and in cold absolute alcohol. It is soluble in ether, in solvent hexane, and in most fixed and volatile oils, and is freely soluble in carbon disulfide, in chloroform, and in turpentine oil. It may contain any antioxidant permitted by the U.S. Food and Drug Administration, in an amount not greater than that required to produce its intended effect.

Functional Use in Foods   Defoaming agent; lubricant; protective coating; release agent.

REQUIREMENTS

Acidity or Alkalinity   Passes test.
Color   Passes test.
Consistency   Passes test (between 100 and 275).
Fixed Oils, Fats, and Rosin   Passes test.
Melting Range   Between 38° and 60°.
Organic Acids   Passes test.
Residue on Ignition   Passes test.
Specific Gravity   Between 0.815 and 0.880 at 60°.
Ultraviolet Absorption (polynuclear hydrocarbons)   Passes test.

TESTS

Acidity or Alkalinity   Introduce 35 g of the sample into a 250-mL separator, add 100 mL of boiling water, and shake vigorously for 5 min. After the Petrolatum and water have separated, draw off the water into a casserole, wash the sample in the separator with two 50-mL portions of boiling water, and add the washings to the casserole. To the accumulated 200 mL of water add 1 drop of phenolphthalein TS, and boil. The solution does not acquire a pink color. If the addition of phenolphthalein produces no pink color, add 0.1 mL of methyl orange TS. No red or pink color is produced.

Color Melt about 10 g on a steam bath, and pour about 5 mL of the liquid into a 150- × 50-mm, clear-glass, bacteriological test tube, keeping the sample melted. The Petrolatum is not darker than a solution made by mixing 3.8 mL of ferric chloride CS and 1.2 mL of cobaltous chloride CS in a similar tube, the comparison of the two being made in reflected light against a white background, holding the sample tube directly against the background at such an angle that there is no fluorescence.

Consistency

Apparatus Determine the consistency of Petrolatum by means of a penetrometer fitted with a polished, cone-shaped, metal plunger weighing 150 g, having a detachable steel tip of the following dimensions: The tip of the cone has an angle of 30°, the point being truncated to a diameter of 0.38 ± 0.03 mm; the base of the tip is 8.38 ± 0.05 mm in diameter; and the length of the tip is 15 ± 0.25 mm. The remaining portion of the cone has an angle of 90°, is 28.2 mm in height, and has a maximum diameter at the base of 65.1 mm. The containers for the test are flat-bottomed metal or glass cylinders that are 102 ± 6 mm in diameter and not less than 60 mm in height.

Procedure Melt a quantity of the sample at 82° ± 2.5°, and pour into one or more of the containers, filling to within 6 mm of the rim. Cool at 25° ± 2.5° over a period of not less than 16 h, protecting from drafts. Two h before the test, place the containers in a water bath at 25° ± 0.5°. If the room temperature is below 23.5° or above 26.5°, adjust the temperature of the cone to 25° ± 0.5° by placing it in the water bath.

Without disturbing the surface of the sample, place the container on the penetrometer table, and lower the cone until the tip just touches the top surface of the sample at a spot 25 to 38 mm from the edge of the container. Adjust the zero setting, and quickly release the plunger, then hold it free for 5 s. Secure the plunger, and read the total penetration from the scale. Make three or more trials, each so spaced that there is no overlapping of the areas of penetration. When the penetration exceeds 20 mm, use a separate container of the sample for each trial. Read the penetration to the nearest 0.1 mm. Calculate the average of the three or more readings, and conduct further trials to a total of 10 if the individual results differ from the average by more than ± 3%. The final average of the trials is not less than 10.0 mm and not more than 27.5 mm, indicating a consistency value between 100 and 275.

Fixed Oils, Fats, and Rosin Digest 10 g of the sample at 100° with 10 g of sodium hydroxide and 50 mL of water for 30 min. Separate the water layer, and add to it an excess of 2 N sulfuric acid. No oily or solid matter separates.

Melting Range Determine as directed for *Melting Range or Temperature, Procedure for Class III*, Appendix IIB.

Organic Acids Weigh 20 g of the sample, add 50 mL of alcohol, previously neutralized to phenolphthalein TS with sodium hydroxide, and 50 mL of water, agitate thoroughly, and heat to boiling. Add 1 mL of phenolphthalein TS, and titrate rapidly, with vigorous agitation, to the production of a sharp pink endpoint, noting the change in the alcohol–water layer. Not more than 0.4 mL of 0.1 N sodium hydroxide is required.

Residue on Ignition Heat 4 g of the sample in an open porcelain or platinum dish over a Bunsen flame. It volatilizes without emitting any acrid odor, and on ignition yields not more than 0.05% of residue.

Specific Gravity Determine by any reliable method (see *General Provisions*).

Ultraviolet Absorption It meets the ultraviolet absorbance specifications required by the U.S. Food and Drug Administration for Petrolatum.

Packaging and Storage Store in tight containers.

Petroleum Wax

Refined Paraffin Wax; Refined Microcrystalline Wax

INS: 905c

DESCRIPTION

A refined mixture of solid hydrocarbons, paraffinic in nature, obtained from petroleum. It occurs as a translucent, tasteless, and odorless wax. It may be prepared as "refined paraffin wax" or as "refined microcrystalline wax." The refined paraffin wax is usually obtained from a lower molecular weight fraction of petroleum and has lower viscosities when molten than the refined microcrystalline wax. The refined microcrystalline wax is usually higher in molecular weight, in flash point, and in melting point than the refined paraffin wax. These waxes are graded and sold according to melting point, which ranges from about 120° to 200°F (48° to 93°C), and color, which varies from amber to almost white. They exhibit a low order of solubility in organic solvents, but are most soluble in aromatic hydrocarbons and least soluble in ketones, in esters, and in alcohols.

Functional Use in Foods Masticatory substance in chewing gum base; protective coating; defoaming agent; microcapsules for spices and flavoring agents.

REQUIREMENTS

Identification Identify refined paraffin wax and refined microcrystalline wax by comparing their infrared absorption spectra with the respective typical spectra as shown in the section on *Infrared Spectra (Series C: Other Substances)*. The samples are melted and prepared for analysis on potassium bromide plates.

Heavy Metals (as Pb) Not more than 10 mg/kg.

Lead Not more than 1 mg/kg.

Ultraviolet Absorbance (polynuclear hydrocarbons) 280 to 289 nm, not more than 0.15; 290 to 299 nm, not more than 0.12; 300 to 359 nm, not more than 0.08; 360 to 400 nm, not more than 0.02.

The following additional specifications, where applicable, should conform to the representations of the vendor: **Color**, **Melting Point**, and **Odor**.

TESTS

Color Determine by any suitable procedure, such as ASTM D 1500.

Heavy Metals Prepare and test a 2-g sample as directed in *Method II* under the *Heavy Metals Test*, Appendix IIIB, using 20 µg of lead ion (Pb) as the control (*Solution A*).

Lead Prepare a *Sample Solution* from a 5-g sample as directed in the general method under *Chewing Gum Base*, Appendix IV. This solution meets the requirements of the *Lead Limit Test*, Appendix IIIB, using 5 µg of lead ion (Pb) in the control.

Melting Point Determine by any suitable procedure, such as ASTM D 127.

Odor Determine by any suitable procedure, such as ASTM D 1833.

Ultraviolet Absorbance Determine as directed in the U.S. Food and Drug Administration regulation for Petroleum Wax (**21** *CFR* 172.886).

Packaging and Storage Store in well-closed containers that are properly vented for liquid materials.

Petroleum Wax, Synthetic

Synthetic Wax (Ethylene Polymer or Ethylene Copolymer with Alpha-Olefins)

DESCRIPTION

A refined mixture of solid hydrocarbons, paraffinic in nature, prepared by the catalytic polymerization of ethylene or copolymer of ethylene with linear (C_3–C_{12}) alpha-olefins. Synthetic Petroleum Wax ranges in melting point from about 170° to 240°F (77° to 116°C). The color of this wax varies from off-white to white. It is most soluble in aromatic hydrocarbons and least soluble in ketones, in esters, and in alcohols.

Functional Use in Foods Masticatory substance in chewing gum base; protective coating; defoaming agent.

REQUIREMENTS

Identification The sample is melted and prepared for analysis on a potassium bromide plate. The infrared absorption spectrum of the sample exhibits maxima at the same wavelengths as the typical spectrum, as shown in the section on *Infrared Spectra* (*Series C: Other Substances*).

Heavy Metals (as Pb) Not more than 10 mg/kg.
Lead Not more than 1 mg/kg.
Molecular Weight (average) Between 500 and 1200.
Ultraviolet Absorbance (polynuclear hydrocarbons) 280 to 289 nm, not more than 0.15; 290 to 299 nm, not more than 0.12; 300 to 359 nm, not more than 0.08; 360 to 400 nm, not more than 0.02.

The following additional specifications, where applicable, should conform to the representations of the vendor: **Color, Melting Point**, and **Odor**.

TESTS

Color Determine by any suitable procedure, such as ASTM D 1500.

Heavy Metals Prepare and test a 2-g sample as directed in *Method II* under the *Heavy Metals Test*, Appendix IIIB, using 20 µg of lead ion (Pb) in the control (*Solution A*).

Lead Prepare a *Sample Solution* from a 5-g sample as directed in the general method under *Chewing Gum Base*, Appendix IV. This solution meets the requirements of the *Lead Limit Test*, Appendix IIIB, using 5 µg of lead ion (Pb) in the control.

Melting Point Determine by any suitable procedure, such as ASTM D 127.

Molecular Weight

Apparatus Use a suitable vapor pressure osmometer, such as the Hewlett-Packard Model 302A, or equivalent, equipped with dual thermistor beads.

Calibration Standards Dissolve accurately weighed amounts of benzil ($C_6H_5COCOC_6H_5$) in *o*-dichlorobenzene to produce solutions containing approximately 3, 7, 10, and 15 mg of benzil, respectively, per g of solution, and heat to 100° on a steam bath.

Sample Preparations Dissolve accurately weighed amounts of the sample in *o*-dichlorobenzene to produce solutions containing approximately 10, 20, 35, and 50 mg of sample, respectively, per g of solution, and heat to 100° on a steam bath. (Other suitable concentrations that give ΔR readings between 5 and 25 may be used in the *Procedure* below.)

Procedure Following the manufacturer's instructions, balance the osmometer to zero with *o*-dichlorobenzene on both thermistor beads, and establish the calibration constant K_S at 100°, using the four *Calibration Standards*. When the temperature within the osmometer has re-equilibrated to 100°, place an aliquot of the most concentrated *Sample Preparation* on the sample thermistor bead. After 4.0 min, balance the instrument to zero with the potentiometer, and record the ΔR value. Repeat this procedure with the same *Sample Preparation* two or three times, and average the ΔR values for that concentration. In a similar manner, obtain the average ΔR values for each of the other three concentrations of the *Sample Preparation*. Plot the four average ΔR values for the *Sample Preparations* as a function of ΔR/concentration, and extrapolate the line to zero to obtain the constant K_U for the sample. Divide K_S by K_U to obtain the molecular weight of the sample tested.

Odor Determine by any suitable procedure, such as ASTM D 1833.

Ultraviolet Absorbance Determine as directed in the U.S. Food and Drug Administration regulation for Petroleum Wax (**21** *CFR* 172.886).

Packaging and Storage Store in well-closed containers.

DL-Phenylalanine

DL-α-Amino-β-phenylpropionic Acid

$C_6H_5CH_2CH(NH_2)COOH$

$C_9H_{11}NO_2$ Formula wt 165.19

CAS: [150-30-1]

DESCRIPTION

White, odorless, crystalline platelets. It is soluble in water, in dilute mineral acids, and in solutions of alkali hydroxides. It is very slightly soluble in alcohol. It is optically inactive.

Functional Use in Foods Nutrient; dietary supplement.

REQUIREMENTS

Identification
A. Heat 5 mL of a 1 in 1000 solution with 1 mL of triketohydrindene hydrate TS. A reddish purple color is produced.
B. Heat 5 mL of a 1 in 100 solution with a few drops of potassium dichromate TS. A characteristic odor is evolved.
C. To a 10-mg sample add 500 mg of potassium nitrate and 2 mL of sulfuric acid, and heat the mixture on a water bath for 20 min. Cool, add 2 mL of hydroxylamine TS, immerse in ice water for 10 min, and then add 10 mL of 1 N sodium hydroxide. A reddish violet color is produced.
Assay Not less than 98.5% and not more than 101.5% of $C_9H_{11}NO_2$ after drying.
Heavy Metals (as Pb) Not more than 0.002%.
Lead Not more than 10 mg/kg.
Loss on Drying Not more than 0.2%.
Residue on Ignition Not more than 0.3%.

TESTS

Assay Transfer about 500 mg, previously dried at 105° for 3 h and accurately weighed, into a 250-mL flask. Dissolve the sample in 75 mL of glacial acetic acid, add 2 drops of crystal violet TS, and titrate with 0.1 N perchloric acid to a bluish green endpoint. Perform a blank determination (see *General Provisions*) and make any necessary correction. Each mL of 0.1 N perchloric acid is equivalent to 16.52 mg of $C_9H_{11}NO_2$.
Heavy Metals Prepare and test a 1-g sample as directed in *Method II* under the *Heavy Metals Test*, Appendix IIIB, using 20 μg of lead ion (Pb) in the control (*Solution A*).
Lead A *Sample Solution* prepared as directed for organic compounds meets the requirements of the *Lead Limit Test*, Appendix IIIB, using 10 μg of lead ion (Pb) in the control.
Loss on Drying, Appendix IIC Dry at 105° for 3 h.
Residue on Ignition, Appendix IIC Ignite 1 g as directed in the general method.

Packaging and Storage Store in well-closed containers.

L-Phenylalanine

L-α-Amino-β-phenylpropionic Acid

$C_6H_5CH_2\overset{\displaystyle |}{\underset{\displaystyle H}{C}}COOH$
       NH$_2$

$C_9H_{11}NO_2$ Formula wt 165.19

CAS: [63-91-2]

DESCRIPTION

Colorless or white, platelike crystals or a white, crystalline powder having a slight characteristic odor and a slightly bitter taste. One g is soluble in about 35 mL of water. It is slightly soluble in alcohol, in dilute mineral acids, and in alkali hydroxide solutions. It melts with decomposition at about 283°. The pH of a 1 in 100 solution is between 5.4 and 6.0.

Functional Use in Foods Nutrient; dietary supplement.

REQUIREMENTS

Identification
A. Heat 5 mL of a 1 in 1000 solution with 1 mL of triketohydrindene hydrate TS. A reddish purple color is produced.
B. Heat 5 mL of a 1 in 100 solution with a few drops of potassium dichromate TS. A characteristic odor is evolved.
C. To a 10 mg sample add 500 mg of potassium nitrate and 2 mL of sulfuric acid, and heat the mixture on a water bath for 20 min. Cool, add 2 mL of hydroxylamine TS, immerse in ice water for 10 min, and then add 10 mL of 1 N sodium hydroxide. A reddish violet color is produced.
Assay Not less than 98.5% and not more than 101.5% of $C_9H_{11}NO_2$, calculated on the dried basis.
Heavy Metals (as Pb) Not more than 0.002%.
Lead Not more than 10 mg/kg.
Loss on Drying Not more than 0.2%.
Residue on Ignition Not more than 0.1%.
Specific Rotation $[\alpha]_D^{20°}$: Between −33.2° and −35.2° after drying; or $[\alpha]_D^{25°}$: Between −32.7° and −34.7° after drying.

TESTS

Assay Dissolve about 300 mg, accurately weighed, in 3 mL of formic acid and 50 mL of glacial acetic acid, add 2 drops of crystal violet TS, and titrate with 0.1 N perchloric acid to a bluish green endpoint. Perform a blank determination (see *General Provisions*), and make any necessary correction. Each mL of 0.1 N perchloric acid is equivalent to 16.52 mg of $C_9H_{11}NO_2$.
Heavy Metals Prepare and test a 1-g sample as directed in *Method II* under the *Heavy Metals Test*, Appendix IIIB, using 20 μg of lead ion (Pb) in the control (*Solution A*).
Lead A *Sample Solution* prepared as directed for organic compounds meets the requirements of the *Lead Limit Test*, Appendix IIIB, using 10 μg of lead ion (Pb) in the control.
Loss on Drying, Appendix IIC Dry at 105° for 3 h.

Residue on Ignition, Appendix IIC Ignite 1 g as directed in the general method.
Specific Rotation, Appendix IIB Determine in a solution containing 2 g of previously dried sample in sufficient water to make 100 mL.

Packaging and Storage Store in well-closed, light-resistant containers.

Phosphoric Acid

Orthophosphoric Acid

H_3PO_4 Formula wt 98.00
INS: 338   CAS: [7664-38-2]

DESCRIPTION

A colorless, odorless solution of H_3PO_4, usually available in concentrations ranging from 75.0% to 85.0%. It is miscible with water and with alcohol.

Functional Use in Foods Acid; sequestrant.

REQUIREMENTS

Labeling Indicate the percent, or the percentage range, of Phosphoric Acid, H_3PO_4.
Identification A 1 in 10 solution gives positive tests for *Phosphate*, Appendix IIIA.
Assay Not less than the minimum or within the range of percentage claimed by the vendor.
Arsenic (as As) Not more than 3 mg/kg.
Fluoride Not more than 10 mg/kg.
Heavy Metals (as Pb) Not more than 5 mg/kg.

TESTS

Assay Weigh accurately about 1.5 g in a tared, glass-stoppered flask, and dilute to 120 mL with water. Add 0.5 mL of thymolphthalein TS, mix, and titrate with 1 N sodium hydroxide to the first appearance of a blue color. Each mL of 1 N sodium hydroxide is equivalent to 49.00 mg of H_3PO_4.
Arsenic A solution of 1 g in 35 mL of water meets the requirements of the *Arsenic Test*, Appendix IIIB.
Fluoride Determine on a 1-g sample as directed in *Method IV* under the *Fluoride Limit Test*, Appendix IIIB, using *Buffer Solution B* and 0.1 mL of *Fluoride Standard Solution*.
Heavy Metals A solution of 4 g in 10 mL of water meets the requirements of the *Heavy Metals Test*, Appendix IIIB, using 20 μg of lead ion (Pb) in the control (*Solution A*).

Packaging and Storage Store in tight containers.

Pimenta Leaf Oil

Pimento Leaf Oil

DESCRIPTION

The volatile oil obtained by steam distillation from the leaves of the evergreen shrub *Pimenta officinalis* Lindl. (Fam. *Myrtaceae*). It is a pale yellow to light brownish yellow liquid when freshly distilled, becoming darker with age. In contact with iron, it acquires a blue shade, turning to dark brown on extended contact. It has a spicy odor. It is soluble in propylene glycol, and it is soluble, with slight opalescence, in most fixed oils. It is relatively insoluble in glycerin and in mineral oil.

Functional Use in Foods Flavoring agent.

REQUIREMENTS

Identification The infrared absorption spectrum of the sample exhibits relative maxima (that may vary in intensity) at the same wavelengths (or frequencies) as those shown in the respective spectrum in the section on *Infrared Spectra (Series A: Essential Oils)*, using the same test conditions as specified therein.
Assay Not less than 80.0% and not more than 91.0%, by volume, of phenols.
Angular Rotation Between −2° and +0.5°.
Heavy Metals (as Pb) Passes test.
Refractive Index Between 1.531 and 1.536 at 20°.
Solubility in Alcohol Passes test.
Specific Gravity Between 1.037 and 1.050.

TESTS

Assay Shake a suitable quantity of the oil with about 2% of powdered tartaric acid for about 2 min, then filter. Using a sample of the filtered oil, proceed as directed under *Phenols*, Appendix VI, modified by heating the flask on a boiling water bath for 10 min and cooling, after shaking the mixture of oil and 1 N potassium hydroxide.
Angular Rotation Determine in a 100-mm tube as directed under *Optical (Specific) Rotation*, Appendix IIB.
Heavy Metals Shake 10 mL of the oil with an equal volume of water to which 1 drop of hydrochloric acid has been added, and pass hydrogen sulfide through the mixture until it is saturated. No darkening in color is produced in either the oil or the water.
Refractive Index, Appendix IIB Determine with an Abbé or other refractometer of equal or greater accuracy.
Solubility in Alcohol Proceed as directed in the general method, Appendix VI. One mL dissolves in 2 mL of 70% alcohol; a slight opalescence may occur when additional solvent is added.
Specific Gravity Determine by any reliable method (see *General Provisions*).

Packaging and Storage Store in full, tight, preferably glass, aluminum, stainless steel, or tin-lined containers in a cool place protected from light.

Pimenta Oil

Pimenta Berries Oil; Pimento Oil; Allspice Oil

DESCRIPTION

The volatile oil distilled from the fruit of *Pimenta officinalis*, Lindley (Fam. *Myrtaceae*). It is a colorless, yellow, or reddish yellow liquid, which becomes darker with age. It has the characteristic odor and taste of allspice. It is affected by light.

Functional Use in Foods Flavoring agent.

REQUIREMENTS

Identification The infrared absorption spectrum of the sample exhibits relative maxima (that may vary in intensity) at the same wavelengths (or frequencies) as those shown in the respective spectrum in the section on *Infrared Spectra* (*Series A: Essential Oils*), using the same test conditions as specified therein.
Assay Not less than 65.0%, by volume, of phenols.
Angular Rotation Between −4° and 0°.
Heavy Metals (as Pb) Passes test.
Refractive Index Between 1.527 and 1.540 at 20°.
Solubility in Alcohol Passes test.
Specific Gravity Between 1.018 and 1.048.

TESTS

Assay Proceed as directed under *Phenols*, Appendix VI, modified by heating on a steam bath for 10 min, after shaking for 5 min. Then cool and let stand overnight, or until the liquids are clear.
Angular Rotation Determine in a 100-mm tube as directed under *Optical (Specific) Rotation*, Appendix IIB.
Heavy Metals Shake 10 mL of the oil with an equal volume of water to which 1 drop of hydrochloric acid has been added, and pass hydrogen sulfide through the mixture until it is saturated. No darkening in color is produced in either the oil or the water.
Refractive Index, Appendix IIB Determine with an Abbé or other refractometer of equal or greater accuracy.
Solubility in Alcohol Proceed as directed in the general method, Appendix VI. One mL dissolves in 2 mL of 70% alcohol.
Specific Gravity Determine by any reliable method (see *General Provisions*).

Packaging and Storage Store in full, tight, preferably glass, aluminum, or tin-lined containers in a cool place protected from light.

Pine Needle Oil, Dwarf

Pine Needle Oil

CAS: [8000-26-8]

DESCRIPTION

The volatile oil obtained by steam distillation from the fresh leaves of *Pinus mugo* Turra var. *pumilio* (Haenke) Zenari (Fam. *Pinaceae*). It is a colorless or yellow liquid having a pleasant aromatic odor and a bitter, pungent taste.

Functional Use in Foods Flavoring agent.

REQUIREMENTS

Identification The infrared absorption spectrum of the sample exhibits relative maxima (that may vary in intensity) at the same wavelengths (or frequencies) as those shown in the respective spectrum in the section on *Infrared Spectra* (*Series A: Essential Oils*), using the same test conditions as specified therein.
Assay Not less than 3.0% and not more than 10.0% of esters, calculated as bornyl acetate ($C_{12}H_{20}O_2$).
Angular Rotation Between −5° and −15°.
Distillation Range Not more than 10% distills below 165°.
Heavy Metals (as Pb) Passes test.
Refractive Index Between 1.475 and 1.480 at 20°.
Solubility in Alcohol Passes test.
Specific Gravity Between 0.853 and 0.871.

TESTS

Assay Weigh accurately about 10 g, and proceed as directed under *Ester Determination*, Appendix VI, using 98.15 as the equivalence factor (*e*) in the calculation.
Angular Rotation Determine in a 100-mm tube as directed under *Optical (Specific) Rotation*, Appendix IIB.
Distillation Range Proceed as directed in the general method, Appendix IIB.
Heavy Metals Shake 10 mL of the oil with an equal volume of water to which 1 drop of hydrochloric acid has been added, and pass hydrogen sulfide through the mixture until it is saturated. No darkening in color is produced in either the oil or the water.
Refractive Index, Appendix IIB Determine with an Abbé or other refractometer of equal or greater accuracy.
Solubility in Alcohol Proceed as directed in the general method, Appendix VI. One mL dissolves in 10 mL of 90% alcohol, often with turbidity.
Specific Gravity Determine by any reliable method (see *General Provisions*).

Packaging and Storage Store in full, tight, preferably glass, aluminum, or tin-lined containers in a cool place protected from light.

Pine Needle Oil, Scotch Type

CAS: [8023-99-2]

DESCRIPTION

A volatile oil obtained by steam distillation from the needles of *Pinus sylvestris* L. (Fam. *Pinaceae*). It is a colorless or yellowish liquid with an aromatic, turpentine-like odor. It is soluble in most fixed oils; soluble, with faint opalescence, in mineral oil; and slightly soluble in propylene glycol. It is practically insoluble in glycerin.

Functional Use in Foods Flavoring agent.

REQUIREMENTS

Identification The infrared absorption spectrum of the sample exhibits relative maxima (that may vary in intensity) at the same wavelengths (or frequencies) as those shown in the respective spectrum in the section on *Infrared Spectra (Series A: Essential Oils)*, using the same test conditions as specified therein.
Assay Not less than 1.5% and not more than 5.0% of esters, calculated as bornyl acetate ($C_{12}H_{20}O_2$).
Angular Rotation Between −4° and +10°.
Heavy Metals (as Pb) Passes test.
Refractive Index Between 1.473 and 1.479 at 20°.
Solubility in Alcohol Passes test.
Specific Gravity Between 0.857 and 0.885.

TESTS

Assay Weigh accurately about 10 g, and proceed as directed under *Ester Determination*, Appendix VI, using 98.15 as the equivalence factor (*e*) in the calculation.
Angular Rotation Determine in a 100-mm tube as directed under *Optical (Specific) Rotation*, Appendix IIB.
Heavy Metals Shake 10 mL of the oil with an equal volume of water to which 1 drop of hydrochloric acid has been added, and pass hydrogen sulfide through the mixture until it is saturated. No darkening in color is produced in either the oil or the water.
Refractive Index, Appendix IIB Determine with an Abbé or other refractometer of equal or greater accuracy.
Solubility in Alcohol Proceed as directed in the general method, Appendix VI. One mL dissolves in 6 mL of 90% alcohol, occasionally with slight opalescence.
Specific Gravity Determine by any reliable method (see *General Provisions*).

Packaging and Storage Store in full, tight, preferably glass, aluminum, or tin-lined containers in a cool place protected from light.

Poloxamer 331

α-Hydro-*omega*-hydroxy-poly(oxyethylene)-poly(oxypropylene)(51–57 moles)poly(oxyethylene) Block Copolymer, Avg formula wt 3800

DESCRIPTION

Poloxamer 331 is a block copolymer condensate of ethylene oxide and propylene oxide having an average formula weight of 3800. It occurs as a practically colorless liquid having a specific gravity of about 1.02 and a refractive index of about 1.452. It is very slightly soluble in water at 25°, but is freely soluble at 0°; it is freely soluble in alcohol, but is insoluble in propylene glycol and in ethylene glycol.

Functional Use in Foods Solubilizing and stabilizing agent in flavor concentrates.

REQUIREMENTS

Cloud Point of a 1 in 10 Solution Between 9° and 12°.
Ethylene Oxide, Propylene Oxide, and 1,4-Dioxane Not more than 5 mg/kg of each.
Heavy Metals (as Pb) Not more than 5 mg/kg.
Hydroxyl Value Between 27.2 and 32.1.
Molecular Weight Between 3500 and 4125.
pH of a 2.5% Solution Between 6.0 and 7.4.

TESTS

Cloud Point of a 1 in 10 Solution Prepare a 10% solution of the sample in water at a temperature below the expected cloud point, and transfer about 100 mL of this solution into a 50- × 120-mm test tube. Immerse the tube in a water bath, previously cooled to at least 10° below the expected cloud point, so that the water level is a few mm above that of the test solution. Place a suitable thermometer (see Appendix I) in the test solution, and position it so that the immersion line will be at the surface of the liquid. Stir the solution slowly with a mechanical stirrer (about 200 rpm), and heat gradually so that the test solution is heated at a rate of about 1°/min. Do not allow the temperature of the water bath to rise more than 10° above that of the test solution at any time. Continue heating in this manner, and record the temperature (cloud point) at which the test solution becomes cloudy.
Ethylene Oxide, Propylene Oxide, and 1,4-Dioxane
Stripped Poloxamer Into a suitable 4-neck, round-bottom flask, equipped with a stirrer, a thermometer, a gas dispersion tube, a dry ice trap, a vacuum outlet, and a heating mantle, place between 100 and 3000 g of Poloxamer 331. At room temperature, evacuate the flask carefully to a pressure of less than 1 mm of mercury, applying the vacuum slowly while observing for excessive foaming due to entrapped gases. After any foaming has subsided, sparge with nitrogen, allowing the pressure to rise to 10 mm of mercury. Heat the flask to 130° while increasing the pressure to about 60 mm of mercury. Continue stripping for 4 h, then cool to room temperature. Shut off the

vacuum pump, and bring the flask pressure back to atmospheric while maintaining nitrogen sparging. Remove the sparging tube with the gas still flowing, then turn off the gas flow. Transfer the *Stripped Poloxamer* to a suitable nitrogen-filled container.

Standard Preparations (**Caution:** Ethylene oxide, propylene oxide, and 1,4-dioxane are toxic and flammable. Prepare these solutions in a well-ventilated fume hood.) To about 50 g of *Stripped Poloxamer*, accurately weighed, in a vial that can be sealed, add suitable quantities of propylene oxide and 1,4-dioxane. Determine the amounts added by weight difference after each addition. Using the special handling described in the following, add ethylene oxide to complete the preparation. Ethylene oxide is a gas at room temperature. It is usually stored in a lecture-type gas cylinder or small metal pressure bomb. Chill the cylinder in a refrigerator before use. Transfer about 5 mL of the liquid ethylene oxide to a 100-mL beaker chilled in wet ice. Using a gas-tight gas chromatographic syringe that has been chilled in a refrigerator, transfer a suitable amount of the liquid ethylene oxide into the mixture. Immediately seal the vial, and shake. Determine the amount added by weight difference. By appropriate dilution with *Stripped Poloxamer*, prepare four solutions, covering the range from 1 to 20 mg/kg for the three components added to the matrix (e.g., 5, 10, 15, and 20 mg/kg). Transfer 1 ± 0.01 g of each of these solutions to separate 22-mL pressure headspace vials, seal each with a silicone septum, star spring, and pressure-relief safety aluminum sealing cap, and crimp the cap closed with a cap-sealing tool.

Test Preparation Transfer 1 ± 0.01 g of poloxamer to a 22-mL pressure headspace vial, and seal, cap, and crimp as directed for the *Standard Preparations*.

Chromatographic System, Appendix IIA The gas chromatograph is equipped with a balanced pressure automatic headspace sampler and a flame-ionization detector and contains a 50-m × 0.32-mm fused silica capillary column bonded with a 5-μm film of 5% phenyl–95% methylsiloxane. The column temperature is programmed from 70° to 250° at 10°/min, with the transfer line at 140° and the detector at 250°. The carrier gas is helium, flowing at a rate of about 0.8 mL/min. The resolution, R, of ethylene oxide and propylene oxide, when the *Standard Preparations* are chromatographed as directed under *Calibration*, is not less than 2.0. On the three *Calibration* plots, no point digresses from its line by more than 10%.

Calibration Place the vials containing the *Standard Preparations* in the automated sampler, and start the sequence so that each vial is heated at a temperature of 110° for 30 min before a suitable portion of its headspace is injected into the chromatograph. Set the automatic sampler for a needle withdrawal time of 0.3 min, a pressurization time of 1 min, an injection time of 0.08 min, and a vial pressure of 22 psig with the vial vent off. Obtain the peak areas for ethylene oxide, propylene oxide, and 1,4-dioxane, which have relative retention times of about 1.0, 1.3, and 3.1, respectively. Plot the area versus mg/kg on linear graph paper, and draw the best straight line through the points.

Procedure Place the vial containing the *Test Preparation* in the automatic sampler, and chromatograph its headspace as done for the *Standard Preparations*. Obtain the peak areas of each of the components and read the concentrations directly from the *Calibration* plots. Not more than 5 mg/kg of ethylene oxide, propylene oxide, or 1,4-dioxane is found.

Heavy Metals Prepare and test a 4-g sample as directed in *Method II* under the *Heavy Metals Test*, Appendix IIIB, using 20 μg of lead ion (Pb) in the control (*Solution A*).

Hydroxyl Value

Distilled Pyridine Distill pyridine over phthalic anhydride (about 60 g for each 1000 mL), discarding the first 25 mL and the last 50 mL of distillate from each 1000 mL distilled.

Phenolphthalein Indicator Prepare a 1% solution of phenolphthalein in undistilled pyridine.

Phthalation Reagent Dissolve 14.4 g of phthalic anhydride in sufficient *Distilled Pyridine* to make 100 mL, mix vigorously, and store in a brown bottle. Allow to stand at least 2 h before use. Determine the suitability of the reagent as follows: Mix 10.0 mL of the reagent with 25 mL of undistilled pyridine and 50 mL of water, allow to stand for 15 min, then add a few drops of *Phenolphthalein Indicator*, and titrate with 0.5 N sodium hydroxide. Multiply the volume, in mL, of the alkali solution required by its exact normality; if the result is not within the range 18.8 to 20.0, adjust the concentration of the reagent accordingly.

Procedure Transfer about 15 g of the sample, accurately weighed, into a 250-mL hydroxyl flask, and add 25.0 mL of *Phthalation Reagent* to the flask, using a pipet previously rinsed with the reagent and touching the tip of the pipet against the protrusion of the flask approximately 15 s after the pipet has drained. In the same manner, transfer the same volume of the reagent into a second flask to serve as the blank. Add a few glass beads to each flask, swirl to dissolve the sample, and reflux for 1 h. Cool the flasks to room temperature, and wash each condenser with two 10-mL portions of undistilled pyridine. Disconnect the condensers, add 10 mL of water to each flask, stopper, swirl, and allow to stand for 10 min. To each flask add 50.0 mL of approximately 0.66 N sodium hydroxide, then add 0.5 mL of *Phenolphthalein Indicator*, and titrate with 0.5 N sodium hydroxide to the first pink color that persists for at least 15 s. Calculate the uncorrected hydroxyl value, h, by the formula

$$(B - S) \times (N \times 56.1/W),$$

in which B is the volume, in mL, of 0.5 N sodium hydroxide required for the blank; S is the volume required for the sample; N is the exact normality of the sodium hydroxide solution; and W is the weight of the sample, in g. Correct the results, if the sample contains significant acidity or alkalinity, as follows: Dissolve approximately the same amount of the sample, accurately weighed, as used above in 40 mL of undistilled pyridine, and add 60 mL of water and 0.5 mL of *Phenolphthalein Indicator*. If the solution is colorless, titrate with 0.1 N sodium hydroxide to a light pink endpoint, recording the volume, in mL, required as v. If the solution is pink, titrate with 0.1 N hydrochloric acid to the disappearance of the pink color, recording the volume, in mL, required as v'. Calculate the acidity correction factor, A, by the formula

$$v \times N \times 56.1/w,$$

in which N is the exact normality of the sodium hydroxide solution, and w is the weight of the sample, in g. Calculate the alkalinity correction factor, A', by the formula

$$v' \times N' \times 56.1/w,$$

in which N' is the exact normality of the hydrochloric acid solution. Finally, calculate the corrected hydroxyl value, H, by the formula

$$h + A, \text{ or } h - A',$$

whichever is appropriate.
Molecular Weight Determine the *Hydroxyl Value*, H, and then calculate the molecular weight by the formula

$$56{,}100 \times (2/H).$$

pH of a 2.5% Solution Prepare a 2.5% solution of the sample in water, and determine the pH by the *Potentiometric Method*, Appendix IIB.

Packaging and Storage Store in tight containers.

Poloxamer 407

α-Hydro-*omega*-hydroxy-poly(oxyethylene)-poly(oxypropylene)(63–71 moles)poly(oxyethylene) Block Copolymer, Avg formula wt 12,500

DESCRIPTION

Poloxamer 407 is a copolymer condensate of ethylene oxide and propylene oxide having an average formula weight of 12,500. It occurs as a white solid having a melting range of about 52° to 56°. It is freely soluble in alcohol and in water, but is insoluble in propylene glycol and in ethylene glycol.

Functional Use in Foods Solubilizing and stabilizing agent in flavor concentrates.

REQUIREMENTS

Cloud Point of a 1 in 10 Solution Above 100°.
Ethylene Oxide, Propylene Oxide, and 1,4-Dioxane Not more than 5 mg/kg of each.
Heavy Metals (as Pb) Not more than 5 mg/kg.
Hydroxyl Value Between 8.5 and 11.5.
Molecular Weight Between 9760 and 13,200.
pH of a 2.5% Solution Between 6.0 and 7.4.

TESTS

Cloud Point of a 1 in 10 Solution Dissolve about 5 g of the sample in 50 mL of water in a test tube, place the tube in a boiling water bath, and heat to 100°. The solution does not become cloudy.
Ethylene Oxide, Propylene Oxide, and 1,4-Dioxane Determine as directed in the monograph for *Poloxamer 331*.
Heavy Metals Prepare and test a 4-g sample as directed in *Method II* under the *Heavy Metals Test*, Appendix IIIB, using 20 μg of lead ion (Pb) in the control (*Solution A*).

Hydroxyl Value
Distilled Pyridine, Phenolphthalein Indicator, and *Phthalation Reagent* Prepare as directed in the monograph for *Poloxamer 331*.
Procedure Proceed as directed for *Procedure* under *Hydroxyl Value* in the monograph for *Poloxamer 331*, using a sample of about 45 g, accurately weighed, and adding 25 mL of *Distilled Pyridine* to both the sample flask and the blank flask before refluxing.
Molecular Weight Determine the *Hydroxyl Value*, H, and then calculate the molecular weight by the formula

$$56{,}100 \times (2/H).$$

pH of a 2.5% Solution Prepare a 2.5% solution of the sample in water, and determine the pH by the *Potentiometric Method*, Appendix IIB.

Packaging and Storage Store in tight containers.

Polydextrose

INS: 1200   CAS: [68424-04-4]

DESCRIPTION

A randomly bonded condensation polymer of D-glucose with some bound sorbitol and citric acid. The 1,6-glycosidic linkage predominates in the polymer, but all other possible bonds are present. The product contains small quantities of free glucose, sorbitol, and 1,6-anhydro-D-glucose (levoglucosan) with a trace of citric acid. It may be neutralized and/or decolorized with potassium hydroxide. It is an off-white to light tan colored solid, very soluble in water.

Functional Use in Foods Bulking agent; formulation aid; humectant; texturizer.

REQUIREMENTS

Identification
A. To 1 drop of a 1 in 10 aqueous solution of the sample, add 4 drops of 5% aqueous phenol solution, then rapidly add 15 drops of sulfuric acid. A deep yellow to orange color is produced.
B. With vigorous swirling (vortex mixer), add 1.0 mL of acetone to 1.0 mL of a 1 in 10 aqueous solution of the sample. The solution remains clear.
C. With vigorous swirling, add 2.0 mL of acetone to the solution from *B*. A heavy, milky turbidity develops immediately.
D. To 1 mL of a 1 in 50 aqueous solution of sample, add 4 mL of alkaline cupric citrate TS. Boil vigorously for 2 to 4 min. Remove from heat, and allow the precipitate (if any) to settle. The supernatant liquid is blue or blue-green.
Assay Not less than 90.0% polymer, calculated on the anhydrous, ash-free basis.

Heavy Metals (as Pb)   Not more than 5 mg/kg.
5-Hydroxymethylfurfural   Not more than 0.1%, calculated on the anhydrous, ash-free basis.
Lead   Not more than 0.5 mg/kg.
Molecular Weight Limit   Passes test.
Monomers

1,6-Anhydro-D-Glucose   Not more than 4.0%, calculated on the anhydrous, ash-free basis.

Glucose and Sorbitol   Not more than 6.0%, calculated on the anhydrous, ash-free basis.

pH of a 10% Solution   Not less than 2.5 for untreated Polydextrose and between 5.0 and 6.0 for neutralized/decolorized Polydextrose.

Residue on Ignition   Not more than 0.3% for untreated Polydextrose and not more than 2.0% for neutralized/decolorized Polydextrose.

Water   Not more than 4.0%.

TESTS

Assay

Glucose Standard Solutions   Weigh accurately 100 mg of α-D-glucose (National Institute of Standards and Technology) into a 500-mL volumetric flask, and make up to volume with distilled water. Dilute 5 aliquots of the solution with distilled water to obtain the following concentrations of standard: 50, 40, 20, 10, and 5 μg/mL.

Phenol Solution   Add 20 mL of water to 80 g of phenol.

Procedure   Weigh accurately approximately 250 mg of the sample into a 250-mL volumetric flask, and make up to volume with distilled water. Transfer a 10.0-mL aliquot to a 250-mL volumetric flask and dilute to volume with distilled water. Proceed as directed in *Standard Curve*. Calculate the percentage of polymer (P) by the formula

$$P = 1.05 [100(A - Y)/(S \times C)] - P_G - 1.11 P_L,$$

in which A is the sample absorbance; Y is the y intercept of the standard curve; S is the slope of absorbance versus glucose concentration, in g/mL, obtained from the *Standard Curve* (S is approximately 0.02); C is the concentration of the sample solution, in g/mL (adjusted for ash and moisture); and P_G and P_L are the percentages of glucose and levoglucosan determined respectively in the *Assay for Monomers*.

Standard Curve   On a daily basis, pipet 2.0 mL of each of the *Glucose Standard Solutions* into 15-mL, acetone-free, screw-cap vials. Add 0.12 mL of the phenol solution, and mix gently. Uncap each vial and add rapidly 5.0 mL of sulfuric acid. Immediately recap each vial, and shake vigorously.

Caution: Use rubber gloves and a safety shield in the sulfuric acid addition step.

Let the vials stand at room temperature for 45 min, then determine the absorbance of each sample at 490 nm in a suitable spectrophotometer, using a *Phenol Solution* sulfuric acid reagent blank in the reference cell. Plot mean absorbances versus concentrations, in μg/mL, obtained from triplicate samples.

Heavy Metals   Prepare and test a 4.0-g sample as directed in *Method II* under the *Heavy Metals Test*, Appendix IIIB, using 20 μg of lead ion (Pb) in the control (*Solution A*).

5-Hydroxymethylfurfural   Transfer approximately 1 g of the sample, accurately weighed, to a 100-mL volumetric flask, and make up to volume with distilled water. Read the absorbance of this solution against a water blank at 283 nm in a 1-cm quartz cell in a spectrophotometer. Calculate the percentage of 5-hydroxymethylfurfural by the formula

$$\% \text{ HMF} = (0.749 \times A)/C,$$

in which A is the absorbance of the sample solution, and C is the concentration of the sample solution, in mg/mL, corrected for ash and moisture.

Lead

Apparatus   Use a suitable spectrophotometer (Perkin-Elmer Model 6000, or equivalent), a graphite furnace containing a L'vov platform (Perkin-Elmer Model HGA-500, or equivalent), and an autosampler (Perkin-Elmer Model AS-40, or equivalent). Use a lead hollow cathode lamp (lamp current of 10 mA), a slit width of 0.7 mm (set low), the wavelength set at 283.3 nm, and a deuterium arc lamp for background correction.

Note: For this test, use reagent-grade chemicals with as low a lead content as is practicable, as well as high-purity water and gases. Before use in this analysis, rinse all glassware and plasticware twice with 10% nitric acid and twice with 10% hydrochloric acid, and then rinse them thoroughly with high-purity water, preferably obtained from a mixed-bed strong-acid, strong-base ion-exchange cartridge capable of producing water with an electrical resistivity of 12 to 15 megohms.

Lead Nitrate Stock Solution   Dissolve 159.8 mg of ACS reagent-grade lead nitrate (alternatively, use NIST Standard Reference Material containing 10 mg of lead per kg, or equivalent) in 100 mL of water containing 1 mL of nitric acid. Dilute to 1000.0 mL with water, and mix. Prepare and store this solution in glass containers that are free from lead salts. Each mL of this solution contains the equivalent of 100 μg of lead ion.

Standard Lead Solution   On the day of use, dilute 10.0 mL of *Lead Nitrate Stock Solution* with water to 100.0 mL, and mix. Each mL of *Standard Lead Solution* contains the equivalent of 10 μg of lead ion.

Standard Solutions   Prepare a series of lead standard solutions serially diluted from the *Standard Lead Solution* in water. Into separate 100-mL volumetric flasks pipet 0.2, 0.5, 1, and 2 mL, respectively, of *Standard Lead Solution*, dilute to volume with water, and mix. The *Standard Solutions* contain, respectively, 0.02, 0.05, 0.1, and 0.2 μg of lead per mL.

Matrix Modifier   Transfer 100.0 mg of ammonium phosphate, dibasic ((NH_4)$_2$$HPO_4$) to a 10-mL volumetric flask, dilute to volume with water, and mix.

Sample Solution   Transfer about 1 g of the sample, accurately weighed, to a 10-mL volumetric flask, add 5 mL of water, and mix. Dilute to volume with water, and mix.

Spiked Sample Solution   Prepare a solution as directed under *Sample Solution*, but add 100 μL of the *Standard Lead Solution*, dilute to volume with water, and mix. This solution contains 0.1 μg/mL of added lead.

Procedure   With the use of an autosampler, atomize 10-μL aliquots of the four *Standard Solutions*, using the following sequence of conditions: step (1) dry at 130° with a 20-s ramp period, a 40-s hold time, and a 300-mL/min argon flow rate; step (2) then char at 800° with a 20-s ramp period, a 40-s hold time, and a 300- mL/min argon flow rate; step (3) atomize at 2400° for 6 s with a 50-mL/min argon flow rate, and read; step (4) clean at 2600° with a 1-s ramp period, a 5-s hold time, and a 300-mL/min argon flow rate; and step (5) recharge at 20° with a 2-s ramp period, a 20-s hold time, and a 300-mL/min argon flow rate. Atomize 10 μL of the *Matrix Modifier* in combination with either 10 μL of the *Sample Solution* or 10 μL of the *Spiked Sample Solution* under identical conditions used for the *Standard Solutions*.

Plot a standard curve using the concentration, in μg/mL, of each *Standard Solution* versus its maximum absorbance value compensated for background correction, and draw the best straight line. From the *Standard Curve*, determine the concentrations C_S and C_A, in μg/mL, of the *Sample Solution* and the *Spiked Sample Solution*, respectively. Calculate the quantity, in mg/kg, of lead in the sample by the formula

$$10C_S/W,$$

in which W is the weight, in g, of the sample taken. Calculate the recovery by the formula

$$[(C_S - C_A)/0.1]100,$$

in which 0.1 is the amount of lead, in μg/mL, added to the *Spiked Sample Solution*.

Molecular Weight Limit

Apparatus Use a suitable high-pressure liquid chromatograph (HPLC) equipped with a differential refractometer, either a loop injector or suitable autosampler, a column heating block or oven and a computing integrator, or computer data handling system with molecular weight determination capabilities. Use a Waters Ultrahydrogel 250 A size exclusion column, or equivalent. The column is maintained at 45°, and the HPLC pump supplies eluent to it at 0.8 mL/min reproducible to 0.5%. The differential refractometer should be set at a sensitivity of 4 × 10^{-6} refractive index units full scale, and the plotter of the integrator should be set to 64 millivolts full scale. Maintain the detector cell at 35° ± 0.1°. Noise attributable to the detector and electronics should be less than 0.1% full scale.

Eluent The eluent is 0.1 N sodium nitrate containing 0.025% sodium azide. Dissolve 35.0 g of sodium nitrate and 1.0 g of sodium azide in 100 mL of HPLC-grade water. Filter through a 0.45-μm filter into a 4-L flask. Dilute to volume with HPLC-grade water. Degas by applying an aspirator vacuum for 30 min.

Standard Solution Transfer 20 mg each of dextrose, stachyose, 5800, 23,700, and 100,000 MW pullulan standards into a 10-mL volumetric flask. Dissolve in and dilute to volume with *Eluent*. Filter through a 0.45-μm syringe filter into a suitable autosampler vial, and seal. All components of the *Standard Solution* are available from Polymer Laboratories, Inc., Technical Center, Amherst Fields Research Park, 160 Old Farm Road, Amherst, MA 01002.

Column Equilibration After installation of a new column in the HPLC, pump *Eluent* through it overnight at 0.3 mL/min. Before calibration or analysis, increase the flow slowly to 0.8 mL/min over a 1-min period, then pump at 0.8 mL/min for at least 1 h before the first injection. Check the flow gravimetrically, and adjust it if necessary. Reduce the flow to 0.1 mL/min when the system is not in use.

Data System Setup Set the integrator or computerized data handling system as their respective manuals instruct for normal gel permeation chromatographic determinations. Set the integration time to 15 min.

Column Standardization After the HPLC system has been equilibrated at a flow rate of 0.8 mL/min for at least 1 h, inject 50 μL of the *Standard Solution* five times. Allow 15 min between injections. Record the retention times of the various components in the *Standard Solution*. Retention times for each component should agree within ± 2 s. Insert the average retention time along with the molecular weight of each component into the calibration table of the molecular weight distribution software.

System Suitability Check the regression results for a cubic fit of the calibration points. They should have an R^2 value of 0.9999+. Dextrose and stachyose should be baseline resolved from one another and from the 5800 MW pullulan standard. Elevated valleys are usually observed between the 5800, 23,700, and 100,000 MW pullulan standards.

Sample Preparation Transfer 50 mg of sample, accurately weighed, into a 10-mL volumetric flask. Dissolve in and dilute to volume with *Eluent*. Filter through a 0.45-μm syringe filter into a suitable autosampler vial.

Procedure Inject 50 μL of the *Sample Preparation*, following the same conditions and procedure as described under *Column Standardization*. Using the Molecular Weight Distribution software of the data reduction system, generate a molecular weight distribution curve of the sample. There is no measurable peak above a molecular weight of 22,000.

Monomers

Apparatus Use a suitable gas chromatograph equipped with a 250-cm × 2-mm (id) glass column packed with 3% OV-1 stationary phase on 100/120-mesh Gas Chrom Q and a flame ionization detector. Maintain the column at 175°, the injection port at 210°, and the detector at 230°. Relative retention times (min): 1,6-anhydro-D-glucose (levoglucosan), pyranose form (3.7), furanose form (not present in standard) (4.3); n-octadecane (5.1); α-D-glucose (8.7); D-sorbitol 11.3; β-D-glucose (13.3).

Standard Solution Weigh accurately 50 mg of α-D-glucose (National Institute of Standards and Technology), 40 mg of anhydrous D-sorbitol, and 35 mg of 1,6-anhydro-D-glucose into a 100-mL volumetric flask; dissolve in and dilute to volume with pyridine.

Octadecane Solution Transfer 50 mg of n-octadecane, accurately weighed, to a 100-mL volumetric flask; dissolve in and dilute to volume with pyridine.

Silylation of Standard Solution Transfer 1.0 mL of *Standard Solution* to a screw-cap vial, and add 1.0 mL of *Octadecane Solution* and 0.5 mL of N-trimethylsilylimidazole. Cap the vial, and immerse it in an ultrasonic bath at 70° for 60 min.

Procedure Accurately weigh 20 mg of the sample into a screw-cap vial, and add 1.0 mL of *Octadecane Solution*, 1 mL of pyridine, and 0.5 mL of N-trimethylsilylimidazole. Cap the vial, and immerse it in an ultrasonic bath at 70° for 60 min. Before sample analysis, inject 3 μL of the silylated *Standard Solution* into the gas chromatograph. Repeat twice, then inject duplicate 3-µL portions of the sample solution. Calculate the percentage of each monomer (P_M) by the formula

$$P_M = (R \times W_S)/(R_S \times W),$$

in which W is the weight, in mg, of the sample, adjusted for ash and moisture; W_S is the weight, in mg, of the respective monomer in the *Standard Solution*; R is the ratio of the area of the monomer peak to the area of the octadecane peak in the sample injection; and R_S is the mean ratio of the area of the monomer peak to the area of the octadecane peak in the standard injections.

pH of a 10% Solution   Determine by the *Potentiometric Method*, Appendix IIB.

Residue on Ignition   Determine as directed in *Method I* under *Residue on Ignition*, Appendix IIC.

Water   Determine by the *Karl Fischer Titrimetric Method*, Appendix IIB, using pyridine instead of methanol in the titration vessel.

Packaging and Storage   Store in tight, light-resistant containers.

Polydextrose Solution

DESCRIPTION

A neutralized, decolorized water solution of polydextrose. It is a clear, straw-colored liquid.

Functional Use in Foods   Bulking agent; formulation aid; humectant; texturizer.

REQUIREMENTS

Identification

A. To 1 drop of a 1 in 10 aqueous solution of sample, add 4 drops of 5% aqueous phenol solution, then rapidly add 15 drops of concentrated sulfuric acid. A deep yellow to orange color is produced.

B. With vigorous swirling (vortex mixer), add 1.0 mL of acetone to 1.0 mL of a 1 in 10 aqueous solution of sample. The solution remains clear.

C. With vigorous swirling, add 2.0 mL of acetone to the solution from *B*. A heavy, milky turbidity develops immediately.

D. To 1 mL of a 1 in 50 aqueous solution of sample, add 4 mL of alkaline cupric citrate TS. Boil vigorously for 2 to 4 min. Remove from heat, and allow the precipitate (if any) to settle. The supernatant liquid is blue or blue-green.

Assay   Not less than 90.0% polymer, calculated on the anhydrous, ash-free basis.

Heavy Metals (as Pb)   Not more than 10 mg/kg.

5-Hydroxymethylfurfural   Not more than 0.1%, calculated on the anhydrous, ash-free basis.

Molecular Weight Limit   Passes test.

Monomers

1,6-Anhydro-D-Glucose   Not more than 4.0%, calculated on the anhydrous, ash-free basis.

Glucose and *Sorbitol*   Not more than 6.0%, calculated on the anhydrous, ash-free basis.

pH of a 10% Solution   Between 5.0 and 6.0.

Residue on Ignition   Not more than 2.0%.

Water   Within the range 27.5% to 32.5%.

TESTS

Assay   Weigh accurately approximately 360 mg of *Polydextrose Solution*, and proceed as directed under *Assay* in the monograph for *Polydextrose*.

Heavy Metals   Prepare and test a 2.0-g sample as directed in *Method II* in the *Heavy Metals Test*, Appendix IIIB, using 20 µg of lead ion (Pb) in the control (*Solution A*).

5-Hydroxymethylfurfural   Proceed as directed in the monograph for *Polydextrose*, using a 1.4-g sample, accurately weighed.

Molecular Weight Limit   Weigh accurately 720 mg of the sample, and proceed as directed under *Molecular Weight Limit* in the monograph for *Polydextrose*.

Monomers

1,6-Anhydro-D-glucose; *Glucose*; and *Sorbitol*   Weigh accurately 30 mg of the sample into a screw-cap vial, and add about 2 mL of pyridine. While flushing the vial with a stream of dry air or nitrogen, heat at 80° to 90° until the solution volume is reduced to 0.2 to 0.5 mL. Add a second portion of pyridine, and repeat the evaporation procedure. Continue as described under *Monomers* in the monograph for *Polydextrose*.

pH of a 10% Solution   Dilute a 1.4-g sample to 10 mL with water, and mix. Proceed as directed under the *Potentiometric Method*, Appendix IIB.

Residue on Ignition   Determine as directed in *Method II* under *Residue on Ignition*, Appendix IIC.

Water   Transfer 1 to 2 mL of the sample into a dropper vial, and accurately weigh the dropper, vial, and sample combined. Add 50 mL of pyridine to a clean, dry reaction jar previously flushed with dry air for 1 min. Titrate the pyridine with Karl Fischer reagent to the endpoint to consume any water present. Transfer one drop of sample (50 to 100 mg) from the weighed sample vial to the reaction jar. Accurately reweigh the dropper, vial, and remaining sample. Stir the pyridine–sample mixture for 5 to 10 min. Titrate with Karl Fischer reagent to the endpoint. For each determination, calculate the percentage water (W) in the sample by the formula

$$W = (V \times F \times 100)/S,$$

in which S is the sample weight, in mg, equal to the difference between the initial and final weighings of the dropper, vial, and sample combination; V is the volume of Karl Fischer reagent consumed in the second titration, in mL; and F is the Karl Fischer reagent standardization factor, in mg/mL. Calculate the water content of the sample as the average of two determinations.

Packaging and Storage   Store in tight, light-resistant containers.

Polyethylene

CAS: [9002-88-4]

DESCRIPTION

A white, translucent, partially crystalline and partially amorphous resin produced by the direct polymerization of ethylene at high temperatures and high pressure. Various grades and types, differing from one another in molecular weight, molecular weight distribution, degree of chain branching, and extent of crystallinity, are available. It is insoluble in water.

Functional Use in Foods Masticatory substance in chewing gum base.

REQUIREMENTS

Identification Prepare the sample by dissolving it in hot toluene and evaporating on a potassium bromide plate. The infrared absorption spectrum of the sample exhibits maxima at the same wavelengths as the typical spectrum, as shown in the section on *Infrared Spectra (Series C: Other Substances)*.
Heavy Metals (as Pb) Not more than 0.002%.
Lead Not more than 3 mg/kg.
Molecular Weight Between 2000 and 21,000.
Volatiles Not more than 0.5%.

TESTS

Heavy Metals Prepare and test a 1-g sample as directed in *Method II* under the *Heavy Metals Test*, Appendix IIIB, using 20 μg of lead ion (Pb) in the control (*Solution A*).
Lead Prepare a *Sample Solution* as directed in the general method under *Chewing Gum Base*, Appendix IV. This solution meets the requirements of the *Lead Limit Test*, Appendix IIIB, using 10 μg of lead ion (Pb) in the control.
Molecular Weight Determine as directed in the general method, Appendix IV.
Volatiles (**Caution:** To reduce explosion hazard, pass carbon dioxide or nitrogen into the lower part of the drying oven at a rate of about 100 mL/min.) Dry a 4-g sample for 45 min at 105° as directed under *Loss on Drying*, Appendix IIC.

Packaging and Storage Store in well-closed containers.

Polyethylene Glycols

PEG

INS: 1521   CAS: [25322-68-3]

DESCRIPTION

Polyethylene Glycols are addition polymers of ethylene oxide and water, ranging in molecular weight from about 200 to about 9500 and having the general formula $HOCH_2(CH_2OCH_2)_n\text{-}CH_2OH$, in which n represents the average number of oxyethylene groups. Commercially available Polyethylene Glycols are usually designated by a number that roughly corresponds to the nominal molecular weight. Polyethylene Glycols having a nominal molecular weight of 600 or below occur as clear to slightly hazy, colorless or practically colorless, viscous, slightly hygroscopic liquids that are miscible with water. Polyethylene Glycols having a nominal molecular weight of 1000 or above are freely soluble in water and occur as creamy white, waxy solids or as flakes resembling paraffin. The Polyethylene Glycols are soluble in many organic solvents, including aliphatic ketones and alcohols, chloroform, glycol ethers, esters, and aromatic hydrocarbons; they are insoluble in ether and in most aliphatic hydrocarbons. As their molecular weight increases, water solubility, vapor pressure, hygroscopicity, and solubility in organic solvents decrease, while solidification point, specific gravity, flash point, and viscosity increase. They may contain a suitable antioxidant.

Functional Use in Foods Dispersing, coating, binding, plasticizing agent; lubricant; flavoring adjuvant.

REQUIREMENTS

Labeling Indicate the nominal average molecular weight.
Completeness and Color of Solution Passes test.
Ethylene Glycol and Diethylene Glycol Not more than 0.25%, combined.
Ethylene Oxide and 1,4-Dioxane Not more than 10 mg/kg singly.
Heavy Metals (as Pb) Not more than 5 mg/kg.
Nominal Molecular Weight *PEGs having nominal molecular weights below 1000*: not less than 95.0% and not more than 105.0% of the labeled value; *PEGs having nominal molecular weights between 1000 and 7000*: not less than 90.0% and not more than 110.0% of the labeled value; *PEGs having nominal molecular weights above 7000*: not less than 87.5% and not more than 112.5% of the labeled value.
pH Between 4.5 and 7.5.
Residue on Ignition Not more than 0.2%.
Viscosity Passes test.

TESTS

Completeness and Color of Solution A solution of 5 g of the sample in 50 mL of water is colorless. It is clear for liquid grades and not more than slightly hazy for solid grades.
Ethylene Glycol and Diethylene Glycol
POLYETHYLENE GLYCOLS HAVING NOMINAL MOLECULAR WEIGHTS BELOW 450

Apparatus Use a suitable gas chromatograph (see Appendix IIA) equipped with a hydrogen flame ionization detector (Varian Aerograph 600D, or equivalent), containing a 1.5-m × 3-mm (id) stainless-steel column packed with sorbitol 12%, by weight, on 60/80-mesh nonacid-washed diatomaceous earth (Chromosorb W, or equivalent).

Operating Conditions The operating parameters may vary, depending upon the particular instrument used, but a suitable chromatogram may be obtained using the following conditions: *column temperature*, 165°; *inlet temperature*, 260°; *carrier gas*, nitrogen (or other suitable inert gas), flowing at a rate of 70 mL/min; *recorder*, 0.5 to +1.05 mV, full span, 1-s full-response time; *hydrogen and air flow to burner*, optimize to give maximum sensitivity.

Standard Solutions Prepare chromatographic standards by dissolving accurately weighed amounts of commercial ethylene glycol and diethylene glycol, previously purified by distillation if necessary, in water. Suitable concentrations range from 0.2 to 1 mg of each glycol per mL.

Sample Preparation Transfer about 4 g of the sample, accurately weighed, into a 10-mL volumetric flask, dilute to volume with water, and mix.

Procedure Inject a 2-μL portion of each of the *Standard Solutions* into the chromatograph, and obtain the chromatogram for each solution. Under the stated conditions, the elution time is approximately 2.0 min for ethylene glycol and 6.5 min for diethylene glycol. Measure the peak heights, and record the values as follows: A = height, in mm, of the ethylene glycol peak; B = weight, in mg, of ethylene glycol per mL of the *Standard Solution*; C = height, in mm, of the diethylene glycol peak; and D = weight, in mg, of diethylene glycol per mL of the *Standard Solution*.

Similarly, inject a 2-μL portion of the *Sample Preparation* into the chromatograph, and obtain the chromatogram, recording the height of the ethylene glycol peak as E and that of the diethylene glycol peak as F. Calculate the percentage of ethylene glycol in the sample by the formula

$$(E \times B)/(A \times \text{sample weight, in g});$$

calculate the percentage of diethylene glycol in the sample by the formula

$$(F \times D)/(C \times \text{sample weight, in g}).$$

Not more than 0.25% of total ethylene and diethylene glycols is found.

POLYETHYLENE GLYCOLS HAVING NOMINAL MOLECULAR WEIGHTS OF 450 OR HIGHER

Sample Preparation Dissolve 50.0 g of the sample in 75 mL of diphenyl ether in a 250-mL distillation flask. Warm the mixture, if necessary, just enough to melt the crystals. Slowly distill at a pressure of 1 to 2 mm of Hg into a receiver graduated to 100 mL in 1-mL subdivisions, until 25 mL of distillate has been collected. Add 20.0 mL of water to the distillate, shake vigorously, and allow the layers to separate. Cool in an ice bath to solidify the diphenyl ether and facilitate its removal. Filter the water layer, wash the diphenyl ether with 5.0 mL of ice-cold water, pass the washings through the filter, and collect the filtrate and washings in a 25-mL volumetric flask. Warm to room temperature, dilute to volume with water, if necessary, and mix. Mix this solution with 25.0 mL of freshly distilled acetonitrile in a 125-mL glass-stoppered flask.

Standard Preparation Transfer 62.5 mg of diethylene glycol to a 25-mL volumetric flask, dilute to volume with a 1:1 mixture of freshly distilled acetonitrile and water, and mix.

Procedure Transfer 10.0 mL each of the *Sample Preparation* and of the *Standard Preparation* into separate 50-mL flasks, each containing 15 mL of ceric ammonium nitrate TS, and mix.

Within 2 to 5 min, using a suitable spectrophotometer, determine the absorbance of each solution in a 1-cm cell at the wavelength of maximum absorbance occurring between 400 and 600 nm, using a blank consisting of 15 mL of ceric ammonium nitrate TS and 10 mL of a 1:1 mixture of acetonitrile and water. The absorbance of the solution from the *Sample Preparation* does not exceed that from the *Standard Preparation*.

Ethylene Oxide and 1,4-Dioxane

Stripped Polyethylene Glycol 400 Into a 5000-mL, 4-neck, round-bottom flask equipped with a stirrer, a thermometer, a gas dispersion tube, a dry ice trap, a vacuum outlet, and a heating mantle, place 3000 g of polyethylene glycol 400. At room temperature, evacuate the flask carefully to a pressure of less than 1 mm of mercury, applying the vacuum slowly while observing for excessive foaming due to entrapped gases. After any foaming has subsided, sparge with nitrogen, allowing the pressure to rise to 10 mm of mercury. Heat the flask to 60° while increasing the pressure to about 60 mm of mercury. Continue stripping for 4 h, then cool to room temperature. Shut off the vacuum pump, and bring the flask pressure back to atmospheric while maintaining nitrogen sparging. Remove the sparging tube with the gas still flowing, then turn off the gas flow. Transfer the *Stripped Polyethylene Glycol 400* to a suitable nitrogen-filled container.

Standard Preparations (**Caution**: Ethylene oxide and 1,4-dioxane are toxic and flammable. Prepare these solutions in a well-ventilated fume hood.) To a known weight of organic-free water in a vial that can be sealed, add a suitable quantity of 1,4-dioxane. Determine the amount added by weight difference. Using the special handling described in the following, complete the preparation. Ethylene oxide is a gas at room temperature. It is usually stored in a lecture-type gas cylinder or small metal pressure bomb. Chill the cylinder in a refrigerator before use. Transfer about 5 mL of the liquid ethylene oxide to a 100-mL beaker chilled in wet ice. Using a gas-tight gas chromatographic syringe that has been chilled in a refrigerator, transfer a suitable amount of the liquid ethylene oxide into the mixture. Immediately seal the vial, and shake. Determine the amount added by weight difference. By appropriate dilution with *Stripped Polyethylene Glycol 400*, prepare four solutions, covering the range from 1 to 20 mg/kg for the two components added to the matrix (e.g., 5, 10, 15, and 20 mg/kg). Transfer 10 mL of each of these solutions to separate 22-mL pressured headspace vials, seal each with a silicone septum, star spring, and pressure-relief safety aluminum sealing cap, and crimp the cap closed with a cap-sealing tool. Shake for 2 min.

Test Preparation Transfer 10 ± 0.01 g of polyethylene glycol to a 22-mL pressure headspace vial, and seal, cap, and crimp as directed for the *Standard Preparations*.

Chromatographic System, Appendix IIA The gas chromatograph is equipped with a balanced pressure automatic headspace sampler and a flame-ionization detector and contains a 50-m × 0.32-mm fused silica capillary column bonded with a 5-μm film of 5% phenyl–95% methylsiloxane. The column temperature is programmed from 70° to 250° at 10°/min, with the transfer line at 140° and the detector at 250°. The carrier gas is helium, flowing at a rate of about 0.8 mL/min. On the two *Calibration* plots, no point digresses from its line by more than 10%.

Calibration Place the vials containing the *Standard Preparations* in the automated sampler, and start the sequence so that each vial is heated at a temperature of 50° for 30 min before a suitable portion of its headspace is injected into the chromatograph. Set the automatic sampler for a needle withdrawal time of 0.3 min, a pressurization time of 1 min, an injection time of 0.08 min, and a vial pressure of 22 psig with the vial vent off. Obtain the peak areas for ethylene oxide and 1,4-dioxane, which have relative retention times of about 1.0 and 3.1, respectively. Plot the area versus mg/kg on linear graph paper, and draw the best straight line through the points.

Procedure Place the vial containing the *Test Preparation* in the automatic sampler, and chromatograph its headspace as done for the *Standard Preparations*. Obtain the peak areas of each of the components, and read the concentrations directly from the *Calibration* plots. Not more than 10 mg/kg of ethylene oxide or 1,4-dioxane is found.

Heavy Metals A solution of 4 g of the sample mixed with 5.0 mL of 0.1 N hydrochloric acid, diluted with water to 25 mL, meets the requirements of the *Heavy Metals Test*, Appendix IIIB, using 20 µg of lead ion (Pb) in the control (*Solution A*).

Nominal Molecular Weight

Phthalic Anhydride Solution Place 49.0 g of phthalic anhydride in an amber bottle, and dissolve it in 300 mL of pyridine that has been freshly distilled over phthalic anhydride. Shake the bottle vigorously until solution is effected, and allow to stand overnight before using.

Sample Preparation for Liquid Polyethylene Glycols Carefully introduce 25.0 mL of the *Phthalic Anhydride Solution* into a clean, dry, heat-resistant pressure bottle. To the bottle add an accurately weighed amount of the sample equivalent to its expected average molecular weight divided by 160. (Thus, a sample of about 1.3 g would be taken for PEG 200, or about 3.8 g for PEG 600.) Stopper the bottle, and wrap it securely in a fabric bag.

Sample Preparation for Solid Polyethylene Glycols Carefully introduce 25.0 mL of the *Phthalic Anhydride Solution* into a clean, dry, heat-resistant pressure bottle. To the bottle add an accurately weighed amount of the sample, previously melted, equivalent to its expected molecular weight divided by 160; because of limited solubility, however, do not use more than 25 g of any sample. Add 25 mL of pyridine, freshly distilled over phthalic anhydride, swirl to effect solution, stopper the bottle, and wrap it securely in a fabric bag.

Procedure Immerse the sample bottle in a water bath, maintained between 96° and 100°, to the same depth as that of the mixture in the bottle. Heat in the water bath for 30 to 60 min, using 60 min for PEGs having molecular weights of 3000 or higher, then remove the bottle from the bath and allow it to cool to room temperature. Uncap the bottle carefully to release any pressure, remove the bottle from the fabric bag, add 5 drops of a 1 in 100 solution of phenolphthalein in pyridine, and titrate with 0.5 N sodium hydroxide to the first pink color that persists for 15 s, recording the volume, in mL, of 0.5 N sodium hydroxide required as S. Perform a blank determination on 25.0 mL on the *Phthalic Anhydride Solution* plus any additional pyridine added to the sample bottle, and record the volume, in mL, of 0.5 N sodium hydroxide required as B. Calculate the nominal molecular weight of the sample by the formula

$$2000W/(B - S)N,$$

in which W is the weight of the sample, in g; $(B - S)$ is the difference between the volume of 0.5 N sodium hydroxide consumed by the blank and by the sample; and N is the exact normality of the sodium hydroxide solution.

pH Determine potentiometrically on a solution prepared by dissolving 5 g of the sample in 100 mL of carbon dioxide-free water and adding 0.3 mL of saturated potassium chloride solution.

Residue on Ignition Ignite 25 g as directed in *Method I*, Appendix IIC.

Viscosity Determine as directed under *Viscosity of Dimethylpolysiloxane*, Appendix IIB, maintaining the constant-temperature bath at 100° ± 0.3° and using a capillary viscometer having a flow time of at least 200 s for the sample being tested. The viscosity is within the limits specified in the table below. (For PEGs not listed in the table, calculate the limits by interpolation.)

Nominal Average Mol Wt	Viscosity Range (centistokes)	Nominal Average Mol Wt	Viscosity Range (centistokes)
200	3.9–4.8	2400	49–65
300	5.4–6.4	2500	51–70
400	6.8–8.0	2600	54–74
500	8.3–9.6	2700	57–78
600	9.9–11.3	2800	60–83
700	11.5–13.0	2900	64–88
800	12.5–14.5	3000	67–93
900	15.0–17.0	3250	73–105
1000	16.0–19.0	3350	76–110
1100	18.0–22.0	3500	87–123
1200	20.0–24.5	3750	99–140
1300	22.0–27.5	4000	110–158
1400	24–30	4250	123–177
1450	25–32	4500	140–200
1500	26–33	4750	155–228
1600	28–36	5000	170–250
1700	31–39	5500	206–315
1800	33–42	6000	250–390
1900	35–45	6500	295–480
2000	38–49	7000	350–590
2100	40–53	7500	405–735
2200	43–56	8000	470–900
2300	46–60		

Packaging and Storage Store in tight containers.

Polyglycerol Esters of Fatty Acids

INS: 475

DESCRIPTION

Polyglycerol Esters of Fatty Acids are mixed partial esters formed by reacting polymerized glycerols with edible fats, oils, or fatty acids. Minor amounts of mono-, di-, and triglycerides; free glycerol and polyglycerols; free fatty acids; and sodium salts of fatty acids may be present. The polyglycerols vary in degree of polymerization, which is specified by a number (such as tri-, penta-, deca-, etc.) that is related to the average number of glycerol residues per polyglycerol molecule. A specified polyglycerol consists of a distribution of molecular species characteristic of its nominal degree of polymerization. By varying the proportions as well as the nature of the fats or fatty acids to be reacted with the polyglycerols, a large and diverse class of products may be obtained. They include light yellow to amber, oily to very viscous liquids; light tan to medium brown, plastic or soft solids; and light tan to brown, hard, waxy solids. The esters range from very hydrophilic to very lipophilic, but as a class tend to be dispersible in water and soluble in organic solvents and oils.

Functional Use in Foods Emulsifier.

REQUIREMENTS

Heavy Metals (as Pb)   Not more than 10 mg/kg.
The following specifications should conform to the representations of the vendor: **Acid Value, Hydroxyl Value**, Iodine Value, **Residue on Ignition, Saponification Value**, and Sodium Salts of Fatty Acids.

TESTS

Acid Value   Determine as directed for *Acid Value, Method II*, under *Fats and Related Substances*, Appendix VII.
Heavy Metals   Prepare and test a 2-g sample as directed in *Method II* under the *Heavy Metals Test*, Appendix IIIB, using 20 μg of lead ion (Pb) in the control (*Solution A*).
Hydroxyl Value   Transfer to a 300-mL Erlenmeyer flask an accurately weighed sample approximately equivalent to 561 divided by the expected hydroxyl value, ± 10%. Mix 9 volumes of pyridine with 1 volume of acetic anhydride, pipet 25.0 mL of this solution into the sample flask, and pipet 25.0 mL into a separate 300-mL Erlenmeyer flask to serve as the blank. Add boiling stones to each flask, and fit the flasks with air condensers, lubricating the joints only with a few drops of pyridine. Reflux the sample solution gently by heating on a hot plate, confining the vapors in the lower portion of the condenser, and continue refluxing for 45 min. Do not heat the blank. Cool the sample flask to room temperature, and rinse the condenser, the condenser tip, and the sides of the flask with 25 mL of pyridine. Add about 50 mL of 0.55 N sodium hydroxide to each flask, mix by swirling for about 45 s, then add 1 mL of phenolphthalein TS and 75 mL of isopropanol, and continue the titration with stirring to the first pink color that persists for at least 30 s. Calculate the hydroxyl value by the formula

$$AV + [(56.1)(B - S)(N/W)],$$

in which AV is the *Acid Value*, determined as directed above; $(B - S)$ is the difference between the volume, in mL, of 0.55 N sodium hydroxide required for the blank and for the sample, respectively; N is the exact normality of the sodium hydroxide solution; and W is the weight of the sample, in g.

Iodine Value   Determine by the *Modified Wijs Method*, Appendix VII.
Residue on Ignition   Determine as directed in the general method, Appendix IIC.
Saponification Value   Determine as directed in the general method, Appendix VII, using an accurately weighed amount of the sample approximately equivalent to 700 divided by the expected saponification value.
Sodium Salts of Fatty Acids   Dissolve about 5 g of the sample, accurately weighed, in 75 mL of glacial acetic acid, add 2 drops of crystal violet TS, and titrate with 0.1 N perchloric acid in glacial acetic acid to an emerald green endpoint. Perform a blank determination and make any necessary correction, recording the net volume of perchloric acid consumed as V. Calculate the number of mg of potassium hydroxide equivalent to the sodium salts per g of the sample by the formula

$$56.1 \times (VN/W),$$

in which N is the exact normality of the perchloric acid, and W is the weight of the sample, in g.

Packaging and Storage   Store in well-closed containers.

Polyisobutylene

CAS: [9003-27-4]

DESCRIPTION

A synthetic polymer produced by the low-temperature polymerization of isobutylene in liquid ethylene, methyl chloride, or hexane, using an aluminum chloride or boron trifluoride catalyst. After completion of polymerization, volatile components are removed by raising the temperature of the reaction mixture or by kneading in a degassing extruder. Low-molecular-weight grades can be prepared by thermo-mechanical degradation of high-molecular-weight grades. Low-molecular-weight grades are soft and gummy; high-molecular-weight grades are tough and elastic. All grades are light in color, odorless, and tasteless and are soluble in diisobutylene, but insoluble in water.

Functional Use in Foods   Masticatory substance in chewing gum base.

REQUIREMENTS

Identification   Prepare the sample by dissolving it in hot toluene and evaporating on a potassium bromide plate. The infrared absorption spectrum of the sample exhibits maxima at the same wavelengths as the typical spectrum, as shown in the section on *Infrared Spectra (Series C: Other Substances)*.
Arsenic (as As)   Not more than 3 mg/kg.
Heavy Metals (as Pb)   Not more than 0.002%.
Lead   Not more than 3 mg/kg.

Molecular Weight (Flory)   Not less than 37,000.
Volatiles   Not more than 0.3%.

TESTS

Arsenic   Prepare a *Sample Solution* as directed in the general method under *Chewing Gum Base*, Appendix IV. This solution meets the requirements of the *Arsenic Test*, Appendix IIIB.

Heavy Metals   Prepare and test a 1-g sample as directed in *Method II* under the *Heavy Metals Test*, Appendix IIIB, using 20 µg of lead ion (Pb) in the control (*Solution A*).

Lead   Prepare a *Sample Solution* as directed in the general method under *Chewing Gum Base*, Appendix IV. This solution meets the requirements of the *Lead Limit Test*, Appendix IIIB, using 10 µg of lead ion (Pb) in the control.

Molecular Weight   Determine as directed in the general method, Appendix IV.

Volatiles   (**Caution**: To reduce explosion hazard, pass carbon dioxide or nitrogen into the lower part of the drying oven at a rate of about 100 mL/min.) Dry a 5-g sample for 2 h at 105° as directed under *Loss on Drying*, Appendix IIC.

Packaging and Storage   Store low-molecular-weight grades in boxes or drums with a release liner or coating; store high-molecular-weight grades wrapped in polyethylene film.

Polypropylene Glycol

CAS: [25322-69-4]

DESCRIPTION

Polypropylene Glycol is an addition polymer of propylene glycol and water, represented by the formula $HO(C_3H_6O)_nC_3H_6OH$, in which n represents the average number of oxypropylene groups. It is a clear, colorless or practically colorless, viscous liquid. It is soluble in water and in such organic solvents as aliphatic ketones and alcohols, but it is insoluble in ether and in most aliphatic hydrocarbons.

Functional Use in Foods   Defoaming agent.

REQUIREMENTS

Heavy Metals (as Pb)   Not more than 5 mg/kg.
Nominal Molecular Weight   Not less than 90.0% and not more than 110.0% of the labeled value.
pH   Between 6.0 and 9.0.
Propylene Oxide   Not more than 0.02%.
Residue on Ignition   Not more than 0.01%.
Viscosity   Passes test.

TESTS

Heavy Metals   A solution of 4 g of the sample mixed with 5.0 mL of 0.1 N hydrochloric acid, diluted with water to 25 mL, meets the requirements of the *Heavy Metals Test*, Appendix IIIB, using 20 µg of lead ion (Pb) in the control (*Solution A*).

Nominal Molecular Weight

Phthalic Anhydride Solution   Place 49.0 g of phthalic anhydride in an amber bottle, and dissolve it in 300 mL of pyridine that has been freshly distilled over phthalic anhydride. Shake the bottle vigorously until solution is effected, and allow to stand overnight before using.

Test Preparation   Carefully introduce 25.0 mL of the *Phthalic Anhydride Solution* into a clean, dry, heat-resistant pressure bottle. To the bottle add an accurately weighed amount of the sample equivalent to its expected nominal molecular weight divided by 160.

Procedure   Immerse the sample bottle in a water bath, maintained between 96° and 100°, to the same depth as that of the mixture in the bottle. Heat the water bath for 30 min, then remove the bottle from the bath, and allow it to cool to room temperature. Uncap the bottle carefully to release any pressure, remove the bottle from the fabric bag, add 5 drops of a 1 in 100 solution of phenolphthalein in pyridine, and titrate with 0.5 N sodium hydroxide to the first pink color that persists for 15 s, recording the volume, in mL, of 0.5 N sodium hydroxide required as S. Perform a blank determination on 25.0 mL of the *Phthalic Anhydride Solution* plus any additional pyridine added to the sample bottle, and record the volume, in mL, of 0.5 N sodium hydroxide required as B. Calculate the nominal molecular weight of the sample by the formula

$$2000W/N(B - S),$$

in which W is the weight of the sample, in g; $(B - S)$ is the difference between the volume of 0.5 N sodium hydroxide consumed by the blank and by the sample; and N is the exact normality of the sodium hydroxide solution.

pH   Determine potentiometrically on a solution prepared by dissolving 10 mL of the sample in 100 mL of methanol neutralized with either 0.1 N hydrochloric acid or 0.1 N sodium hydroxide.

Propylene Oxide

Magnesium Chloride Solution   Add 100 mL of 10 N hydrochloric acid to 950 g of magnesium chloride, $MgCl_2.6H_2O$, dissolve in water, dilute to 1000 mL with water, and mix. Carefully add 400 mL of anhydrous methanol to 100 mL of the stock solution, and allow the mixture to come to room temperature before using.

Indicator   Dissolve 100 mg of bromocresol green in 100 mL of anhydrous methanol.

Procedure   Place 150 mL of anhydrous methanol into each of two 500-mL glass-stoppered conical flasks, the second of which is used as the reagent blank. Into each flask, pressure-pipet 25.0 mL of the *Magnesium Chloride Solution*, allowing the same drainage time for each transfer, and mix thoroughly. To the first flask add about 50 g of the sample, accurately weighed, and dissolve by swirling. Add about 1 mL of the *Indicator* to each flask, and titrate with 0.1 N alcoholic potassium hydroxide to a brilliant blue endpoint, recording the volume required as S, in mL. Titrate the reagent blank flask in the same manner, and record the volume required as B.

To correct for the alkalinity in the sample, place 150 mL of anhydrous methanol in a 500-mL conical flask, add about 50 g of the sample, accurately weighed, and swirl to effect solution. Add 1 mL of the *Indicator*, and titrate with 0.1 N hydrochloric acid to a yellow endpoint, recording the volume required as C, in mL. Calculate the percentage of propylene oxide in the sample by the formula

$$5.81 \times \{[(B - S)N_1]/W_1\} - (CN_2/W_2),$$

in which N_1 is the exact normality of the potassium hydroxide; N_2 is the exact normality of the hydrochloric acid; and W_1 and W_2 are the weights, in g, of the sample taken for the reaction and the alkalinity correction, respectively.

Residue on Ignition Ignite 25 g as directed in the general method, Appendix IIC.

Viscosity Determine as directed under *Viscosity of Dimethylpolysiloxane*, Appendix IIB, maintaining the constant-temperature bath at $37.8° \pm 0.2°$ and using a capillary viscometer having a flow time of at least 200 s for the sample being tested. The viscosity of a sample having a nominal molecular weight of 1000 is between 85 and 97 centistokes, and that of a sample having a nominal molecular weight of 2000 is between 150 and 175 centistokes.

Packaging and Storage Store in tight containers.

Polysorbate 20

Polyoxyethylene (20) Sorbitan Monolaurate; Sorbitan, Monododecanoate; Poly(oxy-1,2-ethanediyl) Derivative

CAS: [9005-64-5]

DESCRIPTION

Polysorbate 20 is a mixture of laurate partial esters of sorbitol and sorbitol anhydrides condensed with approximately 20 moles of ethylene oxide (C_2H_4O) for each mole of sorbitol and its mono- and dianhydrides. It is a lemon- to amber-colored liquid having a faint, characteristic odor and a warm, somewhat bitter taste. It is soluble in water, in alcohol, in ethyl acetate, in methanol, and in dioxane, but is insoluble in mineral oil and in mineral spirits.

Functional Use in Foods Emulsifier; stabilizer.

REQUIREMENTS

Identification To 5 mL of a 1 in 20 solution add 5 mL of 1 N sodium hydroxide, boil for a few min, cool, and acidify with 2.7 N hydrochloric acid. The solution is strongly opalescent.

Assay for Oxyethylene Content Not less than 70.0% and not more than 74.0% of oxyethylene groups (—C_2H_4O—), equivalent to between 97.3% and 103.0% of Polysorbate 20, calculated on the anhydrous basis.

Acid Value Not more than 2.
1,4-Dioxane Not more than 10 mg/kg.
Heavy Metals (as Pb) Not more than 10 mg/kg.
Hydroxyl Value Between 96 and 108.
Lauric Acid Between 15 and 17 g/100 g of sample.
Residue on Ignition Not more than 0.25%.
Saponification Value Between 40 and 50.
Water Not more than 3.0%.

TESTS

Assay for Oxyethylene Content Weigh accurately a 65-mg sample, and proceed as directed in the general method, Appendix VII.

Acid Value Determine as directed for *Acid Value, Method II*, under *Fats and Related Substances*, Appendix VII.

1,4-Dioxane

Stripped Polysorbate Prepare an appropriate quantity of 1,4-dioxane-free Polysorbate 20 by stripping it at 10 mm of mercury, with nitrogen sparge at 130° for 4 h or until, when tested as directed under *Procedure*, no 1,4-dioxane is detected.

1,4-Dioxane Standard Preparation By appropriate quantitative dilutions using *Stripped Polysorbate*, prepare a standard preparation containing 10 μg/mL of 1,4-dioxane. Transfer 5.0 g, accurately weighed, of the *1,4-Dioxane Standard Preparation* to a 22-mL pressure headspace vial; seal with a silicone septum, star spring, and pressure-relief safety aluminum sealing cap; and crimp the cap closed with a cap sealing tool.

Test Preparation Transfer 5.0 g, accurately weighed, of the Polysorbate sample to a 22-mL pressure headspace vial, and seal the cap and crimp as directed for the *1,4-Dioxane Standard Preparation*.

Chromatographic System Chromatographic conditions may vary depending on the type of headspace unit used. The gas chromatograph is equipped with a headspace sampler, flame ionization detector, backflush valve, and 1-mL gas sample loop, and it contains a 9-m × 3.2-mm nickel precolumn and a 55-m × 3.2-mm nickel analytical column containing TENAX TA support (60/80-mesh), or equivalent. The column temperature is maintained at 190°, the detector at 250°, and the injector at 250°. The carrier gas is helium at a flow rate of about 30 mL/min. The backflush valve is programmed to initiate backflushing after 1,4-dioxane elutes into the analytical column.

Procedure Place the vial containing the *1,4-Dioxane Standard Preparation* in the automated sampler, and start the operating sequence so that each vial is heated at 90° for a minimum of 30 min. Using appropriate headspace sampler settings, inject the *1,4-Dioxane Standard Preparation*, and measure the peak area for 1,4-dioxane. Similarly, place the vial containing the *Test Preparation* in the automatic sampler, chromatograph its headspace as done for the *1,4-Dioxane Standard Preparation*, and measure the peak area for 1,4-dioxane. The sample passes the test if the peak area of the *Test Preparation* is not greater than that of the *1,4-Dioxane Standard Preparation*.

Heavy Metals Prepare and test a 2-g sample as directed in *Method II* under the *Heavy Metals Test*, Appendix IIIB, using 20 μg of lead ion (Pb) in the control (*Solution A*).

Hydroxyl Value Determine as directed under *Method II* in the general method, Appendix VII.

Lauric Acid Transfer about 25 g of the sample, accurately weighed, into a 500-mL, round-bottom boiling flask, add 250 mL of alcohol and 7.5 g of potassium hydroxide, and mix. Connect a suitable condenser to the flask, reflux the mixture for 1 to 2 h, then transfer to an 800-mL beaker, rinsing the flask with about 100 mL of water and adding it to the beaker. Heat on a steam bath to evaporate the alcohol, adding water occasionally to replace the alcohol, and evaporate until the odor of alcohol can no longer be detected. Adjust the final volume to about 250 mL with hot water. Neutralize the soap solution with dilute sulfuric acid (1 in 2), add 10% in excess, and heat, while stirring, until the fatty acid layer separates. Transfer the fatty acids into a 500-mL separator, wash with three or four 20-mL portions of hot water, and combine the washings with the original aqueous layer from the saponification. Extract the combined aqueous layer with three 50-mL portions of petroleum ether, add the extracts to the fatty acid layer, evaporate to dryness in a tared dish, cool, and weigh. The lauric acid so obtained has an *Acid Value* between 250 and 275 (*Method I*, Appendix VII).

Residue on Ignition Ignite 5 g as directed in the general method, Appendix IIC.

Saponification Value Determine as directed in the general method, Appendix VII, using about 8 g, accurately weighed.

Water Determine by the *Karl Fischer Titrimetric Method*, Appendix IIB.

Packaging and Storage Store in tight containers.

Polysorbate 60

Polyoxyethylene (20) Sorbitan Monostearate; Sorbitan, Monooctadecanoate; Poly(oxy-1,2-ethanediyl) Derivative

CAS: [9005-67-8]

DESCRIPTION

Polysorbate 60 is a mixture of stearate and palmitate partial esters of sorbitol and sorbitol anhydrides condensed with approximately 20 moles of ethylene oxide (C_2H_4O) for each mole of sorbitol and its mono- and dianhydrides. It is a lemon- to orange-colored, oily liquid or semi-gel having a faint characteristic odor and a warm, somewhat bitter taste. It is soluble in water, in aniline, in ethyl acetate, and in toluene but is insoluble in mineral oil and in vegetable oils.

Functional Use in Foods Emulsifier; stabilizer.

REQUIREMENTS

Identification

A. To 5 mL of a 1 in 20 solution add 5 mL of 1 *N* sodium hydroxide, boil for a few min, cool, and acidify with 2.7 *N* hydrochloric acid. The solution is strongly opalescent.

B. A mixture of 60 volumes of Polysorbate 60 with 40 volumes of water at 25° or below yields a gelatinous mass.

Assay for Oxyethylene Content Not less than 65.0% and not more than 69.5% of oxyethylene groups (—C_2H_4O—), equivalent to between 97.0% and 103.0% of Polysorbate 60, calculated on the anhydrous basis.

Acid Value Not more than 2.

1,4-Dioxane Not more than 10 mg/kg.

Heavy Metals (as Pb) Not more than 10 mg/kg.

Hydroxyl Value Between 81 and 96.

Residue on Ignition Not more than 0.25%.

Saponification Value Between 45 and 55.

Stearic and Palmitic Acids Between 21.5 and 26.0 g/100 g of sample.

Water Not more than 3.0%.

TESTS

Assay for Oxyethylene Content Weigh accurately a 65-mg sample, and proceed as directed in the general method, Appendix VII.

Acid Value Determine as directed for *Acid Value, Method II*, under *Fats and Related Substances*, Appendix VII.

1,4-Dioxane Determine as directed under *1,4-Dioxane* in the monograph for *Polysorbate 20*.

Heavy Metals Prepare and test a 2-g sample as directed in *Method II* under the *Heavy Metals Test*, Appendix IIIB, using 20 µg of lead ion (Pb) in the control (*Solution A*).

Hydroxyl Value Determine as directed under *Method II* in the general method, Appendix VII.

Residue on Ignition Ignite 5 g as directed in the general method, Appendix IIC.

Saponification Value Determine as directed in the general method, Appendix VII, using about 8 g, accurately weighed.

Stearic and Palmitic Acids Isolate the fatty acids as directed under *Lauric Acid* in the monograph for *Polysorbate 20*, and determine the weight of the acids. The product so obtained has an *Acid Value* between 200 and 212 (*Method I*, Appendix VII) and a *Solidification Point*, Appendix IIB, not below 52°.

Water Determine by the *Karl Fischer Titrimetric Method*, Appendix IIB.

Packaging and Storage Store in tight containers.

Polysorbate 65

Polyoxyethylene (20) Sorbitan Tristeara

DESCRIPTION

Polysorbate 65 is a mixture of stearate and palmitate partial esters of sorbitol and its anhydrides condensed with approximately 20 moles of ethylene oxide (C_2H_4O) for each mole of sorbitol and its mono- and dianhydrides. It is a tan, waxy solid having a faint, characteristic odor and a warm, somewhat bitter taste. It is soluble in mineral oil and in vegetable oils, mineral spirits, acetone, ether, dioxane, alcohol, and methanol, and it is dispersible in water.

Functional Use in Foods Emulsifier; stabilizer.

REQUIREMENTS

Identification To 5 mL of a 1 in 20 solution add 5 mL of 1 N sodium hydroxide, boil for a few min, cool, and acidify with 2.7 N hydrochloric acid. The solution is strongly opalescent.
Assay for Oxyethylene Content Not less than 46.0% and not more than 50.0% of oxyethylene groups (—C_2H_4O—), equivalent to between 96.0% and 104.0% of Polysorbate 65, calculated on the anhydrous basis.
Acid Value Not more than 2.
1,4-Dioxane Not more than 10 mg/kg.
Heavy Metals (as Pb) Not more than 10 mg/kg.
Hydroxyl Value Between 44 and 60.
Residue on Ignition Not more than 0.25%.
Saponification Value Between 88 and 98.
Stearic and Palmitic Acids Between 42 and 44 g/100 g of sample.
Water Not more than 3.0%.

TESTS

Assay for Oxyethylene Content Weigh accurately a 90-mg sample, and proceed as directed in the general method, Appendix VII.
Acid Value Determine as directed for *Acid Value, Method II*, under *Fats and Related Substances*, Appendix VII.
1,4-Dioxane Determine as directed under *1,4-Dioxane* in the monograph for *Polysorbate 20*.
Heavy Metals Prepare and test a 2-g sample as directed in *Method II* under the *Heavy Metals Test*, Appendix IIIB, using 20 µg of lead ion (Pb) in the control (*Solution A*).
Hydroxyl Value Determine as directed under *Method II* in the general method, Appendix VII.
Residue on Ignition Ignite 5 g as directed in the general method, Appendix IIC.
Saponification Value Determine as directed in the general method, Appendix VII, using about 6 g, accurately weighed.
Stearic and Palmitic Acids Isolate the fatty acids as directed under *Lauric Acid* in the monograph for *Polysorbate 20*, and determine the weight of the acids. The product so obtained has an *Acid Value* between 200 and 212 (*Method I*, Appendix VII) and a *Solidification Point*, Appendix IIB, not below 52°.
Water Determine by the *Karl Fischer Titrimetric Method*, Appendix IIB.

Packaging and Storage Store in tight containers.

Polysorbate 80

Polyoxyethylene (20) Sorbitan Monooleate; Sorbitan, Mono-9-octadecenoate; Poly(oxy- 1,2-ethanediyl) Derivative

CAS: [9005-65-6]

DESCRIPTION

Polysorbate 80 is a mixture of oleate partial esters of sorbitol and sorbitol anhydrides condensed with approximately 20 moles of ethylene oxide (C_2H_4O) for each mole of sorbitol and its mono- and dianhydrides. It is a yellow- to orange-colored, oily liquid having a faint, characteristic odor and a warm, somewhat bitter taste. It is very soluble in water, producing an odorless, nearly colorless solution, and is soluble in alcohol, in fixed oils, in ethyl acetate, and in toluene. It is insoluble in mineral oil.

Functional Use in Foods Emulsifier; stabilizer.

REQUIREMENTS

Identification
A. To 5 mL of a 1 in 20 solution add 5 mL of 1 N sodium hydroxide, boil for a few min, cool, and acidify with 2.7 N hydrochloric acid. The solution is strongly opalescent.
B. To a 1 in 20 solution add bromine TS, dropwise. The bromine is decolorized.
C. A mixture of 60 volumes of Polysorbate 80 with 40 volumes of water at 25° or below yields a gelatinous mass.
Assay for Oxyethylene Content Not less than 65.0% and not more than 69.5% of oxyethylene groups (—C_2H_4O—), equivalent to between 96.5% and 103.5% of Polysorbate 80, calculated on the anhydrous basis.
Acid Value Not more than 2.
1,4-Dioxane Not more than 10 mg/kg.
Heavy Metals (as Pb) Not more than 10 mg/kg.
Hydroxyl Value Between 65 and 80.
Oleic Acid Between 22 and 24 g/100 g of sample.
Residue on Ignition Not more than 0.25%.
Saponification Value Between 45 and 55.
Water Not more than 3.0%.

TESTS

Assay for Oxyethylene Content Weigh accurately a 65-mg sample, and proceed as directed in the general method, Appendix VII.
Acid Value Determine as directed for *Acid Value, Method II*, under *Fats and Related Substances*, Appendix VII.
1,4-Dioxane Determine as directed under *1,4-Dioxane* in the monograph for *Polysorbate 20*.
Heavy Metals Prepare and test a 2-g sample as directed in *Method II* under the *Heavy Metals Test*, Appendix IIIB, using 20 µg of lead ion (Pb) in the control (*Solution A*).
Hydroxyl Value Determine as directed under *Method II* in the general method, Appendix VII.

Oleic Acid Isolate the fatty acids as directed under *Lauric Acid* in the monograph for *Polysorbate 20*, and determine the weight of the acid. The product so obtained has an *Acid Value* between 193 and 206 (*Method I*, Appendix VII) and an *Iodine Value* between 80 and 92 (*Modified Wijs Method*, Appendix VII).
Residue on Ignition Ignite 5 g as directed in the general method, Appendix IIC.
Saponification Value Determine as directed in the general method, Appendix VII, using about 8 g, accurately weighed.
Water Determine by the *Karl Fischer Titrimetric Method*, Appendix IIB.

Packaging and Storage Store in tight containers.

Polyvinyl Acetate

CAS: [9003-20-7]

DESCRIPTION

A clear, water white to pale yellow, solid resin prepared by the polymerization of vinyl acetate. After completion of polymerization, the resin is freed of traces of residual catalyst (usually a peroxide), monomer, and/or solvent by vacuum drying, steam sparging, washing, or any combination of these treatments. The resin is soluble in acetone, but is insoluble in water.

Functional Use in Foods Masticatory substance in chewing gum base.

REQUIREMENTS

Identification The sample is melted and prepared for analysis on a potassium bromide plate. The infrared absorption spectrum of the sample exhibits maxima at the same wavelengths as the typical spectrum, as shown in the section on *Infrared Spectra* (*Series C: Other Substances*).
Free Acetic Acid Not more than 0.05%.
Heavy Metals (as Pb) Not more than 0.002%.
Lead Not more than 3 mg/kg.
Loss on Drying Not more than 1.0%.
Molecular Weight Not less than 2000.

TESTS

Free Acetic Acid Transfer 10.0 g of the sample into a 250-mL, glass-stoppered Erlenmeyer flask, dissolve in 75 mL of ethylene dichloride, add 60 mL of specially denatured ethanol formula 2B, and mix. Add phenolphthalein TS, and titrate with 0.02 N methanolic potassium hydroxide to a faint pink endpoint. Perform a blank determination (see *General Provisions*), and make any necessary correction. Each mL of 0.02 N methanolic potassium hydroxide is equivalent to 1.201 mg of $C_2H_4O_2$.

Heavy Metals Prepare and test a 1-g sample as directed in *Method II* under the *Heavy Metals Test*, Appendix IIIB, using 20 μg of lead ion (Pb) in the control (*Solution A*).
Lead Prepare a *Sample Solution* as directed in the general method under *Chewing Gum Base*, Appendix IV. This solution meets the requirements of the *Lead Limit Test*, Appendix IIIB, using 10 μg of lead ion (Pb) in the control.
Loss on Drying Dry a sample of about 1.5 g at 100° for 2 h in vacuum as directed under *Loss on Drying*, Appendix IIC.
Molecular Weight Determine as directed in the general method, Appendix IV.

Packaging and Storage Store in well-closed containers.

Polyvinylpolypyrrolidone

Crospovidone; PVPP; 1-Vinyl-2-pyrrolidone Crosslinked Insoluble Polymer

INS: 1202

DESCRIPTION

A crosslinked homopolymer of purified vinylpyrrolidone, produced catalytically. It is insoluble in water and in other common solvents. It occurs as a white to off-white, hygroscopic, free-flowing powder having a faint, bland odor.

Functional Use in Foods Clarifying agent; stabilizer.

REQUIREMENTS

Identification Add 0.1 mL of iodine TS to a suspension of 1 g of the sample in 10 mL of water. The reagent is discolored after the mixture is shaken for 30 s (distinction from polyvinylpyrrolidone, which produces a red color). When 1 mL of starch TS is added and the mixture is shaken, no blue color is formed.
Heavy Metals (as Pb) Not more than 10 mg/kg.
Nitrogen Not less than 11.0% and not more than 12.8%.
pH of a 1 in 100 Suspension Between 5.0 and 11.0.
Residue on Ignition Not more than 0.4%.
Soluble Substances Not more than 0.5% in water; not more than 1.0% in an acid–alcohol medium.
Unsaturation (as vinylpyrrolidinone) Not more than 0.1%.
Water Not more than 6.0%.

TESTS

Heavy Metals Prepare and test a 2-g sample as directed in *Method II* under the *Heavy Metals Test*, Appendix IIIB, using 20 μg of lead ion (Pb) in the control (*Solution A*).
Nitrogen Determine as directed in *Method II* under *Nitrogen Determination*, Appendix IIIC, using a 100-mg sample. In the wet-digestion step, repeat the addition of hydrogen peroxide (usually three to six times) until a clear, light green solution is

obtained, then heat for an additional 4 h, and continue as directed, beginning with "Cautiously add 2 mL of water. . . ."

pH of a 1 in 100 Suspension Determine as directed in the general method, Appendix IIB, using a 1-g sample suspended in 100 mL of water.

Residue on Ignition Ignite a 2-g sample as directed in the general method, Appendix IIC.

Soluble Substances

Solubility in Water Place 10 g of Polyvinylpolypyrrolidone in a 200-mL flask containing 100 mL of water. Shake the flask, and allow the contents to rest for 24 h. Filter on filter screen with a porosity of 2.5 μm, then on a filter screen with a porosity of 0.8 μm. The residue left by evaporating the filtrate over a water bath until dry must be less than 50 mg.

Solubility in an Acid–Alcohol Medium Place 1 g of Polyvinylpolypyrrolidone in a flask containing 500 mL of the following mixture: 3 g of acetic acid, 10 mL of ethanol, and sufficient water to make up the volume to 100 mL. Allow the contents of the flask to rest for 24 h. Filter on a filter screen with a porosity of 2.5 μm, then on a filter screen with a porosity of 0.8 μm. Concentrate the filtrate over a water bath. Finish evaporation over the water bath in a 70-mm diameter tared silica capsule. The dry residue remaining after evaporation must be less than 10 mg, taking account of any residue left by the evaporation of 500 mL of the acetic acid–ethanol mixture.

Unsaturation (as vinylpyrrolidinone) Suspend a 4-g sample in 30 mL of water, stir for 15 min, and filter through a sintered-glass filter having a porosity between 9 and 15 μm, collecting the filtrate in a 250-mL flask. Wash the residue with 100 mL of water, add 500 mg of sodium acetate to the combined filtrates, and titrate with 0.1 N iodine until the color of iodine no longer fades. Add an additional 3.0 mL of 0.1 N iodine, allow to stand for 10 min, and titrate the excess iodine with 0.1 N sodium thiosulfate, adding 3 mL of starch TS as the endpoint is approached. Perform a blank determination (see *General Provisions*), and make any necessary correction. Not more than 0.72 mL is consumed.

Water Determine by the *Karl Fischer Titrimetric Method*, Appendix IIB.

Packaging and Storage Store in tight containers.

Polyvinylpyrrolidone

PVP; Povidone; Poly[1-(2-oxo-1-pyrrolidinyl)ethylene]

INS: 1201 CAS: [9003-39-8]

DESCRIPTION

Polyvinylpyrrolidone is a polymer of purified 1-vinyl-2-pyrrolidinone produced catalytically. It occurs as a white to tan powder, free from objectionable odor. It is soluble in water, in alcohol, and in chloroform, and is insoluble in ether. The pH of a 1 in 20 solution is between 3 and 7.

Functional Use in Foods Clarifying agent; separation/filtration aid; stabilizer; bodying agent; tableting aid; dispersant.

REQUIREMENTS

Labeling Indicate the K-value or the K-value range.

Identification

A. To 10 mL of a 1 in 50 solution of the sample add 20 mL of 1 N hydrochloric acid and 5 mL of potassium dichromate TS. An orange yellow precipitate is produced.

B. Add 5 mL of a 1 in 50 solution of the sample to 75 mg of cobalt nitrate and 300 mg of ammonium thiocyanate dissolved in 2 mL of water, mix, and then make the resulting solution acid with 2.7 N hydrochloric acid. A pale blue precipitate forms.

C. To 5 mL of a 1 in 200 solution of the sample add a few drops of iodine TS. A deep red color is produced.

Aldehydes (as acetaldehyde) Not more than 0.05%.

Heavy Metals (as Pb) Not more than 10 mg/kg.

Hydrazine Not more than 1 mg/kg.

K-Value Between 27 and 32 for the low-molecular-weight product, and between 81 and 97 for the high-molecular-weight product.

Nitrogen Not less than 11.5% and not more than 12.8%, calculated on the anhydrous basis.

Residue on Ignition Not more than 0.1%.

Unsaturation (as vinylpyrrolidinone) Not more than 0.1%.

Water Not more than 5.0%.

TESTS

Aldehydes

Phosphate Buffer Transfer 50.0 g of potassium pyrophosphate to a 500-mL volumetric flask, and dissolve in 400 mL of water. Adjust, if necessary, with 1 N hydrochloric acid to a pH of 9.0, dilute with water to volume, and mix.

Aldehyde Dehydrogenase Solution Transfer a quantity of lyophilized aldehyde dehydrogenase (Sigma A550, or equivalent) to 70 units to a glass vial, dissolve in 10.0 mL of water, and mix.

Note: This solution is stable for 8 h at 4°.

NAD Solution Transfer 40 mg of nicotinamide adenine dinucleotide (B-NAD, Grade III-C, from Sigma Chemical Co.) to a glass vial, dissolve in 10.0 mL of *Phosphate Buffer*, and mix.

Note: This solution is stable for 4 weeks at 4°.

Standard Solution Add about 2 mL of water to a glass weighing bottle, and weigh accurately. Add about 100 mg (about 0.13 mL) of freshly distilled acetaldehyde, and weigh accurately. Transfer this solution to a 100-mL volumetric flask. Rinse the weighing bottle with several portions of water, transferring each rinsing to the 100-mL volumetric flask. Dilute the solution in the 100-mL flask with water to volume, and mix. Store at 4° for about 20 h. Pipet 1 mL of this solution into a 100-mL volumetric flask, dilute with water to volume, and mix.

Test Preparation Transfer about 2 g of Polyvinylpyrrolidone, accurately weighed, to a 100-mL volumetric flask, dissolve in 50 mL of *Phosphate Buffer*, dilute with *Phosphate Buffer* to volume, and mix. Insert a stopper into the flask, heat at 60° for 1 h, and cool to room temperature.

Procedure Pipet 0.5 mL each of the *Standard Solution*, the *Test Preparation*, and water (to provide the reagent blank) into separate 1-cm cells. Add 2.5 mL of *Phosphate Buffer* and 0.2 mL of *NAD Solution* to each cell. Cover the cells to exclude oxygen. Mix by inversion, and allow to stand for 2 to 3 min at 22° ± 2°. Determine the absorbances of the solutions at a wavelength of 340 nm, using water as the reference. Calculate the percentage of aldehydes, expressed as acetaldehyde, in the Polyvinylpyrrolidone taken by the formula

$$10(C/W)\{[(A_{U2} - A_{U1}) - (A_{B2} - A_{B1})]/[(A_{S2} - A_{S1}) - (A_{B2} - A_{B1})]\},$$

in which C is the concentration, in mg/mL, of acetaldehyde in the *Standard Solution*; W is the weight, in g, of Polyvinylpyrrolidone taken; A_{U1}, A_{S1}, and A_{B1} are the absorbances of the solutions obtained from the *Test Preparation*, the *Standard Solution*, and the water reagent blank, respectively, before addition of the *Aldehyde Dehydrogenase Solution*; and A_{U2}, A_{S2}, and A_{B2} are the absorbances of the solutions obtained from the *Test Preparation*, the *Standard Solution*, and the water reagent blank, respectively, after addition of the *Aldehyde Dehydrogenase Solution*. Not more than 0.05% is found.

Heavy Metals Prepare and test a 2-g sample as directed in *Method II* under the *Heavy Metals Test*, Appendix IIIB, using 20 μg of lead ion (Pb) in the control (*Solution A*).

Hydrazine

Salicylaldazine Standard Solution Dissolve 300 mg of hydrazine sulfate in 5 mL of water, add 1 mL of glacial acetic acid and 2 mL of a freshly prepared 20% (v/v) solution of salicylaldehyde in isopropyl alcohol, mix, and allow to stand until a yellow precipitate forms. Extract the mixture with two 15-mL portions of methylene chloride. Combine the methylene chloride extracts, and dry over anhydrous sodium sulfate. Decant the methylene chloride solution, and evaporate it to dryness. Recrystallize the residue of salicylaldazine from a mixture of warm toluene and methanol (60:40) with cooling. Filter, and dry the crystals in a vacuum. The crystals have a melting range of 213° to 219°, but the range between the beginning and end of melting is not to exceed 1°. Prepare a salicylaldazine solution containing 9.38 μg/mL of toluene.

Procedure Transfer 2.5 g to a 50-mL centrifuge tube, add 25 mL of water, and mix to dissolve. Add 500 μL of a 1 in 20 solution of salicylaldehyde in methanol, swirl, and heat in a water bath at 60° for 15 min. Allow to cool, add 2.0 mL of toluene, insert a stopper in the tube, shake vigorously for 2 min, and centrifuge. On a suitable thin-layer chromatographic plate coated with a 0.25-mm layer of dimethylsilanized chromatographic silica gel mixture, apply 10 μL of the clear upper toluene layer in the centrifuge tube and 10 μL of the *Standard Solution* of salicylaldazine. Allow the spots to dry, and develop the chromatogram in a solvent system of methanol and water (2:1) until the solvent front has moved about three-fourths of the length of the plate. Remove the plate from the developing chamber, mark the solvent front, and allow the solvent to evaporate. Locate the spots on the plate by examination under UV light at a wavelength of 365 nm: Salicylaldazine appears as a fluorescent spot having an R_f value of about 0.3, and the fluorescence of any salicylaldazine spot from the test specimen is not more intense than that produced by the spot obtained from the *Standard Solution*.

K-Value The molecular weight of the sample is characterized by its viscosity in aqueous solution, relative to that of water, expressed as a K-value. Determine the relative viscosity, z, as follows: Transfer an accurately weighed portion of the as-is sample, equivalent to approximately 1 g on the anhydrous basis, to a 100-mL volumetric flask, dissolve in about 50 mL of water, dilute to volume, mix thoroughly, and allow to stand for 1 h, then pipet 15 mL of filtrate into a clean, dry Ubbelholde-type viscometer, and place the viscometer in a water bath maintained at 25° ± 0.2°. After allowing the viscometer and the sample solution to warm in the water bath for 10 min, draw the solution by means of very gentle suction up through the capillary until the meniscus is above the upper etched mark. Release suction and begin timing the flow through the capillary. After the meniscus reaches the upper etched mark, record the exact time when the meniscus reaches the lower etched mark, and calculate the flow time to the nearest 0.01 s. Repeat this operation until at least three readings are obtained. The readings must agree within 0.1 s; if not, repeat the determination with additional 15-mL portions of the sample solution after recleaning the viscometer with sulfuric acid–dichromate cleaning solution or with a suitable laboratory cleaning compound that will remove oils, greases, waxes, and other impurities. Calculate the average flow time for the sample solution, and then obtain the flow time in a similar manner for 15 mL of water for the same viscosity pipet. Calculate the relative viscosity, z, of the sample by dividing the average flow time of the sample solution by that of the water sample, and then calculate the K-value by the formula

$$[\sqrt{300c \log z + (c + 1.5c \log z)^2} + 1.5c \log z - c]/(0.15c + 0.003c^2),$$

in which c is the weight, in g, on the anhydrous basis, of the sample in each 100.0 g of solution, and z is as defined above.

Nitrogen Determine as directed in *Method II* under *Nitrogen Determination*, Appendix IIIC, using a 100-mg sample. In the wet-digestion step, omit the use of hydrogen peroxide, and use 5 g of a mixture of potassium sulfate, cupric sulfate, and titanium dioxide (33:1:1) instead of the potassium sulfate and cupric sulfate mixture (10:1). Heat until a clear, light green solution is obtained, heat for an additional 45 min, and continue as directed, beginning with "Cautiously add 20 mL of water, cool, then...," except use 70 mL of water instead of 20.

Residue on Ignition Weigh accurately about 2 g, and proceed as directed in the general method, Appendix IIC.

Unsaturation Dissolve about 10 g of the sample, accurately weighed, in 80 mL of water in a 125-mL, round-bottom flask, add 1.0 g of sodium acetate, mix, and begin titrating with 0.1 N iodine. When the iodine color no longer fades, add 3 additional mL of the titrant, and allow the solution to stand for 5 to 10 min. Add starch TS, and titrate the excess iodine with 0.1 N sodium thiosulfate. Perform a blank determination (see *General Provisions*), using the same volume of 0.1 N iodine, accurately measured, as was used for the sample. Each mL of 0.1 N iodine is equivalent to 5.556 mg of vinylpyrrolidinone.

Potassium Acid Tartrate

Potassium Bitartrate; Cream of Tartar

KOOCCH(OH)CH(OH)COOH

$C_4H_5KO_6$ Formula wt 188.18

CAS: [868-14-4]

DESCRIPTION

Potassium Acid Tartrate is a salt of L(+)-tartaric acid. It occurs as colorless or slightly opaque crystals, or as a white, crystalline powder, having a pleasant, acid taste. A saturated solution is acid to litmus. One g dissolves in 165 mL of water at 25°, in 16 mL of boiling water, and in 8820 mL of alcohol.

Functional Use in Foods   Acid; buffer.

REQUIREMENTS

Identification

A. When sufficiently heated, it chars and emits flammable vapors having an odor resembling that of burning sugar. At a higher temperature and with free access to air, the carbon of the black residue is consumed, and there remains a white, fused mass of potassium carbonate that imparts a reddish purple color to a nonluminous flame.

B. A saturated solution yields a yellowish orange precipitate with sodium cobaltinitrite TS.

C. Neutralize a saturated solution with 1 N sodium hydroxide in a test tube, add silver nitrate TS, then just sufficient 6 N ammonium hydroxide to dissolve the white precipitate, and boil the solution. Silver is deposited on the inner surface of the tube, forming a mirror.

Assay   Not less than 99.0% and not more than 101.0% of $C_4H_5KO_6$ after drying.
Ammonia   Passes test.
Heavy Metals (as Pb)   Not more than 0.002%.
Insoluble Matter   Passes test.
Lead   Not more than 10 mg/kg.

TESTS

Assay   Weigh accurately about 6 g, previously dried at 105° for 3 h, dissolve it in 100 mL of boiling water, add phenolphthalein TS, and titrate with 1 N sodium hydroxide. Each mL of 1 N sodium hydroxide is equivalent to 188.2 mg of $C_4H_5KO_6$.

Ammonia   Heat 500 mg with 5 mL of 1 N sodium hydroxide. No odor of ammonia is detected.
Heavy Metals   Prepare and test a 1-g sample as directed in *Method II* under the *Heavy Metals Test*, Appendix IIIB, using 20 μg of lead ion (Pb) in the control (*Solution A*).
Insoluble Matter   Agitate 500 mg with 3 mL of 6 N ammonium hydroxide. No undissolved residue remains.
Lead   Dissolve 1 g in 3 mL of dilute nitric acid (1 in 2), boil for 1 min, cool, and dilute to 20 mL with water. This solution meets the requirements of the *Lead Limit Test*, Appendix IIIB, using 10 μg of lead ion (Pb) in the control.

Packaging and Storage   Store in tight containers.

Potassium Alginate

Algin

$(C_6H_7O_6K)_n$ Equiv wt, calculated, 214.22
                                     Equiv wt, actual (avg), 238.00

INS: 402   CAS: [9005-36-1]

DESCRIPTION

The potassium salt of alginic acid (see the monograph for *Alginic Acid*) occurs as a white to yellowish, fibrous or granular powder. It is nearly odorless and tasteless. It dissolves in water to form a viscous, colloidal solution. It is insoluble in alcohol and in hydroalcoholic solutions in which the alcohol content is greater than 30% by weight. It is insoluble in chloroform, in ether, and in acids having a pH lower than about 3.

Functional Use in Foods   Stabilizer; thickener; emulsifier.

REQUIREMENTS

Identification

A. To 5 mL of a 1 in 100 solution add 1 mL of calcium chloride TS. A voluminous, gelatinous precipitate is formed.

B. To 10 mL of a 1 in 100 solution add 1 mL of 2 N sulfuric acid. A heavy gelatinous precipitate is formed.

C. Potassium alginate meets the requirements of *Identification Test C* in the monograph for *Alginic Acid*.

Assay   It yields not less than 16.5% and not more than 19.5% of carbon dioxide (CO_2), corresponding to between 89.2 and 105.5% of potassium alginate (equiv wt 238.00), calculated on the dried basis.
Arsenic (as As)   Not more than 3 mg/kg.
Heavy Metals (as Pb)   Not more than 0.002%.
Lead   Not more than 5 mg/kg.
Loss on Drying   Not more than 15.0%.

TESTS

Assay Proceed as directed in the *Alginates Assay*, Appendix IIIC. Each mL of 0.25 N sodium hydroxide consumed in the assay is equivalent to 28.75 mg of potassium alginate (equiv wt 238.00).

Arsenic A *Sample Solution* prepared as directed for organic compounds meets the requirements of the *Arsenic Test*, Appendix IIIB.

Heavy Metals Prepare and test a 1-g sample as directed in *Method II* under the *Heavy Metals Test*, Appendix IIIB, but use nitric acid instead of sulfuric acid to wet the sample prior to ignition, and cautiously ignite in a platinum crucible. Any color does not exceed that produced in a control (*Solution A*) containing 20 µg of lead ion (Pb).

Lead A *Sample Solution* prepared as directed for organic compounds meets the requirements of the *Lead Limit Test*, Appendix IIIB, using 5 µg lead ion (Pb) in the control.

Loss on Drying, Appendix IIC Dry at 105° for 4 h.

Packaging and Storage Store in well-closed containers.

Potassium Benzoate

$C_7H_5KO_2$ Formula wt 160.22

INS: 212 CAS: [582-25-2]

DESCRIPTION

White, odorless or nearly odorless granules, crystalline powder, or flakes. One g dissolves in 2 mL of water, in 75 mL of alcohol, and in 50 mL of 90% alcohol.

Functional Use in Foods Preservative; antimicrobial agent.

REQUIREMENTS

Identification A 1 in 5 solution responds to the flame test for *Potassium* and gives positive tests for *Benzoate*, Appendix IIIA.

Assay Not less than 99.0% and not more than 100.5% of $C_7H_5KO_2$, calculated on the anhydrous basis.

Alkalinity (as KOH) Not more than 0.06%.

Heavy Metals (as Pb) Not more than 10 mg/kg.

Water Not more than 1.5%.

TESTS

Assay Transfer about 600 mg, accurately weighed, to a 250-mL beaker, add 100 mL of glacial acetic acid, and stir until the sample is completely dissolved. Add crystal violet TS, and titrate with 0.1 N perchloric acid in glacial acetic acid. Each mL of 0.1 N perchloric acid is equivalent to 16.02 mg of $C_7H_5KO_2$.

Alkalinity Dissolve 2 g in 20 mL of hot water, and add 2 drops of phenolphthalein TS. If a pink color is produced, not more than 0.2 mL of 0.1 N sulfuric acid is required to discharge it.

Heavy Metals Dissolve 4 g in 40 mL of water, add dropwise, with vigorous stirring, 10 mL of 2.7 N hydrochloric acid, and filter. A 25-mL portion of the filtrate meets the requirements of the *Heavy Metals Test*, Appendix IIIB, using 20 µg of lead ion (Pb) in the control (*Solution A*).

Water Determine by the *Karl Fischer Titrimetric Method*, Appendix IIB.

Packaging and Storage Store in well-closed containers.

Potassium Bicarbonate

$KHCO_3$ Formula wt 100.12

INS: 501(ii) CAS: [298-14-6]

DESCRIPTION

Colorless, transparent, monoclinic prisms or a white, granular powder. It is odorless and stable in air. Its solutions are neutral or alkaline to phenolphthalein TS. One g dissolves in 2.8 mL of water. It is almost insoluble in alcohol.

Functional Use in Foods Alkali; leavening agent.

REQUIREMENTS

Identification A 1 in 10 solution gives positive tests for *Potassium* and for *Bicarbonate*, Appendix IIIA.

Assay Not less than 99.0% and not more than 101.5% of $KHCO_3$, calculated on the dried basis.

Heavy Metals (as Pb) Not more than 10 mg/kg.

Loss on Drying Not more than 0.25%.

Normal Carbonate Passes test.

TESTS

Assay Weigh accurately about 4 g of sample, and dissolve it in 100 mL of water. Add 2 drops of methyl red TS, and titrate with 1 N hydrochloric acid. Add the acid slowly, with constant stirring, until the solution becomes faintly pink. Heat the solution to boiling, cool, and continue the titration until the pink color no longer fades after boiling. Each mL of 1 N hydrochloric acid is equivalent to 100.1 mg of $KHCO_3$.

Heavy Metals Dissolve 2 g in 5 mL of water and 8 mL of 2.7 N hydrochloric acid, boil gently for 1 min, and dilute to 25 mL with water. This solution meets the requirements of the *Heavy Metals Test*, Appendix IIIB, using 20 µg of lead ion (Pb) in the control (*Solution A*).

Loss on Drying, Appendix IIC Dry over silica gel for 4 h.

Normal Carbonate Dissolve 1 g of the sample without agitation in 20 mL of water at a temperature not above 5°, and add 2 mL of 0.1 N hydrochloric acid and 2 drops of phenolphthalein TS. The solution does not assume more than a faint pink color immediately.

Packaging and Storage Store in well-closed containers.

Potassium Bromate

$KBrO_3$ Formula wt 167.00

INS: 924a   CAS: [7758-01-2]

DESCRIPTION

White crystals or a granular powder. It is soluble in water and slightly soluble in alcohol. The pH of a 1 in 20 solution is between 5 and 9.

Functional Use in Foods Maturing agent; oxidizing agent.

REQUIREMENTS

Identification
   A. A 1 in 20 solution imparts a violet color to a nonluminous flame.
   B. To a 1 in 20 solution add sulfurous acid dropwise. A yellow color is produced that disappears upon the addition of an excess of sulfurous acid.
Assay Not less than 99.0% and not more than 101.0% of $KBrO_3$ after drying.
Chloride Not more than 0.05%.
Heavy Metals (as Pb) Not more than 5 mg/kg.
Loss on Drying Not more than 0.1%.
Sulfate Not more than 0.01%.

TESTS

Assay Dissolve about 100 mg, previously dried to constant weight over a suitable desiccant and accurately weighed, in 50 mL of water contained in a 250-mL glass-stoppered Erlenmeyer flask. Add 3 g of potassium iodide, followed by 3 mL of hydrochloric acid. Allow the mixture to stand for 5 min, add 100 mL of cold water, and titrate the liberated iodine with 0.1 N sodium thiosulfate, adding starch TS as the endpoint is approached. Perform a blank determination and make any necessary correction. Each mL of 0.1 N sodium thiosulfate consumed is equivalent to 2.783 mg of $KBrO_3$.
Chloride, Appendix IIIB Any turbidity produced by a 100-mg sample is not greater than that shown in a control containing 50 μg of chloride (Cl) ion.
Heavy Metals Dissolve 4 g in 10 mL of water, add 10 mL of hydrochloric acid, and evaporate to dryness on a steam bath. Dissolve the residue in 10 mL of hydrochloric acid, again evaporate to dryness, and then dissolve the residue in 25 mL of water.

This solution meets the requirements of the *Heavy Metals Test*, Appendix IIIB, using 20 μg of lead ion (Pb) in the control (*Solution A*).
Loss on Drying, Appendix IIC Dry over a suitable desiccant to constant weight.
Sulfate, Appendix IIIB Any turbidity produced by a 100-mg sample is not greater than that shown in a control containing 10 μg of sulfate (SO_4) ion.

Packaging and Storage Store in well-closed containers.

Potassium Carbonate

K_2CO_3 Formula wt 138.21

INS: 501(i)   CAS: [584-08-7]

DESCRIPTION

Potassium Carbonate is anhydrous or contains 1.5 molecules of water of crystallization. The anhydrous form occurs as a white, granular powder and the hydrate form as small, white, translucent crystals or granules. It is odorless, has a strongly alkaline taste, and is very deliquescent. Its solutions are alkaline. One g dissolves in 1 mL of water at 25° and in about 0.7 mL of boiling water. It is insoluble in alcohol.

Functional Use in Foods Alkali.

REQUIREMENTS

Identification A 1 in 10 solution gives positive tests for *Potassium* and for *Carbonate*, Appendix IIIA.
Assay Not less than 99.0% and not more than 100.5% of K_2CO_3 after drying.
Heavy Metals (as Pb) Not more than 0.002%.
Insoluble Substances Passes test.
Lead Not more than 10 mg/kg.
Loss on Drying K_2CO_3 (anhydrous): not more than 1%; $K_2CO_3 \cdot 1\frac{1}{2}H_2O$ (hydrated): between 10.0% and 16.5%.

TESTS

Assay Weigh accurately in a stoppered weighing bottle about 1 g of the dried sample obtained as directed under *Loss on Drying*, and dissolve it in 50 mL of water. Add 2 drops of methyl red TS, and titrate with 1 N hydrochloric acid. Add the acid slowly, with constant stirring, until the solution becomes faintly pink. Head the solution to boiling, cool, and continue the titration until the faint pink color no longer fades after boiling. Each mL of 1 N hydrochloric acid is equivalent to 69.11 mg of K_2CO_3.
Heavy Metals To 1 g add 2 mL of water and 6 mL of 2.7 N hydrochloric acid, boil for 1 min, and dilute to 25 mL with water. This solution meets the requirements of the *Heavy Metals*

Test, Appendix IIIB, using 20 µg of lead ion (Pb) in the control (*Solution A*).

Insoluble Substances No residue is left upon dissolving 1 g in 20 mL of water.

Lead A solution of 1 g, cautiously dissolved in 2.7 N hydrochloric acid (about 5 mL), meets the requirements of the *Lead Limit Test*, Appendix IIIB, using 10 µg of lead ion (Pb) in the control.

Loss on Drying, Appendix IIC Dry about 3 g, accurately weighed, at 180° for 4 h.

Packaging and Storage Store in tight containers.

Potassium Carbonate Solution

INS: 501(i)

DESCRIPTION

A clear or slightly turbid, colorless, alkaline solution that absorbs carbon dioxide when exposed to the air, forming potassium bicarbonate. It is available as solutions with concentrations, weight in weight, of about 50.0%.

Functional Use in Foods Alkali.

REQUIREMENTS

Identification It gives positive tests for *Potassium* and for *Carbonate*, Appendix IIIA.
Assay Not less than 97.0% and not more than 103.0%, by weight, of the labeled amount of K_2CO_3.
Heavy Metals (as Pb) Not more than 0.002%, calculated on the basis of K_2CO_3 determined in the *Assay*.
Lead Not more than 10 mg/kg, calculated on the basis of K_2CO_3 determined in the *Assay*.

TESTS

Assay Based on the stated or labeled percentage of K_2CO_3, weigh accurately a volume of the solution equivalent to about 1 g of Potassium Carbonate, and add it to 50.0 mL of 1 N sulfuric acid. Add methyl orange TS, and titrate the excess acid with 1 N sodium hydroxide. Each mL of 1 N sulfuric acid is equivalent to 69.11 mg of K_2CO_3.
Heavy Metals Dilute the equivalent of 1 g of K_2CO_3, calculated on the basis of the *Assay*, with a mixture of 5 mL of water and 6 mL of 2.7 N hydrochloric acid, and heat to boiling. Cool, and dilute to 25 mL with water. This solution meets the requirements of the *Heavy Metals Test*, Appendix IIIB, using 20 µg of lead ion (Pb) in the control (*Solution A*).
Lead Dilute the equivalent of 1 g of K_2CO_3, calculated on the basis of the *Assay*, with a mixture of 6 mL of 2.7 N hydrochloric acid and 5 mL of water. This solution meets the requirements of the *Lead Limit Test*, Appendix IIIB, using 10 µg of lead ion (Pb) in the control.

Packaging and Storage Store in tight containers.

Potassium Chloride

KCl Formula wt 74.55
INS: 508 CAS: [7447-40-7]

DESCRIPTION

Colorless, elongated, prismatic, or cubical crystals, or a white, granular powder. It is odorless, has a saline taste, and is stable in air. Its solutions are neutral to litmus. It may contain up to 1.0% (total) of suitable food-grade anti-caking, free-flowing, or conditioning agents such as calcium stearate or silicon dioxide, either singly or in combination. One g dissolves in 2.8 mL of water at 25°, and in about 2 mL of boiling water. Potassium Chloride containing anti-caking, free-flowing, or conditioning agents may produce cloudy solutions or dissolve incompletely. It is insoluble in alcohol.

Functional Use in Foods Nutrient; dietary supplement; gelling agent; salt substitute; yeast food.

REQUIREMENTS

Labeling Indicate the name and quantity of any added substance(s) if material contains such substances.
Identification A 1 in 20 solution gives positive tests for *Potassium* and for *Chloride*, Appendix IIIA.
Assay Not less than 99.0% of KCl after drying; and not less than 98.0% of KCl after drying when article contains added substance(s).
Acidity or Alkalinity Article containing no added substance(s) passes test.
Heavy Metals (as Pb) Not more than 5 mg/kg.
Iodide and/or Bromide Passes test.
Loss on Drying Not more than 1.0%.
Sodium Passes test.

TESTS

Assay Dry about 250 mg at 105° for 2 h, weigh accurately, and dissolve in 150 mL of water. Add 1 mL of nitric acid, and immediately titrate with 0.1 N silver nitrate, determining the endpoint potentiometrically, using silver-calomel electrodes and a salt bridge containing 4% agar in a saturated potassium nitrate solution. Perform a blank determination, and make any necessary correction. Each mL of 0.1 N silver nitrate is equivalent to 7.455 mg of KCl.
Acidity or Alkalinity To a solution of 5 g in 50 mL of recently boiled and cooled water add 3 drops of phenolphthalein TS. No pink color is produced. Then add 0.3 mL of 0.02 N sodium hydroxide. A pink color is produced.

Heavy Metals Prepare and test a 4-g sample as directed under the *Heavy Metals Test*, Appendix IIIB, using 20 μg of lead ion (Pb) in the control (*Solution A*).

Iodide and/or Bromide Dissolve 2 g of the sample in 6 mL of water, add 1 mL of chloroform, and then add, dropwise and with constant agitation, 5 mL of a mixture of equal parts of chlorine TS and water. The chloroform is free from even a transient violet or permanent orange color.

Loss on Drying, Appendix IIC Dry at 105° for 2 h.

Sodium A 1 in 20 solution, tested on a platinum wire, does not impart a pronounced yellow color to a nonluminous flame.

Use the following test for material containing added substances:

Preparation of Standard Curve Dissolve 1.2711 g, accurately weighed, of reagent grade sodium chloride in water in a 500-mL volumetric flask, and dilute to volume to obtain a solution containing 1000 μg of sodium per mL. Prepare *Standard Dilutions* from this solution to cover the range 0 to 10 μg/mL of sodium at intervals of 2 μg/mL using 500-mL volumetric flasks and adding 0.5 g of reagent-grade potassium chloride to each volumetric flask before diluting to volume. Atomize portions of the *Standard Dilutions* as described under *Procedure* until readings for the series are reproducible, adjusting the instrument so that the solution containing 10 μg/mL gives a full-scale reading. Prepare a standard curve by plotting the absorbance against the concentration.

Procedure Weigh accurately a 2.50 g sample and transfer to a 250-mL volumetric flask, dilute to volume with water, and mix. Pipet 25 mL of this solution into a 250-mL volumetric flask, dilute to volume with water and mix. This is the *Sample Solution*. Determine the absorbances of the *Standard Dilutions* and the *Sample Solution* at the sodium emission line of 589.6 nm with a flame atomic absorption spectrophotometer equipped with a sodium hollow-cathode lamp and an air–acetylene flame, using water as the blank. Determine concentration C, in μg/mL, of sodium from the *Standard Curve* and calculate the percent of sodium in the sample taken by the formula

$$C/10.$$

Not more than 0.5% is present.

Packaging and Storage Store in well-closed containers.

Potassium Citrate

KOOCCH$_2$C(OH)(COOK)CH$_2$COOK.H$_2$O

C$_6$H$_5$K$_3$O$_7$.H$_2$O Formula wt 324.41

INS: 332(ii) CAS: [6100-05-6]

DESCRIPTION

Transparent crystals or a white, granular powder. It is odorless; has a cooling, saline taste; and is deliquescent when exposed to moist air. One g dissolves in about 0.5 mL of water. It is almost insoluble in alcohol.

Functional Use in Foods Miscellaneous and general purpose; buffer; sequestrant.

REQUIREMENTS

Identification A 1 in 20 solution gives positive tests for *Potassium* and for *Citrate*, Appendix IIIA.

Assay Not less than 99.0% and not more than 100.5% of C$_6$H$_5$K$_3$O$_7$ after drying.

Alkalinity Passes test.

Heavy Metals (as Pb) Not more than 10 mg/kg.

Loss on Drying Between 3.0% and 6.0%.

TESTS

Assay Dissolve about 250 mg, previously dried at 180° to constant weight and accurately weighed, in 40 mL of glacial acetic acid, warming slightly to effect solution. Cool the solution to room temperature, add 2 drops of crystal violet TS, and titrate with 0.1 N perchloric acid. Perform a blank determination (see *General Provisions*), and make any necessary correction. Each mL of 0.1 N perchloric acid is equivalent to 10.213 mg of C$_6$H$_5$K$_3$O$_7$.

Alkalinity A 1 in 20 solution is alkaline to litmus, but after the addition of 0.2 mL of 0.1 N sulfuric acid to 10 mL of this solution, no pink color is produced by the addition of 1 drop of phenolphthalein TS.

Heavy Metals A solution of 2 g in 25 mL meets the requirements of the *Heavy Metals Test*, Appendix IIIB, using 20 μg of lead ion (Pb) in the control (*Solution A*).

Loss on Drying, Appendix IIC Dry at 180° to constant weight.

Packaging and Storage Store in tight containers.

Potassium Gibberellate

C$_{19}$H$_{21}$KO$_6$ Formula wt 384.47

DESCRIPTION

A white to slightly off-white, odorless or practically odorless, crystalline powder. It is soluble in water, in alcohol, and in acetone. The pH of a 1 in 20 solution is about 6. It is deliquescent.

Functional Use in Foods Enzyme activator.

REQUIREMENTS

Identification

A. Dissolve a few mg of the sample in 2 mL of sulfuric acid. A reddish solution having a green fluorescence is formed.

B. A 1 in 10 solution of the sample gives positive tests for *Potassium*, Appendix IIIA.

Assay Not less than 80.0% and not more than 87.0% of $C_{19}H_{21}KO_6$, equivalent to between 72.1% and 78.4% of $C_{19}H_{22}O_6$ (gibberellic acid).
Heavy Metals (as Pb) Not more than 0.002%.
Lead Not more than 10 mg/kg.
Loss on Drying Between 5.0% and 13.0%.
Residue on Ignition Between 19.0% and 23.0%.
Specific Rotation $[\alpha]_D^{20°}$: Between +43.0° and +60.0°.

TESTS

Assay

Standard Preparation Transfer an accurately weighed quantity of FCC Gibberellic Acid Reference Standard, equivalent to about 25 mg of pure gibberellic acid (corrected for phase purity and volatiles content), into a 50-mL volumetric flask, dissolve in methanol, dilute to volume with methanol, and mix. Transfer 10.0 mL of this solution into a second 50-mL volumetric flask, dilute to volume with methanol, and mix.

Assay Preparation Transfer about 65 mg of the sample, accurately weighed, into a 50-mL volumetric flask, dissolve in methanol, dilute to volume with methanol, and mix. Transfer 10.0 mL of this solution into a 100-mL volumetric flask, dilute to volume with methanol, and mix.

Procedure Transfer 5.0 mL of the *Assay Preparation* into a 25- × 200-mm glass-stoppered tube, and transfer 4.0-mL and 5.0-mL portions of the *Standard Preparation* into separate, similar tubes. Place the tubes in a boiling water bath, evaporate to dryness, and then dry in an oven at 90° for 10 min. Remove the tubes from the oven, stopper, and allow to cool to room temperature. Dissolve the residue in each tube in 10.0 mL of dilute sulfuric acid (8 in 10), heat in a boiling water bath for 10 min, and then cool in a 10° water bath for 5 min. Determine the absorbance of the solutions in 1-cm cells at 535 nm with a suitable spectrophotometer, using the dilute sulfuric acid as the blank. Record the absorbance of the solution from the *Assay Preparation* as A_U. Note the absorbance of the two solutions prepared from the 4.0-mL and 5.0-mL aliquots of the *Standard Preparation*, and record the absorbance of the final solution giving the value nearest to that of the sample as A_S; record as V the volume of the aliquot used in preparing this solution. Calculate the quantity, in mg, of $C_{19}H_{21}KO_6$ in the sample taken by the formula

$$500 \times (C/0.8983) \times (V/5) \times (A_U/A_S),$$

in which C is the exact concentration, in mg/mL, of the *Standard Preparation*, and 0.8983 is the ratio of the molecular weight of Potassium Gibberellate to that of gibberellic acid.

Heavy Metals Prepare and test a 1-g sample as directed in *Method II* under the *Heavy Metals Test*, Appendix IIIB, using 20 μg of lead ion (Pb) in the control (*Solution A*).

Lead A *Sample Solution* prepared as directed for organic compounds meets the requirements of the *Lead Limit Test*, Appendix IIIB, using 10 μg of lead ion (Pb) in the control.
Loss on Drying, Appendix IIC Dry at 100° in vacuum for 4 h.
Residue on Ignition Ignite a 1-g sample as directed in the general method, Appendix IIC.
Specific Rotation, Appendix IIB Determine in a solution containing 50 mg in each mL.

Packaging and Storage Store in tight containers protected from light.

Potassium Gluconate

D-Gluconic Acid, Monopotassium Salt; Monopotassium D-Gluconate

$$CH_2OH(CHOH)_4COOK$$

$C_6H_{11}KO_7$ Formula wt, anhydrous 234.25
Formula wt, monohydrate 252.26

INS: 577 CAS: anhydrous [299-27-4]
CAS: monohydrate [35398-15-3]

DESCRIPTION

White or yellowish white, crystalline powder or granules. It is anhydrous or the monohydrate. It is odorless and has a slightly bitter taste. It is freely soluble in water and in glycerin, slightly soluble in alcohol, and insoluble in ether.

Functional Use in Foods Nutrient; dietary supplement; sequestrant.

REQUIREMENTS

Labeling Indicate whether it is anhydrous or the monohydrate.
Identification

A. It responds to the flame test for *Potassium*, Appendix IIIA.

B. It meets the requirements of *Identification Test B* in the monograph for *Calcium Gluconate*.

C. The infrared absorption spectrum of a mineral oil dispersion of the sample, previously dried, exhibits maxima only at the same wavelengths as that of a similar preparation of USP Potassium Gluconate Reference Standard.

Assay Not less than 98.0% of $C_6H_{11}KO_7$, calculated on the dried basis.
Heavy Metals (as Pb) Not more than 0.002%.
Lead Not more than 10 mg/kg.
Loss on Drying Not more than 3.0% for the anhydrous; between 6.0% and 7.5% for the monohydrate.
Reducing Substances Not more than 1.0%.

TESTS

Assay Transfer about 175 mg, accurately weighed, into a clean, dry 200-mL Erlenmeyer flask, add 75 mL of glacial acetic acid, and dissolve by heating on a hot plate. Cool, add quinaldine red TS, and titrate with 0.1 N perchloric acid in glacial acetic acid, using a 10-mL microburet, to a colorless endpoint. Each mL of 0.1 N perchloric acid is equivalent to 23.42 mg of $C_6H_{11}KO_7$.

Heavy Metals A solution of 1 g in 25 mL of water meets the requirements of the *Heavy Metals Test*, Appendix IIIB, using 20 µg of lead ion (Pb) in the control (*Solution A*).

Lead A solution of 1 g in 25 mL of water meets the requirements of the *Lead Limit Test*, Appendix IIIB, using 10 µg of lead ion (Pb) in the control.

Loss on Drying *Appendix IIC* Dry in vacuum at 105° for 4 h.

Reducing Substances Determine as directed under *Reducing Substances* in the monograph for *Copper Gluconate*.

Packaging and Storage Store in well-closed containers.

Potassium Glycerophosphate

$C_3H_7K_2O_6P \cdot 3H_2O$ Formula wt 302.30

CAS: [1319-70-6]

DESCRIPTION

Potassium Glycerophosphate is a pale yellow, syrupy liquid containing three molecules of water of hydration, or it is prepared as a colorless to pale yellow, syrupy solution having a concentration of 50% to 75%. It is very soluble in water, and its solutions are alkaline to litmus paper.

Functional Use in Foods Nutrient; dietary supplement.

REQUIREMENTS

Identification A 1 in 10 solution gives positive tests for *Potassium*, Appendix IIIA.

Assay $C_3H_7K_2O_6P \cdot 3H_2O$: not less than 80.0% of $C_3H_7K_2O_6P$; potassium glycerophosphate solutions: not less than 95.0% and not more than 105.0% of the labeled concentration of $C_3H_7K_2O_6P$.

Heavy Metals (as Pb) Not more than 0.002%.

Lead Not more than 5 mg/kg.

TESTS

Assay Weigh accurately a portion of the sample equivalent to about 4 g of $C_3H_7K_2O_6P$, dissolve it in 30 mL of water, add methyl orange TS, and titrate with 0.5 N hydrochloric acid. Each mL of 0.5 N hydrochloric acid is equivalent to 124.13 mg of $C_3H_7K_2O_6P$.

Heavy Metals A solution of 1 g in 25 mL of water meets the requirements of the *Heavy Metals Test*, Appendix IIIB, using 20 µg of lead ion (Pb) in the control (*Solution A*).

Lead A *Sample Solution* prepared as directed for organic compounds meets the requirements of the *Lead Limit Test*, Appendix IIIB, using 5 µg of lead ion (Pb) in the control.

Packaging and Storage Store in tight containers.

Potassium Hydroxide

Caustic Potash

KOH   Formula wt 56.11

INS: 525   CAS: [1310-58-3]

DESCRIPTION

White or nearly white pellets, flakes, sticks, fused masses, or other forms. Upon exposure to air, it readily absorbs carbon dioxide and moisture, and deliquesces. One g dissolves in 1 mL of water, in about 3 mL of alcohol, and in about 2.5 mL of glycerin. It is very soluble in boiling alcohol.

Functional Use in Foods Alkali.

REQUIREMENTS

Identification A 1 in 25 solution gives positive tests for *Potassium*, Appendix IIIA.

Assay Not less than 85.0% and not more than 100.5% of total alkali, calculated as KOH.

Carbonate (as K_2CO_3) Not more than 3.5%.

Heavy Metals (as Pb) Not more than 0.002%.

Insoluble Substances Passes test.

Lead Not more than 10 mg/kg.

Mercury Not more than 0.1 mg/kg.

TESTS

Assay Dissolve about 1.5 g, accurately weighed, in 40 mL of recently boiled and cooled water, cool to 15°, add phenolphthalein TS, and titrate with 1 N sulfuric acid. At the discharge of the pink color, record the volume of acid required, then add methyl orange TS, and continue the titration until a persistent pink color is produced. Record the total volume of acid required for the titration. Each mL of 1 N sulfuric acid is equivalent to 56.11 mg of total alkali, calculated as KOH.

Carbonate Each mL of 1 N sulfuric acid required between the phenolphthalein and methyl orange endpoints in the *Assay* is equivalent to 138.2 mg of K_2CO_3.

Heavy Metals Dissolve 1 g in a mixture of 5 mL of water and 8 mL of 2.7 N hydrochloric acid. Heat to boiling, cool, dilute to 25 mL with water, and filter. This solution meets the requirements of the *Heavy Metals Test*, Appendix IIIB, using 20 μg of lead ion (Pb) in the control (*Solution A*).

Insoluble Substances   A 1 in 20 solution is complete, clear, and colorless.

Lead   Dissolve 1 g in a mixture of 5 mL of water and 11 mL of 2.7 N hydrochloric acid, and cool. This solution meets the requirements of the *Lead Limit Test*, Appendix IIIB, using 10 μg of lead ion (Pb) in the control.

Mercury   Determine as directed under *Mercury Limit Test*, Appendix IIIB, preparing the *Standard Preparation* and the *Sample Preparation* as follows:

Standard Preparation   Prepare the stock solution and the dilutions, as directed in the test, to obtain a solution containing 1 μg of mercury per mL. Transfer 1.0 mL of the final solution (1 μg of mercury) to a 50-mL beaker, and add 20 mL of water, 1 mL of dilute sulfuric acid solution (1 in 5), and 1 mL of potassium permanganate solution (1 in 25). Cover the beaker with a watch glass, boil for a few s, and cool.

Sample Preparation   Transfer 10.0 g of the sample into a 100-mL beaker, dissolve in 15 mL of water, add 2 drops of phenolphthalein TS, and slowly neutralize, with constant stirring, with dilute hydrochloric acid solution (1 in 2). Add 1 mL of dilute sulfuric acid solution (1 in 5) and 1 mL of potassium permanganate solution (1 in 25), cover the beaker with a watch glass, boil for a few s, and cool.

Packaging and Storage   Store in tight containers.

Potassium Hydroxide Solution

INS: 525

DESCRIPTION

A clear or slightly turbid, colorless or slightly colored, strongly caustic, hygroscopic solution that absorbs carbon dioxide when exposed to the air, forming potassium carbonate. It is available as solutions with nominal concentrations, weight in weight, of 50%.

Functional Use in Foods   Alkali.

REQUIREMENTS

Labeling   Indicate the percent of Potassium Hydroxide, KOH.
Identification   It gives positive tests for *Potassium*, Appendix IIIA.
Assay   Not less than 97.0% and not more than 103.0%, by weight, of the labeled amount of KOH calculated as total alkali.
Carbonate (as K_2CO_3)   Not more than 3.5% of the KOH determined in the *Assay*.
Heavy Metals (as Pb)   Not more than 0.003%, calculated on the basis of KOH determined in the *Assay*.
Lead   Not more than 10 mg/kg, calculated on the basis of KOH determined in the *Assay*.
Mercury   Not more than 1 mg/kg, calculated on the basis of KOH determined in the *Assay*.

TESTS

Assay   Based on the stated or labeled percentage of KOH, weigh accurately a volume of the solution equivalent to about 1.5 g of Potassium Hydroxide, and dilute it to 40 mL with recently boiled and cooled water. Continue as directed under *Assay* in the monograph for *Potassium Hydroxide*, beginning with "...cool to 15°...."

Carbonate   Each mL of 1 N sulfuric acid required between the phenolphthalein and methyl orange endpoints in the *Assay* is equivalent to 138.2 mg of K_2CO_3.

Heavy Metals   Dilute the equivalent of 670 mg of KOH, calculated on the basis of the *Assay*, with a mixture of 5 mL of water and 5 mL of 2.7 N hydrochloric acid, and heat to boiling. Cool, dilute to 25 mL with water, and filter. This solution meets the requirements of the *Heavy Metals Test*, Appendix IIIB, using 20 μg of lead ion (Pb) in the control (*Solution A*).

Lead   Dilute the equivalent of 1 g of KOH, calculated on the basis of the *Assay*, with a mixture of 5 mL of water and 11 mL of 2.7 N hydrochloric acid. This solution meets the requirements of the *Lead Limit Test*, Appendix IIIB, using 10 μg of lead ion (Pb) in the control.

Mercury   Determine as directed under *Mercury Limit Test*, Appendix IIIB, using the following as the *Sample Preparation*: Transfer an accurately weighed amount of the sample, equivalent to 2.0 g of KOH, into a 50-mL beaker, add 10 mL of water and 2 drops of phenolphthalein TS, and slowly neutralize, with constant stirring, with dilute hydrochloric acid solution (1 in 2). Add 1 mL of dilute sulfuric acid solution (1 in 5) and 1 mL of potassium permanganate solution (1 in 25), cover the beaker with a watch glass, boil for a few s, and cool.

Packaging and Storage   Store in tight containers.

Potassium Iodate

KIO_3 Formula wt 214.00
INS: 917   CAS: [7758-05-6]

DESCRIPTION

A white, odorless, crystalline powder. One g dissolves in about 15 mL of water. It is insoluble in alcohol. The pH of a 1 in 20 solution is between 5 and 8.

Functional Use in Foods   Maturing agent; dough conditioner.

REQUIREMENTS

Identification   To 1 mL of a 1 in 10 solution of the sample add 1 drop of starch TS and a few drops of 20% hypophosphorous acid. A transient blue color appears.

Assay Not less than 99.0% and not more than 101.0% of KIO_3 after drying.
Chlorate Passes test (limit about 0.01%).
Heavy Metals (as Pb) Not more than 10 mg/kg.
Iodide Passes test (limit about 0.002%).
Loss on Drying Not more than 0.5%.

TESTS

Assay Weigh accurately about 1.2 g, previously dried at 105° for 3 h, dissolve it in about 50 mL of water in a 100-mL volumetric flask, dilute to volume with water, and mix. Transfer 10.0 mL into a 250-mL glass-stoppered flask, add 40 mL of water, 3 g of potassium iodide, and 10 mL of dilute hydrochloric acid (3 in 10), and stopper the flask. Allow to stand for 5 min, add 100 mL of cold water, and titrate the liberated iodine with 0.1 N sodium thiosulfate, adding starch TS near the endpoint. Perform a blank determination (see *General Provisions*), and make any necessary correction. Each mL of 0.1 N sodium thiosulfate is equivalent to 3.567 mg of KIO_3.
Chlorate To 2 g in a beaker add 2 mL of sulfuric acid. The sample remains white, and no odor or gas is evolved.
Heavy Metals, Appendix IIIB Mix 2 g of the sample with 10 mL of hydrochloric acid (*Solution B*). Prepare a standard (*Solution A*) by adding a few mg of potassium chloride to 10 mL of hydrochloric acid. Cautiously evaporate both solutions to dryness, then repeat the acid treatment on both the sample and standard residues twice, using 5-mL portions of hydrochloric acid and evaporating to dryness each time. Dissolve the residues in 10 mL of water, heat on a steam bath, and discharge any yellow color remaining in the sample solution with hydrazine sulfate. Cool each solution and neutralize to phenolphthalein with 0.1 N sodium hydroxide. Transfer the solutions to 50-mL Nessler tubes, add 2.0 mL of *Standard Lead Solution* (20 μg Pb) to the standard, and dilute both solutions with water to 25 mL. To each tube add 6 mL of 1 N acetic acid and 10 mL of freshly prepared hydrogen sulfide TS, allow to stand for 5 min, and view downward over a white surface. The color of *Solution B* is no darker than that of *Solution A*.
Iodide Dissolve 1 g in 10 mL of water and add 1 mL of 2 N sulfuric acid and 1 drop of starch TS. No blue color is formed.
Loss on Drying, Appendix IIC Dry at 105° for 3 h.

Packaging and Storage Store in well-closed containers.

Potassium Iodide

KI Formula wt 166.00

CAS: [7681-11-0]

DESCRIPTION

Hexahedral crystals, either transparent and colorless or somewhat opaque and white, or a white, granular powder. It is stable in dry air but slightly hygroscopic in moist air. One g is soluble in 0.7 mL of water at 25°, in 0.5 mL of boiling water, in 2 mL of glycerin, and in 22 mL of alcohol. The pH of a 1 in 20 solution is between 6 and 10.

Functional Use in Foods Nutrient; dietary supplement.

REQUIREMENTS

Identification A 1 in 10 solution responds to the tests for *Potassium* and for *Iodide*, Appendix IIIA.
Assay Not less than 99.0% and not more than 101.5% of KI after drying.
Heavy Metals (as Pb) Not more than 10 mg/kg.
Iodate Not more than 4 mg/kg.
Loss on Drying Not more than 1%.
Nitrate, Nitrite, and Ammonia Passes test.
Thiosulfate and Barium Passes test.

TESTS

Assay Dissolve about 500 mg, previously dried at 105° for 4 h and accurately weighed, in about 10 mL of water, and add 35 mL of hydrochloric acid and 5 mL of chloroform. Titrate with 0.05 M potassium iodate until the purple color of iodine disappears from the chloroform. Add the last portions of the iodate solution dropwise, agitating vigorously and continuously. After the chloroform has been decolorized, allow the mixture to stand for 5 min. If the chloroform develops a purple color, titrate further with the iodate solution. Each mL of 0.05 M potassium iodate is equivalent to 16.60 mg of KI.
Heavy Metals A solution of 2 g in 25 mL of water meets the requirements of the *Heavy Metals Test*, Appendix IIIB, using 20 μg of lead ion (Pb) in the control (*Solution A*).
Iodate Dissolve 1.1 g of the sample in sufficient ammonia- and carbon dioxide-free water to make 10 mL of solution, and transfer to a color-comparison tube. Add 1 mL of starch TS and 0.25 mL of 1 N sulfuric acid, mix, and compare the color with that of a control containing, in each 10 mL, 100 mg of Potassium Iodide, 1 mL of standard iodate solution (prepared by diluting 1 mL of a 1 in 2500 solution of potassium iodate to 100 mL with water), 1 mL of starch TS, and 0.25 mL of 1 N sulfuric acid. Any color produced in the solution of the sample does not exceed that in the control.
Loss on Drying, Appendix IIC Dry at 105° for 4 h.
Nitrate, Nitrite, and Ammonia Dissolve 1 g of the sample in 5 mL of water in a 40-mL test tube, and add 5 mL of 1 N sodium hydroxide and about 200 mg of aluminum wire. Insert a plug of cotton in the upper portion of the tube, and place a piece of moistened red litmus paper over the mouth of the tube. Heat in a steam bath for about 15 min. No blue coloration of the paper is discernible.
Thiosulfate and Barium Dissolve 500 mg of the sample in 10 mL of ammonia- and carbon dioxide-free water, and add 2 drops of diluted sulfuric acid. No turbidity develops within 1 min.

Packaging and Storage Store in well-closed containers.

Potassium Lactate Solution

2-Hydroxy-Propanoic Acid, Monopotassium Salt

$C_3H_5KO_3$ Formula wt 128.17

DESCRIPTION

Clear, colorless, or practically colorless, viscous liquid, odorless or having a slight, not unpleasant, odor. It is miscible with water. It is available as solutions with concentrations ranging from about 50% to 70%.

Functional Use in Foods Emulsifier; flavor enhancer; flavoring agent or adjuvant; humectant; pH control agent.

REQUIREMENTS

Labeling Indicate its content, by weight, of Potassium Lactate, $C_3H_5KO_3$.
Identification It gives positive tests for *Potassium* and for *Lactate*, Appendix IIIA.
Assay Not less than 50.0%, by weight, and not less than 95.0% and not more than 105.0%, by weight, of the labeled amount of Potassium Lactate, $C_3H_5KO_3$.
Chloride Not more than 0.05%.
Citrate, Oxalate, Phosphate, or Tartrate Passes test.
Cyanide Not more than 0.5 mg/kg.
Heavy Metals (as Pb) Not more than 10 mg/kg.
Lead Not more than 5 mg/kg.
Methanol and Methyl Esters Not more than 0.025%.
pH Between 5.0 and 9.0.
Sodium Not more than 0.1%.
Sugars Passes test.
Sulfate Not more than 0.005%.

TESTS

Assay Weigh accurately into a suitable flask a volume of Potassium Lactate Solution equivalent to about 500 mg of potassium lactate, add 60 mL of a 1 in 5 mixture of acetic anhydride in glacial acetic acid, mix, and allow to stand for 20 min. Titrate with 0.1 N perchloric acid in glacial acetic acid, determining the endpoint potentiometrically. Perform a blank determination, and make any necessary correction. Each mL of 0.1 N perchloric acid is equivalent to 12.82 mg of $C_3H_5KO_3$.
Chloride, Appendix IIIB Any turbidity produced by a quantity of the Solution containing the equivalent of 40 mg of potassium lactate does not exceed that shown in a control containing 20 μg of chloride ion (Cl).
Citrate, Oxalate, Phosphate, or Tartrate Dilute 5 mL with recently boiled and cooled water to 50 mL. To 4 mL of this solution, add 6 N ammonium hydroxide or 3 N hydrochloric acid, if necessary, to bring the pH to between 7.3 and 7.7. Add 1 mL of calcium chloride TS, and heat in a boiling water bath for 5 min: The solution remains clear.
Cyanide (**Caution**: Because of the extremely poisonous nature of potassium cyanide, conduct this test in a fume hood, and exercise great care to prevent skin contact and inhaling particles or vapors of solutions of the material. Under no conditions pipet solutions by mouth.)

p-Phenylenediamine–Pyridine Mixed Reagent Dissolve 200 mg of *p*-phenylenediamine hydrochloride in 100 mL of water, warming to aid dissolution. Cool, allow the solids to settle, and use the supernatant liquid to make the mixed reagent. Dissolve 128 mL of pyridine in 365 mL of water, add 10 mL of hydrochloric acid, and mix. To prepare the mixed reagent, mix 30 mL of the *p*-phenylenediamine solution with all of the pyridine solution, and allow to stand for 24 h before using. The mixed reagent is stable for about 3 weeks when stored in an amber bottle.

Sample Solution Transfer an accurately weighed quantity of the Solution, equivalent to 20.0 g of potassium lactate, into a 100-mL volumetric flask, dilute to volume with water, and mix.

Cyanide Standard Solution Dissolve 250 mg of potassium cyanide, accurately weighed, in 10 mL of 0.1 N sodium hydroxide in a 100-mL volumetric flask, dilute to volume with 0.1 N sodium hydroxide, and mix. Transfer a 10-mL aliquot into a 1000-mL volumetric flask, dilute to volume with 0.1 N sodium hydroxide, and mix. Each mL of this solution contains 10 μg of cyanide.

Procedure Pipet a 10-mL aliquot of the *Sample Solution* into a 50-mL beaker. Into a second 50-mL beaker, pipet 0.10 mL of the *Cyanide Standard Solution*, and add 10 mL of water. Place the beakers in an ice bath, and adjust the pH to between 9 and 10 with 20% sodium hydroxide, stirring slowly and adding the reagent slowly to avoid overheating. Allow the solutions to stand for 3 min, and then slowly add 10% phosphoric acid to a pH between 5 and 6, measured with a pH meter. Transfer the solutions into 100-mL separators containing 25 mL of cold water, and rinse the beakers and pH meter electrodes with a few mL of cold water, collecting the washings in the respective separator. Add 2 mL of bromine TS, stopper, and mix. Add 2 mL of 2% sodium arsenite solution, stopper, and mix. Add 10 mL of *n*-butanol to the clear solutions, stopper, and mix. Finally, add 5 mL of *p-Phenylenediamine–Pyridine Mixed Reagent*, mix, and allow to stand for 15 min. Remove and discard the aqueous phases, and filter the alcohol phases into 1-cm cells. The absorbance of the solution from the *Sample Solution*, determined at 480 nm with a suitable spectrophotometer, is not greater than that from the *Cyanide Standard Solution*.

Heavy Metals Dilute a quantity of the Solution, equivalent to 2.0 g of potassium lactate, to 25 mL with water. This solution meets the requirements of the *Heavy Metals Test*, Appendix IIIB, using 20 μg of lead ion (Pb) in the control (*Solution A*).
Lead Dilute a quantity of the Solution, equivalent to 2.0 g of potassium lactate, to 25 mL with water. This solution meets the requirements of the *Lead Limit Test*, Appendix IIIB, using 10 μg of lead ion (Pb) in the control.
Methanol and Methyl Esters

Potassium Permanganate and Phosphoric Acid Solution Dissolve 3 g of potassium permanganate in a mixture of 15 mL of phosphoric acid and 70 mL of water. Dilute with water to 100 mL.

Oxalic Acid and Sulfuric Acid Solution Cautiously add 50 mL of sulfuric acid to 50 mL of water, mix, cool, add 5 g of oxalic acid, and mix to dissolve.

Standard Preparation Prepare a solution containing 10.0 mg of methanol in a 100-mL volumetric flask, dilute to volume with dilute alcohol (1 in 10), and mix.

Test Preparation Place 40.0 g of the Solution in a glass-stoppered, round-bottom flask, add 10 mL of water, and cautiously add 30 mL of 5 N potassium hydroxide. Connect a condenser to the flask, and steam-distill, collecting the distillate in a suitable 100-mL graduated vessel containing 10 mL of alcohol. Continue the distillation until the volume in the receiver reaches approximately 95 mL, and dilute the distillate with water to 100.0 mL.

Procedure Transfer 10.0 mL each of the *Standard Preparation* and the *Test Preparation* to separate 25-mL volumetric flasks. To each, add 5.0 mL of *Potassium Permanganate and Phosphoric Acid Solution*, and mix. After 15 min, add 2.0 mL of *Oxalic Acid and Sulfuric Acid Solution* to each, stir with a glass rod until the solution is colorless, add 5.0 mL of fuchsin-sulfurous acid TS (prepared as directed in *Solutions and Indicators*), dilute with water to volume, and mix. After 2 h, concomitantly determine the absorbances of both solutions in 1-cm cells at the wavelength of maximum absorbance at about 575 nm, with a suitable spectrophotometer and using water as the blank: The absorbance of the solution from the *Test Preparation* is not greater than that from the *Standard Preparation*.

pH Determine the pH of the Solution by the *Potentiometric Method*, Appendix IIB.

Sodium

Potassium Chloride Solution Dissolve 100 g of potassium chloride in water and dilute to 1000 mL.

Standard Solutions Transfer 127.1 mg of sodium chloride, previously dried at 105° for 2 h and accurately weighed, to a 500-mL volumetric flask, dilute with water to volume, and mix. Transfer 10.0 mL of this solution to a 100-mL volumetric flask, dilute with water to volume, and mix to obtain a *Stock Solution* containing 10 µg of sodium per mL. Into separate 100-mL volumetric flasks, pipet 1-, 2-, 5-, and 10-mL aliquots of the *Stock Solution*; add 1.0 mL of *Potassium Chloride Solution* followed by 1.0 mL of nitric acid; dilute with water to volume; and mix to obtain *Standard Solutions* containing 0.1, 0.2, 0.5, and 1.0 µg of sodium per mL, respectively.

Test Solution Transfer an accurately weighed quantity of the Solution equivalent to about 4 g of potassium lactate to a 50-mL volumetric flask, dilute to volume with water, and mix. Pipet 1 mL of this solution into a 100-mL volumetric flask, add 1.0 mL of *Potassium Chloride Solution* followed by 1.0 mL of nitric acid, dilute with water to volume, and mix.

Blank Solution Transfer 1.0 mL of *Potassium Chloride Solution* to a 100-mL volumetric flask, add 1.0 mL of nitric acid, dilute with water to volume, and mix.

Procedure Concomitantly determine the absorbances of the *Standard Solutions* and the *Test Solution* at the sodium emission line of 589 nm with a suitable atomic absorption spectrophotometer equipped with a sodium hollow-cathode lamp and an oxidizing air–acetylene flame, using the *Blank Solution* to zero the instrument. Plot the absorbances of the *Standard Solutions* versus concentration, in µg/mL, of sodium, and draw the straight line that best fits the plotted points. From the graph so obtained, determine the concentration C, in µg/mL, of sodium in the *Test Solution*. Calculate the percentage of sodium in the portion of potassium lactate taken by the formula

$$CD/10{,}000W,$$

in which W is the quantity, in g, of potassium lactate taken to prepare the *Test Solution*, and D is the dilution factor for the *Test Solution*.

Sugars To 10 mL of hot alkaline cupric tartrate TS add 5 drops of Potassium Lactate Solution: No red precipitate is formed.

Sulfate, Appendix IIIB Any turbidity produced by a quantity of the Solution containing the equivalent of 4.0 g of potassium lactate does not exceed that shown in a control containing 200 µg of sulfate ion (SO_4).

Packaging and Storage Store in tight containers.

Potassium Metabisulfite

Potassium Pyrosulfite

$K_2S_2O_5$	Formula wt 222.33
INS: 224	CAS: [16731-55-8]

DESCRIPTION

White or colorless, free-flowing crystals, crystalline powder, or granules, usually having an odor of sulfur dioxide. It gradually oxidizes in air to the sulfate. It is soluble in water, and is insoluble in alcohol. Its solutions are acid to litmus.

Functional Use in Foods Preservative; antioxidant; bleaching agent.

REQUIREMENTS

Identification A 1 in 10 solution gives positive tests for *Potassium* and for *Sulfite*, Appendix IIIA.
Assay Not less than 90.0% of $K_2S_2O_5$.
Heavy Metals (as Pb) Not more than 10 mg/kg.
Iron Not more than 10 mg/kg.
Selenium Not more than 0.003%.

TESTS

Assay Weigh accurately about 250 mg, add it to exactly 50 mL of 0.1 N iodine contained in a glass-stoppered flask, and stopper the flask. Allow to stand for 5 min, add 1 mL of hydrochloric acid, and titrate the excess iodine with 0.1 N sodium thiosulfate, using starch TS as the indicator. Each mL of 0.1 N iodine is equivalent to 5.558 mg of $K_2S_2O_5$.

Heavy Metals Dissolve 2 g in 20 mL of water, add 5 mL of hydrochloric acid, evaporate to about 1 mL on a steam bath, and dissolve the residue in 25 mL of water. This solution meets the requirements of the *Heavy Metals Test*, Appendix IIIB, using 20 µg of lead ion (Pb) in the control (*Solution A*).

Iron To 1 g of the sample add 2 mL of hydrochloric acid, and evaporate to dryness on a steam bath. Dissolve the residue in 2 mL of hydrochloric acid and 20 mL of water, add a few drops of bromine TS, and boil the solution to remove the bromine. Cool, dilute with water to 25 mL, then add 50 mg of ammonium persulfate and 5 mL of ammonium thiocyanate TS. Any red or pink color does not exceed that produced in a control containing 1.0 mL of *Iron Standard Solution* (10 μg Fe).
Selenium Determine as directed in *Method I* under the *Selenium Limit Test*, Appendix IIIB, using 100 mg of sample and 100 mg of magnesium oxide.

Packaging and Storage Store in well-filled, tight containers, and avoid exposure to excessive heat.

Potassium Nitrate

KNO_3 Formula wt 101.10
INS: 252 CAS: [7757-79-1]

DESCRIPTION

Colorless, transparent prisms, white granules, or white crystalline powder. It is odorless, has a salty taste, and produces a cooling sensation in the mouth. It is slightly hygroscopic in moist air. Its solutions are neutral to litmus. One g dissolves in 3 mL of water at 25°, in 0.5 mL of boiling water, and in about 620 mL of alcohol.

Functional Use in Foods Antimicrobial agent; preservative.

REQUIREMENTS

Identification A 1 in 10 solution gives positive tests for *Potassium* and *Nitrate*, Appendix IIIA.
Assay Not less than 99.0% and not more than 100.5% of KNO_3 after drying.
Chlorate Passes test.
Heavy Metals (as Pb) Not more than 0.002%.
Lead Not more than 10 mg/kg.
Loss on Drying Not more than 1%.

TESTS

Assay Weigh accurately about 400 mg, previously dried at 105° for 4 h, dissolve in 10 mL of hydrochloric acid in a small beaker or porcelain dish, and evaporate to dryness on a steam bath. Dissolve the residue in 10 mL of hydrochloric acid, and again evaporate to dryness, continuing the heat until the residue, when dissolved in water, is neutral to litmus. Transfer the residue with the aid of 25 mL of water to a glass-stoppered flask, add exactly 50 mL of 0.1 *N* silver nitrate, then add 3 mL of nitric acid and 3 mL of nitrobenzene, and shake vigorously. Add ferric ammonium sulfate TS, and titrate the excess silver nitrate with 0.1 *N* ammonium thiocyanate. Each mL of 0.1 *N* silver nitrate is equivalent to 10.11 mg of KNO_3.
Chlorate Sprinkle about 100 mg of a dry sample upon 1 mL of sulfuric acid. The mixture does not become yellow.
Heavy Metals A solution of 2 g in 25 mL of water meets the requirements of the *Heavy Metals Test*, Appendix IIIB, using 20 μg of lead ion (Pb) and 1 g of the sample in the control (*Solution A*).
Lead A solution of 1 g in 10 mL of water meets the requirements of the *Lead Limit Test*, Appendix IIIB, using 10 μg of lead ion (Pb) in the control.
Loss on Drying, Appendix IIC Dry at 105° for 4 h.

Packaging and Storage Store in tight containers.

Potassium Nitrite

KNO_2 Formula wt 85.10
INS: 249 CAS: [7758-09-0]

DESCRIPTION

Small, white or yellowish, deliquescent granules or cylindrical sticks. It is very soluble in water, but is sparingly soluble in alcohol.

Functional Use in Foods Color fixative in meat and meat products; antimicrobial agent.

REQUIREMENTS

Identification A 1 in 10 solution is alkaline to litmus and gives positive tests for *Potassium* and for *Nitrite*, Appendix IIIA.
Assay Not less than 90.0% and not more than 100.5% of KNO_2.
Heavy Metals (as Pb) Not more than 0.002%.
Lead Not more than 10 mg/kg.

TESTS

Assay Transfer about 1.2 g, accurately weighed, into a 100-mL volumetric flask, dissolve in water, dilute to volume, and mix. Pipet 10 mL of this solution into a mixture of 50.0 mL of 0.1 *N* potassium permanganate, 100 mL of water, and 5 mL of sulfuric acid, keeping the tip of the pipet well below the surface of the liquid. Warm the solution to 40°, allow it to stand for 5 min, and add 25.0 mL of 0.1 *N* oxalic acid. Heat the mixture to about 80°, and titrate with 0.1 *N* potassium permanganate. Each mL of 0.1 *N* potassium permanganate is equivalent to 4.255 mg of KNO_2.
Heavy Metals Dissolve 1 g in 15 mL of 2.7 *N* hydrochloric acid, and evaporate to dryness on a steam bath. To the residue add 2 mL of hydrochloric acid, again evaporate to dryness, and dissolve the residue in 25 mL of water. This solution meets the

requirements of the *Heavy Metals Test*, Appendix IIIB, using 20 μg of lead ion (Pb) in the control (*Solution A*).

Lead A solution of 1 g in 10 mL of water meets the requirements of the *Lead Limit Test*, Appendix IIIB, using 10 μg of lead ion (Pb) in the control.

Packaging and Storage Store in tight containers.

Potassium Phosphate, Dibasic

Dipotassium Monophosphate; Dipotassium Phosphate

K_2HPO_4 Formula wt 174.18

INS: 340(ii) CAS: [7758-11-4]

DESCRIPTION

A colorless or white, granular salt that is deliquescent when exposed to moist air. One g is soluble in about 3 mL of water. It is insoluble in alcohol. The pH of a 1% solution is about 9.

Functional Use in Foods Buffer; sequestrant; yeast food; dietary supplement.

REQUIREMENTS

Identification A 1 in 20 solution gives positive tests for *Potassium* and for *Phosphate*, Appendix IIIA.
Assay Not less than 98.0% of K_2HPO_4 after drying.
Arsenic (as As) Not more than 3 mg/kg.
Fluoride Not more than 10 mg/kg.
Heavy Metals (as Pb) Not more than 0.0015%.
Insoluble Substances Not more than 0.2%.
Lead Not more than 5 mg/kg.
Loss on Drying Not more than 2.0%.

TESTS

Assay Transfer about 6.5 g of the sample, previously dried at 105° for 4 h and accurately weighed, into a 250-mL beaker, add 50.0 mL of 1 *N* hydrochloric acid and 50 mL of water, and stir until the sample is completely dissolved. Place the electrodes of a suitable pH meter in the solution, and titrate the excess acid with 1 *N* sodium hydroxide to the inflection point occurring at about pH 4. Record the buret reading, and calculate the volume (*A*) of 1 *N* hydrochloric acid consumed by the sample. Continue the titration with 1 *N* sodium hydroxide until the inflection point occurring at about pH 8.8 is reached, record the buret reading, and calculate the volume (*B*) of 1 *N* sodium hydroxide required in the titration between the two inflection points (pH 4 to pH 8.8). When *A* is equal to or less than *B*, each mL of the volume *A* of 1 *N* hydrochloric acid is equivalent to 174.2 mg of K_2HPO_4. When *A* is greater than *B*, each mL of the volume 2*B* − *A* of 1 *N* sodium hydroxide is equivalent to 174.2 mg of K_2HPO_4.

Arsenic A solution of 1 g in 10 mL of water meets the requirements of the *Arsenic Test*, Appendix IIIB.
Fluoride Determine as directed under *Fluoride* in the monograph for *Calcium Phosphate, Dibasic*.
Heavy Metals A solution of 1.33 g in 25 mL of water meets the requirements of the *Heavy Metals Test*, Appendix IIIB, using 20 μg of lead ion (Pb) in the control (*Solution A*), using glacial acetic acid to adjust the pH of the sample solution.
Insoluble Substances Dissolve 10 g in 100 mL of hot water, and filter through a tared filtering crucible. Wash the insoluble residue with hot water, dry at 105° for 2 h, cool, and weigh.
Lead A 10-g sample using a 5-μg/mL *Standard Lead Solution* meets the requirements of the *APDC Extraction Method for Lead*, Appendix IIIB.
Loss on Drying, Appendix IIC Dry at 105° for 4 h.

Packaging and Storage Store in tight containers.

Potassium Phosphate, Monobasic

Potassium Biphosphate; Potassium Dihydrogen Phosphate; Monopotassium Phosphate

KH_2PO_4 Formula wt 136.09

INS: 340(i) CAS: [7778-77-0]

DESCRIPTION

Colorless crystals or a white granular or crystalline powder. It is odorless, and is stable in air. It is freely soluble in water, but is insoluble in alcohol. The pH of a 1 in 100 solution is between 4.2 and 4.7.

Functional Use in Foods Buffer; sequestrant; yeast food; dietary supplement.

REQUIREMENTS

Identification A 1 in 20 solution gives positive tests for *Potassium* and for *Phosphate*, Appendix IIIA.
Assay Not less than 98.0% of KH_2PO_4 after drying.
Arsenic (as As) Not more than 3 mg/kg.
Fluoride Not more than 10 mg/kg.
Heavy Metals (as Pb) Not more than 0.0015%.
Insoluble Substances Not more than 0.2%.
Lead Not more than 5 mg/kg.
Loss on Drying Not more than 1%.

TESTS

Assay Transfer about 5 g of the sample, previously dried at 105° for 4 h and accurately weighed, into a 250-mL beaker, add 100 mL of water and 5.0 mL of 1 *N* hydrochloric acid, and stir until the sample is completely dissolved. Place the electrodes of a suitable pH meter in the solution, and slowly titrate the excess acid, stirring constantly, with 1 N sodium hydroxide to the inflection point occurring at about pH 4. Record the buret reading, and calculate the volume (A), if any, of 1 N hydrochloric acid consumed by the sample. Continue the titration with 1 N sodium hydroxide until the inflection point occurring at about pH 8.8 is reached, record the buret reading, and calculate the volume (B) of 1 N sodium hydroxide required in the titration between the two inflection points (pH 4 and pH 8.8). Each mL of the volume B − A of 1 N sodium hydroxide is equivalent to 136.1 mg of KH_2PO_4.

Arsenic A solution of 1 g in 10 mL of water meets the requirements of the *Arsenic Test*, Appendix IIIB.

Fluoride Determine on a 2-g sample as directed in *Method IV* under the *Fluoride Limit Test*, Appendix IIIB, using *Buffer Solution B* and 0.1 mL of *Fluoride Standard Solution*.

Heavy Metals A solution of 1.33 g in 25 mL of water meets the requirements of the *Heavy Metals Test*, Appendix IIIB, using 20 μg of lead ion (Pb) in the control (*Solution A*).

Insoluble Substances Dissolve 10 g in 100 mL of hot water, and filter through a tared filtering crucible. Wash the insoluble residue with hot water, dry at 105° for 2 h, cool, and weigh.

Lead A 10-g sample using a 5-μg/mL *Standard Lead Solution* meets the requirements of the *APDC Extraction Method for Lead*, Appendix IIIB.

Loss on Drying, Appendix IIC Dry at 105° for 4 h.

Packaging and Storage Store in tight containers.

Potassium Phosphate, Tribasic

Tripotassium Phosphate

K_3PO_4 Formula wt 212.27

INS: 340(iii)   CAS: [7778-53-2]

DESCRIPTION

Tribasic Potassium Phosphate is anhydrous or may contain one molecule of water of hydration. It occurs as white, odorless, hygroscopic crystals or granules. It is freely soluble in water, but is insoluble in alcohol. The pH of a 1 in 100 solution is about 11.5.

Functional Use in Foods Emulsifier.

REQUIREMENTS

Identification A 1 in 20 solution gives positive tests for *Potassium* and for *Phosphate*, Appendix IIIA.

Assay Not less than 97.0% of K_3PO_4, calculated on the ignited basis.

Arsenic (as As) Not more than 3 mg/kg.

Fluoride Not more than 10 mg/kg.

Heavy Metals (as Pb) Not more than 0.0015%.

Insoluble Substances Not more than 0.2%.

Lead Not more than 2 mg/kg.

Loss on Ignition K_3PO_4 (anhydrous): not more than 3.0%; $K_3PO_4 \cdot H_2O$ (monohydrate): between 8.0% and 20.0%.

TESTS

Assay Dissolve an accurately weighed quantity of the sample, equivalent to about 8 g of anhydrous K_3PO_4, in 40 mL of water in a 400-mL beaker, and add 100.0 mL of 1 N hydrochloric acid. Pass a stream of carbon dioxide-free air, in fine bubbles, through the solution for 30 min to expel carbon dioxide, covering the beaker loosely to prevent any loss by spraying. Wash the cover and sides of the beaker with a few mL of water, and place the electrodes of a suitable pH meter in the solution. Titrate the solution with 1 N sodium hydroxide to the inflection point occurring at about pH 4, then calculate the volume (A) of 1 N hydrochloric acid consumed. Protect the solution from absorbing carbon dioxide, and continue the titration with 1 N sodium hydroxide until the inflection point occurring at about pH 8.8 is reached. Calculate the volume (B) of 1 N sodium hydroxide consumed in this titration. When A is equal to or greater than 2B, each mL of the volume B of 1 N sodium hydroxide is equivalent to 212.3 mg of K_3PO_4. When A is less than 2B, each mL of the volume A − B of 1 N sodium hydroxide is equivalent to 212.3 mg of K_3PO_4.

Arsenic A solution of 1 g in 10 mL of water meets the requirements of the *Arsenic Test*, Appendix IIIB.

Fluoride Determine on a 2-g sample as directed in *Method IV* under the *Fluoride Limit Test*, Appendix IIIB, using *Buffer Solution A* and 0.1 mL of *Fluoride Standard Solution*.

Heavy Metals A solution of 1.33 g in 25 mL of water meets the requirements of the *Heavy Metals Test*, Appendix IIIB, using 20 μg of lead ion (Pb) in the control (*Solution A*).

Insoluble Substances Dissolve 10 g in 100 mL of hot water, and filter through a tared filtering crucible. Wash the insoluble residue with hot water, dry at 105° for 2 h, cool, and weigh.

Lead A 10-g sample meets the requirements of the *APDC Extraction Method for Lead*, Appendix IIIB.

Loss on Ignition, Appendix IIC Ignite at about 800° for 30 min.

Packaging and Storage Store in tight containers.

Potassium Polymetaphosphate

Potassium Metaphosphate; Potassium Kurrol's Salt

$(KPO_3)x$

                                  CAS: [7990-53-6]

DESCRIPTION

Potassium Polymetaphosphate is a straight-chain polyphosphate having a high degree of polymerization. It occurs as a white, odorless powder. It is insoluble in water, but is soluble in dilute solutions of sodium salts.

Functional Use in Foods Fat emulsifier; moisture-retaining agent.

REQUIREMENTS

Identification

A. Finely powder about 1 g of the sample, and add it slowly to 100 mL of a 1 in 50 solution of sodium chloride while stirring vigorously. A gelatinous mass is formed.

B. Mix 500 mg with 10 mL of nitric acid and 50 mL of water, boil for about 30 min, and cool. The resulting solution gives positive tests for *Potassium* and for *Phosphate*, Appendix IIIA.

Assay Not less than 59.0% and not more than 61.0% of P_2O_5.
Arsenic (as As) Not more than 3 mg/kg.
Fluoride Not more than 10 mg/kg.
Heavy Metals (as Pb) Not more than 0.0015%.
Lead Not more than 2 mg/kg.
Viscosity Between 6.5 and 15 centipoises.

TESTS

Assay Transfer about 1 g of the sample, accurately weighed, into a 400-mL beaker, add 100 mL of water and 25 mL of nitric acid, cover with a watch glass, and boil the solution for 10 min on a hot plate. Rinse any condensate on the watch glass into the beaker, cool the solution to room temperature, transfer it quantitatively to a 500-mL volumetric flask, dilute to volume with water, and mix thoroughly. Pipet 20.0 mL of this solution into a 500-mL Erlenmeyer flask, add 100 mL of water, and heat just to boiling. Add with stirring 50 mL of quimociac TS, then cover with a watch glass, and boil for 1 min in a well-ventilated hood. Cool to room temperature, swirling occasionally while cooling, then filter through a tared, sintered-glass crucible of medium porosity, and wash with five 25-mL portions of water. Dry at about 225° for 30 min, cool, and weigh. Each mg of precipitate thus obtained is equivalent to 32.074 μg of P_2O_5.

Arsenic A solution of 1 g in 15 mL of 2.7 *N* hydrochloric acid meets the requirements of the *Arsenic Test*, Appendix IIIB.

Fluoride Place 5 g of the sample, 25 mL of water, 50 mL of sulfuric acid, 5 drops of silver nitrate solution (1 in 2), and a few glass beads in a 250-mL distilling flask connected with a condenser and carrying a thermometer and a capillary tube, both of which must extend into the liquid. Connect a small dropping funnel, filled with water, or a steam generator to the capillary tube. Support the flask on a fireproof mat with a hole that exposes about one-third of the flask to the flame. Distill into a 250-mL volumetric flask until the temperature reaches 135°. Add water from the funnel or introduce steam through the capillary to maintain the temperature between 135° and 140°. Continue the distillation until 225 to 240 mL has been collected, then dilute to 250 mL with water, and mix.

Place a 50-mL aliquot of this solution in a 100-mL Nessler tube. In another, similar Nessler tube, place 50 mL of water as a control. Add to each tube 0.1 mL of a filtered solution of sodium alizarinsulfonate (1 in 1000) and 1 mL of freshly prepared hydroxylamine hydrochloride solution (1 in 4000), and mix well. Add, dropwise, and with stirring, 0.05 *N* sodium hydroxide to the tube containing the distillate until its color just matches that of the control, which is faintly pink. Then add to each tube exactly 1.0 mL of 0.1 *N* hydrochloric acid, and mix well. From a buret, graduated in 0.05-mL units, add slowly to the tube containing the distillate enough thorium nitrate solution (1 in 4000) so that, after mixing, the color of the liquid just changes to a faint pink. Note the volume of the solution added, add exactly the same volume to the control, and mix. Now add to the control sodium fluoride TS (10 μg F per mL) from a buret to make the colors of the two tubes match after dilution to the same volume. Mix well, and allow all air bubbles to escape before making the final color comparison. Check the endpoint by adding 1 or 2 drops of sodium fluoride TS to the control. A distinct color change should take place. Note the volume of sodium fluoride TS added. The volume of sodium fluoride TS required for the control solution should not exceed 1.0 mL.

Heavy Metals Warm 1.33 g with 10 mL of 2.7 *N* hydrochloric acid until no more dissolves, dilute with water to 25 mL, and filter. This solution meets the requirements of the *Heavy Metals Test*, Appendix IIIB, using 20 μg of lead ion (Pb) in the control (*Solution A*).

Lead A 10-g sample meets the requirements of the *APDC Extraction Method for Lead*, Appendix IIIB.

Viscosity Dissolve 300 mg of the sample in 200 mL of a solution of sodium pyrophosphate (3.5 g of $Na_4P_2O_7$ dissolved in 1000 mL of water), using a magnetic stirrer. When solution is complete, or after 30 min, whichever occurs first, transfer 10 mL into an Ostwald-Fenske viscometer, and determine the time, *T*, in seconds, required for the liquid to flow from the upper to the lower mark in the capillary tube. Calculate the viscosity, in centipoises, by the formula

$$Tv/dt,$$

in which *t* is the time, in s, required for a glycerin–water mixture of known viscosity, *v*, and specific gravity, *d*, to flow from the upper to the lower mark of the capillary tube during calibration of the viscometer under similar conditions.

Packaging and Storage Store in well-closed containers.

Potassium Pyrophosphate

Tetrapotassium Pyrophosphate

$K_4P_2O_7$ Formula wt 330.34

INS: 452(ii) CAS: [7320-34-5]

DESCRIPTION

Colorless crystals, or a white, granular solid. It is hygroscopic. It is very soluble in water, but is insoluble in alcohol. The pH of a 1 in 100 solution is about 10.5.

Functional Use in Foods Emulsifier; texturizer.

REQUIREMENTS

Identification

A. A 1 in 20 solution gives positive tests for *Potassium*, Appendix IIIA.

B. Dissolve 100 mg of the sample in 100 mL of 1.7 N nitric acid. Add 0.5 mL of this solution to 30 mL of quimociac TS. A yellow precipitate does not form. Heat the remaining portion of the sample solution for 10 min at 95°, and then add 0.5 mL of the solution to 30 mL of quimociac TS. A yellow precipitate forms immediately.

Assay Not less than 95.0% of $K_4P_2O_7$, calculated on the ignited basis.
Arsenic (as As)   Not more than 3 mg/kg.
Fluoride   Not more than 10 mg/kg.
Heavy Metals (as Pb)   Not more than 0.0015%.
Insoluble Substances   Not more than 0.1%.
Lead   Not more than 2 mg/kg.
Loss on Ignition   Not more than 0.5%.

TESTS

Assay Dissolve about 600 mg, accurately weighed, in 100 mL of water in a 400-mL beaker, and using a pH meter, adjust the pH of the solution to exactly 3.8 with hydrochloric acid. Add 50 mL of a 1 in 8 solution of zinc sulfate (125 g of $ZnSO_4.7H_2O$ dissolved in water, diluted to 1000 mL, filtered, and adjusted to pH 3.8), and allow to stand for 2 min. Titrate the liberated acid with 0.1 N sodium hydroxide until a pH of 3.8 is again reached. After each addition of sodium hydroxide near the endpoint, time should be allowed for any precipitated zinc hydroxide to redissolve. Each mL of 0.1 N sodium hydroxide is equivalent to 16.52 mg of $K_4P_2O_7$.

Arsenic A solution of 1 g in 35 mL of water meets the requirements of the *Arsenic Test*, Appendix IIIB.

Fluoride Determine on a 2-g sample as directed in *Method IV* under the *Fluoride Limit Test*, Appendix IIIB, using *Buffer Solution A* and 0.1 mL of *Fluoride Standard Solution*.

Heavy Metals A solution of 1.33 g in 25 mL of water meets the requirements of the *Heavy Metals Test*, Appendix IIIB, using 20 µg of lead ion (Pb) in the control (*Solution A*).

Insoluble Substances Dissolve 10 g in 100 mL of hot water, and filter through a tared filtering crucible. Wash the insoluble residue with hot water, dry at 105° for 2 h, cool, and weigh.

Lead A 10-g sample meets the requirements of the *APDC Extraction Method for Lead*, Appendix IIIB.

Loss on Ignition Ignite at about 800° for 30 min.

Packaging and Storage   Store in tight containers.

Potassium Sorbate

2,4-Hexadienoic Acid, Potassium Salt

$$CH_3CH = CHCH = CHCOOK$$

$C_6H_7KO_2$	Formula wt 150.22
INS: 202	CAS: [590-00-1]

DESCRIPTION

White crystals, crystalline powder, or pellets. It decomposes at about 270°.

Functional Use in Foods   Preservative.

REQUIREMENTS

Identification

A. A 1 in 10 solution responds to the flame test for *Potassium*, Appendix IIIA.

B. To 2 mL of a 1 in 10 solution add a few drops of bromine TS. The color is discharged.

Assay Not less than 98.0% and not more than 101.0% of $C_6H_7KO_2$, calculated on the dried basis.
Acidity (as sorbic acid)   Passes test (about 1%).
Alkalinity (as K_2CO_3)   Passes test (about 1%).
Heavy Metals (as Pb)   Not more than 10 mg/kg.
Loss on Drying   Not more than 1.0%.

TESTS

Assay Dissolve about 250 mg, accurately weighed, in 40 mL of glacial acetic acid in a 250-mL, glass-stoppered Erlenmeyer flask, warming if necessary to effect solution. Cool to room temperature, add 2 drops of crystal violet TS, and titrate with 0.1 N perchloric acid in glacial acetic acid to a blue-green endpoint that persists for at least 30 s. Perform a blank determination (see *General Provisions*), and make any necessary correction. Each mL of 0.1 N perchloric acid is equivalent to 15.02 mg of $C_6H_7KO_2$.

Acidity or **Alkalinity** Dissolve 1.1 g in 20 mL of water and add 3 drops of phenolphthalein TS. If the solution is colorless, titrate with 0.1 N sodium hydroxide to a pink color that persists for 15 s. Not more than 1.1 mL is required. If the solution is pink in color, titrate with 0.1 N hydrochloric acid. Not more than 0.8 mL is required to discharge the pink color.

Heavy Metals Prepare and test a 2-g sample as directed in *Method II* under the *Heavy Metals Test*, Appendix IIIB, using 20 µg of lead ion (Pb) in the control (*Solution A*).

Loss on Drying, Appendix IIC   Dry at 105° for 3 h.

Packaging and Storage   Store in tight containers.

Potassium Sulfate

K$_2$SO$_4$ Formula wt 174.26

INS: 515 CAS: [7778-80-5]

DESCRIPTION

Colorless or white crystals or crystalline powder having a bitter, saline taste. One g dissolves in about 8.5 mL of water. It is insoluble in alcohol. The pH of a 1 in 20 solution is about 5.5 to 8.5.

Functional Use in Foods Water corrective; miscellaneous and general purpose.

REQUIREMENTS

Identification A 1 in 10 solution gives positive tests for *Potassium*, Appendix IIIA.
Assay Not less than 99.0% and not more than 100.5% of K$_2$SO$_4$.
Heavy Metals (as Pb) Not more than 10 mg/kg.
Selenium Not more than 30 mg/kg.

TESTS

Assay Dissolve about 500 mg, accurately weighed, in 200 mL of water, add 1 mL of hydrochloric acid, and heat to boiling. Gradually add, in small portions and while stirring constantly, an excess of hot barium chloride TS (about 8 or 9 mL), and heat the mixture on a steam bath for 1 h. Collect the precipitate on a retentive, ashless filter paper, wash until free from chloride, and place the filter in a suitable tared crucible. Carefully burn away the paper, and ignite at 800° ± 25° to constant weight. The weight of the barium sulfate so obtained, multiplied by 0.7466, indicates its equivalent of K$_2$SO$_4$.
Heavy Metals A solution of 3 g in 25 mL of water meets the requirements of the *Heavy Metals Test*, Appendix IIIB, using 20 µg of lead ion (Pb) and 1 g of the sample in the control (*Solution A*).
Selenium Determine as directed in *Method II* under the *Selenium Limit Test*, Appendix IIIB, using 200 mg of sample.

Packaging and Storage Store in well-closed containers.

Potassium Sulfite

K$_2$SO$_3$ Formula wt 158.26

INS: 225 CAS: [10117-38-1]

DESCRIPTION

A white, odorless, granular powder. It undergoes oxidation in air. One g dissolves in about 3.5 mL of water. It is slightly soluble in alcohol.

Functional Use in Foods Preservative; antioxidant.

REQUIREMENTS

Identification A 1 in 20 solution gives positive tests for *Potassium* and for *Sulfite*, Appendix IIIA.
Assay Not less than 90.0% and not more than 100.5% of K$_2$SO$_3$.
Alkalinity (as K$_2$CO$_3$) Between 0.25% and 0.45%.
Heavy Metals (as Pb) Not more than 10 mg/kg.
Selenium Not more than 30 mg/kg.

TESTS

Assay Dissolve about 750 mg, accurately weighed, in a mixture of 100.0 mL of 0.1 N iodine and 5 mL of 2.7 N hydrochloric acid, and titrate the excess iodine with 0.1 N sodium thiosulfate, adding starch TS as the indicator. Each mL of 0.1 N iodine is equivalent to 7.912 mg of K$_2$SO$_3$.
Alkalinity Dissolve 1 g in 20 mL of water, add 25 mL of 3% hydrogen peroxide, previously neutralized to methyl red TS, mix thoroughly, cool to room temperature, and titrate with 0.02 N hydrochloric acid. Perform a blank determination (see *General Provisions*) using 25 mL of the neutralized hydrogen peroxide solution. Each mL of 0.02 N hydrochloric acid is equivalent to 1.38 mg of K$_2$CO$_3$.
Heavy Metals Dissolve 2 g in 10 mL of water, add 4 mL of hydrochloric acid, and evaporate to dryness on a steam bath. To the residue add 5 mL of hot water and 1 mL of hydrochloric acid, and again evaporate to dryness. Dissolve the residue in water and dilute to 25 mL. This solution meets the requirements of the *Heavy Metals Test*, Appendix IIIB, using 20 µg of lead ion (Pb) in the control (*Solution A*).
Selenium Determine as directed in *Method I* under the *Selenium Limit Test*, Appendix IIIB, using 200 mg of sample and 100 mg of magnesium oxide.

Packaging and Storage Store in tight containers.

Potassium Tripolyphosphate

Pentapotassium Triphosphate; Potassium Triphosphate

K$_5$P$_3$O$_{10}$ Formula wt 448.41

INS: 450(v) CAS: [13845-36-8]

DESCRIPTION

White granules or a white powder. It is hygroscopic and is very soluble in water. The pH of a 1 in 100 solution is between 9.2 and 10.1.

Functional Use in Foods Texturizer.

REQUIREMENTS

Identification
A. A 1 in 20 solution gives positive tests for *Potassium*, Appendix IIIA.
B. To 1 mL of a 1 in 100 solution add a few drops of silver nitrate TS. A white precipitate is formed that is soluble in 1.7 N nitric acid.
Assay Not less than 85.0% of $K_5P_3O_{10}$.
Arsenic (as As) Not more than 3 mg/kg.
Fluoride Not more than 10 mg/kg.
Heavy Metals (as Pb) Not more than 10 mg/kg.
Insoluble Substances Not more than 2.0%.
Lead Not more than 2 mg/kg.
Loss on Drying Not more than 0.7%.

TESTS

Assay Proceed as directed in the *Assay* in the monograph for *Sodium Tripolyphosphate*. Calculate the quantity, in mg, of $K_5P_3O_{10}$ in the sample taken by the formula

$$0.650 \times 25V.$$

Arsenic A solution of 1 g in 35 mL of water meets the requirements of the *Arsenic Test*, Appendix IIIB.
Fluoride Determine on a 2-g sample as directed in *Method IV* under the *Fluoride Limit Test*, Appendix IIIB, using *Buffer Solution A* and 0.1 mL of *Fluoride Standard Solution*.
Heavy Metals A solution of 2 g in 25 mL of water meets the requirements of the *Heavy Metals Test*, Appendix IIIB, using 20 µg of lead ion (Pb) in the control (*Solution A*).
Insoluble Substances Dissolve 10 g in 100 mL of hot water, and filter through a tared filtering crucible. Wash the insoluble residue with hot water, dry at 105° for 2 h, cool, and weigh.
Lead A 10-g sample meets the requirements of the *APDC Extraction Method for Lead*, Appendix IIIB.
Loss on Drying, Appendix IIC Dry at about 105° for 1 h.

Packaging and Storage Store in tight containers.

L-Proline

L-2-Pyrrolidinecarboxylic Acid

$C_5H_9NO_2$ Formula wt 115.13

CAS: [147-85-3]

DESCRIPTION

White crystals or a crystalline powder. It is odorless or has a slight, characteristic odor with a slightly sweet taste. It is very soluble in water and in alcohol, but is insoluble in ether.

Functional Use in Foods Nutrient; dietary supplement.

REQUIREMENTS

Identification To 5 mL of a 1 in 1000 solution of the sample add 1 mL of triketohydrindene hydrate TS. A yellow color is produced.
Assay Not less than 98.5% and not more than 101.5% of $C_5H_9NO_2$ after drying.
Heavy Metals (as Pb) Not more than 0.002%.
Lead Not more than 10 mg/kg.
Loss on Drying Not more than 0.3%.
Residue on Ignition Not more than 0.1%.
Specific Rotation $[\alpha]_D^{20°}$: Between $-84.0°$ and $-86.3°$ after drying.

TESTS

Assay Dissolve about 220 mg of the sample, previously dried at 105° for 3 h and accurately weighed, in 3 mL of formic acid and 50 mL of glacial acetic acid, add 2 drops of crystal violet TS, and titrate with 0.1 N perchloric acid to a bluish green endpoint. Perform a blank determination (see *General Provisions*), and make any necessary correction. Each mL of 0.1 N perchloric acid is equivalent to 11.51 mg of $C_5H_9NO_2$.
Heavy Metals Prepare and test a 1-g sample as directed in *Method II* under the *Heavy Metals Test*, Appendix IIIB, using 20 µg of lead ion (Pb) in the control (*Solution A*).
Lead A *Sample Solution* prepared as directed for organic compounds meets the requirements of the *Lead Limit Test*, Appendix IIIB, using 10 µg of lead ion (Pb) in the control.
Loss on Drying, Appendix IIC Dry at 105° for 3 h.
Residue on Ignition, Appendix IIC Ignite 1 g as directed in the general method.
Specific Rotation, Appendix IIB Determine in a solution containing 4 g of a previously dried sample in sufficient water to make 100 mL.

Packaging and Storage Store in well-closed, light-resistant containers.

Propane

$$CH_3CH_2CH_3$$

C_3H_8 Formula wt 44.10

CAS: [74-98-6]

DESCRIPTION

A colorless, odorless, flammable gas (boiling temperature is about $-42°$). One hundred volumes of water dissolves 6.5 volumes at 17.8° and a pressure of 753 mm of mercury; 100 volumes of anhydrous alcohol dissolves 790 volumes at 16.6° and a pressure of 754 mm of mercury; 100 volumes of ether dissolves 926 volumes at 16.6° and a pressure of 757 mm of mercury; 100 volumes of chloroform dissolves 1299 volumes at 21.6° and a pressure of 757 mm of mercury. Vapor pressure at 21° is about 10,290 mm of mercury (108 psi).

Functional Use in Foods Propellant; aerating agent.

REQUIREMENTS

Caution: Propane is highly flammable and explosive. Observe precautions and perform sampling and analytical operations in a well-ventilated fume hood.

Identification
A. The infrared absorption spectrum exhibits maxima, among others, at about the following wavelengths, in μm: 3.4 (vs), 6.8 (s), and 7.2 (m).
B. The vapor pressure of a test specimen, obtained as directed in the *Sampling Procedure*, and determined at 21° by means of a suitable pressure gauge, is between 820 and 875 kPa absolute (119 and 127 psia, respectively).
Assay Not less than 98.0% of Propane (C_3H_8).
Acidity of Residue Passes test.
High-Boiling Residue Not more than 5 mg/kg.
Sulfur Compounds Passes test.
Water Not more than 10 mg/kg.

TESTS

Sampling Procedure, Assay, Acidity of Residue, High-Boiling Residue, Sulfur Compounds, and **Water** Proceed as directed for these **TESTS** in the monograph for *Butane*, except substitute Propane wherever Butane is specified in the text.

Packaging and Storage Store in tight cylinders protected from excessive heat.

Propionic Acid

CH_3CH_2COOH

$C_3H_6O_2$ Formula wt 74.08
INS: 280 CAS: [79-09-4]

DESCRIPTION

An oily liquid having a slightly pungent, rancid odor. It is miscible with water and with alcohol and various other organic solvents.

Functional Use in Foods Preservative; mold inhibitor.

REQUIREMENTS

Assay Not less than 99.5% and not more than 100.5% of $C_3H_6O_2$, calculated on the anhydrous basis.
Aldehydes (as propionaldehyde) Passes test (limit about 0.05%).
Distillation Range Between 138.5° and 142.5°.
Heavy Metals (as Pb) Not more than 10 mg/kg.
Nonvolatile Residue Not more than 0.01%.
Readily Oxidizable Substances (as formic acid) Passes test (limit about 0.05%).
Specific Gravity Between 0.993 and 0.997 at 20°/20°.
Water Not more than 0.15%.

TESTS

Assay Mix about 1.5 g, accurately weighed, with 100 mL of recently boiled and cooled water in a 250-mL Erlenmeyer flask, add phenolphthalein TS, and titrate with 0.5 N sodium hydroxide to the first appearance of a faint pink endpoint that persists for at least 30 s. Each mL of 0.5 N sodium hydroxide is equivalent to 37.04 mg of $C_3H_6O_2$.
Aldehydes Transfer 10.0 mL of the sample into a 250-mL, glass-stoppered Erlenmeyer flask containing 50 mL of water and 10.0 mL of a 1 in 8 solution of sodium bisulfite, stopper the flask, and shake vigorously. Allow the mixture to stand for 30 min, then titrate with 0.1 N iodine to the same brownish yellow endpoint obtained with a blank treated with the same quantities of the same reagents (see *General Provisions*). The difference between the volume of 0.1 N iodine required for the blank and that required for the sample is not more than 1.75 mL.
Distillation Range Determine as directed in the general method, Appendix IIB.
Heavy Metals A solution of 2 g in 25 mL of water meets the requirements of the *Heavy Metals Test*, Appendix IIIB, using 20 μg of lead ion (Pb) in the control (*Solution A*).
Nonvolatile Residue Transfer 100 mL into a tared, 125-mL, platinum evaporating dish, previously heated at 105° to constant weight, and evaporate the sample to dryness on a steam bath. Heat the dish at 105° for 30 min or to constant weight, cool in a desiccator, and weigh.
Readily Oxidizable Substances Dissolve 15 g of sodium hydroxide in 50 mL of water, cool, add 6 mL of bromine, stirring to effect complete solution, and dilute to 2000 mL with water. Transfer 25.0 mL of this solution into a 250-mL, glass-stoppered Erlenmeyer flask containing 100 mL of water, and add 10 mL of a 1 in 5 solution of sodium acetate and 10.0 mL of the sample. Allow to stand for 15 min, add 5 mL of a 1 in 4 solution of potassium iodide and 10 mL of hydrochloric acid, and titrate with 0.1 N sodium thiosulfate just to the disappearance of the brown color. Perform a blank determination (see *General Provisions*). The difference between the volume of 0.1 N sodium thiosulfate required for the blank and that required for the sample is not more than 2.2 mL.
Specific Gravity Determine by any reliable method (see *General Provisions*).

Water Determine by the *Karl Fischer Titrimetric Method*, Appendix IIB.

Packaging and Storage Store in well-closed containers.

Propylene Glycol

1,2-Propanediol; 1,2-Dihydroxypropane; Methyl Glycol

$$CH_3CH(OH)CH_2OH$$

$C_3H_8O_2$	Formula wt 76.10
INS: 1520	CAS: [57-55-6]

DESCRIPTION

A clear, colorless, viscous liquid having a slight, characteristic taste. It is practically odorless. It absorbs moisture when exposed to moist air. It is miscible with water, with acetone, and with chloroform in all proportions. It is soluble in ether and will dissolve many essential oils, but is immiscible with fixed oils.

Functional Use in Foods Solvent; wetting agent; humectant.

REQUIREMENTS

Identification The infrared absorption spectrum of a thin film of the sample exhibits maxima only at the same wavelengths as that of a similar preparation of USP Propylene Glycol Reference Standard.
Assay Not less than 99.5%, by weight, of $C_3H_8O_2$.
Acidity Passes test.
Distillation Range Between 185° and 189°.
Heavy Metals (as Pb) Not more than 5 mg/kg.
Residue on Ignition Not more than 0.007%.
Specific Gravity Between 1.035 and 1.037.
Water Not more than 0.2%.

TESTS

Assay Inject a 10-µL portion of the sample into a suitable gas chromatograph in which the detector is the thermal conductivity type and the column is 1-m × 8-mm stainless steel tubing packed with 4% Carbowax compound 20 M on 40/60-mesh Chromosorb T, or equivalent materials. The carrier gas is helium flowing at 75 mL/min. The injection port temperature is 240°; the column temperature is 120° to 200°, programmed at a rate of 5°/min; and the block temperature is 250°. Under the conditions described, the approximate retention time for Propylene Glycol is 5.7 min, and 8.2, 9.0, and 10.2 min for the three isomers of dipropylene glycol. Measure the area under all peaks by any convenient means, calculate the normalized area percentage of Propylene Glycol, and report as weight percentage.

Acidity Add 3 to 6 drops of phenol red TS to 50 mL of water, then add 0.01 N sodium hydroxide until the solution remains red for 30 s. To this solution add 50 g of the sample, accurately weighed. Titrate with 0.01 N sodium hydroxide until the original red color returns and remains for 15 s: Not more than 1.67 mL of 0.01 N sodium hydroxide is required.
Distillation Range Determine as directed in the general method, Appendix IIB.
Heavy Metals A solution of 4 g in 25 mL of water meets the requirements of the *Heavy Metals Test*, Appendix IIIB, using 20 µg of lead ion (Pb) in the control (*Solution A*).
Residue on Ignition, Appendix IIC Heat a 50-g sample in a tared, 100-mL, shallow dish until it ignites, and allow it to burn without further application of heat in a place free from drafts. Cool, moisten the residue with 5 mL of sulfuric acid, and ignite at about 800° for 15 min.
Specific Gravity Determine by any reliable method (see *General Provisions*).
Water Determine by the *Karl Fischer Titrimetric Method*, Appendix IIB.

Packaging and Storage Store in tight containers.

Propylene Glycol Alginate

Hydroxypropyl Alginate; Algin Derivative

$(C_9H_{14}O_7)_n$	Equiv wt, *Calculated*, 234.21
INS: 405	CAS: [9005-37-2]

DESCRIPTION

The propylene glycol ester of alginic acid (see the monograph for *Alginic Acid*) varies in composition according to its degree of esterification and the percentages of free and neutralized carboxyl groups in the molecule. It occurs as a white to yellowish, fibrous or granular powder. It is practically odorless and tasteless. It dissolves in water, in solutions of dilute organic acids, and depending upon the degree of esterification, in hydroalcoholic mixtures containing up to 60% by weight of alcohol to form stable, viscous colloidal solutions at a pH of 3.

Functional Use in Foods Stabilizer; thickener; emulsifier.

REQUIREMENTS

Identification Transfer 20 mL of the saponified solution obtained in the determination of *Esterified Carboxyl Groups* into a 250-mL Erlenmeyer flask, add 50 mL of 0.1 M periodic acid, swirl, and allow to stand for 30 min. Add 2 g of potassium iodide, titrate with 0.1 N sodium thiosulfate to a faint yellow color, and then dilute the mixture to 200 mL with water. To 10 mL of this solution add 5 mL of hydrochloric acid and 10 mL of modified Schiff's reagent. A blue to blue violet color develops in about 20 min (*formaldehyde*). To another 10-mL portion of the solution add 1 mL of a saturated solution of piperazine hydrate and 0.5 mL of sodium nitroferricyanide TS. A green color develops (*acetaldehyde*).

Note: Oxidation of Propylene Glycol Alginate yields formaldehyde and acetaldehyde.

Assay It yields not less than 16.0% and not more than 20.0% of carbon dioxide (CO_2), calculated on the dried basis.
Arsenic (as As) Not more than 3 mg/kg.
Esterified Carboxyl Groups Not less than 40.0%.
Free Carboxyl Groups Not more than 35.0%, calculated on the dried basis.
Heavy Metals (as Pb) Not more than 0.002%
Lead Not more than 5 mg/kg.
Loss on Drying Not more than 20.0%.
Neutralized Carboxyl Groups Not more than 45.0%.

TESTS

Assay Proceed as directed in the *Alginates Assay*, Appendix IIIC.
Arsenic A *Sample Solution* prepared as directed for organic compounds meets the requirements of the *Arsenic Test*, Appendix IIIB.
Esterified Carboxyl Groups Transfer quantitatively the solution obtained in the determination of *Free Carboxyl Groups* into a 1-L Erlenmeyer flask, add a few drops of phenolphthalein TS and 50.0 mL of 0.1 N sodium hydroxide. Stopper the flask, swirl the solution, and then allow it to stand for 30 min at room temperature. Titrate the excess sodium hydroxide to a faint pink endpoint with 0.1 N hydrochloric acid. Transfer the solution to a 600-mL beaker and complete the titration to a pH of 7.0, determining the endpoint potentiometrically. Calculate the percentage of esterified carboxyl groups by the formula

$$(V \times 44)/(\% \ CO_2 \times W),$$

in which W is the weight, in g, of the sample taken, calculated on the dried basis, for the *Free Carboxyl Groups* test, below; V is the volume, in mL, of 0.1 N sodium hydroxide consumed in this test; and % CO_2 is the percentage of carbon dioxide in the sample as determined by the *Assay*.
Free Carboxyl Groups Transfer about 1 g, accurately weighed, into a 600-mL beaker. Dissolve the sample in 200 mL of water, stirring mechanically for a minimum of 30 min, and titrate with 0.1 N sodium hydroxide to a pH of 7.0, determining the endpoint potentiometrically. Calculate the percentage of free carboxyl groups by the formula

$$(V \times 44)/(\% \ CO_2 \times W),$$

in which V is the volume, in mL, of 0.1 N sodium hydroxide consumed; % CO_2 is the percentage of carbon dioxide in the sample as determined by the *Assay*; and W is the weight, in g, of the sample taken, calculated on the dried basis.
Heavy Metals Determine as directed under *Heavy Metals* in the monograph for *Alginic Acid*.
Lead A *Sample Solution* prepared as directed for organic compounds meets the requirements of the *Lead Limit Test*, Appendix IIIB, using 5 µg of lead ion (Pb) in the control.
Loss on Drying, Appendix IIC Dry at 105° for 4 h.
Neutralized Carboxyl Groups Calculate the percentage of neutralized carboxyl groups by subtracting the sum of the percentage of *Free Carboxyl Groups* and the percentage of *Esterified Carboxyl Groups* from 100%.

Packaging and Storage Store in well-closed containers.

Propylene Glycol Mono- and Diesters

Propylene Glycol Mono- and Diesters of Fatty Acids;
Propylene Glycol Monostearate (or other appropriate ester)

INS: 477

DESCRIPTION

A mixture of Propylene Glycol Mono- and Diesters of fats and/or fatty acids. It has a bland odor and taste and occurs as a clear liquid or as white to yellowish white beads, flakes, or other solid material. It is insoluble in water, but is soluble in alcohol, in ethyl acetate, and in chloroform and other chlorinated hydrocarbons.

Functional Use in Foods Emulsifier; stabilizer.

REQUIREMENTS

Acid Value Not more than 4.
Free Propylene Glycol Not more than 1.5%.
Heavy Metals (as Pb) Not more than 10 mg/kg.
Hydroxyl Value, Iodine Value, and Saponification Value Not greater than the values stated or within the range claimed by the vendor.
Residue on Ignition Not more than 0.5%.
Soap (as potassium stearate) Not more than 7.0%.
Total Monoester Content Not less than the minimum percentage claimed by the vendor.

TESTS

Acid Value Determine as directed for *Acid Value, Method II*, under *Fats and Related Substances*, Appendix VII.
Free Propylene Glycol Determine as directed under *Free Glycerin or Propylene Glycol*, Appendix VII.
Heavy Metals Prepare and test a 2-g sample as directed in *Method II* under the *Heavy Metals Test*, Appendix IIIB, using 20 µg of lead ion (Pb) in the control (*Solution A*).
Hydroxyl Value Determine as directed under *Method II* in the general method, Appendix VII, using about 2 g, accurately weighed.
Iodine Value Determine by the *Modified Wijs Method*, Appendix VII.
Residue on Ignition Ignite 5 g as directed in the general method, Appendix IIC.

Saponification Value Weigh accurately about 4 g, and proceed as directed in the general method, Appendix VII.

Soap Weigh accurately about 5 g, and proceed as directed in the general method, Appendix VII, using 31.0 as the equivalence factor (e) in the calculation.

Total Monoester Content Determine the percentage of free propylene glycol (G) in the sample as directed under *Free Glycerin or Propylene Glycol*, Appendix VII, and determine the *Hydroxyl Value* (H) as directed in *Method II* of the general method, Appendix VII. Calculate the hydroxyl equivalent of free propylene glycol (F) by the formula

$$561.1G/38.$$

Separate the fatty acids as described under *Lauric Acid* in the monograph for *Polysorbate 20*, and determine the *Acid Value* (AV) of the acids as directed in *Method I* of the general method, Appendix VII.

Calculate the average molecular weight (*mol wt*) of the monoester by the formula

$$(56,109/AV) + 76.10 - 18.02.$$

Finally, calculate the percentage of total monoester in the original sample by the formula

$$(H - F) \times mol\ wt/561.$$

Packaging and Storage Store in well-closed containers.

Propyl Gallate

Gallic Acid, Propyl Ester

$C_{10}H_{12}O_5$ Formula wt 212.20

INS: 310   CAS: [121-79-9]

DESCRIPTION

A fine, white to nearly white, odorless powder having a slightly bitter taste. It is slightly soluble in water and freely soluble in alcohol and in ether.

Functional Use in Foods Antioxidant.

REQUIREMENTS

Identification Place about 5 g of the sample and several boiling chips in a 500-mL round-bottom flask, connect a water-cooled condenser to the flask, and introduce a steady stream of nitrogen into the flask, maintaining the flow of nitrogen at all times during the remainder of the procedure. Pour 100 mL of 1 N sodium hydroxide through the top of the condenser, heat the solution to boiling, boil for 30 min, and cool. Place the reaction flask in an ice bath, and slowly, with occasional swirling, add dilute sulfuric acid (10%) until a pH of 2 to 3 is obtained (using pH paper). Filter the precipitate through a sintered-glass crucible, wash with a minimum amount of water, and dry at 110° for 2 h. The gallic acid so obtained melts at about 240° with decomposition (see Appendix IIB, *Melting Range or Temperature*).

Assay Not less than 98.0% and not more than 102.0% of $C_{10}H_{12}O_5$ after drying.

Heavy Metals (as Pb) Not more than 10 mg/kg.

Loss on Drying Not more than 0.5%.

Melting Range Between 146° and 150°.

Residue on Ignition Not more than 0.1%.

TESTS

Assay Transfer about 200 mg, previously dried at 110° for 4 h and accurately weighed, to a 400-mL beaker, dissolve it in 150 mL of water, and heat to boiling. With constant and vigorous stirring, add 50 mL of bismuth nitrate TS, continue stirring and heating until precipitation is complete, and cool. Filter the yellow precipitate on a tared sintered-glass crucible, wash it with cold dilute nitric acid (1 in 300), and dry at 110° to constant weight. The weight of the precipitate so obtained, multiplied by 0.4866, represents its equivalent of $C_{10}H_{12}O_5$.

Heavy Metals Prepare and test a 2-g sample as directed in *Method II* under the *Heavy Metals Test*, Appendix IIIB, using 20 µg of lead ion (Pb) in the control (*Solution A*).

Loss on Drying, Appendix IIC Dry at 110° for 4 h.

Melting Range Determine as directed for *Melting Range or Temperature*, Appendix IIB, after drying at 110° for 4 h.

Residue on Ignition, Appendix IIC Ignite 2 g as directed in the general method.

Packaging and Storage Store in well-closed containers.

Propylparaben

Propyl *p*-Hydroxybenzoate

$C_{10}H_{12}O_3$ Formula wt 180.20

                                        CAS: [94-13-3]

DESCRIPTION

Small, colorless crystals or a white powder. One g dissolves in about 2500 mL of water at 25°, in about 400 mL of boiling water, in about 1.5 mL of alcohol, and in about 3 mL of ether.

Functional Use in Foods Preservative; antimicrobial agent.

REQUIREMENTS

Identification Dissolve about 500 mg in 10 mL of 1 N sodium hydroxide, and boil for 30 min, allowing the solution to evaporate to a volume of about 5 mL. Cool the mixture, and carefully acidify with 2 N sulfuric acid. Collect the precipitate on a filter when cool, wash it several times with small portions of water, and dry in a desiccator over silica gel. The liberated *p*-hydroxybenzoic acid melts between 212° and 217° (see Appendix IIB, *Melting Range or Temperature*).
Assay Not less than 99.0% and not more than 100.5% of $C_{10}H_{12}O_3$, calculated on the dried basis.
Acidity Passes test.
Heavy Metals (as Pb) Not more than 10 mg/kg.
Loss on Drying Not more than 0.5%.
Melting Range Between 95° and 98°.
Residue on Ignition Not more than 0.05%.

TESTS

Assay Place in a flask about 2 g, accurately weighed, add 40.0 mL of 1 N sodium hydroxide, and rinse the sides of the flask with water. Cover with a watch glass, boil gently for 1 h, cool, and titrate the excess sodium hydroxide with 1 N sulfuric acid to pH 6.5. Perform a blank determination with the same quantities of the same reagents in the same manner, and make any necessary correction (see *General Provisions*). Each mL of 1 N sodium hydroxide is equivalent to 180.2 mg of $C_{10}H_{12}O_3$, calculated on the dried basis.
Acidity Heat 750 mg with 15 mL of water at 80° for 1 min, cool, and filter. The filtrate is acid or neutral to litmus. To 10 mL of the filtrate add 0.2 mL of 0.1 N sodium hydroxide and 2 drops of methyl red TS. The solution is yellow, without even a light cast of pink.
Heavy Metals, Appendix IIIB Dissolve 2 g in 23 mL of acetone, and add 2 mL of 1 N acetic acid, 2 mL of water, and 10 mL of hydrogen sulfide TS. Any color does not exceed that produced in a control made with 23 mL of acetone, 2 mL of 1 N acetic acid, 2 mL of *Standard Lead Solution* (20 μg Pb ion), and 10 mL of hydrogen sulfide TS.
Loss on Drying, Appendix IIC Dry over silica gel for 5 h.
Melting Range Determine as directed for *Melting Range or Temperature*, Appendix IIB.
Residue on Ignition Ignite 4 g as directed in the general method, Appendix IIC.

Packaging and Storage Store in well-closed containers.

Pyridoxine Hydrochloride

5-Hydroxy-6-methyl-3,4-pyridinedimethanol Hydrochloride; Vitamin B_6 Hydrochloride; Vitamin B_6

$C_8H_{11}NO_3 \cdot HCl$ Formula wt 205.64

CAS: [58-56-0]

DESCRIPTION

Colorless or white crystals or a white, crystalline powder. It is stable in air, but is slowly affected by sunlight. Its solutions are acid to litmus, having a pH of about 3. One g dissolves in 5 mL of water and in about 100 mL of alcohol. It is insoluble in ether. It melts at about 206° with some decomposition.

Functional Use in Foods Nutrient; dietary supplement.

REQUIREMENTS

Identification

A. Place 1 mL of a solution containing about 100 μg in each mL into each of two test tubes marked A and B, and add to each tube 2 mL of a 1 in 5 sodium acetate solution. To tube A add 1 mL of water, and to tube B add 1 mL of a 1 in 25 boric acid solution, and mix. Cool both tubes to about 20°, and rapidly add to each tube 1 mL of a 1 in 200 solution of 2,6-dichloroquinonechlorimide in alcohol. A blue color is produced in A, which fades rapidly and becomes red in a few min, but no blue color is produced in B.

B. To 2 mL of a 1 in 200 solution add 0.5 mL of phosphotungstic acid TS. A white precipitate is formed.

C. It gives positive tests for *Chloride*, Appendix IIIA.
Assay Not less than 98.0% and not more than 100.5% of $C_8H_{11}NO_3 \cdot HCl$, calculated on the dried basis.
Chloride Content Not less than 16.9% and not more than 17.6% of Cl, calculated on the dried basis.
Heavy Metals (as Pb) Not more than 0.002%.
Loss on Drying Not more than 0.5%.
Residue on Ignition Not more than 0.1%.

TESTS

Assay Dissolve about 400 mg, accurately weighed, in a mixture of 10 mL of glacial acetic acid and 10 mL of mercuric acetate TS, warming slightly to effect solution. Cool to room temperature, add 2 drops of crystal violet TS, and titrate with 0.1 N perchloric acid. Perform a blank determination (see *General Provisions*), and make any necessary correction. Each mL of 0.1 N perchloric acid is equivalent to 20.56 mg of $C_8H_{11}NO_3 \cdot HCl$.
Chloride Content Dissolve about 500 mg of the sample, accurately weighed, in 50 mL of methanol in a glass-stoppered flask.

Add 5 mL of glacial acetic acid and 2 to 3 drops of eosin Y TS, and titrate with 0.1 N silver nitrate. Each mL of 0.1 N silver nitrate is equivalent to 3.545 mg of Cl.

Heavy Metals Prepare and test a 1-g sample as directed in *Method II* under the *Heavy Metals Test*, Appendix IIIB, using 20 µg of lead ion (Pb) in the control (*Solution A*).

Loss on Drying, Appendix IIC   Dry in a vacuum over silica gel for 4 h.

Residue on Ignition   Ignite 2 g as directed in the general method, Appendix IIC.

Packaging and Storage   Store in tight, light-resistant containers, and avoid exposure to sunlight.

Quinine Hydrochloride

$C_{20}H_{24}N_2O_2 \cdot HCl \cdot 2H_2O$ .HCl.2H$_2$O

Formula wt 396.91

CAS: [130-89-2]

DESCRIPTION

White, silky, glistening needles. It is odorless, has a very bitter taste, and effloresces when exposed to warm air. Its solutions are neutral or alkaline to litmus. One g dissolves in 16 mL of water, in 1 mL of alcohol, in about 7 mL of glycerin, and in about 1 mL of chloroform. It is very slightly soluble in ether.

Functional Use in Foods   Flavoring agent.

REQUIREMENTS

Identification
A. To 5 mL of a 1 in 1000 solution of the sample add 1 or 2 drops of bromine TS followed by 1 mL of 6 N ammonium hydroxide. The liquid acquires an emerald green color due to the formation of thalleioquin.

B. A 1 in 100 solution is levorotatory (see Appendix IIB).

C. It gives positive tests for *Chloride*, Appendix IIIA.

Assay   Not less than 99.0% and not more than 101.5% of $C_{20}H_{24}N_2O_2 \cdot HCl$, calculated on the dried basis.

Barium   Passes test.

Chloroform–Alcohol Insoluble Substances   Passes test.

Heavy Metals (as Pb)   Not more than 10 mg/kg.

Loss on Drying   Not more than 10.0%.

Other Cinchona Alkaloids   Passes test.

Readily Carbonizable Substances   Passes test.

Residue on Ignition   Not more than 0.15%.

Specific Rotation $[\alpha]_D^{25°}$: Between −247° and −252°.

Sulfate   Not more than 0.05%.

TESTS

Assay   Dissolve about 150 mg of the sample, accurately weighed, in 20 mL of acetic anhydride, add 2 drops of malachite green TS and 5.5 mL of mercuric acetate TS, and titrate with 0.1 N perchloric acid from a microburet to a yellow endpoint. Perform a blank determination (see *General Provisions*). Each mL of 0.1 N perchloric acid is equivalent to 17.99 mg of $C_{20}H_{24}N_2O_2 \cdot HCl$.

Barium   To 10 mL of a hot solution of the sample (1 in 20) add 1 mL of 2 N sulfuric acid. No turbidity is produced.

Chloroform–Alcohol Insoluble Substances   One g dissolves completely in 7 mL of a mixture of 2 volumes of chloroform and 1 volume of absolute alcohol.

Heavy Metals   Prepare and test a 2-g sample as directed in *Method II* under the *Heavy Metals Test*, Appendix IIIB, using 20 µg of lead ion (Pb) in the control (*Solution A*).

Loss on Drying, Appendix IIC   Dry at 120° for 3 h.

Other Cinchona Alkaloids   Dissolve about 2.5 g in 60 mL of water in a separator, add 10 mL of 6 N ammonium hydroxide, extract the mixture successively with 30 mL and 20 mL of chloroform, and evaporate the combined chloroform extracts to dryness on a steam bath. Dissolve 1.5 g of the residue in 25 mL of alcohol, dilute the solution with 50 mL of hot water, add 1 N sulfuric acid (about 5 mL) until the solution is acid, using 2 drops of methyl red TS as the indicator, and neutralize the excess of acid with 1 N sodium hydroxide. Evaporate the solution to dryness on a steam bath, powder the residue, and agitate it in a test tube with 20 mL of water at 65° for 30 min. Cool the mixture to 15°, macerate it at this temperature for 2 h with occasional shaking, and then filter it through a filter paper (8 to 10 cm). Transfer 5 mL of the filtrate, at a temperature of 15°, to a test tube, and mix it gently, with shaking, with 6 mL of 6 N ammonium hydroxide (which must contain between 10% and 10.2% of NH_3, have a temperature of 15°, and be added at once). A clear liquid is produced.

Readily Carbonizable Substances, Appendix IIB   Dissolve 100 mg in 2 mL of 95% sulfuric acid. The solution is no darker than *Matching Fluid M*.

Residue on Ignition   Ignite 1 g as directed in the general method, Appendix IIC.

Specific Rotation, Appendix IIB   Determine in a solution containing 200 mg in 10 mL of 0.1 N hydrochloric acid.

Sulfate, Appendix IIIB   Any turbidity produced by a 500-mg sample does not exceed that shown in a control containing 250 µg of sulfate (SO_4).

Packaging and Storage   Store in tight, light-resistant containers.

Quinine Sulfate

$(C_{20}H_{24}N_2O_2)_2 \cdot H_2SO_4 \cdot 2H_2O$

$(C_{20}H_{24}N_2O_2)_2 \cdot H_2SO_4$ Formula wt 782.96

CAS: anhydrous [804-63-7]

DESCRIPTION

Fine, white, needlelike crystals, usually lusterless, making a light and readily compressible mass. It is odorless and has a persistent, very bitter taste. It darkens on exposure to light. Its saturated solution is neutral or alkaline to litmus. One g dissolves in about 500 mL of water and in about 120 mL of alcohol at 25°, in about 35 mL of water at 100°, and in about 10 mL of alcohol at 80°. It is slightly soluble in chloroform and in ether, but is freely soluble in a mixture of 2 volumes of chloroform and 1 volume of absolute alcohol.

Functional Use in Foods   Flavoring agent.

REQUIREMENTS

Identification

   A. Acidify a saturated solution of the sample with 2 N sulfuric acid. The resulting solution has a vivid blue fluorescence and is levorotatory (see Appendix IIB).

   B. To 5 mL of a 1 in 1000 solution add 1 or 2 drops of bromine TS followed by 1 mL of 6 N ammonium hydroxide. The liquid acquires an emerald green color due to the formation of thalleioquin.

   C. A 1 in 50 solution made with the aid of a few drops of hydrochloric acid gives positive tests for *Sulfate*, Appendix IIIA.

Assay   Not less than 99.0% and not more than 101.0% of $(C_{20}H_{24}N_2O_2)_2 \cdot H_2SO_4$, calculated on the dried basis.

Chloroform–Alcohol Insoluble Substances   Not more than 0.1%.

Heavy Metals (as Pb)   Not more than 10 mg/kg.

Loss on Drying   Not more than 5.0%.

Other Cinchona Alkaloids   Passes test.

Readily Carbonizable Substances   Passes test.

Residue on Ignition   Not more than 0.05%.

Specific Rotation $[\alpha]_D^{25°}$: Between −240° and −244°.

TESTS

Assay   Dissolve about 200 mg of the sample, accurately weighed, in 20 mL of acetic anhydride, add 2 drops of malachite green TS, and titrate with 0.1 N perchloric acid from a microburet to a yellow endpoint. Perform a blank determination (see *General Provisions*). Each mL of 0.1 N perchloric acid is equivalent to 24.90 mg of $(C_{20}H_{24}N_4O_2)_2 \cdot H_2SO_4$.

Chloroform–Alcohol Insoluble Substances   Warm 2 g of the sample with 15 mL of a mixture of 2 volumes of chloroform and 1 volume of absolute alcohol at 50° for 10 min. Filter through a tared, sintered-glass filter, using gentle suction, and wash the filter with five 10-mL portions of the chloroform–alcohol mixture. Dry at 105° for 1 h, cool, and weigh.

Heavy Metals   Prepare and test a 2-g sample as directed in *Method II* under the *Heavy Metals Test*, Appendix IIIB, using 20 μg of lead ion (Pb) in the control (*Solution A*).

Loss on Drying, Appendix IIC   Dry at 120° for 3 h.

Other Cinchona Alkaloids   Agitate 1.8 g, previously dried at 50° for 2 h, with 20 mL of water at 65° for 30 min. Cool the mixture to 15°, macerate it at this temperature for 2 h with occasional shaking, and then filter it through a filter paper (8 to 10 cm). Transfer 5 mL of the filtrate, at a temperature of 15°, to a test tube, and mix it gently, without shaking, with 6 mL of 6 N ammonium hydroxide (which must contain between 10% and 10.2% of NH_3, have a temperature of 15°, and be added at once). A clear liquid is produced.

Readily Carbonizable Substances, Appendix IIB   Dissolve 200 mg in 5 mL of 95% sulfuric acid. The solution is no darker than *Matching Fluid M*.

Residue on Ignition   Ignite 2 g as directed in the general method, Appendix IIC.

Specific Rotation, Appendix IIB   Determine in a solution containing 200 mg in 10 mL of 0.1 N hydrochloric acid.

Packaging and Storage   Store in well-closed, light-resistant containers.

Rapeseed Oil, Fully Hydrogenated

Fully Hydrogenated Rapeseed Oil

INS: 441   CAS: [84681-71-0]

DESCRIPTION

A white, waxy, odorless solid that is a mixture of triglycerides. The saturated fatty acids are found in the same proportions that result from the full hydrogenation of fatty acids occurring in natural high erucic acid rapeseed oil. The rapeseed oil is obtained from *Brassica napus* and *Brassica campestris* of the family *Cruciferae*. It is made by hydrogenating high erucic acid rapeseed oil in the presence of a nickel catalyst at temperatures not exceeding 245°.

Functional Use in Foods   Coating agent; emulsifying agent; stabilizer; thickener; formulation aid; texturizer.

REQUIREMENTS

Labeling Rapeseed oil products that have been fully hydrogenated should be labeled as Fully Hydrogenated Rapeseed Oil. Label to indicate 1-monoglyceride content.

Identification Fully Hydrogenated Rapeseed Oil exhibits the following composition profile of fatty acids as determined under *Fatty Acid Composition*, Appendix VII.

Fatty Acid:	14:0	16:0	18:0	18:1	18:2
Weight % (Range):	<1.0	3–5	38–42	1.0	<1.0

Fatty Acid:	20:0	20:1	22:0	22:1	24:0
Weight % (Range):	8–10	<1.0	42–50	<1.0	1.0–2.0

Acid Value Not more than 6.
Color (AOCS-Wesson) Not more than 1.5 red/15 yellow.
Erucic Acid Not more than 1.0%.
Free Fatty Acids (as oleic acid) Not more than 2.0%.
Heavy Metals (as Pb) Not more than 5 mg/kg.
Iodine Value Not more than 4.
Lead Not more than 0.1 mg/kg.
Peroxide Value Not more than 2.0 meq/kg.
Residue on Ignition Not more than 0.5%.
Unsaponifiable Matter Not more than 1.5%.
Water Not more than 0.05%.

ADDITIONAL REQUIREMENTS

The following specification should conform to the representations of the vendor: *1-Monoglyceride Content*.

TESTS

Acid Value Determine as directed for *Acid Value, Method II*, under *Fats and Related Substances*, Appendix VII.
Color Proceed as directed for *Color* (AOCS-Wesson) under *Fats and Related Substances*, Appendix VII. Use a 13.34-cm cell.
Erucic Acid Determine as part of *Fatty Acid Composition*, Appendix VII.
Free Fatty Acids Proceed as directed under *Free Fatty Acids*, Appendix VII, using the following equivalence factor (e) in the formula given in the procedure:

Free fatty acids as oleic acid, $e = 28.2$.

Heavy Metals Prepare and test a 2-g sample as directed in *Method II* under the *Heavy Metals Test*, Appendix IIIB, using 10 μg of lead ion (Pb) in the control (*Solution A*).
Iodine Value Proceed as directed under *Modified Wijs Method*, Appendix VII.
Lead Determine as directed under *Method II* in the *Atomic Absorption Spectrophotometric Graphite Furnace Method* under the *Lead Limit Test*, Appendix IIIB, using a 1-g sample.
1-Monoglyceride Content Determine as directed in the general method, Appendix VII.
Peroxide Value Proceed as directed under *Peroxide Value* in the monograph for *Hydroxylated Lecithin*. However, after the addition of saturated potassium iodide and mixing, mix the solution for only 1 min and begin the titration immediately instead of allowing the solution to stand for 10 min.
Residue on Ignition Ignite 5 g as directed in the general method, Appendix IIC.
Unsaponifiable Matter Proceed as directed under *Unsaponifiable Matter*, Appendix VII.
Water Proceed as directed under *Water Determination* by the *Karl Fischer Titrimetric Method*, Appendix IIB. However, in place of 35 to 40 mL of methanol, use 50 mL of a 1:1 chloroform–methanol mixture to dissolve the sample.

Packaging and Storage Store in well-closed containers.

Rapeseed Oil, Superglycerinated

Superglycerinated Fully Hydrogenated Rapeseed Oil

DESCRIPTION

A white solid that is a mixture of mono-, di-, and triglycerides, with triglycerides as a minor component. The saturated fatty acids are found in the same proportions that result from the full hydrogenation of fatty acids occurring in natural high erucic acid rapeseed oil. The rapeseed oil is typically obtained by *n*-hexane extraction from *Brassica napus* and *Brassica campestris* of the family *Cruciferae*. It is made by adding excess glycerin to fully hydrogenated rapeseed oil and heating, in the presence of sodium hydroxide catalyst, to about 165° under partial vacuum and steam sparging agitation.

Functional Use in Foods Coating agent; emulsifying agent; formulation aid; texturizer.

REQUIREMENTS

Labeling Indicate Rapeseed Oil products that have added glycerin (glycerol) and are fully hydrogenated as fully hydrogenated and superglycerinated Rapeseed Oil. The *1-Monoglyceride Content* and *Hydroxyl Value* should conform to the representations of the vendor, and this should be indicated as well.
Identification Superglycerinated Rapeseed Oil exhibits the same fatty acid composition as fully hydrogenated rapeseed oil. It exhibits the following composition profile of fatty acids as determined under *Fatty Acid Composition*, Appendix VII.

Fatty Acid:	14:0	16:0	18:0	18:1	18:2
Weight % (Range):	<1.0	3–5	38–42	1.0	<1.0

Fatty Acid:	20:0	20:1	22:0	22:1	24:0
Weight % (Range):	8–10	<1.0	42–50	<1.0	1.0–2.0

Acid Value Not more than 6.
Color (AOCS-Wesson) Not more than 1.5 red/15 yellow.

Erucic Acid Not more than 1.0%.
Free Fatty Acids (as oleic acid) Not more than 2.0%.
Free Glycerin Not more than 1%.
Heavy Metals Not more than 5 mg/kg.
Iodine Value Not more than 4.
Lead Not more than 0.1 mg/kg.
Peroxide Value Not more than 2.0 meq/kg.
Residue on Ignition Not more than 0.5%.
Unsaponifiable Matter Not more than 1.5%.
Water Not more than 0.05%.

TESTS

1-Monoglyceride Content Determine as directed in the general method, Appendix VII.
Acid Value Determine as directed for *Acid Value, Method II*, under *Fats and Related Substances*, Appendix VII.
Color Proceed as directed for *Color* (AOCS-Wesson) under *Fats and Related Substances*, Appendix VII. Use a 13.34-cm cell.
Erucic Acid Determine as part of *Fatty Acid Composition*, Appendix VII.
Free Fatty Acids Proceed as directed under *Free Fatty Acids*, Appendix VII, using the following equivalence factor (*e*) in the formula given in the procedure:

Free fatty acids as oleic acid, *e* = 28.2.

Free Glycerin Determine as directed in the general method, Appendix VII.
Heavy Metals Prepare and test a 2-g sample as directed in *Method II* under the *Heavy Metals Test*, Appendix IIIB, using 10 μg of lead ion (Pb) in the control (*Solution A*).
Hydroxyl Value Determine as directed under *Method II* in the general method, Appendix VII.
Iodine Value Proceed as directed under the *Modified Wijs Method*, Appendix VII.
Lead Determine as directed under *Method II* in the *Atomic Absorption Spectrophotometric Graphite Furnace Method* under the *Lead Limit Test*, Appendix IIIB, using a 1-g sample.
Peroxide Value Proceed as directed under *Peroxide Value* in the monograph for *Hydroxylated Lecithin*. However, after the addition of saturated potassium iodide and mixing, mix the solution for only 1 min and begin the titration immediately instead of allowing the solution to stand for 10 min.
Residue on Ignition Ignite 5 g as directed in the general method, Appendix IIC.
Total Monoglycerides Determine as directed in the general method, Appendix VII.
Unsaponifiable Matter Proceed as directed under *Unsaponifiable Matter*, Appendix VII.
Water Proceed as directed under *Water Determination* using the *Karl Fischer Titrimetric Method*, Appendix IIB. However, in place of 35 to 40 mL of methanol, use 50 mL of a 1:1 chloroform–methanol mixture to dissolve the sample.

Packaging and Storage Store in well-closed containers.

Riboflavin

Vitamin B_2

$C_{17}H_{20}N_4O_6$ Formula wt 376.37
INS: 101(i) CAS: [83-88-5]

DESCRIPTION

A yellow to orange yellow, crystalline powder having a slight odor. It melts at about 280° with decomposition, and its saturated solution is neutral to litmus. When dry, it is not affected by diffused light, but when in solution, light induces deterioration. One g dissolves in from 3000 to about 20,000 mL of water, the variations being due to differences in the internal crystalline structure. It is less soluble in alcohol than in water. It is insoluble in ether and in chloroform, but is very soluble in dilute solutions of alkalies.

Functional Use in Foods Nutrient; dietary supplement.

REQUIREMENTS

Identification A solution of 1 mg in 100 mL of water is pale greenish yellow by transmitted light and has an intense yellowish green fluorescence that disappears upon the addition of mineral acids or alkalies.
Assay Not less than 98.0% and not more than 102.0% of $C_{17}H_{20}N_4O_6$, calculated on the dried basis.
Loss on Drying Not more than 1.5%.
Lumiflavin Passes test.
Residue on Ignition Not more than 0.3%.
Specific Rotation $[\alpha]_D^{25°}$: Between +56.5° and +59.5°, calculated on the dried basis.

TESTS

Assay (Note: Conduct this assay so that the solutions are protected from direct sunlight at all stages.) Place about 50 mg, accurately weighed, in a 1000-mL volumetric flask containing about 50 mL of water. Add 5 mL of 6 *N* acetic acid and sufficient water to make about 800 mL. Heat on a steam bath, protected from light, with frequent agitation until dissolved. Cool to about 25°, add water to volume, and mix. Dilute this solution with water, quantitatively and stepwise, to bring it within the operating sensitivity of the fluorometer used.

In the same manner, prepare a standard solution to contain, in each mL, a quantity of USP Riboflavin Reference Standard, accurately weighed, equivalent to that of the solution prepared as directed in the preceding paragraph, and measure the intensity of its fluorescence in a fluorometer at about 530 nm, using an excitation wavelength of about 440 nm. Immediately after the reading, add to the solution about 10 mg of sodium hydrosulfite, stirring with a glass rod until dissolved, and at once measure the fluorescence again. The difference between the two readings represents the intensity of the fluorescence due to the standard.

Similarly, measure the intensity of the fluorescence of the final solution of the Riboflavin being assayed, before and after the addition of sodium hydrosulfite. Calculate the quantity of $C_{17}H_{20}N_4O_6$ in the final solution of Riboflavin by the formula

$$C(I_U/I_S),$$

in which C is the concentration, in μg/mL, of USP Riboflavin Reference Standard in the final solution of the standard and I_U and I_S are the corrected fluorescence values observed for the solutions of the Riboflavin and the standard, respectively.

Loss on Drying, Appendix IIC Dry at 105° for 2 h.

Lumiflavin Prepare alcohol-free chloroform as follows: Shake 20 mL of chloroform gently but thoroughly with 20 mL of water for 3 min, draw off the chloroform layer, and wash twice more with 20-mL portions of water. Finally, filter the chloroform through a dry filter paper, shake it well for 5 min with 5 g of powdered anhydrous sodium sulfate, allow the mixture to stand for 2 h, and decant or filter the clear chloroform.

Shake 25 mg of the sample with 10 mL of the alcohol-free chloroform for 5 min, and filter. The absorbance of the filtrate, determined in a 1-cm cell at 440 nm with a suitable spectrophotometer, using alcohol-free chloroform as the blank, does not exceed 0.025.

Residue on Ignition, Appendix IIC Ignite 1 g as directed in the general method.

Specific Rotation, Appendix IIB Determine in a solution of hydrochloric acid containing 50 mg in each 10 mL.

Packaging and Storage Store in tight, light-resistant containers.

Riboflavin 5'-Phosphate Sodium

Riboflavin 5'-Phosphate Ester Monosodium Salt

$C_{17}H_{20}N_4NaO_9P \cdot 2H_2O$ Formula wt 514.36
INS: 101(ii)   CAS: [130-40-5]

DESCRIPTION

A fine, orange yellow, crystalline powder having a slight odor. One g dissolves in about 30 mL of water. When dry, it is not affected by diffused light, but when in solution, light induces deterioration rapidly. It is hygroscopic.

Functional Use in Foods Nutrient; dietary supplement.

REQUIREMENTS

Identification A solution of 1.5 mg in 100 mL of water responds to the *Identification Test* in the monograph for *Riboflavin*.

Assay Not less than the equivalent of 73.0% and not more than the equivalent of 79.0% of riboflavin ($C_{17}H_{20}N_4O_6$), calculated on the dried basis.

Free Phosphate Not more than 1.0%, calculated as PO_4.

Free Riboflavin Not more than 6.0%, calculated on the dried basis.

Loss on Drying Not more than 7.5%.

pH of a 1 in 100 Solution Between 5.0 and 6.5.

Residue on Ignition Not more than 25.0%.

Riboflavin Diphosphate Not more than 6.0% as riboflavin, calculated on the dried basis.

Specific Rotation $[\alpha]_D^{25°}$: Between +37.0° and +42.0°, calculated on the dried basis.

TESTS

Assay (**Note:** Use low-actinic glassware, and conduct this assay so that all solutions are protected from direct sunlight at all stages.)

Assay Preparation Transfer about 50 mg of the sample, accurately weighed, into a 250-mL Erlenmeyer flask, add 20 mL of pyridine and 75 mL of water, and dissolve the sample by frequent shaking. Transfer the solution to a 1000-mL volumetric flask, dilute to volume with water, and mix. Transfer 10.0 mL of this solution into a second 1000-mL volumetric flask, add sufficient 0.1 N sulfuric acid (about 4 mL) so that the final pH of the solution is between 5.9 and 6.1, dilute to volume with water, and mix.

Standard Preparation Transfer about 35 mg of USP Riboflavin Reference Standard, accurately weighed, into a 250-mL Erlenmeyer flask, add 20 mL of pyridine and 75 mL of water, and dissolve the riboflavin by frequent shaking. Transfer the solution to a 1000-mL volumetric flask, dilute to volume with water, and mix. Transfer 10.0 mL of this solution into a second 1000-mL volumetric flask, add sufficient 0.1 N sulfuric acid (about 4 mL) so that the final pH of the solution is between 5.9 and 6.1, dilute to volume with water, and mix.

Procedure With a suitable fluorometer, determine the intensity of the fluorescence of each solution at about 530 nm, using an excitation wavelength of about 440 nm. Record the fluorescence of the *Assay Preparation* as I_U, and that of the *Standard Preparation* as I_S. Calculate the quantity, in mg of $C_{17}H_{20}N_4O_6$ in the sample taken, by the formula

$$100C \times I_U/I_S,$$

in which C is the exact concentration, in µg/mL, of the *Standard Preparation*, corrected for loss on drying.

Free Phosphate

Standard Preparation Transfer 220.0 mg of monobasic potassium phosphate, KH_2PO_4, into a 1000-mL volumetric flask, dissolve in and dilute to volume with water, and mix. Transfer 20.0 mL of this solution into a 100-mL volumetric flask, dilute to volume with water, and mix.

Test Preparation Transfer 300.0 mg of the sample into a 100-mL volumetric flask, dissolve in and dilute to volume with water, and mix.

Acid Molybdate Solution Dilute 25 mL of ammonium molybdate solution (7 g of $(NH_4)_6Mo_7O_{24}\cdot 4H_2O$ in sufficient water to make 100 mL) to 200 mL with water, and then add slowly 25 mL of 7.5 N sulfuric acid.

Ferrous Sulfate Solution Just before use, prepare a 10% aqueous ferrous sulfate solution containing 2 mL of 7.5 N sulfuric acid per 100 mL of final solution.

Procedure Transfer 10.0 mL each of the *Standard Preparation* and of the *Test Preparation* into separate 50-mL Erlenmeyer flasks, add 10.0 mL of *Acid Molybdate Solution* and 5.0 mL of *Ferrous Sulfate Solution* to each flask, and mix. Determine the absorbance of each solution in a 1-cm cell at 700 nm with a suitable spectrophotometer, using as the blank a mixture of 10.0 mL of water, 10.0 mL of *Acid Molybdate Solution*, and 5.0 mL of *Ferrous Sulfate Solution*. The absorbance of the solution from the *Test Preparation* is not greater than that of the *Standard Preparation*.

Free Riboflavin and Riboflavin Diphosphate (Note: Conduct this test so that all solutions are protected from actinic light at all stages, preferably by using low-actinic glassware.)

Mobile Phase Mix 850 mL of 0.054 M monobasic potassium phosphate with 150 mL of methanol, filter, and degas the solution. Make adjustments if necessary.

Standard Preparation Transfer 60 mg of USP Riboflavin Reference Standard, accurately weighed, to a 250-mL volumetric flask, dissolve carefully in 1 mL of hydrochloric acid, dilute with water to volume, and mix. Pipet a 4-mL aliquot into a 100-mL volumetric flask, dilute with *Mobile Phase* to volume, and mix.

Test Preparation Transfer 100.0 mg of the sample to a 100-mL volumetric flask, dissolve in 50 mL of water, dilute with *Mobile Phase* to volume, and mix. Pipet 8 mL of this solution into a 50-mL volumetric flask, dilute with *Mobile Phase* to volume, and mix.

System Suitability Preparation Dissolve USP Phosphated Riboflavin Reference Standard in water to obtain a solution containing 2 mg/mL. Add an equal volume of *Mobile Phase*, and mix. Dilute 8 mL of this solution with *Mobile Phase* to 50 mL, and mix.

Chromatographic System, Appendix IIA The liquid chromatograph is equipped with a fluorometric detector set at 440 nm excitation wavelength and provided with a 470-nm emission filter or set at about 530 nm for a fluorescence detector that uses a monochromator for emission wavelength selection, and a 3.9-mm × 30-cm column that is packed with µBondapak, or equivalent. The flow rate is about 2.0 mL/min. Chromatograph the *System Suitability Preparation*, and record the peak responses. The retention time for riboflavin 5'-monophosphate is about 20 to 25 min, and the approximate relative retention times for the components are

Riboflavin 3'4'-diphosphate:	0.23
Riboflavin 3'5'-diphosphate:	0.39
Riboflavin 4'5'-diphosphate:	0.58
Riboflavin 3'-monophosphate:	0.70
Riboflavin 4'-monophosphate:	0.87
Riboflavin 5'-monophosphate:	1.00
Riboflavin:	1.63

The resolution, R, between the peaks for riboflavin 4'-monophosphate and riboflavin 5'-monophosphate is not less than 1.0, and the relative standard deviation of the response for riboflavin 5'monophosphate in replicate injections is not more than 1.5%.

Procedure Separately inject equal volumes (about 100 µL) of the *Standard Preparation*, the *Test Preparation*, and the *System Suitability Preparation* into the chromatograph. Measure the peak responses obtained from the *Standard Preparation* and the *Test Preparation*, identifying the peaks to be measured in the chromatogram of the *Test Preparation* by comparison of retention times with those of the peaks in the chromatogram of the *System Suitability Preparation*. Calculate the percentage of free riboflavin by the formula

$$625C(r_F/r_S),$$

and calculate the percentage of riboflavin in the form of riboflavin diphosphates by the formula

$$625C(r_D/r_S),$$

in which C is the concentration, in mg/mL, of USP Riboflavin Reference Standard in the *Standard Preparation*; r_F is the riboflavin peak response, if any, obtained from the *Test Preparation*; r_D is the sum of the responses for any of the three riboflavin diphosphate peaks obtained from the *Test Preparation*; and r_S is the riboflavin peak response obtained from the *Standard Preparation*.

Loss on Drying, Appendix IIC Dry at 100° in a vacuum over phosphorus pentoxide for 5 h.
pH of a 1 in 100 Solution Determine by the *Potentiometric Method*, Appendix IIB.
Residue on Ignition Ignite a 1-g sample as directed in the general method, Appendix IIC.
Specific Rotation, Appendix IIB Transfer about 750 mg, accurately weighed, into a 50-mL volumetric flask, dissolve in and dilute to volume with 20% hydrochloric acid, and mix. Determine the rotation in a 1-dm tube within 15 min.

Packaging and Storage Store in tight, light-resistant containers.

Rice Bran Wax

INS: 908   CAS: [8016-60-2]

DESCRIPTION

A refined wax obtained from rice bran. It is hard, slightly crystalline, and ranges in color from tan to light brown. It is soluble in chloroform, but is insoluble in water.

Functional Use in Foods Masticatory substance in chewing gum base; coating agent.

REQUIREMENTS

Identification The sample is melted and prepared for analysis on a potassium bromide plate. The infrared absorption spectrum of the sample exhibits maxima at the same wavelengths as the typical spectrum, as shown in the section on *Infrared Spectra (Series C: Other Substances)*.
Free Fatty Acids Not more than 10.0%.
Heavy Metals (as Pb) Not more than 0.002%.
Iodine Value Not more than 20.
Lead Not more than 3 mg/kg.
Melting Range Between 75° and 80°.
Saponification Value Between 75 and 120.

TESTS

Free Fatty Acids Determine as directed in the general method procedure, Appendix VII.
Heavy Metals Prepare and test a 1-g sample as directed in *Method II* under *Heavy Metals Test*, Appendix IIIB, using 20 μg of lead ion (Pb) in the control (*Solution A*).
Iodine Value Determine by the *Modified Wijs Method*, Appendix VII.
Lead Prepare a *Sample Solution* as directed in the general method under *Chewing Gum Base*, Appendix IV. This solution meets the requirements of the *Lead Limit Test*, Appendix IIIB, using 10 μg of lead ion (Pb) in the control.

Melting Range Determine as directed for *Melting Range or Temperature, Procedure for Class II*, Appendix IIB.
Saponification Value Determine as directed in the general method, Appendix VII.

Packaging and Storage Store in well-closed containers.

Rosemary Oil

CAS: [8000-25-7]

DESCRIPTION

The volatile oil obtained by steam distillation from the fresh flowering tops of *Rosemarinus officinalis* L. (Fam. *Labiatae*). It is a colorless or pale yellow liquid having the characteristic odor of rosemary and a warm, camphoraceous taste.

Functional Use in Foods Flavoring agent.

REQUIREMENTS

Identification The infrared absorption spectrum of the sample exhibits relative maxima (that may vary in intensity) at the same wavelengths (or frequencies) as those shown in the respective spectrum in the section on *Infrared Spectra (Series A: Essential Oils)*, using the same test conditions as specified therein.
Assay for Esters Not less than 1.5% of esters, calculated as bornyl acetate ($C_{12}H_{20}O_2$).
Assay for Total Borneol Not less than 8.0% of borneol ($C_{10}H_{18}O$).
Angular Rotation Between −5° and +10°.
Heavy Metals (as Pb) Passes test.
Refractive Index Between 1.464 and 1.476 at 20°.
Solubility in Alcohol Passes test.
Specific Gravity Between 0.894 and 0.912.

TESTS

Assay for Esters Weigh accurately about 10 mL, and proceed as directed under *Ester Determination*, Appendix VI, using 98.15 as the equivalence factor (*e*) in the calculation.
Assay for Total Borneol Proceed as directed under *Total Alcohols*, Appendix VI, using 5 mL of the dried, acetylated oil, accurately weighed, for the saponification. Calculate the percentage of total borneol by the formula

$$7.712A(1 - 0.0021E)/(B - 0.021A),$$

in which A is the difference between the number of mL of 0.5 N hydrochloric acid required for the sample and the number of mL of 0.5 N hydrochloric acid required for the residual blank titration, B is the weight of the acetylated oil taken, and E is the percentage of esters calculated as bornyl acetate ($C_{12}H_{20}O_2$).

Angular Rotation Determine in a 100-mm tube as directed under *Optical (Specific) Rotation*, Appendix IIB.
Heavy Metals Shake 10 mL of the oil with an equal volume of water to which 1 drop of hydrochloric acid has been added, and pass hydrogen sulfide through the mixture until it is saturated. No darkening in color is produced in either the oil or the water.
Refractive Index, Appendix IIB Determine with an Abbé or other refractometer of equal or greater accuracy.
Solubility in Alcohol Proceed as directed in the general method, Appendix VI. One mL dissolves in 1 mL of 90% alcohol. Upon further dilution, the solution may become turbid.
Specific Gravity Determine by any reliable method (see *General Provisions*).

Packaging and Storage Store in full, tight containers. Avoid exposure to excessive heat.

Rose Oil

CAS: [8007-01-0]

DESCRIPTION

The volatile oil obtained by steam distillation from the fresh flowers of *Rosa gallica* L., *Rosa damascena* Miller, *Rosa alba* L., *Rosa centifolia* L., and varieties of these species (Fam. *Rosaceae*). It is a colorless or yellow liquid having the characteristic odor and taste of rose. At 25° it is a viscous liquid. Upon gradual cooling it changes to a translucent, crystalline mass, which may be liquefied by warming.

Functional Use in Foods Flavoring agent.

REQUIREMENTS

Identification The infrared absorption spectrum of the sample exhibits relative maxima (that may vary in intensity) at the same wavelengths (or frequencies) as those shown in the respective spectrum in the section on *Infrared Spectra (Series A: Essential Oils)*, using the same test conditions as specified therein.
Angular Rotation Between −1° and −4°.
Heavy Metals (as Pb) Passes test.
Refractive Index Between 1.457 and 1.463 at 30°.
Solubility Passes test.
Specific Gravity Between 0.848 and 0.863 at 30°/15°.

TESTS

Angular Rotation Determine in a 100-mm tube as directed under *Optical (Specific) Rotation*, Appendix IIB.
Heavy Metals Shake 10 mL of the oil with an equal volume of water to which 1 drop of hydrochloric acid has been added, and pass hydrogen sulfide through the mixture until it is saturated. No darkening in color is produced in either the oil or the water.
Refractive Index, Appendix IIB Determine with an Abbé or other refractometer of equal or greater accuracy.
Solubility One mL is miscible with 1 mL of chloroform without turbidity. Add 20 mL of 90% alcohol to this mixture. The resulting liquid is neutral or acid to moistened litmus paper and, upon standing at 20°, deposits crystals within 5 min.
Specific Gravity Determine by any reliable method (see *General Provisions*).

Packaging and Storage Store in full, tight containers in a cool place protected from light.

Rue Oil

CAS: [8014-29-7]

DESCRIPTION

The volatile oil obtained by steam distillation from the fresh blossoming plants *Ruta graveolens* L., *Ruta montana* L., or *Ruta bracteosa* L. (Fam. *Rutaceae*). It is a yellow to yellow amber liquid having a characteristic fatty odor. It is soluble in most fixed oils and in mineral oil, but it is relatively insoluble in glycerin and in propylene glycol.

Functional Use in Foods Flavoring agent.

REQUIREMENTS

Identification The infrared absorption spectrum of the sample exhibits relative maxima (that may vary in intensity) at the same wavelengths (or frequencies) as those shown in the respective spectrum in the section on *Infrared Spectra (Series A: Essential Oils)*, using the same test conditions as specified therein.
Assay Not less than 90.0% of ketones, calculated as methyl nonyl ketone ($C_{11}H_{22}O$).
Angular Rotation Between −1° and +3°.
Heavy Metals (as Pb) Passes test.
Refractive Index Between 1.430 and 1.440 at 20°.
Solidification Point Between 7.5° and 10.5°.
Solubility in Alcohol Passes test.
Specific Gravity Between 0.826 and 0.838.

TESTS

Assay Weigh accurately about 1 g, and proceed as directed under *Aldehydes and Ketones—Hydroxylamine Method*, Appendix VI, using 85.10 as the equivalence factor (e) in the calculation.
Angular Rotation Determine in a 100-mm tube as directed under *Optical (Specific) Rotation*, Appendix IIB.

Heavy Metals Shake 10 mL of the oil with an equal volume of water to which 1 drop of hydrochloric acid has been added, and pass hydrogen sulfide through the mixture until it is saturated. No darkening in color is produced in either the oil or the water.
Refractive Index, Appendix IIB Determine with an Abbé or other refractometer of equal or greater accuracy.
Solidification Point Determine as directed in the general method, Appendix IIB.
Solubility in Alcohol Proceed as directed in the general method, Appendix VI. One mL dissolves in 4 mL of 70% alcohol, occasionally with opalescence or precipitation of solids.
Specific Gravity Determine by any reliable method (see *General Provisions*).

Packaging and Storage Store in full, tight, preferably glass, tin-lined, or aluminum containers in a cool place protected from light.

Saccharin

o-Benzosulfimide; Gluside; 1,2-Benzisothiazolin-3-one 1,1-Dioxide

$C_7H_5NO_3S$ Formula wt 183.19
INS: 954   CAS: [81-07-2]

DESCRIPTION

White crystals or a white, crystalline powder. It is odorless or has a faint, aromatic odor. It is intensely sweet. Its solutions are acid to litmus. One g is soluble in 290 mL of water at 25°, in 25 mL of boiling water, and in 30 mL of alcohol. It is slightly soluble in chloroform and in ether, and is readily dissolved by dilute solutions of ammonia, solutions of alkali hydroxides, or solutions of alkali carbonates with the evolution of carbon dioxide.

Functional Use in Foods Nonnutritive sweetener.

REQUIREMENTS

Identification
A. Dissolve about 100 mg in 5 mL of sodium hydroxide solution (1 in 20), evaporate to dryness, and gently fuse the residue over a small flame until it no longer evolves ammonia. After the residue has cooled, dissolve it in 20 mL of water, neutralize the solution with 2.7 *N* hydrochloric acid, and filter. The addition of a drop of ferric chloride TS to the filtrate produces a violet color.
B. Mix 20 mg with 40 mg of resorcinol, add 10 drops of sulfuric acid, and heat the mixture in a liquid bath at 200° for 3 min. After cooling, add 10 mL of water and an excess of 1 *N* sodium hydroxide. A fluorescent green liquid results.
Assay Not less than 98.0% and not more than 101.0% of $C_7H_5NO_3S$ after drying.
Benzoic and Salicylic Acids Passes test.
Heavy Metals (as Pb) Not more than 10 mg/kg.
Loss on Drying Not more than 1%.
Melting Range Between 226° and 230°.
Readily Carbonizable Substances Passes test.
Residue on Ignition Not more than 0.2%.
Selenium Not more than 0.003%.
Toluenesulfonamides Not more than 0.0025%.

TESTS

Assay Dissolve about 500 mg, previously dried at 105° for 2 h and accurately weighed, in 75 mL of hot water, cool quickly, add phenolphthalein TS, and titrate with 0.1 *N* sodium hydroxide. Perform a blank determination and make any necessary correction. Each mL of 0.1 *N* sodium hydroxide is equivalent to 18.32 mg of $C_7H_5NO_3S$.
Benzoic and Salicylic Acids To 10 mL of a hot, saturated solution add ferric chloride TS, dropwise. No precipitate or violet color appears in the liquid.
Heavy Metals Prepare and test a 2-g sample as directed in *Method II* under the *Heavy Metals Test*, Appendix IIIB, using 20 µg of lead ion (Pb) in the control (*Solution A*).
Loss on Drying, Appendix IIC Dry at 105° for 2 h.
Melting Range Determine as directed for *Melting Range or Temperature, Procedure for Class Ia*, Appendix IIB.
Readily Carbonizable Substances, Appendix IIB Dissolve 200 mg in 5 mL of 95% sulfuric acid, and keep at a temperature of 48° to 50° for 10 min. The color is no darker than *Matching Fluid A*.
Residue on Ignition, Appendix IIC Ignite 1 g as directed in the general method.
Selenium Determine as directed in *Method I* under the *Selenium Limit Test*, Appendix IIIB, using 200 mg of sample.
Toluenesulfonamides Determine as directed in the monograph for *Sodium Saccharin*, but use 8.0 mL of sodium carbonate TS to dissolve the sample for the *Test Preparation*.

Packaging and Storage Store in well-closed containers.

Safflower Oil (Unhydrogenated)

DESCRIPTION

A light yellow oil obtained from the plant *Carthamus tinctorius* by mechanical expression or solvent extraction. It is refined, bleached, and deodorized to substantially remove free fatty acids, phospholipids, color, odor and flavor components, and miscellaneous other non-oil materials. It is a liquid at 21° to 27°, but traces of wax may cause the oil to cloud unless removed by winterization. Safflower Oil has the highest linoleic acid [(Z),(Z)-9,12-octadecadienoic acid] content (typically about 78% of total fatty acids) of any known oil. It is free from visible foreign matter at 21° to 27°.

Functional Use in Foods Coating agent; emulsifying agent; formulation aid; texturizer.

REQUIREMENTS

Identification Safflower Oil exhibits the following composition profile of fatty acids as determined under *Fatty Acid Composition*, Appendix VII:

Fatty Acid:	<14	14:0	16:0	16:1	18:0	18:1
Weight % (Range):	<0.1	<1.0	2–10	<0.5	1–10	7.0–42

Fatty Acid:	18:2	18:3	20:0	20:1
Weight % (Range):	72–81	<1.5	<0.5	<0.5

Cold Test Passes test.
Color (AOCS-Wesson) Not more than 1.0 red.
Free Fatty Acids (as oleic acid) Not more than 0.1%.
Iodine Value Between 135 and 150.
Lead Not more than 0.1 mg/kg.
Linoleic Acid Not less than 72% of total fatty acids.
Linolenic Acid Not more than 1.5%.
Peroxide Value Not more than 10 meq/kg.
Unsaponifiable Matter Not more than 1.5%.
Water Not more than 0.1%.

TESTS

Cold Test Proceed as directed under *Cold Test*, Appendix VII.
Color Proceed as directed for *Color* (AOCS-Wesson) under *Fats and Related Substances*, Appendix VII.
Free Fatty Acids Proceed as directed under *Free Fatty Acids*, Appendix VII, using the following equivalence factor (*e*) in the formula given in the procedure:

Free fatty acids as oleic acid, *e* = 28.2.

Iodine Value Proceed as directed under *Iodine Value*, Appendix VII.
Lead Determine as directed under *Method II* in the *Atomic Absorption Spectrophotometric Graphite Furnace Method* under the *Lead Limit Test*, Appendix IIIB, using a 3-g sample.
Linoleic Acid Proceed as directed under *Fatty Acid Composition*, Appendix VII.
Linolenic Acid Proceed as directed under *Fatty Acid Composition*, Appendix VII.
Peroxide Value Proceed as directed under *Peroxide Value* in the monograph for *Hydroxylated Lecithin*. However, after the addition of saturated potassium iodide and mixing, instead of allowing the solution to stand for 10 min, mix the solution for 1 min and begin the titration immediately.
Unsaponifiable Matter Proceed as directed under *Unsaponifiable Matter*, Appendix VII.
Water Proceed as directed under *Water Determination*, Appendix IIB. However, in place of 35 to 40 mL of methanol use 50 mL of chloroform to dissolve the sample.

Packaging and Storage Store in well-closed containers.

Sage Oil, Dalmatian Type

DESCRIPTION

The oil obtained by steam distillation from the partially dried leaves of the plant *Salvia officinalis* L. It is a yellowish or greenish yellow liquid having a warm, camphoraceous and thujonelike odor and flavor. It is soluble in most fixed oils and in mineral oil. Frequently the solutions in mineral oil are opalescent. It is slightly soluble in propylene glycol, but it is practically insoluble in glycerin.

Functional Use in Foods Flavoring agent.

REQUIREMENTS

Identification The infrared absorption spectrum of the sample exhibits relative maxima (that may vary in intensity) at the same wavelengths (or frequencies) as those shown in the respective spectrum in the section on *Infrared Spectra (Series A: Essential Oils)*, using the same test conditions as specified therein.
Assay Not less than 50.0% of ketones, calculated as thujone ($C_{10}H_{16}O$).
Angular Rotation Between +2° and +29°.
Ester Value after Acetylation Between 25 and 60.
Heavy Metals (as Pb) Passes test.
Refractive Index Between 1.457 and 1.469 at 20°.
Saponification Value Between 5 and 20.
Solubility in Alcohol Passes test.
Specific Gravity Between 0.903 and 0.925.

TESTS

Assay Weigh accurately about 1 g, and proceed as directed under *Aldehydes and Ketones—Hydroxylamine Method*, Appendix VI, using 76.12 as the equivalence factor (*e*) in the calculation.
Angular Rotation Determine in a 100-mm tube as directed under *Optical (Specific) Rotation*, Appendix IIB.
Ester Value after Acetylation Proceed as directed under *Total Alcohols*, Appendix VI, using about 2.5 g of the acetylated oil. Calculate the *Ester Value after Acetylation* by the formula

$$A \times 28.05/B,$$

in which A is the number of mL of 0.5 N alcoholic potassium hydroxide consumed in the saponification of the acetylated oil, and B is the weight, in g, of the acetylated oil taken as the sample.
Heavy Metals Shake 10 mL of the oil with an equal volume of water to which 1 drop of hydrochloric acid has been added, and pass hydrogen sulfide through the mixture until it is saturated. No darkening in color is produced in either the oil or the water.
Refractive Index, Appendix IIB Determine with an Abbé or other refractometer of equal or greater accuracy.
Saponification Value Determine as directed for *Saponification Value* under *Essential Oils and Flavors*, Appendix VI, using about 5 g, accurately weighed.
Solubility in Alcohol Proceed as directed in the general method, Appendix VI. One mL dissolves in 1 mL of 80% alcohol.
Specific Gravity Determine by any reliable method (see *General Provisions*).

Packaging and Storage Store in full, tight, preferably glass, tin-lined, or galvanized iron containers in a cool place protected from light.

Sage Oil, Spanish Type

DESCRIPTION

The volatile oil obtained by distillation from the plants of *Salvia lavandulaefolia* Vahl. or *Salvia hispanorium* Lag. (Fam. *Labiatae*). It is a colorless to slightly yellow oil having a camphoraceous odor with a cineole top note. It is soluble in most fixed oils and in glycerin. It is soluble, usually with opalescence, in mineral oil and in propylene glycol.

Functional Use in Foods Flavoring agent.

REQUIREMENTS

Identification The infrared absorption spectrum of the sample exhibits relative maxima (that may vary in intensity) at the same wavelengths (or frequencies) as those shown in the respective spectrum in the section on *Infrared Spectra (Series A: Essential Oils)*, using the same test conditions as specified therein.
Angular Rotation Between –3° and +24°.
Heavy Metals (as Pb) Passes test.
Refractive Index Between 1.468 and 1.473 at 20°.
Saponification Value Between 14 and 57.
Saponification Value after Acetylation Between 56 and 98.
Solubility in Alcohol Passes test.
Specific Gravity Between 0.909 and 0.932.

TESTS

Angular Rotation Determine in a 100-mm tube as directed under *Optical (Specific) Rotation*, Appendix IIB.
Heavy Metals Shake 10 mL of the oil with an equal volume of water to which 1 drop of hydrochloric acid has been added, and pass hydrogen sulfide through the mixture until it is saturated. No darkening in color is produced in either the oil or the water.
Refractive Index, Appendix IIB Determine with an Abbé or other refractometer of equal or greater accuracy.
Saponification Value Determine as directed for *Saponification Value* under *Essential Oils and Flavors*, Appendix VI, using about 5 g, accurately weighed.
Saponification Value after Acetylation Acetylate a 10-mL sample as directed under *Total Alcohols*, Appendix VI. Weigh accurately about 2.5 g of the dried, acetylated oil, and proceed as directed under *Saponification Value*, Appendix VI, using the weight, in g, of the acetylated oil for W in the calculation formula.
Solubility in Alcohol Proceed as directed in the general method, Appendix VI. One mL dissolves in 2 mL of 80% alcohol. The solution may become opalescent upon dilution.
Specific Gravity Determine by any reliable method (see *General Provisions*).

Packaging and Storage Store in full, tight, preferably glass or tin-lined containers in a cool place protected from light.

Sandalwood Oil, East Indian Type

DESCRIPTION

The volatile oil obtained by steam distillation from the dried, ground roots and wood of *Santalum album* L. (Fam. *Santalaceae*). It is a pale yellow to yellow, somewhat viscous, oily liquid having a strong, persistent characteristic odor. It is soluble in most fixed oils, in propylene glycol, and in mineral oil, sometimes with haziness. It is insoluble in glycerin.

Functional Use in Foods Flavoring agent.

REQUIREMENTS

Identification The infrared absorption spectrum of the sample exhibits relative maxima (that may vary in intensity) at the same wavelengths (or frequencies) as those shown in the respective spectrum in the section on *Infrared Spectra (Series A: Essential Oils)*, using the same test conditions as specified therein.
Assay Not less than 90.0% of alcohol, calculated as santalol ($C_{15}H_{24}O$).
Angular Rotation Between –15° and –20°.
Heavy Metals (as Pb) Passes test.
Refractive Index Between 1.500 and 1.510 at 20°.

Solubility in Alcohol Passes test.
Specific Gravity Between 0.965 and 0.980.

TESTS

Assay Proceed as directed under *Total Alcohols*, Appendix VI. Weigh accurately about 1.2 g of the acetylated alcohol for the saponification, and use 110.2 as the equivalence factor (*e*) in the calculation.
Angular Rotation Determine in a 100-mm tube as directed under *Optical (Specific) Rotation*, Appendix IIB.
Heavy Metals Shake 10 mL of the oil with an equal volume of water to which 1 drop of hydrochloric acid has been added, and pass hydrogen sulfide through the mixture until it is saturated. No darkening in color is produced in either the oil or the water.
Refractive Index, Appendix IIB Determine with an Abbé or other refractometer of equal or greater accuracy.
Solubility in Alcohol Proceed as directed in the general method, Appendix VI. One mL dissolves in 5 mL of 70% alcohol and remains in solution on dilution to 10 mL.
Specific Gravity Determine by any reliable method (see *General Provisions*).

Packaging and Storage Store in full, tight, preferably glass, aluminum, or suitably lined containers in a cool place protected from light.

Savory Oil (Summer Variety)

Summer Savory Oil

CAS: [8016-68-0]

DESCRIPTION

The volatile oil obtained by steam distillation from the whole dried plant *Satureia hortensis* L. (Fam. *Labiatae*). It is a light yellow to dark brown liquid having a spicy, aromatic note suggestive of thyme or origanum. It is soluble in most fixed oils and in mineral oil, but it is practically insoluble in glycerin and in propylene glycol.

Functional Use in Foods Flavoring agent.

REQUIREMENTS

Identification The infrared absorption spectrum of the sample exhibits relative maxima (that may vary in intensity) at the same wavelengths (or frequencies) as those shown in the respective spectrum in the section on *Infrared Spectra* (Series A: Essential Oils), using the same test conditions as specified therein.
Assay Not less than 20.0% and not more than 57.0% of phenols as carvacrol ($C_{10}H_{14}O$).
Angular Rotation Between −5° and +4°.

Heavy Metals (as Pb) Passes test.
Refractive Index Between 1.486 and 1.505 at 20°.
Saponification Value Not more than 6.
Solubility in Alcohol Passes test.
Specific Gravity Between 0.875 and 0.954.

TESTS

Assay Proceed as directed under *Phenols*, Appendix VI.
Angular Rotation Determine in a 100-mm tube as directed under *Optical (Specific) Rotation*, Appendix IIB.
Heavy Metals Shake 10 mL of the oil with an equal volume of water to which 1 drop of hydrochloric acid has been added, and pass hydrogen sulfide through the mixture until it is saturated. No darkening in color is produced in either the oil or the water.
Refractive Index, Appendix IIB Determine with an Abbé or other refractometer of equal or greater accuracy.
Saponification Value Determine as directed for *Saponification Value* under *Essential Oils and Flavors*, Appendix VI, using 5 g, accurately weighed.
Solubility in Alcohol Proceed as directed in the general method, Appendix VI. One mL usually dissolves in 2 mL of 80% alcohol. Some oils may be slightly hazy in 10 mL of 90% alcohol.
Specific Gravity Determine by any reliable method (see *General Provisions*).

Packaging and Storage Store in full, tight, preferably glass, aluminum, tin-lined, or other suitably lined containers in a cool place protected from light.

DL-Serine

DL-2-Amino-3-hydroxypropanoic Acid

$$H_2C - CH - COOH$$
$$ | |$$
$$ HO NH_2$$

$C_3H_7NO_3$　　　　　　　　　　　　　　　Formula wt 105.09

CAS: [302-84-1]

DESCRIPTION

White crystals or a crystalline powder. It is soluble in water, but insoluble in alcohol and in ether. It melts with decomposition at about 246° using a closed capillary tube and a bath preheated to 225°. It is optically inactive.

Functional Use in Foods Nutrient; dietary supplement.

REQUIREMENTS

Identification To 5 mL of a 1 in 1000 solution add 1 mL of triketohydrindene hydrate TS. A bluish purple or purple color is produced.
Assay Not less than 98.5% and not more than 101.5% of $C_3H_7NO_3$, calculated on the dried basis.
Heavy Metals (as Pb) Not more than 0.002%.
Lead Not more than 10 mg/kg.
Loss on Drying Not more than 0.3%.
Residue on Ignition Not more than 0.1%.

TESTS

Assay Dissolve about 200 mg, accurately weighed, in 3 mL of formic acid and 50 mL of glacial acetic acid. Titrate with 0.1 N perchloric acid in glacial acetic acid, determining the endpoint potentiometrically. Perform a blank determination (see *General Provisions*), and make any necessary correction. Each mL of 0.1 N perchloric acid is equivalent to 10.51 mg of $C_3H_7NO_3$.
Heavy Metals Prepare and test a 1-g sample as directed in *Method II* under the *Heavy Metals Test*, Appendix IIIB, using 20 µg of lead ion (Pb) in the control (*Solution A*).
Lead A *Sample Solution* prepared as directed for organic compounds meets the requirements of the *Lead Limit Test*, Appendix IIIB, using 10 µg of lead ion (Pb) in the control.
Loss on Drying, Appendix IIC Dry at 105° for 3 h.
Residue on Ignition, Appendix IIC Ignite 1 g as directed in the general method.

Packaging and Storage Store in well-closed containers.

L-Serine

L-2-Amino-3-hydroxypropanoic Acid

$$H_2C-C-COOH$$
$$| \quad /\backslash$$
$$HO \quad H \quad NH_2$$

$C_3H_7NO_3$ Formula wt 105.09

CAS: [56-45-1]

DESCRIPTION

A white crystalline powder without odor and having a sweet taste. It is soluble in water, but is insoluble in alcohol and in ether. It melts with decomposition at about 228°.

Functional Use in Foods Nutrient; dietary supplement.

REQUIREMENTS

Identification
  A. To 5 mL of a 1 in 1000 solution add 1 mL of triketohydrindene hydrate TS. A reddish purple or purple color is produced.
  B. Dissolve about 500 mg in 10 mL of water, add 200 mg of periodic acid, and heat. The odor of formaldehyde is produced.
Assay Not less than 98.5% and not more than 101.5% of $C_3H_7NO_3$, calculated on the dried basis.
Heavy Metals (as Pb) Not more than 0.002%.
Lead Not more than 10 mg/kg.
Loss on Drying Not more than 0.3%.
Residue on Ignition Not more than 0.1%.
Specific Rotation $[\alpha]_D^{20°}$: Between +13.6° and +16.0° after drying; or $[\alpha]_D^{25°}$: Between +13.2° and +15.6° after drying.

TESTS

Assay Dissolve about 200 mg, accurately weighed, in 3 mL of formic acid and 50 mL of glacial acetic acid. Titrate with 0.1 N perchloric acid in glacial acetic acid, determining the endpoint potentiometrically. Perform a blank determination (see *General Provisions*), and make any necessary correction. Each mL of 0.1 N perchloric acid is equivalent to 10.51 mg of $C_3H_7NO_3$.
Heavy Metals Prepare and test a 1-g sample as directed in *Method II* under the *Heavy Metals Test*, Appendix IIIB, using 20 µg of lead ion (Pb) in the control (*Solution A*).
Lead A *Sample Solution* prepared as directed for organic compounds meets the requirements of the *Lead Limit Test*, Appendix IIIB, using 10 µg of lead ion (Pb) in the control.
Loss on Drying, Appendix IIC Dry at 105° for 3 h.
Residue on Ignition, Appendix IIB Ignite 1 g as directed in the general method.
Specific Rotation, Appendix IIB Determine in a solution containing 10 g of a previously dried sample in sufficient 2 N hydrochloric acid to make 100 mL.

Packaging and Storage Store in well-closed containers.

Shellac, Bleached

White Shellac; Regular Bleached Shellac

INS: 904 CAS: [9000-59-3]

DESCRIPTION

Shellac is obtained from lac, the resinous secretion of the insect *Laccifer* (*Tachardia*) *lacca* Kerr (Fam. *Coccidae*). Bleached Shellac is obtained by dissolving the lac in aqueous sodium carbonate, followed by bleaching with sodium hypochlorite, precipitation of the bleached lac with dilute sulfuric acid solution, and drying. It occurs as an off-white, amorphous, granular resin. It is freely (though very slowly) soluble in alcohol, insoluble in water, and slightly soluble in acetone and in ether. Bleached Shellac is usually dissolved in a suitable solvent for application to food products.

Functional Use in Foods Coating agent; surface-finishing agent; glaze.

REQUIREMENTS

Identification To 50 mg of the sample add a few drops of a solution of 1 g of ammonium molybdate in 3 mL of sulfuric acid. A green color is produced, changing to lilac when the solution is neutralized with 6 N ammonium hydroxide.
Acid Value Between 73 and 89.
Heavy Metals (as Pb) Not more than 10 mg/kg.
Loss on Drying Not more than 6.0%.
Rosin Passes test.
Wax Not more than 5.5%.

TESTS

Acid Value Dissolve about 2 g of finely ground sample, accurately weighed, in 50 mL of alcohol previously neutralized to phenolphthalein with sodium hydroxide. Add additional phenolphthalein TS, if necessary, and titrate with 0.1 N sodium hydroxide to a pink endpoint. Calculate the acid value by the formula

$$56.1V \times N/W,$$

in which V is the exact volume, in mL; N is the exact normality of the sodium hydroxide solution; and W is the weight, in g, of sample taken, calculated on the dried basis.
Heavy Metals Prepare and test a 2-g sample as directed in *Method II* under the *Heavy Metals Test*, Appendix IIIB, using 20 μg of lead ion (Pb) in the control (*Solution A*).
Loss on Drying, Appendix IIC Dry at 41° ± 2° to constant weight.
Rosin Dissolve 2 g of the sample in 10 mL of dehydrated alcohol, and add slowly, with shaking, 50 mL of solvent hexane. Transfer to a separator, wash with two 50-mL portions of water, and discard the washings. Filter the solvent layer, evaporate it to dryness, and add to the residue 2 mL of a mixture of 1 volume of liquefied phenol and 2 volumes of methylene chloride. Stir, and transfer a portion of the mixture to a cavity of a color-reaction plate. Fill an adjacent cavity with a mixture of 1 volume of bromine and 4 volumes of methylene chloride, and cover both cavities with an inverted watch glass. No purple or deep indigo blue color is produced in or above the liquid containing the sample residue.
Wax Transfer about 10 g of finely ground sample, accurately weighed, and 2.5 g of sodium carbonate to a 200-mL tall-form beaker. Add 150 mL of hot water, immerse the beaker in a boiling water bath, and stir until the sample is dissolved. Cover the beaker with a watch glass, heat for 3 h without agitation, and cool in a cold water bath. When the wax has floated to the surface, filter the mixture through medium-speed, quantitative, ashless filter paper, transferring the wax to the paper, and wash the filter with water. Pour 5 to 10 mL of alcohol onto the filter to accelerate drying. Wrap the paper loosely in a larger piece of filter paper, bind with a piece of fine wire, and dry with the aid of gentle heat. Extract with chloroform in a suitable continuous extraction apparatus for 2 h, using a previously dried and accurately weighed flask to receive the extracted wax and solvent. Evaporate the solvent, dry the wax at 105° to constant weight, and calculate the percentage of wax.

Packaging and Storage Store in well-closed containers in a cool place protected from heat.

Shellac, Bleached, Wax-Free

Refined Bleached Shellac

DESCRIPTION

Shellac is obtained from lac, the resinous secretion of the insect *Laccifer* (*Tachardia*) *lacca* Kerr (Fam. *Coccidae*). Wax-free Bleached Shellac is obtained by the same process as that described in the monograph for *Bleached Shellac*, except that, in addition, wax is removed by filtration. It occurs as an amorphous, light yellow, granular resin. Its solubility is the same as that of *Bleached Shellac*. Wax-free Bleached Shellac is usually dissolved in a suitable solvent for application to food products.

Functional Use in Foods Coating agent; surface-finishing agent; glaze.

REQUIREMENTS

Identification Wax-free Bleached Shellac responds to the *Identification Test* in the monograph for *Bleached Shellac*.
Acid Value Between 75 and 91.
Heavy Metals (as Pb) Not more than 10 mg/kg.
Loss on Drying Not more than 6.0%.
Rosin Passes test.
Wax Not more than 0.2%.

TESTS

Perform as directed in the monograph for *Bleached Shellac*.

Packaging and Storage Store in well-closed containers in a cool place protected from heat.

Silicon Dioxide

Synthetic Amorphous Silica

SiO_2 Formula wt 60.08

INS: 551   CAS: [7631-86-9]

DESCRIPTION

Silicon Dioxide for food use is an amorphous substance that shows a noncrystalline pattern when examined by X-ray diffraction. It is produced synthetically by either a vapor-phase hydrolysis process, yielding *fumed silica*, or by a wet process, yielding *precipitated silica*, *silica gel*, *colloidal silica*, or *hydrous silica*. Fumed silica is produced in essentially an anhydrous state, whereas the wet-process products are obtained as hydrates or contain surface-adsorbed water.

Fumed silica occurs as a white, fluffy, nongritty powder of extremely fine particle size and is hygroscopic. The wet-process silicas occur as white, fluffy powders or as white, microcellular beads or granules and are hygroscopic or absorb moisture from the air in varying amounts. All of these forms of Silicon Dioxide are insoluble in water and in organic solvents, but are soluble in hydrofluoric acid and in hot, concentrated solutions of alkalies.

Functional Use in Foods Anticaking agent; defoaming agent; carrier; conditioning agent; chillproofing agent in malt beverages; filter aid.

REQUIREMENTS

Identification

A. Place about 5 mg of the sample in a platinum crucible, mix with 200 mg of anhydrous potassium carbonate, and ignite at a red heat for about 10 min over a burner. Cool, dissolve the melt in 2 mL of freshly distilled water, warming if necessary, and slowly add 2 mL of ammonium molybdate TS. A deep yellow color is produced.

B. Place 1 drop of the solution from *Identification Test A* on a filter paper, and evaporate the solvent. Add 1 drop of a saturated solution of *o*-tolidine in glacial acetic acid, and place the paper over ammonium hydroxide. A greenish blue spot is produced.

Assay *Fumed silica*: not less than 99.0% of SiO_2 after ignition; *precipitated silica*, *silica gel*, and *hydrous silica*: not less than 94.0% of SiO_2 after ignition.

Arsenic (as As)   Not more than 3 mg/kg.

Heavy Metals (as Pb)   Not more than 10 mg/kg.

Lead   Not more than 5 mg/kg.

Loss on Drying   *Fumed silica*: not more than 2.5%; *precipitated silica* and *silica gel*: not more than 7.0%; *hydrous silica*: not more than 70.0%; *colloidal silica*: not more than 85.0%.

Loss on Ignition   *Fumed silica*: not more than 2.0% after drying; *silica gel*, *hydrous silica gel*, and *precipitated silica*: not more than 8.5% after drying.

Soluble Ionizable Salts (as Na_2SO_4)   *Precipitated silica*, *silica gel*, and *hydrous silica*: not more than 5.0%.

TESTS

Assay   Transfer about 1 g of the sample, previously dried at 105° for 2 h and accurately weighed, into a tared platinum crucible, ignite as directed in the **TEST** for *Loss on Ignition*, cool in a desiccator, and weigh to obtain the ignited sample weight (W). Moisten the residue with 3 or 4 drops of alcohol, add 2 drops of sulfuric acid, and then add enough hydrofluoric acid to cover the wetted sample. Evaporate to dryness on a hot plate, using medium heat (95° to 105°), then add 5 mL of hydrofluoric acid, swirl the dish carefully to wash down the sides, and again evaporate to dryness. Ignite the dried residue to a red heat over a Meker burner, cool in a desiccator, and weigh to obtain the residual weight (w). The difference between the ignited sample weight and the residual weight ($W - w$) represents the weight of SiO_2 in the ignited sample.

Sample Solution for the Determination of Arsenic, Heavy Metals, and Lead   Transfer 5.0 g of the sample into a 250-mL beaker, add 50 mL of 0.5 N hydrochloric acid, cover with a watch glass, and heat slowly to boiling. Boil gently for 15 min, cool, and let the undissolved material settle. Decant the supernatant liquid through a Whatman No. 3 filter paper, or equivalent, into a 100-mL volumetric flask, retaining as much as possible of the insoluble material in the beaker. Wash the slurry and beaker with three 10-mL portions of hot water, decanting each washing through the filter into the flask. Finally, wash the filter paper with 15 mL of hot water, cool the filtrate to room temperature, dilute to volume with water, and mix.

Arsenic   A 20-mL portion of the *Sample Solution* meets the requirements of the *Arsenic Test*, Appendix IIIB.

Heavy Metals   A 40-mL portion of the *Sample Solution* meets the requirements of the *Heavy Metals Test*, Appendix IIIB, using 20 µg of lead ion (Pb) in the control (*Solution A*).

Lead   Determine as directed under *Lead* in the monograph for *Calcium Silicate*.

Loss on Drying, Appendix IIC   Dry at 105° for 2 h.

Loss on Ignition   Transfer into a suitable tared crucible about 1 g of an accurately weighed sample that has been previously dried at 105° for 2 h. Place the crucible in a cold muffle furnace, and bring the temperature to 900° to 1000° during a 1-h period. Ignite at this temperature for 1 h, cool in a desiccator, and weigh.

Soluble Ionizable Salts   Weigh accurately 5 g of the sample, previously dried at 105° for 2 h, and stir it with 150 mL of water for at least 5 min in a high-speed mixer. Filter with the aid of suction, and wash the mixer and filter with 100 mL of water in divided portions, adding the washings to the filtrate. Dilute the filtrate to 250 mL with water, and determine its conductance with a suitable conductance bridge assembly. The conductance is not greater than that produced by a control containing 250 mg of anhydrous sodium sulfate in each 250 mL.

Packaging and Storage   Store in well-closed containers.

Sodium Acetate

$C_2H_3NaO_2 \cdot 3H_2O$	Formula wt 136.08
INS: 262	CAS: [127-09-3]

DESCRIPTION

Colorless, transparent crystals or a granular, crystalline powder. It is odorless or has a faint, acetous odor. It effloresces in warm, dry air. One g dissolves in about 0.8 mL of water and in about 19 mL of alcohol.

Functional Use in Foods   Buffer.

REQUIREMENTS

Identification   A 1 in 20 solution gives positive tests for *Sodium* and for *Acetate*, Appendix IIIA.
Assay   Not less than 99.0% and not more than 101.0% of $C_2H_3NaO_2$ after drying.
Alkalinity (as Na_2CO_3)   Not more than 0.05%.
Heavy Metals (as Pb)   Not more than 10 mg/kg.
Loss on Drying   Between 36.0% and 41.0%.
Potassium Compounds   Passes test.

TESTS

Assay   Weigh accurately about 400 mg of the sample obtained in the test for *Loss on Drying*, dissolve it in 40 mL of glacial acetic acid, add 2 drops of crystal violet TS, and titrate with 0.1 N perchloric acid in glacial acetic acid. Perform a blank determination (see *General Provisions*), and make any necessary correction. Each mL of 0.1 N perchloric acid is equivalent to 8.203 mg of $C_2H_3NaO_2$.
Alkalinity   Dissolve 2 g in about 20 mL of water, and add 3 drops of phenolphthalein TS. If a pink color is produced, not more than 0.1 mL of 0.1 N sulfuric acid is required to discharge it.
Heavy Metals   A solution of 2 g in 25 mL of water meets the requirements of the *Heavy Metals Test*, Appendix IIIB, using glacial acetic acid to adjust the pH of *Solution B*, and using 20 μg of lead ion (Pb) in the control (*Solution A*).
Loss on Drying, Appendix IIC   Dry at 80° overnight and follow by drying at 120° for 4 h.
Potassium Compounds   Mix a few drops of sodium bitartrate TS with 5 mL of a clear, saturated solution of the sample. No turbidity is produced within 5 min.

Packaging and Storage   Store in tight containers.

Sodium Acetate, Anhydrous

$C_2H_3NaO_2$ Formula wt 82.03

DESCRIPTION

A white, odorless, granular powder. It is hygroscopic. One g dissolves in about 2 mL of water.

Functional Use in Foods   Buffer.

REQUIREMENTS

Identification   A 1 in 20 solution gives positive tests for *Sodium* and for *Acetate*, Appendix IIIA.
Assay   Not less than 99.0% and not more than 101.0% of $C_2H_3NaO_2$ after drying.
Alkalinity (as NaOH)   Not more than 0.2%.
Heavy Metals (as Pb)   Not more than 10 mg/kg.
Loss on Drying   Not more than 1.0%.
Potassium Compounds   Passes test.

TESTS

Assay   Weigh accurately about 400 mg of the sample obtained in the test for *Loss on Drying*, dissolve it in 40 mL of glacial acetic acid, add 2 drops of crystal violet TS, and titrate with 0.1 N perchloric acid in glacial acetic acid. Perform a blank determination (see *General Provisions*), and make any necessary correction. Each mL of 0.1 N perchloric acid is equivalent to 8.203 mg of $C_2H_3NaO_2$.
Alkalinity   Dissolve 2 g in 20 mL of water, and add phenolphthalein TS. If a pink color is produced, not more than 1.0 mL of 0.1 N sulfuric acid is required to discharge it.
Heavy Metals   A solution of 2 g in 25 mL of water meets the requirements of the *Heavy Metals Test*, Appendix IIIB, using glacial acetic acid to adjust the pH of *Solution B*, and using 20 μg of lead ion (Pb) in the control (*Solution A*).
Loss on Drying, Appendix IIC   Dry at 80° overnight and follow by drying at 120° to constant weight.
Potassium Compounds   Mix a few drops of sodium bitartrate TS with 5 mL of a clear, saturated solution of the sample. No turbidity is produced within 5 min.

Packaging and Storage   Store in tight containers.

Sodium Acid Pyrophosphate

Disodium Pyrophosphate; Disodium Dihydrogen Pyrophosphate

$Na_2H_2P_2O_7$ Formula wt 221.94

CAS: [7758-16-9]

DESCRIPTION

White, crystalline powder. It is soluble in water. The pH of a 1 in 100 solution is about 4. It may contain a suitable aluminum and/or calcium salt to control the rate of reaction in leavening systems.

Functional Use in Foods   Buffer; leavening agent; sequestrant.

REQUIREMENTS

Identification

A. A 1 in 20 solution gives positive tests for *Sodium*, Appendix IIIA.

B. Dissolve 100 mg of the sample in 100 mL of 1.7 N nitric acid. Add 0.5 mL of this solution to 30 mL of quimociac TS. A yellow precipitate does not form. Heat the remaining portion of the sample solution for 10 min at 95°, and then add 0.5 mL

of the solution to 30 mL of quimociac TS. A yellow precipitate forms immediately.
Assay Not less than 95.0% of $Na_2H_2P_2O_7$.
Arsenic (as As) Not more than 3 mg/kg.
Fluoride Not more than 0.005%.
Heavy Metals (as Pb) Not more than 0.0015%.
Insoluble Substances Not more than 1%.
Lead Not more than 2 mg/kg.

TESTS

Assay Dissolve about 500 mg, accurately weighed, in 100 mL of water in a 400-mL beaker. Using a pH meter, adjust the pH of the solution to 3.8 with hydrochloric acid, then add 50 mL of a 1 in 8 solution of zinc sulfate (125 g of $ZnSO_4 \cdot 7H_2O$ dissolved in water, diluted to 1000 mL, filtered, and adjusted to pH 3.8), and allow to stand for 2 min. Titrate the liberated acid with 0.1 N sodium hydroxide until a pH of 3.8 is again reached. After each addition of sodium hydroxide near the endpoint, time should be allowed for any precipitated zinc hydroxide to redissolve. Each mL of 0.1 N sodium hydroxide is equivalent to 11.10 mg of $Na_2H_2P_2O_7$.
Arsenic A solution of 1 g in 10 mL of water meets the requirements of the *Arsenic Test*, Appendix IIIB.
Fluoride Determine on a 2-g sample as directed in *Method IV* under the *Fluoride Limit Test*, Appendix IIIB, using *Buffer Solution B* and 0.1 mL of *Fluoride Standard Solution*.
Heavy Metals A solution of 1.33 g in 25 mL of water meets the requirements of the *Heavy Metals Test*, Appendix IIIB, using 20 µg of lead ion (Pb) in the control (*Solution A*).
Insoluble Substances Dissolve 10 g in 100 mL of hot water, and filter through a tared filtering crucible. Wash the insoluble residue with hot water, dry at 105° for 2 h, cool, and weigh.
Lead A 10-g sample meets the requirements of the *APDC Extraction Method for Lead*, Appendix IIIB.

Packaging and Storage Store in tight containers.

Sodium Alginate

Algin

$(C_6H_7O_6Na)_n$ Equiv wt, *Calculated*, 198.11
     Equiv wt, *Actual* (Avg), 222.00

INS: 401   CAS: [9005-38-3]

DESCRIPTION

The sodium salt of alginic acid (see the monograph for *Alginic Acid*) occurs as a white to yellowish, fibrous or granular powder. It is nearly odorless and tasteless. It dissolves in water to form a viscous, colloidal solution. It is insoluble in alcohol and in hydroalcoholic solutions in which the alcohol content is greater than about 30% by weight. It is insoluble in chloroform, in ether, and in acids having a pH lower than about 3.

Functional Use in Foods Stabilizer; thickener; emulsifier.

REQUIREMENTS

Identification
  A. To 5 mL of a 1 in 100 solution add 1 mL of calcium chloride TS. A voluminous, gelatinous precipitate is formed.
  B. To 10 mL of a 1 in 100 solution add 1 mL of 2 N sulfuric acid. A heavy gelatinous precipitate is formed.
  C. Sodium Alginate meets the requirements of *Identification Test C* in the monograph for *Alginic Acid*.
Assay It yields not less than 18.0% and not more than 21.0% of carbon dioxide (CO_2), corresponding to between 90.8% and 106.0% of Sodium Alginate (equiv wt 222.00).
Arsenic (as As) Not more than 3 mg/kg.
Heavy Metals (as Pb) Not more than 0.002%.
Lead Not more than 5 mg/kg.
Loss on Drying Not more than 15.0%.

TESTS

Assay Proceed as directed in the *Alginates Assay*, Appendix IIIC. Each mL of 0.25 N sodium hydroxide consumed in the assay is equivalent to 27.75 mg of Sodium Alginate (equiv wt 222.00), calculated on the dried basis.
Arsenic A *Sample Solution* prepared as directed for organic compounds meets the requirements of the *Arsenic Test*, Appendix IIIB.
Heavy Metals Prepare and test a 1-g sample as directed in *Method II* under the *Heavy Metals Test*, Appendix IIIB, but use nitric acid instead of sulfuric acid to wet the sample prior to ignition, and cautiously ignite in a platinum crucible. Any color does not exceed that produced in a control (*Solution A*) containing 20 µg of lead ion (Pb).
Lead A *Sample Solution* prepared as directed for organic compounds meets the requirements of the *Lead Limit Test*, Appendix IIIB, using 5 µg of lead ion (Pb) in the control.
Loss on Drying, Appendix IIC Dry at 105° for 4 h.

Packaging and Storage Store in well-closed containers.

Sodium Aluminosilicate

Sodium Silicoaluminate

INS: 554   CAS: [1344-00-9]

DESCRIPTION

A series of hydrated sodium aluminum silicates having an $Na_2O/Al_3O_3/SiO_2$ mol ratio of approximately 1/1/13, respectively. It occurs as a fine, white, amorphous powder, or as beads. It is odorless and tasteless. It is insoluble in water and in alcohol and other organic solvents, but at 80° to 100°, it is partially soluble in strong acids and solutions of alkali hydroxides. The pH of a 20% slurry, prepared with carbon dioxide-free water, is between 6.5 and 10.5.

Functional Use in Foods Anticaking agent.

REQUIREMENTS

Identification

A. Mix 500 mg of the sample with 2.5 g of anhydrous potassium carbonate, and heat the mixture in a platinum or nickel crucible until it melts completely. Cool, add 5 mL of water, and allow to stand for 3 min. Heat the bottom of the crucible gently, detach the melt, and transfer it to a beaker with the aid of about 50 mL of water. Add gradually hydrochloric acid until no effervescence is observed, then add 10 mL more of the acid, and evaporate to dryness on a steam bath. Cool, add 20 mL of water, boil, and filter through ash-free filter paper. An insoluble residue of silica remains.

Note: Retain the filtrate for *Identification Test B*.

Transfer the gelatinous residue to a platinum dish, and cautiously add 5 mL of hydrofluoric acid. The precipitate dissolves. (If it does not dissolve, repeat the treatment with hydrofluoric acid.) Heat and hold in the vapors a glass stirring rod with a drop of water on the tip. The drop becomes turbid.

B. Portions of the filtrate obtained in *Identification Test A* give positive tests for *Aluminum* and for *Sodium*, Appendix IIIA.

Assay

Silicon Dioxide Not less than 66.0% and not more than 76.0% of SiO_2 after drying.

Aluminum Oxide Not less than 9.0% and not more than 13.0% of Al_2O_3 after drying.

Sodium Oxide Not less than 4.0% and not more than 7.0% of Na_2O after drying.

Heavy Metals (as Pb) Not more than 10 mg/kg.

Lead Not more than 5 mg/kg.

Loss on Drying Not more than 8.0%.

Loss on Ignition Between 8.0% and 13.0% after drying.

TESTS

Assay

Silicon Dioxide Transfer about 500 mg, previously dried at 105° for 2 h and accurately weighed, into a 250-mL beaker, wash the sides of the beaker with a few mL of water, and then add 30 mL of sulfuric acid and 15 mL of hydrochloric acid. Heat on a hot plate in a hood until dense, white fumes are evolved, cool, add 15 mL of hydrochloric acid, and heat again to dense, white fumes. Cool, add 70 mL of water, and filter through Whatman No. 40, or equivalent, filter paper. Wash the filter paper and precipitate with hot water until the filter paper is free of sulfuric acid. Retain the filtrate for the determination of *Aluminum Oxide*.

Transfer the filter paper and precipitate into a tared platinum crucible, char, and ignite at 900° to constant weight. Moisten the residue with a few drops of water, then add 15 mL of hydrofluoric acid and 8 drops of sulfuric acid, and heat on a hot plate until white fumes of sulfur trioxide are evolved. Cool, add 5 mL of water, 10 mL of hydrofluoric acid, and 3 drops of sulfuric acid, and evaporate to dryness on the hot plate. Heat cautiously over an open flame until sulfur trioxide fumes have ceased, and ignite at 900° to constant weight. The weight loss after the addition of hydrofluoric acid represents the weight of SiO_2 in the sample taken.

Aluminum Oxide Transfer about 500 mg of the sample, previously dried at 105° for 2 h and accurately weighed, into a tared platinum dish, and moisten with 8 to 10 drops of water. Add 25 mL of 70% perchloric acid and 10 mL of hydrofluoric acid, and heat on a hot plate in a hood until dense, white fumes of perchloric acid appear. Cool, add 10 mL of hydrofluoric acid, and heat again to dense, white fumes. Cool, dissolve the residue in sufficient water, quantitatively transfer with the aid of additional water to a 250-mL volumetric flask, and dilute to volume. Retain this solution for sodium oxide analysis.

Transfer by pipet a 10.0-mL aliquot of this solution into a 100-mL volumetric flask, fill to volume with water, and mix.

Set a suitable atomic absorption spectrophotometer to a wavelength of 309.3 nm. Adjust the instrument to zero absorbance against water. Read the absorbance of four standard solutions containing 5, 10, 20, and 50 µg/mL of aluminum, in the form of the chloride, and plot the standard curve as absorbance versus concentration of aluminum.

Aspirate the 1 in 10 diluted sample solution into the spectrophotometer, read the absorbance in the same manner, and by reference to the standard curve, determine the concentration (C) of aluminum, in µg/mL, in the sample solution.

Calculate the quantity, in mg, of Al_2O_3 in the sample taken by the formula

$$250C \times 10 \times 1.8895/1000.$$

Sodium Oxide Set a suitable flame photometer to a wavelength of 589 nm. Adjust the instrument to zero transmittance against water, then adjust it to 100.0% transmittance with a standard solution containing 200 µg of sodium, in the form of the chloride, per mL. Read the percent transmittance of three other standard solutions containing 50, 100, and 150 µg each of sodium per mL, and plot the standard curve as percent transmittance versus concentration of sodium.

Place a portion of the sample solution prepared for the aluminum oxide determination in the photometer, read the percent transmittance in the same manner, and by reference to the standard curve, determine the concentration (C) of sodium, in µg/mL, in the sample solution. Calculate the quantity, in mg, of Na_2O in the sample taken by the formula

$$(250C \times 1.348/1000) - F,$$

in which F, as determined below, is the quantity of sodium oxide equivalent to any sodium sulfate present in the sample.

Correction for Sodium Sulfate Content Transfer about 1 g of the sample, previously dried at 105° for 2 h and accurately weighed, into a tared platinum dish, and moisten with 8 to 10 drops of water. Add 25 mL of 70% perchloric acid and 10 mL of hydrofluoric acid, and heat on a hot plate in a hood until dense, white fumes of perchloric acid appear. Add 10 mL of hydrofluoric acid, and heat again to dense, white fumes. Quantitatively transfer the solution to a 400-mL beaker, add 200 mL of water, and heat to boiling. Gradually add, in small portions at a time and while stirring constantly, an excess of hot barium chloride TS (about 10 mL), and heat the mixture on a steam bath for 1 h. Collect the precipitate on a filter, wash until free from chloride, dry, ignite, and weigh. The weight, in g, of the barium sulfate so obtained, multiplied by 0.6086, indicates its equivalent of Na_2SO_4 (C'). Calculate the correction factor (F) by the formula

$$0.437 \, (C' \times w/W),$$

in which w is the weight, in mg, of the sample taken for the sodium oxide determination, and W is the weight, in mg, of the sample taken for the sodium sulfate determination.

Sample Solution for the Determination of Lead and Heavy Metals Transfer 10.0 g of the sample into a 250-mL beaker, add 50 mL of 0.5 N hydrochloric acid, cover with a watch glass, and heat slowly to boiling. Boil gently for 15 min, cool, and let the undissolved material settle. Decant the supernatant liquid through Whatman No. 4, or equivalent, filter paper into a 100-mL volumetric flask, retaining as much as possible of the insoluble material in the beaker. Wash the slurry and beaker with three 10-mL portions of hot water, decanting each washing through the filter into the flask. Finally, wash the filter paper with 15 mL of hot water, cool the filtrate to room temperature, dilute to volume with water, and mix.

Heavy Metals A 20-mL portion of the *Sample Solution* meets the requirements of the *Heavy Metals Test*, Appendix IIIB, using 20 μg of lead ion (Pb) in the control (*Solution A*).
Lead Determine as directed under *Lead* in the monograph for *Calcium Silicate*, except use a *Standard Lead Solution* containing 0.50 μg/mL.
Loss on Drying, Appendix IIC Dry at 105° for 2 h.
Loss on Ignition Transfer about 5 g, previously dried at 105° for 2 h and accurately weighed, into a suitable tared crucible, and ignite at 900° to constant weight.

Packaging and Storage Store in well-closed containers.

Sodium Aluminum Phosphate, Acidic

SALP

$NaAl_3H_{14}(PO_4)_8 \cdot 4H_2O$	Formula wt 949.88
or	or
$Na_3Al_2H_{15}(PO_4)_8$	Formula wt 897.82
INS: 541(i)	CAS: [7785-88-8]

DESCRIPTION

A white, odorless powder. It is insoluble in water, but is soluble in hydrochloric acid.

Functional Use in Foods Leavening agent.

REQUIREMENTS

Identification A 1 in 10 solution in dilute hydrochloric acid (1 in 2) gives positive tests for *Aluminum* and for *Phosphate*, Appendix IIIA, and it responds to the flame test for *Sodium*, Appendix IIIA.
Assay Not less than 95.0% of $NaAl_3H_{14}(PO_4)_8 \cdot 4H_2O$, or not less than 95.0% of $Na_3Al_2H_{15}(PO_4)_8$.
Arsenic (as As) Not more than 3 mg/kg.
Fluoride Not more than 0.0025%.
Heavy Metals (as Pb) Not more than 0.002%.
Lead Not more than 2 mg/kg.
Loss on Ignition $NaAl_3H_{14}(PO_4)_8 \cdot 4H_2O$: between 19.5% and 21.0%; $Na_3Al_2H_{15}(PO_4)_8$: between 15.0% and 16.0%.

TESTS

Assay Proceed as directed under *Assay* in the monograph for *Sodium Polyphosphates, Glassy*, except replace the final sentence with "Each mg of precipitate thus obtained is equivalent to 53.66 μg of $NaAl_3H_{14}(PO_4)_8 \cdot 4H_2O$ or 50.72 μg of $Na_3Al_2H_{15}(PO_4)_8$."
Arsenic A solution of 1 g in 10 mL of dilute hydrochloric acid (1 in 2) meets the requirements of the *Arsenic Test*, Appendix IIIB.
Fluoride Weigh accurately 2.0 g, and proceed as directed in the *Fluoride Limit Test*, Appendix IIIB.
Heavy Metals Dissolve 1 g in 2.5 mL of 2.7 N hydrochloric acid, and add water to make 25 mL. This solution meets the requirements of the *Heavy Metals Test*, Appendix IIIB, using 20 μg of lead ion (Pb) in the control (*Solution A*).
Lead A 10-g sample meets the requirements of the *APDC Extraction Method for Lead*, Appendix IIIB.
Loss on Ignition Ignite at 750° to 800° for 2 h.

Packaging and Storage Store in well-closed containers.

Sodium Aluminum Phosphate, Basic

Kasal

INS: 541(ii)	CAS: [7785-88-8]

DESCRIPTION

A white, odorless powder consisting of an autogenous mixture of an alkaline sodium aluminum phosphate [approximately $Na_8Al_2(OH)_2(PO_4)_4$] with about 30% dibasic sodium phosphate. It is soluble in hydrochloric acid; the sodium phosphate moiety is soluble in water, whereas the sodium aluminum phosphate moiety is only sparingly soluble in water.

Functional Use in Foods Emulsifier.

REQUIREMENTS

Identification A 1 in 10 solution in dilute hydrochloric acid (1 in 2) gives positive tests for *Aluminum* and for *Phosphate*, Appendix IIIA, and it responds to the flame test for *Sodium*, Appendix IIIA.
Assay Not less than 9.5% and not more than 12.5% of Al_2O_3, calculated on the ignited basis.
Arsenic (as As) Not more than 3 mg/kg.
Fluoride Not more than 0.0025%.
Heavy Metals (as Pb) Not more than 0.0015%.
Lead Not more than 2 mg/kg.
Loss on Ignition Not more than 9.0%.

TESTS

Assay Transfer about 2.5 g of the sample, accurately weighed, into a 400-mL beaker, add 15 mL of hydrochloric acid and one glass bead, cover with a watch glass, and boil gently for about 5 min. Rinse any condensate on the watch glass into the beaker, cool the solution to room temperature, transfer it quantitatively to a 250-mL volumetric flask, dilute to volume with water, and mix thoroughly. Transfer 10.0 mL of this solution to a 250-mL beaker, add phenolphthalein TS, and neutralize with 6 N ammonium hydroxide. Add dilute hydrochloric acid (1 in 2) until the precipitate just dissolves, then dilute to 100 mL with water and heat to 70° to 80°. Add 10 mL of 8-hydroxyquinoline TS and sufficient ammonium acetate TS until a yellow precipitate forms, then add 30 mL in excess. Digest at 70° for 30 min, filter through a previously dried and weighed sintered-glass filter crucible, and wash thoroughly with hot water. Dry at 105° for 2 h, cool, and weigh. Each mg of the precipitate so obtained corresponds to 0.111 mg of Al_2O_3.
Arsenic A solution of 1 g in 10 mL of dilute hydrochloric acid (1 in 2) meets the requirements of the *Arsenic Test*, Appendix IIIB.
Fluoride Weigh accurately 2.0 g, and proceed as directed in the *Fluoride Limit Test*, Appendix IIIB.
Heavy Metals Dissolve 1.33 g in 2.5 mL of 2.7 N hydrochloric acid, and add water to make 25 mL. This solution meets the requirements of the *Heavy Metals Test*, Appendix IIIB, using 20 µg of lead ion (Pb) in the control (*Solution A*).
Lead A 10-g sample meets the requirements of the *APDC Extraction Method for Lead*, Appendix IIIB.
Loss on Ignition Ignite at 750° to 800° for 2 h.

Packaging and Storage Store in well-closed containers.

Sodium Ascorbate

Vitamin C Sodium; Sodium L-Ascorbate

$C_6H_7NaO_6$ Formula wt 198.11
INS: 301   CAS: [134-03-2]

DESCRIPTION

A white to yellowish, crystalline powder. One g is soluble in 2 mL of water. The pH of a 1 in 10 solution is about 7.5.

Functional Use in Foods Antioxidant; meat curing aid; nutrient; dietary supplement.

REQUIREMENTS

Identification
  A. A 1 in 50 solution slowly reduces alkaline cupric tartrate TS at 25°, but more readily upon heating.
  B. To 2 mL of a 1 in 50 solution of the sample acidified with 0.5 mL of 0.1 N hydrochloric acid add 4 drops of methylene blue TS, and warm to 40°. The deep blue color is practically discharged within 3 min.
  C. Dissolve 15 mg of the sample in 15 mL of a 1 in 20 solution of trichloroacetic acid, add about 200 mg of activated charcoal, shake vigorously for 1 min, and filter through a small fluted filter, returning the filtrate, if necessary, until clear. To 5 mL of the filtrate add 1 drop of pyrrole, agitate gently until dissolved, and then heat in a water bath at 50°. A blue color develops.
  D. It gives positive tests for *Sodium*, Appendix IIIA.
Assay Not less than 99.0% and not more than 101.0% of $C_6H_7NaO_6$ after drying.
Heavy Metals (as Pb) Not more than 10 mg/kg.
Loss on Drying Not more than 0.25%.
Specific Rotation $[\alpha]_D^{25°}$: Between +103° and +108°.

TESTS

Assay Dissolve about 400 mg, previously dried over phosphorus pentoxide for 24 h and accurately weighed, in a mixture of 100 mL of water, recently boiled and cooled, and 25 mL of 2 N sulfuric acid, and titrate with 0.1 N iodine, adding starch TS near the endpoint. Each mL of 0.1 N iodine is equivalent to 9.905 mg of $C_6H_7NaO_6$.
Heavy Metals Prepare and test a 2-g sample as directed in *Method II* under the *Heavy Metals Test*, Appendix IIIB, using 20 µg of lead ion (Pb) in the control (*Solution A*).
Loss on Drying Dry in vacuum over phosphorus pentoxide at 60° for 4 h.

Specific Rotation, Appendix IIB Determine in a solution containing 1 g in each 10 mL.

Packaging and Storage Store in tight, light-resistant containers.

Sodium Benzoate

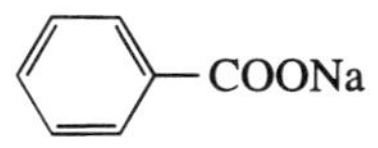

⟨⟩—COONa

C₇H₅NaO₂ Formula wt 144.11
INS: 211 CAS: [532-32-1]

DESCRIPTION

White, odorless or nearly odorless granules, crystalline powder, or flakes. One g dissolves in 2 mL of water, in 75 mL of alcohol, and in 50 mL of 90% alcohol.

Functional Use in Foods Preservative; antimicrobial agent.

REQUIREMENTS

Identification It gives positive tests for *Sodium* and for *Benzoate*, Appendix IIIA.
Assay Not less than 99.0% and not more than 100.5% of C₇H₅NaO₂, calculated on the anhydrous basis.
Alkalinity (as NaOH) Not more than 0.04%.
Heavy Metals (as Pb) Not more than 10 mg/kg.
Water Not more than 1.5%.

TESTS

Assay Transfer about 600 mg, accurately weighed, to a 250-mL beaker, add 100 mL of glacial acetic acid, and stir until the sample is completely dissolved. Add crystal violet TS, and titrate with 0.1 N perchloric acid in glacial acetic acid. Each mL of 0.1 N perchloric acid is equivalent to 14.41 mg of C₇H₅NaO₂.
Alkalinity Dissolve 2 g in 20 mL of hot water, and add 2 drops of phenolphthalein TS. If a pink color is produced, not more than 0.2 mL of 0.1 N sulfuric acid is required to discharge it.
Heavy Metals Dissolve 4 g in 40 mL of water, add dropwise, with vigorous stirring, 10 mL of 2.7 N hydrochloric acid, and filter. A 25-mL portion of the filtrate meets the requirements of the *Heavy Metals Test*, Appendix IIIB, using 20 μg of lead ion (Pb) in the control (*Solution A*).
Water Determine by the *Karl Fischer Titrimetric Method*, Appendix IIB.

Packaging and Storage Store in well-closed containers.

Sodium Bicarbonate

Baking Soda

NaHCO₃ Formula wt 84.01
INS: 500(ii) CAS: [144-55-8]

DESCRIPTION

A white, crystalline powder. It is stable in dry air, but slowly decomposes in moist air. Its solutions, when freshly prepared with cold water without shaking, are alkaline to litmus. The alkalinity increases as the solutions stand, are agitated, or are heated. One g dissolves in 10 mL of water. It is insoluble in alcohol.

Functional Use in Foods Alkali; leavening agent.

REQUIREMENTS

Identification A 1 in 10 solution gives positive tests for *Sodium* and for *Bicarbonate*, Appendix IIIA.
Assay Not less than 99.0% and not more than 100.5% of NaHCO₃ after drying.
Ammonia Passes test.
Heavy Metals (as Pb) Not more than 5 mg/kg.
Insoluble Substances Passes test.
Loss on Drying Not more than 0.25%.

TESTS

Assay Weigh accurately about 3 g of sample, previously dried over silica gel for 4 h, and dissolve it in 100 mL of water. Add 2 drops of methyl red TS, and titrate with 1 N hydrochloric acid. Add the acid slowly, with constant stirring, until the solution becomes faintly pink. Heat the solution to boiling, cool, and continue the titration until the faint pink color no longer fades after boiling. Each mL of 1 N hydrochloric acid is equivalent to 84.01 mg of NaHCO₃.
Ammonia Heat 1 g in a test tube. No odor of ammonia is detected.
Heavy Metals Dissolve 4 g in 10 mL of 2.7 N hydrochloric acid, boil gently for 1 min, and dilute to 25 mL with water. This solution meets the requirements of the *Heavy Metals Test*, Appendix IIIB, using 20 μg of lead ion (Pb) in the control (*Solution A*).
Insoluble Substances One g dissolves completely in 20 mL of water to give a clear solution.
Loss on Drying, Appendix IIC Dry over silica gel for 4 h.

Packaging and Storage Store in well-closed containers.

Sodium Bisulfate

Sodium Acid Sulfate; Nitre Cake

$NaHSO_4$ Formula wt 120.06

CAS: [7681-38-1]

DESCRIPTION

White, odorless crystals or granules. It is soluble in water, and its solutions are strongly acid. It is decomposed by alcohol into sodium sulfate and free sulfuric acid.

Functional Use in Foods   Acid.

REQUIREMENTS

Identification   It gives positive tests for *Sodium* and for *Sulfate*, Appendix IIIA.
Assay   Not less than 35.0% and not more than 39.0% of available H_2SO_4, equivalent to not less than 85.4% and not more than 95.2% of $NaHSO_4$.
Heavy Metals (as Pb)   Not more than 0.003%.
Lead   Not more than 10 mg/kg.
Loss on Drying   Not more than 0.8%.
Selenium   Not more than 0.003%.
Water-Insoluble Substances   Not more than 0.05%.

TESTS

Assay   Dissolve about 5 g, accurately weighed, in about 125 mL of water, add phenolphthalein TS, and titrate with 1 N sodium hydroxide. Each mL of 1 N sodium hydroxide is equivalent to 49.04 mg of H_2SO_4, or to 120.06 mg of $NaHSO_4$.
Heavy Metals   A solution of 670 mg in 25 mL of water meets the requirements of the *Heavy Metals Test*, Appendix IIIB, using 20 µg of lead ion (Pb) in the control (*Solution A*).
Lead   A solution of 1 g in 25 mL of water meets the requirements of the *Lead Limit Test*, Appendix IIIB, using 10 µg of lead ion (Pb) in the control.
Loss on Drying, Appendix IIC   Dry in a desiccator over phosphorus pentoxide for 24 h.
Selenium   Determine as directed in *Method II* under the *Selenium Limit Test*, Appendix IIIB, using 200 mg of sample.
Water-Insoluble Substances   Dissolve 50 g in 300 mL of hot water in a 600-mL beaker, allow the insoluble matter to settle, and filter by decanting through a tared, sintered-glass filter crucible, washing the insoluble matter into the crucible with additional hot water. Dry at 100° to 110° for 1 h, cool in a desiccator, and weigh.

Packaging and Storage   Store in tight containers.

Sodium Bisulfite

Sodium Acid Sulfite; Sodium Hydrogen Sulfite

CAS: [7631-90-5]

DESCRIPTION

Sodium Bisulfite consists of sodium bisulfite ($NaHSO_3$) and sodium metabisulfite ($Na_2S_2O_5$) in varying proportions, and for all practical purposes possesses properties of the true bisulfite. It occurs as white or yellowish white crystals or granular powder having an odor of sulfur dioxide. It is unstable in air. One g dissolves in 4 mL of water. It is slightly soluble in alcohol.

Functional Use in Foods   Preservative.

REQUIREMENTS

Identification   A 1 in 10 solution gives positive tests for *Sodium* and for *Sulfite*, Appendix IIIA.
Assay   Not less than 58.5% and not more than 67.4% of SO_2.
Heavy Metals (as Pb)   Not more than 10 mg/kg.
Iron   Not more than 0.005%.
Selenium   Not more than 0.003%.

TESTS

Assay   Weigh accurately about 200 mg, add it to exactly 50 mL of 0.1 N iodine contained in a glass-stoppered flask, and stopper the flask. Allow to stand for 5 min, add 1 mL of hydrochloric acid, and titrate the excess iodine with 0.1 N sodium thiosulfate, adding starch TS as the indicator. Each mL of 0.1 N iodine is equivalent to 3.203 mg of SO_2.
Heavy Metals   Dissolve 2 g in 10 mL of water, add 5 mL of hydrochloric acid, evaporate to dryness on a steam bath, and dissolve the residue in 25 mL of water. This solution meets the requirements of the *Heavy Metals Test*, Appendix IIIB, using 20 µg of lead ion (Pb) in the control (*Solution A*).
Iron   To 500 mg of the sample add 2 mL of hydrochloric acid, and evaporate to dryness on a steam bath. Dissolve the residue in 2 mL of hydrochloric acid and 20 mL of water, add a few drops of bromine TS, and boil the solution to remove the bromine. Cool, dilute with water to 25 mL, then add 50 mg of ammonium persulfate and 5 mL of ammonium thiocyanate TS. Any red or pink color does not exceed that produced in a control containing 2.5 mL of *Iron Standard Solution* (25 µg Fe).
Selenium   Determine as directed in *Method I* under the *Selenium Limit Test*, Appendix IIIB, using 200 mg of sample.

Packaging and Storage   Store in well-filled, tight containers, and avoid exposure to excessive heat.

Sodium Carbonate

Na$_2$CO$_3 \cdot x$H$_2$O   Formula wt, anhydrous 105.99

INS: 500   CAS: [497-19-8]

DESCRIPTION

Sodium Carbonate is anhydrous or may contain 1 or 10 molecules of water of hydration. It occurs as colorless crystals or as a white, granular or crystalline powder. It is freely soluble in water, and its solutions are alkaline to litmus. The anhydrous salt is hygroscopic, and the two hydrates are efflorescent. The decahydrate melts at about 32°.

Functional Use in Foods   Alkali.

REQUIREMENTS

Identification   It gives positive tests for *Sodium* and for *Carbonate*, Appendix IIIA.
Assay   Not less than 99.5% and not more than 100.5% of Na$_2$CO$_3$ after drying.
Heavy Metals (as Pb)   Not more than 10 mg/kg.
Loss on Drying   Na$_2$CO$_3$ (anhydrous): not more than 1%; Na$_2$CO$_3 \cdot$H$_2$O: between 12.0% and 15.0%; Na$_2$CO$_3 \cdot$10H$_2$O: between 55.0% and 65.0%.

TESTS

Assay   Weigh accurately about 2 g of the dried salt, obtained as directed under *Loss on Drying*, and dissolve it in 50 mL of water. Add 2 drops of methyl red TS, and titrate with 1 N hydrochloric acid. Add the acid slowly, with constant stirring, until the solution becomes faintly pink. Heat the solution to boiling, cool, and continue the titration until the faint pink color no longer fades after boiling. Each mL of 1 N hydrochloric acid is equivalent to 53.00 mg of Na$_2$CO$_3$.
Heavy Metals   Mix 2 g with 5 mL of water and 10 mL of 2.7 N hydrochloric acid, boil for 1 min, cool, and dilute to 25 mL with water. This solution meets the requirements of the *Heavy Metals Test*, Appendix IIIB, using 20 µg of lead ion (Pb) in the control (*Solution A*).
Loss on Drying, Appendix IIC   Dry about 3 g of the anhydrous salt or the monohydrate, accurately weighed, at 275° to 300° to constant weight. For the decahydrate, weigh accurately about 8 g, heat it first at 70°, then gradually raise the temperature, and finally dry at 275° to 300° to constant weight.

Packaging and Storage   Store the anhydrous salt and the decahydrate in tight containers; the monohydrate may be stored in well-closed containers.

Sodium Carboxymethylcellulose

CMC; Cellulose Gum; Modified Cellulose

INS: 466   CAS: [9004-32-4]

DESCRIPTION

It occurs as a white- to cream-colored powder or as granules. The powder is hygroscopic. A 1 in 100 aqueous suspension has a pH between 6.5 and 8.5. It is readily dispersed in water to form colloidal solutions. It is insoluble in most solvents.

Functional Use in Foods   Thickener; stabilizer.

REQUIREMENTS

Identification   Add about 1 g of powdered sample to 50 mL of warm water, while stirring, to produce a uniform dispersion. Continue the stirring until a colloidal solution is produced, and then cool to room temperature.
   A. To about 10 mL of the solution add 10 mL of cupric sulfate TS. A fluffy, bluish white precipitate is formed.
   B. The filtrate from *Identification Test A* gives positive tests for *Sodium*, Appendix IIIA.
Assay   Not less than 99.5% and not more than 100.5% of Sodium Carboxymethylcellulose, calculated on the dried basis.
Degree of Substitution   Not more than 0.95 carboxymethyl groups (—CH$_2$COOH) per anhydroglucose unit after drying.
Heavy Metals (as Pb)   Not more than 10 mg/kg.
Lead   Not more than 3 mg/kg.
Loss on Drying   Not more than 10.0%.
Sodium   Not more than 9.5% after drying.
Viscosity of a 2%, Weight in Weight, Solution   Not less than 25 centipoises.

TESTS

Assay   Calculate the percentage of Sodium Carboxymethylcellulose by subtracting from 100 the percentages of *Sodium Chloride* and *Sodium Glycolate* determined as follows:
   Sodium Chloride   Weigh accurately about 5 g of the sample into a 250-mL beaker, add 50 mL of water and 5 mL of 30% hydrogen peroxide, and heat on a steam bath for 20 min, stirring occasionally to ensure complete dissolution. Cool, add 100 mL of water and 10 mL of nitric acid, and titrate with 0.05 N silver nitrate to a potentiometric endpoint, using silver and mercurous sulfate–potassium sulfate electrodes and stirring constantly. Calculate the percentage of sodium chloride in the sample by the formula

$$584.4 \times V \times N/(100 - b)W,$$

in which V and N represent the volume, in mL, and the normality, respectively, of the silver nitrate; b is the percentage of *Loss on Drying*, determined separately; W is the weight, in g, of the sample; and 584.4 is an equivalence factor for sodium chloride.
   Sodium Glycolate   Transfer about 500 mg, accurately weighed, of the sample into a 100-mL beaker, moisten thor-

oughly with 5 mL of acetic acid, followed by 5 mL of water, and stir with a glass rod until solution is complete (usually about 15 min). Slowly add 50 mL of acetone, with stirring, then add 1 g of sodium chloride, and stir for several min to ensure complete precipitation of the carboxymethylcellulose. Filter through a soft, open-textured paper, previously wetted with a small amount of acetone, and collect the filtrate in a 100-mL volumetric flask. Use an additional 30 mL of acetone to facilitate transfer of the solids and to wash the filter cake, then dilute to volume with acetone, and mix.

Prepare a series of standard solutions as follows: Transfer 100 mg of glycolic acid, previously dried in a desiccator at room temperature overnight and accurately weighed, into a 100-mL volumetric flask, dissolve in water, dilute to volume with water, and mix. Use this solution within 30 days. Transfer 1.0 mL, 2.0 mL, 3.0 mL, and 4.0 mL of the solution into separate 100-mL volumetric flasks, add sufficient water to each flask to make 5 mL, then add 5 mL of glacial acetic acid, and dilute to volume with acetone.

Transfer 2.0 mL of the sample solution and 2.0 mL of each standard solution into separate 25-mL volumetric flasks, and prepare a blank flask containing 2.0 mL of a solution containing 5% each of glacial acetic acid and water in acetone. Place the uncovered flasks in a boiling water bath for exactly 20 min to remove the acetone, remove from the bath, and cool. Add to each flask 5.0 mL of 2,7-dihydroxynaphthalene TS, mix thoroughly, add an additional 15 mL, and again mix thoroughly. Cover the mouth of each flask with a small piece of aluminum foil. Place the flasks upright in a boiling water bath for 20 min, then remove from the bath, cool, dilute to volume with sulfuric acid, and mix.

Using a suitable spectrophotometer, determine the absorbance of each solution at 540 nm against the blank, and prepare a standard curve using the absorbances obtained from the standard solutions. From the standard curve and the absorbance of the sample, determine the weight (w), in mg, of glycolic acid in the sample, and calculate the percentage of sodium glycolate in the sample by the formula

$$12.9 \times w/(100 - b)W,$$

in which 12.9 is a factor converting glycolic acid to sodium glycolate; b is the percentage of *Loss on Drying*, determined separately; and W is the weight of the sample, in g.

Degree of Substitution Weigh accurately about 200 mg of the sample, previously dried at 105° to constant weight, and transfer it into a 250-mL, glass-stoppered Erlenmeyer flask. Add 75 mL of glacial acetic acid, connect the flask with a water-cooled condenser, and reflux gently on a hot plate for 2 h. Cool, transfer the solution to a 250-mL beaker with the aid of 50 mL of glacial acetic acid, and titrate with 0.1 N perchloric acid in dioxane while stirring with a magnetic stirrer. Determine the endpoint potentiometrically with a pH meter equipped with a standard glass electrode and a calomel electrode modified as follows: Discard the aqueous potassium chloride solution contained in the electrode, rinse and fill with the supernatant liquid obtained by shaking thoroughly 2 g each of potassium chloride and silver chloride (or silver oxide) with 100 mL of methanol, then add a few crystals of potassium chloride and silver chloride (or silver oxide) to the electrode.

Record the mL of 0.1 N perchloric acid versus mV (0- to 700-mV range), and continue the titration to a few mL beyond the endpoint. Plot the titration curve, and read the volume (A), in mL, of 0.1 N perchloric acid at the inflection point.

Calculate the degree of substitution (DS) by the formula

$$(16.2A/G)/[1.000 - (8.0\ A/G)],$$

in which A is the volume, in mL, of 0.1 N perchloric acid required; G is the weight, in mg, of the sample taken; 16.2 is one-tenth of the molecular weight of one anhydroglucose unit; and 8.0 is one-tenth of the molecular weight of one sodium carboxymethyl group.

Heavy Metals Prepare and test a 2-g sample as directed in *Method II* under the *Heavy Metals Test*, Appendix IIIB, adding 1 mL of hydroxylamine hydrochloride solution (1 in 5) to the solution of the residue. Any color does not exceed that produced in a control (*Solution A*) containing 20 µg of lead ion (Pb) and 1 mL of the hydroxylamine hydrochloride solution.

Lead A *Sample Solution* prepared from a 2-g sample as directed for organic compounds meets the requirements of the *Lead Limit Test*, Appendix IIIB, using 6 µg of lead ion (Pb) in the control.

Loss on Drying, Appendix IIC Dry to constant weight at 105°.

Sodium From the weight of the sample and the number of mL of 0.1 N perchloric acid consumed in the determination of *Degree of Substitution*, calculate the percentage of sodium. Each mL of 0.1 N perchloric acid is equivalent to 2.299 mg of Na.

Viscosity of a 2%, Weight in Weight, Solution Determine as directed under *Viscosity of Sodium Carboxymethylcellulose*, Appendix IIB.

Packaging and Storage Store in well-closed containers.

Sodium Chloride

Salt

NaCl   Formula wt 58.44

                                               CAS: [7647-14-5]

DESCRIPTION

Salt is a generic term applied to commercially produced Sodium Chloride. It is available in various crystalline forms, referred to as evaporated salt, rock salt, solar salt, or simply salt. It may contain up to 2% (total) of suitable food-grade anticaking, free-flowing, or conditioning agents, either singly or in combination. It may contain not more than 13 mg/kg of sodium ferrocyanide, or not more than 25 mg/kg of green ferric ammonium citrate as crystal-modifying and anticaking agents. If labeled as iodized, it contains not less than 0.006% and not more than 0.010% of potassium iodide.

Sodium Chloride is a transparent to opaque, white crystalline solid of variable particle size. (Rock salt may be white to off-white in color.) It remains dry in air at a relative humidity below 75%, but becomes deliquescent at higher humidities. One g is soluble in 2.8 mL of water at 25°, in 2.7 mL of boiling water, and in about 10 mL of glycerin. Sodium chloride containing water-insoluble anticaking, free-flowing, and conditioning agents may produce cloudy solutions, or may dissolve incompletely. A 1 in 20 solution usually has a pH between 5.5 and 8.5 (the pH may be higher if alkaline conditioning agents have been added).

Functional Use in Foods Nutrient; preservative; flavoring agent and intensifier; curing agent; dough conditioner.

REQUIREMENTS

Labeling Indicate whether the article is iodized.
Identification It gives positive tests for *Sodium* and for *Chloride*, Appendix IIIA.
Assay *Evaporated salt with up to 2% of suitable free-flowing or conditioning agents and anticaking agents such as sodium ferrocyanide*: Not less than 97.5% or more than 100.5% of NaCl after drying at 625° for 2 h.

Evaporated salt with only anticaking agents such as sodium ferrocyanide: Not less than 99.0% or more than 100.5% after drying at 625° for 2 h.

Rock or solar salt: Not less than 97.5% or more than 100.5% of NaCl after drying at 625° for 2 h, the remainder consisting chiefly of minor amounts of naturally occurring components such as alkaline and/or alkaline earth sulfates and chlorides.
Arsenic (as As) Not more than 1 mg/kg.
Calcium and Magnesium Not more than 2.0%.
Heavy Metals (as Pb) Not more than 2 mg/kg.
Iodine Not less than 0.006% and not more than 0.010% of potassium iodide.

Note: This specification applies only to iodized salt.

Iron Not more than 0.0016% of Fe.

Note: This specification applies only to products to which green ferric ammonium citrate has been added.

Loss on Drying Not more than 0.5%.
Sodium Ferrocyanide Not more than 0.0013% of anhydrous $Na_4Fe(CN)_6$.

Note: This specification applies only to products to which sodium ferrocyanide has been added.

TESTS

Note: In the following procedures, it may be necessary to filter the sample solutions to avoid interference by insoluble or suspended anticaking, free-flowing, or conditioning agents.

Assay Weigh accurately about 250 mg of the sample, previously dried at 625° for 2 h, and dissolve it in 50 mL of water in a glass-stoppered flask. Add, while agitating, 3 mL of nitric acid, 5 mL of nitrobenzene, 50.0 mL of 0.1 N silver nitrate, and 2 mL of ferric ammonium sulfate TS. Shake well, and titrate the excess silver nitrate with 0.1 N ammonium thiocyanate. Each mL of 0.1 N silver nitrate is equivalent to 5.844 mg of NaCl.
Arsenic A solution of 3 g of the sample in 25 mL of water meets the requirements of the *Arsenic Test*, Appendix IIIB.
Calcium and Magnesium

Standard EDTA Solution Dissolve 4.0 g of disodium EDTA, $C_{10}H_{14}N_2Na_2O_8 \cdot 2H_2O$, in sufficient water to make 1000 mL.

Magnesium Sulfate Solution Dissolve 2.6 g of magnesium sulfate, $MgSO_4 \cdot 7H_2O$, in sufficient water to make 1000 mL.

Buffer Solution (*a*) Initial Preparation: Transfer 67.5 g of ammonium chloride into a 1000-mL volumetric flask, and dissolve in 570 mL of ammonium hydroxide. Use 2 mL of this solution as directed below under *Titer Determination*. (*b*) Final Preparation: Pipet 50.0 mL of *Magnesium Sulfate Solution* into the flask, add exactly the volume *T*, in mL, of *Standard EDTA Solution*, determined as directed below under *Titer Determination*, then dilute to volume with water, and mix.

Titer Determination Pipet 50.0 mL of *Magnesium Sulfate Solution* into a 400-mL beaker, and add 200 mL of water, 2 mL of *Buffer Solution* (initial preparation), 1.0 mL of potassium cyanide solution (1 in 20), and 5 drops of eriochrome black TS or other suitable indicator. Titrate with the *Standard EDTA Solution*, while stirring with a magnetic stirrer, to a true blue endpoint. Record the volume *T*, in mL, of *Standard EDTA Solution* equivalent to 50.0 mL of the *Magnesium Sulfate Solution*.

Standardization of EDTA Solution Transfer about 1 g, accurately weighed, of primary standard calcium carbonate, $CaCO_3$, into a 1000-mL volumetric flask, dissolve in 800 mL of water containing 5 mL of hydrochloric acid, dilute to volume with water, and mix. Pipet 25.0 mL of this solution into a 400-mL beaker, and add 200 mL of water, 2 mL of *Buffer Solution* (final preparation), 1.0 mL of potassium cyanide solution (1 in 20), and 20 drops of eriochrome black TS or other suitable indicator. Titrate with the *Standard EDTA Solution*, stirring with a magnetic stirrer, to a true blue endpoint. Calculate the factor *F*, giving the number of mg of Ca equivalent to 1.0 mL of *Standard EDTA Solution*, by the formula

$$10.011 w/v,$$

in which *w* is the exact weight, in g, of the primary standard calcium carbonate taken, and *v* is the volume, in mL, of the *Standard EDTA Solution* required in the titration.

Sample Preparation for Rock and Solar Salt Transfer 50.0 g of the sample into a 500-mL volumetric flask, dissolve in 400 mL of water containing 2 mL of hydrochloric acid, dilute to volume with water, and mix. Filter a 50-mL aliquot, then pipet 10.0 mL of the filtrate into a 400-mL beaker, and add 190 mL of water.

Sample Preparation for Evaporated Salt Transfer 10.0 g of the sample into a 400-mL beaker, and dissolve in 100 mL of water. If free-flowing agents are present, filter and rinse quantitatively. Dilute the solution or filtrate to 200 mL with water.

Procedure To the *Sample Preparation* add 5 mL of *Buffer Solution* (final preparation), 1 mL of potassium cyanide solution

(1 in 20), and 5 drops of eriochrome black TS or other suitable indicator. Begin stirring with a magnetic stirrer, and titrate with *Standard EDTA Solution* to a true blue endpoint, recording the volume, in mL, required as *V*. Calculate the mg/kg of total calcium and magnesium (both expressed as Ca) in the sample by the formula

$$V \times F \times 1000/W,$$

in which *W* is the weight, in g, of salt sample in the final solution titrated.

Heavy Metals A solution of 10 g of the sample in 35 mL of water meets the requirements of the *Heavy Metals Test*, Appendix IIIB, using 20 μg of lead ion (Pb) in the control (*Solution A*).

Iodine Transfer about 20 g of the sample, accurately weighed, into a 600-mL beaker, and dissolve in about 300 mL of water. Add a few drops of methyl orange TS, neutralize the solution with phosphoric acid (85%), and then add 1 mL excess of the acid. Add 25 mL of bromine TS and a few glass beads, boil until the solution is clear, then boil for an additional 5 min. Add about 50 mg of salicylic acid crystals, 1 mL of phosphoric acid, and 10 mL of potassium iodide solution (1 in 20), and titrate to a pale yellow color with 0.01 *N* sodium thiosulfate. Add 1 mL of starch TS, and continue the titration to the disappearance of the blue color. Each mL of 0.01 *N* sodium thiosulfate is equivalent to 0.2767 mg of KI.

Iron Dissolve 625.0 mg of the sample in 10 mL of 2.7 *N* hydrochloric acid, and dilute to 50 mL with water. Add about 40 mg of ammonium persulfate crystals and 10 mL of ammonium thiocyanate TS. Any red or pink color does not exceed that produced by 1.0 mL of *Iron Standard Solution* (10 μg Fe) in an equal volume of solution containing 2 mL of hydrochloric acid and the quantities of ammonium persulfate and ammonium thiocyanate used in the test.

Loss on Drying, Appendix IIC Dry at 110° for 2 h.

Sodium Ferrocyanide Dissolve 9.62 g of the sample in 80 mL of water in a 150-mL glass-stoppered cylinder or flask. Prepare a standard solution containing 125 μg of $Na_4Fe(CN)_6$ in each mL by dissolving 99.5 mg of $Na_4Fe(CN)_6 \cdot 10H_2O$ in 500.0 mL of water, then transfer 1.0 mL of this solution into a similar 150-mL container for the control. To each container add 2 mL of ferrous sulfate TS and 1 mL of 2 *N* sulfuric acid, dilute to 100 mL with water, and mix. Transfer 50-mL portions of the respective solutions into matched color-comparison tubes. The sample solution shows no more blue color than the control.

Packaging and Storage Store in well-closed containers.

Sodium Citrate

$C_6H_5Na_3O_7 \cdot 2H_2O$　　　　　　Formula wt 294.10

INS: 331(i)　　　　　　CAS: [68-04-2]

DESCRIPTION

Sodium Citrate is anhydrous or contains two molecules of water of crystallization. It occurs as colorless crystals or as a white, crystalline powder. One g of the dihydrate dissolves in 1.5 mL of water at 25° and in 0.6 mL of boiling water. It is insoluble in alcohol.

Functional Use in Foods Buffer; sequestrant; nutrient for cultured buttermilk.

REQUIREMENTS

Identification A 1 in 20 solution gives positive tests for *Sodium* and for *Citrate*, Appendix IIIA.

Assay Not less than 99.0% and not more than 100.5% of $C_6H_5Na_3O_7$, calculated on the anhydrous basis.

Alkalinity Passes test.

Heavy Metals (as Pb) Not more than 10 mg/kg.

Water *Anhydrous sodium citrate*: not more than 1%; *sodium citrate dihydrate*: between 10.0% and 13.0%.

TESTS

Assay Transfer about 350 mg, accurately weighed, to a 250-mL beaker. Add 100 mL of glacial acetic acid, stir until completely dissolved, and titrate with 0.1 *N* perchloric acid, using crystal violet TS as the indicator. Each mL of 0.1 *N* perchloric acid is equivalent to 8.602 mg of $C_6H_5Na_3O_7$.

Alkalinity A solution of 1 g in 20 mL of water is alkaline to litmus paper, but after the addition of 0.2 mL of 0.1 *N* sulfuric acid, no pink color is produced by 1 drop of phenolphthalein TS.

Heavy Metals A solution of 2 g in 25 mL of water meets the requirements of the *Heavy Metals Test*, Appendix IIIB, using 20 μg of lead ion (Pb) in the control (*Solution A*).

Water Determine by the *Karl Fischer Titrimetric Method*, Appendix IIB.

Packaging and Storage Store in tight containers.

Sodium Dehydroacetate

Sodium 3-(1-Hydroxyethylidene)-6-methyl-1,2-pyran-2,4(3H)-dione

$C_8H_7NaO_4 \cdot H_2O$ Formula wt 208.15
INS: 266   CAS: [4418-26-2]

DESCRIPTION

A white or nearly white, odorless powder having a slight, characteristic taste. One g dissolves in about 3 mL of water, in 2 mL of propylene glycol, and in 7 mL of glycerin.

Functional Use in Foods Preservative.

REQUIREMENTS

Identification Dissolve about 1.5 g in 10 mL of water, add 5 mL of 2.7 N hydrochloric acid, collect the crystals with suction, wash with 10 mL of water, and dry between 75° and 80° for 4 h. The crystals melt between 109° and 111° (see Appendix IIB).
Assay Not less than 98.0% and not more than 100.5% of $C_8H_7NaO_4$, calculated on the anhydrous basis.
Heavy Metals (as Pb) Not more than 10 mg/kg.
Water Between 8.5% and 10.0%.

TESTS

Assay Transfer about 500 mg, accurately weighed, to a 125-mL Erlenmeyer flask, dissolve it in 25 mL of glacial acetic acid containing 1 drop of a 1 in 100 solution of *p*-naphtholbenzein in glacial acetic acid that has been previously neutralized to a blue color, and titrate with 0.1 N perchloric acid to the original blue color. Each mL of 0.1 N perchloric acid is equivalent to 19.01 mg of $C_8H_7NaO_4$.
Heavy Metals Prepare and test a 2-g sample as directed in *Method II* under the *Heavy Metals Test*, Appendix IIIB, using 20 μg of lead ion (Pb) in the control (*Solution A*).
Water Determine by the *Karl Fischer Titrimetric Method*, Appendix IIB.

Packaging and Storage Store in well-closed containers.

Sodium Diacetate

Sodium Hydrogen Diacetate

$CH_3COONa \cdot CH_3COOH \cdot xH_2O$

$C_4H_7NaO_4 \cdot xH_2O$ Formula wt, anhydrous 142.09
INS: 262   CAS: [126-96-5]

DESCRIPTION

Sodium Diacetate is a molecular compound of sodium acetate and acetic acid. It is a white, hygroscopic, crystalline solid having an odor of acetic acid. One g is soluble in about 1 mL of water. The pH of a 1 in 10 solution is between 4.5 and 5.0.

Functional Use in Foods Sequestrant; preservative; mold inhibitor.

REQUIREMENTS

Identification A 1 in 10 solution gives positive tests for *Acetate* and for *Sodium*, Appendix IIIA.
Assay Not less than 39.0% and not more than 41.0% of free acetic acid (CH_3COOH), and not less than 58.0% and not more than 60.0% of sodium acetate (CH_3COONa), calculated on the anhydrous basis.
Heavy Metals (as Pb) Not more than 10 mg/kg.
Readily Oxidizable Substances (as formic acid) Not more than 0.2%.
Water Not more than 2.0%.

TESTS

Assay
Free Acetic Acid Weigh accurately about 4 g, dissolve it in 50 mL of water, add phenolphthalein TS, and titrate with 1 N sodium hydroxide. Each mL of 1 N sodium hydroxide is equivalent to 60.05 mg of CH_3COOH.
Sodium Acetate Content Weigh accurately about 500 mg, dissolve it in 50 mL of glacial acetic acid, and titrate with 0.1 N perchloric acid, determining the endpoint potentiometrically. Each mL of 0.1 N perchloric acid is equivalent to 8.203 mg of CH_3COONa.
Heavy Metals A solution of 2 g in 25 mL of water meets the requirements of the *Heavy Metals Test*, Appendix IIIB, using 20 μg of lead ion (Pb) in the control (*Solution A*).
Readily Oxidizable Substances Dissolve 1.0 g in about 50 mL of water, add 10 mL of 2 N sulfuric acid, and heat the solution to between 80° and 90°. Titrate the hot solution with 0.1 N potassium permanganate to a faint pink color that persists for at least 15 s. Each mL of 0.1 N potassium permanganate is equivalent to 2.301 mg of CH_2O_2.
Water Determine by the *Karl Fischer Titrimetric Method*, Appendix IIB.

Packaging and Storage Store in tight containers.

Sodium Erythorbate

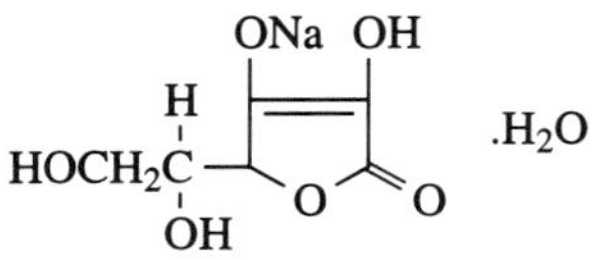

$C_6H_7NaO_6 \cdot H_2O$ Formula wt 216.12

CAS: [6381-77-7]

DESCRIPTION

White, odorless, crystalline powder or granules. In the dry state it is reasonably stable in air, but in solution it deteriorates in the presence of air, trace metals, heat, and light. One g dissolves in about 7 mL of water. The pH of a 1 in 20 solution is between 5.5 and 8.0.

Functional Use in Foods Preservative; antioxidant.

REQUIREMENTS

Identification
A. A 1 in 50 solution slowly reduces alkaline cupric tartrate TS at 25°, but more readily upon heating.
B. To 2 mL of a 1 in 50 solution acidified with 0.5 mL of 0.1 N hydrochloric acid add a few drops of sodium nitroferricyanide TS, followed by 1 mL of 0.1 N sodium hydroxide. A transient blue color is produced immediately.
C. It gives positive tests for *Sodium*, Appendix IIIA.
Assay Not less than 98.0% and not more than 100.5% of $C_6H_7NaO_6 \cdot H_2O$.
Heavy Metals (as Pb) Not more than 10 mg/kg.
Lead Not more than 5 mg/kg.
Oxalate Passes test.
Specific Rotation $[\alpha]_D^{25°}$: Between +95.5° and +98.0°.

TESTS

Assay Dissolve about 400 mg, accurately weighed, in a mixture of 100 mL of water, recently boiled and cooled, and 25 mL of 2 N sulfuric acid, and immediately titrate with 0.1 N iodine, adding starch TS near the endpoint. Each mL of 0.1 N iodine is equivalent to 10.81 mg of $C_6H_7NaO_6 \cdot H_2O$.
Heavy Metals Prepare and test a 2-g sample as directed in *Method II* under the *Heavy Metals Test*, Appendix IIIB, using 20 µg of lead ion (Pb) in the control (*Solution A*).
Lead A *Sample Solution* prepared from a 2-g sample as directed for organic compounds meets the requirements of the *Lead Limit Test*, Appendix IIIB, using 10 µg of lead ion (Pb) in the control.
Oxalate To a solution of 1 g in 10 mL of water add 2 drops of glacial acetic acid and 5 mL of a 1 in 10 solution of calcium acetate. The solution remains clear.
Specific Rotation, Appendix IIB Determine in a solution containing 1 g in each 10 mL.

Packaging and Storage Store in tight, light-resistant containers.

Sodium Ferric Pyrophosphate

Sodium Iron Pyrophosphate

$Na_8Fe_4(P_2O_7)_5 \cdot xH_2O$ Formula wt, anhydrous 1277.02

CAS: [1332-96-3]

DESCRIPTION

A white to tan, odorless powder. It is insoluble in water, but is soluble in hydrochloric acid.

Functional Use in Foods Nutrient; dietary supplement.

REQUIREMENTS

Identification Dissolve 500 mg in 5 mL of dilute hydrochloric acid (1 in 2), and add an excess of 1 N sodium hydroxide. A reddish brown precipitate forms. Age the solution for several min, and then filter, discarding the first few mL. To 5 mL of the clear filtrate add 1 drop of bromophenol blue TS, and titrate with 1 N hydrochloric acid to a green color. Add 10 mL of a 1 in 8 solution of zinc sulfate, and readjust the pH to 3.8 (green color). A white precipitate forms (distinction from orthophosphates).
Assay Not less than 14.5% and not more than 16.0% of Fe.
Fluoride Not more than 0.005%.
Lead Not more than 10 mg/kg.
Loss on Ignition Not more than 8.0%.
Mercury Not more than 3 mg/kg.

TESTS

Assay Proceed as directed under *Assay* in the monograph for *Ferric Phosphate*.
Fluoride Weigh accurately 1.0 g, and proceed as directed in the *Fluoride Limit Test*, Appendix IIIB.
Lead Proceed as directed under *Lead* in the monograph for *Ferric Phosphate*.
Loss on Ignition Ignite at 800° for 1 h.
Mercury Determine as directed under *Mercury* in the monograph for *Ferric Phosphate*.

Packaging and Storage Store in well-closed containers.

Sodium Ferrocyanide

Yellow Prussiate of Soda

$Na_4Fe(CN)_6 \cdot 10H_2O$ Formula wt 484.07

INS: 535 CAS: [13601-19-9]

DESCRIPTION

Yellow crystals or crystalline powder. It is soluble in water, but is practically insoluble in most organic solvents.

Functional Use in Foods Anticaking agent for sodium chloride.

REQUIREMENTS

Identification To 10 mL of a 1% solution of the sample add 1 mL of ferric chloride TS. A dark blue precipitate forms.
Assay Not less than 99.0% of $Na_4Fe(CN)_6 \cdot 10H_2O$.
Chloride Not more than 0.2%.
Cyanide Passes test.
Ferricyanide Passes test.
Free Moisture Not more than 1%.
Insoluble Matter Not more than 0.03%.
Sulfate Not more than 0.07%.

TESTS

Assay Transfer about 3 g, accurately weighed, into a 400-mL beaker, dissolve in 225 mL of water, and add cautiously about 25 mL of 95% sulfuric acid. Add, with stirring, 1 drop of orthophenanthroline TS, and titrate with 0.1 N ceric sulfate until the color changes sharply from orange to pure yellow. Each mL of 0.1 N ceric sulfate is equivalent to 96.81 mg of $Na_4Fe(CN)_6 \cdot 10H_2O$.
Chloride, Appendix IIIB Dissolve 100 mg of the sample in 100 mL of water. Any turbidity produced by a 10-mL portion of this solution does not exceed that shown in a control containing 20 μg of chloride ion (Cl).
Cyanide Dissolve 10 mg of copper sulfate in a mixture of 8 mL of water and 2 mL of 6 N ammonium hydroxide. Wet a strip of filter paper with this solution, and place the wet paper in a stream of hydrogen sulfide. When 1 drop of a 1% solution of the sample is placed on the brown reagent paper, a white circle is not produced.
Ferricyanide Dissolve about 10 mg of the sample in 10 mL of water, and place 1 drop of this solution on a spot plate. Add 1 drop of a 1% solution of lead nitrate followed by a few drops of a solution prepared by saturating cold 2 N acetic acid with benzidine. No blue precipitate or blue coloration appears.
Free Moisture Heat 20 g of the sample at 105° for 6 h, cool in a desiccator, and weigh. Grind the dried sample rapidly, then heat 3 g of the powder to constant weight at 105°, and calculate the total water content (W). Calculate the percentage of free moisture in the sample by the formula

$$W - 0.3721A,$$

in which A is the percentage of $Na_4Fe(CN)_6 \cdot 10H_2O$ found in the *Assay*.
Insoluble Matter Dissolve 50 g of the sample in 300 mL of hot water, and filter off the insoluble matter on a tared, sintered-glass filter crucible. Wash the residue thoroughly with hot water, dry the crucible in an oven at 105° for 1 h, cool in a desiccator, and weigh.
Sulfate, Appendix IIIB Any turbidity produced by a 500-mg sample does not exceed that shown in a control containing 350 μg of sulfate (SO_4).

Packaging and Storage Store in tight containers.

Sodium Gluconate

$CH_2OH(CHOH)_4COONa$

$C_6H_{11}NaO_7$ Formula wt 218.14

INS: 576 CAS: [527-07-1]

DESCRIPTION

A white to tan, granular to fine, crystalline powder. It is very soluble in water, and is sparingly soluble in alcohol. It is insoluble in ether.

Functional Use in Foods Nutrient; dietary supplement; sequestrant.

REQUIREMENTS

Identification

A. A 1 in 20 solution gives positive tests for *Sodium*, Appendix IIIA.

B. It meets the requirements of *Identification Test B* in the monograph for *Calcium Gluconate*.
Assay Not less than 98.0% and not more than 102.0% of $C_6H_{11}NaO_7$.
Heavy Metals (as Pb) Not more than 0.002%.
Lead Not more than 10 mg/kg.
Reducing Substances Not more than 0.5%.

TESTS

Assay Transfer about 150 mg, accurately weighed, into a clean, dry 200-mL Erlenmeyer flask, add 75 mL of glacial acetic acid, and dissolve by heating on a hot plate. Cool, add quinaldine red TS, and titrate with 0.1 N perchloric acid in glacial acetic acid, using a 10-mL microburet, to a colorless endpoint. Each mL of 0.1 N perchloric acid is equivalent to 21.81 mg of $C_6H_{11}NaO_7$.
Heavy Metals A solution of 1 g in 25 mL of water meets the requirements of the *Heavy Metals Test*, Appendix IIIB, using 20 μg of lead ion (Pb) in the control (*Solution A*).

Lead A solution of 1 g in 25 mL of water meets the requirements of the *Lead Limit Test*, Appendix IIIB, using 10 µg of lead ion (Pb) in the control.

Reducing Substances Determine as directed under *Reducing Substances* in the monograph for *Copper Gluconate*.

Packaging and Storage Store in well-closed containers.

Sodium Hydroxide

Caustic Soda

NaOH Formula wt 40.00
INS: 524 CAS: [1310-73-2]

DESCRIPTION

White or nearly white pellets, flakes, sticks, fused masses, or other forms. Upon exposure to air it readily absorbs carbon dioxide and moisture. One g dissolves in 1 mL of water. It is freely soluble in alcohol.

Functional Use in Foods Alkali.

REQUIREMENTS

Identification A 1 in 25 solution gives positive tests for *Sodium*, Appendix IIIA.
Assay Not less than 95.0% and not more than 100.5% of total alkali, calculated as NaOH.
Arsenic (as As) Not more than 3 mg/kg.
Carbonate (as Na_2CO_3) Not more than 3.0%.
Heavy Metals (as Pb) Not more than 0.002%.
Insoluble Substances and Organic Matter Passes test.
Lead Not more than 10 mg/kg.
Mercury Not more than 0.1 mg/kg.

TESTS

Assay Dissolve about 1.5 g, accurately weighed, in 40 mL of recently boiled and cooled water, cool to 15°, add phenolphthalein TS, and titrate with 1 N sulfuric acid. At the discharge of the pink color, record the volume of acid required, then add methyl orange TS, and continue the titration until a persistent pink color is produced. Record the total volume of acid required for the titration. Each mL of 1 N sulfuric acid is equivalent to 40.00 mg of total alkali, calculated as NaOH.

Arsenic Dissolve 1 g in about 10 mL of water, cautiously neutralize to litmus paper with sulfuric acid, and cool. This solution meets the requirements of the *Arsenic Test*, Appendix IIIB.

Carbonate Each mL of 1 N sulfuric acid required between the phenolphthalein and methyl orange endpoints in the *Assay* is equivalent to 106.0 mg of Na_2CO_3.

Heavy Metals Dissolve 1 g in a mixture of 5 mL of water and 11 mL of 2.7 N hydrochloric acid. Heat to boiling, cool, dilute to 25 mL with water, and filter. This solution meets the requirements of the *Heavy Metals Test*, Appendix IIIB, using 20 µg of lead ion (Pb) in the control (*Solution A*).

Insoluble Substances and Organic Matter A 1 in 20 solution is complete, clear, and colorless to slightly colored.

Lead Dissolve 1 g in a mixture of 5 mL of water and 11 mL of 2.7 N hydrochloric acid, and cool. This solution meets the requirements of the *Lead Limit Test*, Appendix IIIB, using 10 µg of lead ion (Pb) in the control.

Mercury Determine as directed under *Mercury Limit Test*, Appendix IIIB, preparing the *Standard Preparation* and the *Sample Preparation* as follows:

Standard Preparation Prepare the stock solution and dilutions to obtain a solution containing 1 µg of Hg per mL, as directed in Appendix IIIB. Transfer 1.0 mL of the final solution (1 µg of Hg) to a 50-mL beaker, and add 20 mL of water, 1 mL of dilute sulfuric acid solution (1 in 5), and 1 mL of potassium permanganate solution (1 in 25). Cover the beaker with a watch glass, boil for a few s, and cool.

Sample Preparation Transfer 10.0 g of the sample into a 100-mL beaker, dissolve in 15 mL of water, add 2 drops of phenolphthalein TS, and slowly neutralize, with constant stirring, with dilute hydrochloric acid solution (1 in 2). Add 1 mL of dilute sulfuric acid solution (1 in 5) and 1 mL of potassium permanganate solution (1 in 25), cover the beaker with a watch glass, boil for a few s, and cool.

Packaging and Storage Store in tight containers.

Sodium Hydroxide Solution

DESCRIPTION

Sodium Hydroxide Solutions are usually available in nominal concentrations of 50% and 73% of NaOH, weight in weight, having freezing points of about 15° and 63°, respectively. These solutions are clear or slightly turbid, colorless or slightly colored, strongly caustic, and hygroscopic, and when exposed to the air, they absorb carbon dioxide, forming sodium carbonate.

Functional Use in Foods Alkali.

REQUIREMENTS

Labeling Indicate the percent of Sodium Hydroxide, NaOH.
Identification Solutions of sodium hydroxide give positive tests for *Sodium*, Appendix IIIA.
Assay Not less than 97.0% and not more than 103.0%, by weight, of the labeled amount of NaOH, calculated as total alkalinity.
Arsenic (as As) Not more than 3 mg/kg, calculated on the basis of NaOH determined in the *Assay*.

Carbonate (as Na_2CO_3) Not more than 3.0%, calculated on the basis of NaOH determined in the *Assay*.
Heavy Metals (as Pb) Not more than 0.002%, calculated on the basis of NaOH determined in the *Assay*.
Lead Not more than 10 mg/kg, calculated on the basis of NaOH determined in the *Assay*.
Mercury Not more than 1 mg/kg, calculated on the basis of NaOH determined in the *Assay*.

TESTS

Assay Based on the stated or labeled percentage of NaOH, weigh accurately a volume of the solution equivalent to about 1.5 g of sodium hydroxide, and dilute it to 40 mL with recently boiled and cooled water. Continue as directed under *Assay* in the monograph for *Sodium Hydroxide*, beginning with "...cool to 15°...."
Arsenic Dilute the equivalent of 1 g of NaOH, calculated on the basis of the *Assay*, to 10 mL with water, cautiously neutralize to litmus paper with sulfuric acid, and cool. This solution meets the requirements of the *Arsenic Test*, Appendix IIIB.
Carbonate Each mL of 1 N sulfuric acid required between the phenolphthalein and methyl orange endpoints in the *Assay* is equivalent to 106.0 mg of Na_2CO_3.
Heavy Metals Dilute the equivalent of 1 g of NaOH, calculated on the basis of the *Assay*, with a mixture of 5 mL of water and 11 mL of 2.7 N hydrochloric acid, and heat to boiling. Cool, dilute to 25 mL with water, and filter. This solution meets the requirements of the *Heavy Metals Test*, Appendix IIIB, using 20 μg of lead ion (Pb) in the control (*Solution A*).
Lead Dilute the equivalent of 1 g of NaOH, calculated on the basis of the *Assay*, with a mixture of 5 mL of water and 11 mL of 2.7 N hydrochloric acid. This solution meets the requirements of the *Lead Limit Test*, Appendix IIIB, using 10 μg of lead ion (Pb) in the control.
Mercury Determine as directed under *Mercury Limit Test*, Appendix IIIB, using the following as the *Sample Preparation*: Transfer an accurately weighed amount of the sample, equivalent to 2.0 g of NaOH, into a 50-mL beaker, add 10 mL of water and 2 drops of phenolphthalein TS, and slowly neutralize, with constant stirring, with dilute hydrochloric acid solution (1 in 2). Add 1 mL of dilute sulfuric acid solution (1 in 5) and 1 mL of potassium permanganate solution (1 in 25), cover the beaker with a watch glass, boil for a few s, and cool.

Packaging and Storage Store in tight containers.

Sodium Hypophosphite

$NaH_2PO_2 \cdot H_2O$ Formula wt 105.99

CAS: [7681-53-0]

DESCRIPTION

Sodium Hypophosphite occurs as a white crystalline powder, as white granules, or as colorless, pearly crystalline plates. It is very deliquescent. One mL of water dissolves about 1 g at 25° and about 6 g at 100°. It is slightly soluble in alcohol.

Caution: Care should be observed in mixing Sodium Hypophosphite with nitrates, chlorates, or other oxidizing agents, as an explosion may occur if triturated or heated.

Functional Use in Foods Preservative; antioxidant.

REQUIREMENTS

Identification A 1 in 20 solution gives positive tests for *Sodium* and for *Hypophosphites*, Appendix IIIA.
Assay Not less than 97.0% and not more than 103.0% of $NaH_2PO_2 \cdot H_2O$.
Fluoride Not more than 10 mg/kg.
Heavy Metals (as Pb) Not more than 10 mg/kg.
Insoluble Substances Not more than 0.1%.

TESTS

Assay Dissolve about 100 mg of the sample, accurately weighed, in 20 mL of water, add 40.0 mL of 0.1 N ceric sulfate, mix well, and add 2 mL of silver sulfate solution (5 g of Ag_2SO_4 dissolved in 95 mL of concentrated sulfuric acid). Cover, heat nearly to boiling, and continue heating for 30 min. Cool to room temperature, and titrate with 0.1 N ferrous sulfate to a pale yellow color. Add 2 drops of orthophenanthroline TS, and continue the titration to a salmon-colored endpoint, recording the volume required, in mL, as S. Perform a residual blank titration (see *General Provisions*), and record the volume required as B. Each mL of the volume $B - S$ is equivalent to 2.650 mg of $NaH_2PO_2 \cdot H_2O$.
Fluoride Proceed as directed in the *Fluoride Limit Test*, Appendix IIIB.
Heavy Metals A solution of 2 g in 25 mL of water meets the requirements of the *Heavy Metals Test*, Appendix IIIB, using 20 μg of lead ion (Pb) in the control (*Solution A*).
Insoluble Substances Dissolve 10 g of the sample in 100 mL of hot water, and filter through a tared filtering crucible. Wash the residue with hot water, dry at 105° for 2 h, cool, and weigh.

Packaging and Storage Store in tight containers.

Sodium Lactate Solution

2-Hydroxy-Propanoic Acid, Monosodium Salt

$C_3H_5NaO_3$ Formula wt 112.06

DESCRIPTION

Sodium Lactate Solution is a clear, colorless or practically colorless, slightly viscous liquid, odorless or having a slight, not unpleasant odor. It is miscible with water, and it is normally available in a concentration range of 60% to about 80% of $C_3H_5NaO_3$, by weight.

Functional Use in Foods Emulsifier; flavor enhancer; flavoring agent or adjuvant; humectant; pH control agent.

REQUIREMENTS

Labeling Indicate its content, by weight, of sodium lactate, $C_3H_5NaO_3$.
Identification It gives positive tests for *Sodium* and for *Lactate*, Appendix IIIA.
Assay Not less than 50.0%, by weight, and not less than 98.0% and not more than 102.0%, by weight, of the labeled amount of $C_3H_5NaO_3$.
Chloride Not more than 0.05%.
Citrate, Oxalate, Phosphate, or Tartrate Passes test.
Cyanide Not more than 0.5 mg/kg.
Heavy Metals (as Pb) Not more than 10 mg/kg.
Lead Not more than 5 mg/kg.
Methanol and Methyl Esters Not more than 0.025%.
pH Between 5.0 and 9.0.
Sugars Passes test.
Sulfate Not more than 0.005%.

TESTS

Assay Weigh accurately into a suitable flask a volume of Sodium Lactate Solution, equivalent to about 300 mg of sodium lactate. Add 60 mL of a 1 in 5 mixture of acetic anhydride in glacial acetic acid, mix, and allow to stand for 20 min. Titrate with 0.1 N perchloric acid in glacial acetic acid, determining the endpoint potentiometrically. Perform a blank determination, and make any necessary correction. Each mL of 0.1 N perchloric acid is equivalent to 11.21 mg of $C_3H_5NaO_3$.
Chloride, Appendix IIIB Any turbidity produced by a quantity of the solution containing the equivalent of 40 mg of sodium lactate does not exceed that shown in a control containing 20 μg of chloride ion (Cl).
Citrate, Oxalate, Phosphate, or Tartrate Dilute 5 mL with recently boiled and cooled water to 50 mL. To 4 mL of this solution add 6 N ammonium hydroxide or 3 N hydrochloric acid, if necessary, to bring the pH to between 7.3 and 7.7. Add 1 mL of calcium chloride TS, and heat in a boiling water bath for 5 min: The solution remains clear.
Cyanide (*Caution*: Because of the extremely poisonous nature of potassium cyanide, conduct this test in a fume hood, and exercise great care to prevent skin contact and inhaling particles or vapors of solutions of the material. Under no conditions pipet solutions by mouth.)

p-Phenylenediamine–Pyridine Mixed Reagent Dissolve 200 mg of *p*-phenylenediamine hydrochloride in 100 mL of water, warming to aid dissolution. Cool, allow the solids to settle, and use the supernatant liquid to make the mixed reagent. Dissolve 128 mL of pyridine in 365 mL of water, add 10 mL of hydrochloric acid, and mix. To prepare the mixed reagent, mix 30 mL of the *p*-phenylenediamine solution with all of the pyridine solution and allow to stand for 24 h before using. The mixed reagent is stable for about 3 weeks when stored in an amber bottle.

Sample Solution Transfer an accurately weighed quantity of the Solution, equivalent to 20.0 g of sodium lactate, into a 100-mL volumetric flask, dilute to volume with water, and mix.

Cyanide Standard Solution Dissolve 250 mg of potassium cyanide, accurately weighed, in 10 mL of 0.1 N sodium hydroxide in a 100 mL volumetric flask, dilute to volume with 0.1 N sodium hydroxide, and mix. Transfer a 10-mL aliquot into a 1000-mL volumetric flask, dilute to volume with 0.1 N sodium hydroxide, and mix. Each mL of this solution contains 10 μg of cyanide.

Procedure Pipet a 10-mL aliquot of the *Sample Solution* into a 50-mL beaker. Into a second 50-mL beaker, pipet 0.1 mL of the *Cyanide Standard Solution*, and add 10 mL of water. Place the beakers in an ice bath, and adjust the pH to between 9 and 10 with 20% sodium hydroxide, stirring slowly and adding the reagent slowly to avoid overheating. Allow the solutions to stand for 3 min, and then slowly add 10% phosphoric acid to a pH between 5 and 6, measured with a pH meter.

Transfer the solutions into 100-mL separators containing 25 mL of cold water, and rinse the beakers and pH meter electrodes with a few mL of cold water, collecting the washings in the respective separator. Add 2 mL of bromine TS, stopper, and mix. Add 2 mL of 2% sodium arsenite solution, stopper, and mix. To the clear solutions add 10 mL of *n*-butanol, stopper, and mix. Finally, add 5 mL of *p-Phenylenediamine–Pyridine Mixed Reagent*, mix, and allow to stand for 15 min. Remove and discard the aqueous phases, and filter the alcohol phases into 1-cm cells. The absorbance of the solution from the *Sample Solution*, determined at 480 nm with a suitable spectrophotometer, is no greater than that from the *Cyanide Standard Solution*.

Heavy Metals Dilute a quantity of the Solution, equivalent to 2.0 g of sodium lactate, to 25 mL with water. This solution meets the requirements of the *Heavy Metals Test*, Appendix IIIB, using 20 μg of lead ion (Pb) in the control (*Solution A*).

Lead Dilute a quantity of the Solution, equivalent to 2.0 g of sodium lactate, to 25 mL with water. This solution meets the requirements of the *Lead Limit Test*, Appendix IIIB, using 10 μg of lead ion (Pb) in the control.

Methanol and Methyl Esters

Potassium Permanganate and Phosphoric Acid Solution Dissolve 3 g of potassium permanganate in a mixture of 15 mL of phosphoric acid and 70 mL of water. Dilute with water to 100 mL.

Oxalic Acid and Sulfuric Acid Solution Cautiously add 50 mL of sulfuric acid to 50 mL of water, mix, cool, add 5 g of oxalic acid, and mix to dissolve.

Standard Preparation Prepare a solution containing 10.0 mg of methanol in 100 mL of dilute alcohol (1 in 10).

Test Preparation Place 40.0 g of the Solution in a glass-stoppered, round-bottom flask, add 10 mL of water, and add cautiously 30 mL of 5 N potassium hydroxide. Connect a condenser to the flask, and steam-distill, collecting the distillate in a suitable 100-mL graduated vessel containing 10 mL of alcohol. Continue the distillation until the volume in the receiver reaches approximately 95 mL, and dilute the distillate with water to 100.0 mL.

Procedure Transfer 10.0 mL each of the *Standard Preparation* and the *Test Preparation* to 25-mL volumetric flasks. To each add 5.0 mL of *Potassium Permanganate and Phosphoric Acid Solution*, and mix. After 15 min, to each add 2.0 mL of *Oxalic Acid and Sulfuric Acid Solution*, stir with a glass rod until the solution is colorless, add 5.0 mL of fuchsin–sulfurous acid TS (prepared as directed under *Solutions and Indicators*), dilute with water to volume, and mix. After 2 h, concomitantly determine the absorbances of both solutions in 1-cm cells at the wavelength of maximum absorbance at about 575 nm, with a suitable spectrophotometer, using water as the blank: The absorbance of the solution from the *Test Preparation* is not greater than that from the *Standard Preparation*.

pH Determine the pH of the Solution by the *Potentiometric Method*, Appendix IIB.

Sugars To 10 mL of hot alkaline cupric tartrate TS add 5 drops of Sodium Lactate Solution: No red precipitate is formed.

Sulfate, Appendix IIIB Any turbidity produced by a quantity of the Solution containing the equivalent of 4.0 g of sodium lactate does not exceed that shown in a control containing 200 µg of sulfate ion (SO_4).

Packaging and Storage Store in tight containers.

Sodium Lauryl Sulfate

INS: 487 CAS: [151-21-3]

DESCRIPTION

Sodium Lauryl Sulfate is a mixture of sodium alkylsulfates consisting chiefly of Sodium Lauryl Sulfate [$CH_3(CH_2)_{10}CH_2O-SO_3Na$]. It occurs as small, white or light yellow crystals having a slight, characteristic odor. One g dissolves in 10 mL of water, forming an opalescent solution.

Functional Use in Foods Surface-active agent.

REQUIREMENTS

Identification A 1 in 10 solution gives positive tests for *Sodium*, Appendix IIIA and, after acidification with hydrochloric acid and boiling gently for 20 min, responds to the tests for *Sulfate*, Appendix IIIA.

Assay Not less than 59.0% of total alcohols.
Alkalinity (as NaOH) Passes test (about 0.25%).
Combined Sodium Chloride and Sodium Sulfate Not more than 8.0%.
Heavy Metals (as Pb) Not more than 0.002%.
Lead Not more than 5 mg/kg.
Unsulfated Alcohols Not more than 4.0%.

TESTS

Assay Transfer about 5 g, accurately weighed, to an 800-mL Kjeldahl flask, and add 150 mL of water, 50 mL of hydrochloric acid, and a few boiling chips. Attach a reflux condenser to the flask, heat carefully to avoid excessive frothing, and then boil for about 4 h. Cool the flask, rinse the condenser with ether, collecting the ether in the flask, and transfer the contents to a 500-mL separator, rinsing the flask twice with ether and adding the washings to the separator. Extract the solution with two 75-mL portions of ether, evaporate the combined ether extracts in a tared beaker on a steam bath, dry the residue at 105° for 30 min, cool, and weigh. The residue represents the total alcohols.

Alkalinity Dissolve 1 g in 100 mL of water, add phenol red TS, and titrate with 0.1 N hydrochloric acid. Not more than 0.5 mL is required for neutralization.

Combined Sodium Chloride and Sodium Sulfate

Sodium Chloride Dissolve about 5 g, accurately weighed, in about 50 mL of water. Neutralize the solution with dilute nitric acid (1 in 20), using litmus paper as the indicator, add 2 mL of potassium chromate TS, and titrate with 0.1 N silver nitrate. Each mL of 0.1 N silver nitrate is equivalent to 5.844 mg of sodium chloride.

Sodium Sulfate Transfer about 1 g, accurately weighed, to a 400-mL beaker, add 10 mL of water, heat the mixture, and stir until completely dissolved. To the hot solution add 100 mL of alcohol, cover, and digest at a temperature just below the boiling point for 2 h. Filter while hot through a sintered-glass filter crucible, and wash the precipitate with 100 mL of hot alcohol. Dissolve the precipitate in the crucible by washing with about 150 mL of water, collecting the washings in a beaker. Acidify with 10 mL of hydrochloric acid, heat to boiling, add 25 mL of barium chloride TS, and allow to stand overnight. Collect the precipitate of barium sulfate on a suitable tared, porous-bottom porcelain filter crucible, wash until free from chloride, dry, and ignite to constant weight at 800° ± 25°. The weight of barium sulfate so obtained, multiplied by 0.6086, represents the weight of Na_2SO_4.

Note: Avoid exposing the crucible to sudden temperature changes.

Heavy Metals A solution of 1 g in 25 mL of water meets the requirements of the *Heavy Metals Test*, Appendix IIIB, using 20 µg of lead ion (Pb) in the control (*Solution A*).

Lead A *Sample Solution* prepared as directed for organic compounds meets the requirements of the *Lead Limit Test*, Appendix IIIB, using 5 µg of lead ion (Pb) in the control.

Unsulfated Alcohols Dissolve about 10 g, accurately weighed, in 100 mL of water, and add 100 mL of alcohol. Transfer the solution to a separator, and extract with three 50-mL portions of solvent hexane. If an emulsion forms, sodium chloride may be added to promote separation of the two layers. Wash the combined solvent hexane extracts with three 50-mL portions of water, and dry with anhydrous sodium sulfate. Filter the solvent hexane extract into a tared beaker, evaporate on a steam bath until the odor of solvent hexane no longer is perceptible, dry the residue at 105° for 30 min, cool, and weigh. The residue represents the unsulfated alcohols.

Packaging and Storage Store in well-closed containers.

Sodium Magnesium Aluminosilicate

CAS: [12040-43-6]

DESCRIPTION

A series of hydrated sodium magnesium aluminosilicates having $Na_2O:MgO:Al_2O_3:SiO_2$ molar ratios of approximately 2:1:2:24, respectively. They are synthetic, amorphous, food-grade coprecipitates that are fine, white powders or beads with a specific gravity of about 2. They are odorless, tasteless, and insoluble in water, in alcohol, and in other organic solvents, but are partially soluble in strongly acidic and alkaline solutions.

Functional Use in Foods Anticaking agent.

REQUIREMENTS

Identification

A. Mix 500 mg of the sample with 2.5 g of anhydrous potassium carbonate, and heat the mixture in a platinum or nickel crucible until it melts completely. Cool, add 5 mL of water, and allow to stand for 3 min. Heat the bottom of the crucible gently, detach the melt, and transfer it to a beaker with the aid of about 50 mL of water. Gradually add hydrochloric acid until no effervescence is observed, add 10 mL more of the acid, and evaporate to dryness on a steam bath. Cool, add 20 mL of water, boil, and filter through ash-free paper. An insoluble residue of silica remains.

Note: Retain the filtrate for *Identification Test B*.

Transfer the gelatinous residue to a platinum dish, and cautiously add 5 mL of hydrofluoric acid. The precipitate dissolves. (If it does not dissolve, repeat the treatment with hydrofluoric acid.) Heat, and introduce into the resulting vapors a glass stirring rod with a drop of water on the tip. The drop becomes turbid.

B. Portions of the filtrate obtained in *Identification Test A* give positive tests for *Aluminum*, for *Sodium*, and for *Magnesium*, Appendix IIIA.

Assay

Silicon Dioxide Not less than 65.0% and not more than 75.0% of SiO_2 after drying.

Aluminum Oxide Not less than 9.0% and not more than 13.0% of Al_2O_3 after drying.

Magnesium Oxide Not less than 1.0% and not more than 3.0% of MgO after drying.

Sodium Oxide Not less than 3.0% and not more than 9.0% of Na_2O after drying.

Heavy Metals (as Pb) Not more than 10 mg/kg.
Lead Not more than 5 mg/kg.
Loss on Drying Not more than 8.0%.
Loss on Ignition Between 8.0% and 11.0% after drying.
pH Between 6.5 and 11.0.
Soluble Salt (as Na_2SO_4) Not more than 7.5%.

TESTS

Assay

Silicon Dioxide Transfer about 500 mg, previously dried at 105° for 2 h and accurately weighed, into a 250-mL beaker, wash the sides of the beaker with a few mL of water, and then add 30 mL of sulfuric acid and 15 mL of hydrochloric acid. Heat on a hot plate in a hood until dense, white fumes are evolved, cool, add 15 mL of hydrochloric acid, and heat again to dense, white fumes. Cool, add 70 mL of water, and filter through Whatman No. 40 (or an equivalent) filter paper. Wash the filter paper and precipitate thoroughly with hot water to remove the sulfuric acid residue.

Transfer the filter paper and precipitate into a tared platinum crucible, char, and ignite at 900° to constant weight. Moisten the residue with a few drops of water, add 15 mL of hydrofluoric acid and 8 drops of sulfuric acid, and heat on a hot plate in a hood until white fumes of sulfur trioxide are evolved. Cool; add 5 mL of water, 10 mL of hydrofluoric acid, and 3 drops of sulfuric acid; and evaporate to dryness on the hot plate. Heat cautiously over an open flame until sulfur trioxide fumes have ceased, and ignite at 900° to constant weight. The weight loss after the addition of hydrofluoric acid represents the weight of SiO_2 in the sample taken.

Aluminum Oxide Transfer about 500 mg of the sample, previously dried at 105° for 2 h and accurately weighed, into a tared platinum dish, and moisten with 8 to 10 drops of water. Add 25 mL of 70% perchloric acid and 10 mL of hydrofluoric acid, and heat on a hot plate in a hood until dense, white fumes of perchloric acid appear. Cool, add 10 mL of hydrofluoric acid, and heat again to dense, white fumes. Cool, dissolve the residue in sufficient water, quantitatively transfer with the aid of additional water to a 250-mL volumetric flask, and dilute to volume with water. Retain this solution for analysis under *Magnesium Oxide* and *Sodium Oxide*.

Transfer by pipet a 10.0-mL aliquot of this solution into a 100-mL volumetric flask, dilute to volume with water, and mix.

Set a suitable atomic absorption spectrophotometer to a wavelength of 309.3 nm. Adjust the instrument to zero absorbance against water. Read the absorbance of four standard solutions containing 5, 10, 20, and 50 µg/mL of aluminum, in the form of the chloride, and plot the standard curve as absorbance versus concentration of aluminum.

Aspirate a 1 in 10 diluted sample solution into the spectrophotometer, read the absorbance in the same manner, and by reference to the standard curve, determine the concentration (C) of aluminum, in µg/mL, in the sample solution.

Calculate the quantity, in mg, of Al_2O_3 in the sample taken by the formula

$$250C \times 10 \times 1.8895/1000.$$

Magnesium Oxide Set a suitable atomic absorption spectrophotometer to a wavelength of 285.2 nm. Adjust the instrument to zero absorbance against water. Read the absorbance of four standard solutions containing 5, 10, 25, and 50 µg/mL of magnesium, in the form of the chloride, and plot the standard curve as absorbance versus concentration of magnesium.

Aspirate the sample solution prepared for the aluminum determination into the spectrophotometer, read the absorbance in the same manner, and by reference to the standard curve, determine the concentration (C) of magnesium, in μg/mL, in the sample solution. Calculate the quantity, in mg, of MgO in the sample taken by the formula

$$250C \times 1.6579/1000.$$

Sodium Oxide Set a suitable flame photometer to a wavelength of 589 nm. Adjust the instrument to zero transmittance against water, and then adjust it to 100.0% transmittance with a standard solution containing 200 μg/mL of sodium, in the form of the chloride. Read the percent transmittance of three other standard solutions containing 50, 100, and 150 μg/mL each of sodium, and plot the standard curve as percent transmittance versus concentration of sodium.

Aspirate the sample solution prepared for the aluminum determination into the photometer, read the percent transmittance in the same manner, and by reference to the standard curve, determine the concentration (C) of sodium, in μg/mL, in the sample solution. Calculate the quantity, in mg, of Na_2O in the sample taken by the formula

$$(250C \times 1.348/1000) - F,$$

in which F, as determined below, is the quantity of sodium oxide equivalent to any sodium sulfate present in the sample.

Correction for Sodium Sulfate Content Transfer about 1 g of the sample, previously dried at 105° for 2 h and accurately weighed, into a tared platinum dish, and moisten with 8 to 10 drops of water. Add 25 mL of 70% perchloric acid and 10 mL of hydrofluoric acid, and heat on a hot plate in a hood until dense, white fumes of perchloric acid appear. Add 10 mL of hydrofluoric acid, and heat again to dense, white fumes. Quantitatively transfer the solution to a 400-mL beaker, add 200 mL of water, and heat to boiling. Gradually add, in small portions at a time and while stirring constantly, an excess of hot barium chloride TS (about 10 mL), and heat the mixture on a steam bath for 1 h. Collect the precipitate on a filter, wash until free from chloride, dry, ignite, and weigh. The weight, in mg, of the barium sulfate so obtained, multiplied by 0.6086, indicates its equivalent of Na_2SO_4 (C'). Calculate the correction factor (F) by the formula

$$0.437(C' \times w/W),$$

in which w is the weight, in mg, of the sample taken for the sodium oxide determination, and W is the weight, in mg, of the sample taken for the sodium sulfate determination.

Sample Solution for the Determination of Lead and Heavy Metals Transfer 10.0 g of the sample into a 250-mL beaker, add 50 mL of 0.5 N hydrochloric acid, cover with a watch glass, and heat slowly to boiling. Boil gently for 15 min, cool, and let the undissolved material settle. Decant the supernatant liquid through Whatman No. 4 (or an equivalent) filter paper into a 100-mL volumetric flask, retaining as much as possible of the insoluble material in the beaker. Wash the slurry and beaker with three 10-mL portions of hot water, decanting each washing through the filter into the flask. Finally, wash the filter paper with 15 mL of hot water, cool the filtrate to room temperature, dilute to volume with water, and mix.

Heavy Metals A 20-mL portion of the *Sample Solution* meets the requirements of the *Heavy Metals Test*, Appendix IIIB, using 20 μg of lead ion (Pb) in the control (*Solution A*).

Lead Determine as directed under *Lead* in the monograph for *Calcium Silicate*, except use a *Standard Lead Solution* containing 0.50 μg/mL.

Loss on Drying, Appendix IIC Dry at 105° for 2 h.

Loss on Ignition Transfer about 5 g, previously dried at 105° for 2 h and accurately weighed, into a suitable tared crucible, and ignite at 900° to constant weight.

pH Determine using a 1 in 5 slurry by the *Potentiometric Method*, Appendix IIB.

Soluble Salt Calculate the percent sodium sulfate from the weight of barium sulfate obtained in the *Correction for Sodium Sulfate Content* in the *Assay*, by the formula

$$N \times 60.86/W,$$

in which N is the weight, in mg, of barium sulfate, and W is the weight, in mg, of the sample taken for sodium sulfate determination.

Packaging and Storage Store in well-closed containers.

Sodium Metabisulfite

Sodium Pyrosulfite

$Na_2S_2O_5$ Formula wt 190.11

INS: 223   CAS: [7681-57-4]

DESCRIPTION

Colorless crystals or a white to yellowish crystalline powder having an odor of sulfur dioxide. It is freely soluble in water and slightly soluble in alcohol. Its solutions are acid to litmus.

Functional Use in Foods Preservative; antioxidant.

REQUIREMENTS

Identification A 1 in 10 solution gives positive tests for *Sodium* and for *Sulfite*, Appendix IIIA.

Assay Not less than 90.0% and not more than 100.5% of $Na_2S_2O_5$.

Heavy Metals (as Pb) Not more than 10 mg/kg.

Iron Not more than 0.002%.

Selenium Not more than 0.003%.

TESTS

Assay Weigh accurately about 200 mg, add it to exactly 50 mL of 0.1 N iodine contained in a glass-stoppered flask, and stopper the flask. Allow to stand for 5 min, add 1 mL of hydrochloric acid, and titrate the excess iodine with 0.1 N sodium thiosulfate, adding starch TS as the indicator. Each mL of 0.1 N iodine is equivalent to 4.752 mg of $Na_2S_2O_5$.

Heavy Metals Dissolve 2 g in 10 mL of water, add 5 mL of hydrochloric acid, evaporate to dryness on a steam bath, and dissolve the residue in 25 mL of water. This solution meets the requirements of the *Heavy Metals Test*, Appendix IIIB, using 20 μg of lead ion (Pb) in the control (*Solution A*).

Iron To 500 mg of the sample add 2 mL of hydrochloric acid, and evaporate to dryness on a steam bath. Dissolve the residue in 2 mL of hydrochloric acid and 20 mL of water, add a few drops of bromine TS, and boil the solution to remove the bromine. Cool, dilute with water to 25 mL, then add 50 mg of ammonium persulfate and 5 mL of ammonium thiocyanate TS. Any red or pink color does not exceed that produced in a control containing 1.0 mL of *Iron Standard Solution* (10 μg Fe).

Selenium Determine as directed in *Method I* under the *Selenium Limit Test*, Appendix IIIB, using 200 mg of sample and 100 mg of magnesium oxide.

Packaging and Storage Store in well-filled, tight containers, and avoid exposure to excessive heat.

Sodium Metaphosphate, Insoluble

Insoluble Sodium Polyphosphate; IMP; Maddrell's Salt

CAS: [50813-16-6]

DESCRIPTION

A high-molecular-weight sodium polyphosphate composed of two long metaphosphate chains ($NaPO_3$) that spiral in opposite directions about a common axis. The Na_2O/P_2O_5 ratio is about 1.0. It occurs as a white, crystalline powder. It is practically insoluble in water but dissolves in mineral acids and in solutions of potassium and ammonium (but not sodium) chlorides. The pH of a 1 in 3 slurry in water is about 6.5.

Functional Use in Foods Emulsifier; sequestrant; texturizer.

REQUIREMENTS

Identification

A. Finely powder about 1 g of the sample, and add it slowly to 100 mL of a 1 in 20 solution of potassium chloride while stirring vigorously. A gelatinous mass is formed.

B. Mix 500 mg of the sample with 10 mL of nitric acid and 50 mL of water, boil for about 30 min, and cool. The resulting solution gives positive tests for *Sodium* and for *Phosphates*, Appendix IIIA.

Assay Not less than 68.7% and not more than 70.0% of P_2O_5.
Arsenic (as As) Not more than 3 mg/kg.
Fluoride Not more than 0.005%.
Heavy Metals (as Pb) Not more than 10 mg/kg.

TESTS

Assay Transfer about 800 mg of the sample, accurately weighed, into a 400-mL beaker, add 100 mL of water and 25 mL of nitric acid, cover with a watch glass, and boil for 10 min on a hot plate. Rinse any condensate from the watch glass into the beaker, cool the solution to room temperature, transfer it quantitatively to a 500-mL volumetric flask, dilute to volume with water, and mix thoroughly. Pipet 20.0 mL of this solution into a 500-mL Erlenmeyer flask, add 100 mL of water, and heat just to boiling. Add with stirring 50 mL of quimociac TS, then cover with a watch glass, and boil for 1 min in a well-ventilated hood. Cool to room temperature, swirling occasionally while cooling, then filter through a tared, sintered-glass filter crucible of medium porosity, and wash with five 25-mL portions of water. Dry at about 225° for 30 min, cool, and weigh. Each mg of precipitate thus obtained is equivalent to 32.074 μg of P_2O_5.

Arsenic A solution of 1 g in 15 mL of 2.7 N hydrochloric acid meets the requirements of the *Arsenic Test*, Appendix IIIB.

Fluoride Determine on a 1-g sample, dissolved in 5 mL of a 1 to 1 hydrochloric acid solution, as directed in *Method IV* under the *Fluoride Limit Test*, Appendix IIIB, using *Buffer Solution B* and 0.1 mL of *Fluoride Standard Solution*.

Heavy Metals Warm 2 g with 25 mL of 2.7 N hydrochloric acid until no more dissolves, dilute with water to 25 mL, and filter if necessary. This solution meets the requirements of the *Heavy Metals Test*, Appendix IIIB, using 20 μg of lead ion (Pb) in the control (*Solution A*).

Packaging and Storage Store in tight containers.

Sodium Metasilicate

$Na_2O \cdot SiO_2 \cdot xH_2O$ Formula wt, anhydrous 122.06

INS: 550   CAS: [6834-92-0]

DESCRIPTION

A hydrous (pentahydrate) or anhydrous silicate having a 1:1 molar ratio of SiO_2 to Na_2O. It occurs as a white, free-flowing, granular material. At 30°, the anhydrous Sodium Metasilicate is easily soluble in water (270 g/L) as is its pentahydrate (610 g/L). The pH values of 1% solutions of anhydrous sodium metasilicate and of its pentahydrate are about 12.6 and 12.4, respectively.

Functional Use in Foods Saponifying agent; boiler water additive.

REQUIREMENTS

Caution: Sodium Metasilicate and its solutions are caustic materials. Use proper protective equipment and avoid contact with the eyes, skin, and clothing. Sodium Metasilicate causes eye and skin burns. Do not inhale vapors from Sodium Metasilicate solutions.

Labeling   Indicate the percent, each, of SiO_2 and Na_2O, and whether it is anhydrous or the pentahydrate.

Identification

A. Dissolve about 200 mg in 10 mL of water. Place a drop of this solution on a spot plate. Add to this 1 drop of 4 M sodium hydroxide and 1 drop of a solution prepared by dissolving 0.5 g of ammonium molybdate in 10 mL of water, followed by the addition of 3 mL of sulfuric acid. A deep-yellow color indicates the presence of silicate.

B. Dip a clean nichrome wire into the Sodium Metasilicate solution prepared in *Identification Test A*, and place the wire in the flame of a Bunsen burner. A bright-yellow color indicates the presence of sodium.

Assay for Silicon Dioxide and Sodium Oxide   Not less than 90.0% and not more than 110.0% of the percents claimed on the label.

Heavy Metals (as Pb)   Not more than 10 mg/kg.

Loss on Drying   Not more than 2.0% for the anhydrous and more than 42.0% for the pentahydrate.

Loss on Ignition   Not more than 0.5% for the anhydrous and between 40.5% and 42.5% for the pentahydrate.

TESTS

Assay for Silicon Dioxide and Sodium Oxide

Silicon Dioxide   In a beaker, acidify 1 g of the sample, accurately weighed, with 5 mL of hydrochloric acid, and evaporate to dryness on a steam bath. Repeat the treatment with an additional 5 mL of hydrochloric acid, and mix the residue with 1 mL of the acid and 20 mL of water. Digest the residue on the steam bath to dissolve the soluble salts, filter the contents of the beaker through an ashless filter paper, and quantitatively transfer the residue to the paper. Wash the paper and residue thoroughly with hot water, transfer the paper to a platinum crucible, dry for 1 h at 105°, and carefully char it at low heat. Gradually increase the heat to burn away the paper, and finally ignite the crucible and its contents to constant weight at 1000°, cool in a desiccator, and weigh. Moisten the ignited residue with few drops of water, add 15 mL of hydrofluoric acid and 5 drops of sulfuric acid (1:3). Heat the crucible with caution in a fume hood on a hot plate until all of the acid is driven off, and then ignite the residue to constant weight at a temperature of 1000°. Cool the crucible in a desiccator, and weigh. The loss in weight is equivalent to the weight of SiO_2 in the sample taken.

Sodium Oxide   Disperse 500 mg of the sample, accurately weighed, in 150 mL of water, and heat to ensure its dissolution. Add 2 to 3 drops of phenolphthalein TS and 100.0 mL of 0.1 N sulfuric acid. Titrate with 0.1 N sodium hydroxide until a permanent pink color first appears. Subtract the volume of 0.1 N sodium hydroxide from the volume of 0.1 N sulfuric acid: Each mL of 0.1 N sulfuric acid is equivalent to 3.099 mg of sodium oxide.

Sample Solution for the Determination of Heavy Metals
Transfer 10.0 g of the sample to a 250-mL beaker, add 50 mL of 0.5 N hydrochloric acid, cover with a watch glass, and heat slowly to boiling. Boil gently for 15 min, cool, and let the undissolved material settle. Decant the supernatant liquid through Whatman No. 4 (or an equivalent) filter paper into a 100-mL volumetric flask, retaining as much as possible of the insoluble material in the beaker. Wash the slurry and beaker with three 10-mL portions of hot water, decanting each washing through the filter into the flask. Finally, wash the filter paper with 15 mL of hot water, cool the filtrate to room temperature, dilute to volume with water, and mix to obtain the *Sample Solution*.

Heavy Metals   A 20-mL portion of the *Sample Solution* meets the requirements of the *Heavy Metals Test*, Appendix IIIB, using 20 µg of lead ion (Pb) in the control (*Solution A*).

Loss on Drying   Appendix IIC Dry at 105° for 2 h. Retain the sample for the *Loss on Ignition Test*.

Loss on Ignition   Ignite the sample retained from *Loss on Drying*, accurately weighed, in a suitable tared crucible at 1000° for 20 min.

Packaging and Storage   Store in tight containers.

Sodium Methylate

Sodium Methoxide

CH_3ONa 　　　　　　　　　　　　Formula wt 54.02

　　　　　　　　　　　　　　　　　CAS: [124-41-4]

DESCRIPTION

A white, amorphous, hygroscopic, free-flowing powder. It is soluble in fats, in esters, and in alcohols. It decomposes without melting above 127°.

Caution: Sodium Methylate and its solutions are caustic and flammable. Avoid contact with the eyes, skin, and clothing, and do not inhale vapors from Sodium Methylate solutions.

Functional Use in Foods   Catalyst for the transesterification of fats.

REQUIREMENTS

Identification   It reacts with oxygen and carbon dioxide and is decomposed by water. The resulting solution gives positive tests for *Sodium*, Appendix IIIA.

Assay   Not less than 97.0% of CH_3NaO.

Arsenic (as As)   Not more than 3 mg/kg.

Heavy Metals (as Pb)   Not more than 0.0025%.

Lead   Not more than 10 mg/kg.

Mercury   Not more than 1 mg/kg.

Sodium Carbonate   Not more than 0.4%.

Sodium Hydroxide   Not more than 1.7%.

TESTS

Note: The tests in the following section must be conducted with a minimum exposure of the sample to air. Preferably the tests should be conducted in a nitrogen hood.

Assay, Sodium Carbonate, and Sodium Hydroxide

Sample Preparation Select two tared weighing bottles, each approximately 30 mm in diameter and 80 mm high, nearly fill each with the sample, which should weigh between 12 and 15 g, securely fit the covers, and weigh.

Determination of Alkalinity as CH_3ONa Remove the top from one of the sample bottles, and quickly drop the bottle into a 500-mL Erlenmeyer flask containing 200 mL of ice-cold, carbon dioxide-free water, sliding the sample bottle gently down the side of the flask to prevent splashing. Immediately stopper the flask with a rubber stopper, and swirl until the sample dissolves. Wash the sample solution into a 250-mL volumetric flask, and nearly dilute to volume with carbon dioxide-free water. Allow the solution to reach room temperature, then dilute to volume with water, and mix. Transfer 50.0 mL of this solution into a 500-mL glass-stoppered Erlenmeyer flask, add 150 mL of carbon dioxide-free water and 5 mL of barium chloride TS, stopper the flask, mix, and allow to stand for 5 min. Add 3 drops of phenolphthalein TS, and titrate with 1 N hydrochloric acid to the disappearance of the pink color. Retain the titrated solution for the *Determination of Sodium Carbonate*. Calculate the percentage of alkalinity as CH_3ONa (% A) by the formula

$$(V_1 \times N \times 5.403)/(W \times 0.2),$$

in which V_1 is the volume, in mL, and N the exact normality of the hydrochloric acid used, and W is the weight, in g, of the sample.

Determination of Sodium Carbonate Add 2 drops of methyl orange TS to the solution retained above, and continue the titration with 1 N hydrochloric acid to a permanent pink color. Calculate the percentage of Na_2CO_3 by the formula

$$(V_2 \times N \times 5.30)/(W \times 0.2),$$

in which V_2 is the volume, in mL, of 1 N hydrochloric acid consumed in the second titration, and N and W are as defined above.

Determination of Sodium Hydroxide The *Karl Fischer Titrimetric Method*, Appendix IIB, may be adapted for this determination at the discretion of the analyst, or the following procedure may be used:

Solution A Add 400 mL of colorless pyridine, containing no more than 0.05% of water, to a 500-mL Florence flask filled with a two-hole rubber stopper and a 7-mm glass tube extending nearly to the bottom of the flask. Place the flask in a cooling bath of running water, and pass dry sulfur dioxide from an upright cylinder until 80 ± 0.5 g has been added. Disconnect the hose from the delivery tube before closing the gas valve. Transfer the solution into a dry, glass-stoppered bottle, add 400 mL of absolute methanol, and mix. Store in a dark place.

Solution B Add 75 g of iodine to 900 mL of absolute methanol contained in a dry, glass-stoppered bottle, and shake until the iodine dissolves. Transfer to a dry automatic buret protected by drying tubes.

Standardization of Solution B Add 15 mL of *Solution A*, measured with a dry graduate, to a dry 125-mL iodine flask, and titrate with *Solution B* to a brownish yellow color. Stopper the flask immediately to prevent moisture contamination and endpoint fading. Disregard the volume of *Solution B* added. To the flask add 50.0 mL of methanol–water standard solution, containing 1.0 mg of H_2O per mL, and immediately titrate with *Solution B* to the same brownish yellow endpoint. Calculate the equivalence factor, F, in mg of water per mL of *Solution B*. Restandardize on each day of use.

Procedure Select two 120-mL wide-mouth glass jars with plastic screw caps, wash with hydrochloric acid, rinse with water followed by isopropanol, and dry with a current of air. Bore a hole through an extra screw cap to accommodate the tip from the automatic buret. Place a magnetic stirring bar in each jar, and flush the jars with dry nitrogen to remove carbon dioxide. To each jar add 30 mL of *Solution A* and 15 mL of absolute methanol, and screw on the caps. Replace the screw cap of one of the jars with the extra cap, insert the buret tip, and begin stirring. Titrate with *Solution B* to a distinct brownish yellow color that persists for at least 5 min. Replace the original cap, and titrate the other solution in the same manner. Remove the caps from both jars, and to one of the jars add about 2 g of the sample, accurately weighed, from the remaining weighing bottle prepared as directed under *Sample Preparation*. Replace the cap on the second jar, and titrate the sample solution with *Solution B* to the brownish yellow color. Titrate the blank to the same color. Calculate the percentage of H_2O in the sample by the formula

$$(F \times V \times 100)/(W \times 1000),$$

in which F is the equivalence factor, in mg/mL, of *Solution B*; V is the net volume, in mL, of *Solution B* required in the titration of the sample; and W is the weight, in g, of the sample taken.

Calculations Calculate the percentage of NaOH by the formula

$$2.222 \times [\% \ H_2O - (\% \ Na_2CO_3 \times 0.170)].$$

Finally, calculate the percentage of CH_3ONa by the formula

$$\% \ A - (\% \ NaOH \times 1.350).$$

Arsenic Cautiously dissolve 1 g of the sample in 10 mL of water, neutralize to litmus paper with 2 N sulfuric acid, and dilute to 35 mL with water. This solution meets the requirements of the *Arsenic Test*, Appendix IIIB.

Heavy Metals Cautiously dissolve 800 mg of the sample in 10 mL of water, add 10 mL of 2.7 N hydrochloric acid, and heat to boiling. Cool, and dilute to 25 mL with water. This solution meets the requirements of the *Heavy Metals Test*, Appendix IIIB, using 20 µg of lead ion (Pb) in the control (*Solution A*).

Lead Cautiously dissolve 1 g of the sample in 10 mL of water, add 10 mL of diluted hydrochloric acid, and heat to boiling. Cool, and dilute to 25 mL with water. This solution meets the requirements of the *Lead Limit Test*, Appendix IIIB, using 10 µg of lead ion (Pb) in the control.

Mercury Determine as directed under *Mercury Limit Test*, Appendix IIIB, using the following as the *Sample Preparation*: Cautiously dissolve 2 g of the sample in 10 mL of water in a small beaker, add 2 drops of phenolphthalein TS, and slowly neutralize, with constant stirring, with dilute sulfuric acid solution (1 in 5). Add 1 mL of dilute sulfuric acid solution (1 in 5) and 1 mL of potassium permanganate solution (1 in 25), and mix.

Packaging and Storage Store in air-tight containers, and take all necessary precautions to prevent combustion during handling.

Sodium Nitrate

$NaNO_3$ Formula wt 84.99

INS: 251 CAS: [7631-99-4]

DESCRIPTION

Colorless, odorless crystals or crystalline granules. It is moderately deliquescent in moist air. It is freely soluble in water, and is sparingly soluble in alcohol.

Functional Use in Foods Antimicrobial agent; preservative.

REQUIREMENTS

Identification A 1 in 5 solution is neutral to litmus and gives positive tests for *Sodium* and for *Nitrate*, Appendix IIIA.
Assay Not less than 99.0% and not more than 100.5% of $NaNO_3$ after drying.
Heavy Metals (as Pb) Not more than 10 mg/kg.
Total Chlorine Passes test (approximately 0.2%).

TESTS

Assay Weigh accurately about 350 mg, previously dried at 105° for 4 h, dissolve in 10 mL of hydrochloric acid in a small beaker or porcelain dish, and evaporate to dryness on a steam bath. Dissolve the residue in 10 mL of hydrochloric acid, and again evaporate to dryness, continuing the heating until the residue, when dissolved in water, is neutral to litmus. Transfer the residue with the aid of 25 mL of water to a glass-stoppered flask, add exactly 50 mL of 0.1 N silver nitrate, then add 3 mL of nitric acid and 3 mL of nitrobenzene, and shake vigorously. Add ferric ammonium sulfate TS, and titrate the excess silver nitrate with 0.1 N ammonium thiocyanate. Each mL of 0.1 N silver nitrate is equivalent to 8.50 mg of $NaNO_3$.
Heavy Metals A solution of 2 g in 25 mL of water meets the requirements of the *Heavy Metals Test*, Appendix IIIB, using 20 µg of lead ion (Pb) and 1 g of the sample in the control (*Solution A*).
Total Chlorine Dissolve 1 g in 100 mL of water, add enough 6% sulfurous acid to give the solution a distinct odor of sulfur dioxide, boil gently until the odor of the sulfur dioxide is no longer apparent, and adjust the volume to 100 mL by the addition of water. Add 1.0 mL of 0.1 N silver nitrate followed by 3 mL of nitric acid and 3 mL of nitrobenzene, and shake vigorously. Add ferric ammonium sulfate TS, and titrate the excess silver nitrate with 0.1 N ammonium thiocyanate. No more than 0.6 mL of the 0.1 N silver nitrate is consumed.

Packaging and Storage Store in tight containers.

Sodium Nitrite

$NaNO_2$ Formula wt 69.00

INS: 250 CAS: [7632-00-0]

DESCRIPTION

It occurs as a white to slightly yellow, granular powder, or as white or nearly white, opaque, fused masses or sticks. It has a mild, saline taste, and is deliquescent in air. Its solutions are alkaline to litmus. One g of Sodium Nitrite dissolves in 1.5 mL of water, but it is sparingly soluble in alcohol.

Functional Use in Foods Color fixative in meat and meat products; antimicrobial agent; preservative.

REQUIREMENTS

Identification Its solutions give positive tests for *Sodium* and for *Nitrite*, Appendix IIIA.
Assay Not less than 97.0% and not more than 100.5% of $NaNO_2$ after drying.
Heavy Metals (as Pb) Not more than 0.002%.
Lead Not more than 10 mg/kg.
Loss on Drying Not more than 0.25%.

TESTS

Assay Dissolve about 3 g, previously dried over silica gel for 4 h and accurately weighed, in water to make 100 mL. Pipet 10 mL of this solution into a mixture of 100.0 mL of 0.1 N potassium permanganate, 50 mL of water, and 5 mL of sulfuric acid, keeping the tip of the pipet well below the surface of the liquid. Warm the solution to 40°, allow it to stand for 5 min, and add 25.0 mL of 0.1 N oxalic acid. Heat the mixture to about 80°, and titrate with 0.1 N potassium permanganate. Each mL of 0.1 N potassium permanganate is equivalent to 3.450 mg of $NaNO_2$.
Heavy Metals Dissolve 1 g in a mixture of 10 mL of water and 2 mL of hydrochloric acid, and evaporate to dryness on a steam bath. Add another 2-mL portion of hydrochloric acid, again evaporate to dryness, and dissolve the residue in 25 mL of water. This solution meets the requirements of the *Heavy Metals Test*, Appendix IIIB, using 20 µg of lead ion (Pb) in the control (*Solution A*).
Lead A solution of 1 g in 10 mL of water meets the requirements of the *Lead Limit Test*, Appendix IIIB, using 10 µg of lead ion (Pb) in the control.

Sodium Phosphate, Dibasic

Disodium Monohydrogen Phosphate; Disodium Phosphate

Na_2HPO_4 Formula wt 141.96

INS: 339(ii)   CAS: [7558-79-4]

DESCRIPTION

Dibasic Sodium Phosphate is anhydrous or contains two molecules of water of hydration. It occurs as a white, crystalline powder or as granules. The anhydrous form is hygroscopic. Both forms are freely soluble in water and insoluble in alcohol.

Functional Use in Foods   Emulsifier; texturizer; buffer; nutrient; dietary supplement.

REQUIREMENTS

Identification   A 1 in 20 solution gives positive tests for *Phosphate* and for *Sodium*, Appendix IIIA.
Assay   Not less than 98.0% of Na_2HPO_4 after drying.
Arsenic (as As)   Not more than 3 mg/kg.
Fluoride   Not more than 0.005%.
Heavy Metals (as Pb)   Not more than 10 mg/kg.
Insoluble Substances   Not more than 0.2%.
Loss on Drying Na_2HPO_4 (anhydrous): not more than 5.0%; $Na_2HPO_4 \cdot 2H_2O$ (dihydrate): between 18.0% and 22.0%.

TESTS

Assay   Transfer about 6.5 g of the sample, previously dried at 105° for 4 h and accurately weighed, into a 250-mL beaker, add 50.0 mL of 1 N hydrochloric acid and 50 mL of water, and stir until the sample is completely dissolved. Place the electrodes of a suitable pH meter in the solution, and titrate the excess acid with 1 N sodium hydroxide to the inflection point occurring at about pH 4. Record the buret reading, and calculate the volume (*A*) of 1 N hydrochloric acid consumed by the sample. Continue the titration with 1 N sodium hydroxide until the inflection point occurring at about pH 8.8 is reached, record the buret reading, and calculate the volume (*B*) of 1 N sodium hydroxide required in the titration between the two inflection points (pH 4 to pH 8.8). When *A* is equal to or less than *B*, each mL of the volume *A* of 1 N hydrochloric acid is equivalent to 142.0 mg of Na_2HPO_4. When *A* is greater than *B*, each mL of the volume 2*B* − *A* of 1 N sodium hydroxide is equivalent to 142.0 mg of Na_2HPO_4.
Arsenic   A solution of 1 g in 35 mL of water meets the requirements of the *Arsenic Test*, Appendix IIIB.
Fluoride   Determine on a 2-g sample as directed in *Method IV* under the *Fluoride Limit Test*, Appendix IIIB, using *Buffer Solution A* and 0.1 mL of *Fluoride Standard Solution*.
Heavy Metals   A solution of 2 g in 25 mL of water meets the requirements of the *Heavy Metals Test*, Appendix IIIB, using 20 μg of lead ion (Pb) in the control (*Solution A*).
Insoluble Substances   Dissolve 10 g in 100 mL of hot water, and filter through a tared filtering crucible (not glass). Wash the insoluble residue with hot water, dry at 105° for 2 h, cool, and weigh.
Loss on Drying, Appendix IIC   Dry at 120° for 4 h.

Packaging and Storage   Store in tight containers.

Sodium Phosphate, Monobasic

Monosodium Phosphate; Sodium Biphosphate; Monosodium Dihydrogen Phosphate

NaH_2PO_4 Formula wt 119.98

INS: 339(i)   CAS: [7558-80-7]

DESCRIPTION

Monobasic Sodium Phosphate is anhydrous or contains one or two molecules of water of hydration. It is odorless and is slightly hygroscopic. The anhydrous form is a white, crystalline powder or granules. The hydrated forms occur as white or transparent crystals or granules. All forms are freely soluble in water, but are insoluble in alcohol. The pH of a 1 in 100 solution is between 4.1 and 4.7.

Functional Use in Foods   Buffer; emulsifier; nutrient; dietary supplement.

REQUIREMENTS

Identification   A 1 in 20 solution gives positive tests for *Phosphate* and for *Sodium*, Appendix IIIA.
Assay   Not less than 98.0% and not more than 103.0% of NaH_2PO_4 after drying.
Arsenic (as As)   Not more than 3 mg/kg.
Fluoride   Not more than 0.005%.
Heavy Metals (as Pb)   Not more than 10 mg/kg.
Insoluble Substances   Not more than 0.2%.
Loss on Drying NaH_2PO_4 (anhydrous): not more than 2.0%; $NaH_2PO_4 \cdot H_2O$ (monohydrate): between 10.0% and 15.0%; $NaH_2PO_4 \cdot 2H_2O$ (dihydrate): between 20.0% and 25.0%.

TESTS

Assay   Transfer about 5 g of the sample, previously dried at 105° for 4 h and accurately weighed, into a 250-mL beaker, add 100 mL of water and 50.0 mL of 1 N hydrochloric acid, and stir until the sample is completely dissolved. Place the electrodes of a suitable pH meter in the solution, and slowly titrate the excess acid, stirring constantly, with 1 N sodium hydroxide to the inflection point occurring at about pH 4. Record the buret reading, and calculate the volume (A), if any, of 1 N hydrochloric acid consumed by the sample. Continue the titration with 1 N sodium hydroxide until the inflection point occurring at about pH 8.8 is reached, record the buret reading, and calculate the volume (B) of 1 N sodium hydroxide required in the titration between the two inflection points (pH 4 and pH 8.8). Each mL of the volume $B - A$ of 1 N sodium hydroxide is equivalent to 120.0 mg of NaH_2PO_4.

Arsenic   A solution of 1 g in 35 mL of water meets the requirements of the *Arsenic Test*, Appendix IIIB.

Fluoride   Determine on a 2-g sample as directed in *Method IV* under the *Fluoride Limit Test*, Appendix IIIB, using *Buffer Solution B* and 0.2 mL of *Fluoride Standard Solution*.

Heavy Metals   A solution of 2 g in 25 mL of water meets the requirements of the *Heavy Metals Test*, Appendix IIIB, using 20 µg of lead ion (Pb) in the control (*Solution A*).

Insoluble Substances   Dissolve 10 g in 100 mL of hot water, and filter through a tared filtering crucible (not glass). Wash the insoluble residue with hot water, dry at 105° for 2 h, cool, and weigh.

Loss on Drying, Appendix IIC   Dry first at 60° for 1 h, then at 105° for 4 h.

Packaging and Storage   Store in tight containers.

Sodium Phosphate, Tribasic

Trisodium Phosphate

Na_3PO_4 Formula wt, anhydrous 163.94

INS: 339(iii)   CAS: [7601-54-9]

DESCRIPTION

Tribasic Sodium Phosphate is anhydrous or contains 1 to 12 molecules of water of hydration. The formula for a crystalline material is approximately $4(Na_3PO_4 \cdot 12H_2O)NaOH$. It occurs as white, odorless crystals or granules or as a crystalline powder. It is freely soluble in water, but is insoluble in alcohol. The pH of a 1 in 100 solution is between 11.5 and 12.0.

Functional Use in Foods   Buffer; emulsifier; nutrient; dietary supplement.

REQUIREMENTS

Identification   A 1 in 20 solution gives positive tests for *Sodium* and for *Phosphate*, Appendix IIIA.

Assay Na_3PO_4 (anhydrous) and $Na_3PO_4 \cdot H_2O$ (monohydrate): not less than 97.0% of Na_3PO_4, calculated on the ignited basis; $4(Na_3PO_4 \cdot 12H_2O)NaOH$ (dodecahydrate): not less than 92.0% of Na_3PO_4, calculated on the ignited basis.

Arsenic (as As)   Not more than 3 mg/kg.

Fluoride   Not more than 0.005%.

Heavy Metals (as Pb)   Not more than 10 mg/kg.

Insoluble Substances   Not more than 0.2%.

Loss on Ignition Na_3PO_4 (anhydrous): not more than 2.0%; $Na_3PO_4 \cdot H_2O$ (monohydrate): between 8.0% and 11.0%; $4(Na_3PO_4 \cdot 12H_2O)NaOH$ (dodecahydrate): between 45.0% and 57.0%.

TESTS

Assay   Dissolve an accurately weighed quantity of the sample, equivalent to between 5.5 and 6 g of anhydrous Na_3PO_4, in 40 mL of water in a 400-mL beaker, and add 100.0 mL of 1 N hydrochloric acid. Pass a stream of carbon dioxide-free air, in fine bubbles, through the solution for 30 min to expel carbon dioxide, covering the beaker loosely to prevent any loss by spraying. Wash the cover and sides of the beaker with a few mL of water, and place the electrodes of a standard pH meter in the solution. Titrate the solution with 1 N sodium hydroxide to the inflection point occurring at about pH 4, then calculate the volume (A) of 1 N hydrochloric acid consumed. Protect the solution from absorbing carbon dioxide from the air, and continue the titration with 1 N sodium hydroxide until the inflection point occurring at about pH 8.8 is reached. Calculate the volume (B) of 1 N sodium hydroxide consumed in the titration. When A is equal to or greater than $2B$, each mL of the volume B of 1 N sodium hydroxide is equivalent to 163.9 mg of Na_3PO_4. When A is less than $2B$, each mL of the volume $A - B$ of 1 N sodium hydroxide is equivalent to 163.9 mg of Na_3PO_4.

Arsenic   A solution of 1 g in 35 mL of water meets the requirements of the *Arsenic Test*, Appendix IIIB.

Fluoride   Determine on a 2-g sample as directed in *Method IV* under the *Fluoride Limit Test*, Appendix IIIB, using *Buffer Solution A* and 0.2 mL of *Fluoride Standard Solution*.

Heavy Metals   A solution of 2 g in 25 mL of water meets the requirements of the *Heavy Metals Test*, Appendix IIIB, using 20 µg of lead ion (Pb) in the control (*Solution A*).

Insoluble Substances   Dissolve 10 g in 100 mL of hot water, and filter through a tared filtering crucible (not glass). Wash the insoluble residue with hot water, dry at 105° for 2 h, cool, and weigh.

Loss on Ignition   Ignite at about 800° for 30 min after drying at 110° for 5 h.

Packaging and Storage   Store in well-closed containers.

Sodium Polyphosphates, Glassy

Sodium Hexametaphosphate; Sodium Tetrapolyphosphate; Graham's Salt

INS: 452   CAS: [68915-31-1]

DESCRIPTION

A class consisting of several amorphous, water-soluble polyphosphates composed of linear chains of metaphosphate units,

$(NaPO_3)_x$ where $x \geq 2$, terminated by Na_2PO_4— groups. These substances occur as colorless or white, transparent platelets, granules, or powders. They are usually identified by their Na_2O/P_2O_5 ratio or their P_2O_5 content. The Na_2O/P_2O_5 ratios vary from about 1.3 for sodium tetrapolyphosphate, where x = approximately 4; through about 1.1 for Graham's salt, commonly called sodium hexametaphosphate, where x = 13 to 18; to about 1.0 for the higher molecular weight sodium polyphosphates, where x = 20 to 100 or more. The pH of their solutions varies from about 3.0 to 9.0. The Glassy Sodium Polyphosphates are very soluble in water.

Functional Use in Foods Emulsifier; sequestrant; texturizer.

REQUIREMENTS

Identification

A. A 1 in 20 solution gives positive tests for *Sodium*, Appendix IIIA.

B. Dissolve about 100 mg in 5 mL of hot 1.7 N nitric acid, warm on a steam bath for 10 min, and cool. Neutralize to litmus paper with 1 N sodium hydroxide, and add silver nitrate TS. A yellow precipitate is formed that is soluble in 1.7 N nitric acid.

Assay Between 60.0% and 71.0% of P_2O_5.
Arsenic (as As) Not more than 3 mg/kg.
Fluoride Not more than 0.005%.
Heavy Metals (as Pb) Not more than 10 mg/kg.
Insoluble Substances Not more than 0.1%.

TESTS

Assay Transfer about 800 mg of the sample, accurately weighed, into a 400-mL beaker, add 100 mL of water and 25 mL of nitric acid, cover with a watch glass, and boil for 10 min on a hot plate. Rinse any condensate from the watch glass into the beaker, cool the solution to room temperature, transfer it quantitatively to a 500-mL volumetric flask, dilute to volume with water, and mix thoroughly. Pipet 20.0 mL of this solution into a 500-mL Erlenmeyer flask, add 100 mL of water, and heat just to boiling. Add with stirring 50 mL of quimociac TS, then cover with a watch glass, and boil for 1 min in a well-ventilated hood. Cool to room temperature, swirling occasionally while cooling, then filter through a tared, sintered-glass filter crucible of medium porosity, and wash with five 25-mL portions of water. Dry at about 225° for 30 min, cool, and weigh. Each mg of precipitate thus obtained is equivalent to 32.074 μg of P_2O_5.

Arsenic A solution of 1 g in 35 mL of water meets the requirements of the *Arsenic Test*, Appendix IIIB.

Fluoride Determine on a 2-g sample as directed in *Method IV* under the *Fluoride Limit Test*, Appendix IIIB, using *Buffer Solution B* and 0.1 mL of *Fluoride Standard Solution*.

Heavy Metals Dissolve 20 g of the sample in 80 mL of water in a 250-mL beaker, cautiously add 20 mL of sulfuric acid, and boil for 1 h. Cool the solution, dilute it to 100 mL with water, mix, and filter through a fritted-disk funnel. Dilute a 10-mL aliquot to 25 mL with water, and adjust the pH to between 3.0 and 4.0 with ammonium hydroxide. Dilute to 40 mL with water, mix, and add 10 mL of freshly prepared hydrogen sulfide TS. Allow to stand for 5 min, and view downward over a white surface. The color of the sample solution is no darker than that of a standard prepared with 20 μg of lead ion (2.0 mL of *Standard Lead Solution*, see *Heavy Metals Test*, Appendix IIIB), treated in the same manner as the sample.

Insoluble Substances Dissolve 10 g in 100 mL of hot water, and filter through a tared filtering crucible. Wash the insoluble residue with hot water, dry at 105° for 2 h, cool, and weigh.

Packaging and Storage Store in tight containers.

Sodium Potassium Tartrate

Rochelle Salt

$C_4H_4KNaO_6 \cdot 4H_2O$ Formula wt 282.22
INS: 337 CAS: [304-59-6]

DESCRIPTION

Sodium Potassium Tartrate is a salt of L(+)-tartaric acid. It occurs as colorless crystals or as a white, crystalline powder and has a cooling, saline taste. As it effloresces slightly in warm, dry air, the crystals are often coated with a white powder. One g dissolves in 1 mL of water. It is practically insoluble in alcohol.

Functional Use in Foods Buffer; sequestrant.

REQUIREMENTS

Identification

A. Upon ignition, it emits the odor of burning sugar and leaves a residue that is alkaline to litmus and that effervesces with acids.

B. To 10 mL of a 1 in 20 solution add 10 mL of acetic acid. A white, crystalline precipitate is formed within 15 min.

C. A 1 in 10 solution gives positive tests for *Tartrate*, Appendix IIIA.

Assay Not less than 99.0% and not more than 102.0% of $C_4H_4KNaO_6 \cdot 4H_2O$.
Alkalinity Passes test.
Heavy Metals (as Pb) Not more than 10 mg/kg.
Water Between 21.0% and 26.0%.

TESTS

Assay Weigh accurately 0.5 g, and mix with 50 mL of glacial acetic acid, 30 mL of 96% formic acid, and 45 mL of acetic anhydride. Heat and stir until dissolution is complete, and titrate with 0.1 N perchloric acid in glacial acetic acid to a green endpoint with crystal violet indicator. Perform a blank determination and make any necessary correction. Each mL of 0.1 N perchloric acid is equivalent to 14.11 mg of $C_4H_4KNaO_6 \cdot 4H_2O$.

Alkalinity A 1 in 20 solution is alkaline to litmus, but after the addition of 0.2 mL of 0.1 N sulfuric acid to 10 mL of this solution, no pink color is produced by the addition of 1 drop of phenolphthalein TS.
Heavy Metals Prepare and test a 2-g sample as directed in *Method II* under the *Heavy Metals Test*, Appendix IIIB, using 20 µg of lead ion (Pb) in the control (*Solution A*).
Water Determine by the *Karl Fischer Titrimetric Method*, Appendix IIB, using a 200-mg sample and 35 mL of methanol in the *Procedure*.

Packaging and Storage Store in tight containers.

Sodium Potassium Tripolyphosphate

Trisodium Dipotassium Tripolyphosphate

$Na_3K_2P_3O_{10}$ Formula wt 400.1

CAS: [24315-83-1]

DESCRIPTION

Sodium Potassium Tripolyphosphate is anhydrous. It occurs as white, slightly hygroscopic granules, or as a powder. It is freely soluble in water. The pH of a 1 in 100 solution is about 10.

Functional Use in Foods Texturizer; sequestrant.

REQUIREMENTS

Identification A 1 in 20 solution gives positive tests for *Sodium*, for *Potassium*, and for *Phosphate*, Appendix IIIA.
Assay Not less than 85.0% and not more than 100.5% of $Na_3K_2P_3O_{10}$.
Arsenic (as As) Not more than 3 mg/kg.
Fluoride Not more than 0.005%.
Heavy Metals (as Pb) Not more than 10 mg/kg.
Insoluble Substances Not more than 0.1%.
Lead Not more than 2 mg/kg.

TESTS

Perform the **TESTS** as described in the monograph for *Sodium Tripolyphosphate*.

Packaging and Storage Store in tight containers.

Sodium Propionate

CH_3CH_2COONa

$C_3H_5NaO_2$ Formula wt 96.06

INS: 281 CAS: [137-40-6]

DESCRIPTION

Colorless, transparent crystals or a granular, crystalline powder. It is odorless or has a faint acetic–butyric odor. It is deliquescent in moist air. One g is soluble in about 1 mL of water at 25°, in about 0.65 mL of boiling water, and in about 24 mL of alcohol. The pH of a 1 in 10 solution is between 8.0 and 10.5.

Functional Use in Foods Preservative; mold inhibitor.

REQUIREMENTS

Identification
  A. A 1 in 20 solution gives positive tests for *Sodium*, Appendix IIIA.
  B. Upon ignition, it yields an alkaline residue that effervesces with acids.
  C. Warm a small sample with sulfuric acid. Propionic acid, recognized by its odor, is evolved.
Assay Not less than 99.0% and not more than 100.5% of $C_3H_5NaO_2$ after drying.
Alkalinity (as Na_2CO_3) Passes test (about 0.15%).
Heavy Metals (as Pb) Not more than 10 mg/kg.
Iron Not more than 0.003%.
Water Not more than 1%.

TESTS

Assay Weigh accurately about 250 mg, previously dried at 105° for 1 h, and dissolve it in 40 mL of glacial acetic acid, warming if necessary to effect solution. Cool to room temperature, add 2 drops of crystal violet TS, and titrate with 0.1 N perchloric acid. Perform a blank determination (see *General Provisions*), and make any necessary correction. Each mL of 0.1 N perchloric acid is equivalent to 9.606 mg of $C_3H_5NaO_2$.
Alkalinity Dissolve 4 g in 20 mL of water, and add 3 drops of phenolphthalein TS. If a pink color is produced, not more than 0.6 mL of 0.1 N sulfuric acid is required to discharge it.
Heavy Metals A solution of 2 g in 25 mL of water meets the requirements of the *Heavy Metals Test*, Appendix IIIB, using 20 µg of lead ion (Pb) in the control (*Solution A*).
Iron Dissolve 300 mg in 40 mL of water, and add 2 mL of hydrochloric acid, about 40 mg of ammonium persulfate, and 10 mL of ammonium thiocyanate TS. Any red or pink color does not exceed that produced by 0.9 mL of *Iron Standard Solution* (9 µg Fe) in an equal volume of solution containing the quantities of reagents used in the test.
Water Determine by the *Karl Fischer Titrimetric Method*, Appendix IIB.

Packaging and Storage Store in tight containers.

Sodium Pyrophosphate

Tetrasodium Diphosphate; Tetrasodium Pyrophosphate

$Na_4P_2O_7$ Formula wt 265.90

INS: 452(i) CAS: [7722-88-5]

DESCRIPTION

Sodium Pyrophosphate is anhydrous or contains 10 molecules of water of hydration. It occurs as a white, crystalline or granular powder. The decahydrate effloresces slightly in dry air. It is soluble in water, but is insoluble in alcohol. The pH of a 1 in 100 solution is about 10.

Functional Use in Foods Emulsifier; buffer; nutrient; dietary supplement; sequestrant; texturizer.

REQUIREMENTS

Identification

A. A 1 in 20 solution gives positive tests for *Sodium*, Appendix IIIA.

B. Dissolve 100 mg of the sample in 100 mL of 1.7 N nitric acid. Add 0.5 mL of this solution to 30 mL of quimociac TS. A yellow precipitate does not form. Heat the remaining portion of the sample solution for 10 min at 95°, and then add 0.5 mL of the solution to 30 mL of quimociac TS. A yellow precipitate forms immediately.

Assay Not less than 95.0% and not more than 100.5% of $Na_4P_2O_7$, calculated on the ignited basis.

Arsenic (as As) Not more than 3 mg/kg.

Fluoride Not more than 0.005%.

Heavy Metals (as Pb) Not more than 10 mg/kg.

Insoluble Substances Not more than 0.2%.

Loss on Ignition $Na_4P_2O_7$ (anhydrous): not more than 0.5%; $Na_4P_2O_7 \cdot 10H_2O$ (decahydrate): between 38.0% and 42.0%.

TESTS

Assay Dissolve an accurately weighed quantity of the sample, equivalent to 500 mg of anhydrous $Na_4P_2O_7$, in 100 mL of water in a 400-mL beaker. Using a pH meter, adjust the pH of the solution to 3.8 with hydrochloric acid, then add 50 mL of a 1 in 8 solution of zinc sulfate (125 g of $ZnSO_4 \cdot 7H_2O$ dissolved in water, diluted to 1000 mL, filtered, and adjusted to pH 3.8), and allow to stand for 2 min. Titrate the liberated acid with 0.1 N sodium hydroxide until a pH of 3.8 is again reached. After each addition of sodium hydroxide near the endpoint, time should be allowed for any precipitated zinc hydroxide to redissolve. Each mL of 0.1 N sodium hydroxide is equivalent to 13.30 mg of $Na_4P_2O_7$.

Arsenic A solution of 1 g in 35 mL of water meets the requirements of the *Arsenic Test*, Appendix IIIB.

Fluoride Determine on a 2-g sample as directed under *Fluoride* in the monograph for *Dibasic Calcium Phosphate*.

Heavy Metals A solution of 2 g in 25 mL of water meets the requirements of the *Heavy Metals Test*, Appendix IIIB, using 20 μg of lead ion (Pb) in the control (*Solution A*).

Insoluble Substances Dissolve 10 g in 100 mL of hot water, and filter through a tared filtering crucible. Wash the insoluble residue with hot water, dry at 105° for 2 h, cool, and weigh.

Loss on Ignition Dry at 110° for 4 h, and then ignite at about 800° for 30 min.

Packaging and Storage Store in tight containers.

Sodium Saccharin

1,2-Benzisothiazolin-3-one 1,1-Dioxide Sodium Salt; Sodium *o*-Benzosulfimide; Soluble Saccharin

$C_7H_4NNaO_3S \cdot 2H_2O$ Formula wt 241.20

INS: 954 CAS: [128-44-9]

DESCRIPTION

White crystals or a white, crystalline powder. It is odorless or has a faint, aromatic odor. It has an intensely sweet taste, even in dilute solutions. In powdered form, it effloresces to the extent that it usually contains only about one-third the amount of water indicated in its molecular formula. One g is soluble in 1.5 mL of water and in about 50 mL of alcohol.

Functional Use in Foods Nonnutritive sweetener.

REQUIREMENTS

Identification

A. Dissolve about 100 mg in 5 mL of sodium hydroxide solution (1 in 20), evaporate to dryness, and gently fuse the residue over a small flame until it no longer evolves ammonia. After the residue has cooled, dissolve it in 20 mL of water, neutralize the solution with 2.7 N hydrochloric acid, and filter. The addition of a drop of ferric chloride TS to the filtrate produces a violet color.

B. Mix 20 mg with 40 mg of resorcinol, add 10 drops of sulfuric acid, and heat the mixture in a liquid bath at 200° for 3 min. After cooling, add 10 mL of water and an excess of 1 N sodium hydroxide. A fluorescent green liquid results.

C. The residue obtained by igniting a 2-g sample gives positive tests for *Sodium*, Appendix IIIA.

D. Add 1 mL of hydrochloric acid to 10 mL of a 1 in 10 solution of the sample, wash the crystalline precipitate well with cold water, and dry at 105° for 2 h. The saccharin thus obtained

melts between 226° and 230° (see Appendix IIB, *Melting Range or Temperature*).
Assay Not less than 98.0% and not more than 101.0% of $C_7H_4NNaO_3S$, calculated on the anhydrous basis.
Alkalinity Passes test.
Benzoate and Salicylate Passes test.
Heavy Metals (as Pb) Not more than 10 mg/kg.
Readily Carbonizable Substances Passes test.
Selenium Not more than 0.003%.
Toluenesulfonamides Not more than 25 mg/kg.
Water Not more than 15.0%.

TESTS

Assay Transfer about 500 mg of the sample, accurately weighed, into a separator with the aid of 10 mL of water, add 2 mL of 2.7 N hydrochloric acid, and extract the precipitated saccharin first with 30 mL, then with five 20-mL portions, of a solvent composed of 9 volumes of chloroform and 1 volume of alcohol. Filter each extract through a small filter paper moistened with the solvent mixture, and evaporate the combined filtrates on a steam bath to dryness with the aid of a current of air. Dissolve the residue in 40 mL of alcohol, add 40 mL of water, mix, add 3 drops of phenolphthalein TS, and titrate with 0.1 N sodium hydroxide. Perform a blank determination on a mixture of 40 mL of alcohol and 40 mL of water (see *General Provisions*). Each mL of 0.1 N sodium hydroxide is equivalent to 20.52 mg of $C_7H_4NNaO_3S$.

Alkalinity A 1 in 10 solution is neutral or alkaline to litmus, but produces no red color with phenolphthalein TS.

Benzoate and Salicylate To 10 mL of a 1 in 20 solution, previously acidified with 5 drops of acetic acid, add 3 drops of ferric chloride TS. No precipitate or violet color appears.

Heavy Metals Prepare and test a 2-g sample as directed in *Method II* under the *Heavy Metals Test*, Appendix IIIB, using 20 μg of lead ion (Pb) in the control (*Solution A*).

Readily Carbonizable Substances, Appendix IIB Dissolve 200 mg in 5 mL of 95% sulfuric acid, and keep at a temperature of 48° to 50° for 10 min. The color is no darker than *Matching Fluid A*.

Selenium Determine as directed in *Method I* under the *Selenium Limit Test*, Appendix IIIB, using 200 mg of sample.

Toluenesulfonamides

Methylene Chloride Use a suitable grade (such as that obtainable from Burdick & Jackson Laboratories, Inc.), equivalent to the product obtained by distillation in all-glass apparatus.

Internal Standard Stock Solution Transfer 100.0 mg of *n*-tricosane (95%, obtainable from Chemical Samples Co.) into a 10-mL volumetric flask, dissolve in *n*-heptane, dilute to volume with the same solvent, and mix.

Stock Standard Preparation Transfer 20.0 mg each of reagent-grade *o*-toluenesulfonamide and *p*-toluenesulfonamide into a 10-mL volumetric flask, dissolve in methylene chloride, dilute to volume with the same solvent, and mix.

Diluted Standard Preparations Pipet into five 10-mL volumetric flasks 0.1, 0.25, 1.0, 2.5, and 5.0 mL, respectively, of the *Stock Standard Preparation*. Pipet 0.25 mL of the *Internal Standard Stock Solution* into each flask, dilute each to volume with methylene chloride, and mix. These solutions contain, respectively, 20, 50, 200, 500, and 1000 μg/mL of each toluenesulfonamide, plus 250 μg of *n*-tricosane.

Test Preparation Dissolve 2.00 g of the sample in 8.0 mL of 5% sodium bicarbonate solution, and mix the solution thoroughly with 10.0 g of chromatographic siliceous earth (Celite 545, Johns-Manville, or equivalent). Transfer the mix into a 25- × 250-mm chromatographic tube having a fritted-glass disk and a Teflon stopcock at the bottom and a reservoir at the top. Pack the contents of the tube by tapping the column on a padded surface, and then by tamping firmly from the top. Place 100 mL of methylene chloride in the reservoir, and adjust the stopcock so that 50 mL of eluate is collected in 20 to 30 min. To the eluate add 25 μL of *Internal Standard Stock Solution*, mix, and then concentrate the solution to a volume of 1.0 mL in a suitable concentrator tube fitted with a modified Snyder column, using a Kontes tube heater maintained at 90°.

Procedure Inject 2.5 μL of the *Test Preparation* into a suitable gas chromatograph equipped with a flame-ionization detector. The column is of glass, approximately 3 m in length and 2 mm in inside diameter, and it is packed with 3% phenyl methyl silicone (OV-17, Applied Science Laboratories, Inc., or equivalent) on 100- to 120-mesh silanized, calcined, diatomaceous silica (Gas-Chrom Q, Applied Science, or equivalent).

Caution: The glass column should extend into the injector for on-column injection and into the detector base to avoid contact with metal.

The carrier is helium, flowing at a rate of 30 mL/min. The injection port, column, and detector are maintained at 225°, 180°, and 250°, respectively. The instrument attenuation setting should be such that 2.5 μL of the *Diluted Standard Preparation* containing 200 μg/mL of each toluenesulfonamide gives a response of 40% to 80% of full-scale deflection. Record the chromatogram, note the peaks for *o*-toluenesulfonamide, *p*-toluenesulfonamide, and the *n*-tricosane internal standard, and calculate the areas for each peak by suitable means. The retention times for *o*-toluenesulfonamide, *p*-toluenesulfonamide, and *n*-tricosane are about 5, 6, and 15 min, respectively.

In a similar manner, obtain the chromatograms for 2.5-μL portions of each of the five *Diluted Standard Preparations*, and for each solution determine the areas of the *o*-toluenesulfonamide, *p*-toluenesulfonamide, and *n*-tricosane peaks. From the values thus obtained, prepare standard curves by plotting concentration of each toluenesulfonamide, in μg/mL, versus the ratio of the respective toluenesulfonamide peak area to that of *n*-tricosane. From the standard curve determine the concentration, in μg/mL, of each toluenesulfonamide in the *Test Preparation*. Divide each value by 2 to convert the result to mg/kg of the toluenesulfonamide in the 2-g sample taken for analysis.

Note: If the toluenesulfonamide content of the sample is greater than about 500 mg/kg, the impurity may crystallize out of the methylene chloride concentrate (see *Test Preparation*). Although this level of impurity exceeds that permitted by the specification, the analysis may be completed by diluting the concentrate with methylene chloride containing 250 μg of *n*-tricosane per mL, and by applying appropriate dilution factors in the calculation. Care must be taken to redissolve completely any crystalline toluenesulfonamide to give a homogeneous solution.

Water Determine by the *Karl Fischer Titrimetric Method*, Appendix IIB.

Packaging and Storage Store in well-closed containers.

Sodium Sesquicarbonate

$Na_2CO_3 \cdot NaHCO_3 \cdot 2H_2O$ Formula wt 226.03

INS: 500(iii)   CAS: [533-96-0]

DESCRIPTION

White crystals, flakes, or a crystalline powder. It is soluble in water, and its solutions are alkaline to litmus.

Functional Use in Foods Alkali; neutralizer in dairy products.

REQUIREMENTS

Identification A 1 in 10 solution gives positive tests for *Sodium* and for *Carbonate*, Appendix IIIA.
Assay *Sodium bicarbonate*: not less than 35.0% and not more than 38.6% of $NaHCO_3$; *sodium carbonate*: not less than 46.4% and not more than 50.0% of Na_2CO_3.
Heavy Metals (as Pb) Not more than 10 mg/kg.
Iron Not more than 0.002%.
Sodium Chloride Not more than 0.5%.
Water Between 13.8% and 16.7%.

TESTS

Assay for Sodium Bicarbonate Dissolve about 3 g of the sample, accurately weighed, in 150 mL of carbon dioxide-free water in a 600-mL beaker containing 50.0 mL of 0.5 N sodium hydroxide. While stirring, add 200 mL of 0.48 M barium chloride that has been adjusted to a pH of 8.0 with a pH meter. Using a pH meter that has been standardized to pH 9.0, titrate the solution with 0.5 N hydrochloric acid until a pH of 8.8 remains for 1 min, and record the volume, in mL, of 0.5 N hydrochloric acid required as *S*. Perform a blank determination using 2.1 g of primary standard sodium carbonate, and record the volume, in mL, of 0.5 N hydrochloric acid required as *B*. Each mL of the volume *B* − *S* of 0.5 N hydrochloric acid is equivalent to 42.00 mg of $NaHCO_3$.

Assay for Sodium Carbonate Determine the total alkalinity (as Na_2O) of the sample as follows: Dissolve about 4.2 g, accurately weighed, in 100 mL of water in a 250-mL beaker, add methyl orange TS, and titrate with 1 N sulfuric acid, stirring vigorously near the endpoint to expel carbon dioxide. Each mL of 1 N sulfuric acid is equivalent to 30.99 mg of Na_2O. Calculate the percentage of sodium oxide (% Na_2O) in the sample taken. Calculate the percentage of sodium carbonate in the sample by the formula

[% Na_2O − (% $NaHCO_3$ × 0.3689)] × 1.7099,

in which % $NaHCO_3$ is the percentage of sodium bicarbonate determined in the *Assay for Sodium Bicarbonate*, 0.3689 is a factor converting $NaHCO_3$ to Na_2O, and 1.7099 is a factor converting Na_2O to Na_2CO_3.

Heavy Metals Mix 2 g with 5 mL of water and 10 mL of 2.7 N hydrochloric acid, boil for 1 min, cool, and dilute to 25 mL with water. This solution meets the requirements of the *Heavy Metals Test*, Appendix IIIB, using 20 µg of lead ion (Pb) in the control (*Solution A*).

Iron Dissolve 500 mg in 10 mL of 2.7 N hydrochloric acid, and dilute to 50 mL with water. Add about 40 mg of ammonium persulfate crystals and 10 mL of ammonium thiocyanate TS. Any red or pink color does not exceed that produced by 1.0 mL of *Iron Standard Solution* (10 µg Fe) in an equal volume of solution containing 2 mL of hydrochloric acid and the quantities of ammonium persulfate and ammonium thiocyanate used in the test.

Sodium Chloride Dissolve about 10 g, accurately weighed, in 50 mL of water in a 250-mL beaker, add sufficient nitric acid to make the solution slightly acid, then add 1 mL of ferric ammonium sulfate TS and 1.00 mL of 0.05 N ammonium thiocyanate, and titrate with 0.05 N silver nitrate, stirring constantly, until the red color is completely discharged. Finally, back titrate with 0.05 N ammonium thiocyanate until a faint reddish color is obtained. Subtract the total volume of 0.05 N ammonium thiocyanate added from the volume of 0.05 N silver nitrate required. Each mL of 0.05 N silver nitrate is equivalent to 2.922 mg of NaCl. Calculate the percentage of sodium chloride in the sample taken.

Water Calculate the percentage of water by subtracting from 100 the sum of the percentages of *Sodium Bicarbonate*, *Sodium Carbonate*, and *Sodium Chloride* found in the sample.

Packaging and Storage Store in well-closed containers.

Sodium Stearoyl Lactylate

INS: 481(i)   CAS: [25383-99-7]

DESCRIPTION

A mixture of sodium salts of stearoyl lactylic acids and minor proportions of other sodium salts of related acids, manufactured by the reaction of stearic acid and lactic acid, neutralized to the sodium salts. It is a slightly hygroscopic, cream-colored powder having a mild, caramel-like odor. It is soluble in hot oil or fat, and is dispersible in warm water.

Functional Use in Foods Emulsifier; dough conditioner; stabilizer; whipping agent.

REQUIREMENTS

Identification

A. Heat 1 g with a mixture of 25 mL of water and 5 mL of hydrochloric acid. Fatty acids are liberated, floating as an oily layer on the surface of the liquid. The water layer gives positive tests for *Sodium*, Appendix IIIA.

B. Mix 25 g of the sample with 50 g of a 15% alcoholic potassium hydroxide solution in an Erlenmeyer flask, and reflux for 1 h or until saponification is complete. Cool, add 150 mL of water, and mix. After complete solution of the soap, add 60 mL of 2 N sulfuric acid, and heat the mixture, with frequent stirring, until the fatty acids separate cleanly as a transparent layer. Wash the fatty acids with boiling water until free from sulfate, collect them in a small beaker, and warm on a steam bath until the water has separated and the fatty acids are clear. Allow the acids to cool, pour off the water layer, then melt the acids, filter into a dry beaker, and dry at 105° for 20 min. The solidification point of the fatty acids so obtained is not below 54° (see Appendix IIB, *Melting Range or Temperature*).

Acid Value Between 60 and 80.
Ester Value Between 150 and 190.
Heavy Metals (as Pb) Not more than 10 mg/kg.
Sodium Content Between 3.5% and 5.0%.
Total Lactic Acid Between 31.0% and 34.0%.

TESTS

Acid Value Transfer about 1 g, accurately weighed, to a 125-mL Erlenmeyer flask, add 25 mL of alcohol, previously neutralized to phenolphthalein TS, and heat on a hot plate until the sample is dissolved. Cool, add 5 drops of phenolphthalein TS, and titrate rapidly with 0.1 N sodium hydroxide to the first pink color that persists for at least 30 s. Calculate the acid value by the formula

$$56.1 V \times N/W,$$

in which V is the volume, in mL, and N is the normality, respectively, of the sodium hydroxide solution, and W is the weight, in g, of the sample taken. Retain the neutralized solution for the determination of *Ester Value*.

Ester Value To the neutralized solution retained in the test for *Acid Value* add 10.0 mL of alcoholic potassium hydroxide solution prepared by dissolving 11.2 g of potassium hydroxide in 250 mL of alcohol and diluting with 25 mL of water. Add 5 drops of phenolphthalein TS, connect a suitable condenser, and reflux for 2 h. Cool, add 5 additional drops of phenolphthalein TS, and titrate the excess alkali with 0.1 N sulfuric acid. Perform a blank determination using 10.0 mL of the alcoholic potassium hydroxide solution. Calculate the ester value by the formula

$$56.1(B - S)N/W,$$

in which $B - S$ represents the difference between the volumes of 0.1 N sulfuric acid required for the blank and the sample, respectively; N is the normality of the sulfuric acid; and W is the weight, in g, of the sample taken.

Heavy Metals Prepare and test a 2-g sample as directed in *Method II* under the *Heavy Metals Test*, Appendix IIIB, using 20 µg of lead ion (Pb) in the control (*Solution A*).

Sodium Content [Note: Ordinary glassware should not be used in this test because of possible contamination by sodium; use suitable plastic (e.g., polyethylene) vessels where necessary.]

Stock Lanthanum Solution Transfer 5.86 g of lanthanum oxide, La_2O_3, to a 100-mL volumetric flask, wet with a few mL of water, slowly add 25 mL of hydrochloric acid, and swirl until the material is completely dissolved. Dilute to volume with water, and mix.

Stock Sodium Solution Use a solution containing 1 mg of Na in each mL (1000 mg/kg Na). The solution may be obtained commercially or prepared as follows: Transfer 1.271 g of sodium chloride, previously dried at 105° for 2 h and accurately weighed, to a 500-mL volumetric flask, dilute to volume with water, and mix.

Standard Preparations Transfer 10.0 mL of the *Stock Lanthanum Solution* to each of three 100-mL volumetric flasks. Using a microliter syringe, transfer 0.20 mL of the *Stock Sodium Solution* to the first flask, 0.40 mL to the second flask, and 0.50 mL to the third flask. Dilute each flask to volume with water, and mix. The flasks contain 2.0, 4.0, and 5.0 µg of Na per mL, respectively. Prepare these solutions fresh daily.

Sample Preparation Transfer about 250 mg of the sample, accurately weighed, to a 30-mL beaker, dissolve with heating in 10 mL of alcohol, and quantitatively transfer the solution into a 25-mL volumetric flask. Wash the beaker with two 5-mL portions of alcohol, adding the washings to the flask, dilute to volume with alcohol, and mix. Transfer 2.5 mL of the *Stock Lanthanum Solution* to a second 25-mL volumetric flask. Using a microliter syringe, transfer 0.25 mL of the alcoholic solution of the sample to the second flask, dilute to volume with water, and mix.

Procedure Concomitantly determine the absorbance of each *Standard Preparation* and of the *Sample Preparation* at 589 nm, with a suitable atomic absorption spectrophotometer, following the operating parameters as recommended by the manufacturer of the instrument. Plot the absorbances of the *Standard Preparations* versus concentration of Na, in µg/mL, and from the curve so obtained determine the concentration, C, in µg/mL, of Na in the *Sample Preparation*. Calculate the quantity, in mg, of Na in the sample taken by the formula

$$2.5C.$$

Total Lactic Acid

Standard Curve Dissolve 1.067 g of lithium lactate in sufficient water to make 1000.0 mL. Transfer 10.0 mL of this solution into a 100-mL volumetric flask, dilute to volume with water, and mix. Transfer 1.0, 2.0, 4.0, 6.0, and 8.0 mL of the diluted standard solution into separate 100-mL volumetric flasks, dilute each flask to volume with water, and mix. These standards represent 1, 2, 4, 6, and 8 µg of lactic acid per mL, respectively. Transfer 1.0 mL of each solution into separate test tubes, and continue as directed in the *Procedure*, beginning with "Add 1 drop of cupric sulfate TS...." After color development and reading the absorbance values, construct a *Standard Curve* by plotting absorbance versus µg of lactic acid.

Test Preparation Transfer about 200 mg of the sample, accurately weighed, into a 125-mL Erlenmeyer flask, add 10 mL of 0.5 N alcoholic potassium hydroxide and 10 mL of water, attach an air condenser, and reflux gently for 45 min. Wash the sides of the flask and the condenser with about 40 mL of water, and heat on a steam bath until no odor of alcohol remains. Add 6 mL of dilute sulfuric acid (1 in 2), heat until the fatty acids are melted, then cool to about 60°, and add 25 mL of petroleum ether. Swirl the mixture gently, and transfer quantitatively to a separator. Collect the water layer in a 100-mL volumetric flask, and wash the petroleum ether layer with two 20-mL portions of water, adding the washings to the volumetric flask. Dilute to volume with water, and mix. Transfer 1.0 mL of this solution into a second 100-mL volumetric flask, dilute to volume with water, and mix.

Procedure Transfer 1.0 mL of the *Test Preparation* into a test tube, and transfer 1.0 mL of water to a second test tube to serve as the blank. Treat each tube as follows: Add 1 drop of cupric sulfate TS, swirl gently, and then add rapidly from a buret 9.0 mL of sulfuric acid. Loosely stopper the tube, and heat in a water bath at 90° for exactly 5 min. Cool immediately to below 20° in an ice bath for 5 min, add 3 drops of *p*-phenylphenol TS, shake immediately, and heat in a water bath at 30° for 30 min, shaking the tube twice during this time to disperse the reagent. Heat the tube in a water bath at 90° for exactly 90 s, and then cool immediately to room temperature in an ice water bath. Determine the absorbance of the solution in a 1-cm cell, at 570 nm, with a suitable spectrophotometer, using the blank to set the instrument. Obtain the weight, in μg, of lactic acid in the portion of the *Test Preparation* taken for the *Procedure* by means of the *Standard Curve*.

Packaging and Storage Store in tight containers in a cool, dry place.

Sodium Stearyl Fumarate

$$\text{NaOOCCH} \atop \| \atop \text{HCCOOC}_{18}\text{H}_{37}$$

$C_{22}H_{39}NaO_4$ Formula wt 390.54

INS: 1169   CAS: [4070-80-8]

DESCRIPTION

A fine, white powder. It is slightly soluble in methanol, but is practically insoluble in water.

Functional Use in Foods Dough conditioner.

REQUIREMENTS

Identification The infrared absorption spectrum of a 1 in 300 potassium bromide dispersion of the sample exhibits maxima only at the same wavelengths as that of a similar preparation of USP Sodium Stearyl Fumarate Reference Standard.

Assay Not less than 99.0% and not more than 101.5% of $C_{22}H_{39}NaO_4$, calculated on the anhydrous basis.

Heavy Metals (as Pb) Not more than 0.002%.

Lead Not more than 10 mg/kg.

Saponification Value Between 142.2 and 146.0, calculated on the anhydrous basis.

Sodium Stearyl Maleate Not more than 0.25%.

Stearyl Alcohol Not more than 0.5%.

Water Not more than 5.0%.

TESTS

Assay Transfer about 250 mg, accurately weighed, into a 50-mL Erlenmeyer flask, mix with 1 mL of chloroform, and add 20 mL of glacial acetic acid to dissolve the sample. Add quinaldine red TS, and titrate with 0.1 N perchloric acid in glacial acetic acid. Each mL of 0.1 N perchloric acid is equivalent to 39.05 mg of $C_{22}H_{39}NaO_4$.

Heavy Metals Prepare and test a 1-g sample as directed in *Method II* under the *Heavy Metals Test*, Appendix IIIB, using 20 μg of lead ion (Pb) in the control.

Lead A *Sample Solution* prepared as directed for organic compounds meets the requirements of the *Lead Limit Test*, Appendix IIIB, using 10 μg of lead ion (Pb) in the control.

Saponification Value Transfer about 450 mg of Sodium Stearyl Fumarate, accurately weighed, into a 300-mL Erlenmeyer flask, and add 50.0 mL of ethanolic potassium hydroxide solution, rinsing down the inside of the flask during the addition. (Prepare the ethanolic potassium hydroxide solutions as follows: Dissolve about 5.5 g of potassium hydroxide in absolute ethanol, heating if necessary to effect solution, and dilute to 1000 mL with absolute ethanol. Prepare fresh daily, and filter if necessary to remove carbonate.) Reflux the mixture gently on a steam bath for at least 2 h, occasionally swirling gently but avoiding splashing the mixture up into the condenser. Rinse the condenser with 10 mL of 70% alcohol, followed by three 10-mL portions of water, collecting the rinsings in the flask. Cool, rinse the sides of the flask with two 10-mL portions of 70% alcohol, add phenolphthalein TS, and titrate with 0.1 N hydrochloric acid to the disappearance of any pink color. Perform a blank determination using the same amount of ethanolic potassium hydroxide solution (see *General Provisions*). Calculate the saponification value by the formula

$$56.1(B - S) \times N/W,$$

in which $B - S$ represents the difference between the volumes of 0.1 N hydrochloric acid required for the blank and for the sample, respectively; N is the exact normality of the hydrochloric acid; and W is the weight, in g, of the sample taken.

Sodium Stearyl Maleate and **Stearyl Alcohol**

Apparatus Assemble a suitable apparatus for ascending thin-layer chromatography (see Appendix IIA). Prepare a slurry of 24 g of chromatographic silica gel G in 75 mL of water, apply a uniformly thin layer to 23-cm square glass plates, or other convenient size, and dry in the air at room temperature for 2 h.

Sample Solution Weigh accurately 200 mg of the sample into a glass-stoppered, 10-mL volumetric flask, dilute to volume with a solution of 10% acetic acid in chloroform, and mix. The mixture may be heated carefully, if necessary, to dissolve the sample, and then cooled before diluting to volume with the solvent mixture.

Standard Solution A Weigh accurately 10 mg of USP Sodium Stearyl Maleate Reference Standard into a 100-mL volumetric flask, dilute to volume with 10% acetic acid in chloroform, and shake well.

Standard Solution B Weigh accurately 20 mg of stearyl alcohol (Aldrich or equivalent) into a 100-mL volumetric flask, dilute to volume with 10% acetic acid in chloroform, and shake well.

Standard Solution C Mix 25.0 mL of *Standard Solution A* with 25.0 mL of *Standard Solution B*, and shake well. This mixture represents 0.25% of sodium stearyl maleate and 0.5% of stearyl alcohol, based upon the weight (200 mg) of the sample taken.

Procedure Spot 10 μL each of the *Sample Solution* and of *Standard Solution C* at the bottom of the plate. Allow the spots to dry, then place the plate in a suitable chromatographic chamber containing a mixture of 5 volumes of toluene, 5 volumes of hexane, and 1 volume of acetic acid, previously equilibrated, and develop by ascending chromatography for 30 min to effect one pass. Remove the plate from the tank, dry in the air for 10 min, and then heat in an oven at 90° for 2 min. After cooling to room temperature, replace the plate in the chamber for a second pass of 30 min. After the second pass, remove the plate from the chamber and dry in the air for 15 to 20 min. Spray evenly with a mixture consisting of 0.5% of potassium permanganate and 0.3% of sodium carbonate in water. Maleate and fumarate appear as yellow spots against a pink background. Spray with sulfuric acid and heat in an oven at 150° for the detection of stearyl alcohol.

Visually compare any spots from the sample against the R_f of the spots from the standards. The spots from the sample do not appear to be stronger than the respective spots from the standards.

Water Determine by the *Karl Fischer Titrimetric Method*, Appendix IIB.

Packaging and Storage Store in well-closed containers.

Sodium Sulfate

Na_2SO_4 Formula wt 142.04

INS: 514 CAS: [7757-82-6]

DESCRIPTION

Sodium Sulfate is anhydrous or contains 10 molecules of water of crystallization. It occurs as colorless crystals or as a fine, white, crystalline powder. The decahydrate is efflorescent. Sodium Sulfate is freely soluble in water and practically insoluble in alcohol. A 1 in 20 solution is neutral or slightly alkaline to litmus paper.

Functional Use in Foods Agent in caramel production.

REQUIREMENTS

Labeling Indicate whether it is anhydrous or the decahydrate.
Identification A 1 in 20 solution gives positive tests for *Sodium* and for *Sulfate*, Appendix IIIA.
Assay Not less than 99.0% and not more than 100.5% of Na_2SO_4 after drying.
Heavy Metals (as Pb) Not more than 10 mg/kg.
Loss on Drying *Anhydrous form*: not more than 1%; *decahydrate*: between 51.0% and 57.0%.
Selenium Not more than 0.003%.

TESTS

Assay Dissolve about 500 mg, previously dried at 105° for 4 h and accurately weighed, in 200 mL of water, add 1 mL of hydrochloric acid, and heat to boiling. Gradually add, in small portions at a time and while stirring constantly, an excess of hot barium chloride TS (about 10 mL), and heat the mixture on a steam bath for 1 h. Collect the precipitate on a retentive, ashless filter paper, wash until free from chloride, and place the filter in a suitable tared crucible. Carefully burn away the paper, and ignite at 800° ± 25° to constant weight. The weight of the barium sulfate so obtained, multiplied by 0.6086, indicates its equivalent of Na_2SO_4.
Heavy Metals A solution of 3 g in 25 mL of water meets the requirements of the *Heavy Metals Test*, Appendix IIIB, using 20 μg of lead ion (Pb) and 1 g of the sample in the control (*Solution A*).
Loss on Drying, Appendix IIC Dry at 105° for 4 h.
Selenium Determine as directed in *Method II* under the *Selenium Limit Test*, Appendix IIIB, using 200 mg of sample.

Packaging and Storage Store in well-closed containers.

Sodium Sulfite

Exsiccated Sodium Sulfite

Na_2SO_3 Formula wt 126.04

INS: 221 CAS: [7757-83-7]

DESCRIPTION

A white or tan to slightly pink, odorless or nearly odorless powder having a cooling, saline, sulfurous taste. It undergoes oxidation in air. Its solutions are alkaline to litmus and to phenolphthalein. One g dissolves in about 4 mL of water. It is sparingly soluble in alcohol.

Functional Use in Foods Preservative; antioxidant; bleaching agent.

REQUIREMENTS

Identification A 1 in 20 solution gives positive tests for *Sodium* and for *Sulfite*, Appendix IIIA.
Assay Not less than 95.0% of Na_2SO_3.
Heavy Metals (as Pb) Not more than 10 mg/kg.
Selenium Not more than 0.003%.

TESTS

Assay Weigh accurately about 250 mg, add it to exactly 50 mL of 0.1 N iodine contained in a glass-stoppered flask, and stopper the flask. Allow to stand for 5 min, add 1 mL of hydrochloric acid, and titrate the excess iodine with 0.1 N sodium thiosulfate, adding starch TS as the indicator. Each mL of 0.1 N iodine is equivalent to 6.302 mg of Na_2SO_3.
Heavy Metals Dissolve 2 g in 10 mL of water, add 4 mL of hydrochloric acid, and evaporate to dryness on a steam bath. To the residue add 5 mL of hot water and 1 mL of hydrochloric acid, and again evaporate to dryness. Dissolve the residue in water, and dilute to 25 mL. This solution meets the requirements of the *Heavy Metals Test*, Appendix IIIB, using 20 μg of lead ion (Pb) in the control (*Solution A*).
Selenium Determine as directed in *Method I* under the *Selenium Limit Test*, Appendix IIIB, using 200 mg of sample and 100 mg of magnesium oxide.

Packaging and Storage Store in tight containers.

Sodium Tartrate

Disodium Tartrate; Disodium D-Tartrate

NaOOCCH(OH)CH(OH)COONa.2H$_2$O

$C_4H_4Na_2O_6 \cdot 2H_2O$ Formula wt 230.08
INS: 335(i)   CAS: [868-18-8]

DESCRIPTION

Sodium Tartrate is the disodium salt of L(+)-tartaric acid. It occurs as transparent, colorless, odorless crystals. One g dissolves in 3 mL of water. It is insoluble in alcohol. The pH of a 1 in 20 solution is between 7 and 9. Upon ignition, it emits the odor of burning sugar and leaves a residue that is alkaline to litmus and that effervesces with acids.

Functional Use in Foods Sequestrant.

REQUIREMENTS

Identification It gives positive tests for *Sodium* and for *Tartrate*, Appendix IIIA.
Assay Not less than 99.0% and not more than 100.5% of $C_4H_4Na_2O_6$ after drying.
Heavy Metals (as Pb) Not more than 10 mg/kg.
Loss on Drying Between 14.0% and 17.0%.
Oxalate Passes test (limit about 0.1%).

TESTS

Assay Weigh accurately about 250 mg, previously dried at 150° for 3 h, and transfer it to a 250-mL beaker. Add 150 mL of glacial acetic acid, heat to near boiling, stir until the sample is dissolved (preferably with a magnetic stirrer), and cool to room temperature. Titrate with 0.1 N perchloric acid in glacial acetic acid, determining the endpoint potentiometrically. Each mL of 0.1 N perchloric acid is equivalent to 9.703 mg of $C_4H_4Na_2O_6$.
Heavy Metals A solution of 2 g in 25 mL of water meets the requirements of the *Heavy Metals Test*, Appendix IIIB, using 20 μg of lead ion (Pb) in the control (*Solution A*).
Loss on Drying, Appendix IIC Dry at 150° for 3 h.
Oxalate Dissolve 1 g in 10 mL of water, and add 5 drops of 1 N acetic acid and 2 mL of calcium chloride TS. No turbidity develops within 1 h.

Packaging and Storage Store in tight containers.

Sodium Thiosulfate

Sodium Hyposulfite

$Na_2S_2O_3 \cdot 5H_2O$ Formula wt 248.19
INS: 539   CAS: [10102-17-7]

DESCRIPTION

Large, colorless crystals or a coarse, crystalline powder. It is deliquescent in moist air and effloresces in dry air at a temperature above 33°. Its solutions are neutral or faintly alkaline to litmus. One g dissolves in 0.5 mL of water. It is insoluble in alcohol.

Functional Use in Foods Sequestrant; antioxidant.

REQUIREMENTS

Identification
A. To a 1 in 10 solution add a few drops of iodine TS. The color is discharged.
B. A 1 in 20 solution gives positive tests for *Sodium* and for *Thiosulfate*, Appendix IIIA.
Assay Not less than 99.0% and not more than 100.5% of $Na_2S_2O_3$ after drying.

Heavy Metals (as Pb)   Not more than 0.002%.
Lead   Not more than 10 mg/kg.
Selenium   Not more than 0.003%.
Water   Between 32.0% and 37.0%.

TESTS

Assay   Weigh accurately about 500 mg of the dried sample obtained in the test for *Water*, dissolve it in 30 mL of water, and titrate with 0.1 N iodine, using starch TS as the indicator. Each mL of 0.1 N iodine is equivalent to 15.81 mg of $Na_2S_2O_3$.

Sample Solution for the Determination of Heavy Metals and Lead   Dissolve 5.0 g in 40 mL of water, slowly add 25 mL of 2.7 N hydrochloric acid, and evaporate to dryness on a steam bath. Add 30 mL of water to the residue, boil gently for 2 min, and filter. Heat the filtrate to boiling, add sufficient bromine TS to produce a clear solution, then add a slight excess of bromine. Boil to expel the excess bromine, cool, and dilute to 50.0 mL with water.

Heavy Metals   Dilute 10.0 mL of the *Sample Solution* (1-g sample) to 25 mL with water, and proceed as directed under the *Heavy Metals Test*, Appendix IIIB, using 20 µg of lead ion (Pb) in the control (*Solution A*).
Lead   A 10.0-mL portion of the *Sample Solution* (1-g sample) meets the requirements of the *Lead Limit Test*, Appendix IIIB, using 10 µg of lead ion (Pb) in the control.
Selenium   Determine as directed in *Method I* under the *Selenium Limit Test*, Appendix IIIB, using 200 mg of sample.
Water   Dry about 1 g, accurately weighed, in a vacuum at 40° to 45° for 16 h, cool, and weigh.

Packaging and Storage   Store in tight containers.

Sodium Trimetaphosphate

$(NaPO_3)_3$ Formula wt 305.89

CAS: [7785-84-4]

DESCRIPTION

A cyclic polyphosphate composed of three metaphosphate units. It occurs as white crystals or as a white, crystalline powder. It is freely soluble in water. The pH of a 1 in 100 solution is about 6.0.

Functional Use in Foods   Starch-modifying agent.

REQUIREMENTS

Identification
A.   A 1 in 20 solution gives positive tests for *Sodium*, Appendix IIIA.

B.   Dissolve about 100 mg in 5 mL of hot 1.7 N nitric acid, warm on a steam bath for 10 min, and cool. Neutralize to litmus paper with 1 N sodium hydroxide, and add silver nitrate TS. A yellow precipitate is formed that is soluble in 1.7 N nitric acid.
Assay   Between 68.0% and 70.0% of P_2O_5.
Arsenic (as As)   Not more than 3 mg/kg.
Fluoride   Not more than 0.005%.
Heavy Metals (as Pb)   Not more than 10 mg/kg.
Insoluble Substances   Not more than 0.1%.

TESTS

Assay   Transfer about 800 mg of the sample, accurately weighed, into a 400-mL beaker, add 100 mL of water and 25 mL of nitric acid, cover with a watch glass, and boil for 10 min on a hot plate. Rinse any condensate from the watch glass into the beaker, cool the solution to room temperature, transfer it quantitatively to a 500-mL volumetric flask, dilute to volume with water, and mix thoroughly. Pipet 20.0 mL of this solution into a 500-mL Erlenmeyer flask, add 100 mL of water, and heat just to boiling. Add with stirring 50 mL of quimociac TS, then cover with a watch glass, and boil for 1 min in a well-ventilated hood. Cool to room temperature, swirling occasionally while cooling, then filter through a tared, sintered-glass filter crucible of medium porosity, and wash with five 25-mL portions of water. Dry at about 225° for 30 min, cool, and weigh. Each mg of precipitate thus obtained is equivalent to 32.074 µg of P_2O_5.
Arsenic   A solution of 1 g in 35 mL of water meets the requirements of the *Arsenic Test*, Appendix IIIB.
Fluoride   Determine on a 2-g sample as directed in *Method IV* under the *Fluoride Limit Test*, Appendix IIIB, using *Buffer Solution A* and 0.1 mL of *Fluoride Standard Solution*.
Heavy Metals   A solution of 2 g in 25 mL of water meets the requirements of the *Heavy Metals Test*, Appendix IIIB, using 20 µg of lead ion (Pb) in the control (*Solution A*).
Insoluble Substances   Dissolve 10 g in 100 mL of hot water, and filter through a tared filtering crucible. Wash the insoluble residue with hot water, dry at 105° for 2 h, cool, and weigh.

Packaging and Storage   Store in tight containers.

Sodium Tripolyphosphate

Pentasodium Triphosphate; Triphosphate; Sodium Triphosphate

$Na_5P_3O_{10}$ Formula wt 367.86

INS: 451(i)   CAS: [7758-29-4]

DESCRIPTION

Sodium Tripolyphosphate is anhydrous or contains six molecules of water of hydration. It occurs as white, slightly hygroscopic granules, or as a powder. It is freely soluble in water. The pH of a 1 in 100 solution is about 9.5.

Functional Use in Foods Texturizer; sequestrant.

REQUIREMENTS

Identification

A. A 1 in 20 solution gives positive tests for *Sodium*, Appendix IIIA.

B. To 1 mL of a 1 in 100 solution add a few drops of silver nitrate TS. A white precipitate is formed that is soluble in 1.7 N nitric acid.

Assay *Anhydrous*: Not less than 85.0% of $Na_5P_3O_{10}$; *hexahydrate*: not less than 65.0% of $Na_5P_3O_{10}$.

Arsenic (as As) Not more than 3 mg/kg.

Fluoride Not more than 0.005%.

Heavy Metals (as Pb) Not more than 10 mg/kg.

Insoluble Substances Not more than 0.1%.

Lead Not more than 5 mg/kg.

TESTS

Assay

Potassium Acetate Buffer (pH 5.0) Dissolve 78.5 g of potassium acetate in 1000 mL of water, and adjust the pH of the solution to 5.0 with acetic acid. Add a few mg of mercuric iodide to inhibit mold growth.

0.3 M Potassium Chloride Solution Dissolve 22.35 g of potassium chloride in water, add 5 mL of *Potassium Acetate Buffer*, dilute with water to 1000 mL, and mix. Add a few mg of mercuric iodide.

0.6 M Potassium Chloride Solution Dissolve 44.7 g of potassium chloride in water, add 5 mL of *Potassium Acetate Buffer*, dilute with water to 1000 mL, and mix. Add a few mg of mercuric iodide.

1 M Potassium Chloride Solution Dissolve 74.5 g of potassium chloride in water, add 5 mL of *Potassium Acetate Buffer*, dilute to 1000 mL with water, and mix. Add a few mg of mercuric iodide.

Chromatographic Column Use a standard chromatographic column, 20- to 40-cm in length, 20- to 28-mm in inside diameter, with a sealed-in, coarse-porosity fritted disk. If a stopcock is not provided, attach a stopcock having a 3- to 4-mm diameter bore to the outlet of the column with a short length of flexible vinyl tubing.

Procedure Close the column stopcock, fill the space between the fritted disk and the stopcock with water, and connect a vacuum line to the stopcock. Prepare a 1:1 water slurry of Dowex 1 × 8, chloride form, 100- to 200- or 200- to 400-mesh, or a comparable grade of styrene–divinylbenzene ion exchange resin, and decant off any very fine particles and any foam. Do this two or three times or until no more finely suspended material or foaming is observed. Fill the column with the slurry, and open the stopcock to allow the vacuum to pack the resin bed until the water level is slightly above the top of the resin, then immediately close the stopcock. Do not allow the liquid level to fall below the resin level at any time. Repeat this procedure until the packed resin column is 15 cm (about 6 in.) above the fritted disk. Place one circle of tightly fitting glass-fiber filter paper on top of the resin bed, then place a perforated polyethylene disk on top of the paper. Alternatively, a loosely packed plug of glass wool may be placed on top of the bed. Close the top of the column with a rubber stopper in which a 7.6-cm length of capillary tubing (1.5-mm id, 7-mm od) has been inserted through the center, so that about 12 mm of the tubing extends through the bottom of the stopper. Connect the top of the capillary tubing to the stem of a 500-mL separator with flexible vinyl tubing, and clamp the separator to a ring stand above the column. Wash the column by adding 100 mL of water to the separator with all stopcocks closed. First open the separator stopcock, then open the column stopcock. The rate of flow should be about 5 mL/min. When the separator is empty, close the stopcock on the column, then close the separator stopcock.

Transfer about 500 mg of the sample, accurately weighed, into a 250-mL volumetric flask, dissolve and dilute to volume with water, and mix. Transfer 10.0 mL of this solution into the separator, open both stopcocks, and allow the solution to drain into the column, rinsing the separator with 20 mL of water. Discard the eluate.

Add 370 mL of *0.3 M Potassium Chloride Solution* to the separator, and allow this solution to pass through the column, discarding the eluate. Add 250 mL of *0.6 M Potassium Chloride Solution* to the column, allow the solution to pass through the column, and receive the eluate in a 400-mL beaker. (To ensure a clean column for the next run, pass 100 mL of *1 M Potassium Chloride Solution* through the column, and then follow with 100 mL of water. Discard all washings.) To the beaker add 15 mL of nitric acid, mix, and boil for 15 to 20 min. Add methyl orange TS, and neutralize the solution with ammonium hydroxide. Add 1 g of ammonium nitrate crystals, stir to dissolve, and cool. Add 15 mL of ammonium molybdate TS, with stirring, and stir vigorously for 3 min, or allow to stand with occasional stirring for 10 to 15 min. Filter the contents of the beaker with suction through a 6- to 7-mm paper-pulp filter pad supported in a 25-mm porcelain disk. The filter pad should be covered with a suspension of infusorial earth. After the contents of the beaker have been transferred to the filter, wash the beaker with five 10-mL portions of a 1 in 100 solution of sodium or potassium nitrate, passing the washings through the filter, then wash the filter with five 5-mL portions of the wash solution. Return the filter pad and the precipitate to the beaker, wash the funnel thoroughly with water into the beaker, and dilute to about 150 mL. Add 0.1 N sodium hydroxide from a buret until the yellow precipitate is dissolved, then add 5 to 8 mL in excess. Add phenolphthalein TS, and titrate the excess alkali with 0.1 N nitric acid. Finally, titrate with 0.1 N sodium hydroxide to the first appearance of the pink color. The difference between the total volume of 0.1 N sodium hydroxide added and the volume of nitric acid required represents the volume, V, in mL, of 0.1 N sodium hydroxide consumed by the phosphomolybdate complex. Calculate the quantity, in mg, of $Na_5P_3O_{10}$ in the sample taken by the formula

$$0.533 \times 25V.$$

Arsenic A solution of 1 g in 35 mL of water meets the requirements of the *Arsenic Test*, Appendix IIIB.

Fluoride Determine on a 2-g sample as directed in *Method IV* under the *Fluoride Limit Test*, Appendix IIIB, using *Buffer Solution A* and 0.1 mL of *Fluoride Standard Solution*.

Heavy Metals A solution of 2 g in 25 mL of water meets the requirements of the *Heavy Metals Test*, Appendix IIIB, using 20 µg of lead ion (Pb) in the control (*Solution A*).

Insoluble Substances Dissolve 10 g in 100 mL of hot water, and filter through a tared filtering crucible. Wash the insoluble residue with hot water, dry at 105° for 2 h, cool, and weigh.

Lead A solution of 1 g in 20 mL of water meets the requirements of the *Lead Limit Test*, Appendix IIIB, using 5 µg of lead ion (Pb) in the control.

Packaging and Storage Store in tight containers.

Heavy Metals Prepare and test a 2-g sample as directed in *Method II* under the *Heavy Metals Test*, Appendix IIIB, using 20 µg of lead ion (Pb) in the control (*Solution A*).

Melting Range, Appendix IIB Determine as directed for *Melting Range or Temperature, Procedure for Class Ia*, but heat at a rate of rise of 1°/min until the melting is complete.

Residue on Ignition, Appendix IIC Ignite 2 g as directed in the general method.

Water Determine by the *Karl Fischer Titrimetric Method*, Appendix IIB.

Packaging and Storage Store in tight containers protected from light, preferably at a temperature not exceeding 38°.

Sorbic Acid

2,4-Hexadienoic Acid

$$CH_3CH=CHCH=CHCOOH$$

$C_6H_8O_2$ Formula wt 112.13

INS: 200 CAS: [110-44-1]

DESCRIPTION

A white, free-flowing powder with a characteristic odor. It is slightly soluble in water. One g dissolves in about 10 mL of alcohol and in about 20 mL of ether.

Functional Use in Foods Preservative; mold inhibitor.

REQUIREMENTS

Identification

A. To 2 mL of a 1 in 10 solution of the sample in alcohol add a few drops of bromine TS. The color is discharged.

B. A 1 in 400,000 solution in isopropanol exhibits an absorbance maximum at 254 ± 2 nm.

Assay Not less than 99.0% and not more than 101.0% of $C_6H_8O_2$, calculated on the anhydrous basis.

Heavy Metals (as Pb) Not more than 10 mg/kg.

Melting Range Between 132° and 135°.

Residue on Ignition Not more than 0.2%.

Water Not more than 0.5%.

TESTS

Assay Dissolve about 250 mg, accurately weighed, in 50 mL of anhydrous methanol that previously has been neutralized with 0.1 N sodium hydroxide, add phenolphthalein TS, and titrate with 0.1 N sodium hydroxide to the first pink color that persists for at least 30 s. Each mL of 0.1 N sodium hydroxide is equivalent to 11.21 mg of $C_6H_8O_2$.

Sorbitan Monostearate

INS: 491 CAS: [1338-41-6]

DESCRIPTION

A mixture of partial stearic and palmitic acid esters of sorbitol and its mono- and dianhydrides. It is manufactured by reacting edible commercial stearic acid (usually containing associated fatty acids, chiefly palmitic) with sorbitol. It is a light cream to tan colored, hard, waxy solid with a bland odor and taste. It is soluble at temperatures above its melting point in toluene, dioxane, ether, ethanol, methanol, and aniline. It is insoluble in cold water, and in mineral spirits and acetone, but is dispersible in warm water and soluble, with haze, above 50° in mineral oil and in ethyl acetate.

Functional Use in Foods Emulsifier; stabilizer; defoaming agent.

REQUIREMENTS

Identification

A. The fatty acid residue obtained in the *Assay* has an *Acid Value* between 200 and 215 (*Method I*, Appendix VII) and an *Iodine Value* not greater than 4.

B. *Sample Solution* Transfer 500 mg of the polyols obtained in the *Assay* into a 2-mL volumetric flask, dissolve in and dilute to volume with water, and mix.

Standard Solution Transfer 25 mg each of sorbitol, of USP 1,4-Sorbitan Reference Standard, and of USP Isosorbide Reference Standard into a 1-mL volumetric flask, dissolve in and dilute to volume with water, and mix.

Procedure Using micropipets, separately spot 2 µL each of the *Sample Solution* and of the *Standard Solution* on a thin-layer chromatographic plate coated with a 0.25-mm layer of chromatographic silica gel. Allow the spots to dry, then place the plate in a suitable chromatographic chamber containing a mixture of 100 volumes of acetone and 2 volumes of acetic acid as the developing solvent, and develop by ascending chromatography until the solvent front has traveled about 15 cm.

Remove the plate from the chamber, dry thoroughly in air, and spray evenly with sulfuric acid solution (1 in 2) until the surface is uniformly wet.

Caution: Do not overspray.

Immediately place the sprayed plate on a hot plate maintained at 200° in a hood. Char until white fumes of sulfur trioxide cease, and cool to room temperature. The spots from the sample are located at the same R_f values as those of the polyols from the standard. The approximate R_f values are sorbitol, 0.07; 1,4-sorbitan, 0.40; and isosorbide, 0.77.

Assay Not less than 27.0 g and not more than 34.0 g of polyols (as sorbitol and its mono- and dianhydrides) per 100 g of sample, and not less than 68 g and not more than 76 g of fatty acids per 100 g of sample, calculated on the anhydrous basis.

Acid Value Between 5 and 10.

Heavy Metals (as Pb) Not more than 10 mg/kg.

Hydroxyl Value Between 235 and 260.

Saponification Value Between 147 and 157.

Water Not more than 1.5%.

TESTS

Assay Transfer about 25 g of the sample, accurately weighed, into a 500-mL round-bottom boiling flask, add 250 mL of alcohol and 7.5 g of potassium hydroxide, and mix. Connect a suitable condenser to the flask, reflux the mixture for 1 to 2 h, then transfer to an 800-mL beaker, rinsing the flask with about 100 mL of water and adding it to the beaker. Heat on a steam bath to evaporate the alcohol, adding water occasionally to replace the alcohol, and evaporate until the odor of alcohol can no longer be detected. Adjust the final volume to about 250 mL with hot water. Neutralize the soap solution with dilute sulfuric acid (1 in 2), add 10% in excess, and heat, while stirring, until the fatty acid layer separates. Transfer the fatty acids to a 500-mL separator, wash with three or four 20-mL portions of hot water to remove polyols, and combine the washings with the original aqueous polyol layer from the saponification. Extract the combined aqueous layer with three 20-mL portions of petroleum ether, add the extracts to the fatty acid layer, evaporate to dryness in a tared dish, cool, and weigh.

Neutralize the polyol solution with a 1 in 10 solution of potassium hydroxide to pH 7 using a suitable pH meter. Evaporate this solution to a moist residue, and separate the polyols from the salts by several extractions with hot alcohol. Evaporate the alcohol extracts on a steam bath to dryness in a tared dish, cool, and weigh. Avoid excessive drying and heating.

Acid Value Determine as directed for *Acid Value, Method II*, under *Fats and Related Substances*, Appendix VII.

Heavy Metals Prepare and test a 2-g sample as directed in *Method II* under the *Heavy Metals Test*, Appendix IIIB, using 20 µg of lead ion (Pb) in the control (*Solution A*).

Hydroxyl Value Determine as directed under *Method II* in the general method, Appendix VII.

Saponification Value Determine as directed for *Saponification Value* under *Fats and Related Substances*, Appendix VII, using about 4 g, accurately weighed.

Water Determine by the *Karl Fischer Titrimetric Method*, Appendix IIB.

Packaging and Storage Store in well-closed containers.

Sorbitol

D-Sorbitol; D-Glucitol; D-Sorbite; 1,2,3,4,5,6-Hexanehexol

$C_6H_{14}O_6$ Formula wt 182.17

INS: 420 CAS: [50-70-4]

DESCRIPTION

White, hygroscopic powder, flakes, or granules having a sweet taste. Its density is about 1.49. One g dissolves in about 0.45 mL of water. It is slightly soluble in alcohol, in methanol, and in acetic acid. Sorbitol can exist in any of several crystalline forms with melting points ranging from 89° to 101°.

Functional Use in Foods Humectant; texturizing agent; sequestrant; nutritive sweetener.

REQUIREMENTS

Identification The retention time of the major peak in the chromatogram of the *Assay Solution* corresponds to that in the chromatogram of the *Standard Preparation* obtained in the *Assay*.

Assay Not less than 91.0% and not more than 100.5% of Sorbitol ($C_6H_{14}O_6$), calculated on the anhydrous basis.

Arsenic (as As) Not more than 3 mg/kg.

Chloride Not more than 0.005%.

Heavy Metals (as Pb) Not more than 5 mg/kg.

Lead Not more than 1 mg/kg.

Reducing Sugars Not more than 0.30%.

Residue on Ignition Not more than 0.1%.

Sulfate Not more than 0.01%.

Total Sugars Not more than 1.0%.

Water Not more than 1.0%.

TESTS

Assay

Mobile Phase Use degassed water.

Standard Solution Dissolve an accurately weighed quantity of USP Sorbitol Reference Standard in water to obtain a solution with a known concentration of about 4.8 mg/mL.

Resolution Solution Dissolve USP Mannitol Reference Standard and USP Sorbitol Reference Standard in water to obtain a solution with concentrations of about 4.8 mg/mL of each.

Assay Solution Transfer about 240 mg of the sample, accurately weighed, to a 50-mL volumetric flask, dilute with water to volume, and mix.

Chromatographic System Use a high-pressure liquid chromatograph equipped with a refractive index detector that is maintained at constant temperature and a 7.5-mm × 30-cm column packed with a strong cation-exchange resin consisting of sulfonated, cross-linked styrene–divinylbenzene copolymer in the calcium form that is about 9 μm in diameter (Sugar Pak, or equivalent) and that is maintained at 30° ± 2°. The flow rate is about 0.2 mL/min. Chromatograph the *Standard Solution*, and record the peak responses. The relative standard deviation for three replicate injections is not more than 2.0%. Chromatograph the *Resolution Solution*: The resolution, R, between the sorbitol and mannitol peaks is not less than 2.0.

Procedure Separately inject equal volumes, about 20 μL each, of the *Assay Solution* and the *Standard Solution* into the chromatograph, and record the peak responses for the major peaks. Calculate the quantity, in mg, of $C_6H_{14}O_6$ in the Sorbitol taken by the formula

$$50C(r_U/r_S),$$

in which C is the concentration, in mg/mL, of USP Sorbitol Reference Standard in the *Standard Solution*, and r_U and r_S are the peak responses obtained with the *Assay Solution* and the *Standard Solution*, respectively.

Arsenic A *Sample Solution* prepared as directed for organic compounds meets the requirements of the *Arsenic Test*, Appendix IIIB.

Chloride, Appendix IIIB A 1.5-g portion shows no more chloride than corresponds to 0.10 mL of 0.020 N hydrochloric acid.

Heavy Metals A solution of 4 g in 25 mL of water meets the requirements of the *Heavy Metals Test*, Appendix IIIB, using 20 μg of lead ion (Pb) in the control (*Solution A*).

Lead Determine as directed for *Flame Atomic Absorption Spectrophotometric Method* under the *Lead Limit Test*, Appendix IIIB, using a 10-g sample.

Reducing Sugars Dissolve 7.0 g in 35 mL of water in a 400-mL beaker, and mix. Add 50 mL of alkaline cupric tartrate TS, cover the beaker with glass, heat the mixture at such a rate that it comes to a boil in about 4 min, and boil for exactly 2 min. Collect the precipitated cuprous oxide in a tared, sintered-glass crucible previously washed with hot water, alcohol, and ether and dried at 100° for 30 min. Thoroughly wash the collected cuprous oxide on the filter with hot water, then with 10 mL of alcohol, and finally with 10 mL of ether, and dry at 100° for 30 min. The weight of the cuprous oxide does not exceed 50 mg.

Residue on Ignition, Appendix IIC Ignite 2 g as directed in the general method.

Sulfate, Appendix IIIB A 1.0-g portion shows no more sulfate than corresponds to 0.10 mL of 0.020 N sulfuric acid.

Total Sugars Transfer 2.1 g into a 250-mL flask fitted with a ground-glass joint, add 40 mL of approximately 0.1 N hydrochloric acid, attach a reflux condenser, and reflux for 4 h. Transfer the solution to a 400-mL beaker, rinsing the flask with about 10 mL of water, neutralize with 6 N sodium hydroxide, and continue as directed under *Reducing Sugars*, beginning with "Add 50 mL of alkaline cupric tartrate TS. . . ." The weight of the cuprous oxide does not exceed 50 mg.

Water Determine by the *Karl Fischer Titrimetric Method*, Appendix IIB.

Packaging and Storage Store in tight containers.

Sorbitol Solution

INS: 420

DESCRIPTION

A water solution of sorbitol ($C_6H_{14}O_6$) containing a small amount of mannitol and other isomeric polyhydric alcohols. It is a clear, colorless, syrupy liquid having a sweet taste. It is neutral to litmus. It is miscible with water, with glycerin, and with propylene glycol.

Functional Use in Foods Humectant; texturizing agent; sequestrant; nutritive sweetener.

REQUIREMENTS

Identification The retention time of the major peak in the chromatogram of the *Assay Solution* corresponds to that in the chromatogram of the *Standard Preparation* obtained in the *Assay*.

Assay Not less than 64.0% of sorbitol ($C_6H_{14}O_6$).
Arsenic (as As) Not more than 3 mg/kg.
Chloride Not more than 0.0035%.
Heavy Metals (as Pb) Not more than 5 mg/kg.
Lead Not more than 1 mg/kg.
Reducing Sugars Not more than 0.21%.
Refractive Index Between 1.455 and 1.465 at 20°.
Residue on Ignition Not more than 0.1%.
Specific Gravity Not less than 1.285.
Sulfate Not more than 0.008%.
Total Sugars Not more than 0.70%.
Water Between 28.5% and 31.5%.

TESTS

Assay

Mobile Phase, Standard Solution, Resolution Solution, and *Chromatographic System* Proceed as directed under *Assay* in the monograph for *Sorbitol*.

Assay Solution Transfer an accurately weighed portion of Sorbitol Solution, equivalent to about 240 mg of sorbitol, to a 50-mL volumetric flask, dilute with water to volume, and mix.

Procedure Proceed as directed for *Procedure* under *Assay* in the monograph for *Sorbitol*, and calculate the quantity, in mg, of $C_6H_{14}O_6$ in the portion of Solution taken by the formula therein given.

Arsenic A *Sample Solution* prepared as directed for organic compounds meets the requirements of the *Arsenic Test*, Appendix IIIB.

Chloride, Appendix IIIB A 2.0-g portion shows no more chloride than corresponds to 0.10 mL of 0.020 *N* hydrochloric acid.

Heavy Metals A mixture of 4 g with 25 mL of water meets the requirements of the *Heavy Metals Test*, Appendix IIIB, using 20 μg of lead ion (Pb) in the control (*Solution A*).

Lead Proceed as directed under *Lead* in the monograph for *Sorbitol*.

Reducing Sugars Transfer 10.0 g, accurately weighed, to a 400-mL beaker, with the aid of 35 mL of water, and mix. Add 50 mL of alkaline cupric tartrate TS, cover the beaker with a watch glass, heat the mixture at such a rate that it comes to a boil in approximately 4 min, and boil for exactly 2 min. Collect the precipitated cuprous oxide in a tared, sintered-glass filter crucible previously washed successively with hot water, alcohol, and ether and dried at 100° for 30 min. Thoroughly wash the collected cuprous oxide on the filter with hot water, then with 10 mL of alcohol, and finally with 10 mL of ether, and dry at 100° for 1 h. The weight of the cuprous oxide does not exceed 50 mg.

Refractive Index, Appendix IIB Determine with an Abbé or other refractometer of equal or greater accuracy.

Residue on Ignition, Appendix IIC Ignite 2 g as directed under *Method II* (for liquids).

Specific Gravity Determine by any reliable method (see *General Provisions*).

Sulfate, Appendix IIIB A 1.37-g portion shows no more sulfate than corresponds to 0.10 mL of 0.020 *N* sulfuric acid.

Total Sugars Transfer 3.0 g into a 250-mL flask fitted with a ground-glass joint, add 40 mL of approximately 0.1 *N* hydrochloric acid, attach a reflux condenser, and reflux for 4 h. Transfer the solution to a 400-mL beaker, rinsing the flask with about 10 mL of water, neutralize with 6 *N* sodium hydroxide, and continue as directed under *Reducing Sugars*, beginning with "Add 50 mL of alkaline cupric tartrate TS. . . ." The weight of the cuprous oxide does not exceed 50 mg.

Water Determine by the *Karl Fischer Titrimetric Method*, Appendix IIB.

Packaging and Storage Store in well-closed containers.

Soybean Oil (Unhydrogenated)

DESCRIPTION

A light amber-colored oil that is obtained from the seed of the legume *Glycine max*, usually by solvent extraction. It is refined, bleached, and deodorized to substantially remove free fatty acids, phospholipids, color, odor and flavor components, and miscellaneous other non-oil materials. It is a liquid at 21° to 27° and remains so even at refrigerator temperatures (2° to 4°). It is free from visible foreign matter at 21° to 27°.

Functional Use in Foods Coating agent; emulsifying agent; formulation aid; texturizer.

REQUIREMENTS

Identification Soybean oil exhibits the following composition profile of fatty acids as determined under *Fatty Acid Composition*, Appendix VII.

Fatty Acid:	<14	14:0	16:0	16:1	18:0	18:1	18:2
Weight % (Range):	<0.1	<0.5	7.0–12	<0.5	2.0–5.5	19–30	48–58

Fatty Acid:	18:3	20:0	20:1	22:0	22:1	24:0
Weight % (Range):	5–10	<1.0	<1.0	<0.5	<0.1	<0.3

Arsenic (as As) Not more than 0.5 mg/kg.
Cold Test Passes test.
Color (AOCS-Wesson) Not more than 20 yellow/2.0 red.
Free Fatty Acids (as oleic acid) Not more than 0.1%.
Iodine Value Between 120 and 143.
Lead Not more than 0.1 mg/kg.
Peroxide Value Not more than 10 meq/kg.
Stability (Active Oxygen Method) Not less than 7 h.
Unsaponifiable Matter Not more than 1.5%.
Water Not more than 0.1%.

TESTS

Arsenic A *Sample Solution* prepared using 2 g of sample, accurately weighed, meets the requirements of the *Arsenic Test*, Appendix IIIB. The absorbance due to any red color from the solution of the sample does not exceed that produced by 1.0 mL of *Standard Arsenic Solution* (1 μg As) when treated in the same manner and under the same conditions as the sample.

Cold Test Proceed as directed under *Cold Test*, Appendix VII.

Color Proceed as directed for *Color* (AOCS-Wesson) under *Fats and Related Substances*, Appendix VII.

Free Fatty Acids Proceed as directed under *Free Fatty Acids*, Appendix VII, using the following equivalence factor (*e*) in the formula given in the procedure:

Free fatty acids as oleic acid, *e* = 28.2.

Iodine Value Proceed as directed under *Iodine Value*, Appendix VII.

Lead Determine as directed under *Method II* in the *Atomic Absorption Spectrophotometric Graphite Furnace Method* under the *Lead Limit Test*, Appendix IIIB, using a 3-g sample.

Peroxide Value Proceed as directed under *Peroxide Value*, in the monograph for *Hydroxylated Lecithin*. However, after the addition of saturated potassium iodide and mixing, instead of allowing the solution to stand for 10 min, mix the solution for 1 min and begin the titration immediately.

Stability (Active Oxygen Method) Proceed as directed under *Stability*, Appendix VII.

Unsaponifiable Matter   Proceed as directed under *Unsaponifiable Matter*, Appendix VII.
Water   Proceed as directed under *Water Determination*, Appendix IIB. However, in place of 35 to 40 mL of methanol, use 50 mL of chloroform to dissolve the sample.

Packaging and Storage   Store in well-closed containers.

Spearmint Oil

CAS: [8008-79-5]

DESCRIPTION

The volatile oil obtained by steam distillation from the fresh overground parts of the flowering plant *Mentha spicata* L. (Common Spearmint), or of *Mentha cardiaca* Gerard ex Baker (Scotch Spearmint) (Fam. *Labiatae*). It is a colorless, yellow, or greenish yellow liquid having the characteristic odor and taste of spearmint. It may be rectified by distillation.

Functional Use in Foods   Flavoring agent.

REQUIREMENTS

Labeling   Indicate whether it is natural or rectified.
Identification   The infrared absorption spectrum of the sample exhibits relative maxima (that may vary in intensity) at the same wavelengths (or frequencies) as those shown in the respective spectrum in the section on *Infrared Spectra (Series A: Essential Oils)*, using the same test conditions as specified therein.
Assay   Not less than 55.0%, by volume, of ketones.
Angular Rotation   Between −48° and −59°.
Heavy Metals (as Pb)   Passes test.
Reaction   Passes test.
Refractive Index   Between 1.484 and 1.491 at 20°.
Solubility in Alcohol   Passes test.
Specific Gravity   Between 0.917 and 0.934.

TESTS

Assay   Proceed as directed under *Aldehydes and Ketones—Neutral Sulfite Method*, Appendix VI.
Angular Rotation   Determine in a 100-mm tube as directed under *Optical (Specific) Rotation*, Appendix IIB.
Heavy Metals   Shake 10 mL of the oil with an equal volume of water to which 1 drop of hydrochloric acid has been added, and pass hydrogen sulfide through the mixture until it is saturated. No darkening in color is produced in either the oil or the water.
Reaction   A recently prepared solution of the sample in 80% alcohol is neutral or only slightly acid to moistened litmus paper.
Refractive Index, Appendix IIB   Determine with an Abbé or other refractometer of equal or greater accuracy.

Solubility in Alcohol   Proceed as directed in the general method, Appendix VI. One mL dissolves in 1 mL of 80% alcohol. On further dilution, the solution may become turbid.
Specific Gravity   Determine by any reliable method (see *General Provisions*).

Packaging and Storage   Store in full, tight containers in a cool place protected from light.

Spice Oleoresins

DESCRIPTION

Spice Oleoresins used in foods are derived from spices and contain the total sapid, odorous, and related characterizing principles normally associated with the respective spices. The oleoresins are produced by one of the following processes: (1) by extraction of the spice with any suitable solvent or solvents, in combination or sequence, followed by removal of the solvent or solvents in conformance with applicable residual solvent regulations (see *General Requirements* below), or (2) by removal of the volatile portion of the spice by distillation, followed by extraction of the nonvolatile portion, which after solvent removal is combined with the total volatile portion.

Spice Oleoresins are frequently used in commerce with added suitable food-grade diluents, preservatives, antioxidants, and other substances consistent with good manufacturing practice, as provided for under Added Substances, see *General Provisions*. When added substances are used, they must be declared on the label in accordance with current U.S. regulations or with the regulations of other countries that recognize the *Food Chemicals Codex*.

The Spice Oleoresins covered by this monograph are

Oleoresin Angelica Seed   Obtained by the solvent extraction of the dried seed of *Angelica archangelica* Linnaeus as a dark brown or green liquid.
Oleoresin Anise   Obtained by the solvent extraction of the dried ripe fruit of *anise, Pimpinella anisum* Linnaeus, or star anise, *Illicium verum* Hooker as a dark brown or green liquid.
Oleoresin Basil   Obtained by the solvent extraction of the dried plant of *Ocimum basilicum* Linnaeus as a dark brown or green semisolid.
Oleoresin Black Pepper   Obtained by the solvent extraction of the dried fruit of *Piper nigrum* Linnaeus as a dark green, olive green, or olive drab extract usually consisting of an upper oily layer and a lower crystalline layer. It may appear as a homogeneous emulsion if examined shortly after the oleoresin has been homogenized, but the product separates on standing. It may be decolorized by partial removal of chlorophyll.
Oleoresin Capsicum   Obtained by the solvent extraction of dried pods of *Capsicum frutescens* Linnaeus or *Capsicum annum* Linnaeus as a clear red to dark red, somewhat viscous liquid of characteristic odor, flavor, and bite. It may be decolorized through good manufacturing practice. It is partly soluble in

alcohol (with oily separation and/or sediment) and is soluble in most fixed oils. The bite is usually standardized according to the label declaration.

Oleoresin Caraway Obtained by the solvent extraction of the dried seeds of *Carum carvi* Linnaeus as a green yellow to brown liquid.

Oleoresin Cardamom Obtained by the solvent extraction of the dried seeds of *Elettaria cardamomum* Maton as a dark brown or green liquid.

Oleoresin Celery Obtained by the solvent extraction of the dried ripe seed of *Apium graveolens* Linnaeus as a dark green, somewhat viscous, nonhomogeneous liquid with the characteristic odor and flavor of celery. It may be decolorized by the partial removal of chlorophyll. It is partly soluble in alcohol (with oily separation), and is soluble in most fixed oils.

Oleoresin Coriander Obtained by the solvent extraction of the dried seeds of *Coriandrum sativum* Linnaeus as a brown yellow to green liquid.

Oleoresin Cubeb Obtained by the solvent extraction of the dried fruit of *Piper cubeba* Linnaeus as a green or green brown liquid.

Oleoresin Cumin Obtained by the solvent extraction of the dried seeds of *Cuminum cyminum* Linnaeus as a brown to yellow green liquid.

Oleoresin Dillseed Obtained by the solvent extraction of the dried seeds of *Anethum graveolens* Linnaeus as a brown or green liquid.

Oleoresin Fennel Obtained by the solvent extraction of the dried fruit of *Foeniculum vulgare* P. Miller as a brown green liquid.

Oleoresin Ginger Obtained by the solvent extraction of the dried rhizomes of *Zingiber officinale* Roscoe as a dark brown, viscous to highly viscous liquid with the characteristic odor and flavor of ginger. It is soluble in alcohol (with sediment).

Oleoresin Hop Obtained by the solvent extraction of the dried membranous cones of the female hop plants of *Humulus lupulus* L. or *Humulus americanus* Nutt. (Fam. *Moraceae*), using a food-grade solvent such as liquid carbon dioxide. It is light golden to black in color, and in consistency, a liquid to semisolid having a characteristic odor. It is soluble in methanol and is slightly soluble in acidified water. It may be reduced with sodium borohydride or with hydrogen and palladium catalyst. It conforms to U.S. Food and Drug Administration regulations pertaining to the specifications for extraction solvents for modified hop extract.

Oleoresin Laurel Leaf Obtained by the solvent extraction of the dried leaves of *Laurus nobilis* Linnaeus as a dark brown or green semisolid.

Oleoresin Marjoram Sweet Obtained by the solvent extraction of the dried herb of the marjoram shrub *Majorana hortensis* Moench as a dark green to brown viscous liquid or semisolid.

Oleoresin Origanum Obtained by the solvent extraction of the dried flowering herb *Origanum* spp. as a dark brown green semisolid.

Oleoresin Paprika Obtained by the solvent extraction of the pods of *Capsicum annuum* Linnaeus as a deep red to deep purplish red, somewhat viscous liquid of characteristic odor and flavor. It frequently occurs as a two-phase mixture. The color is usually standardized according to the label declaration. It is partly soluble in alcohol (with oily separation), and is soluble in most fixed oils.

Oleoresin Parsley Leaf Obtained by the solvent extraction of the dried herb of *Petroselinum crispum* (P. Miller) Nyman ex A.W. Hill as a brown to green liquid.

Oleoresin Parsley Seed Obtained by the solvent extraction of the dried seeds of *Petroselinum crispum* (P. Miller) Nyman ex A.W. Hill as a deep green semiviscous liquid.

Oleoresin Pimenta Berries Obtained by the solvent extraction of the dried fruit of *Pimenta officinalis* Lindl as a brown green to dark green liquid.

Oleoresin Thyme Obtained by the solvent extraction of the dried flowering plant *Thymus vulgaris* Linnaeus or *Thymus zygis* Linnaeus and its var. *gracelis* Boissier as a dark brown to green, viscous semisolid.

Oleoresin Turmeric Obtained by the solvent extraction of the dried rhizomes of *Curcuma longa* Linnaeus as a yellow orange to red brown viscous liquid with a characteristic odor and flavor. The content of curcumin normally varies, and the product is generally standardized according to the label declaration.

Functional Use in Foods Flavoring agent; color (oleoresins paprika and turmeric only).

GENERAL REQUIREMENTS

Identification The volatile oil distilled from an oleoresin is similar in its physical and chemical properties, including its infrared spectrum to that distilled from the spice of the same origin. To obtain the volatile oil from the oleoresin, proceed as directed for *Volatile Oil Content* under *Oleoresins*, Appendix VIII.

Heavy Metals (as Pb) Not more than 0.002%.

Lead Not more than 5 mg/kg.

Residual Solvent *Chlorinated hydrocarbons (total)*: not more than 0.003%; *acetone*: not more than 0.003%; *isopropanol*: not more than 0.003%; *methanol*: not more than 0.005%; *hexane*: not more than 0.0025%.

ADDITIONAL REQUIREMENTS

Oleoresin Angelica Seed *Volatile Oil Content*: between 2 mL and 7 mL/100 g.

Oleoresin Anise *Volatile Oil Content*: between 9 mL and 22 mL/100 g.

Oleoresin Basil *Volatile Oil Content*: between 4 mL and 17 mL/100 g.

Oleoresin Black Pepper *Piperine*: not less than 36%; *Volatile Oil Content*: between 15 mL and 35 mL/100 g.

Oleoresin Capsicum *Scoville Heat Units*: between 100,000 and 2,000,000, as specified on the label.

Oleoresin Caraway *Volatile Oil Content*: between 10 mL and 20 mL/100 g.

Oleoresin Cardamom *Volatile Oil Content*: between 50 mL and 80 mL/100 g.

Oleoresin Celery *Volatile Oil Content*: between 7 mL and 20 mL/100 g.

Oleoresin Coriander *Volatile Oil Content*: between 2 mL and 12 mL/100 g.

Oleoresin Cubeb *Volatile Oil Content*: between 50 mL and 80 mL/100 g.
Oleoresin Cumin *Volatile Oil Content*: between 10 mL and 30 mL/100 g.
Oleoresin Dillseed *Volatile Oil Content*: between 10 mL and 20 mL/100 g.
Oleoresin Fennel *Volatile Oil Content*: between 3 mL and 20 mL/100 g.
Oleoresin Ginger *Volatile Oil Content*: between 18 mL and 35 mL/100 g.
Oleoresin Hop *Volatile Oil Content*: not more than 30 mL/100 g.
Oleoresin Laurel Leaf *Volatile Oil Content*: between 5 mL and 25 mL/100 g.
Oleoresin Marjoram Sweet *Volatile Oil Content*: between 8 mL and 20 mL/100 g.
Oleoresin Origanum *Volatile Oil Content*: between 20 mL and 45 mL/100 g.
Oleoresin Paprika *Color Value*: between 500 and 4500 units, as specified on the label (according to the method of analysis); *Scoville Heat Units (pungency)*: not more than 3,000.
Oleoresin Parsley Leaf *Volatile Oil Content*: between 2 mL and 10 mL/100 g.
Oleoresin Parsley Seed *Volatile Oil Content*: between 2 mL and 7 mL/100 g.
Oleoresin Pimenta Berries *Volatile Oil Content*: between 20 mL and 50 mL/100 g.
Oleoresin Thyme *Volatile Oil Content*: between 5 mL and 12 mL/100 g.
Oleoresin Turmeric *Curcumin (or Color Value equivalent)*: between 1% and 45%, as specified on the label.

TESTS (GENERAL REQUIREMENTS)

Heavy Metals Prepare and test a 1-g sample as directed in *Method II* under the *Heavy Metals Test*, Appendix IIIB, using 20 μg of lead ion (Pb) in the control (*Solution A*).
Lead A *Sample Solution* prepared as directed for organic compounds meets the requirements of the *Lead Limit Test*, Appendix IIIB, using 5 μg of lead ion (Pb) in the control.
Residual Solvent Determine as directed in the general method, Appendix VIII.

TESTS (ADDITIONAL REQUIREMENTS)

Color Value Determine as directed for *Color Value* under *Oleoresins*, Appendix VIII.
Curcumin Determine as directed in the general method, Appendix VIII.
Piperine Determine as directed in the general method, Appendix VIII.
Scoville Heat Units Determine as directed in the general method, Appendix VIII.
Volatile Oil Content Determine as directed for *Volatile Oil Content* under *Oleoresins*, Appendix VIII.

Packaging and Storage Store in full, tight, preferably glass or other suitably lined containers in a cool place protected from light.

Spike Lavender Oil

DESCRIPTION

The volatile oil obtained by steam distillation from the flowers of *Lavandula latifolia*, Vill. (*Lavandula spica*, D.C.) (Fam. *Labiatae*). It is a pale yellow to yellow liquid having a camphoraceous, lavender-like odor. It is soluble in most fixed oils and in propylene glycol. It is slightly soluble in glycerin and in mineral oil.

Functional Use in Foods Flavoring agent.

REQUIREMENTS

Identification The infrared absorption spectrum of the sample exhibits relative maxima (that may vary in intensity) at the same wavelengths (or frequencies) as those shown in the respective spectrum in the section on *Infrared Spectra (Series A: Essential Oils)*, using the same test conditions as specified therein.
Assay Not less than 40.0% and not more than 50.0% of total alcohols, calculated as linalool ($C_{10}H_{18}O$).
Angular Rotation Between −5° and +5°.
Esters Not less than 1.5% and not more than 4.0% of esters, calculated as linalyl acetate ($C_{12}H_{20}O_2$).
Heavy Metals (as Pb) Passes test.
Refractive Index Between 1.463 and 1.468 at 20°.
Solubility in Alcohol Passes test.
Specific Gravity Between 0.893 and 0.909.

TESTS

Assay Proceed as directed under *Linalool Determination*, Appendix VI, using about 1.5 g of the acetylated oil, accurately weighed, for the saponification.
Angular Rotation Determine in a 100-mm tube as directed under *Optical (Specific) Rotation*, Appendix IIB.
Esters Weigh accurately about 10 g, and proceed as directed under *Ester Determination*, Appendix VI, using 98.15 as the equivalence factor (*e*) in the calculation.
Heavy Metals Shake 10 mL of the oil with an equal volume of water to which 1 drop of hydrochloric acid has been added, and pass hydrogen sulfide through the mixture until it is saturated. No darkening in color is produced in either the oil or the water.
Refractive Index, Appendix IIB Determine with an Abbé or other refractometer of equal or greater accuracy.
Solubility in Alcohol Proceed as directed in the general method, Appendix VI. One mL dissolves in 3 mL of 70% alcohol. The solution frequently becomes hazy upon further dilution.
Specific Gravity Determine by any reliable method (see *General Provisions*).

Packaging and Storage Store in full, tight, preferably glass, aluminum, tin-lined, or other suitably lined containers in a cool place protected from light.

Stannous Chloride

SnCl$_2$ Formula wt 189.62
INS: 512   CAS: [7772-99-8]

DESCRIPTION

Stannous Chloride is anhydrous or contains two molecules of water of hydration. It occurs as white or colorless crystals having no odor or a slight odor of hydrochloric acid. It is very soluble in water, and is soluble in alcohol and in glacial acetic acid.

Functional Use in Foods   Reducing agent; antioxidant.

REQUIREMENTS

Identification
     A. To a 1 in 20 solution of the sample in 2.7 N hydrochloric acid add mercuric chloride TS dropwise. A white or grayish white precipitate is formed.
     B. A 1 in 20 solution of the sample gives positive tests for *Chloride*, Appendix IIIA.
Assay   Not less than 99.0% and not more than 101.0% of SnCl$_2$, or not less than 98.0% and not more than 102.2% of SnCl$_2 \cdot$2H$_2$O.
Other Heavy Metals (as Pb)   Not more than 0.01%.
Iron   Not more than 0.005%.
Solubility in Hydrochloric Acid   Passes test.
Substances Not Precipitated by Sulfide   Not more than 0.05%.
Sulfate   Passes test (about 0.003%).

TESTS

Assay   Transfer about 2 g of the sample, accurately weighed, into a 250-mL volumetric flask, dissolve in 15 mL of hydrochloric acid, dilute to volume with water, and mix. Transfer 50.0 mL of this solution into a 500-mL flask, add 5 g of sodium potassium tartrate, and mix. Make the solution alkaline to litmus with a cold saturated solution of sodium bicarbonate, and titrate at once with 0.1 N iodine, using starch TS as the indicator. Each mL of 0.1 N iodine is equivalent to 9.48 mg of SnCl$_2$ or 11.28 mg of SnCl$_2 \cdot$2H$_2$O.
Other Heavy Metals   Dissolve 1 g of the sample in a mixture of 2 mL of hydrochloric acid and 3 mL of nitric acid, and boil until solution is complete and brown fumes are no longer evolved. Cool, and dilute to 50 mL with water. To 10 mL of this solution add 8 mL of sodium hydroxide solution (1 in 10), then cool, and dilute to 40 mL with water. Prepare a control containing 2.0 mL of *Standard Lead Solution* (20 µg Pb), 8 mL of the sodium hydroxide solution, and 30 mL of water. Add 10 mL of hydrogen sulfide TS to each solution. Any color produced in the solution of the sample does not exceed that in the control.
Iron   Add 3 mL of dilute hydrochloric acid (1 in 2) to the residue obtained in the test for *Substances Not Precipitated by Sulfide*, cover with a watch glass, and digest on a steam bath for 15 min. Remove the cover, and evaporate to dryness on the steam bath. Dissolve the residue in a few mL of water and 8 mL of hydrochloric acid, dilute to 100 mL with water, and mix. To 2.0 mL of this solution add 2 mL of hydrochloric acid, 46 mL of water, 40 mg of ammonium persulfate crystals, and 3 mL of ammonium thiocyanate TS. Any red or pink color does not exceed that produced by 2.0 mL of *Iron Standard Solution* (20 µg of Fe) in an equal volume of solution containing the quantities of the reagents used in the test.
Solubility in Hydrochloric Acid   A 5-g portion of the sample dissolves completely in a mixture of 5 mL of hydrochloric acid and 5 mL of water, heating to 40°, if necessary, to effect solution.
Substances Not Precipitated by Sulfide   Transfer about 20 g of the sample, accurately weighed, to a 250-mL beaker, and add 50 mL of a solution prepared by adding 75 mL of bromine carefully to 425 mL of 48% hydrobromic acid. Add 1 mL of sulfuric acid, and mix to effect complete solution. Place the beaker on a hot plate, and volatilize the tin slowly, with gentle boiling, to fumes of sulfur trioxide. Cool, add 30 mL of water, and pass hydrogen sulfide gas through the solution for about 5 min. Filter through Whatman No. 42 filter paper, or equivalent, into a weighed platinum dish, and wash with three small portions of a 1% solution of sulfuric acid saturated with hydrogen sulfide. Evaporate carefully to dryness on a hot plate, and heat in a furnace at 800° ± 25° for 13 min. Cool in a desiccator for at least 30 min, and weigh. Calculate the percentage of substances not precipitated by sulfide by the formula

$$100A/B,$$

in which A and B are the respective weights of the residue and of the sample taken, in g. Retain the residue for the *Iron* test.
Sulfate   Dissolve 5 g of the sample in 5 mL of hydrochloric acid, dilute to 50 mL with water, filter if not clear, and heat the filtrate or clear solution to boiling. Add 5 mL of barium chloride TS, digest in a covered beaker on a steam bath for 2 h, and allow to stand overnight. No precipitate forms.

Packaging and Storage   Store in well-closed containers.

Starter Distillate

Butter Starter Distillate

DESCRIPTION

Starter Distillate is the steam distillate of a culture of one or more species of *Lactococcus lactis* subsp. *diacetylactis* and/or *Leuconostoc cremoris* grown in a medium of skimmed milk that has been fortified with citric acid. It contains more than 97% water and a mixture of organic flavor compounds, principally diacetyl. It is a clear, yellowish, water-soluble liquid.

Functional Use in Foods   Flavoring agent.

REQUIREMENTS

Labeling Indicate the diacetyl content, in mg/mL.
Assay Not less than 90.0% and not more than 110.0% of the labeled amount of diacetyl.
Heavy Metals (as Pb) Not more than 20 mg/kg.
Lead Not more than 5 mg/kg.
Microbial Limits:
    Aerobic Plate Count Not more than 10 per mL.
    Coliform Not more than 10 per mL.
    Listeria monocytogenes Negative in 25 mL.
    Salmonella Negative in 25 mL.
    Yeasts and Molds Not more than 10 per mL.
pH Between 2.8 and 3.8.

TESTS

Assay Transfer an accurately measured volume of Starter Distillate, equivalent to about 25 mg of diacetyl, to a suitable flask. Add 3 drops of phenolphthalein TS, and neutralize the acidity by titrating with 0.05 N sodium hydroxide to a faint pink endpoint. Add 0.25 mL of 30% hydrogen peroxide solution and 3 drops of 0.01% osmic acid. This is prepared by dissolving 1 g of osmium tetroxide in 1 L of water and then making a 1 in 10 dilution.

> *Caution*: Osmium tetroxide and its solutions are toxic. Use proper protective equipment, and avoid contact with the eyes, skin, and clothing.

Mix, cover the flask, and allow it to stand in an incubator held at about 38° for not less than 4 h. Cool to room temperature, and titrate with 0.05 N sodium hydroxide to a faint pink endpoint. Each mL of 0.05 N sodium hydroxide is equivalent to 8.6 mg of diacetyl.
Heavy Metals A solution of 1 g in 25 mL of water meets the requirements of the *Heavy Metals Test*, Appendix IIIB, using 20 μg of lead ion (Pb) in the control (*Solution A*).
Lead A solution of 1 g in 20 mL of water meets the requirements of the *Lead Limit Test*, Appendix IIIB, using 5 μg of lead ion (Pb) in the control.
Microbial Limits:
    Aerobic Plate Count Proceed as directed in Chapter 3 of the FDA *Bacteriological Analytical Manual*, Seventh Edition, 1992.
    Coliform Proceed as directed in Chapter 4 of the FDA *Bacteriological Analytical Manual*, Seventh Edition, 1992.
    Listeria monocytogenes Proceed as directed in Chapter 10 of the FDA *Bacteriological Analytical Manual*, Seventh Edition, 1992.
    Salmonella Proceed as directed in Chapter 5 of the FDA *Bacteriological Analytical Manual*, Seventh Edition, 1992.
    Yeasts and Molds Proceed as directed in Chapter 18 of the FDA *Bacteriological Analytical Manual*, Seventh Edition, 1992.
pH Determine by the *Potentiometric Method*, Appendix IIB.

Packaging and Storage Store in tight containers in a cool place.

Stearic Acid

Octadecanoic Acid

$C_{18}H_{36}O_2$ Formula wt 284.48

CAS: [57-11-4]

DESCRIPTION

A mixture of solid organic acids obtained from fats consisting chiefly of Stearic Acid ($C_{18}H_{36}O_2$) and palmitic acid ($C_{16}H_{32}O_2$). It occurs as a hard, white or faintly yellowish, somewhat glossy and crystalline solid, or as a white or yellowish white powder. It has a slight characteristic odor and taste resembling tallow. Stearic Acid is practically insoluble in water. One g dissolves in about 20 mL of alcohol, in 2 mL of chloroform, and in about 3 mL of ether.

Functional Use in Foods Component in the manufacture of other food-grade additives; lubricant; defoaming agent.

REQUIREMENTS

Acid Value Between 196 and 211.
Heavy Metals (as Pb) Not more than 10 mg/kg.
Iodine Value Not more than 7.
Residue on Ignition Not more than 0.1%.
Saponification Value Between 197 and 212.
Titer (Solidification Point) Between 54.5° and 69°.
Unsaponifiable Matter Not more than 1.5%.
Water Not more than 0.2%.

TESTS

Acid Value Determine as directed for *Acid Value, Method I*, under *Fats and Related Substances*, Appendix VII.
Heavy Metals Prepare and test a 2-g sample as directed in *Method II* under the *Heavy Metals Test*, Appendix IIIB, using 20 μg of lead ion (Pb) in the control (*Solution A*).
Iodine Value Determine by the *Modified Wijs Method*, Appendix VII.
Residue on Ignition, Appendix IIC Ignite 2 g as directed in the general method.
Saponification Value Determine as directed for *Saponification Value* under *Fats and Related Substances*, Appendix VII, using about 3 g, accurately weighed.
Titer (Solidification Point) Determine as directed under *Solidification Point*, Appendix IIB.
Unsaponifiable Matter, Appendix VII Determine as directed in the general method.
Water Determine by the *Karl Fischer Titrimetric Method*, Appendix IIB.

Packaging and Storage Store in well-closed containers.

Stearyl Monoglyceridyl Citrate

DESCRIPTION

A soft, practically tasteless, off-white to tan, waxy solid having a lardlike consistency. It is insoluble in water, but is soluble in chloroform and in ethylene glycol. It is prepared by a controlled chemical reaction from citric acid, monoglycerides of fatty acids (obtained by the glycerolysis of edible fats and oils or derived from fatty acids), and stearyl alcohol.

Functional Use in Foods Emulsion stabilizer.

REQUIREMENTS

Acid Value Between 40 and 52.
Heavy Metals (as Pb) Not more than 10 mg/kg.
Residue on Ignition Not more than 0.1%.
Saponification Value Between 215 and 255.
Total Citric Acid Between 15.0% and 18.0%.
Water Not more than 0.25%.

TESTS

Acid Value Determine as directed for *Acid Value, Method II*, under *Fats and Related Substances*, Appendix VII.
Heavy Metals Prepare and test a 2-g sample as directed in *Method II* under the *Heavy Metals Test*, Appendix IIIB, using 20 μg of lead ion (Pb) in the control (*Solution A*).
Residue on Ignition Ignite 2 g as directed in the general method, Appendix IIC.
Saponification Value Transfer about 1 g, accurately weighed, into a 250-mL Erlenmeyer flask, and add 25 mL of ethylene glycol, 35.0 mL of 0.5 N alcoholic potassium hydroxide, and a few glass beads. Reflux for 1 h, using a water condenser, then rinse the condenser with water, and cool. Add 1 mL of phenolphthalein TS, and titrate with 0.5 N hydrochloric acid. Perform a blank determination (see *General Provisions*), but do not reflux. The difference between the volumes, in mL, of 0.5 N hydrochloric acid consumed in the actual test and in the blank titration, multiplied by 28.05 and divided by the weight, in g, of the sample taken, is the saponification value.

Total Citric Acid

Brominating Solution Dissolve 19.84 g of potassium bromide, 5.44 g of potassium bromate, and 12 g of sodium metavanadate, $NaVO_3$, in water by warming, and dilute to 1000 mL with water. Filter if necessary.

Ferrous Sulfate Solution Dissolve 44 g of ferrous sulfate, $FeSO_4 \cdot 7H_2O$, in 1 N sulfuric acid, dilute to 100 mL with 1 N sulfuric acid, and mix. Use within 5 days of preparation.

Sulfide Solution On the day of use, dissolve 4 g of thiourea in 100 mL of a 1 in 50 solution of sodium borate, $Na_2B_4O_7 \cdot 10H_2O$, and add 2 mL of sodium sulfide TS. Wait 30 min after the addition of the sodium sulfide TS before using.

Standard Solution Transfer about 50 mg of sodium citrate dihydrate, accurately weighed, into a 500-mL volumetric flask, dissolve and dilute to volume with water, and mix. Transfer 15.0 mL of this solution into a 100-mL volumetric flask, dilute to volume with water, and mix. Calculate the concentration (C), in μg/mL, of citric acid in the final solution by the formula

$$(15 \times 1000 \times 0.6533W)/(100 \times 500),$$

in which W is the weight, in mg, of the sodium citrate taken, and 0.6533 is the factor converting sodium citrate dihydrate to citric acid.

Sample Solution Transfer about 250 mg of the sample, accurately weighed, to a 250-mL extraction flask, and add 15 mL of 0.5 N sodium hydroxide, 5 mL of alcohol, and a few glass beads. Connect the flask with a water-cooled condenser, and reflux for 3 h. Immediately cool and neutralize to phenolphthalein TS with 0.5 N hydrochloric acid, then place the flask in an ice bath and add 5 mL of 95% sulfuric acid. Transfer the solution to a 125-mL separator, extract with three 40-mL portions of chloroform, and then extract the combined chloroform extracts in a 250-mL separator with three 10-mL portions of 0.5 N sulfuric acid, adding the acid extracts to a second 250-mL separator. Wash the combined acid extracts with two 60-mL portions of chloroform, and discard the chloroform washes. Filter the acid solution into a 500-mL volumetric flask, neutralize slowly with 6 N sodium carbonate, and dilute to volume with water. Transfer 10.0 mL of this solution into a 100-mL volumetric flask, dilute to volume with water, and mix. Each mL of the final solution contains approximately 10 μg of citric acid.

Procedure Pipet 2 mL each of the *Standard Solution* and of the *Sample Solution* into separate 40- or 45-mL, glass-stoppered centrifuge tubes, and add 3 mL of water to each tube. Place 5 mL of water in a third tube for the reagent blank. Place the tubes in an ice bath, add 5 mL of 95% sulfuric acid, mix thoroughly, and allow to stand for exactly 5 min. Remove the tubes from the ice bath, and allow them to come to room temperature during the next 5 min. To each tube add 5 mL of the *Brominating Solution*, then insert the stoppers, invert the tubes once or twice, and heat in a water bath at 30° for 20 min. Remove the tubes, add 1.5 mL of *Ferrous Sulfate Solution*, invert again, and allow to stand for 5 min, shaking occasionally to ensure complete reduction of the excess free bromine in the tubes. Add 6.5 mL of petroleum ether, shake for 2 or 3 min, and remove the water layer with a syringe. Wash the ether solutions with 15 mL of water, then remove the water and filter the ether extracts into the original centrifuge tubes, which have been previously rinsed with the *Sulfide Solution*. Filter each ether extract through a tight plug of glass wool onto which has been placed a sufficient amount of anhydrous sodium sulfate to remove the last traces of water from the ether. Place 5.0 mL of the filtrate in a clean, dry centrifuge tube, add 3 mL of *Sulfide Solution*, shake vigorously for 1.5 min, and centrifuge. Decant about 0.5 mL of the supernatant ether layer from each tube, then carefully transfer the ether solutions into 1-cm cells and determine the absorbance of the extracts obtained from the *Standard Solution* and the *Sample Solution* at 500 nm with a suitable spectrophotometer, using the reagent blank in the reference cell. Calculate the quantity, in mg, of citric acid in the sample taken by the formula

$$5C \times A_U/A_S,$$

in which C is the exact concentration, in μg/mL, of citric acid in the *Standard Solution*; A_U is the absorbance of the solution from the *Sample Solution*; and A_S is the absorbance of the solution from the *Standard Solution*.

Water   Determine by the *Karl Fischer Titrimetric Method*, Appendix IIB.

Packaging and Storage   Store in well-closed containers.

Succinic Acid

Butanedioic Acid

$$\text{HOOCCH}_2\text{CH}_2\text{COOH}$$

$C_4H_6O_4$ Formula wt 118.09
INS: 363   CAS: [110-15-6]

DESCRIPTION

Colorless or white, odorless crystals. One g dissolves in 13 mL of water at 25°, in 1 mL of boiling water, in 18.5 mL of alcohol, and in 20 mL of glycerin.

Functional Use in Foods   Buffer; neutralizing agent; miscellaneous and general purpose.

REQUIREMENTS

Identification   Place a drop of a saturated solution of the sample in a micro test tube, and add a drop of a 0.5% solution of ammonium chloride and several mg of zinc powder. Cover the mouth of the tube with a disk of filter paper moistened with a solution in hexane of 5% *p*-dimethylaminobenzaldehyde and 20% trichloroacetic acid. Heat with a small flame for about 1 min. A pink to red violet stain appears on the paper.
Assay   Not less than 99.0% and not more than 100.5% of $C_4H_6O_4$.
Heavy Metals (as Pb)   Not more than 10 mg/kg.
Melting Range   Between 185.0° and 190.0°.
Residue on Ignition   Not more than 0.025%.

TESTS

Assay   Dissolve about 250 mg, accurately weighed, in 25 mL of recently boiled and cooled water, add phenolphthalein TS, and titrate with 0.1 N sodium hydroxide to the first appearance of a faint pink color that persists for at least 30 s. Each mL of 0.1 N sodium hydroxide is equivalent to 5.905 mg of $C_4H_6O_4$.
Heavy Metals   A solution of 1.5 g in 25 mL of water meets the requirements of the *Heavy Metals Test*, Appendix IIIB, using 15 μg of lead ion (Pb) in the control (*Solution A*).
Melting Range   Determine as directed for *Melting Range or Temperature*, Appendix IIB.
Residue on Ignition   Ignite 8 g as directed in the general method, Appendix IIC.

Packaging and Storage   Store in well-closed containers.

Succinylated Monoglycerides

INS: 472g

DESCRIPTION

A mixture of succinic acid esters of mono- and diglycerides produced by the succinylation of a product obtained by the glycerolysis of edible fats and oils, or by the direct esterification of glycerol with edible fat-forming fatty acids. It occurs as a waxy solid having an off-white color and a bland taste, melting at about 60°. It is soluble in warm methanol, in ethanol, and in *n*-propanol.

Functional Use in Foods   Emulsifier; dough conditioners.

REQUIREMENTS

Acid Value   Between 70 and 120.
Bound Succinic Acid   Not less than 14.8%.
Free Succinic Acid   Not more than 3%.
Heavy Metals (as Pb)   Not more than 10 mg/kg.
Hydroxyl Value   Between 138 and 152.
Iodine Value   Not more than 3.
Total Succinic Acid   Between 14.8% and 25.6%.

TESTS

Acid Value   Determine as directed for *Acid Value, Method I*, under *Fats and Related Substances*, Appendix VII.
Free and Bound Succinic Acid
0.02 N Sodium Hydroxide in Methanol   Dissolve 4.0 g of sodium hydroxide in 1000 mL of anhydrous methanol. Transfer 200.0 mL of this solution to a 1000-mL volumetric flask, dilute to volume with anhydrous methanol, and mix. Standardize the solution against dried succinic acid, using phenolphthalein TS as the indicator.
Procedure   Transfer about 125 mg of the sample, accurately weighed, into a 250-mL separator containing 100 mL of benzene, and dissolve the sample by heating the separator with warm water. Treat the sample and a blank, consisting of 100 mL of benzene in another separator, in the same manner as follows: Cool the contents of the separator, add 50 mL of water, and mix by inverting the separator about 20 times. Allow to stand for about 15 min, and then transfer the aqueous layer into a 125-mL Erlenmeyer flask. Add 10 mL of water to the separator, wash the benzene layer by inverting the separator five times, and add the washings to the 125-mL flask. To the flask add five drops of phenolphthalein TS, and titrate with *0.02 N Sodium Hydroxide in Methanol*. Perform a blank determination, and record the net volume of alkali, in mL, as V_1.

Transfer the benzene layer into a 500-mL round-bottom flask, and rinse the separator with 10 mL of benzene. Add a few boiling chips to the flask, and evaporate the benzene, preferably on a thin-film evaporator, under partial vacuum at about 60°. Dissolve the residue in the flask in 10 mL of methanol, add 10 mL of water and five drops of phenolphthalein TS, and titrate with *0.02 N Sodium Hydroxide in Methanol*. Perform a blank determination, and record the net volume of alkali, in mL, as V_2.

Calculate the weight, in mg, of free succinic acid in the sample by the formula

$$118.1 \times N \times V_1/2,$$

and calculate the weight, in mg, of bound succinic acid in the sample by the formula

$$118.1 \times N \times V_2/2,$$

in which N is the exact normality of the sodium hydroxide solution.

Heavy Metals Prepare and test a 2-g sample as directed in *Method II* under the *Heavy Metals Test*, Appendix IIIB, using 20 µg of lead ion (Pb) in the control (*Solution A*).

Hydroxyl Value Determine as directed under *Method II* in the general method, Appendix VII.

Iodine Value Determine by the *Modified Wijs Method*, Appendix VII.

Total Succinic Acid The sum of the *Free Succinic Acid* and the *Bound Succinic Acid* represents the *Total Succinic Acid*.

Packaging and Storage Store in well-closed containers.

Sucralose[1]

1,6-Dichloro-1,6-dideoxy-β-D-fructofuranosyl-4-chloro-4-deoxy-α-D-galactopyranoside; 4,1′,6′-Trichlorogalactosucrose

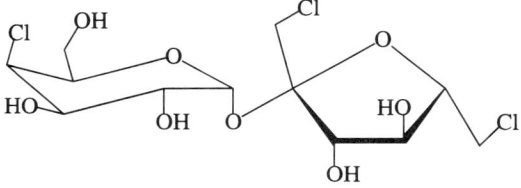

$C_{12}H_{19}Cl_3O_8$ Formula wt 397.64

INS: 955 CAS: [56038-13-2]

DESCRIPTION

A white to off-white, practically odorless, crystalline powder having a sweet taste. It is freely soluble in water, in methanol, and in alcohol and is slightly soluble in ethyl acetate.

Functional Use in Foods Non-nutritive sweetener; flavor enhancer.

---

[1] At press time, Sucralose had not been approved for use in food in the United States; however, since September 25, 1991, the use of Sucralose in food has been permitted in Canada.

REQUIREMENTS

Identification

A. The infrared absorption spectrum of a potassium bromide dispersion sample exhibits relative maxima (that may vary in intensity) at the same wavelengths as that of a similar preparation of Sucralose Standard for analytical use.[2]

B. The retention time of the major peak (excluding the solvent peak) in the liquid chromatogram of the *Sample Preparation* is the same as that of the *Standard Preparation* obtained in the *Assay*.

C. The R_f value of the major spot in the thin-layer chromatogram of the *Test Preparation* is the same as that of the *Standard Preparation* obtained in the test for *Related Substances*.

Assay Not less than 98.0% and not more than 102.0% of $C_{12}H_{19}Cl_3O_8$, calculated on the anhydrous basis.

Heavy Metals (as Pb) Not more than 10 mg/kg.

Hydrolysis Products Passes test.

Methanol Not more than 0.1%.

Related Substances Passes test.

Residue on Ignition Not more than 0.7%.

Specific Rotation $[\alpha]_D^{20°}$: Between +84.0° to +87.5°, calculated on the anhydrous basis.

Water Not more than 2.0%.

TESTS

Assay

Mobile Phase Add 150 mL of acetonitrile (HPLC grade and filtered through a 0.45-µm filter) to 850 mL of water (glass-distilled or equivalent and filtered through a 0.45-µm filter). Mix, and de-gas thoroughly.

Chromatographic System Fit a high-pressure liquid chromatographic system, operated at room temperature, with an 8-mm × 10-cm, 5-µm RadPakC$_{18}$ (or equivalent) reverse-phase column. Maintain the *Mobile Phase* at a pressure and flow rate (typically 1.5 mL/min) capable of giving the required elution time (see *System Suitability Test*). Use a refractive index detector.

Standard Preparation Weigh accurately about 25 mg of Sucralose Standard for analytical use into a 25-mL volumetric flask. Dissolve, and dilute to volume with *Mobile Phase*. Filter the solution through a 0.45-µm filter.

Sample Preparation Weigh accurately about 25 mg of test sample into a 25-mL volumetric flask. Dissolve, and dilute to volume with *Mobile Phase*. Filter the solution through a 0.45-µm filter.

System Suitability Test Chromatograph duplicate 20-µL injections of the *Standard Preparation*. Ensure that the retention time of Sucralose is approximately 9 min. It may be necessary to adjust the *Mobile Phase* composition to obtain the desired retention time. Ensure that the relative standard deviation (100 × standard deviation/mean peak area) does not exceed 2.0%.

Procedure Analyze the *Standard Preparation* and *Sample Preparation* under the conditions described above, making du-

---

[2] Available from McNeil Specialty Products Company, Regulatory Affairs Department, 501 George Street, New Brunswick, NJ 08903-2400.

plicate 20-µL injections, and calculate the mean peak areas. Calculate the percent Sucralose from the peak areas of the *Sample Preparation* (A_U) and *Standard Preparation* (A_S) according to the following formula:

$$\% \text{ Sucralose} = 100(A_U W_S)/(A_S W_U),$$

in which W_S is the weight, in mg, of the standard, and W_U is the weight, in mg, of the sample.

Heavy Metals Prepare and test a 2-g sample as directed in *Method II* under the *Heavy Metals Test*, Appendix IIIB, using 20 µg of lead ion (Pb) in the control (*Solution A*).

Hydrolysis Products

Spray Reagent Dissolve 1.23 g of p-anisidine and 1.66 g of phthalic acid in 100 mL of methanol. Store the solution in darkness, and refrigerate to prevent it from becoming decolorized. Discard if the solution becomes discolored.

Caution: p-Anisidine is toxic if inhaled or absorbed through the skin and should be used with due caution.

Standard Solution A Dissolve 10.0 g of mannitol, weighed to 0.001 g, in water in a 100-mL volumetric flask, and dilute to volume with water.

Standard Solution B Dissolve 40 mg of fructose and 10 g of mannitol, accurately weighed, in 25 mL of water in a 100-mL volumetric flask, and dilute to volume with water.

Sample Solution Dissolve 2.5 g of the sample in 5 mL of methanol in a 10-mL volumetric flask, and dilute to volume with methanol.

Procedure Use a thin-layer chromatographic (TLC) plate coated with a 0.25-mm layer of Merck-silica gel 60 or equivalent. Spot 5 µL of *Standard Solution A* and of *Standard Solution B* onto the plate, applying the solution slowly in 1-µL aliquots and allowing the plate to dry between applications. Spot 5 µL of the *Sample Solution* onto the plate in a similar manner. The three spots should be of similar size. Spray the plate with the *Spray Reagent*, and heat it at 100° ± 2° for 15 min. Immediately after heating, view the plate against a dark background. The spot from the *Sample Solution* is not more intense in color than the spot from *Standard Solution B* (0.1% limit).

Note: Darkening of the mannitol spot (the spot from *Standard Solution A*) indicates that the plate has been held too long in the oven, and a second plate should be prepared.

Methanol

Apparatus Use a suitable gas chromatographic system equipped with a hydrogen flame ionization detector and a 2.1-m × 4-mm (id) glass column packed with 80- to 100-mesh Porapak P.S. or equivalent.

Operating Conditions The operating conditions may vary, depending on the particular instrument used, but a suitable chromatogram using the above-mentioned materials may be obtained under the following conditions: Column temperature: 150° (isothermal); inlet temperature: 200°; detector temperature: 250°; and carrier gas: helium, flowing at a rate of 20 mL/min. The retention time for methanol is about 2 min.

Internal Standard Solution Pipet 1.0 mL of n-propanol into a 100-mL volumetric flask, dilute to volume with pyridine, and mix. Transfer 5.0 mL of this solution to a 500-mL volumetric flask, dilute to volume with pyridine, and mix.

Standard Solution Pipet 2.0 mL of methanol into a 100-mL volumetric flask, dilute to volume with *Internal Standard Solution*, and mix. Transfer 1.0 mL of this solution to a 100-mL volumetric flask, dilute to volume with *Internal Standard Solution*, and mix.

Sample Solution Weigh accurately about 2 g of the sample into a 10-mL volumetric flask, dilute to volume with *Internal Standard Solution*, and mix.

Procedure Inject a 1-µL portion of the *Standard Solution* onto the gas chromatographic column, obtain the chromatogram, and measure the area of the peak produced. The relative standard deviation for replicate injections is not more than 2.0%. Calculate the mean peak areas for the *Standard Solution*. Similarly, inject a 1-µL portion of the *Sample Solution* into the gas chromatograph, and measure the areas of the peaks produced by methanol. Calculate the mean peak areas, and determine the percent of methanol in the portion of Sucralose taken using the following formula:

$$\% \text{ methanol} = (R_U/R_S)(0.158/W_S),$$

in which R_U is the ratio of the peak areas of methanol to that of the internal standard obtained from the *Sample Solution*; R_S is the ratio of the peak areas of methanol to the internal standard obtained from the *Standard Solution*; the factor 0.158 is equal to the volume of methanol in the standard × dilution factor × density of methanol × 100%; and W_S is the weight, in g, of the sample.

Related Substances

Chromatographic Plates Use Whatman LKC$_{18}$ thin-layer chromatographic plates coated with a 0.20-mm layer of silica gel absorbent (or equivalent).

Mobile Phase Mix 70 volumes of 5.0% (w/v) sodium chloride in water, and add 30 volumes of acetonitrile.

Spray Reagent Use a 15% (v/v) solution of sulfuric acid in methanol.

Standard Preparations Dissolve 500.0 mg of Sucralose Standard for analytical use[2] in 5.0 mL of methanol (*Solution A*). Dilute 0.5 mL of *Solution A* with methanol to 100 mL (*Solution B*).

Test Preparation Dissolve 1.0 g of the sample in 10 mL of methanol.

Procedure Apply 5 µL each of *Solution A*, *Solution B*, and *Test Preparation* to the bottom of the chromatographic plate. Place the plate in a suitable chromatographic chamber containing freshly prepared *Mobile Phase*, and allow the solvent front to ascend approximately 15 cm. Remove the plate, allow it to dry, and spray it with the *Spray Reagent*. Heat the plate in an oven at 125° for 10 min. The main spot in the *Test Preparation* is at the same R_f value as the main spot in *Solution A*, and any other single spot in the *Test Preparation* is not more intense than the 0.5% spot in *Solution B*.

Residue on Ignition Ignite a 1- to 2-g sample as directed in *Method I*, Appendix IIC.

Specific Rotation, Appendix IIB Determine in an aqueous solution containing 1.0 g/100 mL, calculated on the anhydrous basis.

Water Determine by the *Karl Fischer Titrimetric Method*, Appendix IIB.

Packaging and Storage Store in well-closed containers in a cool, dry place at less than 21°.

Sucrose

Sugar; Granulated Sugar; Cane Sugar; Beet Sugar

Formula wt 342.30

CAS: [57-50-1]

DESCRIPTION

Sucrose, β-D-fructofuranosyl-α-D-glucopyranoside, is obtained for commercial use from sugar cane and sugar beets. The processed form is a white, crystalline, odorless solid with a sweet taste. It is very soluble in water, in formamide, and in dimethyl sulfoxide and slightly soluble in ethanol.

Functional Use in Foods Nutritive sweetener; formulation and texturizing aid.

REQUIREMENTS

Identification Meets the requirements under *Specific Rotation*.
Assay Not less than 99.8 and not more than 100.2 International Sugar Degrees (°Z).
Arsenic (as As) Not more than 1 mg/kg.
Color Not more than 75 IU.
Heavy Metals (as Pb) Not more than 5 mg/kg.
Invert Sugar Not more than 0.1%.
Lead Not more than 0.5 mg/kg.
Loss on Drying Not more than 0.1%.
Residue on Ignition Not more than 0.15%.
Specific Rotation $[\alpha]_D^{20°}$: Between +65.9° and +66.7°.

TESTS

Note: Consult ICUMSA[1] rules for further details. This applies to *Assay*, *Color*, and *Invert Sugar*.

Assay
Apparatus Use a saccharimeter calibrated with a certified quartz plate according to the directions of the instrument manufacturer and a 20-cm polarimeter tube with cover glasses. The tube and glasses should conform to ICUMSA specifications. Have ready 100-mL flasks accurate to within 0.01 mL. Maintain a water bath at 20° ± 0.1°.
Procedure Quantitatively transfer 26.000 g ± 0.002 g of the sample to the flask, and add about 80 mL of water. Without heating, dissolve the sample by agitation, and add water to the flask to just below the calibration mark. Place the flask in the water bath to adjust the solution to 20° ± 0.1° (degrees Centigrade unless otherwise specified). Dry the inside wall of the flask neck above the calibration mark with filter paper, and using either a hypodermic syringe or a pipet with a drawn out point, adjust to the exact volume with water. Seal the flask with a clean, dry stopper and mix the contents thoroughly by shaking. Carefully rinse the polarimeter tube twice with two-thirds of its volume of sugar solution, and fill it with sugar solution at 20° ± 0.1° in such a way that no air bubbles are trapped. Place the tube in the saccharimeter, and polarize it at 20°. Determine five values to 0.05°Z, and average these values.

Arsenic A *Sample Solution* prepared using a 1-g sample, accurately weighed, meets the requirements of the *Arsenic Test*, Appendix IIIB, using 1 mL of the *Standard Arsenic Solution* in the control (1 μg As).

Color
Apparatus Use a suitable variable wavelength spectrophotometer capable of measuring percent transmittance at 420 nm or a photometer with a 420- ± 10-nm band width filter. The instrument should be designed to permit the use of a 10-cm cell. When an instrument with a reference cell is used, the two cells should be identical with distilled water within ±0.2% when the instrument is set at 100% transmittance on one of the cells.
Procedure Prepare a 50% (w/w) sample solution in water. The pH is adjusted to 7.0 ± 0.2 with 1% sodium hydroxide or 1% hydrochloric acid. Filter through a 0.45-μm pore-size membrane filter, using a vacuum and a diatomaceous earth filter aid (1% on solids) if necessary. Discard the first portion of the filtrate if it is cloudy. Determine the density and concentration of solids (g/mL) refractometrically. Rinse the measuring cell three times with the sample solution, and then fill the cell. Measure absorbency (A_S) at 420 nm. Calculate the color in ICUMSA units (*IU*) as follows:

$$IU = (A_S/bc) \times 1000,$$

in which b is the cell length, in cm, and c is the concentration, in g/mL, of total solids determined refractometrically and calculated from density.

---

[1] International Commission for Uniform Methods of Sugar Analysis (ICUMSA), c/o British Sugar plc, Technical Centre, Colney, Norwich, England.

Heavy Metals A solution of 4 g in 25 mL of water meets the requirements of the *Heavy Metals Test*, Appendix IIIB, using 20 µg of lead ion (Pb) in the control (*Solution A*).

Invert Sugar

Apparatus Use a water bath with vigorously boiling water to ensure that the immersion of flasks does not interrupt the boiling. Place the flasks in the water bath so that the water level is 2 cm above the liquid surface in the flasks.

Muller's Solution Dissolve 35 g of cupric sulfate pentahydrate in 400 mL of boiling water. In a separate beaker, dissolve 173 g of potassium sodium tartrate tetrahydrate and 68 g of anhydrous sodium carbonate in 400 mL of boiling water. Cool both solutions, and while stirring, pour the sodium carbonate–potassium sodium tartrate solution into the cupric sulfate solution. Transfer the combined solutions to a 1000-mL volumetric flask, dilute to volume, and mix. Add 2 g of activated carbon, shake vigorously, and filter through hardened filter paper under vacuum. If cuprous oxide precipitates on storage, refilter the solution.

Standardized Iodine Solution Dissolve about 4.7 g of iodine in a solution of 6 g of iodate-free potassium iodide in 100 mL of water, add 3 drops of hydrochloric acid, and dilute with water to 1000 mL. Standardize to 0.0333 N as directed in *Iodine, 0.1 N* (*Volumetric Solutions* under *Solutions and Indicators*). Adjust normality repeatedly, if necessary.

Standardized Sodium Thiosulfate Solution Dissolve about 8.7 g of sodium thiosulfate ($Na_2S_2O_3 \cdot 5H_2O$) and 67 mg of sodium carbonate in 1000 mL of freshly boiled and cooled water. Add 3 mL of 1.0 N sodium hydroxide. This solution contains 5.54 g of $Na_2S_2O_3$. Standardize to 0.0333 N as directed in *Sodium Thiosulfate, 0.1 N* (*Volumetric Solutions* under *Solutions and Indicators*). Adjust normality repeatedly, if necessary.

Starch Indicator Solution Dissolve 1 g of soluble starch in 100 mL of saturated sodium chloride solution.

Procedure Transfer about 25 g of the sample, accurately weighed, into a 250-mL Erlenmeyer flask, and add 100 mL of water. Dissolve, add 10 mL of *Muller's Solution*, and mix well. Place the flask in a boiling water bath for 10 min ± 5 s. Remove the flask, place a small beaker over its neck, and cool rapidly, without agitation, under cold running water. Acidify the solution with 5 mL of 5 N acetic acid, and immediately add an excess of *Standardized Iodine Solution* (about 20 to 40 mL). Make both of these additions without agitation to avoid the oxidation of cuprous oxide by air. Mix well, and when the precipitate is completely dissolved, titrate the excess iodine with *Standardized Sodium Thiosulfate Solution*, adding a few drops of *Starch Indicator Solution* as the endpoint is approached.

Determine a *Water Blank* by the same procedure, eliminating the sample, and a *Cold Blank* by the same procedure, but allowing the sample solution flask to stand at room temperature for 10 min rather than placing it in the boiling water bath. Calculate the percent Invert Sugar by the formula

$$\% = [(V_I - V_S - B_W - B_S - 0.2W) \times 100]/W,$$

in which V_I is the volume, in mL, of the *Standardized Iodine Solution*; V_S is the volume, in mL, of the *Standardized Sodium Thiosulfate Solution*; B_W is the volume, in mL, of the *Standardized Iodine Solution* in the *Water Blank*; B_S is the volume, in mL, of the *Standardized Iodine Solution* in the *Cold Blank*; 0.2 is a volume correction factor, in mL, used to correct for the reducing value of Sucrose; and W is the sample weight, in g.

Lead Determine as directed under *Method I* in the *Atomic Absorption Spectrophotometric Graphite Furnace Method* under the *Lead Limit Test*, Appendix IIIB, using a 5-g sample.

Loss on Drying Dry about 5 g of the sample in a forced-draft air oven at 105° for 3 h, based on the general method, Appendix IIC.

Residue on Ignition Ignite 1 g as directed in the general method, Appendix IIC.

Specific Rotation, Appendix IIB Dissolve 26 g of the sample in water, and dilute to 100 mL at 20°. Determine the specific rotation using a 20-cm polarimeter tube.

Packaging and Storage Store in tight containers in a dry place.

Sulfur Dioxide

SO_2 Formula wt 64.06

INS: 220   CAS: [7446-09-5]

DESCRIPTION

A colorless, nonflammable gas, under normal conditions of temperature and pressure, having a sharp, pungent odor. It is shipped as a liquid under pressure in containers approved by the U.S. Department of Transportation. Its vapor density is 2.26 times that of air at atmospheric pressure and 0°. The specific gravity of the liquid is about 1.436 at 0°/4°. At 20° the solubility is about 10 g of SO_2 per 100 g of solution.

> **Caution**: Sulfur dioxide gas is intensely irritating to the eyes, throat, and upper respiratory system. Liquid sulfur dioxide may cause skin burns, which result from the freezing effect of the liquid on tissue. Safety precautions to be observed in handling of the material are specified in "Pamphlet G-3" published by the Compressed Gas Association, Suite 1004, 1725 Jefferson Davis Highway, Arlington, VA 22202.

Functional Use in Foods Bleaching agent; preservative.

REQUIREMENTS

Identification A saturated solution of sulfur dioxide in water gives positive tests for *Sulfite*, Appendix IIIA.
Assay Not less than 99.9% of SO_2 by weight.
Heavy Metals (as Pb) Not more than 0.003% by weight.
Lead Not more than 10 mg/kg by weight.
Nonvolatile Residue Not more than 0.05% by weight.
Selenium Not more than 0.002% by weight.
Water Not more than 0.05% by weight.

TESTS

Sampling Samples of sulfur dioxide may be safely withdrawn from a tank or transfer lines, either of which should be equipped with a $\frac{3}{8}$-in. nozzle and valve. Samples should be taken in bombs constructed of 316 stainless steel, designed to withstand 1000 psig and equipped with 316 stainless steel needle valves on both ends. To draw a sample, the bomb is first flushed with dry air to remove any sulfur dioxide, remaining from previous sample drawings, and then attached to the tank or transfer lines with a solid pipe connection. A hose is connected to the other end of the bomb and submerged in either a weak caustic solution or water. Any gas in the bomb is discharged into the caustic or water by first opening the valve at the pipe end, followed by slowly opening the valve at the hose end. When all of the gas is dispelled and liquid sulfur dioxide begins to emerge into the solution, the valve at the hose end is blocked off. The other valves are then tightly closed, and the bomb is detached from the pipe connecting it to the tank or transfer line. Approximately 15% of the liquid sulfur dioxide in the bomb is then discharged into the water or caustic solution. The bomb is then capped at its end and transferred to the laboratory for analysis.

Caution: The bomb should never be stored with more than 85% of the total water capacity of the bomb.

Assay Subtract from 100 the percentages of *Nonvolatile Residue* and of *Water*, as determined herein, to obtain the percentage of SO_2.

Sample Solution for the Determination of Heavy Metals, Lead, and Selenium Measure out 100 mL of sulfur dioxide (144 g) into a 125-mL Erlenmeyer flask, and determine the weight of sample taken by the loss in weight of the sample bomb. Evaporate to dryness on a steam bath, add 3 mL of nitric acid and 10 mL of water to the dry flask, and warm gently on a hot plate for 15 min. Transfer the contents of the flask to a 100-mL volumetric flask, dilute to volume with water, and mix. Transfer a 10.0-mL aliquot into a second 100-mL volumetric flask, dilute to volume with water, and mix.

Note: The tests in which this solution is to be used will be accurate assuming a 144-g sample has been taken; if not, the weight of sample actually taken must be considered in the calculations.

Heavy Metals A 5.0-mL portion of the *Sample Solution*, diluted to 25 mL with water, meets the requirements of the *Heavy Metals Test*, Appendix IIIB, using 20 µg of lead ion (Pb) in the control (*Solution A*).

Lead A 7.0-mL portion of the *Sample Solution*, diluted to 40 mL with water, meets the requirements of the *Lead Limit Test*, Appendix IIIB, using 10 µg of lead ion (Pb) in the control.

Nonvolatile Residue Measure out 200 mL of sulfur dioxide (288 g) into a 250-mL Erlenmeyer flask, and determine the weight of sample taken by the loss in weight of the sample bomb. Evaporate to dryness on a steam bath, and displace the residual vapors with dry air. Wipe the flask dry, cool in a desiccator, and weigh.

Selenium A 2.0-mL portion of the *Sample Solution* meets the requirements of the *Selenium Limit Test, Method II*, Appendix IIIB.

Water Transfer about 50 mL of liquid sulfur dioxide into a Karl Fischer titration jar, determine the weight of sample taken, and determine the water content by the *Karl Fischer Titrimetric Method*, Appendix IIB.

Packaging and Storage Store in suitable pressure containers, observing applicable federal regulations pertaining to shipping containers.

Sulfuric Acid

H_2SO_4 Formula wt 98.08

INS: 513 CAS: [7664-93-9]

DESCRIPTION

A clear, colorless, oily liquid. It is very caustic and corrosive. It is miscible with water and with alcohol with the generation of much heat and contraction in volume. When mixed with other liquids, Sulfuric Acid should be added cautiously to the diluent. Some concentrations of Sulfuric Acid commercially available are expressed in Baumé degrees (Bé°) and others (above 93.0%) as percentage of H_2SO_4. The usually available concentrations are 60° and 66°Bé, equivalent to 77.67% and 93.19% of H_2SO_4, respectively, and 98.0% of H_2SO_4. Its specific gravity varies with the concentration of H_2SO_4 (see *Sulfuric Acid Table*, Appendix IIC).

Functional Use in Foods Acid.

REQUIREMENTS

Identification It responds to the tests for *Sulfate*, Appendix IIIA.
Assay Not less than the minimum or within the range of Bé°, or the percentage of H_2SO_4, claimed or implied by the vendor.
Arsenic (as As) Not more than 3 mg/kg.
Chloride Not more than 0.005%.
Heavy Metals (as Pb) Not more than 0.002%.
Iron Not more than 0.02%.
Lead Not more than 5 mg/kg.
Nitrate Not more than 10 mg/kg.
Reducing Substances (as SO_2) Passes test (approximately 0.004%).
Selenium Not more than 0.002%.

TESTS

Assay Transfer a 1-mL sample into a small, tared, glass-stoppered Erlenmeyer flask, insert the stopper, weigh accurately, and cautiously add about 30 mL of water. Cool the mixture, add methyl orange TS, and titrate with 1 N sodium hydroxide.

Each mL of 1 N sodium hydroxide is equivalent to 49.04 mg of H_2SO_4.

For concentrations of Sulfuric Acid below 93.0%, expressed in Baumé degrees, transfer about 200 mL, previously cooled to a temperature below 15°, into a 250-mL hydrometer cylinder. Insert a suitable Baumé hydrometer graduated at 0.1° intervals, adjust the temperature to exactly 15.6° (60°F), and note the reading at the bottom of the meniscus, estimating it to the nearest 0.05°. The percentage of H_2SO_4 may be obtained by reference to the *Sulfuric Acid Table*, Appendix IIC.

Arsenic A solution of 1 g in 35 mL of water meets the requirements of the *Arsenic Test*, Appendix IIIB, using as a control a mixture of 3 mL of the *Standard Arsenic Solution* (3 μg As) and 1 g of ACS reagent-grade Sulfuric Acid.

Chloride, Appendix IIIB Transfer a volume equivalent to 5 g of the acid into about 25 mL of water contained in a 50-mL volumetric flask, cool, and dilute to volume. Retain the unused portion for the *Heavy Metals*, *Iron*, and *Lead* tests. Any turbidity produced by 4 mL of this solution (400-mg sample) does not exceed that shown in a control containing 20 μg of chloride ion (Cl).

Heavy Metals Dilute 10 mL of the solution (1-g sample) prepared for the *Chloride* test to 25 mL with water. This solution meets the requirements of the *Heavy Metals Test*, Appendix IIIB, using 20 μg of lead ion (Pb) in the control (*Solution A*).

Iron Dilute 1 mL of the solution (100-mg sample) prepared for the *Chloride* test to 40 mL. Add about 30 mg of ammonium persulfate crystals and 10 mL of ammonium thiocyanate TS. Any red color does not exceed that produced by 2.0 mL of *Iron Standard Solution* (20 μg Fe) in an equal volume of solution containing the same quantities of the reagents used in the test.

Lead Dilute 10 mL of the solution (1-g sample) prepared for the *Chloride* test to 40 mL. This solution meets the requirements of the *Lead Limit Test*, Appendix IIIB, using 5 μg of lead ion (Pb) in the control.

Nitrate

Standard Nitrate Solution Transfer 8.022 g of potassium nitrate, KNO_3, previously dried at 105° for 1 h, into a 500-mL volumetric flask, dissolve it in water, dilute to volume, and mix well. Slowly add from a buret 5.0 mL of this solution to 400 mL of ACS reagent-grade Sulfuric Acid, previously cooled to 5°, keeping the tip of the buret below the surface of the acid. After the solution has reached room temperature, transfer it into a 500-mL volumetric flask, and dilute to volume with reagent-grade Sulfuric Acid. Each mL contains 100 μg of HNO_3.

Procedure Into each of two 100-mL Nessler tubes transfer 50 mL of ACS reagent-grade Sulfuric Acid, add slowly 5 mL of a freshly prepared 1 in 10 solution of ferrous sulfate, $FeSO_4 \cdot 7H_2O$, mix with a glass rod, and cool in an ice bath to between 10° and 15°. To one tube of the cooled mixture add a 10-mL sample, previously cooled to between 10° and 15°, and dilute to the 100-mL mark with ACS reagent-grade Sulfuric Acid chilled to about the same temperature. Add the *Standard Nitrate Solution*, dropwise, from a microburet to the other tube, with frequent mixing, until the color of the control nearly matches that of the sample solution. Dilute the control solution to 100 mL, and continue adding the *Standard Nitrate Solution* to as exact a match in color intensity as possible when compared with the sample solution by looking down through the solutions against a white background illuminated by diffused light. Compute the weight of H_2SO_4 in the weight of the sample from the specific gravity and the volume taken (see *Sulfuric Acid Table*, Appendix IIC). Not more than 0.1 mL of the *Standard Nitrate Solution* is required for each g of H_2SO_4.

Reducing Substances Carefully dilute 8 g with about 50 mL of ice-cold water, keeping the solution cool during the addition. To the dilution add 0.1 mL of 0.1 N potassium permanganate. The solution remains pink for not less than 5 min.

Selenium Determine as directed in *Method II* under the *Selenium Limit Test*, Appendix IIIB, using 300 mg of sample. The absorbance of the extract from the *Sample Preparation* is not greater than that from the *Standard Preparation*.

Packaging and Storage Store in tight containers.

Sunflower Oil (Unhydrogenated)

CAS: [8008-31-9]

DESCRIPTION

A light amber-colored oil obtained from the seed of the sunflower plant *Helianthus annuus* by mechanical expression or solvent extraction. It is refined, bleached, and deodorized to substantially remove free fatty acids, phospholipids, color, odor and flavor components, and miscellaneous other non-oil materials. It is a liquid at 21° to 27°, but traces of wax may cause the oil to cloud, unless removed by winterization.

Functional Use in Foods Coating agent; emulsifying agent; formulation aid; texturizer.

REQUIREMENTS

Identification Sunflower oil exhibits the following composition profile of fatty acids as determined under *Fatty Acid Composition*, Appendix VII.

Fatty Acid: Weight %	<14	14:0	16:0	16:1	18:0	18:1	18:2
(Range)	<0.1	<0.5	3.0–10	<1.0	1.0–10	14–65	20–75

Fatty Acid: Weight %	18:3	20:0	20:1	22:0	22:1	24:0
(Range)	<0.5	<1.0	<0.5	<1.0	<0.1	<0.4

Cold Test Passes test.
Color (AOCS-Wesson) Not more than 1.3 red.
Free Fatty Acids (as oleic acid) Not more than 0.1%.
Iodine Value Between 110 and 143.
Lead Not more than 0.1 mg/kg.
Linolenic Acid Not more than 1.5%.
Peroxide Value Not more than 10 meq/kg.

Unsaponifiable Matter Not more than 1.5%.
Water Not more than 0.1%.

TESTS

Cold Test Proceed as directed under *Cold Test*, Appendix VII.

Color Proceed as directed for *Color* (AOCS-Wesson) under *Fats and Related Substances*, Appendix VII.

Free Fatty Acids Proceed as directed under *Free Fatty Acids*, Appendix VII, using the following equivalence factor (e) in the formula given in the procedure:

Free fatty acids as oleic acid, $e = 28.2$.

Iodine Value Proceed as directed under *Iodine Value*, Appendix VII.

Lead Determine as directed under *Method II* in the *Atomic Absorption Spectrophotometric Graphite Furnace Method* under the *Lead Limit Test*, Appendix IIIB, using a 3-g sample.

Linolenic Acid Proceed as directed under *Fatty Acid Composition*, Appendix VII.

Peroxide Value Proceed as directed under *Peroxide Value* in the monograph for *Hydroxylated Lecithin*. However, after the addition of saturated potassium iodide and mixing, instead of allowing the solution to stand for 10 min, mix the solution for 1 min, and begin the titration immediately.

Unsaponifiable Matter Proceed as directed under *Unsaponifiable Matter*, Appendix VII.

Water Proceed as directed under *Water Determination*, Appendix IIB. However, in place of 35 to 40 mL of methanol, use 50 mL of chloroform to dissolve the sample.

Packaging and Storage Store in well-closed containers.

Talc

INS: 553(iii) CAS: [14807-96-6]

DESCRIPTION

A naturally occurring form of hydrous magnesium silicate containing varying proportions of such associated minerals as alpha-quartz, calcite, chlorite, dolomite, kaolin, magnesite, and phlogopite. *Talc derived from deposits that are known to contain associated asbestos is not food grade.* It occurs as a white to grayish white, odorless, tasteless, unctuous powder that is insoluble in water and in solutions of alkali hydroxides but is slightly soluble in dilute mineral acids.

Functional Use in Foods Anticaking agent; coating agent; lubricating and release agent; surface-finishing agent; texturizing agent.

REQUIREMENTS

Identification

A. The X-ray diffraction pattern of a random powder sample exhibits intense reflections at the following d values of 9.34 Å, 4.66 Å, and 3.12 Å.

B. The infrared absorption spectrum of a potassium bromide dispersion of the sample exhibits major peaks at approximately 1015 cm^{-1} and 450 cm^{-1}.

Acid-Soluble Substances (as SO$_4$) Not more than 2.5%.
Arsenic (as As) Not more than 3 mg/kg.
Free Alkali (as NaOH) Not more than 1%.
Heavy Metals (as Pb) Not more than 0.004%.
Lead Not more than 10 mg/kg.
Loss on Drying Not more than 0.5%.
Loss on Ignition Not more than 6.0%.
Soluble Salts Not more than 0.2%.

TESTS

Acid-Soluble Substances Digest 1.00 g with 20 mL of 3 N hydrochloric acid at 50° for 15 min, add water to restore the original volume, mix, and filter. To 10 mL of the filtrate add 1 mL of 2 N sulfuric acid, evaporate to dryness, and ignite to constant weight. The weight of the residue does not exceed 12.5 mg.

Sample Solution for the Determination of Arsenic, Heavy Metals, and Lead Transfer 10.0 g of the sample into a 250-mL flask, and add 50 mL of 0.5 N hydrochloric acid. Attach a reflux condenser to the flask, heat on a steam bath for 30 min, cool, and let the undissolved material settle. Decant the supernatant liquid through Whatman No. 3 filter paper, or equivalent, into a 100-mL volumetric flask, retaining as much as possible of the insoluble material in the beaker. Wash the slurry and beaker with three 10-mL portions of hot water, decanting each washing through the filter into the flask. Finally, wash the filter paper with 15 mL of hot water, cool the filtrate to room temperature, dilute to volume with water, and mix.

Arsenic A 10-mL portion of the *Sample Solution* meets the requirements of the *Arsenic Test*, Appendix IIIB.

Free Alkali Add 2 drops of phenolphthalein TS to 20 mL of the diluted filtrate prepared in the test for *Soluble Salts*, representing 1 g of Talc. If a pink color is produced, not more than 2.5 mL of 0.1 N hydrochloric acid is required to discharge it.

Heavy Metals A 5-mL portion of the *Sample Solution* diluted to 25 mL with water meets the requirements of the *Heavy Metals Test*, Appendix IIIB, using 20 µg of lead ion (Pb) in the control (*Solution A*).

Lead A 10-mL portion of the *Sample Solution* meets the requirements of the *Lead Limit Test*, Appendix IIIB, using 10 µg of lead ion (Pb) in the control.

Loss on Drying, Appendix IIC Dry 10 g at 105° for 1 h.

Loss on Ignition Weigh accurately about 1 g in a tared platinum crucible provided with a cover, initially apply heat gradually, and then ignite to constant weight.

Soluble Salts Boil 10 g of the sample with 150 mL of water for 15 min. Cool to room temperature, and add water to restore the original volume. Allow the mixture to stand for 15 min, and filter until clear. To 75 mL of the clear filtrate add 25 mL of water. Evaporate 50 mL of this solution, representing 2.5 g of Talc, in a tared platinum dish on a steam bath to dryness, and ignite gently to constant weight. The weight of the residue does not exceed 5 mg.

Packaging and Storage Store in well-closed containers.

Tallow

DESCRIPTION

An off-white fat obtained by heat rendering of tissues (cuttings and trimmings) from beef and, to a lesser degree, mutton shortly after slaughter. Rendered Tallow may be alkali refined and bleached, or bleached and deodorized without prior refining. It is a firm fat containing a high proportion of saturated fatty acids and exhibiting greater flavor stability than lard or unhydrogenated vegetable oils.

Rendered, alkali-refined, and bleached-deodorized Tallow are white to off-white solids at 21° to 27°. Alkali-refined and bleached-deodorized Tallow, which are pale yellow to colorless and free of extraneous matter at 54°, differ from rendered Tallow, which is clear to hazy and may contain extraneous matter.

Functional Use in Foods Coating agent; emulsifying agent; formulation aid; texturizer.

REQUIREMENTS

	Rendered Tallow	Alkali Refined Tallow	Bleached and Deodorized Tallow
Color (AOCS-Wesson)	Not more than 3.0 red	Not more than 1.5 red	Not more than 1.5 red
Free Fatty Acids (as oleic acid)	Not more than 1.5%	Not more than 0.5%	Not more than 0.1%
Insoluble Matter	Not more than 0.1%	Not more than 0.01%	Not more than 0.01%
Iodine Value	Between 37 and 50	Between 37 and 50	Between 37 and 50
Water	Not more than 0.5%	Not more than 0.2%	Not more than 0.1%

Identification Tallow exhibits the following composition profile of fatty acids as determined under *Fatty Acid Composition*, Appendix VII.

Fatty Acid Weight % (Range):	<14:0	14:0	14:1	15:0	15:0 iso	16:0
	<0.1	1.4–6.3	0.5–1.5	0.5–1.0	<1.5	20–37

Fatty Acid Weight % (Range):	16:0 iso	16:1	16:2	17:0	17:1	18:0
	<0.5	0.7–8.8	<1.0	0.5–2.0	<1.0	6–40

Fatty Acid Weight % (Range):	18:1	18:2	18:3	20:0	20:1	20:4
	26–50	0.5–5.0	<2.5	<0.5	<0.5	<0.5

Arsenic (as As) Not more than 0.5 mg/kg.
Lead Not more than 0.1 mg/kg.
Peroxide Value Not more than 10 meq/kg.

TESTS

Arsenic A *Sample Solution* prepared using 2 g of sample, accurately weighed, meets the requirements of the *Arsenic Test*, Appendix IIIB. The absorbance due to any red color from the solution of the sample does not exceed that produced by 1.0 mL of *Standard Arsenic Solution* (1 μg As) when treated in the same manner and under the same conditions as the sample.

Color Proceed as directed for *Color* (AOCS-Wesson) under *Fats and Related Substances*, Appendix VII.

Free Fatty Acids Proceed as directed under *Free Fatty Acids*, Appendix VII, using the following equivalence factor (e) in the formula given in the procedure:

Free fatty acids as oleic acid, $e = 28.2$.

Iodine Value Proceed as directed under *Iodine Value*, Appendix VII.

Lead Proceed as directed in the monograph for *Coconut Oil*.

Melting Range Proceed as directed for *Melting Range* under *Fats and Related Substances*, Appendix VII.

Peroxide Value Proceed as directed under *Peroxide Value* in the monograph for *Hydroxylated Lecithin*. However, after the addition of saturated potassium iodide and mixing, instead of allowing the solution to stand for 10 min, mix the solution for 1 min, and begin the titration immediately.

Unsaponifiable Matter Proceed as directed under *Unsaponifiable Matter*, Appendix VII.

Water Proceed as directed under *Water Determination*, Appendix IIB. However, in place of 35 to 40 mL of methanol, use 50 mL of chloroform to dissolve the sample.

Packaging and Storage Store in well-closed containers.

Tangerine Oil, Coldpressed

Tangerine Oil, Expressed

DESCRIPTION

The oil obtained by expression from the peels of the ripe fruit of the Dancy tangerine, and from some other closely related varieties. It is a reddish orange to brownish orange liquid with a pleasant, orangelike odor. Oils produced from unripe fruit often show a green color. It is soluble in most fixed oils and in mineral oil, slightly soluble in propylene glycol, and relatively insoluble in glycerin. It may contain a suitable antioxidant.

Functional Use in Foods Flavoring agent.

REQUIREMENTS

Identification The infrared absorption spectrum of the sample exhibits relative maxima (that may vary in intensity) at the same wavelengths (or frequencies) as those shown in the respective spectrum in the section on *Infrared Spectra (Series A: Essential Oils)*, using the same test conditions as specified therein.
Aldehydes Between 0.8% and 1.9% of aldehydes, calculated as decyl aldehyde ($C_{10}H_{22}O$).
Angular Rotation Between +88° and +96°.
Heavy Metals (as Pb) Not more than 0.004%.
Lead Not more than 10 mg/kg.
Refractive Index Between 1.473 and 1.476 at 20°.
Residue on Evaporation Between 2.3% and 5.8%.
Specific Gravity Between 0.844 and 0.854.

TESTS

Aldehydes Weigh accurately about 10 g, and proceed as directed under *Aldehydes and Ketones—Hydroxylamine/Tert-Butyl Alcohol Method*, Appendix VI, using 78.13 as the equivalence factor (*e*) in the calculation. Allow the samples and the blank to stand at room temperature for 30 min after adding the hydroxylamine hydrochloride solution.
Angular Rotation Determine in a 100-mm tube as directed under *Optical (Specific) Rotation*, Appendix IIB.
Heavy Metals Prepare and test a 500-mg sample as directed in *Method II* under the *Heavy Metals Test*, Appendix IIIB, using 20 µg of lead ion (Pb) in the control (*Solution A*).
Lead A *Sample Solution* prepared as directed for organic compounds meets the requirements of the *Lead Limit Test*, Appendix IIIB, using 10 µg of lead ion (Pb) in the control.
Refractive Index, Appendix IIB Determine with an Abbé or other refractometer of equal or greater accuracy.
Residue on Evaporation Proceed as directed in the general method, Appendix VI, using 5 g, and heat for 5 h.
Specific Gravity Determine by any reliable method (see *General Provisions*).

Packaging and Storage Store in full, tight, preferably glass, tin-lined, galvanized, or other suitably lined containers in a cool place protected from light.

Tannic Acid

CAS: [1401-55-4]

DESCRIPTION

Tannic Acid or hydrolyzable gallotannin is a complex polyphenolic organic structure that yields gallic acid and either glucose or quinic acid as hydrolysis products. Tannic Acid is obtained by solvent extraction from the nutgalls or the excrescences that form on the young twigs of *Quercus infectoria* Olivier and allied species of *Quercus* L. (Fam. *Fagaceae*); from the seed pods of Tara (*Caesalpinia spinosa*); or from the nutgalls of various sumac species, including *Rhus semialata*, *R. coriaria*, *R. galabra*, and *R. typhia*. It occurs as an amorphous powder, as glistening scales, or as spongy masses, varying in color from yellowish white to light brown. It is odorless or has a faint, characteristic odor and an astringent taste. Tannic acid is very soluble in water, in acetone, and in alcohol, but only slightly soluble in absolute alcohol. It is practically insoluble in chloroform, in ether, and in solvent hexane. One g dissolves in about 1 mL of warm glycerin.

Functional Use in Foods Clarifying agent; flavoring agent; flavor enhancer; flavoring adjuvant; pH control agent; boiler water additive.

REQUIREMENTS

Identification
  A. To a 1 in 10 solution add a small quantity of ferric chloride TS. A bluish black color or precipitate forms.
  B. A solution of tannic acid when added to a solution of either an alkaloidal salt, albumin, or gelatin produces a precipitate.
Gums or Dextrin Passes test.
Heavy Metals Not more than 0.002%.
Lead Not more than 10 mg/kg.
Loss on Drying Not more than 7.0%.
Residue on Ignition Not more than 1.0%.
Resinous Substances Passes test.

TESTS

Gums or Dextrin Dissolve 1 g in 5 mL of water, filter, and to the filtrate add 10 mL of alcohol. No turbidity is produced within 15 min.
Heavy Metals Transfer a 1-g sample into a 150-mL beaker, and cautiously add 15 mL of nitric acid and 5 mL of 70% perchloric acid. Evaporate the mixture to dryness on a hot plate under a suitable hood, cool slightly, add 2 mL of hydrochloric acid, and wash down the sides of the beaker with water. Carefully evaporate the solution to dryness on a hot plate, rotating the beaker to avoid spattering. Repeat the addition of 2 mL of hydrochloric acid, washing down the sides of the beaker with water, and evaporate to dryness. Cool the residue, dissolve it in 10 mL of water, and add 1 drop of phenolphthalein TS and sufficient 1 *N* sodium hydroxide, dropwise, to produce a pink color. To this solution add 1 *N* hydrochloric acid, dropwise, until the pink color just disappears, then add 2 mL of 1 *N* acetic acid, and transfer the solution into a 50-mL Nessler tube. Dilute to 25 mL with water, and add 10 mL of hydrogen sulfide TS. After 10 min the color of the solution of the sample is no darker than that produced in a control of equal volume containing 20 µg of lead ion (Pb) and carried through the same procedure as the sample.
Lead A *Sample Solution* prepared as directed for organic compounds meets the requirements of the *Lead Limit Test*, Appendix IIIB, using 10 µg of lead ion (Pb) in the control.

Tarragon Oil

Estragon Oil

DESCRIPTION

The volatile oil obtained by steam distillation from the leaves, stems, and flowers of the plant *Artemesia dracunculus* L. It is a pale yellow to amber liquid having a delicate, spicy odor similar to licorice and sweet basil but characteristic of Tarragon Oil. It is soluble in most fixed oils and in an equal volume of mineral oil, occasionally becoming hazy on further dilution. It is relatively insoluble in propylene glycol, and is insoluble in glycerin.

Functional Use in Foods Flavoring agent.

REQUIREMENTS

Identification The infrared absorption spectrum of the sample exhibits relative maxima (that may vary in intensity) at the same wavelengths (or frequencies) as those shown in the respective spectrum in the section on *Infrared Spectra* (*Series A: Essential Oils*), using the same test conditions as specified therein.
Acid Value Not more than 2.0.
Angular Rotation Between +1.5° and +6.5°.
Heavy Metals (as Pb) Passes test.
Refractive Index Between 1.504 and 1.520 at 20°.
Saponification Value Not more than 18.
Solubility in Alcohol Passes test.
Specific Gravity Between 0.914 and 0.956.

TESTS

Acid Value Determine as directed for *Acid Value* under *Essential Oils and Flavors*, Appendix VI.
Angular Rotation Determine in a 100-mm tube as directed under *Optical (Specific) Rotation*, Appendix IIB.
Heavy Metals Shake 10 mL of the oil with an equal volume of water to which 1 drop of hydrochloric acid has been added, and pass hydrogen sulfide through the mixture until it is saturated. No darkening in color is produced in either the oil or the water.
Refractive Index, Appendix IIB Determine with an Abbé or other refractometer of equal or greater accuracy.
Saponification Value Proceed as directed for *Saponification Value* under *Essential Oils and Flavors*, Appendix VI, using about 5 g, accurately weighed.
Solubility in Alcohol Proceed as directed in the general method, Appendix VI. One mL is soluble in 1 mL of 90% alcohol.
Specific Gravity Determine by any reliable method (see *General Provisions*).

Packaging and Storage Store in full, tight, preferably glass containers in a cool place protected from light.

Tartaric Acid

L(+)-Tartaric Acid

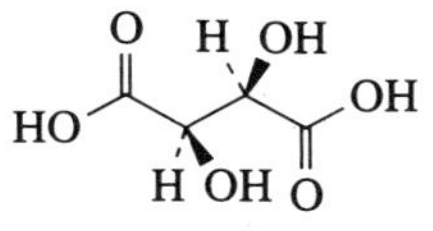

$C_4H_6O_6$ Formula wt 150.09
INS: 334 CAS: [87-69-4]

DESCRIPTION

Colorless or translucent crystals, or a white, fine to granular, crystalline powder. It is odorless, has an acid taste, and is stable in air. One g dissolves in 0.8 mL of water at 25°, in about 0.5 mL of boiling water, and in about 3 mL of alcohol. Its solutions are dextrorotatory.

Functional Use in Foods Acid; sequestrant.

REQUIREMENTS

Identification Its solutions give positive tests for *Tartrate*, Appendix IIIA.
Assay Not less than 99.7% and not more than 100.5% of $C_4H_6O_6$ after drying.
Heavy Metals (as Pb) Not more than 10 mg/kg.
Loss on Drying Not more than 0.5%.
Oxalate Passes test.
Residue on Ignition Not more than 0.05%.
Specific Rotation $[\alpha]_D^{25°}$: Between +12.0° and +13.0°.
Sulfate Passes test.

TESTS

Assay Weigh accurately about 2 g, previously dried over phosphorus pentoxide for 3 h, dissolve it in 40 mL of water, add phenolphthalein TS, and titrate with 1 N sodium hydroxide. Each mL of 1 N sodium hydroxide is equivalent to 75.04 mg of $C_4H_6O_6$.

(from left column, preceding Tarragon Oil:)

Loss on Drying, Appendix IIC Dry at 105° for 2 h.
Residue on Ignition Ignite 1 g as directed in the general method, Appendix IIC.
Resinous Substances Dissolve 1 g in 5 mL of water, filter, and dilute the filtrate to 15 mL. No turbidity is produced.

Packaging and Storage Store in tight, light-resistant containers.

Heavy Metals Prepare and test a 2-g sample as directed in *Method II* under the *Heavy Metals Test*, Appendix IIIB, using 20 μg of lead ion (Pb) in the control (*Solution A*).

Loss on Drying, Appendix IIC Dry over phosphorus pentoxide for 3 h.

Oxalate Nearly neutralize 10 mL of a 1 in 10 solution with 6 N ammonium hydroxide, and add 10 mL of calcium sulfate TS. No turbidity is produced.

Residue on Ignition Ignite 4 g as directed in the general method, Appendix IIC.

Specific Rotation, Appendix IIB Determine in a solution containing 2 g in each 10 mL.

Sulfate To 10 mL of a 1 in 100 solution add 3 drops of hydrochloric acid and 1 mL of barium chloride TS. No turbidity is produced.

Packaging and Storage Store in well-closed containers.

TBHQ

tert-Butylhydroquinone; Mono-*tert*-butylhydroquinone

$C_{10}H_{14}O_2$ Formula wt 166.22
INS: 319 CAS: [1948-33-0]

DESCRIPTION

A white, crystalline solid having a characteristic odor. It is soluble in alcohol and in ether, but is practically insoluble in water.

Functional Use in Foods Antioxidant.

REQUIREMENTS

Identification Dissolve a few mg of the sample in 1 mL of methanol, and add a few drops of a 25% solution of dimethylamine in water. A pink to red color is produced.

Assay Not less than 99.0% of $C_{10}H_{14}O_2$.

***t*-Butyl-*p*-benzoquinone** Not more than 0.2%.

2,5-Di-*t*-butylhydroquinone Not more than 0.2%.

Heavy Metals (as Pb) Not more than 10 mg/kg.

Hydroquinone Not more than 0.1%.

Melting Range Between 126.5° and 128.5°.

Toluene Not more than 0.0025%.

Ultraviolet Absorbance (polynuclear hydrocarbons) Passes test.

TESTS

Assay Transfer about 170 mg of the sample, previously ground to a fine powder and accurately weighed, into a 250-mL, wide-mouth Erlenmeyer flask, and dissolve in 10 mL of methanol. Add 150 mL of water, 1 mL of 1 N sulfuric acid, and 4 drops of diphenylamine indicator (3 mg of *p*-diphenylaminesulfonic acid sodium salt per mL of 0.1 N sulfuric acid), and titrate with 0.1 N ceric sulfate to the first complete color change from yellow to red violet. Record the volume, in mL, of 0.1 N ceric sulfate required as *V*. Calculate the percentage of $C_{10}H_{14}O_2$ in the sample, uncorrected for hydroquinone (HQ) and 2,5-di-*tert*-butylhydroquinone (DTBHQ), by the formula

$$8.311 N(V - 0.1 \text{ mL})/W,$$

in which 0.1 mL represents the volume of ceric sulfate solution consumed by the primary oxidation products of *tert*-Butylhydroquinone ordinarily present in the sample; *N* is the exact normality of the ceric sulfate solution; and *W* is the weight of the sample taken, in g. Record the uncorrected percentage thus calculated as *A*. If HQ and DTBHQ are present in the sample, they will be included in the titration. Calculate the corrected percentage of $C_{10}H_{14}O_2$ in the sample by the formula

$$A - (\% \text{ HQ} \times 1.51) - (\% \text{ DTBHQ} \times 0.75),$$

using the respective values for percentage of HQ and percentage of DTBHQ as determined under *2,5-Di-t-butylhydroquinone* and *Hydroquinone*.

***t*-Butyl-*p*-benzoquinone**

Apparatus Use a suitable double-beam infrared spectrophotometer and matched 0.4-mm liquid sample cells with calcium fluoride windows.

Standard Preparation Transfer about 10 mg of FCC Mono-tertiary-butyl-*p*-benzoquinone Reference Standard, accurately weighed, into a 10-mL volumetric flask, dissolve in chloroform, dilute to volume with the same solvent, and mix.

Sample Preparation Transfer about 1 g of the sample, previously reduced to a fine powder in a high-speed blender and accurately weighed, into a 10-mL volumetric flask, dilute to volume with chloroform, and shake for 5 min to extract the *t*-butyl-*p*-benzoquinone. Filter through a Millipore filter (UHWP01300), or equivalent, before use in the *Procedure* below.

Procedure Fill the reference cell with chloroform and the sample cell with the *Standard Preparation*, place the cells in the respective reference and sample beams of the spectrophotometer, and record the infrared spectrum from 1600 to 1775 cm^{-1}. On the spectrum draw a background line from 1612 to 1750 cm^{-1}, and determine the net absorbance (A_S) of the *Standard Preparation* at 1659 cm^{-1}. Similarly, obtain the spectrum of the *Sample Preparation*, and determine its net absorbance (A_U) at 1659 cm^{-1}. Calculate the percentage of *t*-butyl-*p*-benzoquinone in the sample by the formula

$$100 \times (A_U/A_S) \times (W_S/W_U),$$

in which W_S is the exact weight, in mg, of the Reference Standard taken, and W_U is the exact weight, in mg, of the sample taken.

2,5-Di-*t*-butylhydroquinone and Hydroquinone

Apparatus Use a suitable gas chromatograph (Appendix IIA, *Chromatography*) equipped with a thermal conductivity detector (F and M Model 810, or equivalent), containing a 0.61-m (2-ft) × 6.35-mm (od) stainless steel column packed with 20% Silicone SE-30, by weight, and 80% Diatoport S (60/80-mesh), or equivalent materials.

Operating Conditions The operating parameters may vary, depending upon the particular instrument used, but a suitable chromatogram may be obtained using the following conditions: *column temperature*, programmed from 100° to 270°, at 15°/min; *injection port temperature*, 300°; *carrier gas*, helium, flowing at a rate of 100 mL/min; *bridge current*, 140 ma; *sensitivity*, 1× for integrator (Infotronics CRS-100), 2× for recorder.

Stock Solutions Weigh accurately about 50 mg each of hydroquinone (HQ), 2,5-di-*t*-butylhydroquinone (DTBHQ), and methyl benzoate (internal standard), transfer into separate 50-mL volumetric flasks, dilute to volume with pyridine, and mix.

Calibration Standards Into separate 10-mL volumetric flasks add 0.50, 1.00, 2.00, and 3.00 mL of the HQ stock solution, then to each flask add 2.00 mL of the methyl benzoate (internal standard) stock solution, dilute each to volume with pyridine, and mix. In the same manner prepare four DTBHQ calibrating solutions. Prepare the trimethylsilyl derivative of each solution as follows: Add 9 drops of calibration solution to a 2-mL serum vial, cap the vial, evacuate with a 50-mL gas syringe, add 250 μL of *N,O*-bis-trimethylsilylacetamide, and heat at about 80° for 10 min. Chromatograph 10-μL portions of each standard in duplicate, and plot the concentration ratio of HQ to internal standard (X-axis) against the response ratio of HQ to internal standard (Y-axis). Plot the same relationships between DTBHQ and the internal standard.

Sample Preparation and Procedure Transfer about 1 g of the sample, accurately weighed, into a 10-mL volumetric flask, add 2.00 mL of the methyl benzoate internal standard stock solution, dilute to volume with pyridine, and mix. Prepare the trimethylsilyl derivative as described above under *Calibration Standards*, and then chromatograph duplicate 10-μL portions to obtain the chromatogram. The approximate peak times, in min, are methyl benzoate, 2.5; TMS derivative of HQ, 5.5; TMS derivative of *tert*-Butylhydroquinone, 7.3; TMS derivative of DTBHQ, 8.4.

Calculation Determine the peak areas (response) of interest by automatic integration or manual triangulation. Calculate the response ratio of HQ and DTBHQ to internal standard. From the calibration curves determine the concentration ratio of HQ and DTBHQ to internal standard, and calculate the percentage of HQ and the percentage of DTBHQ in the sample by the formula

$$A = Y \times I \times 10/S,$$

in which A is the percentage of HQ or the percentage of DTBHQ in the sample; Y is the concentration ratio (X-axis on calibration curve); I is the percentage (w/v) of internal standard in the *Sample Preparation*; and S is the weight of sample taken, in g.

Heavy Metals Prepare and test a 2-g sample as directed in *Method II* under *Heavy Metals Test*, Appendix IIIB, using 20 μg of lead ion (Pb) in the control (*Solution A*).

Melting Range Determine as directed for *Melting Range or Temperature*, Appendix IIB.

Toluene

Apparatus Use a suitable gas chromatograph (Appendix IIA, *Chromatography*) equipped with a flame ionization detector (F and M Model 810, or equivalent), containing a 3.66-m (12-ft) × 3.18-mm (od) stainless steel column packed with 10% Silicone SE-30, by weight, and 90% Diatoport S (60/80-mesh), or equivalent materials.

Operating Conditions The operating parameters may vary, depending upon the particular instrument used, but a suitable chromatogram may be obtained using the following conditions: *column temperature*, programmed from 70° to 280° at 15°/min and held; *injection port temperature*, 275°; *carrier gas*, helium, flowing at a rate of 50 mL/min; *cell temperature*, 300°; *hydrogen and air settings*, 20 psi each; *sensitivity*, 1×10^2.

Standard Solution Prepare a solution of toluene in octyl alcohol containing approximately 50 μg/mL, and calculate the exact concentration (C_R) in percent (w/v).

Sample Solution Transfer about 2 g of the sample, accurately weighed, into a 10-mL volumetric flask, dissolve in octyl alcohol, dilute to volume with the same solvent, and mix. Calculate the exact concentration of the solution (C_S) in percent (w/v).

Procedure Inject a 5-μL portion of the *Standard Solution* into the chromatograph, and measure the height of the toluene peak (H_R) on the chromatogram. The toluene retention time is 3.3 min; other peaks are of no interest in this analysis. Similarly, obtain the chromatogram on a 5-μL portion of the *Sample Solution*, and measure the height of the toluene peak (H_S). Calculate the percentage of toluene in the sample by the formula

$$(H_S/H_R) \times (C_R/C_S) \times 100.$$

Ultraviolet Absorbance Dissolve 1 g of L-ascorbic acid in 100 mL of ethanol and 100 mL of water contained in a 500-mL separator (S-1). Transfer about 50 g of the sample, accurately weighed, into the separator, shake to dissolve, then add 50 mL of isooctane, and extract for 3 min. After the phases have separated, drain the lower, aqueous phase into a second 500-mL separator (S-2), then after 1 min of further separating, drain the lower layer into the separator (S-2). Add a second 50-mL portion of isooctane to the aqueous solution in S-2, and repeat the extraction procedure as previously described, drawing off the lower, aqueous layer into a third 500-mL separator (S-3). Add a third 50-mL portion of isooctane to the aqueous solution in S-3, and repeat the extraction procedure as previously described, drawing off and discarding the lower aqueous layer.

Extract each isooctane solution (i.e., the solutions in S-1, -2, -3) with two 100-mL portions of a 0.5% solution of ascorbic acid in ethanol–water (25/75). Shake each mixture for 1 min, allow the phases to separate, and discard the lower, aqueous layers. Next, extract each isooctane solution with two 100-mL portions of a 5% solution of ethanol in water, and discard the lower, aqueous layers. Finally, wash each solution twice with 100 mL of water, and discard the washes.

Lightly pack a standard-size chromatographic tube with 100 g of anhydrous sodium sulfate, and wash the packed column with 75 mL of isooctane, discarding the wash. Filter the isooctane solution from S-1 through the column, and collect the filtrate in a 500-mL distillation flask. Wash S-1 with the isooctane solution contained in S-2, and then pour the solution onto the column, collecting the filtrate in the flask. Wash S-2 and S-1, successively, with the isooctane solution in S-3, and filter the solution through the column as before. Wash S-3, S-2, and S-1 in that order and in tandem with two successive 25-mL portions of isooctane, and pass the washings individually through the column and into the flask. Let the column drain completely.

Add 2 mL of hexadecane and 2 boiling stones to the 500-mL distillation flask containing the combined isooctane extracts, and attach the flask to a suitable vacuum distillation assembly. Evacuate the assembly to about one-third atmosphere, then immerse the flask in a steam bath, and distill the solvent. When isooctane stops dripping into the receiver, turn off the vacuum, wash down the walls of the flask with 5 mL of isooctane added through the top of the distillation head, then replace the thermometer and again evacuate. The isooctane should distill over in about 1 min. At the end of this distillation, add another 5-mL portion of isooctane, and repeat the stripping procedure.

Quantitatively wash the residue from the distillation flask into a 50-mL volumetric flask with isooctane, dilute to volume with isooctane, and mix. Determine the ultraviolet absorption spectrum of the solution in a 5-cm silica cell from 400 nm to 250 nm, with a suitable spectrophotometer, using isooctane as the blank. Determine the absorbance of a solvent control by following the above procedure in every detail, but with the sample omitted. From the sample spectrum determine the maximum absorbance per cm pathlength in each of the following wavelength intervals: (a) 280 to 289 nm; (b) 290 to 299 nm; (c) 300 to 359 nm; and (d) 360 to 400 nm. Calculate the maximum net absorbance per cm in each interval by subtracting from the sample absorbance the corresponding absorbance per cm of the solvent control. The following net absorbance values are not exceeded at the indicated intervals: (a) 0.15; (b) 0.12; (c) 0.08; and (d) 0.02.

Packaging and Storage Store in well-closed containers.

REQUIREMENTS

Acid Value Less than 8.
Arsenic (as As) Not more than 3 mg/kg.
Heavy Metals (as Pb) Not more than 0.002%.
Lead Not more than 3 mg/kg.
Melting Point Not less than 155°.

TESTS

Acid Value Dissolve about 3 g of the sample, accurately weighed, in 100 mL of a mixture of 75 mL of benzene and 36 mL of alcohol previously neutralized to phenolphthalein TS with sodium hydroxide. Add 25 mL of a saturated solution of sodium chloride, then add 10 g in addition of sodium chloride and a few drops of phenolphthalein TS, and titrate with 0.1 N alcoholic potassium hydroxide to the first pink color that persists for at least 30 s. Calculate the acid value by the formula

$$56.1 \times V \times N/W,$$

in which V is the volume, in mL, and N is the normality, respectively, of the alcoholic potassium hydroxide solution, and W is the weight, in g, of the sample.

Arsenic Prepare a *Sample Solution* as directed in the general method under *Chewing Gum Base*, Appendix IV. This solution meets the requirements of the *Arsenic Test*, Appendix IIIB.

Heavy Metals Prepare and test a 1-g sample as directed in *Method II* under the *Heavy Metals Test*, Appendix IIIB, using 20 μg of lead ion (Pb) in the control (*Solution A*).

Lead Prepare a *Sample Solution* as directed in the general method under *Chewing Gum Base*, Appendix IV. This solution meets the requirements of the *Lead Limit Test*, Appendix IIIB, using 10 μg of lead ion (Pb) in the control.

Melting Point Determine as directed for *Class Ib* substances in the general method, Appendix IIB.

Packaging and Storage Store in well-closed containers.

Terpene Resin, Natural

DESCRIPTION

A natural terpene occurring in some coal seams. The resin is separated from coal in froth flotation cells. The crude resin is leached with hexane, and the solution produced is freed of suspended matter by pressure filtration. The resin is concentrated in a pre-evaporator, and most of the solvent is removed in a melter-evaporator. The remaining solvent is removed in a spray dryer.

Functional Use in Foods Masticatory substance in chewing gum base.

Terpene Resin, Synthetic

DESCRIPTION

A synthetic resin composed essentially of polymers of alpha-pinene, beta-pinene, and/or dipentene. The polymer is prepared by a batch or continuous process and is usually purified by steam and water washings. It is insoluble in water. Its color is less than 4 on the Gardner scale (measured in 50% mineral spirit solution).

Functional Use in Foods Masticatory substance in chewing gum base.

REQUIREMENTS

Acid Value Less than 5.
Arsenic (as As) Not more than 3 mg/kg.
Heavy Metals (as Pb) Not more than 0.002%.
Lead Not more than 3 mg/kg.
Saponification Value Less than 5.

TESTS

Acid Value Determine as directed for *Acid Value* under *Fats and Related Substances*, Appendix VII.
Arsenic Prepare a *Sample Solution* as directed in the general method under *Chewing Gum Base*, Appendix IV. This solution meets the requirements of the *Arsenic Test*, Appendix IIIB.
Heavy Metals Prepare and test a 1-g sample as directed in *Method II* under the *Heavy Metals Test*, Appendix IIIB, using 20 μg of lead ion (Pb) in the control (*Solution A*).
Lead Prepare a *Sample Solution* as directed in the general method under *Chewing Gum Base*, Appendix IV. This solution meets the requirements of the *Lead Limit Test*, Appendix IIIB, using 10 μg of lead ion (Pb) in the control.
Saponification Value Determine as directed for *Saponification Value* under *Fats and Related Substances*, Appendix VII.

Packaging and Storage Store in well-closed containers.

Thiamine Hydrochloride

Aneurine Hydrochloride; Thiamine Chloride; Vitamin B_1; Vitamin B_1 Hydrochloride; Thiamine Hydrochloride

$C_{12}H_{17}ClN_4OS \cdot HCl$ Formula wt 337.27

CAS: [67-03-8]

DESCRIPTION

Small, white to yellowish white crystals, or crystalline powder, usually having a slight, characteristic odor. When exposed to air, the anhydrous product rapidly absorbs about 4% of water. It melts at about 248° with some decomposition. One g dissolves in about 1 mL of water and in about 100 mL of alcohol. It is soluble in glycerin, and is insoluble in ether.

Functional Use in Foods Nutrient; dietary supplement.

REQUIREMENTS

Identification
A. The infrared absorption spectrum of a potassium bromide dispersion of the sample, previously dried at 105° for 2 h, exhibits maxima only at the same wavelengths as that of a similar preparation of USP Thiamine Hydrochloride Reference Standard.
B. A 1 in 50 solution gives positive tests for *Chloride*, Appendix IIIA.
Assay Not less than 98.0% and not more than 102.0% of $C_{12}H_{17}ClN_4OS \cdot HCl$, calculated on the anhydrous basis.
Color of Solution Passes test.
Heavy Metals Not more than 10 mg/kg.
Nitrate Passes test.
pH of a 1 in 100 Solution Between 2.7 and 3.4.
Residue on Ignition Not more than 0.2%.
Water Not more than 5.0%.

TESTS

Assay

Solution A Prepare a 0.005 M solution of sodium 1-octanesulfonate in dilute glacial acetic acid (1 to 100).

Solution B Prepare a mixture of methanol and acetonitrile (3:2).

Mobile Phase Prepare a mixture of *Solution A* and *Solution B* (60:40), filter, and degas. Make adjustments to the *Mobile Phase*, if necessary, to obtain baseline separation of thiamine hydrochloride and methyl benzoate.

Internal Standard Solution Transfer 2.0-mL of methyl benzoate to a 100-mL volumetric flask, dilute with methanol to volume, and mix.

Standard Preparation Dissolve an accurately weighed quantity of USP Thiamine Hydrochloride Reference Standard in *Mobile Phase* to obtain a solution having a known concentration of about 1 mg/mL. Transfer 20.0 mL of this solution to a 50-mL volumetric flask, add 5.0 mL of *Internal Standard Solution*, dilute with *Mobile Phase* to volume, and mix to obtain a *Standard Preparation* having a known concentration of about 400 μg/mL.

Assay Preparation Transfer an accurately weighed quantity of about 200 mg of the sample to a 100-mL volumetric flask, dissolve in *Mobile Phase*, dilute with *Mobile Phase* to volume, and mix. Transfer 10.0 mL of this solution to a 50-mL volumetric flask, add 5.0 mL of *Internal Standard Solution*, dilute with *Mobile Phase* to volume, and mix.

Procedure Use a high-pressure liquid chromatograph (see *Chromatography*, Appendix IIA) equipped with an ultraviolet detector that measures at 254 nm. Under typical conditions, the instrument contains a 300- × 4-mm column packed with 10-μm octadecylsilanized silica (μBondapak C 18, or equivalent). The flow rate is about 1 mL/min. The resolution between the thiamine and methyl benzoate peaks is not less than 4.0, the tailing factor for the thiamine peak is not more than 2.0, the column efficiency determined from the thiamine peak is not less than 1500 theoretical plates. Three replicate injections of the *Standard Preparation* show a relative standard deviation of not more than 2.0%. Separately inject about 10-μL portions of the *Standard Preparation* and the *Assay Preparation* into the chromatograph, and record the chromatograms. Measure the peak area responses of the major peaks. Calculate the quantity, in mg, of $C_{12}H_{17}ClN_{14}OS \cdot HCl$ in the sample taken by the formula

$$0.5C(R_U/R_S),$$

in which C is the concentration, in µg/mL, of USP Thiamine Hydrochloride Reference Standard in the *Standard Preparation*, and R_U and R_S are the ratios of the peak area ratios of thiamine to methyl benzoate obtained from the *Assay Preparation* and the *Standard Preparation*, respectively.

Color of Solution Dissolve 1.0 g in water to make 10 mL. This solution exhibits no more color than a dilution of 1.5 mL of 0.1 N potassium dichromate in water to make 1000 mL.

Heavy Metals A solution of 2 g in 25 mL of water meets the requirements of the *Heavy Metals Test*, Appendix IIIB, using 20 µg of lead ion (Pb) in the control (*Solution A*).

Nitrate To 2 mL of a 1 in 50 solution add 2 mL of sulfuric acid, cool, and superimpose 2 mL of ferrous sulfate TS. No brown ring is produced at the junction of the two layers.

pH Determine by the *Potentiometric Method*, Appendix IIB.

Residue on Ignition, Appendix IIC Ignite 1 g as directed in the general method.

Water Proceed as directed under *Water Determination* in the *Karl Fischer Titrimetric Method*, Appendix IIB.

Packaging and Storage Store in tight, light-resistant containers.

Thiamine Mononitrate

Thiamine Nitrate; Vitamin B_1; Vitamin B_1 Mononitrate; Thiamine Mononitrate

$C_{12}H_{17}N_5O_4S$ Formula wt 327.36

CAS: [532-43-4]

DESCRIPTION

White to yellowish white crystals, or crystalline powder, usually having a slight characteristic odor. One g dissolves in about 35 mL of water. It is slightly soluble in alcohol and in chloroform.

Functional Use in Foods Nutrient; dietary supplement.

REQUIREMENTS

Identification

A. To 2 mL of a 1 in 50 solution add 2 mL of sulfuric acid, cool, and superimpose 2 mL of ferrous sulfate TS. A brown ring is produced at the junction of the two liquids.

B. Dissolve about 5 mg of the sample in a mixture of 1 mL of lead acetate TS and 1 mL of sodium hydroxide solution (1 in 10). A yellow color is produced. Heat the mixture for several min on a steam bath. The color changes to brown, and, on standing, a precipitate of lead sulfide separates.

C. A 1 in 10 solution of the sample yields a white precipitate with mercuric chloride TS, and a red brown precipitate with iodine TS. It is precipitated also by mercuric–potassium iodide TS and by trinitrophenol TS.

D. Dissolve about 5 mg of the sample in 5 mL of 0.5 N sodium hydroxide, add 0.5 mL of potassium ferricyanide TS and 5 mL of isobutyl alcohol, shake vigorously for 2 min, and allow the layers to separate. When illuminated from above by a vertical beam of ultraviolet light and viewed at a right angle to this beam, the air–liquid meniscus shows a vivid blue fluorescence, which disappears when the mixture is slightly acidified but reappears when it is again made alkaline.

Assay Not less than 98.0% and not more than 102.0% of $C_{12}H_{17}N_5O_4S$, calculated on the dried basis.

Chloride Not more than 0.06%.

Heavy Metals Not more than 10 mg/kg.

Loss on Drying Not more than 1.0%.

pH of a 1 in 50 Solution Between 6.0 and 7.5.

Residue on Ignition Not more than 0.2%.

TESTS

Assay Proceed as directed in the *Assay* in the monograph for *Thiamine Hydrochloride*. Calculate the quantity, in mg, of $C_{12}H_{17}N_5O_4S$ in the sample taken by the formula

$$(327.36/337.27)0.5C(R_U/R_S),$$

in which 327.36 and 337.27 are the formula weights of Thiamine Mononitrate and Thiamine Hydrochloride, respectively, C is the concentration, in µg/mL, of USP Thiamine Hydrochloride Reference Standard in the *Standard Preparation*, and R_U and R_S are the peak area ratios of thiamine to methyl benzoate obtained from the *Assay Preparation* and the *Standard Preparation*, respectively.

Chloride, Appendix IIIB Any turbidity produced by a 25-mg sample does not exceed that shown in a control containing 15 µg of chloride ion (Cl).

Heavy Metals Prepare and test a 2-g sample as directed in *Method II* under the *Heavy Metals Test*, Appendix IIIB, using 20 µg of lead ion (Pb) in the control (*Solution A*).

Loss on Drying, Appendix IIC Dry about 500 mg, accurately weighed, at 105° for 2 h.

pH Determine by the *Potentiometric Method*, Appendix IIB.

Residue on Ignition, Appendix IIC Ignite 1 g as directed in the general method.

Packaging and Storage Store in tight, light-resistant containers.

L-Threonine

L-2-Amino-3-hydroxybutyric Acid

$$CH_3CHCHCOOH$$
$$\quad\ |\quad\ |$$
$$\ HO\ \ NH_2$$

$C_4H_9NO_3$ Formula wt 119.12

CAS: [72-19-5]

DESCRIPTION

A white, odorless, crystalline powder having a slightly sweet taste. It is freely soluble in water, but insoluble in alcohol, in ether, and in chloroform. It melts with decomposition at about 256°.

Functional Use in Foods   Nutrient; dietary supplement.

REQUIREMENTS

Identification
  A. To 5 mL of a 1 in 1000 solution add 1 mL of triketohydrindene hydrate TS. A reddish purple or purple color is produced.
  B. To 5 mL of a 1 in 10 solution add 5 mL of a saturated solution of potassium periodate, and heat. Ammonia is evolved.
Assay   Not less than 98.5% and not more than 101.5% of $C_4H_9NO_3$, calculated on the dried basis.
Heavy Metals (as Pb)   Not more than 0.002%.
Lead   Not more than 10 mg/kg.
Loss on Drying   Not more than 0.2%.
Residue on Ignition   Not more than 0.1%.
Specific Rotation $[\alpha]_D^{20°}$: Between $-26.5°$ and $-29.0°$ after drying; or $[\alpha]_D^{25°}$: Between $-25.8°$ and $-28.8°$ after drying.

TESTS

Assay   Dissolve about 200 mg, accurately weighed, in 3 mL of formic acid and 50 mL of glacial acetic acid, add 2 drops of crystal violet TS, titrate with 0.1 N perchloric acid to a green endpoint or until the blue color disappears completely. Perform a blank determination (see *General Provisions*), and make any necessary correction. Each mL of 0.1 N perchloric acid is equivalent to 11.91 mg of $C_4H_9NO_3$.
Heavy Metals   Prepare and test a 1-g sample as directed in *Method II* under the *Heavy Metals Test*, Appendix IIIB, using 20 µg of lead ion (Pb) in the control (*Solution A*).
Lead   A *Sample Solution* prepared as directed for organic compounds meets the requirements of the *Lead Limit Test*, Appendix IIIB, using 10 µg of lead ion (Pb) in the control.
Loss on Drying, Appendix IIC   Dry at 105° for 3 h.
Residue on Ignition, Appendix IIC   Ignite 1 g as directed in the general method.
Specific Rotation, Appendix IIB   Determine in a solution containing 6 g of a previously dried sample in sufficient water to make 100 mL.

Packaging and Storage   Store in well-closed containers.

Thyme Oil

CAS: [8007-46-3]

DESCRIPTION

The volatile oil obtained by distillation from the flowering plant *Thymus vulgaris* L., or *Thymus zygis* L., and its var. *gracilis* Boissier (Fam. *Labiatae*). It is a colorless, yellow, or red liquid with a characteristic, pleasant odor and a pungent, persistent taste. It is affected by light.

Functional Use in Foods   Flavoring agent.

REQUIREMENTS

Identification   The infrared absorption spectrum of the sample exhibits relative maxima (that may vary in intensity) at the same wavelengths (or frequencies) as those shown in the respective spectrum in the section on *Infrared Spectra (Series A: Essential Oils)*, using the same test conditions as specified therein.
Assay   Not less than 40%, by volume, of phenols.
Angular Rotation   Levorotatory, but not more than $-3°$.
Heavy Metals (as Pb)   Passes test.
Refractive Index   Between 1.495 and 1.505 at 20°.
Solubility in Alcohol   Passes test.
Specific Gravity   Between 0.915 and 0.935.
Water-Soluble Phenols   Passes test.

TESTS

Assay   Proceed as directed under *Phenols*, Appendix VI, allowing the mixture to stand overnight, then adding sufficient 1 N potassium hydroxide to raise the lower limit of the oily layer into the graduated portion of the neck of the flask. After the solution has become clear, adjust the temperature and read the volume of the residual liquid.
Angular Rotation   Determine in a 100-mm tube as directed under *Optical (Specific) Rotation*, Appendix IIB.
Heavy Metals   Shake 10 mL of the oil with an equal volume of water to which 1 drop of hydrochloric acid has been added, and pass hydrogen sulfide through the mixture until it is saturated. No darkening in color is produced in either the oil or the water.
Refractive Index, Appendix IIB   Determine with an Abbé or other refractometer of equal or greater accuracy.
Solubility in Alcohol   Proceed as directed in the general method, Appendix VI. One mL dissolves in 2 mL of 80% alcohol.
Specific Gravity   Determine by any reliable method (see *General Provisions*).

Water-Soluble Phenols Shake a 1-mL sample with 10 mL of hot water, and after cooling, pass the water layer through a moistened filter. Not even a transient blue or violet color is produced in the filtrate upon the addition of 1 drop of ferric chloride TS.

Packaging and Storage Store in full, tight, light-resistant containers in a cool place.

Titanium Dioxide

TiO$_2$ Formula wt 79.88

INS: 171   CAS: [13463-67-7]

DESCRIPTION

Titanium Dioxide occurs as a white, odorless, tasteless, amorphous powder that is prepared synthetically. It is insoluble in water, in hydrochloric acid, in dilute sulfuric acid, and in alcohol and other organic solvents. It dissolves slowly in hydrofluoric acid and in hot sulfuric acid.

Functional Use in Foods Color.

REQUIREMENTS

Identification Add 5 mL of sulfuric acid to 500 mg of the sample, heat gently until fumes of sulfur trioxide appear, continue heating for at least 10 s, and cool. Cautiously dilute to about 100 mL with water, and filter. When a few drops of hydrogen peroxide TS are added to 5 mL of the clear filtrate, a yellow-red to orange-red color appears immediately.

Assay Not less than 99.0% and not more than 100.5% of TiO$_2$, after drying, after correcting for any aluminum oxide and/or silicon dioxide found to be present in the Titanium Dioxide.

Acid-Soluble Substances Not more than 0.5%.

Aluminum Oxide and/or Silicon Dioxide Not more than 2.0%, either singly or combined.

Antimony Not more than 2 mg/kg.

Arsenic (as As) Not more than 1 mg/kg.

Lead Not more than 10 mg/kg.

Loss on Drying Not more than 0.5%.

Loss on Ignition Not more than 0.5%, after drying.

Mercury Not more than 1 mg/kg.

Water-Soluble Substances Not more than 0.3%.

TESTS

Assay Transfer about 300 mg of the sample, previously dried at 105° for 3 h and accurately weighed, into a 250-mL beaker, add 20 mL of sulfuric acid and 7 to 8 g of ammonium sulfate, and mix. Heat on a hot plate until fumes of sulfur trioxide appear, and continue heating over a strong flame until the sample dissolves or it is apparent that the undissolved residue is siliceous matter. Cool, cautiously dilute with 100 mL of water, and stir. Heat carefully to boiling while stirring, allow the insoluble matter to settle, and filter. Transfer the entire residue to the filter, and wash thoroughly with cold 2 N sulfuric acid. Dilute the filtrate with water to 200 mL, and cautiously add about 10 mL of ammonium hydroxide to reduce the acid concentration to about 5%, by volume, of sulfuric acid.

Prepare a zinc amalgam column in a 25-cm Jones reductor tube, placing a pledget of glass wool in the bottom of the tube and filling the constricted portion of the tube with zinc amalgam prepared as follows: Add 20- to 30-mesh zinc to a 2% mercuric chloride solution, using about 100 mL of the solution for each 100 g of zinc. After about 10 min, decant the solution from the zinc, then wash the zinc with water by decantation. Transfer the zinc amalgam to the reductor tube, and wash the column with 100-mL portions of 2 N sulfuric acid until 100 mL of the washing does not decolorize 1 drop of 0.1 N potassium permanganate.

Place 50 mL of ferric ammonium sulfate TS in a 500-mL suction flask, and add 0.1 N potassium permanganate until a faint pink color persists for 5 min. Attach the Jones reductor tube, containing the zinc amalgam column, to the neck of the flask, and pass 50 mL of 2 N sulfuric acid through the tube at a rate of about 30 mL/min. Pass the prepared titanium solution through the column at the same rate, followed by 100 mL each of 2 N sulfuric acid and water. During these operations, keep the tube filled with solution or water above the upper level of the amalgam column. Gradually release the suction, wash down the outlet tube and the sides of the receiver, and titrate immediately with 0.1 N potassium permanganate. Perform a blank determination (see *General Provisions*), substituting 200 mL of dilute sulfuric acid (1 in 20) for the sample solution, and make any necessary correction. Each mL of 0.1 N potassium permanganate is equivalent to 7.990 mg of TiO$_2$.

Acid-Soluble Substances Suspend 5.0 g of the sample in 100 mL of 0.5 N hydrochloric acid, and heat on a steam bath for 30 min with occasional stirring. Filter through a suitable tared, porous-bottom porcelain filter crucible. Wash with three 10-mL portions of 0.5 N hydrochloric acid, evaporate the combined filtrate and washings to dryness, and ignite at 450° ± 25° to constant weight.

Note: Avoid exposing the crucible to sudden temperature changes.

Aluminum Oxide

0.01 M Zinc Sulfate Dissolve 2.90 g of zinc sulfate, ZnSO$_4$·7H$_2$O, in sufficient water to make 1000 mL. Standardize the solution as follows: Dissolve 500 mg of high-purity (99.9%) aluminum wire, accurately weighed, in 20 mL of hydrochloric acid, heating gently to effect solution, then transfer into a 1000-mL volumetric flask, dilute to volume with water, and mix. Transfer a 10.0-mL aliquot of this solution into a 500-mL Erlenmeyer flask containing 90 mL of water and 3 mL of hydrochloric acid, add 1 drop of methyl orange TS and 25.0 mL of 0.02 M disodium EDTA, and continue as directed below under *Sample Solution B*, beginning with "Add dropwise ammonium hydroxide solution (1 in 5) until. . . ." Calculate the titer T of the zinc sulfate solution by the formula

$$T = (18.896 \times W)/V,$$

in which T is in terms of mg of Al_2O_3 per mL of zinc sulfate solution; W is the weight, in g, of the aluminum wire taken; V is the volume, in mL, of the zinc sulfate solution consumed in the second titration; and 18.896 is a factor derived as follows:

$$(\text{mol wt } Al_2O_3/\text{mol wt Al}) \times (1000 \text{ mg/g}) \times (10 \text{ mL}/2).$$

Sample Solution A Fuse 1 g of the sample, accurately weighed, with 10 g of sodium bisulfate ($NaHSO_4 \cdot H_2O$) contained in a 250-mL high-silica glass Erlenmeyer flask.

Caution: Do not use more sodium bisulfate than specified, since an excess concentration of salt will interfere with the EDTA titration later on in the procedure.

Begin heating at low heat on a hot plate, and then gradually raise the temperature until full heat is reached. When spattering has stopped and light fumes of SO_3 appear, heat in the full flame of a Meker burner, with the flask tilted so that the fusion is concentrated at one end of the flask. Swirl constantly until the melt is clear (except for silica content), but guard against prolonged heating to avoid precipitation of Titanium Dioxide. Cool, add 25 mL of sulfuric acid solution (1 in 2), and then heat until the mass has dissolved and a clear solution results. Cool, and dilute to 120 mL with water.

Sample Solution B Measure out 200 mL of approximately 6.25 M sodium hydroxide, and add 65 mL of it to *Sample Solution A* while stirring constantly with a magnetic stirrer; pour the remaining 135 mL of the alkali solution into a 500-mL volumetric flask. Slowly and with constant stirring add the sample mixture to the alkali solution in the 500-mL volumetric flask, then dilute to volume with water, and mix.

Note: If the procedure is delayed at this point for more than 2 h, store the contents of the volumetric flask in a polyethylene bottle.

Allow most of the precipitate to settle out (or centrifuge for 5 min), and then filter the supernatant liquid through a very fine filter paper. Label the filtrate *Sample Solution B*.

Sample Solution C Transfer 100.0 mL of *Sample Solution B* into a 500-mL Erlenmeyer flask, add 1 drop of methyl orange TS, acidify with hydrochloric acid solution (1 in 2), and then add about 3 mL in excess. Add 25.0 mL of 0.02 M disodium EDTA, and mix.

Note: If the approximate Al_2O_3 content is known, calculate the optimum volume of EDTA solution to be added by the formula

$$(4 \times \% \ Al_2O_3) + 5.$$

Add dropwise ammonium hydroxide solution (1 in 5) until the color is just completely changed from red to orange-yellow, then add 10 mL of ammonium acetate buffer solution (77 g of ammonium acetate plus 10 mL of glacial acetic acid, diluted to 1000 mL with water) and 10 mL of dibasic ammonium phosphate solution (150 g of dibasic ammonium phosphate in 700 mL of water, adjusted to pH 5.5 with a 1 in 2 solution of hydrochloric acid, then diluted to 1000 mL with water.) Boil for 5 min, cool quickly to room temperature in a stream of running water, add 3 drops of xylenol orange TS, and mix. If the solution is purple, yellow-brown, or pink, bring the pH to 5.3 to 5.7 by the addition of acetic acid; at the desired pH a pink color indicates that not enough of the EDTA solution has been added, in which case another 100 mL of *Sample Solution B* should be taken and treated as directed from the beginning of this paragraph, except that 50.0 mL, rather than 25.0 mL, of 0.02 M disodium EDTA should be used.

Procedure Titrate *Sample Solution C* with 0.01 M zinc sulfate to the first yellow-brown or pink endpoint color that persists for 5 to 10 s.

Caution: This titration should be performed quickly near the endpoint by adding rapidly 0.2-mL increments of the titrant until the first color change occurs; although the color will fade in 5 to 10 s, it is the true endpoint. Failure to observe the first color change will result in an incorrect titration. The fading endpoint does not occur at the second endpoint. This first titration should require more than 8 mL of titrant, but for more accurate work a titration of 10 to 15 mL is desirable.

Add 2 g of sodium fluoride, boil the mixture for 2 to 5 min, and cool in a stream of running water. Titrate the EDTA (which is released by fluoride from its aluminum complex) with 0.01 M zinc sulfate to the same fugitive yellow-brown or pink endpoint as described above.

Calculation Calculate the percentage of aluminum oxide (Al_2O_3) in the sample taken by the formula

$$(V \times T)/(2 \times S),$$

in which V is the volume, in mL, of 0.01 M zinc sulfate consumed in the second titration; T is the titer of the zinc sulfate solution, determined previously; and S is the weight, in g, of the sample taken.

Antimony

Stock Standard Solution Transfer 274.28 mg of antimony potassium tartrate, $C_4H_4KO_7Sb \cdot \frac{1}{2}H_2O$, into a 100-mL volumetric flask, dissolve in 6 N hydrochloric acid, dilute to volume with the same solvent, and mix. Each mL contains 1 mg of Sb.

Diluted Standard Solution Pipet 2.00 mL of the *Stock Standard Solution* into a 100-mL volumetric flask, dilute to volume with 6 N hydrochloric acid, and mix. Each mL contains 20 µg of Sb. Prepare this solution fresh weekly.

Standard Preparation Transfer 1.00 mL of the *Diluted Standard Solution* into a 250-mL separator, and add 25 mL of mercuric chloride solution (6% in hydrochloric acid) and 25 mL of hydrochloric acid.

Sample Preparation Transfer 10.00 g of the sample into a 250-mL beaker, add 50 mL of hot 0.5 N hydrochloric acid, cover the beaker, and boil the slurry for 15 min with occasional stirring. Remove from heat, allow to settle for a few s, and decant through a double Whatman No. 42 filter paper plus a No. 12 fluted paper, all previously washed with 0.5 N hydrochloric acid. Evaporate the filtrate slowly on a hot plate until the volume is slightly less than 20 mL, then cool, and transfer to a 25-mL graduate. Rinse the beaker with 5 mL of 0.5 N hydrochloric acid, and add to the graduate. Dilute to the 25-mL mark with 0.5 N hydrochloric acid, and transfer into a 250-mL separator. Rinse the beaker and graduate with a total of 25 mL of mercuric chloride solution (6% in hydrochloric acid) and with a total of 25 mL of hydrochloric acid, adding the washings to the separator.

Procedure To the *Sample Preparation* contained in the separator, add 1.0 mL of 0.1 N ceric sulfate (prepared by dissolving 3.3 g of $Ce(SO_4)_2$ in 100 mL of aqueous solution containing 3 mL of sulfuric acid), and start a stopwatch at the moment of first addition. Mix the solutions together, and pass a stream of clean air over the mixture. At exactly 1.0 min, add 75 mL of water, mix, and continue passing air over the solution. At 2.0 min, add 8 drops of a 1% solution of hydroxylamine hydrochloride, mix, and continue passing air over the solution. At 3.0 min, add 5.0 mL of a 0.2% solution of rhodamine B. Pipet 50.00 mL of benzene into the separator, shake for 1 min, and allow the layers to separate for 90 s. Discard the aqueous layer and a small portion of the organic phase. Transfer about 15 mL of the organic phase to a centrifuge tube, and centrifuge at high speed for 1 min. Determine the absorbance of the clarified solution in a 1-cm cell at 565 nm, with a suitable spectrophotometer, referring to water after having compared benzene in the sample cell to water in the reference cell. The color is stable for several min; it should be measured within 15 to 20 min after starting the stopwatch.

> Note: The colloidal antimony color complex may resist rinsing from the cell with benzene, in which case the cell should be rinsed in succession with dilute nitric acid, hot water, acetone, and benzene. Check the absorbance of the cell with benzene against water contained in the reference cell.

The absorbance produced by the solution from the *Sample Preparation*, after correction for a reagent blank, is not greater than that produced by the *Standard Preparation*, treated in the same manner in the above procedure as the *Sample Preparation*, beginning with "...add 1.0 mL of 0.1 N ceric sulfate...."

Sample Solution for the Determination of Arsenic and Lead Transfer 10.00 g of the sample into a 250-mL beaker, add 50 mL of 0.5 N hydrochloric acid, cover with a watch glass, and heat to boiling on a hot plate. Boil gently for 15 min, then pour the slurry into a 100- to 150-mL centrifuge bottle, and centrifuge for 10 to 15 min, or until undissolved material settles. Decant the supernatant extract through a Whatman No. 4 filter paper, collecting the filtrate in a 100-mL volumetric flask and retaining as much as possible of the undissolved material in the centrifuge bottle. Add 10 mL of hot water to the original beaker, washing off the watch glass with the water, and pour the slurry into the centrifuge bottle. Form a slurry, using a glass stirring rod, and centrifuge. Decant through the same filter paper, and collect the washings in the volumetric flask containing the initial extract. Repeat the entire washing process with two additional 10-mL portions of hot water. Finally, wash the filter paper with 10 to 15 mL of hot water. Cool the contents of the flask to room temperature, dilute to volume with water, and mix.

Arsenic A 30-mL portion of the *Sample Solution* meets the requirements of the *Arsenic Test*, Appendix IIIB.

Lead A 20-mL portion of the *Sample Solution* meets the requirements of the *Lead Limit Test*, Appendix IIIB, using 20 µg of lead ion (Pb) in the control.

Loss on Drying, Appendix IIC Dry at 105° for 3 h.

Loss on Ignition Ignite a 2-g sample, previously dried at 105° for 3 h, at 800° ± 25° to constant weight.

Mercury

Apparatus The apparatus consists of a source of nitrogen (supplied through a regulator or flowmeter capable of measuring a flow rate of 1 L/min) connected to a suitable quartz combustion tube contained in a hinged furnace (Type 70 T, Arthur H. Thomas Co., or equivalent) in which the sample is pyrolyzed at 650°. The exit end of the combustion tube is connected to the optical cell of a suitable mercury vapor meter (Beckman Model K-23, or equivalent), the microammeter of which is connected in parallel through an attenuator to a 1-mV strip chart recorder. The quartz combustion tube (48.3 cm in length and 18.3 mm in outside diameter) is fitted at each end with Pyrex ball-joint adaptors and is packed near the exit end with 40 g of copper oxide (held in place by small wads of quartz wool), the inlet end being used to hold an 88- × 12- × 8-mm combustion boat for the sample.

Standard Mercury Solution Prepare a stock solution by dissolving 1.353 g of mercuric chloride in water and diluting to volume in a 1000-mL volumetric flask. Pipet 1.0 mL of this solution into a 100-mL volumetric flask, dilute to volume with water, and mix to obtain a working solution containing 0.01 µg of Hg per µL.

Standardization of Mercury Vapor Meter Preheat the combustion tube furnace to 650°, and adjust the nitrogen flow to 1 L/min. Standardize the meter in accordance with the manufacturer's instructions, using the internal standard with which the instrument is equipped. Adjust the attenuator so that the scale on the recorder is 200 mV. Under these conditions, a meter reading of 0.078 mg/m^3, obtained with the internal standard, is 50% full scale on the recorder. Check the standardization of the instrument periodically and adjust as necessary.

Calibration of Mercury Vapor Meter Prepare a set of mercury standards containing 0.01, 0.02, and 0.03 µg of Hg by pipetting the required amount of the working standard onto 1- × 0.5- × 0.1-cm pieces of asbestos, previously ignited at 800° for 1 h, contained in separate combustion boats. Cover the asbestos pads with 1 to 2 g of fine granular anhydrous sodium carbonate, previously checked for absence of mercury by this procedure. Place a standard in the tube furnace, and close the inlet with a ball joint sealed at one end and held in place by a clamp. After 1 min, start the gas flow by connecting the nitrogen supply tube to the inlet port of the combustion tube. Record the maximum response from either the observed meter deflection or the chart record for each standard, and prepare a standard curve of the response versus amount of mercury added. The mercury vapor meter should be calibrated each time a series of samples is run, and the calibration should be checked periodically by running a single standard.

Sample Analysis Place 25 mg of the sample in a combustion boat, and cover the sample with 1 to 2 g of the sodium carbonate as described above. Ignite the sample as described above for the standards, record the maximum response, and determine the amount of mercury in the sample by reference to the standard curve.

Silicon Dioxide Fuse 1 g of the sample, accurately weighed, with 10 g of sodium bisulfate (NaHSO$_4$·H$_2$O) contained in a 250-mL high-silica glass Erlenmeyer flask. Heat gently over a Meker burner, while swirling the flask, until decomposition and fusion are complete and the melt is clear, except for the silica content, and then cool.

> **Caution:** Do not overheat the contents of the flask at the beginning, and heat cautiously during fusion to avoid spattering.

To the cold melt add 25 mL of sulfuric acid solution (1 in 2), and heat very carefully and very slowly until the melt is dissolved. Cool, and carefully add 150 mL of water, pouring very small portions down the sides of the flask, with frequent swirling, to avoid overheating and spattering. Allow the contents of the flask to cool, and then filter through fine ashless filter paper, using a 60° gravity funnel. Wash out all of the silica from the flask onto the filter paper with sulfuric acid solution (1 in 10). Transfer the filter paper and its contents into a tared platinum crucible, dry in an oven at 120°, and then heat the partly covered crucible over a Bunsen burner. To prevent flaming of the filter paper, heat first the cover from above, and then the crucible from below. When the filter paper is consumed, transfer the crucible to a muffle furnace and ignite at 1000° for 30 min. Cool in a desiccator, and weigh. Add 2 drops of sulfuric acid (1 in 2) and 5 mL of hydrofluoric acid (sp. gr. 1.15), and carefully evaporate to dryness, first on a low-heat hot plate (to remove the HF) and then over a Bunsen burner (to remove the H$_2$SO$_4$). Take precautions to avoid spattering, especially after removal of the HF. Ignite at 1000° for 10 min, cool in a desiccator, and weigh again. Record the difference between the two weights as the content of SiO$_2$ in the sample.

Water-Soluble Substances Suspend 4.0 g of the sample in 50 mL of water, mix, and allow to stand overnight. Transfer to a 200-mL volumetric flask, add 2 mL of ammonium chloride TS, and mix. If the Titanium Dioxide does not settle, add another 2-mL portion of ammonium chloride TS, then allow the suspension to settle, dilute to volume with water, and mix. Filter through a double thickness of filter paper, discarding the first 10 mL of filtrate, and collect 100 mL of the subsequent clear filtrate. Transfer into a tared platinum dish, evaporate on a hot plate to dryness, and ignite at a dull red heat to constant weight.

Packaging and Storage Store in well-closed containers.

DL-α-Tocopherol

C$_{29}$H$_{50}$O$_2$ Formula wt 430.71

INS: 307   CAS: [2074-53-5]

DESCRIPTION

A form of vitamin E. It occurs as a yellow to amber, nearly odorless, clear, viscous oil that oxidizes and darkens in air and on exposure to light. It is insoluble in water, is freely soluble in alcohol, and is miscible with acetone, chloroform, ether, fats, and vegetable oils.

Functional Use in Foods Nutrient; dietary supplement; antioxidant.

REQUIREMENTS

Labeling Label claims in terms of former International Units (IU) should be based on the following: 1 mg DL-α-tocopherol = 1.1 IU.

Identification

A. Dissolve about 10 mg of the sample in 10 mL of absolute alcohol, add with swirling 2 mL of nitric acid, and heat at about 75° for 15 min. A bright red to orange color develops.

B. The retention time of the major peak (excluding the solvent peak) in the chromatogram of the *Assay Preparation* is the same as that of the *Standard Preparation*, both relative to the internal standard, as obtained in the *Assay*.

C. If the isomeric form is not otherwise known, determine the optical rotation (see Appendix IIB) on a 1 in 10 solution of the sample in chloroform. The specific rotation is not appreciable (approximately ± 0.05°).

Assay Not less than 96.0% and not more than 102.0% of C$_{29}$H$_{50}$O$_2$.

Acidity Passes test.

Heavy Metals (as Pb) Not more than 10 mg/kg.

TESTS

> **Note:** In the following *Assay*, use low-actinic glassware for all solutions containing tocopherols.

Assay

Internal Standard Solution Prepare a solution in *n*-hexane containing 3 mg of hexadecyl hexadecanoate, accurately weighed, in each mL.

Standard Preparation Dissolve about 30 mg of USP Alpha Tocopherol Reference Standard, accurately weighed, in 10.0 mL of the *Internal Standard Solution*.

Assay Preparation Dissolve about 30 mg of the sample, accurately weighed, in 10.0 mL of the *Internal Standard Solution*.

Chromatographic System Use a gas chromatograph equipped with a flame-ionization detector and a glass-lined sample-introduction system or on-column injection. Under typical conditions, the instrument contains a 2-m × 4-mm borosilicate glass column packed with 2% to 5% methylpolysiloxane gum on 80- to 100-mesh acid-base washed silanized chromatographic diatomaceous earth. The column is maintained isothermally between 240° and 260°, the injection port at about 290°, and the detector block at about 300°. The flow rate of dry carrier gas is adjusted to obtain a hexadecyl hexadecanoate peak approximately 18 to 20 min after sample introduction when a 2% column is used, or 30 to 32 min when a 5% column is used.

Note: Cure and condition the column as necessary.

System Suitability Chromatograph a sufficient number of injections of a mixture in *n*-hexane of 1 mg/mL each of USP Alpha Tocopherol Reference Standard and USP Alpha Tocopheryl Acetate Reference Standard, as directed under *Calibration*, to ensure that the resolution factor, R, is not less than 1.0 (see Appendix IIA).

Calibration Chromatograph successive 2- to 5-μL portions of the *Standard Preparation* until the relative response factor, F, is constant (i.e., within a range of approximately 2%) for three consecutive injections. If graphic integration is used, adjust the instrument to obtain at least 70% maximum recorder response for the hexadecyl hexadecanoate peak. Measure the areas under the major peaks occurring at relative retention times of approximately 0.51 (α-tocopherol) and 1.00 (hexadecyl hexadecanoate), and record the values as A_S and A_I, respectively. Calculate the relative response factor, F, by the formula

$$(A_S/A_I) \times (C_I/C_S),$$

in which C_I and C_S are the exact concentrations, in mg/mL, of hexadecyl hexadecanoate and of USP Alpha Tocopherol Reference Standard in the *Standard Preparation*, respectively.

Procedure Inject a suitable portion (2 to 5 μL) of the *Assay Preparation* into the chromatograph, and record the chromatogram. Measure the areas under the major peaks occurring at relative retention times of approximately 0.51 (α-tocopherol) and 1.00 (hexadecyl hexadecanoate), and record the values as a_U and a_I, respectively. Calculate the weight, in mg, of DL-α-tocopherol in the sample by the formula

$$(10C_I/F) \times (a_U/a_I).$$

Acidity Dissolve 1.0 g of the sample in 25 mL of a mixture of equal volumes of alcohol and ether that has been neutralized to phenolphthalein TS with 0.1 *N* sodium hydroxide, add 0.5 mL of phenolphthalein TS, and titrate with 0.1 *N* sodium hydroxide until the solution remains faintly pink after shaking for 30 s. Not more than 1.0 mL of 0.1 *N* sodium hydroxide is required.

Heavy Metals Place a 2-g sample in a silica crucible, and proceed as directed in *Method II* under the *Heavy Metals Test*, Appendix IIIB, using 20 μg of lead ion (Pb) in the control (*Solution A*).

Packaging and Storage Store in tight containers blanketed by inert gas and protected from heat and light.

D-α-Tocopherol Concentrate

DESCRIPTION

A form of vitamin E obtained by the vacuum steam distillation of edible vegetable oil products, comprising a concentrated form of D-α-tocopherol. It occurs as a brownish red to light yellow, nearly odorless, clear viscous oil. It oxidizes and darkens slowly in air and on exposure to light. It may contain an edible vegetable oil added to adjust the required amount of total tocopherols, and the content of D-α-tocopherol may be adjusted by suitable physical and chemical means. It is insoluble in water; is soluble in alcohol; and is miscible with acetone, chloroform, ether, and vegetable oils.

Functional Use in Foods Nutrient; dietary supplement; antioxidant.

REQUIREMENTS

Labeling Indicate the mg/g of D-α-tocopherol present. Label claims in terms of former International Units (IU) should be based on the following: 1 mg D-α-tocopherol = 1.49 IU.
Identification
A. Dissolve about 50 mg of the sample in 10 mL of absolute alcohol, add with swirling 2 mL of nitric acid, and heat at about 75° for 15 min. A bright red to orange color develops.
B. The retention time of the major peak (excluding the solvent peak) in the chromatogram of the *Assay Preparation* is the same as that of the *Standard Preparation*, both relative to the internal standard, as obtained in the *Assay*.
Assay Not less than 40.0% of total tocopherols, of which not less than 95.0% consists of D-α-tocopherol ($C_{29}H_{50}O_2$).
Acidity Passes test.
Heavy Metals (as Pb) Not more than 10 mg/kg.
Specific Rotation $[\alpha]_D^{25°}$: Not less than +24°.

TESTS

Assay, Acidity, Heavy Metals, and **Specific Rotation** Proceed as directed in the monograph for *Tocopherols Concentrate, Mixed*.

Packaging and Storage Store in tight containers blanketed by inert gas and protected from heat and light.

Tocopherols Concentrate, Mixed

INS: 306

DESCRIPTION

This monograph establishes specifications for two types of mixed tocopherols concentrate. Both types are obtained by the vacuum steam distillation of edible vegetable oil products, and both contain a specified minimum amount of total tocopherols (see **REQUIREMENTS**), differing only in the levels of the D-tocopherol forms.

The *high-alpha* type contains a relatively high proportion of D-α-tocopherol and is recognized as a form of vitamin E and also as an antioxidant. The *low-alpha* type contains a relatively high proportion of D-β-, D-γ-, and D-δ-tocopherols, with a minor level of D-α-tocopherol, and thus is not considered to be a form of vitamin E but rather an antioxidant. Both types may contain an edible vegetable oil added to adjust the required amount of total tocopherols, and the tocopherol forms may be adjusted by suitable physical or chemical means.

Mixed tocopherols concentrate occurs as a brownish red to light yellow, clear, viscous oil having a mild, characteristic odor and taste. It may show a slight separation of waxlike constituents in microcrystalline form. It oxidizes and darkens slowly in air and on exposure to light, particularly when in alkaline media. It is insoluble in water, is soluble in alcohol, and is miscible with acetone, chloroform, ether, and vegetable oils.

Functional Use in Foods *High-alpha type*: nutrient; dietary supplement; antioxidant. *Low-alpha type*: antioxidant.

REQUIREMENTS

Labeling *High-alpha type*: Indicate the mg/g of total tocopherols and of D-α-tocopherol present. Label claims in terms of former International Units (IU) should be based on the following: 1 mg D-α-tocopherol = 1.49 IU. *Low-alpha type*: Indicate the mg/g of total tocopherols and of D-β- plus D-γ- plus D-δ-tocopherols present.

Identification

A. Dissolve about 50 mg of the sample in 10 mL of absolute alcohol, add, with swirling, 2 mL of nitric acid, and heat at about 75° for 15 min. A bright red to orange color develops.

B. *High-alpha type*: The retention time of the major peak (excluding the solvent peak) in the chromatogram of the *Assay Preparation* is the same as that of the *Standard Preparation*, both relative to the internal standard, as obtained in the *Assay*. *Low-alpha type*: The retention time of the third major peak (i.e., the peak occurring just before that of the internal standard) in the chromatogram of the *Assay Preparation* is the same as that of the *Standard Preparation*, both relative to the internal standard, as obtained in the *Assay*.

Assay *High-alpha type*: not less than 50.0% of total tocopherols, of which not less than 50.0% consists of D-α-tocopherol ($C_{29}H_{50}O_2$) and not less than 20.0% consists of D-β- plus D-γ- ($C_{28}H_{48}O_2$) plus D-δ-tocopherols ($C_{27}H_{46}O_2$). *Low-alpha type*: not less than 50.0% of total tocopherols, of which not less than 80.0% consists of D-β- plus D-γ- plus D-δ-tocopherols.

Acidity Passes test.

Heavy Metals (as Pb) Not more than 10 mg/kg.

Specific Rotation $[\alpha]_D^{25°}$: *High-alpha type*: not less than +24°. *Low-alpha type*: not less than +20°.

TESTS

Note: In the tests for *Assay* and for *Specific Rotation*, use low-actinic glassware for all solutions containing tocopherols.

Assay

Internal Standard Solution Transfer about 600 mg of hexadecyl hexadecanoate, accurately weighed, to a 200-mL volumetric flask, dissolve in a solution containing 2 parts of pyridine and 1 part of propionic anhydride, dilute to volume with the solution, and mix.

Standard Preparations Transfer 12-, 25-, 37-, and 50-mg portions of USP Alpha Tocopherol Reference Standard, accurately weighed, to separate 50-mL Erlenmeyer flasks having 19/38 standard-taper, ground-glass necks. Pipet 25.0 mL of the *Internal Standard Solution* into each flask, mix, and reflux for 10 min under water-cooled condensers.

Assay Preparation Transfer about 60 mg of the sample, accurately weighed, to another 50-mL Erlenmeyer flask, pipet 10.0 mL of the *Internal Standard Solution* into the flask, mix, and reflux for 10 min under a water-cooled condenser.

Chromatographic System Use the *System* described under *Assay* in the monograph for DL-α-*Tocopherol*.

System Suitability Chromatograph a suitable number of injections of the *Assay Preparation*, as directed under *Calibration*, to ensure that the resolution factor, R, between the major peaks occurring at retention times of approximately 0.50 (δ-tocopheryl propionate) and 0.63 (β- plus γ-tocopheryl propionates), relative to hexadecyl hexadecanoate at 1.00, is not less than 2.5 (see *Chromatography*, Appendix IIA).

Calibration Chromatograph successive 2- to 5-μL portions of each *Standard Preparation* until the relative response factor, F, for each is constant (i.e., within a range of approximately 2%) for three consecutive injections. If graphic integration is used, adjust the instrument to obtain at least 70% maximum recorder response for the hexadecyl hexadecanoate peak. Measure the areas under the first (α-tocopheryl propionate) and second (hexadecyl hexadecanoate) major peaks (excluding the solvent peak), and record the values as A_S and A_I, respectively. Calculate the factor, F, for each concentration of *Standard Preparation* by the formula

$$(A_S/A_I) \times (C_I/C_S),$$

in which C_I and C_S are the exact concentrations, in mg/mL, of hexadecyl hexadecanoate and of USP Alpha Tocopherol Reference Standard in the *Standard Preparation*, respectively. Prepare a relative response factor curve by plotting area of α-tocopheryl propionate versus relative response factor.

Procedure Inject a suitable portion (2 to 5 μL) of the *Assay Preparation* into the chromatograph, and record the chromatogram. Measure the areas under the four major peaks occurring at relative retention times of 0.50, 0.63, 0.76, and 1.00, and record the values as a_δ, $a_{\beta+\gamma}$, a_α, and a_I, corresponding to δ-tocopheryl propionate, β- plus γ-tocopheryl propionates, α-tocopheryl propionate, and hexadecyl hexadecanoate, respectively. Calculate the weight, in mg, of each tocopherol form in the sample by the following formulas:

$$\delta\text{-tocopherol} = (10C_I/F) \times (a_\delta/a_I);$$
$$\beta\text{- plus }\gamma\text{-tocopherols} = (10C_I/F) \times (a_{\beta+\gamma}/a_I);$$
$$\alpha\text{-tocopherol} = (10C_I/F) \times (a_\alpha/a_I),$$

in which F is obtained from the relative response factor curve (see *Calibration*) for each of the corresponding areas under the δ-, β- plus γ-, and α-tocopheryl propionate peaks produced by the *Assay Preparation*.

Note: The relative response factor for δ-tocopheryl propionate and for β- plus γ-tocopheryl propionates has been determined empirically to be the same as for α-tocopheryl propionate.

Acidity Dissolve 1 g of the sample in 25 mL of a mixture of equal volumes of alcohol and ether that has been neutralized to phenolphthalein TS with 0.1 N sodium hydroxide, add 0.5 mL of phenolphthalein TS, and titrate with 0.1 N sodium hydroxide until the solution remains faintly pink after shaking for 30 s. Not more than 1.0 mL of 0.1 N sodium hydroxide is required.

Heavy Metals Place a 2-g sample in a silica crucible, and proceed as directed in *Method II* under the *Heavy Metals Test*, Appendix IIIB, using 20 μg of lead ion (Pb) in the control (*Solution A*).

Specific Rotation Transfer an accurately weighed amount of sample, equivalent to about 100 mg of total tocopherols, to a separator, and dissolve it in 50 mL of ether. To the separator add 20 mL of a 10% solution of potassium ferricyanide in sodium hydroxide solution (1 in 125), and shake for 3 min. Wash the ether solution with four 50-mL portions of water, discard the washings, and dry over anhydrous sodium sulfate. Evaporate the dried ether solution on a water bath under reduced pressure or in an atmosphere of nitrogen until about 7 or 8 mL remain, and then complete the evaporation, removing the last traces of ether without the application of heat. Immediately dissolve the residue in 5.0 mL of isooctane, and determine the optical rotation. Calculate the specific rotation (see Appendix IIB), using as c the concentration expressed as the number of g of total tocopherols, determined in the *Assay*, in 100 mL of the solution.

Packaging and Storage Store in tight containers blanketed by inert gas and protected from heat and light.

D-α-Tocopheryl Acetate

$C_{31}H_{52}O_3$ Formula wt 472.75

CAS: [58-95-7]

DESCRIPTION

A form of vitamin E obtained by the vacuum steam distillation and acetylation of edible vegetable oil products. It occurs as a colorless to yellow, nearly odorless, clear viscous oil. It may solidify on standing, and melts at about 25°. It is unstable in the presence of alkalies. It is insoluble in water; is freely soluble in alcohol; and is miscible with acetone, chloroform, ether, and vegetable oils.

Functional Use in Foods Nutrient; dietary supplement.

REQUIREMENTS

Labeling Label claims in terms of former International Units (IU) should be based on the following: 1 mg D-α-Tocopheryl Acetate = 1.36 IU.

Identification

A. To 10 mL of the *Test Solution* obtained as directed under *Specific Rotation* add with swirling 2 mL of nitric acid, and heat at about 75° for 15 min. A bright red to orange color develops.

B. The retention time of the major peak (excluding the solvent peak) in the chromatogram of the *Assay Preparation* is the same as that of the *Standard Preparation*, both relative to the internal standard, as obtained in the *Assay*.

Assay Not less than 96.0% and not more than 102.0% of $C_{31}H_{52}O_3$.

Acidity Passes test.

Heavy Metals (as Pb) Not more than 10 mg/kg.

Specific Rotation $[\alpha]_D^{25°}$: Not less than +24°.

TESTS

Note: In the **TESTS** for *Assay* and for *Specific Rotation*, use low-actinic glassware for all solutions containing tocopherols.

Assay

Internal Standard Solution Prepare a solution in *n*-hexane containing about 3 mg of hexadecyl hexadecanoate, accurately weighed, in each mL.

Standard Preparation Dissolve about 30 mg of USP Alpha Tocopheryl Acetate Reference Standard, accurately weighed, in 10.0 mL of the *Internal Standard Solution.*

Assay Preparation Dissolve about 30 mg of the sample, accurately weighed, in 10.0 mL of the *Internal Standard Solution.*

Chromatographic System Use the *System* described under *Assay* in the monograph for DL-α-Tocopherol.

System Suitability Chromatograph a suitable number of injections of a mixture in *n*-hexane of 1 mg/mL each of USP Alpha Tocopherol Reference Standard and USP Alpha Tocopheryl Acetate Reference Standard, as directed under *Calibration*, to ensure that the resolution factor, R, is not less than 1.0 (see Appendix IIA).

Calibration Chromatograph successive 2- to 5-µL portions of the *Standard Preparation* until the relative response factor, F, is constant (i.e., within a range of approximately 2%) for three consecutive injections. If graphic integration is used, adjust the instrument to obtain at least 70% maximum recorder response for the hexadecyl hexadecanoate peak. Measure the areas under the major peaks occurring at relative retention times of approximately 0.60 (α-tocopheryl acetate) and 1.00 (hexadecyl hexadecanoate), and record the values as A_S and A_I, respectively. Calculate the relative response factor, F, by the formula

$$(A_S/A_I) \times (C_I/C_S),$$

in which C_I and C_S are the exact concentrations, in mg/mL, of hexadecyl hexadecanoate and of USP Alpha Tocopheryl Acetate Reference Standard in the *Standard Preparation*, respectively.

Procedure Inject a suitable portion (2 to 5 µL) of the *Assay Preparation* into the chromatograph, and record the chromatogram. Measure the areas under the major peaks occurring at relative retention times of approximately 0.60 (α-tocopheryl acetate) and 1.00 (hexadecyl hexadecanoate), and record the values as a_U and a_I, respectively. Calculate the weight, in mg, of D-α-Tocopheryl Acetate in the sample by the formula

$$(10C_I/F) \times (a_U/a_I).$$

Acidity Dissolve 1.0 g of the sample in 25 mL of a mixture of equal volumes of alcohol and ether that has been neutralized to phenolphthalein TS with 0.1 N sodium hydroxide, add 0.5 mL of phenolphthalein TS, and titrate with 0.1 N sodium hydroxide until the solution remains faintly pink after shaking for 30 s. Not more than 1.0 mL of 0.1 N sodium hydroxide is required.

Heavy Metals Place a 2-g sample in a silica crucible, and proceed as directed in *Method II* under the *Heavy Metals Test*, Appendix IIIB, using 20 µg of lead ion (Pb) in the control (*Solution A*).

Specific Rotation

Test Solution Transfer an accurately weighed amount of the sample, equivalent to about 200 mg of α-tocopherol, to a 150-mL, round-bottomed, glass-stoppered flask, and dissolve it in 25 mL of absolute alcohol. Add 20 mL of 2 N sulfuric acid in alcohol (1 in 7), and reflux in an all-glass apparatus for 3 h, protected from sunlight. Cool, transfer to a 200-mL volumetric flask, dilute to volume with 2 N sulfuric acid in alcohol (1 in 72), and mix.

Procedure Transfer an accurately measured volume of the *Test Solution*, equivalent to about 100 mg of α-tocopherol, to a separator, and add 200 mL of water. Extract first with 75 mL, then with two 25-mL portions of ether, and combine the ether extracts in another separator. To the ether solution add 20 mL of a 10% solution of potassium ferricyanide in sodium hydroxide solution (1 in 125), and shake for 3 min. Wash the ether solution with four 50-mL portions of water, discard the washings, and dry over anhydrous sodium sulfate. Evaporate the dried ether solution on a water bath under reduced pressure or in an atmosphere of nitrogen until about 7 or 8 mL remain, and then complete the evaporation, removing the last traces of ether without the application of heat. Immediately dissolve the residue in 5.0 mL of isooctane, and determine the optical rotation. Calculate the specific rotation (see Appendix IIB), using as c the concentration of D-α-Tocopheryl Acetate, determined in the *Assay*, in 100 mL of solution.

Packaging and Storage Store in tight, light-resistant containers.

DL-α-Tocopheryl Acetate

$C_{31}H_{52}O_3$ Formula wt 472.75

CAS: [7695-91-2]

DESCRIPTION

A form of vitamin E. It occurs as a colorless to yellow or greenish yellow, nearly odorless, clear viscous oil. It is unstable in the presence of alkalies. It is insoluble in water, is freely soluble in alcohol, and is miscible with acetone, chloroform, ether, and vegetable oils.

Functional Use in Foods Nutrient; dietary supplement.

REQUIREMENTS

Labeling Label claims in terms of former International Units (IU) should be based on the following: 1 mg DL-α-tocopheryl acetate = 1 IU.

Identification It meets the requirements of *Identification Tests A, B,* and *C* in the monograph for D-α-*Tocopheryl Acetate.*

Assay Not less than 96.0% and not more than 102.0% of $C_{31}H_{52}O_3$.

Acidity Passes test.

Heavy Metals (as Pb) Not more than 10 mg/kg.

TESTS

Note: In the following *Assay*, use low-actinic glassware for all solutions containing tocopherols.

Assay   Proceed as directed in the *Assay* in the monograph for D-α-*Tocopheryl Acetate*, using the following as the *Assay Preparation*: Dissolve an accurately weighed amount of the sample equivalent to about 30 mg of DL-α-tocopheryl acetate in 10.0 mL of the *Internal Standard Solution*.

Acidity   Proceed as directed in the monograph for D-α-*Tocopheryl Acetate*.

Heavy Metals   Place a 2-g sample in a silica crucible, and proceed as directed in *Method II* under the *Heavy Metals Test*, Appendix IIIB, using 20 μg of lead ion (Pb) in the control (*Solution A*).

Packaging and Storage   Store in tight, light-resistant containers.

TESTS

Note: In the **TESTS** for *Assay* and for *Specific Rotation*, use low-actinic glassware for all solutions containing tocopherols.

Assay   Proceed as directed under *Assay* in the monograph for D-α-*Tocopheryl Acetate*, using the following as the *Assay Preparation*: Dissolve an accurately weighed amount of the sample equivalent to about 30 mg of D-α-tocopheryl acetate, in 10.0 mL of the *Internal Standard Solution*.

Acidity and **Heavy Metals**   Proceed as directed in the monograph for D-α-*Tocopheryl Acetate*.

Specific Rotation   Proceed as directed in the monograph for D-α-*Tocopheryl Acetate*, using for the *Test Solution* an accurately weighed sample equivalent to about 200 mg of α-tocopherol.

Packaging and Storage   Store in tight, light-resistant containers.

D-α-Tocopheryl Acetate Concentrate

D-α-Tocopheryl Acetate Preparation

DESCRIPTION

A form of vitamin E obtained by the vacuum steam distillation and acetylation of edible vegetable oil products. The content of D-α-tocopheryl acetate may be adjusted by suitable physical or chemical means. It occurs as a light brown to light yellow, nearly odorless, clear viscous oil. It is unstable in the presence of alkalies. It is insoluble in water; is soluble in alcohol; and is miscible with acetone, chloroform, ether, and vegetable oils.

Functional Use in Foods   Nutrient; dietary supplement.

REQUIREMENTS

Labeling   Indicate the mg/g of D-α-tocopheryl acetate present. Label claims in terms of former International Units (IU) should be based on the following: 1 mg D-α-tocopheryl acetate = 1.36 IU.

Identification   It meets the requirements of *Identification Tests A* and *B* in the monograph for D-α-*Tocopheryl Acetate*.

Assay   Not less than 40.0% of D-α-tocopheryl acetate ($C_{31}H_{52}O_3$).

Acidity   Passes test.

Heavy Metals (as Pb)   Not more than 10 mg/kg.

Specific Rotation $[\alpha]_D^{25°}$: Not less than +24°.

D-α-Tocopheryl Acid Succinate

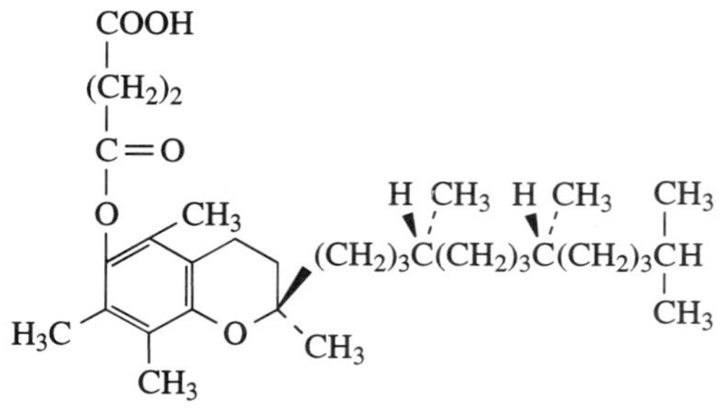

$C_{33}H_{54}O_5$ Formula wt 530.79

CAS: [4345-03-3]

DESCRIPTION

A form of vitamin E obtained by the vacuum steam distillation and succinylation of edible vegetable oil products. It occurs as a white to off-white, crystalline powder having little or no taste or odor. It is stable in air, but is unstable to alkali and to heat. It is insoluble in water; soluble in acetone, alcohol, ether, and vegetable oils; and very soluble in chloroform. It melts at about 75°.

Functional Use in Foods   Nutrient; dietary supplement.

REQUIREMENTS

Labeling   Label claims in terms of former International Units (IU) should be based on the following: 1 mg D-α-Tocopheryl Acid Succinate = 1.21 IU.

Identification

A. To 10 mL of the *Test Solution*, obtained as directed under *Specific Rotation*, add, with swirling, 2 mL of nitric acid, and heat at about 75° for 15 min. A bright red to orange color develops.

B. The retention time of the major peak (excluding the solvent peak) in the chromatogram of the *Assay Preparation* is the same as that of the *Standard Preparation*, both relative to the internal standard, as obtained in the *Assay*.

Assay Not less than 96.0% and not more than 102.0% of $C_{33}H_{54}O_5$.

Acidity Passes test.

Heavy Metals (as Pb) Not more than 10 mg/kg.

Specific Rotation $[\alpha]_D^{25°}$: Not less than +24°.

TESTS

Assay

Internal Standard Solution Prepare a solution in *n*-hexane containing about 3 mg of hexadecyl hexadecanoate, accurately weighed, in each mL.

Standard Preparation Transfer about 30 mg of USP Alpha Tocopheryl Acid Succinate Reference Standard, accurately weighed, to a 4-dram (approximately 15 mL), screw-cap vial. Pipet 2.0 mL of absolute methanol, 1.0 mL of 2,2-dimethoxypropane, and 0.1 mL of concentrated hydrochloric acid into the vial, cap, mix well, and allow to stand for 1 h in the dark. Evaporate just to dryness on a steam bath with the aid of a stream of nitrogen. Pipet 10.0 mL of the *Internal Standard* into the vial, cap, and shake vigorously.

Assay Preparation Prepare as directed for the *Standard Preparation*, using an accurately weighed amount of the sample equivalent to about 30 mg of D-α-Tocopheryl Acid Succinate.

Chromatographic System Use a gas chromatograph equipped with a flame-ionization detector and a glass-lined sample-introduction system or on-column injection. Under typical conditions, the instrument contains a 2-m × 4-mm borosilicate glass column packed with 2% to 5% methylpolysiloxane gum on 80- to 100-mesh, acid-base, washed, silanized, chromatographic diatomaceous earth. The column is maintained isothermally between 260° and 280°, the injection port at about 290°, and the detector block at about 300°. The flow rate of dry carrier gas is adjusted to obtain a hexadecyl hexadecanoate peak 12 to 14 min after sample introduction.

Note: Cure and condition the column as necessary (see *Gas Chromatography*, Appendix IIA).

System Suitability Chromatograph a suitable number of injections of a mixture in *n*-hexane of 1 mg/mL each of USP Alpha Tocopherol Reference Standard and USP Alpha Tocopheryl Acetate Reference Standard, as directed under *Calibration*, to ensure that the resolution factor, *R*, is not less than 1.0 (see *Gas Chromatography*, Appendix IIA).

Calibration Chromatograph successive 2- to 5-μL portions of the upper layer of the *Standard Preparation* until the relative response factor, *F*, is constant (i.e., within a range of approximately 2%) for three consecutive injections. If graphic integration is used, adjust the instrument to obtain 70% maximum recorder response for the hexadecyl hexadecanoate peak. Measure the areas under the major peaks occurring at relative retention times of approximately 1.00 (hexadecyl hexadecanoate) and 1.99 (methyl α-tocopheryl succinate), and record the values as A_I and A_S, respectively. Calculate the relative response factor, *F*, by the formula

$$(A_S/A_I) \times (C_I/C_S),$$

in which C_I and C_S are the exact concentrations, in mg/mL, of hexadecyl hexadecanoate and of USP Alpha Tocopheryl Acid Succinate Reference Standard in the *Standard Preparation*, respectively.

Procedure Inject a suitable portion (2 to 5 μL) of the *Assay Preparation* into the chromatograph, and record the chromatogram. Measure the areas under the major peaks occurring at relative retention times of approximately 1.00 (hexadecyl hexadecanoate) and 1.99 (methyl α-tocopheryl succinate), and record the values as a_I and a_U, respectively. Calculate the weight, in mg, of D-α-Tocopheryl Acid Succinate in the sample by the formula

$$(10C_I/F) \times (a_U/a_I).$$

Acidity Dissolve 1.0 g of the sample in 25 mL of a mixture of equal volumes of alcohol and ether that has been neutralized to phenolphthalein TS with 0.1 *N* sodium hydroxide, add 0.5 mL of phenolphthalein TS, and titrate with 0.1 *N* sodium hydroxide until the solution remains faintly pink after shaking for 30 s. Between 18.0 mL and 19.3 mL of 0.1 *N* sodium hydroxide is required.

Heavy Metals Place a 2-g sample in a silica crucible, and proceed as directed in *Method II* under the *Heavy Metals Test*, Appendix IIIB, using 20 μg of lead ion (Pb) in the control (*Solution A*).

Specific Rotation (Note: Use low-actinic glassware.)

Test Solution Transfer an accurately weighed amount of the sample, equivalent to about 200 mg of α-tocopherol, to a 250-mL, round-bottomed, glass-stoppered flask, dissolve in 50 mL of absolute alcohol, and reflux for 1 min. While the solution is boiling, add through the condenser 1 g of potassium hydroxide pellets, one at a time, to avoid overheating.

Caution: Wear safety goggles.

Continue refluxing for 20 min, and then, without cooling, add 2 mL of hydrochloric acid, dropwise, through the condenser. (This technique is essential to prevent oxidative action by air while the sample is in an alkaline medium.) Cool, and transfer the contents of the flask to a 500-mL separator, rinsing the flask with 100 mL each of water and of ether and adding the rinsings to the separator. Shake vigorously, allow the layers to separate, and collect each of the two layers in separate separators. Extract the aqueous layer with two 50-mL portions of ether, and add these extracts to the main ether extract. Wash the combined ether extracts with four 100-mL portions of water, and then evaporate the solutions on a water bath under reduced pressure or in an atmosphere of nitrogen until about 7 or 8 mL remain. Complete the evaporation, removing the last traces of ether without the application of heat. Immediately dissolve the residue in 2 *N* sulfuric acid in alcohol (1 in 72), transfer to a 200-mL volumetric flask, dilute to volume with the alcoholic sulfuric acid, and mix.

Procedure Transfer an accurately measured volume of the *Test Solution*, equivalent to about 100 mg of α-tocopherol, to a separator, and add 200 mL of water. Extract first with 75 mL, then with two 25-mL portions of ether, and combine the ether extracts in another separator. To the ether solution add 20 mL of a 10% solution of potassium ferricyanide in sodium hydroxide solution (1 in 125), and shake for 3 min. Wash the ether solution with four 50-mL portions of water, discard the washings, and dry over anhydrous sodium sulfate. Evaporate the dried ether solution on a water bath under reduced pressure or in an atmosphere of nitrogen until about 7 or 8 mL remain, and then complete the evaporation, removing the last traces of ether without the application of heat. Immediately dissolve the residue in 5.0 mL of isooctane, and determine the optical rotation. Calculate the specific rotation (see Appendix IIB), using as c the concentration expressed as the number of g of D-α-Tocopheryl Acid Succinate, determined in the *Assay*, in 100 mL of solution.

Packaging and Storage Store in tight, light-resistant containers.

Tragacanth

INS: 413 CAS: [9000-65-1]

DESCRIPTION

A dried gummy exudation obtained from *Astragalus gummifer* Labillardiere, or other Asiatic species of *Astragalus* (Fam. *Leguminosae*). *Unground Tragacanth* occurs as flattened, lamellated, frequently curved fragments or straight or spirally twisted linear pieces from 0.5 to 2.5 mm in thickness. It is white to weak yellow in color, translucent, horny in texture, and having a short fracture. It is odorless and has an insipid, mucilaginous taste. It is rendered more easily pulverizable if heated to a temperature of 50°. *Powdered Tragacanth* is white to yellowish white in color.

Functional Use in Foods Stabilizer; thickener; emulsifier.

REQUIREMENTS

Identification When examined microscopically in water mounts, it shows numerous angular fragments with circular or irregular lamellae, and starch grains up to 25 µm in diameter. It should show very few or no fragments of lignified vegetable tissue. One g in 50 mL of water swells to form a smooth, stiff, opalescent mucilage free from cellular fragments.
Arsenic (as As) Not more than 3 mg/kg.
Ash (Total) Not more than 3.0%.
Ash (Acid-Insoluble) Not more than 0.5%.
Heavy Metals (as Pb) Not more than 0.002%.
Karaya Gum Passes test.
Lead Not more than 5 mg/kg.
Viscosity of a 1% Solution Not less than 250 centipoises.

TESTS

Arsenic A *Sample Solution* prepared as directed for organic compounds meets the requirements of the *Arsenic Test*, Appendix IIIB.
Ash (Total) Determine as directed in the general method, Appendix IIC.
Ash (Acid-Insoluble) Determine as directed in the general method, Appendix IIC.
Heavy Metals Prepare and test a 1-g sample as directed in *Method II* under the *Heavy Metals Test*, Appendix IIIB, using 20 µg of lead ion (Pb) in the control (*Solution A*).
Karaya Gum Boil 1 g with 20 mL of water until a mucilage is formed, add 5 mL of hydrochloric acid, and again boil the mixture for 5 min. No permanent pink or red color develops.
Lead A *Sample Solution* prepared as directed for organic compounds meets the requirements of the *Lead Limit Test*, Appendix IIIB, using 5 µg of lead ion (Pb) in the control.
Viscosity Transfer a 4.0-g sample, finely powdered, into the container of a stirring apparatus equipped with blades capable of revolving at 10,000 rpm. Add 10 mL of alcohol to the sample, swirl to wet the gum uniformly, and then add 390 mL of water, avoiding the formation of lumps. Immediately stir the mixture for 7 min, pour the resulting dispersion into a 500-mL bottle, insert a stopper, and allow to stand for about 24 h in a water bath at 25°. Determine the apparent viscosity at this temperature with a Model LVF Brookfield or equivalent type viscometer (see *Viscosity of Sodium Carboxymethylcellulose*, Appendix IIB) using Spindle No. 2 at 30 rpm and a factor of 10.

Packaging and Storage Store in well-closed containers.

Triacetin

Glyceryl Triacetate

$$\begin{array}{c} H \\ | \\ HCOOCCH_3 \\ | \\ HCOOCCH_3 \\ | \\ HCOOCCH_3 \\ | \\ H \end{array}$$

$C_9H_{14}O_6$ Formula wt 218.21
INS: 1518 CAS: [102-76-1]

DESCRIPTION

A colorless, somewhat oily liquid with a slight, fatty odor and a bitter taste. It is soluble in water, and is miscible with alcohol, with ether, and with chloroform. It distills between 258° and 270°.

Functional Use in Foods Humectant; solvent.

REQUIREMENTS

Identification
A. Heat a few drops in a test tube with about 500 mg of potassium bisulfate. Pungent vapors of acrolein are evolved.

B. The solution resulting from the *Assay* gives positive tests for *Acetate*, Appendix IIIA.

Assay Not less than 98.5% of $C_9H_{14}O_6$.
Acidity Passes test.
Heavy Metals (as Pb) Not more than 5 mg/kg.
Refractive Index Between 1.429 and 1.431 at 25°.
Specific Gravity Between 1.154 and 1.158.
Unsaturated Compounds Passes test.
Water Not more than 0.2%.

TESTS

Assay Transfer about 1 g of the sample, accurately weighed, into a suitable pressure bottle, add 25.0 mL of 1 *N* potassium hydroxide and 15 mL of isopropanol, stopper the bottle, and wrap securely in a canvas bag. Place in a water bath maintained at 98° ± 2°, and heat for 1 h, allowing the water in the bath to just cover the liquid in the bottle. Remove the bottle from the bath, cool in air to room temperature, then loosen the wrapper, uncap the bottle to release any pressure, and remove the wrapper. Add 6 to 8 drops of phenolphthalein TS, and titrate the excess alkali with 0.5 *N* sulfuric acid just to the disappearance of the pink color. Perform a blank determination (see *General Provisions*). Each mL of 0.5 *N* sulfuric acid is equivalent to 36.37 mg of $C_9H_{14}O_6$.

Acidity Transfer about 25 g of the sample, accurately weighed, into a 125-mL conical flask, add 50 mL of toluene and 2 drops of thymol blue TS, and titrate rapidly with 0.02 *M* sodium methoxide in toluene. Swirl the flask continuously until the yellow color changes to a dark color, and then continue the titration without stopping but slowing the addition of titrant until a single drop changes the solution to a clear blue color. The endpoint is stable for about 8 to 15 s. Not more than 1.0 mL of 0.02 *M* sodium methoxide is required.

Heavy Metals Prepare and test a 4-g sample as directed in *Method II* under the *Heavy Metals Test*, Appendix IIIB, using 20 μg of lead ion (Pb) in the control (*Solution A*).

Refractive Index, Appendix IIB Determine at 25° with an Abbé or other refractometer of equal or greater accuracy.

Specific Gravity Determine by any reliable method (see *General Provisions*).

Unsaturated Compounds To 10 mL of the sample in a glass-stoppered tube add, dropwise, a solution of bromine in carbon tetrachloride (1 mL in 100 mL) until a permanent yellow color is produced, and allow to stand in a dark place for 18 h. No turbidity or precipitate appears.

Water Determine by the *Karl Fischer Titrimetric Method*, Appendix IIB.

Packaging and Storage Store in well-closed containers.

Trichloroethylene

Ethylene Trichloride; Trichloroethene; 1,1,2-Trichloroethylene

$$Cl-C(H)=C(Cl)-Cl$$

C_2HCl_3 Formula wt 131.39

CAS: [79-01-6]

DESCRIPTION

A clear, colorless, mobile liquid having a sweet, chloroform-like odor. It is immiscible with water, but is miscible with alcohol, ether, and acetone. Its refractive index at 20° is about 1.477. It may contain a suitable stabilizer.

Functional Use in Foods Extraction solvent.

REQUIREMENTS

Acidity (as HCl) Not more than 10 mg/kg.
Alkalinity (as NaOH) Not more than 10 mg/kg.
Distillation Range Between 86° and 88°.
Free Halogens Passes test.
Heavy Metals (as Pb) Not more than 1 mg/kg.
Nonvolatile Residue Not more than 10 mg/kg.
Specific Gravity Between 1.454 and 1.458.
Water Not more than 0.05%.

TESTS

Acidity or Alkalinity Transfer 25 mL of water and 2 drops of phenolphthalein TS to a 250-mL glass-stoppered flask, and add 0.01 *N* sodium hydroxide to the first appearance of a slight pink color. Add 25 mL (about 36 g) of the sample, and shake for 30 s. If the pink color persists, titrate with 0.01 *N* hydrochloric acid, shaking repeatedly, until the pink color just disappears; not more than 0.9 mL is required. If the pink color is discharged when the sample is added, titrate with 0.01 *N* sodium hydroxide until the faint pink color is restored; not more than 1.0 mL is required.

Distillation Range Determine as directed in the general method, Appendix IIB.

Free Halogens Shake 10 mL of the sample vigorously for 2 min with 10 mL of potassium iodide solution (1 in 10) and 1 mL of starch TS. A blue color does not appear in the water layer.

Heavy Metals Evaporate 14 mL (about 20 g) of the sample to dryness on a steam bath in a glass evaporating dish.

Caution: Use hood.

Cool, add 2 mL of hydrochloric acid, and slowly evaporate to dryness again on the steam bath. Moisten the residue with 1 drop of hydrochloric acid, add 10 mL of hot water, and digest for 2 min. Filter if necessary through a small filter, wash the evaporating dish and the filter with about 10 mL of water, and dilute to 25 mL with water. This solution meets the requirements of the *Heavy Metals Test*, Appendix IIIB, using 20 μg of lead ion (Pb) in the control (*Solution A*).

Nonvolatile Residue   Evaporate 69 mL (about 100 g) of the sample to dryness in a tared dish on a steam bath, dry the residue at 105° for 30 min, cool, and weigh.

   Caution: Use hood.

Specific Gravity   Determine by any reliable method (see *General Provisions*).

Water   Determine by the *Karl Fischer Titrimetric Method*, Appendix IIB.

Packaging and Storage   Store in tight containers.

Triethyl Citrate

Ethyl Citrate

$$\begin{array}{c} CH_2COOC_2H_5 \\ | \\ HO-C-COOC_2H_5 \\ | \\ CH_2COOC_2H_5 \end{array}$$

$C_{12}H_{20}O_7$ Formula wt 276.29
INS: 1505   CAS: [77-93-0]

DESCRIPTION

An odorless, practically colorless, oily liquid. It is slightly soluble in water, but is miscible with alcohol and with ether.

Functional Use in Foods   Solvent.

REQUIREMENTS

Assay   Not less than 99.0% and not more than 100.5% of $C_{12}H_{20}O_7$, on the anhydrous basis.
Acidity (as citric acid)   Not more than 0.02%.
Heavy Metals (as Pb)   Not more than 10 mg/kg.
Refractive Index   Between 1.439 and 1.443 at 25°; or, between 1.440 and 1.444 at 20°.
Specific Gravity   Between 1.135 and 1.139 at 25°.
Water   Not more than 0.25%.

TESTS

Assay   Weigh accurately about 1.5 g of the sample into a 500-mL flask equipped with a standard-taper ground joint, and add 25 mL of isopropyl alcohol and 25 mL of water. Pipet 50 mL of 0.5 *N* sodium hydroxide into the mixture, add a few boiling chips, and attach a suitable water-cooled condenser. Reflux for 1.5 h, then cool, wash down the condenser with about 20 mL of water, add 5 drops of phenolphthalein TS, and titrate the excess alkali with 0.5 *N* sulfuric acid. Perform a blank determination (see *General Provisions*). Each mL of 0.5 *N* sodium hydroxide is equivalent to 46.05 mg of $C_{12}H_{20}O_7$.

Acidity   Dissolve 32 g, accurately weighed, in 30 mL of neutralized alcohol, add bromothymol blue TS, and titrate with 0.1 *N* sodium hydroxide. Not more than 1.0 mL is required.

Heavy Metals   Prepare and test a 2-g sample as directed in *Method II* under the *Heavy Metals Test*, Appendix IIIB, using 20 μg of lead ion (Pb) in the control (*Solution A*).

Refractive Index, Appendix IIB   Determine with an Abbé or other refractometer of equal or greater accuracy.

Specific Gravity   Determine by any reliable method (see *General Provisions*).

Water   Determine by the *Karl Fischer Titrimetric Method*, Appendix IIB.

Packaging and Storage   Store in well-closed containers.

DL-Tryptophan

DL-α-Amino-3-indolepropionic Acid

$C_{11}H_{12}N_2O_2$ Formula wt 204.23
   CAS: [54-12-6]

DESCRIPTION

White crystals or a crystalline powder, odorless or with a slight odor. It is soluble in water and in dilute acids and alkalies. It is sparingly soluble in alcohol. It is optically inactive.

Functional Use in Foods   Nutrient; dietary supplement.

REQUIREMENTS

Identification   To a solution of DL-tryptophan in water add bromine TS. A red color appears that may be extracted with amyl alcohol.
Assay   Not less than 98.5% and not more than 101.5% of $C_{11}H_{12}N_2O_2$ after drying.
Heavy Metals (as Pb)   Not more than 0.002%.
Lead   Not more than 10 mg/kg.
Loss on Drying   Not more than 0.3%.
Residue on Ignition   Not more than 0.1%.

TESTS

Assay Dissolve about 300 mg of the sample, previously dried at 105° for 3 h and accurately weighed, in 3 mL of formic acid and 50 mL of glacial acetic acid, add 2 drops of crystal violet TS, and titrate with 0.1 N perchloric acid to a green endpoint or until the blue color disappears completely. Perform a blank determination (see *General Provisions*), and make any necessary correction. Each mL of 0.1 N perchloric acid is equivalent to 20.42 mg of $C_{11}H_{12}N_2O_2$.

Heavy Metals Prepare and test a 1-g sample as directed under the *Heavy Metals Limit Test*, Appendix IIIB, using 20 µg of lead ion (Pb) in the control (*Solution A*).

Lead A *Sample Solution* prepared as directed for organic compounds meets the requirements of the *Lead Limit Test*, Appendix IIIB, using 10 µg of lead ion (Pb) in the control.

Loss on Drying, Appendix IIC   Dry at 105° for 3 h.

Residue on Ignition   Ignite 2 g as directed in the general method, Appendix IIC.

Packaging and Storage   Store in well-closed containers.

L-Tryptophan

L-α-Amino-3-indolepropionic Acid

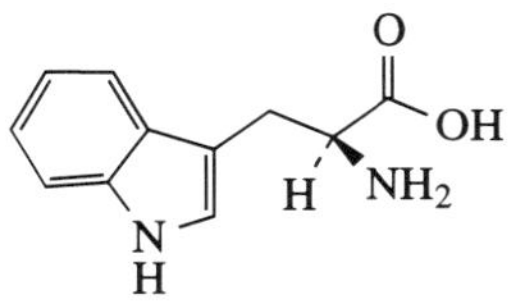

$C_{11}H_{12}N_2O_2$ Formula wt 204.23

CAS: [73-22-3]

DESCRIPTION

White to yellowish white crystals or a crystalline powder. It is odorless or has a slight odor with a slightly bitter taste. One g dissolves in about 100 mL of water. It is soluble in hot alcohol, in dilute hydrochloric acid, and in alkali hydroxide solutions.

Functional Use in Foods   Nutrient; dietary supplement.

REQUIREMENTS

Identification   Dissolve about 1 g in 100 mL of hydrochloric acid solution (1 in 5). To 1 mL of this solution add 1 mL of sodium sulfite solution (1 in 20). A yellow color is produced.

Assay   Not less than 98.5% and not more than 101.5% of $C_{11}H_{12}N_2O_2$, calculated on the dried basis.

Heavy Metals (as Pb)   Not more than 0.002%.

Lead   Not more than 10 mg/kg.

Loss on Drying   Not more than 0.3%.

Residue on Ignition   Not more than 0.1%.

Specific Rotation $[\alpha]_D^{20°}$: Between −30.0° and −33.0° after drying; or $[\alpha]_D^{25°}$: Between −29.7° and −32.7° after drying.

TESTS

Assay   Dissolve about 300 mg, accurately weighed, in 3 mL of formic acid and 50 mL of glacial acetic acid, add 2 drops of crystal violet TS, and titrate with 0.1 N perchloric acid to a green endpoint or until the blue color disappears completely. Perform a blank determination (see *General Provisions*), and make any necessary correction. Each mL of 0.1 N perchloric acid is equivalent to 20.42 mg of $C_{11}H_{12}N_2O_2$.

Heavy Metals   Prepare and test a 1-g sample as directed in *Method II* under the *Heavy Metals Test*, Appendix IIIB, using 20 µg of lead ion (Pb) in the control (*Solution A*).

Lead   A *Sample Solution* prepared as directed for organic compounds meets the requirements of the *Lead Limit Test*, Appendix IIIB, using 10 µg of lead ion (Pb) in the control.

Loss on Drying, Appendix IIC   Dry at 105° for 3 h.

Residue on Ignition, Appendix IIC   Ignite 1 g as directed in the general method.

Specific Rotation, Appendix IIB   Determine in a solution containing 1 g of a previously dried sample in sufficient water to make 100 mL.

Packaging and Storage   Store in well-closed, light-resistant containers.

L-Tyrosine

L-β-(p-Hydroxyphenyl)alanine

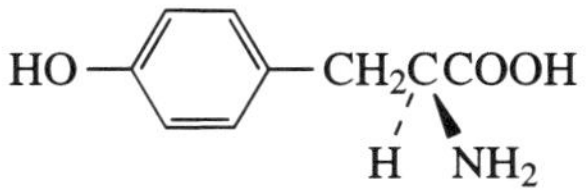

$C_9H_{11}NO_3$ Formula wt 181.19

CAS: [60-18-4]

DESCRIPTION

Colorless, silky needles, or a white crystalline powder. One g is soluble in about 230 mL of water. It is soluble in dilute mineral acids and in alkaline solutions. It is very slightly soluble in alcohol.

Functional Use in Foods   Nutrient; dietary supplement.

REQUIREMENTS

Identification   Heat 5 mL of a 1 in 1000 solution with 1 mL of triketohydrindene hydrate TS. A reddish purple color is produced.

Assay Not less than 98.5% and not more than 101.5% of $C_9H_{11}NO_3$ after drying.
Heavy Metals (as Pb) Not more than 0.003%.
Lead Not more than 10 mg/kg.
Loss on Drying Not more than 0.3%.
Residue on Ignition Not more than 0.1%.
Specific Rotation $[\alpha]_D^{20°}$: Between −11.3° and −12.3° after drying; or $[\alpha]_D^{25°}$: Between −10.0° and −11.0° after drying.

TESTS

Assay Transfer about 400 mg, previously dried at 105° for 3 h and accurately weighed, into a 250-mL flask. Dissolve the sample in about 50 mL of acetic acid, add 2 drops of crystal violet TS, and titrate with 0.1 N perchloric acid to a bluish green endpoint. Perform a blank determination (see *General Provisions*), and make any necessary correction. Each mL of 0.1 N perchloric acid is equivalent to 18.12 mg of $C_9H_{11}NO_3$.
Heavy Metals Prepare and test a 670-mg sample as directed in *Method II* under the *Heavy Metals Test*, Appendix IIIB, using 20 μg of lead ion (Pb) in the control (*Solution A*).
Lead A *Sample Solution* prepared as directed for organic compounds meets the requirements of the *Lead Limit Test*, Appendix IIIB, using 10 μg of lead ion (Pb) in the control.
Loss on Drying, Appendix IIC Dry at 105° for 3 h.
Residue on Ignition Ignite 2 g as directed in the general method, Appendix IIC.
Specific Rotation, Appendix IIB Determine in a solution containing 5 g of a previously dried sample in sufficient 1 N hydrochloric acid to make 100 mL.

Packaging and Storage Store in well-closed containers.

Urea

Carbamide

$$H_2N-\underset{\underset{O}{\|}}{C}-NH_2$$

CH_4N_2O Formula wt 60.06

CAS: [57-13-6]

DESCRIPTION

A colorless to white, prismatic, crystalline powder or small, white pellets. It is practically odorless, but upon standing may develop a slight odor of ammonia. It is freely soluble in water and in boiling alcohol and practically insoluble in chloroform and in ether. It is commonly produced from CO_2 by ammonolysis or from cyanamide by hydrolysis. It melts at a range of 132° to 135°.

Functional Use in Foods Formulation aid; fermentation aid.

REQUIREMENTS

Identification
A. Heat about 500 mg in a test tube until it liquifies. Ammonia vapor is produced. Continue heating until the liquid becomes turbid, and then cool. Dissolve the fused mass in a mixture of 10 mL of water and 1 mL of sodium hydroxide solution (1 in 10). Add 1 drop of cupric sulfate TS. A reddish violet-colored solution is produced.
B. Dissolve 100 mg in 1 mL of water, and add 1 mL of nitric acid. A white precipitate of urea nitrate is produced.
Assay Not less than 99.0% and not more than 100.5% of CH_4N_2O.
Alcohol-Insoluble Matter Not more than 0.04%.
Chloride Not more than 0.007%.
Heavy Metals (as Pb) Not more than 10 mg/kg.
Lead Not more than 5 mg/kg.
Loss on Drying Not more than 1.0%.
Residue on Ignition Not more than 0.1%.
Sulfate Not more than 0.01%.

TESTS

Assay Transfer about 500 mg, accurately weighed, to a 200-mL volumetric flask, and dissolve in 100 mL of water, dilute to volume with water, and mix. Pipet 2 mL of this solution into a semimicro Kjeldahl digestion flask, and proceed as directed under *Nitrogen Determination (Method II)*, Appendix IIIC. Heat the sample until it begins to fume, and then heat for 1 additional h. Each mL of 0.01 N acid is equivalent to 0.3003 mg of CH_4N_2O.
Alcohol-Insoluble Matter Dissolve about 5 g in 50 mL of warm alcohol. If any residue remains, filter the solution through a tared filter, wash the residue, and filter with 20 mL of warm alcohol. Dry at 105° for 1 h. Cool in a desiccator, and weigh.
Chloride A 0.2-g sample meets the requirements of the *Chloride Limit Test*, Appendix IIIB, using 14 μg of chloride ion (Cl) in the control.
Heavy Metals Prepare and test a 2-g sample as directed under the *Heavy Metals Test*, Appendix IIIB, using 20 μg of lead ion (Pb) in the control (*Solution A*).
Lead Prepare and test a 2-g sample as directed under the *Lead Limit Test*, Appendix IIIB, using 10 μg of lead ion (Pb) in the control.
Loss on Drying, Appendix IIC Dry at 105° for 3 h.
Residue on Ignition Ignite 1 g as directed under *Method I* in the general method, Appendix IIC.
Sulfate A 2-g sample meets the requirements of the *Sulfate Limit Test*, Appendix IIIB, using 200 μg of sulfate ion (SO_4) in the control.

Packaging and Storage Store in a well-closed container.

L-Valine

L-2-Amino-3-methylbutyric Acid

$$\text{CH}_3\text{CH}-\text{CCOOH}$$
$$\phantom{\text{CH}_3}|\phantom{\text{CH}}\diagup\backslash$$
$$\phantom{\text{CH}_3}\text{H}_3\text{C}\text{H}\text{NH}_2$$

$C_5H_{11}NO_2$ \hfill Formula wt 117.15

\hfill CAS: [72-18-4]

DESCRIPTION

A white, odorless, crystalline powder having a characteristic taste. It is freely soluble in water, but is practically insoluble in alcohol and in ether. The pH of a 1 in 20 solution is between 5.5 and 7.0. In a closed capillary tube it melts at about 315°.

Functional Use in Foods Nutrient; dietary supplement.

REQUIREMENTS

Identification To 5 mL of a 1 in 1000 solution add 1 mL of triketohydrindene hydrate TS. A reddish purple or bluish color is produced.
Assay Not less than 98.5% and not more than 101.5% of $C_5H_{11}NO_2$, calculated on the dried basis.
Heavy Metals (as Pb) Not more than 0.002%.
Lead Not more than 10 mg/kg.
Loss on Drying Not more than 0.3%.
Residue on Ignition Not more than 0.1%.
Specific Rotation $[\alpha]_D^{20°}$: Between +26.7° and +29.0° after drying; or $[\alpha]_D^{25°}$: Between +26.6° and +28.9° after drying.

TESTS

Assay Dissolve about 200 mg, accurately weighed, in 3 mL of formic acid and 50 mL of glacial acetic acid, add 2 drops of crystal violet TS, and titrate with 0.1 N perchloric acid to a green endpoint or until the blue color disappears completely. Perform a blank determination (see *General Provisions*), and make any necessary correction. Each mL of 0.1 N perchloric acid is equivalent to 11.72 mg of $C_5H_{11}NO_2$.
Heavy Metals (as Pb) Prepare and test a 1-g sample as directed in *Method II* under the *Heavy Metals Test*, Appendix IIIB, using 20 μg of lead ion (Pb) in the control (*Solution A*).
Lead A *Sample Solution* prepared as directed for organic compounds meets the requirements of the *Lead Limit Test*, Appendix IIIB, using 10 μg of lead ion (Pb) in the control.
Loss on Drying, Appendix IIC Dry at 105° for 3 h.
Residue on Ignition, Appendix IIC Ignite 1 g as directed in the general method.
Specific Rotation, Appendix IIB Determine in a solution containing 8 g of a previously dried sample in sufficient 6 N hydrochloric acid to make 100 mL.

Packaging and Storage Store in well-closed containers.

Vitamin A

DESCRIPTION

A suitable form or derivative of retinol ($C_{20}H_{30}O$; Vitamin A alcohol). It usually consists of retinol or esters of retinol formed from edible fatty acids, principally acetic and palmitic acids, or mixtures of these. It may be diluted with edible oils, or it may be incorporated in solid edible carriers, extenders, or excipients. It may contain suitable preservatives, dispersants, and antioxidants, providing it is not to be used in foods in which such substances are prohibited. In liquid form it is a light yellow to red oil that may solidify on refrigeration. In solid form it may have the appearance of the diluent that has been added to it. It may be nearly odorless, or have a mild fishy odor, but it has no rancid odor or taste. In liquid form it is very soluble in chloroform and in ether, it is soluble in absolute alcohol and in vegetable oils, but it is insoluble in glycerin and in water. In solid form it may be dispersible in water. It is unstable to air and light.

Functional Use in Foods Nutrient; dietary supplement.

REQUIREMENTS

Labeling Indicate the form of the Vitamin A, to declare the presence of any preservative, dispersant, antioxidant, or other added substance, and to declare the Vitamin A activity in terms of the equivalent amount of retinol in mg/g and in former International Units.
Identification
A. Dissolve an amount equivalent to about 6 μg of retinol in 1 mL of chloroform, and add 10 mL of antimony trichloride TS. A transient blue color appears at once.
B. Assemble an apparatus for *Thin-layer Chromatography* (see Appendix IIA), using chromatographic silica gel as the adsorbent and a mixture of 4 parts of cyclohexane and 1 part of ether as the solvent system. Prepare a *Standard Solution* by dissolving the contents of 1 capsule of USP Vitamin A Reference Standard in sufficient chloroform to make 25.0 mL.

If the Vitamin A is in liquid form, dissolve a volume representing approximately 15,000 former International Units in sufficient chloroform to make 10 mL. If the Vitamin A is in solid form, weigh a quantity representing approximately 15,000 former International Units, place in a separator, add 75 mL of water, heat, if necessary, to dissolve the carrier, and cool. Shake vigorously for 1 min, extract with 10 mL of chloroform by shaking for 1 min, and centrifuge to clarify the chloroform extract.

Apply at the starting point of the chromatogram 0.015 mL of the *Standard Solution* and 0.01 mL of the chloroform solution of the Vitamin A sample. Develop the chromatogram in the chromatographic chamber, lined with filter paper dipping into the solvent mixture. When the solvent has ascended for a distance of 10 cm, remove the plate, allow it to dry in air, and spray it with antimony trichloride TS. The blue spot formed is indicative of the presence of retinol. The approximate R_f values of the predominant spots, corresponding to the different forms of retinol, are 0.1 for the alcohol, 0.45 for the acetate, and 0.7 for the palmitate.

Assay Not less than 95.0% and not more than 100.5% of the Vitamin A activity declared on the label.

Absorbance Ratio (corrected/observed at 325 nm) Not less than 0.85.

Note: One former International Vitamin A Unit is the specific biologic activity of 0.3 μg of the all-*trans* isomer of retinol.

TESTS

Assay

If Vitamin A is in the form of an ester (acetate or palmitate), use the following procedure:

Assay for Vitamin A Ester (Acetate or Palmitate)   (**Note**: Use low-actinic glassware throughout this procedure.)

Mobile Phase   Use *n*-hexane.

Standard Preparation   Dissolve an accurately weighed quantity of USP Vitamin A Reference Standard (all-*trans* retinyl acetate) in *Mobile Phase*, and dilute quantitatively, and stepwise if necessary, to obtain a solution having a known concentration of about 40 μg/mL.

System Suitability Preparation   Dissolve an accurately weighed quantity of retinyl palmitate[1] in *Mobile Phase* to obtain a solution having a known concentration of about 40 μg/mL. Mix equal volumes of this solution and the *Standard Preparation* to obtain a solution having a concentration of about 20 μg/mL each of retinyl acetate and retinyl palmitate.

Assay Preparation   Dissolve an accurately weighed quantity of sample in *Mobile Phase*, and dilute quantitatively, and stepwise if necessary, to obtain a solution having a concentration of vitamin A ester (acetate or palmitate) of about 40 μg/mL.

Chromatographic System   The liquid chromatograph is equipped with a 325-nm detector and a 4.6-mm × 15-cm column that contains 3-μm silica (Supelcosil LC-Si, or the equivalent). The flow rate is about 1 mL/min. Chromatograph the *System Suitability Preparation*, and measure the peak areas as directed under *Procedure*: The resolution, R, between the all-*trans* retinyl acetate and the all-*trans* retinyl palmitate peaks is not less than 10, and the relative standard deviation for replicate injections is not more than 3.0%.

Procedure   Separately inject equal volumes (about 40 μL) of the *Standard Preparation* and the *Assay Preparation* into the chromatograph, record the chromatograms, and measure the peak areas for the all-*trans* retinyl acetate (or palmitate) and the 13-*cis* retinyl acetate (or palmitate), if present, obtained from the *Standard Preparation* and the peak areas for the all-*trans* retinyl acetate (or palmitate) and the 13-*cis* retinyl acetate (or palmitate), if present, in the chromatogram of the *Assay Preparation*. The relative retention times are about 0.7 for 13-*cis* retinyl acetate and 1.0 for all-*trans* retinyl acetate; or the relative retention times are about 0.8 for 13-*cis* retinyl palmitate and 1.0 for all-*trans* retinyl palmitate.

If Vitamin A is in the form of the acetate, calculate the quantity, in mg/g, of vitamin A acetate in the sample taken by the formula

$$(C/D)(r_U/r_S),$$

in which C is the concentration, in mg/mL, of USP Vitamin A Reference Standard in the *Standard Preparation*; D is the concentration, in mg/mL, of the sample in the *Assay Preparation*; and r_U and r_S are the summed peak areas of the 13-*cis* and all-*trans* retinyl acetate obtained from the *Assay Preparation* and the *Standard Preparation*, respectively.

If Vitamin A is in the form of the palmitate, calculate the quantity, in mg/g, of vitamin A palmitate in the sample taken by the formula

$$(524.96/328.54)(C/D)(r_U/r_S),$$

in which 524.96 is the formula weight of vitamin A palmitate; 328.54 is the formula weight of vitamin A acetate; C is the concentration, in mg/mL, of USP Vitamin A Reference Standard in the *Standard Preparation*; D is the concentration, in mg/mL, of the sample in the *Assay Preparation*; r_U is the summed peak areas of the 13-*cis* and all-*trans* retinyl palmitate obtained from the *Assay Preparation*; and r_S is the summed peak areas of the 13-*cis* and all-*trans* retinyl acetate obtained from the *Standard Preparation*.

Use the following procedure for other forms of Vitamin A:

Complete the assay promptly and exercise care throughout the procedure to keep to a minimum exposure to atmospheric oxygen and other oxidizing agents and to actinic light, preferably by the use of an atmosphere of an inert gas and nonactinic glassware.

Isopropyl Alcohol   Use reagent-grade isopropyl alcohol. Redistill, if necessary, to meet the following requirements for spectral purity: When measured in a 1-cm quartz cell against water it shows an absorbance not greater than 0.05 at 300 nm and not greater than 0.01 between 320 and 350 nm.

Ether   Use freshly redistilled reagent-grade ether, discarding the first and last 10% portions.

Procedure   Transfer into a saponification flask an accurately weighed portion of the sample containing the equivalent of not less than 0.15 mg of retinol, but containing not more than 1 g of fat. If in solid form, heat the portion taken in 10 mL of water on a steam bath for about 10 min, crush the remaining solid with a blunt glass rod, and warm for about 5 min.

Add 30 mL of alcohol if the sample is liquid, or 23 mL of alcohol and 7 mL of glycerin if the sample is solid, followed by 3 mL of potassium hydroxide solution (9 in 10). Reflux under an all-glass condenser for 30 min. Cool the solution, add 30 mL of water, and transfer to a separator. Add 2 g of finely powdered sodium sulfate. Extract by shaking for 2 min with one 150-mL portion of ether and, if an emulsion forms, with three additional 25-mL portions of ether. Combine the ether extracts, if necessary, and wash by swirling gently with 50 mL of water. Repeat the washing more vigorously with three additional 50-mL portions of water. Transfer the washed ether extract to a 250-mL volumetric flask, and add ether to volume.

---

[1] A suitable grade of retinyl palmitate may be obtained either from Sigma Chemical Company, P.O. Box 14508, St. Louis, MO 63178-9916 or from Fluka Chemical Corporation, 980 South Second Street, Ronkonkoma, NY 11779-7238.

Evaporate a 25.0-mL portion of the ether extract to about 5 mL. *Without applying heat and with the aid of a stream of inert gas or vacuum,* continue the evaporation to about 3 mL. Dissolve the residue in sufficient isopropyl alcohol to give an expected concentration of the equivalent of 3 to 5 μg of retinol per mL or such that it will give an absorbance in the range of 0.5 to 0.8 at 325 nm. Determine the absorbances of the resulting solution at the wavelengths 310, 325, and 334 nm, with a suitable spectrophotometer fitted with matched quartz cells.

Calculation Calculate the retinol content as follows:

$$\text{Content (in mg)} = 0.549 A_{325}/LC,$$

in which A_{325} is the observed absorbance at 325 nm; L is the length, in cm, of the absorption cell; and C is the amount of sample expressed as g in each 100 mL of the final isopropyl alcohol solution, provided that A_{325} has a value not less than $[A_{325}]/1.030$ and not more than $[A_{325}]/0.970$, where $[A_{325}]$ is the corrected absorbance at 325 nm and is given by the equation

$$[A_{325}] = 6.815 A_{325} - 2.555 A_{310} - 4.260 A_{334},$$

in which A designates the absorbance at the wavelength indicated by the subscript.

Where $[A_{325}]$ has a value less than $A_{325}/1.030$, apply the following equation:

$$\text{Content (in mg)} = 0.549 [A_{325}]/LC,$$

in which the values are as defined herein.

Confidence Interval The range of the limits of error, indicating the extent of discrepancy to be expected in the results of different laboratories at $P = 0.05$, is approximately ±8%.

Absorbance Ratio Determine by the formula

$$[A_{325}]/A_{325},$$

the terms of which are defined under *Calculation* in the *Assay*, as given above.

Packaging and Storage Store in a cool place in tight containers, preferably under an atmosphere of an inert gas, protected from light.

Vitamin B$_{12}$

Cyanocobalamin

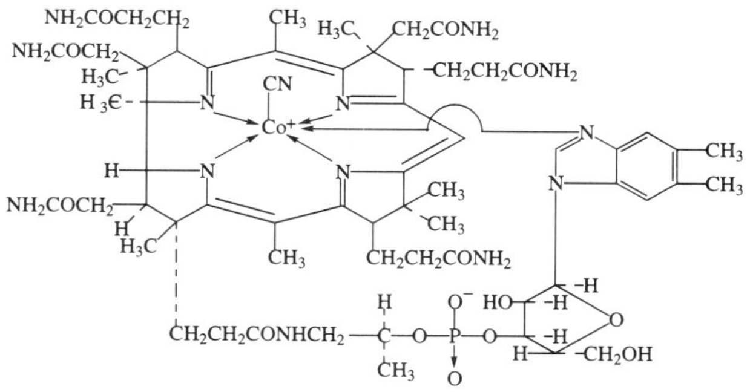

$C_{63}H_{88}CoN_{14}O_{14}P$ Formula wt 1355.38

CAS: [68-19-9]

DESCRIPTION

Dark red crystals, or amorphous or crystalline powder. In the anhydrous form it is very hygroscopic, and when exposed to air it may absorb about 12% of water. It is sparingly soluble in water, soluble in alcohol, and insoluble in acetone, in chloroform, and in ether.

Functional Use in Foods Nutrient; dietary supplement.

REQUIREMENTS

Identification

A. The absorption spectrum of the solution of the sample employed for measurement of absorbance in the *Assay* exhibits maxima within ± 1 nm at 278 nm and 361 nm, and within ± 2 nm at 550 nm. The ratio A_{361}/A_{278} is between 1.70 and 1.90, and the ratio A_{361}/A_{550} is between 3.15 and 3.40.

B. Fuse about 1 mg of the sample with about 50 mg of potassium pyrosulfate in a porcelain crucible, cool, and break up the mass with a glass rod. Add 3 mL of water, and dissolve by boiling. Add 1 drop of phenolphthalein TS, mix, and then add sodium hydroxide solution (1 in 10), dropwise, until just pink. Add 500 mg of sodium acetate, 0.5 mL of 1 N acetic acid, and 0.5 mL of nitroso R salt solution (1 in 500). A red or orange-red color appears at once. Add 0.5 mL of hydrochloric acid, and boil for 1 min. The red color persists.

C. Dissolve about 5 mg of the sample in 5 mL of water in a 50-mL distilling flask connected with a short, water-cooled, vertical condenser, the tip of which dips into a test tube containing 1 mL of sodium hydroxide solution (1 in 50). To the flask add 2.5 mL of hypophosphorous acid, then close the flask, heat at simmering for 10 min, and distill 1 mL into the test tube. To the tube add 4 drops of cold, saturated ferrous ammonium sulfate solution, shake gently, then add about 30 mg of sodium fluoride, and bring the contents to a boil. Immediately add, dropwise, dilute sulfuric acid (1 in 7) until a clear solution results, and then add 3 to 5 drops more of the acid. A blue or blue green color develops within a few min.

Assay Not less than 96.0% and not more than 100.5% of $C_{63}H_{88}CoN_{14}O_{14}P$, calculated on the dried basis.

Loss on Drying Not more than 12.0%.

Pseudo Cyanocobalamin Passes test.

TESTS

Assay With the aid of water, transfer about 30 mg of the sample, accurately weighed, to a 1000-mL volumetric flask, dilute with water to volume, and mix. Dissolve an accurately weighed quantity of USP Cyanocobalamin Reference Standard in water, and dilute quantitatively and stepwise with water to obtain a standard solution having a known concentration of about 30 μg/mL. Concomitantly determine the absorbances of both solutions in 1-cm cells at the wavelength of maximum absorption at about 361 nm, with a suitable spectrophotometer, using water as the blank. Calculate the quantity, in mg, of $C_{63}H_{88}CoN_{14}O_{14}P$ in the sample taken by the formula

$$C \times A_U/A_S,$$

in which C is the concentration of the reference standard solution, in μg/mL, and A_U and A_S are the absorbances of the sample solution and the reference standard solution, respectively.

Loss on Drying, *Appendix IIC* Dry about 25 mg at 105° in a vacuum at a pressure of not more than 5 mm of Hg for 2 h.

Pseudo Cyanocobalamin Dissolve 1.0 mg of the sample in 20 mL of water contained in a small separator, add 5 mL of a mixture of equal volumes of carbon tetrachloride and cresol, and shake well for about 1 min. Allow the layers to separate, and draw off the lower layer into a second small separator. Add 5 mL of dilute sulfuric acid (1 in 7), shake well, and allow to separate completely, centrifuging if necessary. The separated upper layer is colorless or has no more color than a mixture of 0.15 mL of 0.1 N potassium permanganate in 250 mL of water.

Packaging and Storage Store in tight containers.

Vitamin D$_2$

Ergocalciferol; Vitamin D

$C_{28}H_{44}O$ Formula wt 396.66

CAS: [50-14-6]

DESCRIPTION

White, odorless crystals. It is affected by air and by light. It is insoluble in water, but is soluble in alcohol, in chloroform, in ether, and in fatty oils.

Functional Use in Foods Nutrient; dietary supplement.

REQUIREMENTS

Identification

A. Prepare without heating, and handle without delay, a 1 in 100 solution of squalane in chloroform containing 50 mg of the sample per mL, and prepare a solution of USP Ergocalciferol Reference Standard in the same solvent and of the same concentration. Prepare a 1 in 100 solution of squalane in chloroform containing 100 μg of USP Ergosterol Reference Standard per mL. Spot 10 μL each of the sample solution, the ergocalciferol standard solution, and the ergosterol standard solution on a line parallel to and about 2.5 cm from the bottom edge of a thin-layer chromatographic plate coated with a 0.25-mm layer of chromatographic silica gel containing a suitable fluorescing substance. Place the plate in a developing chamber containing, and equilibrated with, a mixture of equal volumes of cyclohexane and ether. Develop the chromatogram until the solvent front has moved about 15 cm above the line of application. Perform the development and subsequent operations in the dark. Remove the plate from the chamber, allow the solvent to evaporate, and spray with a 1 in 50 solution of acetyl chloride in antimony trichloride TS. The chromatogram obtained with the sample solution shows a yellowish orange area (ergocalciferol) having the same R_f value as the area of the ergocalciferol standard and may show a violet area below the ergocalciferol area (see also *Ergosterol* under **TESTS**, below).

B. The infrared absorption spectrum of a potassium bromide dispersion of the sample, in the range of 2 to 12 μm, exhibits maxima only at the same wavelengths as that of a similar preparation of USP Ergocalciferol Reference Standard.

C. The ultraviolet absorption spectrum of the sample in alcohol solution exhibits inflections at the same wavelengths as that of USP Ergocalciferol Reference Standard, similarly measured, and the respective absorptivities at 265 nm do not differ by more than 3.0%.

Assay Not less than 97.0% and not more than 103.0% of $C_{28}H_{44}O$.

Ergosterol Passes test.

Melting Range Between 115° and 119°.

Reducing Substances Passes test.

Specific Rotation $[\alpha]_D^{25°}$: Between +103° and +106°.

TESTS

Assay

Standard Preparation (**Note**: Use low-actinic glassware, and prepare solutions fresh daily.) Transfer about 30 mg of USP Ergocalciferol Reference Standard, accurately weighed, into a 50-mL volumetric flask, dissolve without heating in toluene, add toluene to volume, and mix. Pipet 10 mL of this solution into a second 50-mL volumetric flask, dilute to volume with the *Mobile Phase*, and mix.

Assay Preparation (**Note**: Use low-actinic glassware, and prepare solutions fresh daily.) Transfer about 30 mg of the sample, accurately weighed, into a 50-mL volumetric flask, dissolve without heating in toluene, add toluene to volume, and mix. Pipet 10 mL of this solution into a second 50-mL volumetric flask, dilute to volume with the *Mobile Phase*, and mix.

Mobile Phase Prepare a 3 in 1000 mixture of n-amyl alcohol in ACS Reagent-Grade Hexanes (suitable for use in ultraviolet spectrophotometry) that has been dried by passing through a 60- × 8-cm column containing 500 g of 50- to 250-μm chromatographic siliceous earth. The ratio of components and the flow rate may be varied to meet the requirements of the *System Suitability Test*.

Chromatographic System Use a suitable high-pressure liquid chromatograph, operated at room temperature, fitted with a 25-cm × 4.6-mm stainless steel column packed with porous silica microparticles, 5 to 10 μm in diameter, and an ultraviolet detector that monitors absorption at 254 nm.

System Suitability Preparation Dissolve about 250 mg of USP Vitamin D Assay System Suitability Reference Standard in 10 mL of a mixture of equal volumes of toluene and the *Mobile Phase*. Reflux this solution at 90° for 45 min, and cool. (This solution contains cholecalciferol, pre-cholecalciferol, and *trans*-cholecalciferol.)

System Suitability Test Chromatograph five injections of the *System Suitability Preparation*, and measure the peak responses as directed under the *Procedure*. The relative standard deviation for the peak response does not exceed 2.0%, and the resolution between *trans*-cholecalciferol and pre-cholecalciferol is not less than 1.0. The chromatograms obtained in this test exhibit relative retention times of approximately 0.4, 0.5, and 1.0, for pre-cholecalciferol, *trans*-cholecalciferol, and cholecalciferol, respectively.

Procedure By means of a suitable sampling valve, introduce equal volumes (5 to 10 μL) of the *Standard Preparation* and the *Assay Preparation* into the chromatograph. Measure the peak responses for the major peaks, at corresponding retention times, obtained with the *Assay Preparation* and the *Standard Preparation*. Calculate the quantity, in mg, of $C_{28}H_{44}O$ in the sample taken by the formula

$$0.25C \times A_U/A_S,$$

in which C is the exact concentration, in μg/mL, of USP Ergocalciferol Reference Standard in the *Standard Preparation*, and A_U and A_S are the peak responses for ergocalciferol obtained with the *Assay Preparation* and the *Standard Preparation*, respectively.

Ergosterol The color of any violet area in the chromatogram of the sample solution, obtained as directed under *Identification Test B*, is not more intense than that of the violet area in the chromatogram of the ergosterol standard.

Melting Range Proceed as directed for *Melting Range or Temperature, Procedure for Class Ib*, Appendix IIB.

Reducing Substances To 10 mL of a 1 in 100 solution of the sample in absolute alcohol add 0.5 mL of a 1 in 200 solution of blue tetrazolium in absolute alcohol. Add 0.5 mL of a solution prepared by diluting 1 volume of 10% tetramethylammonium hydroxide with 9 volumes of absolute alcohol. Allow the mixture to stand for 5 min, then add 1 mL of glacial acetic acid, and mix. Prepare a blank by treating 10 mL of absolute alcohol in the same manner. The absorbance of the sample solution, determined at 525 nm with a suitable spectrophotometer against the blank, is not greater than that obtained with a solution containing 0.2 μg/mL of hydroquinone in absolute alcohol, similarly treated.

Specific Rotation, Appendix IIB Determine in a solution in alcohol containing 50 mg in each 10 mL. Prepare the solution without delay, using a sample from a container opened not longer than 30 min, and determine the rotation within 30 min after the solution has been prepared.

Packaging and Storage Store in hermetically sealed containers under nitrogen, in a cool place, and protected from light.

Vitamin D₃

Cholecalciferol; Vitamin D

$C_{27}H_{44}O$

Formula wt 384.65

CAS: [67-97-0]

DESCRIPTION

White, odorless crystals. It is affected by air and by light. It is insoluble in water. It is soluble in alcohol, in chloroform, and in fatty oils.

Functional Use in Foods Nutrient; dietary supplement.

REQUIREMENTS

Identification

A. Prepare without heating, and handle without delay, a 1 in 100 solution of squalane in chloroform containing 50 mg of the sample per mL, and prepare a solution of USP Cholecalciferol Reference Standard in the same solvent and having the same concentration. Spot 10 μL each of the sample solution and of the standard solution on a line parallel to and about 2.5 cm from the bottom edge of a thin-layer chromatographic plate coated with a 0.25-mm layer of chromatographic silica gel containing a suitable fluorescing substance. Place the plate in a developing chamber containing, and equilibrated with, a mixture of equal volumes of cyclohexane and ether. Develop the chromatogram until the solvent front has moved about 15 cm above the line of application. Perform the development and subsequent operations in the dark. Remove the plate from the chamber, allow the solvent to evaporate, and spray with a 1 in 50 solution of acetyl chloride in antimony trichloride TS. The chromatogram obtained with the sample solution shows a yellowish orange area (cholecalciferol) having the same R_f value as the area of the cholecalciferol standard and may show a violet area, attributed to 7-dehydrocholesterol, below the cholecalciferol area.

B. The infrared absorption spectrum of a potassium bromide dispersion of the sample, in the range of 2 to 12 μm, exhibits maxima only at the same wavelengths as that of a similar preparation of USP Cholecalciferol Reference Standard.

C. The ultraviolet absorption spectrum of the sample in a 1 in 100,000 solution in ethanol exhibits inflections at the same wavelengths as that of USP Cholecalciferol Reference Standard, similarly measured, and the respective absorptivities at the point of maximum absorbance occurring at about 265 nm do not differ by more than 3.0%.

Assay Not less than 97.0% and not more than 103.0% of $C_{27}H_{44}O$.
Melting Range Between 84° and 89°.
Specific Rotation $[\alpha]_D^{25°}$: Between +105° and +112°.

TESTS

Assay

Standard Preparation (**Note**: Use low-actinic glassware, and prepare solutions fresh daily.) Transfer about 30 mg of USP Cholecalciferol Reference Standard, accurately weighed, into a 50-mL volumetric flask, dissolve without heating in toluene, add toluene to volume, and mix. Pipet 10 mL of this solution into a second 50-mL volumetric flask, dilute to volume with the *Mobile Phase*, and mix.

Assay Preparation (**Note**: Use low-actinic glassware, and prepare solutions fresh daily.) Transfer about 30 mg of the sample, accurately weighed, into a 50-mL volumetric flask, dissolve without heating in toluene, add toluene to volume, and mix. Pipet 10 mL of this solution into a second 50-mL volumetric flask, dilute to volume with the *Mobile Phase*, and mix.

Mobile Phase, Chromatographic System, System Suitability Preparation, and *System Suitability Test* Proceed as directed in the *Assay* in the monograph for *Vitamin D₂*.

Procedure By means of a suitable sampling valve, introduce equal volumes (5 to 10 μL) of the *Standard Preparation* and the *Assay Preparation* into the chromatograph. Measure the peak responses for the major peaks, at corresponding retention times, obtained with the *Assay Preparation* and the *Standard Preparation*. Calculate the quantity, in mg, of $C_{27}H_{44}O$ in the sample taken by the formula

$$0.25C \times A_U/A_S,$$

in which C is the exact concentration, in μg/mL, of USP Cholecalciferol Reference Standard in the *Standard Preparation*, and A_U and A_S are the peak responses for cholecalciferol obtained with the *Assay Preparation* and the *Standard Preparation*, respectively.

Melting Range, Appendix IIB Proceed as directed for *Melting Range or Temperature, Procedure for Class Ib*, Appendix IIB.
Specific Rotation, Appendix IIB Determine in a solution in alcohol containing 50 mg in each 10 mL. Prepare the solution without delay, using a sample from a container opened not longer than 30 min, and determine the rotation within 30 min after the solution has been prepared.

Packaging and Storage Store in hermetically sealed containers under nitrogen, in a cool place, and protected from light.

Wheat Gluten

Vital Wheat Gluten; Devitalized Wheat Gluten

CAS: [8002-80-0]

DESCRIPTION

Wheat Gluten is the water-insoluble complex protein obtained by water extraction of wheat or wheat flour. It is soluble in alkalies and partly soluble in alcohol and dilute acids. It is a cream to light-tan, free-flowing powder. Vital Wheat Gluten is characterized by high viscoelasticity when hydrated, while devitalized Wheat Gluten has lost this character due to denaturation by heat.

Functional Use in Foods Dough strengthener; formulation aid; nutrient supplement; processing aid; stabilizer and thickener; surface-finishing agent; and texturizing agent.

REQUIREMENTS

Identification Add 40 mL of room-temperature water to 20 g of the sample, and stir. Vital Wheat Gluten will form a cohesive, viscoelastic mass, which can be lifted with the stirring rod without breaking apart. Devitalized Wheat Gluten will not form such a mass.
Assay Not less than 71.0% protein, calculated on the dried basis.
Ash (Total) Not more than 2.0%, calculated on the dried basis.
Crude Fat Not more than 2.0%.
Heavy Metals Not more than 5 mg/kg.
Lead Not more than 1 mg/kg.
Loss on Drying Not more than 10.0%.
Starch Not more than 21.0%.

TESTS

Assay Proceed as directed under *Nitrogen Determination*, Appendix IIIC. Calculate the percent protein (P) by the formula

$$P = 5.7N,$$

in which N is the percent nitrogen.
Ash (Total) Proceed as directed under *Ash (Total)*, Appendix IIC.
Crude Fat Proceed as directed under *Crude Fat*, in *Carbohydrates*, Appendix X.
Heavy Metals Prepare and test a 4-g sample as directed in *Method II* under the *Heavy Metals Test*, Appendix IIIB, using 20 µg of lead ion (Pb) in the control (*Solution A*) and 500° as the ignition temperature.
Lead Determine as directed under *Method I* in the *Atomic Absorption Spectrophotometric Graphite Furnace Method* under the *Lead Limit Test*, Appendix IIIB, using a 1-g sample.
Loss on Drying, Appendix IIC Dry a 2-g sample at 105° for 2 h.

Starch The remainder, after subtracting from 100.0% the sum of the percentages of *Ash (Total)*, *Loss on Drying*, and *Protein* (under *Assay*), represents the percent starch in the sample.

Packaging and Storage Store in well-closed containers.

Whey

DESCRIPTION

Whey is the material obtained as the by-product of cheesemaking resulting after the removal of curds (the solidified casein and butterfat). The process may lead either to a sweet- or acid-type whey. The final product is pasteurized and is available as a liquid or dry product. The acidity of Whey may be adjusted by the addition of safe and suitable pH adjusting ingredients. Whey has a dairy flavor.

Functional Use in Foods Formulation aid; processing aid; flavor enhancer; texturizer; nutritional extender.

REQUIREMENTS

Calculate all analyses except those for *Loss On Drying* and *Microbial Limits* on the dried (moisture-free) basis. Evaporate liquid samples to dryness on a steam bath, then as for the powdered form, dry to constant weight at 65° under vacuum.
Labeling State whether product is sweet or acid and the concentration if it is a liquid product.
Identification Whey exhibits the compositional profile specified below with respect to **Ash, Fat, Lactose, Loss on Drying** and **Protein**.
Ash (Total) Between 7.0% and 14.0%.
Fat Not more than 1.25%.
Free Acid Not more 3.0 mL of 0.1 N sodium hydroxide for sweet Whey and not more than 13.0 mL of 0.1 N sodium hydroxide for acid Whey, based on a sample containing 6.5 g of total solids.
Heavy Metals (as Pb) Not more than 5 mg/kg.
Lactose Between 61.0% and 75.0%.
Lead Not more than 0.5 mg/kg.
Loss on Drying Not more than 8.0% for the dry product and not more than 92.0% for the liquid product.
Microbial Limits:
   Aerobic Plate Count Not more than 50,000 CFU per g.
   Coliform Not more than 10 per g.
   Salmonella Negative in 25 g.
Protein Between 10.0% and 15.0%.

TESTS

Ash (Total) Proceed as directed in the general method, Appendix IIC, to a final gray to white residue.
Fat Transfer to a fat-extraction flask 1 g of the sample, accurately weighed; add 10 mL of water, and shake until homoge-

neous (warm if necessary). Add approximately 1 mL of ammonium hydroxide and heat in a water bath for 15 min at 60° to 70°, shaking occasionally. Add 10 mL of alcohol and mix well. Add 25 mL of peroxide-free ether, stopper, and shake vigorously for 1 min; allow to cool if necessary; add 25 mL of petroleum ether and repeat vigorous shaking. Allow the layers to separate and clarify or centrifuge at 600 rpm to expedite the process. Decant the organic layer into a suitable flask or dish and repeat the extraction twice with 15 mL each of ether and petroleum ether for each extraction. Evaporate the combined ether extractions on a steam bath and dry the residue to a constant weight at 102°, or 70° to 75° at less than 50 mm Hg. Calculate the percent of fat in the sample taken by the formula

$$(R \times 100)/S,$$

in which R is the weight, in mg, of the residue and S is the weight, in mg, of the sample taken.

Free Acid Accurately weigh a portion of the finely ground sample or liquid equivalent to 6.5 g of total solids based on the value obtained under *Loss on Drying*, and transfer to a 500-mL conical flask. Add 100 mL of carbon dioxide-free water, and stir for 1 min. Allow to stand for 1 h at room temperature. Add 0.5 mL of phenolphthalein TS, and titrate with 0.1 N sodium hydroxide to a pink endpoint that persists for 30 s.

Heavy Metals Prepare and test a 4-g sample as directed in *Method II* under the *Heavy Metals Test*, Appendix IIIB, using 20 µg of lead ion (Pb) in the control (*Solution A*).

Lactose

Mobile Phase Use a filtered and degassed acetonitrile–water (80:20) mixture at a flow rate of 2 mL/min.

Internal Standard Solution Prepare an aqueous solution of USP Fructose Reference Standard having a concentration of 100 mg/mL.

Standard Solution Using an accurately weighed quantity of USP Lactose Reference Standard prepare a solution in water having a concentration of 20 mg/mL. Dilute 9 volumes of this solution with 1 volume of the *Internal Standard Solution* to obtain the *Standard Solution* having a known concentration of 18 mg of USP Lactose Reference Standard per mL. Prepare fresh daily.

Assay Preparation Transfer an accurately weighed quantity of Whey containing about 180 mg of lactose to a 10-mL volumetric flask, add 1 mL of the *Internal Standard Solution*, dilute with water to volume, and mix.

Chromatographic System Use a suitable high-performance liquid chromatographic system (see Appendix IIA) operated at room temperature, equipped with a differential refractometer detector, a 250-mm × 4.6-mm microparticle silica gel with siloxane bonded cyano-amino moieties (Whatman P-10 carbohydrate or equivalent) column. Inject a 25-µL portion of the *Standard Solution* and record the peak responses. Replicate injections show a relative standard deviation of not more than 2.0% for the ratio of the response of lactose to that of the internal standard.

Procedure Separately inject 25-µL portions of the *Assay Solution* and the *Standard Preparation* into the chromatograph and record the responses. Calculate the percent of lactose in the Whey sample taken by the formula

$$(10\ C_S/W)(R_U/R_S),$$

in which C_S is the concentration, in mg/mL, of USP Lactose Reference Standard in the *Standard Solution*, W is the weight, in mg, of the Whey sample taken, and R_U and R_S are the response ratios of lactose to the *Internal Standard Solution* obtained with the *Assay Preparation* and the *Standard Solution*, respectively. Correct to the dried basis using the value obtained under *Loss on Drying*.

Lead (Based on AOAC method 973.35) (**Note**: For this test use reagent chemicals with as low a lead content as practicable.)

Ammonium Pyrrolidinedithiocarbamate (APDC), 2% Dissolve 2 g of APDC (Aldrich or equivalent) in 100 mL of water. Remove any insoluble free acid and other impurities by extracting the solution with three 10-mL portions of butylacetate and discarding the organic phase.

Citric Acid Solution, 10% Transfer 10.0 g of lead-free citric acid, accurately weighed, to a 100-mL volumetric flask, dilute to volume with water, and mix.

Standard Lead Solutions

Lead Nitrate Stock Solution Dissolve 159.8 mg, accurately weighed, of ACS reagent-grade lead nitrate in 100 mL of 1 N nitric acid to obtain a stock solution having a concentration of lead of 1 mg/mL.

Standard Dilutions Quantitatively and serially dilute accurately measured volumes of the *Lead Nitrate Stock Solution* with 1 N nitric acid to obtain *Standard Dilutions* having known concentrations of 1.0, 0.5, and 0.25 µg of lead per mL, respectively.

Sample Preparation Transfer a previously dried, accurately weighed 25.0 g sample to an ashing vessel. Place in a furnace held at 250° and slowly and gradually raise the temperature in 50° increments to 350°, and hold at this temperature until smoking ceases. Taking care not to ignite the sample, raise the temperature to 500° in 75° increments, and hold overnight. Remove from the furnace and allow to cool to room temperature. The ash must be white and essentially carbon-free. If it is gray containing excess carbon particles, destroy this by wetting the sample with a minimum amount of water followed by the dropwise addition of about 3 mL of nitric acid. Dry on a hot plate and then transfer to the furnace held at 250°, slowly increase the temperature to 500° and maintain it for 2 h. Repeat this procedure until a white, carbon-free residue is obtained. Dissolve the residue in 5 mL of nitric acid with warming on a steam bath to aid solution. Transfer the solution, filtering if necessary, into a 50-mL volumetric flask, and dilute to volume with 1 N nitric acid.

Procedure Transfer 20.0-mL aliquots of the *Sample Preparation*, each *Standard Dilution*, and water to serve as a blank, to separate 60-mL separations. Treat each solution as follows: add 4 mL of *Citric Acid Solution*, 10% and 3 drops bromocresol green indicator. Adjust the pH to about 5.4 using 6 N ammonium hydroxide. Add 4 mL of APDC, stopper, and shake for 60 s. Add 5.0 mL of butyl acetate, stopper, and shake vigorously for 60 s. Allow the layers to separate and discard the lower, aqueous phase. If an emulsion forms, use centrifugation to break the emulsion.

Using an atomic absorption spectrophotometer with an air–acetylene flame and set at the 283.3 nm lead line, separately aspirate the *Standard Dilutions*. Plot a standard curve of each corrected absorbance versus the concentration, in µg/mL. Aspirate the *Sample Preparation*, determine its corrected absorbance A and its corresponding lead concentration, C_s, from the standard curve. Calculate the concentration, in mg/kg, of lead in the sample taken by the formula

$$12.5\ C_s/W_s,$$

in which W_s is the weight, in g, of the sample.

Loss on Drying Dry at 65° at a pressure less than 100 mm of mercury for 16 h.

Microbial Limits:

Aerobic Plate Count Proceed as directed in chapter 3 of the *FDA Bacteriological Analytical Manual*, Seventh Edition, 1992.

Coliforms Proceed as directed in chapter 4 of the *FDA Bacteriological Analytical Manual*, Seventh Edition, 1992.

Salmonella Proceed as directed in chapter 5 of the *FDA Bacteriological Analytical Manual*, Seventh Edition, 1992.

Protein Determine the percent of nitrogen as directed in the *Nitrogen Determination (Kjeldahl Method)*, Appendix IIIC. The percent protein equals percent N × 6.38.

Packaging and Storage Store in tight containers.

Wintergreen Oil

Gaultheria Oil

CAS: [68917-75-9]

DESCRIPTION

Wintergreen Oil is obtained by maceration and subsequent distillation with steam from the leaves of *Gualtheria procumbens* L. (Fam. *Ericaceae*) or from the bark of *Betula lenta* L. (Fam. *Betulaceae*). It is a colorless, yellowish, or reddish liquid having the characteristic odor and taste of wintergreen. It boils, with decomposition, between 219° and 224°. It is soluble in alcohol and in glacial acetic acid, and it is very slightly soluble in water.

Functional Use in Foods Flavoring agent.

REQUIREMENTS

Identification

A. Shake 1 drop with about 5 mL of water, and add 1 drop of ferric chloride TS. A deep violet color is produced.

B. The infrared absorption spectrum of the sample exhibits relative maxima (that may vary in intensity) at the same wavelengths (or frequencies) as those shown in the respective spectrum in the section on *Infrared Spectra* (*Series A: Essential Oils*), using the same test conditions as specified therein.

Assay Not less than 98.0% and not more than 100.5% of methyl salicylate ($C_8H_8O_3$).

Acid Value Not more than 1.0.

Angular Rotation Slightly levorotatory, exhibiting a rotation of not more than −1.5°.

Heavy Metals (as Pb) Passes test.

Refractive Index Between 1.535 and 1.538 at 20°.

Solubility in Alcohol Passes test.

Specific Gravity Between 1.176 and 1.182.

TESTS

Assay Weigh accurately about 2 g, and proceed as directed for *Ester Determination*, Appendix VI, using 76.08 as the equivalence factor (*e*) in the calculation. Modify the procedure by using 50.0 mL of 0.5 N alcoholic potassium hydroxide and by refluxing on the steam bath for 2 h.

Acid Value Determine as directed for *Acid Value* under *Essential Oils and Flavors*, Appendix VI, using bromocresol purple TS instead of phenolphthalein TS as the indicator.

Angular Rotation Determine in a 100-mm tube as directed under *Optical (Specific) Rotation*, Appendix IIB.

Heavy Metals Shake 10 mL of the oil with an equal volume of water to which 1 drop of hydrochloric acid has been added, and pass hydrogen sulfide through the mixture until it is saturated. No darkening in color is produced in either the oil or the water.

Refractive Index, Appendix IIB Determine with an Abbé or other refractometer of equal or greater accuracy.

Solubility in Alcohol Proceed as directed in the general method, Appendix VI. One mL dissolves in 7 mL of 70% alcohol. The solution may have not more than a slight cloudiness.

Specific Gravity Determine by any reliable method (see *General Provisions*).

Packaging and Storage Store in full, tight containers in a cool place protected from light.

Xanthan Gum

INS: 415 CAS: [11138-66-2]

DESCRIPTION

A high-molecular-weight polysaccharide gum produced by a pure-culture fermentation of a carbohydrate with *Xanthomonas campestris*, purified by recovery with isopropyl alcohol, dried, and milled. It contains D-glucose and D-mannose as the dominant hexose units, along with D-glucuronic acid, and is prepared as the sodium, potassium, or calcium salt. It occurs as a cream-colored powder that is readily soluble in hot or cold water. Its solutions are neutral.

Functional Use in Foods Stabilizer; thickener; emulsifier; suspending agent; bodying agent; foam enhancer.

REQUIREMENTS

Identification To 300 mL of water, previously heated to 80° and stirred rapidly with a mechanical stirrer in a 400-mL beaker, add, at the point of maximum agitation, a dry blend of 1.5 g of the sample and 1.5 g of locust bean gum. Stir until the mixture goes into solution, and then continue stirring for 30 min longer. Do not allow the water temperature to drop below 60° during stirring. Discontinue stirring, and allow the mixture to cool at room temperature for at least 2 h. A firm rubbery gel forms after the temperature drops below 40°, but no such gel forms in a 1% control solution of the sample prepared in the same manner but omitting the locust bean gum.

Assay It yields, on the dry basis, not less than 4.2% and not more than 5.4% carbon dioxide (CO_2), corresponding to between 91.0% and 117.0% of Xanthan Gum.

Arsenic (as As)   Not more than 3 mg/kg.

Heavy Metals (as Pb)   Not more than 0.002%.

Isopropyl Alcohol   Not more than 0.075%.

Lead   Not more than 5 mg/kg.

Loss on Drying   Not more than 15.0%.

Pyruvic Acid   Not less than 1.5%.

Viscosity   Passes test.

TESTS

Assay Proceed as directed under *Alginates Assay*, Appendix IIIC, but use an undried sample of about 1.2 g, accurately weighed.

Arsenic A *Sample Solution* prepared as directed for organic compounds meets the requirements of the *Arsenic Test*, Appendix IIIB.

Heavy Metals Prepare and test a 1-g sample as directed in *Method II* under the *Heavy Metals Test*, Appendix IIIB, using a platinum crucible for the ignition and 20 μg of lead ion (Pb) in the control (*Solution A*).

Isopropyl Alcohol

IPA Standard Solution Transfer 500.0 mg of chromatographic-quality isopropyl alcohol into a 50-mL volumetric flask, dilute to volume with water, and mix. Pipet 10 mL of this solution into a 100-mL volumetric flask, dilute to volume with water, and mix.

TBA Standard Solution Transfer 500.0 mg of chromatographic-quality *tert*-butyl alcohol into a 50-mL volumetric flask, dilute to volume with water, and mix. Pipet 10 mL of this solution into a 100-mL volumetric flask, dilute to volume with water, and mix.

Mixed Standard Solution Pipet 4 mL each of the *IPA Standard Solution* and of the *TBA Standard Solution* into a 125-mL graduated Erlenmeyer flask, dilute to about 100 mL with water, and mix. This solution contains approximately 40 μg each of isopropyl alcohol and of *tert*-butyl alcohol per mL.

Sample Preparation Disperse 1 mL of a suitable antifoam emulsion, such as Dow-Corning G-10 or equivalent, in 200 mL of water contained in a 1000-mL, 24/40, round-bottom distilling flask. Add about 5 g of the sample, accurately weighed, and shake for 1 h on a wrist-action mechanical shaker. Connect the flask to a fractionating column, and distill about 100 mL, adjusting the heat so that foam does not enter the column. Add 4.0 mL of *TBA Standard Solution* to the distillate to obtain the *Sample Preparation*.

Procedure Inject about 5 μL of the *Mixed Standard Solution* into a suitable gas chromatograph equipped with a flame-ionization detector and a 1.8-m × 3.2-mm stainless steel column packed with 80/100-mesh Porapak QS, or equivalent. The carrier is helium flowing at 80 mL/min. The injection port temperature is 200°, the column temperature is 165°, and the detector temperature is 200°. The retention time of isopropyl alcohol is about 2 min, and that of *tert*-butyl alcohol is about 3 min.

Determine the areas of the IPA and TBA peaks, and calculate the response factor, f, by the formula

$$A_{IPA}/A_{TBA},$$

in which A_{IPA} is the area of the isopropyl alcohol peak, and A_{TBA} is the area of the *tert*-butyl alcohol peak.

Similarly, inject about 5 μL of the *Sample Preparation*, and determine the peak areas, recording the area of the isopropyl alcohol peak as a_{IPA}, and that of the *tert*-butyl alcohol peak as a_{TBA}. Calculate the isopropyl alcohol content, in mg/kg, in the sample taken by the formula

$$(a_{IPA} \times 4000)/(f \times a_{TBA} \times W),$$

in which W is the weight, in g, of the sample taken.

Lead A *Sample Solution* prepared as directed for organic compounds meets the requirements of the *Lead Limit Test*, Appendix IIIB, using 5 μg of lead ion (Pb) in the control.

Loss on Drying, Appendix IIC   Dry at 105° for 2.5 h.

Pyruvic Acid

Sample Preparation Dissolve 600.0 mg of the sample, accurately weighed, in sufficient water to make 100.0 mL, and transfer 10.0 mL of the solution into a 50-mL glass-stoppered flask. Pipet 20 mL of 1 N hydrochloric acid into the flask, weigh the flask, and reflux for 3 h, taking precautions to prevent loss of vapors. Cool to room temperature, and add water to make up for any weight loss during refluxing. Pipet 1.0 mL of a 1 in 200 solution of 2,4-dinitrophenylhydrazine in 2 N hydrochloric acid into a 30-mL separator, then add 2.0 mL of the sample solution, mix, and allow to stand at room temperature for 5 min. Extract the mixture with 5 mL of ethyl acetate, and discard the aqueous layer. Extract the hydrazone from the ethyl acetate with three 5-mL portions of sodium carbonate TS, collecting the extracts in a 50-mL volumetric flask. Dilute to volume with sodium carbonate TS, and mix.

Standard Preparation Transfer 45.0 mg of pyruvic acid, accurately weighed, into a 500-mL volumetric flask, dilute to volume with water, and mix. Transfer 10.0 mL of this solution into a 50-mL glass-stoppered flask, and continue as directed under *Sample Preparation*, beginning with "Pipet 20 mL of 1 N hydrochloric acid into the flask. . . ."

Procedure Determine the absorbance of each solution with a suitable spectrophotometer in 1-cm cells at the maximum at about 375 nm, using sodium carbonate TS as the blank. The absorbance of the *Sample Preparation* is equal to or greater than that of the *Standard Preparation*.

Viscosity Prepare two identical solutions, each containing 1% of the sample and 1% of potassium chloride in water, and stir for 2 h. Determine the viscosity of one solution at 23.9° (75°F) as directed under *Viscosity of Sodium Carboxymethylcellulose*,

Appendix IIB, using a No. 3 spindle rotating at 60 rpm (Brookfield or equivalent). The viscosity (V_1) thus determined is not less than 600 centipoises. Determine the viscosity (V_2) of the other solution in the same manner, but maintain the temperature at 65.6° (150°F). The ratio of the viscosities, V_1/V_2, is between 1.02 and 1.45.

Packaging and Storage Store in well-closed containers.

Xylitol

1,2,3,4,5-Pentahydroxypentane

$C_5H_{12}O_5$ Formula wt 152.15
INS: 967   CAS: [87-99-0]

DESCRIPTION

White crystals or crystalline powder. Xylitol is found in most fruits and berries, as well as in vegetables. It has a sweet taste and produces a cooling sensation in the mouth. One g dissolves in about 0.65 mL of water. It is sparingly soluble in alcohol. Crystalline Xylitol has a melting range between 92° and 96°.

Functional Use in Foods Nutritive sweetener.

REQUIREMENTS

Identification

A. Dissolve 5 g of the sample in 10 mL of a mixture of equal volumes of hydrochloric acid and formalin, and allow to react at 50° for 2 h. Add 25 mL of ethanol, collect the crystals produced, and dissolve them in 10 mL of water by heating. Add 50 mL of ethanol to crystallize, collect the separated crystals by filtration, recrystallize them from ethanol twice, and dry at 105° for 2 h. The crystals melt between 195° and 201°.

B. The infrared absorption spectrum of a potassium bromide dispersion of Xylitol exhibits maxima only at the same wavelengths as those of a similar preparation of USP Xylitol Reference Standard. If a difference appears, dissolve portions of both the sample and the reference standard in a suitable solvent, evaporate the solutions to dryness, and repeat the test on the residues.

Assay Not less than 98.5% and not more than 101.0% of $C_5H_{12}O_5$, calculated on the anhydrous basis.
Arsenic (as As) Not more than 3 mg/kg.
Heavy Metals (as Pb) Not more than 10 mg/kg.
Lead Not more than 1 mg/kg.
Other Polyols Not more than 2.0%.
Reducing Sugars (as glucose) Not more than 0.2%.
Residue on Ignition Not more than 0.5%.
Water Not more than 0.5%.

TESTS

Assay

Internal Standard Solution Transfer about 500 mg of erythritol, accurately weighed, into a 25-mL volumetric flask, dilute to volume with water, and mix.

Standard Solution Transfer about 25 mg each of L-arabinitol, galactitol, mannitol, and sorbitol, accurately weighed, to a 100-mL volumetric flask, dilute to volume with water, and mix. To an accurately measured volume of this solution, add an accurately weighed amount of USP Xylitol Reference Standard to obtain a solution with a known concentration of about 49 mg/mL.

Assay Preparation Transfer about 5 g of the sample, accurately weighed, into a 100-mL volumetric flask, dilute to volume with water, and mix.

Chromatographic System Use a gas chromatograph equipped with a flame-ionization detector and a 2-m × 2-mm glass column packed with 3% liquid phase of 25% phenyl–25% cyanopropylmethylsilicone (OV-225 or equivalent) on silanized siliceous earth support (Chromosorb W-HP or equivalent). The carrier gas is nitrogen flowing at about 30 mL/min. The injector port temperature is 250°, the column temperature is 200°, and the detector temperature is 250°. Chromatograph the derivatized *Standard Solution* prepared as directed under *Procedure*, and record the peak responses. The relative retention times corresponding to erythritol, L-arabinitol, Xylitol, galactitol, mannitol, and sorbitol are usually about 1.0, 2.77, 3.90, 6.96, 7.63, and 8.43, respectively. The relative standard deviation of the response ratios of the derivatized Xylitol to the derivatized erythritol from three replicate injections does not exceed 2.0%.

Procedure Pipet 1-mL portions of the *Standard Solution* and the *Assay Preparation* into separate 100-mL, round-bottom boiling flasks. To each flask, add 1.0 mL of *Internal Standard Solution*, and evaporate the respective mixtures to dryness on a water bath at 60° with the aid of a rotary evaporator. Dissolve each dry residue in 1 mL of pyridine, and add 1 mL of acetic anhydride to each flask. Boil each solution under reflux for 1 h to complete the acetylation. Separately inject 1-μL portions of the derivatized solutions from the *Assay Preparation* and the *Standard Solution* into the gas chromatograph and measure the peak responses. Calculate the percentage of Xylitol, on the as-is basis, by the formula

$$100(W_S/W_U)(R_U/R_S),$$

in which W_S is the weight, in mg, of USP Xylitol Reference Standard used for the *Standard Solution*; W_U is the weight, in mg, of the sample taken for the *Assay Preparation*; and R_U and R_S are the ratios of peak responses of the derivatized analyte to the derivatized erythritol from the *Internal Standard Solution* obtained from the *Assay Preparation* and the *Standard Solution*, respectively. Using the value obtained in the *Water* determination, correct the percentage to the anhydrous basis.

Arsenic A *Sample Solution* prepared as directed for organic compounds meets the requirements of the *Arsenic Test*, Appendix IIIB.

Heavy Metals A solution of 2 g in 25 mL of water meets the requirements of the *Heavy Metals Test*, Appendix IIIB, using 20 μg of lead ion (Pb) in the control (*Solution A*).

Lead Determine as directed for *Flame Atomic Absorption Spectrophotometric Method* under the *Lead Limit Test*, Appendix IIIB, using a 10-g sample.

Other Polyols

Internal Standard Solution, Standard Solution, Assay Preparation, and *Chromatographic System* Proceed as directed under *Assay*.

Procedure Proceed as described under *Assay*. Calculate the percentage of each polyol—L-arabinitol, galactitol, mannitol, and sorbitol—by the formula therein given, in which W_S refers to the weight, in mg, of the respective polyol taken for the *Standard Solution*; R_S is the peak response ratio of the corresponding polyol obtained from the *Standard Solution*; and R_U is the peak response ratio of the corresponding polyol obtained from the *Assay Preparation*. Sum the four individual polyol percentages to obtain the total.

Reducing Sugars Dissolve about 500 mg of the sample, accurately weighed, in 2 mL of water in a 10-mL conical flask, and add 2 mL of a dextrose solution, containing 0.5 mg/mL, to another conical flask. Add 1 mL of *Fehling's Solution A* and of *Fehling's Solution B* (see *Cupric Tartrate TS, Alkaline* under *Solutions and Indicators*) to each flask, heat to boiling, and cool. The sample solution is less turbid than the dextrose solution, which forms a reddish brown precipitate.

Residue on Ignition, Appendix IIC Ignite 2 g as directed in the general method.

Water Determine water content as directed under the *Karl Fischer Titrimetric Method*, Appendix IIB.

Packaging and Storage Store in well-closed containers in a dry place.

Dried Yeast

Brewer's Yeast; Torula Yeast

DESCRIPTION

Dried Yeast is the comminuted, washed, dried, and pasteurized cell walls from *Saccharomyces cerevisiae, Saccharomyces fragilis,* or *Torula utilis*. It is a light brown to buff powder, granules, or flakes with the characteristic odor of yeast. It contains no added substances.

Functional Use in Foods Carrier; flavor enhancer.

REQUIREMENTS

Identification When examined under a microscope, it exhibits numerous irregular masses and isolated yeast cells—the latter ovate, elliptical, spheroidal, or elliptic-elongate in shape, some with one or more attached buds—up to 12 μm in length and up to 7.5 μm in width. Each has a wall of cellulose surrounding a protoplast containing refractile glycogen vacuoles and oil globules.

Assay Not less than 45.0% protein.
Ash (Total) Not more than 8.0%.
Folic Acid Not more than 0.04 mg/g.
Lead Not more than 1 mg/kg.
Loss on Drying Not more than 7.0%.
Microbial Limits:
  Aerobic Plate Count Not more than 7500 CFU per g.
  E. coli Negative in 25 g.
  Salmonella Negative in 25 g.

TESTS

Assay Determine the percent of nitrogen by the *Kjeldahl Method*, under *Nitrogen Determination*, Appendix IIIC, and multiply by 6.25 to obtain the percent protein.

Ash (Total) Proceed as directed in Appendix IIC.

Folic Acid (Note: In the microbiological assay of folic acid, the microorganism is highly sensitive to minute amounts of growth factors and to many cleansing agents. Meticulously cleanse 20- × 150-mm test tubes and other necessary glassware with a suitable detergent, sodium lauryl sulfate, or an equivalent substitute. Follow cleansing by heating for 1 to 2 h at approximately 250°.) This method is based on AOAC method 960.46.

Vitamin-Free, Acid Hydrolyzed Casein Solution Prepare the solution by mixing 400 g of vitamin-free casein with 2 L of boiling 5 N hydrochloric acid. Autoclave for 10 h at 121°. Concentrate the mixture by distillation under reduced pressure until a thick paste remains. Redissolve the paste in water, adjust the solution to pH 3.5 ± 0.1 with a 10% solution of sodium hydroxide, and dilute with water to a final volume of 4 L. Add 80 g of activated charcoal, stir for 1 h, and filter. Repeat the treatment with activated charcoal. Filter the solution if a precipitate forms on storage.

Adenine–Guanine–Uracil Solution Dissolve 1.0 g each of adenine sulfate, guanine hydrochloride, and uracil in 50 mL of warm dilute hydrochloric acid (1 in 2), cool, and dilute with water to 1 L.

Asparagine Solution Dissolve 10 g of L-asparagine monohydrate in approximately 500 mL of water, and dilute with water to 1 L.

Manganese Sulfate Solution Dissolve 2.0 g of manganese sulfate monohydrate in water, and dilute with water to 200 mL.

Polysorbate 80 Solution Dissolve 25 g of polysorbate 80 (polyoxyethylene sorbitan monooleate) in ethyl alcohol, and dilute with ethyl alcohol to make 250 mL.

Salt Solution Dissolve 20 g of magnesium sulfate heptahydrate, 1 g of sodium chloride, 1 g of ferrous sulfate heptahydrate, and 1 g of manganese sulfate monohydrate in water, dilute with water to 1 L, add 10 drops of hydrochloric acid, and mix.

Tryptophan Solution Suspend 2.0 g of L-tryptophan in 800 mL of water, heat to 80°, and add dilute hydrochloric acid (1 in 2), dropwise, with stirring, until the suspension dissolves. Cool, and dilute with water to 1 L.

Vitamin Solution Dissolve 10 mg of *p*-aminobenzoic acid, 8 mg of calcium pantothenate, 40 mg of pyridoxine hydrochloride, 4 mg of thiamine hydrochloride, 8 mg of niacin, and 0.2 mg of biotin in approximately 300 mL of water. Add 10 mg of riboflavin dissolved in approximately 200 mL of 0.02 N acetic acid. Add a solution containing 1.9 g of anhydrous sodium acetate and 1.6 mL of glacial acetic acid in approximately 40 mL of water. Dilute the solution with water to a final volume of 2 L.

Xanthine Solution Suspend 1.0 g of xanthine in 200 mL of water, heat to approximately 70°, add 30 mL of dilute ammonium hydroxide (2 in 5), and stir until the suspension dissolves. Cool, and dilute with water to 1 L.

Basal Medium Stock Solution Prepare by adding, with mixing, in the following order, 25 mL of the *Vitamin-Free, Acid-Hydrolyzed Casein Solution*, 25 mL of the *Tryptophan Solution*, 2.5 mL of the *Adenine–Guanine–Uracil Solution*, 5 mL of the *Xanthine Solution*, 15 mL of the *Asparagine Solution*, 50 mL of the *Vitamin Solution*, and 5 mL of the *Salt Solution*. Add approximately 50 mL of water, and add, with mixing, 0.19 g of L-cysteine monohydrochloride monohydrate, 10 g of anhydrous glucose, 13 g of sodium citrate dihydrate, 1.6 g of anhydrous dipotassium hydrogen phosphate, and 0.0013 g of glutathione. When solution is complete, adjust to pH 6.8 with 10% sodium hydroxide solution, and add, with mixing, 0.25 mL of the *Polysorbate 80 Solution* and 5 mL of the *Manganese Sulfate Solution*. Dilute to a final volume of 250 mL with water.

Liquid Culture Medium Dissolve 15 g of peptonized milk, 5 g of water-soluble yeast extract, 10 g of anhydrous glucose, and 2 g of anhydrous potassium dihydrogen phosphate in about 600 mL of water. Add 100 mL of filtered tomato juice (filtered through Whatman No. 1 filter paper, or equivalent), and adjust to pH 6.5 by the dropwise addition of 1.0 N sodium hydroxide. Add, with mixing, 10 mL of the *Polysorbate 80 Solution* (see above). Dilute with water to a final volume of 1000 mL. Add 10-mL portions of this *Liquid Culture Medium* to test tubes, cover to prevent contamination, and sterilize by heating in an autoclave at 121° for 15 min. Cool the tubes rapidly to keep color formation to a minimum, and store at 10° in the dark.

Agar Culture Medium To 500 mL of *Liquid Culture Medium*, add 6.0 g of agar, and heat with stirring on a steam bath until the agar dissolves. Add approximately 10-mL portions of the hot solution to test tubes, cover to prevent contamination, sterilize by heating in an autoclave at 121° for 15 min, and cool tubes in an upright position to keep color formation to a minimum. Store at 10° in the dark.

Suspension Medium Dilute an appropriate volume of the *Basal Medium Stock Solution* with an equal volume of water. Distribute 10-mL portions of the *Suspension Medium* to test tubes, cover to prevent contamination, sterilize by heating in an autoclave at 121° for 15 min, and cool tubes rapidly to keep color formation to a minimum. Store at 10° in the dark.

Assay Organism Maintain *Enterococcus* (*Streptococcus*) *faecalis* ATCC 8043 by subculturing in stab cultures of *Agar Culture Medium* and incubating at 37° for 24 h. Stab cultures may be stored in the dark at 10° for a maximum of 7 days until use. Prepare fresh stab cultures at least on a weekly basis. Before using a new culture in the assay, make several successive transfers of the culture over a 1- to 2-week period. Transfer cells from the stab culture of *Assay Organism* to a sterile tube containing 10 mL of *Liquid Culture Medium*. Incubate for 18 h at 37°. Under aseptic conditions, centrifuge the culture, and decant the supernate. Wash the cells with three 10-mL portions of sterile *Suspension Medium*. Resuspend cells in 10 mL of sterile *Suspension Medium*—these cells serve as the inoculum.

Folic Acid Stock Solutions Accurately weigh, in a closed system, 50 to 60 mg of USP Folic Acid Reference Standard that has been dried to constant weight and stored in the dark over phosphorus pentoxide in a desiccator. Dissolve in approximately 30 mL of 0.01 N sodium hydroxide, add approximately 300 mL of water, adjust to pH 7.5 with hydrochloric acid (1 in 2), and dilute with additional water to a final folic acid concentration of exactly 100 μg/mL. Store under toluene in the dark at 10°.

Prepare an intermediate *Folic Acid Stock Solution* containing 1 μg/mL by placing 10 mL of the 100 μg/mL *Folic Acid Stock Solution* in a flask, adding approximately 500 mL of water, adjusting to pH 7.5 with dilute hydrochloric acid or sodium hydroxide as necessary, and diluting with additional water to a final volume of 1 L. Store under toluene in the dark at 10°.

Prepare the final *Folic Acid Stock Solution* by taking 100 mL of the intermediate *Folic Acid Stock Solution*, adding approximately 500 mL of water, adjusting to pH 7.5 with dilute hydrochloric acid or sodium hydroxide as necessary, and diluting with additional water to a final volume of 1 L. Store under toluene in the dark at 10°. This final *Folic Acid Stock Solution* has a concentration of 100 ng/mL.

Preparation of the Standard Curve Dilute the *Folic Acid Stock Solution* with water to a measured volume such that after incubation, as described below, response at the 5.0-mL level of this solution is equivalent to a titration volume of 8 to 12 mL. This concentration is usually 1 to 4 ng of folic acid per mL but can vary with the culture used in the assay. Designate this solution as the *Folic Acid Working Standard Solution*. To duplicate test tubes, add 0.0 (for uninoculated blanks), 0.0 (for inoculated blanks), 1.0, 2.0, 3.0, 4.0, and 5.0 mL, respectively, of the *Folic Acid Working Standard Solution*. To each tube, add water to make a final volume of 5.0 mL. Add 5.0 mL of the *Basal Medium Stock Solution* to each tube, and mix. Cover the tubes suitably to prevent bacterial contamination, and sterilize by heating in an autoclave at 121° for 10 min. Cool tubes rapidly to keep color formation to a minimum.

Note: Sterilizing and cooling conditions must be kept uniform to obtain reproducible results.

Aseptically inoculate each tube with 1 drop of the *Assay Organism* inoculum, except for one set of duplicate tubes containing 0.0 mL of the *Folic Acid Working Standard Solution*, which serve as the uninoculated blanks. Incubate the tubes for 72 h at 37°.

Note: Contamination of assay tubes with any foreign organism invalidates the assay.

Titrate the contents of each tube with 0.1 N sodium hydroxide, using bromothymol blue as the indicator. Disregard the results of the assay if the titration volume for the inoculated blank is more than 1.5 mL greater than that for the uninoculated blank. The titration volume for the 5.0-mL level of the *Folic Acid Working Standard Solution* should be approximately 8 to 12 mL. Prepare a standard curve by plotting the titration values, expressed in mL of 0.1 N sodium hydroxide for each level of the *Folic Acid Working Standard Solution* used, against the amount of folic acid contained in that tube.

Assay Solution Weigh and suspend 1.0 g of the sample in 100 mL of water. Add 2 mL of ammonium hydroxide (2 in 5). If the sample is not readily soluble, comminute to disperse it evenly in the liquid, then agitate vigorously and wash down the sides of the flask with 0.1 N ammonium hydroxide. Heat the mixture in an autoclave at 121° for 15 min. If lumping occurs, agitate the sample until the particles are evenly dispersed. Dilute the mixture with water to 200 mL. Filter through Whatman No. 1 filter paper, or equivalent, if necessary, to remove any undissolved particles. Adjust the filtered mixture to pH 6.8 and dilute to 1000 mL with water. The final *Assay Solution* is prepared by diluting with water 1.0 mL of the intermediate solution to a final volume of 50.0 mL.

Procedure To duplicate test tubes, add 0.0 (for uninoculated blanks), 0.0 (for inoculated blanks), 1.0, 2.0, 3.0, 4.0, and 5.0 mL, respectively, of the *Assay Solution*. To each tube, add water to make a final volume of 5.0 mL. Proceed as directed above for *Preparation of the Standard Curve*. Determine the amount of folic acid for each level of the *Assay Solution* by interpolation from the standard curve. Discard any observed titration values equivalent to less than 0.5 mL or more than 4.5 mL of the *Folic Acid Working Standard Solution*. If necessary, the *Assay Solution* can be diluted to achieve the ideal concentration range of folic acid. For each level of *Assay Solution* used, calculate the vitamin content per mL of *Assay Solution*. Calculate the average of values obtained from tubes that do not vary by greater than 10% from this average. More than two-thirds of the original number of tubes must be within 10% of the average folic acid value, or the data cannot be used to calculate the folic acid concentration in the sample. If the data are acceptable, the folic acid concentration in the sample can be determined by multiplying the average folic acid concentration, in ng/mL, of the *Assay Solution* by 0.025 to give the mg of folic acid per g of sample.

Lead Determine as directed under the *Dithizone Method* under the *Lead Limit Test*, Appendix IIIB, using a 4-g sample and 4 μg of lead ion (Pb) in the control.

Loss on Drying, Appendix IIC Dry a 1-g sample at 105° for 4 h.

Microbial Limits:
  Aerobic Plate Count Proceed as directed in Chapter 3 of the *FDA Bacteriological Analytical Manual*, Seventh Edition, 1992.
  E. coli Proceed as directed in Chapter 4 of the *FDA Bacteriological Analytical Manual*, Seventh Edition, 1992.
  Salmonella Proceed as directed in Chapter 5 of the *FDA Bacteriological Analytical Manual*, Seventh Edition, 1992.

Packaging and Storage Store in tight containers in a cool, dry place.

Yeast Extract

Autolyzed Yeast Extract

DESCRIPTION

Yeast Extract comprises the water-soluble components of the yeast cell, the composition of which is primarily amino acids, peptides, carbohydrates, and salts. Yeast Extract is produced through the hydrolysis of peptide bonds by the naturally occurring enzymes present in edible yeasts or by the addition of food-grade enzymes. Food-grade salts may be added during processing. Individual products may be in liquid, paste, powder, or granular form.

Functional Use in Foods Flavoring agent; flavor enhancer.

REQUIREMENTS

Calculate all analyses on the dried basis. In a suitable tared container, evaporate liquid and paste samples to dryness on a steam bath, then, as for the powdered and granular forms, dry to constant weight at 105° (see *General Provisions*).
Assay (Protein) Not less than 42.0% protein.
α-Amino Nitrogen/Total Nitrogen (AN/TN) Percent Ratio Not less than 15.0% or more than 55.0%.
Ammonia Nitrogen Not more than 2.0%, calculated on a dry, salt-free basis.
Glutamic Acid Not more than 12.0% as $C_5H_9NO_4$ and not more than 28.0% of the total amino acids.
Heavy Metals (as Pb) Not more than 10 mg/kg.
Insoluble Matter Not more than 2%.
Lead Not more than 3 mg/kg.
Mercury Not more than 3 mg/kg.
Microbial Limits:
  Aerobic Plate Count Not more than 50,000 CFU per g.
  Coliforms Not more than 10 per g.
  Yeast and Mold Not more than 50 CFU per g.
  Salmonella Negative in 25 g.
Potassium Not more than 13.0%.
Sodium Not more than 20.0%.

TESTS

Assay (Protein) Proceed as directed under *Nitrogen Determination*, Appendix IIIC. Calculate the percent protein (P) by the formula

$$P = 6.25N,$$

in which N is the percent nitrogen.
α-Amino Nitrogen/Total Nitrogen (AN/TN) Percent Ratio Determine α-*Amino Nitrogen* as directed in the test in the mono-

graph for *Acid Hydrolysates of Proteins*. Determine Total Nitrogen as directed under *Nitrogen Determination*, Appendix IIIC. Calculate the AN/TN percent ratio, in which AN is the percent α-amino nitrogen and TN is the percent total nitrogen.

Ammonia Nitrogen Proceed as directed in the test for *Ammonia Nitrogen* in the monograph for *Acid Hydrolysates of Proteins*.

Glutamic Acid Proceed as directed in the test for *Glutamic Acid* in the monograph for *Acid Hydrolysates of Proteins*.

Heavy Metals Prepare and test a 2-g sample as directed in *Method II* under the *Heavy Metals Test*, Appendix IIIB, using 20 µg of lead ion (Pb) in the control (*Solution A*).

Insoluble Matter Transfer about 5 g, accurately weighed, into a 250-mL Erlenmeyer flask, add 75 mL of water, cover the flask with a watch glass, and boil gently for 2 min. Filter the solution through a tared filtering crucible, dry at 105° for 1 h, cool, and weigh.

Lead A *Sample Solution* prepared as directed for organic compounds meets the requirements of the *Lead Limit Test*, Appendix IIIB, using 3 µg of lead ion (Pb) in the control.

Mercury Determine as directed in the *Mercury Limit Test*, Appendix IIIB.

Microbial Limits:

  Aerobic Plate Count Proceed as directed in chapter 3 of the *FDA Bacteriological Analytical Manual*, Seventh Edition, 1992.

  Coliforms Proceed as directed in chapter 4 of the *FDA Bacteriological Analytical Manual*, Seventh Edition, 1992.

  Yeast and Mold Proceed as directed in chapter 18 of the *FDA Bacteriological Analytical Manual*, Seventh Edition, 1992.

  Salmonella Proceed as directed in chapter 5 of the *FDA Bacteriological Analytical Manual*, Seventh Edition, 1992.

Potassium Proceed as directed in the test for *Potassium* in the monograph for *Acid Hydrolysates of Proteins*.

Sodium Proceed as directed in the test for *Sodium* in the monograph for *Acid Hydrolysates of Proteins*.

Packaging and Storage Store in well-closed containers.

Zein

CAS: [9010-66-6]

DESCRIPTION

Zein is a very light yellow- to straw-colored, water-insoluble, granular or fine powder comprising the prolamine protein component of corn (*Zea mays* Linne'). It is produced commercially by extraction from corn gluten with alkaline aqueous isopropyl alcohol. The extract is then cooled, which causes the Zein to precipitate.

Functional Use in Foods Surface-finishing agent; texturizing agent.

REQUIREMENTS

Identification

  A. Dissolve about 0.1 g in 10 mL of 0.1 N sodium hydroxide, and add a few drops of cupric sulfate TS. Warm in a water bath: A purple color develops.

  B. To a test tube containing 25 mg of the sample, add 1 mL of nitric acid. Agitate vigorously: The solution becomes light yellow. Further addition of about 10 mL of 6 N ammonium hydroxide produces an orange color.

Assay Not less than 88.0% and not more than 96.0% protein, calculated on the dried basis.

Heavy Metals (as Pb) Not more than 20 mg/kg.

Lead Not more than 5 mg/kg.

Loss on Drying Not more than 8.0%.

Loss on Ignition Not more than 2%.

TESTS

Assay Proceed as directed under *Nitrogen Determination*, Appendix IIIC. Calculate the percent protein (P) by the formula

$$P = 6.25N,$$

in which N is the percent nitrogen.

Heavy Metals Prepare and test a 1-g sample as directed under *Method II* under the *Heavy Metals Test*, Appendix IIIB, using 20 µg of lead ion (Pb) in the control (*Solution A*).

Lead A *Sample Solution* prepared as directed for organic compounds meets the requirements of the *Lead Limit Test*, Appendix IIIB, using 5 µg of lead ion (Pb) in the control.

Loss on Drying, Appendix IIC Dry a 2-g sample in an air oven at 105° for 2 h.

Loss on Ignition Proceed as directed under *Ash (Total)*, Appendix IIC, using a 2-g sample.

Packaging and Storage Store in well-closed containers.

Zinc Gluconate

[CH$_2$OH(CHOH)$_4$COO]$_2$Zn

C$_{12}$H$_{22}$O$_{14}$Zn   Formula wt 455.69

CAS: [4468-02-4]

DESCRIPTION

Zinc Gluconate is a white or nearly white, granular or crystalline powder. It occurs as a mixture of various states of hydration, up to the trihydrate, depending on the method of isolation. It is freely soluble in water and very slightly soluble in alcohol.

Functional Use in Foods Nutrient; dietary supplement.

REQUIREMENTS

Labeling Indicate the powder or granular form of the product.
Identification

A. A 1 in 10 solution gives positive tests for *Zinc*, Appendix IIIA.

B. It meets the requirements of *Identification Test B* in the monograph for *Calcium Gluconate*.

Assay Not less than 97.0% and not more than 102.0% of $C_{12}H_{22}O_{14}Zn$, calculated on the anhydrous basis.
Cadmium Not more than 5 mg/kg.
Chloride Not more than 0.05%.
Lead Not more than 10 mg/kg.
Reducing Substances Not more than 1.0%.
Sulfate Not more than 0.05%.
Water Not more than 11.6%.

TESTS

Assay Dissolve about 700 mg of the sample, accurately weighed, in 100 mL of water, warming if necessary, and add 5 mL of ammonia–ammonium chloride buffer TS and 0.1 mL of eriochrome black TS. Titrate with 0.05 M disodium EDTA until the solution is blue in color. Each mL of 0.05 M disodium EDTA is equivalent to 22.78 mg of $C_{12}H_{22}O_{14}Zn$.

Cadmium

Standard Preparation Transfer 137.2 mg of cadmium nitrate to a 1000-mL volumetric flask, dissolve in water, dilute with water to volume, and mix. Pipet 25 mL of the resulting solution into a 100-mL volumetric flask, add 1 mL of hydrochloric acid, dilute with water to volume, and mix. Each mL of this *Standard Preparation* contains 12.5 µg of Cd.

Test Preparation Transfer 10.0 g of Zinc Gluconate to a 50-mL volumetric flask, dissolve in water, dilute with water to volume, and mix.

Procedure To three separate 25-mL volumetric flasks add 0 mL, 2.0 mL, and 4.0 mL of *Standard Preparation*, respectively. To each flask add 5.0 mL of *Test Preparation*, dilute with water to volume, and mix. These test solutions contain, respectively, 0, 1.0 and 2.0 µg/mL of cadmium from the *Standard Preparation*. Concomitantly determine the absorbances of the test solutions at the cadmium emission line at 228.8 nm, with a suitable atomic absorption spectrophotometer equipped with a cadmium hollow-cathode lamp and an air–acetylene flame, using water as the blank. Plot the absorbances of the test solutions versus their contents of cadmium, in µg/mL, as furnished by the *Standard Preparation*, draw the straight line best fitting the three points, and extrapolate the line until it intercepts the concentration axis. From the intercept determine the amount, in µg, of cadmium in each mL of the test solution containing 0 mL of the *Standard Preparation*. Calculate the quantity, in mg/kg, of Cd in the specimen by multiplying this value by 25.

Chloride, Appendix IIIB Any turbidity produced by a 40-mg sample does not exceed that shown in a control containing 20 µg of chloride ion (Cl).

Lead (Note: For the preparation of all aqueous solutions and for the rinsing of glassware before use, employ water that has been passed through a strong-acid, strong-base, mixed-bed ion-exchange resin before use. Select all reagents to have as low a content of lead as practicable, and store all reagent solutions in containers of borosilicate glass. Cleanse glassware before use by soaking in warm 8 N nitric acid for 30 min and by rinsing with deionized water.)

Standard Preparation Transfer 10.0 mL of Lead Nitrate Stock Solution, (see *Heavy Metals Test*, Appendix IIIB), to a 100-mL volumetric flask, add 40 mL of water and 5 mL of nitric acid, dilute with water to volume, and mix. Transfer 0.40 mL of this solution to a second 100-mL volumetric flask, add 50 mL of water and 1 mL of nitric acid, dilute with water to volume, and mix. This solution contains 0.04 µg of lead per mL.

Test Preparation Transfer about 4 g of Zinc Gluconate, accurately weighed, to a 100-mL volumetric flask. Add 50 mL of water and 5 mL of nitric acid, and sonicate to dissolve the sample. Dilute with water to volume, and mix. Transfer 4.0 mL of this solution to a second 100-mL volumetric flask, add 50 mL of water and 1 mL of nitric acid, dilute with water to volume and mix.

Blank Transfer 1.2 mL of nitric acid to a 100-mL volumetric flask, dilute with water to volume, and mix.

Test Solutions Prepare mixtures of the *Test Preparation*, the *Standard Preparation*, and the *Blank* with the following proportional compositions, by volume: 10.0:0:10.0, 10.0:4.0:6.0, 10.0:7.0:3.0, and 10.0:10.0:0. These *Test Solutions* contain, respectively, 0, 0.008, 0.014, and 0.020 µg/mL of lead from the *Standard Preparation*.

Procedure Separately inject equal volumes (about 20 µL) of the *Test Solutions* and the *Blank* into the graphite tube of a suitable graphite furnace atomic absorption spectrophotometer, temperature-programmed as follows to reach 2000° in about 2 min, using an argon gas flow of about 3 L/min, except where indicated: 70° for 10 s, 90° for 60 s, 120° for 10 s, 250° for 2 s (no gas flow), and 2000° for 3.2 s. When the temperature reaches 2000°, determine the absorbance at the lead emission line at 283.3 nm, corrected for background absorption. Correct the absorbance values obtained from the *Test Solutions* by subtracting from each the absorbance value obtained from the *Blank*. Plot the corrected absorbances of the *Test Solutions* versus their content of lead, in µg/mL, as furnished by the *Standard Preparation*, draw the straight line best fitting the four points, and extrapolate the line until it intercepts the concentration axis. From the intercept determine the concentration, C, in µg/mL, of lead in each mL of the *Test Solution* containing 0 µg/mL of lead from the *Test Preparation*. Calculate the mg/kg of lead in the sample by the formula

$$5000 C/W,$$

in which W is the weight, in g, of Zinc Gluconate taken to prepare the *Test Preparation*.

Reducing Substances Transfer about 1 g of the sample, accurately weighed, into a 250-mL Erlenmeyer flask, dissolve in 10 mL of water, and add 25 mL of alkaline cupric citrate TS. Cover the flask with a small beaker, boil gently for exactly 5 min, and cool rapidly to room temperature. Add 25 mL of a 1 in 10 solution of acetic acid, 10.0 mL of 0.1 N iodine, 10 mL of 2.7 N hydrochloric acid, and 3 mL of starch TS, and titrate with 0.1 N sodium thiosulfate to the disappearance of the blue color.

Calculate the weight, in mg, of reducing substances (as D-glucose) by the formula

$$27 \times (V_1 N_1 - V_2 N_2),$$

in which 27 is an empirically determined equivalence factor for D-glucose; V_1 and N_1 are the volume, in mL, and the normality of the iodine solution, respectively; and V_2 and N_2 are the volume, in mL, and the normality of the sodium thiosulfate solution, respectively.

Sulfate, Appendix IIIB Any turbidity produced by a 500-mg sample does not exceed that in a control containing 250 μg of sulfate ion (SO_4).

Water Determine by the *Karl Fischer Titrimetric Method*, Appendix IIB, using the *Residual Titration* procedure.

Packaging and Storage Store in well-closed containers.

Zinc Oxide

ZnO   Formula wt 81.39

CAS: [1314-13-2]

DESCRIPTION

A fine, white, odorless, amorphous powder. It gradually absorbs carbon dioxide from the air. It is insoluble in water and in alcohol, and is soluble in dilute acids and in strong bases.

Functional Use in Foods Nutrient; dietary supplement.

REQUIREMENTS

Identification
  A. When strongly heated, it assumes a yellow color that disappears on cooling.
  B. A solution of the sample in a slight excess of 3 N hydrochloric acid gives positive tests for *Zinc*, Appendix IIIA.
Assay Not less than 99.0% of ZnO after ignition.
Alkalinity Passes test.
Cadmium Not more than 10 mg/kg.
Lead Not more than 10 mg/kg.
Loss on Ignition Not more than 1.0%.
Substances Not Precipitated by Sulfide Not more than 0.5%.

TESTS

Assay Dissolve about 1.5 g of freshly ignited sample, accurately weighed, and 2.5 g of ammonium chloride in 50.0 mL of 1 N sulfuric acid with the aid of gentle heat, if necessary. When solution is complete, add methyl orange TS, and titrate the excess sulfuric acid with 1 N sodium hydroxide. Each mL of 1 N sulfuric acid is equivalent to 40.69 mg of ZnO.
Alkalinity Suspend 2 g in 20 mL of water, boil for 1 min, filter, and add 0.1 mL of phenolphthalein TS to the filtrate. No red color is produced.

Cadmium Determine as directed under *Cadmium* in the monograph for *Zinc Sulfate*, using the following as the *Sample Solution*: Transfer 5 g of the sample, accurately weighed, into a 50-mL volumetric flask, dissolve in a minimum volume of dilute hydrochloric acid (2 in 3), dilute to volume with water, and mix.
Lead, Appendix IIIB Suspend 2 g in 5 mL of water, add just enough hydrochloric acid to dissolve, and add a few drops of nitric acid. Add ammonium hydroxide dropwise until a faint but permanent precipitate forms. Clear the turbidity with a few drops of 10% nitric acid, and dilute to 40 mL with water. Add 20 mL of this solution to 25 mL of a 10% sodium cyanide solution, stirring constantly during the addition. If a precipitate persists, add solid sodium cyanide until the solution remains clear. Dilute to 50 mL with 10% sodium cyanide solution, add 0.2 mL of sodium sulfide TS, mix, and transfer into a Nessler tube. After 5 min, any color does not exceed that produced by 1.0 mL of *Standard Lead Solution* (10 μg Pb ion) in an equal volume of solution containing the quantities of reagents used in the test.
Loss on Ignition Weigh accurately about 2 g, and ignite at 800° ± 25° to constant weight.
Substances Not Precipitated by Sulfide Transfer 2 g, accurately weighed, to a 200-mL volumetric flask, dissolve in 20 mL of dilute acetic acid (1 in 4), dilute to volume with water, and mix. Precipitate the zinc completely with ammonium sulfide TS, dilute to volume with water, and mix. Filter through a dry filter, discarding the first portion of filtrate, and collect 100 mL of the subsequent filtrate. Add a few drops of sulfuric acid, and evaporate to dryness on a steam bath in a tared dish. Ignite cautiously until the ammonium salts are volatilized, ignite to constant weight at 800° ± 25°, cool, and weigh. The weight of the residue does not exceed 5 mg.

Packaging and Storage Store in well-closed containers.

Zinc Sulfate

$ZnSO_4 \cdot xH_2O$ Formula wt, anhydrous 161.45

CAS: [7733-02-0]

DESCRIPTION

Zinc Sulfate contains one or seven molecules of water of hydration. It occurs as colorless, transparent prisms or small needles, or as a granular, crystalline powder. It is odorless. The monohydrate loses water at temperatures above 238°, whereas the heptahydrate effloresces in dry air at room temperature. Its solutions are acid to litmus. The monohydrate is soluble in water and practically insoluble in alcohol. One g of the heptahydrate dissolves in about 0.6 mL of water and in about 2.5 mL of glycerin; it is insoluble in alcohol.

Functional Use in Foods Nutrient; dietary supplement.

REQUIREMENTS

Identification A 1 in 20 solution gives positive tests for *Zinc* and for *Sulfate*, Appendix IIIA.
Assay *Monohydrate*: not less than 98.0% and not more than 100.5% of $ZnSO_4 \cdot H_2O$; *heptahydrate*: not less than 99.0% and not more than 108.7% of $ZnSO_4 \cdot 7H_2O$.
Acidity Passes test.
Alkalies and Alkaline Earths Not more than 0.5%.
Cadmium Not more than 5 mg/kg.
Lead Not more than 10 mg/kg.
Mercury Not more than 5 mg/kg.
Selenium Not more than 0.003%.

TESTS

Assay Dissolve about 175 mg of the monohydrate, or about 300 mg of the heptahydrate, accurately weighed, in 100 mL of water, add 5 mL of ammonia–ammonium chloride buffer TS and 0.1 mL of eriochrome black TS, and titrate with 0.05 M disodium EDTA until the solution is deep blue in color. Each mL of 0.05 M disodium EDTA is equivalent to 8.973 mg of $ZnSO_4 \cdot H_2O$, or 14.38 mg of $ZnSO_4 \cdot 7H_2O$.

Acidity A 1 in 20 solution is not colored pink by methyl orange TS.

Alkalies and Alkaline Earths Transfer a 2-g sample to a 200-mL volumetric flask, dissolve in about 150 mL of water, and precipitate the zinc completely with ammonium sulfide TS. Dilute to volume with water, and mix. Filter through a dry filter, rejecting the first portion of the filtrate, and add a few drops of sulfuric acid to 100 mL of the subsequent filtrate. Evaporate to dryness in a tared dish, ignite to constant weight, cool, and weigh. The weight of the residue does not exceed 5 mg.

Cadmium

Spectrophotometer Use any suitable atomic absorption spectrophotometer equipped with a Boling-type burner, an air–acetylene flame, and a hollow-cathode cadmium lamp. The instrument should be capable of operating within the sensitivity necessary for the determination.

Standard Solution Transfer 100 mg of cadmium chloride crystals ($CdCl_2 \cdot 2\frac{1}{2}H_2O$), accurately weighed, into a 1000-mL volumetric flask, dissolve in and dilute to volume with water, and mix. Pipet 25 mL of this solution into a 100-mL volumetric flask, add 1 mL of hydrochloric acid, dilute to volume with water, and mix. Each mL contains 12.5 µg of Cd.

Sample Solution Transfer 10 g of the sample, accurately weighed, into a 50-mL volumetric flask, dissolve in and dilute to volume with water, and mix.

Procedure Transfer 5.0 mL of the *Sample Solution* into each of five separate 25-mL volumetric flasks. Dilute *Flask 1* to volume with water, and mix. To *Flasks 2, 3, 4,* and *5* add 1.00, 2.00, 3.00, and 4.00 mL of the *Standard Solution*, respectively, then dilute each flask to volume with water, and mix. Determine the absorbance of each solution at 228.8 nm, setting the instrument to previously established optimum conditions. Plot absorbance versus µg of added Cd per 25 mL, and extrapolate to zero absorbance. Read the (negative) concentration of Cd in *Flask 1*, which equals the mg/kg of Cd in the sample taken.

Lead Dissolve 500 mg of the sample in 5 mL of water, transfer to a color-comparison tube (*A*), add 10 mL of potassium cyanide solution (1 in 10), mix, and allow the mixture to become clear. In a similar matched color-comparison tube (*B*) place 5 mL of water, and add 0.50 mL of Standard Lead Solution (5 µg of Pb) and 10 mL of potassium cyanide solution (1 in 10). To each solution add 0.1 mL of sodium sulfide TS, mix, and allow to stand for 5 min. When viewed downward over a white surface, the solution in tube *A* is no darker than that in tube *B*.

Mercury Determine as directed under *Mercury Limit Test*, Appendix IIIB, using the following as the *Sample Preparation*: Dissolve 400 mg of the sample in 10 mL of water in a small beaker, add 1 mL of dilute sulfuric acid solution (1 in 5) and 1 mL of potassium permanganate solution (1 in 25), cover the beaker, boil for a few s, and cool.

Selenium Determine as directed in *Method I* under the *Selenium Limit Test*, Appendix IIIB, using 200 mg of sample.

Packaging and Storage Store in tight containers.

3 / *Flavor Chemicals*

Contents

Specifications for Flavor Chemicals ... 449
Test Methods for Flavor Chemicals ... 564
Gas Chromatographic (GC) Assay of Flavor Chemicals ... 569

SPECIFICATIONS FOR FLAVOR CHEMICALS

Specifications for flavoring agents other than the essential oils are presented in tabular form rather than as separate monographs. Specifications for such ingredients from the *Food Chemicals Codex*, third edition, together with a number of new specifications, are provided on the following pages of this section. Following the tabular specifications are the *Test Methods for Flavor Chemicals* (M-1 through M-18) and the *Gas Chromatographic Assay of Flavor Chemicals* used to determine the assay values and various other physico-chemical properties of the flavors. The infrared spectra, used for identification and comparison purposes, are provided in the section entitled *Infrared Spectra*.

Explanatory Notes to Tabular Specifications

The *Food Chemicals Codex*, fourth edition, uses the nomenclature convention (Z) and (E) to specify the structure of acyclic double bonds in organic chemicals. The "(Z)" notation of the fourth edition replaces the "*cis*" notation of the third edition, and the "(E)" notation replaces the "*trans*" notation.

The FCC name of the substance is followed, where available, by the number assigned to the substance by the Flavor and Extract Manufacturers Association (FEMA) and by its synonym(s). The explanatory notes to the specifications in this section apply throughout the tabular series and are as follows:

Note 1 (Solubility) Approximate solubilities (see *General Provisions*) are indicated by the following abbreviations: vs = very soluble; s = soluble; ss = slightly soluble; vss = very slightly soluble; m = miscible; ins = insoluble or practically insoluble. Other abbreviations are as follows: alc = alcohol; gly = glycerin; org = organic; prop = propylene (as in propylene glycol); veg = vegetable (as in vegetable oil); vol = volatile.

Note 2 (B.P.) Boiling points (B.P.) are expressed in °C. They are approximate values given for information only and not as requirements.

Note 3 (Solubility in Alcohol) Determine the solubility in alcohol at 25° as directed in the general method, Appendix VI, *Essential Oils and Flavors*.

Note 4 (I.D.) The notation "IR" in the identification (I.D.) column indicates that an infrared absorption spectrum is provided for the particular substance in the section entitled *Infrared Spectra*. Where the IR requirement is specified, the infrared absorption spectrum of the sample shall exhibit maxima at the same wavelengths (or frequencies) as those shown in the respective spectrum, using the test conditions as specified therein.

Note 5 (Assay) Assay requirements are specified as *minimum* values (unless a range of assay values is given) and are stated in weight percent unless otherwise indicated. References to assay methods are indicated by citations in parentheses, e.g., "(M-1a)," to methods provided under *Test Methods for Flavor Chemicals*.

Note 6 (A.V.) Unless otherwise indicated, determine the acid value (A.V.) as directed in M-16, using phenolphthalein TS as indicator unless another indicator is specified for an individual substance. Where *Method II* is specified, determine the acid value as directed in the general method, Appendix VII, *Fats and Related Substances*.

Note 7 (Ref. Index) Refractive index (Ref. Index) determinations are made at 20° unless another temperature is specified, according to the general method, Appendix IIB, *Physical Tests and Determinations*.

Note 8 (Sp. Gr.) Specific gravity (Sp. Gr.) determinations are made at 25° unless another temperature is specified by any reliable method (see *General Provisions*).

Note 9 (Other Requirements) Numerical limits for other requirements are specified as *maximum* values unless otherwise indicated (max = maximum; NLT = not lower than or not less than, as appropriate). Test methods are indicated by citations in parentheses, which refer either to methods given in the section that follows this tabular section or to general methods given in Appendix VI, *Essential Oils and Flavors*.

Unless specifically prohibited by a notation in this column, the flavor chemicals listed herein may contain a suitable antioxidant. If an antioxidant is used, it shall be named on the label of the substance to which it is added, and its percentage shall be indicated.

General Information and Description

Name of Substance/ Synonyms	Formula Wt/Formula/ Structure	Physical Form/ Odor	Solubility[1]/ B.P.[2]	Solubility in Alcohol[3]
Acetaldehyde FEMA No. 2003 Ethanal; Acetic Aldehyde	44.05/C_2H_4O/ CH_3CHO	flammable, colorless liq/ pungent, ethereal	m—water, alc, org solvents/ 21°	
Acetanisole FEMA No. 2005 p-Methoxyacetophenone; 4-Acetylanisole	150.18/$C_9H_{10}O_2$/	colorless to pale yel fused solid/ hawthorn-like	s—most fixed oils, prop glycol; ins—gly/ 153° (26 mm Hg)	1 g in 5 mL 50% alc
Acetoin FEMA No. 2008 Acetyl Methyl Carbinol; Dimethyl-ketol; 3-Hydroxy-2-butanone	88.11/$C_4H_8O_2$/ $CH_3CH(OH)COCH_3$	colorless to pale yel liq (monomer), or white cryst powder (dimer)/ buttery	m—alc, prop glycol, water; ins—veg oils/ 148°	
Acetophenone FEMA No. 2009 Methyl Phenyl Ketone; Acetylbenzene	120.15/C_8H_8O/	practically colorless liq above 20°/ very sweet, pungent	vs—prop glycol, most fixed oils; s—alc, chloroform, ether; ss—water; ins—gly/ 202°	1 mL in 5 mL 50% alc
3-Acetyl-2,5-dimethyl Furan FEMA No. 3391 2,5-Dimethyl-3-acetylfuran	138.17/$C_8H_{10}O_2$/	yel liq/ powerful, slightly roasted, nut-like	s—alc, prop glycol, most fixed oils; ss—water/ 83° (11 mm Hg)	
2-Acetylpyrazine FEMA No. 3126 Methyl Pyrazinyl Ketone	122.13/$C_6H_6N_2O$/	colorless to pale yel cryst; m.p. 78°/ popcorn-like		1 g in 20 mL 95% alc
3-Acetylpyridine FEMA No. 3424 Methyl Pyridyl Ketone	121.14/C_7H_7NO/	colorless to yel liq/ sweet, nutty, popcorn-like	s—acids, alc, ether, water/ 230°	
2-Acetylpyrrole FEMA No. 3202 Methyl 2-Pyrrolyl Ketone	109.13/C_6H_7NO/	light beige to yellowish, fine cryst/ bready		

Requirements

I.D. Test[4]	Assay Min. %[5]	A.V. Max.[6]	Ref. Index[7]	Sp. Gr.[8]	Other Requirements[9]
IR	99.0% of C_2H_4O (M-2b)	5.0		0.804–0.811 (0°/20°)	**Res. on Evap.**—0.006% (M-17)
IR	98.0% of $C_9H_{10}O_2$ (M-1b)				**Chlorinated Cmpds.**—passes test (Appendix VI) **Heavy Metals**—0.004% (M-7) **Lead**—10 mg/kg (M-10)
IR	96.0% of $C_4H_8O_2$ (M-1b)		1.417–1.420	1.005–1.019	
IR	98.0% of C_8H_8O (M-1b)		1.533–1.535	1.025–1.028	**Chlorinated Cmpds.**—passes test (Appendix VI) **Solidification Pt.**—NLT 19° (Appendix IIB)
IR	99.0% of $C_8H_{10}O_2$ (M-1a)		1.475–1.496 (25°)	1.027–1.048	**Angular Rotation**—between -1° and +1° (Appendix IIB)
IR	99.0% of $C_6H_6N_2O$ (M-1a)				**Melting Range**—between 75° and 78° (Appendix IIB)
IR	98.0% of C_7H_7NO (M-1a)		1.530–1.540 (25°)	1.100–1.115 (20°)	**Water**—0.5% (Appendix IIB, KF)
	98.0% of C_6H_7NO (M-1a)				**Heavy Metals**—10 mg/kg (M-7) **Melting Range**—between 85° and 90° (Appendix IIB) **Res. on Ignit.**—0.3% (Appendix IIC) **Water**—0.5% (Appendix IIB, KF)

General Information and Description

Name of Substance (Synonyms)	Formula Wt/Formula/ Structure	Physical Form/ Odor	Solubility[1]/ B.P.[2]	Solubility in Alcohol[3]
Allyl Cyclohexanepropionate FEMA No. 2026 Allyl-3-cyclohexanepropionate	196.29/$C_{12}H_{20}O_2$/ ⟨hexyl⟩—$CH_2CH_2COOCH_2CH=CH_2$	colorless liq/ pineapple-like	m—alc, chloroform, ether; ins—gly, water	1 mL in 4 mL 80% alc
Allyl Heptanoate FEMA No. 2031 Allyl Heptoate	170.25/$C_{10}H_{18}O_2$/ $CH_3(CH_2)_5COOC_3H_5$	colorless to pale yel liq/ sweet, pineapple	210°	1 mL in 1 mL 95% alc
Allyl Hexanoate FEMA No. 2032 Allyl Caproate	156.22/$C_9H_{16}O_2$/ $CH_3(CH_2)_4COOCH_2CH=CH_2$	colorless to light yel liq/ strong, pineapple-like	m—alc, most fixed oils; ins—prop glycol, water/ 185°	1 mL in 6 mL 70% alc
Allyl α-Ionone FEMA No. 2033 Allyl Ionone	232.37/$C_{16}H_{24}O$/ (structure)—$CH_2CH_2CH=CH_2$	colorless to yel liq/ fruity, woody	s—alc; ins—water/ 265°	1 mL in 1 mL 90% alc gives clear soln
Allyl Isothiocyanate FEMA No. 2034	99.16/C_3H_5NCS/ $CH_2=CH-CH_2-N=C=S$	colorless to pale yel, strongly refractive liq/ very pungent irritating odor, acrid taste, mustard-like [caution: lachrymator]	m—alc, carbon disulfide, ether/ 88°	
Allyl Isovalerate FEMA No. 2045 Allyl Isopentanoate	142.20/$C_8H_{14}O_2$/ $(CH_3)_2CHCH_2CO_2CH_2CH=CH_2$	colorless to pale yel liq/ fruit-like, apple aroma	155°	1 mL in 1 mL 95% alc

Requirements

I.D. Test[4]	Assay Min. %[5]	A.V. Max.[6]	Ref. Index[7]	Sp. Gr.[8]	Other Requirements[9]
IR	98.0% of $C_{12}H_{20}O_2$ (M-1b)	5.0	1.457–1.462	0.945–0.950	**Allyl Alcohol**—NMT 0.1% (M-1b)
IR	97.0% of $C_{10}H_{18}O_2$ (M-1b)	1.0	1.426–1.430	0.880–0.885	**Allyl Alcohol**—NMT 0.1% (M-1b)
IR	98.0% of $C_9H_{16}O_2$ (M-1b)	1.0	1.422–1.426	0.884–0.890	**Allyl Alcohol**—NMT 0.1% (M-1b)
IR	88.0% of $C_{16}H_{24}O$ (M-1b)		1.502–1.507	0.926–0.932	**Allyl Alcohol**—NMT 0.1% (M-1b)
IR	93.0% of C_3H_5NCS (M-1a)		1.527–1.531	1.013–1.020	**Dist. Range**— between 148° and 154°(Appendix IIB) **Phenols**—passes test (M-18) **Allyl Alcohol**—NMT 0.1% (M-1b)
IR	98.0% of $C_8H_{14}O_2$ (M-1b)	1.0	1.413–1.418	0.879–0.884	**Allyl Alcohol**—NMT 0.1% (M-1b)

General Information and Description

Name of Substance (Synonyms)	Formula Wt/Formula/ Structure	Physical Form/ Odor	Solubility[1]/ B.P.[2]	Solubility in Alcohol[3]
1-Amyl Alcohol FEMA No. 2056 1-Pentanol	88.15/$C_5H_{12}O$/ $CH_3(CH_2)_4OH$	colorless to pale yel liq	m—alc/ 136°	
Amyl Butyrate FEMA No. 2059 1-Pentyl Butyrate	158.23/$C_9H_{18}O_2$/ $CH_3CH_2CH_2COOCH_2(CH_2)_3CH_3$	colorless to pale yel liq		
α-Amylcinnamaldehyde FEMA No. 2061 Amylcinnamaldehyde	202.30/$C_{14}H_{18}O$/ Ph−CH=C(−CHO)−$(CH_2)_4CH_3$	yel liq/ strong, floral, like jasmine on dilution, spicy	s—most fixed oils; ins—gly, prop glycol/ 285°	1 mL in 5 mL 80% alc
Amyl Cinnamate FEMA No. 2063 Isoamyl Cinnamate; Isoamyl 3-Phenyl Propenate	218.28/$C_{14}H_{18}O_2$/ Ph−CH=CHCOO$(CH_2)_4CH_3$	colorless to pale yel liq/ faint, balsamic, cocoa-like	s—most fixed oils; ss—prop glycol; ins—gly/ 310°	1 mL in 7 mL 80% alc may be opalescent
Amyl Formate FEMA No. 2068 1-Pentyl Formate	116.16/$C_6H_{12}O_2$/ $CH_3(CH_2)_4OCHO$	colorless to pale yel liq	m—alc/ 128°–130°	
Amyl Heptanoate FEMA No. 2073 Pentyl Heptanoate	200.32/$C_{12}H_{24}O_2$/ $CH_3(CH_2)_5COO(CH_2)_4CH_3$	colorless to pale yel liq/ fruity	245°	1 mL in 1 mL 95% alc
Amyl Octanoate FEMA No. 2079 Isoamyl Octanoate; Isoamyl Caprylate; Amyl Caprylate	214.35/$C_{13}H_{26}O_2$/ $CH_3(CH_2)_6COOC_5H_{11}$	colorless liq/ fruity	s—alc, most fixed oils; ss—prop glycol; ins—gly, water/ 260°	1 mL in 7 mL 80% alc remains clear to 10 mL
Amyl Propionate FEMA No. 2082 Isoamyl Propionate	144.21/$C_8H_{16}O_2$/ $CH_3CH_2COOC_5H_{11}$	colorless liq/ fruity, apricot-pineapple	s—alc, most fixed oils; ins—gly, prop glycol, water/ 160°	1 mL in 3 mL 70% alc

Requirements

I.D. Test[4]	Assay Min. %[5]	A.V. Max.[6]	Ref. Index[7]	Sp. Gr.[8]	Other Requirements[9]
	98.0% of $C_5H_{12}O$ (M-1b)		1.407–1.412	0.810–0.816	
	98.0% of $C_9H_{18}O_2$ (sum of isomers) (M-1b)	1.0	1.409–1.414	0.863–0.866	
IR	97.0% of $C_{14}H_{18}O$ (M-1b)	5.0	1.554–1.559	0.963–0.968	**Chlorinated Cmpds.**—passes test (Appendix VI)
IR	96.0% of $C_{14}H_{18}O_2$ (M-1b)	1.0	1.535–1.539	0.992–0.997	
	92.0% of $C_6H_{12}O_2$ (sum of isomers) (M-1b)	5.0 add ice to soln	1.396–1.402	0.881–0.887	
	93.0% of $C_{12}H_{24}O_2$ (sum of isomers) (M-1a)	1.0	1.422–1.426	0.859–0.863	
IR	98.0% of $C_{13}H_{26}O_2$ (M-1b)	1.0	1.425–1.429	0.855–0.861	
IR	98.0% of $C_8H_{16}O_2$ (M-1b)	1.0	1.405–1.409	0.866–0.871	

General Information and Description

Name of Substance (Synonyms)	Formula Wt/Formula/ Structure	Physical Form/ Odor	Solubility[1]/ B.P.[2]	Solubility in Alcohol[3]
Anethole FEMA No. 2086 p-Propenylanisole; trans-Anethole; Isoestragole	148.20/$C_{10}H_{12}O$/ CH_3O—⟨⟩—$CH=CHCH_3$	colorless to faintly yel liq at or above 23°; sweet taste; affected by light/ anise-like	ss—water; m—chloroform, ether/ 234°	1 mL in 2 mL alc
Anisole FEMA No. 2097 Methylphenyl Ether	108.14/C_7H_8O/ ⟨⟩—OCH_3	colorless liq/ phenolic, anise-like	s—alc, ether; ins—water/ 154°	
Anisyl Acetate FEMA No. 2098 p-Methoxybenzyl Acetate	180.20/$C_{10}H_{12}O_3$/ CH_3O—⟨⟩—CH_2OOCCH_3	colorless to slightly yel liq/ floral, fruity, balsamic	s—alc, most fixed oils; ins—gly, prop glycol/ 235°	1 mL in 6 mL 60% alc remains in soln to 10 mL
Anisyl Alcohol FEMA No. 2099 Anisic Alcohol; p-Methoxybenzyl Alcohol	138.17/$C_8H_{10}O_2$/ CH_3O—⟨⟩—CH_2OH	colorless to slightly yel liq/ floral	s—most fixed oils; ss—gly/ 259°	1 mL in 1 mL 50% alc remains in soln to 10 mL
Anisyl Formate FEMA No. 2101 p-Methoxybenzyl Formate	166.18/$C_9H_{10}O_3$/ CH_3O—⟨⟩—CH_2-O-CH (=O)	colorless to pale yel liq/ sweet, floral, tonka-like	100°	1 mL in 1 mL 95% alc
Benzaldehyde FEMA No. 2127	106.12/C_7H_6O/ ⟨⟩—CHO	colorless liq, burning taste/ bitter almond oil	ss—water; m—alc, ether, vol oils, fixed oils/ 178°	
Benzaldehyde Glyceryl Acetal FEMA No. 2129 Mixture of 1,2- and 1,3- Benzaldehyde Cyclic Acetals of Glycerin	180.20/$C_{10}H_{12}O_3$/ (a) 1,3-dioxolane with phenyl and CH₂OH (b) 1,3-dioxane with phenyl and OH	colorless to pale yel liq/ mild almond odor	185°	1 mL in 1 mL 95% alc

Requirements

I.D. Test[4]	Assay Min. %[5]	A.V. Max.[6]	Ref. Index[7]	Sp. Gr.[8]	Other Requirements[9]
IR	99.0% of $C_{10}H_{12}O$ (M-1b)		1.557–1.562	0.983–0.988	**Angular Rotation**—between -0.15° and +0.15° (Appendix IIB, 100-mm tube) **Dist. Range**— between 231° and 237° (Appendix IIB) **Phenols**—passes test (M-18) **Solidification Pt.**—NLT 20° (Appendix IIB)
IR			1.515–1.518	0.990–0.993	**Phenols**—passes test (M-18) **Dist. Range**—within a 2° range (Appendix IIB)
IR	97.0% of $C_{10}H_{12}O_3$ (M-1b)	1.0	1.511–1.516	1.104–1.111	
IR	97.0% of $C_8H_{10}O_2$ (M-1b)	1.0	1.542–1.547	1.110–1.115	**Aldehydes**—1.0% as anisaldehyde (M-1b) **Solidification Pt.**—min 23.5° (Appendix IIB)
IR	90.0% of $C_9H_{10}O_3$ (M-1b)	3.0	1.521–1.525	1.138–1.142	
IR	98.0% of C_7H_6O (M-1b)		1.544–1.547	1.041–1.046	**Chlorinated Cmpds.**—passes test (Appendix VI) **Hydrocyanic Acid**—passes test (M-9)
IR	95.0% of $C_{10}H_{12}O_3$ (sum of isomers) (M-1a)	2.0	1.535–1.541	1.181–1.191	

General Information and Description

Name of Substance (Synonyms)	Formula Wt/Formula/ Structure	Physical Form/ Odor	Solubility[1]/ B.P.[2]	Solubility in Alcohol[3]
1,2-Benzodihydropyrone FEMA No. 2381 Dihydrocoumarin	148.16/$C_9H_8O_2$/	colorless to pale yel liq/ coconut-like	272°	1 mL in 1 mL 95% alc
Benzophenone FEMA No. 2134 Diphenyl Ketone; Benzoylbenzene	182.22/$C_{13}H_{10}O$/	white rhombic cryst or flaky solid; m.p. 48.5°/ delicate, persistent, rose-like	s—most fixed oils; ss—prop glycol; ins—gly/ 305°	1 g in 10 mL 80% alc
Benzyl Acetate FEMA No. 2135	150.18/$C_9H_{10}O_2$/	colorless liq/ sweet, floral, fruity	s—alc, most fixed oils, prop glycol; ins—gly, water/ 214°	1 mL in 5 mL 60% alc
Benzyl Alcohol FEMA No. 2137 Phenyl Carbinol	108.14/C_7H_8O/	colorless liq with a sharp burning taste/ faint, aromatic	m—alc, chloroform, ether, 1 mL in 30 mL water/ 206° (decomp)	
Benzyl Benzoate FEMA No. 2138	212.25/$C_{14}H_{12}O_2$/	colorless, oily liq/ slight, aromatic odor	m—alc, chloroform, ether; ins—gly, water/ 323°	
Benzyl Butyrate FEMA No. 2140 Benzyl n-Butyrate	178.23/$C_{11}H_{14}O_2$/	colorless liq/ floral, fruity, plum-like	s—alc, most fixed oils; ins—gly, prop glycol, water/ 239°	1 mL in 2 mL 80% alc
Benzyl Cinnamate FEMA No. 2142	238.29/$C_{16}H_{14}O_2$/	white to pale yel solid/ sweet, balsamic	s—most fixed oils; ins—gly, prop glycol/ 195° (5 mm Hg)	1 g in 8 mL 90% alc

Requirements

I.D. Test[4]	Assay Min. %[5]	A.V. Max.[6]	Ref. Index[7]	Sp. Gr.[8]	Other Requirements[9]
IR	99.0% of $C_9H_8O_2$		1.555–1.559	1.186–1.192	**Solidification Pt.**—NLT 22° (Appendix IIB)
IR	97.0% of $C_{13}H_{10}O$				**Chlorinated Cmpds.**—passes test (Appendix VI) **Heavy Metals**—0.004% (M-7) **Lead**—10 mg/kg (M-10) **Solidification Pt.**—NLT 47° (Appendix IIB)
IR	98.0% of $C_9H_{10}O_2$ (M-1b)	1.0 (phenol red TS)	1.501–1.504	1.052–1.056	**Chlorinated Cmpds.**—passes test (Appendix VI)
IR	99.0% of C_7H_8O (M-1a)		1.539–1.541	1.042–1.047	**Aldehydes**—0.2% (M-1b) **Chlorinated Cmpds.**—passes test (Appendix VI) **Dist. Range**—NLT 95% between 202.5° and 206.5° (Appendix IIB)
IR	99.0% of $C_{14}H_{12}O_2$ (M-1b)	1.0	1.568–1.570	1.116–1.120	**Chlorinated Cmpds.**—passes test (Appendix VI) **Solidification Pt.**—NLT 18° (Appendix IIB)
IR	98.0% of $C_{11}H_{14}O_2$ (M-1b)	1.0	1.492–1.496	1.006–1.009	
IR	98.0% of $C_{16}H_{14}O_2$ (M-1b)	1.0			**Chlorinated Cmpds.**—passes test (Appendix VI) **Heavy Metals**—0.004% (M-7) **Lead**—10 mg/kg (M-10) **Solidification Pt.**—between 33.0° and 35.0° (Appendix IIB)

General Information and Description

Name of Substance (Synonyms)	Formula Wt/Formula/ Structure	Physical Form/ Odor	Solubility[1]/ B.P.[2]	Solubility in Alcohol[3]
Benzyl Formate FEMA No. 2145	136.15/$C_8H_8O_2$/ $\text{C}_6\text{H}_5\text{-CH}_2\text{-O-CHO}$	colorless to pale yel liq/ sweet, balsamic, floral	203°	1 mL in 1 mL 95% alc
Benzyl Isobutyrate FEMA No. 2141 Benzyl 2-Methyl Propionate	178.23/$C_{11}H_{14}O_2$/ $\text{C}_6\text{H}_5\text{-CH}_2\text{OOCCH(CH}_3)_2$	colorless liq/ floral, fruity, jasmine-like	s—alc, most fixed oils; ss—prop glycol; ins—gly/ 229°	1 mL in 6 mL 70% alc
Benzyl Isovalerate FEMA No. 2152 Benzyl 3-Methyl Butyrate	192.26/$C_{12}H_{16}O_2$/ $\text{C}_6\text{H}_5\text{-CH}_2\text{OOCCH}_2\text{CH(CH}_3)_2$	colorless liq/ fruity, herbaceous, apple-like	s—alc, most fixed oils; ss—prop glycol; ins—gly, water/ 246°	1 mL in 3 mL 80% alc remains in soln on dilution
Benzyl Phenylacetate FEMA No. 2149	226.27/$C_{15}H_{14}O_2$/ $\text{C}_6\text{H}_5\text{-CH}_2\text{COOCH}_2\text{-C}_6\text{H}_5$	colorless liq/ sweet, floral, honey undertone	m—alc, chloroform, ether/ 317°	1 mL in 3 mL 90% alc gives clear soln
Benzyl Propionate FEMA No. 2150 Benzyl Propanoate	164.20/$C_{10}H_{12}O_2$/ $\text{C}_6\text{H}_5\text{-CH}_2\text{OOCCH}_2\text{CH}_3$	colorless liq/ sweet, floral fruity	s—alc, most fixed oils; ss—prop glycol; ins—gly, water/ 222°	1 mL in 3 mL 70% alc remains clear to 10 mL
Benzyl Salicylate FEMA No. 2151	228.25/$C_{14}H_{12}O_3$/ 2-OH-$\text{C}_6\text{H}_4\text{-COOCH}_2\text{-C}_6\text{H}_5$	almost colorless liq/ faint, sweet odor	s—most fixed oils; ins—gly, prop glycol/ 300°	1 mL in 5 mL 95% alc
Bornyl Acetate FEMA No. 2159 L-Bornyl Acetate	196.29/$C_{12}H_{20}O_2$/ (bornyl acetate structure)	colorless liq, semicryst mass, or white cryst solid/ sweet herbaceous, piney	s—alc, most fixed oils; ss—water; ins—gly, prop glycol/ 226°	1 mL in 3 mL 70% alc remains in soln to 10 mL

Requirements

I.D. Test[4]	Assay Min. %[5]	A.V. Max.[6]	Ref. Index[7]	Sp. Gr.[8]	Other Requirements[9]
IR	95.0% of $C_8H_8O_2$ (M-1b)	3.0	1.508–1.515	1.082–1.092	
IR	97.0% of $C_{11}H_{14}O_2$ (M-1b)	1.0	1.488–1.492	1.000–1.005	
IR	98.0% of $C_{12}H_{16}O_2$ (M-1b)	1.0	1.486–1.490	0.983–0.989	
IR	98.0% of $C_{15}H_{14}O_2$ (M-1b)	1.0	1.553–1.558	1.095–1.099	
IR	98.0% of $C_{10}H_{12}O_2$ (M-1b)	1.0	1.496–1.500	1.028–1.032	
IR	98.0% of $C_{14}H_{12}O_3$ (M-1b)	1.0 (phenol red TS)	1.573–1.582	1.176–1.180	**Solidification Pt.**—NLT 23.5° (Appendix IIB)
IR	98.0% of $C_{12}H_{20}O_2$ (M-1b)	1.0	1.462–1.466	0.981–0.985	**Angular Rotation**— between -39.5° and -45.0° (Appendix IIB, 100-mm tube) **Solidification Pt.**—NLT 25° (Appendix IIB)

General Information and Description

Name of Substance (Synonyms)	Formula Wt/Formula/ Structure	Physical Form/ Odor	Solubility[1]/ B.P.[2]	Solubility in Alcohol[3]
2-Butanone FEMA No. 2170 Methyl Ethyl Ketone	72.11/C_4H_8O/ $CH_3COCH_2CH_3$	colorless, mobile liq/ ethereal, nauseating	*m*—alc, ether, most fixed oils, 1 mL in 4 mL water/ 78.6°–80°	
Butan-3-one-2-yl Butanoate FEMA No. 3332	158.20/$C_8H_{14}O_3$/	white to slightly yel liq/ sweet, red berry character	*s*—alc, prop glycol, most fixed oils; *ins*—water	
Butyl Acetate FEMA No. 2174 *n*-Butyl Acetate	116.16/$C_6H_{12}O_2$/ $CH_3COO(CH_2)_3CH_3$	colorless, mobile liq/ strong, fruity	*m*—alc, ether, prop glycol, 1 mL in 145 mL water/ 126°	
Butyl Alcohol FEMA No. 2178 1-Butanol	74.12/$C_4H_{10}O$/ $CH_3(CH_2)_2CH_2OH$	colorless, mobile liq/ vinous	*m*—alc, ether, other org solvents, 1 mL in 15 mL water/ 117.7°	
Butyl Butyrate FEMA No. 2186 *n*-Butyl *n*-Butyrate	144.21/$C_8H_{16}O_2$/ $CH_3CH_2CH_2COOC_4H_9$	colorless liq/ fruity, pineapple-like on dilution	*ss*—prop glycol, water, 1 mL in 3 mL 70% alc; *m*—alc, ether, most veg oils/ 165°	
Butyl Butyryllactate FEMA No. 2190 Butyryllactic Acid; Butyl Ester; Lactic Acid Butyl Ester, Butyrate	216.28/$C_{11}H_{20}O_4$/ $CH_3CHCOOC_4H_9$ $\quad\quad\mid$ $CH_3CH_2CH_2COO$	colorless liq/ mild, buttery, cream-like	*s*—prop glycol; *m*—alc, most fixed oils; *ins*—water	1 mL in 3mL 70% alc

Requirements

I.D. Test[4]	Assay Min. %[5]	A.V. Max.[6]	Ref. Index[7]	Sp. Gr.[8]	Other Requirements[9]
IR	99.5% of C_4H_8O (M-1b)	2.0 (M-16)		0.801–0.803	**Dist. Range**—within 1.5° (Appendix IIB) **Water**—0.2% (Appendix IIB, KF; use freshly dist. pyridine as solvent)
IR	98.0% of $C_8H_{14}O_3$ (M-1a)		1.408–1.429	0.972–0.992	
IR	98.0% of $C_6H_{12}O_2$ (M-1b)	2.0 (M-16)	1.393–1.396	0.876–0.880	**Dist. Range**—between 120° and 128° (Appendix IIB)
IR	99.5% of $C_4H_{10}O$ (M-1b)	2.0 (M-16)		0.807–0.809	**Butyl Ether**—0.15% (M-1b) **Dist. Range**—max. 1.5° between beginning and end (Appendix IIB)
IR	98.0% of $C_8H_{16}O_2$ (M-1b)	1.0	1.405–1.407	0.867–0.871	
IR	95.0% of $C_{11}H_{20}O_4$ (M-1b)	1.0	1.420–1.423	0.970–0.974	

General Information and Description

Name of Substance (Synonyms)	Formula Wt/Formula/ Structure	Physical Form/ Odor	Solubility[1]/ B.P.[2]	Solubility in Alcohol[3]
Butyl Isobutyrate FEMA No. 2188	144.21/$C_8H_{16}O_2$/ $(CH_3)_2CHCOOC_4H_9$	colorless liq/ fresh, fruity, apple-pineapple	m—alc, ether, most fixed oils; ins—gly, prop glycol, water/ 166°	1 mL in 7 mL 60% alc
Butyl Isovalerate FEMA No. 2218	158.24/$C_9H_{18}O_2$/ $(CH_3)_2CHCH_2COOC_4H_9$	colorless to pale yel liq/ fruity	175°	1 mL in 1 mL 95% alc
Butyl Phenylacetate FEMA No. 2209	196.26/$C_{12}H_{16}O_2$/	colorless to pale yel liq/ honey, rose-like	260°	1 mL in 1 mL 95% alc
Butyl Stearate FEMA No. 2214 Butyl Octadecanoate	340.59/$C_{22}H_{44}O_2$/ $CH_3(CH_2)_{16}COO(CH_2)_3CH_3$	colorless, waxy solid/ odorless to faintly fatty	223°	
Butyraldehyde FEMA No. 2219 Butyl Aldehyde	72.11/C_6H_8O/ $CH_3(CH_2)_2CHO$	colorless, mobile liq/ pungent, nutty	s—1 mL in 15 mL water; m—alc, ether/ 74.8°	
Butyric Acid FEMA No. 2221	88.11/$C_4H_8O_2$/ $CH_3(CH_2)_2COOH$	colorless liq/ strong, rancid, butter-like	m—alc, most fixed oils, prop glycol water/ 164°	
γ-Butyrolactone FEMA No. 3291	86.09/$C_4H_6O_2$/	colorless to slightly yel liq/ faint, sweet, caramel-like	s—water; m—alc/ 204°	

Requirements

I.D. Test[4]	Assay Min. %[5]	A.V. Max.[6]	Ref. Index[7]	Sp. Gr.[8]	Other Requirements[9]
IR	97.0% of $C_8H_{16}O_2$ (M-1b)	1.0	1.401–1.404	0.859–0.864	
IR	97.0% of $C_9H_{18}O_2$ (M-1b)	1.0	1.407–1.411	0.856–0.859	
IR	98.0% of $C_{12}H_{16}O_2$ (M-1a)	1.0	1.488–1.492	0.990–0.997	
					Melting Range—between 17° and 21° (Appendix IIB) **Iodine Value**—1 max (Appendix VII) **Saponification Value**—between 165° and 180° (Appendix VI)
IR	98.0% of C_6H_8O (M-2c)	5.0		0.797–0.802	**Dist. Range**—between 72° and 80° (first 95%, Appendix IIB) *p*-**Butyraldehyde**—2.5% (M-1b) **Water**—0.5% (Appendix IIB, KF)
IR	99.0% of $C_4H_8O_2$ (M-3a)		1.397–1.399	0.952–0.956	**Heavy Metals**—0.004% (M-7) **Lead**—10 mg/kg (M-10) **Reducing Subs.**—passes test (M-15)
IR	98.0% of $C_4H_6O_2$ (M-1a)		1.434–1.454 (25°)	1.120–1.130	

General Information and Description

Name of Substance (Synonyms)	Formula Wt/Formula/ Structure	Physical Form/ Odor	Solubility[1]/ B.P.[2]	Solubility in Alcohol[3]
Camphene FEMA No. 2229	136.24/$C_{10}H_{16}$/	colorless cryst mass; m.p. 52°/ camphoraceous-oily odor	s—alc; m—most fixed oils; ins—water	
***d*-Camphor** FEMA No. 2230	152.24/$C_{10}H_{16}O$/	white to gray translucent cryst or fused mass/ characteristic	204°	1 mL in 1 mL 95% alc
Carvacrol FEMA No. 2245	150.22/$C_{10}H_{14}O$/	colorless to pale yel liq/ pungent, spicy, thymol-like	s—alc, ether; ins—water/ 238°	1 mL in 4 mL 60% alc gives clear soln
***l*-Carveol** FEMA No. 2247 *p*-Mentha-6,8-dien-2-ol	152.24/$C_{10}H_{16}O$/	colorless to pale yel liq/ spearminty	226°–227° (751 mm Hg)	1 mL in 1 mL 95% alc
***d*-Carvone** FEMA No. 2249 dextro-Carvone; *d*-1-Methyl-4-isopropenyl-6-cyclohexen-2-one	150.22/$C_{10}H_{14}O$/	colorless to light yel liq/ caraway-like	s—prop glycol, most fixed oils; m—alc; ins—gly/ 230°	1 mL in 5 mL 60% alc

Requirements

I.D. Test[4]	Assay Min. %[5]	A.V. Max.[6]	Ref. Index[7]	Sp. Gr.[8]	Other Requirements[9]
	80.0% of $C_{10}H_{16}$ (M-1a)				**Solidification Pt.**—$40°$ (Appendix IIB)
IR					**Melting Range**—between $174°$ and $179°$ (Appendix IIB) **Angular Rotation**—between $+41°$ and $+43°$ (Appendix IIB)
IR	98.0% of $C_{10}H_{14}O$ (M-1b)		1.521–1.526	0.974–0.980	
IR	96.0% of $C_{10}H_{16}O$ [(Z) isomer 45% +/- 5%; (E) isomer 55% +/- 5%] (M-1b)		1.493–1.497	0.947–0.953	**Angular Rotation**— between $-117°$ and $-130°$ (Appendix IIB)
IR	95.0% of $C_{10}H_{14}O$ (M-1b)		1.496–1.499	0.955–0.960	**Angular Rotation**—between $+50°$ and $+60°$ (Appendix IIB, 100-mm tube)

General Information and Description

Name of Substance (Synonyms)	Formula Wt/Formula/ Structure	Physical Form/ Odor	Solubility[1]/ B.P.[2]	Solubility in Alcohol[3]
l-Carvone FEMA No. 2249 levo-Carvone; *l*-1-Methyl-4-isopropenyl-6-cyclohexen-2-one	150.22/$C_{10}H_{14}O$/	colorless to pale strawberry colored liq/ spearmint-like	s—prop glycol, most fixed oils; m—alc; ins—gly/ 231°	1 mL in 2 mL 70% alc
l-Carvyl Acetate FEMA No. 2250 *p*-Mentha-6,8-dien-2-yl Acetate	194.27/$C_{12}H_{18}O_2$/	colorless to pale yel liq/ spearminty	s—alc/ 77°–79° (0.1 mm Hg)	
β-Caryophyllene FEMA No. 2252	204.36/$C_{15}H_{24}$/	colorless to slightly yel, oily liq/ clove-like	s—alc, ether; ins—water/ 256°	1 mL in 6 mL 95% alc gives clear soln
Cinnamaldehyde FEMA No. 2286 Cinnamic Aldehyde; Cinnamal	132.16/C_9H_8O/ ⌬—CH=CHCHO	yel, strongly refractive liq/ cinnamon-like, burning aromatic taste	m—alc, chloroform, ether, fixed and vol oils, 1 g in 700 mL water	1 mL in 5 mL 60% alc
Cinnamic Acid FEMA No. 2288 3-Phenylpropenoic Acid	148.16/$C_9H_8O_2$/ ⌬—CH=CHCOOH	white cryst scales; m.p. 133°/ honey-floral	s—acetic acid, acetone, benzene, most fixed oils, 1 g in 2000 mL water	
Cinnamyl Acetate FEMA No. 2293	176.22/$C_{11}H_{12}O_2$/ ⌬—CH=CHCH₂OOCCH₃	colorless to slightly yel liq/ sweet, balsamic, floral	m—alc, chloroform, ether, most fixed oils; ins—gly, water/ 264°	1 mL in 5 mL 70% alc

Requirements

I.D. Test[4]	Assay Min. %[5]	A.V. Max.[6]	Ref. Index[7]	Sp. Gr.[8]	Other Requirements[9]
IR	97.0% of $C_{10}H_{14}O$ (M-1b)		1.495–1.499	0.956–0.960	**Angular Rotation**—between -57° and -62° (Appendix IIB, 100-mm tube)
IR	98.0% of $C_{12}H_{18}O_2$ (M-1b)	1.0	1.473–1.479	0.964–0.970	**Angular Rotation**— between -90° and -120° (Appendix IIB)
IR			1.498–1.504	0.897–0.910	**Angular Rotation**— between -5° and -10° (Appendix IIB, 100-mm tube) **Phenols**—3.0% (M-1b)
IR	98.0% of C_9H_8O (M-1b)	10.0	1.619–1.623	1.046–1.050	**Chlorinated Cmpds.**—passes test (Appendix VI)
IR	99.0% of $C_9H_8O_2$ (after drying) (M-3b)				**Melting Range**—NLT 130° (Appendix IIB) **Res. on Ignit.**—0.05% (Appendix IIC)
IR	98.0% of $C_{11}H_{12}O_2$ (M-1b)	1.0	1.539–1.543	1.050–1.054	

General Information and Description

Name of Substance (Synonyms)	Formula Wt/Formula/ Structure	Physical Form/ Odor	Solubility[1]/ B.P.[2]	Solubility in Alcohol[3]
Cinnamyl Alcohol FEMA No. 2294 Cinnamic Alcohol	134.18/$C_9H_{10}O$/ ⌬—CH=CHCH$_2$OH	white to slightly yel cryst solid; m.p. 33°/ balsamic	s—most fixed oils, prop glycol; ins—gly/ 258°	1 g in 1 mL 70% alc remains in soln to 10 mL
Cinnamyl Butyrate FEMA No. 2296	204.27/$C_{13}H_{16}O_2$/ ⌬—CH=CHCH$_2$OOC(CH$_2$)$_2$CH$_3$	colorless to pale yel liq/ fruity, balsamic	300°	1 mL in 1 mL 95% alc
Cinnamyl Cinnamate FEMA No. 2298	264.32/$C_{18}H_{16}O_2$/ ⌬—CH=CHCOOCH$_2$CH=CH—⌬	mixture of (Z) and (E) isomers; low-melting solid	370°	1 mL in 1 mL 95% alc
Cinnamyl Formate FEMA No. 2299	162.19/$C_{10}H_{10}O_2$/ ⌬—CH=CHCH$_2$OOCH	colorless to slightly yel liq/ green, herbaceous, balsamic odor	m—alc, chloroform, ether, most fixed oils; ins—water/ 250°	1 mL in 2 mL 80% alc gives clear soln
Cinnamyl Isobutyrate FEMA No. 2297	204.27/$C_{13}H_{16}O_2$/ ⌬—CH=CHCH$_2$OOCH(CH$_3$)$_2$	colorless to pale yel liq/ sweet, balsamic, fruity	254°	1 mL in 1 mL 95% alc
Cinnamyl Isovalerate FEMA No. 2302	218.30/$C_{14}H_{18}O_2$/ ⌬—CH=CHCH$_2$OOCCH$_2$CH(CH$_3$)CH$_3$	colorless to slightly yel liq/ spicy, floral, fruity	m—alc, chloroform, most fixed oils, ether; ins—gly, prop glycol, water/ 313°	1 mL in 1 mL 90% alc

Requirements

I.D. Test[4]	Assay Min. %[5]	A.V. Max.[6]	Ref. Index[7]	Sp. Gr.[8]	Other Requirements[9]
IR	98.0% of $C_9H_{10}O$ (M-1b)				**Aldehydes**—1.5% (M-1b) **Chlorinated Cmpds.**—passes test (Appendix VI) **Solidification Pt.**—NLT 31° (Appendix IIB)
IR	96.0% of $C_{13}H_{16}O_2$ (M-1b)	1.0	1.525–1.530	1.010–1.015	
IR	95.0% of $C_{18}H_{16}O_2$ (M-1b)	2.0			
IR	92.0% of $C_{10}H_{10}O_2$ (M-1a)	3.0	1.550–1.556	1.077–1.082	**Cinnamyl Alcohol**—8.0% (M-1a)
IR	96.0% of $C_{13}H_{16}O_2$ (M-1b)	3.0	1.523–1.528	1.006–1.009	
IR	95.0% of $C_{14}H_{18}O_2$ (M-1b)	3.0	1.518–1.524	0.991–0.996	

General Information and Description

Name of Substance (Synonyms)	Formula Wt/Formula/ Structure	Physical Form/ Odor	Solubility[1]/ B.P.[2]	Solubility in Alcohol[3]
Cinnamyl Propionate FEMA No. 2301	190.24/$C_{12}H_{14}O_2$/ 〈phenyl〉—CH=CHCH$_2$OOCCH$_2$CH$_3$	colorless to pale yel liq/ spicy, fruity, balsamic	m—alc, chloroform, ether, most fixed oils; ins—gly, prop glycol, water/ 289°	
Citral FEMA No. 2303 [Mixture of Geranial (E)-3,7-dimethyl-2,6-octadien-1-al) and Neral(the (Z) isomer)]	152.24/$C_{10}H_{16}O$/ (a) Geranial (b) Neral	pale yel liq/ strong, lemon-like	s—fixed oils, min oil, prop glycol; ins—gly/ 228°	1 mL in 7 mL 70% alc
Citronellal FEMA No. 2307 3,7-Dimethyl-6-octen-1-al	154.25/$C_{10}H_{18}O$/	colorless to slightly yel liq/ intense lemon-citronella-rose	s—alc, most fixed oils; ss—prop glycol; ins—gly, water/ 206°	1 mL in 5 mL 70% alc remains clear on dilution
Citronellol FEMA No. 2309 3,7-Dimethyl-6-octen-1-ol	156.27/$C_{10}H_{20}O$/ —CH$_2$OH	colorless, oily liq/ rose-like	s—most fixed oils, prop glycol; ss—water; ins—gly/ 225°	1 mL in 2 mL 70% alc remains in soln to 10 mL
Citronellyl Acetate FEMA No. 2311 3,7-Dimethyl-6-octen-1-yl Acetate	198.31/$C_{12}H_{22}O_2$/ —CH$_2$OC(=O)CH$_3$	colorless liq/ fruity	s—alc, most fixed oils; ins—gly, prop glycol, water/ 229°	1 mL in 9 mL 70% alc
Citronellyl Butyrate FEMA No. 2312 3,7-Dimethyl-6-octen-1-yl Butyrate	226.36/$C_{14}H_{26}O_2$/ —CH$_2$OC(=O)CH$_2$CH$_2$CH$_3$	colorless liq/ strong, fruity-rosy	m—alc, ether, most fixed oils, chloroform; ins—water/ 245°	1 mL in 6 mL 80% alc gives clear soln

Requirements

I.D. Test[4]	Assay Min. %[5]	A.V. Max.[6]	Ref. Index[7]	Sp. Gr.[8]	Other Requirements[9]
IR	98.0% of $C_{12}H_{14}O_2$ (M-1b)	3.0	1.532–1.537	1.029–1.035	
IR	96.0% of $C_{10}H_{16}O$ (M-1b)		1.486–1.490	0.885–0.891	
IR	85.0% of aldehydes as $C_{10}H_{18}O$ (M-1b)	3.0	1.446–1.456	0.850–0.860	**Angular Rotation**— between -1° and +11° (Appendix IIB, 100-mm tube)
IR	90.0% of total alcohols as $C_{10}H_{20}O$ (Appendix VI; 1.2g/78.13)		1.454–1.462	0.850–0.860	**Aldehydes**—1.0% as citronellal (M-2d; 5g/66.08) **Esters**—1.0% as citronellyl acetate (Appendix VI; 5g/99.15)
IR	92.0% of total esters as $C_{12}H_{22}O_2$ (Appendix VI; 1.4g/ 99.15)	1.0	1.440–1.450	0.883–0.893	
IR	90.0% of total esters as $C_{14}H_{26}O_2$ (Appendix VI; 1.5g/113.2)	1.0	1.444–1.448	0.873–0.883	

General Information and Description

Name of Substance (Synonyms)	Formula Wt/Formula/ Structure	Physical Form/ Odor	Solubility[1]/ B.P.[2]	Solubility in Alcohol[3]
Citronellyl Formate FEMA No. 2314 3,7-Dimethyl-6-octen-1-yl Formate	184.28/$C_{11}H_{20}O_2$/	colorless liq/ strong, fruity, floral	s—alc, most fixed oils; ss—prop glycol; ins—gly, water/ 235°	1 mL in 3 mL 80% alc remains in soln to 10 mL
Citronellyl Isobutyrate FEMA No. 2313 3,7-Dimethyl-6-octen-1-yl Isobutyrate	226.36/$C_{14}H_{26}O_2$/	colorless liq/ rosy-fruity	m—alc, chloroform, ether, most fixed oils; ins—water/ 249°	1 mL in 6 mL 80% alc gives clear soln
Citronellyl Propionate FEMA No. 2316 Citronellyl Propanoate; 3,7-Dimethyl-6-octen-1-yl Propionate	212.33/$C_{13}H_{24}O_2$/	colorless liq/ fruity-rosy	m—alc, most fixed oils; ins—water/ 242°	1 mL in 4 mL 80% alc gives clear soln
p-Cresyl Acetate FEMA No. 3073 p-Tolyl Acetate; p-Methylphenyl Acetate	150.18/$C_9H_{10}O_2$/	colorless liq/ strong, floral	s—most fixed oils, prop glycol; ins—gly/ 212°	1 mL in 2 mL 70% alc
Cuminic Aldehyde FEMA No. 2341 p-Cuminic Aldehyde; Cuminaldehyde; p-Isopropylbenzaldehyde; Cuminal	148.20/$C_{10}H_{12}O$/	colorless to pale yel liq/ strong, pungent odor of cumin oil	s—alc, ether; ins—water/ 236°	1 mL in 4 mL 70% alc
Cyclamen Aldehyde FEMA No. 2743 2-Methyl-3-(p-isopropylphenyl) propionaldehyde	190.29/$C_{13}H_{18}O$/	colorless to pale yel liq/ strong, floral	s—most fixed oils; ins—prop glycol, gly/ 270°	1 mL in 3 mL 80% alc
Cyclohexyl Acetate FEMA No. 2349	121.1/$C_8H_{14}O_2$/	colorless to pale yel liq	s—alc/ 174°	

Requirements

I.D. Test[4]	Assay Min. %[5]	A.V. Max.[6]	Ref. Index[7]	Sp. Gr.[8]	Other Requirements[9]
IR	86.0% of total esters as $C_{11}H_{20}O_2$ (Appendix VI; 1.0g/92.14)	3.0	1.443–1.452	0.890–0.903	
IR	92.0% of total esters as $C_{14}H_{26}O_2$ (Appendix VI; 1.5g/113.2)	1.0	1.440–1.448	0.870–0.880	
IR	90.0% of total esters as $C_{13}H_{24}O_2$ (Appendix VI; 1.2g/95.12)	1.0	1.443–1.449	0.877–0.886	
IR	98.0% of $C_9H_{10}O_2$ (M-1b)	1.0 (phenol red TS)	1.499–1.502	1.044–1.050	**Free Cresol**—1.0% (M-18)
IR	95.0% of $C_{10}H_{12}O$ (M-2a)	5.0	1.529–1.534	0.976–0.980	**Chlorinated Cmpds.**—passes test (Appendix VI)
IR	90.0% of $C_{13}H_{18}O$ (M-1b)	5.0	1.503–1.508	0.946–0.952	
	98.0% of $C_8H_{14}O_2$ (M-1b)	1.0	1.436–1.441	0.966–0.970	

General Information and Description

Name of Substance (Synonyms)	Formula Wt/Formula/ Structure	Physical Form/ Odor	Solubility[1]/ B.P.[2]	Solubility in Alcohol[3]
p-Cymene FEMA No. 2356	134.22/$C_{10}H_{14}$/ [structure: benzene ring with CH$_3$ and CH(CH$_3$)$_2$ para substituents]	colorless to pale yel liq/ no odor	177°	1 mL in 1 mL 95% alc
(E),(E)-2,4-Decadienal FEMA No. 3135 *trans,trans*-2,4-Decadienal	152.24/$C_{10}H_{16}O$/ CH$_3$(CH$_2$)$_4$\C=C/H ... C=C ... CHO	yel liq/ powerful, oily, like chicken fat	s—alc, fixed oils; *ins*—water/ 104° (7 mm Hg)	
δ-Decalactone FEMA No. 2361	170.25/$C_{10}H_{18}O_2$/ CH$_3$(CH$_2$)$_4$ [δ-lactone ring]	colorless liq/ coconut-fruity, butter-like on dilution	*vs*—alc, prop glycol, veg oils; *ins*—water/ 281°	
γ-Decalactone FEMA No. 2360 4-Hydroxydecanoic Acid Lactone	170.25/$C_{10}H_{18}O_2$/ CH$_3$(CH$_2$)$_5$CHCH$_2$CH$_2$C=O └────────O	colorless to pale yel liq/ fruity, peach-like	281°	1 mL in 1 mL 95% alc
Decanal FEMA No. 2362 Aldehyde C-10; Capraldehyde	156.27/$C_{10}H_{20}O$/ CH$_3$(CH$_2$)$_8$CHO	colorless to light yel liq/ fatty, floral-orange on dilution	*m*—alc, fixed oils, prop glycol (may be turbid); *ins*—gly, water/ 209°	
Decyl Alcohol FEMA No. 2365 1-Decanol; Alcohol C-10	158.28/$C_{10}H_{22}O$/ CH$_3$(CH$_2$)$_8$CH$_2$OH	colorless liq/ floral, waxy, fruity	s—alc, ether, min oil, prop glycol most fixed oils; *ins*—gly, water/ 233°	1 mL in 3 mL 60% alc
(E)-2-Decenal FEMA No. 2366 *trans*-2-Decenal	154.25/$C_{10}H_{18}O$/ CH$_3$(CH$_2$)$_6$\C=C/H ... CHO	slightly yel liq/ orange, wax-like odor	s—alc, most fixed oils; *ins*—water/ 229°	

Requirements

I.D. Test[4]	Assay Min. %[5]	A.V. Max.[6]	Ref. Index[7]	Sp. Gr.[8]	Other Requirements[9]
IR	97.0% (M-1a)		1.489–1.491	0.853–0.855	
IR	89.0% of $C_{10}H_{16}O$ (M-1a)		1.514–1.516	0.866–0.876	
IR	98.0% of $C_{10}H_{18}O_2$ (M-1b)	5.0 (Appendix VII)	1.456–1.459		**Saponification Value**—between 323 and 333 (Appendix VI, 1-g sample)
IR	95.0% of $C_{10}H_{18}O_2$ (M-1a)	1.0	1.447–1.451	0.950–0.955	
IR	92.0% of $C_{10}H_{20}O$ (M-1b)	10.0	1.426–1.430	0.823–0.832	
IR	98.0% of $C_{10}H_{22}O$ (M-1b)	1.0	1.435–1.439	0.826–0.831	**Solidification Pt.**—NLT 5° (Appendix IIB)
IR	92.0% of $C_{10}H_{18}O$ (M-1a)		1.452–1.457	0.836–0.846	

General Information and Description

Name of Substance (Synonyms)	Formula Wt/Formula/ Structure	Physical Form/ Odor	Solubility[1]/ B.P.[2]	Solubility in Alcohol[3]
(Z)-4-Decenal FEMA No. 3264 *cis*-4-Decenal	154.25/$C_{10}H_{18}O$/ $CH_3(CH_2)_4\text{-CH=CH-}(CH_2)_2CHO$ (cis)	colorless to slightly yel liq/ orange-like, fatty	*s*—alc, most fixed oils; *ins*—water	
Diacetyl FEMA No. 2370 2,3-Butanedione; Dimethyldiketone; Dimethylglyoxal	86.09/$C_4H_6O_2$/ $CH_3-\overset{O}{\underset{\|}{C}}-\overset{O}{\underset{\|}{C}}-CH_3$	yel to yel-green liq/ powerful, buttery in very dilute soln	*s*—gly, water; *m*—alc, most fixed oils, prop glycol/ 88°	
Dibenzyl Ether FEMA No. 2371	198.26/$C_{14}H_{14}O$/ (C₆H₅CH₂-O-CH₂C₆H₅)	colorless to pale yel liq/ earthy	298°	1 mL in 1 mL 95% alc
1,2-Di[(1′-ethoxy)ethoxy]propane FEMA No. 3534	220.31/$C_{11}H_{24}O_4$ $CH_2-O-\overset{CH_3}{\underset{\|}{CH}}-O-CH_2CH_3$ $\|$ $CH-O-\overset{CH_3}{\underset{\|}{CH}}-O-CH_2CH_3$ $\|$ CH_3	colorless to pale yel liq		
Diethyl Malonate FEMA No. 2375 Ethyl Malonate; Malonic Ester	160.17/$C_7H_{12}O_4$/ $CH_2\begin{smallmatrix}\diagup COOCH_2CH_3\\\diagdown COOCH_2CH_3\end{smallmatrix}$	colorless liq/ slight, fruit-like	*s*—most fixed oils, prop glycol; *ss*—alc, water; *ins*—gly, min oil/ 200°	1 mL in 1.5 mL 60% alc
Diethyl Sebacate FEMA No. 2376 Ethyl Sebacate	258.36/$C_{14}H_{26}O_4$/ $C_2H_5OOC(CH_2)_8COOC_2H_5$	colorless to slightly yel liq/ faint, winy, fruity	*m*—alc, ether, other org solvents most fixed oils; *ins*—water/ 302°	
Diethyl Succinate FEMA No. 2377 Ethyl Succinate	174.20/$C_8H_{14}O_4$/ $C_2H_5OOCCH_2CH_2COOC_2H_5$	colorless, mobile liq/ faint, pleasant	1 mL in 50 mL water; *m*—alc, ether, most fixed oils/ 217°	

Requirements

I.D. Test[4]	Assay Min. %[5]	A.V. Max.[6]	Ref. Index[7]	Sp. Gr.[8]	Other Requirements[9]
IR	90.0% of $C_{10}H_{18}O$ (M-1a)		1.442–1.444	0.847–0.848	
IR	95.0% of $C_4H_6O_2$ (M-1b)		1.393–1.397	0.979–0.985	**Solidification Pt.**— between -2.0° and -4.0° (Appendix IIB)
IR	98.0% of $C_{14}H_{14}O$ (M-1b)		1.557–1.565	1.039–1.044	**Chlorinated Cmpds.**—passes test (Appendix VI)
	97.0% of $C_{11}H_{24}O_4$ (M-1b)	0.1	1.409–1.413	0.915–0.925	
IR	98.0% of $C_7H_{12}O_4$ (M-1b)	1.0	1.413–1.416	1.053–1.056	
IR	98.0% of $C_{14}H_{26}O_4$ (M-1b)	1.0	1.435–1.438	0.960–0.965	
IR	99.0% of $C_8H_{14}O_4$ (M-1a)	2.0			**Diethyl Maleate**—0.03% (M-1b) **Water**—0.05% (Appendix IIB, KF)

General Information and Description

Name of Substance (Synonyms)	Formula Wt/Formula/ Structure	Physical Form/ Odor	Solubility[1]/ B.P.[2]	Solubility in Alcohol[3]
Dihydrocarveol FEMA No. 2379	154.25/$C_{10}H_{18}O$/ (structure: 2-methyl-5-(isopropenyl)cyclohexanol)	almost colorless, oily liq/ spearmint-like	s—alc, most fixed oils; ins—water/ 225°	
***d*-Dihydrocarvone** FEMA No. 3565 *d*-2-Methyl-5-(1-methylethenyl)-cyclohexanone	154.24/$C_{10}H_{16}O$/ (structure: 2-methyl-5-(isopropenyl)cyclohexanone)	almost colorless liq/ herbaceous, spearmint-like	s—alc, most fixed oils; ins—water/ 222°	
Dimethyl Anthranilate FEMA No. 2718 Methyl *N*-Methyl Anthranilate	165.19/$C_9H_{11}NO_2$/ (structure: benzene with -NHCH$_3$ and -COOCH$_3$)	pale yel liq with bluish fluorescence/ grape-like	s—most fixed oils; ss—prop glycol; ins—gly, water/ 256°	1 mL in 3 mL 80% alc remains in soln to 10 mL
Dimethyl Benzyl Carbinol FEMA No. 2393 α,α-Dimethylphenethyl Alcohol	150.22/$C_{10}H_{14}O$/ $\mathrm{C_6H_5-CH_2C(CH_3)_2OH}$	white cryst solid that melts readily; may exist in supercooled form as colorless to pale yel liq/ floral	s—most fixed oils, min oil, prop glycol; ins—gly	1 mL in 3 mL 50% alc remains in soln to 10 mL
Dimethyl Benzyl Carbinyl Acetate FEMA No. 2392 α,α-Dimethylphenethyl Acetate	192.26/$C_{12}H_{16}O_2$/ $\mathrm{C_6H_5-CH_2C(CH_3)_2OOCCH_3}$	colorless liq; solidifies at room temp/ floral, fruity	s—most fixed oils; ss—prop glycol; ins—water/ 250°	1 mL in 4 mL 70% alc
Dimethyl Benzyl Carbinyl Butyrate FEMA No. 2394 α,α-Dimethylphenethyl Butyrate	220.31/$C_{14}H_{20}O_2$/ $\mathrm{C_6H_5-CH_2C(CH_3)_2OOC(CH_2)_2CH_3}$	almost colorless liq/ prune-like	s—alc, most fixed oils; ins—water, prop glycol	
2,6-Dimethyl-5-heptenal FEMA No. 2389	140.23/$C_9H_{16}O$/ $\mathrm{CH_3C(CH_3)=CH(CH_2)_2CH(CH_3)CHO}$	pale yel liq/ melon-like	116°–124° (100 mm Hg)	1 mL in 1 mL 95% alc

Requirements

I.D. Test[4]	Assay Min. %[5]	A.V. Max.[6]	Ref. Index[7]	Sp. Gr.[8]	Other Requirements[9]
	96.0% of $C_{10}H_{18}O$ (sum of isomers) (M-1a)		1.477–1.481	0.921–0.926	
	92.0% of $C_{10}H_{16}O$ (sum of isomers) (M-1a)		1.470–1.474	0.923–0.928	**Angular Rotation**—min. +14.0° (Appendix IIB)
	98.0–101.3% of total esters as $C_9H_{11}NO_2$ (Appendix VI; 1.1g/82.60)		1.578–1.581	1.126–1.132	**Solidification Pt.**—NLT 14° (Appendix IIB)
IR	97.0% of $C_{10}H_{14}O$ (M-1b)	1.0	1.514–1.517 (20°, as supercooled liq)	0.972–0.977	**Chlorinated Cmpds.**—passes test (Appendix VI) **Solidification Pt.**—NLT 22° (Appendix IIB)
IR	98.0% of $C_{12}H_{16}O_2$ (M-1b)	1.0	1.490–1.495	0.995–1.002	**Chlorinated Cmpds.**—passes test (Appendix VI) **Solidification Pt.**—NLT 28° (Appendix IIB)
IR	95.0% of $C_{14}H_{20}O_2$ (M-1b)		1.473–1.493 (25°)	0.960–0.981	**Angular Rotation**— between -1° and +2° (Appendix IIB)
IR	85.0% of $C_9H_{16}O$ (M-2d; 1g/14.01)	5.0	1.442–1.447	0.848–0.854	

General Information and Description

Name of Substance (Synonyms)	Formula Wt/Formula/ Structure	Physical Form/ Odor	Solubility[1]/ B.P.[2]	Solubility in Alcohol[3]
3,7-Dimethyl-1-octanol FEMA No. 2391 Dimethyl Octanol; Tetrahydrogeraniol	158.28/$C_{10}H_{22}O$/ CH$_2$OH	colorless liq/ sweet, rose-like	s—most fixed oils, prop glycol; ins—gly/ 213°	1 mL in 3 mL 70% alc
2,3-Dimethylpyrazine FEMA No. 3271	108.14/$C_6H_8N_2$/	colorless to slightly yel liq/ nutty, cocoa-like	m—water, org solvents/ 156°	
2,5-Dimethylpyrazine FEMA No. 3272	108.14/$C_6H_8N_2$/	colorless to slightly yel liq/ earthy, potato-like	m—water, org solvents/ 155°	
2,6-Dimethylpyrazine FEMA No. 3273	108.14/$C_6H_8N_2$/	white to yel, lumpy cryst, m.p. 48°/ nutty, coffee-like	s—water, org solvents/ 155°	
2,5-Dimethylpyrrole	95.14/C_6H_9N/	colorless to yellowish, oily liq	vs—alc, ether; vss—water/ 165°	
Dimethyl Succinate FEMA No. 2396	146.14/$C_6H_{10}O_4$/ CH$_3$OC(CH$_2$)$_2$COCH$_3$	colorless to pale yel liq	196°	1 mL in 1 mL 95% alc
Dimethyl Sulfide FEMA No. 2746 Methyl Sulfide; Thiobismethane	62.14/C_2H_6S/ CH$_3$SCH$_3$	colorless to pale yel liq/ disagreeable, intense boiled cabbage	109°	1 mL in 1 mL 95% alc

Requirements

I.D. Test[4]	Assay Min. %[5]	A.V. Max.[6]	Ref. Index[7]	Sp. Gr.[8]	Other Requirements[9]
IR	90.0% of total alcohols as $C_{10}H_{22}O$ (Appendix VI; 1.2g/79.15)	1.0	1.435–1.445	0.826–0.842	
IR	95.0% of $C_6H_8N_2$ (M-1a)		1.506–1.509	1.000–1.022 (20°)	**Dist. Range**—between 152° and 157° (Appendix IIB) **Heavy Metals**—10 mg/kg (M-7) **Solidification Pt.**—between 11° and 13° (Appendix IIB) **Tri- and Tetrapyrazines**—5% (by GC assay) **Water**—0.5% (Appendix IIB, KF; use freshly dist. pyridine as solvent)
IR	99.0% of $C_6H_8N_2$ (M-1a)		1.497–1.501	0.980–1.000	**Heavy Metals**—10 mg/kg (M-7) **Solidification Pt.**—between 12° and 17° (Appendix IIB) **Water**—0.5% (Appendix IIB, KF; use freshly dist. pyridine as solvent)
IR	98.0% of $C_6H_8N_2$ (M-1a)			0.965 (50°)	**Melting Range**—between 35° and 40° (Appendix IIB) **Res. on Ignit.**—0.1% (Appendix IIC) **Water**—0.5% (Appendix IIB, KF; use freshly dist. pyridine as solvent)
IR	98.0% of C_6H_9N (M-1a)		1.503–1.506	0.935–0.945 (20°)	**Water**—0.5% (Appendix IIB, KF; use freshly dist. pyridine as solvent)
	98.0% of $C_6H_{10}O_4$ (M-1b)	1.0	1.418–1.421	1.114–1.118	
IR	99.0% of C_2H_6S (M-1a)		1.431–1.441	0.842–0.847	

General Information and Description

Name of Substance (Synonyms)	Formula Wt/Formula/ Structure	Physical Form/ Odor	Solubility[1]/ B.P.[2]	Solubility in Alcohol[3]
δ-Dodecalactone FEMA No. 2401	198.31/$C_{12}H_{22}O_2$/ $CH_3(CH_2)_6$–[lactone ring]–O	colorless to yel liq/ coconut-fruity, butter-like on dilution	vs—alc, prop glycol, veg oils; ins—water	
γ-Dodecalactone FEMA No. 2400 4-Hydroxydodecanoic Acid Lactone	198.31/$C_{12}H_{22}O_2$/ $CH_3(CH_2)_6CH(CH_2)_3C=O$ $\quad\quad\quad |_____O$	colorless to pale yel liq/ fruity, peach-like, pear-like	131° (1.5 mm Hg)	1 mL in 1 mL 95% alc
(E)-2-Dodecen-1-al FEMA No. 2402 trans-2-Dodecen-1-al	182.31/$C_{12}H_{22}O$/ $CH_3(CH_2)_8$\ /H $\quad\quad$ C=C H/ \CHO	slightly yel liq/ fatty, citrus-like	s—alc, most fixed oils; ins—water/ 272°	
Estragole FEMA No. 2411 p-Allylanisole; Methyl Chavicol	148.20/$C_{10}H_{12}O$/ CH_3O–⟨⟩–$CH_2CH=CH_2$	colorless to light yel liq/ anise-like	s—alc; ins—water/ 216°	1 mL in 6 mL 80% alc gives clear soln
Ethone FEMA No. 2673 1-(p-Methoxyphenyl)-1-penten-3-one	190.24/$C_{12}H_{14}O_2$/ CH_3O–⟨⟩–$CH=CHCOCH_2CH_3$	white to pale yel cryst; m.p. 60°		1 g in 7 ml 95% alc
Ethyl Acetate FEMA No. 2414	88.11/$C_4H_8O_2$/ $CH_3COOC_2H_5$	colorless liq; vol at low temp; flammable/ fragrant, acetous, ethereal	m—alc, ether, gly, fixed oils, vol oils, 1 mL in 10 mL water/ 54°	
Ethyl Acetoacetate FEMA No. 2415 Acetoacetic Ester; Ethyl 3-Oxybutanoate	130.14/$C_6H_{10}O_3$/ $CH_3COCH_2COOC_2H_5$ ⇅ $CH_3C=CHCOOC_2H_5$ $\quad\mid$ $\quad OH$	colorless to very light yel, mobile liq/ agreeable odor	m—alc, ether, ethyl acetate, 1 mL in 12 mL water/ 181°	

Requirements

I.D. Test[4]	Assay Min. %[5]	A.V. Max.[6]	Ref. Index[7]	Sp. Gr.[8]	Other Requirements[9]
IR	98.0% of $C_{12}H_{22}O_2$ (M-1a)	8.0 (Appendix VII)	1.458–1.461		**Saponification Value**—between 278 and 286 (Appendix VI, 1-g sample)
	97.0% of $C_{12}H_{22}O_2$ (M-1a)	1.0	1.451–1.456	0.933–0.938	
IR	93.0% of $C_{12}H_{22}O$ (M-1a)		1.462–1.464	0.839–0.849	
IR	95.0% of $C_{10}H_{12}O$ (M-1a)		1.519–1.524	0.960–0.968	
IR	98.0% of $C_{12}H_{14}O_2$ (M-1b)				**Solidification Pt.**—min 59.0° (Appendix IIB)
IR	99.0% of $C_4H_8O_2$ (M-1b)	5.0		0.894–0.898	**Dist. Range**—between 76° and 77.5° (Appendix IIB) **Methyl Cmpds.**—passes test (M-11) **Readily Carb. Subs.**—passes test (M-13) **Res. on Evap.**—0.02% (M-17, 10-g sample, 105°)
IR	97.5% of $C_6H_{10}O_3$ (M-1b)	5.0	1.418–1.421	1.022–1.027	

General Information and Description

Name of Substance (Synonyms)	Formula Wt/Formula/ Structure	Physical Form/ Odor	Solubility[1]/ B.P.[2]	Solubility in Alcohol[3]
Ethyl Acrylate FEMA No. 2418	100.12/$C_5H_8O_2$/ $CH_2=CHCOOC_2H_5$	colorless, mobile liq; lachrymator/ intense, harsh, fruity	m—alc, ether, 1 mL in 50 mL water/ 99°	
Ethyl p-Anisate FEMA No. 2420 Ethyl p-Methoxybenzoate	180.20/$C_{10}H_{12}O_3$/ $CH_3O-\langle\rangle-COOC_2H_5$	colorless to slightly yel liq/ light, fruity, anise-like	s—alc, chloroform, ether; ins—water/ 270°	1 mL in 7 mL 60% alc gives clear soln
Ethyl Anthranilate FEMA No. 2421 Ethyl o-Aminobenzoate	165.19/$C_9H_{11}NO_2$/ NH_2 $\langle\rangle-COOC_2H_5$	colorless to amber-colored liq/ floral, orange blossom-like	s—alc, most fixed oils, prop glycol/ 267°	1 mL in 2 mL 70% alc
Ethyl Benzoate FEMA No. 2422	150.18/$C_9H_{10}O_2$/ $\langle\rangle-COOC_2H_5$	colorless liq/ heavy, floral, fruity	s—alc, most fixed oils, prop glycol; ins—gly, water/ 212°	1 mL in 6 mL 60% alc
Ethyl Benzoyl Acetate FEMA No. 2423	192.21/$C_{11}H_{12}O_3$/ $\langle\rangle-COCH_2CO_2CH_2CH_3$	light yel liq	265°	
Ethyl-(E)-2-butenoate FEMA No. 3486 Ethyl Crotonate; Ethyl-*trans*-2-butenoate	114.14/$C_6H_{10}O_2$ $CH_3CH=CHOOCH_2CH_3$	colorless to pale yel liq	136°	
2-Ethylbutyraldehyde FEMA No. 2426	100.16/$C_6H_{12}O$/ C_2H_5 $\|$ CH_3CH_2CHCHO	colorless, mobile liq/ pungent	m—alc, ether, 1 mL in 50 mL water/ 117°	
Ethyl Butyrate FEMA No. 2427	116.16/$C_6H_{12}O_2$/ $CH_3CH_2CH_2COOC_2H_5$	colorless liq/ banana-pineapple	s—fixed oils, prop glycol; ins—gly/ 121°	1 mL in 3 mL 60% alc

Requirements

I.D. Test[4]	Assay Min. %[5]	A.V. Max.[6]	Ref. Index[7]	Sp. Gr.[8]	Other Requirements[9]
IR	99.5% of $C_5H_8O_2$ (M-1b)	5.0		0.916–0.919	**Antioxidants**—0.022% (M-6) **Water**—0.05% (Appendix IIB, KF)
IR	97.0% of $C_{10}H_{12}O_3$ (M-1b)	1.0	1.522–1.526	1.101–1.104	
IR	96.0% of total esters as $C_9H_{11}NO_2$ (Appendix VI; 1.5g/82.6)	1.0	1.563–1.566	1.115–1.120	**Solidification Pt.**—NLT 13° (Appendix IIB)
IR	98.0% of $C_9H_{10}O_2$ (M-1b)	1.0	1.502–1.506	1.043–1.046	**Chlorinated Cmpds.**—passes test (Appendix VI)
	88.0% of $C_{11}H_{12}O_3$ (M-1b)	2.0	1.528–1.533	1.107–1.120	
	97.0% of $C_6H_{10}O_2$ (M-1b)	1.0	1.422–1.427	0.913–0.920	
IR	95.0% of $C_6H_{12}O$ (M-1b)	2.0		0.808–0.814	**Dist. Range**—NLT 95% between 100° and 120° (Appendix IIB)
IR	98.0% of $C_6H_{12}O_2$ (M-1b)	1.0	1.391–1.394	0.870–0.877	

General Information and Description

Name of Substance (Synonyms)	Formula Wt/Formula/ Structure	Physical Form/ Odor	Solubility[1]/ B.P.[2]	Solubility in Alcohol[3]
2-Ethylbutyric Acid FEMA No. 2429	116.16/$C_6H_{12}O_2$/ $CH_3CH_2\overset{\overset{\displaystyle C_2H_5}{\|}}{C}HCOOH$	colorless liq/ mildly rancid odor	*m*—alc, ether, 1 mL in 65 mL water/ 99° (18 mm Hg)	
Ethyl Cinnamate FEMA No. 2430 Ethyl 3-Phenylpropenate	176.22/$C_{11}H_{12}O_2$/ ⌬—CH=CHCOOC$_2$H$_5$	colorless, oily liq/ faint, cinnamon-like	*m*—alc, ether, most fixed oils; *ins*—gly, water/ 272°	1 mL in 5 mL 70% alc
Ethyl Decanoate FEMA No. 2432 Ethyl Caprate	200.32/$C_{12}H_{24}O_2$/ $CH_3(CH_2)_8COOC_2H_5$	colorless liq/ oily, brandy-like odor	*s*—most fixed oils; *ins*—gly, prop glycol/ 243°	1 mL in 4 mL 80% alc
2-Ethyl-3,5(6)-dimethylpyrazine FEMA No. 3149	136.20/$C_8H_{12}N_2$/ (pyrazine ring with H$_3$C, CH$_2$CH$_3$, CH$_3$ substituents)	colorless to slightly yel liq/ roasted cocoa	*s*—water, org solvents	
Ethylene Brassylate FEMA No. 3543	270.37/$C_{15}H_{26}O_4$/ $CH_2-\overset{\overset{\displaystyle O}{\|\|}}{C}-O-CH_2$ | | $(CH_2)_9$ | | | $CH_2-\underset{\underset{\displaystyle O}{\|\|}}{C}-O-CH_2$	colorless to pale yel liq/ sweet, musky	138°–142° (1 mm Hg)	1 mL in 1 mL 95% alc
2-Ethyl Fenchol FEMA No. 3491	182.31/$C_{12}H_{22}O$/ (bicyclic structure with CH$_3$, OH, CH$_2$CH$_3$, CH$_3$, CH$_3$)	pale yel liq/ sharp, camphoraceous, earthy character	*s*—alc, prop glycol, most fixed oils; *ins*—water	
Ethyl Formate FEMA No. 2434	74.08/$C_3H_6O_2$/ $HCOOC_2H_5$	colorless, flammable liq/ sharp, rum-like	*s*—most fixed oils, prop glycol, water (decomp); *ss*—min oil; *ins*—gly/ 54°	1 mL in 5 mL 50% alc

Requirements

I.D. Test[4]	Assay Min. %[5]	A.V. Max.[6]	Ref. Index[7]	Sp. Gr.[8]	Other Requirements[9]
IR	98.0% of $C_6H_{12}O_2$ (M-3b)			0.917–0.922	**Dist. Range**—between 190° and 200° (Appendix IIB) **Water**—0.2% (Appendix IIB, KF)
IR	98.0% of $C_{11}H_{12}O_2$ (M-1b)	1.0	1.558–1.561	1.045–1.051	
IR	98.0% of $C_{12}H_{24}O_2$ (M-1b)	1.0	1.424–1.427	0.863–0.868	
IR	95.0% of $C_8H_{12}N_2$ (M-1a)		1.500–1.503	0.950–0.980 (20°)	**Water**—0.1% (Appendix IIB, KF; use freshly dist. pyridine as solvent)
IR	97.0% of $C_{15}H_{26}O_4$ (M-1b)	1.0	1.468–1.473	1.040–1.045	
IR	95.0% of $C_{12}H_{22}O$ (M-1a)		1.470–1.491	0.946–0.967	
IR	95.0% of $C_3H_6O_2$ (M-1b)		1.359–1.363	0.916–0.921	**Acidity**—0.2% (M-5)

General Information and Description

Name of Substance (Synonyms)	Formula Wt/Formula/ Structure	Physical Form/ Odor	Solubility[1]/ B.P.[2]	Solubility in Alcohol[3]
4-Ethyl Guaiacol FEMA No. 2436 4-Hydroxy-3-methylethylbenzene	152.19/$C_9H_{12}O_2$/ OH ⌬—OCH₃ CH₂CH₃	colorless to pale yel liq/ warm, spicy, medicinal	235°	1 mL in 1 mL 95% alc
Ethyl Heptanoate FEMA No. 2437 Ethyl Heptoate	158.24/$C_9H_{18}O_2$/ $CH_3(CH_2)_5COOC_2H_5$	colorless liq/ winy-brandy	ss—prop glycol; m—alc, chloroform, most fixed oils; ins—gly/ 189° (72% water azeotrope, 98.5°)	1 mL in 3 mL 70% alc
Ethyl Hexanoate FEMA No. 2439 Ethyl Caproate; Ethyl Capronate	144.21/$C_8H_{16}O_2$/ $CH_3(CH_2)_4COOC_2H_5$	colorless liq/ winy	s—most fixed oils; ss—prop glycol; ins—gly/ 166°	1 mL in 2 mL 70% alc
2-Ethyl Hexanol FEMA No. 3151 2-Ethyl-1-hexanol	130.23/$C_8H_{18}O$/ $CH_3(CH_2)_3CH(C_2H_5)CH_2OH$	colorless to pale yel liq	183°	
Ethyl Isobutyrate FEMA No. 2428	116.16/$C_6H_{12}O_2$/ $(CH_3)_2CHCOOC_2H_5$	colorless liq/ fruity	112°–113°	1 mL in 1 mL 95% alc
Ethyl Isovalerate FEMA No. 2463 Ethyl 3-Methylbutyrate	130.19/$C_7H_{14}O_2$/ $(CH_3)_2CHCH_2COOC_2H_5$	colorless liq/ strong, fruity, vinous, apple-like on dilution	s—prop glycol, 1 mL in 350 mL water; m—alc, most fixed oils/ 135°	
Ethyl Lactate FEMA No. 2440 Ethyl 2-Hydroxypropionate	118.13/$C_5H_{10}O_3$/ $CH_3CHOHCOOC_2H_5$	colorless liq/ characteristic odor	vs—alc, ether, chloroform, water/ 154°	

Requirements

I.D. Test[4]	Assay Min. %[5]	A.V. Max.[6]	Ref. Index[7]	Sp. Gr.[8]	Other Requirements[9]
IR	98.0% of $C_9H_{12}O_2$ (M-1a)		1.525–1.530	1.061–1.064	
IR	98.0% of $C_9H_{18}O_2$ (M-1b)	1.0	1.411–1.415	0.867–0.872	
IR	98.0% of $C_8H_{16}O_2$ (M-1b)	1.0	1.406–1.409	0.867–0.871	
	97.0% of $C_8H_{18}O$ (M-1b)		1.429–1.434	0.830–0.834	
IR	98.0% of $C_6H_{12}O_2$ (M-1b)	1.0	1.385–1.391	0.862–0.868	
IR	98.0% of $C_7H_{14}O_2$ (M-1b)	2.0	1.395–1.399	0.862–0.866	
IR	98.0% of $C_5H_{10}O_3$ (M-1b)	1.0	1.410–1.420	1.029–1.032	

General Information and Description

Name of Substance (Synonyms)	Formula Wt/Formula/ Structure	Physical Form/ Odor	Solubility[1]/ B.P.[2]	Solubility in Alcohol[3]
Ethyl Laurate FEMA No. 2441 Ethyl Dodecanoate	228.38/$C_{14}H_{28}O_2$/ $CH_3(CH_2)_{10}COOC_2H_5$	colorless, oily liq/ fruity-floral	m—alc, chloroform, ether; ins—water/ 269°	1 mL in 9 mL 80% alc gives clear soln
Ethyl Levulinate FEMA No. 2442	144.17/$C_7H_{12}O_3$/ $CH_3COCH_2CH_2COOC_2H_5$	colorless to pale yel liq	93°–94°	1 mL in 1 mL 95% alc
Ethyl 2-Methylbutyrate FEMA No. 2443	130.19/$C_7H_{14}O_2$/ $CH_3CH_2OOCCHCH_2CH_3$ $\quad\quad\quad\quad\quad\; | \;$ $\quad\quad\quad\quad\quad CH_3$	colorless liq/ strong, green-fruity, apple-like	s—alc, prop glycol; vss—water; m—most fixed oils/ 133°	
Ethyl 2-Methylpentanoate FEMA No. 3488	144.21/$C_8H_{16}O_2$/ $CH_3CH_2CH_2CH(CH_3)CO_2CH_2CH_3$	colorless to pale yel liq		
Ethyl Methylphenylglycidate FEMA No. 2444 Aldehyde C-16; Strawberry Aldehyde	206.24/$C_{12}H_{14}O_3$/ Ph—C(CH$_3$)(—O—)CHCOOCH$_2$CH$_3$ (epoxide)	colorless to pale yel liq/ strong, fruity, strawberry-like	s—most fixed oils, prop glycol; ins—gly	1 mL in 3 mL 70% alc
2-Ethyl-3-methylpyrazine FEMA No. 3155	122.17/$C_7H_{10}N_2$/ pyrazine with CH$_2$CH$_3$ and CH$_3$	colorless to slightly yel liq/ strong, raw potato	s—water, org solvents/ 57° (10 mm Hg)	
Ethyl 3-Methylthiopropionate FEMA No. 3343	148.23/$C_6H_{12}O_2S$/ $CH_3SCH_2CH_2COOCH_2CH_3$	colorless to pale yel liq/ onion-like, fruity, sweet	89°–91° (15 mm Hg)	1 mL in 1 mL 95% alc
Ethyl Myristate FEMA No. 2445	256.43/$C_{16}H_{32}O_2$/ $CH_3(CH_2)_{12}COOC_2H_5$	colorless to pale yel liq/ waxy	178°–180° (12 mm Hg)	1 mL in 1 mL 95% alc

Requirements

I.D. Test[4]	Assay Min. %[5]	A.V. Max.[6]	Ref. Index[7]	Sp. Gr.[8]	Other Requirements[9]
IR	98.0% of $C_{14}H_{28}O_2$ (M-1b)	1.0	1.430–1.434	0.858–0.863	
IR	98.0% of $C_7H_{12}O_3$ (M-1b)	2.0	1.420–1.425	1.009–1.014	
	95.0% of $C_7H_{14}O_2$ (M-1b)	2.0	1.393–1.400	0.863–0.870	
	98.0% of $C_8H_{16}O_2$ (M-1b)	1.0	1.401–1.404	0.859–0.865	
IR	98.0% of $C_{12}H_{14}O_3$ (M-1b)	2.0	1.504–1.513	1.086–1.096	
IR	98.0% of $C_7H_{10}N_2$ (M-1a)		1.502–1.505	0.980–0.999 (20°)	**Water**—0.1% (Appendix IIB, KF; use freshly dist. pyridine as solvent)
IR	99.0% of $C_6H_{12}O_2S$ (M-3)		1.457–1.463	1.030–1.035	
IR	98.0% of $C_{16}H_{32}O_2$ (M-1b)	1.0	1.434–1.438	0.857–0.862	

General Information and Description

Name of Substance (Synonyms)	Formula Wt/Formula/ Structure	Physical Form/ Odor	Solubility[1]/ B.P.[2]	Solubility in Alcohol[3]
Ethyl Nonanoate FEMA No. 2447 Ethyl Pelargonate	186.29/$C_{11}H_{22}O_2$/ $CH_3(CH_2)_7COOC_2H_5$	colorless liq/ fatty, fruity, cognac-like	m—alc, prop glycol; ins—water, 1 mL in 10 mL 70% alc/ 229°	
Ethyl Octanoate FEMA No. 2449 Ethyl Caprylate; Ethyl Octoate	172.27/$C_{10}H_{20}O_2$/ $CH_3(CH_2)_6COOC_2H_5$	colorless liq/ winy-brandy, fruity-floral	s—most fixed oils; ss—prop glycol; ins—gly, water/ 209°	1 mL in 4 mL 70% alc
Ethyl Oleate FEMA No. 2450 Ethyl 9-Octadecenoate	310.52/$C_{20}H_{38}O_2$/ $CH_3(CH_2)_7CH=CH(CH_2)_7COOCH_2CH_3$	colorless to pale yel liq/ floral	205°–208°	
Ethyl Oxyhydrate (so-called) FEMA No. 2996 Rum Ether, So-Called		colorless liq/ sharp rum-like	m—alc, gly, prop glycol	
Ethyl Phenylacetate FEMA No. 2452	164.20/$C_{10}H_{12}O_2$/ ⟨⟩—$CH_2COOC_2H_5$	colorless or nearly colorless liq/ sweet, honey-like	s—alc, most fixed oils; ins—gly, prop glycol, water/ 228°	1 mL in 3 mL 70% alc
Ethyl Phenylglycidate FEMA No. 2454	192.21/$C_{11}H_{12}O_3$/ ⟨⟩—CH—CHCOOCH$_2$CH$_3$ $\quad\quad$ \O/	colorless to slightly yel liq/ strong, strawberry-like	s—alc, chloroform, ether; ins—water	1 mL in 6 mL 70% alc, and in 1 mL 80% alc gives clear solns
Ethyl Propionate FEMA No. 2456	102.13/$C_5H_{10}O_2$/ $CH_3CH_2COOC_2H_5$	colorless liq/ fruity, rum-like, ethereal	s—most fixed oils, 1 mL in 42 mL water; m—alc, prop glycol/ 99°	

Requirements

I.D. Test[4]	Assay Min. %[5]	A.V. Max.[6]	Ref. Index[7]	Sp. Gr.[8]	Other Requirements[9]
IR	98.0% of $C_{11}H_{22}O_2$ (M-1b)	3.0	1.420–1.424	0.863–0.867	
IR	98.0% of $C_{10}H_{20}O_2$ (M-1b)	1.0	1.417–1.419	0.865–0.868	
		1.0	1.448–1.453	0.868–0.873	**Saponification Value**—between 175 and 190 (Appendix VI)
					Alcohol Content—min 14.0% by vol, at 15.56° (M-4)
					Ester Value—min 25 (Appendix VI, 1- to 3-g sample)
IR	98.0% of $C_{10}H_{12}O_2$ (M-1b)	1.0	1.496–1.500	1.027–1.032	**Chlorinated Cmpds.**—passes test (Appendix VI)
IR	98.0% of $C_{11}H_{12}O_3$ (M-1b)		1.516–1.521	1.120–1.125	
IR	97.0% of $C_5H_{10}O_2$ (M-1b)	2.0	1.383–1.385	0.886–0.889	

General Information and Description

Name of Substance (Synonyms)	Formula Wt/Formula/ Structure	Physical Form/ Odor	Solubility[1]/ B.P.[2]	Solubility in Alcohol[3]
Ethyl Salicylate FEMA No. 2458	166.18/$C_9H_{10}O_3$/ (structure: benzene ring with COOC$_2$H$_5$ and OH)	colorless liq/ wintergreen-like	s—alc, acetic acid, most fixed oils; ss—gly, water/ 234°	1 mL in 4 mL 80% alc gives clear soln
Ethyl 10-Undecanoate FEMA No. 2461	212.33/$C_{13}H_{24}O_2$/ $H_2C=CH(CH_2)_8CO_2C_2H_5$	colorless to pale yel liq	258°–259° (761 mm Hg)	
Ethyl Valerate FEMA No. 2462 Ethyl *n*-Pentanoate	130.19/$C_7H_{14}O_2$/ $CH_3(CH_2)_3COOCH_2CH_3$	colorless to pale yel liq	145°	
Ethyl Vanillin FEMA No. 2464 3-Ethoxy-4-hydroxybenzaldehyde	166.18/$C_9H_{10}O_3$/ (structure: benzene with CHO, OC$_2$H$_5$, OH)	fine white or slightly yellowish cryst; affected by strong light; m.p. 78°/ strong, vanilla-like	s—alc, chloroform, ether, prop glycol solns of alkali hydroxides, 1 g in 100 mL water at 50°	
Eucalyptol FEMA No. 2465 1,8-Cineol; Anhydride of Menthane 1,8 Diole; 1:8 Oxido-*p*-menthane; 1,8 Epoxy-*p*-menthane	154.25/$C_{10}H_{18}O$/ (bicyclic structure with CH$_3$ groups)	colorless liq/ characteristic odor; pungent, cooling taste	s—alc, most fixed oils, gly, prop glycol/ 176°	1 mL in 5 mL 60% alc
Eugenol FEMA No. 2467 4-Allyl-2-methoxyphenol; Eugenic Acid; 4-Allylguaiacol	164.20/$C_{10}H_{12}O_2$/ (structure: benzene with OH, OCH$_3$, CH$_2$CH=CH$_2$)	colorless to pale yel liq with a pungent, spicy taste; darkens and thickens on exposure to air/ strong aromatic odor of clove	ss—water; m—alc, chloroform, ether, fixed oils/ 256°	1 mL in 2 mL 70% alc

Requirements

I.D. Test[4]	Assay Min. %[5]	A.V. Max.[6]	Ref. Index[7]	Sp. Gr.[8]	Other Requirements[9]
IR	99.0% of $C_9H_{10}O_3$ (M-1b)	1.0 (phenol red TS)	1.520–1.523	1.127–1.129	
	98.0% of $C_{13}H_{24}O_2$ (M-1b)	1.0	1.436–1.440	0.877–0.879	
	98.0% of $C_7H_{14}O_2$ (M-1b)	1.0	1.399–1.404	0.870–0.875	
IR	98.0% of $C_9H_{10}O_3$ (M-1b)				**Heavy Metals**—10 mg/kg (M-7) **Melting Range**—between 76° and 78° (Appendix IIB, dry over P_2O_5/4h) **Loss on Drying**—0.5% (Appendix IIC, P_2O_5/4h) **Res. on Ignit.**—0.05% (Appendix IIC, 2-g sample)
IR			1.455–1.460	0.921–0.924	**Angular Rotation**— between -0.5° and +0.5° (Appendix IIB, 100-mm tube) **Solidification Pt.**—NLT 0° (Appendix IIB)
IR	98.0% of $C_{10}H_{12}O_2$ (M-1b)		1.540–1.542	1.064–1.070	**Hydrocarbons**—passes test (M-8)

General Information and Description

Name of Substance (Synonyms)	Formula Wt/Formula/ Structure	Physical Form/ Odor	Solubility[1]/ B.P.[2]	Solubility in Alcohol[3]
Eugenyl Acetate FEMA No. 2469 4-Allyl-2-methoxy-phenyl Acetate; Eugenol Acetate; Acetyl Eugenol; Aceteugenol	206.24/$C_{12}H_{14}O_3$/	fused solid, melts at warm room temp to a pale yel liq/ mild, clove-like	s—alc, ether; ins—water	1 mL in 5 mL 70% alc
Farnesol FEMA No. 2478 3,7,11-Trimethyl-2,6,10-dodecatrien-1-ol	222.37/$C_{15}H_{26}O$/	slightly yel liq/ mild, oily odor	ins—water/ 263°	
Furfural FEMA No. 2489 2-Furaldehyde; Pyromucic Aldehyde	96.09/$C_5H_4O_2$/	colorless to yel oily liq, turns reddish brown on long storage/ typical of cyclic aldehydes	s—water; m—alc	
Fusel Oil, Refined FEMA No. 2497		colorless to pale yel liq/ no odor	128°–130°	1 mL in 1 mL 95% alc
Geraniol FEMA No. 2507 trans-3,7-Dimethyl-2,6-octadien-1-ol; E-3,7-Dimethyl-2,6-octadien-1-ol	154.25/$C_{10}H_{18}O$/	colorless liq/ rose-like	s—most fixed oils, prop glycol; ss—water; ins—gly/ 230°	1 mL in 3 mL 70% alc remains in soln to 10 mL
Geranyl Acetate FEMA No. 2509 3,7-Dimethyl-2,6-octadien-1-yl Acetate	196.29/$C_{12}H_{20}O_2$/	colorless liq/ floral	s—alc, most fixed oils; ss—prop glycol; ins—gly, water/ 245°	1 mL in 9 mL 70 % alc
Geranyl Benzoate FEMA No. 2511 3,7-Dimethyl-2,6-octadien-1-yl Benzoate	258.36/$C_{17}H_{22}O_2$/	slightly yellowish liq/ floral, resembling ylang ylang oil	m—alc, chloroform; ins—water/ 305°	1 mL in 4 mL 90% alc gives clear soln

Requirements

I.D. Test[4]	Assay Min. %[5]	A.V. Max.[6]	Ref. Index[7]	Sp. Gr.[8]	Other Requirements[9]
IR	98.0% of $C_{12}H_{14}O_3$ (M-1b)	1.0 (phenol red TS)		1.077–1.082 (melted, supercooled liq)	**Solidification Pt.**—NLT 25° (Appendix IIB)
IR	NLT 96.0% of $C_{15}H_{26}O$ (sum of isomers) (M-1a)		1.487–1.492	0.884–0.889 (20°)	
IR	96.0% of $C_5H_4O_2$ (M-1b)	1.0	1.522–1.528	1.154–1.158	
IR	95.0% of 2 & 3 methyl butanol (M-1a)		1.405–1.410	0.807–0.813	**Angular Rotation**— between -0.5° and -2.0° (Appendix IIB)
IR	88.0% of total alcohols as $C_{10}H_{18}O$ (Appendix VI; 1.2g/77.13)		1.469–1.478	0.870–0.885	**Aldehydes**—1.0% as citronellal (M-2d; 5g/77.13) **Esters**—1.0% as geranyl acetate (Appendix VI; 5g/98.15)
IR	90.0% of total esters as $C_{12}H_{20}O_2$ (Appendix VI; 1.0g/98.15)		1.458–1.464	0.900–0.914	
IR	95.0% of total esters as $C_{17}H_{22}O_2$ (Appendix VI; 1.5g/129.2)	1.0	1.513–1.518	0.978–0.984	

General Information and Description

Name of Substance (Synonyms)	Formula Wt/Formula/ Structure	Physical Form/ Odor	Solubility[1]/ B.P.[2]	Solubility in Alcohol[3]
Geranyl Butyrate FEMA No. 2512 3,7-Dimethyl-2,6-octadien-1-yl Butyrate	224.34/$C_{14}H_{24}O_2$/	colorless to pale yel liq/ fruity, rose-like	s—alc, most fixed oils; ins—gly, prop glycol, water/ 253°	1 mL in 6 mL 80% alc
Geranyl Formate FEMA No. 2514 3,7-Dimethyl-2,6-octadien-1-yl Formate	182.26/$C_{11}H_{18}O_2$/	colorless to pale yel liq/ fresh, leafy, rose-like	s—alc, most fixed oils; ins—gly, prop glycol, water/ 216°	1 mL in 3 mL 80% alc
Geranyl Phenylacetate FEMA No. 2516 3,7-Dimethyl-2,6-octadien-1-yl Phenylacetate	279.39/$C_{18}H_{24}O_2$/	yel liq/ honey-rose	m—alc, chloroform, ether; ins—water	1 mL in 4 mL 90% alc gives clear soln
Geranyl Propionate FEMA No. 2517 3,7-Dimethyl-2,6-octadien-1-yl Propionate	210.32/$C_{13}H_{22}O_2$/	colorless liq/ rosy, fruity	s—alc, most fixed oils; ins—gly, prop glycol, water/ 253°	1 mL in 4 mL 80% alc
Glyceryl Tripropanoate FEMA No. 3286 Tripropionin	260.29/$C_{12}H_{20}O_6$/	colorless to pale yel liq/ odorless with a bitter taste	175°–176° (20 mm Hg)	
(E),(E)-2,4-Heptadienal FEMA No. 3164 trans,trans-2,4-Heptadienal	110.16/$C_7H_{10}O$/	slightly yel liq/ fatty, green odor	s—alc, most fixed oils; ins—water	

Requirements

I.D. Test[4]	Assay Min. %[5]	A.V. Max.[6]	Ref. Index[7]	Sp. Gr.[8]	Other Requirements[9]
IR	92.0% of total esters as $C_{14}H_{24}O_2$ (Appendix VI; 1.0g/112.2)	1.0	1.455–1.462	0.889–0.904	
IR	85.0% of total esters as $C_{11}H_{18}O_2$ (Appendix VI; 1.0g/91.13)	3.0 (Appendix VI)	1.457–1.466	0.906–0.920	
IR	97.0% of total esters as $C_{18}H_{24}O_2$ (Appendix VI; 1.6g/136.2)	2.0	1.506–1.511	0.971–0.978	
IR	92.0% of total esters as $C_{13}H_{22}O_2$ (Appendix VI; 1.6g/105.2)	1.0	1.456–1.464	0.896–0.913	
	97.1% of $C_{12}H_{20}O_6$ (M-1b)	2.0	1.431–1.435	1.078–1.082	
IR	92.0% of $C_7H_{10}O$ (sum of isomers) (M-1a)		1.478–1.480		

General Information and Description

Name of Substance (Synonyms)	Formula Wt/Formula/ Structure	Physical Form/ Odor	Solubility[1]/ B.P.[2]	Solubility in Alcohol[3]
γ-Heptalactone FEMA No. 2539	128.17/$C_7H_{12}O_2$/ $CH_3(CH_2)_2$—(lactone ring)=O	colorless, slightly oily liq/ coconut, sweet, malty, caramel	m—alc, most fixed oils	
Heptanal FEMA No. 2540 Aldehyde C-7; Heptaldehyde	114.19/$C_7H_{14}O$/ $CH_3(CH_2)_5CHO$	colorless to slightly yel liq/ penetrating, oily odor	ss—water; m—alc, ether, fixed oils/ 153°	1 mL in 2 mL 70% alc gives clear soln
2-Heptanone FEMA No. 2544 Methyl Amyl Ketone	114.19/$C_7H_{14}O$/ $CH_3CO(CH_2)_4CH_3$	colorless, mobile liq/ fruity, spicy	m—alc, ether, 1 mL in 250 mL water/ 151°	
3-Heptanone FEMA No. 2545 Ethyl Butyl Ketone	114.19/$C_7H_{14}O$/ $CH_3(CH_2)_3COCH_2CH_3$	colorless, mobile liq/ fruity, green, fatty odor	m—alc, ether, 1 mL in 70 mL water/ 149°	
(Z)-4-Hepten-1-al FEMA No. 3289 cis-4-Hepten-1-al	112.17/$C_7H_{12}O$/ CH_3CH_2\C=C/$(CH_2)_2CHO$	slightly yel liq/ fatty, green odor	s—alc, most fixed oils; ins—water	
Heptyl Alcohol FEMA No. 2548 Enanthic Alcohol	116.20/$C_7H_{16}O$/ $CH_3(CH_2)_5CH_2OH$	colorless liq/ citrus	ss—water; m—alc, ether, most fixed oils/ 175°	1 mL in 2 mL 60% alc gives clear soln
γ-Hexalactone FEMA No. 2556 4-Hydroxyhexanoic Acid Lactone	114.14/$C_6H_{10}O_2$/ $CH_3CH_2CHCH_2CH_2C$=O (lactone)	colorless to pale yel liq/ herbaceous, sweet	220°	1 mL in 1 mL 95% alc

Requirements

I.D. Test[4]	Assay Min. %[5]	A.V. Max.[6]	Ref. Index[7]	Sp. Gr.[8]	Other Requirements[9]
IR	98.0% of $C_7H_{12}O_2$ (M-1a)		1.439–1.445	0.997–1.004 (20°)	
IR	92.0% of $C_7H_{14}O$ (M-1b)	10.0	1.412–1.420	0.814–0.819	
IR	95.0% of $C_7H_{14}O$ (M-1b)	2.0		0.814–0.819	**Dist. Range**—between 147° and 154° (Appendix IIB) **Res. on Evap.**—5 mg/100mL (M-17, 100-mL sample) **Water**—0.3% (Appendix IIB, KF; use freshly dist. pyridine as solvent)
IR	97.0% of $C_7H_{14}O$ (M-1b)	2.0		0.813–0.818	**Dist. Range**—between 143° and 151° (Appendix IIB) **Water**—0.3% (Appendix IIB, KF; use freshly dist. pyridine as solvent)
IR	98.0% of $C_7H_{12}O$ (sum of isomers) (M-1a)		1.432–1.436		
IR	97.0% of $C_7H_{16}O$ (M-1b)	1.0	1.423–1.427	0.820–0.824	**Aldehydes**—1.0% as heptanal (M-1b)
IR	98.0% of $C_6H_{10}O_2$ (M-1a)	1.0	1.437–1.442	1.020–1.025	

General Information and Description

Name of Substance (Synonyms)	Formula Wt/Formula/ Structure	Physical Form/ Odor	Solubility[1]/ B.P.[2]	Solubility in Alcohol[3]
Hexanal FEMA No. 2557 Caproic Aldehyde; Hexaldehyde; Aldehyde C-6	$100.16/C_6H_{12}O/$ $CH_3(CH_2)_4CHO$	almost colorless liq/ fatty-green, grassy odor	vss—water; m—alc, prop glycol, most fixed oils/ 131°	
Hexanoic Acid FEMA No. 2559 Caproic Acid	$116.16/C_6H_{12}O_2/$ $CH_3(CH_2)_4COOH$	colorless to very pale yel, oily liq/ cheesy, sweat-like	m—alc, most fixed oils, ether, 1 mL in 250 mL water/ 223°	
(E)-2-Hexen-1-al FEMA No. 2560 trans-2-Hexen-1-al	$98.14/C_6H_{10}O/$ $CH_3(CH_2)_2\text{C=C}(H)(CHO)$ (H)	pale yel liq/ strong, fruity-green, vegetable-like	s—alc, prop glycol, most fixed oils; vss—water/ 47° (17 mm Hg)	
(E)-2-Hexen-1-ol FEMA No. 2562 trans-2-Hexen-1-ol	$100.16/C_6H_{12}O/$ $CH_3(CH_2)_2\text{C=C}(H)(CH_2OH)$ (H)	almost colorless liq/ strong, fruity-green	s—alc, prop glycol, most fixed oils; vss—water/ 158°	
(Z)-3-Hexen-1-ol FEMA No. 2563 cis-3-Hexen-1-ol	$100.16/C_6H_{12}O/$ $CH_3CH_2\text{C=C}(H)((CH_2)_2OH)$ (H)	colorless liq/ powerful, grassy-green	s—alc, prop glycol, most fixed oils; vss—water/ 156°	
(E)-2-Hexenyl Acetate FEMA No. 2564 trans-2-Hexenyl Acetate	$142.20/C_8H_{14}O_2/$ $CH_3COOCH_2\text{C=C}(H)((CH_2)_2CH_3)$ (H)	colorless to pale yel liq/ green note	166°	1 mL in 1 mL 95% alc
(Z)-3-Hexenyl Acetate FEMA No. 3171 cis-3-Hexen-1-yl Acetate	$142.20/C_8H_{14}O_2/$ $CH_3COOCH_2\text{C=C}(H)((CH_2)_2CH_3)$ (H)	colorless to pale yel liq/ powerful green note	198°	1 mL in 1 mL 95% alc

Requirements

I.D. Test[4]	Assay Min. %[5]	A.V. Max.[6]	Ref. Index[7]	Sp. Gr.[8]	Other Requirements[9]
	97.0% of $C_6H_{12}O$ (M-1a)	10.0	1.402–1.407	0.808–0.817	
IR	98.0% of $C_6H_{12}O_2$ (M-3a)		1.415–1.418	0.923–0.928	**Solidification Pt.**—NLT -4.5° (Appendix IIB)
	92.0% of $C_6H_{10}O$ (M-1a)		1.445–1.449	0.841–0.848	
IR	95.0% of $C_6H_{12}O$ (M-1a)		1.437–1.442	0.836–0.841	
IR	98.0% as sum of (Z) and (E) isomers; min 92% (Z) (M-1a)		1.439–1.441	0.846–0.850	
IR	98.0% as sum of (Z) and (E) isomers; min 90% (E) (M-1a)		1.425–1.430	0.890–0.897	
IR	98.0% as sum of (Z) and (E) isomers; min 92% (Z) (M-1a)	1.0	1.425–1.429	0.896–0.901	

General Information and Description

Name of Substance (Synonyms)	Formula Wt/Formula/ Structure	Physical Form/ Odor	Solubility[1]/ B.P.[2]	Solubility in Alcohol[3]
(Z)-3-Hexenyl Isovalerate FEMA No. 3498 cis-3-Hexenyl Isovalerate	184.28/$C_{11}H_{20}O_2$/ CH_3CH_2 \C=C/ H H/ \\$(CH_2)_2OOCCH_2CH(CH_3)_2$	colorless liq/ sweet, apple-like	s—alc, prop glycol, most fixed oils; ins—water/ 199°	
(Z)-3-Hexenyl 2-Methylbutyrate FEMA No. 3497 cis-3-Hexenyl 2-Methylbutyrate	184.28/$C_{11}H_{20}O_2$/ CH_3CH_2 \C=C/ H H/ \\$(CH_2)_2OOCCHCHCH_3$ CH_3	almost colorless liq/ powerful, fruity, like unripe apples	s—alc, most fixed oils; ins—water	
n-Hexyl Acetate FEMA No. 2565	144.21/$C_8H_{16}O_2$/ $CH_3(CH_2)_5OOCCH_3$	colorless liq/ fruity	168°–170°	1 mL in 1 mL 95% alc
Hexyl Alcohol FEMA No. 2567 1-Hexanol; Alcohol C-6	102.18/$C_6H_{14}O$/ $CH_3(CH_2)_4CH_2OH$	colorless, mobile liq/ mild, sweet, green	m—alc, ether, 1 mL in 175 mL water/ 157°	
Hexyl-2-butenoate FEMA No. 3354	170.25/$C_{10}H_{18}O_2$/ $CH_3(CH_2)_5OOCCH=CHCH_3$	colorless liq/ fruity	s—alc, most fixed oils; ins—water, prop glycol	
α-Hexylcinnamaldehyde FEMA No. 2569	216.32/$C_{15}H_{20}O$/ ⌬—CH=C$(CH_2)_5CH_3$ CHO	pale yel liq/ jasmine-like	s—most fixed oils; ins—gly, prop glycol/ 174° (15 mm Hg)	1 mL in 1 mL 90% alc
Hexyl Isovalerate FEMA No. 3500	186.29/$C_{11}H_{22}O_2$/ $(CH_3)_2CHCH_2COOCH_2(CH_2)_4CH_3$	colorless liq/ pungent, fruity	s—alc, most fixed oils; ins—water/ 215°	
Hexyl 2-Methylbutyrate FEMA No. 3499	186.29/$C_{11}H_{22}O_2$/ $CH_3(CH_2)_5OOCCHCH_2CH_3$ CH_3	colorless liq/ strong, fresh-green, fruity	s—alc, most fixed oils; ins—water	

Requirements

I.D. Test[4]	Assay Min. %[5]	A.V. Max.[6]	Ref. Index[7]	Sp. Gr.[8]	Other Requirements[9]
	95.0% of $C_{11}H_{20}O_2$ (M-1a)	2.0	1.429–1.435	0.876–0.874	
	90.0% of $C_{11}H_{20}O_2$ (M-1a)	2.0	1.430–1.434	0.876–0.880	
IR	98.0% of $C_8H_{16}O_2$ (M-1b)	1.0	1.407–1.411	0.868–0.872	
IR	96.5% of $C_6H_{14}O$ (M-1b)	2.0		0.816–0.821	
IR	95.0% of $C_{10}H_{18}O_2$ (M-1a)		1.428–1.449	0.880–0.900	
IR	95.0% of $C_{15}H_{20}O$ (M-1b)	5.0	1.548–1.552	0.953–0.959	**Chlorinated Cmpds.**—passes test (Appendix VI)
	95.0% of $C_{11}H_{22}O_2$ (M-1b)	2.0	1.417–1.421	0.853–0.857	
	95.0% of $C_{11}H_{22}O_2$ (M-1a)	2.0	1.416–1.421	0.854–0.859	

General Information and Description

Name of Substance (Synonyms)	Formula Wt/Formula/ Structure	Physical Form/ Odor	Solubility[1]/ B.P.[2]	Solubility in Alcohol[3]
Hydroxycitronellal FEMA No. 2583 7-Hydroxy-3,7-dimethyl Octanal	172.27/$C_{10}H_{20}O_2$/	colorless liq/ sweet, floral, lily-like	s—most fixed oils, prop glycol; ins—gly/ 241°	1 mL in 1 mL 50% alc
Hydroxycitronellal Dimethyl Acetal FEMA No. 2585 7-Hydroxy-3,7-dimethyl Octanal: Acetal	218.34/$C_{12}H_{26}O_3$/	colorless liq/ floral	s—most fixed oils, prop glycol; ins—gly/ 252°	1 mL in 2 mL 50% alc
4-Hydroxy-2,5-dimethyl-3(2H)-furanone FEMA No. 3174	128.13/$C_6H_8O_3$/	white to pale yel solid/ fruity, caramel, burnt sugar		1 g in 1 mL 95% alc
6-Hydroxy-3,7-dimethyloctanoic Acid Lactone FEMA No. 3355	170.25/$C_{10}H_{18}O_2$/	colorless, low-melting solid/ maple syrup or brown sugar	vs—water; s—alc	
4-(p-Hydroxyphenyl)-2-butanone FEMA No. 2588	164.20/$C_{10}H_{12}O_2$/	white solid/ raspberry		1 g in 1 mL 98% alc
Indole FEMA No. 2593	117.15/C_8H_7N/	white, lustrous, flaky, cryst solid/ unpleasant odor in high conc, but free from fecal quality; odor becomes floral in higher dilutions	s—alc, most fixed oils, prop glycol; ins—gly	1 g in 3 mL 70% alc

Requirements

I.D. Test[4]	Assay Min. %[5]	A.V. Max.[6]	Ref. Index[7]	Sp. Gr.[8]	Other Requirements[9]
IR	95.0% of $C_{10}H_{20}O_2$ (M-1b)	5.0	1.447–1.450	0.918–0.923	
IR	95.0% of $C_{12}H_{26}O_3$ (M-1b)	1.0	1.441–1.444	0.925–0.930	**Free Hydroxy Citronellal**—3.0% (M-1b)
IR	98.0% of $C_6H_8O_3$ in a suitable solvent (M-1a)				
	90.0% of $C_{10}H_{18}O_2$ (M-1a)		1.457–1.461	0.966–0.973	
IR	98.0% of $C_{10}H_{12}O_2$ (M-1b)				**Melting Range**—between 82° and 84° (Appendix IIB)
IR					**Solidification Pt.**—NLT 51° (Appendix IIB, dry over H_2SO_4)

General Information and Description

Name of Substance (Synonyms)	Formula Wt/Formula/ Structure	Physical Form/ Odor	Solubility[1]/ B.P.[2]	Solubility in Alcohol[3]
α-Ionone FEMA No. 2594 4(2,6,6-Trimethyl-2-cyclohexenyl)-3-butene-2-one	192.30/$C_{13}H_{20}O$/	colorless to pale yel liq/ warm, woody, violet-floral	s—alc, most fixed oils, prop glycol; ins—gly, water/ 237°	1 mL in 10 mL 60% alc
β-Ionone FEMA No. 2595 4(2,6,6-Trimethyl-1-cyclohexenyl)-3-butene-2-one	192.30/$C_{13}H_{20}O$/	colorless to pale straw-colored liq/ warm, woody, dry	s—alc, most fixed oils, prop glycol; ins—gly, water/ 239°	
Isoamyl Acetate FEMA No. 2055 Amyl Acetate; β-Methyl Butyl Acetate	130.19/$C_7H_{14}O_2$/ $CH_3COOCH_2CH_2HC(CH_3)_2$	colorless liq/ fruity, pear-like, banana-like	ss—water; m—alc, ether, ethyl acetate, most fixed oils; ins—gly, prop glycol/ 145°	1 mL in 3 mL 60% alc gives clear soln
Isoamyl Alcohol FEMA No. 2057	88.15/$C_5H_{12}O$/ $(CH_3)_2CHCH_2CH_2OH$	colorless to pale yel liq	m—alc/ 130°	
Isoamyl Benzoate FEMA No. 2058	196.26/$C_{12}H_{16}O_2$/ $CH_2OOCCH_2CH(CH_3)_2$	colorless to pale yel liq/ pungent fruit-like odor	261° (746 mm Hg)	
Isoamyl Butyrate FEMA No. 2060 Amyl Butyrate	158.24/$C_9H_{18}O_2$/ $CH_3(CH_2)_2COOCH_2CH_2CH(CH_3)_2$	colorless liq/ fruity	s—alc, most fixed oils; ins—gly, prop glycol, water/ 179°	1 mL in 4 mL 70% alc

Requirements

I.D. Test[4]	Assay Min. %[5]	A.V. Max.[6]	Ref. Index[7]	Sp. Gr.[8]	Other Requirements[9]
IR	98.0% of $C_{13}H_{20}O$ (M-1b)		1.497–1.502	0.927–0.933	
IR	97.0% of $C_{13}H_{20}O$ (M-1b)		1.517–1.522	0.940–0.947	
IR	95.0% of $C_7H_{14}O_2$ (M-1b)	1.0	1.400–1.404	0.868–0.878	
	98.0% of $C_5H_{12}O$ (sum of isomers) (M-1b)		1.405–1.410	0.807–0.813	
IR	98.0% of $C_{12}H_{16}O_2$ (sum of isomers) (M-1a)	1.0	1.492–1.496	0.986–0.992	
IR	98.0% of $C_9H_{18}O_2$ (M-1b)	1.0	1.409–1.414	0.861–0.866	

General Information and Description

Name of Substance (Synonyms)	Formula Wt/Formula/ Structure	Physical Form/ Odor	Solubility[1]/ B.P.[2]	Solubility in Alcohol[3]
Isoamyl Formate FEMA No. 2069 Amyl Formate	116.16/$C_6H_{12}O_2$/ HCOOCH$_2$CH$_2$CH(CH$_3$)$_2$	colorless liq/ plum-like	s—alc, most fixed oils, prop glycol; ss—water; ins—gly/ 124°	1 mL in 4 mL 60% alc remains in soln to 10 mL
Isoamyl Hexanoate FEMA No. 2075 Amyl Hexanoate; Isoamyl Caproate; Pentyl Hexanoate	186.29/$C_{11}H_{22}O_2$/ CH$_3$(CH$_2$)$_4$COOCH$_2$CH$_2$CH(CH$_3$)$_2$	colorless liq/ fruity	s—alc, fixed oils; ins—gly, prop glycol, water/ 222°	1 mL in 3 mL 80% alc gives clear soln
Isoamyl Isovalerate FEMA No. 2085 Amyl Valerate; Amyl Isovalerate	172.27/$C_{10}H_{20}O_2$/ (CH$_3$)$_2$CHCH$_2$COOCH$_2$CH$_2$CH(CH$_3$)$_2$	colorless liq/ fruity, apple-like	ss—prop glycol; m—alc, most fixed oils; ins—water, 1 mL in 6 mL 70% alc/ 192°	
Isoamyl Phenyl Acetate FEMA No. 2081	206.29/$C_{13}H_{18}O_2$/ ⟨benzene⟩—CH$_2$COOCH$_2$CH$_2$CH(CH$_3$)$_2$	colorless to pale yel liq	268°	1 mL in 1 mL 95% alc
Isoamyl Salicylate FEMA No. 2084 Amyl Salicylate	208.26/$C_{12}H_{16}O_3$/ ⟨benzene-OH⟩—COOCH$_2$CH$_2$CH(CH$_3$)$_2$	colorless liq/ pleasant odor	m—alc, chloroform, ether, most fixed oils; ins—gly, prop glycol, water/ 277°	1 mL in 3 mL 90% alc remains in soln on dilution
Isoborneol FEMA No. 2158	154.25/$C_{10}H_{18}O$/ ⟨bornyl structure⟩—OH	white cryst solid/ piney, camphoraceous	214°	1 g in 1 mL 95% alc

Requirements

I.D. Test[4]	Assay Min. %[5]	A.V. Max.[6]	Ref. Index[7]	Sp. Gr.[8]	Other Requirements[9]
IR	92.0% of $C_6H_{12}O_2$ (M-1b)	3.0	1.396–1.400	0.881–0.889	
IR	98.0% of $C_{11}H_{22}O_2$ (M-1b)	1.0	1.418–1.422	0.858–0.863	
	98.0% of $C_{10}H_{20}O_2$ (M-1b)	2.0	1.411–1.414	0.851–0.857	
IR	98.0% of $C_{13}H_{18}O_2$ (sum of isomers) (M-1b)	1.0	1.485–1.490	0.975–0.981	
IR	98.0% of $C_{12}H_{16}O_3$ (M-1b)	1.0 (phenol red TS)	1.505–1.509	1.047–1.053	
IR					**Melting Range**—between 212° and 214° (Appendix IIB)

General Information and Description

Name of Substance (Synonyms)	Formula Wt/Formula/ Structure	Physical Form/ Odor	Solubility[1]/ B.P.[2]	Solubility in Alcohol[3]
Isobornyl Acetate FEMA No. 2160	196.29/$C_{12}H_{20}O_2$/	colorless liq when fresh, develops very pale straw shade on storage/ camphoraceous, piney, balsamic	s—alc, most fixed oils; ss—prop glycol; ins—gly, water/ 227°	1 mL in 3 mL 70% alc
Isobutyl Acetate FEMA No. 2175	116.16/$C_6H_{12}O_2$/ $CH_3COOCH_2CH(CH_3)_2$	colorless liq/ fruity, banana-like on dilution	s—alc, most fixed oils, prop glycol, 1 mL in 180 mL water/ 116°	
Isobutyl Alcohol FEMA No. 2179	74.12/$C_4H_{10}O$/ $(CH_3)_2CHCH_2OH$	colorless, mobile liq/ penetrating, winy	m—alc, ether, 1 mL in 140 mL water/ 108°	
Isobutyl-2-butenoate FEMA No. 3432	142.20/$C_8H_{14}O_2$/ $(CH_3)_2CHCH_2OOCCH=CHCH_3$	colorless liq/ powerful, fruity	s—alc, prop glycol, most fixed oils; ss—water	
Isobutyl Butyrate FEMA No. 2187 2-Methyl Propanyl Butyrate	144.21/$C_8H_{16}O_2$/ $C_3H_7COOCH_2CH(CH_3)_2$	colorless liq/ sweet, fruity, apple-like, pineapple-like	s—alc, most fixed oils; ss—water; ins—gly/ 157°	1 mL in 8 mL 60% alc
Isobutyl Cinnamate FEMA No. 2193	204.27/$C_{13}H_{16}O_2$/	colorless liq/ sweet, fruity, balsamic	m—alc, chloroform, ether, most fixed oils; ins—water/ 271°	1 mL in 3 mL 80% alc gives clear soln
Isobutyl Phenylacetate FEMA No. 2210	192.26/$C_{12}H_{16}O_2$/	colorless liq/ rose, honey-like	s—alc, most fixed oils; ins—gly, prop glycol, water/ 247°	1 mL in 2 mL 80% alc remains in soln to 10 mL

Requirements

I.D. Test[4]	Assay Min. %[5]	A.V. Max.[6]	Ref. Index[7]	Sp. Gr.[8]	Other Requirements[9]
IR	97.0% of $C_{12}H_{20}O_2$ (M-1b)	1.0	1.462–1.465	0.979–0.984	**Angular Rotation**— between -4° and 0° (Appendix IIB, 100-mm tube)
IR	90.0% of $C_6H_{12}O_2$ (M-1b)	1.0	1.389–1.392	0.862–0.871	
IR	98.0% of $C_4H_{10}O$ (M-1a)	2.0		0.799–0.801	
IR	95.0% of $C_8H_{14}O_2$ (M-1a)		1.426–1.430	0.880–0.900	
IR	98.0% of $C_8H_{16}O_2$ (M-1b)	1.0	1.402–1.405	0.858–0.863	
IR	98.0% of $C_{13}H_{16}O_2$ (M-1b)	1.0	1.539–1.541	1.001–1.004	
IR	98.0% of $C_{12}H_{16}O_2$ (M-1b)	1.0	1.486–1.488	0.984–0.988	

General Information and Description

Name of Substance (Synonyms)	Formula Wt/Formula/ Structure	Physical Form/ Odor	Solubility[1]/ B.P.[2]	Solubility in Alcohol[3]
Isobutyl Salicylate FEMA No. 2213	194.23/$C_{11}H_{14}O_3$/ benzene ring with OH and $COOCH_2CHCH_3$ with CH_3	colorless liq/ orchid-like	s—most fixed oils; ins—gly, prop glycol/ 260°	1 mL in 9 mL 80% alc remains in soln to 10 mL
Isobutyraldehyde FEMA No. 2220	72.11/C_4H_8O/ $(CH_3)_2CHCHO$	colorless, mobile liq/ sharp, pungent	m—alc, ether, 1 mL in 125 mL water/ 64°	
Isobutyric Acid FEMA No. 2222 2-Methyl Propanoic Acid; Isopropylformic Acid	88.11/$C_4H_8O_2$/ $(CH_3)_2CHCOOH$	colorless liq/ strong, penetrating odor of rancid butter	m—alc, most fixed oils, gly, prop glycol; ins—water/ 155°	
Isoeugenol FEMA No. 2468 2-Methoxy-4-propenylphenol	164.20/$C_{10}H_{12}O_2$/ CH_3O, HO-benzene-$CH=CHCH_3$	pale yel, viscous liq/ floral, carnation-like	s—most fixed oils, ether; ins—gly/ 266°	1 mL in 5 mL 50% alc
Isoeugenyl Acetate FEMA No. 2470 2-Methoxy-4-propenyl Phenyl Acetate	206.24/$C_{12}H_{14}O_3$/ $OOCCH_3$, OCH_3 on benzene with $CH=CHCH_3$	white cryst/ spicy, clove-like	s—alc, most fixed oils, chloroform; ins—water	1 g in 27 mL 95% alc gives clear soln
Isopropyl Acetate FEMA No. 2926	102.13/$C_5H_{10}O_2$/ $CH_3COOCH(CH_3)_2$	colorless, mobile liq/ characteristic odor	m—alc, ether, fixed oils, 1 g in 72 mL water/ 88°	
Isopulegol FEMA No. 2962 p-Menth-4-en-3-ol	154.25/$C_{10}H_{18}O$/ cyclohexane with CH_3, OH, and $H_3C-C=CH_2$	colorless liq/ harsh, camphoraceous, mint-like, with rose leaf and geranium background	91° (12 mm Hg)	1 mL in 4 mL 60% alc gives clear soln

Requirements

I.D. Test[4]	Assay Min. %[5]	A.V. Max.[6]	Ref. Index[7]	Sp. Gr.[8]	Other Requirements[9]
IR	98.0% of $C_{11}H_{14}O_3$ (M-1b)	1.0 (phenol red TS)	1.507–1.510	1.062–1.066	
IR	98.0% of C_4H_8O (M-5c)	5.0		0.783–0.788	
IR	99.0–101.1% of $C_4H_8O_2$ (M-3a)		1.392–1.395	0.944–0.948	**Reducing Subs.**—passes test (M-15)
IR	99.0% of phenols by vol (M-1b)		1.572–1.577	1.079–1.085	**Solidification Pt.**—NLT 12° (Appendix IIB)
IR	98.0% of $C_{12}H_{14}O_3$ (M-1b)	2.0 (phenol red TS)			**Solidification Pt.**—NLT 76° (Appendix IIB)
IR	99.0% of $C_5H_{10}O_2$ (M-1b)	2.0		0.866–0.869	
IR	95.0% of total alcohols as $C_{10}H_{18}O$ (Appendix VI; 1.2g/77.12 with 2h reflux)	1.0	1.470–1.475	0.904–0.913	**Aldehydes**—1.0% as citronellal (M-2d; 10g/77.13) **Angular Rotation**—between 0° and -7° (Appendix IIB, 100-mm tube)

General Information and Description

Name of Substance (Synonyms)	Formula Wt/Formula/ Structure	Physical Form/ Odor	Solubility[1]/ B.P.[2]	Solubility in Alcohol[3]
Isovaleric Acid FEMA No. 3102 Isopropylacetic Acid	102.13/$C_5H_{10}O_2$/ $(CH_3)_2CHCH_2COOH$	colorless to pale yel liq/ disagreeable, rancid, cheese-like	s—alc, chloroform, ether, water/ 175°	
Lauryl Alcohol FEMA No. 2617 1-Dodecanol; Alcohol C-12	186.34/$C_{12}H_{26}O$/ $CH_3(CH_2)_{10}CH_2OH$	colorless liq above 21°/ fatty odor	s—most fixed oils, prop glycol; ins—gly, water/ 259°	1 mL in 3 mL 70% alc remains clear to 10 mL
Lauryl Aldehyde FEMA No. 2615 Aldehyde C-12; Dodecanal	184.32/$C_{12}H_{24}O$/ $CH_3(CH_2)_{10}CHO$	colorless to light yel liq/ fatty odor	s—alc, most fixed oils, prop glycol (may be turbid) ins—gly, water/ 249°	
Levulinic Acid FEMA No. 2627	116.12/$C_5H_8O_3$/ $CH_3COCH_2CH_2COOH$	yel to brown liq; may congeal	245°	1 mL in 1 mL 95% alc
d-**Limonene** FEMA No. 2633 *d-p*-Mentha-1,8-diene; Cinene	136.24/$C_{10}H_{16}$/	colorless liq/ mildly citrus odor, free from camphoraceous and terpene-like notes	ss—gly; m—alc, most fixed oils; ins—prop glycol, water/ 177°	
l-**Limonene** *l-p*-Mentha-1,8-diene	136.24/$C_{10}H_{16}$/	colorless liq/ refreshing, light, clean odor	m—alc, most fixed oils; ins—water/ 177°	
Linalool FEMA No. 2635 3,7-Dimethyl-1,6-octadien-3-ol	154.25/$C_{10}H_{18}O$/	colorless liq/ pleasant, floral odor	s—fixed oils, prop glycol; ins—gly/ 198°	1 mL in 4 mL 60% alc

Requirements

I.D. Test[4]	Assay Min. %[5]	A.V. Max.[6]	Ref. Index[7]	Sp. Gr.[8]	Other Requirements[9]
IR	99.0% of $C_5H_{10}O_2$ (M-3a)		1.401–1.405	0.923–0.928	
IR	97.0% of $C_{12}H_{26}O$ (M-1b)	1.0	1.440–1.444	0.830–0.836	**Solidification Pt.**—NLT 21° (Appendix IIB)
IR	92.0% of $C_{12}H_{24}O$ (M-1b)	10.0	1.433–1.439	0.826–0.836	
IR	97.0% of $C_5H_8O_3$ (M-3a)		1.440–1.445	1.136–1.142	**Solidification Pt.**—min 27° (Appendix IIB)
IR	93.0% of $C_{10}H_{16}$ (M-1a)		1.471–1.474	0.838–0.843	**Angular Rotation**—between +96° and +104° (Appendix IIB, 100-mm tube) **Peroxide Value**—5.0 (M-12)
	95.0% of $C_{10}H_{16}$ (M-1a)		1.469–1.473	0.837–0.841	**Angular Rotation**—min -90° (Appendix IIB, 100-mm tube) **Peroxide Value**—5.0 (M-12)
IR	92.0% of $C_{10}H_{18}O$ (M-1b)		1.461–1.465	0.858–0.867	**Esters**— 0.5% as linalylacetate (Appendix VI; 10g/98.15)

General Information and Description

Name of Substance (Synonyms)	Formula Wt/Formula/ Structure	Physical Form/ Odor	Solubility[1]/ B.P.[2]	Solubility in Alcohol[3]
Linalyl Acetate FEMA No. 2636 3,7-Dimethyl-1,6-octadien-3-yl Acetate	196.29/$C_{12}H_{20}O_2$/	colorless liq/ floral, fruity	ss—prop glycol; m—alc, fixed oils; ins—gly, water/ 220°	1 mL in 5 mL 70% alc
Linalyl Benzoate FEMA No. 2638 3,7-Dimethyl-1,6-octadien-3-yl Benzoate	258.36/$C_{17}H_{22}O_2$/	yellowish to brownish yel liq/ tuberose-like	s—alc, chloroform, ether; ins—water/ 263°	1 mL in 1 mL 90% alc gives clear soln
Linalyl Formate FEMA No. 2642 3,7-Dimethyl-1,6-octadien-3-yl Formate	182.26/$C_{11}H_{18}O_2$/	colorless liq/ fresh, citrus, green, herbaceous, bergamot-like	s—alc, most fixed oils; ss—prop glycol, water; ins—gly/ 202°	1 mL in 6 mL 70% alc
Linalyl Isobutyrate FEMA No. 2640 3,7-Dimethyl-6-octadien-3-yl Isobutyrate	224.34/$C_{14}H_{24}O_2$/	colorless to slightly yel liq/ sweet, fresh, rosy	m—alc, chloroform, ether; ins—water/ 230°	1 mL in 3 mL 80% alc gives clear soln
Linalyl Propionate FEMA No. 2645 3,7-Dimethyl-6-octadien-3-yl Propionate	210.32/$C_{13}H_{22}O_2$/	colorless or almost colorless liq/ fresh, floral, sweet, fruity, pear-like	s—alc, most fixed oils; ss—prop glycol; ins—gly/ 226°	1 mL in 2 mL 80% alc

Requirements

I.D. Test[4]	Assay Min. %[5]	A.V. Max.[6]	Ref. Index[7]	Sp. Gr.[8]	Other Requirements[9]
IR	90.0% of total esters as $C_{12}H_{20}O_2$ (M-1b)	1.0	1.449–1.457	0.895–0.914	**Angular Rotation**— between -1° and +1° (Appendix IIB, 100-mm tube)
IR	75.0% of $C_{17}H_{22}O_2$ (M-1b)	5.0	1.505–1.520	0.980–0.999	
IR	90.0% of $C_{11}H_{18}O_2$ (M-1b)	3.0	1.453–1.458	0.910–0.918	
IR	95.0% of $C_{14}H_{24}O_2$ (M-1b)	1.0	1.446–1.451	0.882–0.888	
IR	92.0% of $C_{13}H_{22}O_2$ (M-1b)	1.0	1.449–1.454	0.893–0.902	**Angular Rotation**— between -1° and +1° (Appendix IIB, 100-mm tube)

General Information and Description

Name of Substance (Synonyms)	Formula Wt/Formula/ Structure	Physical Form/ Odor	Solubility[1]/ B.P.[2]	Solubility in Alcohol[3]
Menthol FEMA No. 2665 3-*p*-Menthanol [NOTE: *l*-Menthol is obtained from natural sources or by synthetic processes; *dl*-Menthol is produced synthetically]	156.27/$C_{10}H_{20}O$/	colorless, hexagonal crysts, usually needle-like; fused masses or cryst powder; m.p. 43°/ pleasant, peppermint-like	*vs*—alc, vol oils; *ss*—water/ 212°	
***l*-Menthone** FEMA No. 2667 *l-p*-Menthan-3-one	154.25/$C_{10}H_{18}O$/	almost colorless liq/ mint-like	*s*—alc, most fixed oils; *vss*—water/ 207°	
***dl*-Menthyl Acetate** FEMA No. 2668 *dl-p*-Menthan-3-yl Acetate	198.31/$C_{12}H_{22}O_2$/	colorless liq/ mild, minty	*s*—alc, prop glycol, most fixed oils; *ss*—water, gly	
***l*-Menthyl Acetate** FEMA No. 2668 *l-p*-Menthan-3-yl Acetate	198.31/$C_{12}H_{22}O_2$/	colorless liq/ mild, minty	*s*—alc, prop glycol, most fixed oils; *ss*—water	
2-Mercaptopropionic Acid FEMA No. 3180	106.16/$C_3H_6O_2S$/ $CH_3CH(SH)COOH$	colorless to pale yel liq/ roasted, meaty odor	*m*—water, alc, ether, acetone/ 117°	1 mL in 1 mL 95% alc

Requirements

I.D. Test[4]	Assay Min. %[5]	A.V. Max.[6]	Ref. Index[7]	Sp. Gr.[8]	Other Requirements[9]
IR	95.0% of $C_{10}H_{20}O$ (sum of isomers) (M-1b)				**Heavy Metals**—0.004% (M-7) **Lead**—10 mg/kg (M-10) **Melting Range (*l*-menthol)**—between 41° and 43° (Appendix IIB) **Nonvol. Res**—0.05% (M-17) **Readily Ox. Subs. (*dl*-menthol)**— passes test (M-14) **Specific Rotation (*l*-menthol)**— between -45° and -51° (Appendix IIB) **Specific Rotation (*dl*-menthol)**— between -2° and +2° (Appendix IIB)
	96.0% of $C_{10}H_{18}O$ (M-1b)	1.0	1.448–1.453	0.888–0.895	**Angular Rotation**—min -20° (Appendix IIB, 100-mm tube)
	97.0% of $C_{12}H_{22}O_2$ (sum of isomers) (M-1b)	2.0	1.443–1.450	0.919–0.924	
	98.0% of $C_{12}H_{22}O_2$ (M-1b)	2.0	1.443–1.447	0.919–0.924	**Angular Rotation**—min -69.0° (Appendix IIB, 100-mm tube)
IR	98.0% of $C_3H_6O_2S$ (M-3a)		1.479–1.484	1.192–1.200	

General Information and Description

Name of Substance (Synonyms)	Formula Wt/Formula/ Structure	Physical Form/ Odor	Solubility[1]/ B.P.[2]	Solubility in Alcohol[3]
p-Methoxybenzaldehyde FEMA No. 2670 Anisic Aldehyde; p-Anisaldehyde	136.15/$C_8H_8O_2$/	colorless to slightly yel liq/ hawthorn-like	s—prop glycol; m—alc, ether, most fixed oils; ins—alc, water/ 248°	1 mL in 3 mL 60% alc gives clear soln
2-Methoxy-3(5)-methylpyrazine FEMA No. 3183	124.14/$C_6H_8N_2O$/	colorless liq/ roasted, like hazelnut	s—water, org solvents	
4-p-Methoxyphenyl-2-butanone FEMA No. 2672 Anisyl Acetone	178.23/$C_{11}H_{14}O_2$/	colorless to pale yel liq/ sweet, floral, fruity	277°	1 mL in 1 mL 95% alc
2-Methoxypyrazine FEMA No. 3302	110.12/$C_5H_6N_2O$/	colorless to yellowish liq/ nutty, cocoa-like	s—alc; ins—water/ 61° (29 mm Hg)	
Methyl Acetate FEMA No. 2676	74.08/$C_3H_6O_2$/ CH_3COOCH_3	colorless liq	57.5°	1 mL in 1 mL 95% alc
4-Methyl Acetophenone FEMA No. 2677 Methyl p-Tolyl Ketone	134.18/$C_9H_{10}O$/	colorless or nearly colorless liq/ fruity floral odor resembling acetophenone	s—most fixed oils, prop glycol; ins—gly/ 226°	1 mL in 10 mL 50% alc
p-Methyl Anisole FEMA No. 2681 p-Cresyl Methyl Ether; Methyl p-Cresol	122.17/$C_8H_{10}O$/	colorless liq/ ylang-ylang	s—most fixed oils; ins—gly, prop glycol/ 174°	1 mL in 3 mL 80% alc remains in soln on dilution

Requirements

I.D. Test[4]	Assay Min. %[5]	A.V. Max.[6]	Ref. Index[7]	Sp. Gr.[8]	Other Requirements[9]
IR	97.5% of $C_8H_8O_2$ (M-1b)	6.0	1.571–1.574	1.119–1.123	**Chlorinated Cmpds.**—passes test (Appendix VI)
IR	99.0% of $C_6H_8N_2O$ (sum of isomers) (M-1b)		1.506–1.510	1.060–1.090 (20°)	
IR	98.0% of $C_{11}H_{14}O_2$ (M-1a)		1.517–1.521	1.042–1.048	
IR	99.0% of $C_5H_6N_2O$ (M-1b)		1.508–1.511	1.110–1.140 (20°)	
IR	98.0% of $C_3H_6O_2$ (M-1b)	1.0	1.358–1.363	0.927–0.932	
IR	95.0% of $C_9H_{10}O$ (M-1b)		1.530–1.535	0.996–1.004	**Chlorinated Cmpds.**—passes test (Appendix VI)
IR			1.510–1.513	0.966–0.970	**Cresol**—0.5% (M-1b)

General Information and Description

Name of Substance (Synonyms)	Formula Wt/Formula/ Structure	Physical Form/ Odor	Solubility[1]/ B.P.[2]	Solubility in Alcohol[3]
Methyl Anthranilate FEMA No. 2682	151.16/$C_8H_9NO_2$/ COOCH$_3$, NH$_2$ (on benzene ring)	colorless to pale yel liq with bluish fluorescence/ grape-like	s—most fixed oils, prop glycol; ins—gly/ 256°	1 mL in 5 mL 60% alc remains in soln to 10 mL
Methyl Benzoate FEMA No. 2683	136.15/$C_8H_8O_2$/ benzene-COOCH$_3$	colorless liq/ deep, pungent, floral	s—alc, most fixed oils, prop glycol; ins—gly/ 198°	1 mL in 4 mL 60% alc
α-Methylbenzyl Acetate FEMA No. 2684 Tolyl Acetate So-Called	164.20/$C_{10}H_{12}O_2$/ (structure with CH$_3$ on ring and CH$_2$OOCCH$_3$)	colorless liq/ sweet, nutty	s—most fixed oils; ss—prop glycol; ins—gly	1 mL in 2 mL 70% alc remains clear on dilution
α-Methylbenzyl Alcohol FEMA No. 2685 Methyl Phenylcarbinol; α-Phenethyl Alcohol	122.17/$C_8H_{10}O$/ benzene-CHCH$_3$, OH	colorless liq above room temp/ mild, hyacinth-like	vs—gly; s—most fixed oils, prop glycol/ 204°	1 mL in 3 mL 50% alc
2-Methyl Butanal FEMA No. 2691	86.13/$C_5H_{10}O$/ $CH_3CH_2CH(CH_3)CHO$	colorless to pale yel liq	93°	
3-Methyl Butanal FEMA No. 2692	86.13/$C_5H_{10}O$/ $(CH_3)_2CHCH_2CHO$	colorless to pale yel liq	93°	
2-Methylbutyl Acetate FEMA No. 3644	130.18/$C_7H_{14}O_2$/ $CH_3CH_2CH(CH_3)CH_2OOCCH_3$	colorless to pale yel liq		
2-Methylbutyl Isovalerate FEMA No. 2753 2-Methylbutyl-3-methylbutanoate	172.27/$C_{10}H_{20}O_2$/ CH_3 | $CH_3CH_2CHCH_2OOCCH_2CH(CH_3)_2$	colorless liq/ herbaceous, fruity	s—alc, most fixed oils; ins—water	

Requirements

I.D. Test[4]	Assay Min. %[5]	A.V. Max.[6]	Ref. Index[7]	Sp. Gr.[8]	Other Requirements[9]
IR	98.0% of total esters as $C_8H_9NO_2$ (Appendix VI; 1.0g/75.59)		1.582–1.584 (as supercooled liq)	1.161–1.169	**Solidification Pt.**—NLT 23.8° (Appendix IIB)
IR	98.0% of $C_8H_8O_2$ (M-1b)	1.0	1.514–1.518	1.082–1.088	**Chlorinated Cmpds.**—passes test (Appendix VI)
IR	98.0% of $C_{10}H_{12}O_2$ (M-1b)	1.0	1.501–1.504	1.030–1.035	**Chlorinated Cmpds.**—passes test (Appendix VI)
IR	99.0% of $C_8H_{10}O$ (M-1b)		1.525–1.529	1.009–1.014	**Ketones**—1.0% as acetophenone (M-2d; 10.0g/60.07) **Solidification Pt.**—NLT 19° (Appendix IIB)
	97.0% of $C_5H_{10}O$ (M-1b)	10.0	1.388–1.393	0.799–0.804	
	97.0% of $C_5H_{10}O$ (M-1b)	10.0	1.388–1.391	0.795–0.802	
	97.0% of $C_7H_{14}O_2$ (M-1b)	1.0	1.399–1.404	0.872–0.877	
	98.0% of $C_{10}H_{20}O_2$ (M-1a)	2.0	1.413–1.416	0.852–0.857	

General Information and Description

Name of Substance (Synonyms)	Formula Wt/Formula/ Structure	Physical Form/ Odor	Solubility[1]/ B.P.[2]	Solubility in Alcohol[3]
Methyl Butyrate FEMA No. 2693	102.13/$C_5H_{10}O_2$/ $CH_3(CH_2)_2COOCH_3$	colorless liq	102°	
2-Methylbutyric Acid FEMA No. 2695	102.13/$C_5H_{10}O_2$/ $CH_3CH_2\underset{\underset{CH_3}{\mid}}{CH}COOH$	colorless to pale yel liq	176°	1 mL in 1 mL 95% alc
α-Methylcinnamaldehyde FEMA No. 2697	146.19/$C_{10}H_{10}O$/ Ph—CH=C(CH₃)CHO	yel liq/ cinnamon-like	s—most fixed oils, prop glycol; ins—gly/ 148° (27 mm Hg)	1 mL in 3mL 70% alc remains clear on dilution
Methyl Cinnamate FEMA No. 2698	162.19/$C_{10}H_{10}O_2$/ Ph—CH=CHCOOCH₃	white to slightly yel cryst mass/ fruity, balsamic	s—alc, most fixed oils, gly, prop glycol; ins—water/ 260°	1 mL in 4 mL 80% alc
6-Methylcoumarin FEMA No. 2699	160.17/$C_{10}H_8O_2$/ H₃C-coumarin structure	white cryst solid/ coconut-like	303° (725 mm Hg)	1 g in 20 mL 95% alc
Methyl Cyclopentenolone FEMA No. 2700 3-Methylcyclopentane-1,2-dione	112.13/$C_6H_8O_2$/ cyclopentanedione with CH₃	white, cryst powder/ nutty odor, maple-licorice aroma in dilute soln	s—alc, prop glycol; ss—most fixed oils, 1 g in 72 mL water	1 g in 5 mL 90% alc
Methyl Eugenol FEMA No. 2475 Eugenyl Methyl Ether; 1,2-Dimethoxy-4-allylbenzene	178.23/$C_{11}H_{14}O_2$/ dimethoxyallylbenzene	colorless to pale yel liq/ delicate, clove-carnation	s—most fixed oils; ins—gly, prop glycol/ 249°	1 mL in 2 mL 70% alc remains clear to 10 mL
6-Methyl-5-hepten-2-one FEMA No. 2707 Methyl Heptenone	126.20/$C_8H_{14}O$/ $CH_3\underset{\underset{CH_3}{\mid}}{C}=CHCH_2CH_2COCH_3$	slightly yel liq/ sharp, citrus-lemongrass	m—alc, most fixed oils, ether; ins—water/ 73° (18 mm Hg)	1 mL in 2 mL 70% alc gives clear soln

Requirements

I.D. Test[4]	Assay Min. %[5]	A.V. Max.[6]	Ref. Index[7]	Sp. Gr.[8]	Other Requirements[9]
	98.0% of $C_5H_{10}O_2$ (M-1b)	1.0	1.386–1.390	0.892–0.897	
IR	98.0% of $C_5H_{10}O_2$ (M-3a)		1.404–1.408	0.932–0.936	
IR	97.0% of $C_{10}H_{10}O$ (M-1b)	5.0	1.602–1.607	1.035–1.039	
IR	98.0% of $C_{10}H_{10}O_2$ (M-1b)	2.0			**Chlorinated Cmpds.**—passes test (Appendix VI) **Heavy Metals**—0.004% (M-7) **Lead**—10 mg/kg (M-10)
IR	99.0% of $C_{10}H_8O_2$ (M-1a)				**Melting Range**—between 73° and 76° (Appendix IIB)
					Heavy Metals—0.004% (M-7) **Melting Range**—between 104° and 108° (Appendix IIB)
IR	98.0% of $C_{11}H_{14}O_2$ (M-1b)		1.532–1.536	1.032–1.036	**Eugenol**—1.0% (M-1b)
IR	98.0% of $C_8H_{14}O$ (M-1b)		1.438–1.442	0.846–0.851	

General Information and Description

Name of Substance (Synonyms)	Formula Wt/Formula/ Structure	Physical Form/ Odor	Solubility[1]/ B.P.[2]	Solubility in Alcohol[3]
Methyl Hexyl Ketone FEMA No. 2802 2-Octanone	128.21/$C_8H_{16}O$/ $CH_3(CH_2)_5COCH_3$	colorless to pale yel liq/ apple-like	175°	1 mL in 1 mL 95% alc
Methyl Ionones Mixture of α-, β-, γ- or α-iso, and δ- isomers	206.3/$C_{14}H_{22}O$/	clear to pale yel to yel liq		
Methyl Isobutyrate FEMA No. 2694	102.13/$C_5H_{10}O_2$/ $CH_3COOCH(CH_3)_2$	colorless liq	90°	1 mL in 1 mL 95% alc
Methyl Isoeugenol FEMA No. 2476 4-Allyl-1,2-dimethoxy Benzene; Isoeugenyl Methyl Ether; 4-Propenyl Veratrole	178.23/$C_{11}H_{14}O_2$/ $CH_3O-\text{(ring)}-CH=CHCH_3$ CH_3O	colorless to pale yel liq/ delicate, clove carnation	s—most fixed oils; ins—gly, prop glycol/ 270°	1 mL in 2 mL 70% alc remains in soln to 10 mL
5-Methyl-2-isopropyl-2-hexenal FEMA No. 3406	154.25/$C_{10}H_{18}O$/ CH_3 $CH_3CHCH_2CH=CCHO$ $CH(CH_3)_2$	slightly yel liq/ herbaceous, woody, fruity, chocolate	s—alc, most fixed oils; ins—water, prop glycol	
Methyl 2-Methylbutyrate FEMA No. 2719 Methyl 2-Methyl-butanoate	116.16/$C_6H_{12}O_2$/ $CH_3OOCCHCH_2CH_3$ CH_3	almost colorless liq/ sweet, fruity, apple-like	s—alc, most fixed oils; ins—water/ 115°	
Methyl-3-methylthiopropionate FEMA No. 2720	134.19/$C_5H_{10}O_2S$/ $CH_3SCH_2CH_2CO_2CH_3$	colorless to pale yel liq/ onion-like	74°–78°	
Methyl β-Naphthyl Ketone FEMA No. 2723 2-Acetonaphthone	170.21/$C_{12}H_{10}O$/ (naphthyl)-C($=O$)-CH_3	white or nearly white cryst solid/ orange blossom	s—most fixed oils; ss—prop glycol; ins—gly/ 300°	1 g in 5 mL 95% alc

Requirements

I.D. Test[4]	Assay Min. %[5]	A.V. Max.[6]	Ref. Index[7]	Sp. Gr.[8]	Other Requirements[9]
IR	95.0% of $C_8H_{16}O$ (M-1a)	1.0	1.414–1.418	0.813–0.818	
	88.0% of $C_{14}H_{22}O$ (M-1b)	5.0	1.497–1.507	0.925–0.934	
IR	97.0% of $C_5H_{10}O_2$ (M-1b)	1.0	1.382–1.386	0.884–0.888	
IR			1.566–1.569	1.047–1.053	**Isoeugenol**—1.0% (M-1b)
IR	90.0% of $C_{10}H_{18}O$ (sum of isomers) (M-1a)		1.448–1.453	0.845–0.860	
	92.0% of $C_6H_{12}O_2$ (M-1b)	2.0	1.393–1.397	0.879–0.883	
IR	97.0% of $C_5H_{10}O_2S$ (M-1a)	1.0	1.462–1.468	1.069–1.078	
IR	99.0% of $C_{12}H_{10}O$ (M-1b)				**Solidification Pt.**—NLT 53° (Appendix IIB)

General Information and Description

Name of Substance (Synonyms)	Formula Wt/Formula/ Structure	Physical Form/ Odor	Solubility[1]/ B.P.[2]	Solubility in Alcohol[3]
Methyl 2-Octynoate FEMA No. 2729 Methyl Heptine Carbonate	154.21/$C_9H_{14}O_2$/ $CH_3(CH_2)_4C{\equiv}CCOOCH_3$	colorless to slightly yel liq/ powerful, unpleasant, violet-like when diluted	s—most fixed oils; ss—prop glycol; ins—gly/ 217°	1 mL in 5 mL 70% alc
2-Methylpentanoic Acid FEMA No. 2754	116.16/$C_6H_{12}O_2$/ $CH_3CH_2CH_2\overset{\underset{\mid}{CH_3}}{C}HCOOH$	colorless to pale yel liq/ caramel, pungent	196°–197°	1 mL in 1 mL 95% alc
4-Methylpentanoic Acid FEMA No. 3463	116.16/$C_6H_{12}O_2$/ $CH_3\overset{\underset{\mid}{CH_3}}{C}HCH_2CH_2COOH$	colorless to pale yel liq/ sour, penetrating	199°–201°	1 mL in 1 mL 95% alc
4-Methyl-2-pentanone FEMA No. 2731 Methyl Isobutyl Ketone	100.16/$C_6H_{12}O$/ $CH_3COCH_2CH(CH_3)_2$	colorless, mobile liq/ fruity, ethereal	m—alc, ether, 1 mL in 50 mL water/ 117°	
2-Methyl-2-pentenoic Acid FEMA No. 3195	114.14/$C_6H_{10}O_2$/ $CH_3CH_2CH{=}\overset{\underset{\mid}{CH_3}}{C}COOH$	colorless to pale yel liq	123° (30 mm Hg)	1 mL in 1 mL 95% alc
Methyl Phenylacetate FEMA No. 2733	150.18/$C_9H_{10}O_2$/ ⟨C₆H₅⟩—CH_2COOCH_3	colorless or nearly colorless liq/ honey-like, jasmine-like	s—alc, most fixed oils; ins—gly, prop glycol, water/ 215°	1 mL in 6 mL 60% alc
Methyl Phenylcarbinyl Acetate FEMA No. 2684 α-Phenyl Ethyl Acetate	164.20/$C_{10}H_{12}O_2$/ ⟨C₆H₅⟩—$\overset{\underset{\mid}{OCOCH_3}}{C}HCH_3$	colorless liq/ gardenia-like	s—most fixed oils, gly; ins—water/ 214°	1 mL in 7 mL 60% alc
Methyl Propyl 3-Methyl Butyrate FEMA No. 3369 Isobutyl Isovalerate	158.24/$C_9H_{18}O_2$/ $CH_3CH(CH_3)CH_2COOCH_2CH(CH_3)CH_3$	colorless to pale yel liq	m—alc/ 170°	

Requirements

I.D. Test[4]	Assay Min. %[5]	A.V. Max.[6]	Ref. Index[7]	Sp. Gr.[8]	Other Requirements[9]
IR	96.0% of $C_9H_{14}O_2$ (M-1b)	1.0	1.446–1.449	0.919–0.924	**Chlorinated Cmpds.**—passes test (Appendix VI)
	98.0% of $C_6H_{12}O_2$ (M-3a)		1.411–1.416	0.919–0.922	
IR	98.0% of $C_6H_{12}O_2$ (M-3a)		1.412–1.417	0.919–0.926	
	99.0% of $C_6H_{12}O$ (M-2d)	2.0		0.796–0.799	**Dist. Range**—between 114° and 117° (Appendix IIB) **Water**—0.1% (Appendix IIB, KF; use freshly dist. pyridine as solvent)
IR	98.0% of $C_6H_{10}O_2$ (M-3a)		1.450–1.460	0.976–0.982	
IR	98.0% of $C_9H_{10}O_2$ (M-1b)	1.0	1.503–1.509	1.061–1.067	**Chlorinated Cmpds.**—passes test (Appendix VI)
IR	97.0% of $C_{10}H_{12}O_2$ (M-1b)	2.0	1.493–1.497	1.023–1.026	**Chlorinated Cmpds.**—passes test (Appendix VI)
	98.0% of $C_9H_{18}O_2$ (M-1b)	1.0	1.404–1.408	0.850–0.854	

General Information and Description

Name of Substance (Synonyms)	Formula Wt/Formula/ Structure	Physical Form/ Odor	Solubility[1]/ B.P.[2]	Solubility in Alcohol[3]
2-Methylpyrazine FEMA No. 3309	94.12/$C_5H_6N_2$/	colorless to slightly yel liq/ nutty, cocoa-like	m—water, alc, acetone, most fixed oils/ 137°	
Methyl Salicylate FEMA No. 2745	152.15/$C_8H_8O_3$/	colorless, yellowish, or reddish liq: optically inactive/ odor and taste of wintergreen	s—alc, glacial acetic acid; ss—water/ 222° (decomp)	1 mL in 7 mL 70% alc may be slightly cloudy
4-Methyl-5-thiazole Ethanol FEMA No. 3204 Sulfurol	143.20/C_9H_9NOS/	colorless to pale yel liq; may darken upon aging	135° (7 mm Hg)	
3-Methylthiopropionaldehyde FEMA No. 2747 Methional	104.17/C_4H_8OS/ $CH_3S(CH_2)_2CH{=}O$	colorless to pale yel liq/ meaty potato	165°–166°	1 mL in 1 mL 95% alc
2-Methylundecanal FEMA No. 2749 Aldehyde C-12 MNA; Methyl n-Nonyl Acetaldehyde	184.32/$C_{12}H_{24}O$/ $CH_3(CH_2)_8CH(CH_3)CHO$	colorless to slightly yel liq/ fatty odor	s—fixed oils, alc, prop glycol (may be turbid); ins—gly/ 171°	
Myrcene FEMA No. 2762 7-Methyl-3-methylene-1,6-octadiene	136.24/$C_{10}H_{16}$/	colorless to pale yel liq/ sweet, balsamic	s—alc, most fixed oils; ins—water/ 167°	
Myristaldehyde FEMA No. 2763 Tetradecanal	212.38/$C_{14}H_{28}O$/ $CH_3(CH_2)_{12}CHO$	colorless to pale yel liq/ fatty, orris-like	260°	

Requirements

I.D. Test[4]	Assay Min. %[5]	A.V. Max.[6]	Ref. Index[7]	Sp. Gr.[8]	Other Requirements[9]
IR	99.0% of $C_5H_6N_2$ (M-1a)		1.504–1.506	1.010–1.030 (20°)	**Water**—0.5% (Appendix IIB, KF; use freshly dist. pyridine as solvent)
IR	98.0% of $C_8H_8O_3$ (M-1b)	1.0 (phenol red TS)	1.535–1.538	1.180–1.185	
	98.0% of C_9H_9NOS (M-1b)		1.548–1.552	1.196–1.210	
	98.0% of C_4H_8OS (M-1a)		1.484–1.493	1.038–1.048	
IR	94.0% of $C_{12}H_{24}O$ (M-1b)	10.0	1.431–1.436	0.822–0.830	
	90.0% of $C_{10}H_{16}$ (M-1a)		1.466–1.471	0.789–0.793	**Peroxide Value**—50.0 (M-12)
	85.0% of $C_{14}H_{28}O$ (M-2a)	5.0	1.438–1.445	0.825–0.830	

General Information and Description

Name of Substance (Synonyms)	Formula Wt/Formula/ Structure	Physical Form/ Odor	Solubility[1]/ B.P.[2]	Solubility in Alcohol[3]
Myristyl Alcohol 1-Tetradecanol; Tetradecyl Alcohol	214.38/$C_{14}H_{30}O$/ $CH_3(CH_2)_{12}CH_2OH$	colorless to white, waxy, solid flakes/ waxy	s—ether; ss—alc; ins—water/ 289°	
Nerol FEMA No. 2770 cis-3,7-Dimethyl-2,6-octadien-1-ol	154.25/$C_{10}H_{18}O$/	colorless liq/ fresh, sweet, rose-like	m—alc, chloroform, ether; ins—water/ 227°	1 mL in 9 mL 50% alc gives clear soln
Nerolidol FEMA No. 2772 3,7,11-Trimethyl-1,6,10-dodecatrien-3-ol	227.37/$C_{15}H_{26}O$/	colorless to straw-colored liq/ faint, floral, rose-like, apple-like	s—most fixed oils, prop glycol; ins—gly/ 276°	1 mL in 4 mL 70% alc
Neryl Acetate FEMA No. 2773 cis-3,7-Dimethyl-2,6-octadien-1-yl Acetate	196.29/$C_{12}H_{20}O_2$/	colorless to pale yel liq/ sweet, floral	134° (25 mm Hg)	1 mL in 1 mL 95% alc
(E),(E)-2,4-Nonadienal FEMA No. 3212 trans,trans-2,4-Nonadienal	138.21/$C_9H_{14}O$/	slightly yel liq/ strong, fatty, floral	s—alc, most fixed oils; ins—water/ 97° (10 mm Hg)	
(E),(Z)-2,6-Nonadienal FEMA No. 3377 trans,cis-2,6-Nonadienal	138.21/$C_9H_{14}O$/	slightly yel liq/ powerful, violet, cucumber	s—alc, most fixed oils; ins—water/ 94° (18 mm Hg)	
(E),(Z)-2,6-Nonadienol FEMA No. 2780 trans,cis-2,6-Nonadienol	140.22/$C_9H_{16}O$/	white to yellowish liq/ powerful, green, vegetable	ins—water/ 196°	

Requirements

I.D. Test[4]	Assay Min. %[5]	A.V. Max.[6]	Ref. Index[7]	Sp. Gr.[8]	Other Requirements[9]
	98.0% of $C_{14}H_{30}O$ (M-1b)	1.0			**Melting Range**—between 38° and 41° (Appendix IIB) **Iodine Value**—3.0 max. (Appendix VII) **Saponification Value**—1.0 max. (Appendix VI)
IR	95.0% of total alcohols as $C_{10}H_{18}O$ (Appendix VI; 1.2g/77.13)		1.467–1.478	0.875–0.880	
IR	90.0% of total alcohols as $C_{15}H_{26}O$ (M-1b)		1.478–1.483	0.870–0.880	**Angular Rotation (Natural)**—between +11° and +14° (Appendix IIB, 100-mm tube) **Esters**—0.5% as nerolidyl acetate (Appendix VI; 10g/132.7)
IR	96.0% of $C_{12}H_{20}O_2$; predominantly (Z) isomer by M-1a (M-1b)	1.0	1.458–1.464	0.905–0.914	
IR	89.0% of $C_9H_{14}O$ (M-1a)		1.522–1.525	0.850–0.870	
IR	92.0% of $C_9H_{14}O$ (M-1a)		1.470–1.475	0.850–0.870	
IR	92.0% of $C_9H_{16}O$ (M-1a)		1.463–1.465	0.860–0.880	

General Information and Description

Name of Substance (Synonyms)	Formula Wt/Formula/ Structure	Physical Form/ Odor	Solubility[1]/ B.P.[2]	Solubility in Alcohol[3]
δ-Nonalactone FEMA No. 3356 5-Hydroxynonanoic Acid, Lactone	156.22/$C_9H_{16}O_2$/ $CH_3(CH_2)_3$— (δ-lactone ring)	colorless to pale yel liq/ coconut-like	250°	1 mL in 1 mL 95% alc
γ-Nonalactone FEMA No. 2781 Aldehyde C-18, So-Called	156.22/$C_9H_{16}O_2$/ $CH_3(CH_2)_4$— (γ-lactone ring)	colorless to slightly yel liq/ coconut-like	s—alc, most fixed oils, prop glycol; ins—water	1 mL in 5 mL 60% alc
Nonanal FEMA No. 2782 Aldehyde C-9; Pelargonic Aldehyde	142.24/$C_9H_{18}O$/ $CH_3(CH_2)_7CHO$	colorless to light yel liq/ fatty odor, citrus-rose on dilution	s—alc, most fixed oils, prop glycol; ins—gly/ 93° (23 mm Hg)	
Nonanoic Acid FEMA No. 2784	158.24/$C_9H_{18}O_2$/ $CH_3(CH_2)_7COOH$	colorless to pale yel liq	254°	
2-Nonanone FEMA No. 2785 Methyl Heptyl Ketone	142.24/$C_9H_{18}O$/ $CH_3CO(CH_2)_6CH_3$	colorless to pale yel liq/ fruity, floral, fatty, herbaceous	195°	1 mL in 1 mL 95% alc
(E)-2-Nonenal FEMA No. 3213 trans-2-Nonenal	140.22/$C_9H_{16}O$/ $CH_3(CH_2)_5$CH=CH-CHO	white to slightly yellowish liq/ fatty, violet	s—alc, most fixed oils; ins—water/ 88° (12 mm Hg)	
(E)-2-Nonen-1-ol FEMA No. 3379 trans-2-Nonenol	142.23/$C_9H_{18}O$/ $CH_3(CH_2)_5$CH=CH-CH_2OH	white liq/ fatty, violet	ins—water	
(Z)-6-Nonen-1-ol FEMA No. 3465 cis-6-Nonen-1-ol	142.24/$C_9H_{18}O$/ CH_3CH_2CH=CH-$(CH_2)_4CH_2OH$	white to slightly yel liq/ powerful, melon-like	ins—water	

FCC IV / Flavor Chemicals / 539

Requirements

I.D. Test[4]	Assay Min. %[5]	A.V. Max.[6]	Ref. Index[7]	Sp. Gr.[8]	Other Requirements[9]
IR	98.0% of $C_9H_{16}O_2$ (M-1a)		1.452–1.458	0.994–0.999	
IR	98.0% of $C_9H_{16}O_2$ (M-1b)	2.0	1.446–1.450	0.958–0.966	
IR	92.0% of $C_9H_{18}O$ (M-1b)	10.0	1.422–1.429	0.820–0.830	
	98.0% of $C_9H_{18}O_2$ (M-3a)		1.431–1.435	0.901–0.906	
IR	97.0% of $C_9H_{18}O$ (M-1a)		1.418–1.423	0.817–0.823	
IR	92.0% of $C_9H_{16}O$ (M-1a)		1.457–1.460	0.850–0.870	
IR	96.0% of $C_9H_{18}O$ (M-1a)		1.444–1.448	0.830–0.850	
IR	95.0% of $C_9H_{18}O$ (M-1a)		1.448–1.450	0.850–0.870	

540 / FCC IV / Flavor Chemicals

General Information and Description

Name of Substance (Synonyms)	Formula Wt/Formula/ Structure	Physical Form/ Odor	Solubility[1]/ B.P.[2]	Solubility in Alcohol[3]
Nonyl Acetate FEMA No. 2788	186.29/$C_{11}H_{22}O_2$/ $CH_3COO(CH_2)_8CH_3$	colorless liq/ floral, fruity	s—alc, ether; ins—water/ 212°	1 mL in 6 mL 70% alc gives clear soln
Nonyl Alcohol FEMA No. 2789 1-Nonanol; Alcohol C-9	144.26/$C_9H_{20}O$/ $CH_3(CH_2)_7CH_2OH$	colorless liq/ rose-citrus	m—alc, chloroform, ether; ins—water/ 213°	1 mL in 3 mL 60% alc gives clear soln
δ-Octalactone FEMA No. 3214 5-Hydroxyoctanoic Acid Lactone	142.20/$C_8H_{14}O_2$/	colorless to pale yel liq/ coconut-like		
γ-Octalactone FEMA No. 2796	142.20/$C_8H_{14}O_2$/	colorless to slightly yel liq/ sweet, coconut, fruity	s—alc; ss—water	
Octanal FEMA No. 2797 Aldehyde C-8; Caprylic Aldehyde	128.21/$C_8H_{16}O$/ $CH_3(CH_2)_6CHO$	colorless to light yel liq/ fatty-orange odor	s—alc, most fixed oils, prop glycol; ins—gly/ 171°	
3-Octanol FEMA No. 3581	130.23/$C_8H_{18}O$/ $CH_3(CH_2)_4\overset{\underset{\mid}{OH}}{C}HCH_2CH_3$	colorless liq/ strong, oily-nutty, herbaceous	s—alc, most fixed oils; ins—water/ 174°	
(E)-2-Octen-1-al FEMA No. 3215 trans-2-Octen-1-al	126.20/$C_8H_{14}O$/	slightly yel liq/ fatty, green odor	s—alc, most fixed oils; ss—water/ 84° (19 mm Hg)	
1-Octen-3-ol FEMA No. 2805 Amyl Vinyl Carbinol	128.21/$C_8H_{16}O$/ $CH_3(CH_2)_4CHCH=CH_2$ $\mid$ OH	colorless to pale yel liq/ mushroom-like, herbaceous	175°	1 mL in 1 mL 95% alc

Requirements

I.D. Test[4]	Assay Min. %[5]	A.V. Max.[6]	Ref. Index[7]	Sp. Gr.[8]	Other Requirements[9]
IR	97.0% of $C_{11}H_{22}O_2$ (M-1b)	1.0	1.422–1.426	0.864–0.868	
IR	97.0% of $C_9H_{20}O$ (M-1b)	1.0	1.431–1.435	0.824–0.830	
IR	98.0% of $C_8H_{14}O_2$ (M-1a)		1.452–1.458	0.994–0.999	
IR	95.0% of $C_8H_{14}O_2$ (M-1a)	8.0	1.443–1.447	0.970–0.980	
IR	92.0% of $C_8H_{16}O$ (M-1b)	10.0	1.417–1.425	0.810–0.830	
	97.0% of $C_8H_{18}O$ (M-1a)		1.425–1.429	0.816–0.821	
IR	92.0% of $C_8H_{14}O$ [as (E) isomer] (M-1a)		1.450–1.455	0.830–0.850	
IR	97.0% of $C_8H_{16}O$ (M-1a)		1.434–1.442	0.831–0.839	

General Information and Description

Name of Substance (Synonyms)	Formula Wt/Formula/ Structure	Physical Form/ Odor	Solubility[1]/ B.P.[2]	Solubility in Alcohol[3]
(Z)-3-Octen-1-ol FEMA No. 3467 cis-3-Octen-1-ol	128.21/$C_8H_{16}O$/ $CH_3(CH_2)_3$\C=C/$(CH_2)_2OH$ (H, H)	white to slightly yellowish liq/ musty, mushroom-like	ins—water/ 174°	
1-Octen-3-yl Acetate FEMA No. 3582	170.25/$C_{10}H_{18}O_2$/ $CH_3(CH_2)_3CH_2\underset{\underset{CH=CH_2}{\mid}}{C}HOOCCH_3$	almost colorless liq/ metallic, mushroom-like	s—alc, most fixed oils; ins—water, prop glycol	
1-Octen-3-yl Butyrate FEMA No. 3612	198.31/$C_{12}H_{22}O_2$/ $CH_3(CH_2)_3CH_2\underset{\underset{CH=CH_2}{\mid}}{C}HOOC(CH_2)_2CH_3$	almost colorless liq/ metallic, mushroom-like	s—alc, most fixed oils; ss—prop glycol; ins—water	
Octyl Acetate FEMA No. 2806	172.27/$C_{10}H_{20}O_2$/ $CH_3COO(CH_2)_7CH_3$	colorless liq/ fruity, orange-like, jasmine-like	m—alc, most fixed oils, org solvents; ins—water/ 208°	1 mL in 4 mL 70% alc gives clear soln
3-Octyl Acetate FEMA No. 3583	172.27/$C_{10}H_{20}O_2$/ $CH_3(CH_2)_3CH_2\underset{\underset{CH_2CH_3}{\mid}}{C}HOOCCH_3$	colorless liq/ rosy-minty	s—alc, prop glycol, most fixed oils; ss—water/ 187°	
Octyl Alcohol FEMA No. 2800 Alcohol C-8; 1-Octanol; Capryl Alcohol	130.23/$C_8H_{18}O$/ $CH_3(CH_2)_6CH_2OH$	colorless liq/ sharp fatty-citrus	s—most fixed oils, prop glycol; ins—gly/ 195°	1 mL in 5 mL 50% alc
Octyl Formate FEMA No. 2809	158.24/$C_9H_{18}O_2$/ $HCOO(CH_2)_7CH_3$	colorless liq/ fruity	s—most fixed oils, min oil, prop glycol; ins—gly/ 200°	1 mL in 5 mL 70% alc remains in soln to 10 mL
Octyl Isobutyrate FEMA No. 2808 Octyl 2-Methylpropanoate	200.32/$C_{12}H_{24}O_2$/ $CH_3(CH_2)_7OOCCH(CH_3)_2$	colorless to pale yel liq/ refreshing, herbaceous	245°	1 mL in 1 mL 95% alc

Requirements

I.D. Test[4]	Assay Min. %[5]	A.V. Max.[6]	Ref. Index[7]	Sp. Gr.[8]	Other Requirements[9]
IR	95.0% of $C_8H_{16}O$ [as (Z) isomer] (M-1a)		1.440–1.446	0.830–0.850	
IR	95.0% of $C_{10}H_{18}O_2$ (M-1b)		1.414–1.434 (25°)	0.865–0.886	
IR	95.0% of $C_{12}H_{22}O_2$ (M-1b)		1.416–1.437 (25°)	0.859–0.880	
IR	98.0% of $C_{10}H_{20}O_2$ (M-1b)	1.0	1.418–1.421	0.865–0.868	
	98.0% of $C_{10}H_{20}O_2$ (M-1b)	2.0	1.414–1.419	0.856–0.860	
IR	98.0% of $C_8H_{18}O$ (M-1b)	1.0	1.428–1.431	0.822–0.830	
IR	96.0% of $C_9H_{18}O_2$ (M-1b)	1.0	1.418–1.420	0.869–0.874	
	98.0% of $C_{12}H_{24}O_2$ (M-1a)	1.0	1.420–1.425	0.853–0.858	

General Information and Description

Name of Substance (Synonyms)	Formula Wt/Formula/ Structure	Physical Form/ Odor	Solubility[1]/ B.P.[2]	Solubility in Alcohol[3]
2,3-Pentanedione FEMA No. 2841 Acetyl Propionyl	100.12/$C_5H_8O_2$/ $CH_3CH_2\overset{O}{\overset{\|\|}{C}}-\overset{O}{\overset{\|\|}{C}}CH_3$	yel to yel green liq/ penetrating, buttery on dilution	*m*—alc, prop glycol, fixed oils; *ins*—gly, water/ 108°	1 mL in 3 mL 50% alc
2-Pentanone FEMA No. 2842 Methyl Propyl Ketone	86.13/$C_5H_{10}O$/ $CH_3COCH_2CH_2CH_3$	colorless, mobile liq/ fruity, ethereal	*m*—alc, ether, 1 mL in 25 mL water/ 102°	
α-Phellandrene FEMA No. 2856 *p*-Mentha-1,5-diene	136.24/$C_{10}H_{16}$/ (cyclohexadiene with CH$_3$ and CH(CH$_3$)$_2$ substituents)	colorless to slightly yel liq/ herbaceous odor, mint-like background	*s*—alc; *ins*—water	1 mL in 1 mL 95% alc gives clear soln
Phenethyl Acetate FEMA No. 2857 2-Phenethyl Acetate	164.20/$C_{10}H_{12}O_2$/ C$_6$H$_5$—CH$_2$CH$_2$OOCCH$_3$	colorless liq/ sweet, rosy, honey-like	*s*—alc, most fixed oils, prop glycol; *ins*—gly, water/ 232°	1 mL in 2 mL 70% alc remains clear to 10 mL
Phenethyl Alcohol FEMA No. 2858 2-Phenylethyl Alcohol	122.17/$C_8H_{10}O$/ C$_6$H$_5$—CH$_2$CH$_2$OH	colorless liq/ rose-like	*s*—most fixed oils, water, prop glycol/ 219°	1 mL in 2 mL 50% alc remains clear to 10 mL
Phenethyl Isobutyrate FEMA No. 2862	192.26/$C_{12}H_{16}O_2$/ C$_6$H$_5$—CH$_2$CH$_2$OOCCH(CH$_3$)CH$_3$	colorless to slightly yel liq/ fruity, rosy	*s*—alc, most fixed oils; *ins*—water/ 230°	1 mL in 3 mL 80% alc gives clear soln
Phenethyl Isovalerate FEMA No. 2871	206.28/$C_{13}H_{18}O_2$/ C$_6$H$_5$—CH$_2$CH$_2$OOCCH$_2$CH(CH$_3$)CH$_3$	colorless to slightly yel liq/ fruity, rosy	*s*—alc, most fixed oils; *ins*—water/ 263°	1 mL in 3 mL 80% alc gives clear soln

Requirements

I.D. Test[4]	Assay Min. %[5]	A.V. Max.[6]	Ref. Index[7]	Sp. Gr.[8]	Other Requirements[9]
IR	93.0% of $C_5H_8O_2$ (M-1b)		1.402–1.406	0.952–0.962	
IR	95.0% of $C_5H_{10}O$ (M-1b)	2.0		0.801–0.806	
IR			1.471–1.477	0.835–0.865	**Angular Rotation**—between -80° and -120° (Appendix IIB, 100-mm tube)
IR	98.0% of $C_{10}H_{12}O_2$ (M-1b)	1.0	1.497–1.501	1.030–1.034	
IR			1.531–1.534	1.017–1.020	**Chlorinated Cmpds.**—passes test (Appendix VI)
IR	98.0% of $C_{12}H_{16}O_2$ (M-1b)	1.0	1.486–1.490	0.987–0.990	
IR	98.0% of $C_{13}H_{18}O_2$ (M-1b)	1.0	1.484–1.486	0.973–0.976	

General Information and Description

Name of Substance (Synonyms)	Formula Wt/Formula/ Structure	Physical Form/ Odor	Solubility[1]/ B.P.[2]	Solubility in Alcohol[3]
2-Phenethyl 2-Methylbutyrate FEMA No. 3632	206.28/$C_{13}H_{18}O_2$/ Ph—$CH_2CH_2OOCCHCH_2CH_3$ | CH_3	colorless liq/ floral-fruity	s—alc, most fixed oils; ins—water	
Phenethyl Phenylacetate FEMA No. 2866	240.30/$C_{16}H_{16}O_2$/ Ph—$CH_2CH_2OOCCH_2$—Ph	colorless to slightly yel liq above 26°/ rosy, hyacinth-like	s—alc; ins—water/ 325°	1 mL in 4 mL 90% alc gives clear soln
Phenethyl Salicylate FEMA No. 2868	242.27/$C_{15}H_{14}O_3$/ Ph—CH_2CH_2OOC—(o-HO-Ph)	white cryst/ balsamic	s—alc; ins—water	1 g in 20 mL 95% alc gives clear soln
Phenoxyethyl Isobutyrate FEMA No. 2873	208.26/$C_{12}H_{16}O_3$/ Ph—$OCH_2CH_2OOCCHCH_3$ | CH_3	colorless liq/ honey, rose-like	m—alc, chloroform, ether; ins—water	1 mL in 3 mL 70% alc gives clear soln
Phenylacetaldehyde FEMA No. 2874 α-Toluic Aldehyde	120.15/C_8H_8O/ Ph—CH_2CHO	colorless to slightly yel, oily liq; becomes more viscous on aging/ harsh odor, hyacinth-like on dilution	s—most fixed oils, prop glycol; ins—gly/ 195°	1 mL in 2 mL 80% alc
Phenylacetaldehyde Dimethyl Acetal FEMA No. 2876	166.22/$C_{10}H_{14}O_2$/ Ph—$CH_2CH(OCH_3)_2$	colorless liq/ strong odor	s—most fixed oils, prop glycol; ins—gly/ 219°	1 mL in 2 mL 70% alc remains clear to 10 mL
Phenylacetic Acid FEMA No. 2878 α-Toluic Acid	136.15/$C_8H_8O_2$/ Ph—CH_2COOH	glistening white cryst solid/ persistent, disagreeable, suggestive of geranium leaf and rose when diluted	s—most fixed oils, gly; ss—water/ 265°	

Requirements

I.D. Test[4]	Assay Min. %[5]	A.V. Max.[6]	Ref. Index[7]	Sp. Gr.[8]	Other Requirements[9]
	95.0% of $C_{13}H_{18}O_2$ (M-1b)	2.0	1.484–1.488	0.973–0.977	
IR	98.0% of $C_{16}H_{16}O_2$ (M-1b)	1.0		1.079–1.082	**Solidification Pt.**—NLT 26° (Appendix IIB)
IR	98.0% of $C_{15}H_{14}O_3$ (M-1b)	1.0 (phenol red TS)			**Solidification Pt.**—NLT 41° (Appendix IIB)
IR	97.0% of $C_{12}H_{16}O_3$ (M-1b)	1.0	1.492–1.495	1.044–1.048	
IR	90.0% of C_8H_8O (M-1b)	5.0	1.525–1.545	1.025–1.045	
IR	95.0% of $C_{10}H_{14}O_2$ (M-1b)	1.0	1.493–1.496	1.000–1.006	**Chlorinated Cmpds.**—passes test (Appendix VI) **Free Phenyl Acetaldehyde**—1.0% (M-1b)
IR	99.0% of $C_8H_8O_2$ (after drying) (M-3b)				**Heavy Metals**—0.004% (M-7) **Melting Range**—between 76° and 78° (Appendix IIB, Class Ia) **Lead**—10 mg/kg (M-10)

General Information and Description

Name of Substance (Synonyms)	Formula Wt/Formula/ Structure	Physical Form/ Odor	Solubility[1]/ B.P.[2]	Solubility in Alcohol[3]
Phenylethyl Anthranilate FEMA No. 2859	241.29/$C_{15}H_{15}NO_2$/ Ph-CH$_2$CH$_2$OC=O / Ph-NH$_2$	colorless to pale yel cryst mass/ neroli-like, grape undertone	s—alc/ 324°	
Phenylethyl Butyrate FEMA No. 2861	192.26/$C_{12}H_{16}O_2$/ Ph-CH$_2$CH$_2$OOCCH$_2$CH$_2$CH$_3$	colorless to pale yel liq/ green, hay-like	238°	1 mL in 1 mL 95% alc
3-Phenyl-1-propanol FEMA No. 2885 Phenylpropyl Alcohol; Hydrocinnamyl Alcohol	136.19/$C_9H_{12}O$/ Ph-CH$_2$CH$_2$CH$_2$OH	colorless, slightly viscous liq/ sweet, hyacinth-mignonette	s—most fixed oils, prop glycol; ins—gly/ 236°	1 mL in 1 mL 70% alc
2-Phenylpropionaldehyde FEMA No. 2886 Hydratropic Aldehyde; α-Methyl Phenylacetaldehyde	134.18/$C_9H_{10}O$/ Ph-CHCHO / CH$_3$	water-white to pale yel liq/ floral	s—most fixed oils; ss—prop glycol; ins—gly/ 222°	
3-Phenylpropionaldehyde FEMA No. 2887 Hydrocinnamaldehyde; Phenylpropyl Aldehyde	134.18/$C_9H_{10}O$/ Ph-CH$_2$CH$_2$CHO	colorless to slightly yel liq/ strong, pungent, floral, hyacinth-like	m—alc, ether; ins—water	1 mL in 7 mL 60% alc remains clear on dilution
2-Phenylpropionaldehyde Dimethyl Acetal FEMA No. 2888 Hydratropic Aldehyde Dimethyl Acetal	180.25/$C_{11}H_{16}O_2$/ Ph-CH-CH(OCH$_3$) / CH$_3$ OCH$_3$	colorless to slightly yel liq/ mushroom-like	s—alc, ether; ins—water/ 241°	1 mL in 7 mL 60% alc, and in 3 mL 70% alc gives clear solns
3-Phenylpropyl Acetate FEMA No. 2890	178.23/$C_{11}H_{14}O_2$/ Ph-CH$_2$CH$_2$CH$_2$OOCCH$_3$	colorless liq/ spicy, floral	s—alc; ins—water	1 mL in 3 mL 70% alc gives clear soln

Requirements

I.D. Test[4]	Assay Min. %[5]	A.V. Max.[6]	Ref. Index[7]	Sp. Gr.[8]	Other Requirements[9]
IR	98.0% of $C_{15}H_{15}NO_2$ (M-1b)	1.0			**Solidification Pt.**—NLT 40° (Appendix IIB)
IR	98.0% of $C_{12}H_{16}O_2$ (M-1b)	1.0	1.487–1.492	0.991–0.995	
IR	98.0% of $C_9H_{12}O$ (M-1b)		1.524–1.528	0.998–1.002	**Free 3-Phenyl Propionaldehyde**—0.5% (M-1b)
IR	95.0% of $C_9H_{10}O$ (M-1b)	5.0	1.515–1.520	0.998–1.006	
IR	90.0% of aldehydes (M-1b)	10.0	1.520–1.532	1.010–1.020	**Chlorinated Cmpds.**—passes test (Appendix VI)
IR	95.0% of $C_{11}H_{16}O_2$ (M-1b)		1.492–1.497	0.989–0.994	**Free 2-Phenyl-Propionaldehyde**—3.0% (M-1b)
IR	98.0% of $C_{11}H_{14}O_2$ (M-1b)	1.0	1.494–1.497	1.012–1.015	

General Information and Description

Name of Substance (Synonyms)	Formula Wt/Formula/ Structure	Physical Form/ Odor	Solubility[1]/ B.P.[2]	Solubility in Alcohol[3]
α-Pinene FEMA No. 2902 2,6,6-Trimethylbicyclo(3.1.1)hept-2-ene; 2-Pinene; l-α-Pinene	136.24/$C_{10}H_{16}$/	colorless liq/ fresh, piney	s—alc, most fixed oils; ins—water/ 155°	
β-Pinene FEMA No. 2903 6,6-Dimethyl-2-methylenebicyclo-[3.1.1]heptane	136.24/$C_{10}H_{16}$/	colorless liq/ resinous-piney	s—most fixed oils; ins—water, prop glycol/ 165°	
Piperidine FEMA No. 2908 Hexahydropyridine	85.15/$C_5H_{11}N$/	colorless to pale yel liq/ ammoniacal, fishy, nauseating	106°	
Piperonal FEMA No. 2911 3,4-(Methylenedioxy)-benzaldehyde; Heliotropine; Piperonyl Aldehyde	150.13/$C_8H_6O_3$/	white cryst substance; m.p. 37°/ floral odor, like heliotrope, free from safrole by-odor	vs—alc; s—most fixed oils, prop glycol; ins—gly, water/ 264°	1 g in 4 mL 70% alc
Propenylguaethol FEMA No. 2922 1-Ethoxy-2-hydroxy-4-propenylbenzene	178.23/$C_{11}H_{14}O_2$/	white cryst powder/ vanilla-like	s—veg oils; ins—water, 1 g in 20 mL 95% alc	
Propionaldehyde FEMA No. 2923	58.08/C_3H_6O/ CH_3CH_2CHO	colorless, mobile liq/ sharp, pungent	m—alc, ether, water/ 49°	
Propyl Acetate FEMA No. 2925 n-Propyl Acetate	102.13/$C_5H_{10}O_2$/ $CH_3CH_2CH_2OOCCH_3$	colorless liq	102°	1 mL in 1 mL 95% alc

Requirements

I.D. Test[4]	Assay Min. %[5]	A.V. Max.[6]	Ref. Index[7]	Sp. Gr.[8]	Other Requirements[9]
	97.0% of $C_{10}H_{16}$ (M-1a)		1.464–1.468	0.855–0.860	**Angular Rotation**— between -35° and -50° (Appendix IIB)
	97.0% of $C_{10}H_{16}$ (M-1a)		1.477–1.481	0.867–0.871	**Angular Rotation**— between -15° and -30° (Appendix IIB)
IR	98.0% of $C_5H_{11}N$ (M-1a)		1.450–1.454	0.858–0.862	
IR	99.0% of $C_8H_6O_3$ (M-1b)				**Heavy Metals**—0.004% (M-7) **Lead**—10 mg/kg (M-10) **Solidification Pt.**—NLT 35° (Appendix IIB)
IR					**Heavy Metals**—10 mg/kg (M-7) **Melting Range**—between 85° and 88° (Appendix IIB) **Res. on Ignit.**—0.1% (Appendix IIC, 2-g sample)
IR	97.0% of C_3H_6O (M-2c)	5.0 (M-16)		0.800–0.805	**Dist. Range**—between 46° and 50° (first 97%, Appendix IIB) **Water**—2.5% (Appendix IIB, KF; use freshly dist. pyridine as solvent)
IR	97.0% of $C_5H_{10}O_2$ (M-1b)	1.0	1.382–1.387	0.880–0.886	

General Information and Description

Name of Substance (Synonyms)	Formula Wt/Formula/ Structure	Physical Form/ Odor	Solubility[1]/ B.P.[2]	Solubility in Alcohol[3]
Propyl Alcohol FEMA No. 2928 *n*-Propanol	60.09/C_3H_8O/ $CH_3CH_2CH_2OH$	colorless liq	97°	1 mL in 1 mL 95% alc
***p*-Propyl Anisole** FEMA No. 2930 Dihydroanethole	150.22/$C_{10}H_{14}O$/ CH_3O—⟨⟩—$CH_2CH_2CH_3$	colorless to pale yel liq/ anisetype odor, with sassafras background	*s*—most fixed oils; *ins*—gly, prop glycol	1 mL in 5 mL 80% alc remains in soln on dilution
Propyl Propionate FEMA No. 2958 *n*-Propyl Propionate	116.16/$C_6H_{12}O_2$/ $CH_3CH_2CH_2OOCCH_2CH_3$	colorless to pale yel liq	123°	1 mL in 1 mL 95% alc
Pyrrole FEMA No. 3386	67.09/C_4H_5N/ ⟨N-H⟩	colorless to yellowish liq, darkens on aging/ nutty, sweet, warm, ethereal	*s*—alc, most fixed oils; *ss*—water/ 130° (decomp)	
Rhodinol FEMA No. 2980	[see *Citronellol, Geraniol,* and *Nerol*]	colorless liq/ pronounced, rose-like	*s*—most fixed oils, prop glycol; *ins*—gly	1 mL in 1.2 mL 70% alc
Rhodinyl Acetate FEMA No. 2981	[see *Citronellyl Acetate* and *Geranyl Acetate*]	colorless to slightly yel liq/ light, fresh, rose-like	*s*—alc, most fixed oils; *ins*—gly, prop glycol, water/ 237°	1 mL in 2 mL 80% alc remains in soln to 10 mL
Rhodinyl Formate FEMA No. 2984	[see *Citronellyl Formate*]	colorless to slightly yel liq/ leafy, rose-like	*s*—alc, most fixed oils; *ins*—gly, prop glycol, water/ 220°	1 mL in 2 mL 80% alc gives clear soln

Requirements

I.D. Test[4]	Assay Min. %[5]	A.V. Max.[6]	Ref. Index[7]	Sp. Gr.[8]	Other Requirements[9]
IR	99.0% of C_3H_8O (M-1b)		1.383–1.388	0.800–0.805	
IR			1.502–1.506	0.940–0.943	
IR	98.0% of $C_6H_{12}O_2$ (M-1b)	1.0	1.391–1.396	0.873–0.879	
IR	98.0% of C_4H_5N (M-1a)		1.507–1.510	0.950–0.980 (20°)	**Dist. Range**—between 125° and 130° (Appendix IIB) **Water**—0.5% (Appendix IIB, KF; use freshly dist. pyridine as solvent)
IR	82.0% of total alcohols as $C_{10}H_{20}O$ (Appendix VI; 1.2g/78.14)		1.463–1.473	0.860–0.880	**Angular Rotation**— between -4° and -9° (Appendix IIB, 100-mm tube) **Esters**—1.0% as citronellyl acetate (Appendix VI; 5g/99.15)
IR	87.0% of total esters as $C_{12}H_{22}O_2$ (Appendix VI; 1.3g/99.15)	1.0	1.450–1.458	0.895–0.908	**Angular Rotation**— between -2° and -6° (Appendix IIB, 100-mm tube)
IR	85.0% of total esters as $C_{11}H_{20}O_2$ (Appendix VI; 1.3g/92.14)	2.0	1.453–1.458	0.901–0.908	

General Information and Description

Name of Substance (Synonyms)	Formula Wt/Formula/ Structure	Physical Form/ Odor	Solubility[1]/ B.P.[2]	Solubility in Alcohol[3]
Santalol FEMA No. 3006 [Mixture of α- and β-isomers]	220.35/$C_{15}H_{24}O$/	colorless to slightly yel, viscous liq/ sandalwood-like	vs—alc, fixed oils, prop glycol; ins—gly, water/ 302°	1 mL in 4 mL 70% alc gives clear soln
Santalyl Acetate FEMA No. 3007	[mixture of α- and β-isomers from acetylation of *Santalol*]	colorless to slightly yel liq/ sandalwood-like	s—alc; ins—water/ 315°	1 mL in 9 mL 80% alc gives clear soln
α-Terpinene FEMA No. 3558 1-Methyl-4-(1-methylethyl)-1,3-cyclohexadiene	136.24/$C_{10}H_{16}$/	colorless liq/ lemon-like	s—alc, most fixed oils; ins—water/ 173°	
γ-Terpinene FEMA No. 3559 1-Methyl-4-(1-methylethyl)-1,4-cyclohexadiene	136.24/$C_{10}H_{16}$/	colorless liq/ herbaceous, citrusy	s—alc, most fixed oils; ins—water/ 182°	
Terpinen-4-ol FEMA No. 2248 4-Carvomenthenol	154.25/$C_{10}H_{18}O$	colorless to pale yel liq	s—alc/ 88° (6 mm Hg)	
α-Terpineol FEMA No. 3045 *p*-Menth-1-en-8-ol	154.25/$C_{10}H_{18}O$/	colorless, viscous liq/ lilac-like	ss—gly, water/ 217°	1 mL in 2 mL 70% alc, 4 mL 60% alc, 8 mL 50% alc

Requirements

I.D. Test[4]	Assay Min. %[5]	A.V. Max.[6]	Ref. Index[7]	Sp. Gr.[8]	Other Requirements[9]
IR	95.0% of total alcohols as $C_{15}H_{24}O$ (Appendix VI; 1.6g/110.18)		1.505–1.509	0.965–0.975	**Angular Rotation**— between -11° and -19° (Appendix IIB, 100-mm tube)
IR	95.0% of total esters as $C_{17}H_{26}O_2$ (Appendix VI; 1.6g/131.2)	1.0	1.488–1.491	0.980–0.986	
	89.0% of $C_{10}H_{16}$ (M-1a)		1.475–1.480	0.833–0.838	
	95.0% of $C_{10}H_{16}$ (M-1a)		1.473–1.477	0.841–0.845	
	92.0% of $C_{10}H_{18}O$ (M-1b)		1.476–1.480	0.928–0.934	
IR	96.0% of $C_{10}H_{18}O$ (sum of isomers) (M-1a)		1.482–1.485	0.930–0.936	

General Information and Description

Name of Substance (Synonyms)	Formula Wt/Formula/ Structure	Physical Form/ Odor	Solubility[1]/ B.P.[2]	Solubility in Alcohol[3]
Terpinyl Acetate FEMA No. 3047 Menthen-1-yl-8 Acetate	196.29/$C_{12}H_{20}O_2$/	colorless liq/ sweet, refreshing, herbaceous	s—alc, most fixed oils, min oil, prop glycol; ss—gly; ins—water/ 220°	1 mL in 5 mL 70% alc remains in soln to 10 mL
Terpinyl Propionate FEMA No. 3053 Menthen-1-yl-8 Propionate	210.32/$C_{13}H_{22}O_2$/	colorless to slightly yel liq/ sweet, floral, herbaceous, lavender-like	s—gly; ss—prop glycol; m—alc, chloroform, ether, most fixed oils; ins—water/ 240°	1 mL in 2 mL 80% alc gives clear soln
Tetrahydrofurfuryl Alcohol FEMA No. 3056	102.13/$C_5H_{10}O_2$/	colorless liq/ mild, warm, oily, caramel	178°	
Tetrahydrolinalool FEMA No. 3060 3,7-Dimethyl-3-octanol	158.28/$C_{10}H_{22}O$/	colorless liq/ distinct floral odor	s—alc, most fixed oils; ins—water/ 71° (6 mm Hg)	
2,3,5,6-Tetramethylpyrazine FEMA No. 3237	136.20/$C_8H_{12}N_2$/	white cryst or powder/ fermented soybeans	s—alc, prop glycol, most fixed oils; ss—water/ 190°	
Thymol FEMA No. 3066	150.22/$C_{10}H_{14}O$/	white cryst/ phenol-like	232°	1 g in 1 mL 95% alc

Requirements

I.D. Test[4]	Assay Min. %[5]	A.V. Max.[6]	Ref. Index[7]	Sp. Gr.[8]	Other Requirements[9]
IR	97.0% of $C_{12}H_{20}O_2$ (M-1b)		1.464–1.467	0.953–0.962	
IR	95.0% of $C_{13}H_{22}O_2$ (M-1b)	1.0	1.462–1.468	0.947–0.952	
IR	99.0% of $C_5H_{10}O_2$ (M-1a)		1.452–1.453	1.050–1.052	
	95.0% of $C_{10}H_{22}O$ (M-1a)		1.431–1.435	0.823–0.829	
IR	95.0% of $C_8H_{12}N_2$ (M-1a)				**Melting Range**—between 85° and 90° (Appendix IIB) **Water**—0.2% (Appendix IIB, KF; use freshly dist. pyridine as solvent)
IR					**Melting Range**—between 49° and 51° (Appendix IIB)

General Information and Description

Name of Substance (Synonyms)	Formula Wt/Formula/ Structure	Physical Form/ Odor	Solubility[1]/ B.P.[2]	Solubility in Alcohol[3]
Tolualdehyde FEMA No. 3068 mixed isomers; Tolyl Aldehyde, mixed isomers; Methylbenzaldehyde	120.15/C_8H_8O/ (CHO, CH₃ on benzene ring)	colorless liq/ cherry-like	198°	1 mL in 1 mL 95% alc
p-Tolualdehyde FEMA No. 3068 p-Tolyl Aldehyde; p-Methylbenzaldehyde	120.15/C_8H_8O/ (CHO, CH₃ para on benzene ring)	colorless liq/ cherry-like	83°–85° (11 mm Hg)	1 mL in 1 mL 95% alc
p-Tolyl Isobutyrate FEMA No. 3075 p-Cresyl Isobutyrate	178.23/$C_{11}H_{14}O_2$/ H_3C—⟨⟩—$OOCCHCH_3$ $\quad\quad\quad\quad\quad\quad CH_3$	colorless liq	s—alc; ins—water	1 mL in 7 mL 70% alc gives clear soln
Tributyrin FEMA No. 2223 Glyceryl Tributyrate; Butyrin	302.37/$C_{15}H_{26}O_6$/ H–COCOC₃H₇ H–COCOC₃H₇ H–COCOC₃H₇ (glyceryl structure)	colorless, somewhat oily liq	s—alc, chloroform, ether; ins—water/ 308°	
2-Tridecenal FEMA No. 3082	196.33/$C_{13}H_{24}O$/ $CH_3(CH_2)_9CH=CHCHO$	white or slightly yellowish liq/ oily, citrus odor	s—alc, most fixed oils; ins—water	
Trimethylamine FEMA No. 3241	59.11/C_3H_9N/ $(CH_3)_3N$	gas/ pungent, fishy, ammoniacal	2.9°	
3,5,5-Trimethyl Hexanal FEMA No. 3524	142.24/$C_9H_{18}O$/ $(CH_3)_3CCH_2CH(CH_3)CH_2CHO$	colorless to pale yel liq	67° (2.5 mm Hg)	

Requirements

I.D. Test[4]	Assay Min. %[5]	A.V. Max.[6]	Ref. Index[7]	Sp. Gr.[8]	Other Requirements[9]
IR	94.0% of C_8H_8O (M-1b)	5.0	1.540–1.548	1.019–1.029	
IR	97.0% of C_8H_8O (M-1b)	5.0	1.542–1.548	1.012–1.018	
IR	95.0% of $C_{11}H_{14}O_2$ (M-1b)	1.0	1.485–1.489	0.990–0.996	
IR	99.0% of $C_{15}H_{26}O_6$ (M-1b)	5.0		1.034–1.037	
IR	92.0% of $C_{13}H_{24}O$ (M-1a)		1.457–1.460	0.842–0.862	
	98.0% of C_3H_9N in a suitable solvent (M-1a)		1.432		
	97.0% (M-1b)	5.0	1.419–1.424	0.817–0.823	

General Information and Description

Name of Substance (Synonyms)	Formula Wt/Formula/ Structure	Physical Form/ Odor	Solubility[1]/ B.P.[2]	Solubility in Alcohol[3]
2,4,5-Trimethyl δ-3-Oxazoline FEMA No. 3525	113.16/$C_6H_{11}NO$/	yel orange liq/ powerful, musty, slight green, wood, nut-like	*s*—alc, prop glycol, water; *ins*—most fixed oils	
2,3,5-Trimethylpyrazine FEMA No. 3244	122.17/$C_7H_{10}N_2$/	colorless to slightly yel liq/ baked potato, peanut	*s*—water, org solvents/ 171°	
δ-Undecalactone FEMA No. 3294 5-Hydroxyundecanoic Acid Lactone	184.28/$C_{11}H_{20}O_2$/ $CH_3(CH_2)_5$—	colorless to pale yel liq/ creamy, peach-like	152°–155° (10.5 mm Hg)	1 mL in 1 mL 95% alc
γ-Undecalactone FEMA No. 3091 Aldehyde C-14 Pure, So-Called; Peach Aldehyde	184.28/$C_{11}H_{20}O_2$/ $CH_3(CH_2)_6CHCH_2CH_2$ $\quad\quad\quad\quad\;\; |\quad\quad\;\;\;|$ $\quad\quad\quad\quad\;\; O\!\!-\!\!\!-\!\!C\!\!=\!\!O$	colorless to slightly yel liq/ fruity, peach-like	*s*—alc, most fixed oils, prop glycol; *ins*—gly, water/ 297°	1 mL in 5 mL 60% alc
Undecanal FEMA No. 3092 Aldehyde C-11 Undecyclic; *n*-Undecyl Aldehyde	170.30/$C_{11}H_{22}O$/ $CH_3(CH_2)_9CHO$	colorless to slightly yel liq/ sweet, fatty, floral odor	*s*—most fixed oils, prop glycol; *ins*—gly, water/ 223°	
2-Undecanone FEMA No. 3093 Methyl Nonyl Ketone	170.30/$C_{11}H_{22}O$/ $CH_3CO(CH_2)_9CH_3$	colorless to pale yel liq/ citrus, fatty, rue-like	231°–232°	1 mL in 1 mL 95% alc
10-Undecenal FEMA No. 3095 Aldehyde C-11 Undecylenic; Undecen-10-al	168.28/$C_{11}H_{20}O$/ $CH_2=CH(CH_2)_8CHO$	colorless to light yel liq/ fatty; rose-like odor on dilution	*s*—most fixed oils, prop glycol; *ins*—gly, water/ 235°	
2-Undecenol	170.30/$C_{11}H_{22}O$/ $CH_3(CH_2)_7CH=CHCH_2OH$	white to slightly yel liq/ oily, sweet, floral	*ins*—water	

Requirements

I.D. Test[4]	Assay Min. %[5]	A.V. Max.[6]	Ref. Index[7]	Sp. Gr.[8]	Other Requirements[9]
IR	94.0% of $C_6H_{11}NO$ (M-1a)		1.414–1.435	0.911–0.932	
IR	98.0% of $C_7H_{10}N_2$ (M-1a)		1.503–1.507	0.960–0.990 (20°)	**Water**—0.2% (Appendix IIB, KF; use freshly dist. pyridine as solvent)
IR	98.0% of $C_{11}H_{20}O_2$ (M-1a)		1.457–1.461	0.956–0.961	
IR	98.0% of $C_{11}H_{20}O_2$ (M-1b)	5.0	1.448–1.453	0.942–0.945	
IR	92.0% of $C_{11}H_{22}O$ (M-1b)	10.0	1.430–1.435	0.825–0.832	
IR	96.0% of $C_{11}H_{22}O$ (M-1a)	5.0	1.428–1.432	0.822–0.826	
IR	90.0% of $C_{11}H_{20}O$ (M-1b)	6.0	1.441–1.447	0.840–0.850	
IR	92.0% of $C_{11}H_{22}O$ (M-1a)		1.450–1.452 (22°)	0.847–0.848	

General Information and Description

Name of Substance (Synonyms)	Formula Wt/Formula/ Structure	Physical Form/ Odor	Solubility[1]/ B.P.[2]	Solubility in Alcohol[3]
Undecyl Alcohol FEMA No. 3097 Alcohol C-11	172.31/$C_{11}H_{24}O$/ $CH_3(CH_2)_9CH_2OH$	colorless liq/ fatty-floral	s—most fixed oils; ins—water/ 146° (30 mm Hg)	
Valeraldehyde FEMA No. 3098	86.13/$C_5H_{10}O$/ $CH_3CH_2CH_2CH_2CHO$	colorless to pale yel liq/ characteristic	103°	1 mL in 1 mL 95% alc
Valeric Acid FEMA No. 3101 Pentanoic Acid	102.13/$C_5H_{10}O_2$/ $CH_3(CH_2)_3COOH$	colorless to pale yel, mobile liq/ unpleasant, penetrating, rancid odor	m—alc, ether, 1 mL in 40 mL water/ 186°	
γ-Valerolactone FEMA No. 3103	100.12/$C_5H_8O_2$/	colorless to slightly yel liq/ warm, sweet, herbaceous	m—alc, most fixed oils, water/ 207°	
Vanillin FEMA No. 3107 4-Hydroxy-3-methoxybenzaldehyde	152.15/$C_8H_8O_3$/	fine, white to slightly yel cryst, usually needle-like; affected by light/ odor and taste of vanilla	s—alc, chloroform, ether, 1 g in 100 mL water at 25°, in 20 mL gly, in 20 mL water at 80°	
Zingerone FEMA No. 3124	194.23/$C_{11}H_{14}O_3$	yel to yel-brown liq	290°	

Requirements

I.D. Test[4]	Assay Min. %[5]	A.V. Max.[6]	Ref. Index[7]	Sp. Gr.[8]	Other Requirements[9]
IR	97.0% of $C_{11}H_{24}O$ (M-1a)		1.437–1.443	0.820–0.840	
IR	97.0% of $C_5H_{10}O$ (M-2a)	5.0	1.390–1.395	0.805–0.809	
IR	99.0% of $C_5H_{10}O_2$ (M-3b)		1.405–1.414 (25°)	0.935–0.940	
IR	95.0% of $C_5H_8O_2$ (M-1b)		1.431–1.434	1.047–1.054	
IR	97.0% of $C_8H_8O_3$ (on dried basis) (M-1b)				**Heavy Metals**—10 mg/kg (M-7) **Loss on Drying**—0.5% (Appendix IIC, silica gel/4h) **Melting Range**—between 81° and 83° (Appendix IIB) **Res. on Ignit.**—0.05% (Appendix IIC, 2-g sample)
	95.0% of $C_{11}H_{14}O_3$ (M-1b)		1.538–1.545	1.136–1.140	

TEST METHODS FOR FLAVOR CHEMICALS

This section provides the test methods by which certain flavor chemicals listed in the above tabular section are to be analyzed.

M-1 ASSAY BY GAS CHROMATOGRAPHY

M-1a General Method, Polar Column

Proceed as directed in *GC Assay of Flavor Chemicals*. The composition of the polar column and the conditions of analysis may be varied at the discretion of the analyst, provided that such changes would result in equal or improved separations and/or quantification as would be obtained by use of the particular column material and test conditions specified therein.

M-1b General Method, Nonpolar Column

Proceed as directed in *GC Assay of Flavor Chemicals*. The composition of the nonpolar column and the conditions of analysis may be varied at the discretion of the analyst, provided that such changes would result in equal or improved separations and/or quantification as would be obtained by use of the particular column material and test conditions specified therein.

M-2 ASSAYS FOR CERTAIN ALDEHYDES AND KETONES

M-2a Aldehydes—Hydroxylamine/Tert-Butyl Alcohol Method

Hydroxylamine Solution Dissolve 45 g of reagent-grade hydroxylamine hydrochloride in 130 mL of water, add 850 mL of *tert*-butyl alcohol, mix, and using a pH meter, neutralize to a pH of 3.0 to 3.5 with sodium hydroxide.

Caution: Do not heat the solution.

Procedure Weigh accurately the quantity of sample specified below, and transfer it into a 250-mL glass-stoppered flask. Add 50 mL of the *Hydroxylamine Solution*, mix thoroughly, and allow to stand at room temperature for the time specified. Titrate with 0.5 N sodium hydroxide to the same pH as the *Hydroxylamine Solution* used. Calculate the percentage of aldehyde or ketone by the formula

$$AK = (S)(100e)/W,$$

in which AK is the percentage of aldehyde or ketone; S is the number of mL of 0.5 N sodium hydroxide consumed in the titration of the sample; e is the equivalence factor given below; and W is the weight, in mg, of the sample.

Substance	Sample Weight (g)	Reaction Time (min)	1 mL of 0.5 N NaOH Equivalent to
Cuminic Aldehyde	1	60	74.11 mg of $C_{10}H_{12}O$
Myristaldehyde	1.5	60	106.18 mg of $C_{14}H_{28}O$
Valeraldehyde	1	60	43.07 mg of $C_5H_{10}O$

M-2b Procedure Requiring the Use of Sealed Glass Vials or Ampules

Transfer 65 mL of 0.5 N hydroxylamine hydrochloride and 50.0 mL of 0.5 N triethanolamine into a suitable heat-resistant pressure bottle provided with a tight closure that can be fastened securely. Replace the air in the bottle by passing a gentle stream of nitrogen for 2 min through a glass tube positioned so that the end is just above the surface of the liquid. To the mixture in the pressure bottle add the quantity of sample specified below, contained in a sealed glass ampule. Introduce several pieces of 8-mm glass rod, cap the bottle, and shake vigorously to break the ampule. Allow the bottle to stand at room temperature for the time specified, swirling occasionally. Cool, if necessary, and uncap the bottle cautiously to prevent any loss of the contents. Titrate with 0.5 N sulfuric acid to a greenish blue endpoint. Perform a residual blank titration (see *General Provisions*). Each mL of 0.5 N sulfuric acid is equivalent to the amount specified below.

Substance	Sample Weight (g)	Reaction Time (min)	1 mL of 0.5 N H_2SO_4 Equivalent to
Acetaldehyde	600	30	22.03 mg of C_2H_4O

M-2c Aldehydes—Hydroxylamine Method

Hydroxylamine Hydrochloride Solution Dissolve 50 g of hydroxylamine hydrochloride (preferably reagent grade or freshly recrystallized before using) in 90 mL of water, and dilute to 1000 mL with aldehyde-free alcohol. Adjust the solution to a pH of 3.4 with 0.5 N alcoholic potassium hydroxide.

Procedure Weigh accurately the quantity of sample specified in the table below, and transfer it into a 125-mL Erlenmeyer flask. Add 30 mL of *Hydroxylamine Hydrochloride Solution*, mix thoroughly, and allow to stand at room temperature for the time specified below. Titrate with 0.5 N alcoholic potassium hydroxide to a greenish yellow endpoint that matches the color of 30 mL of *Hydroxylamine Hydrochloride Solution* in a 125-mL flask when the same volume of bromophenol blue TS has been added to each flask, or preferably titrate to a pH of 3.4 using a suitable pH meter. Calculate the percentage of aldehyde (A) by the formula

$$A = (S - b)(100e)/W,$$

in which S is the number of mL of 0.5 N alcoholic potassium hydroxide consumed in the titration of the sample; b is the number of mL of 0.5 N alcoholic potassium hydroxide consumed in the titration of the blank; e is the equivalence factor given below; and W is the weight, in mg, of the sample.

Substance	Sample Weight (g)	Reaction Time (min)	1 mL of 0.5 N KOH Equivalent to
Butyraldehyde	900	60	36.06 mg of C_4H_8O
Isobutyraldehyde	900	60	36.06 mg of C_4H_8O
Propionaldehyde	750	30	29.04 mg of C_3H_6O

M-2d Ketones—Hydroxylamine Method

Hydroxylamine Solution Dissolve 20 g of hydroxylamine hydrochloride (reagent grade or, preferably, freshly crystallized) in 40 mL of water, and dilute to 400 mL with alcohol. Add, with stirring, 300 mL of 0.5 N alcoholic potassium hydroxide, and filter. Use this solution within 2 days.

Procedure Weigh accurately the quantity of sample specified below, and transfer it into a 250-mL glass-stoppered flask. Add 75.0 mL of *Hydroxylamine Solution* to this flask and to a similar flask for a residual blank titration (see *General Provisions*). Attach the flask to a suitable condenser, reflux the mixture for the time specified, and then cool to room temperature. Titrate both flasks with 0.5 N hydrochloric acid to the same greenish yellow endpoint using bromophenol blue TS as the indicator or, preferably, to a pH of 3.4 using a pH meter. (If the indicator is used, the endpoint color must be the same as that produced when the blank is titrated to a pH of 3.4.) Calculate the percentage of ketone by the formula

$$K = (b - S)(100e)/W,$$

in which K is the percentage of ketone; b is the number of mL of 0.5 N hydrochloric acid consumed in the residual blank titration; S is the number of mL of 0.5 N hydrochloric acid consumed in the titration of the sample; e is the equivalence factor given below; and W is the weight, in mg, of the sample.

Substance	Sample Weight (g)	Reaction Time (min)	1 mL of 0.5 N HCl Equivalent to
4-Methyl-2-pentanone	1200	60	50.08 mg of $C_6H_{12}O$

M-3 ASSAY BY TITRIMETRIC PROCEDURES

M-3a Direct Aqueous Acid–Base Titrations

Transfer an accurately weighed amount of the sample, as specified below, to a 250-mL Erlenmeyer flask containing 75 to 100 mL of water, add phenolphthalein TS, and titrate with 0.5 N sodium hydroxide to the first pink color that persists for 15 s. Each mL of 0.5 N sodium hydroxide is equivalent to the amount of substance as specified below.

Substance	Sample Weight (g)	1 mL of 0.5 N NaOH Equivalent to
Butyric Acid	1.5	44.06 mg of $C_4H_8O_2$
Hexanoic Acid	2.0	58.08 mg of $C_6H_{12}O_2$
Isobutyric Acid	1.5	44.06 mg of $C_4H_8O_2$
Isovaleric Acid	1.5	51.07 mg of $C_5H_{10}O_2$
Levulinic Acid	1.0	58.08 mg of $C_5H_8O_3$
2-Mercaptopropionic Acid	1.0	53.08 mg of $C_3H_6O_2S$
2-Methyl-2-pentenoic Acid	2.0	57.02 mg of $C_6H_{10}O_2$
2-Methylbutyric Acid	1.0	51.06 mg of $C_5H_{10}O_2$
4-Methylpentanoic Acid	2.0	58.05 mg of $C_6H_{12}O_2$
2-Methylpentanoic Acid	2.0	58.05 mg of $C_6H_{12}O_2$
Nonanoic Acid	1.0	79.12 mg of $C_9H_{18}O_2$

M-3b Direct Aqueous–Alcoholic Acid–Base Titrations

Dissolve 1 g, accurately weighed, of the sample in 50% ethanol/water, which has been previously neutralized to phenolphthalein TS with 0.1 N sodium hydroxide. Titrate with 0.5 N sodium hydroxide to a pink color. Each mL of titrant is equivalent to the amount of substance specified.

Conditions for Direct Aqueous–Alcoholic Acid–Base Titrations

Substance	1 mL of 0.5 N NaOH Equivalent to
Cinnamic Acid (dried in desiccator 3 h over silica gel)	74.08 mg of $C_9H_8O_2$
2-Ethylbutyric Acid	58.08 mg of $C_6H_{12}O_2$
Phenylacetic Acid (dried 3 h over H_2SO_4)	68.08 mg of $C_8H_8O_2$
Valeric Acid	51.07 mg of $C_5H_{10}O_2$

M-4 ALCOHOL CONTENT OF ETHYL OXYHYDRATE

Mix 25.0 mL of the sample with an equal volume of water in a separator, saturate with sodium chloride, and extract with three 25-mL portions of solvent hexane. Extract the combined solvent hexane extracts with three 10-mL portions of a saturated solution of sodium chloride, and then discard the solvent hexane solutions. Combine the saline solutions in a suitable distillation flask, and distill, collecting 25 mL of distillate. The specific gravity of the distillate is not greater than 0.9814, indicating an alcohol content of not less than 14.0% by volume.

M-5 ACIDITY DETERMINATION BY IODOMETRIC METHOD

Ethyl Formate (Acidity as Formic Acid) Transfer about 5 g, accurately weighed, into a glass-stoppered flask containing a solution of 500 mg of potassium iodate and 2 g of potassium iodide in 50 mL of water. Titrate the liberated iodine with 0.1 N sodium thiosulfate, using starch TS as the indicator. Each mL of 0.1 N sodium thiosulfate is equivalent to 4.603 mg of CH_2O_2.

M-6 LIMIT TEST FOR ANTIOXIDANTS IN ETHYL ACRYLATE

Preliminary Examination of the Sample Wash a 25-mL portion of the sample with 25 mL of sodium hydroxide solution (1 in 10). Any yellow or brown coloration in the extract indicates the presence of hydroquinone, in which case both of the procedures below (*A* and *B*) must be followed to determine the antioxidant content. If the sodium hydroxide extract remains colorless, the first procedure (*A*) need not be run, and the antioxidant content is determined by the second procedure (*B*) alone.

A. Determination of Hydroquinone

Carbonyl-Free Methanol To 500 mL of anhydrous methanol add 5 g of 2,4-dinitrophenylhydrazine, heat the mixture under a reflux condenser for 2 h, and then recover the methanol by distillation. Store the carbonyl-free methanol in tight containers.

2,4-Dinitrophenylhydrazine Solution Dissolve 100 mg of 2,4-dinitrophenylhydrazine in 50 mL of *Carbonyl-Free Methanol*, add 4 mL of hydrochloric acid, and dilute to 100 mL with water.

Sodium Carbonate Solution Dissolve 530 mg of sodium carbonate in sufficient water to make 100 mL.

Pyridine–Diethanolamine Solution Mix 5 mL of diethanolamine with 500 mL of freshly distilled pyridine.

Calibration Curve Transfer 25 mg of hydroquinone, accurately weighed, into a 100-mL volumetric flask, add sufficient butyl acetate to volume, and mix thoroughly (250 μg/mL). Prepare a series of standards by transferring 1.0-, 2.0-, 3.0-, 4.0-, and 6.0-mL portions of this solution into separate 50-mL volumetric flasks, and diluting each aliquot to 50.0 mL with butyl acetate. One mL of each of these standards contains 5, 10, 15, 20, and 30 μg, respectively, of hydroquinone. Transfer 1.0 mL of each solution into separate 25-mL glass-stoppered graduates, and continue as directed in the *Procedure*, beginning with "... add 2.0 mL of water...." Plot a calibration curve of absorbance versus μg of hydroquinone. Fifteen μg of hydroquinone should be equivalent to approximately 0.30 units of absorbance, and the curve should intersect the origin.

Procedure Using a hypodermic syringe, transfer 0.2 mL of the sample, accurately weighed, into a 25-mL glass-stoppered graduate, add 2.0 mL of water, stopper the graduate, and mix the contents well without allowing contact between the liquid and the stopper. Add to the mixture 0.5 mL of the *Sodium Carbonate Solution*, and immediately shake gently for 5 s, avoiding contact between the solution and the stopper. Immediately add 1.0 mL of a 15%, volume in volume, solution of sulfuric acid, shake as previously directed, and add 1-mL of the *Dinitrophenylhydrazine Solution*. Stopper the graduate and place it in a water bath, maintained at a temperature between 70° and 72°, for 1 h. Shake samples three times during the heating period. Cool the graduate to room temperature, dilute the contents to 15 mL with water, add 5.8 mL of benzene, stopper, shake vigorously, and then allow the phases to separate. Transfer 2.0 mL of the benzene layer, using a suitable pipet, into a test tube, add 10.0 mL of *Pyridine–Diethanolamine Solution*, and mix. Transfer a portion of this solution into a 2-cm cell, and determine the absorbance at 620 nm with a suitable spectrophotometer, using as a blank 1.0 mL of butyl acetate treated in the same manner as the sample except that 5.0 mL of benzene is used for the extraction instead of 5.8 mL. From the previously prepared *Calibration Curve*, read the μg of hydroquinone and/or benzoquinone corresponding to the absorbance of the solution from the sample, and record this value as w. Calculate the mg/kg of hydroquinone (mg/kg HQ) in the sample by the formula

$$1000w/W,$$

in which W is the weight, in mg, of the sample taken.

B. Determination of Hydroquinone Monomethyl Ether

Antioxidant-Free Ethyl Acrylate Wash a suitable volume of the sample with three separate, similar-sized volumes of sodium hydroxide solution (1 in 10). After the last washing, add a small amount of sodium chloride, if necessary, to remove any turbidity that may be present.

Calibration Curve Transfer 25.0 mg of hydroquinone monomethyl ether, accurately weighed, into a 100-mL volumetric flask, add *Antioxidant-Free Ethyl Acrylate* to volume, and shake to effect complete solution (250 μg/mL). Prepare a series of standards by transferring 1.0-, 5.0-, 10.0-, and 20.0-mL portions of this solution into separate 25-mL volumetric flasks, diluting to volume with *Antioxidant-Free Ethyl Acrylate*, and mixing. One mL of each of the standards contains 10, 50, 100, and 200 μg, respectively, of hydroquinone monomethyl ether. Transfer 5.0 mL of each solution into separate 50-mL volumetric flasks, dilute each to volume with isooctane, and mix. Determine the absorbance of each solution in a 1-cm silica cell at 292 nm with a suitable spectrophotometer, using a 1 in 10 dilution of *Antioxidant-Free Ethyl Acrylate* as the blank. Plot a calibration curve of absorbance versus μg of hydroquinone monomethyl ether. The curve should be linear and should intersect the origin.

Procedure Transfer 5.0 mL of the sample, accurately weighed, into a 50-mL volumetric flask, dilute to volume with isooctane, and mix. Determine the absorbance of this solution in a 1-cm silica cell at 292 nm with a suitable spectrophotometer, using a 1 in 10 dilution of *Antioxidant-Free Ethyl Acrylate* in isooctane as the blank. From the previously prepared *Calibration Curve* read the μg of hydroquinone monomethyl ether corresponding to the absorbance of the sample solution, and record this value as w. Calculate the mg/kg of hydroquinone monomethyl ether (mg/kg HMME) in the sample by the formula

$$w/W,$$

in which W is the weight, in g, of the sample taken.

Note: If the first sodium hydroxide extract obtained under *Preliminary Examination of the Sample* (or under *Antioxidant-Free Ethyl Acrylate*) showed a yellow coloration, the true mg/kg HMME is obtained by subtracting the mg/kg HQ, obtained under section *A*, from the apparent mg/kg HMME.

M-7 LIMIT TEST FOR HEAVY METALS

Prepare and test, as directed in *Method II* under the *Heavy Metals Test*, Appendix IIIB, an accurately weighed sample corresponding to the following.

Heavy Metals Limit	Sample Weight (mg)
0.004%	500
0.002%	1000
10 mg/kg	2000
5 mg/kg	4000

Use 20 μg of lead ion (Pb) in the control (*Solution A*).

M-8 LIMIT TEST FOR HYDROCARBONS IN EUGENOL

Dissolve 1 mL of the sample in 20 mL of 0.5 N sodium hydroxide in a stoppered 50-mL tube, add 18 mL of water, and mix. A clear mixture results immediately, but it may become turbid when exposed to air.

M-9 LIMIT TEST FOR HYDROCYANIC ACID IN BENZALDEHYDE

Shake 0.5 mL with 5 mL of water, add 0.5 mL of 1 N sodium hydroxide and 0.1 mL of ferrous sulfate TS, and warm the mixture gently. Upon the addition of a slight excess of hydrochloric acid, no greenish blue color or blue precipitate is produced within 15 min.

M-10 LIMIT TEST FOR LEAD

A *Sample Solution* prepared as directed for organic compounds using a 1-g sample meets the requirements of the *Lead Limit Test*, Appendix IIIB, using 10 μg of lead ion (Pb) in the control.

M-11 LIMIT TEST FOR METHYL COMPOUNDS IN ETHYL ACETATE

Place 20 mL of sample in a 500-mL separator, add a solution of 20 g of sodium hydroxide in 50 mL of water, stopper the separator, and wrap it securely in a towel for protection against the heat of the reaction. Shake the mixture vigorously for about 5 min, cautiously opening the stopcock from time to time to permit the escape of air. Continue the shaking vigorously until a homogeneous liquid results, then distill, and collect about 25 mL of the distillate. To 1 drop of the distillate add 1 drop of dilute phosphoric acid (1 in 20) and 1 drop of potassium permanganate solution (1 in 20). Mix, allow to stand for 1 min, and add sodium bisulfite solution (1 in 20) dropwise until the color is discharged. If a brown color remains, add 1 drop of the dilute phosphoric acid. To the colorless solution add 5 mL of a freshly prepared solution of chromotropic acid (1 in 2000) in 75% sulfuric acid, and heat on a steam bath for 10 min at 60°. No violet color appears.

M-12 LIMIT TEST FOR PEROXIDE VALUE

To 50 mL of a mixture of 3 volumes of glacial acetic acid and 2 volumes of chloroform add 10 mL of the sample. To this solution add 1 mL of a saturated solution of potassium iodide, allow to stand for exactly 1 min with gentle shaking, and then introduce 100 mL of water and a few drops of starch TS. Titrate immediately with 0.1 N sodium thiosulfate. Each mL of 0.1 N sodium thiosulfate, multiplied by 5, equals the peroxide value, expressed in millimoles of peroxide per liter of the sample.

M-13 LIMIT TEST FOR READILY CARBONIZABLE SUBSTANCES IN ETHYL ACETATE

Carefully pour 2 mL of the sample on 10 mL of 95% sulfuric acid so as to form separate layers. No discoloration is developed within 15 min.

M-14 LIMIT TEST FOR READILY OXIDIZABLE SUBSTANCES IN *dl*-MENTHOL

Transfer 500 mg of *dl*-menthol into a clean, dry test tube, and add 10 mL of potassium permanganate solution prepared by diluting 3 mL of 0.1 N potassium permanganate with water to 100 mL. Place the test tube in a beaker of water maintained between 45° and 50°. At 30-s intervals, quickly remove the test tube from the bath and shake. The color of potassium permanganate is still apparent after 5 min.

M-15 LIMIT TEST FOR REDUCING SUBSTANCES

Dilute 2 mL of the sample in a glass-stoppered flask with 50 mL of water and 5 mL of sulfuric acid, shaking the flask during the addition. While the solution is still warm, titrate with 0.1 N potassium permanganate. Not more than 1 mL is required to produce a pink color that persists for 30 min.

M-16 ACID VALUE

Dissolve about 10 g of the sample, accurately weighed, in 50 mL of alcohol, previously neutralized to phenolphthalein with 0.1 N sodium hydroxide. (Add 50 g of ice when testing cinnamyl formate, citronellyl formate, geranyl formate, isoamyl formate, and linalyl formate.) Add 1 mL of phenolphthalein TS, and titrate with 0.1 N sodium hydroxide until the solution remains faintly pink after shaking for 10 s, unless otherwise directed in the individual monograph. Calculate the acid value (AV) by the formula

$$AV = (5.61 \times S)/W,$$

in which S is the number of mL of 0.1 N sodium hydroxide consumed in the titration of the sample, and W is the weight, in g, of the sample.

M-17 RESIDUE ON EVAPORATION

Weigh accurately the quantity of sample specified in the monograph, and transfer it into a suitable evaporating dish that has previously been heated on a steam bath, cooled to room temperature in a desiccator, and accurately weighed. Weigh the sample in the dish. Heat the evaporating dish containing the sample on the steam bath for 1 h. Cool the dish and its contents to room temperature in a desiccator, and weigh accurately. Calculate the residue as percentage of the sample used.

M-18 QUALITATIVE TEST FOR PHENOLS USING FERRIC CHLORIDE

Allyl Isothiocyanate Dilute 1 mL of the sample with 5 mL of alcohol, and add 1 drop of ferric chloride TS. A blue color is not produced immediately.

Anethole Shake 1 mL with 20 mL of water, and allow the liquids to separate. Filter the water layer through a filter paper previously moistened with water, and to 10 mL of the filtrate add 3 drops of ferric chloride TS. No purplish color is produced.

Anisole Shake 1 mL with about 20 mL of water, allow the layers to separate, collect the water layer in a test tube, and add to it a few drops of ferric chloride TS. No greenish, bluish, or purplish color is produced.

Cresyl Acetate (Test for Free Cresol)
  Ferric Chloride Solution Add 1.5 g of anhydrous ferric chloride to 850 mL of chloroform in a 2-L beaker. Add 100 mL of ethylene glycol monobutyl ether. When the ferric chloride has dissolved, add 50 mL of pyridine, mix, and filter through a Büchner funnel.
  Procedure Transfer a 5-mL sample to a 15-mm test tube, and add 10 mL of the *Ferric Chloride Solution*. The color of the solution is no darker green than a solution of 5 mL of a 1% solution of cresol in cresol-free methyl *p*-cresol mixed with 10 mL of the *Ferric Chloride Solution*.

GAS CHROMATOGRAPHIC (GC) ASSAY OF FLAVOR CHEMICALS

This procedure applies both to the assay of flavor chemicals and to the quantitation of minor components in flavor chemicals. Analysts following this procedure and performing the test should obtain sufficient resolution of major and even trace components of a mixture to calculate accurately the concentration of the desired component; should be familiar with the general principles, usual techniques, and instrumental variables normally met in gas chromatographic analysis; and should pay particular attention to the following:

1. Stability of baseline, return to baseline before and after each peak of interest, and minimum use of recorder attenuation.
2. Any incompatibility between a sensitive sample component and column support, liquid substrate, or construction material.
3. The response to different components of the same or different detectors. Since sizable errors may be encountered in correlating area percent directly to weight percent, the methods for calculating response factors should be known.
4. Where limits for minor components are specified in the column entitled *Other Requirements* in the above tabular specifications for flavor chemicals, analysts should use authentic materials to confirm the retention times of minor components. Determine the quantity of components following the instructions below under *Calculations and Methods*.

I GC CONDITIONS FOR ANALYSIS

Column: open tubular capillary column of fused silica or deactivated glass 30 m long × 0.25 to 0.53 mm id.

Stationary phase:
1. For a **nonpolar column**:   methyl silicone gum, or equivalent (preferably a bonded and cross-linked dimethyl polysiloxane);
2. For a **polar column**:   polyethylene glycol, or equivalent (preferably a bonded and cross-linked polyethylene glycol);
3. The stationary phase coating should have a thickness of 1 to 3 μm.

Carrier gas: helium flowing at a linear velocity of 20 to 40 mL/s.

Sample size: 0.1 to 1.0 μL.

Split ratio: (for 0.25 mm to 0.35 mm id columns only) 50:1 to 200:1, typically, making sure that no one component exceeds the capacity of the column.

Inlet temperature: 225° to 275°.

Detector temperature: 250° to 300°.

Detectors: use either a thermal conductivity or flame ionization detector, operating both as recommended by the manufacturer.

Oven program: 50° to 240°, increasing the temperature by 5°/min; hold at 240° for 5 min.

Analysts can also use any GC conditions providing separations equal to (or better than) those obtained with the above method, but in the case of a dispute, the above method must stand.

II CALCULATIONS AND METHODS

A. Peak area integration with total area detected normalized to 100%, using electronic integrators: Use an electronic peak integrator in accordance with the manufacturer's recommendations, ensuring that the integration parameters permit proper integration of the peaks of a variety of shapes and magnitudes and do not interpret baseline shifts and noise spikes as area contributed by the sample. Use internal or external standards as needed to confirm that the total GC peak area corresponds to 100% of the components present in the sample.

B. Results obtained as described above are based on the assumption that the entire sample has eluted and the peaks of all of the components have been included in the calculation. They will be incorrect if any part of the sample does not elute or if all the peaks are not measured. In such cases, and in all methods described above, the internal standard method may be used to determine percentages based on the total sample. For this method, measurements are required of the peaks of the component(s) being assayed and of the internal standard.

An accurately weighed mixture of the internal standard and the sample is prepared and chromatographed, the area ratio(s) of the component(s) to the standard is computed, and the percentage(s) of the component(s) is calculated.

If this calculation is to be applied, the substance used as the standard should be one that meets the following criteria:

a. Its detector response is similar to that of the component(s) to be determined. In general, the more nearly the chemical structure of the component resembles that of the standard, the closer the response will be.

b. Its retention time is close to, but not identical with, that of the component(s).

c. Its elution time is different from that of any other component in the sample so that its peak does not superimpose on any other.

The weight ratio of the internal standard to the sample should be such that the internal standard and the component sought produce approximately equal peaks. This is, of course, not possible if several components of interest are at different levels of concentration.

If the internal standard method is applied properly, it may be assumed that the ratio of the weight of component to the weight of internal standard is exactly proportional to the peak area ratio, and under these conditions no correction factor is needed. The sample is first run by itself to determine whether the internal standard would mask any component by peak superposition. If there is no interference, a mixture is prepared of the sample and of the internal standard in the specified weight ratio, and the percentages of the internal standard and of the sample in the mixture are calculated. The mixture is chromatographed, and the areas of the component peak and the internal standard peak are calculated by one of the methods described above.

The calculations are as follows:

1a. % Component in Mixture / % Internal Standard in Mixture = Component Area / Internal Standard Area

or

1b. % Component in Mixture = % Int. Std. in Mixture × (Component Area / Internal Standard Area)

2. % Component in Sample = (% Component in Mixture × 100) / % Sample in Mixture

Should calibration be necessary, mixtures should be prepared of internal standard and component of either 100% or of known purity. The number of mixtures and the weight ratios to be used depend on the component being analyzed. Usually, three mixtures will be required. The weight ratio of one is chosen so that the heights of component and standard are equal. The ratios of the other two may be two-thirds and four-thirds of this value. Each mixture should be chromatographed at least three times, and areas calculated. The factor for each chromatograph should be calculated as specified below, and the averages taken for each mixture. An overall average factor is calculated from them. The calibration should be performed periodically.

1. Factor = [(Wt. Component × % Purity) / (Wt. Int. Std. × % Purity)] × [(Int. Std. Area) / Component Area]

2. % Component in Sample Mixture = (Component Area × Factor × % Int. Std. in Sample Mixture) / Int. Std. Area

3. % Component in Sample = (% Component in Sample Mixture × 100) / % Sample in Sample Mixture

III GC SYSTEM SUITABILITY TEST SAMPLE

The GC system suitability test sample consists of an equal-weight mixture of FCC-quality acetophenone, benzyl alcohol, benzyl acetate, linalool, and hydroxycitronellal.

Using the test sample described below, periodically test the performance of and resolution provided by the gas chromatograph employed. The test sample must display results comparable in quantitative composition, peak shape, and elution order to those specified herein. The quantitative composition should not deviate from the results listed below by more than ±10%. Analyze the GC test sample using the *GC Conditions for Analysis* given above.

Component in Test Sample	Order of Elution		Normalized % Area (FID)	
	Nonpolar	Polar	Nonpolar	Polar
Benzyl Alcohol	1	4	22.0	21.3
Acetophenone	2	2	21.1	21.4
Linalool	3	1	20.8	21.0
Benzyl Acetate	4	3	18.6	19.1
Hydroxycitronellal	5	5	16.7	16.7

4 / Infrared Spectra

INTRODUCTION

The infrared absorption spectra contained in this section are provided in conjunction with the requirement for *Identification* as specified for a number of substances in this edition.

ORGANIZATION

This section contains reproductions of infrared spectra for three major groups of substances.

Series A (pages 572–600), for essential oils
Series B (pages 601–711), for the flavor chemicals
Series C (pages 713–722), for miscellaneous other substances

Most of the spectra were prepared especially for use in the *Food Chemicals Codex* by the Technical Committee of the Flavor and Extract Manufacturers Association. It was not feasible for them to be recorded in a single uniform format, however. Consequently, spectra of several shapes and sizes, contributed by different laboratories using various types of infrared spectrophotometers, will be found. Within each series, however, spectra of the same format have been grouped together in uniform subseries. An alphabetical listing, with page references, is provided at the beginning of each series to aid the reader in locating the desired spectrum.

SAMPLE PREPARATION

Most of the substances for which spectra are provided are liquids at or near room temperature. Unless otherwise noted in the caption for an individual spectrum, the spectra for substances in *Series A* and *B* were obtained on the neat liquids contained in fixed-volume sodium chloride cells or between salt plates. For substances in *Series A* and *B* that are not liquids, the sample was prepared as a potassium bromide pellet or a mineral oil (Nujol or equivalent) dispersion, as indicated in the individual spectrum caption. For substances in *Series C*, the samples were prepared as directed under *Identification* in the individual monographs.

Series A: Essential Oils

This series is divided into two subseries depending on the format of the spectra involved. The substances are listed below alphabetically.

Almond Oil, Bitter, FFPA ... 573
Ambrette Seed Oil ... 596
Amyris Oil, West Indian Type ... 573
Angelica Root Oil ... 573
Angelica Seed Oil ... 574
Anise Oil ... 574
Balsam Peru Oil ... 574
Basil Oil, Comoros Type ... 574
Basil Oil, European Type ... 597
Bay Oil ... 575
Bergamot Oil, Coldpressed ... 575
Birch Tar Oil, Rectified ... 575
Black Pepper Oil ... 575
Bois de Rose Oil ... 576
Cananga Oil ... 576
Caraway Oil ... 576
Cardamom Oil ... 576
Carrot Seed Oil ... 577
Cascarilla Oil ... 577
Cassia Oil ... 577
Cedar Leaf Oil ... 577
Celery Seed Oil ... 578
Chamomile Oil, English Type ... 578
Chamomile Oil, German Type ... 578
Cinnamon Bark Oil, Ceylon Type ... 578
Cinnamon Leaf Oil ... 579
Clary Oil ... 579
Clove Leaf Oil ... 579
Clove Oil ... 579
Clove Stem Oil ... 580
Cognac Oil, Green ... 580
Copaiba Oil ... 580
Coriander Oil ... 580
Costus Root Oil ... 581

Cubeb Oil ... 581
Cumin Oil ... 581
Dill Seed Oil, European Type ... 581
Dill Seed Oil, Indian Type ... 597
Dillweed Oil, American Type ... 582
Eucalyptus Oil ... 582
Fennel Oil ... 582
Fir Needle Oil, Canadian Type ... 582
Fir Needle Oil, Siberian Type ... 583
Garlic Oil ... 583
Geranium Oil, Algerian Type ... 583
Ginger Oil ... 583
Grapefruit Oil, Coldpressed ... 584
Hops Oil ... 584
Juniper Berries Oil ... 584
Labdanum Oil ... 584
Laurel Leaf Oil ... 585
Lavandin Oil, Abrial Type ... 585
Lavender Oil ... 585
Lemongrass Oil ... 585
Lemon Oil, Coldpressed ... 586
Lemon Oil, Desert Type, Coldpressed ... 598
Lemon Oil, Distilled ... 586
Lime Oil, Coldpressed (Mexican Type) ... 598
Lime Oil, Coldpressed (Tahitian Type) ... 586
Lime Oil, Distilled ... 586
Linaloe Wood Oil ... 587
Lovage Oil ... 587
Mace Oil ... 587
Mandarin Oil, Coldpressed ... 587
Marjoram Oil, Spanish Type ... 588
Marjoram Oil, Sweet ... 588

Mentha Arvensis Oil, Partially Dementholized ... 588
Myrrh Oil ... 588
Nutmeg Oil ... 589
Olibanum Oil ... 589
Onion Oil ... 589
Orange Oil, Coldpressed ... 589
Orange Oil, Bitter, Coldpressed ... 590
Orange Oil, Distilled ... 590
Origanum Oil, Spanish Type ... 590
Orris Root Oil ... 590
Palmarosa Oil ... 591
Parsley Herb Oil ... 591
Parsley Seed Oil ... 591
Pennyroyal Oil ... 591
Peppermint Oil ... 592
Petitgrain Oil, Paraguay Type ... 592
Pimenta Leaf Oil ... 592
Pimenta Oil ... 592
Pine Needle Oil, Dwarf ... 599
Pine Needle Oil, Scotch Type ... 599
Rosemary Oil ... 593
Rose Oil ... 593
Rue Oil ... 593
Sage Oil, Dalmatian Type ... 593
Sage Oil, Spanish Type ... 594
Sandalwood Oil, East Indian Type ... 594
Savory Oil (Summer Variety) ... 594
Spearmint Oil ... 600
Spike Lavender Oil ... 594
Tangerine Oil, Coldpressed ... 595
Tarragon Oil ... 595
Thyme Oil ... 595
Wintergreen Oil ... 600

SERIES A-1

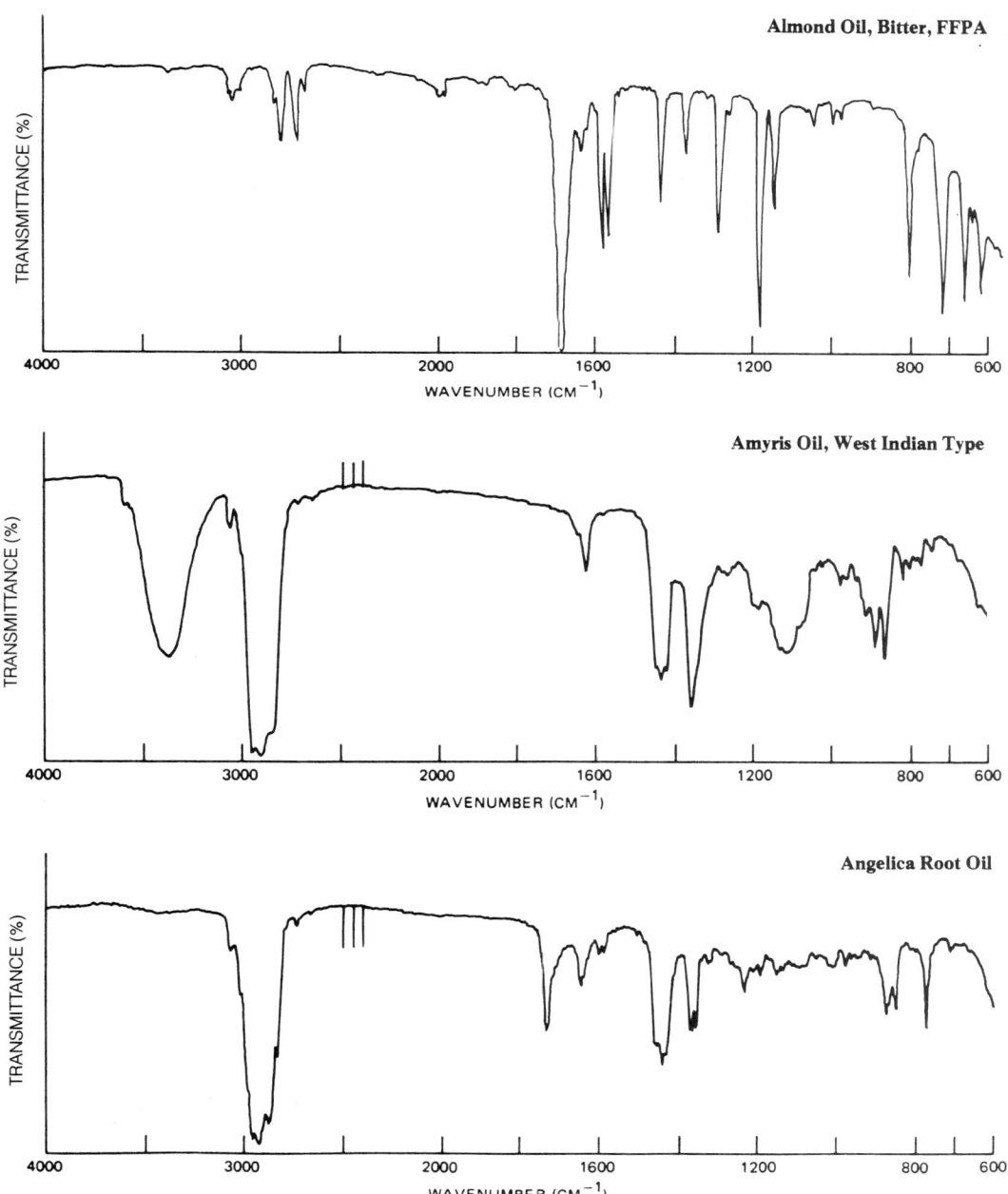

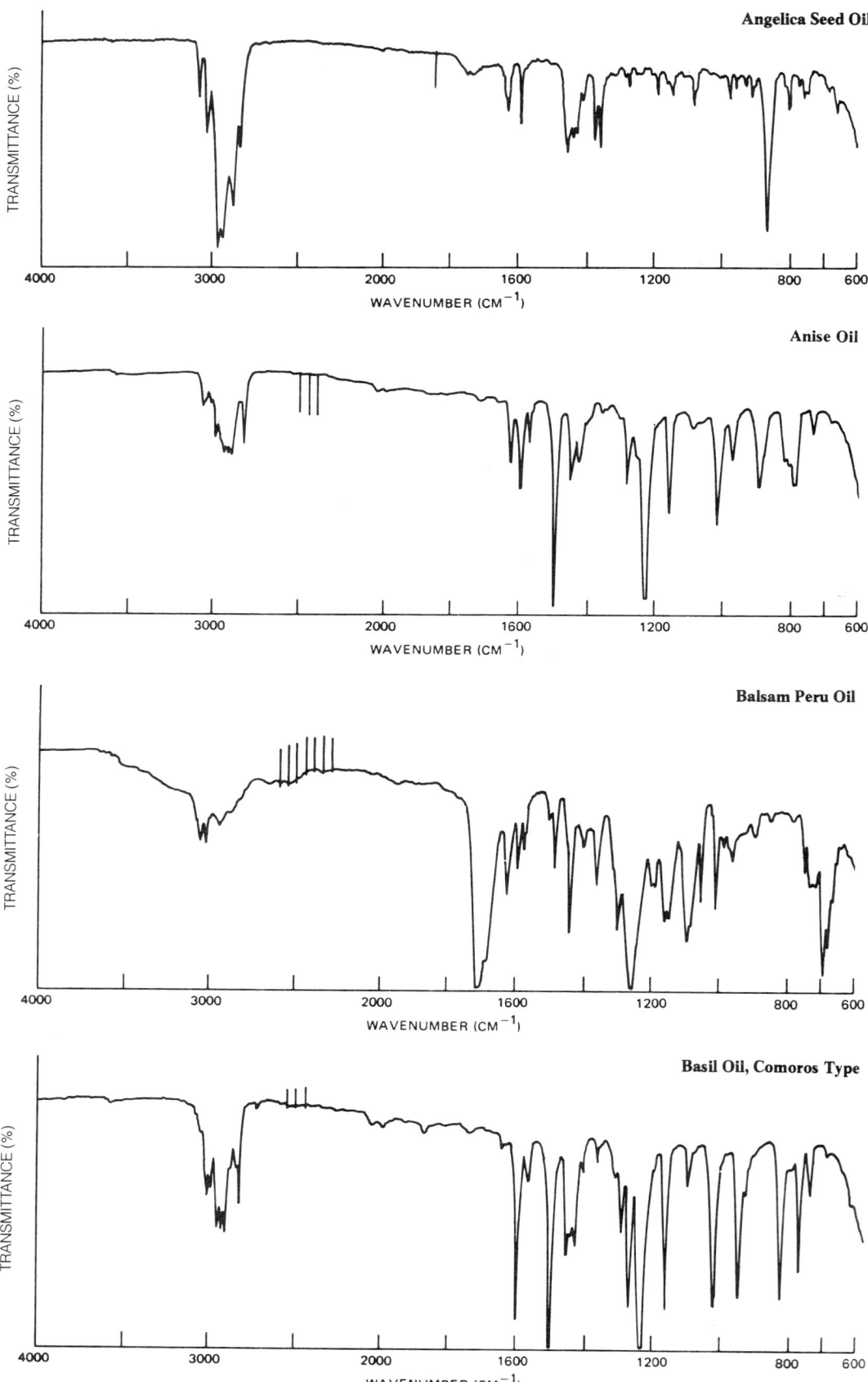

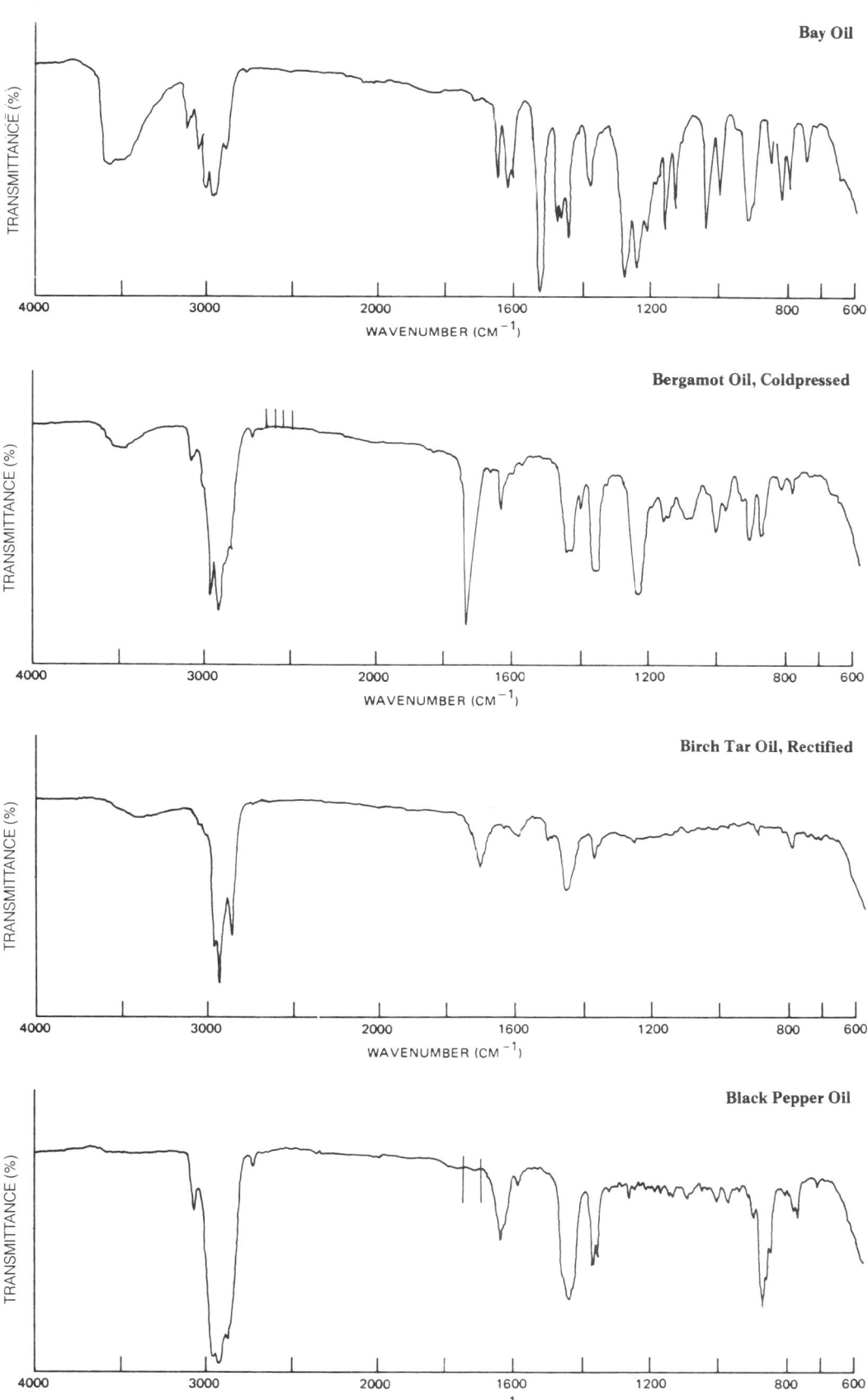

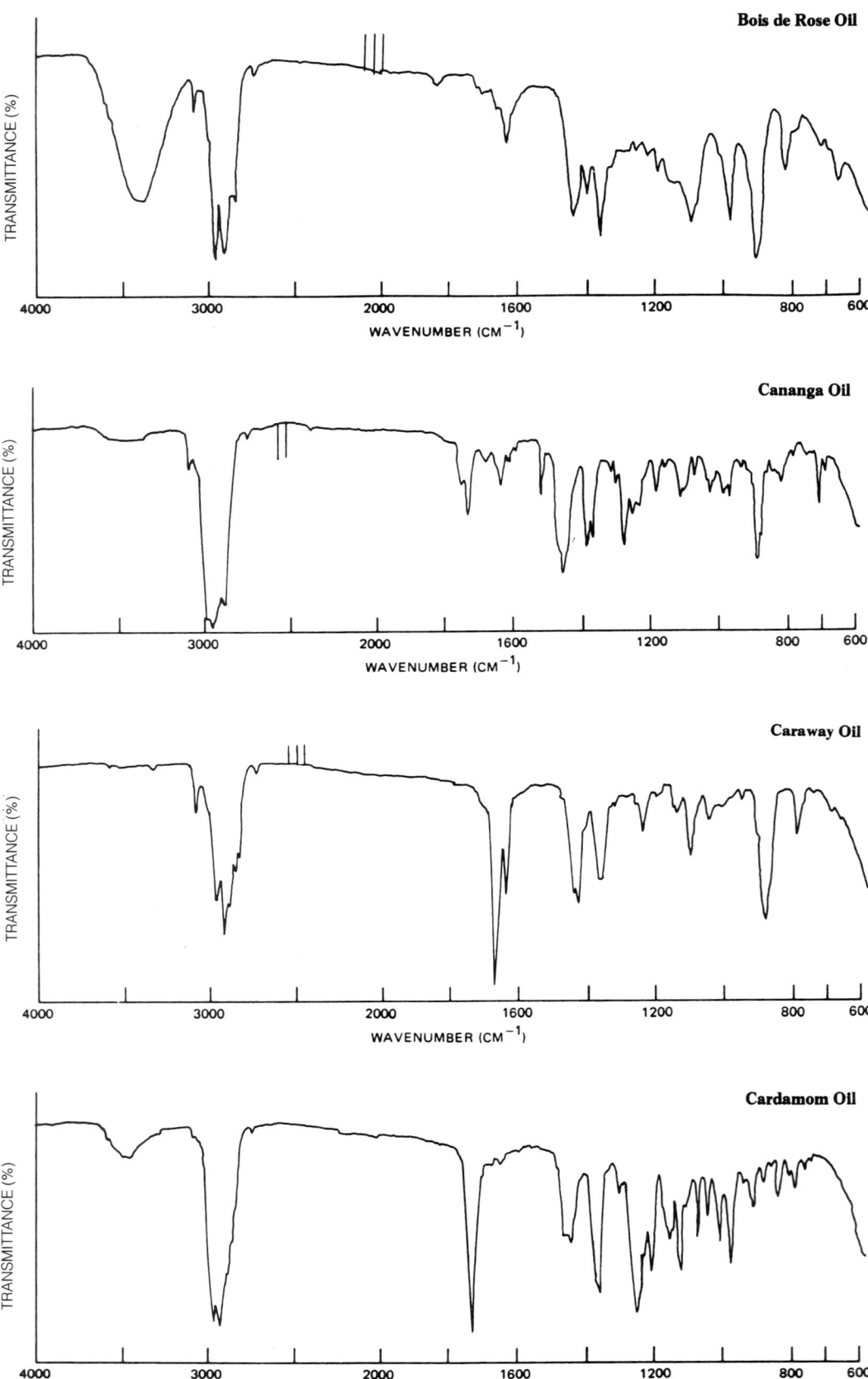

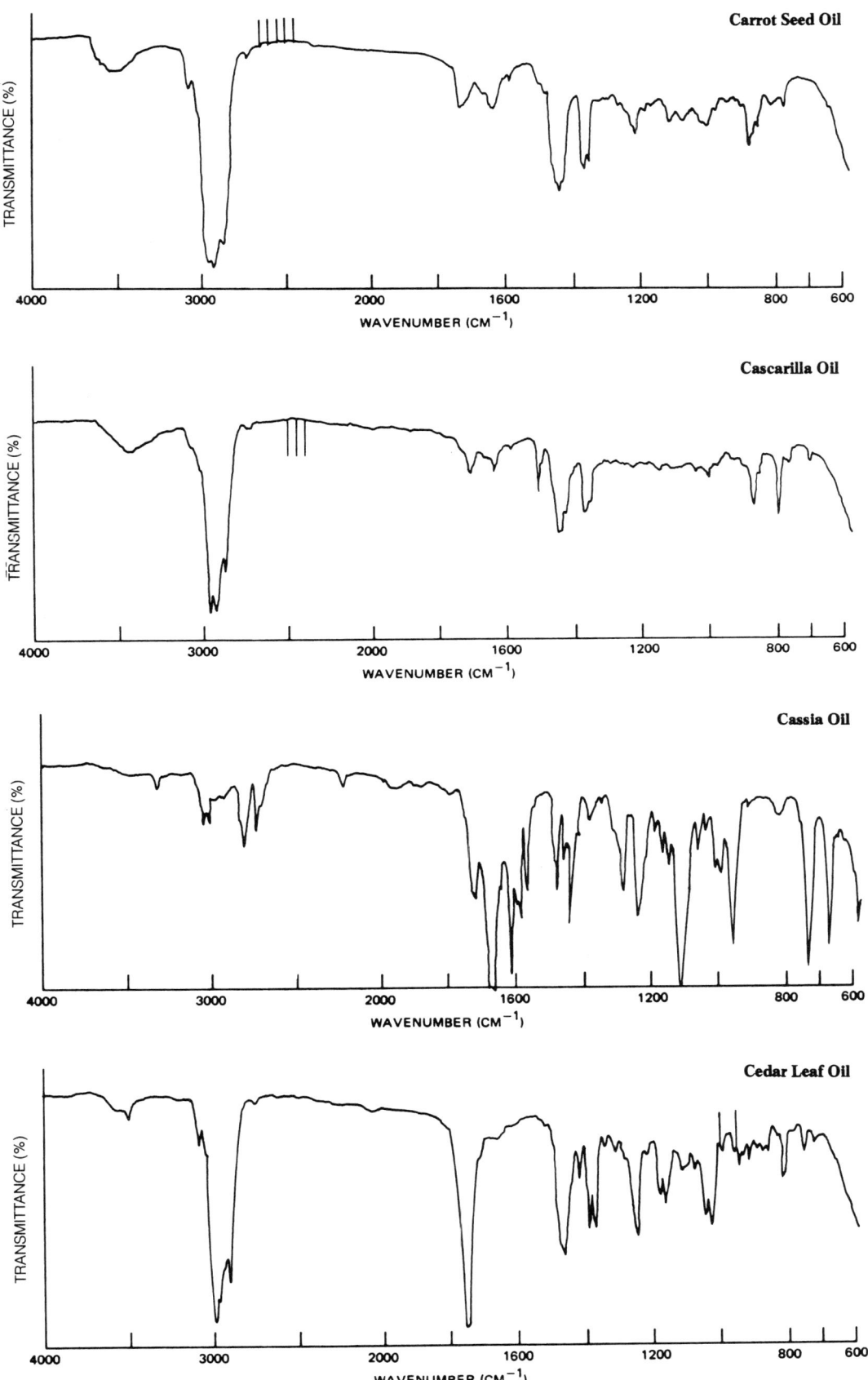

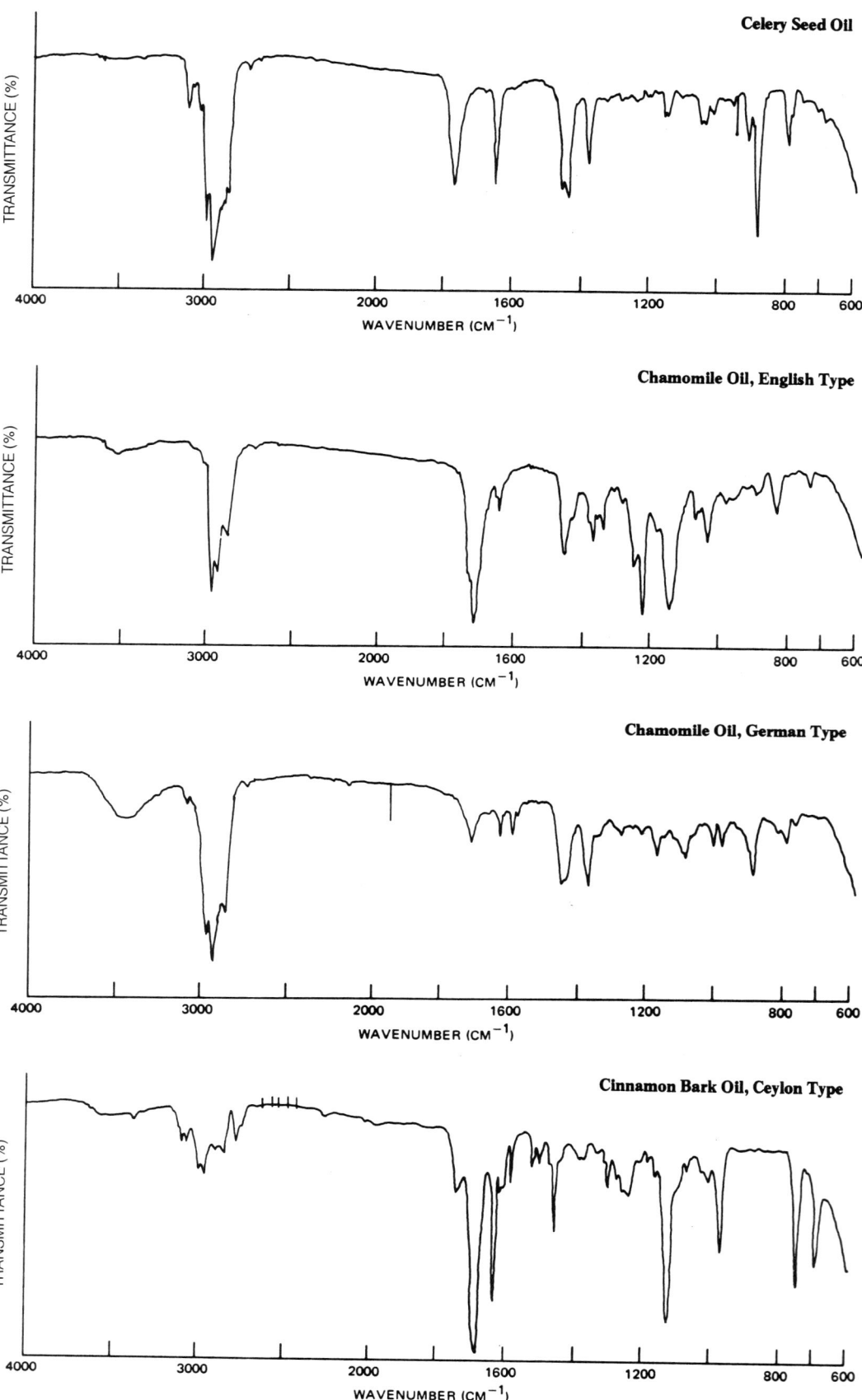

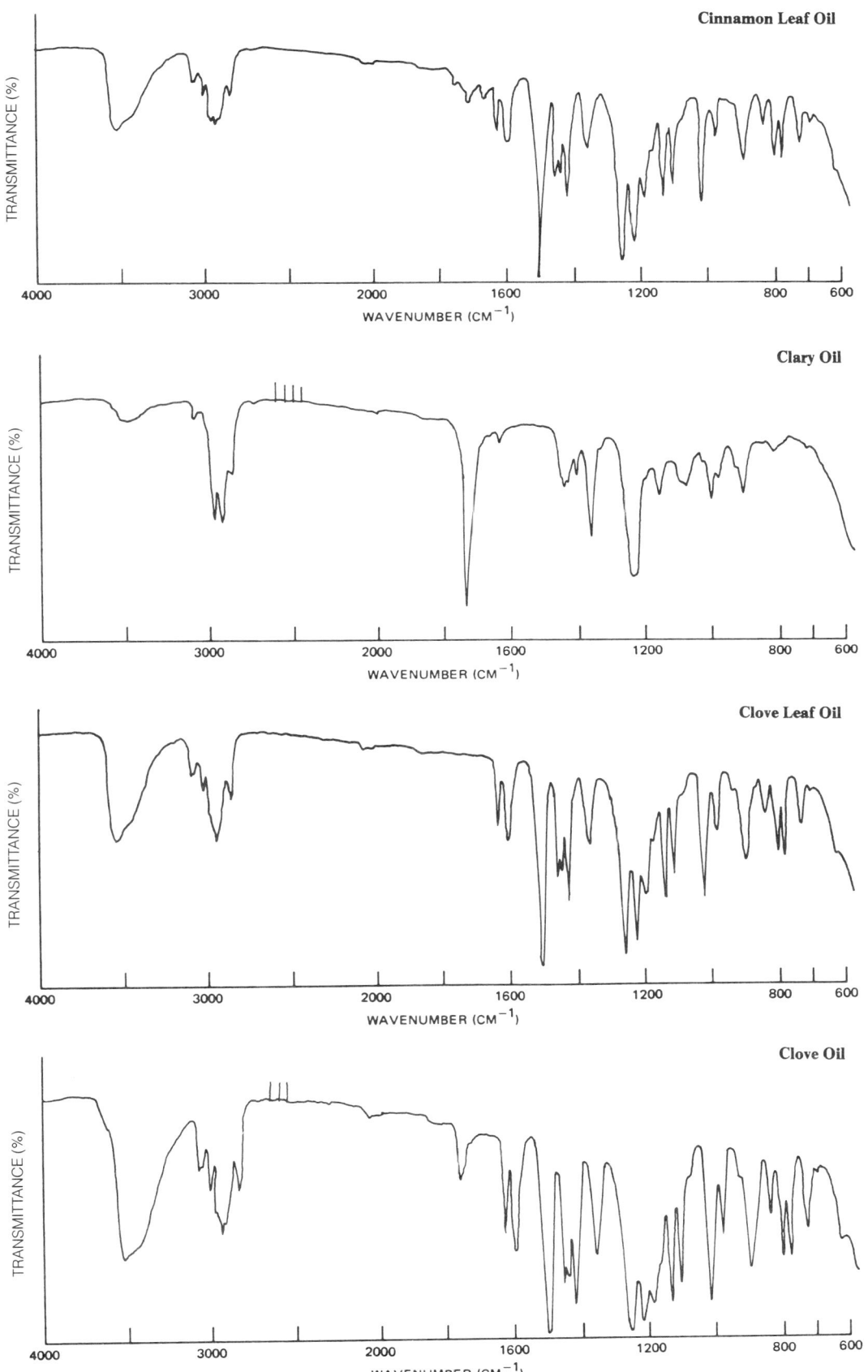

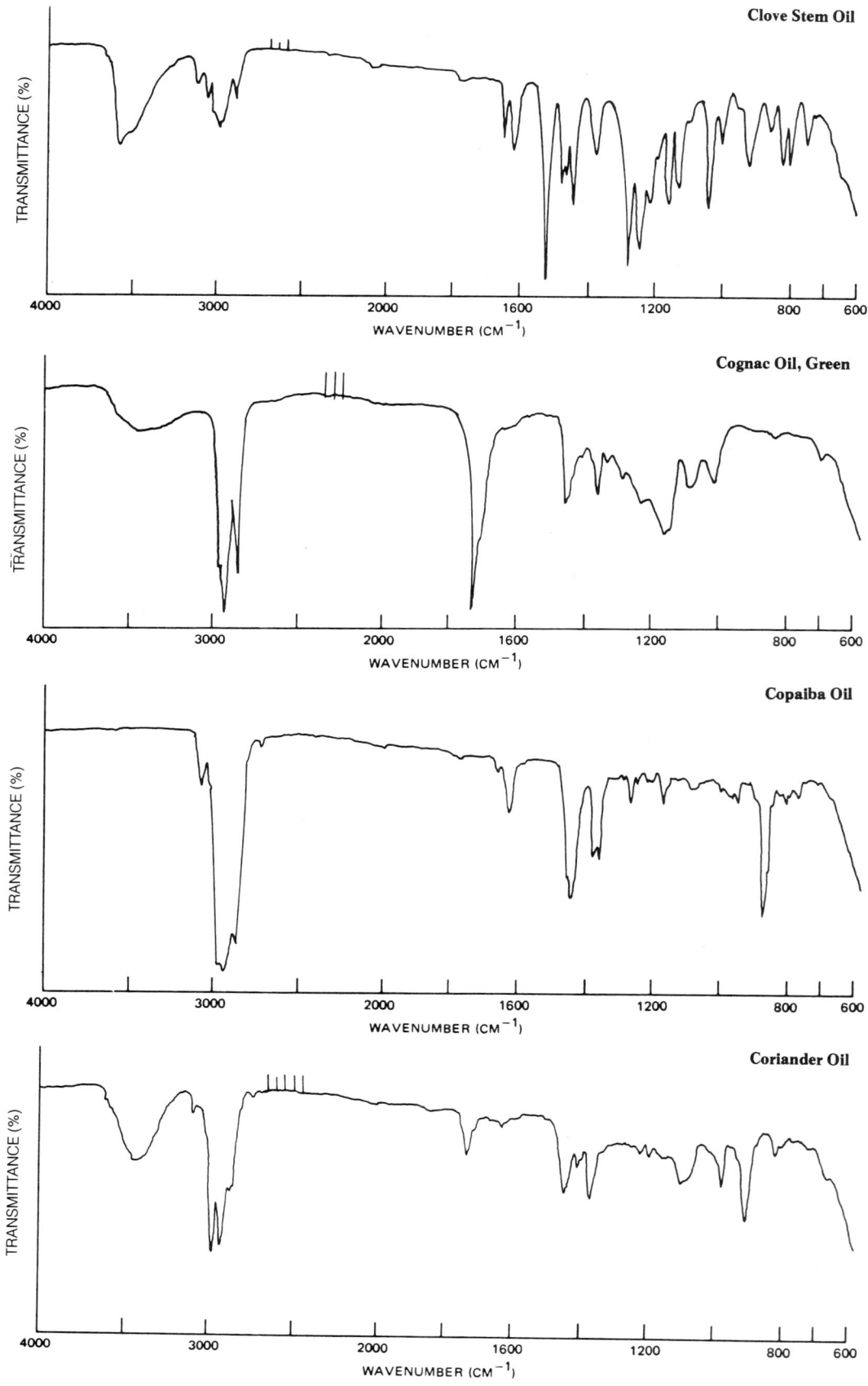

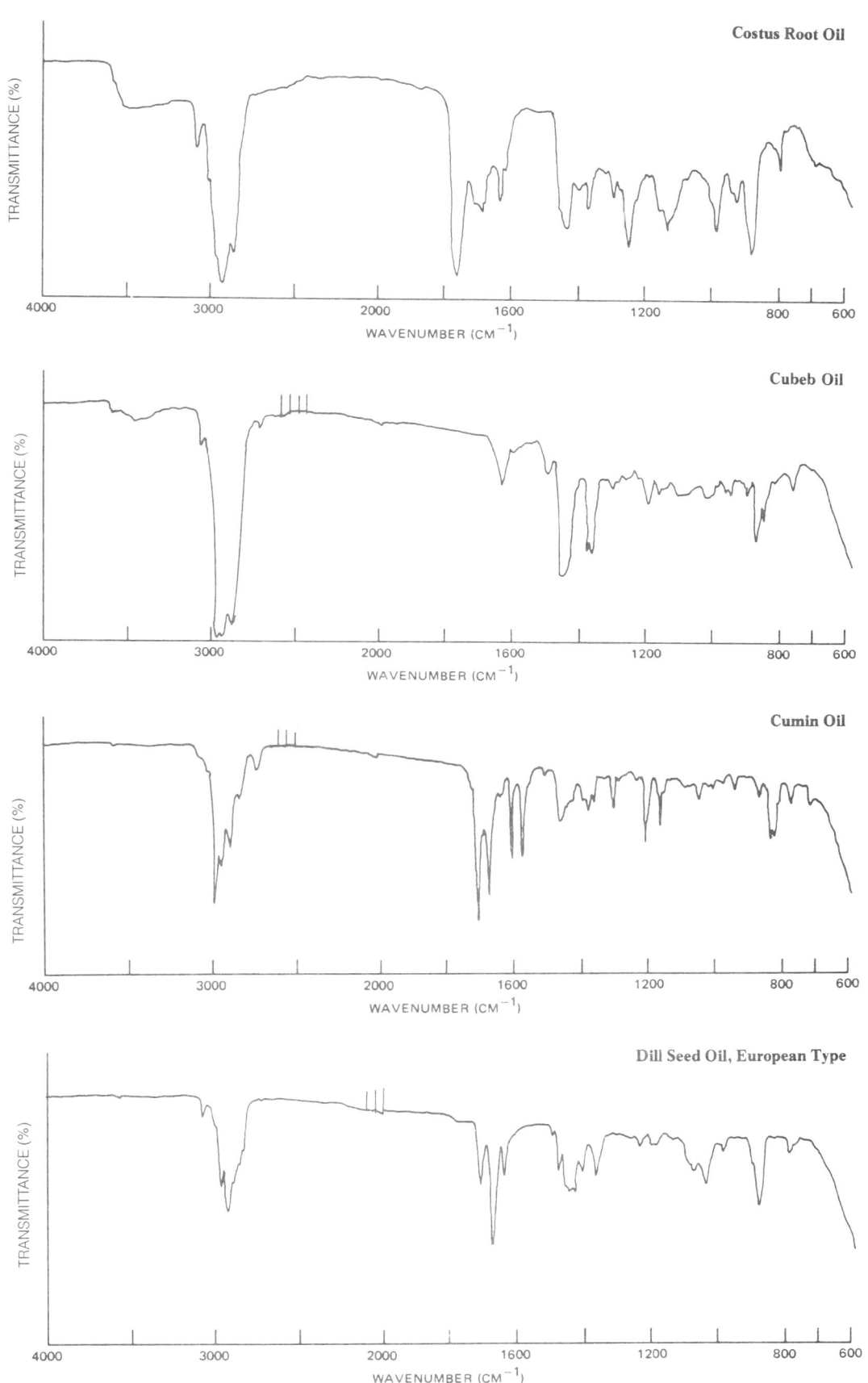

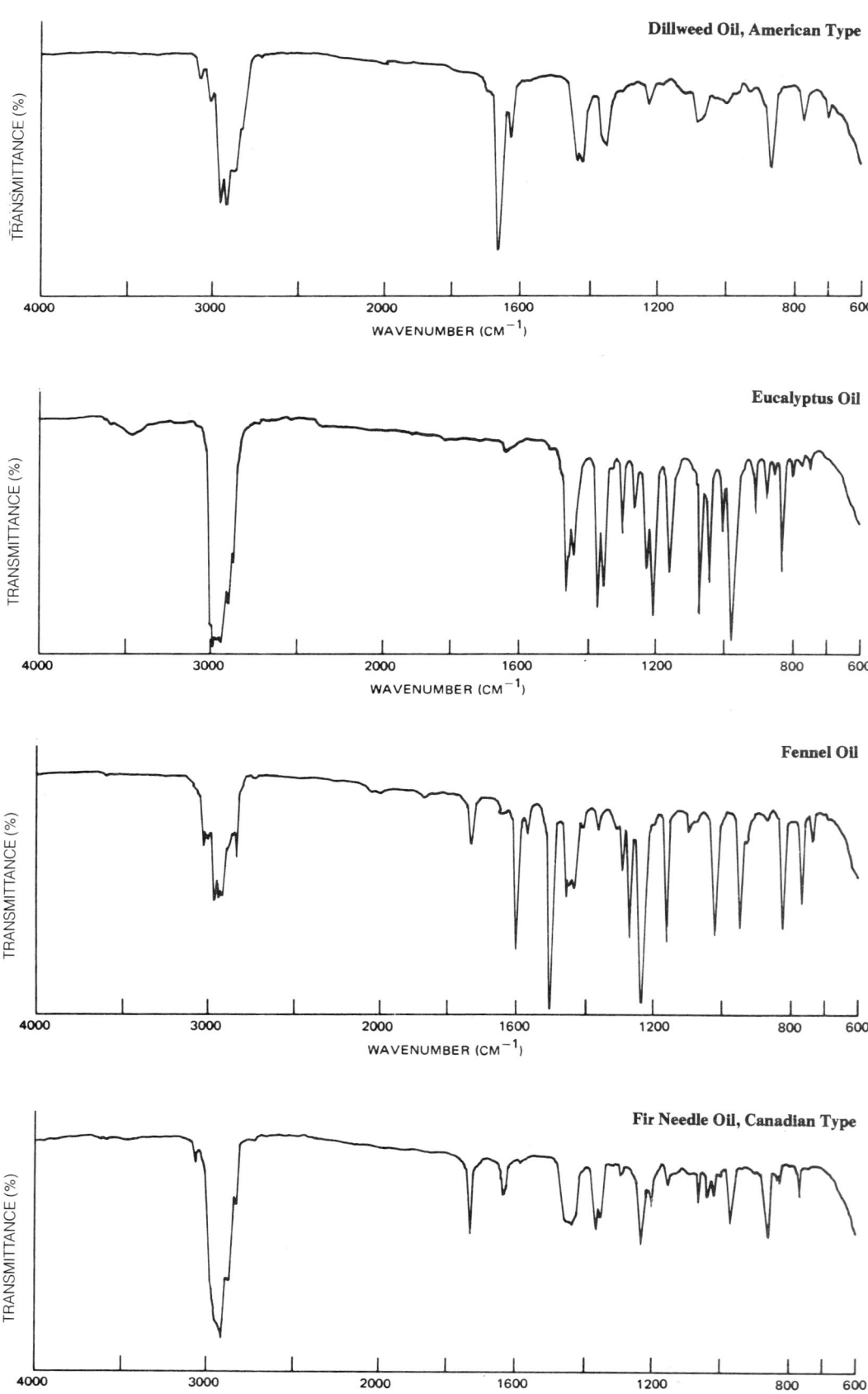

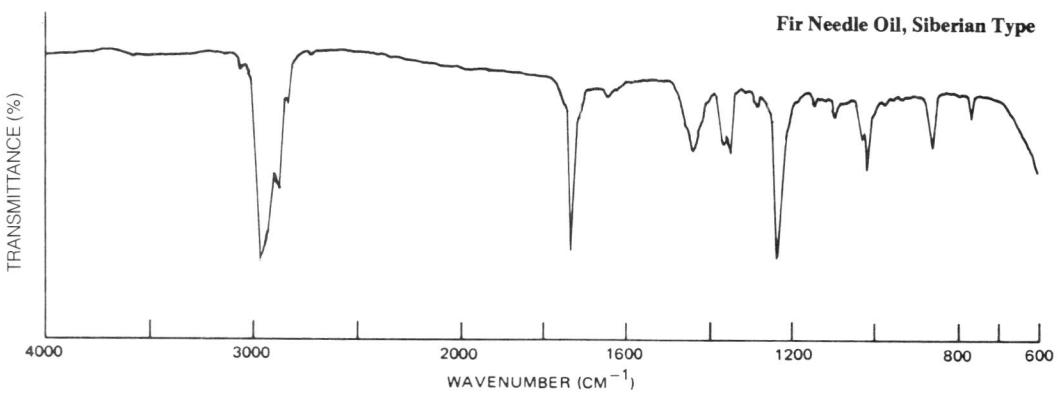
Fir Needle Oil, Siberian Type

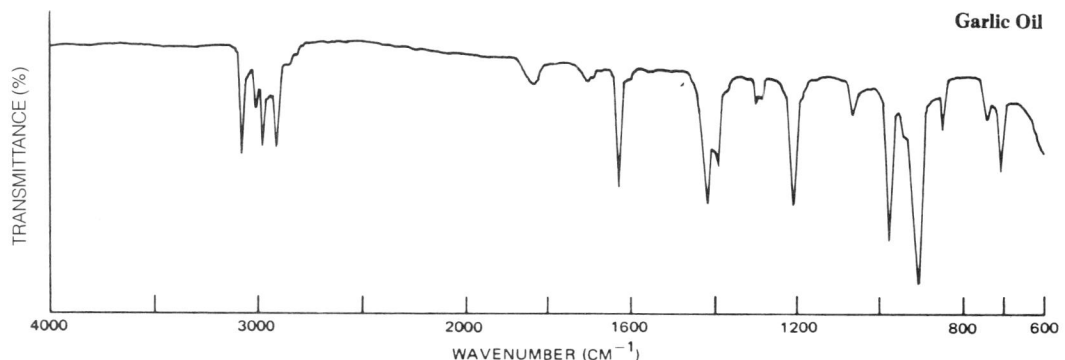
Garlic Oil

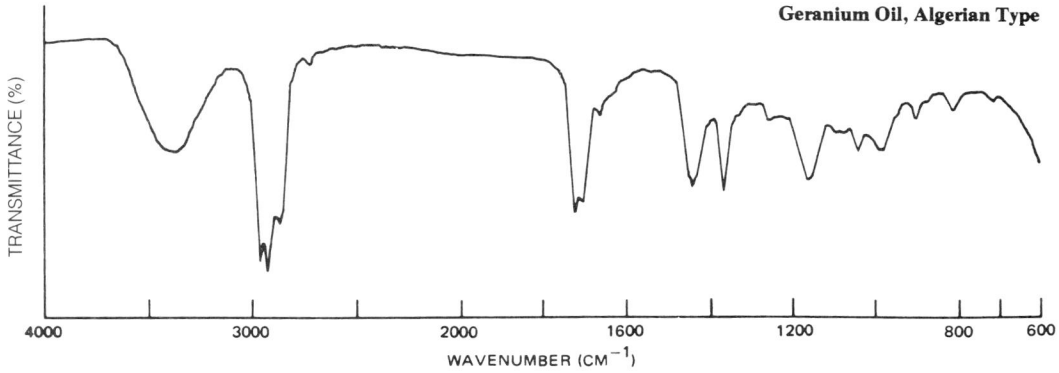
Geranium Oil, Algerian Type

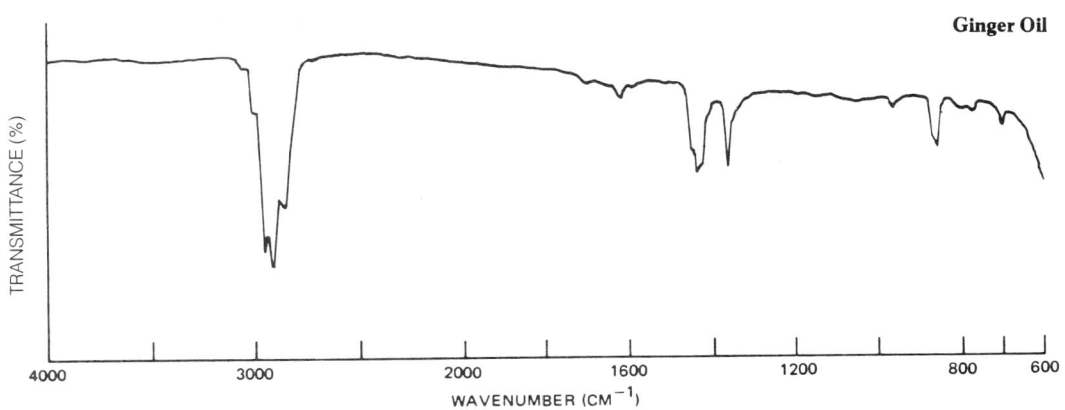
Ginger Oil

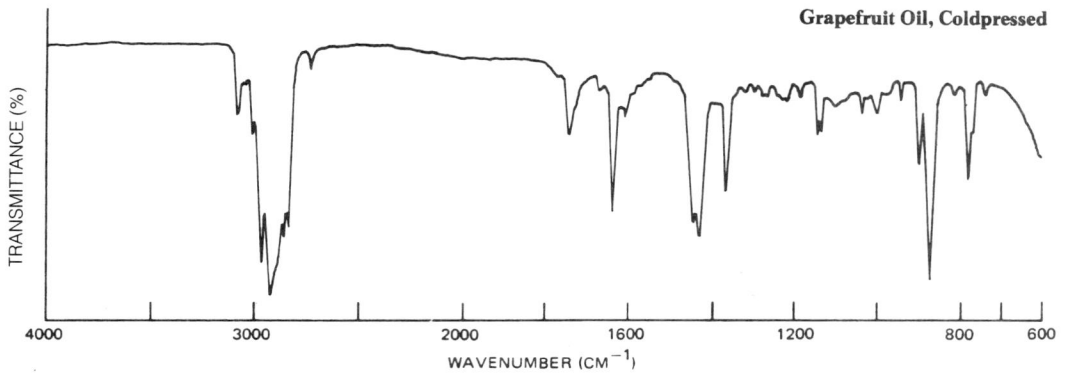

Grapefruit Oil, Coldpressed

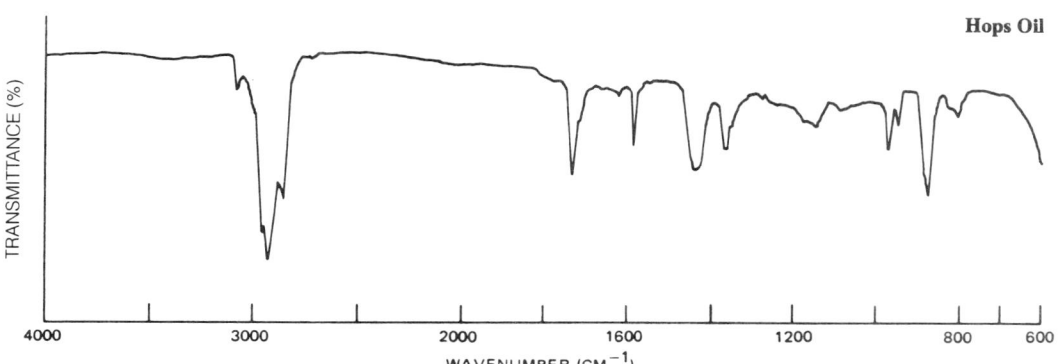

Hops Oil

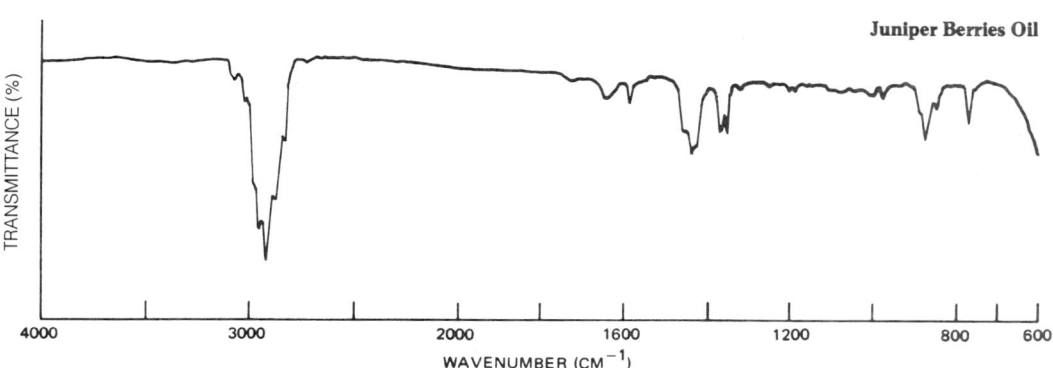

Juniper Berries Oil

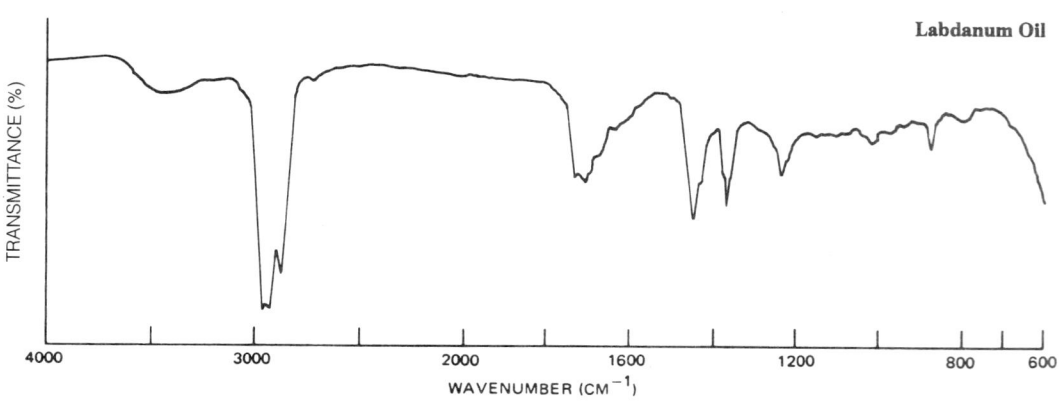

Labdanum Oil

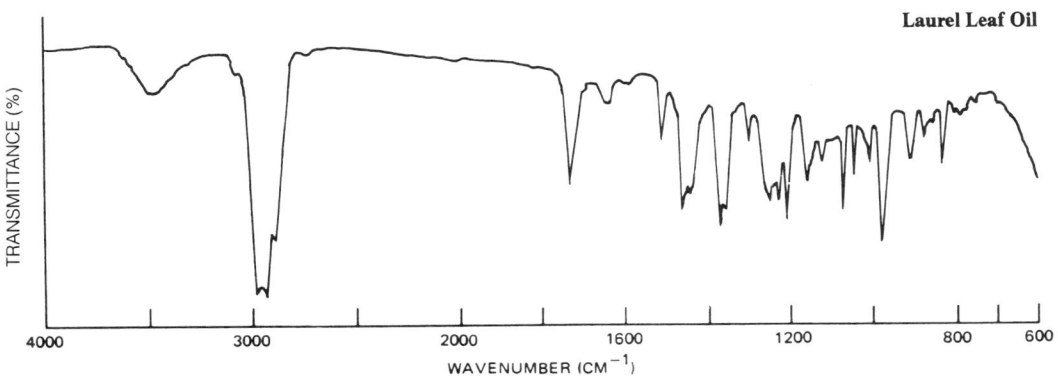
Laurel Leaf Oil

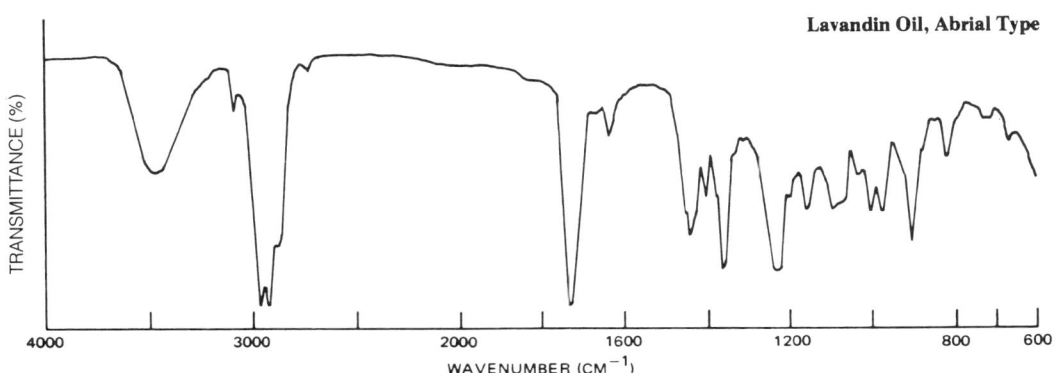
Lavandin Oil, Abrial Type

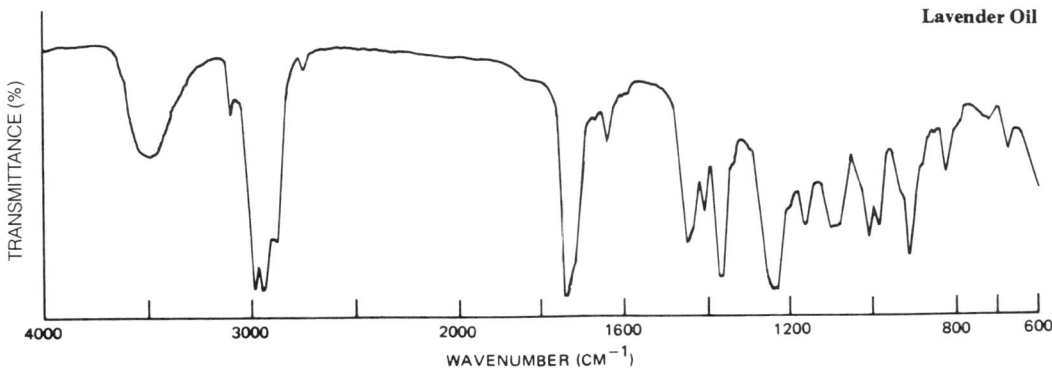
Lavender Oil

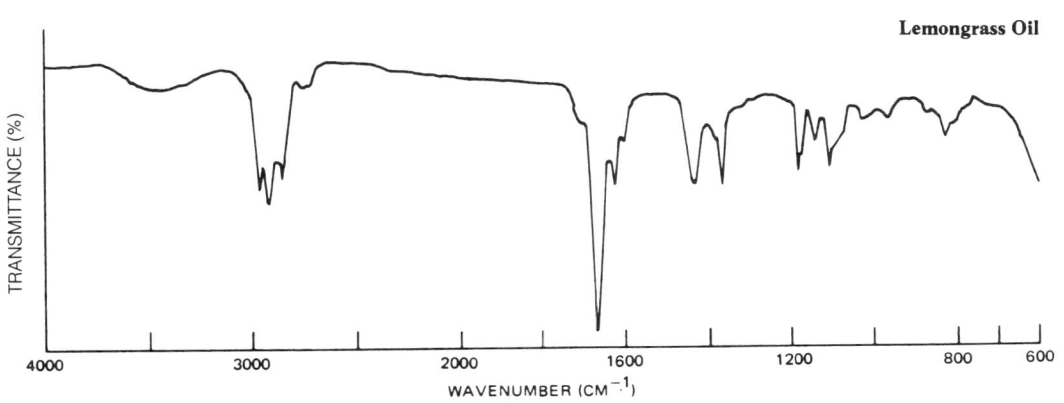
Lemongrass Oil

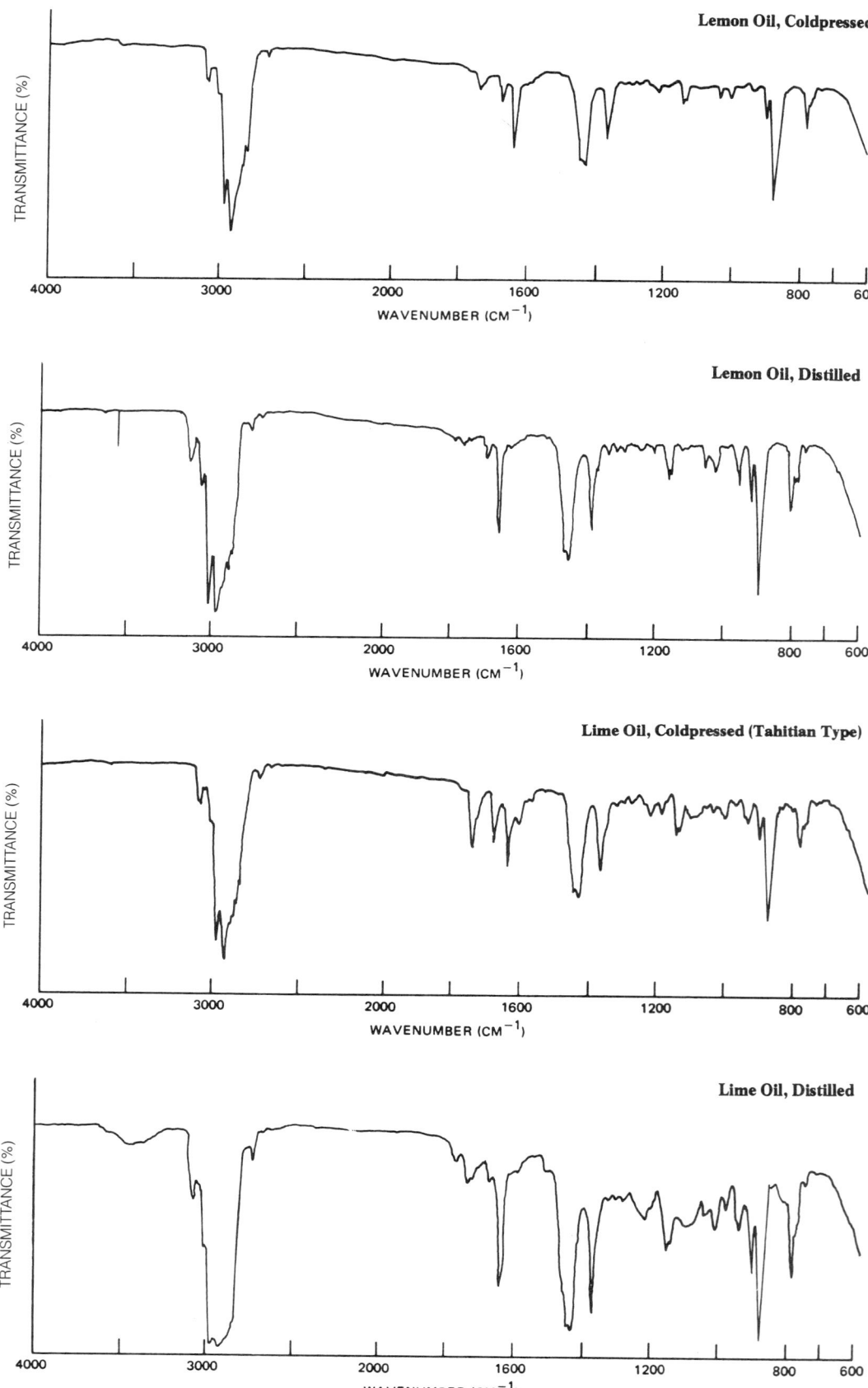

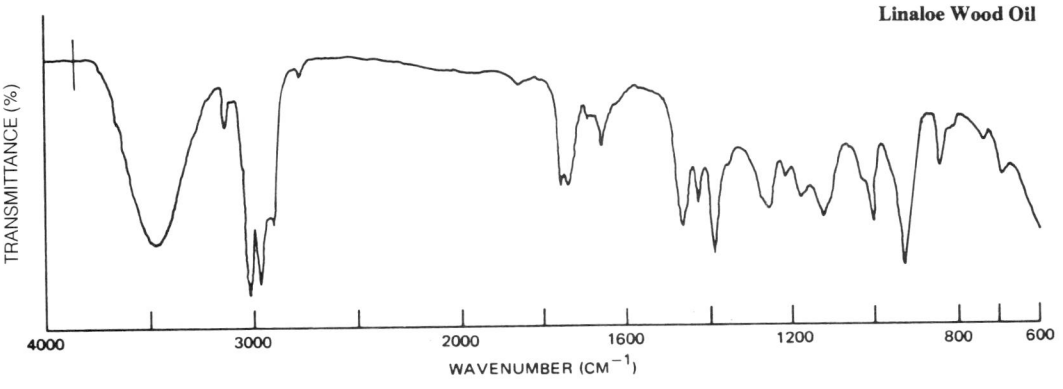
Linaloe Wood Oil

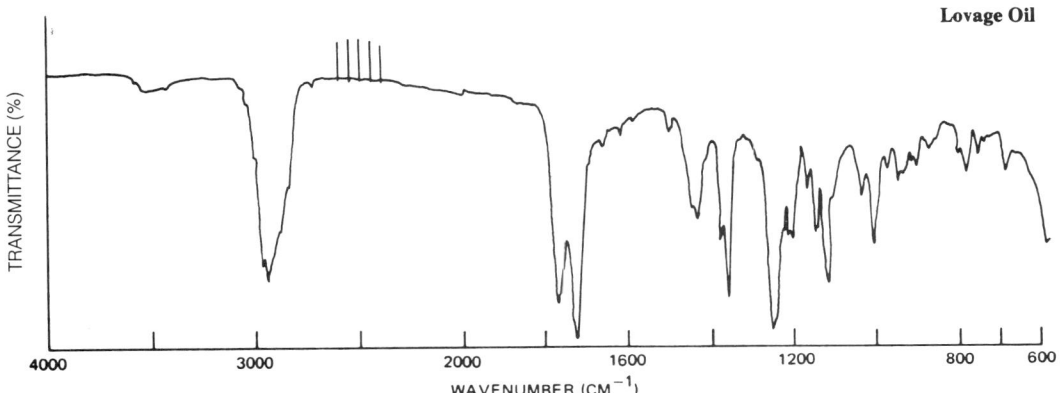
Lovage Oil

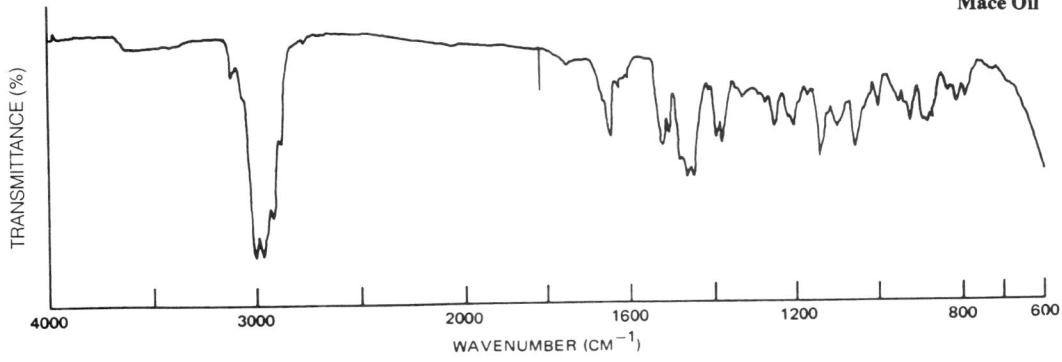
Mace Oil

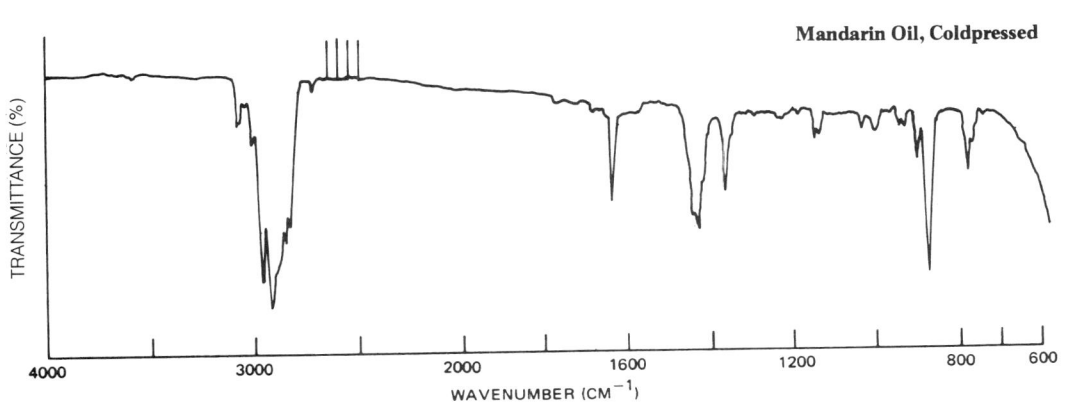
Mandarin Oil, Coldpressed

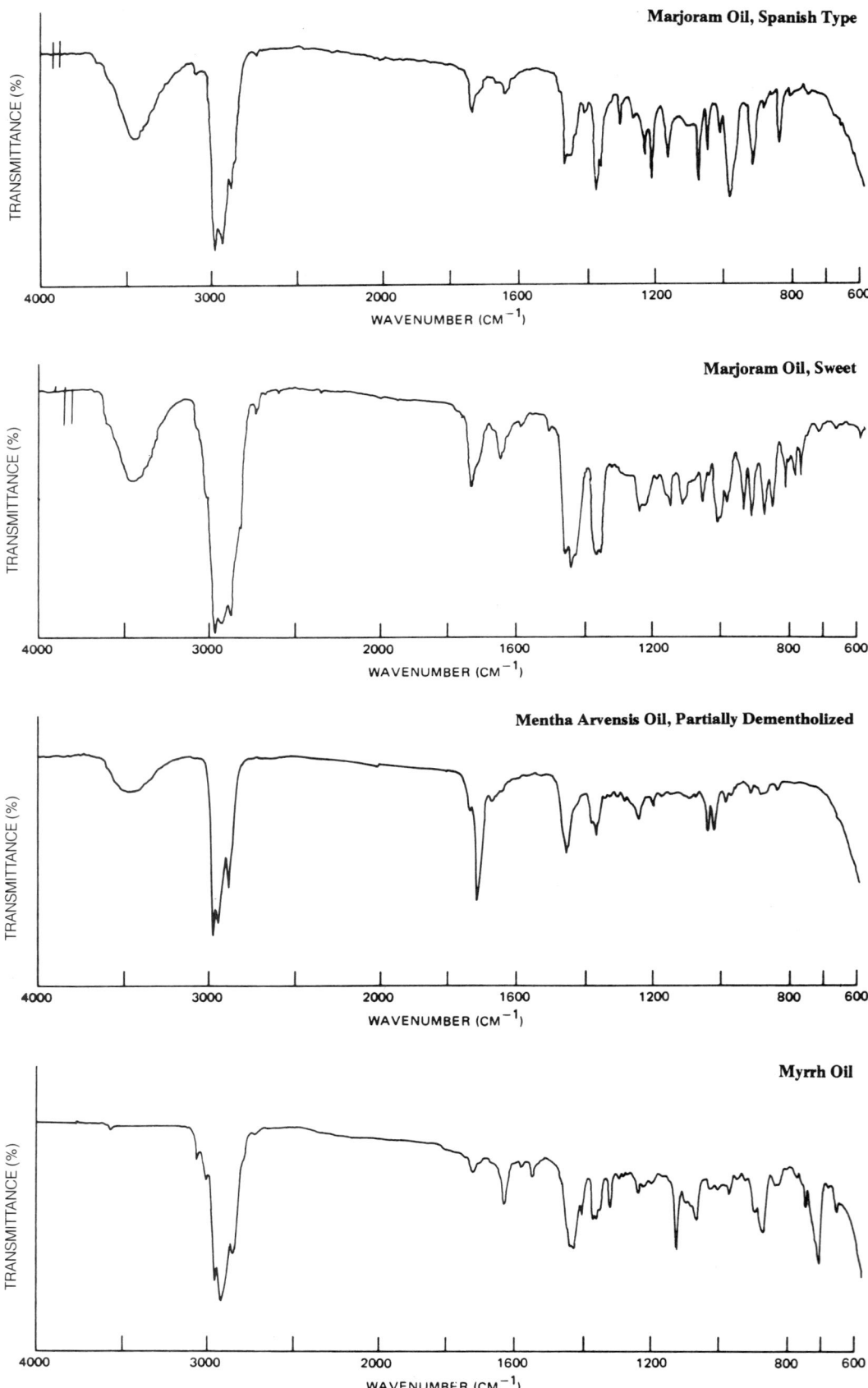

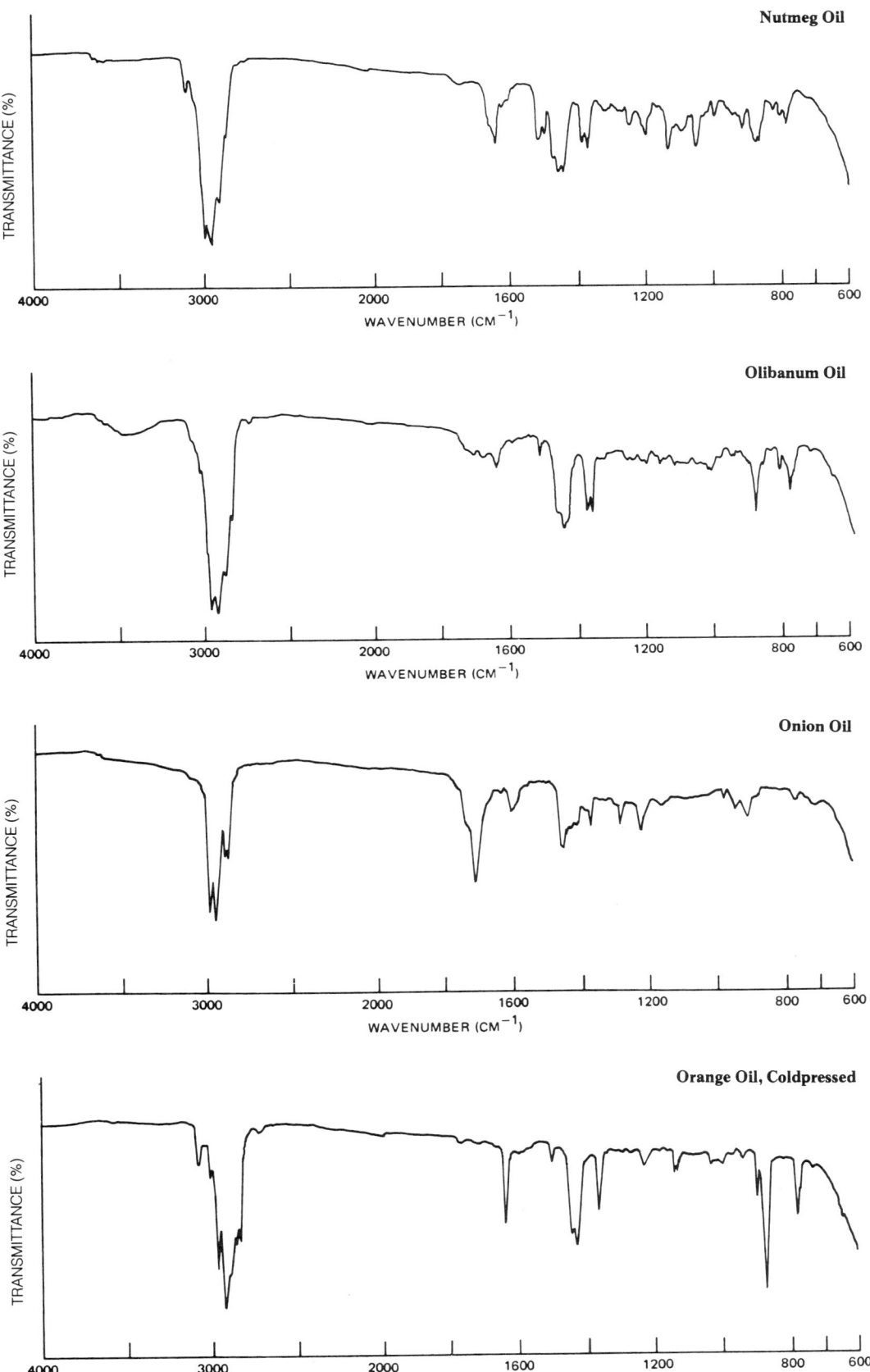

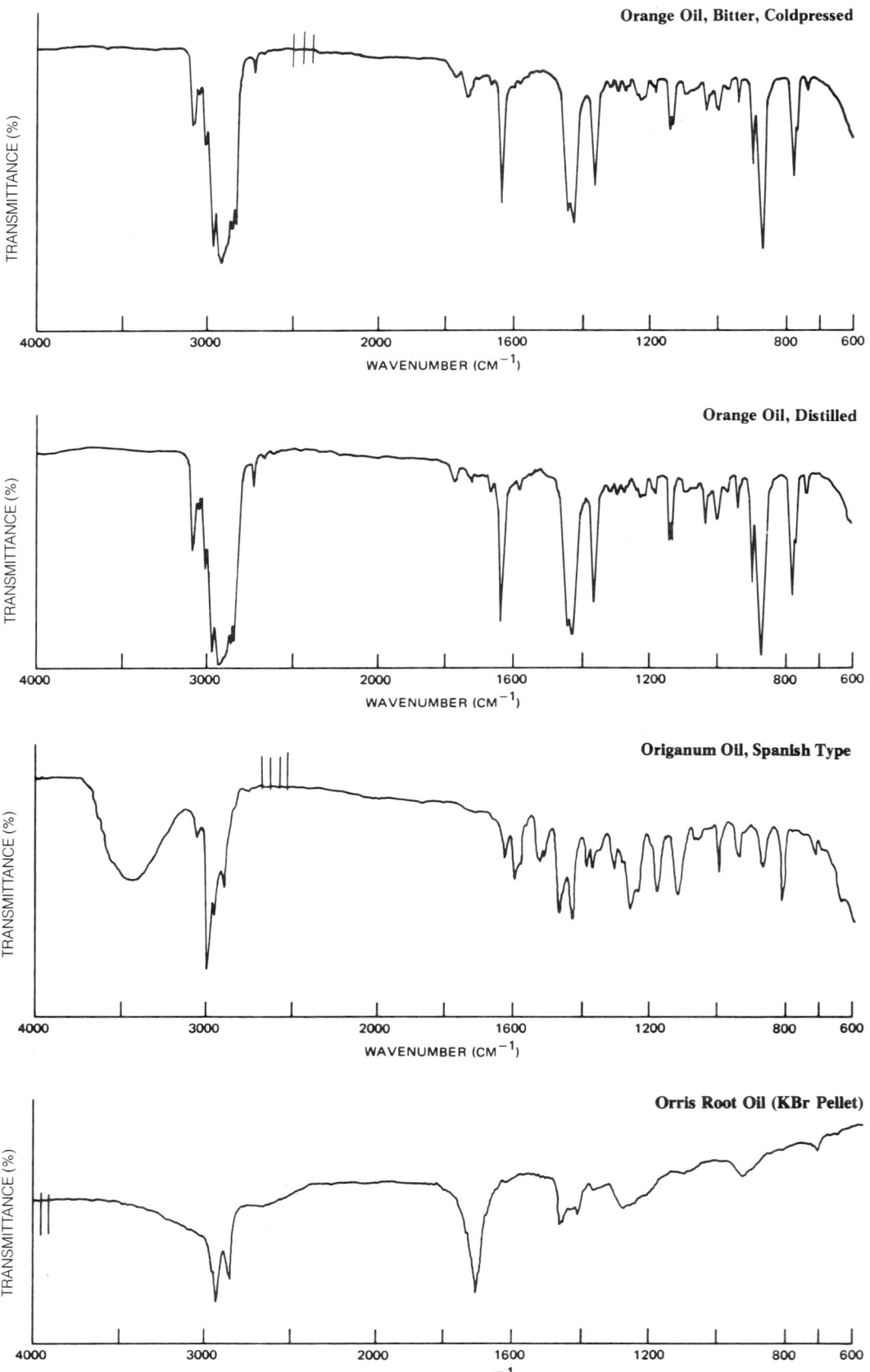

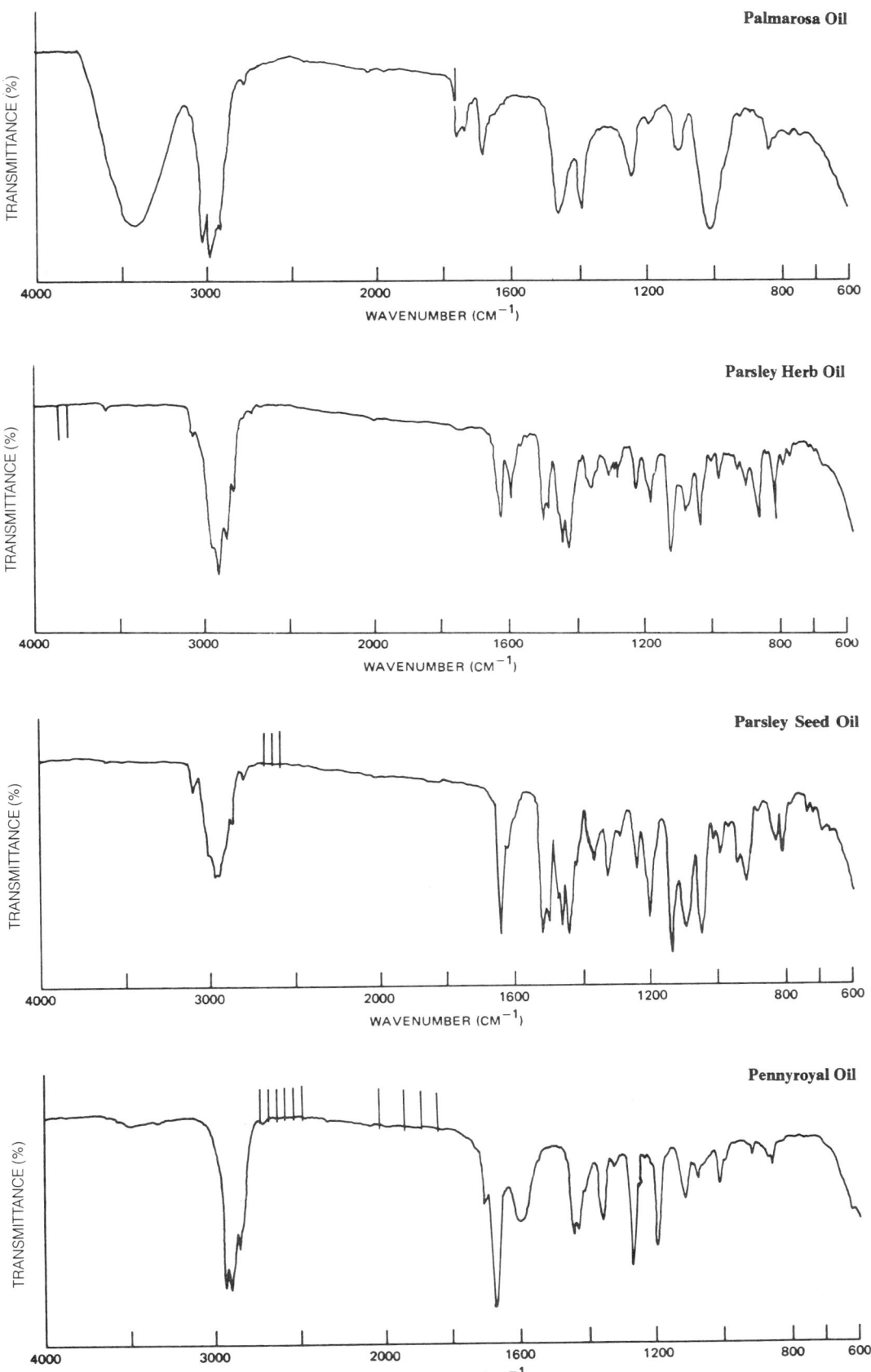

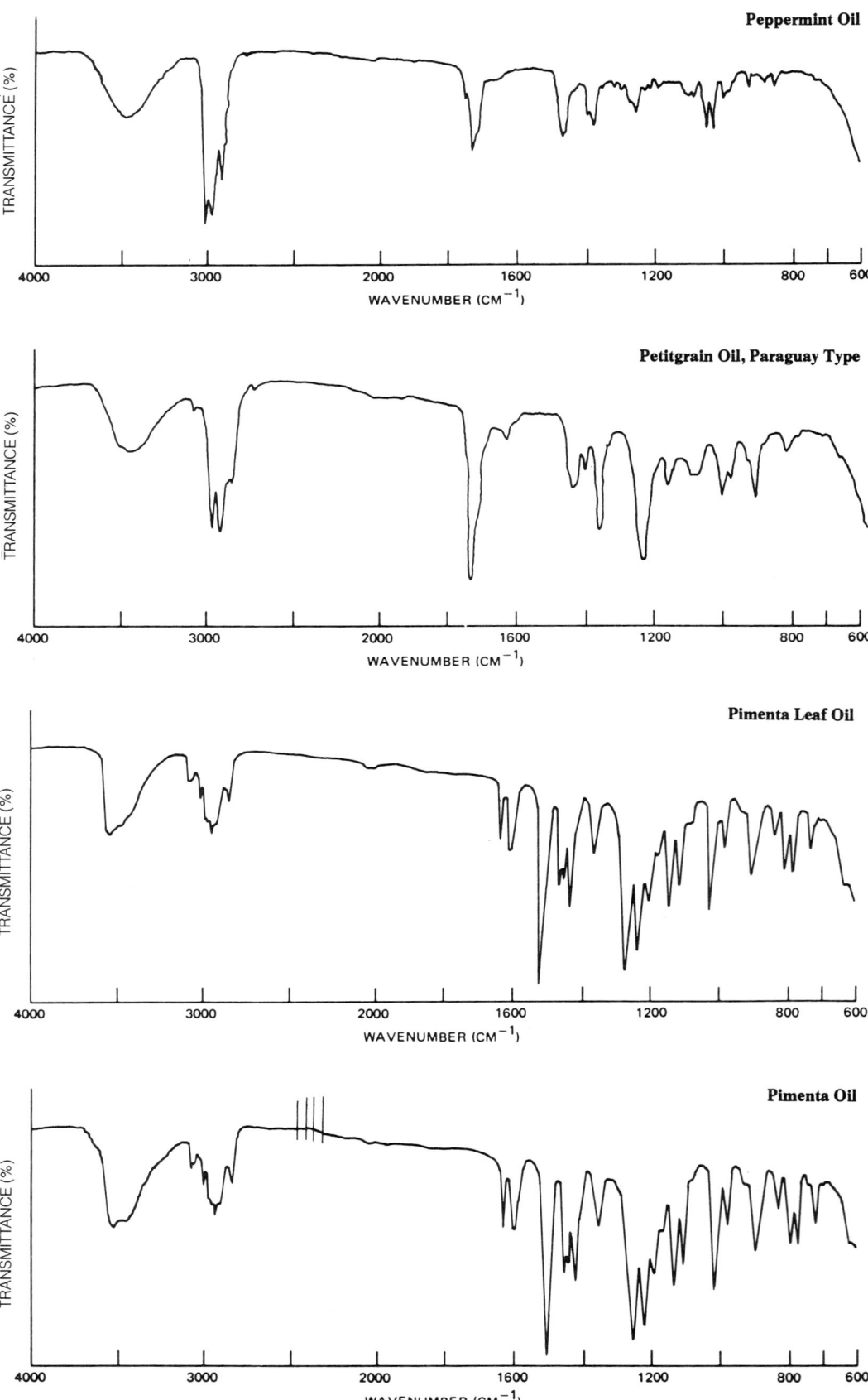

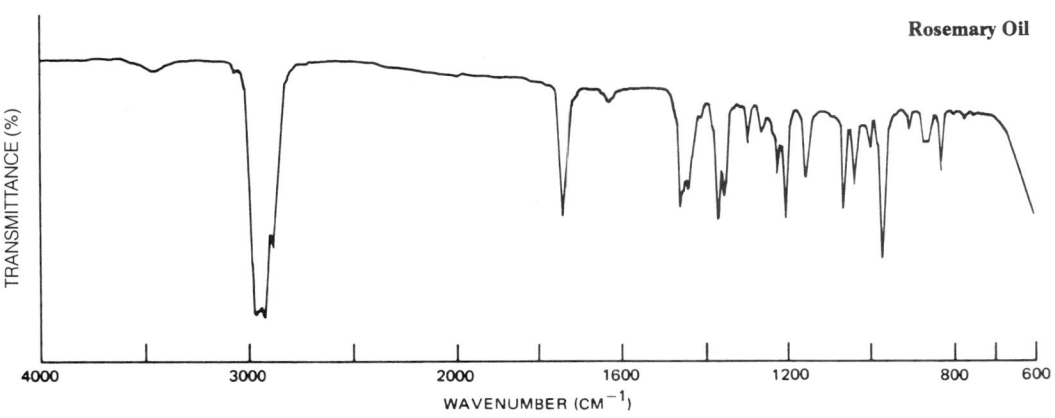

Rosemary Oil

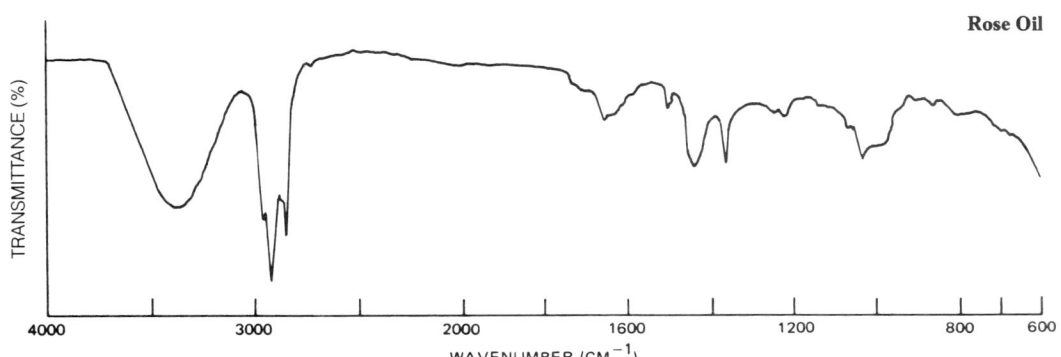

Rose Oil

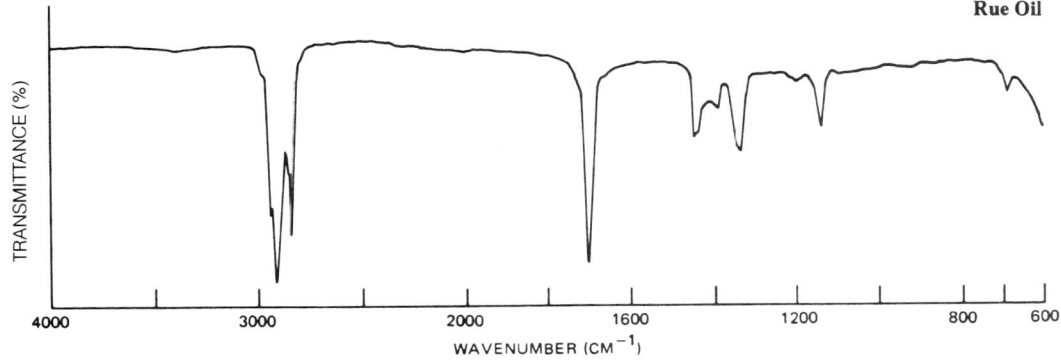

Rue Oil

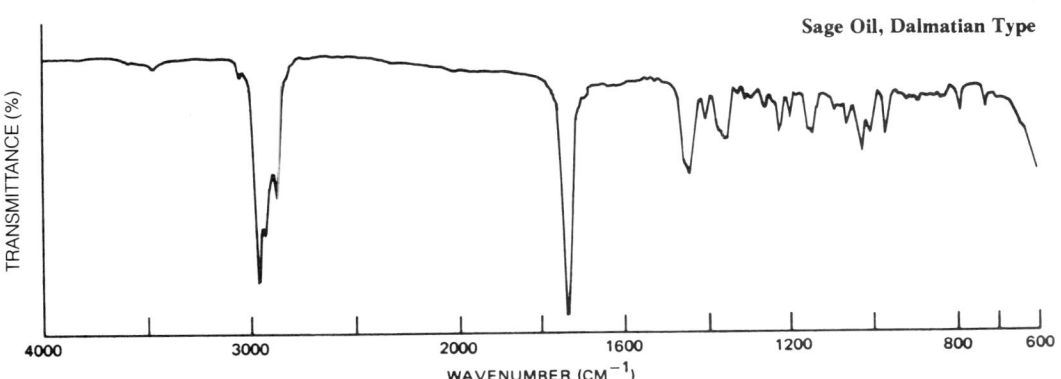

Sage Oil, Dalmatian Type

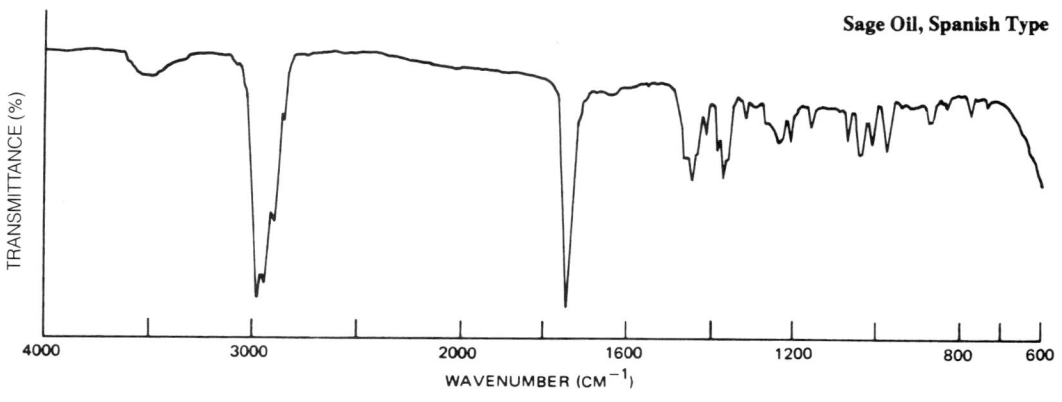
Sage Oil, Spanish Type

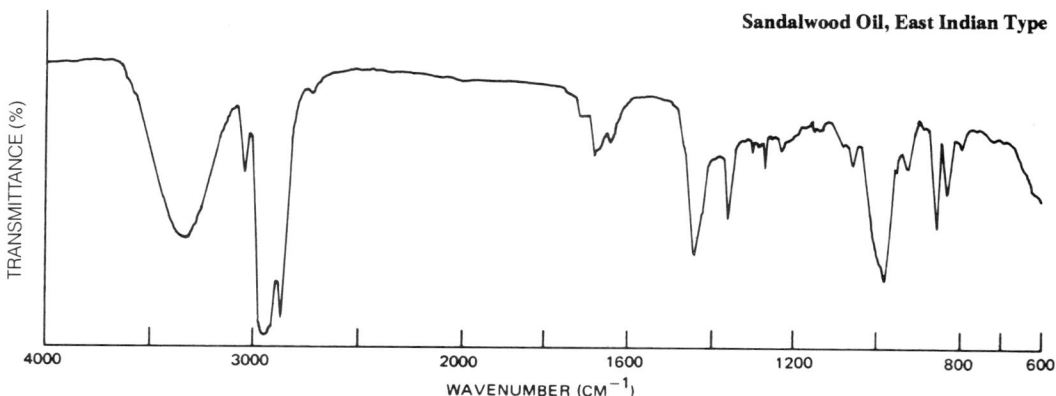
Sandalwood Oil, East Indian Type

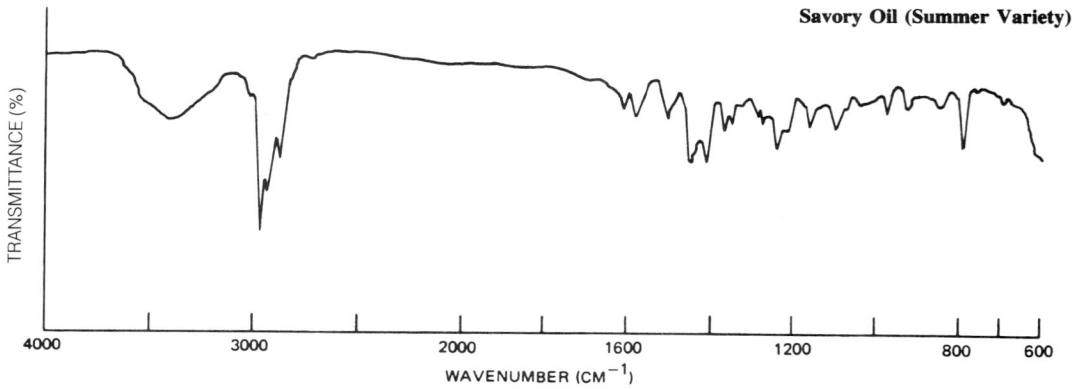
Savory Oil (Summer Variety)

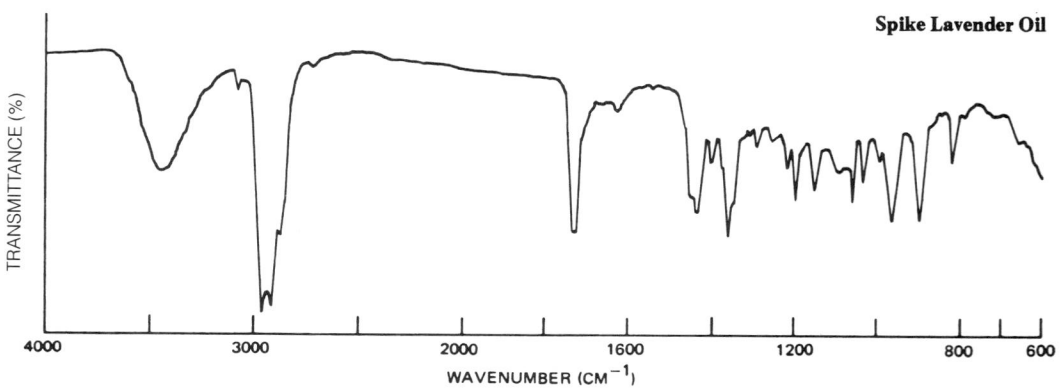
Spike Lavender Oil

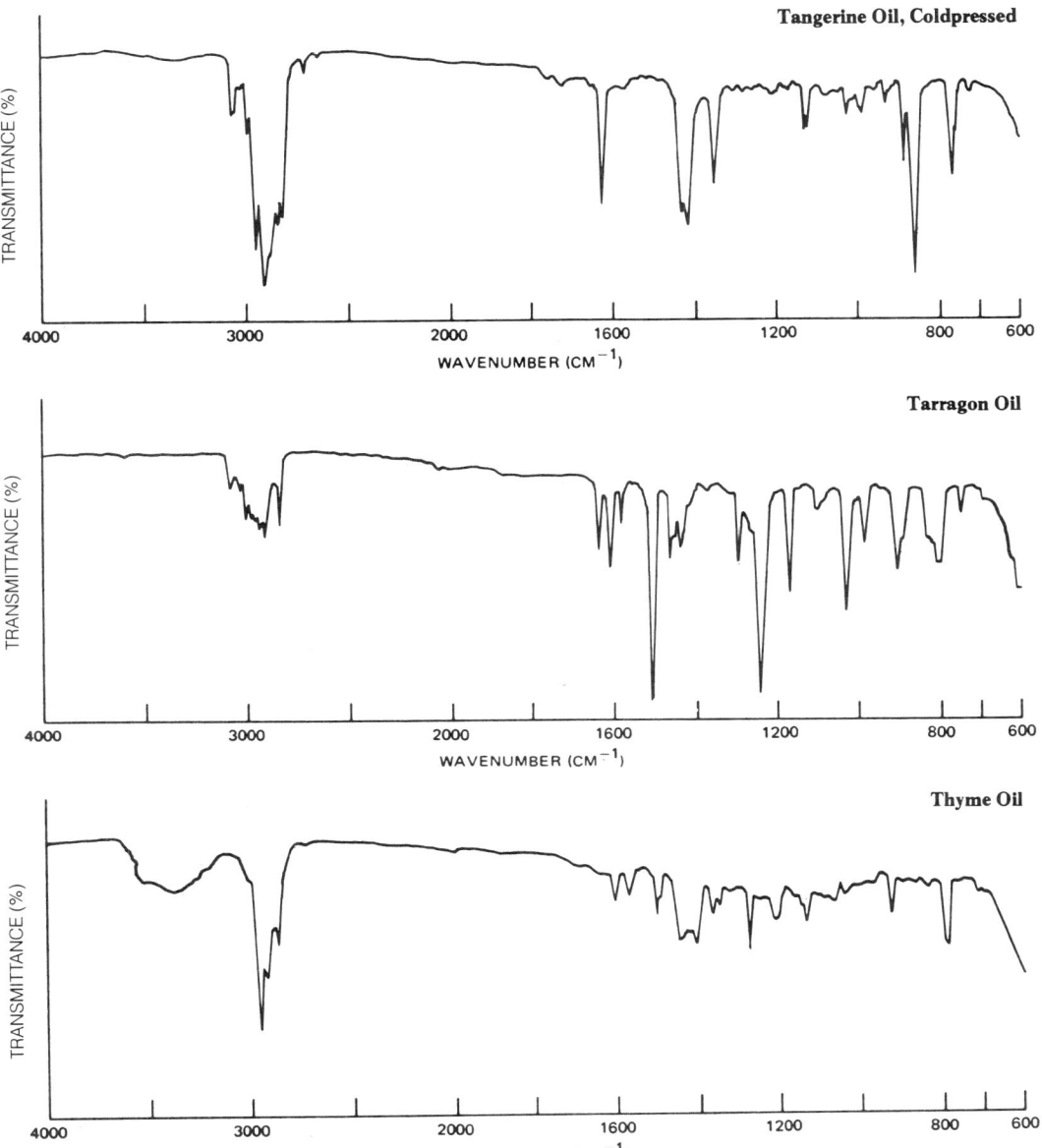

SERIES A-2

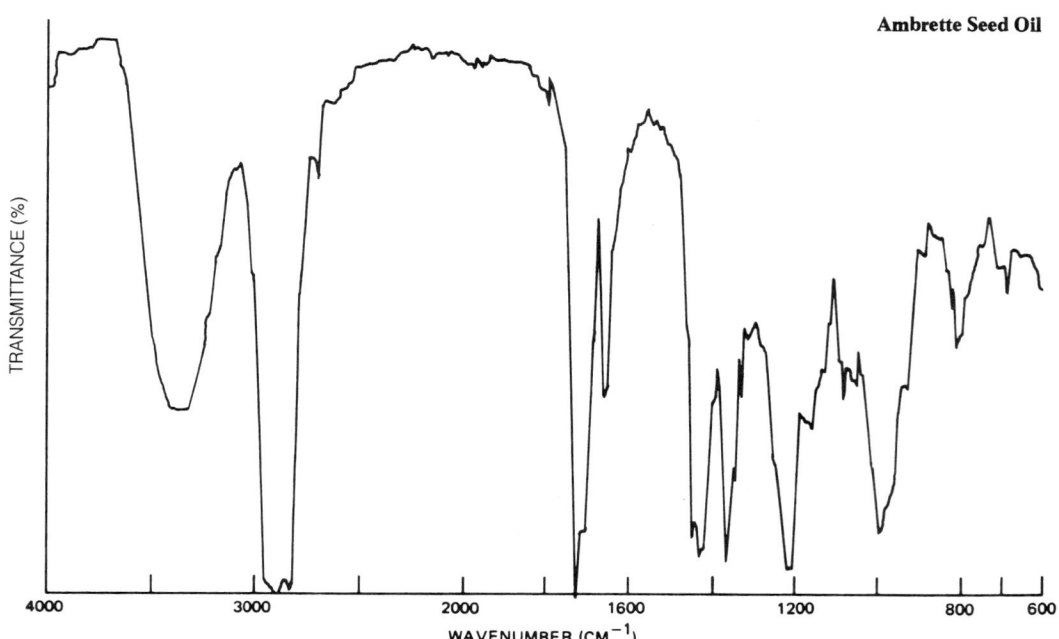

Ambrette Seed Oil

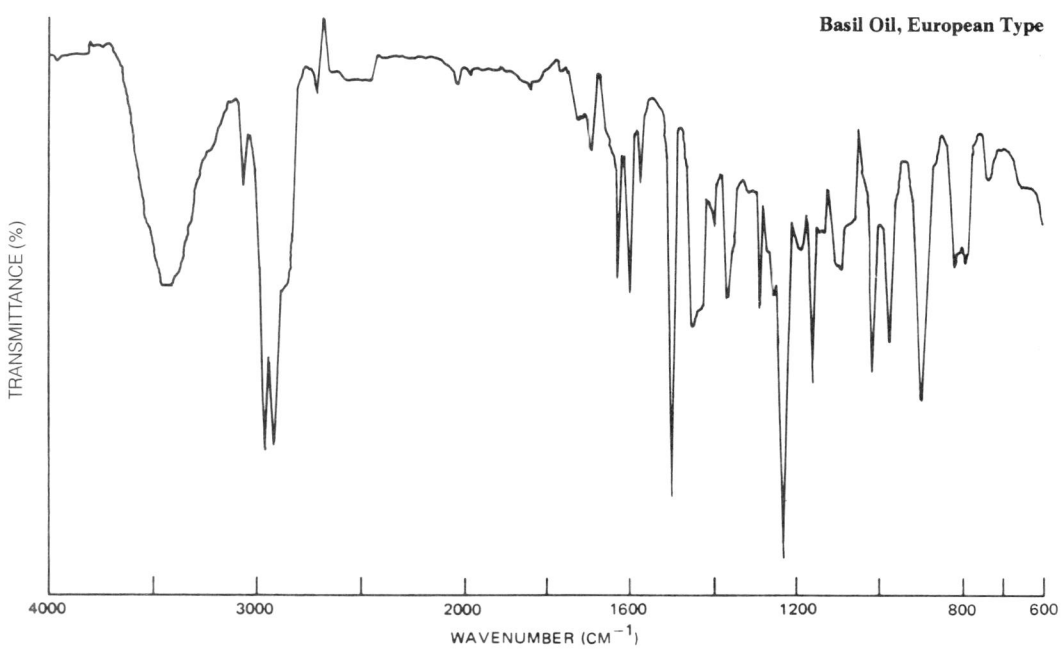
Basil Oil, European Type

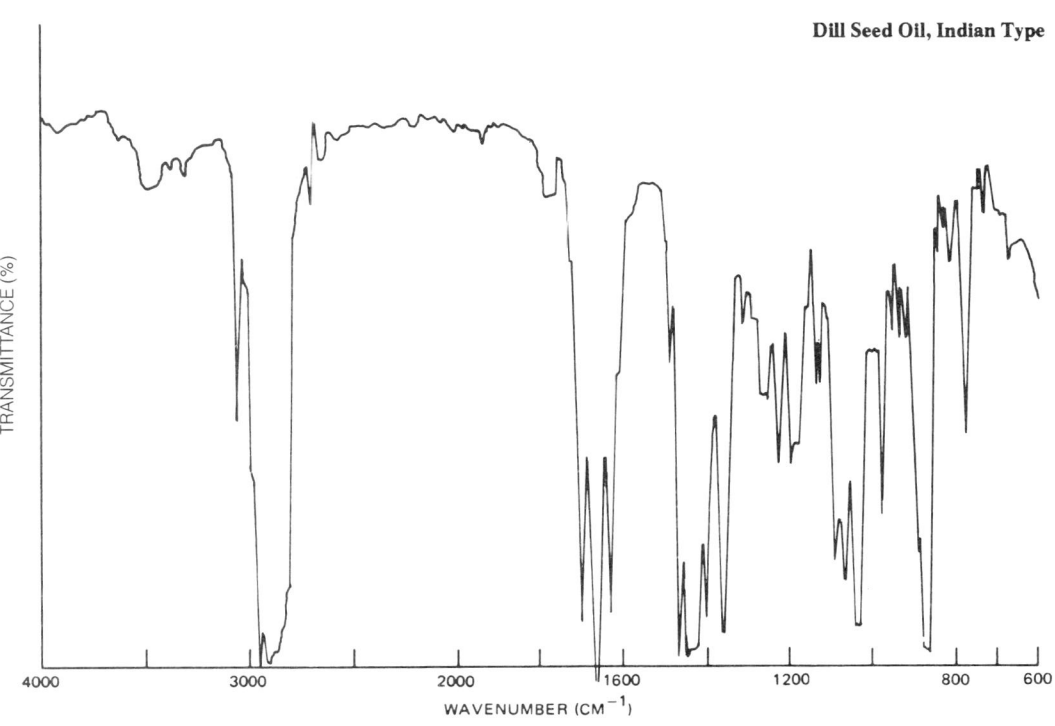
Dill Seed Oil, Indian Type

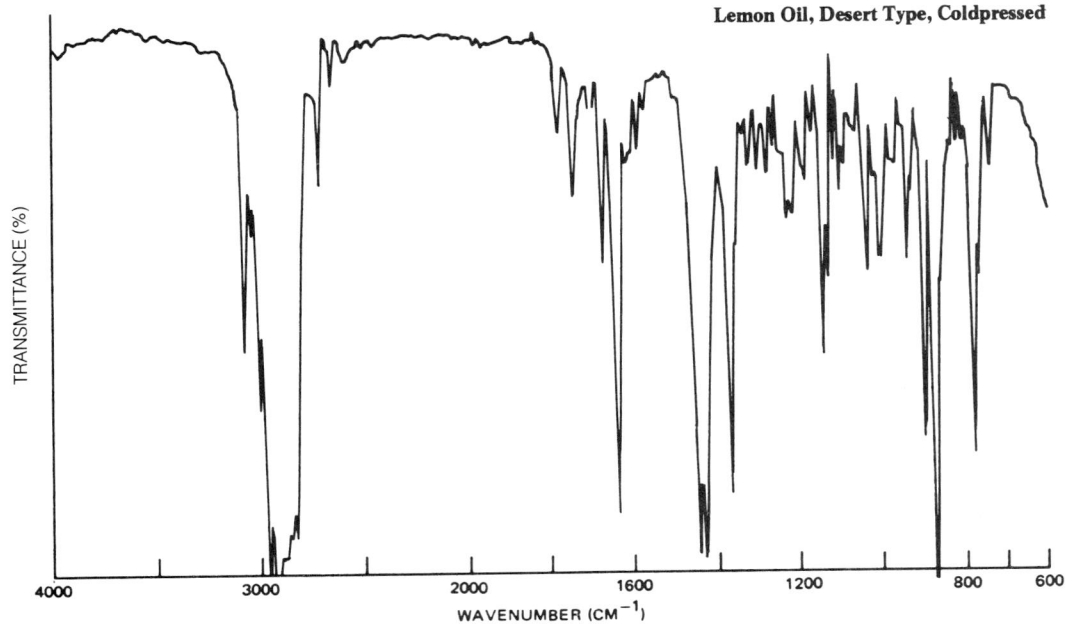

Lemon Oil, Desert Type, Coldpressed

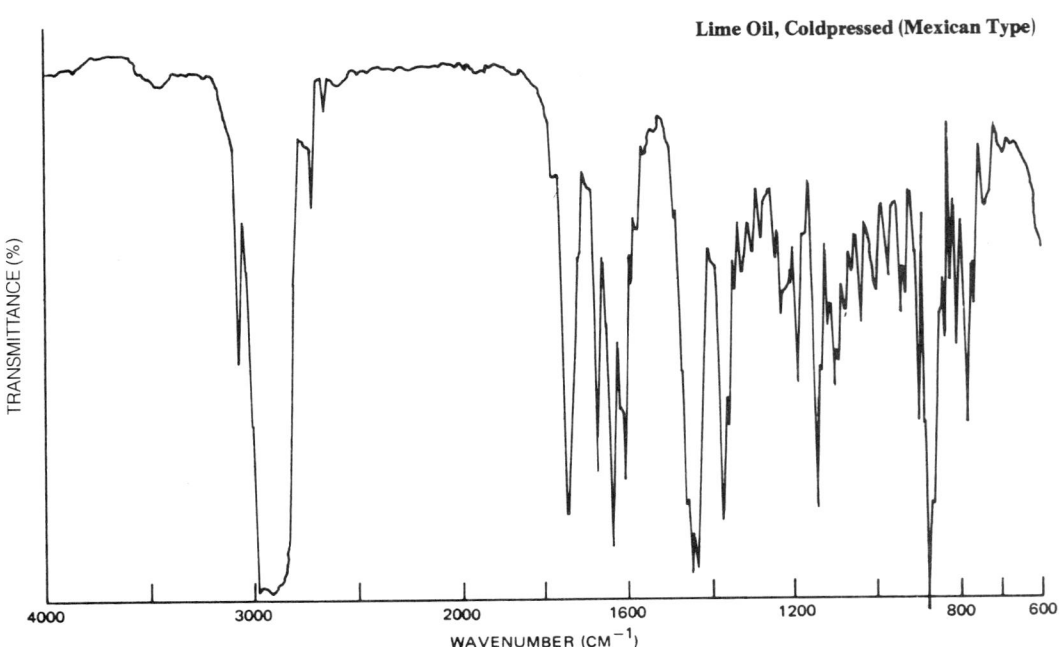

Lime Oil, Coldpressed (Mexican Type)

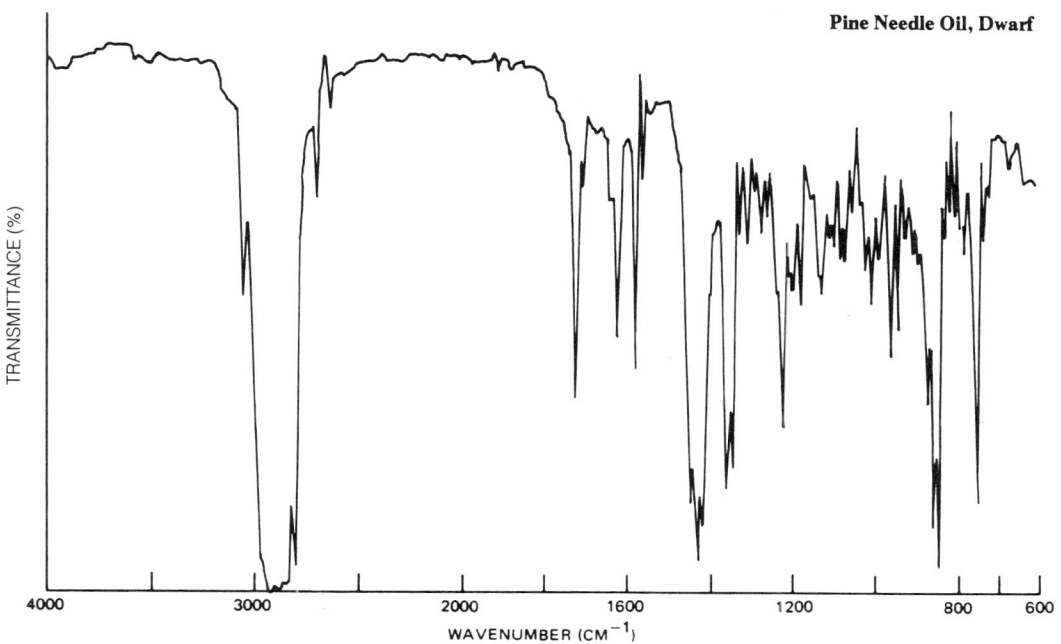

Pine Needle Oil, Dwarf

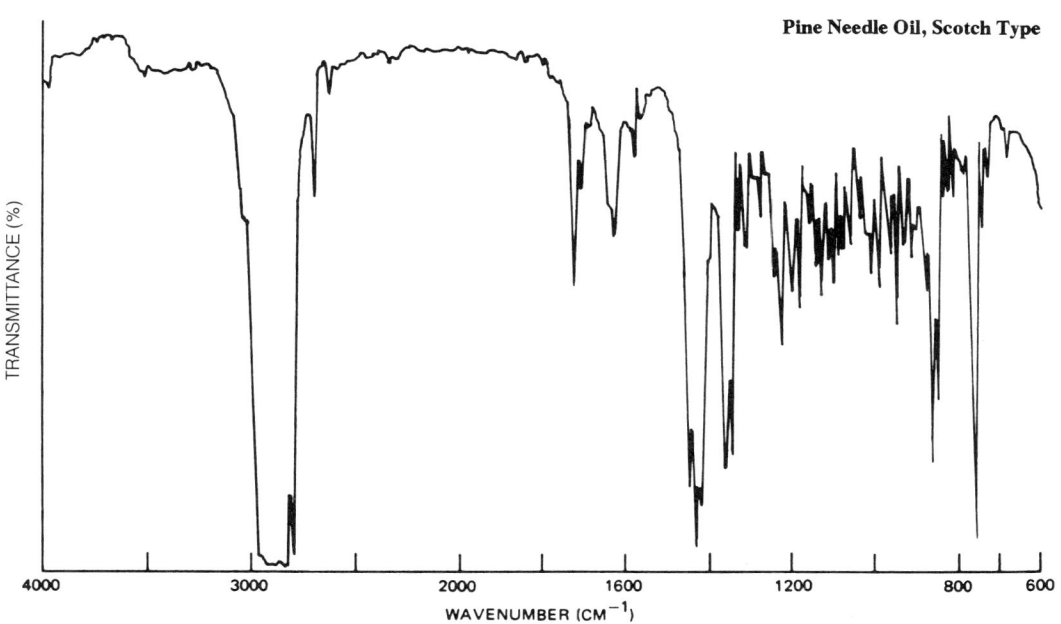

Pine Needle Oil, Scotch Type

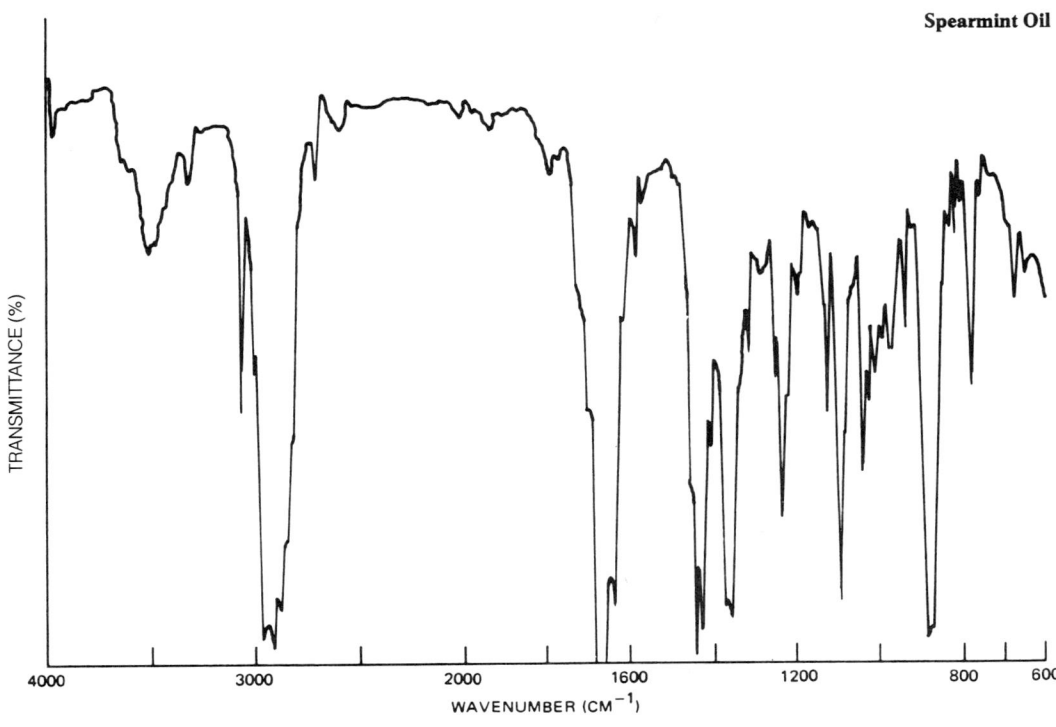

Spearmint Oil

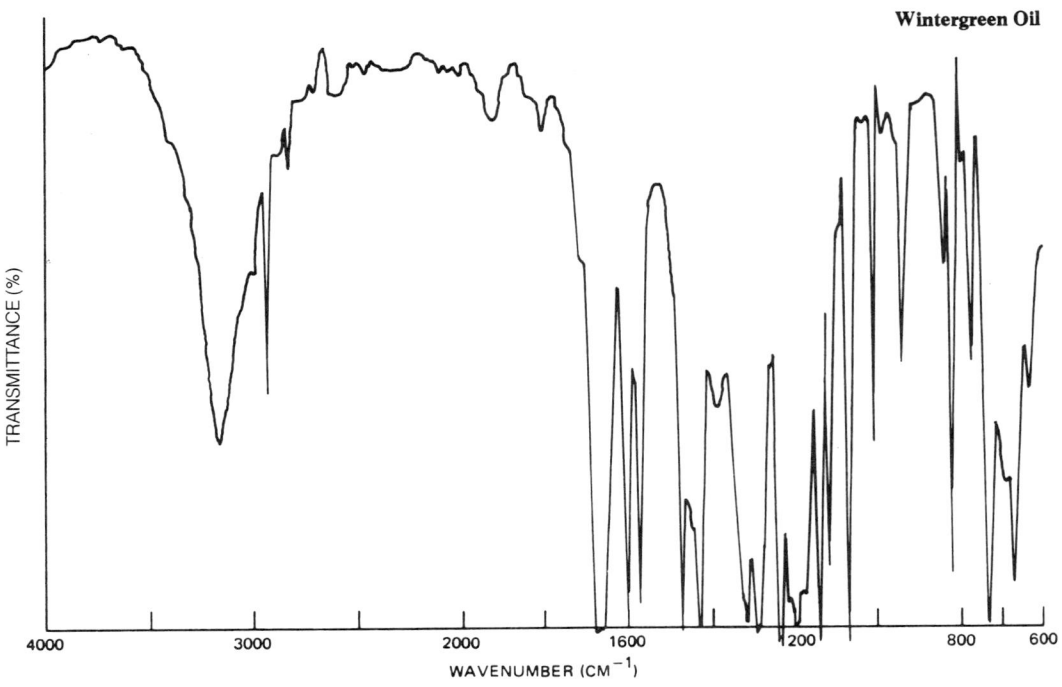

Wintergreen Oil

Series B: Flavor Chemicals

This series is divided into three subseries depending on the format of the spectra involved. The substances are listed below alphabetically.

Acetaldehyde ... 604
Acetanisole ... 680
Acetoin ... 681
Acetophenone ... 644
3-Acetyl-2,5-dimethylfuran ... 644
2-Acetylpyrazine ... 604
3-Acetylpyridine ... 604
Allyl Cyclohexanepropionate ... 645
Allyl Heptanoate ... 605
Allyl Hexanoate ... 645
Allyl α-Ionone ... 645
Allyl Isothiocyanate ... 681
Allyl Isovalerate ... 605
α-Amylcinnamaldehyde ... 605
Amyl Cinnamate ... 646
Amyl Octanoate ... 605
Amyl Propionate ... 606
Anethole ... 682
Anisole ... 606
Anisyl Acetate ... 606
Anisyl Alcohol ... 606
Anisyl Formate ... 607
Benzaldehyde ... 607
Benzaldehyde Glyceryl Acetal ... 607
1,2-Benzodihydropyrone ... 607
Benzophenone ... 646
Benzyl Acetate ... 608
Benzyl Alcohol ... 646
Benzyl Benzoate ... 647
Benzyl Butyrate ... 647
Benzyl Cinnamate ... 647
Benzyl Formate ... 608
Benzyl Isobutyrate ... 608
Benzyl Isovalerate ... 648
Benzyl Phenylacetate ... 608
Benzyl Propionate ... 648
Benzyl Salicylate ... 682
Bornyl Acetate ... 648
2-Butanone ... 609
Butan-3-one-2-yl Butanoate ... 649
Butyl Acetate ... 649
Butyl Alcohol ... 649
Butyl Butyrate ... 650
Butyl Butyryllactate ... 609
Butyl Isobutyrate ... 650

Butyl Isovalerate ... 609
Butyl Phenylacetate ... 609
Butyraldehyde ... 610
Butyric Acid ... 610
γ-Butyrolactone ... 650
d-Camphor ... 610
Carvacrol ... 610
l-Carveol ... 611
d-Carvone ... 651
l-Carvone ... 651
l-Carvyl Acetate ... 611
β-Caryophyllene ... 611
Cinnamaldehyde ... 611
Cinnamic Acid ... 612
Cinnamyl Acetate ... 612
Cinnamyl Alcohol ... 612
Cinnamyl Butyrate ... 612
Cinnamyl Cinnamate ... 613
Cinnamyl Formate ... 613
Cinnamyl Isobutyrate ... 613
Cinnamyl Isovalerate ... 613
Cinnamyl Propionate ... 614
Citral ... 651
Citronellal ... 652
Citronellol ... 683
Citronellyl Acetate ... 652
Citronellyl Butyrate ... 614
Citronellyl Formate ... 683
Citronellyl Isobutyrate ... 614
Citronellyl Propionate ... 652
p-Cresyl Acetate ... 684
Cuminic Aldehyde ... 653
Cyclamen Aldehyde ... 653
p-Cymene ... 614
(E),(E)-2,4-Decadienal ... 615
δ-Decalactone ... 615
γ-Decalactone ... 615
Decanal ... 653
(E)-2-Decenal ... 615
(Z)-4-Decenal ... 616
Decyl Alcohol ... 616
Diacetyl ... 616
Dibenzyl Ether ... 616
Diethyl Malonate ... 684
Diethyl Sebacate ... 654

Diethyl Succinate ... 685
Dimethyl Benzyl Carbinol ... 617
Dimethyl Benzyl Carbinyl Acetate ... 654
Dimethyl Benzyl Carbinyl Butyrate ... 654
2,6-Dimethyl-5-heptenal ... 617
3,7-Dimethyl-1-octanol ... 655
2,3-Dimethylpyrazine ... 617
2,5-Dimethylpyrazine ... 617
2,6-Dimethylpyrazine ... 618
2,5-Dimethylpyrrole ... 618
Dimethyl Sulfide ... 618
δ-Dodecalactone ... 618
(E)-2-Dodecen-1-al ... 619
Estragole ... 619
Ethone ... 619
Ethyl Acetate ... 685
Ethyl Acetoacetate ... 655
Ethyl Acrylate ... 686
Ethyl p-Anisate ... 686
Ethyl Anthranilate ... 687
Ethyl Benzoate ... 655
2-Ethylbutyraldehyde ... 619
Ethyl Butyrate ... 656
2-Ethylbutyric Acid ... 620
Ethyl Cinnamate ... 620
Ethyl Decanoate ... 687
2-Ethyl-3,5(6)-dimethylpyrazine ... 656
Ethylene Brassylate ... 620
2-Ethyl Fenchol ... 656
Ethyl Formate ... 657
4-Ethyl Guaiacol ... 620
Ethyl Heptanoate ... 657
Ethyl Hexanoate ... 657
Ethyl Isobutyrate ... 621
Ethyl Isovalerate ... 658
Ethyl Lactate ... 688
Ethyl Laurate ... 621
Ethyl Levulinate ... 621
Ethyl Methylphenylglycidate ... 658
2-Ethyl-3-methylpyrazine ... 658
Ethyl-3-methylthiopropionate ... 621
Ethyl Myristate ... 622
Ethyl Nonanoate ... 622

Ethyl Octanoate ... 659
Ethyl Phenylacetate ... 688
Ethyl Phenylglycidate ... 689
Ethyl Propionate ... 659
Ethyl Salicylate ... 659
Ethyl Vanillin ... 660
Eucalyptol ... 622
Eugenol ... 689
Eugenyl Acetate ... 622
Farnesol ... 690
Furfural ... 660
Fusel Oil, Refined ... 623
Geraniol ... 660
Geranyl Acetate ... 661
Geranyl Benzoate ... 690
Geranyl Butyrate ... 691
Geranyl Formate ... 661
Geranyl Phenylacetate ... 661
Geranyl Propionate ... 662
(E),(E)-2,4-Heptadienal ... 623
γ-Heptalactone ... 662
Heptanal ... 691
2-Heptanone ... 623
3-Heptanone ... 692
(Z)-4-Hepten-1-al ... 623
Heptyl Alcohol ... 692
γ-Hexalactone ... 624
Hexanoic Acid ... 662
(E)-2-Hexen-1-ol ... 624
(Z)-3-Hexen-1-ol ... 624
(E)-2-Hexenyl Acetate ... 624
(Z)-3-Hexenyl Acetate ... 625
n-Hexyl Acetate ... 625
Hexyl Alcohol ... 693
Hexyl-2-butenoate ... 663
α-Hexylcinnamaldehyde ... 625
Hydroxycitronellal ... 693
Hydroxycitronellal Dimethyl Acetal ... 663
4-Hydroxy-2,5-dimethyl-3(2H)-furanone ... 625
4-(p-Hydroxyphenyl)-2-butanone ... 626
Indole ... 694
α-Ionone ... 663
β-Ionone ... 694
Isoamyl Acetate ... 664
Isoamyl Benzoate ... 626
Isoamyl Butyrate ... 664
Isoamyl Formate ... 664
Isoamyl Hexanoate ... 695

Isoamyl Phenyl Acetate ... 626
Isoamyl Salicylate ... 695
Isoborneol ... 626
Isobornyl Acetate ... 696
Isobutyl Acetate ... 627
Isobutyl Alcohol ... 627
Isobutyl-2-butenoate ... 665
Isobutyl Butyrate ... 665
Isobutyl Cinnamate ... 627
Isobutyl Phenylacetate ... 665
Isobutyl Salicylate ... 627
Isobutyraldehyde ... 628
Isobutyric Acid ... 628
Isoeugenol ... 666
Isoeugenyl Acetate ... 696
Isopropyl Acetate ... 628
Isopulegol ... 666
Isovaleric Acid ... 628
Lauryl Alcohol ... 629
Lauryl Aldehyde ... 697
Levulinic Acid ... 629
d-Limonene ... 666
Linalool ... 667
Linalyl Acetate ... 667
Linalyl Benzoate ... 697
Linalyl Formate ... 629
Linalyl Isobutyrate ... 698
Linalyl Propionate ... 698
Menthol ... 667
2-Mercaptopropionic Acid ... 629
p-Methoxybenzaldehyde ... 699
2-Methoxy-3(5)-methylpyrazine ... 668
4-p-Methoxyphenyl-2-butanone ... 630
2-Methoxypyrazine ... 630
Methyl Acetate ... 630
4'-Methyl Acetophenone ... 668
p-Methyl Anisole ... 668
Methyl Anthranilate ... 669
Methyl Benzoate ... 630
α-Methylbenzyl Acetate ... 699
α-Methylbenzyl Alcohol ... 669
2-Methylbutyric Acid ... 631
α-Methylcinnamaldehyde ... 631
Methyl Cinnamate ... 669
6-Methylcoumarin ... 631
Methyl Cyclopentenolone ... 631
Methyl Eugenol ... 700
6-Methyl-5-hepten-2-one ... 700
Methyl Hexyl Ketone ... 632
Methyl Isobutyrate ... 632
Methyl Isoeugenol ... 670

5-Methyl-2-isopropyl-2-hexenal ... 670
Methyl-3-(methylthio)propionate ... 632
Methyl β-Naphthyl Ketone ... 670
Methyl 2-Octynoate ... 671
4-Methylpentanoic Acid ... 632
2-Methyl-2-pentenoic Acid ... 633
Methyl Phenylacetate ... 671
Methyl Phenylcarbinyl Acetate ... 671
2-Methylpyrazine ... 633
Methyl Salicylate ... 701
2-Methylundecanal ... 633
Nerol ... 701
Nerolidol ... 633
Neryl Acetate ... 634
(E),(E)-2,4-Nonadienal ... 634
(E),(Z)-2,6-Nonadienal ... 634
(E),(Z)-2,6-Nonadienol ... 634
δ-Nonalactone ... 635
γ-Nonalactone ... 702
Nonanal ... 635
2-Nonanone ... 635
(E)-2-Nonenal ... 635
(E)-2-Nonen-1-ol ... 636
(Z)-6-Nonen-1-ol ... 636
Nonyl Acetate ... 702
Nonyl Alcohol ... 703
δ-Octalactone ... 636
γ-Octalactone ... 672
Octanal ... 636
(E)-2-Octen-1-al ... 637
1-Octen-3-ol ... 637
(Z)-3-Octen-1-ol ... 637
1-Octen-3-yl Acetate ... 672
1-Octen-3-yl Butyrate ... 672
Octyl Acetate ... 703
Octyl Alcohol ... 704
Octyl Formate ... 704
2,3-Pentanedione ... 673
2-Pentanone ... 673
α-Phellandrene ... 705
Phenethyl Acetate ... 705
Phenethyl Alcohol ... 706
Phenethyl Isobutyrate ... 637
Phenethyl Isovalerate ... 638
Phenethyl Phenylacetate ... 673
Phenethyl Salicylate ... 674
Phenoxyethyl Isobutyrate ... 674
Phenylacetaldehyde ... 674
Phenylacetaldehyde Dimethyl Acetal ... 706

Phenylacetic Acid ... 638
Phenylethyl Anthranilate ... 638
Phenylethyl Butyrate ... 638
3-Phenyl-1-propanol ... 707
2-Phenylpropionaldehyde ... 707
3-Phenylpropionaldehyde ... 675
2-Phenylpropionaldehyde Dimethyl Acetal ... 708
3-Phenylpropyl Acetate ... 639
Piperidine ... 639
Piperonal ... 675
Propenylguaethol ... 708
Propionaldehyde ... 639
n-Propyl Acetate ... 639
n-Propyl Alcohol ... 640
p-Propyl Anisole ... 675

n-Propyl Propionate ... 640
Pyrrole ... 640
Rhodinol ... 676
Rhodinyl Acetate ... 709
Rhodinyl Formate ... 709
Santalol ... 710
Santalyl Acetate ... 676
α-Terpineol ... 710
Terpinyl Acetate ... 676
Terpinyl Propionate ... 711
Tetrahydrofurfuryl Alcohol ... 640
2,3,5,6-Tetramethylpyrazine ... 641
Thymol ... 641
Tolualdehyde, Mixed Isomers ... 641
p-Tolyl Isobutyrate ... 641

Tributyrin ... 677
2-Tridecenal ... 642
2,4,5-Trimethyl δ-3-Oxazoline ... 677
2,3,5-Trimethylpyrazine ... 642
δ-Undecalactone ... 642
γ-Undecalactone ... 642
Undecanal ... 677
2-Undecanone ... 643
10-Undecenal ... 643
2-Undecenol ... 678
Undecyl Alcohol ... 678
Valeraldehyde ... 643
Valeric Acid ... 678
γ-Valerolactone ... 679
Vanillin ... 679

SERIES B-1

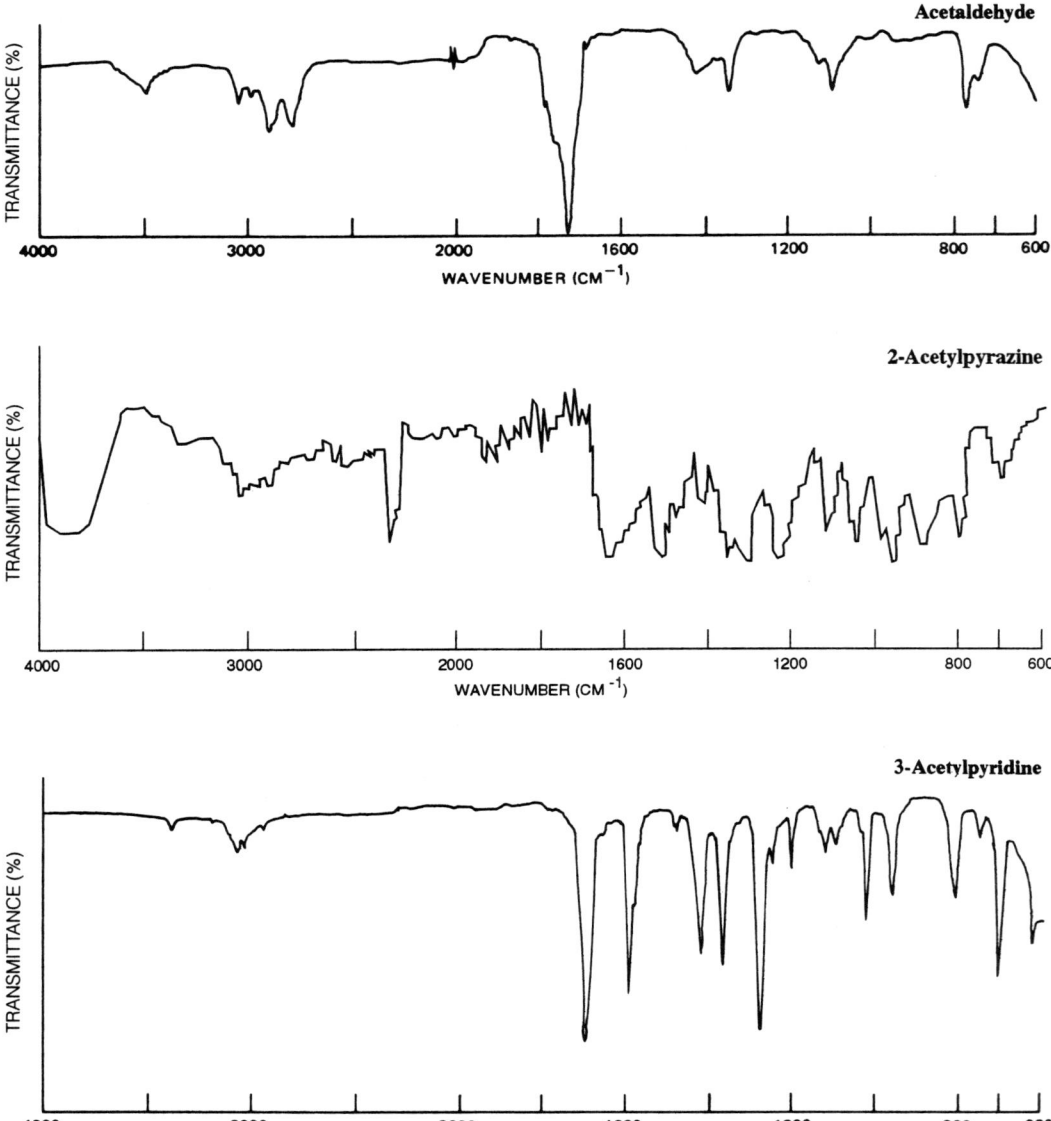

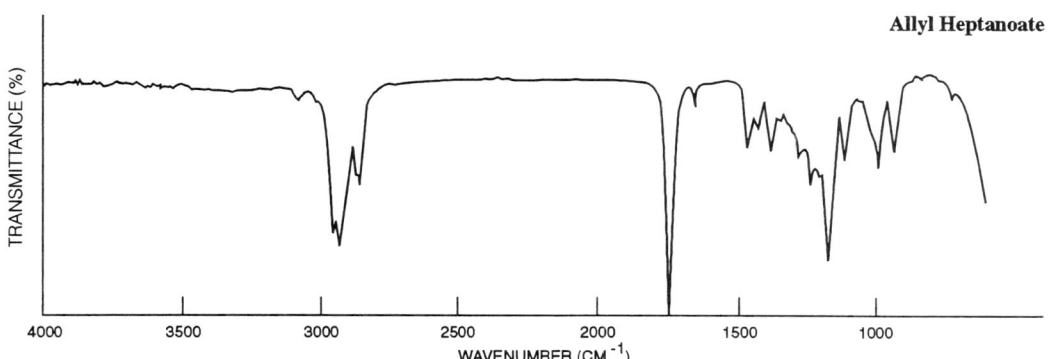

Allyl Heptanoate

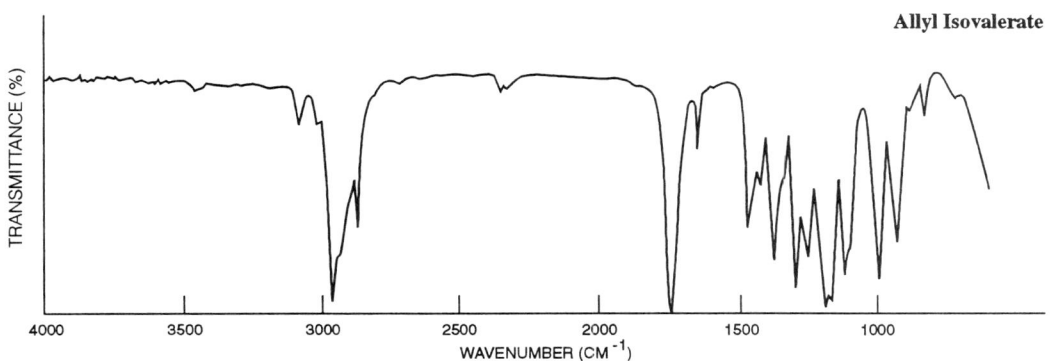

Allyl Isovalerate

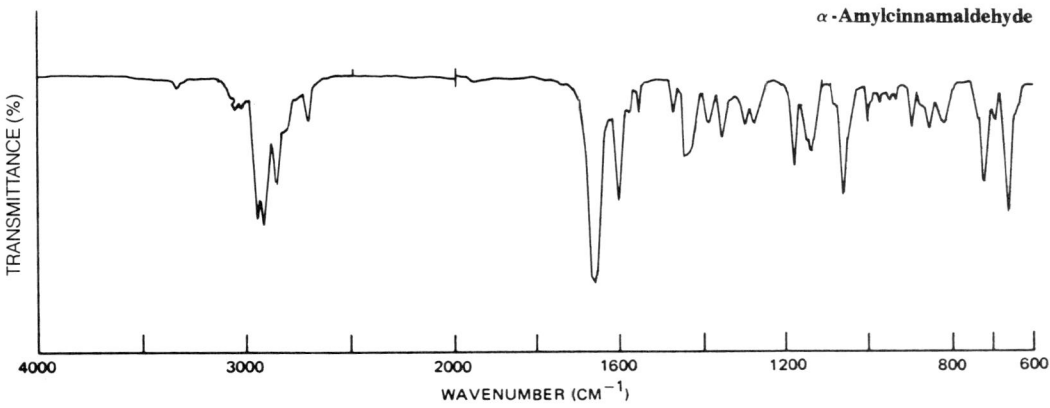

α-Amylcinnamaldehyde

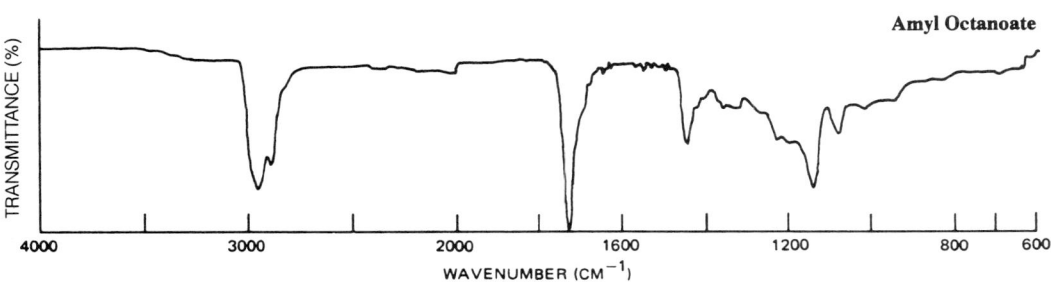

Amyl Octanoate

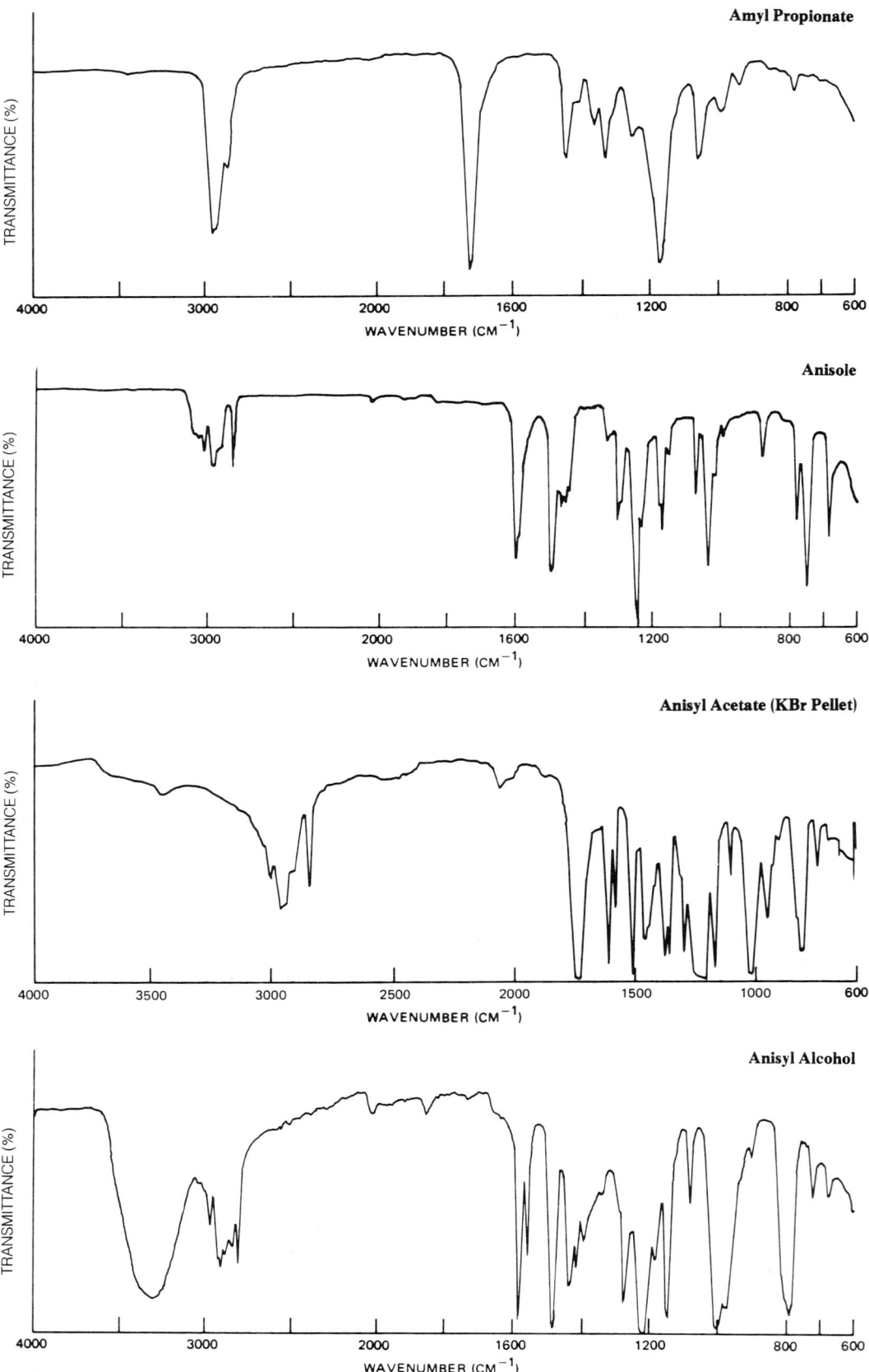

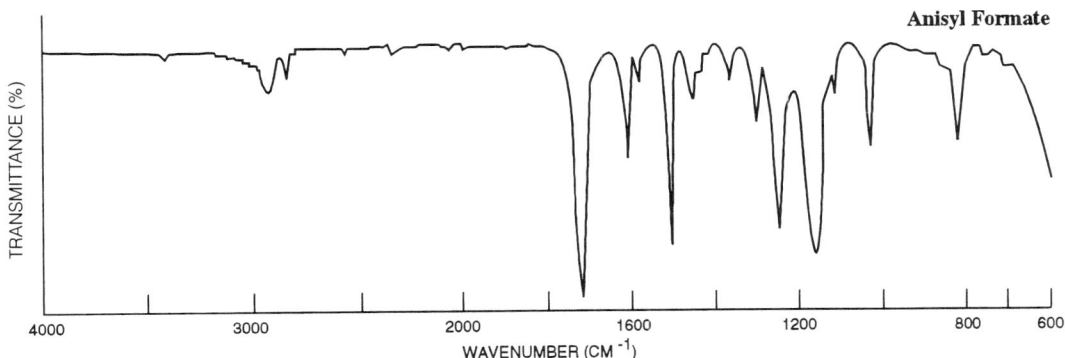

Anisyl Formate

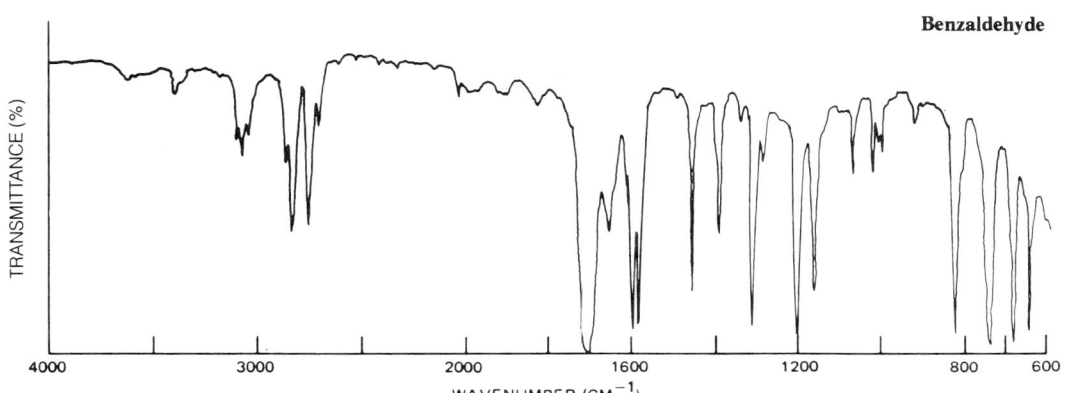

Benzaldehyde

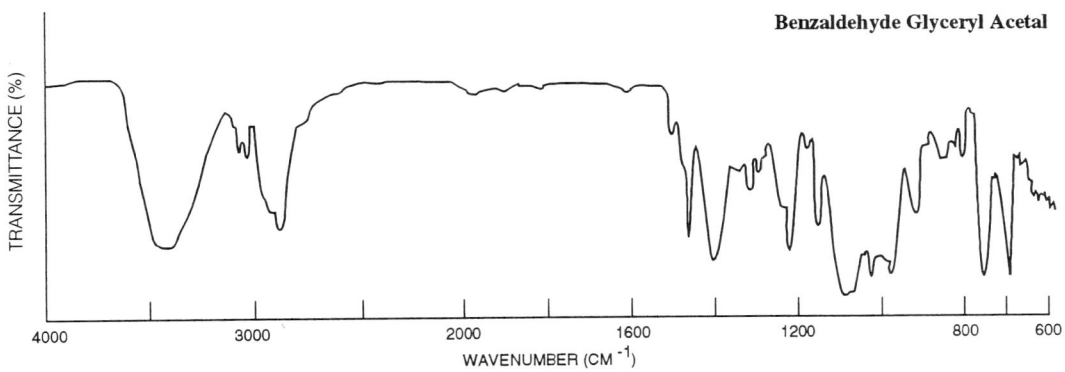

Benzaldehyde Glyceryl Acetal

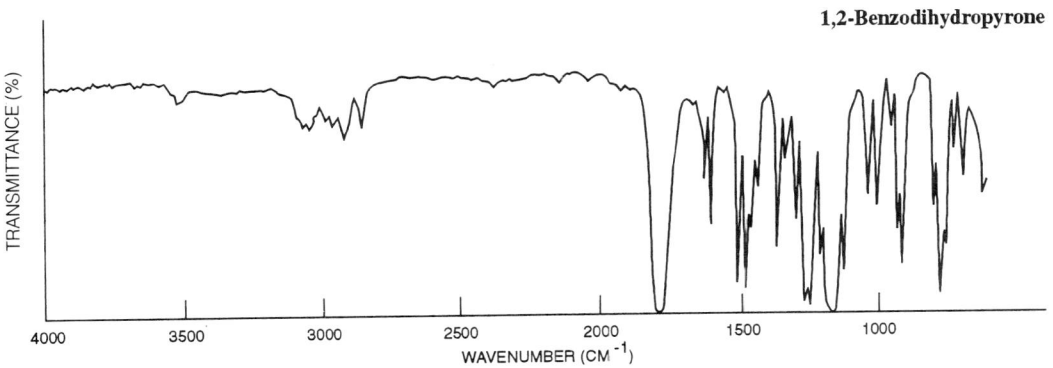

1,2-Benzodihydropyrone

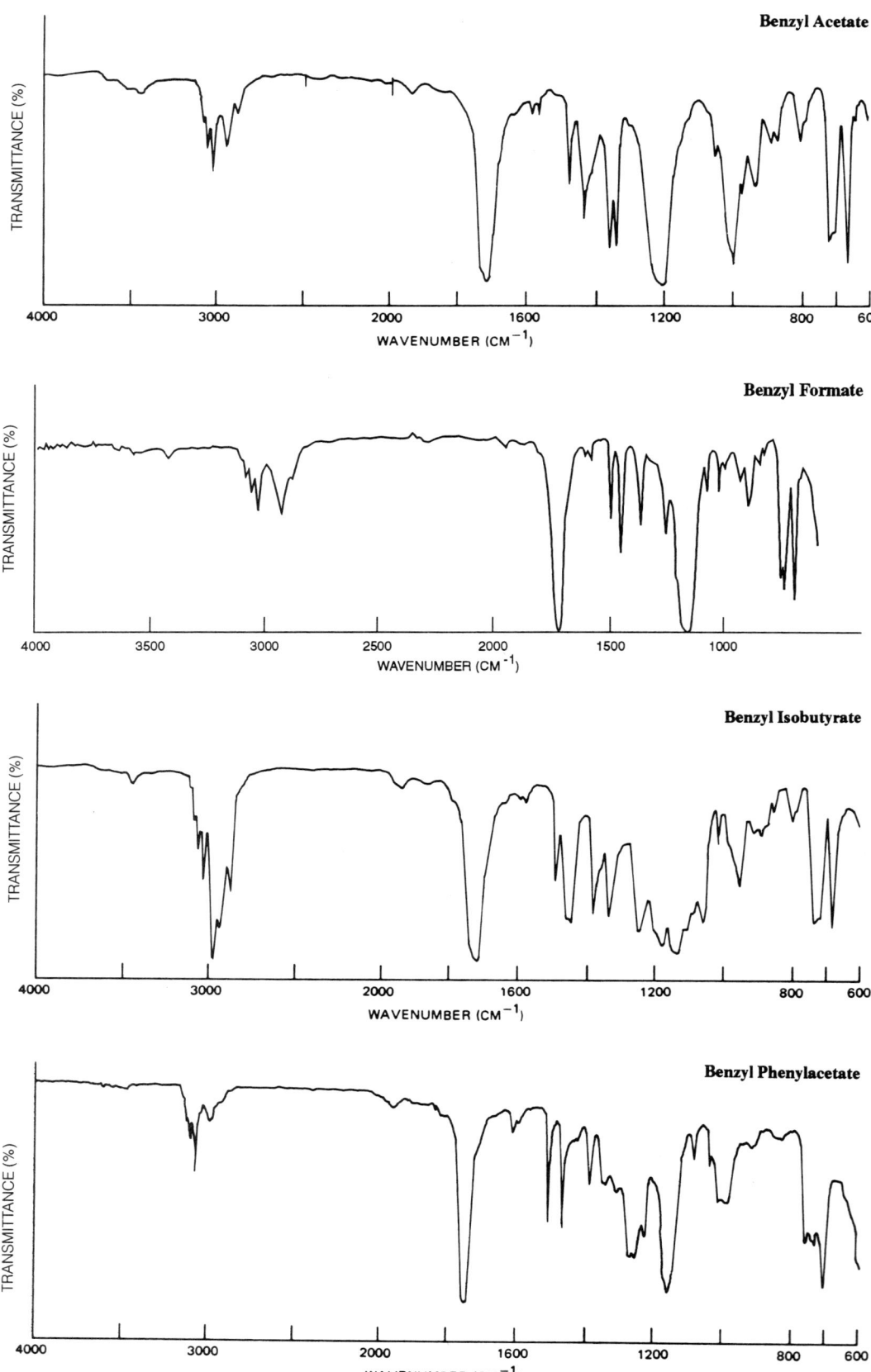

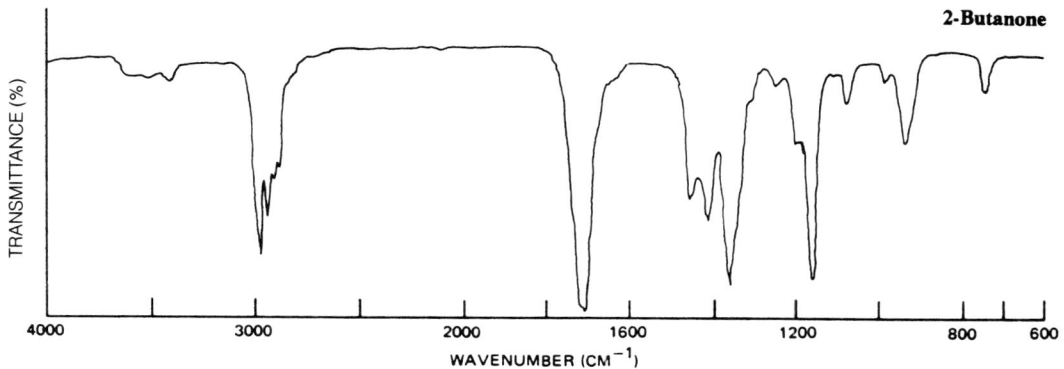

2-Butanone

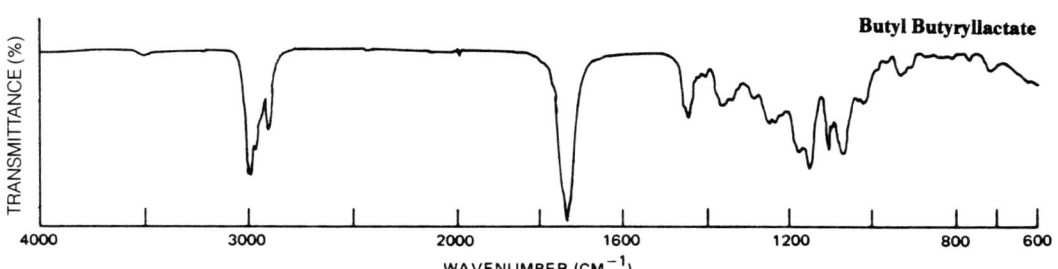

Butyl Butyryllactate

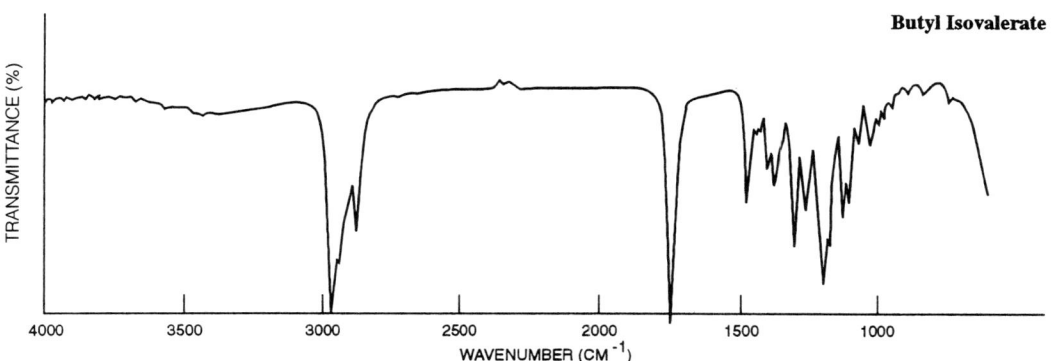

Butyl Isovalerate

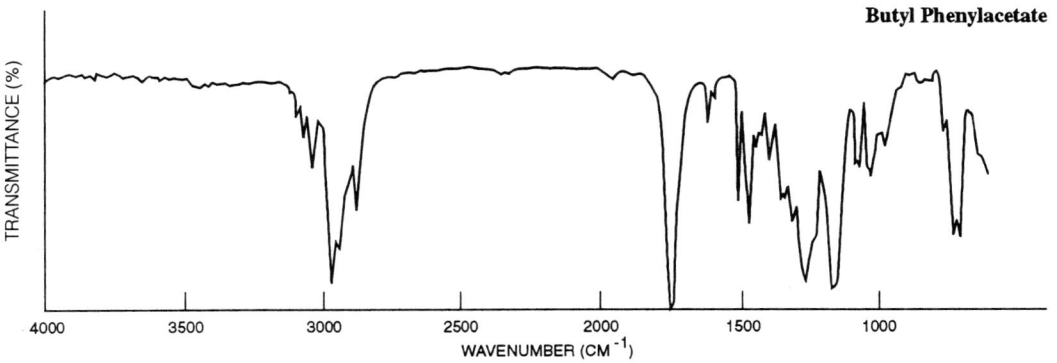

Butyl Phenylacetate

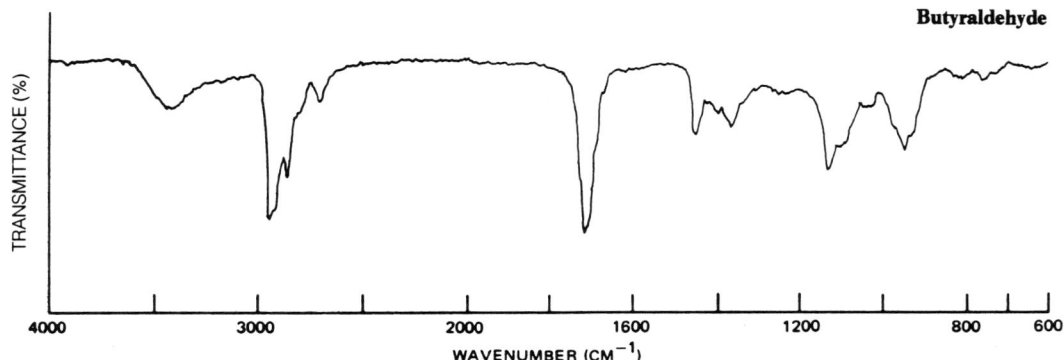
Butyraldehyde

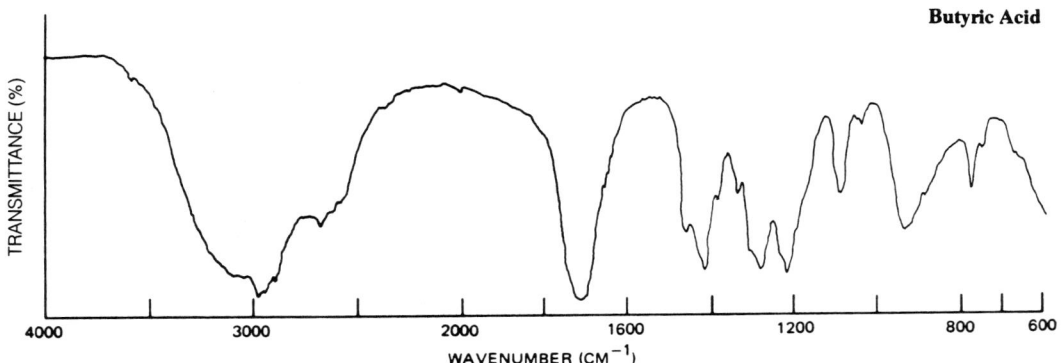
Butyric Acid

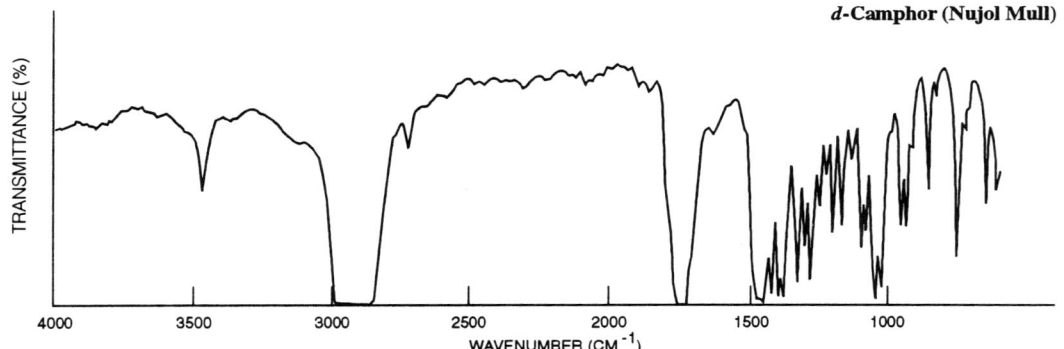
d-Camphor (Nujol Mull)

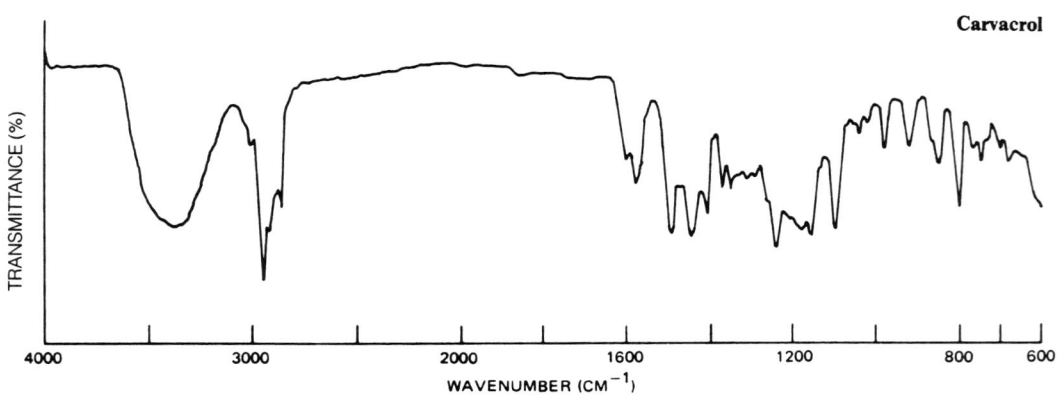
Carvacrol

l-Carveol

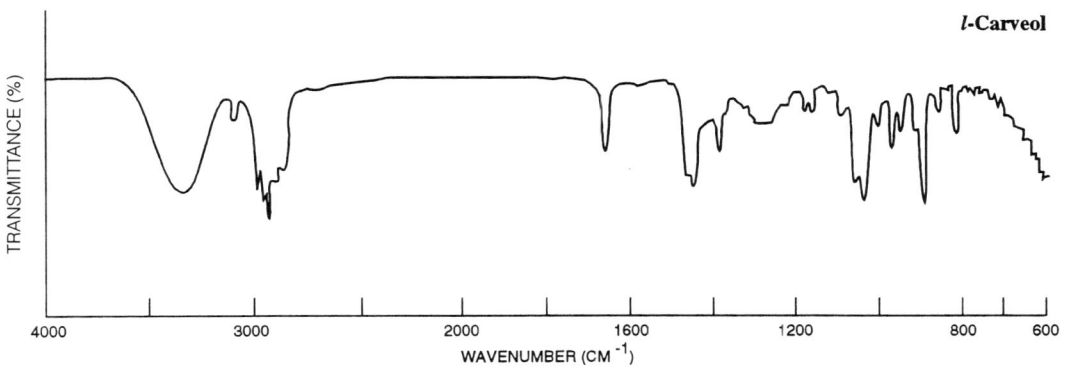

l-Carvyl Acetate

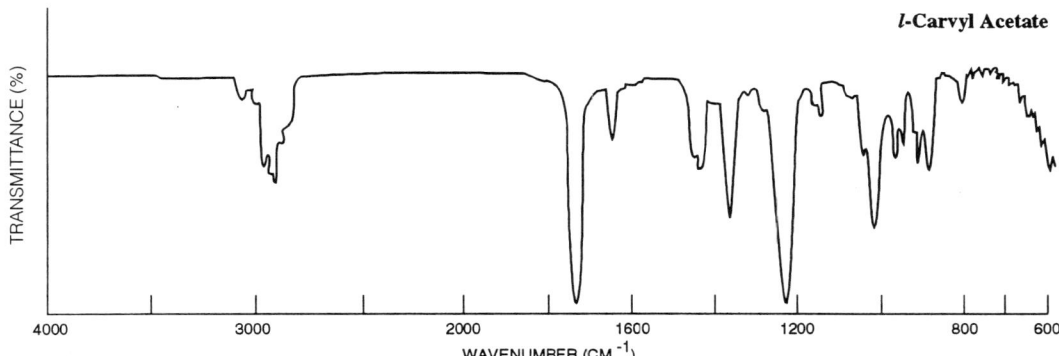

β-Caryophyllene

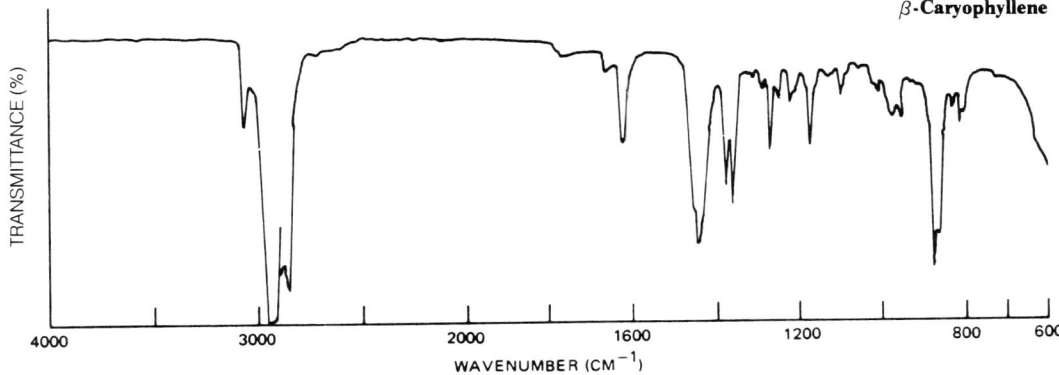

Cinnamaldehyde

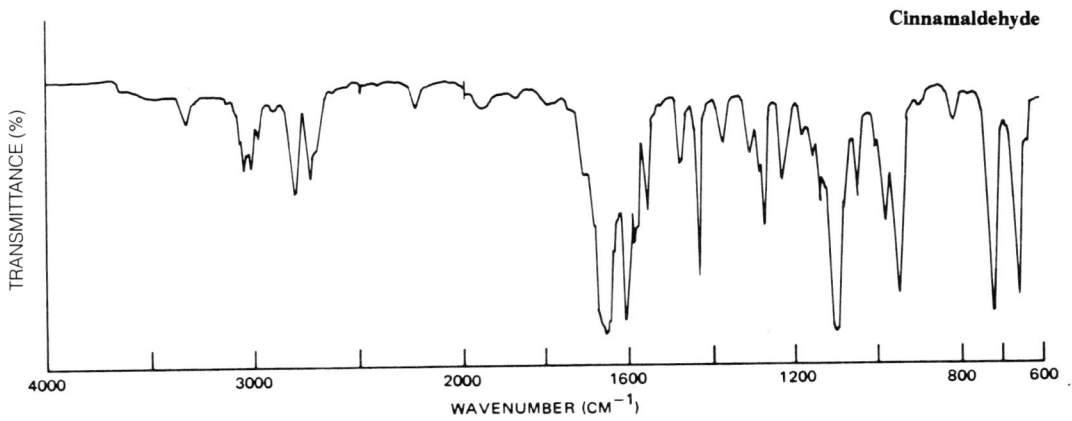

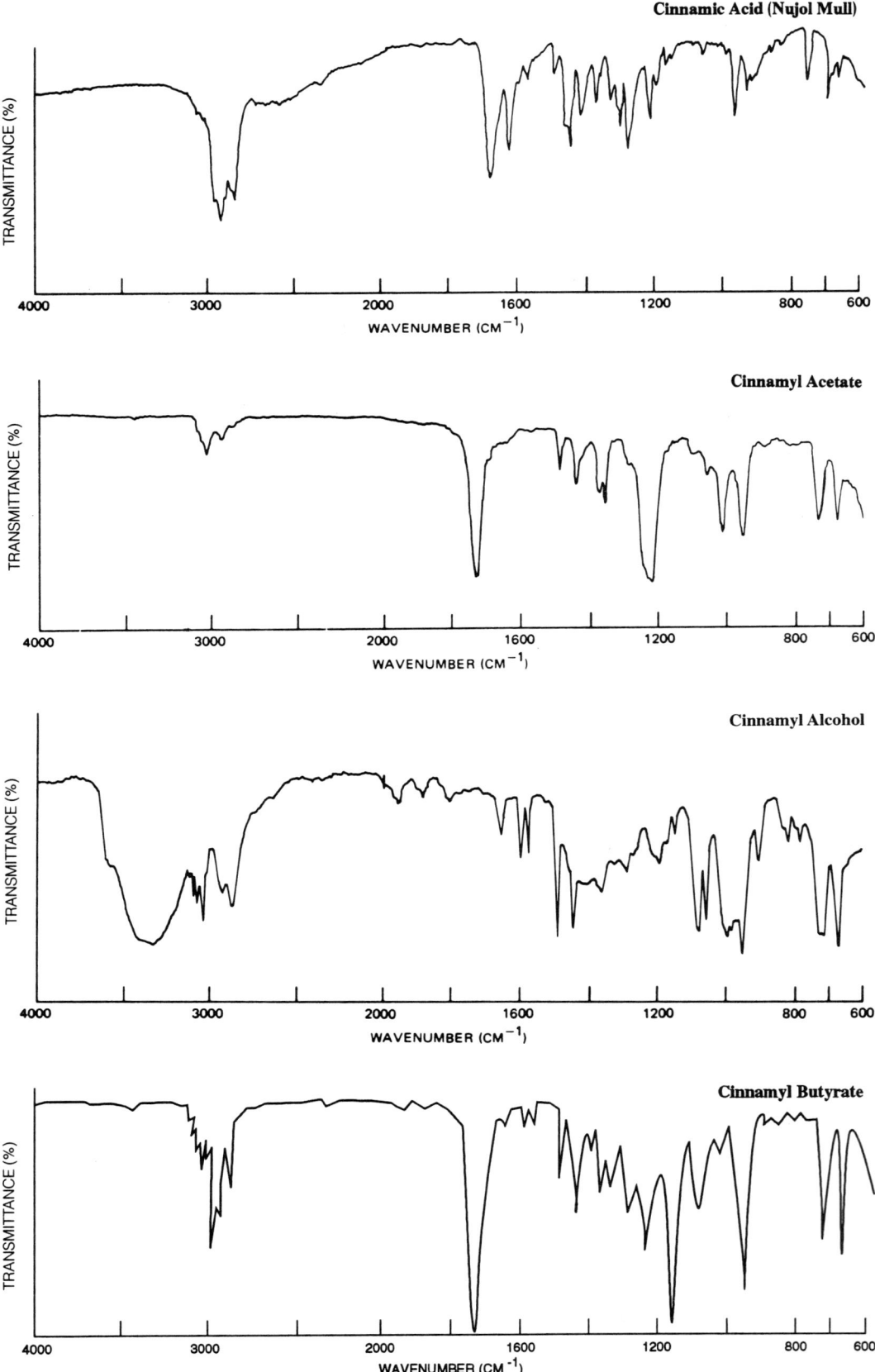

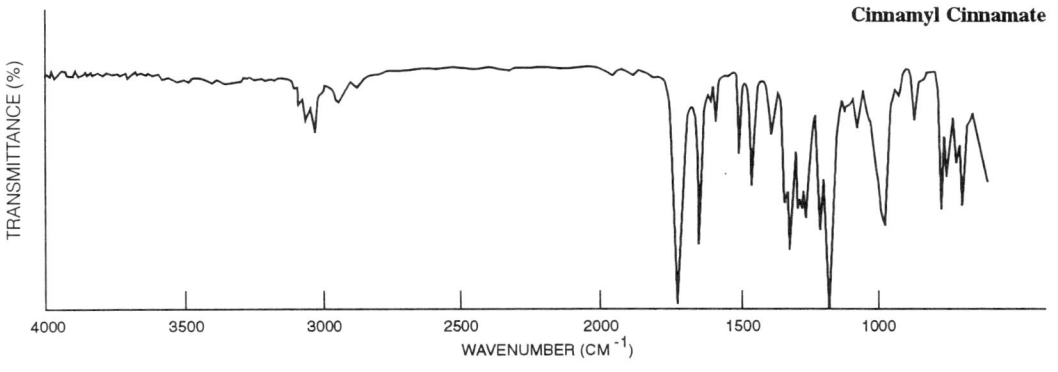

Cinnamyl Cinnamate

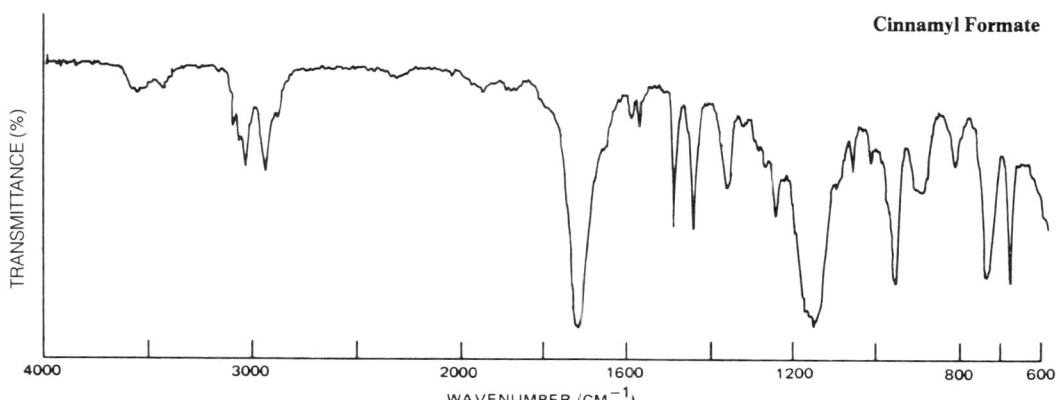

Cinnamyl Formate

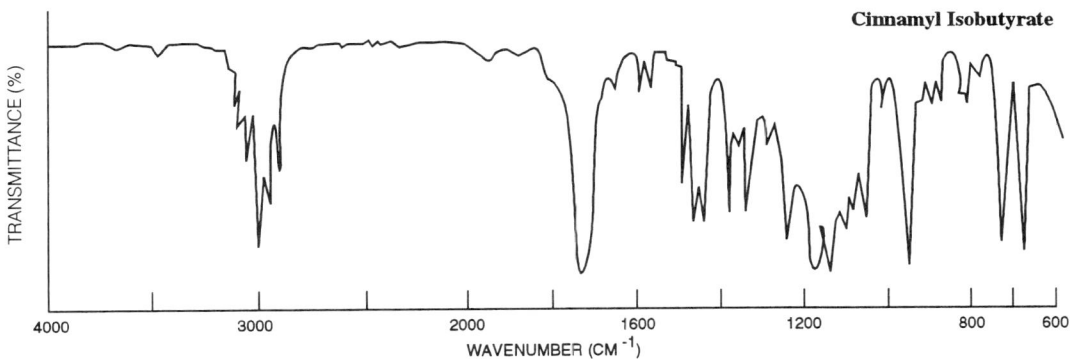

Cinnamyl Isobutyrate

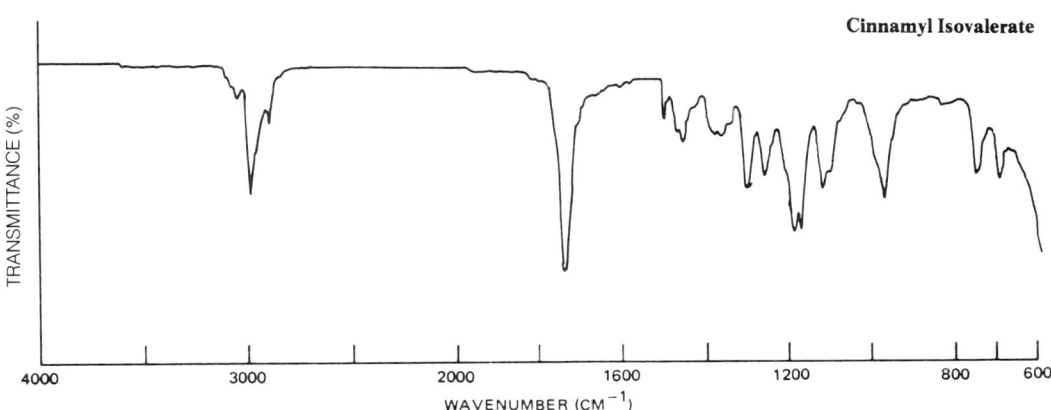

Cinnamyl Isovalerate

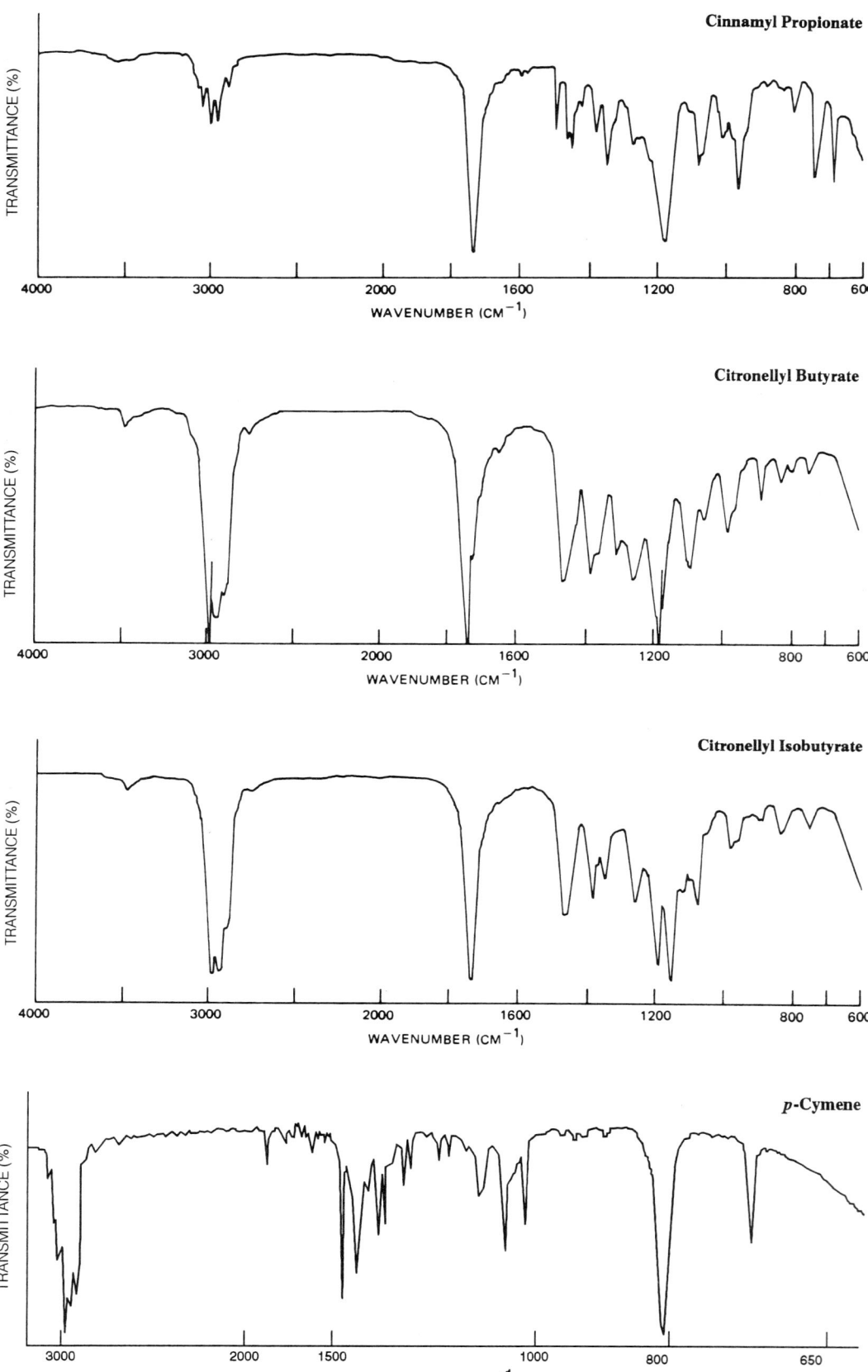

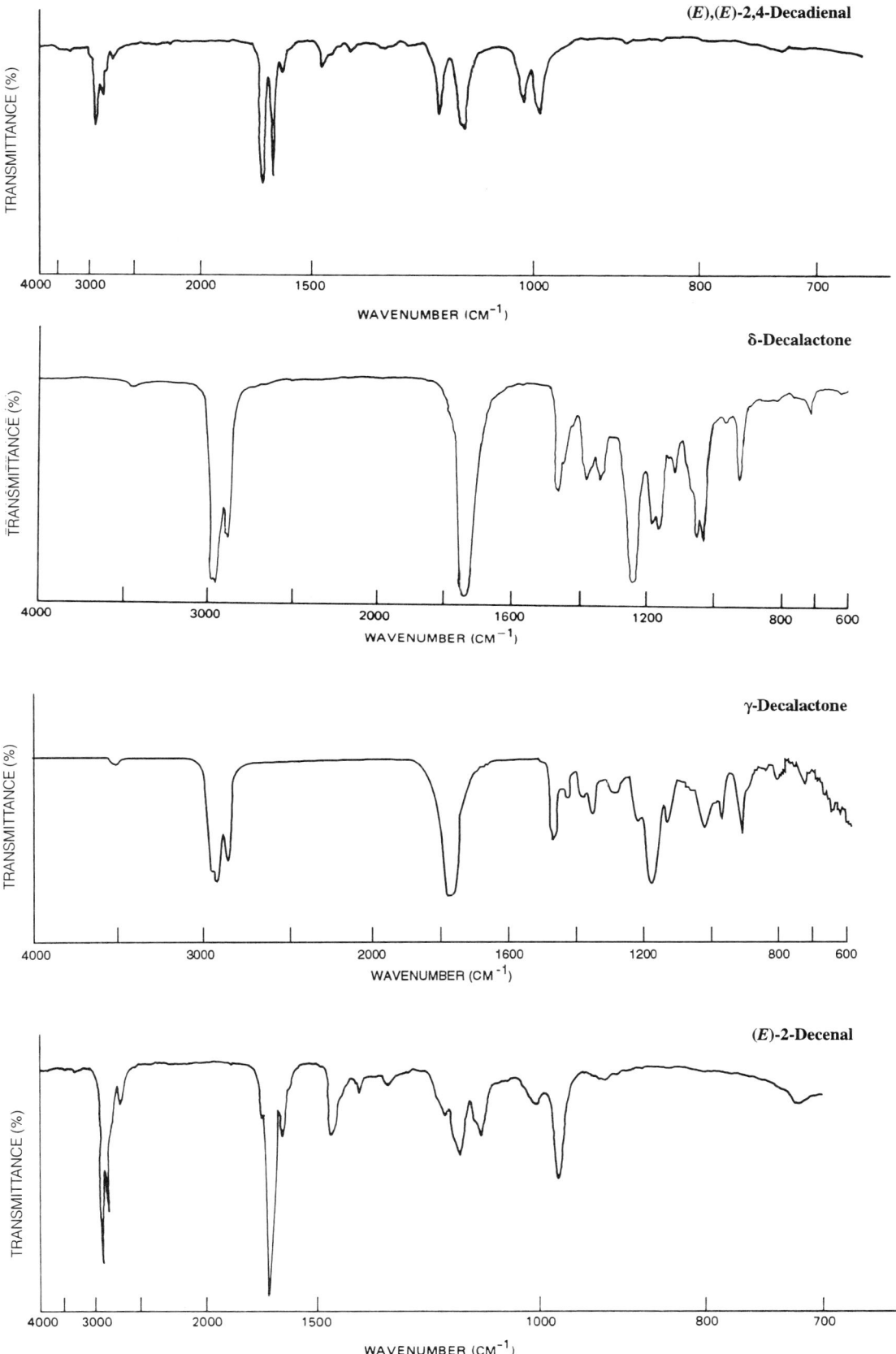

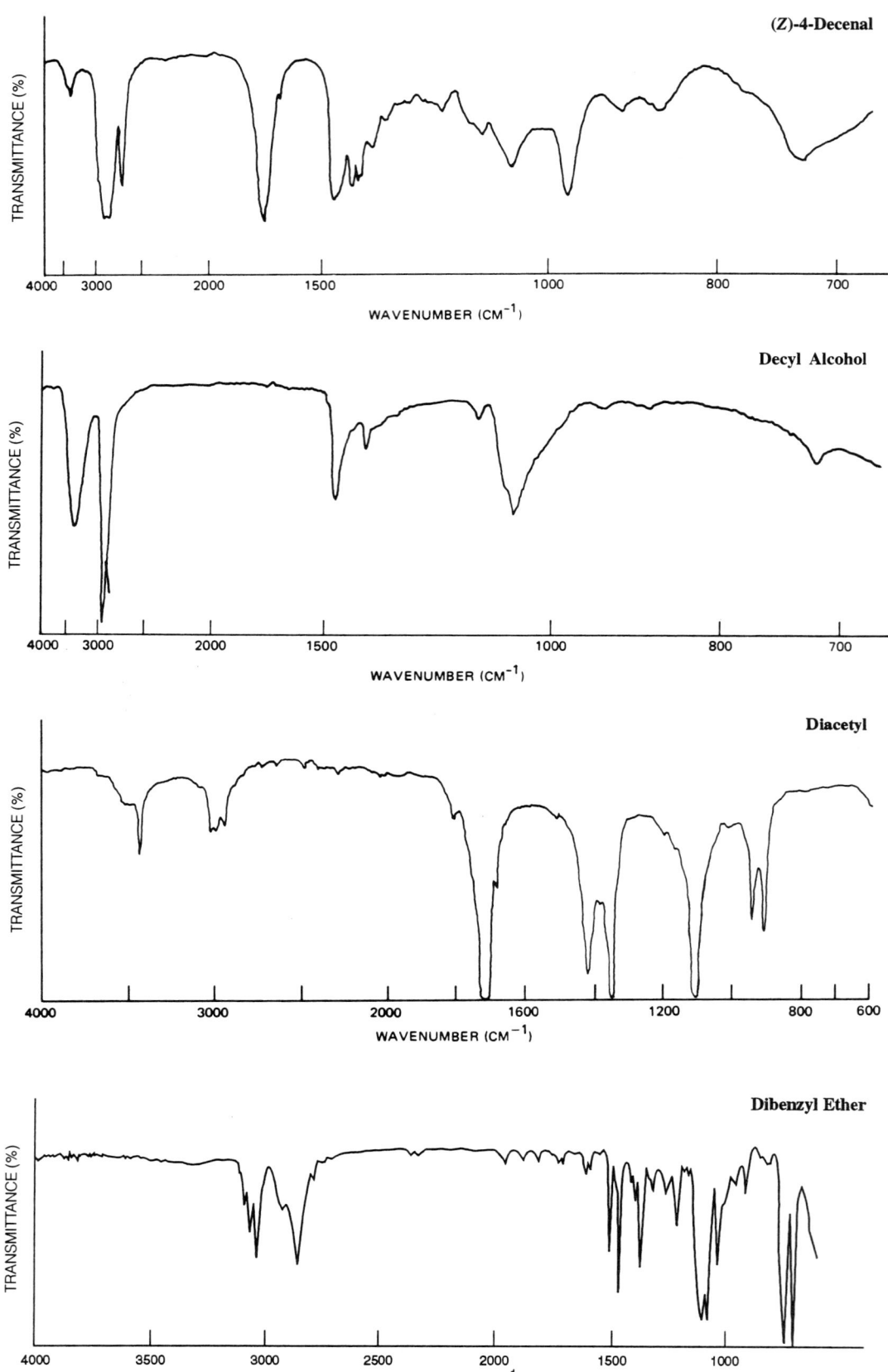

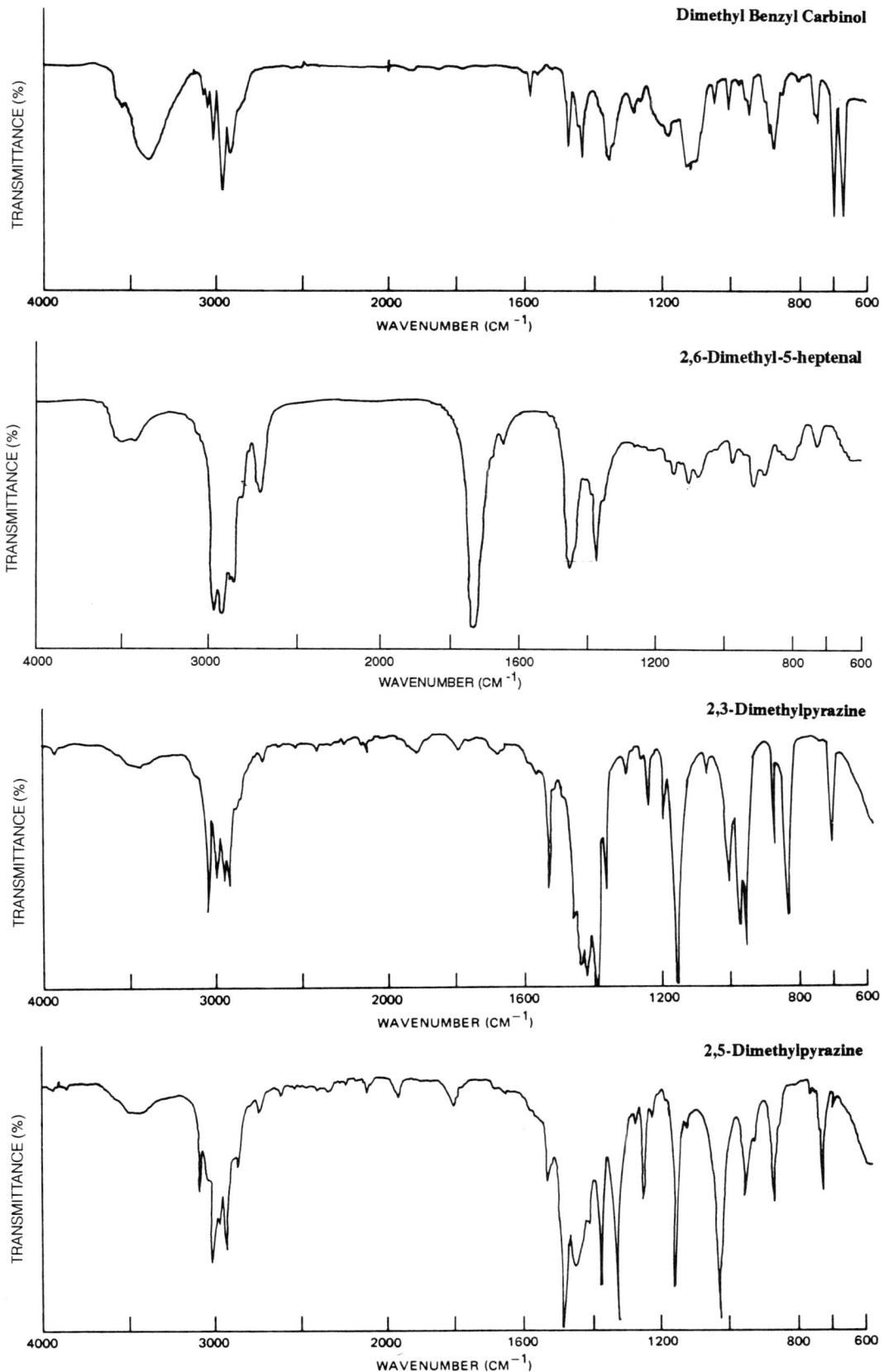

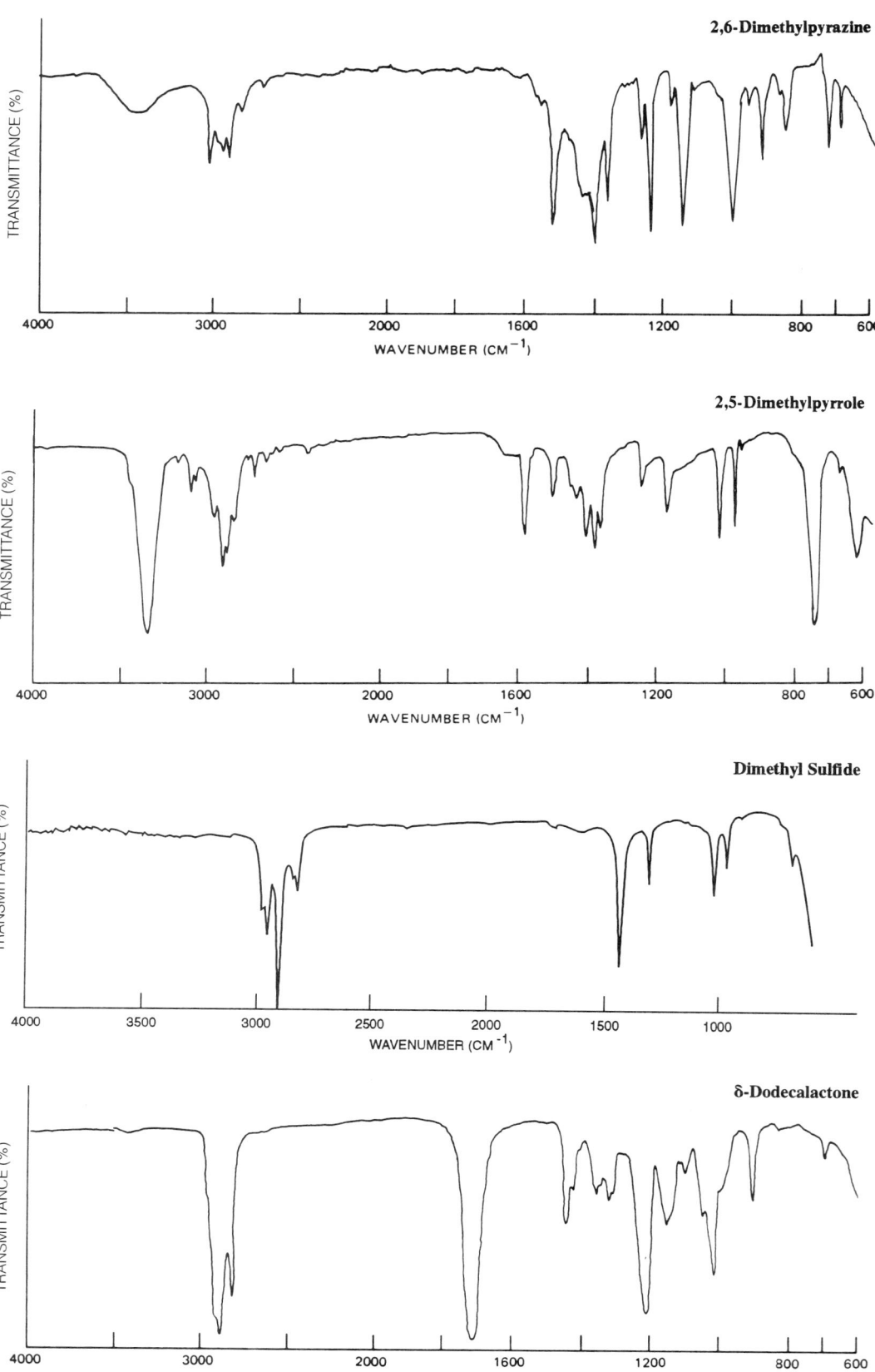

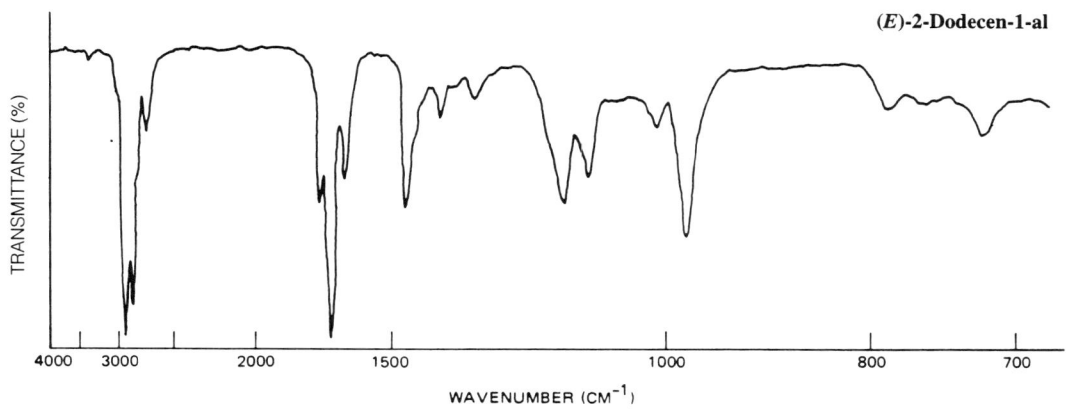
(E)-2-Dodecen-1-al

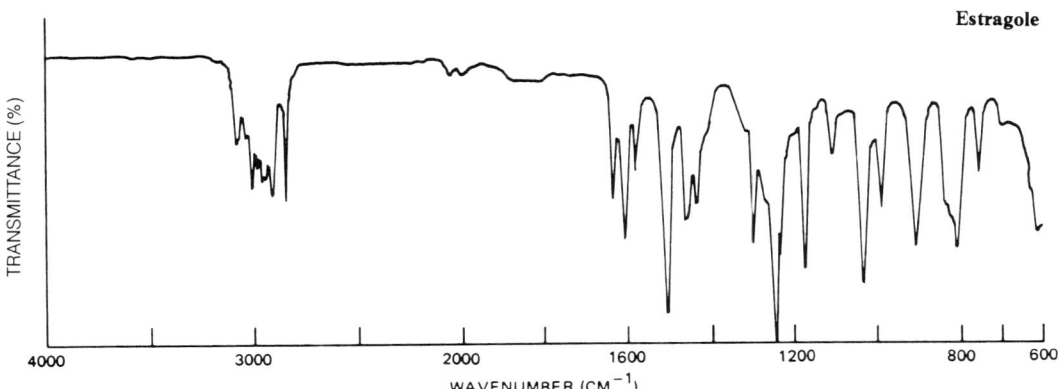
Estragole

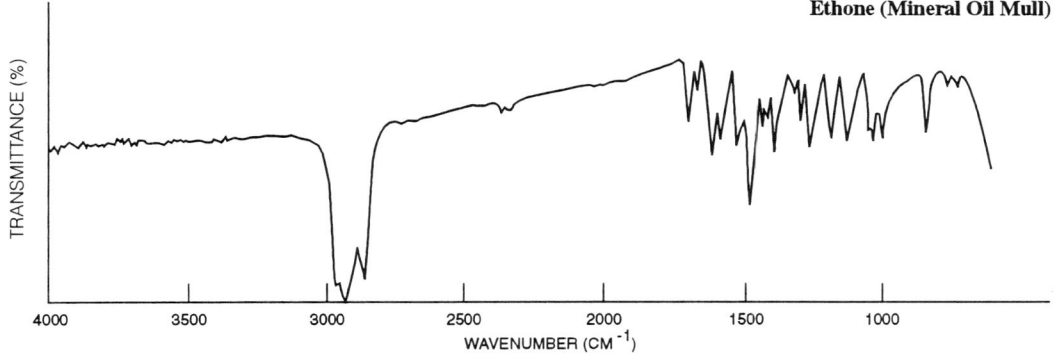
Ethone (Mineral Oil Mull)

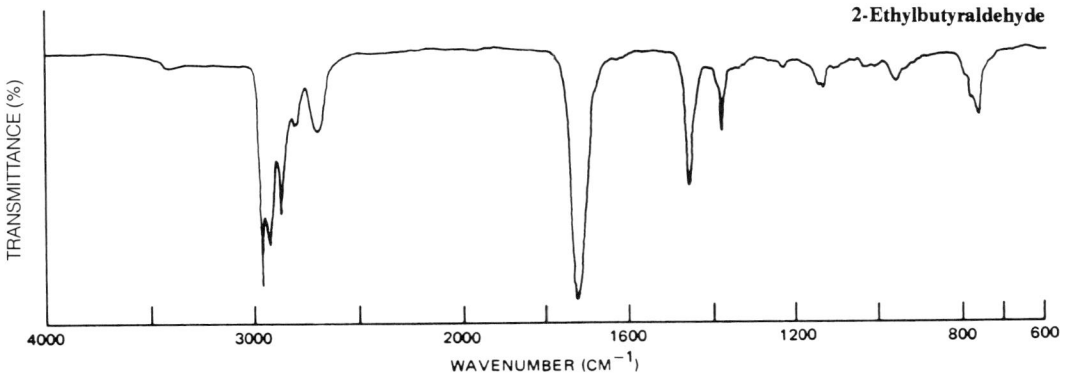
2-Ethylbutyraldehyde

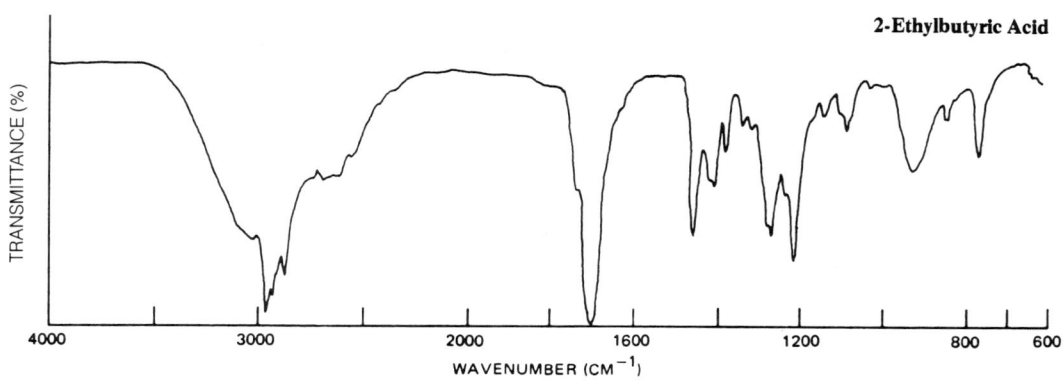

2-Ethylbutyric Acid

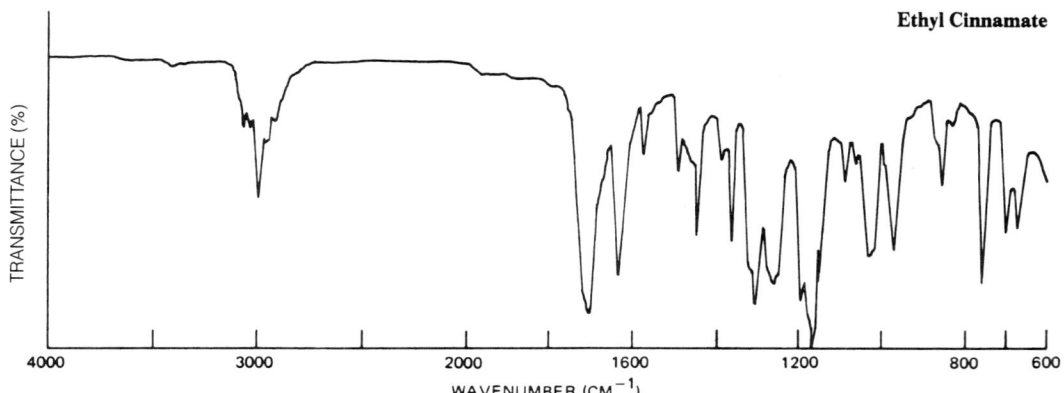

Ethyl Cinnamate

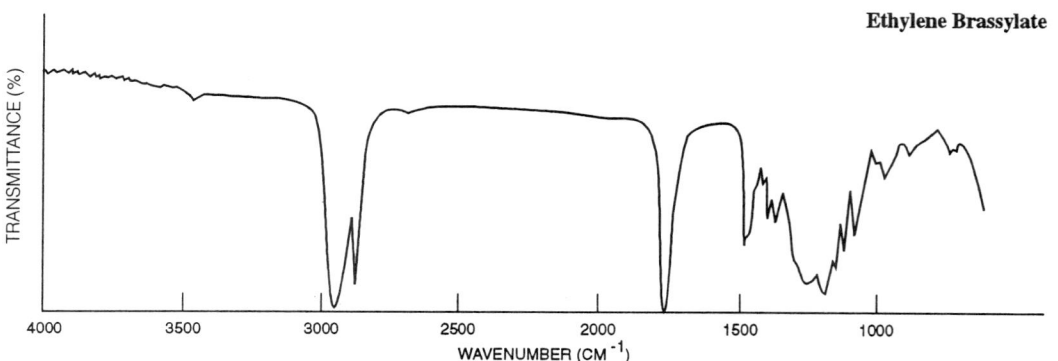

Ethylene Brassylate

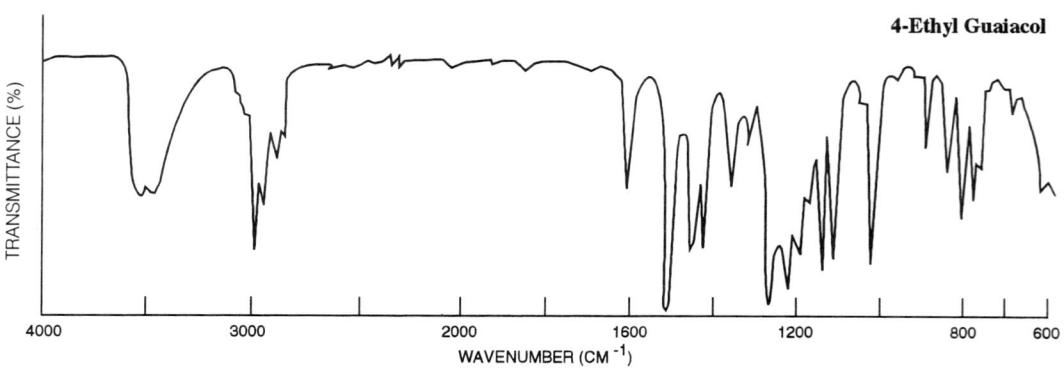

4-Ethyl Guaiacol

Ethyl Isobutyrate

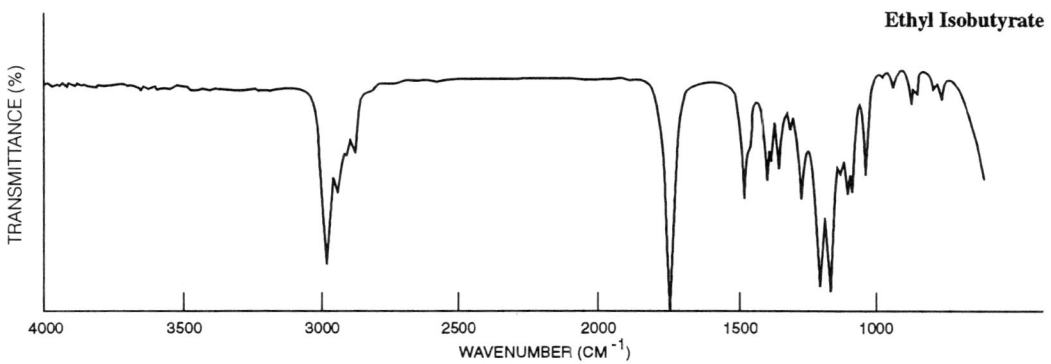

Ethyl Laurate

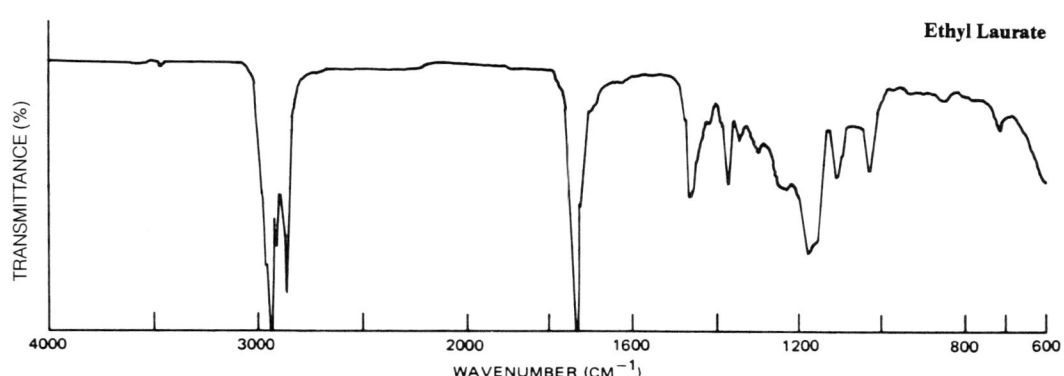

Ethyl Levulinate

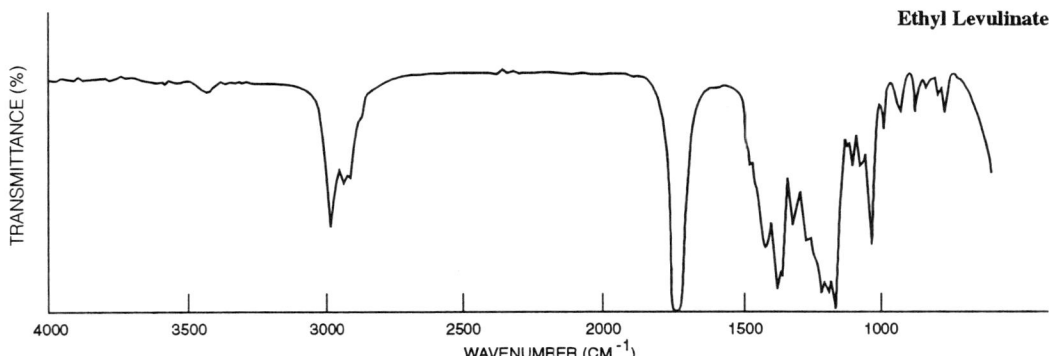

Ethyl-3-methylthiopropionate

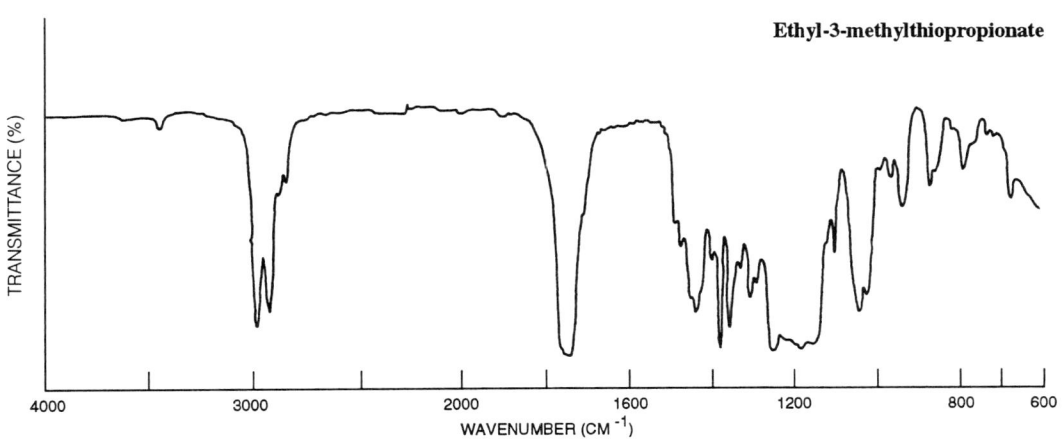

Ethyl Myristate

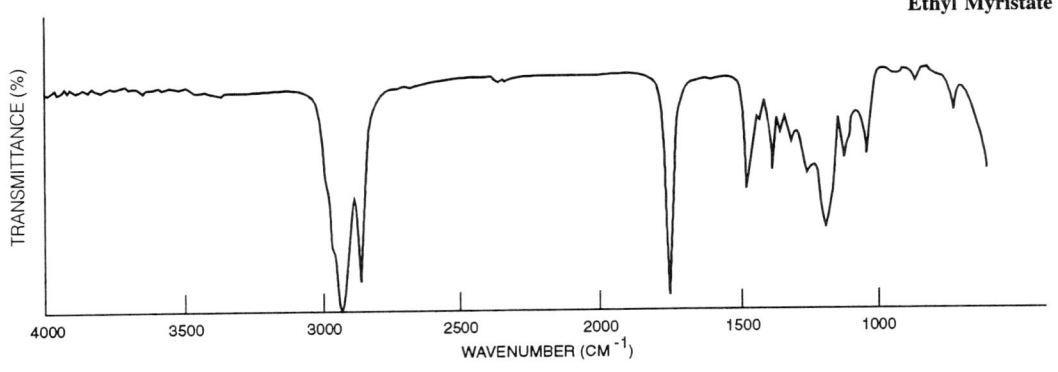

Ethyl Nonanoate

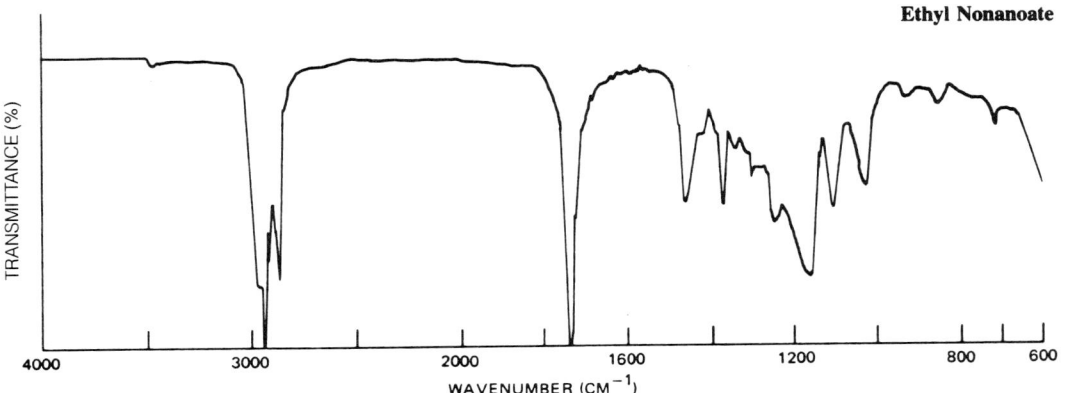

Eucalyptol

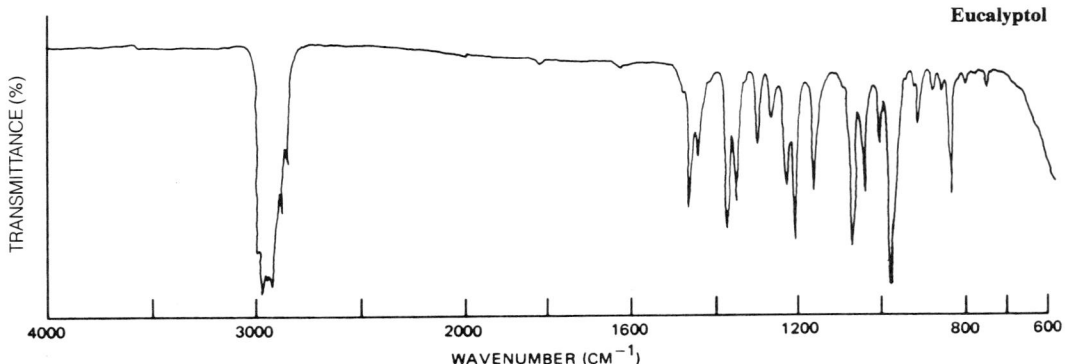

Eugenyl Acetate

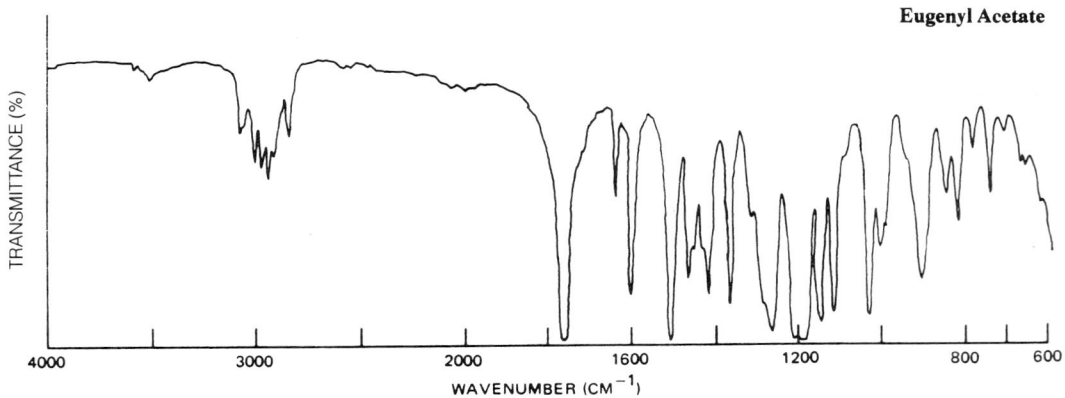

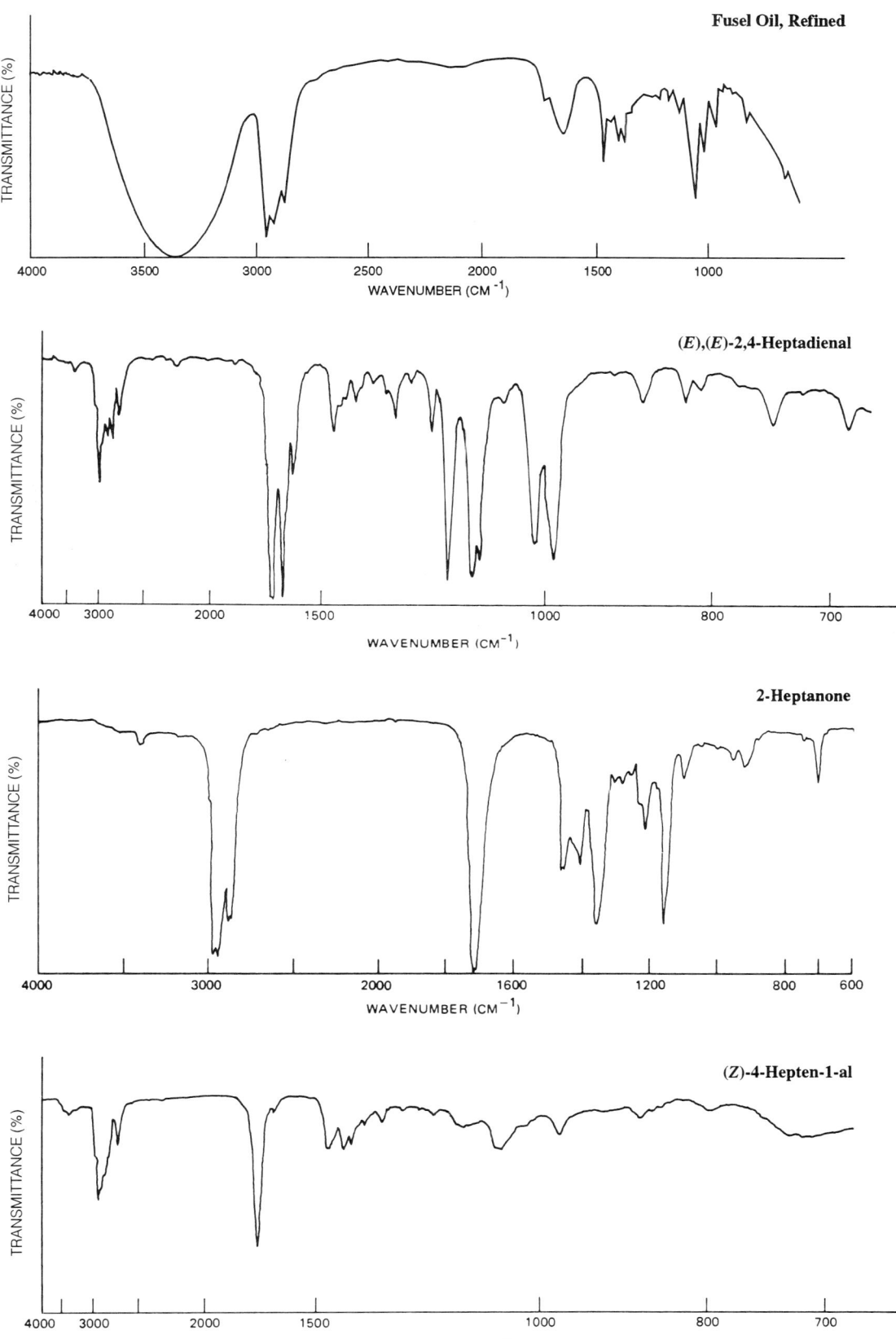

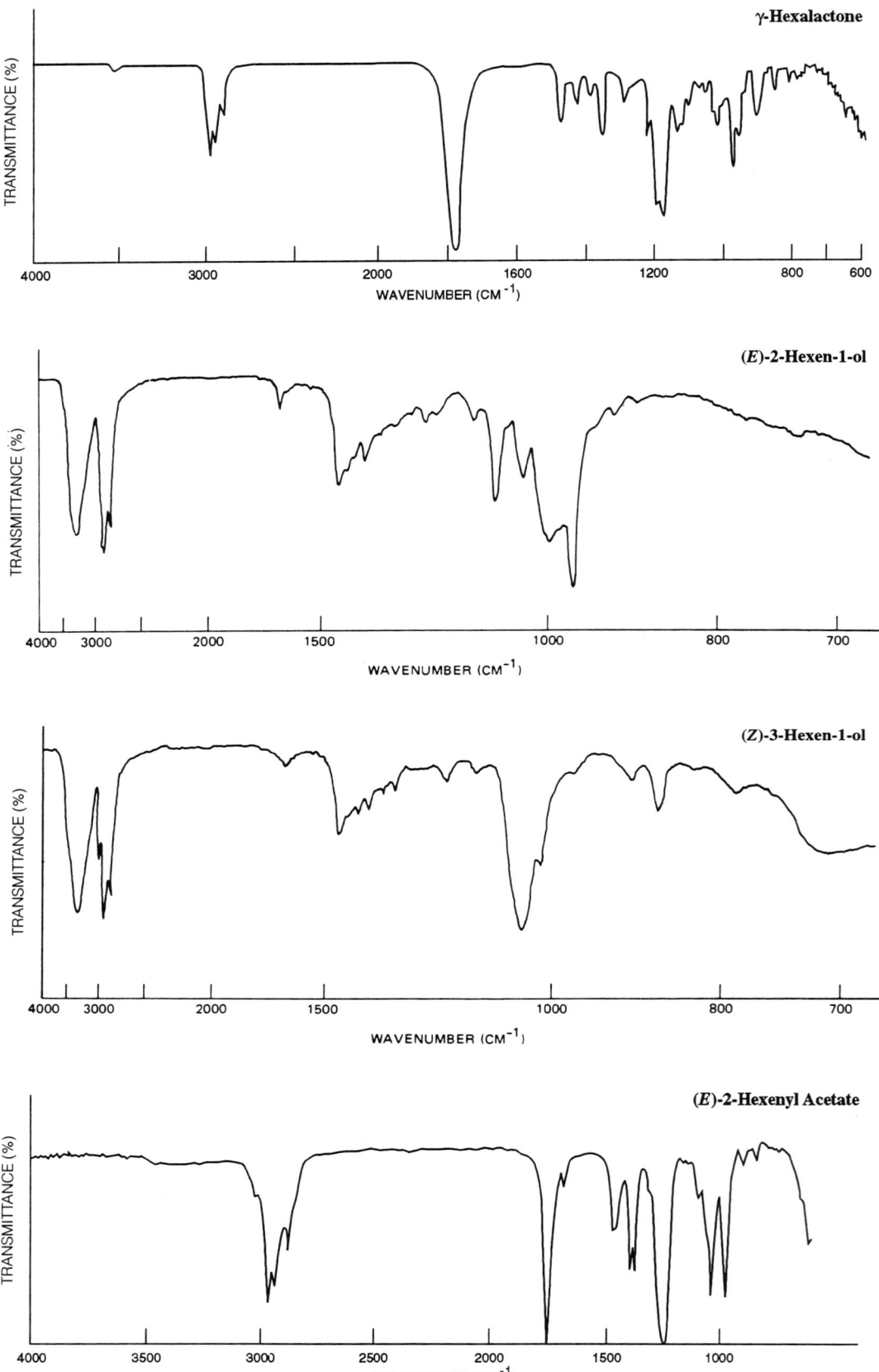

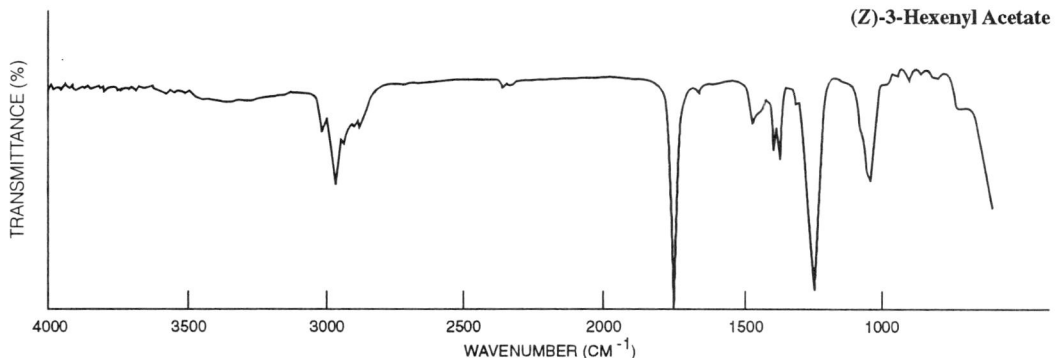

(Z)-3-Hexenyl Acetate

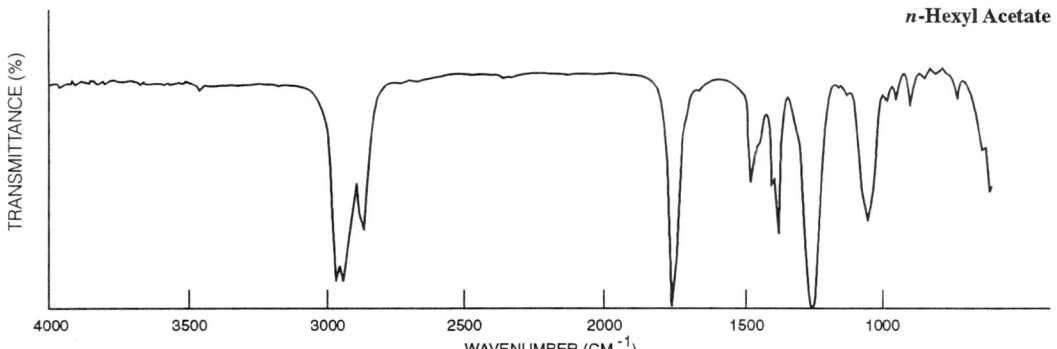

n-Hexyl Acetate

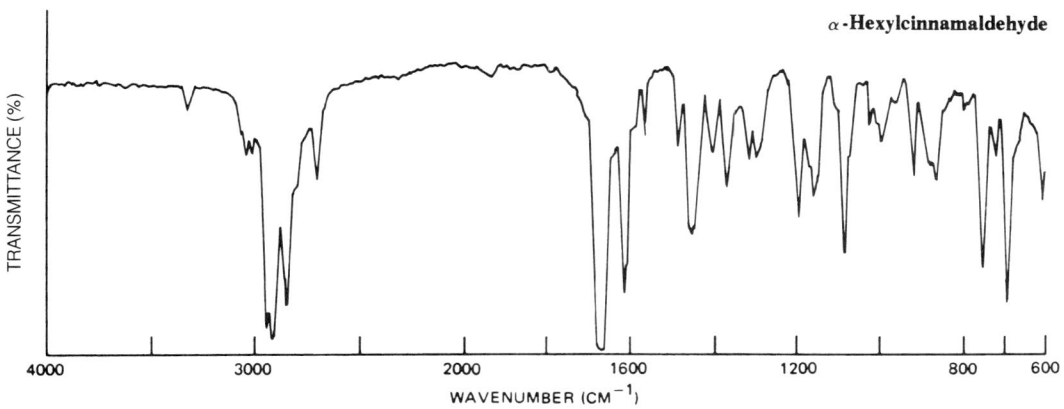

α-Hexylcinnamaldehyde

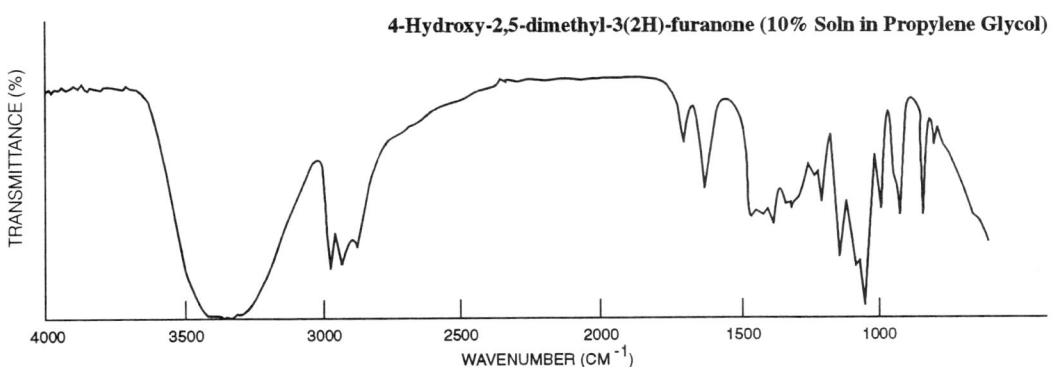

4-Hydroxy-2,5-dimethyl-3(2H)-furanone (10% Soln in Propylene Glycol)

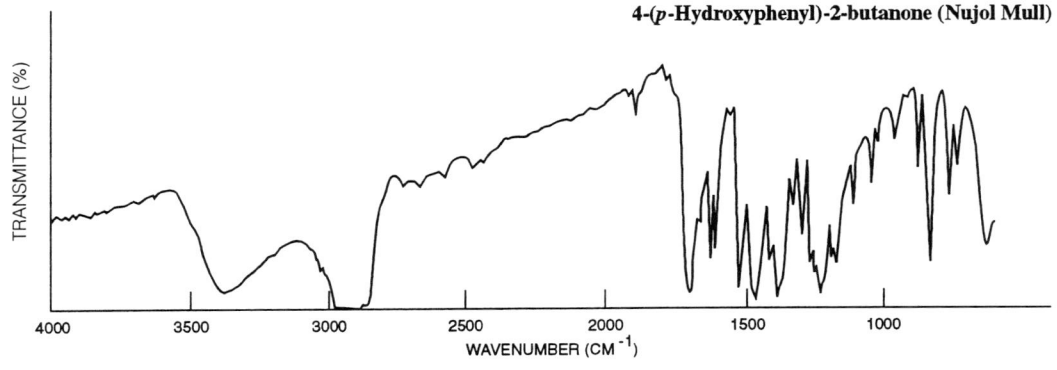

4-(*p*-Hydroxyphenyl)-2-butanone (Nujol Mull)

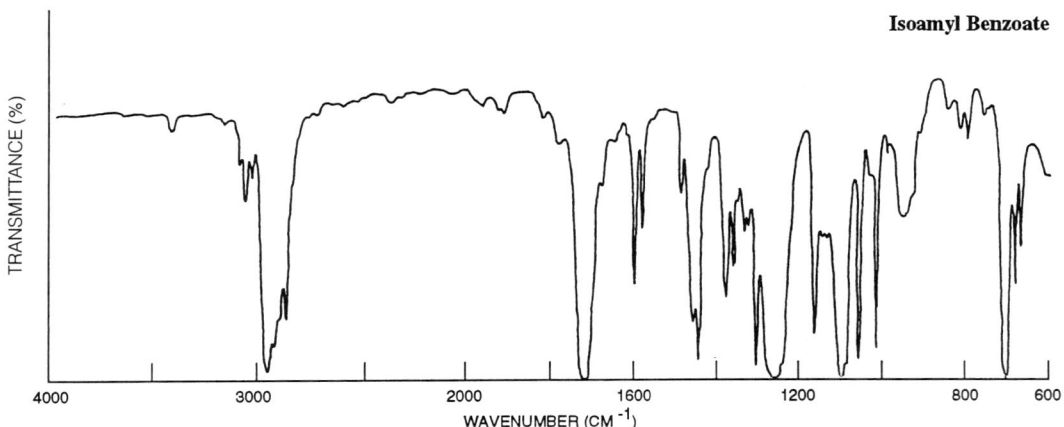

Isoamyl Benzoate

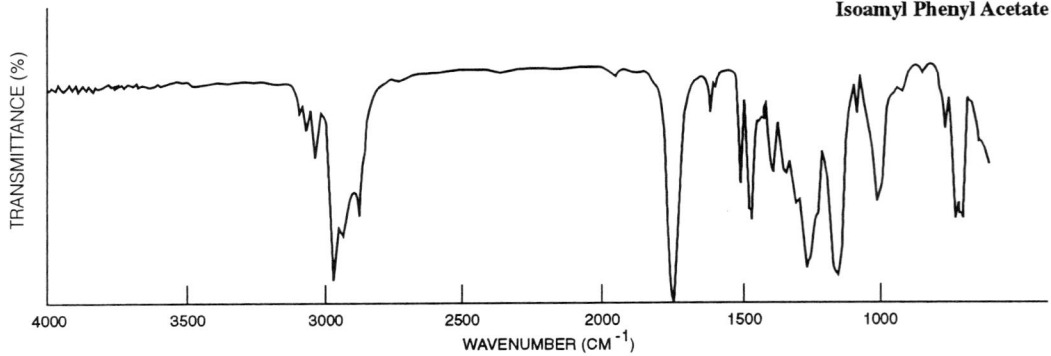

Isoamyl Phenyl Acetate

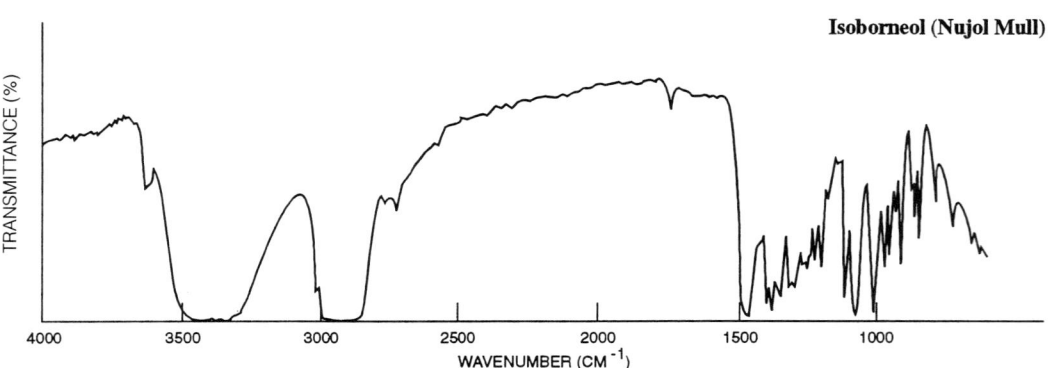

Isoborneol (Nujol Mull)

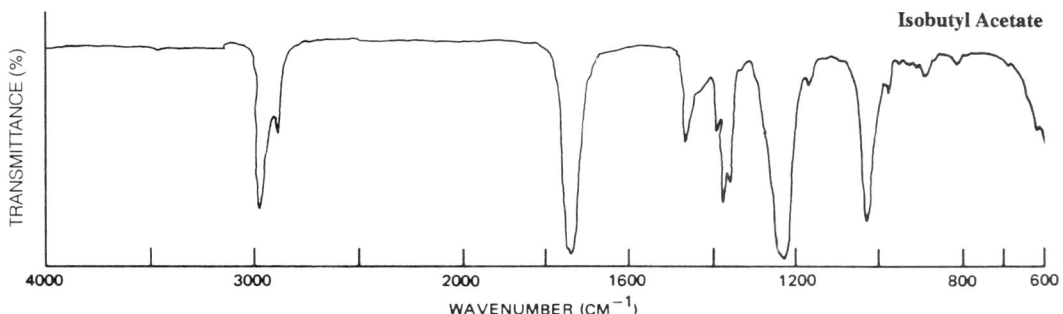

Isobutyl Acetate

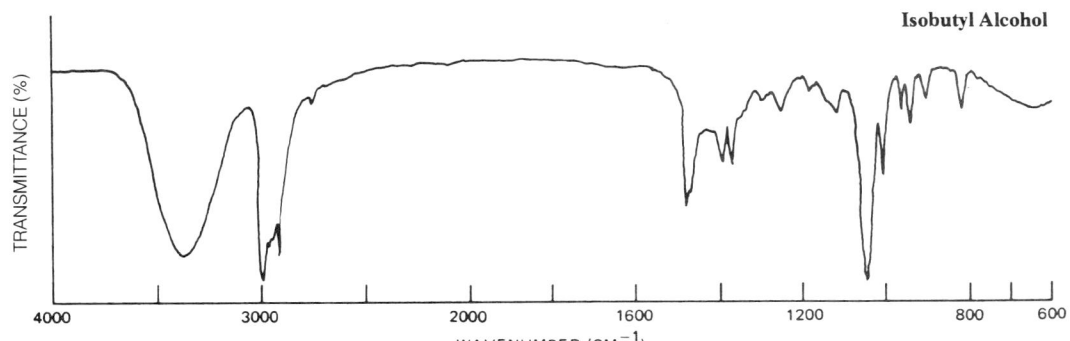

Isobutyl Alcohol

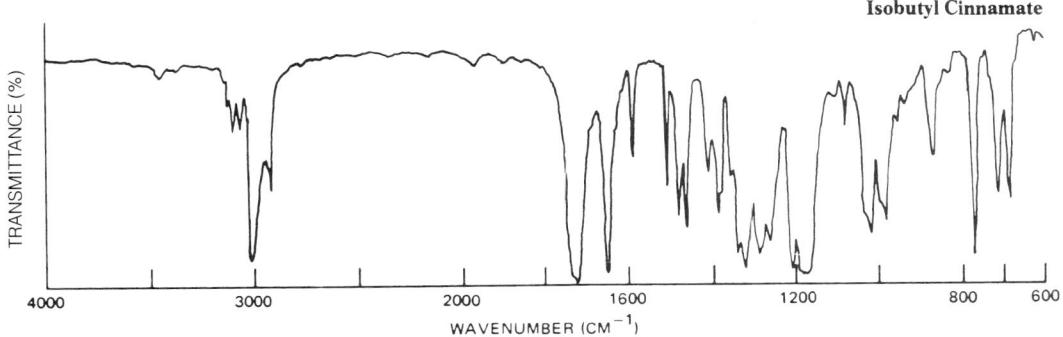

Isobutyl Cinnamate

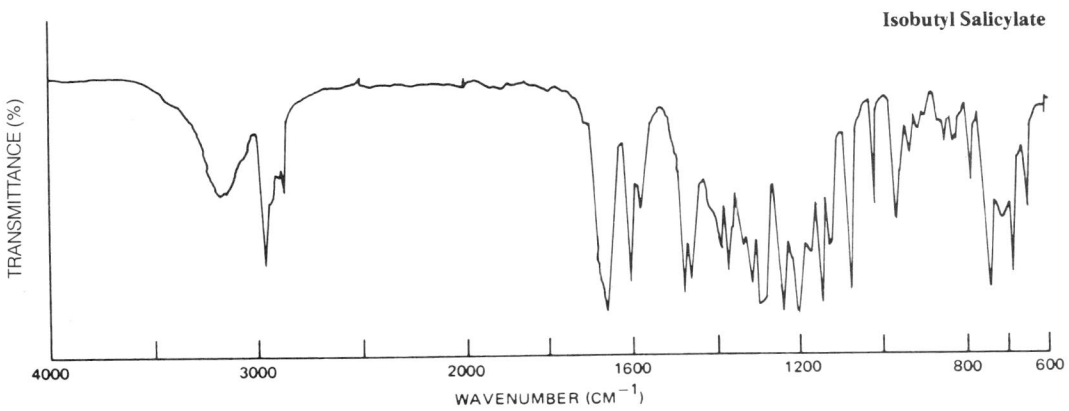

Isobutyl Salicylate

Isobutyl Acetate

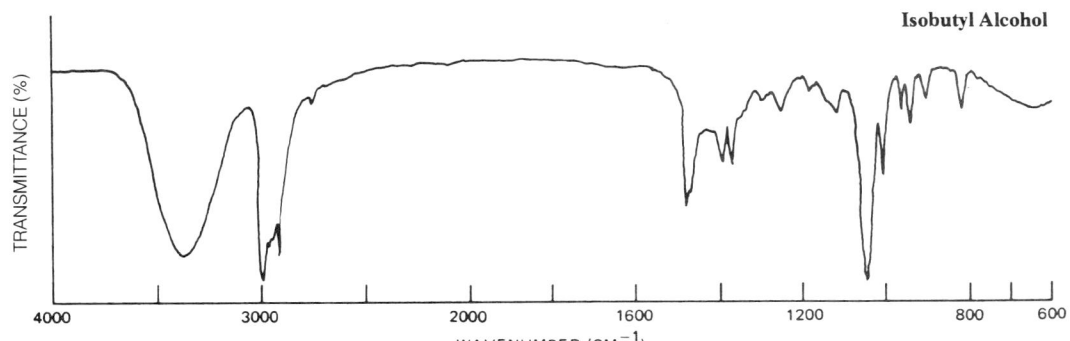

Isobutyl Alcohol

Isobutyl Cinnamate

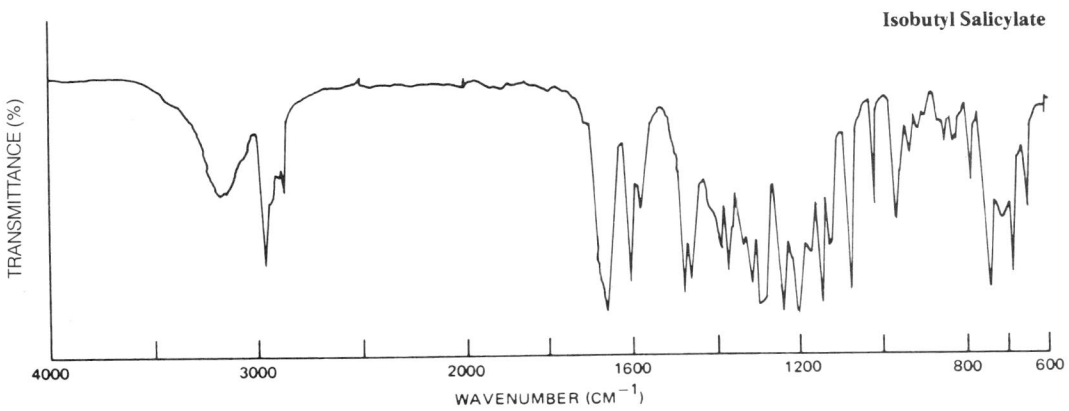

Isobutyl Salicylate

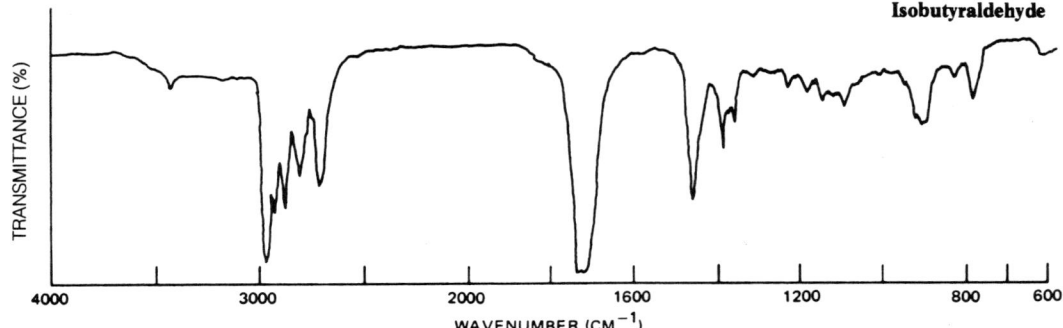

Isobutyraldehyde

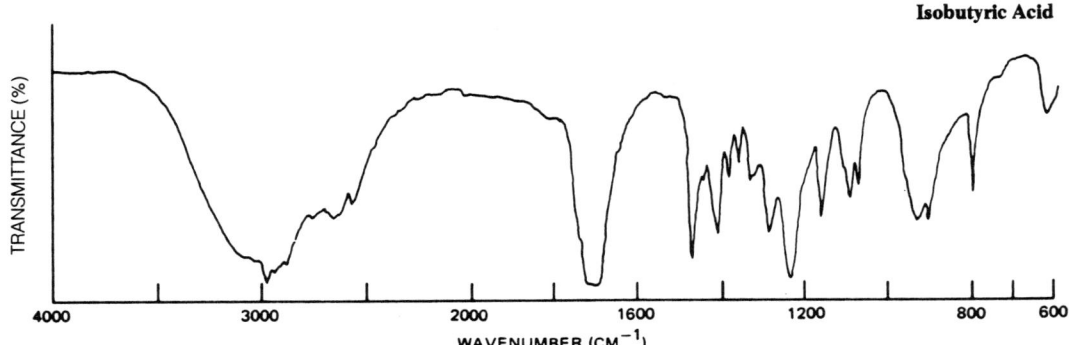

Isobutyric Acid

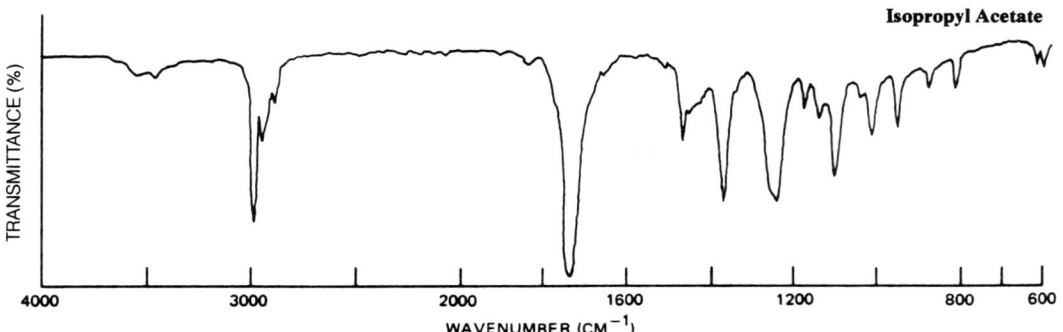

Isopropyl Acetate

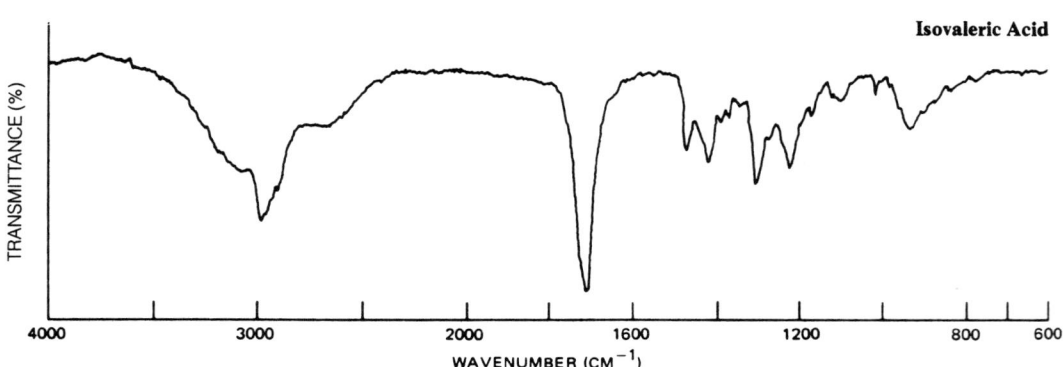

Isovaleric Acid

Lauryl Alcohol

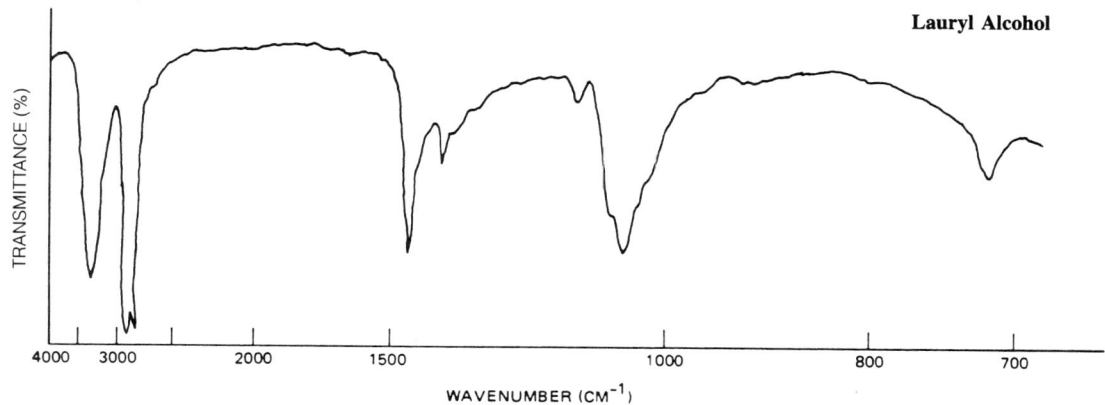

Levulinic Acid

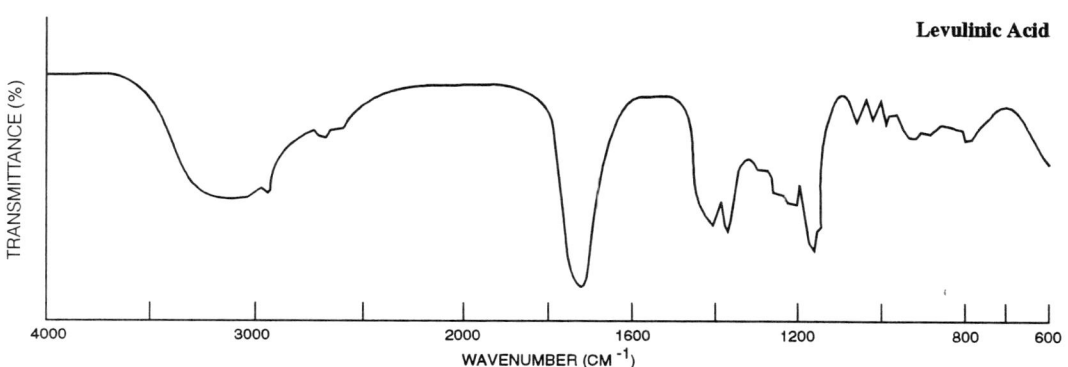

Linalyl Formate

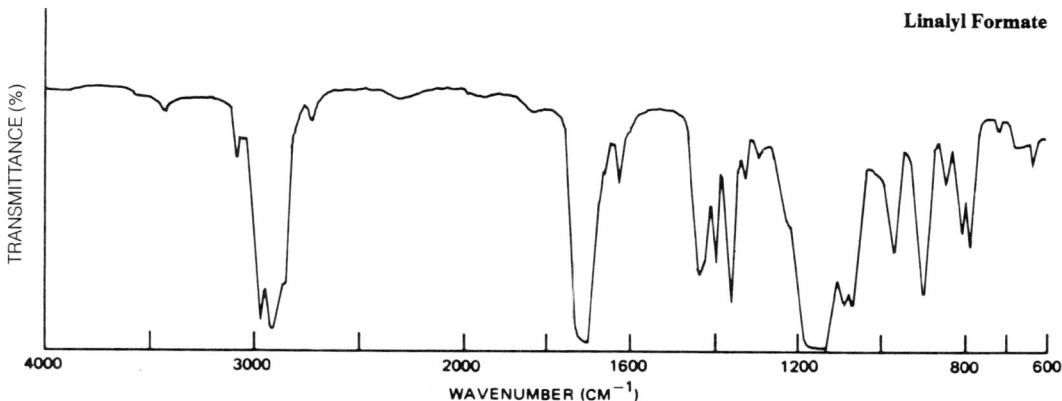

2-Mercaptopropionic Acid

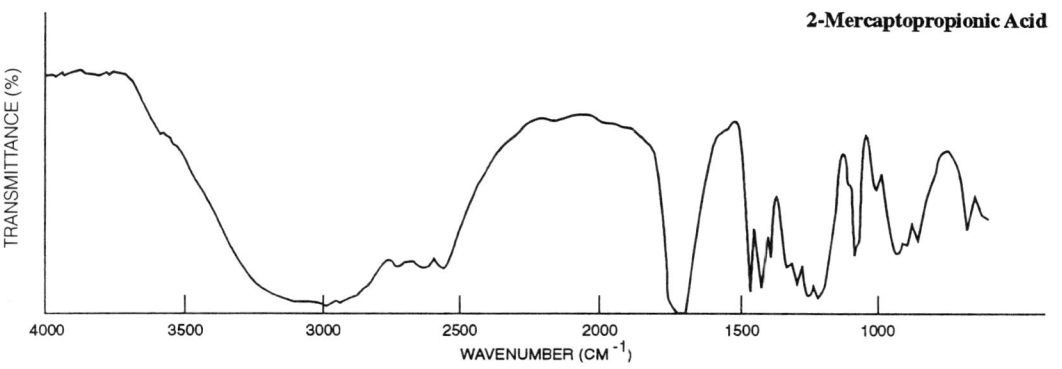

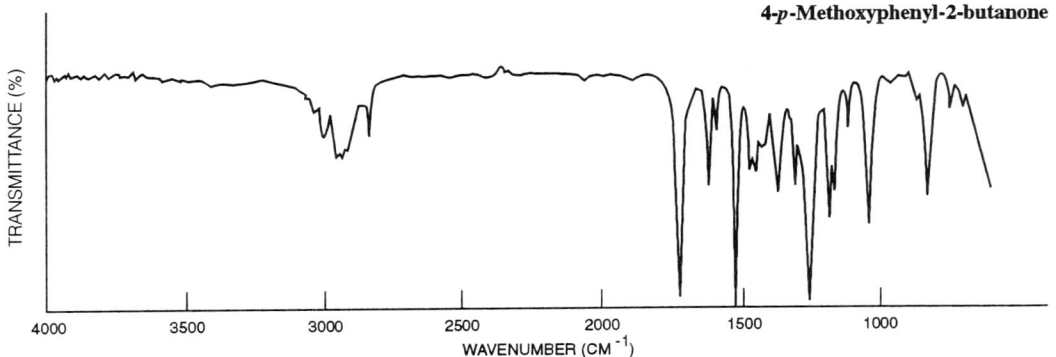

4-p-Methoxyphenyl-2-butanone

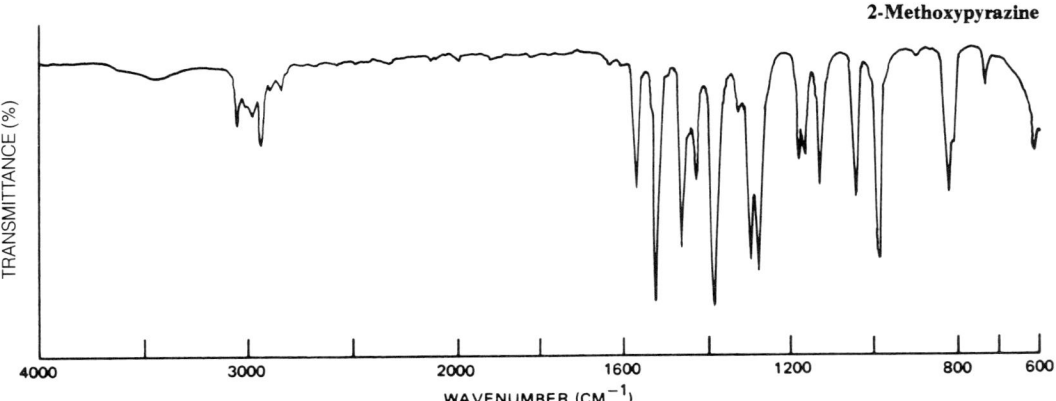

2-Methoxypyrazine

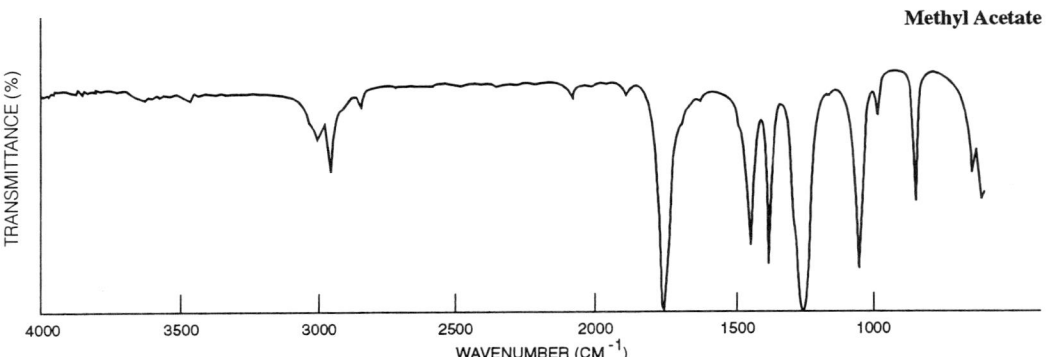

Methyl Acetate

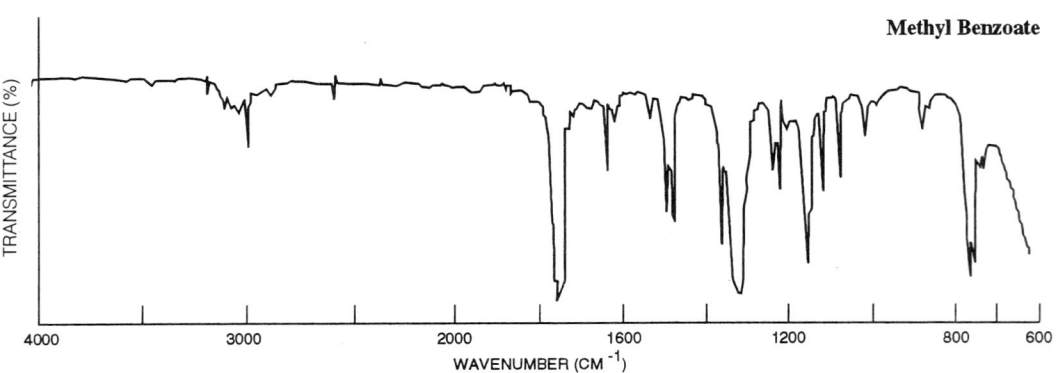

Methyl Benzoate

2-Methylbutyric Acid

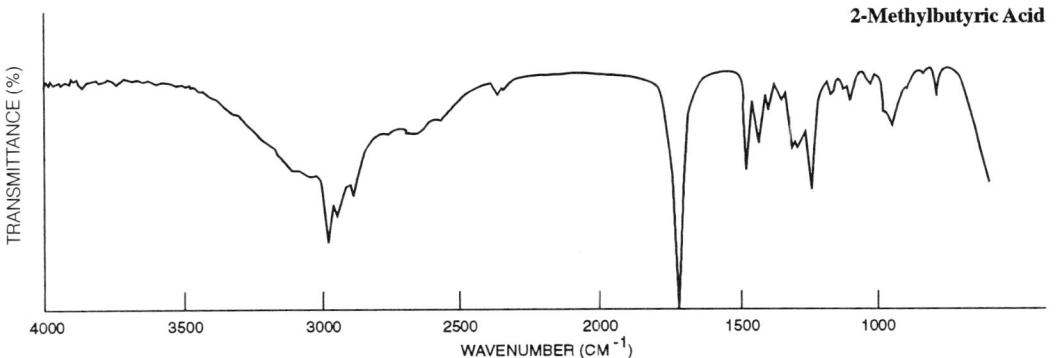

α-Methylcinnamaldehyde

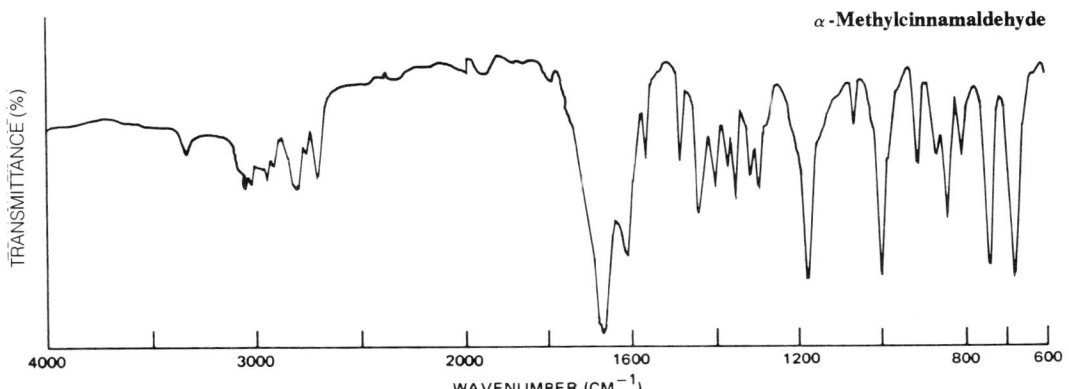

6-Methylcoumarin

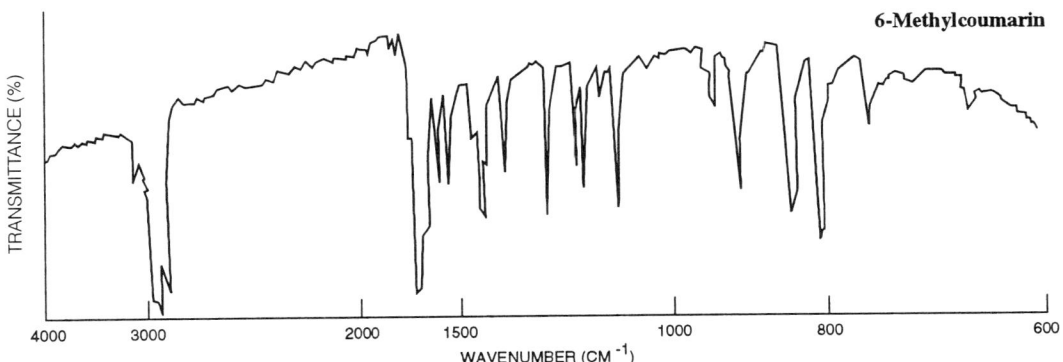

Methyl Cyclopentenolone (KBr Pellet)

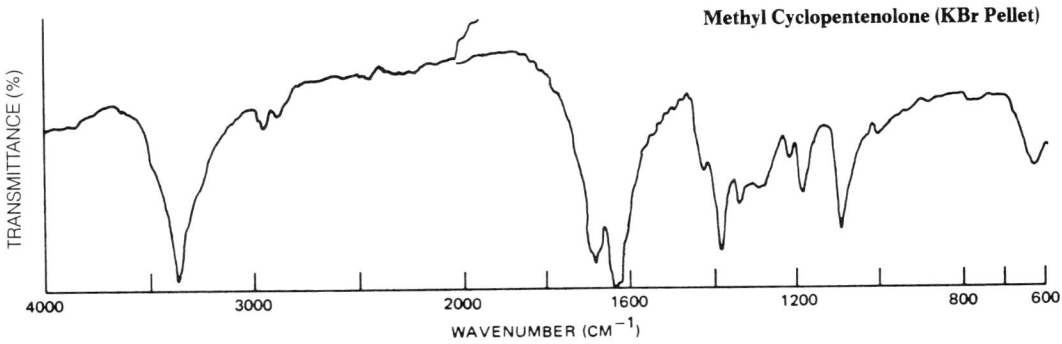

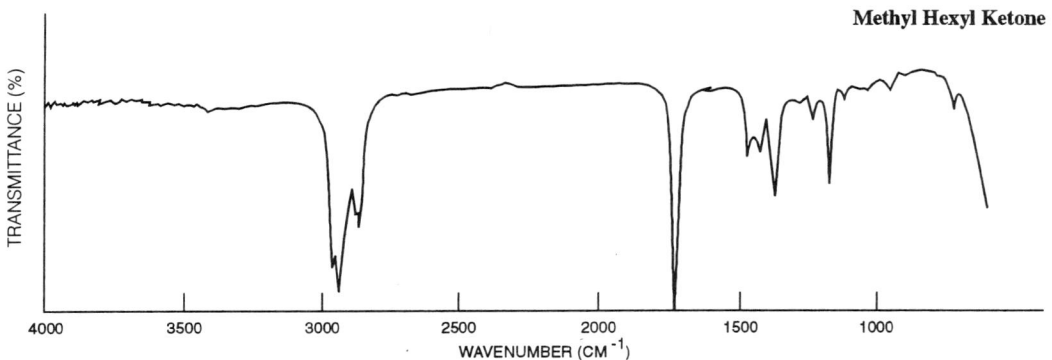

Methyl Hexyl Ketone

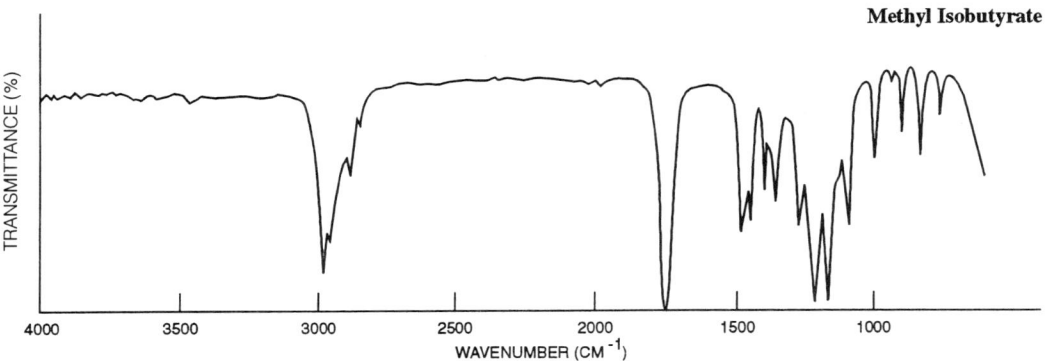

Methyl Isobutyrate

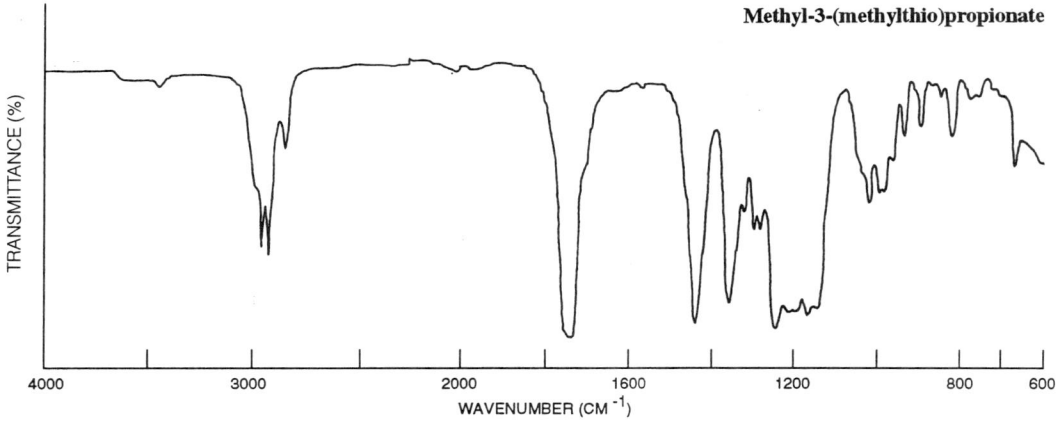

Methyl-3-(methylthio)propionate

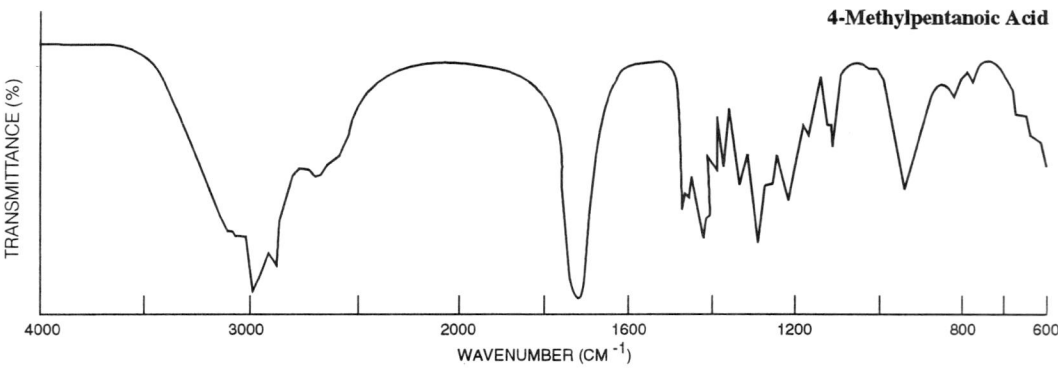

4-Methylpentanoic Acid

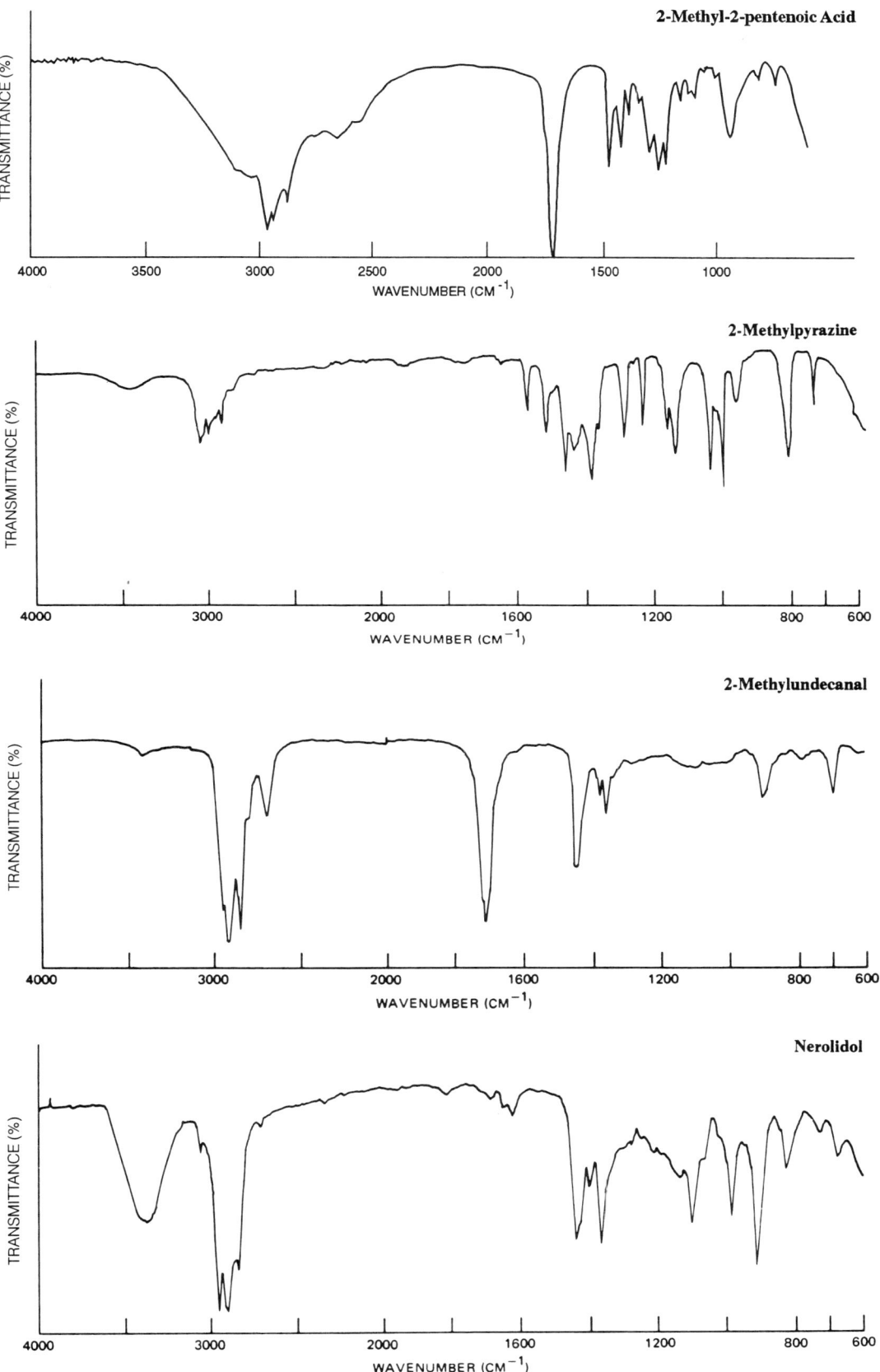

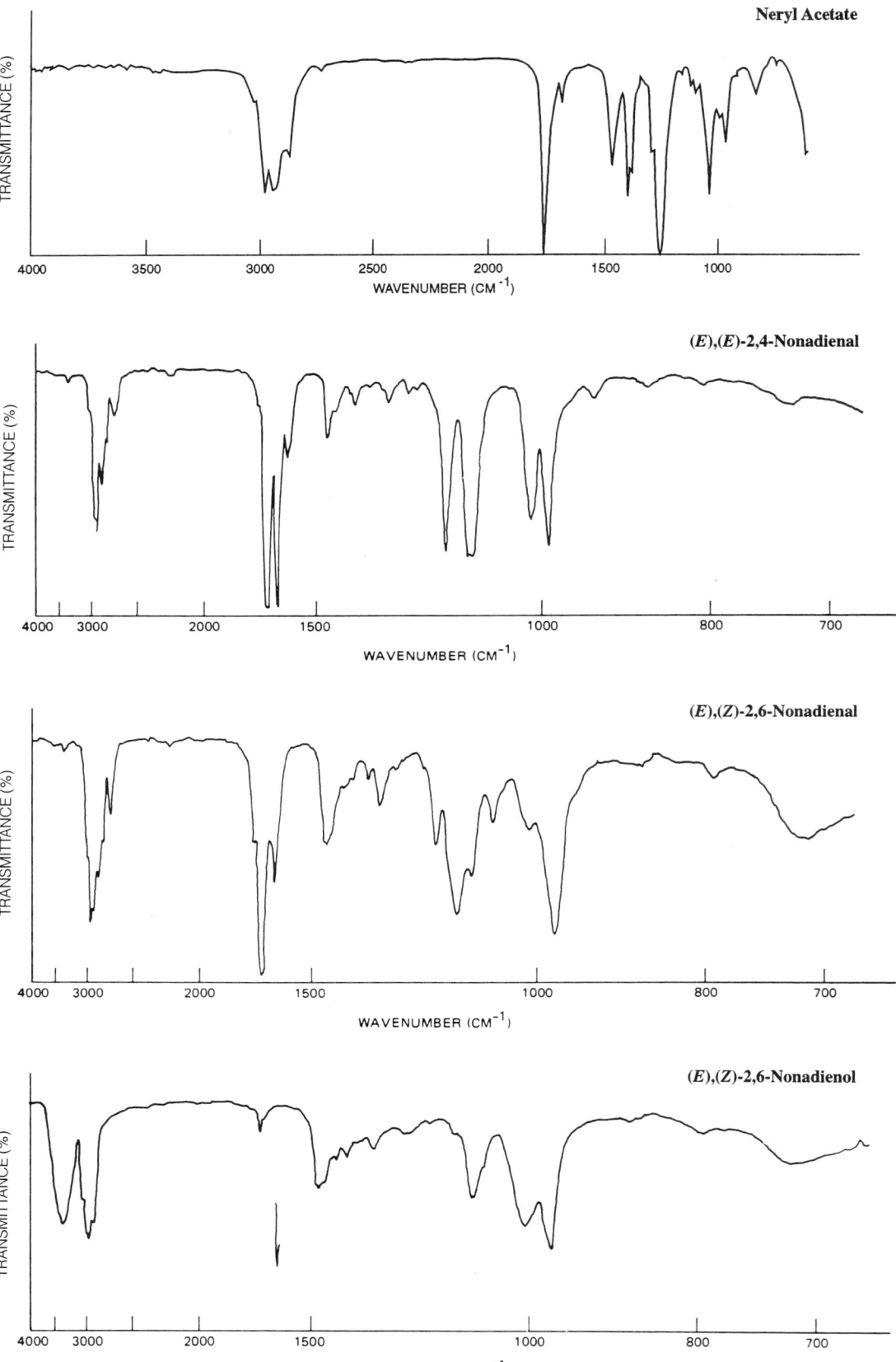

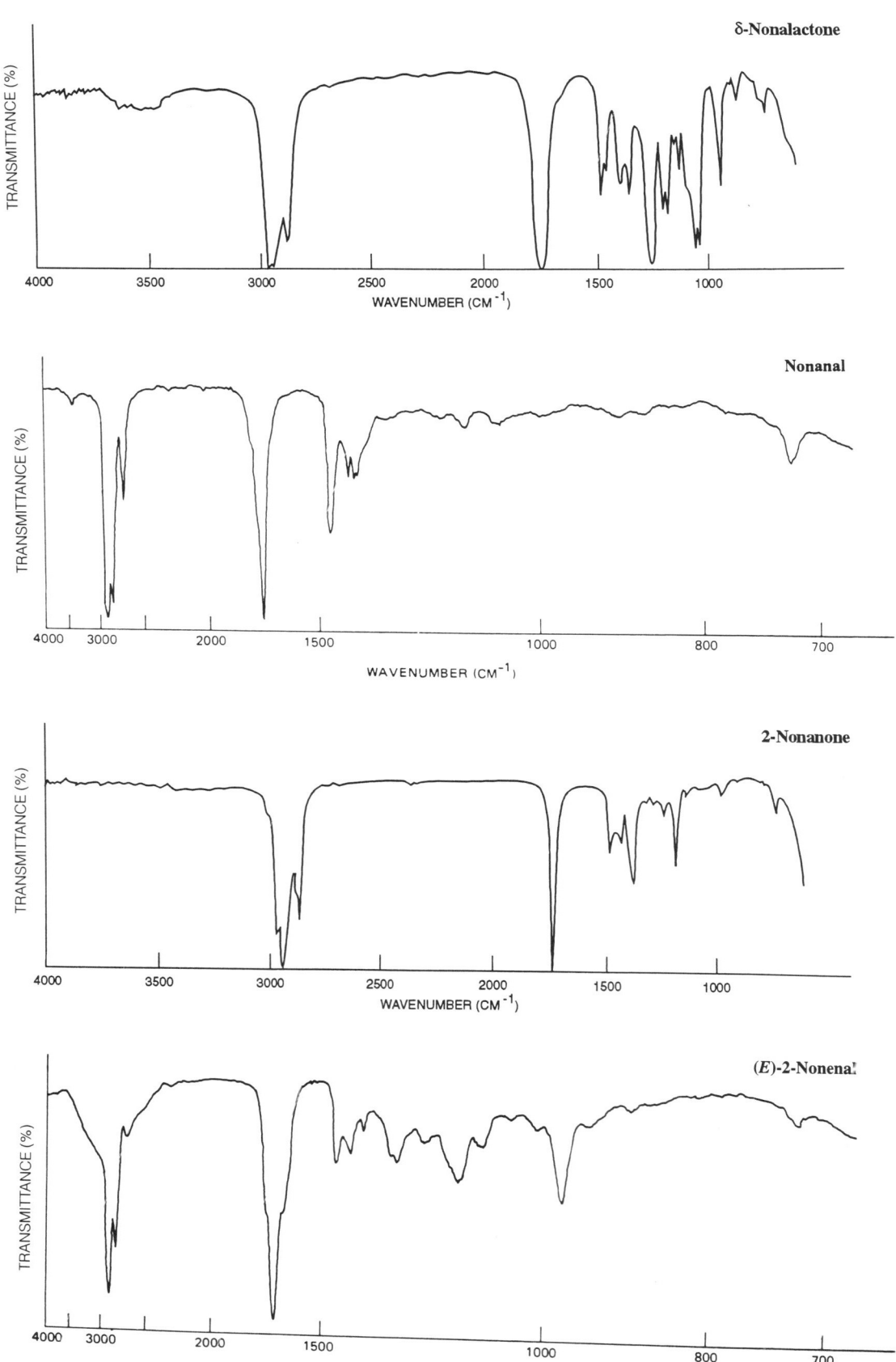

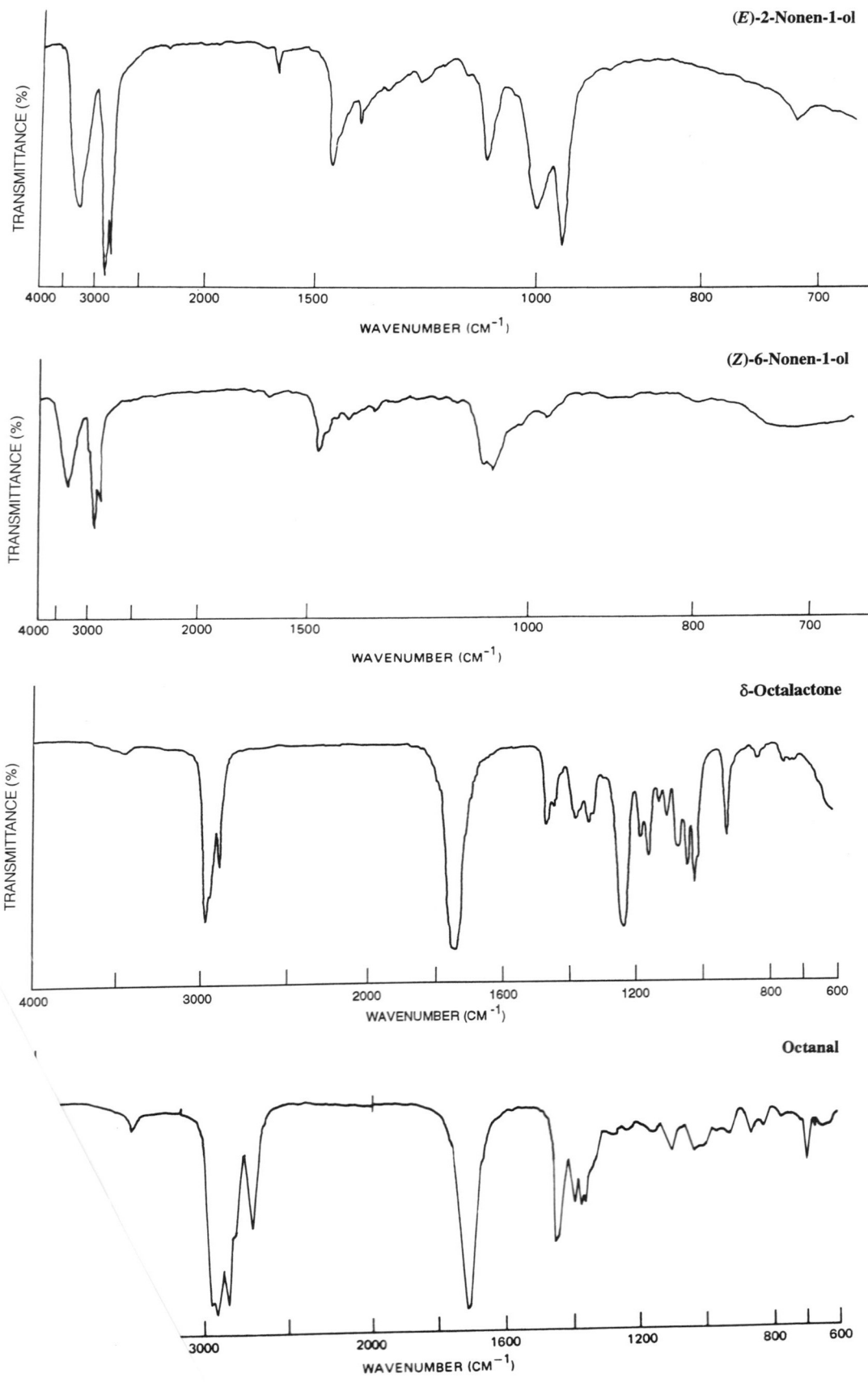

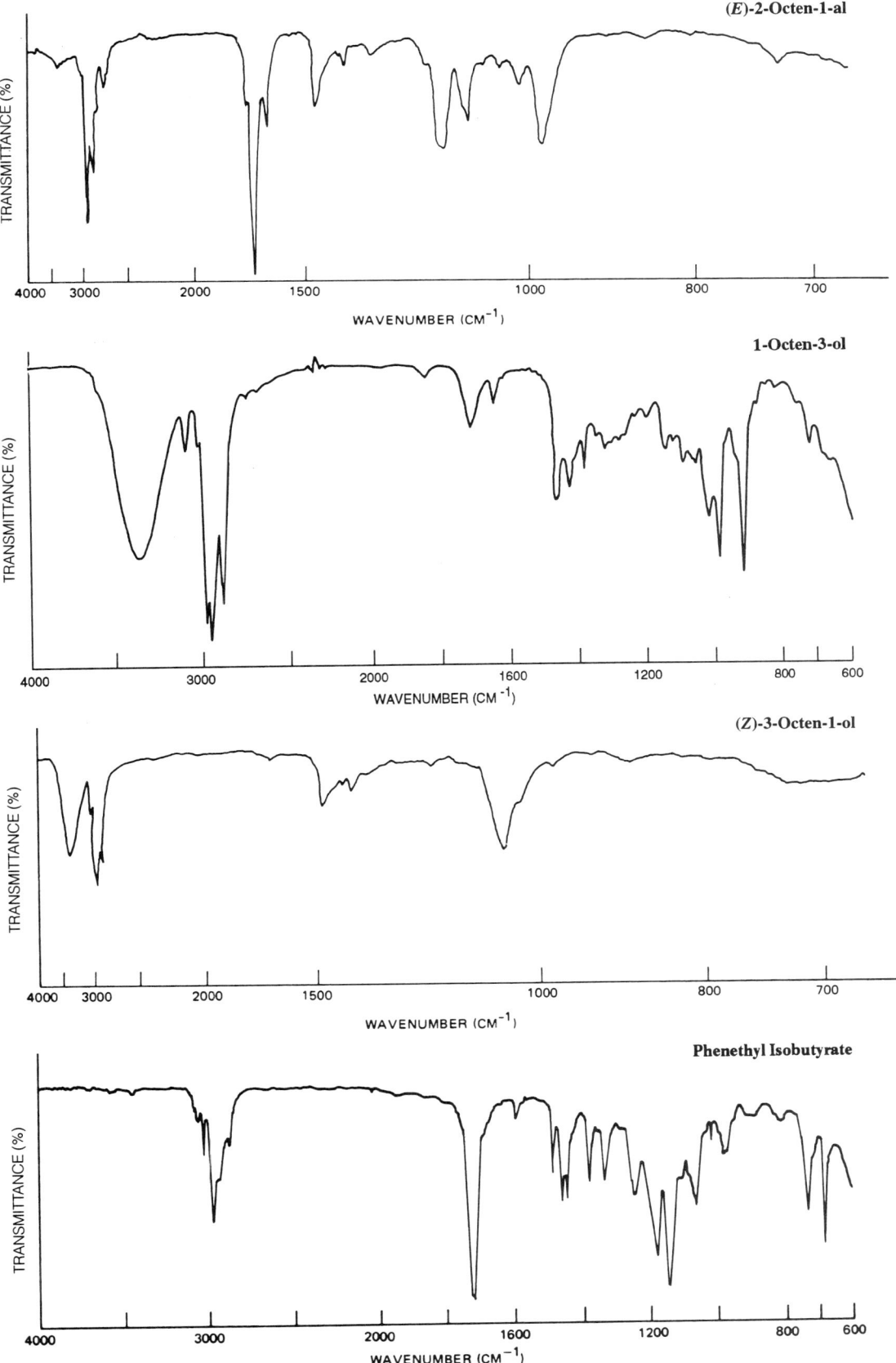

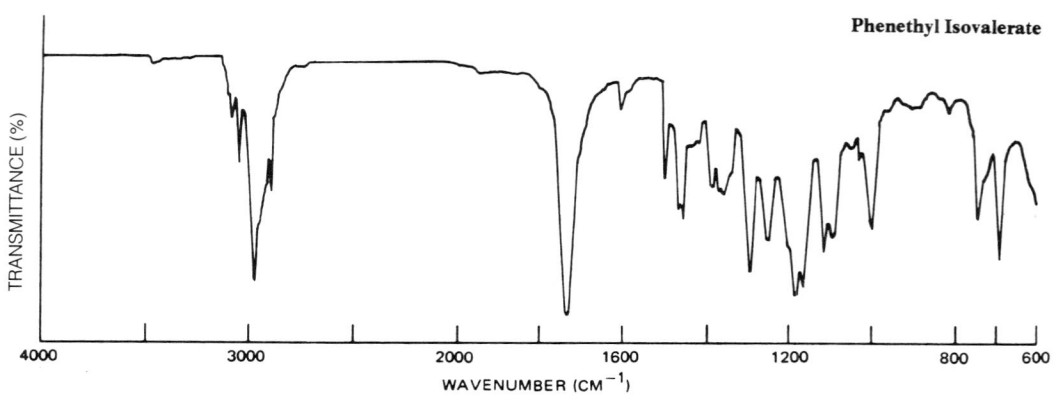

Phenethyl Isovalerate

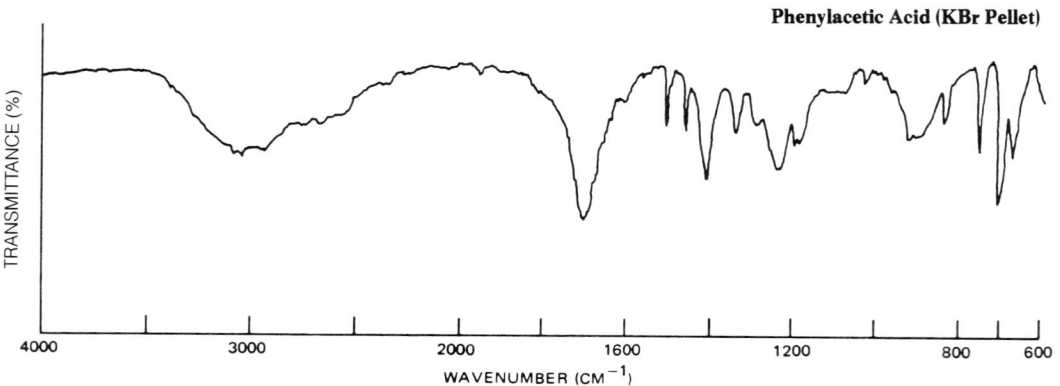

Phenylacetic Acid (KBr Pellet)

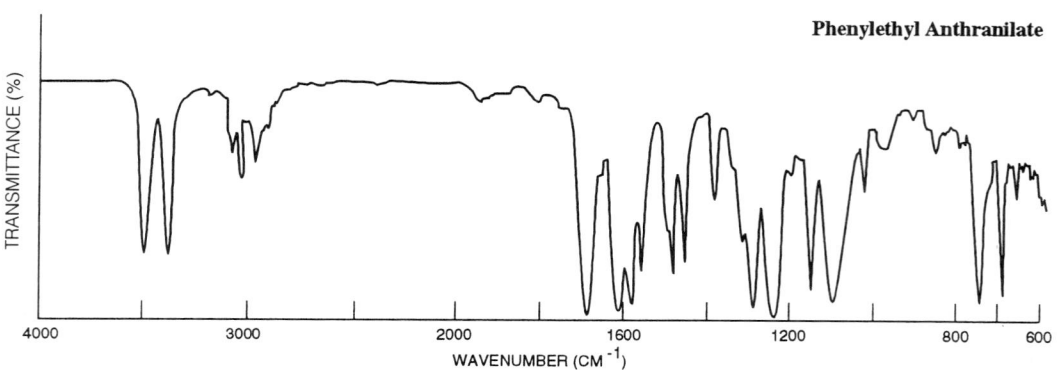

Phenylethyl Anthranilate

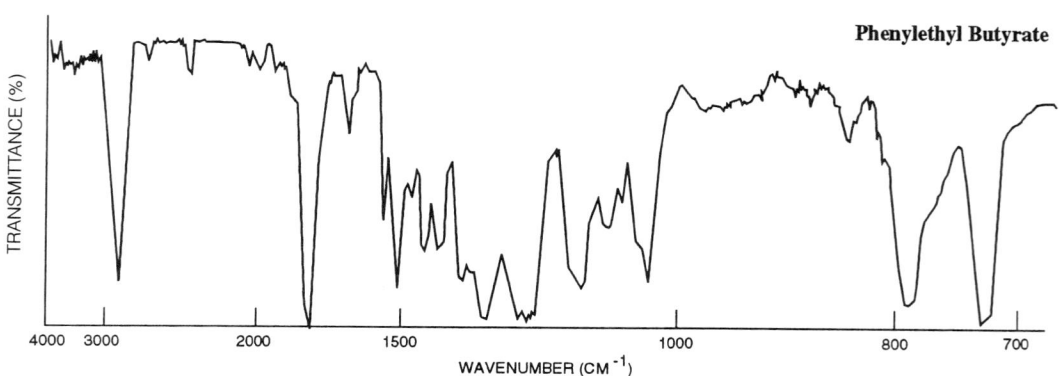

Phenylethyl Butyrate

3-Phenylpropyl Acetate

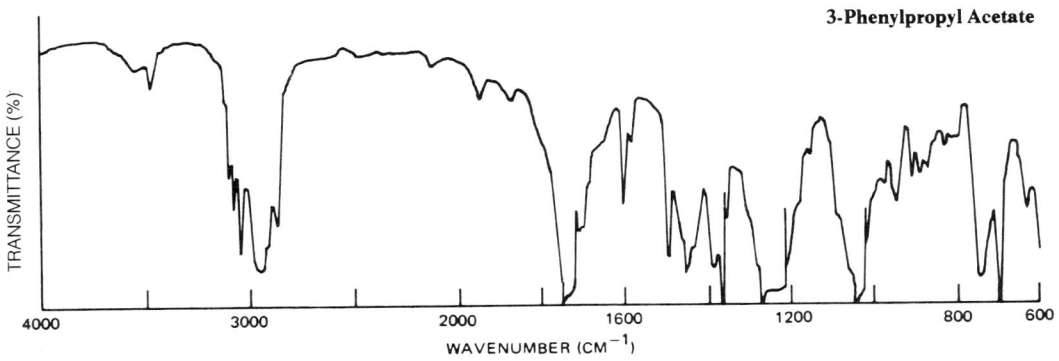

Piperidine

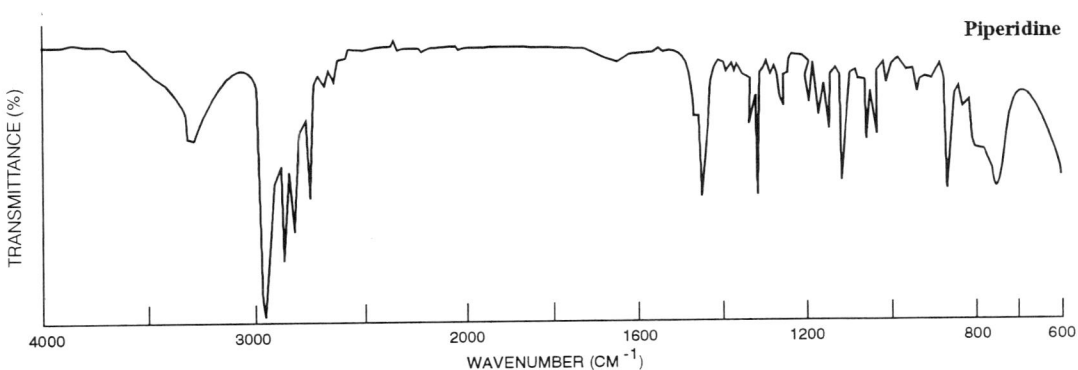

Propionaldehyde

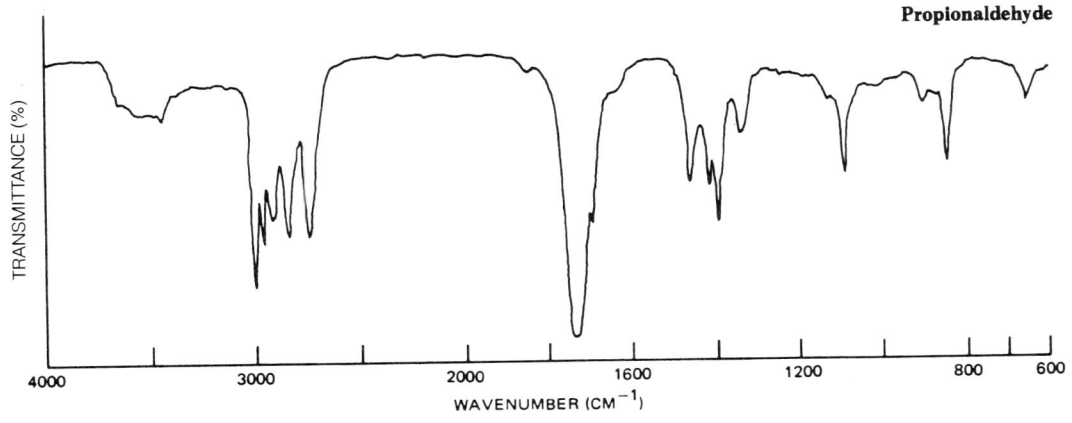

n-Propyl Acetate

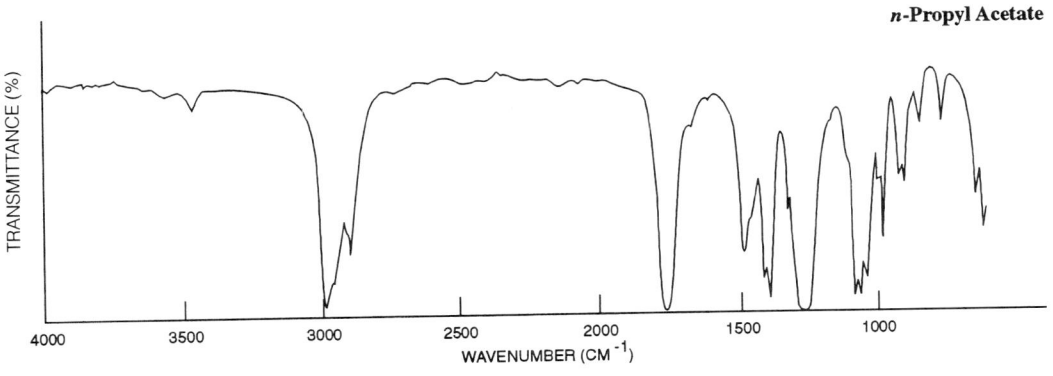

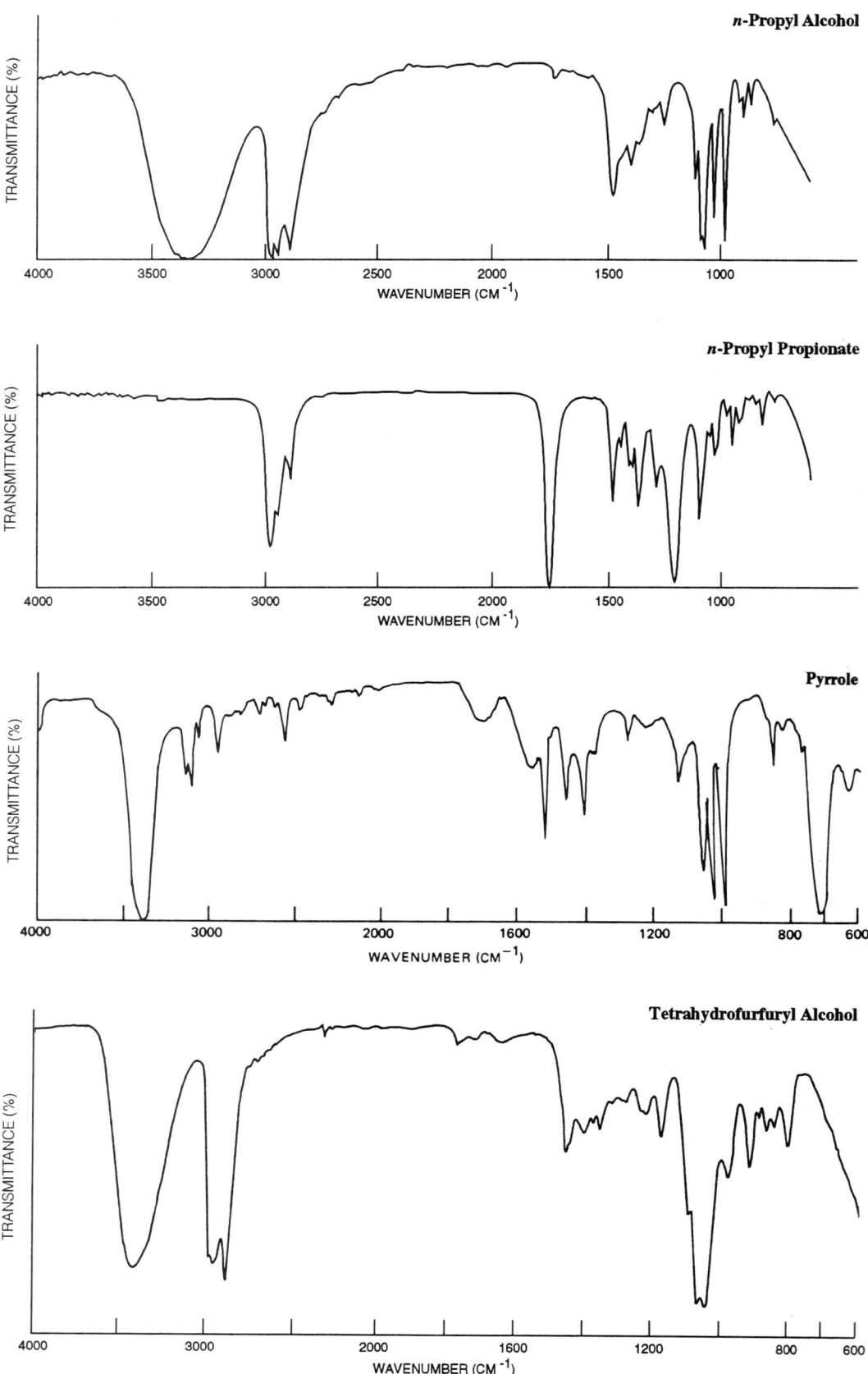

2,3,5,6-Tetramethylpyrazine (Nujol Mull)

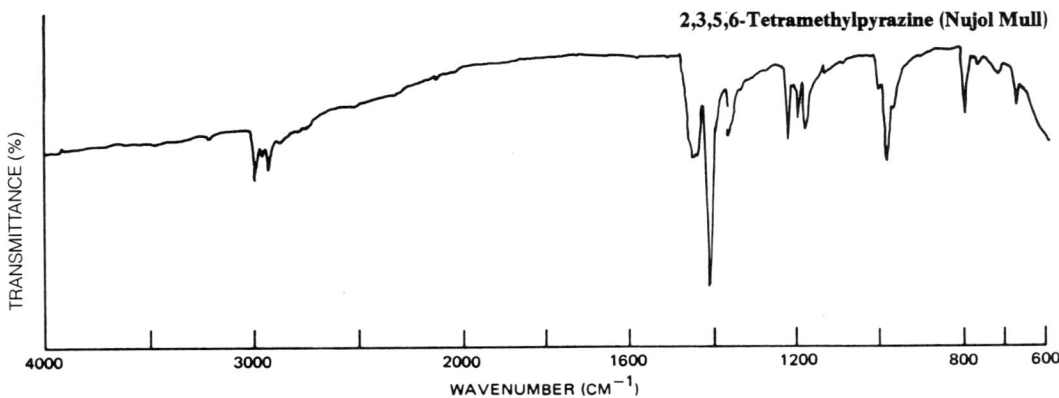

Thymol (Nujol Mull)

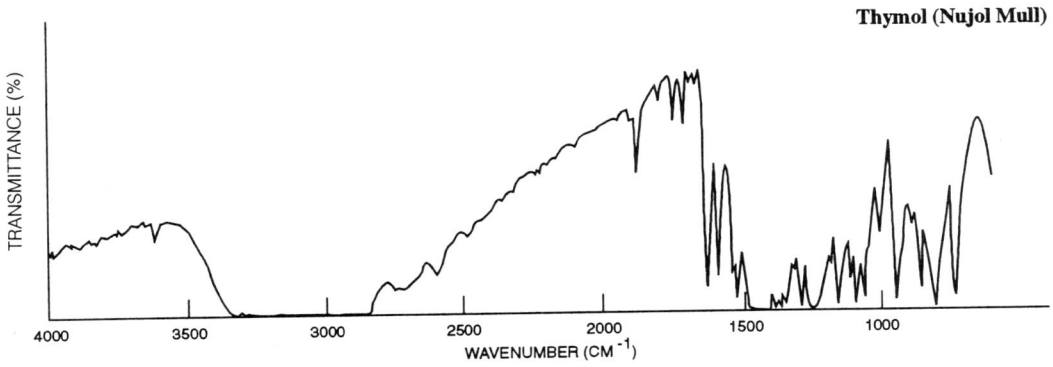

Tolualdehyde, Mixed Isomers

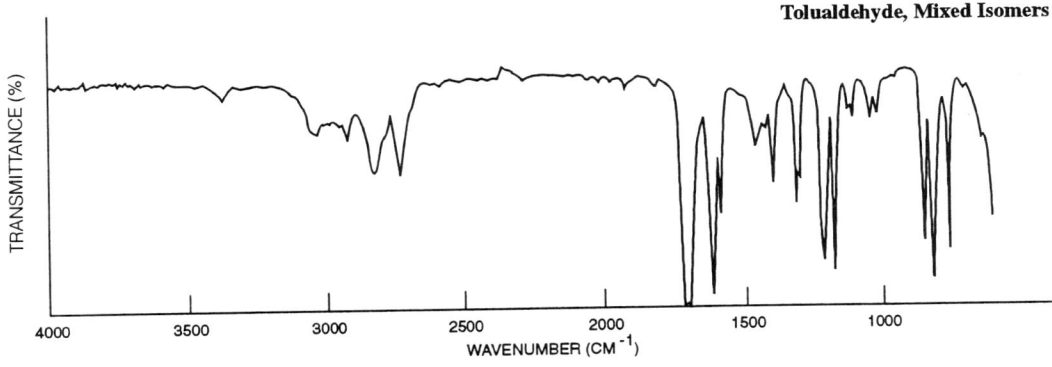

***p*-Tolyl Isobutyrate**

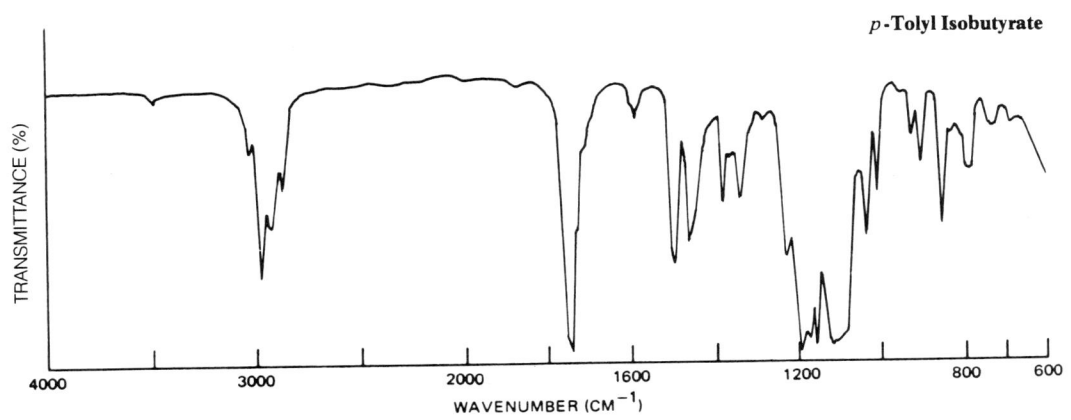

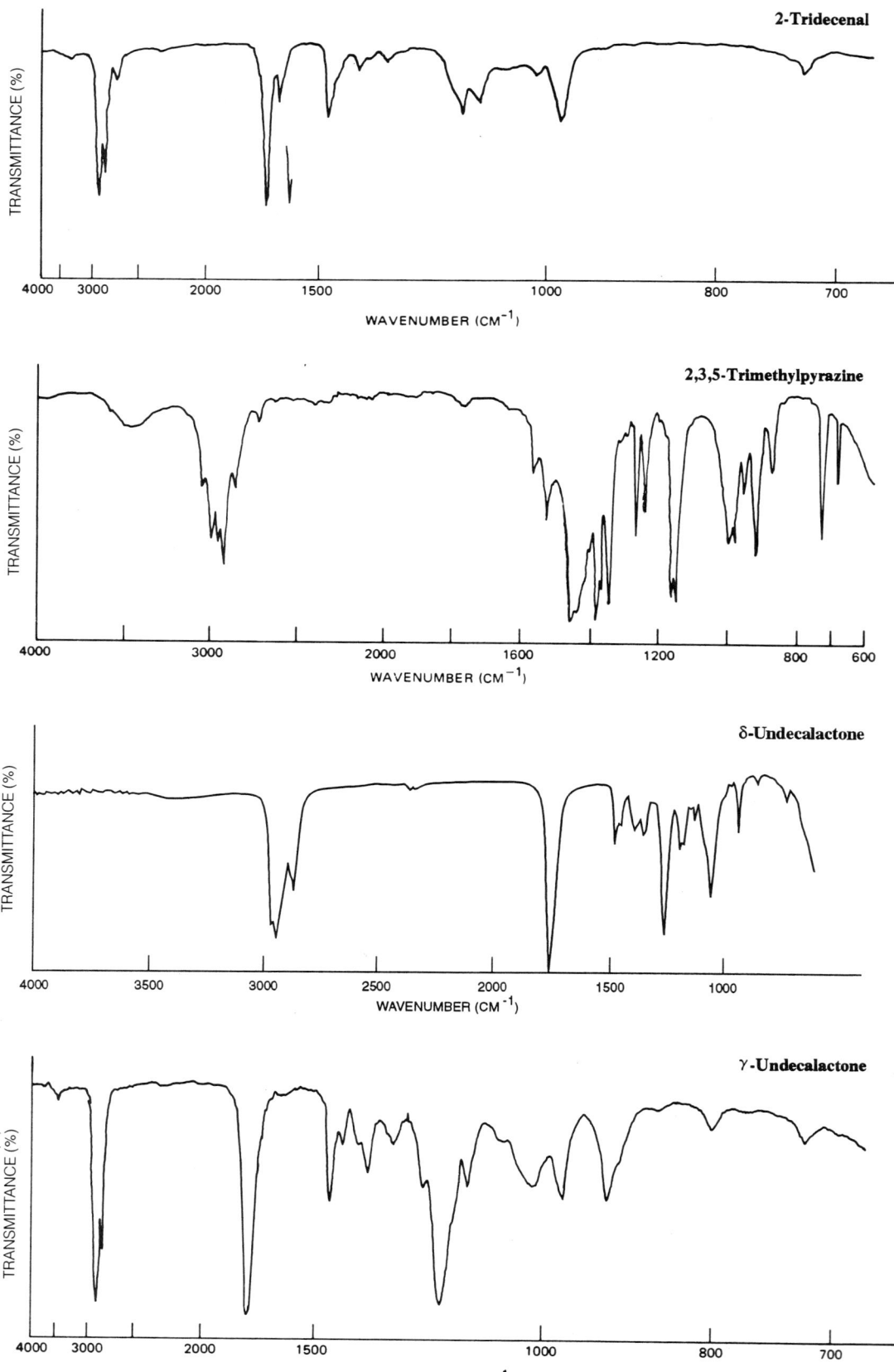

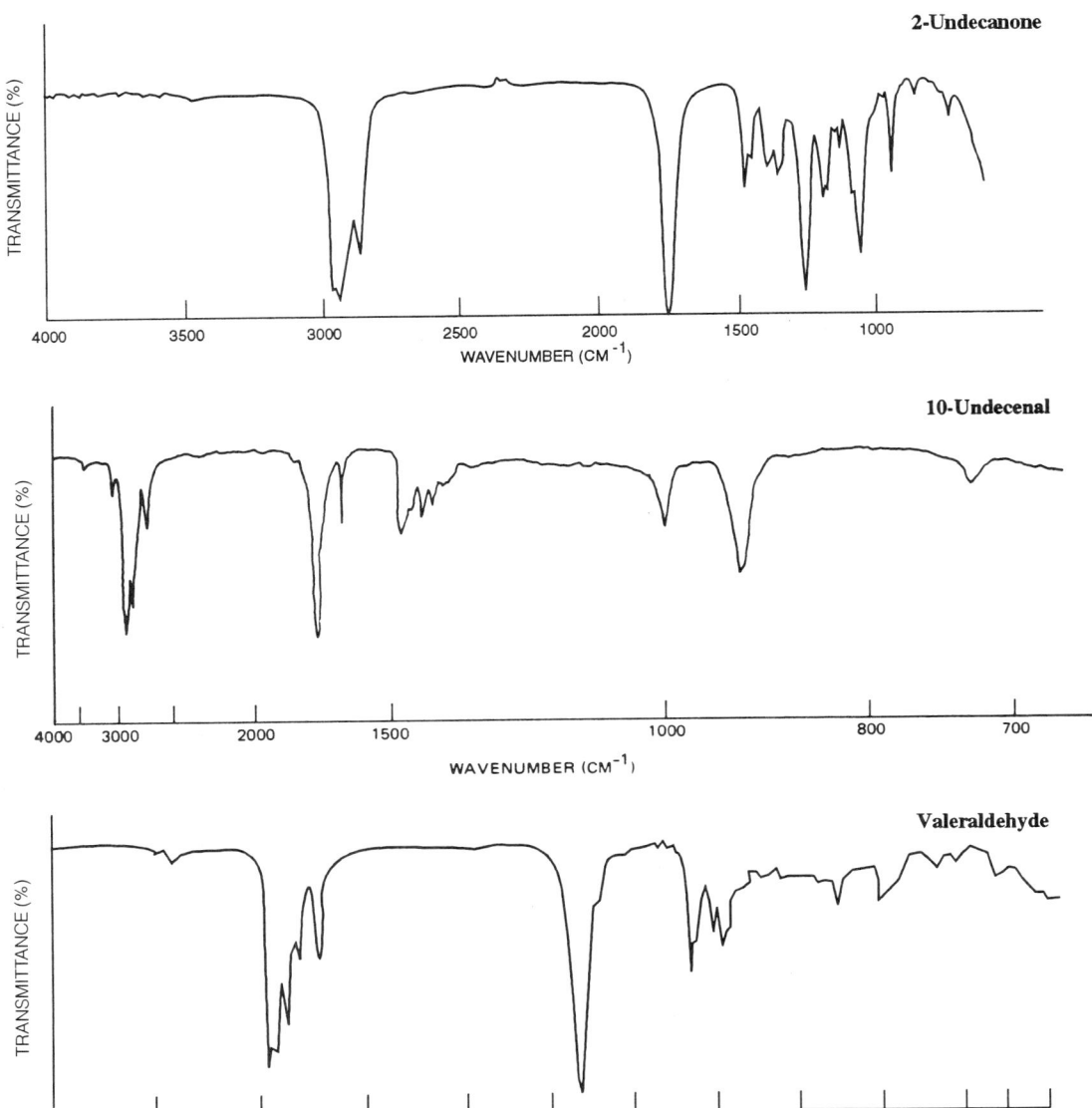

SERIES B-2

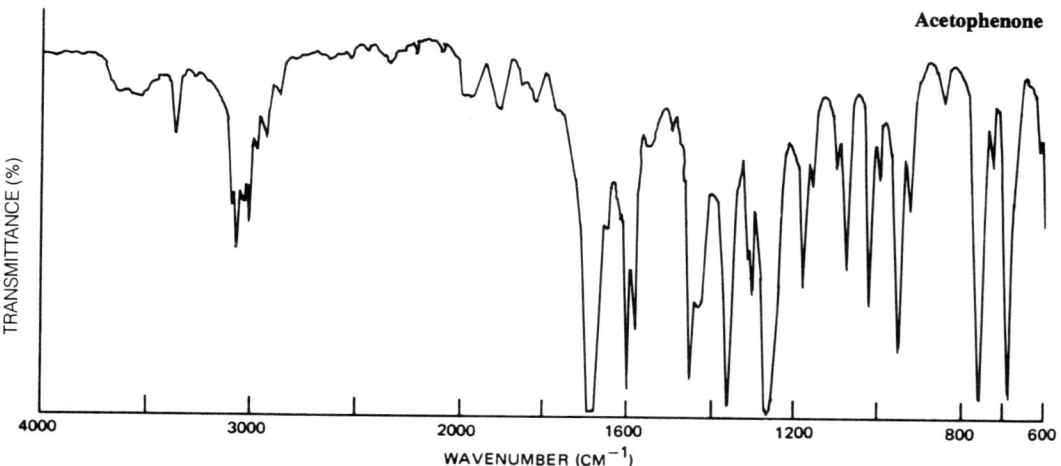
Acetophenone

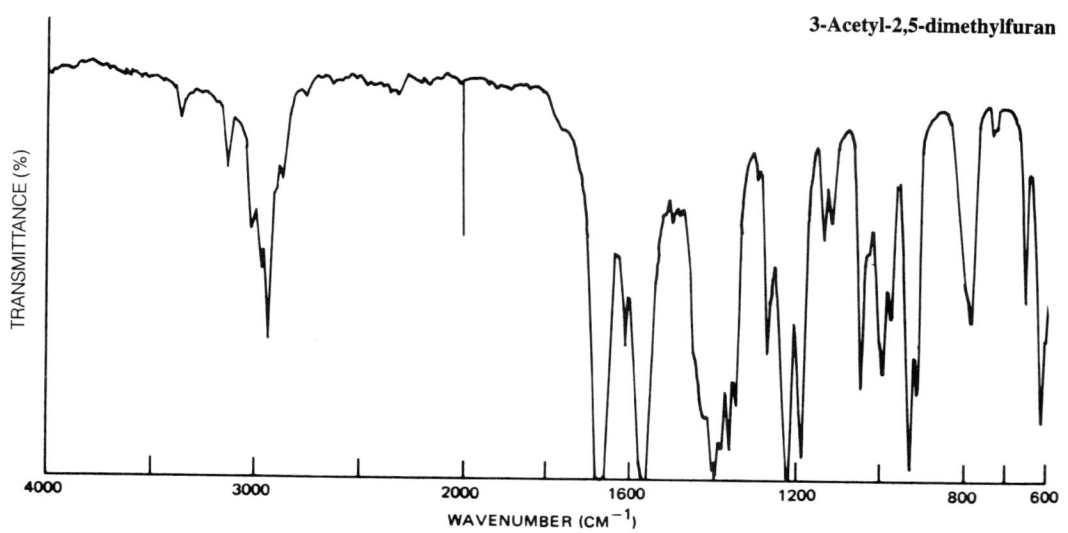
3-Acetyl-2,5-dimethylfuran

Allyl Cyclohexanepropionate

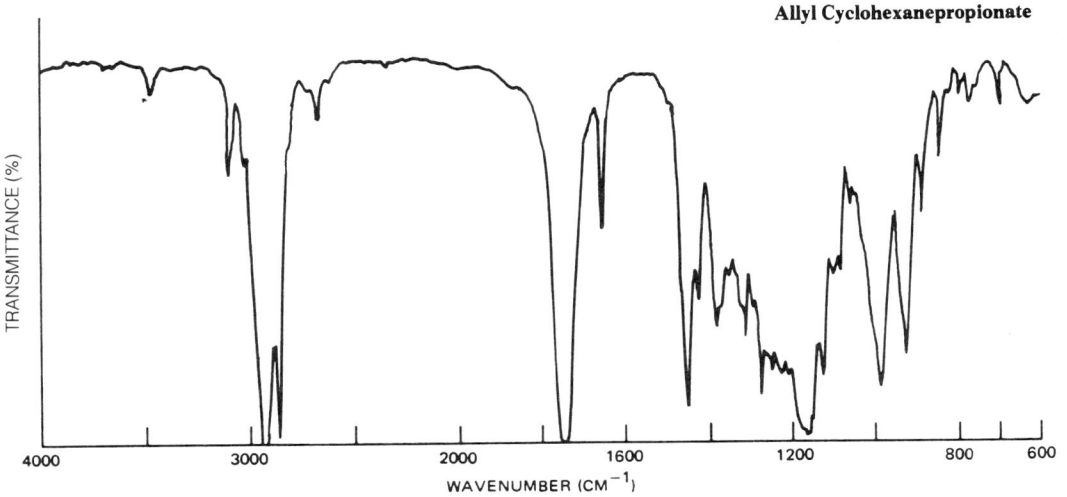

Allyl Hexanoate

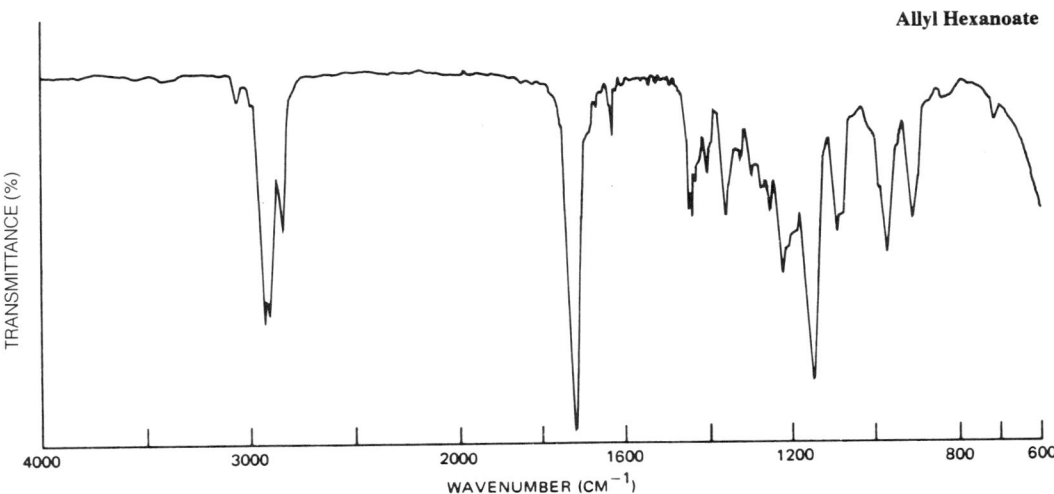

Allyl α-Ionone

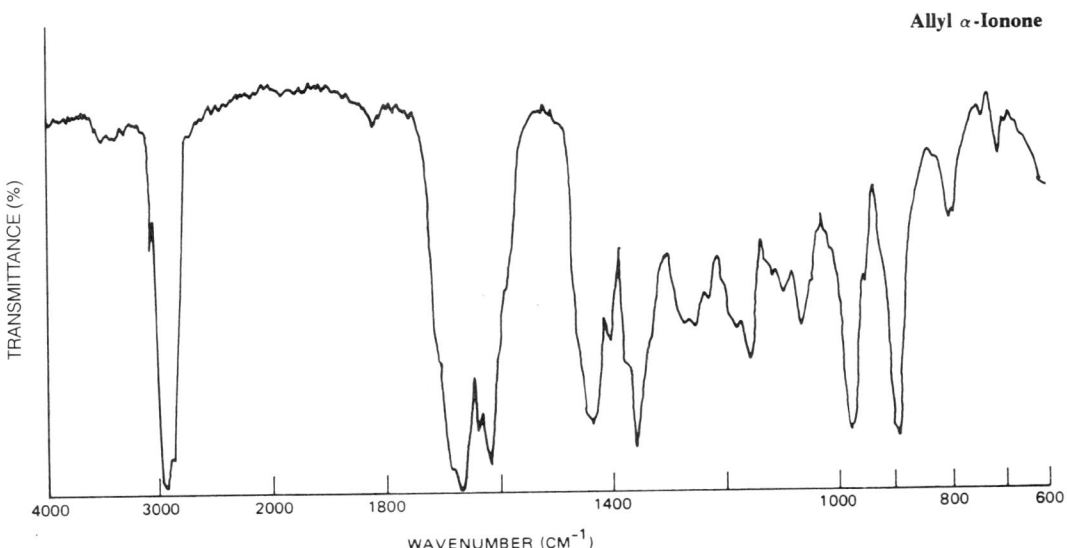

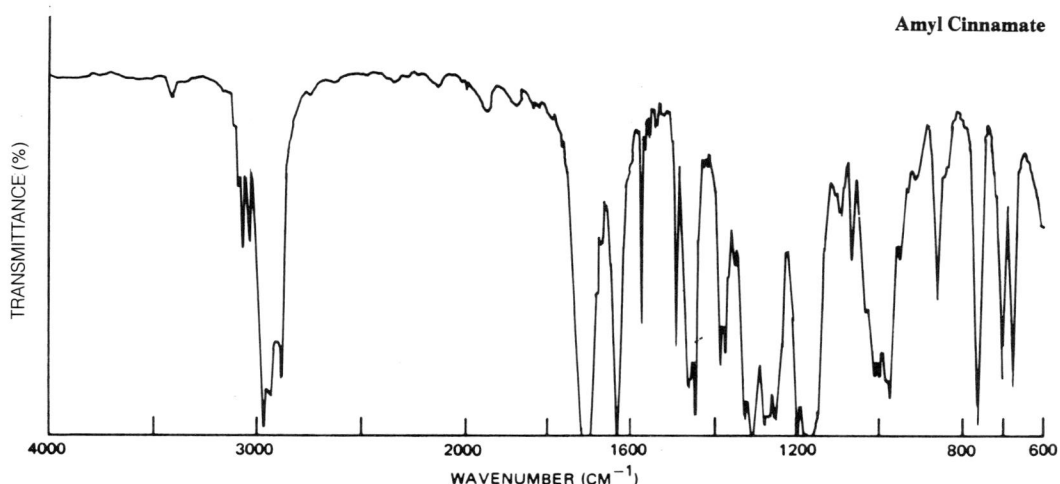

Amyl Cinnamate

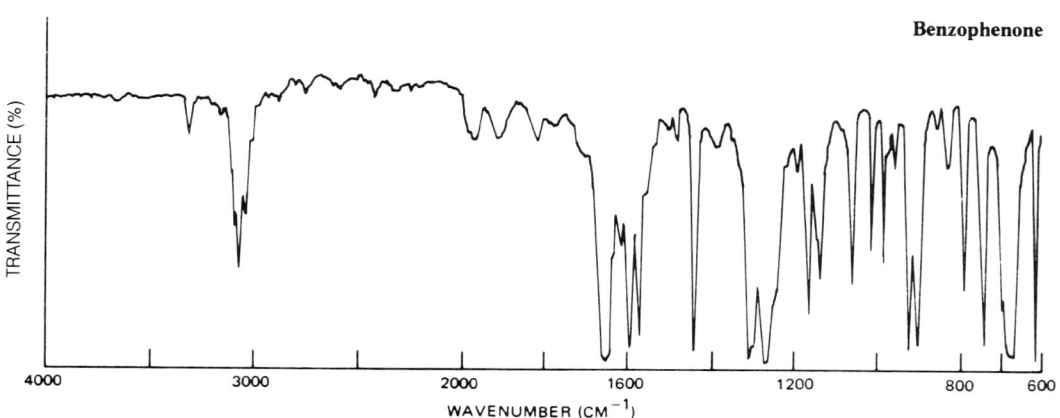

Benzophenone

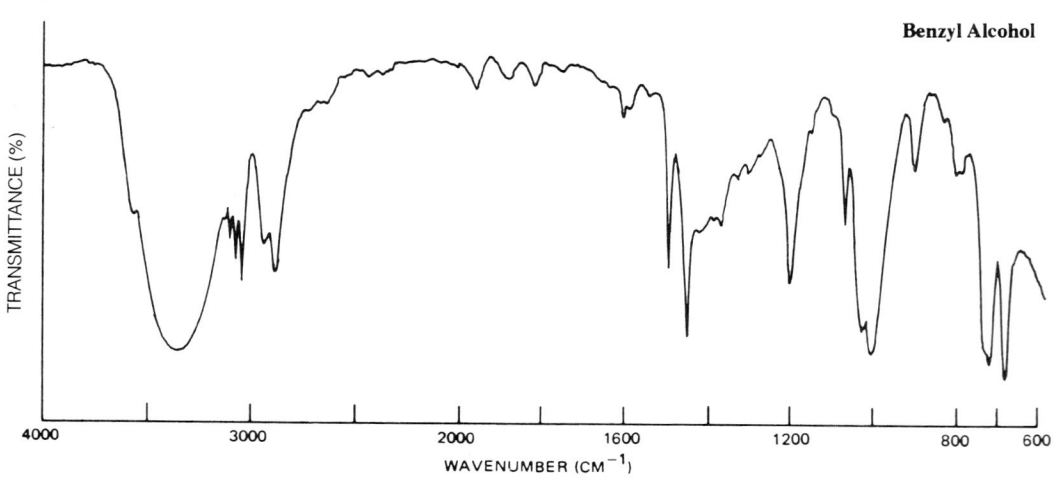

Benzyl Alcohol

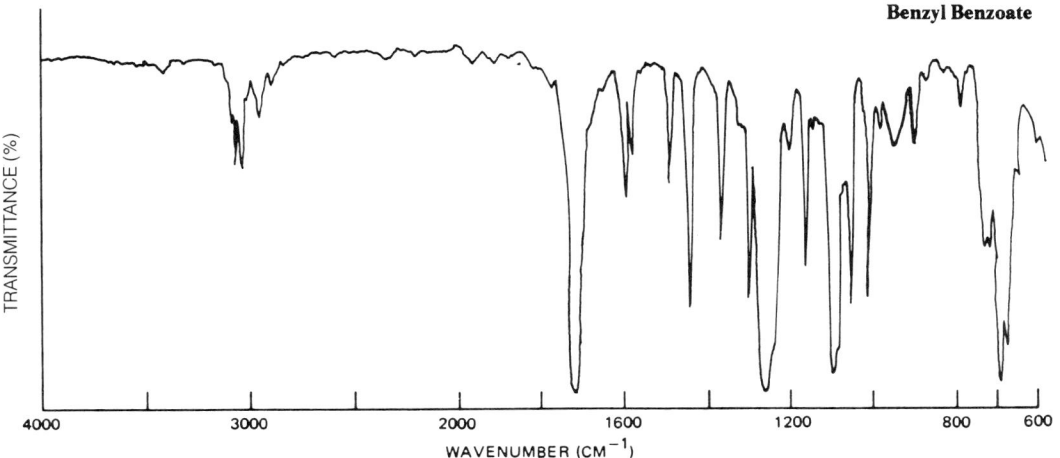

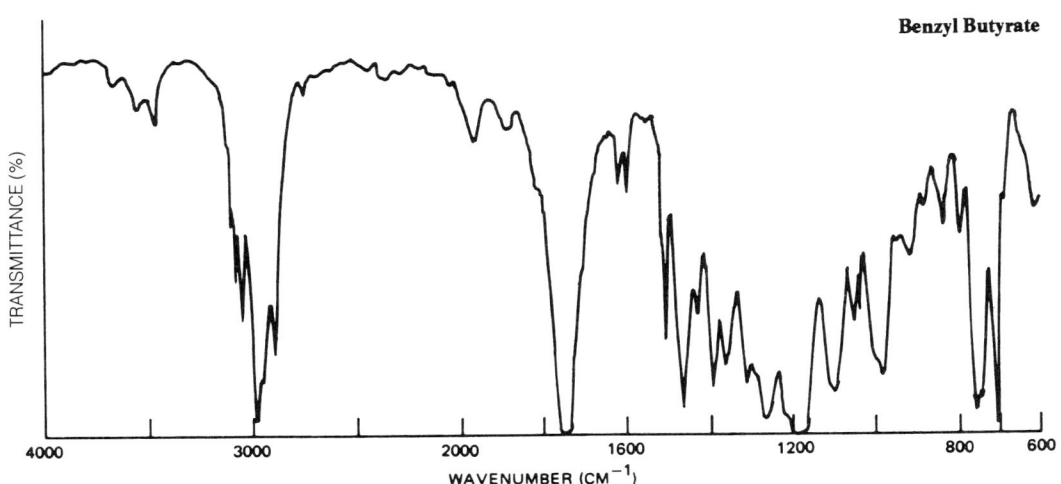

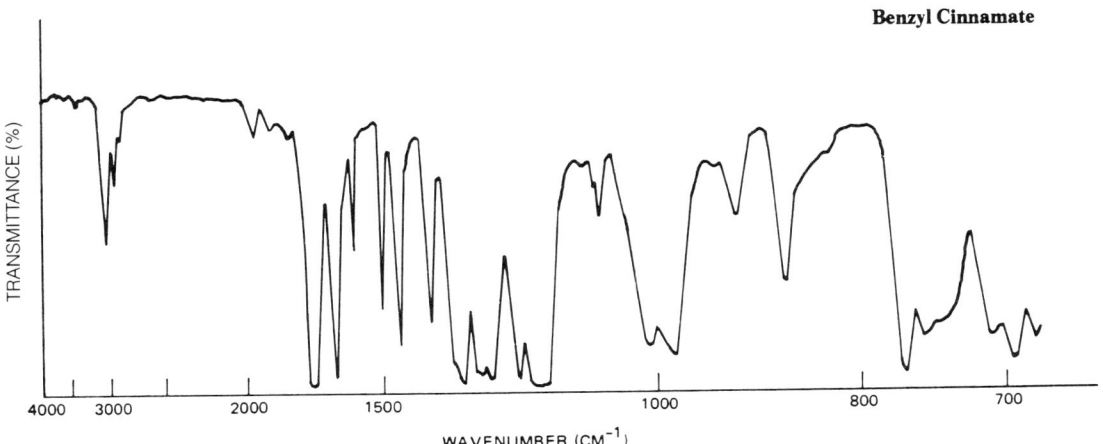

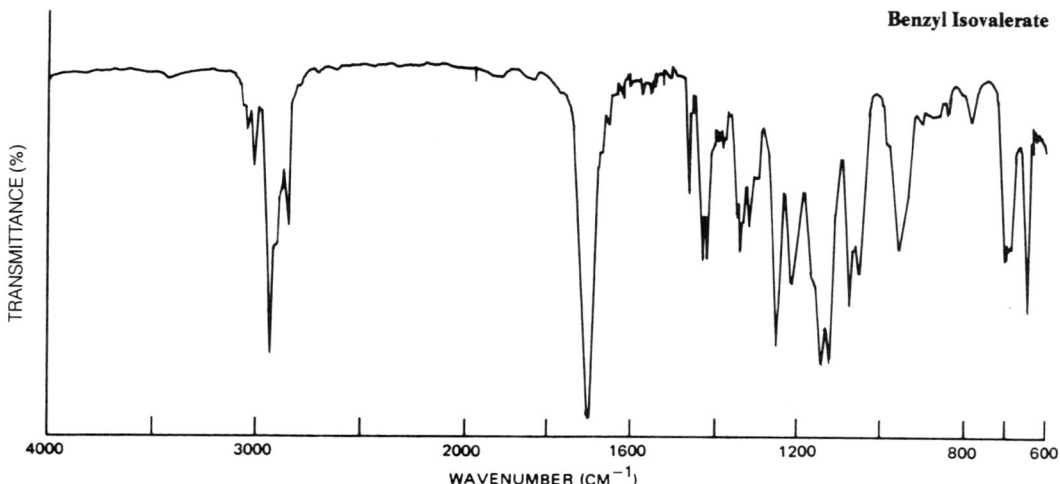

Benzyl Isovalerate

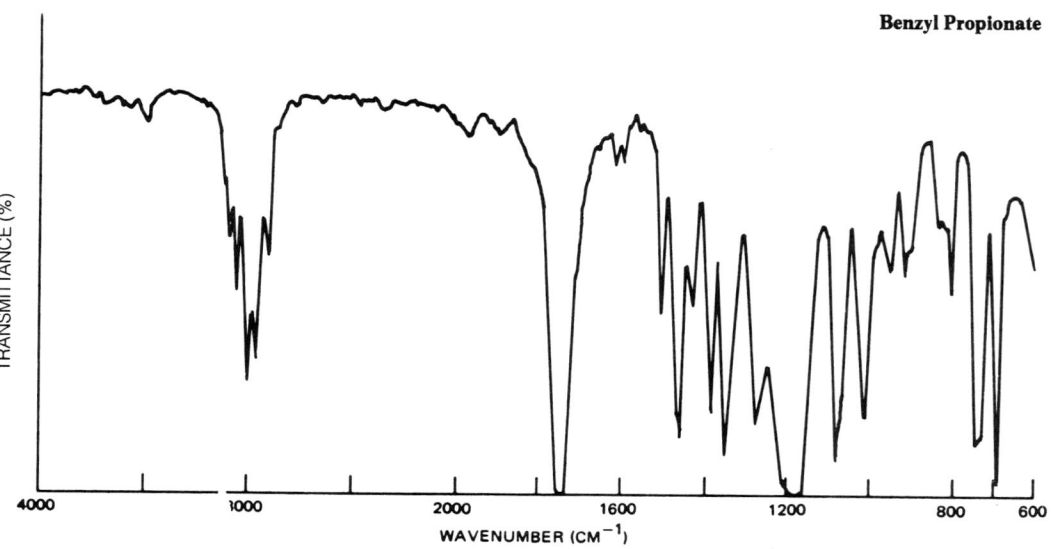

Benzyl Propionate

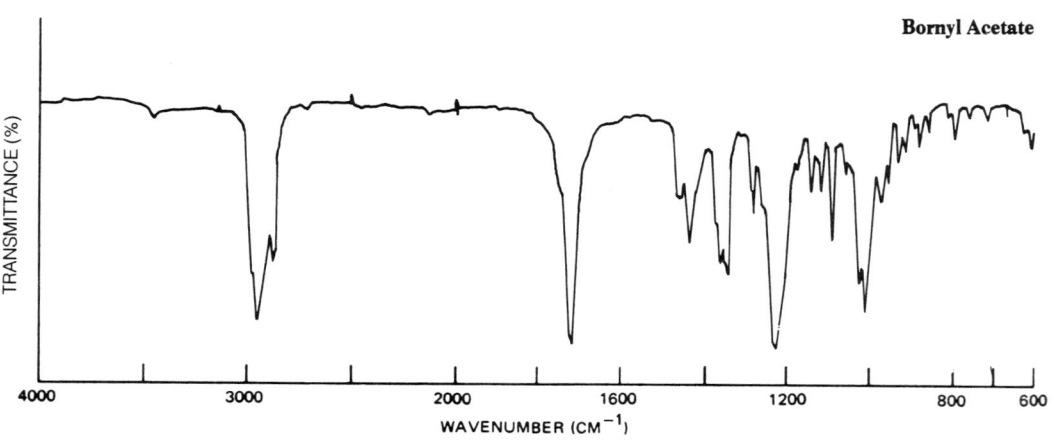

Bornyl Acetate

Butan-3-one-2-yl Butanoate

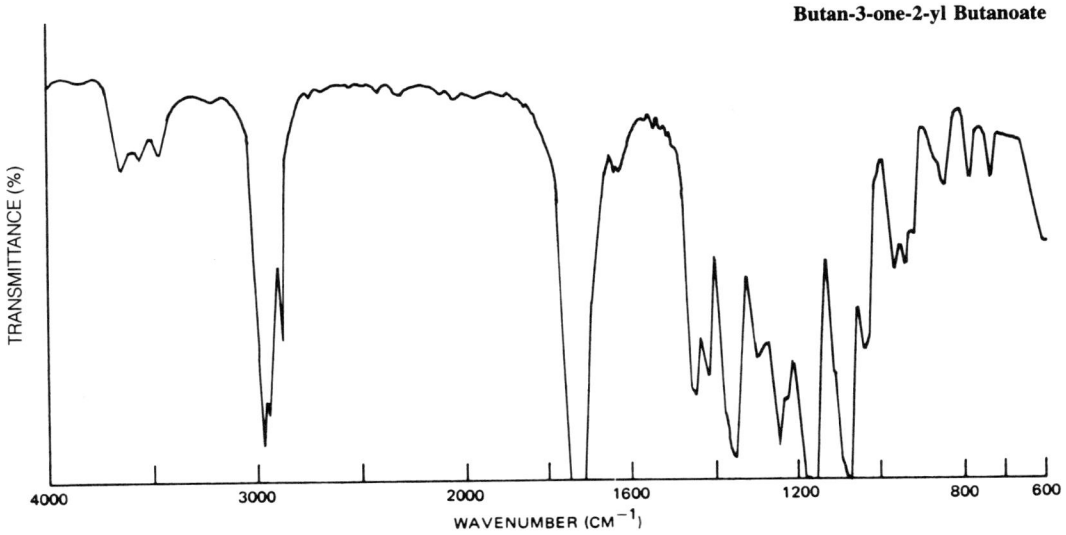

Butyl Acetate

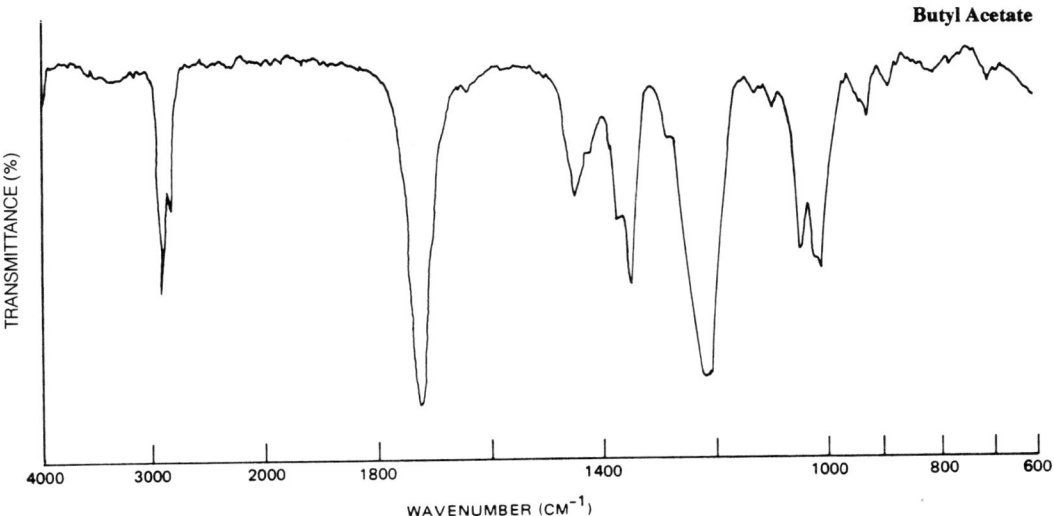

Butyl Alcohol

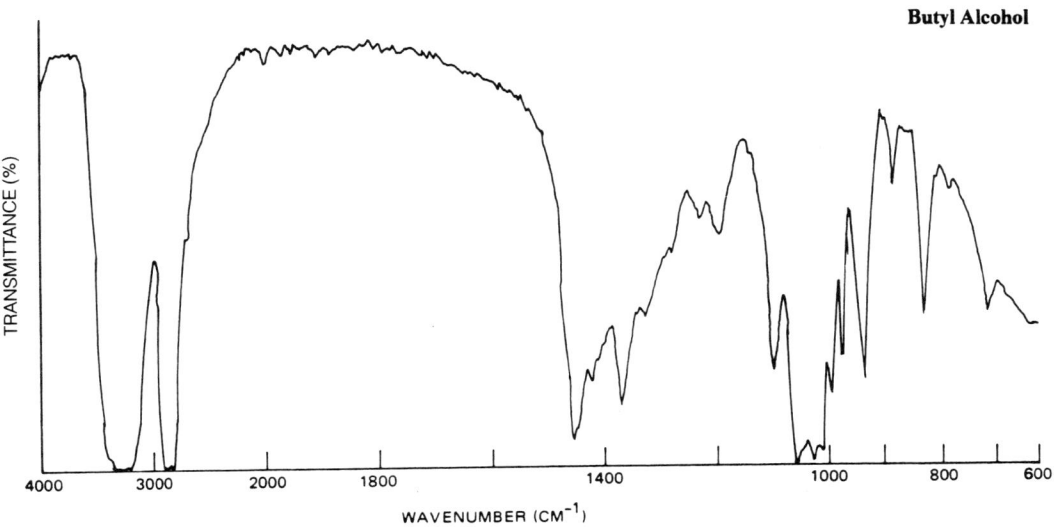

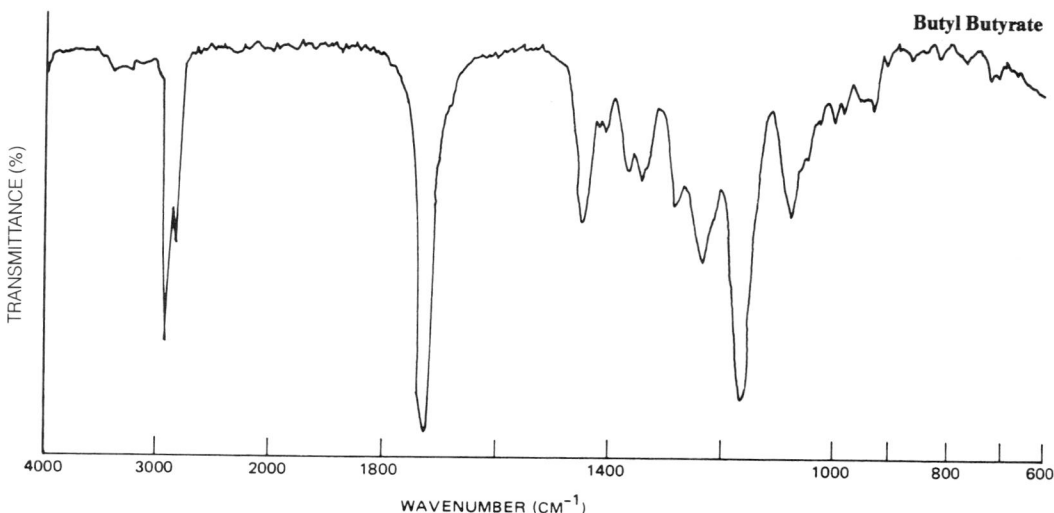

Butyl Butyrate

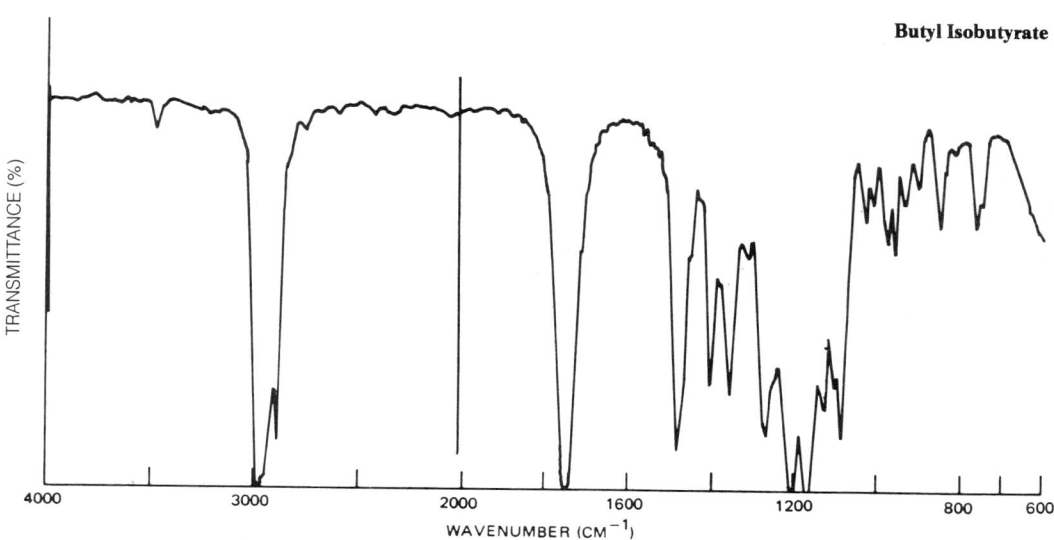

Butyl Isobutyrate

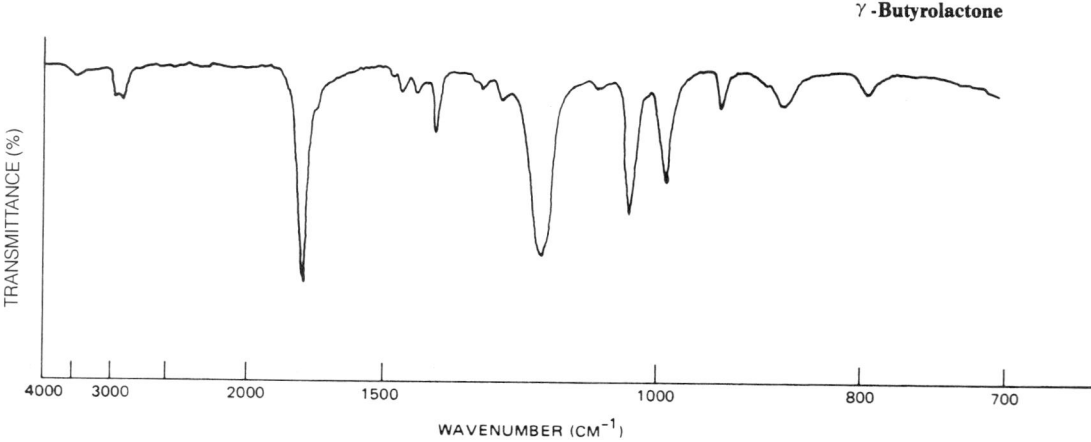

γ-Butyrolactone

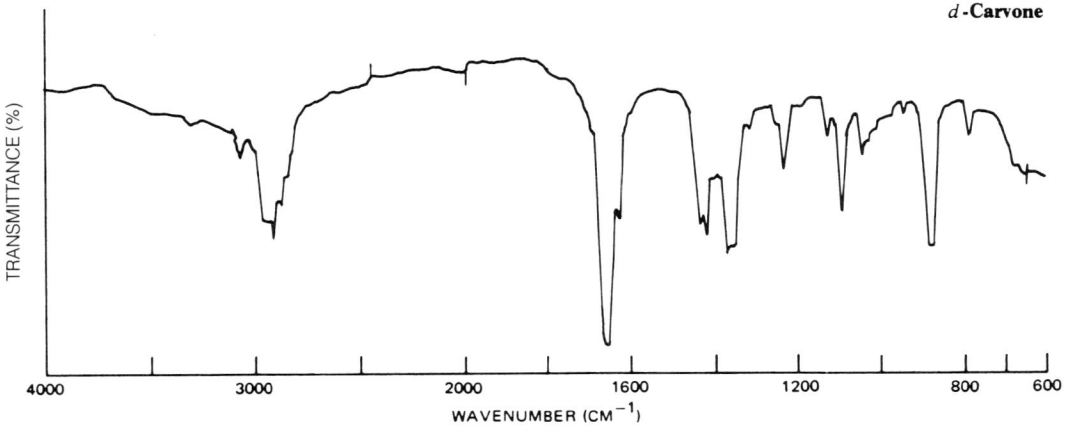

d-Carvone

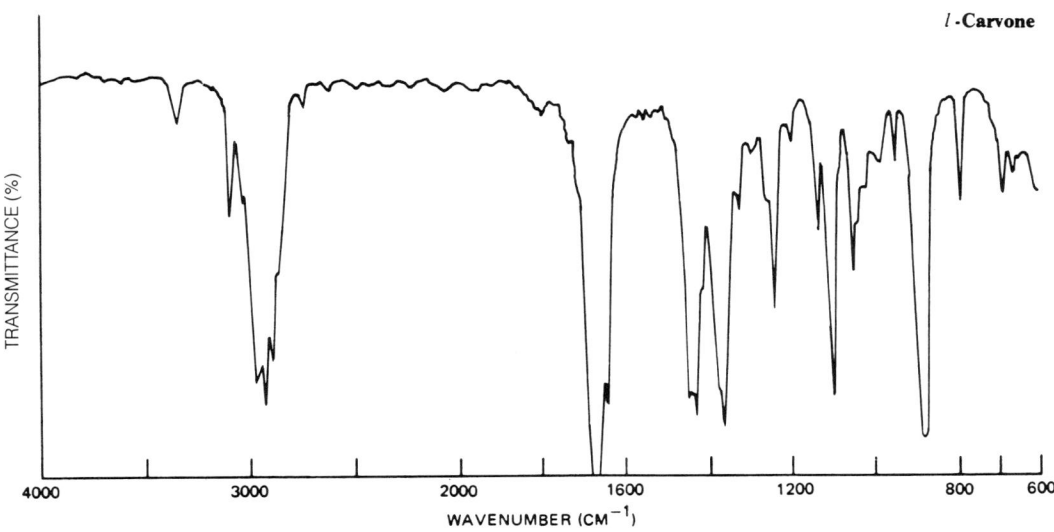

l-Carvone

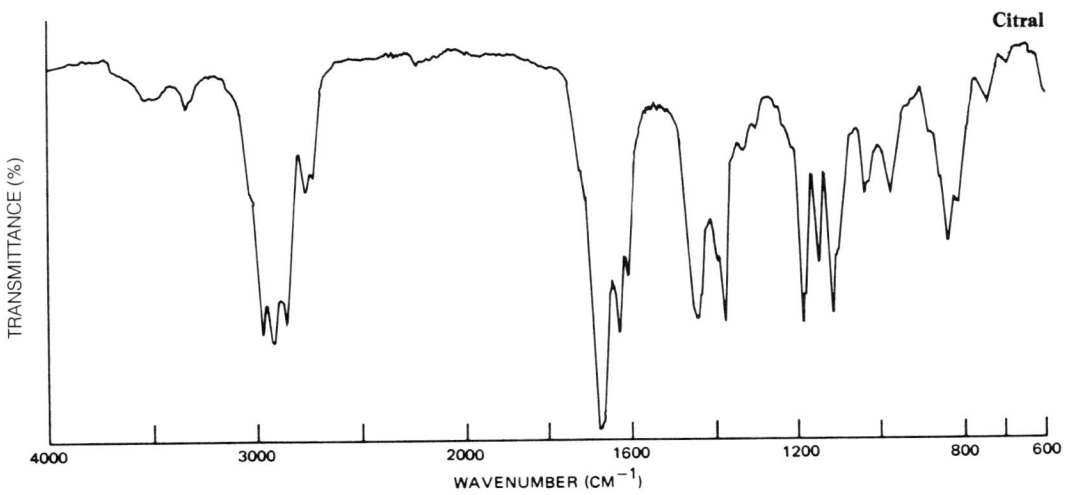

Citral

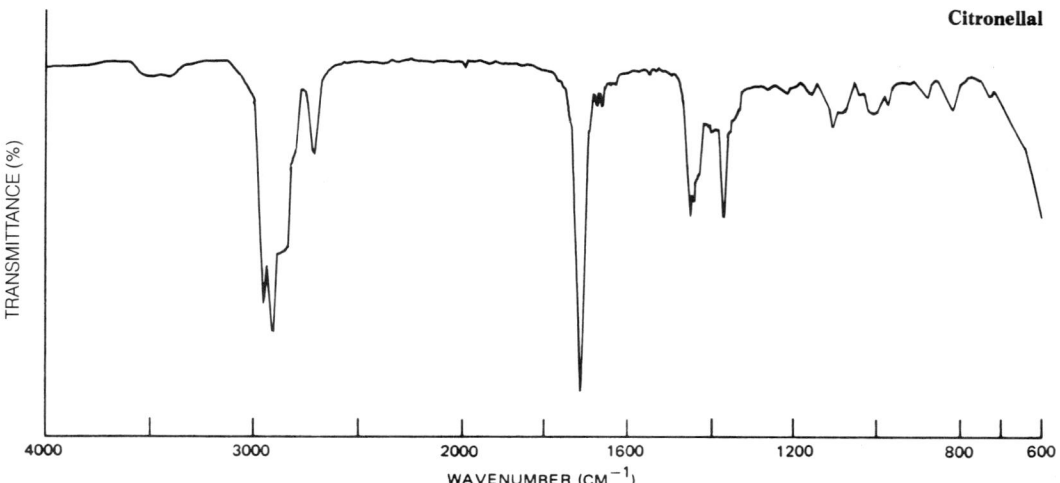

Citronellal

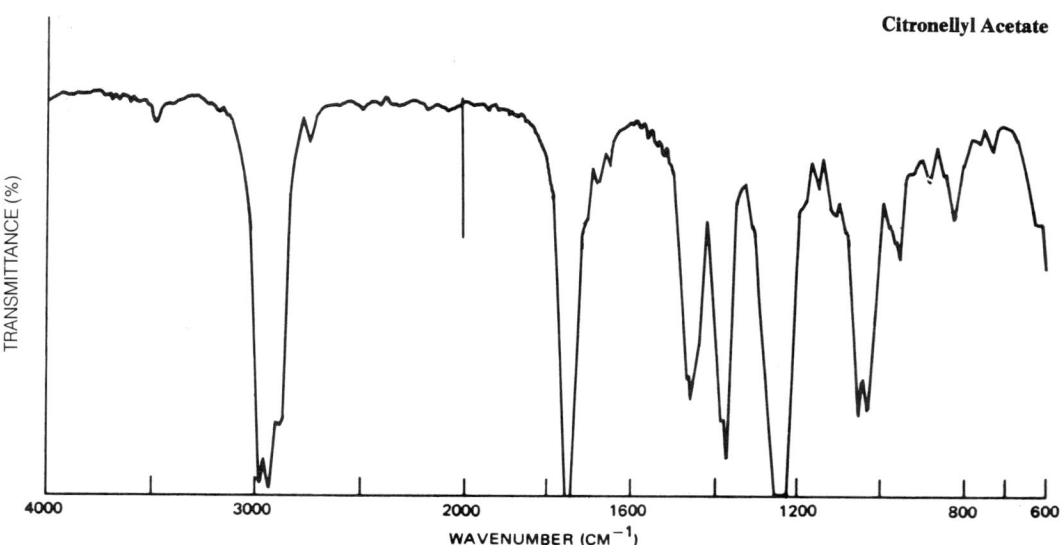

Citronellyl Acetate

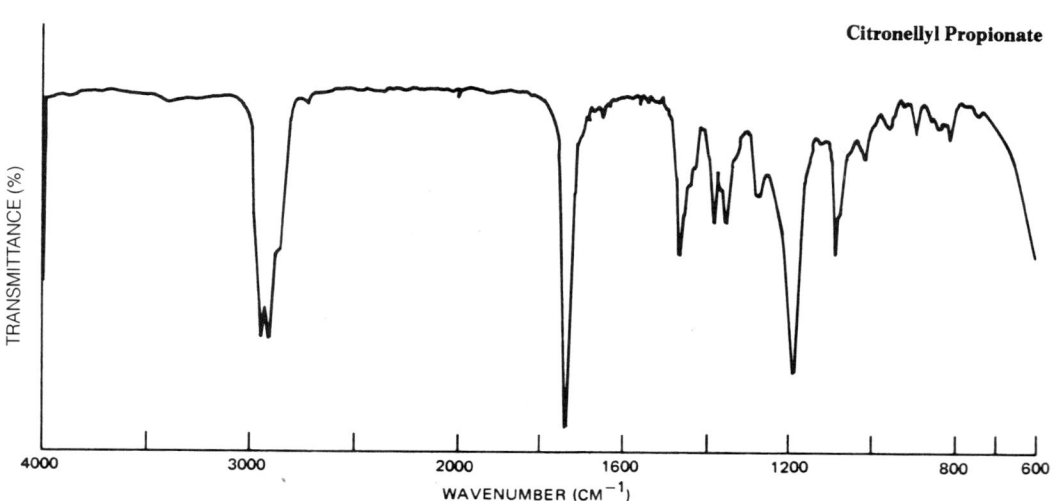

Citronellyl Propionate

Cuminic Aldehyde

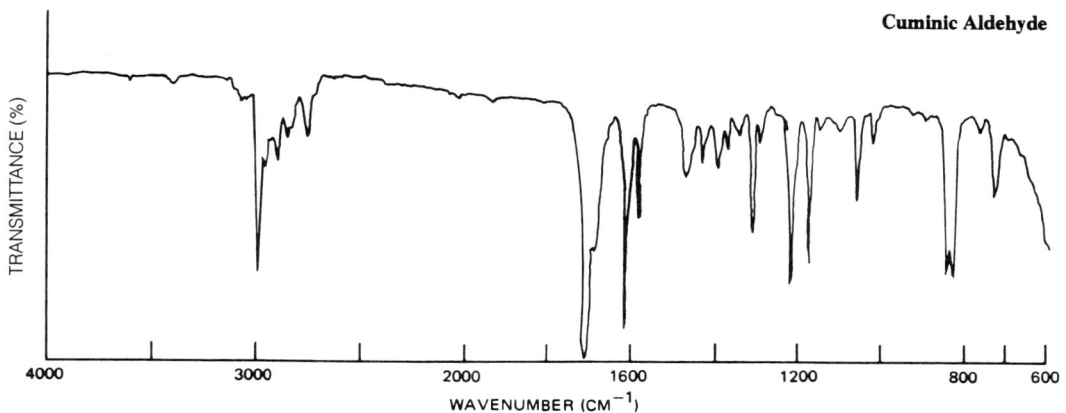

Cyclamen Aldehyde

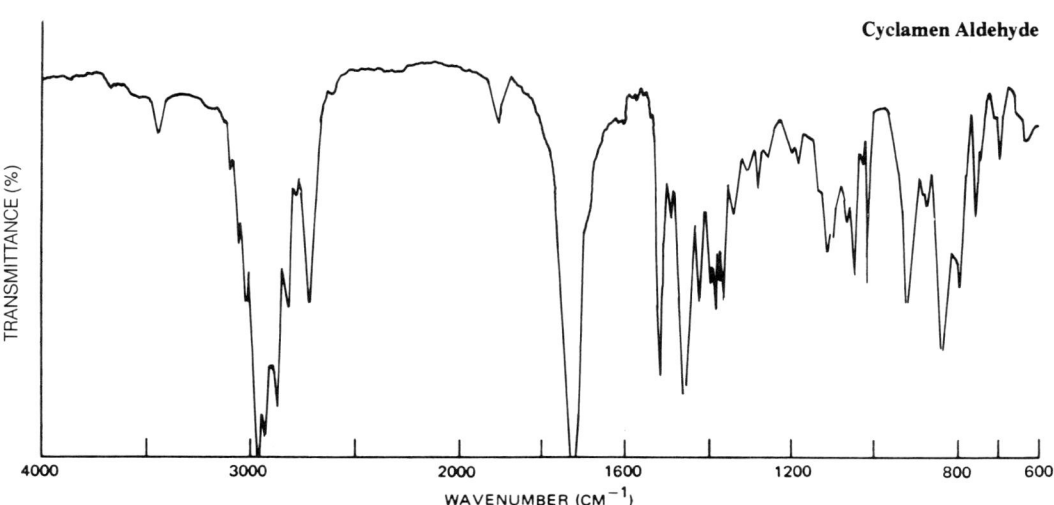

Decanal

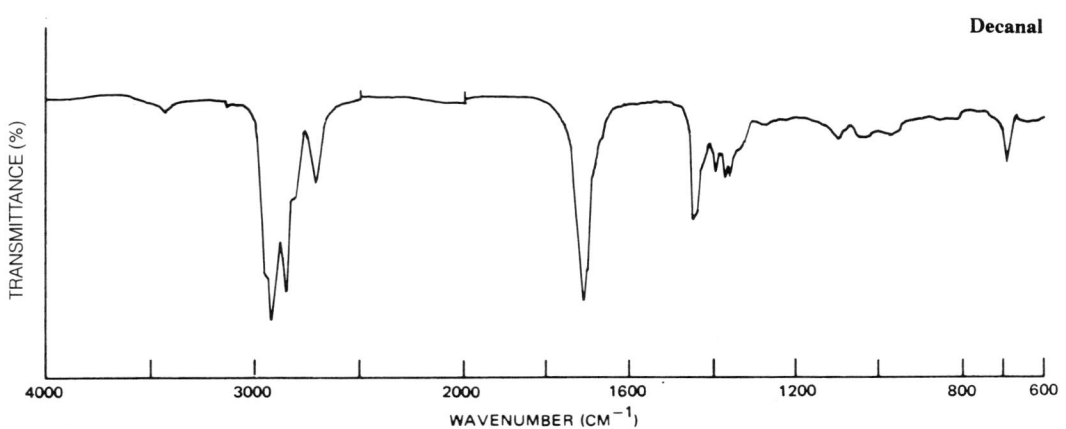

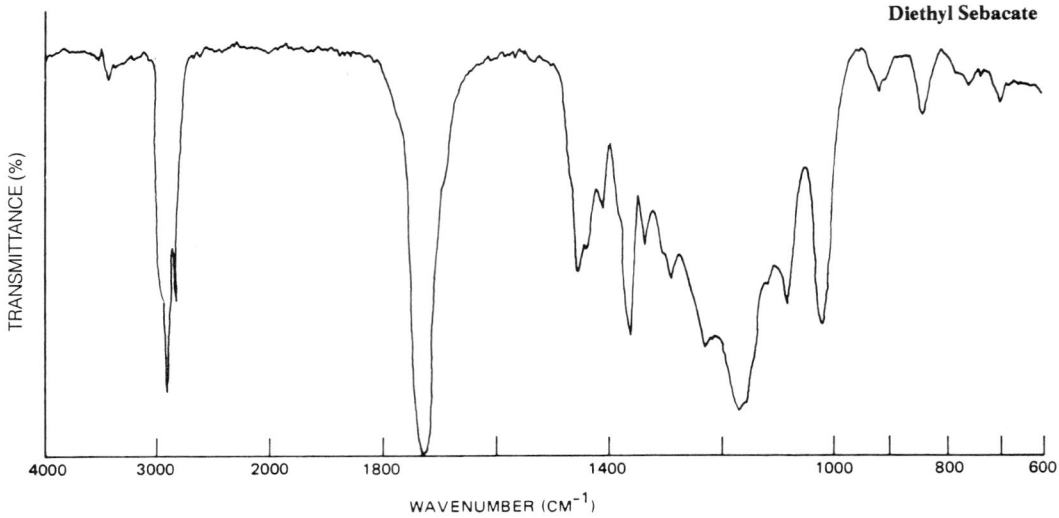

Diethyl Sebacate

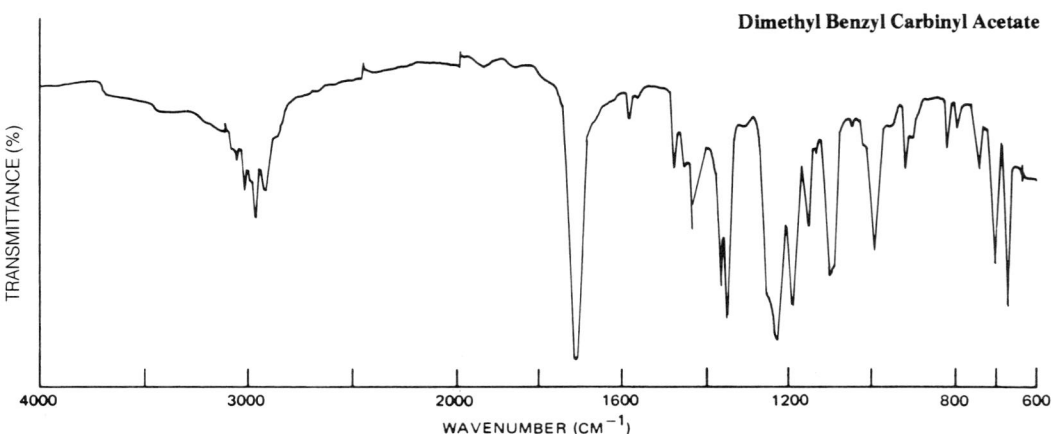

Dimethyl Benzyl Carbinyl Acetate

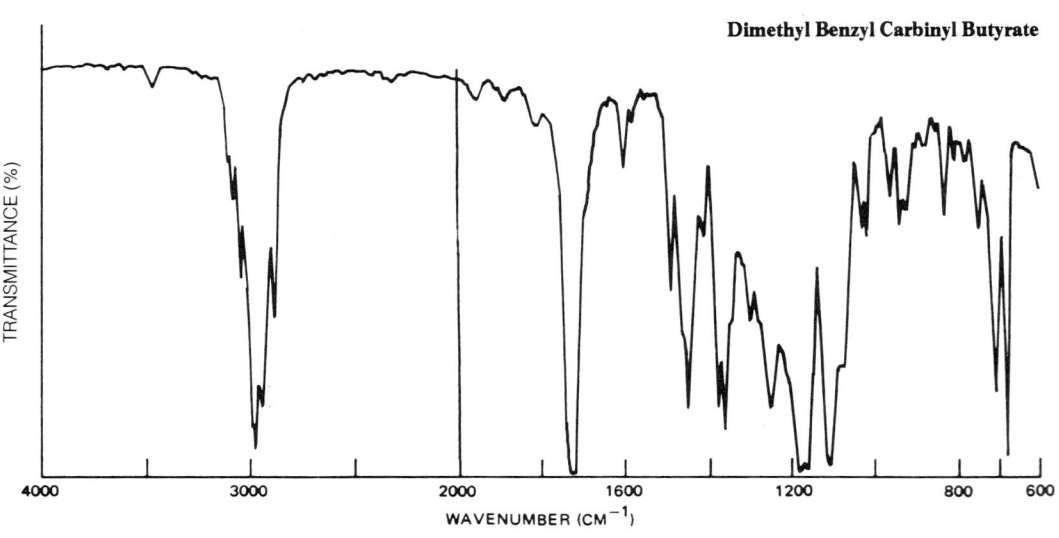

Dimethyl Benzyl Carbinyl Butyrate

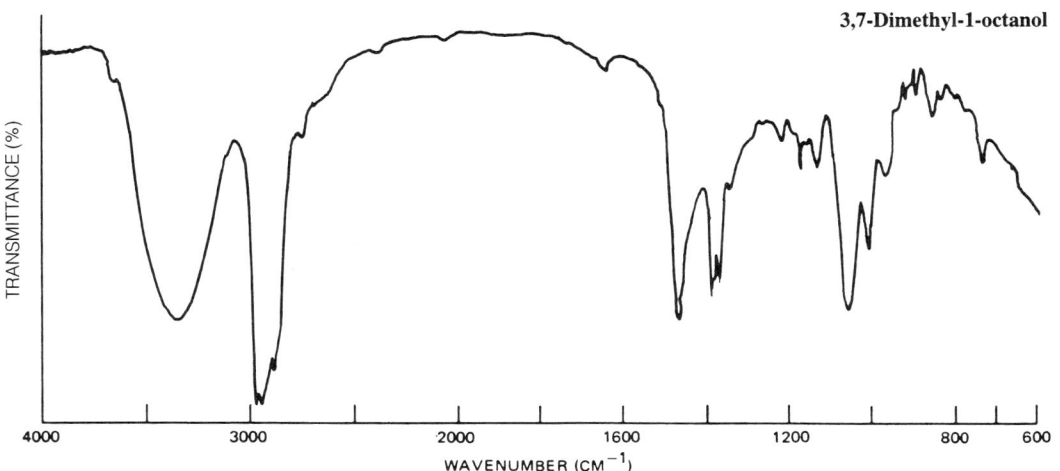

3,7-Dimethyl-1-octanol

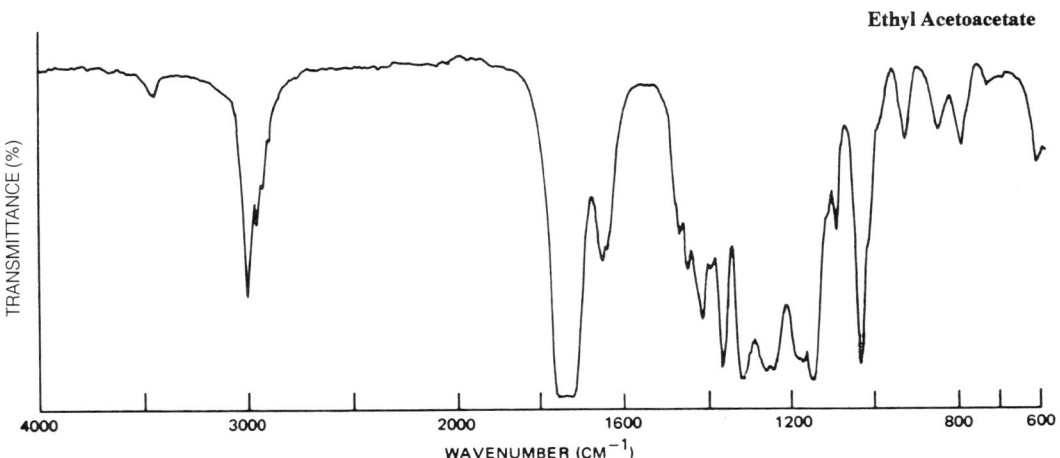

Ethyl Acetoacetate

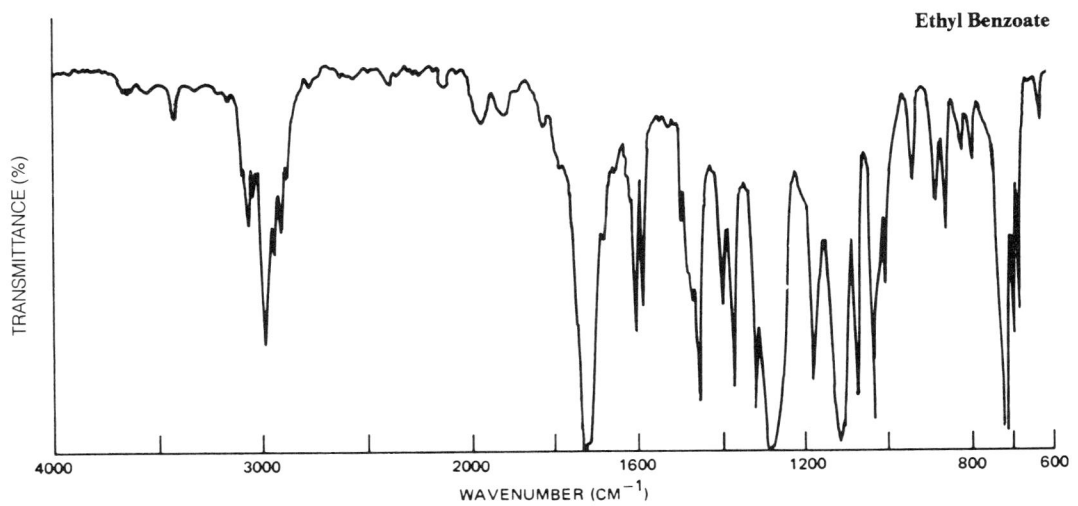

Ethyl Benzoate

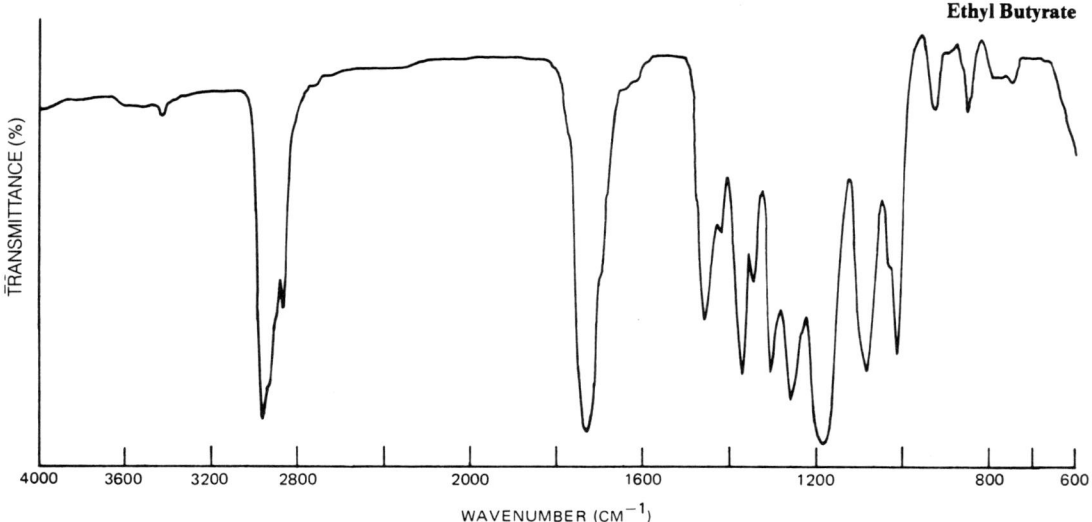

Ethyl Butyrate

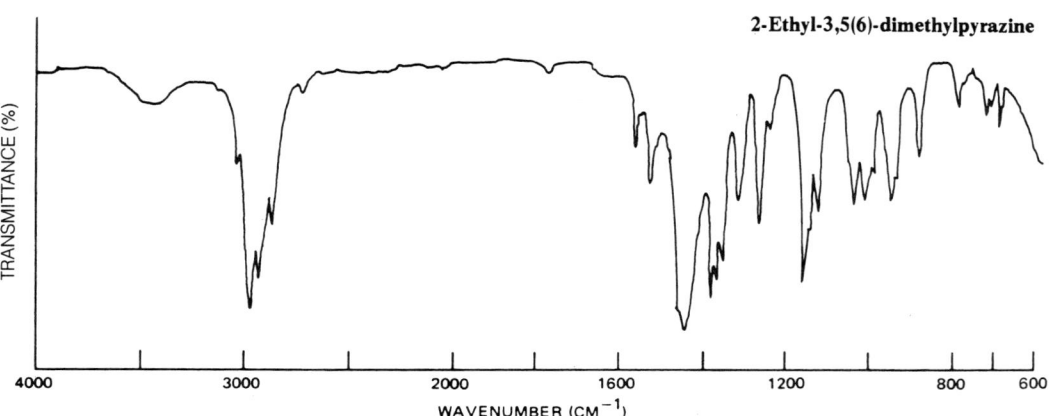

2-Ethyl-3,5(6)-dimethylpyrazine

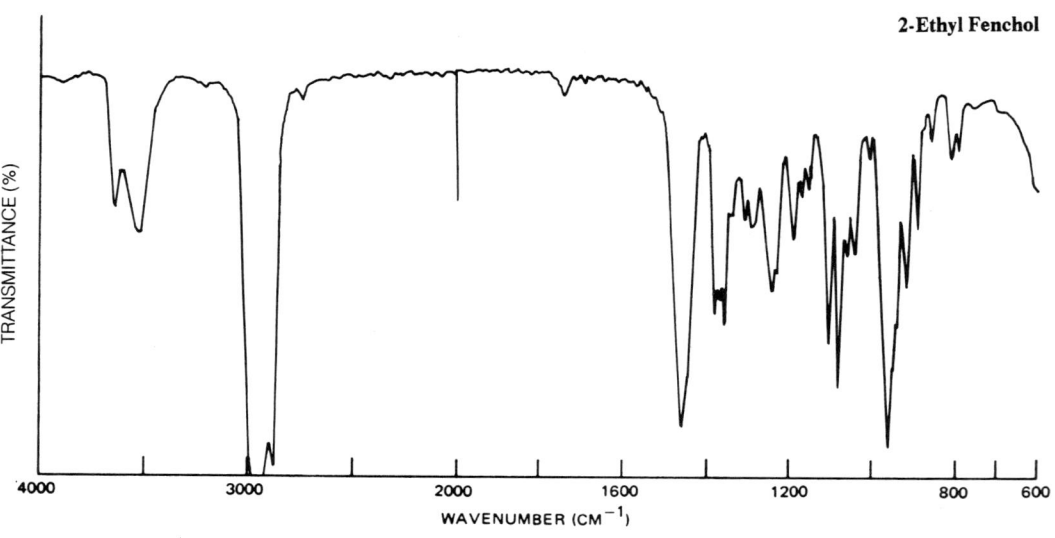

2-Ethyl Fenchol

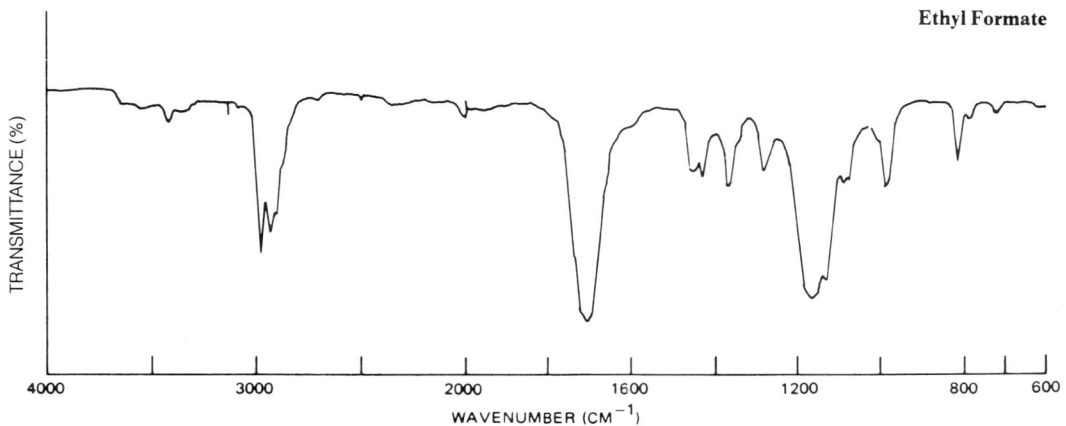

Ethyl Formate

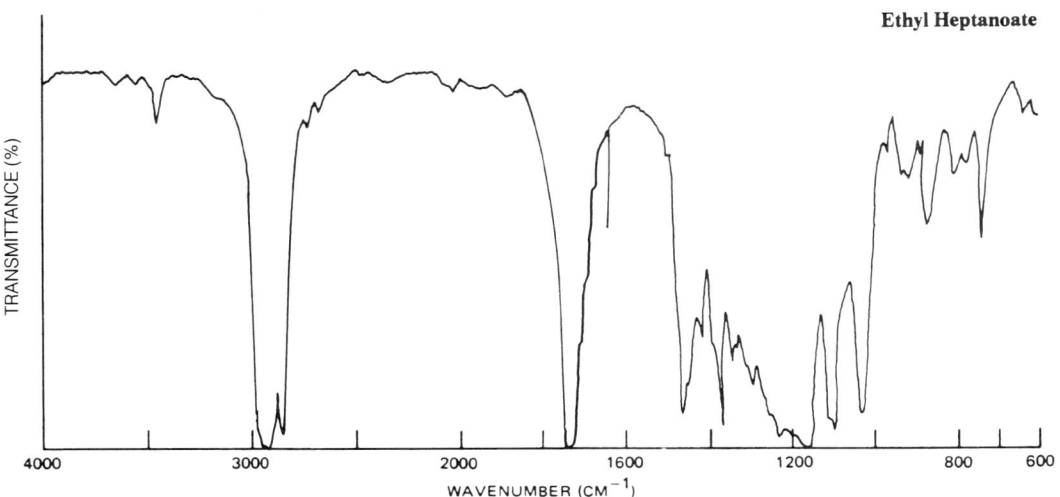

Ethyl Heptanoate

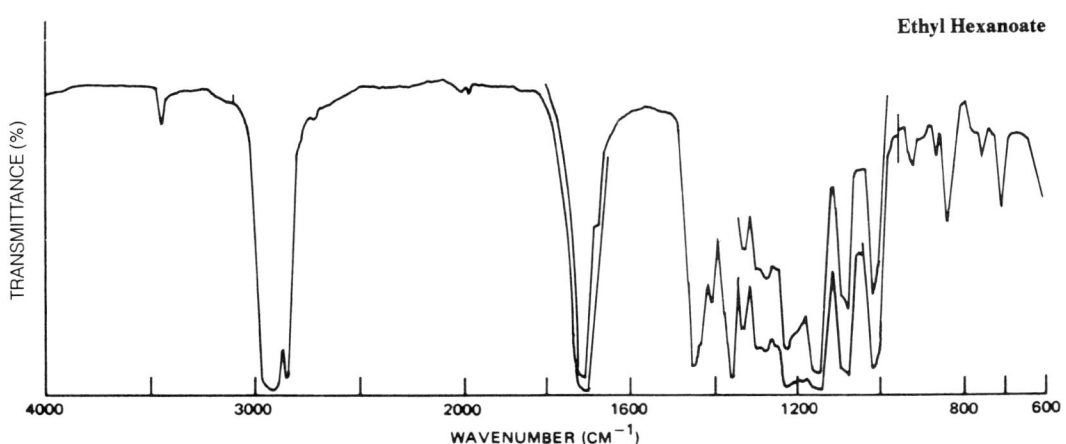

Ethyl Hexanoate

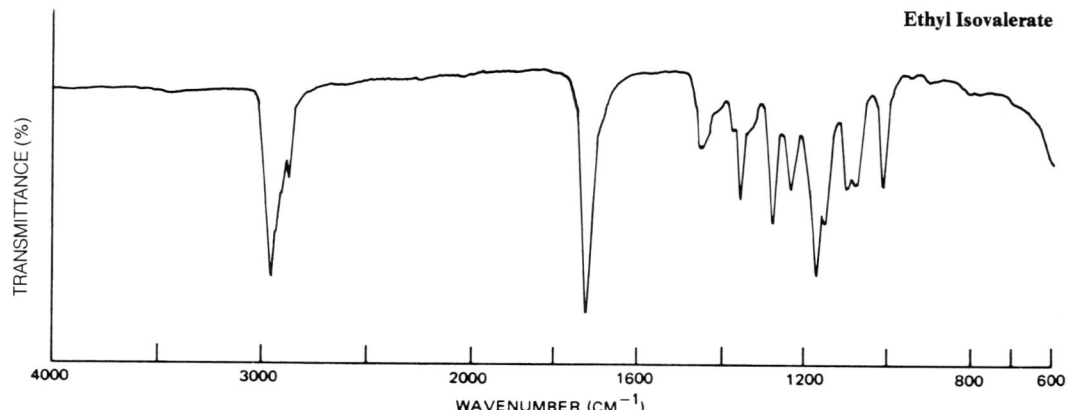

Ethyl Isovalerate

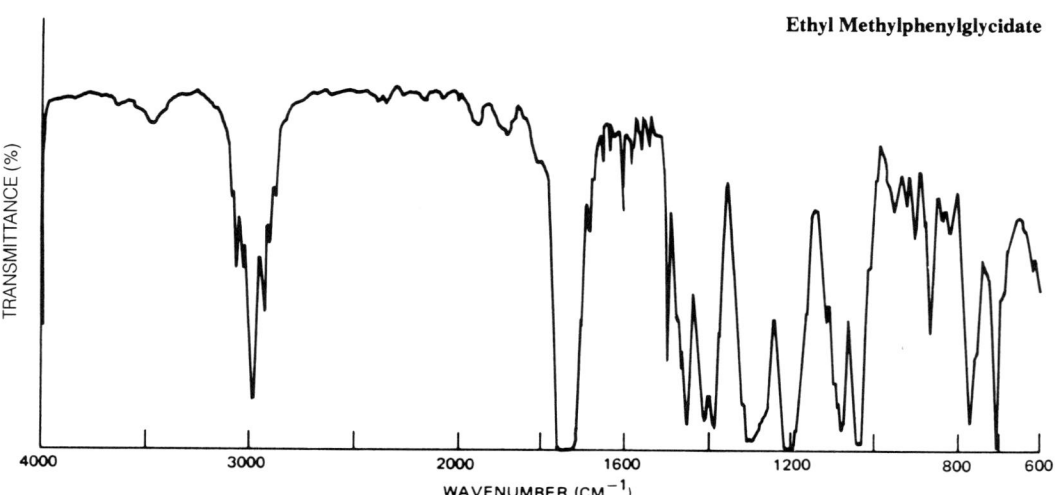

Ethyl Methylphenylglycidate

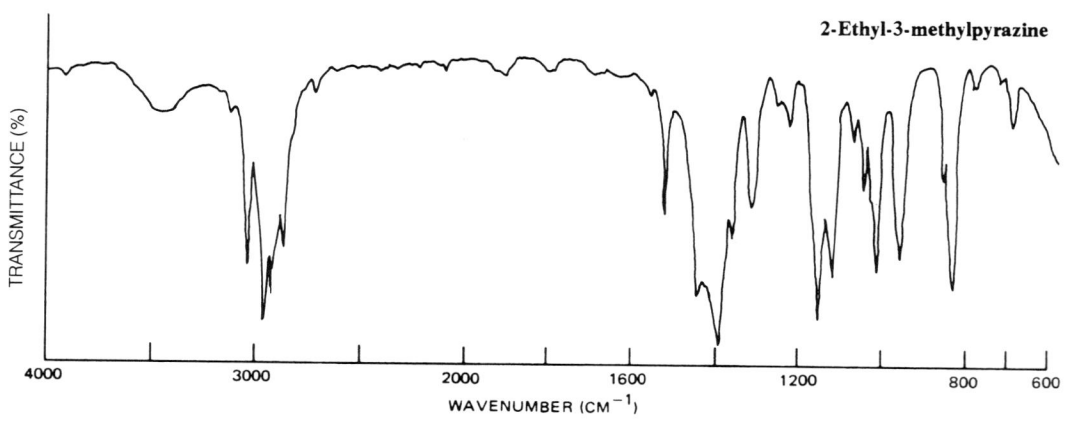

2-Ethyl-3-methylpyrazine

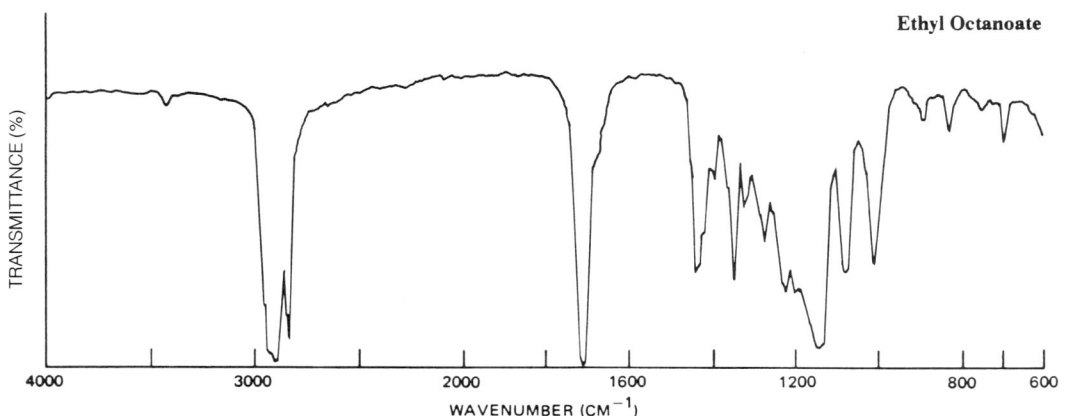

Ethyl Octanoate

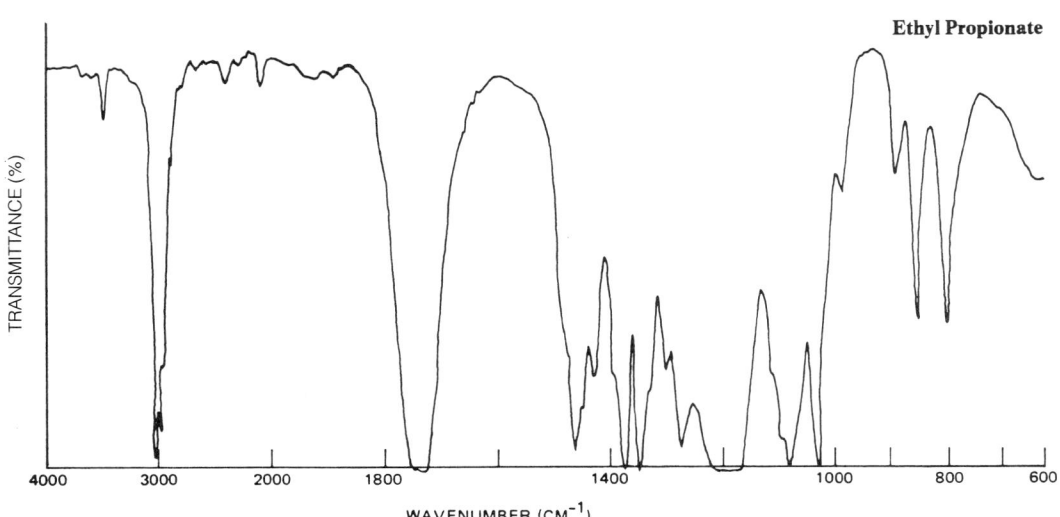

Ethyl Propionate

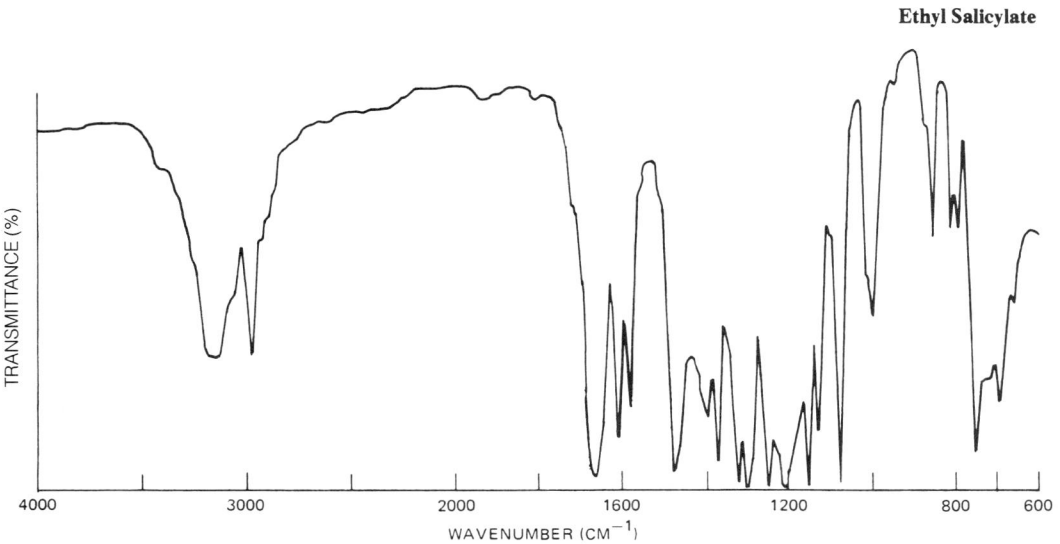

Ethyl Salicylate

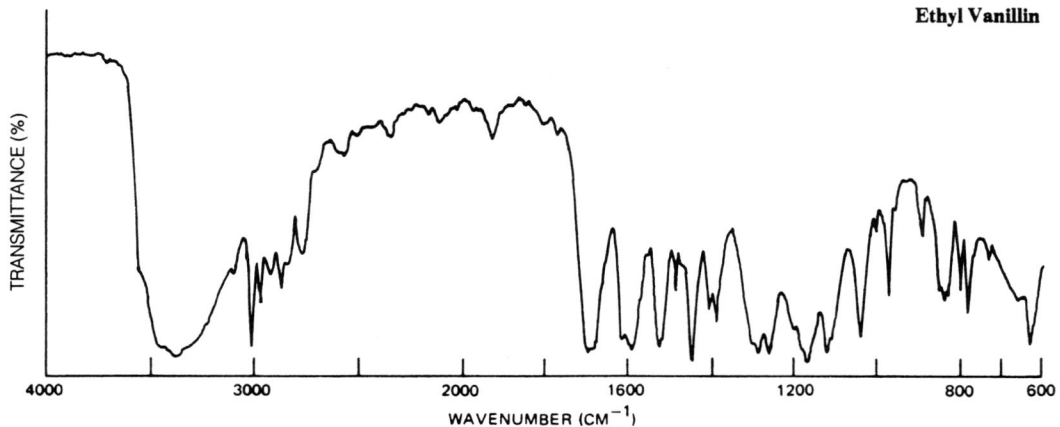

Ethyl Vanillin

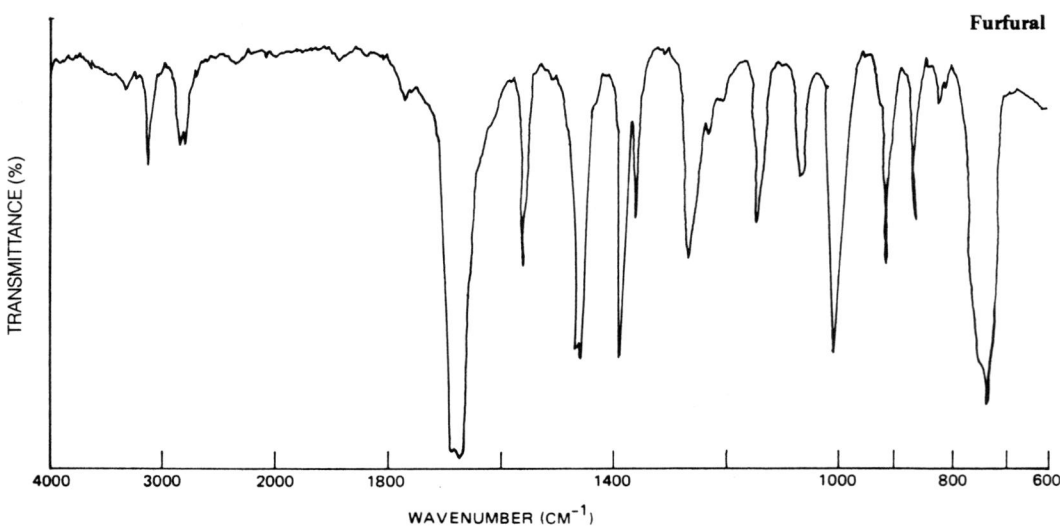

Furfural

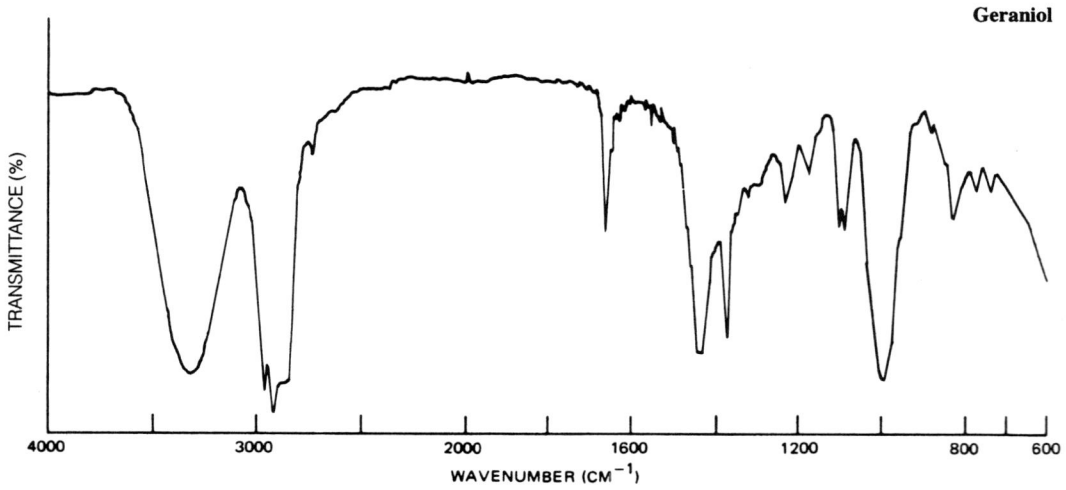

Geraniol

Geranyl Acetate

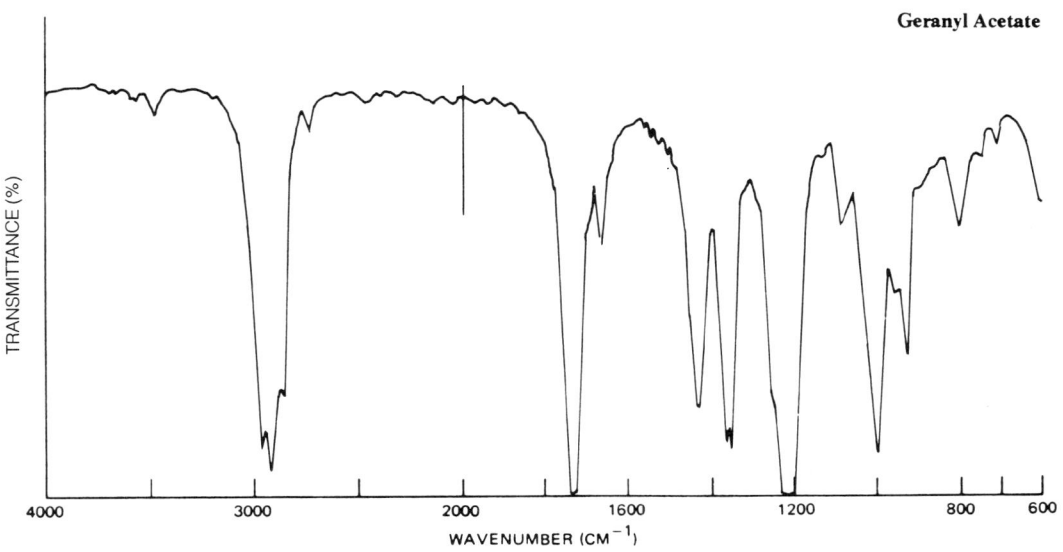

Geranyl Formate

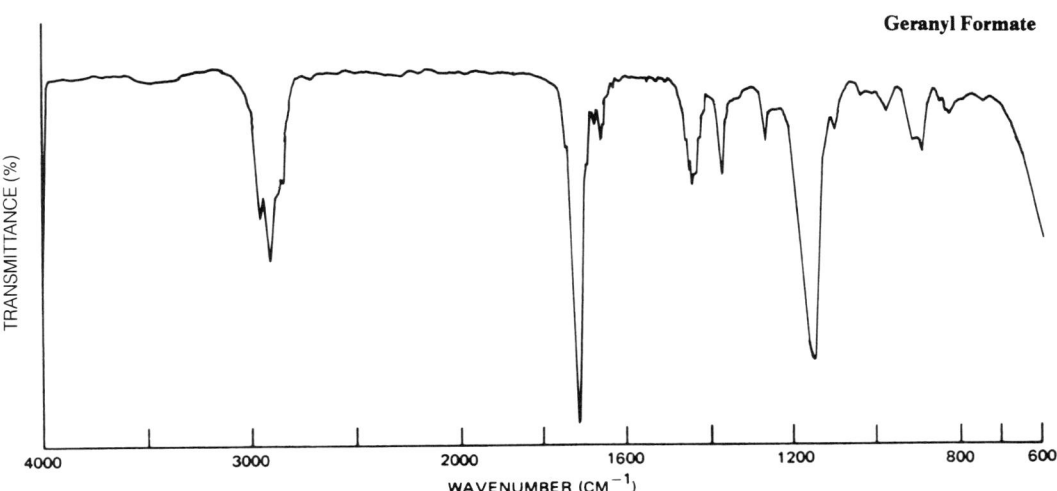

Geranyl Phenylacetate

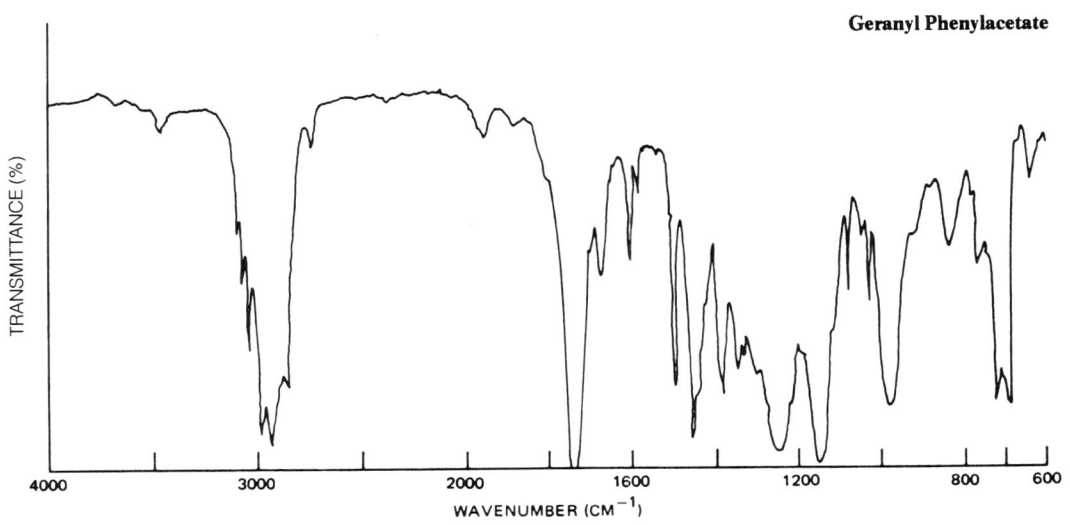

Geranyl Propionate

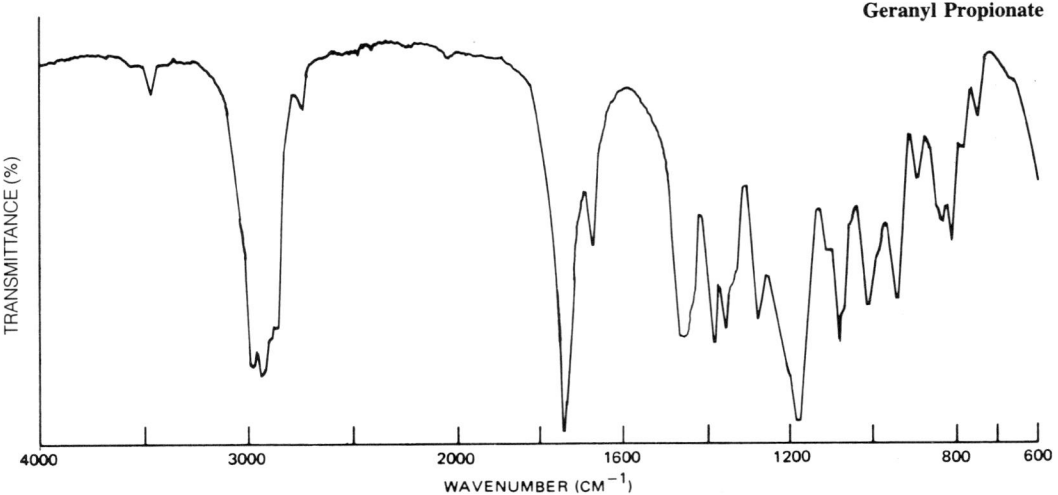

γ-Heptalactone

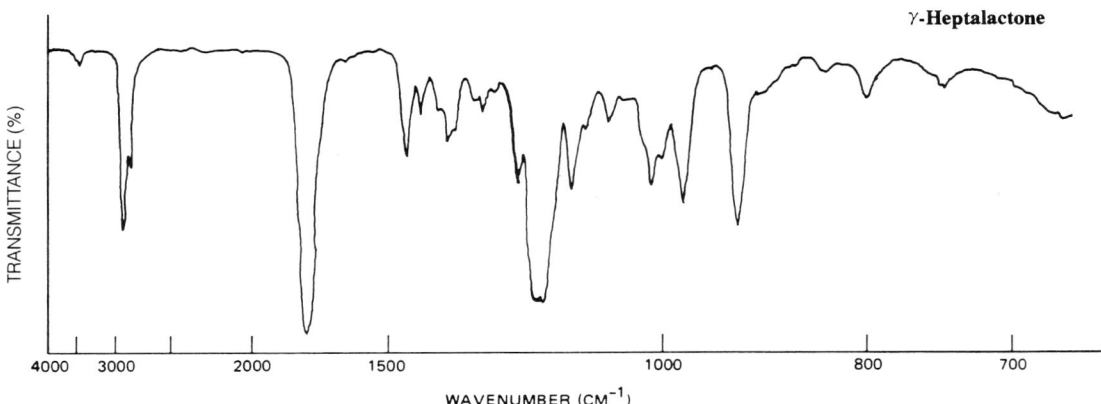

Hexanoic Acid

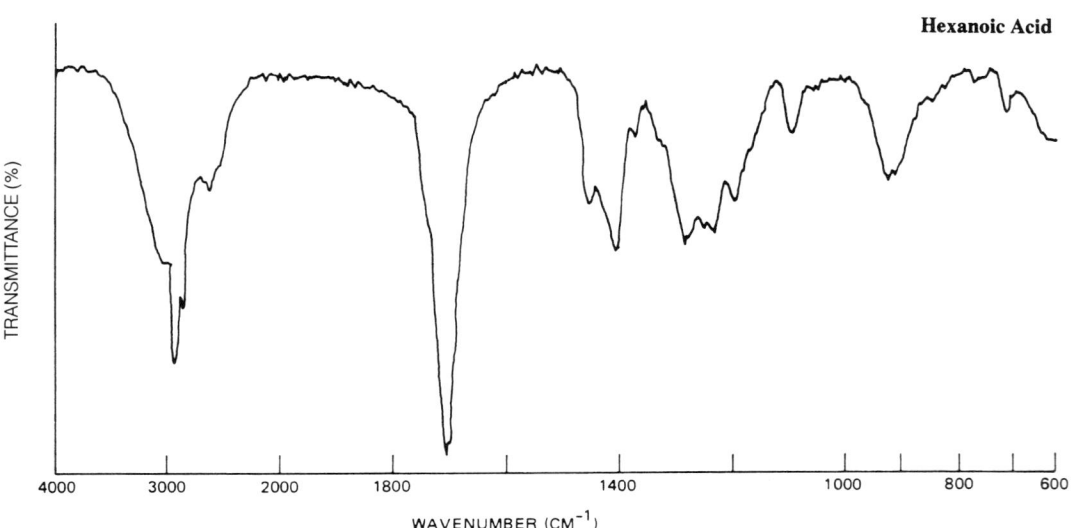

Hexyl-2-butenoate

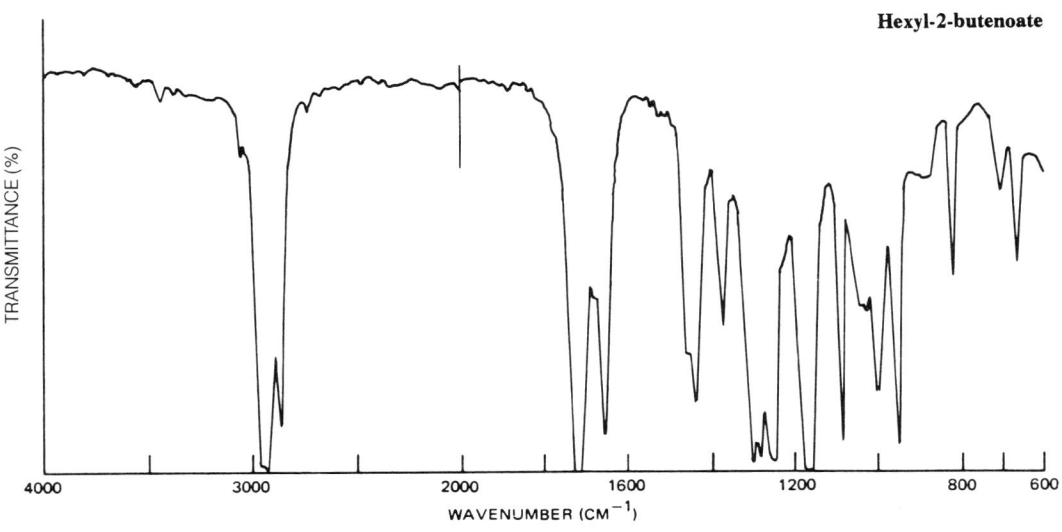

Hydroxycitronellal Dimethyl Acetal

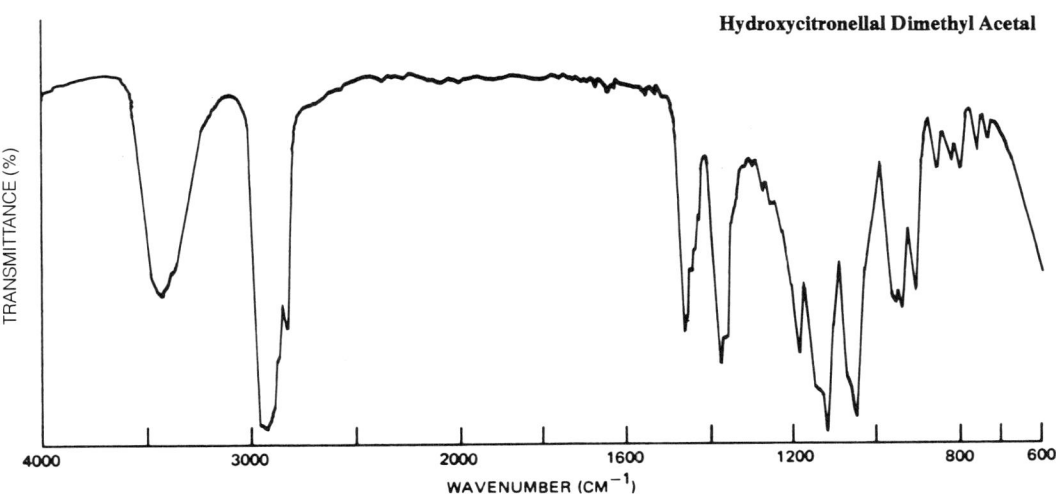

α-Ionone

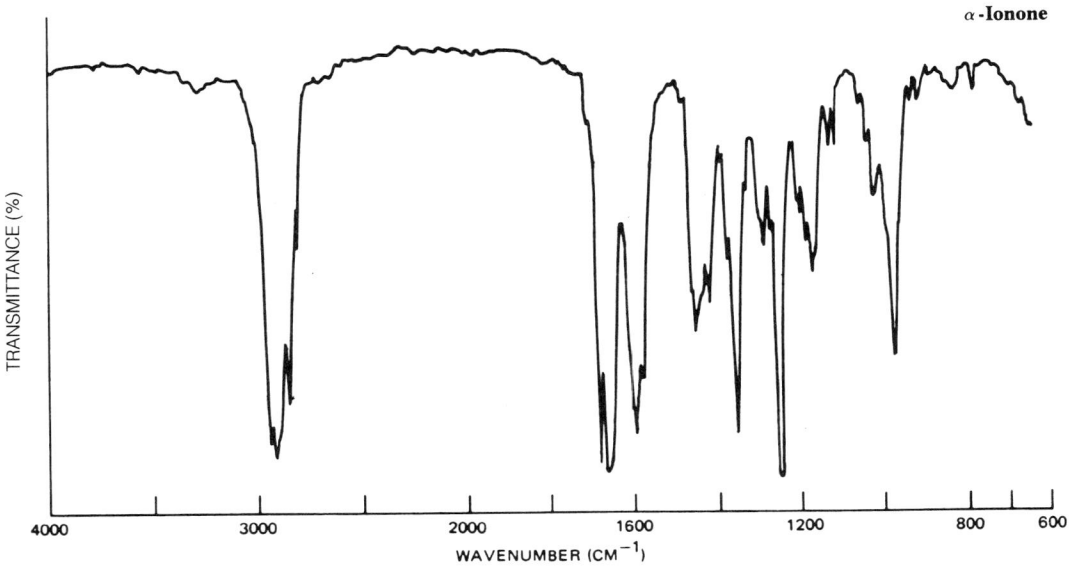

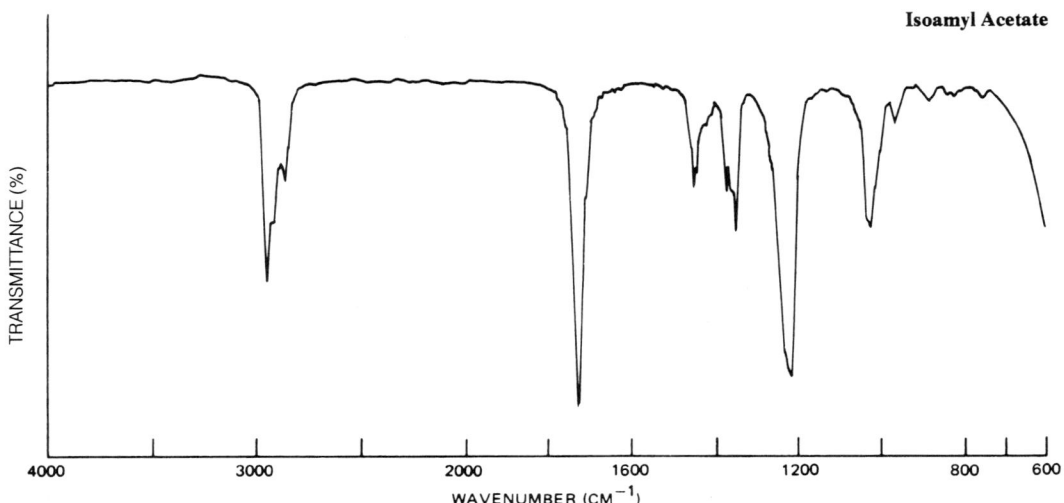

Isoamyl Acetate

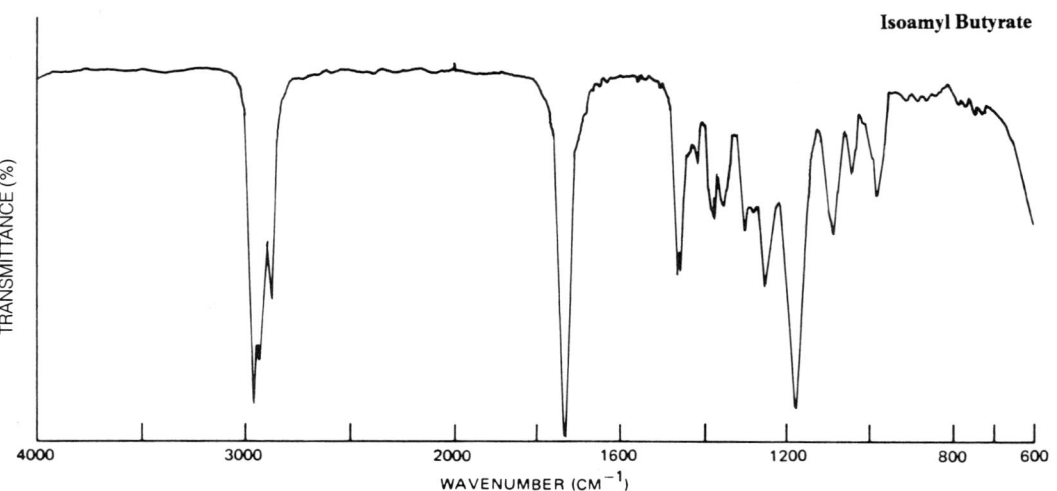

Isoamyl Butyrate

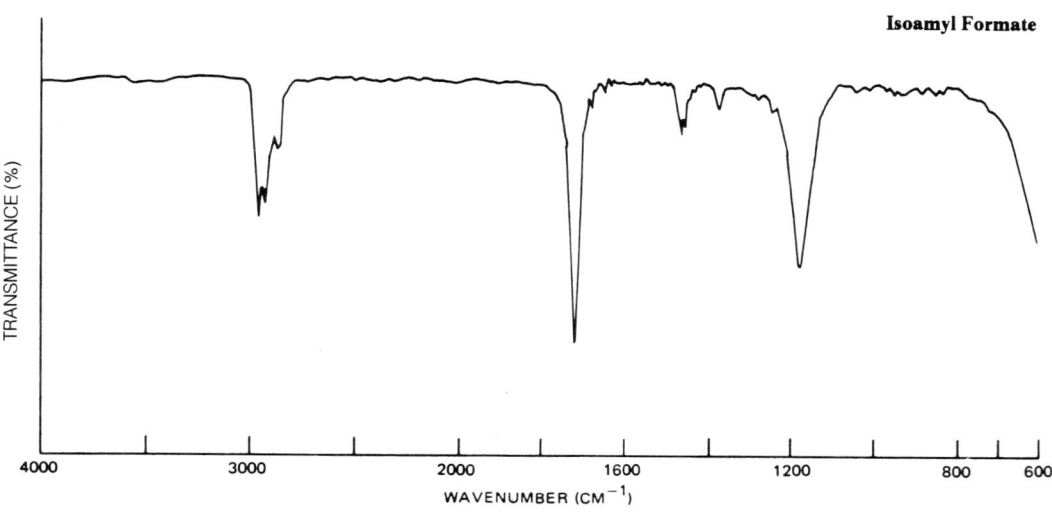

Isoamyl Formate

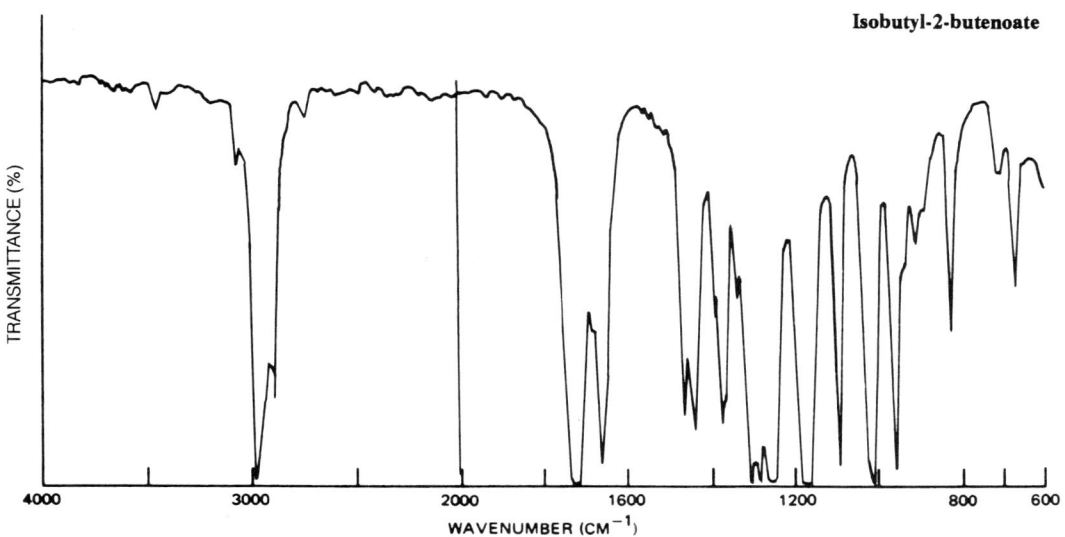

Isobutyl-2-butenoate

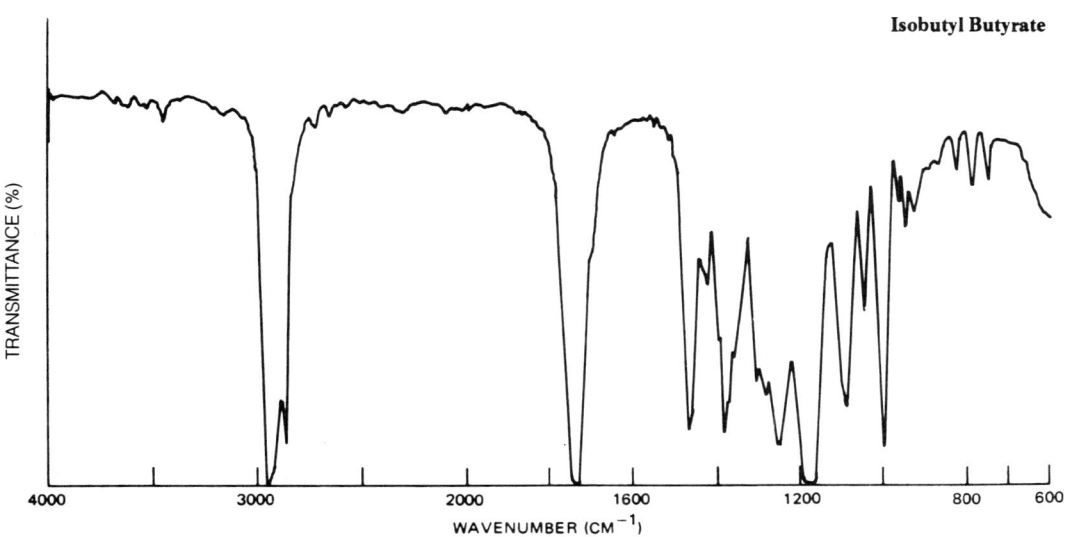

Isobutyl Butyrate

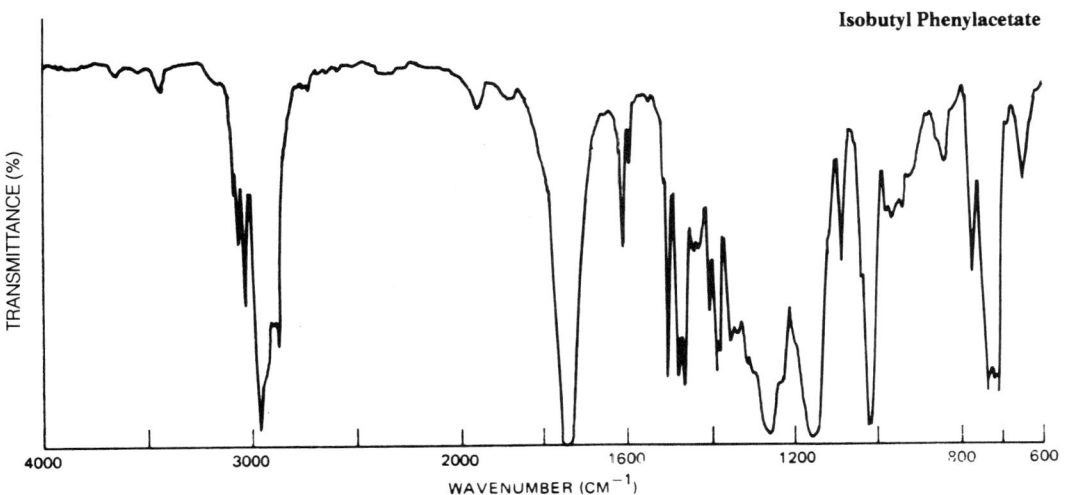

Isobutyl Phenylacetate

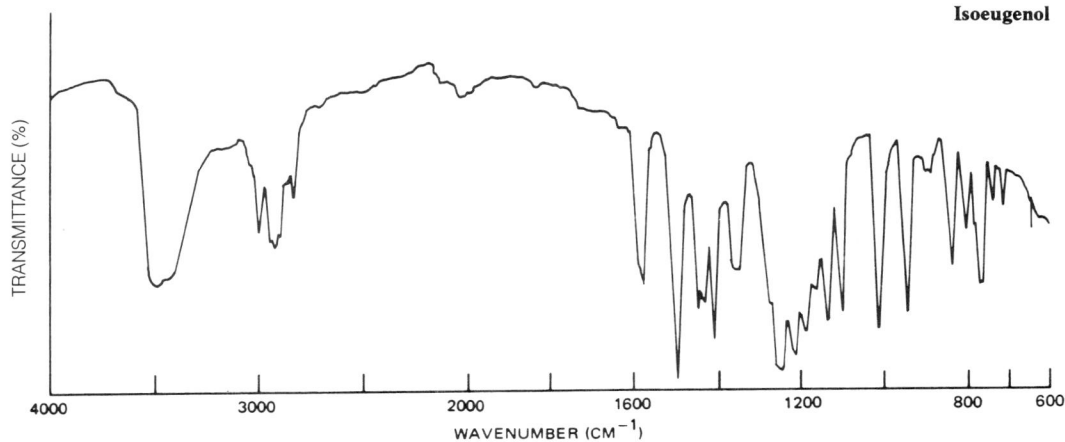

Isoeugenol

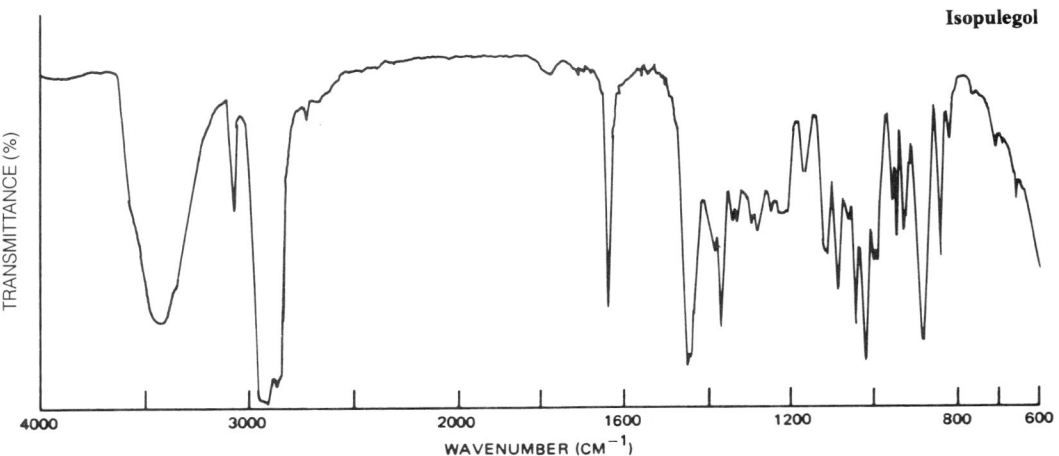

Isopulegol

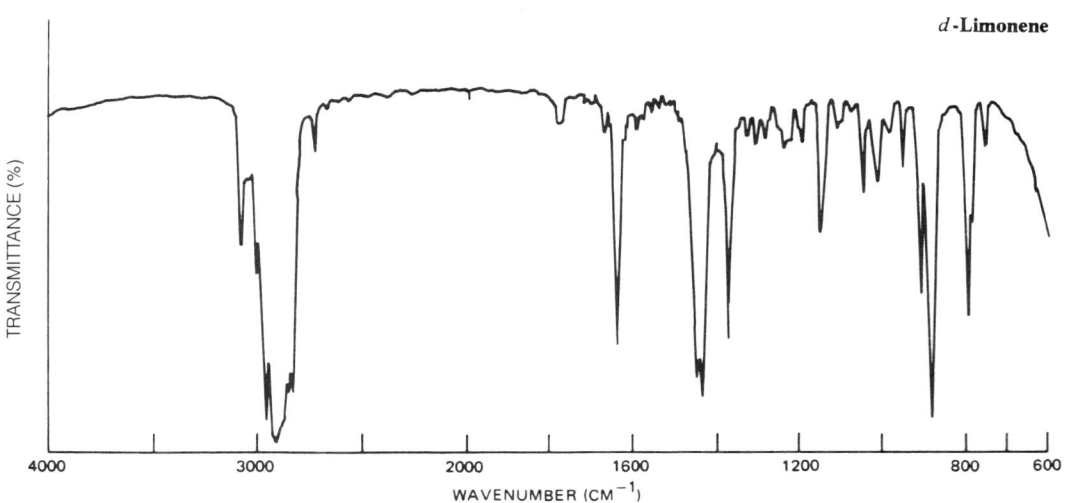

***d*-Limonene**

Linalool

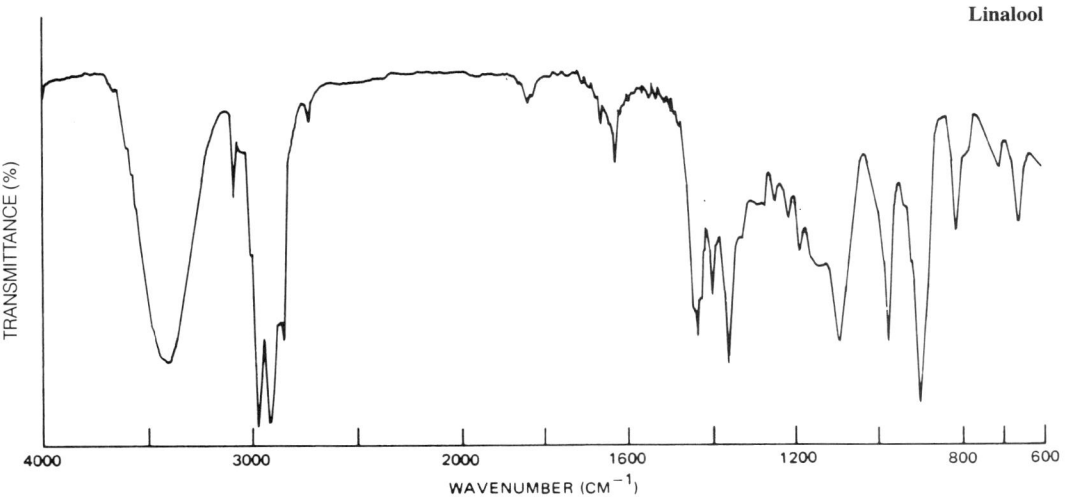

Linalyl Acetate

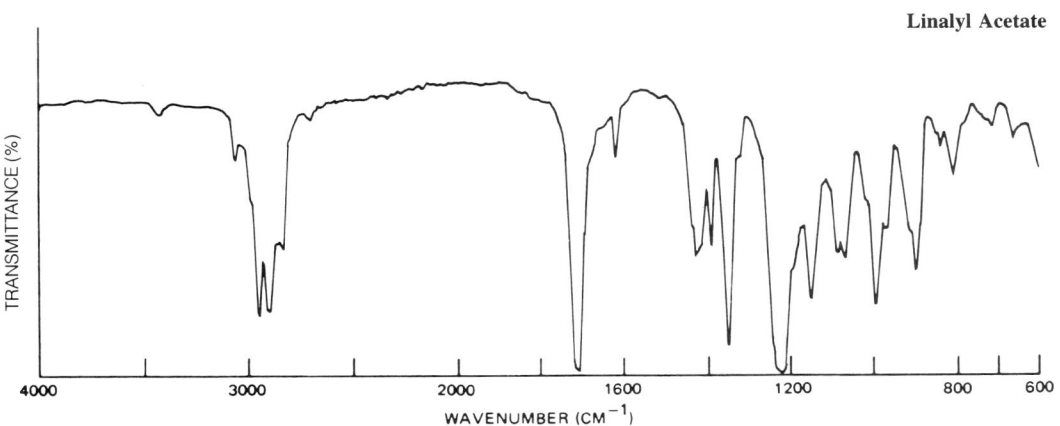

Menthol (*l*-form)

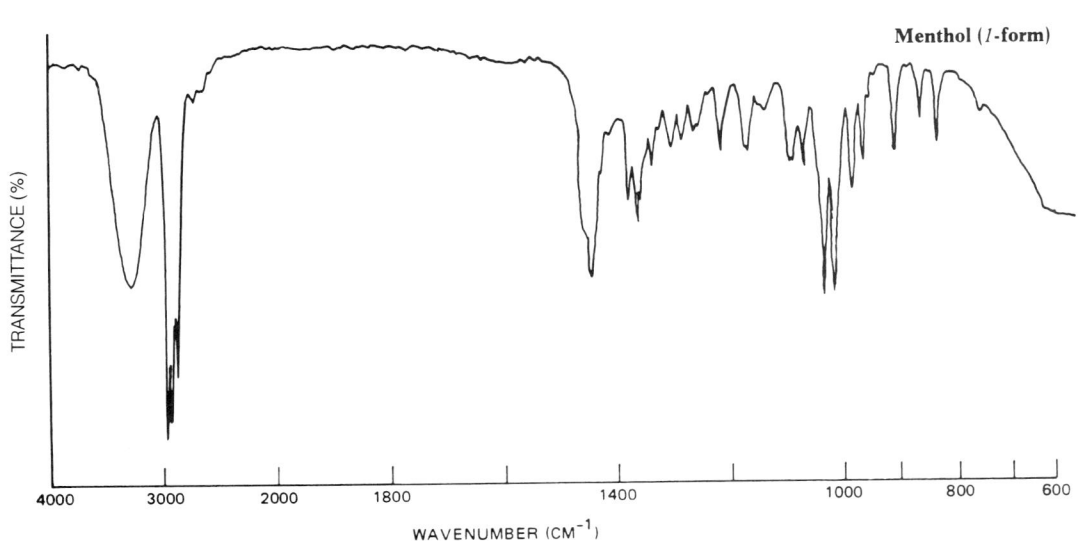

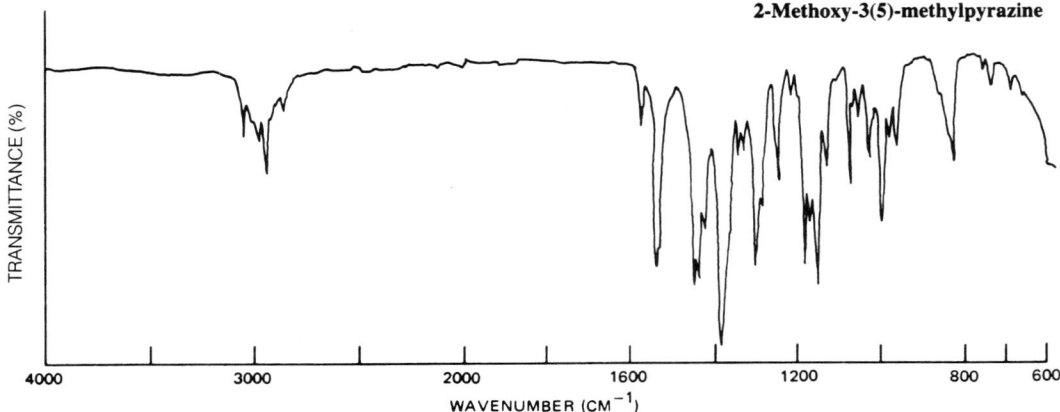

2-Methoxy-3(5)-methylpyrazine

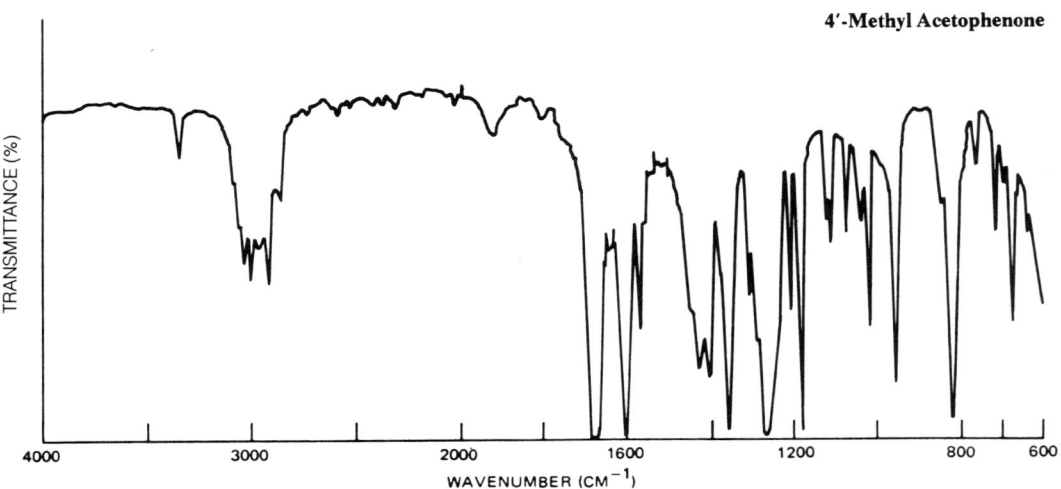

4'-Methyl Acetophenone

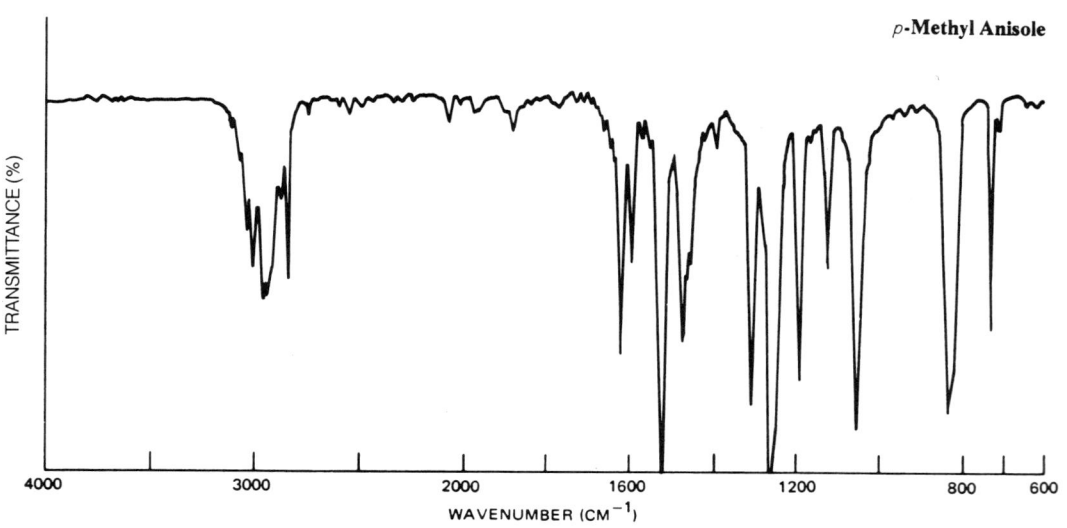

p-Methyl Anisole

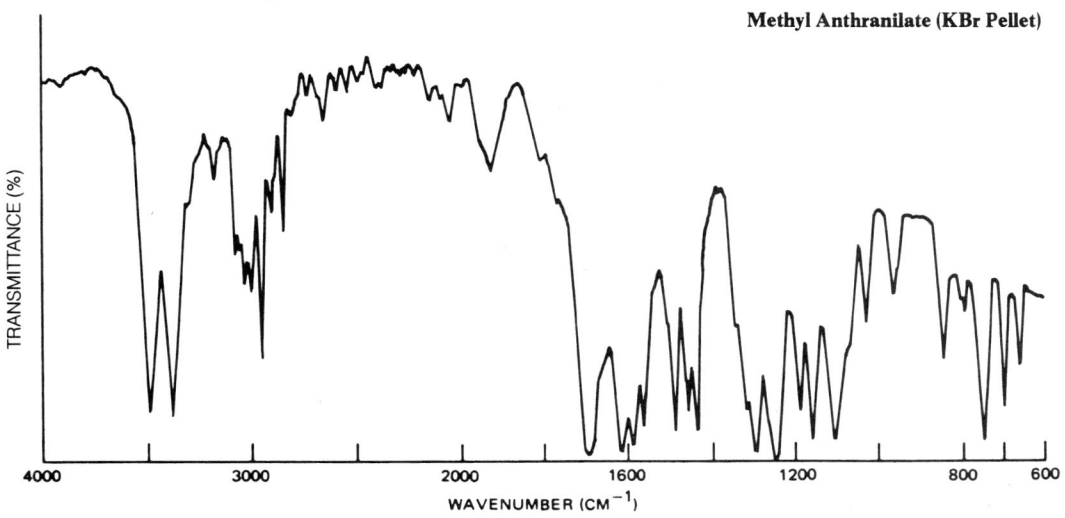
Methyl Anthranilate (KBr Pellet)

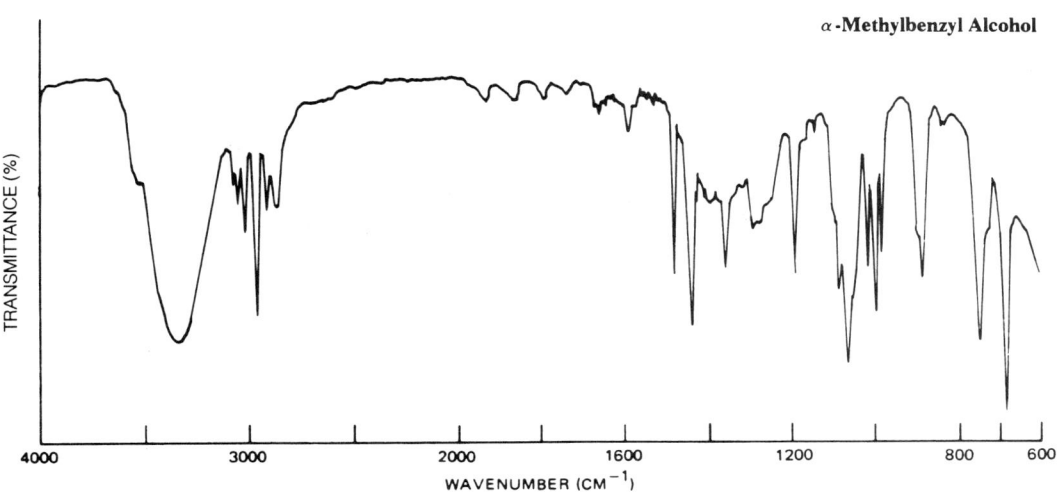
α-Methylbenzyl Alcohol

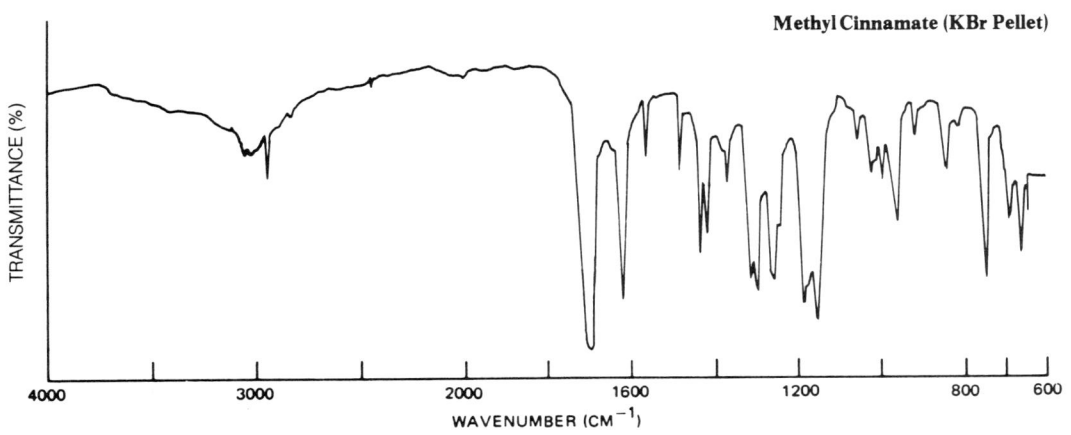
Methyl Cinnamate (KBr Pellet)

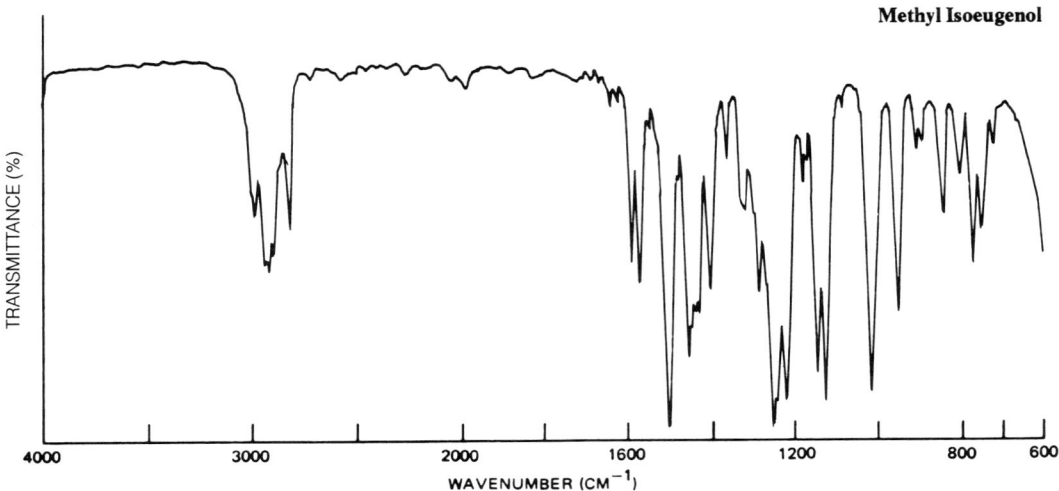

Methyl Isoeugenol

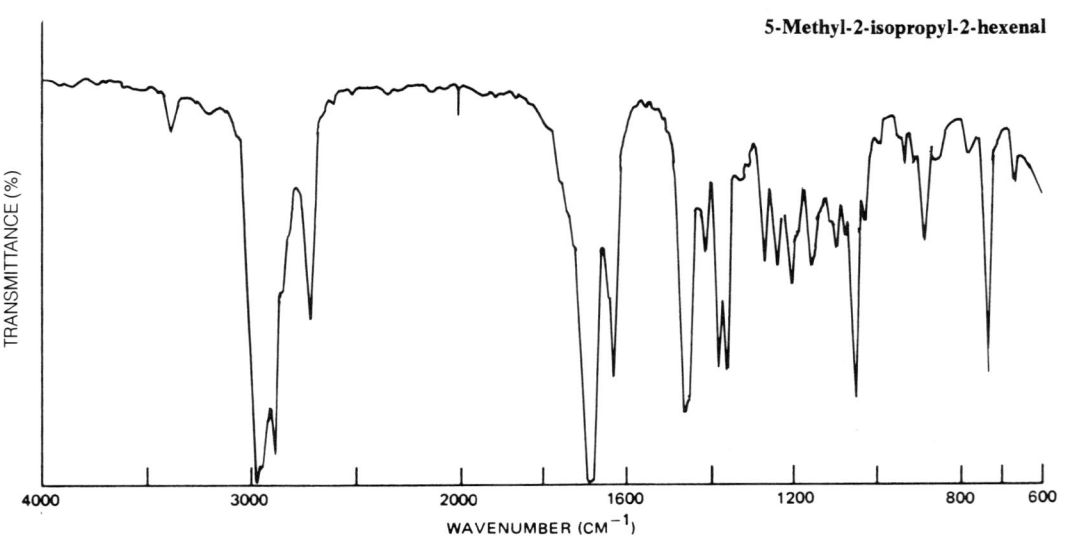

5-Methyl-2-isopropyl-2-hexenal

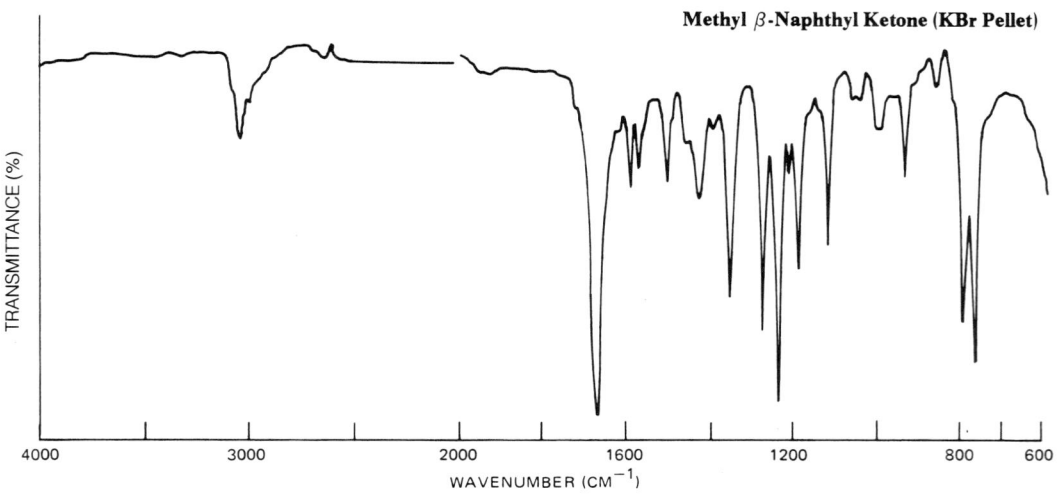

Methyl β-Naphthyl Ketone (KBr Pellet)

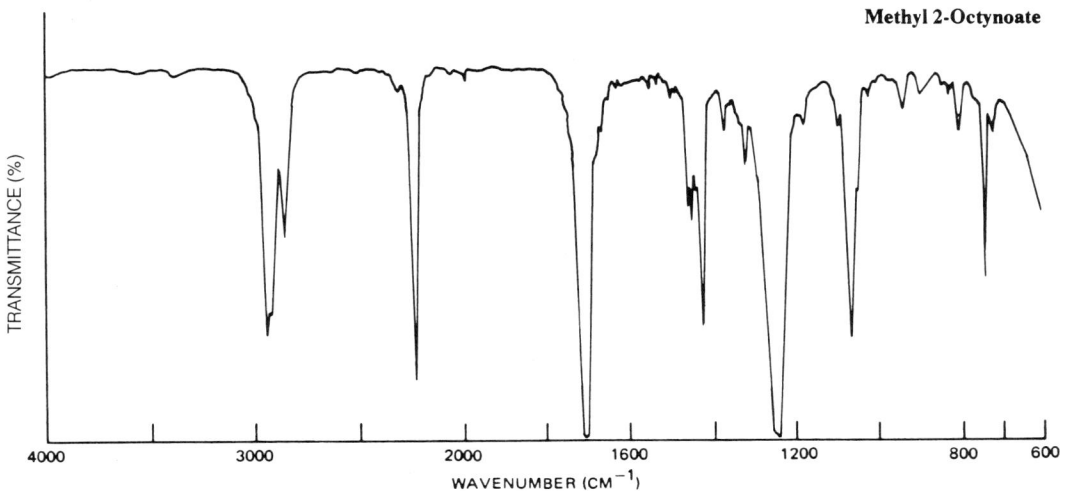

Methyl 2-Octynoate

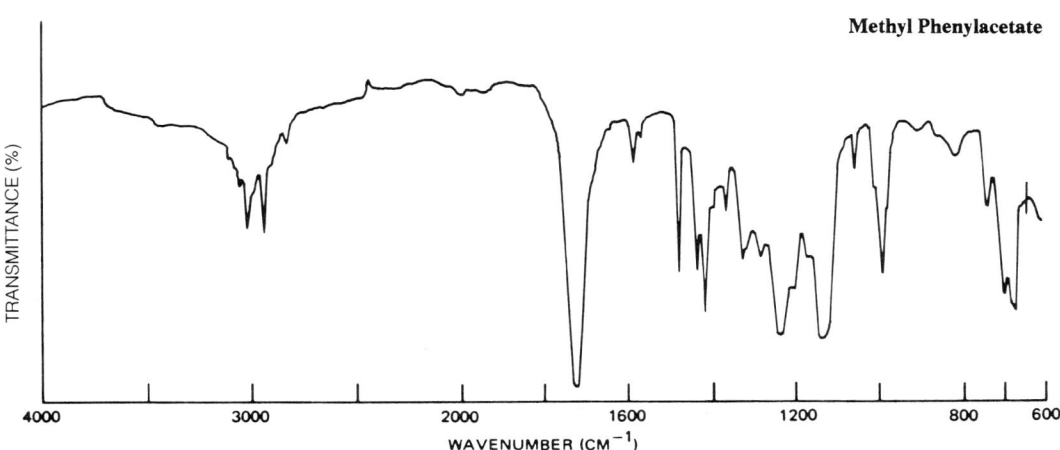

Methyl Phenylacetate

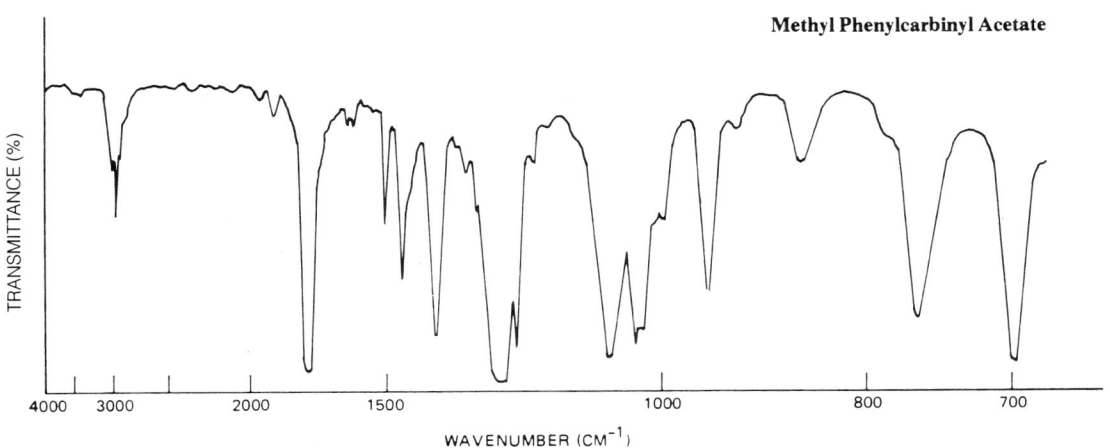

Methyl Phenylcarbinyl Acetate

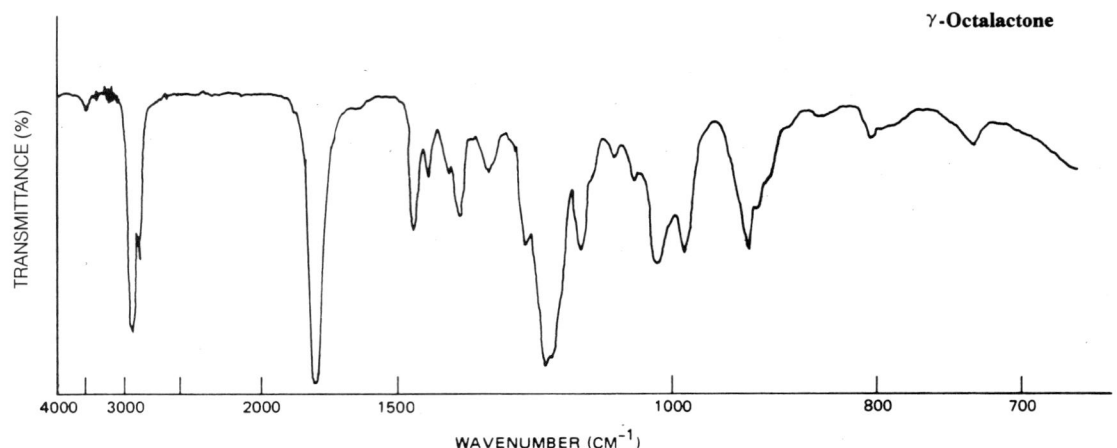

γ-Octalactone

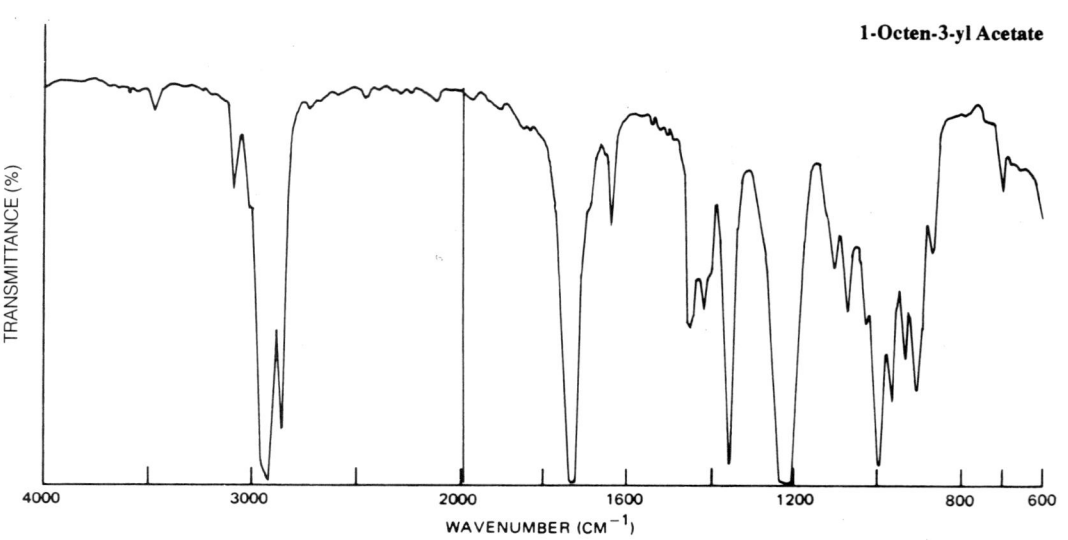

1-Octen-3-yl Acetate

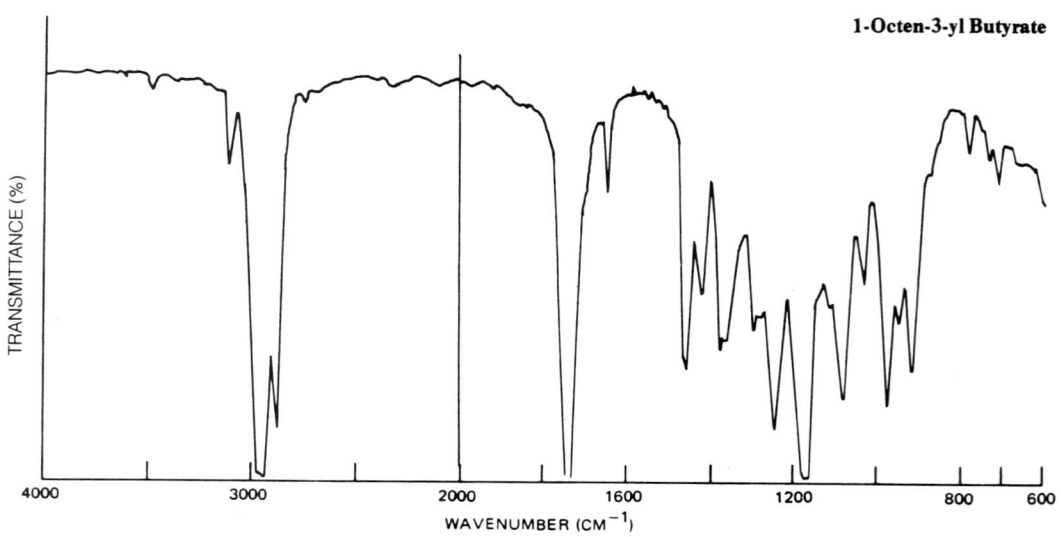

1-Octen-3-yl Butyrate

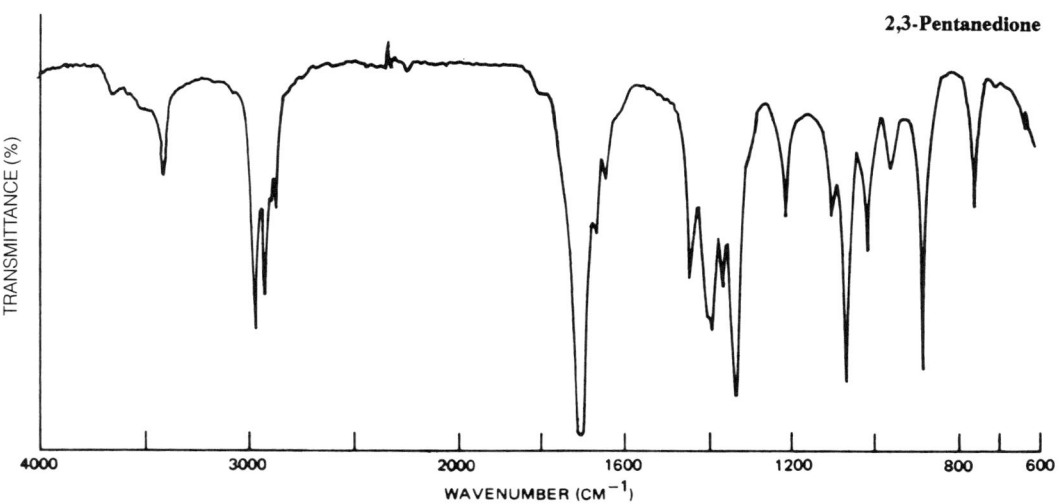

2,3-Pentanedione

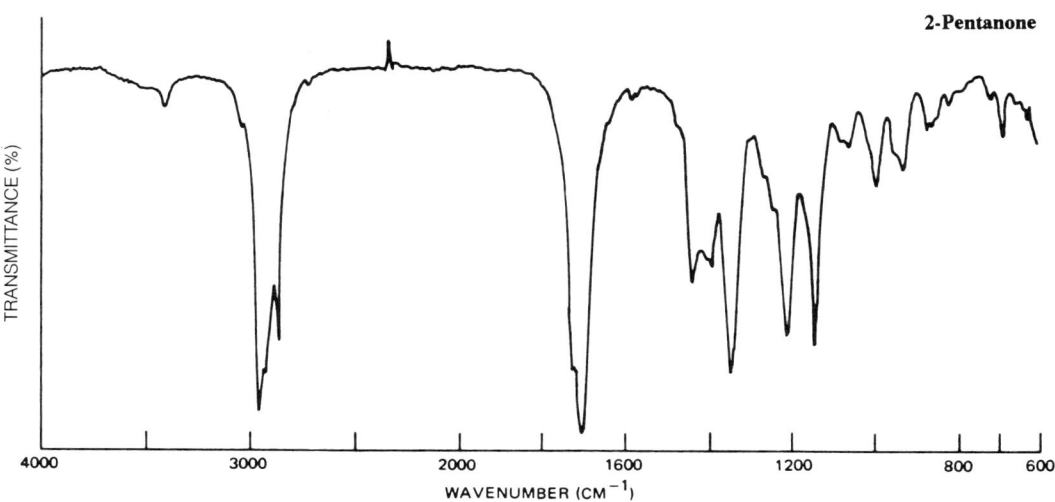

2-Pentanone

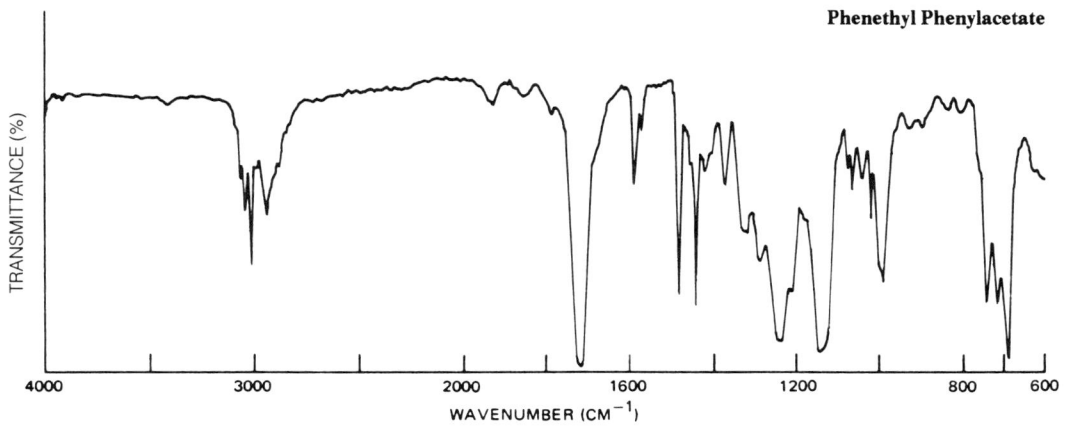

Phenethyl Phenylacetate

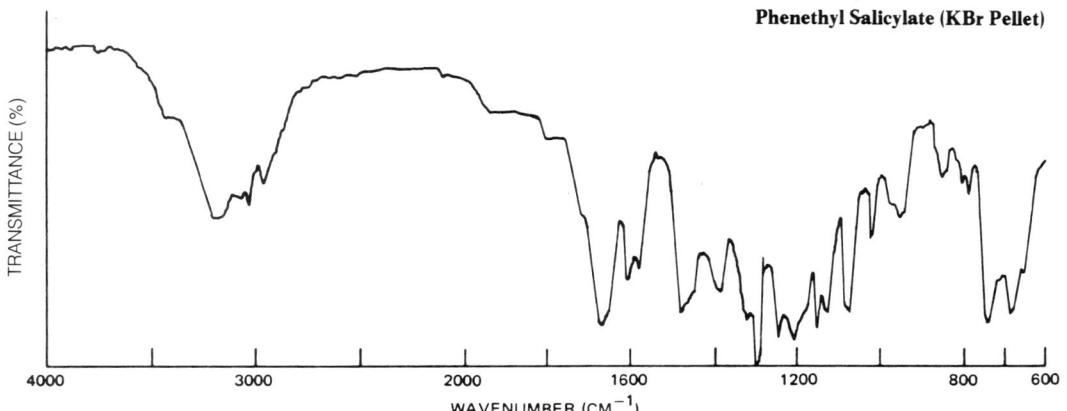
Phenethyl Salicylate (KBr Pellet)

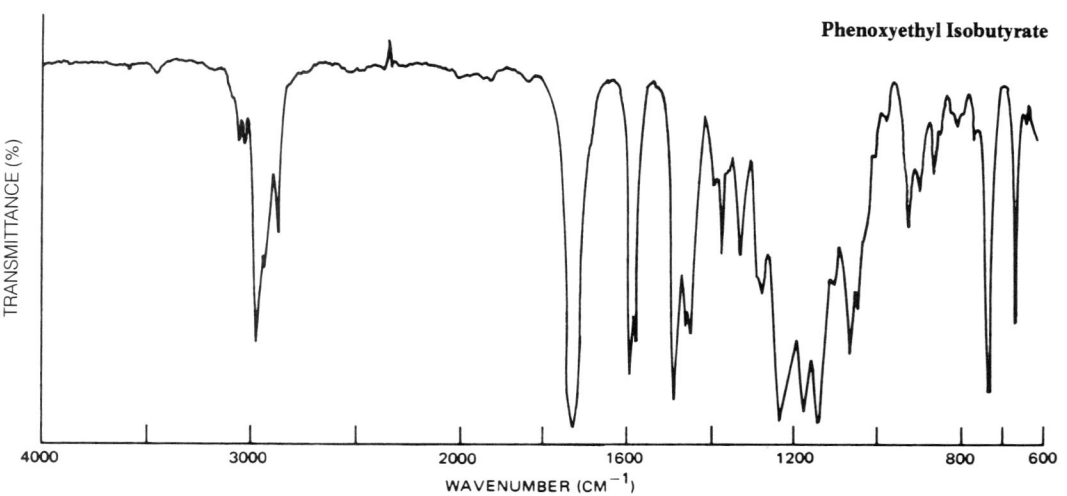
Phenoxyethyl Isobutyrate

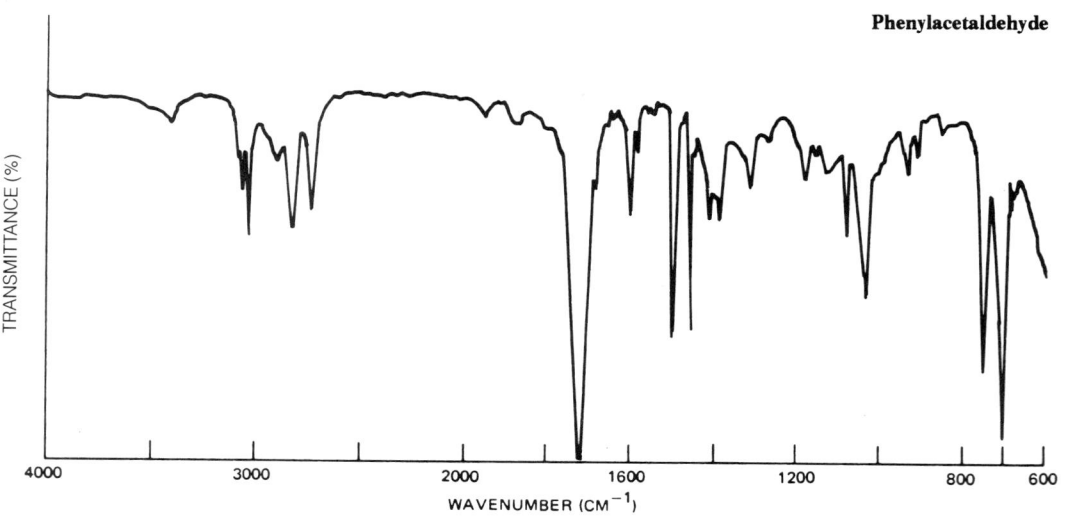
Phenylacetaldehyde

3-Phenylpropionaldehyde

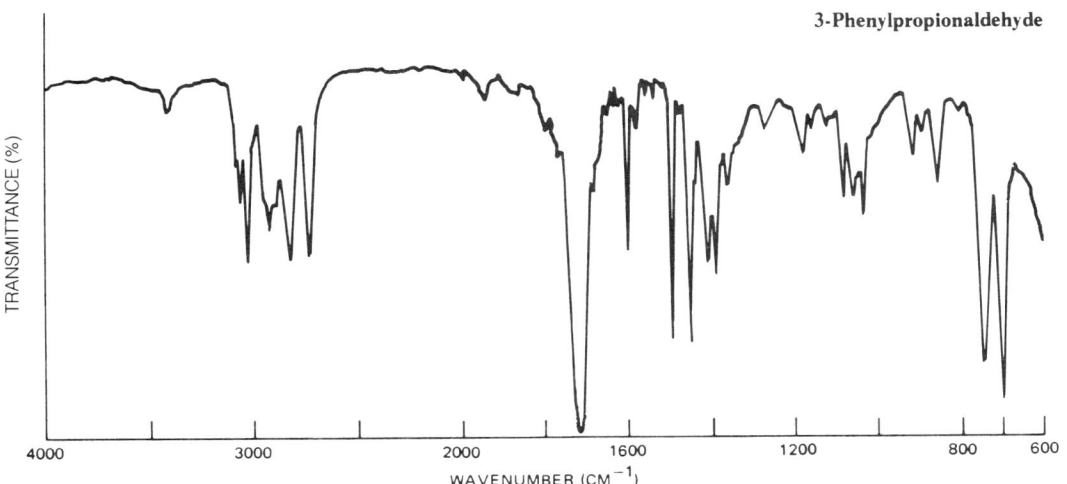

Piperonal (KBr Pellet)

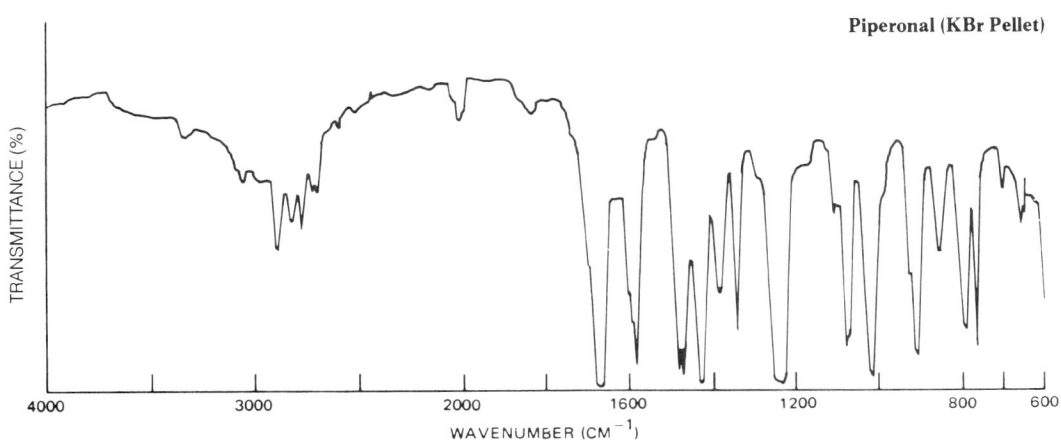

***p*-Propyl Anisole**

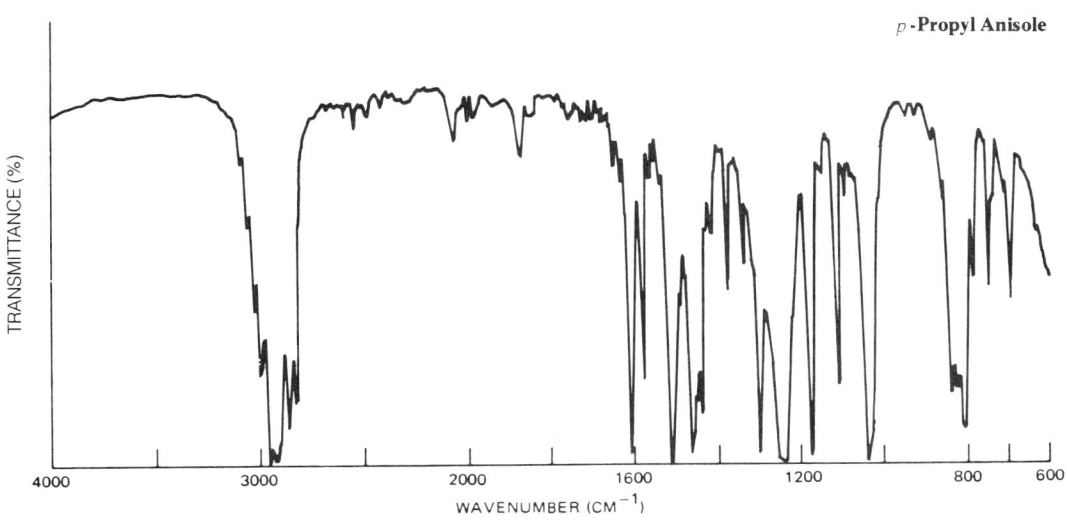

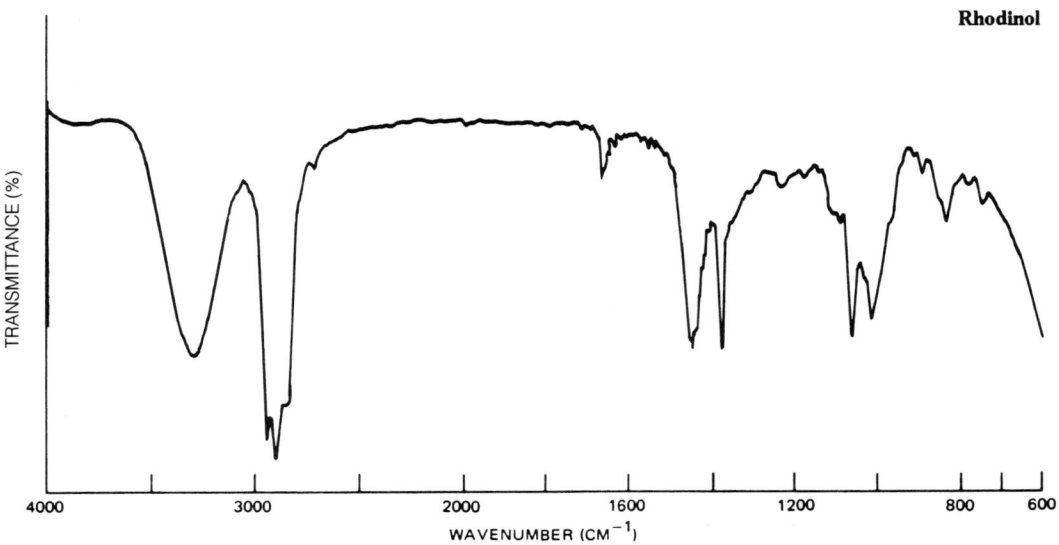

Rhodinol

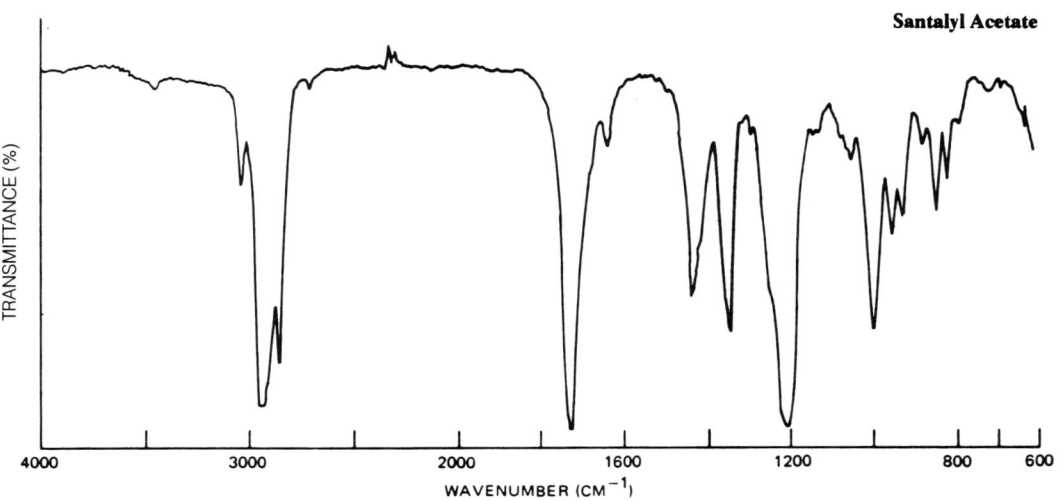

Santalyl Acetate

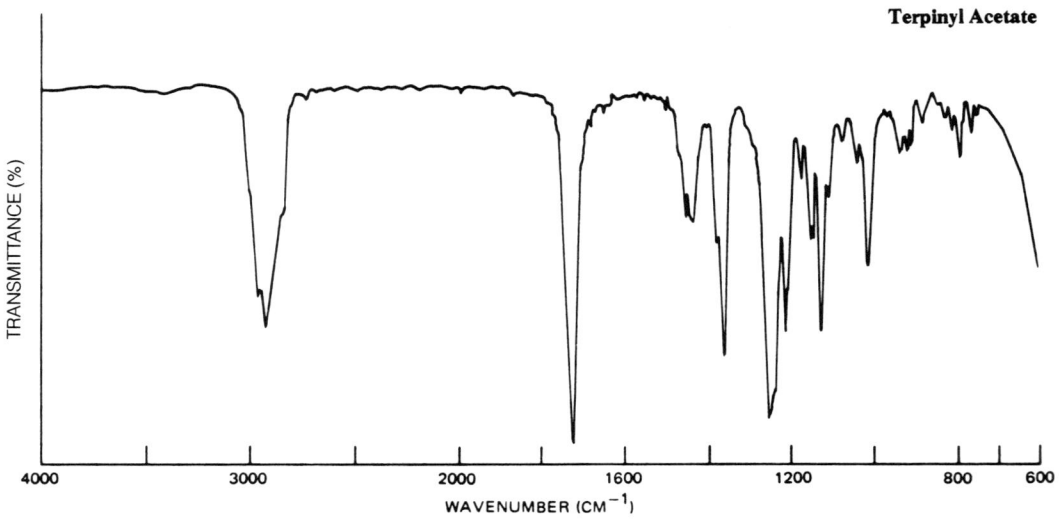

Terpinyl Acetate

Tributyrin

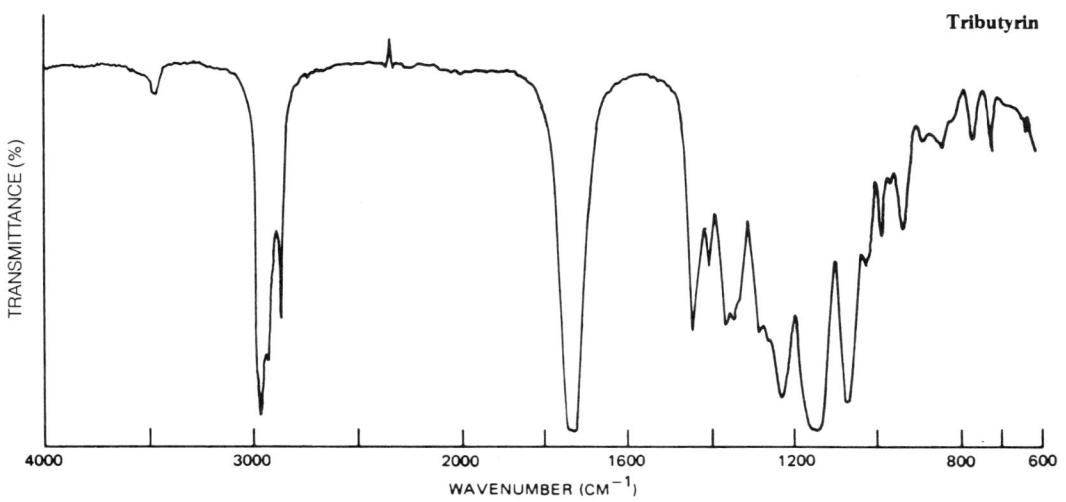

2,4,5-Trimethyl δ-3-Oxazoline

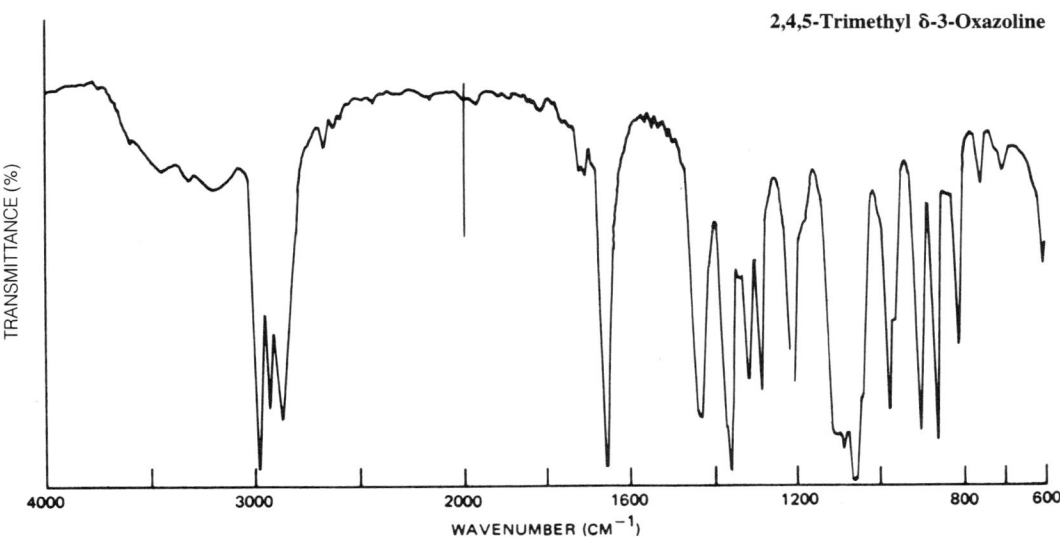

Undecanal

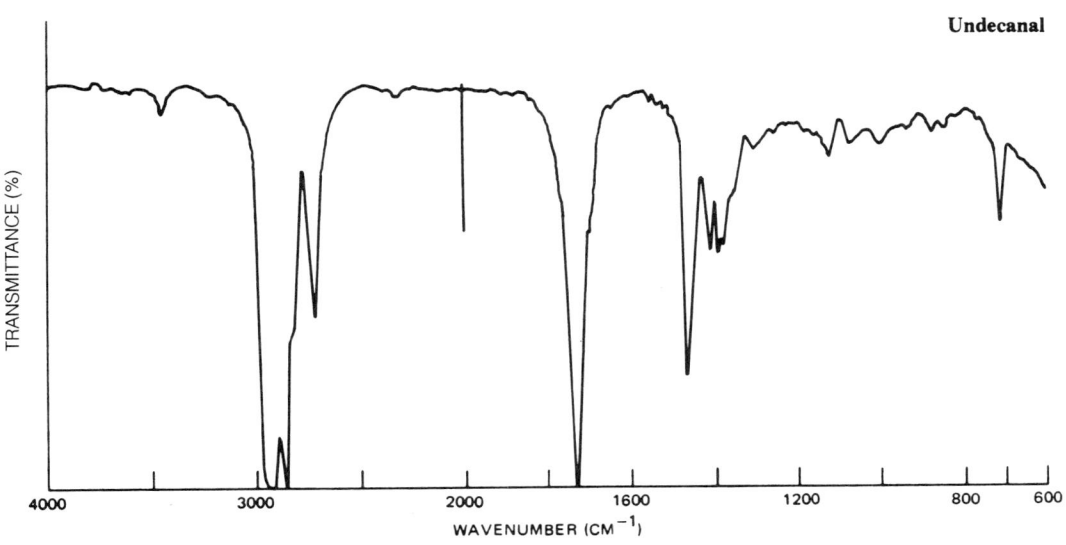

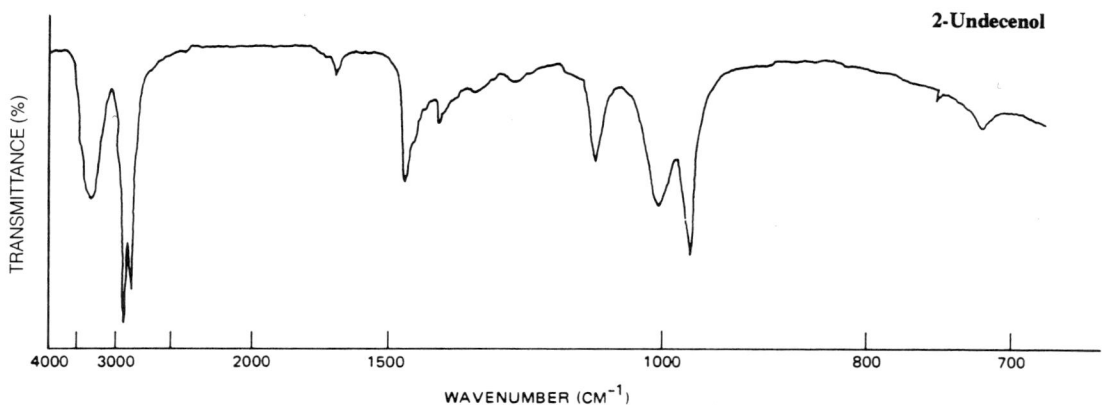

2-Undecenol

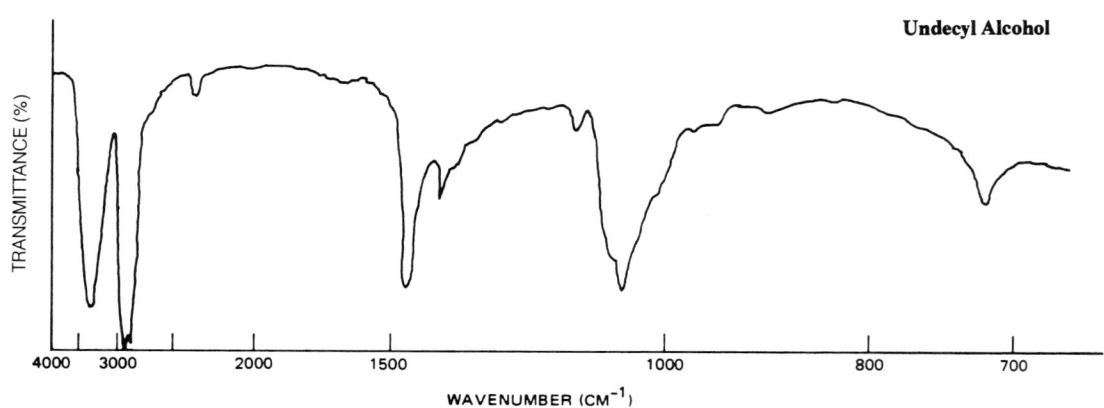

Undecyl Alcohol

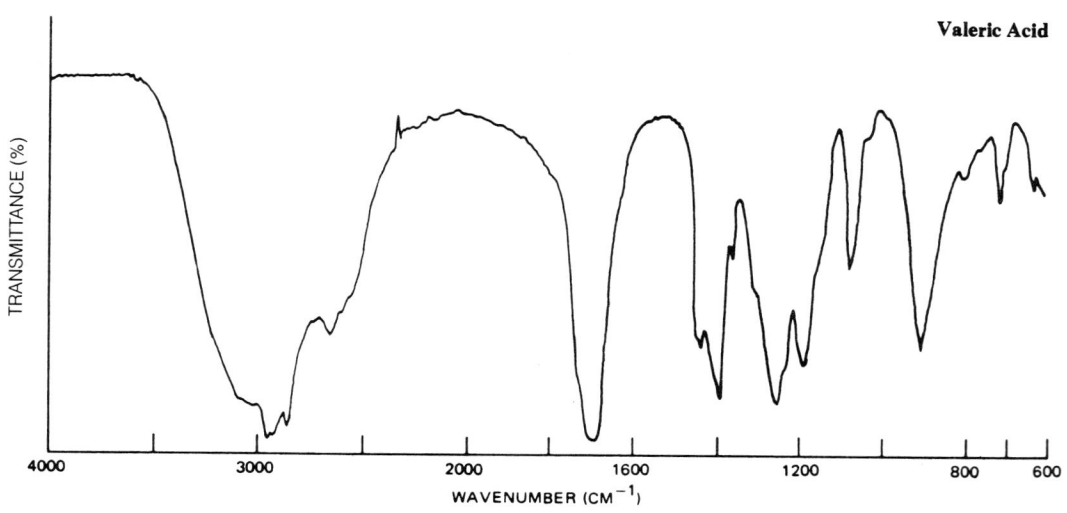

Valeric Acid

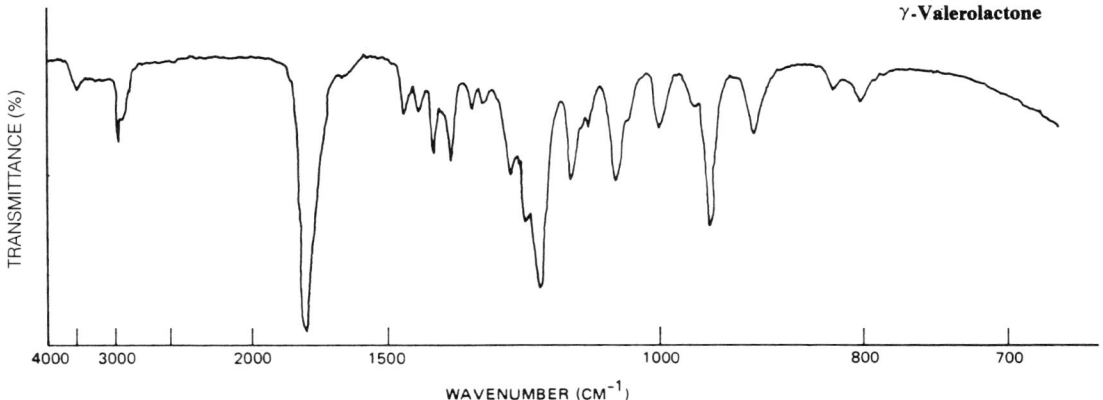

γ-Valerolactone

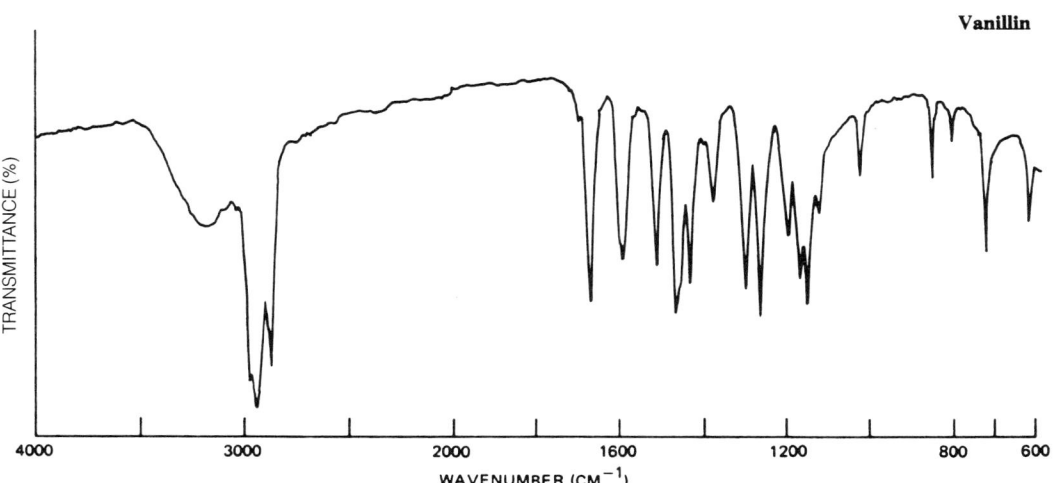

Vanillin

SERIES B-3

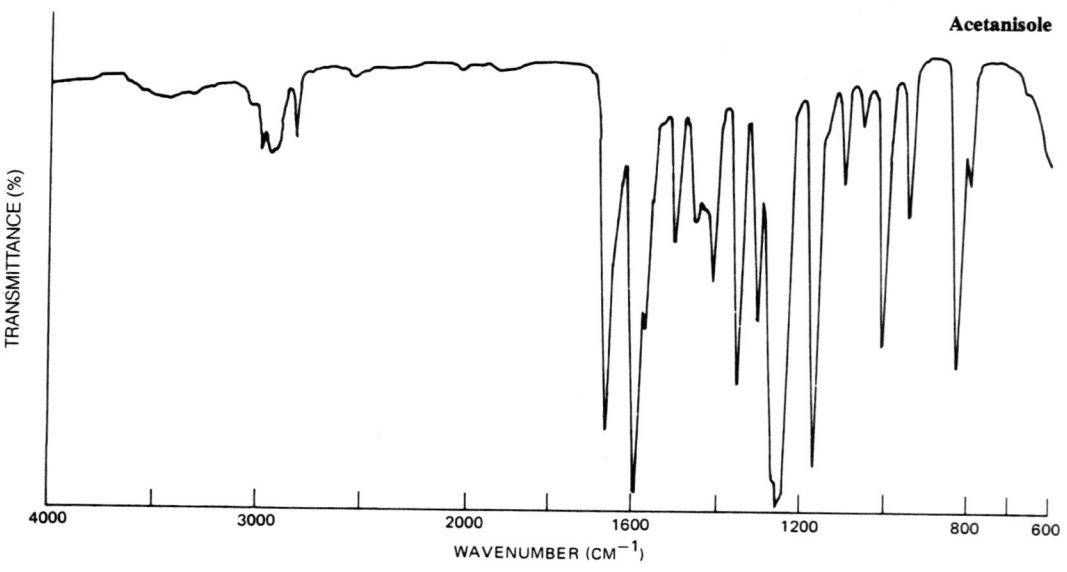

Acetanisole

Acetoin

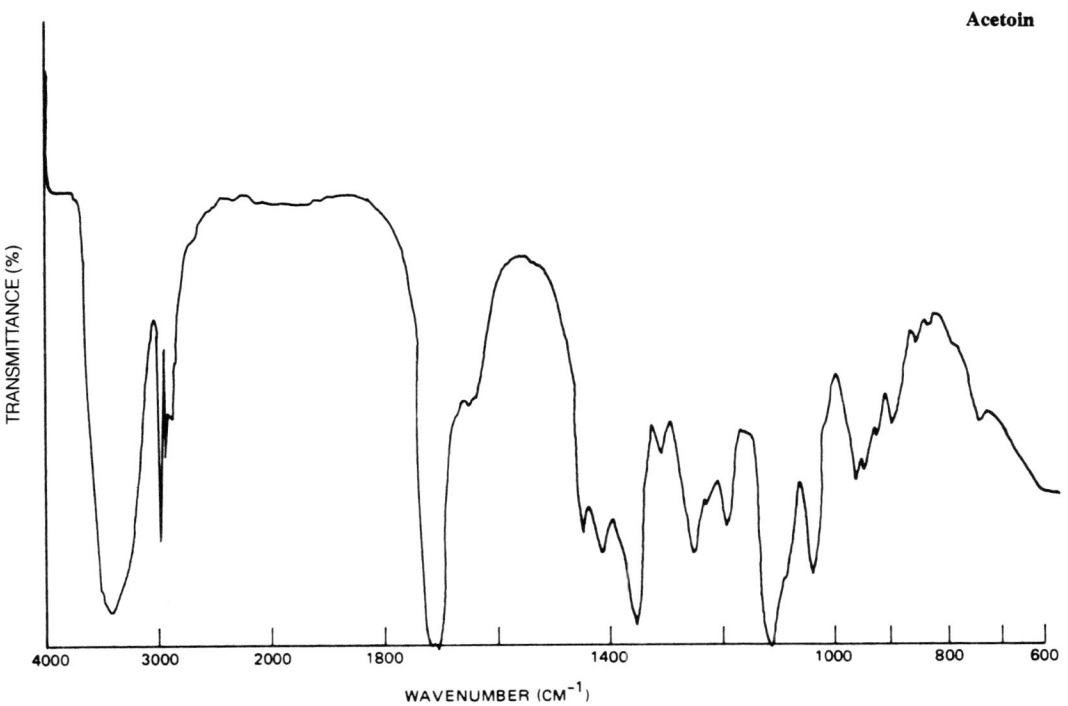

Allyl Isothiocyanate

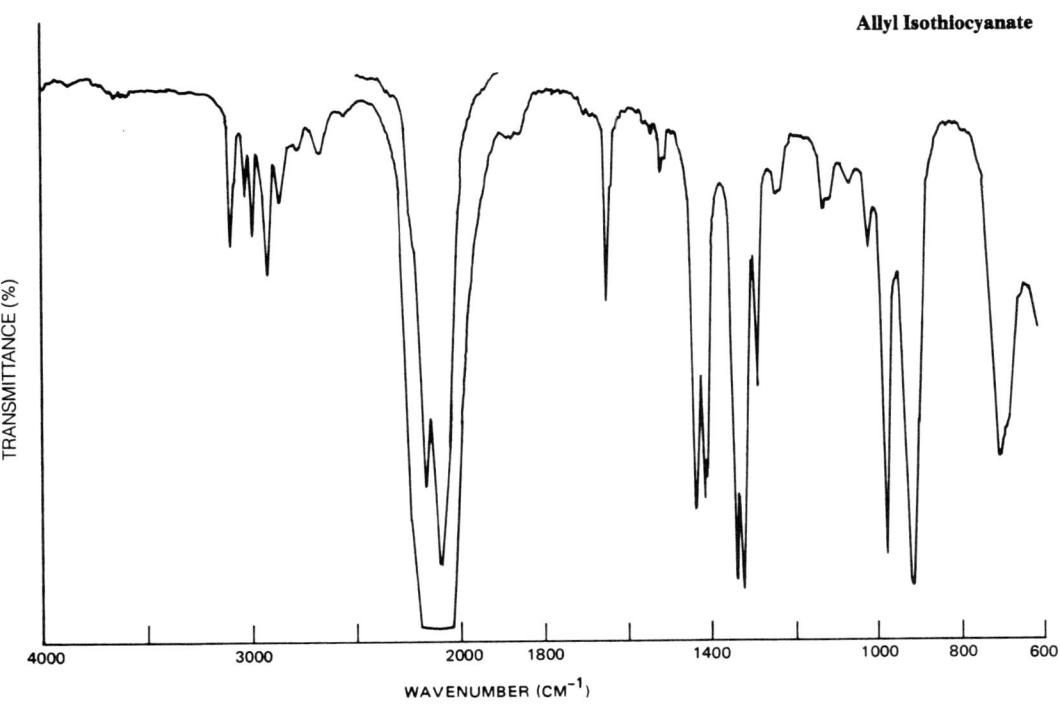

Acetoin

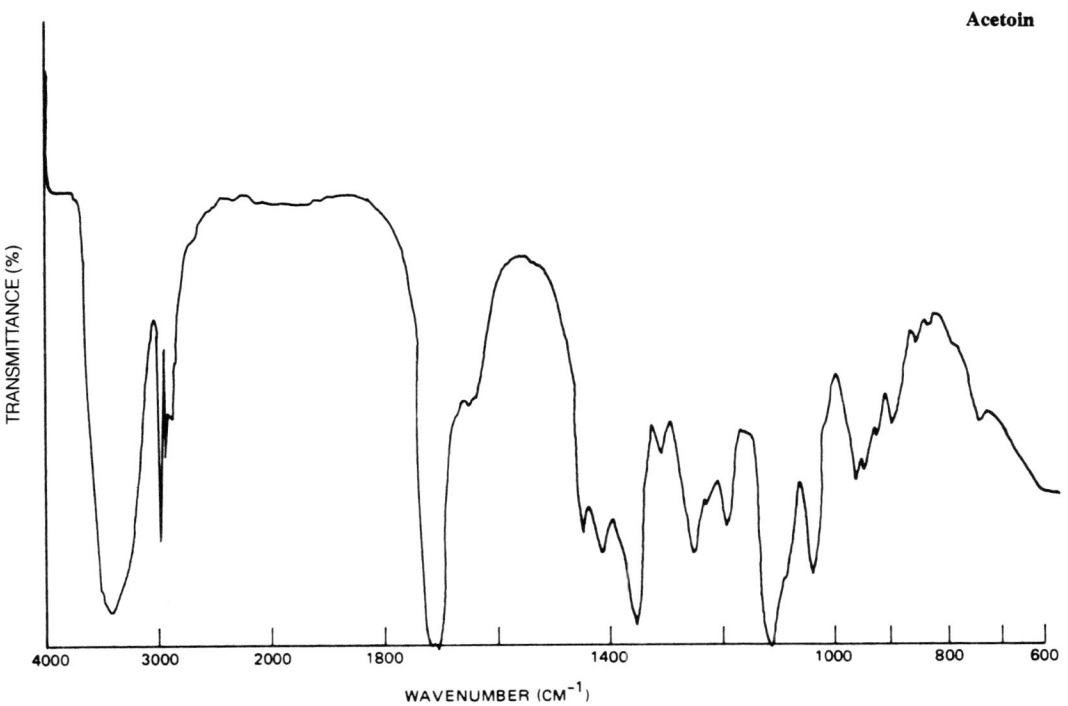

Allyl Isothiocyanate

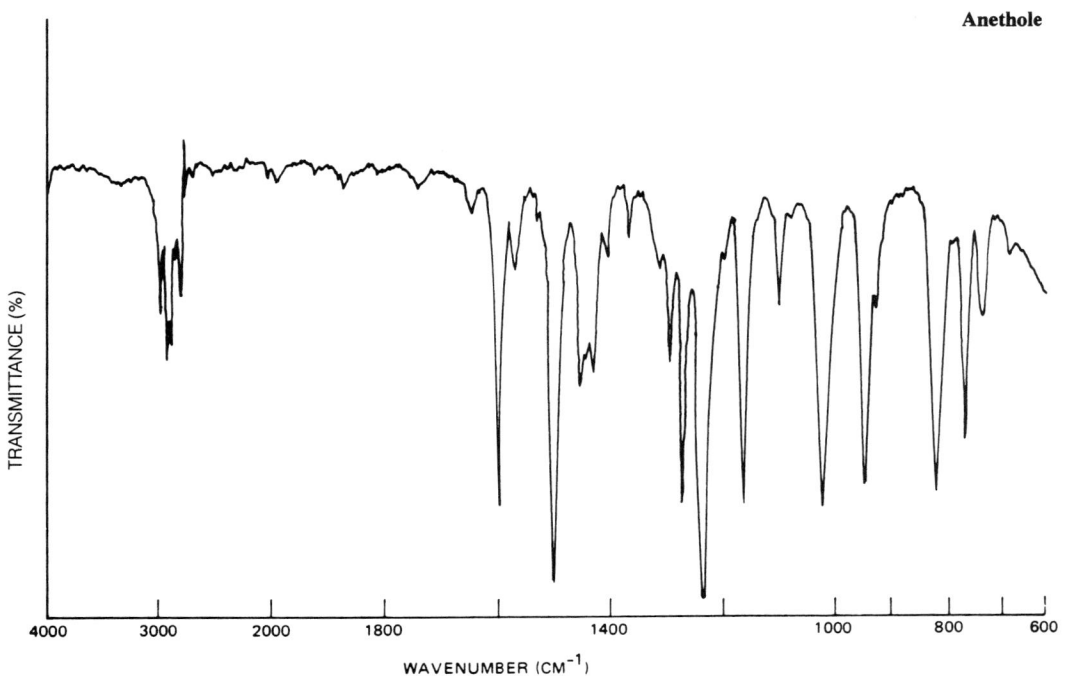

Anethole

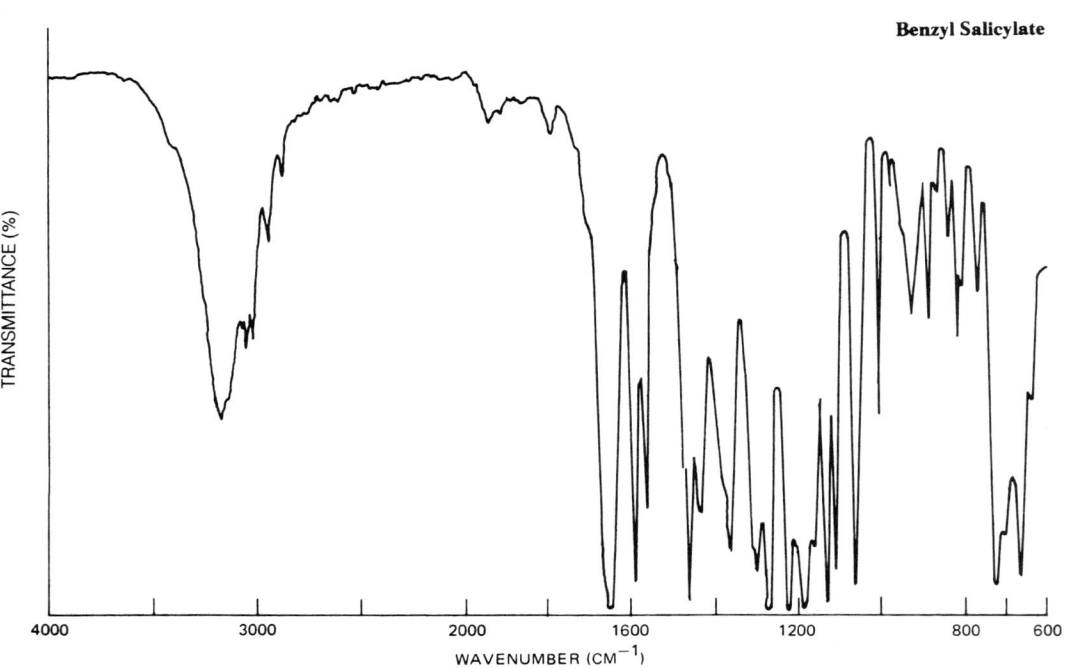

Benzyl Salicylate

Citronellol

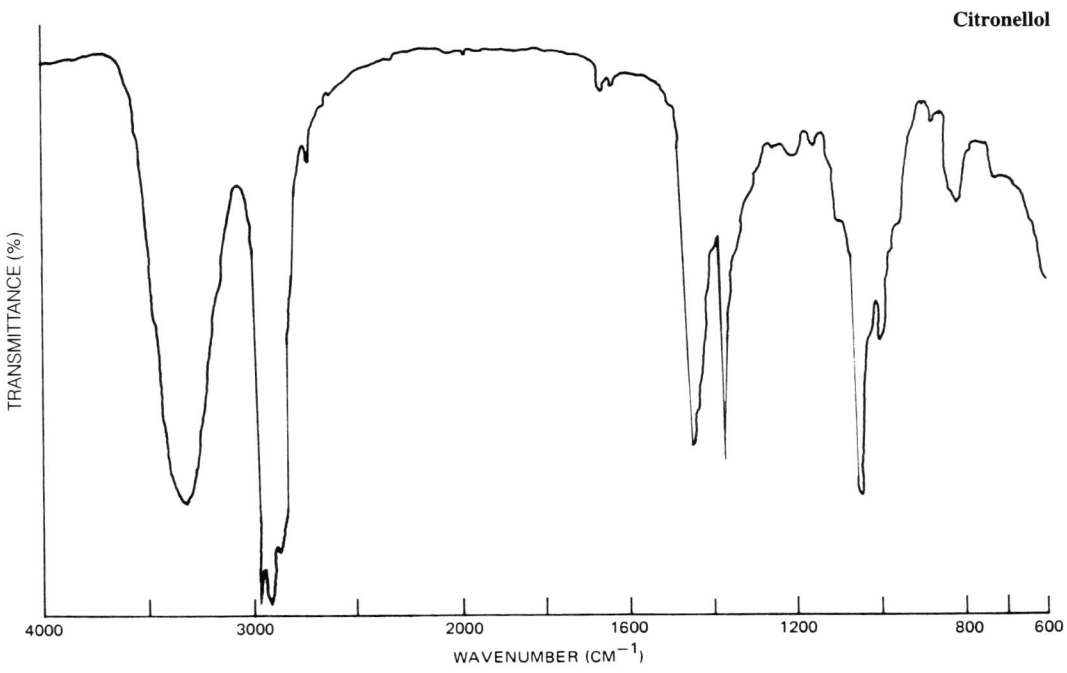

Citronellyl Formate

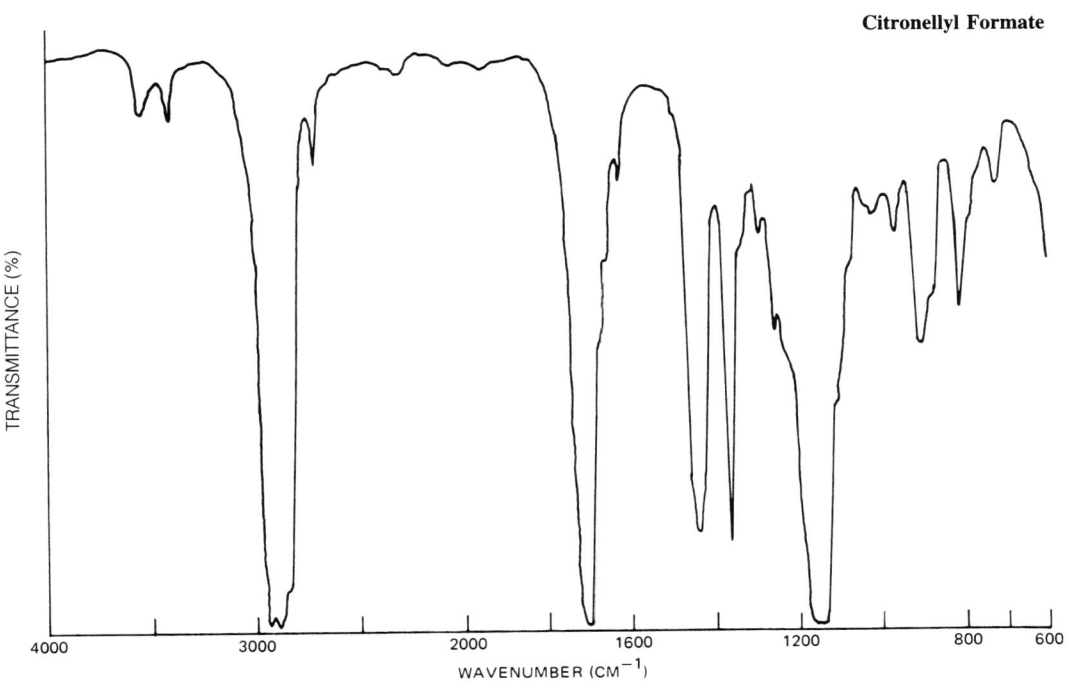

p-Cresyl Acetate

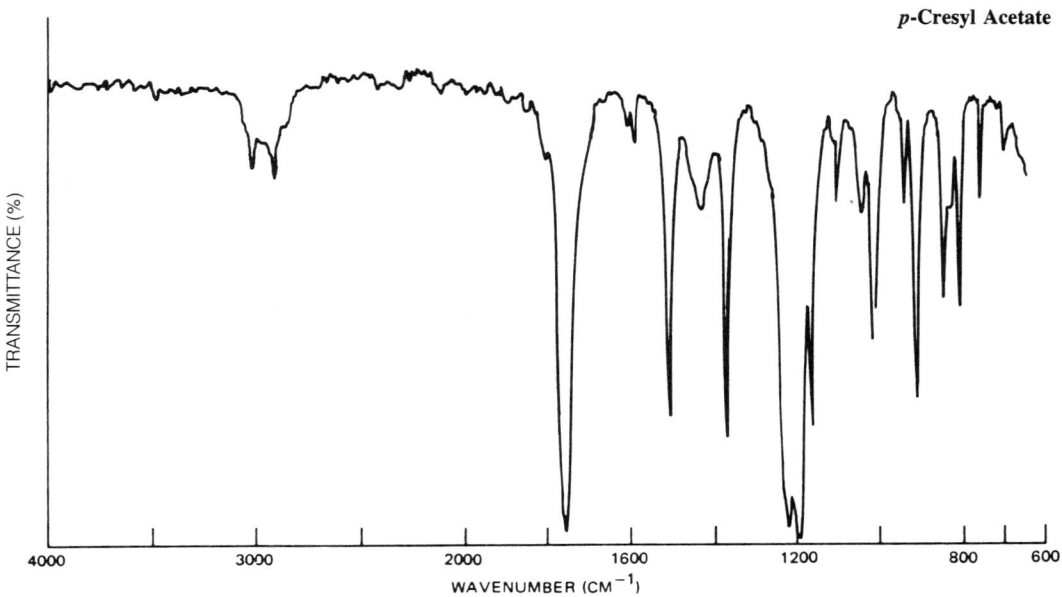

Diethyl Malonate

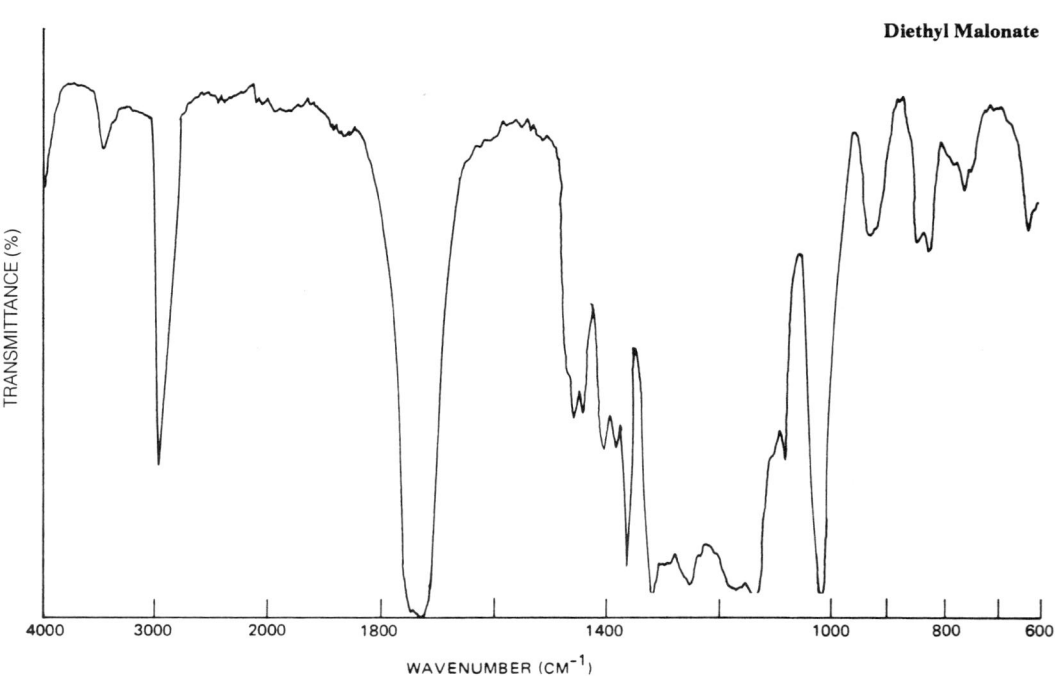

Diethyl Succinate

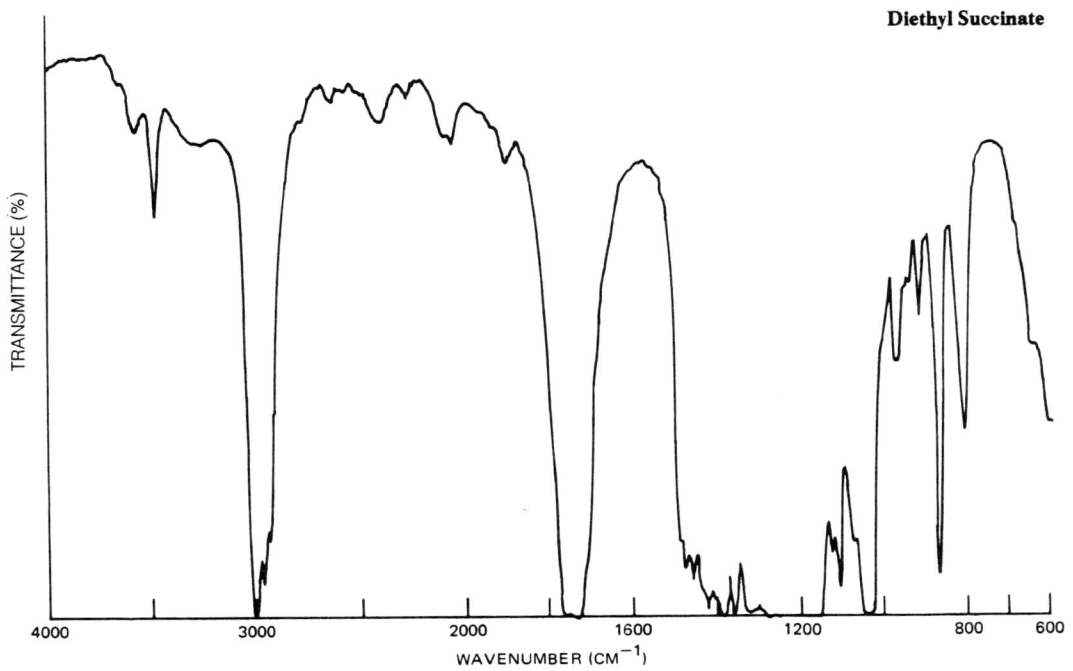

Ethyl Acetate

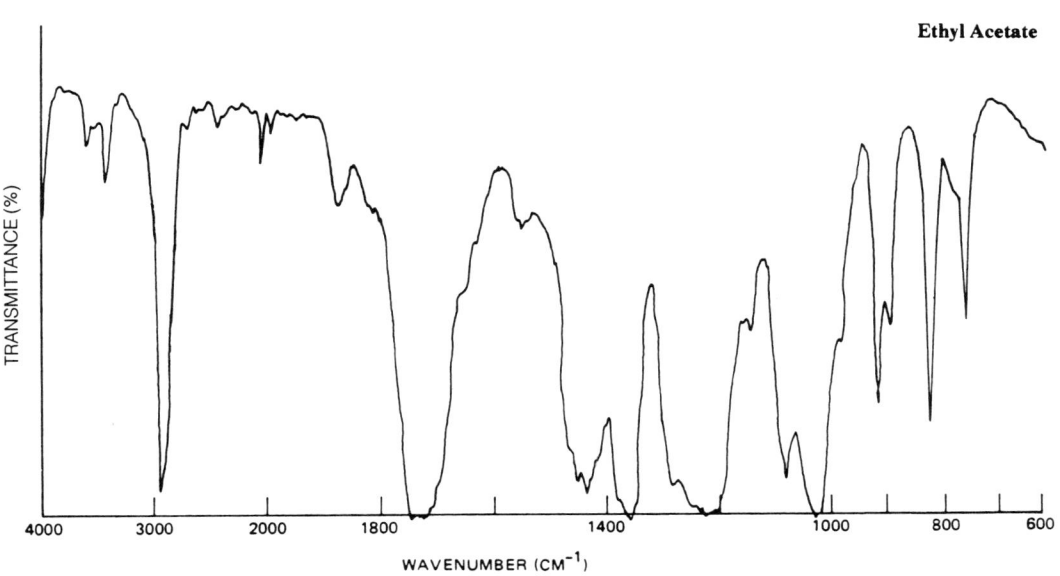

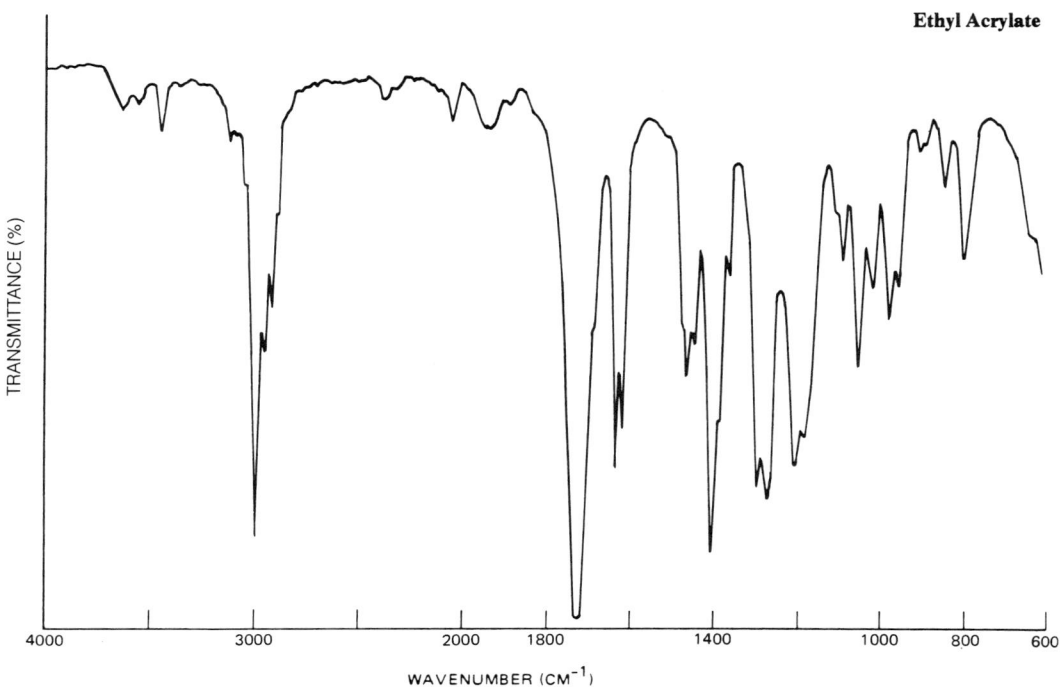

Ethyl Acrylate

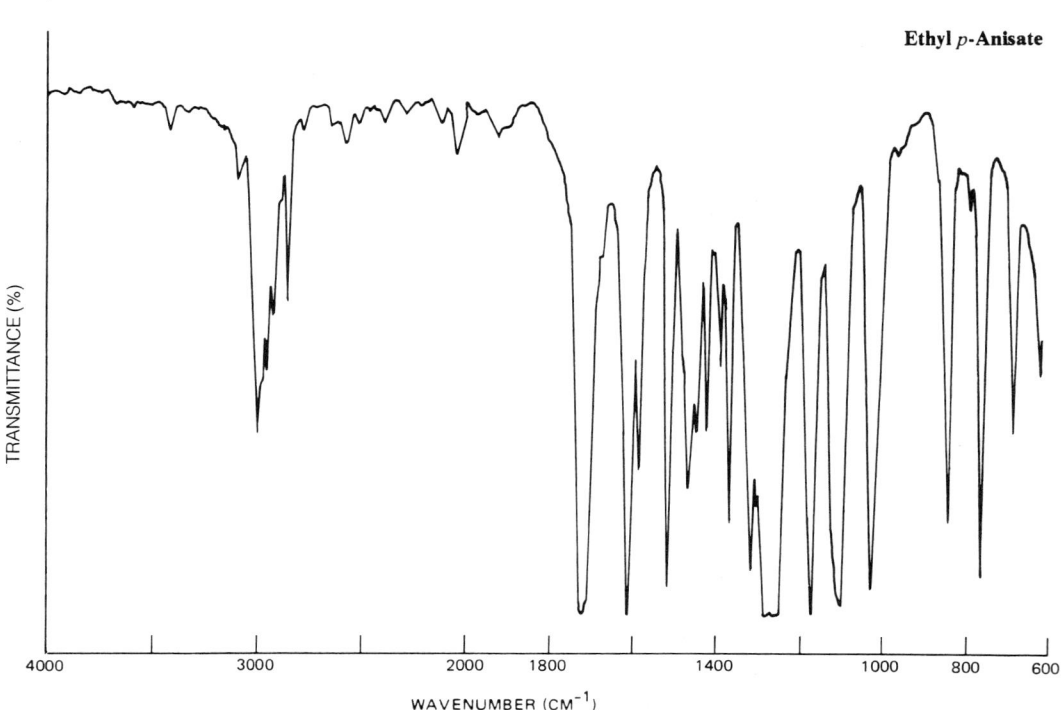

Ethyl p-Anisate

Ethyl Anthranilate

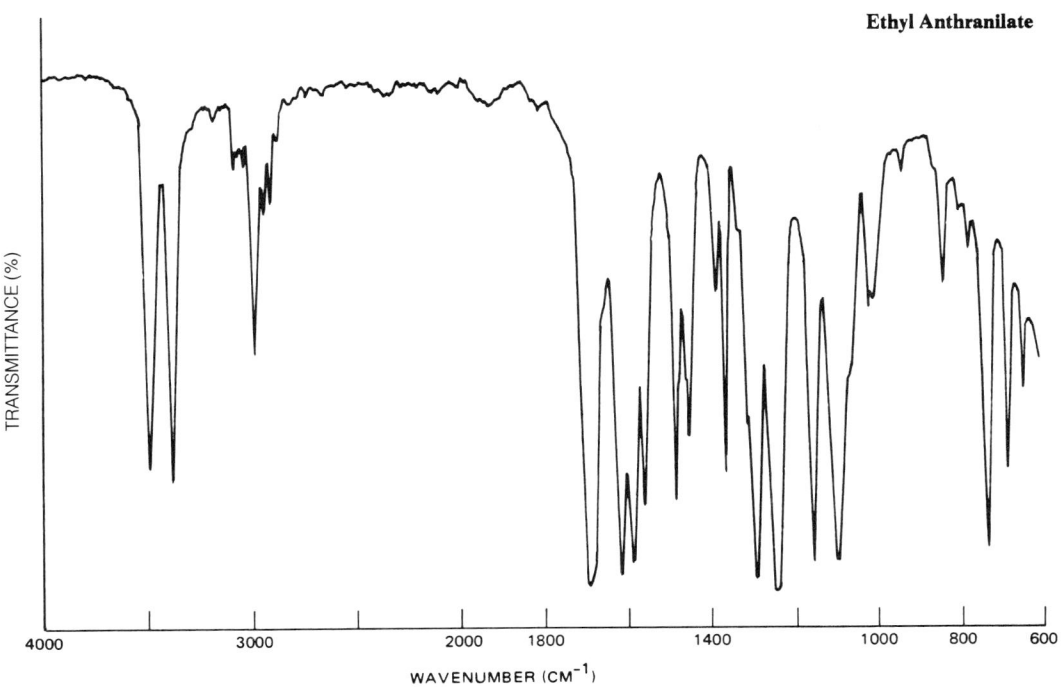

Ethyl Decanoate

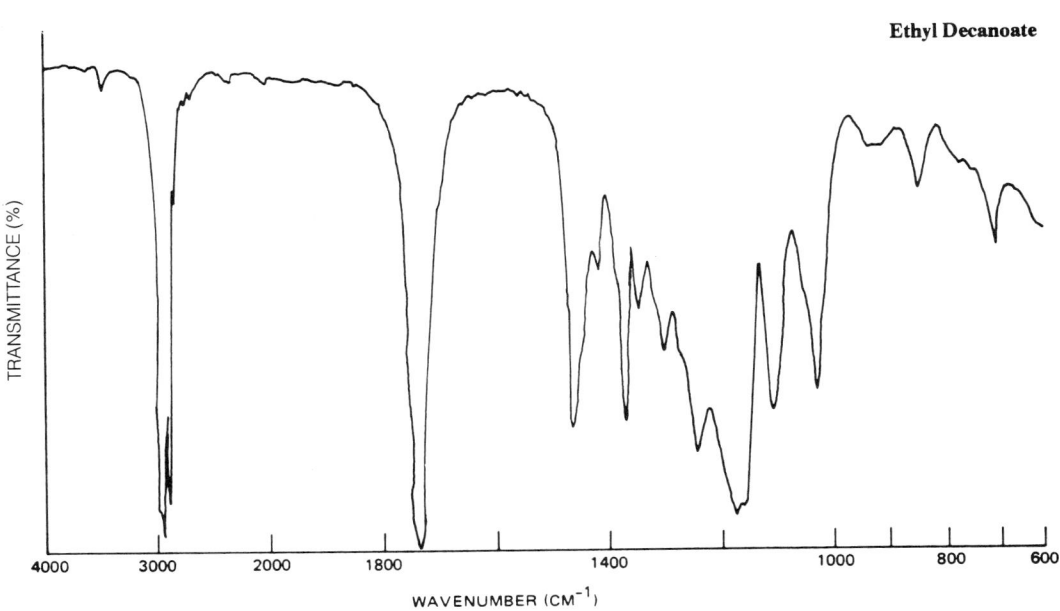

Ethyl Lactate

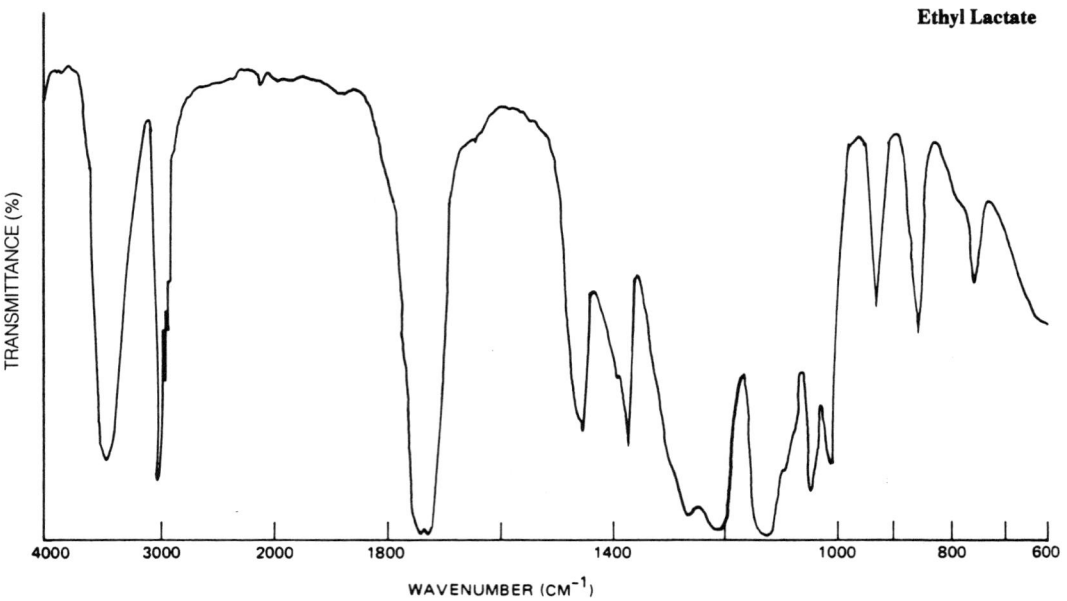

Ethyl Phenylacetate

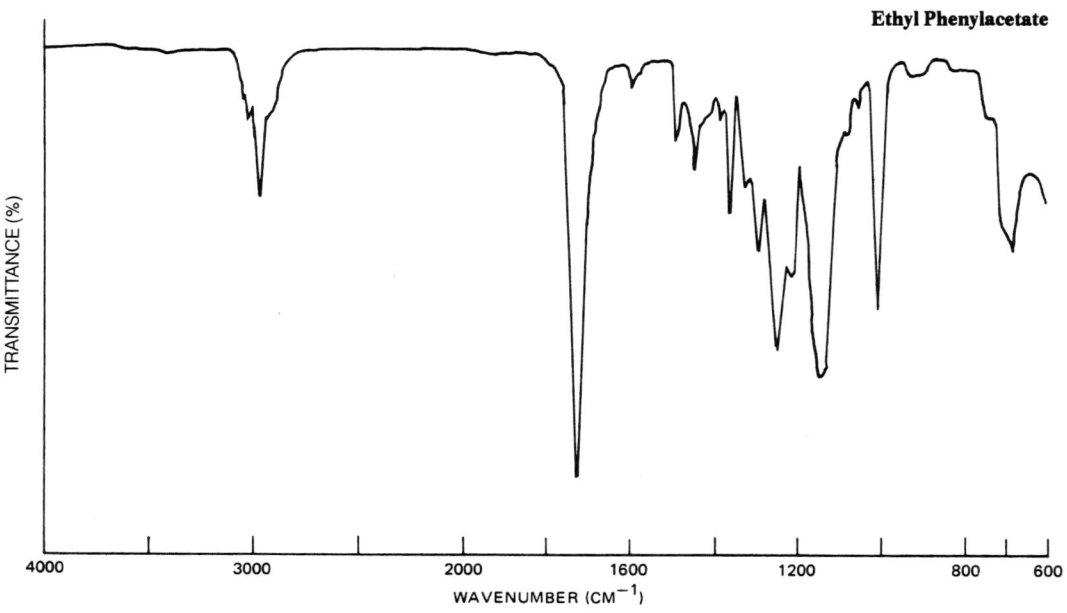

Ethyl Phenylglycidate

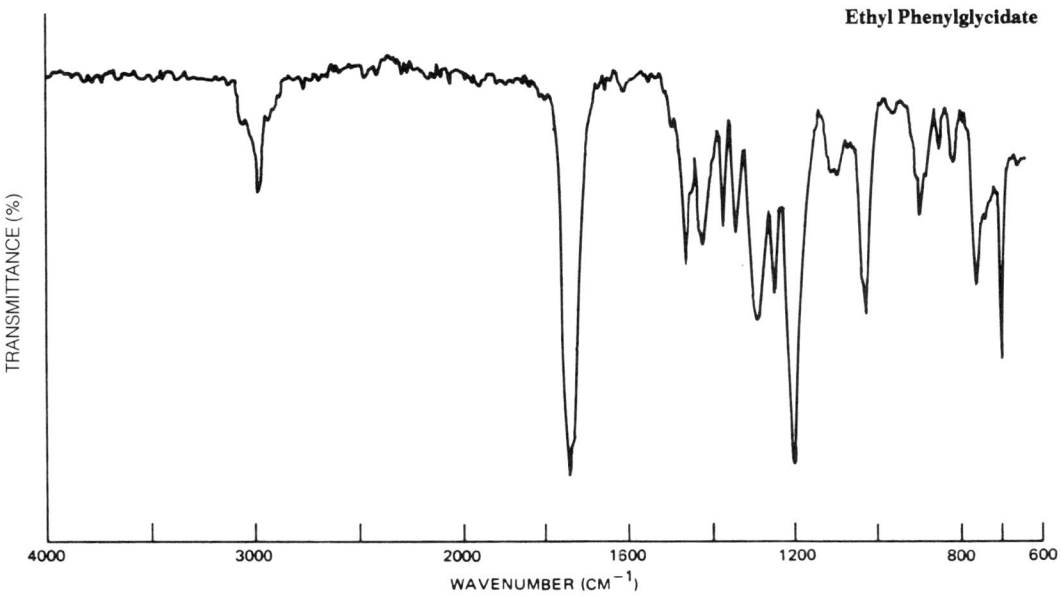

Eugenol

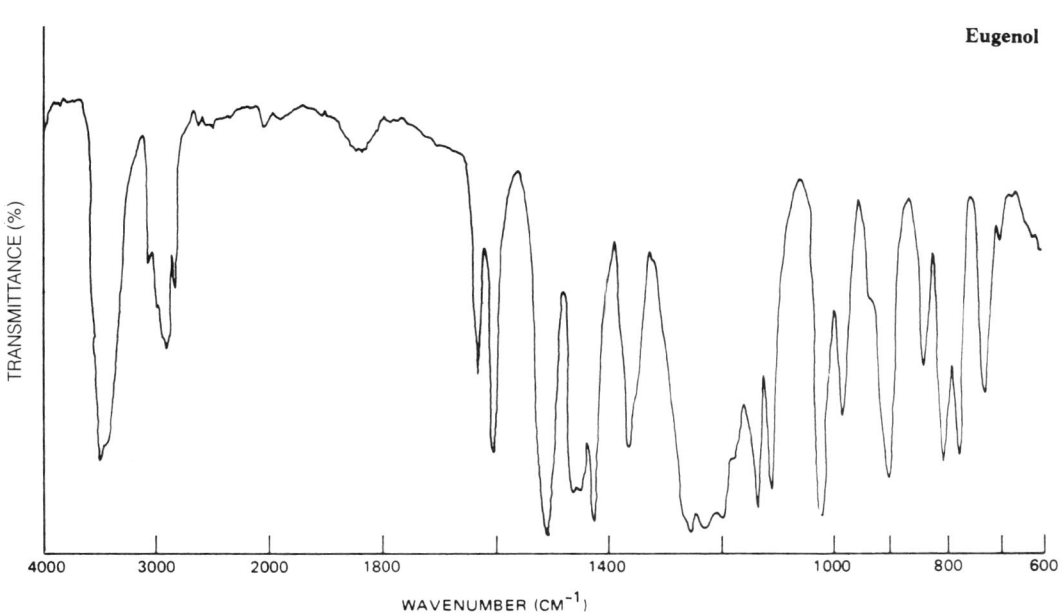

Farnesol

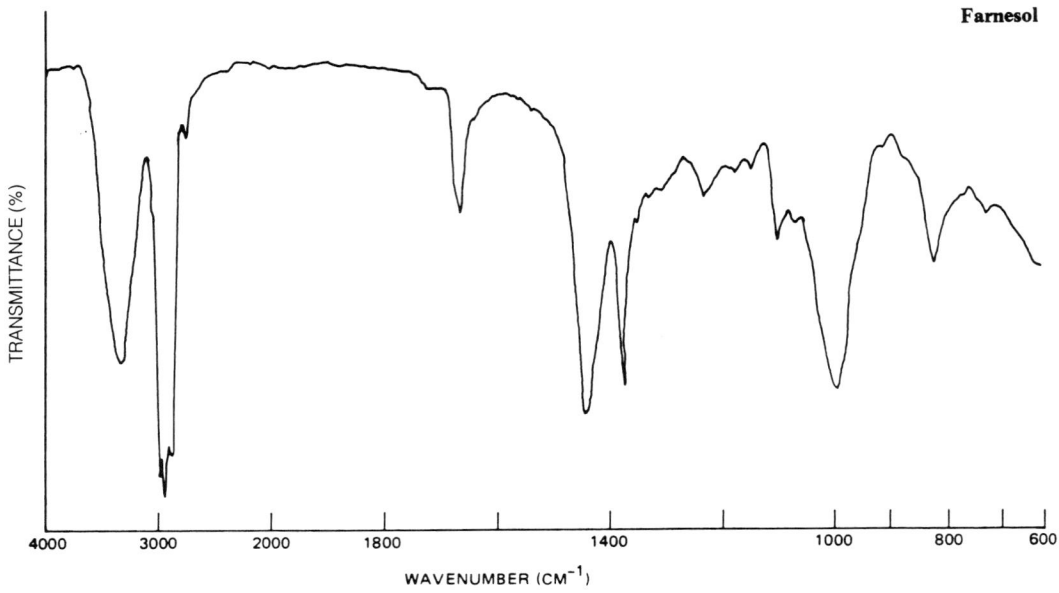

Geranyl Benzoate

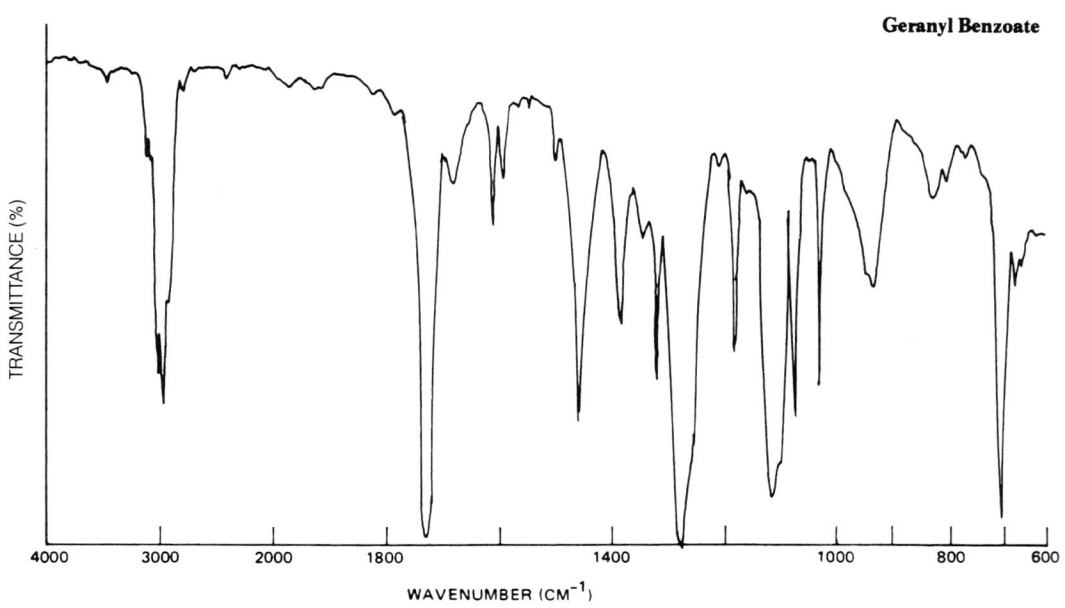

Geranyl Butyrate

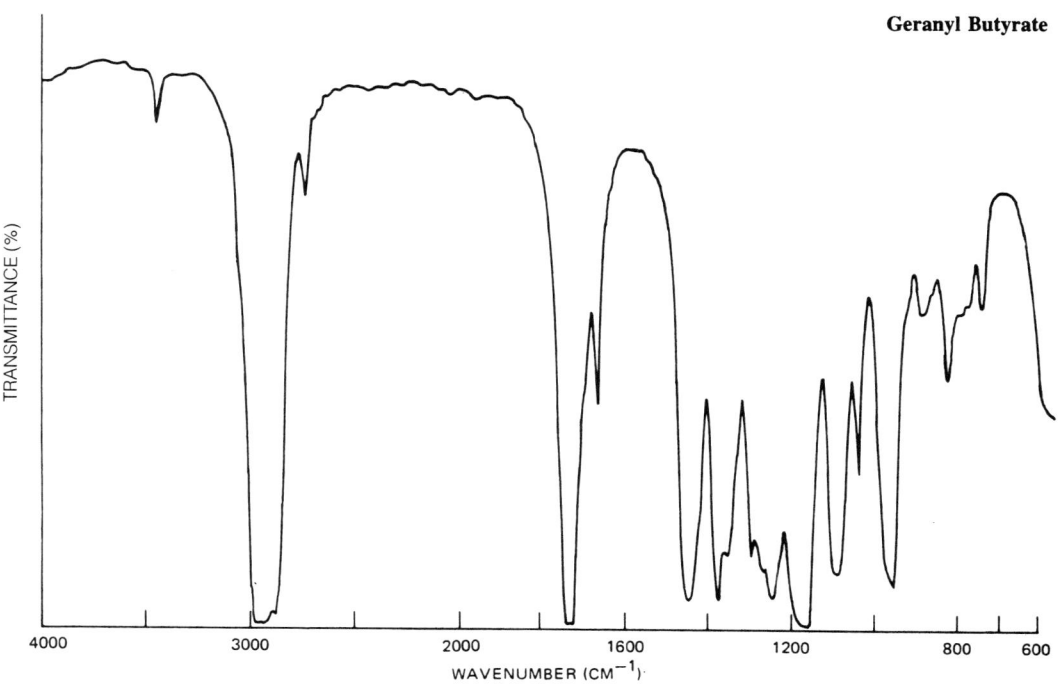

Heptanal

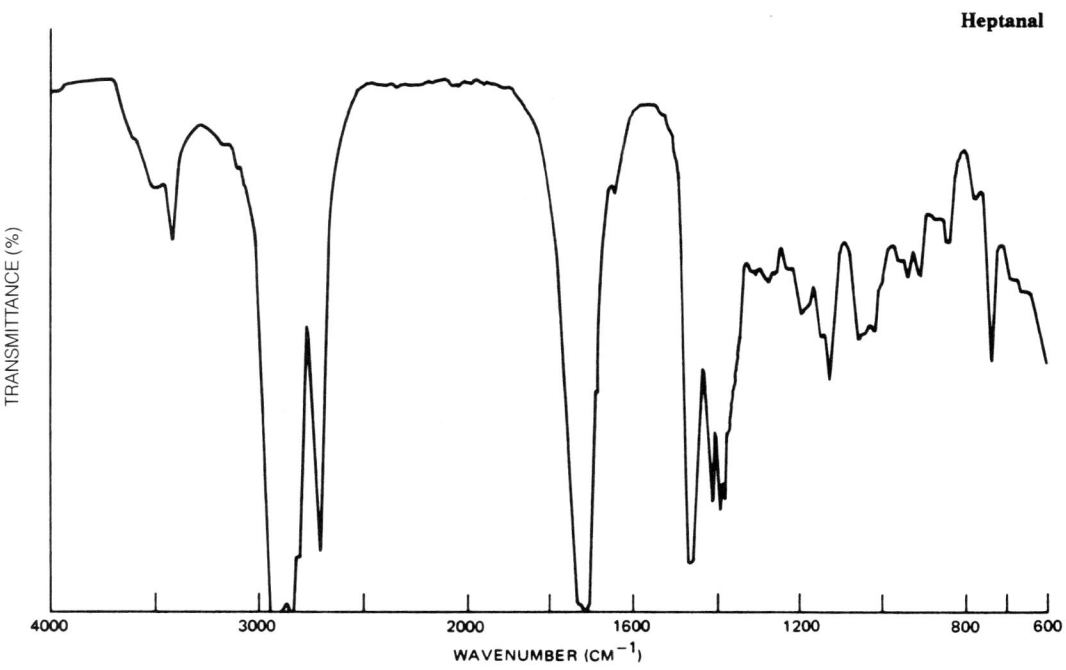

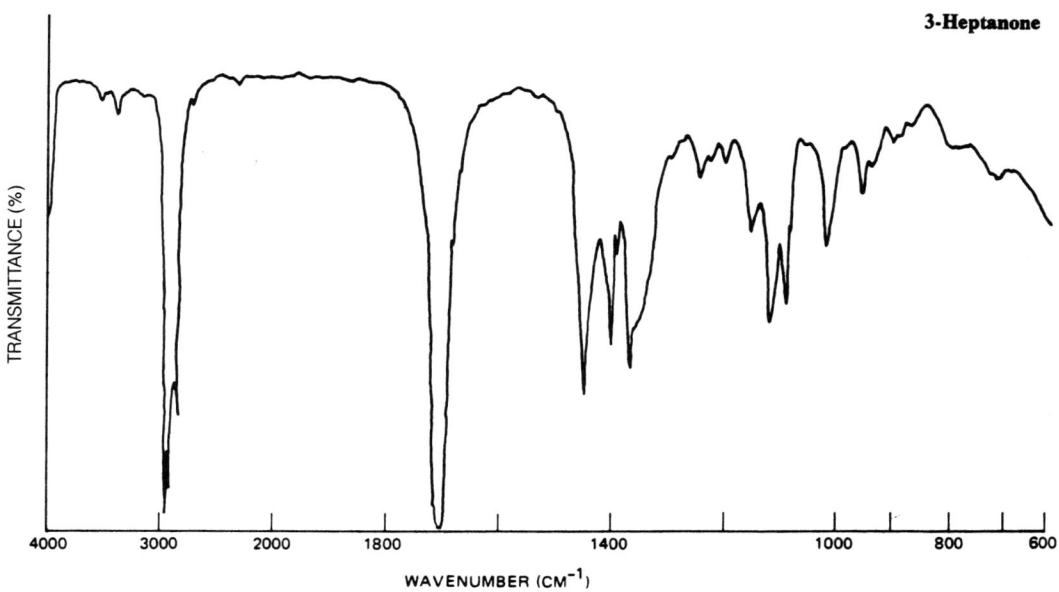

3-Heptanone

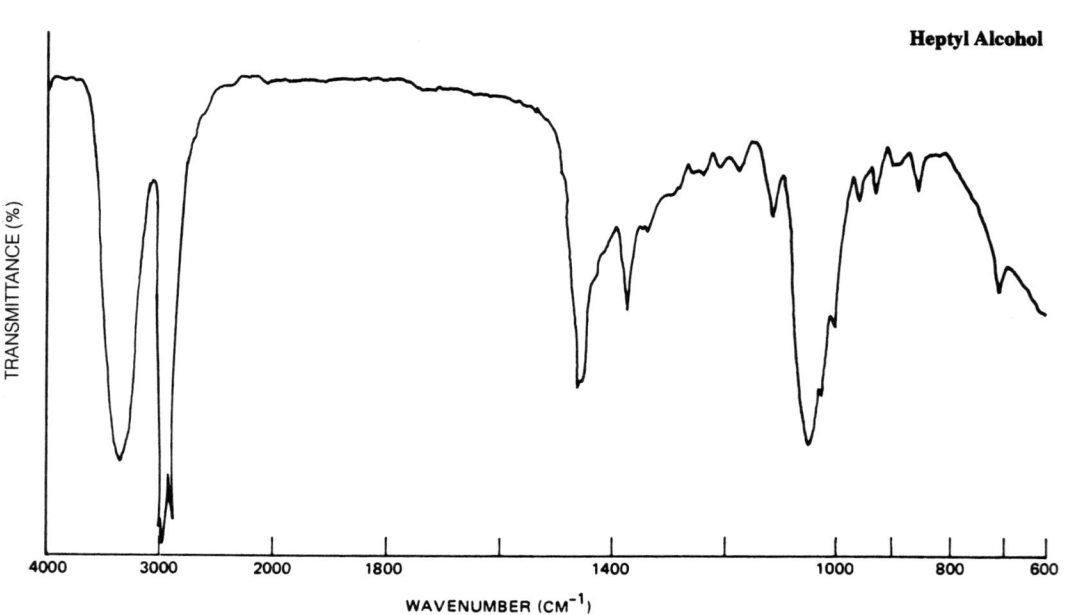

Heptyl Alcohol

Hexyl Alcohol

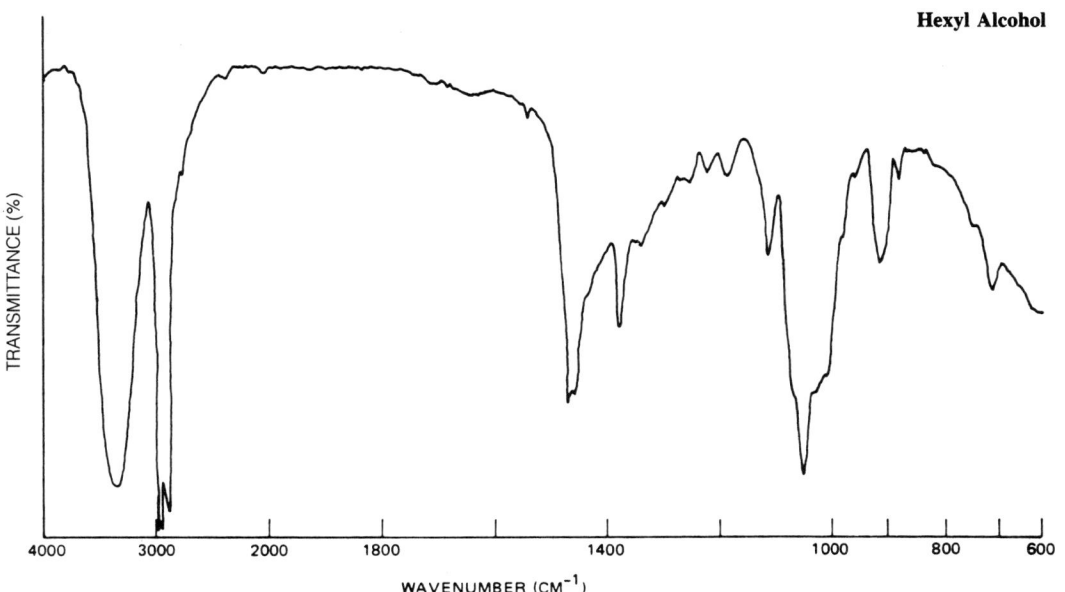

Hydroxycitronellal

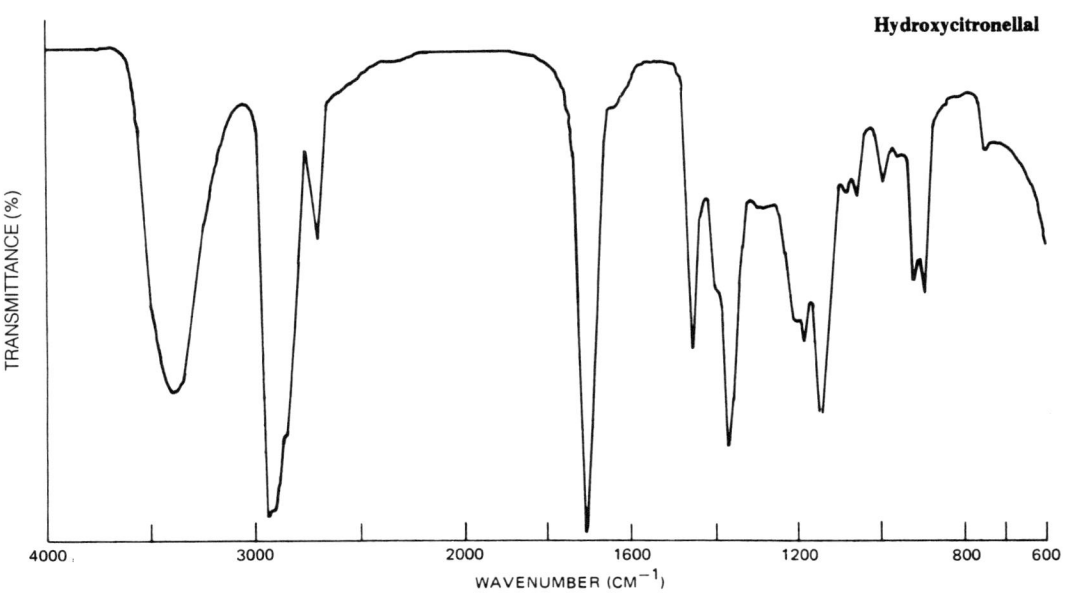

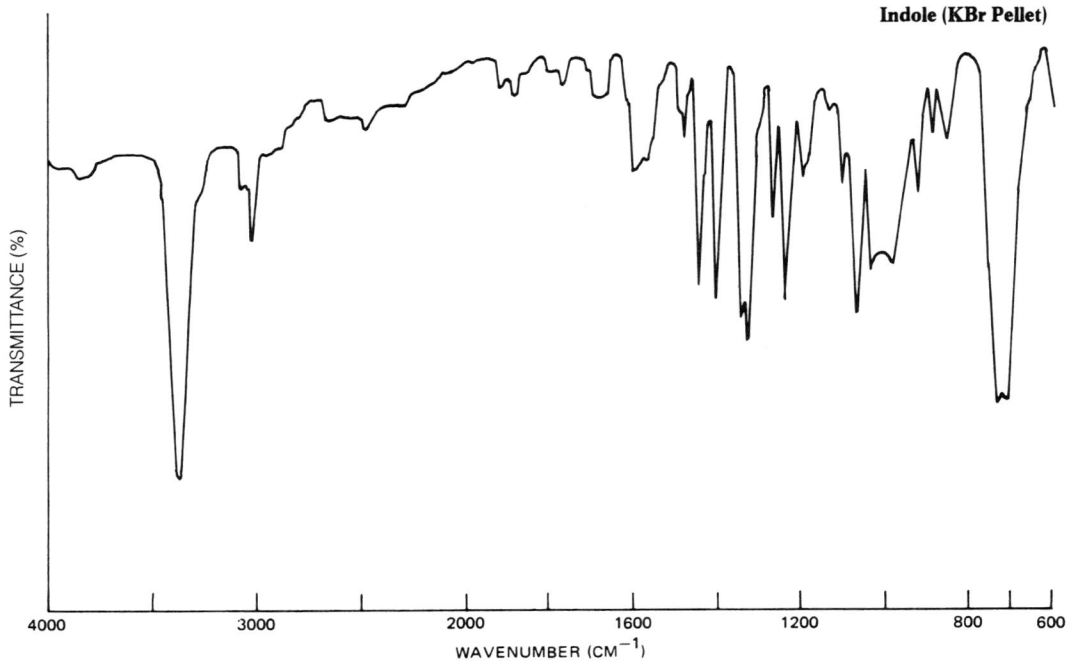

Indole (KBr Pellet)

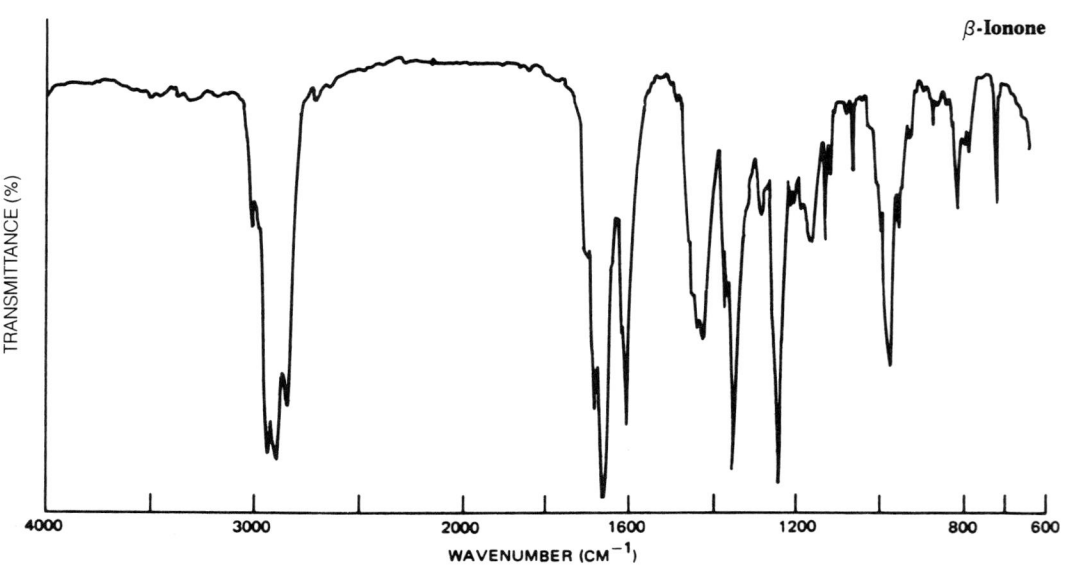

β-Ionone

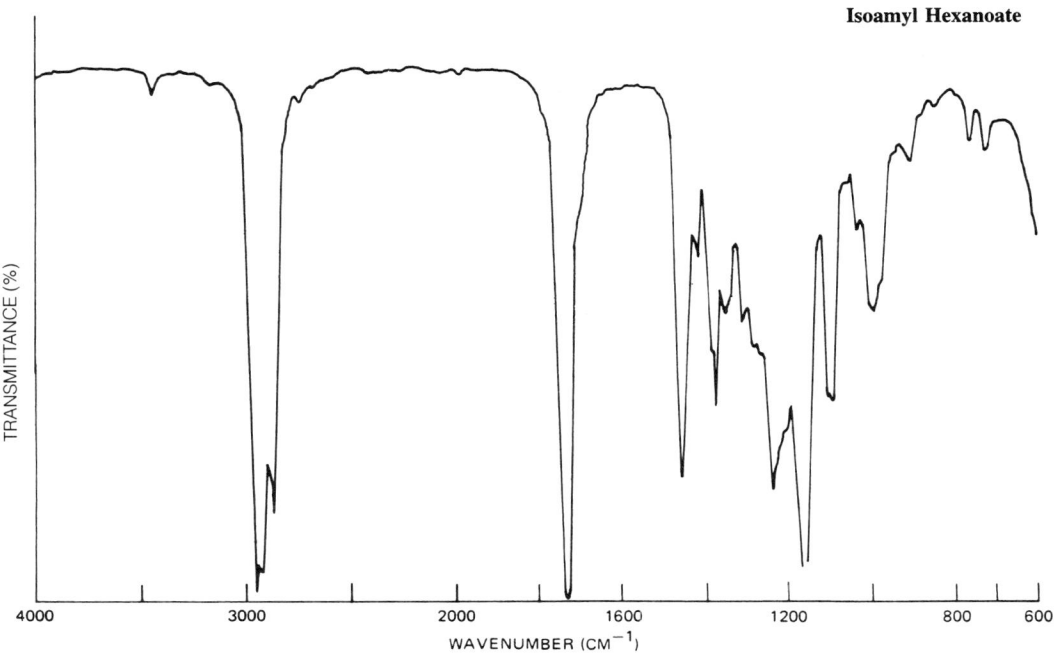

Isoamyl Hexanoate

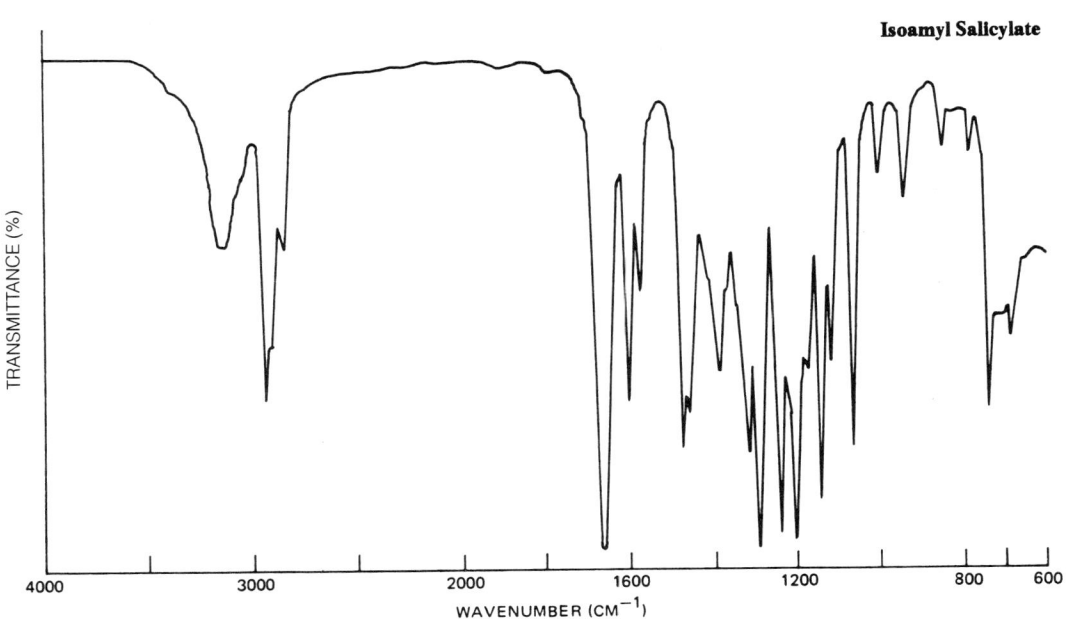

Isoamyl Salicylate

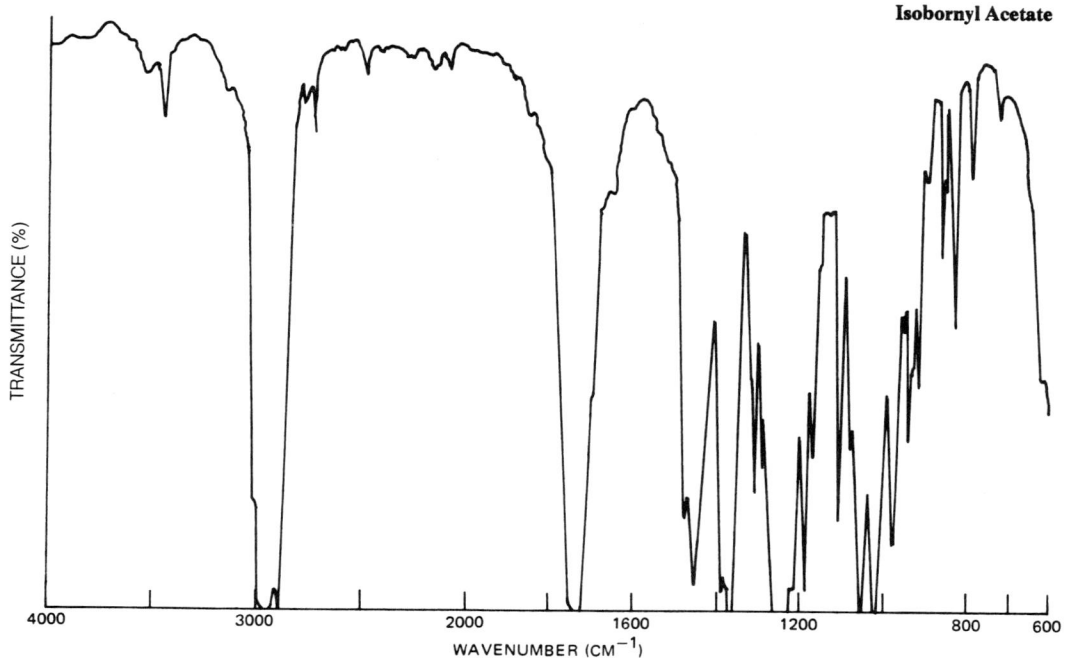

Isobornyl Acetate

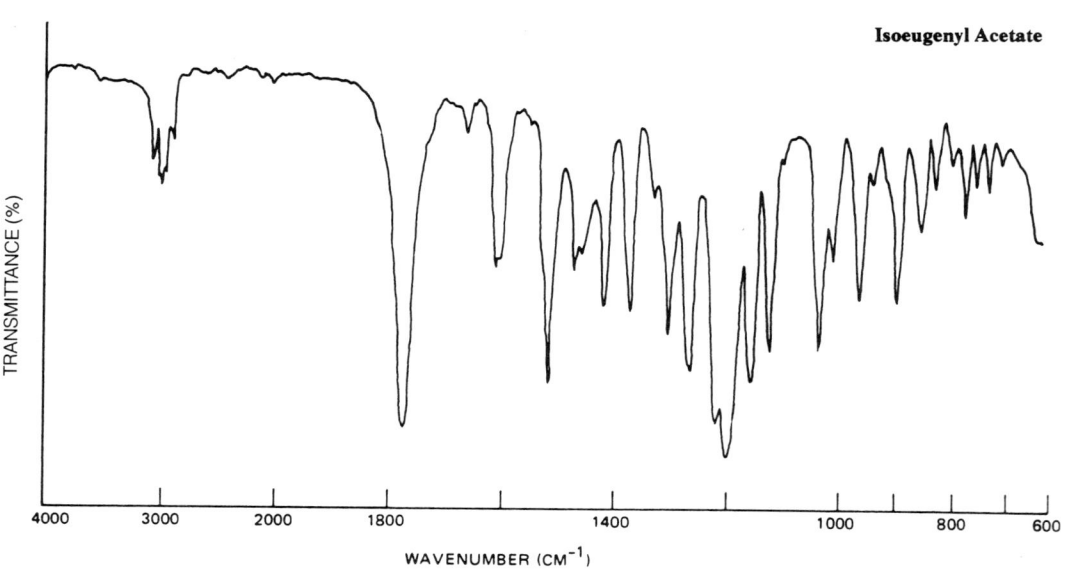

Isoeugenyl Acetate

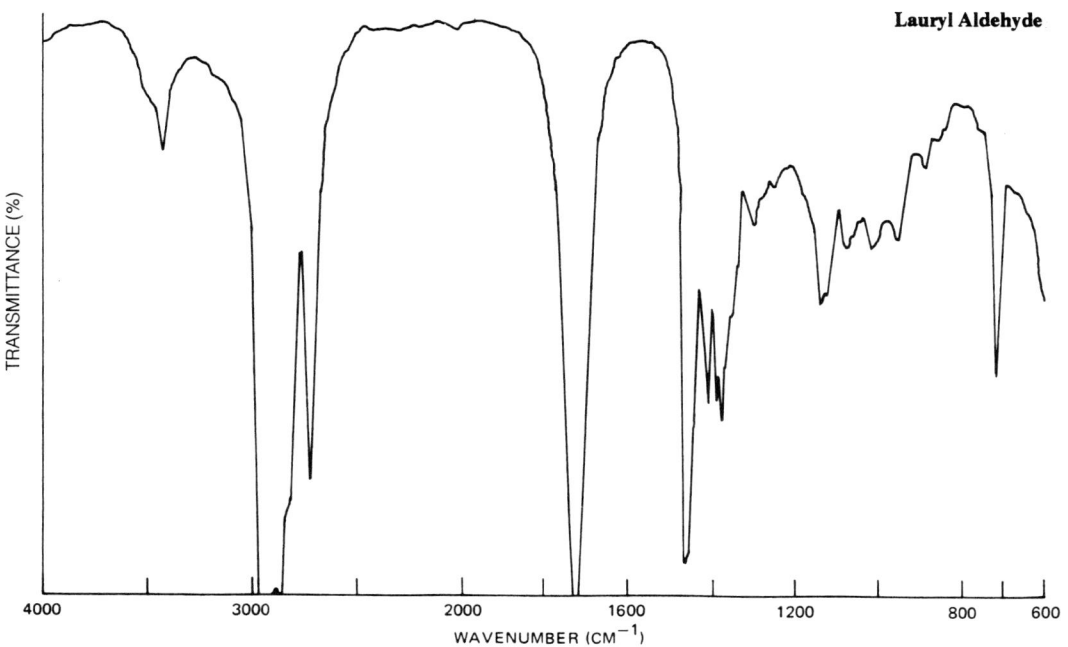

Lauryl Aldehyde

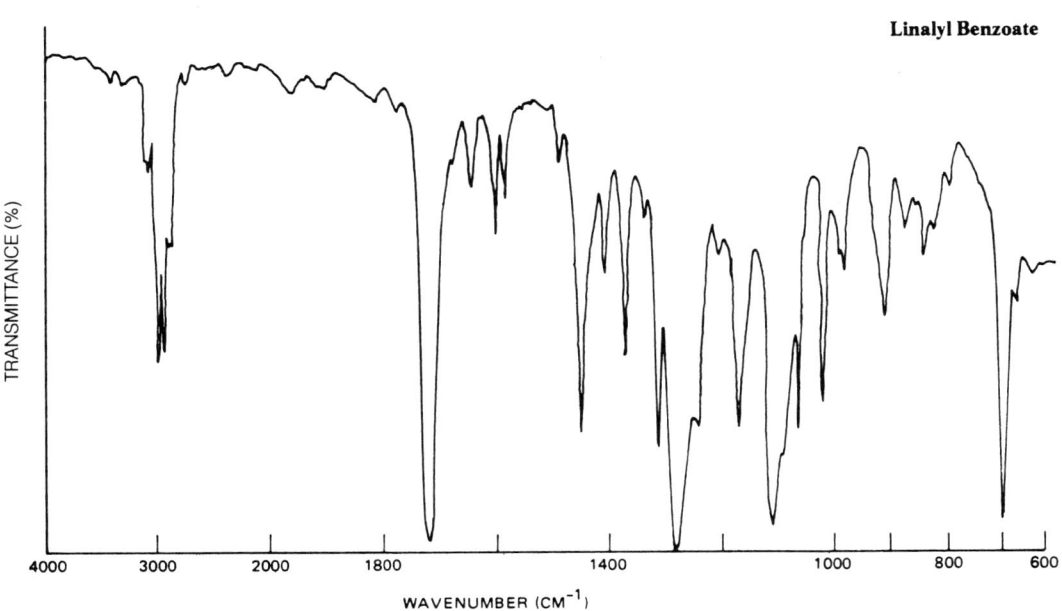

Linalyl Benzoate

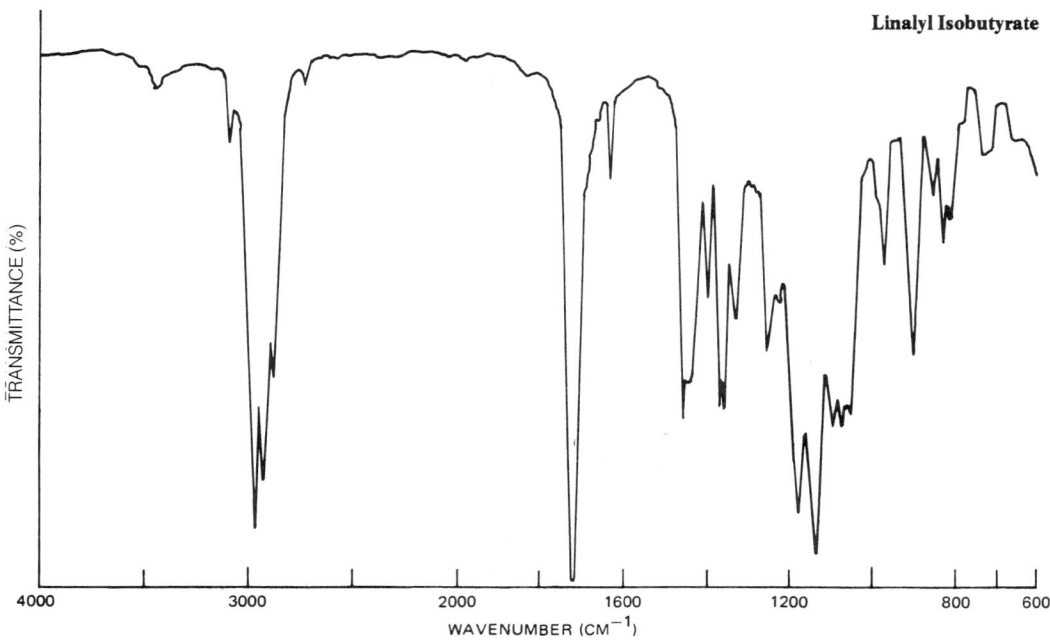

Linalyl Isobutyrate

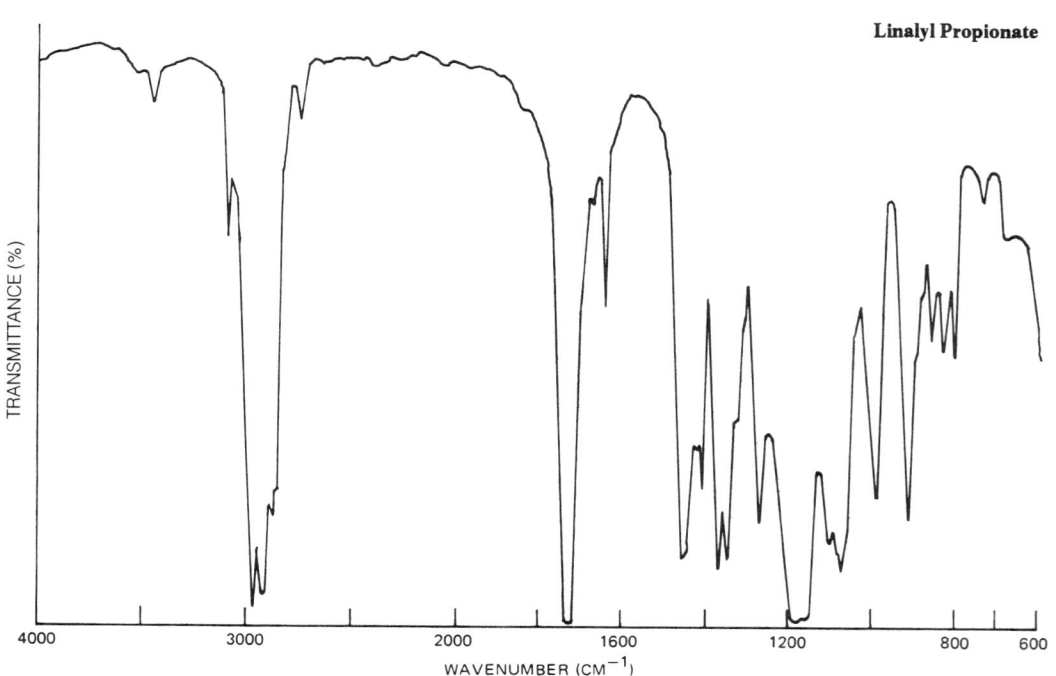

Linalyl Propionate

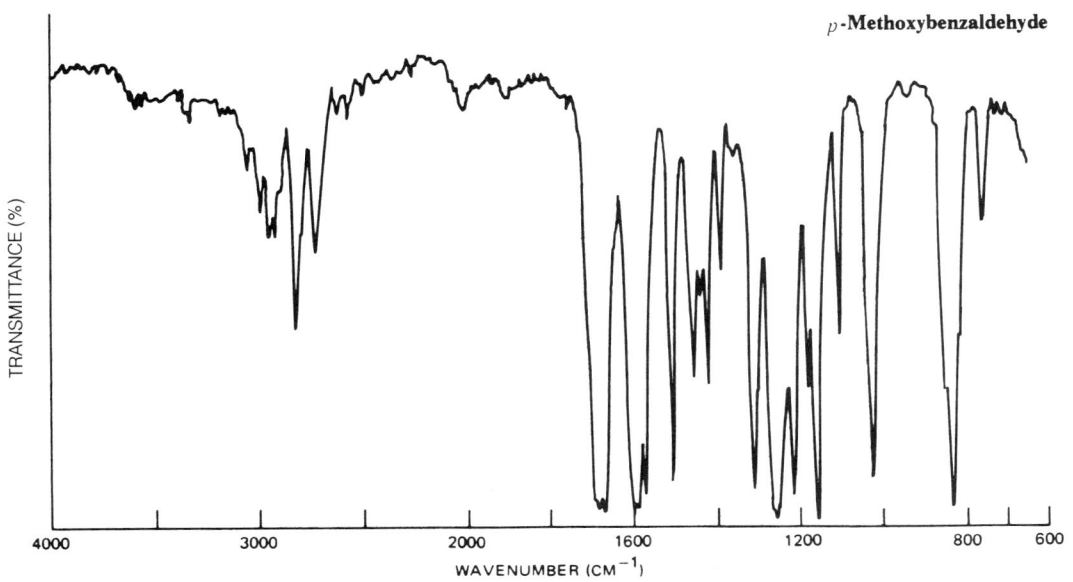

p-Methoxybenzaldehyde

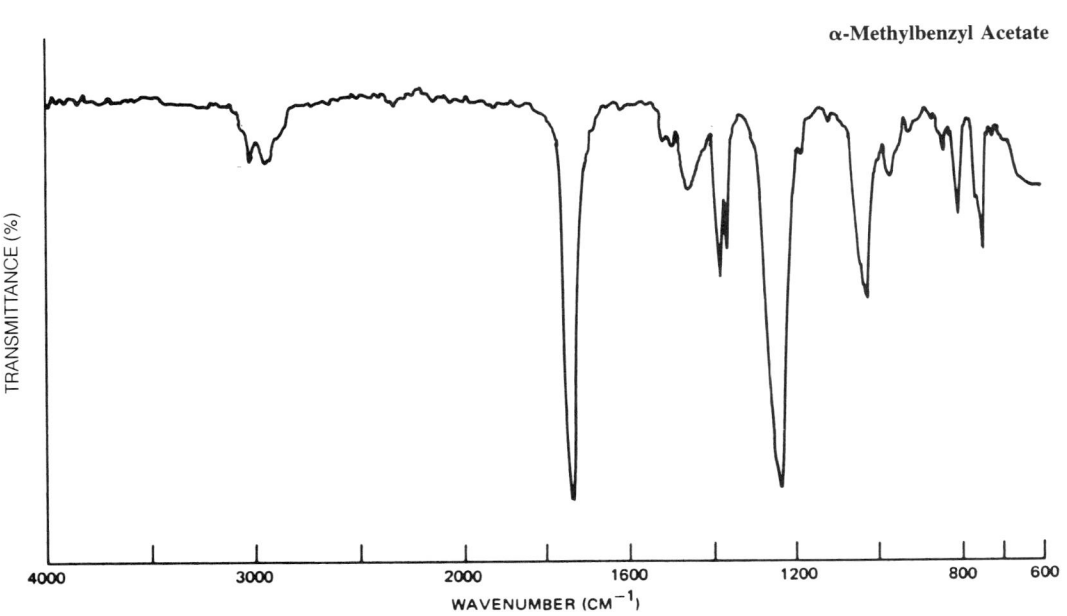

α-Methylbenzyl Acetate

Methyl Eugenol

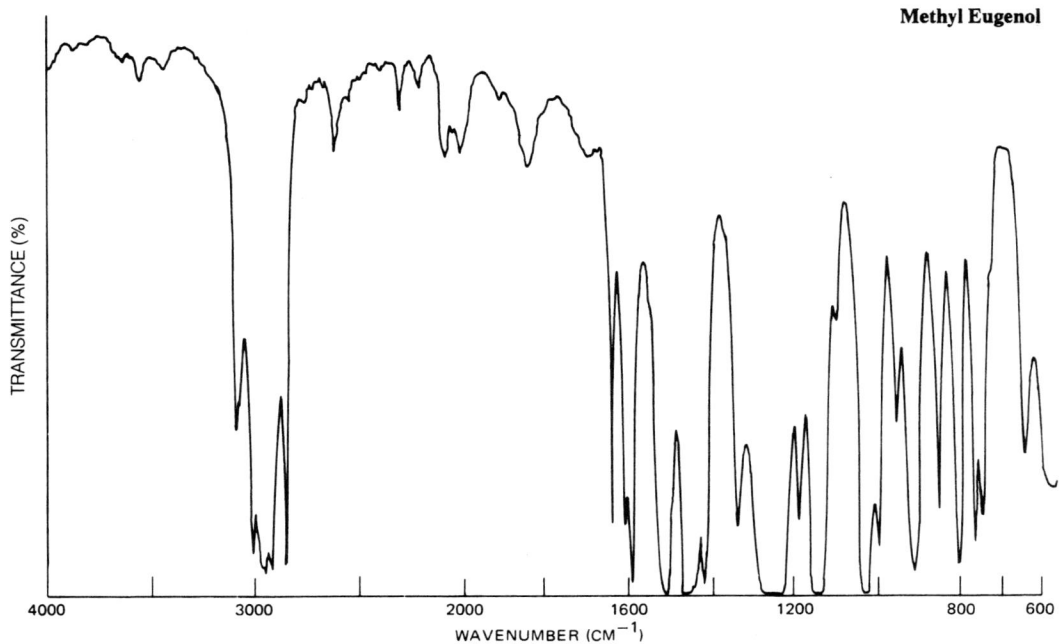

6-Methyl-5-hepten-2-one

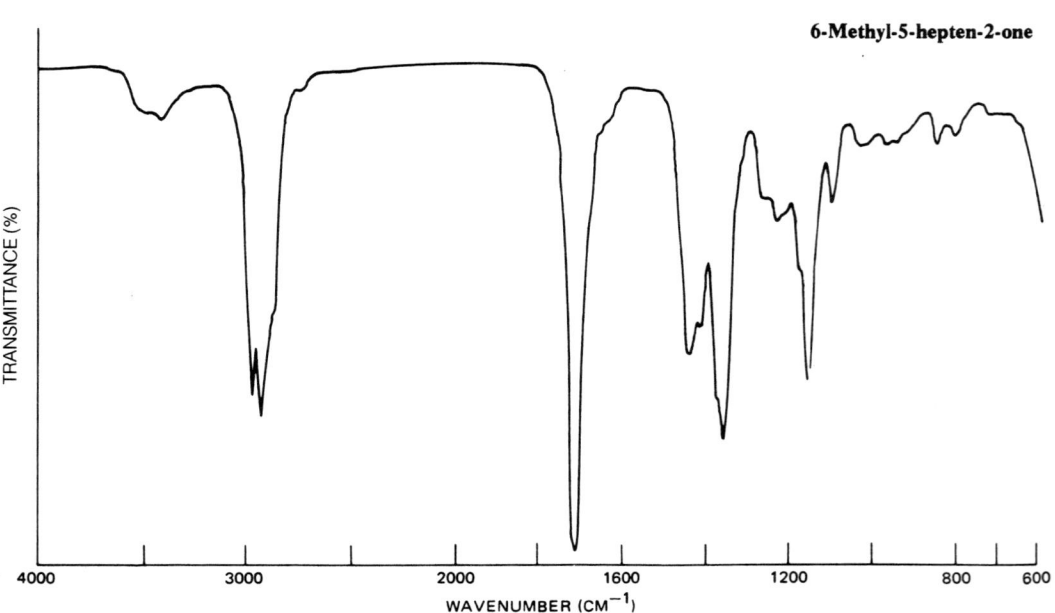

Methyl Salicylate

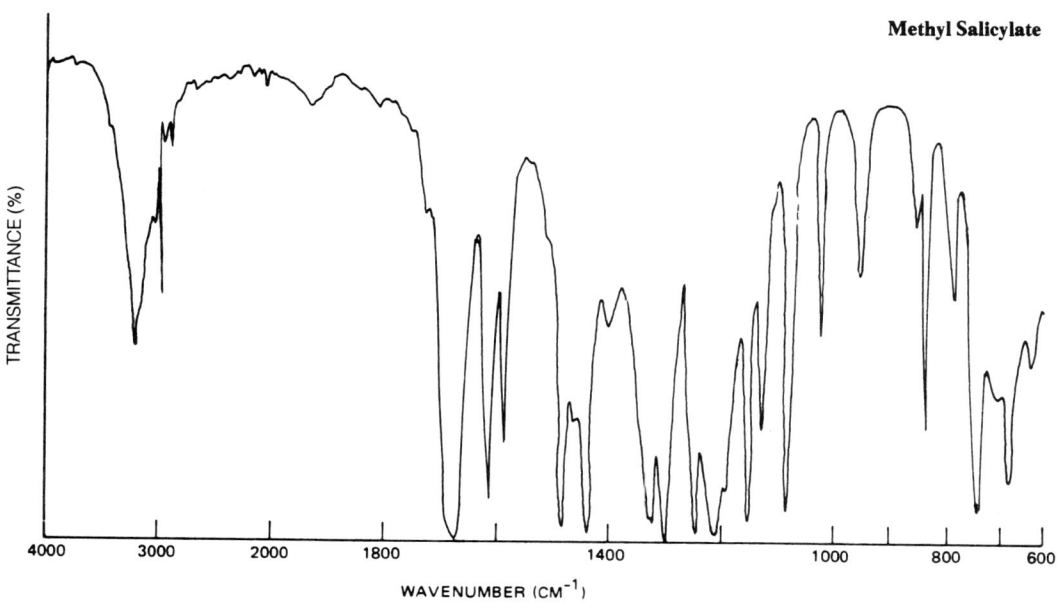

Nerol

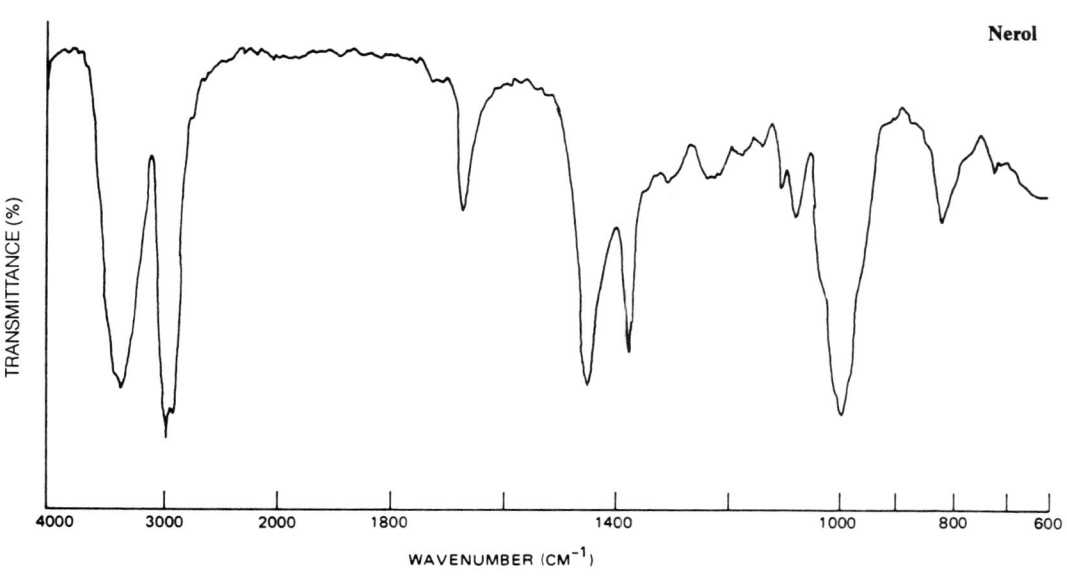

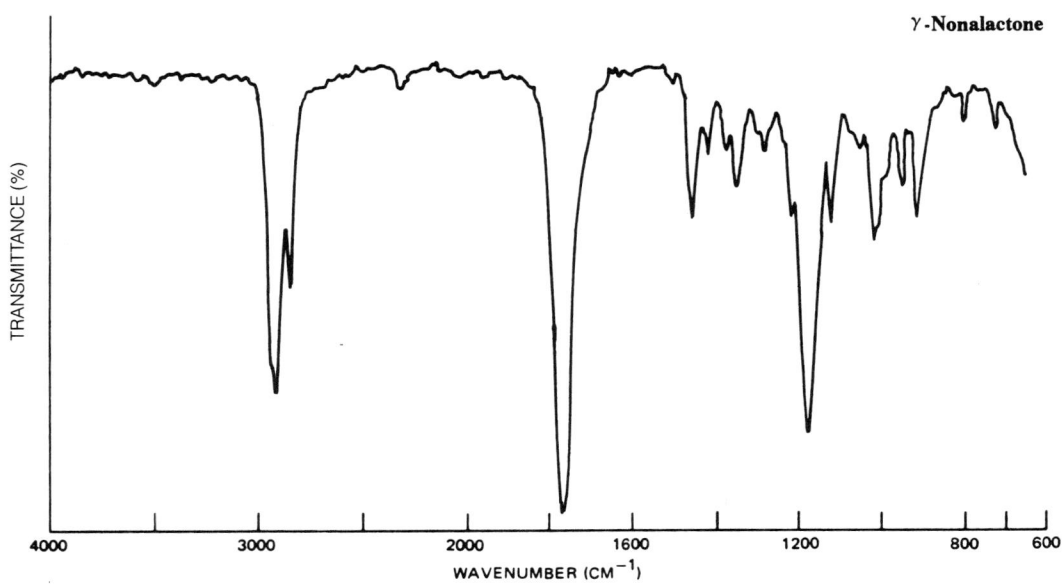

γ-Nonalactone

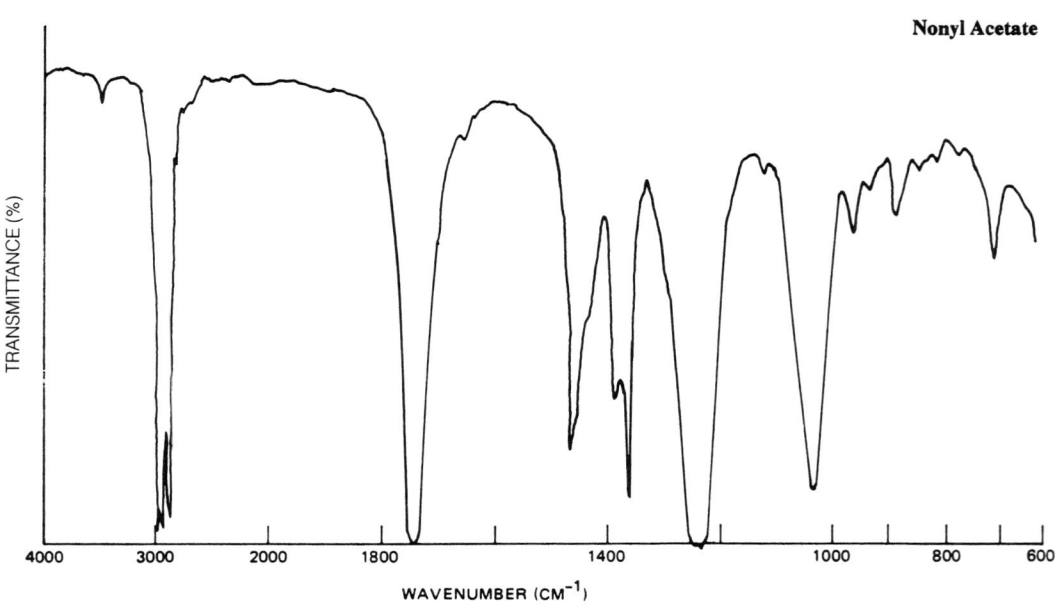

Nonyl Acetate

Nonyl Alcohol

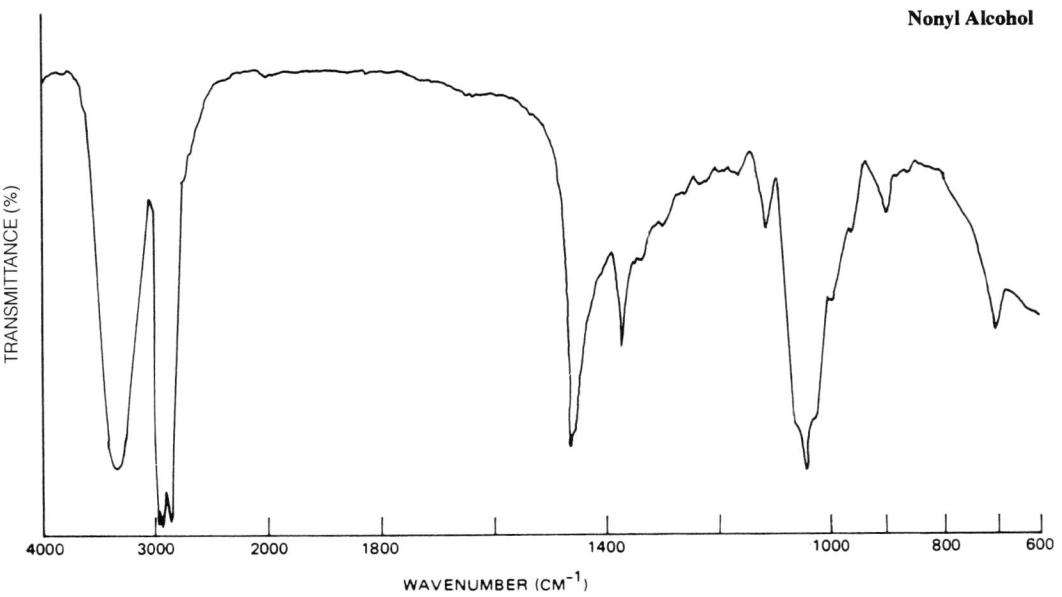

Octyl Acetate

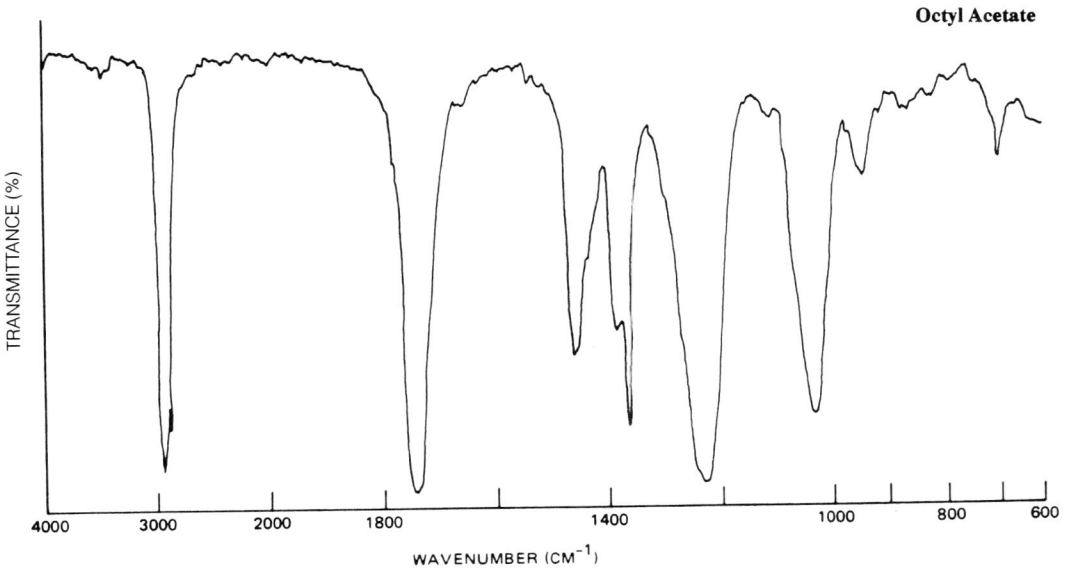

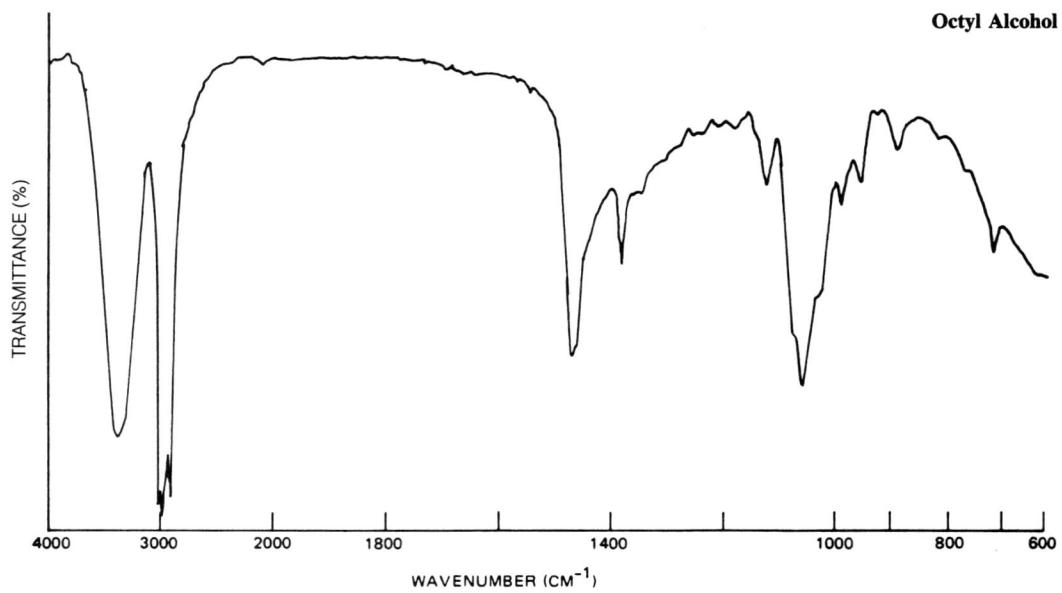

Octyl Alcohol

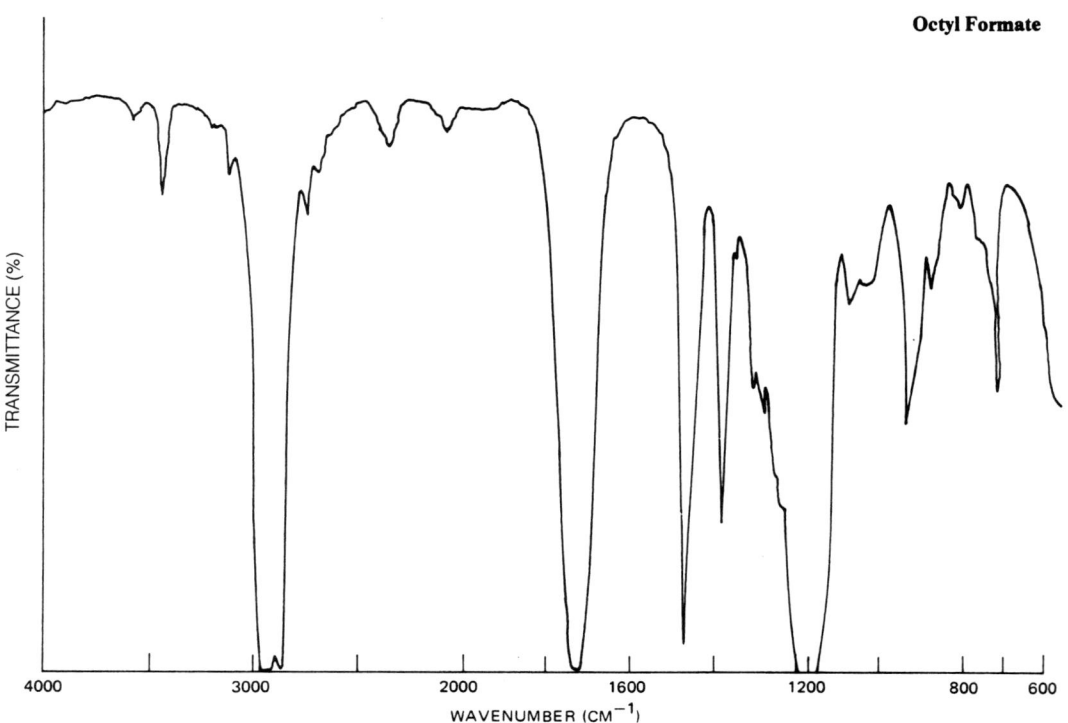

Octyl Formate

α-Phellandrene

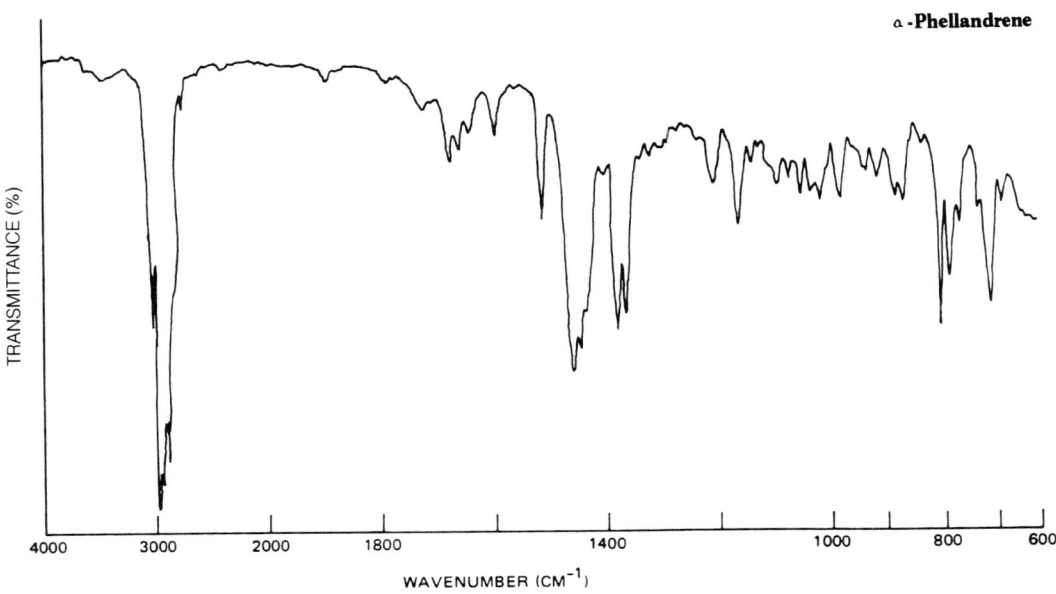

Phenethyl Acetate

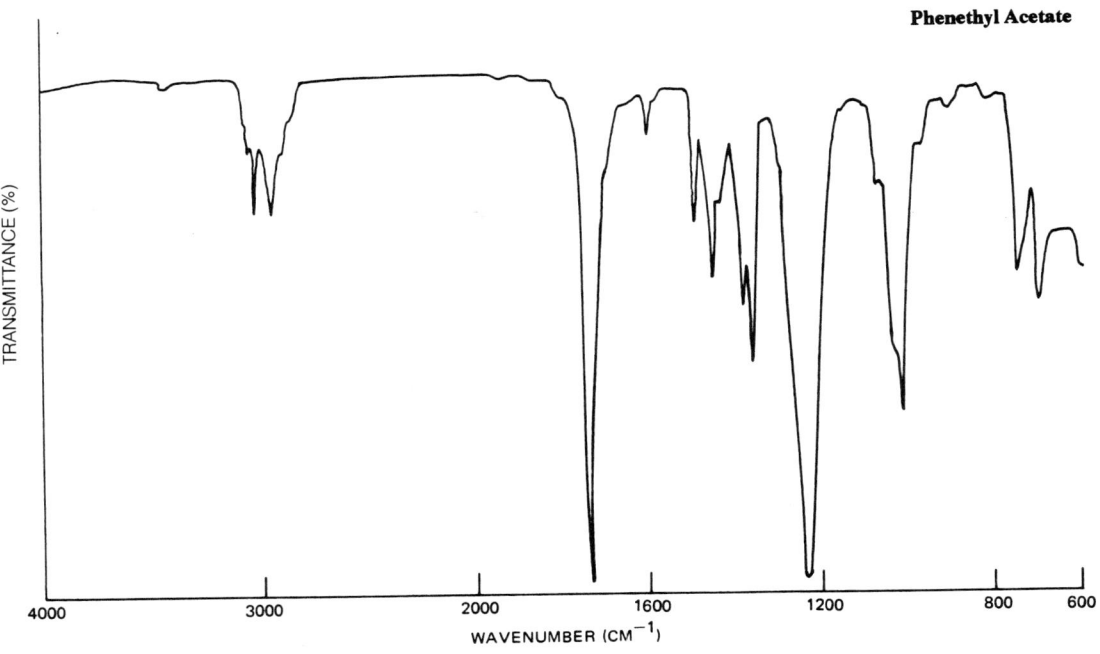

Phenethyl Alcohol

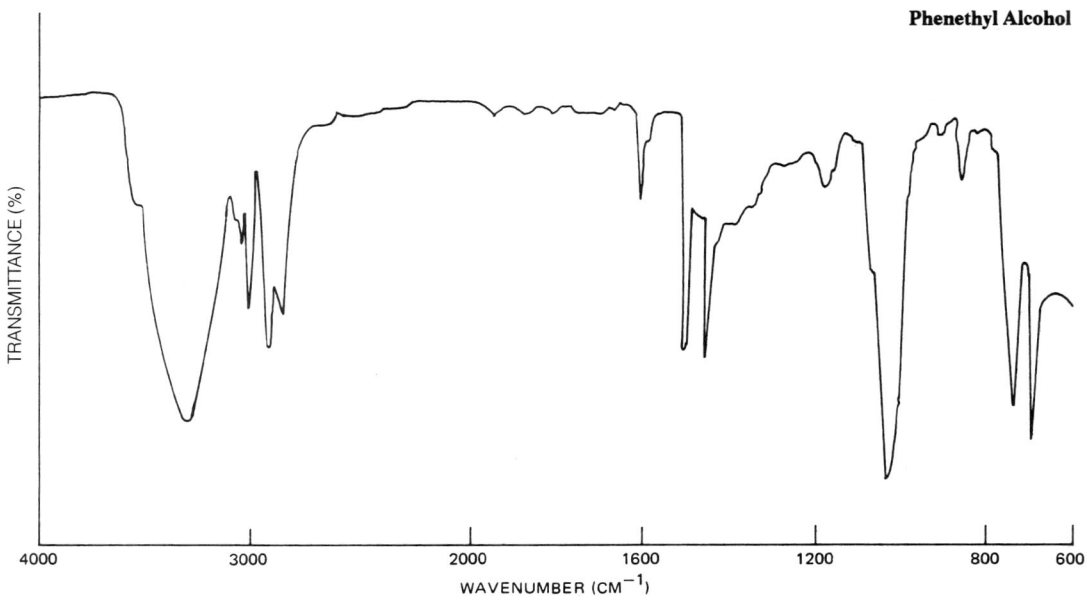

Phenylacetaldehyde Dimethyl Acetal

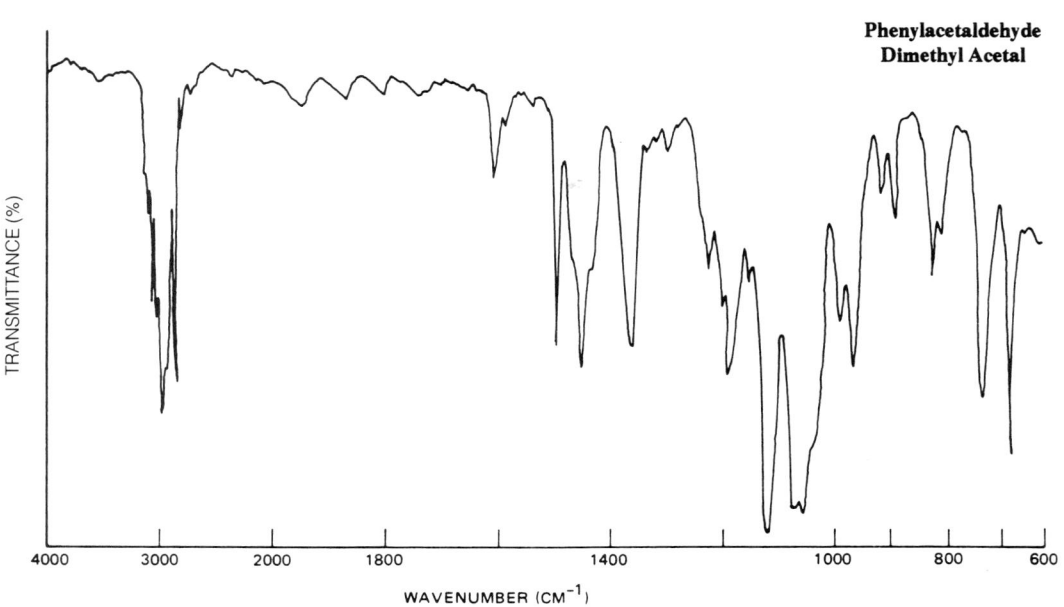

3-Phenyl-1-propanol

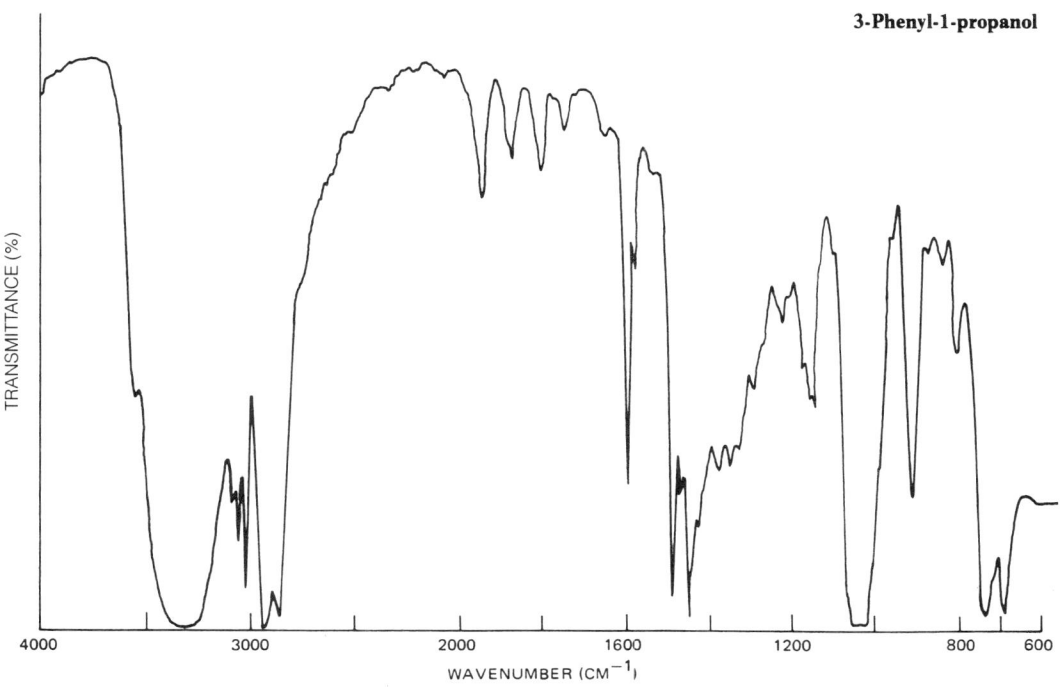

2-Phenylpropionaldehyde

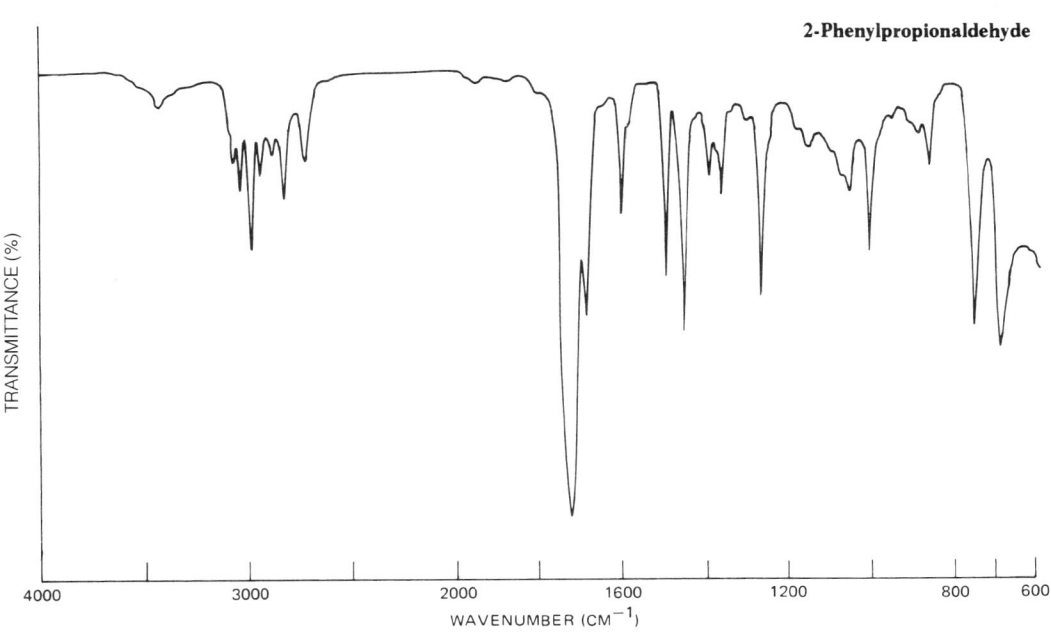

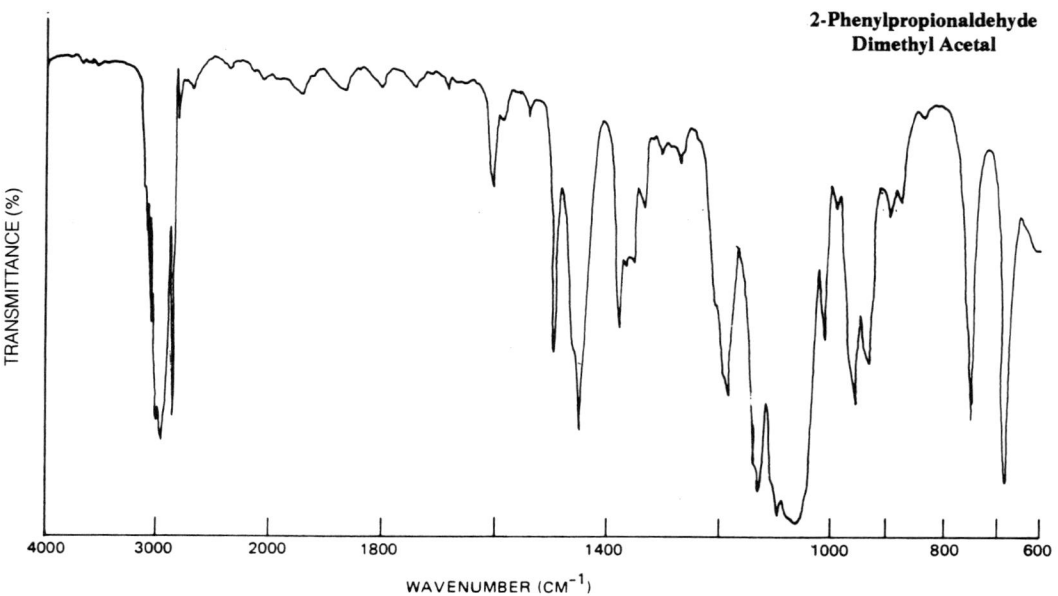

2-Phenylpropionaldehyde Dimethyl Acetal

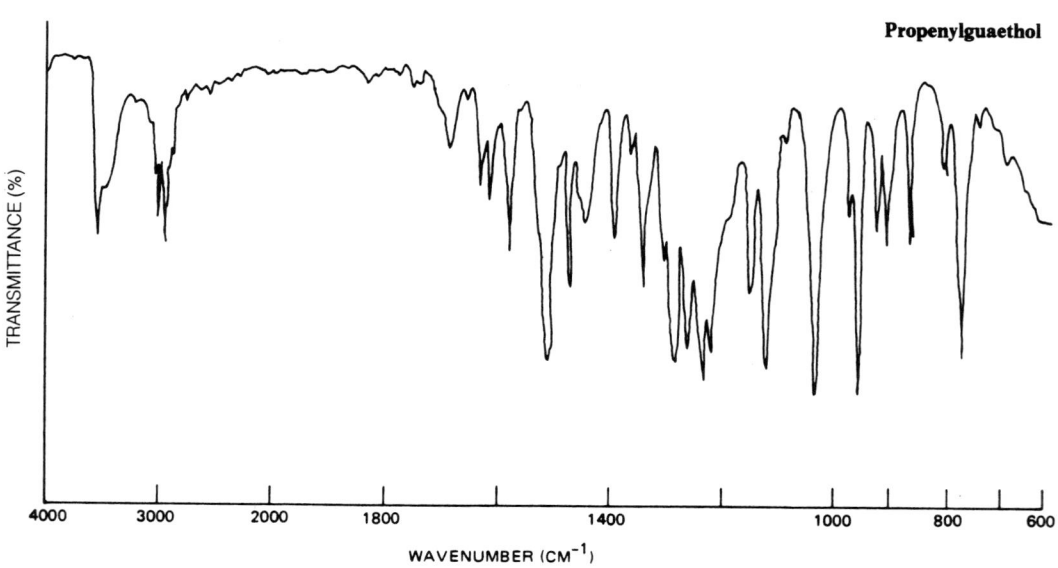

Propenylguaethol

Rhodinyl Acetate

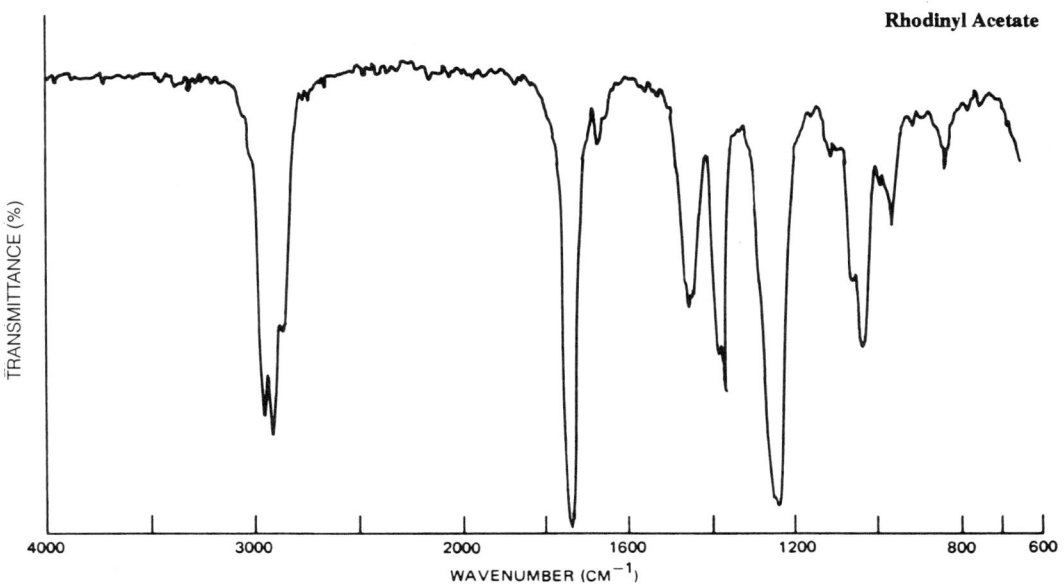

Rhodinyl Formate

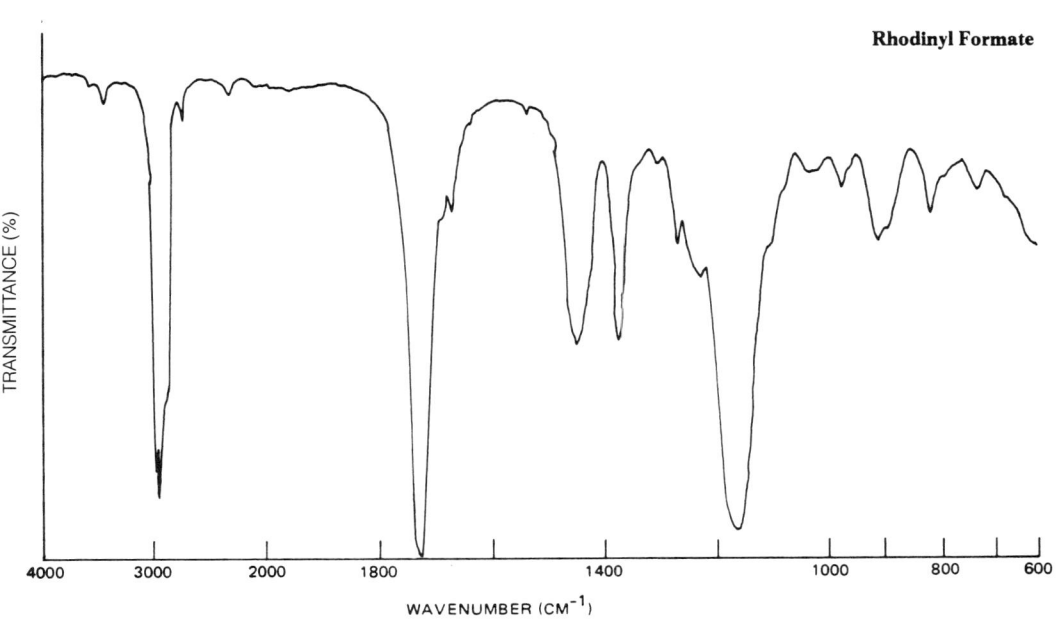

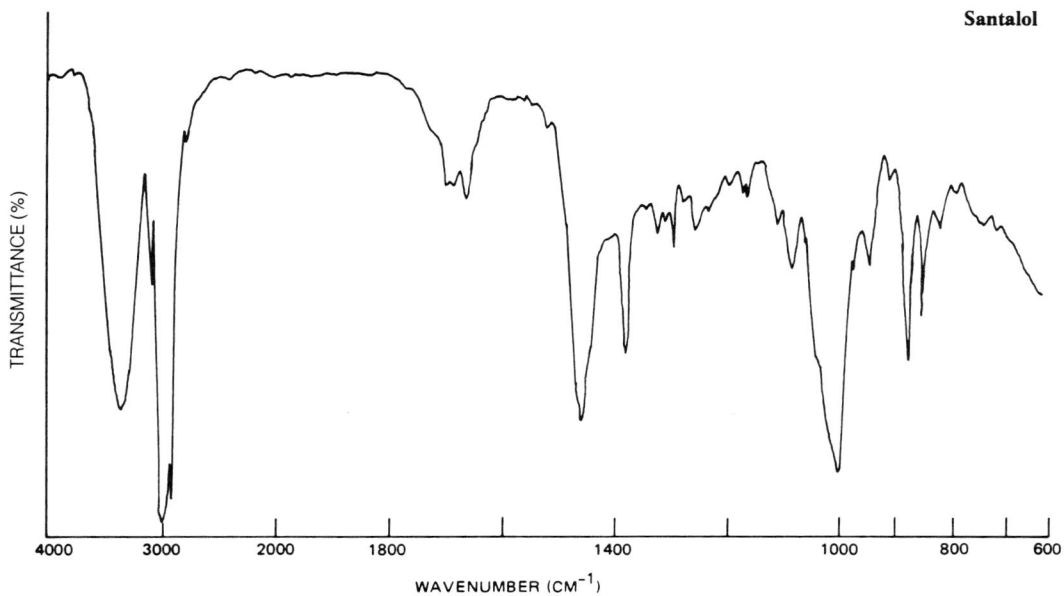

Santalol

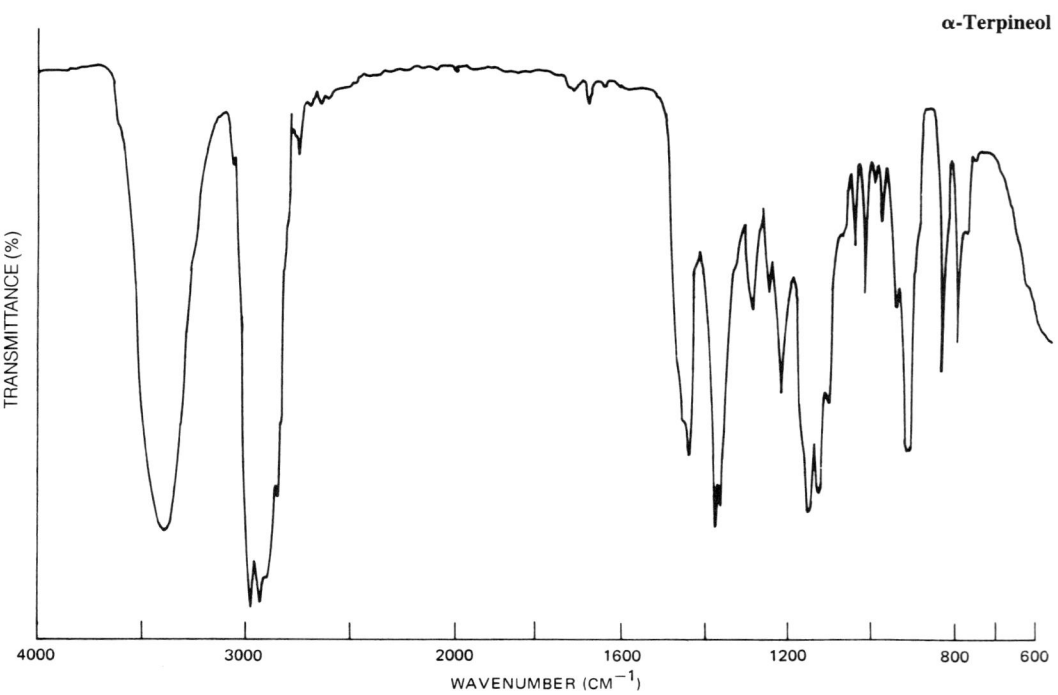

α-Terpineol

Terpinyl Propionate

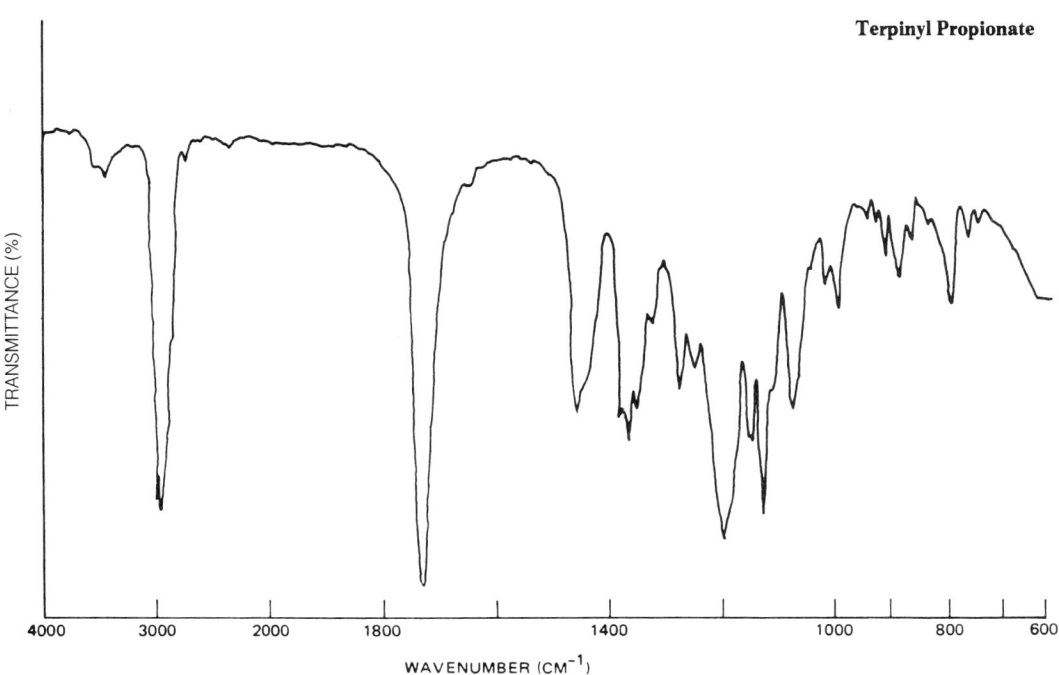

Series C: Other Substances

The spectra in this series are listed below alphabetically.

Butadiene-Styrene 75/25 Rubber (Emulsion-Polymerized Latex) ... 714
Butadiene-Styrene 75/25 Rubber (Emulsion-Polymerized Solid) ... 714
Butadiene-Styrene 50/50 Rubber (Latex) ... 714
Butadiene-Styrene 50/50 Rubber (Solid) ... 715
Candelilla Wax ... 715
Glycerol Ester of Gum Rosin ... 715
Glycerol Ester of Partially Dimerized Rosin ... 716

Glycerol Ester of Partially Hydrogenated Gum Rosin ... 716
Glycerol Ester of Partially Hydrogenated Wood Rosin ... 716
Glycerol Ester of Polymerized Rosin ... 717
Glycerol Ester of Tall Oil Rosin ... 717
Glycerol Ester of Wood Rosin ... 717
Isobutylene-Isoprene Copolymer ... 718
Methyl Ester of Rosin, Partially Hydrogenated ... 718
Morpholine ... 718

Paraffin, Synthetic ... 719
Pentaerythritol Ester of Partially Hydrogenated Wood Rosin ... 719
Pentaerythritol Ester of Wood Rosin ... 719
Petroleum Wax (Refined) ... 720
Petroleum Wax (Microcrystalline) ... 720
Petroleum Wax, Synthetic ... 720
Polyethylene ... 721
Polyisobutylene ... 721
Polyvinyl Acetate ... 721
Rice Bran Wax ... 722

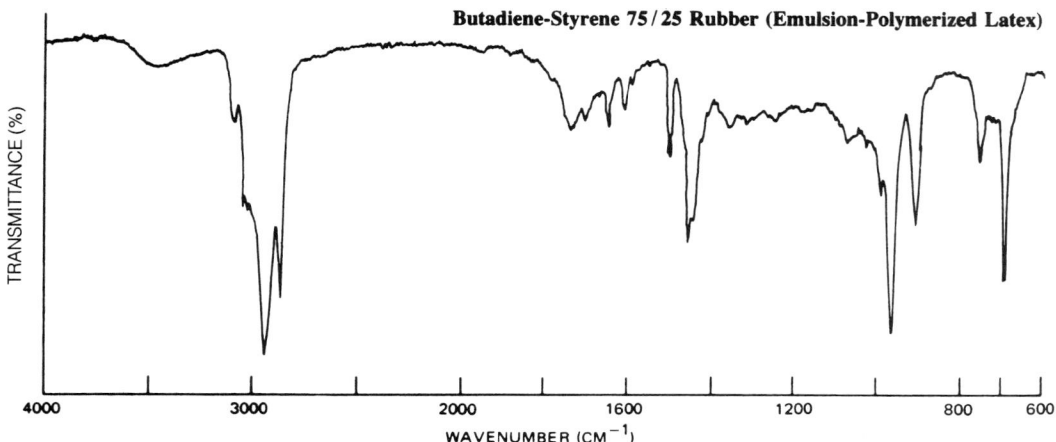

Butadiene-Styrene 75/25 Rubber (Emulsion-Polymerized Latex)

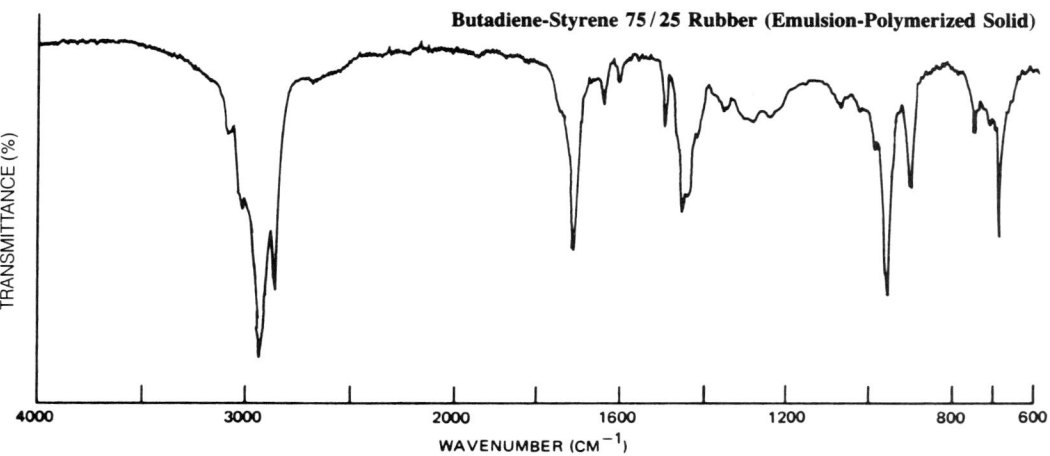

Butadiene-Styrene 75/25 Rubber (Emulsion-Polymerized Solid)

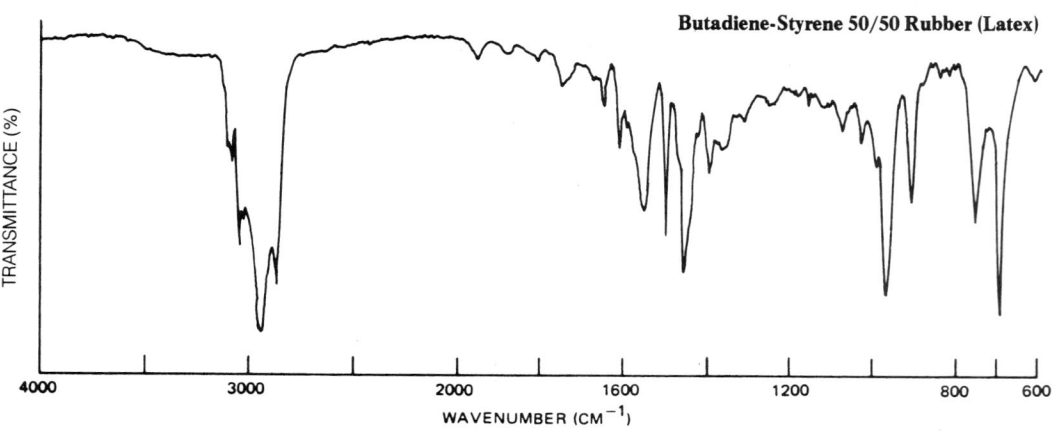

Butadiene-Styrene 50/50 Rubber (Latex)

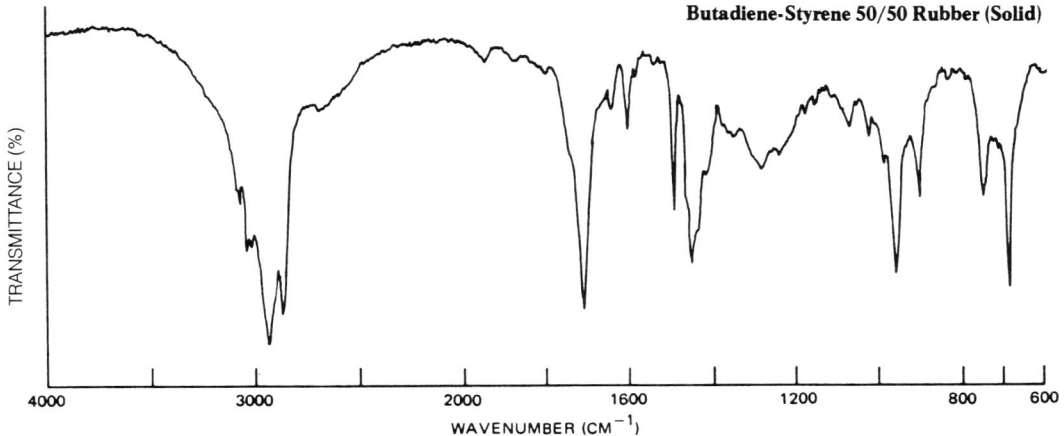
Butadiene-Styrene 50/50 Rubber (Solid)

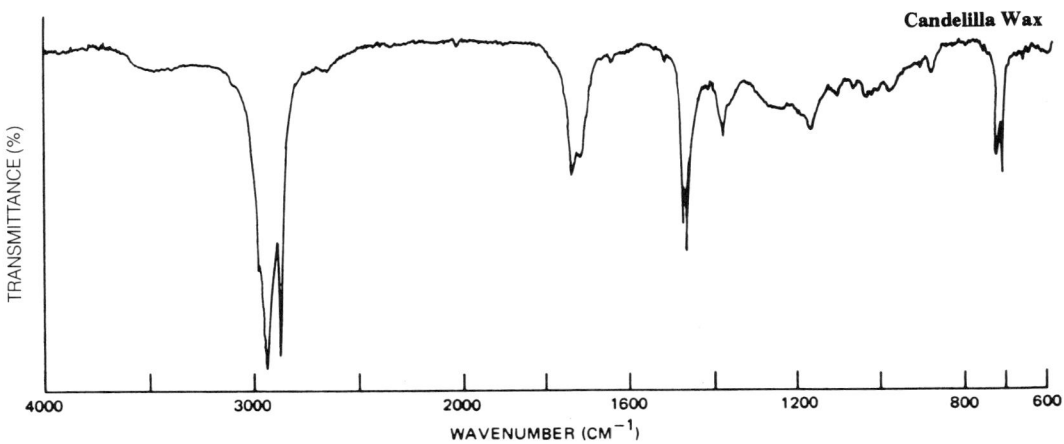
Candelilla Wax

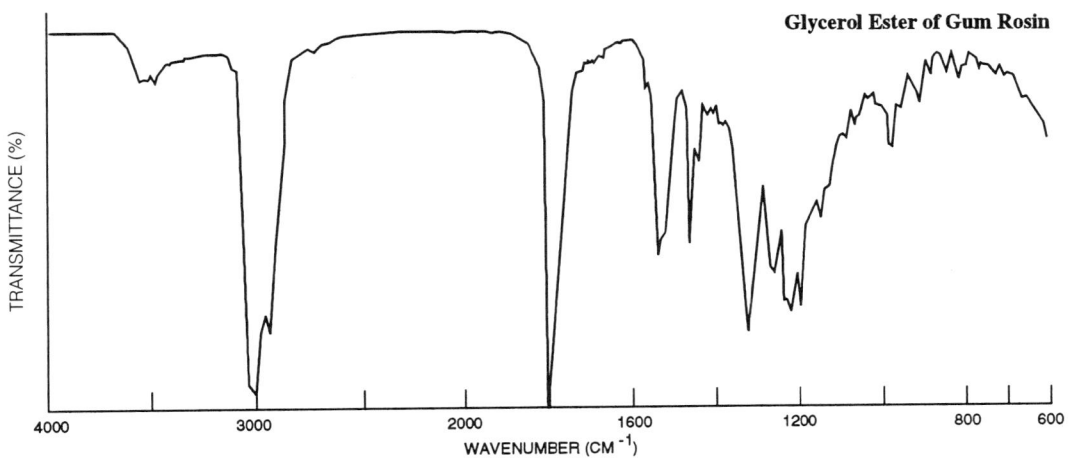
Glycerol Ester of Gum Rosin

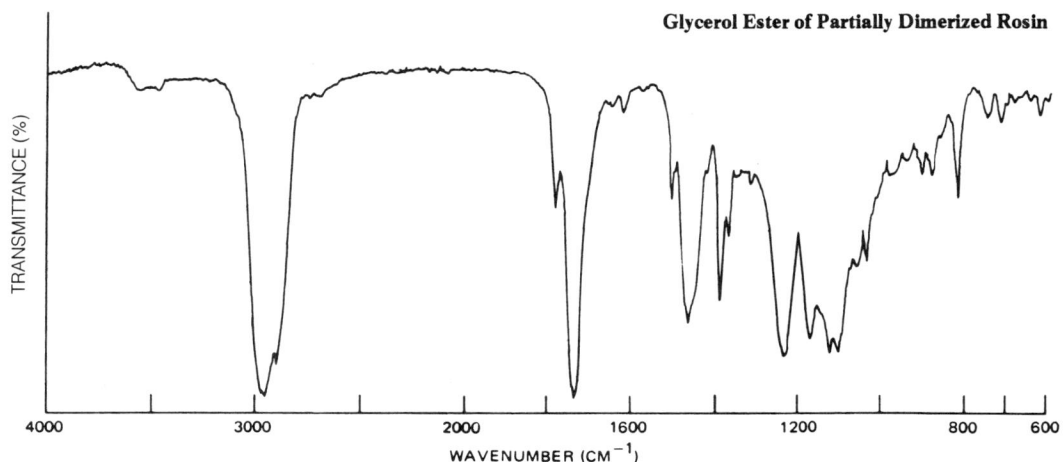
Glycerol Ester of Partially Dimerized Rosin

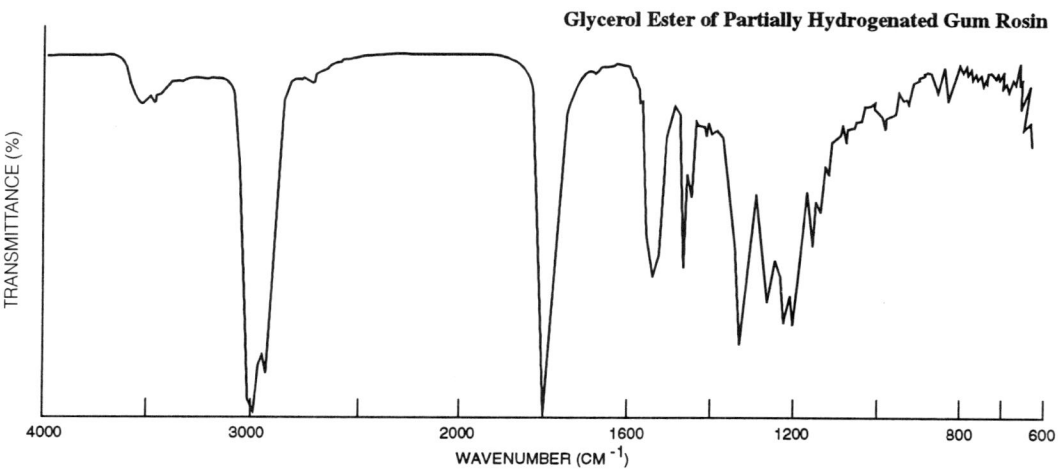
Glycerol Ester of Partially Hydrogenated Gum Rosin

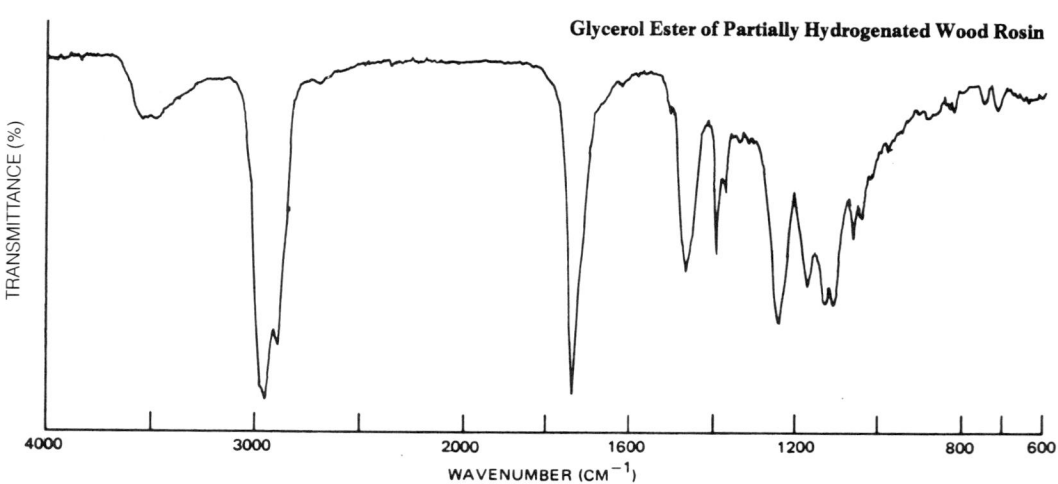
Glycerol Ester of Partially Hydrogenated Wood Rosin

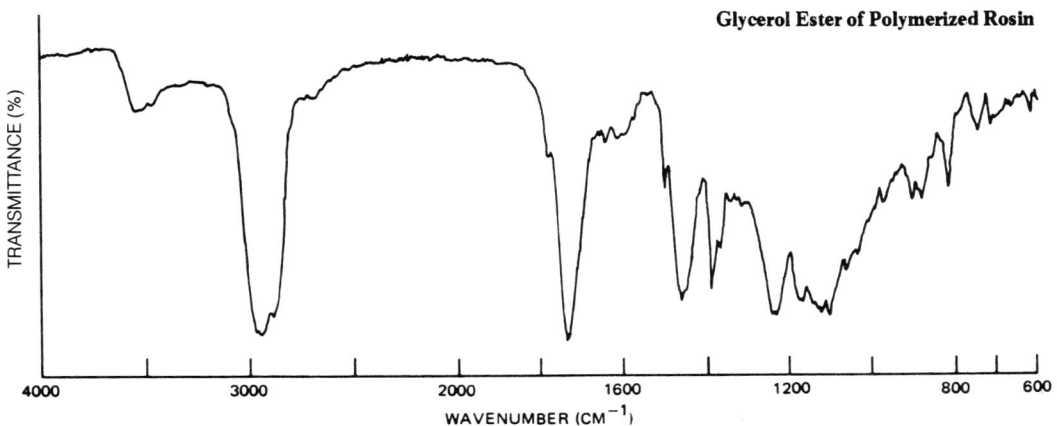
Glycerol Ester of Polymerized Rosin

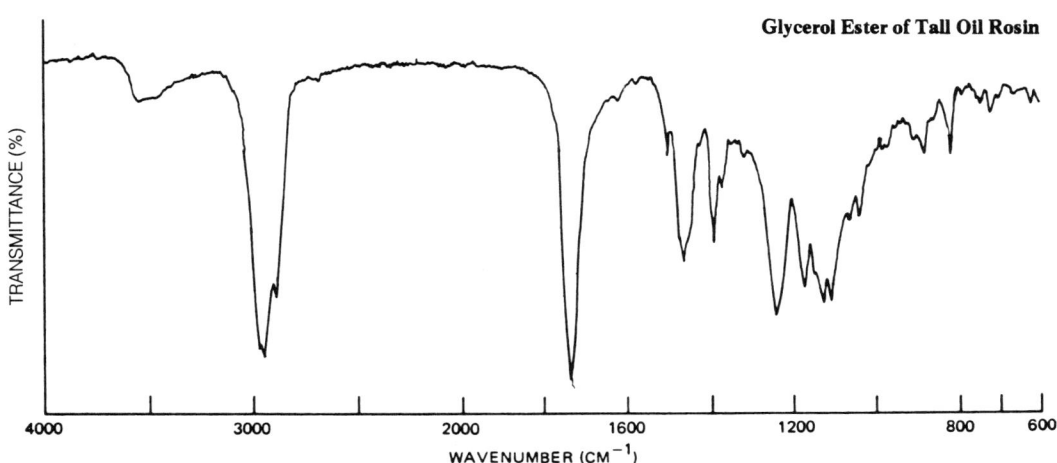
Glycerol Ester of Tall Oil Rosin

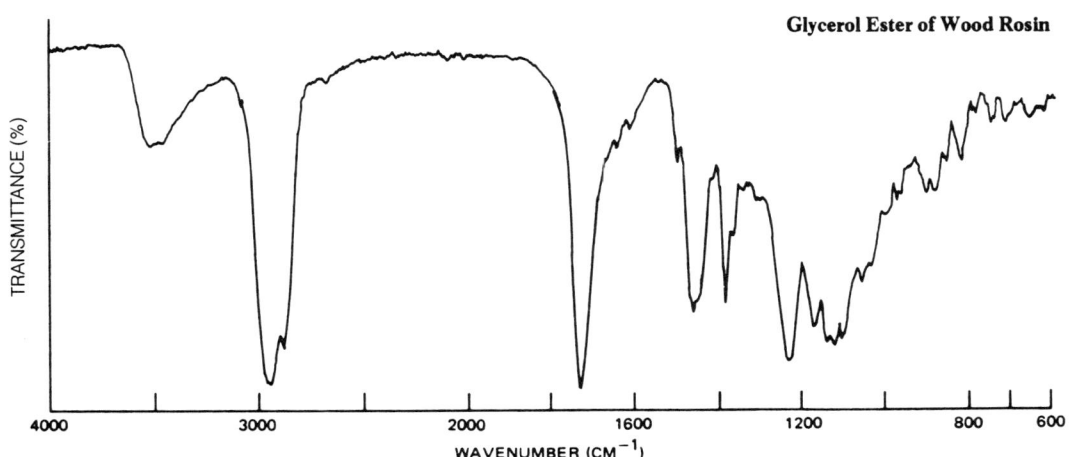
Glycerol Ester of Wood Rosin

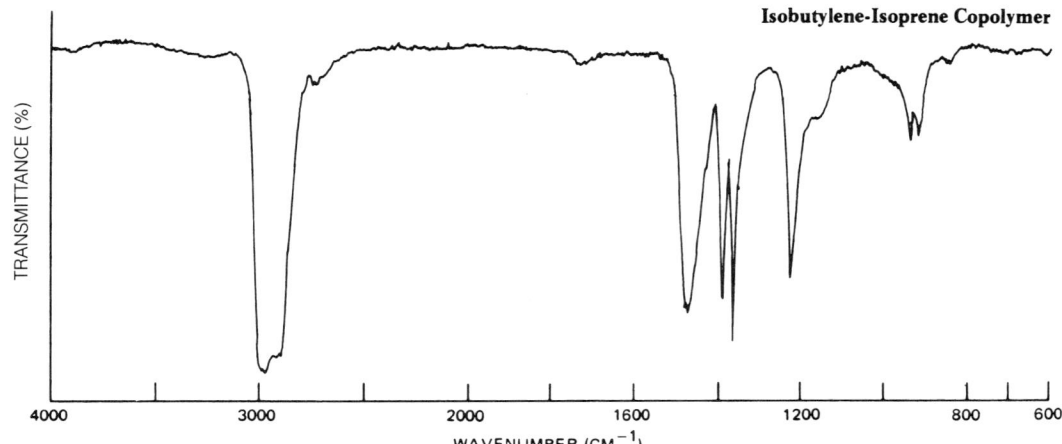

Isobutylene-Isoprene Copolymer

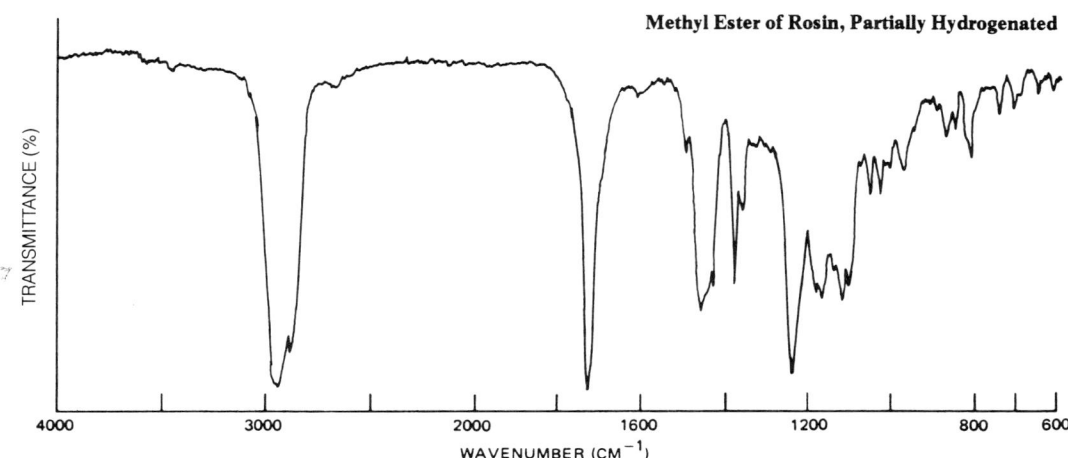

Methyl Ester of Rosin, Partially Hydrogenated

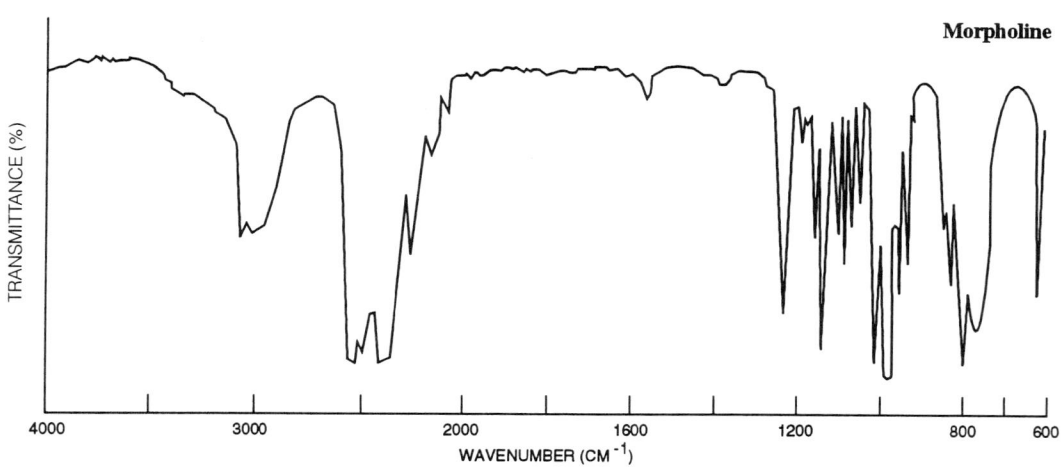

Morpholine

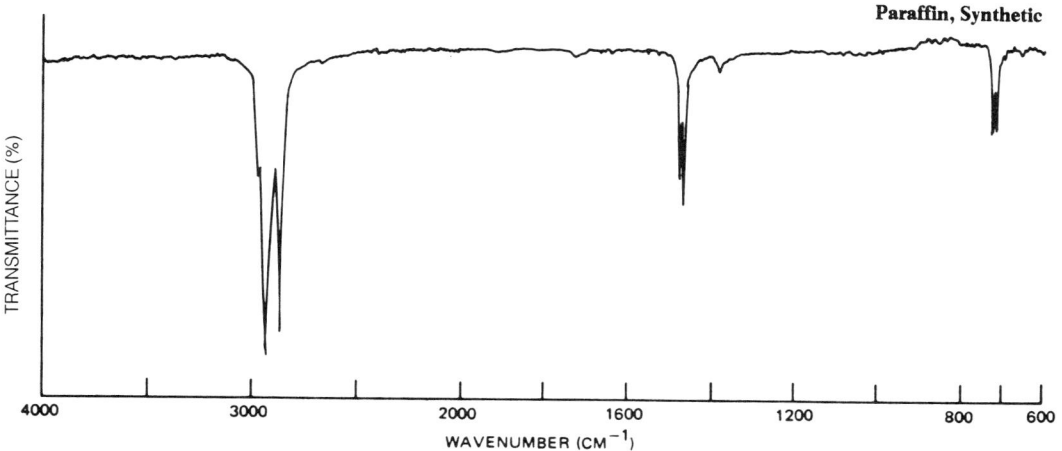

Paraffin, Synthetic

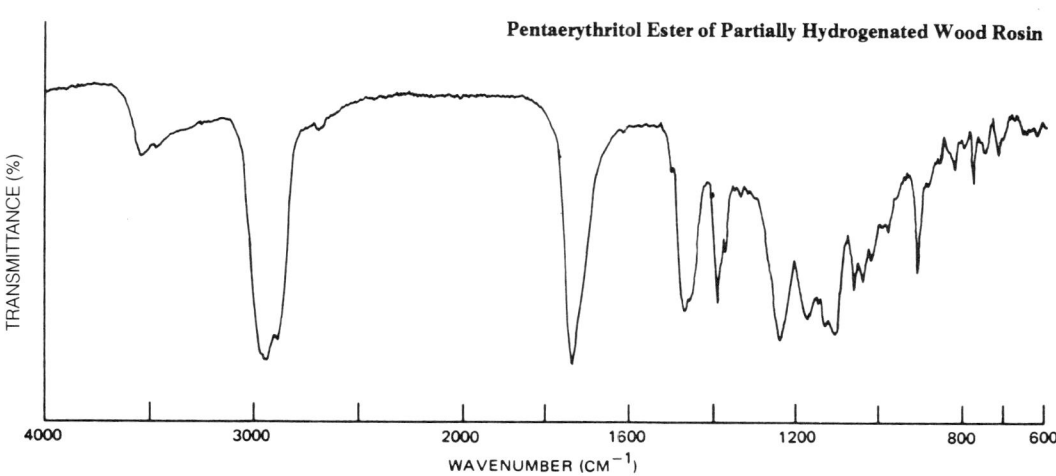

Pentaerythritol Ester of Partially Hydrogenated Wood Rosin

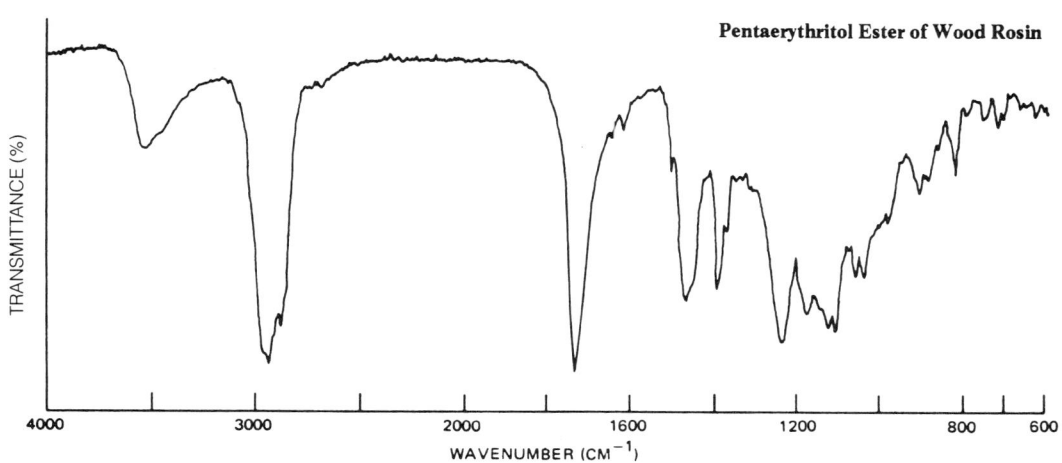

Pentaerythritol Ester of Wood Rosin

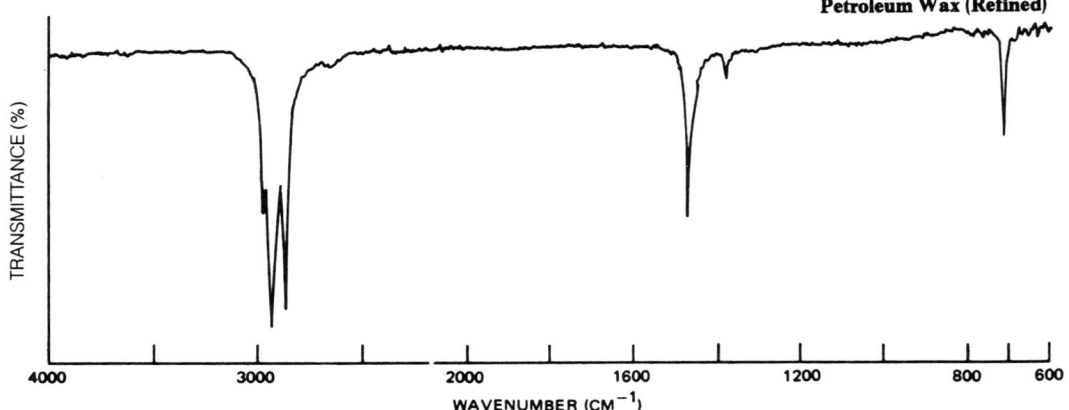
Petroleum Wax (Refined)

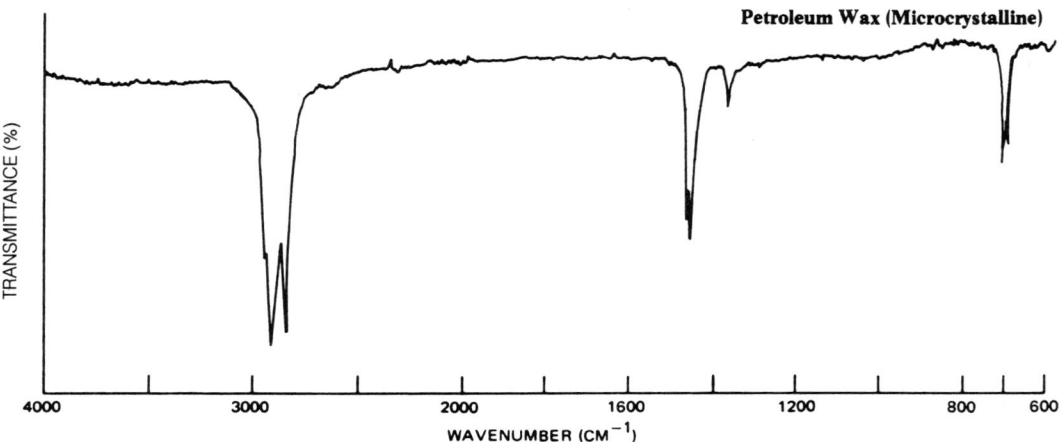
Petroleum Wax (Microcrystalline)

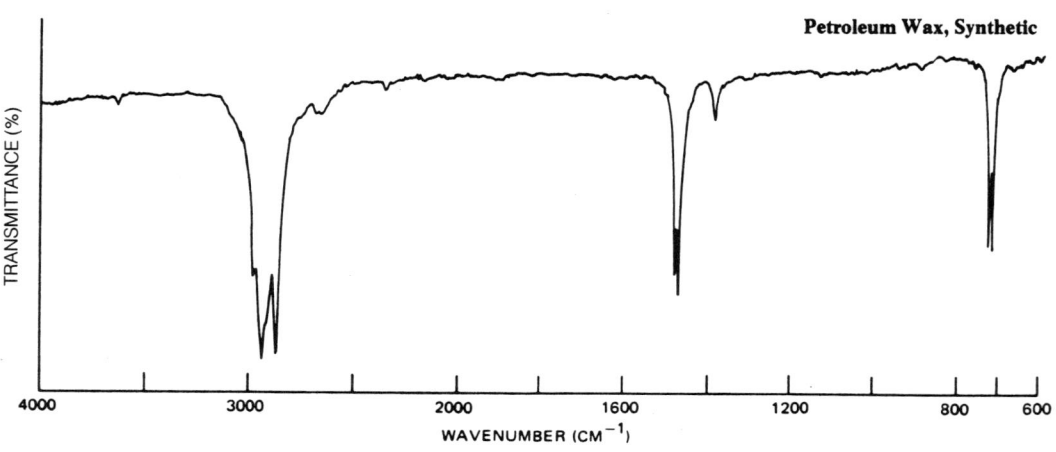
Petroleum Wax, Synthetic

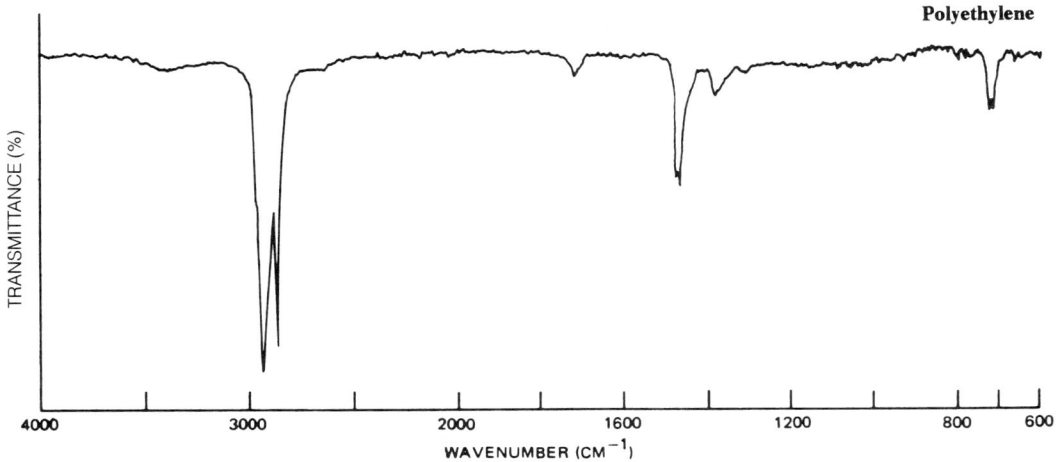

Polyethylene

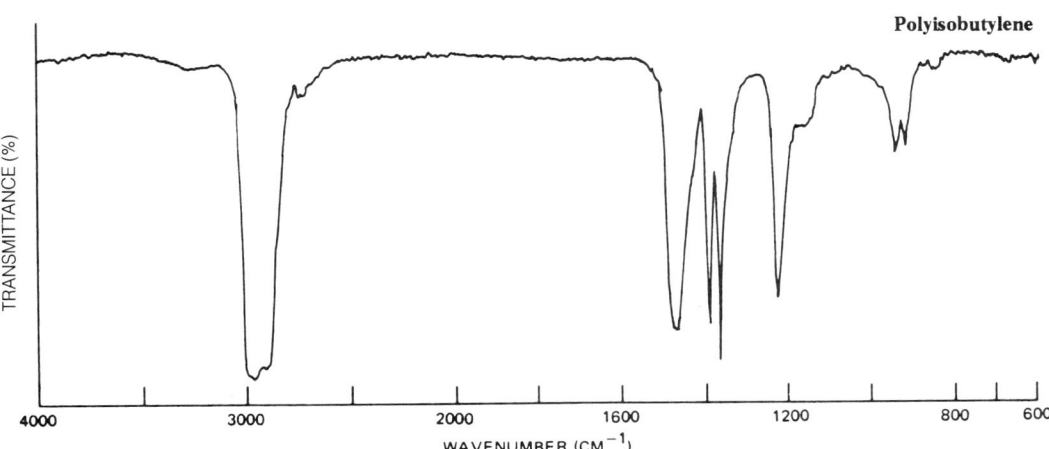

Polyisobutylene

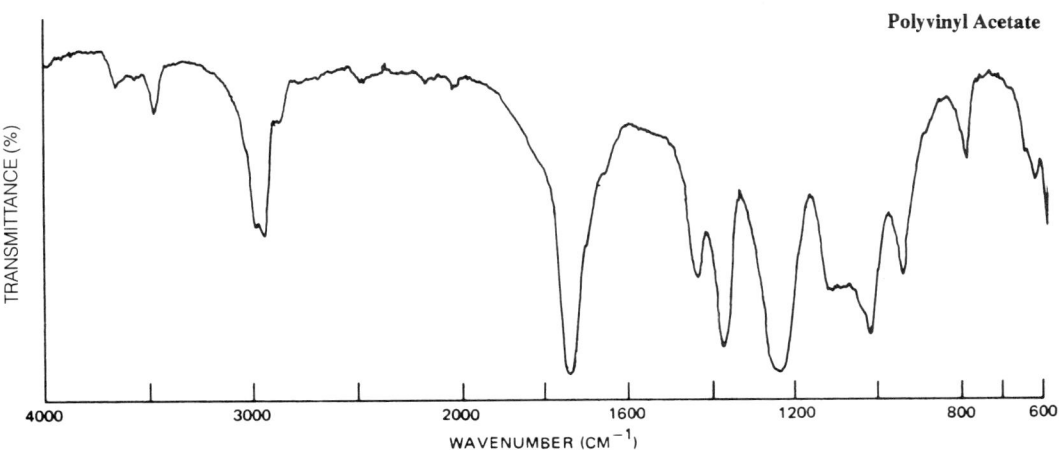

Polyvinyl Acetate

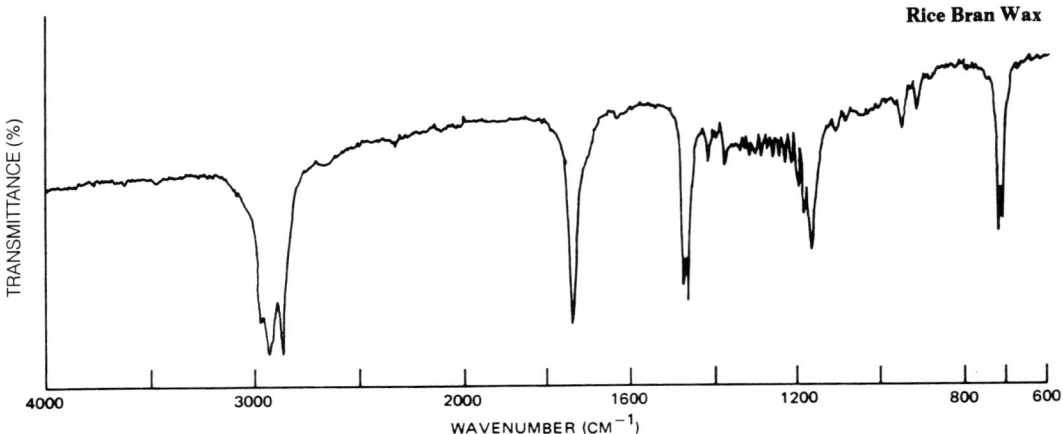
Rice Bran Wax

5 / General Tests and Assays

Contents

General Tests and Assays

APPENDIX I:
APPARATUS FOR TESTS AND ASSAYS ... 727
- Oxygen Flask Combustion ... 727
- Thermometers ... 727
- Volumetric Apparatus ... 728
- Weights and Balances ... 729

APPENDIX II:
PHYSICAL TESTS AND DETERMINATIONS ... 729
- A Chromatography ... 729
  - Column Chromatography ... 730
  - Thin-Layer Chromatography ... 731
  - Gas Chromatography ... 732
  - High-Pressure Liquid Chromatography ... 734
- B Physicochemical Properties ... 737
  - Distillation Range ... 737
  - Melting Range or Temperature ... 738
  - Optical (Specific) Rotation ... 739
  - pH Determination ... 740
  - Readily Carbonizable Substances ... 740
  - Refractive Index ... 741
  - Solidification Point ... 742
  - Viscosity of Dimethylpolysiloxane ... 743
  - Viscosity of Methylcellulose ... 743
  - Viscosity of Sodium Carboxymethylcellulose ... 744
  - Water Determination ... 745
    - Method I: Karl Fischer Titrimetric Method ... 745
    - Method II: Toluene Distillation Method ... 747
- C Others ... 748
  - Ash (Acid-Insoluble) ... 748
  - Ash (Total) ... 748
  - Hydrochloric Acid Table ... 748
  - Loss on Drying ... 749
  - Oil Content of Synthetic Paraffin ... 750
  - Residue on Ignition ... 751
  - Sieve Analysis of Granular Metal Powders ... 752
  - Sulfuric Acid Table ... 752

APPENDIX III:
CHEMICAL TESTS AND DETERMINATIONS ... 753
- A Identification Tests ... 753
- B Limit Tests ... 755
  - Arsenic Test ... 755
  - Chloride and Sulfate Limit Tests ... 757
  - 1,4-Dioxane Limit Test ... 757
  - Fluoride Limit Test ... 758
  - Heavy Metals Test ... 760
  - Lead Limit Test ... 762
    - Dithizone Method ... 762
    - Flame Atomic Absorption Spectrophotometric Method ... 763
    - Atomic Absorption Spectrophotometric Graphite Furnace Method ... 763
      - Method I ... 763
      - Method II ... 765
    - APDC Extraction Method ... 766
  - Mercury Limit Test ... 766
  - Selenium Limit Test ... 767

C Others ... 768
    Alginates Assay ... 768
    Benzene (in Paraffinic Hydrocarbon Solvents) ... 769
    Colors, F D & C ... 771
    Hydroxypropoxyl Determination ... 777
    Methoxyl Determination ... 778
    Nitrogen Determination ... 779
    Sulfur (by Oxidative Microcoulometry) ... 780

APPENDIX IV:
CHEWING GUM BASE ... 782
    Bound Styrene ... 782
    Molecular Weight ... 783
        Polyethylene ... 783
        Polyisobutylene ... 783
        Polyvinyl Acetate ... 783
    Quinones ... 784
    Residual Styrene ... 784
    Sample Solution for Arsenic Test ... 785
    Sample Solution for Lead Limit Test ... 785
    Total Unsaturation ... 785

APPENDIX V:
ENZYME ASSAYS ... 786
    Acid Phosphatase Activity ... 788
    Alpha-Amylase Activity (Nonbacterial) ... 789
    Alpha-Amylase Activity (Bacterial) ... 790
    Catalase Activity ... 791
    Cellulase Activity ... 791
    Chymotrypsin Activity ... 793
    Diastase Activity ... 793
    α-Galactosidase Activity ... 794
    β-Glucanase Activity ... 795
    Glucoamylase Activity ... 795
    Glucose Isomerase Activity ... 796
    Glucose Oxidase Activity ... 798
    β-D-Glucosidase Activity ... 798
    Hemicellulase Activity ... 799
    Invertase Activity ... 800
    Lactase (Neutral) (β-Galactosidase) Activity ... 801
    Lactase (Acid) (β-Galactosidase) Activity ... 802
    Lipase Activity ... 803
    Lipase/Esterase (Forestomach) Activity ... 804
    Maltogenic Amylase Activity ... 804
    Milk-Clotting Activity ... 805
    Pancreatin Activity ... 805
    Pepsin Activity ... 807
    Phospholipase A_2 Activity ... 808
    Phytase Activity ... 808
    Plant Proteolytic Activity ... 810
    Proteolytic Activity, Bacterial (PC) ... 811
    Proteolytic Activity, Fungal (HUT) ... 812
    Proteolytic Activity, Fungal (SAP) ... 813
    Pullulanase Activity ... 814
    Trypsin Activity ... 814

APPENDIX VI:
ESSENTIAL OILS AND FLAVORS ... 815
    Acetals ... 815
    Acid Value ... 815
    Alcohols, Total ... 815
    Aldehydes ... 816
    Aldehydes and Ketones ... 816
        Hydroxylamine Method ... 816
        Hydroxylamine/Tert-Butyl Alcohol Method ... 816
        Neutral Sulfite Method ... 817
    Chlorinated Compounds ... 817
    Esters ... 817
        Ester Determination ... 817
        Ester Determination (High-Boiling Solvent) ... 817
        Saponification Value ... 817
        Ester Value ... 817
    Linalool Determination ... 818
    Percentage of Cineole ... 818
    Phenols ... 818
    Phenols, Free ... 818
    Residue on Evaporation ... 819
    Solubility in Alcohol ... 819
    Ultraviolet Absorbance of Citrus Oils ... 819
    Volatile Oil Content ... 819

APPENDIX VII:
FATS AND RELATED SUBSTANCES ... 820
    Acetyl Value ... 820
    Acid Value ... 820
    Chlorophyll ... 821
    Cold Test ... 821
    Color (AOCS-Wesson) ... 821
    Fatty Acid Composition ... 821
    Fatty Acids, Free ... 822
    Glycerin or Propylene Glycol, Free ... 822
    Hydroxyl Value ... 822
    Iodine Value ... 823
        Wijs Method, Modified ... 823
    Melting Range ... 824
    1-Monoglycerides ... 824
    Monoglycerides, Total ... 824
    Oxyethylene Determination ... 825
    Reichert-Meissl Value ... 826
    Saponification Value ... 827
    Soap ... 827
    Specific Gravity ... 827
    Stability ... 828
    Unsaponifiable Matter ... 828
    Volatile Acidity ... 829

APPENDIX VIII:
OLEORESINS ... 829
    Color Value ... 829
    Curcumin Content ... 829
    Piperine Content ... 830
    Residual Solvent ... 830
    Scoville Heat Units ... 831
    Volatile Oil Content ... 832

APPENDIX IX:
ROSINS AND RELATED SUBSTANCES ... 832
    Acid Number ... 832
    Softening Point ... 832
        Drop Method ... 832
        Ring-and-Ball Method ... 834
    Viscosity ... 835

APPENDIX X:
CARBOHYDRATES (STARCHES, SUGARS, AND RELATED SUBSTANCES) ... 836
    Acetyl Groups ... 836
    Crude Fat ... 836
    Invert Sugar ... 836
        Assay ... 836
    Lactose ... 837
    Manganese ... 838
    Phosphorus ... 838
    Propylene Chlorohydrin ... 839
    Reducing Sugars Assay ... 840
    Sulfur Dioxide Determination ... 841
    Total Solids ... 842
        Glucose Syrup (Corn Syrup) ... 843
        High-Fructose Corn Syrup Solids ... 844
        Maltodextrin ... 845
        Invert Sugar ... 846
        Sucrose ... 846

SOLUTIONS AND INDICATORS ... 848
    Colorimetric Solutions (CS) ... 848
    Standard Buffer Solutions ... 848
    Standard Solutions for the Preparation of Controls and Standards ... 849
    Test Solutions (TS) and Other Reagents ... 849
    Volumetric Solutions ... 855
    Indicators ... 860
    Indicator Papers and Test Papers ... 861
    Detector Tubes ... 862

List of Figures

No.

1. Chromatographic Separation of Two Substances ... 736
2. Asymmetrical Chromatographic Peak ... 736
3. Apparatus for Determination of Solidification Point ... 742
4. Stirrer for Solidification Point Determination ... 742
5. Ubbelohde Viscometer for Dimethylpolysiloxane ... 743
6. Methylcellulose Viscometers ... 744
7. Agitator for Viscosity of Sodium Carboxymethylcellulose ... 745
8. Moisture Distillation Apparatus ... 748
9. Assembly for Checking Pore Diameter of Filter Sticks ... 750
10. Filtration Assembly for Determination of Oil Content ... 750
11. General Apparatus for Arsenic Test ... 755
12. Modified Bethge Apparatus for the Distillation of Arsenic Tribromide ... 755
13. Special Apparatus for the Distillation of Arsenic Trichloride ... 756
14. Special Apparatus for the Determination of Inorganic Arsenic ... 756
15. Closed-System Vacuum Distillation Apparatus for 1,4-Dioxane ... 757
16. Aeration Apparatus for Mercury Limit Test ... 766
17. Apparatus for Alginates Assay ... 768
18. Typical Chromatogram for the Determination of Benzene in Hexanes Using Column No. 5 ... 769
19. Illustration of *A/B* Ratio ... 769
20. Illustration of *A/B* Ratio for a Small Component Peak on the Tail of a Large Peak ... 769
21. Upward Displacement-Type Liquid–Liquid Extractor with Sintered-Glass Diffuser ... 771
22. (a) Schematic Diagram of Apparatus for Photometric Mercury Vapor M Method; (b) Quartz Combustion Tube with Boat and Copper Oxide Packing; (c) Schematic Diagram of Trap Used To Contain Ascarite, Dehydrite, and Aluminum Oxide ... 772
23. Titanous Chloride Titration Apparatus ... 774
24. F D & C Red No. 40 Top and Bottom Traces ... 776
25. F D & C Yellow No. 5 Top and Bottom Traces ... 776
26. F D & C Yellow No. 6 Top and Bottom Traces ... 777
27. Apparatus for Hydroxypropoxyl Determination ... 777
28. Chart for Converting Percentage of Substitution, by Weight, of Hydroxypropoxyl Groups to Molecular Substitution per Glucose Unit ... 778
29. Distillation Apparatus for Methoxyl Determination ... 778
30. Microcoulometric Titrating System for the Determination of Sulfur in Hexanes ... 780
31. Raney Nickel Reduction Apparatus ... 781
32. Column Assembly for Assay of Immobilized Glucose Isomerase ... 797
33. Typical Spectrogram of Lemon Oil ... 819
34. Apparatus for Determination of Volatile Oil Content ... 819
35. Apparatus for Oxyethylene Distribution ... 825
36. Reichert-Meissl Distillation Apparatus ... 827
37. Modified Hortvet-Sellier Distillation Apparatus ... 829
38. Clevenger Traps ... 830
39. Apparatus for Drop Softening Point Determination ... 833
40. Shouldered Ring, Ring Holder, Ball-Centering Guide, and Assembly of Apparatus Showing Two Rings ... 834
41. Assembly of Apparatus Showing Stirrer and Single-Shouldered Ring ... 835
42. Butt-Type Extractor for Crude Fat Determination ... 836
43. The Optimized Monier-Williams Apparatus ... 841
44. Diagram of Bubbler ... 842

APPENDIX I: APPARATUS FOR TESTS AND ASSAYS

OXYGEN FLASK COMBUSTION

Apparatus The apparatus consists of a heavy-walled, deeply lipped or cupped, conical flask of a volume suitable for the complete combustion of the sample in which the particular element is being determined (e.g., see *Selenium Limit Test*, Appendix IIIB). The flask is fitted with a ground-glass stopper to which is fused a sample carrier consisting of heavy-gauge platinum wire and a piece of welded platinum gauze measuring about 1.5 × 2 cm. A suitable apparatus may be obtained as Catalog Nos. 6513-C20 (500-mL capacity) and 6513-C30 (1000-mL capacity) from the Arthur H. Thomas Co., P.O. Box 779, Philadelphia, PA 19105. Equivalent apparatus available from other sources, or other suitable apparatus embodying the principles described herein, may also be used.

Procedure (Caution: The analyst should wear safety glasses and should use a suitable safety shield between himself and the apparatus. Further safety measures should be observed as necessary to ensure maximum protection of the analyst. Furthermore, the flask must be scrupulously clean and free from even traces of organic solvents. Samples containing water of hydration or more than 1% of moisture should be dried at 140° for 2 h before combustion, unless otherwise directed.) Accurately weigh the amount of sample specified in the monograph or general test. Solids should be weighed on a 4-cm square piece of halide-free paper, which should be folded around the sample. Liquid samples not exceeding 0.2 mL in volume should be weighed in tared cellulose acetate capsules [available as Catalog Nos. 6513-C80 (100 capsules) and 6513-C82 (1000 capsules) from the Arthur H. Thomas Co.]; gelatin capsules are satisfactory for liquid samples exceeding 0.2 mL in volume.

> Note: Gelatin capsules may contain significant amounts of combined halide or sulfur, in which case a blank determination should be made as necessary.

Place the sample, together with a filter paper fuse-strip, in the platinum gauze sample holder. Place the absorbing liquid, as specified in the individual monograph or general test, in the flask, moisten the joint of the stopper with water, and flush the air from the flask with a stream of rapidly flowing oxygen, swirling the liquid to facilitate its taking up oxygen.

> Note: Saturation of the liquid with oxygen is essential for successful performance of this procedure.

Ignite the fuse-strip by suitable means. If the strip is ignited outside the flask, immediately plunge the sample holder into the flask, invert the flask so that the absorption solution makes a seal around the stopper, and hold the stopper firmly in place. If the ignition is carried out in a closed system, the inversion of the flask may be omitted. After combustion is complete, shake the flask vigorously, and allow to stand for not less than 10 min with intermittent shaking. Then continue as directed in the individual monograph or general test chapter.

THERMOMETERS

Thermometers suitable for *Food Chemicals Codex* use conform to the specifications of the American Society for Testing and Materials, ASTM Standards E 1, and are standardized in accordance with ASTM Method E 77.

The thermometers are of the mercury in glass type, and the column above the liquid is filled with nitrogen. They may be standardized for *total immersion* or for *partial immersion* and should be used as near as practicable under the same condition of immersion.

"Total immersion" means standardization with the thermometer immersed to the top of the mercury column, with the remainder of the stem and the upper expansion chamber exposed to the ambient temperature. "Partial immersion" means standardization with the thermometer immersed to the indicated immersion line etched on the front of the thermometer, with the remainder of the stem exposed to the ambient temperature. If used under any other condition of immersion, an emergent-stem correction is necessary to obtain correct temperature readings.

Thermometer Specifications

| ASTM | Range | | Subdivisions | | Immersion |
No. E 1	°C	°F	°C	°F	(mm)
For General Use					
1 C	−20 to +150	—	1	—	76
1 F	—	0 to 302	—	2	76
2 C	−5 to +300	—	1	—	76
2 F	—	20 to 580	—	2	76
3 C	−5 to +400	—	1	—	76
3 F	—	20 to 760	—	2	76
For Determination of Softening Point					
15 C	−2 to +80	—	0.2	—	total
15 F	—	30 to 180	—	0.5	total
16 C	30 to 200	—	0.5	—	total
16 F	—	85 to 392	—	1	total
For Determination of Kinematic Viscosity					
44 F	—	66.5 to 71.5	—	0.1	total
45 F	—	74.5 to 79.5	—	0.1	total
28 F	—	97.5 to 102.5	—	0.1	total
46 F	—	119.5 to 124.5	—	0.1	total
29 F	—	127.5 to 132.5	—	0.1	total
47 F	—	137.5 to 142.5	—	0.1	total
48 F	—	177.5 to 182.5	—	0.1	total
30 F	—	207.5 to 212.5	—	0.1	total
For Determination of Distillation Range					
37 C	−2 to +52	—	0.2	—	100
38 C	24 to 78	—	0.2	—	100
39 C	48 to 102	—	0.2	—	100
40 C	72 to 126	—	0.2	—	100
41 C	98 to 152	—	0.2	—	100
102 C	123 to 177	—	0.2	—	100

Thermometer Specifications (*continued*)

ASTM No. E 1	Range °C	°F	Subdivisions °C	°F	Immersion (mm)
103 C	148 to 202	—	0.2	—	100
104 C	173 to 227	—	0.2	—	100
105 C	198 to 252	—	0.2	—	100
106 C	223 to 277	—	0.2	—	100
107 C	248 to 302	—	0.2	—	100
For Determination of Solidification Point					
89 C	−20 to +10	—	0.1	—	76
90 C	0 to 30	—	0.1	—	76
91 C	20 to 50	—	0.1	—	76
92 C	40 to 70	—	0.1	—	76
93 C	60 to 90	—	0.1	—	76
94 C	80 to 110	—	0.1	—	76
95 C	100 to 130	—	0.1	—	76
96 C	120 to 150	—	0.1	—	76
100 C	145 to 205	—	0.2	—	76
101 C	195 to 305	—	0.5	—	76
For Special Use					
14 C[a]	38 to 82	—	0.1	—	79
38 C[b]	−2 to +68	—	0.2	—	45
18 C[c]	34 to 42	—	0.1	—	total
18 F[c]	—	94 to 108	—	0.2	total
22 C[c]	95 to 103	—	0.1	—	total
22 F[c]	—	204 to 218	—	0.2	total
23 C[d]	18 to 28	—	0.2	—	90
24 C[d]	30 to 54	—	0.2	—	90
54 F[e]	—	68 to 213	—	0.5	total
71 F[f]	—	−35 to +70	—	1	76

[a]For determination of melting range of Class III solids.
[b]For determination of the titer of fatty acids.
[c]For determination of Saybolt viscosity.
[d]For determination of Engler viscosity.
[e]For determination of congealing point.
[f]For determination of oil in wax.

In selecting a thermometer, careful consideration should be given to the conditions under which it is to be used. The preceding table lists several ASTM thermometers, together with their usual conditions of use, which may be required in *Food Chemicals Codex* tests. Complete specifications for these thermometers are given in "ASTM Standards on Thermometers."

VOLUMETRIC APPARATUS

Most of the volumetric apparatus available in the United States is calibrated at 20°, although the temperatures generally prevailing in laboratories more nearly approach 25°, which is the temperature specified generally for tests and assays. This discrepancy is inconsequential provided the room temperature is reasonably constant and the apparatus has been calibrated accurately prior to and under the conditions of its intended use.

Before use, all volumetric ware must be cleaned in such a manner that, when rinsed with water, no droplet of water can be seen on the inside walls. Many kinds of "degreasing" solutions are available, and the user should consult the manufacturer's literature for the system of choice.

Use To attain the degree of precision required in many assays involving volumetric measurements and directing that a quantity be "accurately measured," the apparatus must be chosen and used with exceptional care. Where less than 10 mL of titrant is to be measured, a 10-mL buret or microburet generally is required.

The design of volumetric apparatus is an important factor in ensuring accuracy. For example, the length of the graduated portions of graduated cylinders should be not less than five times the inside diameter, and the tips of burets should permit an outflow of not more than 0.5 mL/s.

Pipets and burets must be allowed to drain properly in use. Usually, transfer pipets for dilute aqueous solutions should drain for the time specified by the manufacturer before the tip is touched to the wall of the vessel. Buret volumes should not be read immediately upon delivery of the titrant. A suitable length of time should elapse to allow the titrant retained on the walls to drain down. A time interval of 5 to 10 s is usually sufficient.

Standards of Accuracy The capacity tolerances for volumetric flasks, transfer pipets, and burets are those accepted by the National Institute of Standards and Technology (Class A),[1] as indicated in the accompanying tables. Use Class A volumetric apparatus unless otherwise specified in the individual monograph. For plastic volumetric apparatus, the accepted capacity tolerances are Class B.[2]

Volumetric Flasks

	Designated Volume (mL)						
	10	25	50	100	250	500	1000
Limit of error (mL)	0.02	0.03	0.05	0.08	0.12	0.15	0.30
Limit of error (%)	0.20	0.12	0.10	0.08	0.05	0.03	0.03

Transfer Pipets

	Designated Volume (mL)						
	1	2	5	10	25	50	100
Limit of error (mL)	0.006	0.006	0.01	0.02	0.03	0.05	0.08
Limit of error (%)	0.6	0.30	0.20	0.20	0.12	0.10	0.08

Burets

	Designated Volume (mL)		
	10 ("micro" type)	25	50
Subdivisions (mL)	0.02	0.10	0.10
Limit of error (mL)	0.02	0.03	0.05

[1]See "Testing of Glass Volumetric Apparatus," NBS Circ. 602, April 1, 1959. Apparatus meeting the specifications of NBSIR 74-461 ("The Calibration of Small Volumetric Laboratory Glassware"), as well as of ANSI/ASTM E 694-79 ("Specifications for Volumetric Ware"), is also acceptable.

[2]See ASTM E 288, Fed. Spec. NNN-F-289, and ISO Standard 284.

The capacity tolerances for measuring (i.e., "graduated") pipets of up to and including 10-mL capacity are somewhat larger than those for the corresponding sizes of transfer pipets, namely, 0.01, 0.02, and 0.03 mL for the 2-, 5-, and 10-mL sizes, respectively.

Transfer and measuring pipets calibrated "to deliver" should be drained in a vertical position and then touched against the wall of the receiving vessel to drain the tips. Volume readings on burets should be estimated to the nearest 0.01 mL for 25- and 50-mL burets, and to the nearest 0.005 mL for 5- and 10-mL burets. Pipets calibrated "to contain" may be called for in special cases, generally for measuring viscous fluids. In such cases, the pipet should be washed clean, after draining, and the washings added to the measured portion.

WEIGHTS AND BALANCES

Codex tests and assays are designed for use with three types of analytical balances, known as macro-, semimicro-, and micro-.

By custom, microbalances weigh objects with a sensitivity down to the μg range (or lower); semimicrobalances down to the 0.01-mg range; and analytical macrobalances down to the 0.1-mg range.

Tolerances The analytical weights meet the tolerances of the American National Standard ANSI/ASTM E 617, "Laboratory Weights and Precision Mass Standards." This standard is incorporated by reference and should be consulted for full description and information on the tolerances and construction of weights.[3] Where quantities of 25 mg or less are to be "weighed accurately," any applicable corrections for weights should be used.

Class 1.1 weights are used for calibration of low-capacity, high-sensitivity balances. They are available in various denominations from 1 to 500 mg. The tolerance for any denomination in this class is 5 μg. They are recommended for calibration of balances using optical or electrical methods for accurately weighing quantities below 20 mg.

Class 1 weights are designated as high-precision standards for calibration. They may be used for weighing accurately quantities below 20 mg.

Class 2 weights are used as working standards for calibration, built-in weights for analytical balances, and laboratory weights for routine analytical work.

Class 3 and Class 4 weights are used with moderate-precision laboratory balances.

Use Where substances are to be "accurately weighed" in an assay or a test, the weighing is to be performed in such manner as to limit the error to 0.1% or less. For example, a quantity of 50 mg is to be weighed to the nearest 0.05 mg; a quantity of 0.1 g is to be weighed to the nearest 0.1 mg; and a quantity of 10 g is to be weighed to the nearest 10 mg.

Calibration All precision balances and weights should be calibrated periodically (preferably at least once a year) and a record kept of the calibration date and results. The user may have a set of weights calibrated by the nearest Department of Weights and Measurements (or its equivalent). This is usually done for little or no charge. Alternatively, an independent, outside company may be retained for the purpose of performing such calibrations.

Buoyancy Effect When a weighing is to be performed with an accuracy of 0.1% or better, the buoyancy effect should not be neglected. The equation to be used in correcting for this effect is

$$M_v = M_a[1 + 0.0012(1/D_o - D_w)],$$

in which M_v is the mass in vacuum; M_a is the mass in air; 0.0012 is the density of air; D_o is the density of the weighed object; and D_w is the density of the calibrated weights.

APPENDIX II: PHYSICAL TESTS AND DETERMINATIONS

A. CHROMATOGRAPHY

For the purposes of the *Food Chemicals Codex*, chromatography is defined as an analytical technique whereby a mixture of chemicals may be separated by virtue of their differential affinities for two immiscible phases. One of these, the stationary phase, consists of a fixed bed of small particles with a large surface area, while the other, the mobile phase, is a gas or liquid that moves constantly through, or over the surface of, the fixed phase. Chromatographic systems achieve their ability to separate mixtures by selectively retarding the passage of some compounds through the stationary phase while permitting others to move more freely. Therefore, the chromatogram may be evaluated qualitatively by determining the R_f, or retardation factor, for each of the eluted substances. The R_f is a measure of that fraction of its total elution time that any compound spends in the mobile phase. Since this fraction is directly related to the fraction of the total amount of the solute that is in the mobile phase, the R_f can be expressed as

---

[3]Copies of ASTM Standard E 617-81 (Reapproved 1985) may be obtained from the American Society for Testing and Materials, 1916 Race Street, Philadelphia, PA 19103.

$$R_f = V_m C_m/(V_m C_m + V_s C_s),$$

in which V_m and V_s are the volumes of the mobile and stationary phase, respectively, and C_m and C_s are the concentrations of the solute in either phase at any time. This can be simplified to

$$R_f = V_m/(V_m + KV_s),$$

in which $K = C_s/C_m$ and is an equilibrium constant that indicates this differential affinity of the solute for the phases. Alternatively, a new constant, k', the capacity factor, may be introduced, giving another form of the expression:

$$R_f = 1/(1 + k'),$$

in which $k' = KV_s/V_m$. The capacity factor, k', which is normally constant for small samples, is a parameter that expresses the ability of a particular chromatographic system to interact with a solute. The larger the k' value, the more the sample is retarded.

Both the retardation factor and the capacity factor may be used for qualitative identification of a solute or for developing strategies for improving separation. In terms of parameters easily obtainable from the chromatogram, the R_f is defined as the ratio of the distance traveled by the solute band to the distance traveled by the mobile solvent in a particular time. The capacity factor, k', can be evaluated by the expression

$$k' = (t_r - t_o)/t_o,$$

in which t_r, the retention time, is the elapsed time from the start of the chromatogram to the elution maximum of the solute, and t_o is the retention time of a solute that is not retained by the chromatographic system.

Retardation of the solutes by the stationary phase may be achieved by one or a combination of mechanisms. Certain substances, such as alumina or silica gel, interact with the solutes primarily by *adsorption*, either *physical adsorption*, in which the binding forces are weak and easily reversible, or *chemisorption*, in which strong bonding to the surface can occur. Another important mechanism of retardation is *partition*, which occurs when the solute dissolves in the stationary phase, usually a liquid coated as a thin layer on the surface of an inert particle or chemically bonded to it. If the liquid phase is a polar substance (e.g., polyethylene glycol) and the mobile phase is nonpolar, the process is termed *normal-phase chromatography*. When the stationary phase is nonpolar (e.g., octadecylsilane) and the mobile phase is polar, the process is *reversed-phase chromatography*. For the separation of mixtures of ionic species, insoluble polymers called *ion exchangers* are used as the stationary phase. Ions of the solutes contained in the mobile phase are adsorbed onto the surface of the ion exchanger while at the same time displacing an electrically equivalent amount of less strongly bound ions in order to maintain the electroneutrality of both phases. The chromatographic separation of mixtures of large molecules such as proteins may be accomplished by a mechanism called *size exclusion chromatography*. The stationary phases used are highly cross-linked polymers that have imbibed a sufficient amount of solvent to form a gel. The separation is based on the physical size of the solutes; those that are too large to fit within the interstices of the gel are eluted rapidly, while the smaller molecules follow an irregular path through the pores of the gel and are eluted later. In any chromatographic separation, more than one of the above mechanisms may be occurring simultaneously.

Chromatographic separations may also be characterized according to the type of instrumentation or apparatus used. The types of chromatography that may be used in the FCC are column, thin-layer, gas, and high-pressure or high-performance liquid chromatography.

COLUMN CHROMATOGRAPHY

Apparatus The equipment needed for column chromatography is not elaborate, consisting only of a cylindrical glass or Teflon tube that has a restricted outflow orifice. The dimensions of the tube are not critical and may vary from 10 to 40 mm in inside diameter and from 100 to 600 mm in length. For a given separation, greater efficiency may be obtained with a long narrow column, but the resultant flow rate will be lower. A fritted-glass disk may be seated in the end of the tube to act as a support for the packing material. The column is fitted at the end with a stopcock or other flow-restriction device in order to control the rate of delivery of the eluant.

Procedure The stationary phase is introduced into the column either as a dry powder or as a slurry in the mobile phase. Since a homogeneous bed free of void spaces is necessary to achieve maximum separation efficiency, the packing material is introduced in small portions and allowed to settle before further additions are made. Settling may be accomplished by allowing the mobile phase to flow through the bed, by tapping or vibrating the column if a dry powder is used, or by compressing each added portion using a tamping rod. The rod can be a solid glass, plastic, or metal cylinder whose diameter is slightly smaller than the column, or it can be a thinner rod onto the end of which has been attached a disk of suitable diameter. Ion-exchange resins and exclusion polymers are never packed as dry powders since after introduction of the mobile phase they will swell and create sufficient pressure to shatter the column. When the packing has been completed, the sample is introduced onto the top of the column. If the sample is soluble, it is dissolved in a minimum amount of the mobile phase, pipetted onto the column, and allowed to percolate into the top of the bed. If it is not soluble or if the volume of solution is too large, it may be mixed with a small amount of the column packing. This material is then transferred to the chromatographic tube to form the top of the bed.

The chromatogram is then developed by adding the mobile phase to the column in small portions and allowing it to percolate through the packed bed either by gravity or under the influence of pressure or vacuum. Development of the chromatogram takes place by selective retardation of the components of the mixture as a result of their interaction with the stationary phase. In column chromatography, the stationary phase may act by adsorption, partition, ion exchange, exclusion of the solutes, or a combination of these effects.

When the development is complete, the components of the sample mixture may be detected and isolated by either of two procedures. The entire column may be extruded carefully from the tube, and if the compounds are colored or fluorescent under ultraviolet light, the appropriate segments may be cut from the column using a razor blade. If the components are colorless, they may be visualized by painting or spraying a thin longitudinal section of the surface of the chromatogram with color-developing reagents. The chemical may then be separated from the stationary phase by extraction with a strong solvent such as methanol and subsequently quantitated by suitable methods.

In the second procedure, the mobile phase may be allowed to flow through the column until the components of the mixture successively appear in the effluent. This eluate may be collected in fractions and the mobile phase evaporated if desired. The chemicals present in each fraction may then be determined by suitable analytical techniques.

THIN-LAYER CHROMATOGRAPHY

In thin-layer chromatography (TLC), the stationary phase is a uniform layer of a finely divided powder that has been coated on the surface of a glass or plastic sheet and that is held in place by a binder. The capacity of the system is dependent on the thickness of the layer, which may range from 0.1 to 2.0 mm. The thinner layers are used primarily for analytical separations, while the thicker layers, because of their greater sample-handling ability, are useful for preparative work.

Substances that are used as coatings in TLC include silica gel, alumina, or cellulose. Separations occur due to adsorption of the solutes from the mobile phase onto the surface of the thin layer. However, adsorption of water from the air or solvent components from the mobile phase can give rise to partition or liquid–liquid chromatography. Specially coated plates are available that permit ion exchange or reversed-phase separations.

Apparatus Acceptable apparatus and materials for thin-layer chromatography consist of the following:

Glass Plates Flat glass plates of uniform thickness throughout their areas. The most common sizes are 20, 10, and 5 cm × 20 cm.

Aligning Tray An aligning tray or other suitable flat surface is used to align and hold plates during application of the adsorbent.

Adsorbent The adsorbent may consist of finely divided adsorbent materials for chromatography. It can be applied directly to the glass plate, or it can be bonded to the plate by means of plaster of Paris or with starch paste. Pretreated chromatographic plates are available commercially.

Spreader A suitable spreading device that when moved over the glass plate applies a uniform layer of adsorbent of desired thickness over the entire surface of the plate.

Storage Rack A rack of convenient size to hold the prepared plates during drying and transportation.

Developing Chamber A glass chamber that can accommodate one or more plates and can be properly closed and sealed. It is fitted with a plate-support rack that can support the plates when the lid of the chamber is in place.

Note: Pre-formed TLC plates available commercially may also be utilized.

Procedure Clean the plates scrupulously, as by immersion in a chromic acid cleansing mixture, and rinse them with copious quantities of water until the water runs off the plates without leaving any visible water or oily spots, and then dry.

Arrange the plate or plates on the aligning tray, and secure them so that they will not slip during the application of the adsorbent. Mix an appropriate quantity of adsorbent and liquid, usually water, which when shaken for 30 s gives a smooth slurry that will spread evenly with the aid of a spreader. Transfer the slurry to the spreader, and apply the coating at once before the binder begins to harden. Move the spreader smoothly over the plates from one end of the tray to the other. Remove the spreader, and wipe away excess slurry. Allow the plates to set for 10 min, and then place them in the storage rack and dry at 105° for 30 min or as directed in the monograph. Store the finished plates in a desiccator.

Equilibrate the atmosphere in the *Developing Chamber*, by placing a volume of the mobile phase in excess of that required for complete development of the chromatogram, cover the chamber with its lid and allow to stand for at least 30 min.

Apply the *Sample Solution* and the *Standard Solution* at points about 1.5 cm apart and about 2 cm from the lower edge of the plate (the lower edge is the first part over which the spreader moves in the application of the adsorbent layer), and allow to dry. A template will aid in determining the spot points and the 10- to 15-cm distance through which the solvent front should move.

Arrange the plate on the supporting rack (sample spots on the bottom), and introduce the rack into the developing chamber. The solvent in the chamber must be deep enough to reach the lower edge of the adsorbent, but must not touch the spot points. Seal the cover in place, and maintain the system until the solvent ascends to a point 10 to 15 cm above the initial spots, this usually requiring from 15 min to 1 h. Remove the plates, and dry them in air. Measure and record the distance of each spot from the point of origin. If so directed, spray the spots with the reagent specified, observe, and compare the sample with the standard chromatogram.

Detection and Identification Detection and identification of solute bands is done by methods essentially the same as those described in *Column Chromatography*. However, in TLC an additional method called *fluorescence quenching* is also used. In this procedure, an inorganic phosphor is mixed with the adsorbent before it is coated on the plate. When the developed chromatogram is irradiated with ultraviolet light, the surface of the plate fluoresces with a characteristic color, except in those places where ultraviolet-absorbing solutes are situated. These quench the fluorescence and are detectable as dark spots.

Detection with an ultraviolet light source suitable for observations with short (254 nm) and long (360 nm) ultraviolet wavelengths may be called for in some cases.

Quantitative Analysis Two methods are available if quantitation of the solute is necessary. In the first, the bands are detected and their positions marked. Those areas of adsorbent containing the compounds of interest are scraped from the surface of the plate into a centrifuge tube. The chemicals are extracted from the adsorbent with the aid of a suitable strong solvent, the suspension is centrifuged, and the supernatant layer is subjected to appropriate methods of quantitative analysis.

The second method involves the use of a scanning densitometer. This is a spectrophotometric device that directs a beam of monochromatic radiation across the surface of the plate. After interaction with the solutes in the adsorbent layer, the radiation is detected as transmitted or reflected light and a recording of light intensity versus distance traveled is produced. The concentration of a particular species is proportional to the area under its peak and can be determined accurately by comparison with standards.

GAS CHROMATOGRAPHY

The distinguishing features of gas chromatography are a gaseous mobile phase and a solid or immobilized liquid stationary phase. Liquid stationary phases are available in packed or capillary columns. In the packed columns, the liquid phase is deposited on a finely divided, inert solid support, such as diatomaceous earth, porous polymer, or graphitized carbon, which is packed into a column that is typically 2 to 4 mm in internal diameter and 1 to 3 m in length. In capillary columns, which contain no packing, the liquid phase is deposited on the inner surface of the fused silica column and may be chemically bonded to it. In gas–solid chromatography, the solid phase is an active adsorbent, such as alumina, silica, or carbon, packed into a column. Polyaromatic porous resins, which are sometimes used in packed columns, are not coated with a liquid phase.

When a volatile compound is introduced into the carrier gas and carried into the column, it is partitioned between the gas and stationary phases by a dynamic countercurrent distribution process. The compound is carried down the column by the carrier gas, retarded to a greater or lesser extent by sorption and desorption in the stationary phase. The elution of the compound is characterized by the partition ratio, k', a dimensionless quantity also called the capacity factor. It is equivalent to the ratio of the time required for the compound to flow through the column (the retention time) to the retention time of a nonretarded compound. The value of the capacity factor depends on the chemical nature of the compound, the nature, amount, and surface area of the liquid phase, and the column temperature. Under a specified set of experimental conditions, a characteristic capacity factor exists for every compound. Separation by gas chromatography occurs only if the compounds concerned have different capacity factors.

Apparatus A gas chromatograph consists of a carrier gas source, an injection port, column, detector, and recording device. The injection port, column, and detector are temperature controlled. The typical carrier gas is helium, nitrogen, or hydrogen, depending on the column and detector in use. The gas is supplied from a high-pressure cylinder and passes through suitable pressure-reducing valves to the injection port and column. Compounds to be chromatographed, either in solution or as gases, are injected into the gas stream at the injection port. Depending upon the configuration of the apparatus, the test mixture may be injected directly into the column or be vaporized in the injection port and mixed into the flowing carrier gas prior to entering the column.

Once in the column, compounds in the test mixture are separated by virtue of differences in their capacity factors, which in turn depend upon their vapor pressure and degree of interaction with the stationary phase. The capacity factor, which governs resolution and retention times of components of the test mixture, is also temperature dependent. The use of temperature-programmable column ovens takes advantage of this dependence to achieve efficient separation of compounds differing widely in vapor pressure.

As resolved compounds emerge separately from the column, they pass through a detector, which responds to the amount of each compound present. The type of detector to be used depends upon the nature of the compounds to be analyzed, and is specified in the individual monograph. Detectors are heated to prevent condensation of the eluting compounds.

Detector output is recorded as a function of time, producing a chromatogram, which consists of a series of peaks on a time axis. Each peak represents a compound in the vaporized test mixture, although some peaks may overlap. The elution time is characteristic of the individual compounds, and the peak area is a function of the amount present.

Injectors Sample injection devices range from simple syringes to fully programmable automatic injectors. The amount of sample that can be injected into a capillary column without overloading is small compared to the amount that can be injected into a packed column, and may be less than the smallest amount that can be manipulated satisfactorily by syringe. Capillary columns are therefore used with injectors able to split samples into two fractions, a small one that enters the column and a large one that goes to waste (split injector). Such injectors may also be used in a splitless mode for analyses of trace or minor components.

Purge and trap injectors are equipped with a sparging device by which volatile compounds in solution are carried into a low-temperature trap. When sparging is complete, trapped compounds are thermally desorbed into the carrier gas by rapid heating of the temperature-programmable trap.

Headspace injectors are equipped with a thermostatically controlled sample heating chamber. Solid or liquid samples in tightly closed containers are heated in the chamber for a fixed period of time, allowing the volatile components in the sample to reach an equilibrium between the non-gaseous phase and the gaseous or headspace phase.

After this equilibrium has been established, the injector automatically introduces a fixed amount of the headspace in the sample container into the gas chromatograph.

Columns Capillary columns, which are usually made of fused silica, are 0.2 to 0.53 mm in internal diameter and 5 to

30 m in length. The liquid or stationary phase is 0.1 to 1.0 μm thick, although nonpolar stationary phases may be up to 5 μm thick.

Packed columns, made of glass or metal, are 1 to 3 m in length with internal diameters of 2 to 4 mm. Those used for analysis typically have liquid phase loadings of about 5% (w/w) on a solid support.

Supports for analysis of polar compounds on low-capacity, low-polarity liquid phase columns must be inert to avoid peak tailing. The reactivity of support materials can be reduced by silanizing prior to coating with liquid phase. Acid-washed, flux-calcined diatomaceous earth is often used for drug analysis. Support materials are available in various mesh sizes, with 80- to 100-mesh and 100- to 120-mesh being most commonly used with 2- to 4-mm columns.

Retention time and the peak efficiency depend on the carrier gas flow rate; retention time is also directly proportional to column length, while resolution is proportional to the square root of the column length. For packed columns, the carrier gas flow rate is usually expressed in mL/min at atmospheric pressure and room temperature. It is measured at the detector outlet with a bubble tube while the column is at operating temperature. The linear flow rate through a packed column is inversely proportional to the square of the column diameter for a given flow volume. Flow rates of 60 mL/min in a 4-mm column and 15 mL/min in a 2-mm column give identical linear flow rates and thus similar retention times. Unless otherwise specified in the individual monograph, flow rates for packed columns are 60 to 75 mL/min for 4-mm id columns and ~30 mL/min for 2-mm id columns. For capillary columns, linear flow velocity is often used instead of flow rate. This is conveniently determined from the length of the column and the retention time of a dilute methane sample, provided a flame-ionization detector is in use. Typical linear velocities are 20 to 60 cm/s for helium and 35 to 70 cm/s for hydrogen. At high operating temperatures there is sufficient vapor pressure to result in a gradual loss of liquid phase, a process called bleeding.

Detectors Flame-ionization detectors are used for most analyses, with lesser use made of thermal conductivity, electron-capture, nitrogen-phosphorus, and mass spectrometric detectors. For quantitative analyses, detectors must have a wide linear dynamic range: the response must be directly proportional to the amount of compound present in the detector over a wide range of concentrations. Flame-ionization detectors have a wide linear range ($\sim 10^6$) and are sensitive to organic compounds. Unless otherwise specified in individual monographs, flame-ionization detectors with either helium or nitrogen carrier gas are to be used for packed columns, and helium or hydrogen is used for capillary columns.

The thermal conductivity detector detects changes in the thermal conductivity of the gas stream as solutes are eluted. Although its linear dynamic range is smaller than that of the flame-ionization detector, it is quite rugged and occasionally used with packed columns, especially for compounds that do not respond to flame-ionization detectors.

The alkali flame-ionization detector, sometimes called an NP or nitrogen–phosphorus detector, contains a thermionic source, such as an alkali–metal salt or a glass element containing rubidium or other metal, that results in the efficient ionization of organic nitrogen and phosphorus compounds. It is a selective detector that shows little response to hydrocarbons.

The electron-capture detector contains a radioactive source (usually ^{63}Ni) of ionizing radiation. It exhibits an extremely high response to compounds containing halogens and nitro groups but little response to hydrocarbons. The sensitivity increases with the number and atomic weight of the halogen atoms.

Data Collection Devices Modern data stations receive the detector output, calculate peak areas, and print chromatograms, complete with run parameters and peak data. Chromatographic data may be stored and reprocessed, with integration and other calculation variables being changed as required. Data stations are used also to program the chromatograph, controlling most operational variables and providing for long periods of unattended operation.

Data can also be collected for manual measurement on simple recorders or on integrators whose capabilities range from those providing a printout of peak areas to those providing chromatograms with peak areas and peak heights calculated and data stored for possible reprocessing.

Procedure Capillary columns must be tested to ensure that they comply with the manufacturers' specifications before they are used. These tests consist of the following injections: a dilute methane sample to determine the linear flow velocity; a mixture of alkanes (e.g., C_{14}, C_{15}, and C_{16}) to determine resolution; and a polarity test mixture to check for active sites on the column. The latter mixture may include a methyl ester, an unsaturated compound, a phenol, an aromatic amine, a diol, a free carboxylic acid, and a polycyclic aromatic compound, depending upon the samples to be analyzed.

Packed columns must be conditioned before use until the baseline and other characteristics are stable. This may be done by operation at a temperature above that called for by the method or by repeated injections of the compound or mixture to be chromatographed. A suitable test for support inertness should be done. Very polar molecules (like free fatty acids) may require a derivatization step.

Before any column is used for assay purposes, a calibration curve should be constructed to verify that the instrumental response is linear over the required range and that the curve passes through the origin. If the compound to be analyzed is adsorbed within the system, the calibration curve will intersect the abscissa at a nonzero value. This may result in error, particularly for compounds at low concentrations determined by a procedure based on a single reference point. At high concentrations, the liquid phase may be overloaded, leading to loss of peak height and symmetry.

Assays require quantitative comparison of one chromatogram with another. A major source of error is irreproducibility in the amount of sample injected, notably, when manual injections are made with a syringe. The effects of variability can be minimized by addition of an internal standard, a non-interfering compound present at the same concentration in sample and standard solutions. The ratio of peak response of the analyte to that of the internal standard is compared from one chromatogram to another. Where the internal standard is chemically similar to the substance being determined, there is also compensation for minor variations in column and detector characteristics. In some cases, the internal standard may be carried through the sample preparation procedure prior to gas chromatography to control other quantitative aspects of the assay. Automatic injectors greatly improve the reproducibility of sample injections and reduce the need for internal standards.

Many monographs require that system suitability requirements be met before samples are analyzed, see *System Suitability* below.

HIGH-PRESSURE LIQUID CHROMATOGRAPHY

High-pressure liquid chromatography (HPLC), sometimes called high-performance liquid chromatography, is a separation technique based on a solid stationary phase and a liquid mobile phase. Separations are achieved by partition, adsorption, exclusion, or ion-exchange processes, depending upon the type of stationary phase used. HPLC has distinct advantages over gas chromatography for the analysis of organic compounds. Compounds to be analyzed are dissolved in a liquid, and most separations take place at room temperature.

As in gas chromatography, the elution time of a compound can be described by the capacity factor, k', which depends on the chemical nature of the composition and flow rate of the mobile phase, and the composition and surface area of the stationary phase. Column length is an important determinant of resolution. Only compounds having different capacity factors can be separated by HPLC.

Apparatus A liquid chromatograph consists of a reservoir containing the mobile phase, a pump to force the mobile phase through the system at high pressure, an injector to introduce the sample into the mobile phase, a chromatographic column, a detector, and a data collection device such as a computer, integrator, or recorder. Short, small-bore columns containing densely packed particles of stationary phase provide for the rapid exchange of compounds between the mobile and stationary phases. In addition to receiving and reporting detector output, computers are used to control chromatographic settings and operations, thus providing for long periods of unattended operation.

Pumping Systems HPLC pumping systems deliver metered amounts of mobile phase from the solvent reservoirs to the column through high-pressure tubing and fittings. Modern systems consist of one or more computer-controlled metering pumps that can be programmed to vary the ratio of mobile phase components, as is required for gradient chromatography, or to mix isocratic mobile phases (i.e., mobile phases having a fixed ratio of solvents). However, the proportion of ingredients in premixed isocratic mobile phases can be more accurately controlled than in those delivered by most pumping systems. Operating pressures up to 5000 psi with delivery rates up to about 10 mL/min are typical. Pumps used for quantitative analysis should be constructed of materials inert to corrosive mobile phase components and be capable of delivering the mobile phase at a constant rate with minimal fluctuations over extended periods of time.

Injectors After dissolution in mobile phase or other suitable solution, compounds to be chromatographed are injected into the mobile phase, either manually by syringe or loop injectors, or automatically by autosamplers. The latter consist of a carousel or rack to hold sample vials with tops that have a pierceable septum or stopper and an injection device to transfer sample from the vials to a calibrated, fixed-volume loop from which it is loaded into the chromatograph. Some autosamplers can be programmed to control sample volume, the number of injections and loop rinse cycles, the interval between injections, and other operating variables.

Some valve systems incorporate a calibrated sample loop that is filled with test solution for transfer to the column in the mobile phase. In other systems, test solution is transferred to a cavity by syringe and then switched into the mobile phase.

Columns For most analyses, separation is achieved by partition of compounds in the test solution between the mobile and stationary phases. Systems consisting of polar stationary phases and nonpolar mobile phases are described as normal phase, while the opposite arrangement, polar mobile phases and nonpolar stationary phases, is called reversed-phase chromatography. Partition chromatography is almost always used for hydrocarbon-soluble compounds of molecular weight less than 1000. The affinity of a compound for the stationary phase, and thus its retention time on the column, is controlled by making the mobile phase more or less polar. Mobile phase polarity can be varied by the addition of a second, and sometimes a third or even a fourth, component.

Stationary phases for modern, reversed-phase liquid chromatography typically consist of an organic phase chemically bound to silica or other materials. Particles are usually 3 or 5 or 10 μm in diameter, but sizes may range up to 50 μm for preparative columns. Small particles thinly coated with organic phase provide for fast mass transfer and hence rapid transfer of compounds between the stationary and mobile phases. Column polarity depends on the polarity, of the bound functional groups, which range from relatively nonpolar octadecyl silane to very polar nitrile groups.

Columns used for analytical separations usually have internal diameters of 2 to 5 mm; larger diameter columns are used for preparative chromatography. Columns may be heated to give more efficient separations, but only rarely are they used at temperatures above 60° because of potential stationary phase degradation or mobile phase volatility. Unless otherwise specified in the individual monograph, columns are used at ambient temperature.

Ion-exchange chromatography is used to separate water-soluble, ionizable compounds of molecular weight less than 1500. The stationary phases are usually synthetic organic resins; cation-exchange resins contain negatively charged active sites and are used to separate basic substances such as amines, while anion-exchange resins have positively charged active sites for separation of compounds with negatively charged groups, such as phosphate, sulfonate, or carboxylate groups. Water-soluble ionic or ionizable compounds are attracted to the resins, and differences in affinity bring about the chromatographic separation. The pH of the mobile phase, temperature, ion type, ionic concentration, and organic modifiers affect the equilibrium, and these variables can be adjusted to obtain the desired degree of separation.

In size-exclusion chromatography, columns are packed with a porous stationary phase. Molecules of the compounds being chromatographed are filtered according to size. Those too large to enter the pores pass unretained through the column (total exclusion). Smaller molecules enter the pores and are increasingly retained as molecular size decreases. These columns are typically used to remove high molecular weight matrices or to characterize the molecular weight distribution of a polymer.

Detectors Many compendial HPLC methods require the use of spectrophotometric detectors. Such a detector consists of a flow-through cell mounted at the end of the column. A beam of ultraviolet radiation passes through the flow cell and into the detector. As compounds elute from the column, they pass through the cell and absorb the radiation, resulting in measurable energy level changes.

Fixed, variable, and photodiode array (PDA) detectors are widely available. Fixed wavelength detectors operate at a single wavelength, typically 254 nm, emitted by a low-pressure mercury lamp. Variable wavelength detectors contain a continuous source, such as a deuterium or high-pressure xenon lamp, and a monochromator or an interference filter to generate monochromatic radiation at a wavelength selected by the operator. Modern variable wavelength detectors can be programmed to change wavelength while an analysis is in progress. Multi-wavelength detectors measure absorbance at two or more wavelengths simultaneously. In diode array multi-wavelength detectors, continuous radiation is passed through the sample cell, then resolved into its constituent wavelengths, which are individually detected by the photodiode array. These detectors acquire absorbance data over the entire UV-visible range, thus providing the analyst with chromatograms at multiple, selectable wavelengths and spectra of the eluting peaks. Diode array detectors usually have lower signal-to-noise ratios than fixed or variable wavelength detectors, and thus are less suitable for analysis of compounds present at low concentrations.

Differential refractometer detectors measure the difference between the refractive index of the mobile phase alone and that of the mobile phase containing chromatographed compounds as it emerges from the column. Refractive index detectors are used to detect non-UV absorbing compounds, but they are less sensitive than UV detectors. They are sensitive to small changes in solvent composition, flow rate, and temperature, so that a reference column may be required to obtain a satisfactory baseline.

Fluorometric detectors are sensitive to compounds that are inherently fluorescent or that can be converted to fluorescent derivatives either by chemical transformation of the compound or by coupling with fluorescent reagents at specific functional groups. If derivatization is required, it can be done prior to chromatographic separation or, alternatively, the reagent can be introduced into the mobile phase just prior to its entering the detector.

Potentiometric, voltammetric, or polarographic electrochemical detectors are useful for the quantitation of species that can be oxidized or reduced at a working electrode. These detectors are selective, sensitive, and reliable, but require conducting mobile phases free of dissolved oxygen and reducible metal ions. A pulseless pump must be used, and care must be taken to ensure that the pH, ionic strength, and temperature of the mobile phase remain constant. Working electrodes are prone to contamination by reaction products with consequent variable responses.

Electrochemical detectors with carbon-paste electrodes may be used advantageously to measure nanogram quantities of easily oxidized compounds, notably phenols and catechols.

Data Collection Devices Modern data stations receive and store detector output and print out chromatograms complete with peak heights, peak areas, sample identification, and method variables. They are also used to program the liquid chromatograph, controlling most variables and providing for long periods of unattended operation.

Data also may be collected on simple recorders for manual measurement or on stand-alone integrators, which range in complexity, from those providing a printout of peak areas to those providing a printout of peak areas and peak heights calculated and data stored for possible subsequent reprocessing.

Procedure The mobile phase composition significantly influences chromatographic performance and the resolution of compounds in the mixture being chromatographed. Composition has a much greater effect than temperature on the capacity factor, k'.

In partition chromatography, the partition coefficient, and hence the separation, can be changed by addition of another component to the mobile phase. In ion-exchange chromatography, pH and ionic strength, as well as changes in the composition of the mobile phase, affect capacity factors. The technique of continuously changing the solvent composition during the chromatographic run is called gradient elution or solvent programming. It is sometimes used to chromatograph complex mixtures of components differing greatly in their capacity factors. Detectors that are sensitive to change in solvent composition, such as the differential refractometer, are more difficult to use with the gradient elution technique.

For accurate quantitative work, high-purity, "HPLC-grade" solvents and reagents must be used. The detector must have a broad linear dynamic range, and compounds to be measured must be resolved from any interfering substances. The linear dynamic range of a compound is the range over which the detector signal response is directly proportional to the amount of the compound. For maximum flexibility in quantitative work, this range should be about three orders of magnitude. HPLC systems are calibrated by plotting peak responses in comparison with known concentrations of a reference standard, using either an external or an internal standardization procedure.

Reliable quantitative results are obtained by external calibration if automatic injectors or autosamplers are used. This method involves direct comparison of the peak responses obtained by separately chromatographing the test and reference standard solutions. If syringe injection, which is irreproducible at the high pressures involved, must be used, better quantitative results are obtained by the internal calibration procedure where a known amount of a noninterfering compound, the internal standard, is added to the test and reference standard solutions, and the ratios of peak responses of the analyte and internal standard are compared.

Because of normal variations in equipment, supplies, and techniques, a system suitability test is required to ensure that a given operating system may be generally applicable. The main features of *System Suitability* tests are described below.

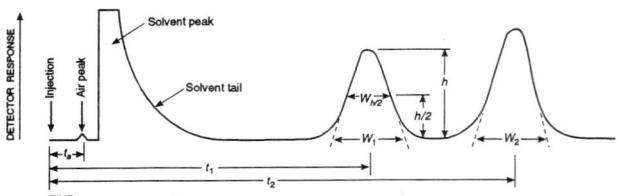

FIGURE 1 Chromatographic Separation of Two Substances.

For information on the interpretation of results, see the section *Interpretation of Chromatograms*.

Interpretation of Chromatograms Fig. 1 represents a typical chromatographic separation of two substances, 1 and 2, in which t_1 and t_2 are the respective retention times; h, $h/2$, and $W_{h/2}$ are the height, the half-height, and the width at half-height, respectively, for peak 1; and W_1 and W_2 are the respective widths of peaks 1 and 2 at the baseline. Air peaks are a feature of gas chromatograms and correspond to the solvent front in liquid chromatography.

Chromatographic retention times are characteristic of the compounds they represent but are not unique. Coincidence of retention times of a test and a reference substance can be used as a feature in construction of an identity profile but is insufficient on its own to establish identity. Absolute retention times of a given compound vary from one chromatogram to the next. Comparisons are normally made in terms of relative retention, which is calculated by the equation

$$\alpha = (t_2 - t_a)/(t_1 - t_a)$$

in which t_2 and t_1 are the retention times, measured from the point of injection, of the test and reference substances, respectively, determined under identical experimental conditions on the same column, and t_a is the retention time of a nonretained substance, such as methane in this case, of gas chromatography.

In this and the following expressions, the corresponding retention volumes or linear separations on the chromatogram, both of which are directly proportional to retention time, may be substituted in the equations. Where the value of t_a is small, R_r may be estimated from the retention times measured from the point of injection (t_2/t_1).

The number of theoretical plates, N, is a measure of column efficiency. For Gaussian peaks, it is calculated by the equations

$$N = 16(t/W)^2 \quad \text{or} \quad N = 5.54(t/W_{1/2})^2$$

in which t is the retention time of the substance and W is the width of the peak at its base, obtained by extrapolating the relatively straight sides of the peak to the baseline. $W_{1/2}$ is the peak width at half-height, obtained directly by electronic integrators. The value of N depends upon the substance being chromatographed as well as the operating conditions such as mobile phase or carrier gas flow rates and temperature, the quality of the packing, the uniformity of the packing within the column and, for capillary columns, the thickness of the stationary phase film, and the internal diameter and length of the column.

The separation of two components in a mixture, the resolution, R, is determined by the equation

$$R = 2(t_2 - t_1)/(W_2 + W_1)$$

in which t_2 and t_1 are the retention times of the two components, and W_2 and W_1, are the corresponding widths at the bases of the peaks obtained by extrapolating the relatively straight sides of the peaks to the baseline.

Peak areas and peak heights are usually proportional to the quantity of compound eluting. These are commonly measured by electronic integrators but may be determined by more classical approaches. Peak areas are generally used but may be less accurate if peak interference occurs. For manual measurements, the chart should be run faster than usual, or a comparator should be used to measure the width at half-height and the width at the base of the peak, to minimize error in these measurements. For accurate quantitative work, the components to be measured should be separated from any interfering components. Peak tailing and fronting and the measurement of peaks on solvent tails are to be avoided (see Fig. 2). The calculation is expressed by the equation

$$S_R\ (\%) = (100/\overline{X})\left\{\left[\sum_{i=1}^{N} (X_i - \overline{X})^2\right]/(N - 1)\right\}^{1/2},$$

in which S_R is the relative standard deviation in percent, $\overline{X}$ is the mean of the set of N measurements, and X_i is an individual measurement. When an internal standard is used, the measurement X_i usually refers to the measurement of relative area, A_s,

$$X_i = A_s = a_r/a_i,$$

in which a_r is the area of the peak corresponding to the standard substance and a_i is the area of the peak corresponding to the internal standard. When peak heights are used, the measurement X_i refers to the measurement of relative heights, H_s,

$$X_i = H_s = h_r/h_i,$$

in which h_r is the height of the peak corresponding to the standard substance and h_i is the height of the peak corresponding to the internal standard.

System Suitability Such tests are an integral part of gas and liquid chromatographic methods. They are used to verify that the resolution and reproducibility of the chromatographic system are adequate for the analysis to be done. The tests are based on the concept that the equipment, electronics, analytical operations, and samples to be analyzed constitute an integral system that can be evaluated as such.

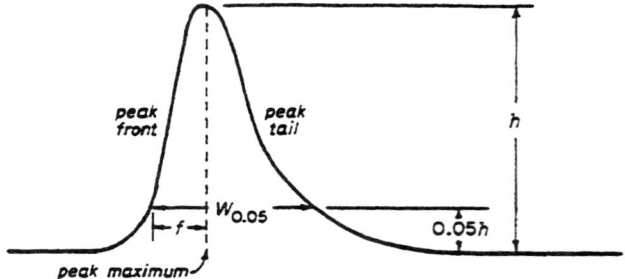

FIGURE 2 Asymmetrical Chromatographic Peak.

The resolution, R, is a function of column efficiency, N, and is specified to ensure that closely eluting compounds are resolved from each other, to establish the general resolving power of the system, and to ensure that internal standards are resolved from the analyte. Column efficiency may be specified also as a system suitability requirement, especially if there is only one peak of interest in the chromatogram; however, it is a less reliable means to ensure resolution than direct measurement. Column efficiency is a measure of peak sharpness, which is important for the detection of trace components.

Replicate injections of a standard preparation used in the assay or other standard solution are compared to ascertain whether requirements for precision are met. Unless otherwise specified in the individual monograph, data from five replicate injections of the analyte are used to calculate the relative standard deviation if the requirement is 2.0% or less; data from six replicate injections are used if the relative standard deviation requirement is more than 2.0%.

The tailing factor, T, a measure of peak symmetry, is unity for perfectly symmetrical peaks and its value increases as tailing becomes more pronounced. In some cases, values less than unity may be observed. As peak asymmetry increases, integration, and hence precision, becomes less reliable. The calculation is expressed by the equation

$$\text{tailing factor} = T = W_{0.05}/2f.$$

These tests are performed by collecting data from replicate injections of standard or other solutions as specified in the individual monograph. The specification of definitive parameters in a monograph does not preclude the use of other suitable operating conditions (see *Procedures* under *Tests and Assays* in *General Provisions*). Adjustments of operating conditions to meet system suitability requirements may be necessary.

Unless otherwise directed in the monograph, system suitability parameters are determined from the analyte peak.

To ascertain the effectiveness of the final operating system, it should be subjected to a suitability test prior to use and during testing whenever there is a significant change in equipment, or in a critical reagent, or when a malfunction is suspected.

B. PHYSICOCHEMICAL PROPERTIES

DISTILLATION RANGE

Scope This method is to be used for determining the distillation range of pure or nearly pure compounds or mixtures having a relatively narrow distillation range of about 40° or less. The result so determined is an indication of purity, not necessarily of identity. Products having a distillation range of greater than 40° may be determined by this method if a wide-range thermometer, such as ASTM E 1, 2C, or 3C, is specified in the individual monograph.

Definitions

Distillation Range The difference between the temperature observed at the start of a distillation and that observed at which a specified volume has distilled, or at which the dry point is reached.

Initial Boiling Point The temperature indicated by the distillation thermometer at the instant the first drop of condensate leaves the end of the condenser tube.

Dry Point The temperature indicated at the instant the last drop of liquid evaporates from the lowest point in the distillation flask, disregarding any liquid on the side of the flask.

Apparatus

Distillation Flask A 200-mL, round-bottom distilling flask of heat-resistant glass is preferred when sufficient sample (in excess of 100 mL) is available for the test. If a sample of less than 100 mL must be used, a smaller flask having a capacity of at least double the volume of the liquid taken may be employed. The 200-mL flask has a total length of 17 to 19 cm, and the inside diameter of the neck is 20 to 22 mm. Attached about midway on the neck, approximately 12 cm from the bottom of the flask, is a side arm 10 to 12.7 cm long and 5 mm in internal diameter, which forms an angle of 70° to 75° with the lower portion of the neck.

Condenser Use a straight glass condenser of heat-resistant tubing, 56 to 60 cm long and equipped with a water jacket so that about 40 cm of the tubing is in contact with the cooling medium. The lower end of the condenser may be bent to provide a delivery tube or it may be connected to a bent adapter that serves as the delivery tube.

Note: All-glass apparatus with standard-taper ground joints may be used alternatively if the assembly employed provides results equal to those obtained with the flask and condenser described above.

Receiver The receiver is a 100-mL cylinder that is graduated in 1-mL subdivisions and calibrated "to contain." It is used for measuring the sample as well as for receiving the distillate.

Thermometer An accurately standardized partial-immersion thermometer having the smallest practical subdivisions (not greater than 0.2°) is recommended in order to avoid the necessity for an emergent stem correction. Suitable thermometers are available as the ASTM E 1 Series 37C through 41C, and 102C through 107C, or as the MCA types R-1 through R-4 (see *Thermometers*, Appendix I).

Source of Heat A Bunsen burner is the preferred source of heat. An electric heater may be used, however, if it is shown to give results comparable to those obtained with the gas burner.

Shield The entire burner and flask assembly should be protected from external air currents. Any efficient shield may be employed for this purpose.

Flask Support A heat-resistant board, 5 to 7 mm in thickness and having a 10-cm circular hole, is placed on a suitable ring or platform support and fitted loosely inside the shield to ensure that hot gases from the source of heat do not come in contact with the sides or neck of the flask. A second 5- to 7-mm thick heat-resistant board, 14- to 16-cm square and provided with a 30- to 40-mm circular hole, is placed on top of the first board. This board is used to hold the 200-mL distillation flask, which should be fitted firmly on the board so that direct heat is applied to the flask only through the opening in the board.

Procedure (Note: For materials boiling below 50°, cool the liquid to below 10° before sampling, receive the distillate in a water bath cooled to below 10°, and use water cooled to below 10° in the condenser.) Measure 100 ± 0.5 mL of the liquid in the 100-mL graduate, and transfer the sample together with an efficient antibumping device to the distilling flask. Do not use a funnel in the transfer or allow any of the sample to enter the side arm of the flask. Place the flask on the heat-resistant boards, which are supported on a ring or platform, and place in position the shield for the flask and burner. Connect the flask and condenser, place the graduate under the outlet of the condenser tube, and insert the thermometer. The thermometer should be located in the center of the neck so that the top of the contraction chamber (or bulb, if 37C or 38C is used) is level with the bottom of the outlet to the side arm. Regulate the heating so that the first drop of liquid is collected within 5 to 10 min. Read the thermometer at the instant the first drop of distillate falls from the end of the condenser tube, and record as the initial boiling point. Continue the distillation at the rate of 4 or 5 mL of distillate per min, noting the temperature as soon as the last drop of liquid evaporates from the bottom of the flask (dry point) or when the specified percentage has distilled over. Correct the observed temperature readings for any variation in the barometric pressure from the normal (760 mm) by allowing 0.1° for each 2.7 mm of variation, adding the correction if the pressure is lower, or subtracting if higher than 760 mm.

When a total-immersion thermometer is used, correct for the temperature of the emergent stem by the formula

$$0.00015 \times N(T - t),$$

in which N represents the number of degrees of emergent stem from the bottom of the stopper, T, the observed temperatures of the distillation, and t, the temperature registered by an auxiliary thermometer, the bulb of which is placed midway of the emergent stem, adding the correction to the observed readings of the main thermometer.

MELTING RANGE OR TEMPERATURE

For purposes of this Codex, the melting range or temperature of a solid is defined as those points of temperature within which or the point at which the solid coalesces and is completely melted, when determined as directed below. Any apparatus or method capable of equal accuracy may be used. The accuracy should be checked frequently by the use of one or more of the six USP Melting Point Reference Standards, preferably the one that melts nearest the melting temperature of the compound to be tested.

Five procedures for the determination of melting range or temperature are given herein, varying in accordance with the nature of the substance. When no class is designated in the monograph, use the procedure for *Class I*.

The procedure known as the mixed melting point determination, whereby the melting range of a solid under test is compared with that of an intimate mixture of equal parts of the solid and an authentic specimen of it, may be used as a confirmatory identification test. Agreement of the observations on the original and the mixture usually constitutes reliable evidence of chemical identity.

Apparatus The melting range apparatus consists of a glass container for a bath of colorless fluid, a suitable stirring device, an accurate thermometer (see Appendix I), and a controlled source of heat. The bath fluid is selected consistent with the temperature required, but light paraffin is used generally, and certain liquid silicones are well adapted to the higher temperature ranges. The fluid is deep enough to permit immersion of the thermometer to its specified immersion depth so that the bulb is still about 2 cm above the bottom of the bath. The heat may be supplied electrically or by an open flame. The capillary tube is about 10 cm long, with an internal diameter of 0.8 to 1.2 mm, and with walls 0.2 to 0.3 mm thick.

The thermometer is preferably one that conforms to the specifications provided under *Thermometers*, Appendix I, selected for the desired accuracy and range of temperature.

Procedure for Class I Reduce the sample to a very fine powder, and unless otherwise directed, render it anhydrous when it contains water of hydration by drying it at the temperature specified in the monograph, or when the substance contains no water of hydration, dry it over a suitable desiccant for 16 to 24 h.

Charge a capillary glass tube, one end of which is sealed, with a sufficient amount of the dry powder to form a column in the bottom of the tube 2.5 to 3.5 mm high when packed down as closely as possible by moderate tapping on a solid surface.

Heat the bath until a temperature approximately 30° below the expected melting point is reached, attach the capillary tube to the thermometer, and adjust its height so that the material in the capillary is level with the thermometer bulb. Return the thermometer to the bath, continue the heating, with constant stirring, at a rate of rise of approximately 3°/min until a temperature 3° below the expected melting point is attained, then carefully regulate the rate to about 1° to 2°/min until melting is complete.

The temperature at which the column of the sample is observed to collapse definitely against the side of the tube at any point is defined as the beginning of melting, and the temperature at which the sample becomes liquid throughout is defined as the end of melting. The two temperatures fall within the limits of the melting range.

Procedure for Class Ia Prepare the sample and charge the capillary glass tube as directed for *Class I*. Heat the bath until a temperature 10° ± 1° below the expected melting range is reached, then introduce the charged tube, and heat at a rate of rise of 3° ± 0.5°/min until melting is complete. Record the melting range as for *Class I*.

Procedure for Class Ib Place the sample in a closed container, and cool to 10° or lower for at least 2 h. Without previous powdering, charge the cooled material into the capillary tube as directed for *Class I*, then immediately place the charged tube in a vacuum desiccator, and dry at a pressure not exceeding 20 mm of Hg for 3 h. Immediately upon removal from the desiccator, fire-seal the open end of the tube, and as soon as practicable, proceed with the determination of the melting range as follows: Heat the bath until a temperature of 10° ± 1° below the expected melting range is reached, then introduce the charged tube, and heat at a rate of rise of 3° ± 0.5°/min until melting is complete. Record the melting range as directed in *Class I*.

If the particle size of the material is too large for the capillary, precool the sample as directed above, then with as little pressure as possible gently crush the particles to fit the capillary, and immediately charge the tube.

Procedure for Class II Carefully melt the material to be tested at as low a temperature as possible, and draw it into a capillary tube that is left open at both ends to a depth of about 10 mm. Cool the charged tube at 10°, or lower, for 24 h, or in contact with ice for at least 2 h. Then attach the tube to the thermometer by means of a rubber band, adjust it in a water bath so that the upper edge of the material is 10 mm below the water level, and heat as directed for *Class I* except within 5° of the expected melting temperature, regulate the rate of rise of temperature to 0.5° to 1.0°/min. The temperature at which the material is observed to rise in the capillary tube is the melting temperature.

Procedure for Class III Melt a quantity of the substance slowly, while stirring, until it reaches a temperature of 90° to 92°. Remove the source of heat, and allow the molten substance to cool to a temperature of 8° to 10° above the expected melting point. Chill the bulb of an ASTM 14C thermometer (see Appendix I) to 5°, wipe it dry, and while it is still cold, dip it into the molten substance so that approximately the lower half of the bulb is submerged. Withdraw it immediately, and hold it vertically away from the heat until the wax surface dulls, then dip it for 5 min into a water bath having a temperature not higher than 16°.

Fix the thermometer securely in a test tube so that the lower point is 15 mm above the bottom of the test tube. Suspend the test tube in a water bath adjusted to about 16°, and raise the temperature of the bath at the rate of 2°/min to 30°, then change to a rate of 1°/min, and note the temperature at which the first drop of melted substance leaves the thermometer. Repeat the determination twice on a freshly melted portion of the sample. If the variation of three determinations is less than 1°, take the average of the three as the melting point. If the variation of three determinations is greater than 1°, make two additional determinations and take the average of the five.

OPTICAL (SPECIFIC) ROTATION

Many chemicals in a pure state or in solution are optically active in the sense that they cause incident polarized light to emerge in a plane forming a measurable angle with the plane of the incident light. When this effect is large enough for precise measurement, it may serve as the basis for an assay or an identity test. In this connection, the optical rotation is expressed in degrees, as either *angular rotation* (observed) or *specific rotation* (calculated with reference to the specific concentration of 1 g of solute in 1 mL of solution, measured under stated conditions).

Specific rotation usually is expressed by the term $[\alpha]_x^t$, in which t represents, in degrees centigrade, the temperature at which the rotation is determined, and x represents the characteristic spectral line or wavelength of the light used. Spectral lines most frequently employed are the D line of sodium (doublet at 589.0 and 589.6 nm) and the yellow–green line of mercury at 546.1 nm. The specific gravity and the rotatory power vary appreciably with the temperature.

The accuracy and precision of optical rotatory measurements will be increased if they are carried out with due regard for the following general considerations.

Supplement the source of illumination with a filtering system capable of transmitting light of a sufficiently monochromatic nature. Precision polarimeters generally are designed to accommodate interchangeable disks to isolate the D line from sodium light or the 546.1-nm line from the mercury spectrum. With polarimeters not thus designed, cells containing suitably colored liquids may be employed as filters (see also A. Weissberger and B. W. Rossiter, *Techniques of Chemistry*, Vol. I: *Physical Methods of Chemistry*, Part 3, Wiley-Interscience, New York, 1972).

Pay special attention to temperature control of the solution and of the polarimeter. Make accurate and reproducible observations to the extent that differences between replicates, or between observed and true values of rotation (the latter value having been established by calibration of the polarimeter scale with suitable standards), calculated in terms of either specific rotation or angular rotation, whichever is appropriate, do not exceed one-fourth of the range given in the individual monograph for the rotation of the article being tested. Generally, a polarimeter accurate to 0.05° of angular rotation, and capable of being read with the same precision, suffices for FCC purposes; in some cases, a polarimeter accurate to 0.01°, or less, of angular rotation, and read with comparable precision, may be required.

Fill polarimeter tubes in such a way as to avoid creating or leaving air bubbles, which interfere with the passage of the beam of light. Interference from bubbles is minimized with tubes in which the bore is expanded at one end. However, tubes of uniform bore, such as semimicro- or micro-tubes, require care for proper filling. At the time of filling, the tubes and the liquid or solution should be at a temperature not higher than that specified for the determination to guard against the formation of a bubble upon cooling and contraction of the contents.

In closing tubes having removable end plates fitted with gaskets and caps, the latter should be tightened only enough to ensure a leak-proof seal between the end plate and the body of the tube. Excessive pressure on the end plate may set up strains that result in interference with the measurements. In determining the specific rotation of a substance of low rotatory power, loosen the caps and tighten them again between successive readings in the measurement of both the rotation and the zero point. Differences arising from end plate strain thus generally will be revealed and appropriate adjustments to eliminate the cause may be made.

Procedure In the case of a solid, dissolve the substance in a suitable solvent, reserving a separate portion of the latter for a blank determination. Make at least five readings of the rotation of the solution, or of the substance itself if liquid, at 25° or the temperature specified in the individual monograph. Replace the solution with the reserved portion of the solvent (or, in the case of a liquid, use the empty tube), make the same number of readings, and use the average as the zero point value. Subtract the zero point value from the average observed rotation if the two figures are of the same sign, or add if opposite in sign, to obtain the corrected observed rotation.

Calculation Calculate the specific rotation of a liquid substance, or of a solid in solution, by application of one of the following formulas: (I) for liquid substances,

$$[\alpha]_x^t = a/ld,$$

(II) for solutions of solids,

$$[\alpha]_x^t = 100a/lpd = 100a/lc,$$

in which a is the corrected observed rotation, in degrees, at temperature t; x is the wavelength of the light used; l is the length of the polarimeter tube, in dm; d is the specific gravity of the liquid or solution at the temperature of observation; p is the concentration of the solution expressed as the number of g of substance in 100 g of solution; and c is the concentration of the solution expressed as the number of g of substance in 100 mL of solution. The concentrations p and c should be calculated on the dried or anhydrous basis, unless otherwise specified.

pH DETERMINATION

Principle The definition of pH is the negative log of the hydrogen ion concentration in moles per liter of aqueous solutions. Measure pH potentiometrically by using a pH meter or colorimetrically by using pH indicator paper.

Scope This method is suitable to determine the pH of aqueous solutions. While pH meters, calibrated with aqueous solutions, are sometimes used to make measurements in semiaqueous solutions or in nonaqueous polar solutions, the value obtained is the apparent pH value only and should not be compared with the pH of aqueous solutions. For nonpolar solutions, pH has no meaning, and pH electrodes may be damaged by direct contact with these solutions. References to the pH of nonpolar solutions or liquids usually indicate the pH of a water extract of the nonpolar liquid or the apparent pH of a mixture of the nonpolar liquid in a polar liquid such as alcohol or alcohol–water mixtures.

Procedure

Potentiometric Method (pH Meter)
Calibration Select two standard buffers to bracket, if possible, the anticipated pH of the unknown substances. These commercially available standards and the sample should be at the same temperature, within two degrees. Set the temperature compensator of the pH meter to the temperature of the samples and standards. Follow the manufacturer's instructions for setting temperature compensation and for adjusting the output during calibration. Rinse the electrodes with distilled or deionized water, and blot them dry with clean, absorbent laboratory tissue. Place the electrode(s) in the first standard buffer solution, and adjust the standardization control so that the pH reading matches the stated pH of the standard buffer. Repeat this procedure with fresh portions of the first buffer solution until two successive readings are within ± 0.02 pH units with no further adjustment. Rinse the electrodes, blot them dry, and place them in a portion of the second standard buffer solution. Following the manufacturer's instructions, adjust the slope control (not the standardization control) until the output displays the pH of the second standard buffer.

Repeat the sequence of standardization with both buffers until pH readings are within ± 0.02 pH units for both buffers without adjustments to either the slope or standardization controls. The pH of the unknown may then be measured, using either a pH electrode in combination with a reference electrode or a single combination electrode. Select electrodes made of chemically resistant glass when measuring samples of either low or high pH.

pH Indicator Paper Test papers impregnated with acid–base indicators, although less accurate than pH meters, offer a convenient way to determine the pH of an aqueous solution. They may be purchased in rolls or strips covering all or part of the pH range; papers covering a narrow part of the pH range can be sensitive to differences of 0.2 pH units. Some test papers comprise a plastic strip with small squares of test paper attached. The different squares are sensitive to different pH ranges. When using this type of test paper, wet all of the squares with the test sample to ensure a correct pH reading.

Test paper can contaminate the sample being tested; therefore, do not dip it into the sample. Either use a clean glass rod to remove a drop of the test solution and place it on the test paper, or transfer a small amount of the sample to a small container, dip the test paper into this portion, and compare the developed color with the color comparison chart provided with the test paper to determine the pH of the sample.

READILY CARBONIZABLE SUBSTANCES

Reagents
Sulfuric Acid, 95% Add a quantity of sulfuric acid of known concentration to sufficient water to adjust the final concentration to between 94.5% and 95.5% of H_2SO_4. Because the acid concentration may change upon standing or upon intermittent use, check the concentration frequently and either adjust solutions assaying more than 95.5% or less than 94.5% by adding either diluted or fuming sulfuric acid, as required, or discard them.

Cobaltous Chloride CS Dissolve about 65 g of cobaltous chloride ($CoCl_2.6H_2O$) in enough of a mixture of 25 mL of hydrochloric acid and 975 mL of water to make 1000 mL. Pipet 5 mL of this solution into a 250-mL iodine flask, add 5 mL hydrogen peroxide TS (3%) and 15 mL of sodium hydroxide solution (1 in 5), boil for 10 min, cool, and add 2 g of potassium iodide and 20 mL of dilute sulfuric acid (1 in 4). When the precipitate has dissolved, titrate the liberated iodine with 0.1 N sodium thiosulfate. The titration is sensitive to air oxidation and should be blanketed with carbon dioxide. Each mL of 0.1 N sodium thiosulfate is equivalent to 23.79 mg of $CoCl_2 \cdot 6H_2O$. Adjust the final volume of the solution by adding enough of the mixture of hydrochloric acid and water to make each mL contain 59.5 mg of $CoCl_2 \cdot 6H_2O$.

Cupric Sulfate CS Dissolve about 65 g of cupric sulfate ($CuSO_4 \cdot 5H_2O$) in enough of a mixture of 25 mL of hydrochloric acid and 975 mL of water to make 1000 mL. Pipet 10 mL of this solution into a 250-mL iodine flask; add 40 mL of water, 4 mL of acetic acid, and 3 g of potassium iodide; and titrate the liberated iodine with 0.1 N sodium thiosulfate, adding starch TS as the indicator. Each mL of 0.1 N sodium thiosulfate is equivalent to 24.97 mg of $CuSO_4 \cdot 5H_2O$. Adjust the final volume of the solution by adding enough of the mixture of hydrochloric acid and water to make each mL contain 62.4 mg of $CuSO_4 \cdot 5H_2O$.

Ferric Chloride CS Dissolve about 55 g of ferric chloride ($FeCl_3 \cdot 6H_2O$) in enough of a mixture of 25 mL of hydrochloric acid and 975 mL of water to make 1000 mL. Pipet 10 mL of this solution into a 250-mL iodine flask; add 15 mL of water, 5 mL of hydrochloric acid, and 3 g of potassium iodide; and allow the mixture to stand for 15 min. Dilute with 100 mL of water, and titrate the liberated iodine with 0.1 N sodium thiosulfate, adding starch TS as the indicator. Perform a blank determination with the same quantities of the same reagents and in the same manner, and make any necessary correction. Each mL of 0.1 N sodium thiosulfate is equivalent to 27.03 mg of $FeCl_3 \cdot 6H_2O$. Adjust the final volume of the solution by adding the mixture of hydrochloric acid and water to make each mL contain 45.0 mg of $FeCl_3 \cdot 6H_2O$.

Platinum–Cobalt CS Transfer 1.246 g of potassium chloroplatinate, K_2PtCl_6, and 1.00 g of crystallized cobaltous chloride, $CoCl_2 \cdot 6H_2O$, into a 1000-mL volumetric flask, dissolve in about 200 mL of water and 100 mL of hydrochloric acid, dilute to volume with water, and mix. This solution has a color of 500 APHA units.

Note: Use this solution only when specified in an individual monograph.

Procedure Unless otherwise directed, add the specified quantity of the substance, finely powdered if in solid form, in small portions to the comparison container, which is made of colorless glass resistant to the action of sulfuric acid and contains the specified volume of *95% Sulfuric Acid*.

Stir the mixture with a glass rod until solution is complete, allow the solution to stand for 15 min, unless otherwise directed, and compare the color of the solution with that of the specified matching fluid in a comparison container that also is of colorless glass and has the same internal and cross-section dimensions, viewing the fluids transversely against a background of white porcelain or white glass.

When heat is directed to effect solution of the substance in the *95% Sulfuric Acid*, mix the sample and the acid in a test tube, heat as directed, cool, and transfer the solution to the comparison container for matching.

Matching Fluids For purposes of comparison, a series of 20 matching fluids, each designated by a letter of the alphabet, is provided, the composition of each being as indicated in the accompanying table. To prepare the matching fluid specified, pipet the prescribed volumes of the colorimetric test solutions (CS) and water into one of the matching containers, and mix the solutions in the container.

Matching Fluids[a]

Matching Fluid	Parts of Cobaltous Chloride CS	Parts of Ferric Chloride CS	Parts of Cupric Sulfate CS	Parts of Water
A	0.1	0.4	0.1	4.4
B	0.3	0.9	0.3	8.5
C	0.1	0.6	0.1	4.2
D	0.3	0.6	0.4	3.7
E	0.4	1.2	0.3	3.1
F	0.3	1.2	0.0	3.5
G	0.5	1.2	0.2	3.1
H	0.2	1.5	0.0	3.3
I	0.4	2.2	0.1	2.3
J	0.4	3.5	0.1	1.0
K	0.5	4.5	0.0	0.0
L	0.8	3.8	0.1	0.3
M	0.1	2.0	0.1	2.8
N	0.0	4.9	0.1	0.0
O	0.1	4.8	0.1	0.0
P	0.2	0.4	0.1	4.3
Q	0.2	0.3	0.1	4.4
R	0.3	0.4	0.2	4.1
S	0.2	0.1	0.0	4.7
T	0.5	0.5	0.4	3.6

[a] Solutions A–D, very light brownish yellow.
Solutions E–L, yellow through reddish yellow.
Solutions M–O, greenish yellow.
Solutions P–T, light pink.

REFRACTIVE INDEX

The refractive index of a transparent substance is the ratio of the velocity of light in air to its velocity in that material under like conditions. It is equal to the ratio of the sine of the angle of incidence made by a ray in air to the sine of the angle of refraction made by the ray in the material being tested. The refractive index values specified in this Codex are for the D line of sodium (589 nm) unless otherwise specified. The determination should be made at the temperature specified in the individual monograph, or at 25° if no temperature is specified. This physical constant is used as a means for identification of, and detection of impurities in, volatile oils and other liquid substances. The Abbé refractometer, or other refractometers of equal or greater accuracy, may be employed at the discretion of the operator.

SOLIDIFICATION POINT

Scope This method is designed to determine the solidification point of food-grade chemicals having appreciable heats of fusion. It is applicable to chemicals having solidification points between −20° and +150°. Necessary modifications will be noted in individual monographs.

Definition *Solidification Point* is an empirical constant defined as the temperature at which the liquid phase of a substance is in approximate equilibrium with a relatively small portion of the solid phase. It is measured by noting the maximum temperature reached during a controlled cooling cycle after the appearance of a solid phase.

The solidification point is distinguished from the freezing point in that the latter term applies to the temperature of equilibrium between the solid and liquid state of pure compounds.

Some chemical compounds have two temperatures at which there may be a temperature equilibrium between the solid and liquid state depending upon the crystal form of the solid that is present.

Apparatus The apparatus illustrated in Figs. 3 and 4 consists of the components described in the following paragraphs.

Thermometer A thermometer having a range not exceeding 30°, graduated in 0.1° divisions, and calibrated for 76-mm immersion should be employed. A satisfactory series of thermometers, covering a range from −20° to +150°, is available as ASTM E 1 89C through 96C (see *Thermometers*, Appendix I). A thermometer should be chosen such that the solidification point is not obscured by the cork stopper of the sample container.

Sample Container Use a standard glass 25- × 150-mm test tube with a lip, fitted with a two-hole cork stopper to hold the

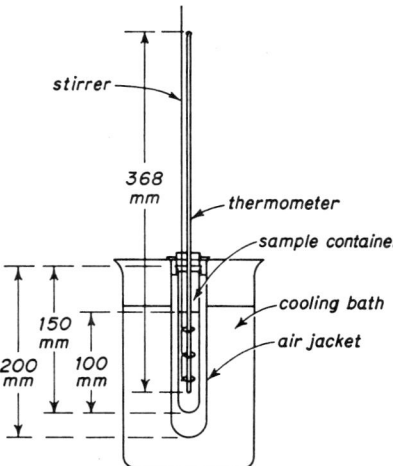

FIGURE 3 Apparatus for Determination of Solidification Point.

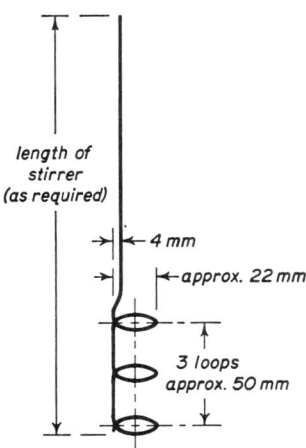

FIGURE 4 Stirrer for Solidification Point Determination.

thermometer in place and to allow adequate stirring with a stirrer.

Air Jacket For the air jacket, use a standard glass 38- × 200-mm test tube with a lip and fitted with a cork or rubber stopper bored with a hole into which the sample container can easily be inserted up to the lip.

Cooling Bath Use a 2000-mL beaker or a similar, suitable container as a cooling bath. Fill it with an appropriate cooling medium such as glycerin, mineral oil, water, water and ice, or alcohol–dry ice.

Stirrer The stirrer (Fig. 4) consists of a 1-mm diameter (B & S gauge 18), corrosion-resistant wire bent into a series of three loops about 25 mm apart. It should be made so that it will move freely in the space between the thermometer and the inner wall of the sample container. The shaft of the stirrer should be of a convenient length designed to pass loosely through a hole in the cork holding the thermometer. Stirring may be hand operated or mechanically activated at 20 to 30 strokes/min.

Assembly Assemble the apparatus in such a way that the cooling bath can be heated or cooled to control the desired temperature ranges. Clamp the air jacket so that it is held rigidly just below the lip, and immerse it in the cooling bath to a depth of 160 mm.

Sample Preparation The solidification point of chemicals is usually determined as they are received. Some may be hygroscopic, however, and will require special drying. If this is necessary, it will be noted in the individual monographs.

Products that are normally solid at room temperature must be carefully melted at a temperature about 10° above the expected solidification point. Care should be observed to avoid heating in such a way as to decompose or distill any portion of a sample.

Procedure Adjust the temperature of the cooling bath to about 5° below the expected solidification point. Fit the thermometer and stirrer with a cork stopper so that the thermometer is centered and the bulb is about 20 mm from the bottom of the sample container. Transfer a sufficient amount of the sample, previously melted if necessary, into the sample container to fill it to a depth of about 90 mm when in the molten state. Place the thermometer and stirrer in the sample container, and adjust the thermometer so that the immersion line will be at the surface of the liquid and the end of the bulb is 20 ± 4 mm from the bottom of the sample container. When the temperature of the sample is about 5° above the expected solidification point, place the assembled sample tube in the air jacket.

Allow the sample to cool while stirring, at the rate of 20 to 30 strokes/min, in such a manner that the stirrer does not touch the thermometer. Stir the sample continuously during the remainder of the test.

The temperature at first will gradually fall, then will become constant as crystallization starts and continues under equilibrium conditions, and finally will start to drop again. Some chemicals may supercool slightly below (0.5°) the solidification point; as crystallization begins, the temperature will rise and remain constant as equilibrium conditions are established. Other products may cool more than 0.5° and cause deviation from the normal pattern of temperature change. If the temperature rise exceeds 0.5° after the initial crystallization begins, repeat the test, and seed the melted compound with small crystals of the sample at 0.5° intervals as the temperature approaches the expected solidification point. Crystals for seeding may be obtained by freezing a small sample in a test tube directly in the cooling bath. It is preferable that seed of the stable phase be used from a previous determination.

Observe and record the temperature readings at regular intervals until the temperature rises from a minimum, due to supercooling, to a maximum and then finally drops. The maximum temperature reading is the solidification point. Readings 10 s apart should be taken to establish that the temperature is at the maximum level and should continue until the drop in temperature is established.

VISCOSITY OF DIMETHYLPOLYSILOXANE

Apparatus The Ubbelohde suspended level viscometer, shown in Fig. 5, is preferred to determine the viscosity of dimethylpolysiloxane. Alternatively, a Cannon-Ubbelohde viscometer may be used.

Select a viscometer having a minimum flow time of at least 200 s. Use a No. 3 size Ubbelohde, or a No. 400 size Cannon-Ubbelohde, viscometer for the range of 300 to 600 centistokes. The viscometer should be fitted with holders that satisfy the dimensional positions of the separate tubes as shown in the diagram and that hold the viscometer vertically. Filling lines in bulb A indicate the minimum and maximum volumes of liquid to be used for convenient operation. The volume of bulb B is approximately 5 mL.

Calibration of the Viscometer Determine the viscosity constant, C, for each viscometer by using an oil of known viscosity.[1]

---

[1]Oils of known viscosities may be obtained from the Cannon Instrument Co., P.O. Box 812, State College, PA 16801. For determining the viscosity of dimethylpolysiloxane, choose an oil whose viscosity is as close as possible to that of the type of sample to be tested.

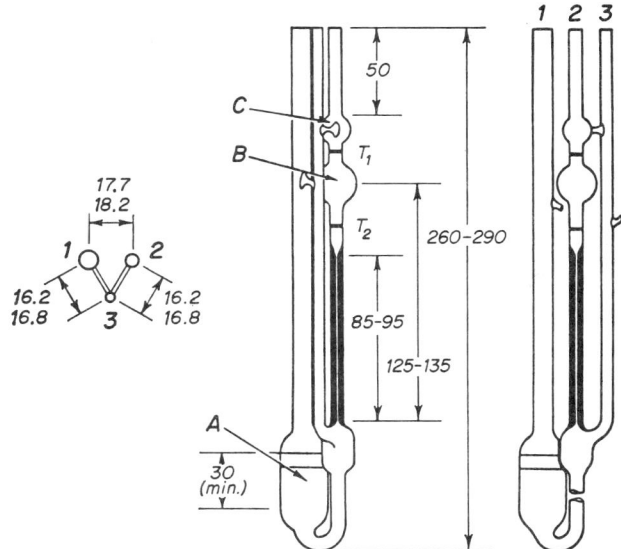

FIGURE 5 Ubbelohde Viscometer for Dimethylpolysiloxane (all dimensions are in mm).

Charge the viscometer by tilting the instrument about 30 degrees from the vertical, with bulb A below the capillary, and then introduce enough of the sample into tube 1 to bring the level up to the lower filling line. The level should not be above the upper filling line when the viscometer is returned to the vertical position and the sample has drained from tube 1. Charge the viscometer in such a manner that the U-tube at the bottom fills completely without trapping air.

After the viscometer has been in a constant-temperature bath (25° ± 0.2°) long enough for the sample to reach temperature equilibrium, place a finger over tube 3, and apply suction to tube 2 until the liquid reaches the center of bulb C. Remove suction from tube 2, then remove the finger from tube 3, and place it over tube 2 until the sample drops away from the lower end of the capillary. Remove the finger from tube 2, and measure the time, to the nearest 0.1 s, required for the meniscus to pass from the first timing mark (T_1) to the second (T_2).

Calculate the viscometer constant, C, by the equation $C = cs/t_1$, in which cs is the viscosity, in centistokes, and t_1 is the efflux time, in s, for the standard liquid.

Determination of the Viscosity of Dimethylpolysiloxane
Charge the viscometer with the sample in the same manner as described for the calibration procedure; determine the efflux time, t_2; and calculate the viscosity of the dimethylpolysiloxane by the formula

$$C \times t_2.$$

VISCOSITY OF METHYLCELLULOSE

Apparatus Viscometers used to determine the viscosity of methylcellulose and some related compounds are illustrated in

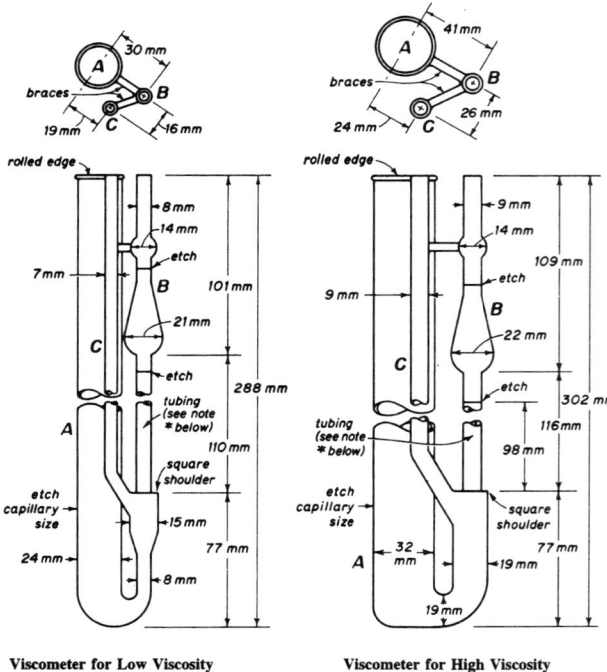

FIGURE 6 Methylcellulose Viscometers.

Fig. 6 and consist of three parts: a large filling tube, A; an orifice tube, B; and an air vent to the reservoir, C.

There are two basic types of viscometers—one for cellulose derivatives of a range between 1500 and 4000 centipoises, and the other for less viscous ones. Each type of viscometer is modified slightly for the different viscosities.

Calibration of the Viscometer Determine the viscometer constant, K, for each viscometer by using an oil of known viscosity.[2] Place an excess of the liquid that is to be tested (adjusted to 20° ± 0.1°) in the filling tube, A, and transfer it to the orifice tube, B, by gentle suction, taking care to keep the liquid free from air bubbles by closing the air vent tube, C. Adjust the column of liquid in tube B so it is even with the top graduation line. Open both tubes B and C to permit the liquid to flow into the reservoir against atmospheric pressure.

Note: Failure to open air vent tube C before determining the viscosity will yield false values.

Record the time, in s, for the liquid to flow from the upper mark to the lower mark in tube B.

Calculate the viscometer constant, K, from the equation

$$K = V/dt,$$

---

[2]Oils of known viscosities may be obtained from The Cannon Instrument Co., P.O. Box 812, State College, PA 16801. For determining the viscosity of methylcellulose, choose an oil whose viscosity is as close as possible to that of the type of sample to be tested.

in which V is the viscosity of the liquid, in centipoises; K is the viscometer constant; d is the specific gravity of the liquid tested at 20°/20°; and t is the time, in s, for the liquid to pass from the upper to the lower mark.

For the calibration, all values in the equation are known or can be determined except K, which must be solved. If a tube is repaired, it must be recalibrated to avoid obtaining significant changes in the value of K.

Determination of the Viscosity of Methylcellulose Prepare a 2% solution of methylcellulose or other cellulose derivative, by weight, as directed in the monograph. Place the solution in the proper viscometer and determine the time, t, required for the solution to flow from the upper mark to the lower mark in orifice tube B. Separately determine the specific gravity, d, at 20°/20°. Viscosity, $V = Kdt$.

VISCOSITY OF SODIUM CARBOXYMETHYLCELLULOSE

Apparatus Use a Viscometer A Model LVG Brookfield or equivalent type viscometer for the determination of viscosity of aqueous solutions of sodium carboxymethylcellulose within the range of 25 to 10,000 centipoises at 25°. Instruments of this type have spindles for use in determining the viscosity of different viscosity types of sodium carboxymethylcellulose. The spindles and speeds for determining viscosity within different ranges are tabulated below.

Viscometer Spindles Required for Given Speeds

Viscosity Range (centipoises)	Spindle No.	Speed (rpm)	Scale	Factor
10–100	1	60	100	1
100–200	1	30	100	2
200–1000	2	30	100	10
1000–4000	3	30	100	40
4000–10,000	4	30	100	200

Mechanical Stirrer Use an agitator essentially as shown in Fig. 7 that can be attached to a variable-speed motor capable of turning up to 1500 rpm.

Note: Stirrers equipped with 1½-in., 3-blade type, stainless steel propellers, such as A. H. Thomas Co. Catalog No. 9240-K, are also satisfactory.

Sample Container Use a glass jar about 5¼ in. deep (133 mm) having an outside diameter of approximately 2⅜ in. (60 mm) and a capacity of about 8 oz (236 mL).

Sample Preparation Weigh accurately an amount of sample equivalent to 4.8 g of sodium carboxymethylcellulose on the dried basis, and record the actual quantity required as S, in g. Transfer to the sample container an accurately measured volume of water equivalent to $240 - S$ g. Position the stirrer in the

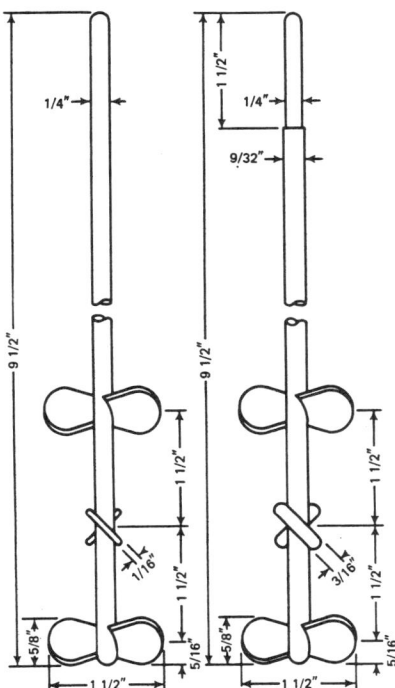

FIGURE 7   Agitator for Viscosity of Sodium Carboxymethylcellulose.

sample container, allowing minimum clearance between the stirrer and the bottom of the container, and begin stirring. Slowly add the sample, and adjust the stirring speed to approximately 800 ± 100 rpm. Mix for exactly 2 h. Do not allow the stirring speed to exceed 900 rpm. Remove the stirrer, and transfer the sample container to a constant-temperature water bath maintained at 25° ± 0.2°. Check the sample temperature at the end of 1 h to ensure that the test temperature has been reached.

Procedure   Remove the sample container from the water bath, and measure the viscosity with the Brookfield viscometer, using the proper spindle and speed indicated in the accompanying table. Allow the spindle to rotate until a constant reading is obtained. Calculate the viscosity, in centipoises, by multiplying the reading observed by the appropriate factor from the table.

WATER DETERMINATION

Method I (Karl Fischer Titrimetric Method)   Determine the water by *Method Ia*, unless otherwise specified in the individual monograph.

Method Ia (Direct Titration)
Principle   The titrimetric determination of water is based on the quantitative reaction of water with an anhydrous solution of sulfur dioxide and iodine in the presence of a buffer that reacts with hydrogen ions.

In the original titrimetric solution, known as *Karl Fischer Reagent*, the sulfur dioxide and iodine are dissolved in pyridine and methanol. Pyridine-free reagents are more commonly used now. The test specimen may be titrated with the *Karl Fischer Reagent* directly, or the analysis may be carried out by a residual titration procedure. The stoichiometry of the reaction is not exact, and the reproducibility of the determination depends on such factors as the relative concentrations of the *Karl Fischer Reagent* ingredients, the nature of the inert solvent used to dissolve the test specimen, the apparent pH of the final mixture, and the technique used in the particular determination. Therefore, an empirically standardized technique is used to achieve the desired accuracy. Precision in the method is governed largely by the extent to which atmospheric moisture is excluded from the system. The titration of water is usually carried out with the use of anhydrous methanol as the solvent for the test specimen; however, other suitable solvents may be used for special or unusual test specimens.

Substances that may interfere with the test results are ferric ion, chlorine, and similar oxidizing agents, as well as significant amounts of strong acids or bases, phosgene, or anything that will reduce iodide to iodine, poison the reagent, and show the sample to be bone dry when water may be present (false negative). 8-Hydroxyquinoline may be added to the vessel to eliminate interference from ferric ion. Chlorine interference can be eliminated with SO_2 or unsaturated hydrocarbon. Excess pyridine or other amines may be added to the vessel to eliminate the interference of strong acids. Excess acetic acid or other carboxylic acid can be added to reduce the interference of strong bases. Aldehydes and ketones may react with the solution, showing the sample to be wet while the detector never reaches an endpoint (false positive).

Apparatus   Any apparatus may be used that provides for adequate exclusion of atmospheric moisture and determination of the endpoint. In the case of a colorless solution that is titrated directly, the endpoint may be observed visually as a change in color from canary yellow to amber. The reverse is observed in the case of a test specimen that is titrated residually. More commonly, however, the endpoint is determined electrometrically with an apparatus employing a simple electrical circuit that serves to impress about 200 mV of applied potential between a pair of platinum electrodes (about 5 mm^2 in area and about 2.5 cm apart) immersed in the solution to be titrated. At the endpoint of the titration, a slight excess of the reagent increases the flow of current to between 50 and 150 microamperes for 30 s to 30 min, depending on the solution being titrated. The time is shortest for substances that dissolve in the reagent. The longer times are required for solid materials that do not readily go into solution in the *Karl Fischer Reagent*. With some automatic titrators, the abrupt change in current or potential at the endpoint serves to close a solenoid-operated valve that controls the buret delivering the titrant. A commercially available apparatus generally comprises a closed system consisting of one or two automatic burets and a tightly covered titration vessel fitted with the necessary electrodes and a magnetic stirrer. The air in the system is kept dry with a suitable desiccant such as phosphorus pentoxide, and the titration vessel may be purged by means of a stream of dry nitrogen or a current of dry air.

Reagent The *Karl Fischer Reagent* may be prepared as follows: Add 125 g of iodine to a solution containing 670 mL of methanol and 170 mL of pyridine, and cool. Place 100 mL of pyridine in a 250 mL graduated cylinder, and keeping the pyridine cold in an ice bath, pass in dry sulfur dioxide until the volume reaches 200 mL. Slowly add this solution, with shaking, to the cooled iodine mixture. Shake to dissolve the iodine, transfer the solution to the apparatus, and allow the solution to stand overnight before standardizing. One mL of this solution when freshly prepared is equivalent to approximately 5 mg of water, but it deteriorates gradually; therefore, standardize it within 1 h before use, or daily in continual use. Protect the solution from light while in use. Store any bulk stock of the solution in a suitably sealed, glass-stoppered container, fully protected from light and under refrigeration.

A commercially available, stabilized solution of a Karl Fischer-type reagent may be used. Commercially available reagents containing solvents or bases other than pyridine and/or alcohols other than methanol also may be used. These may be single solutions or reagents formed in situ by combining the components of the reagents present in two discrete solutions. The diluted *Karl Fischer Reagent* called for in some monographs should be diluted as directed by the manufacturer. Either methanol, or another suitable solvent such as ethylene glycol monomethyl ether, may be used as the diluent.

Test Preparation Unless otherwise specified in the individual monograph, use an accurately weighed or measured amount of the specimen under test estimated to contain 10 to 250 mg of water.

Where the monograph specifies that the specimen under test is hygroscopic, accurately weigh a sample of the specimen into a suitable container. Use a dry syringe to inject an appropriate volume of methanol, or other suitable solvent, accurately measured, into the container and shake to dissolve the specimen. Dry the syringe, and use it to remove the solution from the container and transfer it to a titration vessel prepared as directed under *Procedure*. Repeat the procedure with a second portion of methanol, or other suitable solvent, accurately measured; add this washing to the titration vessel; and immediately titrate. Determine the water content, in mg, of a portion of solvent of the same total volume as that used to dissolve the specimen and to wash the container and syringe, as directed under *Standardization of Water Solution for Residual Titrations*, and subtract this value from the water content, in mg, obtained in the titration of the specimen under test.

Standardization of the Reagent Place enough methanol or other suitable solvent in the titration vessel to cover the electrodes, and add sufficient *Karl Fischer Reagent* to give the characteristic color, or 100 ± 50 microamperes of direct current at about 200 mV of applied potential. Pure methanol can make the detector overly sensitive, particularly at low ppm levels of water, causing it to deflect to dryness and slowly recover with each addition of reagent. This slows down the titration and may allow the system to actually pick up ambient moisture during the resulting long titration. Adding chloroform or a similar nonconducting solvent will retard this sensitivity and can improve the analysis.

For determination of trace amounts of water (less than 1%), quickly add 25 μL (25 mg) of pure water, using a 25- or 50- μL syringe, and titrate to the endpoint. The water equivalence factor F, in mg of water per mL of reagent, is given by the formula

$$25/V,$$

in which V is the volume, in mL, of the *Karl Fischer Reagent* consumed in the second titration.

For the precise determination of significant amounts of water (more than 1%), quickly add between 25 and 250 mg (25 to 250 μL) of pure water, accurately weighed by difference from a weighing pipet or from a precalibrated syringe or micropipet, the amount of water used being governed by the reagent strength and the buret size, as referred to under *Volumetric Apparatus*. Titrate to the endpoint. Calculate the water equivalence factor, F, in mg of water per mL of reagent by the formula

$$W/V,$$

in which W is the weight, in mg, of the water, and V is the volume, in mL, of the *Karl Fischer Reagent* required.

Procedure Unless otherwise specified, transfer 35 to 40 mL of methanol or other suitable solvent to the titration vessel, and titrate with the *Karl Fischer Reagent* to the electrometric or visual endpoint to consume any moisture that may be present. (Disregard the volume consumed since it does not enter into the calculations.) Quickly add the *Test Preparation*, mix, and again titrate with the *Karl Fischer Reagent* to the electrometric or visual endpoint. Calculate the water content of the specimen, in mg, by the formula

$$SF,$$

in which S is the volume, in mL, of the *Karl Fischer Reagent*, consumed in the second titration, and F is the water equivalence factor of the *Karl Fischer Reagent*.

Method 1b (Residual Titration)

Principle See the information in the section entitled *Principle* under *Method Ia*. In the residual titration, add excess *Karl Fischer Reagent* to the test specimen, allow sufficient time for the reaction to reach completion, and titrate the unconsumed *Karl Fischer Reagent* with a standard solution of water in a solvent such as methanol. The residual titration procedure is generally applicable and avoids the difficulties that may be encountered in the direct titration of substances from which the bound water is released slowly.

Apparatus, Reagent, and Test Preparation Use *Method Ia*.

Standardization of Water Solution for Residual Titration Prepare a *Water Solution* by diluting 2 mL of pure water with methanol or another suitable solvent to 1000 mL. Standardize this solution by titrating 25.0 mL with the *Karl Fischer Reagent*, previously standardized as directed under *Standardization of the Reagent*. Calculate the water content, in mg/mL, of the *Water Solution* with the formula

$$VF/25,$$

in which V is the volume of the *Karl Fischer Reagent* consumed, and F is the water equivalence factor of the *Karl Fischer Reagent*. Determine the water content of the *Water Solution* weekly, and standardize the *Karl Fischer Reagent* against it periodically as needed. Store the *Water Solution* in a tightly capped container.

Procedure Where the individual monograph specifies the water content is to be determined by *Method Ib*, transfer 35 to 40 mL of methanol or other suitable solvent to the titration vessel, and titrate with the *Karl Fischer Reagent* to the electrometric or visual endpoint. Quickly add the *Test Preparation*, mix, and add an accurately measured excess of the *Karl Fischer Reagent*. Allow sufficient time for the reaction to reach completion, and titrate the unconsumed *Karl Fischer Reagent* with standardized *Water Solution* to the electrometric or visual endpoint. Calculate the water content of the specimen, in mg, with the formula

$$F(X' - XR),$$

in which F is the water equivalence factor of the *Karl Fischer Reagent*; X' is the volume, in mL, of the *Karl Fischer Reagent*, added after introduction of the specimen; X is the volume, in mL, of standardized *Water Solution* required to neutralize the unconsumed *Karl Fischer Reagent*; and R is the ratio $V/25$ (mL *Karl Fischer Reagent*/mL *Water Solution*), determined from the *Standardization of Water Solution for Residual Titration*.

Method Ic (Coulometric Titration)

Principle Use the Karl Fischer reaction in the coulometric determination of water. In this determination, iodine is not added in the form of a volumetric solution, but is produced in an iodide-containing solution by anodic oxidation. The reaction cell usually consists of a large anode compartment and a small cathode compartment that are separated by a diaphragm. Other suitable types of reaction cells (e.g., without diaphragms) may be used. Each compartment has a platinum electrode that conducts current through the cell. Iodine, which is produced at the anode electrode, immediately reacts with the water present in the compartment. When all the water has been consumed, an excess of iodine occurs, which can be detected potentiometrically, thus indicating the endpoint. Pre-electrolysis, which can take several hours, eliminates moisture from the system. Therefore, changing the *Karl Fischer Reagent* after each determination is not practical. Individual determinations may be carried out in succession in the same reagent solution. A requirement for this method is that each component of the test specimen be compatible with the other components and that no side reactions take place. Samples may be transferred into the vessel as solids or as solutions by means of injection through a septum. Gases can be introduced into the cell by means of a suitable gas inlet tube. For the water determination of solids, another common technique is to dissolve the solid in a suitable solvent and then inject a portion of this solution into the cell. In the case of insoluble solids, water may be extracted using suitable solvents, and then the extracts injected into the coulometric cell. Alternatively, an evaporation technique may be used in which the sample is heated in a tube and the water is evaporated and carried into the cell by means of a stream of dry, inert gas. Precision in the method is predominantly governed by the extent to which atmospheric moisture is excluded from the system. Control of the system may be monitored by measuring the amount of baseline drift. The titration of water in solid test specimens is usually carried out with the use of anhydrous methanol as the solvent. Other suitable solvents may be used for special or unusual test specimens. This method is particularly suited to chemically inert substances such as hydrocarbons, alcohols, and ethers. In comparison with the volumetric Karl Fischer titration, coulometry is a micromethod. The method uses extremely small amounts of current. It is predominantly used for substances with a very low water content (0.1% to 0.0001%).

Apparatus Any commercially available apparatus consisting of an absolutely tight system fitted with the necessary electrodes and a magnetic stirrer is appropriate. The instrument's microprocessor controls the analytical procedure and displays the results. Calibration of the instrument is not necessary as the current consumed can be measured absolutely. Proper operation of the instrument can be confirmed by injecting 1 μL of water into the vessel. The instrument should read 1000 μg of water on reaching the endpoint.

Reagent See *Reagent* under *Method Ia*.

Test Preparation Using a dry syringe, inject an appropriate volume of test specimen estimated to contain 0.5 to 5 mg of water, accurately measured, into the anolyte solution. The sample may also be introduced as a solid, accurately weighed, into the anolyte solution. Perform coulometric titration and determine the water content of the specimen under test.

Alternatively, when the specimen is a suitable solid, dissolve an appropriate quantity, accurately weighed, in anhydrous methanol or another suitable solvent, and inject a suitable portion into the anolyte solution.

When the specimen is an insoluble solid, extract the water by using a suitable anhydrous solvent from which an appropriate quantity, accurately weighed, may be injected into the anolyte solution. Alternatively use an evaporation technique.

Procedure Quickly inject the *Test Preparation*, or transfer the solid sample, into the anolyte, mix, and perform the coulometric titration to the electrometric endpoint. Read the water content of the *Test Preparation* directly from the instrument's display, and calculate the percentage that is present in the substance.

Method II (Toluene Distillation Method)

Principle This method determines water by distillation of a sample with an immiscible solvent, usually toluene.

Apparatus Use a glass distillation apparatus (see Fig. 8) provided with 24/40 ground-glass connections. The components consist of a 500-mL short-neck, round-bottom flask connected by means of a trap to a 400-mm, water-cooled condenser. The lower tip of the condenser should be about 7 mm above the surface of the liquid in the trap after distillation conditions have been established (see *Procedure*).

The trap should be constructed of well-annealed glass, the receiving end of which is graduated to contain 5 mL and subdivided into 0.1-mL divisions, with each 1-mL line numbered from 5 mL beginning at the top. Calibrate the receiver by adding

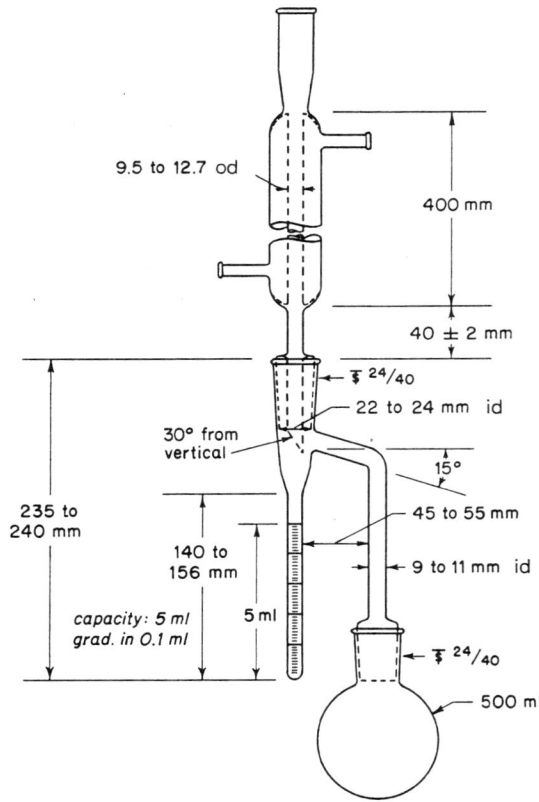

FIGURE 8 Moisture Distillation Apparatus.

1 mL of water, accurately measured, to 100 mL of toluene contained in the distillation flask. Conduct the distillation, and calculate the volume of water obtained as directed in the *Procedure*. Add another mL of water to the cooled apparatus, and repeat the distillation. Continue in this manner until five 1-mL portions of water have been added. The error at any indicated capacity should not exceed 0.05 mL. The source of heat is either an oil bath or an electric heater provided with a suitable means of temperature control. The distillation may be better controlled by insulating the tube leading from the flask to the receiver. It is also advantageous to protect the flask from drafts. Clean the entire apparatus with potassium dichromate–sulfuric acid cleaning solution, rinse thoroughly, and dry completely before using.

Procedure Place in the previously cleaned and dried flask a quantity of the substance, weighed accurately to the nearest 0.01 g, that is expected to yield from 1.5 to 4 mL of water. If the substance is of a pastelike consistency, weigh it in a boat of metal foil that will pass through the neck of the flask. If the substance is likely to cause bumping, take suitable precautions to prevent it. Transfer about 200 mL of ACS reagent-grade toluene into the flask, and swirl to mix it with the sample. Assemble the apparatus, fill the receiver with toluene by pouring it through the condenser until it begins to overflow into the flask, and insert a loose cotton plug in the top of the condenser. Heat the flask so that the distillation rate will be about 200 drops/min, and continue distilling until the volume of water in the trap remains constant for 5 min. Discontinue the heating, use a copper or nichrome wire spiral to dislodge any drops of water that may be adhering to the inside of the condenser tube or receiver, and wash down with about 5 mL of toluene. Disconnect the receiver, immerse it in water at 25° for at least 15 min or until the toluene layer is clear, and then read the volume of water. Conduct a blank determination using the same volume of toluene as used when distilling the sample mixture, and make any necessary correction (see *General Provisions*).

C. OTHERS

ASH (Acid-Insoluble)

Boil the ash obtained as directed under *Ash (Total)*, below, with 25 mL of 2.7 N hydrochloric acid for 5 min, collect the insoluble matter on a tared, porous-bottom porcelain filter crucible or ashless filter, wash it with hot water, ignite to constant weight at 675° ± 25°, and weigh. Calculate the percentage of acid-insoluble ash from the weight of the sample taken.

Note: Avoid exposing the crucible to sudden temperature changes.

ASH (Total)

Unless otherwise directed, accurately weigh about 3 g of the sample in a tared crucible, ignite it at a low temperature (about 550°), not to exceed a very dull redness, until it is free from carbon, cool it in a desiccator, and weigh. If a carbon-free ash is not obtained, wet the charred mass with hot water, collect the insoluble residue on an ashless filter paper, and ignite the residue and filter paper until the ash is white or nearly so. Finally, add the filtrate, evaporate it to dryness, and heat the whole to a dull redness. If a carbon-free ash is still not obtained, cool the crucible, add 15 mL of ethanol, break up the ash with a glass rod, then burn off the ethanol, again heat the whole to a dull redness, cool it in a desiccator, and weigh.

HYDROCHLORIC ACID TABLE

°Bé	Sp. Gr.	Percent HCl	°Bé	Sp. Gr.	Percent HCl
1.00	1.0069	1.40	5.25	1.0375	7.52
2.00	1.0140	2.82	5.50	1.0394	7.89
3.00	1.0211	4.25	5.75	1.0413	8.26
4.00	1.0284	5.69	6.00	1.0432	8.64
5.00	1.0357	7.15	6.25	1.0450	9.02

°Bé	Sp. Gr.	Percent HCl	°Bé	Sp. Gr.	Percent HCl
6.50	1.0469	9.40	17.8	1.1399	27.58
6.75	1.0488	9.78	17.9	1.1408	27.75
7.00	1.0507	10.17	18.0	1.1417	27.92
7.25	1.0526	10.55	18.1	1.1426	28.09
7.50	1.0545	10.94	18.2	1.1435	28.26
7.75	1.0564	11.32	18.3	1.1444	28.44
8.00	1.0584	11.71	18.4	1.1453	28.61
8.25	1.0603	12.09	18.5	1.1462	28.78
8.50	1.0623	12.48	18.6	1.1471	28.95
8.75	1.0642	12.87	18.7	1.1480	29.13
9.00	1.0662	13.26	18.8	1.1489	29.30
9.25	1.0681	13.65	18.9	1.1498	29.48
9.50	1.0701	14.04	19.0	1.1508	29.65
9.75	1.0721	14.43	19.1	1.1517	29.83
10.00	1.0741	14.83	19.2	1.1526	30.00
10.25	1.0761	15.22	19.3	1.1535	30.18
10.50	1.0781	15.62	19.4	1.1544	30.35
10.75	1.0801	16.01	19.5	1.1554	30.53
11.00	1.0821	16.41	19.6	1.1563	30.71
11.25	1.0841	16.81	19.7	1.1572	30.90
11.50	1.0861	17.21	19.8	1.1581	31.08
11.75	1.0881	17.61	19.9	1.1590	31.27
12.00	1.0902	18.01	20.0	1.1600	31.45
12.25	1.0922	18.41	20.1	1.1609	31.64
12.50	1.0943	18.82	20.2	1.1619	31.82
12.75	1.0964	19.22	20.3	1.1628	32.01
13.00	1.0985	19.63	20.4	1.1637	32.19
13.25	1.1006	20.04	20.5	1.1647	32.38
13.50	1.1027	20.44	20.6	1.1656	32.56
13.75	1.1048	20.86	20.7	1.1666	32.75
14.00	1.1069	21.27	20.8	1.1675	32.93
14.25	1.1090	21.68	20.9	1.1684	33.12
14.50	1.1111	22.09	21.0	1.1694	33.31
14.75	1.1132	22.50	21.1	1.1703	33.50
15.00	1.1154	22.92	21.2	1.1713	33.69
15.25	1.1176	23.33	21.3	1.1722	33.88
15.50	1.1197	23.75	21.4	1.1732	34.07
15.75	1.1219	24.16	21.5	1.1741	34.26
16.0	1.1240	24.57	21.6	1.1751	34.45
16.1	1.1248	24.73	21.7	1.1760	34.64
16.2	1.1256	24.90	21.8	1.1770	34.83
16.3	1.1265	25.06	21.9	1.1779	35.02
16.4	1.1274	25.23	22.0	1.1789	35.21
16.5	1.1283	25.39	22.1	1.1798	35.40
16.6	1.1292	25.56	22.2	1.1808	35.59
16.7	1.1301	25.72	22.3	1.1817	35.78
16.8	1.1310	25.89	22.4	1.1827	35.97
16.9	1.1319	26.05	22.5	1.1836	36.16
17.0	1.1328	26.22	22.6	1.1846	36.35
17.1	1.1336	26.39	22.7	1.1856	36.54
17.2	1.1345	26.56	22.8	1.1866	36.73
17.3	1.1354	26.73	22.9	1.1875	36.93
17.4	1.1363	26.90	23.0	1.1885	37.14
17.5	1.1372	27.07	23.1	1.1895	37.36
17.6	1.1381	27.24	23.2	1.1904	37.58
17.7	1.1390	27.41	23.3	1.1914	37.80
23.4	1.1924	38.03	24.5	1.2033	40.55
23.5	1.1934	38.26	24.6	1.2043	40.78
23.6	1.1944	38.49	24.7	1.2053	41.01
23.7	1.1953	38.72	24.8	1.2063	41.24
23.8	1.1963	38.95	24.9	1.2073	41.48
23.9	1.1973	39.18	25.0	1.2083	41.72
24.0	1.1983	39.41	25.1	1.2093	41.99
24.1	1.1993	39.64	25.2	1.2103	42.30
24.2	1.2003	39.86	25.3	1.2114	42.64
24.3	1.2013	40.09	25.4	1.2124	43.01
24.4	1.2023	40.32	25.5	1.2134	43.40

Source: Courtesy of the Manufacturing Chemists Association.

Specific gravity determinations were made at 60°F, compared with water at 60°F.

From the specific gravities, the corresponding degrees Baumé were calculated by the following formula:

$$\text{Baumé} = 145 - (145/\text{sp. gr.})$$

Baumé hydrometers for use with this table must be graduated by the above formula, which should always be printed on the scale.

Allowance for Temperature
10° to 15°Bé: $\frac{1}{40}$°Bé or 0.0002 sp. gr. for 1°F
15° to 22°Bé: $\frac{1}{30}$°Bé or 0.0003 sp. gr. for 1°F
22° to 25°Bé: $\frac{1}{28}$°Bé or 0.00035 sp. gr. for 1°F

LOSS ON DRYING

This procedure is used to determine the amount of volatile matter expelled under the conditions specified in the monograph. Since the volatile matter may include material other than adsorbed moisture, this test is designed for compounds in which the loss on drying may not definitely be attributable to water alone. For substances appearing to contain water as the only volatile constituent, the *Direct (Karl Fischer) Titration Method*, provided under *Water*, Appendix IIC, is usually appropriate.

Procedure Unless otherwise directed in the monograph, conduct the determination on 1 to 2 g of the substance, previously mixed and accurately weighed. If the sample is in the form of large crystals, reduce the particle size to about 2 mm, quickly crushing the sample to avoid absorption or loss of moisture. Tare a glass-stoppered, shallow weighing bottle that has been dried for 30 min under the same conditions to be used in the determination. Transfer the sample to the bottle, replace the cover, and weigh the bottle and its contents. By gentle sideways shaking, distribute the sample as evenly as possible to a depth of about 5 mm for most substances and not over 10 mm in the case of bulky materials. Place the loaded bottle in the drying chamber, removing the stopper and leaving it also in the chamber, and dry at the temperature and for the length of time specified in the monograph. Upon opening the chamber, close the bottle promptly and allow it to come to room temperature, preferably in a desiccator, before weighing.

Where drying in vacuum is specified in the monograph, use a pressure as low as that obtainable by an aspirating water pump (not higher than 20 mm of Hg).

If the test substance melts at a temperature lower than that specified for the determination, preheat the bottle and its contents for 1 to 2 h at a temperature 5° to 10° below the melting range, then continue drying at the specified temperature for the determination. When drying the sample in a desiccator, ensure that the desiccant is kept fully effective by replacing it frequently.

OIL CONTENT OF SYNTHETIC PARAFFIN

Apparatus

Filter Stick Use either a 10-mm diameter sintered-glass filter stick of 10- to 15-μm maximum pore diameter, or a filter stick made of stainless steel and having a 0.5-in. disk of 10- to 15-μm maximum pore diameter. Determine conformance with the pore diameter specified as follows: Clean sintered-glass filter sticks by soaking in hydrochloric acid, or stainless steel sticks by soaking in nitric acid, wash with water, rinse with acetone, and dry in air followed by drying in an oven at 105° for 30 min.

Thoroughly wet the clean filter stick by soaking in water, and then connect it with an apparatus (see Fig. 9) consisting of a mercury-filled manometer, readable to 0.5 mm; a clean and filtered air supply; a drying bulb filled with silica gel; and a needle-valve type air pressure regulator. Apply pressure slowly from the air source, and immerse the filter just below the surface of water contained in a beaker.

Note: If a head of liquid is noted above the surface of the filter after it is inserted into the water, the back pressure thus

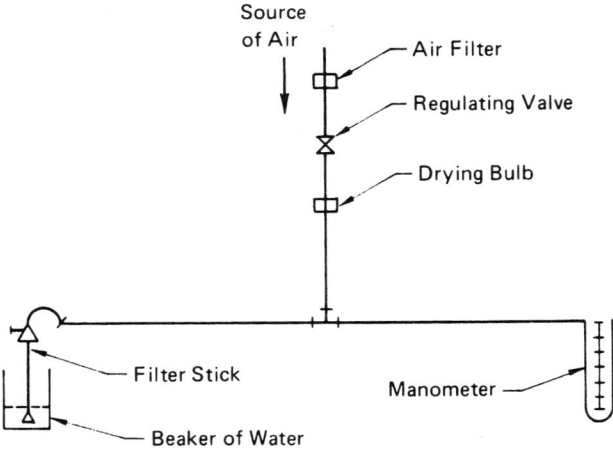

FIGURE 9 Assembly for Checking Pore Diameter of Filter Sticks.

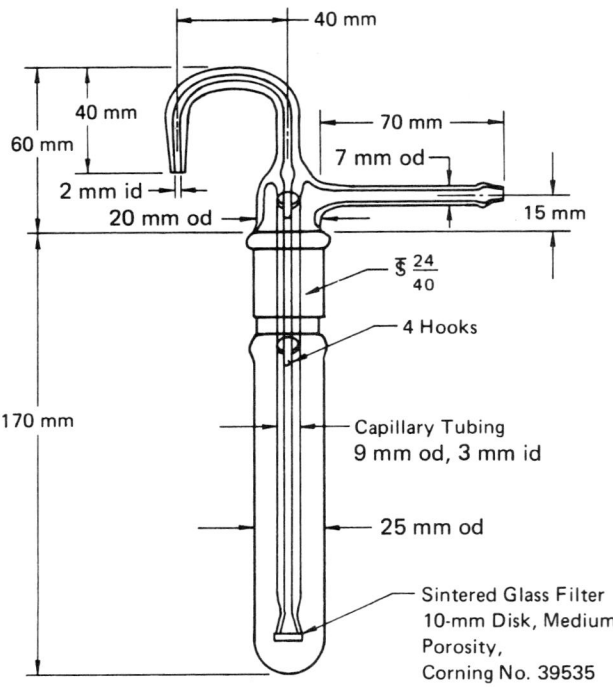

FIGURE 10 Filtration Assembly for Determination of Oil Content.

produced should be subtracted from the observed pressure when the pore diameter is calculated as directed below.

Increase the air pressure to 10 mm below the acceptable pressure limit, and then increase the pressure at a slow, uniform rate of about 3 mm of Hg per min until the first bubble passes through the filter. This can be conveniently observed by placing the beaker over a mirror. Read the manometer when the first bubble passes off the underside of the filter. Calculate the pore diameter, in μm, by the formula

$$2180/p,$$

in which p is the observed pressure, in mm, corrected for any back pressure as mentioned above.

Filtration Assembly Connect the *Filter Stick* with an air pressure inlet tube and delivery nozzle and ground-glass joint to fit a 25- × 170-mm test tube as shown in Fig. 10. If a stainless steel *Filter Stick* is used, make the connection to the test tube by means of a cork.

Cooling Bath Use a suitable insulated box having 1-in. holes in the center to accommodate any desired number of test tubes. The bath may be filled with a suitable medium such as kerosene and may be cooled by circulating a refrigerant through coils, or by using solid carbon dioxide, to produce a temperature of −30° ± 2°F.

Air Pressure Regulator Use a suitable pressure-reduction valve, or other suitable regulator, that will supply air to the *Filtration Assembly* at the volume and pressure required to give an even flow of filtrate (see *Procedure*). Connect the regulator with rubber tubing to the end of the *Filter Stick* in the *Filtration Assembly*.

Thermometer Use an ASTM Oil in Wax Thermometer having the range of −35° to +70°F and conforming to the requirements for an ASTM 71 F thermometer (see *Thermometers*, Appendix I).

Weighing Bottles Use glass-stoppered conical bottles having a capacity of 15 mL. The bottles are used as evaporating flasks in the *Procedure*.

Evaporation Assembly The assembly consists of an evaporating cabinet capable of maintaining a temperature of 95° ± 2°F around the evaporation flasks, and air jets (4 ± 0.2 mm id) for delivering a stream of clean, dry air vertically downward into the flasks. In the *Procedure* below, support each jet so that the tip is 15 ± 5 mm above the surface of the liquid at the start of the evaporation. Supply the air (purified by passage through a tube of 1-cm bore packed loosely to a height of 20 cm with absorbent cotton) at the rate of 2 to 3 L/min per jet. The cleanliness of the air should be checked periodically to ensure that not more than 0.1 mg of residue is obtained when 4 mL of methyl ethyl ketone is evaporated as directed in the *Procedure*.

Wire Stirrer Use a 250-mm length of stiff iron or nichrome wire of about No. 20 B & S gauge. Form a 10-mm diameter loop at each end, and bend the loop at the bottom end so that the plane of the loop is perpendicular to the length of the wire.

Sample Selection If the sample is about 1 kg or less, obtain a representative portion by melting the entire sample and stirring thoroughly. For samples greater than about 1 kg, exercise special care to ensure that a truly representative portion is obtained, noting that the oil may not be distributed uniformly throughout the sample and that mechanical operations may have expressed some of the oil.

Procedure Melt a representative portion of the sample in a beaker, using a water bath or oven maintained at 160° to 210°F. As soon as the sample is completely melted, thoroughly mix it by stirring. Preheat a dropper pipet, provided with a rubber bulb and calibrated to deliver 1 ± 0.05 g of molten sample, and withdraw a 1-g portion of the sample as soon as possible after it has melted. Hold the pipet in a vertical position, and carefully transfer its contents into a clean, dry test tube previously weighed to the nearest mg. Evenly coat the bottom of the tube by swirling, allow the tube to cool, and weigh to the nearest mg. Calculate the sample weight, in g, and record it as B (see *Calculation*). Pipet 15 mL of methyl ethyl ketone (ASTM Specification D 740 or equivalent) into the tube, and immerse the tube up to the top of the liquid in a hot water or steam bath. Stir with an up-and-down motion with the wire stirrer, and continue heating and stirring until a homogeneous solution is obtained, exercising care to avoid loss of solvent by prolonged boiling.

> **Note**: If it appears that a clear solution will not be obtained, stir until any undissolved material is well dispersed so as to produce a slightly cloudy solution.

After the sample solution is prepared, plunge the test tube into an 800-mL beaker of ice water, and continue to stir until the contents are cold. Remove the stirrer, then remove the test tube from the bath, dry the outside of the tube with a cloth, and weigh to the nearest 100 mg. Calculate the weight, in g, of solvent in the test tube, and record it as C (see *Calculation*).

Place the tube in the cooling bath, maintained at −30° ± 2°F, and stir continuously with the thermometer until the temperature reaches −25° ± 0.5°F, maintaining the slurry at a uniform consistency and taking precautions to prevent the sample from setting up on the walls of the tube or forming crystals.

Place the filter stick in a test tube and cool at −30° ± 2°F in the cooling bath for a minimum of 10 min. Immerse the cooled filter stick in the sample, then connect the filtration assembly, seating the ground-glass joint of the filter so as to make an airtight seal. Placed an unstoppered weighing bottle, previously weighed together with the glass stopper to the nearest 0.1 mg, under the delivery nozzle of the filtration assembly.

> **Note**: Suitable precautions and proper analytical technique should be applied to ensure the accuracy of the weight of the bottle. Prior to determining its weight, the bottle and its stopper should have been cleaned and dried, then rinsed with methyl ethyl ketone, wiped dry on the outside, dried in the evaporation assembly for about 5 min, and cooled. Then allow it to stand for about 10 min near the balance before weighing.

Apply air pressure to the filtration assembly, immediately collect about 4 mL of filtrate in the weighing bottle, and release the air pressure to permit the liquid to drain back slowly from the delivery nozzle. Stopper the bottle, and weigh it to the nearest 10 mg without waiting for it to come to room temperature. Remove the stopper, transfer the bottle to the evaporation assembly maintained at 95° ± 2°F, and place it under an air jet centered inside the neck, with the tip 15 ± 5 mm above the surface of the liquid. After the solvent has evaporated (usually less than 30 min time), stopper the bottle, and allow it to stand near the balance for about 10 min before it is weighed to the nearest 0.1 mg. Repeat the evaporation procedure for 5-min periods until the loss between successive weighings is not more than 0.2 mg. Determine the weight of the oil residue, in g, by subtracting the weight of the empty stoppered bottle from the weight of the stoppered bottle plus the oil residue after the evaporation procedure, and record the results as A (see *Calculation*). Determine the weight of solvent evaporated, in g, by subtracting the weight of the bottle plus oil residue from the weight of the bottle plus filtrate, and record the result as D (see *Calculation*).

Calculation Calculate the percentage, by weight, of oil in the sample by the formula

$$(100\ AC/BD) - 0.15,$$

in which 0.15 is a factor to correct for solubility of the sample in the solvent at −25°F.

RESIDUE ON IGNITION (Sulfated Ash)

Method I (for Solids) Transfer the quantity of the sample directed in the individual monograph to a tared 50- to 100-mL platinum dish or other suitable container, and add sufficient 2

N sulfuric acid to moisten the entire sample. Heat gently, using a hot plate, an Argand burner, or an infrared heat lamp, until the sample is dry and thoroughly charred, then continue heating until all of the sample has been volatilized or nearly all of the carbon has been oxidized, and cool. Moisten the residue with 0.1 mL of sulfuric acid, and heat in the same manner until the remainder of the sample and any excess sulfuric acid have been volatilized. Finally, ignite to constant weight in a muffle furnace at 800° ± 25° for 15 min, or longer if necessary to complete ignition, cool in a desiccator, and weigh.

Method II (for Liquids) Unless otherwise directed, transfer the required weight of the sample to a tared 75- to 100-mL platinum dish. Heat gently, using an Argand or Meker burner, until the sample ignites, then allow the sample to burn until it self-extinguishes. Cool, then wet the residue with 2 mL of concentrated sulfuric acid, and heat the sample over a low flame until dry. Ignite to constant weight in a muffle furnace at 800° ± 25° for 30 min, or longer if necessary for complete ignition, cool in a desiccator, and weigh.

SIEVE ANALYSIS OF GRANULAR METAL POWDERS
(Based on ASTM Designation: B 214)

Apparatus
Sieves Use a set of standard sieves, ranging from 80 mesh to 325 mesh, conforming to the specifications in ASTM Designation: E 11 (Sieves for Testing Purposes).
Sieve Shaker Use a mechanically operated sieve shaker that imparts to the set of sieves a horizontal rotary motion of between 270 and 300 rotations/min and a tapping action of between 140 and 160 taps/min. The sieve shaker is fitted with a plug to receive the impact of the tapping device. The entire apparatus is rigidly mounted—bolted to a solid foundation, preferably of concrete. Preferably a time switch is provided to ensure the accuracy of test duration.

Procedure Assemble the sieves in consecutive order by opening size, with the coarsest sieve (80-mesh) at the top, and place a solid-collecting pan below the bottom sieve (325 mesh). Place 100.0 g of the test sample, W, on the top sieve, and close the sieve with a solid cover. Securely fasten the assembly to the sieve shaker, and operate the shaker for 15 min. Remove the most coarse sieve from the nest, gently tap its contents to one side, and pour the contents onto a tared, glazed paper. Using a soft brush, transfer onto the next finer sieve any material adhering to the bottom of the sieve and frame. Place the sieve just removed upside down on the paper containing the retained portion, and tap the sieve. Accurately weigh the paper and its contents, and record the net weight of the fraction, F, obtained. Repeat this process for each sieve in the nest and for the portion of the sample that has been collected in the bottom pan. Record the total of the fractions retained on the sieves as T and that portion collected in the pan as t. The combined total, S, of $T + t$ is the amount of the sample, W, recovered in the test. Calculate the percent recovery by the formula

$$S/W \times 100.$$

If the percent recovery is less than 99.0%, check the condition of the sieves and for possible errors in weighing, and repeat the test. If the percent recovery is not less than 99.0%, calculate the percent retained on each sieve by the formula

$$F/W \times 100.$$

Calculate the percent through the smallest mesh sieve from the portion collected in the pan by the formula

$$[(100 - t)/W] \times 100.$$

SULFURIC ACID TABLE

°Bé	Sp. Gr.	Percent H_2SO_4	°Bé	Sp. Gr.	Percent H_2SO_4
0	1.0000	0.00	36	1.3303	42.63
1	1.0069	1.02	37	1.3426	43.99
2	1.0140	2.08	38	1.3551	45.35
3	1.0211	3.13	39	1.3679	46.72
4	1.0284	4.21	40	1.3810	48.10
5	1.0357	5.28	41	1.3942	49.47
6	1.0432	6.37	42	1.4078	50.87
7	1.0507	7.45	43	1.4216	52.26
8	1.0584	8.55	44	1.4356	53.66
9	1.0662	9.66	45	1.4500	55.07
10	1.0741	10.77	46	1.4646	56.48
11	1.0821	11.89	47	1.4796	57.90
12	1.0902	13.01	48	1.4948	59.32
13	1.0985	14.13	49	1.5104	60.75
14	1.1069	15.25	50	1.5263	62.18
15	1.1154	16.38	51	1.5426	63.66
16	1.1240	17.53	52	1.5591	65.13
17	1.1328	18.71	53	1.5761	66.63
18	1.1417	19.89	54	1.5934	68.13
19	1.1508	21.07	55	1.6111	69.65
20	1.1600	22.25	56	1.6292	71.17
21	1.1694	23.43	57	1.6477	72.75
22	1.1789	24.61	58	1.6667	74.36
23	1.1885	25.81	59	1.6860	75.99
24	1.1983	27.03	60	1.7059	77.67
25	1.2083	28.28	61	1.7262	79.43
26	1.2185	29.53	62	1.7470	81.30
27	1.2288	30.79	63	1.7683	83.34
28	1.2393	32.05	64	1.7901	85.66
29	1.2500	33.33	64.25	1.7957	86.33
30	1.2609	34.63	64.50	1.8012	87.04
31	1.2719	35.93	64.75	1.8068	87.81
32	1.2832	37.26	65	1.8125	88.65
33	1.2946	38.58	65.25	1.8182	89.55
34	1.3063	39.92	65.50	1.8239	90.60
35	1.3182	41.27	66	1.8354	93.19

Source: Courtesy of the Manufacturing Chemists Association.

Specific gravity determinations were made at 60°F, compared with water at 60°F. The values given above for aqueous sulfuric acid solutions were adopted as standard in 1904 by the Manufacturing Chemists Association.

From the specific gravities, the corresponding degrees Baumé were calculated by the following formula:

Baumé = 145 − (145/sp. gr.).

Baumé hydrometers for use with this table must be graduated by the above formula, which should always be printed on the scale. Acids stronger than 66°Bé should have their percentage compositions determined by chemical analysis.

APPENDIX III: CHEMICAL TESTS AND DETERMINATIONS

A. IDENTIFICATION TESTS

The identification tests described in *Section A* of this Appendix are frequently referred to in the Codex for the presumptive identification of FCC-grade chemicals taken from labeled containers. These tests are not intended to be applied to mixtures unless so specified.

Acetate
Acetic acid or acetates, when warmed with sulfuric acid and alcohol, form ethyl acetate, recognizable by its characteristic odor. With neutral solutions of acetates, ferric chloride TS produces a deep red color that is destroyed by the addition of a mineral acid.

Aluminum
Solutions of aluminum salts yield with 6 N ammonia a white, gelatinous precipitate that is insoluble in an excess of the 6 N ammonia. The same precipitate is produced by 1 N sodium hydroxide, but it dissolves in an excess of this reagent.

Ammonium
Ammonium salts are decomposed by 1 N sodium hydroxide with the evolution of ammonia, recognizable by its alkaline effect upon moistened red litmus paper. The decomposition is accelerated by warming.

Benzoate
Neutral solutions of benzoates yield a salmon-colored precipitate with ferric chloride TS. From moderately concentrated solutions of benzoate, 2 N sulfuric acid precipitates free benzoic acid, which is readily soluble in ether.

Bicarbonate
See *Carbonate*.

Bisulfite
See *Sulfite*.

Bromide
Free bromine is liberated from solutions of bromides upon the dropwise addition of chlorine TS. When shaken with chloroform, the bromine dissolves, coloring the chloroform red to reddish brown. A yellowish white precipitate, which is insoluble in nitric acid and slightly soluble in 6 N ammonia, is produced when solutions of bromides are treated with silver nitrate TS.

Calcium
Insoluble oxalate salts are formed when solutions of calcium salts are treated in the following manner: Using 2 drops of methyl red TS as indicator, neutralize a solution of a calcium salt (1 in 20) with 6 N ammonia, then add 2.7 N hydrochloric acid, dropwise, until the solution is acid. A white precipitate of calcium oxalate forms upon the addition of ammonium oxalate TS. This precipitate is insoluble in acetic acid but dissolves in hydrochloric acid.

Calcium salts moistened with hydrochloric acid impart a transient yellowish red color to a nonluminous flame.

Carbonate
Carbonates and bicarbonates effervesce with acids, yielding a colorless gas that produces a white precipitate immediately when passed into calcium hydroxide TS. Cold solutions of soluble carbonates are colored red by phenolphthalein TS, whereas solutions of bicarbonates remain unchanged or are slightly changed.

Chloride
Solutions of chlorides yield with silver nitrate TS a white, curdy precipitate that is insoluble in nitric acid but soluble in a slight excess of 6 N ammonia.

Citrate
To 15 mL of pyridine add a few mg of a citrate salt, dissolved or suspended in 1 mL of water, and shake. To this mixture add 5 mL of acetic anhydride, and shake: A light red color is produced.

Cobalt

Solutions of cobalt salts (1 in 20) in 2.7 N hydrochloric acid yield a red precipitate when heated on a steam bath with an equal volume of a hot, freshly prepared 1 in 10 solution of 1-nitroso-2-naphthol in 9 N acetic acid. Solutions of cobalt salts yield a yellow precipitate when saturated with potassium chloride and treated with potassium nitrite and acetic acid.

Copper

When solutions of cupric compounds are acidified with hydrochloric acid, a red film of metallic copper is deposited upon a bright untarnished surface of metallic iron. An excess of 6 N ammonia, added to a solution of a cupric salt, produces first a bluish precipitate and then a deep blue-colored solution. Solutions of cupric salts yield with potassium ferrocyanide TS a reddish brown precipitate, insoluble in diluted acids.

Hypophosphite

Hypophosphites evolve spontaneously flammable phosphine when strongly heated. Solutions of hypophosphites yield a white precipitate with mercuric chloride TS. This precipitate becomes gray when an excess of hypophosphite is present. Hypophosphite solutions, acidified with sulfuric acid and warmed with copper sulfate TS, yield a red precipitate.

Iodide

Solutions of iodides, upon the dropwise addition of chlorine TS, liberate iodine, which colors the solution yellow to red. Chloroform is colored violet when shaken with this solution. The iodine thus liberated gives a blue color with starch TS. Silver nitrate TS produces in solutions of iodides a yellow, curdy precipitate that is insoluble in nitric acid and in 6 N ammonia.

Iron

Solutions of ferrous and ferric compounds yield a black precipitate with ammonium sulfide TS. This precipitate is dissolved by cold 2.7 N hydrochloric acid with the evolution of hydrogen sulfide.

Ferric Salts   Potassium ferrocyanide TS (10%) produces a dark blue precipitate in acid solutions of ferric salts. With an excess of 1 N sodium hydroxide, a reddish brown precipitate is formed. Solutions of ferric salts produce with ammonium thiocyanate TS (1.0 N) a deep red color that is not destroyed by diluted mineral acids.

Ferrous Salts   Potassium ferricyanide TS (10%) produces a dark blue precipitate in solutions of ferrous salts. This precipitate, which is insoluble in dilute hydrochloric acid, is decomposed by 1 N sodium hydroxide. Solutions of ferrous salts yield with 1 N sodium hydroxide a greenish white precipitate, the color rapidly changing to green and then to brown when shaken.

Lactate

When solutions of lactates are acidified with sulfuric acid, potassium permanganate TS (0.1 N) is added, and the mixture is heated, acetaldehyde is evolved. This can be detected by allowing the vapor to come into contact with a filter paper that has been moistened with a freshly prepared mixture of equal volumes of 20% aqueous morpholine and sodium nitroferricyanide TS: A blue color is produced.

Magnesium

Solutions of magnesium salts in the presence of ammonium chloride yield no precipitate with ammonium carbonate TS, but a white crystalline precipitate, which is insoluble in 6 N ammonium hydroxide, is formed upon the subsequent addition of sodium phosphate TS (6%).

Manganese

Solutions of manganous salts yield with ammonium sulfide TS a salmon-colored precipitate that dissolves in acetic acid.

Nitrate

When a solution of a nitrate is mixed with an equal volume of sulfuric acid, the mixture cooled, and a solution of ferrous sulfate superimposed, a brown color is produced at the junction of the two liquids. Brownish red fumes are evolved when a nitrate is heated with sulfuric acid and metallic copper. Nitrates do not decolorize acidified potassium permanganate TS (0.1 N) (distinction from nitrites).

Nitrite

Nitrites yield brownish red fumes when treated with diluted mineral acids or acetic acid. A few drops of potassium iodide TS (15%) and a few drops of 2 N sulfuric acid added to a solution of nitrite liberate iodine, which colors starch TS blue.

Peroxide

Solutions of peroxides slightly acidified with sulfuric acid yield a deep blue color upon the addition of potassium dichromate TS. On shaking the mixture with an equal volume of diethyl ether and allowing the liquids to separate, the blue color is transferred to the ether layer.

Phosphate

Neutral solutions of orthophosphates yield with silver nitrate TS (0.1 N) a yellow precipitate, which is soluble in 1.7 N nitric acid or in 6 N ammonium hydroxide. With ammonium molybdate TS, a yellow precipitate, which is soluble in 6 N ammonium hydroxide, is formed.

Potassium

Potassium compounds impart a violet color to a nonluminous flame if not masked by the presence of small quantities of sodium. In neutral, concentrated or moderately concentrated solutions of potassium salts, sodium bitartrate TS (10%) slowly produces a white, crystalline precipitate that is soluble in 6 N ammonium hydroxide and in solutions of alkali hydroxides or carbonates. The precipitation may be accelerated by stirring or rubbing the inside of the test tube with a glass rod or by the addition of a small amount of glacial acetic acid or alcohol.

Sodium

Dissolve 0.1 g of the sodium compound in 2 mL of water, evaporate to dryness, and dissolve the residue in 2 mL of water. Add 2 mL of 15% potassium carbonate, and heat to boiling.

No precipitate is formed. Add 4 mL of potassium pyroantimonate TS, and heat to boiling. Allow to cool in ice water, and if necessary, rub the inside of the test tube with a glass rod. A dense precipitate is formed. Sodium compounds impart an intense yellow color to a nonluminous flame.

Sulfate
Solutions of sulfates yield with barium chloride TS (10%) a white precipitate that is insoluble in hydrochloric and nitric acids. Sulfates yield with lead acetate TS (8%) a white precipitate that is soluble in ammonium acetate solution. Hydrochloric acid produces no precipitate when added to solutions of sulfates (distinction from thiosulfates).

Sulfite
When treated with 2.7 N hydrochloric acid, sulfites and bisulfites yield sulfur dioxide, recognizable by its characteristic odor. This gas blackens filter paper moistened with mercurous nitrate TS.

Tartrate
When a few mg of a tartrate are added to a mixture of 15 mL of pyridine and 5 mL of acetic anhydride, an emerald green color is produced.

Thiosulfate
Solutions of thiosulfates yield with hydrochloric acid a white precipitate that soon turns yellow, liberating sulfur dioxide, recognizable by its odor. The addition of ferric chloride TS to solutions of thiosulfates produces a dark violet color that quickly disappears.

Zinc
Zinc salts, in the presence of sodium acetate, yield a white precipitate with hydrogen sulfide. This precipitate, which is insoluble in acetic acid, is dissolved by 2.7 N hydrochloric acid. A similar precipitate is produced by ammonium sulfide TS in neutral or alkaline solutions. Solutions of zinc salts yield with potassium ferrocyanide TS (10%) a white precipitate that is insoluble in 2.7 N hydrochloric acid.

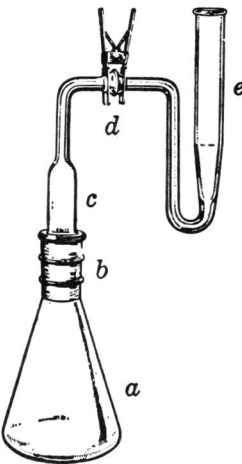

FIGURE 11 General Apparatus for Arsenic Test. (Courtesy of the Fisher Scientific Co., Pittsburgh, PA.)

Apparatus Use the general apparatus shown in Fig. 11 unless otherwise specified in an individual monograph. It consists of a 125-mL arsine generator flask (*a*) fitted with a scrubber unit (*c*) and an absorber tube (*e*), with a 24/40 standard-taper joint (*b*) and a ball-and-socket joint (*d*), secured with a No. 12 clamp, connecting the units. The tubing between *d* and *e* and between *d* and *c* is a capillary having an id of 2 mm and an od of 8 mm. Alternatively, an apparatus embodying the principle of the general assembly described and illustrated may be used.

Note: The special assemblies shown in Figs. 12, 13, and 14 are to be used only when specified in certain monographs.

Standard Arsenic Solution Accurately weigh 132.0 mg of arsenic trioxide that has been previously dried at 105° for 1 h, and dissolve it in 5 mL of sodium hydroxide solution (1 in 5). Neutralize the solution with 2 N sulfuric acid, add 10 mL in excess, and dilute to 1000.0 mL with recently boiled water.

B. LIMIT TESTS

ARSENIC TEST

Silver Diethyldithiocarbamate Colorimetric Method
(*Note*: All reagents used in this test should be very low in arsenic content.)

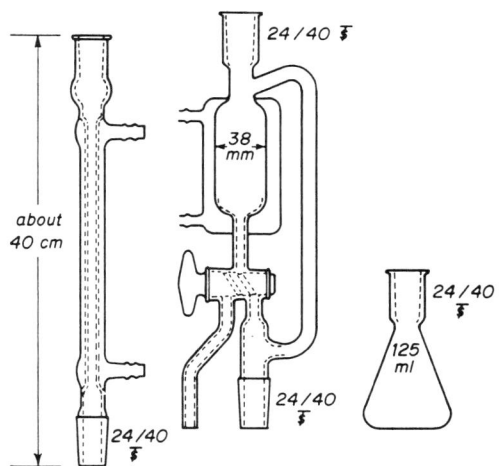

FIGURE 12 Modified Bethge Apparatus for the Distillation of Arsenic Tribromide.

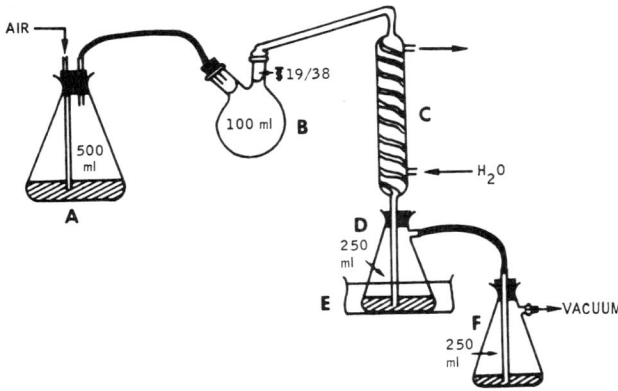

FIGURE 13 Special Apparatus for the Distillation of Arsenic Trichloride. (Flask *A* contains 150 mL of hydrochloric acid; flasks *D* and *F* contain 20 mL of water. Flask *D* is placed in an ice water bath, *E*.)

Transfer 10.0 mL of this solution into a 1000-mL volumetric flask, add 10 mL of 2 N sulfuric acid, dilute to volume with recently boiled water, and mix. Use this final solution, which contains 1 μg of arsenic (As) in each mL, within 3 days.

Silver Diethyldithiocarbamate Solution Dissolve 1 g of recrystallized silver diethyldithiocarbamate in 200 mL of recently distilled pyridine. Store this solution in a light-resistant container and use within 1 month.

Stannous Chloride Solution Dissolve 40 g of stannous chloride dihydrate, $SnCl_2 \cdot 2H_2O$, in 100 mL of hydrochloric acid. Store the solution in glass containers and use within 3 months.

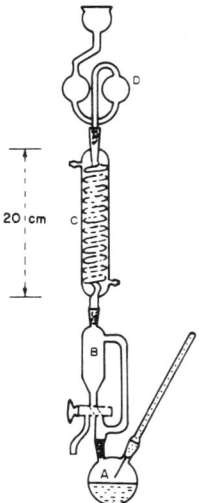

FIGURE 14 Special Apparatus for the Determination of Inorganic Arsenic. (*A*, 250-mL distillation flask; *B*, receiver chamber, approximately 50-mL capacity; *C*, reflux condenser; *D*, splash head.)

Lead Acetate-Impregnated Cotton Soak cotton in a saturated solution of lead acetate trihydrate, squeeze out the excess solution, and dry in a vacuum at room temperature.

Sample Solution Use directly as the *Sample Solution* in the *Procedure* the solution obtained by treating the sample as directed in an individual monograph. Prepare sample solutions of organic compounds in the generator flask (*a*), unless otherwise directed, according to the following general procedure:

Caution: Some substances may react unexpectedly with explosive violence when digested with hydrogen peroxide. Use appropriate safety precautions at all times.

Note: If halogen-containing compounds are present, use a lower temperature while heating the sample with sulfuric acid, do not boil the mixture, and add the peroxide, with caution, before charring begins to prevent loss of trivalent arsenic.

Transfer 1.0 g of the sample into the generator flask, add 5 mL of sulfuric acid and a few glass beads, and digest at a temperature not exceeding 120° until charring begins, preferably using a hot plate in a fume hood. (Additional sulfuric acid may be necessary to completely wet some samples, but the total volume added should not exceed about 10 mL.) After the acid has initially decomposed the sample, cautiously add, dropwise, hydrogen peroxide (30%), allowing the reaction to subside and reheating the sample between drops. Add the first few drops very slowly with sufficient mixing to prevent a rapid reaction, and discontinue heating if foaming becomes excessive. Swirl the solution in the flask to prevent unreacted substance from caking on the walls or bottom of the flask during digestion. *Maintain oxidizing conditions at all times during the digestion by adding small quantities of the peroxide whenever the mixture turns brown or darkens.* Continue the digestion until the organic matter is destroyed, gradually raising the temperature of the hot plate to 250° to 300° until fumes of sulfur trioxide are copiously evolved and the solution becomes colorless or retains only a light straw color. Cool, cautiously add 10 mL of water, heat again to strong fuming, and cool. Cautiously add 10 mL of water, mix, wash the sides of the flask with a few mL of water, and dilute to 35 mL.

Procedure If the *Sample Solution* was not prepared in the generator flask, transfer to the flask a volume of the solution, prepared as directed, equivalent to 1.0 g of the substance being tested, and add water to make 35 mL. Add 20 mL of dilute sulfuric acid (1 in 5), 2 mL of potassium iodide TS, 0.5 mL of *Stannous Chloride Solution*, and 1 mL of isopropyl alcohol, and mix. Allow the mixture to stand for 30 min at room temperature. Pack the scrubber unit (*c*) with two plugs of *Lead Acetate-Impregnated Cotton*, leaving a small air space between the two plugs, lubricate joints *b* and *d* with stopcock grease, if necessary, and connect the scrubber unit with the absorber tube (*e*). Transfer 3.0 mL of *Silver Diethyldithiocarbamate Solution* to the absorber tube, add 3.0 g of granular zinc (20-mesh) to the mixture in the flask, and immediately insert the standard-taper joint (*b*) in the flask. Allow the evolution of hydrogen and color development to proceed at room temperature (25° ± 3°) for 45 min, swirling the flask gently at 10-min intervals. Disconnect the absorber tube from the generator and scrubber units, and transfer the *Silver Diethyldithiocarbamate Solution* to a 1-cm absorption cell. Determine the absorbance at the wavelength of maximum absorption between 535 nm and 540 nm, with a suitable spectrophotometer or colorimeter, using *Silver Diethyldithiocarbamate Solution* as the blank. The absorbance due to any red color from the solution of the sample does not exceed that produced by 3.0 mL of *Standard Arsenic Solution* (3 μg As) when treated in the same manner and under the same conditions as the sample. The room temperature during the generation of arsine from the standard should be held to within ± 2° of that observed during the determination of the sample.

Interferences Metals or salts of metals such as chromium, cobalt, copper, mercury, molybdenum, nickel, palladium, and silver may interfere with the evolution of arsine. Antimony, which forms stibine, is the only metal likely to produce a positive interference in the color development with the silver diethyldithiocarbamate. Stibine forms a red color with silver diethyldithiocarbamate that has a maximum absorbance at 510 nm, but at 535 to 540 nm, the absorbance of the antimony complex is so diminished that the results of the determination would not be altered significantly.

CHLORIDE AND SULFATE LIMIT TESTS

Where limits for chloride and sulfate are specified in the individual monograph, compare the sample solution and control in appropriate glass cylinders of the same dimensions and matched as closely as practicable with respect to their optical characteristics.

If the solution is not perfectly clear after acidification, filter it through filter paper that has been washed free of chloride and sulfate. Add identical quantities of the precipitant (silver nitrate TS or barium chloride TS) in rapid succession to both the sample solution and the control solution.

Experience has shown that visual turbidimetric comparisons are best made between solutions containing from 10 to 20 μg of chloride ion (Cl) or from 200 to 400 μg of sulfate ion (SO_4) in 50 mL. Weights of samples are specified on this basis in the individual monographs in which these limits are included.

Chloride Test
Standard Chloride Solution Dissolve 165 mg of sodium chloride in water and dilute to 100.0 mL. Transfer 10.0 mL of this solution to a 1000-mL volumetric flask, dilute to volume with water, and mix. Each mL of the final solution contains 10 μg of chloride ion (Cl).

Procedure Unless otherwise directed, dissolve the specified amount of the test substance in 30 to 40 mL of water, neutralize to litmus external indicator with nitric acid, if necessary, then add 1 mL in excess. To the clear solution or filtrate add 1 mL of silver nitrate TS, dilute to 50 mL with water, mix, and allow to stand for 5 min protected from direct sunlight. Compare the turbidity, if any, with that produced similarly in a control solution containing the required volume of *Standard Chloride Solution* and the quantities of the reagents used for the sample.

Sulfate Test
Standard Sulfate Solution Dissolve 148 mg of anhydrous sodium sulfate in water and dilute to 100.0 mL. Transfer 10.0 mL of this solution to a 1000-mL volumetric flask, dilute to volume with water, and mix. Each mL of the final solution contains 10 μg of sulfate (SO_4).

Procedure Unless otherwise directed, dissolve the specified amount of the test substance in 30 to 40 mL of water, neutralize to litmus external indicator with hydrochloric acid, if necessary, then add 1 mL of 2.7 N hydrochloric acid. To the clear solution or filtrate add 3 mL of barium chloride TS, dilute to 50 mL with water, and mix. After 10 min compare the turbidity, if any, with that produced in a solution containing the required volume of *Standard Sulfate Solution* and the quantities of the reagents used for the sample.

1,4-DIOXANE LIMIT TEST

Vacuum Distillation Apparatus Assemble a closed-system vacuum distillation apparatus, employing glass vacuum stopcocks (*A*, *B*, and *C*), as shown in Fig. 15. The concentrator tube (*D*) is made of borosilicate or quartz (not flint) glass, graduated precisely enough to measure the 0.9 mL or more of distillate and marked so that the analyst can dilute accurately to 2.0 mL (available as Chromaflex concentrator tube, Kontes Glass Co., Vineland, NJ, Catalog No. K42560-0000).

Standard Preparation Prepare a solution of 1,4-dioxane in water containing 100 μg/mL. Keep the solution refrigerated, and prepare fresh weekly.

Sample Preparation Transfer 20 g of the sample, accurately weighed, into a 50-mL round-bottom flask (*E*) having a 24/40 ground-glass neck. Semi-solid or waxy samples should be

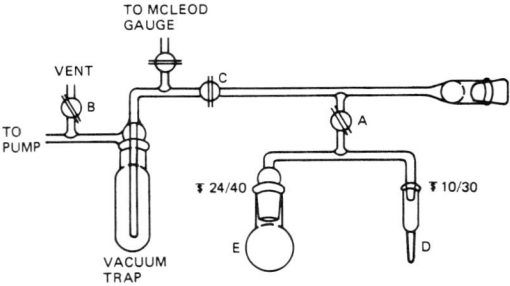

FIGURE 15 Closed-System Vacuum Distillation Apparatus for 1,4-Dioxane.

liquefied by heating on a steam bath before making the transfer. Add 2.0 mL of water to the flask for crystalline samples, and 1.0 mL for liquid, semi-solid, or waxy samples. Place a small Teflon-covered stirring bar in the flask, stopper, and stir to mix. Immerse the flask in an ice bath, and chill for about 1 min.

Wrap heating tape around the tube connecting the Chromaflex tube (*D*) and the round-bottom flask (*E*), and apply about 10 V to the tape. Apply a light coating of high-vacuum silicone grease to the ground-glass joints, and connect the Chromaflex tube to the 10/30 joint and the round-bottom flask to the 24/40 joint. Immerse the vacuum trap in a Dewar flask filled with liquid nitrogen, close stopcocks *A* and *B*, open stopcock *C*, and begin evacuating the system with a vacuum pump. Prepare a slush bath from powdered dry ice and methanol, and raise the bath to the neck of the round-bottom flask. After freezing the contents of the flask for about 10 min, and when the vacuum system is operating at 0.05 mm pressure or lower, open stopcock *A* for 20 s, and then close it. Remove the slush bath, and allow the flask to warm in air for about 1 min. Immerse the flask in a water bath at 20° to 25°, and after about 5 min warm the water in the bath to 35° to 40° (sufficient to liquefy most samples) while stirring slowly but constantly with the magnetic bar. Cool the water in the bath by adding ice, and chill for about 2 min. Replace the water bath with the slush bath, freeze the contents of the flask for about 10 min, then open stopcock *A* for 20 s, and close it. Remove the slush bath, and repeat the heating steps as before, this time reaching a final temperature of 45° to 50° or a temperature necessary to melt the sample completely. If there is any condensation in the tube connecting the round-bottom flask to the Chromaflex tube, slowly increase the voltage to the heating tape and heat until condensation disappears.

Stir with the magnetic stirrer throughout the following steps: Very slowly immerse the Chromaflex tube in the Dewar flask containing liquid nitrogen.

Caution: When there is liquid distillate in the Chromaflex tube, the tube must be immersed in the nitrogen *very slowly*, or the tube will break.

Water will begin to distill into the tube. As ice forms in the tube, raise the Dewar flask to keep the liquid nitrogen level only slightly below the level of ice in the tube. When water begins to freeze in the neck of the 10/30 joint, or when liquid nitrogen reaches the 2.0-mL graduation mark on the Chromaflex tube, remove the Dewar flask and let the ice melt without heating. After the ice has melted, check the volume of water that has distilled, and repeat the sequence of chilling and thawing until at least 0.9 mL of water has been collected. Freeze the tube once again for about 2 min, and release the vacuum first by opening stopcock *B*, followed by stopcock *A*. Remove the Chromaflex tube from the apparatus, close it with a greased stopper, and let the ice melt without heating. Mix the contents of the tube by swirling, note the volume of distillate, and dilute to 2.0 mL with water, if necessary. Use this *Sample Preparation* as directed under *Chromatography*.

Chromatography Use a gas chromatograph equipped with a flame-ionization detector. Under typical conditions, the instrument contains a 4-mm (id) × 6-ft glass column packed with 80/100- or 100/120-mesh Chromosorb 104 or equivalent. The column is maintained isothermally at about 140°, the injection port at 200°, and the detector at 250°. Nitrogen is the carrier gas, flowing at a rate of about 35 mL/min. Install an oxygen scrubber between the carrier gas line and the column. The column should be conditioned for about 72 h at 250° with 30 to 40 mL/min carrier flow.

Note: Chromosorb 104 is oxygen sensitive. Both new and used columns should be flushed with carrier gas for 30 to 60 min before heating each time they are installed in the gas chromatograph.

Inject a volume of the *Standard Preparation*, accurately measured, to give about 20% of maximum recorder response. Where possible, keep the injection volume in the range of 2 to 4 µL, and use the solvent-flush technique to minimize errors associated with injection volumes. In the same manner, inject an identical volume of the *Sample Preparation*. The height of the peak produced by the *Sample Preparation* does not exceed that produced by the *Standard Preparation*.[1]

FLUORIDE LIMIT TEST

Method I (Thorium Nitrate Colorimetric Method)
Use this method unless otherwise directed in the individual monograph.

Caution: When applying this test to organic compounds, rigidly control at all times the temperature at which the distillation is conducted to the recommended range of 135° to 140° to avoid the possibility of explosion.

Note: To minimize the distillation blank resulting from fluoride leached from the glassware, treat the distillation apparatus as follows: Treat the glassware with hot 10% sodium hydroxide solution, followed by flushing with tap water and rinsing with distilled water. At least once daily, treat in addition by boiling down 15 to 20 mL of dilute sulfuric acid (1 in 2) until the still is filled with fumes; cool, pour off the acid, treat again with 10% sodium hydroxide solution, and rinse thoroughly. For further details, see AOAC method 944.08.

Unless otherwise directed, place a 5.0-g sample and 30 mL of water in a 125-mL Pyrex distillation flask having a side arm and trap. The flask is connected with a condenser and carries a thermometer and a capillary tube, both of which must extend into the liquid. Slowly add, with continuous stirring, 10 mL of 70% perchloric acid, and then add 2 or 3 drops of silver nitrate solution (1 in 2) and a few glass beads. Connect a small dropping funnel or a steam generator to the capillary tube. Support the flask on a flame-resistant mat or shielding board, with a hole

---
[1] If the sample fails the test because of known or suspected interference, another aliquot may be run on a 6-ft × 2-mm (id) column of 0.2% Carbowax 1500 on Carbopak C, column operating at 100° isothermal, with 20 mL/min of helium carrier flow. Under these conditions, the 1,4-dioxane elutes in about 4 min.

that exposes about one-third of the flask to the low, "clean" flame of a Bunsen burner.

Note: The shielding is essential to prevent the walls of the flask from overheating above the level of its liquid contents.

Distill until the temperature reaches 135°. Add water from the funnel or introduce steam through the capillary, maintaining the temperature between 135° and 140° at all times. Continue the distillation until 100 mL of distillate has been collected. After the 100-mL portion (*Distillate A*) is collected, collect an additional 50-mL portion of distillate (*Distillate B*) to ensure that all of the fluorine has been volatilized.

Place 50 mL of *Distillate A* in a 50-mL Nessler tube. In another similar Nessler tube place 50 mL of water distilled through the apparatus as a control. Add to each tube 0.1 mL of a filtered solution of sodium alizarinsulfonate (1 in 1000) and 1 mL of freshly prepared hydroxylamine hydrochloride solution (1 in 4000), and mix well. Add, dropwise and with stirring, either 1 N or 0.05 N sodium hydroxide, depending upon the expected volume of volatile acid distilling over, to the tube containing the distillate until its color just matches that of the control, which is faintly pink. Then add to each tube 1.0 mL of 0.1 N hydrochloric acid, and mix well. From a buret, graduated in 0.05 mL, add slowly to the tube containing the distillate enough thorium nitrate solution (1 in 4000) so that, after mixing, the color of the liquid just changes to a faint pink. Note the volume of the solution added, then add exactly the same volume to the control, and mix. Now add to the control solution sodium fluoride TS (10 μg F per mL) from a buret to make the colors of the two tubes match after dilution to the same volume. Mix well, and allow all air bubbles to escape before making the final color comparison. Check the endpoint by adding 1 or 2 drops of sodium fluoride TS to the control. A distinct change in color should take place. Note the volume of sodium fluoride TS added.

Dilute *Distillate B* to 100 mL, and mix well. Place 50 mL of this solution in a 50-mL Nessler tube, and follow the procedure used for *Distillate A*. The total volume of sodium fluoride TS required for the solutions from both *Distillate A* and *Distillate B* should not exceed 2.5 mL.

Method II (Ion-Selective Electrode Method A)

Buffer Solution   Dissolve 36 g of cyclohexylenedinitrilotetraacetic acid (CDTA) in sufficient 1 N sodium hydroxide to make 200 mL. Transfer 20 mL of this solution (equivalent to 4 g of disodium CDTA) into a 1000-mL beaker containing 500 mL of water, 57 mL of glacial acetic acid, and 58 g of sodium chloride, and stir to dissolve. Adjust the pH of the solution to between 5.0 and 5.5 by the addition of 5 N sodium hydroxide, then cool to room temperature, dilute to 1000 mL with water, and mix.

Procedure   Unless otherwise directed in the individual monograph, place an 8.0-g sample and 20 mL of water in a 250-mL distilling flask, cautiously add 20 mL of perchloric acid, and then add 2 or 3 drops of silver nitrate solution (1 in 2) and a few glass beads. Following the directions, and observing the *Caution* and *Notes*, as given under *Method I*, distill the solution until 200 mL of distillate has been collected.

Transfer a 25.0-mL aliquot of the distillate into a 250-mL plastic beaker, and dilute to 100 mL with the *Buffer Solution*. Place the fluoride ion and reference electrodes (or a combination fluoride electrode) of a suitable ion-selective electrode apparatus in the solution, and adjust the calibration control until the indicator needle points to the center on the logarithmic concentration scale, allowing sufficient time for equilibration (about 20 min) and stirring constantly during the equilibration period and throughout the remainder of the procedure. Pipet 1.0 mL of a solution containing 100 μg of fluoride ion (F) per mL (prepared by dissolving 22.2 mg of sodium fluoride, previously dried at 200° for 4 h, in sufficient water to make 100.0 mL) into the beaker, allow the electrode to come to equilibrium, and record the final reading on the logarithmic concentration scale.

Note: Follow the instrument manufacturer's instructions regarding precautions and interferences, electrode filling and check, temperature compensation, and calibration.

Calculations   Calculate the fluoride content, in mg/kg, of the sample taken by the formula

$$[IA/(R - I)] \times 100 \times (200/25W),$$

in which I is the initial scale reading before the addition of the sodium fluoride solution; A is the concentration, in μg/mL, of fluoride in the sodium fluoride solution added to the sample solution; R is the final scale reading after addition of the sodium fluoride solution; and W is the original weight, in g, of the sample.

Method III (Ion-Selective Electrode Method B)

Sodium Fluoride Solution   (5 μg F per mL) Transfer 2.210 g of sodium fluoride, previously dried at 200° for 4 h and accurately weighed, into a 400-mL plastic beaker, add 200 mL of water, and stir until dissolved. Quantitatively transfer this solution into a 1000-mL volumetric flask with the aid of water, dilute to volume with water, and mix. Store this stock solution in a plastic bottle. On the day of use, transfer 5.0 mL of the stock solution into a 1000-mL volumetric flask, dilute to volume with water, and mix.

Calibration Curve   Transfer into separate 250-mL plastic beakers 1.0, 2.0, 3.0, 5.0, 10.0, and 15.0 mL of the *Sodium Fluoride Solution*, add 50 mL of water, 5 mL of 1 N hydrochloric acid, 10 mL of 1 M sodium citrate, and 10 mL of 0.2 M disodium EDTA to each beaker, and mix. Transfer each solution into a 100-mL volumetric flask, dilute to volume with water, and mix. Transfer a 50-mL portion of each solution into a 125-mL plastic beaker, and measure the potential of each solution with a suitable ion-selective electrode apparatus (such as the Orion Model No. 94-09, with solid-state membrane), using a suitable reference electrode (such as the Orion Model No. 90-01, with single junction). Plot the calibration curve on two-cycle semilogarithmic paper (such as K & E No. 465130), with μg F per 100 mL solution on the logarithmic scale.

Procedure   Transfer 1.00 g of the sample into a 150-mL glass beaker, add 10 mL of water, and, while stirring continuously, add 20 mL of 1 N hydrochloric acid slowly to dissolve the sample. Boil rapidly for 1 min, then transfer into a 250-mL plastic beaker, and cool rapidly in ice water. Add 15 mL of 1 M sodium citrate and 10 mL of 0.2 M disodium EDTA, and mix. Adjust the pH to 5.5 ± 0.1 with 1 N hydrochloric acid or 1 N sodium hydroxide, if necessary, then transfer into a 100-mL volumetric flask, dilute to volume with water, and mix. Transfer a 50-mL portion of this solution into a 125-mL plastic beaker, and measure the potential of the solution with the apparatus described under *Calibration Curve*. Determine the fluoride content, in μg, of the sample from the *Calibration Curve*.

Method IV (Ion-Selective Electrode Method C)

Buffer Solution A Add 2 volumes of 6 N acetic acid to 1 volume of water, and adjust the pH to 5.0 with 50% potassium hydroxide solution.

Buffer Solution B Dissolve 150 g of sodium citrate dihydrate and 10.3 g of disodium EDTA dihydrate in 800 mL of water, adjust the pH to 8.0 with 50% sodium hydroxide solution, and dilute to 1000 mL with water.

Buffer Solution C Dissolve 36 g of cyclohexylene dinitrilotetraacetic acid (CDTA) in sufficient 1 N sodium hydroxide to make 200 mL by boiling, then cool, and filter through glass-fiber filter paper. Pipet 30 mL of this solution into a mixture consisting of 750 mL of water, 87 g of sodium chloride, and 85.5 mL of glacial acetic acid. Adjust the pH to between 5.0 and 5.5 by the addition of 50% sodium hydroxide solution, then cool, and dilute to 3000 mL with water.

Fluoride Standard Solution Use a solution containing 100 μg of fluoride ion (F) per mL (100 mg/kg), prepared by dissolving 22.2 mg of sodium fluoride, previously dried at 200° for 4 h, in sufficient water to make 100.0 mL.

Sample Preparation Weigh accurately the amount of sample specified in the monograph, transfer it into a 100-mL volumetric flask, and dissolve it in a minimum amount of water or in the volume of hydrochloric acid solution specified in the monograph. Add 50.0 mL of the appropriate buffer solution, A, B, or C, as specified in the monograph, dilute to volume with water, and mix.

Procedure Pipet a 50-mL aliquot of the *Sample Preparation* into a plastic beaker, and place in the solution the fluoride ion and reference electrodes (or a combination fluoride electrode) of a suitable ion-selective electrode apparatus with magnetic stirrer. Begin stirring slowly, and set the slope of the meter to 100% and the temperature control to room temperature, which should be the temperature of the solution. Adjust the calibration control to read infinity on the increment logarithmic scale, and allow the instrument to equilibrate.

> Note: The ion-selective electrode responds much more slowly than does a pH electrode, and a stable reading may not be obtained until 2 or 3 min have passed. The reading should not change for 30 to 60 s.

Add the volume, accurately measured, of *Fluoride Standard Solution* specified in the monograph, allow the electrode to equilibrate with continued stirring, and take the final reading on the increment logarithmic scale, recording the value obtained as S. Perform a blank determination using 50 mL of the same buffer solution as used for the sample under analysis, and record the value obtained as B.

Calculation Determine the Δ value by the formula

$$(V \times C)/50,$$

in which V is the volume, in mL, of *Fluoride Standard Solution* added; C is the exact concentration, in mg/kg, of the *Fluoride Standard Solution*; and 50 is the mL of *Sample Preparation* used. Calculate the mg/kg of fluoride (F) in the sample by the formula

$$[(S \times \Delta) - B] \times (100/W),$$

in which W is the weight, in g, of sample taken.

HEAVY METALS TEST

This test is designed to limit the content of common metallic impurities colored by sulfide ion (Ag, As, Bi, Cd, Cu, Hg, Pb, Sb, Sn) by comparing the color with a standard containing lead ion (Pb) under the specified test conditions. It demonstrates that the test substance is not grossly contaminated by such heavy metals and, within the precision of the test, that it does not exceed the *Heavy Metals* limit given in the individual monograph as determined by concomitant visual comparison with a control solution. In the specified pH range, the optimum concentration of lead ion (Pb) for matching purposes by this method is 20 μg in 50 mL of solution.

The most common limitation of the *Heavy Metals Test* is that the color the sulfide ion produces in the *Sample Solution* depends on the metals present and may not match the color in the *Lead Solution* used for matching purposes. Lead sulfide is brown, as are Ag, Bi, Cu, Hg, and Sn sulfides. While it is possible that ions not mentioned here may also yield nonmatching colors, among the nine common metallic impurities listed above, the sulfides with different colors are those of As and Cd, which are yellow, and that of Sb, which is orange. If a yellow or orange color is observed, the following action is indicated: If the monograph does not include an arsenic requirement, As should be determined. Any As found should not exceed the requirement in the monograph, or 3 mg/kg if there is no requirement. If these criteria are met, Cd may be a contributor to the yellow color, so the Cd content should be determined. If an orange color is observed, the Sb content should be determined. These additional tests are in accord with the section on *Trace Impurities* in the *General Provisions* of this book, as follows: "... if other possible impurities may be present, additional tests may be required, and should be applied, as necessary, by the manufacturer, vendor, or user to demonstrate that the substance is suitable for its intended application in foods or in food processing."

Determine the amount of heavy metals by *Method I* unless otherwise specified in the individual monograph. Use *Method I* for substances that yield clear, colorless solutions before adding sulfide ion. Use *Method II* for those substances that do not yield clear, colorless solutions under the test conditions specified for *Method I* or for substances that by virtue of their complex nature, interfere with the precipitation of metals by sulfide ion. Use *Method III*, a wet digestion method, only in those cases where neither *Method I* nor *Method II* can be used.

Special Reagents

Lead Nitrate Stock Solution Dissolve 159.8 mg of ACS reagent-grade lead nitrate [Pb(NO$_3$)$_2$] in 100 mL of water containing 1 mL of nitric acid, dilute with water to 1000.0 mL, and mix. Prepare and store this solution in glass containers that are free from lead salts.

Standard Lead Solution On the day of use, dilute 10.0 mL of *Lead Nitrate Stock Solution* with water to 100.0 mL. Each mL of *Standard Lead Solution* contains the equivalent of 10 μg of lead ion (Pb).

Procedure (Note: In the following procedures, failure to accurately adjust the pH of the solution within the specified limits may result in a significant loss of test sensitivity.)

Method I

Solution A Pipet 2.0 mL of *Standard Lead Solution* (20 μg of Pb) into a 50-mL color-comparison tube, and add water to make 25 mL. Adjust the pH to between 3.0 and 4.0 (using short-range pH indicator paper) by adding 1 *N* acetic acid or 6 *N* ammonia, dilute with water to 40 mL, and mix.

Solution B Place 25 mL of the solution, prepared as directed in the individual monograph, into a 50-mL color-comparison tube that matches the one used for *Solution A*, adjust the pH to between 3.0 and 4.0 (using short-range pH indicator paper) by adding 1 *N* acetic acid or 6 *N* ammonia, dilute with water to 40 mL, and mix.

Solution C Place 25 mL of the solution prepared as directed in the individual monograph into a third color-comparison tube that matches those used for *Solutions A* and *B*, and add 2.0 mL of *Standard Lead Solution*. Adjust the pH to between 3.0 and 4.0 (using short-range pH indicator paper) by adding 1 *N* acetic acid or 6 *N* ammonia, dilute with water to 40 mL, and mix.

Add 10 mL of freshly prepared hydrogen sulfide TS to each tube, mix, allow to stand for 5 min, and view downward over a white surface. The color of *Solution B* is not darker than that of *Solution A*, and the intensity of the color of *Solution C* is equal to or greater than that of *Solution A*. If the color of *Solution C* is lighter than that of *Solution A*, the test substance is providing an interference with the test procedure, and *Method II* must be used for the substance under examination.

Method II

Solution A Prepare as directed under *Method I*.

Solution B Place the quantity, accurately weighed, of sample specified in the individual monograph in a suitable crucible, add sufficient sulfuric acid to wet the sample, and carefully ignite at a low temperature until thoroughly charred, covering the crucible loosely with a suitable lid during the ignition. After the substance is thoroughly carbonized, add 2 mL of nitric acid and 5 drops of sulfuric acid, cautiously heat until white fumes evolve, then ignite, preferably in a muffle furnace, at 500° to 600° until all of the carbon is burned off. Cool, add 4 mL of dilute hydrochloric acid (1 in 2), cover, and digest on a steam bath for 10 to 15 min. Uncover, and slowly evaporate on a steam bath to dryness. Moisten the residue with 1 drop of hydrochloric acid, add 10 mL of hot water, and digest for 2 min. Add 6 *N* ammonia dropwise until the solution is just alkaline to litmus paper, dilute with water to 25 mL, and adjust the pH to between 3.0 and 4.0 (using short-range pH indicator paper) by adding 1 *N* acetic acid. Filter if necessary, rinse the crucible and the filter with 10 mL of water, transfer the solution and rinsings to a 50-mL color-comparison tube, dilute with water to 40 mL, and mix.

Add 10 mL of freshly prepared hydrogen sulfide TS to each tube, mix, allow to stand for 5 min, and view downward over a white surface. The color of *Solution B* is not darker than that of *Solution A*.

Method III

Solution A Transfer a mixture of 8 mL of sulfuric acid and 10 mL of nitric acid into a 100-mL Kjeldahl flask, clamp the flask at an angle of 45°, and then add, in small increments, an additional volume of nitric acid equal to that added in the preparation of *Solution B*, below. Heat the solution to dense, white fumes, cool, and cautiously add 10 mL of water. Add a volume of hydrogen peroxide (30%) equal to that added in the preparation of *Solution B*, below, then boil gently to dense, white fumes, and cool. Cautiously add 5 mL of water, mix, and boil gently to dense, white fumes. Continue boiling until the volume is reduced to about 2 or 3 mL, then cool, and dilute cautiously with a few mL of water. Into this solution pipet 2.0 mL of *Standard Lead Solution*, and mix. Transfer into a 50-mL color-comparison tube, rinse the flask with water, adding the rinsings to the tube until the volume is 25 mL, and mix. Adjust the pH to between 3.0 and 4.0 (using short-range pH indicator paper) with ammonium hydroxide initially, and then with 6 *N* ammonia as the desired range is neared, dilute with water to 40 mL, and mix.

Solution B Transfer the quantity, accurately weighed, of sample specified in the individual monograph into a 100-mL Kjeldahl flask (or into a 300-mL flask if the reaction foams excessively) clamp the flask at an angle of 45°, and then add a sufficient amount of a mixture of 8 mL of sulfuric acid and 10 mL of nitric acid to moisten the sample thoroughly.

Note: For liquid samples, use 3 mL of the acid mixture.

Warm gently until the reaction commences, allow the reaction to subside, and then add additional portions of the acid mixture, heating after each addition, until all of the 18 mL of acid mixture has been added. Increase the heat, and boil gently until the reaction mixture darkens. Remove the flask from the heat, add 2 mL of nitric acid, and heat to boiling again. Continue the intermittent heating and addition of 2-mL portions of nitric acid until no further darkening occurs, then heat strongly to dense, white fumes, and cool. Cautiously add 5 mL of water, mix, boil gently to dense, white fumes, and continue heating until the volume is reduced to about 2 or 3 mL. Cool, cautiously add 5 mL of water, and examine. If the solution is yellow, cautiously add 1 mL of hydrogen peroxide (30%), and again evaporate to dense, white fumes and to a volume of about 2 or 3 mL. Cool, dilute cautiously with a few mL of water, and mix. Transfer into a 50-mL color-comparison tube, rinse the flask with water, adding the rinsings to the tube until the volume is 25 mL, and mix. Adjust the pH to between 3.0 and 4.0 (using short-range pH indicator paper) with ammonium hydroxide initially, and then with 6 *N* ammonia as the desired range is neared, dilute with water to 40 mL, and mix.

To each tube add 10 mL of freshly prepared hydrogen sulfide TS, mix, allow to stand for 5 min, and view downward over a white surface. The color of *Solution B* is not darker than that of *Solution A*.

LEAD LIMIT TEST

Note: Unless otherwise specified in the monograph, use the *Dithizone Method* to determine lead levels.

Dithizone Method

Special Reagents   Select reagents having as low a lead content as practicable, and store all solutions in containers of borosilicate glass. Rinse all glassware thoroughly with warm dilute nitric acid (1 in 2) followed by water.

Ammonia–Cyanide Solution   Dissolve 2 g of potassium cyanide in 15 mL of ammonium hydroxide, and dilute with water to 100 mL.

Ammonium Citrate Solution   Dissolve 40 g of citric acid in 90 mL of water, add 2 or 3 drops of phenol red TS, then cautiously add ammonium hydroxide until the solution acquires a reddish color. Extract it with 20-mL portions of *Dithizone Extraction Solution* until the dithizone solution retains its green color or remains unchanged.

Diluted Standard Lead Solution (1 μg Pb in 1 mL)   Immediately before use, transfer 10.0 mL of *Standard Lead Solution* (*Heavy Metals Test*, Appendix IIIB) containing 10 μg of lead per mL, to a 100-mL volumetric flask, dilute to volume with dilute nitric acid (1 in 100), and mix.

Dithizone Extraction Solution   Dissolve 30 mg of dithizone in 1000 mL of chloroform, add 5 mL of alcohol, and mix. Store in a refrigerator. Before use, shake a suitable volume of the solution with about half its volume of dilute nitric acid (1 in 100), discarding the nitric acid. Do not use if more than 1 month old.

Hydroxylamine Hydrochloride Solution   Dissolve 20 g of hydroxylamine hydrochloride in sufficient water to make about 65 mL, transfer the solution to a separator, add a few drops of thymol blue TS, then add ammonium hydroxide until the solution assumes a yellow color. Add 10 mL of sodium diethyldithiocarbamate solution (1 in 25), mix, and allow to stand for 5 min. Extract the solution with successive 10- to 15-mL portions of chloroform until a 5-mL test portion of the chloroform extract does not assume a yellow color when shaken with cupric sulfate TS. Add 2.7 *N* hydrochloric acid until the extracted solution is pink, adding 1 or 2 drops more of thymol blue TS if necessary, then dilute with water to 100 mL, and mix.

Potassium Cyanide Solution   Dissolve 50 g of potassium cyanide in sufficient water to make 100 mL. Remove the lead from the solution by extraction with successive portions of *Dithizone Extraction Solution* as described under *Ammonium Citrate Solution*, then extract any dithizone remaining in the cyanide solution by shaking with chloroform. Finally, dilute the cyanide solution with sufficient water so that each 100 mL contains 10 g of potassium cyanide.

Standard Dithizone Solution   Dissolve 10 mg of dithizone in 1000 mL of chloroform, keeping the solution in a glass-stoppered, lead-free bottle suitably wrapped to protect it from light and stored in a refrigerator.

Sample Solution   The solution obtained by treating the sample as directed in an individual monograph is used directly as the *Sample Solution* in the *Procedure*. Sample solutions of organic compounds are prepared, unless otherwise directed, according to the following general method:

> **Caution**: Some substances may react unexpectedly with explosive violence when digested with hydrogen peroxide. Appropriate safety precautions must be employed at all times.

Transfer 1.0 g of the sample into a suitable flask, add 5 mL of sulfuric acid and a few glass beads, and digest at a temperature not exceeding 120° until charring begins, using, preferably, a hot plate in a fume hood. (Additional sulfuric acid may be necessary to completely wet some samples, but the total volume added should not exceed about 10 mL.) After the sample has been initially decomposed by the acid, add with caution, dropwise, hydrogen peroxide (30%), allowing the reaction to subside and reheating between drops. The first few drops must be added very slowly with sufficient mixing to prevent a rapid reaction, and heating should be discontinued if foaming becomes excessive. Swirl the solution in the flask to prevent unreacted substance from caking on the walls or bottom of the flask during the digestion. *Add small quantities of the peroxide when the solution begins to darken*, and continue the digestion until the organic matter is destroyed, gradually raising the temperature of the hot plate to 250° to 300° until fumes of sulfur trioxide are copiously evolved and the solution becomes colorless or retains only a light straw color. Cool, add cautiously 10 mL of water, again evaporate to strong fuming, and cool. Quantitatively transfer the solution into a separator with the aid of small quantities of water.

Procedure   Transfer the *Sample Solution*, prepared as directed in the individual monograph, into a separator, and, unless otherwise directed, add 6 mL of *Ammonium Citrate Solution* and 2 mL of *Hydroxylamine Hydrochloride Solution*. (Use 10 mL of the citrate solution when determining lead in iron salts.) To the separator add 2 drops of phenol red TS, and make the solution just alkaline (red in color) by the addition of ammonium hydroxide. Cool the solution, if necessary, under a stream of tap water, then add 2 mL of *Potassium Cyanide Solution*. Immediately extract the solution with 5-mL portions of *Dithizone Extraction Solution*, draining each extract into another separator, until the dithizone solution retains its green color. Shake the combined dithizone solutions for 30 s with 20 mL of dilute nitric acid (1 in 100), discard the chloroform layer, add to the acid solution 5.0 mL of *Standard Dithizone Solution* and 4 mL of *Ammonia–Cyanide Solution*, and shake for 30 s. The purplish hue in the chloroform solution of the sample due to any lead dithizonate present does not exceed that in a control, containing the volume of *Diluted Standard Lead Solution* equivalent to the amount of lead specified in the monograph, when treated in the same manner as the sample.

Flame Atomic Absorption Spectrophotometric Method
Select reagents having as low a lead content as practicable, and store all solutions in high-density polyethylene containers. Rinse all plastic and glassware thoroughly with warm dilute nitric acid (1 in 2) followed by water.

Lead Nitrate Stock Solution (100 µg/mL) Dissolve 159.8 mg of reagent-grade lead nitrate, $Pb(NO_3)_2$, in 100 mL of water containing 1 mL of nitric acid in a 1000-mL volumetric flask, and dilute to volume with water.

Standard Lead Solution (10 µg/mL) On the day of use, transfer 10 mL of *Lead Nitrate Stock Solution* to a 100-mL volumetric flask, and dilute to volume with water.

Diluted Standard Lead Solutions On the day of use, prepare a set of standard lead solutions that corresponds to the lead limit specified in the monograph:
1-mg/kg Lead Limit (0.5-, 1.0-, and 1.5-µg/mL standards) On the day of use, transfer 5.0, 10.0, and 15.0 mL of *Standard Lead Solution* to three separate 100-mL volumetric flasks, add 10 mL of 3 *N* hydrochloric acid to each, and dilute to volume with water.
5-mg/kg Lead Limit (1.0-, 5.0-, and 10.0-µg/mL standards) On the day of use, transfer 1.0. and 50.0 mL of *Standard Lead Solution* to two separate 100-mL volumetric flasks, add 10 mL of 3 *N* hydrochloric acid to each, and dilute to volume with water. The final standard, 10.0 µg/mL, is taken directly from the *Standard Lead Solution*.
10-mg/kg Lead Limit (5.0-, 10.0-, and 15.0-µg/mL standards) On the day of use, transfer 5.0, 10.0, and 15.0 mL of *Lead Nitrate Stock Solution* to three separate 100-mL volumetric flasks, add 10 mL of 3 *N* hydrochloric acid to each, and dilute to volume with water.

25% Sulfuric Acid Solution (by volume) Cautiously add 100 mL of sulfuric acid to 300 mL of water with constant stirring while cooling in an ice bath.

Sample Preparation Weigh to the nearest 0.1 mg the sample weight specified in the monograph, and transfer to an evaporating dish. Add 5 mL of *25% Sulfuric Acid Solution*, and distribute the sulfuric acid uniformly through the sample. Within a hood, place the dish on a steam bath to evaporate most of the water. Place the dish on a burner, and slowly pre-ash the sample by expelling most of the sulfuric acid. Place the dish in a muffle furnace that has been set at 525°, and ash the sample until the residue appears free from carbon. Prepare a *Sample Blank* by ashing 5 mL of 25% sulfuric acid. Cool and cautiously wash down the inside of each evaporation dish with water.
Add 5 mL of 1 *N* hydrochloric acid. Place the dish on a steam bath, and evaporate to dryness. Add 1.0 mL of 3 *N* hydrochloric acid and approximately 5 mL of water, and heat briefly on a steam bath to dissolve any residue. Transfer each solution quantitatively to a 10-mL volumetric flask, dilute to volume, and mix.

Procedure Concomitantly determine the absorbances of the *Sample Blank*, the *Diluted Standard Lead Solutions*, and the *Sample Preparation* at the lead emission line of 283.3 nm, using a slit-width of 0.7 nm. Use a suitable atomic absorption spectrophotometer equipped with a lead electrodeless discharge lamp (EDL), an air–acetylene flame, and a 4-in. burner head. Use water as the blank.

Calculations Determine the corrected absorbance values by subtracting the *Sample Blank* absorbance from each of the *Diluted Standard Lead Solutions* and from the *Sample Preparation* absorbances. Prepare a standard curve by plotting the corrected *Diluted Standard Lead Solutions* absorbance values versus their corresponding concentrations expressed as µg/mL. Determine the *Sample Preparation* lead concentration by reference to the calibration curve. Calculate the quantity of lead, in mg/kg, in the sample taken by the formula

$$10C/W_S,$$

in which C is the concentration, in µg/mL, of lead from the standard curve, and W_S is the weight, in g, of the sample taken.

Atomic Absorption Spectrophotometric Graphite Furnace Method
The following methods are primarily intended for the analysis of applicable substances containing less than 1 mg/kg of lead.

Method I
This method is intended for the quantitation of lead in substances that are soluble in water, such as sugars and sugar syrups, at levels as low as 0.03 mg/kg. The method detection limit is approximately 5 ng/kg.

Apparatus Use a suitable modern graphite furnace atomic absorption spectrophotometer set at 283.3 nm and equipped with an autosampler, pyrolytically coated graphite tubes, solid pyrolytic graphite platforms, and an adequate means of background correction. Zeeman effect or Smith-Hieftje background correction are preferred, but deuterium arc background correction should be acceptable. (This method was developed on a Perkin-Elmer Model Z5100, 0.7-nm slit, HGA-600 furnace, AS-60 autosampler with Zeeman background correction.) If the instrument does not have a well-defined calibration function, a separate calculator or computer is required for linear least squares, nonlinear, or quadratic calibrations. Use either a hollow cathode lamp or an electrodeless discharge lamp as the source, and use argon as the purge gas and breathing-quality air (for oxygen ashing to avoid residue build up during the char step) as the alternate gas. Set up the instrument according to manufacturer's specifications, with consideration of current good GFAAS practice, addressing such factors as line voltage, cooling water temperature, graphite part specifications, and furnace temperature. If an optical pyrometer or thermocouple is not available to check the furnace controller temperature calibration, dim the room lights, and observe the furnace emission through the sample introduction port while increasing the furnace temperature. A characteristic cherry-red glow should begin to appear at 800°. If it glows at a lower temperature (typical with Perkin-Elmer ZL furnaces), then the furnace is hotter, and temperatures must be adjusted downwards accordingly.

Use acid-cleaned [in a mixture of 5% sub-boiling, distilled nitric acid and 5% sub-boiling, distilled hydrochloric acid made up in deionized, distilled water (18 megohm), and thoroughly rinsed with deionized, distilled water (18 megohm)] autosampler cups (PE B008-7600 Teflon, or equivalent) to avoid contamination. Use micropipets with disposable tips free of lead contamination for dilution. Ensure accuracy and precision of micropipets and tips by dispensing and weighing 5 to 10 replicate portions of water onto a microbalance. Use acid-cleaned volumetric glassware to prepare standards and dilute samples to a final volume. For digestion, use acid-cleaned, high-density polyethylene tubes, polypropylene tubes, Teflon tubes, or quartz tubes. Store final diluted samples in plastic tubes.

Standard Solutions Prepare all lead solutions in 5% sub-boiling distilled nitric acid. Use a single-element 1000- or 10,000-μg/mL lead stock to prepare (weekly) an intermediate 10-μg/mL standard in 5% nitric acid. Prepare (daily) a *Lead Standard Solution* (1 μg/mL) by diluting the intermediate 10-μg/mL stock solution 1:10. Prepare *Working Calibration Standards* of 100.0, 50.0, 25.0, and 10.0 ng/mL from this, using appropriate dilutions. Store standards in acid-cleaned polyethylene test tubes or bottles. If the GFAAS autosampler is used to automatically dilute standards, ensure calibration accuracy by pipetting volumes of 3 μL or greater.

Modifier Stock Solution Weigh 20 g of ultrapure magnesium nitrate hexahydrate and dilute to 100 mL. Just before use, prepare a *Modifier Working Solution* by diluting stock solution 1:10. A volume of 5 μL will provide 0.06 mg of magnesium nitrate.

Sample Digestion (**Caution**: Perform the procedure in a fume hood, and wear safety glasses.) Obtain a representative subsample to be analyzed. For liquid samples such as sugar syrups, ultrasonicate and/or vortex mix before weighing. For solid samples such as crystalline sucrose, make a sugar solution using equal weights of sample (5-g minimum) and deionized, distilled (18 megohm) water. Mix samples until completely dissolved. Accurately weigh approximately 1.5 g (record to nearest mg) of sample (or 3.0 g of sugar solution) into a digestion tube. Run a *Sample Preparation Blank* of 1.5 g of deionized distilled (18 megohm) water through the entire procedure with each batch of samples. Add 0.75 mL of sub-boiling, distilled nitric acid. Heat plastic tubes in a water bath, quartz tubes in a water bath or heating block, warming slowly to between 90° and 95° to avoid spattering. Monitor the temperature by using a "dummy" sample. Heat until all brown vapors have dissipated and any rust-colored tint is gone (20 to 30 min). Cool. Add 0.5 mL of 50% hydrogen peroxide dropwise, heat at 90° to 95° for 5 min, and cool. Add a second 0.5-mL portion of 50% hydrogen peroxide dropwise, and heat at 90° to 100° for 5 to 10 min until clear. Cool, and dilute to a final volume of 10 mL.

Procedure The furnace program is as follows: (1) dry at 200°, using a 20-s ramp and a 30-s hold and a 300-mL/min argon flow; (2) char the sample at 750°, using a 40-s ramp and a 40-s hold and a 300-mL/min air flow; (3) cool down, and purge the air from the furnace for 60 s, using a 20° set temperature and a 300-mL/min argon flow; (4) atomize at 1800°, using a 0-s ramp and a 10-s hold with the argon flow stopped; (5) clean out at 2600°, with a 1-s ramp and a 7-s hold; (6) cool down the furnace (if necessary) at 20°, with a 1-s ramp and a 5-s hold with a 300-mL/min argon flow. Use the autosampler to inject 20 μL of blanks, calibration standards, and sample solutions and 5 μL of *Modifier Working Solution*. Inject each respective solution in triplicate, and average results. Use peak area measurements for all quantitation. After ensuring that the furnace is clean by running a 5% nitric acid blank, check the instrument sensitivity by running the 25-ng/mL calibration standard. If the integrated absorbance is less than 0.14 abs-sec for a standard, 28-mm × 6-mm, end-heated furnace tube, correct the cause of insufficient sensitivity before proceeding. (Please note that the newer Perkin-Elmer THGA furnaces are smaller and typically provide an integrated absorbance of 0.05 to 0.07 abs-sec.) If the integrated absorbance is greater than 0.25 abs-sec, contamination is likely, and the source should be investigated. Calculate the characteristic mass (m_o) (mass of Pb pg necessary to produce an integrated absorbance of 0.0044 abs-sec) as follows:

$$m_o = (0.0044 \text{ abs-sec})(25 \text{ pg}/\mu\text{L})(20 \text{ }\mu\text{L})/$$
$$(\text{measured 25 pg}/\mu\text{L abs-sec}).$$

Record and track the integrated absorbance and m_o for reference and quality assurance.

Standard Curve Inject each calibration standard in triplicate. Normal instrument linearity extends to 25 ng/mL. If nonlinear calibration capability is not available, limit the working calibration curve to ≤25 ng/mL. Use the calibration algorithms provided in the instrument software. Recheck calibration periodically (≤15 samples) by running a 25- or 50-ng/mL calibration standard interspersed with samples. If recheck differs from calibration by >10%, recalibrate the instrument. The instrumental detection limit (DL) and quantitation limit (QL), in picograms, may be based on 7 to 10 replicates of the *Sample Preparation Blank* and calculated as follows:

$$DL = (3)(\text{s.d. blank abs-sec})(10 \text{ pg}/\mu\text{L})(20 \text{ }\mu\text{L})/$$
$$(\text{abs-sec 10 ng/mL std}),$$

$$QL = (10)(\text{s.d. blank abs-sec})(10 \text{ pg}/\mu\text{L})(20 \text{ }\mu\text{L})/$$
$$(\text{abs-sec 10 ng/mL std}).$$

During method development, detection limits were typically 10 to 14 pg, corresponding to 0.5 to 0.7 ng/mL for 20 μL. This corresponds to a method detection limit of 3.3 to 4.7 ng/g of sugar.

Sample Analyses Inject each sample digest in triplicate, and record the integrated absorbance. If instrument response exceeds that of the calibration curve, dilute with 5% nitric acid to bring the sample response into working range, and note the dilution factor (DF). Sample solutions having a final concentration of >25 ng/mL should be diluted 1:10 to facilitate analysis in the linear range for systems not equipped with non-linear calibration. All sample analyses should be blank corrected using the sample preparation blank. This can typically be done automatically by the software after identifying and running a representative sample preparation blank. Use the calibration algorithm provided in the instrument software to calculate a blank-corrected, digest lead concentration (in ng/mL).

Calculation of Lead Content Calculate the lead level in the original sample as follows:

Pb(ng/g) = (blank-corrected Pb ng/mL)(DF)[sample vol (10 mL)]/[sample weight (approx. 1.5 g)][2]

Quality Assurance To ensure analytical accuracy, NIST SRM 1643c acidified water or a similar material should be analyzed before the unknown samples are. The certified content of SRM 1643c is 35.3 ± 0.9 ng/mL. If the concentration determined is not within 10% of the mean reference value (31.8 to 38.8 ng/mL), the reason for inaccuracy should be evaluated, and unknown samples should not be analyzed until acceptable accuracy is achieved. Also prepare an in-house control solution made from uncontaminated table sugar or reagent-grade sucrose (or other appropriate substance with a Pb content <5 ng/g as received) mixed with an equal volume of water. Spike this solution with Pb to produce a concentration of 100 ng/g. Analyze with each batch of samples. Recoveries should be 100% ± 20%, and the precision for complete replicate digestions should be <5% RSD. Periodically, a sample digest should be checked using the method of standard additions to ensure that there are no multiplicative or chemical interferences. Spiking samples and checking recoveries is always a good practice.

Method II

This method is primarily intended for the determination of lead at levels of less than 1 mg/kg in substances immiscible with water, such as edible oils.

Apparatus Use a suitable atomic absorption spectrophotometer (Perkin-Elmer Model 3100 or equivalent) fitted with a graphite furnace (Perkin-Elmer HGA 600 or equivalent). Use a lead hollow-cathode lamp (Perkin-Elmer or equivalent) with argon as the carrier gas. Follow the manufacturers' directions for setting the appropriate instrument parameters for lead determination.

Note: For this test, use reagent-grade chemicals with as low a lead content as is practicable, as well as high-purity water and gases. Before use in this analysis, rinse all glassware and plasticware twice with 10% nitric acid and twice with 10% hydrochloric acid, and then rinse them thoroughly with high-purity water, preferably obtained from a mixed-bed strong-acid, strong-base ion-exchange cartridge capable of producing water with an electrical resistivity of 12 to 15 megohms.

Hydrogen Peroxide–Nitric Acid Solution Dissolve equal volumes of 10% hydrogen peroxide and 10% nitric acid.

Note: Use caution.

---

[2]If a sample solution was prepared initially to ensure sample homogeneity, this is the weight of the original sugar digested (not the weight of the solution).

Lead Nitrate Stock Solution Dissolve 159.8 mg of ACS reagent-grade lead nitrate (alternatively, use NIST Standard Reference Material, containing 10 mg of lead per kg, or equivalent) in 100 mL of *Hydrogen Peroxide–Nitric Acid Solution*. Dilute to 1000.0 mL with *Hydrogen Peroxide–Nitric Acid Solution*, and mix. Prepare and store this solution in glass containers that are free from lead salts. Each mL of this solution contains the equivalent of 100 µg of lead ion.

Standard Lead Solution On the day of use, dilute 10.0 mL of *Lead Nitrate Stock Solution* with *Hydrogen Peroxide–Nitric Acid Solution* to 100.0 mL, and mix. Each mL of *Standard Lead Solution* contains the equivalent of 10 µg of lead ion.

Butanol–Nitric Acid Solution Slowly add 50 mL of nitric acid to approximately 500 mL of butanol in a 1000-mL volumetric flask. Dilute to volume with butanol, and mix.

Standard Solutions Prepare a series of lead standard solutions serially diluted from the *Standard Lead Solution* in *Butanol–Nitric Acid Solution*. Into separate 100-mL volumetric flasks pipet 0.2, 0.5, 1, and 2 mL, respectively, of *Standard Lead Solution*, dilute to volume with *Butanol–Nitric Acid Solution*, and mix. The *Standard Solutions* contain, respectively, 0.02, 0.05, 0.1, and 0.2 µg of lead per mL. (For lead limits greater than 1 mg/kg, prepare a series of standard solutions in a range encompassing the expected lead concentration in the sample.)

Sample Solution (Note: Perform this procedure in a fume hood, and wear safety glasses.) Accurately weigh 1 g of the sample, and place it in a large test tube. Add 1 mL of nitric acid. Place the test tube in a rack in a boiling water bath. As soon as the rusty tint is gone, add 1 mL of 30% hydrogen peroxide dropwise to avoid a vigorous reaction, and wait for bubbles to form. Stir with an acid-washed plastic spatula if necessary. Remove the test tube from the water bath, and let it cool. Transfer the solution to a 10-mL volumetric flask, and dilute to volume with *Butanol–Nitric Acid Solution*, and mix. Use this solution for analysis.

Procedure

Tungsten Solution Transfer 0.1 g of tungstic acid (H_2WO_4) and 5 g of sodium hydroxide pellets into a 50-mL plastic bottle. Add 5.0 mL of high-purity water, and mix. Heat the mixture in a hot water bath until complete solution is achieved. Cool, and store at room temperature.

Procedure Place the graphite tube in the furnace. Inject a 20-µL aliquot of the *Tungsten Solution* into the graphite tube, using a 300-mL/min argon flow and the following sequence of conditions: Dry at 110° for 20 s, char at 700° to 900° for 20 s, and, with the argon flow stopped, atomize at 2700° for 10 s; repeat this procedure once more using a second 20-µL aliquot of the *Tungsten Solution*. Clean the quartz windows.

Standard Curve (Note: The sample injection technique is the most crucial step in controlling the precision of the analysis; the volume of the sample must remain constant. Rinse the µL pipet tip (Eppendorf or equivalent) three times with either the *Standard Solutions* or *Sample Solution* before injection. Use a fresh pipet tip for each injection, and start the atomization process immediately after injecting the sample. Between injections, flush the graphite tube of any residual lead by purging at a high temperature as recommended by the manufacturer.) With the hollow cathode lamp properly aligned for maximum absorbance and the wavelength set at 283.3 nm, atomize 20-μL aliquots of the four *Standard Solutions*, using a 300-mL/min argon flow and the following sequence of conditions: Dry at 110° for 30 s, with a 20-s ramp period and a 10-s hold time, then char at 700° for 42 s, with a 20-s ramp period and a 22-s hold time, and then, with the argon flow stopped, atomize at 2300° for 7 s.

Plot a standard curve using the concentration, in μg/mL, of each *Standard Solution* versus its maximum absorbance value compensated for background correction as directed for the particular instrument, and draw the best straight line.

Atomize 20 μL of the *Sample Solution* under identical conditions, and measure its corrected maximum absorbance. From the *Standard Curve*, determine the concentration C, in μg/mL, of the *Sample Solution*. Calculate the quantity, in mg/kg, of lead in the sample by the formula

$$10C/W,$$

in which W is the weight, in g, of the sample taken.

APDC Extraction Method
Select reagents having as low a lead content as practicable, and store all solutions in containers of high-density polyethylene containers. Rinse all plastic and glassware thoroughly with warm, dilute nitric acid (1 in 2) followed by water.

2% APDC Solution Dissolve 2.0 g of ammonium pyrrolidinedithiocarbamate (APDC) in 100 mL of water. Filter any slight residue of insoluble APDC from the solution before use.

Lead Nitrate Stock Solution (100 μg/mL) Dissolve 159.8 mg of reagent-grade lead nitrate, $Pb(NO_3)_2$, in 100 mL of water containing 1 mL of nitric acid in a 1000-mL volumetric flask, and dilute to volume with water.

Standard Lead Solution (2 μg/mL) On the day of use, transfer 2.0 mL of *Lead Nitrate Stock Solution* to a 100-mL volumetric flask, and dilute to volume with water.

Sample Preparation Transfer a 10.0-g sample to a clean 150-mL beaker, and 10 mL of water to a second 150-mL beaker to serve as the blank. To each add 30 mL of water and the minimum amount of hydrochloric acid needed to dissolve the sample, plus an additional 1 mL of hydrochloric acid to ensure the dissolution of any lead present. Heat to boiling, and boil for several min. Allow to cool, and dilute to about 100 mL with deionized water. Adjust the pH of the resulting solution to between 1.0 and 1.5 with 25% NaOH. Quantitatively transfer the pH-adjusted solution to a clean 250-mL separatory funnel, and dilute to about 200 mL with water. Add 2 mL of *2% APDC Solution*, and mix. Extract with two 20-mL portions of chloroform, collecting the extracts in a clean 50-mL beaker. Evaporate to dryness on a steam bath. Add 3 mL of nitric acid to the residue, and heat to near dryness. Then add 0.5 mL of nitric acid and 10 mL of deionized water to the beaker, and heat until the volume is reduced to about 3 to 5 mL. Transfer the digested extract to a clean 10-mL volumetric flask, and dilute to volume with water.

Procedure Concomitantly determine the absorbances of the *Standard Lead Solution* and the *Sample Preparation* against the blank at the lead emission line of 283.3 nm, using a slit-width of 0.7 nm. Use a suitable atomic absorption spectrophotometer equipped with a lead electrodeless discharge lamp (EDL), or equivalent; an air–acetylene flame; and a 4-in. burner head. Use water as the blank. The absorbance of the *Sample Preparation* is not greater than that of the *Standard Lead Solution*.

MERCURY LIMIT TEST

Mercury Detection Instrument Use any suitable atomic absorption spectrophotometer equipped with a fast-response recorder and capable of measuring the radiation absorbed by mercury vapors at the mercury resonance line of 253.6 nm. A simple mercury vapor meter or detector equipped with a variable span recorder is also satisfactory.

Note: Wash all glassware associated with the test with nitric acid, and rinse thoroughly with water before use.

Aeration Apparatus The apparatus, shown in Fig. 16, consists of a flowmeter (*a*), capable of measuring flow rates from 500 to 1000 mL/min, connected via a three-way stopcock (*b*), with a Teflon plug, to 125-mL gas washing bottles (*c* and *d*), followed by a drying tube (*e*), and finally a suitable quartz liquid absorption cell (*f*), terminating with a vent (*g*) to a fume hood.

Note: The absorption cell will vary in optical pathlength depending upon the type of mercury detection instrument used.

Bottle *c* is fitted with an extra-coarse fritted bubbler (Corning 31770 125 EC, or equivalent), and the bottle is marked with a 60-mL calibration line. The drying tube *e* is lightly packed with magnesium perchlorate. Bottle *c* is used for the test solution, and bottle *d*, which remains empty throughout the procedure, is used to collect water droplets. Alternatively, an apparatus embodying the principle of the assembly described and illustrated may be used. The aerating medium may be either compressed air or compressed nitrogen.

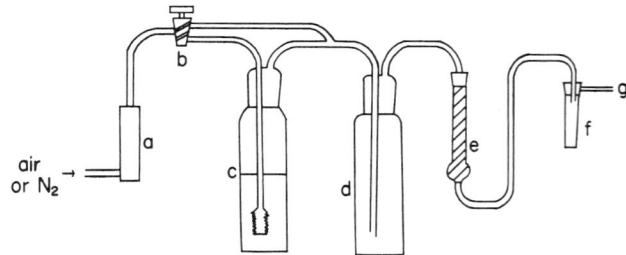

FIGURE 16 Aeration Apparatus for Mercury Limit Test.

Standard Preparation Transfer 1.71 g of mercuric nitrate, Hg(NO$_3$).H$_2$O, to a 1000-mL volumetric flask, dissolve in a mixture of 100 mL of water and 2 mL of nitric acid, dilute to volume with water, and mix. Discard after 1 month. Transfer 10.0 mL of this solution to a second 1000-mL volumetric flask, acidify with 5 mL of dilute sulfuric acid solution (1 in 5), dilute to volume with water, and mix. Discard after 1 week. On the day of use, transfer 10.0 mL of the second solution to a 100-mL volumetric flask, acidify with 5 mL of dilute sulfuric acid (1 in 5), dilute to volume with water, and mix. Each mL of this solution contains 1 μg of Hg. Transfer 2.0 mL of this solution (2 μg of Hg) to a 50-mL beaker, and add 20 mL of water, 1 mL of dilute sulfuric acid solution (1 in 5), and 1 mL of potassium permanganate solution (1 in 25). Cover the beaker with a watch glass, boil for a few s, and cool.

Sample Preparation Prepare as directed in the individual monograph.

Procedure Assemble the aerating apparatus as shown in Fig. 16, with bottles *c* and *d* empty and stopcock *b* in the bypass position. Connect the apparatus to absorption cell *f* in the instrument, and adjust the air or nitrogen flow rate so that, in the following procedure, maximum absorption and reproducibility are obtained without excessive foaming in the test solution. Obtain a baseline reading at 253.6 nm, following the manufacturer's instructions for operating the instrument. Treat the *Standard Preparation* as follows: Destroy the excess permanganate by adding a 1 in 10 solution of hydroxylamine hydrochloride, dropwise, until the solution is colorless. Immediately wash the solution into bottle *c* with water, and dilute to the 60-mL mark with water. Add 2 mL of 10% stannous chloride solution (prepared fresh each week by dissolving 10 g of SnCl$_2$.2H$_2$O in 20 mL of warm hydrochloric acid and diluting with 80 mL of water), and immediately reconnect bottle *c* to the aerating apparatus. Turn stopcock *b* from the bypass to the aerating position, and continue the aeration until the absorption peak has been passed and the recorder pen has returned to the baseline. Disconnect bottle *c* from the aerating apparatus, discard the *Standard Preparation* mixture, wash bottle *c* with water, and repeat the foregoing procedure using the *Sample Preparation*; any absorbance produced by the *Sample Preparation* does not exceed that produced by the *Standard Preparation*.

SELENIUM LIMIT TEST

Reagents and Solutions

2,3-Diaminonaphthalene Solution On the day of use, dissolve 100 mg of 2,3-diaminonaphthalene (C$_{10}$H$_{10}$N$_2$) and 500 mg of hydroxylamine hydrochloride (NH$_2$OH.HCl) in sufficient 0.1 *N* hydrochloric acid to make 100 mL.

Selenium Stock Solution Transfer 40.0 mg of powdered metallic selenium into a 1000-mL volumetric flask, and dissolve in 100 mL of dilute nitric acid (1 in 2), warming gently on a steam bath to effect solution. Cool, dilute with water to volume, and mix.

Selenium Standard Solution Pipet 5.0 mL of *Selenium Stock Solution* into a 200-mL volumetric flask, dilute to volume with water, and mix. Each mL of this solution contains the equivalent of 1 μg of selenium (Se).

Method I

Standard Preparation Pipet 6.0 mL of *Selenium Standard Solution* into a 150-mL beaker, add 50 mL of 0.25 *N* nitric acid, and mix.

Sample Preparation Using a 1000-mL combustion flask and 25 mL of 0.5 *N* nitric acid as the absorbing liquid, proceed as directed under *Oxygen Flask Combustion*, Appendix I, using the amount of sample specified in the individual monograph (and the magnesium oxide or other reagent, where specified).

Note: If the sample contains water of hydration or more than 1% of moisture, dry it at 140° for 2 h before combustion, unless otherwise directed.

Upon completion of combustion, place a few mL of water in the cup or lip of the combustion flask, loosen the stopper of the flask, and rinse the stopper, sample holder, and sides of the flask with about 10 mL of water. Transfer the solution, with the aid of about 20 mL of water, into a 150-mL beaker, heat gently to boiling, boil for 10 min, and cool.

Procedure Treat the *Sample Preparation*, the *Standard Preparation*, and 50 mL of 0.25 *N* nitric acid, to serve as the blank, similarly and in parallel as follows: Add dilute ammonium hydroxide (1 in 2) to adjust the pH of the solution to 2.0 ± 0.2. Dilute with water to 60.0 mL, and transfer to a low-actinic separator with the aid of 10.0 mL of water, adding the 10.0 mL of rinsings to the separator. Add 200 mg of hydroxylamine hydrochloride, swirl to dissolve, immediately add 5.0 mL of *2,3-Diaminonaphthalene Solution*, insert the stopper, and swirl to mix. Allow the solution to stand at room temperature for 100 min. Add 5.0 mL of cyclohexane, shake vigorously for 2 min, and allow the layers to separate. Discard the aqueous phases, and centrifuge the cyclohexane extracts to remove any traces of water. Determine the absorbance of each extract in a 1-cm cell at the maximum at about 380 nm with a suitable spectrophotometer, using the extract from the blank to set the instrument. The absorbance of the extract from the *Sample Preparation* is not greater than that from the *Standard Preparation* when a 200-mg sample is tested, or not greater than one-half the absorbance of the extract from the *Standard Preparation* when a 100-mg sample is tested.

Method II

Standard Preparation Pipet 6.0 mL of *Selenium Standard Solution* into a 150-mL beaker, add 50 mL of 2 *N* hydrochloric acid, and mix.

Sample Preparation Transfer the amount of sample specified in the individual monograph into a 150-mL beaker, dissolve in 25 mL of 4 *N* hydrochloric acid, swirling if necessary to effect solution, heat gently to boiling, and digest on a steam bath for 15 min. Remove from heat, add 25 mL of water, and allow to cool to room temperature.

Procedure Place the beakers containing the *Standard Preparation* and the *Sample Preparation* in a fume hood, and to a third beaker add 50 mL of 2 N hydrochloric acid to serve as the blank. Cautiously add 5 mL of ammonium hydroxide to each beaker, mix, and allow the solution to cool. Treat each solution, similarly and in parallel, as directed under *Procedure* in *Method I*, beginning with "Add dilute ammonium hydroxide (1 in 2). . . ."

C. OTHERS

ALGINATES ASSAY

In a suitable closed system, liberate the carbon dioxide from the uronic acid groups of about 250 mg of the test sample by heating with hydrochloric acid, and sweep the carbon dioxide by means of an inert gas into a titration vessel containing excess standardized sodium hydroxide. Any suitable system may be used as long as it provides precautions against leakage and overheating of the reaction mixture, adequate sweeping time, avoidance of entrainment of hydrochloric acid, and meets the requirements of the *System Suitability Test*. One suitable system, with accompanying procedure is given below.

Apparatus The apparatus is shown in Fig. 17. It consists essentially of a soda lime column, A, a mercury valve, B, connected through a side arm, C, to a reaction flask, D, by means of a rubber connection. Flask D is a 100-mL round-bottomed, long-neck boiling flask, resting in a suitable heating mantle, E.

The reaction flask is provided with a reflux condenser, F, to which is fitted a delivery tube, G, of 40-mL capacity, having a stopcock, H. The reflux condenser terminates in a trap, I, containing 25 g of 20-mesh zinc or tin, which can be connected with an absorption tower, J.

The absorption tower consists of a 45-cm tube fitted with a medium-porosity fritted glass disk sealed to the inner part above the side arm and having a delivery tube sealed to it extending down to the end of the tube. A trap, consisting of a bulb of approximately 100-mL capacity, is blown above the fritted disk and the outer portion of a ground spherical joint is sealed on above the bulb. A 250-mL Erlenmeyer flask, K, is connected to the bottom of the absorption tower. The top of the tower is connected to a soda lime tower, L, which is connected to a suitable pump to provide vacuum and air supply, the choice of which is made by a three-way stopcock, M. The volume of air or vacuum is controlled by a capillary-tube regulator or needle valve, N.

All joints are size 35/25, ground spherical type.

Standard D-Glucurono-6,3-lactone This chemical ($C_6H_8O_6$) is available as a reference standard with an assay of 100.0 ± 1.0% (24.99 ± 0.25% CO_2) from Aldrich Chemical Co., Catalog No. 85,145-0.

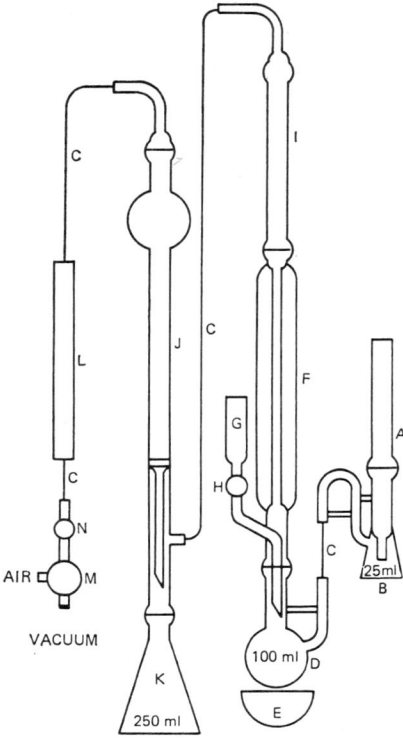

FIGURE 17 Apparatus for Alginates Assay.

System Suitability Test Transfer about 250.0 mg of standard D-glucurono-6,3-lactone, accurately weighed, to the reaction flask, D, and carry out the *Procedure* described below. The system is considered suitable when the net titration results in a calculation of % CO_2 in a range of 24.73 to 25.26, which is equivalent to a range of 98.95% to 101.06% D-glucurono-6,3-lactone.

Procedure Transfer about 250 mg of the sample, accurately weighed, into the reaction flask, D, add 25 mL of 0.1 N hydrochloric acid, insert several boiling chips, and connect the flask to the reflux condenser, F, using syrupy phosphoric acid as a lubricant.

Note: Stopcock grease may be used for the other connections.

Check the system for air leaks by forcing mercury up into the inner tube of the mercury valve, B, to a height of about 5 cm. Turn off the pressure using the stopcock, M. If the mercury level does not fall appreciably after 1 to 2 min, the apparatus may be considered to be free from leaks. Draw carbon dioxide-free air through the apparatus at a rate of 3000 to 6000 mL/h. Raise the heating mantle, E, to the flask, heat the sample to boiling, and boil gently for 2 min. Turn off and lower the mantle, and allow the sample to cool for 15 min. Charge the delivery tube, G, with 23 mL of hydrochloric acid. Disconnect the absorption tower, J, rapidly transfer 25.0 mL of 0.25 N sodium hydroxide into the tower, add 5 drops of n-butanol, and again connect the absorption tower. Draw carbon dioxide-free air through the apparatus at the rate of about 2000 mL/h, add the hydrochloric acid to the reaction flask through the delivery tube, raise the heating mantle, and heat the reaction mixture to boiling. After 2 h, discontinue the current of air and heating. Force the sodium hydroxide solution down into the flask, K, using gentle air pressure, and then rinse down the absorption tower with three 15-mL portions of water, forcing each washing into the flask with air pressure. Remove the flask, and add to it 10 mL of a 10% solution of barium chloride ($BaCl_2 \cdot 2H_2O$). Stopper the flask, shake gently for about 2 min, add phenolphthalein TS, and titrate with 0.1 N hydrochloric acid. Perform a blank determination (see *General Provisions*). Each mL of 0.25 N sodium hydroxide consumed is equivalent to 5.5 mg of carbon dioxide (CO_2). Calculate the results on the dried basis.

BENZENE (in Paraffinic Hydrocarbon Solvents)

Apparatus Use a suitable gas chromatograph (see Appendix IIA), equipped with a column that will elute n-decane before benzene under the conditions of the *System Suitability Test* (see below). Column materials and conditions that have been found suitable for this method are listed in the accompanying Tables. See Fig. 18 for a typical chromatogram obtained with column No. 5.

Reagents
Isooctane 99 mole percent minimum containing less than 0.05 mole percent aromatic material.
Benzene 99.5 mole percent minimum.

Internal Standard n-Decane and either n-undecane or n-dodecane according to the requirement of the *System Suitability Test*.

Reference Solution A Prepare a standard solution containing 0.5% by weight each of the *Internal Standard* and benzene in isooctane.

Reference Solution B Prepare a standard solution containing about 0.5% by weight each of n-decane, *Internal Standard*, and benzene in isooctane.

Calibration Select the instrument conditions necessary to give the desired sensitivity. Inject a known volume of *Reference Solution A* and change the attenuation, if necessary, so that the benzene peak is measured with a chart deflection of not less than 25% nor more than 95% of full scale. When choosing the attenuation, consider all unresolved peaks to represent a single compound. There may be tailing of the nonaromatic peak, but do not use any conditions that lead to a depth of the valley ahead of the benzene peak (A) less than 50% of the weight of the benzene peak (B) as depicted in Fig. 19.

If there is tailing of the nonaromatic material, construct a baseline by drawing a line from the bottom of the valley ahead of the benzene peak to the point of tangency after the peak (see Fig. 20). Measure the areas of the benzene peak and the internal standard peak by any of the following means: triangulation, planimeter, paper cutout, or mechanical or electronic integrator.

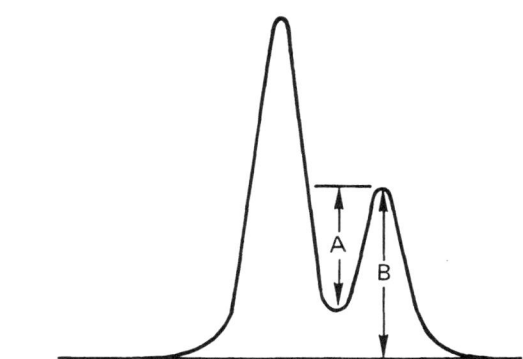

FIGURE 19 Illustration of A/B Ratio.

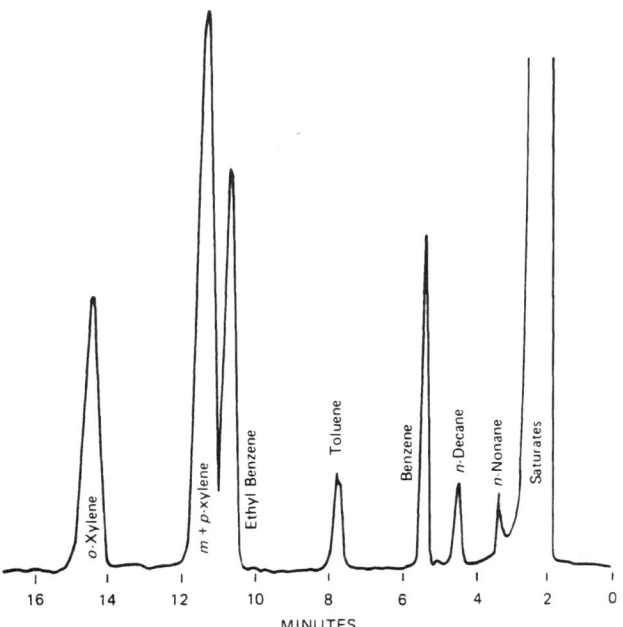

FIGURE 18 Typical Chromatogram for the Determination of Benzene in Hexanes Using Column No. 5.

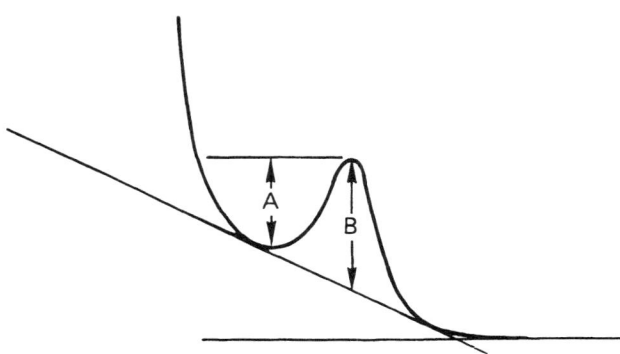

FIGURE 20 Illustration of A/B Ratio for a Small Component Peak on the Tail of a Large Peak.

Do not use integrators on peaks without a constant baseline, unless the integrator has provision for making baseline corrections with accuracy at least as good as that of manual methods.

Calculate a response factor for benzene (R_b) relative to the internal standard by the formula

$$R_b = A_i/W_i \times W_b/A_b,$$

in which A_i is the area of the internal standard peak in arbitrary units corrected for attenuation; W_i is the weight percent of internal standard in *Reference Solution A*; A_b is the area of the benzene peak in arbitrary units corrected for attenuation; and W_b is the weight percent of benzene in *Reference Solution A*.

Procedure Place approximately 0.1 mL of *Internal Standard* in a tared 25-mL volumetric flask, weigh on an analytical balance; dissolve in and dilute to volume with the sample to be analyzed.

Using the exact instrumental conditions that were used in the calibration, inject the same volume of sample containing the *Internal Standard*. Before measuring the area of the *Internal Standard* and benzene peaks change the attenuation to ensure at least 25% chart deflection.

Measure the area of the internal standard and benzene peaks in the same manner as was used for the calibration. Calculate the weight percent of benzene in the sample (W_B) by the formula

$$W_B = (A_b \times R_b \times W_i \times 100)/(A_i \times S),$$

in which A_b is the area of the benzene peak corrected for attenuation; R_b is the relative response factor for benzene; W_i is the weight of *Internal Standard* added, in g; A_i is the area of the *Internal Standard* peak corrected for attenuation; and S is the weight of the sample, in g.

System Suitability Test Inject the same volume of *Reference Solution B* as in the *Calibration* and record the chromatogram. *n*-Decane must be eluted before benzene, and the ratio of A to B (Fig. 19) must be at least 0.5 where A is equal to the depth of the valley between the *n*-decane and benzene peaks and B is equal to the height of the benzene peak.

Column Materials and Conditions for the Determination of Benzene in Hexanes

Column No.	1	2	3	4	5	6	7
Liquid phase	CEF	PEF 200	CEF	DEGS	TCEPE	TCEPE	DEGS
Length, ft	15	6	16	10	15	100	12
m	—	4.5	2	5	3.1	—	313.7
Diameter, in. (mm)							
Inside	0.07(1.8)	—	0.07	0.18(4.5)	0.06(1.5)	0.01(.254)	—
Outside	1/8(3.2)	1/4(6.4)	1/8	—	—	—	1/8
Weight, percent	17	30	20	20	10	—	20
Solid support	Chromosorb P	Chromosorb P	Chromosorb P	Chromosorb P	Chromosorb P	Capillary	Chromosorb P
Mesh	60–80	60–80	60–80	80–100	60–80	—	80–100
Treatment	AW	AW	AW	none	AW	none	AW Sil
Inlet, deg	200	210	250	260	250	275	260
Detector, deg	200	155	250	200	175	250	240
Column, deg	115	95	90	100	115	95	65
Carrier gas	N_2	He	He	He	N_2	N_2	He
Flow rate, cm^3/min	30	60	60	60	1	3	52
Detector	FI	TC	FI	FI	FI	FI	FI
Recorder, mV	5	1	1	1	10	1	1
Sample, l	5	10	1	2	5	0.8	5
Split	9 + 1	—	—	—	100 + 1	100 − 1	—
Area	Tri	EI	DI	Tri Plan	EI	EI	Tri

Abbreviations Used in Table
AW—Acid washed; CEF—*N,N*-Bis(2-cyanoethyl)formamide; DEGS—Diethylene Glycol Succinate; DI—Disk integrator; EI—Electronic integrator; FI—Flame ionization; Sil—Silanized; TC—Thermal conductivity; TCEPE—Tetracyanoethylated Pentaerythritrol; Tri—Triangulation.

Retention Times in Minutes for Selected Hydrocarbons Under the Conditions for the Determination of Benzene in Hexanes

Column No.	1	2	3	4	5	6	7
Benzene	3.4	2.0	6.5	6.7	5.4	6.1	6.7
Toluene	4.4	3.2	9.0	10.3	7.8	7.0	10.3
Ethylbenzene	5.4	5.2	11.5	14.8	10.8	8.0	14.8
p-m-Xylenes	5.8	—	12.5	—	11.4	8.5	—
o-Xylene	7.5	6.8	17.0	16.1	14.5	10.0	—
n-Undecane	3.0	2.8	3.5	—	—	—	—
n-Dodecane	—	—	—	12.8	8.5	6.5	—

COLORS, F D & C[3]

Chromium
Standards

Standard Chromium Solution (1000 mg/kg) Transfer 2.829 g of $K_2Cr_2O_7$, accurately weighed (National Institute of Standards and Technology No. 136) to a 1-L volumetric flask; dissolve in and dilute to volume with water.

Standard Colorant Solution Transfer 62.5 g of colorant previously shown to be free of chromium to a 1-L volumetric flask; dissolve in and dilute to volume with water.

Apparatus Use any suitable atomic absorption spectrophotometer equipped with a fast response recorder and capable of measuring the radiation absorbed at 357.9 nm.

Instrument Parameters *Wavelength setting*: 357.9 nm; *optical passes*: 5; *lamp current*: 8 mA; *lamp voltage*: 500 v; *fuel*: hydrogen; *oxidant*: air; *recorder*: 1 mv with a scale expansion of 5 or 10. Alternatively, follow the instructions supplied with the instrument.

Procedure Set the instrument at the optimum conditions for measuring chromium as directed by the manufacturer's instructions. Prepare a series of seven standard chromium solutions containing Cr at approximately 5, 10, 15, 20, 40, 50, and 60 mg/kg by appropriate dilutions of the *Standard Chromium Solution* into 100-mL volumetric flasks; add 80 mL of the *Standard Colorant Solution*, and dilute each flask to volume with water.

Transfer 5 g of the colorant to be analyzed to a 100-mL volumetric flask; dissolve in and dilute to volume with water. Prepare a calibration curve using the series of standards, and using this curve, determine the chromium content of the colorant samples.

Ether Extracts

Caution: Isopropyl ether forms explosive peroxides. To ensure the absence of peroxides, perform the following test: Prepare a colorless solution of ferrous thiocyanate by mixing equal volumes of 0.1 N ferrous sulfate and 0.1 N ammonium thiocyanate. Using titanous chloride, carefully discharge any red coloration due to ferric ions. To 50 mL of the solution add 10 mL of ether, and shake vigorously for 2 to 3 min. A red color indicates the presence of peroxides. If redistillation is necessary, the usual precautions against peroxide detonation should be observed. Immediately before use, the ether should be passed through a 30-cm column of chromatography-grade aluminum oxide to remove peroxides and inhibitors.

---

[3]For the most current analytical procedures, refer to the latest *Color Certification Manual of Analytical Procedures* available from the Office of Cosmetics and Colors, Food and Drug Administration, Washington, D.C. 20204.

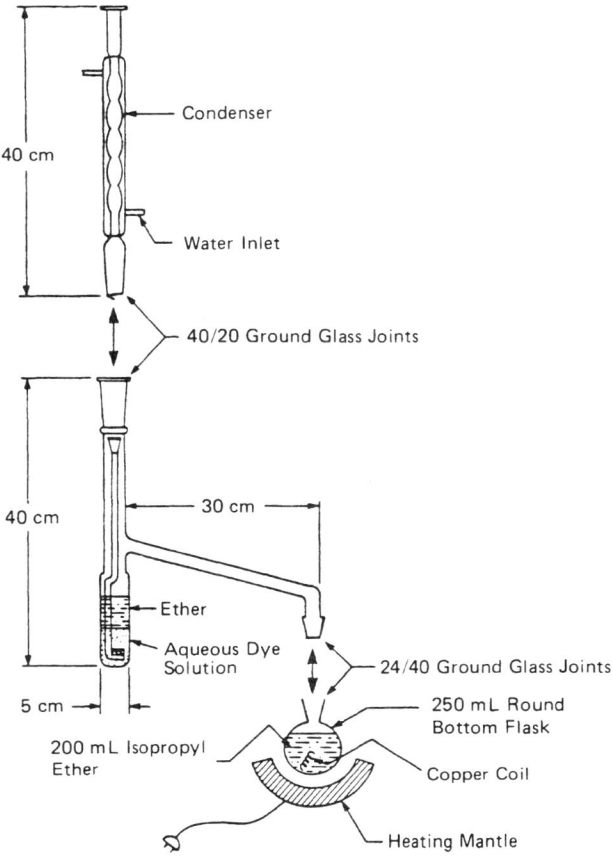

FIGURE 21 Upward Displacement Type Liquid–Liquid Extractor with Sintered-Glass Diffuser.

Apparatus Use an upward displacement-type liquid–liquid extractor, as shown in Fig. 21, with a sintered-glass diffuser and a working capacity of 200 mL. Suspend a piece of bright copper wire through the condenser, and place a small coil of copper wire (about 0.5 g) in the distillation flask.

Alkaline Ether Extract Transfer 5 g of the colorant to a beaker, and dissolve in 150 mL of water. Add 2 mL of 2.5 N NaOH solution, and transfer the solution to the extractor; dilute to approximately 200 mL with water. Add 200 mL of ether to the distillation flask, and extract for 2 h with a reflux rate of about 15 mL/min. Set the extracted colorant solution aside. Transfer the ether extract to a separatory funnel, and wash with two 25-mL portions of 0.1 N NaOH followed by two 25-mL portions of water. Reduce the volume of the ether extract to about 5 mL by distillation (in portions) from a tared flask containing a small piece of clean copper coil.

Acid Ether Extract To the extracted colorant solution set aside in the alkaline ether extract procedure above, add 5 mL of 3 N hydrochloric acid, mix, and extract with ether as directed above. Wash the ether extract with two 25-mL portions of 0.1 N hydrochloric acid and water. Transfer the washed ether in portions to the flask containing the evaporated alkaline extract, and carefully remove all the ether by distillation. Dry the residue in an oven at 85° for 20 min. Then allow the flask to cool in a desiccator for 30 min, and weigh. Repeat drying and cooling until a constant weight is obtained. The increase in weight of the tared flask, expressed as a percentage of the sample weight, is the combined ether extract.

Leuco Base

Reagents and Solutions

Cupric Chloride Solution Transfer 10.0 g of $CuCl_2.2H_2O$ to a 1-L volumetric flask; dissolve in and dilute to volume with dimethylformamide (DMF).

Sample Solution Prepare as directed in the individual monograph.

Procedure

Solution 1 Pipet 50 mL of DMF into a 250-mL volumetric flask, cover, and place in the dark.

Solution 2 Pipet 10 mL of the sample solution into a 250-mL volumetric flask, add 50 mL of DMF, and place in the dark.

Solution 3 Pipet 50 mL of *Cupric Chloride Solution* into a 250-mL volumetric flask, and gently bubble air through the solution for 30 min.

Solutions 4a and 4b Pipet 10 mL of the sample solution into each of two 250-mL volumetric flasks, add 50 mL of *Cupric Chloride Solution* to each, and bubble air gently through the solutions for 30 min. Dilute all the solutions nearly to volume with water; incubate for 5 to 10 min, but no longer, in a water bath cooled with tap water; and dilute to volume. Record the spectrum for each solution between 500 nm and 700 nm using an absorbance range of 0 to 1 and a 1-cm pathlength cell; record all spectra on the same spectrogram.

Curve No.	Solution in Sample Cell	Solution in Reference Cell
I	1	1
II	1	2
III	3	3
IVa	3	4a
IVb	3	4b

Calculation

$$\% \text{ Leuco Base} = \frac{[(IV - III) - (II - I)] \times 2500}{a \times W \times r},$$

in which the Roman numerals I through IV represent the absorbance readings for solutions of the corresponding Arabic numerals (above) at the wavelength maximum; a is the absorptivity (for F D & C Green No. 3, $a = 0.156$ at 625 nm; for F D & C Blue No. 1, $a = 0.164$ at 630 nm); W is the weight, in g, of the sample; and r is the ratio of the molecular weights of colorant and leuco base (for F D & C Green No. 3, $r = 0.9712$; for F D & C Blue No. 1, $r = 0.9706$).

Mercury

Apparatus The apparatus used for the direct microdetermination of mercury is shown in Fig. 22. It consists of a quartz combustion tube designed to hold a porcelain combustion boat (60 × 10 × 8 mm) and a small piece of copper oxide wire. The combustion tube is placed in a heavy-duty hinged combustion tube furnace (Lindburg Type 70T, or equivalent) and connected by clamped ball-joints at one end to a source of nitrogen and at the other, to a series of three traps. The traps are constructed of a linear array of 18- × 2-mm Pyrex tubes connected by clamped ball-joints and extend from the connection at the combustion tube. Trap I contains anhydrous calcium sulfate packed between quartz-wool plugs, trap II contains Ascarite packed between cotton plugs, and trap III contains aluminum oxide packed between cotton plugs. The nitrogen flow forces the mercury through the combustion tube, the three traps, and a section of Tygon tube to a mercury vapor meter (Beckman model K-23, or equivalent). The mercury released from a sample during combustion is quantitated by comparing the recorder response with that given by a series of mercury standards.

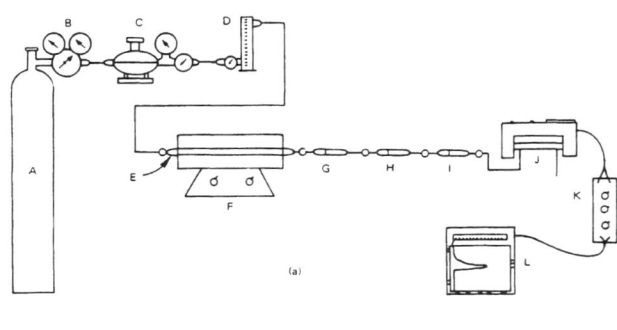

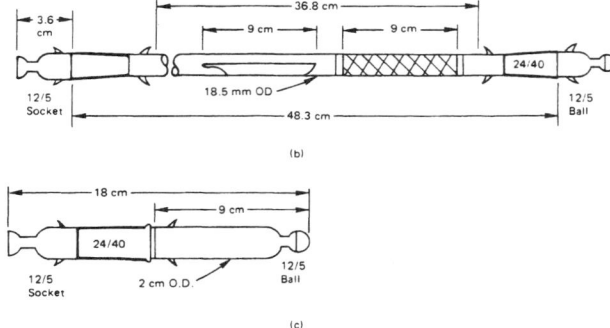

FIGURE 22 (a) Schematic Diagram of Apparatus for Photometric Mercury Vapor M Method:

A. Tank of nitrogen
B. Two-stage pressure regulator
C. Low-pressure regulator
D. Flowmeter
E. Combustion tube
F. Combustion-tube furnace
G. Dehydrite trap
H. Ascarite trap
I. Aluminum oxide trap
J. Mercury vapor meter
K. Attenuator
L. Recorder

(b) Quartz Combustion Tube with Boat and Copper Oxide Packing;
(c) Schematic Diagram of Trap Used to Contain Ascarite, Dehydrite, and Aluminum Oxide.

Reagents and Equipment
Absorbent Cotton
Aluminum Oxide   Anhydrous.
Calcium Sulfate   Anhydrous, dehydrite, or equivalent.
Asbestos Pads, (1 × 0.5 × 1 cm)   Preheated at 800° for 1 h.
Ascarite   20- to 30-mesh.
Copper Oxide Wire   Preheated at 850° for 2 h.
Nitrogen   Purified grade.
Quartz Wool
Sodium Carbonate   Anhydrous, fine granular.
Standard Solution   Transfer approximately 1.35 g of reagent-grade mercurous chloride, accurately weighed, to a 1-L volumetric flask. Dissolve in and dilute to volume with water. When diluted 100-fold, the solution contains 0.01 μg Hg per μL (*Diluted Standard Solution*).

Procedure   Preheat the furnace to 650°, and adjust the nitrogen flow to 1 L/min.
Blank Analysis   Place a square piece of preheated asbestos pad in the combustion boat, and cover it with sodium carbonate. Stop the nitrogen flow, disconnect the ball-joint, quickly insert the boat into the combustion tube with large forceps, and reconnect the joint. Note the time, allow the boat to sit in the tube with no nitrogen flow for exactly 1 min, then restart the flow of nitrogen. Mercury elutes almost immediately with the reinstated nitrogen flow; note the recorder response. Allow about 30 s between runs.
Calibration   Determine the recorder response after the application to the asbestos pad of 1, 2, and 3 μL of the *Diluted Standard Solution*.
Sample Analysis   Transfer 25 mg of colorant, accurately weighed, to the combustion boat, and cover the sample completely with sodium carbonate. Follow the procedure used for the blank analysis above, and calculate the mercury content using the standard curve.

Trap Problems   (1) Some colorants (e.g., F D & C Blue No. 1 and F D & C Green No. 3) may give a response that is symmetrically dissimilar to the Hg peak. If such a response "carries over" to the next sample, then the aluminum oxide trap may need to be changed. (2) If the recorder response is of inadequate sensitivity (peak height induced by 0.01 μg less than 0.5 cm), then the traps are packed too tightly. Remove or redistribute packing first in the aluminum oxide trap, then try the other traps. (3) The traps will need changing periodically as indicated by a change in the physical appearance of the trap material or by chart responses of different retention times or different symmetry from that of mercury standards. (4) If two or more standards are run in succession, a later sample might give an erroneous mercury response. Run blanks and then repeat the sample analysis to confirm the validity of the response.

Sodium Chloride   Dissolve approximately 2 g of colorant, accurately weighed, in 100 mL of water, and add 10 g of activated carbon that is free of chloride and sulfate. Boil gently for 2 to 3 min. Cool to room temperature, add 1 mL of 6 N nitric acid, and stir. Dilute to volume with water in a 200-mL volumetric flask, and then filter through dry paper. Repeat the treatment with 2-g portions of carbon until no color is adsorbed onto filter paper dipped into the filtrate.

Transfer 50 mL of filtrate to a 250-mL flask. Add 2 mL of 6 N nitric acid, 5 mL of nitrobenzene, and 10 mL of standardized 0.1 N silver nitrate solution. Shake the flask until the silver chloride coagulates. Prepare a saturated solution of ferric ammonium sulfate, and add just enough concentrated nitric acid to discharge the red color; add 1 mL of this solution to the 250-mL flask to serve as the indicator. Titrate with 0.1 N ammonium thiocyanate solution that has been standardized against the silver nitrate solution until the color persists after shaking for 1 min. Calculate the weight percent of the sodium chloride, P, by the formula

$$P = [(V \times N)/W] \times 22.79,$$

in which V is the net volume, in mL, of silver nitrate solution required; N is the normality of the silver nitrate solution; and W is the weight, in g, of the sample. The factor 22.79 incorporates a total volume of 195 mL because 10 g of activated carbon occupies 5 mL.

Sodium Sulfate   Place 25 mL of the decolorized filtrate obtained from the *Sodium Chloride* test above in a 125 mL Erlenmeyer flask, and add 1 drop of a 0.5% phenolphthalein solution in 50% ethanol. Add 0.05 N sodium hydroxide dropwise until the solution is alkaline to pH paper, and then add 0.002 N hydrochloric acid until the indicator is decolorized. Add 25 mL of ethanol and about 0.2 g of tetrahydroquinone sulfate indicator. Titrate with 0.03 N barium chloride solution to a red endpoint. Make a blank determination.

Calculate the weight percent, P, of sodium sufate by the formula

$$P = [(V - B) \times N/W] \times 55.4,$$

in which V is the volume, in mL, of barium chloride solution required to titrate the sample; B is the volume, in mL, of barium chloride solution required for the blank; N is the normality of the barium chloride solution; and W is the sample weight, in g. The factor 55.4 incorporates a total volume of 195 mL because 10 g of activated carbon occupies 5 mL.

Total Color
Method I (Spectrophotometric)
Pipet 10.0 mL of the dissolved colorant into a 250-mL Erlenmeyer flask containing 90 mL of 0.04 N ammonium acetate, and mix well. Determine the net absorbance of the solution relative to water at the wavelength maximum given for each color. Calculate the percentage of colorant present using the following formula, which presumes a 1-cm pathlength cell:

$$\% \text{ total color} = (A \times 100)/(a \times W),$$

in which A is the absorbance; a is the absorptivity; and W is the weight, in g, of the sample.

Method II (Titration with Titanium Chloride)
Apparatus   The apparatus for determining total color by titration with $TiCl_3$ is shown in Fig. 23. It consists of a storage bottle, A, of 0.1 N $TiCl_3$ titrant maintained under hydrogen produced by a Kipp generator; an Erlenmeyer flask, B, equipped

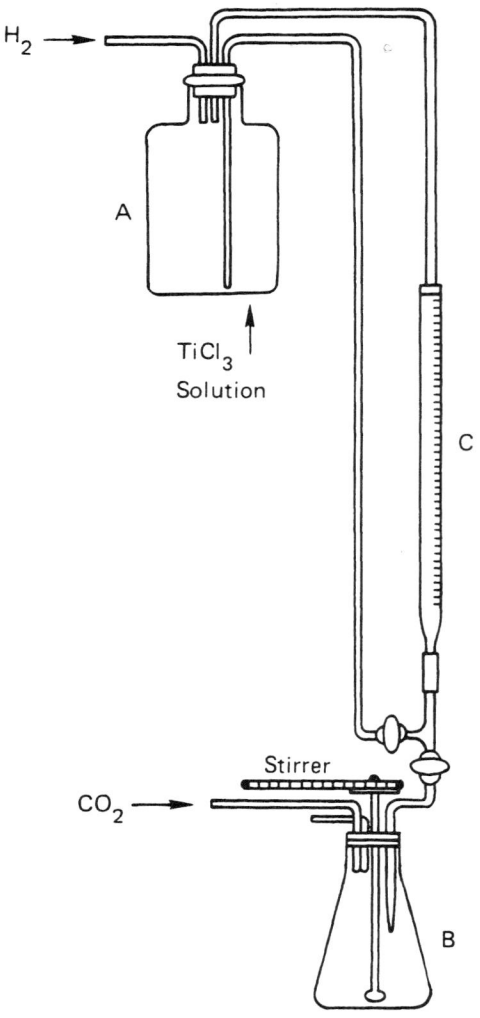

FIGURE 23 Titanous Chloride Titration Apparatus.

with a source of CO_2 or N_2 to maintain an inert atmosphere in which the reaction takes place; a stirrer; and the buret, C.

Reagents and Solutions

Titanium Chloride Solution (0.1 N) Transfer to a storage bottle 73 mL of commercially prepared 20% $TiCl_3$ solution, and carefully add 82 mL of concentrated HCl per L of final solution. Mix well, and bubble CO_2 or N_2 through the solution for 1 h. Before standardizing, maintain the solution under a hydrogen atmosphere for at least 16 h using a Kipp generator.

Potassium Dichromate (0.1 N, primary standard) Transfer 4.9032 g of $K_2Cr_2O_7$ (National Institute of Standards and Technology No. 136) to a 1-L volumetric flask; dissolve in and dilute to volume with water.

Ammonium Thiocyanate (50%) Transfer 500 g of NH_4SCN (ACS certified) to a 1-L volumetric flask; dissolve in about 600 mL of water, warming if necessary; and dilute to volume.

Ferrous Ammonium Sulfate $Fe(NH_4)_2(SO_4)_2 \cdot 6H_2O$, ACS certified.

Sodium Bitartrate

Standardization of the Titanium Chloride Solution Drain any standing $TiCl_3$ from the feed lines and buret, and refill with fresh solution. Add 3.0 g of ferrous ammonium sulfate to a wide-mouth Erlenmeyer flask followed by 200 mL of water, 25 mL of 50% H_2SO_4, 25 mL of 0.1 N $K_2Cr_2O_7$ (by pipet), and 2 or 3 boiling chips. Boil the solution vigorously on a hot plate for 30 s to remove dissolved air, then quickly transfer the flask to the titration apparatus, securely connect the stopper assembly, and start the CO_2 flow and stirrer. Pass CO_2 over the solution for 1 min before beginning the titration.

Add the $TiCl_3$ solution at a fast, steady drip to within 1 mL of the estimated endpoint (about 20 mL). Reduce the CO_2 flow, remove the solid glass rod from the stopper assembly, pipet 10 mL of *Ammonium Thiocyanate* (50%) into the flask, insert the glass rod, and increase the CO_2 flow. Continue titrating slowly until the endpoint: A color change from brown-red to light green is observed. Perform a blank determination using the same reagents and quantities, and calculate the normality, N, of the $TiCl_3$ solution on the basis of three titrations by the formula

$$N = (V_r \times N_r/V_t - V_b),$$

in which V_r is the volume, in mL, of $K_2Cr_2O_7$ solution used; N_r is the normality of the $K_2Cr_2O_7$ solution; V_t is the volume, in mL, of $TiCl_3$ solution used; and V_b is the volume, in mL, of $TiCl_3$ used in the blank titration.

Procedure Transfer the quantity of colorant prescribed in the individual monograph to a 500-mL, wide-mouth Erlenmeyer flask and add 21 to 22 g of sodium bitartrate (sodium citrate for F D & C Yellow No. 6), 275 mL of water, and two or three boiling chips. Boil the solution vigorously on a hot plate for 30 s to remove dissolved air, then quickly transfer the flask to the titration apparatus, securely connect the stopper assembly, and start the CO_2 flow and stirrer. Pass CO_2 over the solution for 1 min before beginning the titration.

Titrate the sample until the color lightens, wait 20 s, and then continue the addition with about 2 s between drops. When the color is almost completely bleached, wait 20 s, then continue the addition with 5 s between drops. A complete color change indicates the endpoint. Perform a blank determination using the same reagents and quantities, and calculate the total color, T, in percent, on the basis of three titrations by the formula

$$T = [(V_t - V_b)/(W \times F_s)] \times 100 \times N,$$

in which V_t is the volume of titrant used; V_b is the volume of titrant required to produce the endpoint in a blank; N is the normality of the titrant; W is the sample weight, in g; and F_s is a factor derived from the stoichiometry of the reaction characteristics of each colorant and is given in the individual monograph.

Method III (Gravimetric)

Transfer approximately 0.5 g of colorant, accurately weighed, to a 400-mL beaker, add 100 mL of water, and heat to boiling. Add 25 mL of dilute HCl (1 in 50), and bring to a boil. Wash down the sides of the beaker with water, and cover and keep on a steam bath for several hours or overnight. Cool to room temperature, and quantitatively transfer the precipitate to a tared filtering crucible with dilute HCl (1 in 100). Wash the precipitate with two 15-mL portions of water, and dry the crucible for 3 h at 135°. Cool in a desiccator, and weigh. Calculate the total color, P, in weight percent, by the formula

$$P = [(W_p \times F)/W_s] \times 100,$$

in which W_p is the weight, in g, of the precipitate; F is the gravimetric conversion factor given in the individual monograph; and W_s is the original sample weight, in g.

Uncombined Intermediates and Products of Side Reactions
Method I
Sample Solution Transfer approximately 2 g of colorant to a 100-mL volumetric flask; dissolve in and dilute to volume with water.

Apparatus Pack a 2.5- × 45-cm glass column with approximately 20 g of cellulose (Whatman CF-11 grade, or the equivalent) that has been slurried in the eluant and from which the fines have been removed by decantation. Equilibrate the column thoroughly with the eluant, 35% ammonium sulfate.

Procedure Pipet 5 mL of *Sample Solution* into a beaker containing 5 g of cellulose that has been slurried in eluant and from which the fines have been removed by decantation. Stir the mixture thoroughly, add 10 g of ammonium sulfate, and stir until uniformly mixed. Mix the slurry with 15 mL of eluant, and apply it to the column. Allow the fluid to enter the column, and wash the beaker with eluant until the sample is quantitatively transferred. Elute the column with approximately 500 mL of 35% ammonium sulfate, and collect a total of eight 60-mL fractions. Divide each collected fraction in half and add 0.5 mL of NH$_4$OH to one half and 0.5 mL of HCl to the other.

Calculation After identifying each intermediate and side product by comparing spectra of the fractions with commercial standards, calculate the concentration, C, of each using the formula

$$C = A/(a \times b),$$

in which A is the absorbance at the wavelength of maximal absorption; b is the cell pathlength, in cm; and a is the absorptivity given in the individual monograph.

Method II
Apparatus Use a suitable high-performance liquid chromatography system (see Appendix IIA) equipped with a dual wavelength detector system such that the effluent can be monitored serially at 254 nm and 325 to 385 nm (wide-band pass). Use a 1-m × 2.1-mm (id) column packed with a strong anion-exchange resin (Dupont No. 830950405, or the equivalent).

Operating Conditions The operating conditions required may vary depending on the system used. The following conditions have been shown to give suitable results for F D & C Red No. 40, F D & C Yellow No. 5, and F D & C Yellow No. 6.

F D & C Red No. 40
  Primary eluant: 0.01 M aqueous Na$_2$B$_4$O$_7$.
  Secondary eluant: 0.20 M NaClO$_4$ in aqueous 0.01 M Na$_2$B$_4$O$_7$.
  Sample size: 20 μL of a 0.25% solution.
  Flow rate: 0.60 mL/min.
  Gradient: Linear, in two phases: 0% to 18% in 40 min, 18% to 62% in 8 min more, then hold for 18 min more at 62%.
  Temperature: 50°.
  Pressure: 1000 psi.
  Order of elution: (1) Cresidinesulfonic acid (CSA), (2) unknown, (3) Schaeffer's salt (SS), (4) unknown, (5) 4,4'-diazoaminobis(5-methoxy-2-methylbenzenesulfonic acid) (DMMA), (6) unknown, (7) F D & C Red No. 40, (8) 6,6'-oxybis(2-naphthalenesulfonic acid) (DONS).

F D & C Yellow No. 5
  Primary eluant: 0.01 M aqueous Na$_2$B$_4$O$_7$.
  Secondary eluant: 0.10 M NaClO$_4$ in aqueous 0.01 M Na$_2$B$_4$O$_7$.
  Sample Size: 50 μL of a 0.15% solution, prepared within 13 min of injection.
  Flow rate: 1.00 mL/min.
  Gradient: Exponential at 4%/min: 0.95%.
  Temperature: 50°.
  Pressure: 1000 psi.
  Order of elution: (1) Phenylhydrazine-*p*-sulfonic acid (PHSA), (2) sulfanilic acid (SA), (3) 1-(4-sulfophenyl)-3-ethylcarboxy-5-hydroxypyrazolone (PY-T), (4) 1-(4-sulfophenyl)-3-carboxy-5-hydroxypyrazolone (EEPT), (5) 4,4'-(diazoamino)-dibenzenesulfonic acid (DAADBSA).

F D & C Yellow No. 6
  Primary eluant: 0.01 M aqueous Na$_2$B$_4$O$_7$.
  Secondary eluant: 0.20 M NaClO$_4$ in aqueous 0.01 M Na$_2$B$_4$O$_7$.
  Sample size: 5 μL of a 1% solution.
  Flow rate: 0.50 mL/min.
  Gradient: Linear in four phases: 0% to 11% in 10 min; hold 25 min; 11% to 38% in 10 min; 38% to 42% in 10 min; 42% to 98% in 20 min; hold 20 min.
  Temperature: 50°.
  Pressure: 1000 psi.
  Order of elution: (1) Sulfanilic acid (SA), (2) Schaeffer's salt (SS), (3) 4,4'-(diazoamino)-dibenzenesulfonic acid (DAADBSA), (4) *R*-salt dye, (5) F D & C Yellow No. 6, (6) 6,6'-oxybis(2-naphthalenesulfonic acid) (DONS).

Standard Solutions
  F D & C Red No. 40 Prepare a solution containing 0.25 g of colorant, 0.5 mg of CSA, 0.75 mg of SS, 0.25 mg of DMMA, and 1.25 mg of DONS in a 100-mL volumetric flask. Dissolve in and dilute to volume with 0.1 M Na$_2$B$_4$O$_7$.
  F D & C Yellow No. 5 Prepare a solution containing 0.15 g of colorant and 0.3 mg each of PHSA, SA, PY-T, EEPT, and DAADBSA in a 100-mL volumetric flask. Dissolve in and dilute to volume with 0.1 M Na$_2$B$_4$O$_7$.
  F D & C Yellow No. 6 Prepare a solution containing 0.25 g of colorant, 0.5 mg of SA, 0.75 mg of SS, 0.25 mg of DAADBSA, and 1.25 mg of DONS in a 100-mL volumetric flask. Dissolve in and dilute to volume with 0.1 M Na$_2$B$_4$O$_7$.

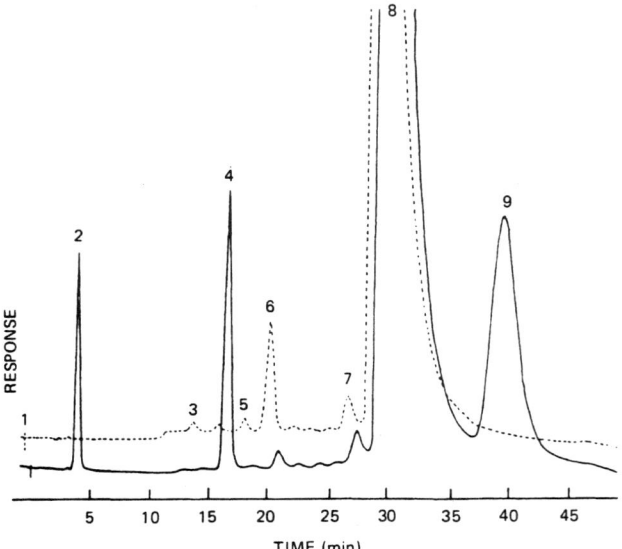

FIGURE 24 F D & C Red No. 40 Top Trace: Eluant Monitored at 254 nm; Bottom Trace: Eluant Monitored at 375 to 385 nm.

Test Solutions Prepare at least four test solutions, each containing the colorant, and one impurity, accurately weighed, dissolved in 0.1 M $Na_2B_4O_7$, and diluted to volume in a 100-mL volumetric flask. The solutions should encompass the range of concentrations, evenly spaced, given below for each constituent:

F D & C Red No. 40 (250 mg): CSA (0.05 to 0.5 mg); SS (0.05 to 0.75 mg); DONS (0.5 to 2.5 mg); DMMA (0.025 to 0.25 mg). Inject 20 µL of each solution.

F D & C Yellow No. 5 (150 mg): SA (7.5 to 300 µg); PY-T (7.5 to 300 µg); EEPT (7.5 to 300 µg); DAADBSA (7.5 to 300 µg). Inject 50 µL of each solution.

F D & C Yellow No. 6 (250 mg): SA (0.05 to 0.5 mg); SS (0.05 to 0.75 mg); DONS (0.5 to 2.5 mg); DAADBSA (0.05 to 0.25 mg). Inject 20 µL of each solution.

System Suitability

Resolution Elute the column with the gradient specified under *Operating Conditions* until a smooth baseline is obtained. Inject an aliquot of the *Standard Solution*. The resolution of the eluted components matches or exceeds that shown for the corresponding colorant (see Figs. 24, 25, and 26). After determining that the column will give the required resolution, allow it to rest for 2 weeks before use.

Calibration Inject the designated volume of each *Test Solution* onto a conditioned column, and prepare a standard curve corresponding to each unreacted intermediate and side reaction product. Determine the area, A, for each peak from the integrator if an automated system is used or by multiplying the peak height by the width at one-half the height. The peak height alone may be used for EEPT, PY-T, and DAADBSA. Calculate the concentration, C_i, of each intermediate or side product using the formula

$$C_i = mA_i + b,$$

in which A_i is the area of its corresponding chromatographic peak. Calculate the slope, m, and intercept, b, using the following linear regression formulas:

$$m = [N\Sigma C_iA_i - \Sigma C_i\Sigma A_i]/[N\Sigma A_i^2 - (\Sigma A_i)^2],$$

$$b = [\overline{A}]_i - m[\overline{C}]_i,$$

in which $\overline{C}$ and $\overline{A}$ are the calculated averages of the concentrations and peak areas, respectively, used to construct the standard curve for one intermediate or side reaction product. Calculate the correlation coefficient, r, from the following formula:

$$r = [\Sigma(C_i - \overline{C})(A_i - A)]/[\Sigma(C_i - \overline{C}^2) \times \Sigma(A_i - \overline{A})^2].$$

Each time the system is calibrated, add the new data to those accumulated from previous analyses. The correlation coefficient must be between 0.95 and 1.00 for any single experiment or from accumulated data.

Recalibrate the system after every ten determinations or 2 days, whichever occurs first.

Sample Preparation Prepare as directed in the individual monograph.

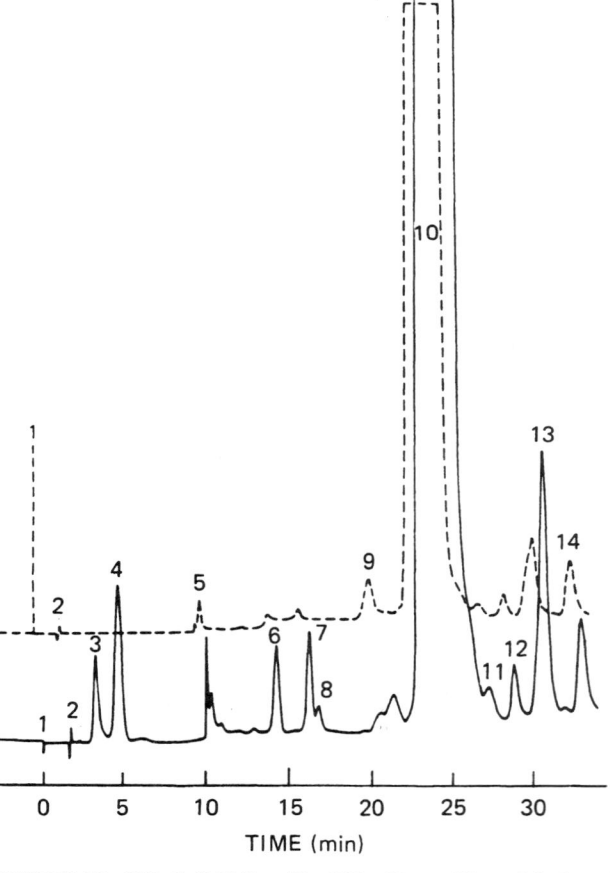

FIGURE 25 F D & C Yellow No. 5 Top Trace: Eluant Monitored at 254 nm; Bottom Trace: Eluant Monitored at 375 to 385 nm.

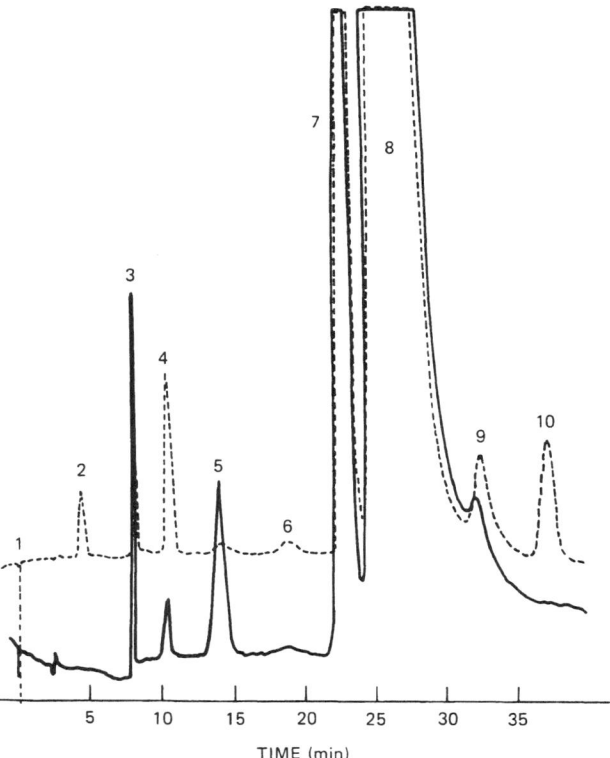

FIGURE 26 F D & C Yellow No. 6 Top Trace: Eluant Monitored at 54 nm; Bottom Trace: Eluant Monitored at 375 to 385 nm.

Procedure Inject the volume of *Sample Preparation* as designated in the monograph into the column. Determine the concentration of intermediates and side reaction products from the peak areas using the slope, m, and intercept, b, calculated under *Calibration* by the equation

$$C_s = mA_s + b,$$

in which C_s is the concentration of the unknown in the *Sample Preparation* and A_s its corresponding peak area.

Volatile Matter Transfer 1.5 to 2.5 g of colorant, accurately weighed, to a tared crucible. Heat in a vacuum oven at 135° for 12 to 15 h. Lower the pressure in the oven to −125 mm Hg, and continue heating for an additional 2 h. Cover the crucible, and allow to cool in a desiccator. Reweigh the crucible when cool. The loss of weight is defined as the volatile matter.

Water-Insoluble Matter Transfer about 1 g of colorant, accurately weighed, to a 250-mL beaker, and add 200 mL of boiling water. Stir to facilitate dissolution of the color.

Tare a filtering crucible equipped with a glass fiber filter (Reeve Angel, No. 5270, or equivalent). Filter the solution with the aid of suction when it has cooled to ambient temperature. Rinse the beaker three times, pouring the rinse through the crucible. Wash the filter with water until the filtrate is colorless.

Dry the crucible and filter in an oven at 135° for at least 3 h, cool them in a desiccator and reweigh to the nearest 0.1 mg. Calculate the percent water-insoluble matter, I, by the formula

$$I = (W_c/W_s) \times 100,$$

in which W_c is the difference in crucible weight and W_s is the sample weight.

HYDROXYPROPOXYL DETERMINATION

Apparatus The apparatus for hydroxypropoxyl group determination is shown in Fig. 27. The boiling flask, D, is fitted with an aluminum foil-covered Vigreaux column, E, on the side arm and with a bleeder tube through the neck and to the bottom of the flask for the introduction of steam and nitrogen. A steam generator, B, is attached to the bleeder tube through tube C, and a condenser, F, is attached to the Vigreaux column. The boiling flask and steam generator are immersed in an oil bath, A, equipped with a thermoregulator such that a temperature of 155° and the desired heating rate may be maintained. The distillate is collected in a 150-mL beaker, G, or other suitable container.

Procedure Unless otherwise directed, transfer about 100 mg of the sample, previously dried at 105° for 2 h and accurately weighed, into the boiling flask, and add 10 mL of chromium trioxide solution (60 g in 140 mL of water). Immerse the steam generator and the boiling flask in the oil bath (at room temperature) to the level of the top of the chromium trioxide solution. Start cooling water through the condenser, and pass nitrogen gas through the boiling flask at the rate of one bubble per s. Starting at room temperature, raise the temperature of the oil bath to 155° over a period of not less than 30 min, and maintain this temperature until the end of the determination. Distill until 50 mL of distillate is collected. Detach the condenser from the Vigreaux column, and wash it with water, collecting the washings in the distillate container. Titrate the combined washings and distillate with 0.02 N sodium hydroxide to a pH of 7.0, using a pH meter set at the expanded scale.

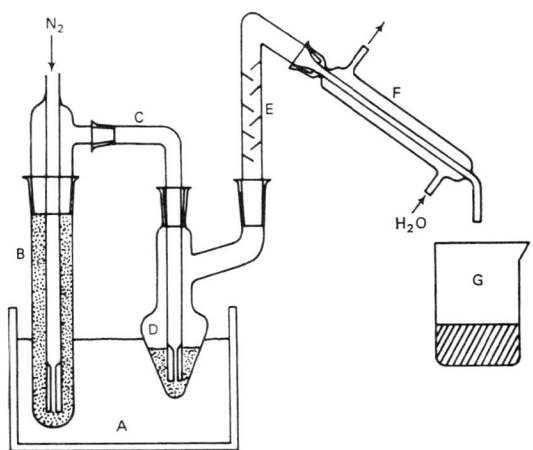

FIGURE 27 Apparatus for Hydroxypropoxyl Determination.

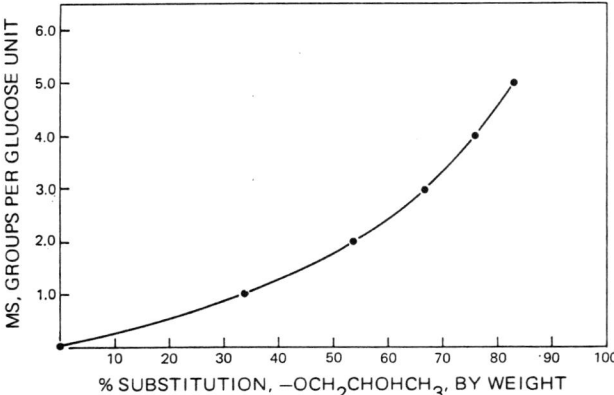

FIGURE 28 Chart for Converting Percentage of Substitution, by Weight, of Hydroxypropoxyl Groups to Molecular Substitution per Glucose Unit.

Note: Phenolphthalein TS may be used for this titration if it is also used for all standards and blanks.

Record the volume, V_a, of the 0.02 N sodium hydroxide used. Add 500 mg of sodium bicarbonate and 10 mL of 2 N sulfuric acid, and then after evolution of carbon dioxide has ceased, add 1 g of potassium iodide. Stopper the flask, shake the mixture, and allow it to stand in the dark for 5 min. Titrate the liberated iodine with 0.02 N sodium thiosulfate to the sharp disappearance of the yellow color, confirming the endpoint by the addition of a few drops of starch TS. Record the volume of 0.02 N sodium thiosulfate required as Y_a.

Make several reagent blank determinations, using only the chromium trioxide solution in the above procedure. The ratio of the sodium hydroxide titration (V_b) to the sodium thiosulfate titration (Y_b), corrected for variation in normalities, will give the acidity-to-oxidizing ratio, $V_b/Y_b = K$, for the chromium trioxide carried over in the distillation. The factor K should be constant for all determinations.

Make a series of blank determinations using 100 mg of methylcellulose (containing no foreign material) in place of the sample, recording the average volume of 0.02 N sodium hydroxide required as V_m and the average volume of 0.02 N sodium thiosulfate required as Y_m.

Calculate the hydroxypropoxyl content of the sample, in mg, by the formula

$$75.0 \times [N_1(V_a - V_m) - kN_2(Y_a - Y_m)],$$

in which N_1 is the exact normality of the 0.02 N sodium hydroxide solution, N_2 is the exact normality of the 0.02 N sodium thiosulfate solution, and $k = V_bN_1/Y_bN_2$.

The percentage of substitution, by weight, of hydroxypropoxyl groups, determined as directed above, may be converted to molecular substitution, per glucose unit, by reference to Fig. 28.

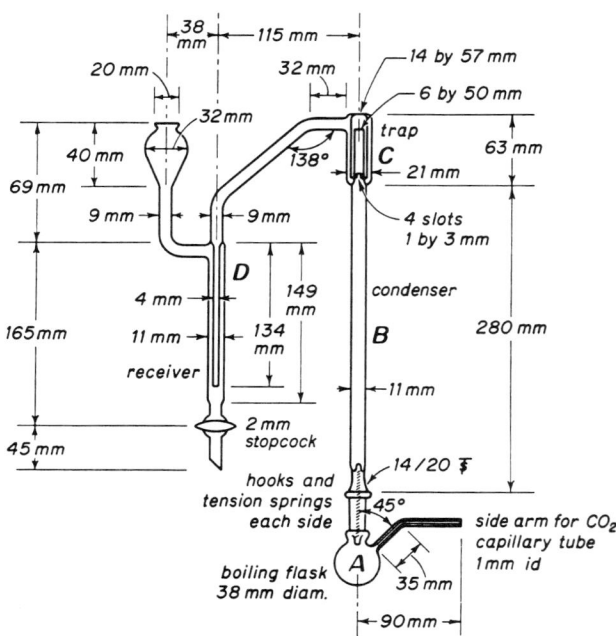

FIGURE 29 Distillation Apparatus for Methoxyl Determination.

METHOXYL DETERMINATION

Apparatus The apparatus for methoxyl determination, as shown in Fig. 29, consists of a boiling flask, A, fitted with a capillary side arm to provide an inlet for carbon dioxide and connected to a column, B, which separates aqueous hydriodic acid from the more volatile methyl iodide. After the methyl iodide passes through a suspension of aqueous red phosphorus in the scrubber trap, C, it is absorbed in the bromine–acetic acid absorption tube, D. The carbon dioxide is introduced from a device arranged to minimize pressure fluctuations and connected to the apparatus by a small capillary containing a small plug of cotton.

Reagents

Acetic Potassium Acetate Dissolve 100 g of potassium acetate in 1000 mL of a mixture consisting of 900 mL of glacial acetic acid and 100 mL of acetic anhydride.

Bromine–Acetic Acid Solution On the day of use, dissolve 5 mL of bromine in 145 mL of the *Acetic Potassium Acetate* solution.

Hydriodic Acid Use special-grade hydriodic acid suitable for alkoxyl determinations, or purify reagent grade as follows: Distill over red phosphorus in an all-glass apparatus, passing a slow stream of carbon dioxide through the apparatus until the distillation is terminated and the receiving flask has completely cooled.

Caution: Use a safety shield and conduct the distillation in a hood.

Collect the colorless, or almost colorless, constant-boiling acid distilling between 126° and 127°. Store the acid in a cool, dark place in small, brown, glass-stoppered bottles previously flushed with carbon dioxide and finally sealed with paraffin.

Procedure Fill trap *C* half full with a suspension of about 60 mg of red phosphorus in 100 mL of water, introduced through the funnel on tube *D* and the side arm that connects with the trap at *C*. Rinse tube *D* and the side arm with water, collecting the rinsings in trap *C*, then charge absorption tube *D* with 7 mL of *Bromine–Acetic Acid Solution*. Place the sample, accurately weighed in a tared gelatin capsule, in the boiling flask *A*, along with a few glass beads or boiling stones, then add 6 mL of *Hydriodic Acid*. Connect the flask to the condenser, using a few drops of the acid to seal the junction, and begin passing the carbon dioxide through the apparatus at the rate of about two bubbles per s. Heat the flask in an oil bath at 150°, continue the reaction for 40 min, and drain the contents of absorption tube *D* into a 500-mL Erlenmeyer flask containing 10 mL of sodium acetate solution (1 in 4). Rinse tube *D* with water, collecting the rinsings in the flask, and dilute to about 125 mL with water. Discharge the reddish brown color of bromine by adding formic acid dropwise, with swirling, then add 3 drops in excess. Usually a total of 12 to 15 drops of formic acid is required. Allow the flask to stand for 3 min, add 15 mL of 2 *N* sulfuric acid and 3 g of potassium iodide, and titrate immediately with 0.1 *N* sodium thiosulfate, adding starch TS near the endpoint. Perform a blank determination with the same quantities of the same reagents, including the gelatin capsule, and in the same manner, and make any necessary correction. Each mL of 0.1 *N* sodium thiosulfate is equivalent to 0.517 mg (517 μg) of methoxyl groups (—OCH$_3$).

NITROGEN DETERMINATION
(Kjeldahl Method)

Caution: Provide adequate ventilation in the laboratory, and do not permit accumulation of exposed mercury.

Note: All reagents should be nitrogen free, where available, or otherwise very low in nitrogen content.

Method I
This method should be used unless otherwise directed in the individual monograph. It is not applicable to certain nitrogen-containing compounds that do not yield their entire nitrogen upon digestion with sulfuric acid.

A. Nitrites and Nitrates Absent Unless otherwise directed, transfer from 700 mg to 2.2 g of the sample into a 500- to 800-mL Kjeldahl digestion flask of hard, moderately thick, well-annealed glass, wrapping the sample, if solid or semi-solid, in nitrogen-free filter paper to facilitate the transfer if desired. Add 700 mg of mercuric oxide or 650 mg of metallic mercury, 15 g of powdered potassium sulfate or anhydrous sodium sulfate, and 25 mL of 93% to 98% sulfuric acid. (If a sample weight greater than 2.2 g is used, increase the sulfuric acid by 10 mL for each additional gram of sample.) Place the flask in an inclined position, and heat gently until frothing ceases, adding a small amount of paraffin, if necessary, to reduce frothing.

Caution: The digestion should be conducted in a fume hood, or the digestion apparatus should be equipped with a fume exhaust system.

Boil briskly until the solution clears, and then continue boiling for 30 min longer (or for 2 h for samples containing organic material). Cool, add about 200 mL of water, mix, and then cool to below 25°. Add 25 mL of sulfide or thiosulfate solution (40 g of K$_2$S, 40 g of Na$_2$S, or 80 g of Na$_2$S$_2$O$_3$.5H$_2$O in 1000 mL of water), and mix to precipitate the mercury. Add a few granules of zinc to prevent bumping, tilt the flask, and cautiously pour sodium hydroxide pellets, or a 2 in 5 sodium hydroxide solution, down the inside of the flask so that it forms a layer under the acid solution, using a sufficient amount (usually about 25 g of solid NaOH) to make the mixture strongly alkaline. Immediately connect the flask to a distillation apparatus consisting of a Kjeldahl connecting bulb and a condenser, the delivery tube of which extends well beneath the surface of a measured excess of 0.5 *N* hydrochloric or sulfuric acid contained in a 500-mL flask. Add from 5 to 7 drops of methyl red indicator (1 g of methyl red in 200 mL of alcohol) to the receiver flask. Rotate the Kjeldahl flask to mix its contents thoroughly, and then heat until all of the ammonia has distilled, collecting at least 150 mL of distillate. Wash the tip of the delivery tube, collecting the washings in the receiving flask, and titrate the excess acid with 0.5 *N* sodium hydroxide. Perform a blank determination, substituting 2 g of sucrose for the sample, and make any necessary correction (see *General Provisions*). Each mL of 0.5 *N* acid consumed is equivalent to 7.003 mg of nitrogen.

Note: If the substance to be determined is known to have a low nitrogen content, 0.1 *N* acid and alkali may be used, in which case each mL of 0.1 *N* acid consumed is equivalent to 1.401 mg of nitrogen.

B. Nitrites and Nitrates Present (**Note**: This procedure is not applicable to liquids or to materials having a high chlorine-to-nitrate ratio.) Unless otherwise directed, transfer from 700 mg to 2.2 g of the sample into a Kjeldahl flask, and add 40 mL of 93% to 98% sulfuric acid containing 2 g of salicylic acid. Mix thoroughly by shaking, and then allow to stand for 30 min or more, with occasional shaking. Add 5 g of Na$_2$S$_2$O$_3$.5H$_2$O, or 2 g of zinc dust (as an impalpable powder, not granules or filings), shake, and allow to stand for 5 min. Heat over a low flame until frothing ceases, then remove the heat, add 700 mg of mercuric oxide (or 650 mg of metallic mercury) and 15 g of powdered potassium sulfate (or anhydrous sodium sulfate), and boil briskly until the solution clears. Continue boiling for 30 min longer (or for 2 h for samples containing organic material), and then continue as directed under *A*, beginning with "Cool, add about 200 mL of water...."

Method II (Semimicro)

Note: Automated instruments may be used in place of this manual method, provided the automated equipment has been properly calibrated.

Transfer an accurately weighed or measured quantity of the sample, equivalent to about 2 or 3 mg of nitrogen, to the digestion flask of a semimicro Kjeldahl apparatus. Add 1 g of a powdered mixture of potassium sulfate and cupric sulfate (10 to 1), using a fine jet of water to wash down any material adhering to the neck of the flask, and then pour 7 mL of sulfuric acid down the inside wall of the flask to rinse it. Cautiously add down the inside of the flask 1 mL of 30% hydrogen peroxide, swirling the flask during the addition.

Caution: Do not add any peroxide during the digestion.

Heat over a free flame or an electric heater until the solution has attained a clear blue color and the walls of the flask are free from carbonized material. Cautiously add 20 mL of water, cool, then add through a funnel 30 mL of sodium hydroxide solution (2 in 5), and rinse the funnel with 10 mL of water. Connect the flask to a steam distillation apparatus, and immediately begin the distillation with steam. Collect the distillate in 15 mL of boric acid solution (1 in 25) to which has been added 3 drops of methyl red–methylene blue TS and enough water to cover the end of the condensing tube. Continue passing the steam until 80 to 100 mL of distillate has been collected, then remove the absorption flask, rinse the end of the condenser tube with a small quantity of water, and titrate with 0.01 N sulfuric acid. Each mL of 0.01 N acid is equivalent to 140 μg of nitrogen.

When more than 2 or 3 mg of nitrogen is present in the measured quantity of the substance to be determined, 0.02 or 0.1 N sulfuric acid may be used in the titration if at least 15 mL of titrant is required. If the total dry weight of the material taken is greater than 100 mg, increase proportionately the quantities of sulfuric acid and sodium hydroxide added before distillation.

SULFUR (by Oxidative Microcoulometry) (Based on ASTM D3120)

Note: All reagents used in this test should be reagent grade; water should be of high purity, and gases must be high-purity grade.

Apparatus The Dohrmann Microcoulometric Titrating System (MCTS-30), or its equivalent, shown in Fig. 30, is to be used unless otherwise specified in an individual monograph. It consists of a constant rate injector (A), a pyrolysis furnace (B), a quartz pyrolysis tube (C), a granular tin scrubber (D), a titration cell (E), and a microcoulometer with a digital readout (F).

Granular-Tin Scrubber Place 5 g of 20/30-mesh granular reagent grade tin between quartz-wool plugs in an elongated 18/9-12/5 standard-taper adaptor that connects the pyrolysis tube and the titration cell.

Microcoulometer Must have variable attenuation, gain control, and be capable of measuring the potential of the sensing-reference electrode pair, and comparing this potential with a bias potential, amplifying the potential difference, and applying the amplified difference to the working-auxiliary electrode pair to generate a titrant. Also the microcoulometer output voltage signal must be proportional to the generating current.

Pyrolysis Furnace The sample should be pyrolyzed in an electric furnace having at least two separate and independently controlled temperature zones, the first being an inlet section that can maintain a temperature sufficient to volatilize all the organic sample. The second zone shall be a pyrolysis section that can maintain a temperature sufficient to pyrolyze the organic matrix and oxidize all the organically bound sulfur. A third outlet temperature zone is optional.

Pyrolysis Tube Must be fabricated from quartz and constructed in such a way that a sample, which is vaporized completely in the inlet section, is swept into the pyrolysis zone by an inert gas where it mixes with oxygen and is burned. The inlet end of the tube shall hold a septum for syringe entry of the sample and side arms for the introduction of oxygen and inert gases. The center or pyrolysis section should be of sufficient volume to ensure complete pyrolysis of the sample.

Sampling Syringe A microliter syringe of 10-μL capacity capable of accurately delivering 1 to 10 μL of sample into the pyrolysis tube. Three-in. × 24-gauge needles are recommended to reach the inlet zone of the pyroloysis furnace.

Titration Cell Must contain a sensor-reference pair of electrodes to detect changes in triiodide ion concentration and a generator anode–cathode pair of electrodes to maintain constant triiodide ion concentration and an inlet for a gaseous sample

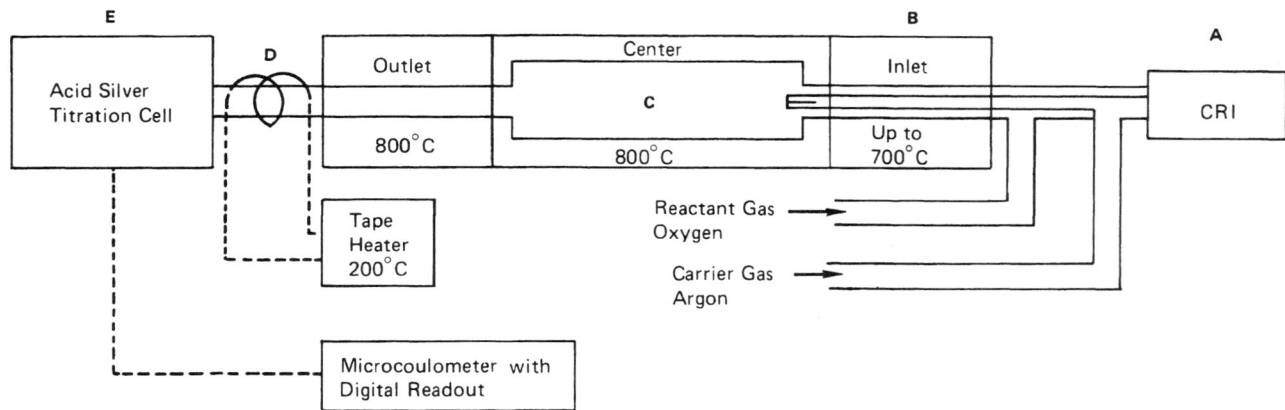

FIGURE 30 Microcoulometric Titrating System for the Determination of Sulfur in Hexanes.

from the pyrolysis tube. The sensor electrode shall be platinum foil and the reference electrode platinum wire in saturated triiodide half-cell. The generator anode and cathode half-cell shall also be platinum. The titration cell shall be placed on a suitable magnetic stirrer.

Preparation of Apparatus Carefully insert the quartz pyrolysis tube into the furnace, attach the tin scrubber, and connect the reactant and carrier-gas lines. Add the *Cell Electrolyte Solution* (see below) to the titration cell, and flush the cell several times. Maintain an electrolyte level of 3.2 to 6.6 mm above the platinum electrodes. Place the titration cell on a magnetic stirrer, and connect the cell inlet to the tin scrubber outlet. Position the platinum-foil electrodes (mounted on the movable cell head) so that the gas-inlet flow is parallel to the electrodes with the generator anode adjacent to the generator cathode. Assemble and connect the coulometer in accordance with the manufacturer's instructions. Double-wrap the adaptor containing the tin scrubber with heating tape and turn the heating tape on. Adjust the flow of the gases, the pyrolysis furnace temperature, the titration cell, and the coulometer to the desired operating conditions. Typical operating conditions are as follows:

Reactant gas flow (oxygen), cm³/min	200
Carrier-gas flow (Ar, He), cm³/min	40
Furnace temperature, °C	
Inlet zone	700 (maximum)
Pyrolysis zone	800–1000
Outlet zone	800 (maximum)
Tin-Scrubber temperature, °C	200
Titration cell	Stirrer speed set to produce slight vortex
Coulometer	
Bias voltage, mV	160
Gain	50
Constant Rate Injector, µL/s	0.25

The tin scrubber must be conditioned to sulfur, nitrogen, and chlorine before quantitative analysis can be achieved. A solution containing 10 mg/kg butyl sulfide, 100 mg/kg pyridine, and 200 mg/kg chlorobenzene in isooctane has proven an effective conditioning agent. With a fresh scrubber installed and heated, two 30-µL samples of this conditioning agent injected at a flow rate of 0.5 µL/s produces a steadily increasing response, with final conditioning indicated by a constant reading from the offset during the second injection.

Reagents

Argon or Helium (Argon preferred) High-purity grade, used as the carrier gas. Two-stage gas regulators must be used.

Cell Electrolyte Solution Dissolve 0.5 g of potassium iodide and 0.6 g of sodium azide in 500 mL of high-purity water, add 5 mL of glacial acetic acid and dilute to 1 L. Store in a dark bottle or in a dark place and prepare fresh at least every 3 months.

Oxygen High-purity grade, used as the reactant gas.

Iodine Resublimed, 20-mesh or less, for saturated reference electrode.

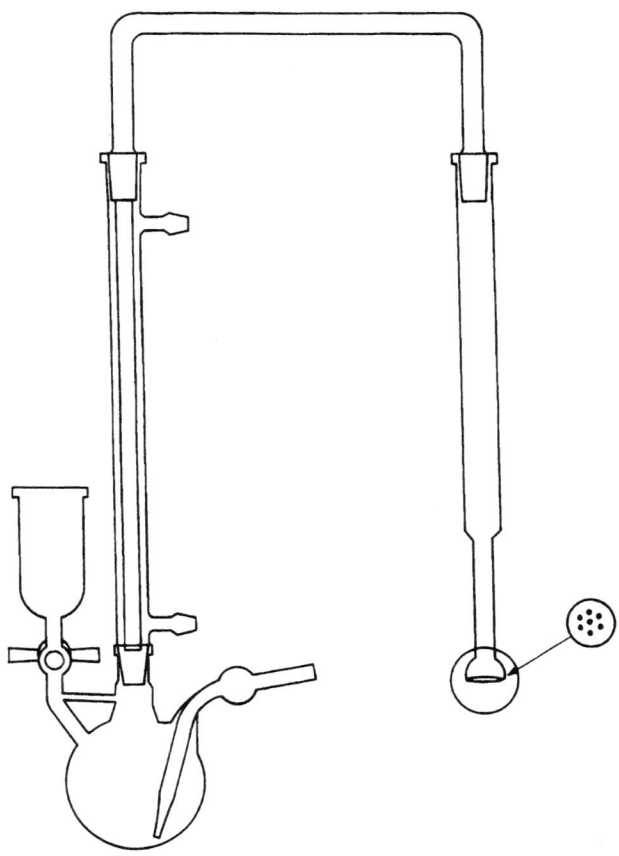

FIGURE 31 Raney Nickel Reduction Apparatus.

Sulfur Standard (approximately 100 mg/kg) Weigh accurately 0.1569 g of *n*-butyl sulfide into a tared 500-mL volumetric flask. Dilute to the mark with isooctane, and reweigh. Calculate the sulfur concentration (S), in percent, by the formula

$$S = W_b/W_s \times 2.192 \times 10^5,$$

in which W_b is the weight of *n*-butyl sulfide and W_s is the weight of the solution.

Calibration Prepare a calibration standard (approximately 5 mg/kg) by pipetting 5 mL of *Sulfur Standard* into a 10-mL volumetric flask and diluting to volume with isooctane. Fill and clamp the syringe onto the constant rate injector, push the sliding carriage forward to penetrate the septum with the needle, and zero the meter in case of long-term drift in the automatic baseline zero circuitry. Switch S_1 automatically starts the stepper-motor syringe drive and initiates the analysis cycle. At 2.5 min (before the 3-min meter hold point) set the digital meter with the scan potentiometer to correspond to the sulfur content of the known standard to the nearest 0.01 mg/kg. At the 3-min point, the number displayed on the meter stops, the plunger drive block is retracted to its original position, as preset by switch S_2, and a baseline re-equilibration period equal to the injection period elapses before a ready light and a beeper indicate that a new sample may be injected. Repeat the *Calibration* step a total of at least four times.

Procedure Rinse the syringe several times with sample; then fill it, clamp it onto the constant-rate injector, push the sliding carriage forward to penetrate the septum with the needle, and zero the meter. Turn on switch S_1 to start the stepper-motor syringe drive automatically and initiate the analysis cycle. After the 3-min hold point, the number displayed on the meter corresponds to the sulfur content of the injected sample.

APPENDIX IV: CHEWING GUM BASE

BOUND STYRENE

Abbé-Type Refractometer Use an instrument, having fourth decimal place accuracy, that can be placed in a nearly horizontal position for measurement of the refractive index of solids. An Amici-type compensating prism for achromatization is necessary unless a sodium vapor lamp is used as a light source.

Ethanol–Toluene Azeotrope Mix 70 volumes of ethanol or of formula 2B ethanol with 30 volumes of toluene, reflux for 4 h over calcium oxide, and then distill, discarding the first and last portions and collecting only that portion distilling within a range of 1°.

Note: Refluxing and distilling are not necessary if anhydrous 2B ethanol or absolute grain alcohol is used.

Sample Preparation Sheet out a sample of the polymer to a thickness of 0.5 mm, and cut the sheeted sample into strips approximately 13 mm wide and 25 mm long. Fasten one strip to each leg of a "spider," consisting of a 13-mm square of sheet aluminum or stainless steel having a Nichrome wire leg about 38 mm long attached to each corner. Place the spider and strips in a 400-mL flask containing 60 mL of *Ethanol–Toluene Azeotrope*, positioning the spider so that each sample strip is in contact on all sides with the solvent. Extract for 1 h at a temperature at which the solvent boils gently, then replace the solvent with another 60-mL portion of *Ethanol–Toluene Azeotrope*, and extract for an additional hour. Remove the spider and sample strips from the flask, and dry them at 100° to constant weight in a vacuum oven at a pressure of about 10 mm of Hg.

Caution: The samples must be extracted and dried thoroughly, but overheating, which would cause plastication, must be avoided.

Remove the extracted and dried strips from the spider, and press the strips between aluminum foil (0.025 to 0.08 mm thick, having good tear strength) at 100° for 3 to 10 min (preferably not more than 5 min), using any suitable apparatus to produce a uniform thickness not exceeding 0.5 mm. If the pressing is done between flat platens, without a cavity, use a force between about 500 and 1500 lb, increasing the applied force proportionally if several strips are pressed at one time; if cavity pressing plates are used, close the platens without applying pressure and preheat for 1 min, then apply a force of about 11 tons for 3 min, remove the specimens from the press, and allow them to cool.

Procedure Cut the pressed sample in half with sharp scissors, and peel off one piece of the foil. Cut off a strip about 6 mm wide and 12 mm long with the scissors so that one of the narrower ends is freshly cut.

Check the adjustment of the refractometer by means of a glass test plate pressed firmly against the prism, using a drop of α-bromo-naphthalene as the contact liquid. The small light source should be collimated. The best readings are obtained with the glass test piece if the light is diffused through crumpled tissue paper. After this adjustment, clean the prism well with lens paper moistened with alcohol. The refractive index of the glass test piece and of the test specimen must be measured at a known constant temperature, preferably 25°.

Place the test sample on the prism with the cut edge toward the light source approximately where the edge of the glass test piece was positioned. Remove the tissue paper from the light source, press the specimen firmly on the prism, and wait at least 1 min for the sample to attain temperature equilibrium. The upper prism may be closed lightly on the specimen if adequate light can still be focused on the end of the specimen. Unless the specimen is very thin, however, this operation can damage the prism or its mounting. Adjust the compensating prism until a sharp dividing line between light and dark fields with minimum color is obtained. Test the contact between the specimen and the prism by pressing the test specimen against the prism: There should be no change in the position of the boundary line during this test. Move the hand control from the light into the dark field until the boundary line just reaches the cross hairs, and make at least three readings. If the readings differ by more than 0.0001 refractive index unit, make further readings. Repeat the process of obtaining readings with another portion of the sample having a freshly cut edge, and average the mean values of the two sets of readings thus obtained. If the two mean values do not differ by more than 0.0002 refractive index unit, report the average as the results of the calculations. If necessary, correct the refractive index measurements to 25° using the formula

$$n_{25} = n_t + 0.00037(t - 25),$$

in which n_{25} is the refractive index at 25°, and n_t is the refractive index at the temperature, t, of measurement.

Calculate the percentage of bound styrene in emulsion-polymerized samples by the formula

$$23.50 + 1164(n_{25} - 1.53456) - 3497(n_{25} - 1.53456)^2.$$

Calculate the percentage of bound styrene in solution-polymerized samples by the formula

$$(1212.1212)(n_{25}) - 1838.3636.$$

Alternatively, the percentage of bound styrene may be determined by reference to suitable tables.

MOLECULAR WEIGHT

Polyethylene
Sample Solutions Dissolve 1 g of the sample, accurately weighed, in 95 mL of tetrahydronaphthalene, filter into a 100-mL volumetric flask, dilute to volume with the solvent, and mix (*Solution 1*). Transfer 50.0 mL of *Solution 1* into a tared dish, evaporate on a steam bath for about 1 h, and then complete the evaporation to dryness by heating in a vacuum oven at 70° for 2 h or to constant weight. Calculate the concentration, C_1, in g per 100 mL, of *Solution 1*. Prepare *Solutions 2* and *3*, respectively, by diluting 5.0-mL and 10.0-mL portions of *Solution 1* to 50.0 mL with the solvent, and then calculate the concentration of each (C_2 and C_3, respectively).

Procedure Determine the flow time, in s, of the solvent (t_0) and of the three *Sample Solutions* (t_1, t_2, and t_3, respectively) in a Cannon-Fenske viscometer immersed in a constant-temperature bath maintained at 130°. Calculate the specific viscosity, η_{sp}, of each *Sample Solution* by the formula

$$(t/t_0) - 1,$$

and then calculate the reduced viscosity of each by the formula

$$\eta_{sp}/C.$$

Plot the reduced viscosity of each solution against concentration, and extrapolate to zero concentration to obtain the intrinsic viscosity, $[\eta]$. Finally, calculate the molecular weight of the polyethylene by the formula

$$([\eta]/K)^{1/a},$$

in which K is 5.1×10^{-4}, and a is 0.725.

Polyisobutylene (Flory Method)
Sample Solutions Dissolve 1 g of the sample, accurately weighed, in 95 mL of diisobutylene, filter into a 100-mL volumetric flask, dilute to volume with the solvent, and mix (*Solution 1*). Transfer 50.0 mL of *Solution 1* into a tared dish, evaporate on a steam bath for about 1 h, and then complete the evaporation to dryness by heating in a vacuum oven at 70° for 2 h or to constant weight. Calculate the concentration, C_1, in g per 100 mL, of *Solution 1*. Prepare *Solutions 2* and *3*, respectively, by diluting 5.0-mL and 10.0-mL portions of *Solution 1* to 50.0 mL with the solvent, and then calculate the concentration of each (C_2 and C_3, respectively).

Procedure Determine the flow time, in s, of the solvent (t_0) and of the three *Sample Solutions* (t_1, t_2, and t_3, respectively) in a Cannon-Fenske viscometer immersed in a constant-temperature bath maintained at 20°. Calculate the specific viscosity, η_{sp}, of each *Sample Solution* by the formula

$$(t/t_0) - 1,$$

and then calculate the reduced viscosity of each by the formula

$$\eta_{sp}/C.$$

Plot the reduced viscosity of each solution against concentration, and extrapolate to zero concentration to obtain the intrinsic viscosity, $[\eta]$. Finally, calculate the molecular weight of the polyisobutylene by the formula

$$([\eta]/K)^{1/a},$$

in which K is 3.60×10^{-4}, and a is 0.64.

Polyvinyl Acetate
Sample Solutions Dissolve 1 g of the sample, accurately weighed, in 95 mL of acetone, filter into a 100-mL volumetric flask, dilute to volume with the solvent, and mix (*Solution 1*). Transfer 50.0 mL of *Solution 1* into a tared dish, evaporate on a steam bath for about 1 h, and then complete the evaporation to dryness by heating in a vacuum oven at 70° for 2 h or to constant weight. Calculate the concentration, C_1, in g per 100 mL, of *Solution 1*. Prepare *Solutions 2* and *3*, respectively, by diluting 5.0-mL and 10.0-mL portions of *Solution 1* to 50.0 mL with the solvent, and then calculate the concentration of each (C_2 and C_3, respectively).

Procedure Determine the flow time, in s, of the solvent (t_0) and of the three *Sample Solutions* (t_1, t_2, and t_3, respectively) in a Cannon-Fenske viscometer immersed in a constant-temperature bath maintained at 25°. Calculate the specific viscosity, η_{sp}, of each *Sample Solution* by the formula

$$(t/t_0) - 1,$$

and then calculate the reduced viscosity of each by the formula

$$\eta_{sp}/C.$$

Plot the reduced viscosity of each solution against concentration, and extrapolate to zero concentration to obtain the intrinsic viscosity, $[\eta]$. Finally, calculate the molecular weight of the polyvinyl acetate by the formula

$$([\eta]/K)^{1/a},$$

in which K is 1.88×10^{-4}, and a is 0.69.

QUINONES

Standard Preparations Transfer 25.0 mg of hydroquinone into a 100-mL volumetric flask, dissolve in water, dilute to volume with water, and mix. Transfer 1.0-, 2.0-, 3.0-, 4.0-, and 6.0-mL aliquots of this solution into a series of 100-mL volumetric flasks, dilute each to volume with water, and mix. Transfer 2.0 mL of each of these solutions and 3.0 mL of water into a series of 25-mL graduates, add 0.5 mL of 0.1 N sodium carbonate to each, and continue as directed under *Sample Preparations*, beginning with "... shake immediately, then add 1.0 mL of 15% sulfuric acid...."

Sample Preparations Place 30 g of freshly coagulated and washed sample in a two-necked, 250-mL flask, add 100 mL of water, and heat at 66° for 2 h.

Caution: Do not boil.

Cool to room temperature, and transfer 5.0 mL of the extract into a 25-mL glass-stoppered graduate. Transfer 5.0 mL of water into a second graduate to serve as the blank. To each graduate add 1.0 mL of 15% sulfuric acid. To the graduate containing the sample extract add 0.5 mL of 0.1 N sodium carbonate, shake immediately, and then add 1.0 mL of 15% sulfuric acid.

Note: The elapsed time for this operation should not exceed 15 s.

To each graduate add 1.0 mL of 2,4-dinitrophenylhydrazine solution (dissolve 100 mg of 2,4-dinitrophenylhydrazine in 50 mL of carbonyl-free methanol, add 4 mL of hydrochloric acid, and dilute to 100 mL with water), stopper, and heat at 70° in a water bath for 1 h. Cool to room temperature, then add to each graduate 13 mL of water and 5.0 mL of benzene, stopper, and shake vigorously. Allow the phases to separate, and pipet 2.0 mL of the benzene layer from each graduate into corresponding test tubes containing 10 mL of a 1 in 100 solution of diethanolamine in pyridine. Shake each tube, and allow the color to develop for 10 min.

Procedure Determine the absorbance of the *Sample Preparation* in a 1-cm cell at 620 nm, with a suitable spectrophotometer, against the reagent blank. Determine the absorbance of each *Standard Preparation* in the same manner. Prepare a *Standard Curve* by plotting absorbance of each *Standard Preparation* against µg of quinone. From the *Standard Curve*, read the quantity, in µg, of quinones (as benzoquinone) in the *Sample Preparation*, and record the value thus obtained as Q. Calculate the quantity of quinones (as benzoquinone), in ppm, in the sample by the formula

$$20Q/W,$$

in which W is the weight, in g, of the sample taken.

RESIDUAL STYRENE

Standard Preparation Place 25 mL of carbon disulfide in a 100-mL volumetric flask, cap with a serum stopper, and tare the flask to the nearest 0.1 mg. Inject, with 50-µg syringes, 15 µL each of styrene and of alpha-methylstyrene (AMS), reweighing after each addition to obtain the weight of each solution injected. Record the weight, in mg, of styrene as w_1 and that of AMS as w_2. Dilute to volume with carbon disulfide, and mix. Pipet 2 mL of this solution into a second 100-mL volumetric flask, dilute to volume with carbon disulfide, and mix. Finally, pipet 25 mL of the diluted solution into a third 100-mL volumetric flask, dilute to volume with carbon disulfide, and mix.

AMS–Solvent Solution Place 25 mL of carbon disulfide in a 100-mL volumetric flask, cap with a serum stopper, and tare the flask to the nearest 0.1 mg. Using a 50-µL syringe, inject 15 µL of AMS, and reweigh to obtain the weight of AMS injected. Dilute to volume with carbon disulfide, and mix. Pipet 2 mL of this solution into a second 100-mL volumetric flask, dilute to volume with carbon disulfide, and mix. Finally, pipet 25 mL of the diluted solution into a third 100-mL volumetric flask, dilute to volume, and mix. Calculate the weight, in g, of AMS in each mL of the final solution, and record the result as w' (approximately 7.5×10^{-7}).

Sample Preparation

Latex Samples Add, with agitation, 100 mL of the latex to a mixture consisting of 15 mL of glacial acetic acid and 10 g of sodium chloride in 500 mL of hot water. Coagulation starts almost immediately. When coagulation is complete, collect the coagulum on a coarse filter or cheesecloth, and wash with 1000 mL of a hot solution prepared with 5.6 g of sodium hydroxide and 1000 mL of water. Wash with hot water until the wash water is free of alkali, then cut the coagulum into small pieces, and dry at 105° for 4 h. Continue as directed under *Solid Samples*, beginning with "Transfer 1.5 g, accurately weighed...."

Solid Samples Cut a piece approximately 2 in. × 3 in. × 5 in. from the corner of a polymer bale, and pass it through a cold mill, set at least $\frac{1}{4}$ in. open, four times, reversing the sample on each pass. Cut the sample into two pieces at least 1 in. from the edge to expose clean polymer, and then dice or cut into small strips approximately 2 g of the clean polymer. Transfer 1.5 g, accurately weighed, into a 4-oz bottle fitted with a polyethylene cap, add 25.0 mL of the *AMS–Solvent Solution*, cap tightly, and agitate on a mechanical shaker until the polymer dissolves.

Note: Some polymers tend to swell and form viscous cements instead of dissolving cleanly. If this occurs, add 5- to 10-mL increments of carbon disulfide to obtain a mobile slurry, and in the next step increase the volume of methanol by a proportional amount.

Add 25 mL of methanol, cap the bottle, and shake vigorously on the shaker for 30 min. After the contents have settled, decant

10 mL of the coagulant serum into a 1-oz bottle, add 10 mL of water, and stopper with a serum cap. Shake vigorously for 1 min, then turn the bottle upside down, and allow the layers to separate. Withdraw by syringe 1 to 2 mL of the lower (carbon disulfide) layer, and transfer it into a 10-dram vial filled $\frac{1}{4}$ in. with anhydrous sodium sulfate. Seal with a polyethylene cap, shake to mix, and allow to settle.

Procedure Inject a 10-µL portion of the *Sample Preparation* into a suitable gas chromatograph in which the detector is the hydrogen flame-ionization type and the column is 3-m × 5-mm stainless steel tubing packed with 25% Ucon 50 HB 2000 on 60/80-mesh acid-washed DMCS Chromosorb W, or with equivalent packing materials. The carrier is nitrogen or helium flowing at 40 mL/min. The injection port temperature is 240°; the column temperature, 170° isothermal; and the detector temperature, 250°. Adjust the sensitivity of the instrument to give as large a signal as possible for styrene and AMS as is consistent with an acceptable background level. Measure the styrene and AMS peaks by any convenient method, recording the area of the styrene peak as A_1 and that of the AMS peak as A_2.

In the same manner, inject a 10-µL portion of the *Standard Preparation* into the chromatograph, obtain the chromatogram, and record the area of the styrene peak as a_1 and that of the AMS peak as a_2. Calculate the styrene factor, F, by the formula

$$(w_1/w_2) \times (a_2/a_1).$$

Calculate the content of residual styrene in the sample taken, in ppm, by the formula

$$(A_1/A_2) \times F \times 25 \times (w'/W) \times 10^6,$$

in which W is the weight, in g, of the sample taken.

SAMPLE SOLUTION FOR ARSENIC TEST

Transfer a 1-g sample, accurately weighed, into a Kjeldahl flask, rest the open end of the flask in a Kjeldahl fume bulb attached to a water aspirator, add 5 mL of sulfuric acid and 4 mL of 30% hydrogen peroxide, and digest over a small flame. (See *Caution* statement under *Arsenic Test*, Appendix IIIB.) Continue adding the peroxide in 2-mL portions, allowing the reaction to subside between additions, until all organic matter is destroyed, fumes of sulfuric acid are copiously evolved, and the solution becomes colorless. *Maintain oxidizing conditions at all times during the digestion by adding peroxide whenever the mixture turns brown or darkens.* (The amount of peroxide required to completely digest the samples will vary, but as much as 200 mL may be required in some cases, depending upon the nature of the material.) Cool, cautiously add 10 mL of water, again evaporate to strong fuming, and cool. Transfer the solution into an arsine generator flask, wash the Kjeldahl flask and bulb with water, adding the washings to the generator flask, and dilute to 35 mL with water.

SAMPLE SOLUTION FOR LEAD LIMIT TEST

Transfer a 3.3-g sample, accurately weighed, into a porcelain dish or casserole, heat on a hot plate until completely charred, then heat in a muffle furnace at 480° for 8 h or overnight, and cool. Cautiously add 5 mL of nitric acid, evaporate to dryness on a hot plate, then heat again in the muffle furnace at 480° for exactly 15 min, and cool. Extract the ash with two 10-mL portions of water, filtering each extract into a separator. Leach any insoluble material on the filter with 6 mL of *Ammonium Citrate Solution*, 2 mL of *Hydroxylamine Hydrochloride Solution*, and 5 mL of water (see *Lead Limit Test*, Appendix IIIB, for preparation of these solutions), adding the filtered washings to the separator. Continue as directed under *Procedure* in the *Lead Limit Test*, Appendix IIIB, beginning with "To the separator add 2 drops of phenol red TS...."

TOTAL UNSATURATION

Transfer about 500 mg of the sample, accurately weighed, into a 500-mL iodine flask containing 100 mL of filtered 1,1,1-trichloroethane. Stopper the flask securely, cover to protect the mixture from light, and shake in a mechanical shaker until the sample is completely dissolved (about 1.5 h). Remove the flask from the shaker and the cover from the flask, and then add 5 mL of a 1 in 5 solution of trichloroacetic acid in 1,1,1-trichloroethane, 25.0 mL of 0.1 N iodine in 1,1,1-trichloroethane, and 25 mL of a 3 in 10 solution of mercuric acetate in glacial acetic acid. Stopper the flask, mix the contents thoroughly by shaking, and allow to stand in a dark place for exactly 30 min. Add 50 mL of potassium iodide TS, immediately titrate with 0.1 N sodium thiosulfate, using starch TS as the indicator, and record the volume, in mL, required as S. Perform a blank determination (see *General Provisions*), and record the volume of 0.1 N sodium thiosulfate as B. Calculate the percent total unsaturation by the formula

$$(B - S) \times (N/W) \times 1.87,$$

in which N is the exact normality of the 0.1 N sodium thiosulfate solution; W is the weight, in g, of the sample taken; and 1.87 is an equivalence factor for isobutylene, which is the chief contributor of olefin linkages in the copolymer.

Note: The percentage thus obtained will be high if the copolymer contains an antioxidant that reacts as an unsaturated compound in this test procedure; in this case, the weight percent of the antioxidant must be determined by appropriate means and subtracted from the percent total unsaturation as determined herein.

APPENDIX V: ENZYME ASSAYS

A list of the enzymes covered by the general monograph on *Enzyme Preparations*, is shown in the accompanying table. Also incorporated in the table are the trivial names by which each is commonly known, as well as the systematic names of the major components or of the enzyme for which the preparation is standardized, in accordance with the *Recommendations (1992) of the Nomenclature Committee of the International Union of Biochemistry and Molecular Biology on the Nomenclature and Classification of Enzymes.*

Enzyme Preparations Used in Food Processing

TRIVIAL NAME	CLASSIFICATION	SOURCE	SYSTEMATIC NAMES (IUB)[a]	IUB NO.[a]
α-Amylase	carbohydrase	(1) *Aspergillus niger* var. (2) *Aspergillus oryzae* var. (3) *Rhizopus oryzae* var. (4) *Bacillus subtilis* var. (5) barley malt (6) *Bacillus licheniformis* var. (7) *Bacillus stearothermophilus* (8) *Bacillus subtilis** 　d-*Bacillus megaterium* (9) *Bacillus subtilis** 　d-*Bacillus stearothermophilus* (10) *Bacillus licheniformis** 　d-*Bacillus stearothermophilus*	1,4-α-D-glucan glucanohydrolase	3.2.1.1
β-Amylase	carbohydrase	(1) barley malt (2) barley	1,4-α-D-glucan maltohydrolase	3.2.1.2
Bromelain	protease	pineapples: *Ananas comosus* *Ananas bracteatus* (L)	none	3.4.22.32 3.4.22.33
Catalase	oxidoreductase	(1) *Aspergillus niger* var. (2) bovine liver (3) *Micrococcus lysodeikticus*	hydrogen peroxide: hydrogen peroxide oxidoreductase	1.11.1.6
Cellulase	carbohydrase	(1) *Aspergillus niger* var. (2) *Trichoderma longibrachiatum* (formerly *reesei*)	Endo-1,4-(1,3;1,4)-β-D-glucan 4-glucanohydrolase	3.2.1.4
Chymosin	protease	(1) *Aspergillus niger* var. *awamori** 　d-calf prochymosin gene (2) *Escherichia coli* K-12* 　d-calf prochymosin gene (3) *Kluyveromyces marxianus** 　d-calf prochymosin gene	cleaves a single bond in *kappa* casein	3.4.23.4
Chymotrypsin	protease	bovine or porcine pancreatic extract	none	3.4.21.1
Ficin	protease	figs: *Ficus* sp.	none	3.4.22.3
α-Galactosidase	carbohydrase	(1) *Mortierella vinacea* var. *raffinoseutilizer* (2) *Aspergillus niger*	α-D-galactoside galactohydrolase	3.2.1.22
β-Glucanase	carbohydrase	(1) *Aspergillus niger* var. (2) *Bacillus subtilis* var. (3) *Trichoderma longibrachiatum*	1,3-(1,3;1,4)-β-D-glucan 3(4)-glucanohydrolase	3.2.1.6
Glucoamylase (Amyloglucosidase)	carbohydrase	(1) *Aspergillus niger* var. (2) *Aspergillus oryzae* var. (3) *Rhizopus oryzae* var. (4) *Rhizopus niveus*	1,4-α-D-glucan glucohydrolase	3.2.1.3

TRIVIAL NAME	CLASSIFICATION	SOURCE	SYSTEMATIC NAMES (IUB)[a]	IUB NO.[a]
Glucose Isomerase	isomerase	(1) *Actinoplanes missouriensis* (2) *Bacillus coagulans* (3) *Streptomyces olivaceus* (4) *Streptomyces olivochromogenus* (5) *Streptomyces rubiginosus* (6) *Streptomyces murinus* (7) *Microbacterium arborescens*	D-xylose ketoisomerase	5.3.1.5
Glucose Oxidase	oxidoreductase	*Aspergillus niger* var.	β-D-glucose: oxygen 1-oxidoreductase	1.1.3.4
β-D-Glucosidase	carbohydrase	(1) *Aspergillus niger* var. (2) *Trichoderma longibrachiatum*	β-D-glucoside glucohydrolase	3.2.1.21
Hemicellulase	carbohydrase	(1) *Aspergillus niger* var. (2) *Trichoderma longibrachiatum*	(1) α-L-arabinofuranoside arabinofuranohydrolase (2) 1,4-β-D-mannan mannanohydrolase (3) 1,3-β-D-xylan xylanohydrolase (4) 1,5-α-L-arabinan arabinanohydrolase	3.2.1.55 3.2.1.78 3.2.1.32 3.2.1.99
Invertase	carbohydrase	*Saccharomyces* sp. (*Kluyveromyces*)	β-D-fructofuranoside fructohydrolase	3.2.1.26
Lactase	carbohydrase	(1) *Aspergillus niger* var. (2) *Aspergillus oryzae* var. (3) *Saccharomyces* sp. (4) *Candida pseudotropicalis* (5) *Kluyveromyces marxianus* var. *lactis*	β-D-galactoside galactohydrolase	3.2.1.23
Lipase	lipase	(1) edible forestomach tissue of calves, kids, and lambs (2) animal pancreatic tissues (3) *Aspergillus oryzae* var. (4) *Aspergillus niger* var. (5) *Rhizomucor miehei* (6) *Candida rugosa*	(1) carboxylic-ester hydrolase (2) triacylglycerol acylhydrolase	3.1.1.1 3.1.1.3
Maltogenic Amylase	carbohydrase	*Bacillus subtilis** d-*Bacillus stearothermophilus*	1,4-α-D-glucan α-maltohydrolase	3.2.1.133
Pancreatin	mixed carbohydrase, protease, and lipase	bovine and porcine pancreatic tissue	(1) 1,4-α-D-glucan glucanohydrolase (2) triacylglycerol acylhydrolase (3) protease	3.2.1.1 3.1.1.3 3.4.21.4
Papain	protease	papaya: *Carica papaya* (L)	none	3.4.22.2 3.4.22.6
Pectinase[b]	carbohydrase	(1) *Aspergillus niger* var. (2) *Rhizopus oryzae* var.	(1) poly(1,4-α-D-galacturonide) glycanohydrolase (2) pectin pectylhydrolase (3) poly(1,4-α-D-galacturonide)lyase (4) poly(methoxyl-L-galacturonide)lyase	3.2.1.15 3.1.1.11 4.2.2.2 4.2.2.10

TRIVIAL NAME	CLASSIFICATION	SOURCE	SYSTEMATIC NAMES (IUB)[a]	IUB NO.[a]
Pepsin	protease	porcine or other animal stomach tissue	none	3.4.23.1 3.4.23.2
Phospholipase A_2	lipase	animal pancreatic tissue	phosphatidylcholine 2-acylhydrolase	3.1.1.4
Phytase	phosphatase	*Aspergillus niger* var.	(1) *myo*-inositol-hexakisphosphate-3-phosphohydrolase	3.1.3.8
			(2) orthophosphoric-mono ester phosphohydrolase	3.1.3.2
Protease (general)	protease	(1) *Aspergillus niger* var. (2) *Aspergillus oryzae* var (3) *Bacillus subtilis* var. (4) *Bacillus licheniformis* var.	none	3.4.23.18 3.4.24.28 3.4.21.62
Pullulanase	carbohydrase	*Bacillus acidopullulyticus*	α-Dextrin-6-glucanohydrolase	3.2.1.41
Rennet	protease	(1) fourth stomach of ruminant animals (2) *Endothia parasitica* (3) *Rhizomucor miehei* (4) *Rhizomucor pusillus* (Lindt) (5) *Bacillus cereus*	none	3.4.23.1 3.4.23.4 3.4.23.22 3.4.23.23
Trypsin	protease	animal pancreas	none	3.4.21.4

[a] *Enzyme Nomenclature: Recommendations (1992) of the Nomenclature Committee of the International Union of Biochemistry*, Academic Press, New York, 1992.
[b] Usually a mixture of pectin depolymerase, pectin methylesterase, pectin lyase, and pectate lyase.
*The asterisk indicates a genetically modified organism. The donor organism is listed after "d-."

The following procedures are provided for application as necessary in determining compliance with the vendor's declared representations for enzyme activity. For all of the procedures use filtered, ultra-high purity water with a resistivity of 16 to 18 megohms.

ACID PHOSPHATASE ACTIVITY

Application and Principle This procedure is used to determine acid phosphatase activity in preparations derived from *Aspergillus niger* var. The test is based on the enzymatic hydrolysis of *p*-nitrophenyl phosphate, followed by the measurement of the released inorganic phosphate.

Reagents and Solutions
Glycine Buffer (0.2 M, pH 2.5) Dissolve 15.014 g of glycine (Merck, Catalog No. 4201) in about 800 mL of water. Adjust the pH to 2.5 with 1 M hydrochloric acid (consumption should be about 80 mL), and dilute to 1000 mL with water.

Substrate (30 mM) Dissolve 1.114 g of *p*-nitrophenyl phosphate (Boehringer, Catalog No. 738 352) in glycine buffer, and adjust the volume to 100 mL with the buffer. Prepare fresh substrate solution daily.

TCA Solution Dissolve 15 g of trichloroacetic acid in water, and dilute to 100 mL.

Ascorbic Acid Solution Dissolve 10 g of ascorbic acid in water, and dilute to 100 mL. Store under refrigeration. The solution is stable for 7 days.

Ammonium Molybdate Solution Dissolve 2.5 g of ammonium molybdate [$(NH_4)_6MoO_{24}\cdot 4H_2O$] (Merck, Catalog No. 1182) in water, and dilute to 100 mL.

1 M Sulfuric Acid Stir 55.6 mL of concentrated sulfuric acid (H_2SO_4) (Merck, Catalog No. 731) into about 800 mL of water. Allow to cool, and make up to 1000 mL with water.

Reagent C Mix 3 volumes of *1 M Sulfuric Acid* with 1 volume of *Ammonium Molybdate Solution*, then add 1 volume of *Ascorbic Acid Solution*, and mix well. Prepare fresh daily.

Standard Phosphate Solution Prepare a 9.0-mM phosphate stock solution. Dissolve and dilute 612.4 mg of potassium dihydrogen phosphate (KH_2PO_4) (dried in desiccator with silica) to 500 mL with water in a volumetric flask. Make the following dilutions in water from the stock solution, and use these as standards.

Dilution	Phosphorus Concentration (nmol/mL)	Acid Phosphatase Activity (HFU/mL)
1:100	90	2400
1:200	45	1200
1:400	22.5	600

Pipet 4.0 mL of each dilution into two test tubes. Also pipet 4.0 mL of water into one tube (reagent blank). Add 4.0 mL of *Reagent C*, and mix. Incubate at 50° for 20 min, and cool to room temperature. Measure the absorbances at 820 nm against that of reagent blank. Prepare a standard curve by plotting the absorbances against acid phosphatase activity [HFU (acid phosphatase unit)/mL]. Construct a new standard curve with each series of assays.

Test Preparation Prepare a solution of the enzyme preparation in the *Glycine Buffer* so that 1 mL will contain between 600 and 2400 HFU/mL.

Procedure Pipet 1.9 mL of *Substrate* in two test tubes. Add 2.0 mL of *TCA Solution* to one of the tubes (blank), and mix. Put the tubes without *TCA Solution* in a water bath at 37° and let them equilibrate for 5 min. While using a stopwatch, start the hydrolysis by adding sequentially at proper intervals 0.1 mL of *Test Preparation* to each tube, and mix. After exactly 15 min of incubation, stop the reaction by adding 2.0 mL of *TCA Solution* to each tube. Mix, and cool to room temperature. Add 0.1 mL of *Test Preparation* to the reagent blank tube (kept at room temperature), and mix. If precipitate occurs, separate it by centrifugation for 10 min at 2000 g.

Pipet 0.4 mL of each sample after hydrolysis into separate test tubes. Add 3.6 mL of water to each tube. Add 4.0 mL of *Reagent C*, and mix. Incubate at 50° for 20 min, and cool to room temperature. Determine the absorbance against that of reagent blank at 820 nm.

Calculation One acid phosphatase unit (HFU) is the amount of enzyme that liberates, under the conditions of the assay, inorganic phosphate from *p*-nitrophenyl phosphate at the rate of 1 nmol/min.

Subtract the blank absorbance from the sample absorbance (the difference should be between 0.100 and 1.000). Determine the acid phosphatase activity (HFU/mL) from the standard curve, and multiply by the dilution factor. For the activity of solid samples, use the following formula:

$$\text{HFU/g} = (\text{HFU/mL} \times f)/g,$$

in which f is the dilution factor and g is the weight, in g, of the sample.

ALPHA-AMYLASE ACTIVITY (NONBACTERIAL)

Application and Principle This procedure is used to determine the α-amylase activity of enzyme preparations derived from *Aspergillus niger* var.; *Aspergillus oryzae* var.; *Rhizopus oryzae* var.; and barley malt. The assay is based on the time required to obtain a standard degree of hydrolysis of a starch solution at 30° ± 0.1°. The degree of hydrolysis is determined by comparing the iodine color of the hydrolysate with that of a standard.

Apparatus

Reference Color Standard Use a special Alpha-Amylase Color Disk (Orbeco Analytical Systems, 185 Marine Street, Farmingdale, NY 11735, Catalog No. 620-S5). Alternatively, prepare a color standard by dissolving 25.0 g of cobaltous chloride ($CoCl_2 \cdot 6H_2O$) and 3.84 g of potassium dichromate in 100 mL of 0.01 N hydrochloric acid. This standard is stable indefinitely when stored in a stoppered bottle or comparator tube.

Comparator Use either the standard Hellige comparator (Orbeco, Catalog No. 607) or the pocket comparator with prism attachment (Orbeco, Catalog No. 605AHT). The comparator should be illuminated with a 100-W frosted lamp placed 6 in. from the rear opal glass of the comparator and mounted so that direct rays from the lamp do not shine into the operator's eyes.

Comparator Tubes Use the precision-bored square tubes with a 13-mm viewing depth that are supplied with the Hellige comparator. Suitable tubes are also available from other apparatus suppliers (e.g., Thomas Scientific).

Reagents and Solutions

Buffer Solution (pH 4.8) Dissolve 164 g of anhydrous sodium acetate in about 500 mL of water, add 120 mL of glacial acetic acid, and adjust the pH to 4.8 with glacial acetic acid. Dilute to 1000 mL with water, and mix.

β-Amylase Solution Dissolve into 5 mL of water a quantity of β-amylase, free from α-amylase activity (Sigma Chemical Co., Catalog No. A7005), equivalent to 250 mg of β-amylase with 2000° diastatic power.

Special Starch Use starch designated as "Starch (Lintner) Soluble," (Baker Analyzed Reagent, Catalog No. 4010). Before using new batches, test them in parallel with previous lots known to be satisfactory. Variations of more than ± 3° diastatic power in the averages of a series of parallel tests indicate an unsuitable batch.

Buffered Substrate Solution Disperse 10.0 g (dry-weight basis) of *Special Starch* in 100 mL of cold water, and slowly pour the mixture into 300 mL of boiling water. Boil and stir for 1 to 2 min, then cool, and add 25 mL of *Buffer Solution*, followed by all of the *β-Amylase Solution*. Quantitatively transfer the mixture into a 500-mL volumetric flask with the aid of water saturated with toluene, dilute to volume with the same solvent, and mix. Store the solution at 30° ± 2° for not less than 18 h nor more than 72 h before use. (This solution is also known as "buffered limit dextrin substrate.")

Stock Iodine Solution Dissolve 5.5 g of iodine and 11.0 g of potassium iodide in about 200 mL of water, dilute to 250 mL with water, and mix. Store in a dark bottle, and make a fresh solution every 30 days.

Dilute Iodine Solution Dissolve 20 g of potassium iodide in 300 mL of water, and add 2.0 mL of *Stock Iodine Solution*. Quantitatively transfer the mixture into a 500-mL volumetric flask, dilute to volume with water, and mix. Prepare daily.

Sample Preparation Prepare a solution of the sample so that 5 mL of the final dilution will give an endpoint between 10 and 30 min under the conditions of the assay.

For barley malt, finely grind 25 g of the sample in a Miag-Seck mill, (Buhler-Miag, Inc., P.O. Box 9497, Minneapolis, MN 55440). Quantitatively transfer the powder into a 1000-mL Erlenmeyer flask, add 500 mL of a 0.5% solution of sodium chloride, and allow the infusion to stand for 2.5 h at $30° \pm 0.2°$, agitating the contents by gently rotating the flask at 20-min intervals.

Caution: Do not mix the infusion by inverting the flask. The quantity of the grist left adhering to the inner walls of the flask as a result of agitation must be as small as possible.

Filter the infusion through a 32-cm fluted filter of Whatman No. 1, or equivalent, paper on a 20-cm funnel, returning the first 50 mL of filtrate to the filter. Collect the filtrate until 3 h have elapsed from the time the sodium chloride solution and the sample were first mixed. Pipet 20.0 mL of the filtered infusion into a 100-mL volumetric flask, dilute to volume with the 0.5% sodium chloride solution, and mix.

Procedure Pipet 5.0 mL of *Dilute Iodine Solution* into a series of 13- × 100-mm test tubes, and place them in a water bath maintained at $30° \pm 0.1°$, allowing 20 tubes for each assay.

Pipet 20.0 mL of the *Buffered Substrate Solution*, previously heated in the water bath for 20 min, into a 50-mL Erlenmeyer flask, and add 5.0 mL of 0.5% sodium chloride solution, also previously heated in the water bath for 20 min. Place the flask in the water bath.

At zero time, rapidly pipet 5.0 mL of the *Sample Preparation* into the equilibrated substrate, mix immediately by swirling, stopper the flask, and place it back in the water bath. After 10 min, transfer 1.0 mL of the reaction mixture from the 50-mL flask into one of the test tubes containing the *Dilute Iodine Solution*, shake the tube, then pour its contents into a *Comparator Tube*, and immediately compare with the *Reference Color Standard* in the *Comparator*, using a tube of water behind the color disk.

Note: Be certain that the pipet tip does not touch the iodine solution; carryback of iodine to the hydrolyzing mixture will interfere with enzyme action and will affect the results of the determination.

In the same manner, repeat the transfer and comparison procedure at accurately timed intervals until the α-amylase color is reached, at which time record the elapsed time. In cases where two comparisons 30 s apart show that one is darker and the other lighter than the *Reference Color Standard*, record the endpoint to the nearest quarter min. Shake out the 13-mm *Comparator Tube* between successive readings. Minimize slight differences in color discrimination between operators by using a prism attachment and by maintaining a 6- to 10-in. distance between the *Comparator* and the operator's eye.

Calculation One α-amylase dextrinizing unit (DU) is defined as the quantity of α-amylase that will dextrinize soluble starch in the presence of an excess of β-amylase at the rate of 1 g/h at 30°.

Calculate the α-amylase dextrinizing units in the sample as follows:

$$DU \text{ (solution)} = 24/(W \times T),$$

and

$$DU \text{ (dry basis)} = DU \text{ (solution)} \times 100/(100 - M),$$

in which W is the weight, in g, of the enzyme sample added to the incubation mixture in the 5-mL aliquot of the *Sample Preparation* used; T is the elapsed dextrinizing time, in min; 24 is the product of the weight of the starch substrate (0.4 g) and 60 min; and M is the percent moisture in the sample, determined by suitable means.

ALPHA-AMYLASE ACTIVITY (BACTERIAL)

Application and Principle This procedure is used to determine the α-amylase activity, expressed as bacterial amylase units (BAU), of enzyme preparations derived from *Bacillus subtilis* var., *Bacillus licheniformis* var., and *Bacillus stearothermophilus*. It is not applicable to products that contain β-amylase. The assay is based on the time required to obtain a standard degree of hydrolysis of a starch solution at $30° \pm 0.1°$. The degree of hydrolysis is determined by comparing the iodine color of the hydrolysate with that of a standard.

Apparatus Use the *Reference Color Standard*, the *Comparator*, and the *Comparator Tubes* as described under *Alpha-Amylase Activity (Nonbacterial)*, described in this Appendix, but use either daylight or daylight-type fluorescent lamps as the light source for the *Comparator*. (Incandescent lamps give slightly lower results.)

Reagents and Solutions

pH 6.6 Buffer Dissolve 9.1 g of potassium dihydrogen phosphate (KH_2PO_4) in sufficient water to make 1000 mL (*Solution A*). Dissolve 9.5 g of dibasic sodium phosphate (Na_2HPO_4) in sufficient water to make 1000 mL (*Solution B*). Add 400 mL of *Solution A* to 600 mL of *Solution B*, mix, and adjust the pH to 6.6, if necessary, by the addition of *Solution A* or *Solution B* as required.

Dilute Iodine Solution Prepare as directed under *Alpha-Amylase Activity (Nonbacterial)*.

Special Starch Use the material described under *Alpha-Amylase Activity (Nonbacterial)*.

Starch Substrate Solution Disperse 10.0 g (dry-weight basis) of *Special Starch* in 100 mL of cold water, and slowly pour the mixture into 300 mL of boiling water. Boil and stir for 1 to 2 min, and then cool while continuously stirring. Quantitatively transfer the mixture into a 500-mL volumetric flask with the aid of water, add 10 mL of *pH 6.6 Buffer*, dilute to volume with water, and mix.

Sample Preparation Prepare a solution of the sample so that 10 mL of the final dilution will give an endpoint between 15 and 35 min under the conditions of the assay.

Procedure Pipet 5.0 mL of *Dilute Iodine Solution* into a series of 13- × 100-mm test tubes, and place them in a water bath maintained at 30° ± 0.1°, allowing 20 tubes for each assay.

Pipet 20.0 mL of the *Starch Substrate Solution* into a 50-mL Erlenmeyer flask, stopper, and allow to equilibrate for 20 min in the water bath at 30°.

At zero time, rapidly pipet 10.0 mL of the *Sample Preparation* into the equilibrated mixture, and continue as directed in the *Procedure* under *Alpha-Amylase Activity (Nonbacterial)*, beginning with ". . . mix immediately by swirling, stopper the flask. . . ."

Calculation One bacterial amylase unit (BAU) is defined as that quantity of enzyme that will dextrinize starch at the rate of 1 mg/min under the specified test conditions.

Calculate the α-amylase activity of the sample, expressed as BAU, by the formula

$$BAU/g = 40F/T,$$

in which 40 is a factor (400/10) derived from the 400 mg of starch (20 mL of a 2% solution) and the 10-mL aliquot of *Sample Preparation* used; F is the dilution factor (total dilution volume/sample weight, in g); and T is the dextrinizing time, in min.

CATALASE ACTIVITY

Application and Principle This procedure is used to determine the catalase activity, expressed as Baker Units, of preparations derived from *Aspergillus niger* var., bovine liver, or *Micrococcus lysodeikticus*. The assay is an exhaustion method based on the breakdown of hydrogen peroxide by catalase and the simultaneous breakdown of the catalase by the peroxide under controlled conditions.

Reagents and Solutions

Ammonium Molybdate Solution (1%) Dissolve 1.0 g of ammonium molybdate [$(NH_4)_6MoO_{24}\cdot 4H_2O$] (Merck, Catalog No. 1182) in water, and dilute to 100 mL.

0.250 N Sodium Thiosulfate Dissolve 62.5 g of sodium thiosulfate ($Na_2S_2O_3\cdot 5H_2O$) in 750 mL of recently boiled and cooled water, add 3.0 mL of 0.2 N sodium hydroxide as a stabilizer, dilute to 1000 mL with water, and mix. Standardize as directed for *0.1 N Sodium Thiosulfate*, (see *Solutions and Indicators*), and if necessary, adjust to exactly 0.250 N.

Peroxide Substrate Solution Dissolve 25.0 g of anhydrous dibasic sodium phosphate (Na_2HPO_4), or 70.8 g of $Na_2HPO_4\cdot 12H_2O$, in about 1500 mL of water, and adjust to pH 7.0 ± 0.1 with 85% phosphoric acid. Cautiously add 100 mL of 30% hydrogen peroxide, dilute to 2000 mL in a graduate, and mix. Store in a clean amber bottle, loosely stoppered. The solution is stable for more than 1 week if kept at 5° in a full container. (With freshly prepared substrate, the blank will require about 16 mL of *0.250 N Sodium Thiosulfate*. If the blank requires fewer than 14 mL, the substrate solution is unsuitable and should be prepared fresh again. The sample titration must be between 50% and 80% of that required for the blank.)

Procedure Pipet an aliquot of not more than 1.0 mL of the sample, previously diluted to contain approximately 3.5 Baker Units of catalase, into a 200-mL beaker. Rapidly add 100 mL of *Peroxide Substrate Solution*, previously adjusted to 25°, and stir immediately for 5 to 10 s. Cover the beaker, and incubate at 25° ± 1° until the reaction is completed. Stir vigorously for 5 s, and then pipet 4.0 mL from the beaker into a 50-mL Erlenmeyer flask. Add 5 mL of 2 N sulfuric acid to the flask, mix, then add 5.0 mL of 40% potassium iodide solution, freshly prepared, and 1 drop of *Ammonium Molybdate Solution* (1%), and mix. While continuing to mix, titrate rapidly to a colorless endpoint with *0.250 N Sodium Thiosulfate*, recording the volume, in mL, required as S. Perform a blank determination with 4.0 mL of *Peroxide Substrate Solution*, and record the volume required, in mL, as B.

Note: When preparations derived from beef liver are tested, the reaction is complete within 30 min. Preparations derived from *Aspergillus* and other sources may require up to 1 h. In assaying an enzyme of unknown origin, run a titration after 30 min and then at 10-min intervals thereafter. The reaction is complete when two consecutive titrations are the same.

Calculation One Baker Unit is defined as the amount of catalase that will decompose 264 mg of hydrogen peroxide under the conditions of the assay.

Calculate the activity of the sample by the formula

$$\text{Baker Units/g or mL} = 0.4(B - S) \times (1/C),$$

in which C is the mL of aliquot of original enzyme preparation added to each 100 mL of *Peroxide Substrate Solution*, or when 1 mL of diluted enzyme is used, C is the dilution factor; B is the volume, in mL, as defined above; and S is the mL of *0.250 N Sodium Thiosulfate*, as defined above.

CELLULASE ACTIVITY

Application and Principle This assay is based on the enzymatic hydrolysis of the interior β-1,4-glucosidic bonds of a defined carboxymethyl cellulose substrate at pH 4.5 and at 40°. The corresponding reduction in substrate viscosity is determined with a calibrated viscometer.

Apparatus

Calibrated Viscometer Use a size 100 Calibrated Cannon-Fenske Type Viscometer, or its equivalent (Scientific Products, Catalog No. P2885-100).

Constant-Temperature Glass Water Bath (40° ± 0.1°) Use a constant-temperature glass water bath, or its equivalent (Scientific Products, Catalog No. W3520-10).

Stopwatches Use two stopwatches—*Stopwatch No. 1*, calibrated in $1/10$ min for determining the reaction time (T_r), and

Stopwatch No. 2, calibrated in $\frac{1}{5}$ s for determining the efflux time (T_t).

Waring Blender Use a two-speed Waring blender, or its equivalent (Scientific Products, Catalog No. 58350-1).

Reagents and Solutions

Acetic Acid Solution (2 N) While agitating a 1-L beaker containing 800 mL of water, carefully add 116 mL of glacial acetic acid. Cool to room temperature. Quantitatively transfer the solution to a 1-L volumetric flask, and dilute to volume with water.

Sodium Acetate Solution (2 N) Dissolve 272.16 g of sodium acetate trihydrate in approximately 800 mL of water contained in a 1-L beaker. Quantitatively transfer to a 1-L volumetric flask, and dilute to volume with water.

Acetic Acid Solution (0.4 N) Transfer 200 mL of *Acetic Acid Solution (2 N)* into a 1-L volumetric flask, and dilute to volume with water.

Sodium Acetate Solution (0.4 N) Transfer 200 mL of *Sodium Acetate Solution (2 N)* into a 1-L volumetric flask, and dilute to volume with water.

Acetate Buffer (pH 4.5) Using a standardized pH meter, add *Sodium Acetate Solution (0.4 N)* with continuous agitation to 400 mL of *Acetic Acid Solution (0.4 N)* in a suitable flask until the pH is 4.5 ± 0.05.

Sodium Carboxymethyl Cellulose Use sodium carboxymethyl cellulose (Hercules, Inc., CMC Type 7HP).

Sodium Carboxymethyl Cellulose Substrate (0.2% w/v) Transfer 200 mL of water into the bowl of the *Waring Blender*. With the blender on low speed, slowly disperse 1.0 g (moisture-free basis) of the *Sodium Carboxymethyl Cellulose* into the bowl, being careful not to splash out any of the liquid. Using a rubber policeman, wash down the sides of the glass bowl with water. Place the top on the bowl and blend at high speed for 1 min. Quantitatively transfer to a 500-mL volumetric flask, and dilute to volume with water. Filter the substrate through gauze before use.

Sample Preparation Prepare an enzyme solution so that 1 mL of the final dilution will produce a relative fluidity change between 0.18 and 0.22 in 5 min under the conditions of the assay. Weigh the enzyme, and quantitatively transfer it to a glass mortar. Triturate with water and quantitatively transfer the mixture to an appropriate volumetric flask. Dilute to volume with water, and filter the enzyme solution through Whatman No. 1 filter paper before use.

Procedure Place the *Calibrated Viscometer* in the 40° ± 0.1° water bath in an exactly vertical position. Use only a scrupulously clean viscometer. (To clean the viscometer, draw a large volume of detergent solution followed by water through the viscometer by using an aspirator with a rubber tube connected to the narrow arm of the viscometer.)

Pipet 20 mL of filtered *Sodium Carboxymethyl Cellulose Substrate* and 4 mL of *Acetate Buffer* into a 50-mL Erlenmeyer flask. Allow at least two flasks for each enzyme sample and one flask for a substrate blank. Stopper the flasks, and equilibrate them in the water bath for 15 min.

At zero time, pipet 1 mL of the enzyme solution into the equilibrated substrate. Start *Stopwatch No. 1*, and mix the solution thoroughly. Immediately pipet 10 mL of the reaction mixture into the wide arm of the viscometer.

After approximately 2 min, apply suction with a rubber tube connected to the narrow arm of the viscometer, drawing the reaction mixture above the upper mark into the driving fluid head. Measure the efflux time by allowing the reaction mixture to freely flow down past the upper mark. As the meniscus of the reaction mixture falls past the upper mark, start *Stopwatch No. 2*. At the same time, record the reaction time, in min, from *Stopwatch No. 1* (T_R). As the meniscus of the reaction mixture falls past the lower mark, record the time, in s, from *Stopwatch No. 2* (T_T).

Repeat the final step until a total of four determinations is obtained over a reaction time (T_R) of not more than 15 min.

Prepare a substrate blank by pipetting 1 mL of water into 24 mL of buffered substrate. Pipet 10 mL of the reaction mixture into the wide arm of the viscometer. Determine the time (T_S) in s required for the meniscus to fall between the two marks. Use an average of five determinations for (T_S).

Prepare a water blank by pipetting 10 mL of equilibrated water into the wide arm of the viscometer. Determine the time (T_W) in s required for the meniscus to fall between the two marks. Use an average of five determinations for (T_W).

Calculations One Cellulase Unit (CU) is defined as the amount of activity that will produce a relative fluidity change of 1 in 5 min in a defined carboxymethyl cellulose substrate under the conditions of the assay.

Calculate the relative fluidities (F_R) and the (T_N) values for each of the four efflux times (T_T) and reaction times (T_R) as follows:

$$F_R = (T_S - T_W)/(T_T - T_W),$$

$$T_N = \frac{1}{2}(T_T/60 \text{ s/min}) + T_R = (T_T/120) + T_R,$$

in which F_R is the relative fluidity for each reaction time; T_S is the average efflux time, in s, for the substrate blank; T_W is the average efflux time, in s, for the water blank; T_T is the efflux time, in s, of reaction mixture; T_R is the elapsed time, in min, from zero time, that is, the time from addition of the enzyme solution to the buffered substrate until the beginning of the measurement of efflux time (T_T); and T_N is the reaction time, in min, (T_R) plus one-half of the efflux time (T_T) converted to min.

Plot the four relative fluidities (F_R) as the ordinate against the four reaction times (T_N) as the abscissa. A straight line should be obtained. The slope of this line corresponds to the relative fluidity change per min and is proportional to the enzyme concentration. The slope of the best line through a series of experimental points is a better criterion of enzyme activity than is a single relative fluidity value. From the graph, determine the F_R values at 10 and 5 min. They should have a difference in fluidity of not more than 0.22 nor less than 0.18. Calculate the activity of the enzyme unknown as follows:

$$\text{CU/g} = [1000(F_{R10} - F_{R5})]/W,$$

in which F_{R5} is the relative fluidity at 5 min of reaction time; F_{R10} is the relative fluidity at 10 min of reaction time; 1000 is the mg/g; and W is the weight, in mg, of enzyme added to the reaction mixture in a 1-mL aliquot of enzyme solution.

CHYMOTRYPSIN ACTIVITY

Application and Principle This procedure is used to determine chymotrypsin activity in chymotrypsin preparations derived from purified extracts of porcine or bovine pancreas.

Reagents and Solutions

0.15 M Phosphate Buffer (pH 7.0) Dissolve 4.54 g of monobasic potassium phosphate in water, and dilute to 500 mL. Dissolve 4.73 g of anhydrous dibasic sodium phosphate in water, and dilute to 500 mL. Mix 38.0 mL of the monobasic potassium phosphate solution with 61.1 mL of the dibasic sodium phosphate solution. Adjust the pH of the mixture to 7.0 by the dropwise addition of the dibasic sodium phosphate solution, if necessary.

Substrate Solution Dissolve 23.7 mg of N-acetyl-L-tyrosine ethyl ester in about 50 mL of the *0.15 M Phosphate Buffer* with warming. When the solution has cooled, dilute to 100.0 mL with the *0.15 M Phosphate Buffer*.

Sample Preparation Dissolve a sufficient amount of sample, accurately weighed, in 0.001 N hydrochloric acid to produce a solution containing between 12 and 16 USP Chymotrypsin Units per mL. This solution should cause a change in absorbance between 0.008 and 0.012 in a 30-s interval.

Procedure Conduct the assay in a suitable spectrophotometer equipped to maintain a temperature of 24° ± 0.1° in the cell compartment. Determine the temperature before and after measuring the absorbance to ensure that the temperature does not change more than 0.5° during the assay. Pipet 0.2 mL of the 0.001 N hydrochloric acid and 3.0 mL of the *Substrate Solution* into a 1-cm cell. Place the cell in the spectrophotometer, and adjust the instrument so that the absorbance will read 0.200 at 237 nm. Pipet 0.2 mL of the *Sample Preparation* into a second cell, add 3.0 mL of the *Substrate Solution*, and place the cell in the spectrophotometer. Begin timing the reaction from the addition of the *Substrate Solution*. Read the absorbance at 30-s intervals for at least 5 min. Repeat the procedure at least once. If the rate of change fails to remain constant for at least 3 min, repeat the test, and if necessary, use a lower sample concentration. The duplicate determinations at the same sample concentration should match the first determination in rate of absorbance change.

Calculations One USP Chymotrypsin Unit is defined as the activity causing a change in absorbance at the rate 0.0075/min under the conditions of the assay. Determine the average absorbance change per min using only those values within the 3-min portion of the curve where the rate of change is constant. Plot a curve of absorbance against time.

Calculate the number of Chymotrypsin Units per mg by the formula

$$(A_2 - A_1)/(0.0075TW),$$

in which A_2 is the straight-line initial absorbance reading; A_1 is the straight-line final absorbance reading; T is the elapsed time, in min; and W is the weight, in mg, of the sample in the volume of solution used to determine the absorbance.

DIASTASE ACTIVITY (DIASTATIC POWER)

Application and Principle This procedure is used to determine the amylase activity of barley malt and other enzyme preparations. The assay is based on a 30-min hydrolysis of a starch substrate at pH 4.6 and 20°. The reducing sugar groups produced on hydrolysis are measured in a titrimetric procedure using alkaline ferricyanide.

Apparatus

Mill Use a laboratory mill of the type Miag-Seck, for fine grinding of malt (Buhler Miag, Inc.).

Reagents and Solutions

Acetate Buffer Solution Dissolve 68 g of sodium acetate ($NaC_2H_3O_2 \cdot 3H_2O$) in 500 mL of 1 N acetic acid in a 1000-mL volumetric flask, dilute to volume with water, and mix.

Special Starch Use the material described under *Alpha-Amylase Activity (Nonbacterial)*.

Starch Substrate Solution Disperse 20.0 g (dry-weight basis) of *Special Starch* in 50 mL of water, mix to a fine paste, and pour slowly into 750 mL of boiling water. Boil and stir for 2 min, cool, add 20 mL of *Acetate Buffer Solution*, and mix. Quantitatively transfer into a 1000-mL volumetric flask, dilute to volume with water, and mix.

Acetic Acid–Potassium Chloride–Zinc Sulfate Solution (A-P-Z) Dissolve 70 g of potassium chloride and 20 g of zinc sulfate ($ZnSO_4 \cdot 7H_2O$) in 700 mL of water in a 1000-mL volumetric flask, add 200 mL of glacial acetic acid, dilute to volume with water, and mix.

Alkaline Ferricyanide Solution (0.05 N) Dissolve 16.5 g of potassium ferricyanide [$K_3Fe(CN)_6$] and 22 g of anhydrous sodium carbonate in 800 mL of water in a 1000-mL volumetric flask, dilute to volume with water, and mix.

Potassium Iodide Solution Dissolve 50 g of potassium iodide in 50 mL of water in a 100-mL volumetric flask, dilute to volume with water, and mix. Add 2 drops of 50% sodium hydroxide solution, and mix. The solution should be colorless.

Sample Preparation

Malt Samples Grind 30 g of the sample to a fine powder in a *Maig-Seck Mill*. Accurately weigh 25 g of the powder, and transfer it into a 1000-mL Erlenmeyer flask. Add 500 mL of a 0.5% sodium chloride solution, and allow the infusion to stand for 2.5 h at 20° ± 0.2°, agitating the contents by gently rotating the flask at 20-min intervals.

Note: Do not mix the infusion by inverting the flask. The quantity of grist left adhering to the inner walls of the flask as a result of agitation must be as small as possible. Gently swirl the contents of the flask without splashing them against the walls to mix sufficiently.

Filter the infusion through a 32-cm fluted filter of Whatman No. 1, or equivalent, paper on a 20-cm funnel, returning the first 50 mL of filtrate to the filter. Place a watch glass over the funnel, and use a suitable cover around the stem and over the receiver to reduce evaporation losses during filtration. Collect the filtrate until 30 min of filtration time have elapsed. Pipet 20.0 mL of the filtrate into a 100-mL volumetric flask, dilute to volume with 0.5% sodium chloride solution, and mix.

Other Enzyme Preparations Prepare a solution so that 10 mL of the final dilution will give a diastatic power (DP) value between 2° and 150°.

Procedure Pipet 10.0 mL of the *Sample Preparation* into a 250-mL volumetric flask, and at zero time, add 200 mL of *Starch Substrate Solution*, previously equilibrated for 30 min in a water bath maintained at 20° ± 0.2°. Start the stopwatch at zero time.

Place the mixture in the water bath at 20°, and allow it to cool for exactly 30 min, then add 20.0 mL of 0.5 N sodium hydroxide, dilute to volume with water, and mix.

Prepare a blank by adding 20.0 mL of 0.5 N sodium hydroxide to a 250-mL volumetric flask, followed by 10.0 mL of the *Sample Preparation*. Swirl to mix, add 200 mL of *Starch Substrate Solution*, dilute to volume with water, and mix.

Pipet 5.0 mL of the sample digestion mixture into a 125-mL Erlenmeyer flask, add 10.0 mL of *Alkaline Ferricyanide Solution*, and swirl to mix. Heat the flask for exactly 20 min in a boiling water bath, and then cool to room temperature. Add 25 mL of *A-P-Z Solution*, followed by 1 mL of *Potassium Iodide Solution*, and swirl to mix. Titrate with 0.05 N sodium thiosulfate to the complete disappearance of the blue color, recording the volume, in mL, of 0.05 N sodium thiosulfate required as S.

Treat the blank solution in the same manner as described for the sample, recording the volume, in mL, of 0.05 N sodium thiosulfate required as B.

Calculation One unit of diastase activity, expressed as degrees diastatic power (DP°), is defined as that amount of enzyme contained in 0.1 mL of a 5% solution of the sample enzyme preparation that will produce sufficient reducing sugars to reduce 5 mL of Fehling's solution when the sample is incubated with 100 mL of the substrate for 1 h at 20°.

Note: The definition of the unit does not correspond to the method of the determination.

Calculate the diastase activity, expressed as DP°, of the sample by the formulas

$$DP°, \text{ as-is basis} = (B - S) \times 23,$$

and

$$DP°, \text{ dry basis} = DP°, \text{ as-is basis} \times 100/(100 - M),$$

in which 23 is a factor, determined by collaborative study, required to convert to the units of the definition, and M is the percent moisture of the sample, determined by suitable means.

α-GALACTOSIDASE ACTIVITY

Application and Principle This procedure is used to determine α-galactosidase activity in enzyme preparations derived from *Aspergillus niger* var. The assay is based on a 15-min hydrolysis of *p*-nitrophenyl-α-D-galactopyranoside followed by spectrophotometric measurement of the liberated 4-nitrophenol.

Reagents and Solutions

Acetate Buffer Dissolve 11.55 mL of glacial acetic acid in water, and dilute to 1 L (*Solution A*). Dissolve 16.4 g of sodium acetate in water, and dilute to 1 L (*Solution B*). Mix 7.5 mL of *Solution A* and 42.5 mL of *Solution B*, and dilute to 200 mL. Adjust the pH of this solution to 5.5 with either *Solution A* or *B* as necessary.

Substrate Solution Dissolve 0.0383 g of *p*-nitrophenyl-α-D-galactopyranoside (Sigma Chemical Co., Catalog No. 877, or equivalent) in *Acetate Buffer*, and dilute to 100 mL.

Borax Buffer Dissolve 47.63 g of sodium borate decahydrate in warm water. Cool to room temperature. Add 20 mL of 4 N sodium hydroxide solution, adjust the pH of the solution to 9.7 with 4 N sodium hydroxide, and dilute to 2 L.

4-Nitrophenol Stock Solution Dissolve 0.0334 g of 4-nitrophenol (Aldrich Chemical Co., Catalog No. 24,132-6) in water, and dilute to 1 L. This solution contains 0.24 μmol/mL.

Preparation of Standards and Samples

Standards Prepare the following dilutions of *4-Nitrophenol Stock Solution* with water: 100 mL plus 50 mL of water (0.16 μmol/mL), 50 mL plus 100 mL of water (0.08 μmol/mL); and 25 mL plus 125 mL of water (0.04 μmol/mL). Transfer 2.0 mL of the substrate solution to each of five separate test tubes. Add 1 mL of the *4-Nitrophenol Stock Solution* to the first tube, 1.0 mL of each dilution to the next three tubes, and 1.0 mL of water to the fifth tube. Add 5.0 mL of *Borax Buffer* to each tube, and mix.

Samples Prepare a solution that contains between 0.001 and 0.003 galactosidase units of activity per mL.

Procedure Equilibrate the *Substrate Solution* in a water bath at 37° ± 0.2° for at least 15 min. For active samples, transfer 1.0 mL of each sample to separate test tubes and equilibrate in the 37° ± 0.2° water bath. At zero time, add 2.0 mL of *Substrate Solution*, mix, and return to the water bath. After exactly 15.0 min, add 5.0 mL of *Borax Buffer* to each tube, mix, and remove from the water bath.

For sample blanks, transfer in sequence 1.0 mL of each sample to separate test tubes, add 5.0 mL of *Borax Buffer*, and mix. Add 2.0 mL of *Substrate Solution* to each tube, and mix.

Measure the absorbance of each standard sample and blank at 405 nm versus that of water. Determine the absorbances of all solutions within 30 min of completing the tests.

Calculations One galactosidase activity unit (GalU) is defined as the quantity of the enzyme that will liberate *p*-nitrophenol at the rate of 1 μmol/min under the conditions of the assay.

Calculate the millimolar extinction of 4-nitrophenol standards using the following formula:

$$\epsilon = A_N/C,$$

in which A_N is the absorbance of the p-nitrophenol standards at 405 nm and C is the concentration of 4-nitrophenol, in μmol/mL.

The averaged millimolar extinction coefficient should be approximately 18.3.

$$\text{GalU/g} = [(A_S - A_B) \times F]/(\epsilon \times T \times M),$$

in which A_S is the sample absorbance; A_B is the blank absorbance; F is the appropriate dilution factor; T is the reaction time, in min; M is the sample weight, in g; and ϵ is the millimolar extinction coefficient.

β-GLUCANASE ACTIVITY

Application and Principle This procedure is used to determine β-glucanase activity of enzyme preparations derived from *Aspergillus niger* var. and *Bacillus subtilis* var. The assay is based on a 15-min hydrolysis of lichenin substrate at 40° and at pH 6.5. The increase in reducing power due to liberated reducing groups is measured by the neocuproine method.

Reagents and Solutions

Phosphate Buffer Dissolve 13.6 g of monobasic potassium phosphate in about 1900 mL of water, add 70% sodium hydroxide solution until the pH is 6.5 ± 0.05, then transfer the solution into a 2000-mL volumetric flask, dilute to volume with water, and mix.

Neocuproine Solution A Dissolve 40.0 g of anhydrous sodium carbonate, 16.0 g of glycine, and 450 mg of cupric sulfate pentahydrate in about 600 mL of water. Transfer the solution into a 1000-mL volumetric flask, dilute to volume with water, and mix.

Neocuproine Solution B Dissolve 600 mg of neocuproine hydrochloride in about 400 mL of water, transfer the solution into a 500-mL volumetric flask, dilute to volume with water, and mix. Discard when a yellow color develops.

Lichenin Substrate Grind 150 mg of lichenin (Sigma Chemical Co., Catalog No. L-6133, or equivalent) to a fine powder in a mortar, and dissolve it in about 50 mL of water at about 85°. After solution is complete (20 to 30 min), add 90 mg of sodium borohydride and continue heating below the boiling point for 1 h. Add 15 g of Amberlite MB-3, or an equivalent ion-exchange resin, and stir continuously for 30 min. Filter with the aid of a vacuum through Whatman No. 1 filter paper, or equivalent, in a Buchner funnel, and wash the paper with about 20 mL of water. Add 680 mg of monobasic potassium phosphate to the filtrate, and refilter through a 0.22-μm Millipore filter pad, or equivalent. Wash the pad with 10 mL of water, and adjust the pH of the filtrate to 6.5 ± 0.05 with 1 N sodium hydroxide or 1 N hydrochloric acid. Transfer the filtrate into a 100-mL volumetric flask, dilute to volume with water, and mix. Store at 2° to 4° for not more than 3 days.

Glucose Standard Solution Dissolve 36.0 mg of anhydrous dextrose in *Phosphate Buffer* in a 1000-mL volumetric flask, dilute to volume with water, and mix.

Test Preparation Prepare a solution from the enzyme preparation sample so that 1 mL of the final dilution will contain between 0.01 and 0.02 β-glucanase units. Weigh the sample, transfer it into a volumetric flask of appropriate size, dilute to volume with *Phosphate Buffer*, and mix.

Procedure Pipet 2 mL of *Lichenin Substrate* into each of four separate test tubes graduated at 25 mL, and heat the tubes in a water bath at 40° for 10 to 15 min to equilibrate.

After equilibration, add 1 mL of *Phosphate Buffer* to tube 1 (substrate blank), 1 mL of *Glucose Standard Solution* to tube 2 (glucose standard), 4 mL of *Neocuproine Solution A* and 1 mL of the *Test Preparation* to tube 3 (enzyme blank), and 1 mL of the *Test Preparation* to tube 4 (sample). Prepare a fifth tube for the buffer blank, and add 3 mL of *Phosphate Buffer*.

Incubate the five tubes at 40° for exactly 15 min, and then add 4 mL of *Neocuproine Solution A* to tubes 1, 2, 4, and 5. Add 4 mL of *Neocuproine Solution B* to all five tubes, and cap each with a suitably sized glass marble.

Caution: Do not use rubber stoppers.

Heat the tubes in a vigorously boiling water bath for exactly 12 min to develop color, then cool to room temperature in cold water, and adjust the volume of each to 25 mL with water. Cap the tubes with Parafilm, or other suitable closure, and mix by inverting several times.

Determine the absorbance of each solution at 450 nm in 1-cm cells, with a suitable spectrophotometer, against the buffer blank in tube 5.

Calculation One β-glucanase unit (BGU) is defined as that quantity of enzyme that will liberate reducing sugar (as glucose equivalence) at a rate of 1 μmol/min under the conditions of the assay.

Calculate the activity of the enzyme preparation taken for analysis as follows:

$$\text{BGU} = [(A_4 - A_3) \times 36 \times 10^6]/\\ [(A_2 - A_1) \times 180 \times 15 \times \mu\text{g sample}],$$

in which A_4 is the absorbance of the sample (tube 4), A_3 is the absorbance of the enzyme blank (tube 3), A_2 is the absorbance of the glucose standard (tube 2), A_1 is the absorbance of the substrate blank (tube 1), 36 is the μg of glucose in the *Glucose Standard Solution*, 10^6 is the factor converting μg to g, 180 is the weight of 1 μmol of glucose, and 15 is the reaction time.

GLUCOAMYLASE ACTIVITY (AMYLOGLUCOSIDASE ACTIVITY)

Application and Principle This procedure is used to determine the glucoamylase activity of preparations derived from *Aspergillus niger* var., but it may be modified to determine preparations derived from *Aspergillus oryzae* var. and *Rhizopus oryzae* var. (as indicated by the variations in the text below). The sample hydrolyzes *p*-nitrophenyl-α-D-glucopyranoside (PNPG) to *p*-nitrophenol (PNP) and glucose at pH 4.3 and 50°.

Use the quantity of PNP liberated per unit of time to calculate the enzyme activity. Measure the PNP liberated against a quantity of a standard preparation of PNP by measuring the absorbance of the solutions at 400 nm after adjusting the pH of the reaction mixture to pH 8.0.

Note: Use a pH of 5.0 when testing preparations derived from *Aspergillus oryzae* var. or *Rhizopus oryzae* var.

Apparatus

Water Bath Use an open, circulating water bath with control accuracy of at least ± 0.1°.

Spectrophotometer Use a spectrophotometer suitable for measuring absorbances at 400 nm.

Cuvettes Use 10-mm light-path fused quartz.

Thermometer Use a partial immersion thermometer with a suitable range, graduated in $\frac{1}{10}$°.

Timer Use a solid-state timer, model 69240 (GCS Corporation, Precision Scientific Group), or equivalent, accurate to ± 0.01 min in 240 min.

Vortex Mixer Use a standard variable-speed mixer.

Reagents and Solutions

p-Nitrophenol Stock Solution (PNP) (0.001 M) Dissolve 139.11 mg of *p*-nitrophenol previously dried (60°, maximum 4 h) into water, and dilute to 1000 mL.

Caution: Avoid contact with skin. If contact occurs, wash the affected area with water. Work in a well-ventilated area.

Acetate Buffer Solution (0.1 M) Dissolve 4.4 g of sodium acetate trihydrate ($NaC_2H_3O_2 \cdot 3H_2O$) in approximately 800 mL of water, add 4.5 mL of acetic acid ($C_2H_4O_2$). Adjust to pH 4.5 ± .05 by adding either sodium acetate or glacial acetic acid as required. Dilute to 1 L.

Note: Use a pH of 5.0 when testing preparations derived from *Aspergillus oryzae* var. or *Rhizopus oryzae* var.

The pH optimum is 5.0 for *Aspergillus oryzae* var.- or *Rhizopus oryzae* var.-derived preparations.

Sodium Carbonate Solution (0.3 M) Dissolve 15.9 g of sodium carbonate (Na_2CO_3) in water, and dilute to 500 mL.

p-Nitrophenyl-α-D-glucopyranoside Solution (PNPG) Dissolve 100.0 mg of PNPG (Sigma Chemical Co., Catalog No. N1377) in acetate buffer and dilute to 100 mL.

Preparation of Standards and Samples

Standards Dilute three portions of *PNP Stock Solution* to produce standards for the standard curve. Add 3 mL of the *PNP Stock Solution* to 125 mL of *Sodium Carbonate Solution*, and dilute to 500 mL with water to produce the first standard, containing 0.006 μmol/mL. Add 2 mL of *PNP Stock Solution* to 25 mL of *Sodium Carbonate Solution*, and dilute to 100 mL with water to produce the second standard containing 0.02 μmol/mL. Add 5 mL of *PNP Stock Solutions* to 25 mL of *Sodium Carbonate Solution*, and dilute to 100 mL with water to produce the third standard, containing 0.05 μmol/mL.

Sample Solution Dilute 1.00 ± 0.01 g of sample in sufficient *Acetate Buffer Solution* to produce a solution that contains between 0.1 and 0.3 glucoamylase units of activity per mL.

Procedure Measure absorbances of each of the three *PNP Standard Solutions* to calculate the molar extinction coefficient. Equilibrate the *PNPG Solution* in a 50° water bath for at least 15 min. For active samples, transfer 2.0 mL of the *Sample Solution* to a test tube. Loosely stopper, and place the tube in the water bath to equilibrate for 5 min. At zero time, add 2.0 mL of *PNPG Solution*, and mix at moderate speed on a vortex mixer. Return the mixture to the water bath. Exactly 10.0 min later, add 3.0 mL of the *Sodium Carbonate Solution*, mix on the vortex, and remove from the water bath.

For sample blanks, transfer 2.0 mL of the *Sample Solution* and 3.0 mL of the *Sodium Carbonate Solution* into a test tube, and mix. Add 2.0 mL of *PNPG Solution*, and mix. Measure the absorbance of each sample and the blank versus water in a 10-mm cell.

Note: Determine the absorbance of the sample and blank solutions not more than 20 min after adding *Sodium Carbonate Solution*.

Calculations One unit of glucoamylase activity is defined as the amount of glucoamylase that will liberate 0.1 μmol/min of *p*-nitrophenol from the *PNPG Solution* under the conditions of the assay.

Calculate the millimolar extinction of the PNP standards using the following formula:

$$\epsilon = A_N/C,$$

in which A_N is the absorbance of the *p*-nitrophenol standard, at 400 nm, and C is concentration, in μmol/mL, of *p*-nitrophenol.

The averaged millimolar extinction coefficient ϵ should be approximately 18.2.

$$\text{Glucoamylase } \mu/g = [(A_S - A_B) \times 7 \times F]/\epsilon \times 10 \times 0.10 \times W \times 2,$$

in which A_S is the sample absorbance; A_B is the blank absorbance; F is the appropriate dilution factor; W is the weight of sample, in g; 7 is the final volume of the test solutions; 10 is the reaction time, in min; 0.10 is the amount of PNP liberated, in μmol/min/unit of enzyme; 2 is the sample aliquot, in mL; and ϵ is the millimolar extinction coefficient.

GLUCOSE ISOMERASE ACTIVITY

Note: Glucose isomerase activity of the commercial enzyme is usually determined on the enzyme that has been immobilized by binding with a polymer matrix or other suitable material. The following method is designed for use with such preparations.

Application and Principle Use this procedure to determine glucose isomerase preparations derived from *Actinoplanes missouriensis*, *Bacillus coagulans*, *Microbacterium arborescens*, *Streptomyces murinus*, *Streptomyces olivaceus*, *Streptomyces olivochromogenes*, and *Streptomyces rubiginosus*. It is based on measurement of the rate of conversion of glucose to fructose in a packed-bed reactor. The procedure as outlined approximates an initial velocity assay method. Specific conditions are glucose concentration, 45% w/w; pH (inlet), measured at room temperature in the 7.0 to 8.5 range, as specified; temperature, 60.0°; and magnesium concentration, 4×10^{-3} M.

The optimum conditions for enzymes from different microbial sources and methods of preparation may vary; therefore, if the manufacturer recommends different pH conditions, buffering systems, or methods of sample preparation, use such variations in the instructions given in the text.

Apparatus

Column Assembly and Apparatus (**Note**: Make all connections with inert tubing, glass, or plastic as appropriate.) The column assembly is shown in Fig. 32. Use a 2.5- × 40-cm glass column provided with a coarse, sintered-glass bottom and a water jacket connected to a constant-temperature water bath, maintained at 60.0°, by means of a circulating pump. Connect the top of the column to a variable-speed peristaltic pump having a maximum flow rate of 800 mL/h. The diameter of the tubing with which the peristaltic pump is fitted should permit variation of the pumping volume from 60 to 150 mL/h. Connect the outlet of the column with a collecting vessel.

Reagents and Solutions

Glucose Substrate Dissolve 539 g of anhydrous glucose and 1.0 g of magnesium sulfate (MgSO$_4$.7H$_2$O) in 700 mL of water or the manufacturer's recommended buffer, previously heated to 50° to 60°. Cool the solution to room temperature, and adjust the pH as specified by the enzyme manufacturer. Transfer the solution to a 1000-mL volumetric flask, dilute to volume with water or the specified buffer, and mix. Transfer to a vacuum flask, and de-aerate for 30 min.

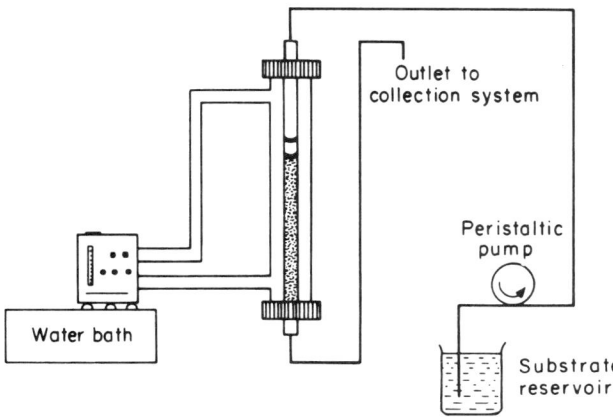

FIGURE 32 Column Assembly for Assay of Immobilized Glucose Isomerase.

Magnesium Sulfate Solution Dissolve 1.0 g of magnesium sulfate (MgSO$_4$.7H$_2$O) in 700 mL of water. Adjust the pH to 7.5 to 8.0 as specified by the manufacturer, using 1 N sodium hydroxide, dilute to 1000 mL with water, and mix.

Sample Preparation Transfer to a 500-mL vacuum flask an amount of the sample, accurately weighed in g or measured in mL, as appropriate, sufficient to obtain 2000 to 8000 glucose isomerase units (GI$_c$U). Add 200 mL of *Glucose Substrate*, stir gently for 15 s, and repeat the stirring every 5 min for 40 min. De-aerate by vacuum for 30 min.

Column Preparation Quantitatively transfer the *Sample Preparation* to the column with the aid of *Magnesium Sulfate Solution* as necessary. Allow the enzyme granules to settle, and then place a porous disk so that it is even with, and in contact with, the top of the enzyme bed. Displace all of the air from the disk. Place a cotton plug about 1 or 2 cm above the disk. (This plug acts as a filter. It ensures proper heating of the solution and traps dissolved gases that may be present in the *Glucose Substrate*.) Connect the tubing from the peristaltic pump with the top of the column, and seal the connection by suitable means to protect the column contents from the atmosphere. Place the inlet tube of the peristaltic pump into the *Glucose Substrate* solution, and begin a downward flow of the *Glucose Substrate* into the column at a rate of at least 80 mL/h. Maintain the flow rate for 1 h at room temperature.

Assay Adjust the flow of the *Glucose Substrate* to such a rate that a fractional conversion of 0.2 to 0.3 will be produced, based on the estimated activity of the sample. Calculate the fractional conversion from optical rotation values obtained on the starting *Glucose Substrate* and the sample effluent, as specified under *Calculations*, below. After establishing the correct flow rate, run the column overnight (16 h minimum), then check the pH of the *Glucose Substrate*, and readjust if necessary to the specified pH. Measure the flow rate, and collect a sample of the column effluent. Cover the effluent sample, allow it to stand for 30 min at room temperature, and then determine the fractional conversion of glucose to fructose (see *Calculations*, below). If the conversion is less than 0.2 or more than 0.3, adjust the flow rate to bring the conversion into this range. If a flow rate adjustment is required, collect an additional effluent sample after allowing the column to re-equilibrate for at least 2 h, and then determine the fractional conversion. Measure the flow rate, and collect an effluent sample. Cover the sample, let it stand at room temperature for 30 min, and determine the fractional conversion.

Calculations

Specific Rotation Measure the optical rotation of the effluent sample and of the starting *Glucose Substrate* at 25.0°, and calculate their specific rotations [see *Optical (Specific) Rotation*, Appendix IIB] by the formula

$$[\alpha] = 100a/lpd,$$

in which a is the corrected observed rotation, in degrees; l is the length of the polarimeter tube, in dm; p is the concentration of the test solution, expressed as g of solute per 100 g of solution; and d is the specific gravity of the solution at 25°.

Fractional Conversion Calculate the fractional conversion, X, by the formula

$$X = (\alpha_E - \alpha_S)/(\alpha_F - \alpha_S),$$

in which α_E is the specific rotation of the column effluent, α_S is the specific rotation of the *Glucose Substrate*, and α_F is the specific rotation of fructose (which, in this case, has been calculated to be −94.54).

Activity The enzyme activity is expressed in glucose isomerase units (GI_cU, the subscript c signifying column process). One GI_cU is defined as the amount of enzyme that converts glucose to fructose at an initial rate of 1 μmol/min, under the conditions specified.

Calculate the glucose isomerase activity by the formula:

$$GI_cU/g \text{ or } mL = (FS/W) \times X_E \times \ln[X_E/(X_E - X)],$$

in which F is the flow rate, in mL/min; S is the concentration of the *Glucose Substrate*, in μmol/mL; X is the fractional conversion, as determined above; X_E is the fractional conversion at equilibrium, or 0.51; and W is the weight or volume of the sample taken, in g or mL, respectively.

GLUCOSE OXIDASE ACTIVITY

Application and Principle This procedure is used to determine glucose oxidase activity in preparations derived from *Aspergillus niger* var. The assay is based on the titrimetric measurement of gluconic acid produced in the presence of excess substrate and excess air.

Reagents and Solutions

Chloride–Acetate Buffer Solution Dissolve 2.92 g of sodium chloride and 4.10 g of sodium acetate in about 900 mL of water. Adjust the pH to 5.1 with either dilute acetic acid or dilute sodium hydroxide solution and dilute to 1000.0 mL.

Sodium Hydroxide Solution (0.1 N)

Hydrochloric Acid Solution (0.05 N) Standardized.

Phenolphthalein Solution (2% w/v) Solution in methanol.

Octadecanol Solution Prepare a saturated solution in methanol.

Substrate Solution Dissolve 30.00 g of anhydrous glucose in 1000 mL of the *Chloride–Acetate Buffer Solution*.

Sample Preparation Dissolve an accurately weighed amount of enzyme preparation in the *Chloride–Acetate Buffer Solution*, and dilute in the buffer solution to obtain an enzyme activity of 5 to 7 activity units per mL.

Procedure Transfer 25.0 mL of the *Substrate Solution* to a 32- × 200-mm test tube. To a second 32- × 200-mm test tube transfer 25.0 mL of the *Chloride–Acetate Buffer Solution* (blank). Equilibrate both tubes in a 35° ± 0.1° water bath for 20 min. Add 3.0 mL of the *Sample Preparation* to each test tube, mix, and insert a glass sparger into each tube with a preadjusted air flow of 700 to 750 mL/min. If excessive foaming occurs, add 3 drops of the *Octadecanol Solution* to each tube. After exactly 15 min, remove the sparge and rinse any adhering reaction mixture back into the tube with water. Immediately add 10 mL of the *Sodium Hydroxide Solution* and 3 drops of the *Phenolphthalein Solution* to each tube. Insert a small magnetic stirrer bar, stir, and titrate to the phenolphthalein endpoint with the standardized 0.05 N *Hydrochloric Acid Solution*.

Calculation One Glucose Oxidase Titrimetric unit of activity (GOTu) is the quantity of enzyme that will oxidize 3 mg of glucose to gluconic acid under the conditions of the assay. Determine the enzyme activity using the following equation:

$$GOTu/g = [(B - T) \times N \times 180 \times F]/[3 \times W],$$

in which B is the titration volume, in mL, of the blank; T is the titration volume, in mL, of the sample; N is the normality of the titrant; 180 is the molecular weight of glucose; F is the sample dilution factor; 3 is from the unit definition; and W is the weight, in g, of the enzyme preparation contained in each mL of the sample solution.

β-D-GLUCOSIDASE ACTIVITY

Application and Principle This procedure is used to determine β-D-glucosidase activity of preparations derived from *Aspergillus niger* and *Trichoderma longibrachiatum*. The reaction is based on the release of glucose from cellobiose substrate; the glucose is measured using a glucose oxidase assay. The unit of activity is the amount of enzyme required to release exactly 1.0 mg of glucose from cellobiose under the conditions of the assay.

Apparatus

Glucose Reagent Kit (Sigma Chemical Co., Catalog No. 16-10, or equivalent) Contains all reagents required to determine the glucose liberated in the *Procedure*.

Reagents and Solutions

Citrate Buffer (pH 4.8, 0.05 M) Dissolve 9.6 g of citric acid in about 700 mL of water. Adjust the pH to 4.8 with 1 N sodium hydroxide, and dilute to 1000 mL.

Substrate Solution Dissolve 513 mg of cellobiose (Sigma Chemical Co., Catalog No. C-7252) in *Citrate Buffer*, and dilute to 100.0 mL with *Citrate Buffer*. Prepare fresh daily.

Procedure Dilute 1.0 mL of enzyme sample solution in the *Citrate Buffer*, and pipet the mixture into small test tubes. Make at least two dilutions for each sample investigated such that one of these dilutions would release slightly more than 1 mg of glucose in the reaction and the other, slightly less. Equilibrate the solution at 50° ± 0.1° in a water bath, add 1 mL of substrate solution preequilibrated at 50°, and incubate the reaction mixture at 50° for 30 min. Terminate the reaction by immersing the tubes in boiling water for exactly 5 min. Transfer the tubes to a cold water bath, and determine the amount of glucose released by a standard procedure, which is based on the hexokinase reaction, recommended by the manufacturer of the above *Glucose Reagent Kit*. Include all the necessary enzyme reagent blanks, and subtract them from the absorbance obtained in the glucose hexokinase reaction.

Calculation The unit of activity is defined as the amount of enzyme required to release exactly 1.0 mg of glucose from cellobiose under the conditions of the assay. Calculate the unit as follows:

1. Determine the glucose concentrations (mg/mL) in the cellobiose reaction mixture by using at least two different enzyme dilutions.
2. Multiply the glucose concentrations obtained in the previous step by 2 to convert glucose concentrations into absolute amounts (mg).
3. Convert the enzyme dilutions into concentrations:

$$\text{Concentration} = 1/\text{dilution}.$$

4. Obtain the concentration of enzyme that would have released exactly 1.0 mg of glucose by plotting glucose liberated (as in 2, above) against enzyme concentration (as in 3, above).
5. Calculate the cellobiose activity:

Units/mL = 0.0926/enzyme concentration to release 1.0 mg glucose.

The value 0.0926 is derived as follows:

$$1 \text{ IU} = 1 \text{ μmol/min of cellobiose converted}$$
$$= 2 \text{ μmol/min of glucose formed.}$$

HEMICELLULASE ACTIVITY

Application and Principle This procedure is used to determine hemicellulase activity of preparations derived from *Aspergillus niger* var. The test is based on the enzymatic hydrolysis of the interior glucosidic bonds of a defined locust (carob) bean gum substrate at 40° and pH 4.5. Determine the corresponding reduction in substrate viscosity with a calibrated viscometer.

Apparatus

Viscometer Use a size 100 calibrated Cannon-Fenske Type Viscometer, or its equivalent (Scientific Products, Catalog No. 2885-100).

Glass Water Bath Use a constant-temperature glass water bath, maintained at 40° ± 0.1° (Scientific Products, Catalog No. W3520-10).

Stopwatches Use two stopwatches—*Stopwatch No. 1*, calibrated in $\frac{1}{10}$ min for determining the reaction time (T_R), and *Stopwatch No. 2*, calibrated in $\frac{1}{5}$ s for determining the efflux time (T_T).

Reagents and Solutions

Acetate Buffer (pH 4.5) Add 0.2 N sodium acetate, with continuous agitation, to 400 mL of 0.2 N acetic acid until the pH is 4.5 ± 0.05, as determined by a pH meter.

Locust Bean Gum Use Powdered Type D-200 locust bean gum, or its equivalent, (Meer Corp.). Because the substrate may vary from lot to lot, test each lot in parallel with a previous lot known to be satisfactory. Variations of more than ± 5% viscosity in the average of a series of parallel tests indicate an unsuitable lot.

Substrate Solution Place 12.5 mL of 0.2 N hydrochloric acid and 250 mL of warm water (72° to 75°) in the bowl of a power blender (Waring two-speed, or its equivalent, Scientific Products, Catalog No. 58350-1), and set the blender on low speed. Slowly disperse 2.0 g of *Locust Bean Gum*, on a moisture-free basis, into the bowl, taking care not to splash out any of the liquid in the bowl. Wash down the sides of the bowl with warm water, using a rubber policeman, cover the bowl, and blend at high speed for 5 min. Quantitatively transfer the mixture to a 1000-mL beaker, and cool to room temperature. Using a pH meter, adjust the mixture to pH 6.0 with 0.2 N sodium hydroxide. Quantitatively transfer the mixture to a 1000-mL volumetric flask, dilute to volume with water, and mix. Filter the substrate through gauze before use.

Sample Preparation Prepare a solution of the sample in water so that 1 mL of the final dilution will produce a change in relative fluidity between 0.18 and 0.22 in 5 min under the conditions specified in the *Procedure*.

Weigh the enzyme preparation, quantitatively transfer it to a glass mortar, and triturate with water. Quantitatively transfer the mixture to an appropriately sized volumetric flask, dilute to volume with water, and mix. Filter through Whatman No. 1 filter paper, or equivalent, before use.

Procedure Scrupulously clean the viscometer by drawing a large volume of detergent solution, followed by water, through the instrument, and place the viscometer, previously calibrated, in the glass water bath in an exactly vertical position. Pipet 20.0 mL of *Substrate Solution* and 4.0 mL of *Acetate Buffer* into a 50-mL Erlenmeyer flask, allowing at least two flasks for each enzyme sample and one flask for a substrate blank. Stopper the flasks, and equilibrate them in the water bath for 15 min. At zero time, pipet 1.0 mL of the *Sample Preparation* into the equilibrated substrate, start timing with *Stopwatch No. 1*, and mix thoroughly. Immediately pipet 10.0 mL of this mixture into the wide arm of the viscometer. After about 2 min, draw the reaction mixture above the upper mark into the driving fluid head by applying suction with a rubber tube connected to the narrow arm of the instrument. Measure the efflux time by allowing the reaction mixture to flow freely down past the upper mark. As the meniscus falls past the upper mark, start *Stopwatch No. 2*, and at the same time, record the reaction time (T_R), in min, from *Stopwatch No. 1*. As the meniscus of the reaction mixture falls past the lower mark, record the time (T_T), in s, from *Stopwatch No. 2*. Immediately re-draw the reaction mixture above the upper mark and into the driving fluid head. As the meniscus falls freely past the upper mark, restart *Stopwatch No. 2*, and at the same time record the reaction time (T_R), in min, from *Stopwatch No. 1*. As the meniscus falls past the lower mark, record the time (T_T), in s, from *Stopwatch No. 2*.

Repeat the latter operation, beginning with "Immediately re-draw the reaction mixture, . . ." until a total of four determinations is obtained over a reaction time (T_R) of not more than 15 min.

Prepare a substrate blank by pipetting 1.0 mL of water into a mixture of 20.0 mL of *Substrate Solution* and 4.0 mL of

Acetate Buffer, and then immediately pipet 10.0 mL of this mixture into the wide arm of the viscometer. Determine the time (T_S), in s, required for the meniscus to fall between the two marks. Use an average of five determinations as T_S.

Prepare a water blank by pipetting 10.0 mL of water, previously equilibrated to 40° ± 0.1°, into the wide arm of the viscometer. Determine the time (T_W), in s, required for the meniscus to fall between the two marks. Use an average of five determinations as T_W.

Calculation One hemicellulase unit (HCU) is defined as that activity that will produce a relative fluidity change of 1 over a period of 5 min in a locust bean gum substrate under the conditions specified. Calculate the relative fluidities (F_R) and T_N values (see definition below) for each of the four efflux times (T_T) and reaction times (T_R) as follows:

$$F_R = (T_S - T_W)/(T_T - T_W),$$

and

$$T_N = \tfrac{1}{2}(T_T/60 \text{ s/min}) + T_R = (T_T/120) + T_R,$$

in which F_R is the relative fluidity for each reaction time; T_S is the average efflux time, in s, for the substrate blank; T_W is the average efflux time, in s, for the water blank; T_T is the efflux time, in s, of the sample reaction mixture; T_R is the elapsed time, in min, from zero time, that is, the time from addition of the enzyme solution to the buffered substrate until the beginning of the measurement of the efflux time (T_T); and T_N is the reaction time (T_R), in min, plus one half of the efflux time (T_T) converted to min.

Plot the four relative fluidities (F_R) as the ordinate against the four reaction times (T_N) as the abscissa. This should result in a straight line. The slope of the line corresponds to the relative fluidity change per min and is proportional to the enzyme concentration. The slope of the best line through a series of experimental points is a better criterion of enzyme activity than is a single relative fluidity value. From the curve, determine the F_R values at 10 and 5 min. They should have a difference in fluidity of not more than 0.22 and not less than 0.18. Calculate the activity of the enzyme sample as follows:

$$HCU/g = 1000(F_{R10} - F_{R5}/W),$$

in which F_{R10} is the relative fluidity at 10 min reaction time; F_{R5} is the relative fluidity at 5 min reaction time; 1000 is mg/g; and W is the weight, in mg, of the enzyme sample contained in the 1.0-mL aliquot of *Sample Preparation* added to the equilibrated substrate in the *Procedure*.

INVERTASE ACTIVITY

Application and Principle This procedure is used to determine the invertase activity of enzyme preparations from yeast *Saccharomyces* sp (*Kluyveromyces*). The assay is based on a 30-min hydrolysis of sucrose at 30° ± 0.1° and at pH 4.62. The degree of hydrolysis is determined by measuring the optical rotation of the solution with a polarimeter.

Reagents and Solutions

Acetate Buffer Dissolve 4.0 g of sodium hydroxide (NaOH) in about 900 mL of water, and carefully neutralize with 12.0 g of acetic acid 98% to 100% (CH_3COOH). Cool to room temperature. Transfer the solution into a 1000-mL volumetric flask, dilute to volume with water, and mix. The pH should be 4.62 ± 0.05.

Sucrose Substrate Solution Dissolve 82.152 g of sucrose in about 900 mL of water. Transfer the solution into a 1000-mL volumetric flask, dilute to volume with water, and mix. Use a freshly prepared solution only.

Sodium Carbonate Solution Dissolve 53.0 g of sodium carbonate (Na_2CO_3) in about 400 mL of water, then transfer the solution into a 500-mL volumetric flask, dilute to volume with water, and mix.

Test Preparation Using a 100-mL volumetric flask, prepare a solution from the starting enzyme preparation by weighing a minimum of 1 g of sample accurately to within 1 mg. Dilute with water so that the final solution will contain between 1.3 and 5.3 Invertase Units per 20 mL. Pipet 20.0 mL of this solution into a 100-mL Erlenmeyer flask.

Blank Preparation Pipet 20.0 mL water in a 100-mL Erlenmeyer flask.

Procedure To the flasks containing 20.0 mL of each *Test Preparation* and to the *Blank Preparation*, add 5.00 mL of *Acetate Buffer*. At zero time, and at regular time intervals so that each test sample is analyzed in the same elapsed time, place the flasks containing the *Test Preparations* and the *Blank Preparation* in a circulating water bath maintained at 30.0° ± 0.1°. Equilibrate the samples for 10 min in the water bath. In the same order and with the same time intervals, rapidly pipet 25.00 mL of equilibrated *Sucrose Substrate Solution* into the test flasks. Incubate for 30.0 min, and stir continuously. Terminate the reaction by adding 10.00 mL of *Sodium Carbonate Solution*, and swirl to mix. Place the flasks containing the *Test Preparations* and the *Blank Preparation* in a water bath maintained at 20.0° ± 0.1° for 30 min. Use a polarimeter with an accuracy of at least 0.001 degrees of arc. With the same precision, determine the optical rotation of each solution at 589 nm (sodium lamp), using a 10-cm path-length cell with the thermostat set at 20.0° ± 0.1°. Use water to blank the polarimeter initially.

Calculation One Invertase Unit (INVU) is defined as the quantity of enzyme that will hydrolyze 1.142 μmol sucrose per min under the conditions of the assay. Calculate the invertase units per 20 mL as follows:

$$\frac{(R_{BT} - R_{TEST}) \times 2 \times 1{,}000{,}000/[66.77 - (Glu + Fru)]}{(342.3)(1.142)} = INVU/20 \text{ mL},$$

where Glu is (0.525)(52.5) and Fru is (0.525)(−91.315), or simplified:

$$(R_{BL} - R_{TEST}) \times 58.71 = INVU/20 \text{ mL},$$

in which R_{BL} is the rotation of *Blank Preparation*; R_{TEST} is the rotation of the *Test Preparation*; 66.77 is the specific rotation of sucrose; 52.5 is the specific rotation of glucose; −91.315 is the specific rotation of fructose; 0.525 is 0.5 corrected for 5% weight increase by hydrolysis; 342.3 is the molecular mass (g/mol) of sucrose; and 1.142 is the unit definition factor. Specific rotations are valid at the average concentrations in this test.

Invertase Activity in Weighed Samples:

$$INVU/20 \text{ mL} \times d/w = INVU/g,$$

Invertase Activity in Pipetted Samples:

$$INVU/20 \text{ mL} \times d = INVU/mL,$$

in which d is the total dilution factor and w is the weight of the sample.

LACTASE (NEUTRAL) (β-GALACTOSIDASE) ACTIVITY

Application and Principle This procedure is used to determine the neutral lactase activity of enzyme preparations derived from *Kluyveromyces marxianus* var. *lactis* and *Saccharomyces* sp. The assay is based on a 10-min hydrolysis of an *o*-nitrophenyl-β-D-galactopyranoside (ONPG) substrate at $30.0° \pm 0.1°$ and at pH 6.50.

Reagents and Solutions

Magnesium Solution Dilute 24.65 g of magnesium sulphate heptahydrate ($MgSO_4.7H_2O$) in about 950 mL of water. Transfer the solution into a 1000-mL volumetric flask, dilute to volume with water, and mix.

EDTA Solution Dissolve 1.86 g of disodium EDTA dihydrate ($C_{10}H_{14}N_2Na_2O_8.2H_2O$) in about 950 mL of water. Transfer the solution into a 1000-mL volumetric flask, dilute to volume with water, and mix.

P-E-M Buffer Dissolve 8.8 g of potassium dihydrogen phosphate (KH_2PO_4) and 8.0 g of dipotassium hydrogen phosphate trihydrate ($K_2HPO_4.3H_2O$) in about 900 mL of water. Add 10.0 mL of *Magnesium Solution* and 10.0 mL of *EDTA Solution*. Transfer the solution into a 1000-mL volumetric flask, dilute to volume with water, and mix. The pH should be 6.50 ± 0.05.

Lactase Reference Preparation (Highly concentrated lactase preparation) This preparation can be obtained from Gist-Brocades, Delft, The Netherlands.

ONPG (*o*-nitrophenyl-β-D-galactopyranoside) is validated according to the following procedure:

Validation of New ONPG Transfer 150, 250, and 375 mg of the new *ONPG* into separate 100-mL volumetric flasks, dilute to volume with *P-E-M Buffer*, and mix. Prepare solutions of the *Lactase Reference Preparation* by weighing an amount of *Lactase Reference Preparation* corresponding to 5000 ± 250 Neutral Lactase Units (NLU) accurately to within 1 mg in duplicate in 50-mL volumetric flasks, dissolve in *P-E-M Buffer*, dilute to volume with the same, and mix. Prepare dilutions of this initial solution with *P-E-M Buffer* so that 1 mL of the final dilution will contain 0.0375, 0.0750, and 0.1125 NLU of activity. In duplicate, determine the enzyme activity of the three enzyme concentrations using each of the new *ONPG Substrate* solutions corresponding to 150, 250, and 375 mg and the old *ONPG Substrate* at 250 mg by following the steps in the *Procedure*, below.

Calculation Calculate the enzyme activity following the steps indicated under *Calculation for NLU Activity*, below. Determine the average of the duplicates for each enzyme concentration at each level of *ONPG Substrate* (the maximum allowable difference between these duplicates is 6.5%). Determine the overall average for the three enzyme concentrations (0.0375, 0.0750, and 0.1125) for each *ONPG Substrate* level (150, 250, and 375 mg of *ONPG*).

To determine the overall average of three enzyme concentrations at 150 mg of *ONPG*:

$$X = (A + B + C)/3,$$

in which A is the average result of 0.0375 at 150 mg of *ONPG*, B is the average result of 0.0750 at 150 mg of *ONPG*, and C is the average result of 0.1125 at 150 mg of *ONPG*.

To determine the overall average of three enzyme concentrations at 250 mg of *ONPG*:

$$Y = (D + E + F)/3,$$

in which D is the average result of 0.0375 at 250 mg of *ONPG*, E is the average result of 0.0750 at 250 mg of *ONPG*, and F is the average result of 0.1125 at 250 mg of *ONPG*.

To determine the overall average of three enzyme concentrations at 375 mg of *ONPG*:

$$Z = (G + H + I)/3,$$

in which G is the average result of 0.0375 at 375 mg of *ONPG*, H is the average result of 0.0750 at 375 mg of *ONPG*, and I is the average result of 0.1125 at 375 mg of *ONPG*.

The *ONPG* analyzed is suitable for use when the following specifications are met for each *ONPG* concentration:

1. The average result of each enzyme concentration for each *ONPG* level does not deviate more than 3% from the overall average of the three enzyme concentrations for that level of *ONPG*. For example, A or B or C should not deviate more than 3% from X; D or E or F should not deviate more than 3% from Y; G or H or I should not deviate more than 3% from Z.

2. The overall average of the three enzyme concentrations found for 150 mg of *ONPG* (X) should not vary more than 81% to 99% of the overall average of the three enzymes concentrations found for 250 mg of *ONPG* (Y). The overall average of the three enzyme concentrations found for 375 mg of *ONPG* (Z) should not vary more than 96% to 114% of the overall average of the three enzyme concentrations of 250 mg of *ONPG* (Y).

3. The absorbance of each blank is less than 0.050.

4. For each new lot of *ONPG*, the overall average of the three enzyme concentrations found for 250 mg of *ONPG* (Y) per 100 mL should be within 5% of the overall average of the three enzyme concentrations found for 250 mg of *ONPG* of the lot in use at that moment.

ONPG Substrate Dissolve 250.0 mg *ONPG* (use lot currently in use) in about 80 mL of *P-E-M Buffer*. Transfer the solution to a 100-mL volumetric flask, dilute to volume with *P-E-M Buffer*, and mix. Prepare, at most, 2 h before incubation.

Sodium Carbonate Solution Dissolve 50.0 g of sodium carbonate anhydrous (Na_2CO_3) and 37.2 g of disodium EDTA dihydrate ($C_{10}H_{14}N_2Na_2O_8 \cdot 2H_2O$) in about 900 mL of water. Transfer the solution into a 1000-mL volumetric flask, dilute to volume with water, and mix.

Standard o-Nitrophenol Solution Transfer 139.0 mg of *o*-nitrophenol into a 1000-mL volumetric flask, dissolve in 10 mL of 96% ethanol, dilute to volume with water, and mix. Pipet 2-, 4-, 6-, 8-, 10-, 12-, and 14-mL portions of this solution into a series of 100-mL volumetric flasks, add 25 mL of *Sodium Carbonate Solution* to each, dilute each to volume with *P-E-M Buffer*, and mix. The dilutions contain, respectively, 0.02, 0.04, 0.06, 0.08, 0.10, 0.12 and 0.14 μmol/mL of *o*-nitrophenol.

Determine the absorbance of each dilution at 420 nm in a 1-cm path-length cell, with a suitable spectrophotometer, using water as the blank. For each dilution, plot absorbance against μmol of *o*-nitrophenol (this must result in a straight line through the origin). Divide the absorbance of each dilution by μmol of *o*-nitrophenol to obtain the extinction coefficient (ϵ) at that dilution (the slope of the line is the extinction coefficient). Average the seven values thus calculated (this should result in a value of 4.60 ± 0.05).

Test Preparation Using a volumetric flask, prepare a test solution from the starting enzyme preparation by accurately weighing out a minimum of 1 g of sample to the nearest mg. Dissolve in *P-E-M Buffer* so that 1 mL of the final dilution will contain between 0.027 and 0.095 NLU. Transfer 1 mL of this final dilution to a 15- × 150-mm test tube as the *Test Preparation*. Perform in duplicate.

Procedure Equilibrate the test tubes containing each *Test Preparation* in a water bath maintained at 30.0° ± 0.1° for at least 5 but not more than 15 min. At zero time, in the order of the series and at regular time intervals, rapidly pipet 5.00 mL of *ONPG Substrate*, equilibrated at 30.0° ± 0.1°, into the test tubes, and mix by shaking. After a 10.0-min incubation (reaction) time, in the same order and with the same regular intervals, pipet 2.00 mL of *Sodium Carbonate Solution* into each, mix by shaking, and hold at room temperature. Determine the absorbance of each solution within 30 min at 420 nm in a 1-cm cell with a suitable spectrophotometer, using as the blank a solution prepared in the same manner as for the sample except adding *ONPG Substrate* and *Sodium Carbonate Solution* in reverse order.

Calculation for NLU Activity One Neutral Lactase Unit (NLU) is defined as that quantity of enzyme that will liberate 1.30 μmol/min of *o*-nitrophenol under the conditions of the assay. Calculate the activity of the enzyme preparation taken for the analysis as follows:

$$NLU/g = [(A \times 8 \times f)/(\epsilon \times 10 \times W)]/1.30,$$

in which A is the average of the absorbance readings for the sample, corrected for the sample blank; 8 is the volume, in mL, of the incubation mixture after termination; f is the total dilution factor of the sample; ϵ is the extinction coefficient, determined as directed under *Standard o-Nitrophenol Solution*; 10 is the incubation time, in min; W is the sample weight, in g; and 1.30 is the factor used in the unit definition.

LACTASE (ACID) (β-GALACTOSIDASE) ACTIVITY

Application and Principle This procedure is used to determine lactase activity of enzyme preparations derived from *Aspergillus oryzae* var. The assay is based on a 15-min hydrolysis of an *o*-nitrophenyl-β-D-galactopyranoside substrate at 37° and pH 4.5.

Reagents and Solutions

2.0 N Acetic Acid Dilute 57.5 mL of glacial acetic acid to 500 mL with water. Mix well, and store in a refrigerator.

4.0 N Sodium Hydroxide Dissolve 40.0 g of sodium hydroxide in sufficient water to make 250 mL.

Acetate Buffer Combine 50 mL of *2.0 N Acetic Acid* and 11.3 mL of *4.0 N Sodium Hydroxide* in a 1000-mL volumetric flask, and dilute to volume with water. Verify that the pH is 4.50 ± 0.05, using a pH meter, and adjust, if necessary, with *2.0 N Acetic Acid* or *4.0 N Sodium Hydroxide*.

2.0 mM o-Nitrophenol Stock Transfer 139.0 mg of *o*-nitrophenol to a 500-mL volumetric flask, dissolve in 10 mL of USP alcohol (95% ethanol) by swirling, and dilute to volume with 1% sodium carbonate.

o-Nitrophenol Standards

0.10 mM Standard Solution Pipet 5.0 mL of the *2.0 mM o-Nitrophenol Stock* solution into a 100-mL volumetric flask, and dilute to volume with 1% sodium carbonate solution.

0.14 mM Standard Solution Pipet 7.0 mL of the *2.0 mM o-Nitrophenol Stock* solution into a 100-mL volumetric flask, and dilute to volume with 1% sodium carbonate solution.

0.18 mM Standard Solution Pipet 9.0 mL of the *2.0 mM o-Nitrophenol Stock* solution into a 100-mL volumetric flask, and dilute to volume with 1% sodium carbonate solution.

Substrate Transfer 370.0 mg of *o*-nitrophenyl-β-D-galactopyranoside to a 100-mL volumetric flask, and add 50 mL of *Acetate Buffer*. Swirl to dissolve, and dilute to volume with *Acetate Buffer*.

Note: Perform the assay procedure within 2 h of *Substrate* preparation.

Test Preparation Prepare a solution from the test sample preparation such that 1 mL of the final dilution will contain between 0.15 and 0.65 lactase unit. Weigh, and quantitatively transfer the enzyme to a volumetric flask of appropriate size. Dissolve the enzyme in water, swirling gently, and dilute with water if necessary.

Note: Perform the assay procedure within 2 h of dissolution of the *Test Preparation*.

System Suitability Determine the absorbance of the three *o-Nitrophenol Standards* at 420 nm in a 1-cm cell, using a suitable spectrophotometer. Use water to zero the instrument.

Calculate the millimolar extinction, ϵ, for each of the *o-Nitrophenol Standards* (0.10, 0.14, and 0.18 mM) by the formula

$$\epsilon = A_N/C,$$

in which A_N is the absorbance of each *o*-nitrophenol standard at 420 nm and C is the corresponding concentration of *o*-nitrophenol in the standard. ϵ for each standard should be approximately 4.60/mM. Perform a linear regression analysis of the absorbance readings of the three *o-Nitrophenol Standards* versus the *o*-nitrophenol concentration in each (0.10, 0.14, and 0.18 mM). The r^2 should not be less than 0.99. Determine the mean ϵ of the three *o-Nitrophenol Standards* for use in the calculations below.

Procedure For each sample or blank, pipet 2.0 mL of the *Substrate* solution into a 25- × 150-mm test tube, and equilibrate in a water bath maintained at 37.0° ± 0.1° for approximately 10 min. At zero time, rapidly pipet 0.5 mL of the *Test Preparation* (or 0.5 mL of water as a blank) into the equilibrated substrate, mix by brief (1 s) vortex, and immediately return the tubes to the water bath. After exactly 15 min of incubation, rapidly add 2.5 mL of 10% sodium carbonate solution, and vortex the tube to stop the enzyme reaction. Dilute the samples and blanks to 25.0 mL by adding 20.0 mL of water, and thoroughly mix. Determine the absorbance of the diluted samples and blanks at 420 nm in a 1-cm cell, using a suitable spectrophotometer. Use water to zero the instrument.

Calculation One lactase unit (ALU) is defined as that quantity of enzyme that will liberate *o*-nitrophenol at a rate of 1 μmol/min under the conditions of the assay.

Calculate the activity (lactase activity per g) of the enzyme preparation taken for analysis as follows:

$$\text{ALU/g} = [(A_S - B)(25)]/[(\epsilon)(15)(W)],$$

in which A_S is the average of absorbance readings for the *Test Preparation*; B is the average of absorbance readings for the blank; 25 is the final volume, in mL, of the diluted incubation mixture; ϵ is the mean absorptivity of the *o-Nitrophenol Standards* per μM, 15 is the incubation time, in min, and W is the weight, in g, of original enzyme preparation contained in the 0.5-mL aliquot of *Test Preparation* used in the incubation.

LIPASE ACTIVITY

Application and Principle This procedure is used to determine the lipase activity in preparations derived from microbial sources and animal pancreatic tissues. The assay is based on the potentiometric measurement of the rate at which the preparations will catalyze the hydrolysis of tributyrin.

Apparatus
Automatic Recording Titrimeter Use an instrument operating in the pH stat mode and equipped with a jacketed titration cell (Radiometer Titralab, or equivalent).

Constant Temperature Bath Operated at 30° ± 0.1°.
Blender

Reagents and Solutions
0.05 N Sodium Hydroxide Dissolve 2.0 g of sodium hydroxide in water, and dilute to 100 mL. Standardize with NBS grade potassium hydrogen phthalate.

Emulsification Reagent Dissolve 17.9 g of sodium chloride and 0.41 g of monobasic potassium phosphate in about 400 mL of water. Add 540 mL of glycerol and, with vigorous stirring, add 6.0 g of gum arabic (Sigma, Catalog No. G 9752). Stir until dissolved. Dilute to 1000 mL.

Glycine Buffer (0.1 M) Dissolve 7.50 g of glycine (Sigma, Catalog No. G 126) and 3.8 g of sodium hydroxide in about 900 mL water. Adjust the pH to 10.8, and dilute to 1000 mL.

Note: Instead of the *Glycine Buffer*, some enzyme preparations may require the use of 0.01 M pH 8.0 *Tris Buffer* prepared as directed for *Tris Buffer* under *Proteolytic Activity, Bacterial (PC)*, except to titrate with 1 N hydrochloric acid to pH 8.0.

Substrate Emulsion Transfer 15.9 mL of tributyrin (Sigma, Catalog No. T 8626) to a blender, add 50 mL *Emulsification Reagent* and 235 mL water. Blend for 15 min at maximum speed. Equilibrate in the 30° constant temperature bath for at least 15 min before use. Use within 4 h.

Sample Preparation Dissolve an accurately weighed amount of the enzyme preparation in *Glycine Buffer* (or pH 8.0 *Tris Buffer* is specified) so that each mL contains between 2000 and 5000 lipase units per mL. Accurately dilute a portion of this solution with water to obtain a final solution containing between 0.5 and 1.5 lipase units per mL.

Procedure Fill the titrator buret with the *0.05 N Sodium Hydroxide* solution, and following the manufacturer's instructions, set the temperature to 30° and the pH set point to 7.0. Transfer 15.0 mL of the *Substrate Emulsion* to the titration cell, and add a small stirrer bar. Add 1.0 mL of the diluted *Sample Preparation*, and actuate the titrator. Record the rate of *0.05 N Sodium Hydroxide* addition. Stop the titration after a constant (linear) rate of addition has been observed for 5 min. Determine the addition rate, in mL/min, from the linear portion of the recording and record this value as R.

Calculation One lipase unit (LU) is defined as the quantity of enzyme that will liberate 1 μmol of butyric acid per min under the conditions of the test.

Calculate the activity of the enzyme preparation by the formula

$$\text{LU/g} = R \times N \times 1000/W,$$

in which R is the addition rate, in mL/min; N is the normality of the *Sodium Hydroxide* solution; 1000 converts mM to μM; and W is the weight, in g, of the enzyme preparation contained in 1 mL of the diluted *Sample Preparation*.

LIPASE/ESTERASE (FORESTOMACH) ACTIVITY

Application and Principle This procedure is primarily applicable to lipases from animal forestomach sources. The analysis is performed by potentiometric titration.

Apparatus Use a suitable automatic recording titrator equipped with thermostatic control (Thermostatic Recording pH Stat, Sargent-Welch, or equivalent).

Reagents and Solutions
Sodium Caseinate Use the hydrophilic powder, soluble form (Erie Casein, Inc.).
Hydroxylated Lecithin Solution Use the material available from Food Technology, Inc., and prepare a 10% solution in light mineral oil (FCC or USP grade).
Tri-n-butyrin Use the material available from Aldrich Chemical Company, Inc., or equivalent.
Substrate Preparation Disperse an amount of *Sodium Caseinate*, equivalent to 600 mg of casein, in 95 mL of water contained in a one-half pint freezer jar (Ball Mason, or equivalent) that fits the head of a suitable high-speed blender. Add 0.5 mL of *Hydroxylated Lecithin Solution* and 5.0 mL of *Tri-n-butyrin*, and mix for 60 s at low speed. Adjust the temperature of the mixture to 42°, and use within 4 h.
Sample Preparation Suspend or dissolve an accurately weighed amount of the enzyme preparation in water, and dilute to obtain an enzyme activity of 10 to 12 lipase (forestomach) units per mL.

Procedure Fill the buret of the titrator with 0.025 N sodium hydroxide, and calibrate the instrument following the manufacturer's instructions, setting the temperature at 42° and the pH at 6.20. Mix the *Substrate Preparation* for about 15 s with a magnetic stirrer, then pipet 10.0 mL into the reaction vessel of the titrator, and add a small stirring bar. Place the vessel on the titrator, add 1.0 mL of the *Sample Preparation*, and equilibrate the mixture for 15 min. Actuate the recorder, and record the titration curve for 15 min.

Note: The recorder trace reflecting delivery of the titrant must be linear.

Determine the rate, in mL/min, at which the titrant was delivered during the titration, and record this value as R.

Calculation One lipase (forestomach) unit (LFU) is the activity that releases butyric acid at the rate of 1.25 μmol/min under the conditions of the test.

Calculate the activity of the enzyme preparation by the formula

$$\text{LFU/g} = (R \times 0.025 \times 10^3)/(W \times 1.25),$$

in which R is the rate, in mL/min, of titrant delivery; 0.025 is the normality of the titrant; 10^3 is the conversion from mg to μg; and W is the weight, in g, of the enzyme preparation contained in the 1.0 mL of *Sample Preparation* taken for analysis.

MALTOGENIC AMYLASE ACTIVITY

Application and Principle This procedure is used to determine maltogenic amylase activity in preparations derived from *Bacillus subtilis* containing a *Bacillus stearothermophilus* amylase gene. The test is based on a 30-min hydrolysis of maltotriose under controlled conditions and measurement of the glucose formed by high-pressure liquid chromatography (HPLC).

Reagents and Solutions
Citrate Buffer, 0.1 M Dissolve 5.255 g of citric acid ($C_6H_8O_7 \cdot H_2O$) in about 150 mL of water. Adjust the pH to 5.0 with 1 N sodium hydroxide, and dilute to 250 mL.
Substrate Solution Dissolve 1.00 g of maltotriose (Sigma Chemical Co., Catalog No. M 8378) in *Citrate Buffer* in a 50-mL volumetric flask, and dilute to volume with *Citrate Buffer*.
Sodium Chloride Solution, 1 M Dissolve 29.22 g of sodium chloride in water, and dilute to 500 mL.
Amberlite MB-1 Ion Exchange Resin Air dry at room temperature for about 1 week. Protect from contamination.
Glucose Standards Dissolve 1.80 g of anhydrous glucose in water, and dilute to 1000 mL. Transfer 20.0, 50.0, 75.0, and 100.0 mL to separate 100-mL volumetric flasks, and dilute to volume with water. These solutions contain 0.36, 0.9, 1.35, and 1.80 mg of glucose per mL. Using filtered, degassed water as the mobile phase, equilibrate an HPX 87C column, or equivalent, in a high-pressure liquid chromatograph equipped with a differential refractometer. Chromatograph 5-μL portions of the glucose standards, and record the chromatograms. Prepare a standard curve of the glucose concentration versus the peak height.
Sample Preparation Prepare a solution of each sample to contain approximately 7.5 Maltogenic Amylase Units (MANU) per mL. Further dilute an aliquot of each sample so that the final dilution contains 1% by volume of the *Sodium Chloride Solution, 1 M* and contains between 0.150 and 0.600 MANU per mL.

Procedure Transfer 2.00 mL of each sample to separate test tubes, and equilibrate in the 37° water bath for at least 10 min. At the same time, equilibrate the *Substrate Solution* in the same water bath. At zero time, transfer 2.0 mL of the equilibrated *Substrate Solution* to the first sample tube, mix thoroughly, and return the tube to the 37° bath. Repeat the process for each sample. After exactly 30.0 min, transfer the test tube to a boiling water bath for 15 min, then remove and cool to room temperature. Add approximately 100 mg of *Amberlite MB-1 Ion Exchange Resin* to each tube, place the tubes on the shaker, and mix for at least 15 min. Filter the treated solution through a 0.45-μm filter. Use a separate filter for each sample. Inject a 5-μL portion of each filtered sample into a previously equilibrated high-pressure liquid chromatograph equipped with an HPX 87C column (Biorad, or equivalent) and a differential refractometer. Filtered, degassed water is the mobile phase. Record the elution curve.

Calculation One Maltogenic Amylase Unit (MANU) is defined as the amount of enzyme that will cleave maltotriose at a rate of 1 μmol/min under the conditions of the test. From the elution curve of each sample, determine the glucose concentration (G) in the sample from the previously prepared standard curve. Calculate the MANU/g by the formula

$$\text{MANU/g} = G \times 4 \times F/180.1 \times 30 \times W,$$

in which G is the glucose concentration in the test solution; 4 is the total test solution volume; 30 is the reaction time, in min; F is the dilution factor; and W is the sample weight, in g.

MILK-CLOTTING ACTIVITY

Application and Principle This procedure is to be applied to enzyme preparations derived from either animal or microbial sources.

Apparatus

Bottle-Rotating Apparatus Use a suitable assembly designed to rotate at a rate of 16 to 18 rpm.

Sample Bottles Use 125-mL, squat, round, wide-mouth bottles (such as Scientific Products, Catalog No. B-7545-125).

Substrate Solution Dissolve 60 g of low-heat, nonfat dry milk (such as Galloway West, Peake Grade A) in 500 mL of a solution, adjusted to pH 6.3 if necessary, containing in each mL 2.05 mg of sodium acetate ($NaC_2H_3O_2$) and 1.11 mg of calcium chloride ($CaCl_2$).

Standard Preparation Use a standard-strength rennet, bovine rennet, microbial rennet (*E. parasitica*), or microbial rennet (*Mucor* species), as appropriate for the preparation to be assayed. Such standards, which are available from commercial coagulant manufacturers, should be of known activity. Dilute the standard-strength material 1 to 200 with water, and mix. Equilibrate to 30° before use, and prepare no more than 2 h before use.

Sample Preparation Prepare aqueous solutions or dilutions of the sample to produce a final concentration such that the clotting time, as determined in the *Procedure* below, will be within 1 min of that of the *Standard Preparation*. Prepare no more than 1 h before use.

Procedure Transfer 50.0 mL of the *Substrate Solution* into each of four 125-mL *Sample Bottles*. Place the bottles on the *Bottle-Rotating Apparatus*, and suspend the apparatus in a water bath, maintained at 30° ± 0.5°, so that the bottles are at an angle of approximately 20° to 30° to the horizontal. Immerse the bottles so that the water level in the bath is about equal to the substrate level in the bottles. Begin rotating the apparatus at 16 to 18 rpm, then add 1.0 mL of the *Sample Preparation* to each of two bottles, and record the exact time of addition. Add 1.0 mL of the *Standard Preparation* to each of the other two bottles, recording the exact time.

Observe the rotating bottles, and record the exact time of the first evidence of clotting (i.e., when fine granules or flecks adhere to the sides of the bottle). Variations in the response of different lots of the substrate may cause variations in clotting time; therefore, measure the test samples and standards simultaneously on the same substrate. Average the clotting time, in s, of the duplicate samples, recording the time for the *Standard Preparation* as T_S and that for the *Sample Preparation* as T_U.

Calculation Calculate the activity of the enzyme preparation by the formula

$$\text{Milk-clotting units/mL} = 100 \times (T_S/T_U) \times (D_S/D_U),$$

in which 100 is the activity assigned to the *Standard Preparation*, D_S is the dilution factor for the *Standard Preparation*, and D_U is the dilution factor for the *Sample Preparation*.

Note: The dilution factors should be expressed as fractions; for example, a dilution of 1 to 200 would be expressed as $1/200$.

PANCREATIN ACTIVITY

Application and Principle These procedures are used to determine the primary enzyme activities in pancreatin preparations.

Reference Standards

USP Sodium Taurocholate Reference Standard (**Caution**: Avoid inhaling airborne particles.) Keep container tightly closed. Dry at 105° for 4 h before using.

USP Pancreatin Reference Standard Keep container tightly closed, and store in a refrigerator. Do not open while cold, and do not dry before using.

Amylase Activity

pH 6.8 Phosphate Buffer On the day of use, dissolve 13.6 g of monobasic potassium phosphate in water to make 500 mL of solution. Dissolve 14.2 g of anhydrous dibasic sodium phosphate in water to make 500 mL of solution. Mix 51 mL of the monobasic potassium phosphate solution with 49 mL of the dibasic sodium phosphate solution. If necessary, adjust by the dropwise addition of the appropriate solution to a pH of 6.8.

Substrate Solution On the day of use, stir a portion of purified soluble starch equivalent to 2.0 g of dried substance with 10 mL of water, and add this mixture to 160 mL of water, add it to the hot solution, and heat to boiling, with continuous mixing. Cool to room temperature, and add water to make 200 mL.

Standard Preparation Weigh accurately about 20 mg of USP Pancreatin Reference Standard into a suitable mortar. Add about 30 mL of *pH 6.8 Phosphate Buffer*, and triturate for 5 to 10 min. Transfer the mixture with the aid of *pH 6.8 Phosphate Buffer* to a 50-mL volumetric flask, dilute with *pH 6.8 Phosphate Buffer* to volume, and mix. Calculate the activity, in USP Units of amylase activity per mL, of the resulting solution from the declared potency on the label of the Reference Standard.

Assay Preparation For Pancreatin having about the same amylase activity as the USP Pancreatin Reference Standard, weigh accurately about 40 mg of Pancreatin into a suitable mortar.

Note: For Pancreatin having a different amylase activity, weigh accurately the amount necessary to obtain an *Assay Preparation* having amylase activity per mL corresponding approximately to that of the *Standard Preparation*.

Add about 3 mL of *pH 6.8 Phosphate Buffer*, and triturate for 5 to 10 min. Transfer the mixture with the aid of *pH 6.8 Phosphate Buffer* to a 100-mL volumetric flask, dilute with *pH 6.8 Phosphate Buffer* to volume, and mix.

Procedure Prepare four stoppered, 250-mL conical flasks, and mark them *S*, *U*, *BS*, and *BU*. Pipet into each flask 25 mL of *Substrate Solution*, 10 mL of *pH 6.8 Phosphate Buffer*, and 1 mL of sodium chloride solution (11.7 in 1000), insert the stoppers, and mix. Place the flasks in a water bath maintained at 25° ± 0.1°, and allow them to equilibrate. To flasks *BU* and *BS* add 2 mL of 1 *N* hydrochloric acid, mix, and return the flasks to the water bath. To flasks *U* and *BU* add 1.0-mL portions of the *Assay Preparation*, and to flasks *S* and *BS* add 1.0 mL of the *Standard Preparation*. Mix each, and return the flasks to the water bath. After 10 min, accurately timed from the addition of the enzyme, add 2-mL portions of 1 *N* hydrochloric acid to flasks *S* and *U*, and mix. To each flask, with continuous stirring, add 10.0 mL of 1 *N* iodine volumetric solution, and immediately add 45 mL of 0.1 *N* sodium hydroxide. Place the flasks in the dark at a temperature between 15° and 25° for 15 min. To each flask add 4 mL of 2 *N* sulfuric acid, and titrate with 0.1 *N* sodium thiosulfate volumetric solution to the disappearance of the blue color. Calculate the amylase activity, in USP Units per mg, by the formula

$$100(C_S/W_U)(V_{BU} - V_U)/(V_{BS} - V_S),$$

in which C_S is the amylase activity of the *Standard Preparation*, in USP Units per mL; W_U is the amount, in mg, of Pancreatin taken; and V_U, V_S, V_{BU}, and V_{BS} are the volumes, in mL, of 0.1 *N* sodium thiosulfate consumed in the titration of the solutions in flasks, *U*, *S*, *BU*, and *BS*, respectively.

Lipase Activity
Acacia Solution Centrifuge a solution of acacia (1 in 10) until clear. Use only the clear solution.

Olive Oil Substrate Combine 165 mL of the *Acacia Solution*, 20 mL of olive oil, and 15 g of crushed ice in the cup of an electric blender. Cool the mixture in an ice bath to 5°, and homogenize at high speed for 15 min, intermittently cooling in an ice bath to prevent the temperature from exceeding 30°. Test for suitability of mixing as follows: Place a drop of the homogenate on a microscope slide and gently press a cover slide in place to spread the liquid. Examine the entire field under high power (43× magnification objective lens and 5× magnification ocular), using an eyepiece equipped with a calibrated micrometer. The substrate is satisfactory if 90% of the particles do not exceed 2 μm in diameter and none exceeds 10 μm in diameter.

Buffer Solution Dissolve 60 mg of tris(hydroxymethyl) aminomethane and 234 mg of sodium chloride in water to make 100 mL.

Sodium Taurocholate Solution Prepare a solution to contain 80.0 mg of USP Sodium Taurocholate Reference Standard in each mL.

Standard Test Dilution Suspend about 200 mg of USP Pancreatin Reference Standard, accurately weighed, in about 3 mL of cold water in a mortar, triturate for 10 min, and add cold water to a volume necessary to produce a concentration of 8 to 16 USP Units of lipase activity per mL, based on the declared potency on the label of the Reference Standard. Maintain the suspension at 4°, and mix before using. For each determination, withdraw 5 to 10 mL of the cold suspension, and allow the temperature to rise to 20° before pipeting the exact volume.

Assay Test Dilution Suspend about 200 mg of the Pancreatin sample, accurately weighed, in about 3 mL of cold water in a mortar, triturate for 10 min, and add cold water to a volume necessary to produce a concentration of 8 to 16 USP Units of lipase activity per mL, based on the estimated potency of the test material. Maintain the suspension at 4°, and mix before using. For each determination, withdraw 5 to 10 mL of the cold suspension, and allow the temperature to rise to 20° before pipeting the exact volume.

Procedure Mix 10.0 mL of *Olive Oil Substrate*, 8.0 mL of *Buffer Solution*, 2.0 mL of *Sodium Taurocholate Solution*, and 9.0 mL of water in a jacketed glass vessel of about 50-mL capacity, the outer chamber of which is connected to a thermostatically controlled water bath. Cover the mixture, and stir continuously with a mechanical stirring device.

With the mixture maintained at a temperature of 37° ± 0.1°, add 0.1 *N* sodium hydroxide, from a microburet inserted through an opening in the cover, to adjust potentiometrically the pH to 9.20, using a calomel-glass electrode system. Add 1.0 mL of *Assay Test Dilution*, and then continue adding the 0.1 *N* sodium hydroxide for 5 min to maintain the pH at 9.0. Determine the volume of 0.1 *N* sodium hydroxide added after each min.

In the same manner, titrate 1.0 mL of *Standard Test Dilution*.

Calculation From the *Standard Test Dilution*, plot the volume of 0.1 *N* sodium hydroxide titrated against time. Using only the points that fall on the straight-line segment of the curve, calculate the mean acidity released per min by the *Assay Test Dilution*. Taking into consideration dilution factors, calculate the lipase activity of the *Standard Test Dilution*, using the lipase activity of the USP Pancreatin Reference Standard stated on the label.

Protease Activity
Casein Substrate Place 1.25 g of finely powdered casein in a 100-mL conical flask containing 5 mL of water, shake to form a suspension, add 10 mL of 0.1 *N* sodium hydroxide, shake for 1 min, add 50 mL of water, and shake for about 1 h to dissolve the casein. Adjust the pH to about 8.0 ± 0.1, using 1 *N* sodium

hydroxide or 1 N hydrochloric acid. Transfer the solution to a 100-mL volumetric flask, dilute with water to volume, and mix. Use this substrate on the day it is prepared.

Buffer Solution Dissolve 6.8 g of monobasic potassium phosphate and 1.8 g of sodium hydroxide in 950 mL of water in a 1000-mL volumetric flask, adjust to a pH of 7.5 ± 0.2, using 0.2 N sodium hydroxide, dilute with water to volume, and mix. Store this solution in a refrigerator.

Trichloroacetic Acid Solution Dissolve 50 g of trichloroacetic acid in 1000 mL of water. Store this solution at room temperature.

Filter Paper Determine the suitability of the filter paper by filtering a 5-mL portion of *Trichloroacetic Acid Solution* through the paper and measuring the absorbance of the filtrate at 280 nm, using an unfiltered portion of the same *Trichloroacetic Acid Solution* as the blank: The absorbance is not more than 0.04. If the absorbance is more than 0.04, the filter paper may be washed repeatedly with *Trichloroacetic Acid Solution* until the absorbance of the filtrate, determined as above, is not more than 0.04.

Standard Test Dilution Add about 100 mg of USP Pancreatin Reference Standard, accurately weighed, to 100.0 mL of *Buffer Solution*, and mix by shaking intermittently at room temperature for about 25 min. Dilute quantitatively with *Buffer Solution* to produce a concentration of about 2.5 USP Units of protease activity per mL, based on the potency declared on the label of the Reference Standard.

Assay Test Dilution Add about 100 mg of USP Pancreatin Reference Standard, accurately weighed, to 100.0 mL of *Buffer Solution*, and mix by shaking intermittently at room temperature for 25 min. Dilute quantitatively with *Buffer Solution* to obtain a dilution that corresponds in activity to the *Standard Test Dilution*.

Procedure Label test tubes in duplicate S_1, S_2, and S_3 for the standard series, and U for the sample. Pipet into tubes S_1 2.0 mL, into S_2 and U 1.5 mL, and into S_3 1.0 mL of *Buffer Solution*. Pipet into tubes S_1 1.0 mL, into S_2 1.5 mL, and into S_3 2.0 mL of the *Standard Test Dilution*. Pipet into tubes U 1.5 mL of the *Assay Test Dilution*. Pipet into one tube each of S_1, S_2, S_3, and U 5.0 mL of *Trichloroacetic Acid Solution*, and mix. Designate these tubes as S_{1B}, S_{2B}, S_{3B}, and U_B, respectively. Prepare a blank by mixing 3 mL of *Buffer Solution* and 5 mL of *Trichloroacetic Acid Solution* in a separate test tube marked B. Place all the tubes in a 40° water bath, insert a glass stirring rod into each tube, and allow temperature equilibration. At zero time, add to each tube, at timed intervals, 2.0 mL of the *Casein Substrate*, preheated to the bath temperature, and mix. Accurately timed, 60 min after the addition of the *Casein Substrate*, stop the reaction in tubes S_1, S_2, S_3, and U by adding 5.0 mL of *Trichloroacetic Acid Solution* at the corresponding time intervals, stir, and remove all the tubes from the bath. Allow to stand for 10 min at room temperature to complete protein precipitation, and filter. The filtrates must be free from haze. Determine the absorbances of each filtrate, in a 1-cm cell, at 280 nm, with a suitable spectrophotometer, using the intake from the blank (tube B) to set the instrument.

Calculation Correct the absorbance values for the filtrates from tubes S_1, S_2, and S_3 by subtracting the absorbance values for the filtrates from tubes S_{1B}, S_{2B}, and S_{3B}, respectively, and plot the corrected absorbance values against the corresponding volumes of the *Standard Test Dilution* used. From the curve, using the corrected absorbance value ($U - U_B$ for the USP Pancreatin Reference Standard taken), and taking into consideration the dilution factors, calculate the protease activity, in USP Units, of the USP Pancreatin Reference Standard taken by comparison with that of the standard, using the protease activity stated on the label of USP Pancreatin Reference Standard.

PEPSIN ACTIVITY

Application This procedure is to be applied to preparations derived from porcine or other animal stomachs.

Apparatus

Measuring Vessels Use 100-mL, conically shaped measuring vessels complying with the following descriptions: (1) diameters not exceeding 1 cm at the bottom; (2) comply in other respects with the water and sediment tube ASTM Standard Method D96-68; (3) graduated from 0 to 0.5 mL in 0.05-mL graduations, from 2 to 3 mL in 0.1-mL graduations, from 3 to 5 mL in 0.2-mL graduations, from 5 to 10 mL in 1-mL graduations, from 10 to 25 mL in 5-mL graduations, and with graduation marks at 50, 75, and 100 mL.

Note: Measuring vessels other than the type described herein may be used if they are of such design and graduation to permit measurement of the residue with equivalent accuracy.

Reagents and Solutions

Hydrochloric Acid Solution Mix 35 mL of 1.0 N hydrochloric acid with 385 mL of water.

Substrate Boil one or more hen eggs for 15 min to provide coagulated albumen, (Miles, Inc.) and cool rapidly by immersion in cold water. Remove the shell and pellicle and all of the yolk, and at once rub the albumen through a clean, dry No. 40 sieve, rejecting the first portion that passes through the sieve.

Substrate Preparation Place 10 g of the *Substrate* in each of as many 100-mL wide-mouth bottles as needed for the test, and immediately add 35 mL of *Hydrochloric Acid Solution* (all at one time or in portions). By suitable means, thoroughly disintegrate the particles of albumen. Equilibrate to 52° before use in the *Procedure*, below.

Standard Preparation Dissolve 100 mg of USP Pepsin Reference Standard in 150 mL of *Hydrochloric Acid Solution*. Use this solution within 1 h.

Sample Preparation Dissolve 100 mg of the pepsin sample, or an amount of the enzyme preparation that will provide a solution similar to or slightly stronger than the *Standard Preparation*, in 150 mL of *Hydrochloric Acid Solution*. Use this solution within 1 h.

Procedure Pipet 5.0 mL of the *Standard Preparation* into each of two bottles containing the *Substrate Preparation*. To two or more additional substrate bottles add graduated aliquots of the *Sample Preparation* so that one bottle will contain approximately the same amount, and the others will contain successively lesser amounts, of pepsin as is contained in the 5.0 mL of the *Standard Preparation*, using, for example, 5.0, 4.9, and 4.8 mL. When less than 5.0 mL of the *Sample Preparation* is used, add sufficient *Hydrochloric Acid Solution* to make 5.0 mL of combined *Sample Preparation* plus acid added. At once stopper the bottles securely, invert them three times, and heat in a water bath, maintained at 52° ± 0.5°, for 2.5 h, agitating the contents equally every 10 min by inverting the bottles once. Remove the bottles from the bath, and pour the contents of each into separate measuring vessels.

Transfer the undigested albumen that adheres to the sides of the bottles into the respective measuring vessel with the aid of small portions of water until 50 mL has been used for each. Mix the contents of each vessel, allow them to stand for 30 min, and then read for each the volume of undigested albumen. Average the sediment volumes in the two standard vessels, and note which of the sample vessels contains undigested albumen closest to the average for the standards. Finally, record as v the volume, in mL, of *Sample Preparation* that produced the undigested albumen closest to the average produced by the *Standard Preparations*.

Calculation One pepsin unit is defined as that quantity of enzyme that digests 3000 times its weight of coagulated egg albumen under the conditions of the assay.

Calculate the activity of the enzyme preparation by the formula

$$\text{Pepsin units/mg} = 3000 \times (S/u) \times (5.0/v),$$

in which S is the weight, in mg, of USP Pepsin Reference Standard used to make the *Standard Preparation*; u is the weight, in mg, of enzyme preparation taken for analysis; and v is as defined in the *Procedure*.

PHOSPHOLIPASE A_2 ACTIVITY

Application and Principle This procedure is used to determine the phospholipase A_2 activity from extracts of porcine pancreatic tissue. The analysis is performed by potentiometric titration.

Apparatus
Automatic Titrator Use a suitable automatic recording titrator equipped with a stirred, thermostated, controlled-atmosphere titration cell (e.g., Radiometer Autotitrator).

Homogenizer Use a suitable homogenizer (e.g., Biomixer; Fisher Scientific, Catalog No. 11-504-2-4, or equivalent).
Constant-Temperature Water Bath Set at 40° ± 0.1°.

Reagents and Solutions
Calcium Chloride Solution (0.3 M) Transfer 4.41 g of calcium chloride dihydrate to a 100-mL volumetric flask, dissolve in, and dilute to volume with water.

Sodium Deoxycholate Solution (0.016 M) Dissolve 0.67 g of sodium deoxycholate (Sigma Chemical Co., Catalog. No. D6750) in 100 mL of water.

Sodium Hydroxide Solution (0.1 N) Use a standardized solution.

Substrate Solution Add the yolk of one fresh egg to 100 mL of deionized water and homogenize until a stable emulsion is obtained. Add 5 mL of the *Calcium Chloride Solution*, and mix.

Sample Preparation Dissolve an accurately weighed amount of enzyme preparation in 0.001 N hydrochloric acid, and dilute to obtain an enzyme activity of 10 to 80 units of activity per mL.

Procedure Pre-equilibrate the *Substrate Solution*, the *Sodium Deoxycholate Solution*, and about 50 mL of water to 40° in the water bath. Transfer 10 mL of the *Substrate Solution* to the thermostated titration vessel. Add 5 mL of the *Sodium Deoxycholate Solution* and 10 mL of deionized water. Blanket the cell with nitrogen and equilibrate for approximately 5 min. Using the *Automatic Titrator* filled with *0.1 N Sodium Hydroxide Solution*, adjust the pH of the solution to 8.0 ± 0.05. Monitor the consumption (if any) of sodium hydroxide for 5 min as a blank. Refill the *Automatic Titrator*. Add 0.1 mL of *Sample Solution* containing between 1 and 8 units of activity and start the *Automatic Titrator*. Record the sodium hydroxide consumption for at least 5 min.

Calculation One phospholipase unit is defined as the quantity of enzyme that produces 1 microequivalent of free fatty acid per min under the conditions of the test. Determine the rate, R, of titrant consumption during 0 to 3 min of the reaction.

Note: The recorder trace must be linear during the first 3 min of the reaction.

Determine the rate of titrant consumption (if any) during equilibration (blank) (R_B):

$$\text{Units/g} = (R \times N) - (R_B \times N)/W,$$

in which R and R_B are the rates of titrant consumption of the sample and blank, in µL/min; N is the normality of the titrant; and W is the weight, in g, contained in 0.1 mL of the *Sample Preparation* taken for the test.

PHYTASE ACTIVITY

Application and Principle This procedure is used to determine the activity of enzymes releasing phosphate from phytate.

The assay is based on enzymatic hydrolysis of sodium phytate under controlled conditions by measurement of the amount of ortho phosphate released.

Reagents and Solutions (Note: All glassware must be acid washed, rinsed, and scrupulously cleaned to ensure the absence of phosphate.)

Acetate Buffer (pH 5.5) Dissolve 1.76 g of 100% acetic acid ($C_2H_4O_2$), 30.02 g of sodium acetate trihydrate ($C_2H_3O_2Na \cdot 3H_2O$), and 0.147 g of calcium chloride dihydrate in about 900 mL of water. Transfer the solution into a 1000-mL volumetric flask, dilute to volume with water, and mix. The pH should be 5.50 ± 0.05.

Substrate Solution Dissolve 8.40 g of sodium phytate decahydrate ($C_6H_6O_{24}P_6Na_{12} \cdot 10H_2O$) (Sigma Chemical Co.) in 900 mL of *Acetate Buffer*. Adjust the pH to 5.50 ± 0.05 at 37.0° ± 0.1° by adding 4 M acetic acid. Cool to ambient temperature. Quantitatively transfer the mixture to a 1000-mL volumetric flask, dilute to volume with *Acetate Buffer*, and mix. Prepare fresh daily.

Nitric Acid Solution (27%) While stirring, slowly add 70 mL of 65% nitric acid to 130 mL of water.

Ammonium Heptamolybdate Solution Dissolve 100 g of ammonium heptamolybdate tetrahydrate [$(NH_4)_6Mo_7O_{24} \cdot 4H_2O$] in 900 mL of water in a 1000-mL volumetric flask. Add 10 mL of 25% ammonia solution, dilute to volume with water, and mix. This solution is stable for 4 weeks when stored at ambient temperature and shielded from light.

Ammonium Vanadate Solution Dissolve 2.35 g of ammonium monovanadate (NH_4VO_3) in 400 mL of warm (60°) water. While stirring, slowly add 20 mL of *Nitric Acid Solution* (27%). Cool to ambient temperature. Quantitatively transfer the mixture to a 1000-mL volumetric flask, dilute to volume with water, and mix. This solution is stable for 4 weeks when stored at ambient temperature and shielded from light.

Color/Stop Solution While stirring, add 250 mL of *Ammonium Vanadate Solution* to 250 mL of *Ammonium Heptamolybdate Solution*. Slowly add 165 mL of 65% nitric acid. Cool to ambient temperature. Quantitatively transfer the mixture to a 1000-mL volumetric flask, dilute to volume with water, and mix. Prepare fresh daily.

Potassium Dihydrogen Phosphate Solution Dry a sufficient amount of potassium dihydrogen phosphate (KH_2PO_4) in a vacuum oven at 100° to 104° for 2 h. Cool to ambient temperature in a desiccator over dried silica gel.

In duplicate (solutions A and B), weigh approximately 0.245 g of dried potassium dihydrogen phosphate accurately to within 1 mg and dilute with *Acetate Buffer* to 1 L to obtain solutions containing 1.80 mmol/L of potassium dihydrogen phosphate.

Phytase Reference Preparation (Highly concentrated phytase preparation) This preparation can be obtained from Gist-Brocades, Delft, The Netherlands, with an assigned activity (by collaborative assay), or the activity of the reference preparation can be determined according to *Procedure 2*.

Phytase Reference Solutions, Procedure 1 Weigh an amount of *Phytase Reference Preparation* corresponding with 20,000 phytase units accurately to within 1 mg in duplicate in 200-mL volumetric flasks. Dissolve in and dilute to volume with *Acetate Buffer*, and mix. Dilute with *Acetate Buffer* to obtain dilutions containing approximately 0.01, 0.02, 0.04, 0.06, and 0.08 phytase units per 2.0 mL of the final dilution.

Sample Preparation, Procedure 1 Suspend or dissolve and dilute accurately weighed amounts of sample in *Acetate Buffer* so that 2.0 mL of the final dilution will contain between 0.02 and 0.08 phytase units.

Sample Preparation, Procedure 2 In duplicate, accurately weigh amounts of *Phytase Reference Preparation* and dissolve and dilute in *Acetate Buffer* to obtain dilutions containing 0.06 ± 0.006 phytase units per 2.0 mL of the final dilution.

Procedures

Procedure 1 (Determination of the phytase activity) Transfer 2.00 mL of the *Sample Preparation, Procedure 1*, and the *Phytase Reference Solutions, Procedure 1*, into separate 20- × 150-mm glass test tubes. Using a stopwatch and starting at time equals zero, in the order of the series and within regular time intervals, place the tubes into a 37.0° ± 0.1° water bath and allow their contents to equilibrate for 5 min. At time equals 5 min, in the same order of the series and with the same time intervals, add 4.0 mL of *Substrate Solution* (previously equilibrated to 37.0° ± 0.1°) to each test tube. Mix, and replace in the 37.0° ± 0.1° water bath. At time equals 65 min, in the same order and within the same time intervals, terminate the incubation by adding 4.0 mL of *Color/Stop Solution*. Mix, and cool to ambient temperature.

Prepare blanks by transferring 2.00 mL of the *Sample Preparation, Procedure 1*, and the *Phytase Reference Solutions, Procedure 1*, into separate 20- × 150-mm glass test tubes. Using a stopwatch and starting at time equals zero, in the order of the series and within regular time intervals, place the tubes into a 37.0° ± 0.1° water bath and allow them to equilibrate for 5 min. At time equal 5 min, in the same order of the series and within the same time intervals, add 4.0 mL of *Color/Stop Solution*. Mix, and cool to ambient temperature. Next add 4.00 mL of *Substrate Solution* to the blank tubes, and mix.

Centrifuge all test tubes for 5 min at 3000 × g. Determine the absorbance of each solution at 415 nm in a 1-cm path-length cell with a suitable spectrophotometer, using water to zero the instrument.

Procedure 2 (Determination of the phytase activity of the *Phytase Reference Preparation*) Transfer 2.00 mL of *Sample Preparation, Procedure 2*, and 2.00 mL (three times from *Potassium Dihydrogen Phosphate Solution A* and two times from *B*) of *Potassium Dihydrogen Phosphate Solutions* into separate 20- × 150-mm glass test tubes. Using a stopwatch and starting at time equals zero, in the order of the series and within regular time intervals, place the tubes into a 37.0° ± 0.1° water bath and allow their contents to equilibrate for 5 min. At time equals 5 min, in the same order of the series and within the same time intervals, add 4.0 mL of *Substrate Solution* (previously equilibrated to 37.0° ± 0.1°) to the test tubes. Mix, and replace in the 37.0° ± 0.1° water bath. At time equals 35 min, in the same order and within the same time intervals, terminate the incubation by adding 4.0 mL of *Color/Stop Solution*. Mix, and cool to ambient temperature.

Prepare blanks by transferring 2.00 mL of *Sample Preparation, Procedure 2*, into separate 20- × 150-mm glass test tubes.

Prepare *Reagent Blanks* by transferring 2.00 mL of water into a series of five separate 20- × 150-mm glass test tubes. Add 4.0 mL of *Color/Stop Solution* to all blank tubes, and mix. Next add 4.0 mL of *Substrate Solution*, and mix. Determine the absorbance of each solution at 415 nm in a 1-cm path-length cell with a suitable spectrophotometer, using water to zero the instrument.

Calculations

Calculation, Procedure 1   One Phytase (fytase) unit (FTU) is the amount of enzyme that liberates inorganic phosphate at 1 μmol/min from sodium phytate 0.0051 mol/L at 37.0° at pH 5.50 under the conditions of the test. Calculate the corrected absorbance (sample minus blank) for each *Sample Preparation* and *Phytase Reference Solution*. Plot the accurately calculated phytase activity (FTU per 2 mL) of each *Phytase Reference Solution* against the corresponding absorbance. From the curve, determine the phytase activity in each *Sample Preparation* (FTU per 2 mL):

Activity (FTU/g) = (FTU per 2 mL × dilution)/ sample weight (g).

Calculation, Procedure 2   Calculate the corrected absorbances, A_R, for each *Sample Preparation* (absorbance *Phytase Reference Solution* minus corresponding absorbance blank) and for each *Potassium Dihydrogen Phosphate Solution*, A_P (absorbance *Potassium Dihydrogen Phosphate Solution* minus average absorbance reagent blank). Calculate C, the phosphate concentration of each *Potassium Dihydrogen Phosphate Solution*:

$(W \times 1000 \times 2)/MW = C$ (mmol/2 mL).

Calculate the absorbances, D, for each *Potassium Dihydrogen Phosphate Solution* after correction for the amount of potassium dihydrogen phosphate weighed:

$A_P/C = D$ (absorbance units/mmol of phosphate per 2 mL).

Calculate the average of results D, giving E (maximum allowable difference, 5%). Calculate the activity for each *Phytase Reference Preparation*:

$(A_R \times f)/(30 \times R \times E) = $ FTU/g,

in which A_R equals the corrected absorbance of the *Phytase Standard Solution*; f equals the total dilution factor of the reference preparation; 30 equals the incubation time, in min; R equals sample weight, in g; E equals average of D factors; W equals the weight of potassium dihydrogen phosphate, in g; and MW equals the molecular weight of potassium dihydrogen phosphate, 136.09 (g/mol).

PLANT PROTEOLYTIC ACTIVITY

Application and Principle   This procedure is used to determine the proteolytic activity of papain, ficin, and bromelain. The assay is based on a 60-min proteolytic hydrolysis of a casein substrate at pH 6.0 and 40°. Unhydrolyzed substrate is precipitated with trichloroacetic acid and removed by filtration; solubilized casein is then measured spectrophotometrically.

Reagents and Solutions

Sodium Phosphate Solution (0.05 M)   Transfer 7.1 g of anhydrous dibasic sodium phosphate into a 1000-mL volumetric flask, dissolve in about 500 mL of water, dilute to volume with water, and mix. Add 1 drop of toluene as a preservative.

Citric Acid Solution (0.05 M)   Transfer 10.5 g of citric acid monohydrate into a 1000-mL volumetric flask, dissolve in about 500 mL of water, dilute to volume with water, and mix. Add 1 drop of toluene as a preservative.

Phosphate–Cysteine–EDTA Buffer Solution   Dissolve 7.1 g of anhydrous dibasic sodium phosphate in about 800 mL of water, and then dissolve in this solution 14.0 g of disodium EDTA dihydrate and 6.1 g of cysteine hydrochloride monohydrate.

Adjust to pH 6.0 ± 0.1 with 1 N hydrochloric acid or 1 N sodium hydroxide, then transfer into a 1000-mL volumetric flask, dilute to volume with water, and mix.

Trichloroacetic Acid Solution   Dissolve 30 g of trichloroacetic acid in 100 mL of water.

Casein Substrate Solution   Disperse 1 g (moisture-free basis) of Hammarsten-grade casein (United States Biochemical Corp., Catalog No. 12840, or equivalent) in 50 mL of *Sodium Phosphate Solution*, and heat for 30 min in a boiling water bath, with occasional agitation. Cool to room temperature, and with rapid and continuous agitation, adjust to pH 6.0 ± 0.1 by the addition of *Citric Acid Solution*.

Note: Rapid and continuous agitation during the addition prevents casein precipitation.

Quantitatively transfer the mixture into a 100-mL volumetric flask, dilute to volume with water, and mix.

Stock Standard Solution   Transfer 100.0 mg of USP Papain Reference Standard into a 100-mL volumetric flask, dissolve, and dilute to volume with *Phosphate–Cysteine–EDTA Buffer Solution*, and mix.

Diluted Standard Solutions   Pipet 2, 3, 4, 5, 6, and 7 mL of *Stock Standard Solution* into a series of 100-mL volumetric flasks, dilute each to volume with *Phosphate–Cysteine–EDTA Buffer Solution*, and mix by inversion.

Test Solution   Prepare a solution from the enzyme preparation so that 2 mL of the final dilution will give a ΔA in the *Procedure* between 0.2 and 0.5. Weigh the sample accurately, transfer it quantitatively to a glass mortar, and triturate with *Phosphate–Cysteine–EDTA Buffer Solution*. Transfer the mixture quantitatively into a volumetric flask of appropriate size, dilute to volume with *Phosphate–Cysteine–EDTA Buffer Solution*, and mix.

Procedure   Pipet 5 mL of *Casein Substrate Solution* into each of a series of 25- × 150-mm test tubes, allowing three tubes for the enzyme unknown, six for a papain standard curve, and nine for enzyme blanks. Equilibrate the tubes for 15 min in a water bath maintained at 40° ± 0.1°. Starting the stopwatch at zero time, rapidly pipet 2 mL of each of the *Diluted Standard Solutions*, and 2-mL portions of the *Test Solution*, into the equilibrated substrate. Mix each by swirling, stopper, and place the tubes back in the water bath. After 60.0 min, add 3 mL of *Trichloroacetic Acid Solution* to each tube. Immediately mix each tube by swirling.

Prepare enzyme blanks containing 5.0 mL of *Casein Substrate Solution*, 3.0 mL of *Trichloroacetic Acid Solution*, and 2.0 mL of one of the appropriate *Diluted Standard Solutions* or the *Test Solution*.

Return all tubes to the water bath, and heat for 30.0 min, allowing the precipitated protein to coagulate completely. Filter each mixture through Whatman No. 42, or equivalent, filter paper, discarding the first 3 mL of filtrate. The subsequent filtrate must be perfectly clear. Determine the absorbance of each filtrate in a 1-cm cell at 280 nm, with a suitable spectrophotometer, against its respective blank.

Calculation One papain unit (PU) is defined in this assay as that quantity of enzyme that liberates the equivalent of 1 μg of tyrosine per h under the conditions of the assay.

Prepare a standard curve by plotting the absorbances of filtrates from the *Diluted Standard Solutions* against the corresponding enzyme concentrations, in mg/mL. By interpolation from the standard curve, obtain the equivalent concentration of the filtrate from the *Test Solution*.

Calculate the activity of the enzyme preparation taken for analysis as follows:

$$PU/mg = (A \times C \times 10)/W,$$

in which A is the activity of USP Papain Reference Standard, in PU per mg; C is the concentration, in mg/mL, of Reference Standard from the standard curve, equivalent to the enzyme unknown; 10 is the total volume, in mL, of the final incubation mixture; and W is the weight, in mg, of original enzyme preparation in the 2-mL aliquot of *Test Solution* added to the incubation mixture.

PROTEOLYTIC ACTIVITY, BACTERIAL (PC)

Application and Principle This procedure is used to determine protease activity, expressed as PC units, of preparations derived from *Bacillus subtilis* var. and *Bacillus licheniformis* var. The assay is based on a 30-min proteolytic hydrolysis of casein at 37° and pH 7.0. Unhydrolyzed casein is removed by filtration, and the solubilized casein is determined spectrophotometrically.

Reagents and Solutions

Casein Use Hammarsten-grade casein (United States Biochemical Corp., Catalog No. 12840, or equivalent).

Tris Buffer (pH 7.0) Dissolve 12.1 g of enzyme-grade (or equivalent) tris(hydroxymethyl)aminomethane in 800 mL of water, and titrate with 1 N hydrochloric acid to pH 7.0. Transfer into a 1000-mL volumetric flask, dilute to volume with water, and mix.

TCA Solution Dissolve 18 g of trichloroacetic acid and 19 g of sodium acetate trihydrate in 800 mL of water in a 1000-mL volumetric flask, add 20 mL of glacial acetic acid, dilute to volume with water, and mix.

Substrate Solution Dissolve 6.05 g of enzyme-grade tris(hydroxymethyl)aminomethane in 500 mL of water, add 8 mL of 1 N hydrochloric acid, and mix. Dissolve 7 g of *Casein* in this solution, and heat for 30 min in a boiling water bath, stirring occasionally.

Cool to room temperature, and adjust to pH 7.0 with 0.2 N hydrochloric acid, adding the acid slowly, with vigorous stirring, to prevent precipitation of the casein. Transfer the mixture into a 1000-mL volumetric flask, dilute to volume with water, and mix.

Sample Preparation Using *Tris Buffer*, prepare a solution of the sample enzyme preparation so that 2 mL of the final dilution will contain between 10 and 44 bacterial protease units.

Procedure Pipet 10.0 mL of the *Substrate Solution* into each of a series of 25- × 150-mm test tubes, allowing one tube for each enzyme test, one tube for each enzyme blank, and one tube for a substrate blank. Equilibrate the tubes for 15 min in a water bath maintained at 37° ± 0.1°.

Starting the stopwatch at zero time, rapidly pipet 2.0 mL of the *Sample Preparation* into the equilibrated substrate. Mix, and replace the tubes in the water bath. Add 2 mL of *Tris Buffer* (instead of the *Sample Preparation*) to the substrate blank. After exactly 30 min, add 10 mL of *TCA Solution* to each enzyme incubation and to the substrate blank to stop the reaction. Heat the tubes in the water bath for an additional 30 min to allow the protein to coagulate completely.

At the end of the second heating period, shake each tube vigorously, and filter through 11-cm Whatman No. 42, or equivalent, filter paper, discarding the first 3 mL of filtrate.

Note: The filtrate must be perfectly clear.

Determine the absorbance of each sample filtrate in a 1-cm cell, at 275 nm, with a suitable spectrophotometer, using the filtrate from the substrate blank to set the instrument at zero. Correct each reading by subtracting the appropriate enzyme blank reading, and record the value so obtained as A_U.

Standard Curve Transfer 100.0 mg of L-tyrosine, chromatographic-grade or equivalent (Aldrich Chemical Co.), previously dried to constant weight, to a 1000-mL volumetric flask. Dissolve in 60 mL of 0.1 N hydrochloric acid. When completely dissolved, dilute the solution to volume with water, and mix thoroughly. This solution contains 100 μg of tyrosine in 1.0 mL. Prepare three more dilutions from this stock solution to contain 75.0, 50.0, and 25.0 μg of tyrosine per mL. Determine the absorbance of the four solutions at 275 nm in a 1-cm cell on a suitable spectrophotometer versus 0.006 N hydrochloric acid. Prepare a plot of absorbance versus tyrosine concentration.

Calculation One bacterial protease unit (PC) is defined as that quantity of enzyme that produces the equivalent of 1.5 μg/mL of L-tyrosine per min under the conditions of the assay.

From the *Standard Curve*, and by interpolation, determine the absorbance of a solution having a tyrosine concentration of 60 μg/mL. A figure close to 0.0115 should be obtained. Divide the interpolated value by 40 to obtain the absorbance equivalent to that of a solution having a tyrosine concentration of 1.5 µg/mL, and record the value thus derived as A_S.

Calculate the activity of the sample enzyme preparation by the formula

$$PC/g = (A_U/A_S) \times (22/30W),$$

in which 22 is the final volume, in mL, of the reaction mixture; 30 is the time, in min, of the reaction; and W is the weight, in g, of the original sample taken.

PROTEOLYTIC ACTIVITY, FUNGAL (HUT)

Application and Principle This procedure is used to determine the proteolytic activity, expressed as hemoglobin units on the tyrosine basis (HUT), of preparations derived from *Aspergillus oryzae* var. and *Aspergillus niger* var., and it may be used to determine the activity of other proteases at pH 4.7. The test is based on the 30-min enzymatic hydrolysis of a hemoglobin substrate at pH 4.7 and 40°. Unhydrolyzed substrate is precipitated with trichloroacetic acid and removed by filtration. The quantity of solubilized hemoglobin in the filtrate is determined spectrophotometrically.

Reagents and Solutions

Hemoglobin Use Hemoglobin Substrate Powder (Sigma Chemical Co., Catalog No. H2625) or a similar high-grade material that is completely soluble in water.

Acetate Buffer Solution Dissolve 136 g of sodium acetate ($NaC_2H_3O_2 \cdot 3H_2O$) in sufficient water to make 500 mL. Mix 25.0 mL of this solution with 50.0 mL of 1 M acetic acid, dilute to 1000 mL with water, and mix. The pH of this solution should be 4.7 ± 0.02.

Substrate Solution Transfer 4.0 g of the *Hemoglobin* into a 250-mL beaker, add 100 mL of water, and stir for 10 min to dissolve. Immerse the electrodes of a pH meter in the solution, and while stirring continuously, adjust the pH to 1.7 by adding 0.3 N hydrochloric acid. After 10 min, adjust the pH to 4.7 by adding 0.5 M sodium acetate. Transfer the solution into a 200-mL volumetric flask, dilute to volume with water, and mix. This solution is stable for about 5 days when refrigerated.

Trichloroacetic Acid Solution Dissolve 140 g of trichloroacetic acid in about 750 mL of water. Transfer the solution to a 1000-mL volumetric flask, dilute to volume with water, and mix thoroughly.

Sample Preparation Dissolve an amount of the sample in the *Acetate Buffer Solution* to produce a solution containing, in each mL, between 9 and 22 HUT. (Such a concentration will produce an absorbance reading, in the procedure below, within the preferred range of 0.2 to 0.5.)

Procedure Pipet 10.0 mL of the *Substrate Solution* into each of a series of 25- × 150-mm test tubes: One for each enzyme test and one for the substrate blank. Heat the tubes in a water bath at 40° for about 5 min. To each tube, except the substrate blank, add 2.0 mL of the *Sample Preparation*, and begin timing the reaction at the moment the solution is added; add 2.0 mL of the *Acetate Buffer Solution* to the substrate blank tube. Close the tubes with No. 4 rubber stoppers, and tap each tube gently for 30 s against the palm of the hand to mix. Heat each tube in a water bath at 40° for exactly 30 min, and then rapidly pipet 10.0 mL of the *Trichloroacetic Acid Solution* into each tube. Shake each tube vigorously against the stopper for about 40 s, and then allow to cool to room temperature for 1 h, shaking each tube against the stopper at 10- to 12-min intervals during this period. Prepare enzyme blanks as follows: Heat, in separate tubes, 10.0 mL of the *Substrate Solution* and about 5 mL of the *Sample Preparation* in the water bath for 30 min, then add 10.0 mL of the *Trichloroacetic Acid Solution* to the *Substrate Solution*, shake well for 40 s, and to this mixture add 2.0 mL of the preheated *Sample Preparation*. Shake again, and cool at room temperature for 1 h, shaking at 10- to 12-min intervals.

At the end of 1 h, shake each tube vigorously, and filter through 11-cm Whatman No. 42, or equivalent, filter paper, refiltering the first half of the filtrate through the same paper. Determine the absorbance of each filtrate in a 1-cm cell, at 275 nm, with a suitable spectrophotometer, using the filtrate from the substrate blank to set the instrument to zero. Correct each reading by subtracting the appropriate enzyme blank reading, and record the value so obtained as A_U.

Note: If a corrected absorbance reading between 0.2 and 0.5 is not obtained, repeat the test using more or less of the enzyme preparation as necessary.

Standard Curve Transfer 100.0 mg of L-tyrosine, chromatographic-grade, or equivalent (Aldrich Chemical Co.), previously dried to constant weight, to a 1000-mL volumetric flask. Dissolve in 60 mL of 0.1 N hydrochloric acid. When the L-tyrosene is completely dissolved, dilute the solution to volume with water, and mix thoroughly. This solution contains 100 µg of tyrosine in 1.0 mL. Prepare three more dilutions from this stock solution to contain 75.0, 50.0, and 25.0 µg of tyrosine per mL. Determine the absorbance of the four solutions at 275 nm in a 1-cm cell on a suitable spectrophotometer versus 0.006 N hydrochloric acid. Prepare a plot of absorbance versus tyrosine concentration. Determine the slope of the curve in terms of absorbance per µg of tyrosine. Multiply this value by 1.10, and record it as A_S. A value of approximately 0.0084 should be obtained.

Calculation One HUT unit of proteolytic (protease) activity is defined as that amount of enzyme that produces, in 1 min under the specified conditions, a hydrolysate whose absorbance at 275 nm is the same as that of a solution containing 1.10 µg/mL of tyrosine in 0.006 N hydrochloric acid.

Calculate the HUT per g of the original enzyme preparation by the formula

$$HUT/g = (A_U/A_S) \times (22/30W),$$

in which 22 is the final volume of the test solution; 30 is the reaction time, in min; and W is the weight, in g, of the original sample taken.

Note: The value for A_S, under carefully controlled and standardized conditions, is 0.0084; this value may be used for routine work in lieu of the value obtained from the standard curve, but the exact value calculated from the standard curve should be used for more accurate results and in cases of doubt.

PROTEOLYTIC ACTIVITY, FUNGAL (SAP)

Application and Principle This procedure is used to determine proteolytic activity, expressed in spectrophotometric acid protease units (SAP), of preparations derived from *Aspergillus niger* var. and *Aspergillus oryzae* var. The test is based on a 30-min enzymatic hydrolysis of a Hammarsten Casein Substrate at pH 3.0 and 37°. Unhydrolyzed substrate is precipitated with trichloroacetic acid and removed by filtration. The quantity of solubilized casein in the filtrate is determined spectrophotometrically.

Reagents and Solutions

Casein Use Hammarsten-grade casein (United States Biochemical Corp., Catalog No. 12840, or equivalent).

Glycine–Hydrochloric Acid Buffer (0.05 M) Dissolve 3.75 g of glycine in about 800 mL of water. Add 1 N hydrochloric acid until the solution is pH 3.0, determined with a pH meter. Quantitatively transfer the solution to a 1000-mL volumetric flask, dilute to volume with water, and mix.

TCA Solution Dissolve 18.0 g of trichloroacetic acid and 11.45 g of anhydrous sodium acetate in about 800 mL of water, and add 21.0 mL of glacial acetic acid. Quantitatively transfer the solution to a 1000-mL volumetric flask, dilute to volume with water, and mix.

Substrate Solution Pipet 8 mL of 1 N hydrochloric acid into about 500 mL of water, and with continuous agitation, disperse 7.0 g (moisture-free basis) of *Casein* into this solution. Heat for 30 min in a boiling water bath, stirring occasionally, and cool to room temperature. Dissolve 3.75 g of glycine in the solution, and using a pH meter, adjust to pH 3.0 with 0.1 N hydrochloric acid. Quantitatively transfer the solution to a 1000-mL volumetric flask, dilute to volume with water, and mix.

Sample Preparation Using *Glycine–Hydrochloric Acid Buffer*, prepare a solution of the sample enzyme preparation so that 2 mL of the final dilution will give a corrected absorbance of enzyme incubation filtrate at 275 nm (ΔA, as defined in the *Procedure*) between 0.200 and 0.500. Weigh the enzyme preparation, quantitatively transfer it to a glass mortar, and triturate with *Glycine–Hydrochloric Acid Buffer*. Quantitatively transfer the mixture to an appropriately sized volumetric flask, dilute to volume with *Glycine–Hydrochloric Acid Buffer*, and mix.

Procedure Pipet 10.0 mL of *Substrate Solution* into each of a series of 25- × 150-mm test tubes, allowing at least two tubes for each sample, one for each enzyme blank, and one for a substrate blank. Stopper the tubes, and equilibrate them for 15 min in a water bath maintained at 37° ± 0.1°.

At zero time, start the stopwatch, and rapidly pipet 2.0 mL of the *Sample Preparation* into the equilibrated substrate. Mix by swirling, and replace the tubes in the water bath.

Note: Keep the tubes stoppered during incubation.

Add 2 mL of *Glycine–Hydrochloric Acid Buffer* (instead of the *Sample Preparation*) to the substrate blank. After exactly 30 min, add 10 mL of *TCA Solution* to each enzyme incubation and to the substrate blank to stop the reaction. In the following order, prepare an enzyme blank containing 10 mL of *Substrate Solution*, 10 mL of *TCA Solution*, and 2 mL of the *Sample Preparation*. Heat all tubes in the water bath for 30 min, allowing the precipitated protein to coagulate completely.

At the end of the second heating period, cool the tubes in an ice bath for 5 min, and filter through Whatman No. 42 filter paper, or equivalent. The filtrates must be perfectly clear. Determine the absorbance of each filtrate in a 1-cm cell at 275 nm with a suitable spectrophotometer, against the substrate blank. Correct each absorbance by subtracting the absorbance of the respective enzyme blank.

Standard Curve Transfer 181.2 mg of L-tyrosine, chromatographic-grade or equivalent (Sigma Chemical Co.), previously dried to constant weight, to a 1000-mL volumetric flask. Dissolve in 60 mL of 0.1 N hydrochloric acid. When the L-tyrosine is completely dissolved, dilute the solution to volume with water, and mix thoroughly. This solution contains 1.00 μmol of tyrosine per 1.0 mL. Prepare dilutions from this stock solution to contain 0.10, 0.20, 0.30, 0.40, and 0.50 μmol/mL. Determine against a water blank the absorbance of each dilution in a 1-cm cell at 275 nm. Prepare a plot of absorbance versus μmol of tyrosine per mL. A straight line must be obtained. Determine the slope and intercept for use in the *Calculation* below. A value close to 1.38 should be obtained. The slope and intercept may be calculated by the least squares method as follows:

$$\text{Slope} = [n\Sigma(MA) - \Sigma(M)\Sigma(A)]/[n\Sigma(M^2) - (\Sigma M)^2],$$

$$\text{Intercept} = [\Sigma(A)\Sigma(M^2) - \Sigma(M)\Sigma(MA)]/[n\Sigma(M^2) - (\Sigma M)^2],$$

in which n is the number of points on the standard curve, M is the μmol of tyrosine per mL for each point on the standard curve, and A is the absorbance of the sample.

Calculation One spectrophotometric acid protease unit is that activity that will liberate 1 μmol of tyrosine per min under the conditions specified. The activity is expressed as follows:

$$\text{SAP}/g = (\Delta A - I) \times 22/(S \times 30 \times W),$$

in which ΔA is the corrected absorbance of the enzyme incubation filtrate; I is the intercept of the *Standard Curve*; 22 is the final volume of the incubation mixture, in mL; S is the slope of the *Standard Curve*; 30 is the incubation time, in min; and W is the weight, in g, of the enzyme sample contained in the 2.0-mL aliquot of *Sample Preparation* added to the incubation mixture in the *Procedure*.

PULLULANASE ACTIVITY

Application and Principle This procedure is used to determine pullulanase activity derived from *Bacillus acidopullulyticus*. The method is based on measuring the increase in reducing sugars formed by a 30-min hydrolysis of pullulan at 40° and pH 5.0. The increase in reducing sugars is measured spectrophotometrically at 520 nm using a modified Nelson–Somogyi procedure.

Reagents and Solutions

Citrate Buffer (pH 5.0) Dissolve 10.5 g of citric acid monohydrate in 950 mL of water, adjust the pH to 5.0 ± 0.05 using 5 N sodium hydroxide, and dilute to 1000 mL.

Nelson's Color Reagent Dissolve 25.0 g of ammonium molybdate tetrahydrate in 300 mL of water. Carefully add 20.0 mL of concentrated sulfuric acid while stirring. Dissolve 3.0 g of sodium arsenate heptahydrate in 25 mL of water. Slowly add this solution to the ammonium molybdate solution with stirring. Dilute this solution to 500 mL with water.

Somogyi's Copper Reagent Dissolve 14.0 g of anhydrous dibasic sodium phosphate and 20.0 g of potassium sodium tartrate tetrahydrate into 250 mL of water. Add 60.0 g of 1 M sodium hydroxide solution. Dissolve 4.0 g of cupric sulfate pentahydrate into 25 mL of water. Add this solution to the tartrate solution. Add 90.0 g of anhydrous sodium sulfate while stirring. Dilute the final solution to 500 mL.

Glucose Standards Dissolve 800 mg of previously dried anhydrous D-glucose in 100 mL of the *Citrate Buffer*. Prepare glucose standards containing 16, 40, 80, and 120 μg/mL of glucose.

Pullulan Substrate Dissolve 150 mg of pullulan (Sigma Chemical Co., Catalog No. P-4516, or equivalent) in 49.80 g of the *Citrate Buffer*. Prepare daily.

Sample Preparation Dissolve an accurately weighed amount of the enzyme preparation in *Citrate Buffer* and dilute in *Citrate Buffer* to obtain an enzyme activity of 0.01 to 0.03 activity units per mL.

Procedure Transfer 1.0-mL aliquots of *Pullulan Substrate* to separate 15- × 150-mm test tubes. Insert a one-hole stopper in each tube, and equilibrate for 15 min in a 40° ± 0.1° water bath. At time equals zero and at 30-s intervals, add 1 mL of the respective samples, and mix. Exactly 30.0 min after addition of the samples, add 2.0 mL of *Somogyi's Copper Reagent* to each tube to terminate the reaction. Mix thoroughly, and allow the tubes to come to room temperature. To obtain sample blanks, add the *Somogyi's Copper Reagent* to the sample before adding the substrate.

Prepare a standard curve by adding 2.0 mL of each glucose standard and 2.0 mL of *Somogyi's Copper Reagent* to a test tube, and mix. The blanks and *Glucose Standards* should not be incubated at 40°.

Loosely stopper samples, blanks, and *Glucose Standards* containing *Somogyi's Copper Reagent*, and place them in a vigorously boiling water bath for exactly 25.0 min. Cool in an ice-water bath for approximately 1 min.

Add 2.00 mL of *Nelson's Color Reagent* to each tube, and mix thoroughly to dissolve any red precipitate that might be present. Let the solutions stand for 5 min. Add 2.0 mL of water to each tube, and mix.

Measure the absorbance of all solutions at 520 nm, using water as the reference. Mix the contents of each tube before transferring them to the cuvette.

Calculations One pullulanase unit (PUN) is the amount of activity that under the conditions of the test, will liberate reducing sugars equivalent to 1 μmol of glucose per min. Determine the linear regression line for absorbance versus 2 times the glucose concentration (μg/mL) in the standards. Use the slope, α, in the following equation to determine the activity in the enzyme preparation:

$$\text{PUN/g} = (A_S - A_B)/\alpha \times W \times 180 \times 30,$$

in which A_S is the absorbance of the sample; A_B is the absorbance of the blank; W is the weight, in g, of the enzyme preparation contained in the 1.0 mL of *Sample Preparation* taken for analysis; 180 is the molecular weight of glucose; and 30 is the incubation time, in min.

TRYPSIN ACTIVITY

Application and Principle This procedure is used to determine the trypsin activity of trypsin preparations derived from purified extracts of porcine or bovine pancreas.

Reagents and Solutions

Fifteenth Molar Phosphate Buffer (pH 7.6) Dissolve 4.54 g of monobasic potassium phosphate in sufficient water to make 500 mL of solution. Dissolve 4.73 g of anhydrous dibasic sodium phosphate in sufficient water to make 500 mL of solution.

Mix 13 mL of the monobasic potassium phosphate solution with 87 mL of the anhydrous dibasic sodium phosphate solution.

Substrate Solution Dissolve 85.7 mg of *N*-benzoyl-L-arginine ethyl ester hydrochloride, suitable for use in assaying trypsin, in sufficient water to make 100 mL.

Note: Determine the suitability of the substrate and check the adjustment of the spectrophotometer by performing the assay using USP Trypsin Reference Standard.

Dilute 10.0 mL of this solution to 100.0 mL with *Fifteenth Molar Phosphate Buffer*. Determine the absorbance of this solution at 253 nm in a 1-cm cell, with a suitable spectrophotometer, using water as the blank and maintaining the cell temperature at 25° ± 0.1°. Adjust the absorbance of the solution, if necessary, by the addition of *Fifteenth Molar Phosphate Buffer* so that it measures not less than 0.575 and not more than 0.585. Use this solution within a period of 2 h.

Sample Preparations Dissolve a sufficient amount of sample, accurately weighed, in 0.001 N hydrochloric acid to produce a solution containing about 3000 USP trypsin units in each mL. Prepare three dilutions using 0.001 N hydrochloric acid so that the final solutions will contain 12, 18, and 24 USP trypsin units in each 0.2 mL. Use these concentrations in the *Procedure* below.

Procedure Conduct the test in a spectrophotometer equipped to maintain a temperature of 25° ± 0.1° in the cell compartment.

Determine the temperature in the reaction cell before and after the measurement of absorbance to ensure that the temperature does not change by more than 0.5°.

Pipet 0.2 mL of 0.001 N hydrochloric acid and 3.0 mL of *Substrate Solution* into a 1-cm cell. Place this cell in the spectrophotometer, and adjust the instrument so that the absorbance will read 0.050 at 253 nm. Pipet 0.2 mL of the *Sample Preparation* containing 12 USP units into another 1-cm cell. Add 3.0 mL of *Substrate Solution*, and place the cell in the spectrophotometer. At the same time the *Substrate Solution* is added, start a stopwatch, and read the absorbance at 30-s intervals for 5 min. Repeat the procedure with the *Sample Preparations* containing 18 and 24 USP units. Plot curves of absorbance versus time for each concentration, and use only those values that form a straight line to determine the activity of the trypsin. Discard the values on the plateau, and take the average of the results from the three concentration levels as the actual activity of the trypsin.

Calculations One USP trypsin unit is the activity causing a change in the absorbance of 0.003/min under the conditions specified in this assay.

Calculate the number of USP trypsin units per mg at each level by the formula

$$\text{USP trypsin units} = (A_1 - A_2)/(T \times W \times 0.003),$$

in which A_1 is the absorbance straight-line final reading; A_2 is the absorbance straight-line initial reading; T is the elapsed time, in min, between the initial and final readings; and W is the weight, in mg, of trypsin in the volume of solution used in determining the absorbance.

APPENDIX VI: ESSENTIAL OILS AND FLAVORS

ACETALS

Hydroxylamine Hydrochloride Solution Prepare as directed under *Aldehydes*, this Appendix.

Procedure Weigh accurately the quantity of the sample specified in the monograph, and transfer it into a 125-mL Erlenmeyer flask. Add 30 mL of *Hydroxylamine Hydrochloride Solution*, and reflux on a steam bath for exactly 60 min. Allow the condenser to drain into the flask for 5 min after removing the flask from the steam bath. Detach, and rapidly cool the flask to room temperature. Add bromophenol blue TS as the indicator, and titrate with 0.5 N alcoholic potassium hydroxide to pH 3.4, or to the same light color as produced in the original hydroxylamine hydrochloride solution on adding the indicator. Calculate the mL of 0.5 N alcoholic potassium hydroxide consumed per g of sample (A).

Using a separate portion of the sample, proceed as directed under *Aldehydes*, this Appendix. Calculate the mL of 0.5 N alcoholic potassium hydroxide consumed per g of sample (B).

Calculate the percentage of acetals by the formula

$$(A - B) \times f,$$

in which f is the equivalence factor given in the monograph.

ACID VALUE

Dissolve about 10 g of the sample, accurately weighed, in 50 mL of alcohol, previously neutralized to phenolphthalein with 0.1 N sodium hydroxide. (Add 50 g of ice when testing cinnamyl formate, citronellyl formate, geranyl formate, isoamyl formate, and linalyl formate.) Add 1 mL of phenolphthalein TS, and titrate with 0.1 N sodium hydroxide until the solution remains faintly pink after shaking for 10 s, unless otherwise directed in the individual monograph. Calculate the acid value (AV) by the formula

$$AV = (5.61 \times S)/W,$$

in which S is the number of mL of 0.1 N sodium hydroxide consumed in the titration of the sample, and W is the weight, in g, of the sample.

TOTAL ALCOHOLS

Unless otherwise stated in the monograph, transfer 10 g of a solid sample, or 10 mL of a liquid sample, accurately weighed, into a 100-mL flask having a standard-taper neck. Add 10 mL of acetic anhydride and 1 g of anhydrous sodium acetate, mix these materials, attach a reflux condenser to the flask, and reflux the mixture for 1 h. Cool, and through the condenser, add 50 mL of water at a temperature between 50° and 60°. Shake intermittently during a period of 15 min, cool to room temperature, transfer the mixture completely to a separator, allow the layers to separate, and then remove and reject the lower, aqueous layer. Wash the oil layer successively with 50 mL of a saturated sodium chloride solution, 50 mL of a 10% sodium carbonate solution, and 50 mL of saturated sodium chloride solution. If the oil is still acid to moistened litmus paper, wash it with additional portions of sodium chloride solution until it is free from acid. Drain off the oil, dry it with anhydrous sodium sulfate, then filter it.

Weigh the quantity of acetylated oil specified in the monograph into a tared 125-mL Erlenmeyer flask, and add 10 mL of neutral alcohol, 10 drops of phenolphthalein TS, and 0.1 N alcoholic potassium hydroxide, dropwise, until a pink endpoint is obtained. If more than 0.20 mL is needed, reject the sample, and wash and test the remaining acetylated oil until its acid content is below this level. Prepare a blank for residual titration (see *General Provisions*), using the same volume of alcohol and indicator, and add 1 drop of 0.1 N alkali to produce a pink endpoint. Measure 25.0 mL of 0.5 N alcoholic potassium hydroxide into each of the flasks, reflux them simultaneously for 1 h, cool, and titrate the contents of each flask with 0.5 N hydrochloric acid to the disappearance of the pink color. Calculate the percentage of *Total Alcohols* (A) by the formula

$$A = [(b - S)(100e)]/[W - 21(b - S)],$$

in which b is the number of mL of 0.5 N hydrochloric acid consumed in the residual blank titration, S is the number of mL of 0.5 N hydrochloric acid consumed in the titration of the sample, e is the equivalence factor given in the monograph, and W is the weight, in mg, of the sample of the acetylated oil.

ALDEHYDES

Hydroxylamine Hydrochloride Solution Dissolve 50 g of hydroxylamine hydrochloride (preferably reagent grade or freshly recrystallized before using) in 90 mL of water, and dilute to 1000 mL with aldehyde-free alcohol. Adjust the solution to a pH of 3.4 with 0.5 N alcoholic potassium hydroxide.

Procedure Weigh accurately the quantity of sample specified in the monograph, and transfer it into a 125-mL Erlenmeyer flask. Add 30 mL of *Hydroxylamine Hydrochloride Solution*, mix thoroughly, and allow to stand at room temperature for 10 min, unless otherwise specified in the monograph. Titrate with 0.5 N alcoholic potassium hydroxide to a greenish yellow endpoint that matches the color of 30 mL of *Hydroxylamine Hydrochloride Solution* in a 125-mL flask when the same volume of bromophenol blue TS has been added to each flask, or preferably titrate to a pH of 3.4 using a suitable pH meter. Calculate the percentage of aldehyde (A) by the formula

$$A = (S - b)(100e)/W,$$

in which S is the number of mL of 0.5 N alcoholic potassium hydroxide consumed in the titration of the sample, b is the number of mL of 0.5 N alcoholic potassium hydroxide consumed in the titration of the blank, e is the equivalence factor given in the monograph, and W is the weight, in mg, of the sample.

ALDEHYDES AND KETONES

Hydroxylamine Method
Hydroxylamine Solution Dissolve 20 g of hydroxylamine hydrochloride (reagent grade or, preferably, freshly crystallized) in 40 mL of water, and dilute to 400 mL with alcohol. Add, with stirring, 300 mL of 0.5 N alcoholic potassium hydroxide, and filter. Use this solution within 2 days.

Procedure Weigh accurately the quantity of the sample specified in the individual monograph, and transfer it into a 250-mL glass-stoppered flask. Add 75.0 mL of *Hydroxylamine Solution* to this flask and to a similar flask for a residual blank titration (see *General Provisions*). If the component to be determined is an *aldehyde*, stopper the flasks and allow them to stand at room temperature for 1 h unless otherwise stated in the monograph. If the component to be determined is a *ketone*, attach the flask to a suitable condenser, and reflux the mixture for 1 h unless otherwise stated in the monograph, and then cool to room temperature. Titrate both flasks to the same greenish yellow endpoint using bromophenol blue TS as the indicator or, preferably, to a pH of 3.4 using a pH meter. (If the indicator is used, the endpoint color must be the same as that produced when the blank is titrated to a pH of 3.4.) Calculate the percentage of aldehyde or ketone by the formula

$$AK = (b - S)(100e)/W,$$

in which AK is the percentage of aldehyde or ketone, b is the number of mL of 0.5 N hydrochloric acid consumed in the residual blank titration, S is the number of mL of 0.5 N hydrochloric acid consumed in the titration of the sample, e is the equivalence factor given in the monograph, and W is the weight, in mg, of the sample.

Hydroxylamine/Tert-Butyl Alcohol Method
Hydroxylamine Solution Dissolve 45 g of reagent-grade hydroxylamine hydrochloride in 130 mL of water, add 850 mL of *tert*-butyl alcohol, mix, and using a pH meter, neutralize to a pH of 3.0 to 3.5 with sodium hydroxide.

Caution: Do not heat the solution.

Procedure Weigh accurately the quantity of the sample specified in the individual monograph, and transfer it into a 250-mL glass-stoppered flask. Add 50 mL of the *Hydroxylamine Solution*, or the volume specified in the monograph, mix thoroughly, and allow to stand at room temperature for the time specified in the monograph. Titrate with 0.5 N sodium hydroxide to the same pH as the *Hydroxylamine Solution* used. Calculate the percentage of aldehyde or ketone by the formula

$$AK = (S)(100e)/W,$$

in which AK is the percentage of aldehyde or ketone, S is the number of mL of 0.5 N sodium hydroxide consumed in the titration of the sample, e is the equivalence factor given in the monograph, and W is the weight, in mg, of the sample.

Neutral Sulfite Method
Pipet a 10-mL sample into a 100-mL cassia flask fitted with a stopper, and add 50 mL of a freshly prepared 30 in 100 solution of sodium sulfite. Add 2 drops of phenolphthalein TS, and neutralize with 50% (by volume) acetic acid solution. Heat the mixture in a boiling water bath, and shake the flask repeatedly, neutralizing the mixture from time to time by the addition of a few drops of the 50% acetic acid solution, stoppering the flask to prevent loss of volatile material. After no coloration appears upon the addition of a few more drops of phenolphthalein TS and heating for 15 min, cool to room temperature. When the liquids have separated completely, add sufficient sodium sulfite solution to raise the lower level of the oily layer within the graduated portion of the neck of the flask. Calculate the percentage, by volume, of the aldehyde or ketone by the formula

$$AK = 100 - (V \times 10),$$

in which AK is the percentage, by volume, of the aldehyde or ketone in the sample, and V is the number of mL of separated oil in the graduated neck of the flask.

CHLORINATED COMPOUNDS

Wind a 1.5- × 5-cm strip of 20-mesh copper gauze around the end of a copper wire. Heat the gauze in a nonluminous flame of a Bunsen burner until it glows without coloring the flame green. Permit the gauze to cool, and re-ignite it several times until a good coat of oxide has formed. With a medicine dropper, apply 2 drops of the sample to the cooled gauze, ignite, and permit it to burn freely in the air. Again cool the gauze, add 2 more drops, and burn as before. Continue this process until a total of 6 drops have been added and ignited. Then hold the gauze in the outer edge of a Bunsen flame adjusted to a height of 4 cm. Not even a transient green color is imparted to the flame. If at any of the additions the sample appears to be instantly vaporized, the test must be repeated from the beginning.

ESTERS

Ester Determination Weigh accurately the quantity of the sample specified in the monograph, and transfer it into a 125-mL Erlenmeyer flask containing a few boiling stones. Add to this flask and, simultaneously, to a similar flask for a residual blank titration (see *General Provisions*) 25.0 mL of 0.5 N alcoholic potassium hydroxide. Connect each flask to a reflux condenser, and reflux the mixtures on a steam bath for exactly 1 h, unless otherwise directed in the monograph. Allow the mixtures to cool, add 10 drops of phenolphthalein TS to each flask, and titrate the excess alkali in each flask with 0.5 N hydrochloric acid. Calculate the percentage of esters (E) in the sample by the formula

$$E = (b - S)(100e)/W,$$

in which b is the number of mL of 0.5 N hydrochloric acid consumed in the residual blank titration, S is the number of mL of 0.5 N hydrochloric acid consumed in the titration of the sample, e is the equivalence factor given in the monograph, and W is the weight, in mg, of the sample.

Ester Determination (High-Boiling Solvent)
0.5 N Potassium Hydroxide Solution Dissolve about 35 g of potassium hydroxide in 75 mL of water, add 1000 mL of a suitable grade of monoethyl ether of diethylene glycol, and mix.
Procedure Weigh accurately the quantity of the sample specified in the monograph, and transfer it into a 200-mL Erlenmeyer flask having a standard-taper joint. To this flask and to a similar flask for a residual blank titration (see *General Provisions*) add two glass beads and 25.0 mL of *0.5 N Potassium Hydroxide Solution*, allowing exactly 1 min for drainage from the buret or pipet. Attach an air condenser to each flask, reflux gently for 1 h, and cool. Rinse down the condensers with about 50 mL of water, then add phenolphthalein TS to each flask, and titrate the excess alkali with 0.5 N sulfuric acid to the disappearance of the pink color. Calculate the percentage of esters (E) in the sample by the formula

$$E = (b - S)(100e)/W,$$

in which b is the number of mL of 0.5 N sulfuric acid consumed in the blank determination, S is the number of mL of 0.5 N sulfuric acid required in the titration of the sample, e is the equivalence factor given in the monograph, and W is the weight, in mg, of the sample.

Saponification Value Proceed as directed for *Ester Determination* or *Ester Determination (High-Boiling Solvent)*, as specified in the monograph. Calculate the saponification value (SV) by the formula

$$SV = (b - S)(28.05)/W,$$

in which b and S are as defined under *Ester Determination*, and W is the weight, in g, of the sample.

Ester Value If the sample contains no free acids, the saponification value and the ester value are identical. If a determination of the *Acid Value* (AV) is specified in the monograph, calculate the ester value (EV) by the formula

$$EV = SV - AV,$$

in which SV is the saponification value.

LINALOOL DETERMINATION

Transfer a 10-mL sample, previously dried with sodium sulfate, into a 125-mL glass-stoppered Erlenmeyer flask previously cooled in an ice bath. Add to the cooled oil 20 mL of dimethyl aniline (monomethyl-free), and mix thoroughly. To the mixture add 8 mL of acetyl chloride and 5 mL of acetic anhydride, cool for several min, permit to stand at room temperature for another 30 min, then immerse the flask in a water bath maintained at 40° ± 1° for 16 h. Wash the acetylated oil with three 75-mL portions of ice water, followed by successive washes with 25-mL portions of 5% sulfuric acid, until the separated acid layer no longer becomes cloudy or emits an odor of dimethyl aniline when made alkaline. After removal of the dimethyl aniline, wash the acetylated oil first with 10 mL of sodium carbonate TS and then with successive portions of water until the washings are neutral to litmus. Finally, dry the acetylated oil with anhydrous sodium sulfate, and proceed as directed for *Ester Determination* under *Esters*, this Appendix. Calculate the percentage of linalool ($C_{10}H_{18}O$) by the formula

$$L = [7.707(b - S)]/[W - 0.021(b - S)],$$

in which L is the percentage of linalool, b is the the number of mL of 0.5 N hydrochloric acid consumed in the residual blank titration, S is the number of mL of 0.5 N hydrochloric acid consumed in the titration of the sample, and W is the weight, in g, of the sample.

Note: When this method is applied to essential oils containing appreciable amounts of esters, perform an *Ester Determination*, this Appendix, on a sample of the *original oil* and calculate the percentage of total linalool by the formula

$$L = [7.707(b - S)(1 - 0.0021E)]/[W - 0.21(b - S)],$$

in which L is the percentage of linalool, E is the percentage of esters, calculated as linalyl acetate ($C_{12}H_{20}O_2$) in the sample of the original oil, and b, S, and W are as defined in the preceding paragraph.

Note: This entire procedure is applicable only to linalool and linalool-containing oils. It is not intended for the determination of other tertiary alcohols.

PERCENTAGE OF CINEOLE

Temperature	0.0	0.1	0.2	0.3	0.4	0.5	0.6	0.7	0.8	0.9
24	45.6	45.7	45.9	46.0	46.1	46.3	46.4	46.5	46.6	46.8
25	46.9	47.0	47.2	47.3	47.4	47.6	47.7	47.8	47.9	48.1
26	48.2	48.3	48.5	48.6	48.7	48.9	49.0	49.1	49.2	49.4
27	49.5	49.6	49.8	49.9	50.0	50.2	50.3	50.4	50.5	50.7
28	50.8	50.9	51.1	51.2	51.3	51.5	51.6	51.7	51.8	52.0
29	52.1	52.2	52.4	52.5	52.6	52.8	52.9	53.0	53.1	53.3
30	53.4	53.5	53.7	53.8	53.9	54.1	54.2	54.3	54.4	54.6
31	54.7	54.8	55.0	55.1	55.2	55.4	55.5	55.6	55.7	55.9
32	56.0	56.1	56.3	56.4	56.5	56.7	56.8	56.9	57.0	57.2
33	57.3	57.4	57.6	57.7	57.8	58.0	58.1	58.2	58.3	58.5
34	58.6	58.7	58.9	59.0	59.1	59.3	59.4	59.5	59.6	59.8
35	59.9	60.0	60.2	60.3	60.4	60.6	60.7	60.8	60.9	61.1
36	61.2	61.3	61.5	61.6	61.7	61.9	62.0	62.1	62.2	62.4
37	62.5	62.6	62.8	62.9	63.0	63.2	63.3	63.4	63.5	63.7
38	63.8	63.9	64.1	64.2	64.4	64.5	64.6	64.8	64.9	65.1
39	65.2	65.4	65.5	65.7	65.8	66.0	66.2	66.3	66.5	66.6
40	66.8	67.0	67.2	67.3	67.5	67.7	67.9	68.1	68.2	68.4
41	68.6	68.8	69.0	69.2	69.4	69.6	69.7	69.9	70.1	70.3
42	70.5	70.7	70.9	71.0	71.2	71.4	71.6	71.8	71.9	72.1
43	72.3	72.5	72.7	72.9	73.1	73.3	73.4	73.6	73.8	74.0
44	74.2	74.4	74.6	74.8	75.0	75.2	75.3	75.5	75.7	75.9
45	76.1	76.3	76.5	76.7	76.9	77.1	77.2	77.4	77.6	77.8
46	78.0	78.2	78.4	78.6	78.8	79.0	79.2	79.4	79.6	79.8
47	80.0	80.2	80.4	80.6	80.8	81.1	81.3	81.5	81.7	81.9
48	82.1	82.3	82.5	82.7	82.9	83.2	83.4	83.6	83.8	84.0
49	84.2	84.4	84.6	84.8	85.0	85.3	85.5	85.7	85.9	86.1
50	86.3	86.6	86.8	87.1	87.3	87.6	87.8	88.1	88.3	88.6
51	88.8	89.1	89.3	89.6	89.8	90.1	90.3	90.6	90.8	91.1
52	91.3	91.6	91.8	92.1	92.3	92.6	92.8	93.1	93.3	93.6
53	93.8	94.1	94.3	94.6	94.8	95.1	95.3	95.6	95.8	96.1
54	96.3	96.6	96.9	97.2	97.5	97.8	98.1	98.4	98.7	99.0
55	99.3	99.7	100.0							

PHENOLS

Pipet 10 mL of the oil, which has been subjected to any treatment specified in the monograph, into a 100-mL cassia flask, add 75 mL of 1 N potassium hydroxide, and shake vigorously for 5 min to ensure complete extraction of the phenol by the alkali solution. Allow the mixture to stand for about 30 min, then add sufficient 1 N potassium hydroxide to raise the oily layer into the graduated portion of the flask, stopper the flask, and allow it to stand overnight. Read the volume of insoluble oil to 0.05 mL. Calculate the percentage, by volume, of phenols by the formula

$$P = (10 - V) \times 10,$$

in which P is the percentage, by volume, of phenols, and V is the observed volume, in mL, of insoluble oil.

FREE PHENOLS

Transfer about 5 g, accurately weighed, of the sample into a 150-mL flask having a standard-taper neck. Pipet exactly 10 mL of a 1 in 10 solution of acetic anhydride in anhydrous pyridine into the flask, and pipet exactly 10 mL of this solution, preferably measured with the same pipet, into a second 150-mL flask for the residual blank titration (see *General Provisions*). Connect the flasks to condensers, reflux for 1 h, and cool to a temperature below 100°. Add 25 mL of water to each flask through the condensers, and reflux again for 10 min. Cool the flasks, add phenolphthalein TS, and titrate with 0.5 N potassium hydroxide. Calculate the percentage of free phenols by the formula

$$\text{Percentage of Free Phenols} = (b - S) \times 100f/W,$$

in which b is the number of mL of 0.5 N potassium hydroxide consumed in the residual blank titration, S is the number of mL of 0.5 N potassium hydroxide consumed in the titration of the sample, f is the equivalence factor given in the monograph, and W is the weight, in mg, of the sample.

RESIDUE ON EVAPORATION

Weigh accurately the quantity of sample specified in the monograph, and transfer it into a suitable evaporating dish that has previously been heated on a steam bath, cooled to room temperature in a desiccator, and accurately weighed. Weigh the sample in the dish. Heat the evaporating dish containing the sample on the steam bath for the period of time specified in the monograph. Cool the dish and its contents to room temperature in a desiccator, and weigh accurately. Calculate the residue as percentage of the sample used.

SOLUBILITY IN ALCOHOL

Transfer a 1.0-mL sample into a calibrated 10-mL glass-stoppered cylinder graduated in 0.1-mL subdivisions, and add slowly, in small portions, alcohol of the concentration specified in the monograph. Maintain the temperature at 25°, and shake the cylinder thoroughly after each addition of alcohol. When a clear solution is first obtained, record the number of mL of alcohol required. Continue the addition of the alcohol until a total of 10 mL has been added. If opalescence or cloudiness occurs during these subsequent additions of alcohol, record the number of mL of alcohol at which the phenomenon occurs.

ULTRAVIOLET ABSORBANCE OF CITRUS OILS

Transfer the quantity of the sample specified in the monograph into a 100-mL volumetric flask, add alcohol to volume, and mix. Determine the ultraviolet absorption spectrum of the solution in the range of 260 to 400 nm in a 1-cm cell with a suitable recording or manual spectrophotometer, using alcohol as the blank. If a manual instrument is used, read absorbances at 5-nm intervals from 260 nm to a point about 12 nm from the expected maximum absorbance, then at 3-nm intervals for three readings, and at 1-nm intervals to a point about 5 nm beyond the maximum, and then at 10-nm intervals to 400 nm. From these data, plot the absorbances as ordinates against wavelength on the abscissa, and draw the spectrogram. Draw a baseline tangent to the areas of minimum absorbance, as shown in Fig. 33 (which is typical of lemon oil), joining point A in the region of 280 to 300 nm and a second point, B, in the region of 355 to 380 nm. Locate the point of maximum absorbance, C, and from it drop a vertical line, perpendicular to the abscissa, that intersects line AB at D. Read from the ordinate the absorbances corresponding to points C and D, subtract the latter from the former, and correct the difference for the actual weight of oil taken, calculating on the basis of the sample weight specified in the monograph.

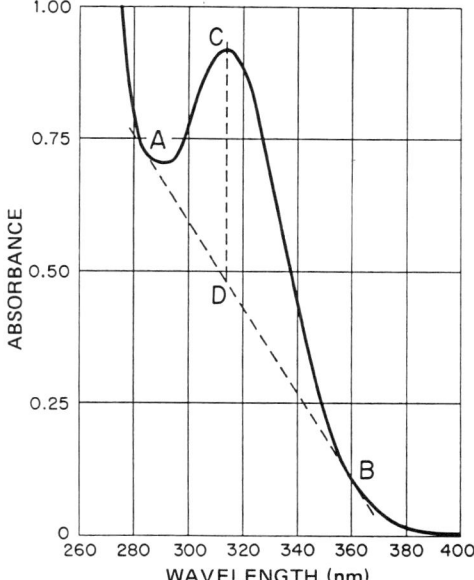

FIGURE 33 Typical Spectrogram of Lemon Oil.

VOLATILE OIL CONTENT

This procedure is used, when specified in the individual monograph, for determining the volatile oil content of gums, resins, and essential oils.

Apparatus The apparatus is shown in Fig. 34. It consists of a 1000-mL boiling flask, A, attached through a trap, D, to a Liebig condenser, C, which is connected to a 25-mL collector tube, B, graduated in 0.10-mL units.

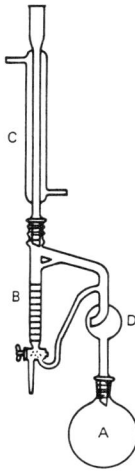

FIGURE 34 Apparatus for Determination of Volatile Oil Content.

Procedure   Place 750 mL of water in the boiling flask, boil for 10 min, and cool to 50°. Transfer the specified volume of the sample, prepared as directed in the monograph, into the flask, then immediately attach the remainder of the apparatus to the flask, and boil until the volume of distilled oil collected in the graduated collector tube remains constant. Avoid splashing the contents of the flask in order to prevent contamination of the distillate with nonvolatile material, and do not continue distillation for an extended time after the volume of distillate becomes constant. If the distilled oil is heavier than water, set the stopcock in the closed position to prevent return of the heavy distillate to the flask.

When distillation is complete, allow the contents of the collection tube to settle until the oil and water layers are separated completely. Allow the distillate to cool to room temperature, read its volume, and calculate therefrom the percentage of volatile oil.

Note: When the volatile oil thus collected is to be used in additional tests, as may be specified in the monograph, the oil should be drained off, dried, and filtered before use.

APPENDIX VII: FATS AND RELATED SUBSTANCES

ACETYL VALUE
(Based on AOCS Method Cd 4-40)

The acetyl value is defined as the number of mg of potassium hydroxide required to neutralize the acetic acid obtained by saponifying 1 g of the acetylated sample.

Acetylation   Boil 50 mL of the oil or melted fat with 50 mL of freshly distilled acetic anhydride for 2 h under a reflux condenser. Pour the mixture into a beaker containing 500 mL of water, and boil for 15 min, bubbling a stream of nitrogen or carbon dioxide through the mixture to prevent bumping. Cool slightly, remove the water, add another 500 mL of water, and boil again. Repeat for a third time with another 500-mL portion of water, and remove the wash water, which should be neutral to litmus. Transfer the acetylated fat to a separator, and wash with two 200-mL portions of warm water, separating as much as possible of the wash water each time. Transfer the washed sample to a beaker, add 5 g of anhydrous sodium sulfate, and let stand for 1 h, agitating occasionally to assist drying. Filter the oil through a dry filter paper, preferably in an oven at 100° to 110°, and keep the filtered oil in the oven until it is completely dry. The acetylated product should be a clear, brilliant oil.

Saponification   Weigh accurately from 2 to 2.5 g each of the acetylated oil and of the original, untreated sample into separate 250-mL Erlenmeyer flasks. Add to each flask 25.0 mL of 0.5 N alcoholic potassium hydroxide, and continue as directed in the *Procedure* under *Saponification Value*, in this Appendix, beginning with "Connect an air condenser...." Record the saponification value of the untreated sample as S, and that of the acetylized oil as S', then calculate the acetyl value of the sample by the formula

$$(S' - S)/(1.000 - 0.00075S).$$

ACID VALUE
(Based on AOCS Methods Te 1a-64 and Cd 3d-63)

The acid value is defined as the number of mg of potassium hydroxide required to neutralize the fatty acids in 1 g of the test substance.

Method I (Commercial Fatty Acids)
Unless otherwise directed, weigh accurately about 5 g of the sample into a 500-mL Erlenmeyer flask, and dissolve it in 75 to 100 mL of hot alcohol, previously boiled and neutralized to phenolphthalein TS with sodium hydroxide. Agitation and further heating may be necessary to effect complete solution of the sample. Add 0.5 mL of phenolphthalein TS, and titrate immediately, while shaking, with 0.5 N sodium hydroxide to the first pink color that persists for at least 30 s. Calculate the acid value by the formula

$$56.1V \times N/W,$$

in which V is the volume, in mL, and N is the normality, respectively, of the sodium hydroxide solution; and W is the weight, in g, of the sample taken.

Method II (Animal Fats and Vegetable and Marine Oils)
Prepare a solvent mixture consisting of equal parts, by volume, of isopropyl alcohol and toluene. Add 2 mL of a 1% solution of phenolphthalein in isopropyl alcohol to 125 mL of the mixture, and neutralize with alkali to a faint but permanent pink color. Weigh accurately the appropriate amount of well-mixed liquid sample indicated in the table below, dissolve it in the neutralized solvent mixture, warming if necessary, and shake vigorously while titrating with 0.1 N potassium hydroxide to the first permanent pink color of the same intensity as that of

the neutralized solvent before mixing with the sample. Calculate the acid value by the formula

$$56.1V \times N/W,$$

in which V is the volume, in mL, and N is the normality, respectively, of the potassium hydroxide solution; and W is the weight, in g, of the sample taken.

Acid Value	Sample Weight (g)
0–1	20
1–4	10
4–15	2.5
15–75	0.5
75 and over	0.1

CHLOROPHYLL
(Based on AOCS Method Cc 13d-55)

Use a reliable spectrophotometer with a sample holder equilibrated at 44° ± 3° to obtain absorbance values at 630, 670, and 710 nm. Calculate the concentration of chlorophyll (C) using the following formula:

$$C = [A_{670} - (A_{630}/2) - (A_{710}/2)]/(K \times b),$$

in which C is the concentration of chlorophyll, in mg/kg; A is the absorbance at the wavelength indicated by the subscript; K is the constant for the specific spectrophotometer being used and is equal to 0.1016 for the Beckman Model DU; and b is the optical pathlength through the sample, in cm.

COLD TEST
(Based on AOCS Method Cc 11-53)

Filter a sample (200 to 300 mL), and transfer to a clean, dry bottle. Fill the bottle completely, and insert a cork stopper. Seal with paraffin, and equilibrate at 25° in a water bath so that it is completely covered. Next, immerse the bottle in an ice and water bath so it is completely covered. Monitor the bath during the test and replenish the ice frequently to keep the bath at 0°.

After 5.5 h remove the bottle from the bath. The sample must be clear; fat crystals or cloudiness must be totally absent.

COLOR (AOCS-Wesson)
(Based on AOCS Method Cc 13b-45)

Apparatus Use a Lovibond tintometer or the equivalent and a set of color comparison glasses that conform to the AOCS-Wesson Tintometer Color Scale (available from the National Institute of Standards and Technology). A minimum set of glasses consists of

Red	0.1	0.2	0.3	0.4	0.5	0.6	0.8	0.9
	1.0	2.0	2.5	3.0	3.5	4.0	5.0	6.0
	7.0	7.6	8.0	9.0	10.0	11.0	12.0	16.0
	20.0
Yellow	1.0	2.0	3.0	5.0	10.0	15.0	20.0	35.0
	50.0	70.0

For making color comparisons, use color tubes of clear, colorless glass with a smooth, flat, polished bottom (length 154 mm; id 19 mm; od 22 mm), and marked to indicate liquid columns of 25.4 and 133.35 mm.

Procedure Add 0.1 g of diatomaceous earth to a 60-g sample, agitate for 2.5 min at room temperature (or 10° to 15° above the melting point if the sample is not liquid), and filter. Adjust the temperature to 25° to 35° (or not more than 10° above the melting point), and fill the color tube to the desired mark. Place the tube in the tintometer (in a dark booth or cabinet), and match the sample color as closely as possible with a standard glass.

FATTY ACID COMPOSITION

Apparatus Use a suitable gas chromatograph (see Appendix IIA) equipped with a thermal conductivity detector, and containing a 3.05-m × 6.4-mm (od) glass or aluminum column packed with preconditioned 10%, by weight, DEGS-PS on 100/120-mesh diatomaceous earth (Chromosorb WHP, or equivalent).

Operating Conditions The operating conditions may vary with the instrument used, but a suitable chromatogram may be obtained using the following conditions: *column temperature*, 215°; *inlet temperature* (injector), 300°; *detector*, 300°; and *carrier gas flow*, 60 mL/min.

Standard Solutions Chromatograph a commercially available standard (such as NuCheck 17A)[1] containing a mixture of fatty-acid methyl esters, including the methyl esters of oleic acid ($C_{18:1}$) and erucic acid ($C_{22:1}$). The calculated concentration should compare to that claimed within ± 2 σ, where σ is the standard deviation calculated from at least 10 replicate determinations, preferably made over a period of several days.

Chromatograph a suitable number of samples of the standard to ensure that the resolution factor R defining the efficiency of the separation between methyl stearate, eluting at approximately 5.6 min, and methyl oleate, eluting at approximately 6.5 min, is 0.9 or greater.

Sample Preparation Transfer approximately 10 to 15 g of sample (melted if necessary) to a 250-mL glass stoppered boiling

---

[1] NuCheck Prep. Inc., PO Box 172, Elysian, MN 56028.

flask, and add 50 mL of sodium methoxide reagent prepared by dissolving 9 g of metallic sodium in 3 L of fresh, reagent-grade methanol (carry out this preparation in a hood; hydrogen gas is evolved). Place a stirring bar in the flask, attach a water-cooled condenser, and reflux with stirring over low heat for 3 to 5 min or until the cloudiness disappears, indicating that methylation is complete. Add 25 mL of 0.5% hydrochloric acid saturated with sodium chloride through the condenser, and continue stirring until the two solutions mix. Remove the sample, and extract with hexane. Decant the hexane extract through reagent-grade sodium sulfate in a filter paper assembly, and evaporate the hexane with a stream of dry nitrogen on a steam bath.

Procedure Inject an appropriate volume (1 μL) of sample into the chromatograph. If an automated system is used, follow the manufacturer's instructions; if calculations are to be done manually, proceed as follows:

Calculate the percent of each component (C_N) by the formula

$$C_N = [(A_N \times M_N)/T_s] \times 100,$$

in which A_N is the area of the peak corresponding to component C_N, M_N is the molecular weight of C_N, and T_s is the total obtained by summing over all ($A_N \times M_N$), calculated for all detected components $T_s = \Sigma(A_N \times M_N)$.

FREE FATTY ACIDS
(Based on AOCS Method Ca 5a-40)

Unless otherwise directed, accurately weigh the appropriate amount of the sample, indicated in the table below, into a 250-mL Erlenmeyer flask or other suitable container. Add 2 mL of phenolphthalein TS to the specified amount of hot alcohol, neutralize with alkali to the first faint, but permanent, pink color, and then add the hot, neutralized alcohol to the sample container. Titrate with the appropriate normality of sodium hydroxide, shaking vigorously, to the first permanent pink color of the same intensity as that of the neutralized alcohol. The color must persist for at least 30 s. Calculate the percentage of free fatty acids (FFA) in the sample by the formula

$$VNe/W,$$

in which V is the volume and N is the normality, respectively, of the sodium hydroxide used; W is the weight of the sample, in g; and e is the equivalence factor given in the monograph.

FFA Range (%)	Grams of Sample	Milliliters of Alcohol	Strength of NaOH
0.00– 0.2	56.4 ± 0.2	50	0.1 N
0.2 – 1.0	28.2 ± 0.2	50	0.1 N
1.0 – 30.0	7.05 ± 0.05	75	0.25 N
30.0 – 50.0	7.05 ± 0.05	100	0.25–1.0 N
50.0 –100	3.525 ± 0.001	100	1.0 N

FREE GLYCERIN OR PROPYLENE GLYCOL
(Based on AOCS Method Ca 14-56)

Reagents and Solutions Use the *Periodic Acid Solution* and *Chloroform* as described under *1-Monoglycerides*, in this Appendix.

Procedure To the combined aqueous extracts obtained as directed under *1-Monoglycerides*, add 50.0 mL of *Periodic Acid Solution*. Run two blanks by adding 50.0 mL of this reagent solution to two 500-mL glass-stoppered Erlenmeyer flasks, each containing 75 mL of water. Continue as directed in the *Procedure* under *1-Monoglycerides*, beginning with "... and allow to stand for at least 30 min but no longer than 90 min."

Calculation Calculate the percentage of free glycerin in the original sample by the formula

$$(b - S) \times N \times 2.30/W,$$

or calculate the percentage of free propylene glycol by the formula

$$(b - S) \times N \times 3.81/W,$$

in which b is the number of mL of sodium thiosulfate consumed in the blank determination; S is the number of mL required in the titration of the aqueous extracts from the sample; N is the exact normality of the sodium thiosulfate; W is the weight, in g, of the original sample taken; 2.30 is the molecular weight of glycerin divided by 40; and 3.81 is the molecular weight of propylene glycol divided by 20.

Note: If the aqueous extract contains more than 20 mg of glycerin or more than 30 mg of propylene glycol, dilute the extract in a volumetric flask and transfer a suitable aliquot into a 500-mL glass-stoppered Erlenmeyer flask before proceeding with the test. The weight of the sample should be corrected in the calculation.

HYDROXYL VALUE
(Based on AOCS Methods Cd 4-40 and Cd 13-60)

The hydroxyl value is defined as the number of mg of potassium hydroxide equivalent to the hydroxyl content of 1 g of the unacetylated sample.

Method I
Proceed as directed under *Acetyl Value*, in this Appendix, but calculate the hydroxyl value by the formula

$$(S' - S)/(1.000 - 0.00075S').$$

Method II

Unless otherwise directed, accurately weigh the appropriate amount of the sample indicated in the table below, transfer it into a 250-mL glass-stoppered Erlenmeyer flask, and add 5.0 mL of pyridine–acetic anhydride reagent (mix 3 volumes of freshly distilled pyridine with 1 volume of freshly distilled acetic anhydride).

Hydroxyl Value	Sample Weight (g)
0– 20	10
20– 50	5
50–100	3
100–150	2
150–200	1.50
200–250	1.25
250–300	1
300–350	0.75

Pipet 5 mL of the pyridine–acetic anhydride reagent into a second 250-mL flask for the reagent blank. Heat the flasks for 1 h on a steam bath under reflux condensers, then add 10 mL of water through each condenser, heat for 10 min longer, and allow the flasks to cool to room temperature. Add 15 mL of n-butyl alcohol, previously neutralized to phenolphthalein TS with 0.5 N alcoholic potassium hydroxide, through the condenser, then remove the condensers, and wash the sides of the flasks with 10 mL of n-butyl alcohol. To each flask add 1 mL of phenolphthalein TS, and titrate to a faint pink endpoint with 0.5 N alcoholic potassium hydroxide, recording the mL required for the sample as S and that for the blank as B. To correct for free acid, mix about 10 g of the sample, accurately weighed, with 10 mL of freshly distilled pyridine, previously neutralized to phenolphthalein, add 1 mL of phenolphthalein TS, and titrate to a faint endpoint with 0.5 N alcoholic potassium hydroxide, recording the mL required as A. Calculate the hydroxyl value by the formula

$$[B + (WA/C) - S] \times 56.1 N/W,$$

in which W and C are the weights, in g, of the samples taken for acetylation and for the free acid determination, respectively; and N is the exact normality of the alcoholic potassium hydroxide.

IODINE VALUE
(Based on AOCS Method Cd 1d-92)

The iodine value is a measure of unsaturation and is expressed as the number of g of iodine absorbed, under the prescribed conditions, by 100 g of the test substance.

Modified Wijs Method (Acetic Acid/Cyclohexane Method)
Wijs Solution Dissolve 13 g of resublimed iodine in 1000 mL of glacial acetic acid. Pipet 10.0 mL of this solution into a 250-mL flask, add 20 mL of potassium iodide TS and 100 mL of water, and titrate with 0.1 N sodium thiosulfate, adding starch TS near the endpoint. Record the volume required as A. Set aside about 100 mL of the iodine–acetic acid solution for future use. Pass chlorine gas, washed and dried with sulfuric acid, through the remainder of the solution until a 10.0-mL portion requires not quite twice the volume of 0.1 N sodium thiosulfate consumed in the titration of the original iodine solution. A characteristic color change occurs when the desired amount of chlorine has been added. Alternatively, *Wijs Solution* may be prepared by dissolving 16.5 g of iodine monochloride, ICl, in 1000 mL of glacial acetic acid. Store the solution in amber bottles sealed with paraffin until ready for use, and use within 30 days.

Total Halogen Content Pipet 10.0 mL of *Wijs Solution* into a 500-mL Erlenmeyer flask containing 150 mL of recently boiled and cooled water and 15 mL of potassium iodide TS. Titrate immediately with 0.1 N sodium thiosulfate, recording the volume required as B.

Halogen Ratio Calculate the I/Cl ratio by the formula

$$A/(B - A).$$

The halogen ratio must be between 1.0 and 1.2. If the ratio is not within this range, the halogen content can be adjusted by adding the original solution or by passing more chlorine through the solution.

Note: *Wijs Solution* is commercially available.

Procedure The appropriate weight of the sample, in g, is calculated by dividing the number 25 by the expected iodine value. Melt the sample, if necessary, and filter it through a dry filter paper. Transfer the accurately weighed quantity of the sample into a clean, dry, 500-mL glass-stoppered bottle or flask containing 20 mL of glacial acetic acid/cyclohexane, 1:1, v/v, and pipet 25.0 mL of *Wijs Solution* into the flask. The excess of iodine should be between 50% and 60% of the quantity added, that is, between 100% and 150% of the quantity absorbed. Swirl, and let stand in the dark for 1.0 h where the iodine value is < 150 and for 2.0 h where the iodine value is ≥ 150. Add 20 mL of potassium iodide TS and 100 mL of recently boiled and cooled water, and titrate the excess iodine with 0.1 N sodium thiosulfate, adding the titrant gradually and shaking constantly until the yellow color of the solution almost disappears. Add starch TS, and continue the titration until the blue color disappears entirely. Toward the end of the titration, stopper the container and shake it violently so that any iodine remaining in solution in the glacial acetic acid/cyclohexane, 1:1 solution, may be taken up by the potassium iodide solution. Concomitantly, conduct two determinations on blanks in the same manner and at the same temperature. Calculate the iodine value by the formula

$$(B - S) \times 12.69 N/W,$$

in which $B - S$ represents the difference between the volumes of sodium thiosulfate required for the blank and for the sample, respectively; N is the normality of the sodium thiosulfate; and W is the weight, in g, of the sample taken.

MELTING RANGE

Fats of animal and vegetable origin do not exhibit a sharp melting point. For the purpose of this test, melting range is defined as the range of temperature in which the sample becomes a perfectly clear liquid after first passing through a stage of gradual softening, during which it may become opalescent.

Apparatus Use any suitable commercial or other apparatus. Use melting-point capillary tubes—id, 1 mm; od, 2 mm; length, 50 to 80 mm; and open at both ends.

Procedure
Capillary Method (Based on AOCS Method Cc 1-25). Melt the sample and filter it through filter paper; the sample must be absolutely dry. Dip three capillary tubes in the liquid sample so that the oil stands approximately 10 mm high in the tubes, and fuse the end of the tube containing the sample without burning it. Place the tubes containing the liquid sample in a beaker, and equilibrate them at least 16 h at 4° to 10° in a refrigerator. Determine the melting range, using a temperature increase of 0.5° per min when within 10° of the anticipated melting point. The melting ranges of the three samples should be no more than 0.5° apart.

1-MONOGLYCERIDES
(Based on AOCS Method Cd 11-57)

Reagents and Solutions
Periodic Acid Solution Dissolve 5.4 g of periodic acid, H_5IO_6, in 100 mL of water, add 1900 mL of glacial acetic acid, and mix. Store in a light-resistant, glass-stoppered bottle or in a clear, glass-stoppered bottle protected from light.

Chloroform Use chloroform meeting the following test: To each of three 500-mL flasks add 50.0 mL of *Periodic Acid Solution*, then add 50 mL of chloroform and 10 mL of water to two of the flasks and 50 mL of water to the third. To each flask add 20 mL of potassium iodide TS, mix gently, and continue as directed in the *Procedure*, beginning with "... allow to stand at least 1 min. . . ." The difference between the volume of 0.1 N sodium thiosulfate required in the titrations with and without the chloroform is not greater than 0.5 mL.

Procedure Melt the sample, if not liquid, at a temperature not higher than 10° above its melting point, and mix thoroughly. Transfer an accurately weighed portion of the sample, equivalent to about 150 mg of 1-monoglycerides, into a 100-mL beaker (or weigh a sample equivalent to 20 mg of glycerin or 30 mg of propylene glycol if only *Free Glycerin or Propylene Glycol* is to be determined), and dissolve in 25 mL of chloroform. Transfer the solution, with the aid of an additional 25 mL of chloroform, into a separator, wash the beaker with 25 mL of water, and add the washing to the separator. Stopper the separator tightly, shake vigorously for 30 to 60 s, and allow the layers to separate. (Add 1 to 2 mL of glacial acetic acid to break emulsions formed due to the presence of soap.) Collect the aqueous layer in a 500-mL glass-stoppered Erlenmeyer flask, and extract the chloroform solution again using two 25-mL portions of water. Retain the combined aqueous extracts for the determination of *Free Glycerin or Propylene Glycol* (in this Appendix). Transfer the chloroform to a 500-mL glass-stoppered Erlenmeyer flask, and add 50.0 mL of *Periodic Acid Solution* to this flask and to each of two blank flasks containing 50 mL of chloroform and 10 mL of water. Swirl the flasks during the addition of the reagent, and allow to stand for at least 30 min, but no longer than 90 min. To each flask, add 20 mL of potassium iodide TS, and allow to stand at least 1 min, but no longer than 5 min, before titrating. Add 100 mL of water, and titrate with 0.1 N sodium thiosulfate, using a magnetic stirrer to keep the solution thoroughly mixed, to the disappearance of the brown iodine color, then add 2 mL of starch TS and continue the titration to the disappearance of the blue color. Calculate the percentage of 1-monoglycerides[2] in the sample by the formula

$$(B - S) \times N \times 17.927/W,$$

in which B is the number of mL of sodium thiosulfate consumed in the blank determination; S is the number of mL required in the titration of the sample; N is the exact normality of the sodium thiosulfate; W is the weight, in g, of the sample taken; and 17.927 is the molecular weight of glyceryl monostearate divided by 20.

TOTAL MONOGLYCERIDES

Preparation of Silica Gel Place about 10 g of 100- to 200-mesh silica gel of a grade suitable for chromatographic work in a tared weighing bottle, cap immediately, and weigh accurately. Remove the cap, dry at 200° for 2 h, cap immediately, and cool for 30 min. Raise the cap momentarily to equalize the pressure, then weigh again, reheat for 5 min at 200°, cool, and reweigh. Repeat this 5-min drying cycle until two consecutive weights agree within 10 mg. Calculate the percentage of water in the original silica gel (A) by the formula

$$(\text{loss in wt/sample wt}) \times 100,$$

then calculate the amount of water required to adjust the water content to 5% by the formula

$$W \times (5 - A)/95,$$

in which W is the weight, in g, of the undried sample to be used.

Accurately weigh the appropriate amount of the undried silica gel to be used in the determination, transfer to a suitable blender or mixer, and add the calculated amount of water to give a final

---

[2]The monoglyceride may be calculated to some monoester other than glyceryl monostearate by dividing the molecular weight of the monoglyceride by 20 and substituting the value so obtained for 17.927 in the formula, using 17.80, for example, in calculating to the monooleate.

water content of 5% ± 0.1%. Blend for 1 h to ensure complete water distribution, and store in a sealed container. Determine the water content of the adjusted silica gel as directed above, and readjust if necessary.

Note: Each new lot of silica gel should be checked for suitability by the analysis of a monoglyceride of known composition.

Sample Preparation (**Caution**: To avoid rearrangement of partial glycerides, use extreme caution in applying heat to samples, and do not heat above 50°.)

Samples Melting Below 50° Melt the sample, if necessary, by warming for short periods below 50°, not exceeding a total of 30 min.

Samples Melting Above 50° Grind about 10 g in a mortar and pestle, chilling solid samples, if necessary, in carbon dioxide.

Weigh accurately about 1 g of the prepared sample into a 100-mL beaker, add 15 mL of chloroform, and warm, if necessary, to effect solution. Use only minimal heat, and do not heat above 40°.

Preparation of Chromatographic Column Connect a 19- × 290-mm chromatographic tube, equipped with an outer 19/22 standard-taper joint at the top and a coarse, fritted-glass disk and inner 19/22 standard-taper joint at the bottom, with an adapter consisting of an outer 19/22 joint connected to a Teflon stopcock. Do not grease the joints. Weigh 30 g of the prepared silica gel into a 150-mL beaker, add 50 to 60 mL of petroleum ether, and stir slowly with a glass rod until all air bubbles are expelled. Transfer the slurry to the column through a powder funnel, and open the stopcock, allowing the liquid level to drop to about 2 cm above the silica gel. Transfer any silica gel slurry remaining in the beaker into the column with a minimum amount of petroleum ether, then rinse the funnel and sides of the column. Drain the solvent through the stopcock until the level drops to 2 cm above the silica gel, and remove the powder funnel.

Procedure Carefully add the *Sample Preparation* to the prepared column. Open the stopcock, and adjust the flow rate to about 2 mL/min, discarding the eluate. Rinse the sample beaker with 5 mL of chloroform, and add the rinsing to the column when the level drops to 2 cm above the silica gel. Never allow the column to become dry on top, and maintain a flow rate of 2 mL/min throughout the elution. Avoid interruptions during elution as they may cause pressure buildup and result in leakage through the stopcock or cracks in the silica gel packing.

Attach a 250-mL reservoir separator, provided with a Teflon stopcock and a 19/22 standard-taper drip tip inner joint, to the column. Add 200 mL of benzene, elute, and discard the eluate, which contains the triglycerides fraction. When the level of benzene drops to 2 cm above the silica gel, add 200 mL of a 1 in 10 mixture of ether in benzene, elute, and discard the eluate, which contains the diglycerides and the free fatty acid fraction. When all of the ether–benzene solvent has been added from the separator and the level in the column drops to 2 cm above the silica gel, add from 250 to 300 mL of ether, and collect the monoglyceride fraction in a tared flask. Rinse the tip of the column into the flask with a few mL of ether, and evaporate to dryness on a steam bath under a stream of nitrogen or dry air.

Cool for at least 15 min, weigh, then reheat on the steam bath for 5 min in the same manner. Cool, reweigh, and repeat the 5-min evaporation, cooling, and reweighing procedures until two consecutive weights agree within 2 mg. The weight of the residue represents the total monoglycerides in the sample taken.

OXYETHYLENE DETERMINATION

Apparatus The apparatus for oxyethylene group determination is shown in Fig. 35. It consists of a boiling flask, A, fitted with a capillary side tube to provide an inlet for carbon dioxide and connected by a condenser with trap B, which contains an aqueous suspension of red phosphorus. The first absorption tube, C, contains a silver nitrate solution to absorb ethyl iodide. Absorption tube D is fitted with a 1.75-mm spiral rod (23 turns, 8.5-mm rise per turn), which is required to provide a longer contact of the evolved ethylene with the bromine solution. A standard-taper adapter and stopcock are connected to tube D to permit the transfer of the bromine solution into a titration flask without loss. A final trap, E, containing a potassium iodide solution, collects any bromine swept out by the flow of carbon dioxide.

Dimensions of the apparatus not readily determined from Fig. 35 are as follows: carbon dioxide inlet capillary, 1-mm id; flask A, 28-mm diameter, 12/18 standard-taper joint; condenser,

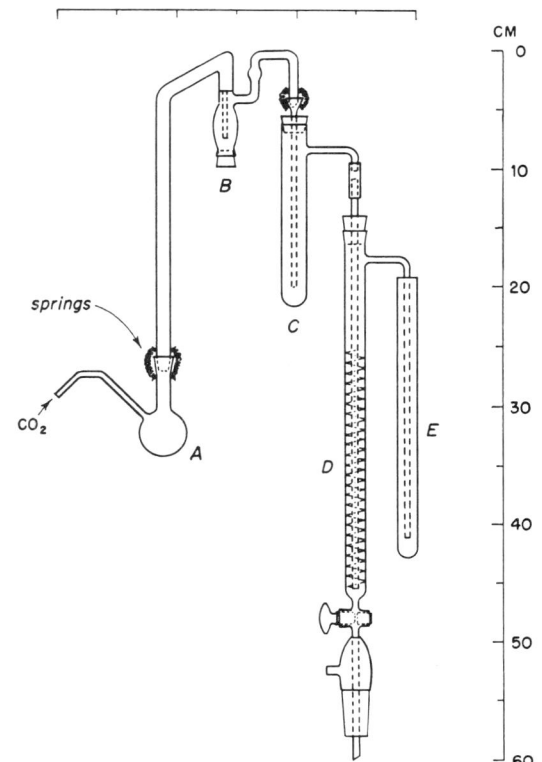

FIGURE 35 Apparatus for Oxyethylene Determination.

9-mm id; inlet to trap B, 2-mm id; inlet to trap C, 7/15 standard-taper joint, 2-mm id; trap C, 14-mm id; trap D, inner tube, 8-mm od, 2-mm opening at bottom of spiral; outer tube, approximately 12.5-mm id; side arm 7 cm from top of inserted spiral, 3.5-mm id, 2-mm opening at bottom.

Reagents

Hydriodic Acid Use special-grade hydriodic acid suitable for alkoxyl determinations, or purify reagent-grade as follows: Distill over red phosphorus in an all-glass apparatus, passing a slow stream of carbon dioxide through the apparatus until the distillation is terminated and the receiving flask has completely cooled.

Caution: Use a safety shield, and conduct the distillation in a hood.

Silver Nitrate Solution Dissolve 15 g of silver nitrate in 50 mL of water, mix with 400 mL of alcohol, and add a few drops of nitric acid.

Bromine–Bromide Solution Add 1 mL of bromine to 300 mL of glacial acetic acid saturated with dry potassium iodide (about 5 g). Fifteen mL of this solution requires about 40 mL of 0.05 N sodium thiosulfate. Store in a brown bottle in a dark place, and standardize at least once a day during use.

Procedure Fill trap B with enough of a suspension of 60 mg of red phosphorus in 100 mL of water to cover the inlet tube. Pipet 10 mL of the *Silver Nitrate Solution* into tube C and 15 mL of the *Bromine–Bromide Solution* into tube D, and place 10 mL of a 1 in 10 solution of potassium iodide in trap E. Transfer an accurately weighed quantity of the sample specified in the monograph into the reaction flask, A, and add 10 mL of *Hydriodic Acid* along with a few glass beads or boiling stones. Connect the flask to the condenser, and begin passing carbon dioxide through the apparatus at the rate of about one bubble per s. Heat the flask in an oil bath at 140° to 145°, and continue the reaction at this temperature for at least 40 min. Heating should be continued until the cloudy reflux in the condenser becomes clear and until the supernatant liquid in the silver nitrate tube, C, is almost completely clarified. Five min before the reaction is terminated, heat the *Silver Nitrate Solution* in tube C in a hot water bath at 50° to 60° to expel any dissolved olefin. At the completion of the decomposition, cautiously disconnect tubes D and C in the order named, then disconnect the carbon dioxide source and remove the oil bath. Connect tube D to a 500-mL iodine flask containing 150 mL of water and 10 mL of a 1 in 10 solution of potassium iodide, run the *Bromine–Bromide Solution* into the flask, and rinse the tube and spiral with water. Add the potassium iodide solution from trap E to the flask, rinsing the side arm and tube with a few mL of water, stopper the flask, and allow to stand for 5 min. Add 5 mL of 2 N sulfuric acid, and titrate immediately with 0.05 N sodium thiosulfate, using 2 mL of starch TS for the endpoint. Transfer the silver nitrate solution from tube C into a flask, rinsing the tube with water, dilute to 150 mL with water, and heat to boiling. Cool, and titrate with 0.05 N ammonium thiocyanate, using 3 mL of ferric ammonium sulfate TS as the indicator. Perform a blank determination. Calculate the percentage of oxyethylene groups (—CH$_2$CH$_2$O—), as ethylene, by the formula

$$(B - S) \times N \times 2.203/W,$$

in which $B - S$ represents the difference between the volumes of sodium thiosulfate required for the blank and the sample solution, respectively; N is the normality of the sodium thiosulfate; W is the weight, in g, of the sample taken; and 2.203 is an equivalence factor for oxyethylene. Calculate the percentage of oxyethylene groups, as ethyl iodide, by the formula

$$(B' - S') \times N' \times 4.405/W,$$

in which $B' - S'$ represents the difference between the volumes of ammonium thiocyanate required for the blank and the sample solution, respectively; N' is the normality of the ammonium thiocyanate; and 4.405 is an equivalence factor for oxyethylene. The sum of the values so obtained represents the percentage of oxyethylene groups in the sample taken.

REICHERT-MEISSL VALUE
(Based on AOCS Method Cd 5-40)

The Reichert-Meissl value is a measure of soluble volatile fatty acids (chiefly butyric and caproic). It is expressed in terms of the number of mL of 0.1 N sodium hydroxide required to neutralize the fatty acids obtained from a 5-g sample under the specified conditions of the method.

Apparatus Use a glass distillation apparatus of the same dimensions and construction as that shown in Fig. 36.

Reagents

Sodium Hydroxide Solution Prepare a solution containing 50.0% by weight of NaOH, and protect from contact with carbon dioxide. Allow the solution to settle, and use only the clear liquid.

Glycerin–Sodium Hydroxide Mixture Add 20 mL of the *Sodium Hydroxide Solution* to 180 mL of glycerin.

Procedure Unless otherwise directed, accurately weigh about 5 g of the sample, previously melted if necessary, into the 300-mL distillation flask. Add 20.0 mL of the *Glycerin–Sodium Hydroxide Mixture*, and heat until the sample is completely saponified, as indicated by the mixture becoming perfectly clear. Shake the flask gently if any foaming occurs. Add 135 mL of recently boiled and cooled water, dropwise at first to prevent foaming, then add 6 mL of dilute sulfuric acid (1 in 5) and a few pieces of pumice stone or silicon carbide. Rest the flask on a piece of heat-proof board having a center hole 5 cm in diameter, and begin the distillation, regulating the flame so as to collect 110 mL of distillate in 30 ± 2 min (measure time from the passage of the first drop of distillate from the condenser to the receiving flask), letting the distillate drip into the flask at a temperature not higher than 20°.

When 110 mL has distilled, disconnect the receiving flask, and remove the flame. Mix the contents of the flask with gentle shaking, and immerse almost completely for 15 min in water cooled to 15°. Filter the distillate through dry, 9-cm, moderately

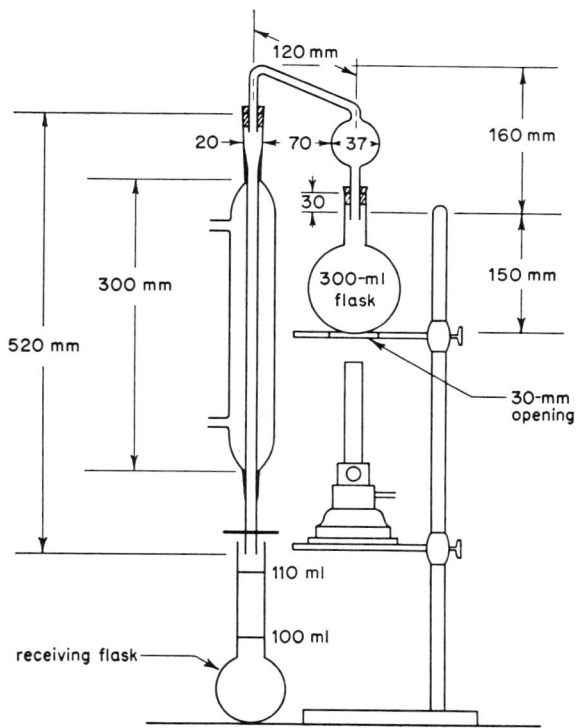

FIGURE 36 Reichert-Meissl Distillation Apparatus. [**Note**: A suitable heating mantle may be substituted for the burner.]

retentive paper (S & S No. 589 White Ribbon, or equivalent), add phenolphthalein TS, and titrate 100 mL of the filtrate with 0.1 N sodium hydroxide to the first pink color that remains unchanged for 2 to 3 min. Perform a blank determination using the same quantities of the same reagents, and calculate the Reichert-Meissl value by the formula

$$1.1 \times (S - B),$$

in which S is the volume of 0.1 N sodium hydroxide required for the sample, and B is the volume required for the blank.

SAPONIFICATION VALUE
(Based on AOCS Methods Tl 1a-64 and Cd 3-25)

The saponification value is defined as the number of mg of potassium hydroxide required to neutralize the free acids and saponify the esters in 1 g of the test substance.

Procedure Melt the sample, if necessary, and filter it through a dry filter paper to remove any traces of moisture. Unless otherwise directed, weigh accurately into a 250-mL flask a sample of such size that the titration of the sample solution after saponification will require between 45% and 55% of the volume of 0.5 N hydrochloric acid required for the blank, and add to the flask 50.0 mL of 0.5 N alcoholic potassium hydroxide.

Connect an air condenser, at least 65 cm in length, to the flask, and reflux gently until the sample is completely saponified (usually 30 min to 1 h). Cool slightly, wash the condenser with a few mL of water, add 1 mL of phenolphthalein TS, and titrate the excess potassium hydroxide with 0.5 N hydrochloric acid. Heat the contents of the flask to boiling, again titrate to the disappearance of any pink color that may have developed, and record the total volume of acid required. Perform a blank determination using the same amount of 0.5 N alcoholic potassium hydroxide. Calculate the saponification value by the formula

$$56.1(B - S) \times N/W,$$

in which $B - S$ represents the difference between the volumes of 0.5 N hydrochloric acid required for the blank and the sample, respectively; N is the normality of the hydrochloric acid; and W is the weight, in g, of the sample taken.

Note: A "masked phenolphthalein indicator" may be used with off-color materials. Prepare the indicator by dissolving 1.6 g of phenolphthalein and 2.7 g of methylene blue in 500 mL of alcohol, and adjust the pH with alcoholic alkali solution so that the greenish blue color is faintly tinged with purple. The color change, when going from acid to alkali, is from green to purple.

SOAP

Prepare a solvent mixture consisting of equal parts, by volume, of benzene and methanol, add bromophenol blue TS, and neutralize with 0.5 N hydrochloric acid, or use neutralized acetone as the solvent. Accurately weigh the amount of sample specified in the individual monograph, dissolve it in 100 mL of the neutralized solvent mixture, and titrate with 0.5 N hydrochloric acid to a definite yellow endpoint. Calculate the percentage of soap in the sample by the formula

$$VNe/W,$$

in which V and N are the volume and normality, respectively, of the hydrochloric acid; W is the weight of the sample, in g; and e is the equivalence factor given in the monograph.

SPECIFIC GRAVITY

The specific gravity of a fat or oil is determined at 25°, except when the substance is a solid at that temperature, in which case the specific gravity is determined at the temperature specified in the monograph, and is referred to water at 25°.

Clean a suitable pycnometer by filling it with a saturated solution of chromic acid (CrO_3) in sulfuric acid and allowing it to stand for at least 4 h. Empty the pycnometer, rinse it thoroughly, then fill it with recently boiled water, previously cooled to about 20°, and place in a constant-temperature bath at 25°. After 30 min, adjust the level of water to the proper point on the pycnometer, and stopper. Remove the pycnometer from the bath, wipe dry with a clean cloth free from lint, and weigh. Empty the pycnometer, rinse several times with alcohol and then with ether, allow to dry completely, remove any ether vapor, and weigh. Determine the weight of the contained water at 25° by subtracting the weight of the pycnometer from its weight when full.

Filter the oil or melted sample through filter paper to remove any impurities and the last traces of moisture, and cool to a few degrees below the temperature at which the determination is to be made. Fill the clean, dry pycnometer with the sample, and place it in the constant-temperature bath at the specified temperature. After 30 min, adjust the level of the oil to the mark on the pycnometer, insert the stopper, wipe dry, and weigh. Subtract the weight of the empty pycnometer from its weight when filled with the sample, and divide the difference by the weight of the water contained at 25°. The quotient is the specific gravity at the temperature of observation, referred to water at 25°.

STABILITY (Active Oxygen Method)
(Based on AOCS Method Cd 12-57)

Fat stability is the time, in h, required for a sample of fat or oil to attain a peroxide value of 100. This period of time is determined by interpolation between two measurements and is assumed to be an index of resistance to rancidity.

Caution: All equipment must be scrupulously clean (for an acceptable cleaning procedure, see AOCS Official Method Cd 12-57). Do not use chromic acid or other acidic cleaning agents. All receptacles in the heater must be calibrated for temperature under the exact conditions of the test. During the test, the temperature must be monitored in a sample tube containing the recommended quantity of oil.

Apparatus Use a suitable heating block and aeration apparatus, such as shown in the Official and Tentative Methods of the AOCS or in JAOCS 33 (1956), pp. 628–630.

Sampling Remove samples from large containers or processing equipment with sampling devices only of stainless steel, aluminum, nickel, or glass. Solid fat samples should be taken at least 5 cm from the walls of large containers and 2.5 cm from the walls of small containers. If liquid oil is to be poured from a container, clean the spout or lip with an acetone-moistened cloth. Under no circumstances should samples be taken from containers equipped with plastic or enameled tops or paper or wax liners.

Procedure Unless already completely liquid, the sample should be melted at a temperature not more than 10° above its melting point. Pour 20 mL into each of two or more sample tubes ensuring that the sample does not contact the tube where the stopper will later fit. Insert the aeration tube assembly so that the end of the air delivery tube is 5 cm below the surface of the sample. Place the sample tube in a container of vigorously boiling water for 5 min (during this time adjust the air flow rate from the manifold). Remove the tube, wipe dry, and transfer immediately to the constant-temperature heater, maintained at 97.8° ± 0.2°, and connect the aeration tube to the manifold. Determine to the nearest h the time required for the sample to attain a *Peroxide Value* (determine as directed for *Peroxide Value* in the monograph for *Hydroxylated Lecithin*) of 100 milliequivalents (meq) as follows: With 1-g samples determine when the peroxide value is approximately 75 meq and 125 meq, then perform the test on four 5-g samples determining the peroxide value in duplicate at the times corresponding to 75 and 125 meq. Make a second determination on two 5-g samples exactly 1 h after the first pair. Plot these values against aeration time; the AOM stability value in h is given where the line crosses 100 meq.

UNSAPONIFIABLE MATTER
(Based on AOCS Method Ca 6a-40)

This procedure determines those substances frequently found dissolved in fatty materials that cannot be saponified by alkali hydroxides but that are soluble in the ordinary fat solvents.

Procedure Accurately weigh 5.0 g of the sample into a 250-mL flask, add a solution of 2 g of potassium hydroxide in 40 mL of alcohol, and boil gently under a reflux condenser for 1 h or until saponification is complete. Transfer the contents of the flask to a glass-stoppered extraction cylinder (approximately 30 cm in length, 3.5 cm in diameter, and graduated at 40, 80, and 130 mL). Wash the flask with sufficient alcohol to make a volume of 40 mL in the cylinder, and complete the transfer with warm and then cold water until the total volume is 80 mL. Finally, wash the flask with a few mL of petroleum ether, add the washings to the cylinder, cool the contents of the cylinder to room temperature, and add 50 mL of petroleum ether.

Insert the stopper, shake the cylinder vigorously for at least 1 min, and allow both layers to become clear. Siphon the upper layer as completely as possible without removing any of the lower layer, collecting the ether fraction in a 500-mL separator. Repeat the extraction and siphoning at least six times with 50-mL portions of petroleum ether, shaking vigorously each time. Wash the combined extracts, with vigorous shaking, with 25-mL portions of 10% alcohol until the wash water is neutral to phenolphthalein, and discard the washings. Transfer the ether extract to a tared beaker, and rinse the separator with 10 mL of ether, adding the rinsings to the beaker. Evaporate the ether on a steam bath just to dryness, and dry the residue to constant weight, preferably at 75° to 80° under a vacuum of not more than 200 mm of Hg, or at 100° for 30 min. Cool in a desiccator, and weigh to obtain the uncorrected weight of unsaponifiable matter.

Determine the quantity of fatty acids in the residue as follows: Dissolve the residue in 50 mL of warm alcohol (containing phenolphthalein TS and previously neutralized with sodium hydroxide to a faint pink color), and titrate with 0.02 N sodium hydroxide to the same color. Each mL of 0.02 N sodium hydroxide is equivalent to 5.659 mg of fatty acids, calculated as oleic acid.

Subtract the calculated weight of fatty acids from the weight of the residue to obtain the corrected weight of unsaponifiable matter in the sample.

VOLATILE ACIDITY

Modified Hortvet-Sellier Method

Apparatus Assemble a modified Hortvet-Sellier distillation apparatus as shown in Fig. 37, using a sufficiently large (approximately 38- × 203-mm) inner Sellier tube and large distillation trap.

Procedure Transfer the amount of sample, accurately weighed, specified in the monograph into the inner tube of the assembly, and insert the tube in the outer flask containing about 300 mL of recently boiled hot water. To the sample add 10 mL of approximately 4 N perchloric acid [35 mL (60 g) of 70% perchloric acid in 100 mL of water], and connect the inner tube to a water-cooled condenser through the distillation trap. Distill by heating the outer flask so that 100 mL of distillate is collected within 20 to 25 min. Collect the distillate in 100-mL portions, add phenolphthalein TS to each portion, and titrate with 0.5 N sodium hydroxide. Continue the distillation until a 100-mL portion of the distillate requires no more than 0.5 mL of 0.5 N sodium hydroxide for neutralization.

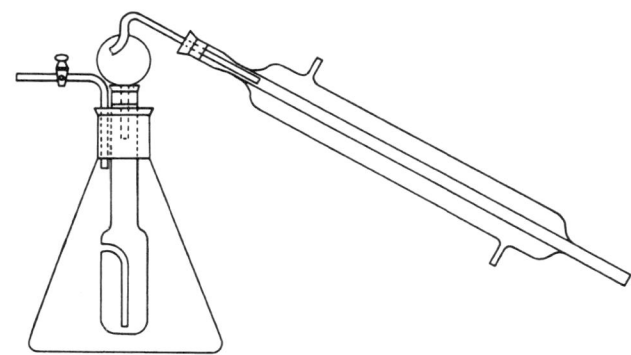

FIGURE 37   Modified Hortvet-Sellier Distillation Apparatus.

Caution: Do not distill to dryness.

Calculate the weight, in mg, of volatile acids in the sample taken by the formula

$$V \times e,$$

in which V is the total volume, in mL, of 0.5 N sodium hydroxide consumed in the series of titrations and e is the equivalence factor given in the monograph.

APPENDIX VIII: OLEORESINS

COLOR VALUE

Sample Preparation Transfer 70 to 100 mg of the sample, previously mixed well by shaking and accurately weighed, into a 100-mL volumetric flask, dissolve in acetone, dilute to volume with acetone, and mix. Allow the solution to stand for 2 min, then pipet 10 mL into a second 100-mL volumetric flask, dilute to volume with acetone, and mix.

Procedure Determine the absorbance of the *Sample Preparation* with a suitable spectrophotometer in a 1-cm cell at 460 nm, using acetone as the blank. Record the value obtained as A_S. In the same manner, determine the absorbance of a National Institute of Standards and Technology Standard Glass Filter 930, and record the value obtained as A_F.

   Note: The recommended range for absorbance values is between 0.30 and 0.70. Solutions having absorbances greater than 0.70 should be diluted with acetone to one-half the original concentration, and those having absorbances less than 0.30 should be discarded and the *Sample Preparation* prepared with a larger sample. Appropriate adjustments should be made in the sample weight (W) used in the *Calculation* below.

Calculation Determine the instrument correction factor, F, by the formula

$$A_N/A_F,$$

in which A_N is the absorbance of the filter as stated by the National Institute of Standards and Technology. Calculate the color value of the sample by the formula

$$(A_S \times 164 \times F)/W,$$

in which W is the weight, in g, of sample taken.

CURCUMIN CONTENT

Standard Preparation Transfer about 250 mg of pure curcumin, accurately weighed, into a 100-mL volumetric flask, and record the weight as W, in mg. Dissolve in acetone, dilute to volume with acetone, and mix. Pipet a 1-mL portion of this solution into a second 100-mL volumetric flask, dilute to volume with acetone, and mix. Finally, pipet a 5-mL portion of the last solution into a 50-mL volumetric flask, dilute to volume with acetone, and mix.

Sample Preparation Transfer an accurately weighed amount of the sample, equivalent to about 250 mg of curcumin, into a 100-mL volumetric flask, and record the weight as w, in mg. Dissolve in acetone, dilute to volume with acetone, and mix. Pipet a 1-mL portion of this solution into a second 100-mL volumetric flask, dilute to volume with acetone, and mix. Finally, pipet a 5-mL portion of the last solution into a 50-mL volumetric flask, dilute to volume with acetone, and mix.

Procedure Determine the absorbance of each solution in 1-cm cells at the wavelength of maximum absorption at about 421 nm with a suitable spectrophotometer, using acetone as the blank. Calculate the percentage of curcumin in the sample by the formula

$$100 \times (W/w) \times (A_U/A_S),$$

in which A_U is the absorbance of the *Sample Preparation*, and A_S is the absorbance of the *Standard Preparation*.

Note: The absorbance readings should be made as soon as possible after the solutions are prepared to avoid color loss.

PIPERINE CONTENT

Stock Standard Solution Purify piperine by repeated crystallization from isopropanol until a product having a melting range of 129° to 130° is obtained. Transfer 100.0 mg of the crystals, accurately weighed, into a 100-mL volumetric flask, dissolve in ethylene dichloride, dilute to volume with ethylene dichloride, and mix. Pipet 10.0 mL of this solution into a second 100-mL volumetric flask, dilute to volume with ethylene dichloride, and mix.

Standard Dilutions Pipet 1.0, 3.0, 5.0, and 10.0 mL of the *Stock Standard Solution* (corresponding to 0.1, 0.3, 0.5, and 1.0 mg of piperine, respectively) into separate 100-mL volumetric flasks, dilute each flask to volume with ethylene dichloride, and mix. Determine the absorbance of each dilution at once, as directed in the *Procedure*.

Sample Preparation Heat a portion of the sample to 100° on a steam bath or in an oven (but not on a hot plate), mix with a glass stirring rod, and transfer 100 mg, accurately weighed, into a 100-mL volumetric flask. Dissolve in ethylene dichloride, dilute to volume with ethylene dichloride, and mix. Pipet 1.0 mL of this solution into a second 100-mL volumetric flask, dilute to volume with ethylene dichloride, and mix. Determine the absorbance of the solution at once, as directed in the *Procedure*.

Procedure Determine the absorbance of the *Sample Preparation* and of each of the *Standard Dilutions* in 1-cm cells at the wavelength of maximum absorption at about 342 nm with a suitable spectrophotometer, using ethylene dichloride as the blank. Prepare a standard curve of concentration, in mg per 100 mL, versus absorbance for the four *Standard Dilutions*, including the absorbance at zero concentration obtained with the blank. From the standard curve, determine the concentration of piperine in the *Sample Preparation*, and record the value as C, in mg per 100 mL. Calculate the percentage of piperine in the sample by the formula

$$100 \times (100C/W),$$

in which W is the weight, in mg, of sample taken.

RESIDUAL SOLVENT

This procedure is for the determination of acetone, ethylene dichloride, hexane, isopropanol, methanol, methylene chloride, and trichloroethylene residues.

Distilling Head Use a Clevenger trap designed for use with oils heavier than water. A suitable design is shown in Fig. 38a.

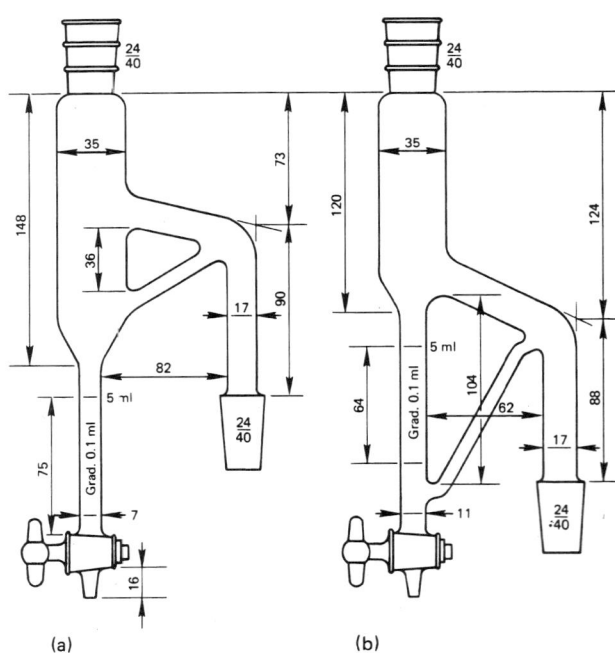

FIGURE 38 Clevenger Traps (all measurements are in mm) (a) Oils Heavier Than Water; (b) Oils Lighter Than Water.

Toluene The toluene used for this analysis should not contain any of the solvents determined by this method. The purity may be determined by gas chromatographic analysis, using one of the following columns or their equivalent: (1) 17% by weight of Ucon 75-H-90,000 on 35/80-mesh Chromosorb W; (2) 20% Ucon LB-135 on 35/80-mesh Chromosorb W; (3) 15% Ucon LB-1715 on 60/80-mesh Chromosorb W; or (4) Porapak Q 50/60 mesh. Follow the conditions described under *Procedure*, and inject the same amount of toluene as will be injected in the analysis of the solvents. If impurities interfering with the test are present, they will appear as peaks occurring before the toluene peak and should be removed by fractional distillation.

Benzene The benzene used for this analysis should be free from interfering impurities. The purity may be determined as described under *Toluene*.

Detergent and **Antifoam** Any such products that are free from volatile compounds may be used. If volatile compounds are present, they may be removed by prolonged boiling of the aqueous solutions of the products.

Reference Solution A Prepare a solution in *Toluene* containing 2500 ppm of benzene. If the toluene available contains benzene as the only impurity, the benzene level can be determined by gas chromatography and sufficient benzene added to bring the level to 2500 ppm.

Reference Solution B Prepare a solution containing 0.63% v/w of acetone in water.

Sample Preparation A (all solvents except methanol) Place 50.0 g of the sample, 1.00 mL of *Reference Solution A*, 10 g of anhydrous sodium sulfate, 50 mL of water, and a small amount each of *Detergent* and *Antifoam* in a 250-mL round-bottom flask with a 24/40 ground-glass neck. Attach the *Distilling Head*, a 400-mm water-cooled condenser, and a receiver, and collect approximately 15 mL of distillate. Add 15 g of anhydrous potassium carbonate to the distillate, cool while shaking, and allow the phases to separate. All of the solvents except methanol will be present in the toluene layer, which is used in the *Procedure*. Draw off the aqueous layer for use in *Sample Preparation B*.

Sample Preparation B (methanol only) Place the aqueous layer obtained from *Sample Preparation A* in a 50-mL round-bottom distilling flask with a 24/40 ground-glass neck, add a few boiling chips and 1.00 mL of *Reference Solution B*, and collect approximately 1 mL of distillate, which will contain any methanol from the sample, together with acetone as the internal standard. The distillate is used in the *Procedure*.

Procedure Use a gas chromatograph equipped with a hot-wire detector and a suitable sample-injection system or on-column injection. Under typical conditions, the instrument contains a $^1\!/_4$-in. (od) × 6- to 8-ft column maintained isothermally at 70° to 80°. The flow rate of dry carrier gas is 50 to 80 mL/min, and the sample size is 15 to 20 μL (for the hot-wire detector). The column selected for use in the chromatograph depends on the components to be analyzed and, to a certain extent, on the preference of the analyst. The columns 1, 2, 3, and 4, as described under *Toluene*, may be used as follows: (1) This column separates acetone and methanol from their aqueous solution. It may be used for the separation and analysis of hexane, acetone, and trichloroethylene in the toluene layer from *Sample Preparation A*. The elution order is acetone, methanol, and water, or hexane, acetone, isopropanol plus methylene chloride, benzene, trichloroethylene, and ethylene dichloride plus toluene. (2) This column separates methylene chloride and isopropanol, and ethylene dichloride. The elution order is hexane plus acetone, methylene chloride, isopropanol, benzene, ethylene dichloride, trichloroethylene, and toluene. (3) This is the best general-purpose column, except for the determination of methanol. The elution order is hexane, acetone, benzene, ethylene dichloride, and toluene. (4) This column is used for the determination of methanol, which elutes just after the large water peak.

Calibration Determine the response of the detector for known ratios of solvents by injecting known mixtures of solvents and benzene in toluene. The levels of the solvents and benzene in toluene should be of the same magnitude as they will be present in the sample under analysis.

Calculate the areas of the solvents with respect to benzene, and then calculate the calibration factor, F, as follows:

$$F \text{ (solvent)} = (\text{wt \% solvent/wt \% benzene}) \times (\text{area of benzene/area of solvent}).$$

The recovery of the various solvents from the oleoresin sample, with respect to the recovery of benzene, is as follows: hexane, 52%; acetone, 85%; isopropanol, 100%; methylene chloride, 87.5%; trichloroethylene, 113%; ethylene dichloride, 102%; and methanol, 87%.

Calculation Calculate the ppm of residual solvent (except methanol) by the formula

$$\text{Res. solv.} = \{[43.4 \times F \text{ (solvent)} \times 100]/[\% \text{ recovery of solvent}]\} \times (\text{area of solvent/area of benzene}),$$

in which 43.4 is the ppm of benzene internal standard, related to the 50-g oleoresin sample taken for analysis. Calculate the ppm of residual methanol by the formula

$$\text{Methanol} = \{[100 \times F \text{ (methanol)}]/0.87\} \times \text{area of methanol/area of benzene}),$$

in which 100 is the ppm of acetone internal standard, related to the 50-g oleoresin sample taken for analysis.

SCOVILLE HEAT UNITS

Sample Preparation Transfer 200 mg of the sample into a 50-mL volumetric flask, dilute to volume with alcohol, and mix thoroughly by shaking. Allow the insolubles to settle before use.

Sucrose Solution Prepare a suitable volume of a 10% w/v solution of sucrose in water.

Standard Solution Add 0.15 mL of the *Sample Preparation* to 140 mL of the *Sucrose Solution*, and mix. This solution contains the equivalent of 240,000 Scoville Heat Units.

Test Solutions If the oleoresin sample is claimed to contain more than 240,000 Scoville Heat Units, prepare one or more dilutions according to the following table:

Scoville Heat Units	Standard Solution (mL)	Sucrose Solution (mL)
360,000	20	10
480,000	20	20
600,000	20	30
720,000	20	40
840,000	20	50
960,000	20	60
1,080,000	20	70
1,200,000	20	80
1,320,000	20	90
1,440,000	20	100
1,560,000	20	110
1,680,000	20	120
1,800,000	20	130
1,920,000	20	140
2,040,000	20	150

If the oleoresin sample is claimed to contain less than 240,000 Scoville Heat Units, prepare one or more dilutions according to the following table:

Scoville Heat Units	Sample Preparation (mL)	Sucrose Solution (mL)
100,000	0.15	60
117,500	0.15	70
170,000	0.15	100
205,000	0.15	120

Procedure Select five panel members who are thoroughly experienced with this method. Instruct the panelists to swallow 5 mL of the solution corresponding to the claimed content of Scoville Heat Units. The sample passes the test if three of the five panel members perceive a pungent or stinging sensation in the throat.

VOLATILE OIL CONTENT

Weigh accurately an amount of sample sufficient to yield 2 to 5 mL of volatile oil, and transfer with the aid of water into a 1000- or 2000-mL round-bottom shortneck flask with a 24/40 ground-glass neck. Add a magnetic stirring bar and about 500 mL of water, and connect a Clevenger trap of the proper type (see Figs. 38a and 38b) and a 400-mm water-cooled condenser. Heat the flask with stirring, and distill at a rate of 1 to 1.5 drops per s until two consecutive readings taken at 1-h intervals show no change of oil volume in the trap. Cool to room temperature, allow to stand until the oil layer is clear, and read the volume of oil collected, estimating to the nearest 0.02 mL. Calculate the percentage (v/w) of volatile oil in the sample by the formula

$$100(V/W),$$

in which V is the volume, in mL, of oil collected, and W is the weight, in g, of sample taken.

APPENDIX IX: ROSINS AND RELATED SUBSTANCES

ACID NUMBER

The acid number is the number of mg of potassium hydroxide required to neutralize the free acids in 1 g of the test substance.

Procedure Unless otherwise directed in the individual monograph, transfer about 4 g of the sample, previously crushed into small lumps and accurately weighed, into a 250-mL Erlenmeyer flask, and add 100 mL of a 1 to 3 mixture of toluene–isopropyl alcohol, previously neutralized to phenolphthalein TS with sodium hydroxide. Dissolve the sample by shaking or heating gently, if necessary, then add about 0.5 mL of phenolphthalein TS, and titrate with 0.5 N or 0.1 N alcoholic potassium hydroxide to the first pink color that persists for 30 s. Calculate the acid number by the formula

$$56.1V \times N/W,$$

in which V is the exact volume, in mL, and N is the exact normality, respectively, of the potassium hydroxide solution, and W is the weight, in g, of the sample.

SOFTENING POINT

Drop Method The *drop softening point* is that temperature at which a given weight of rosin or rosin derivative begins to

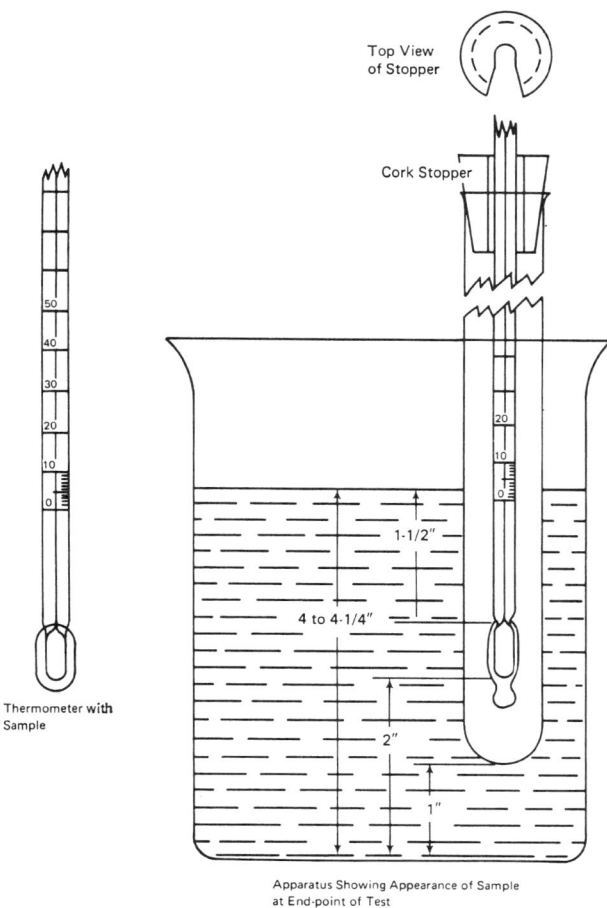

FIGURE 39 Apparatus for Drop Softening Point Determination.

drop from the bulb of a special thermometer mounted in a test tube that is immersed in a constant-temperature bath.

Apparatus The apparatus illustrated in Fig. 39 consists of the components described in the following paragraphs.

Thermometer Use a special total-immersion softening point thermometer,[1] covering the range from 0° to 250° and graduated in 1° divisions. The bulb should be 15.9 ± 0.8 mm in length and 6.35 ± 0.4 mm in diameter.

Heating Bath Use an 800- to 2000-mL beaker containing a suitable heating medium. For rosins having a softening point below 80°, use water; for those having softening points above 80°, use glycerin or silicone oil, depending upon the temperature range required. Maintain the temperature of the heating medium within ± 1° of the temperature specified in the individual monograph. Stir the bath medium constantly during the test with a suitable mechanical stirrer to ensure uniform heating of the medium.

Test Tube Use a standard 22-mm od × 200- to 250-mm test tube with a rim, fitted with a cork stopper as shown in Fig. 39.

---
[1]Available from the Walter K. Kessler Co., Inc.

Sample Preparation Place about 20 g of the sample in a 50-mL beaker, and heat it in an oven, on a sand bath or hot plate, or in an oil bath until the sample becomes soft enough to mold on the thermometer bulb. Tare the softening point thermometer, and cautiously warm the bulb over a hot plate until it registers 15° to 20° above the expected softening point of the sample. Immediately dip the thermometer bulb into the melted sample, withdraw, and rotate it to deposit a uniform film of the molten sample over the surface of the bulb, taking care not to extend the film higher than the top of the bulb. Quickly place the thermometer on a balance, and weigh. The weight of the sample on the thermometer bulb should be between 0.5 and 0.55 g. If the weight is low, again dip the bulb in the molten sample; if the weight is high, pull off some of the sample with the fingers. When the correct sample weight has been obtained, mold the sample uniformly around the bulb by rolling it on the palm of the hand or between the fingers. The sample must be of uniform thickness over the bulb, and it must not extend up onto the thermometer stem (see Fig. 39). (If the film of the sample is not uniform when cooled, remove it completely from the bulb and apply a new one. Do not reheat the film and try to remold it.) Allow the film and thermometer to cool to approximately 35° or lower, allowing about 15 min for cooling.

Note: If samples having high softening points crack or "check" on the thermometer bulb upon cooling to room temperature, prepare another sample film and cool only to about 50° below the expected softening point.

Procedure Fill the glass beaker to a depth of not less than 101.6 mm or more than 108 mm with a suitable heating medium; support the beaker over a Bunsen burner, hot plate, or other suitable source of heat; and insert the bath stirrer and a bath temperature thermometer. Place the stirrer to one side so that the impeller clears the side of the beaker and is about 12.7 mm above the bottom of the beaker. Start the stirrer, heat the bath to the temperature specified in the individual monograph, and maintain this temperature within ± 1° throughout the test.

Insert the prepared sample thermometer in the test tube, supporting it with a notched cork stopper so that the lower end of the bulb is 25.4 mm from the bottom of the test tube. Place the test tube in the bath so that the bottom of the thermometer bulb is 50.8 mm from the bottom of the beaker; the top of the bulb should be about 25.4 to 38.1 mm below the liquid level of the bath. Stir the bath to keep its temperature uniform throughout. Observe the sample thermometer, and record as the softening point the reading at which the elongated drop of sample on the end of the bulb first becomes constricted (see Fig. 39). Report the softening point to the nearest 1.0°.

Caution: If the rosin crystallizes, thus making it difficult to obtain the correct softening point, prepare a new sample by heating the rosin rapidly, yet cautiously, over a flame to a temperature of 160° to 170° to destroy all crystal nuclei. Dip the thermometer bulb into the molten resin, remove it momentarily, and rotate the thermometer to provide a uniform resin film on the bulb as it partially cools in the air. Dip the bulb in the melted sample repeatedly until the proper amount of resin is deposited on the bulb. Do not report results if a crystal-free sample cannot be obtained.

Ring-and-Ball Method The *ring-and-ball softening point* is the temperature at which a disk of the sample held within a horizontal ring is forced downward a distance of 25.4 mm under the weight of a steel ball as the sample is heated at a prescribed rate in a water, glycerin, or silicone oil (Dow Corning 200 fluid 50 cs or an equivalent is suitable) bath.

Apparatus Ring-and-ball softening point may be determined manually using the apparatus described below. Automated apparatus may be used provided equivalent results are obtained. The calibration of any automated apparatus should be monitored on a regular basis because accurate temperature control is required. The apparatus illustrated in Figs. 40 and 41 consists of the components described in the following paragraphs.

Ring Use a brass-shouldered ring conforming to the dimensions shown in Fig. 40a. If desired, the ring may be attached by brazing or other convenient manner to a brass wire of about 13 B & S gauge (1.52 to 2.03 mm in diameter) as shown in Fig. 41a.

Ball Use a steel ball, 9.53 mm in diameter, weighing between 3.45 and 3.55 g.

Ball-Centering Guide If desired, center the ball by using a guide constructed of brass and having the general shape and dimensions illustrated in Fig. 40c.

Container Use a heat-resistant glass vessel, such as an 800-mL low-form Griffin beaker, not less than 85 mm in diameter and not less than 127 mm in depth from the bottom of the flare.

Support for Ring and Thermometer Use any convenient device for supporting the ring and thermometer, provided that it meets the following requirements: (1) the ring is supported in a substantially horizontal position; (2) when the apparatus shown in Fig. 40d is used, the bottom of the ring is 25.4 mm above the horizontal plate below it, the bottom surface of the horizontal plate is 12.7 to 19 mm above the bottom of the container, and the depth of the liquid in the container is not less than 101.6 mm; (3) if the apparatus shown in Fig. 41e is used, the bottom of the ring is 25.4 mm above the bottom of the container, with the bottom end of the rod resting on the bottom of the container, and the depth of the liquid in the container is not less than 101.6 mm, as shown in Figs. 41a, b, and c; and (4) in both assemblies, the thermometer is suspended so that the bottom of the bulb is level with the bottom of the ring and within 12.7 mm of, but not touching, the ring.

Thermometers Depending on the expected softening point of the sample, use either an ASTM 15C or 15F low-softening-point thermometer (−2° to 80°) or an ASTM 16C or 16F high-softening-point thermometer (30° to 200°), as described under *Thermometers*, Appendix I.

Stirrer Use a suitable mechanical stirrer rotating between 500 and 700 rpm. To ensure uniform heat distribution in the heating medium, the direction of the shaft rotation should move the liquid upward. (See Fig. 41d for recommended dimensions.)

Sample Preparation Select a representative sample of the material under test consisting of freshly broken lumps free of oxidized surfaces. Immediately before use, scrape off the surface layer of samples received as lumps, avoiding inclusion of finely

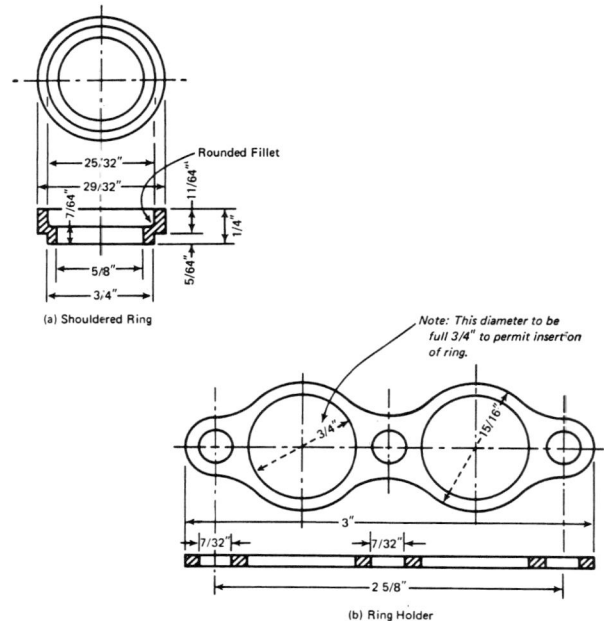

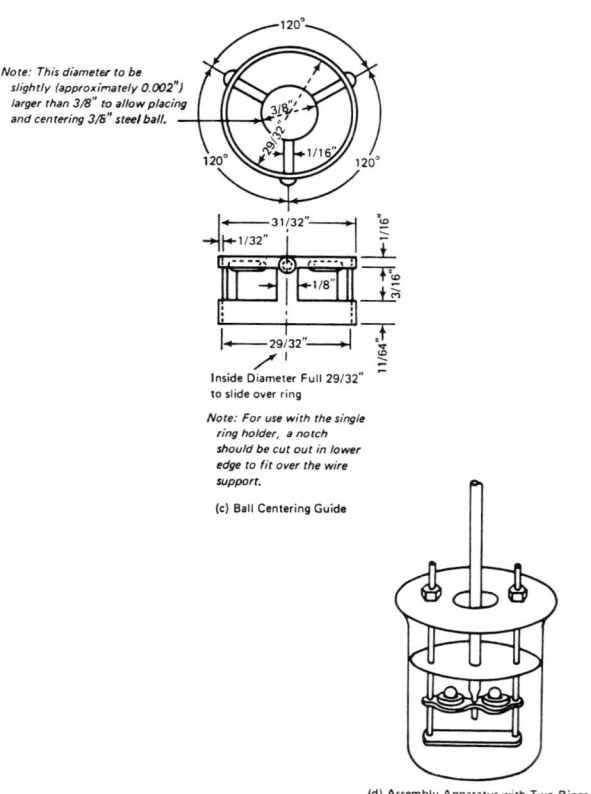

FIGURE 40 Shouldered Ring, Ring Holder, Ball-Centering Guide, and Assembly of Apparatus Showing Two Rings.

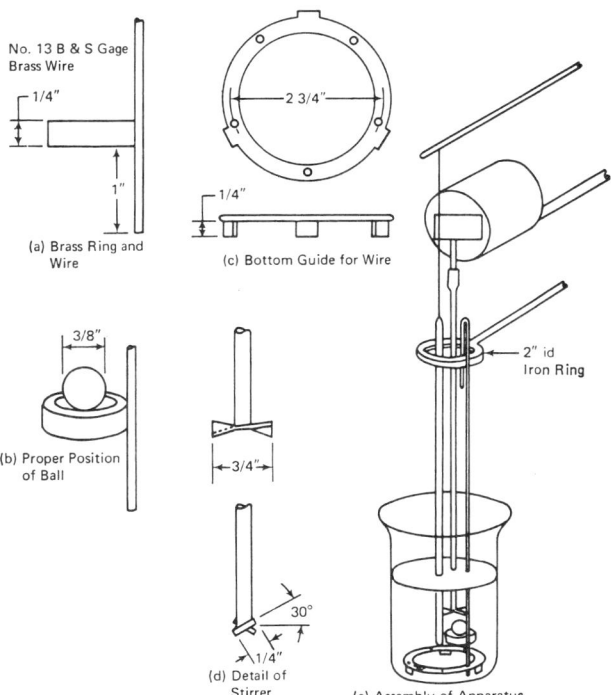

FIGURE 41 Assembly of Apparatus Showing Stirrer and Single-Shouldered Ring.

divided material or dust. The amount of sample taken should be at least twice that necessary to fill the desired number of rings, but in no case less than 40 g. Immediately melt the sample in a clean container, using an oven, hot plate, or sand or oil bath to prevent local overheating. Avoid incorporating air bubbles in the melting sample, which must not be heated above the temperature necessary to pour the material readily without inclusion of air bubbles. The time from the beginning of heating to the pouring of the sample shall not exceed 15 min. Immediately before filling the rings, preheat them to approximately the same temperature at which the sample is to be poured. While being filled, the rings should rest on an aluminum or steel plate. Pour a sufficient amount of the sample into the rings to leave an excess on cooling. Cool for at least 30 min, and then cut the excess material off cleanly with a slightly heated knife or spatula. Use a clean container and a fresh sample if the test is repeated.

Procedure

Materials Having Softening Points above 80° Fill the glass vessel with glycerin to a depth of not less than 101.6 mm and not more than 107.95 mm. The starting temperature of the bath shall be 32°. For resins (including rosin), cool the bath liquid to not less than 27° below the anticipated softening point, but in no case lower than 35°. Position the axis of the stirrer shaft near the back wall of the container, with the blades clearing the wall and with the bottom of the blades 19 mm above the top of the ring. Unless the ball-centering guide is used, make a slight indentation in the center of the sample by pressing the ball or a rounded rod, slightly heated for hard materials, into the sample at this point. Suspend the ring containing the sample in the bath so that the lower surface of the filled ring is 25.4 mm above the upper surface of the lower horizontal plate (see Fig. 40d), which is at least 12.7 mm and not more than 19 mm above the bottom of the glass vessel, or 25.4 mm above the bottom of the container (see Fig. 41e). Place the ball in the bath but not on the test specimen. Suspend an ASTM high-softening-point thermometer (16C or 16F) in the bath so that the bottom of its bulb is level with the bottom of the ring and within 12.7 mm of, but not touching, the ring. Maintain the initial temperature of the bath for 15 min. Begin stirring, and continue stirring at 500 to 700 rpm until the determination is complete. Apply heat in such a manner that the temperature of the bath liquid is raised 5° per min, avoiding the effects of drafts by using shields if necessary.

Note: The rate of rise of the temperature should be uniform and should not be averaged over the test period. Reject all tests in which the rate of rise exceeds ± 0.5° for any min period after the first three.

Record as the softening point the temperature of the thermometer at the instant the sample touches the lower horizontal plate (see Fig. 40d) or the bottom of the container (see Fig. 41e). Make no correction for the emergent stem of the thermometer.

Materials Having Softening Points of 80° or Below Follow the above procedure, except use an ASTM low-softening-point thermometer (15C or 15F) and use freshly boiled water cooled to 5° as the heating medium. For resins (including rosins), use water cooled to not less than 27° below the anticipated softening point, but in no case lower than 5°. Report the softening point to the nearest 1.0°.

VISCOSITY

Unless otherwise directed in the individual monograph, transfer the prepared sample into an 8-oz wide-mouth glass jar, 10.8 cm high and 7 cm in inside diameter, equipped with a screw lid. Condition the sample in a water bath at 25° ± 0.2° for 30 min (± 5 min), taking care to prevent water from coming into contact with the sample. Insert a No. 4 spindle in a Brookfield Model RVF viscometer,[2] or equivalent, and move the jar into place under the spindle, adjusting the elevation of the jar so that the upper surface of the sample is in the center of the shaft indentation and the spindle is in the center of the jar.

Note: Keep the viscometer level at all times during the test procedure.

Set the viscometer to rotate at 20 rpm, and allow the spindle to rotate until a constant dial reading is obtained. The viscosity, in centipoise, is the dial reading on the 0 to 100 scale multiplied by the appropriate factor (for a Brookfield RVF, spindle No. 4, 20 rpm, the factor is 100).

---

[2]Available from Brookfield Engineering Laboratories, Inc., Stoughton, MA.

APPENDIX X: CARBOHYDRATES (STARCHES, SUGARS, AND RELATED SUBSTANCES)

ACETYL GROUPS

Transfer about 5 g of the sample, accurately weighed, into a 250-mL Erlenmeyer flask, suspend in 50 mL of water, add a few drops of phenolphthalein TS, and titrate with 0.1 N sodium hydroxide to a permanent pink endpoint. Add 25.0 mL of 0.45 N sodium hydroxide, stopper the flask, and shake vigorously for 30 min, preferably with a mechanical shaker. Remove the stopper, wash the stopper and sides of the flask with a few mL of water, and titrate the excess alkali with 0.2 N hydrochloric acid to the disappearance of the pink color, recording the volume, in mL, of 0.2 N hydrochloric acid required as S. Perform a blank titration of 25.0 mL of 0.45 N sodium hydroxide, and record the volume, in mL, of 0.2 N hydrochloric acid required as B. Calculate the percentage of acetyl groups by the formula

% Acetyl Groups = $(B - S) \times N \times 0.043 \times 100/W$,

in which N is the exact normality of the hydrochloric acid solution, and W is the weight, in g, of the sample.

CRUDE FAT

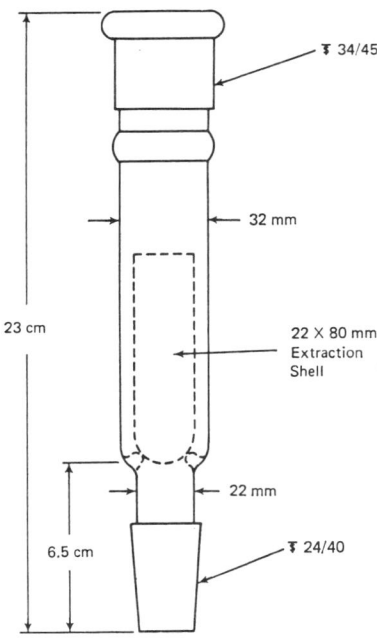

FIGURE 42   Butt-Type Extractor for Crude Fat Determination.

Apparatus   The apparatus consists of a Butt-type extractor,[1] as shown in Fig. 42, having a standard-taper 34/45 female joint at the upper end, to which is attached a Friedrichs- or Hopkins-type condenser, and a 24/40 male joint at the lower end, to which is attached a 125-mL Erlenmeyer flask.

Procedure   Transfer about 10 g of the sample, previously ground to 20-mesh or finer and accurately weighed, to a 15-cm filter paper, roll the paper tightly around the sample, and place it in a suitable extraction shell. Plug the top of the shell with cotton previously extracted with hexane, and place the shell in the extractor. Attach the extractor to a dry 125-mL Erlenmeyer flask containing about 50 mL of hexane and to a water-cooled condenser, apply heat to the flask to produce 150 to 200 drops of condensed solvent per min, and extract for 16 h. Disconnect the flask, and filter the extract to remove any insoluble residue. Rinse the flask and filter with a few mL of hexane, combine the washings and filtrate in a tared flask, and evaporate on a steam bath until no odor of solvent remains. Dry in a vacuum for 1 h at 100°, cool in a desiccator, and weigh.

INVERT SUGAR

Assay
Apparatus   Mount a ring support on a ringstand 1 to 2 in. above a gas burner, and mount a second ring 6 to 7 in. above the first. Place a 6-in. open-wire gauze on the lower ring to support a 250-mL Erlenmeyer flask, and place a 4-in. watch glass with a center hole on the upper ring to deflect heat. Attach a 50-mL buret to the ringstand so that the tip just passes through the watch glass centered above the flask. Alternatively, a buret with an offset tip may be used in place of a buret with a straight tip extending through the hole in the watch glass. Place an indirectly lighted white surface behind the assembly for observing the endpoint.

Soxhlet Solution
Copper Sulfate Solution   Dissolve 34.639 g of $CuSO_4.5H_2O$ in water, dilute to 500 mL, and filter.
Alkaline Tartrate Solution   Dissolve 173 g of potassium sodium tartrate ($KNaC_4H_4O_6.4H_2O$) and 50 g of NaOH in water, and dilute to 500 mL. Allow to stand 2 days, and filter before use.
Just before use, prepare the *Soxhlet Solution* by mixing equal volumes of *Copper Sulfate Solution* and *Alkaline Tartrate Solution*.

Standard Solution   Transfer approximately 9.5 g of NF-grade sucrose, accurately weighed, to a 1000-mL volumetric flask; dissolve in 100 mL of water; add 5 mL of hydrochloric acid; and store 3 days at 20° to 25°. Dilute to volume with water. This solution is stable for several months.

Sample Solution   Transfer 10 g of sample, accurately weighed, to a 1-L volumetric flask; dissolve in and dilute to volume with water.

---
[1] Available from H. S. Martin & Co., Evanston, IL.

Standardized Soxhlet Solution Pipet 25.0 mL of *Soxhlet Solution* into a 400-mL Erlenmeyer flask containing a few boiling chips, and titrate with the *Standard Solution* as directed under *Procedure*. Accurately dilute the *Standard Solution* so that the titration requires more than 15 but less than 50 mL.

Procedure

Invert Sugar Pipet 25.0 mL of *Standardized Soxhlet Solution* into a 400-mL Erlenmeyer flask containing a few boiling chips. Rapidly add *Sample Solution* from a buret to within 0.5 mL of the endpoint (determined by a preliminary titration). Immediately place the flask on the wire gauze of the *Apparatus*, and adjust the burner so that the boiling point will be reached in 2 min. As boiling proceeds, add 1 mL of a 0.2% aqueous solution of methylene blue, and complete the titration within 1 min by adding the *Sample Solution* dropwise or in small increments until the blue color disappears.

Find the invert sugar factor (F_I) in the table of conversion factors, immediately following this *Assay*, corresponding to the volume of titrant (V) used. When using the table, interpolation might be required to obtain correct factors corresponding to titrant volumes not shown and for solutions of invert sugar containing between 0 and 1 g of sucrose per 100 mL of solution. Calculate the concentration (C_I) of invert sugar, in mg/mL, used in the titration by using the following formula:

$$C_I = F_I/V.$$

Calculate the percent of invert sugar (P_I) in the sample by using the following formula:

$$P_I = (C_I \times 100)/C_S,$$

in which C_I is the concentration of invert sugar in mg/mL, as defined above, and C_S is the concentration of sample in mg/mL, which is calculated from the quantity of sample used to prepare the *Sample Solution*.

Sucrose Pipet 100 mL of *Sample Solution* into a 200-mL volumetric flask, and add slowly 10 mL of 2.7 N hydrochloric acid, diluted 1:1, while gently swirling the solution; place in a constant-temperature bath maintained at 60°; agitate continuously for 3 min; and allow to sit in the bath for an additional 7 min. Remove the flask from the bath, and cool to 20° as rapidly as possible; dilute to volume with water, and mix well. Continue as directed in the *Procedure* above under *Invert Sugar*. Calculate the percent invert sugar present after hydrolysis (P_H) using the formulas

$$C_H = F_S/V,$$

$$P_H = 100C_H/C_S,$$

in which C_H is the concentration of invert sugar, in mg/mL, after hydrolysis; F_S is the sucrose factor from the conversion table corresponding to the volume of titrant V; and C_S is the concentration of sample, in mg/mL, in the sample solution, as defined above. Calculate the percent sucrose by using the following formula:

$$P_S = 0.95(P_H - P_I),$$

in which P_H and P_I are the percentages of invert sugar determined after and before hydrolysis, respectively.

Conversion Factors for the Determination of Invert Sugar in the Presence of Sucrose

V(mL)	F_S^a	F_I^b	V(mL)	F_S^a	F_I^b
15	123.6	122.6	33	124.4	123.2
16	123.6	122.7	34	124.5	123.3
17	123.6	122.7	35	124.5	123.3
18	123.7	122.7	36	124.6	123.3
19	123.7	122.8	37	124.6	123.4
20	123.8	122.8	38	124.7	123.4
21	123.8	122.8	39	124.7	123.4
22	123.9	122.9	40	124.8	123.4
23	123.9	122.9	41	124.8	123.5
24	124.0	122.9	42	124.9	123.5
25	124.0	123.0	43	124.9	123.5
26	124.1	123.0	44	125.0	123.6
27	124.1	123.0	45	125.0	123.6
28	124.2	123.1	46	125.1	123.6
29	124.2	123.1	47	125.1	123.7
30	124.3	123.1	48	125.2	123.7
31	124.3	123.2	49	125.2	123.7
32	124.4	123.2	50	125.3	123.8

[a] Invert sugar only
[b] Invert sugar containing 1 g of sucrose per 100 mL

LACTOSE

Assay

Apparatus Use a suitable high-performance liquid chromatographic system (see *Chromatography*, Appendix IIA) equipped with a differential refractometer detector, a precolumn, an online 0.45-μm filter, and a 250-mm × 4.6-mm (id) stainless steel column.

Solid Phase Microparticle silica gel with siloxane bonded cyano-amino moieties (Whatman P-10 carbohydrate or equivalent) equilibrated and operated at room temperature.

Mobile Phase Acetonitrile–water (80:20) at a flow rate of 2 mL/min.

Reagents

Acetonitrile An appropriate grade for liquid chromatography.

Fructose Prepare a solution of fructose to be used as an internal standard by transferring 50 g of commercial grade β-D(−)fructose powder to a 500-mL volumetric flask, and dissolve in and dilute to volume with water.

Standard Solution Transfer about 2 g of NF-grade anhydrous lactose, accurately weighed, to a 100-mL volumetric flask, add 10 mL of fructose internal standard solution, and dilute to volume with water. Prepare fresh daily.

Water An appropriate grade for liquid chromatography.

System Suitability (see *Chromatography*, Appendix IIA).

Repeatability Allow the chromatographic system to equilibrate at a flow rate of 2 mL/min, then inject 25-μL aliquots of the *Standard Solution*. The chromatogram should show baseline resolution and a retention time for water of 1 to 2 min; fructose, 2 to 3 min; and lactose, 5 to 6 min. The coefficient of variation for the relative peak heights (lactose peak height/fructose peak height) for ten injections should be ≤0.6% when column equilibration is complete.

Linearity of Detector Response On a monthly basis (or when changes in the system are made), monitor the linearity of detector response by injecting standard lactose solutions containing 1.4%, 1.8%, 2.0%, 2.2%, and 2.6% lactose. Linear regression of the curve generated by plotting peak height versus concentration should give a correlation coefficient of at least 0.999.

Sample Preparation Prepare the sample as directed in the individual monograph. Analysis must be performed within 24 h.

Procedure Inject triplicate 25-μL aliquots of sample and standard solutions. If more than one sample is to be analyzed, inject the standard solution after every third sample. Calculate results using average standard response factors bracketing every three samples (see *Chromatography*, Appendix IIA).

Calculation Calculate the % Lactose (dried basis) by the formula

$$(R_L/R_F) \times (W_L/W_S) \times (100 - M_L/100 - M_S) \times P,$$

in which R_L and R_F are the response factors for lactose and fructose; W_S and W_L are the weights, in g, of the sample and lactose standard in their respective solutions; M_S and M_L are the percentages of moisture in the sample and lactose standard; and P is the purity, in percent, of the lactose standard. Determine the moisture content by drying at 120° for 16 h.

MANGANESE

Manganese Detection Instrument Use any suitable atomic absorption spectrophotometer equipped with a fast-response recorder or other readout device and capable of measuring the radiation absorbed by manganese atoms at the manganese resonance line of 279.5 nm.

Standard Preparations Transfer 1000 mg, accurately weighed, of manganese metal powder into a 1000-mL volumetric flask, dissolve by warming in a mixture of 10 mL of water and 10 mL of 0.5 N hydrochloric acid, cool, dilute to volume with water, and mix. Pipet 5.0 mL of this solution into a 50-mL volumetric flask, dilute to volume with water, and mix. Finally, pipet 5.0, 10.0, 15.0, and 25.0 mL of this solution into separate 1000-mL volumetric flasks, dilute each flask to volume with water, and mix. The final solutions contain 0.5, 1.0, 1.5, and 2.5 mg/kg of Mn, respectively.

Sample Preparation Transfer 10.000 g of the sample into a 200-mL Kohlrausch volumetric flask, previously rinsed with 0.5 N hydrochloric acid, add 140 mL of 0.5 N hydrochloric acid, and shake vigorously for 15 min, preferably with a mechanical shaker. Dilute to volume with 0.5 N hydrochloric acid, and shake. Centrifuge approximately 100 mL of the sample mixture in a heavy-walled centrifuge tube at 2000 rpm for 5 min, and use the clear supernatant liquid in the following *Procedure*.

Procedure Aspirate 0.5 N hydrochloric acid through the air–acetylene burner for 5 min, and obtain a baseline reading at 279.5 nm, following the manufacturer's instructions for operating the atomic absorption spectrophotometer being used for the analysis. Aspirate a portion of each *Standard Preparation* in the same manner, note the readings, then aspirate a portion of the *Sample Preparation*, and note the reading. Prepare a standard curve by plotting the mg/kg of Mn in each *Standard Preparation* against the respective readings. From the graph determine the mg/kg of Mn in the *Sample Preparation*, and multiply this value by 20 to obtain the mg/kg of Mn in the original sample taken for analysis.

PHOSPHORUS

Reagents

Ammonium Molybdate Solution (5%) Dissolve 50 g of ammonium molybdate tetrahydrate, $(NH_4)_6Mo_7O_{24}.4H_2O$, in 900 mL of warm water, cool to room temperature, dilute to 1000 mL with water, and mix.

Ammonium Vanadate Solution (0.25%) Dissolve 2.5 g of ammonium metavanadate, NH_4VO_3, in 600 mL of boiling water, cool to 60° to 70°, and add 20 mL of nitric acid. Cool to room temperature, dilute to 1000 mL with water, and mix.

Zinc Acetate Solution (10%) Dissolve 120 g of zinc acetate dihydrate, $Zn(C_2H_3O_2)_2.2H_2O$, in 880 mL of water, and filter through Whatman No. 2V or equivalent filter paper before use.

Nitric Acid Solution (29%) Add 300 mL of nitric acid (sp. gr. 1.42) to 600 mL of water, and mix.

Standard Phosphorus Solution (100 μg P in 1 mL) Dissolve 438.7 mg of monobasic potassium phosphate, KH_2PO_4, in water in a 1000-mL volumetric flask, dilute to volume with water, and mix.

Standard Curve Pipet 5.0, 10.0, and 15.0 mL of the *Standard Phosphorus Solution* into separate 100-mL volumetric flasks. To each of these flasks, and to a fourth, blank flask, add in the order stated 10 mL of *Nitric Acid Solution*, 10 mL of *Ammonium Vanadate Solution*, and 10 mL of *Ammonium Molybdate Solution*, mixing thoroughly after each addition. Dilute to volume with water, mix, and allow to stand for 10 min. Determine the absorbance of each standard solution in a 1-cm cell at 460 nm, with a suitable spectrophotometer, using the blank to set the instrument to zero. Prepare a standard curve by plotting the absorbance of each solution versus its concentration, in mg of P per 100 mL.

Treated Sample Place 20 to 25 g of the starch sample in a 250-mL beaker, add 200 mL of a 7 to 3 methanol–water mixture, disperse the sample, and agitate mechanically for 15 min. Recover the starch by vacuum filtration in a 150-mL medium-porosity fritted-glass or Büchner funnel, and wash the wet cake with 200 mL of the methanol–water mixture. Reslurry the wet cake in the solvent, and wash it a second time in the same manner. Dry the filter cake in an air oven at a temperature below 50°, then grind the sample to 20-mesh or finer, and blend thoroughly. Determine the amount of dry substance by drying a 5-g portion in a vacuum oven, not exceeding 100 mm of Hg, at 120° for 5 h.

Note: The treatment outlined above is satisfactory for starch products that are insoluble in cold water. For pregelatinized starch and other water-soluble starches, prepare a 1% to 2% aqueous paste, place it in a cellophane tube, and dialyze against running distilled water for 30 h to 40 h. Precipitate the starch by pouring the solution into 4 volumes of acetone per volume of paste, while stirring. Recover the starch by vacuum filtration in a medium-porosity fritted-glass or Büchner funnel, and wash the filter cake with absolute ethanol. Dry the filter cake, and determine the amount of dry substance as directed for water-insoluble starches.

Sample Preparation Transfer about 10 g of the *Treated Sample*, calculated on the dry-substance basis and accurately weighed, into a Vycor dish, and add 10 mL of *Zinc Acetate Solution* in a fine stream, distributing the solution uniformly in the sample. Carefully evaporate to dryness on a hot plate, then increase the heat, and carbonize the sample on the hot plate or over a gas flame. Ignite in a muffle furnace at 550° until the ash is free from carbon (about 1 to 2 h), and cool. Wet the ash with 15 mL of water, and slowly wash down the sides of the dish with 5 mL of *Nitric Acid Solution*. Heat to boiling, cool, and quantitatively transfer the mixture into a 200-mL volumetric flask, rinsing the dish with three 20-mL portions of water and adding the rinsings to the flask. Dilute to volume with water, and mix. Transfer an accurately measured aliquot (V, in mL) of this solution, containing not more than 1.5 mg of phosphorus, into a 100-mL volumetric flask, and add 50 mL of water to a second flask to serve as a blank. To each flask add in the order stated 10 mL of *Nitric Acid Solution*, 10 mL of *Ammonium Vanadate Solution*, and 10 mL of *Ammonium Molybdate Solution*, mixing thoroughly after each addition. Dilute to volume with water, mix, and allow to stand for 10 min.

Procedure Determine the absorbance of the *Sample Preparation* in a 1-cm cell at 460 nm, with a suitable spectrophotometer, using the blank to set the instrument at zero. From the *Standard Curve*, determine the mg of phosphorus in the aliquot taken, recording this value as a. Calculate the amount, in mg/kg, of phosphorus (P) in the original sample by the formula

$$\text{mg/kg P} = (a \times 200 \times 1000)/(V \times W),$$

in which W is the weight, in g, of the sample taken.

PROPYLENE CHLOROHYDRIN

Special Apparatus

Gas Chromatograph Use a suitable gas chromatograph. A dual-column instrument equipped with a flame-ionization detector and an integrator is preferred.

Concentrator Use a Kuderna-Danish concentrator having a 500-mL flask, available from Kontes Glass Co., Vineland, NJ (Catalog No. K-57000), or equivalent.

Pressure Bottles Use 200-mL pressure bottles, with a Neoprene washer, glass stopper, and attached wire clamp, available from Fisher Scientific Co. (Vitro 400, Catalog No. 3-100), or equivalent.

Gas Chromatography Column Use a stainless steel column, 3 m × 3.2 mm (od), packed with 10% Carbowax 20 M on 80/100-mesh Gas Chrom 2, or equivalent. After packing and before use, condition the column overnight at 200°, using a helium flow of 25 mL/min.

Reagents

Diethyl Ether Use anhydrous, analytical reagent-grade diethyl ether, available from Fisher Scientific Co. or J. T. Baker Co., or other suitable sources.

Note: Some lots of diethyl ether contain foreign residues that interfere with the analysis and/or the interpretation of the chromatograms. If the ether quality is unknown or suspect, concentrate 50 mL to a volume of about 1 mL in the concentrator, and then chromatograph a 2.0-µL portion using the conditions outlined under the *Procedure*. If the chromatogram is excessively noisy and contains signal peaks that overlap or interfere in the measurement of the peaks produced by the propylene chlorohydrin isomers, the ether should be redistilled.

Florisil PR Use 60/100-mesh material, available from Floridin Co., 3 Penn Center, Pittsburgh, PA 15235, or an equivalent product available from Supelco, Bellefonte, PA 16823.

Propylene Chlorohydrins Use 1-Chloro-2-propanol Practical Grade, containing 25% 2-chloro-1-propanol, available from Aldrich Chemical Company, Milwaukee, WI 53233.

Standard Preparation Draw 25 µL of *Propylene Chlorohydrins* into a 50-µL syringe, weigh accurately, and discharge the contents into a 500-mL volumetric flask partially filled with water. Reweigh the syringe, and record the weight of the chlorohydrins taken. Dilute to volume with water, and mix. This solution contains about 27.5 mg of mixed chlorohydrins, or about 55 µg/mL. Prepare this solution fresh daily.

Sample Preparation Transfer a blended representative 50.0-g sample into a pressure bottle, and add 125 mL of 2 N sulfuric acid. Clamp the top in place, and swirl the contents until the sample is completely dispersed. Place the bottle in a boiling water bath, heat for 10 min, then swirl the bottle to mix the contents, and heat in the bath for an additional 15 min. Cool in air to room temperature, then neutralize the hydrolyzed sample to pH 7 with 25% sodium hydroxide solution, and filter through Whatman No. 1 paper, or equivalent, in a Büchner funnel, using suction. Wash the bottle and filter paper with 25 mL of water, and combine the washings with the filtrate. Add 30 g of anhydrous sodium sulfate, and stir with a magnetic stirring bar for 5 to 10 min, or until the sodium sulfate is completely dissolved. Transfer the solution into a 500-mL separator equipped with a Teflon plug, rinse the flask with 25 mL of water, and combine the washings with the sample solution. Extract with five

50-mL portions of *Diethyl Ether*, allowing at least 5 min in each extraction for adequate phase separation. Transfer the combined ether extracts in a concentrator, place the graduated receiver of the concentrator in a water bath maintained at 50° to 55°, and concentrate the extract to a volume of 4 mL.

> Note: Ether extracts of samples may contain foreign residues that interfere with the analysis and/or interpretation of the chromatograms. These residues are believed to be degradation products arising during the hydrolysis treatment. Analytical problems created by their presence can be avoided through application of a cleanup treatment performed as follows: Concentrate the ether extract to about 8 mL, instead of 4 mL specified above. Add 10 g of *Florisil PR*, previously heated to 130° for 16 h just before use, to a chromatographic tube of suitable size, then tap gently, and add 1 g of anhydrous sodium sulfate to the top of the column. Wet the column with 25 mL of *Diethyl Ether*, and quantitatively transfer the concentrated extract to the column with the aid of small portions of the ether. Elute with three 25-mL portions of the ether, collect all of the eluate, transfer it to a concentrator, and concentrate to a volume of 4 mL.

Cool the extract to room temperature, transfer it quantitatively to a 5.0-mL volumetric flask with the aid of small portions of *Diethyl Ether*, dilute to volume with the ether, and mix.

Control Preparations Transfer 50.0-g portions of unmodified (underivatized) waxy corn starch into five separate pressure bottles, and add 125 mL of 2 N sulfuric acid to each bottle. Add 0.0, 0.5, 1.0, 2.0, and 5.0 mL of the *Standard Preparation* to the bottles, respectively, giving propylene chlorohydrin concentrations, on the starch basis, of 0, 0.5, 1, 2, and 5 mg/kg, respectively. Calculate the exact concentration in each bottle from the weight of *Propylene Chlorohydrins* used in making the *Standard Preparation*. Clamp the tops in place, swirl until the contents of each bottle are completely dissolved, and proceed with the hydrolysis, neutralization, filtration, extraction, extract concentration, and final dilution as directed under *Sample Preparation*.

Procedure Perform the analysis by gas chromatography with the gas chromatograph and gas chromatography column previously described. The operating conditions may be varied, depending on the column and instrument used. A suitable chromatogram was obtained using a column oven temperature of 110°, isothermal; injection port temperature of 210°; detector temperature of 240°; and hydrogen (30 mL/min), air (350 mL/min), or helium (25 mL/min), as the carrier gas.

Inject 2.0-µL aliquots of each of the concentrated extracts, prepared as directed under *Control Preparations*, allowing sufficient time between injections for signal peaks corresponding to the two chlorohydrin isomers to be recorded (and integrated) and for the column to be purged. Record and sum the signal areas (integrator outputs) from the two chlorohydrin isomers for each of the controls.

Using identical operating conditions, inject a 2.0-µL aliquot of the concentrated extract prepared as directed under *Sample Preparation*, and record and sum the signal areas (integrator outputs) from the sample.

Calculation Prepare a standard curve for the summed signal areas for each of the controls against the calculated propylene chlorohydrin concentrations, in mg/kg, derived from the actual weight of chlorohydrin isomers used. Using the summed signal areas corresponding to the 1-chloro-2-propanol and 2-chloro-1-propanol from the sample, determine the concentration of mixed propylene chlorohydrins, in mg/kg, in the sample by reference to the calibration plot.

> Note: After gaining experience with the procedure and demonstrating that the calibration plot derived from the control samples is linear and reproducible, the number of controls can be reduced to one containing about 5 mg/kg of mixed propylene chlorohydrin isomers. The propylene chlorohydrin level in the sample can then be calculated as follows:
>
> Propylene chlorohydrins, mg/kg = $(C \times a)/A$,
>
> in which C is the concentration, in mg/kg, of propylene chlorohydrins (sum of isomers) in the control; a is the sum of the signal areas produced by the propylene chlorohydrin isomers in the sample; and A is the sum of the signal areas produced by the propylene chlorohydrin isomers in the control.

REDUCING SUGARS ASSAY

Apparatus Mount a ring support on a ringstand 1 to 2 in. above a gas burner, and mount a second ring 6 to 7 in. above the first. Place a 6-in. open-wire gauze on the lower ring to support a 250-mL Erlenmeyer flask, and place a 4-in. watch glass with a center hole on the upper ring to deflect heat. Attach a 25-mL buret to the ringstand so that the tip just passes through the watch glass centered above the flask. Place an indirectly lighted white surface behind the assembly for observing the endpoint.

Standardized Fehling's Solution Measure a quantity of *Fehling's Solution A*, add an equal quantity of *Fehling's Solution B*, and mix (see *Cupric Tartrate TS, Alkaline* in the section on *Solutions and Indicators*). Immediately before use, standardize as follows: Transfer 3.000 g of primary standard dextrose (NIST Standard Reference Material, or equivalent), previously dried in vacuum at 100° for 2 h, into a 500-mL volumetric flask, dissolve in and dilute to volume with water, and mix. Pipet 25 mL of the mixed Fehling's solution into a 200-mL Erlenmeyer flask containing a few glass beads, and titrate with the standard dextrose solution as directed under *Procedure*. Adjust the concentration of *Fehling's Solution A* by dilution or the addition of copper sulfate, so that the titration requires 20.0 mL of the standard dextrose solution.

Procedure Transfer about 3 g of the sample, accurately weighed, into a 500-mL volumetric flask, dissolve in and dilute to volume with water, and mix. Pipet 25.0 mL of *Standardized Fehling's Solution* into a 200-mL Erlenmeyer flask containing a few glass beads, and add the sample solution from a buret to within 0.5 mL of the anticipated endpoint (determined by preliminary titration). Immediately place the flask on the wire gauze of the *Apparatus*, and adjust the burner so that the boiling point will be reached in about 2 min. Bring to a boil, and boil gently for 2 min. As boiling continues, add 2 drops of a 1% aqueous solution of methylene blue, and complete the titration within 1 min by adding the sample solution dropwise or in small increments until the blue color disappears. Record the volume, in mL, of sample solution required as V. Calculate the percentage of reducing sugars, as D-glucose on the dried basis, by the formula

$$\% \text{ Reducing Sugars} = (500 \times 0.12 \times 100)/(V \times W),$$

in which W is the weight, in g, of the sample of dry substance.

SULFUR DIOXIDE DETERMINATION
(Based on AOAC Method 962.16)

Reagents

3% Hydrogen Peroxide Solution Dilute 30% hydrogen peroxide to 3% with water. Just before use, add 3 drops of methyl red TS, and titrate to a yellow endpoint using 0.01 N sodium hydroxide. If the endpoint is exceeded, discard the solution and prepare another 3% hydrogen peroxide solution.

Standardized Titrant Prepare a solution of 0.01 N sodium hydroxide.

Nitrogen A source of high-purity nitrogen is required with a flow regulator that will maintain a flow of 200 ± 10 mL/min. To guard against the presence of oxygen in the nitrogen, an oxygen scrubbing apparatus or solution such as an alkaline pyrogallol trap may be used. Prepare the pyrogallol trap as follows: Add 4.5 g of pyrogallol to the trap, purge the trap with nitrogen for 2 to 3 min, and add potassium hydroxide solution (65 g of potassium hydroxide added to 85 mL of water) to the trap while maintaining an atmosphere of nitrogen in the trap.

Caution: Exothermic reaction.

Sample Preparation (for solids) Transfer 50 g of the sample, or a quantity of the sample with a known quantity of sulfur dioxide (500 to 1500 μg of SO_2), to a food processor or blender, if necessary. Add 50 mL of 5% ethanol in water, and briefly grind the mixture, reserving another 50 mL of 5% ethanol in water to rinse the blender jar. Grinding or blending should be continued only until the food is chopped into pieces small enough to pass through the 24/40 joint of a flask (see Fig. 43).

Sample Preparation (for liquids) Mix 50 g of the sample, or a quantity with a known amount of sulfur dioxide (500 to 1500 μg of SO_2), with 100 mL of 5% ethanol in water.

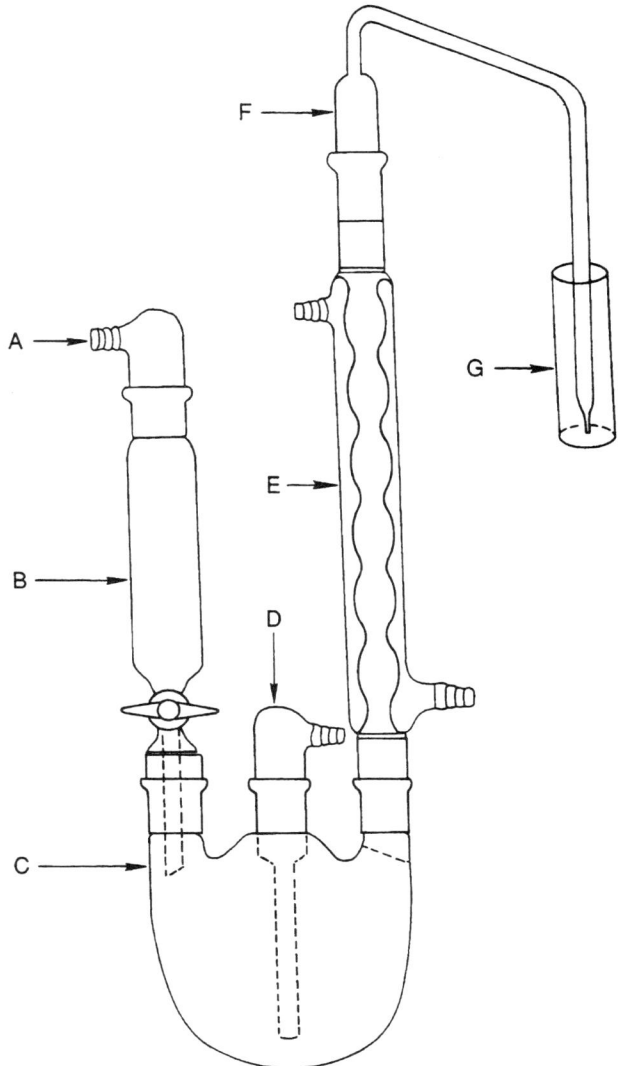

FIGURE 43 The Optimized Monier-Williams Apparatus; Component Identification is Given in Text (component F is depicted in FIGURE 44).

Apparatus The apparatus shown diagrammatically (Fig. 43) is designed to accomplish the selective transfer of sulfur dioxide from the sample in boiling aqueous hydrochloric acid to the *3% Hydrogen Peroxide Solution*. This apparatus is easier to assemble than the official AOAC apparatus, and the back-pressure inside the apparatus is limited to the unavoidable pressure due to the height of the *3% Hydrogen Peroxide Solution* above the tip of the bubbler, F. Keeping the back-pressure as low as possible reduces the likelihood that sulfur dioxide will be lost through leaks.

Note: Tygon and silicon tubing should be preboiled before use in this procedure.

The apparatus should be assembled as shown in Fig. 43 with a thin film of stopcock grease on the sealing surfaces of all the joints except the joint between the separatory funnel and the

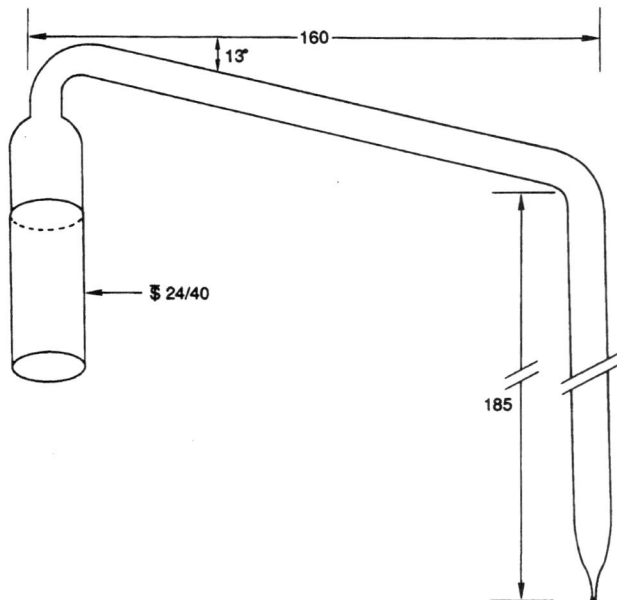

FIGURE 44 Diagram of Bubbler (*F* in FIGURE 43) (lengths are given in mm).

flask. Each joint should be clamped together to ensure a complete seal throughout the analysis. The separatory funnel, *B*, should have a capacity of 100 mL or greater. An inlet adapter, *A*, with a hose connector (Kontes K-183000, or equivalent) is required to provide a means of applying a head of pressure above the solution. (A pressure-equalizing dropping funnel is not recommended because condensate, perhaps with sulfur dioxide, is deposited in the funnel and the side arm.) The round-bottom flask, *C*, is a 1000-mL flask with three 24/40 tapered joints. The gas inlet tube, *D* (Kontes K-179000, or equivalent), should be of sufficient length to permit introduction of the nitrogen within 2.5 cm of the bottom of the flask. The Allihn condenser, *E* (Kontes K-431000-2430, or equivalent), has a jacket length of 300 mm. The bubbler, *F*, is fabricated from glass according to the dimensions given in Fig. 44, and it has the same dimensions as a 50-mL graduated cylinder (see Fig. 44). The *3% Hydrogen Peroxide Solution* can be contained in a receiving vessel, *G*, with an id of about 2.5 cm and a depth of 18 cm.

Buret Use a 10-mL buret with overflow tube and hose connections for an Ascarite tube or equivalent air-scrubbing apparatus. This will permit the maintenance of a carbon dioxide-free atmosphere over the *Standardized Titrant*.

Chilled Water Circulator The condenser must be chilled with a coolant, such as 20% methanol–water, at a flow rate so that the condenser outlet temperature is maintained at 5°. A circulating pump equivalent to the Neslab Coolflow 33 is suitable.

Determination Assemble the apparatus as shown in Fig. 43. The flask must be positioned in a heating mantle that is controlled by a power-regulating device such as Variac or equivalent. Add 400 mL of distilled water to the flask. Close the stopcock of the separatory funnel, and add 90 mL of 4 *N* hydrochloric acid to the separatory funnel. Begin the flow of nitrogen at a rate of 200 ± 10 mL/min. The condenser coolant flow must be initiated at this time. Add 30 mL of *3% Hydrogen Peroxide Solution*, which has been titrated to a yellow endpoint with the *Standardized Titrant*, to the receiving vessel, *G*. After 15 min, the apparatus and the water will be thoroughly deoxygenated, and the apparatus will be ready for sample introduction.

Sample Introduction and Distillation Remove the separatory funnel, and quantitatively transfer the food sample in aqueous ethanol to the flask. Wipe the tapered joint clean with a laboratory tissue, apply stopcock grease to the outer joint of the separatory funnel, and return the separatory funnel to the tapered joint flask. The nitrogen flow through the *3% Hydrogen Peroxide Solution* should resume as soon as the funnel is reinserted into the appropriate joint in the flask. Examine each joint to ensure that it is sealed.

Apply a head pressure above the hydrochloric acid solution in the separatory funnel with a rubber bulb equipped with a valve. Open the stopcock in the separatory funnel, and permit the hydrochloric acid solution to flow into the flask. Continue to maintain sufficient pressure above the acid solution to force the solution into the flask. The stopcock may temporarily be closed, if necessary, to pump up the pressure above the acid. To guard against the escape of sulfur dioxide into the separatory funnel, close the stopcock before the last few mL drain out of the separatory funnel.

Apply the power to the heating mantle. Use a power setting that will cause 80 to 90 drops of condensate to return to the flask from the condenser per min. After 1.75 h of boiling, cool the contents of the 1000-mL flask at the condensation rate stated above, and remove the contents of the receiving vessel, *G*.

Titration Add 3 drops of *Methyl Red Indicator*, and titrate the above-mentioned contents with the *Standardized Titrant* to a yellow endpoint that persists for at least 20 s. Calculate the sulfur dioxide content, expressed as μg of sulfur dioxide per g of food (μg/g or mg/kg) as follows:

$$\text{mg/kg} = (32.03 \times V_B \times N \times 1000)/Wt,$$

in which 32.03 is the milliequivalent weight, in mg, of sulfur dioxide; V_B is the volume, in mL, of sodium hydroxide titrant of normality, *N*, required to reach the endpoint; the factor 1000 converts mg to μg; and *Wt* is the weight, in g, of food sample introduced into the 1000-mL flask.

TOTAL SOLIDS

Note: The refractive index, RI, of solutions of various carbohydrates at specific temperatures is directly correlated with the solutions' concentrations (in g/100 g or percent dried solids). The following tables, as required in some monographs in this edition, are provided for the user's convenience.

Apparatus Use a suitable refractometer (see *Refractive Index*, Appendix IIB) equipped with a jacket for water circulation or some other mechanism for maintaining the sample at 20.0° ± 0.1° or some other fixed temperature. Before proceeding with measurements, ensure that the prism has reached the equilibrium temperature.

Standardization To achieve the theoretical accuracy of ± 0.0001, calibrate the instrument daily by determining the refractive index of distilled water, which is 1.3330 at 20°, and 1.3325 at 25°.

Procedure Determine the refractive index after ensuring that the sample and prism have reached the equilibrium temperature.

For *Corn Syrups, High-Fructose Corn Syrups, Liquid Fructose,* and *Maltodextrin*, convert the refractive index to approximate percent solids using the accompanying tables.

Note: These tables cover the approximate total solids levels of these products in commerce. If the ash or dextrose equivalent of the sample differs from the product in the table, use the accompanying ash and dextrose equivalent correction table.

Glucose Syrup (Corn Syrup)

28 DE[a] Glucose Syrup—0.3% Ash

%DS[b]	RI[c] 20°C	RI 45°C	°Baumé at 140°F (60°C) + 1
76.0	1.4888	1.4837	40.98
77.0	1.4915	1.4864	41.49
78.0	1.4943	1.4892	42.00
79.0	1.4971	1.4919	42.51
80.0	1.4999	1.4947	43.01

[a]Dextrose Equivalent
[b]Dry Substance
[c]Refractive Index

36 DE Glucose Syrup—0.3% Ash

%DS	RI 20°C	RI 45°C	°Baumé at 140°F (60°C) + 1
78.4	1.4938	1.4887	42.01
79.4	1.4965	1.4914	42.52
80.4	1.4993	1.4941	43.02
81.4	1.5021	1.4969	43.52
82.4	1.5049	1.4997	44.02

34 DE High-Maltose Glucose Syrup—0.3% Ash

%DS	RI 20°C	RI 45°C	°Baumé at 140°F (60°C) + 1
78.6	1.4933	1.4882	41.99
79.6	1.4960	1.4909	42.49
80.6	1.4988	1.4936	42.99
81.6	1.5015	1.4964	43.49
82.6	1.5043	1.4992	43.99

43 DE High-Maltose Glucose Syrup—0.3% Ash

%DS	RI 20°C	RI 45°C	°Baumé at 140°F (60°C) + 1
78.9	1.4934	1.4883	42.00
79.9	1.4961	1.4910	42.51
80.9	1.4988	1.4937	43.01
81.9	1.5016	1.4964	43.51
82.9	1.5044	1.4992	44.01

43 DE Glucose Syrup—0.3% Ash

%DS	RI 20°C	RI 45°C	°Baumé at 140°F (60°C) + 1
78.7	1.4933	1.4882	42.01
79.7	1.4960	1.4909	42.51
80.7	1.4988	1.4936	43.02
81.7	1.5015	1.4964	43.52
82.7	1.5043	1.4992	44.01

43 DE (Ion-Exchanged) Glucose Syrup—0.03% Ash

%DS	RI 20°C	RI 45°C	°Baumé at 140°F (60°C) + 1
78.8	1.4935	1.4884	41.99
79.8	1.4962	1.4911	42.50
80.8	1.4990	1.4938	43.00
81.8	1.5018	1.4966	43.50
82.8	1.5045	1.4994	43.99

53 DE Glucose Syrup—0.3% Ash

%DS	RI 20°C	RI 45°C	°Baumé at 140°F (60°C) + 1
80.5	1.4962	1.4911	42.64
81.5	1.4989	1.4938	43.14
82.5	1.5016	1.4965	43.64
83.5	1.5044	1.4992	44.13
84.5	1.5072	1.5020	44.63

63 DE Glucose Syrup—0.3% Ash

%DS	RI 20°C	RI 45°C	°Baumé at 140°F (60°C) + 1
81.0	1.4955	1.4904	42.53
82.0	1.4982	1.4931	43.02
83.0	1.5009	1.4958	43.52
84.0	1.5037	1.4985	44.01
85.0	1.5064	1.5012	44.50

63 DE (Ion-Exchanged) Glucose Syrup—0.03% Ash

%DS	RI 20°C	RI 45°C	°Baumé at 140°F (60°C) + 1
81.3	1.4963	1.4912	42.60
82.3	1.4990	1.4939	43.10
83.3	1.5017	1.4965	43.59
84.3	1.5044	1.4993	44.09
85.3	1.5072	1.5020	44.58

66 DE Glucose Syrup—0.3% Ash

%DS	RI 20°C	RI 45°C	°Baumé at 140°F (60°C) + 1
81.0	1.4949	1.4898	42.36
82.0	1.4975	1.4924	42.86
83.0	1.5002	1.4951	43.36
84.0	1.5029	1.4978	43.85
85.0	1.5056	1.5005	44.35

95 DE Glucose Syrup—0.3% Ash

%DS	RI 20°C	RI 45°C	°Baumé at 140°F (60°C) + 1
69.0	1.4598	1.4550	35.46
70.0	1.4621	1.4573	35.96
71.0	1.4644	1.4596	36.46
72.0	1.4668	1.4619	36.96
73.0	1.4692	1.4643	37.45

95 DE (Ion-Exchanged) Glucose Syrup—0.03% Ash

%DS	RI 20°C	RI 45°C	°Baumé at 140°F (60°C) + 1
69.0	1.4597	1.4549	35.39
70.0	1.4620	1.4572	35.89
71.0	1.4644	1.4595	36.39
72.0	1.4667	1.4619	36.89
73.0	1.4691	1.4642	37.38

High-Fructose Corn Syrup Solids

42% High-Fructose Corn Syrup—0.03% Ash

%DS[a]	RI[b] 20°C	RI 45°C
69.0	1.4597	1.4543
70.0	1.4620	1.4565
71.0	1.4643	1.4589
72.0	1.4667	1.4612
73.0	1.4691	1.4635

[a] Dry Substance
[b] Refractive Index

55% High-Fructose Corn Syrup—0.05% Ash

%DS	RI 20°C	RI 45°C
75.0	1.4738	1.4680
76.0	1.4762	1.4704
77.0	1.4786	1.4728
78.0	1.4811	1.4752
79.0	1.4835	1.4776

Liquid Fructose

%DS	RI 20°C	RI 45°C
75.0	1.4732	1.4667
76.0	1.4756	1.4691
77.0	1.4780	1.4715
78.0	1.4805	1.4739
79.0	1.4829	1.4763

Maltodextrin

12 DE[a] Maltodextrin—0.3% Ash

%DS[b]	RI[c] 20°C	RI 45°C	Commercial °Baumé 140°F (60°C) + 1
45.0	1.4149	1.4105	24.57
46.0	1.4171	1.4126	25.13
47.0	1.4193	1.4148	25.68
48.0	1.4215	1.4170	26.24
49.0	1.4237	1.4192	26.79
50.0	1.4260	1.4214	27.34
51.0	1.4282	1.4237	27.89
52.0	1.4305	1.4259	28.44
53.0	1.4328	1.4282	28.99
54.0	1.4351	1.4305	29.53
55.0	1.4375	1.4328	30.08
56.0	1.4398	1.4351	30.62
57.0	1.4422	1.4375	31.16
58.0	1.4446	1.4399	31.71
59.0	1.4470	1.4422	32.24
60.0	1.4494	1.4446	32.78
61.0	1.4519	1.4471	33.32
62.0	1.4544	1.4495	33.85
63.0	1.4569	1.4520	34.39
64.0	1.4594	1.4545	34.92
65.0	1.4619	1.4570	35.45
66.0	1.4644	1.4595	35.98
67.0	1.4670	1.4621	36.51
68.0	1.4696	1.4646	37.04
69.0	1.4722	1.4672	37.56
70.0	1.4748	1.4698	38.08
71.0	1.4775	1.4724	38.61
72.0	1.4801	1.4751	39.13
73.0	1.4828	1.4778	39.65
74.0	1.4855	1.4805	40.16
75.0	1.4883	1.4832	40.68
76.0	1.4910	1.4859	41.19
77.0	1.4938	1.4887	41.71
78.0	1.4966	1.4915	42.22
79.0	1.4994	1.4943	42.73
80.0	1.5023	1.4971	43.24
81.0	1.5051	1.4999	43.74
82.0	1.5080	1.5028	44.25
83.0	1.5110	1.5057	44.75
84.0	1.5139	1.5086	45.26
85.0	1.5168	1.5116	45.76
86.0	1.5198	1.5145	46.26
87.0	1.5228	1.5175	46.76
88.0	1.5259	1.5206	47.25
89.0	1.5289	1.5236	47.75
90.0	1.5320	1.5267	48.24
91.0	1.5351	1.5298	48.73
92.0	1.5382	1.5329	49.23
93.0	1.5414	1.5360	49.72
94.0	1.5446	1.5392	50.21
95.0	1.5478	1.5424	50.69

[a]Dextrose Equivalent
[b]Dry Substance
[c]Refractive Index

Ash and DE[a] Corrections for Corn Syrup and Maltodextrin:[b] Changes in Refractive Index for an increase of ...

%DS[c]	1% Ash	1 DE
2	0.000000	−0.000001
4	0.000000	−0.000003
6	0.000001	−0.000005
8	0.000002	−0.000007
10	0.000003	−0.000010
12	0.000004	−0.000012
14	0.000006	−0.000015
16	0.000008	−0.000017
18	0.000010	−0.000020
20	0.000013	−0.000023
22	0.000016	−0.000026
24	0.000019	−0.000029
26	0.000022	−0.000033
28	0.000026	−0.000036
30	0.000030	−0.000040
32	0.000034	−0.000044
34	0.000039	−0.000048
36	0.000044	−0.000052
38	0.000049	−0.000057
40	0.000055	−0.000061
42	0.000061	−0.000066
44	0.000068	−0.000071
46	0.000074	−0.000076
48	0.000082	−0.000081
50	0.000089	−0.000087
52	0.000097	−0.000093
54	0.000105	−0.000099
56	0.000114	−0.000105
58	0.000123	−0.000112
60	0.000133	−0.000118
62	0.000143	−0.000125
64	0.000153	−0.000132
66	0.000164	−0.000140
68	0.000175	−0.000147
70	0.000187	−0.000155
72	0.000199	−0.000163
74	0.000212	−0.000172
76	0.000225	−0.000181
78	0.000239	−0.000190
80	0.000253	−0.000199
82	0.000268	−0.000208
84	0.000283	−0.000218

[a]Dextrose Equivalent
[b]Wartman, A. M., et al. J. Chemical and Engineering Data 21:467, 1976.
[c]Dry Substance

Invert Sugar

For invert sugar, convert the refractive index to approximate percent solids (uncorrected for invert sugar) using the accompanying sucrose table. Correct for invert sugar by using the following formula:

$$D = (S + C) + (P_1 \times 0.022),$$

in which S is the approximate percent solids determined from the refractive index table for sucrose, C is the temperature correction derived from the accompanying temperature correction table if the refractometer was operated at other than 20°, and P_1 is the percent invert sugar determined as directed under *Assay* for *Invert Sugar* in this Appendix.

Sucrose

International Refractive Index Scale of ICUMSA[a] (1974) for Pure Sucrose Solutions at 20°C and 589 nm[b]

Sucrose g/100 g	0.0	0.1	0.2	0.3	0.4	0.5	0.6	0.7	0.8	0.9
56	1.4329	4332	4334	4336	4338	4340	4343	4345	4347	4349
57	1.4352	4354	4356	4358	4360	4363	4365	4367	4369	4372
58	1.4374	4376	4378	4380	4383	4385	4387	4389	4392	4394
59	1.4396	4398	4401	4403	4405	4407	4410	4412	4414	4417
60	1.4419	4421	4423	4426	4428	4430	4432	4435	4437	4439
61	1.4442	4444	4446	4448	4451	4453	4455	4458	4460	4462
62	1.4464	4467	4469	4471	4474	4476	4478	4481	4483	4485
63	1.4488	4490	4492	4495	4497	4499	4502	4504	4506	4509
64	1.4511	4513	4516	4518	4520	4523	4525	4527	4530	4532
65	1.4534	4537	4539	4541	4544	4546	4548	4551	4553	4556
66	1.4558	4560	4563	4565	4567	4570	4572	4575	4577	4579
67	1.4582	4584	4586	4589	4591	4594	4596	4598	4601	4603
68	1.4606	4608	4610	4613	4615	4618	4620	4623	4625	4627
69	1.4630	4632	4635	4637	4639	4642	4644	4647	4649	4652
70	1.4654	4657	4659	4661	4664	4666	4669	4671	4674	4676
71	1.4679	4681	4683	4686	4688	4691	4693	4696	4698	4701
72	1.4703	4706	4708	4711	4713	4716	4718	4721	4723	4726
73	1.4728	4730	4733	4735	4738	4740	4743	4745	4748	4750
74	1.4753	4756	4758	4761	4763	4766	4768	4771	4773	4776
75	1.4778	4781	4783	4786	4788	4791	4793	4796	4798	4801
76	1.4804	4806	4809	4811	4814	4816	4819	4821	4824	4826
77	1.4829	4832	4834	4837	4839	4842	4844	4847	4850	4852
78	1.4855	4857	4860	4862	4865	4868	4870	4873	4875	4878
79	1.4881	4883	4886	4888	4891	4894	4896	4899	4901	4904
80	1.4907	4909	4912	4914	4917	4920	4922	4925	4928	4930
81	1.4933	4935	4938	4941	4943	4946	4949	4951	4954	4957
82	1.4959	4962	4964	4967	4970	4972	4975	4978	4980	4983
83	1.4986	4988	4991	4994	4996	4999	5002	5004	5007	5010
84	1.5012	5015	5018	5020	5023	5026	5029	5031	5034	5037
85	1.5039									

[a] Adapted from "Refractometry and Tables—Official" (ICUMSA SPS-3 1994), International Commission for Uniform Methods of Sugar Analysis (ICUMSA), c/o British Sugar Technical Centre, Colney, Norwich NR4 7UB, England.
[b] No rounding has been carried out; therefore, values given may be too low by a maximum of 1×10^{-4}.

Temperature Corrections for Refractometric Sucrose Solutions with Measurements at 20° and 589 nm

Temperature (°C)	Measured Sucrose (% solids)																	
	0	5	10	15	20	25	30	35	40	45	50	55	60	65	70	75	80	85
	Subtract from the measured value																	
15	0.29	0.30	0.32	0.33	0.34	0.35	0.36	0.37	0.37	0.38	0.38	0.38	0.38	0.38	0.38	0.38	0.37	0.37
16	0.24	0.25	0.26	0.27	0.28	0.28	0.29	0.30	0.30	0.30	0.31	0.31	0.31	0.31	0.31	0.30	0.30	0.30
17	0.18	0.19	0.20	0.20	0.21	0.21	0.22	0.22	0.23	0.23	0.23	0.23	0.23	0.23	0.23	0.23	0.23	0.22
18	0.12	0.13	0.13	0.14	0.14	0.14	0.15	0.15	0.15	0.15	0.15	0.15	0.15	0.15	0.15	0.15	0.15	0.15
19	0.06	0.06	0.07	0.07	0.07	0.07	0.07	0.08	0.08	0.08	0.08	0.08	0.08	0.08	0.08	0.08	0.08	0.07
	Add to the measured value																	
21	0.06	0.07	0.07	0.07	0.07	0.07	0.08	0.08	0.08	0.08	0.08	0.08	0.08	0.08	0.08	0.08	0.08	0.07
22	0.13	0.14	0.14	0.14	0.15	0.15	0.15	0.15	0.16	0.16	0.16	0.16	0.16	0.16	0.15	0.15	0.15	0.15
23	0.20	0.21	0.21	0.22	0.22	0.23	0.23	0.23	0.23	0.24	0.24	0.24	0.24	0.23	0.23	0.23	0.23	0.22
24	0.27	0.28	0.29	0.29	0.30	0.30	0.31	0.31	0.31	0.32	0.32	0.32	0.32	0.31	0.31	0.31	0.30	0.30
25	0.34	0.35	0.36	0.37	0.38	0.38	0.39	0.39	0.40	0.40	0.40	0.40	0.40	0.39	0.39	0.38	0.38	0.37
26	0.42	0.43	0.44	0.45	0.46	0.46	0.47	0.47	0.48	0.48	0.48	0.48	0.48	0.47	0.47	0.46	0.46	0.45
27	0.50	0.51	0.52	0.53	0.54	0.55	0.55	0.56	0.56	0.56	0.56	0.56	0.56	0.55	0.55	0.54	0.53	0.52
28	0.58	0.59	0.60	0.61	0.62	0.63	0.64	0.64	0.64	0.65	0.65	0.64	0.64	0.63	0.63	0.62	0.61	0.60
29	0.66	0.67	0.68	0.70	0.71	0.71	0.72	0.73	0.73	0.73	0.73	0.73	0.72	0.72	0.71	0.70	0.69	0.67
30	0.74	0.76	0.77	0.78	0.79	0.80	0.81	0.81	0.82	0.82	0.81	0.81	0.80	0.80	0.79	0.78	0.76	0.75
31	0.83	0.84	0.85	0.87	0.88	0.89	0.89	0.90	0.90	0.90	0.90	0.89	0.89	0.88	0.87	0.86	0.84	0.82
32	0.92	0.93	0.94	0.96	0.97	0.98	0.98	0.99	0.99	0.99	0.99	0.98	0.97	0.96	0.95	0.93	0.92	0.90
33	1.01	1.02	1.03	1.05	1.06	1.07	1.07	1.08	1.08	1.08	1.07	1.07	1.06	1.04	1.03	1.01	1.00	0.98
34	1.10	1.11	1.13	1.14	1.15	1.16	1.16	1.17	1.17	1.16	1.16	1.15	1.14	1.13	1.11	1.09	1.07	1.05
35	1.19	1.21	1.22	1.23	1.24	1.25	1.25	1.26	1.26	1.25	1.25	1.24	1.23	1.21	1.19	1.17	1.15	1.13
36	1.29	1.30	1.31	1.33	1.34	1.34	1.35	1.35	1.35	1.34	1.34	1.33	1.31	1.29	1.28	1.25	1.23	1.20
37	1.39	1.40	1.41	1.42	1.43	1.44	1.44	1.44	1.44	1.43	1.43	1.41	1.40	1.38	1.36	1.33	1.31	1.28
38	1.49	1.50	1.51	1.52	1.53	1.53	1.54	1.54	1.53	1.53	1.52	1.50	1.48	1.46	1.44	1.42	1.39	1.36
39	1.59	1.60	1.61	1.62	1.63	1.63	1.63	1.63	1.63	1.62	1.61	1.59	1.57	1.55	1.52	1.50	1.47	1.43
40	1.69	1.70	1.71	1.72	1.73	1.73	1.73	1.73	1.72	1.71	1.70	1.68	1.66	1.63	1.61	1.58	1.54	1.51

SOURCE: Adapted from "Refractometry and Tables—Official" (ICUMSA SPS-3 1994), International Commission for Uniform Methods of Sugar Analysis (ICUMSA), c/o British Sugar Technical Centre, Colney, Norwich NR4 7UB, England.

SOLUTIONS AND INDICATORS

COLORIMETRIC SOLUTIONS (CS)

Colorimetric solutions are used in the preparation of colorimetric standards for certain chemicals, and for the carbonization tests with sulfuric acid that are specified in several monographs. Directions for the preparation of the primary colorimetric solutions and *Matching Fluids* are given under the test for *Readily Carbonizable Substances*, Appendix IIB. Store the solutions in suitably resistant, tight containers.

Comparison of colors as directed in the *Food Chemicals Codex* tests preferably is made in matched color-comparison tubes or in a suitable colorimeter under conditions that ensure that the colorimetric reference solution and that of the specimen under test are treated alike in all respects.

STANDARD BUFFER SOLUTIONS

Reagent Solutions Before mixing, dry the crystalline reagents, except the boric acid, at 110° to 120°, and use water that has been previously boiled and cooled in preparing the solutions. Store the prepared reagent solutions in chemically resistant glass or polyethylene bottles, and use within 3 months. Discard if molding is evident.

Potassium Chloride, 0.2 M Dissolve 14.91 g of potassium chloride (KCl) in sufficient water to make 1000.0 mL.

Potassium Biphthalate, 0.2 M Dissolve 40.84 g of potassium biphthalate [$KHC_6H_4(COO)_2$] in sufficient water to make 1000.0 mL.

Potassium Phosphate, Monobasic, 0.2 M Dissolve 27.22 g of monobasic potassium phosphate (KH_2PO_4) in sufficient water to make 1000.0 mL.

Boric Acid–Potassium Chloride, 0.2 M Dissolve 12.37 g of boric acid (H_3BO_3) and 14.91 g of potassium chloride (KCl) in sufficient water to make 1000.0 mL.

Hydrochloric Acid, 0.2 M, and *Sodium Hydroxide, 0.2 M* Prepare and standardize as directed under *Volumetric Solutions* in this section.

Procedure To prepare 200 mL of a standard buffer solution having a pH within the range 1.2 to 10.0, place 50.0 mL of the appropriate 0.2 M salt solution, prepared above, in a 200-mL volumetric flask, add the volume of 0.2 M hydrochloric acid or a sodium hydroxide specified for the desired pH in the accompanying table, dilute to volume with water, and mix.

Composition of Standard Buffer Solutions

Hydrochloric Acid Buffer		Acid Phthalate Buffer		Neutralized Phthalate Buffer		Phosphate Buffer		Alkaline Borate Buffer	
To 50.0 mL of 0.2 M KCl add the mL of HCl specified		To 50.0 mL of 0.2 M $KHC_6H_4(COO)_2$ add the mL of HCl specified		To 50.0 mL of 0.2 M $KHC_6H_4(COO)_2$ add the mL of NaOH specified		To 50.0 mL of 0.2 M KH_2PO_4 add the mL of NaOH specified		To 50.0 mL of 0.2 M H_3BO_3-KCl add the mL of NaOH specified	
pH	0.2 M HCl (mL)	pH	0.2 M HCl (mL)	pH	0.2 M NaOH (mL)	pH	0.2 M NaOH (mL)	pH	0.2 M NaOH (mL)
1.2	85.0	2.2	49.5	4.2	3.0	5.8	3.6	8.0	3.9
1.3	67.2	2.4	42.2	4.4	6.6	6.0	5.6	8.2	6.0
1.4	53.2	2.6	35.4	4.6	11.1	6.2	8.1	8.4	8.6
1.5	41.4	2.8	28.9	4.8	16.5	6.4	11.6	8.6	11.8
1.6	32.4	3.0	22.3	5.0	22.6	6.6	16.4	8.8	15.8
1.7	26.0	3.2	15.7	5.2	28.8	6.8	22.4	9.0	20.8
1.8	20.4	3.4	10.4	5.4	34.1	7.0	29.1	9.2	26.4
1.9	16.2	3.6	6.3	5.6	38.8	7.2	34.7	9.4	32.1
2.0	13.0	3.8	2.9	5.8	42.3	7.4	39.1	9.6	36.9
2.1	10.2	4.0	0.1	—	—	7.6	42.4	9.8	40.6
2.2	7.8	—	—	—	—	7.8	44.5	10.0	43.7
						8.0	46.1	—	—

Dilute all final solutions to 200.0 mL (see *Procedure*). The standard pH values given in this table are considered to be reproducible to within ± 0.02 of the pH unit specified at 25°.

STANDARD SOLUTIONS FOR THE PREPARATION OF CONTROLS AND STANDARDS

The following solutions are used in tests for impurities that require the comparison of the color or turbidity produced in a solution of the test substance with that produced by a known amount of the impurity in a control. Directions for the preparation of other standard solutions are given in the monographs or under the general tests in which they are required (see also *Index*).

Ammonium Standard Solution (10 µg NH_4 in 1 mL) Dissolve 296.0 mg of ammonium chloride (NH_4Cl) in sufficient water to make 100.0 mL, and mix. Transfer 10.0 mL of this solution into a 1000-mL volumetric flask, dilute to volume with water, and mix.

Barium Standard Solution (100 µg Ba in 1 mL) Dissolve 177.9 mg of barium chloride ($BaCl_2 \cdot 2H_2O$) in water in a 1000-mL volumetric flask, dilute to volume with water, and mix.

Iron Standard Solution (10 µg Fe in 1 mL) Dissolve 702.2 mg of ferrous ammonium sulfate [$Fe(NH_4)_2(SO_4)_2 \cdot 6H_2O$] in 10 mL of 2 N sulfuric acid in a 100-mL volumetric flask, dilute to volume with water, and mix. Transfer 10.0 mL of this solution into a 1000-mL volumetric flask, add 10 mL of 2 N sulfuric acid, dilute to volume with water, and mix.

Magnesium Standard Solution (50 µg Mg in 1 mL) Dissolve 50.0 mg of magnesium metal (Mg) in 1 mL of hydrochloric acid in a 1000-mL volumetric flask, dilute to volume with water, and mix.

Phosphate Standard Solution (10 µg PO_4 in 1 mL) Dissolve 143.3 mg of monobasic potassium phosphate (KH_2PO_4) in water in a 100-mL volumetric flask, dilute to volume with water, and mix. Transfer 10.0 mL of this solution into a 1000-mL volumetric flask, dilute to volume with water, and mix.

TEST SOLUTIONS (TS) AND OTHER REAGENTS

Certain of the following test solutions are intended for use as acid–base indicators in volumetric analyses. Such solutions should be adjusted so that when 0.15 mL of the indicator solution is added to 25 mL of carbon dioxide-free water, 0.25 mL of 0.02 N acid or alkali, respectively, will produce the characteristic color change.

In general, the directive to prepare a solution "fresh" indicates that the solution is of limited stability and must be prepared on the day of use.

Acetic Acid (approximately 17.5 N) Use ACS reagent-grade *Acetic Acid, Glacial* (99.7% of CH_3COOH).

Acetic Acid TS, Diluted (1 N) A solution containing about 6% (w/v) of CH_3COOH. Prepare by diluting 60.0 mL of glacial acetic acid, or 166.6 mL of 36% acetic acid (6 N), with sufficient water to make 1000 mL.

Alcohol (*Ethanol; Ethyl Alcohol;* C_2H_5OH) Use ACS reagent-grade *Ethyl Alcohol* (not less than 95.0%, by volume, of C_2H_5OH).

Note: For use in assays and tests involving ultraviolet spectrophotometry, use ACS reagent-grade *Ethyl Alcohol Suitable for Use in Ultraviolet Spectrophotometry*.

Alcohol, Absolute (*Anhydrous Alcohol; Dehydrated Alcohol*) Use ACS reagent-grade *Ethyl Alcohol, Absolute* (not less than 99.5%, by volume, of C_2H_5OH).

Alcohol, Diluted A solution containing 41.0% to 42.0%, by weight, corresponding to 48.4% to 49.5%, by volume, at 15.56°, of C_2H_5OH.

Alcohol, 70% (at 15.56°) A 38.6:15 mixture (v/v) of 95% alcohol and water, having a specific gravity of 0.884 at 25°. To prepare 100 mL, dilute 73.7 mL of alcohol to 100 mL with water at 25°.

Alcohol, 80% (at 15.56°) A 45.5:9.5 mixture (v/v) of 95% alcohol and water, having a specific gravity of 0.857 at 25°. To prepare 100 mL, dilute 84.3 mL of alcohol to 100 mL with water at 25°.

Alcohol, 90% (at 15.56°) A 51:3 mixture (v/v) of 95% alcohol and water, having a specific gravity of 0.827 at 25°. To prepare 100 mL, dilute 94.8 mL of alcohol to 100 mL with water at 25°.

Alcohol, Aldehyde-Free Dissolve 2.5 g of lead acetate in 5 mL of water, add the solution to 1000 mL of alcohol contained in a glass-stoppered bottle, and mix. Dissolve 5 g of potassium hydroxide in 25 mL of warm alcohol, cool, and add slowly, without stirring, to the alcoholic solution of lead acetate. Allow to stand for 1 h, then shake the mixture vigorously, allow to stand overnight, decant the clear liquid, and recover the alcohol by distillation. Ethyl Alcohol FCC, Alcohol USP, or USSD #3A or #30 may be used. If the titration of a 250-mL sample of the alcohol by hydroxylamine hydrochloride TS (in this section) does not exceed 0.25 mL of 0.5 N alcoholic potassium hydroxide, the above treatment may be omitted.

Alcoholic Potassium Hydroxide TS See *Potassium Hydroxide TS, Alcoholic*.

Alkaline Cupric Tartrate TS (*Fehling's Solution*) See *Cupric Tartrate TS, Alkaline*.

Alkaline Mercuric-Potassium Iodide TS (*Nessler's Reagent*)   See *Mercuric-Potassium Iodide TS, Alkaline*.

Ammonia–Ammonium Chloride Buffer TS (approximately pH 10)   Dissolve 67.5 g of ammonium chloride, NH_4Cl, in water, add 570 mL of ammonium hydroxide (28%), and dilute with water to 1000 mL.

Ammonia TS (6 N in NH_3)   A solution containing between 9.5% and 10.5% of NH_3. Prepare by diluting 400 mL of ammonium hydroxide (28%) with sufficient water to make 1000 mL.

Ammonia TS, Stronger (15.2 N in NH_3) (*Ammonium Hydroxide, Stronger Ammonia Water*)   Use ACS reagent-grade *Ammonium Hydroxide*, which is a practically saturated solution of ammonia in water, containing between 28% and 30% of NH_3.

Ammoniacal Silver Nitrate TS   Add 6 N ammonium hydroxide, dropwise, to a 1 in 20 solution of silver nitrate until the precipitate that first forms is almost, but not entirely, dissolved. Filter the solution, and place in a dark bottle.

> **Caution**: Ammoniacal silver nitrate TS forms explosive compounds on standing. Do not store this solution, but prepare a fresh quantity for each series of determinations. Neutralize the excess reagent and rinse all glassware with hydrochloric acid immediately after completing a test.

Ammonium Acetate TS   Dissolve 10 g of ammonium acetate $(NH_4C_2H_3O_2)$ in sufficient water to make 100 mL.

Ammonium Carbonate TS   Dissolve 20 g of ammonium carbonate and 20 mL of ammonia TS in sufficient water to make 100 mL.

Ammonium Chloride TS   Dissolve 10.5 g of ammonium chloride (NH_4Cl) in sufficient water to make 100 mL.

Ammonium Molybdate TS   Dissolve 6.5 g of finely powdered molybdic acid (85%) in a mixture of 14 mL of water and 14.5 mL of ammonium hydroxide. Cool the solution, and add it slowly, with stirring, to a well-cooled mixture of 32 mL of nitric acid and 40 mL of water. Allow to stand for 48 h, and filter through a fine-porosity, sintered-glass crucible lined at the bottom with a layer of glass wool. This solution deteriorates upon standing and is unsuitable for use if, upon the addition of 2 mL of sodium phosphate TS to 5 mL of the solution, an abundant yellow precipitate does not form at once or after slight warming. Store it in the dark. If a precipitate forms during storage, use only the clear, supernatant solution.

Ammonium Oxalate TS   Dissolve 3.5 g of ammonium oxalate $[(NH_4)_2C_2O_4.H_2O]$ in sufficient water to make 100 mL.

Ammonium Sulfanilate TS   To 2.5 g of sulfanilic acid add 15 mL of water and 3 mL of 6 N ammonium hydroxide, and mix. Add, with stirring, more 6 N ammonium hydroxide, if necessary, until the acid dissolves, adjust the pH of the solution to about 4.5 with 2.7 N hydrochloric acid, using bromocresol green TS as an outside indicator, and dilute to 25 mL.

Ammonium Sulfide TS   Saturate 6 N ammonium hydroxide with hydrogen sulfide (H_2S) and add two-thirds of its volume of 6 N ammonium hydroxide. Residue upon ignition: not more than 0.05%. The solution is not rendered turbid either by magnesium sulfate TS or by calcium chloride TS (*carbonate*). This solution is unsuitable for use if an abundant precipitate of sulfur is present. Store it in small, well-filled, dark amber-colored bottles in a cold, dark place.

Ammonium Thiocyanate TS (1 N)   Dissolve 8 g of ammonium thiocyanate (NH_4SCN) in sufficient water to make 100 mL.

Antimony Trichloride TS   Dissolve 20 g of antimony trichloride $(SbCl_3)$ in chloroform to make 100 mL. Filter if necessary.

Barium Chloride TS   Dissolve 12 g of barium chloride $(BaCl_2.2H_2O)$ in sufficient water to make 100 mL.

Barium Diphenylamine Sulfonate TS   Dissolve 300 mg of *p*-diphenylamine sulfonic acid barium salt in 100 mL of water.

Benedict's Qualitative Reagent   See *Cupric Citrate TS, Alkaline*.

Benzidine TS   Dissolve 50 mg of benzidine in 10 mL of glacial acetic acid, dilute to 100 mL with water, and mix.

Bismuth Nitrate TS   Reflux 5 g of bismuth nitrate $[Bi(NO_3)_3.5H_2O]$ with 7.5 mL of nitric acid and 10 mL of water until dissolved, cool, filter, and dilute to 250 mL with water.

Bromine TS (*Bromine Water*)   Prepare a saturated solution of bromine by agitating 2 to 3 mL of bromine (Br_2) with 100 mL of cold water in a glass-stoppered bottle, the stopper of which should be lubricated with petrolatum. Store it in a cold place protected from light.

Bromocresol Blue TS   Use *Bromocresol Green TS*.

Bromocresol Green TS   Dissolve 50 mg of bromocresol green in 100 mL of alcohol, and filter if necessary.

Bromocresol Purple TS   Dissolve 250 mg of bromocresol purple in 20 mL of 0.5 N sodium hydroxide, and dilute with water to 250 mL.

Bromophenol Blue TS   Dissolve 100 mg of bromophenol blue in 100 mL of dilute alcohol (1 in 2), and filter if necessary.

Bromothymol Blue TS   Dissolve 100 mg of bromothymol blue in 100 mL of dilute alcohol (1 in 2), and filter if necessary.

Calcium Chloride TS   Dissolve 7.5 g of calcium chloride $(CaCl_2.2H_2O)$ in sufficient water to make 100 mL.

Calcium Hydroxide TS   A solution containing approximately 140 mg of $Ca(OH)_2$ in each 100 mL. To prepare, add 3 g of calcium hydroxide $[Ca(OH)_2]$ to 1000 mL of water, and agitate the mixture vigorously and repeatedly for 1 h. Allow the excess calcium hydroxide to settle, and decant or draw off the clear, supernatant liquid.

Calcium Sulfate TS   A saturated solution of calcium sulfate in water.

Carr-Price Reagent   See *Antimony Trichloride TS*.

Ceric Ammonium Nitrate TS   Dissolve 6.25 g of ceric ammonium nitrate $[(NH_4)_2Ce(NO_3)_6]$ in 100 mL of 0.25 N nitric acid. Prepare the solution fresh every third day.

Chlorine TS (*Chlorine Water*)   A saturated solution of chlorine in water. Place the solution in small, completely filled, light-resistant containers. *Chlorine TS,* even when kept from light and air, is apt to deteriorate. Store it in a cold, dark place. For full strength, prepare this solution fresh.

Chromotropic Acid TS   Dissolve 50 mg of chromotropic acid or its sodium salt in 100 mL of 75% sulfuric acid (made by cautiously adding 75 mL of 95% to 98% sulfuric acid to 33.3 mL of water).

Cobaltous Chloride TS   Dissolve 2 g of cobaltous chloride $(CoCl_2.6H_2O)$ in 1 mL of hydrochloric acid and sufficient water to make 100 mL.

Cobalt–Uranyl Acetate TS   Dissolve, with warming, 40 g of uranyl acetate $[UO_2(C_2H_3O_2)_2.2H_2O]$ in a mixture of 30 g of glacial acetic acid and sufficient water to make 500 mL. Similarly, prepare a solution containing 200 g of cobaltous acetate $[Co(C_2H_3O_2)_2.4H_2O]$ in a mixture of 30 g of glacial acetic acid and sufficient water to make 500 mL. Mix the two solutions while still warm, and cool to 20°. Maintain the temperature at 20° for about 2 h to separate the excess salts from solution, and then filter through a dry filter.

Congo Red TS   Dissolve 500 mg of congo red in a mixture of 10 mL of alcohol and 90 mL of water.

Cresol Red TS   Triturate 100 mg of cresol red in a mortar with 26.2 mL of 0.01 N sodium hydroxide until solution is complete, then dilute the solution with water to 250 mL.

Cresol Red–Thymol Blue TS   Add 15 mL of *Thymol Blue TS* to 5 mL of *Cresol Red TS*, and mix.

Crystal Violet TS   Dissolve 100 mg of crystal violet in 10 mL of glacial acetic acid.

Cupric Citrate TS, Alkaline (*Benedict's Qualitative Reagent*)   With the aid of heat, dissolve 173 g of sodium citrate $(C_6H_5Na_3O_7.2H_2O)$ and 117 g of sodium carbonate $(Na_2CO_3.H_2O)$ in about 700 mL of water, and filter through paper, if necessary. In a separate container, dissolve 17.3 g of cupric sulfate $(CuSO_4.5H_2O)$ in about 100 mL of water, and slowly add this solution, with constant stirring, to the first solution. Cool the mixture, dilute to 1000 mL, and mix.

Cupric Nitrate TS   Dissolve 2.4 g of cupric nitrate $[Cu(NO_3)_2.3H_2O]$ in sufficient water to make 100 mL.

Cupric Sulfate TS   Dissolve 12.5 g of cupric sulfate $(CuSO_4.5H_2O)$ in sufficient water to make 100 mL, and mix.

Cupric Tartrate TS, Alkaline (*Fehling's Solution*)   *The Copper Solution* (*A*): Dissolve 34.66 g of carefully selected, small crystals of cupric sulfate, $CuSO_4.5H_2O$, showing no trace of efflorescence or of adhering moisture, in sufficient water to make 500 mL. Store this solution in small, tight containers. *The Alkaline Tartrate Solution* (*B*): Dissolve 173 g of crystallized potassium sodium tartrate $(KNaC_4H_4O_6.4H_2O)$ and 50 g of sodium hydroxide (NaOH) in sufficient water to make 500 mL. Store this solution in small, alkali-resistant containers. For use, mix exactly equal volumes of solutions *A* and *B* at the time required.

Cyanogen Bromide TS   Dissolve 5 g of cyanogen bromide in water to make 50 mL.

   Caution: Prepare this solution under a hood, as cyanogen bromide volatilizes at room temperature and the vapor is highly irritating and poisonous.

Denigès' Reagent   See *Mercuric Sulfate TS*.

2,7-Dihydroxynaphthalene TS   Dissolve 100 mg of 2,7-dihydroxynaphthalene in 1000 mL of sulfuric acid, and allow the solution to stand until the initial color disappears. If the solution is very dark, discard it and prepare a new solution from a different supply of sulfuric acid. This solution is stable for approximately 1 month if stored in a dark bottle.

Diphenylamine TS   Dissolve 1 g of diphenylamine in 100 mL of sulfuric acid. The solution should be colorless.

Diphenylcarbazone TS   Dissolve about 1 g of diphenylcarbazone $(C_{13}H_{12}N_4O)$ in sufficient alcohol to make 100 mL. Store this solution in a brown bottle.

α,α′-Dipyridyl TS   Dissolve 100 mg of α,α′-dipyridyl $(C_{10}H_8N_2)$ in 50 mL of absolute alcohol.

Dithizone TS   Dissolve 25.6 mg of dithizone in 100 mL of alcohol.

Eosin Y TS (adsorption indicator)   Dissolve 50 mg of eosin Y in 10 mL of water.

Eriochrome Black TS   Dissolve 200 mg of eriochrome black T and 2 g of hydroxylamine hydrochloride $(NH_2OH.HCl)$ in sufficient methanol to make 50 mL, and filter. Store the solution in a light-resistant container and use within 2 weeks.

***p*-Ethoxychrysoidin TS**   Dissolve 50 mg of *p*-ethoxychrysoidin monohydrochloride in a mixture of 25 mL of water and 25 mL of alcohol, add 3 drops of hydrochloric acid, stir vigorously, and filter if necessary to obtain a clear solution.

Fehling's Solution See *Cupric Tartrate TS, Alkaline*.

Ferric Ammonium Sulfate TS Dissolve 8 g of ferric ammonium sulfate [FeNH$_4$(SO$_4$)$_2$.12H$_2$O] in sufficient water to make 100 mL.

Ferric Chloride TS Dissolve 9 g of ferric chloride (FeCl$_3$.6H$_2$O) in sufficient water to make 100 mL.

Ferric Chloride TS, Alcoholic Dissolve 100 mg of ferric chloride (FeCl$_3$.6H$_2$O) in 50 mL of absolute alcohol. Prepare this solution fresh.

Ferric Sulfate TS, Acid Add 7.5 mL of sulfuric acid to 100 mL of water, and dissolve 80 g of ferrous sulfate in the mixture with the aid of heat. Mix 7.5 mL of nitric acid and 20 mL of water, warm, and add to this the ferrous sulfate solution. Concentrate the mixture until, upon the sudden disengagement of ruddy vapors, the black color of the liquid changes to red. Test for the absence of ferrous iron, and, if necessary, add a few drops of nitric acid and heat again. When the solution is cold, add sufficient water to make 110 mL.

Ferrous Sulfate TS Dissolve 8 g of clear crystals of ferrous sulfate (FeSO$_4$.7H$_2$O) in about 100 mL of recently boiled and thoroughly cooled water. Prepare this solution fresh.

Formaldehyde TS A solution containing approximately 37.0% (w/v) of HCHO. It may contain methanol to prevent polymerization.

Fuchsin–Sulfurous Acid TS Dissolve 200 mg of basic fuchsin in 120 mL of hot water, and allow the solution to cool. Add a solution of 2 g of anhydrous sodium sulfite in 20 mL of water, and then add 2 mL of hydrochloric acid. Dilute the solution with water to 200 mL, and allow to stand for at least 1 h. Prepare this solution fresh.

Hydrochloric Acid (approximately 12 N) Use ACS reagent-grade *Hydrochloric Acid* (36.5% to 38.0% of HCl).

Hydrochloric Acid TS, Diluted (2.7 N) A solution containing 10% (w/v) of HCl. Prepare by diluting 226 mL of hydrochloric acid (36%) with sufficient water to make 1000 mL.

Hydrogen Peroxide TS A solution containing between 2.5 and 3.5 g of H$_2$O$_2$ in each 100 mL. It may contain suitable preservatives, totaling not more than 0.05%.

Hydrogen Sulfide TS A saturated solution of hydrogen sulfide made by passing H$_2$S into cold water. Store it in small, dark, amber-colored bottles, filled nearly to the top. It is unsuitable unless it possesses a strong odor of H$_2$S, and unless it produces at once a copious precipitate of sulfur when added to an equal volume of ferric chloride TS. Store in a cold, dark place.

Hydroxylamine Hydrochloride TS Dissolve 3.5 g of hydroxylamine hydrochloride (NH$_2$OH.HCl) in 95 mL of 60% alcohol, and add 0.5 mL of bromophenol blue solution (1 in 1000) and 0.5 N alcoholic potassium hydroxide until a greenish tint develops in the solution. Then add sufficient 60% alcohol to make 100 mL.

8-Hydroxyquinoline TS Dissolve 5 g of 8-hydroxyquinoline (oxine) in sufficient alcohol to make 100 mL.

Indigo Carmine TS (*Sodium Indigotindisulfonate TS*) Dissolve a quantity of sodium indigotindisulfonate, equivalent to 180 mg of C$_{16}$H$_8$N$_2$O$_2$(SO$_3$Na)$_2$, in sufficient water to make 100 mL. Use within 60 days.

Iodine TS Dissolve 14 g of iodine (I$_2$) in a solution of 36 g of potassium iodide (KI) in 100 mL of water, add 3 drops of hydrochloric acid, dilute with water to 1000 mL, and mix.

Isopropanol [*Isopropyl Alcohol; 2-Propanol; (CH$_3$)$_2$CHOH*] Use ACS reagent-grade *Isopropyl Alcohol*.

  Note: For use in assays and tests involving ultraviolet spectrophotometry, use ACS reagent-grade *Isopropyl Alcohol Suitable for Use in Ultraviolet Spectrophotometry*.

Isopropanol, Anhydrous (*Dehydrated Isopropanol*) Use isopropanol that has been previously dried by shaking with anhydrous calcium chloride, followed by filtering.

Lead Acetate TS Dissolve 9.5 g of clear, transparent crystals of lead acetate [Pb(C$_2$H$_3$O$_2$)$_2$.3H$_2$O] in sufficient recently boiled water to make 100 mL. Store in well-stoppered bottles.

Lead Subacetate TS Triturate 14 g of lead monoxide (PbO) to a smooth paste with 10 mL of water, and transfer the mixture to a bottle, using an additional 10 mL of water for rinsing. Dissolve 22 g of lead acetate [Pb(C$_2$H$_3$O$_2$)$_2$.3H$_2$O] in 70 mL of water, and add the solution to the lead oxide mixture. Shake it vigorously for 5 min, then set it aside, shaking it frequently during 7 days. Finally filter, and add enough recently boiled water through the filter to make 100 mL.

Lead Subacetate TS, Diluted Dilute 3.25 mL of *Lead Subacetate TS* with sufficient water, recently boiled and cooled, to make 100 mL. Store in small, well-fitted, tight containers.

Litmus TS Digest 25 g of powdered litmus with three successive 100-mL portions of boiling alcohol, continuing each extraction for about 1 h. Filter, wash with alcohol, and discard the alcohol filtrate. Macerate the residue with about 25 mL of cold water for 4 h, filter, and discard the filtrate. Finally, digest the residue with 125 mL of boiling water for 1 h, cool, and filter.

Magnesia Mixture TS Dissolve 5.5 g of magnesium chloride (MgCl$_2$.6H$_2$O) and 7 g of ammonium chloride (NH$_4$Cl) in 65 mL of water, add 35 mL of 6 N ammonium hydroxide, set the mixture aside for a few days in a well-stoppered bottle, and filter. If the solution is not perfectly clear, filter it before using.

Magnesium Sulfate TS Dissolve 12 g of crystals of magnesium sulfate (MgSO$_4$.7H$_2$O) selected for freedom from efflorescence, in water to make 100 mL.

Malachite Green TS Dissolve 1 g of malachite green oxalate in 100 mL of glacial acetic acid.

Mayer's Reagent See *Mercuric-Potassium Iodide TS*.

Mercuric Acetate TS Dissolve 6 g of mercuric acetate [$Hg(C_2H_3O_2)_2$] in sufficient glacial acetic acid to make 100 mL. Store in tight containers protected from direct sunlight.

Mercuric Chloride TS Dissolve 6.5 g of mercuric chloride ($HgCl_2$) in water to make 100 mL.

Mercuric-Potassium Iodide TS (*Mayer's Reagent*) Dissolve 1.358 g of mercuric chloride ($HgCl_2$) in 60 mL of water. Dissolve 5 g of potassium iodide (KI) in 10 mL of water. Mix the two solutions, and add water to make 100 mL.

Mercuric-Potassium Iodide TS, Alkaline (*Nessler's Reagent*) Dissolve 10 g of potassium iodide (KI) in 10 mL of water, and add slowly, with stirring, a saturated solution of mercuric chloride until a slight red precipitate remains undissolved. To this mixture add an ice-cold solution of 30 g of potassium hydroxide (KOH) in 60 mL of water, then add 1 mL more of the saturated solution of mercuric chloride. Dilute with water to 200 mL. Allow the precipitate to settle, and draw off the clear liquid. A 2-mL portion of this reagent, when added to 100 mL of a 1 in 300,000 solution of ammonium chloride in ammonia-free water, produces at once a yellowish brown color.

Mercuric Sulfate TS (*Denigès' Reagent*) Mix 5 g of yellow mercuric oxide (HgO) with 40 mL of water, and while stirring, slowly add 20 mL of sulfuric acid, then add another 40 mL of water, and stir until completely dissolved.

Mercurous Nitrate TS Dissolve 15 g of mercurous nitrate in a mixture of 90 mL of water and 10 mL of 2 N nitric acid. Store in dark, amber-colored bottles in which a small globule of mercury has been placed.

Methanol (*Methyl Alcohol*) Use ACS reagent-grade *Methanol*.

Methanol, Anhydrous (*Dehydrated Methanol*) Use *Methanol*.

p-Methylaminophenol Sulfate TS Dissolve 2 g of p-methylaminophenol sulfate [$(HOC_6H_4NHCH_3)_2 \cdot H_2SO_4$] in 100 mL of water. To 10 mL of this solution add 90 mL of water and 20 g of sodium bisulfite. Confirm the suitability of this solution by the following test: Add 1 mL of the solution to each of four tubes containing 25 mL of 0.5 N sulfuric acid and 1 mL of ammonium molybdate TS. Add 5 μg of phosphate (PO_4) to one tube, 10 μg to a second, and 20 μg to a third, using 0.5, 1.0, and 2.0 mL, respectively, of *Phosphate Standard Solution*, and allow to stand for 2 h. The solutions in the three tubes should show readily perceptible differences in blue color corresponding to the relative amounts of phosphate added, and the one to which 5 μg of phosphate was added should be perceptibly bluer than the blank.

Methylene Blue TS Dissolve 125 mg of methylene blue in 100 mL of alcohol, and dilute with alcohol to 250 mL.

Methyl Orange TS Dissolve 100 mg of methyl orange in 100 mL of water, and filter if necessary.

Methyl Red TS Dissolve 100 mg of methyl red in 100 mL of alcohol, and filter if necessary.

Methyl Red–Methylene Blue TS Add 10 mL of *Methyl Red TS* to 10 mL of *Methylene Blue TS*, and mix.

Methylrosaniline Chloride TS See *Crystal Violet TS*.

Methyl Violet TS See *Crystal Violet TS*.

Millon's Reagent To 2 mL of mercury in an Erlenmeyer flask add 20 mL of nitric acid. Shake the flask under a hood to break the mercury into small globules. After about 10 min add 35 mL of water, and, if a precipitate or crystals appear, add sufficient dilute nitric acid (1 in 5, prepared from nitric acid from which the oxides have been removed by blowing air through it until it is colorless) to dissolve the separated solid. Add sodium hydroxide solution (1 in 10), dropwise, with thorough mixing, until the curdy precipitate that forms after the addition of each drop no longer redissolves but is dispersed to form a suspension. Add 5 mL more of the dilute nitric acid, and mix well. Prepare this solution fresh.

α-Naphtholbenzein TS Dissolve 0.2 g of α-naphtholbenzein in glacial acetic acid to make 100 mL. *Sensitivity*: Add 100 mL of freshly boiled and cooled water to 0.2 mL of a solution of α-naphtholbenzein in ethanol (1 in 1000), and add 0.1 mL of 0.1 N sodium hydroxide: a green color develops. Add subsequently 0.2 mL of 0.1 N hydrochloric acid: the color of the solution changes to yellow-red.

Naphthol Green TS Dissolve 500 mg of naphthol green B in water to make 1000 mL.

Nessler's Reagent See *Mercuric-Potassium Iodide TS, Alkaline*.

Neutral Red TS Dissolve 100 mg of neutral red in 100 mL of 50% alcohol.

Ninhydrin TS See *Triketohydrindene Hydrate TS*.

Nitric Acid (approximately 15.7 N) Use ACS reagent-grade *Nitric Acid* (69.0% to 71.0% of HNO_3).

Nitric Acid TS, Diluted (1.7 N) A solution containing about 10% (w/v) of HNO_3. Prepare by diluting 105 mL of nitric acid (70%) with water to make 1000 mL.

Orthophenanthroline TS Dissolve 150 mg of orthophenanthroline ($C_{12}H_8N_2 \cdot H_2O$) in 10 mL of a solution of ferrous sulfate, prepared by dissolving 700 mg of clear crystals of ferrous sulfate ($FeSO_4 \cdot 7H_2O$) in 100 mL of water. The ferrous sulfate solution must be prepared immediately before dissolving the orthophenanthroline. Store the solution in well-closed containers.

Oxalic Acid TS Dissolve 6.3 g of oxalic acid ($H_2C_2O_4.2H_2O$) in water to make 100 mL.

Phenol Red TS (*Phenolsulfonphthalein TS*) Dissolve 100 mg of phenolsulfonphthalein in 100 mL of alcohol, and filter if necessary.

Phenolphthalein TS Dissolve 1 g of phenolphthalein in 100 mL of alcohol.

Phenolsulfonphthalein TS See *Phenol Red TS*.

***p*-Phenylphenol TS** On the day of use, dissolve 750 mg of *p*-phenylphenol in 50 mL of *Sodium Hydroxide TS*.

Phosphoric Acid Use ACS reagent-grade *Phosphoric Acid* (not less than 85.0% of H_3PO_4).

Phosphotungstic Acid TS Dissolve 1 g of phosphotungstic acid (approximately $24WO_3.2H_3PO_4.48H_2O$) in water to make 100 mL.

Picric Acid TS See *Trinitrophenol TS*.

Potassium Acetate TS Dissolve 10 g of potassium acetate ($KC_2H_3O_2$) in water to make 100 mL.

Potassium Chromate TS Dissolve 10 g of potassium chromate (K_2CrO_4) in water to make 100 mL.

Potassium Dichromate TS Dissolve 7.5 g of potassium dichromate ($K_2Cr_2O_7$) in water to make 100 mL.

Potassium Ferricyanide TS (10%) Dissolve 1 g of potassium ferricyanide [$K_3Fe(CN)_6$] in 10 mL of water. Prepare this solution fresh.

Potassium Ferrocyanide TS Dissolve 1 g of potassium ferrocyanide [$K_4Fe(CN)_6.3H_2O$] in 10 mL of water. Prepare this solution fresh.

Potassium Hydroxide TS (1 *N*) Dissolve 6.5 g of potassium hydroxide (KOH) in water to make 100 mL.

Potassium Hydroxide TS, Alcoholic Use *0.5 N Alcoholic Potassium Hydroxide* (see *Volumetric Solutions* in this section).

Potassium Iodide TS Dissolve 16.5 g of potassium iodide (KI) in water to make 100 mL. Store in light-resistant containers.

Potassium Permanganate TS Use *0.1 N Potassium Permanganate* (see *Volumetric Solutions* in this section).

Potassium Pyroantimonate TS Dissolve 2 g of potassium pyroantimonate in 95 mL of hot water. Cool quickly, and add a solution containing 2.5 g of potassium hydroxide in 50 mL of water and 1 mL of sodium hydroxide solution (8.5 in 100). Allow to stand for 24 h, filter, and dilute with water to 150 mL.

Potassium Sulfate TS Dissolve 1 g of potassium sulfate (K_2SO_4) in sufficient water to make 100 mL.

Quimociac TS Dissolve 70 g of sodium molybdate ($Na_2MoO_4.2H_2O$) in 150 mL of water (*Solution A*). Dissolve 60 g of citric acid in a mixture of 85 mL of nitric acid and 150 mL of water, and cool (*Solution B*). Gradually add *Solution A* to *Solution B*, with stirring, to produce *Solution C*. Dissolve 5.0 mL of synthetic quinoline in a mixture of 35 mL of nitric acid and 100 mL of water (*Solution D*). Gradually add *Solution D* to *Solution C*, mix well, and allow to stand overnight. Filter the mixture, add 280 mL of acetone to the filtrate, dilute to 1000 mL with water, and mix. Store in a polyethylene bottle.

  Caution: This reagent contains acetone. Do not use it near an open flame. Operations involving heating or boiling should be conducted in a well-ventilated hood.

Quinaldine Red TS Dissolve 100 mg of quinaldine red in 100 mL of glacial acetic acid.

Schiff's Reagent, Modified Dissolve 200 mg of rosaniline hydrochloride ($C_{20}H_{20}ClN_3$) in 120 mL of hot water. Cool, add 2 g of sodium bisulfite ($NaHSO_3$) followed by 2 mL of hydrochloric acid, and dilute to 200 mL with water. Store in a brown bottle at 15° or lower.

Silver Nitrate TS Use *0.1 N Silver Nitrate* (see *Volumetric Solutions* in this section).

Sodium Bisulfite TS Dissolve 10 g of sodium bisulfite ($NaHSO_3$) in water to make 30 mL. Prepare this solution fresh.

Sodium Bitartrate TS Dissolve 1 g of sodium bitartrate ($NaHC_4H_4O_6.H_2O$) in water to make 10 mL. Prepare this solution fresh.

Sodium Borate TS Dissolve 2 g of sodium borate ($Na_2B_4O_7.10H_2O$) in water to make 100 mL.

Sodium Carbonate TS Dissolve 10.6 g of anhydrous sodium carbonate (Na_2CO_3) in water to make 100 mL.

Sodium Cobaltinitrite TS Dissolve 10 g of sodium cobaltinitrite [$Na_3Co(NO_2)_6$] in water to make 50 mL, and filter if necessary.

Sodium Fluoride TS Dry about 500 mg of sodium fluoride (NaF) at 200° for 4 h. Weigh accurately 222 mg of the dried sodium fluoride, and dissolve it in sufficient water to make exactly 100 mL. Transfer 10.0 mL of this solution into a 1000-mL volumetric flask, dilute to volume with water, and mix. Each mL of this final solution corresponds to 10 μg of fluorine (F).

Sodium Hydroxide TS (1 *N*) Dissolve 4.3 g of sodium hydroxide (NaOH) in water to make 100 mL.

Sodium Indigotindisulfonate TS   See *Indigo Carmine TS*.

Sodium Nitroferricyanide TS   Dissolve 1 g of sodium nitroferricyanide [$Na_2Fe(NO)(CN)_5 \cdot 2H_2O$] in water to make 20 mL. Prepare this solution fresh.

Sodium Phosphate TS   Dissolve 12 g of clear crystals of dibasic sodium phosphate ($Na_2HPO_4 \cdot 7H_2O$) in water to make 100 mL.

Sodium Sulfide TS   Dissolve 1 g of sodium sulfide ($Na_2S \cdot 9H_2O$) in water to make 10 mL. Prepare this solution fresh.

Sodium Thiosulfate TS   Use *0.1 N Sodium Thiosulfate* (see *Volumetric Solutions* in this section).

Stannous Chloride TS   Dissolve 40 g of reagent-grade stannous chloride dihydrate ($SnCl_2 \cdot 2H_2O$) in 100 mL of hydrochloric acid.

Starch TS   Mix 1 g of a suitable starch with 10 mg of red mercuric oxide and sufficient cold water to make a thin paste. Add 20 mL of boiling water, boil for 1 min with continuous stirring, and cool. Use only the clear solution.

Starch Iodide Paste TS   Heat 100 mg of water in a 250-mL beaker to boiling, add a solution of 750 mg of potassium iodide (KI) in 5 mL of water, then add 2 g of zinc chloride ($ZnCl_2$) dissolved in 10 mL of water, and, while the solution is boiling, add with stirring a smooth suspension of 5 g of potato starch in 30 mL of cold water. Continue to boil for 2 min, then cool. Store in well-closed containers in a cool place. This mixture must show a definite blue streak when a glass rod dipped in a mixture of 1 mL of 0.1 M sodium nitrite, 500 mL of water, and 10 mL of hydrochloric acid is streaked on a smear of the paste.

Sulfanilic Acid TS   Dissolve 800 mg of sulfanilic acid (p-$NH_2C_6H_4SO_3H \cdot H_2O$) in 100 mL of acetic acid. Store in tight containers.

Sulfuric Acid (approximately 36 N)   Use ACS reagent-grade *Sulfuric Acid* (95.0% to 98.0% of H_2SO_4).

Sulfuric Acid TS (95%)   Add a quantity of sulfuric acid of known concentration to sufficient water to adjust the final concentration to between 94.5% and 95.5% of H_2SO_4. Since the acid concentration may change upon standing or upon intermittent use, the concentration should be checked frequently and solutions assaying more than 95.5% or less than 94.5% discarded or adjusted by adding either diluted or fuming sulfuric acid, as required.

Sulfuric Acid TS, Diluted (2 N)   A solution containing 10% (w/v) of H_2SO_4. Prepare by cautiously adding 57 mL of sulfuric acid (95% to 98%) or *Sulfuric Acid TS* to about 100 mL of water, then cool to room temperature, and dilute with water to 1000 mL.

Tannic Acid TS   Dissolve 1 g of tannic acid (tannin) in 1 mL of alcohol, and add water to make 10 mL. Prepare this solution fresh.

Thymol Blue TS   Dissolve 100 mg of thymol blue in 100 mL of alcohol, and filter if necessary.

Thymolphthalein TS   Dissolve 100 mg of thymolphthalein in 100 mL of alcohol, and filter if necessary.

Triketohydrindene Hydrate TS (*Ninhydrin TS*)   Dissolve 200 mg of triketohydrindene hydrate ($C_9H_4O_3 \cdot H_2O$) in water to make 100 mL. Prepare this solution fresh.

Trinitrophenol TS (*Picric Acid TS*)   Dissolve the equivalent of 1 g of anhydrous trinitrophenol in 100 mL of hot water. Cool the solution, and filter if necessary.

Xylenol Orange TS   Dissolve 100 mg of xylenol orange in 100 mL of alcohol.

VOLUMETRIC SOLUTIONS

Normal Solutions   A normal solution contains 1 g equivalent weight of the solute per L of solution. The normalities of solutions used in volumetric determinations are designated as 1 N, 0.1 N, 0.05 N, etc., in this Codex.

Molar Solutions   A molar solution contains 1 g molecular weight of the solute per L of solution. The molarities of such solutions are designated as 1 M, 0.1 M, 0.05 M, etc., in this Codex.

Preparation and Methods of Standardization   The details for the preparation and standardization of solutions used in several normalities are usually given only for the one most frequently required. Solutions of other normalities are prepared and standardized in the same general manner as described. Solutions of lower normalities may be prepared accurately by making an exact dilution of a stronger solution, but solutions prepared in this way should be restandardized before use.

Dilute solutions that are not stable, such as 0.01 N potassium permanganate and sodium thiosulfate, are preferably prepared by diluting exactly the higher normality with thoroughly boiled and cooled water on the same day they are to be used.

All volumetric solutions should be prepared, standardized, and used at the standard temperature of 25°, if practicable. When a titration must be carried out at a markedly different temperature, the volumetric solution should be standardized at that same temperature, or a suitable temperature correction should be made. Since the strength of a standard solution may change upon standing, the normality or molarity factor should be redetermined frequently.

Although the directions provide only one method of standardization, other methods of equal or greater accuracy may be used. For substances available as certified primary standards, or of comparable quality, the final standard solution may be prepared by weighing accurately a suitable quantity of the substance and dissolving it to produce a specific volume solution of known concentration. Hydrochloric and sulfuric acids may be standardized against a certified primary standard.

In volumetric assays described in this Codex, the number of mg of the test substance equivalent to 1 mL of the primary volumetric solution is given. In general, these equivalents may be derived by simple calculation (see also *Solutions*, in the *General Provisions*).

Ammonium Thiocyanate, 0.1 N (7.612 g NH$_4$SCN per 1000 mL) Dissolve about 8 g of ammonium thiocyanate (NH$_4$SCN) in 1000 mL of water, and standardize by titrating the solution against *0.1 N Silver Nitrate* as follows: Transfer about 30 mL of *0.1 N Silver Nitrate*, accurately measured, into a glass-stoppered flask. Dilute with 50 mL of water, then add 2 mL of *Ferric Ammonium Sulfate TS* and 2 mL of nitric acid, and titrate with the ammonium thiocyanate solution to the first appearance of a red-brown color. Calculate the normality, and, if desired, adjust the solution to exactly 0.1 N. If desired, *0.1 N Ammonium Thiocyanate* may be replaced by 0.1 N potassium thiocyanate where the former is directed in various tests and assays.

Barium Hydroxide, 0.2 N [17.14 g Ba(OH)$_2$ per 1000 mL] Dissolve about 36 g of barium hydroxide [Ba(OH)$_2$.8H$_2$O] in 1 L of recently boiled and cooled water, and quickly filter the solution. Keep this solution in bottles with well-fitted rubber stoppers with a soda–lime tube attached to each bottle to protect the solution from carbon dioxide in the air. Standardize as follows: Transfer quantitatively about 60 mL of 0.1 N hydrochloric acid, accurately measured, to a flask; add 2 drops of *Phenolphthalein TS*; and slowly titrate with the barium hydroxide solution, with constant stirring, until a permanent pink color is produced. Calculate the normality of the barium hydroxide solution and, if desired, adjust to exactly 0.2 N with freshly boiled and cooled water.

> **Note**: Solutions of alkali hydroxides absorb carbon dioxide when exposed to air. Connect the buret used for titrations with barium hydroxide solution directly to the storage bottle and provide the bottle with a soda–lime tube so that air entering must pass through this tube, which will absorb carbon dioxide. Frequently restandardize standard solutions of barium hydroxide.

Bromine, 0.1 N (7.990 g Br per 1000 mL) Dissolve 3 g of potassium bromate (KBrO$_3$) and 15 g of potassium bromide (KBr) in sufficient water to make 1000 mL, and standardize the solution as follows: Transfer about 25 mL of the solution, accurately measured, into a 500-mL iodine flask, and dilute with 120 mL of water. Add 5 mL of hydrochloric acid, stopper the flask, and shake it gently. Then add 5 mL of *Potassium Iodide TS*, restopper, shake the mixture, allow it to stand for 5 min, and titrate the liberated iodine with *0.1 N Sodium Thiosulfate*, adding *Starch TS* near the end of the titration. Calculate the normality. Store this solution in dark, amber-colored, glass-stoppered bottles.

Ceric Sulfate, 0.1 N [33.22 g Ce(SO$_4$)$_2$ per 1000 mL] Transfer 59 g of ceric ammonium nitrate [Ce(NO$_3$)$_4$.2NH$_4$NO$_3$.2H$_2$O] to a beaker, add 31 mL of sulfuric acid, mix, and cautiously add water, in 20-mL portions, until solution is complete. Cover the beaker, let stand overnight, filter through a sintered-glass crucible of fine porosity, add water to make 1000 mL, and mix. Standardize the solution as follows: Weigh accurately 200 mg of primary standard arsenic trioxide (As$_2$O$_3$) previously dried at 100° for 1 h, and transfer to a 500-mL Erlenmeyer flask. Wash down the inner walls of the flask with 25 mL of sodium hydroxide solution (2 in 25), swirl to dissolve the sample, and when solution is complete, add 100 mL of water, and mix. Add 10 mL of dilute sulfuric acid (1 in 3) and 2 drops each of *Orthophenanthroline TS* and a solution of osmium tetroxide in 0.1 N sulfuric acid (1 in 400), and slowly titrate with the ceric sulfate solution until the pink color is changed to a very pale blue. Calculate the normality. Each 4.946 mg of As$_2$O$_3$ is equivalent to 1 mL of *0.1 N Ceric Sulfate*.

Ceric Sulfate, 0.01 N [3.322 g Ce(SO$_4$)$_2$ per 1000 mL] Dissolve 4.2 g of ceric sulfate [Ce(SO$_4$)$_2$.4H$_2$O] or 5.5 g of the acid sulfate [Ce(HSO$_4$)$_4$] in about 500 mL of water containing 28 mL of sulfuric acid, and dilute to 1000 mL. Allow the solution to stand overnight, and filter. Standardize this solution daily as follows: Weigh accurately about 275 mg of hydroquinone (C$_6$H$_6$O$_2$), dissolve it in sufficient *0.5 N Alcoholic Sulfuric Acid* to make 500.0 mL, and mix. To 25.0 mL of this solution add 75 mL of 0.5 N sulfuric acid, 20 mL of water, and 2 drops of *Diphenylamine TS*. Titrate with the ceric sulfate solution at a rate of about 25 drops per 10 s until an endpoint is reached that persists for 10 s. Perform a blank determination using 100 mL of *0.5 N Alcoholic Sulfuric Acid*, 20 mL of water, and 2 drops of *Diphenylamine TS*, and make any necessary correction. Calculate the normality of the ceric sulfate solution by the formula

$$0.05W/55.057V,$$

in which W is the weight, in mg, of the hydroquinone sample taken, and V is the volume, in mL, of the ceric sulfate solution consumed in the titration.

Disodium EDTA, 0.05 M (16.81 g C$_{10}$H$_{14}$N$_2$Na$_2$O$_8$ per 1000 mL) Dissolve 18.6 g of disodium ethylenediaminetetraacetate (C$_{10}$H$_{14}$N$_2$Na$_2$O$_8$.2H$_2$O) in sufficient water to make 1000 mL, and standardize the solution as follows: Weigh accurately about 200 mg of chelometric standard calcium carbonate (CaCO$_3$), transfer to a 400-mL beaker, add 10 mL of water, and swirl to form a slurry. Cover the beaker with a watch glass, and introduce 2 mL of 2.7 N hydrochloric acid from a pipet inserted between the lip of the beaker and the edge of the watch glass. Swirl the contents of the beaker to dissolve the calcium carbonate. Wash down the sides of the beaker, the outer surface of the pipet, and the watch glass, and dilute to about 100 mL with water. While stirring, preferably with a magnetic stirrer, add about 30 mL of the disodium EDTA solution from a 50-mL buret, then add 15 mL of *1 N Sodium Hydroxide* and 300 mg of *Hydroxy Naphthol Blue Indicator*, and continue the titration to a blue endpoint. Calculate the molarity by the formula

$$W/100.09V,$$

in which W is the weight, in mg, of $CaCO_3$ in the sample of calcium carbonate taken, and V is the volume, in mL, of disodium EDTA solution consumed. Each 5.004 mg of $CaCO_3$ is equivalent to 1 mL of *0.05 M Disodium EDTA*.

For the determination of aluminum in its salts, use *0.05 M Disodium EDTA* standardized as follows: Transfer 2 g, accurately weighed, of aluminum wire to a 1000-mL volumetric flask, and add 50 mL of a 1:1 hydrochloric acid–water mixture. Swirl the flask to ensure complete wetting of the wire, and allow the reaction to proceed. When dissolution is complete, dilute with water to volume, and mix. Transfer 10.0 mL of this solution to a 250-mL beaker, add 25.0 mL of the disodium EDTA solution, boil gently for 5 min, and cool. Add in the order given, and with continuous stirring, 20 mL of pH 4.5 buffer solution (77.1 g of ammonium acetate and 57 mL of glacial acetic acid in 1000 mL of solution), 50 mL of alcohol, and 2 mL of *Dithizone TS*. Titrate with *0.05 M Zinc Sulfate* to a bright rose pink color, and perform a blank determination, substituting 10 mL of water for the 10.0 mL of aluminum solution. Each mL of disodium EDTA solution is equivalent to 1.349 mg of aluminum (Al).

Ferrous Ammonium Sulfate, 0.1 N [39.21 g $Fe(NH_4)_2(SO_4)_2 \cdot 6H_2O$ per 1000 mL] Dissolve 40 g of ferrous ammonium sulfate hexahydrate in a previously cooled mixture of 40 mL of sulfuric acid and 200 mL of water, dilute to 1000 mL with water, and mix. On the day of use, standardize the solution as follows: Transfer from 25 to 30 mL of the solution, accurately measured, into a flask, add 2 drops of *Orthophenanthroline TS*, and titrate with *0.1 N Ceric Sulfate* until the red color is changed to pale blue. From the volume of *0.1 N Ceric Sulfate* consumed, calculate the normality.

Hydrochloric Acid, 1 N (36.46 g HCl per 1000 mL) Dilute 85 mL of hydrochloric acid with water to make 1000 mL, and standardize the solution as follows: Accurately weigh about 1.5 g of primary standard anhydrous sodium carbonate (Na_2CO_3) that has been heated at a temperature of about 270° for 1 h. Dissolve it in 100 mL of water, and add 2 drops of *Methyl Red TS*. Add the acid slowly from a buret, with constant stirring, until the solution becomes faintly pink. Heat the solution to boiling, and continue the titration until the faint pink color is no longer affected by continued boiling. Calculate the normality. Each 52.99 mg of Na_2CO_3 is equivalent to 1 mL of *1 N Hydrochloric Acid*.

Hydroxylamine Hydrochloride, 0.5 N (35 g $NH_2OH \cdot HCl$ per 1000 mL) Dissolve 35 g of hydroxylamine hydrochloride in 150 mL of water, and dilute to 1000 mL with anhydrous methanol. To 500 mL of this solution add 15 mL of a 0.04% solution of bromophenol blue in alcohol, and titrate with *0.5 N Triethanolamine* until the solution appears greenish blue by transmitted light. Prepare this solution fresh before each series of analyses.

Iodine, 0.1 N (12.69 g I per 1000 mL) Dissolve about 14 g of iodine (I) in a solution of 36 g of potassium iodide (KI) in 100 mL of water, add 3 drops of hydrochloric acid, dilute with water to 1000 mL, and standardize as follows: Weigh accurately about 150 mg of primary standard arsenic trioxide (As_2O_3) previously dried at 105° for 1 h, and dissolve it in 20 mL of *1 N Sodium Hydroxide* by warming if necessary. Dilute with 40 mL of water, add 2 drops of *Methyl Orange TS*, and follow with 2.7 N hydrochloric acid until the yellow color is changed to pink. Then add 2 g of sodium bicarbonate ($NaHCO_3$), dilute with 50 mL of water, add 3 mL of *Starch TS*, and slowly add the iodine solution from a buret until a permanent blue color is produced. Calculate the normality. Each 4.946 mg of As_2O_3 is equivalent to 1 mL of *0.1 N Iodine*. Store this solution in glass-stoppered bottles.

Lithium Methoxide, 0.1 N (3.797 g CH_3OLi per 1000 mL) Dissolve 600 mg of freshly cut lithium metal in a mixture of 150 mL of anhydrous methanol and 850 mL of benzene. Filter the resulting solution if it is cloudy, and standardize it as follows: Dissolve about 80 mg of benzoic acid (National Institute of Standards and Technology primary standard), accurately weighed, in 35 mL of dimethylformamide, add 5 drops of *Thymol Blue TS*, and titrate with the lithium methoxide solution to a dark blue endpoint.

Caution: Protect the solution from absorption of carbon dioxide and moisture by covering the titration vessel with aluminum foil while dissolving the benzoic acid sample and during the titration.

Each mL of *0.1 N Lithium Methoxide* is equivalent to 12.21 mg of benzoic acid.

Mercuric Nitrate, 0.1 M [32.46 g $Hg(NO_3)_2$ per 1000 mL] Dissolve about 35 g of mercuric nitrate [$Hg(NO_3)_2 \cdot H_2O$] in a mixture of 5 mL of nitric acid and 500 mL of water, and dilute with water to 1000 mL. Standardize the solution as follows: Transfer an accurately measured volume of about 20 mL of the solution into an Erlenmeyer flask, and add 2 mL of nitric acid and 2 mL of *Ferric Ammonium Sulfate TS*. Cool to below 20°, and titrate with *0.1 N Ammonium Thiocyanate* to the first appearance of a permanent brownish color. Calculate the molarity.

Oxalic Acid, 0.1 N (4.502 g $H_2C_2O_4$ per 1000 mL) Dissolve 6.45 g of oxalic acid ($H_2C_2O_4 \cdot 2H_2O$) in sufficient water to make 1000 mL. Standardize by titration against freshly standardized *0.1 N Potassium Permanganate* as directed under *Potassium Permanganate, 0.1 N*. Store this solution in glass-stoppered bottles, protected from light.

Perchloric Acid, 0.1 N (10.046 g $HClO_4$ per 1000 mL) Mix 8.5 mL of perchloric acid (70%) with 500 mL of glacial acetic acid and 30 mL of acetic anhydride. Cool, and add glacial acetic acid to make 1000 mL. Allow the prepared solution to stand for 1 day for the excess acetic anhydride to be combined, and determine the water content by the *Karl Fischer Titrimetric Method*, Appendix IIB. If the water content exceeds 0.05%, add more acetic anhydride, but if the solution contains no titratable water, add sufficient water to make the content between 0.02% and 0.05% of water. Allow to stand for 1 day, and again determine the water content by titration. Standardize the solution as follows: Weigh accurately about 700 mg of primary standard potassium biphthalate [$KHC_6H_4(COO)_2$], previously dried at

105° for 2 h, and dissolve it in 50 mL of glacial acetic acid in a 250-mL flask. Add 2 drops of *Crystal Violet TS*, and titrate with the perchloric acid solution until the violet color changes to emerald green. Deduct the volume of the perchloric acid consumed by 50 mL of the glacial acetic acid, and calculate the normality. Each 20.42 mg of $KHC_6H_4(COO)_2$ is equivalent to 1 mL of *0.1 N Perchloric Acid*.

Perchloric Acid, 0.1 N, in Dioxane Mix 8.5 mL of perchloric acid (70%) with sufficient dioxane, which has been especially purified by adsorption, to make 1000 mL. Standardize the solution as follows: Weigh accurately about 700 mg of primary standard potassium biphthalate [$KHC_6H_4(COO)_2$], previously dried at 105° for 2 h, and dissolve in 50 mL of glacial acetic acid in a 250-mL flask. Add 2 drops of *Crystal Violet TS*, and titrate with the perchloric acid solution until the violet color changes to bluish green. Deduct the volume of the perchloric acid consumed by 50 mL of the glacial acetic acid, and calculate the normality. Each 20.42 mg of $KHC_6H_4(COO)_2$ is equivalent to 1 mL of *0.1 N Perchloric Acid*.

Potassium Acid Phthalate, 0.1 N [20.42 g $KHC_6H_4(COO)_2$ per 1000 mL] Dissolve 20.42 g of primary standard potassium biphthalate [$KHC_6H_4(COO)_2$], previously dried at 105° for 2 h, in glacial acetic acid in a 1000-mL volumetric flask, warming on a steam bath if necessary to effect solution and protecting the solution from contamination by moisture. Cool to room temperature, dilute to volume with glacial acetic acid, and mix.

Potassium Dichromate, 0.1 N (4.903 g $K_2Cr_2O_7$ per 1000 mL) Dissolve about 5 g of potassium dichromate ($K_2Cr_2O_7$) in 1000 mL of water, transfer quantitatively 25 mL of this solution to a 500-mL glass-stoppered flask, add 2 g of potassium iodide (free from iodate) (KI), dilute with 200 mL of water, add 5 mL of hydrochloric acid, and mix. Allow to stand for 10 min in a dark place, and titrate the liberated iodine with *0.1 N Sodium Thiosulfate*, adding *Starch TS* as the endpoint is approached. Correct for a blank run on the same quantities of the same reagents, and calculate the normality.

Potassium Hydroxide, 1 N (56.11 g KOH per 1000 mL) Prepare and standardize 1 N potassium hydroxide by the procedure set forth for *Sodium Hydroxide, 1 N*, using 74 g of the potassium hydroxide (KOH) to prepare the solution. Each 204.2 mg of $KHC_6H_4(COO)_2$ is equivalent to 1 mL of 1 N potassium hydroxide.

Potassium Hydroxide, 0.5 N, Alcoholic Dissolve about 35 g of potassium hydroxide (KOH) in 20 mL of water, and add sufficient aldehyde-free alcohol to make 1000 mL. Allow the solution to stand in a tightly stoppered bottle for 24 h. Then quickly decant the clear supernatant liquid into a suitable, tight container, and standardize as follows: Transfer quantitatively 25 mL of 0.5 N hydrochloric acid into a flask, dilute with 50 mL of water, add 2 drops of *Phenolphthalein TS*, and titrate with the alcoholic potassium hydroxide solution until a permanent, pale pink color is produced. Calculate the normality. Store this solution in tightly stoppered bottles protected from light.

Potassium Iodate, 0.05 M (10.70 g KIO_3 per 1000 mL) Dissolve 10.700 g of potassium iodate of primary standard quality (KIO_3), previously dried at 110° to constant weight, in sufficient water to make 1000.0 mL.

Potassium Permanganate, 0.1 N (3.161 g $KMnO_4$ per 1000 mL) Dissolve about 3.3 g of potassium permanganate ($KMnO_4$) in 1000 mL of water in a flask, and boil the solution for about 15 min. Stopper the flask, allow it to stand for at least 2 days, and filter through a fine-porosity, sintered-glass crucible. If necessary, the bottom of the crucible may be lined with a pledget of glass wool. Standardize the solution as follows: Weigh accurately about 200 mg of sodium oxalate of primary standard quality ($Na_2C_2O_4$), previously dried at 100° to constant weight, and dissolve it in 250 mL of water. Add 7 mL of sulfuric acid, heat to about 70°, and then slowly add the permanganate solution from a buret, with constant stirring, until a pale pink color that persists for 15 s is produced. The temperature at the conclusion of the titration should be not less than 60°. Calculate the normality. Each 6.700 mg of $Na_2C_2O_4$ is equivalent to 1 mL of 0.1 N potassium permanganate. Potassium permanganate is reduced on contact with organic substances such as rubber; therefore, the solution must be handled in apparatus entirely of glass or other suitably inert material. Store it in glass-stoppered, amber-colored bottles, and restandardize frequently.

Silver Nitrate, 0.1 N (16.99 g $AgNO_3$ per 1000 mL) Dissolve about 17.5 g of silver nitrate ($AgNO_3$) in 1000 mL of water, and standardize the solution as follows: Weigh accurately 100 mg of primary standard sodium chloride, previously dried at 120° for 16 h, into a 150-mL beaker, and dissolve it in 5 mL of water. Add 5 mL of acetic acid, 50 mL of methanol, and 2 or 3 drops of *Eosin Y TS*, and titrate with the silver nitrate solution to the endpoint. Calculate the normality.

Sodium Acetate, 0.1 N (8.203 g CH_3COONa per 1000 mL) Dissolve 8.20 g of anhydrous sodium acetate in glacial acetic acid to make 1000 mL, and standardize the solution as follows: To 25.0 mL of the prepared sodium acetate solution, add 50 mL of glacial acetic acid and 1 mL of α-*Naphtholbenzein TS*. Titrate with *0.1 N Perchloric Acid* until a yellow-brown color changes through yellow to green. Perform a blank determination, and make any necessary correction. Calculate the normality factor.

Sodium Arsenite, 0.05 N (3.248 g $NaAsO_2$ per 1000 mL) Transfer 2.4725 g of arsenic trioxide, which has been pulverized and dried at 100° to constant weight, to a 1000-mL volumetric flask, dissolve it in 20 mL of *1 N Sodium Hydroxide*, and add *1 N Sulfuric Acid* or *1 N Hydrochloric Acid* until the solution is neutral or only slightly acid to litmus. Add 15 g of sodium bicarbonate, dilute to volume with water, and mix.

Sodium Hydroxide, 1 N (40.00 g NaOH per 1000 mL) Dissolve about 45 g of sodium hydroxide (NaOH) in about 950 mL of water, and add a freshly prepared saturated solution of barium hydroxide until no more precipitate forms. Shake the mixture thoroughly, and allow it to stand overnight in a stoppered bottle. Decant or filter the solution, and standardize the clear liquid as follows: Transfer about 5 g of primary standard potassium biphthalate [$KHC_6H_4(COO)_2$], previously dried at 105° for 2 h and accurately weighed, to a flask, and dissolve it in 75 mL of carbon dioxide-free water. If the potassium biphthalate is in the form of large crystals, crush it before drying. To the flask add 2 drops of *Phenolphthalein TS*, and titrate with the sodium hydroxide solution to a permanent pink color. Calculate the normality. Each 204.2 mg of potassium biphthalate is equivalent to 1 mL of 1 N sodium hydroxide.

> Note: Solutions of alkali hydroxides absorb carbon dioxide when exposed to air. Therefore store them in bottles with well-fitted, suitable stoppers, provided with a tube filled with a mixture of sodium hydroxide and lime so that air entering the container must pass through this tube, which will absorb the carbon dioxide. Frequently restandardize standard solutions of sodium hydroxide.

Sodium Methoxide, 0.1 N, in Pyridine (5.40 g CH_3ONa per 1000 mL) Weigh 14 g of freshly cut sodium metal, and cut into small cubes. Place about 0.5 mL of anhydrous methanol in a round-bottom 120-mL flask equipped with a ground-glass joint, add 1 cube of the sodium metal, and when the reaction subsides, add the remaining sodium metal to the flask. Connect a water-cooled condenser to the flask, and slowly add 100 mL of anhydrous methanol, in small portions, through the top of the condenser. Regulate the addition of the methanol so that the vapors are condensed and do not escape through the top of the condenser. After addition of the methanol is complete, connect a drying tube to the top of the condenser, and allow the solution to cool. Transfer 17.5 mL of this solution (approximately 6 N) into a 1000-mL volumetric flask containing 70 mL of anhydrous methanol, and dilute to volume with freshly distilled pyridine. Store preferably in the reservoir of an automatic buret suitably protected from carbon dioxide and moisture. Standardize the solution as follows: Weigh accurately about 400 mg of primary standard benzoic acid, transfer it into a 250-mL wide-mouth Erlenmeyer flask, and dissolve it in 50 mL of freshly distilled pyridine. Add a few drops of *Thymolphthalein TS*, and titrate immediately with the sodium methoxide solution to a blue endpoint. During the titration, direct a gentle stream of nitrogen into the flask through a short piece of 6-mm glass tubing fastened near the tip of the buret. Perform a blank determination (see the *General Provisions*), correct for the volume of sodium methoxide solution consumed by the blank, and calculate the normality. Each 12.21 mg of benzoic acid is equivalent to 1 mL of *0.1 N Sodium Methoxide in Pyridine*.

Sodium Methoxide, 0.02 N, in Toluene (1.08 g CH_3ONa per 1000 mL) Weigh 2.5 g of freshly cut sodium metal, and cut into small cubes. Place about 200 mL of anhydrous methanol in a 1000-mL volumetric flask, chill in an ice bath, and add the cubes one at a time to the methanol. When the last cube is dissolved, dilute to the mark with toluene, and mix. Standardize the solution as follows: Weigh accurately about 20 mg of primary standard benzoic acid, transfer it into a 50-mL conical flask, and dissolve it in 25 mL of dimethylformamide. Add 2 drops of a solution of 100 mg of thymol blue in 10 mL of dimethylformamide, and titrate immediately with the sodium methoxide solution to a blue endpoint. Titrate a blank solution of dimethylformamide in the same manner, correct the volume of sodium methoxide solution consumed by the blank, and calculate the normality. Each 2.442 mg of benzoic acid is equivalent to 1 mL of *0.02 N Sodium Methoxide in Toluene*.

Sodium Thiosulfate, 0.1 N (15.81 g $Na_2S_2O_3$ per 1000 mL) Dissolve about 26 g of sodium thiosulfate ($Na_2S_2O_3 \cdot 5H_2O$) and 200 mg of sodium carbonate (Na_2CO_3) in 1000 mL of recently boiled and cooled water. Standardize the solution as follows: Weigh accurately about 210 mg of primary standard potassium dichromate, previously pulverized and dried at 120° for 4 h, and dissolve in 100 mL of water in a 500-mL glass-stoppered flask. Swirl to dissolve the sample, remove the stopper, and quickly add 2 g of sodium bicarbonate, 3 g of potassium iodide, and 5 mL of hydrochloric acid. Stopper the flask, swirl to mix, and let stand in the dark for 10 min. Rinse the stopper and inner walls of the flask with water, and titrate the liberated iodine with the sodium thiosulfate solution until the solution is only faint yellow. Add *Starch TS*, and continue the titration to the discharge of the blue color. Calculate the normality.

Sulfuric Acid, 1 N (49.04 g H_2SO_4 per 1000 mL) Add slowly, with stirring, 30 mL of sulfuric acid to about 1020 mL of water, allow to cool to 25°, and standardize by titration against primary standard sodium carbonate (Na_2CO_3) as directed under *Hydrochloric Acid, 1 N*. Each 52.99 mg of Na_2CO_3 is equivalent to 1 mL of *1 N Sulfuric Acid*.

Sulfuric Acid, Alcoholic, 5 N (245.2 g H_2SO_4 per 1000 mL) Add cautiously, with stirring, 139 mL of sulfuric acid to a sufficient quantity of absolute alcohol to make 1000.0 mL.

Sulfuric Acid, Alcoholic, 0.5 N Add cautiously, with stirring, 13.9 mL of sulfuric acid to a sufficient quantity of absolute alcohol to make 1000.0 mL. Alternatively, prepare this solution by diluting 100.0 mL of 5 N sulfuric acid with absolute alcohol to make 1000.0 mL.

Thorium Nitrate, 0.1 M [48.01 g $Th(NO_3)_4$ per 1000 mL] Weigh accurately 55.21 g of thorium nitrate [$Th(NO_3)_4 \cdot 4H_2O$], dissolve it in water, dilute to 1000.0 mL, and mix. Standardize the solution as follows: Transfer 50.0 mL into a 500-mL volumetric flask, dilute to volume with water, and mix. Transfer 50.0 mL of the diluted solution into a 400-mL beaker, add 150 mL of water and 5 mL of hydrochloric acid, and heat to boiling. While stirring, add 25 mL of a saturated solution of oxalic acid, then digest the mixture for 1 h just below the boiling point, and allow to stand overnight. Decant through Whatman No. 42, or equivalent, filter paper, and transfer the precipitate to the filter using about 100 mL of a wash solution consisting of 70 mL of the saturated oxalic acid solution, 430 mL of water, and 5 mL of hydrochloric acid. Transfer the precipitate and filter paper to a tared tall-form porcelain crucible, dry, char the paper, and ignite at 950° for 1.5 h or to constant weight. Cool in a desiccator, weigh, and calculate the molarity of the solution by the formula

$$200W/264.04,$$

in which W is the weight, in g, of thorium oxide obtained.

Triethanolamine, 0.5 N [74 g N(CH$_2$CH$_2$OH)$_3$ per 1000 mL] Transfer 65 mL (74 g) of 98% triethanolamine into a 1000-mL volumetric flask, dilute to volume with water, stopper the flask, and mix thoroughly.

Zinc Sulfate, 0.05 M (8.072 g ZnSO$_4$ per 1000 mL) Dissolve about 15 g of zinc sulfate (ZnSO$_4$.7H$_2$O) in sufficient water to make 1000 mL, and standardize the solution as follows: Dilute about 35 mL, accurately measured, with 75 mL of water, add 5 mL of *Ammonia–Ammonium Chloride Buffer TS* and 0.1 mL of *Eriochrome Black TS*, and titrate with *0.05 M Disodium EDTA* until the solution is deep blue. Calculate the molarity.

INDICATORS

The necessary solutions of indicators may be prepared as directed under *Test Solutions (TS) and Other Reagents*. The sodium salts of many indicators are commercially available and may be used interchangeably in water solutions with the alcohol solutions specified for the free indicators.

Useful pH indicators, listed in ascending order of the lower limit of their range, are methyl yellow (pH 2.9 to 4.0), bromophenol blue (pH 3.0 to 4.6), bromocresol green (pH 4.0 to 5.4), methyl red (pH 4.2 to 6.2), bromocresol purple (pH 5.2 to 6.8), bromothymol blue (pH 6.0 to 7.6), phenol red (pH 6.8 to 8.2), thymol blue (pH 8.0 to 9.2), and thymolphthalein (pH 9.3 to 10.5).

Alphazurine 2G Use a suitable grade.

Azo Violet [*4-(p-Nitrophenylazo) Resorcinol*] A red powder, melting at about 193° with decomposition.

Bromocresol Blue Use *Bromocresol Green*.

Bromocresol Green (*Bromocresol Blue; Tetrabromo-m-cresolsulfonphthalein*) A white or pale buff-colored powder; slightly soluble in water; soluble in alcohol and in solutions of alkali hydroxides. Transition interval: from pH 3.8 (yellow) to 5.4 (blue).

Bromocresol Purple (*Dibromo-o-cresolsulfonphthalein*) A white to pink, crystalline powder; insoluble in water; soluble in alcohol and in solutions of alkali hydroxides. Transition interval: from pH 5.2 (yellow) to 6.8 (purple).

Bromophenol Blue (*Tetrabromophenolsulfonphthalein*) Pinkish crystals, soluble in alcohol. Insoluble in water; soluble in solutions of alkali hydroxides. Transition interval: from pH 3.0 (yellow) to 4.6 (blue).

Bromothymol Blue (*Dibromothymolsulfonphthalein*) A rose red powder. Insoluble in water; soluble in alcohol and in solutions of alkali hydroxides. Transition interval: from pH 6.0 (yellow) to 7.6 (blue).

Cresol Red (*o-Cresolsulfonphthalein*) A red brown powder. Slightly soluble in water; soluble in alcohol and in dilute solutions of alkali hydroxides. Transition interval: from pH 7.2 (yellow) to 8.8 (blue).

Crystal Violet (*Hexamethyl-p-rosaniline Chloride*) Dark green crystals. Slightly soluble in water; sparingly soluble in alcohol and in glacial acetic acid. Its solutions are deep violet.

Sensitiveness Dissolve 100 mg in 100 mL of glacial acetic acid, and mix. Pipet 1 mL of the solution into a 100-mL volumetric flask, and dilute with glacial acetic acid to volume. The solution is violet blue and does not show a reddish tint. Pipet 20 mL of the diluted solution into a beaker, and titrate with *0.1 N Perchloric Acid*, adding the perchloric acid slowly from a microburet. Not more than 0.1 mL of *0.1 N Perchloric Acid* is required to produce an emerald green color.

Dithizone (*Diphenylthiocarbazone*) A bluish black powder. Insoluble in water; soluble in alcohol and in chloroform, yielding intensely green solutions even in high dilutions.

Eriochrome Black T [*Sodium 1-(1-Hydroxy-2-naphthylazo)-5-nitro-2-naphthol-4-sulfonate*] A brownish black powder having a faint metallic sheen. Soluble in alcohol, in methanol, and in hot water.

Sensitiveness To 10 mL of a 1 in 200,000 solution in a mixture of equal parts of methanol and water add sodium hydroxide solution (1 in 100) until the pH is 10. The solution is pure blue and free from cloudiness. Add 0.2 mL of *Magnesium Standard Solution* (10 µg Mg ion). The color of the solution changes to red violet, and with the continued addition of magnesium ion, it becomes wine red.

p-Ethoxychrysoidin Monohydrochloride [*4-(p-Ethoxyphenylazo)-m-phenylenediamine Monohydrochloride; 4′-Ethoxy-2,4-diaminoazobenzene Monohydrochloride*] A reddish powder, insoluble in water. Transition interval: from pH 3.5 (red) to 5.5 (yellow).

Hydroxy Naphthol Blue The disodium salt of 1-(2-naphtholazo-3,6-disulfonic acid)-2-naphthol-4-sulfonic acid deposited on crystals of sodium chloride. Small blue crystals, freely soluble in water. In the pH range between 12 and 13, its solution is reddish pink in the presence of calcium ion and deep blue in the presence of excess disodium EDTA.

Suitability for Calcium Determinations Dissolve 300 mg in 100 mL of water, add 10 mL of *1 N Sodium Hydroxide* and 1.0 mL of calcium chloride solution (1 in 200), and dilute with water to 165 mL. The solution is reddish pink. Add 1.0 mL of *0.05 M Disodium EDTA*. The solution becomes deep blue.

Litmus A blue powder, cubes, or pieces. Partly soluble in water and in alcohol. Transition interval: from approximately pH 4.5 (red) to 8 (blue). Litmus is unsuitable for determining the pH of solutions of carbonates or bicarbonates.

Methylene Blue [*3,7-Bis(dimethylamino)phenazathionium Chloride*] Dark green crystals or a crystalline powder having a bronzelike luster. Soluble in water and in chloroform; sparingly soluble in alcohol.

Methyl Orange (*Helianthin; Tropaeolin D; 4′-Dimethyl-amino-azobenzene-4-sodium Sulfonate*)   An orange-yellow powder or crystalline scales. Slightly soluble in cold water; readily soluble in hot water; insoluble in alcohol. Transition interval: from pH 3.2 (pink) to 4.4 (yellow).

Methyl Red (*o-Carboxybenzeneazodimethylaniline Hydrochloride*)   A dark red powder or violet crystals. Sparingly soluble in water; soluble in alcohol. Transition interval: from pH 4.2 (red) to 6.2 (yellow).

Methyl Red Sodium   The sodium salt of *o*-carboxybenzene-azo-dimethylaniline. An orange-brown powder. Freely soluble in cold water and in alcohol. Transition interval: from pH 4.2 (red) to 6.2 (yellow).

Methyl Yellow (*p-Dimethylaminoazobenzene*)   Yellow crystals, melting between 114° and 117°. Insoluble in water; soluble in alcohol, in benzene, in chloroform, in ether, in dilute mineral acids, and in oils. Transition interval: from pH 2.9 (red) to 4.0 (yellow).

Murexide Indicator Preparation   Add 400 mg of murexide to 40 g of powdered potassium sulfate (K_2SO_4), and grind in a glass mortar to a homogeneous mixture. Alternatively, use tablets containing 0.4 mg of murexide admixed with potassium sulfate or potassium chloride, available commercially.

Naphthol Green B   The ferric salt of 6-sodium sulfo-1-isonitroso-1,2-naphthoquinone. A dark green powder, insoluble in water.

Neutral Red (*3-Amino-7-dimethylamino-2-methylphenazine Chloride*)   A coarse, reddish to olive green powder. Sparingly soluble in water and in alcohol. Transition interval: from pH 6.8 (red) to 8.0 (orange).

Phenol Red (*Phenolsulfonphthalein*)   A bright to dark red, crystalline powder. Very slightly soluble in water; sparingly soluble in alcohol; soluble in solutions of alkali hydroxides. Transition interval: from pH 6.8 (yellow) to 8.2 (red).

Phenolphthalein   White or yellowish white crystals. Practically insoluble in water; soluble in alcohol and in solutions of alkali hydroxides. Transition interval: from pH 8.0 (colorless) to 10.0 (red).

Quinaldine Red (*5-Dimethylamino-2-strylethylquinolinium Iodide*)   A dark, blue-black powder, melting at about 260° with decomposition. Sparingly soluble in water; freely soluble in alcohol. Transition interval: from pH 1.4 (colorless) to 3.2 (red).

Thymol Blue (*Thymolsulfonphthalein*)   A dark, brownish green, crystalline powder. Slightly soluble in water; soluble in alcohol and in dilute alkali solutions. Acid transition interval: from pH 1.2 (red) to 2.8 (yellow). Alkaline transition interval: from pH 8.0 (yellow) to 9.2 (blue).

Thymolphthalein   A white to slightly yellow, crystalline powder. Insoluble in water; soluble in alcohol and in solutions of alkali hydroxides. Transition interval: from pH 9.3 (colorless) to 10.5 (blue).

Xylenol Orange   [*3,3′-Bis-di(carboxymethyl)aminomethyl-o-cresolsulfonphthalein*]   An orange powder. Soluble in water and in alcohol. In acid solution it is lemon yellow, and its metal complexes are intensely red. It gives a distinct endpoint in the direct EDTA titration of metals such as bismuth, thorium, scandium, lead, zinc, lanthanum, cadmium, and mercury.

INDICATOR PAPERS AND TEST PAPERS

Indicator papers and test papers are strips of paper of suitable dimension and grade (usually Swedish O filter paper or other makes of like surface, quality, and ash) impregnated with a sufficiently stable indicator solution or reagent.

Treat strong, white filter paper with hydrochloric acid, and wash with water until the last washing shows no acid reaction to *Methyl Red TS*. Then treat with 6 *N* ammonium hydroxide, wash again with water until the last washing is not alkaline toward *Phenolphthalein TS*, and dry thoroughly. Saturate the dry paper with the appropriate indicator solution prepared as directed below, and dry carefully by suspending from glass rods or other inert material in still air free from acid, alkali, and other fumes. Cut the paper into strips of convenient size, and store in well-closed containers protected from light and moisture.

Indicator papers and test papers that are available commercially may be used, if desired.

Acetaldehyde Test Paper   Use a solution prepared by mixing equal volumes of a 20% solution of morpholine and a 5% solution of sodium nitroferricyanide. Saturate the prepared filter paper in the mixture, and use the moistened paper without drying.

Cupric Sulfate Test Paper   Use *Cupric Sulfate TS*.

Lead Acetate Test Paper   Usually about 6 × 80 mm in size. Use *Lead Acetate TS*, and dry the paper at 100°, avoiding contact with metal.

Litmus Paper, Blue   Usually about 6 × 50 mm in size. It meets the requirements of the following tests.
Phosphate   Place 10 strips in 10 mL of water to which have been added 1 mL of nitric acid and 0.5 mL of 6 *N* ammonium hydroxide. Allow to stand for 10 min, then decant the solution, warm, and add 5 mL of *Ammonium Molybdate TS*. Shake at about 40° for 5 min. No precipitate of phosphomolybdate is formed.

Residue on Ignition Ignite carefully 10 strips of the paper to constant weight. The weight of the residue corresponds to not more than 400 µg per strip of about 3 cm^2.

Rosins, Acids, etc. Immerse a strip of the blue paper in a solution of 100 mg of silver nitrate (AgNO$_3$) in 50 mL of water. The color of the paper does not change in 30 s.

Sensitiveness Drop a 10- to 12-mm strip in 100 mL of 0.0005 N hydrochloric acid contained in a beaker, and stir continuously. The color of the paper is changed within 45 s.

Litmus Paper, Red Usually about 6 × 50 mm in size. Red litmus meets the requirements for *Phosphate, Residue on Ignition*, and *Rosins, Acids, etc.*, under *Litmus Paper, Blue*.

Sensitiveness Drop a 10- × 12-mm strip into 100 mL of 0.0005 N sodium hydroxide contained in a beaker, and stir continuously. The color of the paper changes within 30 s.

Phenolphthalein Paper Use a 1 in 1000 solution of phenolphthalein in dilute alcohol (1 in 2).

Starch Iodate Paper Use a mixture of equal volumes of *Starch TS* and potassium iodate solution (1 in 20).

Starch Iodide Paper Use a solution of 500 mg of potassium iodide (KI) in 100 mL of freshly prepared *Starch TS*.

DETECTOR TUBES

Ammonia Detector Tube A fuse-sealed glass tube (Draeger or equivalent) that is designed to allow gas to be passed through it and that contains suitable absorbing filters and support media for the indicator bromophenol blue. The Draeger Reference Number is CH 20501; the measuring range is 5 to 70 ppm.

Note: Suitable detector tubes are available from National Draeger, Inc., P.O. Box 120, Pittsburgh, PA 15205-0120. Tubes other than those specified in the monograph may be used in accordance with the section entitled *Codex Specifications* in the *General Provisions*.

Carbon Dioxide Detector Tube A fuse-sealed glass tube (Draeger or equivalent) that is designed to allow gas to be passed through it and that contains suitable absorbing filters and support media for the indicators hydrazine and crystal violet. The Draeger Reference Number is CH 30801; the measuring range is 0.01% to 0.30%.

Carbon Monoxide Detector Tube A fuse-sealed glass tube (Draeger or equivalent) that is designed to allow gas to be passed through it and that contains suitable absorbing filters and support media for the indicators iodine pentoxide, selenium dioxide, and fuming sulfuric acid. The Draeger Reference Number is CH 25601; the measuring range is 5 to 150 ppm.

Chlorine Detector Tube A fuse-sealed glass tube (Draeger or equivalent) that is designed to allow gas to be passed through it and that contains suitable absorbing filters and support media for the indicator *o*-toluidine. The Draeger Reference Number is CH 24301; the measuring range is 0.2 to 3 ppm.

Hydrogen Sulfide Detector Tube A fuse-sealed glass tube (Draeger or equivalent) that is designed to allow gas to be passed through it and that contains suitable absorbing filters and support media for the indicator, which is a suitable lead salt. The Draeger Reference Number is 6719001; the measuring range is 1 to 20 ppm.

Nitric Oxide–Nitrogen Dioxide Detector Tube A fuse-sealed glass tube (Draeger or equivalent) that is designed to allow gas to be passed through it and that contains suitable absorbing filters and support media for an oxidizing layer and the indicator diphenylbenzidine. The Draeger Reference Number is CH 29401; the measuring range is 0.5 to 10 ppm.

Sulfur Dioxide Detector Tube A fuse-sealed glass tube (Draeger or equivalent) that is designed to allow gas to be passed through it and that contains suitable absorbing filters and support media for an iodine–starch indicator. The Draeger Reference Number is CH 31701; the measuring range is 1 to 25 ppm.

Water Vapor Detector Tube A fuse-sealed glass tube (Draeger or equivalent) that is designed to allow gas to be passed through it and that contains suitable absorbing filters and support media for the indicator, which consists of a selenium sol in suspension in sulfuric acid. The Draeger Reference Number is CH 67 28531; the measuring range is 5 to 200 mg/m^3.

Index

Titles of monographs in *Section 2* and those of flavor chemicals in *Section 3* are shown in **boldface type**. An asterisk indicates a new monograph for the fourth edition. For titles of new monographs introduced in the four supplements to the third edition, consult the listings under *General Information*, pages xxxi–xxxii.

A

Abbreviations, 6–7
Absolute Alcohol, 849
Acacia, 9
"Accurately," Defined, xxiii
Acesulfame K, 10
*****Acesulfame Potassium**, 10
Acetaldehyde, 450, 604
Acetaldehyde Test Paper, 861
Acetals, 815
Acetanisole, 450, 680
Acetate Identification Test, 753
Aceteugenol, 498
Acetic Acid, Glacial, 10, 849
Acetic Acid TS, Diluted, 849
Acetic Aldehyde, 450
Acetoacetic Ester, 484
Acetoin, 450, 681
2-Acetonaphthone, 530
Acetone, 11
Acetone Peroxides, 12
Acetophenone, 450, 570, 644
N-Acetyl-L-2-amino-4-(methylthio)butyric Acid, 13
4-Acetylanisole, 450
Acetylated Distarch Adipate, 159
Acetylated Distarch Phosphate, 160
Acetylated Mono- and Diglycerides, 12
Acetylated Monoglycerides, 12
Acetylbenzene, 450
3-Acetyl-2,5-dimethylfuran, 450, 644
Acetyl Eugenol, 498

Acetyl Groups, 836
***N*-Acetyl-L-Methionine**, 13
Acetyl Methyl Carbinol, 450
3-Acetyl-6-methyl-1,2-pyran-2,4(3H)-dione, 115
Acetyl Propionyl, 544
2-Acetylpyrazine, 450, 604
3-Acetylpyridine, 450, 604
2-Acetylpyrrole, 450
Acetyl Value, 820
Achilleic Acid, 15
Acid Calcium Phosphate, 68
Acid Ferric Sulfate TS, 852
Acid Hydrolysates of Proteins, 13
Acid-Hydrolyzed Milk Protein, 13
Acid-Hydrolyzed Proteins, 13
Acid-Insoluble Ash, 748
Acidity Determination by Iodometric Method (Flavor Chemicals), 565
Acid-Modified Starch, 159
Acid Number (Rosins and Related Substances), 832
Acid Phosphatase Activity, 788
Acid Phthalate Buffer, 848
Acid Value (Essential Oils and Flavors), 815
Acid Value (Fats and Related Substances), 820
Acid Value (Flavor Chemicals), 568
Acknowledgments, xvii
Aconitic Acid, 15
Added Substances Policy, 2
Adipic Acid, 16
Agar, 17

DL-Alanine, 18
L-Alanine, 18
Alcohol, 136, 849
Alcohol, Absolute, 849
Alcohol, Aldehyde-Free, 849
Alcohol, Anhydrous, 849
Alcohol, Dehydrated, 849
Alcohol, Diluted, 849
Alcohol, 70%, 849
Alcohol, 80%, 849
Alcohol, 90%, 849
Alcohol C-6, 506
Alcohol C-8, 542
Alcohol C-9, 540
Alcohol C-10, 476
Alcohol C-11, 562
Alcohol C-12, 518
Alcohol Content of Ethyl Oxyhydrate (Flavor Chemicals), 585
Alcoholic Ferric Chloride TS, 852
Alcoholic Potassium Hydroxide, 0.5 N, 854
Alcoholic Potassium Hydroxide TS, 849, 854
Alcoholic Sulfuric Acid, 0.5 N, 859
Alcoholic Sulfuric Acid, 5 N, 859
Alcohols, Total, 815
Aldehyde C-6, 504
Aldehyde C-7, 502
Aldehyde C-8, 540
Aldehyde C-9, 538
Aldehyde C-10, 476
Aldehyde C-11 Undecyclic, 560
Aldehyde C-11 Undecylenic, 560

Aldehyde C-12, 518
Aldehyde C-12 MNA, 534
Aldehyde C-14 Pure, So-Called, 560
Aldehyde C-16, 492
Aldehyde C-18, So-Called, 538
Aldehyde-Free Alcohol, 849
Aldehydes and Ketones, 816
    Hydroxylamine Method, 816
    Hydroxylamine/Tert-Butyl Alcohol Method, 816
    Neutral Sulfite Method, 817
Algin, 24, 54, 312, 351
Alginates Assay, 768
Algin Derivative, 331
Alginic Acid, 19
Alkaline Borate Buffer, 848
Alkaline Cupric Citrate TS, 851
Alkaline Cupric Tartrate TS, 849
Alkaline Mercuric-Potassium Iodide TS, 850
Allspice Oil, 294
Allura Red AC, 144
p-Allylanisole, 484
Allyl Caproate, 452
Allyl Cyclohexanepropionate, 452, 645
Allyl-3-cyclohexanepropionate, 452
4-Allyl-1,2-dimethoxy Benzene, 530
4-Allylguaiacol, 496
Allyl Heptanoate, 452, 605
Allyl Heptoate, 452
Allyl Hexanoate, 452, 645
Allyl Ionone, 452
Allyl α-Ionone, 452, 645
Allyl Isopentanoate, 452
Allyl Isothiocyanate, 452, 568, 681
Allyl Isovalerate, 452, 605
4-Allyl-2-methoxyphenol, 496
4-Allyl-2-methoxyphenol Acetate, 498
Almond Oil, Bitter, FFPA, 19, 573
Alpha-Amylase Activity (Bacterial), 789
Alpha-Amylase Activity (Nonbacterial), 790
Alphazurine 2G, 860
Aluminum Ammonium Sulfate, 20
Aluminum Identification Test, 753
Aluminum Magnesium Silicate, 41
Aluminum Potassium Sulfate, 21
Aluminum Sodium Sulfate, 21
Aluminum Sulfate, 22
Ambrette Seed Liquid, 23
Ambrette Seed Oil, 23, 596
Aminoacetic Acid, 186
N-[4-[[(2-Amino-1,4-dihydro-4-oxo-6-pteridinyl)methyl]amino]benzoyl]-L-glutamic Acid, 157
3-Amino-7-dimethylamino-2-methylphenazine Chloride, 861
L-2-Aminoglutaramic Acid, 175

L-2-Amino-5-guanidinovaleric Acid, 32
L-2-Amino-5-guanidinovaleric Acid Monohydrochloride, 33
L-2-Amino-3-hydroxybutyric Acid, 413
DL-2-Amino-3-hydroxypropanoic Acid, 346
L-2-Amino-3-hydroxypropanoic Acid, 347
N-[p-[[(2-Amino-4-hydroxy-6-pteridinyl)methyl]amino]benzoyl]glutamic Acid, 157
L-α-Amino-4(or 5)-imidazolepropionic Acid, 192
L-α-Amino-4(or 5)-imidazolepropionic Acid Monohydrochloride, 193
DL-α-Amino-3-indolepropionic Acid, 426
L-α-Amino-3-indolepropionic Acid, 427
L-2-Amino-3-mercaptopropanoic Acid Monohydrochloride, 113
L-2-Amino-3-methylbutyric Acid, 429
DL-2-Amino-4-(methylthio)butyric Acid, 251
L-2-Amino-4-(methylthio)butyric Acid, 250
DL-2-Amino-3-methylvaleric Acid, 206
L-2-Amino-3-methylvaleric Acid, 206
DL-2-Amino-4-methylvaleric Acid, 224
L-2-Amino-4-methylvaleric Acid, 224
L-2-Aminopentanedioic Acid, 174
2-Aminopentanedioic Acid Hydrochloride, 174
DL-α-Amino-β-phenylpropionic Acid, 292
L-α-Amino-β-phenylpropionic Acid, 292
DL-2-Aminopropanoic Acid, 18
L-2-Aminopropanoic Acid, 18
L-α-Aminosuccinamic Acid, 35
DL-Aminosuccinic Acid, 36
L-Aminosuccinic Acid, 37
Ammonia–Ammonium Chloride Buffer TS, 850
Ammoniacal Silver Nitrate TS, 850
Ammonia Detector Tube, 862
Ammoniated Glycyrrhizin, 23
Ammonia TS, 850
Ammonia TS, Stronger, 850
Ammonia Water, Stronger, 850
Ammonium Acetate TS, 850
Ammonium Alginate, 24
Ammonium Alum, 20
Ammonium Bicarbonate, 25
Ammonium Carbonate, 25
Ammonium Carbonate TS, 850
Ammonium Chloride, 26
Ammonium Chloride TS, 850
Ammonium Glutamate, 257
Ammonium Glycyrrhizinate, 257
Ammonium Glycyrrhizinate, Pentahydrate, 257

Ammonium Hydroxide, 26
Ammonium Hydroxide, 850
Ammonium Identification Test, 753
Ammonium Molybdate TS, 850
Ammonium Oxalate TS, 850
Ammonium Phosphate, Dibasic, 27
Ammonium Phosphate, Monobasic, 27
Ammonium Saccharin, 28
Ammonium Standard Solution, 849
Ammonium Sulfanilate TS, 850
Ammonium Sulfate, 28
Ammonium Sulfide TS, 850
Ammonium Thiocyanate, 0.1 N, 856
Ammonium Thiocyanate TS, 850
Amyl Acetate, 510
*1-Amyl Alcohol, 454
Amylase, 130
α-Amylase, 132, 786
β-Amylase, 132, 786
*Amyl Butyrate, 454, 510
Amyl Caprylate, 454
Amylcinnamaldehyde, 454
α-Amylcinnamaldehyde, 454, 605
Amyl Cinnamate, 454, 646
*Amyl Formate, 454, 512
Amyl Heptanoate, 454
Amyl Hexanoate, 512
Amyl Isovalerate, 512
Amyl Octanoate, 454, 605
Amyloglucosidase, 798
Amyl Propionate, 454, 606
Amyl Salicylate, 512
Amyl Valerate, 512
Amyl Vinyl Carbinol, 540
Amyris Oil, West Indian Type, 29, 573
Analytical Performance Parameters, xxiii
Anethole, 456, 568, 682
trans-Anethole, 456
Aneurine Hydrochloride, 411
Angelica Root Oil, 29, 573
Angelica Seed Oil, 30, 574
Angelica Seed Oleoresin, 391, 392
Anhydride of Menthane 1:8 Diole, 496
Anhydrous Alcohol, 849
Anhydrous Isopropanol, 852
Anhydrous Methanol, 853
Animal Lipase, 787
p-Anisaldehyde, 524
Anise Oil, 30, 572
Anise Oleoresin, 391, 392
Anisic Alcohol, 456
Anisic Aldehyde, 524
Anisole, 456, 568, 606
Anisyl Acetate, 456, 606
Anisyl Acetone, 524
Anisyl Alcohol, 456, 606
Anisyl Formate, 456, 607
Annatto Extracts, 31
Antimony Trichloride TS, 850

APDC Extraction Method, 766
APM, 35
APO, 32
Apocarotenal, 32
β-Apo-8′-Carotenal, 32
Apparatus for Tests and Assays, 4, 727
D-Araboascorbic Acid, 134
L-Arginine, 32
L-Arginine Monohydrochloride, 33
Arsenic Specification, Requirements for Keeping, xv
Arsenic Specifications Policy, 2
Arsenic Test, 755
Ascorbic Acid, 33
L-Ascorbic Acid, 33
Ascorbyl Palmitate, 34
Ash (Acid-Insoluble), 748
Ash (Total), 748
L-Asparagine, 35
Aspartame, 35
DL-Aspartic Acid, 36
L-Aspartic Acid, 37
N-L-α-Aspartyl-L-phenylalanine 1-Methyl Ester, 35
Assay by Gas Chromatography (Flavor Chemicals), 564
  General Method, Nonpolar Column, 564
  General Method, Polar Column, 564
Assay by Titrimetric Procedures (Flavor Chemicals), 565
  Direct Aqueous Acid–Base Titrations, 565
  Direct Aqueous–Alcoholic Acid–Base Titrations, 565
Assays for Certain Aldehydes and Ketones (Flavor Chemicals), 564
  Aldehydes—Hydroxylamine Method, 564
  Aldehydes—Hydroxylamine/Tert-Butyl Alcohol Method, 564
  Ketones—Hydroxylamine Method, 565
  Procedure Requiring the Use of Sealed Glass Vials or Ampules, 564
Assay Tolerances, Maximum, 6
Atomic Absorption Spectrophotometric Graphite Furnace Method, 763
Atomic Weights and Chemical Formulas, 3
Autolyzed Yeast Extract, 442
Azodicarbonamide, 37
Azodicarboxylic Acid Diamide, 37
Azo Violet, 860

B

Bacterial Alpha-Amylase Activity, 789
Bacterial (PC) Proteolytic Activity, 811
Baking Soda, 355

Balances and Weights, 729
Balsam Fir Oil, 156
Balsam Peru Oil, 38, 574
Barium Chloride TS, 850
Barium Diphenylamine Sulfonate TS, 850
Barium Hydroxide, 0.2 N, 856
Barium Standard Solution, 849
Basil Oil, Comoros Type, 39, 574
Basil Oil, European Type, 39, 597
Basil Oil, Italian Type, 39
Basil Oil, Réunion Type, 39
Basil Oil Exotic, 39
Basil Oleoresin, 391, 392
Bay Leaf Oil, 217
Bay Oil, 40, 575
Beeswax, White, 40
Beeswax, Yellow, 41
Beet Sugar, 400
Benedict's Qualitative Reagent, 850, 851
*Bentonite, 41
Benzaldehyde, 456, 607
Benzaldehyde Glyceryl Acetal, 456, 607
Benzene (in Paraffinic Hydrocarbon Solvents), 769
Benzidine TS, 850
1,2-Benzisothiazolin-3-one 1,1-Dioxide, 343
1,2-Benzisothiazolin-3-one 1,1-Dioxide Ammonium Salt, 28
1,2-Benzisothiazolin-3-one 1,1-Dioxide Calcium Salt, 71
1,2-Benzisothiazolin-3-one 1,1-Dioxide Sodium Salt, 378
Benzoate Identification Test, 753
1,2-Benzodihydropyrone, 458, 607
Benzoic Acid, 43
Benzophenone, 458, 646
o-**Benzosulfimide**, 343
Benzoylbenzene, 458
Benzoyl Peroxide, 43
Benzyl Acetate, 458, 570, 608
Benzyl Alcohol, 458, 570, 646
Benzyl Benzoate, 458, 647
Benzyl Butyrate, 458, 647
Benzyl *n*-Butyrate, 458
Benzyl Cinnamate, 458, 647
Benzyl Formate, 460, 608
Benzyl Isobutyrate, 460, 608
Benzyl Isovalerate, 460, 648
Benzyl 3-Methyl Butyrate, 460
Benzyl 2-Methyl Propionate, 460
Benzyl Phenylacetate, 460, 608
Benzyl Propanoate, 460
Benzyl Propionate, 648
Benzyl Salicylate, 460, 682
Bergamot Oil, Coldpressed, 44, 575
BHA, 44
BHT, 45

Bicarbonate Identification Test, 753
Biotin, 46
D-Biotin, 46
Birch Tar Oil, Rectified, 46, 575
3,3′-Bis-di(carboxymethyl)aminomethyl-*o*-cresolsulfonphthalein, 861
3,7-Bis(dimethylamino)phenazathionium Chloride, 860
Bismuth Nitrate TS, 850
Bisulfite Identification Test, 753
Bitter Almond Oil Free from Prussic Acid, 19
Black Pepper Oil, 47, 572
Black Pepper Oleoresin, 391, 392
Blank Tests, 4
Blank Titration, Residual, 4
Bleached Starch, 159
Blue Litmus Paper, 861
Bois de Rose Oil, 47, 576
Boric Acid–Potassium Chloride, 0.2 M, 848
Bornyl Acetate, 460, 648
l-Bornyl Acetate, 460
Bound Styrene, 782
Bovine Rennet, 133
Brewer's Yeast, 440
Brilliant Blue FCF, 139
Bromelain, 132, 786
Bromide Identification Test, 753
Brominated Vegetable Oil, 48
Bromine, 0.1 N, 856
Bromine TS, 850
Bromine Water, 850
Bromocresol Blue, 860
Bromocresol Blue TS, 850
Bromocresol Green, 860
Bromocresol Green TS, 850
Bromocresol Purple, 860
Bromocresol Purple TS, 850
Bromophenol Blue, 860
Bromophenol Blue TS, 850
Bromothymol Blue, 860
Bromothymol Blue TS, 850
Buffer Solutions, Standard, 848
Butadiene–Styrene 50/50 Rubber, 49
Butadiene–Styrene 75/25 Rubber, 48
Butane, 50
n-Butane, 50
1,4-Butanedicarboxylic Acid, 16
Butanedioic Acid, 397
2,3-Butanedione, 478
1-Butanol, 462
2-Butanone, 462, 609
Butan-3-one-2-yl Butanoate, 462, 649
(*E*)-Butenedioic Acid, 164
Butter Starter Distillate, 394
Butyl Acetate, 462, 649
n-Butyl Acetate, 462
Butyl Alcohol, 462, 649

Butyl Aldehyde, 462
Butylated Hydroxyanisole, 44
Butylated Hydroxymethylphenol, 51
Butylated Hydroxytoluene, 45
Butyl Butyrate, 462, 650
n-Butyl n-Butyrate, 462
Butyl Butyryllactate, 462, 609
1,3-Butylene Glycol, 52
tert-Butylhydroquinone, 408
Butyl Isobutyrate, 464, 650
Butyl Isovalerate, 464, 609
Butyl Octadecanoate, 464
Butyl Phenylacetate, 464, 609
Butyl Rubber, 205
Butyl Stearate, 464
Butyraldehyde, 464, 565, 610
Butyric Acid, 464, 565, 610
Butyrin, 558
γ-Butyrolactone, 464, 650
Butyryllactic Acid, Butyl Ester, 462

C

Caffeine, 52
Calcium Acetate, 53
*****Calcium Acid Pyrophosphate**, 53
Calcium Alginate, 54
Calcium Ascorbate, 54
Calcium Biphosphate, 68
Calcium Bromate, 55
Calcium Carbonate, 55
Calcium Chloride, 56
Calcium Chloride, Anhydrous, 56
Calcium Chloride Double Salt of DL- or D-Calcium Pantothenate, 66
Calcium Chloride Solution, 57
Calcium Chloride TS, 850
Calcium Citrate, 58
Calcium Disodium Edetate, 59
Calcium Disodium EDTA, 59
Calcium Disodium Ethylenediaminetetraacetate, 59
Calcium Disodium (Ethylenedinitrilo)tetraacetate, 59
Calcium 4-(β,D-Galactosido)-D-gluconate, 63
Calcium Gluconate, 60
Calcium Glycerophosphate, 60
Calcium Hydroxide, 61
Calcium Hydroxide TS, 850
Calcium Hydroxyapatite, 69
Calcium Identification Test, 753
Calcium Iodate, 62
Calcium Lactate, 62
Calcium Lactobionate, 63
Calcium Oxide, 64
Calcium Pantothenate, 64
D-Calcium Pantothenate, 64

Calcium Pantothenate, Calcium Chloride Double Salt, 66
Calcium Pantothenate, Racemic, 65
Calcium Peroxide, 67
Calcium Phosphate, Dibasic, 67
Calcium Phosphate, Monobasic, 68
Calcium Phosphate, Tribasic, 69
Calcium Propionate, 70
Calcium Pyrophosphate, 70
Calcium Saccharin, 71
Calcium Silicate, 72
Calcium Sorbate, 73
Calcium Stearate, 74
Calcium Stearoyl Lactylate, 74
Calcium Sulfate, 76
Calcium Sulfate TS, 851
Calf Rennet, 129, 788
Camphene, 466
d-**Camphor**, 466, 610
Cananga Oil, 76, 572
Candelilla Wax, 77, 715
Cane Sugar, 400
Canola Oil, 77
Cantha, 79
Canthaxanthin, 79
Capraldehyde, 476
Capric Acid, 115
Caproic Acid, 504
Caproic Aldehyde, 504
Capryl Alcohol, 542
Caprylic Acid, 271
Caprylic Aldehyde, 540
Capsicum Oleoresin, 391, 392
Caramel, 80
Caramel Color, 80
Caraway Oil, 85, 576
Caraway Oleoresin, 392
Carbamide, 428
Carbohydrase (Aspergillus niger var.), 786
Carbohydrase (Aspergillus oryzae var.), 786
Carbohydrase (Bacillus acidopullulyticus var.), 786
Carbohydrase (Bacillus stearothermophilus), 787
Carbohydrase (Bacillus subtilis), 786, 787
Carbohydrase (Candida pseudotropicalis), 786
Carbohydrase (Kluyveromyces marxianus var. lactis), 786
Carbohydrase (Martierella vinaceae var. raffinoseutilizer), 786
Carbohydrase (Rhizopus niveus), 786
Carbohydrase (Rhizopus oryzae var.), 786
Carbohydrase (Saccharomyces species), 787

Carbohydrase (Trichoderma longibrachiatum var.), 787
Carbohydrase and Protease, Mixed (Bacillus licheniformis), 131
Carbohydrase and Protease, Mixed (Bacillus subtilis), 131
Carbohydrates (Starches, Sugars, and Related Substances), 836
Carbon, Activated, 85
Carbonate Identification Test, 753
*****Carbon Dioxide**, 87
Carbon Dioxide Detector Tube, 862
Carbon Monoxide Detector Tube, 862
o-Carboxybenzeneazodimethylaniline Hydrochloride, 861
Cardamom Oil, 88, 576
Cardamom Oleoresin, 392
Carmine, 89
Carminic Acid, 89
Carnauba Wax, 80
Carob Bean Gum, 228
Carotene, 90
β-Carotene, 90
Carrageenan, xiv, xxxii
Carrot Seed Oil, 91, 577
Carr-Price Reagent, 851
Carvacrol, 466, 610
l-Carveol, 466, 611
4-Carvomenthenol, 554
dextro-Carvone, 466
d-Carvone, 466, 651
l-Carvone, 468, 651
levo-Carvone, 468
l-Carvyl Acetate, 468, 611
β-Caryophyllene, 468, 611
Cascarilla Oil, 91, 577
Casein and Caseinate Salts, 92
Cassia Oil, 93, 577
Castor Oil, 94
Catalase, 132, 786
Catalase (Aspergillus niger var.), 786
Catalase (bovine liver), 786
Catalase (Micrococcus lysodeikticus), 786
Catalase Activity, 791
Caustic Potash, 318
Caustic Soda, 364
Cedar Leaf Oil, 94, 577
Cedar Leaf Oil, White, 94
Celery Oleoresin, 392
Celery Seed Oil, 95, 578
Cellulase, 132, 786
Cellulase Activity, 791
Cellulose, Microcrystalline, 95
Cellulose, Powdered, 96
Cellulose Gel, 95
Cellulose Gum, 357
"Centrifuge," Defined, 4
Ceric Ammonium Nitrate TS, 851
Ceric Sulfate, 0.01 N, 856

Ceric Sulfate, 0.1 N, 856
Chamomile Oil, English Type, 98, 578
Chamomile Oil, German Type, 98, 578
Chamomile Oil, Hungarian Type, 98
Changes in Format to the *Food Chemicals Codex*, Fourth Edition, xv
Chemical Formulas and Atomic Weights, 3
Chemical Tests and Determinations, 753
Chewing Gum Base, 782
　Bound Styrene, 782
　Molecular Weight, 783
　Quinones, 784
　Residual Styrene, 784
　Total Unsaturation, 785
Chicle, 249
Chicle, Venezuelan, 249
Chilte, 249
China Clay, 207
Chiquibul, 249
Chloride and Sulfate Limit Tests, 757
Chloride Identification Test, 753
Chlorinated Compounds, 817
Chlorine, 99
Chlorine Detector Tube, 862
Chlorine TS, 851
Chlorophyll, 821
Cholalic Acid, 99
Cholecalciferol, 434
Cholic Acid, 99
Choline Bitartrate, 100
Choline Chloride, 101
Chromatography, 729
　Column, 730
　Gas, 732
　High-Pressure Liquid, 734
　Thin-Layer, 731
Chromium, 771
Chromotropic Acid TS, 851
Chymosin, 132
Chymosin (*Aspergillus niger* var.), 786
Chymotrypsin, 786
Chymotrypsin Activity, 793
Cinene, 518
1,8-Cineol, 496
Cineole, Percentage of, 818
Cinnamal, 468
Cinnamaldehyde, 468, 611
Cinnamic Acid, 468, 565, 612
Cinnamic Alcohol, 470
Cinnamic Aldehyde, 468
Cinnamon Bark Oil, Ceylon Type, 101, 578
Cinnamon Leaf Oil, 102, 579
Cinnamon Oil, 93
Cinnamyl Acetate, 468, 612
Cinnamyl Alcohol, 470, 612
Cinnamyl Anthranilate, xiv, xxxii
Cinnamyl Butyrate, 470, 612

Cinnamyl Cinnamate, 470, 613
Cinnamyl Formate, 470, 613
Cinnamyl Isobutyrate, 470, 613
Cinnamyl Isovalerate, 470, 613
Cinnamyl Propionate, 472, 614
cis and *trans* changed to (Z) and (E), xvi
Citral, 472, 651
Citrate Identification Test, 753
Citric Acid, 102
Citridic Acid, 15
Citronellal, 472, 652
Citronellol, 472, 683
Citronellyl Acetate, 472, 652
Citronellyl Butyrate, 472, 614
Citronellyl Formate, 474, 683
Citronellyl Isobutyrate, 474, 614
Citronellyl Propanoate, 474
Citronellyl Propionate, 474, 652
Citrus Oils, Ultraviolet Absorbance of, 819
Clary Oil, 103, 579
Clary Sage Oil, 103
Clove Bud Oil, 104
Clove Leaf Oil, 104, 579
Clove Oil, 104, 579
Clove Stem Oil, 105, 580
CMC, 357
Coagulated or Concentrated Latices of Vegetable Origin, 249
Cobalt Identification Test, 754
Cobaltous Chloride CS, 741
Cobaltous Chloride TS, 851
Cobalt–Uranyl Acetate TS, 851
Cocoa Butter Substitute, 105
Coconut Oil (Unhydrogenated), 107
Cognac Oil, Green, 108, 580
Cold Test, 821
Color (AOCS-Wesson), 821
Colorimetric Solutions, 848
Colors, F D & C (also see specific color under F D & C), 771
　Chromium, 771
　Ether Extracts, 771
　Leuco Base, 772
　Mercury, 772
　Total Color, 773
　Uncombined Intermediates and Products of Side Reactions, 775
Column Chromatography, 730
Compliance with Federal Statutes, ii
Congo Red TS, 851
"Constant Weight," Defined, 6
Containers, 6, 7
　Light-Resistant, 7
　Tared, 6
　Tight, 7
　Well-Closed, 7
"Cool Place," Defined, 7
Copaiba Oil, 108, 580

Copper Gluconate, 109
Copper Identification Test, 754
Copper Sulfate, 109
Coriander Oil, 110, 580
Coriander Oleoresin, 392
Cornmint Oil, Partially Dementholized, 250
Corn Oil (Unhydrogenated), 110
Corn Sugar, 118
Corn Syrup, 173, 843
Costus Root Oil, 111, 581
Cottonseed Oil (Unhydrogenated), 111
Cream of Tartar, 312
Cresol Red, 860
Cresol Red–Thymol Blue TS, 851
Cresol Red TS, 851
o-Cresolsulfonphthalein, 860
Cresyl Acetate, 474, 568, 684
p-Cresyl Isobutyrate, 558
p-Cresyl Methyl Ether, 524
Criteria for *Food Chemicals Codex* Grade, xx
Crospovidone, 309
Crown Gum, 249
Crude Fat, 836
Crystal Violet, 860
Crystal Violet TS, 851, 853
"CS," Defined, 848
Cubeb Oil, 112, 581
Cubeb Oleoresin, 392, 393
Cuminal, 474
Cuminaldehyde, 474
Cuminic Aldehyde, 474, 564, 653
p-Cuminic Aldehyde, 474
Cumin Oil, 113, 581
Cumin Oleoresin, 392, 393
Cupric Citrate TS, Alkaline, 850, 851
Cupric Nitrate TS, 851
Cupric Sulfate, 109
Cupric Sulfate CS, 741
Cupric Sulfate Test Paper, 861
Cupric Sulfate TS, 851
Cupric Tartrate TS, Alkaline, 849, 851, 852
Curcumin Content, 829
Cyanocobalamin, 431
Cyanogen Bromide TS, 851
Cyclamen Aldehyde, 474, 653
1,2,3,5/4,6-Cyclohexanehexol, 201
*Cyclohexyl Acetate, 474
p-Cymene, 476, 614
L-Cysteine Monohydrochloride, 113
L-Cystine, 114

D

Damar Gum, 114
Damar Resin, 114

Dammar, 114
Dammar Gum, 114
Dammar Resin, 114
Danish Agar, 164
Data Elements Required for Assay Validation, xxvi, xxvii
D.E., 120
"Deaerated Water," Defined, 5
(*E*),(*E*)-2,4-Decadienal, 476, 615
trans,trans-2,4-Decadienal, 476
δ-Decalactone, 474, 615
γ-Decalactone, 474, 615
d-Decanal, 474
Decanoic Acid, 115
1-Decanol, 476
(*E*)-2-Decenal, 476, 615
trans-2-Decenal, 476
(*Z*)-4-Decenal, 478, 616
cis-4-Decenal, 478
1-Decyl Alcohol, 476
Dehydrated Alcohol, 849
Dehydrated Isopropanol, 852
Dehydrated Methanol, 853
Dehydroacetic Acid, 115
Deleted Monographs, xxxii
Denigès' Reagent, 851, 853
Deoxycholic Acid, 116
"Desiccators and Desiccants," Defined, 4
Desoxycholic Acid, 116
Detector Tubes, 862
  Ammonia, 862
  Carbon Dioxide, 862
  Carbon Monoxide, 862
  Chlorine, 862
  Hydrogen Sulfide, 862
  Nitric Oxide–Nitrogen Dioxide, 862
  Sulfur Dioxide, 862
  Water Vapor, 862
Devitalized Wheat Gluten, 435
Dexpanthenol, 116
Dextrin, 117
Dextro Calcium Pantothenate, 64
Dextrose, 118
DHA, 115
Diacetyl, 478, 616
Diacetyl Tartaric Acid Esters of Mono- and Diglycerides, 119
2,6-Diaminohexanoic Acid Hydrochloride, 229
Diammonium Phosphate, 27
Diastase Activity (Diastatic Power), 793
Diatomaceous Earth, 120
Diatomaceous Silica, 120
Diatomite, 120
Dibenzyl Ether, 478, 616
Dibromo-*o*-cresolsulfonphthalein, 860
Dibromothymolsulfonphthalein, 860
2,6-Di-*tert*-butyl-*p*-cresol, 45
Dicalcium Phosphate, 67

1,6-Dichloro-1,6-dideoxy-β-D-fructofuranosyl-4-chloro-4-deoxy-α-D-galactopyranoside, 398
1,2-Dichloroethane, 136
Dichloromethane, 253
1,2-Di[(1'-ethoxy)ethoxy]propane, 478
Diethylene Imidoxide, 261
Diethylene Oximide, 261
Diethyl Malonate, 478, 684
Diethyl Sebacate, 478, 654
Diethyl Succinate, 478, 685
Dihydroanethole, 552
Dihydrocarveol, 480
d-Dihydrocarvone, 480
Dihydrocoumarin, 458
13α,12α-Dihydroxycholanic Acid, 116
2,7-Dihydroxynaphthalene TS, 851
1,2-Dihydroxypropane, 331
4,4'-Diketo-β-carotene, 79
Dilauryl Thiodipropionate, 121
Dill Herb Oil, American Type, 123
Dill Oil, 123
Dill Oil, Indian Type, 122
Dill Seed Oil, European Type, 122, 581
Dill Seed Oil, Indian, 122, 597
Dill Seed Oil, Indian Type, 122
Dillseed Oleoresin, 392, 393
Dillweed Oil, American Type, 123, 582
Diluted Acetic Acid TS, 849
Diluted Alcohol, 849
Diluted Hydrochloric Acid TS, 852
Diluted Lead Subacetate TS, 852
Diluted Nitric Acid TS, 853
Dimagnesium Phosphate, 233
1,2-Dimethoxy-4-allylbenzene, 528
2,5-Dimethyl-3-acetylfuran, 450
p-Dimethylaminoazobenzene, 861
4'-Dimethylaminoazobenzene-4-sodium Sulfonate, 861
5-Dimethylamino-2-strylethylquinolinium Iodide, 861
Dimethyl Anthranilate, 480
Dimethyl Benzyl Carbinol, 480, 617
Dimethyl Benzyl Carbinyl Acetate, 480, 654
Dimethyl Benzyl Carbinyl Butyrate, 480, 654
Dimethyldiketone, 478
Dimethylglyoxal, 478
2,6-Dimethyl-5-heptenal, 480, 617
Dimethylketol, 450
Dimethyl Ketone, 11
6,6-Dimethyl-2-methylenebicyclo-[3.1.1]heptane, 550
(*E*)-3,7-Dimethyl-2,6-octadien-1-al, 498
trans-3,7-Dimethyl-2,6-octadien-1-al, 498
3,7-Dimethyl-1,6-octadien-3-ol, 518
(*E*)-3,7-Dimethyl-2,6-octadien-1-ol, 498

trans-3,7-Dimethyl-2,6-octadien-1-ol, 498
3,7-Dimethyl-1,6-octadien-3-yl Acetate, 860
3,7-Dimethyl-2,6-octadien-1-yl Acetate, 498
3,7-Dimethyl-1,6-octadien-1-yl Benzoate, 498
3,7-Dimethyl-2,6-octadien-1-yl Butyrate, 500
3,7-Dimethyl-1,6-octadien-3-yl Formate, 520
3,7-Dimethyl-2,6-octadien-1-yl Formate, 500
3,7-Dimethyl-2,6-octadien-3-yl Isobutyrate, 860
3,7-Dimethyl-2,6-octadien-1-yl Phenylacetate, 500
3,7-Dimethyl-2,6-octadien-1-yl Propionate, 500
3,7-Dimethyl-2,6-octadien-3-yl Propionate, 520
Dimethyl Octanol, 482
3,7-Dimethyl-1-octanol, 482, 655
3,7-Dimethyl-3-octanol, 556
3,7-Dimethyl-6-octen-1-ol, 472
3,7-Dimethyl-6-octen-1-yl Acetate, 472
3,7-Dimethyl-6-octen-1-yl Butyrate, 472
3,7-Dimethyl-6-octen-1-yl Formate, 474
3,7-Dimethyl-6-octen-1-yl Isobutyrate, 474
3,7-Dimethyl-6-octen-1-yl Propionate, 474
α,α-Dimethylphenethyl Acetate, 480
α,α-Dimethylphenethyl Alcohol, 480
α,α-Dimethylphenethyl Butyrate, 480
Dimethylpolysiloxane, 123
Dimethylpolysiloxane, Viscosity of, 743
2,3-Dimethylpyrazine, 482, 617
2,5-Dimethylpyrazine, 482, 617
2,6-Dimethylpyrazine, 482, 618
2,5-Dimethylpyrrole, 482, 618
Dimethyl Silicone, 123
*Dimethyl Succinate, 482
Dimethyl Sulfide, 482, 618
Dioctyl Sodium Sulfosuccinate, 124
1,4-Dioxane Limit Test, 757
Diphenylamine TS, 851
Diphenylcarbazone TS, 851
Diphenyl Ketone, 458
Diphenylthiocarbazone, 860
Dipotassium Monophosphate, 324
Dipotassium Phosphate, 324
α,α'-Dipyridyl TS, 851
Disodium Dihydrogen Pyrophosphate, 350
Disodium Edetate, 125
Disodium EDTA, 125
Disodium EDTA, 0.05 *M*, 856

Disodium Ethylenediaminetetraacetate, 125
Disodium (Ethylenedinitrilo)tetraacetate, 125
Disodium Guanosine-5′-monophosphate, 126
Disodium Guanylate, 126
Disodium Inosinate, 127
Disodium Inosine-5′-monophosphate, 127
Disodium Monohydrogen Phosphate, 374
Disodium Phosphate, 374
Disodium Pyrophosphate, 350
Disodium Tartrate, 384
Disodium D-Tartrate, 384
Distarch Phosphate, 160
Distillation Range, 727, 737
3,3′-Dithiobis(2-aminopropanoic acid), 114
Dithizone, 860
Dithizone Method, 762
Dithizone TS, 851
δ-Dodecalactone, 484, 618
γ-Dodecalactone, 484
Dodecanal, 518
Dodecanoic Acid, 218
1-Dodecanol, 518
(E)-2-Dodecen-1-al, 484, 619
trans-2-Dodecen-1-al, 484
Dried Glucose Syrup, 173
"Dried to Constant Weight," Defined, 6
Dried Yeast, 440
Drop Method, 832
DSS, 124

E

Edible Gelatin, 166
Enanthic Alcohol, 502
Enocianina, 187
Enzyme Activity
　Acid Phosphatase Activity, 788
　Alpha-Amylase Activity (Bacterial), 789
　Alpha-Amylase Activity (Nonbacterial), 790
　Catalase Activity, 791
　Cellulose Activity, 791
　Chymotrypsin Activity, 793
　Diastase Activity (Diastic Power), 793
　α-Galactosidase Activity, 794
　β-Glucanase Activity, 795
　Glucoamylase Activity (Amyloglucosidase Activity), 798
　Glucose Isomerase Activity, 796
　Glucose Oxidase Activity, 798
　β-D-Glucosidase Activity, 798
　Hemicellulase Activity, 799
　Invertase Activity, 800
　Lactase (Acid) (β-Galactosidase) Activity, 802
　Lactase (Neutral) (β-Galactosidase) Activity, 801
　Lipase Activity, 803
　Lipase/Esterase (Forestomach) Activity, 804
　Maltogenic Amylase Activity, 804
　Milk-Clotting Activity, 805
　Pancreatin Activity, 805
　Pepsin Activity, 807
　Phospholipase A_2 Activity, 808
　Phytase Activity, 808
　Plant Proteolytic Activity, 810
　Proteolytic Activity, Bacterial (PC), 811
　Proteolytic Activity, Fungal (HUT), 812
　Proteolytic Activity, Fungal (SAP), 813
　Pullulanase Activity, 814
　Trypsin Activity, 814
Enzyme Assays, 786
Enzyme-Hydrolyzed (Source) Protein, 282
Enzyme-Modified Fat, 128
Enzyme-Modified Milkfat, 128
Enzyme-Modified (Source) Protein, 282
Enzyme Preparations, 129
　Animal-Derived Preparations, 129
　　Catalase (bovine liver), 129, 786
　　Chymotrypsin, 129, 786
　　Lipase, Animal, 129, 787
　　Pancreatin, 129, 787
　　Pepsin, 129, 788
　　Phospholipase A_2, 129, 788
　　Rennet, Bovine, 129, 133
　　Rennet, Calf, 129
　　Trypsin, 129, 788
　Plant-Derived Preparations, 130
　　Amylase, 130, 786
　　Bromelain, 130, 786
　　Ficin, 130, 132, 786
　　Malt, 130
　　Papain, 130, 787
　Microbially Derived Preparations, 130
　　Carbohdrase (Aspergillus niger var.), 130, 786
　　Carbohdrase (Aspergillus oryzae var.), 130, 786
　　Carbohdrase (Bacillus acidopullulyticus var.), 130, 788
　　Carbohdrase (Bacillus stearothermophilus), 130, 786
　　Carbohdrase (Bacillus subtilis), 130, 786
　　Carbohdrase (Candida pseudotropicalis), 130, 787
　　Carbohdrase (Kluyveromyces marxianus var. lactis), 130, 787
　　Carbohdrase (Martierella vinaceae var. raffinoseutilizer), 130, 786
　　Carbohdrase (Rhizopus niveus), 130, 786
　　Carbohdrase (Rhizopus oryzae var.), 131, 786
　　Carbohdrase (Saccharomyces species), 131, 787
　　Carbohdrase (Trichoderma longibrachiatum var.), 131, 786
　　Carbohdrase and Protease, Mixed (Bacillus licheniformis), 131
　　Carbohdrase and Protease, Mixed (Bacillus subtilis), 131, 787
　　Catalase (Aspergillus niger var.), 131, 786
　　Catalase (Micrococcus lysodeikticus), 131, 786
　　Chymosin (Aspergillus niger var.), 131, 786
　　Glucose Isomerase, 131, 787
　　Glucose Oxidase (Aspergillus niger var.), 131, 787
　　Lipase (Aspergillus niger var.), 131, 787
　　Lipase (Aspergillus oryzae var.), 132, 787
　　Lipase (Candida rugosa), 132, 787
　　Lipase (Rhizomucor (Mucor) miehei), 132, 787
　　Phytase (Aspergillus niger var.), 132, 788
　　Protease (Aspergillus niger var.), 132, 786, 788
　　Protease (Aspergillus oryzae var.), 132, 788
　　Rennet, Microbial (Bacillus cereus), 132, 788
　　Rennet, Microbial (Endothia parasitica), 132, 788
　　Rennet, Microbial (Mucor species), 132
Enzyme Preparations Used in Food Processing (Table), 786–788
Eosin Y TS, 851
1:8-Epoxy-p-menthane, 496
Epsom Salt, 237
Equisetic Acid, 15
Ergocalciferol, 432
Eriochrome Black T, 860
Eriochrome Black TS, 851
Erythorbic Acid, 134
Erythrosine, 143
Essential Oils, Infrared Spectra of, 572
Essential Oils and Flavors, 815
Esters, 817
　Ester Determination, 817

Ester Determination (High-Boiling Solvent), 817
Ester Value, 817
Saponification Value, 817
Estragole, 484, 619
Estragon Oil, 407
Ethanal, 450
Ethanol, 136, 849
Ether Extracts, 771
*****Ethone**, 484, 619
p-Ethoxychrysoidin Monohydrochloride, 860
p-Ethoxychrysoidin TS, 851
4'-Ethoxy-2,4-diaminoazobenzene Monohydrochloride, 860
6-Ethoxy-1,2-dihydro-2,2,4-trimethylquinoline, 135
3-Ethoxy-4-hydroxybenzaldehyde, 496
1-Ethoxy-2-hydroxy-4-propenylbenzene, 550
Ethoxylated Mono- and Diglycerides, 134
4-(*p*-Ethoxyphenylazo)-*m*-phenylenediamine Monohydrochloride, 860
Ethoxyquin, 135
Ethyl Acetate, 484, 685
Ethyl Acetoacetate, 484, 655
Ethyl Acrylate, 486, 686
Ethyl Alcohol, 136, 849
Ethyl Alcohol, 136
Ethyl *o*-Aminobenzoate, 486
Ethyl *p*-Anisate, 486, 686
Ethyl Anthranilate, 486, 687
Ethyl Benzoate, 486, 655
*****Ethyl Benzoyl Acetate**, 486
*****Ethyl-(*E*)-2-butenoate**, 486
Ethyl-*trans*-2-butenoate, 486
Ethyl Butyl Ketone, 502
2-Ethylbutyraldehyde, 486, 619
Ethyl Butyrate, 486, 656
2-Ethylbutyric Acid, 488, 565, 620
Ethyl Caprate, 488
Ethyl Caproate, 490
Ethyl Capronate, 490
Ethyl Caprylate, 494
Ethyl Cellulose, 136
Ethyl Cinnamate, 488, 620
Ethyl Citrate, 426
Ethyl Crotonate, 486
Ethyl Decanoate, 488, 687
2-Ethyl-3,5(6)-dimethylpyrazine, 488, 656
Ethyl Dodecanoate, 492
Ethylene Brassylate, 488, 620
trans-1,2-Ethylenedicarboxylic Acid, 164
Ethylene Dichloride, 137
Ethylene Trichloride, 425
2-Ethyl Fenchol, 488, 656

Ethyl Formate, 488, 565, 657
4-Ethyl Guaiacol, 490, 620
Ethyl Heptanoate, 490, 657
Ethyl Heptoate, 490
Ethyl Hexanoate, 490, 657
*****2-Ethyl Hexanol**, 490
2-Ethyl-1-hexanol, 490
Ethyl 2-Hydroxypropionate, 490
Ethyl Isobutyrate, 490, 621
Ethyl Isovalerate, 490, 658
Ethyl Lactate, 490, 688
Ethyl Laurate, 492, 621
*****Ethyl Levulinate**, 492, 621
Ethyl Malonate, 478
Ethyl Maltol, 138
Ethyl *p*-Methoxybenzoate, 486
Ethyl 2-Methylbutyrate, 492
Ethyl 3-Methylbutyrate, 490
*****Ethyl 2-Methylpentanoate**, 492
Ethyl Methylphenylglycidate, 492, 658
2-Ethyl-3-methylpyrazine, 492, 658
Ethyl 3-Methylthiopropionate, 492, 621
Ethyl Myristate, 492, 622
Ethyl Nonanoate, 494, 622
Ethyl 9-Octadecenoate, 494
Ethyl Octanoate, 494, 659
Ethyl Octoate, 494
Ethyl Oleate, 494
Ethyl 3-Oxybutanoate, 484
Ethyl Oxyhydrate, 494
Ethyl Pelargonate, 494
Ethyl *n*-Pentanoate, 496
Ethyl Phenylacetate, 494, 688
Ethyl Phenylglycidate, 494, 689
Ethyl 3-Phenylpropenate, 488
Ethyl Propionate, 494, 659
Ethyl Salicylate, 496, 659
Ethyl Sebacate, 478
Ethyl Succinate, 478
Ethyl 10-Undecanoate, 496
*****Ethyl Valerate**, 496
Ethyl Vanillin, 496, 660
Eucalyptol, 496, 622
Eucalyptus Oil, 138, 582
Eugenic Acid, 496
Eugenol, 496, 689
Eugenol Acetate, 498
Eugenyl Acetate, 498, 622
Eugenyl Methyl Ether, 528
"Excessive Heat," Defined, 7
Expanded Perlite, 288
Exsiccated Sodium Sulfite, 383

F

Farnesol, 498, 690
Fast Green FCF, 141
Fat, Crude, 836
Fats and Related Substances, 820

Fatty Acid Composition, 821
Fatty Acids, Free, 822
F D & C Blue No. 1, 139
F D & C Blue No. 2, 140
F D & C Colors, 771
F D & C Green No. 3, 141
F D & C Red No. 3, 143
F D & C Red No. 40, 144
F D & C Yellow No. 5, 145
F D & C Yellow No. 6, 146
Federal Statutes, Compliance with, ii
Fehling's Solution, 849, 851, 852
FEMA Numbers, 449
Fennel Oil, 147, 572
Fennel Oleoresin, 392, 393
Ferric Ammonium Citrate, Brown, 147
Ferric Ammonium Citrate, Green, 148
Ferric Ammonium Sulfate TS, 852
Ferric Chloride CS, 741
Ferric Chloride TS, 852
Ferric Chloride TS, Alcoholic, 852
Ferric Orthophosphate, 149
Ferric Phosphate, 149
Ferric Pyrophosphate, 150
Ferric Sulfate TS, Acid, 852
Ferrous Ammonium Sulfate, 0.1 *N*, 857
Ferrous Fumarate, 151
Ferrous Gluconate, 153
*****Ferrous Lactate**, 154
Ferrous Sulfate, 155
Ferrous Sulfate, Dried, 155
Ferrous Sulfate TS, 852
Ficin, 130, 132, 786
Figures, Significant, 5
"Filtration," Defined, 4
Fir Needle Oil, Canadian Type, 156, 582
Fir Needle Oil, Siberian Type, 156, 583
Fischer-Tropsch Paraffin, 280
Flame Atomic Absorption Spectrophotometric Method, 763
Flavor Chemicals, Gas Chromatographic (GC) Assay of, 569
Flavor Chemicals, Infrared Spectra of, 571
Flavor Chemicals, Specifications for, 449
Flavor Chemicals, Test Methods for, 564
Flavor Chemicals Policy, 2
Fluoride Limits Guidelines, 2
Fluoride Limit Test, 758
Folacin, 157
Folic Acid, 157
Food Chemicals Codex, Operating Procedures of, xix–xxiii
Food Chemicals Codex, Organization of, v–vii
Food Chemicals Codex, Scope of, xiv
Food-Grade Gelatin, 166
Food Starch, Modified, 158

Food Starch–Modified, 158
Food Starch, Unmodified, 161
Formaldehyde TS, 852
Former and Current Titles of *Food Chemicals Codex* Monographs, xxxi
Formic Acid, 162, 565
Formula Weight, xvi
Free Fatty Acids, 822
Free Glycerin or Propylene Glycol, 822
Free Phenols, 818
"Fresh," Defined, for Solutions, 849
Fructose, 162
d-Fructose, 162
Fruit Sugar, 162
Fuchsin–Sulfurous Acid TS, 852
Fully Hydrogenated Rapeseed Oil, 336
Fumaric Acid, 161
Functional Use in Foods, 7
Functions of the Committee on Food Chemicals Codex, xiv, xix
Fungal (HUT) Proteolytic Activity, 812
Fungal (SAP) Proteolytic Activity, 813
2-Furaldehyde, 498
Furcelleran, 164
Furfural, 498, 660
Fusel Oil, Refined, 498, 623

G

4-*O*-β-Galactopyranosyl-D-glucose, 212
α-Galactosidase, 132, 786
α-Galactosidase Activity, 794
Gallic Acid, Propyl Ester, 333
Garlic Oil, 166, 583
Gas Chromatographic (GC) Assay of Flavor Chemicals, 449, 569
  Calculations and Methods, 569
  GC Conditions for Analysis, 569
  GC System Suitability Test Sample, 570
Gas Chromatography, 732
Gaultheria Oil, 437
Gelatin, 166
Gelatinized Starch, 159
Gellan Gum, 169
General Good Manufacturing Practices Guidelines for Food Chemicals, xxvii–xxx
General Information, xix
General Provisions and Requirements, xiv, xv, xx, xxii, 1
General Tests and Assays, 723
Geranial, 472
Geraniol, 498, 660
Geranium Oil, Algerian Type, 170, 583
Geranium Oil, East Indian Type, 277

Geranium Oil, Turkish Type, 277
Geranyl Acetate, 498, 661
Geranyl Benzoate, 498, 690
Geranyl Butyrate, 500, 691
Geranyl Formate, 500, 661
Geranyl Phenylacetate, 500, 661
Geranyl Propionate, 500, 622
Gibberellic Acid, 171
Ginger Oil, 171, 583
Ginger Oleoresin, 392, 393
β-Glucanase, 132, 786
β-Glucanase Activity, 795
D-Glucitol, 388
Glucoamylase, 132, 786
Glucoamylase Activity (Amyloglucosidase Activity), 795
D-Gluconic Acid, Monopotassium Salt, 317
Glucono Delta-Lactone, 172
Glucose, 118
D-Glucose, 118
Glucose Isomerase, 132, 787
Glucose Isomerase Activity, 796
Glucose Oxidase, 133
Glucose Oxidase (*Aspergillus niger* var.), 131, 787
Glucose Oxidase Activity, 798
Glucose Syrup, 173, 843
Glucose Syrup, Dried, 173
Glucose Syrup Solids, 173
β-D-Glucosidase, 133, 787
β-D-Glucosidase Activity, 798
Gluside, 343
Glutamic Acid, 174
L-Glutamic Acid, 174
L-Glutamic Acid Hydrochloride, 174
L-Glutamine, 175
Glutaral, 175
Glutaraldehyde, 175
Glycerin, 176
Glycerin or Propylene Glycol, Free, 822
Glycerol, 176
Glycerol Ester of Gum Rosin, 177, 715
Glycerol Ester of Partially Dimerized Rosin, 178, 716
Glycerol Ester of Partially Hydrogenated Gum Rosin, 178, 716
Glycerol Ester of Partially Hydrogenated Wood Rosin, 178, 716
Glycerol Ester of Polymerized Rosin, 179, 717
Glycerol Ester of Tall Oil Rosin, 179, 717
Glycerol Ester of Wood Rosin, 180, 717
Glyceryl Behenate, 180
Glyceryl-Lacto Esters of Fatty Acids, 181

Glyceryl Monooleate, 182
Glyceryl Monostearate, 183
Glyceryl Triacetate, 424
Glyceryl Tribehenate, 180
Glyceryl Tributyrate, 558
Glyceryl Tridocosanoate, 180
Glyceryl Tripropanoate, 500
Glyceryl Tristearate, 185
Glycine, 186
Glycocoll, 186
Good Manufacturing Practices Guidelines for Food Chemicals, xxvii–xxx
Graham's Salt, 375
Granular Metal Powders, Sieve Analysis of, 752
Granulated Sugar, 400
Grapefruit Oil, Coldpressed, 186, 584
Grapefruit Oil, Expressed, 186
Grape Skin Extract, 187
Guaiac Resin, 189
Guar Gum, 187
Gum Arabic, 9
Gum Ghatti, 188
Gum Guaiac, 189
Gutta hang kang, 249
Gutta Katiau, 249

H

Harmonization, International, xv
Heavy Metals Limit, Reduced, 3
Heavy Metals Test, 760
Helianthin, 861
Heliotropine, 550
Helium, 189
Hemicellulase, 133, 787
Hemicellulase Activity, 799
(*E*),(*E*)-2,4-Heptadienal, 500, 623
trans,trans-2,4-Heptadienal, 500
γ-Heptalactone, 502, 662
Heptaldehyde, 502
Heptanal, 502, 691
2-Heptanone, 502, 623
3-Heptanone, 502, 692
(*Z*)-4-Hepten-1-al, 502, 623
cis-4-Hepten-1-al, 502
Heptyl Alcohol, 502, 692
n-Heptyl-*p*-hydroxybenzoate, 190
Heptylparaben, 190
Hexadecanoic Acid, 278
2,4-Hexadienoic Acid, 387
2,4-Hexadienoic Acid, Calcium Salt, 73
2,4-Hexadienoic Acid, Potassium Salt, 327
cis-Hexahydro-2-oxo-1H-thieno[3,4]imidazole-4-valeric Acid, 46
Hexahydropyridine, 550

γ-**Hexalactone**, 502, 624
Hexaldehyde, 504
Hexamethyl-*p*-rosaniline Chloride, 860
Hexanal, 504
Hexanedioic Acid, 16
1,2,3,4,5,6-Hexanehexol, 247, 388
Hexanes, 190
Hexanoic Acid, 504, 565, 662
1-Hexanol, 506
(*E*)-**2-Hexen-1-al**, 504
trans-2-Hexen-1-al, 504
(*E*)-**2-Hexen-1-ol**, 504, 624
trans-2-Hexen-1-ol, 504
(*Z*)-**3-Hexen-1-ol**, 504, 624
cis-3-Hexen-1-ol, 504
(*E*)-**2-Hexenyl Acetate**, 504, 624
trans-2-Hexenyl Acetate, 504
(*Z*)-**3-Hexen-1-yl Acetate**, 504, 625
cis-3-Hexen-1-yl Acetate, 504
(*Z*)-**3-Hexenyl Isovalerate**, 506
cis-3-Hexenyl Isovalerate, 506
(*Z*)-**3-Hexenyl 2-Methylbutyrate**, 506
cis-3-Hexenyl 2-Methylbutyrate, 506
n-**Hexyl Acetate**, 506, 625
Hexyl Alcohol, 506, 693
Hexyl-2-Butenoate, 506, 663
α-**Hexylcinnamaldehyde**, 506, 625
Hexyl Isovalerate, 506
Hexyl 2-Methylbutyrate, 506
High-Fructose Corn Syrup, 191
High-Pressure Liquid Chromatography, 734
L-**Histidine**, 192
L-**Histidine Monohydrochloride**, 193
Hop Oleoresin, 392, 393
Hops Oil, 193, 584
Hortvet-Sellier Method, Modified, 829
Hydratropic Aldehyde, 548
Hydratropic Aldehyde Dimethyl Acetal, 548
Hydrochloric Acid, 194, 852
Hydrochloric Acid (reagent), 848
Hydrochloric Acid, 1 *N*, 857
Hydrochloric Acid Buffer, 848
Hydrochloric Acid Table, 748
Hydrochloric Acid TS, Diluted, 852
Hydrocinnamaldehyde, 548
Hydrocinnamyl Alcohol, 548
Hydrogenated Glucose Syrup, 238
Hydrogen Peroxide, 197
Hydrogen Peroxide TS, 852
Hydrogen Sulfide Detector Tube, 862
Hydrogen Sulfide TS, 852
α-Hydro-*omega*-hydroxy-poly(oxyethylene)-poly(oxypropylene)(51–57 moles)poly(oxyethylene) Block Copolymer, 295

α-Hydro-*omega*-hydroxy-poly(oxyethylene)-poly(oxypropylene)(63–71 moles)poly(oxyethylene) Block Copolymer, 297
Hydrolyzed Plant Protein (HPP), 13
Hydrolyzed (Source) Protein Extract, 13
Hydrolyzed Vegetable Protein (HVP), 13
3-Hydroxy-2-butanone, 138, 450
Hydroxycitronellal, 508, 570, 693
Hydroxycitronellal Dimethyl Acetal, 508, 663
4-Hydroxydecanoic Acid Lactone, 474
4-Hydroxy-2,5-dimethyl-3(2H)-furanone, 508, 625
7-Hydroxy-3,7-dimethyl Octanal, 508
7-Hydroxy-3,7-dimethyl Octanal: Acetal, 508
6-Hydroxy-3,7-dimethyloctanoic Acid Lactone, 508
4-Hydroxydodecanoic Acid Lactone, 484
3-Hydroxy-2-ethyl-4-pyrone, 138
(2-Hydroxyethyl)trimethylammonium Bitartrate, 100
(2-Hydroxyethyl)trimethylammonium Chloride, 101
Hydroxylamine Hydrochloride, 0.5 *N*, 857
Hydroxylamine Hydrochloride TS, 852
Hydroxylamine Method, 564, 565
Hydroxylamine/Tert-Butyl Alcohol Method, 564
Hydroxylated Lecithin, 198
Hydroxyl Value, 822
4-Hydroxy-3-methoxybenzaldehyde, 562
4-Hydroxy-3-methylethylbenzene, 490
5-Hydroxy-6-methyl-3,4-pyridinedimethanol Hydrochloride, 334
3-Hydroxy-2-methyl-4-pyrone, 240
Hydroxy Naphthol Blue, 860
5-Hydroxynonanoic Acid Lactone, 538
5-Hydroxyoctanoic Acid Lactone, 540
L-β-(*p*-Hydroxyphenyl)alanine, 427
4-(*p*-Hydroxyphenyl)-2-butanone, 508, 626
4-Hydroxyphexanoic Acid Lactone, 502
2-Hydroxypropanoic Acid, Monopotassium Salt, 321
2-Hydroxypropanoic Acid, Monosodium Salt, 365
2-Hydroxypropionic Acid, 211
α-Hydroxypropionic Acid, 211
Hydroxypropoxyl Determination, 777
Hydroxypropyl Alginate, 331
Hydroxypropyl Cellulose, 199
Hydroxypropyl Distarch Phosphate, 159
Hydroxypropyl Methylcellulose, 200
Hydroxypropyl Starch, 160
8-Hydroxyquinoline TS, 852
Hydroxysuccinic Acid, 237

5-Hydroxyundecanoic Acid Lactone, 560
Hypophosphite Identification Test, 754

I

Identification Specifications, 4
Identification Tests, 753
  Acetate, 753
  Aluminum, 753
  Ammonium, 753
  Benzoate, 753
  Bicarbonate, 753
  Bisulfite, 753
  Bromide, 753
  Calcium, 753
  Carbonate, 753
  Chloride, 753
  Citrate, 753
  Cobalt, 754
  Copper, 754
  Hypophosphite, 754
  Iodide, 754
  Iron, 754
  Lactate, 754
  Magnesium, 754
  Manganese, 754
  Nitrate, 754
  Nitrite, 754
  Peroxide, 754
  Phosphate, 754
  Potassium, 754
  Sodium, 754, 755
  Sulfate, 755
  Sulfite, 755
  Tartrate, 755
  Thiosulfate, 755
  Zinc, 755
"Ignite to Constant Weight," Defined, 6
IMP, 370
Indian Gum, 188
Indicator Papers and Test Papers, 861
Indicators, 4, 860
Indicators, Quantity Used, 4
Indigo Carmine, 140
Indigo Carmine TS, 852, 855
Indigotine, 140
Indigotine Disulfonate, 140
Indole, 508, 694
Infrared Spectra, 571
  Essential Oils, 572
  Flavor Chemicals, 601
  Other Substances, 713
Inositol, 201
i-Inositol, 201
meso-Inositol, 201
Insoluble Sodium Polyphosphate, 370
International Harmonization, xv
Invertase Activity, 800
Invert Sugar, 202, 846

Invert Sugar (Carbohydrates), 846, 836
Invert Sugar Syrup, 202
Iodide Identification Test, 754
Iodine, 0.1 N, 857
Iodine TS, 852
Iodine Value, 823
α-Ionone, 510, 663
β-Ionone, 510, 694
Iron Ammonium Citrate, 147, 148
Iron, Carbonyl, 202
Iron, Electrolytic, 203
Iron, Reduced, 204
Iron (II) 2-Hydroxypropionate, 154
Iron Identification Test, 754
Iron (II) Lactate, 154
Iron Phosphate, 149
Iron Pyrophosphate, 150
Iron Standard Solution, 849
Isoamyl Acetate, 510, 664
*****Isoamyl Alcohol**, 510
Isoamyl Benzoate, 510, 626
Isoamyl Butyrate, 510, 664
Isoamyl Caproate, 512
Isoamyl Caprylate, 454
Isoamyl Cinnamate, 454
Isoamyl Formate, 512, 664
Isoamyl Hexanoate, 512, 695
Isoamyl Isovalerate, 512
Isoamyl Octanoate, 454
*****Isoamyl Phenyl Acetate**, 512, 626
Isoamyl 3-Phenyl Propenate, 454
Isoamyl Propionate, 454
Isoamyl Salicylate, 512, 695
Isoborneol, 512, 626
Isobornyl Acetate, 514, 696
Isobutane, 205
Isobutyl Acetate, 514, 627
Isobutyl Alcohol, 514, 627
Isobutyl-2-butenoate, 514, 665
Isobutyl Butyrate, 514, 665
Isobutyl Cinnamate, 514, 627
Isobutylene–Isoprene Copolymer, 205, 718
Isobutyl Isovalerate, 532
Isobutyl Phenylacetate, 514, 665
Isobutyl Salicylate, 516, 627
Isobutyraldehyde, 516, 565, 628
Isobutyric Acid, 516, 565, 628
Isoestragole, 456
Isoeugenol, 516, 666
Isoeugenyl Acetate, 516, 696
Isoeugenyl Methyl Ether, 530
DL-Isoleucine, 206
L-Isoleucine, 206
Isopropanol, 207, 852
Isopropanol, Anhydrous, 852
Isopropanol, Dehydrated, 852
Isopropyl Acetate, 516, 628
Isopropylacetic Acid, 518

Isopropyl Alcohol, 207
Isopropyl Alcohol, 852
p-Isopropylbenzaldehyde, 474
Isopropylformic Acid, 516
Isopulegol, 516, 666
Isovaleric Acid, 518, 565, 628

J

Jelutong, 249
Juniper Berries Oil, 207, 584

K

Kaolin, 208
Karaya Gum, 208
Karl Fischer Titrimetric Method for Water, 5, 6, 745
Kasal, 353
Kelp, 209
Kjeldahl Method for Nitrogen Determination, 779
Konjac, 210
Konjac Flour, 210
Konjac Gum, 210
Konnyaku, 210

L

Labdanum Oil, 211, 584
Labeling, Use of "FCC" in, 2
Labeling Policy, 2
Lactase, 133, 787
Lactase (Acid) (β-Galactosidase) Activity, 802
Lactase (Neutral) (β-Galactosidase) Activity, 801
Lactated Mono-Diglycerides, 181
Lactate Identification Test, 754
Lactic Acid, 211
Lactic Acid Butyl Ester, Butyrate, 462
Lactose, 212, 837
Lactose (Carbohydrates), 837
Lactylated Fatty Acid Esters of Glycerol and Propylene Glycol, 213
Lactylic Esters of Fatty Acids, 215
Lanolin, Anhydrous, 216
Lard (Unhydrogenated), 217
Laurel Leaf Oil, 217, 585
Laurel Leaf Oleoresin, 392, 393
Lauric Acid, 218
Lauryl Alcohol, 518, 629
Lauryl Aldehyde, 518, 697
Lavandin Oil, Abrial Type, 218, 585
Lavender Oil, 219, 585
Lead Acetate Test Paper, 861
Lead Acetate TS, 852

Lead (and Heavy Metals) Limits Policy, 3
Lead Limit, Reduced, 3
Lead Limit Test, 762
  APDC Extraction Method, 766
  Atomic Absorption Spectrophotometric Graphite Furnace Method, 763
  Dithizone Method, 762
  Flame Atomic Absorption Spectrophotometric Method, 763
Lead Subacetate TS, 852
Lead Subacetate TS, Diluted, 852
LEAR, 77
Leche caspi (sorva), 249
Leche de vaca, 249
Lecithin, 220
Legal Status of the *Food Chemicals Codex*, xvi
Lemongrass Oil, 221, 585
Lemon Oil, Coldpressed, 222, 586
Lemon Oil, Desert Type, Coldpressed, 223, 598
Lemon Oil, Distilled, 222, 586
Lemon Oil, Expressed, 222
Lemon Oil Arizona, 223
DL-Leucine, 224
L-Leucine, 224
Leuco Base, 772
*****Levulinic Acid**, 518, 565, 629
Levulose, 162
Light-Resistant Container, 7
Lime, 64
Lime Oil, Coldpressed, 225
Lime Oil, Distilled, 225, 586
Lime Oil, Expressed, 225
Limestone, Ground, 225
"Limit of Quantitation," Defined, xxv
Limit Tests
  Chloride and Sulfate, 757
  1,4-Dioxane, 757
  Fluoride, 758
  Lead, 762
  Mercury, 766
  Selenium, 767
Limit Tests (Flavor Chemicals), 566
  Antioxidants in Ethyl Acrylate, 566
  Heavy Metals, 567
  Hydrocarbons in Eugenol, 567
  Hydrocyanic Acid in Benzaldehyde, 567
  Lead, 567
  Methyl Compounds in Ethyl Acetate, 567
  Peroxide Value, 567
  Readily Carbonizable Substances in Ethyl Acetate, 567
  Readily Oxidizable Substances in *dl*-Menthol, 567
  Reducing Substances, 568

d-Limonene, 518, 666
l-Limonene, 518
Linaloe Wood Oil, 227, 587
Linalool, 518, 570, 667
Linalool Determination, 818
Linalyl Acetate, 520, 667
Linalyl Benzoate, 520, 697
Linalyl Formate, 510, 629
Linalyl Isobutyrate, 520, 698
Linalyl Propionate, 520, 698
"Linearity," Defined, xxv
Linoleic Acid, 227
Lipase, 133, 787
Lipase (*Aspergillus niger* var.), 787
Lipase (*Aspergillus oryzae* var.), 787
Lipase (*Candida rugosa*), 787
Lipase (*Rhizomucor (Mucor) miehei*), 787
Lipase, Animal, 129, 787
Lipase Activity, 803
Lipase/Esterase (Forestomach) Activity, 804
Liquid Petrolatum, 256
Listing Substances in the *Food Chemicals Codex*, Requirements for, xx
Lithium Methoxide, 0.1 N, 857
Litmus, 860
Litmus Paper, Blue, 861
Litmus Paper, Red, 862
Litmus TS, 852
Locust Bean Gum, 228
Locust (Carob) Bean Gum, 228
Loss on Drying, 4, 6, 749
Lovage Oil, 228, 587
Low Erucic Acid Rapeseed Oil, 77
L-**Lysine Monohydrochloride**, 229

M

Mace Oil, 230, 587
Maddrell's Salt, 370
Magnesia Mixture TS, 852
Magnesium Carbonate, 230
Magnesium Chloride, 231
*****Magnesium Gluconate**, 231
Magnesium Hydroxide, 232
Magnesium Identification Test, 754
Magnesium Oxide, 233
Magnesium Phosphate, Dibasic, 233
Magnesium Phosphate, Tribasic, 234
Magnesium Silicate, 235
Magnesium Silicate, Synthetic, 235
Magnesium Standard Solution, 849
Magnesium Stearate, 236
Magnesium Sulfate, 237
Magnesium Sulfate TS, 852
Malachite Green TS, 852
Malic Acid, 237

DL-Malic Acid, 237
Malonic Ester, 478
Malt, 130
Malt Extract, 241
*****Maltitol Syrup**, 238
Maltodextrin, 239, 845
Maltogenic Amylase, 133, 787
Maltogenic Amylase Activity, 804
Maltol, 240
Malt Syrup, 241
Mandarin Oil, Coldpressed, 243, 587
Mandarin Oil, Expressed, 243
Manganese (Carbohydrates), 838
Manganese Chloride, 244
Manganese Gluconate, 245
Manganese Glycerophosphate, 245
Manganese Hypophosphite, 246
Manganese Identification Test, 754
Manganese Sulfate, 246
Mannite, 247
Mannitol, 247
D-Mannitol, 247
Marjoram Oil, Spanish Type, 248, 588
Marjoram Oil, Sweet, 248, 588
Marjoram Oleoresin, Sweet, 392, 393
Massaranduba balata, 249
Massaranduba chocolate, 249
Masticatory Substances, Natural, 249
Mayer's Reagent, 853
Maximum Assay Tolerances, 6
Melting Range (Fats and Related Substances), 824
Melting Range or Temperature (Physical Tests and Determinations), 728, 738
Mentha Arvensis Oil, Partially Dementholized, 250, 588
p-Mentha-1,5-diene, 544
d-p-Mentha-1,8-diene, 518
l-p-Mentha-1,8-diene, 518
p-Mentha-6,8-dien-2-ol, 466
p-Mentha-6,8-dien-2-yl Acetate, 468
Menthane 1:8 Diole, Anhydride of, 496
3-*p*-Menthanol, 522
l-p-Menthan-3-one, 522
dl-p-Menthan-3-yl Acetate, 522
l-p-Menthan-3-yl Acetate, 522
p-Menth-1-en-8-ol, 554
p-Menth-4-en-3-ol, 516
Menthen-1-yl-8 Acetate, 556
Menthen-1-yl-8 Propionate, 556
Menthol, 522, 667
l-**Menthone**, 522
dl-**Menthyl Acetate**, 522
l-**Menthyl Acetate**, 522
2-Mercaptopropionic Acid, 522, 565, 629
Mercuric Acetate TS, 853
Mercuric Chloride TS, 853
Mercuric Nitrate, 0.1 M, 857

Mercuric-Potassium Iodide TS, 853
Mercuric-Potassium Iodide TS, Alkaline, 850, 853
Mercuric Sulfate TS, 851, 853
Mercurous Nitrate TS, 853
Mercury Limit Test, 767
Methanol, 251, 853
Methanol, Anhydrous, 853
Methanol, Dehydrated, 853
Methional, 534
DL-**Methionine**, 250
L-**Methionine**, 251
p-Methoxyacetophenone, 450
***p*-Methoxybenzaldehyde**, 524, 699
p-Methoxybenzyl Acetate, 456
p-Methoxybenzyl Alcohol, 456
p-Methoxybenzyl Formate, 456
Methoxyl Determination, 778
2-Methoxy-3(5)-methylpyrazine, 524, 668
4-*p*-Methoxyphenyl-2-butanone, 524, 630
1-(*p*-Methoxyphenyl)-1-penten-3-one, 484
2-Methoxy-4-propenylphenol, 516
2-Methoxy-4-propenyl Phenyl Acetate, 516
2-Methoxypyrazine, 524, 630
*****Methyl Acetate**, 524, 630
4'-Methyl Acetophenone, 524, 668
Methylacetopyronone, 115
Methyl Alcohol, 251
Methyl Alcohol, 853
p-Methylaminophenol Sulfate TS, 853
Methyl Amyl Ketone, 502
***p*-Methyl Anisole**, 524, 668
Methyl Anthranilate, 526, 669
Methylbenzaldehyde, 558
p-Methylbenzaldehyde, 558
Methyl Benzoate, 526, 630
α-**Methylbenzyl Acetate**, 526, 699
α-**Methylbenzyl Alcohol**, 526, 669
*****2-Methyl Butanal**, 526
*****3-Methyl Butanal**, 526
β-Methyl Butyl Acetate, 510
*****2-Methylbutyl Acetate**, 526
2-Methylbutyl Isovalerate, 526
2-Methylbutyl-3-methylbutanoate, 526
*****Methyl Butyrate**, 528
*****2-Methylbutyric Acid**, 528, 565, 631
Methylcellulose, 252
Methylcellulose, Viscosity of, 743
Methyl Chavicol, 484
α-**Methylcinnamaldehyde**, 528, 631
Methyl Cinnamate, 528, 669
6-Methylcoumarin, 528, 631
Methyl *p*-Cresol, 524
3-Methylcyclopentane-1,2-dione, 528
Methyl Cyclopentenolone, 528, 631

Methylene Blue, 860
Methylene Blue TS, 853
Methylene Chloride, 253
3,4-(Methylenedioxy)benzaldehyde, 550
Methyl Ester of Rosin, Partially Hydrogenated, 254, 718
Methyl Ethyl Cellulose, 254
Methyl Ethyl Ketone, 462
Methyl Eugenol, 528, 700
Methyl Formate, xiv, xxxii
Methyl Glycol, 331
Methyl Heptenone, 528
6-Methyl-5-hepten-2-one, 528, 700
Methyl Heptine Carbonate, 532
Methyl Heptyl Ketone, 538
Methyl Hexyl Ketone, 530, 632
Methyl p-Hydroxybenzoate, 255
*Methyl Ionones, 530
Methyl Isobutyl Ketone, 532
*Methyl Isobutyrate, 530, 632
Methyl Isoeugenol, 530, 670
d-1-Methyl-4-isopropenyl-6-cyclohexen-2-one, 466
l-1-Methyl-4-isopropenyl-6-cyclohexen-2-one, 468
5-Methyl-2-isopropyl-2-hexenal, 530, 670
2-Methyl-3-(p-isopropylphenyl)-propionaldehyde, 474
Methyl N-Methyl Anthranilate, 480
Methyl 2-Methyl-butanoate, 530
Methyl 2-Methylbutyrate, 530
7-Methyl-3-methylene-1,6-octadiene, 534
d-2-Methyl-5-(1-methylethenyl)-cyclohexanone, 480
1-Methyl-4-(1-methylethyl)-1,3-cyclohexadiene, 554
1-Methyl-4-(1-methylethyl)-1,4-cyclohexadiene, 554
Methyl-3-methylthiopropionate, 530
Methyl β-Naphthyl Ketone, 530, 670
Methyl n-Nonyl Acetaldehyde, 534
Methyl Nonyl Ketone, 560
Methyl 2-Octynoate, 532, 671
Methyl Orange, 861
Methyl Orange TS, 853
Methylparaben, 255
2-Methylpentanoic Acid, 532, 565
4-Methylpentanoic Acid, 532, 565, 632
4-Methyl-2-pentanone, 532, 565
2-Methyl-2-pentenoic Acid, 532, 565, 633
α-Methyl Phenylacetaldehyde, 548
Methyl Phenylacetate, 532, 671
p-Methylphenyl Acetate, 474
Methyl Phenylcarbinol, 526
Methyl Phenylcarbinyl Acetate, 532, 671
Methylphenyl Ether, 456

Methyl Phenyl Ketone, 450
2-Methyl Propanoic Acid, 516
2-Methyl Propanyl Butyrate, 514
Methyl Propyl Ketone, 544
*Methyl Propyl 3-Methyl Butyrate, 532
2-Methylpyrazine, 534, 633
Methyl Pyrazinyl Ketone, 450
Methyl Pyridyl Ketone, 450
Methyl 2-Pyrrolyl Ketone, 450
Methyl Red, 861
Methyl Red–Methylene Blue TS, 853
Methyl Red Sodium, 861
Methyl Red TS, 853
Methylrosaniline Chloride TS, 853
Methyl Salicylate, 534, 701
Methyl Sulfide, 482
*4-Methyl-5-thiazole Ethanol, 534
3-Methylthiopropionaldehyde, 534
Methyl p-Tolyl Ketone, 524
2-Methylundecanal, 534, 633
Methyl Violet TS, 853
Methyl Yellow, 861
Mg/Kg and Percent Policy, 3
Microbial Rennet (*Bacillus cereus*), 132, 788
Microbial Rennet (*Endothia parasitica*), 132, 788
Microbial Rennet (*Mucor* species), 132
Microbiological Attributes Policy, 3
Microcrystalline Wax, Refined, 290
Milk-Clotting Activity, 805
Millon's Reagent, 853
Mineral Oil, White, 256
Mixed Paraffinic Hydrocarbons, 190
"mm of Mercury," Defined, 4
Modified Cellulose, 199, 254, 357
Modified Cellulose, EC, 136
Modified Cellulose, HPMC, 200
Modified Cellulose, MC, 252
Modified Food Starch, 158
Modified Hortvet–Sellier Method, 829
Modified Schiff's Reagent, 854
Molar Solutions, 855
Molecular Weight, xvi
Monoammonium L-Glutamate, 257
Monoammonium Glutamate Monohydrate, 257
Monoammonium Glycyrrhizinate, 257
Monoammonium Phosphate, 27
Mono- and Diglycerides, 258
Monobasic Potassium Phosphate, 0.2 M, 828
Mono-*tert*-butylhydroquinone, 408
Monocalcium Phosphate, 68
Monoglyceride Citrate, 259
1-Monoglycerides, 824
Monoglycerides, Total, 824
Monographs, Codex Specifications, 9

Monographs Added to the *Food Chemicals Codex*, Third Edition, xxxi
Monographs Deleted, xxxii
Monographs New to the *Food Chemicals Codex*, Fourth Edition, xxxii
Monopotassium D-Gluconate, 317
Monopotassium L-Glutamate, 259
Monopotassium Glutamate Monohydrate, 259
Monopotassium Phosphate, 324
Monosodium Dihydrogen Phosphate, 374
Monosodium L-Glutamate, 260
Monosodium Glutamate Monohydrate, 260
Monosodium Phosphate, 374
Monostearin, 183
Morpholine, 261, 718
MPG, 259
MSG, 260
Murexide Indicator Preparation, 861
Mustard Oil, 261
Myrcene, 534
Myrcia Oil, 40
Myristaldehyde, 534, 564
Myristic Acid, 262
Myristica Oil, 270
Myristyl Alcohol, 536
Myrrh Oil, 262, 588

N

α-Naphtholbenzein TS, 853
Naphthol Green B, 853, 861
Naphthol Green TS, 853
Natamycin, 263
Natural Rubber (Latex Solids), 249
"Negligible," Defined, 4
Neral, 472
Nerol, 536, 701
Nerolidol, 536, 633
Neryl Acetate, 536, 634
Nessler's Reagent, 850, 853
Neutralized Phthalate Buffer, 848
Neutral Red, 861
Neutral Red TS, 853
Neutral Sulfite Method, 817
New Monographs in the *Food Chemicals Codex*, Fourth Edition, xxxii
Niacin, 264
Niacinamide, 264
Niacinamide Ascorbate, 265
Nickel, 265
Nickel Catalysts, 265
Nicotinamide, 264
Nicotinic Acid, 264
Niger Gutta, 249
Ninhydrin TS, 853, 855
Nisin Preparation, 266
Nispero, 249

Nitrate Identification Test, 754
Nitre Cake, 356
Nitric Acid, 853
Nitric Acid TS, Diluted, 853
Nitric Oxide–Nitrogen Dioxide Detector Tube, 862
Nitrite Identification Test, 754
Nitrogen, 268
Nitrogen Determination, 779
Nitrogen Enriched Air, 268
Nitrogen Oxide, 269
4-(p-Nitrophenylazo) Resorcinol, 860
Nitrous Oxide, 269
(E),(E)-2,4-Nonadienal, 536, 634
trans,trans-2,4-Nonadienal, 536
(E),(Z)-2,6-Nonadienal, 536, 634
trans,cis-2,6-Nonadienal, 536
(E),(Z)-2,6-Nonadienol, 536, 634
trans,cis-2,6-Nonadienol, 536
δ-Nonalactone, 538, 635
γ-Nonalactone, 538, 702
Nonanal, 538, 635
*Nonanoic Acid, 538, 565
1-Nonanol, 540
2-Nonanone, 538, 635
Nonbacterial Alpha-Amylase Activity, 789
(E)-2-Nonenal, 538, 635
trans-2-Nonenal, 538
(E)-2-Nonen-1-ol, 538, 636
trans-2-Nonen-1-ol, 538
(Z)-6-Nonen-1-ol, 538, 636
cis-6-Nonen-1-ol, 538
Nonyl Acetate, 540, 702
Nonyl Alcohol, 540, 703
Normal Solutions, 855
Nutmeg Oil, 270, 589

O

(Z),(Z)-9,12-Octadecadienoic Acid, 227
Octadecanoic Acid, 395
Octadecanoic Acid, 1,2,3-Propane Triyl Ester, 185
(Z)-9-Octadecenoic Acid, 271
δ-Octalactone, 540, 636
γ-Octalactone, 540, 672
Octanal, 540, 636
Octanoic Acid, 271
1-Octanol, 542
3-Octanol, 540
2-Octanone, 530
(E)-2-Octen-1-al, 540, 637
trans-2-Octen-1-al, 540
1-Octen-3-ol, 540, 637
(Z)-3-Octen-1-ol, 542, 637
cis-3-Octen-1-ol, 542
1-Octen-3-yl Acetate, 542, 672
1-Octen-3-yl Butyrate, 542, 672

Octyl Acetate, 542, 703
3-Octyl Acetate, 542
Octyl Alcohol, 542, 704
Octyl Formate, 542, 704
Octyl Isobutyrate, 542
Octyl 2-Methylpropanoate, 542
"Odorless," Defined, 4
Oil Content of Synthetic Paraffin, 750
Oil of Frankincense, 272
Oil of Muscatel, 103
Oil of Shaddock, 186
Oleic Acid, 271
Oleoresin Angelica Seed, 391, 392
Oleoresin Anise, 391, 392
Oleoresin Basil, 391, 392
Oleoresin Black Pepper, 391, 392
Oleoresin Capsicum, 391, 392
Oleoresin Caraway, 392
Oleoresin Cardamom, 392
Oleoresin Celery, 392
Oleoresin Coriander, 392
Oleoresin Cubeb, 392, 393
Oleoresin Cumin, 392, 393
Oleoresin Dillseed, 392, 393
Oleoresin Fennel, 392, 393
Oleoresin Ginger, 392, 393
Oleoresin Hop, 392, 393
Oleoresin Laurel Leaf, 392, 393
Oleoresin Marjoram Sweet, 392, 393
Oleoresin Origanum, 392, 393
Oleoresin Paprika, 392, 393
Oleoresin Parsley Leaf, 392, 393
Oleoresin Parsley Seed, 392, 393
Oleoresin Pimenta Berries, 392, 393
Oleoresin Thyme, 392, 393
Oleoresin Turmeric, 392, 393
Oleoresins, 829
    Color Value, 829
    Curcumin Content, 829
    Piperine Content, 830
    Residual Solvent, 830
    Scoville Heat Units, 831
    Volatile Oil Content, 832
Olibanum Oil, 272, 589
Onion Oil, 272, 589
Operating Procedures of the *Food Chemicals Codex*, xix–xxii
Optical (Specific) Rotation, 739
Orange Oil, Bitter, Coldpressed, 273, 590
Orange Oil, Coldpressed, 273, 589
Orange Oil, Distilled, 274, 590
Organization, 571
Organization of the Food Chemicals Codex, 1981–1995, v
Origanum Oil, Spanish Type, 274, 590
Origanum Oleoresin, 392, 393
Orris Root Oil, 275, 590

Orthophenanthroline TS, 853
Orthophosphoric Acid, 293
Others Who Provided Assistance, 1981–1995, x, xi
Oxalic Acid, 0.1 N, 857
Oxalic Acid TS, 854
Ox Bile Extract, 275
Oxidized Hydroxypropyl Starch, 159
Oxidized Starch, 159
1:8-Oxido-p-menthane, 496
Oxyethylene Determination, 825
Oxygen Flask Combustion, 727
Oxystearin, 276
Ozone, 277

P

Packaging and Storage, 7
Palmarosa Oil, 277, 591
Palmitic Acid, 278
Palmitoyl L-Ascorbic Acid, 34
Palm Kernel Oil (Unhydrogenated), 278
Palm Oil (Unhydrogenated), 279
Pancreatin, 133, 787
Pancreatin Activity, 805
Panthenol, 116
DL-Panthenol, 280
D(+)-Pantothenyl Alcohol, 116
DL-Pantothenyl Alcohol, 280
Pantothenyl Alcohol, Racemic, 280
Papain, 787
Paprika Oleoresin, 392, 393
Paraffin, Synthetic, 280, 719
Paraffin, Synthetic, Oil Content of, 750
Paraffin Wax, Refined, 290
Parsley Herb Oil, 281, 591
Parsley Leaf Oleoresin, 392, 393
Parsley Seed Oil, 282, 591
Parsley Seed Oleoresin, 392, 393
Partial Acid Digest of (Source) Protein, 282
Partial Enzymatic Digest of (Source) Protein, 282
*Partially Hydrolyzed Proteins, 282
Partially Hydrolyzed (Source) Protein, 282
Participants in Committee Activities and Other Programs, viii, ix
Peach Aldehyde, 560
Peanut Oil (Unhydrogenated), 283
Pectinase, 133, 787
Pectins, 283
PEG, 301
Pelargonic Aldehyde, 538
Pendare, 249
Pennyroyal Oil, 286, 591
Pentaerythritol Ester of Partially Hydrogenated Wood Rosin, 286, 719

Pentaerythritol Ester of Wood Rosin, 287, 719
1,2,3,4,5-Pentahydroxypentane, 439
1,5-Pentanedial, 175
2,3-Pentanedione, 544, 673
Pentanoic Acid, 562
1-Pentanol, 454
2-Pentanone, 544, 673
Pentapotassium Triphosphate, 328
Pentasodium Triphosphate, 385
1-Pentyl Butyrate, 454
1-Pentyl Formate, 454
Pentyl Heptanoate, 454
Pentyl Hexanoate, 512
Peppermint Oil, 287, 592
Pepsin, 133, 788
Pepsin Activity, 807
(Source) Peptone, 282
Percentage of Cineole, 818
Perchloric Acid, 0.1 N, 857
Perchloric Acid, 0.1 N, in Dioxane, 858
Perillo, 249
Perlite, 288
Perlite, Expanded, 288
Peroxide Identification Test, 754
Petitgrain Oil, Paraguay Type, 289, 592
Petrolatum, 289
Petrolatum, White, 289
Petrolatum, Yellow, 289
Petroleum Wax, 290
Petroleum Wax, Synthetic, 291, 720
pH Determination, 740
α-Phellandrene, 544, 705
Phenethyl Acetate, 544, 705
2-Phenethyl Acetate, 544
Phenethyl Alcohol, 544, 706
α-Phenethyl Alcohol, 526
Phenethyl Isobutyrate, 544, 637
Phenethyl Isovalerate, 544, 638
2-Phenethyl 2-Methylbutyrate, 546
Phenethyl Phenylacetate, 546, 673
Phenethyl Salicylate, 546, 674
Phenolphthalein, 861
Phenolphthalein Paper, 862
Phenolphthalein TS, 854
Phenol Red, 861
Phenol Red TS, 854
Phenols, 818
Phenols, Free, 818
Phenolsulfonphthalein, 861
Phenolsulfonphthalein TS, 854
Phenoxyethyl Isobutyrate, 546, 674
Phenylacetaldehyde, 546, 674
Phenylacetaldehyde Dimethyl Acetal, 546, 706
Phenylacetic Acid, 546, 565, 638
DL-Phenylalanine, 292
L-Phenylalanine, 292

Phenyl Carbinol, 458
α-Phenyl Ethyl Acetate, 532
2-Phenylethyl Alcohol, 544
Phenylethyl Anthranilate, 548, 638
Phenylethyl Butyrate, 548, 638
p-Phenylphenol TS, 854
3-Phenyl-1-propanol, 548, 707
3-Phenylpropenoic Acid, 468
2-Phenylpropionaldehyde, 548, 707
3-Phenylpropionaldehyde, 548, 675
2-Phenylpropionaldehyde Dimethyl Acetal, 548, 708
3-Phenylpropyl Acetate, 548, 639
Phenylpropyl Alcohol, 548
Phenylpropyl Aldehyde, 548
pH Indicators, 740
Phosphate Buffer, 848
Phosphated Distarch Phosphate, 160
Phosphate Identification Test, 754
Phosphate Standard Solution, 849
Phospholipase A_2, 133, 788
Phospholipase A_2 Activity, 808
Phosphoric Acid, 293, 854
Phosphoric Acid, 854
Phosphorus (Carbohydrates), 838
Phosphotungstic Acid TS, 854
Physical Tests and Determinations, 729
Physicochemical Properties, 737
Phytase, 133, 788
Phytase (*Aspergillus niger* var.), 788
Phytase Activity, 808
Picric Acid TS, 854, 855
Pimaricin, 263
Pimenta Berries Oil, 294
Pimenta Berries Oleoresin, 392, 393
Pimenta Leaf Oil, 293, 592
Pimenta Oil, 294, 592
Pimento Leaf Oil, 293
Pimento Oil, 294
α-Pinene, 550
β-Pinene, 550
2-Pinene, 550
l-α-Pinene, 550
Pine Needle Oil, 156, 294
Pine Needle Oil, Dwarf, 294, 599
Pine Needle Oil, Scotch Type, 295, 599
Piperidine, 550, 639
Piperine Content, 830
Piperonal, 550, 675
Piperonyl Aldehyde, 550
Plant Proteolytic Activity, 810
Platinum–Cobalt CS, 741
Policies and Guidelines, 1
  Added Substances Policy, 2
  Arsenic Specifications Policy, 2
  FCC Substances Containing Sulfur Dioxide Policy, 2
  Flavor Chemicals Policy, 2
  Fluoride Limits Guidelines, 2

  General Policy, 1
  Labeling Policy, 2
  Lead (and Heavy Metals) Limits Policy, 3
  Mg/Kg and Percent Policy, 3
  Microbiological Attributes Policy, 3
Poloxamer 331, 295
Poloxamer 407, 497
Polydextrose, 297
Polydextrose Solution, 300
Polydimethylsiloxane, 123
Polyethylene, 301, 721, 783
Polyethylene Glycols, 301
Polyglycerate (60), 134
Polyglycerol Esters of Fatty Acids, 303
Polyisobutylene, 304, 721, 783
Poly[1-(2-oxo-1-pyrrolidinyl)ethylene], 310
Poly(oxy-1,2-ethanediyl) Derivative, 306, 307, 308
Polyoxyethylene (20) Mono- and Diglycerides of Fatty Acids, 134
Polyoxyethylene (20) Sorbitan Monolaurate, 306
Polyoxyethylene (20) Sorbitan Monooleate, 308
Polyoxyethylene (20) Sorbitan Monostearate, 307
Polyoxyethylene (20) Sorbitan Tristearate, 307
Polypropylene Glycol, 305
Polysorbate 20, 306
Polysorbate 60, 307
Polysorbate 65, 307
Polysorbate 80, 308
Polyvinyl Acetate, 309, 721, 783
Polyvinylpolypyrrolidone, 309
Polyvinylpyrrolidone, 310
Potassium Acetate TS, 854
Potassium Acid Phthalate, 0.1 N, 858
Potassium Acid Tartrate, 312
Potassium Alginate, 312
Potassium Alum, 21
Potassium Benzoate, 313
Potassium Bicarbonate, 313
Potassium Biphosphate, 324
Potassium Biphthalate, 0.2 M, 848
Potassium Bitartrate, 312
Potassium Bromate, 314
Potassium Carbonate, 314
Potassium Carbonate Solution, 315
Potassium Chloride, 315
Potassium Chloride, 0.2 M, 848
Potassium Chromate TS, 854
Potassium Citrate, 316
Potassium Dichromate, 0.1 N, 858
Potassium Dichromate TS, 854
Potassium Dihydrogen Phosphate, 324
Potassium Ferricyanide TS, 854

Potassium Ferrocyanide TS, 854
Potassium Gibberellate, 316
Potassium Gluconate, 317
Potassium Glutamate, 259
Potassium Glycerophosphate, 318
Potassium Hydroxide, 318, 854
Potassium Hydroxide, 0.5 N, Alcoholic, 849
Potassium Hydroxide, 1 N, 858
Potassium Hydroxide Solution, 319
Potassium Hydroxide TS, 854
Potassium Hydroxide TS, Alcoholic, 854
Potassium Identification Test, 754
Potassium Iodate, 319
Potassium Iodate, 0.05 M, 858
Potassium Iodide, 320
Potassium Iodide TS, 854
Potassium Kurrol's Salt, 325
Potassium Lactate Solution, 321
Potassium Metabisulfite, 322
Potassium Metaphosphate, 325
Potassium Nitrate, 323
Potassium Nitrite, 323
Potassium Permanganate, 0.1 N, 858
Potassium Permanganate TS, 854
Potassium Phosphate, Dibasic, 324
Potassium Phosphate, Monobasic, 324
Potassium Phosphate, Monobasic, 0.2 M, 848
Potassium Phosphate, Tribasic, 325
Potassium Polymetaphosphate, 325
Potassium Pyroantimonate TS, 854
Potassium Pyrophosphate, 326
Potassium Pyrosulfite, 322
Potassium Salt, 7
Potassium Sorbate, 327
Potassium Sulfate, 328
Potassium Sulfate TS, 854
Potassium Sulfite, 328
Potassium Triphosphate, 328
Potassium Tripolyphosphate, 328
Povidone, 310
Precipitated Calcium Phosphate, 69
"Precision," Defined, xxiii
Preface, xiii
Procedure for Revising Specifications, xxii
Procedures for Submission and Development of Specifications, xxi, xxii
L-**Proline**, 329
Propane, 329
1,2-Propanediol, 331
1,2,3-Propanetriol Octadecanoate, 183
2-Propanol, 207, 852
n-Propanol, 552
2-Propanone, 11
p-Propenylanisole, 456
Propenylguaethol, 550, 708

4-Propenyl Veratrole, 530
Propionaldehyde, 550, 565, 639
Propionic Acid, 330
***Propyl Acetate**, 550
n-Propyl Acetate, 550, 639
***Propyl Alcohol**, 552
p-**Propyl Anisole**, 552, 675
Propylene Chlorohydrin (Carbohydrates), 839
Propylene Glycol, 331
Propylene Glycol Alginate, 331
Propylene Glycol Ether of Methylcellulose, 200
Propylene Glycol Lactostearate, 213
Propylene Glycol Mono- and Diesters, 332
Propylene Glycol Mono- and Diesters of Fatty Acids, 332
Propylene Glycol Monostearate, 332
Propyl Gallate, 333
Propyl p-Hydroxybenzoate, 333
Propylparaben, 333
***Propyl Propionate**, 552
n-Propyl Propionate, 552, 640
Protease, 133, 788
Protease (Aspergillus niger var.), 788
Protease (Aspergillus oryzae var.), 788
Proteolytic Activity, Bacterial (PC), 811
Proteolytic Activity, Fungal (HUT), 812
Proteolytic Activity, Fungal (SAP), 813
Proteolytic Activity, Plant, 810
Pteroylglutamic Acid, 157
Pullulanase, 133, 788
Pullulanase Activity, 814
Purified Oxgall, 275
PVP, 310
PVPP, 309
3-Pyridinecarboxylic Acid, 264
Pyridoxine Hydrochloride, 334
Pyromucic Aldehyde, 498
Pyrrole, 552, 640
L-2-Pyrrolidinecarboxylic Acid, 329

Q

Qualitative Test for Phenols Using Ferric Chloride, 568
Quimociac TS, 854
Quinaldine Red, 861
Quinaldine Red TS, 854
Quinine Hydrochloride, 335
Quinine Sulfate, 336
Quinones, 784

R

Racemic Pantothenyl Alcohol, 280
"Range," Defined, xxv

Rapeseed Oil, Fully Hydrogenated, 336
Rapeseed Oil, Superglycerinated, 337
Readily Carbonizable Substances, 740
Reagent Solutions, 5, 848
Reagent Specifications, 4
Reagents, Hazardous or Toxic, 5
Red Litmus Paper, 862
Reduced Heavy Metals Limit, 3
Reduced Lead Limit, 3
Reducing Sugars Assay, 840
Reference Standards, 5
Refined Bleached Shellac, 348
Refined Microcrystalline Wax, 290
Refined Paraffin Wax, 290
Refractive Index, 741
Regular Bleached Shellac, 347
Reichert-Meissl Value, 826
Rennet, 788
Rennet, Bovine, 133
Rennet, Calf, 129, 133
Rennet, Microbial (Bacillus cereus), 132, 788
Rennet, Microbial (Endothia parasitica), 132, 788
Rennet, Microbial (Mucor species), 132
Requirements for Keeping the Arsenic Specification, 2
Requirements for Listing Substances in the Food Chemicals Codex, xx
Residual Blank Titration, 4
Residual Solvent, 830
Residual Styrene, 784
Residue on Evaporation, 819
Residue on Evaporation (Flavor Chemicals), 568
Residue on Ignition (Sulfated Ash), 751
Revising Specifications, Procedure for, xxii
Rhodinol, 552, 676
Rhodinyl Acetate, 522, 709
Rhodinyl Formate, 522, 709
Riboflavin, 338
Riboflavin 5'-Phosphate Ester Monosodium Salt, 339
Riboflavin 5'-Phosphate Sodium, 339
Rice Bran Wax, 341, 722
Ring-and-Ball Method, 834
"Robustness," Defined, xxvi
Rochelle Salt, 376
Rose Geranium Oil, Algerian Type, 170
Rosemary Oil, 341, 593
Rose Oil, 342, 593
Rosidinha, 249
Rosins and Related Substances, 832
Rue Oil, 342, 593
"Ruggedness," Defined, xxvi
Rum Ether, So-Called, 494

S

Saccharin, 343
Safflower Oil (Unhydrogenated), 343
Sage Oil, Dalmatian Type, 344, 593
Sage Oil, Spanish Type, 345, 594
SALP, 353
Salt, 358
Sandalwood Oil, East Indian Type, 345, 594
Sandalwood Oil, West Indian Type, 29
Santalol, 554, 710
Santalyl Acetate, 554, 676
Saponification Value (Esters) (Essential Oils and Flavors), 817
Saponification Value (Fats and Related Substances), 827
Savory Oil (Summer Variety), 346, 594
Schiff's Reagent, Modified, 854
Scope of the *Food Chemicals Codex*, xiv
Scoville Heat Units, 831
Selenium Limit Test, 767
DL-Serine, 346
L-Serine, 347
Shellac, Bleached, 347
Shellac, Bleached, Wax-Free, 348
Shellac, Refined Bleached, 348
Shellac, Regular Bleached, 347
Shellac, White, 347
Sieve Analysis of Granular Metal Powders, 752
Significant Figures, 5
Silica, Synthetic Amorphous, 348
Silicon Dioxide, 348
Silver Nitrate, 0.1 N, 858
Silver Nitrate, Ammoniacal, TS, 850
Silver Nitrate TS, 854
Slaked Lime, 61
Smectite, 41
Soap, 827
Soda Alum, 21
Sodium Acetate, 349
Sodium Acetate, 0.1 N, 858
Sodium Acetate, Anhydrous, 350
Sodium Acid Pyrophosphate, 350
Sodium Acid Sulfate, 356
Sodium Acid Sulfite, 356
Sodium Alginate, 351
Sodium Alum, 21
Sodium Aluminosilicate, 351
Sodium Aluminum Phosphate, Acidic, 353
Sodium Aluminum Phosphate, Basic, 353
Sodium Arsenite, 0.05 N, 858
Sodium Ascorbate, 354
Sodium L-Ascorbate, 354
Sodium Benzoate, 355
Sodium *o*-Benzosulfimide, 378

Sodium Bicarbonate, 355
Sodium Biphosphate, 356, 374
Sodium Bisulfate, 356
Sodium Bisulfite, 346
Sodium Bisulfite TS, 854
Sodium Bitartrate TS, 854
Sodium Borate TS, 854
Sodium Carbonate, 357
Sodium Carbonate TS, 854
Sodium Carboxymethylcellulose, 357
Sodium Carboxymethylcellulose, Viscosity of, 744
Sodium Chloride, 358
Sodium Choleate, 275
Sodium Citrate, 360
Sodium Cobaltinitrite TS, 854
Sodium Dehydroacetate, 361
Sodium Diacetate, 361
Sodium Erythorbate, 362
Sodium Ferric Pyrophosphate, 362
Sodium Ferrocyanide, 363
Sodium Fluoride TS, 854
Sodium Gluconate, 363
Sodium Glutamate, 260
Sodium 5'-Guanylate, 126
Sodium Hexametaphosphate, 375
Sodium Hydrogen Diacetate, 361
Sodium Hydrogen Sulfite, 356
Sodium Hydroxide, 364, 854
Sodium Hydroxide, 1 N, 858
Sodium Hydroxide Solution, 364
Sodium Hydroxide TS, 854
Sodium 3-(1-Hydroxyethylidene)-6-methyl-1,2-pyran-2,4(3H)-dione, 361
Sodium 1-(1-Hydroxy-2-naphthylazo)-5-nitro-2-naphthol-4-sulfonate, 860
Sodium Hypophosphite, 365
Sodium Hyposulfite, 384
Sodium Identification Test, 754
Sodium Indigotindisulfonate TS, 852, 855
Sodium 5'-Inosinate, 127
Sodium Iron Pyrophosphate, 362
Sodium Lactate Solution, 365
Sodium Lauryl Sulfate, 367
Sodium Magnesium Aluminosilicate, 368
Sodium Metabisulfite, 369
Sodium Metaphosphate, Insoluble, 370
Sodium Metasilicate, 370
Sodium Methoxide, 371
Sodium Methoxide, 0.02 N, in Toluene, 859
Sodium Methoxide, 0.1 N, in Pyridine, 859
Sodium Methylate, 371
Sodium Nitrate, 373

Sodium Nitrite, 373
Sodium Nitroferricyanide TS, 855
Sodium Phosphate, Dibasic, 374
Sodium Phosphate, Monobasic, 374
Sodium Phosphate, Tribasic, 375
Sodium Phosphate TS, 855
Sodium Polyphosphate, Insoluble, 370
Sodium Polyphosphates, Glassy, 375
Sodium Potassium Tartrate, 376
*Sodium Potassium Tripolyphosphate, 377
Sodium Propionate, 377
Sodium Pyrophosphate, 378
Sodium Pyrosulfite, 369
Sodium Saccharin, 378
Sodium Sesquicarbonate, 380
Sodium Silicoaluminate, 351
Sodium Stearoyl Lactylate, 380
Sodium Stearyl Fumarate, 382
Sodium Sulfate, 383
Sodium Sulfide TS, 855
Sodium Sulfite, 383
Sodium Sulfite, Exsiccated, 383
Sodium Tartrate, 384
Sodium Tetrapolyphosphate, 375
Sodium Thiosulfate, 384
Sodium Thiosulfate, 0.1 N, 859
Sodium Thiosulfate TS, 855
Sodium Trimetaphosphate, 385
Sodium Triphosphate, 385
Sodium Tripolyphosphate, 385
Softening Point, 727, 832
  Drop Method, 832
  Ring-and-Ball Method, 834
Solidification Point, 728, 742
Solubility in Alcohol, 449, 819
Solubility (Table), 7
Soluble Saccharin, 378
Solutions, Colorimetric (CS), 848
Solutions, Standard, for the Preparation of Controls and Standards, 849
Solutions, Standard Buffer, 848
Solutions, Test (TS) and Other Reagents, 5, 849
Solutions, Volumetric, 855
Solutions and Indicators, 5, 848, 860
Sorbic Acid, 387
Sorbitan, Monododecanoate, 306
Sorbitan, Monooctadecanoate, 307
Sorbitan, Mono-9-octadecenoate, 308
Sorbitan Monostearate, 387
D-Sorbite, 388
Sorbitol, 388
D-Sorbitol, 388
Sorbitol Solution, 389
Soybean Oil (Unhydrogenated), 390
Spearmint Oil, 391, 600
Specifications, Procedure for Revising, xxii

Specifications, Procedures for Submission and Development of, xxi
Specifications for Flavor Chemicals, 449
Specific Gravity, 5, 449
Specific Gravity (Fats and Related Substances), 827
"Specificity," Defined, xxiv
Specific Rotation, 739
Spice Oleoresins, 391
    Angelica Seed, 391, 392
    Anise, 391, 392
    Basil, 391, 392
    Black Pepper, 391, 392
    Capsicum, 391, 392
    Caraway, 392
    Cardamom, 392
    Celery, 392
    Coriander, 392
    Cubeb, 392, 393
    Cumin, 392, 393
    Dillseed, 392, 393
    Fennel, 392, 393
    Ginger, 392, 393
    Hop, 392, 393
    Laurel Leaf, 392, 393
    Marjoram Sweet, 392, 393
    Origanum, 392, 393
    Paprika, 392, 393
    Parsley Leaf, 392, 393
    Parsley Seed, 392, 393
    Pimenta Berries, 392, 393
    Thyme, 392, 393
    Turmeric, 392, 393
Spike Lavender Oil, 393, 594
Stability (Active Oxygen Method), 828
Standard Ammonium Solution, 849
Standard Barium Solution, 849
Standard Buffer Solutions, 848
Standard Iron Solution, 849
Standard Magnesium Solution, 849
Standard Phosphate Solution, 849
Standard Solutions for the Preparation of Controls and Standards, 849
Stannous Chloride, 394
Stannous Chloride TS, 855
Starch, Acid-Modified, 159
Starch, Bleached, 159
Starch, Gelatinized, 159
Starch, Hydroxypropyl, 160
Starch, Oxidized, 159
Starch, Thin-Boiling, 159
Starch Acetate, 159
Starch Aluminum Octenyl Succinate, 160
Starch Esters, 159
Starch Ether-Esters, 159
Starch Ethers-Hemiacetals, or Ethers, 159
Starch Iodate Paper, 862

Starch Iodide Paper, 862
Starch Iodide Paste TS, 855
Starch Octenyl Succinate, 159
Starch Phosphate, 159
Starch Sodium Octenyl Succinate, 159
Starch Sodium Succinate, 169
Starch TS, 855
Starter Distillate, 394
Stearic Acid, 395
Stearin, 185
Stearyl Monoglyceridyl Citrate, 396
Sterculia Gum, 208
Strawberry Aldehyde, 492
Strong Ammonia Solution, 26
Stronger Ammonia TS, 850
Stronger Ammonia Water, 26, 850
Styrene, Bound, 782
Styrene, Residual, 784
Submission and Development of Specifications, Procedures for, xxi
Submissions to the Codex, xxii
Succinic Acid, 397
Succinylated Monoglycerides, 397
Sucralose, 398
Sucrose, 400, 846
Sugar, 400
Sulfanilic Acid TS, 855
Sulfate and Chloride Limit Test, 757
Sulfate Identification Test, 755
Sulfite Identification Test, 755
Sulfur (by Oxidative Microcoulometry), 780
Sulfur Dioxide, 401
Sulfur Dioxide Detector Tube, 862
Sulfur Dioxide Determination, 841
Sulfur Dioxide Policy, 2
Sulfuric Acid, 402, 855
Sulfuric Acid (reagent), 855
Sulfuric Acid, 1 N, 859
Sulfuric Acid, 95%, 855
Sulfuric Acid, Alcoholic, 0.5 N, 859
Sulfuric Acid, Alcoholic, 5 N, 859
Sulfuric Acid Table, 752
Sulfuric Acid TS, 855
Sulfuric Acid TS, Diluted, 855
Sulfurol, 524
Summer Savory Oil, 346
Sunflower Oil (Unhydrogenated), 403
Sunset Yellow FCF, 146
Superglycerinated Fully Hydrogenated Rapeseed Oil, 337
Sweet Basil Oil, 39
Sweet Orange Oil, 273
Sweetwood Bark Oil, 91
Synthetic Amorphous Silica, 348
Synthetic Magnesium Silicate, 235
Synthetic Paraffin, Oil Content of, 750
Synthetic Wax, 291

T

Talc, 404
Tallow, 405
Tangerine Oil, Coldpressed, 405, 595
Tangerine Oil, Expressed, 405
Tannic Acid, 406
Tannic Acid TS, 855
Tared Container, 6
Tarragon Oil, 407, 595
Tartaric Acid, 407
L(+)-Tartaric Acid, 407
Tartrate Identification Test, 755
Tartrazine, 145
TBHQ, 408
Temperatures, 5
Terpene Resin, Natural, 410
Terpene Resin, Synthetic, 410
α-Terpinene, 554
γ-Terpinene, 554
*Terpinen-4-ol, 554
α-Terpineol, 554, 710
Terpinyl Acetate, 556, 676
Terpinyl Propionate, 556, 711
Test Methods for Flavor Chemicals, 449, 564
    Acidity Determination by Iodometric Method, 565
    Acid Value, 568, 815
    Alcohol Content of Ethyl Oxyhydrate, 565
    Assay by Gas Chromatography, 564
        General Method, Nonpolar Method, 564
        General Method, Polar Method, 564
    Assay by Titrimetric Procedures, 565
        Direct Aqueous Acid–Base Titrations, 565
        Direct Aqueous–Alcoholic Acid–Base Titrations, 565
    Assays for Certain Aldehydes and Ketones, 564
        Aldehydes—Hydroxylamine Method, 564
        Aldehydes—Hydroxylamine/Tert-Butyl Alcohol Method, 564
        Ketones—Hydroxylamine Method, 565
        Procedure Requiring the Use of Sealed Glass Vials or Ampules, 564
    Limit Test for Antioxidants in Ethyl Acrylate, 566
    Limit Test for Heavy Metals, 567
    Limit Test for Hydrocarbons in Eugenol, 567
    Limit Test for Hydrocyanic Acid in Benzaldehyde, 567
    Limit Test for Lead, 567

Limit Test for Methyl Compounds in Ethyl Acetate, 567
Limit Test for Peroxide Value, 567
Limit Test for Readily Carbonizable Substances in Ethyl Acetate, 567
Limit Test for Readily Oxidizable Substances in dl-Menthol, 567
Limit Test for Reducing Substances, 568
Qualitative Test for Phenols Using Ferric Chloride, 568
Residue on Evaporation, 568
Test Solutions (TS) and Other Reagents, 5, 849
Tetrabromo-m-cresolsulfonphthalein, 860
Tetrabromophenolsulfonphthalein, 860
Tetradecanal, 534
Tetradecanoic Acid, 262
1-Tetradecanol, 536
Tetradecyl Alcohol, 536
Tetrahydrofurfuryl Alcohol, 556, 640
Tetrahydrogeraniol, 482
Tetrahydrolinalool, 556
Tetrahydro-2H-1,4-oxazine, 261
2,3,5,6-Tetramethylpyrazine, 556, 641
Tetrapotassium Pyrophosphate, 326
Tetrasodium Diphosphate, 378
Tetrasodium Pyrophosphate, 378
Thermometers, 727
Thiamine Chloride, 411
Thiamine Hydrochloride, 411
Thiamine Mononitrate, 412
Thiamine Nitrate, 412
Thiamin Hydrochloride, 411
Thiamin Mononitrate, 412
Thin-Boiling Starch, 159
Thin-Layer Chromatography, 731
Thiobismethane, 482
Thiosulfate Identification Test, 755
Thorium Nitrate, 0.1 M, 859
L-**Threonine**, 413
Thuja Oil, 94
Thyme Oil, 413, 595
Thyme Oleoresin, 392, 393
Thymol, 556, 641
Thymol Blue, 861
Thymol Blue TS, 851, 855
Thymolphthalein, 861
Thymolphthalein TS, 855
Thymolsulfonphthalein, 861
Tight Container, 7
Time Limits, 6
Titanium Dioxide, 414
DL-α-**Tocopherol**, 417
D-α-**Tocopherol Concentrate**, 418
Tocopherols Concentrate, Mixed, 419
D-α-**Tocopheryl Acetate**, 420
DL-α-**Tocopheryl Acetate**, 421

D-α-**Tocopheryl Acetate Concentrate**, 422
D-α-Tocopheryl Acetate Preparation, 422
D-α-**Tocopheryl Acid Succinate**, 422
Tolerances, 6
Tolualdehyde, 558
p-**Tolualdehyde**, 558
Toluene Distillation Method for Water, 747
α-Toluic Acid, 546
α-Toluic Aldehyde, 546
p-Tolyl Acetate, 474
Tolyl Acetate, So-Called, 526
p-Tolyl Aldehyde, 558
Tolyl Aldehyde, mixed isomers, 558
p-Tolyl Isobutyrate, 558, 641
Torula Yeast, 440
Total Alcohols, 815
Total Ash, 748
Total Color, 773, 774
Total Monoglycerides, 824
Total Solids, 842
   Glucose Syrup (Corn Syrup), 843, 843
   High-Fructose Corn Syrup Solids, 844
   Maltodextrin, 845
   Sucrose, 846
Total Unsaturation, 785
Trace Impurities, xx, 6
Tragacanth, 424
"Transfer," Defined, 4
Triacetin, 424
Triatomic Oxygen, 277
Tributyrin, 558, 677
Tricalcium Phosphate, 69
Trichloroethene, 425
Trichloroethylene, 425
1,1,2-Trichloroethylene, 425
4,1',6'-Trichlorogalactosucrose, 398
2-Tridecenal, 558, 642
Triethanolamine, 0.5 N, 860
Triethyl Citrate, 426
3,7,12-Trihydroxycholanic Acid, 99
Triketohydrindene Hydrate TS, 853, 855
Trimagnesium Phosphate, 234
Trimethylamine, 558
2,6,6-Trimethylbicyclo[3.1.1]hept-2-ene, 550
4(2,6,6-Trimethyl-1-cyclohexenyl)-3-butene-2-one, 510
4(2,6,6-Trimethyl-2-cyclohexenyl)-3-butene-2-one, 510
3,7,11-Trimethyl-1,6,10-dodecatrien-3-ol, 536
3,7,11-Trimethyl-2,6,10-dodecatrien-1-ol, 498
*3,5,5-**Trimethyl Hexanal**, 558
2,4,5-Trimethyl δ-3-Oxazoline, 560, 677
2,3,5-Trimethylpyrazine, 560, 642

1,3,7-Trimethylxanthine, 52
Trinitrophenol TS, 854, 855
Triphosphate, 385
Tripotassium Phosphate, 325
Tripropionin, 500
Trisodium Dipotassium Tripolyphosphate, 377
Trisodium Phosphate, 375
Tristearin, 185
Tropaeolin D, 861
Trypsin, 133, 788
Trypsin Activity, 814
DL-**Tryptophan**, 426
L-**Tryptophan**, 427
Tunu, 249
Turmeric Oleoresin, 392, 393
L-**Tyrosine**, 427

U

Ultraviolet Absorbance of Citrus Oils, 819
δ-**Undecalactone**, 560, 642
γ-**Undecalactone**, 560, 642
Undecanal, 560, 677
2-Undecanone, 560, 643
10-Undecenal, 560, 643
Undecen-10-al, 560
2-Undecenol, 560, 678
Undecyl Alcohol, 562, 678
n-Undecyl Aldehyde, 560
Unsaponifiable Matter, 828
Urea, 428

V

"Vacuum," Defined, 6
Valeraldehyde, 562, 564, 643
Valeric Acid, 562, 565, 678
γ-**Valerolactone**, 562, 679
"Validation," Defined, xxiii
Validation of Codex Methods, xxii
L-**Valine**, 429
Vanillin, 562, 679
Venezuelan Chicle, 249
1-Vinyl-2-pyrrolidone Crosslinked Insoluble Polymer, 309
Viscosity (Rosins and Related Substances), 835
Viscosity of Dimethylpolysiloxane, 743
Viscosity of Methylcellulose, 743
Viscosity of Sodium Carboxymethylcellulose, 744
Vital Wheat Gluten, 535
Vitamin A, 429
Vitamin B_1, 411, 412
Vitamin B_1 Hydrochloride, 411
Vitamin B_1 Mononitrate, 412
Vitamin B_2, 338

Vitamin B$_6$, 334
Vitamin B$_6$ Hydrochloride, 336
Vitamin B$_{12}$, 431
Vitamin C, 33
Vitamin C Sodium, 354
Vitamin D, 432, 434
Vitamin D$_2$, 432
Vitamin D$_3$, 434
Volatile Acidity, 829
Volatile Oil Content (Essential Oils and Flavors), 815
Volatile Oil Content (Oleoresins), 832
Volumetric Apparatus, 4, 728
Volumetric Solutions, 855
"VS," Defined, 449

W

Water and Loss on Drying, 4, 6, 749
Water Determination, 745
    Karl Fischer Titrimetric Method, 745
    Toluene Distillation Method, 747
Water Vapor Detector Tube, 862
Weighing Practices, 6
Weights and Balances, 4, 729
Weights and Measures, Symbols, and Abbreviations, 6
Well-Closed Container, 7
Wheat Gluten, 435
*****Whey**, 435
White Cedar Leaf Oil, 94
White Petrolatum, 289
White Shellac, 347
White Wax, 40
Wine Yeast Oil, 108
Wintergreen Oil, 437, 600
Wool Fat, 216

X

Xanthan Gum, 437
Xylenol Orange, 861
Xylenol Orange TS, 855
Xylitol, 439

Y

Yam Flour, 210
*****Yeast, Dried**, 440
Yeast Extract, 442
Yellow Petrolatum, 289
Yellow Prussiate of Soda, 363
Yellow Wax, 41

Z

Zein, 443
Zinc Gluconate, 443
Zinc Identification Test, 755
Zinc Oxide, 445
Zinc Sulfate, 445
Zinc Sulfate, 0.05 M, 860
*****Zingerone**, 562